An Introduction to

Statistical Communication Theory

IEEE Press
445 Hoes Lane, P.O. Box 1331
Piscataway, NJ 08855-1331

IEEE Communications Society, *Sponsor*
C-S Liaison to IEEE Press, Tom Robertazzi

IEEE Information Theory Society, *Sponsor*
IT-S Liaison to IEEE Press, Stuart Schwartz

First printing (1960) by McGraw-Hill, Inc.
Second Printing (1987) by Peninsula Publishing.

Printed in the United States of America

10 9 8 7 6 5 4 3 2 1

ISBN 0-7803-1178-7
IEEE Order Number: PC5648

Library of Congress Cataloging-in-Publication Data

Middleton, David. 1920–
An introduction to statistical communication theory / David Middleton : sponsored by IEEE Communications Society, IEEE Information Theory Society.
p. cm.
Reprint. Originally published: New York, McGraw-Hill, 1960.
Includes bibliographical references and indexes.
ISBN 0-7803-1178-7
1. Statistical communication theory. I. Title.
TK5101.M45 1996
621.382—dc20 96-11241
CIP

An IEEE Press Classic Reissue

An Introduction to

Statistical Communication Theory

DAVID MIDDLETON

IEEE Communications Society, *Sponsor*
IEEE Information Theory Society, *Sponsor*

IEEE PRESS

The Institute of Electrical and Electronics Engineers, Inc., New York

This book and other IEEE PRESS books may be purchased at a discount from the publisher when ordered in bulk quantities. Contact:

IEEE Press Marketing
Attn: Special Sales
445 Hoes Lane, P.O. Box 1331
Piscataway, NJ 08855-1331
Fax: (908) 981-9334

For more information about IEEE PRESS products, visit the IEEE Home Page: http://www.ieee.org/

To
J. B. Wiesner
(1915–1994)

FOREWORD TO THE IEEE PRESS REISSUE

The statistical theory of communication has proven to be a powerful methodology for the design, analysis, and understanding of practical systems for electronic communications and related applications. From its origins in the 1940's to the present day, this theory has remained remarkably vibrant and useful, while continuing to evolve as new modes of communication emerge. The publication in 1960 of *Introduction to Statistical Communication Theory (ISCT)* was a landmark for the field of statistical communication. For the first time, the disciplines comprising statistical communication theory—random processes, modulation and detection, signal extraction, information theory—were combined into a single, unified treatment at a level of depth and degree of completeness which have not been matched in any subsequent comprehensive work. Moreover, *ISCT* introduced a further interdisciplinary feature, in which relevant physical characteristics of communication channels were incorporated into many topics.

Today, some thirty-five years since its first appearance, this book remains a unique sourcebook in statistical communications. Most of the topics treated in this book are still taught in the classroom today, and they continue to arise in the day-to-day work of engineers in fields such as electronic communications, radar, sonar, and radio astronomy, among many others. Although there are a great many excellent books on communications in print, no other book treats the mathematical and physical foundations of the discipline in the comprehensive, interdisciplinary way found here. This unified presentation offers the reader a distinct advantage in terms of understanding, as well as a significant value in terms of compactness of resources.

ISCT is a classic book in the field of communications. As such, it ranks with Davenport and Root's *An Introduction to the Theory of Random Signals and Noise* and Wozencraft and Jacob's *Principles of Communications Engineering*. However, even without its historical stature, this book stands today as a valuable resource for engineers, researchers, and students in communications and related fields. By republishing this book, the IEEE is providing a valuable service to its members and to the electronic communications community in general. This reissue is a very welcome event indeed.

H. Vincent Poor
Princeton University
November 1995

PREFACE TO THE SECOND REPRINT EDITION (1996)

It has been thirty-six years since *An Introduction to Statistical Communication Theory* was originally published, in 1960 by McGraw-Hill in its International Series in Pure and Applied Physics [1]. Nine years have also passed since its republication as a Reprint Edition by the Peninsula Publishing Co., of Los Altos, California, in 1987 [1a]. The present volume is the reincarnation of the previous editions, through the kind offices of the IEEE Press and under the welcome sponsorship of the Communication and Information Theory Societies of the IEEE, with selected changes and additions described below. From a historical viewpoint, *An Introduction* follows, at an advanced level of application, along the path, primarily, of the earlier work of Lawson and Uhlenbeck in the MIT Radiation Laboratory Series [2] and of Davenport and Root [3]. Because so much of the original material of *An Introduction* appears still to be pertinent and useful today and because almost three full generations of engineers and scientists have arrived since the book's inception, it seems appropriate once more to make it available to the concerned technical community.

Unlike the first Reprint Edition (1987–1995), for which no "updating" was attempted (save for the brief list of references cited in the Preface thereof), this edition does provide selected references which connect the earlier work of the original book's period with many new concepts, methods, and results which have appeared since 1960 and which may broadly be considered to follow from and extend the basic ideas and techniques of statistical communication theory (SCT), as exposited in [1, 1a]. These include, for example, (1) developments in such areas as detection and estimation in non-Gaussian noise environments involving propagation and scattering in nonhomogeneous channels [4–9, 14–15]; (2) state-space, model-based signal processing (estimation and control) [10]; (3) digital signal processing [11]; (4) spectrum analysis and array processing [12, 13]; as well as (5) other extensions of classical signal detection and estimation [14–16], including fuzzy logic [17] and neural networks [18]; (6) the rôle of information theory in physical science [20]; and of course, (7) the essential developments in computing technology which have made the practical implementation of SCT more fully possible.

In addition to these important topics, selectively referenced below, a rather extensive list of some 125 references has been added at the end of this Edition, with few exceptions, on a chapter-by-chapter basis. These represent mainly books which in the author's opinion describe the technical progress in the greatly expanded, interdisciplinary field of SCT from 1960 to the present. By and large, they themselves also remain vital and useful today, as well as of historical interest, and serve to connect the still active past with the ongoing present. This list is not intended to be, nor could it be, complete in any practical sense, nor is any slight to omitted worthy work intended. In any case, many useful references are also to be sought and found in these volumes.

Finally, the author has taken the opportunity of this Reprint Edition to correct the errors and misprints (known to him), fortunately few since the many printings (1960–1972) of the original work [1]. The author is deeply indebted to Professor Poor (Princeton University) for his support and interest in the republication of this book and for his generous foreword thereto. The author also wishes to extend his appreciation to the various reviewers and to Mr. Dudley Kay and the excellent editorial staff at the IEEE Press for their essential efforts in bringing this work to publication again.

David Middleton
New York, 1996

[1] D. Middleton, *An Introduction to Statistical Communication Theory*, International Series in Pure and Applied Physics, McGraw-Hill, New York, 1960.

[1a] ———, *An Introduction to Statistical Communication Theory*; (First) Reprint Edition, Peninsula, Los Altos, CA, 1987–1995; see following Preface.

[2] J. L Lawson and G. E. Uhlenbeck, *Threshold Signals*, MIT Radiation Laboratory Series, Vol. 24, McGraw-Hill, New York, 1950.

[3] W. B. Davenport and W. L. Root, *An Introduction to Random Signals and Noise*, McGraw-Hill, New York, 1958; Reprint Edition, IEEE Press, Piscataway, NJ, 1984.

[4] D. Middleton, *Threshold Signal Processing (with Applications to Ocean Acoustics, Non-Gaussian Random Media, and Generalized Telecommunications)*, American Institute of Physics (AIP), Series in ''Modern Acoustics and Signal Processing,'' American Institute of Physics Press, New York, in press (1997).

[5] ———, ''Threshold Detection in Correlated Non-Gaussian Noise Fields,'' *IEEE Trans. Information Theory*, Vol. 41, no. 4, pp. 976–1000, July 1995.

[6] D. Middleton and A. D. Spaulding, ''Optimum Reception in Non-Gaussian Electromagnetic Environments II. Optimum and Suboptimum Threshold Signal Detection in Class A and B Noise,'' NTIA Report 83-120, (NTIS pub. No. PB 83-241141), ITS (NTIA), U.S. Department of Commerce, 325 Broadway, Boulder, CO 80303. May 1983.

[7] ———, *Elements of Weak Signal Detection in Non-Gaussian Noise Environments*, Chapter 5 of *Advances in Statistical Signal Processing,* Volume 2: *Detection*, eds. H. V. Poor and J. B. Thomas, JAI Press, 55 Old Post Road, No. 2, P.O. Box 1678, Greenwich, CT 06836-1678, December 1993.

[8] S. A. Kassam, *Signal Detection in Non-Gaussian Noise*, Springer-Verlag, New York, 1988.

[9] E. J. Wegman, S. C. Schwartz, and J. B. Thomas, eds., *Topics in Non-Gaussian Signal Processing*, Springer-Verlag, New York, 1989. See also
E. J. Wegman and J. G. Smith, eds., *Statistical Signal Processing*, Marcel Dekker, New York, 1984.

[10] J. V. Candy, *Signal Processing: A Model-Based Approach*, McGraw-Hill, New York, 1986.

[11] A. V. Oppenheim and R. W. Schafer, *Digital Signal Processing*, Prentice Hall, Englewood Cliffs, NJ, 1975.

[12] A. M. Yaglom, *Correlation Theory of Stationary and Related Random Functions:* I. *Basic Results* (1987); II. *Supplementary Notes and References* (1987), Springer-Verlag, New York.

[13] S. Haykin, ed., *Advances in Spectrum Analysis and Array Processing*, Vols. 1 and 2, Prentice Hall, Englewood Cliffs, NJ, 1991.

[14] S. M. Rytov, Yu. A. Kravtsov, and V. I. Tatarskii, *Principles of Statistical Radio Physics* (English Edition); Vol. 1, *Elements of Random Process Theory* (1987); Vol. 2, *Correlation Theory of Random Process* (1988); Vol. 3, *Elements of Random Fields* (1989); Vol. 4, *Wave Propagation Through Random Media* (1989), Springer-Verlag, New York. (See also [1] above.)

[15] D. Middleton, "Space-Time Processing for Weak-Signal Detection in Non-Gaussian and Non-Uniform Electromagnetic Interference (EMI) Fields":, Contractor Report 86-36 (NTIS: PB-86-193406), ITS (NTIA), U.S. Department of Commerce, 325 Broadway, Boulder, CO 80303), February 1986.

[16] C. W. Helstrom, *Elements of Signal Detection and Estimation*, Prentice Hall, Englewood Cliffs, NJ, 1995.

[17] L. A. Zadeh, *Fuzzy Sets and Applications: Selected Papers by L. A. Zadeh*, eds. R. R. Yager et al., John Wiley, New York, 1987.

[18] S. Haykin, *Neural Networks*, IEEE Press (Macmillan College Publishing Co.), New York, 1994.

[19] R. Young, *Wavelet Theory and Its Applications*, Kluwer Academic Publishers, Boston, 1993.

[20] L. Brillouin, *Science and Information Theory*, 2nd edition, Academic Press, New York, 1962.

See also the earlier selected references at the end of the Preface to the First Reprint Edition following. Additional references, by chapter, are given at the end of this book.

PREFACE TO THE FIRST REPRINT EDITION (1987–1995)

It has been over a quarter of a century since *An Introduction to Statistical Communication Theory* was first published, in 1960, by McGraw-Hill (New York) in its International Series in Pure and Applied Physics. Since then almost two generations of scientists and engineers have appeared and many specialized works within the broad domain of Statistical Communication Theory (SCT) have been produced. Today, what is often referred to as "signal processing" has become a standard engineering discipline in all areas and applications which require the transmission and reception of "information." Signal processing itself has become an ensemble of techniques whose foundations were introduced and broadly described in the original edition of the present book.

Because so much of the material of *An Introduction* appears to be continuously useful and because this work has been out-of-print since 1972, it seems particularly appropriate to reintroduce the book at this time. Indeed, inasmuch as the concepts and methods, and many of the examples described therein, are *canonical*, that is, have a general, functional *form* independent of specific physical models and analytical detail, they remain relevant to current and future applications. It is in this spirit that this Reprint Edition (Peninsula Publishing) is presented.

No attempt to "update" the book itself has been made here. To do so with the same emphasis on comprehensiveness and detail covered in the original would now require a whole *series* of books. Apart from losing the advantages of self-containment in a single volume, such a task does not appear attractive, in the face of the many excellent books which have subsequently appeared and which treat in more detail many of the topics originally discussed in the author's treatise.

Of course, new methods and ideas in Statistical Communication Theory (SCT) have appeared since the early sixties. In addition to the topics treated generally in the original edition, such as *noise theory*, *noise physics*, *information theory*, *statistical decision theory (SDT)*, and *data processing*, one should add within the domain of SCT important new developments: (1) *nongaussian noise and interference models* [10], [12], [18]; (2) *threshold signal detection and estimation* [11] - [13], [15], [16]; (3) *spatial processing*, as well as temporal sampling, including general arrays and apertures [17]; and (4) *statistical-physical approaches* [10], [17] - [20], to describe the channel itself, including scattering and various types of propagation encountered in real-world environments. Moreover, as predicted at the time (1960), cf. Sec. 23.5, [1], the many areas of then future study noted therein have indeed been (and are being) investigated, as well as others not then known. For example, whole areas of current importance such as Electromagnetic Compatibility (EMC) and Spectrum Management [11], [12], [13], [16], quantum optical signal processing [8], generalized arrays and spatial sampling [17] - [19], as well as the development of powerful and economical computers and programs for handling the vast data loads incurred in practice, were either unknown or barely beginning to emerge. In fact, the recognition (slow even today) that one lives in a "nongaussian" world [10], [12] is perhaps one of the more significant features of the current evolution of SCT and its specific applications.

Finally, to assist the reader in relating the SCT fundamentals of 1960 with the developing practicalities of the present, the author has appended below his own brief, highly personalized list of books and papers. No attempt at completeness is made and no slight to worthy work not mentioned here is intended.

David Middleton
New York, 1987

I. Books

[1]. D. Middleton, *An Introduction to Statistical Communication Theory*, McGraw-Hill (New York), 1960.

[1a]. Д. Миддлтон: Введение в Статистическую Теорию Связи, Том 1, 1961; Том 2, 1962, Советское Радио, Москва, С.С.С.Р. (vol. 1, 1961; vol. 2, 1962, Soviet Radio, Moscow, U.S.S.R.).

[2]. L. A. Wainstein and V. D. Zubakov, *Extraction of Signals from Noise* (translated from the Russian by R. L. Silverman), Prentice Hall (New Jersey), 1962.

[3]. D. Middleton, *Topics in Communication Theory*, McGraw-Hill (New York), 1965.

[3a]. Д. Миддлтон ОЧЕРКИ ТЕОРИИ СВЯЗИ
Издательство «СОВЕТСКОЕ РАДИО» *Москва — 1966*

[4]. C. W. Helstrom, *Statistical Theory of Signal Detection*, Second Edition, Pergamon Press (New York), 1968.

[5]. J. B. Thomas, *An Introduction to Statistical Communication Theory*, John Wiley (New York), 1969.

[6]. H. Van Trees, *Detection, Estimation, and Modulation Theory*, Part I (1968), Part II (1971), Part III (1971), John Wiley (New York).

[7]. B. R. Levin, *Theoretical Bases of Statistical Radio Engineering*, Moscow, "Soviet Radio," Vol. 1, 1974, Vol. 2, 1975, Vol. 3, 1976.
(В.Р. Левин, Теоретиуеские Осиовъи Статистиуеской Радиотехники.)

[8]. C. W. Helstrom, *Quantum Detection and Estimation Theory*, Academic Press (New York), 1976.

[9]. N.H. Blachman, *Noise and Its Effect on Communication,* 2nd Edition, R. E. Krieger Pub. Co. (Malabar, Florida), 1982.

II. Papers

[10]. D. Middleton, "Statistical-Physical Models of Electromagnetic Interference," IEEE Trans. on Electromagnetic Compatibility, Vol. EMC-17, No. 3, pp. 106-127, Aug., 1977.

[11]. A. D. Spaulding and D. Middleton, "Optimum Reception in an Impulsive Interference Environment — Part I: Coherent Detection; Part II: Incoherent Reception," IEEE Trans. Commun., Vol. COM-25, pp. 910-934, Sept. 1977.

[12]. D. Middleton, "Canonical Non-Gaussian Noise Models: Their Implications for Measurement and for Prediction of Receiver Performance," IEEE Trans. Electromag. Compat., Vol. EMC-21, No. 3, pp. 209-220, Aug. 1979.

[13]. D. Middleton, "Threshold Detection in Non-Gaussian Interference Environments: Exposition and Interpretation of New Results for EMC Applications," IEEE Trans. on Electromag. Compat., Vol. EMC-26, No. 1, pp. 19-28, Feb., 1984.

[14]. S. A. Kassam and H. V. Poor, "Robust Techniques for Signal Processing — A Survey," Proc. IEEE, Vol. 73, No. 3, March, 1985, pp. 433-481.

[15]. D. Middleton, "Threshold Signal and Parameter Estimation in Non-Gaussian EMC Environments," pp. 429-435, *Proceedings*, International Symp. on EMC, Zurich, Switz., March 5-7, 1985.

[16]. A. D. Spaulding, "Locally Optimum and Suboptimum Detector Performance in a Non-Gaussian Interference Environment," IEEE Trans. Commun., Vol. COM-33, No. 6, June 1985, pp. 509-517.

[17]. D. Middleton, "Space-Time Processing for Weak Signal Detection in Non-Gaussian and Non-Uniform Electromagnetic Interference (EMI) Fields," Contractor Report 86-36, Feb., 1986, ITS/NTIA — U.S. Dept. of Commerce, 325 Broadway, Boulder, CO 80303. NTIS: PB-86-193406.

[18]. D. Middleton, "Second-Order Non-Gaussian Probability Distributions and Their Application to 'Classical' Nonlinear Processing Problems in Communication Theory," *Proceedings* of 1986 Conf. on Information Sciences and Systems, Princeton University, March 19-21, 1986.

[19]. D. Middleton, "A Statistical Theory of Reverberation and Similar First-Order Scattered Fields," Parts I, II, IEEE Trans. Information Theory, Vol. *IT-13*, 372-392; 393-414 (1967); Parts III, IV, ibid., Vol. *IT-18*, 35-67; 68-90 (1972).

[20]. D. Middleton, "Channel Modeling and Threshold Signal Processing in Underwater Acoustics: An Analytical Overview," IEEE J.of Oceanic Engineering, Vol. OE-12, No. 1, Jan., 1987.

PREFACE TO THE FIRST EDITION (1960)

Statistical communication theory may be broadly described as a theory which applies probability concepts and statistical methods to the communication process. In the wide sense, this includes not only the transmission and reception of messages, the measurement and processing of data, measures of information, coding techniques, and the design and evaluation of decision systems for these purposes, but also the statistical study of language, the bio- and psychophysical mechanisms of communication, the human observer, and his role in the group environment. Here, however, we consider statistical communication theory from the narrower viewpoint of the physical systems that are specifically involved, e.g., radio, radar, etc. For this purpose, it is convenient to regard statistical communication theory as consisting of the following contiguous and, to varying degrees, overlapping areas of interest: (1) *noise theory*, which embraces the mathematical description of random processes and their properties under various linear and nonlinear transformations; (2) *noise physics*, which is concerned mainly with the underlying physical mechanisms of the noise processes encountered in applications; (3) *information theory*, defined here in the strict sense as a theory of information measures and coding; and (4) *statistical decision theory*, as modified and extended to system design, evaluation, and the comparison of optimum and suboptimum systems, where the desired end product is some definite decision—a "yes" or "no" or a measurement. With (3) and (4), it is natural to consider also *data processing*, in the course of transmission and reception, the *physics and the technology of the components*, e.g., tubes, transistors, etc., by which actual systems for the above purposes are to be realized, and, of course, the various concepts and methods of *probability* and *statistics* that are required for a quantitative treatment, subject now to the different constraints imposed by the communication process itself.

The present book is addressed principally to engineers, physicists, and applied mathematicians. Its aims are threefold: (1) to outline a systematic approach to the design of optimal communication systems of various fundamental types, including an evaluation of performance and a comparison with nonoptimum systems for similar purposes; (2) to incorporate within a framework of a unified theoretical approach the principal results of earlier work, as well as to indicate some of the more important new ones; and finally (3) to be used as a text at various levels, as an instrument for the research worker in current problems, and, it is hoped, as a starting point for new developments. The emphasis here is mainly on (1) and (2), although (3) has not been neglected: method as well as results are stressed. About 300 problems have been included, not only as exercises but also as a source of additional results which could not otherwise be presented. While the bibliography is necessarily selective, 500 references (including the supplement) are available, from which more specialized interests may in turn be served. The mathematical exposition is for the most part heuristic, and although a detailed, rigorous treatment is outside the scope and intent here, in this respect reference is made to the appropriate literature. Moreover, little space is devoted to the probabilistic and statistical background, the elements of which the reader is assumed to possess. A knowledge of Fourier- and Laplace-transform methods, contour integration, matrices, simple integral equations, and the usual techniques of advanced calculus courses is also required, along with the elements of circuit theory and the principles of radio, radar, and other types of electronic communication systems.

The book is divided into four main parts. Part 1 introduces and describes some of the statistical techniques required in the analysis of communication systems and concludes with an introductory chapter on information theory. Part 2 considers the normal process and some of the processes derived from it and gives a short account of the physical models of shot and thermal noise. Part 3 is concerned mainly with various nonlinear operations that are common in transmission and reception, such as modulation and demodulation, and the calculation of signal-to-noise ratios. Linear measurement, filtering, and prediction and more general distribution problems, the results of which are needed in the general analysis of Part 4, are also examined here. Finally, Part 4 gives a detailed development of a statistical communication theory for the basic single-link communication system, consisting of message or signal source, transmitter, medium of propagation (or channel), and receiver and decision-making elements. The attention here is on optimization and evaluation of receiver performance, e.g., signal detection and extraction, with a short introduction (in Chap. 23) to the still more general question of simultaneous optimization of both the reception and transmission operations. About a third of the material of Parts 1 to 3 coincides with that in recent publications on noise theory. Part 4 has not appeared in book form before, and much of the rest, Part 3 particularly, is also believed to be new in this respect. While about three-quarters of the subject matter of the book as a whole may be found scattered through the original technical literature, many of the results mentioned in Chaps. 12, 17, 19 to 21, and 23 have not been presented previously.

In spite of its size, this work must still be considered as an introduction, for it has not been possible to treat more than a few important topics, such as detection theory, with any great degree of completeness and still maintain the attempt at a broad coverage which is one aim of the book. In fact, a number of important subjects have been omitted, because they require a separate volume for an adequate treatment, or because they are already so handled in the literature, or in some instances because no theory of any stature is as yet available. Among the topics which have been regretfully dropped are linear and nonlinear communication feedback (and related control) systems, coding methods, sequential detection and estimation, various types of noise-measuring systems, and many more specialized results in the theory of noise, such as the probability distributions of zeros and maxima of random waves. Problems involving large-scale, i.e., many-link, communication networks, game-theory applications, and their relationship to the single-link systems examined in Part 4 are likewise omitted. Some attention is given to Markoff processes, the Fokker-Planck and related equations, with applications to shot and thermal noise, and, later, to questions of optimum linear measurements, their statistical errors, filtering, prediction, and the like. However, a more extensive coverage of the second-moment theory (involving spectra and signal-to-noise ratios) of AM and FM transmission and reception is provided. The same is true for spectra and covariance functions of linearly filtered random and mixed processes, including random pulse trains. The treatment of the normal process, and those processes derived from it, including gaussian and nongaussian functionals, is intermediate, sufficient to satisfy the needs of subsequent chapters. The discussion of optimum systems for detecting and extracting signals from noise backgrounds, on the other hand, is comparatively detailed, though by no means complete, since this whole subject is still in a state of rapid development. Finally, besides the analytical features of system optimization and evaluation, considerable attention is given to the interpretation of system structure in such cases: the representation of optimum systems in terms of ordered sets of physically realizable elements. Threshold or weak-signal systems receive particular emphasis in the present study, since they represent the furthest extent to which performance can be pushed under a given environment. Moreover, optimum design for threshold operation, while not usually optimum for strong signals, is nevertheless almost always satisfactory in the latter instances, so that a system designed to make the most of threshold operation will also function acceptably when the interference is weak or ignorable. Threshold performance is important whenever we are required to operate at the limit of system capabilities. Terrestrial, satellite, and space communications offer many conspicuous examples.

As a text, the selection of material may be made rather arbitrarily, depending on the level and scope of the intended course. For example, one might use Part 1 followed by Chaps. 7, 9, 11 and parts of Chaps. 12 to 16. A more advanced program might use portions of Part 1, all of Part 2, most of Part 3, and, as an introduction to decision-theory methods in communication, Chaps. 18 and 19. Specific topics, such as system optimization, might be based on Chap. 17 and Part 4, with Chap. 6 and selected portions of earlier chapters. A perusal of the table of contents may suggest still other programs. An excellent introduction to Parts 1 to 3, with a more detailed coverage in various instances, may be found in the book by Davenport and Root.† The principal references on which the material of each chapter is based are listed at the chapter ends, with a supplementary bibliography at the end of the book. For a critique of the methods discussed in Part 4 and some possible avenues of future development, see Chap. 23. A glossary of principal symbols is included, with two appendixes on special functions and integral equations that occur frequently in the analysis.

Portions of Parts 1, 2, 3 were developed from material presented by the author during 1949 to 1954 as a series of graduate courses in the Division of Applied Science at Harvard University; also, some of the author's published work referred to in the text stemmed from research originally supported at Harvard and at the Massachusetts Institute of Technology under Department of Defense contracts with those institutions. The book itself was begun in 1949, the final draft of Part 1 was completed during 1955–1956, those of Parts 2 and 3 in 1956–1957, and Part 4 in 1957–1958.

While it is regretfully impracticable to mention here the many who have contributed to this work over the last decade, they will find themselves for the most part recorded in the name index at the end of the book, as well as in the references throughout the text. However, the author wishes specifically to thank Prof. Leon Brillouin for his original encouragement of this project, Dr. Arthur Kohlenberg for his criticism of Chap. 4, and Prof. Peter Elias for his valuable comments concerning Chap. 6. The author also wishes to thank several of his former students for their assistance, and in particular Dr. David Van Meter for his careful reading of the entire final manuscript and Dr. Julian Bussgang for his detailed criticisms of the first 10 chapters. Thanks are also due Dr. James A. Mullen and Dr. D. B. Brick for assistance with proofs. In all instances, the errors of omission and commission remain the author's, as do the opinions and results therein, unless otherwise indicated. Finally, the author is indebted to Drs. Davenport and Root for the opportunity to examine their manuscript when both books were in the final stages of completion and to discuss with them various aspects of the several works. For their heroic efforts in coping with a difficult manuscript, the author's sincere appreciation is also extended to Mrs. Edith Nicholson, who typed the first third, and to Miss Anne Guest, who handled most of the remaining two-thirds.

Finally, the author wishes to thank the editors of various journals for permission to use certain material which appeared originally in their publications. In particular, the author is indebted to the editors of the *Journal*

† W. B. Davenport, Jr., and W. L. Root, "An Introduction to the Theory of Random Signals and Noise," McGraw-Hill, New York, 1958.

of the Society of Industrial and Applied Mathematics for a number of excerpts from the paper by Middleton and Van Meter, "Detection and Extraction of Signals in Noise from the Point of View of Statistical Decision Theory," which appeared in the journal of the society in December, 1955, and June, 1956 (see Chap. 1, Ref. 6). Specifically, most of Chap. 18, about 40 per cent of Chap. 19, a few pages of Chap. 20, most of Secs. 21.1-3, 21.1-4, 21.4, and 21.5, parts of Secs. 21.2-2, 21.2-3, 21.3-1, and all of Chap. 22, except for Secs. 22.1, 22.2, have been taken directly or with minor modifications from the aforementioned paper. The author is also grateful to the Institute of Radio Engineers for permission to excerpt and paraphrase portions of his earlier work, in particular, Middleton and Van Meter, *IRE Trans. on Inform. Theory*, **IT-1**: 1-7 (September, 1955), and to use figs. 1 to 3 of this paper (Figs. 23.3, 23.4, P 23.3 here); also, for figs. 4, 5 (Figs. 13.11, 13.12 here), *Proc. IRE*, **36**: 1467 (1948); for figs. 2, 3, 7 (Fig. 20.4*a*, *b* here), *IRE Trans. on Inform. Theory, Symposium*, **PGIT-4**: 145 (September, 1954) (with Van Meter); for fig. 4, *IRE Trans. on Inform. Theory*, **IT-3**: 86 (1957). Similarly, the author wishes to thank the *Quarterly of Applied Mathematics* for permission to use figs. 3 to 10, *Quart. Appl. Math.*, **5**: 445 (1948) (here embodied in Figs. 5.3, 5.5, 9.1, 9.3, 13.3, 13.5); for figs. 2.1, 2.2, *ibid.*, **9**: 337 (1952), and for fig. 3.2, *ibid.*, **10**: 38 (1952) (here Figs. 12.2, 12.3, 14.9); and for figs. 4, 5, 6, *ibid.*, **7**: 129 (1949) (here Figs. 15.14, 15.15*a*, *b*). The author is likewise indebted to the *Journal of Applied Physics* for permission to use figs. 1, 2, 6, 11, 13, *J. Appl. Phys.*, **17**: 778 (1946) (here embodied in Figs. 12.1, 13.17*a*, *b*, 13.18); for figs. 1, 3, 4, 8, *ibid.*, **20,** 334 (1949) (here as Figs. 15.5, 15.6, 15.8, 15.16); for figs. 1, 4, *ibid.*, **23**: 377 (1952) (with W. B. Davenport and R. A. Johnson) (here as Figs. 16.2, 16.3); and for figs. 4, 5, 7, *ibid.*, **24**: 379 (1953) (here as Figs. 20.10 to 20.12).

David Middleton

नष्टो मोहः स्मृतिर्लब्धा त्वत्प्रसादान्मयाच्युत ।।
स्थितोऽस्मि गतसन्देहः करिष्ये वचनं तव ।। ७६ ।।

Srimad-Bhagavad-Gita
Chapter XVIII, verse 73

CONTENTS

PART 1

AN INTRODUCTION TO STATISTICAL COMMUNICATION THEORY

CHAPTER 1

STATISTICAL PRELIMINARIES

1.1 *Introductory Remarks*

In the large, the subject of this book is statistical communication theory. In detail, it is an application of modern statistical methods to random phenomena which influence the design and operation of communication systems. The specific treatment is confined to certain classes of electrical and electronic devices, e.g., radio, radar, etc., and their component elements, although the general statistical methods are applicable in other areas as well. Among such may be mentioned the theory of the Brownian motion, the turbulent flow of liquids and gases (in hydro- and aerodynamics), corresponding communication problems in acoustical media, and various phenomena in astronomy, in cosmic radiation, and in the actuarial and economic fields.

In the present case, the phenomena we have mainly to deal with are the fluctuating, or random, currents and voltages that appear at different stages of the communication process—i.e., the inherent, or background, *noise* which arises in the various system elements, noise and signals originating in the medium of propagation, and the desired messages, or *signals*, which are generated in or impressed upon the system at various points.† Phenomena, or *processes*, of this type are characterized by unpredictable changes in time: they exhibit variations from observation to observation which no amount of effort or control in the course of a run or trial can remove. However, if they show regularities or stabilized properties as the number of such runs or observations is increased under similar conditions, these regularities are called *statistical properties* and it is for these that a mathematical theory can be constructed. Physical processes in the natural, or real, world which possess wholly or in part a random mechanism in their structure and therefore exhibit this sort of behavior are called *random* or *stochastic*‡ *processes*, the latter term being frequently used to call attention to their time-dependent nature. We shall also refer to their mathematical description as a stochastic process, remembering that the analytical model here is always a more or less precise representation of a corresponding set, or *ensemble*, of possible events in the physical world, possessing properties of regularity in

† For a more general definition of signal and noise, see Sec. 1.3-5.

‡ From στοχος, meaning "chance."

the above sense. Since the theory of probability is the mathematical theory of phenomena which show statistical regularity, our methods of treating random processes are based on such concepts. The necessity of a statistical approach stems, of course, from the fact not only that it is usually impossible to specify the initial states of many physical systems with sufficient accuracy to yield unique descriptions of final states but also that the very laws of nature which we invoke are themselves idealizations, which ignore all but the principal characteristics of the model and of necessity omit the perturbations. Even then, a detailed application is often unworkable because of the inherent complexity of the system, so that a statistical treatment alone is productive.

It has been widely recognized since about 1945 that an effective study of communication systems and devices cannot generally be carried out in terms of an individual message or signal alone,[1,2]† nor can such a study safely neglect the inhibiting effects of the background noise or other interference on system performance.[3] Rather, one must consider the set, or ensemble, of possible signals for which the system is designed, the ensemble appropriate to the accompanying noise, and the manner in which they combine in the communication process itself. For those systems and system inputs, then, which possess statistical regularity (and fortunately these are usually the physically important ones here), we expect probability methods to provide the needed approach. Our concern with the properties of individual signals and associated system behavior, characteristic of earlier treatments, remains, but the emphasis is shifted now to the properties of the ensemble as a whole. It is these which are ultimately significant in analysis, design, and performance. Thus, when we speak of the probability of an event at a single observation, such as the presence (or absence) of a particular signal in a particular noise sample, we recall that this has meaning only in the context of an ensemble of such observations and that system performance is properly judged in terms of this ensemble, and not on the basis of an individual run alone.

Communication systems may be broadly described in terms of the operations which they perform on postulated classes of inputs. Chief among such operations electronically are the typical *linear*‡ ones of amplification, differentiation, smoothing (or integration, e.g., linear filtering) and such *nonlinear*‡ ones as modulation, demodulation, distortion, detection, etc., included under the general term of *rectification*. From the statistical viewpoint, these operations have their mathematical counterparts in corresponding transformations (e.g., translation, multiplication, differentiation, integration, etc.) *of the signal and noise ensembles*. In this way, the study of communication systems in a general sense becomes one of determining the statistical properties associated with the new ensembles that result from

† Superscript numerals are keyed to the References at the end of the chapter.

‡ For a definition of linear and nonlinear operation, see Sec. 2.1-3(2).

these various linear and nonlinear transformations. Just what transformations are important in a given situation depends, of course, on the purpose or intent of the particular communication process. Therefore, from a still broader viewpoint not only have we to consider ensemble properties—i.e., the statistical properties of signals and noise, including those for transformed ensembles—but we have also to determine which ensembles and transformations to select when the consequences of an action or decision are specifically made a part of system function itself. For this purpose, *statistical decision theory*[4] provides necessary concepts and techniques.[5,6]

Before moving on to examine the salient features of signal and noise ensembles, let us consider very briefly a few results of probability theory, after which it will be possible to discuss more precisely the notion of a stochastic process and to proceed thus to the main task of the book.

1.2 *Probability Distributions and Distribution Densities*†

In this section, we summarize in a formal way some of the results of probability theory that are required for our subsequent discussion; for a detailed and rigorous treatment, the reader is referred to standard works.‡ Let us begin, then, with the notion of a *random variable.*§

Consider a definite random experiment **E**, which may be repeated under similar (controlled) conditions a large number of times. Furthermore, let the result of each particular trial, or run, of this experiment be given by a (single real) quantity X. Then let us introduce a corresponding variable point X in a *probability*, or *measure, space* R_1. X is called a (one-dimensional) random variable. Accordingly, if a *value* x_k is associated with a particular event or outcome of the experiment **E** on a particular trial, we say that the random variable X has a *probability* $P(X = x_k) = p_k$ of assuming the realized value x_k. In this way, probability is defined as a number which is associated with a possible outcome of the random experiment **E**; p_k, of course, is equal to or greater than 0 but cannot exceed 1. In fact, $p_k = 0$ is to be interpreted that X takes the value x_j a negligible fraction of the total number of trials in the ensemble of possible runs, while $p_k = 1$ is equivalent to certainty that X has the value x_j on all (but a negligible portion of the total possible number of) runs in the ensemble.

The concept of random variable can be extended. If the outcome of a

† Section 1.2 may be omitted by the reader familiar with probability theory.

‡ Among these, see in particular Cramér.[7] The modern mathematical foundations and development of probability theory stem mainly from the work of Kolmogoroff,[8] Borel,[9] Fréchet,[10] and others. From a more applied point of view, we mention, for example, Uspensky[11] and Feller.[12] We shall use Cramér in this section as our principal reference for the detailed exposition and points of rigor as well as for an extensive bibliography of the pertinent mathematical literature.

§ As an introduction, see Cramér, *op. cit.*, chap. 13, and also Arley and Buch,[13] chaps. 1 and 2; fundamental definitions and axioms are considered in Cramér, chap. 14.

particular trial is the set of values $\mathbf{x} = (x_1, \ldots, x_n)$, we then introduce a corresponding variable *vector* point $\mathbf{X} = (X_1, \ldots, X_n)$ in probability, or measure, space R_n and say that $\mathbf{X}$ is a *vector* random variable, or an n-dimensional random variable. We remark, for both X and $\mathbf{X}$, that the distribution of the points in measure space, or, more suggestively, the "mass" (i.e., the p_k for X, etc., above), or weighting associated with the points of measure space, is equal to the *probability measure*, or, equivalently, the probability, associated with the corresponding set of values of the random variable.

A SINGLE RANDOM VARIABLE

1.2-1 Distribution Functions.† Consider a single random variable X; from the above remarks we write the distribution of X as the function

$$D(x) = P(X \leq x) \tag{1.1}$$

where $P(X \leq x)$ is the probability that X takes any one of the allowed values x, or less, on any given trial of the experiment $\mathbf{E}$; in short form, we write $D(x)$ = d.f. of X. The general properties that the distribution must possess‡ are

$$\begin{array}{lll}
(1) & D(x) \geq 0 & -\infty < x < \infty, \text{ all } x \\
(2) & 0 \leq D(x) \leq 1 & D(\infty) = 1,\ D(-\infty) = 0 \\
(3) & \text{Nondecreasing (point) function as } x \to \infty & \\
(4) & \text{Everywhere continuous to the right} &
\end{array} \tag{1.2}$$

The probability that our random variable X takes a value in the interval $(a < X \leq b)$ is thus

$$D(b) - D(a) = P(a < X \leq b) \tag{1.3}$$

The derivative of $D(x)$, when it exists, is defined by

$$\lim_{\epsilon \to 0} \frac{D(x + \epsilon) - D(x - \epsilon)}{2\epsilon} \to w(x) \tag{1.4}$$

if x is a point of continuity; $w(x)$ is the density of "mass" at x and is called the *probability density* (p.d.), or frequency function (f.f.), of the distribution D. Properties of $w(x)$, corresponding to those of $D(x)$, are easily deduced from the above.

$$\begin{array}{ll}
(1) & w(x) \geq 0 \qquad \text{all } x \\
(2) & \int_{-\infty}^{\infty} w(x)\, dx = 1 \\
(3) & D(x) = \int_{-\infty}^{x+} w(x)\, dx
\end{array} \tag{1.5}$$

The quantity $w(x)\, dx$ is the probability of the random variable X having

† Cramér, *op. cit.*, chap. 15.

‡ *Ibid.*, chap. 6, sec. 6.6.

the value x in the range $(x, x + dx)$ and is often called the *probability element* of x.†

1.2-2 Functions of a Random Variable.‡ Any (real) function $Y = g(X)$ of the (real) random variable X is itself a random variable, with its own probability measure. Thus, its distribution function is

$$D(y) = P(Y \leq y) \tag{1.6}$$

where y are the values corresponding to x, obtained from the transformation $y = g(x)$; $D(y)$ possesses exactly the same general properties [Eq. (1.2)] as does $D(x)$. These notions may be extended to complex functions $X = X_1 + iX_2$ (X_1, X_2 real), $Y = Y_1 + iY_2$ (Y_1, Y_2 real).§ One has now a *pair* of random variables X_1, X_2, or Y_1, Y_2, with their appropriate distributions, etc. (see Secs. 1.2-8 to 1.2-14).

1.2-3 Discrete and Continuous Distributions. Most of our applied problems fall into three classes, those possessing (1) discrete, (2) continuous,

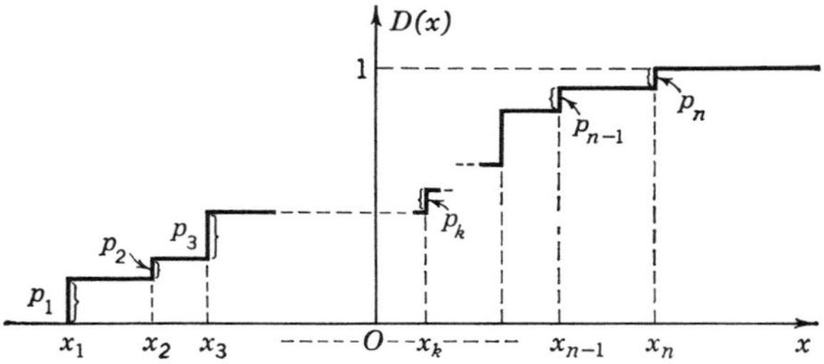

FIG. 1.1. Distribution of X in the discrete case.

or (3) mixed distributions. The discrete distribution (class 1) is one for which the weighting of the distribution of the random variable X is concentrated at a discrete set of points, such that any finite interval in measure space contains at most a finite number of such "mass" points. As an example, consider the mass points $x_1, x_2, \ldots, x_n$, with the corresponding "masses," or probabilities, $p_1, p_2, \ldots, p_n$, such that

$$\sum_{k=1}^{n} p_k = 1$$

(for unity measure), with $p_k \geq 0$ $(k = 1, \ldots, n)$. As shown in Fig. 1.1, the d.f. of X is here

$$D(x) = P(X \leq x) = \sum_{x_k \leq x} p_k \tag{1.7}$$

where $P(X = x_k) = p_k$ and $P(X \neq x_k)$ (all k) vanishes.

† Cf. *ibid.*, sec. 15.1, p. 167.
‡ *Ibid.*, sec. 14.5.
§ *Ibid.*

For random variables possessing a continuous distribution (class 2), we require that the d.f. is everywhere continuous, and that the f.f. $w(x) = D'(x)$ [cf. Eq. (1.4)] exists and is continuous for all x (except possibly for a finite number of points in any finite interval). Here

$$D(x) = P(X \leq x) = \int_{-\infty}^{x} w(x)\,dx \tag{1.8}$$

In the specific example of the *normal frequency function* $w(x) = e^{-x^2/2}/(2\pi)^{1/2}$ (all x), we have from Eq. (1.8) the corresponding d.f.

$$D(x) = \frac{1}{2}\left[1 + \Theta\left(\frac{x}{\sqrt{2}}\right)\right] \qquad \Theta(x) = \frac{2}{\sqrt{\pi}}\int_0^x e^{-y^2}\,dy \tag{1.9}$$

where Θ is the *error function;*† $D(x)$ is sketched in Fig. 1.2. Note that

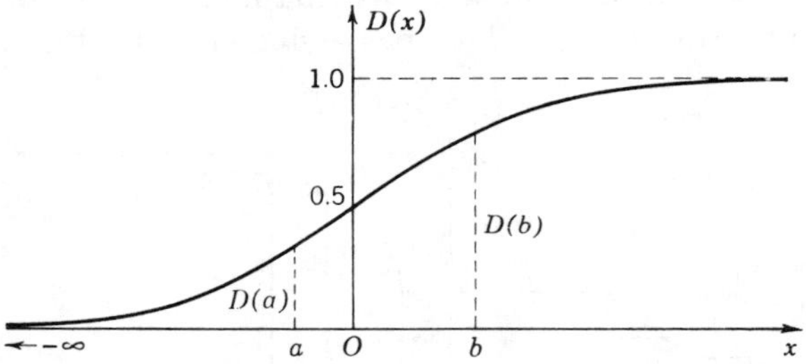

FIG. 1.2. The continuous distribution $D(x) = \frac{1}{2}[1 + \Theta(x/\sqrt{2})]$.

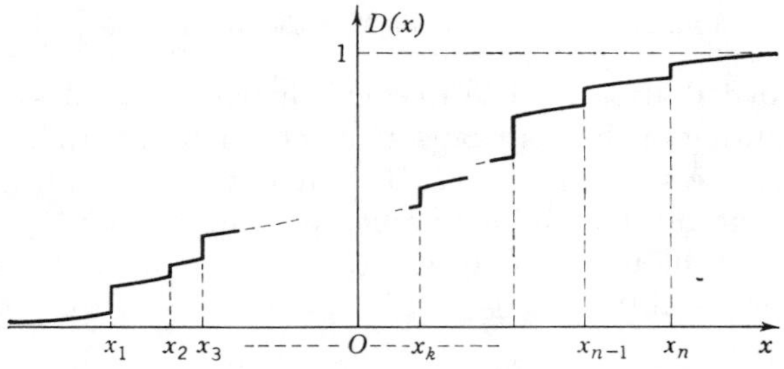

FIG. 1.3. Distribution of X in the mixed case (3).

the probability that X takes a value belonging to the interval (a,b) is $P(a < X \leq b) = D(b) - D(a)$, the difference of the two ordinates indicated in Fig. 1.2.

Finally, we have the case of the mixed distribution (class 3).‡

$$D(x) = d_1 D_1(x) + d_2 D_2(x) \qquad d_1, d_2 \geq 0;\ d_1 + d_2 = 1 \tag{1.10}$$

† For properties of this function, see, for example, W. Magnus and F. Oberhettinger, "Special Functions of Mathematical Physics," chap. 6, par. 4, Chelsea, New York, 1949, and also Tables of the Error Function and Its First Twenty Derivatives, *Ann. Harvard Computation Lab.*, January, 1952.

‡ Cramér, *op. cit.*, sec. 6.6, p. 58.

Here D_1 is a discrete distribution and D_2 a continuous one. In our two examples (classes 1 and 2), we see that $d_1 = 1$, $d_2 = 0$ and $d_1 = 0$, $d_2 = 1$, respectively, representing the two extremes of the mixed case. A typical distribution curve, combining these two examples according to Eq. (1.10), is shown in Fig. 1.3.

1.2-4 Distribution Densities. We have observed above in the case of continuous distributions that the d.f. of X possesses a derivative $D'(x) = w(x)$, called the frequency function, or d.d., of X, expressed in terms of the values x which the random variable X can assume on any one trial of the experiment **E**. For discrete distributions (cf. class 1), no such density function strictly exists. However, probabilities and d.f.'s may still be rigorously expressed in integral form with the help of the Riemann-Stieltjes, or Lebesgue-Stieltjes,† representations,

$$D(x) = \int_{[x]} dD(x) \tag{1.11a}$$

which become for operations on a function g of X, such as integration,

$$Z(a,b) \equiv \int_a^b g(x)\, dD(x) \tag{1.11b}$$

When the distribution is continuous, $dD(x)$ is simply the probability element $w(x)\, dx$ (cf. class 1).

Here, however, because of its convenience in manipulation and its familiarity to the engineer and physicist, we shall introduce the formalism of the Dirac singular ("impulse" or "delta") function δ to write *probability-density operators* [in place of $dD(x)$], which for all practical purposes give results identical with the distribution functions of the continuous cases and, upon integration, for the discrete cases as well.‡ Thus, we include, within the usual representation of a Riemann integral which combines the ordinary integration procedures with the properties (described below) of this density operator, the notion of probability density, extended formally to discrete distributions.§

The salient properties of the δ function needed here are briefly summarized.¶ We remark first that the δ function can be regarded as a limiting form of a sequence of functions:

† *Ibid.*, sec. 7.4 for L-S and sec. 7.5 for R-S representations. For more general discussions of the Lebesgue-Stieltjes integral, see chap. 7; for Lebesgue measure, chaps. 8, 9 (based on chaps. 1–3, 4–7).

‡ In this connection, no other meaning is to be imparted to this p.d. operator; from a probabilistic point of view, it is only the distribution functions in the discrete cases which can have physical significance.

§ Of course, the δ function has other interpretations under other, nonprobabilistic circumstances.

¶ See Van der Pol and Bremmer,[14] chap. 5, especially pp. 56–83. This reference contains a full discussion of the δ function, its interpretation, and its relation to the rigorous treatment. See also a paper by Clavier.[15] In addition, a particularly convenient reference for a systematic treatment is given by M. J. Lighthill.[15a]

$$\lim_{\alpha\to 0} \delta_\alpha(x - x') = \delta(x - x') \begin{cases} \text{when } \delta_\alpha = \dfrac{\alpha}{\pi[\alpha^2 + (x - x')^2]} & (1.12a) \\ \text{or} \quad \delta_\alpha = \dfrac{\sin \alpha^{-1}(x - x')}{\pi(x - x')} & (1.12b) \\ \text{or} \quad \delta_\alpha = \alpha^{-1}e^{-(x-x')/\alpha},\ x - x' > 0 & (1.12c) \end{cases}$$

with the properties that, as $\alpha \to 0$, $x = x'$, $\delta \to \infty$, while, if $x \neq x'$, then $\delta \to 0$, in such a way that

$$\int_a^b \delta(x - x')\,dx = \begin{cases} 1 & a < x' < b \\ \tfrac{1}{2} & a = x' \text{ or } b = x' \\ 0 & x' < a,\ x' > b\ (b > a) \end{cases} \tag{1.13}$$

where a symmetrical representation, like Eqs. (1.12a), (1.12b), is used; that is, $\delta(x - x') = \delta(x' - x)$. The functions are of two general types: in one case, the singularity appears as an infinite "spike" at a single point $x = x'$, with the function smooth elsewhere, while, in the other, the infinite spike still occurs at $x = x'$, but now the function oscillates rapidly everywhere except at this point. Note that the third example [Eq. (1.12c)] is a one-sided delta function in the limit. Here we shall always use the two-sided, or symmetrical, form, with the added convention that if the point $x' = a$ (or b) is to be included, i.e., if the interval (a,b) is to be closed, we write

$$\int_{a-\epsilon}^b \delta(x - x')\,dx = 1 \qquad \epsilon > 0$$

or

$$\int_{a-}^b \delta(x - x')\,dx = 1 \qquad x' = a \tag{1.14}$$

where ϵ is a very small positive number. Otherwise, Eq. (1.13) applies: the average value $\tfrac{1}{2}$ at $x' = a$ (or b) is taken. The principal property possessed by these functions, which justifies its formal use in applications, is

$$\int_a^b F(x)\,\delta(x - x')\,dx = \begin{cases} F(x') & a < x' < b \\ \tfrac{1}{2}F(x') & x' = a \text{ or } b \\ 0 & x' > b \text{ or } < a \end{cases} \tag{1.15}$$

and

$$\int_{a-}^b F(x)\,\delta(x - a)\,dx = F(a), \text{ etc.}$$

Particularly useful forms of the δ function are the integral representations

$$\int_{-\infty}^{\infty} e^{\pm 2\pi i x t}\,dt = \delta(x - 0) \qquad \int_{-\infty}^{\infty} e^{\pm 2\pi i (x-x')t}\,dt = \delta(x - x') \tag{1.16a}$$

which are simpler examples of the many-dimensional *vector* $\boldsymbol{\delta}$ function

$$\int \cdots \int_{-\infty}^{\infty} \exp\,[\pm 2\pi i(\mathbf{x} - \mathbf{x}') \cdot \mathbf{t}]\,dt_1 \cdots dt_n = \boldsymbol{\delta}(\mathbf{x} - \mathbf{x}') = \prod_{j=1}^{n} \delta(x_j - x_j') \tag{1.16b}$$

in which $(\mathbf{x} - \mathbf{x}') \cdot \mathbf{t} = \sum_{j=1}^{n} (x_j - x_j')t_j$, with our usual vector notation.† The delta function $\delta(x - x')$ is also recognized as the derivative of the discontinuous *unit function* $u(x - x') = 1$ $(x > x')$, $u(x - x') = 0$ $(x < x')$.

With the help of the above, we can now consider probability densities for discrete distributions. For class 1 in Sec. 1.2-3, we may write the p.d. of X as

$$w(x) = \sum_{k=1}^{n} p_k \, \delta(x - x_k) \tag{1.17}$$

From the properties (1.13), it is clear that the p_k are the "masses" associated with the values x_k of the random variable X, and $\delta(x - x_k)$ is an oper-

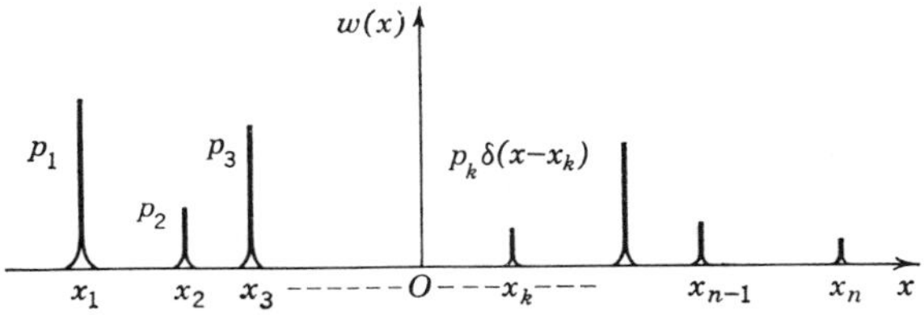

FIG. 1.4. The d.d. of X in the discrete case.

ator which picks out from the continuum of point values $[x]$ only that one, x_k, for which there is a nonzero weighting. Observe that

$$\int_{-\infty}^{\infty} w(x)\, dx = \sum_{k=1}^{n} p_k \int_{-\infty}^{\infty} \delta(x - x_k)\, dx = \sum_{k=1}^{n} p_k = 1 \tag{1.18}$$

as required: the measure is unity. Furthermore, the d.f. of X is now

$$D(x) = \int_{-\infty}^{x} w(x)\, dx = \sum_{k=1}^{n} p_k \int_{-\infty}^{x} \delta(x - x_k)\, dx = \sum_{x_k \le x} p_k \tag{1.19}$$

in agreement with Eq. (1.7). The distribution density [Eq. (1.17)] is sketched in Fig. 1.4; the relative heights of the "spikes" at $x = x_k$ are determined by the relative values of the weightings p_k. (The representation of Fig. 1.4 is purely conventional in that a finite, rather than an infinite, ordinate is used for convenience.) Our δ-function formalism

† For the properties, definitions, etc., of vectors and tensors, see, for example, H. Margenau and G. M. Murphy, "The Mathematics of Physics and Chemistry," chap. 5, Van Nostrand, Princeton, N.J., 1943; L. Page, "Introduction to Theoretical Physics," 2d ed., introduction, Van Nostrand, Princeton, N.J., 1935; and in particular P. M. Morse and H. Feshbach, "Methods of Theoretical Physics," secs. 1.1–1.6, McGraw-Hill, New York, 1953.

enables us also to treat mixed distributions [cf. class 3 in Sec. 1.2-3] in similar fashion: the distribution density is then a mixture of the type

$$w(x) = d_1 \sum_k^n p_k \, \delta(x - x_k) + d_2 w_2(x) \qquad d_1, d_2 > 0; d_1 + d_2 = 1 \quad (1.20)$$

where $w_2(x)$ belongs to the continuous part of the distribution $D(x)$, sketched already in Fig. 1.3.

1.2-5 Mean Values and Moments. Let $w(x)$ be the density function of a random variable X, which may exhibit either continuous or singular properties, or both, as described above. Consider now some real function $g(X)$ of the original random variable, integrable over $(-\infty, \infty)$ with respect to $w(x)$. We interpret†

$$\overline{g(X)} = \boldsymbol{E}\{g(X)\} = \int_{-\infty}^{\infty} g(x) w(x) \, dx \qquad (1.21)$$

as the *weighted mean*, *mean value*, or *expectation* of $g(X)$, with respect to the d.d. $w(x)$. Here $\boldsymbol{E}$ is the expectation operator, defined according to Eq. (1.21). Note that, for purely discrete distributions, from Eqs. (1.17) and (1.15) this becomes

$$\overline{g(X)} = \boldsymbol{E}\{g(X)\} = \sum_{k=1}^{n} g(x_k) p_k \qquad (1.22)$$

while for the mixed distribution [Eq. (1.10)] the weighting is of the type of Eq. (1.20), so that for $\overline{g(X)}$ one gets in addition to an expression like Eq. (1.22) an integral over the continuous portion of $w(x)$.

Letting $g(x) = |x|^\nu$ ($\nu > 0$), we write the *absolute* ν*th moment of* X as‡

$$\overline{|x|^\nu} = \int_{-\infty}^{\infty} |x|^\nu w(x) \, dx \qquad (1.23)$$

In most applications, $\nu = n$ is a positive integer, so that $\overline{g(x)} = \overline{x^n}$ is called the *nth moment* of X, when it exists, i.e., when the integral $\int_{-\infty}^{\infty} |x|^n w(x) \, dx$ is *absolutely convergent*. Moments of particular interest are $\bar{x}$, $\overline{x^2}$, the *mean* and *mean-square* values of X, while

$$\boldsymbol{E}\{(X - \boldsymbol{E}\{X\})^2\} = \overline{(x - \bar{x})^2} = \overline{x^2} - \bar{x}^2 = \int_{-\infty}^{\infty} (x - \bar{x})^2 w(x) \, dx \equiv \sigma_x^2 \quad (1.24)$$

is called the *variance* of X and σ_x its *standard deviation.* Moments of the type $\mu_n = \overline{(x - \bar{x})^n}$ ($n = 1, 2, \ldots$) are known as *central moments*.

† Cramér, *op. cit.*, sec. 15.3.

‡ Here and in the following discussion, we shall not make a special distinction between a random variable X and the values x that it may assume (cf. the beginning of Sec. 1.2), using X or x interchangeably, as the case may be.

1.2-6 The Characteristic Function.† A mean value of particular importance is represented by the *characteristic function*‡

$$E\{e^{i\xi X}\} = \int_{-\infty}^{\infty} w(x)e^{i\xi x}\,dx \equiv F_x(\xi) \tag{1.25}$$

with the properties, following immediately from the definition above,§

$$F_x(0) = 1 \qquad |F_x(\xi)| \le 1 \qquad F_x(\xi)^* = F_x(-\xi) \tag{1.26}$$

where the asterisk denotes the complex conjugate. Furthermore, we have¶

$$\begin{aligned}\int_{-\infty}^{\infty} F_x(\xi)e^{-i\xi x}\frac{d\xi}{2\pi} &= \iint_{-\infty}^{\infty} w(x')e^{i\xi(x'-x)}\,dx'\,\frac{d\xi}{2\pi} \\ &= \int_{-\infty}^{\infty} w(x')\,\delta(x'-x)\,dx' = w(x)\end{aligned} \tag{1.27}$$

from Eq. (1.15). Accordingly, *the characteristic function* (abbreviated c.f.) *and the distribution density are a Fourier-transform pair.*

$$F_x(\xi) = \overline{e^{i\xi x}} = \int_{-\infty}^{\infty} w(x)e^{i\xi x}\,dx \equiv \mathfrak{F}\{w(x)\} \qquad \text{by definition} \tag{1.28a}$$

$$w(x) = \int_{-\infty}^{\infty} F_x(\xi)e^{-i\xi x}\frac{d\xi}{2\pi} \equiv \mathfrak{F}^{-1}\{F_x(\xi)\} \tag{1.28b}$$

as indicated symbolically by the Fourier-transform operator $\mathfrak{F}$ and its inverse $\mathfrak{F}^{-1}$. Moreover, this relation is *unique:*‖ once the c.f. is given, so also is the corresponding d.d., and vice versa; that is, $\mathfrak{F}^{-1}\{\mathfrak{F}\} = \mathfrak{F}\{\mathfrak{F}^{-1}\}$.

The characteristic function is of great importance in applications, because it is usually analytically simpler to deal with than the corresponding p.d. function $w(x)$.

For the discrete distribution [Eqs. (1.7), (1.17)], we see that the c.f. is

$$F_x(\xi) = \sum_{k=1}^{n} p_k \int_{-\infty}^{\infty} e^{i\xi x}\,\delta(x - x_k)\,dx = \sum_{k=1}^{n} p_k e^{i\xi x_k} \tag{1.29}$$

and, conversely, inserting this into Eq. (1.28b), we get Eq. (1.17) once again, as expected. Observe that, for these singular d.d.'s, $F_x(\pm\infty)$ oscillates indefinitely. On the other hand, for continuous distributions, $F_x(\pm\infty)$

† Cramér, *op. cit.*, sec. 15.9. Mathematical proofs and properties are discussed in detail in chap. 10, especially secs. 10.1–10.4.

‡ This is a special case of the *moment-generating* function $E\{e^{\theta x}\}$, where $\theta = i\xi$.

§ We shall also write $F_x(i\xi)$ for $F_x(\xi)$ on occasion, emphasizing the fact that F_x is a function of $i\xi$.

¶ Cf. Cramér, *op. cit.*, sec. 10.3.

‖ *Loc. cit.*

vanishes. As an example, consider the *normal frequency function* $w(x) = e^{-(x-\bar{x})^2/2\sigma_x^2}(2\pi\sigma_x^2)^{-1/2}$; from Eq. (1.28*a*), one has

$$F_x(\xi) = \int_{-\infty}^{\infty} e^{i\xi x-(x-\bar{x})^2/2\sigma_x^2} \frac{dx}{(2\pi\sigma_x^2)^{1/2}} = e^{i\xi\bar{x}-\sigma_x^2\xi^2/2} \tag{1.30}$$

from which it is observed that Eq. (1.30) does indeed possess the general properties of Eq. (1.26), with $F_x(\pm\infty) = 0$.

When the c.f. possesses all derivatives about the point $\xi = 0$, or, equivalently, when all moments $\overline{x^n}$ $(n \geq 0)$ exist, we can write the Taylor's expansion

$$F_x(\xi) = \sum_{n=0}^{\infty} \mu_n' \frac{(i\xi)^n}{n!} = \sum_{n=0}^{\infty} \int_{-\infty}^{\infty} \frac{(i\xi)^n}{n!} x^n w(x)\, dx \tag{1.31a}$$

from the series for $e^{i\xi x}$. Thus,

$$\mu_n' = (-i)^n \frac{d^n}{d\xi^n} F_x \bigg|_{\xi=0} = \overline{x^n} \tag{1.31b}$$

showing that the coefficient of $(i\xi)^n/n!$ in the Taylor's series for the characteristic function is simply the nth moment of X. When $\bar{x} = 0$, then $\mu_n' = \mu_n = \overline{(x - \bar{x})^n}$. Even when the characteristic function is not expandable beyond a certain order n' of derivative, i.e., when moments of the distribution higher than n' do not exist, we can still get the lower-order moments by the above process. An example where *none* of the moments exists (absolutely) is given by the Cauchy distribution,† whose f.f. and c.f. are

$$w(x) = [\pi(1 + x^2)]^{-1} \qquad F_x(\xi) = e^{-|\xi|} \tag{1.32}$$

1.2-7 Generalizations; Semi-invariants. The definition of the characteristic function can be extended. We define the c.f. of any (real) random variable $g(X)$ as the mean value

$$F_g(\xi) = \boldsymbol{E}\{e^{i\xi g(X)}\} = \int_{-\infty}^{\infty} e^{i\xi g(x)} w(x)\, dx \tag{1.33}$$

Now, however, we no longer have the Fourier-transform relations (1.28) *in* x unless $g(x)$ is linear. Of course, with g as the new random variable, the Fourier-transform pair $\{F_g(\xi),W(g)\}$, corresponding to Eq. (1.28) with g instead of x, is defined as before. Accordingly, let us write with the help of Eqs. (1.28*b*) and (1.33)

$$\begin{aligned} W(g) &= \mathfrak{F}^{-1}\{F_g\} = \int_{-\infty}^{\infty} F_g(\xi)e^{-i\xi g} \frac{d\xi}{2\pi} = \int_{-\infty}^{\infty} \frac{d\xi}{2\pi} \int_{-\infty}^{\infty} w(x)e^{-i\xi[g-g(x)]}\, dx \\ &= \int_{-\infty}^{\infty} w(x)\, \delta[g - g(x)]\, dx \end{aligned} \tag{1.34}$$

† One can, however, define a mean, in the sense of a Cauchy principal value, for example, $\lim_{a\to\infty} \int_{-a}^{a} xw(x)\, dx$, which is seen to be zero for Eq. (1.32).

which gives the distribution density $W(g)$, once the transformation $g = g(x)$ and the d.d. $w(x)$ have been specified. The relation (1.34) may be interpreted as a mapping operation, with $\delta[g - g(x)]$ the mapping operator which takes the points on the x axis over into the new set g in the interval $[G]$, when $[G]$ is the portion of the (real) axis g, defined explicitly by the transformation $g = g(x)$. We shall find Eq. (1.34) and its extensions to two or more random variables especially useful in actual manipulations; for an example, see Prob. 1.1.

Closely related to the moments μ'_n (and the central moments μ_n) are the *semi-invariants*,† or cumulants, λ_m, which are obtained from the characteristic function by taking its logarithm and developing it in a Maclaurin series according to

$$\log F_x(\xi) = \sum_{m=1}^{\infty} \frac{\lambda_m (i\xi)^m}{m!} \tag{1.35}$$

By comparing the two series (1.31*a*) and (1.35), we can easily establish the relations between the various moments and semi-invariants (cf. Prob. 1.2).

TWO OR MORE RANDOM VARIABLES

In much of our subsequent work, we shall have to deal with more than one random variable and, in some cases, with a very large number. The extension of the preceding discussion to situations involving two or more random variables is briefly summarized, to indicate some of the important new features which appear under these conditions.

1.2-8 Two Random Variables.‡ If X and Y are two random variables, with x and y their respective values that can be assumed on any one trial, then we can write for the *joint* distribution $D(x,y)$ and the corresponding joint distribution density $W(x,y)$

$$D(x,y) = P(X \le x,\, Y \le y) = \int_{-\infty}^{x} \int_{-\infty}^{y} W(x,y)\, dx\, dy \tag{1.36}$$

with

$$\begin{aligned} D_1(x) &= \int_{-\infty}^{x} dx \int_{-\infty}^{\infty} W(x,y)\, dy = \int_{-\infty}^{x} w_1(x)\, dx \\ w_1(x) &= \int_{-\infty}^{\infty} W(x,y)\, dy \end{aligned} \tag{1.37}$$

and similar expressions for $D_2(y)$, $w_2(y)$. Here a statistical relationship is in general implied between X and Y; $D_1(x)$, $D_2(y)$ are the *marginal distributions* of X and Y, respectively, while $w_1(x)$, $w_2(y)$ are the corresponding *marginal* d.d.'s. [For discrete and mixed distributions, Eqs. (1.36), (1.37) still apply, with the aid of the formalism of Sec. 1.2-4.] The usual properties of Eqs. (1.2), (1.5), etc., are extended, in an obvious fashion, so that

† Cramér, *op. cit.*, sec. 15.10.

‡ *Ibid.*, chap. 21.

$D(x,y) \geq 0$, $D(\infty,\infty) = 1$. $D(b_1,b_2) + D(a_1,a_2) - D(a_1,b_2) - D(b_1,a_2)$ is similarly interpreted as the *joint* probability that the pair of random variables X, Y takes values in the regions $(a_1 < X \leq b_1)$, $(a_2 < Y \leq b_2)$, and so forth. For example, in the purely discrete case we can write

$$P(X = x_k;\ Y = y_l) = p_{kl} \qquad \text{with} \sum_{k,l}^{m,n} p_{kl} = 1 \tag{1.38}$$

for measure unity, (x_k,y_l) being the points at which the "masses" p_{kl} are situated. The marginal distributions of X and Y are here

$$p_{k(\ )} = P(X = x_k) = \sum_l p_{kl} \qquad p_{(\ )l} = P(Y = y_l) = \sum_k p_{kl} \tag{1.39}$$

One also has the characteristic function, with its unique Fourier transform†

$$\begin{aligned} F_{x,y}(\xi_1,\xi_2) &= \boldsymbol{E}\{e^{i\xi_1 X + i\xi_2 Y}\} = \iint\limits_{-\infty}^{\infty} e^{i\xi_1 x + i\xi_2 y} W(x,y)\, dx\, dy \\ &= \mathfrak{F}\{W(x,y)\} \end{aligned} \tag{1.40a}$$

$$\begin{aligned} W(x,y) &= \iint\limits_{-\infty}^{\infty} e^{-i\xi_1 x - i\xi_2 y}\, F_{x,y}(\xi_1,\xi_2)\, \frac{d\xi_1\, d\xi_2}{(2\pi)^2} \\ &= \mathfrak{F}^{-1}\{F_{x,y}(\xi_1,\xi_2)\} \end{aligned} \tag{1.40b}$$

and the *joint moments* (provided that they exist) are

$$\overline{x^k y^l} = \iint\limits_{-\infty}^{\infty} x^k y^l W(x,y)\, dx\, dy = (-i)^{k+l} \left. \frac{\partial^{k+l}}{\partial \xi_1{}^k\, \partial \xi_2{}^l} F_{x,y} \right|_{\xi_1 = \xi_2 = 0} \tag{1.41}$$

as obtained by the double Taylor's-series expansion of $e^{i\xi_1 x + i\xi_2 y}$ in Eq. (1.40*a*), etc. [cf. Eq. (1.31)]. The marginal c.f.'s are

$$F_x(\xi_1) = F_{x,y}(\xi_1,0) \qquad F_y(\xi_2) = F_{x,y}(0,\xi_2) \tag{1.42}$$

corresponding to $w_1(x)$, $w_2(y)$. Complex random variables can also be defined. If $Z = X + iY$ (X, Y real) and X, Y obey a joint distribution $D(x,y)$, then Z is a complex random variable whose distribution and distribution density are taken to be the joint distribution and d.d. of X and Y, since by definition the distribution and d.d. are always real and positive (or zero). Moments of Z, however, may be complex, viz.,

$$\boldsymbol{E}\{Z\} = \boldsymbol{E}\{X\} + i\boldsymbol{E}\{Y\} \tag{1.42a}$$

We shall encounter examples of complex random variables later when deal-

† *Ibid.*, sec. 21.3.

ing with the spectral properties of random processes with the help of Eq. (1.28*b*).

1.2-9 Conditional Distributions;† Mean Values and Moments. We can also define *conditional* distributions and their d.d.'s. Thus, the conditional distribution of a random variable Y relative to the hypothesis that another random variable X has a value in the range $(x, x + \epsilon)$ is‡ for continuous distributions

$$P(Y \leq y|x < X < x + \epsilon) = \int_x^{x+\epsilon} dx \int_{-\infty}^{y} \frac{W(x,y)\, dy}{\int_x^{x+\epsilon} dx \int_{-\infty}^{\infty} W(x,y)\, dy} \tag{1.43}$$

The corresponding frequency function for Y, given $X = x$, is

$$\begin{aligned} W(y|x) &= \frac{d}{dy} \lim_{\epsilon \to 0} P(Y \leq y|x < X < x + \epsilon) \\ &= \frac{d}{dy} \int_{-\infty}^{y} \frac{W(x,y')\, dy'}{w_1(x)} = \frac{W(x,y)}{w_1(x)} \end{aligned} \tag{1.44}$$

and with our formalism (cf. Sec. 1.2-4) we include the discrete and mixed cases as well. Thus, $W(y|x)\, dy$ is the probability that Y takes a value in the range $(y, y + dy)$ when X has the value x. The corresponding expression for $W(x|y)$ is obtained in the same way, so that we can write alternatively

$$W(x,y) = w_1(x)W(y|x) = w_2(y)W(x|y) \tag{1.45}$$

[$W(x|y)$ and $W(y|x)$ are, of course, different functions of their arguments, in general.] These conditional densities must also satisfy the usual conditions imposed by measurability and their probabilistic interpretation, viz.,

$$W(y|x) \geq 0 \qquad \int_{-\infty}^{\infty} W(y|x)\, dy = 1 \qquad \text{etc.} \tag{1.46}$$

Eliminating the joint density $W(x,y)$ with the aid of Eqs. (1.44), (1.45), we can write still another expression for the conditional densities:

$$W(y|x) = \frac{W(x,y)}{w_1(x)} = \frac{w_2(y)W(x|y)}{w_1(x)} \tag{1.47}$$

Conditional *averages* and *moments* can also be defined. Consider the conditional average of $g(Y)$, given $X = x$.

† *Ibid.*, sec. 14.3.

‡ Two notations for conditional distributions, densities, averages, etc., will be used throughout. In the first, the quantity following the vertical bar is the given quantity or represents the stated hypothesis, while that preceding this vertical bar is the argument of the function, for example, $W(y|x)$, x given, y the functional argument. The second notation just reverses that of the first. The former is common in mathematical work, while the latter is often used in physical applications (cf. Sec. 1.4 and Chap. 10). In any case, the proper interpretation will be clear in each instance.

$$\boldsymbol{E}\{g(Y)|X = x\} = \int_{-\infty}^{\infty} g(y)W(y|x)\,dy \tag{1.48a}$$

$$\boldsymbol{E}\{Y^m|X = x\} = \int_{-\infty}^{\infty} y^m W(y|x)\,dy \tag{1.48b}$$

This last is the conditional mth moment of the random variable Y when X takes the value x. $\boldsymbol{E}\{g(X)|Y = y\}$, $\boldsymbol{E}\{x^m|Y = y\}$, etc., are defined in the same way. As will become evident as we proceed, conditional distributions, densities, and averages are of fundamental importance in the study of random processes and in our particular applications of that study to statistical communication theory.

1.2-10 Statistical Independence.† It may happen that the mechanism underlying the random variable X in no way affects that of Y, and vice versa. We say then that X and Y are *statistically independent.* A necessary and sufficient condition that this be true mathematically is that the joint d.d. $W(x,y)$ factors into the two marginal densities $w_1(x)$, $w_2(y)$, viz.,

$$W(x,y) = w_1(x)w_2(y) \tag{1.49}$$

An alternative condition, equivalent to this, is that the c.f. $F_{x,y}(\xi_1,\xi_2)$ factors into the two marginal c.f.'s:

$$F_{x,y}(\xi_1,\xi_2) = F_x(\xi_1)F_y(\xi_2) \tag{1.50}$$

The extension to discrete and mixed d.d.'s is made as before. From Eqs. (1.45) or (1.47) and (1.49), we see that

$$W(y|x) = w_2(y) \qquad W(x|y) = w_1(x) \tag{1.51}$$

since here knowledge of x (or y) in no way influences the distribution of y (or x).

When all moments $\overline{x^k y^l}$ exist, a third condition, equivalent to the above [Eqs. (1.49), (1.50)] for statistical independence, is $\overline{x^k y^l} = \overline{x^k} \cdot \overline{y^l}$ (*all* $k, l \geq 0$). It is not enough that just the joint second moment factors, for example, $\overline{xy} = \bar{x} \cdot \bar{y}$, for the higher moments may not, and then X and Y are not statistically independent. An example is provided by the two random variables $X = \cos\phi$, $Y = \sin\phi$, where ϕ is uniformly distributed over the interval $(0 \leq \phi < 2\pi)$. Then, $\bar{x} = \bar{y} = 0$, $\overline{xy} = \bar{x} \cdot \bar{y} = 0$, but since $\overline{x^2} = \overline{y^2} = \frac{1}{2}$ and $\overline{x^2y^2} = \frac{1}{8} \neq \overline{x^2} \cdot \overline{y^2}$, we can say only that X and Y are *linearly* independent.

1.2-11 The Covariance. A quantity of considerable importance in subsequent applications is the *covariance* K_{xy}, defined by

$$K_{xy} = \boldsymbol{E}\{(X - \bar{x})(Y - \bar{y})\} = \overline{(x - \bar{x})(y - \bar{y})} \tag{1.52}$$

with the properties that $K_{xy} \leq \sigma_x\sigma_y$, $K_{xx} = \sigma_x^2$, $K_{yy} = \sigma_y^2$. If X and Y are statistically independent, K_{xy} vanishes, but if K_{xy} vanishes, the most we can say is that X and Y are linearly independent (cf. the example above).

† Cramér, *op. cit.*, secs. 15.11, 21.3.

In a similar way, for a pair of complex random variables $Z_1 = X_1 + iY_1$, $Z_2 = X_2 + iY_2$ (cf. Sec. 1.2-8) we can define a complex covariance by

$$K_{12} = \boldsymbol{E}\{(Z_1 - \bar{z}_1)(Z_2 - \bar{z}_2)^*\} = \overline{(z_1 - \bar{z}_1)(z_2 - \bar{z}_2)^*} \tag{1.52a}$$

$$K_{12} = \overline{(x_1 - \bar{x}_1)(x_2 + \bar{x}_2)} + \overline{(y_1 - \bar{y}_1)(y_2 - \bar{y}_2)} + i\{\overline{(x_2 - \bar{x}_2)(y_1 - \bar{y}_1)} - \overline{(y_2 - \bar{y}_2)(x_1 - \bar{x}_1)}\} \tag{1.52b}$$

$$K_{12} = K_{x_1x_2} + K_{y_1y_2} + i\{K_{x_2y_1} - K_{x_1y_2}\} \tag{1.52c}$$

Statistical independence of Z_1 and Z_2 requires that $W(x_1,y_1;x_2,y_2) = W(x_1,y_1)W(x_2,y_2)$, while for linear independence it is enough that all $K_{x_1x_2}$, $K_{y_1y_2}$, etc., in Eq. (1.52*c*) vanish.

1.2-12 Multivariate Cases.† The treatment of two random variables is readily extended to greater numbers. Such systems, of two or more random variables, are called *multivariate*, or *vector*, systems. A few of the important extensions are listed below, where it is now convenient to introduce matrix notation.‡ Thus, $\mathbf{y}$ is a column vector of n components; $\mathbf{K}$ is an $n \times n$ matrix; $\tilde{\mathbf{y}}$ is a transposed vector, i.e., a row vector, viz.,

$$\mathbf{y} = \begin{bmatrix} y_1 \\ y_2 \\ \cdots \\ y_k \\ \cdots \\ y_n \end{bmatrix} \qquad \tilde{\mathbf{y}} = [y_1,y_2, \ . \ . \ . \ ,y_k, \ . \ . \ . \ ,y_n] \qquad \mathbf{K} = \begin{bmatrix} K_{11} & K_{12} & \cdots & K_{1n} \\ K_{21} & K_{22} & \cdots & \cdots \\ \cdots & \cdots & \cdots & \cdots \\ K_{n1} & \cdots & \cdots & K_{nn} \end{bmatrix} \tag{1.53}$$

The tilde (∼) indicates the transposed matrix. The nth-order multivariate distribution D_n and distribution density W_n are given in the following relation:

$$D_n(y_1, \ . \ . \ . \ ,y_n) = \int_{-\infty}^{y_1} dy_1 \cdots \int_{-\infty}^{y_n} dy_n \, W_n(y_1, \ . \ . \ . \ ,y_n) \tag{1.54}$$

Similarly, we have a variety of marginal distributions and densities. For example, in the latter instance

$$W_m(y_1, \ . \ . \ . \ ,y_m) = \int_{-\infty}^{\infty} dy_{m+1} \cdots \int_{-\infty}^{\infty} dy_n \, W_n(y_1, \ . \ . \ . \ ,y_n) \qquad 1 \le m < n \tag{1.55}$$

with the usual properties $W_n \geq 0$, $W_m \geq 0$; $\int_{-\infty}^{\infty} dy_1 \cdots \int_{-\infty}^{\infty} dy_n \, W_n = 1$, etc. Various conditional distributions and densities can also be constructed along the lines of Sec. 1.2-9 [cf. Eqs. (1.45) et seq.].§ We shall see examples of these presently, in Sec. 1.3.

† *Ibid.*, chap. 22.

‡ See, for example, *ibid.*, chap. 11; Margenau and Murphy, *op. cit.*, chap. 10; and R. A. Frazer, W. J. Duncan, and A. R. Collar, "Elementary Matrices," chap. 1, Cambridge, New York, 1950; etc.

§ Cramér, *op. cit.*, sec. 22.1.

1.2-13 Characteristic Functions and Semi-invariants. Like the one-dimensional cases, the corresponding characteristic functions for W_n, D_n, etc., are defined as the expectation of $e^{i\tilde{\xi}\mathbf{Y}}$, viz.,

$$F_{y_1,\ldots,y_n}(\xi_1, \ldots, \xi_n) = E\{e^{i\tilde{\xi}\mathbf{Y}}\} = \int_{-\infty}^{\infty} \cdots \int d\mathbf{y}\, W_n(y_1, \ldots, y_n)e^{i\tilde{\xi}\mathbf{y}}$$
$$= F_{\mathbf{y}}(\boldsymbol{\xi}) = \mathfrak{F}\{W_n(y_1, \ldots, y_n)\} \tag{1.56a}$$

$$W_n(y_1, \ldots, y_n) = \int_{-\infty}^{\infty} \cdots \int e^{-i\tilde{\xi}\mathbf{y}} F_{\mathbf{y}}(\boldsymbol{\xi}) \frac{d\boldsymbol{\xi}}{(2\pi)^n} = \mathfrak{F}^{-1}\{F_{\mathbf{y}}(\boldsymbol{\xi})\} \tag{1.56b}$$

where we have used the abbreviation $d\mathbf{y}$ for $dy_1 \cdots dy_n$, $\boldsymbol{\xi}$ for $(\xi_1, \ldots, \xi_n)$, etc. Again $F_{\mathbf{y}}$ and W_n are uniquely related by the Fourier-transform pair (1.56). The moments, when they exist, follow from

$$\mu^{(M)\prime}_{m_1,\ldots,m_n} \equiv \overline{y_1^{m_1} \cdots y_n^{m_n}}$$
$$= (-i)^M \frac{\partial^M}{\partial \xi_1^{m_1} \cdots \partial \xi_n^{m_n}} F_y \bigg|_{\xi=0} \qquad M = \sum_{j=1}^{n} m_j \geq 0 \tag{1.57a}$$

$$\mu^{(M)\prime}_{m_1,\ldots,m_n} = \int_{-\infty}^{\infty} \cdots \int d\mathbf{y}\, y_1^{m_1} \cdots y_n^{m_n} W_n(y_1, \ldots, y_n) \tag{1.57b}$$

The semi-invariants, or cumulants $\lambda^{(L)}$, are found from

$$\log F_{\mathbf{y}}(\boldsymbol{\xi}) = \sum_{L=0}^{\infty} \frac{\lambda^{(L)}_{l_1,\ldots,l_n}(i\xi_1)^{l_1} \cdots (i\xi_n)^{l_n}}{l_1!l_2! \cdots l_n!} \qquad L = \sum_{j}^{n} l_j \geq 0 \tag{1.58}$$
$$\log F_{\mathbf{y}}(\boldsymbol{\xi}) \equiv \mathcal{K}(i\boldsymbol{\xi}) \tag{1.58a}$$

where $\mathcal{K}$ is the *cumulant function* and the cumulants $\lambda^{(L)}$ themselves can be expressed in terms of the moments $\mu^{(M)\prime}$ as in the one-variable case [cf. Sec. 1.2-7, Eq. (1.35)]. A necessary and sufficient condition that the y_k ($k = 1, \ldots, n$) be statistically independent is that the c.f. $F_{\mathbf{y}}(\boldsymbol{\xi})$ (or the d.d. W_n) factors be

$$F_{\mathbf{y}}(\boldsymbol{\xi}) = \prod_{k=1}^{n} F_{y_k}(\xi_k) \qquad \text{or} \qquad W_n(\mathbf{y}) = \prod_{k=1}^{n} w_k(y_k) \tag{1.59}$$

An example of general interest here and elsewhere is provided by the *normal*, or *gaussian*, distribution for the vector random variable $\mathbf{y} = (y_1, \ldots, y_n)$, defined by the cumulant function

$$\mathcal{K}(i\boldsymbol{\xi}) = i\tilde{\boldsymbol{\xi}}\bar{\mathbf{y}} - \tfrac{1}{2}\tilde{\boldsymbol{\xi}}\mathbf{K}\boldsymbol{\xi} \tag{1.60}$$

$\mathbf{K}$ is now the *covariance matrix*,†

† The covariance matrix is *positive definite*, that is, $\tilde{\mathbf{a}}\mathbf{K}\mathbf{a} \geq 0$, with equality only if $\mathbf{a} = 0$ identically, since if we set $\mathbf{y} - \bar{\mathbf{y}} = \mathbf{y}'$, then $\tilde{\mathbf{a}}\mathbf{K}\mathbf{a} = \sum_{kl} a_k a_l \overline{y_k' y_l'} = \overline{\left(\sum_k a_k y_k'\right)^2} \geq 0$.

$$\mathbf{K} = \|K_{kl}\| = \|\overline{(y_k - \bar{y}_k)(y_l - \bar{y}_l)}\| \qquad k, l = 1, \ldots, n \quad (1.60a)$$

The corresponding c.f. and d.d. become in this instance†

$$F_{\mathbf{y}}(\boldsymbol{\xi}) = e^{\mathcal{K}(i\boldsymbol{\xi})} = e^{i\tilde{\boldsymbol{\xi}}\bar{\mathbf{y}} - \frac{1}{2}\tilde{\boldsymbol{\xi}}\mathbf{K}\boldsymbol{\xi}} \quad (1.61a)$$

$$W_n(\mathbf{y}) = (2\pi)^{-n/2} (\det \mathbf{K})^{-\frac{1}{2}} \exp\left[-\tfrac{1}{2}(\tilde{\mathbf{y}} - \bar{\tilde{\mathbf{y}}})\mathbf{K}^{-1}(\mathbf{y} - \bar{\mathbf{y}})\right] \quad (1.61b)$$

with $\mathbf{K}^{-1}$ the inverse of $\mathbf{K}$ and det $\mathbf{K}$ the latter's determinant. In Chap. 7, we shall discuss the normal distribution and *normal process* in more detail.

1.2-14 Generalizations. As in the one- and two-variable cases (cf. Sec. 1.2-7), the concept of the characteristic function may be extended to a new vector random variable $\mathbf{G}$, obtained from $\mathbf{Y}$ by the set of transformations $\mathbf{g} = (g_1, \ldots, g_m)$ $(m \le n)$, where $g_k = g_k(y_1, \ldots, y_n)$. We write

$$F_{\mathbf{g}}(\boldsymbol{\xi}) = E\{e^{i\tilde{\boldsymbol{\xi}}\mathbf{G}(\mathbf{y})}\} = \int_{-\infty}^{\infty} \cdots \int dy_1 \cdots dy_n \, e^{i\xi_1 g_1 + \cdots + i\xi_m g_m} W_n(y_1, \ldots, y_n) \quad (1.62)$$

in which the corresponding p.d. of $\mathbf{G}$ is found to be

$$W_m(g_1, \ldots, g_m) = \int_{-\infty}^{\infty} \cdots \int W_n(y_1, \ldots, y_n) \times \boldsymbol{\delta}[\mathbf{g} - \mathbf{g}(y_1, \ldots, y_n)]\, dy_1 \cdots dy_n \quad (1.63)$$

and where $\boldsymbol{\delta}[\mathbf{g} - \mathbf{g}(\mathbf{y})]$ is the m-dimensional delta function $\delta[g_1 - g_1(\mathbf{y})]$ $\delta[g_2 - g_2(\mathbf{y})] \cdots \delta[g_m - g_m(\mathbf{y})]$ [cf. Eq. (1.16b)]. In conjunction with the integral form (1.16b) for $\boldsymbol{\delta}$, Eq. (1.63) provides a powerful method for dealing explicitly with many types of transformations and mappings.‡ Trans-

† See Cramér, *op. cit.*, chap. 11, sec. 11.12, for the explicit evaluation of Eq. (1.56b), to give Eq. (1.61b) when Eq. (1.61a) is the c.f., and also chap. 24. See Sec. 7.3-1 here.

‡ In the case $m = n$ and the y distribution of the continuous type, with the following conditions on the transformation satisfied:

1. g_k everywhere unique and continuous, with continuous partial derivatives $\partial g_l/\partial x_k$, $\partial x_l/\partial g_k$

2. The inverse transformation $y_k = h_k(g_1, \ldots, g_n)$ existing and unique

it can be shown that the probability element of $\mathbf{g}$ in G-space is obtained from the probability element of $\mathbf{y}$ according to

$$w_n(g_1, \ldots, g_n)dg_1 \ldots dg_n = W_n[h_1(g_1, \ldots, g_n), \ldots, h_n(g_1, \ldots, g_n)]\,|J|dy_1 \ldots dy_n$$

where J is the jacobian $\partial(y_1, \ldots, y_n)/\partial(g_1, \ldots, g_n)$ (Cramér, *op. cit.*, sec. 22.2). Even when $m < n$, we can use this approach as an alternative to the δ-function technique above, introducing $n - m$ additional relations $y_{k'} = h_{k'}(g_1, \ldots, g_n)$ $(k' = m + 1, \ldots, n)$. Apart from the conditions 1, 2, these relations are arbitrary and may be selected to facilitate the (subsequent) integrations and transformations. However, in most cases of interest to us, the transformations implied by Eq. (1.63) are more convenient (cf. Chaps. 17, 19, 21, 22).

formations of this type, where $m < n$, are sometimes called *irreversible transformations*, since they define a unique transformation from **Y**- to **G**-space, but *not* from **G**- to **Y**-space. That is, given $\mathbf{g} = (g_1, \ldots, g_m)$, it is not possible uniquely to find $\mathbf{y} = (y_1, \ldots, y_n)$, $y_k = y_k(g_1, \ldots, g_m)$.

As a very simple, but important, example, consider the problem of determining the d.d. of a random variable G, which is the sum of n *independent* random variables $Y_1, Y_2, \ldots, Y_n$. Thus, $g = \sum_1^n y_k$, and from Eq. (1.62) we write

$$F_g(\xi) = \overline{e^{i\xi g(y_1, \ldots, y_k)}} = \int \cdots \int_{-\infty}^{\infty} e^{i\xi \sum_k y_k} W_n(y_1, \ldots, y_n)\, dy_1 \cdots dy_n \tag{1.64}$$

Since the Y's are statistically independent, however, we can use Eq. (1.59) to write

$$F_g(\xi) = \prod_{k=1}^{n} \int_{-\infty}^{\infty} e^{i\xi y_k} w_k(y_k)\, dy_k = \prod_{k=1}^{n} F_{y_k}(\xi) \tag{1.65a}$$

Therefore
$$W_1(g) = \int_{-\infty}^{\infty} e^{-i\xi g} \prod_k^n F_{y_k}(\xi) \frac{d\xi}{2\pi} \tag{1.65b}$$

which is the desired distribution density.

So far, we have not introduced explicitly the notion of *process*—i.e., dependence on time—which is needed for most of our subsequent operations. This is simply done if we agree in our experiment **E** (and its mathematical counterpart) that the various values $y_1, \ldots, y_n$ that our vector random variable $\mathbf{Y} = (Y_1, \ldots, Y_n)$ can assume represent values observed at particular times $(t_1, \ldots, t_n)$. For example, $y_1 = y(t_1)$, $y_2 = y(t_2)$, etc., and the $y_1, \ldots, y_n$ are thus stochastic variables or functions of the time t, where for the moment t is allowed to have a discrete set of values $(t_1, \ldots, t_n)$. Often, we shall order these values of time, so that $t_1 \leq t_2 \leq \cdots \leq t_n$, and allow the t_k's to be chosen in a finite or infinite interval. The probability densities and characteristic functions (Sec. 1.2-6) are now more fully written

$$W_n(y_1,t_1; \ldots ;y_n,t_n);\ F_{y_1, \ldots, y_n}(\xi_1,t_1; \ldots ;\xi_n,t_n) = E\{e^{i\tilde{\boldsymbol{\xi}}\mathbf{Y}(t)}\} = \overline{e^{i\tilde{\boldsymbol{\xi}}\mathbf{y}(t)}} \tag{1.66}$$

exhibiting the explicit dependence on the n parameters $t_1, \ldots, t_n$. We shall see in the next section how these probability concepts may be used for the mathematical description of a random process.

PROBLEMS

1.1 (*a*) If the random variables X, Y are related by $Y = aX + b$, show that

$$D_Y(y) = \begin{cases} D\left(\frac{y-b}{a}\right) & a > 0 \\ 1 - D\left(\frac{y-b}{a}\right) & a < 0 \end{cases} \qquad W(y) = |a|^{-1} w\left(\frac{y-b}{a}\right) \tag{1}$$

(*b*) If $Y = X^2$, show that

$$D_Y(y) = \begin{cases} 0 & y < 0 \\ D(\sqrt{y}) - D(-\sqrt{y}) & y \geq 0 \end{cases} \tag{2}$$

$$W(y) = \begin{cases} w[x(y)] = 0 & y < 0 \\ \frac{1}{2\sqrt{y}}[w(\sqrt{y}) + w(-\sqrt{y})] & y > 0 \end{cases} \tag{3}$$

(*c*) If $Y = A_0 \cos(\omega_0 t + \phi)$, where ϕ is uniformly distributed in the interval $(0 \leq \phi \leq 2\pi)$, show that

$$F_y(\xi) = J_0(A_0\xi) \qquad W(y) = \begin{cases} (\pi\sqrt{A_0^2 - y^2})^{-1} & y < |A_0| \\ 0 & |y| > |A_0| \end{cases} \tag{4}$$

1.2 (*a*) In terms of the central moments μ_n of the random variable X, obtain the following relations between them and the semi-invariants λ_m:

$$\lambda_1 = \bar{x} \qquad \lambda_2 = \mu_2 = \sigma^2 \qquad \lambda_3 = \mu_3 \qquad \lambda_4 = \mu_4 - 3\mu_2^2$$
$$\lambda_5 = \mu_5 - 10\mu_2\mu_3 \qquad \lambda_6 = \mu_6 - 15\mu_2\mu_4 - 10\mu_3^2 + 30\mu_2^3$$

What are the corresponding expressions for μ_n in terms of the λ_m?

(*b*) For n independent random variables X_k $(k = 1, \ldots, n)$ whose first, second, and third moments, at least, exist, show for the random variable Z that is their sum that its mean, variance, and third central moment are

$$\bar{Z} = \bar{x}_1 + \bar{x}_2 + \cdots + \bar{x}_n \qquad \sigma_Z^2 = \sigma_1^2 + \sigma_2^2 + \cdots + \sigma_n^2$$
$$(\mu_3)_Z = \mu_3^{(1)} + \mu_3^{(2)} + \cdots \mu_3^{(n)} \tag{1}$$

(*c*) For the semi-invariants of Z in (*b*), show also that

$$(\lambda_m)_Z = \lambda_m^{(1)} + \lambda_m^{(2)} + \cdots + \lambda_m^{(n)} \qquad m \geq 1 \tag{2}$$

It is this simple linear combination which makes the semi-invariant so convenient in manipulations.

1.3 Let the random variable X_k take the (real) values $x_k = a$, with probability p, or $x_k = b$ $(\neq a)$, with probability q $(= 1 - p)$ on the kth trial of an experiment **E**, and let each trial be independent of every other. Show that

(*a*) The d.d. of $U = \sum_1^n X_k$ is

$$W(u) = \sum_0^n {}_nC_k p^k q^{n-k}\, \delta\{u - [ak + b(n-k)]\} \qquad {}_nC_k = n!/(n-k)!k! \tag{1}$$

What is the corresponding c.f.? (The distribution of U is called the *binomial* distribution.)

(b) $\mathbf{E}\{U\} = n(ap + bq)$; $\mathbf{E}\{U^2\} = n(a^2p + b^2q) + n(n-1)(ap + bq)^2$, and hence the variance is

$$\sigma_u^2 = npq(a - b)^2 \tag{2}$$

1.4 In Prob. 1.3, consider the case $p = \lambda/n$ [$\lambda > 0$ (fixed)], letting $n \to \infty$.

(a) Show that the d.d. of U, with values $u = a_k$ ($k \geq 0$), is the *Poisson* distribution density

$$W(u) = \sum_0^\infty \frac{\lambda^k e^{-\lambda}}{k!}\,\delta(u - a_k) \tag{1}$$

and that the c.f. when $a_k = ka$ is

$$F_u(\xi) = e^{\lambda(e^{i\xi a}-1)} \tag{2}$$

(b) Verify that $\mathbf{E}\{U\} = \lambda a$; $\mathbf{E}\{(U - \bar{u})^2\} = \sigma_u^2 = \lambda a^2$ ($a_k = ka$).

1.5 (a) Show that the sum of any number of independent, normally distributed random variables is itself normally distributed with

$$\text{Mean } \bar{u} = \sum_k \bar{x}_k$$

$$\text{Variance } \sigma_u^2 = \sum_k \sigma_k^2 \tag{1}$$

that is,

$$W(u) = \frac{e^{-(u-\bar{u})^2/2\sigma_u^2}}{\sqrt{2\pi\sigma_\mu^2}}$$

(b) Show that any linear function of independent normal random variables is itself normal.

1.6 (a) Let X be a normal random variable with zero mean and unity variance. First show that, for $Z = X^2$, $w(z) = e^{-z}/\sqrt{2\pi z}$ ($z > 0$), $w(z) = 0$ ($z < 0$), and then verify that its c.f. is $(1 - 2i\xi)^{-1/2}$.

(b) Now, letting $X_1, \ldots, X_n$ be n independent random variables of the above type, show that the d.d. of $U = \sum_k X_k^2 = \chi^2$ is

$$w_n(u) = \begin{cases} 2^{-n/2}\Gamma(n/2)^{-1}e^{-u/2}u^{n/2-1} & u > 0 \\ 0 & u < 0 \end{cases} \tag{1}$$

Sketch the curves of $w_n(u)$ for $n = 0, 1, 2, 3, 4$. [Equation (1) is closely related to the χ^2 distribution.]

1.7 (a) Show that if one has $n + 1$ independent normal random variables $X, X_1, \ldots, X_n$, each with zero means and variance σ^2, the random variable $U = X/Z$, $Z = \sum_{k=1}^n X_k^2$, has the d.d.

$$S_n(u) = \frac{1}{\sqrt{\pi n}}\,\Gamma\left(\frac{n+1}{2}\right)\left(1 + \frac{u^2}{n}\right)^{-(n+1)/2} \tag{1}$$

Note that $S_n(u)$ does not depend on σ^2. [Equation (1) is known as Student's d.d.]

(b) Obtain the c.f. for $S_n(u)$, viz.,

$$F_u(\xi) = 2^{1-n/2}\Gamma(n/2)^{-1}(n|\xi|)^{n/2}K_{n/2}(\sqrt{n}|\xi|) \tag{2}$$

where K is a modified Bessel function of the second kind.† Note that, for $n = 1$, $F_u(\xi) = e^{-|\xi|}$ of the Cauchy d.d. [Eq. (1.32)].

(c) Verify that

$$\overline{u^{2m}} = \frac{(2m)!n^m}{2^m m!(n-2)(n-4)\cdots(n-2m)} \qquad m \leq \frac{n-1}{2};\ \overline{u^{2m+1}} = 0 \tag{3}$$

1.8 (a) If $F_{y_1,y_2}(\xi_1,\xi_2) = \exp[-\psi(\xi_1^2 + \xi_2^2 + 2\xi_1\xi_2\rho)/2]$ is the characteristic function of the two random variables Y_1, Y_2, show that

$$W_2(y_1,y_2) = \frac{\exp[-(y_1^2 + y_2^2 - 2\rho y_1 y_2)/2\psi(1-\rho^2)]}{2\pi\psi\sqrt{1-\rho^2}} \tag{1}$$

(W_2 is now the *bivariate* normal distribution density of y_1 and y_2.)

(b) Verify that

$$\bar{y}_1 = \bar{y}_2 = 0 \qquad \overline{y_1^2} = \overline{y_2^2} = \psi \qquad \overline{y_1 y_2} = \psi\rho \tag{2}$$

What is the covariance matrix? Find the semi-invariants.

(c) Obtain the conditional probability density and its associated characteristic function.

$$W_2(y_2|y_1) = (\sqrt{2\pi\psi}\sqrt{1-\rho^2})^{-1}\exp[-(y_2 - \rho y_1)^2/2\psi(1-\rho^2)] \tag{3a}$$
$$F_{y_2}(\xi_2|y_1) = \exp[i\xi_2\rho y_1 - \psi(1-\rho^2)\xi_2^2/2] \tag{3b}$$

(d) Verify that the d.d. in the more general case for which

$$F_{y_1,y_2}(\xi_1,\xi_2) = \exp[i\xi_1\bar{y}_1 + i\xi_2\bar{y}_2 - \tfrac{1}{2}(\sigma_1^2\xi_1^2 + \sigma_2^2\xi_2^2 + 2\rho\sigma_1\sigma_2\xi_1\xi_2)] \tag{4a}$$

is

$$W_2(y_1,y_2) = [2\pi\sigma_1\sigma_2(1-\rho^2)^{1/2}]\exp\frac{-\left(\frac{y_1-\bar{y}_1}{\sigma_1}\right)^2 - \left(\frac{y_2-\bar{y}_2}{\sigma_2}\right)^2 + 2\rho\frac{y_1-\bar{y}_1}{\sigma_1}\frac{y_2-\bar{y}_2}{\sigma_2}}{2(1-\rho^2)} \tag{4b}$$

1.3 *Description of Random Processes*[16–18]

As mentioned in Sec. 1.1, a *random* or *stochastic process* from the physical point of view is a process in time, occurring in the real world and governed at least in part by a random mechanism. From the mathematical viewpoint, a stochastic process is an ensemble of events in time for which there exists a probability measure, descriptive of the statistical properties, or regularities, that can be exhibited in the real world in a long series of trials under similar conditions. Of course, in practice one almost never has the complete ensemble physically, but only a few members (i.e., a *subensemble*), from which, however, it is often possible to deduce the statistical properties characteristic of the corresponding mathematical model. This analytical description, like all quantitative accounts of the physical world, is at best an approximation, albeit a good one if sufficient pertinent data can be obtained.

More concisely, then, we may say that a *stochastic process*, $y(t)$, is an

† Cf. G. N. Watson, "Theory of Bessel Functions," 2d ed., pp. 78–80, Macmillan, New York, 1944.

ensemble, or set, of real-valued† functions of time t, for example, $y^{(1)}(t)$, $y^{(2)}(t), \ldots, y^{(j)}(t), \ldots, y^{(M)}(t)$ (M finite or infinite), together with a suitable probability measure by which we may determine the probability that any one member (or group of members) of the ensemble has certain observable properties.‡ By suitable probability measure here is meant one of total measure unity, obeying the usual conditions for measurability (see Sec. 1.2 and references). A function $y^{(j)}(t)$ belonging to the process is called a *member function* of the ensemble, or a *representation*, and the process itself $y(t)$ is a *stochastic* or *random variable*, whose distribution and frequency functions accordingly exhibit the same general properties as the random variables discussed in Sec. 1.2.§ The member functions $y^{(j)}(t)$ may be defined for discrete values of the time t, namely $t_1, t_2, \ldots$, on a finite interval $(0,T)$ or on an infinite one $(-\infty,\infty)$; or they may be defined over a continuum of values of t for such intervals. The former is usually called a *random series* and the latter a *random process* (with the term process applied as well in the more general sense given above).

For example, consider the ensemble defined by the set of sinusoids

$$y(t,\phi) = A_0 \cos(\omega_0 t + \phi) \tag{1.67}$$

where A_0 and ω_0 are fixed and the phase ϕ has random properties. The ensemble is also defined as a collection of different *units*, $y^{(k)}(t)$ ($k = 1, \ldots, N$), and these units (indicated by the superscript k) are taken to be the points of our probability or measure space (sometimes called *index space*), upon which the probability measure characteristic of a stochastic process is defined. Each of these different units may consist of one or more of the member functions $y^{(j)}(t)$ ($j = 1, \ldots, M$) of the ensemble, so that $M \geq N$, which occurs when various of the ensemble members are identical. Then, with each unit (or set of units) is associated a number, which is the probability that on any one trial a member function $y^{(j)}(t)$ is selected which has the particular value $y^{(k)}(t)$ associated with the unit k.

To illustrate this, let us suppose that the ensemble (1.67) contains a finite number N of units; that is, N different values of y, namely, $y^{(k)}(t)$ ($k = 1, \ldots, N$) are possible [for any allowed t in the selected time interval $(0,T)$, $(-\infty,\infty)$, etc.], depending on a discrete set (M) of values $\phi^{(j)}$, say, in the interval $(0 \leq \phi \leq 2\pi)$. If we require further that to every distinct unit k there corresponds a single value of $\phi^{(j)}$, then $j = k$, $M = N$, and all representations are equally important. Consequently, the probability of pick-

† Only real-valued functions are considered here for the time being.

‡ For our purposes, these are to be identified ultimately with measurable quantities in the physical world.

§ Henceforth, we shall not distinguish between the random variable Y, or $Y(t)$, and its values $y(t)$. This identity always implies (at least a possible) experiment whereby the values $y(t)$ can be observed physically. See Sec. 1.2.

ing the jth member on a given trial is the same as that for any other member i; that is, $p_j = p_i = M^{-1} = N^{-1}$. Similarly, the probability that on a given observation a member function is selected belonging to the subensemble $(j, j + l)$, say, is simply $(l + 1)/M$, and so forth.

Usually, however, there are more member functions $y^{(j)}$ in an ensemble than there are units $y^{(k)}$; that is, $M > N$. In our example above, this corresponds to several identical representations for each distinct value of ϕ. Here not every unit is equally significant, and thus different probabilities or weightings p_k (≥ 0) may exist for each $k = 1, \ldots, N$, with $\sum_{k=1}^{N} p_k = 1$ for unity measure. A similar argument can be applied to the cases where ϕ has a continuum of values, and the ensemble an infinite set of distinguishable units, as well as member functions. In many cases, the probability distributions and probability densities are then continuous or can be represented by our formalism of Sec. 1.2-4, the process in question being assumed measurable. Our next example is typical.

We remark, finally, that the representations of the ensemble need *not* be entirely random; members containing definite periodic or aperiodic components are not excluded. In fact, sets of functions exhibiting no random mechanism at all, e.g., Eq. (1.67), where ϕ also has but a single, specified value, are included as limiting cases in the general notion of ensemble. These are ensembles of a single unit, where all members are identical. Of course, not any ensemble can be used to describe a process: we must place some restrictions on the ensemble properties, namely, that the probability distributions for the function values do exist. When this is so, we have a suitable random process from which predictions and estimates can be made.

1.3-1 A Random Noise Process. To show how the various distribution densities which embody the statistical properties of the ensemble may in principle be obtained for a stochastic process, let us consider as our second example the case of an ensemble of random noise voltages $y(t)$. Here $y^{(j)}(t)$ is a typical member function, such as might be observed at the output of a cathode-ray oscilloscope when shot or thermal noise voltages are applied at the input of the device. Assuming the existence of this ensemble, then, and observing that y can take a continuum of values, we exclude for the time being any deterministic—i.e., steady, periodic, or nonrandom aperiodic—components, so that $y(t)$ is a *continuous, entirely random process.* [Later on, we shall deal with *mixed* processes, having deterministic components. These, we may expect, are described in terms of mixed densities (cf. Sec. 1.2-4).]

Consider now the ensemble shown in Fig. 1.5, consisting of representations specified in the interval $(t_0, t_0 + T)$, and let us begin by determining the first-order p.d. $W_1(y_1,t_1)$. First let the ensemble contain a finite number M of these representations, where M is very large. Next, divide y into a set

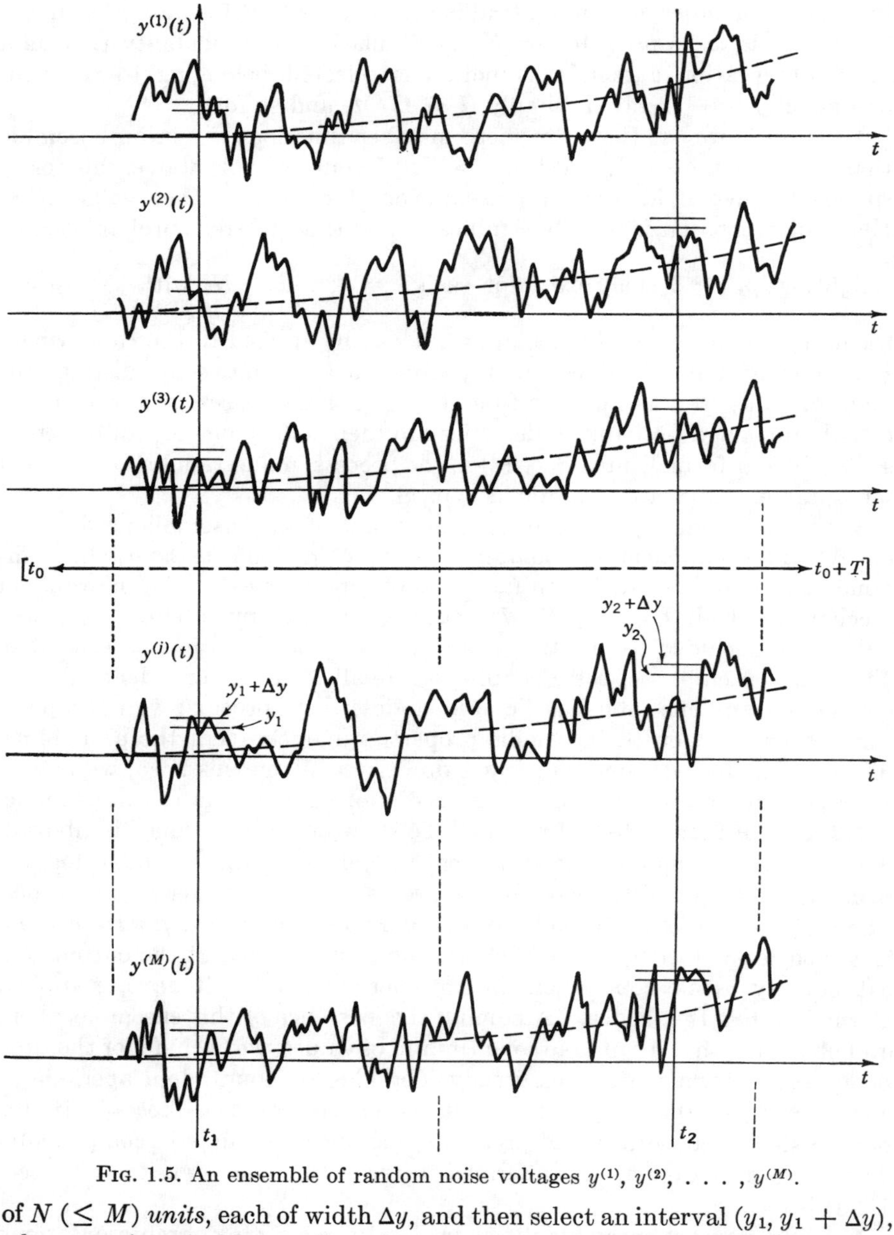

FIG. 1.5. An ensemble of random noise voltages $y^{(1)}, y^{(2)}, \ldots, y^{(M)}$.

of N ($\leq M$) *units*, each of width Δy, and then select an interval $(y_1, y_1 + \Delta y)$, where Δy is suitably small and $y_1 = l\,\Delta y$. We then count the number $n_1(M)$ of member functions $y^{(j)}$ ($j = 1, \ldots, m, \ldots, M$) which at time t_1 fall in the interval $(y_1,y,+\Delta y)$, i.e., which belong to the unit corresponding to $(y_1,y,+\Delta y)$. The probability measure associated with this interval or unit is then simply

$$P(y_1 < y < y_1 + \Delta y; t_1) = [n_1(M)/M]_{t_1} \tag{1.68}$$

Taking the limits $M \to \infty$, $\Delta y \to dy$, we assume that $n_1(M) \to \infty$ in such a way that†

$$\lim_{M,\Delta y} P(y_1 < y < y_1 + \Delta y; t_1) \to W_1(y_1,t_1)\, dy_1$$
$$= \text{probability that } y \text{ at time } t_1 \text{ falls in interval } (y_1, y_1 + dy) \tag{1.69}$$

Repeating this counting operation for each unit $(y_1, y_1 + \Delta y)$, allowing y_1 to range over all possible values, then gives in the limit† the desired probability density $W_1(y_1,t_1)$.

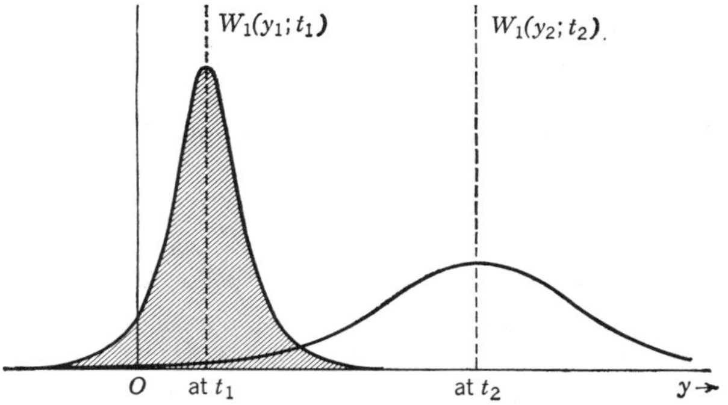

FIG. 1.6. First-order d.d.'s at two different times (cf. Fig. 1.5).

A similar procedure is followed for the second-order density $W_2(y_1,t_1;y_2,t_2)$. Here we select as a typical unit the *joint* interval $(y_1, y_1 + \Delta y)$, $(y_2, y_2 + \Delta y)$ and then count the number of pairs $n_{12}(M)$ of crossings in the two regions (cf. Fig. 1.5) for all possible values of (y_1,y_2). We accordingly define the joint probability $P(y_1 < y < y_1 + \Delta y; t_1; y_2 < y < y_2 + \Delta y; t_2) = n_{12}(M)/M$, which in the limit $M \to \infty$, $\Delta y \to dy_1, dy_2$ is assumed to become

$$\lim_{\substack{M\to\infty \\ \Delta y \to dy_1, dy_2}} P(y_1 < y < \Delta y + y_1; t_1; y_2 < y < y_2 + \Delta y; t_2)$$
$$\to W_2(y_1,t_1;y_2,t_2)\, dy_1\, dy_2$$
$$= \textit{joint} \text{ probability that } y \text{ takes values in range } (y_1, y_1 + dy_1) \text{ at time } t_1 \textit{ and} \text{ values in interval } (y_2, y_2 + dy_2) \text{ at time } t_2 \tag{1.70}$$

The higher-order densities are obtained in a similar fashion. The counting process is laborious, even for W_2, while for W_n $(n \geq 3)$ it may prove entirely impractical in actual physical situations, although conceptually possible in the mathematical model. It is important to note the times $(t_1,t_2, \ldots ,t_n)$

† For the physical systems considered here, this limit will always exist, at least in the sense of Sec. 1.2-4, within our powers of observation. For the corresponding mathematical model above, we *postulate* the existence of these limits.

at which these probabilities are computed. The actual structure of the d.d.'s may be quite different for different times, as shown, for example, in Fig. 1.6.

1.3-2 Mathematical Description of a Random Process. The hierarchy of distribution (densities) $W_1, W_2, \ldots, W_n$, as $n \rightarrow \infty$, accordingly provides the mathematical description of the random process.† Since the $W_1, W_2, \ldots, W_{n-1}$ are marginal d.d.'s of W_n, all statistical information for the process y is essentially contained in W_n, in general as $n \rightarrow \infty$. Knowing W_n, as $n \rightarrow \infty$, we may say that the process is *completely*‡ *described* {provided, of course, that certain conditions of regularity for the realizations $y^{(1)}(t)$, $y^{(2)}(t)$, etc., are imposed, so that these distribution densities exist [Eqs. (1.69) and (1.70)], for all t in the interval $(t_0, t_0 + T)$}. It sometimes happens that knowledge of W_1 or W_2 is sufficient to give us W_n for all n, so that a complete description in the above sense follows. We shall see examples of this in Sec. 1.4.

1.3-3 Some Other Types of Process. Besides the continuous random process (a) discussed above, there are several other categories to be distinguished (cf. Fig. 1.7). For instance, if we allow the stochastic variable $y(t)$ to take a continuum of values but restrict the parameter t to a discrete set $t_1, \ldots, t_n$ (which may be finite or infinite), we call $y(t)$ a *continuous random series* (b). An important example is the classical problem of the *random walk*,[19,20] where the "steps," or displacements, at any one move may take a value $y^{(j)}(t_k)$ from a continuum $(a < y < b)$, say, but these moves are made only at the discrete instants $t_1, t_2, \ldots, t_k$, etc. A third possibility arises when y is restricted as well to a discrete set of values $y(t_k)_1$, $y(t_k)_2$, $y(t_k)_3$, . . . at discrete times t_k. Such processes are often called *discrete random series* (c), one instance of which is provided by a sequence of tossed coins. The t_k represent the times at which the coin is tossed; the two possible values $y(t_k)_1$, $y(t_k)_2$ correspond to heads and tails, respectively. Finally, there are the cases where y again can take only a set of discrete values, but now t represents a continuum such as occurs, for example, in the output of a clamped servomechanism, or counter system, etc. (where time moves on continuously and the output of such systems maintains one of a *discrete* set of possible values for various periods, changing abruptly to new ones). These are *discrete random processes* (d).§

For most of the applications in this book, we shall be concerned with the continuous random series (b) and with the continuous random process (a).

† The description is equally well given in terms of the characteristic functions $F_{y_1}(\xi_1,t_1)$, $F_{y_1y_2}(\xi_1,t_1;\xi_2,t_2)$, etc.

‡ Strictly speaking, we must require, in addition, the topological property of separability of the process (see Doob,[16] chap. 2, sec. 2).

§ In distinguishing the various types, it is helpful to observe that the terms discrete and continuous refer to the stochastic variable y, while process and series are applied to the parameter t.

Besides providing a useful model of the detailed structure of various noise mechanisms (cf. Part 2), processes of type *b* occur frequently when continuous data are quantized (in time) by sampling procedures (cf. Part 4). On the other hand, the continuous process is appropriate for the macroscopic treatment of electronic noise and other fluctuation phenomena, observed at a scale well above the level where the fundamentally discrete nature of the physical mechanism is apparent. It is appropriate, also, to many situations in communication theory where continuous (and discrete) data are processed

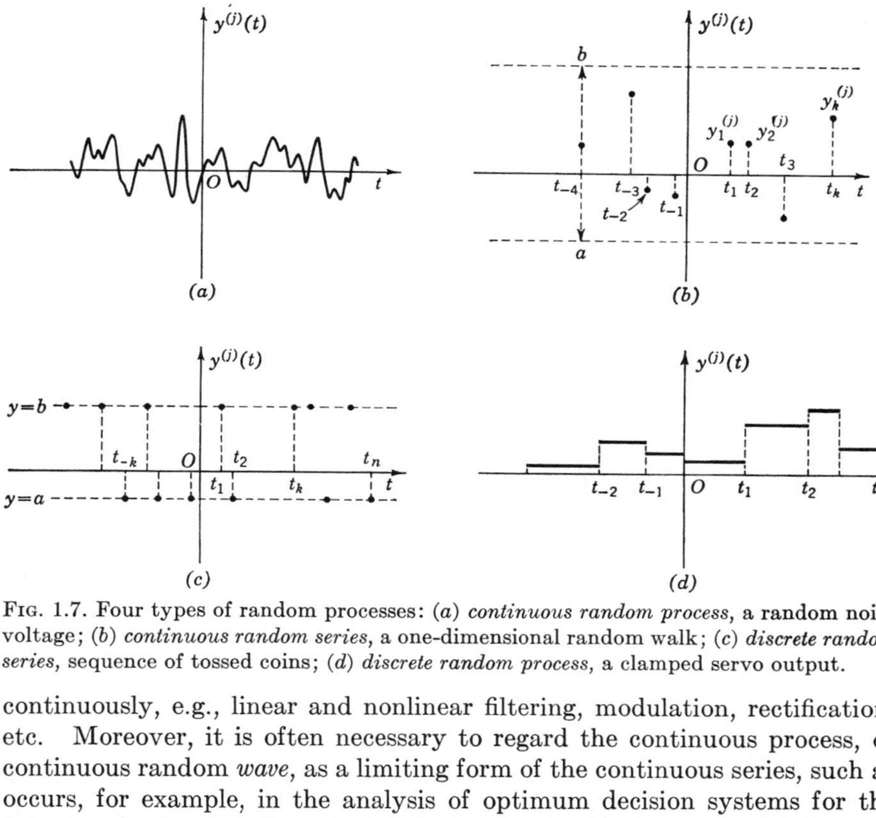

FIG. 1.7. Four types of random processes: (*a*) *continuous random process,* a random noise voltage; (*b*) *continuous random series,* a one-dimensional random walk; (*c*) *discrete random series,* sequence of tossed coins; (*d*) *discrete random process,* a clamped servo output.

continuously, e.g., linear and nonlinear filtering, modulation, rectification, etc. Moreover, it is often necessary to regard the continuous process, or continuous random *wave,* as a limiting form of the continuous series, such as occurs, for example, in the analysis of optimum decision systems for the detection and extraction of signals from noise backgrounds [cf. Part 4, Chaps. 18 to 20]. We remark that one can, of course, equally well discuss these processes in terms of their distributions, instead of their distribution densities, although the δ formalism (Sec. 1.2-4) provides the appropriate d.d. for "pure" processes, and for mixed ones as well, containing deterministic components, e.g., steady periodic terms, etc. The more refined mathematical concepts[16,18,21–25] of a random process, besides the comparatively simple notions of continuity, introduce questions of bounded variation, differentiability, convergence in probability, expandability, etc., to which for our purposes we shall here give no more than the briefest of treatments (cf. Secs. 2.1-1, 2.1-2); the interested reader is referred to the appropriate literature.

1.3-4 Deterministic Processes. Besides the entirely random processes (*a*) to (*d*) (Secs. 1.3-2, 1.3-3), there are degenerate forms of such processes. Consider, for example, the process $y(t) = A_0 \cos(\omega_0 t + \phi)$, where A_0, ω_0 are fixed and ϕ has some d.d. in $(0,2\pi)$. Then, representations or realized values of ϕ [for example, $\phi^{(j)}$ $(j = 1, \ldots)$] can be deduced by measurements of $y^{(j)}(t)$, and the future behavior of each $y^{(j)}(t)$, corresponding to each $\phi^{(j)}$, is accordingly specified exactly. When a finite number (here two) of such measurements determines the future behavior of any representation of the ensemble, the process is *completely deterministic*. Deterministic processes possess a definite functional dependence on time, while their random character appears parametrically. We shall normally reserve the term stochastic or random process for those cases, e.g., thermal noise, shot noise, etc., to which such definite functional structures (in time) cannot be given and shall call those processes "mixed" which contain in addition a deterministic component. A typical example of interest is the additive mixture of noise and signal ensembles, representations of which constitute the received waves in many communication problems. Sometimes, too, the mixture is multiplicative, as well as additive: the desired signal may be a noise-modulated carrier, for example, $N_1(t) \cos(\omega_0 t + \phi)$, appearing in a noise background $N_2(t)$, and so forth. A variety of mixed processes will occur in practical applications, as we shall see in subsequent chapters.

1.3-5 Signals and Noise. We define in general a *signal* to be any *desired* component of a transmitted or received wave, while *noise* is the accompanying *undesired* component. Signals may be deterministic or entirely random, while the noise, usually random, may in many cases possess the deterministic properties frequently associated with signal ensembles. In particular problems, it will be evident which is which.

We may use these ideas to describe the signal types with which we have to deal. Writing

$$y_s(t) = As(t,\epsilon;\boldsymbol{\theta}) \tag{1.71}$$

for a general signal ensemble, we note that A is a measure of the power content (i.e., scale in amplitude) of this signal, while $s(t)$, a normalized waveform, indicates its structure in time, which is assumed given. Here $\boldsymbol{\theta}$ represents all other descriptive parameters, which may or may not possess continuous or discrete p.d.'s, while ϵ is an *epoch*, relating the wave in time to some arbitrary origin, as shown, for example, in the ensemble of Fig. 1.8. The epoch is measured from some definite (i.e., functionally determined) point in the waveform to some chosen origin of time.

Suppose $\boldsymbol{\theta}$ has measure unity for $\boldsymbol{\theta} = \theta_0$. A signal ensemble of (infinitely) many units is produced if we allow ϵ to have a (continuous) d.d. in the interval ($a < \epsilon < b$). For periodic waves, this usually becomes ($0 < \epsilon < T_0$), where T_0 is the period of a typical member. For *aperiodic* waves, a similar situation exists for $0 < \epsilon < T_s$, with a p.d. for ϵ in this interval. We observe also that the descriptive parameters ($A,\epsilon,\boldsymbol{\theta}$) may be themselves functions of time, albeit stochastic ones. For example, our signal may be the triangular wave shown in Fig. 1.8, but with random periods (for example, T_0 possesses a d.d.); or the periods may be fixed [$w(T) = \delta(T - T_0)$], but the amplitude (i.e., scale) of the wave may be subject to random variations, either from member to member (but determined for any one representation) or from time to time (for each member of the ensemble). Still other possibilities

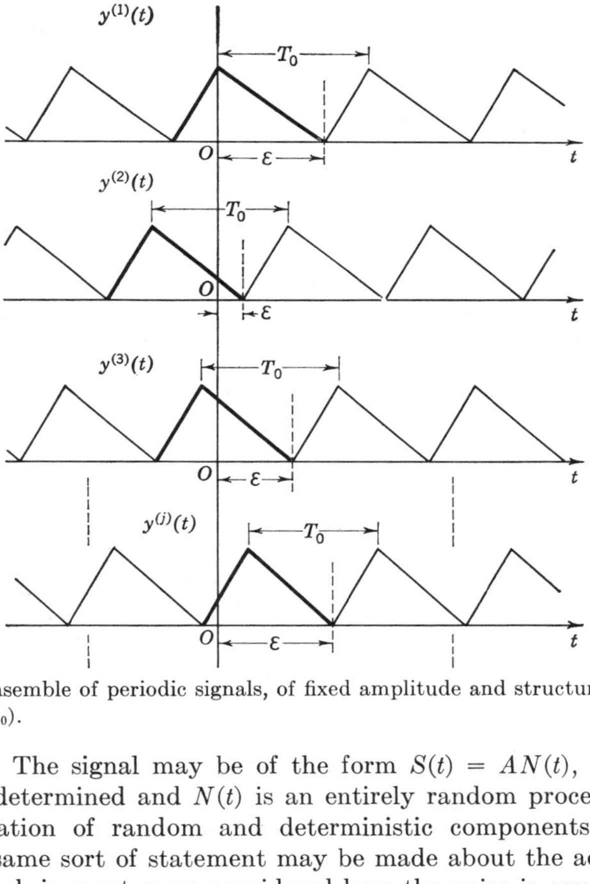

FIG. 1.8. An ensemble of periodic signals, of fixed amplitude and structure, but random epoch ϵ in $(0,T_0)$.

may occur. The signal may be of the form $S(t) = AN(t)$, for example, where A is determined and $N(t)$ is an entirely random process. Clearly, any combination of random and deterministic components is possible; exactly the same sort of statement may be made about the accompanying noise, although in most cases considered here the noise is assumed to contain no deterministic components, at least before it is processed with the signal.

1.3-6 Stationarity and Nonstationarity. The process is said to be *nonstationary* when its ensemble properties, i.e., the descriptive hierarchy of

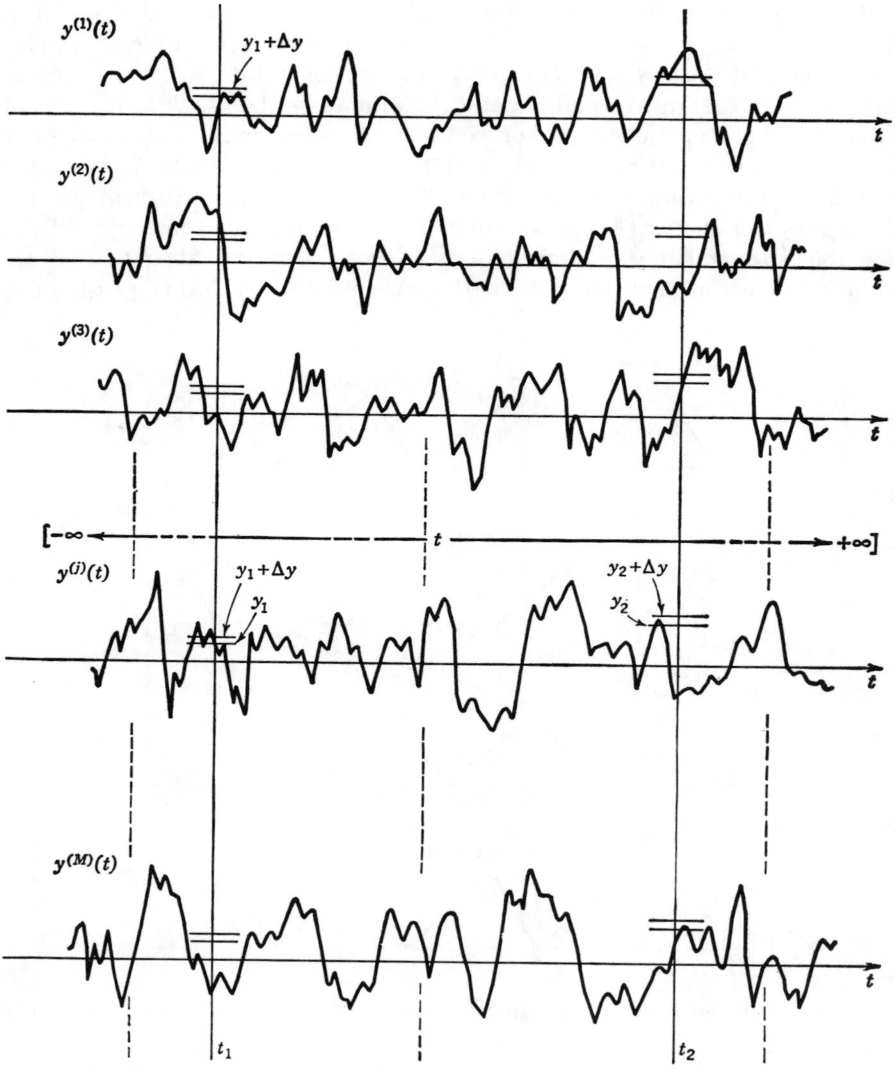

FIG. 1.9. A stationary ensemble of random noise voltages $y^{(1)}, y^{(2)}, \ldots, y^{(M)}, \ldots$.

probability densities (cf. Sec. 1.3-2), depend on the absolute origin of time, through $t_1, t_2, \ldots, t_n$, and not alone on the differences $t_2 - t_1, t_3 - t_1, \ldots$. Expressed somewhat more concisely, a process is nonstationary if its ensemble does not remain invariant under an arbitrary (linear) time displacement. Thus, if the ensemble $y(t)$ goes into a *different* ensemble $y'(t) = y(t + t')$, this is equivalent to a change of measures $(W_1,W_2, \ldots ,W_n, \ldots)$ to a new set $(W_1', \ldots ,W_n', \ldots)$ with t'. Physically, the underlying physical mechanism changes in the course of time, as shown, for example, in Figs. 1.5, 1.6: W_1 at t_1 is quite different from W_1 at t_2 ($\neq t_1$), and so on.

Conversely, a process is said to be *stationary*[26] if the ensemble remains invariant of an arbitrary (linear) time shift, $y(t) \to y(t + t') = y(t)$; the ensemble remains unchanged, or equivalently *all* probability measures remain unaffected by this displacement, viz.,

$$W_n(y_1,t_1; \ . \ . \ . \ ;y_n,t_n) = W_n(y_1, t_1 + t'; \ . \ . \ . \ ; y_n, t_n + t') \qquad n \geq 1 \quad (1.72)$$

The statistical properties of the ensemble are independent of the absolute origin of time and are functions only of the various time differences between observations, so that we may write

$$\begin{aligned} W_n(y_1,t_1;y_2,t_2; \ . \ . \ . \ ;y_n,t_n) &= W_n(y_1, 0; y_2, t_2 - t_1; \ . \ . \ . \ ; y_n, t_n - t_1) \\ &= W_n(y_1;y_2,t_2; \ . \ . \ . \ ;y_n,t_n) \qquad n \geq 1 \qquad (1.73) \end{aligned}$$

letting $t_1 = 0$ for convenience. Equivalently, for every arbitrary (linear) shift $t \to t'$, each representation $y^{(j)}$ of the original ensemble goes over into another member $y^{(k)}$ of the new ensemble in such a way that the weighting or measure associated with corresponding units in the two ensembles remains unchanged. Figure 1.9 shows a typical ensemble of stationary random noise voltage waves, where each representation persists over the interval $(-\infty < t < \infty)$.†

The hierarchy of defining probability densities is determined as before in the nonstationary case (cf. Sec. 1.3-1). We write

$W_1(y_1)\,dy_1 = W_1(y)\,dy =$ probability that y takes values in interval $(y, y + dy)$

$W_2(y_1; y_2, t_2 - t_1)\,dy_1\,dy_2 =$ joint probability that y takes values in range $(y_1, y_1 + dy_1)$ at time t_1 and values in interval $(y_2, y_2 + dy_2)$ at another time t_2

$W_n(y_1; y_2, t_2 - t_1; \ . \ . \ . \ ; y_n, t_n - t_1)\,dy_1, \ . \ . \ . \ , dy_n =$ joint probability of finding y in intervals $(y_1, y_1 + dy_1)$, $(y_2, y_2 + dy_2)$, . . . , $(y_n, y_n + dy_n)$ at an initial time t_1 and times $t_2 - t_1, \ . \ . \ . \ , t_n - t_1$, respectively (1.74)

where usually, to simplify the notation and the discussion, we shall arbitrarily take $t_1 = 0$ as the time origin [cf. Eq. (1.73)]. In contrast with the nonstationary cases, the underlying physical mechanism of these random processes does not change with time.

† Note that, if the interval $(t_0 < t < t_0 + T)$ over which the member functions are in general different from zero is finite while the members are zero outside, the process is nonstationary, even if within that interval the measures remain invariant: one can always find an arbitrary translation $|t'| > T$ which takes one outside the original interval and therefore yields a different ensemble, each member of which is now zero. The process of Fig. 1.5 is nonstationary for this reason, in addition to the obvious variation of its scale in the interval $(t_0, t_0 + T)$.

Processes for which Eq. (1.72) is true are also called *strictly*, or *completely*, *stationary processes*.† A less strict form of stationarity arises when we require only that the covariance function of the process depend on time differences alone.

$$E\{[Y(t_k) - E\{Y(t_k)\}][Y(t_l) - E\{Y(t_l)\}]\} \\ = K_y(|t_l - t_k|) \qquad E\{Y^2\} < \infty, |E\{Y\}| < \infty \quad (1.75)$$

[cf. Eqs. (1.52), (1.60*a*)]. This, of course, is not enough to ensure that the distribution densities W_1, W_2, etc., remain invariant of an arbitrary shift t' [cf. Eq. (1.72)]. For example, consider the process (e.g., series) for which

$$\begin{aligned} E\{Y_k\ (= \overline{y_k})\} &= 0 \\ E\{Y_k^2\ (= \overline{y_k^2})\} &= \sigma^2\ (> 0) \\ K_y(k) &= 0 \qquad k \neq 0 \\ K_y(0) &= \sigma^2 \end{aligned} \quad (1.76)$$

where the y_k are statistically independent (cf. Sec. 1.2-10). Then unless the $w_k(y_k)$ are all alike, higher moments, for example, $\overline{y_k^2 y_l^2 y_m^2}$, $\overline{y_k^3 y_l^3 y_m^3}$, will depend on k, l, m, not on the differences $l - k$, $m - k$, etc., and the process is not completely stationary.

When a process obeys Eq. (1.75), it is said to be *wide-sense stationary*, or *stationary to the second order*.‡ It is clear that processes that are strictly stationary are also wide-sense stationary, but the reverse is not true, except in the very important special case of the *normal*, or *gaussian*, *process*, where the process itself is determined entirely by the structure of the covariance function (or matrix) (and the expectations $\bar{y}$) [cf. Eqs. (1.60), (1.61), and Chap. 7]. The conditions for the existence of processes stationary in the wide sense are thus determined by the conditions for the existence of the covariance function.

Of course, stationarity, either in the wide or in the strict sense, is an idealization. Physically, no process can have started at $t = -\infty$, nor can it continue to $t \to \infty$ without changes in the underlying physical mechanism. In actuality, one always deals with a finite sample, but if the fluctuation time of the process—roughly, the mean spacing between successive "zeros" in time of any member function $y^{(j)}(t)$—is small compared with the observation time, or sample length, and if the underlying random mechanism does not change in this interval, we may say that the process is practically stationary. Then we expect that results calculated on the assumption of true stationarity will agree closely with those determined by taking the finite duration of the ensemble into account. Accordingly, our first applications of probability methods assume strict stationarity, either for the ensemble itself or for the various subensembles comprising the process. A more refined theory follows which considers the effects of finite sample (and ensemble) size in time (cf. Chaps. 18 to 23).

† Doob,[16] chap. 2, sec. 8; Bartlett,[17] sec. 6.4.

‡ Doob, *op. cit.*, chap. 2, sec. 8, pp. 94, 95; Bartlett, *op. cit.*, sec. 6.1.

1.3-7 Examples of Stationary and Nonstationary Processes. To illustrate the remarks of Sec. 1.3-6, let us consider a number of simple examples:

(1) The process $y(t,\phi) = A_0 \cos (\omega_0 t + \phi)$. Here we have an ensemble of sinusoids (A_0 and ω_0 fixed), where ϕ has a p.d. $w(\phi)$ defined in the primary interval $(0,2\pi)$. [Because of the periodic structure of the member functions, $w(\phi)$ is similarly defined in all other period intervals of the *same*

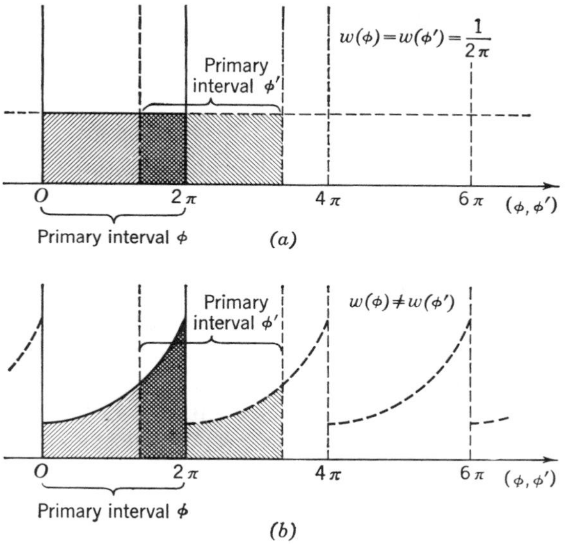

FIG. 1.10. Probability densities of ϕ, ϕ' for the (*a*) stationary and (*b*) nonstationary ensemble [cf. Eq. (1.77)].

ensemble.] Testing the definition of (strict) stationarity above, we see that the arbitrary shift $t \to t' = t + t_0'$ gives

$$y(t + t_0', \phi) = A_0 \cos (\omega_0 t + \omega_0 t_0' + \phi) = A_0 \cos (\omega_0 t + \phi') \qquad (1.77)$$
$$w(\phi) \to w'(\phi')$$

where $w'(\phi')$ is the corresponding p.d. for ϕ'. Unless ϕ is *uniformly* distributed in the primary interval $(0,2\pi)$, ϕ' will have a different distribution in its own primary interval, as can be seen in Fig. 1.10*a* and *b*, and the measures w, w' will not be the same. For the former d.d. w, then, the process is stationary.† For the latter, it is not.

† Strictly speaking, we must show that all W_n $(n \geq 1)$ for y remain invariant [cf. Eq. (1.72)]. However, here it is clear that the transformed ensemble has the same *form* and that, since $w(\phi) = w'(\phi')$, every subensemble [e.g., those y's corresponding to ϕ and to ϕ' in the interval $(0,\pi/2)$] has the same measure, so that the ensemble itself remains unchanged. Using the techniques of Sec. 1.3-8, the reader may find it instructive to show that this is equivalent to the invariance of W_n $(n \geq 1)$ for this particular ensemble y (see Prob. 1.10).

(1a) The process $y(t,\phi) = A_0 \cos(\omega_0 t + \phi) + N(t)$. Here $N(t)$ is a noise ensemble. If $N(t)$ is stationary, then y is also stationary, provided that ϕ is uniformly distributed in the primary interval $(0,2\pi)$, as we have noted above. On the other hand, if either (or both) of the subensembles $A_0 \cos(\omega_0 t + \phi)$, $N(t)$ is nonstationary, so also is y.

(2) The continuous process $y(t,\phi) = A_0[1 + \lambda \cos \omega_m(t + \phi)] \cos(\omega_0 t + \phi + \phi_0)$. This is an ensemble of sinusoidally modulated carriers, A_0, ω_m, ϕ_0, ω_0 fixed, and λ chosen so that $A_0(\ \ ) \geq 0$. To test it for stationarity, let $t \to t' = t_0' + t$, so that

$$y(t + t_0', \phi) = A_0\{1 + \lambda \cos[\omega_m t + \phi' + (\omega_m - \omega_0)t_0']\} \cos(\omega_0 t + \phi_0 + \phi') \qquad \phi' = \phi + \omega_0 t_0' \qquad (1.78)$$

Even if $w(\phi)$ is uniform, so that $w'(\phi')$ is also [in its primary interval (cf. Fig. 1.10*a*)], *the same ensemble does not result for arbitrary* t' [cf. Eq. (1.78)] and so $y(t,\phi)$ here is not stationary. This is an example where the W_1, W_2, W_3, . . . , *for* y do not remain invariant,† although $w(\phi)$ does.‡

1.3-8 Distribution Densities for Deterministic Processes. In many practical cases, the signals which one has to deal with are deterministic. Their ensemble properties are accordingly derived from the probability measures associated with the various descriptive parameters A, $\boldsymbol{\epsilon}$, $\boldsymbol{\theta}$ [cf. Eq. (1.71)] of the signal. From the methods of Sec. 1.2-14, especially Eqs. (1.62), (1.63), we may represent the hierarchy of d.d.'s of $y_s = As(t;\boldsymbol{\epsilon},\boldsymbol{\theta})$ by

$$W_n(y_1,t_1; \ldots ;y_n,t_n)_S = \int_{-\infty}^{\infty} \cdots \int e^{-i\boldsymbol{\xi}\mathbf{y}} F_{y_1,\ldots,y_n}(\xi_1,t_1; \ldots ;\xi_n,t_n)_S \frac{d\xi_1 \cdots d\xi_n}{(2\pi)^n} \qquad n \geq 1 \qquad (1.79)$$

where specifically the characteristic function F_S is

$$F_{y_1,\ldots,y_n}(\xi_1,t_1; \ldots ;\xi_n,t_n)_S = \int_{-\infty}^{\infty} \cdots \int \exp\,[iA\{\xi_1 s(t_1;\boldsymbol{\epsilon},\boldsymbol{\theta}) + \cdots + \xi_n s(t_n;\boldsymbol{\epsilon},\boldsymbol{\theta})\}] w(A,\boldsymbol{\epsilon},\boldsymbol{\theta})\, dA\, d\boldsymbol{\epsilon}\, d\boldsymbol{\theta} \qquad (1.80)$$

In the still more general cases where the defining parameters are also time-dependent, Eq. (1.80) becomes

$$F_S = \int_{-\infty}^{\infty} \cdots \int \exp\left[i \sum_k^n \xi_k A_k s(t_k,\boldsymbol{\epsilon}_k,\boldsymbol{\theta}_k)\right] \times w(A_1, \ldots ,A_n;\epsilon_1, \ldots ,\epsilon_n;\boldsymbol{\theta}_1, \ldots ,\boldsymbol{\theta}_n)\, \mathbf{dA}\, \mathbf{d\epsilon}\, \mathbf{d\theta}_1 \cdots \mathbf{d\theta}_n \qquad (1.81)$$

† See p. 34.

‡ For deterministic processes, it is usually possible to decide at once by inspection whether or not the same ensemble is generated as $t \to t' = t_0' + t$, without having explicitly to demonstrate the invariance or noninvariance of W_n.

As an example, let us consider the simple ensemble of *periodic* waves $y_s = s(t,\epsilon) = s(t + \epsilon)$, with period T_0, so that $s(t + T_0, \epsilon) = s(t,\epsilon)$, etc., where the epoch ϵ is uniformly distributed in a period interval $(0,T_0)$ (cf. Fig. 1.8). This is equivalent to a uniform distribution of the phase $\phi =$

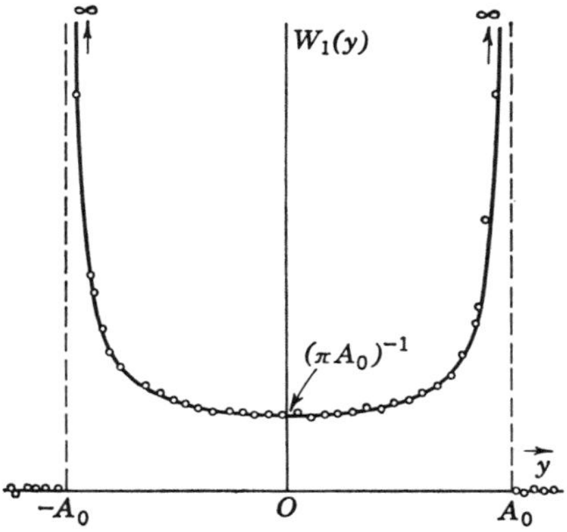

FIG. 1.11. First-order probability density $W_1(y)$ of the ensemble of sine waves $y(t,\phi) = A_0 \cos(\omega_0 t + \phi)$, ϕ uniformly distributed in $(0,2\pi)$. (*Experimental points after N. Knudtzon, Experimental Study of Statistical Characteristics of Filtered Random Noise, MIT Research Lab. Electronics Tech. Rept.* 115, *July* 15, 1949.)

$2\pi\epsilon/T_0$ in $(0,2\pi)$. Applying Eq. (1.80) to Eq. (1.79) for the case $n = 1$, we can write at once

$$\begin{aligned} W_1(y_1,t_1)_S &= \int_{-\infty}^{\infty} e^{-i\xi_1 y_1} \frac{d\xi_1}{2\pi} \int_0^{2\pi} e^{i\xi_1 s(t_1;\phi)} \frac{d\phi}{2\pi} \\ &= \int_{-\infty}^{\infty} e^{-i\xi_1 y_1} \frac{d\xi_1}{2\pi} \int_0^{2\pi} e^{i\xi_1 s(0;\phi)} \frac{d\phi}{2\pi} = W_1(y_1)_S \end{aligned} \tag{1.82}$$

since the process is stationary under these conditions (cf. Prob. 1.9*a*), and hence independent of t_1, which we set equal to zero here for convenience. In a similar way, the higher-order densities depend on t only through the differences $t_2 - t_1, t_3 - t_1, \ldots$. As an example, Fig. 1.11 shows $W_1(y)_S$ for the ensemble of sinusoids $y_s = A_0 \cos(\omega_0 t + \phi)$, where specifically from Eq. (1.82) we get (cf. Prob. 1.11)

$$W_1(y)_S = \begin{cases} \dfrac{1}{\pi\sqrt{A_0{}^2 - y^2}} & -A_0 < y < A_0 \\ 0 & |y| > A_0 \end{cases} \tag{1.82a}$$

An experimental demonstration of this has been given by Knudtzon.[27]

While Eqs. (1.79), (1.80) give these higher-order densities in a general fashion, we can take advantage of the periodic and stationary character of this ensemble $s(t;\phi)$ to obtain $(W_n)_S$ more simply. Let us illustrate with

$(W_2)_S$, introducing the *conditional* density $W_2(y_1|y_2; t_2 - t_1)_S$ through the relation†

$$W_2(y_1; y_2, t_2 - t_1)_S = W_1(y_1)_S W_2(y_1|y_2; t_2 - t_1)_S \tag{1.83}$$

Now, $y(t_1) = s(t_1 + \epsilon)$, and $y(t_2) = s(t_2 + \epsilon)$, where ϵ is uniformly distributed over a period interval, since y is assumed to be stationary. Furthermore, since y is deterministic, we can invert these relations to obtain

$$t_1 + \epsilon = s^{-1}(y_1) \qquad t_2 + \epsilon = s^{-1}(y_2) \tag{1.84}$$

Also because of the deterministic structure of y, once s is known at a time $t_1 + \epsilon$, it is specified (functionally) for all other times, so that the conditional density in Eq. (1.83) becomes simply

$$\begin{aligned} W_2(y_1|y_2; t_2 - t_1)_S &= \delta[y_2 - s(t_2 + \epsilon)] = \delta[y_2 - s(t_2 - t_1 + t_1 + \epsilon)] \\ &= \delta\{y_2 - s[t_2 - t_1 + s^{-1}(y_1)]\} \end{aligned} \tag{1.85}$$

The second-order density [Eq. (1.83)] can then be written

$$W_2(y_1; y_2, t_2 - t_1)_S = W_1(y_1)_S \, \delta\{y_2 - s[t_2 - t_1 + s^{-1}(y_1)]\} \tag{1.86}$$

when $W_1(y_1)_S$ is obtained from Eq. (1.82). Higher-order densities may be constructed in the same way (cf. Prob. 1.13). Thus, with W_n $(n \geq 1)$ for entirely random processes and $(W_n)_S$ $(n \geq 1)$ for deterministic processes, we can now, in principle at least, give a complete description (cf. Sec. 1.3-2) of the mixed processes which occur most frequently in applications.

PROBLEMS

1.9 (*a*) Show that any ensemble of periodic waves with stationary parameters is stationary.

(*b*) Verify that the ensemble of periodically modulated carriers, $f(t) \cos(\omega_0 t + \phi)$, $f(t) = f(t + T_0')$, is in general not stationary. When is it stationary? When is it wide-sense stationary?

1.10 If ϕ is uniformly distributed over the primary interval $(0,2\pi)$, show with the help of Sec. 1.3-8 that the ensemble $y = A_0 \cos(\omega_0 t + \phi)$ $(A_0, \omega_0$ fixed) is stationary; i.e., show the invariance of $W_n(y_1,t_1; \ldots ;y_n,t_n)$ $(n \geq 1)$ under an arbitrary (linear) time shift.

1.11 For the ensemble of Prob. 1.10, show that

$$W_1(y) = \begin{cases} (\pi\sqrt{A_0^2 - y^2})^{-1} & -A_0 < y < A_0 \\ 0 & |y| > A_0 \end{cases} \tag{1}$$

$$F_y(\xi) = J_0(A_0\xi) \qquad \text{Bessel function of first kind} \tag{2}$$

† See Sec. 1.2-9, especially Eqs. (1.45) et seq. Note, however, that here we use the alternative notation: the quantity preceding the vertical bar is assumed given; that following is the argument of the function.

(See Fig. 1.11.) Show also that the distribution function of y is

$$D_1(y) = \begin{cases} 0 & y < -A_0 \\ \dfrac{1}{2} + \dfrac{1}{\pi}\sin^{-1}\dfrac{y}{A_0} & -A_0 \le y \le A_0 \\ 1 & y \ge A_0 \end{cases} \tag{3}$$

1.12 Show for the ensemble $y = A_0 \cos(\omega_0 t + \phi)$ [A_0, ω_0 fixed, ϕ uniformly distributed in $(0,2\pi)$] that

$$F_{y_1,y_2}(\xi_1, \xi_2; t_2 - t_1) = \sum_{m=0}^{\infty} (-1)^m \epsilon_m J_m(A_0\xi_1) J_m(A_0\xi_2) \cos m\omega_0(t_2 - t_1) \tag{1}$$

and

$$W_2(y_1, y_2; t_2 - t_1) = \frac{1}{\pi\sqrt{A_0^2 - y_1^2}}\, \delta\left\{y_2 - A_0 \cos\left[\cos^{-1}\frac{y_1}{A_0} + (t_2 - t_1)\omega_0\right]\right\} \tag{2a}$$

or

$$W_2(y_1, y_2; t_2 - t_1) = \begin{cases} \dfrac{1}{\pi^2} \displaystyle\sum_{m=0}^{\infty} \epsilon_m \cos m\omega_0(t_2 - t_1)\, \dfrac{\cos\left(m \sin^{-1}\dfrac{y_1}{A_0} + \dfrac{m\pi}{2}\right)}{\sqrt{A_0^2 - y_1^2}}\, \dfrac{\cos\left(m \sin^{-1}\dfrac{y_2}{A_0} + \dfrac{m\pi}{2}\right)}{\sqrt{A_0^2 - y_2^2}} & -A_0 < y_1, y_2 < A_0 \\ 0 & |y_1|, |y_2| > A_0 \end{cases} \tag{2b}$$

Since y is stationary, W_2 depends only on the time difference $t_2 - t_1$.

1.13 Show for the ensemble of Prob. 1.12 that

$$W_n(y_1; y_2, t_2; \ldots; y_n, t_n) = \begin{cases} \dfrac{1}{\pi\sqrt{A_0^2 - y_1^2}} \displaystyle\prod_{k=1}^{n-1} \delta\left[y_{k+1} - A_0 \cos\left(\cos^{-1}\frac{y_1}{A_0} + t_{k+1}\omega_0\right)\right] & -A_0 < y_k < A_0 \\ 0 & \text{otherwise} \end{cases} \tag{1}$$

1.14 (*a*) For the ensemble of Prob. 1.12, show that the joint first-order density of y and $\dot{y}$ is

$$W_1(y,\dot{y}) = W_1(y)[\delta(\dot{y} + \omega_0\sqrt{A_0^2 - y^2}) + \delta(\dot{y} - \omega_0\sqrt{A_0^2 - y^2})]/2 \tag{1}$$

where $W_1(y)$ is given in Prob. 1.11, Eq. (1), if we observe that

$$F_{y,\dot{y}}(\xi_1,\xi_2) = J_0(A_0\sqrt{\xi_1^2 + \omega_0^2\xi_2^2}) \tag{2}$$

(*b*) Show also that y and $\dot{y}$ are uncorrelated, that is, $\overline{y\dot{y}} = 0$, but that y and $\dot{y}$ are not statistically independent.

1.15 (*a*) Repeat Prob. 1.14, except that now one is to show that the joint first-order density of y and $\ddot{y}$ is

$$W_1(y,\ddot{y}) = W_1(y)\,\delta(\ddot{y} + \omega_0^2 y) \tag{1}$$

with the characteristic function

$$F_{y,\ddot{y}}(\xi_1,\xi_2) = J_0[A_0(\xi_1 - \omega_0^2\xi_2)] \tag{2}$$

(*b*) Verify that y and $\ddot{y}$ are not statistically independent, and show specifically that

$$\overline{y\ddot{y}} = -A_0^2\omega_0^2/2 \tag{3}$$

1.4 *Further Classification of Random Processes*

We have seen in Sec. 1.3-2 that knowledge of the probability densities W_n for all $n \geq 1$ completely describes the random process. Many physical processes, however, are not so statistically complex that all W_n are needed to provide complete information about the process; it is frequently sufficient to know W_1 or W_2 alone. As we shall observe presently, once these distribution densities have been found to contain all the possible statistical information, we can then determine from them the higher-order densities W_n $(n \to \infty)$, completing the descriptive hierarchy. An example is the periodic deterministic process of Sec. 1.3-8, where $(W_n)_S$ is completely determined once $(W_1)_S$ has been given [cf. Eq. (1.86)]. Other examples of entirely random processes are considered below in Secs. 1.4-2, 1.4-3.

Classification is important, because often, for the solution of particular problems, we need to know whether or not W_1, or W_2, etc., really contains all the information about the process. For example, in the theory of the Brownian motion, $W_1(\mathbf{r},\dot{\mathbf{r}})$† is not sufficient to describe the stochastic process completely. One needs in addition the joint second-order density $W_2(\mathbf{r}_1, \dot{\mathbf{r}}_1; \mathbf{r}_2, \dot{\mathbf{r}}_2; t_2 - t_1)$. Similar remarks apply in the theory of thermal noise (cf. Chap. 11). Furthermore, although a second-order density is adequate for many uses (cf. Chaps. 5 and 12 to 15), all orders of distribution density may be required for both linear and nonlinear cases when the more general methods of statistical decision theory are applied to system analysis and design (cf. Chaps. 18 to 23).

1.4-1 Conditional Distribution Densities. It is convenient at this stage to extend the notion of conditional probability mentioned in Sec. 1.2-9 and very briefly in Sec. 1.3-8. We define the simplest conditional p.d. for a random process $y(t)$ by‡

$$W_2(y_1,t_1|y_2,t_2)\,dy_2 = \text{probability that, if } y \text{ has value } y_1 \text{ at time } t_1, \; y \text{ will have values in range } (y_2, y_2 + dy_2) \text{ at time } t_2 \text{ later} \tag{1.87}$$

Higher-order densities may be defined in a similar way. We have as one generalization

$$W_n(y_1,t_1; \ldots ;y_{n-1},t_{n-1}|y_n,t_n)\,dy_n = \text{probability that, if } y \text{ takes values } y_1, \ldots , y_{n-1} \text{ at respective times } t_1, \ldots , t_{n-1}, \; y \text{ will have values in interval } (y_n, y_n + dy_n) \text{ at time } t_n \text{ later } (t_n \geq t_{n-1}, \ldots) \tag{1.88}$$

† Here $\mathbf{r}$ is the vector displacement and $\dot{\mathbf{r}}$ the velocity of the heavy particle in suspension.

‡ Here the hypothesis precedes the argument, which is the reverse of the notation in Sec. 1.2-9 (cf. second footnote, p. 17).

From the definitions (1.45), etc., we have the following relations for $W_n(|)$ and $W_n(\ )$,

$$W_2(y_1,t_1;y_2,t_2) = W_1(y_1,t_1)W_2(y_1,t_1|y_2,t_2) \quad (1.89)$$

$$\begin{aligned} W_3(y_1,t_1;y_2,t_2;y_3,t_3) &= W_2(y_1,t_1;y_2,t_2)W_3(y_1,t_1;y_2,t_2|y_3,t_3) \\ &= W_1(y_1,t_1)W_2(y_1,t_1|y_2,t_2)W_3(y_1,t_1;y_2,t_2|y_3,t_3) \end{aligned} \quad (1.90)$$

and so on, for all $n \geq 2$. Since $W_n(|)$ is a probability density, it must satisfy the conditions

$$W_n(y_1,t_1; \ldots ;y_{n-1},t_{n-1}|y_n,t_n) \geq 0 \quad (1.91a)$$

$$\int W_n(y_1,t_1; \ldots ;y_{n-1},t_{n-1}|y_n,t_n)\, dy_n = 1 \quad (1.91b)$$

$$\int \cdots \int W_n(y_1,t_1; \ldots ;y_{n-1},t_{n-1})W_n(y_1,t_1; \ldots ;y_{n-1},t_{n-1}|y_n,t_n) \times dy_1 \cdots dy_{n-1} = W_1(y_n,t_n) \quad (1.91c)$$

which follow at once from the definition of conditional probability. Note that here the times are ordered, namely $(t_1 \leq t_2 \leq \cdots \leq t_n)$. Other conditional distribution densities exist also. For instance, one can also define a marginal conditional probability density of the form $W_{m,n} = W_{m,n}(y_1,t_1; \ldots ;y_m,t_m|y_{m+1},t_{m+1}; \ldots ;y_n,t_n)$, where as before the quantities to the left of the vertical bar are given and those to the right are to be predicted. In subsequent applications, however, $W_n(. \ . \ . \ |y_n,t_n)$ is sufficient. Note that other properties of $W_n(. \ . \ . \ | \ . \ . \ .)$ are consequences of the defining relations (1.87), (1.88),

$$\lim_{t_2-t_1\to 0} W_2(y_1,t_1|y_2,t_2) = \delta(y_2 - y_1)$$

therefore with (1.92)

$$\lim_{t_2-t_1\to 0} W_2(y_1,t_1;y_2,t_2) = W_1(y_1,t_1)\ \delta(y_2 - y_1)$$

since it is certain that $y_2 = y_1$, a given quantity, when the times of observation coincide [remember that $W_2(|)$ is a probability *density;* cf. Sec. 1.2-4]. At the other extreme, we have

$$\lim_{t_2-t_1\to \infty} W_2(y_1,t_1|y_2,t_2) = W_1(y_2,t_2)$$

therefore with (1.93)

$$\lim_{t_2-t_1\to \infty} W_2(y_1,t_1;y_2,t_2) = W_1(y_2,t_1)W_1(y_2,t_2)$$

which, for the entirely random processes assumed here, is simply a statement of the fact that there is no "memory," or statistical dependence of values of y observed at two different times sufficiently separated in time.† Similar interpretations follow directly for $W_{m,n}$ as $t_k - t_l \to \infty$ or 0.

1.4-2 The Purely Random Process.‡ We return now to classification. The simplest type of process $y(t)$ is said to be purely random when suc-

† When the process contains a steady or periodic component, then there is memory for $t_2 - t_1 \to \infty$; the values of y at t_2 are dependent in this respect on values at t_1. See Sec. 3.1-5, for example.

‡ Note the distinction between purely random and entirely random (cf. Sec. 1.3-1).

cessive values of y are in no way correlated, i.e., when successive values of the random variable are statistically independent of the preceding ones (no matter how small the interval between observations). This can be written from Eq. (1.88) as

$$W_n(y_1,t_1; \ldots ;y_{n-1},t_{n-1}|y_n,t_n) = W_1(y_n,t_n) \tag{1.94}$$

Substitution into Eqs. (1.89), (1.90), etc., for $n \geq 2$ gives us the hierarchy

$$W_2(y_1,t_1;y_2,t_2) = W_1(y_1,t_1)W_1(y_2,t_2) \tag{1.95a}$$

$$\cdots\cdots\cdots\cdots\cdots\cdots$$

$$W_n(y_1,t_1; \ldots ;y_n,t_n) = \prod_{k=1}^{n} W_1(y_k,t_k) \tag{1.95b}$$

for all orders n. From Eq. (1.59), we may say that the $y_1, \ldots , y_n$ are *statistically independent* for every $t_j \neq t_k$. All information about the process is contained in the first-order d.d. W_1, which now enables us to describe the process completely, according to Eq. (1.95*b*) (cf. Sec. 1.3-2). In practical cases, it is easy to give examples of a random series, i.e., when t is discrete (cf. Sec. 1.3-3). The toss of a coin at stated times, yielding a sequence of heads and tails, is one example. The method of random sampling in classical statistics is another. However, when t is continuous, the purely random process is strictly a limiting class; in physical situations, y_1 and y_2 are always correlated when the corresponding time difference $t_2 - t_1$ is finite (> 0) though small. In fact, it is the nature and degree of such correlations that are of particular interest in physical cases.

1.4-3 The Simple Markoff Process.† The next more complicated process arises when all the information is contained in the second-order probability density $W_2(y_1,t_1;y_2,t_2)$. These are called Markoff processes,‡[28] after the Russian mathematician who first studied them. They form a class of considerable importance, since in one sense or another most noise and many signal processes are of this type. By a simple Markoff process we mean a stochastic process $y(t)$ such that values of the random variable at *any* set of times $(t_1, \ldots ,t_k, \ldots ,t_n)$ depend on values of y at any set of *previous* times $(t_1, \ldots ,t_l, \ldots)$ only through the last available value y_{n-1}. Thus, we write for the conditional density [Eq. (1.88)]

$$W_n(y_1,t_1; \ldots ;y_{n-1},t_{n-1}|y_n,t_n) = W_2(y_{n-1},t_{n-1}|y_n,t_n) \qquad t_1 \leq t_2 \leq \cdots \leq t_n \tag{1.96}$$

† For a discussion of Markoff processes in the discrete case, e.g., *Markoff chains*, or discrete Markoff processes, see Feller, *op. cit.*, chap. 15, and also Bartlett, *op. cit.*, sec. 2.2, and chap. 2 in general (secs. 6.3–6.5 consider continuous Markoff processes). For some applications to random walk, renewal, and queueing problems, see Bartlett, chap. 2. Further references are contained in Bartlett's bibliography.

‡ Here we consider only simple, or first-order, Markoff processes.

and with this and the definition of a conditional d.d., viz.,

$$W_k(y_1,t_1; \ldots ;y_k,t_k) = W_{k-1}(y_1,t_1; \ldots ;y_{k-1},t_{k-1}) W_2(y_{k-1},t_{k-1}|y_k,t_k) \qquad k \geq 2 \tag{1.97}$$

we see that

$$\begin{aligned} W_n(y_1,t_1; \ldots ;y_n,t_n) &= W_1(y_1,t_1) W_2(y_1,t_1|y_2,t_2) W_3(y_1,t_1;y_2,t_2|y_3,t_3) \cdots \\ &= W_1(y_1,t_1) W_2(y_1,t_1|y_2,t_2) W_2(y_2,t_2|y_3,t_3) \\ &\qquad \cdots W_2(y_{n-1},t_{n-1}|y_n,t_n) \\ &= W_1(y_1,t_1) \prod_{k=2}^{n} W_2(y_{k-1},t_{k-1}|y_k,t_k) \qquad \text{all } n \geq 2 \end{aligned} \tag{1.98}$$

and the process is completely described (cf. Sec. 1.3-2), it being assumed, of course, that W_1 and the various $W_2(|)$ exist. The Markoff condition (1.96) also implies that the process y is entirely random. Then, since

$$\begin{aligned} \lim_{t_k - t_{k-1} \to \infty} W_2(y_{k-1},t_{k-1}|y_k,t_k) &= W_1(y_k,t_k) \\ \lim_{t_k - t_{k-1} \to \infty} W_2(y_{k-1},t_{k-1};y_k,t_k) &= W_1(y_{k-1},t_{k-1}) W_1(y_k,t_k) \end{aligned} \tag{1.99}$$

from Eq. (1.93), it is easily seen that our more general expression W_n [Eq. (1.98)] reduces to that for a purely random process [Eq. (1.95*b*)] when all observation times t_k ($k = 1, \ldots, n$) are infinitely far apart. Similarly, at the other extreme where all t_k are identical, we use Eq. (1.92) to write Eq. (1.98) as

$$\lim_{t_n \to t_{n-1} \to \cdots \to t_1} W_n(y_1,t_1; \ldots ;y_n,t_n) = W_1(y_1,t_1) \prod_{k=2}^{n} \delta(y_k - y_{k-1}) \tag{1.100}$$

In the former case, there is no memory between successive values; in the latter, the memory is perfect: $y_n = y_{n-1} = \cdots = y_1$ with probability 1. When the process contains a deterministic component, these relations must be suitably modified. Equations (1.96) to (1.100) then apply only to the entirely random portion of the process. Finally, we observe that more complicated, or higher-order, Markoff processes can easily be constructed by altering the condition (1.96). For example, we can require dependence on *two* previous events, e.g.,

$$W_n(y_1,t_1; \ldots ;y_n,t_n) = W_3(y_{n-2},t_{n-2};y_{n-1},t_{n-1}|y_n,t_n) \qquad t_1 \leq t_2 \leq \cdots \leq t_n \tag{1.101}$$

and so on. Note that, for *stationary* processes (cf. Sec. 1.3-6), the Markoff condition (1.96) reduces to $W_2(y_{n-1}|y_n, t_n - t_{n-1})$, so that Eq. (1.98) is simply

$$W_n(y_1;y_2,t_2; \ldots ;y_n,t_n) = W_1(y_1) \prod_{k=2}^{n} W_2(y_{k-1}|y_k,\Delta t_k) \qquad \Delta t_k \equiv t_k - t_{k-1} \tag{1.102}$$

where it is only the various time *differences* Δt_k that are now significant, not the absolute times $t_1, \ldots, t_n$.

1.4-4 The Smoluchowski Equation. In the treatment of Markoff processes, we cannot take $W_2(|)$ or W_2 as an arbitrary function of its variables, although $W_2(|)$ (or W_2) completely describes the process [cf. Eq. (1.98)]. Not only must the general conditions (1.91) be satisfied, but so also must

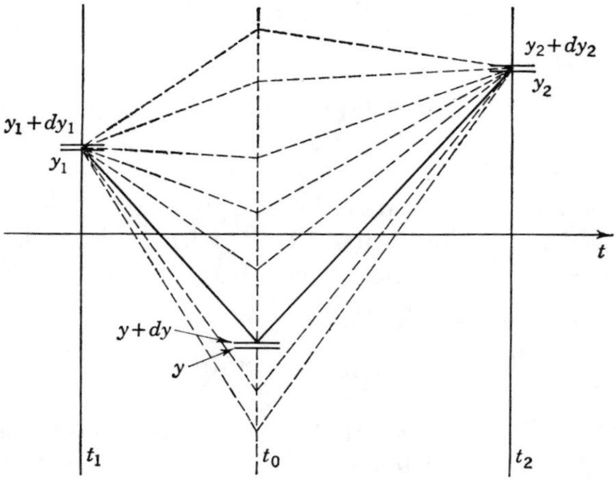

FIG. 1.12. Diagram showing transitions from y_1 at time t_1 to y_2 at time t_2 through possible intermediate values y at time t_0.

the Smoluchowski equation[28–30] hold (or, equivalently, the equation of Chapman and Kolmogoroff[30]), viz.,

$$W_2(y_1,t_1|y_2,t_2) = \int W_2(y_1,t_1|y,t_0)W_2(y,t_0|y_2,t_2)\,dy \qquad t_1 \le t_0 \le t_2 \quad (1.103a)$$

which in the stationary case is

$$W_2(y_1|y_2;t) = \int W_2(y_1|y,t_0)W_2(y|y_2,\ t-t_0)\,dy \qquad 0 \le t_0 \le t \quad (1.103b)$$

The Smoluchowski equation is essentially an expression of the transition (it need not be continuous), or "unfolding," of probability from instant to instant for different choices of $t = t_2 - t_1$. Starting with y_1 at a time t_1, one can go to a value y at some arbitrary later time t_0. Then, given this particular value of y as a new starting point, one has a certain probability of ending finally in the range $(y_2,\ y_2 + dy_2)$ at still later time t_2 $(\ge t_0 \ge t_1)$. The unfolding is repeated for all allowed values of y, and the $W_2(|)$ appearing in the integrand of Eqs. (1.103) are accordingly interpreted as *transition probabilities*. This is illustrated schematically in Fig. 1.12.

To see how Smoluchowski's equation is established, let us start with the third-order density for the process, namely, $W_3(y_1,t_1;y,t_0;y_2,t_2)$ $(t_1 \le t_0 \le t_2)$. From the definition of conditional probability, we write at once

$$W_3(y_1,t_1;y,t_0;y_2,t_0) = W_2(y_1,t_1;y,t_0)W_3(y_1,t_1;y,t_0|y_2,t_2)$$

Integrating over y and applying the Markoff condition (1.96) gives

$$W_2(y_1,t_1;y_2,t_2) = \int W_2(y_1,t_1;y,t_0)W_2(y,t_0|y_2,t_2)\,dy \tag{1.104}$$

With the help of Eq. (1.89), $W_1(y_1,t_1)$ can be factored from both members of Eq. (1.104), and Eq. (1.103*a*) is the result.

1.4-5 Higher-order Processes and Projections.[29] We can continue as in Secs. 1.4-2 and 1.4-3 for the higher-order cases by appropriate extensions of Eq. (1.96). Thus, $W_3(y_1,t_1;y_2,t_2|y_3,t_3)$ or W_3 describes completely the next more complex class of process; $W_4(|)$ or W_4, the next; and so on.

For the majority of purely physical problems, however, the simple Markoff process is the one of chief importance. Often, an apparently non-Markovian process can be transformed into a Markoff system by introducing one or more new random variables, such as $z = \dot{y}$, $u = \ddot{y}$, etc., or by letting z represent a coordinate of another system. Then, it may happen that for y and z *together* we have a simple Markoff process, for which the Smoluchowski equation now is

$$W_2(y_1,z_1,t_1|y_2,z_2,t_2) = \iint W_2(y_1,z_1,t_1|y,z,t_0)W_2(y,z,t_0|y_2,z_2,t_2)\,dy\,dz \qquad t_1 \leq t_0 \leq t_2 \tag{1.105}$$

If we can satisfy Eq. (1.105) with our new distribution density, we have as marginal d.d.

$$W_2(y_1,t_1;y_2,t_2) = \iint W_2(y_1,z_1,t_1;y_2,z_2,t_2)\,dz_1\,dz_2 \tag{1.106}$$

and in general $W_2(y_1,t_1;y_2,t_2)$ no longer describes a Markoff process. If we can find a conditional density $W_2(|)$ which obeys Eq. (1.105), the joint density $W_2(y_1,t_1;y_2,t_2)$ may be regarded as a *projection* of the more complicated conditional Markovian distribution density $W_2(y_1,z_1,t_1|y_2,z_2;t_2)$. That $W_2(y_1,t_1;y_2,t_2)$ or $W_2(y_1,t_1|y_2,t_2)$ by itself does not describe a Markoff process is due to the fact that we have not originally given a complete enough statistical account of the system. Our ability to select appropriate additional random variables $z, u, v, \ldots$ in order to extend the given process to one that is Markovian depends on the physical mechanism of the process itself, as we shall see presently in Chap. 11.

PROBLEMS

1.16 (*a*) Show that for a simple Markoff process one can also write

$$W_n(y_1,t_1; \ldots ;y_n,t_n) = \frac{\prod_{k=2}^{n} W_2(y_{k-1},t_{k-1};y_k,t_k)}{\prod_{k=2}^{n-1} W_1(y_k,t_k)} \qquad n \geq 3 \tag{1}$$

(b) Verify that, if $y(t)$ is a process which is completely described by a third-order density, W_n is given by

$$W_n(y_1,t_1; \ldots ;y_n,t_n) = \frac{\prod_{k=2}^{n-1} W_3(y_{k-1},t_{k-1}; \ldots ;y_{k+1},t_{k+1})}{\prod_{k=2}^{n-2} W_2(y_k,t_k;y_{k+1},t_{k+1})} \qquad n \geq 4 \tag{2}$$

(c) Show that the generalization of the Smoluchowski equation for (b) is

$$W_3(y_1,t_1;y_2,t_2|y_3,t_3) = \int W_3(y_1,t_1;y_2,t_2|y,t_0) W_3(y_2,t_2,y,t_0|y_3,t_3)\, dy \tag{3}$$

for all t_0 such that $t_1 \leq t_0 \leq t_3$ ($t_1 \leq t_2 \leq t_3$).

1.17 Suppose that $W_2(y|x) = W_2(x|y)$, where x, y are continuous random variables in $(-\infty < x, y < \infty)$.

(a) Show that, if $W_2(y|x)$ vanishes nowhere in the finite part of the x, y plane, both x and y must be uniformly distributed.

(b) For the stationary random process $y(t)$, where $x = y(t_1) = y_1$, $y = y(t_2) = y_2$, we have

$$W_2(y_1|y_2;\, t_2 - t_1) = W_2(y_2|y_1;\, t_1 - t_2) \tag{1}$$

We assume in addition that $W_1(y) = W_1(-y)$ and is a monotonically decreasing, nonvanishing continuous function in the interval $(0 \leq y \leq \infty)$. Show that

$$W_2(y_1|y_2; t_2 - t_1) = A\,\delta(y_1 - y_2) + (1 - A)\,\delta(y_1 + y_2) \qquad 0 < A < 1 \tag{2}$$

and A may depend on $t_2 - t_1$, that is, $A = A(t_2 - t_1)$.

1.5 *Time and Ensemble Averages*

We have seen in Sec. 1.3 how from the ensemble $y(t)$ one can obtain statistical information about the process in the form of probability densities (or distributions). It is also natural to ask what can be obtained from a single representation $y^{(j)}(t)$ when this member function is considered over some period of time $(t, t + T)$, since in most physical situations we shall not have the entire ensemble actually at our disposal at any one instant, but only one, or at most several, more or less representative members. In this case, certain useful quantities depending on the given $y^{(j)}(t)$ can be obtained from a suitable measurement over an observation period and are generally known as *time averages.* Analogous quantities can be defined over the ensemble *at specified times* t_1, t_2, . . . (cf. Figs. 1.5, 1.9) and are called *statistical,* or *ensemble, averages.* After defining these average quantities here, we shall examine in the next section (Sec. 1.6) how and under what conditions the two types of averages can be related.

1.5-1 Time Averages. Let us begin rather generally by considering some jth member, $G^{(j)}(y_1^{(j)}, \ldots, y_n^{(j)})$, of the ensemble G, which is itself a function of the ensemble $y(t)$. Here, $y_k = y(t_0 + t_k)$ $(k = 1, \ldots, n)$, and the t_k are parameters representing n times at which observations on the

y ensemble yield the particular members $y_k^{(j)} = y^{(j)}(t_0 + t_k)$. Then we define the *time average* of $G^{(j)}$ as

$$\langle G^{(j)}[y^{(j)}(t_1), \ldots, y^{(j)}(t_n)]\rangle \equiv \lim_{T\to\infty} \frac{1}{T}\int_{-T/2}^{T/2} dt_0\, G^{(j)}[y^{(j)}(t_0 + t_1), \ldots, y^{(j)}(t_0 + t_n)] \quad (1.107)$$

This definition applies for members of entirely random, deterministic, or mixed processes (cf. Sec. 1.3-4), provided that $G^{(j)}$ is suitably bounded so that the limit exists.

It is, of course, possible to introduce other definitions of the time average. Here Eq. (1.107) is called the *principal* limit,† where the interval $(-T/2,T/2)$ is extended (in the limit $T \to \infty$) to include the entire member function $G^{(j)}$ for all $-\infty < t < \infty$, but one can also consider the one-sided limits

$$\langle G^{(j)}\rangle_{+\infty} \equiv \lim_{T\to\infty} \frac{1}{T}\int_{0}^{T} dt_0\, G^{(j)}[y^{(j)}(t_0 + t_1), \ldots, y^{(j)}(t_0 + t_n)] \quad (1.108a)$$

$$\langle G^{(j)}\rangle_{-\infty} \equiv \lim_{T\to\infty} \frac{1}{T}\int_{-T}^{0} dt_0\, G^{(j)}[y^{(j)}(t_0 + t_1), \ldots, y^{(j)}(t_0 + t_n)] \quad (1.108b)$$

which, as noted in the next section, are under certain conditions equal to the corresponding principal limit for nearly all member functions of the ensemble. Still other possibilities, like

$$\langle G^{(j)}\rangle_{t_1} \equiv \lim_{T\to\infty} \frac{1}{T}\int_{t_1}^{t_1\pm T} dt_0\, G^{(j)}[y^{(j)}(t_0 + t_1), \ldots, y^{(j)}(t_0 + t_n)] \quad (1.108c)$$

can be considered, but for nearly all cases in the present work we shall find it convenient to employ the principal limit and in certain cases the one-sided form $\langle G^{(j)}\rangle_{+\infty}$ when the two are equivalent [cf. Sec. 1.6, Eqs. (1.126), (1.127)].

At this point, observe that the existence of the principal limit [Eq. (1.107)] implies that $\langle G^{(j)}\rangle$ depends on the time differences $t_2 - t_1, \ldots, t_n - t_1$ and is independent of the absolute time scale of the process.‡ For if we let $t_0' = t_0 + t_1$ in Eq. (1.107), we can write

$$\langle G^{(j)}\rangle = \lim_{T\to\infty} \frac{1}{T}\int_{t_1-T/2}^{t_1+T/2} dt_0'\, G^{(j)}[y^{(j)}(t_0'), \ldots, y^{(j)}(t_n - t_1 + t_0')] \quad (1.109)$$

† This is also often written $\lim_{T\to\infty} \frac{1}{2T}\int_{-T}^{T} G^{(j)}\, dt_0$. It is assumed that the $y^{(j)}$, and hence the $G^{(j)}$, persist for all time $(-\infty < t < \infty)$ and that these various limits exist.

‡ This is *not* the same thing as saying that $G^{(j)}$ belongs to a stationary process, since it does not necessarily follow that the *statistical* properties W_n $(n \geq 1)$ for G remain invariant under a (linear) time shift, even if we add the condition that $\langle G^j\rangle$ exists for almost all j [cf. Eq. (1.100)].

The integral can now be split into three parts,

$$\frac{1}{T}\int_{-T/2}^{T/2}(\quad)\,dt_0' - \frac{1}{T}\int_{-T/2}^{t_1-T/2}(\quad)\,dt_0' + \frac{1}{T}\int_{T/2}^{T/2+t_1}(\quad)\,dt_0'$$

only the first of which in the limit $T \to \infty$ in general is finite ($\neq 0$), so that Eq. (1.109) becomes, on dropping primes,

$$\langle G^{(j)}\rangle = \lim_{T\to\infty}\frac{1}{T}\int_{-T/2}^{T/2} dt_0\, G^{(j)}[y^{(j)}(t_0), \ldots, y^{(j)}(t_n - t_1 + t_0)] \tag{1.110}$$

which establishes the statement that $\langle G^{(j)}\rangle$ depends only on the time differences $t_2 - t_1, \ldots, t_n - t_1$. From this we might be tempted to say that the limit exists only if $y^{(j)}$ (and hence $G^{(j)}$) belongs to a stationary process, but the limit can exist also when $y(t)$ is not stationary. Consider the ensemble of Sec. 1.3-7(2), and let $G = y^2$, for example. Then we have explicitly in Eq. (1.110)

$$\begin{aligned}\langle G^{(j)}\rangle &= \lim_{T\to\infty}\frac{A_0^2}{T}\int_{-T/2}^{T/2}[1 + \lambda\cos(\omega_m t_0 + \phi^{(j)})]^2 \cos^2(\omega_0 t_0 + \phi^{(j)} + \phi_0)\,dt_0 \\ &= \frac{A_0^2}{2}\lim_{T\to\infty}\frac{1}{T}\int_{-T/2}^{T/2}\Big\{1 + \lambda\cos(\omega_m t_0 + \phi^{(j)}) \\ &\qquad + \frac{\lambda^2}{2}[1 + \cos(2\omega_m t_0 + 2\phi^{(j)})]\Big\}[1 + \cos(2\omega_0 t_0 + 2\phi^{(j)} + 2\phi_0)]\,dt_0 \\ &= \frac{A_0^2}{2}\left(1 + \frac{\lambda^2}{2}\right) \qquad \omega_m,\ \omega_0 \text{ incommensurable}\end{aligned} \tag{1.110a}$$

Of course, *finite* time averages like

$$\frac{1}{T}\int_0^T dt_0\, G^{(j)}[y^{(j)}(t_0 + t_1), \ldots, y^{(j)}(t_0 + t_n)] = \frac{1}{T}\int_{t_1}^{T+t_1} dt_0\, G^{(j)}[y^{(j)}(t_0), \ldots, y^{(j)}(t_n - t_1 + t_0)]$$

do depend on the absolute origin of time, through t_1. We observe also that similar remarks hold for more general situations, e.g., where $G^{(j)}$ depends on several random processes $x(t), y(t), z(t), \ldots$, so that $G^{(j)} = G^{(j)}[x^{(j)}(t_1 + t_0), \ldots; y^{(j)}(t_1 + t_0), \ldots; z^{(j)}(t_1 + t_0), \ldots; \ldots]$.

Time averages of particular interest in practical applications are:

1. The mean value

$$\langle y^{(j)}(t)\rangle = \lim_{T\to\infty}\frac{1}{T}\int_{-T/2}^{T/2} y^{(j)}(t_0 + t)\,dt_0 = \langle y^{(j)}\rangle \tag{1.111}$$

2. The mean square or intensity

$$\langle y^{(j)}(t)^2\rangle = \lim_{T\to\infty}\frac{1}{T}\int_{-T/2}^{T/2} y^{(j)}(t_0 + t)^2\,dt_0 = \langle (y^{(j)})^2\rangle \tag{1.112}$$

3. The autocorrelation function

$$
\begin{aligned}
\langle y^{(j)}(t_1)y^{(j)}(t_2)\rangle &= \lim_{T\to\infty} \frac{1}{T}\int_{-T/2}^{T/2} y^{(j)}(t_0+t_1)y^{(j)}(t_0+t_2)\,dt_0 \\
&\equiv R_y^{(j)}(t_2-t_1) \\
&= R_y^{(j)}(t_1-t_2)
\end{aligned} \tag{1.113}
$$

(cf. Prob. 1.18*c*).

4. The cross-correlation function

$$
\begin{aligned}
\langle y^{(j)}(t_1)z^{(j)}(t_2)\rangle &= \lim_{T\to\infty} \frac{1}{T}\int_{-T/2}^{T/2} y^{(j)}(t_0+t_1)z^{(j)}(t_0+t_2)\,dt_0 \\
&\equiv R_{zy}^{(j)}(t_2-t_1) \\
&= R_{yz}^{(j)}(t_1-t_2)
\end{aligned} \tag{1.114}
$$

The first item (1) gives the steady, or "d-c," component in the wave $y^{(j)}(t)$, if any, while item 2 is a measure of the intensity of the disturbance and is proportional to or equal to the mean power in this wave (depending on the units in which $y^{(j)}$ is given). The auto- and cross-correlation functions (1.113), (1.114) are averages of special importance in practical operation, as we shall see later in Part 3. A discussion of their salient features is reserved for Chap. 3. Note from the argument of Eqs. (1.108) to (1.110) that the first-order time moments or time averages $\langle y^{(j)}\rangle, \langle y^{(j)2}\rangle, \ldots, \langle y^{(j)n}\rangle$, etc., are independent of t, while the *second-order averages depend only on the differences* $t_2 - t_1$ [or $t_1 - t_2$ *in some fashion* regardless of whether $y^{(j)}$ is a member of a stationary or nonstationary ensemble; cf. Eqs. (1.113), (1.114) in the simplest cases]. For *periodic waves*, we observe that if we divide the interval $(t_1 - T/2,\ t_1 + T/2)$ into N periods, plus a possible remainder which never exceeds the contribution of a single period interval T_0, we can write the familiar result

$$
\begin{aligned}
\langle y^{(j)}(t)_{per}\rangle &= \lim_{N\to\infty}\left[\frac{1}{NT_0}\int_{-NT_0/2}^{NT_0/2} y^{(j)}(t_0+t_1)\,dt_0 + O\left(\frac{1}{N}\right)\right] \\
&= \frac{1}{T_0}\int_{(-T_0/2,0)}^{(T_0/2,T_0)} y^{(j)}(t_0)_{per}\,dt_0
\end{aligned} \tag{1.115}
$$

since the integral over N periods is equal to N times the integral over a single period. Generalizations for periodic $y^{(j)}(t)^n$, $G^{(j)}$, etc., are directly made.

1.5-2 Ensemble Averages. The other type of average (as already noted in Secs. 1.2, 1.3) is defined *over the ensemble*, at one or more times $t_1, t_2, \ldots$, as distinct from the time averages above, which are performed on a single member of the ensemble over an infinite period. Let us consider again the general ensemble $G = G(y_1,t_1; \ldots ;y_n,t_n)$ and for the moment restrict it to a finite number of members M. Thus, we define the *ensemble*, or *statistical*, *average* of G as

$$\overline{G(y_1,t_1; \ldots ;y_n,t_n)} \equiv \frac{1}{M} \sum_{j=1}^{M} G^{(j)}[y^{(j)}(t_1), \ldots ,y^{(j)}(t_n)] = \bar{G} \quad (1.116)$$

In terms of the measure, or weighting, $P(G)$ of G, this is equivalent to†

$$\bar{G} = \int_G G\, dP(G) = \int_G GW_1(G)\, dG \quad (1.117)$$

where, as previously, we may use the δ formalism (cf. Sec. 1.2-4) if the distribution of G contains discrete mass points. When, as is usually the case, the number M of representations becomes infinite, we extend the definition to

$$\overline{G(y_1,t_1; \ldots ;y_n,t_n)} = \lim_{M\to\infty} \frac{1}{M} \sum_{j=1}^{M} G^{(j)}[y^{(j)}(t_1), \ldots ,y^{(j)}(t_n)] \quad (1.118a)$$

where it is assumed in addition that this limit exists and converges to the expectation of G, so that Eq. (1.117) applies.‡ In terms of the random variables $y_1, \ldots , y_n$, we may write also

$$\bar{G} = \boldsymbol{E}\{G\} = \int \cdots \int G(y_1,t_1; \ldots ;y_n,t_n)W_n(y_1,t_1; \ldots ;y_n,t_n)\, dy_1 \cdots dy_n \quad (1.118b)$$

and the observation times $t_1, \ldots , t_n$ appear now explicitly as parameters. Equations (1.116), (1.118a) suggest the "counting" operations that one goes through in determining $\bar{G}$ from a given ensemble, just as Eqs. (1.68) to (1.70) indicate how the hierarchy of probability densities may be computed for the process itself.

Setting $G(y_1, \ldots ,y_n)$ equal to $y_1^n = y(t_1)^n$ ($n = 1, \ldots$), we obtain the *n first-order moments*

$$\overline{y_1^n} = \lim_{M\to\infty} \frac{1}{M} \sum_{j}^{M} [y^{(j)}(t_1)]^n = \int y_1^n W_1(y_1,t_1)\, dy_1 \quad (1.119)$$

and, like Eqs. (1.111), (1.112), $\bar{y}_1$, $\overline{y_1^2}$ are analogous measures of the mean and mean intensity, but now with respect to the ensemble at time t_1, as distinguished from the steady component and mean intensity of a particular representation throughout $T \to \infty$. More complicated statistical averages are also possible:

$$\overline{y(t_1)y(t_2)} = \iint y_1y_2W_2(y_1,t_1;y_2,t_2)\, dy_1\, dy_2 \equiv M_y(t_1,t_2) \quad (1.120a)$$

$$\overline{y(t_1)^k y(t_2)^l y(t_3)^m} = \iint y_1^k y_2^l y_3^m W_3(y_1,t_1;y_2,t_2;y_3,t_3)\, dy_1\, dy_2\, dy_3 \quad (1.120b)$$

† Cramér, *op. cit.*, secs. 7.4, 7.5 and chaps. 7–9. See also Sec. 1.2-4 above and comments. Specifically, for moments or statistical averages, see Sec. 1.2-5 above and Cramér, sec. 15.3.

‡ $W_1(G)$ will, of course, be a different density function here, ordinarily.

Table 1.1

Type of average	Ensemble	Time
1. Mean value	$\overline{y(t_1)} = \bar{y}_1$ ($= \bar{y}$, stat. process)	$\langle y^{(j)}(t_1)\rangle$
2. Mean square; intensity	$\overline{y_1^2}$ ($= \overline{y^2}$, stat. process)	$\langle y^{(j)}(t_1)^2\rangle$
3. Second-order joint moments (SJM)	$M_y(t_1,t_2) = \overline{y(t_1)y(t_2)} = \overline{y_1y_2}$ $= M_y(t_1 - t_2)$ (stat. process) $= M_y(t_2 - t_1)$	$R_y^{(j)}(t_1,t_2) = R_y^{(j)}(t_1 - t_2)$ $= \langle y^{(j)}(t_1)y^{(j)}(t_2)\rangle$ Autocorrelation function
4. Cross second moments	$M_{xy}(t_1,t_2) = \overline{x(t_1)y(t_2)} = M_{yx}(t_2,t_1)$ $= \overline{x_1y_2} = M_{yx}(t_2 - t_1)$ (stat. process) $= M_{xy}(t_1 - t_2)$	$R_{xy}^{(j)}(t_1,t_2) = R_{yx}^{(j)}(t_2,t_1) = R_{yx}^{(j)}(t_2 - t_1)$ $= \langle x^{(j)}(t_1)y^{(j)}(t_2)\rangle$ $= R_{xy}^{(j)}(t_1 - t_2)$ Cross-correlation function
5. Covariance functions ($t = t_2 - t_1$) (for statistical averages only)	$K_y(t_1,t_2) = \overline{(y_1 - \bar{y}_1)(y_2 - \bar{y}_2)}$ $= M_y(t_1,t_2) - \bar{y}_1\bar{y}_2$ $= M_y(t) - \bar{y}^2$ (stat. process) Autovariance	$R_y^{(j)}(t_1,t_2) - \langle y^{(j)}(t_1)\rangle\langle y^{(j)}(t_2)\rangle$
	$K_{xy}(t_1,t_2) = \overline{(x_1 - \bar{x}_1)(y_2 - \bar{y}_2)}$ $= M_{xy}(t_1,t_2) - \overline{x_1y_2}$ $= M_{yx}(t) - \bar{x}\bar{y}$ (stat. process) Cross-variance	$R_{xy}^{(j)}(t_1,t_2) - \langle x^{(j)}(t_1)\rangle\langle y^{(j)}(t_2)\rangle$

and still more generally, where y, z, x, etc., are different random variables,

$$\overline{G(y_1,z_2, \ldots ,x_n)} = \int \cdots \int dy_1\, dz_2 \cdots dx_n\, G(y_1,t_1;z_2,t_2; \ldots ;x_n,t_n) \tag{1.120c}$$

Statistical and corresponding time averages of particular interest are summarized in Table 1.1.

For stationary processes (cf. Sec. 1.3-6), we can set $t_1 = 0$, remembering that t_2, t_3, etc., are measured now from $t_1 = 0$, and write Eq. (1.118b) accordingly. Specifically, we have

$$\overline{y_1{}^n} = \overline{y(t_1)^n} = \int y^n W_1(y)\, dy = \overline{y^n} \tag{1.121a}$$

$$\begin{aligned} M_y(t_1,t_2) = \overline{y_1y_2} = \overline{y(t_1)y(t_2)} &= \iint y_1y_2W_2(y_1, y_2; t_2 - t_1)\, dy_1\, dy_2 \\ &= \iint y_1y_2W_2(y_1,y_2;t)\, dy_1\, dy_2 = M_y(t), \text{ etc.} \end{aligned} \tag{1.121b}$$

where we adopt the convention that $t_2 - t_1 = t$, recalling that only the time differences $t_2 - t_1 = t$, $t_3 - t_1$, etc., between observations are important now. In particular, observe that, if in the limit $t_2 - t_1 \to \infty$, we obtain from Eq. (1.99)

$$\begin{aligned} \lim_{t_2-t_1\to\infty} \overline{y_1y_2} &= \iint y_1y_2 \lim_{t_2-t_1\to\infty} W_2(y_1, y_2; t_2 - t_1)\, dy_1\, dy_2 \\ &= \iint y_1y_2W_1(y_1)W_1(y_2)\, dy_1\, dy_2 = \bar{y}^2 \end{aligned} \tag{1.122}$$

for stationary processes, while in nonstationary cases this becomes $\lim_{t_2-t_1\to\infty} \overline{y_1y_2} = \bar{y}_1\bar{y}_2$. For purely random processes (Sec. 1.4-2), this is true for all finite $|t_2 - t_1| > 0$ and for all moments (that exist). Note, finally, that, if we go to the other extreme ($t_2 - t_1 \to 0$) of observations at identical times, we may use Eq. (1.100) to write

$$\begin{aligned} \lim_{t_2-t_1\to 0} \overline{y_1y_2} &= \iint y_1y_2W_1(y_1)\, \delta(y_2 - y_1)\, dy_1\, dy_2 \\ &= \int y_1{}^2W_1(y_1)\, dy_1 = \overline{y_1{}^2} = \overline{y^2} \end{aligned} \tag{1.123}$$

with $\overline{y_1{}^2} = \overline{y(t_1)^2}$ for nonstationary situations. From this and Eq. (1.122), it is clear that we can readily obtain the lower-order moments $\bar{y}^2$, $\overline{y^2}$ directly from $\overline{y_1y_2}$ by the suitable limit on $t_2 - t_1$. This is a consequence, of course, of the circumstance that $W_1(y_1)$, $W_1(y_2)$ are marginal distribution densities of W_2 and reflects the fact that first-order information is completely contained in the second- (and higher-) order distributions.

PROBLEMS

1.18 (a) Does Eq. (1.122) hold for deterministic processes? Discuss, considering Secs. 1.3-4, 1.3-5. Compare stationary and nonstationary cases.

(b) Show, for the ensemble $y(t,\phi) = A \cos (\omega_0 t + \phi)$ [A, ω_0 fixed, ϕ uniformly distributed in the primary interval $(0,2\pi)$], that by an actual calculation of $\overline{y_1y_2}$ and $\langle y_1^{(j)}y_2^{(j)}\rangle$

$$\begin{aligned} R_y^{(j)}(t_2 - t_1) &= \langle y^{(j)}(t_1)y^{(j)}(t_2)\rangle = \overline{y_1y_2} \\ &= \tfrac{1}{2}A_0^2 \cos \omega_0(t_2 - t_1) \qquad \text{probability 1} \end{aligned} \tag{1}$$

(c) Show that $R_y^{(j)}(t_2 - t_1) = R_y^{(j)}(t_1 - t_2)$ and that $K_y(t_2 - t_1) = K_y(t_1 - t_2)$.

(d) Show also that $0 \le |R_y^{(j)}(t_2 - t_1)| \le R_y^{(j)}(0)$ and, for the ensemble, that $0 \le |K_y(t_2 - t_1)| \le K_y(0)$, $0 \le |M_y(t_2 - t_1)| \le M_y(0)$.

1.19 (a) For the ensemble of additive and independent signal and noise processes, $y(t) = A \cos(\omega_0 t + \phi) + N(t)$, where $S(t,\phi) = A\cos(\omega_0 t + \phi)$ has the properties of Prob. 1.18b, show that the autocorrelation function of y is

$$R_y^{(j)}(t_2 - t_1) = R_S^{(j)}(t_2 - t_1) + R_N^{(j)}(t_2 - t_1) \tag{1}$$

and if $N(t)$ is normal, with $W_2(N_1, N_2; t_2 - t_1)$ given by Prob. 1.8a, show that

$$\begin{aligned} K_y(t_2 - t_1) &= \frac{A_0^2}{2} \cos \omega_0(t_2 - t_1) + \overline{N^2}\rho_N(t_2 - t_1) \\ &= K_S(t_2 - t_1) + K_N(t_2 - t_1) \end{aligned} \tag{2}$$

(b) If the noise contains a steady component $\bar{N} \neq 0$, modify Eq. (2) explicitly.

(c) If the noise contains a deterministic component ta (a fixed), what is $R_y^{(j)}$? What now is the form of the covariance function K_y?

(d) In (a), let us suppose that signal and noise are no longer independent. Show that the autocorrelation function of y is now

$$\begin{aligned} R_y^{(j)}(t) &= R_S^{(j)}(t) + R_N^{(j)}(t) + R_{SN}^{(j)}(t) + R_{NS}^{(j)}(t) \\ &= R_S^{(j)}(t) + R_N^{(j)}(t) + R_{SN}^{(j)}(t) + R_{SN}^{(j)}(-t) \end{aligned} \tag{3}$$

1.20 (a) Prove Wiener's lemma:[26,35] If $g(t) \ge 0$ for all $(-\infty < T < \infty)$, and if $\lim_{T\to\infty} \frac{1}{T}\int_{T/2}^{T/2} g(t_0)\,dt_0$ exists, then $\lim_{T\to\infty} \frac{1}{T}\int_{-T/2+t_1}^{T/2+t_1} g(t_0)\,dt_0$ exists for all (real) t_1 and these two limits are equal.

(b) Show that if $g(t)$ is not bounded at least from one side—i.e., if $g(t) \ge 0$ or $g(t) \le 0$ does not hold for all $(-\infty < t < \infty)$—Wiener's lemma fails.

(c) Extend (a) to cases where $g(t) \ge C$ or $g(t) \le C$ for all $(-\infty < t < \infty)$, for some real constant C.

1.6 *Ergodicity and Applications*

Although an effective theory of noise and signals in communication systems is necessarily based on probability methods (cf. Sec. 1.1), it is not usually possible to deal experimentally with the entire ensemble of possible signal and noise waves. One considers, instead, a single member (sometimes several) of the ensemble $y(t)$ and observes this member's behavior in the course of time, obtaining, for example, the time averages $\langle y^{(j)}\rangle$, $\langle (y^{(j)})^2\rangle$, $\langle y^{(j)}(t_1)y^{(j)}(t_1 + t)\rangle$, etc., discussed in Sec. 1.5.

Now, for a theory to be useful in applications, it must permit us to relate the a priori quantities, predicted by the theory from the assumed statistical properties of the ensemble, to the corresponding quantities actually observed experimentally (i.e., a posteriori) from the given member of the ensemble in the course of time. In other words, we wish to know (1) in what sense and under what conditions time and ensemble averages can be related and (2) whether our particular representation $y^{(j)}(t)$ belongs to an ensemble where this is possible. Sets for which a definite relationship between time and ensemble averages exists, so that (1) can be applied, are then said to be *ergodic*, and the conditions for an interchangeability of time and ensemble averages are established by a form of the *ergodic theorem*. The condition 2 may be called an *ergodic hypothesis*. This is the assumption, usually made, that we are actually dealing with a member of an ergodic ensemble and hence can apply the ergodic theorem, although in physical situations this is never strictly true, since it is not possible to maintain the required conditions, as we shall see in more detail presently. The notion of ergodicity appeared first in kinetic theory and classical statistical mechanics, mainly through the work[31] of Maxwell, Boltzmann, Clausius, and Gibbs, with more refined concepts culminating in various forms of ergodic theory in the researches of G. D. Birkhoff,[32] Von Neumann,[33] Hopf,[34] and others during the 1930s and subsequently and with important application to the theory of turbulence, also at about this time.[35] Applications of this concept to communication theory are more recent, dating in large part from the work of Wiener[36] in the early 1940s and appearing since in studies of signals and noise in electronic systems.[37]

1.6-1 The Ergodic Theorem. Let us begin by stating first what is meant by an ergodic ensemble:

> *An ensemble is ergodic if* (*a*) *it is stationary* (in the strict sense; cf. Sec. 1.3-6) and (*b*) *it contains no strictly stationary subensemble*(s) *with measure other than* 1 *or* 0 (1.124)

To illustrate, let us consider again the ensemble $y(t,\phi) = A_0 \cos(\omega_0 t + \phi)$ of Sec. 1.3-7(1). If ϕ is uniformly distributed in $(0,2\pi)$ and A_0, ω_0 are independent of time, then $y(t,\phi)$ is strictly stationary, as we have already shown. Moreover, if A_0, ω_0 are fixed, i.e., have measure 1 at a single value $(\sim A_0,\omega_0)$, then from the above we observe also that $y(t,\phi)$ is an ergodic ensemble since it contains no *stationary* subsets of measure different from 1 or 0. This is easily seen from the following example: Consider the subset for $[0 < \phi < a\ (< 2\pi)]$. Then, with an arbitrary (linear) translation, one gets ϕ', with the ϕ' subset as indicated in Fig. 1.13. The measure of the two subsets (in their respective primary intervals) is now quite different. The measure is no longer invariant, and the subset $(0 \leq \phi \leq a)$ is accordingly not stationary. On the other hand, if now we allow A_0 to have a distribution density of the form $W(A_0) = A_0 e^{-A_0^2/2\sigma^2}/\sigma^2$ $(A_0 \geq 0)$, $W(A_0) = 0$ $(A_0 < 0)$, with ϕ uniformly distributed as before in $(0,2\pi)$, the ensemble is still strictly stationary but it is no longer ergodic. The subset of functions for which A_0 lies between 1 and 2, for example, is also stationary and possesses a measure $\int_1^2 W(A_0)\, dA_0$ which is *not* 1 or 0.

We can now state the ergodic theorem in a form which allows us to interchange time and ensemble averages in almost all cases, when external conditions permit. Given that the ensemble in question is ergodic [Eq. (1.124)], the following has been proved:[32–34,38]

Ergodic Theorem. For an ergodic ensemble, the average of a function of the random variables over the ensemble is equal with probability unity to the average (of the corresponding function) *over all possible time translations*† *of a particular member function of the ensemble, except for a subset of representations of measure zero.*‡

In brief, except for a set of measure zero, time and statistical averages are equal (with probability 1) when the ensemble is ergodic. Thus, if

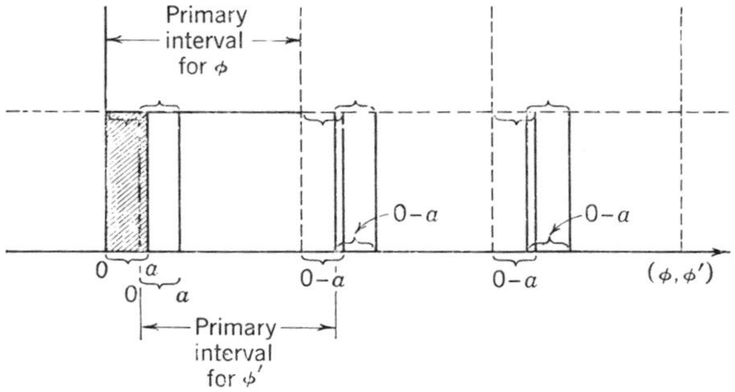

FIG. 1.13. A nonstationary subset $(0,\phi,a)$ of the ensemble $y(t,\phi) = A \cos (\omega_0 t + \phi)$.

$G^{(j)}(y_1, \ldots ,y_n)$ (cf. Sec. 1.5-1) is some (jth) member function of an ergodic ensemble, then the ergodic theorem above says that

$$\langle G^{(j)}[y(t_1), \ldots ,y(t_n)]\rangle = \overline{G(y_1;y_2,t_2; \ldots ;y_n,t_n)} \qquad \text{probability 1} \qquad (1.125a)$$

or

$$\langle G \rangle = \boldsymbol{E}\{G\} = \int_G GW(G)\, dG \qquad \text{probability 1} \qquad (1.125b)$$

if we wish to express the measure in the G rather than in the $(y_1, \ldots ,y_n)$ probability space.

The essential force of the theorem, in cruder language, is that, if the ensemble is ergodic, essentially any one function of the ensemble is typical of the ensemble as a whole and, consequently, operations performed in time on the typical member yield the same results as corresponding operations taken over the entire ensemble at any particular instant. It is easy to see that the condition of (strict) stationarity [Eq. (1.124a)] is necessary, for if it did not hold, ensemble averages at various times $t_1, t_2, \ldots$ would have different values in general and, moreover, would not be equal to the corre-

† The time translations which take the ensemble into itself, i.e., leave its measure invariant in the sense of strict stationarity (cf. Sec. 1.3-6), are said in this case to be *metrically transitive;* i.e., they have here the further property [Eq. (1.124b)] that only subsets in the ensemble which are stationary have measure 1 or 0.

‡ This theorem is also known as the strong law of large numbers for (strictly) stationary processes (cf. Doob[38]).

sponding time average over a particular member function (cf. Figs. 1.5, 1.6 for noise or Fig. 1.10*b* for a deterministic process). Strict stationarity is not enough, however, if we are to equate time and ensemble averages. The condition [Eq. (1.124*a*)] of metrical transitivity is needed to ensure that the jth member is truly representative. For example, in the case above where the amplitudes A_0 have the measure $W(A_0) = A_0 e^{-A_0^2/2\sigma^2}/\sigma^2$ $(A_0 > 0)$, members in the subset $(0 < A_0 < 2\sigma)$ are much more likely to be chosen than are members in the subset $(2\sigma < A_0 < \infty)$. Time averages on members from neither set can then be expected to give the same statistical average, when the whole ensemble is considered, since the members of neither set are representative of the entire ensemble. For some examples where metrical transitivity *does* hold, see the "geometrical" cases discussed by Kampé de Fériet[35] (chap. V, sec. 8).

The time average referred to in the statement (1.125) of the ergodic theorem is the principal limit [Eq. (1.107)]. However, if $G^{(j)}$ is chosen from an ergodic ensemble, then it is also true[39] that the one-sided averages of Eq. (1.108) exist and are equal, with probability 1, that is, for almost every member $G^{(j)}$, so that

$$\langle G^{(j)}\rangle_{+\infty} = \langle G^{(j)}\rangle_{-\infty} \qquad \text{probability 1} \tag{1.126}$$

and from this it follows at once that

$$\begin{aligned}\lim_{T\to\infty} \frac{1}{2T}\int_{-T}^{T} G^{(j)}(t_0)\,dt_0 &= \frac{1}{2}\left(\langle G^{(j)}\rangle_{+\infty} + \langle G^{(j)}\rangle_{-\infty}\right)\\ &= \langle G^{(j)}\rangle_{\pm\infty} = \langle G^{(j)}\rangle \qquad \text{probability 1}\end{aligned} \tag{1.127}$$

since the left-hand member of Eq. (1.127) is equivalent to $\langle G^{(j)}\rangle$ [Eq. (1.107)]. Because of ergodicity, $\langle G^{(j)}\rangle_{\pm\infty} = \langle G^{(j)}\rangle = \bar{G}^{(j)}$ (probability 1), and the limits are identical for (almost) all members of the ensemble. However, even without the additional condition of metrical transitivity required for ergodicity [i.e., Eq. (1.124*a*)], it is still true that Eqs. (1.126), (1.127) hold, provided only that the process G is strictly stationary and measurable with $\boldsymbol{E}\{|G|\} < \infty$.[39] The actual values of these limits vary, of course, with j (i.e., over the parameter of distribution), assuming the common limiting or ensemble value $\bar{G}$ with probability 1 when the process G is also metrically transitive.

As Wiener has emphasized, the equivalence of $\langle G^{(j)}\rangle_{+\infty}$, $\langle G^{(j)}\rangle_{-\infty}$, and $\langle G^{(j)}\rangle$ under the above conditions is a very important result, for it permits us to use the "past" of an individual member (where now we take $t_0 = 0$ in $\langle G^{(j)}\rangle_{-\infty}$ to be the "present" of the observation) to predict its "future" *in some average sense.* Thus, time averages carried out on the past $(-T,0)$ are equivalent (probability 1) to those which are to be carried out $(0,T)$ with respect to the disturbance's future behavior (lim $T \to \infty$). If, moreover, G is ergodic, in almost all cases we can determine beforehand from the given ensemble data just what this average should be, provided that it is possible actually to achieve these limiting operations in a physical system. The precise future structure of individual waveforms cannot, of course, be determined from the past, unless the process is completely deterministic, which is clearly a completely trivial case for communication purposes. The information-bearing members are all essentially unpredictable in this strict sense: it is only the ensemble properties, and the time averages related to them, that are predictable and hence useful quantitatively.

1.6-2 Applications to Signal and Noise Ensembles. In many problems of statistical communication theory, the signal, noise, and noise and signal ensembles together are not ergodic, even for the idealized situations where stationarity is assumed. The theorem [Eq. (1.125)] cannot then be applied directly. However, it often happens that these more complicated ensembles are made up of subensembles which are themselves ergodic, and to these latter the theorem can be applied separately. In other cases, both time and statistical averages may be required to determine the expected behavior of a system. Let us illustrate these remarks with some typical examples.

Example 1. As the first illustration of the interchangeability of time and ensemble averages under suitable conditions, consider the ensemble $y(t) = a(t) \cos [\omega_0 t + \phi(t)]$, where $a(t)$, $\phi(t)$ are, respectively, an *envelope* and *phase* that are entirely random; ω_0 is fixed. If $a(t)$ and $\phi(t)$ are stationary, *and* if $\phi(t)$ is uniformly distributed over $(0,2\pi)$, the ensemble $y(t)$ is also stationary with $\bar{y} = 0$. Furthermore, if both $a(t)$ and $\phi(t)$ are ergodic [cf. Eq. (1.124)], then so also is y. Now suppose that $a(t)$, $\phi(t)$ are statistically independent, that we are given the particular data $y^{(j)}(t)$ over $(-\infty < t < \infty)$, and that we wish to find the autocorrelation function $\langle y^{(j)}(t_0)y^{(j)}(t_0 + t)\rangle$ [cf. Eq. (1.113)], viz.,

$$R_{y^{(j)}}(t) = \lim_{T\to\infty} \frac{1}{T} \int_{-T/2}^{T/2} y^{(j)}(t_0)y^{(j)}(t_0 + t)\, dt_0 \tag{1.128}$$

This we can do directly by performing the indicated operations of Eq. (1.128) on the given data. However, if the statistical properties of the ensemble are known, we can also obtain $R^{(j)}(t)$ directly from the covariance function $K(|t|)$ (cf. Table 1.1) with the help of the ergodic theorem (1.125). We have

$$R_{y^{(j)}}(t) = \langle y_1^{(j)}y_2^{(j)}\rangle = K_y(t) = \overline{y_1y_2} = \overline{y(t_1)y(t_2)} \qquad t = t_2 - t_1,\ \bar{y} = 0, \text{ probability 1} \tag{1.129}$$

Explicitly, this becomes

$$\begin{aligned} R_{y^{(j)}}(t) &= \int \cdots \int a(t_1)a(t_2) \cos [\omega_0 t_1 + \phi(t_2)] \\ &\qquad \times \cos [\omega_0 t_2 + \phi(t_2)] W_2(a_1,\phi_1;a_2,\phi_2;t)\, da_1 \cdots d\phi_2 \\ &= \int \cdots \int a_1a_2 \cos (\omega_0 t_1 + \phi) \\ &\qquad \times \cos (\omega_0 t_2 + \phi) W_2(a_1,a_2;\phi;t)\, da_1\, da_2\, d\phi \qquad \text{probability 1} \end{aligned} \tag{1.130a}$$

since ϕ is uniformly distributed. Also, since the (a,ϕ) processes are stationary and independent, we have $\overline{[\cos(\omega_0 t_1 + \phi)\cos(\omega_0 t_2 + \phi)]}_\phi = \overline{[\cos\phi' \cos(\omega_0 t + \phi')]}_\phi = \frac{1}{2}\cos\omega_0 t$, so that finally

$$R_y^{(j)}(t) = \tfrac{1}{2}\cos\omega_0 t \iint_{-\infty}^{\infty} a_1 a_2 W_2(a_1,a_2;t)\, da_1\, da_2$$

$$= \tfrac{1}{2}\overline{a_1 a_2}\cos\omega_0 t \qquad \text{probability 1} \tag{1.130b}$$

The $y(t)$ may represent an ensemble of narrowband noise voltage or current waves such as are encountered in radio or radar reception or an amplitude modulation of a carrier $\cos(\omega_0 t + \phi)$ (cf. Chaps. 12, 13, 17 to 19 for specific applications).

Example 2. Mixed ensembles, consisting of signal and noise processes in some linear or nonlinear combination, offer further opportunity for the interchange of time and ensemble averages. Even if the combined process is ergodic, the most convenient form of result is often a mixture of both types of average, as our example below indicates. Accordingly, instead of noise alone, let us consider now the ensemble $z = g(y)$, where g is a one-valued function of y and $y = y(t)$ is the sum of a noise ensemble $N(t)$ and a signal set $S(t)$, for example, $y = S + N$, with the additional conditions that S and N are statistically independent and ergodic. The function, or transformation, g, for example, may represent the effect of an amplifier or a second detector in a radio receiver, or some other linear or nonlinear† operation, such that the instantaneous output of the device in question is

$$z^{(j)}(t) = g[y^{(j)}(t)] \qquad y^{(j)}(t) = S^{(j)}(t) + N^{(j)}(t) \tag{1.131}$$

when $y^{(j)}(t)$ is the input wave.

Let us calculate the correlation function of this output wave, viz.,

$$R_z^{(j)}(t) = \lim_{T\to\infty} \frac{1}{T}\int_{-T/2}^{T/2} g[y^{(j)}(t_0)]g[y^{(j)}(t_0 + t)]\, dt_0 \tag{1.132}$$

from the corresponding statistical average $\overline{g_1 g_2}$, with the help of the ergodic theorem. We have

$$R_z^{(j)}(t) = \begin{cases} \overline{g(t_0)g(t_0 + t)} = \boldsymbol{E}\{g(t_0)g(t_0+t)\} & \text{probability 1} \\ \iint g(y_1)g(y_2)W_2(y_1,y_2;t)\, dy_1\, dy_2 = \overline{g_1 g_2} & t = t_2 - t_1 \end{cases} \tag{1.133a}$$

When the signal is at least partially deterministic, usually it is easier to evaluate the integrals in Eq. (1.133a) by taking advantage of the explicit signal structure. Writing

$$R_z^{(j)}(t) = \overline{g(S_1 + N_1)g(S_2 + N_2)} = \int \cdots \int g(S_1 + N_1)g(S_2 + N_2) \times W_2(S_1,N_1;S_2,N_2;t)_{S+N}\, dS_1 \cdots dN_2 \tag{1.133b}$$

† For a definition of linear and nonlinear operation, see Sec. 2.1-3(2).

where $(W_2)_{S+N}$ now is the joint second-order density of S and N, and remembering that S and N are independent, we see that $(W_2)_{S+N}$ factors, namely, $(W_2)_{S+N} = (W_2)_S(W_2)_N$. Next, we employ the ergodic theorem again, now to replace the statistical average over the signal components by the equivalent time average, which makes direct use of the signal's functional form. The final result is a mixed expression containing the statistical average over the noise terms and the time average over the signal elements, viz.,

$$\begin{aligned} R_{z=g}^{(j)}(t) = \overline{g_1 g_2} &= \langle \textstyle\iint g(S_1{}^{(j)} + N_1)g(S_2{}^{(j)} + N_2)W_2(N_1,N_2;t)\, dN_1\, dN_2 \rangle_S \\ &= \langle \overline{g(S_1{}^{(j)} + N_1)g(S_2{}^{(j)} + N_2)^{(N)}} \rangle_S \qquad \text{probability 1} \end{aligned} \tag{1.133c}$$

where, following the convention of Sec. 1.5, the bar refers to the ensemble average and the pointed brackets $\langle\ \rangle_S$ to the time average.

Example 3. In Examples 1 and 2, we considered stationary (and ergodic) systems only. What happens now when we wish to calculate averages for nonstationary systems? Here the ergodic theorem cannot be applied *in toto* to interchange time and ensemble averages. Instead, for useful results we must apply both time and statistical averages. Let us illustrate by modifying Example 2, extending it to include the case of a periodic amplitude modulation of a carrier $\cos \omega_0 t$, so that the signal $S(t)$ has the deterministic structure $S(t) = V_{mod}(t) \cos \omega_0 t$ $[V_{mod}(t) \geq 0]$. Then the ensemble $y(t) = S(t) + N(t)$ for the input to a *nonlinear system* g, and the corresponding ensemble $z = g(y)$ for the output, are no longer stationary (cf. Sec. 1.3-7, also Prob. 1.9*b*) and therefore not ergodic. In this case, we cannot compute the correlation function (for either y or z) a priori from the ensemble properties of y and the transformation g, but only the average of this quantity over the ensemble (when such an average exists). Thus, we have

$$\begin{aligned} \overline{R_z{}^{(j)}(t)} &= \lim_{T\to\infty} \frac{1}{T} \int_{-T/2}^{T/2} \overline{g[y^{(j)}(t_0)]g[y^{(j)}(t_0 + t)]}\, dt_0 \\ &= \langle \overline{g(y_1{}^{(j)})g(y_2{}^{(j)})}^{(C,N)} \rangle_{mod} \end{aligned} \tag{1.134a}$$

$$\begin{aligned} \overline{R_z{}^{(j)}(t)} = \langle \textstyle\int \cdots \int g(S_1{}^{(j)} + N_1)g(S_2{}^{(j)} + N_2)W_2(S_1,N_1;S_2,N_2;t)_{C+N} \\ \times\, dN_1 \cdots dS_2 \rangle_{mod} \end{aligned} \tag{1.134b}$$

$$\overline{R_z{}^{(j)}(t)} = \langle\langle \overline{g(S_1{}^{(j)} + N_1)g(S_2{}^{(j)} + N_2)^{(N)}} \rangle_C \rangle_{mod} \tag{1.134c}$$

where we have used the fact that for fixed phases ($\sim t'$) of the modulation the subensemble formed of modulated carrier plus noise, for example, $y' = V_{mod}(t') \cos (\omega_0 t + \phi) + N(t)$ (t' fixed), is ergodic. Finally, we perform a time (or equivalent statistical) average over the epochs t' of the modulation as indicated by $\langle\ \rangle_{mod}$ in Eqs. (1.134*b*), (1.134*c*); $\langle\ \rangle_C$ is the time average over the carrier, $\cos \omega_0 t$, which with $\langle\ \rangle_{mod}$ takes advantage of the deterministic structure of the signal. For nonstationary systems, then, both types of averaging are needed if one is to obtain results independent of the particular member of the ensemble originally available, unlike the entirely ergodic cases of Examples 1 and 2, where only one averaging process is required.

Usually, we are interested in the (ergodic) signal or modulation subensemble after reception in noise. To calculate its correlation function, for example, or to observe and obtain the latter experimentally from the data of a particular representation, we must be able to separate it from the carrier ensemble. The ergodic theorem may then be applied directly to this subensemble, as in Example 1 or 2. A radio or radar receiver actually performs this separation. The process of detection and filtering embodied in the transformation g enables us to pick out the desired signal (contaminated by noise, of course), after which the various time or ensemble averages can be performed upon it. When the modulation is not ergodic, both types of average are again required. Finally, we emphasize once more that ergodicity here is an *assumption*, an idealization which can give only an approximate account of actual physical situations, because of finite observation times, e.g., finite samples in the statistical sense, and because of the ultimately nonstationary character of all physical processes. Nevertheless, for many applications in statistical communication theory the notion of ergodicity is a useful one and is a convenient starting point for the technically more involved and realistic, but conceptually no more difficult treatment which specifically considers the finite observation period or sample size (cf. Chaps. 17 to 23).

REFERENCES

1. Wiener, N.: "Extrapolation, Interpolation, and Smoothing of Stationary Time Series," Technology Press, Cambridge, Mass., and Wiley, New York, 1949; published originally at Massachusetts Institute of Technology, Feb. 1, 1942, as a report on DIC Contract 6037, National Research Council, sec. D2.
2. Shannon, C. E.: A Mathematical Theory of Communication, *Bell System Tech. J.*, **27**: 379, 623 (1949).
3. Rice, S. O.: Mathematical Analysis of Random Noise, *Bell System Tech. J.*, **23**: 282 (1944), **24**: 46 (1945).
4. Wald, A.: "Statistical Decision Functions," Wiley, New York, 1950.
5. Van Meter, D., and D. Middleton: Modern Statistical Approaches to Reception in Communication Theory, *IRE Trans. on Inform. Theory*, **PGIT-4**: 119 (September, 1954).
6. Middleton, D., and D. Van Meter: Detection and Extraction of Signals in Noise from the Point of View of Statistical Decision Theory, *J. Soc. Ind. Appl. Math.*, **3**(4): 192 (December, 1955), **4**(2): 86 (June, 1956).
7. Cramér, H.: "Mathematical Methods of Statistics," Princeton University Press, Princeton, N.J., 1946.
8. Kolmogoroff, A.: "Grundbegriffe der Wahrscheinlichkeitsrechnung," Berlin, 1933.
9. Borel, E., et al.: "Traité du calcul des probabilités et de ses applications," Paris, 1924ff.

10. Fréchet, M.: "Recherches théoriques modernes sur la théorie des probabilités," Ref. 9, vol. 1, pt. III, 1937, 1938.
11. Uspensky, J. V.: "Introduction to Mathematical Probability," McGraw-Hill, New York, 1937.
12. Feller, W.: "An Introduction to Probability Theory and Its Applications," Wiley, New York, 1950.
13. Arley, N., and K. R. Buch: "Introduction to the Theory of Probability and Statistics," Wiley, New York, 1950.
14. Van der Pol, B., and H. Bremmer: "Operational Calculus, Based on the Two-sided Laplace Transform," Cambridge, New York, 1950.
15. Clavier, P. A.: Some Applications of the Laurent Schwartz Distribution Theory to Network Problems, *Proc. Symposium on Modern Network Synthesis,* 1956, pp. 249–265.
15a. Lighthill, M. J.: "An Introduction to Fourier Analysis and Generalized Functions," Cambridge, New York, 1958.
16. Doob, J. L.: "Stochastic Processes," Wiley, New York, 1950. See chap. 2, secs. 1 and 2, for example.
17. Bartlett, M. S.: "An Introduction to Stochastic Processes," secs. 1–3, Cambridge, New York, 1955.
18. Blanc-Lapierre, A., and R. Fortet: "Théorie des fonctions aléatoires," chaps. 3, 4, Masson et Cie, Paris, 1953.
19. Rayleigh, Lord: "Scientific Papers," vol. 1, p. 491; vol. 3, p. 473; vol. 4, p. 370; vol. 6, p. 604, 1920. For further references, see Chap. 7.
20. Ref. 17, chap. 2, also parts of chap. 3.
21. Wiener, N.: Generalized Harmonic Analysis, *Acta Math.,* **55**: 117 (1930).
22. Kolmogoroff, A.: Korrelations Theorie der stationaren stochastichen Prozesse, *Math. Ann.,* **109**: 604 (1934).
23. Lévy, P.: "Processes stochastiques et mouvement brownien," Paris, 1948.
24. Moyal, J. F.: Stochastic Processes and Statistical Physics, *J. Research Roy. Stat. Soc.,* **B11**: 167 (1949).
25. Ref. 12, chap. 17.
26. For a concise discussion of the notion of a stationary random function of time, from a more rigorous mathematical viewpoint, see J. Kampé de Fériet, Intoduction to the Statistical Theory of Turbulence. III, *Soc. Ind. Appl. Math.,* **2**: 244 (1954), especially chap. V, sec. 6.
27. Knudtzon, N.: Experimental Study of Statistical Characteristics of Filtered Random Noise, *MIT Research Lab. Electronics Tech. Rept.* 115, July 15, 1949; cf. fig. 2-10.
28. Markoff, A.: Extension of the Law of Large Numbers to Dependent Events, *Bull. Soc. Phys. Math. Kazan,* (2)**15**: 135–156 (1906); "Wahrscheinlichkeitsrechnung," Leipzig, 1912.
29. Wang, M. C., and G. E. Uhlenbeck: On the Theory of the Brownian Motion. II, *Revs. Modern Phys.,* **17**: 323 (1945).
30. Ref. 12, sec. 17.9; Ref. 17, sec. 2.2, for example.
31. Ter Haar, D.: Foundations of Statistical Mechanics, *Revs. Modern Phys.,* **27**: 289 (1955); detailed discussion and very extensive bibliography. The interested reader is also referred to Ter Haar, "Elements of Statistical Mechanics," app. I, Rinehart, New York, 1954. See also A. I. Khintchine, "Statistical Mechanics," chaps. II, III, Dover, New York, 1949.
32. Birkhoff, G. D.: Proof of the Ergodic Theorem, *Proc. Natl. Acad. Sci. U.S.,* **17**: 650, 656 (1931).
33. Von Neumann, J.: Proof of the Quasi-ergodic Hypothesis, *Proc. Natl. Acad. Sci. U.S.,* **18**: 70 (1932).
34. Hopf, E.: "Ergodentheorie," Chelsea, New York, 1948.

35. For a detailed discussion, see Ref. 26, I, **2**: 1 (1954); II, **2**: 143 (1954); III, **2**: 244 (1954); IV, **3**: 90 (1955), with an extensive bibliography of earlier work in **2**: 143 (1954); cf. pp. 172–174 and also especially pt. III, chap. V.
36. Ref. 1, secs. 0.8, 1.4.
37. See, for example, Ref. 3; D. Middleton, Some General Results in the Theory of Noise through Non Linear Devices, *Quart. Appl. Math.*, **5**: 445 (1948); also, Chap. 13.
38. Doob, J. L.: "Stochastic Processes," Wiley, New York, 1953; cf. p. 464 and also pp. 515ff.
39. Ref. 38, pp. 515ff.

CHAPTER 2

OPERATIONS ON ENSEMBLES

2.1 *Operations on Ensembles*

As has been mentioned earlier (Sec. 1.1), communication theory may be considered to be a theory involving various types of operations on ensembles. These operations are transformations, like $z = g(y)$ in Examples 2, 3 of Sec. 1.6-2, which convert an initial, or *input*, ensemble into a final, or *output*, ensemble, representing such physical operations as modulation, filtering, rectification, etc. Mathematically, the most common of these are the linear ones of differentiation and integration and the nonlinear ones of product formation,† raising to powers, and mapping by means of special transformations like $z = g(y) = y^{\nu}$ $(y \geq 0, \nu > 0)$, $z = g(y) = 0$ $(y < 0)$, etc. While it is clear how these operations affect an individual member of the input ensemble, it is necessary to extend them to the ensemble as a whole, so that they take on a significance with respect to the probability measures describing the ensemble. Differentiation, for example, is no longer simply defined as a limiting process for a single member, but as a limit which must hold for almost all members (i.e., with probability 1); similarly for integration and for the various nonlinear transformations of interest to us here. A few of the more refined concepts needed to justify formal operations are briefly summarized in the following paragraphs,‡ to remind us of some of the mathematical questions that are raised in the treatment of actual communication processes and to indicate, without any attempt at completeness, something of the existing rigorous apparatus for handling such questions.[2,3]

2.1-1 Stochastic Convergence. Let us consider first a sequence of random variables§ $Y_1, Y_2, \ldots, Y_n$. There are three principal ways in which such a sequence is said to converge to a *stochastic limit* Y.

(1) Convergence in probability (CIP). Y_n converges to Y in probability as $n \to \infty$, if for any $\epsilon > 0$

$$\lim_{n\to\infty} P(|Y_n - Y| > \epsilon) = 0 \tag{2.1a}$$

† See Secs. 2.2, 2.3 for definitions of linear and nonlinear operations.

‡ For a more detailed discussion, see Bartlett,[1] secs. 5.1, 5.11.

§ We remark that Y_1, Y_2, . . . are real; however, the following can be extended directly to complex random variables, where the measure is defined appropriately for both the real and the imaginary parts (Blanc-LaPierre and Fortet,[4] chap. 2, sec. 5).

A necessary and sufficient condition is that for any $(\epsilon,\lambda) > 0$ there exists an n' such that

$$P(|Y_n - Y_m| > \epsilon) < \lambda \qquad \text{for all } n, m \geq n' \tag{2.1b}$$

The limit Y is unique in the following way: If $Y_n \to Y$ (CIP) and $Y_n \to Z$ (CIP), then $Y = Z$ almost certainly, that is, $P(Y - Z = 0) = 1$.

(2) Almost certain, or strong, convergence (ACC). A stronger criterion of convergence, implying convergence in probability, is defined in terms of the *entire* sequence of stochastic variables $Y_1, \ldots, Y_n$. Letting such a sequence be regarded now as a new random variable with the values $y_1, y_2, \ldots, y_n, \ldots$, we say that this sequence of values does or does not converge in the usual sense to the limit y. If $P[Y_n(= y_n) \to Y(= y)] = 1$, then $Y_n \to Y$ almost certainly (ACC). This is equivalent to the condition of strong convergence in probability theory, which is defined by

$$\lim_{n\to\infty} P(|Y_m - Y| > \epsilon \text{ for at least one } m \geq n) = 0 \qquad \text{any } \epsilon > 0 \tag{2.2}$$

The following are sufficient conditions here:

$$\begin{array}{llll} \text{If} & \sum_n \boldsymbol{E}\{|Y_n - Y|^q\} < \infty & q > 0 & \\ \text{then} & Y_n \to Y & \text{ACC} & (2.3) \\ \text{or if} & \sum_n \boldsymbol{E}\{(|Y_{n+1} - Y_n|\epsilon_n^{-1})^q\} < \infty & \text{when } \Sigma_n\epsilon_n < \infty & \\ \text{then} & Y_n \to Y & \text{ACC} & \end{array}$$

(3) Convergence in the mean square (MSC). This is a particularly useful concept in the theory of stochastic processes and is defined in general by

$$\lim_{n\to\infty} \boldsymbol{E}\{|Y_n - Y|^q\} = 0 \qquad q \geq 1 \tag{2.4}$$

where unless otherwise indicated we shall always take $q = 2$, for *mean-square convergence*. This is then analogous to the usual notion of "convergence in the mean" of analysis,† for example, $\underset{n\to\infty}{\text{l.i.m.}}\, Y_n = Y$, but in order to avoid confusion with the statistical usage of "in the mean" we refer to this as mean-square convergence, that is, $\underset{n\to\infty}{\text{MSC}}\, Y_n = Y \ (q = 2)$. The necessary and sufficient condition for MSC, corresponding to Eqs. (2.1a), (2.1b) for CIP and ACC, is

$$\boldsymbol{E}\{|Y_n - Y_m|^2\} < \epsilon \qquad \text{for all } n \text{ with } m > n' \tag{2.5}$$

Observe that such ACC or MSC implies CIP and at the same time the uniform convergence of the d.f. $D_n(y)$ to the limit $D(y)$ at all points of continuity of $D(y)$.

† See footnote †, p. 70.

(4) Random functions. These notions of convergence can be extended from random sequences, where the parameter t takes discrete values only, to random variables, where the parameter t is continuous over a finite interval. Then we may consider the stochastic convergence of the random variable $Y(t)$, where t is continuous in a (nonzero) interval, to $Y(\tau)$ as $t \to \tau$. We say that $Y(t)$ is mean-square *continuous* at $t = \tau$ if

$$\underset{t \to \tau}{\text{MSC}}\ Y(t) = Y(\tau) \qquad \text{or} \qquad \lim_{t \to \tau} \boldsymbol{E}\{|Y(t) - Y(\tau)|^2\} = 0 \tag{2.6}$$

Carrying out the operations indicated in Eq. (2.6) then gives the condition†

$$\begin{aligned} \lim_{t_1,t_2 \to \tau} \boldsymbol{E}\{Y(t_1)Y(t_2)\} &= \lim_{t_1,t_2 \to \tau} M_y(t_1,t_2) = M_y(\tau,\tau) \\ &= \boldsymbol{E}\{|Y(\tau)|^2\} \end{aligned} \tag{2.7a}$$

where M_y is the second-moment function [cf. Eq. (1.120*a*)]. From this, one can show that the expectation and MSC operations commute (cf. Prob. 2.1).

In a similar way, one can show that the limit and expectation operations also commute for other than the mean-square case, provided, of course, that the moments in question are continuous at the limit points. For example, one finds that

$$\begin{aligned} \lim_{t_1 \to \tau_1, \ldots, t_m \to \tau_m} & \boldsymbol{E}\{Y(t_1) \cdots Y(t_m)\} \\ &= \lim_{t_1 \to \tau_1, \ldots} M_y(t_1, \ldots, t_m) \\ &= M_y(\tau_1, \ldots, \tau_m) = \boldsymbol{E}\{Y(\tau_1) \cdots Y(\tau_m)\} \qquad m \geq 1 \end{aligned} \tag{2.7b}$$

when $M_y(\tau_1, \ldots, \tau_m)$ exists.

The extension of the idea of almost certain, or strong, convergence to continuous functions of the parameter t requires that it be applicable to almost all realizations $y(t)$ of the ensemble, as $t \to \tau$, and this in turn implies certain further conditions on the regularity of these possible member functions. Thus, for example, stochastic convergence of $Y(t_n)$ to $Y(t)$ for any arbitrary sequence of $t_n \to \tau$ is no longer sufficient for almost certain convergence, although it is the natural property for examining the convergence in probability or mean-square convergence of $Y(t) \to Y(\tau)$. Mean-square or in-probability continuity at every point t of an interval $(0 \leq t \leq T)$ defines continuity mean square (or in probability) over that interval, but this is not necessarily true any more of almost certain (or strong) convergence over that interval, which is the really useful concept. A sufficient condition for the latter is

$$\boldsymbol{E}\{|Y(t) - Y(t + \epsilon)|^2\} \leq C|\epsilon|^{a+1} \qquad C, a > 0 \text{ for all } t, t + \epsilon \text{ in } (0,T) \tag{2.8}$$

We remark at this point that these more refined questions of convergence frequently appear in our analysis of physical problems, particularly in the

† For a discussion of continuity of a function, see, for example, R. Courant, "Differential and Integral Calculus," vol. I, pp. 49ff., Nordemann, New York, 1938.

description of systems that employ continuous or analogue sampling techniques over a finite observation interval $(0,T)$. While almost certain continuity and convergence can safely be postulated in most cases, it is often necessary to verify that operations carried out on the basis of such assumptions do indeed lead to meaningful results.

2.1-2 Stochastic Differentiation and Integration. In a similar fashion, we may employ the above concepts in differentiation and integration. Accordingly, we say that $\dot{Y}(t)$ is the *mean-square* differential coefficient of $Y(t)$ when†

$$\lim_{\epsilon\to 0} \boldsymbol{E}\left\{\left| \frac{Y(t+\epsilon) - Y(t)}{\epsilon} - \dot{Y}(t) \right|^2\right\} = 0 \tag{2.9}$$

and as may be directly shown from Sec. 2.1-1(4) (cf. Prob. 2.1*b*), the equivalent condition is that the second-moment function $M_y(t_1,t_2)$ of $Y(t)$ has the partial derivatives

$$\frac{\partial M_y}{\partial t_1}, \frac{\partial M_y}{\partial t_2}, \frac{\partial^2 M_y}{\partial t_1\,\partial t_2} = \frac{\partial^2 M_y}{\partial t_2\,\partial t_1} \qquad \text{all at } t_1 = t_2 \tag{2.10a}$$

These partial derivatives exist on the whole line $t_1 = t_2$, and over the entire (t_1,t_2) plane, if $\dot{Y}(t)$ exists at all points of t_1, so that we can also write

$$\begin{aligned}\boldsymbol{E}\{\dot{Y}(t_1)\} &= \boldsymbol{E}\left\{\lim_{\epsilon\to 0} \frac{Y(t_1+\epsilon) - Y(t_1)}{\epsilon}\right\} \\ &= \lim_{\epsilon\to 0} \frac{\boldsymbol{E}\{Y(t_1+\epsilon)\} - \boldsymbol{E}\{Y(t_1)\}}{\epsilon} = \frac{\partial}{\partial t_1}\boldsymbol{E}\{Y(t_1)\}\end{aligned} \tag{2.10b}$$

since by Eq. (2.7*b*) the expectation and limit operations commute (provided that $\boldsymbol{E}\{Y(t_1)\}$ and its derivative at t_1 are continuous in the usual sense). Similarly, we get

$$\boldsymbol{E}\{\dot{Y}(t_1)Y(t_2)\} = \frac{\partial}{\partial t_1} M_y(t_1,t_2) \qquad \boldsymbol{E}\{\dot{Y}(t_1)\dot{Y}(t_2)\} = \frac{\partial^2}{\partial t_1\,\partial t_2} M_y(t_1,t_2) \qquad \text{etc.} \tag{2.10c}$$

Repeating the argument applied above to $\dot{Y}$, we get the conditions for $\ddot{Y}$ to exist stochastically, and so on for $\dddot{Y}$, etc. Then $Y(t)$ can be developed in a mean-square-convergent Taylor's series about t', in the region for t and t',

$$Y(t) = \sum_{n=0}^{\infty} \frac{(t-t')^n}{n!} Y^{(n)}(t') \tag{2.11}$$

† Usually, we deal with real processes, corresponding to the fact that physical inputs to physical systems result in real, rather than complex, output processes. An exception is the analytical situation where we are interested in the spectral properties of these processes (cf. Secs. 3.2, 3.3).

provided, of course, that all derivatives $Y^{(n)}$ exist in the mean-square sense, i.e., provided that the moment function $M_y(t_1,t_2)$ possesses the required higher-order partial derivatives for all (t,t') in the region. Note that if $y(t)$ is a stationary process, so that $M_y(t_1,t_2) = M_y(|t_1 - t_2|)$, it is necessarily differentiable for all t_1, if differentiable at a point. When $\dot{Y}$ exists (mean square), we say that the corresponding process $y(t)$ is differentiable through the first order (mean square) if $\dot{Y}$, $\ddot{Y}$ exist (mean square) through the second order, and so forth. Some models of physical processes are not differentiable at all, such as thermal (or shot) noise through a reasonably narrow RC filter (cf. Sec. 3.3), while others may not be differentiable beyond a given order, e.g., thermal noise through a linear passive lumped-constant network consisting of a finite number of independent LRC meshes (cf. Sec. 2.2-1), the order of differentiability being determined by the number of such meshes.

Integration for random variables may now be defined over the ensemble in a similar way.† If one defines the (Riemann) integrals

$$X \equiv \int_0^T A(t)U(t)\,dt \qquad Y \equiv \int_0^T B(t)V(t)\,dt \tag{2.12a}$$

in the mean-square sense, where A, B are real bounded functions, U, V are random variables, the necessary and sufficient condition that either of the integrals in Eq. (2.12a) exists in this sense mean square is that‡

$$\begin{aligned} \boldsymbol{E}\{X^2\} &= \iint_0^T A(t_1)A(t_2)\boldsymbol{E}\{U(t_1)U(t_2)\}\,dt_1\,dt_2 \\ \text{or} \qquad \boldsymbol{E}\{Y^2\} &= \iint_0^T B(t_1)B(t_2)\boldsymbol{E}\{V(t_1)V(t_2)\}\,dt_1\,dt_2 \end{aligned} \tag{2.12b}$$

exists in the usual way. The operations involved here are

$$\boldsymbol{E}\{X^2\} = \boldsymbol{E}\left\{\iint_0^T A(t_1)A(t_2)U(t_1)U(t_2)\,dt_1\,dt_2\right\} \qquad \text{etc.}$$

and, replacing the integrals by their usual definitions as the limit of a sum, one may use the result above, that the limit and expectation operations commute, to obtain Eq. (2.12b) as indicated. The condition on the interchangeability of integration and expectation is, of course, the existence of the resulting integral.

† Doob.[5] Bartlett, *op. cit.*, sec. 5.11.

‡ The random variables $U(t_1)$, $V(t_1)$ may also be considered as points (for each t) in a (separable) Hilbert space if $\boldsymbol{E}\{U(t_1)U(t_2)\}$, $\boldsymbol{E}\{V(t_1)V(t_2)\}$ are continuous mean square. For a discussion, see Blanc-LaPierre and Fortet, *op. cit.*, p. 89, and also, for differentiability, pp. 90, 91, and, in general, pt. C, chap. 3.

Similarly, the condition for the existence (mean square) of the two-dimensional random variable $Z = (X,Y)$, with X, Y defined by the integrals in Eq. (2.12*a*), is governed by the existence in the usual (Riemann) sense† of

$$\boldsymbol{E}\{XY\} = \iint_0^T A(t_1)B(t_2)\boldsymbol{E}\{U(t_1)V(t_2)\}\,dt_1\,dt_2 \tag{2.12c}$$

For complex variables, one has directly

$$\boldsymbol{E}\{XX^*\} = \iint_0^T A(t_1)A(t_2)^*\boldsymbol{E}\{U(t_1)U(t_2)^*\}\,dt_1\,dt_2 \tag{2.12d}$$

with corresponding expressions for $\boldsymbol{E}\{YY^*\}$, $\boldsymbol{E}\{XY^*\}$, etc. All these notions are readily extended to infinite ranges of integration, random functions of more than one variable [as suggested by Eq. (2.12*c*)], complex random variables, etc.‡

For example, suppose that $x(t)$, $y(t)$ are two *output* ensembles or random processes, determined from the *input* processes $u(t)$, $v(t)$ by

$$x(t) = \iint_{-\infty}^{\infty} A(t,\tau)u(\tau)\,d\tau \qquad y(t) = \iint_{-\infty}^{\infty} B(t,\tau)v(\tau)\,d\tau \tag{2.13a}$$

A, B suitably bounded. Then $x(t)$, $y(t)$ as defined above exist mean square, provided that

$$M_{xy}(t_1,t_2) = \overline{x_1y_2} = \iint_{-\infty}^{\infty} A(t_1,\tau_1)B(t_2,\tau_2)\overline{u(\tau_1)v(\tau_2)}\,d\tau_1\,d\tau_2 \tag{2.13b}$$

exists in the ordinary (Riemann) sense [cf. Eq. (2.12*b*) and footnotes]. If $M_{xy}(t_1,t_2) = M_{xy}(t_1 - t_2) = M_{yx}(t_2 - t_1)$, the output processes are wide-sense stationary (cf. Sec. 1.3-6). Note that for nonlinear operations§ on an ensemble $z(t)$, where, say, $u(t)$ and $v(t)$ are obtained by the transformations $u = g_1(z)$, $v = g_2(z)$, the corresponding condition for the (mean-square) existence of $x(t)$, $y(t)$ as defined by Eq. (2.13*a*) is, like Eq. (2.13*b*), the existence of

$$M_{xy}(t_1,t_2) = \iint_{-\infty}^{\infty} A(t_1,\tau_1)B(t_2,\tau_2)\,\overline{g_1[z(\tau_1)]g_2[z(\tau_2)]}\,d\tau_1\,d\tau_2 \tag{2.14}$$

† Riemann-Stieltjes integrals may be defined (mean square) in similar fashion. For remarks on Lebesgue-Stieltjes integration, see Bartlett, *op. cit.*, p. 142. We remark that, in the ordinary nonprobabilistic sense, *convergence in the mean* for a particular member $x^{(j)}(t,\lambda)$ of an ensemble is defined with respect to an interval (a,b) according to

$$\lim_{\lambda\to 0} \int_a^b |x^{(j)}(t,\lambda) - x^{(j)}(t)|^2\,dt = 0$$

‡ See footnote ‡, p. 69.

§ Cf. Sec. 2.3-2.

in the ordinary sense.† The condition is now on the higher (and possibly on all) joint moments of $z_1 = z(\tau_1)$, $z_2 = z(\tau_2)$, as they appear through $\overline{g_1 g_2}$. In actual practice, we shall always assume that we are dealing with ensembles for which Eqs. (2.13*b*), (2.14) exist (mean square). Physically, Eq. (2.14) might represent the joint moment of the filtered outputs $x(t)$, $y(t)$ of two nonlinear devices—e.g., rectifiers or second detectors—when the input in each case is a noise or signal process $z(t)$, A and B representing the linear filtering operation following the nonlinear one of rectification (or g_1, g_2). Many such cases will be considered in detail later (cf. Chaps. 3, 4, 12 to 21).

We note, finally, that the equivalence of the (mean-square) property for the random variables x, y, z, etc., and the corresponding ordinary analytical property for the moment functions M_x, M_y, M_{xy}, M_{xz}, etc., stems from the *linear* character of differentiation and integration. These remarks are not restricted to mean-square differentiation and integration, however, as the above argument may be directly extended to multiple integrals and derivatives: the operations of differentiation and expectation or integration and expectation commute in general, provided that the appropriate derivatives and integrals that ensue exist. We shall make extensive use of this fact in subsequent chapters, where it is always assumed that the resulting expressions have definite values (see the rest of Chap. 2, Chaps. 3 to 5, Chap. 8, and Parts 3, 4 for specific examples).

2.1-3 Remarks on General Transformations. So far, a number of different linear and nonlinear operations‡ on ensembles have been introduced at various points in our discussion to illustrate the kind of process that we may expect in communication theory. Nothing has been said as yet, however, about the effects on the stationarity, or ergodicity, of the output ensembles, e.g., on $x(t)$, $y(t)$ above.

To illustrate, let $\boldsymbol{T}$ be a general transformation which takes, save for a set of measure zero, almost all members of an *input* ensemble $y(t)$ into a new, or *output*, ensemble $z(t)$, in some fashion according to

$$z(t) = \boldsymbol{T}\{y(t)\} \tag{2.15}$$

Transformations $\boldsymbol{T}$ are said to be either *measure-preserving* or *non-measure-preserving*, according as the transformed process $z = \boldsymbol{T}\{y\}$ does or does not have the same hierarchy of distribution densities as y. The transformations we shall encounter are not restricted to measure-preserving ones alone,§ like the linear time translations for strictly stationary processes, where $\boldsymbol{T}\{y\} = y$ (cf. Sec. 1.3-6). The physically much more common cases are included, which are not measure-preserving, since the probability densities (and distributions) of the transformed output ensemble $z(t)$ are different from those of the input ensemble $y(t)$. [Thus, for instance, $\boldsymbol{T}$ might be

† See footnote †, p. 70.
‡ For a definition of linear and nonlinear, see Sec. 2.1-3(2).
§ Doob, *op. cit.*, chap. 10, sec. 1, p. 452.

$g(\ )$ of Sec. 1.6-2, Examples 2, 3.] The measure for the set $z(t)$ is still defined in the usual way (cf. Sec. 1.2-14), in terms of the measure of $y(t)$, and the measure for a given subset of z, say $(z_1 < z < z_2)$, is equal to that for the subset $(y_1 < y < y_2)$ of y, from which the z subset is generated by the transformation $\boldsymbol{T}$. The hierarchy of probability densities describing the random process z is related to the corresponding hierarchy describing y through $\boldsymbol{T}$, and when this transformation is one to one, the relationship is specified by means of the jacobian of $\boldsymbol{T}$.[6,†] Total measure unity (for z) is preserved.

In most physical situations considered in the present study, $\boldsymbol{T}_L$ corresponds to putting the input ensemble y through a linear filter (e.g., an amplifier, delay line, linear transducer) or a nonlinear filter, such as the second detector, or mixer, of radio receivers, or combinations of such elements, to obtain a desired output ensemble z. These are almost invariably *not* measure-preserving. Thus, for simple linear lumped-constant filters (as we shall see in more detail in Sec. 2.2), $\boldsymbol{T}_L$ becomes the convolution operation with respect to a *weighting function* h, which for the steady-state condition is explicitly

$$z(t) = \int_{-\infty}^{\infty} h(t - \tau)y(\tau)\, d\tau \tag{2.16}$$

[conditions for stochastic integrability are given in Eqs. (2-13*b*), (2.14)].

Clearly, this transformation is not measure-preserving, for the statistics of z are changed from those of y. Suppose y possesses a nonzero moment $\bar{y}$ which this filter removes; then $W_1(z)\, dz \neq W_1(y)\, dy$, etc. Another example is afforded by a half-wave linear rectifier, represented analytically by a $\boldsymbol{T}_g$ such that $z = y\ (y \geq 0)$, $z = 0\ (y \leq 0)$. Here the measure, $b = \int_0^\infty W_1(y)\, dy$, for the subset $(0 < y < \infty)$ of the input is preserved, but that for all $y < 0$ is now concentrated at the point $z = 0$ in the output, i.e.,

$$W_1(z) = \begin{cases} 2a\, \delta(z - 0) + W_1(z)_y & z \geq 0 \\ 0 & z < 0 \end{cases} \qquad (a + b = 1) \tag{2.17}$$

where $W_1(\ )_y$ is the d.d. of y. Since $\int_0^\infty W_1(y)_y\, dy = b$ here, we see that $\int_{-\infty}^{\infty} W_1(z)\, dz = \int_0^\infty 2a\, \delta(z - 0)\, dz + \int_0^\infty W_1(z)_y\, dz = a + b = 1$; the total measure of the output is still unity, as required.

(1) Representation of a communication system. Let us apply these concepts in a general way to communication systems. These can be represented by a sequence of transformations $\boldsymbol{T}_n\boldsymbol{T}_{n-1} \cdots \boldsymbol{T}_2\boldsymbol{T}_1$, where the order is important. Thus, in general $\boldsymbol{T}_j$ and $\boldsymbol{T}_k$ do not commute; that is, $\boldsymbol{T}_j\boldsymbol{T}_k \neq \boldsymbol{T}_k\boldsymbol{T}_j$: linear filtering before rectification is usually not the same as rectifi-

† More generally, the mapping process is given by the Fourier-transform pair [Eqs. (1.62), (1.63)] (cf. Sec. 1.2-14).

cation followed by linear filtering. However, it will be true for most of the linear systems with which we have to deal that the succession of transformations $\boldsymbol{T}_j\boldsymbol{T}_k, \ldots, \boldsymbol{T}_m$ does commute, and for some sequences of nonlinear operations the same will also be true. We shall see a number of examples of this in Secs. 2.2, 2.3. Often, it is possible that application of $\boldsymbol{T}_j$ to $\boldsymbol{T}_k$, say, does *not* change the composite operation $(\boldsymbol{T}_j\boldsymbol{T}_k)\{\ \}$; that is, $\boldsymbol{T}_j\{\boldsymbol{T}_k\{x\}\}$ is equivalent to $(\boldsymbol{T}_j\boldsymbol{T}_k)\{x\}$; in other words, $\boldsymbol{T}_j$ and $\boldsymbol{T}_k$ are *factorable*, in the order indicated.

Physically, each transformation corresponds to a stage† of the communication system. Factorability implies that the output of the first stage does not change if a second stage be connected. The two stages are then said to be *isolated*. (Note, incidentally, that if two transformations commute they also factor.) Similarly, operations $\boldsymbol{T}_j$ which cannot be resolved into a

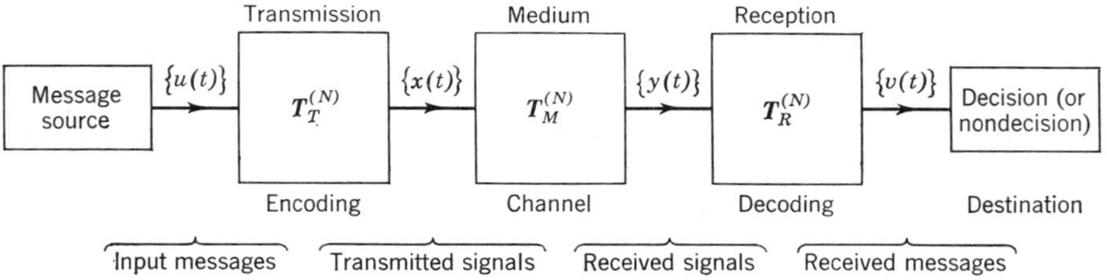

FIG. 2.1. A simple single-link communication system.

sequence of two or more suboperations are said to be *irreducible*, and the system elements comprising $\boldsymbol{T}_j$ may be called the prime elements of this transformation. Many of the operations which we shall discuss in subsequent sections, however, are not irreducible: it is usually possible either to factor or to resolve the transformation in question (often not uniquely) into two or more suboperations. The process of reception in a communication receiver is one example.

A simple communication system is indicated symbolically in Fig. 2.1. Thus, if $u(t)$ represents the ensemble of possible messages and $\boldsymbol{T}'_1 \cdots \boldsymbol{T}'_m \equiv \boldsymbol{T}_{T^{(N)}}$ is the succession of operations on u to produce the signal ensemble x, we have

$$x_1 = \boldsymbol{T}'_1\{u\} \qquad x_2 = \boldsymbol{T}'_2\{x_1\} \qquad \cdots \qquad x = \boldsymbol{T}'_m\{x_{m-1}\}$$
$$\text{or} \qquad x = \boldsymbol{T}'_m\boldsymbol{T}'_{m-1} \cdots \boldsymbol{T}'_2\boldsymbol{T}'_1\{u\} = \boldsymbol{T}_{T^{(N)}}\{u\} \tag{2.18}$$

for the set of possible transmitted waves. In reception, $\boldsymbol{T}_n\boldsymbol{T}_{n-1} \cdots \boldsymbol{T}_1$ are the corresponding operations on the received signal ensemble $y(t)$, yielding finally the received message set $v(t)$. Thus, we have

$$y_n = \boldsymbol{T}_n\{y\} \qquad y_{n-1} = \boldsymbol{T}_{n-1}\{y_n\} \qquad \cdots \qquad v = \boldsymbol{T}_1\{y_1\}$$
$$\text{or} \qquad v = \boldsymbol{T}_1\boldsymbol{T}_2 \cdots \boldsymbol{T}_{n-1}\boldsymbol{T}_n\{y\} = \boldsymbol{T}_{R^{(N)}}\{y\} \tag{2.19}$$

† For "stage," one can also read "sequence of stages."

The transmitted waves $x(t)$ are different from the original input message ensemble $u(t)$, mainly because of (1) the physical conditions imposed by the medium through which transmission takes place, (2) the nature of the reception process (this may or may not be significant), and (3) the introduction of noise in the course of the transformation $\mathbf{T}_T^{(N)}$, in physical systems sometimes an unavoidable concomitant of the *encoding* process, as the operation $\mathbf{T}_T^{(N)}$ is frequently called. The received signal ensemble $y(t)$ is also usually different from the transmitted one, owing to noise generated, in the course of propagation, from external sources and owing to the nature of the medium itself. The received message ensemble $v(t)$ is also modified, often seriously, by the noise inherent in the process of reception (or *decoding*). If $\mathbf{T}_M^{(N)}$ represents the effect of the medium, that is, $y = \mathbf{T}_M^{(N)}\{x\}$, we have from Eqs. (2.18), (2.19)

$$v = \mathbf{T}_R^{(N)}\mathbf{T}_M^{(N)}\mathbf{T}_T^{(N)}\{u\} \tag{2.20}$$

Thus, u and v are related, and in many cases closely for effective systems, more distantly for ineffective ones, so that it is the attempt to bring the original and the received message more fully into identity that is one motivation for our present study. Of course, we are not always interested in bringing u and v into identity or closest correspondence. Sometimes v implies an entirely different kind of quantity from u—for example, in detection v leads to a *decision*, "yes" or "no," a signal is or is not present, while u is the original signal ensemble. In such cases, the most we can say is that u is given and an end result from v is desired with respect to some reasonable criterion of performance.

As mentioned above, noise may be injected into the system in the course of a transformation $\mathbf{T}$. To indicate this explicitly, we may write (1) $\mathbf{T}^{(N)} = \mathbf{T}\mathbf{T}_N$ or (2) $\mathbf{T}^{(N)} = \mathbf{T}_N\mathbf{T}$, where $\mathbf{T}$ represents a noise-free operation, while $\mathbf{T}_N$ indicates only that noise is introduced in some fashion, either additively, as is most common in practice, or in a more complicated way, like the multiplicative processes characteristic of scatter and fading in the course of propagation (cf. Probs. 11.12, 11.13). Here (1) shows that the noise appears *before* the operation $\mathbf{T}$, while (2) indicates that it is introduced following that operation. We may also have the situation (3) $\mathbf{T}^{(N)} = \mathbf{T}_{N2}\mathbf{T}\mathbf{T}_{N1}$, where noise arises both before and after the transformation $\mathbf{T}$, as occurs, for example, in crystal mixers (cf. Chap. 12). However, unless otherwise specified, we shall regard $\mathbf{T}^{(N)}$ in the sense of (1) $\mathbf{T}^{(N)} = \mathbf{T}\mathbf{T}_N$: the injected noise appears before the operation in question and hence is subject to that operation, whether it is linear or nonlinear. An initial step in the analysis of any system, of course, is to determine the nature and location of the various noise sources, as they appear in $\mathbf{T}^{(N)}$.

Actually, all physical devices and systems are ultimately noisy, because of the inherently random character of the conduction and propagation mechanisms which govern their operation. Accordingly, we should also modify the constituent elements of $\boldsymbol{T}_M^{(N)}$, $\boldsymbol{T}_R^{(N)}$, $\boldsymbol{T}_T^{(N)}$ in Eq. (2.20) to indicate the presence of injected noise. However, for practical purposes only the initial stages of $\boldsymbol{T}_T^{(N)}$, $\boldsymbol{T}_R^{(N)}$ effectively exhibit fluctuating voltages or currents, which for the rest may be considered negligible. We can therefore speak in applications of *ideal*, or *noise-free*, transformations ($\sim\boldsymbol{T}$) and noise-generating operations ($\sim\boldsymbol{T}^{(N)}$), where in most instances the latter may be factored into the sequence 1 or 2. In the remaining chapters of Part 1, we shall regard $\boldsymbol{T}$ as noise-free (unless otherwise indicated), while the appearance of noise is assigned to a source external to the ideal element or device in question. In Part 2, on the other hand, we shall attempt in certain cases to describe the physical origins of the injected noise that is symbolized by the operation $\boldsymbol{T}_N$ and, if possible, to obtain the hierarchy of distribution densities which completely describes the process.

A short list of the more common transformations encountered here and in later work follows:

$\boldsymbol{T}_N$ = transformation, or operation, by which noise only is injected

$\boldsymbol{T}^{(N)}$ = operation involving one or more noise sources ($\sim\boldsymbol{T}_N$) and one or more ideal (i.e., noise-free) transformations

$\boldsymbol{T}_g$ = ideal, i.e., noise-free nonlinear operation

$\boldsymbol{T}_L$ = ideal linear operation (included as special case of $\boldsymbol{T}_g$)

$\boldsymbol{T}_g^{(N)}$ = nonlinear operation, with injected noise (includes $\boldsymbol{T}_L^{(N)}$ as special case) (2.20*a*)

A variety of general transformations can be compounded from these, for example, $\boldsymbol{T}_T^{(N)}$, $\boldsymbol{T}_R^{(N)}$, etc., as the need arises. Our convention here, in any case, is to represent linear operations by the subscript L and nonlinear ones by the subscript g.

Observe, incidentally, that for ideal (i.e., noise-free) systems, where the medium introduces no modification of $x(t)$ other than a simple translation, or delay, for example, $\boldsymbol{T}\{x(t)\} = x(t + t')$, apart from this delay t', $\boldsymbol{T}_R$ is the "inverse" of $\boldsymbol{T}_T$, that is, $\boldsymbol{T}_R = \boldsymbol{T}_T^{-1}$ in the sense that perfect recovery of the transmitted process takes place. It is also possible, though not necessary, that $\boldsymbol{T}_k^{-1} = \boldsymbol{T}'_k$ $(k = 1, \ldots, m = n)$ individually. This is not true of actual systems, since $\boldsymbol{T}'^{(N)}_k$, $\boldsymbol{T}_k^{(N)}$ represent the introduction of noise as well, $\boldsymbol{T}'^{(N)}_k$ is only approximately the inverse of, say, $\boldsymbol{T}_j^{(N)}$ $(m \neq n)$, and for some $\boldsymbol{T}'^{(N)}_k$ there may be no direct counterparts in $\boldsymbol{T}_j^{(N)}$ at all.

Of course, the communication link, or system, $\boldsymbol{T}_R^{(N)}\boldsymbol{T}_M^{(N)}\boldsymbol{T}_T^{(N)}$ above is by no means the most general: one can have *feedback*, in the form of a return channel, with receiver and transmitter now sharing each other's functions, or there can be more than one message source applied to the receiver, with this in turn linked to other sources and receivers. In this book, we shall not, however, go beyond the comparatively simple single-link system indicated schematically by Fig. 2.1, as this is sufficiently general to provide an adequate introduction to the concepts and applications of statistical communication theory.

(2) Classification of transformations. With the foregoing remarks and those of Secs. 1.3-6, 1.6 in mind, we can classify the typical transformations encountered in our study. Classification is helpful because it gives us an indication of the statistical complexity to be expected of the output ensemble and, accordingly, some idea of the analytical methods needed for the solution of explicit problems.

(*a*) *A transformation (or operation)* $\mathbf{T}_L$ *is linear if the principle of superposition holds,* i.e.,

$$\begin{array}{ll} \text{If} & z_1 = \mathbf{T}_L\{y_1\} \qquad z_2 = \mathbf{T}_L\{y_2\} \\ \text{then} & c_1z_1 + c_2z_2 = \mathbf{T}_L\{c_1y_1 + c_2y_2\} \end{array} \tag{2.21}$$

where c_1 and c_2 are arbitrary finite constants. The linear lumped-constant filters of circuit theory are conspicuous examples of devices which perform such operations (as we shall see in more detail in Sec. 2.2).

(*b*) *Conversely, a transformation (or operation)* $\mathbf{T}_g$ *is nonlinear if the principle of superposition is not satisfied,* i.e.,

$$\begin{array}{ll} \text{If} & z_1 = \mathbf{T}_g\{y_1\} \qquad z_2 = \mathbf{T}_g\{y_2\} \\ \text{and} & c_1z_1 + c_2z_2 \neq \mathbf{T}_g\{c_1y_1 + c_2y_2\} \end{array} \tag{2.22}$$

Consider, for example, $\mathbf{T}_g$ as a squaring operation. Then, $\mathbf{T}_g\{y_1\} = y_1^2 = z_1$, and $\mathbf{T}_g\{y_2\} = y_2^2 = z_2$, but $\mathbf{T}_g\{y_1 + y_2\} = (y_1 + y_2)^2 = y_1^2 + y_2^2 + 2y_1y_2 \neq z_1 + z_2$ $(c_1 = c_2 = 1)$. The operation of mixers and second detectors, or rectifiers, in radio and radar receivers is a common instance of nonlinear behavior (cf. Sec. 2.3 and Chaps. 12 to 15).

We can also describe a transformation in terms of the time shifts that it may cause in the input ensemble.†

(*c*) *A transformation* $\mathbf{T}$ *is invariant* if an arbitrary linear time shift t' in the input ensemble $y(t)$ merely shifts the output $z(t)$ without introducing additional (fixed or arbitrary) delays or advances, for almost all members of y and for all t'. Thus, if $z(t) = \mathbf{T}\{y(t)\}$, for $t \to t + t'$, we get $z(t + t') = \mathbf{T}\{y(t + t')\}$. As an example, consider the transformation $\mathbf{T}_L$ representing the linear lumped-constant filter of Eq. (2.16). $\mathbf{T}_L$ is invariant, for setting $\tau \to \tau + t'$ in $y(\tau)$ gives

$$\int_{-\infty}^{\infty} h(t - \tau)y(\tau + t')\,d\tau = \int_{-\infty}^{\infty} h(t + t' - \tau')y(\tau')\,d\tau' = z(t + t') \tag{2.23}$$

If (*c*) is not fulfilled, then $\mathbf{T}_L$ is not invariant. Suppose one observes the output of this linear filter, now only for the period $(0,T)$, so that $z_T(t) = \int_0^T h(t - \tau)y(\tau)\,d\tau$ is the new output ensemble. Then, clearly, the oper-

† The question of the effect of the various transformations $\mathbf{T}_k$, $\mathbf{T}'_j$, $\mathbf{T}_{R^{(N)}}$, $\mathbf{T}_{T^{(N)}}$, etc., on the Markovian nature of the input and output ensemble is considered briefly in Sec. 7.6.

ation $T_L = \int_0^T h(t-\tau)\{\quad\}\,d\tau$ is not invariant, for

$$z_T(t,t') = \int_0^T h(t-\tau)y(\tau+t')\,d\tau = \int_{t'}^{T+t'} h(t+t'-\tau')y(\tau')\,d\tau' \neq z(t+t') \quad (2.24)$$

as we might have expected. The process of modulation is also usually not invariant, since an additional delay (or phase shift) is then introduced. Accordingly, if $T = (1 + \lambda \cos \omega_m t)\{\quad\}$ $(1 \geq \lambda \geq 0)$ is the modulation operator for a sinusoidal modulation and the input ensemble is, say, $A_0 \cos(\omega_0 t + \phi)$, we get

$$z(t) = T\{y(t)\} = A_0(1 + \lambda \cos \omega_m t) \cos(\omega_0 t + \phi) \quad (2.25a)$$

while

$$z(t,t') = T\{y(t+t')\} = A_0(1 + \lambda \cos \omega_m t) \cos(\omega_0 t + \omega_0 t' + \phi) \neq z(t+t') \quad (2.25b)$$

(However, restricting t' to be such that $\omega_m t' = 2\pi n$, we can then say that T in this sense is invariant for all translations that are multiples of the period of the modulation.)

The important features of *invariance* are that, if T possesses this property and if $z = T\{y\}$, then:

1. The output ensemble z is *strictly* (or wide-sense) *stationary* if the input ensemble y is similarly strictly (or wide-sense) stationary.
2. The output ensemble z is *ergodic* if the input y is ergodic (cf. Prob. 2.3).

Note that this does *not* mean that an arbitrary invariant transformation T, operating on an ergodic input ensemble y, preserves measure in going from y to z *measure space*, since in general, $W_1[z(y)]_z$ is different from $W_1(y)_y$, and so on, for W_2 [cf. Eq. (2.17)]. The form and scale of the descriptive density functions are changed in passing from y- to z-space. However, in the output, or z measure space, z itself is ergodic—it fulfills the conditions of Sec. 1.6-1 that an ensemble be metrically transitive. In this connection, observe that, if T is measure-preserving, it is also invariant, since $y = T\{y\}$. Many of the systems encountered in the physical communication process and many of their component operations—i.e., the T_j, T'_k, etc., of Sec. 2.1-3(1)—belong ideally (i.e., in the limit of infinite observation periods) to the invariant non-measure-preserving class. On the other hand, in most actual (as opposed to those limiting) cases, the representative transformations are both noninvariant and non-measure-preserving, largely in consequence of finite observation times: messages and signals are of finite length; transmission, reception, measurement, and decision are all necessarily limited to "real times," or finite periods.

PROBLEMS

2.1 (*a*) Show that, if $Y(t)$ is a continuous random variable with continuous parameter t in an interval, the expectation $E\{Y(t_1)Y(t_2)\}$ and MSC operations commute.

(*b*) Show that a random process $Y(t)$ is almost certainly continuous if mean-square differentiable and almost certainly differentiable once if mean-square differentiable twice, etc. Consequently, if $Y(t)$ is differentiable mean square to all orders, so also is it almost certainly differentiable to all orders.

2.2 (*a*) Verify that the necessary and sufficient condition that $Y(t)$ is expandable in a mean-square-convergent Taylor's series in some region of the t plane is that $M_y(t_1,t_2)$ is analytic in the corresponding t_1, t_2 regions.

(*b*) Show that the above series for $Y(t)$ converges almost certainly to $Y(t)$ and hence that $Y(t)$ is also analytic.

(*c*) Discuss the differentiability (mean square and almost certain) of the processes with the following $M_y(t_1,t_2)$:

$$e^{-\alpha|t_1-t_2|^2} \qquad e^{-\beta|t_1-t_2|} \qquad e^{-\gamma|t_1-t_2|^\nu} \quad (\nu > 0) \qquad \cos \eta\,(t_1 - t_2)$$

Discuss also the mean-square continuity of the above.

2.3 (*a*) Prove that, if $y(t)$ is a stationary random process and $\boldsymbol{T}$ is an invariant operation or transformation, $z = \boldsymbol{T}\{y\}$ is also a stationary process.

(*b*) In the above, show that, if y is ergodic, then z is also ergodic.

(*c*) Show that invariance is a necessary and sufficient condition for (*a*), (*b*).

2.2 *Linear Systems; Representations*

So far we have considered operations on ensembles and their corresponding statistical properties from a rather general and descriptive point of view (cf. Secs. 1.3 to 1.6, 2.1). To be more specific, we summarize now for later use some of the salient properties of selected ideal linear and, in Sec. 2.3, ideal nonlinear circuit elements or networks, by which many transformations of ensembles are actually effected. For the time being, we shall regard these networks as ideal noise-free systems. Later, in Part 2, we shall extend the analysis to include the effects of the noise present in nonideal physical systems.

2.2-1 Remarks on General Networks. The transformations mentioned above fall into a number of categories of increasing complexity.†

1. *Linear passive lumped-constant networks*, e.g., ordinary *RLC* networks, bandpass and low-pass filters, etc.
2. *Linear active systems*, e.g., amplifiers.
3. *Time-varying linear devices*, e.g., the linear switch, "on" at $t = 0$, "off" at $t = T$ [used in the process of data sampling in $(0,T)$]; periodic linear switches for essentially infinite observation periods.
4. *Nonlinear systems* (usually active, and without hysteresis or feedback), e.g., mixers, second detectors in radio and radar receivers.
5. *Time-varying nonlinear systems* such as computers, where storage, variable programming, and a variety of nonlinear operations characteristic of data processing, e.g., multiplication, raising to powers, general transformations like $z = g(y)$, occur, either digitally (discrete sampling in time) or by analogue means (e.g., continuous sampling).

† This list is not intended to be exhaustive.

This last, class 5, is the most general class of combined operations considered in this volume, and that mainly from the viewpoint of optimum system design (cf. Part 4). Here we shall be concerned mainly with class 1, with simple versions of class 2, and with class 4. The operations are essentially continuous (cf. Parts 2, 3); some simple examples of class 3 are considered briefly in Part 3, while class 5 is illustrated in Part 4.

A general network is composed of resistances, mutual- and self-inductances, and capacitances, called *network parameters*, connected together in some fashion (cf. Fig. 2.2, where the boxes represent the basic, or elementary, circuit elements connected together in series).† The junctions a, b, c, d, . . . are *branch points*, or *nodes;* sections of network between branch

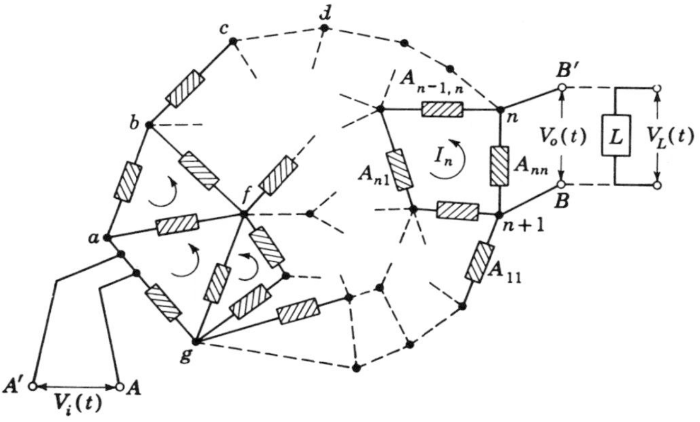

FIG. 2.2. A general network showing branches, branch points, and meshes, with input and output voltages V_i, V_o and load L.

points are *branches*, for example, *ab*, *bc*, *bf*, . . . ; and a succession of branches forming a closed loop is a *mesh*, for example, *abf*, *abfg*, ... While there is no unique way of designating the meshes, it is customary to give currents in the same direction a common sign and to take the counterclockwise direction as positive. The number M of independent meshes (like *abf*, but *not abfg*) is $B - P + 1$, where B is the number of branches and P the number of branch points, or nodes. Node and mesh analysis have equal status in modern treatments. In the former, the variables are the *node voltages*, and the inputs are *current sources*. In the latter, referred to briefly below, the *mesh currents* are the variables, with *voltage sources* as inputs (cf. Fig. 2.2).

Networks can be *active* or *passive*. An active network contains one or more sources or sinks of energy, such as a mesh including a thermionic

† It is assumed that the reader is familiar with the elements of linear filter theory, as well as the basic features of electric- and electronic-circuit analysis. For details, consult any standard reference, e.g., E. A. Guillemin, "Communication Networks," vols. 1, 2. Wiley, New York, 1956.

valve, apart from those sinks attributable to the purely resistive portions of the circuit. Passive networks contain only resistive sinks, for example, a connection of resistances and capacitances. The *input* to the circuit is for our discussion an ensemble of voltages $V_i(t)$, supplied from an ensemble of sources of zero internal impedance, and is applied by insertion in any one of the branches, while the *output* is ordinarily taken to be a corresponding ensemble of open-circuited voltages across some pair of branch points (cf. Fig. 2.2). Frequently, a *load*, i.e., another network or network element, is introduced between these terminals, in the manner of the figure cited. The open-circuited voltages $V_L(t)$ across it are called the load voltages. Two- and four-terminal networks (the circuit AA'-BB' above is a four-terminal one) are types considered here mainly, although examples of higher-order systems will be encountered in later chapters.

Applying Kirchhoff's laws† to the four-terminal network of Fig. 2.2, we observe that the ensemble of mesh currents $[I_1, \ldots, I_n]$, which we write more compactly as the column vector $\mathbf{I}(t)$, the input voltages $\mathbf{V}_{in}(t)$, and the output voltage $\mathbf{V}_o(t)$ across the $(n, n+1)$ branch are determined by the equations

$$\boldsymbol{T}_L\{I(t)\} = \mathbf{T}_L\mathbf{I}(t) = \mathbf{V}_{in}(t) \qquad \mathbf{T}'_L(t)\mathbf{I}(t) = \mathbf{V}_o(t) \tag{2.26}$$

where the transformation $\boldsymbol{T}_L$ is the square matrix‡ $\mathbf{T}_L$,

$$\mathbf{T}_L(t) = \|T_{kl}\| = \left[\frac{d}{dt}\{L_{kl}(t)\} + R_{kl}(t) + \frac{1}{C_{kl}(t)}\int^t (\quad)\,dt\right] \qquad k, l = 1, \ldots, n \tag{2.27a}$$

$$\tilde{\mathbf{T}}'_L(t) = [T_{n+1,1}, T_{n+1,2}, \ldots, T_{n+1,n}] \qquad \text{row vector} \tag{2.27b}$$

$$\mathbf{V}_{in}(t) = [V_{in}(t), 0, \ldots, 0] \qquad \text{column vector} \tag{2.27c}$$

When $k = l$, the mesh parameters are the *self*-inductances, *self*-capacitances, etc., while for $k \neq l$ these are known as *mutual* inductances, mutual capacitances, etc.

2.2-2 The Transient State; Differential Equations and Solutions. Let us now require that the operations $\boldsymbol{T}_L$ above represent a linear system, so

† Namely, the sum of the currents leaving a branch point is zero (conservation of charge), and the sum of the voltages around any *independent* mesh is zero, since no sources or sinks (other than purely resistive ones) are enclosed by the mesh. More generally, the sum of the voltages around *any* closed path (mesh) is algebraically zero, whether or not the path is an independent mesh and whether or not it contains voltage sources or sinks. We restrict the generality by considering effectively lumped-constant systems, where the minimum wavelength of the disturbances excited in the network is always large compared with the dimensions of network itself and where consequently radiation is ignorable. Maxwell's equations, of course, are ultimately the fundamental ones in all cases.

‡ If we are interested instead in the set of mesh *charges* Q_l, the integrodifferential operators $\boldsymbol{T}_{kl}$ are replaced by the differential operators $(d^2/dt^2)[L_{kl}(t)] + (d/dt)[R_{kl}(t)] + [S_{kl}(t)]$, where $S_{kl}(t)$ are the *mesh elastances* (Guillemin, *op. cit.*).

that the principle of superposition [Eq. (2.21)] is satisfied. For the time being, we consider only one member of a possible ensemble of exciting voltages $\mathbf{V}_{in}^{(j)}$ $(j = 1, \ldots)$ and restrict our attention further to the important subclass of linear lumped-constant passive networks. This means not only that the principle of superposition holds, true also for time-varying linear systems, but also that negative values of resistances R, self-inductances L, and self-capacitances C are excluded (i.e., the passivity condition), and that L, C, R are constants, independent of time. While the representation of such a network in terms of the n mesh currents (or n mesh charges) is particularly useful when the currents (charges) and voltages in a given mesh are desired, the analysis of the over-all, or "lumped," behavior of the network as a whole is more conveniently given in terms of the input and output voltages $\mathbf{V}_{in}^{(j)}$ and $\mathbf{V}_o^{(j)}$, which here will be taken as a single input $V_{in}^{(j)}$ and single output $V_o^{(j)}$, across some pair of branch points (cf. Fig. 2.2). Eliminating the mesh currents $\mathbf{I}^{(j)}$ and their derivatives $\dot{\mathbf{I}}^{(j)}$, $\ddot{\mathbf{I}}^{(j)}$, we obtain finally[7] a linear differential equation of the type

$$\sum_{q=0}^{N} a_q \frac{d^q}{dt^q} V_o^{(j)}(t) = \sum_{m=0}^{M} c_m \frac{d^m}{dt^m} V_{in}^{(j)}(t) \tag{2.28}$$

where the a_q, c_m are constants depending only upon the mesh parameters and $M, N \leq 2n$, with n as before the number of independent meshes. Apart from the fact that such differential equations are a natural starting point for the analysis of linear networks, a further justification for introducing them at this stage will become apparent in Chap. 11, where an account of circuit noise, arising from actual nonideal elements, is given. These differential equations then play a significant role in the ensuing discussion of the noise processes.

The relation (2.28) can be solved explicitly if the input, or driving, voltage $V_{in}^{(j)}(t)$ is given. As is well known,[8] the general solution can be expressed as the sum of any solution of the *inhomogeneous equation* (2.28) (called, for example, a *particular* solution) and the general solution of the *homogeneous equation*, obtained from Eq. (2.28) on setting $V_{in}^{(j)} = 0$, subject to the N initial conditions which determine the N arbitrary constants characteristic of this solution. From Eq. (2.28), we can define the *transient response* of a network to the excitation $V_{in}^{(j)}(t)$ as *that response which is the difference between the output* $V_o^{(j)}(t)$ *at time* $t > t_0$ [= *time at which* $V_{in}^{(j)}(t)$ *is first applied*] *and the limiting form*† *that this output approaches as* $t \to \infty$. This limiting form of solution is the *steady-state* output, representing a particular solution of Eq. (2.28), which we shall touch on briefly in Sec. 2.2-5. When the system starts from rest (at $t = 0$ for convenience), so that $V_o^{(j)}(t) = 0$ $(t \leq 0)$, the output $V_o^{(j)}(t)$ $(t > 0)$ is also referred to as the

† It is assumed that this limit exists, which, as noted presently, will always be the case for stable networks and bounded inputs.

normal response of the system. Note that the normal response may include transient effects as well as the steady-state contribution.

The most general solution of the homogeneous equation, known also as the complementary function, is $V_o^{(j)}(t)_{no} = \sum_{q=0}^{N} b_q^{(j)} f_q(t)$, where the $f_q(t)$ are particular solutions of the type $f_q = t^{l-1}e^{-\beta_q t}$ $(q = 1, \ldots, N; l = 1, \ldots, L)$, β_q is in general a complex constant, and the $b_q^{(j)}$ depend on the initial conditions. The f_q and β_q are, respectively, *eigenfunctions* (or normal-mode solutions) and *eigenvalues* (or normal modes) associated with the homogeneous equation. To determine their solutions and eigenvalues, we set $V_o^{(j)} = e^{pt}$ in the usual way, substitute in the homogeneous equation, and observe that we get the polynomial $F(p) = \sum_{q=0}^{N} a_q p^q$. This must vanish, if e^{pt} is a solution, with the result that the normal modes $p_q = \beta_q$ are determined from the modal equation $F(p) = \prod_{q=1}^{N} (p - \beta_q) = 0$. Two cases are distinguished:[8]

1. All roots are distinct: $\beta_l \neq \beta_k$ $(l \neq k)$. The normal-mode solutions are then $F_q^{(j)}(t) = b_q^{(j)} e^{\beta_q t}$.

2. Some (or all) roots are multiple: $\beta_q = \beta_{q+1} = \cdots = \beta_{q+l-1}$. Here the normal-mode solutions take the form $F_q^{(j)}(t) = (b_q^{(j)} + b_{q+1}^{(j)} t + \cdots b_{q+l-1}^{(j)} t^{l-1}) e^{\beta_q t}$.

For the complete solution of the inhomogeneous case [Eq. (2.28)], we need, in addition to the above, particular solutions to Eq. (2.28), and these are readily indicated in terms of the applied $V_{in}^{(j)}(t)$ (cf. Prob. 2.6). It is convenient to write β_q in terms of its real and imaginary parts,

$$\beta_q = \alpha_q + i\omega_q \tag{2.29}$$

where α_q, ω_q are real (and $\omega_q = 0$ on occasion). Note that, if β_q is complex, its complex conjugate β_q^* is also a solution of $F(p) = 0$. The normal-mode solutions are then all complex but occur in the transient solution in linear combinations which are themselves real, for example, $(f_q + f_q^*)/2 = t^{l-1}e^{\alpha_q t} \cos \omega_q t$, $(f_q - f_q^*)/2i = t^{l-1}e^{\alpha_q t} \sin \omega_q t$, and if there is no damping, i.e., if $\alpha_q = 0$, the transients will be purely sinusoidal. Since the filter elements are all constants, the normal response is independent of the time origin and depends only on the past values of the excitation. Furthermore, from Eq. (2.29) it is clear that stability† is achieved only if *all* modal solutions have eigenvalues with negative real parts. On the other hand, critical cases are the ones for which the real parts of one (or more) eigenvalues vanish. If such a mode is excited, a finite excitation will produce an output which (for

† Stability is the condition that a finite input produces a finite output, both in magnitude and in duration.

simple roots) is bounded but which does not die down to zero in the course of time or which (for multiple roots) increases indefinitely. Finally, if the real parts of β_q are *positive*, the output becomes ultimately unbounded. However, since there is always some dissipation in physical systems, stability is assumed for the passive network under consideration here: analytically, the real parts of all normal-mode values are always taken to be negative.

2.2-3 The Weighting Function. While the behavior of a linear filter may be completely described by the differential equation relating the input and output voltages [cf. Eq. (2.28)], it is usually more convenient to use other, more compact expressions which, because they are independent of the particular excitation $V_{in}^{(j)}(t)$, are more general and at the same time more specific as far as the filter itself is concerned. One such description is given by the *weighting* function, $h(t)$, of the filter, defined as the normal response to a unit impulse $\delta(t - 0)$, applied at $t = 0$ [cf. Eq. (2.28)],

$$\sum_{q=0}^{N} a_q \frac{d^q h(t)}{dt^q} = \delta(t - 0) \qquad h(t) = 0,\ t < 0 \tag{2.30}$$

We remark that $h(t)$ may be discontinuous at $t = 0$ and may also contain other delta-function excitations for $t > 0$. Since $V_{in}^{(j)}(t) = \delta(t - 0)$ (all j here) is zero after, as well as before, $t = 0$, the response $h(t)$ satisfies the homogeneous equation of the network and the solution is therefore a linear combination of *all* normal modes of the filter, with possibly a linear sum of delta functions for $t > 0$.

From the principle of superposition [Eq. (2.21)], one can determine (cf. Prob. 2.6) the normal response of the filter to some arbitrary (real) input $V_{in}^{(j)}(t)$, which except for a finite number of delta functions ($t \geq 0$) is bounded in magnitude. This output, for individual members of the ensemble $V_{in}(t)$, is

$$V_o^{(j)}(t) = \int_{0-}^{t+} V_{in}^{(j)}(\tau) h(t - \tau)\, d\tau$$

$$= \int_{(0-)}^{t} V_{in}^{(j)}(t - \tau) h(\tau)\, d\tau \qquad V_{in}^{(j)} = 0,\ t < 0 \tag{2.31a}$$

or

$$V_o^{(j)}(t) = \int_{-\infty}^{t} V_{in}^{(j)}(t - \tau) h(\tau)\, d\tau \qquad h = 0,\ \tau < 0 \tag{2.31b}$$

where plus or minus on the limits indicates that the entire singularity (if any) at $\tau = 0$ (or $\tau = t$) is included [cf. (1.15)]. As an operation on the input ensemble $V_{in}(t)$, $\boldsymbol{T}_L \left[= \int_{0-}^{t} h(\tau) \{\quad\} \, d\tau = \int_{0}^{t+} h(t - \tau) \{\quad\} \, d\tau \right]$ generates a corresponding output ensemble $V_o(t)$, provided, of course, that the filter is stable and that the integrals (2.31) exist mean square, i.e., provided that

$$M_{V_o}(t_1,t_2) = \overline{V_o(t_1) V_o(t_2)} = \int_0^{(t_1+)} \int_0^{(t_2+)} h(t_1 - \tau_1)\, h(t_2 - \tau_2) \overline{V_{in}(\tau_1) V_{in}(\tau_2)}\, d\tau_1\, d\tau_2 \tag{2.32}$$

exists in the ordinary sense (see Sec. 2.1-2). Since $h(\tau)$ is by definition real, as are the $V_{in}(t)$, so also must the output ensemble be real. The operations $\boldsymbol{T}_L$ above are also called *convolutions* of the input with a weighting function, which here considers only the *past* of the input disturbance, since $h(t) = 0$, $t < 0$. Note that $\boldsymbol{T}_L$ is neither measure-preserving nor invariant [cf. Eqs. (2.16), (2.24) and Sec. 2.1-3(2)].

The weighting function provides a complete description of the filter, since it includes all the features contained in the original homogeneous differential equation (2.30). We may also say that the weighting function represents the memory of the filter: if $h(t)$ is slowly varying in time, the memory is "long," while for rapidly varying $h(t)$ it is "short." This memory itself represents the distortion imposed by the filter upon the original input ensemble, often appearing as a *lag* for sudden changes of input. No restriction on the existence of the weighting function is imposed by considerations of stability: weighting functions for unstable filters are included in the definition as well. For these reasons, it is natural to regard the weighting function as the fundamental quantity which fully defines the filter's properties.

2.2-4 The Transfer Function. A useful alternative method of describing the filter's characteristics is provided by the *transfer function* $Y(p)$. To define it, we first introduce the *Laplace* transform,† which enables us to deal with the transient states of operation considered here. Thus, if we let $g(t)$ be a (real) function defined for $t > 0$ and vanishing for $t < 0$, such that $|g(t)| \leq c_0 e^{\alpha_0 t}$ $(c_0, \alpha_0 > 0)$, then the integral

$$G(p) \equiv \int_{0-}^{\infty} e^{-pt} g(t)\, dt = \int_{-\infty}^{\infty} g(t) e^{-pt}\, dt \qquad \text{Re}\, p > \alpha_0,\ g = 0,\ t < 0 \tag{2.33a}$$

is also convergent for all complex values of p for which $\text{Re}\, p > \alpha_0$. As a function of the complex variable p, the quantity $G(p)$ is called the (one-sided) Laplace transform of $g(t)$, more often represented by

$$\mathfrak{L}\{p,g(t)\} = \mathfrak{L}\{g\} = \int_{0-}^{\infty} e^{-pt} g(t)\, dt \qquad \text{Re}\, p > \alpha_0 \tag{2.33b}$$

and now $\boldsymbol{T}_L = \mathfrak{L}$ is the *Laplace transformation*, or operator, $\int_{0-}^{\infty} e^{-pt}\{\quad\}\, dt$, as indicated explicitly by the integral (2.33*b*). Figure 2.3 shows the region

† For comprehensive accounts of the Laplace transformation, from both the mathematical and the applied points of view, see, for example, G. Doetsch, "Theorie und Anwendung der Laplace Transformation," Springer, Berlin, 1937; D. V. Widder, "The Laplace Transform," Princeton University Press, Princeton, N.J., 1941; W. R. Smythe, "Static and Dynamic Electricity," 1st ed., McGraw-Hill, New York, 1939; J. A. Stratton, "Electromagnetic Theory," McGraw-Hill, New York, 1941; M. F. Gardiner and J. L. Barnes, "Transients in Linear Systems," Wiley, New York, 1942; B. Van der Pol and H. Bremmer, "Operational Calculus Based on the Two-sided Laplace Transform," Cambridge, New York, 1950; P. M. Morse and H. Feshbach, "Methods of Theoretical Physics," chap. IV, McGraw-Hill, New York, 1953.

in the complex p plane for which $\mathfrak{L}\{g\}$ converges absolutely, with α_0 the abscissa of absolute convergence. Once $G(p)$ has been found [cf. Eq. (2.33)], this region can be extended by *analytic continuation*† to include the entire p plane, excepting only those points for which $G(p)$ is not analytic.‡

The inverse of $\mathfrak{L}$ is $\mathfrak{L}^{-1}\{G(p)\}$, which is easily obtained by integrating $\mathfrak{L}(g) = G(p)$ with respect to $e^{pt}/2\pi i$ $(t > 0)$ along the contour $\mathbf{C}_2$ (cf. Fig. 2.3), with $c > \alpha_0$ and $p = \alpha + i\omega$, that is,

$$\begin{aligned} \mathfrak{L}^{-1}\{G(p)\} &\equiv \int_{-\infty i+c}^{\infty i+c(>0)} G(p)e^{pt}\,\frac{dp}{2\pi i} \\ &= \int_{0-}^{\infty} g(t')\,dt' \int_{-\infty i+c}^{\infty i+c} e^{p(t-t')}\frac{dp}{2\pi i} \\ &= \int_{0-}^{\infty} g(t')e^{\alpha(t-t')}\,dt' \int_{-\infty}^{\infty} e^{i\omega(t-t')}\,df \qquad \omega = 2\pi f \\ &= \int_{0-}^{\infty} g(t')e^{\alpha(t-t')}\,dt'\;\delta(t-t') = g(t) \qquad t > 0- \\ &= \int_{0-}^{\infty} g(t')e^{\alpha(t-t')}\,dt'\;\delta(t-t') = 0 \qquad t < 0- \end{aligned} \tag{2.34}$$

Like the Fourier transformations $\mathfrak{F}$, $\mathfrak{F}^{-1}$ (Sec. 1.2), $\mathfrak{L}$ and $\mathfrak{L}^{-1}$ are uniquely related,§ so that $\mathfrak{L}\mathfrak{L}^{-1}\{g\} = \mathfrak{L}^{-1}\mathfrak{L}\{g\} = g$. An alternative form which we shall make frequent use of in the study of nonlinear systems (cf. Sec. 2.3 and Chaps. 5, 12 to 15) is obtained by setting $p = i\xi$, a rotation of the contours of Fig. 2.3 clockwise by $\pi/2$. Equations (2.33*a*) and (2.34) become

$$G(i\xi) = \int_{0-}^{\infty} e^{-i\xi t}g(t)\,dt \qquad \operatorname{Im}\xi < -\alpha_0 \tag{2.35a}$$

$$g(t) = \frac{1}{2\pi}\int_{\mathbf{C}_2} G(i\xi)e^{i\xi t}\,d\xi = \frac{1}{2\pi}\int_{\mathbf{C}} G(i\xi)e^{i\xi t}\,d\xi \tag{2.35b}$$

this last provided that there are no singularities to the right of the contour $\mathbf{C}$ in Fig. 2.3. In Eq. (2.35*b*), $\mathbf{C}$ is the real axis, except for an infinitesimal semicircle indented downward about a possible singularity at $z = 0$ if $\alpha_0 = 0$.

A few useful relations which follow directly from Eqs. (2.33), (2.34) are listed below:¶

† See, for example, E. T. Copson, "Theory of Functions of a Complex Variable," sec. 4.6, Oxford, New York, 1935, and K. Knopp, "Theory of Functions," chap. 8, Dover, New York, 1945.

‡ This extension of the domain is assumed in the subsequent discussion. Note in Fig. 2.3 for Re $p < \alpha_1$ $(<\alpha_0)$, where α_1 is an abscissa of convergence [usually finite, and depending on the particular $g(t)$], that although $G(p)$ is determined by analytic continuation, the integral (2.33) from which $G(p)$ is originally obtained no longer represents $G(p)$ for Re $p < \alpha_1$.

§ See, for example, Morse and Feshbach, *op. cit.*, sec. 4.8.

¶ For extensive tables of Laplace transforms, see, for example, A. Erdélyi (ed.), "Tables of Integral Transforms," vol. I, chaps. 4, 5, McGraw-Hill, New York, 1954; Magnus and Oberhettinger, "Special Functions of Mathematical Physics," chap. 8, par. 2, Chelsea, New York, 1949; and Van der Pol and Bremmer, *op. cit.*, chaps. XVII, XVIII.

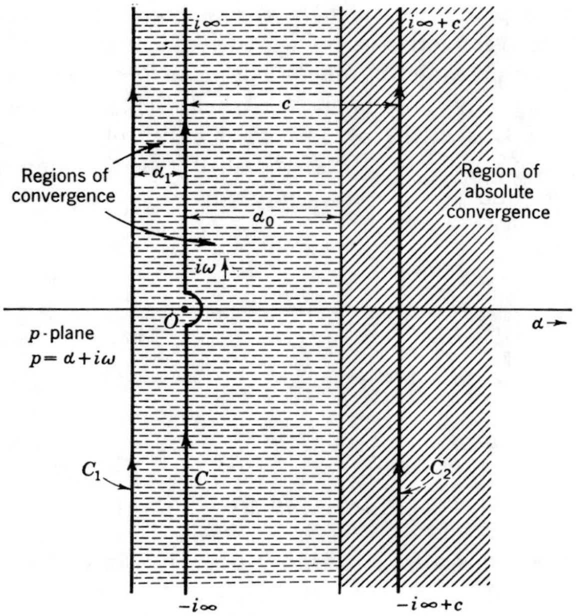

FIG. 2.3. Complex p plane, showing contours $(\mathbf{C}_1,\mathbf{C}_2)$ used in the theory of the Laplace transform and a region of convergence for some $g(t)$.

1. Linearity†

$$\mathfrak{L}\{a_1g_1 + a_2g_2\} = a_1\mathfrak{L}\{g_1\} + a_2\mathfrak{L}\{g_2\} \qquad a_1,\ a_2 \text{ arbitrary constants}$$

2. Convolution. If

$$G(t) = \int_{0-}^{\infty} g_1(\tau)g_2(t-\tau)\,d\tau = \int_{0-}^{t+} g_1(\tau)g_2(t-\tau)\,d\tau$$

then $$\mathfrak{L}\{G\} = \mathfrak{L}\{g_1\}\mathfrak{L}\{g_2\}$$

3. Derivative (2.36)

$$\mathfrak{L}\left\{\frac{dg}{dt}\right\} = p\mathfrak{L}\{g\} - g(0-), = p\mathfrak{L}\{g\} \qquad \text{if } g(0-) = 0$$

$$\mathfrak{L}\left\{\frac{d^ng}{dt^n}\right\} = p^n\mathfrak{L}\{g\} \qquad \text{if } g^{(n)}(0-) = 0, \text{ all } n \geq 0$$

4. Integral

$$\mathfrak{L}\left\{\int_{0-}^{t} g(t)\,dt\right\} = p^{-1}\mathfrak{L}\{g\}, \qquad \operatorname{Re} p > \max(0, \alpha_0)$$

Let us now apply $T_L' = \mathfrak{L}$ to the solution of the basic inhomogeneous differential equation (2.28) for a particular member function $V_{in}^{(j)}(t)$ of the

† It is assumed that $g(t)$, $g_1(t)$, and $g_2(t)$ possess Laplace transforms.

input ensemble $V_{in}(t)$. Taking the Laplace transform of both sides and using (3) [Eq. (2.36)] for the various derivatives [$V_{in}^{(j)}$, $V_o^{(j)} = 0$ $(t \leq 0-)$ since $V_o^{(j)}$ is a *normal response*], we get†

$$(a_n p^n + a_{n-1} p^{n-1} + \cdots + a_0)\mathfrak{L}\{V_o^{(j)}\} = (c_m p^m + c_{m-1} p^{m-1} + \cdots + c_0)\mathfrak{L}\{V_{in}^{(j)}\} \tag{2.37}$$

or

$$\mathfrak{L}\{V_o^{(j)}\} = Y(p)\mathfrak{L}\{V_{in}^{(j)}\}$$

where we have written

$$Y(p) = \frac{c_m p^m + c_{m-1} p^{m-1} + \cdots + c_0}{a_n p^n + a_{n-1} p^{n-1} + \cdots + a_0} = \frac{\prod_{k=1}^{m} (p - \gamma_k)}{\prod_{k=1}^{n} (p - \beta_k)} \tag{2.38}$$

in which γ_k $(k = 1, \ldots, m)$ and β_k $(k = 1, \ldots, n)$ are, respectively, the *zeros* and *poles* (i.e., eigenvalues) of $Y(p)$. Here $Y(p)$ is called the *transfer function of the filter*, which, as we can see from Eq. (2.38), is independent of the input and output members $V_{in}^{(j)}$, $V_o^{(j)}$, and hence $Y(p)$ is independent of the ensemble, operating as it does in the same fashion on every member function. If we let $V_{in}^{(j)}(t) = \delta(t - 0)$, for all j, the corresponding output is the weighting function $V_o^{(j)} = h(t)$, likewise for all j, by definition [cf. Eq. (2.30)]. Since $\mathfrak{L}\{V_{in}\} = 1$ [cf. Eq. (2.33*a*)], we see at once from Eqs. (2.33), (2.34), (2.37) that

$$\mathfrak{L}\{h\} = Y(p) = \int_{0-}^{\infty} h(t)e^{-pt}\,dt$$

$$\mathfrak{L}^{-1}\{Y(p)\} = h(t) = \begin{cases} \displaystyle\int_{-\infty i+c}^{\infty i+c} Y(p)e^{pt}\,\frac{dp}{2\pi i} & t > 0- \\ 0 & t < 0- \end{cases} \tag{2.39}$$

Thus, *the transfer function is the Laplace transform of the filter's weighting function.* Equation (2.39), in conjunction with Eq. (2.38), provides a technique for determining $h(t)$ explicitly by the methods of contour integration.‡ Since $h(t)$ is defined for unstable systems as well as stable ones (Sec. 2.2-3), so also is the corresponding transfer function $Y(p)$. Note that if $m > n$, even if $\mathrm{Re}\,\beta_k < 0$, the filter is unstable. When $m \leq n$ (the usual condition), we can resolve $Y(p)$ into partial fractions, for both single and multiple roots, and calculate $\mathfrak{L}^{-1}\{Y\}$ from Eq. (2.39); here $\mathrm{Re}\,\beta_k < 0$ for stability. If the filter is not stable, some of the poles of $Y(p)$ will lie to the right of the line $p = 0$ (cf. Fig. 2.3).

Since the output ensemble $V_o(t)$ is given by the convolution operation $T_L\{V_{in}(t)\} = \int_0^{t+(\infty)} V_{in}(\tau)h(t - \tau)\,d\tau$, we may apply (2) [Eq. (2.36)] to

† We now replace M, N in Eq. (2.28) by m, n.

‡ Morse and Feshbach, *op. cit.*, chap. IV; N. W. McLachlan, "Complex Variable and Operational Calculus with Technical Applications," pt. I, Cambridge, New York, 1939.

both sides of this expression, obtaining for an ensemble of normal responses

$$\mathfrak{L}\{V_o\} = \mathfrak{L}\{h\}\mathfrak{L}\{V_{in}\}$$

with
$$\mathfrak{L}^{-1}\{\mathfrak{L}\{V_o\}\} = V_o(t) = \int_{-\infty i+c}^{\infty i+c(>0)} Y(p)\mathfrak{L}(V_{in})e^{pt}\,\frac{dp}{2\pi i} \tag{2.40}$$

From this, the transformation T_L above becomes in terms of $\mathfrak{L}$

$$T_L\{\quad\} = \mathfrak{L}^{-1}\{\mathfrak{L}\{h\}\mathfrak{L}\{\quad\}\} \tag{2.41}$$

and the existence of $V_o = T_L\{V_{in}\}$ in the mean-square sense (cf. Sec. 2.1-2) follows from the condition (2.32), or, from Eqs. (2.32) and (2.40), its equivalent exists,

$$M_{V_o}(t_1,t_2) = \iint_{-\infty i+c}^{\infty i+c} Y(p_1)Y(p_2)\,\overline{\mathfrak{L}\{V_{in};p_1\}\mathfrak{L}\{V_{in};p_2\}}\,e^{p_1t_1+p_2t_2}\,\frac{dp_1\,dp_2}{(2\pi i)^2} \tag{2.42}$$

Here $\mathfrak{L}(V_{in},p)$ is a (complex) random variable, since V_{in} is a (real) random variable (cf. Sec. 1.2-2). Note that $\mathfrak{L}$, like T_L in this present transient state, is neither measure-preserving nor invariant.

2.2-5 System Functions and the Steady State. For many applications, it is the behavior of the filter long after the initial excitation that is of interest, when the transients have decayed to zero and the input voltage has established itself as a member of a stationary ensemble. This is called the "steady state," and for it to happen, of course, the initial excitation must occur at a time indefinitely removed in the past, so that the interval of interest is now $(-\infty < t < \infty)$, in contrast to $(0,t)$ above for transient responses. Like the impulsive excitation in the transient case above, any periodic wave, usually a simple sinusoid for convenience, can be used to describe the properties of stable filters here, and just as the Laplace transformation $\mathfrak{L}$ is the natural mathematical tool for handling transient phenomena in these filters, so also the Fourier transformation $\mathfrak{F}$ turns out to be the appropriate analytical device for treating the steady-state condition.

(1) The Fourier transform. To see how this comes about, let us begin with a real function $g(t)$, defined over $(-\infty < t < \infty)$ and for the moment integrable as $|g(t)|$ in $(-\infty,\infty)$. The Fourier-transform pair for $g(t)$ is accordingly†

$$\mathfrak{F}\{f,g(t)\} = \mathfrak{F}\{g\} = \int_{-\infty}^{\infty} e^{-i\omega t}g(t)\,dt \equiv S_g(f) \qquad \omega = 2\pi f \tag{2.43a}$$

with
$$\mathfrak{F}^{-1}\{S_g\} = g(t) = \int_{-\infty}^{\infty} e^{i\omega t}S_g(f)\,df \tag{2.43b}$$

† For a detailed discussion of the properties of $\mathfrak{F}$, for both real and complex arguments, the conditions of convergence, applications, etc., see, for example, H. S. Carslaw, "Fourier Series and Integrals" (1906), 3d rev. ed., Dover, New York, 1930; R. E. A. C. Paley and N. Wiener, Fourier Transforms in the Complex Domain, *Am. Math. Soc. Colloq. Publ.*, vol. 10, 1934; Titchmarsh;[11] Morse and Feshbach, *op. cit.*, sec. 4.8.

where $S_g(f)$ is called the *amplitude spectral density* of $g(t)$ if t has the dimensions of time (seconds), so that f is a *frequency* in cycles per second and $\omega\ (= 2\pi f)$ is an *angular frequency* in radians per second. $S_g(f)$ is in general complex, and $\mathfrak{F}$ and $\mathfrak{F}^{-1}$ are uniquely related, with the result that $\mathfrak{F}\mathfrak{F}^{-1}\{g\} = \mathfrak{F}^{-1}\mathfrak{F}\{g\} = g$, as the reader can verify directly.

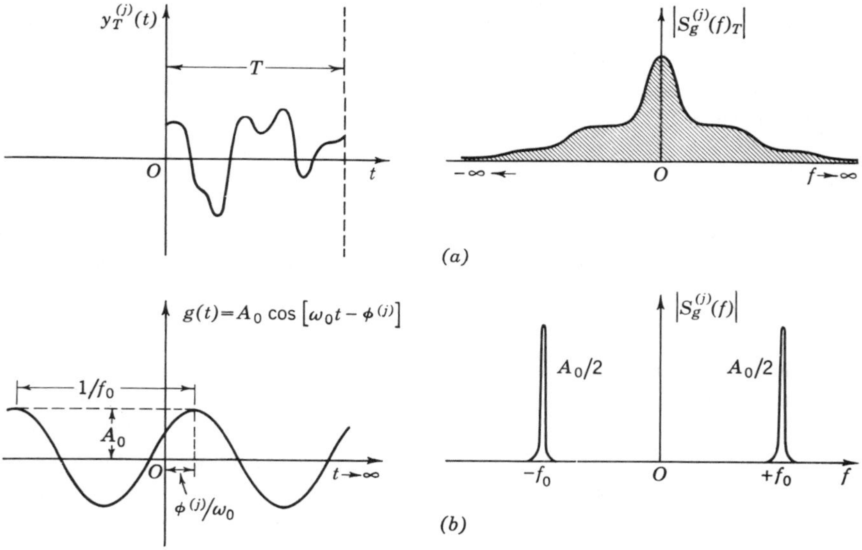

FIG. 2.4. Amplitude spectra (moduli only) of (*a*) a truncated noise wave $y_T{}^{(j)}(t)$ and (*b*) a sinusoid of constant phase $\phi^{(j)}$ and frequency f_0.

Suppose we let $g(t) = y^{(j)}(t)$, some suitably bounded member of an ensemble, in the observation interval $(0,T)$, so that $g(t) = y^{(j)}(t) = y_T{}^{(j)}(t)$ in $(-\infty,\infty)$, with $y_T{}^{(j)}(t)$ zero outside the region $(0,T)$. Then we can apply Eq. (2.43) to get

$$S_g{}^{(j)}(f)_T \equiv \mathfrak{F}\{y_T{}^{(j)}(t)\} = \int_0^T e^{-i\omega t} y^{(j)}(t)\,dt = \int_{-\infty}^{\infty} e^{-i\omega t} y_T{}^{(j)}(t)\,dt \tag{2.44a}$$

with the inverse relation

$$y_T{}^{(j)}(t) = \int_{-\infty}^{\infty} S_g{}^{(j)}(f)_T e^{i\omega t}\,df = \mathfrak{F}^{-1}\{S_g{}^{(i)}(f)_T\} \tag{2.44b}$$

and as long as T is finite, these transforms exist in the usual sense.

The amplitude spectrum is continuous (in f) and bounded in magnitude in the manner of Fig. 2.4*a*, although not limited to finite frequencies: there is always some contribution in the interval $(-\infty < f < \infty)$.

We can also extend Eq. (2.43) to include periodic waves, with the help of Sec. 1.2-4. Suppose now that $g(t) = A_0 \cos(\omega_0 t + \phi^{(j)})$ in $(-\infty,\infty)$.

Then, from Eqs. (2.43) and (1.16), it follows for this singular case that

$$S_g^{(j)}(f) = \int_{-\infty}^{\infty} A_0 e^{-i\omega t} \cos(\omega_0 t + \phi^{(j)})\, dt = \frac{A_0 e^{i\phi^{(j)}}}{2}\, \delta(f - f_0) + \frac{A_0 e^{-\phi^{(j)}}}{2}\, \delta(f + f_0) \qquad \omega_0 = 2\pi f_0 \quad (2.45)$$

and we have an example of a *line spectrum;* the amplitude, or "mass," associated with frequencies f is concentrated at the two frequencies $(\pm f_0)$ and is zero everywhere else, as sketched in Fig. 2.4*b*. Note that $S_g(f)$ is complex $(\phi^{(j)} \neq \pi n)$. Since $g(t)$ is a real function of the time, it follows at once from Eq. (2.43*b*) that

$$S_g(f) = S_g(-f)^* \qquad \text{or} \qquad S_g(f) = |S_g(f)|\, e^{i\phi_g(f)}$$
$$\text{where} \qquad |S_g(-f)| = |S_g(f)| \qquad\qquad \phi_g(f) = -\phi_g(-f) \qquad (2.46)$$

The *modulus* $|S_g(f)|$ is thus an even function of frequency and the *phase* $\phi_g(f)$ an odd one. $|S_g|$ and ϕ_g are, of course, real; and as before, the asterisk denotes the complex conjugate. It should also be observed from Eqs. (2.43), (2.44*a*) that $\lim_{T\to\infty} S_g^{(j)}(f)_T$ does *not* exist, even in the delta-function sense [Eq. (2.45)], if $g(t)$ does not die down sufficiently rapidly as $t \to +\infty$. This means that noise and (nondeterministic) signal waves which are members of stationary ensembles [or nonstationary ones, for the interval $(-\infty < t < \infty)$] do not possess Fourier transforms, i.e., amplitude spectral densities, in the ordinary sense, as defined above. The concept of "spectrum" has to be broadened before it is possible to discuss such waves from the frequency point of view in the steady state $(-\infty < t < \infty)$ (cf. Sec. 3.2).

It will be convenient to describe a spectrum, or any function of frequency, according as it is:

1. Broadband. That is, the width Δf between half-amplitude points of the modulus $|S_g(f)|$ or half-*intensity* points of $|S_g(f)|^2$ (or of the envelope of the modulus, apart from the delta function for all periodic or multiply periodic waves) is comparable with or larger than the frequency at which $|S_g(f)|$ is a maximum, in the manner of Fig. 2.5. The half-amplitude or intensity points are defined as those frequencies for which $|S_g(f)|$ or $|S_g(f)|^2$ is one-half $|S_g(f)|_{\max}$ or $|S_g(f)|^2_{\max}$.

2. Narrowband. That is, the spectral width, as defined above, is much less than the central, or *resonant*, frequency† of the spectrum (cf. Fig. 2.5*b*).

These notions are, of course, only approximate; we shall attempt later, in Chap. 3 and subsequently, to render them more precise and quantitative in dealing with actual spectra. Furthermore, they clearly begin to break down when one attempts to apply them to spectral shapes which are not unimodal (for either $f > 0$ or $f < 0$), although the general concepts of "broad" and "narrow" can still be extended to these more complicated spectral distributions. Broad spectra imply comparatively rapid *changes*

† The frequency for which $|S_g(f)| = |S_g(f)|_{\max}$.

in the waveform, while narrowband spectra correspond to waves which have a decided sinusoidal quality. In these latter, the maximum (and minimum) amplitudes change slowly compared with the instantaneous ones, while the interval between successive "zeros" of the wave also varies gradually about a "period" which is the reciprocal of the central, or resonant,

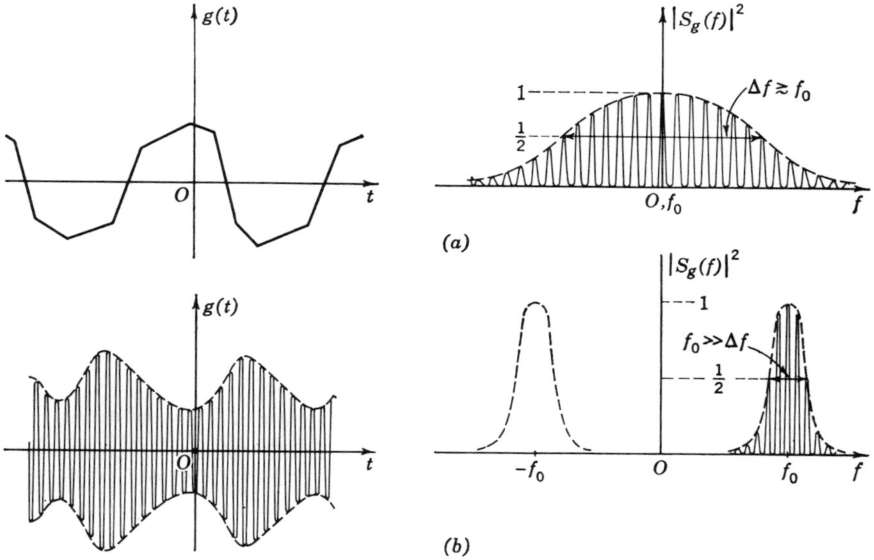

FIG. 2.5. Spectra of (a) broadband and (b) narrowband periodic waves or multiply periodic waves, e.g., a sinusoidally modulated carrier of the type (2.25a), etc. (arbitrary normalization).

frequency f_0 of the band. As we shall see later, these same notions can be carried over to entirely random processes in the steady state as well.

Like $\mathfrak{L}$ in Eq. (2.36), we may expect similar properties† for $\mathfrak{F}$:

1. Linearity‡

$$\mathfrak{F}\{a_1g_1 + a_2g_2\} = a_1\mathfrak{F}\{g_1\} + a_2\mathfrak{F}\{g_2\} \qquad a_1, a_2 \text{ arbitrary constants}$$

2. Convolution. If

$$G(t) = \int_{-\infty}^{\infty} g_1(\tau)g_2(t-\tau)\,d\tau$$

then

$$\mathfrak{F}\{G\} = \mathfrak{F}\{g_1\}\mathfrak{F}\{g_2\}$$

3. Derivative

$$\mathfrak{F}\left\{\frac{dg}{dt}\right\} = i\omega\mathfrak{F}\{g\} = i\omega S_g(f) \qquad \text{if } \int_{-\infty}^{\infty} \omega|S_g(f)|\,df \text{ exists} \tag{2.47}$$

$$\mathfrak{F}\left\{\frac{d^ng}{dt^n}\right\} = (i\omega)^n\mathfrak{F}\{g\} = (i\omega)^nS_g(f) \qquad \text{if } \int_{-\infty}^{\infty} |\omega^n|\,|S_g(f)|\,df \text{ exists}$$

† See Magnus and Oberhettinger, *op. cit.*, chap. 8, sec. 1; Erdélyi (ed.), *op. cit.*, vol. I.

‡ It is assumed, again, that g_1, g_2, g possess Fourier transforms.

4. Integration

$$\mathfrak{F}\left\{\int_{-\infty}^{t} g(t)\,dt\right\} = S_g(f)/i\omega \qquad \text{if } |S_g(0)| = 0$$

(2) The system function. Restricting our attention now to individual *periodic* members $V_{in}^{(j)}(t)$ of a periodic input ensemble, so that Fourier transforms of $V_{in}^{(j)}$ and $V_o^{(j)}$ exist (in the δ-function sense), we take the Fourier transforms of both sides of the inhomogeneous equation (2.28), with the help of (3) [Eq. (2.47)], getting

$$[a_n(i\omega)^n + a_{n-1}(i\omega)^{n-1} + \cdots + a_0]\mathfrak{F}\{V_o^{(j)}\} = [c_m(i\omega)^m + c_{m-1}(i\omega)^{m-1} + \cdots + c_0]\mathfrak{F}\{V_{in}^{(j)}\} \qquad (2.48)$$

or

$$\mathfrak{F}\{V_o^{(j)}\} = Y(i\omega)\mathfrak{F}\{V_{in}^{(j)}\} \qquad 0 \le m \le n$$

where

$$Y(i\omega) = \frac{c_m(i\omega)^m + c_{m-1}(i\omega)^{m-1} + \cdots + c_0}{a_n(i\omega)^n + a_{n-1}(i\omega)^{n-1} + \cdots + a_0} \qquad (2.49)$$

The quantity $Y(i\omega)$ is called the *system*, or *response*, *function* of the filter and is identical with the transfer function $Y(p)$ when $p = i\omega$. Note, however, that, while $Y(p)$ (p complex) can exist for unstable as well as stable filters, $Y(i\omega)$ is defined only for stable ones, since only then can $\mathfrak{F}\{V_o^{(j)}\}$ exist in Eqs. (2.48) when $\mathfrak{F}\{V_{in}^{(j)}\}$ is given. From Eq. (2.49) it is clear that $Y(i\omega)$ is here defined independent of the particular members $V_{in}^{(j)}$, $V_o^{(j)}$ of the input and output ensembles, as long as $V_{in}^{(j)}$ is periodic. [For stable filters, for which $m \le n$, and for less restricted inputs, we can, however, define the system function (2.49) generally by analytic continuation of the transfer function $Y(p)$—cf. Eq. (2.38)—setting $p = i\omega$ (ω real).] Like $S_g(f)$, $Y(i\omega)$ is usually complex, so that we can write [cf. Eq. (2.46)]

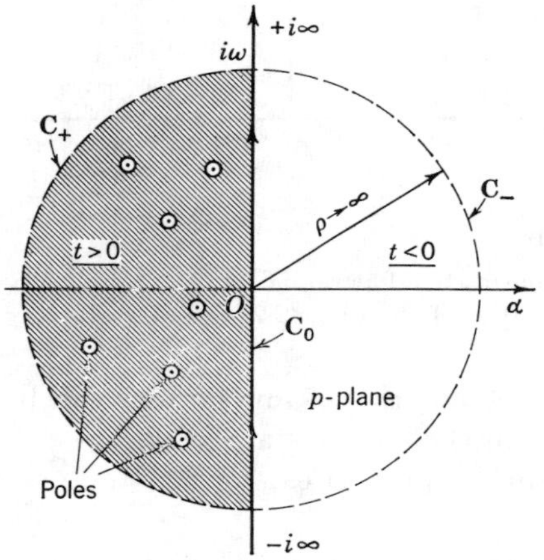

FIG. 2.6. Contours of integration and poles of a linear stable filter in the complex p plane.

$$Y(i\omega) = |Y(i\omega)|e^{i\phi(i\omega)} \qquad \text{with } |Y| = (YY^*)^{1/2}$$
$$\phi(i\omega) = \tan^{-1}\frac{Y - Y^*}{i(Y + Y^*)} \qquad (2.50)$$

where $|Y|$ and ϕ are real functions of angular frequency ω and, like $S_g^{(j)}(f)_T$ of Eq. (2.44*a*), Y is normally a continuous function of f (or ω).

The steady-state output ensemble of our filter, where again $V_{in}(t)$ is a periodic input ensemble, now follows from Eq. (2.31*a*) if we set the upper limit on τ to be ∞. We write accordingly

$$V_o(t) = \int_{-\infty}^{\infty} V_{in}(\tau)h(t-\tau)\,d\tau = \int_{-\infty}^{\infty} V_{in}(t-\tau)h(\tau)\,d\tau$$
$$= \int_{-\infty}^{t} V_{in}(\tau)h(t-\tau)\,d\tau \tag{2.51}$$

Taking the Fourier transform of both sides of Eq. (2.51) gives us

$$\mathfrak{F}\{V_o\} = \mathfrak{F}\{h\}\mathfrak{F}\{V_{in}\} \qquad \text{or} \qquad S_{V_o}(f) = \mathfrak{F}\{h\}S_{V_{in}}(f) \tag{2.52}$$

and comparing this with Eq. (2.48) shows that

$$\mathfrak{F}\{h\} = Y(i\omega) = \int_{-\infty}^{\infty} h(t)e^{-i\omega t}\,dt \qquad \omega = 2\pi f \tag{2.53a}$$

with

$$\mathfrak{F}^{-1}\{Y\} = h(t) = \int_{-\infty}^{\infty} Y(i\omega)e^{i\omega t}\,df \tag{2.53b}$$

Thus, *the system function of a stable* (linear) *filter is the Fourier transform of the weighting function*, analogous to the relation (2.39) for the more general (linear) cases where stability may not exist. We can now write Eq. (2.52) alternatively as

$$S_{V_o}(f) = Y(i\omega)S_{V_{in}}(f) \tag{2.54a}$$

with its inverse

$$V_o(t) = \mathfrak{F}^{-1}\{S_{V_o}\} = \frac{1}{2\pi}\int_{-\infty}^{\infty} Y(i\omega)\mathfrak{F}\{V_{in}\}e^{i\omega t}\,d\omega$$
$$= \int_{-\infty}^{\infty} Y(i\omega)S_{V_{in}}(f)e^{i\omega t}\,df \tag{2.54b}$$

for input and output ensembles of periodic waves† [cf. Eq. (2.40)]. As an example, let $V_{in}(t) = A_0 \cos(\omega_0 t + \phi)$. Then, from Eq. (2.43a), $S_{V_{in}}(f)$ becomes $(A_0/2)e^{i\phi}\,\delta(f - f_0) + (A_0/2)e^{-i\phi}\,\delta(f + f_0)$, and from Eq. (2.54b) it follows at once that the output ensemble is

$$V_o(t) = \frac{A_0}{2}[e^{i\phi+i\omega_0 t}Y(i\omega_0) + e^{-i\phi-i\omega_0 t}Y(-i\omega_0)] = A_0 \operatorname{Re}\{e^{i\omega_0 t+i\phi}Y(i\omega_0)\} \tag{2.55}$$

The output is still sinusoidal at the input frequency f_0, but with a new phase and amplitude, determined by the system function of the network.

To calculate the weighting function, it is usually simplest to use Eq. (2.53b), replacing $i\omega$ by the complex variable $p = \alpha + i\omega$, so that

$$h(t) = \int_{-\infty i}^{\infty i} Y(p)e^{pt}\frac{dp}{2\pi i} = \int_{\mathbf{C}_0} Y(p)e^{pt}\frac{dp}{2\pi i} \tag{2.56}$$

where now $\mathbf{C}_0$ is the imaginary axis (cf. Fig. 2.6). Since the filter is stable [otherwise $Y(i\omega)$ does not exist], all the roots β_q of the modal equation

† These relations apply, of course, to ensembles for which $V_{in}(\pm\infty)$ vanish suitably rapidly and which are suitably behaved in $(-\infty, \infty)$ so that $S_{V_{in}}(f)$ exists, but these are clearly not what we mean by steady-state phenomena.

TABLE 2.1. RESULTS FOR

Types of filter	Network diagrams and differential equations	Weighting functions $h(t)$, $t > 0-$ [$h(t) = 0$, $t < 0-$]	Transfer function $Y(p)$
1. *RC* filter	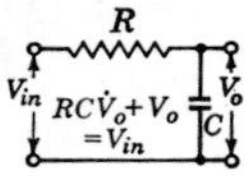 $RC\dot{V}_o + V_o = V_{in}$	$(RC)^{-1}e^{-t/RC}$	$(1 + RCp)^{-1}$
2. *CR* filter	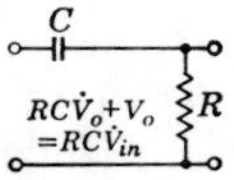 $RC\dot{V}_o + V_o = RC\dot{V}_{in}$	$\delta(t - 0) - (RC)^{-1}e^{-t/RC}$	$\dfrac{RCp}{RCp + 1}$
3. Series *LRC* filter	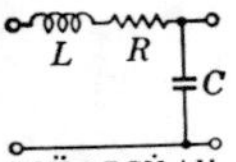 $LC\ddot{V}_o + RC\dot{V}_o + V_o = V_{in}$	$\dfrac{e^{-Rt/2L}}{i\omega_1 LC}\sinh i\omega_1 t$, $\omega_1^2 < 0$ $\dfrac{te^{-Rt/2L}}{LC}$, $\omega_1 = 0$ $\dfrac{e^{-Rt/2L}}{\omega_1 LC}\sin \omega_1 t$, $\omega_1^2 > 0$	$\dfrac{1}{LCp^2 + RCp + 1}$
4. Series *CLR* filter	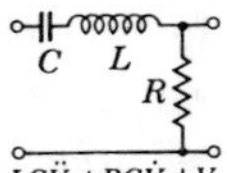 $LC\ddot{V}_o + RC\dot{V}_o + V_o = RC\dot{V}_{in}$	$\dfrac{2\omega_F e^{-\omega_F t}}{i\omega_1}(\omega_F \sinh i\omega_1 t - \omega_1 \cosh i\omega_1 t)$, $\omega_1^2 < 0$ $2\omega_F(1 - \omega_F t)e^{-\omega_F t}$, $\omega_1^2 = 0$ $\dfrac{2\omega_0\omega_F}{\omega_1} e^{-\omega_F t} \cos\left(\omega_1 t + \tan^{-1}\dfrac{\omega_1}{\omega_F}\right)$, $\omega_1^2 > 0$	$\dfrac{RCp}{LCp^2 + RCp + 1}$
5. Parallel *LRC* filter	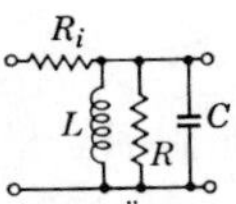 $CLR_i\ddot{V}_o + \left(\dfrac{R_iL}{R} + L\right)\dot{V}_o + R_iV_o = LV_{in}$	$(iCR_i\omega_1)^{-1}e^{-\omega_F t}(\omega_F \sinh \omega_1 t - \omega_1 \cosh \omega_1 t)$, $\omega_1^2 < 0$ $\dfrac{1}{CR_i}(1 - \omega_F t)e^{-\omega_F t}$, $\omega_1^2 = 0$ $\dfrac{e^{-\omega_F t}}{CR_i\omega_1}(-\omega_F \sin \omega_1 t + \omega_1 \cos \omega_1 t)$, $\omega_1^2 > 0$	$\dfrac{Lp}{CLR_ip^2 + L\left(\dfrac{R_i}{R} + 1\right)p + R_i}$

SOME SIMPLE FILTERS

Phase function $\phi(i\omega)$	Modulus² of system function $\lvert Y(i\omega)\rvert^2$	Narrowband relations	
		$Y_0(i\omega')$, $\omega' = \omega - \omega_0$	$h_0(t)$, $\gamma_0(t)$
$\tan^{-1}(-RC\omega)$	$[1 + (RC\omega)^2]^{-1}$		
$\tan^{-1}\frac{1}{RC\omega}$	$\frac{(RC\omega)^2}{(RC\omega)^2 + 1}$		
$\tan^{-1}\frac{RC\omega}{LC\omega^2 - 1}$ High Q: $\phi_0(i\omega') \doteq \tan^{-1}\frac{\omega_F}{\omega'}$	$\frac{1}{(RC\omega)^2\left[1 + \left(\frac{\omega - \omega_0}{\omega_F}\right)^2\left(\frac{\omega + \omega_0}{2\omega}\right)^2\right]}$ $\omega_0^2 = \frac{1}{LC};\ \omega_F = \frac{R}{2L}$ $Q \equiv \frac{\omega_0}{2\omega_F} = \left(\frac{L}{CR^2}\right)^{1/2}$ $(RC)^{-1} = \frac{\omega_0^2}{2\omega_F}$ $\omega_1 = \sqrt{\omega_0^2 - \omega_F^2} = \sqrt{\frac{1}{LC} - \frac{R^2}{4L^2}}$ High Q: $\lvert Y(i\omega)\rvert^2 \doteq \frac{\omega_0^2}{4\omega_F^2}\frac{1}{1 + \left(\frac{\omega_0 - \omega}{\omega_F}\right)^2},\ Q \gg 1$	$\frac{-i\omega_0/2\omega_F}{1 + i(\omega'/\omega_F)}$	$\gamma_0 = -\frac{\pi}{2}$ $h_0(t) = \frac{\omega_0}{2}e^{-\omega_F t}$
$\tan^{-1}\frac{1 - LC\omega^2}{RC\omega}$ $\phi_0(i\omega') \doteq \tan^{-1}\frac{-\omega'}{\omega_F},\ Q \gg 1$	$\frac{1}{1 + \left(\frac{\omega - \omega_0}{\omega_F}\right)^2\left(\frac{\omega + \omega_0}{2\omega}\right)^2}$ $\lvert Y(i\omega)\rvert^2 \doteq \frac{1}{1 + \left(\frac{\omega - \omega_0}{\omega_F}\right)^2},\ Q \gg 1$	$\frac{i}{1 + i(\omega'/\omega_F)}$	$\gamma_0 = 0$ $h_0(t) = \omega_F e^{-\omega_F t}$
$\tan^{-1}\frac{2C^2\omega_F(\omega_0^2 - \omega^2)}{\omega}$ $\phi_0(i\omega') \doteq \tan^{-1}(-4C^2\omega_F^2\omega')$	$\frac{1}{1 + \left(\frac{\omega - \omega_0}{\omega_F}\right)^2\left(\frac{\omega + \omega_0}{2\omega}\right)^2}$ $\omega_F = \left(\frac{2RR_iC}{R_i + R}\right)^{-1}$ ω_0, ω_1 as above (3); high Q as above (4)	$\frac{i}{C^2RR_i\omega_0}\frac{1}{1 + i\omega'(2/CR\omega_0^2)}$	$\gamma_0 = -\frac{\pi}{2}$ $h_0(t) = \frac{e^{-CR\omega_0^2 t/2}}{2CR_i\omega_0}$

$F(p) = \prod_{q=1}^{n} (p - \beta_q)$ possess *negative* real parts $(\alpha_q < 0)$. Observe that for $t < 0$ we can add to $h(t)$ the zero contribution of the integrand along the infinite semicircle $\mathbf{C}_-$ in the right half plane, since $\lim_{\alpha \to \infty} e^{\alpha t} Y(\alpha) \to 0$ $(\alpha \leq 0, t < 0)$. Similarly, since all poles of $Y(p)$ are to the left of $\mathbf{C}_0$, adding the zero contribution of the integrand along the infinite semicircle $\mathbf{C}_+$, for $t > 0$, we may apply Cauchy's theorem† to give us, as required,

$$
\begin{aligned}
h(t) &= 2\pi i \sum_{q}^{n} (\text{residues at } p_q = \beta_q) \qquad t > 0- \\
\text{and} \qquad h(t) &= 0 \qquad t < 0-
\end{aligned}
\tag{2.57}
$$

since there are no singularities in the right-half p plane. Results for some common filters, of interest to us subsequently, are listed in Table 2.1.

(3) Physical realizability. At this point, we observe that the existence of the system function is closely related to the question of *physical realizability*. A lumped-constant network is said to be physically realizable if it can be constructed from a finite number of passive L, R, C circuit elements, where, in addition, ideal amplifiers may be included which are linear and whose sole function is to modify the gain, or scale, of the input disturbance, i.e., without introducing feedback or nonlinear effects.‡ The necessary and sufficient conditions for physical realizability (cf. Prob. 2.6*b*, *c*) may be expressed mathematically by:

1. $h(t) = 0$ $(t < 0-)$. The network operates only on the past of the input. It cannot respond to an input before the input arrives.

2. $\lim_{t \to \infty} h(t) \to 0$ sufficiently rapidly, or, more explicitly, $\int_{-\infty}^{\infty} |h(t)|\, dt < \infty$, so that the filter is stable. An input that is bounded in duration produces an output disturbance which ultimately dies down to zero as $t \to \infty$.

In the frequency domain, these two conditions are equivalent to the requirement that $Y(p)$, as a function of the complex variable p, is analytic everywhere in the half plane $\text{Re}\, p > 0$ and contains no singularities on the imaginary axis (cf. Fig. 2.6).

An alternative criterion for the physical realizability of (linear) systems for which a system function can be specified is provided by the Paley-Wiener relation§

† See, for example, Morse and Feshbach, *op. cit.*, secs. 4.2, 4.5, and McLachlan, *op. cit.*

‡ "Physical realizability" is used here in this rather restricted sense for lumped-constant elements. Distributed elements, such as waveguides and transmission lines, are certainly physically realizable, but the above concepts of weighting and system functions then lose their simple character, and one must in general return to the original Maxwell equations for a description of such systems.

§ See Paley and Wiener, *loc. cit.*, theorem XII. See also G. E. Valley and H. Wallman, "Vacuum Tube Amplifiers," app. A, Massachusetts Institute of Technology Radiation Laboratory Series, vol. 18, McGraw-Hill, New York, 1948.

$$J \equiv \int_{-\infty}^{\infty} \frac{\log |Y(i\omega)|^2 \, d\omega}{1 + \omega^2} \tag{2.58}$$

If $|J| < \infty$, then $|Y(i\omega)|$ is the modulus of a system function of a realizable filter, while if $|J| \to \infty$, the filter is not physically realizable but can only be approximated by a finite number of linear lumped elements. Furthermore, if $|J| < \infty$, then we have $|Y|^2 = YY^*$; $|Y|^2$ can be factored, and it is then possible to choose $Y(i\omega)$ in such a way that $Y(p)$ contains no poles in the right half plane Re $p > 0$. For some applications of this approach, see Chap. 16.

(**4**) **Fourier transform of the ensemble.** So far, we have considered only ensembles of periodic waves as inputs to these linear filters. Let us at this point lift the restriction that the input ensembles in the steady state must be deterministic, with a periodic structure, letting $V_{in}(t)$ now represent an entirely random noise or signal ensemble. Then the steady-state transformation $\boldsymbol{T}_L = \int_{-\infty}^{\infty} h(t - \tau)\{\quad\} \, d\tau$ [cf. Eq. (2.51)] does indeed yield an output ensemble $V_o(t)$ when the input $V_{in}(t)$ is of this more general type, provided that $\boldsymbol{T}_L\{V_{in}\}$ exists mean square (cf. Sec. 2.1-2), i.e., provided that

$$M_{V_o}(t_1,t_2) = \iint\limits_{-\infty}^{\infty} h(t_1 - \tau_1)h(t_2 - \tau_2)\overline{V_{in}(\tau_1)V_{in}(\tau_2)} \, d\tau_1 \, d\tau_2 \tag{2.59a}$$

is defined in the usual Riemann (or Riemann-Stieltjes) sense. *Here, however, it is not true that*

$$M_{V_o}(t_1,t_2) = \iint\limits_{-\infty i}^{\infty i} Y(p_1)Y(p_2)\overline{\mathfrak{F}\left\{V_{in}; \frac{p_1}{2\pi i}\right\} \mathfrak{F}\left\{V_{in}; \frac{p_2}{2\pi i}\right\}} \, e^{p_1t_1+p_2t_2} \frac{dp_1 \, dp_2}{(2\pi i)^2} \tag{2.59b}$$

nor can one write $\boldsymbol{T}_L\{\quad\} = \mathfrak{F}^{-1}\{\mathfrak{F}\{h\}\mathfrak{F}\{\quad\}\}$, since the amplitude spectra $\mathfrak{F}\{V_{in}\}$ are not defined for these inputs, unlike the transient cases considered in Sec. 2.2-4 [cf. Eq. (2.42)]. Before we can give the proper expression for the condition (2.59*a*) in the frequency domain, we shall have to broaden the concept of the spectrum to include partially and entirely random processes. Note, incidentally, that while $\boldsymbol{T}_L$ is non-measure-preserving, as before, it is also invariant [Sec. 2.1-3(2)]: stationary and ergodic input ensembles are transformed into new (and different) stationary and ergodic output ensembles.

(**5**) **Narrowband filters.** When the system function $Y(i\omega)$ is *narrowband* [cf. Sec. 2.2-5(1)], or strongly and unimodally resonant, a useful simplification in the form of the associated weighting function is possible. To show this, we start with Eq. (2.53*a*) and write

$$\begin{aligned} h(t) &= \int_{-\infty}^{0} Y(i\omega)e^{i\omega t} \, df + \int_{0}^{\infty} Y(i\omega)e^{i\omega t} \, df \\ &= \int_{0}^{\infty} Y(i\omega)^*e^{-i\omega t} \, df + \int_{0}^{\infty} Y(i\omega)e^{i\omega t} \, df \end{aligned} \tag{2.60}$$

and since Y is assumed to be narrowband, Y has its resonance at f_0 and Y^* at $-f_0$. Now let $\omega = \omega' + \omega_0$, so that Eq. (2.60) becomes

$$\begin{aligned} h(t) &= e^{-i\omega_0 t} \int_{-\omega_0}^{\infty} Y(-i\omega' - i\omega_0) e^{-i\omega' t}\, df' + e^{i\omega_0 t} \int_{-\omega_0}^{\infty} Y(i\omega' + i\omega_0) e^{i\omega' t}\, df' \\ &\doteq e^{-i\omega_0 t} \int_{-\infty}^{\infty} Y(i\omega' + i\omega_0)^* e^{-i\omega' t}\, df' + e^{i\omega_0 t} \int_{-\infty}^{\infty} Y(i\omega' + i\omega_0) e^{i\omega' t}\, df' \\ &\doteq 2 \,\mathrm{Re} \int_{-\infty}^{\infty} Y(i\omega' + i\omega_0) e^{i\omega' t + i\omega_0 t}\, df' \end{aligned} \tag{2.61}$$

which is a good approximation as long as f_0 is sufficiently large compared with the bandwidth. Setting

$$Y(i\omega' + i\omega_0) \equiv Y_0(i\omega') \tag{2.62}$$

where now $Y_0(i\omega')$ is the system function of a low-pass broadband filter, *with the new origin of frequencies at* f_0, in virtue of $\omega' = \omega - \omega_0$, we obtain from Eq. (2.61) the desired result,

$$\begin{aligned} h(t) &\doteq 2 \,\mathrm{Re} \int_{-\infty}^{\infty} Y_0(i\omega') e^{i\omega' t + i\omega_0 t}\, df' \equiv 2h_0(t) \cos\,[\omega_0 t - \gamma_0(t)] \\ \text{or} \quad h(t) &\doteq 2 \,\mathrm{Re} \int_{-\infty}^{\infty} Y_0(i\omega') e^{i\omega' t + i\omega_0 t}\, df' = 2 \,\mathrm{Re}\, h_0(t) e^{i\omega_0 t - i\gamma_0(t)} \end{aligned} \tag{2.63a}$$

$$\begin{aligned} \text{so that} \quad h_0(t) e^{-i\gamma_0(t)} &= \int_{-\infty}^{\infty} Y_0(i\omega') e^{i\omega' t}\, df' \qquad \omega' \equiv \omega - \omega_0 \\ \text{with} \quad Y_0(i\omega') &= \mathfrak{F}\{h_0(t) e^{-i\gamma_0(t)}\} \end{aligned} \tag{2.63b}$$

and $h_0(t)$, $\gamma_0(t)$ are slowly varying functions of time compared with $\cos \omega_0 t$, $\sin \omega_0 t$.

This weighting function $h_0(t)$ and phase angle $\gamma_0(t)$ may be obtained from the original $h(t)$ by applying the "high-Q," or narrowband, approximation appropriate to the filter in question, or they may be calculated from $\mathfrak{F}^{-1}\{Y_0\}$ as indicated above. Like $h(t)$, $h_0(t)$ and $\gamma_0(t)$ are real, and $h_0(t)$ vanishes for $t < 0-$, so that $Y_0(i\omega')$ is in general complex, like $Y(i\omega)$. Some specific examples are included in Table 2.1. For many of the networks which we shall encounter in communication applications, to a first approximation the phase $\gamma_0(t)$ is independent of t and in many cases is actually zero. As one can easily see from Eq. (2.62) in Eq. (2.54), each member of the output ensemble is always narrowband, even if the original input representation is not.

Consider now a (real) input ensemble of narrowband waves in the steady state. These can be written in the form

$$V_{in}(t) = E_0(t) \cos\,[\omega_c t + \phi - \Phi(t)] = \mathrm{Re}\, E_0(t) e^{i\omega_c t + i\phi - i\Phi(t)} \tag{2.64}$$

where ϕ is assigned an appropriate distribution in the primary interval $(0,2\pi)$. Usually $E_0 \geq 0$,† and in any case $E_0(t)$, $\Phi(t)$ are real and slowly varying compared with $\cos\,(\omega_c t + \phi)$, $\sin\,(\omega_c t + \phi)$, so that $V_{in}(t)$ may truly

† In certain propagation problems, it is possible, though not usual, for $E_0 < 0$.

be considered narrowband,† within the rather loose definition of the term as given in Sec. 2.2-5(1). One calls E_0 and $\Phi - \phi$, or Φ, the *envelope* and *phase*, respectively, of the narrowband disturbance. These quantities may be operationally defined in terms of the specific measuring devices which yield $E_0(t)$ and, separately, $\Phi(t)$—for example, the familiar envelope and phase detectors of communication theory [cf. Part 3, Chaps. 13, 15; see also Sec. 2.2-5(6)].

The output ensemble $V_o(t)$, which follows when $V_{in}(t)$ above is applied to a narrowband filter, is, of course, also an ensemble of narrowband waves. Let us see how $V_o(t)$ and $V_{in}(t)$ are related, taking specifically into account this narrowband property of the filter. We have

$$V_o(t) = \mathbf{T}_L\{V_{in}(t)\}$$

where now (2.65a)

$$\mathbf{T}_L\{\quad\} \doteq 2\int_{-\infty}^{\infty} h_0(\tau)\{\quad\}_{t-\tau} \cos\left[\omega_0\tau - \gamma_0(t)\right] d\tau$$

with the help of Eq. (2.63). This becomes

$$\begin{aligned} V_o(t) &\doteq 2\int_{-\infty}^{\infty} h_0(\tau)E_0(t-\tau)\cos\left[\omega_0\tau - \gamma_0(\tau)\right] \\ &\qquad \cos\left[\omega_c(t-\tau) + \phi - \Phi(t-\tau)\right] d\tau \\ &\doteq \int_{-\infty}^{\infty} h_0(\tau)E_0(t-\tau)\cos\left[\omega_0 t + \omega_D(t-\tau) - \Phi(t-\tau)\right. \\ &\qquad \left. + \phi - \gamma_0(\tau)\right] d\tau \qquad \omega_D \equiv \omega_c - \omega_0 \ll \omega_c, \omega_0 \end{aligned} \tag{2.65b}$$

since E_0, h_0, γ_0, Φ are slowly varying compared with $\cos \omega_0\tau$, $\cos \omega_c\tau$, etc., and ω_D, which measures the extent of "detuning," is required to be small compared with both ω_c and ω_0. Accordingly, if we write

$$\begin{aligned} V_o(t) &\equiv \mathrm{Re}\left[\alpha(t) - i\beta(t)\right]e^{i\omega_0 t + i\phi} = \alpha(t)\cos(\omega_0 t + \phi) + \beta(t)\sin(\omega_0 t + \phi) \\ &= \mathrm{Re}\, e^{i\omega_0 t + i\phi}\int_{-\infty}^{\infty} h_0(\tau)E_0(t-\tau)e^{i\omega_D(t-\tau) - i\Phi(t-\tau) - i\gamma_0(\tau)}\, d\tau \end{aligned} \tag{2.66}$$

defining $\alpha(t)$ and $\beta(t)$ from these relations, and if we further define an $\alpha_0(t)$, $\beta_0(t)$ by

$$V_{in}(t) \equiv \mathrm{Re}\left[\alpha_0(t) - i\beta_0(t)\right]e^{i\omega_c t + i\phi} = \alpha_0(t)\cos(\omega_c t + \phi) + \beta_0(t)\sin(\omega_c t + \phi) \tag{2.67}$$

we observe on comparing Eqs. (2.66) and (2.67) that

$$\alpha(t) - i\beta(t) = \int_{-\infty}^{\infty} \left[\alpha_0(\tau) - i\beta_0(\tau)\right]e^{i\omega_D\tau - i\gamma_0(t-\tau)}h_0(t-\tau)\, d\tau \tag{2.68}$$

† Or, equivalently, having an essentially sinusoidal time structure for intervals less than the mean period of changes in the extremum amplitudes of the instantaneous waveform (cf. Fig. 2.5b). In terms of *bandwidth* and *amplitude spectra*, we can speak so far here only of ensembles of periodic waves *in the steady state*, since it is only for these that amplitude spectra exist (in the δ-function sense above). However, as we shall see in Chap. 3, these same notions of broadband and narrowband can also be applied directly to the more general cases of entirely random or mixed processes when the concept of spectrum has been extended to include such processes. Therefore, in the subsequent discussion we shall use the terms broad- and narrowband without restriction.

A number of useful alternative forms are also possible. For example, by analogy with Eq. (2.64) we can write

$$V_o(t) = \text{Re}\, B_0(t)e^{i\omega_0 t + i\phi - i\Psi(t)} = B_0(t)\cos\,[\omega_0 t + \phi - \Psi(t)] \tag{2.69}$$

where similarly $B_0(t)$ and $\Psi(t)$ are the envelopes and phases of the output ensemble, whose approximate center frequency is now $f_0\ (= f_c - f_D)$. Comparison with Eqs. (2.64) to (2.68) leads to the following relations,

$$\begin{aligned}
\alpha(t) &= \text{Re} \int_{-\infty}^{\infty} E_0(\tau)h_0(t-\tau)e^{i\omega_D\tau - i\Phi(\tau) - i\gamma_0(t-\tau)}\,d\tau \\
\beta(t) &= \text{Im} \int_{-\infty}^{\infty} E_0(\tau)h_0(t-\tau)e^{i\omega_D\tau - i\Phi(\tau) - i\gamma_0(t-\tau)}\,d\tau \\
\sqrt{\alpha_0(t)^2 + \beta_0(t)^2} &= |\alpha_0 - i\beta_0| = E_0(t) \\
\sqrt{\alpha(t)^2 + \beta(t)^2} &= |\alpha - i\beta| = B_0(t) \\
\tan^{-1}[\beta_0(t)/\alpha_0(t)] &= \Phi(t) \\
\tan^{-1}[\beta(t)/\alpha(t)] &= \Psi(t)
\end{aligned} \tag{2.70}$$

from which the output envelopes and phases can be determined by the corresponding input quantities.

In the frequency domain, one can find similar relations if the input ensembles are again restricted to the class of deterministic periodic waves. Defining then

$$S_{V_{in}}(f')_0 \equiv \mathfrak{F}\{\alpha_0 - i\beta_0\} = \int_{-\infty}^{\infty} [\alpha_0(t) - i\beta_0(t)]e^{-i\omega t}\,dt \qquad \omega' = \omega - \omega_0 \tag{2.71a}$$

$$S_{V_o}(f')_0 \equiv \mathfrak{F}\{\alpha - i\beta\} = \int_{-\infty}^{\infty} [\alpha(t) - i\beta(t)]e^{-i\omega t}\,dt \tag{2.71b}$$

where now these amplitude spectra for V_{in} and V_o are taken about the resonant or center frequencies f_c, f_0, respectively, we find that

$$\alpha(t) - i\beta(t) = \int_{-\infty}^{\infty} S_{V_{in}}(f')_0 Y_0(i\omega')e^{i\omega' t}\,df' = \mathfrak{F}^{-1}\{S_{V_o}(f')_0\} \tag{2.72a}$$

and therefore

$$S_{V_o}(f')_0 = S_{V_{in}}(f')_0 Y_0(i\omega') \qquad \omega' = \omega - \omega_0 \tag{2.72b}$$

We shall see some specific examples of narrowband behavior later, in Chap. 3, and in Part 3 when some typical receiving systems are analyzed.

(6) Instantaneous envelope, phase, and frequency, and the analytic signal. In Sec. 2.2-5(5), we have tacitly assumed for the representation (2.64) of an ensemble of narrowband waves, $V_{in}(t)$, that the envelopes $E_0(t)$ and instantaneous phases $\omega_c t + \phi - \Phi(t)$ are uniquely specified and hence uniquely resolvable (at least in principle) by the appropriate physical operation of envelope (AM) and phase detection. Then, if this be true, in terms of the instantaneous phase we can also *define* an *instantaneous frequency*,[9] according to

$$f_{inst} \equiv \frac{1}{2\pi}\frac{d}{dt}\,(\text{instantaneous phase}) = [\omega_c - \dot{\Phi}(t)]/2\pi \tag{2.73}$$

here remembering, of course, that for this concept to be useful, i.e., quantitatively operational, the narrowband condition that $\Phi(t)$ varies slowly compared with $\omega_c t$ must hold, in some average sense, for example, $\omega_c \gg \sqrt{\overline{\dot{\Phi}^2}}$, at least.

Analytically, these assumptions of uniqueness imply a certain definite relationship between the envelope and the phase, $\omega_c t + \phi - \Phi(t)$. Frequently, however, in describing an ensemble of narrowband waves, $V_{in}(t)$, in this way, the "carrier"-frequency term $(\omega_c t/2\pi)$ is not a priori specified; any component in the spectral region where $V_{in}(t)$ is significant may serve as well as any other as a reference frequency, so that one may write $V_{in}(t) = E_0(t) \cos [\omega_c t + \phi - \Phi(t)] = E_0'(t) \cos [\omega_c' t + \phi - \Phi'(t)]$, and it is evident that $E_0' \neq E_0$, $\omega_c' t - \Phi'(t) \neq \omega_c t - \Phi(t)$ if $\omega_c \neq \omega_c'$. Thus, if ω_c (or ω_c') is selected arbitrarily, it is not clear analytically just what constitutes an observable envelope and phase. Of course, if ω_c is initially determined [i.e., it might be the (angular) frequency of a carrier modulated by $E_0(t)$ and $\Phi(t)$], then $E_0(t)$ and $\Phi(t)$ (apart from multiples of π) are uniquely specified. This is easily seen if we write

$$
\begin{aligned}
V_{in}(t) &= (E_0 \cos \Phi) \cos (\omega_c t + \phi) + (E_0 \sin \Phi) \sin (\omega_c t + \phi) \\
&= (E_0' \cos \Phi') \cos (\omega_c t + \phi) + (E_0' \sin \Phi') \sin (\omega_c t + \phi) \qquad (2.74)
\end{aligned}
$$

and since these are equal, the only solution of the simultaneous equations $E_0 \cos \Phi = E_0' \cos \Phi$, $E_0 \sin \Phi = E_0' \sin \Phi'$ for which E_0, E_0', etc., remain slowly varying compared with $\cos (\omega_c t + \phi)$, etc., is $E_0 = E_0'$, $\Phi = \Phi' \pm q\pi$.

To resolve these analytical† ambiguities and, equally important, to provide a definite quantitative interpretation of the instantaneous frequency [Eq. (2.73)] when the h-f component $\omega_c t$ of the phase is not a priori determined, we need a procedure for setting $E_0'(t) = E_0(t)$, $\omega_c' t - \Phi'(t) = \omega_c t - \Phi(t)$, so that E and $\omega_c t - \Phi$ are uniquely specified, and hence coincide with what is to be obtained in an actual detection process. Following an approach suggested originally by Ville,[10] we begin by constructing the complex random variable $Z(t) = X(t) - iY(t)$ (X, Y real), where $X(t) = V_{in}(t)$ here, so that alternatively $Z(t) = |Z|e^{i \arg Z} = (X^2 + Y^2)^{1/2} e^{-i \tan^{-1}(Y/X)}$ and $\operatorname{Re} Z(t) = X(t) = V_{in}(t)$. Now, *provided that* $Y(t)$ *is properly chosen as a suitable function of* $X(t)$, we observe that

$$
V_{in}(t) = \operatorname{Re} Z(t) = \operatorname{Re} |Z(t)| e^{i \arg Z(t)} = E_0(t) \cos [\omega_c t + \phi - \Phi(t)] \qquad (2.75)
$$

Now E_0 and the phase $\omega_c t + \phi - \Phi(t)$ are uniquely specified (apart from multiples of π), since $|Z|$ and $\arg Z$ are functions solely of $X(t)$. Specifically, $E_0(t)$ is identified with $|Z(t)|$ and $\omega_c t + \phi - \Phi(t)$ with $\arg Z(t)$, and we say

† There is no physical ambiguity, since a properly designed AM and FM receiver will automatically determine for us the correct $E_0(t)$ and $\phi - \Phi(t)$ when measurements can be carried out.

that $V_{in}(t)$ is represented by a "carrier," cos [arg $Z(t)$], which is amplitude-modulated by a "signal" $E_0(t) = |Z(t)|$. The instantaneous frequency [Eq. (2.73)] then becomes

$$f_{inst} = \frac{1}{2\pi}\frac{d}{dt}[\arg Z(t)] \tag{2.75a}$$

The next step is to choose $Y = Y\{X(t)\}$ so that Eq. (2.75) is possible. Observe first that Z equals $X - iY$, and hence Y is *in quadrature* with X. A unique representation of Y in terms of X, where this quadrature property is exhibited and preserved, is provided by the *Hilbert-transform* pair[11]

$$Y(t) = -\frac{1}{\pi}\,\mathcal{P}\int_{-\infty}^{\infty}\frac{X(\tau)}{t-\tau}\,d\tau \qquad X(t) = \frac{1}{\pi}\,\mathcal{P}\int_{-\infty}^{\infty}\frac{Y(\tau)\,d\tau}{t-\tau} \tag{2.76}$$

where $\mathcal{P}$ indicates that the principal value of the integral at $t = \tau$ is to be taken, e.g.,

$$\mathcal{P}\int_{-\infty}^{\infty}\frac{g(x)}{a-x}\,dx \equiv \lim_{\epsilon\to 0}\left[\left(\int_{-\infty}^{a-\epsilon} + \int_{a+\epsilon}^{\infty}\right)\frac{g(x)}{a-x}\,dx\right] \tag{2.76a}$$

Writing $\mathfrak{H}$ for $(1/\pi)\,\mathcal{P}\int_{-\infty}^{\infty}\{\quad\}\,d\tau/(t-\tau)$, we may express $Y(t)$ and $X(t)$ more compactly as the Hilbert transforms

$$Y(t) = -\mathfrak{H}\{X(t)\} \qquad X(t) = \mathfrak{H}\{Y(t)\} \tag{2.77}$$

from which it is at once apparent that

$$X(t) = \mathfrak{H}\{Y(t)\} = -\mathfrak{H}^2\{X(t)\} \tag{2.78}$$

The operation $-\mathfrak{H}^2$ brings us back to the original function. Thus, X and Y are skew-reciprocal, i.e., reciprocal apart from a minus sign, and are uniquely related in this fashion.† If X and Y, as random variables, represent random processes, then Eqs. (2.76) to (2.78) apply in the usual mean-square sense (cf. Sec. 2.1-2). Furthermore, $T_L \equiv \mathfrak{H}$ is easily shown to be *invariant* [cf. Sec. 2.1-3(2)]. We remark, also, that the complex variable $Z(t) = X(t) - iY(t)$ above is sometimes called[10,10a] the *analytic signal* associated with $X(t) = V_{in}(t)$, inasmuch as $Z(t) = \lim_{u\to 0} Z(t,u)$, where $Z(t,u) = X(t,u) - iY(t,u)$ is an analytic function of $z = t + iu$. It can be shown (cf. Prob. 2.8c) that the transform relations (2.77), (2.78) do indeed satisfy the Cauchy-Riemann conditions that $Z(t,u)$ be analytic.

Let us illustrate these remarks by determining the quadrature function $Y(t)$ from Eq. (2.76) in a number of simple cases. Suppose that $X(t) =$

† For a proof of the transform properties (2.76) and the details of their derivation from the Fourier-integral formula, see Titchmarsh, *op. cit.*

$\cos \omega_0 t$. Then, from Eq. (2.76) we get

$$\begin{aligned} Y(t) &= \frac{-1}{\pi} \mathcal{P} \int_{-\infty}^{\infty} \frac{\cos \omega_0 \tau}{t - \tau} d\tau \\ &= -\frac{1}{\pi} \lim_{\epsilon \to 0} \left[\int_{\epsilon}^{\infty} \frac{\cos \omega_0 (t - y)}{y} dy + \int_{-\infty}^{-\epsilon} \frac{\cos \omega_0 (t - y)}{y} dy \right] \\ &= -\frac{1}{\pi} \left(\cos \omega_0 t \, \mathcal{P} \int_{-\infty}^{\infty} \frac{\cos \omega_0 y}{y} dy + \sin \omega_0 t \, \mathcal{P} \int_{-\infty}^{\infty} \frac{\sin \omega_0 y}{y} dy \right) \\ &= -\sin \omega_0 t \qquad \text{since } \mathcal{P} \int_{-\infty}^{\infty} \cos x \frac{dx}{x} = 0 \end{aligned} \tag{2.79}$$

Thus, $Y(t) = -\mathcal{H}\{\cos \omega_0 t\} = -\cos (\omega_0 t - \pi/2) = -\sin \omega_0 t$, which is the expected quadrature function. Similarly, one easily shows that $\mathcal{H}\{\sin \omega_0 t\} = \sin (\omega_0 t - \pi/2) = -\cos \omega_0 t$. Other examples are given in Prob. 2.8.

Frequently, the expression (2.67)

$$V_{in}(t) = \alpha_0(t) \cos \omega_c t + \beta_0(t) \sin \omega_c t = X(t) \tag{2.80a}$$

is a useful alternative representation to the envelope-phase form (2.64). For narrowband waves, one shows from Eq. (2.76) (cf. Prob. 2.9) that the quadrature function in this instance becomes (*approximately*)

$$\begin{aligned} Y(t) &\doteq -\mathcal{H}\{\alpha_0(t) \cos \omega_c t + \beta_0(t) \sin \omega_c t\} \\ &\doteq -\alpha_0(t) \sin \omega_c t + \beta_0(t) \cos \omega_c t \qquad \text{all } \omega_c \gg 0 \end{aligned} \tag{2.80b}$$

and, from Eqs. (2.80*a*), (2.80*b*) and the fact that $Y = -\mathcal{H}\{X\}$, we easily find that (approximately)

$$\begin{aligned} \alpha_0(t) &\doteq X(t) \cos \omega_c t + \mathcal{H}\{X\} \sin \omega_c t \\ \beta_0(t) &\doteq X(t) \sin \omega_c t - \mathcal{H}\{X\} \cos \omega_c t \end{aligned} \tag{2.81}$$

In any case, note that, although α_0 and β_0 are not uniquely represented if the choice of ω_c is left open, the envelope $E_0 = \sqrt{\alpha_0^2 + \beta_0^2} = |Z|$ and phase $\phi + \omega_c t - \tan^{-1} (\beta_0/\alpha_0) = \arg Z(t)$ are uniquely determined, by the argument of Eqs. (2.75) et seq., and therefore so is the instantaneous frequency

$$\begin{aligned} f_{inst} &= \frac{1}{2\pi} \frac{d}{dt} [\arg Z(t)] = \frac{-1}{2\pi} \frac{d}{dt} \left[\tan^{-1} \frac{Y(t)}{X(t)} \right] \\ &= \frac{1}{2\pi} \left[\omega_c - \frac{d}{dt} \tan^{-1} \frac{\beta_0(t)}{\alpha_0(t)} \right] \end{aligned} \tag{2.82}$$

with $Z = X - iY = X + i\mathcal{H}\{X\}$.

PROBLEMS

2.4 (*a*) Prove the properties 1 to 4 of Eq. (2.36) for the Laplace transform $\mathcal{L}$.
(*b*) Show that, if the indicated limits exist,

$$\lim_{p \to 0} \{pG(p)\} = \lim_{p \to 0} (p\mathcal{L}\{g(t)\}) = \lim_{t \to \infty} g(t)$$
$$\lim_{p \to \infty} pG(p) = g(0+)$$

if $g(t)$ has no delta function at $t = 0$.

$$\lim_{p \to \infty} p[G(p) - A] = g(0+) \qquad \text{with } \lim_{p \to \infty} G(p) = A$$

when $g(t)$ has a delta function at $t = 0$.

2.5 Obtain the following Laplace transforms when $g(t)$ is given:

$g(t)$	$\mathfrak{L}\{g\}$
$\sin at$	$\dfrac{a}{a^2 + p^2}$
$\cos at$	$\dfrac{p}{a^2 + p^2}$
$\dfrac{\sin at}{t}$	$\tan^{-1} \dfrac{a}{p}$
$e^{-a^2t^2}$	$\sqrt{\dfrac{\pi}{p}} \left[1 - \Theta\left(\dfrac{p}{2a}\right)\right] e^{p^2/4a^2}, \quad \Theta(x) = \dfrac{2}{\sqrt{\pi}} \int_0^x e^{-\lambda^2} \, d\lambda$
$t^\nu e^{at}$, Re $\nu \geq 0$	$\dfrac{\Gamma(\nu + 1)}{(a - p)^{\nu+1}}$
$J_0(at)$	$(a^2 + p^2)^{-1/2}$
$I_0(at)$	$(p^2 - a^2)^{-1/2}$

What are the regions of absolute convergence in each instance?

2.6 (*a*) Show, using the principle of superposition, that, if $V_{in}(t)$ is an input ensemble to a linear system, the corresponding output ensemble is

$$V_o(t) = \boldsymbol{T}_L\{V_{in}(t)\} \qquad \text{where } \boldsymbol{T}_L = \int_{0-}^{t} V_{in}(t - \tau) h(\tau) \, d\tau \tag{1}$$

when $h(\tau)$ is the weighting function of the linear network. Include the effects of possible δ-function responses at $\tau_1, \tau_2, \ldots \geq 0$.

(*b*) Prove that a linear filter is stable if and only if $\int_{-\infty}^{\infty} |h(\tau)| \, d\tau < \infty$.

(*c*) Show that, if the transfer function $Y(p)$ of this linear filter is analytic in the positive half plane Re $p > 0$ and contains no singularities on the imaginary axis, e.g., if $|dY/dp|^2$ is integrable, the filter is stable.

2.7 (*a*) Prove the properties 1 to 4 of Eq. (2.47) for the Fourier transform $\mathfrak{F}$.

(*b*) Obtain the following Fourier transforms when $g(t)$ is given [cf. Eq. (2.43)], and sketch $\mathfrak{F}\{g\}$ in each case:

$g(t)$	$\mathfrak{F}\{g\}$
$\dfrac{\sin at}{t}$	$\left.\begin{matrix} \pi, \ \lvert\omega\rvert < a \\ 0, \ \lvert\omega\rvert > a \end{matrix}\right\} \ \omega = 2\pi f$
$\dfrac{1}{\sqrt{a^2 + t^2}}$	$2K_0(a\lvert\omega\rvert)$
$\dfrac{1}{a^2 + t^2}$	$\dfrac{\pi}{a} e^{-a\lvert\omega\rvert}$
$\dfrac{1}{4b} e^{-b\lvert t\rvert}$, Re $b > 0$	$\dfrac{1}{b^2 + \omega^2}$
$J_0(at)$	$\dfrac{2}{\sqrt{a^2 - \omega^2}}$
$(1 + t^4)^{-1}$	$\pi e^{-\lvert\omega\rvert/\sqrt{2}} \cos\left(\dfrac{\lvert\omega\rvert}{\sqrt{2}} - \dfrac{\pi}{4}\right)$

2.8 (*a*) Show that

$$Y(t) = -\frac{1}{\pi}\,\mathcal{P}\int_{-\infty}^{\infty}\frac{X(\tau)}{t-\tau}\,d\tau \qquad X(t) = \frac{1}{\pi}\int_{-\infty}^{\infty}\mathcal{P}\,\frac{Y(\tau)}{t-\tau}\,d\tau \tag{1}$$

are a Hilbert-transform pair. (HINT: Start with $\psi(z) \equiv \int_{-\infty}^{\infty}[\alpha(\tau) - i\beta(\tau)]e^{i\tau z}\,d\tau$, and set X and Y equal to the real and imaginary parts of $\psi(z)$, $z = x + iy$, in the limit $y \to 0$. Hence, show, for example, that $Y(t) = \lim_{\rho\to\infty}(1/\pi)\int_0^{\rho} d\tau \int_{-\infty}^{\infty}\sin(u-t)\tau X(u)\,du$, and from this deduce the first relation of Eq. (1). Repeat, with appropriate modifications, for $X(t) = \mathcal{H}\{Y(t)\}$.)

(*b*) Given $X(t)$, obtain the following Hilbert transforms:

	$X(t)$	$Y(t)$
(1)	$X(t) = \begin{cases} 1, & -a < t < a \\ 0, & \lvert t\rvert > a \end{cases}$	$-\frac{1}{\pi}\log\left\lvert\frac{a+t}{a-t}\right\rvert$
(2)	$\frac{1}{1+t^2}$	$\frac{-t}{1+t^2}$
(3)	$\frac{\sin 2Bt}{2Bt}$	$\frac{\cos 2Bt - 1}{2Bt}$

If $Y(t)$ is given, as above [(1) to (3)], what is $X(t)$?

(*c*) Show, if $Z(t) = X(t) + i\mathcal{H}\{X\}$, $Y = -\mathcal{H}\{X\}$, that the Hilbert-transform relations between X and Y satisfy the Cauchy-Riemann conditions that $Z(t,u)$ be analytic, where $Z(t) = \lim_{u\to 0} Z(t,u)$, $z = t + iu$.

2.9 (*a*) Show that, if $X(t) = \alpha(t)\cos\omega_0 t + \beta(t)\sin\omega_0 t$ $(\omega_0 \gg 0)$,

$$\mathcal{H}\{X\} \doteq -\alpha(t)\sin\omega_0 t + \beta(t)\cos\omega_0 t \tag{1}$$

approximately.

(*b*) If $X(t)$ is a narrowband noise voltage ensemble, show that $\mathcal{H}\{X\}$ is also an ensemble of narrowband noise voltages, *with the same measure*, provided that $\tan^{-1}[\beta_0(t)/\alpha_0(t)]$ is uniformly distributed in any $(0,2\pi)$ interval.

(*c*) Show that $\mathcal{H}$ is an invariant operator, and discuss the conditions under which it is (1) measure-preserving and (2) non-measure-preserving, giving an example in each case.

2.10 Obtain the results in columns 3 to 8 of Table 2.1.

2.3 *Nonlinear Systems; Representations*

The preceding section has reviewed briefly some of the principal features of ideal linear lumped-constant filters. The communication process, however, is essentially nonlinear. Linear operations enter only as special cases, so that our task now is to consider the distinguishing features of the nonlinear transformations that normally appear in transmission and reception systems. As before, it is with the properties of the transformed signal and noise ensembles that we are ultimately concerned, and not with those of particular member functions alone. Again, for the time being, we restrict our attention to the ideal, or noise-free, part $\boldsymbol{T}_g$ of transformations like $\boldsymbol{T}_g^{(N)}$ [cf. Sec. 2.1-3(1)].

2.3-1 Some Nonlinear Responses. As we have observed already in Eq. (2.22), a nonlinear operator $\boldsymbol{T}_g$ is one for which the superposition principle does not hold, that is, $T_g\{y_1 + y_2\} \neq z_1 + z_2$ if $T_g\{y_1\} = z_1$, $T_g\{y_2\} = z_2$. For instance, if $T_g\{y\} = g(y) = y^\nu \ (y > 0)$, $T_g\{y\} = g(y) = 0 \ (y < 0)$, then $T_g\{ \ \}$ is nonlinear. When the input and output ensembles (y and z) are related by an expression of this kind, we also call z the *rectified* output (ensemble) and the operation T_g *rectification.* Rectification is a common feature of communication systems, since the processes of *modulation,* or encoding, represented by $T_T^{(N)}$ and the process of *demodulation,* or decoding, as indicated by $T_R^{(N)}$ (cf. Fig. 2.1) normally require nonlinear operations of some type. This is particularly true of the latter in the critical situation of threshold reception, where the desired and undesired components of the received wave are comparable in energy. At the extremes of negligible noise backgrounds, $T_T^{(N)}$ and $T_R^{(N)}$ may ideally approach linear operation, as far as the desired or signal component is concerned, although nonlinear elements are mainly used to achieve this effect in practice (cf. Chaps. 12 to 14).

For the electronic systems with which we shall be mainly concerned in the following, $T_g\{y\}$ represents a process of *amplitude modulation,* where the ensemble of desired signals $V_S(t)$ is impressed upon a *carrier*[12,13] $y = \cos(\omega_0 t + \phi)$, ideally according to the operation $T_g = A_0[1 + \lambda V_S(t)]\{ \ \}$ $(\lambda V_S \geq -1)$, so that the modulated carrier ensemble becomes

$$z(t) = T_g\{y\} = A_0[1 + \lambda V_S(t)] \cos(\omega_0 t + \phi) \tag{2.83}$$

The transformation T_g clearly reduces to a linear operation provided that the same signal ensemble is used [T_g, however, is not *invariant,* as Eq. (2.25) shows]. Actually, to achieve this result in practice, a nonlinear operation is employed.

Frequency, or phase, modulation is always nonlinear. Here the carrier is modified according to

$$z(t) = T_g\{y(t)\} = A_0 \cos[\omega_0 t + \Phi(t) + \phi] \tag{2.84}$$

where the desired signal is embodied in the instantaneous phase $\Phi(t)$. For reception, again except in certain special cases, $T_R^{(N)}$ is nonlinear: recovery of the desired signal in some form requires one or more nonlinear operations on the set of possible inputs. In the AM case, a *detector* [often the *second detector* of heterodyne systems (cf. Fig. 5.11), where there is a *mixer,* i.e., first detector, and local oscillator as well (cf. Chaps. 5, 12)] is the essential nonlinear element. In FM receivers, there are several nonlinear devices (apart from the mixer)—a *limiter,* to remove unwanted amplitude variations, and a *discriminator,* to extract the intelligence contained in the phase $\Phi(t)$. The combination of limiter and discriminator is for FM reception the analogue of the second detector in the simpler AM case (cf. Chap. 15). And finally, as we shall see in Part 4, Chap. 20, optimum methods of reception ($T_{R,opt}^{(N)}$) combine in various ways both the amplitude and phase techniques described above.

Accordingly, detectors, mixers, limiters, discriminators, and amplifiers with distortion due to cutoff or saturation effects (cf. Fig. 2.7) are common nonlinear elements that we may expect to find in electronic communication systems. Figure 2.7 shows a number of typical cases; the transformation, or operation, $\mathbf{T}_g$ corresponds in these instances to the *dynamic characteristic, dynamic response, dynamic path,* or *dynamic transfer function* of the element in question.† Here, $\mathbf{T}_g$ may be expressed analytically for each of the examples of Fig. 2.7 as:

1. Half-wave linear detector

$$z = \begin{cases} \beta y & y \geq 0 \\ 0 & y < 0 \end{cases} \qquad (2.85a)$$

where $\beta(>0)$ is a *dynamic transconductance.*

2. Small-signal rectifier

$$z = \sum_{k=0}^{\infty} \alpha_k (y - V_o)^k \qquad (2.85b)$$

with the α_k a set of dynamic transconductances. For large variations about $y = 0$, this might be represented by the half-wave νth-law characteristic

$$z = \begin{cases} \alpha y^\nu & y \geq 0 \\ 0 & y \leq 0 \end{cases} \qquad \nu > 0 \quad (2.85c)$$

3. Amplifier with saturation and cutoff

$$z = \begin{cases} 0 & y < V_o \\ g(y) & V_o < y < V_o' \\ \beta V_o' & y > V_o' \end{cases} \qquad (2.85d)$$

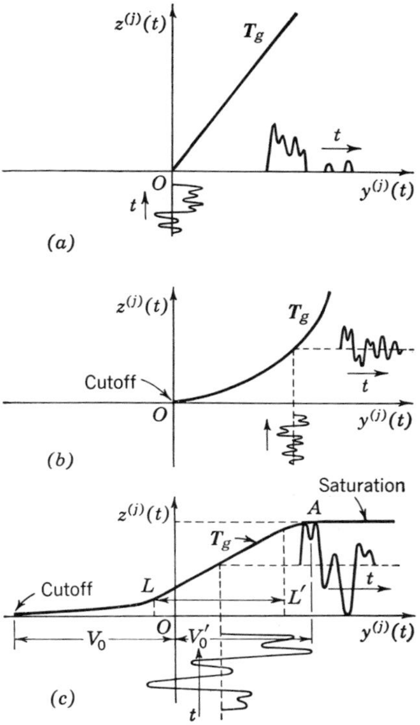

FIG. 2.7. (*a*) $y^{(j)}$, $z^{(j)}$, respectively, input and output members of corresponding ensembles, where $\mathbf{T}_g$ represents a half-wave linear detector, in *class C* operation, typical also of mixers. (*b*) Input and output ($y^{(j)}$, $z^{(j)}$) of a *small-signal* detector, where rectification takes place because of the curvature of the dynamic path $\mathbf{T}_g$. For large inputs, one has again essentially class C operation, with rectification occurring because of cutoff, as well as curvature. (*c*) Typical input and output for an actual amplifier. For sufficiently large input amplitudes, the device saturates (at A), "clipping" the input. Similarly, clipping may also occur because of cutoff. Cutoff A is also a typical dynamic characteristic for a biased saturated rectifier. LL' is the interval of *linear* operation.

Note that these dynamic paths, relating instantaneous input and output, are all one-valued in the above examples. The memory of these nonlinear

† For a derivation in particular cases from the static properties of the element, see Reich,[13] secs. 2.7, 4.6.

elements is zero: the past behavior of the input wave does not influence the present structure of the output disturbance. In some electronic systems, however, this is not true: the single-valued dynamic path is expanded into a loop, or multivalued characteristic, as sketched in Fig. 2.8, whenever the

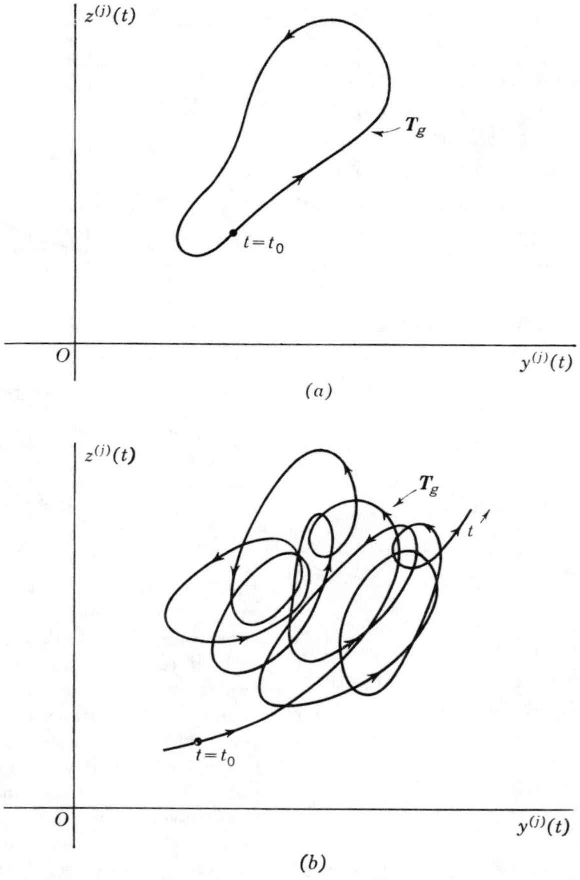

FIG. 2.8. (a) A dynamic transfer characteristic for a highly reactive plate load in an amplifier, as traced out by a periodic wave ($y^{(j)}$). (b) Same as (a) except that now the input is a random noise wave.

load on the element is appreciably reactive, as occurs, for example, in the case of the familiar RC-loaded diode rectifier, as a detector, or in amplifiers with low-Q, or broadband, plate loads whose reactance is at all appreciable. Instead of the simple relations above [Eq. (2.85)], the output ensemble $z(t)$ is now obtained from the input $y(t)$ by some such transformation as

$$z(t) = \mathbf{T}_g\{y\} = g[y(t);y(t')] \qquad -\infty < t' \leq t \tag{2.86a}$$

where the *past* of the input, as well as its present, or instantaneous, value, determines the output z.

Still more generally, one may have *feedback*, so that in place of Eq. (2.86*a*) there results

$$z(t) = \boldsymbol{T}_g\{y\} = g[y(t);g(t')] \qquad -\infty < t' \leq t \tag{2.86b}$$

for example; the past of the *output* influences the instantaneous output, as well as the input itself. Behavior of this type can often be described as a kind of *hysteresis* and the corresponding dynamic characteristic as a *hysteresis loop*.† Except in a number of special cases, no systematic way of handling these more general problems appears to exist as yet, and even for the special cases themselves the analysis turns out to be extremely complicated. Fortunately, the nonlinear element $\boldsymbol{T}_g$ with zero memory such that $z(t) = g[y(t)]$ can be made a satisfactorily close approximation to most of the important nonlinear system functions that appear in ordinary system usage if care is taken to establish the appropriate analytic structure for given operating conditions. Through Part 3 of this book, at any rate, we shall confine our discussion, without exception, to those with zero memory, like Eqs. (2.85*a*) to (2.85*d*).

2.3-2 Representation of Dynamic Characteristics with Zero Memory. Since our interest is mainly in the input and output ensembles as a whole, rather than in a particular representation, it is the properties, i.e., statistical properties, of these ensembles that are significant.

For example, suppose that z is generated from an input y, according to a zero-memory transformation $z = \boldsymbol{T}_g\{y\} = g(y)$, and we wish to determine the average value of z, for example, $\bar{z} = \boldsymbol{E}\{g(y)\} = \int_{-\infty}^{\infty} g(y)W_1(y)\,dy$, where y is a stationary process. To be more specific, let g be a *biased linear rectifier* such that $z = \beta(y - b_0)$ $(y \geq b_0)$, $z = 0$ $(y \leq b_0)$, so that

$$\overline{z(t)} = \boldsymbol{E}\{g[y(t)]\} = \int_{b_0}^{\infty} (y - b_0)W_1(y)\,dy \tag{2.87}$$

Given the first-order density $W_1(y)$, we can certainly find $\bar{z}$ in principle, although the technical features of the integration may be quite involved. To show this more forcibly, let us ask also for the covariance function‡ of the output ensemble, namely,

$$M_z(t_2 - t_1) = \overline{z(t_1)z(t_2)} = \overline{\boldsymbol{T}_g\{y_1\}\boldsymbol{T}_g\{y_2\}} = \boldsymbol{E}\{g[y(t_1)]g[y(t_2)]\} \tag{2.88a}$$

$$M_z(t_2 - t_1) = \beta^2 \iint_{-b_0}^{\infty} (y_1 - b_0)(y_2 - b_0)W_2(y_1; y_2; t_2 - t_1)\,dy_1\,dy_2 \tag{2.88b}$$

† For another illustration, see D. Middleton, An Approximate Theory of Eddy-current Loss in Transformer Cores Excited by Sine Wave or by Random Noise, *Proc. IRE*, **35**: 270 (1947); cf. fig. 3.

‡ We shall use the term covariance interchangeably with the second-order second-moment function M in the following, although our original definition [Eq. (1.52)] is in terms of the random variable $z - \bar{z}$, rather than z.

[cf. Eq. (1.121*b*)]. Observe, incidentally, that T_g in these zero-memory cases is invariant. Hence, the output ensemble $z(t)$ is also stationary, since the input $y(t)$ is assumed to be a stationary process [cf. Sec. 2.1-3(2)*c*]. The explicit evaluation of M_z is now even more difficult than $\bar{z}$ [Eq. (2.87)], again essentially because of the finite region of integration, imposed here by the cutoff of the nonlinear element at $y \leq b_0$.

Higher moments can be represented in the same way. For example, we can write

$$E\{z(t_1) \cdots z(t_n)\} = \overline{T_g\{y_1\}T_g\{y_2\} \cdots T_g\{y_n\}}$$
$$= \int \cdots \int_{-\infty}^{\infty} g(y_1) \cdots g(y_n) W_n(y_1;y_2,t_2; \ldots ;y_n,t_n)\, dy_1 \cdots dy_n \tag{2.88c}$$

for a zero-memory nonlinear device g. In the case of *linear* systems, we can include memory [cf. Eqs. (2.31)] and obtain in place of Eq. (2.88*c*)

$$E\{z(t_1) \cdots z(t_n)\} = \overline{T_L\{y_1\} \cdots T_L\{y_2\}}$$
$$= \int_0^{t_1} \cdots \int_0^{t_n} E\{y(\tau_1) \cdots y(\tau_n)\} h(t_1 - \tau_1) \cdots h(t_n - \tau_n)\, d\tau_1 \cdots d\tau_n \tag{2.88d}$$

All these expressions exhibit progressively greater technical difficulties because of the particular character of the transformation $T_g\{\ \} = g(\ )$. It is important, therefore, to look for an alternative representation of g which removes or at least mitigates these difficulties. Fortunately, this can be done with the help of the Laplace transform $\mathfrak{L}$ (cf. Sec. 2.2-4). Considering for the time being only a single member $y^{(j)}$ of the ensemble, taking the Laplace transform of both sides of $z^{(j)} = g(y^{(j)})$, and then writing $p = i\xi$ (p, ξ complex), we get

$$f(i\xi) \equiv \mathfrak{L}\{g[y^{(j)}(t)];\, p = i\xi\} = \int_{-\infty}^{\infty} e^{-i\xi y^{(j)}(t)} g[y^{(j)}(t)]\, dy^{(j)}$$
$$= \int_{-\infty}^{\infty} e^{-i\xi y^{(j)}} g(y^{(j)})\, dy^{(j)} \qquad \text{Im}\ \xi < -c_0\ (c_0 > 0),\ g(y^{(j)}) = 0\ (y^{(j)} < y_0) \tag{2.89}$$

where c_0 is some positive (real) constant (cf. Fig. 2.9), to ensure convergence of the integral, and $g(y^{(j)})$ does not increase faster than $e^{c_1 y^{(j)}}$ ($c_1 < c_0$). Note that since g is a zero-memory transformation—only the instantaneous input $y^{(j)}(t)$ determines the instantaneous output $z^{(j)}(t)$—we can drop the argument t in Eq. (2.89). For systems with memory, we cannot do this, and the success of this representation breaks down.†

Since $f(i\xi)$ above is the Laplace transform of the dynamic characteristic $g(\ )$ in the p plane, it is also equivalently a *complex* Fourier transform of

† Observe also that, since $g(\ )$ is a transformation applied independent of the particular input $y^{(j)}$, its Laplace transform $f(i\xi)$ is similarly independent of the choice of input; hence, we do not write $g^{(j)}$, or $f^{(j)}(i\xi)$.

$g(\ )$ in the ξ plane, as indicated directly by the form of Eq. (2.89). It is convenient to write Eq. (2.89) alternatively as

$$f(i\xi) = \mathfrak{F}_{C'}\{g(y^{(j)})\} = \int_{-\infty}^{\infty} g(y^{(j)})e^{-i\xi y^{(j)}}\, dy^{(j)} \qquad \text{Im } \xi < -c_0\ (c_0 > 0),$$
$$g(y^{(j)}) = 0\ (y^{(j)} < y_0) \quad (2.90a)$$

with the inverse relation

$$g[y^{(j)}(t)] = \mathfrak{F}_{C'}{}^{-1}\{f(i\xi)\} = \frac{1}{2\pi}\int_{-\infty - ic_1}^{\infty - ic_1} f(i\xi)e^{i\xi y^{(j)}(t)}\, d\xi \qquad c_1 > c_0 > 0 \quad (2.90b)$$

The contour of integration and the region of convergence of $f(i\xi)$ are shown in Fig. 2.9; $\mathfrak{F}_{C'}$ is a *one-sided* complex Fourier transform (and, correspondingly, $\mathfrak{L}$ is a one-sided Laplace transform) because the region of convergence

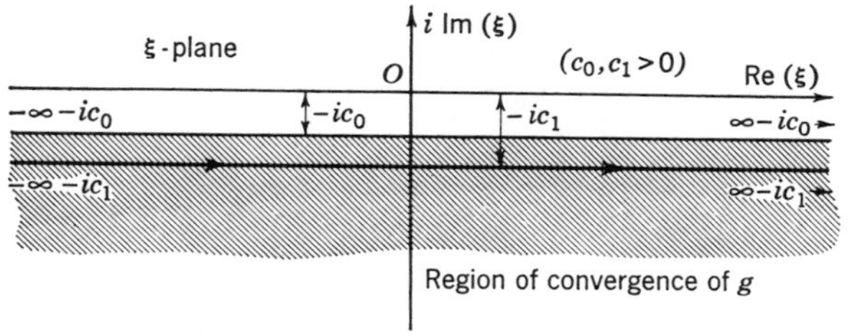

Fig. 2.9. Contour of integration and region of convergence for the transformation g.

of g is a portion of the lower-half ξ plane, reflecting the fact that $|g(y^{(j)})|$ is bounded by $e^{c_0 y}$. The vanishing of $g(y^{(j)})$ for $y^{(j)} < y_0$ is represented by one or more singularities in the ξ plane, for Im $\xi > -c_0$ (and we speak of g itself in this connection as a *one-sided* characteristic).

As an example, consider the (unbiased) half-wave linear rectifier of Eq. (2.85a) (cf. Fig. 2.7a). From Eq. (2.90a), we get directly

$$f(i\xi) = \beta\int_0^{\infty} e^{-i\xi y^{(j)}} y^{(j)}\, dy^{(j)} = \frac{\beta}{(i\xi)^2} \qquad \text{Im } \xi < 0 \qquad (2.91)$$

Inserting this in Eq. (2.90b) gives us again $g[y^{(j)}(t)] = \beta y^{(j)}$ $(y^{(j)} > 0)$, $g[y^{(j)}(t)] = 0$ $(y^{(j)} < 0)$, as expected, since $\mathfrak{F}_{C'}$ and $\mathfrak{F}_{C'}{}^{-1}$ are uniquely related. Observe that $f(i\xi)$ is analytic everywhere except at $\xi = 0$.

If $g(y^{(j)})$ does not, however, vanish for $y^{(j)} < y_0$, that is, if g is a *two-sided* dynamic characteristic, we can use a *two-sided* complex Fourier transform† (or a two-sided Laplace transform in the p plane†), obtained by splitting $g(y^{(j)})$ into two parts $g_-(y^{(j)}) = 0$ $(y > y_0)$, $g_+(y^{(j)}) = 0$ $(y < y_0)$ and writing

$$f(i\xi) = \mathfrak{F}_{C^{(2)}}\{g(y^{(j)})\} = f_+(i\xi) + f_-(i\xi)$$
$$= \int_{y_0}^{\infty} e^{-i\xi y^{(j)}} g_+(y^{(j)})\, dy^{(j)}\Big]_{\text{Im }\xi < -c_+} + \int_{-\infty}^{y_0} e^{-i\xi y^{(j)}} g_-(y^{(j)})\, dy^{(j)}\Big]_{\text{Im }\xi > c_-} \quad (2.92a)$$

† Morse and Feshbach, *op. cit.*, pp. 459ff.; Van der Pol and Bremmer, *op. cit.*, chap. II.

respectively for $f_+(i\xi)$ and $f_-(i\xi)$, where c_+ and c_- are real positive numbers chosen for convergence, with g_+, g_- suitably bounded as $y^{(j)} \to \pm\infty$, that is, $g_+ < a_+e^{c_+y^{(j)}}$, $g_- < a_-e^{-c_-y^{(j)}}$ (a_+, a_- also real). The inverse relation is

$$z^{(j)}(t) = \mathfrak{F}_{C^{(2)}}{}^{-1}\{f(i\xi)\} = \int_{-\infty-ic_1}^{\infty-ic_1} f_+(i\xi)e^{i\xi y^{(j)}}\frac{d\xi}{2\pi} + \int_{-\infty+ic_2}^{\infty+ic_2} f_-(i\xi)e^{i\xi y^{(j)}}\frac{d\xi}{2\pi} \tag{2.92b}$$

with the contours of integration and regions of convergence shown in Fig. 2.10, where $c_1 > c_+ \geq 0$, $c_2 > c_- \geq 0$. Without seriously limiting ourselves in application, we choose g now so that all the singularities of $f_+(i\xi)$ lie above

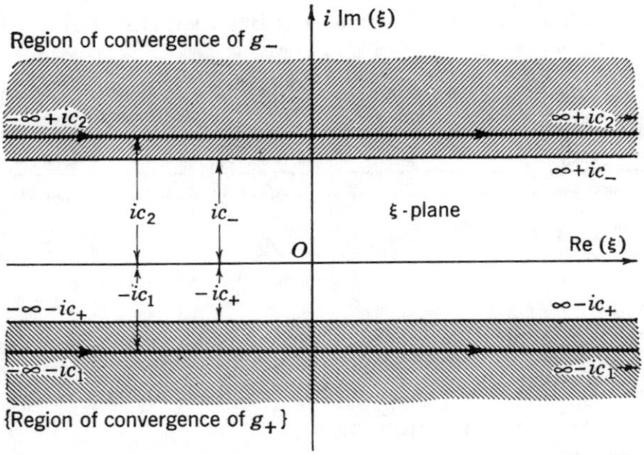

FIG. 2.10. Contours of integration and regions of convergence for $g = g_+ + g_-$ in the ξ plane.

the line Im $\xi = -ic_+$ and all those of $f_-(i\xi)$ below the line Im $\xi = +ic_-$. Note that here $f_+(i\xi)$ is identical with $f(i\xi)$ for the one-sided case, where $g_+ = g$ (cf. above).

To illustrate these remarks, let us calculate the (complex) Fourier transform $f(i\xi)$ of the dynamic characteristic shown in Fig. 2.11. Here, $g(y) = \beta$ $(y > 0)$, $g(y) = -\beta$ $(y < 0)$, the response of an ideal or "super" limiter, where all amplitude variations of the input wave are smoothed out to appear as rectangular pulses of magnitude $\pm\beta$. Consequently, Eq. (2.92a) gives

$$\begin{aligned} f_+(i\xi) &= \beta\int_0^\infty e^{-i\xi y^{(j)}}\,dy^{(j)} = \frac{\beta}{i\xi}\Big]_{\text{Im }\xi<-c_+} \\ f_-(i\xi) &= -\beta\int_{-\infty}^0 e^{-i\xi y^{(j)}}\,dy^{(j)} = +\frac{\beta}{i\xi}\Big]_{\text{Im }\xi>c_-} \end{aligned} \tag{2.93}$$

and $f(i\xi)$ is $f_+(i\xi) + f_-(i\xi)$, by analytic continuation. The singularities in both instances occur at $\xi = 0$, in the interval of the ξ plane between the two regions of convergence (cf. Fig. 2.10), and $f_+(i\xi)$, $f_-(i\xi)$ are analytic everywhere except for the simple pole at $\xi = 0$.

In general, by putting the discontinuity of the transformation g, which is reflected in the limits of integration [cf. Eqs. (2.87), (2.88)], into the integrand, as it were, we can consider *all* allowed values of the original input $y^{(j)}$. As our subsequent examples will show, the actual labor of obtaining explicit results is thereby greatly reduced, particularly if $y^{(j)}$ comes from a mixed process, one of whose components is a time-dependent deterministic wave. The complex Fourier transform is, of course, simply a technical device, whose end product must be the same as that obtained directly without it. However, transform relations like Eqs. (2.90*b*) and (2.92*b*) are often the only tools available by which explicit results can be secured in the more complicated cases, as many of the problems considered in Part 3 will amply illustrate.

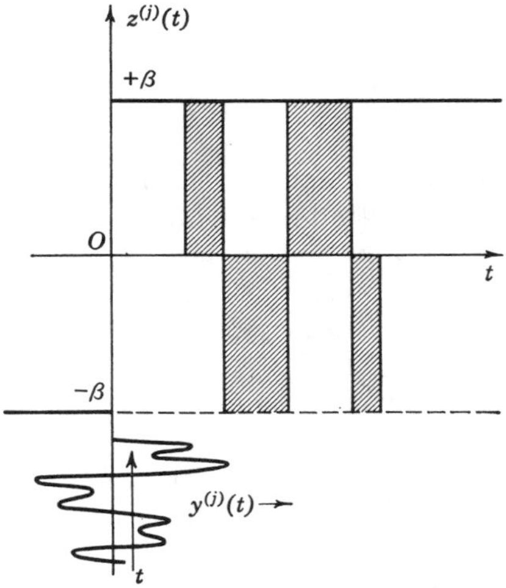

Fig. 2.11. Dynamic response of an ideal or "super" limiter.

Our next step is to extend these results to the input and output ensembles as a whole. This is done with the help of Eq. (2.90*b*), for one-sided transforms, where we now write for the output ensemble

$$z(t) = g[y(t)] = \frac{1}{2\pi} \int_{-\infty - ic}^{\infty - ic} f(i\xi) e^{i\xi y} \, d\xi \tag{2.94}$$

Applying the condition that this integral exists mean square (cf. Sec. 2.1-2), we have the requirement that

$$\boldsymbol{E}\{g(y_1)g(y_2)\} = \frac{1}{(2\pi)^2} \iint_{-\infty - ic}^{\infty - ic} f(i\xi_1) f(i\xi_2) \, \overline{e^{i\xi_1 y_1 + i\xi_2 y_2}} \, d\xi_1 \, d\xi_2 \tag{2.95}$$

and is defined in the ordinary (Riemann or Riemann-Stieltjes) sense. Observe that, since ξ_1, ξ_2 are in general complex, $\overline{e^{i\xi_1 y_1 + i\xi_2 y_2}}$ is *not* the joint characteristic function F_{y_1,y_2} of the random variables y_1, y_2, which is defined for real ξ_1, ξ_2 only, but by analytic continuation it has the same functional form as this characteristic function. Consequently, we may still write $F_{y_1,y_2}(i\xi_1, i\xi_2; t_2 - t_1)$, with ξ_1, ξ_2 complex, remembering that the statistical average

$$\overline{e^{i\xi_1 y_1 + i\xi_2 y_2}} = \iint_{-\infty}^{\infty} e^{i\xi_1 y_1 + i\xi_2 y_2} W_2(y_1, y_2; t_2 - t_1) \, dy_1 \, dy_2 \qquad \xi_1, \xi_2 \text{ complex} \tag{2.96}$$

can be performed (cf. Sec. 1.2-8) as long as the integrals for the real and imaginary parts of Eq. (2.96) exist. Thus, we can rewrite Eq. (2.95) for the existence of Eq. (2.94), or $z = \boldsymbol{T}_g\{y\}$, as a stochastic integral,

$$\boldsymbol{E}\{g(y_1)g(y_2)\} = \frac{1}{(2\pi)^2} \iint_{-\infty - ic}^{\infty - ic} f(i\xi_1)f(i\xi_2)F_{y_1,y_2}(\xi_1, \xi_2; t_2 - t_1)\, d\xi_1\, d\xi_2 \tag{2.97}$$

which, incidentally, is the covariance function $M_z(t_2 - t_1)$ [cf. Eq. (2.88*a*)] of the output ensemble. As before, it is assumed that Eq. (2.97) exists for all input ensembles y and for transformations g which possess complex Fourier transforms $f(i\xi)$. An exactly parallel argument applies for the two-sided transforms, specific examples of which will be encountered in Sec. 13.4. Because $F_y(i\xi)$, F_{y_1,y_2}, etc., appear in expressions like Eqs. (2.97), (2.98), the present technique for calculating moments of the output ensemble is sometimes called the *characteristic-function* or *indirect method*, as distinguished from the equivalent, though much less tractable, method of determining the moments directly from their definition, e.g., from $\boldsymbol{E}\{g(y)\}$, etc.

Returning to our first example [Eq. (2.87)], we can now apply Eq. (2.94) for the ensemble as a whole to get

$$\boldsymbol{E}\{z\} = \boldsymbol{E}\{g(y)\} = \frac{1}{2\pi}\int_{-\infty - ic}^{\infty - ic} f(i\xi)\overline{e^{i\xi y}}\, d\xi = \frac{1}{2\pi}\int_{-\infty - ic}^{\infty - ic} f(i\xi)F_y(i\xi)\, d\xi \tag{2.98}$$

where now $F_y(i\xi)$ is the extension of the first-order characteristic function of y to complex arguments. For the biased linear rectifier specifically mentioned, we find (cf. Prob. 2.11) that $f(i\xi) = \beta e^{-ib_0\xi}/(i\xi)^2$, which may also be inserted in Eq. (2.97) to give us $\boldsymbol{E}\{g(y_1)g(y_2)\}$ of our second example [Eq. (2.88*b*)]. Some explicit results for normal statistics are given in Chap. 5, where we shall pursue this subject in greater detail. We remark, finally, that the complex Fourier transform was first introduced into problems of this type by Bennett and Rice[14] and employed subsequently by Fränz,[15] Rice,[16] Middleton,[17] and others in receiver problems involving signals and noise.

2.3-3 Equivalent Contours. In evaluating $\boldsymbol{E}\{g(y)\}$, $\boldsymbol{E}\{g(y_1)g(y_2)\}$, etc., when $\boldsymbol{T}_g\{\ \} = g(\ )$ can be represented by a one-sided complex Fourier transform of the type of Eq. (2.90*b*), it is convenient to use a contour **C** along the real axis (cf. Fig. 2.12), indented downward about a possible singularity at $\xi = 0$, which is where singularities occur for most of the $f(i\xi)$ encountered in our study, as can be seen from Table 2.2 (case 5 is an exception). We observe that if $g(y)$ is such that $\int_a^\infty g(y)\, dy\, e^{-i\xi y}$ (ξ real) converges, as is normally the case, the contributions of the arcs $A'A$, BB' at $\xi_{\text{real}} \to +\infty$ vanish. Application of Cauchy's theorem allows us to write

$$g(y) = \int_{-\infty - ic}^{\infty - ic} f(i\xi)e^{i\xi y}\frac{d\xi}{2\pi} = \frac{1}{2\pi}\int_{\mathbf{C}} f(i\xi)e^{i\xi y}\, d\xi \tag{2.99}$$

The integration is now to be performed with respect to the real variable ξ along the entire real axis, except for the semicircle about $\xi = 0$, where ξ takes on complex values and the integration is along the downward semicircle. This

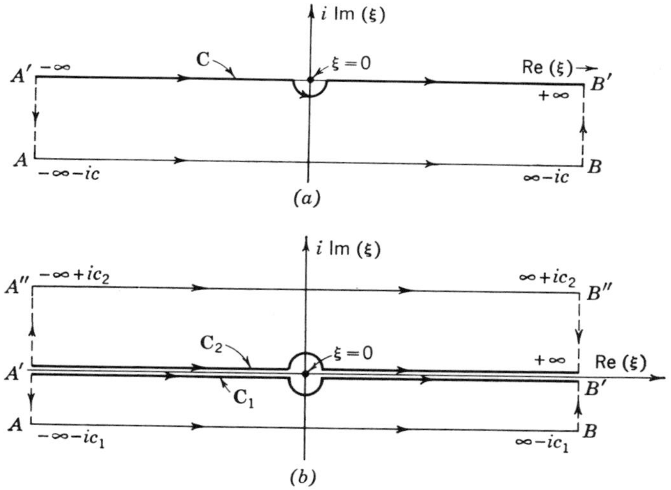

FIG. 2.12. Equivalent contours for one and two-sided dynamic characteristics.

is usually much easier to carry out than integration along the line $\xi = x - ic$, or other equivalent contours with Im $\xi < 0$.

In the same way, one gets two equivalent contours $\mathbf{C}_1$ and $\mathbf{C}_2$ (Fig. 2.12*b*), when g is represented by the two-sided complex Fourier transform (2.92*b*), g_+ and g_- being such that convergence at $\xi \to \pm\infty$ is assured. In Prob. 2.13, an alternative representation of the two-sided transform is given, which is often more convenient in actual use.

When $g(y)$ is a simple polynomial and does not vanish in $(-\infty < y < \infty)$, we can alternatively use

$$g(y) = \frac{1}{2\pi}\int_{\mathbf{C}_0} f_+(i\xi)e^{i\xi y}\,d\xi \qquad (2.100)$$

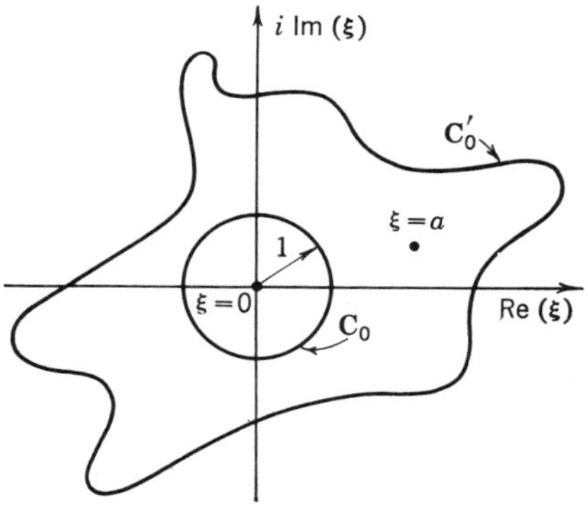

FIG. 2.13. Alternative contours for two-sided polynomial dynamic paths [cf. Eq. (2.100)].

where now $f_+(i\xi)$ [Eq. (2.90*a*) or (2.92*a*)] is extended analytically over the finite ξ plane, except for singularities at $\xi = 0$ (or $\xi = a \neq 0$). The contour $\mathbf{C}_0$ is the unit circle, or is any finite, closed contour $\mathbf{C}_0'$ within which lie all poles of $f_+(i\xi)$ (cf. Fig. 2.13 and Table 2.2).

TABLE 2.2. SOME COMMON NONLINEAR TRANSFORMATIONS

Type of device	$z = T_g\{y\} = g(y)$	$f(i\xi) = \mathfrak{F}_C\{g\}$	Contours C	Dynamic path
1. Half-wave linear rectifier	$z = \begin{cases} \beta y, & y > 0 \\ 0, & y < 0 \end{cases}$	$\frac{\beta}{(i\xi)^2}$	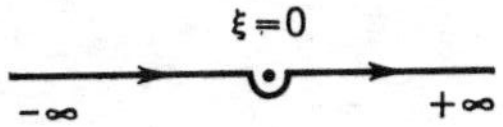	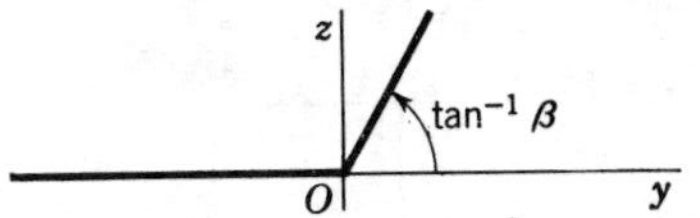
2. Half-wave square-law rectifier	$z = \begin{cases} \alpha y^2, & y > 0 \\ 0, & y < 0 \end{cases}$	$\frac{2\alpha}{(i\xi)^3}$	Same as (1)	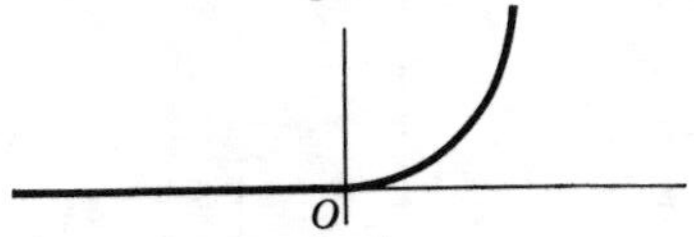
3. Biased νth-law rectifier, $\nu \geq 0$	$z = \begin{cases} \beta(y - b_0)^\nu, & y \geq b_0 \\ 0, & y < b_0 \end{cases}$	$\frac{\beta\Gamma(\nu + 1)e^{-ib_0\xi}}{(i\xi)^{\nu+1}}$	Same as (1)	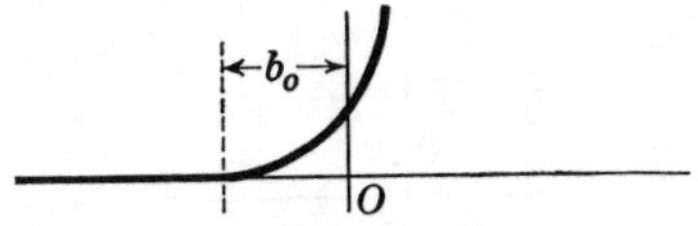
4. Biased linear rectifier with saturation	$z = \begin{cases} \beta(R_0 - b_0), & y \geq R_0 \\ \beta(y - b_0), & b_0 \leq y \leq R_0 \\ 0, & y \leq b_0 \end{cases}$	$\beta \frac{e^{-ib_0\xi} - e^{-iR_0\xi}}{(i\xi)^2}$	Same as (1)	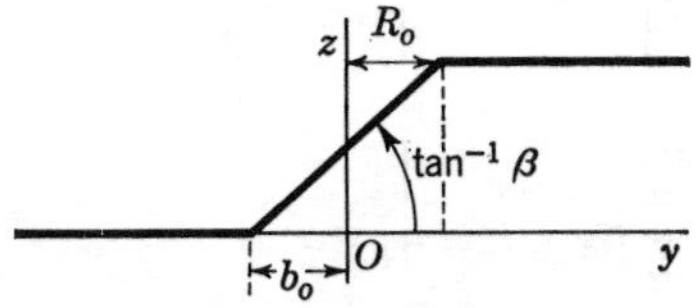
5. Half-wave rectifier with exponential saturation	$z = \begin{cases} \beta(1 - e^{-\alpha y}), & y \geq 0 \\ 0, & y \leq 0 \end{cases}$	$\frac{\alpha\beta}{i\xi(\alpha + i\xi)}, \ \alpha, \beta > 0$	Same as (1)	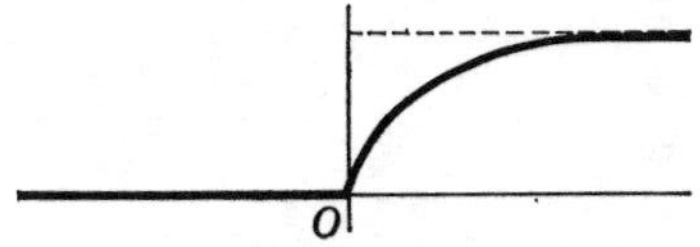

6. Ideal linear limiter	$z = \begin{cases} \beta R_0, & y \geq R_0 \\ \beta y, & -R_0 \leq y \leq R_0 \\ -\beta R_0, & y \leq -R_0 \end{cases}$	$f_+(i\xi) = \dfrac{\beta(1 - e^{-iR_0\xi})}{(i\xi)^2}$ $f_-(i\xi) = \dfrac{-\beta(1 - e^{+iR_0\xi})}{(i\xi)^2}$	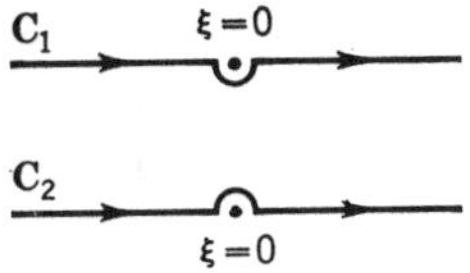	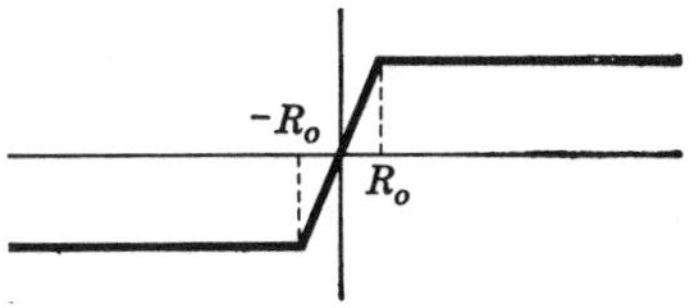
7. nth-power device (n = integer ≥ 0)	$z = \beta y^n, \; -\infty < y < \infty$	$\dfrac{\beta n!}{(i\xi)^{n+1}}$	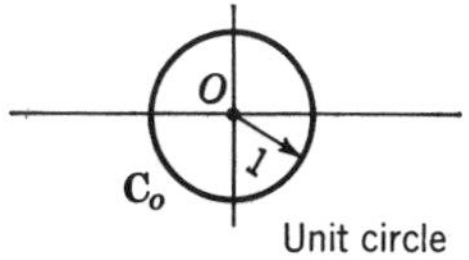	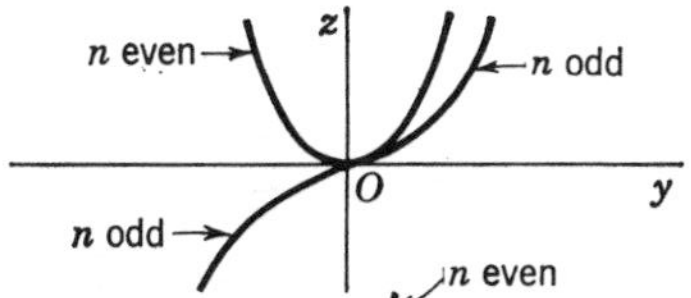
8. nth-power device with bias	$z = \beta(y - b_0)^n, \; -\infty < y < \infty$	$\dfrac{\beta n! e^{-ib_0\xi}}{(i\xi)^{n+1}}$	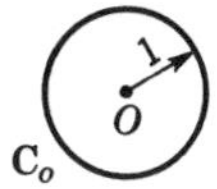	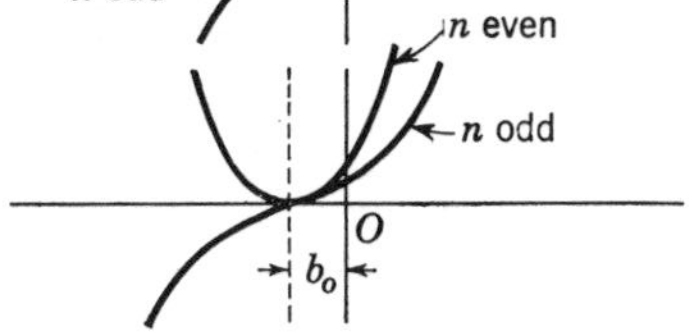
9. General νth-law device, $\nu > 0$ (symmetrical)	$z = \beta\|y\|^\nu, \; \nu > 0, \; -\infty < y < \infty$	$f_+(i\xi) = \dfrac{\beta\Gamma(\nu + 1)}{(i\xi)^{\nu+1}}$ $f_-(i\xi) = \dfrac{\beta\Gamma(\nu + 1)}{(-i\xi)^{\nu+1}}$	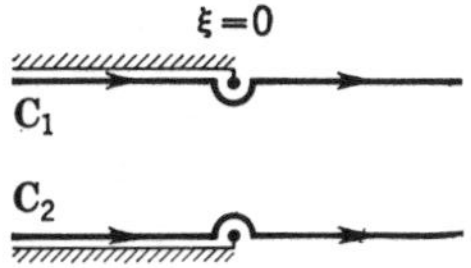	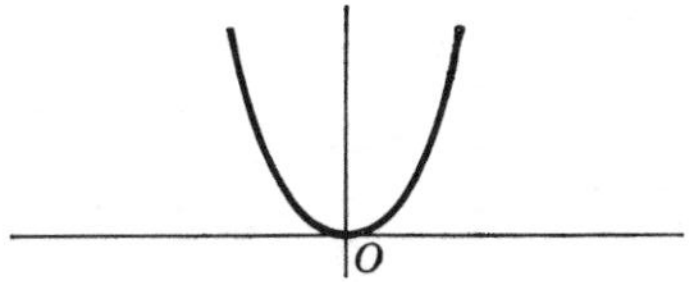
10. General νth-law device, $\nu > 0$ (antisymmetrical)	$z = \begin{cases} \beta y^\nu, & y > 0 \\ -\beta\|y\|^\nu, & y < 0 \end{cases}$	$f_+(i\xi) = \dfrac{\beta\Gamma(\nu + 1)}{(i\xi)^{\nu+1}}$ $f_-(i\xi) = \dfrac{-\beta\Gamma(\nu + 1)}{(-i\xi)^{\nu+1}}$	Same as (9)	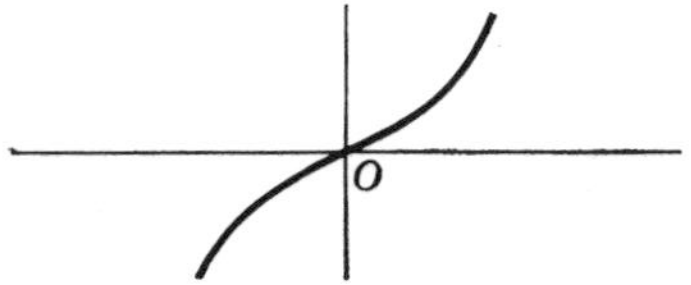

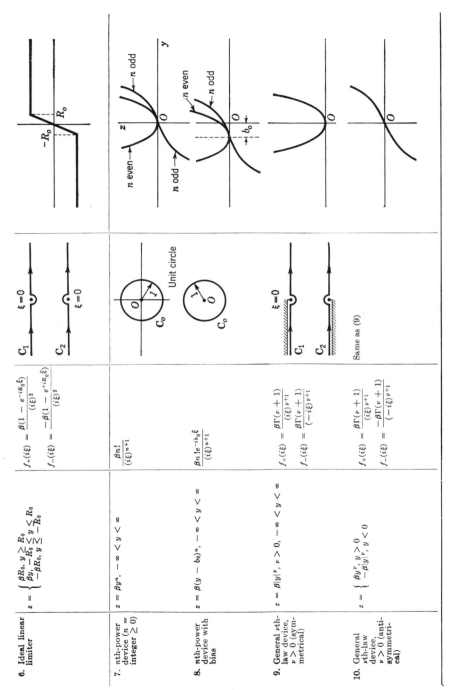

While integrands with branch points are excluded in this representation, they can be handled with contours of the type shown in Fig. 2.14. As an example, consider the half-wave νth-law rectifier ($\nu \geq 0$), $g(y) = \beta y^\nu$ ($y \geq 0$), $g(y) = 0$ ($y < 0$). From Eq. (2.90*a*), we get

$$f(i\xi) = \beta \int_0^\infty e^{-i\xi y} y^\nu \, dy = \frac{\beta\Gamma(\nu+1)}{(i\xi)^{\nu+1}} \tag{2.101a}$$

and the inverse relation (2.90*b*) is alternatively

$$g(y) = \int_C \frac{\beta\Gamma(\nu+1)}{(i\xi)^{\nu+1}} e^{i\xi y} \frac{d\xi}{2\pi} = \int_\Gamma \frac{\beta\Gamma(\nu+1) e^{i\xi y}\, d\xi}{(i\xi)^{\nu+1} 2\pi} \tag{2.101b}$$

which is easily evaluated along the various portions of the real axis with the help of the gamma function and the contours of Fig. 2.14 to give us the

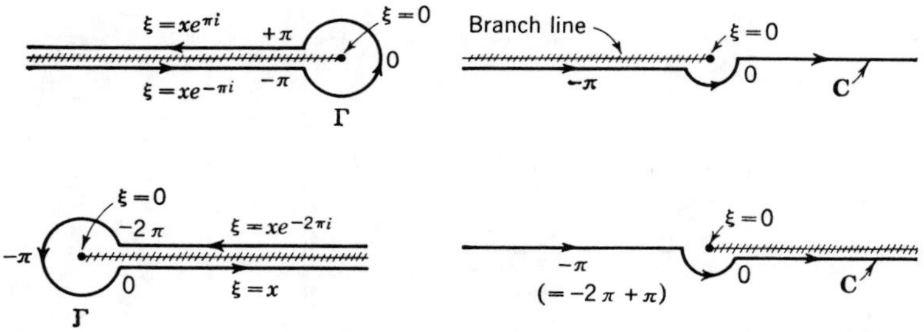

FIG. 2.14. Some equivalent contours of integration when $f(i\xi)$ contains a branch point at $\xi = 0$.

original relation $g(y)$ above. Other integrals of this type will be encountered in our study of noise and signals through nonlinear elements (cf. Chaps. 12 to 14). For a discussion of various possible and equivalent contours, the reader is referred to any of the standard works on complex variables and contour integration.†

The integral representations (2.90*b*) or (2.99), (2.92*b*), (2.100) also enable us to write the output ensemble $z = g(y)$, when y is a narrowband input of the type of Eq. (2.64), in a form useful in the subsequent study of AM and FM receivers. If we note that‡

$$e^{ia\cos\phi} = \sum_{l=0}^{\infty} \epsilon_l i^l J_l(a) \cos l\phi \qquad \epsilon_0 = 1,\ \epsilon_l = 2\ (l \neq 0) \tag{2.102}$$

where J_l is a Bessel function of the first kind, lth order, is the Fourier expansion of $e^{ia\cos\phi}$, we can apply this to Eq. (2.90*b*) when the input is

† For instance, see McLachlan, *op. cit.*, chap. VI, and also appendixes; Morse and Feshbach, *op. cit.*, secs. 4.4, 4.5; and Van der Pol and Bremmer, *op. cit.*

‡ G. N. Watson, "Theory of Bessel Functions," sec. 2.22, Macmillan, New York, 1944 (Cambridge University Press).

narrowband. The result is

$$z = g(y) = \frac{1}{2\pi} \int_C e^{i\xi E_0(t) \cos [\omega_c t - \Phi(t) + \phi]} f(i\xi)\, d\xi$$

$$= \sum_{l=0}^{\infty} B_l(t) \cos l[\omega_c t - \Phi(t) + \phi] \qquad (2.103a)$$

where

$$B_l(t) = \frac{\epsilon_l i^l}{2\pi} \int_C J_l[\xi E_0(t)] f(i\xi)\, d\xi \qquad (2.103b)$$

with a pair of series of similar expressions for two-sided dynamic paths. Since $E_0(t)$ and $\Phi(t)$ are slowly varying compared with $\cos l(\omega_c t + \phi)$ $(l \neq 0)$, we see that the nonlinear operation $\boldsymbol{T}_g\{\ \} = g(\ )$ has in effect generated an output ensemble consisting of the sum of an infinite number of narrowband components $B_l(t) \cos l[\omega_c t - \Phi(t) + \phi]$ $(l \neq 0)$. These components can be resolved with the aid of suitable narrowband (i.e., bandpass) filters (cf. Sec. 2.2-5). For example, if we use a filter whose system function is centered about lf_c and is sufficiently narrow so that the contributions of the other components are negligible, we get from Eqs. (2.65) and (2.103) in the steady state

$$z_{\text{filtered}}(t) = \boldsymbol{T}_L \boldsymbol{T}_g\{y\} = \sum_{l=0}^{\infty} 2 \int_{-\infty}^{\infty} h_0(t - \tau) B_l(\tau)$$
$$\cos l[\omega_c \tau - \Phi(\tau) + \phi] \cos [\omega_0(t - \tau) - \gamma_0(t - \tau)]\, d\tau$$
$$\doteq \int_{-\infty}^{\infty} h_0(t - \tau) B_l(\tau) \cos [l\omega_c t - \Phi(\tau) + \phi - \gamma_0(t - \tau)]\, d\tau$$
$$\omega_0 = l\omega_c \ (l \geq 1) \qquad (2.104)$$

since the rapidly varying terms in the first integrand can be neglected. When $l = 0$, one uses a *low-pass* filter, which is insensitive to all rapidly varying inputs of $O(lf_c, l \geq 1)$; $z(t)_{l=0}$ becomes, instead,

$$z(t)_{l=0,\text{filtered}} \doteq \int_{-\infty}^{\infty} h(t - \tau)_{\text{low-pass}} B_0(\tau)\, d\tau \qquad (2.105)$$

and $z(t)_{0,\text{filtered}}$ is now broadband. Broadband inputs to zero-memory nonlinear devices cannot, of course, be resolved in this way since the output is also broadband; z_{filtered} becomes for this case simply

$$z(t)_{\text{filtered}} = \boldsymbol{T}_L \boldsymbol{T}_g\{y\} = \int_{-\infty}^{\infty} h(t - \tau) g[y(\tau)]\, d\tau \qquad (2.106)$$

where as usual we assume that the nonlinear operation has zero memory, and, subsequently, the linear filters are isolated [cf. Sec. 2.1-3(1)]. See Table 2.2 for some common dynamic responses and their complex Fourier-transform representations.

PROBLEMS

2.11 Obtain the complex Fourier transforms $f(i\xi)$ of the dynamic characteristics 1 to 5 in Table 2.2.

2.12 Work Prob. 2.11 for the dynamic characteristics 6 to 10 of Table 2.2.

2.13 (*a*) By splitting the general (zero-memory) dynamic response $g(y)$ into odd and even parts, according to

$$g(y) = g_e(y) + g_o(y) \tag{1}$$

where $g_e = g_{e1} + g_{e2}$, $g_o = g_{o1} - g_{o2}$, $g_{(1)} = 0\ (V < 0)$, $g_{(2)} = 0\ (V > 0)$, show that

$$z = T_g\{y\} = \frac{1}{\pi}\int_C f_e(i\xi)_1 \cos \xi y\, d\xi + \frac{i}{\pi}\int_C f_o(i\xi)_1 \sin \xi y\, d\xi \tag{2}$$

in which

$$f_{(e,o)}(i\xi)_1 = \int_{-\infty}^{\infty} g_{(e1,o1)}(y)e^{-i\xi y}\, dy \tag{3}$$

(*b*) Accordingly, show for symmetrical and antisymmetrical characteristics (about $y = 0$) that

$$z_{sym}(t) = \frac{1}{\pi}\int_C f_+(i\xi) \cos y(t)\xi\, d\xi \qquad z_{antisym}(t) = \frac{i}{\pi}\int_C f_+(i\xi) \sin y(t)\xi\, d\xi \tag{4}$$

2.14 Show that, if general waves of the form $y = A_0(t) \cos [\omega_0 t - \Phi(t) + \phi]$ (ϕ a random parameter) are introduced to the zero-memory half-wave νth-law rectifier of No. 3, Table 2.2 ($b_0 = 0$), the output ensemble $z(t)$ becomes

$$z(t) = T\{y\} = \beta A_0(t)^\nu \frac{\Gamma(\nu + 1)}{2^{\nu+1}} \sum_{l=0}^{\infty} \frac{\epsilon_l \cos l[\omega_0 t - \Phi(t) + \phi]}{\Gamma[(2 + \nu - l)/2]\Gamma[(2 + \nu + l)/2]} \tag{1}$$

Thus, if $A_0(t)$, $\Phi(t)$ are narrowband and a low-pass filter is used, with a uniform response to frequencies much higher than the highest significant components of $A_0(t)$, $\Phi(t)$, but falling off to zero before frequencies $O(f_0)$ are reached, we see that the l-f output is

$$z(t)_{filt} = T_L\{z\} = \frac{\beta A_0(t)^\nu \Gamma(\nu + 1)}{\Gamma(1 + \nu/2)^2 2^{\nu+1}} \qquad l = 0 \tag{2}$$

independent of the phases $\Phi(t)$ and ϕ. Since $A_0(t) = E_0(t)$ [Eq. (2.64)], this becomes

$$z(t)_{filt} = \frac{\beta|\alpha(t) - i\beta(t)|^\nu \Gamma(\nu + 1)}{\Gamma(1 + \nu/2)^2 2^{\nu+1}} = \begin{cases} \dfrac{\beta}{\pi}|\alpha(t) - i\beta(t)| & \nu = 1 \\ \dfrac{\beta}{4}[\alpha(t)^2 + \beta(t)^2] & \nu = 2 \end{cases} \tag{3}$$

2.15 Show that, if the narrowband ensemble y of Prob. 2.14 is passed into a zero-memory symmetrical or antisymmetrical full-wave rectifier of the type No. 9 or No. 10, respectively (Table 2.2), the output ensemble is

$$z(t)_{sym,antisym} = \beta A_0(t) \frac{\Gamma(\nu + 1)}{2^{\nu+1}} \sum_{l=0}^{\infty} \frac{\epsilon_l[1 \pm (-1)^l] \cos l[\omega_0 t - \Phi(t) + \phi]}{\Gamma[(2 + \nu - l)/2]\Gamma[(2 + \nu + l)/2]} \tag{1}$$

where plus applies for the symmetrical and minus for the antisymmetrical device. As expected, all terms of odd order ($l = 1, 3, 5, \ldots$) vanish in the former, while all terms of even order ($l = 0, 2, 4, \ldots$) drop out in the latter.

2.16 By the methods of contour integration and with the help of analytic continuation, show that

(a) $$\int_C e^{-c^2\xi^2}\xi^{2\mu-1}\,d\xi = \frac{i\pi e^{-\mu\pi i}}{c^{2\mu}\Gamma(1-\mu)} = ic^{-2\mu}\Gamma(\mu)e^{-i\pi\mu}\sin\pi\mu \qquad |\arg c| < \frac{\pi}{4}$$

(b) $$\int_C \xi^{\mu-1}J_\nu(a\xi)e^{-q^2\xi^2}\,d\xi = \frac{\pi i^{1-\nu-\mu}(a/2q)^\nu\,{}_1F_1[(\mu+\nu)/2;\,1+\nu;\,-a^2/4q^2]}{q^\mu\Gamma(\nu+1)\Gamma[1-(\nu+\mu)/2]} \qquad |\arg q| < \frac{\pi}{4}$$

(c) $$\int_C \xi^\mu e^{+ib\xi-c^2\xi^2}\,d\xi = \frac{\pi i^{-\mu}}{c^{\mu+1}}\left\{\frac{{}_1F_1[(\mu+1)/2;\,\tfrac{1}{2};\,-b^2/4c^2]}{\Gamma[(1-\mu)/2]} + \frac{b}{c}\,i\,\frac{{}_1F_1[(\mu+2)/2;\,\tfrac{3}{2};\,-b^2/4c^2]}{\Gamma[(-\mu)/2]}\right\} \qquad |\arg c| < \frac{\pi}{4}$$

(d) $$\int_C J_\alpha(a\xi)J_\beta(b\xi)\frac{d\xi}{\xi^\gamma} = \frac{\pi e^{-\pi i(\alpha+\beta-\gamma)/2}b^\beta 2^{1-\gamma}a^{\gamma-1-\beta}}{\Gamma(\beta+1)\Gamma[(1+\gamma-\alpha-\beta)/2]\Gamma[(1+\gamma+\alpha-\beta)/2]}$$
$$ {}_2F_1\left[\frac{1}{2}(\alpha+\beta-\gamma+1),\frac{1}{2}(\beta-\gamma-\alpha+1);\beta+1;\frac{b^2}{a^2}\right] \qquad 0 \le b \le a,\ \mathrm{Re}\,\gamma > 0$$

Here ${}_1F_1$ and ${}_2F_1$ are hypergeometric functions (see also Appendix 1).

D. Middleton,[17] *Quart. Appl. Math.*, **5**: 445 (1948), app. III.

2.4 *Objectives of the Present Study*

In the preceding sections (2.1 to 2.3), a number of classes of linear and nonlinear transformations commonly used in the communication process have been considered, while much of the statistical apparatus needed to describe the ensembles of signals and noise so transformed has been outlined in Chap. 1. We are now in a position to review from a rather general point of view the problems ahead of us in the remaining chapters. These are symbolized by the relation

$$\{v\} = \mathbf{T}_R^{(N)}\mathbf{T}_M^{(N)}\mathbf{T}_T^{(N)}\{u\} \tag{2.107}$$

[cf. Eq. (2.20)], which connects the ensemble of received messages $\{v\}$ with the original ensemble $\{u\}$ of messages to be transmitted [see Sec. 2.1-3(1)]. With $\mathbf{T}_T^{(N)}$, $\mathbf{T}_R^{(N)}$ specified or adjustable, with $\{u\}$ given or open to choice, a central problem of communication theory is that of finding the $\mathbf{T}_T^{(N)}$ or $\mathbf{T}_R^{(N)}$ or both which in some definite and meaningful sense bring $\{v\}$, or decisions based upon $\{v\}$, into closest correspondence with or otherwise optimum dependence upon $\{u\}$. In other words, this is the problem of determining the structure of the transmission and reception operations in order to yield a received message ensemble $\{v\}$ or a set of decisions based upon $\{v\}$ which comes closest (in some definite sense) to the original ensemble $\{u\}$ or to the original intent of the communication process itself.

Before such problems can be expressed quantitatively, however, it is necessary to examine $\{v\}$ (or decisions based on $\{v\}$) in terms of $\{u\}$ when

$\boldsymbol{T}_R^{(N)}$, $\boldsymbol{T}_M^{(N)}$, and $\boldsymbol{T}_T^{(N)}$ are already determined. This is what we mean by the analysis of system performance when the system in question has already been specified. Part 3 of this book is devoted chiefly to a study of such systems, where $\boldsymbol{T}_T^{(N)}$ and $\boldsymbol{T}_R^{(N)}$ represent coding procedures such as amplitude or frequency (or phase) modulation and demodulation. We shall begin in Chaps. 3ff. with the simplest linear filters and progress to the more involved nonlinear devices, until we reach a position, finally, in Part 3, Chaps. 12 to 15, where we can consider the essential elements of AM and FM transmitters and receivers and in particular their effects on mixed ensembles of signal and noise waves. In almost all cases here, however, we shall require that the original message ensemble $\{u\}$ and the encoding process $\boldsymbol{T}_T^{(N)}$ be selected beforehand. A discussion of some of the latter problems is included in Chap. 6 on information theory. The greater space is devoted to reception (decoding), where the accompanying noise enjoys a critical place in system performance. Our premier problem is thus to find $\{v\}$, given $\boldsymbol{T}_T^{(N)}$, $\boldsymbol{T}_M^{(N)}$, $\boldsymbol{T}_R^{(N)}$, and to compare it with the given $\{u\}$ when $\boldsymbol{T}_T^{(N)}$, $\boldsymbol{T}_M^{(N)}$, $\boldsymbol{T}_R^{(N)}$ may take a variety of different forms and where either one or more of these transformations imply the insertion of noise at some point of the system.

Closely allied to this is the question of the precise structure of $\boldsymbol{T}_M^{(N)}$ and of the component transformations of $\boldsymbol{T}_T^{(N)}$, $\boldsymbol{T}_R^{(N)}$ [cf. Eqs. (2.18), (2.19)], which, apart from their operations on message and signal ensembles, represent the injection of noise into the system. In Part 2, Chaps. 7 to 11, the salient statistical properties of the more common physical sources of noise in electronic devices are briefly examined, in order to determine the hierarchy of distribution densities $W_1, W_2, \ldots, W_n, \ldots$ needed to describe the noise process. Having these, we can at least in principle find the statistical properties of the various transformed ensembles $x = \boldsymbol{T}_T^{(N)}\{u\}$, $y = \boldsymbol{T}_M^{(N)}\boldsymbol{T}_T^{(N)}\{u\}$, etc., comprising Eq. (2.107).

Considered, finally, in Part 4, Chaps. 18 to 23, is the still more general and interesting situation where the precise structure of $\boldsymbol{T}_R^{(N)}$ is left open, with only $y = \boldsymbol{T}_M^{(N)}\boldsymbol{T}_T^{(N)}\{u\}$ given a priori. We then ask for $\boldsymbol{T}_R^{(N)}$ so that $\{v\}$ may represent the original message ensemble $\{u\}$ in some "best" sense, where "best" depends, of course, on the criterion originally selected, such as the cost of a decision or the amount of information lost when $\{v\}$ is received (and acted upon). It is here that the methods of statistical decision theory exhibit their power and enable us to compare actual, suboptimum systems like those of Part 3 with the optimum devices predicted by the decision-theoretical approach. Still more complicated cases, involving multiple sources, receivers, feedback, etc., can be treated by these methods, where simultaneous adjustment of $\boldsymbol{T}_R^{(N)}$ *and* $\boldsymbol{T}_T^{(N)}$ is required for optimum performance. A few of these topics are briefly reviewed in the concluding chapter, along with some important problems in statistical communication theory that are as yet unsolved.

REFERENCES

1. Bartlett, M. S.: "An Introduction to Stochastic Processes," Cambridge, New York, 1955.
2. Fréchet, M.: Recherches théoretiques modernes sur la théorie des probabilités, in Borel et al., "Traité du calcul des probabilités et de ses applications," vol. 1, pt. III, Paris, 1937, 1938; also, chap. 5.
3. Moyal, J. E.: Stochastic Processes and Statistical Physics, *J. Research Roy. Stat. Soc.*, **B11**: 150 (1949).
4. Blanc-LaPierre, A., and R. Fortet: "Théorie des fonctions aléatoires," Masson et Cie, Paris, 1953.
5. Doob, J. L.: "Stochastic Processes," chap. 11, sec. 9, Wiley, New York, 1953.
6. Cramér, H.: "Mathematical Methods of Statistics," sec. 22.2, Princeton University Press, Princeton, N.J., 1946.
7. For details, see, for example, H. M. James, N. B. Nichols, and R. S. Phillips, "Theory of Servomechanisms," chap. 2, Massachusetts Institute of Technology Radiation Laboratory Series, vol. 25, McGraw-Hill, New York, 1947.
8. Ince, E. L.: "Ordinary Differential Equations," chap. 6, Dover, New York, 1944.
9. Van der Pol, B.: Frequency Modulation, *J. IEE (London)*, **93**: 153 (1946).
10. Ville, J.: Théorie et applications de la notion de signal analytique, *Cables et transmissions*, **2**: 61 (1948). See also the discussion in Ref. 4, chap. IX, sec. 10, especially pp. 411, 412.

10a. Dugundji, J.: Envelopes and Preënvelopes of Real Waveforms, *IRE Trans. on Inform. Theory*, **IT-4**: 53 (1958).

11. Titchmarsh, E. C.: "Introduction to the Theory of Fourier Integrals," 2d ed., chap. 5, Oxford, New York, 1948.
12. Chaffee, E. L.: "Theory of Thermionic Vacuum Tubes," McGraw-Hill, New York, 1933.
13. Reich, H. J.: "Theory and Application of Electron Tubes," 2d ed., chaps. 3–9, McGraw-Hill, New York, 1944.
14. Bennett, W. R., and S. O. Rice: Note on Methods of Computing Modulation Products, *Phil. Mag.*, **18**: 422 (1934).
15. Fränz, K.: Die Ubertragung von Rauschspannungen über den linearen Gleichrichter, *Z. Hoch. u. Elek.*, **57**: 146 (1941); see also *Elek. Nachr.*, **19**: 286 (1942).
16. Rice, S. O.: Mathematical Analysis of Random Noise, *Bell System Tech. J.*, **24**: 46 (1945), pt. IV.
17. Middleton, D.: The Response of Biased, Saturated Linear and Quadratic Rectifiers to Random Noise, *J. Appl. Phys.*, **17**: 778 (1946); Some General Results in the Theory of Noise through Non Linear Devices, *Quart. Appl. Math.*, **5**: 445 (1948). Also, for further references, see Chap. 13.

CHAPTER 3

SPECTRA, COVARIANCE, AND CORRELATION FUNCTIONS

In Chap. 1, we have seen in a general way how signal and noise ensembles can be described and how time and ensemble averages can be determined. In Chap. 2, we have indicated some of the more common transformations of such ensembles that we may expect in the communication process. Our next step in the study of system behavior is to apply time and ensemble averages to specific signal and noise ensembles. We shall confine our attention in this chapter to first- and second-order moments (cf. Sec. 1.5) and for the most part to linear systems ($\boldsymbol{T}_L\{\quad\}$), extending the treatment to zero-memory nonlinear devices ($\boldsymbol{T}_g\{\quad\}$) in Chap. 5. Chapters 3 to 5 and 12 to 15 are preliminary to the more powerful methods and more intricate analyses of Part 4 (Chaps. 18 to 23), which employ the full ensemble properties, i.e., the hierarchy of distribution densities, W_n ($n \to \infty$), as well as selected moments and such lower-order densities as W_1, W_2, required in the simpler and less comprehensive criteria of system performance (cf. Chap. 5).

3.1 *The Autocorrelation Function*†

Here we shall begin by considering principally real continuous processes that contain no periodic (or deterministic) components, so that, apart from a possible constant (or d-c) component, the process is entirely random. Thermal and shot noise voltages or currents again provide typical examples. Also, for the time being we shall not attempt to distinguish between linear and nonlinear processes. However, in Chap. 4 we shall give some attention to the important cases of discrete, as distinguished from continuous, data, generated by sampling processes commonly encountered in digital and analogue systems. In any case, our ensemble $y(t)$ of (real) noise voltages may be regarded as the output of some device upon which the sole constraints are that $y(t)$ possesses measurable properties in all time and that $M_y(t_1,t_2)$ or $K_y(t_1,t_2)$ is defined. In addition, a steady state (cf. Sec. 2.2-5) is also assumed: the outputs $y^{(j)}(t)$ are not bounded in duration, only in amplitude, to ensure that the time averages $\langle y^{(j)}\rangle$, $\langle y^{(j)2}\rangle$ exist.

3.1-1 Definition of the Autocorrelation Function. Since in actual practice we deal with only one member of an ensemble at one time or, at most,

† For specific references to the calculation and application of correlation functions in communication problems, see Secs. 3.2 to 3.4 and also Chaps. 4, 5, 13, and 16.

with several such members, the time average is the natural way of obtaining *experimentally* such physically important quantities as the mean or d-c component of $y^{(j)}(t)$, the mean power $[\sim\langle y^{(j)2}\rangle]$, and as we shall see later, the mean power as a function of frequency—the *mean-power spectrum.* The time moment of special interest in this connection is the autocorrelation function $R_{y^{(j)}}(t_2 - t_1)$, already noted in Eq. (1.111), viz.,

$$R_{y^{(j)}}(t) \equiv \lim_{T\to\infty} \frac{1}{T} \int_{-T/2}^{T/2} y^{(j)}(t_0)y^{(j)}(t_0 + t)\, dt_0 \qquad t = t_2 - t_1 \tag{3.1}$$

From this, the desired *mean intensity* of the wave is obtained at once on setting $t = 0$, namely,

$$R_{y^{(j)}}(0) = \lim_{T\to\infty} \frac{1}{T} \int_{-T/2}^{T/2} y^{(j)}(t_0)^2\, dt_0 = \langle y^{(j)2}\rangle \tag{3.1a}$$

Thus, if $y^{(j)}(t)$ is the instantaneous voltage drop of this noise wave across a unit resistance, $\langle y^{(j)2}\rangle$ is a measure of the average power (watts) in $y^{(j)}(t)$.

Although we have assumed a steady-state condition, we observe that the correlation function (3.1), and hence the mean intensity (3.1a), is defined for members of nonstationary, as well as stationary, processes [cf. Eqs. (1.111), (1.112)]; $R_{y^{(j)}}(t)$ is a function only of the difference $t = t_2 - t_1$ of initial and final observation times t_1, t_2 in such cases. The fact that $y^{(j)}(t)$ must be suitably bounded for the time average to exist excludes nonstationarity due to instabilities or drift, in the manner of Fig. 1.5, where the mean value of the noise voltage $y^{(j)}(t)$ increases indefinitely with time (except when these variations are themselves bounded in magnitude and rate). However, for nonstationarity which arises solely because of the lack of invariance of some operation $\mathbf{T}$ on an input ensemble, such as amplitude modulation in the example of Eq. (1.110a), the autocorrelation function is still defined. For these reasons, most of the purely random waves that we shall discuss here are restricted to be stationary, whereas bounded, deterministic processes need not be so restricted.

3.1-2 Ergodic Processes. When y is a real ergodic process, we can equate time and ensemble averages according to the theorem of Sec. 1.6-1, so that the autocorrelation function $R_{y^{(j)}}$ and the autovariance K_y are equal (probability 1), viz.,

$$R_{y^{(j)}}(t) = K_y(t) = \iint_{-\infty}^{\infty} y_1 y_2 W_2(y_1,y_2;t)\, dy_1\, dy_2 \qquad \text{probability 1} \tag{3.2}$$

since $\bar{y} = 0$ here.† Similarly, if $\bar{y} \neq 0$, we get in terms of the second-order

† We retain the superscript (j) to emphasize the fact that the autocorrelation function $R_{y^{(j)}}$ is obtained *experimentally* (in principle) from a time average over a *single* member of the ensemble, while the autovariance K_y is an ensemble relation determined theoretically from the (assumed) statistically known properties of the ensemble as a whole (cf. Table 1.1). For other, equivalent time averages, see Eqs. (1.126), (1.127).

moment function M_y

$$R_y^{(j)}(t) = M_y(t) = \iint_{-\infty}^{\infty} y_1 y_2 W_2(y_1,y_2;t)\, dy_1\, dy_2$$
$$= K_y(t) + \bar{y}^2 \qquad \text{probability 1} \quad (3.2a)$$

With the help of Eqs. (1.92) in Eq. (3.2), the mean intensity is simply

$$R_y^{(j)}(0) = M_y(0) = \int_{-\infty}^{\infty} y^2 W_1(y)\, dy = K_y(0) + \bar{y}^2 \qquad \text{probability 1} \quad (3.3)$$

When the characteristic function of y_1 and y_2 is known, it is usually simpler to determine these moments by differentiation, according to Eqs. (1.57). Equation (3.2a) becomes equivalently†

$$R_y^{(j)}(t) = -\frac{\partial^2}{\partial \xi_1\, \partial \xi_2} F_2(\xi_1,\xi_2;t)\Big|_{\xi_1=\xi_2=0} = \overline{y_1 y_2} \qquad \text{probability 1} \quad (3.4)$$

with $\bar{y} = -i(\partial/\partial\xi_1)F_1(\xi_1)\Big|_{\xi_1=0}$. For example, if in Prob. 1.8b we set $\rho = \rho(t)$, we can easily verify that $R_y^{(j)}(t) = \psi\rho(t) = K_y(t)$ (probability 1), or if we take the more general version [Prob. 1.8d, Eq. (4a)], we get $R_y^{(j)}(t) = \psi\rho(t) + \bar{y}^2$ [cf. Eq. (3.2a)]. We assume, of course, that the autovariance is properly specified; as we shall see in the next section, not any function of t is suitable, but only those which satisfy rather special conditions [Secs. 3.2-3, and 3.2-3(2)].

3.1-3 Some Properties of $R_y^{(j)}(t)$. From the definition (3.1), the following properties of the autocorrelation function of a real wave are easily obtained:

(1) $R_y^{(j)}(0) = \langle y^{(j)2} \rangle$ = mean intensity of the wave.

(2) $R_y^{(j)}(t) = R_y^{(j)}(-t)$. The autocorrelation function is even in $t = t_2 - t_1$, as can be seen from the fact that $\langle y^{(j)}(t_0)y^{(j)}(t_0 + t)\rangle = \langle y^{(j)}(t_0' - t)y^{(j)}(t_0')\rangle$.

(3) $R_y^{(j)}(0) \geq |R_y^{(j)}(t)| \geq 0$. This follows from the fact that $0 \leq \langle [y^{(j)}(t_1) \pm y^{(j)}(t_2)]^2 \rangle = 2R_y^{(j)}(0) \pm 2R_y^{(j)}(t)$. From (3) it is also seen that, if $R_y^{(j)}(0)$ exists, so also must $R_y^{(j)}(t)$.

For a member $y^{(j)}$ of an ergodic process which is entirely random except for a possible steady component, one can use Eqs. (1.89) and (1.93) to show an additional property that

(4) $$R_y^{(j)}(\pm\infty) = M_y(\pm\infty) = \left(\int_{-\infty}^{\infty} y W_1(y)\, dy\right)^2 = \bar{y}^2 \qquad \text{probability 1}$$

All memory, save that contained in $\bar{y}$ (if $\bar{y} \neq 0$), between pairs of observations at time t apart vanishes as $t \to \pm\infty$. If $\bar{y}$ is zero, then $R_y^{(j)}(\pm\infty) = 0$ (probability 1); y behaves like a purely random process (Sec. 1.4-2) for such t, and y_1 and y_2 are said to be *uncorrelated* for $|t| \to \infty$. In fact, somewhat

† For convenience, we make the obvious abbreviations $F_2(\xi_1,\xi_2;t)$ for $F_{y_1,y_2}(\xi_1,\xi_2;t)$, etc.

more broadly we say that, if $\langle y^{(j)}(t_1)y^{(j)}(t_2)\rangle = \langle y^{(j)}\rangle^2$ (all $t_1 \neq t_2$), $y^{(j)}(t_1)$ and $y^{(j)}(t_2)$ are uncorrelated. Observe that this does not necessarily mean that $g(y_1^{(j)})$, $g(y_2^{(j)})$ are uncorrelated, nor does it imply that the ensembles from which $y_1^{(j)}$ and $y_2^{(j)}$ are taken are such that y_1 and y_2 are statistically independent. As was observed in Sec. 1.4-1, the most that can be said is that y_1 and y_2 are wide-sense independent.

When y contains periodic terms, (4) above must be modified; $W_2(y_1,y_2;t)$ no longer factors completely into two first-order densities according to Eqs. (1.93) as $t \to \pm\infty$. The autocorrelation function not only contains a d-c, or steady, component but one or more oscillating terms, which do not die down as $t \to \pm\infty$ (cf. Sec. 3.1-5).

Some related results for correlation functions are:

(5) $$\dot{R}_y^{(j)}(t) = \langle y_1^{(j)}\dot{y}_2^{(j)}\rangle = \overline{y_1\dot{y}_2} = \frac{d}{dt}M_y(t) \qquad \text{probability 1}$$

One easily derives such relations as:

(6) $$\begin{aligned} \langle y_1^{(j)}\dot{y}_2^{(j)}\rangle + \langle \dot{y}_1^{(j)}y_2^{(j)}\rangle &= 0 \\ \langle y_1^{(j)}\ddot{y}_2^{(j)}\rangle + \langle \ddot{y}_1^{(j)}y_2^{(j)}\rangle &= 2\ddot{R}_y^{(j)}(t) \end{aligned}$$

These, for ergodic y, are equal with probability 1 to

$$\overline{y_1\dot{y}_2} + \overline{\dot{y}_1y_2} = 0 \qquad \overline{y_1\ddot{y}_2} + \overline{\ddot{y}_1y_2} = 2\ddot{M}_y(t)$$

respectively, it being assumed, of course, that dM_y/dt, d^2M_y/dt^2 in the above exist (all t), that is, that the process y is differentiable mean square to the required order (Sec. 2.1-2). Higher-order relations can be constructed in a similar way.

We have also the definition:

(7) Normalization. It is often convenient to use instead of $R_y^{(j)}$ and M_y the normalized forms

$$r_y^{(j)}(t) \equiv R_y^{(j)}(t)/R_y^{(j)}(0) \qquad \rho_y(t) \equiv M_y(t)/M_y(0)$$

From (3) it follows at once that $-1 \leq r_y^{(j)} \leq 1$, and since $0 \leq |M_y| \leq M_y(0)$, ρ_y is also restricted to the interval $(-1,1)$ for every bounded correlation and covariance function.

We can use the properties 1 to 4 above to sketch in Fig. 3.1 a typical autocorrelation function for a member of a stationary and ergodic shot noise voltage ensemble $y(t)$. For simplicity, we assume that $\langle y^{(j)}\rangle = 0$, so that $R_y^{(j)}$ ultimately vanishes as $t \to \pm\infty$, and that $y^{(j)}(t)$ represents the output of some stable linear device† with a finite memory (cf. Sec. 2.2-3), so that $R_y^{(j)}(0)$ is also finite. As we shall see specifically in Sec. 3.3, $R_y^{(j)}(t)$ of Fig. 3.1b is characteristic of a moderately narrowband filter acting on a

† Or a combination of finite-memory linear and zero-memory nonlinear elements.

wideband noise voltage input. We remark, finally, that the discussion of Secs. 3.1-2 and 3.1-3 may be extended in a straightforward way to complex waves $y = (y)_{\text{real}} + i(y)_{imag}$.

3.1-4 The Discrete Case. Instead of continuous processes, we may often have to consider a discrete or continuous random *series* (cf. Sec. 1.3-3) whose specific structure as a series is determined by the nature of the underlying physical mechanism. Its structure may also be determined by

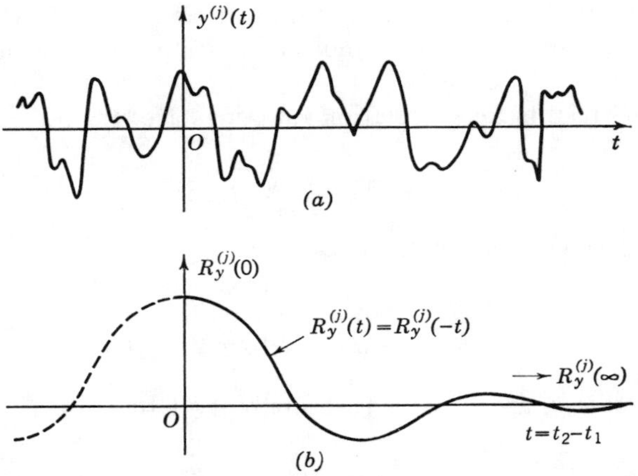

FIG. 3.1. A typical stationary shot noise voltage (a) after a stable filter, with (b) its autocorrelation function.

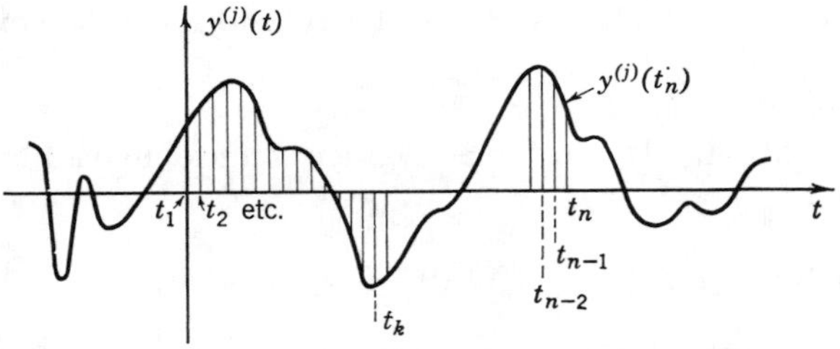

FIG. 3.2. A continuous random series obtained from a continuous process by sampling at fixed times $t_1, t_2, \ldots$.

sampling an originally discrete or continuous *process* at discrete time intervals, in the manner of Fig. 3.2, to generate the sequence $y^{(j)}(t_1), y^{(j)}(t_2), \ldots, y^{(j)}(t_n), \ldots$ $(t_1 < t_2, \ldots, t_n, \ldots)$. The classical one-dimensional random walk, mentioned briefly in Sec. 1.3-3, and the sampled output of, say, the "super clipper" of Fig. 2.11 when a shot noise voltage is the input are typical examples.

Instead of an autocorrelation function $R_y^{(j)}(t)$, one has now an *autocorrelation coefficient* $R_y^{(j)}(t_k) \equiv R_y^{(j)}(k)$ and, similarly for the ensemble, an *auto-*

variance $K_y(t_k) = K_y(k)$ in place of the autovariance function $K_y(t)$, with corresponding analogues of the other moments. These quantities are defined (for the real waves assumed here) by

$$R_y^{(j)}(k) \equiv \lim_{N\to\infty} \left(\frac{1}{2N+1} \sum_{-N}^{N} y_n^{(j)} y_{n+k}^{(j)} \right) = R_y^{(j)}(\Delta_k t) \tag{3.5}$$

$$y_n^{(j)} = y^{(j)}(t_n) \qquad \Delta_n t \equiv t_{n+1} - t_n \,(>0)$$

with

$$\begin{aligned} K_y(k) &\equiv \boldsymbol{E}\{(y_n - \bar{y}_n)(y_{n+k} - \bar{y}_{n+k})\} = K_y(\Delta_k t) \\ M_y(k) &\equiv \boldsymbol{E}\{y_n y_{n+k}\} = \overline{y_n y_{n+k}}, \text{ etc.} \end{aligned} \tag{3.6}$$

where it is postulated that $\boldsymbol{E}\{\quad\}$ exists and the process y may or may not be ergodic. For ergodic processes, we have, of course, $R_y^{(j)}(k) = K_y(k)$ ($\bar{y} = 0$), or $M_y(k)$, with probability 1. Again excluding periodic components, we observe that properties 2, 3, and 7 for continuous waves here have the analogues:

$$\begin{aligned} &(2) \qquad R_y^{(j)}(k) = R_y^{(j)}(-k) \\ &(3) \qquad R_y^{(j)}(0) \geq |R_y^{(j)}(k)| \geq 0 \\ &(7) \qquad r_y^{(j)}(k) = R_y^{(j)}(k)/R_y^{(j)}(0) \leq 1 \end{aligned} \tag{3.7}$$

Specific examples of practical interest are considered presently, in Chap. 4.

3.1-5 Periodic Waves. Let us now consider the cases where $y^{(j)}(t)$ belongs to an ensemble of deterministic waveforms (cf. Sec. 1.3-8), remembering that a deterministic waveform is still a random one, albeit of a special type (the ensemble variable is here a random constant). Since we are assuming a steady-state condition, this means, of course, that the $y^{(j)}(t)$ have some sort of periodic structure. The simplest example arises when $y^{(j)}(t)$ is a pure sinusoid. Applying Eq. (1.115) and the argument leading up to it, with $G^{(j)} = y^{(j)}(t_0)y^{(j)}(t_0 + t)$ here, we obtain for the autocorrelation function of $y^{(j)} = A_0 \cos(\omega_0 t + \phi^{(j)})$

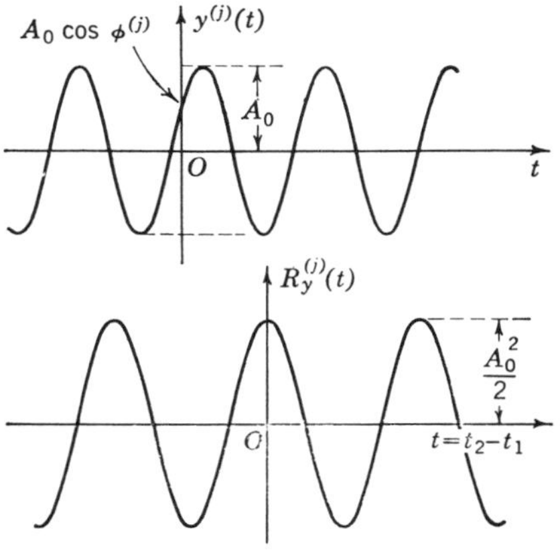

FIG. 3.3. A sinusoid and its autocorrelation function.

$$\begin{aligned} R_y^{(j)}(t) &= \frac{1}{T_0} \int_0^{T_0} y^{(j)}(t_0) y^{(j)}(t_0 + t)\, dt_0 \\ &= \frac{A_0^2}{2\pi} \int_0^{2\pi} \cos\psi \cos(\psi + \omega_0 t)\, d\psi \qquad \psi = \omega_0 t_0 = \frac{2\pi t_0}{T_0} \\ &= \frac{A_0^2}{2} \cos \omega_0 t \end{aligned} \tag{3.8}$$

Note that the (constant) phase $\phi^{(j)}$ does not appear in this expression: phase data are lost. Observe, also, that the autocorrelation function is also periodic, with the same fundamental frequency ($f_0 = 1/T_0$) as the original wave; $R_y^{(j)}(t)$ and $y^{(j)}(t)$ are sketched in Fig. 3.3.

General periodic waves, like that shown in Fig. 3.4, can be handled in several ways. Our first is to expand the real wave $y^{(j)}(t)$ in a Fourier series†

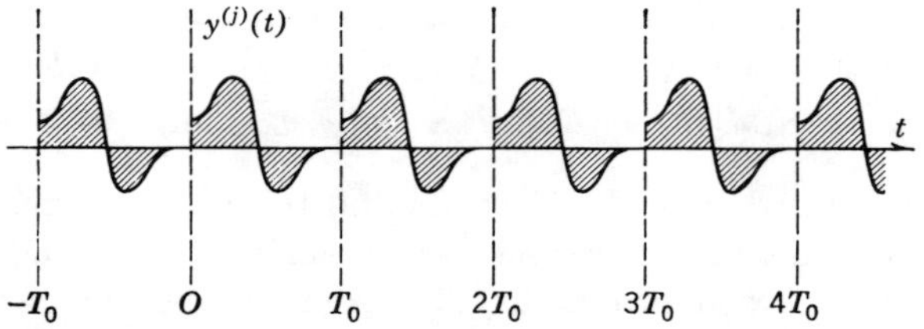

FIG. 3.4. A general periodic wave, with period interval $(0, T_0)$.

on the period interval (0, $T_0 = 1/f_0$), so that we may write the equivalent forms

$$
\begin{aligned}
y^{(j)}(t) &= \sum_{n=0}^{\infty} (a_n^{(j)} \cos n\omega_0 t + b_n^{(j)} \sin n\omega_0 t) \\
&= \sum_{n=0}^{\infty} c_n^{(j)} \cos (n\omega_0 t + \phi_n^{(j)}) \\
&= \sum_{n=-\infty}^{\infty} C_n^{(j)} e^{in\omega_0 t}
\end{aligned}
\tag{3.9}
$$

where the coefficients $a_n^{(j)}$, $b_n^{(j)}$, etc., are determined from

$$
\begin{aligned}
a_n^{(j)} &= \frac{1}{T_0} \int_0^{T_0} y^{(j)}(t_0) \cos n\omega_0 t_0 \, dt_0 \\
b_n^{(j)} &= \frac{1}{T_0} \int_0^{T_0} y^{(j)}(t_0) \sin n\omega_0 t_0 \, dt_0 \\
c_n^{(j)} &= \sqrt{a_n^{(j)2} + b_n^{(j)2}} \\
\phi_n^{(j)} &= -\tan^{-1} [(b_n^{(j)})/(a_n^{(j)})] \\
C_n^{(j)} &= \frac{1}{T_0} \int_0^{T_0} y(t_0) e^{-in\omega_0 t_0} \, dt_0 = |C_n^{(j)}| e^{i\phi_n^{(j)}}
\end{aligned}
\tag{3.10}
$$

with
$$|C_n^{(j)}| = |C_{-n}^{(j)}| \qquad \phi_n^{(j)} = -\phi_{-n}^{(j)}$$

respectively. Since $y^{(j)}(t)$ is a real wave, we have also $C_n^{(j)} = C_{-n}^{(j)*}$, and therefore $|C_n^{(j)}| = c_n^{(j)}/\sqrt{\epsilon_n} = \sqrt{a_n^{(j)2} + b_n^{(j)2}}/\sqrt{\epsilon_n}$, with $\epsilon_0 = 1$, $\epsilon_n = 2$ ($n \neq 0$). Inserting Eq. (3.9) into Eq. (3.8) gives at once for the autocorrelation function

† For a general discussion of Fourier series and their properties, see E. T. Whittaker and G. N. Watson, "Modern Analysis," 4th ed., chap. IX, Cambridge, New York, 1940.

$$R_y^{(j)}(t) = \frac{1}{2}\sum_{n=0}^{\infty}(a_n^{(j)2} + b_n^{(j)2})\cos n\omega_0 t$$

$$= \frac{1}{2}\sum_{n=0}^{\infty} c_n^{(j)2}\cos n\omega_0 t = \sum_{-\infty}^{\infty}|C_n^{(j)}|^2 e^{-in\omega_0 t} \tag{3.11}$$

or

$$R_y^{(j)}(t) = \sum_{n=0}^{\infty}\epsilon_n|C_n^{(j)}|^2\cos n\omega_0 t$$

Again, $R_y^{(j)}(t)$ has the same period as the original wave, and all phase information ($\sim\phi_n^{(j)}$) is suppressed. Our simple sinusoid of Eq. (3.8) is an obvious special case of Eq. (3.11).

The mean total intensity is found from (1) of Sec. 3.1-3; setting $t = 0$ in Eq. (3.11) gives at once

$$R_y^{(j)}(0) = \langle y^{(j)2}\rangle = \sum_{n=0}^{\infty}\epsilon_n|C_n^{(j)}|^2 = \frac{1}{2}\sum_{n=0}^{\infty}c_n^{(j)2} = \frac{1}{2}\sum_{0}^{\infty}(a_n^{(j)2} + b_n^{(j)2}) \tag{3.12}$$

which is, of course, just one form of Parseval's theorem.† The intensity of the nth component of $y^{(j)}(t)$ is immediately exhibited as the coefficient of $\cos n\omega_0 t$ in Eq. (3.11) for $R_y^{(j)}(t)$. We observe also that $R_y^{(j)}(\pm\infty)$ does not vanish but oscillates indefinitely, unlike the entirely random waves discussed above. In fact, Eq. (3.11) applies for all t and is typical of periodic waves. We may accordingly broaden our interpretation of property 4 (Sec. 3.1-3), noting now that, if $y^{(j)}(t)$ consists of a (general) mixture of periodic and random components, $R_y^{(j)}(\pm\infty)$ reveals (1) the d-c, or steady, component in the wave, which may be attributed to *both* the periodic and random terms, and (2) the periodic, nonzero frequency components, which can come only from the original periodic term, since $R_y^{(j)}(\pm\infty)_{\text{noise}}$ dies down to zero ultimately.

While Eq. (3.11) is particularly useful in exhibiting the harmonic structure (apart from phase) of the periodic wave $y^{(j)}(t)$ and the intensity of its individual components, it is not usually easy to see how $R_y^{(j)}(t)$ itself varies with $t\ (= t_2 - t_1)$. Rather than a Fourier-series development, we should like now a closed expression for $R_y^{(j)}(t)$. This is readily found if we take advantage of the periodic structure of $y^{(j)}(t)$ and use Eq. (3.8) for a typical kth-period interval $[kT_0, (k+1)T_0]$ to write

$$R_y^{(j)}(t_k) \equiv \frac{1}{T_0}\int_{kT_0}^{(k+1)T_0} y^{(j)}(t_0)y^{(j)}(t_0 + t_k)\,dt_0$$

$$R_y^{(j)}(t_k) = \begin{cases} \dfrac{1}{T_0}\displaystyle\int_0^{T_0} y^{(j)}(t_0')y^{(j)}(t_0' + t_k)\,dt_0' & kT_0 \le t_k \le (k+1)T_0 \\ 0 & \text{elsewhere} \end{cases} \tag{3.13a}$$

† *Ibid.*

with $t_k = t - kT_0$, and observe that the desired autocorrelation function becomes

$$R_y^{(j)}(t) = \sum_{-\infty}^{\infty} R_y^{(j)}(t_k) \qquad t_k = t - kT_0 \tag{3.13b}$$

In computing $R_y^{(j)}(t_k)$, only the *overlap* of $y^{(j)}(t_0)$ with $y_0^{(j)}(t_0 + t_k)$ that falls within the period interval $[kT_0 \le t \le (k+1)T_0]$ is to be considered. If $y^{(j)}(t)$ is symmetrical in a period interval, we have

$$R_y^{(j)}(t_k) = \begin{cases} \dfrac{1}{T_0} \displaystyle\int_{-T_0/2}^{T_0/2} y^{(j)}(t_0')y^{(j)}(t_0' + t_k)\, dt_0' \\ \qquad\qquad\qquad\qquad (k - \tfrac{1}{2})T_0 \le t \le (k + \tfrac{1}{2})T_0 \\ 0 \qquad\qquad\qquad\qquad t \text{ elsewhere} \end{cases} \tag{3.14}$$

(sometimes a more convenient relation).

As an example illustrating both series approaches, let us consider the periodic train of rectangular pulses of duration τ, shown in Fig. 3.5. From Eqs. (3.10), we obtain at once

$$C_n = A_0 \frac{1 - e^{-in\omega_0\tau}}{2\pi in} \qquad 0 < \tau \le T_0 \tag{3.15a}$$

and therefore

$$|C_n|^2 = A_0{}^2 \frac{1 - \cos n\omega_0\tau}{2\pi^2 n^2} = A_0{}^2 \frac{\sin^2 (n\omega_0\tau/2)}{\pi^2 n^2} \tag{3.15b}$$

whereupon the series form (3.11) of the autocorrelation function becomes

$$R_y^{(j)}(t) = \sum_{n=0} \frac{A_0{}^2\epsilon_n \sin^2 (n\omega_0\tau/2)}{\pi^2 n^2} \cos n\omega_0 t \tag{3.16}$$

revealing the intensity of the various components. On the other hand, to find a "closed" expression for $R_y^{(j)}(t)$ we use, instead, Eqs. (3.13). Assuming first that $\tau \le T_0/2$, with $\Delta_k = 1$, $|t - kT_0| \le \tau$, we compute the overlap according to

$$R_y^{(j)}(t_k) = \begin{cases} \dfrac{1}{T_0} A_0{}^2 \displaystyle\int_{t_k}^{\tau} dt_0 = A_0{}^2 \frac{\tau - t_k}{T_0} & 0 \le t_k \le \tau \\ \dfrac{1}{T_0} A_0{}^2 \displaystyle\int_{0}^{\tau + t_k} dt_0 = A_0{}^2 \frac{\tau + t_k}{T_0} & -\tau \le t_k \le 0 \end{cases} \tag{3.17}$$

and

$$R_y^{(j)}(t) = A_0{}^2 \sum_{-\infty}^{\infty} \frac{\tau - |t - kT_0|}{T_0} \Delta_k \tag{3.18}$$

where $\Delta_k = 0$ for $|t - kT_0| > \tau$. Figure 3.6*a*, *b* illustrates the calculation of Eq. (3.17), while Fig. 3.7 shows the autocorrelation function itself as a function of t. (The case $T_0/2 \le \tau \le T_0$ is left as an exercise for the reader.)

Thus, Eq. (3.18) represents the sum of the series (3.16), while the latter is simply the cosine Fourier-series development of $R_y^{(j)}(t)$ on a period interval. The mean total intensity is simply $A_0^2\tau/T_0$, with a d-c term of intensity $(A_0\tau/T_0)^2$.

3.1-6 Mixed Processes. When $y^{(j)}(t)$ is an *additive* mixture of a deterministic and a random component, so that $y^{(j)}(t) = S^{(j)}(t) + N^{(j)}(t)$ and

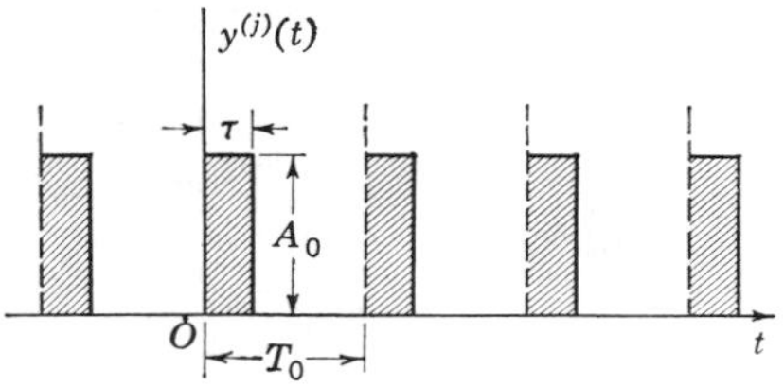

FIG. 3.5. A periodic train of rectangular pulses.

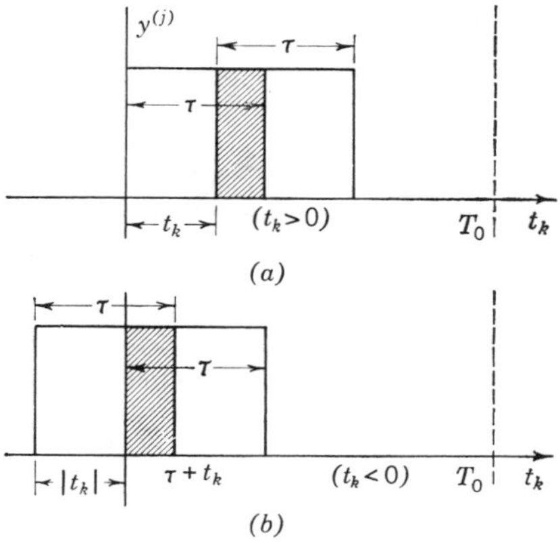

FIG. 3.6. Overlap of $y^{(j)}(t_0 + t_k)$ with $y^{(j)}(t_0)$ in a typical period interval of the pulse train of Fig. 3.5.

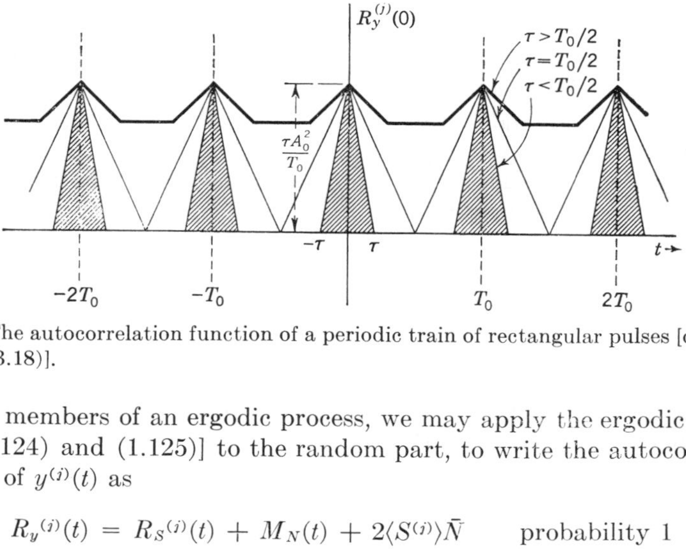

Fig. 3.7. The autocorrelation function of a periodic train of rectangular pulses [cf. Fig. 3.5 and Eq. (3.18)].

both are members of an ergodic process, we may apply the ergodic theorem [Eqs. (1.124) and (1.125)] to the random part, to write the autocorrelation function of $y^{(j)}(t)$ as

$$R_y^{(j)}(t) = R_{S^{(j)}}(t) + M_N(t) + 2\langle S^{(j)}\rangle \bar{N} \qquad \text{probability 1} \qquad (3.19)$$

Here $\langle S^{(j)}N\rangle$ factors into $\langle S^{(j)}\rangle \bar{N}$ since S and N are assumed statistically independent. Usually, either $\langle S^{(j)}\rangle$ or $\bar{N}$ vanishes, and we obtain the simpler relation

$$R_y^{(j)}(t) = R_{S^{(j)}}(t) + M_N(t) = R_{S^{(j)}}(t) + K_N(t) \qquad \text{if } \bar{N} = 0 \text{ (probability 1)} \qquad (3.19a)$$

From the properties 1 to 4 of Sec. 3.1-3 and the results of Sec. 3.1-5, we can sketch a typical autocorrelation function for Eq. (3.19a), indicated in Fig. 3.8. As $t \to \pm\infty$, K_N vanishes, while the periodic structure of the deterministic component reveals itself. This characteristic behavior is taken advantage of in the so-called *correlation detector* as a possible device for detecting the presence of a steady deterministic signal [here $S^{(j)}(t)$] in noise. Multiplicative processes are also of some interest. Suppose that $y^{(j)}(t) = S^{(j)}(t)N^{(j)}(t)$. Then, again applying the ergodic theorem to the random factors, we easily obtain

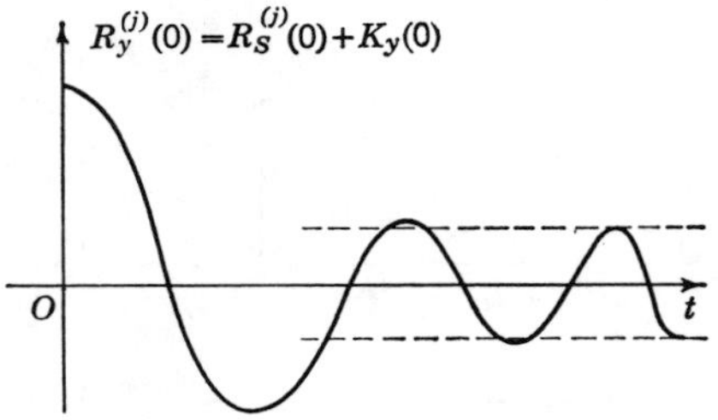

FIG. 3.8. Autocorrelation function of the sum of a periodic and a random wave.

$$R^{(j)}(t) = R_S{}^{(j)}(t)R_N{}^{(j)}(t) = R_S{}^{(j)}(t)M_N(t) \qquad \text{probability 1}$$

with similar relations for more complicated forms of $y^{(j)}(t)$.

Besides additive, multiplicative, and independent mixtures of signal and noise waves, we can easily construct other combinations, as Examples 1 to 3 of Sec. 1.6 show, where nonlinear elements and nonergodic ensembles may also appear. For instance, suppose that $y^{(j)} = a^{(j)} \cos (\omega_0 t + \phi^{(j)})$, where now a, considered over the ensemble, is normally distributed, so that y is no longer ergodic, even if ϕ is uniformly distributed in the primary interval $(0,2\pi)$ [cf. Sec. 1.3-7 and the remarks following Eqs. (1.125)]. An $R_y{}^{(j)}(t)$ can be *measured* in a given experimental situation but cannot be calculated a priori by the usual appeal to the ergodic theorem. We can, however, calculate an *average* autocorrelation function

$$\overline{R_y{}^{(j)}(t)} = \frac{\overline{a^2}}{2} \cos \omega_0 t = K_y(t) = \langle\langle y^{(j)}(t_1)y^{(j)}(t_2)\rangle\rangle_{stat\,av} \tag{3.20}$$

As a second example, where the somewhat more complicated structures of mixed processes are evident, let us consider the correlation function of a narrowband wave (cf. Sec. 2.2-5), modulated in amplitude and phase by a random or deterministic disturbance, e.g.,

$$y^{(j)}(t) = A_0{}^{(j)}(t) \cos [\omega_0 t - \Phi^{(j)}(t)] \tag{3.21}$$

which becomes now specifically

$$R_y{}^{(j)}(t) = \frac{1}{2} \lim_{T\to\infty} \frac{1}{T} \left\{ \int_{-T/2}^{T/2} A_0{}^{(j)}(t_0)A_0{}^{(j)}(t_0 + t) \times \cos [2\omega_0 t_0 + \omega_0 t - \Phi^{(j)}(t_0) - \Phi^{(j)}(t_0 + t)]\, dt_0 + \int_{-T/2}^{T/2} A_0{}^{(j)}(t_0)A_0{}^{(j)}(t_0 + t) \cos [\omega_0 t + \Phi^{(j)}(t_0) - \Phi^{(j)}(t_0 + t)]\, dt_0 \right\} \tag{3.22}$$

Taking advantage now of the narrowband structure of $y^{(j)}$ [for example, $A_0(t)$, $\Phi(t)$ are slowly varying compared with $\cos \omega_0 t$, $\sin \omega_0 t$, etc.], we may neglect the first term in Eq. (3.22) in comparison with the second, to write approximately

$$R_y^{(j)}(t) \doteq \tfrac{1}{2} \operatorname{Re} \langle A_0(t_1) A_0(t_2) e^{-i\omega_0 t + i\Phi_2 - i\Phi_1} \rangle \qquad \Phi_1 = \Phi(t_1), \text{ etc.} \tag{3.23}$$

where the approximation becomes progressively better as A_0, $\cos \Phi$ are more slowly variable with t. Applying Eqs. (2.66) et seq. and writing

$$\begin{aligned} \mathcal{R}^{(j)}_{\alpha_0,\beta_0}(t) &\equiv \langle [\alpha_0(t_1) - i\beta_0(t_1)][\alpha_0(t_2) + i\beta_0(t_2)] \rangle^{(j)} = \mathcal{R}^{(j)}_{\alpha_0,\beta_0}(-t)^* \\ &= \langle A_0(t_1) A_0(t_2) e^{-i\Phi(t_1) + i\Phi(t_2)} \rangle^{(j)} \end{aligned} \tag{3.24}$$

we may express Eq. (3.23) alternatively as

$$R_y^{(j)}(t) \doteq \tfrac{1}{2} \operatorname{Re} [e^{-i\omega_0 t} \mathcal{R}^{(j)}_{\alpha_0,\beta_0}(t)] \tag{3.25}$$

which, as we shall see in Sec. 3.3, is a useful form of the autocorrelation function of a narrowband wave. In general, y is not an ergodic process, so that while an $R_y^{(j)}(t)$ can be observed experimentally, we can only calculate a priori the statistical average $\overline{R_y^{(j)}(t)}$ of the correlation function.

We shall encounter many examples of mixed processes in our subsequent study of communication systems, where additional statistical averages are required to yield useful quantities that can be determined a priori.

PROBLEMS

3.1 (*a*) Show that the autocorrelation function of $y^{(j)}(t) = A_0 \cos \omega_a t \cos (\omega_0 t + \phi^{(j)})$ (ω_0, ω_a incommensurable) is $R_y^{(j)}(t) = (A_0^2/4) \cos \omega_a t \cos \omega_0 t$. Thus, discuss the condition for the existence of the autocorrelation function when $y^{(j)}$ does *not* belong to an ergodic ensemble.

(*b*) If y_1 and y_2 are ergodic ensembles (for a deterministic process), such that $\bar{y}_1 = \bar{y}_2 = 0$, is $y_1 y_2 = z$ an ergodic ensemble? Under what conditions? Show in general that $\overline{R_z^{(j)}(t)} = K_{y_1}(t) \times K_{y_2}(t) = K_z(t)$ if y_1, y_2 are statistically independent here. When does $R_z^{(j)}(t) = K_z(t)$ (probability 1)? Illustrate the discussion with the results of (*a*).

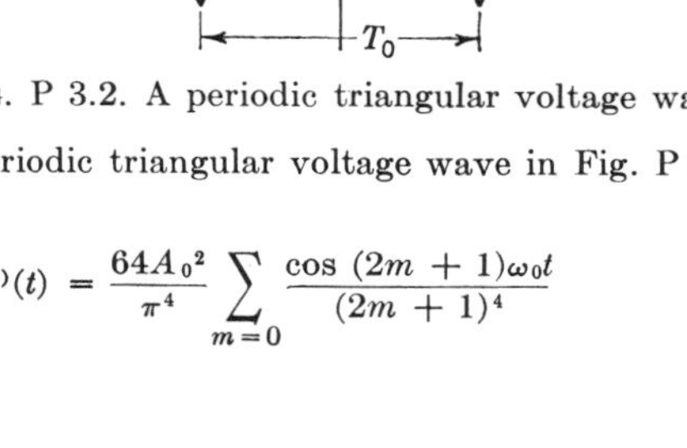

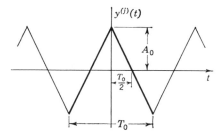

FIG. P 3.2. A periodic triangular voltage wave.

3.2 Show for the periodic triangular voltage wave in Fig. P 3.2 that its autocorrelation function is

$$R_y^{(j)}(t) = \frac{64 A_0^2}{\pi^4} \sum_{m=0} \frac{\cos (2m + 1)\omega_0 t}{(2m + 1)^4} \tag{1}$$

or
$$R_{y^{(j)}}(t) = 2A_0^2 \sum_{-\infty}^{\infty} \left[\frac{1}{6} - 4\left(\frac{t_k}{T_0}\right)^2 + \frac{16}{3}\frac{|t_k|^3}{T_0^3}\right] \Delta_k \tag{2}$$

where, for $-T_0/2 < t_k < T_0/2$, $\Delta_k = 1$; otherwise $\Delta_k = 0$; $t_k = t - kT_0$. Sketch $R_{y^{(j)}}(t)$, and find $\langle y^{(j)2}\rangle$, $\langle y^{(j)}\rangle$.

3.3 Verify that the autocorrelation function of the periodic saw-tooth voltage wave in Fig. P 3.3*a* can be written as

$$R_{y^{(j)}}(t) = \frac{A_0^2}{2}\left(\frac{1}{2} + \sum_{m=1}^{\infty} \frac{\cos m\omega_0 t}{\pi^2 m^2}\right) \tag{1}$$

or
$$R_{y^{(j)}}(t) = A_0^2 \sum_{-\infty}^{\infty} \left[\frac{1}{3} - \frac{1}{2}\frac{|t_k|}{T_0} + \frac{1}{2}\left(\frac{t_k}{T_0}\right)^2\right] \Delta_k \tag{2}$$

with $\Delta_k = 1$ $[kT_0 < t < (k+1)T_0]$, $\Delta_k = 0$ otherwise, and $t_k = t - kT_0$ as before. Sketch $R_{y^{(j)}}(t)$, and obtain $\langle y^{(j)2}\rangle$ and $\langle y^{(j)}\rangle$. What is the autocorrelation function of the wave in Fig. P 3.3*b*?

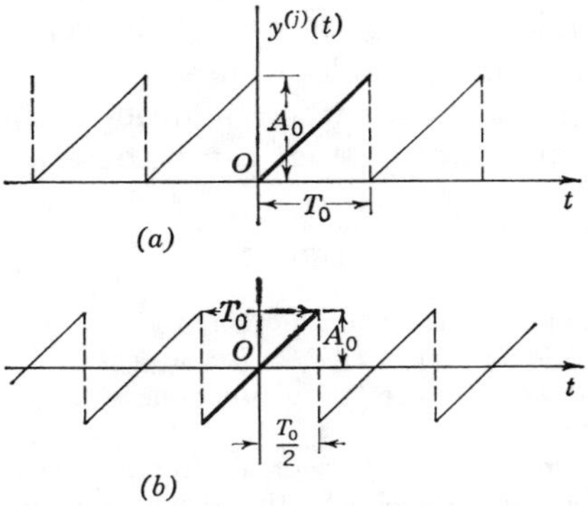

FIG. P 3.3. (*a*) A periodic saw-tooth voltage wave; (*b*) the same, without a d-c component.

3.4 Obtain the autocorrelation function of $z^{(j)}(t)$ when $z^{(j)} = \boldsymbol{T}_g\{y^{(j)}\}$ and $\boldsymbol{T}_g\{\ \}$ is a *half-wave linear rectifier* of dynamic transconductance β for the periodic input voltage $y^{(j)} = A_0 \cos[\omega_0 t + \phi^{(j)}]$. Thus, show that

$$R_{z^{(j)}}(t) = \frac{A_0^2\beta^2}{\pi^2}\left[1 + \frac{\pi^2}{8}\cos\omega_0 t + 2\sum_{m=1}^{\infty} \frac{\cos 2m\omega_0 t}{(4m^2-1)^2}\right] \tag{1}$$

or
$$R_{z^{(j)}}(t) = A_0^2\beta^2 \sum_{-\infty}^{\infty} \left[\left(\frac{1}{4} - \frac{1}{2}\left|\frac{t_k}{T_0}\right|\right)\cos\omega_0 t_k + \frac{\sin\omega_0|t_k|}{4\pi}\right] \Delta_k \tag{2}$$

where $t_k = t - kT_0$, and $\Delta_k = 1$ if $-T_0/2 < t_k < T_0/2$, $\Delta_k = 0$ otherwise. What are $\langle z^{(j)}\rangle$ and $\langle z^{(j)2}\rangle$? Sketch $R_{z^{(j)}}(t)$.

3.5 Repeat Prob. 3.4, now for the *full-wave linear rectifier* $z^{(j)} = \boldsymbol{T}_g\{y\} = \beta|y^{(j)}|$, and show specifically that

$$R_{z^{(j)}}(t) = \frac{4A_0^2\beta^2}{\pi^2}\left[1 + 2\sum_{m=1}^{\infty} \frac{\cos 2m\omega_0 t}{(4m^2-1)^2}\right] \tag{1}$$

with

$$R_z^{(j)}(t) = A_0^2\beta^2 \sum_{-\infty}^{\infty} \left[\left(\left|\frac{t_k}{T_0}\right| - \frac{1}{2\pi}\right) \sin \omega_0|t_k| + \left(\frac{1}{2} - \left|\frac{t_k}{T_0}\right|\right) \cos \omega_0 t_k\right] \Delta_k \qquad (2)$$

where, as before, $t_k = t - kT_0$ and $\Delta_k = 1$ if $-T_0/2 < t_k < T_0/2$, $\Delta_k = 0$ otherwise. Obtain $\langle z^{(j)2}\rangle$ and $\langle z^{(j)}\rangle$, and sketch $R_z^{(j)}(t)$.

3.2 *The Intensity Spectrum*

In our discussion of the steady state (cf. Sec. 2.2-5), we defined the amplitude spectrum $S_y^{(j)}(f)$ of a disturbance $y^{(j)}(t)$ as the Fourier transform

$$S_y^{(j)}(f) = \mathfrak{F}\{y^{(j)}\} = \int_{-\infty}^{\infty} y^{(j)}(t)e^{-i\omega t}\,dt \qquad \omega = 2\pi f \qquad (3.26)$$

when this Fourier transform exists or at most exhibits the δ-function singularities characteristic of a steady-state *periodic* wave $y^{(j)}(t)$ [cf. Eq. (2.45)]. However, as we have already noted when we attempt to deal with entirely random waves under these conditions, $\mathfrak{F}\{y^{(j)}\}$ does not exist in any of the above senses, since $y^{(j)}$ does not die down to zero as $t \to \pm\infty$ with sufficient rapidity to ensure convergence, nor does $y^{(j)}$ possess a definite periodic structure which could be interpreted as a "line" spectrum [cf. Eq. (2.45)] in terms of δ functions. Strictly speaking, then, we observe that the amplitude spectrum of any steady-state aperiodic disturbance does not converge to a finite limit for all frequencies. This does not mean, however, that the notion of "spectrum" is useless in such cases. On the contrary, as Wiener,[1] Khintchine,[2] Bochner,[3] and others[4] have shown in the development of a generalized harmonic analysis, it is possible to extend the usual notions of the spectrum (when $\mathfrak{F}\{y^{(j)}\}$ is defined) to include random functions (of time) where the familiar treatment fails. The present section outlines a few of the pertinent features of such an extension.

3.2-1 Introductory Remarks. Let us begin by considering a particular member $y^{(j)}(t)$ of an ensemble y of (real) voltage waves, where for the moment y may be deterministic, entirely random, or a mixed process, but in any case such that $\mathfrak{F}\{y^{(j)}\}$ converges. This means, of course, that, apart from suitable behavior in any finite interval, $\lim_{t\to\pm\infty} y^{(j)}(t) \to 0$ sufficiently rapidly, in the manner of Fig. 3.9*a*, *b*, or, analytically, that $\int_{-\infty}^{\infty} |y^{(j)}(t)|\,dt < \infty$ and has a definite value.† Then, by Plancherel's theorem,† we have

$$\int_{-\infty}^{\infty} y^{(j)}(t)^2\,dt = \int_{-\infty}^{\infty} |\mathfrak{F}\{y^{(j)}\}|^2\,df = \int_{-\infty}^{\infty} |S_y^{(j)}(f)|^2\,df \equiv E_y^{(j)} < \infty \qquad (3.27)$$

where $E_y^{(j)}$ is the (total) *energy* (in joules) in the voltage wave $y^{(j)}$ if this voltage appears across a unit resistance. The quantity $|S_y^{(j)}(f)|^2$ is corre-

† E. C. Titchmarsh, "Fourier Integrals," 2d ed., Oxford, New York, 1948.

spondingly the (total) *energy density* (in joules per cycle), as is at once apparent from Eq. (3.27). Generally, regardless of the units of $y^{(j)}(t)$, we shall call $E_{y^{(j)}}$ the *total intensity* of the wave and $|S_{y^{(j)}}|^2$ the total intensity

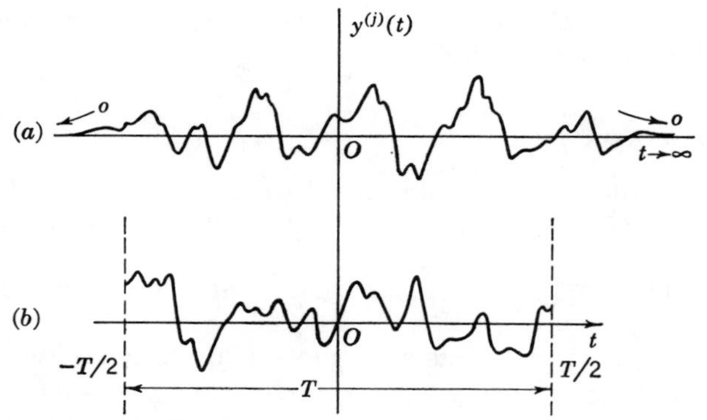

FIG. 3.9. (a) An aperiodic voltage $(-\infty < t < \infty)$. (b) An aperiodic wave which vanishes everywhere outside $(-T/2, T/2)$.

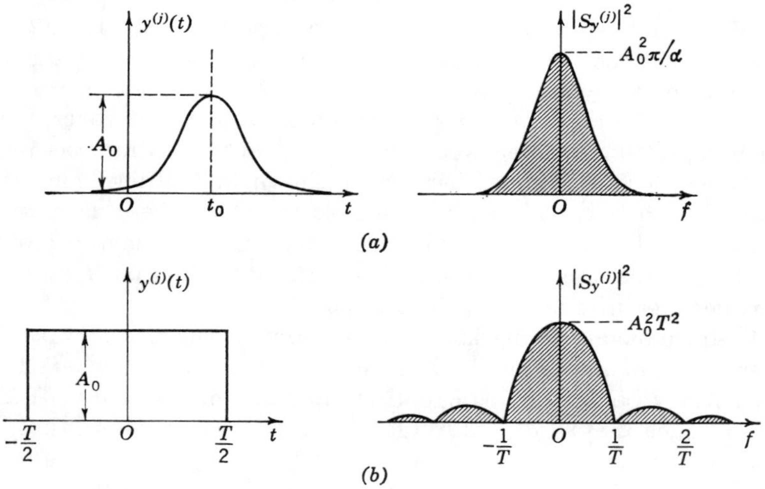

FIG. 3.10. (a) Gaussian pulse. (b) Rectangular pulse. Corresponding total intensity densities.

density. For example, if $y^{(j)}$ is the gaussian pulse $y^{(j)} = A_0e^{-\alpha(t-t_0)^2}$ (cf. Fig. 3.10a), we obtain directly from Eq. (3.26)

$$\begin{aligned} S_{y^{(j)}}(f) &= A_0(\pi/\alpha)^{1/2}e^{-i\omega t_0-\omega^2/4\alpha^2} \\ |S_{y^{(j)}}(f)|^2 &= A_0^2(\pi/\alpha)e^{-\omega^2/2\alpha} \\ E_{y^{(j)}} &= (\pi/2\alpha)^{1/2}A_0^2 \end{aligned} \tag{3.28}$$

while for the wave of Fig. 3.10*b* we get

$$\begin{aligned} S_y^{(j)}(f)_T &= \frac{2A_0}{\omega} \sin \frac{\omega T}{2} \\ |S_y^{(j)}(f)_T|^2 &= \frac{4A_0^2}{\omega^2} \sin^2 \frac{\omega T}{2} \\ E_y^{(j)}(T) &= A_0^2 T \end{aligned} \tag{3.29}$$

The total intensity densities are also sketched in Fig. 3.10*a*, *b*. Note that $|S_y^{(j)}|^2$ depends only on the duration of $y^{(j)}$ and not on where $y^{(j)}$ is located with respect to the time scale.

When the disturbance $y^{(j)}(t)$ vanishes everywhere outside some interval $(t_1, t_1 + T)$ or, equivalently, outside $(-T/2,T/2)$, we can define an *average power*, or *average intensity*, over the interval in question, according to

$$P_y^{(j)}(T) \equiv \frac{E_y^{(j)}(T)}{T} = \frac{1}{T} \int_{t_1}^{t_1+T} y^{(j)}(t_0)^2 \, dt_0 = \frac{1}{T} \int_{-\infty}^{\infty} y_T^{(j)}(t_0)^2 \, dt_0 \tag{3.30}$$

where $y_T^{(j)} = y^{(j)}$ $(-T/2 < t < T/2)$ and is zero outside this interval. From Eq. (3.27), this can be written in still another form,

$$P_y^{(j)}(T) = \int_{-\infty}^{\infty} |S_y^{(j)}(f)_T|^2 \, df/T \qquad \text{where } S_y^{(j)}(f)_T = \int_{-\infty}^{\infty} y_T^{(j)}(t) e^{-i\omega t} \, dt \tag{3.31}$$

Here we shall call $|S_y^{(j)}(f)_T|^2$ the (total) energy or intensity density of $y^{(j)}$ for the interval $(-T/2,T/2)$. Defining

$$\mathcal{W}_y^{(j)}(f)_T \equiv 2|S_y^{(j)}(f)_T|^2/T \tag{3.32}$$

we see at once from Eq. (3.31) and the fact that $|S_y|^2$ is an even function of frequency (for real y) that

$$P_y^{(j)}(T) = \int_0^{\infty} \mathcal{W}_y^{(j)}(f)_T \, df \tag{3.33}$$

so that we may call $\mathcal{W}_y^{(j)}(f)_T$ the *average power density* or *average intensity density* of $y_T^{(j)}(t)$. Note that all *phase* structure associated with the original wave disappears in the representation (3.32) of the intensity density. Furthermore, since $\mathcal{W}_y^{(j)}(f)_T$ is also an even function of f, all distinction between positive and negative frequencies is lost here, corresponding in the time domain to the fact that $P_y^{(j)}(T)$ is independent of the origin of time. Therefore, it is usually convenient in dealing with spectral *intensities* such as $|S_y^{(j)}(f)_T|$, $\mathcal{W}_y^{(j)}(f)_T$ to consider only the positive frequency range $(0 \le f < \infty)$, although for analytical purposes it is sometimes helpful to use the symmetry property inherent in the definition of spectral intensities of real waves [cf. Eqs. (3.46) et seq. and comment].

If $y^{(j)}(t)$ is a member of an ensemble of noise voltages in the steady state so that $y^{(j)}(t)$ does not die down properly to zero as $t \to \pm\infty$, we can still define an *average power* or *average intensity* $P_y^{(j)}$ as the limit $T \to \infty$ of $P_y^{(j)}(T)$, according to Eqs. (3.30), (3.31), in a variety of equivalent ways, viz.:

$$P_y^{(j)} \equiv \lim_{T\to\infty} P_y^{(j)}(T) = \lim_{T\to\infty} \frac{1}{T} \int_{-T/2}^{T/2} y^{(j)}(t_0)^2\, dt_0$$

$$= \lim_{T\to\infty} \frac{1}{T} \int_{-\infty}^{\infty} y_T^{(j)}(t_0)^2\, dt_0 \qquad (3.34a)$$

$$P_y^{(j)} = \lim_{T\to\infty} \frac{1}{T} \int_{-\infty}^{\infty} |S_y^{(j)}(f)_T|^2\, df \qquad (3.34b)$$

or

$$P_y^{(j)} = \lim_{T\to\infty} \int_0^{\infty} \mathcal{W}_y^{(j)}(f)_T\, df \qquad (3.34c)$$

Thus, the average intensity is simply the second time moment of the representation $y^{(j)}(t)$, and we *postulate* that this limit exists. Such an assumption in no way restricts the usefulness of this concept in physical situations, since we never deal with processes possessing or dissipating infinite *power* over any finite interval of time. On the other hand, as we expect, the total energy, $\lim_{T\to\infty} TP_y^{(j)}(T) = \lim_{T\to\infty} E_y^{(j)}(T)$, does become infinite $O(T)$, which is merely the mathematical statement of the fact that the source generating $y^{(j)}(t)$ operates steadily from $-\infty$ in the past to $+\infty$ in the future and produces a nonvanishing output during an infinite number of finite periods in $(-\infty, \infty)$.

Although $\lim_{T\to\infty} P_y^{(j)}(T)$ exists, *it is not true for any steady-state entirely random wave that* [for frequencies f such that $\mathcal{W}_y^{(j)}(f)_T > 0$] $\lim_{T\to\infty} \mathcal{W}_y^{(j)}(f)_T$ [cf. Eq. (3.32)] *approaches a definite limit.* In fact, as indicated below, $\lim_{T\to\infty} \mathcal{W}_y^{(j)}(f)_T \equiv \mathcal{W}_y^{(j)}(f)_\infty$ is usually bounded but oscillates indefinitely in some fashion (with $\mathcal{W}_\infty \geq 0$ always) as $T \to \infty$. To see how this comes about, we observe first that $\mathcal{W}_y(f)_T$ is a random variable when considered over the ensemble of representations $\mathcal{W}_y^{(j)}(f)_T$ corresponding to $y_T(t)$. Similarly, $\mathcal{W}_y(f)_\infty$ is also a random variable corresponding to the ensemble $y(t)$ in the limit. It can be shown† that $\mathcal{W}_y(f)_T$, and $\mathcal{W}_y(f)_\infty$ in the limit for all f in some finite interval where $\mathcal{W}_y(f)_T > 0$, possess the distribution densities

$$W_1(\mathcal{W}_T) \begin{cases} \doteq \dfrac{e^{-\mathcal{W}_T/\overline{\mathcal{W}}_T}}{T} & \mathcal{W}_T > 0 \\ = 0 & \mathcal{W}_T < 0,\ \text{any } fT \gg 1 \end{cases} \qquad (3.35a)$$

$$W_1(\mathcal{W}_\infty) = \begin{cases} \dfrac{e^{-\mathcal{W}_\infty/\overline{\mathcal{W}}_\infty}}{\overline{\mathcal{W}}_\infty} & \mathcal{W}_\infty > 0 \\ 0 & \mathcal{W}_\infty < 0,\ \text{any } f \geq 0 \end{cases} \qquad (3.35b)$$

† See Sec. 17.2–3.

Consequently, the probability that $\mathcal{W}_\infty$ exceeds a value C is found at once to be $e^{-C/\bar{\mathcal{W}}_\infty}$, and provided that $\bar{\mathcal{W}}_\infty$ is finite and greater than zero,† this vanishes as $C \to \infty$, so that $\mathcal{W}_\infty^{(j)}$ also cannot assume an infinite value with finite probability. On the other hand, since $\mathcal{W}_\infty$ has a finite nonzero variance‡ inasmuch as $\overline{\mathcal{W}_\infty^2} - \bar{\mathcal{W}}_\infty^2 = \bar{\mathcal{W}}_\infty^2$ (> 0) from Eq. (3.35*b*), the distribution density $W_1(\mathcal{W}_\infty)$ cannot be represented by $\delta[\mathcal{W}_\infty - \mathcal{W}_\infty^{(j)} (= D)]$. Thus, $\mathcal{W}_\infty^{(j)}$ ($= \lim_{T\to\infty} \mathcal{W}_T^{(j)}$) cannot have a unique value D (> 0) in the limit with probability 1. In fact, the probability that $\mathcal{W}_\infty^{(j)}$ takes some *definite* value D is zero. An immediate consequence is that *we cannot exchange the order of the limit and integration in Eq.* (3.34*c*). Much more important, this result indicates that *the spectral intensity density* $\mathcal{W}_\infty$, in order to be suitably defined in the limit $T \to \infty$, *must be expressed as a property of the ensemble as a whole:* as we shall see presently, in this instance it is in terms of an appropriate ensemble *average* that the concept of "spectral density" can be generalized§ to include the steady-state aperiodic random disturbances encountered so often in statistical communication theory.

3.2-2 The Intensity Spectrum and the Wiener-Khintchine Theorem. Let us accordingly consider an ensemble $y_T(t)$ of real, entirely random noise waves which is obtained from the ensemble y ($-\infty < t < \infty$), with $\boldsymbol{E}\{y^2\} < \infty$ for every finite interval $\{t\}$, by truncation, so that y_T vanishes everywhere outside $(-T/2, T/2)$, in the manner of Fig. 3.9*b*. Then let us define in the mean-square sense (cf. Sec. 2.1) an amplitude spectral density

$$S_y(f)_T = \int_{-T/2}^{T/2} y(t)e^{-i\omega t}\,dt = \int_{-\infty}^{\infty} y_T(t)e^{-i\omega t}\,dt = \mathfrak{F}\{y_T\} \qquad \omega = 2\pi f \quad (3.36a)$$

with the usual inverse relation

$$y_T(t) = \mathfrak{F}^{-1}\{S_y(f)_T\} = \int_{-\infty}^{\infty} S_y(f)_T e^{i\omega t}\,df \quad (3.36b)$$

Definition mean-square means, of course, that, with a characteristic loss of phase information as to the structure of y,

$$\mathcal{W}_y(f)_T \equiv \boldsymbol{E}\left\{\frac{2}{T} S_y(f)_T S_y(f)_T^*\right\} = \frac{2}{T}\,\overline{|\mathfrak{F}\{y_T\}|^2} \quad (3.37)$$

† That $\bar{\mathcal{W}}_\infty$ exceeds zero is ensured by our choice of frequency f; that this is possible follows from $P_y^{(j)} > 0$. The finiteness of $\bar{\mathcal{W}}_\infty$, alternatively, is a direct result of the initial assumption that $P_y^{(j)} < \infty$ and therefore that $\overline{P_y^{(j)}} < \infty$. Note also that, if $\mathcal{W}_\infty \to \infty$, $\lim_{T\to\infty} \mathcal{W}_T^{(j)}$ must also be unlimited in magnitude, as is $P^{(j)}(T)$ [Eqs. (3.34)] under these same circumstances.

‡ See W. B. Davenport, Jr., and W. L. Root, "An Introduction to the Theory of Random Signals and Noise," sec. 6.6, McGraw-Hill, New York, 1958.

§ By a *different* limiting process it is also possible to define a spectral density *individually* (i.e., each j) for almost all members of the ensemble, but this cannot be achieved as the limit of Eq. (3.32) directly [cf. Sec. 3.2-3(3)].

exists [cf. Eq. (3.32)], or explicitly from Eq. (3.36a) that

$$\mathfrak{W}_y(f)_T = \frac{2}{T} \iint_{-T/2}^{T/2} \overline{y(t_1)y(t_2)} e^{-i\omega(t_1-t_2)}\, dt_1\, dt_2$$
$$= \frac{2}{T} \iint_{-T/2}^{T/2} K_y(t_1,t_2) e^{-i\omega(t_1-t_2)}\, dt_1\, dt_2 \tag{3.38}$$

is finite and positive (or zero) with K_y the covariance function (cf. Table 1.1) and $\bar{y} = 0$. The Fourier transform of the intensity density $\mathfrak{W}_y(f)_T$, as defined over the ensemble, is

$$\int_{-\infty}^{\infty} \mathfrak{W}_y(f)_T e^{i\omega t}\, df = \frac{2}{T} \iint_{-T/2}^{T/2} K_y(t_1,t_2)\, dt_1\, dt_2 \int_{-\infty}^{\infty} e^{-i\omega(t_1-t_2-t)}\, df$$
$$= \frac{2}{T} \int_{-T/2}^{T/2} K_y(t_1, t_1 - t)\, dt_1 \tag{3.39}$$

since $\int_{-\infty}^{\infty} e^{i\omega(t_2-t_1+t)}\, df = \delta(t_2 - t_1 + t)$ [cf. Eq. (1.16a)].

At this point, let us require that $y(t)$ be wide-sense stationary (cf. Sec. 1.3-6), so that $K_y(t_1,t_2) = K_y(t_2 - t_1) = K_y(t_1 - t_2)$. Then Eq. (3.39) becomes

$$\mathfrak{W}_y(f)_T = \frac{2}{T} \iint_{-T/2}^{T/2} K_y(t_1 - t_2) e^{-i\omega(t_1-t_2)}\, dt_1\, dt_2$$
$$= 2 \int_{-T}^{T} K_y(x)_T e^{-i\omega x}\, dx \geq 0 \qquad K_y(x)_T \equiv K_y(x)\left(1 - \frac{|x|}{T}\right) \tag{3.40}$$

Now if $K_y(x)$ is continuous, it can be shown† by the following argument that the right-hand member of Eq. (3.40) approaches a definite limit as $T \to \infty$, which is

$$\lim_{T\to\infty} \mathfrak{W}_y(f)_T = 2 \int_{-\infty}^{\infty} K_y(x) e^{-i\omega x}\, dx = 2\mathfrak{F}\{K_y\} \equiv \mathfrak{W}_y(f)_\infty \tag{3.41}$$

or $\mathfrak{W}_y(f)$, as we shall henceforth write it.

To see this, we start with the nonnegative form (3.40) and multiply it by $(1 - |\omega|/\omega_L)e^{i\omega t}$, after which we integrate with respect to ω over the region $(-\omega_L \leq \omega \leq \omega_L)$, to get

$$\int_{-\omega_L/2\pi}^{\omega_L/2\pi} \left(1 - \frac{|\omega|}{\omega_L}\right) e^{i\omega t} \mathfrak{W}_y(f)_T\, df = 2 \int_{-\infty}^{\infty} K_y(x)_T \left\{\frac{\sin[\omega_L(x-t)/2]}{\omega_L(x-t)/2}\right\}^2 \omega_L\, dx \tag{3.41a}$$

† The proof is due to Loève.[5] For a comprehensive discussion of the analytical questions involved, see, for example, Bartlett,[15] secs. 6.1, 6.11; J. L. Doob, "Stochastic Processes," chap. XI, secs. 3–5, Wiley, New York, 1953; and also Blanc-Lapierre and Fortet,[10] chaps. IX, X.

The coefficient of $e^{i\omega t}$ in the left integrand is not negative, and so the integral itself is proportional to a characteristic function (i.e., the integrand is proportional to a proper probability density). By the continuity theorem of Lévy,[6] the right-hand side converges uniformly in every finite x interval to $K_y(x)_T$ as $\omega_L \to \infty$ [provided that $K_y(x)_T$ is continuous, as assumed above]. Thus, the $K_y(x)_T$ for which $K_y(0)_T = K_y(0)$ (> 0) is, except for a scale factor $K_y(0)$, interpretable as a characteristic function. Also from the Lévy continuity theorem it follows that, as $T \to \infty$ in every finite interval of x, $K_y(x)_T$ is also a characteristic function [when divided by $K_y(0)$]. Inasmuch as $\lim_{T\to\infty} K_y(x)_T = K_y(x)$, Eq. (3.41) accordingly follows, and the theorem is proved. Thus, observing that

$$K_y(t) = K_y(-t) \qquad \mathcal{W}_y(f) = \mathcal{W}_y(-f)$$

and then taking the Fourier transform ($\mathfrak{F}^{-1}$) of Eq. (3.41), we have finally the relations first obtained by Wiener[1] and Khintchine,[2]

$$\begin{aligned} \mathcal{W}_y(f) &= 2\mathfrak{F}\{K_y\} = 4\int_0^\infty K_y(t)\cos\omega t\,dt \qquad \omega = 2\pi f \\ K_y(t) &= \tfrac{1}{2}\mathfrak{F}^{-1}\{\mathcal{W}_y\} = \int_0^\infty \mathcal{W}_y(f)\cos\omega t\,df \end{aligned} \tag{3.42}$$

That is, the intensity spectrum and the covariance function of a wide-sense stationary, entirely random (real) process are each other's cosine Fourier transforms.† Note in addition that $K_y(0) = \overline{y^2}$ and that $K_y(\pm\infty) = \bar{y}^2$ [cf. Eqs. (1.122), (1.123)]; here $K_y(\pm\infty) = 0$, since the process is assumed to be entirely random, that is, $\boldsymbol{E}\{y\} = 0$. The relations (3.42) are also sometimes collectively called the Wiener-Khintchine (or W-K) theorem for random processes. We observe from Eqs. (3.37) and (3.40) that the spectral intensity density here is properly defined in terms of the ensemble average

$$\mathcal{W}_y(f) = \lim_{T\to\infty} \mathcal{W}_y(f)_T \equiv \lim_{T\to\infty} \overline{\mathcal{W}_y^{(j)}(f)_T} = \lim_{T\to\infty} \frac{2\overline{|S_y^{(j)}(f)_T|^2}}{T} \tag{3.43}$$

where this average must be carried out *before* the limit is taken.

As we shall note presently in Sec. 3.2-4, the Wiener-Khintchine theorem can be directly extended to include deterministic and mixed processes as well, and the spectral intensity density $\mathcal{W}_y(f) = \lim_{T\to\infty} \boldsymbol{E}\{\mathcal{W}_y^{(j)}(f)_T\}$ can still be interpreted as an average over the ensemble. Note, too, that in the derivation of the Wiener-Khintchine theorem $y(t)$ may be the result of a linear or nonlinear operation, with or without memory, provided only that the process is wide-sense stationary with a continuous covariance function [cf. Eqs. (3.40) et seq.].

† Of course, it is also possible *to define* an intensity density $\mathcal{W}_y(f)$ as $2\mathfrak{F}\{K_y\}$, but this does not relate it in any clear fashion to what is observed physically, or to the amplitude density $S_y^{(j)}(f)_T$ in the limit. [See also the remarks in Sec. 3.2-3(3).]

The foregoing is easily extended to complex processes y. We define (cf. Sec. 1.2-8)

$$K_y(t) \equiv \boldsymbol{E}\{y(t_1)y(t_2)^*\} = K_y(-t)^* \qquad t = t_2 - t_1,\ \boldsymbol{E}\{y\} = 0 \quad (3.44)$$

The covariance function is complex, except at $t = 0$, where, of course, $K_y(0) = \overline{yy^*}$ is real. It can be shown by precisely the same argument [Eqs. (3.36), (3.40)] that the W-K theorem (3.42) becomes in this instance

$$\begin{aligned} \mathcal{W}_y(f) &= 2\mathfrak{F}\{K_y\} = 2\int_{-\infty}^{\infty} K_y(t)e^{-i\omega t}\,dt \qquad \omega = 2\pi f \\ K_y(t) &= \tfrac{1}{2}\mathfrak{F}^{-1}\{\mathcal{W}_y\} = \frac{1}{2}\int_{-\infty}^{\infty} \mathcal{W}_y(f)e^{i\omega t}\,df = K_y(-t)^* \end{aligned} \quad (3.45)$$

Although $\mathcal{W}_y(f)$ is always real and equal to or greater than zero [cf. Eqs. (3.37), (3.42)], $\mathcal{W}_y(f)$ is no longer an even function of frequency since

$$\mathcal{W}_y(-f) = 2\int_{-\infty}^{\infty} K_y(t)^*e^{-i\omega t}\,dt \neq \mathcal{W}_y(f) \quad (3.46)$$

Observe for both real and complex processes y that

$$\frac{1}{2}\int_{-\infty}^{\infty} \mathcal{W}_y(f)\,df = \int_{-\infty}^{\infty} K_y(t)\delta(t-0)\,dt = K_y(0) = \overline{yy^*} < \infty \quad (3.47)$$

The mean intensity of the process is still equal to the integral of the (ensemble) spectral intensity density over all frequencies, as expected.

As we remarked earlier [cf. Eqs. (3.40) et seq.], the continuity condition on $K_y(t)$ (all t) is the necessary and sufficient condition for the existence of an intensity density which must satisfy $\mathcal{W}_y(f) \geq 0$, $\mathcal{W}_y(\pm\infty) = 0$, and $\int_{-\infty}^{\infty} \mathcal{W}_y(f)\,df < \infty$ (all f). Conversely, the existence of $\mathfrak{F}^{-1}\{\mathcal{W}_y\}$, when $\mathcal{W}_y$ has these properties, is necessary and sufficient for the existence of a process with specified autovariance $K_y(t)$. The assumption that $K_y(t)$ is continuous is, of course, equivalent to mean-square continuity† of $y(t)$.

These properties may also be expressed in terms of the *spectral intensity distribution*, defined according to

$$\mathcal{D}_y(f) \equiv \frac{1}{2}\int_{-\infty}^{f} \mathcal{W}_y(f)\,df \quad (3.48)$$

with the features that $\mathcal{D}_y(-\infty) = 0$, $\mathcal{D}_y(\infty) = \overline{yy^*} = K_y(0)$, and $\mathcal{D}_y(f)$ is nondecreasing as $f \to \infty$ (cf. Sec. 1.2-3). For real processes, we may make

† Bartlett, *op. cit.*, sec. 6.11; Bochner.[3] In all cases, we limit ourselves to processes for which the singular component (continuous but with zero derivative almost everywhere) is absent.

the useful alternative definition, since now $\mathcal{W}_y(f) = \mathcal{W}_y(-f)$, that

$$\mathcal{D}_y(f)_+ \equiv \int_0^f \mathcal{W}_y(f)\, df$$

and therefore (3.49)

$$\mathcal{D}_y(f) = \tfrac{1}{2}[\overline{y^2} + \mathcal{D}_y(f)_+]$$

with $\mathcal{D}_y(\infty)_+ = \overline{y^2} < \infty$, $\mathcal{D}_y(f)_+ \geq 0$ (all f), and $\mathcal{D}_y(f)_+$ also nondecreasing as $f \to \infty$. In nearly all applications, we shall deal with real processes alone, since these are the only kind that can be generated by a physical communication system to which a real input is applied, with the result that $\mathcal{D}_y(f)_+$ and $\mathcal{W}_y(f) = \mathcal{W}_y(-f)$ are the natural quantities here.

In the next few sections, we shall examine some of the consequences of the W-K theorem when applied to both stationary and ergodic processes.

3.2-3 Spectra, Correlation Functions, and Spectral Moments. In the physical world, a single member alone of the ensemble $y(t)$ is normally available. But if the process is ergodic, we can in principle use the Wiener-Khintchine theorem above to determine the ensemble spectral intensity of the process *from the appropriate time average on this single member function* [cf. Sec. 1.6 and the discussion in Sec. 3.2-3(3)].

(1) Ergodic processes. To show this, let us start with Eq. (3.36*b*) and determine first the autocorrelation function,

$$\begin{aligned} R_y{}^{(j)}(t) &\equiv \lim_{T\to\infty} \frac{1}{T} \int_{-\infty}^{\infty} dt_0\, y_T{}^{(j)}(t_0) y_T{}^{(j)}(t_0 + t) \\ &= \lim_{T\to\infty} \iint_{-\infty}^{\infty} \frac{S_y{}^{(j)}(f)_T S_y{}^{(j)}(f')_T}{T}\, df\, df' \int_{-\infty}^{\infty} e^{i\omega t_0 + i\omega'(t_0+t)}\, dt_0 \end{aligned} \qquad (3.49a)$$

Since the integral over t_0 is the delta function $\delta(f' + f)$ [cf. Eq. (1.16*a*)], we may write Eq. (3.49*a*) as

$$R_y{}^{(j)}(t) = \lim_{T\to\infty} \frac{1}{2} \int_{-\infty}^{\infty} \mathcal{W}_y{}^{(j)}(f)_T e^{i\omega t}\, df \qquad (3.49b)$$

where Eq. (3.32) has been used, along with the fact that $S_y{}^{(j)}(-f)_T = S_y{}^{(j)}(f)_T^*$ for real waves $y^{(j)}$ and that $\mathcal{W}_y{}^{(j)}$ is an even function of f.

Next, we take the ensemble average of both sides of Eq. (3.49*b*) to get [in virtue of Eqs. (3.43), (3.38) to (3.41)]†

$$\overline{R_y{}^{(j)}(t)} = \frac{1}{2} \int_{-\infty}^{\infty} \lim_{T\to\infty} \overline{\mathcal{W}_y{}^{(j)}(f)_T} e^{i\omega t}\, df = \int_0^{\infty} \mathcal{W}_y(f) \cos \omega t\, df \qquad (3.50)$$

† Since y is entirely random, $(\mathcal{W}_y{}^{(j)})_T$ is absolutely continuous, for both finite T and in the limit $T \to \infty$, while the integral is convergent in both these cases. We are then permitted to interchange the order of the limit and integration. However, for a deterministic process, as we shall see presently (cf. Sec. 3.2-4), the spectral density is not continuous, and the operations of integration and the limit $T \to \infty$ do not commute, so that an alternative approach is required.

If we observe that as long as y is wide-sense stationary,

$$\overline{R_y^{(j)}(t)} = \lim_{T\to\infty} \frac{1}{T} \int_{-T/2}^{T/2} dt_0 \, \overline{y^{(j)}(t_0) y^{(j)}(t_0 + t)}$$
$$= K_y(t) \lim_{T\to\infty} \frac{1}{T} \int_{-T/2}^{T/2} dt_0 = K_y(t) \qquad (3.51a)$$

and we obtain the second relation of Eqs. (3.42). Now, taking advantage of the ergodicity of y that is assumed here, along with the fact that $\boldsymbol{E}\{y\} = 0$ also, we have, by Eq. (1.124),

$$R_y^{(j)}(t) = K_y(t) \qquad \text{probability 1} \qquad (3.51b)$$

so that the Wiener-Khintchine theorem (3.42), in terms of the autocorrelation as distinct from the covariance function, becomes finally from Eqs. (3.50), (3.51)

$$\left.\begin{aligned} R_y^{(j)}(t) &= \tfrac{1}{2}\mathfrak{F}^{-1}\{\mathcal{W}_y\} = \int_0^\infty \mathcal{W}_y(f) \cos \omega t \, df \\ \mathcal{W}_y(f) &= 2\mathfrak{F}\{R_y^{(j)}\} = 4 \int_0^\infty R_y^{(j)}(t) \cos \omega t \, dt \end{aligned}\right\} \quad \text{probability 1} \qquad (3.52)$$

and, as expected,

$$R_y^{(j)}(0) = \langle y^{(j)2} \rangle = \overline{y^2} = K_y(0) = \int_0^\infty \mathcal{W}_y(f) \, df \qquad \text{probability 1} \qquad (3.53)$$

[cf. Eq. (3.47), with $R_y^{(j)}(\pm\infty) = \bar{y}^2 = 0$ (probability 1), corresponding to $K_y(\pm\infty) = 0$].

Thus, in extension of the familiar nonstatistical treatment of deterministic waves in the time and frequency domains (cf. Sec. 2.2), with the help of the Wiener-Khintchine theorem (3.52) we can now construct analogous relations in the case of the ensemble and its representative members. We observe again that the experimental determination of the autocorrelation function $R_y^{(j)}(t)$ can never, of course, be carried out strictly in the limit $T \to \infty$. But if T is sufficiently large compared with the highest frequency f for which $\mathcal{W}_y^{(j)}(f)_T$ is noticeably different from zero, and if $y^{(j)}$ is stable in the observation period $(0,T)$, the result is a good approximation of $R_y^{(j)}$, which when substituted into Eqs. (3.52) yields a reasonable measure of the spectral density $\mathcal{W}_y$ of the process. (The reader is referred to Secs. 16.1-1, 16.1-2 for a discussion of the errors introduced by the finite observation time.)

Let us illustrate the foregoing with some simple examples for real ergodic processes, with $\boldsymbol{E}\{y\} = 0$, representing random noise voltages, so that Eq. (3.51b) applies in the following. Figures 3.11 to 3.13 show representative intensity spectra, along with the corresponding covariance functions K_y.

(*a*) *Band-limited "white" noise.* Here†

$$\mathcal{W}_y(f) = \begin{cases} W_0 & 0 \le f < B \\ 0 & f > B \end{cases} \qquad (3.54a)$$

† Since we are dealing with real waves, $\mathcal{W}_y(f)$ being an even function of frequency, it

and therefore $$K_y(t) = \psi \frac{\sin 2\pi Bt}{2\pi Bt} \qquad \psi \equiv \overline{y^2} = BW_0 \tag{3.54b}$$

from Eqs. (3.42) or (3.52), where W_0 is the constant intensity density (amplitude²/frequency) and B is the bandwidth. Because $\mathcal{W}_y(f)$ is uniform over the band $(0,B)$, the spectrum is called "white." A similar behavior is noted in Example c when the bandwidth B is increased (W_0 fixed). In general, widening the spectrum corresponds in the time domain to a narrowing or more rapid variation of the autovariance (and autocorrelation) function.

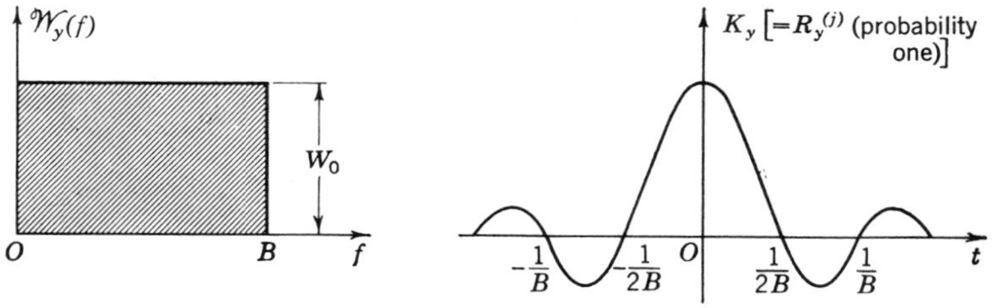

FIG. 3.11. Spectrum and autovariance $K_y(t)$ of band-limited white noise.

(b) *RC noise.* In this instance, the characteristic spectral intensity is

$$\mathcal{W}_y(f) = \frac{W_0}{1 + \omega^2/\alpha^2} \qquad \alpha > 0 \tag{3.55a}$$

and therefore

$$K_y(t) = \frac{W_0\alpha}{4} e^{-\alpha|t|} = \psi e^{-\alpha|t|} \qquad \psi = \overline{y^2} = \frac{W_0\alpha}{4} \tag{3.55b}$$

This spectrum, as we shall see presently in Sec. 3.3, results when very broadband noise is passed through a simple RC filter (cf. type 1 of Table 2.1). It is also characteristic of several types of physical noise process (see Sec. 11.1). Here, all frequencies ($f \geq 0$) contribute to varying degrees according to Eq. (3.55a): the spectrum does not terminate abruptly like that of the preceding example. Note also that, as the bandwidth ($\sim\alpha$) is increased, the covariance function falls off more rapidly as $t \to \infty$ (cf. Fig. 3.12), which we may interpret as a shortening of memory between observation times t ($= t_2 - t_1$) sec apart. The broader-band waves fluctuate more rapidly in a given interval, and hence the less is the second observation at t_2 dependent on, or predictable from, the first, at t_1.

---

is sufficient to consider positive frequencies only for the *intensity* spectrum [cf. the remarks following Eq. (3.33)]. However, both positive and negative frequencies are distinguishable in the case of *amplitude* spectra, when these exist (at most in the δ-function sense), as then phase information is preserved [cf. Eq. (2.45)].

(c) *Unlimited white noise.*† Here the spectrum is constant (W_0) for all frequencies, that is, $B \to \infty$ in case a, so that

$$\mathcal{W}_y(f) = W_0 \qquad \text{all } f \geq 0 \tag{3.56a}$$

and therefore
$$K_y(t) = W_0 \lim_{B\to\infty} \frac{\sin 2\pi Bt}{2\pi t} = \frac{W_0}{2}\,\delta(t-0) \tag{3.56b}$$

which is easily verified by inserting Eq. (3.56a) into Eqs. (3.42). This is a limiting case of considerable utility in the analysis of actual systems,

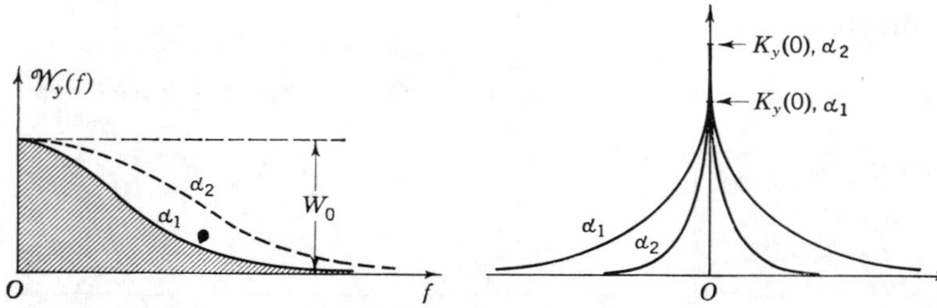

FIG. 3.12. Spectra and autovariances of RC noise, $\alpha = (RC)^{-1}$ ($\alpha_2 > \alpha_1$).

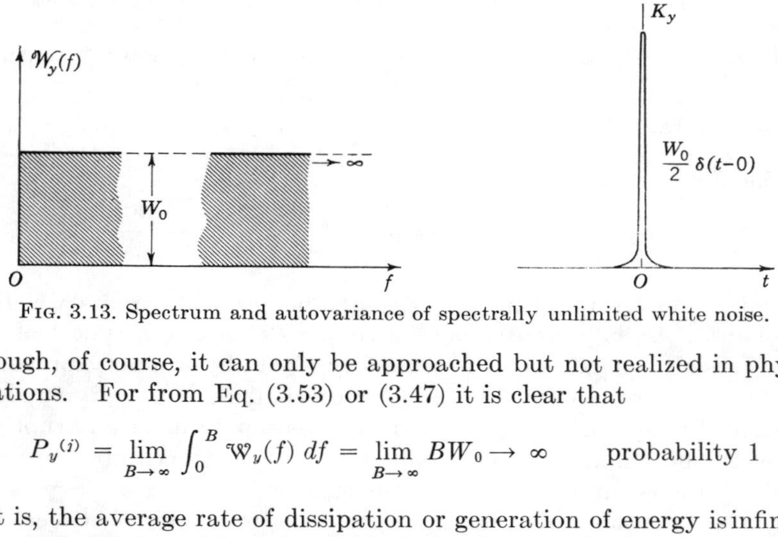

FIG. 3.13. Spectrum and autovariance of spectrally unlimited white noise.

although, of course, it can only be approached but not realized in physical situations. For from Eq. (3.53) or (3.47) it is clear that

$$P_y^{(j)} = \lim_{B\to\infty} \int_0^B \mathcal{W}_y(f)\,df = \lim_{B\to\infty} BW_0 \to \infty \qquad \text{probability 1}$$

That is, the average rate of dissipation or generation of energy is infinite—the so-called "ultraviolet catastrophe" (as $B \to \infty$),‡ which cannot occur in any physical process (with $W_0 > 0$). Spectra and covariance functions like those of Eqs. (3.56) are also characteristic of purely random processes with infinite second moment. Here there is no memory as in the above

† Henceforth, we shall simply call this white noise, reserving the term "band-limited" for the case when B is finite.

‡ See, for example, F. K. Richtmyer and E. H. Kennard, "Introduction to Modern Physics," 2d ed., sec. 88, McGraw-Hill, New York, 1942.

examples: all observations are completely uncorrelated, except of course at $t_2 = t_1$ (or $t = 0$), where the correlation is necessarily total.

(2) The intensity spectrum as a probability density. We observe in Sec. 3.2-2 that the spectral intensity density $\mathcal{W}_y(f)$ has the properties $\mathcal{W}_y(f) \geq 0, \int_{-\infty}^{\infty} \mathcal{W}_y(f)\, df = 2\overline{y^2} < \infty$ of a probability density. Equivalently, $\mathfrak{D}_y(f)$ [cf. Eq. (3.48)] has all the necessary and sufficient properties of a probability *distribution*,† apart from a normalizing factor $\frac{1}{2} \int_{-\infty}^{\infty} \mathcal{W}_y(f)\, df = K_y(0)$. Correspondingly, again apart from a scale factor $K_y(0)$, the autovariance $K_y(t)$ is mathematically identical with a *characteristic function.*‡ In fact, if we write

$$W_y(f) \equiv \mathcal{W}_y(f)/\tfrac{1}{2} \int_{-\infty}^{\infty} \mathcal{W}_y(f)\, df = \mathcal{W}_y(f)/K_y(0) \qquad (3.57)$$
$$F_f(t)_y \equiv K_y(t)/K_y(0) = k_y(t)$$

then $W_y(f)$ and F_f are, respectively, the probability density and characteristic function of a random variable f.

That the frequency f can be considered a random variable here is capable of a direct physical interpretation: $W_y(f)\, df$ is the probability that a random wave possesses an average power or intensity $\mathcal{W}_y(f)\, df$ in the frequency range $(f, f + df)$. Similarly, for real processes, $\mathfrak{D}(f)_+/K_y(0)$ [cf. Eqs. (3.49)] is the cumulative *distribution* of frequencies, and $\mathfrak{D}(f)_+$ itself is simply the average intensity of the ensemble in question for the frequency interval $(0,f)$.

From these relations (3.57), or from the W-K theorem (3.42), it is clear that not any (symmetrical) function of time will serve as a proper autocorrelation or autovariance function, since it is always necessary that they lead to a spectral density which is never negative, or to a spectral distribution $\mathfrak{D}$, or $\mathfrak{D}_+$, which is always monotonically increasing as $f \to \infty$, except for a finite number of (finite) jumps in any finite interval (f_1,f_2) (cf. Prob. 3.9).

(3) Limiting forms and the individual Wiener-Khintchine theorem; interpretation. To see in a general way how the spectral density $\mathcal{W}_y(f)$, by definition a property of the ensemble $y(t)$ as a whole, is related to what is observed in the physical process of an actual spectrum analysis, we observe first that physical analyzers are designed to give intensity (or power) measurements, appropriate to various frequency intervals $(f, f + \Delta f)$ $(0 \leq f)$. These, in turn, because of practical limitations, fluctuate in the course of time about the (infinite-time) average output powers, $P_y^{(j)}(f, f + \Delta f)$, associated with these intervals. Thus, for a given observation or data-acquisition period T and a specified *aperture*, or "resolving power," Δf (that is, a smooth-

† That $\mathcal{W}_y(f)$ and $K_y(t)$ have these interpretations follows at once from the W-K theorem and its proof. For details, see Bartlett, *op. cit.*, pp. 161–163, and for various limit and continuity properties associated with the characteristic function, see Cramér,[6] sec. 10.3.

‡ Bartlett, *op. cit.*

ing, or integration, time $\sim\Delta f^{-1}$), what is actually produced by the analyzer is a quantity with the dimensions of power that fluctuates about a limiting value $P_y^{(j)}(f, f + \Delta f)$. (We have assumed here, of course, that y is a real ensemble, for which this limit exists for almost all members.) Accordingly, for our present purpose it will be sufficient to examine the relationship between intensity spectra and this limiting value $P_y^{(j)}$ for the analyzer output. A more detailed analysis of the operation of an actual analyzer we shall reserve for Sec. 5.2-3.

The discussion is somewhat idealized, for convenience, by our assumption of a rectangular smoothing filter, of width Δf, and with infinite rejection outside the interval $(f, f + \Delta f)$ (cf. Sec. 5.2-3 also). Now, starting with a finite sample $y_T^{(j)}$, we note that if we actually perform a Fourier analysis of the observed quantity, for the interval $(-T/2, T/2)$, we obtain $S_y^{(j)}(f)_T$. Hence, we can *define*, and compute, but still not observe directly on the physical analyzer, an average intensity for the components in the frequency range $(f, f + \Delta f)$ given by [cf. Eq. (3.31)]

$$P_y^{(j)}(f, f + f)_T \equiv \mathfrak{D}_y^{(j)}(f + \Delta f)_T - \mathfrak{D}_y^{(j)}(f)_T \tag{3.58a}$$

where this spectral intensity distribution $\mathfrak{D}_y^{(j)}(f)_T$ [cf. Eqs. (3.48), (3.49)] is in turn specified by

$$\mathfrak{D}_y^{(j)}(f)_T \equiv \frac{1}{T}\int_{-\infty}^{f} |S_y^{(j)}(f)_T|^2\, df = \frac{1}{2}\int_{-\infty}^{f} \mathcal{W}_y^{(j)}(f)_T\, df \tag{3.58b}$$

Since T is finite (and $\Delta f > 0$), it is clear that these expressions for $P_y^{(j)}(f, f + \Delta f)_T$ and the spectral density $\mathcal{W}_y^{(j)}(f)_T$ are mathematically well-defined quantities, which, of course, assume different values for different members of the ensemble, as does the output of the analyzer.

Now, to see what happens in the limit $T \to \infty$, let us assume first that $y^{(j)}(t)$ is a member of a wide-sense stationary ensemble, for which $\boldsymbol{E}\{y^2\} < \infty$ and for which the autocorrelation function approaches a definite limit as $T \to \infty$, and is continuous at $t = 0$. Furthermore, let us require that y be an entirely random process. Then it can be shown[6] that

$$\lim_{T\to\infty} \mathfrak{D}_y^{(j)}(f)_T = \mathfrak{D}_y^{(j)}(f) = \int_{-\infty}^{f} \mathcal{W}_y^{(j)}(f)\, df \qquad \text{or} \qquad \mathcal{W}_y^{(j)}(f) = \frac{d}{df}\mathfrak{D}_y^{(j)} \tag{3.58c}$$

for almost all members of the ensemble; $\mathfrak{D}_y^{(j)}(f)$ is now absolutely continuous,† *so that an individual spectral density* $\mathcal{W}_y^{(j)}(f)$ *is also defined* in the limit $T \to \infty$, according to Eq. (3.58*c*), *for almost all members of the ensemble*. Ideally, then, if we were able to carry out the smoothing process for $T \to \infty$,

† $\mathfrak{D}_y^{(j)}(f)$, of course, is monotone increasing with f, and $\mathfrak{D}_y^{(j)}(-\infty) = 0$, $\mathfrak{D}_y^{(j)}(\infty) < \infty$, while $\mathcal{W}_y^{(j)}(f) \geq 0$. Since y is entirely random here, there are no discontinuities in $\mathfrak{D}_y^{(j)}(f)$. See also the footnote that refers to Eq. (3.48). Note that $\mathcal{W}_y^{(j)}(f) \neq \lim_{T\to\infty} 2|S_T^{(j)}(f)|^2/T$ [cf. Eqs. (3.32) et seq.]. These are two different limiting operations.

we would observe under these conditions that

$$\lim_{T\to\infty} [\mathfrak{D}_y^{(j)}(f+\Delta f)_T - \mathfrak{D}_y^{(j)}(f)_T]\,\Delta f^{-1} = \frac{1}{\Delta f}[\mathfrak{D}_y^{(j)}(f+\Delta f) - \mathfrak{D}_y^{(j)}(f)] \tag{3.59}$$

exists, for almost all members, and that, in addition, if we allow Δf to approach zero, i.e., effectively make the integration or smoothing time infinite, as well as the observation time T, *after* $T \to \infty$,† we get the individual *spectral density*

$$\lim_{\Delta f\to 0}\lim_{T\to\infty} \frac{\mathfrak{D}_y^{(j)}(f+\Delta f)_T - \mathfrak{D}_y^{(j)}(f)_T}{\Delta f} = \lim_{\Delta f\to 0}\frac{1}{\Delta f}\int_f^{f+\Delta f} \mathfrak{W}_y^{(j)}(f)\,df = \mathfrak{W}_y^{(j)}(f) \tag{3.60}$$

again for almost all members of the ensemble. Note, however, that we *cannot* reverse the order of the limits,‡ for from Eq. (3.58*b*)

$$\lim_{\Delta f\to 0} \frac{\mathfrak{D}_y^{(j)}(f+\Delta f)_T - \mathfrak{D}_y^{(j)}(f)_T}{\Delta f} = \mathfrak{W}_y^{(j)}(f)_T \tag{3.61}$$

and we have already seen that $\lim_{T\to\infty} \mathfrak{W}_y^{(j)}(f)_T$ does not exist (probability 1) for any member of the ensemble [Eqs. (3.35) and the subsequent argument].

The limiting individual spectral density $\mathfrak{W}_y^{(j)}(f)$, defined according to Eq. (3.58*c*), varies from member to member of the ensemble and, by the argument used to obtain Eq. (3.58*c*), can be shown[6] to be the Fourier transform of the autocorrelation function $R_y^{(j)}(t)$, which, as we have seen, also takes on different values for different members of the ensemble y. In fact, analogous to the ensemble form of the Wiener-Khintchine theorem (3.42), we can write

$$\begin{aligned} \mathfrak{W}_y^{(j)}(f) &= 2\mathfrak{F}\{R_y^{(j)}\} = 4\int_0^\infty R_y^{(j)}(t)\cos\omega t\,dt \qquad \omega = 2\pi f \\ R_y^{(j)}(t) &= \tfrac{1}{2}\mathfrak{F}^{-1}\{\mathfrak{W}_y^{(j)}\} = \int_0^\infty \mathfrak{W}_y^{(j)}(f)\cos\omega t\,dt \end{aligned} \tag{3.62}$$

which we shall henceforth call the *individual Wiener-Khintchine theorem*, to distinguish it from the Wiener-Khintchine theorem (3.42) defined over the ensemble. Note now that if the process is ergodic in addition, we may say that

$$\lim_{T\to\infty} \overline{\mathfrak{W}_y^{(j)}(f)}_T \equiv \mathfrak{W}_y(f) = \mathfrak{W}_y^{(j)}(f) \qquad \text{probability 1} \tag{3.63a}$$

with the corresponding equivalence

$$R_y^{(j)}(t) = K_y(t) \qquad \text{probability 1} \tag{3.63b}$$

in the time domain. From Eqs. (3.50), (3.51*b*), and (3.63), then, the two Wiener-Khintchine theorems (3.42), (3.62), respectively, for the ensemble

† See footnote †, p. 152.

‡ Of course, if $\Delta f \to 0$ while T is still finite, the actual output of a *physical* system must vanish, since the analyzer is then unable to accept an input in the first place.

and for almost every individual member are equivalent provided that the process is ergodic. In all instances, the order of limits and integration is crucial, since otherwise the spectral density for the truncated wave (as $T \to \infty$) has no definite limiting value.† Also, as noted above, this order has a definite physical interpretation in terms of the way a spectrum is actually measured: in any case, one does not observe $\mathcal{W}_y^{(j)}$ or $\mathcal{W}_y$, or even $(\mathcal{W}_y^{(j)})_T$, directly on an analyzer (cf. Sec. 5.2-3).

To summarize, we observe first that, for all truly physical processes, smoothing and observation times ($\sim\Delta f^{-1}$ and T, respectively) are finite, so that there is no question of the existence of a spectral density, although the latter is never observed directly by an analyzer. If $T\,\Delta f$ is at all large, and if the process from which the individual member under study is taken is also wide-sense stationary, etc., a density can be constructed from measurements which will be close to the individual limiting density $\mathcal{W}_y^{(j)}(f)$ in nearly all cases. Second, this individual limiting density, while usually variable over the ensemble, is equal (probability 1) to the ensemble spectral density $\mathcal{W}_y(f)$, provided now that y is ergodic. Thus, in the analysis, we distinguish two limiting types of spectral density, which are the results of two different limiting operations: the individual limit [Eq. (3.60)], which is not a common property of the ensemble y, and the ensemble density $\mathcal{W}_y(f)$, which is, and which, moreover, can be interpreted as a probability density. It is in terms of the latter, and the autovariance K_y, that we shall usually express the Wiener-Khintchine theorem. Also, it is in terms of the latter as a limiting form that we shall employ the concept of spectral density in theoretical applications, although, experimentally, $\mathcal{W}_y^{(j)}(f)$ and $R_y^{(j)}(t)$ are the appropriate limiting expressions.‡

(4) Spectral moments. In Sec. 2.1-2, we have considered the mean-square differentiability of a process y, which is necessarily expressed in terms of the various derivatives of the second-moment, or covariance,

† This point appears to have been rather generally overlooked in physical and engineering applications, where $\mathcal{W}_y(f)$ is often erroneously expressed as $\lim_{T\to\infty} \mathcal{W}_y^{(j)}(f)_T$, or even $E\{\lim_{T\to\infty} \mathcal{W}_y^{(j)}(f)_T\}$, rather than $\lim_{T\to\infty} E\{\mathcal{W}_y^{(j)}(f)_T\}$. However, final results for spectra are usually correct when either form (3.42) or form (3.62) of the Wiener-Khintchine theorem is used, in place of a direct calculation based on the incorrect definition $\lim_{T\to\infty} [2|S_y^{(j)}(f)_T|^2/T]$. The assumption that the process is ergodic is usually made *implicitly* (wide-sense or even strict stationarity is *not* enough), and the correlation and covariance functions are equated, with Eqs. (3.52) or (3.62) yielding the desired spectrum. No difficulties arise when the spectrum is determined experimentally by analogue means ($0 < \Delta f < \infty$ always), but if an actual calculation of $2|S_y^{(j)}(f)_T|^2/T$ by digital methods is attempted (where now Δf can be infinite—i.e., no inherent smoothing is introduced by the measuring device or process itself), it is found that, as $T \to \infty$, $2|S_y^{(j)}(f)_T|^2/T$ oscillates indefinitely (cf. our earlier remarks in Sec. 3.2-1).

‡ For a more detailed discussion of the mathematical features of the spectrum of random waves, see Kampé de Fériet,[6] especially chap. II.

function of y. If $y(t)$ is wide-sense stationary, we see from the W-K theorem (3.42), (3.45) that this implies corresponding relations in terms of the spectral density. Thus, for real waves $y(t)$, since

$$\begin{aligned} \boldsymbol{E}\{y(t_0)\dot{y}(t_0+t)\} &= \boldsymbol{E}\{\dot{y}(t_0)y(t_0-t)\} = \frac{d}{dt}\boldsymbol{E}\{y(t_0)y(t_0+t)\} \\ &= \frac{dK_y}{dt} \qquad \boldsymbol{E}\{y\} = 0 \text{ here} \end{aligned} \tag{3.64a}$$

we easily obtain

$$\boldsymbol{E}\{\dot{y}(t_0)\dot{y}(t_0+t)\} = -\frac{d}{dt}\boldsymbol{E}\{\dot{y}(t_0)y(t_0-t)\} = -\frac{d^2K_y}{dt^2} \tag{3.64b}$$

which from (3.42) becomes finally

$$\begin{aligned} K_{\dot{y}}(t) &= -\frac{d^2K_y}{dt^2} = \int_0^\infty \omega^2\mathcal{W}_y(f)\cos\omega t\,df = \tfrac{1}{2}\mathfrak{F}^{-1}\{\mathcal{W}_{\dot{y}}\} \\ \text{so that} \qquad \mathcal{W}_{\dot{y}}(f) &= \omega^2\mathcal{W}_y(f) = 2\mathfrak{F}\{K_{\dot{y}}\} \end{aligned} \tag{3.64c}$$

which together are the W-K theorem for $\dot{y}$ when y is wide-sense stationary.

In a similar way, we obtain the other derivatives of various orders:

$$\begin{aligned} K_{y\dot{y}}(t) &= -K_{\dot{y}y}(t) = \frac{dK_y}{dt} = -\frac{1}{2}\int_{-\infty}^{\infty}\omega\sin\omega t\,\mathcal{W}_y(f)\,df = 0 \\ K_{y\ddot{y}}(t) &= K_{\ddot{y}y}(t) = \frac{d^2K_y}{dt^2} = -K_{\dot{y}}(t) = -\int_0^\infty \omega^2\cos\omega t\,\mathcal{W}_y(f)\,df \\ K_{y^{(n)}}(t) &= (-1)^n\frac{d^{2n}}{dt^{2n}}K_y = \int_0^\infty \omega^{2n}\cos\omega t\,\mathcal{W}_y(f)\,df \end{aligned} \tag{3.65}$$

Analogous to the moments of a distribution obtained by differentiating the characteristic function (cf. Sec. 1.2-6), we find that

$$b_n \equiv (-i)^n\frac{d^n}{dt^n}K_y(t)\bigg|_{t=0} = (-i)^n\frac{d^n}{dt^n}\int_{-\infty}^{\infty}\tfrac{1}{2}\mathcal{W}_y(f)e^{i\omega t}\,df\bigg|_{t=0} \tag{3.66a}$$

$$= \int_{-\infty}^{\infty}\frac{\omega^n}{2}\mathcal{W}_y(f)\,df\begin{cases} = 0 & n = 1, 3, 5, \ldots \\ \neq 0 & n = 0, 2, 4, \ldots \end{cases} \quad \text{for real } y \tag{3.66b}$$

The quantities b_{2n} (for real y) are the *spectral moments* of order $2n$. These exist provided that the process y is mean-square differentiable through the nth order (cf. Slutsky[7]). Conversely, if these moments exist, the process is differentiable, order n. For instance, consider the examples (a to c) in Sec. 3.2-3(1). From the above [Eq. (3.66b)], we see at once that band-limited white noise is differentiable to all orders, while neither the RC noise (b) nor the unlimited white noise (c) are differentiable at all. In physical situations, RC noise is differentiable, however, since the spectrum [Eq. (3.55a)]

is no longer an accurate model at very high frequencies: a more rapid falling off is to be expected, from the nature of the noise mechanism itself, where quantum effects become significant (cf. Sec. 11.1-7).

We note, finally, that when the process y is also ergodic, and differentiable in the first order at least, we may obtain the W-K theorem for $\dot{y}$ in terms of $R_{\dot{y}}^{(j)}(t)$, following the procedure used in the similar situation above for $R_y^{(j)}$ [cf. Eqs. (3.49a) to (3.52) and Prob. 3.10]. Higher derivatives may be treated in the same way, provided that the process is differentiable to the appropriate order. This approach illustrates a general technique which we shall exploit in subsequent applications, such as frequency modulation by noise or by noiselike waves, and in the detection of similar disturbances (cf. Chaps. 14, 15).

3.2-4 Periodic Waves and Mixed Processes. The concept of the (ensemble) spectral intensity density $\mathcal{W}_y(f)$, defined above for entirely random processes only, can be extended with the aid of the δ-function formalism (Sec. 1.2-4) to include ensembles of deterministic waves in the steady state, namely, periodic disturbances. The notion of the intensity density is then easily broadened to include the important cases of mixed processes, characteristic of deterministic signals with random noise backgrounds.†

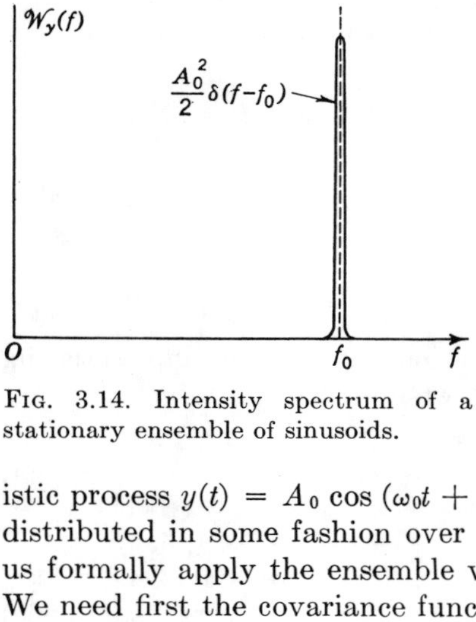

FIG. 3.14. Intensity spectrum of a stationary ensemble of sinusoids.

(1) Periodic waves. As has already been noted [cf. Eq. (2.45)], the amplitude spectral density of a periodic wave consists of pairs of "line" spectra, or delta functions, at both positive and negative frequencies, viz., at $\pm f_0$, $\pm 2f_0$, etc., where f_0 is the fundamental frequency (see Fig. 2.5a). To determine the corresponding intensity spectrum, let us begin with the simple deterministic process $y(t) = A_0 \cos(\omega_0 t + \phi)$, with $f_0\ (= \omega_0/2\pi) > 0$, and where ϕ is distributed in some fashion over the primary interval $(0,2\pi)$, and then let us formally apply the ensemble version of the W-K theorem [Eqs. (3.42)]. We need first the covariance function $K_y(t)$, which is here

$$K_y(t) = A_0^2 \overline{\cos(\omega_0 t_1 + \phi)\cos(\omega_0 t_2 + \phi)} = \frac{A_0^2}{2}\cos\omega_0 t \qquad t = t_2 - t_1,\ \omega_0 > 0 \quad (3.67)$$

provided that ϕ is uniformly distributed in $(0,2\pi)$ [which incidentally ensures that y is also strictly stationary, and ergodic if the subsets of A_0 and ω_0 have measures 1 or 0 only; cf. Eqs. (1.124), (1.125)]. Inserting

† A similar procedure can be carried out for the individual spectral density $\mathcal{W}_y^{(j)}$; however, in the subsequent discussion we shall consider only the ensemble property $\mathcal{W}_y$.

Eq. (3.67) into the first member of Eqs. (3.42), we obtain directly

$$\mathcal{W}_y(f) = \frac{A_0^2}{2} \int_{-\infty}^{\infty} [\cos(\omega + \omega_0)t + \cos(\omega - \omega_0)t]\, dt$$

$$= \frac{A_0^2}{2}\, \delta(f - f_0) \qquad f_0 > 0 \tag{3.68}$$

The term in $\delta(f + f_0)$ is discarded, since only positive frequencies f are allowed in our definition of $\mathcal{W}_y(f)$ for the real process assumed here.†

The result (3.68) agrees with what we expect: the spectral "line" at $f = f_0$ has a weighting, or total intensity, equal to $A_0^2/2$, which is the average intensity of any member function, since here $R_y^{(j)}(0) = K_y(0) = A_0^2/2$ (probability 1). This is also easily observed from

$$P_y = \int_0^\infty \mathcal{W}_y(f)\, df = \frac{A_0^2}{2} \int_0^\infty \delta(f - f_0)\, df = \frac{A_0^2}{2} \qquad f_0 > 0 \tag{3.69}$$

The spectrum is shown in Fig. 3.14. We observe, incidentally, that for d-c waves, for example, $f_0 = 0$, $M_y = A_0^2$, Eq. (3.68) becomes instead (on replacing K_y by M_y wherever $\boldsymbol{E}\{y\} \neq 0$)

$$\mathcal{W}_y(f) = 2A_0^2 \int_{-\infty}^{\infty} \cos \omega t\, dt = 2A_0^2 \delta(f - 0) \tag{3.70a}$$

and

$$\int_0^\infty \mathcal{W}_y(f)\, df = 2A_0^2 \int_0^\infty \delta(f - f_0)\, df = A_0^2 \tag{3.70b}$$

(cf. Sec. 1.2-4), once more the expected result.

We may apply the above directly to ensembles y of general periodic waves. With ϕ uniformly distributed and y ergodic, we have at once from Eq. (3.11)

$$M_y(t) = R_y^{(j)}(t) = \frac{1}{2} \sum_{n=0}^{\infty} (a_n^{(j)2} + b_n^{(j)2}) \cos n\omega_0 t$$

$$= \frac{1}{2} \sum_{n=0}^{\infty} c_n^{(j)2} \cos n\omega_0 t, \text{ etc.} \qquad \text{probability 1} \tag{3.71}$$

† As mentioned earlier in Secs. 3.2-1, 3.2-2, this corresponds physically to the fact that there is no observable distinction between positive and negative frequencies in *intensity* spectra, since all phase information (by which a sign can be assigned to frequency) is lost in the process of measuring the mean square of a quantity. Mathematically, the contribution of the negative frequencies in $\lim_{T\to\infty} \overline{|S_y^{(j)}(f)_T|^2/T}$ is "folded over" and appears as the factor 2 in the definition (3.43). One can, of course, equally well define an intensity density $G_y(f) \equiv \lim_{T\to\infty} \overline{|S_y^{(j)}(f)_T|^2/T} = \frac{1}{2}\mathcal{W}_y(f)$ for real processes, where now negative frequencies are included in parallel with the results [cf. Eqs. (2.45), (2.46), (3.28), (3.29)] for *amplitude* densities when these exist. In such cases, for example, both terms in the integrand of Eq. (3.68) are then used. Throughout the present book, however, we shall always use $\mathcal{W}_y(f)$ [Eq. (3.43)] as the defining relation for the spectral intensity, unless otherwise indicated.

With Eq. (3.70a) for the steady part and Eq. (3.68) for the oscillating part of the correlation function, Eqs. (3.42) give

$$\mathcal{W}_y(f) = a_0^2\delta(f-0) + \frac{1}{2}\sum_{n=0}^{\infty}(a_n^2+b_n^2)\delta(f-nf_0)$$

$$= c_0^2\delta(f-0) + \frac{1}{2}\sum_{n=0}^{\infty}c_n^2\delta(f-nf_0) \tag{3.72}$$

($a_n = a_n^{(j)}$ here). A typical spectral density is sketched in Fig. 3.15. Comparing Eqs. (3.71), (3.72), we observe that the coefficient of $\cos n\omega_0 t$ in the autocorrelation (or autovariance) function is simply the average

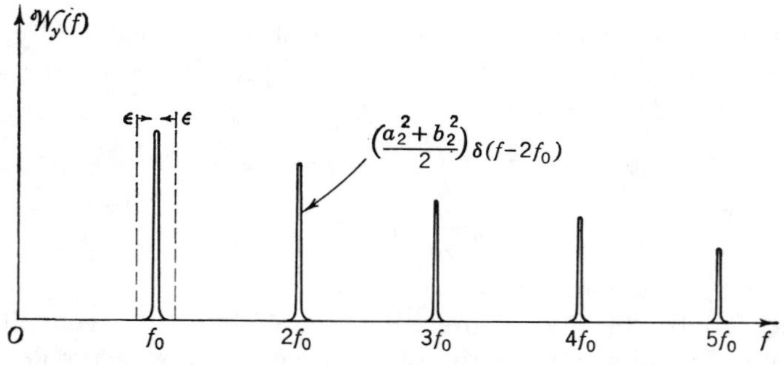

FIG. 3.15. Intensity spectrum of a general periodic ensemble.

intensity P_{nf_0} of the corresponding frequency term of the periodic wave, for, like Eqs. (3.69), (3.70b), we can write

$$(P_y)_{nf_0} \equiv \int_{nf_0-\epsilon}^{nf_0+\epsilon} \mathcal{W}_y(f)\,df = \frac{1}{2}(a_n^2+b_n^2) \qquad \epsilon > 0 \tag{3.73}$$

There remains now to relate the ensemble spectral density $\mathcal{W}_y(f)$ specifically to the amplitude spectral density $S_y^{(j)}(f)_T$ for these purely periodic processes, namely, to obtain the analogue of Eq. (3.43). This is easily accomplished if we start with the defining relation $\mathcal{W}_y^{(j)}(f)_T \equiv 2|S_y^{(j)}(f)_T|^2/T$ and recall from Eq. (3.9) that $y^{(j)}(t) = \sum_{-\infty}^{\infty} C_n^{(j)}e^{in\omega_0 t}$ in the steady state. We assume that the only random element in the $C_n^{(j)}$ are the phases $\phi^{(j)}$, so that we can write at once

$$\overline{\mathcal{W}_y^{(j)}(f)_T} = \frac{2}{T}\sum_{m,n}|C_m|\,|C_n|\overline{e^{i(\phi_m-\phi_n)}}\iint_{-T/2}^{T/2} e^{i(\omega-m\omega_0)t-i(\omega-n\omega_0)t'}\,dt\,dt' \tag{3.74}$$

Integration and division by T then gives

$$\overline{\mathcal{W}_y^{(j)}(f)_T/T} \equiv 2\overline{|S_y^{(j)}(f)_T|^2}/T^2$$
$$= 2\sum_{m,n} |C_m|\,|C_n| \frac{\sin[(\omega - m\omega_0)T/2]}{(\omega - m\omega_0)T/2} \frac{\sin[(\omega - n\omega_0)T/2]}{(\omega - n\omega_0)T/2} \overline{e^{i(\phi_m - \phi_n)}} \tag{3.75}$$

If we now take the limit $T \to \infty$, it is at once clear from this that, if $m \neq n$, the result vanishes $O(T^{-1})$ when $\omega = m\omega_0$ or $\omega = n\omega_0$, while, for all other frequencies ($\sim\omega$), the series is $O(T^{-2})$, vanishing even more rapidly.† However, when $m = n$ *and* $f = mf_0$, we get

$$\lim_{T\to\infty} [2\overline{|S_y^{(j)}(mf_0)_T|^2}/T^2] = 2|C_m|^2 = c_m^2 \tag{3.76}$$

[cf. Eqs. (3.1) to (3.10) et seq.], while for $f \neq mf_0$ the result is again zero, $O(T^{-2})$. But, from Eq. (3.72), we see that $\mathcal{W}_y(f) = C_m^2\delta(f - mf_0)$ for all frequencies in the immediate neighborhood of the particular component mf_0, so that our desired relation between the intensity and amplitude spectra becomes explicitly

$$\mathcal{W}_y(f) = \left[\lim_{T\to\infty} \frac{2\overline{|S_y^{(j)}(mf_0)_T|^2}}{T^2}\right] \delta(f - mf_0) \qquad \text{all } m = 0, 1, 2, \ldots \tag{3.77}$$

This is to be compared with Eq. (3.43) for entirely random processes.

Finally, if we use the inverse relation of the W-K theorem, starting with Eq. (3.72) as a typical intensity density, we get again $M_y(t) = R_y^{(j)}(t)$ (probability 1) [Eq. (3.71)], which is entirely consistent with our interpretation of $\mathcal{W}_y(f)$. This, along with Eqs. (3.69), (3.70*b*), and (3.73), justifies our formal use of the W-K theorem in the singular cases of the "line," or discrete, spectrum. From an alternative viewpoint, $\mathcal{W}_y(f)M_y(0)^{-1}$ may be interpreted as a probability density [cf. Eq. (3.57)], and $M_y(t)/M_y(0)$ as the associated characteristic function, where now one has a distribution of the discrete type (cf. Sec. 1.2-3), corresponding to the fact that the probability density consists of "masses," $\int_{nf_0-}^{nf_0+} \mathcal{W}_y(f)\,df\,M_y(0)^{-1}$, located at the frequencies 0, f_0, $2f_0$, etc. The proof of the W-K relations in the present discrete case is based on considerations of this type (see Sec. 1.2 and the associated references, and also Ref. 6); we shall content ourselves here simply with referring the reader to the pertinent mathematical literature for details.

† $\sum_{m,n} |C_m|\,|C_n|\overline{e^{i(\phi_m-\phi_n)}} < \sum_m |C_m|^2$ $(< \infty)$, since the average intensity of the wave is required to be finite.

(2) **Mixed processes.** Here let us consider the real ensemble $y = N + S$, where S is deterministic and N is an entirely random process;† S and N are assumed statistically independent. We can combine the results of Secs. 3.2-2, 3.2-3, and 3.2-4(1), just as has been done for a probability distribution with discrete and continuous components (cf. Sec. 1.2-4, Fig. 1.3), to write for the W-K theorem in this more general situation

$$M_y(t) = K_N(t) + M_S(t) = \int_0^\infty [\mathcal{W}_N(f) + \mathcal{W}_S(f)] \cos \omega t \, df = \tfrac{1}{2}\mathfrak{F}^{-1}\{\mathcal{W}_y\} \tag{3.78a}$$

$$\mathcal{W}_y(f) = \mathcal{W}_N(f) + \mathcal{W}_S(f) = 4\int_0^\infty [K_N(t) + M_S(t)] \cos \omega t \, dt = 2\mathfrak{F}\{M_y\} \tag{3.78b}$$

where now M_y replaces K_y, since $\boldsymbol{E}\{y\} \neq 0$ if S contains a steady component. The spectral density can be separated into discrete and continuous parts, viz.,

$$\mathcal{W}_y(f) = \mathcal{W}_y(f)_{cont} + \mathcal{W}_y(f)_{\text{discrete}} = \mathcal{W}_N(f) + \sum_{n=0}^{\infty} A_n^2 \delta(f - nf_0) \tag{3.79}$$

the former due to N alone and the latter to S ($\boldsymbol{E}\{N\} = 0$ here). A typical spectral intensity density has the distribution shown in Fig. 3.16, for

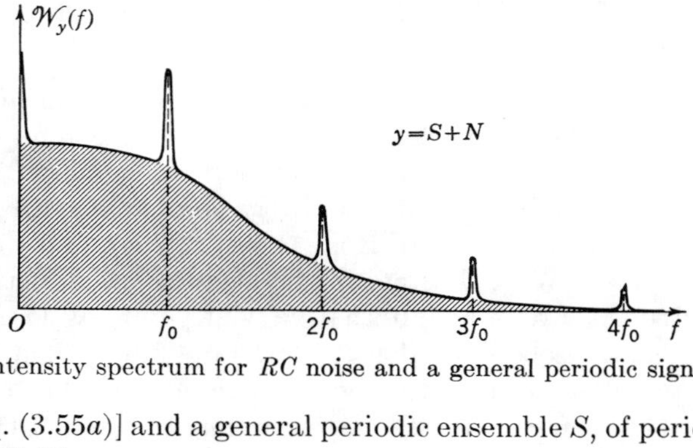

FIG. 3.16. Intensity spectrum for RC noise and a general periodic signal ensemble.

RC noise [Eq. (3.55a)] and a general periodic ensemble S, of period $T_0 = 1/f_0$. We write explicitly

$$\mathcal{W}_y(f) = \frac{W_0}{1 + \omega^2/\alpha^2} + \sum_{n=0}^{\infty} A_n^2 \delta(f - nf_0) \tag{3.80}$$

Note that here the noise makes no contribution to the steady component.

As a second example, let y be an ensemble of narrowband waves,

$$y = \text{Re}\,\{[\alpha_0(t) - i\beta_0(t)]e^{i\omega_c t + i\phi}\} = \text{Re}\,\{A_0(t)e^{i\omega_c t + i\phi - i\Phi_0(t)}\}$$

† Again it is assumed that the singular portion of the integrated operation, which is continuous but with zero derivatives almost everywhere, is absent.

[cf. Eq. (2.64)], where ϕ is uniformly distributed in $(0,2\pi)$, with $\boldsymbol{E}\{y\} = 0$, and α_0, β_0, or $A_0(t)$ and $\Phi_0(t)$ are also wide-sense stationary; A_0 and Φ_0 may be deterministic, entirely random, or mixed processes, statistically related or independent. The analogue of Eq. (3.25) becomes

$$K_y(t) = \tfrac{1}{2}\,\mathrm{Re}\,[e^{-i\omega_c t}\mathcal{K}_{\alpha_0,\beta_0}(t)]$$

where $$\mathcal{K}_{\alpha_0,\beta_0}(t) = \overline{[\alpha_0(t_1) - i\beta_0(t_1)][\alpha_0(t_2) + i\beta_0(t_2)]} \tag{3.81}$$

or $$\mathcal{K}_{\alpha_0,\beta_0}(t) = \boldsymbol{E}\{A_0(t_1)A_0(t_2)e^{-i\Phi_0(t_1)+i\Phi_0(t_2)}\}$$

and the corresponding spectral density is here

$$\mathcal{W}_y(f) = \mathfrak{F}\{\mathrm{Re}\,[e^{-i\omega_c t}\mathcal{K}_{\alpha_0,\beta_0}(t)]\} \tag{3.82}$$

A typical spectrum is sketched in Fig. 3.17; the ensemble average, of course, is over all random variables in A_0, Φ_0 above.

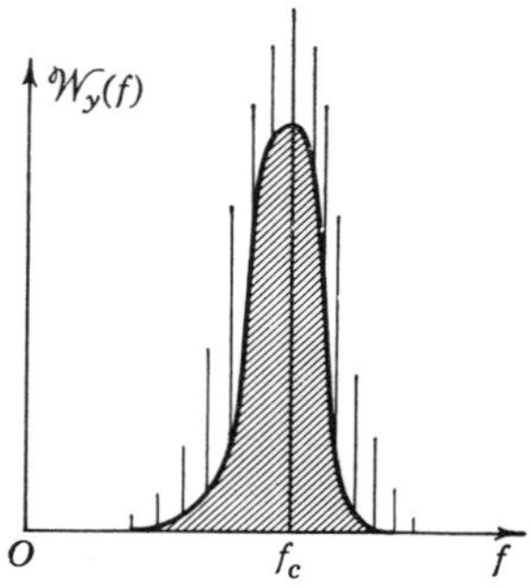

FIG. 3.17. Intensity spectrum of a narrowband mixed process.

The same argument applies to general mixed ensembles of random and deterministic processes: thus, Eq. (3.79) applies equally well here if we understand that $\mathcal{W}_{cont}$ represents the portion of the mixture that possesses a continuous spectrum, corresponding, as we have seen, to the entirely random component, and that $\mathcal{W}_{\text{discrete}}$ applies to the periodic elements in the combined process. Thus, using Eq. (3.77) and our earlier result, Eq. (3.43), we may interpret the ensemble intensity density in all situations where it is capable of definition as

$$\mathcal{W}_y(f) = \begin{cases} \lim\limits_{T\to\infty} \dfrac{2\overline{|S_y^{(j)}(f)_T|^2}}{T} & \text{for continuous part of spectrum} \\ \left[\lim\limits_{T\to\infty} \dfrac{2\overline{|S_y^{(j)}(mf_0)_T|^2}}{T^2}\right]\delta(f - mf_0) & \text{for all spectral lines at } mf_0\ (m = 0, 1, 2, \ldots) \end{cases} \tag{3.83}$$

We shall encounter many spectra of these mixed types in our subsequent treatment of signals and noise in linear and nonlinear systems (cf. Chaps. 4, 5, 12 to 17).

PROBLEMS

3.6 Derive the W-K theorem for complex, entirely random, ergodic processes z, namely,

$$R_z^{(j)}(t) = \tfrac{1}{2}\mathfrak{F}^{-1}\{\mathcal{W}_z\} = \tfrac{1}{2}\int_{-\infty}^{\infty} \mathcal{W}_z(f)e^{i\omega t}\,df$$
$$\mathcal{W}_z(f) = 2\mathfrak{F}\{R_z^{(j)}\} = 2\int_{-\infty}^{\infty} R_z^{(j)}(t)e^{-i\omega t}\,dt \qquad \text{probability 1} \quad (1)$$

where $R_z^{(j)}(t) \equiv \langle z(t_0)z(t_0 + t)^*\rangle$.

3.7 Obtain $\mathcal{W}_z(f)$ for the complex wave $z = x - i\mathcal{H}\{x\}$. What is $K_z(t)$? Work the problem for broad- and narrowband waves.

3.8 (*a*) Which of the following covariance functions are suitable for describing the second-moment properties of a real wide-sense stationary process y?

(1) $e^{-\alpha t}$.
(2) $1(-T/2 < t < T/2)$; 0 (elsewhere).
(3) $t^{2n}e^{-\beta|t|}$.
(4) $\Gamma(1 + |t|)^{-1}$.
(5) $J_0(at)e^{-\beta t^2}$.
(6) $\dfrac{\sin \alpha t}{t} e^{-\beta|t|}$.

Verify your answer in each case.

(*b*) What are sufficient conditions that a wide-sense stationary complex process possess a covariance function?

3.9 (*a*) Obtain the spectral moments of a complex process z, differentiable to the appropriate order [cf. Eqs. (3.66)].

(*b*) Prove that, if the second spectral moment b_2 is finite, then the (real) process $y(t)$ is mean-square differentiable.

(*c*) Show, in addition, for (*b*) that $y(t)$ is also a-c differentiable.

(*d*) For a real process, under suitable conditions, show that the spectral distribution $\mathfrak{D}_y(f)_+$ is

$$\mathfrak{D}_y(f)_+ = 4\int_0^\infty \frac{\sin \omega t}{t} K_y(t)\, dt \tag{1}$$

and that, in general, for a complex process and for all points of continuity of $\mathfrak{D}_z(f)$,

$$\mathfrak{D}_z(f_2) - \mathfrak{D}_z(f_1) = \lim_{T\to\infty} \int_{-T/2}^{T/2} \frac{e^{-i\omega_2 t} - e^{-i\omega_1 t}}{-2\pi i t} K_z(t)\, dt \tag{2}$$

(*e*) **Show that the *time moments* of a real process y, defined by**

$$\langle t_K{}^n\rangle \equiv \int_{-\infty}^{\infty} t^n K_y(t)\, dt \tag{3}$$

are equal to $i^n(d^n/d\omega^n)\mathcal{W}_y(\omega/2\pi)\Big]_{\omega=0}$. What are the conditions for their existence? Extend the concept to a complex process $z(t)$, and discuss the results.

3.10 (*a*) For the ergodic, entirely random (real) process $y(t)$, differentiable in the first order (mean square), show that the W-K theorem for the derived process $\dot{y}$ is

$$\left.\begin{aligned} R_{\dot{y}}{}^{(j)}(t) &= \tfrac{1}{2}\mathfrak{F}^{-1}\{\mathcal{W}_{\dot{y}}\} = \int_0^\infty \omega^2 \mathcal{W}_y(f) \cos \omega t\, df = -\frac{d^2K_y}{dt^2} \\ \mathcal{W}_{\dot{y}}(f) &= 2\mathfrak{F}\{R_{\dot{y}}{}^{(j)}\} = 4\int_0^\infty R_{\dot{y}}{}^{(j)}(t) \cos \omega t\, dt \end{aligned}\right\} \text{probability 1} \tag{1}$$

(*b*) Obtain in a similar way the W-K theorem for $\ddot{y}$ if y is now differentiable through the second order (mean square).

3.11 (*a*) Show that

$$\int_{-\infty}^{\infty} \frac{\sin(x+a)}{x+a}\frac{\sin x}{x} e^{-ibx}\, dx = \frac{\pi}{2}\frac{e^{i\gamma_+ a} - e^{i\gamma_- a}}{ia} \tag{1}$$

$$\left.\begin{aligned} \gamma_+ &= 1 \text{ or } b + 1 > -1, \text{ whichever is smaller} \\ \gamma_- &= -1 \text{ or } b - 1 < 1, \text{ whichever is larger} \end{aligned}\right\} a, b \text{ real}$$

(*b*) Prove with the aid of (*a*), for an entirely deterministic stationary process with a nonrandom structure (e.g., an ensemble of *periodic* waves, whose epochs only are random) starting from Eq. (3.49*b*), e.g.,

$$\overline{R_y^{(i)}(t)} = \lim_{T\to\infty} \frac{1}{2}\int_{-\infty}^{\infty} \overline{\mathcal{W}_y^{(i)}(f)_T}\, e^{i\omega t}\, df \tag{2}$$

that

$$M_y(t) = \sum_{-\infty}^{\infty} e^{-in\omega_0 t}|C_n|^2 \tag{3}$$

(NOTE: Since $\overline{\mathcal{W}_T}$ is not absolutely continuous, we cannot interchange the order of the limit and the integration here.) What is $\mathcal{W}_y(f)$?

3.3 *Autovariance, Autocorrelation, and Spectra of Linearly Filtered Waves*†

In the preceding section, we have established the necessary relations between the time and frequency properties of a random, deterministic, or mixed process y that is at least wide-sense stationary. Let us now examine what happens when such an ensemble is passed through a physically realizable, linear filter operating in the steady state (cf. Sec. 2.2-5). Our purpose here is to consider briefly the general second-moment properties of the new ensemble $z = \boldsymbol{T}_L\{y\}$ generated by this linear filter when the autovariance $K_y(t)$ is given and, in particular instances, the autocorrelation function $R_y^{(i)}(t)$. Attention is directed to the narrowband cases, which occur so frequently in practical communication situations. We assume real inputs (y), so that z is also a set of real time functions.

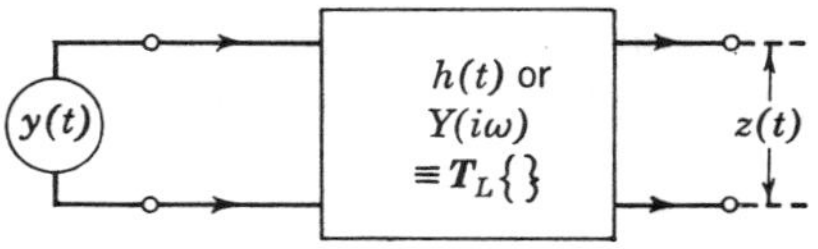

FIG. 3.18. Ensemble output of a physically realizable linear filter when the input ensemble is y; here y is white noise (cf. Fig. 3.13).

3.3-1 General (Linear) Filters (I). Let us consider a general lumped-constant stable filter‡ with weighting and system functions $h(t)$, $Y(i\omega)$, respectively (Sec. 2.2-5), and let the input be an ensemble of white noise voltage waves that are wide-sense stationary [see, for example, case *c* of Sec. 3.2-3(1)]. Then the steady-state output ensemble z (cf. Fig. 3.18) is

$$z(t) = \boldsymbol{T}_L\{y\} = \int_{-\infty}^{\infty} h(t-\tau)y(\tau)\, d\tau \tag{3.84}$$

† Problems of this type have been extensively studied by Y. W. Lee and others;[8] see the references and bibliography of Chaps. 5, 16.

‡ Unless otherwise indicated, we shall assume that these networks are ideal, in the sense that they are noise-free.

which is assumed to exist mean square. This means, of course, that

$$\boldsymbol{E}\{z(t_1)^2\} = K_z(0) = \iint\limits_{-\infty}^{\infty} h(t_1 - \tau_1)h(t_1 - \tau_2)\overline{y(\tau_1)y(\tau_2)}\, d\tau_1\, d\tau_2 < \infty \tag{3.85a}$$

or

$$\boldsymbol{E}\{z(t_1)^2\} = K_z(0) = \frac{W_0}{2}\int_{-\infty}^{\infty} h(t_1 - \tau_1)^2\, d\tau_1 = \frac{W_0}{2}\int_0^{\infty} h(x)^2\, dx < \infty \tag{3.85b}$$

where we have used $K_y(\tau_2 - \tau_1) = (W_0/2)\delta(\tau_2 - \tau_1)$ [Eq. (3.56b)]. The ensemble z is also wide-sense stationary, since $\boldsymbol{T}_L$ is invariant (cf. Secs. 2.1, 3.2). Moreover, note that the condition for the existence mean square of $z(t)$ [Eq. (3.84)] is somewhat less strict than the usual requirement of stability, that is, $\int_{-\infty}^{\infty} |h(x)|\, dx < \infty$ (Prob. 2.6b). One demands here, instead, that the average power output be finite, although in practice the latter guarantees the former.

Using Eq. (3.84) and the fact that $K_y(t) = (W_0/2)\,\delta(t - 0)$, we obtain at once the autovariance of the process $z(t)$. We have

$$K_z(t) = \boldsymbol{E}\{z(t_1)z(t_2)\} = \iint\limits_{-\infty}^{\infty} h(t_1 - \tau_1)\overline{y(\tau_1)y(\tau_2)}h(t_2 - \tau_2)\, d\tau_1\, d\tau_2 \tag{3.86a}$$

$$= \frac{W_0}{2}\int_{-\infty}^{\infty} h(x)h(x + t)\, dx \qquad t = t_2 - t_1 \tag{3.86b}$$

after the substitution $x = t_1 - \tau_1$. This suggests that we may define a function $\rho(t)$ which is analogous to the covariance or autocorrelation function of a (real) process but which is determined solely by the filter structure, according to

$$\rho(t) \equiv \int_{-\infty}^{\infty} h(x)h(x + t)\, dx = 2K_z(t)/W_0 \tag{3.87}$$

Observe that, like an autocorrelation function, $\rho(t)$ is an even function whose maximum value never exceeds that at $t = 0$. Because of its frequent appearance in the analysis of system behavior and its formal similarity to $K(t)$, $R^{(j)}(t)$, we shall call $\rho(t)$ *the autocorrelation function*† *of the filter*,[9] although its dimensions ($\text{second}^{-1} \equiv$ frequency) are different from that of a true autocorrelation function, and although it is quite independent of the input and output ensembles. The noise output of such filters we shall also refer to as *lumped-constant noise.*

Applying the W-K theorem (3.42) now to Eqs. (3.86), with Eq. (3.87) in mind, we find that the spectral intensity density of the output z is

$$\mathcal{W}_z(f) = W_0\int_{-\infty}^{\infty} \rho(t)e^{-i\omega t}\, dt = W_0|Y(i\omega)|^2 \qquad \omega = 2\pi f \tag{3.88}$$

† We could equally well call it the autovariance function of the filter.

where we have used Eq. (2.53*b*), for example, $h(t) = \mathfrak{F}^{-1}\{Y\}$, with h always real for physical networks. Thus, the originally white noise of Fig. 3.13 is "colored" and the spectrum shaped by the system function of the subsequent filter. The analogue of the W-K theorem for the ensemble is for the filter

$$|Y(i\omega)|^2 = \mathfrak{F}\{\rho\} = \int_{-\infty}^{\infty} \rho(t)e^{-i\omega t}\,dt \qquad \omega = 2\pi f \tag{3.89a}$$

$$\rho(t) = \mathfrak{F}^{-1}\{|Y|^2\} = \int_{-\infty}^{\infty} |Y(i\omega)|^2 e^{i\omega t}\,df = \int_{-\infty i}^{\infty i} Y(p)Y(-p)e^{pt}\,\frac{dp}{2\pi i}$$

or (3.89*b*)

$$\rho(t) = \mathfrak{F}^{-1}\{|Y|^2\} = \int_{-\infty}^{\infty} |Y(i\omega)|^2 e^{i\omega t}\,df = \int_{-\infty}^{\infty} h(x)h(x+t)\,dx$$

from Eqs. (3.86). That $\rho(t) = \rho(-t)$ is easily verified from Eq. (3.89*b*), so that ρ and $|Y|^2$ are also each other's Fourier cosine transforms. Note that ρ has the dimensions second^{-1}, while Y is dimensionless, since we are dealing with transformations on *voltage* ensembles that yield in turn other ensembles of voltage waves. Finally, when $Y(i\omega)$ refers to a lumped-constant network, we say in addition that the resulting spectrum $\mathcal{W}_z(f)$ [Eq. (3.88)] is a *rational*, or *lumped-constant, spectrum*, inasmuch as $|Y(i\omega)|^2$ is a rational fraction in $i\omega$ [cf. Eq. (2.49)].

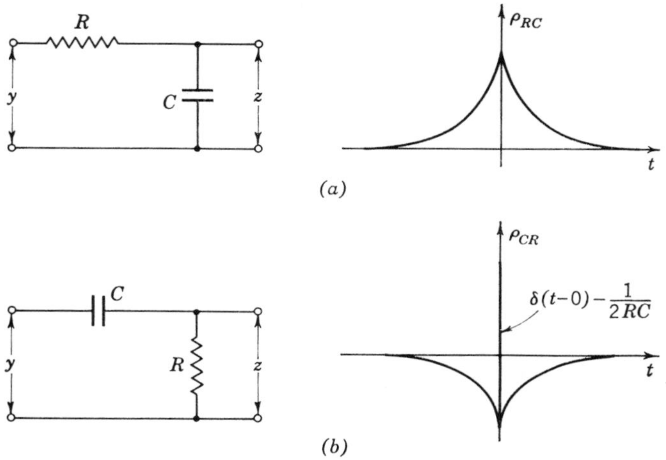

FIG. 3.19. Autocorrelation functions of (*a*) an *RC* filter, (*b*) a *CR* filter.

Let us illustrate Eqs. (3.89) with two simple examples: the *RC* and *CR* filters shown in Fig. 3.19. Starting with either the appropriate system or weighting function (cf. Table 2.1) and using Eq. (3.89*b*), we readily find that:

1. *RC* filter

$$Y(p) = (1 + RCp)^{-1}$$

Therefore,
$$\rho_{RC}(t) = \frac{1}{2RC} e^{-|t|/RC} \tag{3.90a}$$
$$h(t) = (RC)^{-1}e^{-t/RC} \qquad t > 0$$

2. *CR* filter

$$Y(p) = RCp/(1 + RCp)$$

Therefore,
$$\rho_{CR}(t) = \delta(t - 0) - \frac{1}{2RC} e^{-|t|/RC} \tag{3.90b}$$
$$h(t) = \delta(t - 0) - \frac{1}{RC} e^{-t/RC} \qquad t > 0$$

with $\rho(t)$ shown in Fig. 3.19. More complicated networks are handled in precisely similar fashion (cf. Probs. 3.12 to 3.15). Spectra and covariance functions of the output ensemble follow at once from Eqs. (3.90) in Eqs. (3.86) to (3.88) (see Table 3.1).

3.3-2 Narrowband Filters (I). When the filter representing $\boldsymbol{T}_L$ is narrowband, the output is also and we suspect by analogy with the result of Sec. 2.2-6 that the correlation and covariance functions, $\rho(t)$ and $K_z(t)$, of the filter and output ensemble likewise exhibit a characteristic envelope and phase structure. Substitution of Eqs. (2.67), (2.68) into Eqs. (3.86), with $K_y(\tau_2 - \tau_1) = (W_0/2)\delta(\tau_2 - \tau_1)$, gives

$$\begin{aligned} K_z(t) &\doteq 2W_0 \int_{-\infty}^{\infty} h_0(t_1 - \tau_1)h_0(t_2 - \tau_1) \cos [\omega_0 t_1 - \omega_0\tau_1 - \gamma_0(t_1 - \tau_1)] \\ &\qquad \times \cos [\omega_0 t_2 - \omega_0\tau_1 - \gamma_0(t_2 - \tau_1)]\, d\tau_1 \\ &= 2W_0 \int_{-\infty}^{\infty} h_0(x)h_0(x + t) \cos [\omega_0 x - \gamma_0(x)] \\ &\qquad \times \cos [\omega_0(x + t) - \gamma_0(x + t)]\, dx \end{aligned} \tag{3.91}$$

with $t_1 - \tau_1 = x$. Since h_0 and γ_0 are slowly varying as compared with $\cos \omega_0 x$, $\sin \omega_0 x$, etc., we may reduce Eq. (3.91) still further to

$$\begin{aligned} K_z(t) &\doteq W_0 \int_{-\infty}^{\infty} h_0(x)h_0(x + t) \cos [\omega_0 t + \gamma_0(x) - \gamma_0(x + t)]\, dx \\ &\equiv W_0[\rho_0(t) \cos \omega_0 t + \lambda_0(t) \sin \omega_0 t] = W_0(\rho_0^2 + \lambda_0^2)^{1/2} \\ &\qquad \times \cos [\omega_0 t - \tan^{-1} (\lambda_0/\rho_0)] \\ &= K_0(t)_z \cos [\omega_0 t - \phi_0(t)_z] \end{aligned} \tag{3.92}$$

where specifically

$$\rho_0(t) \equiv \int_{-\infty}^{\infty} h_0(x)h_0(x + t) \cos [\gamma_0(x) - \gamma_0(x + t)]\, dx = \rho_0(-t) \tag{3.93a}$$
$$\lambda_0(t) \equiv -\int_{-\infty}^{\infty} h_0(x)h_0(x + t) \sin [\gamma_0(x) - \gamma_0(x + t)]\, dx = -\lambda_0(-t) \tag{3.93b}$$

and

$$\rho(t) = 2[\rho_0(t) \cos \omega_0(t) + \lambda_0(t) \sin \omega_0 t] \tag{3.93c}$$

this last from Eq. (3.87). That ρ_0 is even and λ_0 is odd in t follows directly from Eq. (3.92) and the fact that for these real processes $K_z(t)$ is necessarily an even function of t $(= t_2 - t_1)$. Comparing Eqs. (3.92) and (3.81), we have also

$$K_z(t) = W_0 \operatorname{Re} [e^{-i\omega_0 t}\mathcal{K}_0(t)] \tag{3.93d}$$

where

$$\begin{aligned} \mathcal{K}_0 &\equiv \rho_0 + i\lambda_0 \\ &= \sqrt{\rho_0{}^2 + \lambda_0{}^2}\, e^{i \tan^{-1} (\lambda_0/\rho_0)} \\ &= K_0(t)_z e^{i\phi_0(t)_z} \end{aligned}$$

$\mathcal{K}_0$ may alternatively be expressed in terms of the envelope and phase of the narrowband ensemble z (cf. Sec. 2.2-6).

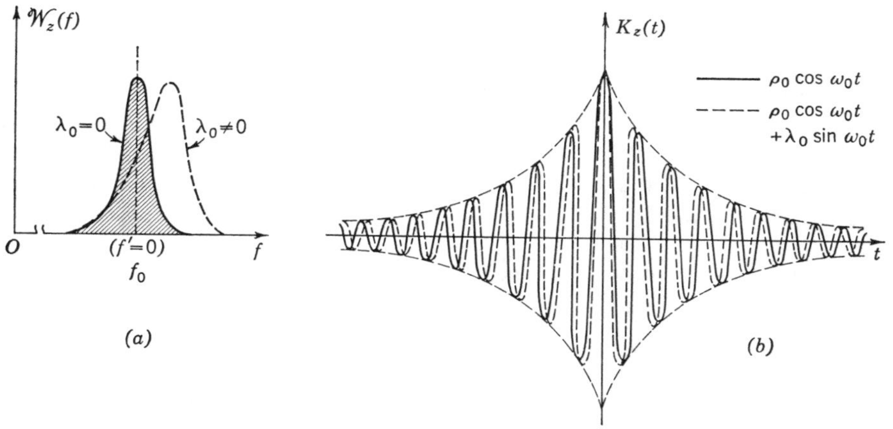

FIG. 3.20. (*a*) Symmetrical and unsymmetrical intensity spectra of a narrowband random noise ensemble. (*b*) The covariance function of the narrowband noise ensemble of Eq. (3.92).

Applying the W-K theorem to Eq. (3.92) or (3.93*d*) gives the corresponding intensity density,

$$\mathcal{W}_z(f) = 2W_0 \int_{-\infty}^{\infty} [\rho_0(t) \cos \omega_0 t + \lambda_0(t) \sin \omega_0 t] \cos \omega t \, dt$$

$$\omega = 2\pi f,\ \omega_0 = 2\pi f_0 \tag{3.94a}$$

$$\mathcal{W}_z(f) \doteq W_0 \int_{-\infty}^{\infty} [\rho_0(t) \cos \omega' t - \lambda_0(t) \sin \omega' t]\, dt \qquad \omega' = \omega - \omega_0 \tag{3.94b}$$

or

$$\mathcal{W}_z(f) = 2W_0 \mathfrak{F}\{\operatorname{Re} [e^{-i\omega_0 t}\mathcal{K}_0(t)]\} \tag{3.94c}$$

When $\lambda_0 \neq 0$, it is clear from Eq. (3.94*b*) that the spectral density is unsymmetrical: while ρ_0 is even, λ_0 is antisymmetrical. A typical spectrum and autovariance are shown in Fig. 3.20*a*, *b*. For an equivalent autocorrelation function (probability 1), obtained experimentally, see Figs. 13.4, 13.13, and 13.14.

We observe also that the total average intensity associated with the ensemble z is

$$P_z = \int_0^\infty \mathcal{W}_z(f)\,df \doteq \int_{-\infty}^{\infty} \mathcal{W}_z(f')\,df' = W_0\rho_0(0) = W_0 \int_{-\infty}^{\infty} |Y_0(i\omega')|^2\,df'$$

or

$$P_z = \int_0^\infty \mathcal{W}_z(f)\,df \doteq \int_{-\infty}^{\infty} \mathcal{W}_z(f')\,df' = W_0 \int_{-\infty}^{\infty} h_0(x)^2\,dx \tag{3.95}$$

since $\lambda_0(0) = 0$, from Eq. (3.93b). In fact, from Eqs. (3.87) and (3.95) we have also

$$\frac{W_0}{2}\rho(0) = W_0\rho_0(0) = \overline{z^2} = K_z(0) \tag{3.96}$$

If $\gamma_0(x)$ is a constant (cf. Table 2.1), then $\lambda_0(t)$ vanishes and the spectrum is symmetrical about f_0. We have

$$\rho_0(t) = \int_{-\infty}^{\infty} h_0(x)h_0(x+t)\,dx = \int_{-\infty i}^{\infty i} Y_0(p')Y_0(-p')e^{p't}\frac{dp'}{2\pi i} \qquad \lambda_0 = 0 \tag{3.97}$$

with the corresponding covariance and spectral density of the output noise ensemble given by

$$K_z(t) = W_0\rho_0(t)\cos\omega_0 t \tag{3.98a}$$

$$\mathcal{W}_z(f) = W_0 \int_{-\infty}^{\infty} \rho_0(t)\cos(\omega_0 - \omega)t\,dt \tag{3.98b}$$

as sketched in Fig. 3.20 also. Characteristically, the spectrum is confined to the immediate vicinity of f_0, while the covariance K_z oscillates rapidly, with an envelope determined by $\rho_0(t)$ (cf. Table 3.1). Since y and z are entirely random processes, $K_z(\pm\infty) = W_0\rho_0(\pm\infty) = \bar{z}^2 = 0$, as expected.

A noise process is said to be "colored" when its spectral distribution (density) is not uniform for all frequencies $f \geq 0$. We shall regard such noise henceforth (unless otherwise indicated) as derived originally from a white noise source and then passed through an appropriate filter to give it the required spectral density $\mathcal{W}$. Note that when the reference frequency f_0 ($= \omega_0/2\pi$) is not given a priori, as far as an observer of the output is concerned, the resulting envelope and phase are then uniquely specified only upon detection with the appropriate device or, mathematically, in terms of the *analytic signal* discussed in Sec. 2.2-6. If $y^{(j)}(t)$ belongs to an ergodic ensemble, so also does $z^{(j)}(t)$, since $\mathbf{T}_L$ is an invariant operator here. By the ergodic theorem (1.125), we may set $R_z^{(j)}(t) = K_z(t)$, $R_y^{(j)}(t) = K_y(t)$ (probability 1) in the usual way, after which the ensemble form of the W-K theorem may be applied for the desired intensity spectrum.

Now that the spectral intensity density has been defined (cf. Sec. 3.2), we can make more precise what we mean by "narrowband" or "broadband" noise in the steady state, where, of course, the usual definition (cf. Sec.

2.2-5) breaks down, since the amplitude spectrum does not exist for such waves. Thus, as suggested by Fig. 3.21, we call a wave narrowband if its *intensity* spectrum is narrow compared with, say, the central frequency in the band, while we say that a disturbance is broadband when the spectrum is comparable in width with the highest significant frequency component in that intensity band. The spectral width Δf is measured between the half-intensity points. The definition is clearly a loose one, depending for its utility on a reasonably wide separation of these two extremes.

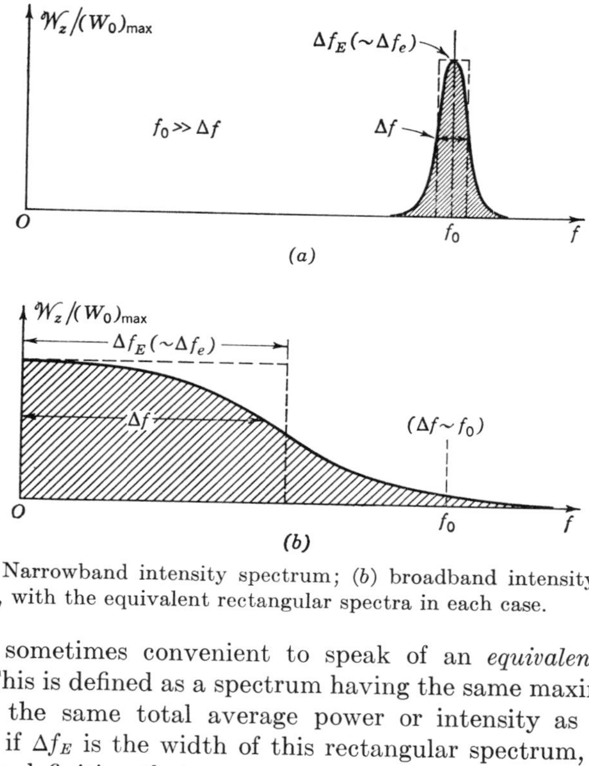

FIG. 3.21. (*a*) Narrowband intensity spectrum; (*b*) broadband intensity spectrum (or filter response), with the equivalent rectangular spectra in each case.

It is also sometimes convenient to speak of an *equivalent rectangular spectrum.* This is defined as a spectrum having the same maximum density $(W_0)_{max}$ and the same total average power or intensity as the original. Accordingly, if Δf_E is the width of this rectangular spectrum, it follows at once from the definition that

$$\Delta f_E \equiv \int_0^\infty \mathcal{W}_z(f)\, df/(W_0)_{max} \tag{3.99a}$$

An equivalent rectangular *filter response* is similarly defined: from Eqs. (3.88), (3.89) in Eq. (3.99*a*), with $(W_0)_{max}$ now set equal to W_0 (the density of the original white noise source), the filter width in this case becomes

$$\Delta f_e \equiv \int_0^\infty |Y(i\omega)|^2\, df = \frac{1}{2}\int_{0-}^\infty h(x)^2\, dx \tag{3.99b}$$

Δf, Δf_E, Δf_e are indicated in Fig. 3-21.

TABLE 3.1. OUTPUT NOISE SPECTRA AND COVARIANCE

Type	Circuit	$\rho(t) = \rho(-t) = \int_{-\infty i}^{\infty i} Y(p)Y(-p)e^{pt}\frac{dp}{2\pi i}$	$\rho(t)$
1. *RC*	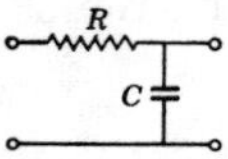	$\frac{1}{2RC}e^{-\lvert t\rvert/RC}$	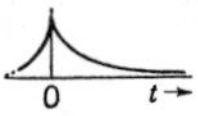
2. *CR*	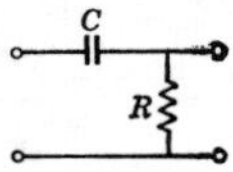	$\delta(t-0) - \frac{1}{2RC}e^{-\lvert t\rvert/RC}$	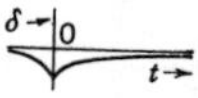
3. Series *LRC*	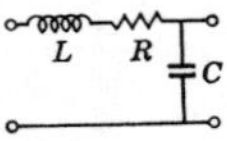	$\frac{\omega_0^2 e^{-\omega_F\lvert t\rvert}}{4\omega_F}\begin{cases}\cosh i\omega_1 t + \frac{\omega_F}{i\omega_1}\sinh i\omega_1\lvert t\rvert, & \omega_1^2<0\\ (1+\omega_F\lvert t\rvert), & \omega_1=0\\ \cos\omega_1 t + \frac{\omega_F}{\omega_1}\sin\omega_1\lvert t\rvert, & \omega_1^2>0\end{cases}$	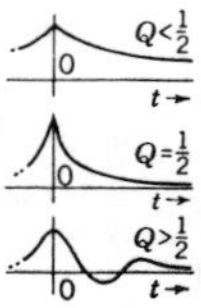
4. Series *CLR*	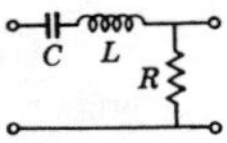	$\omega_F e^{-\omega_F\lvert t\rvert}\begin{cases}\cosh i\omega_1 t - \frac{\omega_F}{i\omega_1}\sinh i\omega_1\lvert t\rvert, & \omega_1^2<0\\ (1-\omega_F\lvert t\rvert), & \omega_1=0\\ \cos\omega_1 t - \frac{\omega_F}{\omega_1}\sin\omega_1\lvert t\rvert, & \omega_1^2>0\end{cases}$	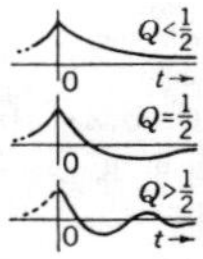
5. Series *RCL*	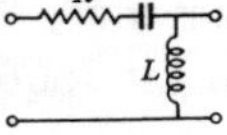	$\delta(t-0) + \omega_F e^{-\omega_F\lvert t\rvert}\begin{cases}(Q^2-1)\cosh\omega_1 it + \frac{\omega_F}{\omega_1}(Q^2+1)\sinh i\omega_1\lvert t\rvert, & \omega_1^2<0\\ (Q^2-1)+\omega_F\lvert t\rvert(Q^2+1), & \omega_1=0\\ (Q^2-1)\cos\omega_1 t + \frac{\omega_F}{\omega_1}(Q^2+1)\sin\omega_1\lvert t\rvert, & \omega_1^2>0\end{cases}$	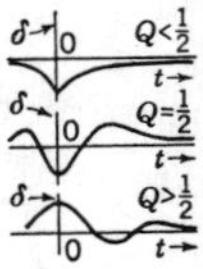
6. Gaussian	Limit of *N* cascaded on-tune filters, type 4, as $N \to \infty$	$\frac{\omega_G e^{-\omega_G^2 t^2/4}}{2\sqrt{\pi}}$	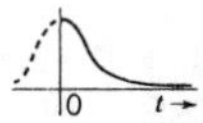
7. Rectangular	Limit of *N* stagger-tuned filters, as $N \to \infty$	$\frac{\omega_R}{\pi}\frac{\sin\omega_R t}{\omega_R t}$	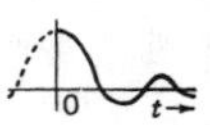

Functions for Some Simple Linear Filters

$\frac{W_0}{2}\rho(0) = W_0\rho_0(0)$			High $Q \gg 1$		
$\mathcal{W}_z(f) = W_0\lvert Y(i\omega)\rvert^2$	$\overline{z^2} = K_z(0) = \int_0^\infty \mathcal{W}_z(f)\,df$		$\rho_0(t)\cos\omega_0 t \doteq \frac{\rho(t)}{2}$		$W_0\lvert Y_0(i\omega')\rvert^2$
$\frac{W_0}{1+(RC\omega)^2}$	$\frac{W_0\omega_F}{4}$ $\omega_F = (RC)^{-1}$				
$\frac{(RC\omega)^2}{1+(RC\omega)^2}$	∞	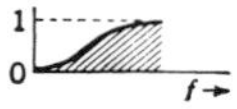			
$\frac{W_0\omega_0^4}{4\omega_F^2}\times\frac{1}{\omega^2\left[1+\left(\frac{\omega-\omega_0}{\omega_F}\right)^2\left(\frac{\omega+\omega_0}{2\omega}\right)^2\right]}$ $\omega_0^2 = (LC)^{-1}$ $\omega_F = \frac{R}{2L}$ $\omega_1 = \sqrt{\omega_0^2 - \omega_F^2}$ $Q \equiv \frac{\omega_0}{2\omega_F},\ RC = \frac{2\omega_F}{\omega_0^2}$	$\frac{W_0\omega_0^2}{8\omega_F}$	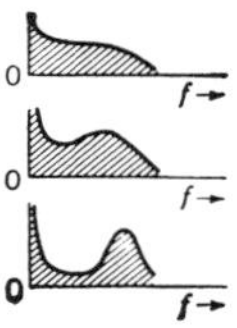	$\frac{\omega_0^2 e^{-\omega_F\lvert t\rvert}}{8\omega_F}\cos\omega_0 t$ $\rho_0 \sim 0(Q^2)$ $\lambda_0 \sim 0(Q)$	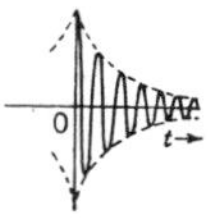	$\frac{W_0\omega_0^2/4\omega_F^2}{1+\omega'^2/\omega_F^2}$ $\omega' \equiv \omega - \omega_0$
$\frac{W_0}{1+\left(\frac{\omega-\omega_0}{\omega_F}\right)^2\left(\frac{\omega+\omega_0}{2\omega}\right)^2}$	$\frac{W_0\omega_F}{2}$	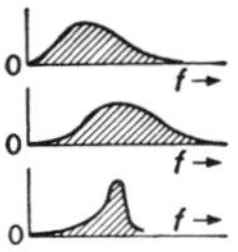	$\frac{\omega_F}{2}e^{-\omega_F\lvert t\rvert}\cos\omega_0 t$ $\rho_0 \sim 0(Q^0)$ $\lambda_0 \sim 0(Q^{-1})$	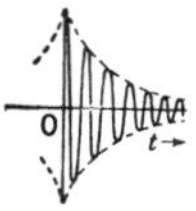	$\frac{W_0}{1+\left(\frac{\omega-\omega_0}{\omega_F}\right)^2}$
$W_0\left[1+\frac{(RC\omega)^{-2}-1}{1+\left(\frac{\omega-\omega_0}{\omega_F}\right)^2\left(\frac{\omega+\omega_0}{2\omega}\right)^2}\right]$	∞	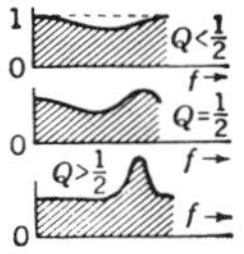	$\delta(t-0) + \frac{Q^2\omega_F e^{-\omega_F\lvert t\rvert}}{2}\cos\omega_0 t$	Similar to above plus $\delta(t-0)$	$\frac{W_0Q^2}{1+\left(\frac{\omega-\omega_0}{\omega_F}\right)^2} + W_0$
$W_0 e^{-\omega^2/\omega_G^2}$	$\frac{W_0\omega_G}{4\sqrt{\pi}}$	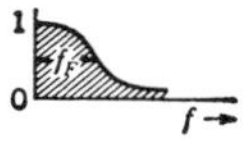	$\frac{\omega_G e^{-\omega_G^2 t^2/4}}{4\sqrt{\pi}}\cos\omega_0 t$	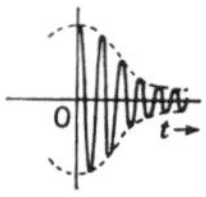	$W_0 e^{-\omega'^2/\omega_G^2}$ $\psi = \frac{W_0\omega_G}{2\sqrt{\pi}}$
$W_0,\ 0 < f < f_R\left(=\frac{\omega_R}{2\pi}\right)$ $0,\ f > f_R$	$\frac{W_0\omega_R}{2\pi}$	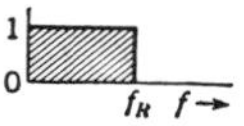	$\frac{\omega_R}{2\pi}\frac{\sin\omega_R t}{\omega_R t}\cos\omega_0 t$	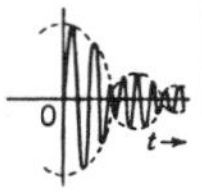	$W_0,\ f_0 - f_R < f < f_0 + f_R$ $0,\ f > f_0 + f_R$ $< f_0 - f_R$

Three simple types of narrowband (and broadband) filters provide useful models of actual noise spectra when the original source is white noise, in the manner of Fig. 3.18. Listed in Table 3.1, these are:

1. The *single-tuned response* [cf. (4) in Table 2.1]; also the characteristic form of pressure-broadened spectral lines and the response of simple dispersive media

2. The *gaussian response*, which is the limit, apart from exceptional cases, of a large number (N) of cascaded on-tune filters of type 1, as $N \to \infty$ †

3. The *rectangular filter*, mentioned above, which is the limit of a large number (N) of judiciously stagger-tuned responses of type 1, as $N \to \infty$,‡ for example, a limiting form of Butterworth filter

Although types 2 and 3 are not strictly realizable (cf. Sec. 2.2-5), they are convenient limiting expressions between which most actual system responses lie so long as we are concerned with spectral behavior not too far out on the "wings" of the spectrum. Then, of course, specific filter structure must be taken into account. If now we let ω_F, ω_G, ω_R be filter (or spectral) parameters of the (angular) frequency width of the single-tuned, gaussian, and rectangular forms above, as indicated in Table 3.1, and if these spectra are equivalent [i.e., contain the same total average intensity, with $(W_0)_{\max}$ the same in each instance], we readily find that

$$\omega_F = \frac{\omega_G}{\sqrt{\pi}} = \frac{2\omega_R}{\pi} \tag{3.100}$$

Included in Table 3.1 are both narrow- and broadband versions of types 1 to 3, as well as a number of other simple responses.

3.3-3 General Linear Filters [II]. Let us now take the further step of considering an ensemble $y(t)$, of non-"white," or colored, noise voltage waves, which is again at least wide-sense stationary and whose autovariance K_y is specified. This is a case of considerable practical importance, since broad- and narrowband waves, and particularly the latter, occur commonly in the various stages of telecommunication receivers and other electronic devices.

Next, we let this ensemble be the input to a general linear passive (stable) filter, represented by the transformation $\mathbf{T}_L$, and ask for the second-moment properties, i.e., the covariance and spectral density functions of the output ensemble $z = \mathbf{T}_L\{y\}$ in steady-state operation. The situation is indicated schematically in Fig. 3.22, where the source $y_0(t)$ and the input $y = \mathbf{T}_{L_0}\{y_0\}$

† See G. E. Valley, Jr., and H. Wallman, "Vacuum Tube Amplifiers," app. A, p. 721, Massachusetts Institute of Technology Radiation Laboratory Series, vol. 18, McGraw-Hill, New York, 1948.

‡ *Ibid.* The idea is that, unless the stagger tuning is adjusted to make the initial terms of the Taylor's-series expansion of the filter response vanish (and the number vanishing must approach ∞ with N), the response will be more like the gaussian, i.e., an exponential with the first nonvanishing term in the exponent.

are assumed to be unaffected by the subsequent network T_L. Without seriously limiting ourselves, let us assume also that y_0, y, and z are all *real* ensembles and represent entirely random processes, so that $E\{y_0,y,z\} = 0$, that is, $M_y = K_y$, etc. Equations (3.84) and (3.85*a*) or (3.86*a*) are again

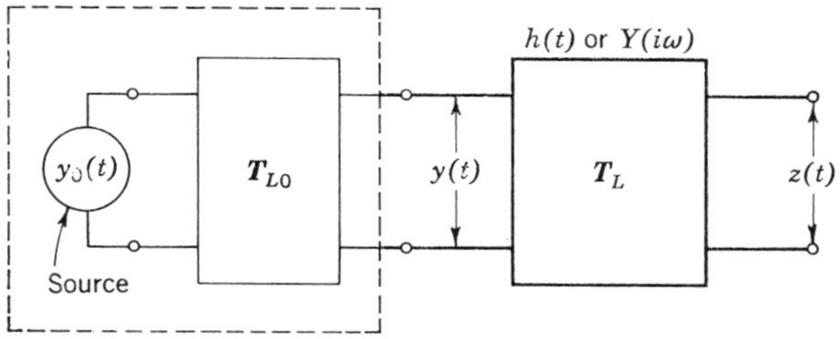

FIG. 3.22. Ensemble output of a linear, physically realizable filter when the input ensemble is colored noise.

applicable, but now K_y is no longer a delta function. We write instead of Eq. (3.86*a*)

$$K_z(t) = \overline{z(t_1)z(t_2)} = \iint_{-\infty}^{\infty} h(t_1 - \tau_1)K_y(\tau_2 - \tau_1)h(t_2 - \tau_2)\,d\tau_1\,d\tau_2$$

$$= \iint_{-\infty}^{\infty} h(x_1)K_y(t + x_1 - x_2)h(x_2)\,dx_1\,dx_2 \qquad t = t_2 - t_1 \tag{3.101}$$

with the substitution $x_1 = t_1 - \tau_1$, $x_2 = t_2 - \tau_2$.

Next, letting $x_2 = x_1 + \tau$, with x_2 and τ the new variables,† we can use the definition (3.87) and write Eq. (3.101) more compactly, in a variety of equivalent forms:

$$K_z(t) = \int_{-\infty}^{\infty} \rho(\tau)K_y(t - \tau)\,d\tau = \int_{-\infty}^{\infty} \rho(\tau)K_y(t + \tau)\,d\tau$$

or

$$K_z(t) = \int_{-\infty}^{\infty} \rho(\tau)K_y(t + \tau)\,d\tau = \int_{-\infty}^{\infty} \rho(t - \tau)K_y(\tau)\,d\tau \tag{3.102}$$

In terms of the system function of the filter, this is alternatively

$$K_z(t) = \int_{-\infty}^{\infty} K_y(\tau) \int_{-\infty i}^{\infty i} Y(p)Y(-p)e^{p(t\pm\tau)}\,\frac{dp}{2\pi i}\,d\tau \tag{3.102a}$$

since ρ and K_y are even functions of their arguments. Thus, K_z is the convolution of K_y and the filter's correlation function, a result which is directly a consequence of the fact that the input ensemble is wide-sense stationary, so that $K_y(\tau_1,\tau_2) = K_y(|\tau_2 - \tau_1|)$. The output z is also wide-sense stationary, since T_L is an invariant operation [cf. Sec. 2.1-3(2)]. If, in addition, y is ergodic, then z is also and the autocorrelation function of the

† The transformation is $x_1 = x_1$, $x_2 = x_1 + \tau$, with x_1, x_2 the old and x_2, τ the new variables, giving a jacobian of unity.

output can be expressed as

$$R_z^{(j)}(t) = \int_{-\infty}^{\infty} \rho(\tau) K_y(t \pm \tau)\, d\tau \qquad \text{probability 1} \tag{3.103a}$$

with the other forms of Eqs. (3.102) following in the same way. In fact, it can be shown that $R_z^{(j)}$ may be expressed in terms of the autocorrelation function of the input wave $y^{(j)}$ by

$$R_z^{(j)}(t) = \int_{-\infty}^{\infty} \rho(\tau) R_y^{(j)}(t \pm \tau)\, d\tau \tag{3.103b}$$

even if y (and hence z) are *not* members of wide-sense stationary ensembles. Of course, Eq. (3.103*a*) does not then apply in such cases.

The intensity spectrum associated with the ensemble z follows at once from the W-K theorem (3.42) applied to Eqs. (3.102). With the help of the relations (3.89) between $\rho(t)$ and $|Y(i\omega)|^2$, we indicate the principal steps:

$$\begin{aligned}
\mathcal{W}_z(f) = 2\mathfrak{F}\{K_z\} &= 2 \iint_{-\infty}^{\infty} \rho(\tau) K_y(t - \tau) e^{-i\omega t}\, d\tau\, dt \\
&= \iint_{-\infty}^{\infty} df'\, df''\, |Y(i\omega')|^2 \mathcal{W}_y(f'') \int_{-\infty}^{\infty} e^{i(\omega'-\omega'')\tau}\, d\tau \int_{-\infty}^{\infty} e^{i(\omega''-\omega)t}\, dt \\
&= \iint_{-\infty}^{\infty} df'\, df''\, |Y(i\omega')|^2 \mathcal{W}_y(f'') \delta(f' - f'') \delta(f'' - f) \\
&= \int_{-\infty}^{\infty} df''\, |Y(i\omega'')|^2 \mathcal{W}_y(f'') \delta(f'' - f)
\end{aligned}$$

or

$$\mathcal{W}_z(f) = |Y(i\omega)|^2 \mathcal{W}_y(f) \tag{3.104}$$

As we expect, the spectrum of the output ensemble is further modified, or shaped, by the intervening filter.† Furthermore, if the spectral density of the original source is W_0 (amperes² per frequency) and $Y_S(i\omega)$ is the system function of the network T_{L_0}, assumed linear and with output $y(t)$, we can write at once from Eq. (3.88)

$$\mathcal{W}_z(f) = W_0 |Y_S(i\omega)|^2 |Y(i\omega)|^2 \tag{3.104a}$$

and so on for any number of properly isolated linear filters in series with $\boldsymbol{T_L}$.

The relations (3.102) and (3.104), taken together, establish the connection between the spectra and autovariance of the input and output ensembles when the former is modified by a linear filter. To summarize, we write

$$\begin{aligned}
K_z(t) &= \int_{-\infty}^{\infty} \rho(\tau) K_y(t \mp \tau)\, d\tau \\
\mathcal{W}_z(f) &= |Y(i\omega)|^2 \mathcal{W}_y(f)
\end{aligned} \tag{3.105}$$

† The corresponding relation for the individual spectral density is

$$\mathcal{W}_z^{(j)}(f) = |Y(i\omega)|^2 \mathcal{W}_y^{(j)}(f)$$

For the definitions of $\mathcal{W}_{y,z}^{(j)}$ see Sec. 3.2-3(3).

The total average intensity of the output is obtained from Eqs. (3.105); as before,

$$\overline{z^2} = K_z(0) = \int_{-\infty}^{\infty} \rho(\tau)K_y(\tau)\,d\tau = \int_0^{\infty} |Y(i\omega)|^2 \mathcal{W}_y(f)\,df \qquad (3.106)$$

with $\langle z^{(j)^2}\rangle = R_z^{(j)}(0)$, etc. [cf. Eq. (3.103*b*)], for individual members of the output ensemble.

As a simple example, consider *RC* noise into an *RC* filter. One has at once for the spectrum $\mathcal{W}_z(f) = [1 + (R_0C_0\omega)^2]^{-1}[1 + (RC\omega)^2]^{-1}$, where R_0C_0 is characteristic of the source and *RC* of the filter. The resulting spectrum is narrower than that of y, so that K_z falls off to zero *less* rapidly as $t \to \infty$.

3.3-4 Narrowband Filters (II). The treatment of Sec. 3.3-2 may be extended to the important cases of narrowband filters and narrowband ensembles. We distinguish these situations as (1) broadband and (2) narrowband inputs into narrowband networks and (3) narrowband ensembles into broadband filters.

(1) Broadband inputs and narrowband filters. Examining (1) first, we use the first of Eqs. (2.93) in Eq. (3.101) to get

$$K_z(t) \doteq 2 \iint_{-\infty}^{\infty} h_0(x_1)h_0(x_1 + \tau)K_y(t - \tau) \times \cos\,[\omega_0\tau + \gamma_0(x_1) - \gamma_0(x_1 + \tau)]\,dx_1\,d\tau \qquad (3.107)$$

where the rapidly varying terms in the cosine product have been neglected. From the definitions (3.93*a*), (3.93*b*) for ρ_0 and λ_0, we see that Eq. (3.107) becomes alternatively

$$K_z(t) \doteq 2 \int_{-\infty}^{\infty} [\rho_0(\tau)\cos\omega_0\tau + \lambda_0(\tau)\sin\omega_0\tau]K_y(t - \tau)\,d\tau \qquad (3.108a)$$

This is the analogue of Eqs. (3.102).

The output is, of course, narrowband, so that K_z can also be written

$$K_z(t) = \rho_{0z}(t)\cos\omega_0 t + \lambda_{0z}(t)\sin\omega_0 t = K_0(t)_z \cos\,[\omega_0 t - \phi_0(t)_z] \qquad (3.108b)$$

where

$$K_0(t)_z = \sqrt{\rho_{0z}(t)^2 + \lambda_{0z}(t)^2} \qquad \phi_0(t)_z = \tan^{-1}(\lambda_{0z}/\rho_{0z}) \qquad (3.108c)$$

and

$$\rho_{0z}(t) \equiv 2 \int_{-\infty}^{\infty} [\rho_0(t + \tau)\cos\omega_0\tau + \lambda_0(t + \tau)\sin\omega_0\tau]K_y(\tau)\,d\tau = \rho_{0z}(-t) \qquad (3.108d)$$

$$\lambda_{0z}(t) \equiv 2 \int_{-\infty}^{\infty} [\lambda_0(t + \tau)\cos\omega_0\tau - \rho_0(t + \tau)\sin\omega_0\tau]K_y(\tau)\,d\tau = -\lambda_{0z}(-t)$$

We also note that, in terms of the spectrum, Eq. (3.108d) can be expressed alternatively as

$$\rho_{0z}(t) = \int_0^\infty \mathcal{W}_z(f) \cos(\omega - \omega_0)t \, df$$
$$\lambda_{0z}(t) = -\int_0^\infty \mathcal{W}_z(f) \sin(\omega - \omega_0)t \, df \tag{3.108e}$$

{This last is readily established with the help of the identity $\mathcal{W}_z(f) \cos \omega t = [\cos \omega_0 t \cos(\omega - \omega_0)t - \sin \omega_0 t \sin(\omega - \omega_0)t]\mathcal{W}_z(f)$ followed by integration over f and comparison with Eq. (3.108b).} Note that $\lambda_{0z}(0) = 0$, since $\lambda_0(\tau)$ is odd and $K_y(\tau)$ is even; moreover, whenever γ_0 is a constant, λ_0 vanishes. However, $\lambda_{0z}(t)$ $(t \neq 0)$ does not vanish, as we can see at once from Eq. (3.108d).

Since $z(t)$ is a narrowband ensemble, we can also represent it by

$$z(t) = \alpha_z(t) \cos(\omega_0 t + \phi) + \beta_z(t) \sin(\omega_0 t + \phi)$$
$$= \operatorname{Re}\{[\alpha_z(t) - i\beta_z(t)]e^{i\omega_0 t + i\phi}\} \tag{3.109}$$

where now α_z, β_z are entirely random, wide-sense stationary processes, not necessarily independent, and ϕ, independent of α_z, β_z, is uniformly distributed in $(0,2\pi)$. Collecting coefficients of $\cos \omega_0 t$, $\sin \omega_0 t$ in $\boldsymbol{E}\{z(t_1)z(t_2)\}$ $(t = t_2 - t_1)$, we get directly

$$K_z(t) = \frac{\overline{\alpha_{1z}\alpha_{2z}} + \overline{\beta_{1z}\beta_{2z}}}{2} \cos \omega_0 t + \frac{\overline{\alpha_{1z}\beta_{2z}} - \overline{\alpha_{2z}\beta_{1z}}}{2} \sin \omega_0 t \qquad \alpha_{1z} = \alpha_z(t_1), \text{ etc.} \tag{3.110}$$

Comparison with Eq. (3.108b) shows that

$$\rho_{0z}(t) = (\overline{\alpha_{1z}\alpha_{2z}} + \overline{\beta_{1z}\beta_{2z}})/2 \qquad \lambda_{0z}(t) = (\overline{\alpha_{1z}\beta_{2z}} - \overline{\alpha_{2z}\beta_{1z}})/2 \tag{3.111a}$$

with

$$K_z(0) = \overline{z^2} = \rho_{0z}(0) = (\overline{\alpha_z^2} + \overline{\beta_z^2})/2 \tag{3.111b}$$

In the frequency domain, we may expect corresponding extensions of the spectral relation (3.104). We indicate various equivalent forms below, leaving their verification to the reader.

$$\mathcal{W}_z(f) = 4 \iint_{-\infty}^{\infty} [\rho_0(\tau) \cos \omega_0 \tau + \lambda_0(\tau) \sin \omega_0 \tau] K_y(t - \tau) e^{-i\omega t} \, d\tau \, dt$$
$$\doteq |Y_0[i(\omega - \omega_0)]|^2 \mathcal{W}_y(f) \tag{3.112}$$

in the manner of Fig. 3.23. The over-all frequency structure is seen to be much less involved than the time representations. For the total average intensity [cf. Eq. (3.111b)], we have alternatively

$$K_z(0) = \overline{z^2} = \int_0^\infty |Y_0(i\omega - i\omega_0)|^2 \mathcal{W}_y(f) \, df$$
$$= \int_{-\omega_0}^\infty |Y_0(i\omega')|^2 \mathcal{W}_y(f' + f_0) \, df'$$
$$\doteq \int_{-\infty}^\infty |Y_0(i\omega')|^2 \mathcal{W}_y(f' + f_0) \, df' \tag{3.113}$$

since $\omega_0 \gg 2\pi \, \Delta f_{\text{filter}}$.

(2) Narrowband inputs and narrowband filters. To illustrate the modification of the preceding relations that occur when the ensemble of input disturbances possesses narrowband structure, as well as the filter $\boldsymbol{T}_L$, let us return to Eqs. (2.67) to (2.70), with ω_c now the characteristic "central"

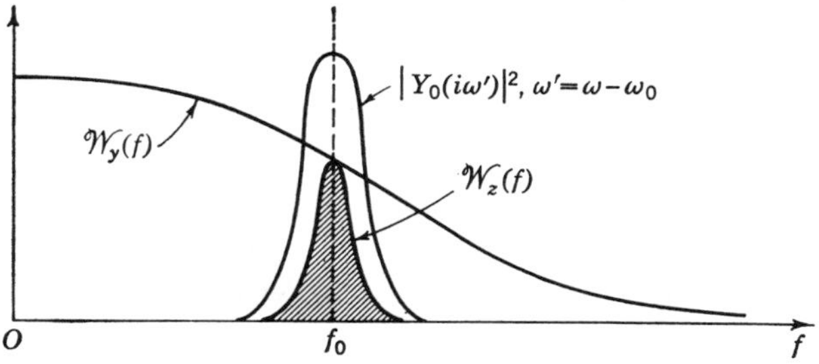

FIG. 3.23. Case 1. Spectral density of a broadband ensemble after passage through a linear narrowband filter.

(angular) frequency of the input, and ω_0 (as above) the "central," or resonant, frequency of the filter in question. Then, we can write for the input

$$y(t) = \text{Re}\,\{[\alpha_{0y}(t) - i\beta_{0y}(t)]e^{i\omega_c t + i\phi}\} \tag{3.114}$$

with ϕ, as above, uniformly distributed in $(0,2\pi)$, etc. We obtain, as for z in Eq. (3.110),

$$\begin{aligned} K_y(t) &= \frac{\overline{(\alpha_{0y})_1(\alpha_{0y})_2} + \overline{(\beta_{0y})_1(\beta_{0y})_2}}{2}\cos\omega_c t + \frac{\overline{(\alpha_{0y})_1(\beta_{0y})_2} - \overline{(\alpha_{0y})_2(\beta_{0y})_1}}{2}\sin\omega_c t \\ &= \rho_{0y}(t)\cos\omega_c t + \lambda_{0y}(t)\sin\omega_c t \end{aligned} \tag{3.115}$$

When this is substituted into Eq. (3.108*a*), we obtain the desired output covariance

$$\begin{aligned} K_z(t) &= 2\int_{-\infty}^{\infty} [\rho_0(\tau)\cos\omega_0\tau + \lambda_0(t)\sin\omega_0\tau] \\ &\qquad \times [\rho_{0y}(t-\tau)\cos\omega_c(t-\tau) + \lambda_{0y}(t-\tau)\sin\omega_c(t-\tau)]\,d\tau \\ &= \rho_{0z}(t)\cos\omega_0 t + \lambda_{0z}\sin\omega_0 t = \sqrt{\rho_{0z}{}^2 + \lambda_{0z}{}^2} \\ &\qquad \times \cos\,[\omega_0 t - \tan^{-1}(\lambda_{0z}/\rho_{0z})] \qquad \omega_D \equiv \omega_c - \omega_0 \end{aligned} \tag{3.116}$$

where now

$$\begin{aligned} \rho_{0z}(t) &= \int_{-\infty}^{\infty} \{\rho_0(\tau)[\rho_{0y}(t-\tau)\cos\omega_D(t-\tau) + \lambda_{0y}(t-\tau)\sin\omega_D(t-\tau)] \\ &\quad + \lambda_0(\tau)[\rho_{0y}(t-\tau)\sin\omega_D(t-\tau) + \lambda_{0y}(t-\tau)\cos\omega_0(t-\tau)]\}\,d\tau \end{aligned} \tag{3.117a}$$

and

$$\begin{aligned} \lambda_{0z}(t) &= \int_{-\infty}^{\infty} \{\rho_0(\tau)[-\rho_{0y}(t-\tau)\sin\omega_D(t-\tau) + \lambda_{0y}(t-\tau)\cos\omega_D(t-\tau)] \\ &\quad - \lambda_0(\tau)[\rho_{0y}(t-\tau)\cos\omega_D(t-\tau) - \lambda_{0y}(t-\tau)\sin\omega_D(t-\tau)]\}\,d\tau \end{aligned} \tag{3.117b}$$

The degree of "off-tuning," ω_D, is always small compared with ω_0, ω_c, of course. Observe that when the input and the filter are *on-tune*, that is, when $\omega_D = 0$, Eqs. (3.117a), (3.117b) simplify somewhat to

$$\rho_{0z}(t) = \text{Re} \int_{-\infty}^{\infty} [\rho_0(\tau) - i\lambda_0(\tau)][\rho_{0y}(t-\tau) + i\lambda_{0y}(t-\tau)]\, d\tau = \rho_{0z}(-t) \tag{3.117c}$$

$$\lambda_{0z}(t) = \text{Im} \int_{-\infty}^{\infty} [\rho_0(\tau) - i\lambda_0(\tau)][\rho_{0y}(t-\tau) + i\lambda_{0y}(t-\tau)]\, d\tau = -\lambda_{0z}(-t)$$

Alternatively [cf. Eq. (3.114)] the narrowband output also may be represented in terms of an envelope and phase (see Sec. 2.2-5), so that

$$z(t) = \alpha_z(t) \cos \omega_0 t + \beta_z(t) \sin \omega_0 t = \text{Re}\, [E_{0z}(t) e^{i\omega_0 t - i\Phi_{0z}(t)}] \tag{3.118}$$

with the result that we have still another equivalent representation of the covariance function K_z, now in terms of the envelope and phase, viz.,

$$\begin{aligned} K_z(t) &= \tfrac{1}{2} \text{Re}\, [\mathcal{K}_{\alpha,\beta}(t)_z e^{-i\omega_0 t}] \\ \text{with} \qquad \mathcal{K}_{\alpha,\beta}(t)_z &= \rho_{0z} + i\lambda_{0z} = \mathcal{K}^*_{\alpha,\beta}(-t)_z \\ &= \boldsymbol{E}\{E_{0z}(t_1) E_{0z}(t_2) e^{i\Phi_{0z}(t_2) - i\Phi_{0z}(t_1)}\} \\ &= \overline{(\alpha_{z1} - i\beta_{z1})(\alpha_{z2} + i\beta_{z2})} \end{aligned} \tag{3.119}$$

like Eqs. (3.93c), (3.93d).

Using Eqs. (3.93), now appropriately modified, e.g.,

$$K_y(t) = \tfrac{1}{2} \text{Re}\, [\mathcal{K}_{\alpha_0,\beta_0}(t)_y e^{-i\omega_c t}] \tag{3.120}$$

in conjunction with Eqs. (3.105), we obtain finally the following relation between input and output covariance,

$$K_z(t) = \int_{-\infty}^{\infty} \rho(t-\tau) K_y(\tau)\, d\tau \doteq \tfrac{1}{2} \text{Re} \int_{-\infty}^{\infty} [\rho_0(\tau) + i\lambda_0(\tau)] e^{-i(\omega_c t - \omega_D \tau)} \times \mathcal{K}_{\alpha_0,\beta_0}(t-\tau)_y\, d\tau$$

which becomes

$$K_z(t) \doteq \tfrac{1}{2} \text{Re}\, [e^{-i\omega_c t} \int_{-\infty}^{\infty} \mathcal{K}_0(\pm\tau) \mathcal{K}_{\alpha_0,\beta_0}(t \mp \tau)_y e^{\pm i\omega_D \tau}\, d\tau] \qquad \omega_D = \omega_c - \omega_0 \tag{3.121}$$

Comparison with Eq. (3.119) shows at once that

$$\mathcal{K}_{\alpha,\beta}(t)_z = e^{-i\omega_D t} \int_{-\infty}^{\infty} \mathcal{K}_0(\pm\tau) \mathcal{K}_{\alpha_0,\beta_0}(t \mp \tau)_y e^{\pm i\omega_D \tau}\, d\tau \tag{3.122a}$$

or

$$\mathcal{K}_{\alpha,\beta}(t)_z = \int_{-\infty}^{\infty} \mathcal{K}_0(t \mp x) \mathcal{K}_{\alpha_0,\beta_0}(\pm x)_y e^{\mp i\omega_D x}\, dx \tag{3.122b}$$

When the filter is on-tune *and* symmetrical, $\mathcal{K}_0(\tau) = \rho_0(\tau)$ and Eqs. (3.122) reduce to the simpler expression

$$\mathcal{K}_{\alpha,\beta}(t)_z = \int_{-\infty}^{\infty} \rho_0(\tau) \mathcal{K}_{\alpha_0,\beta_0}(t \mp \tau)_y\, d\tau \tag{3.122c}$$

For the corresponding intensity spectrum, we again apply the W-K theorem (3.42) to get, as sketched in Fig. 3.24,

$$\mathcal{W}_z(t) = 2\mathfrak{F}\{K_z\} \doteq \mathcal{W}_y(f) |Y_0(i\omega - i\omega_0)|^2 \doteq \mathcal{W}_y(f - f_c) |Y_0(i\omega - i\omega_0)|^2 \tag{3-123}$$

since $\mathcal{W}_y(f) \doteq \mathcal{W}_y(f - f_c)$ for the narrowband ensemble y. The total intensity of the filtered ensemble is

$$P_z \equiv K_z(0) = \tfrac{1}{2} \operatorname{Re} \int_{-\infty}^{\infty} \mathcal{K}_0(\tau)\mathcal{K}_{\alpha_0,\beta_0}(\tau)_y^* e^{i\omega_0\tau}\, d\tau$$
$$= \int_{-\infty}^{\infty} \mathcal{W}_y(f' - f_D)|Y_0(i\omega')|^2\, df' \qquad (3.124)$$

in terms of the narrowband filter's system function, centered at $f' = 0$ (cf. Fig. 3.24). In general, the frequency forms are simpler than the expression for the covariance, as is evident from Eq. (3.123) vs. Eq. (3.121). For the correlation function of a particular $z^{(j)}(t)$, we have only formally to replace $\mathcal{K}_{\alpha_0,\beta_0}$, $\mathcal{K}_{\alpha,\beta}$ by $R^{(j)}_{\alpha_0,\beta_0}R^{(j)}_{\alpha,\beta}$ in the preceding; this can be verified from the definition of $R^{(j)}$ as a time average. The ensemble spectral density $\mathcal{W}$ is likewise replaced by the individual density $\mathcal{W}^{(j)}$ [cf. Eq. (3.112)].

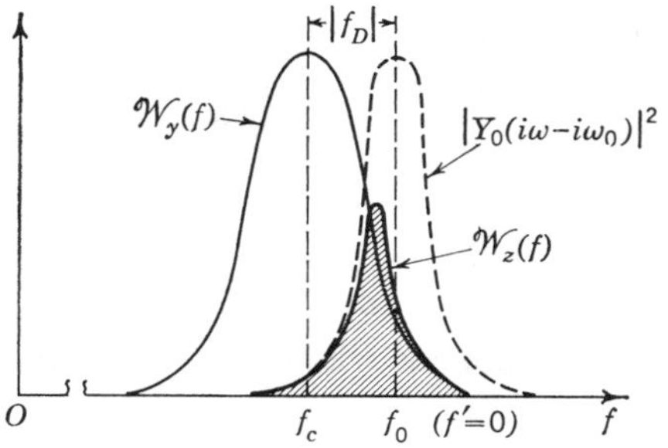

FIG. 3.24. Case 2. Spectral density of a narrowband ensemble after passage through a narrowband filter.

(3) Narrowband inputs and broadband filters. The appropriate results in this case are quickly obtained from the preceding, if we use Eqs. (3.105) and (3.106). The spectral situation is sketched in Fig. 3.25, while for the covariance function we get

$$K_z(t) \doteq \tfrac{1}{2} \operatorname{Re} \int_{-\infty}^{\infty} \rho(\tau)\mathcal{K}_{\alpha_0,\beta_0}(t - \tau)_y e^{-i\omega_c(t-\tau)}\, d\tau$$

or

$$K_z(t) = \tfrac{1}{2} \operatorname{Re} \int_{-\infty}^{\infty} \rho(t - \tau)\mathcal{K}_{\alpha_0,\beta_0}(\tau)_y e^{-i\omega_c\tau}\, d\tau \qquad (3.125)$$

with

$$\mathcal{W}_z(f) \doteq |Y(i\omega)|^2\mathcal{W}_y(f - f_c) \qquad (3.126)$$

and a total average intensity

$$P_z = K_z(0) \doteq \int_{-\infty}^{\infty} |Y(i\omega)|^2\mathcal{W}_y(f - f_c)\, df \qquad (3.127)$$

It should be remembered, of course, that our expressions above for narrowband filters and ensemble outputs are only approximate, their validity

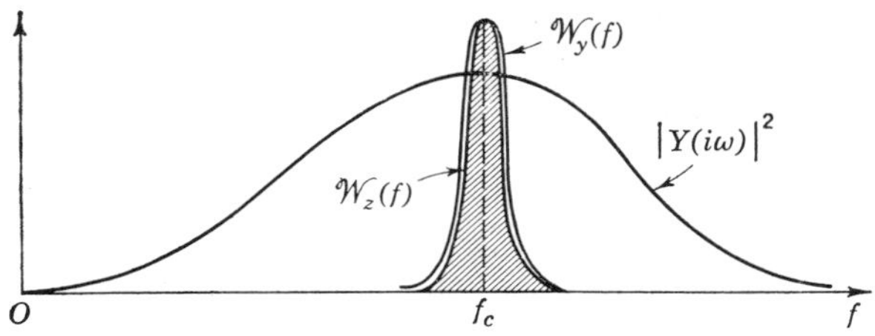

FIG. 3.25. Case 3. Spectral density of a narrowband ensemble after passage through a broadband filter.

improving as the limit of infinitely narrowband structures is approached. This does not, however, preclude their usefulness in practical applications as long as one is dealing with a reasonably narrowband phenomenon. By "reasonable," here, is meant that one may choose a ratio of equivalent filter and intensity spectral widths, that is, Δf_e, Δf_E [cf. Eqs. (3.99a), (3.99b)], to the central frequency of as little as 1:4, although in most system applications the ratio is usually O(1:100) or even smaller.

3.3-5 Approximations; Time Moments. When the input ensemble and the filter differ greatly in spectral width, e.g., when the filter is narrowband and the intensity spectrum of the input is broadband, one can often obtain useful approximate expressions for the autovariance and spectrum of the resulting output by taking advantage of these spectral differences.

To show this, it is convenient to start with the autovariance $K_z(t)$ as given by Eqs. (3.102). Then, if the filter is wide compared with the input y, that is, if $\Delta f_e \gg (\Delta f_E)_y$ [cf. Eqs. (3.99)], $K_y(t - \tau)$ varies slowly (for given t) compared with $\rho(\tau)$ for the filter, so that it is natural in this instance to expand the covariance in a Taylor's series about t, writing Eqs. (3.102) as

$$K_z(t) = \int_{-\infty}^{\infty} \rho(\tau) K_y(t - \tau)\, d\tau = \sum_{n=0}^{N} \frac{(-1)^n}{n!} K_y^{(n)}(t) \int_{-\infty}^{\infty} \rho(\tau)\tau^n\, d\tau + R_{N+1} \qquad \Delta f_e \gg (\Delta f_E)_y \qquad (3.128)$$

where R_{N+1} is the remainder $\int_{-\infty}^{\infty} (-\tau)^{N+1} K^{(N+1)}(t - \theta\tau)\rho(\tau)\, d\tau/(N + 1)!$ $(0 < \theta < 1)$.

On the other hand, if the filter is narrow compared with the input, we use an alternative expression for K_z and develop ρ instead, getting

$$K_z(t) = \int_{-\infty}^{\infty} \rho(t - \tau) K_y(\tau)\, d\tau = \sum_{m=0}^{M} \frac{(-1)^m}{m!} \rho^{(m)}(t) \int_{-\infty}^{\infty} K_y(\tau)\tau^m\, d\tau + R_{M+1} \qquad \Delta f_e \ll (\Delta f_E)_y \qquad (3.129)$$

with R_{M+1} the appropriate remainder in this case; $K_y^{(n)}$, $\rho^{(m)}$ are, respectively, the nth and mth derivatives of the covariance and filter correlation functions. Usually, the first couple of terms are a sufficiently good approximation,† and we have assumed for the moment that an expansion through the remainder term is possible.

The quantities $\langle \tau_\rho{}^n \rangle$ and $\langle \tau_K{}^m \rangle$, defined by

$$\langle \tau_\rho{}^n \rangle \equiv \int_{-\infty}^{\infty} \tau^n \rho(\tau)\, d\tau \qquad \langle \tau_K{}^m \rangle \equiv \int_{-\infty}^{\infty} \tau^m K_y(\tau)\, d\tau \qquad (3.130)$$

† As an estimate of "goodness," one might use the closeness with which the expanded version of $K_z(0)$ comes to the integral $\int_{-\infty}^{\infty} \rho(\tau) K_y(\tau)\, d\tau$, that is, a comparison of total average powers.

we call the *time moments* of the filter and of the (input) ensemble, respectively. In their spectral form, these become

$$\langle \tau_\rho{}^n \rangle = i^n \frac{d^n}{d\omega^n} \int_{-\infty}^{\infty} \rho(\tau) e^{i\omega t}\, d\tau \bigg|_{\omega=0} = i^n \frac{d^n}{d\omega^n} |Y(i\omega)|^2 \bigg|_{\omega=0} \begin{cases} = 0 & n = 1, 3, 5, \ldots \\ \neq 0 & n = 0, 2, 4, \ldots \end{cases} \tag{3.131a}$$

$$\langle \tau_K{}^m \rangle = i^m \frac{d^m}{d\omega^m} \int_{-\infty}^{\infty} K_y(\tau) e^{i\omega\tau}\, d\tau \bigg|_{\omega=0} = i^m \frac{d^m}{d\omega^m} \mathcal{W}_y\left(\frac{\omega}{2\pi}\right) \bigg|_{\omega=0} \begin{cases} = 0 & m = 1, 3, 5, \ldots \\ \neq 0 & m = 0, 2, 4, \ldots \end{cases} \tag{3.131b}$$

which exist, provided that the filter's system function and the spectral density possess derivatives at $\omega = 0$, at least for some m and $n \geq 0$.

In a similar way, we observe that the derivatives of ρ and K_y are related to the spectral moments [cf. Sec. 3.2-3(4)] according to

$$\rho^{(m)}(t) = \frac{d^m}{dt^m} \int_{-\infty}^{\infty} |Y(i\omega)|^2 e^{i\omega t}\, df = i^m \int_{-\infty}^{\infty} \omega^m |Y(i\omega)|^2 e^{i\omega t}\, df \tag{3.132a}$$

and
$$K_y{}^{(n)}(t) = \frac{1}{2} \frac{d^n}{dt^n} \int_{-\infty}^{\infty} \mathcal{W}_y(f) e^{i\omega t}\, df = i^n \int_{-\infty}^{\infty} \omega^n \mathcal{W}_y(f) e^{i\omega t}\, df \tag{3.132b}$$

with the obvious condition for the existence of these derivatives depending on the existence of the appropriate spectral moments $\rho^{(m)}(0)$, $K_y{}^{(n)}(0)$. It is clear that, for m, n an odd integer, $\rho^{(m)}$, $K_y{}^{(n)}$ vanish. Combining these in Eqs. (3.128), (3.129), we see that, for the latter expansions to be applicable, time and spectral moments of the filter and of the input must exist, at least up to some finite order m, $n > 0$. Accordingly, for comparatively wide filters $[\Delta f_e \gg (\Delta f_E)_y]$, we require the existence through order n of the filter's time moments and the corresponding spectral moments of the input ensemble. On the other hand, for comparatively broad inputs $[(\Delta f_E)_y \gg \Delta f_e]$, it is the time moments of the input and the spectral moments of the filter that are significant. Thus, for high-Q LRC noise through an RC filter with a long time constant, we can safely use the first three terms ($n = 0, 1, 2$) in the development, since spectral moments $O(n \leqq 3)$ exist for LRC noise. If the input noise is, for example, comparatively broad RC noise, through a narrow RC filter $[(\Delta f_E)_y \gg \Delta f_e]$, only the first term ($m = 0$) is possible in the development, since, for $m \geq 2$, $\rho^{(m)}(t)$ does not exist, i.e., the (even) spectral moments of $O(m \geq 2)$ of the filter are divergent. If, however, physically unrealizable filters are included, it is possible for filters and spectral moments of *all* orders to remain finite. For suppose the input noise is shaped by the gaussian filter of (6) in Table 3.1, so that $\rho(\tau)$, $K_y(\tau) \sim e^{-\alpha^2\tau^2}$; then from Eqs. (3.130) and (3.132) it is evident at once that $\rho^{(m)}$, $K_y{}^{(n)}$, $\langle \tau_K{}^{(m)} \rangle$, and $\langle \tau_\rho{}^n \rangle$ are defined for all orders m, $n \geq 0$. Incidentally, input processes of this type are also differentiable mean square to all orders. In general, the degree of expandability of the output covariance is closely related to the existence of the appropriate spectral moments of the filter and input process and is perhaps most easily determined by inspection of the associated spectral and system functions. We shall have occasion in subsequent chapters to consider this question again.

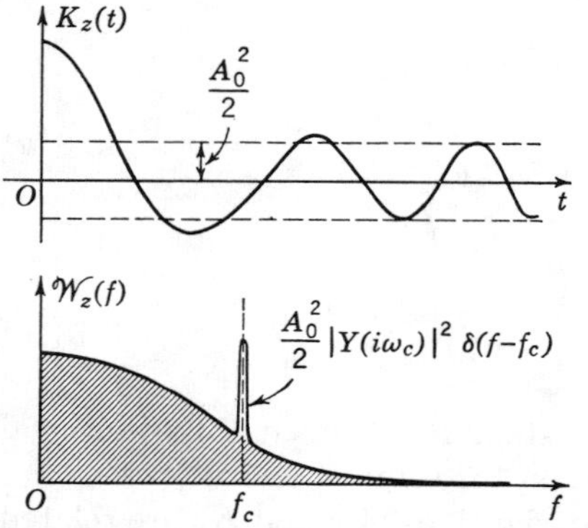

FIG. 3.26. Covariance and spectral density functions at the output of an RC filter, for a sinusoid and input white noise.

3.3-6 Mixed Processes; Examples. It is evident that the remarks of the preceding sections (3.3-1, 3.3-4) can be extended to include deterministic and mixed processes in the steady state as well, since the filter operations T_L are invariant of the various possible input ensembles. We illustrate this with three simple examples of mixed processes as input ensembles:

(1) Sinusoid and white noise through an ***RC*** **filter.** From Eqs. (3.90*a*) and (3.131), (3.132), we have here

$$\rho(t) = (2RC)^{-1}e^{-|t|/RC} \qquad K_y(t) = \frac{W_0}{2}\,\delta(t-0) + \frac{A_0^2}{2}\cos\omega_c t$$

and

$$\mathcal{W}_y(f) = W_0 + \frac{A_0^2}{2}\,\delta(f-f_c) \qquad f_c > 0$$

so that from Eqs. (3.102), (3.104) we obtain

$$\begin{aligned} K_z(t) &= \frac{1}{2RC}\int_{-\infty}^{\infty} e^{-|\tau|/RC}\left[\frac{W_0}{2}\,\delta(\tau - t) + \frac{A_0^2}{2}\cos\omega_c(t-\tau)\right]d\tau \\ &= \frac{W_0}{4RC}\,e^{-|\tau|/RC} + \frac{A_0^2\cos\omega_c t}{2[1+(RC\omega_c)^2]} \end{aligned} \tag{3.133}$$

the second term of which is simply $A_0^2|Y(i\omega_c)|^2\cos(\omega_c t/2)$. The corresponding spectral density may be obtained from Eq. (3.133) with the help of the W-K theorem or more directly from Eq. (3.104), viz.:

$$\mathcal{W}_z(f) = \frac{W_0}{1+(RC\omega)^2} + \frac{A_0^2\delta(f-f_c)}{2(1+R^2C^2\omega_c^2)} \tag{3.134}$$

The time and frequency pictures are shown in Fig. 3.26. Observe that the contribution of the noise in K_z dies out, as expected, leaving the periodic term, which maintains its characteristic oscillations at the period of the original wave. The intensity spectrum has a continuous part, on which is superimposed the delta function representing the effect of the deterministic component. The average total intensity is easily obtained from Eq.

(3.133) on setting $t = 0$, to give

$$\overline{z^2} = \frac{W_0}{4RC} + \frac{A_0^2}{2(1 + R^2C^2\omega_c^2)} \tag{3.135}$$

of which the first term represents the intensity of the filtered noise wave and the second that of the periodic component.

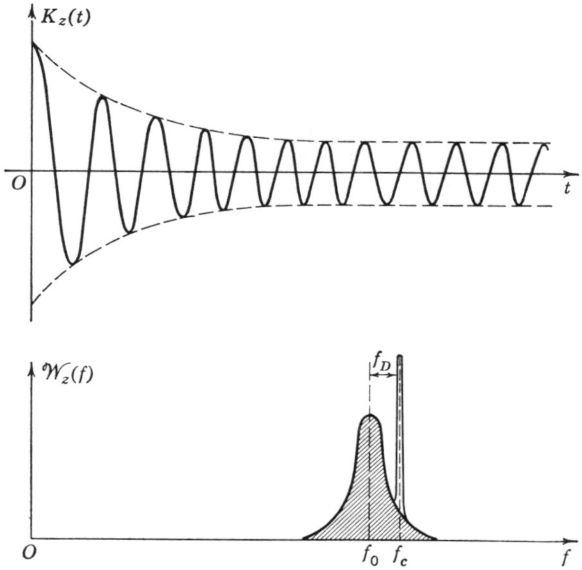

FIG. 3.27. Spectrum and covariance for case 2: high-Q CLR filter.

(2) Sine-wave and white noise through a high-Q CLR filter. Let us take as the filter in this example the CLR filter [(4) of Table 3.1] and require it to have the narrowband structure of a high-Q system. Then we have

$$\rho(t)_{LCR} \doteq \frac{\omega_F}{2} e^{-\omega_F|t|} \cos \omega_0 t \qquad Q = \frac{\omega_0}{2\omega_F} \gg 1 \qquad K_y(t) \text{ as above}$$

and use Eq. (3.108a) to write

$$\begin{aligned} K_z(t) &\doteq \omega_F \int_{-\infty}^{\infty} \cos \omega_0\tau \left[\frac{W_0}{2} \delta(t - \tau) + \frac{A_0^2}{2} \cos \omega_0(t - \tau) \right] e^{-\omega_F|t|}\, d\tau \\ &= \frac{\omega_F W_0}{2} e^{-\omega_F|t|} \cos \omega_0 t + \frac{A_0^2}{2} \cos \omega_c t \left(1 + \frac{\omega_D^2}{\omega_F^2}\right)^{-1} \qquad \omega_D = \omega_c - \omega_0 \end{aligned} \tag{3.136}$$

The corresponding intensity spectrum is obtained directly from Eq. (3.112), viz.,

$$\mathcal{W}_z(f) \doteq \frac{W_0}{1 + (\omega - \omega_0)^2/\omega_F^2} + \frac{A_0^2(f - f_c)}{2[1 + (\omega - \omega_0)^2/\omega_F^2]} \tag{3.137}$$

as sketched in Fig. 3.27. The total average intensity in this instance is

$$K_z(0) = \overline{z^2} = \frac{\omega_F W_0}{2} + \frac{A_0^2}{2}\left(1 + \frac{\omega_D^2}{\omega_F^2}\right)^{-1} \tag{3.138}$$

and the contribution of the discrete component is reduced by the amount of detuning ω_D.

(3) Narrowband input noise; narrowband filter. As a simple example of this case, let us consider, as the input to the high-*Q* *CLR* filter above, noise with the covariance of Eq. (3.120), e.g.,

$$K_y(t) = \tfrac{1}{2}\,\mathrm{Re}\left(\frac{W_0\omega_{Fy}}{2}\, e^{-\omega_{F_y}|t| - i\omega_c t}\right) \tag{3.139}$$

specifically; as before, $\mathcal{K}_0 = (\omega_F/2)e^{-\omega_F|t|}$ here, so that from Eq. (3.121) we obtain

$$K_z(t) = \psi_y \frac{\omega_F}{4}\,\mathrm{Re}\left(e^{-i\omega_c t}\int_{-\infty}^{\infty} e^{-\omega_F|\tau| - \omega_{F_y}|t-\tau| + i\omega_D\tau}\,d\tau\right) \qquad \omega_D = \omega_c - \omega_0 \tag{3.140}$$

which may be readily evaluated; $\psi_y = W_0\omega_{Fy}/2$ is the mean intensity $\overline{y^2}$ of the input noise ensemble. From Eq. (3.123) and the results of Table 3.1, it is a simple matter to write for the corresponding intensity spectrum

$$\mathcal{W}_z(f) = W_0\left[1 + \left(\frac{\omega - \omega_c}{\omega_F}\right)^2\right]^{-1}\left\{1 + \left[\frac{\omega - (\omega_c - \omega_D)}{\omega_F}\right]^2\right\}^{-1} \tag{3.141}$$

with a total average intensity $\overline{z^2}$ [cf. Eq. (3.124)], from Eq. (3.140), which is easily found to be

$$\overline{z^2} = \psi_y(1 + a_y)/[(1 + a_y)^2 + b_D^2] \qquad a_y \equiv \omega_{Fy}/\omega_F \qquad b_D = \omega_D/\omega_F \tag{3.142}$$

The covariance and spectral density have the same type of structure sketched in Fig. 3.27, apart from the "spike" at $f = f_c$ and apart from the periodic term ($\sim \cos \omega_c t$), indicated in the figure as $|t| \to \infty$.

We remark, finally, that, for all the narrowband cases discussed in Secs. 3.3-2, 3.3-4, and above, the representations of K_z, $\mathcal{W}_z$ are approximate, becoming exact in the limit of infinitely narrowband ensembles. The precise forms are of interest, however, only when spectral properties of the ensemble well away from the maximum response are desired, a situation which is not common in our present study. Additional examples appear in Probs. 3.12 to 3.15.

PROBLEMS

3.12 Show for the filter given in Fig. P 3.12 that, if $Z(i\omega)$ is the filter's impedance function $[= Y(i\omega)^{-1}]$ and if $I(t)$ is a stationary ensemble of white noise *currents* (of fixed intensity density W_I), then

$$|Z(i\omega)|^2 = \frac{R^2}{1 + \left(\frac{\omega - \omega_0}{\omega_F}\right)^2 \left(\frac{\omega + \omega_0}{2\omega}\right)^2} \equiv \mathfrak{F}\{\rho_Z\} \tag{1}$$

and hence for all Q $(= \omega_0/2\omega_F)$ obtain ρ_Z, the autocorrelation function of the network under these driving conditions, where $\rho_Z(t) = \mathfrak{F}^{-1}\{|Z|^2\}$ [cf. Eq. (3.89)]. [Remember here that $V(t) = \int_{-\infty}^{\infty} h_I(t - \tau)I(t)\, d\tau \equiv T_L\{I\}$ in the steady state, and so, by analogy with Eq. (3.87), $\rho_Z \equiv \int_{-\infty}^{\infty} h_I(x)h_I(x + t)\, dx$.]

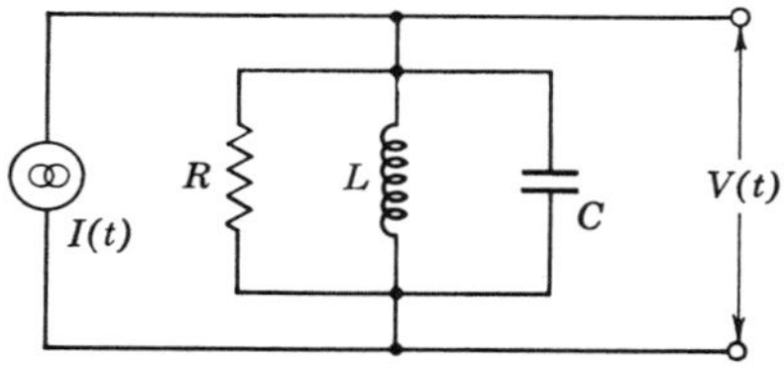

FIG. P 3.12. A parallel circuit with a white-noise current source.

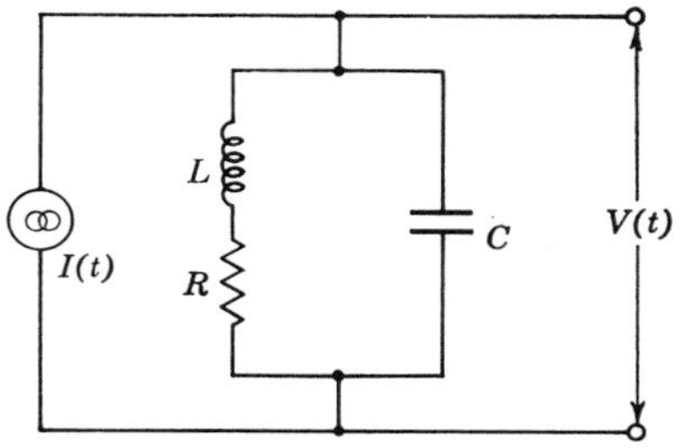

FIG. P 3.13. A parallel circuit with a white-noise current source.

3.13 Repeat Prob. 3.12, but now for the network of Fig. P 3.13, and thus show that

$$|Z(i\omega)|^2 = \frac{R^2\omega_0^4}{4\omega_F^2} \frac{\omega^2 + 4\omega_F^2}{(\omega_0^2 - \omega^2)^2 + 4\omega^2\omega_F^2} \tag{1}$$

which, for high Q $(= \omega_0/2\omega_F) \gg 1$, $\omega_0 = (LC)^{-1}$, $\omega_F = R/2L$, becomes

$$|Z(i\omega)|^2_{Q\gg 1} \doteq \frac{R^2Q^4}{1 + \left(\frac{\omega - \omega_0}{\omega_F}\right)^2} \tag{1a}$$

Hence, show that $\rho_Z = \mathfrak{F}^{-1}\{|Z|^2\}$ is here

$$\rho_Z(t) = \rho_Z(0)e^{-\omega_F|t|} \begin{cases} \cosh i\omega_1 t + \dfrac{\omega_F}{i\omega_1}\dfrac{1 - Q^2}{1 + Q^2}\sinh i\omega_1|t| & \omega_1^2 < 0 \\ 1 + \tfrac{3}{5}\omega_F|t| & \omega_1^2 = 0 \ (Q = \tfrac{1}{2}) \\ \cos \omega_1 y + \dfrac{\omega_F}{\omega_1}\dfrac{1 - Q^2}{1 + Q^2}\sin \omega_1|t| & \omega_1^2 > 0 \end{cases} \tag{2}$$

when, as before (cf. Table 3.1), $\omega_1 = \sqrt{\omega_0^2 - \omega_F^2}$. Verify that

$$\rho_Z(0) = R^2Q^2\omega_F(1 + Q^2)$$

and show that

$$\rho_Z(t)_0 \doteq \frac{R^2Q^4\omega_F}{2} e^{-\omega_F|t|} \cos \omega_0 t \qquad Q \gg 1 \tag{3}$$

3.14 In the case of n synchronous single-tuned stages (that is, CLR) (high Q), where $|Y|^2 = (1 + x^2)^{-n}$, $x = (\omega - \omega_0)/\omega_F$, show that the correlation function of these n

stages, taken together, is $\rho_0 \cos \omega_0 t$, where

$$\rho_0(t) = \omega_F \sqrt{\frac{2}{\pi}} \frac{1}{(m-1)!2^m} |\omega_F t|^{(m-1)/2} K_{(m-1)/2}(\omega_F|t|) \tag{1a}$$

or

$$\rho_0(t) = \frac{\omega_F e^{-\omega_F|t|}}{2^m(m-1)!} \sum_{l=0}^{m-1} \frac{(\omega_F|t|)^{m-1-l}(m-1+l)!}{l!(m-1-l)!2^l} \qquad m \geq 1 \tag{1b}$$

in which $K_{(m-1)/2}$ is a modified Bessel function of the second kind and ω_0 is the (angular) resonant frequency.

3.15 In the case of n staggered-tuned CLR stages, for which $|Y|^2 = (1 + x^{2n})^{-1}$ $x = (\omega - \omega_0)/\omega_F$ in the high-Q condition, show that

$$\rho_0(t) = \frac{\omega_E}{2n} \sum_{l=1}^{n} e^{-\omega_F|t| \sin \phi_{n,l}} \sin [\phi_{n,l} + \omega_F(t) \cos \phi_{n,l}] \qquad \phi_{n,l} \equiv \frac{2l-1}{2n} \pi \tag{1}$$

3.4 *Cross-correlation Functions, Spectra, and Generalizations*

Up to this point in Chap. 3, we have considered the second-moment properties of single ensembles y, z, as represented by their covariance and spectral density functions $\overline{y_1 y_2}$, $\mathcal{W}_y$, etc. Often, however, we have to deal with more than one ensemble at a time, so that it is natural to seek the various second-moment properties in these mixed cases. One important example, which we have already encountered in previous chapters, is the ensemble which is the sum of a signal and noise ensemble. Another case of interest is obtained by forming the product of the two ensembles y and z, at selected times t_1, t_2. More complicated structures are easily generated: $G(yz)$, $G(y,z)$, where G is some nonlinear operation, etc.

Here, however, we shall consider only the single product types $y \times z$, where it is convenient now to regard y and z as the components of a *vector* process, for example, $\mathbf{u} = \hat{\mathbf{i}}_1 y + \hat{\mathbf{i}}_2 z$, in which $\hat{\mathbf{i}}_1$, $\hat{\mathbf{i}}_2$ are unit vectors (usually in the directions of an orthogonal axis system). Operations on products of the type $y_1 z_2$, $y_2 z_1$, as well as on $y_1 y_2$, $z_1 z_2$, are of particular importance in applications where we wish to determine whether or not pairs of member functions $y^{(j)}$, $z^{(j)}$ are correlated, i.e., exhibit a statistical or a coherent relationship, respectively, over the ensemble or in time. Such situations arise, for example, in the detection of signals in noise and in prediction and tracking problems, where the waves in question either are entirely random or contain significant random components. A detailed account of some of these applications is deferred to Chaps. 13 to 16. Here we shall introduce the topic of cross correlation and consider briefly some of the second-moment properties familiarly associated with this concept.

3.4-1 Definitions and Properties. Let us consider a pair of real ensembles, y and z, which we may interpret as the components of a two-dimensional vector process $\mathbf{u} = \hat{\mathbf{i}}_1 y + \hat{\mathbf{i}}_2 z$. These components are, moreover, required to share a common physical mechanism, so that we may expect a statistical relationship of some sort between them. In addition, let y and z for the moment be wide-sense stationary and entirely random, so that $\boldsymbol{E}\{y\} = \boldsymbol{E}\{z\} = 0$, $\overline{y_1 y_2} = K_y(|t_2 - t_1|)$, etc., and let us consider the four components $(y_1y_2, y_1z_2, z_1y_2, z_1z_2)$ of the second-rank tensor $\mathbf{u}(t_1)\mathbf{u}(t_2)$ when the ensemble average is performed over $\mathbf{u}_1\mathbf{u}_2$. It is, of course, assumed that $\overline{\mathbf{u}_1\mathbf{u}_2}$ exists, so that these components become the elements of the *covariance matrix* (or *tensor*)

$$\mathbf{K}_{yz}(t) = \begin{bmatrix} K_y(t) & K_{yz}(t) \\ K_{zy}(t) & K_z(t) \end{bmatrix} = \begin{bmatrix} \overline{y_1y_2} & \overline{y_1z_2} \\ \overline{z_1y_2} & \overline{z_1z_2} \end{bmatrix} = \tilde{\mathbf{K}}_{zy}(-t) \qquad t = t_1 - t_2 \tag{3.143}$$

For the individual members $[\mathbf{u}(t_1)\mathbf{u}(t_2)]^{(j)}$ of the ensemble $\mathbf{u}_1\mathbf{u}_2$, we have analogously the *coherency matrix* (or *tensor*)

$$\mathbf{R}_{yz}{}^{(j)}(t) = \begin{bmatrix} R_y{}^{(j)}(t) & R_{yz}{}^{(j)}(t) \\ R_{zy}{}^{(j)}(t) & R_z{}^{(j)}(t) \end{bmatrix} = \begin{bmatrix} \langle (y_1y_2)^{(j)} \rangle & \langle (y_1z_2)^{(j)} \rangle \\ \langle (z_1y_2)^{(j)} \rangle & \langle (z_1z_2)^{(j)} \rangle \end{bmatrix} = \tilde{\mathbf{R}}_{zy}{}^{(j)}(-t) \tag{3.144}$$

with $t = t_1 - t_2$ again.

The off-diagonal terms in Eqs. (3.143), (3.144) are, respectively, the *cross-variance* and *cross-correlation* functions associated with the components of the vector process $\mathbf{u}(t)$, while the diagonal elements are the more familiar autovariance and autocorrelation functions. The notion of the covariance and coherency matrices is extended in obvious fashion to vector processes of more than two components. Physical examples where these concepts prove useful arise in the theory of linear electrical networks (cf. Chap. 11), in physical optics,[10] and in the theory of turbulent fluid flow,[11,12] to mention a few applications.

Spectral properties may be derived as before (cf. Sec. 3.2). Starting with the ensembles of truncated waves $y_T(t)$, $z_T(t)$, we may define, as for the (auto) spectral densities, a corresponding *cross-spectral* density for each member of the ensemble, viz.:

$$\mathcal{W}_{yz}{}^{(j)}(f)_T \equiv 2S_y{}^{(j)}(f)_T S_z{}^{(j)}(f)_T^*/T \tag{3.145}$$

In the limit,† we have the *ensemble* density

$$\mathcal{W}_{yz}(f) \equiv \lim_{T\to\infty} \boldsymbol{E}\left\{\frac{2}{T} S_y{}^{(j)}(f)_T S_z{}^{(j)}(f)_T^*\right\} \tag{3.146}$$

For these particular processes, we can also define a limiting individual cross-spectral density $\mathcal{W}_{yz}{}^{(j)}(f)$ for almost all members of this product ensemble, according to

$$\lim_{T\to\infty} \mathfrak{D}_{yz}{}^{(j)}(f)_T \equiv \mathfrak{D}_{yz}{}^{(j)}(f) = \int_{-\infty}^{f} \mathcal{W}_{yz}{}^{(j)}(f)\, df \qquad \mathcal{W}_{yz}{}^{(j)}(f) = \frac{d}{df}\mathfrak{D}_{yz}{}^{(j)}(f) \tag{3.147}$$

† See Eq. (3.43) and also the discussion of Sec. 3.2-1.

[see the discussion in Sec. 3.2-3(3); however, since $\mathcal{W}_{yz}^{(j)}$ is neither non-negative nor real (unless $y = z$), the proof of Eq. (3.147) has to be modified[13] from that for $\mathcal{W}_y^{(j)}$.]

Corresponding to the covariance and coherency matrices (3.143), (3.144), we have the spectral matrices

$$\mathbf{W}_{yz}(f) = \begin{bmatrix} \mathcal{W}_y(f) & \mathcal{W}_{yz}(f) \\ \mathcal{W}_{zy}(f) & \mathcal{W}_z(f) \end{bmatrix} \qquad \mathbf{W}_{yz}^{(j)}(f) = \begin{bmatrix} \mathcal{W}_y^{(j)}(f) & \mathcal{W}_{yz}^{(j)}(f) \\ \mathcal{W}_{zy}^{(j)}(f) & \mathcal{W}_z^{(j)}(f) \end{bmatrix} \tag{3.148}$$

the components of which are related to the corresponding elements of $\mathbf{K}_{yz}$ and $\mathbf{R}_{yz}^{(j)}$ through the W-K theorems (3.42), (3.62) for the diagonal terms. Through the extensions† below of these theorems for the off-diagonal terms, we write

$$\mathcal{W}_{yz}(f) = 2\mathfrak{F}\{K_{yz}\} = 2\int_{-\infty}^{\infty} e^{-i\omega t} K_{yz}(t)\,dt = \mathcal{W}_{zy}(f)^* \qquad \omega = 2\pi f \tag{3.149a}$$

$$K_{yz}(t) = \tfrac{1}{2}\mathfrak{F}^{-1}\{\mathcal{W}_{yz}\} = \frac{1}{2}\int_{-\infty}^{\infty} e^{i\omega t}\mathcal{W}_{yz}(f)\,df = K_{zy}(-t) \qquad t = t_1 - t_2 \tag{3.149b}$$

As mentioned above, although y and z are taken to be real here so that the cross variance K_{yz} is also real, $\mathcal{W}_{yz}(f)$ is normally a complex quantity. If y and z are complex processes, then $K_{yz}(t) = K_{zy}(-t)^*$, $R_{yz}(t) = R_{zy}(-t)^*$, and in general $\mathcal{W}_{yz}(f) = \mathcal{W}_{zy}(f)^*$. Similarly, for almost all members of the vector process $\mathbf{u}(t) = \hat{\mathbf{i}}_1 y(t) + \hat{\mathbf{i}}_2 z(t)$, the individual W-K theorem, corresponding to Eqs. (3.62), becomes

$$\mathcal{W}_{yz}^{(j)}(f) = 2\mathfrak{F}\{R_{yz}^{(j)}\} \qquad R_{yz}^{(j)}(t) = \tfrac{1}{2}\mathfrak{F}^{-1}\{\mathcal{W}_{yz}^{(j)}\} \tag{3.149c}$$

where now, of course, $\mathcal{W}_{yz}^{(j)}$ and $R_{yz}^{(j)}$ vary from representation to representation.

When the vector process $\mathbf{u}$ is ergodic, we may equate ensemble and individual spectral densities in the usual fashion (probability 1). Deterministic and mixed processes are handled as before (cf. Sec. 3.2-4): the periodic components, if any, appear as line spectra (Figs. 3.14, 3.15), while the aperiodic terms yield a continuous spectrum, in combination like those sketched in Figs. 3.16, 3.17. Note, however, that the ensemble spectral density $\mathcal{W}_{yz}$ is not now susceptible to interpretation as a probability density, unlike $\mathcal{W}_y$, $\mathcal{W}_z$ (cf. Sec. 3.3-3), since $\mathcal{W}_{yz}$ is usually complex and such that, for some finite (> 0) frequency intervals, its real and imaginary parts can be negative (cf. Sec. 3.4-2).

† The proof here, from Eq. (3.40) on, must again be modified, since $\mathcal{W}_{yz}$ is usually neither real nor always positive (or zero). For one approach, see Kampé de Fériet,[6] sec. 5, for the appropriate intensity conditions and theorems.

The relations K_{yz} and $\mathcal{W}_{yz}$ exhibit a number of features not found in the autovariances and spectral densities of the component processes y and z. For example, if $K_{yz} = 0$ (all t) (or, more generally, if $M_{yz} = 0$), y and z are *wide-sense independent* and, in addition, $\mathcal{W}_{yz}(f)$ also vanishes. However, if K_{yz} does not vanish for all t, then we say that y and z are statistically dependent, at least through the first order.

Wide-sense independence can arise in several ways, as is perhaps most easily seen from the limiting expression for the spectral density of Eq. (3.146). If y and z are both derived from a common source but are subject to (linear or nonlinear) filtering, such that they share no common spectral range of frequencies, then $(S_yS_z^*)_T^{(j)} = 0$ [for all (j) and f] and the cross density vanishes. On the other hand, if y and z share a common spectral region but stem from different, statistically unrelated sources, $E\{2(S_yS_z^*)_T^{(j)}/T\}$ is now zero (since we are assuming y and z to be entirely random processes) and again $\mathcal{W}_{yz}$ vanishes.† Of course, if we relax the assumption that $\bar{y}$ and $\bar{z}$ are zero, statistical interdependence of y and z appears when both $\bar{y}, \bar{z} \neq 0$. Moreover, if the processes y and z are mixed, containing commensurable periodic components, there will be a common line spectrum, as well as a continuum, which occurs if the random components are wide-sense dependent.

These observations have significance, for example, in the detection of signals by methods which use correlation. The received wave often consists of an additive mixture of the desired signal and noise, $V = S + N$, or of noise alone, N. When the structure of the desired signal S is known a priori, one method of determining the presence or absence of the signal is to cross-correlate $V^{(j)}$ and $S^{(j)}$, with a suitable adjustment of the delay $t_1 - t_2$ for maximum response, and observe whether or not $\langle (VS)^{(j)} \rangle$ differs significantly from $\langle (NS)^{(j)} \rangle$. As we shall see in Part 4, optimum detection systems often employ techniques of this type. A few simple cases of the cross-variance and spectral density are examined and illustrated in Sec. 3.4-2.

As in the case of the autovariance, we may obtain the cross-variance function from the appropriate second-order distribution density or characteristic function. For strictly stationary random processes, we write, analogous to Eq. (3.2),

$$K_{yz}(t) = \overline{y(t_1)z(t_2)} = \iint_{-\infty}^{\infty} y_1 z_2 W_2(y_1, z_2; t_1 - t_2)\, dy_1\, dz_2 \qquad \begin{matrix} \bar{y} = \bar{z} = 0 \\ t = t_1 - t_2 \end{matrix}$$

$$= -\frac{\partial^2}{\partial \xi_1\, \partial \xi_2} F_2(\xi_1, \xi_2; t_1 - t_2)\Big|_{\xi_1 = \xi_2 = 0} = R_{yz}^{(j)}(t) \qquad \text{probability } 1 \tag{3.150}$$

† Similarly, for almost all members of the product ensemble y_1z_2 the individual spectral density vanishes, since $R_{yz}^{(j)}(t)$ is zero in these instances: $y^{(j)}$ and $z^{(j)}$ are completely incoherent.

this last if $y \times z$ is ergodic; generally, if $\bar{y}$, $\bar{z}$ are different from zero, Eq. (3.150) becomes instead

$$M_{yz}(t) = \iint\limits_{-\infty}^{\infty} y_1 z_2 W_2(y_1, z_2; t_1 - t_2)\, dy_1\, dz_2 = K_{yz}(t) + \bar{y}\bar{z} \qquad (3.150a)$$

which is replaced by $M_{yz}(t_1,t_2) = K_{yz}(t_1,t_2) + \bar{y}_1\bar{z}_2$ if y and z are not stationary. Unlike the autovariance, $K_{yz}(0)$ may vanish if $\mathcal{W}_{yz}(f) = -\mathcal{W}_{yz}(-f)$; corresponding to case 3 in Sec. 3.1-3, we have, however,

$$|\overline{y_1 z_2}| \leq \sqrt{\overline{y_1^2}\,\overline{z_2^2}} \quad \text{or} \quad |K_{yz}(t)|/\sqrt{K_y(0)K_z(0)} \leq 1 \qquad \bar{y} = \bar{z} = 0 \qquad (3.151)$$

(cf. Prob. 3.18*a*). It is also evident that, if y and z are free of deterministic terms, $K_{yz}(\pm\infty) = 0$, $M_{yz}(\pm\infty) = \bar{y}\bar{z}$, and all memory between pairs of observations separated by sufficiently long intervals of time vanishes if either $\bar{y}$ or $\bar{z}$ is zero. On the other hand, if $y \times z$ contains a periodic component, $K_{yz}(\pm\infty)$ oscillates indefinitely at the frequency of this periodic term in the manner of Fig. 3.3.

The features of the continuous case are readily extended to random series (cf. Sec. 3.1-4), generated, say, by sampling a continuous process at discrete intervals $t_1, t_2, \ldots, t_k, \ldots, t_n, \ldots$. Analogous to Eqs. (3.5), (3.6), we can write for an individual representation

$$R_{yz}^{(j)}(k) \equiv \lim_{N\to\infty} \left[\frac{1}{2N+1} \sum_{-N}^{N} (y_n z_{n+k})^{(j)}\right] \qquad y_n = y(t_n), \ldots \qquad (3.152)$$

and for the ensemble

$$K_{yz}(k) \equiv \mathbf{E}\{(y_n - \bar{y}_n)(z_{n+k} - \bar{z}_{n+k})\} \qquad (3.153)$$

We remark, finally, that we can also define spectral and time moments for these product ensembles $y \times z$ [cf. Sec. 3.2-3(4)]. For example, if $y = y$ and $z = \dot{y}$, we get from Eqs. (3.65) $K_{yz} = K_{y\dot{y}} = dK_y/dt$; or if $z = \ddot{y}$, then we have $K_{yz} = K_{y\ddot{y}} = d^2K_y/dt^2$, etc. One can also have $K_{y\dot{z}}$, $K_{\dot{y}\dot{z}}$ and other combinations, subject, of course, to our ability to differentiate the process to the appropriate order. In terms of the original cross density $\mathcal{W}_{yz}$, these higher-order crossvariances can be written

$$K_{y\dot{z}}(t) = -\frac{i}{2}\int_{-\infty}^{\infty} \omega \mathcal{W}_{yz}(f) e^{i\omega t}\, df = R_{y\dot{z}}^{(j)} \qquad \text{probability 1} \qquad (3.154a)$$

$$K_{\dot{y}\dot{z}}(t) = \frac{1}{2}\int_{-\infty}^{\infty} \omega^2 \mathcal{W}_{yz}(f) e^{i\omega t}\, df = R_{\dot{y}\dot{z}}^{(j)} \qquad \text{probability 1, etc.} \qquad (3.154b)$$

this last provided that $y \times z$ is ergodic (see Probs. 3.18 to 3.21 for further examples).

3.4-2 Linearly Filtered Waves. Let us suppose that $x(t)$ is an entirely random, wide-sense stationary process, representing an ensemble of thermal

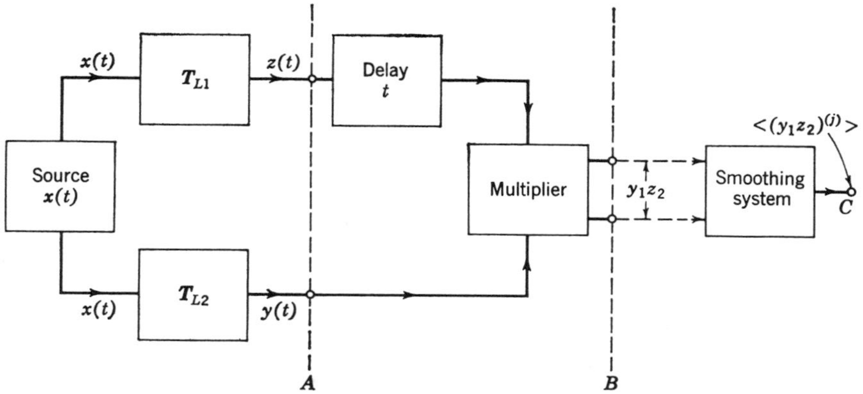

FIG. 3.28. Two ensembles generated by a common source, with a delay and multiplication yielding the product ensemble y_1z_2. Final output of the smoothing system is ideally $\langle y_1z_2\rangle^{(j)}$ for each member of the input ensemble.

noise voltages. Next, let us select two linear filters, $\boldsymbol{T}_{L1}$, $\boldsymbol{T}_{L2}$, through which $x(t)$ is simultaneously passed, in the manner of Fig. 3.28, to generate two new processes $y = \boldsymbol{T}_{L2}\{x\}$, $z = \boldsymbol{T}_{L1}\{x\}$. Specifically, for the steady-state condition assumed here, we write

$$y(t) = \int_{-\infty}^{\infty} h_y(t-\tau)x(\tau)\,d\tau \qquad z(t) = \int_{-\infty}^{\infty} h_z(t-\tau)x(\tau)\,d\tau \quad (3.155)$$

y and z are also wide-sense stationary, since $\boldsymbol{T}_{L1,2}$ are invariant operators. The cross-variance function $\overline{y_1z_2}$ is therefore

$$\begin{aligned} K_{yz}(t) = \overline{y(t_1)z(t_2)} &= \iint_{-\infty}^{\infty} h_y(t_1-\tau_1)\overline{x(\tau_1)x(\tau_2)}h_z(t_2-\tau_2)\,d\tau_1\,d\tau_2 \\ &= \iint_{-\infty}^{\infty} h_y(t_0')K_x(t+t_0'-t_0'')h_z(t_0'')\,dt_0'\,dt_0'' \qquad t = t_1 - t_2 \quad (3.156) \end{aligned}$$

following the steps indicated in Eq. (3.101). Here, just as for the autocorrelation function of a filter, we can define a cross-correlation function for the pair of linear filters $\boldsymbol{T}_{L1}$, $\boldsymbol{T}_{L2}$ above, in terms of their respective weighting and system functions, as

$$\begin{aligned} \rho_{(yz)}(\tau) &\equiv \int_{-\infty}^{\infty} h_y(t_0)h_z(t_0+\tau)\,d\tau \\ &= \int_{-\infty}^{\infty} Y_y(i\omega)Y_z(i\omega)^*e^{i\omega t}\,df = \mathfrak{F}^{-1}\{Y_yY_z^*\} \\ &= \int_{-\infty i}^{\infty i} Y_y(p)Y_z(-p)e^{pt}\frac{dp}{2\pi i} = \rho_{(zy)}(-\tau) \qquad (3.157a) \end{aligned}$$

with the inverse relation

$$Y_y(i\omega)Y_z(i\omega)^* = \mathfrak{F}\{\rho_{(yz)}\} = \int_{-\infty}^{\infty} \rho_{(yz)}(t)e^{-i\omega t}\,dt \tag{3.157b}$$

Making the same transformations as in Eqs. (3.101), (3.102) and using Eqs. (3.157), we obtain for Eq. (3.156) the equivalent forms

$$\begin{aligned} K_{yz}(t) &= \int_{-\infty}^{\infty} \rho_{(yz)}(\tau)K_x(t-\tau)\,d\tau = \int_{-\infty}^{\infty} \rho_{(zy)}(\tau)K_x(t+\tau)\,d\tau \\ &= \int_{-\infty}^{\infty} \rho_{(yz)}(t \mp \tau)K_x(\tau)\,d\tau \end{aligned} \tag{3.158a}$$

or, in terms of the transfer functions of the two filters,

$$K_{yz}(t) = \int_{-\infty}^{\infty} K_x(\tau)\,d\tau \int_{-\infty i}^{\infty i} Y_y(p)Y_z(-p)e^{p(t\mp\tau)}\,\frac{dp}{2\pi i} \tag{3.158b}$$

The cross variance of the component ensembles of the vector process **u** is simply the convolution of the autovariance of the common source ensemble and the cross-correlation function of the two filters. [For the various autovariances $K_y(t)$, $K_z(t)$, we have only to set $\boldsymbol{T}_{L2} = \boldsymbol{T}_{L1}$ or $\boldsymbol{T}_{L1} = \boldsymbol{T}_{L2}$, respectively, in the above; cf. Fig. 3.28.] Similarly, Eqs. (3.158) may be used for individual members of the output ensembles y, z if we replace K_{yz} by $R_{yz}^{(j)}$, K_x by $R_x^{(j)}$; if $x(t)$ is an ergodic process, so also are y and z (and the product y_1z_2 here), since $\boldsymbol{T}_{L1,2}$ are invariant operations, and K_{yz}, K_x, etc., are equal to $R_{yz}^{(j)}$, $R_x^{(j)}$, etc. (probability 1), in the usual way. Applying the W-K theorem (3.149) to Eqs. (3.158), following the steps indicated in Eq. (3.104), we readily show that

$$\mathcal{W}_{yz}(f) = Y_y(i\omega)Y_z(i\omega)^*\mathcal{W}_x(f) \tag{3.159}$$

from which it is immediately evident that $\mathcal{W}_{yz}$ is a complex quantity with a negative imaginary part for some frequencies. Corresponding results obtain for the individual spectral densities. The relation (3.159) is the counterpart in the frequency domain of the covariance expressions (3.158) in the time domain, and these, of course, are connected to the former by the W-K theorem (3.149).

3.4-3 Examples. As our first example, let us consider the problem of determining the weighting function of a given linear (passive) network.[14] In place of the familiar approach using a sine wave (cf. Sec. 2.2-5), we employ instead an entirely random noise $x(t)$ as the input to the filter $\boldsymbol{T}_{L1}$ under test and choose $\boldsymbol{T}_{L2}$ to be a purely resistive (i.e., frequency-insensitive) network whose system function is accordingly a constant, for example, $Y_z(i\omega) = C_z$, with the weighting function $h_z(t) = C_z\delta(t-0)$. Then, from Eq. (3.157*a*), it follows at once that

$$\rho_{(yz)}(t) = C_z \int_{-\infty i}^{\infty i} Y_y(p)e^{pt}\,\frac{dp}{2\pi i} = C_z h_y(t) \tag{3.160a}$$

and so the cross variance observed at the output C in Fig. 3.28 becomes, from Eq. (3.158a),

$$K_{yz}(t) = C_z \int_{-\infty}^{\infty} h_y(t-\tau)K_x(\tau)\,d\tau \tag{3.160b}$$

If now we choose unlimited white noise for the input $x(t)$ [Sec. 3.2-3(1)c], we have $K_x(\tau) = (W_0/2)\delta(\tau - 0)$, which applied in Eq. (3.160b) gives immediately

$$K_{yz}(t) = C_z h_y(t) \tag{3.161a}$$

Correspondingly, in the frequency domain one has, from Eq. (3.159), for the system function

$$Y_y(i\omega) = \mathcal{W}_{yz}(f)/C_z\mathcal{W}_x(f) = \mathcal{W}_{yz}(f)/C_zW_0 \tag{3.161b}$$

so that by observing $K_{yz}(t)$ at the output of the cross-correlation scheme of Fig. 3.28, for various delays t, or, equivalently, by computing $\mathcal{W}_{yz}(f)$ from the observed spectral data at this output [cf. Sec. 3.2-3(3)], we can in principle determine either the weighting or the system function of the filter in question. Of course, this is an idealization (like the sine-wave approach) which assumes infinite observation and smoothing times—to determine each value of $K_{yz}(t)$ precisely—and which makes use of unlimited white noise. Neither of these conditions obtains in practice. We should, however, use noise that is spectrally as broad as possible, to minimize the errors introduced, not only by the assumption that K_x above is a delta function, but also by the necessarily finite observation and smoothing periods available. In actual measurements, these effects should be taken into account, in the manner of Sec. 16.1.

For our second example, we assume now that each filter in Fig. 3.28 is a simple RC network [cf. Eq. (3.90a), Fig. 3.19a] with the respective system functions $Y_y(i\omega) = (1 + i\omega/\alpha_y)^{-1}$, $Y_z(i\omega) = (1 + i\omega/\alpha_z)^{-1}$, where $\alpha_y = (RC)_y^{-1}$, $\alpha_z = (RC)_z^{-1}$. From Eq. (3.157a), the cross-correlation function of this pair of filters is readily found,

$$\rho_{yz}(t) = \int_{-\infty i}^{\infty i} \frac{e^{pt}}{(1+p/\alpha_y)(1-p/\alpha_z)}\frac{dp}{2\pi i} = \begin{cases} \dfrac{\alpha_y\alpha_z}{\alpha_y+\alpha_z}\,e^{-\alpha_y t} & t \geq 0\ (\alpha_y,\ \alpha_z > 0) \\ \dfrac{\alpha_y\alpha_z}{\alpha_y+\alpha_z}\,e^{\alpha_z t} & t \leq 0 \end{cases} \tag{3.162}$$

as sketched in Fig. 3.29. Let us assume in addition that the noise source $x(t)$ produces RC noise, such that $K_x(t) = \psi_x e^{-\alpha_x|t|}$, $\psi_x = W_0\alpha_x/4$ [Eqs. (3.87), (3.90a)], and so Eq. (3.158a) gives

$$K_{yz}(t) = \frac{\psi_x\alpha_y\alpha_z}{\alpha_y+\alpha_z}\left(\int_{-\infty}^{t} e^{-\alpha_y(t-\tau)-\alpha_x|\tau|}\,d\tau + \int_{t}^{\infty} e^{\alpha_z(t-\tau)-\alpha_x|\tau|}\,d\tau\right) \qquad \alpha_x > 0 \tag{3.163a}$$

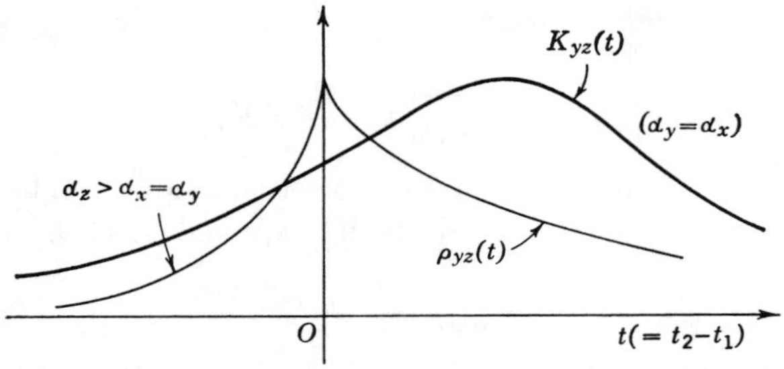

FIG. 3.29. The cross-correlation function ρ_{yz} of two RC filters ($\alpha_z > \alpha_y$), and the cross variance of these filter outputs when the common source is RC noise, with time constant α_x^{-1}.

This integrates finally to

$$K_{yz}(t) = \frac{\psi_x \alpha_y \alpha_z}{\alpha_y + \alpha_z} \begin{cases} \dfrac{e^{-\alpha_y t}}{\alpha_x + \alpha_y} + \dfrac{e^{-\alpha_x t} - e^{-\alpha_y t}}{\alpha_y - \alpha_x} + \dfrac{e^{-\alpha_x t}}{\alpha_x + \alpha_z} & t \geq 0 \\ \dfrac{e^{\alpha_z t}}{\alpha_x + \alpha_z} + \dfrac{e^{\alpha_x t} - e^{\alpha_z t}}{\alpha_z - \alpha_x} + \dfrac{e^{\alpha_x t}}{\alpha_x + \alpha_y} & t \leq 0 \end{cases} \tag{3.163b}$$

Observe, incidentally, that $K_{yz}(t) = K_{zy}(-t)$ and that ρ_{yz} and K_{yz} are not symmetrical about $t = 0$ unless the filters are identical, in which case ρ_{yz} becomes the autocorrelation function ρ_y ($= \rho_z$) and K_{yz} the autovariance K_y. The spectral counterparts of Eqs. (3.162), (3.163*b*) are obtained directly from Eqs. (3.157*a*), (3.159) or by application of the W-K theorem (3.149). For our example above, we get

$$\mathfrak{F}\{\rho_{yz}\} = [(1 + i\omega/\alpha_y)(1 - i\omega/\alpha_z)]^{-1} \tag{3.164a}$$

$$\mathcal{W}_{yz}(f) = \frac{W_0[1 - i\omega(\alpha_z^{-1} - \alpha_y^{-1}) + \omega^2/\alpha_y\alpha_z]}{(1 + \omega^2/\alpha_y^2)(1 + \omega^2/\alpha_z^2)(1 + \omega^2/\alpha_x^2)} \tag{3.164b}$$

Note that $\mathcal{W}_{yz}(f)$ is complex, with an imaginary part that is negative for some frequencies, in accordance with our earlier observations.

3.4-4 Truncated Waves.† As we have seen in the preceding section, it is sometimes convenient to deal with truncated waves (cf. Fig. 3.9*b*). Truncated waves, of course, are what we obtain physically when we consider a finite portion of an existing disturbance, in the interval $(0,T)$ or $(-T/2,T/2)$, say, since in all practical applications observation time T is necessarily finite. Just as in the steady-state cases when $\langle y^{(j)2}\rangle$ exists, we can also define an autocorrelation function and spectral density for the corresponding truncated waves $y_T^{(j)}$. The spectral density $\mathcal{W}_y^{(j)}(f)_T$ has already been specified [cf. Eqs. (3.31), (3.32)]. The autocorrelation function of $y_T^{(j)}$, defined by

$$R_y^{(j)}(t)_T \equiv \frac{1}{T}\int_{-\infty}^{\infty} y_T^{(j)}(t_0)\, y_T^{(j)}(t_0 + t)\, dt_0$$

$$y_T^{(j)} = 0, \ |t| > |T/2| \text{ or } t > T, <0 \tag{3.165}$$

† See, for example, Kampé de Fériet,[6] chap. I.

[cf. Eq. (3.49a)], is easily shown (cf. Prob. 3.22) to be the cosine Fourier transform of the spectral density $\mathcal{W}_y^{(j)}(t)_T$, for example,

$$R_y^{(j)}(t)_T = \int_0^\infty \cos \omega t \, \mathcal{W}_y^{(j)}(f)_T \, df = \tfrac{1}{2}\mathfrak{F}^{-1}\{(\mathcal{W}_y^{(j)})_T\} \qquad (3.166a)$$

with the corresponding relation

$$\begin{aligned} \mathcal{W}_y^{(j)}(f)_T &= 4 \int_0^\infty \cos \omega t \, R_y^{(j)}(f)_T \, df = 2\mathfrak{F}\{(R_y^{(j)})_T\} \\ &\equiv 2|S_y^{(j)}(f)|^2/T \end{aligned} \qquad (3.166b)$$

[cf. Eq. (3.32)]. It also follows (cf. Prob. 3.22b) that $R_y^{(j)}(t)_T$ is an even function of t, for example, $R_y^{(j)}(t)_T = R_y^{(j)}(-t)_T$, and that $R_y^{(j)}(0) \geq |R_y^{(j)}(t)|$.

The ensemble spectral density in this case is given by (Eq. 3.40), provided that the ensemble y_T of truncated waves is derived from an originally *untruncated*, wide-sense stationary ensemble y. We write specifically

$$\mathcal{W}_y(f)_T = 2 \int_{-T}^{T} K_y(\tau)_T e^{-i\omega\tau} \, d\tau = 2\mathfrak{F}\{(K_y)_T\} \qquad (3.167a)$$

where $K_y(\tau)_T = K_y(\tau)(1 - |\tau|/T)$, $K_y(\tau)_T = 0$ ($|\tau| > T$), and observe that the other member of the transform pair is

$$K_y(\tau)_T = \begin{cases} \tfrac{1}{2}\mathfrak{F}^{-1}\{\mathcal{W}_y(f)_T\} = K_y(\tau)\left(1 - \dfrac{|\tau|}{T}\right) & |\tau| < T \\ 0 & |\tau| > T \end{cases} \qquad (3.167b)$$

Of course, $\mathcal{W}_y^{(j)}(f)_T$, in general, is different from $\mathcal{W}_y(f)_T$, even if the original ensemble y from which y_T is taken is ergodic, since y_T is clearly not even wide-sense stationary. The relations (3.166), (3.167) are the analogues of the W-K theorems for the infinite interval ($-\infty < t < \infty$) considered in Sec. 3.2. Corresponding expressions can also be defined and derived for cross-correlation functions and spectra (cf. Prob. 3.23a, b) associated with the truncated ensembles y_T, z_T, etc. In later chapters (cf. Chaps. 16, 17 and Part 4) we shall specifically consider the effects of truncation on the observed statistical properties of random and mixed processes and various decision processes associated with reception.

3.4-5 Nonstationary Processes. We observed earlier, in Sec. 1.4, that nonstationarity can arise mainly from (1) truncation (as above) of an already stationary process or (2) instabilities, systematic drifts and changes, etc., in the underlying physical mechanism of the process itself. In fact, in practical situations we have always to deal with nonstationary, or ultimately nonstationary, situations. However, the much greater emphasis in earlier chapters and sections on the theory of stationary and even ergodic processes in no way attempts to minimize the more realistic questions imposed by such matters as finite observation and data-processing times or the unstable character of the observed data themselves. A very wide class of situations is covered by truncation, due to finite acquisition or observation periods, of *stationary* processes: this is an excellent model when the truncation interval can be considered short compared with the drift time or other time measures of instability. Thus, here we expect the treatment to be based always on the pertinent properties of stationary ensembles. Then, too, as pointed out in Sec. 1.6 (cf. Examples 1 to 3), we are often able to reduce many complicated, nonstationary, and nonergodic cases to suitable functions of stationary ensembles where the techniques of Chaps. 2, 3 can be safely applied without undue strain on realism. Very often, as well, the instability mechanisms which modify otherwise stationary systems can be adequately described by nonstationary deterministic structures together with stationary random processes, so that, again, the starting point of the analysis of random waves is the theory of the stationary process. Examples of this approach will appear in quantity in the later chapters, particularly Chaps. 16 and 17, and in Part 4.

For the moment, let us return to Eq. (3.39) and consider what happens to the spectrum and autovariance function when $y(t)$ is not a stationary ensemble in the steady state. We have

$$\int_{-\infty}^{\infty} \mathcal{W}_y(f)_T e^{i\omega t}\, df = \frac{2}{T}\int_{-T/2}^{T/2} K_y(t_1, t_1 - t)\, dt_1 \tag{3.168}$$

In the limit as $T \to \infty$, it can be shown that the right-hand member of Eq. (3.168) exists,† provided that K_y is suitably continuous at $t_1 = t_2 = 0$ and $y^{(j)}$, of course, possesses a finite time average $\langle y^{2(j)}\rangle$ (for almost all j); y is initially assumed to be a real, entirely random process. Thus, as $T \to \infty$, we have

$$\frac{1}{2}\int_{-\infty}^{\infty} \mathcal{W}_y(f) e^{i\omega t}\, df = \langle K_y(t_1, t_1 - t)\rangle = \tfrac{1}{2}\mathfrak{F}^{-1}\{(\mathcal{W}_y)\} \tag{3.169a}$$

with
$$2\int_{-\infty}^{\infty} \langle K_y(t_1, t_1 - t)\rangle e^{-i\omega t}\, dt = \mathcal{W}_y(f) = 2\mathfrak{F}\{\langle K_y\rangle\} \tag{3.169b}$$

where now $\lim_{T\to\infty} \mathcal{W}_y(f)_T \equiv \mathcal{W}_y(f)_\infty$, and Eqs. (3.169*a*), (3.169*b*) are an extended form of the W-K theorem (3.42). If the nonstationary behavior is such that $|\langle y^{(j)}\rangle^2| \to \infty$, as might occur, for example, with a random wave whose intensity level increases proportionally to time, then the limit $\mathcal{W}_y(f)$ as $T \to \infty$ does not exist and Eqs. (3.169) are meaningless. For suitably bounded nonstationarities, on the other hand, we may expect that Eqs. (3.169) can be applied.

Observe, finally, that, since $K_y(t_1,t_2) = K_y(t_2,t_1)$ for the real processes characteristic of most of our communication applications, $\langle K_y(t_1, t_1 - t)\rangle =$

† The proof of existence follows as in the stationary case above (Sec. 3.2-2) (cf. Bartlett, *op. cit.*, sec. 6.11), where now $\Psi_T(\tau) \equiv (2/T)\int_{-T/2}^{T/2} K_y(t_1, t_1 - \tau)\, dt$ $(|\tau| < T/2)$, $\Psi_T(\tau) = 0$, $(|\tau| > T/2)$ as before.

$\langle K_y(t_1, t_1 + t)\rangle$, with the result that $\mathcal{W}_y(f) = \mathcal{W}_y(-f)$ and $\langle K_y(t_1, t_1 - t)\rangle$ is an even function of t. If the process is stationary, then we get at once $\mathcal{W}_y(f) = \mathcal{W}_y(f)$ [Eqs. (3.42)] as before. The essential analytical complication introduced here by nonstationarity is the fact that the covariance K_y is no longer a function of the difference alone of the observation times t_1, t_2, so that one cannot take advantage of the simplification inherent in the convolution form when filtering is concerned [cf. Eqs. (3.102) et seq.]. From a more profound viewpoint, the requirements on our a priori knowledge of the process are greater: ensemble averages are now always required, in addition to any time average. We can no longer appeal to the ergodic theorem for prediction as to the ensemble behavior from observations on a single member function in the time domain. Corresponding results for the components of the vector process $\mathbf{u}$ considered earlier in Sec. 3.4-1 are obtained in a similar way: analogous to Eq. (3.52), we write

$$\frac{1}{2}\int_{-\infty}^{\infty} \mathcal{W}_{yz}(f)_\infty e^{i\omega t}\, df = \langle K_{yz}(t_1, t_1 - t)\rangle = \tfrac{1}{2}\mathfrak{F}^{-1}\{\langle K_{yz}\rangle\} \tag{3.170a}$$

$$2\int_{-\infty}^{\infty} \langle K_{yz}(t_1, t_1 - t)\rangle e^{-i\omega t}\, df = 2\mathfrak{F}\{\langle K_{yz}\rangle\} = \mathcal{W}_{yz}(f) \tag{3.170b}$$

where now $K_{yz}(t_1, t_1 - t) = K_{zy}(t_1, t_1 + t)$ and $\mathcal{W}_{yz}(f) = \mathcal{W}_{zy}(f)^*$ as before [cf. Eq. (3.149a)]. The details are left to the reader.

Finally, we remark that the concept of the intensity spectrum for stationary stochastic processes discussed above can be extended to nonstationary situations. When they exist, such spectra depend on the interval in question and are often called "instantaneous" spectra. For a detailed discussion, the reader may consult Refs. 16 to 19 at the end of this chapter.

PROBLEMS

3.16 Starting with members of the truncated ensembles and making the appropriate assumptions of stationarity, etc., show that Eq. (3.159) is

$$\mathcal{W}_{yz}(f) = Y_y(i\omega)Y_z(i\omega)^*\mathcal{W}_x(f) \tag{1}$$

where $x(t)$ is the original (steady-state) ensemble at the inputs to the y and z filters.

3.17 White noise, of spectral density W_0, is passed through two separate and different series LCR networks (cf. Table 2.1). If $y(t)$ and $z(t)$ are the ensemble outputs of these networks, respectively,

(*a*) Show that:

For $\omega_1^2 > 0$

$$K_{yz}(t) = \frac{2W_0\omega_{Fy}\omega_{Fz}e^{-\omega_{Fy}t}(A_c \cos \omega_{1y}t + A_S \sin \omega_{1y}t)}{(\omega_{Fy} + \omega_{Fz})^4 + 2(\omega_{1z}^2 + \omega_{1y}^2)(\omega_{Fy} + \omega_{Fz})^2 + (\omega_{1z}^2 - \omega_{1y}^2)^2} \qquad t > 0 \tag{1}$$

where $\omega_{1y} = \sqrt{\omega_{0y}^2 - \omega_{Fy}^2}$, $\omega_{Fy} = \omega_{0y}^2(RC)_y/2$, $\omega_{0y}^2 = 1/(LC)_y$, etc., and

$$\begin{aligned} A_c &= 2\omega_{Fy}[(\omega_{Fy} + \omega_{Fz})^2 + \omega_{1z}^2 - \omega_{1y}^2] - 2(\omega_{Fy} + \omega_{Fz})(\omega_{Fy}^2 - \omega_{1y}^2) \\ A_S &= \omega_{1y}^{-1}[(\omega_{1y}^2 - \omega_{Fy}^2)(\omega_{Fy} + \omega_{Fz})^2 + \omega_{1z}^2 - \omega_{1y}^2] - 4\omega_{Fy}\omega_{1y}^2(\omega_{Fy} + \omega_{Fz}) \end{aligned} \tag{2}$$

For $t < 0$

$$K_{yz}(t) = \frac{2W_0\omega_{Fy}\omega_{Fz}e^{\omega_{Fz}t}(A_c' \cos \omega_{1z}t - A_S' \sin \omega_{1z}t)}{(\omega_{Fy} + \omega_{Fz})^4 + 2(\omega_{1z}^2 + \omega_{1y}^2)(\omega_{Fy} + \omega_{Fz}) + (\omega_{1z}^2 - \omega_{1y}^2)^2} \tag{3}$$

where now A_c', A_S' are obtained from A_c, A_S, respectively, on replacing ω_{Fy} by ω_{Fz}, ω_{Fz} by ω_{Fy}, respectively.

(*b*) Obtain the high-Q forms.

$$K_{yz}(t) = \frac{2W_0\Omega_F}{(1+\Omega_F)(1+\Omega_0^2)} \begin{cases} \omega_{Fy}e^{-\omega_{Fy}t} \cos \omega_{0y}t & t > 0 \\ \omega_{Fz}\Omega_0^2 e^{\omega_{Fz}t} \cos \omega_{0z}t & t < 0 \end{cases} \tag{4}$$

where now $\Omega_F \equiv \omega_{Fz}/\omega_{Fy}$, $\Omega_0 \equiv \omega_{0z}/\omega_{0y}$, and $\omega_{0y} \approx \omega_{0z}$, $\omega_{0y}/2\omega_{Fy} \gg 1$, etc.

(*c*) In the case of low Q, where in both instances $\omega_{1y}^2 < 0$, $\omega_{1z}^2 < 0$, show that Eqs. (1) and (2) may be used if ω_1 is replaced by $i\omega_1'$, $(\omega_2')^2 = \sqrt{\omega_F^2 - \omega_0^2} > 0$.

(*d*) In the case of critical damping in both filters, obtain

$$K_{yz}(t) = \frac{2W_0\omega_{Fy}\Omega_F}{(1+\Omega_F)^2} \begin{cases} \left(\dfrac{2\Omega_F}{1+\Omega_F} - \omega_{Fy}t\right) e^{-\omega_{Fy}t} & t > 0 \\ \Omega_F^2 \left(\dfrac{2\Omega_F}{1+\Omega_F} + \omega_{Fz}t\right) e^{\omega_{Fz}t} & t < 0 \end{cases} \tag{5}$$

3.18 (*a*) Show that $|K_{yz}(t)|/\sqrt{K_y(0)K_z(0)} \le 1$, $\bar{y} = \bar{z} = 0$ for Prob. 3.17. Extend this relation to complex processes.

(*b*) If $x(t)$ is a stationary RC noise ensemble, of spectral density W_0 (at $f = 0$), e.g.,

$$\mathcal{W}_x(f) = W_0/(1 + \omega^2/\alpha_x^2) \qquad \alpha_x = (RC)_x^{-1} \tag{1}$$

and if x is then passed through two RC filters [$\alpha_y = (RC)_y^{-1}$, $\alpha_z = (RC)_z^{-1}$], to generate a y and a z ensemble, show that

$$K_{y\dot{z}}(t) = \frac{W_0\alpha_x^2\alpha_y\alpha_z}{4(\alpha_y+\alpha_z)(\alpha_y-\alpha_x)}\left[\frac{-(\alpha_y+\alpha_z)e^{-\alpha_x t}}{\alpha_x+\alpha_z} + \frac{2\alpha_y e^{-\alpha_y t}}{\alpha_x+\alpha_y}\right] \qquad t > 0 \tag{2a}$$

$(\alpha_x \neq \alpha_y \neq \alpha_z)$

$$K_{y\dot{z}}(t) = \frac{W_0\alpha_x^2\alpha_y\alpha_z}{4(\alpha_x+\alpha_z)(\alpha_z-\alpha_x)}\left[\frac{(\alpha_y+\alpha_z)e^{\alpha_x t}}{\alpha_x+\alpha_y} - \frac{2\alpha_z e^{\alpha_z t}}{\alpha_x+\alpha_z}\right] \qquad t < 0 \tag{2b}$$

Observe that $K_{y\dot{z}}(0+) \neq K_{\dot{y}z}(0-)$ unless $\alpha_y = \alpha_z$. Show also that, if $\alpha_x = \alpha_y$ ($\neq \alpha_z$), $K_{y\dot{z}}(t)$ ($t > 0$) is unchanged from Eq. (2*a*) but that

$$K_{y\dot{z}}(t) = \frac{-W_0\alpha_x\alpha_z e^{-\alpha_x t}}{4(\alpha_x+\alpha_z)}\left[1 - \alpha_x t - \frac{3\alpha_x+\alpha_z}{2(\alpha_x+\alpha_z)}\right] \qquad t < 0,\ \alpha_x = \alpha_y \tag{3}$$

(*c*) Obtain $K_{y\ddot{z}}(t)$ for the waves of (*b*).

3.19 Consider two general linear (passive) networks of m and n independent meshes, respectively (cf. Fig. 2.2), through each of which RC noise is passed. Discuss the general structure of the cross-variance functions $K_{yz}(t), \ldots, K_{y^{(k)}z^{(l)}}(t)$, where $y^{(k)} = d^ky/dt^k$, $z^{(l)} = d^lz/dt^l$ (all k, l), and describe the behavior of k at $t \to \pm\infty, 0$.

3.20 Defining the cross time moments

$$\langle\tau^n\rangle_{\rho_{xy}} \equiv \int_{-\infty}^{\infty} \tau^n\rho_{xy}(\tau)\,d\tau \qquad \langle\tau^m\rangle_{K_{xy}} \equiv \int_{-\infty}^{\infty} \tau^m K_{xy}(\tau)\,d\tau \tag{1}$$

associated with the cross-correlation function of a pair of linear filters and of the cross-variance function of the two processes (x,y), develop the properties analogous to Eqs.

(3.130), (3.131), and obtain the analogues of Eqs. (3.128), (3.129). Give in each instance the appropriate conditions on the processes and on the filters for such developments.

3.21 Two stationary processes, y and z, are obtained from the passage of an original stationary process x through a pair of linear passive *narrowband* filters, so that $y = \boldsymbol{T}_y\{x\}$, $z = \boldsymbol{T}_z\{x\}$. Specifically, we have

$$h_y(t) \doteq 2h_0(t)_y \cos [\omega_{0y}t - \gamma_{0y}(t)] \qquad h_z(t) \doteq 2h_0(y)_z \cos [\omega_{0z}t - \gamma_{0z}(t)] \tag{1}$$

[cf. Eq. (2.63a)].

(a) Defining

$$\begin{aligned}\rho_0(t)_{0yz} &\equiv \int_{-\infty}^{\infty} h_0(\tau)_y h_0(\tau + t)_z \cos [(\omega_{0z} - \omega_{0y})\tau + \gamma_{0y}(\tau) - \gamma_{0z}(\tau + t)]\, d\tau \\ &= \rho_0(-t)_{0yz} \end{aligned} \tag{2a}$$

$$\begin{aligned}\lambda_0(t)_{0yz} &\equiv -\int_{-\infty}^{\infty} h_0(\tau)_y h_0(\tau + t)_z \sin [(\omega_{0z} - \omega_{0y})\tau + \gamma_{0y}(\tau) - \gamma_{0z}(\tau + t)]\, d\tau \\ &= -\lambda_0(-t)_{0yz} \end{aligned} \tag{2b}$$

show that

$$K_{yz}(t) \doteq 2\int_{-\infty}^{\infty} [\rho_0(\tau)_{0yz} \cos \omega_{0z}\tau + \lambda_0(\tau)_{0yz} \sin \omega_{0z}\tau] K_x(t - \tau)\, d\tau \tag{3}$$

$$K_{yz}(t) \equiv \rho_0(t)_{yz} \cos \omega_{0z}(t) + \lambda_0(t)_{yz} \sin \omega_{0z}t \tag{3a}$$

and hence, if we make the further definitions

$$\rho_0(t)_{0yz} + i\lambda_0(t)_{0yz} \equiv \mathcal{K}_0(t)_{0yz} \qquad \rho_0(t)_{yz} + i\lambda_0(t)_{yz} \equiv \mathcal{K}_0(t)_{yz} \tag{4}$$

show that

$$K_{yz}(t) = \text{Re}\, [\mathcal{K}_0(t)_{yz} e^{-i\omega_{0z}t}] \tag{5a}$$

where

$$\mathcal{K}_0(t)_{yz} \doteq 2\int_{-\infty}^{\infty} \mathcal{K}_0(t \mp \tau)_{0yz} e^{\pm i\omega_{0z}} K_x(\tau)\, d\tau \tag{5b}$$

Note that $\mathcal{K}_0(\tau)_{0yz} = \mathcal{K}_0(-\tau)^*_{0zy} e^{i(\omega_{0z} - \omega_{0y})\tau}$, while $\mathcal{K}_0(t)_{yz} = \mathcal{K}_0(-t)_{zy}$.

(b) If the source ensemble $x(t)$ is, moreover, itself narrowband, show that now

$$K_{yz}(t) = \text{Re} \left[e^{-i\omega_{0z}t} \int_{-\infty}^{\infty} \mathcal{K}_0(t \mp \tau)_{0yz} \mathcal{K}_0(\pm\tau)_x e^{\pm i(\omega_{0z} - \omega_{0x})\tau}\, d\tau \right] \tag{6}$$

where

$$\mathcal{K}_0(t)_{yz} = \int_{-\infty}^{\infty} \mathcal{K}_0(t \mp \tau)_{0yz} \mathcal{K}_0(\pm\tau)_x e^{\pm i(\omega_{0z} - \omega_{1x})\tau}\, d\tau$$

$$K_x(\tau) \doteq \text{Re}\, [\mathcal{K}_0(\tau)_x e^{-i\omega_{0x}\tau}] \tag{7}$$

with $\mathcal{K}_0(\tau)_x = \rho_{0x}(\tau) + i\lambda_{0x}(\tau) = \frac{1}{2}\mathcal{K}_{\alpha,\beta}(t)_x$ [cf. Eq. (3.119)] and $\mathcal{K}_0(\tau)_x = \mathcal{K}_0(-\tau)^*_x$, etc.

(c) In the case of a narrowband source x and *broadband* filters, verify that

$$K_{yz}(t) = \text{Re} \int_{-\infty}^{\infty} \rho_{yz}(t \mp \tau) \mathcal{K}_0(\tau)_x e^{\mp i\omega_{0x}\tau}\, d\tau \tag{8}$$

[cf. Eq. (3.125)].

(d) Show that in terms of the narrowband system functions of the y and z filters for (a), (b), the cross-spectral density is given by

$$\mathcal{W}_{yz}(f) \doteq \mathcal{W}_x(f) Y_0(i\omega - i\omega_{0y})_y Y_0(i\omega - i\omega_{0z})^*_z \tag{9}$$

for a general source ensemble, while if the source itself is narrowband, $\mathcal{W}_x(f)$ can be replaced by $\mathcal{W}_x(f - f_{0x})$.

3.22 (a) Given a truncated ensemble $y(t)_T$, show that the autocorrelation function $R_y{}^{(j)}(t)_T$ for almost all members is the Fourier transform of the individual spectral density $\mathcal{W}_y{}^{(j)}(f)_T$.

(*b*) Establish the properties $R_y^{(i)}(t)_T = R_y^{(i)}(-t)_T$, $|R_y^{(i)}(t)_T| \leq R_y^{(i)}(0)_T$. State carefully your assumptions for (*a*), (*b*). J. Kampé de Fériet.[6]

3.23 (*a*) Given a pair of truncated ensembles $y(t)_T$, $z(t)_T$ derived from a common source, show that $\mathcal{W}_{yz}^{(i)}(f)_T = 2\mathfrak{F}^{-1}\{R_{yz}^{(i)}\}$.

(*b*) Establish the properties $R_{yz}^{(i)}(t)_T = R_{zy}^{(i)}(-t)_T^*$ and

$$|R_{yz}^{(i)}(t)_T|^2 \leq R_y^{(i)}(0)_T R_z^{(i)}(0)_T$$

and state the assumptions on the processes y_T and z_T for which (*a*) and (*b*) hold. J. Kampé de Fériet.[6]

REFERENCES

1. Wiener, N.: Generalized Harmonic Analysis, *Acta Math.*, **55:** 117 (1930).
2. Khintchine, A.: Korrelationstheorie der stationären stochastischen Prozesse, *Math. Ann.*, **109:** 604 (1934); The Theory of Correlation of Stationary Chance Processes, *Uspekhi Mat. Nauk*, no. 5, 1938.
3. Bochner, S.: "Lectures on Fourier Analysis," Princeton, N.J., 1936–1937; also, "Harmonic Analysis and the Theory of Probability," University of California Press, Berkeley, Calif., 1955.
4. See, for example, H. Wold, Analysis of Stationary Time Series, thesis, Uppsala University, 1938; M. Loève, Sur les fonctions aléatoires stationnaires de seconde ordre, *Rev. sci.*, **83:** 297 (1945). An extensive bibliography is included in Ref. 10 and in Ref. 15, especially chap. VI.
5. Loève, M.: Ref. 4.
6. Kampé de Fériet, J.: Introduction to the Statistical Theory of Turbulence. II, *J. Soc. Ind. Appl. Math.*, **2:** 143 (1954), chap. II, secs. 9, 10. The proof is based on Lévy's continuity theorem; cf. H. Cramér, "Mathematical Methods of Statistics," sec. 10.4, Princeton University Press, Princeton, N.J., 1946.
7. Slutsky, E.: Sur les fonctions aléatoires presque périodiques et sur la décomposition des fonctions aléatoires stationnaires et composantes, *Actualités sci. et ind.*, **35:** 738 (1938).
8. Lee, Y. W.: Communication Applications of Correlation Analysis, Symposium on Applications of Autocorrelation Analysis to Physical Problems (ONR), Woods Hole, June, 1949. Also, Y. W. Lee and J. B. Wiesner, Correlation Functions and Communication Applications, *Electronics*, **23:** 86 (June, 1950), and Y. W. Lee, Applications of Statistical Methods to Communication Problems, *MIT Research Lab. Electronics Tech. Rept.* 181, September, 1950.
9. Davenport, W. B., R. A. Johnson, and D. Middleton: Statistical Errors in Measurements on Random Time Functions, *J. Appl. Phys.*, **23:** 377 (1952), especially eqs. (2.21), (2.22).
10. Blanc-Lapierre, A., and R. Fortet: "Théorie des fonctions aléatoires," pp. 478–514, Masson et Cie, Paris, 1953.
11. Kampé de Fériet, J.: Ref. 10, chap. 14; see also Ref. 6.
12. Batchelor, G. K.: "The Theory of Homogeneous Turbulence," Cambridge, New York, 1956.
13. Ref. 6, pt. III, p. 244; cf. chap. III, secs. 5, 6.
14. Lee, Y. W.: Applications of Statistical Methods to Communication Problems, *MIT Research Lab. Electronics Tech. Rept.* 181, September, 1950; see also, for example, S. Goldman, "Information Theory," chap. 8, sec. 8.7, Prentice-Hall, Englewood Cliffs, N.J., 1953.

15. Bartlett, M. S.: "An Introduction to Stochastic Processes," Cambridge, New York, 1955.
16. Fano, R. M.: Short-time Autocorrelation Functions and Power Spectra, *J. Acoust. Soc. Am.*, **22**: 546 (1950).
17. Page, C. H.: Instantaneous Power Spectra, *J. Appl. Phys.*, **23**: 103 (1952).
18. Lampard, D. G.: Generalization of the Wiener-Khintchine Theorem to Nonstationary Processes, *J. Appl. Phys.*, **25**, 802 (1954); see also The Response of Linear Networks to Suddenly Applied Stationary Random Noise, *IRE Trans. on Circuit Theory*, **CT-2**(1): 49 (March, 1955).
19. Silverman, R. A.: Locally Stationary Random Processes, *IRE Trans. on Inform. Theory*, **IT-3**: 182 (1957).

CHAPTER 4

SAMPLING, INTERPOLATION, AND RANDOM PULSE TRAINS

In practical applications, it is often necessary to reduce continuous data to a series of numbers, to sequences of voltage or current pulses, and to determine the statistical properties of random wave trains that are generated in the course of the communication process itself. Such situations arise in the reduction of continuous data for digital computation, in feedback and other control systems, in radar, and in various types of message coding, e.g., telegraphy, pulse-time modulation, and time-multiplexing systems. Other examples, e.g., shot and thermal noise, impulsive interference, etc., are provided by the characteristic noise mechanisms of the propagation medium and of the system elements themselves. Methods of sampling and interpolation also offer a number of useful technical devices for the analysis of waveform, as we shall see in more detail in Chap. 6 and in Part 4. In the present chapter as in Chap. 3, we consider again such ensemble properties as the intensity spectrum and the covariance function. A discussion of higher-order statistics, such as the probability densities themselves, is reserved for subsequent chapters (cf. Chaps. 7 to 11 and 17).

As an operation on the ensemble, sampling and interpolation may be regarded as a form of linear switching or as a simple form of multiplication between two time-dependent quantities. We distinguish two main types of operation $\boldsymbol{T}_S$ here:

1. Discrete sampling, and interpolation
2. Pulsed sampling

To complete the discussion, we include a third type:

3. Generation of random pulse trains and continuous waveforms by various random mechanisms, which, however, cannot be regarded as a deterministic operation $\boldsymbol{T}_S$ on the original ensemble

In case 1, examined in Secs. 4.1, 4.2, a continuous random *series*† (or *process*, if interpolation is desired) is produced by sampling a continuous process. On the other hand, Secs. 4.3ff. consider case 2, where a new waveform is derived, exhibiting in modified fashion the random (and

† See Sec. 1.3-3, for example.

deterministic) features of the original, while for case 3 it is the resultant wave that claims our interest. In all these instances, steady-state operation is once more assumed.

SAMPLING AND INTERPOLATION

4.1 *Discrete Sampling (Continuous Random Series)*

Let us begin with the continuous random series $y_n = y(t_n)$ $(-\infty < n < \infty)$, obtained by sampling the steady-state continuous random process $y(t)$,

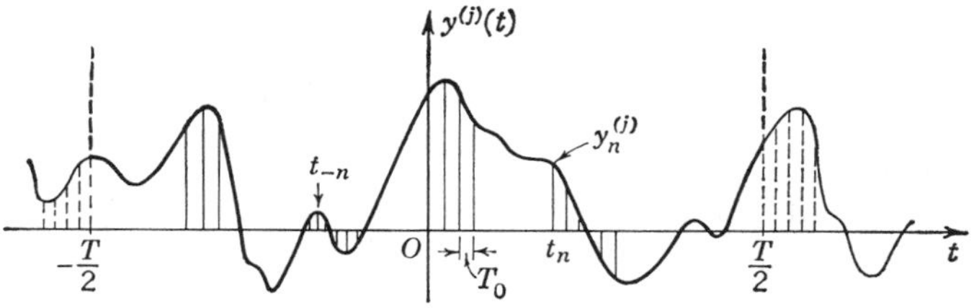

FIG. 4.1. Discrete (periodic) sampling of a continuous random wave in the interval $(-T/2, T/2)$.

in the manner of Fig. 4.1. If $\boldsymbol{T}_S$ is the sampling operation, we write specifically

$$y_n = \boldsymbol{T}_S\{y(t)\} = \int_{-\infty}^{\infty} y(t_0)\delta(t_0 - t_n)\, dt_0 \qquad -\infty < n < \infty \tag{4.1}$$

For the moment, we assume the process $y(t)$ to be wide-sense stationary and entirely random, except for a possible steady component. Considering now truncated samples $(-N \leq n \leq N)$ with periodic sampling, so that $t_n = nT_0$, we may *define* an amplitude spectral density for each member of the series y_n by analogy with the corresponding situation for continuous waves. We write†

$$S_{y_n}{}^{(j)}(f)_N \equiv T_0 \sum_{-N}^{N} y_n{}^{(j)} e^{-i\omega n T_0} \tag{4.2a}$$

$$y_n{}^{(j)} = \mathfrak{F}^{-1}\{S_{y_n}{}^{(j)}\} = \int_{-f_0/2}^{f_0/2} S_{y_n}{}^{(j)}(f)_N e^{i\omega n T_0}\, df \tag{4.2b}$$

where $f_0 = 1/T_0$ and $S_{y_n}{}^{(j)}$ and $y_n{}^{(j)}$ are a Fourier-transform pair. From Eq. (4-2*a*), it is evident that this amplitude spectral density is periodic *in frequency*, with period $T_0{}^{-1}$, and that it is completely specified by its values in the primary interval $(-\frac{1}{2}T_0, \frac{1}{2}T_0)$. Since $y(t)$, and hence y_n, is a real quantity, we have $S_{y_n}{}^{(j)}(f)_N = S_{y_n}{}^{(j)}(-f)_N^*$ and we may further

† See, for example, James, Nichols, and Phillips,[20] sec. 6.7, pp. 281ff. See also J. L. Doob, "Stochastic Processes," chap. 10, secs. 3, 4, 7, and 9, Wiley, New York, 1953, for a rigorous treatment of some of these questions.

define an intensity density, analogous to Eq. (3.32), by

$$\mathcal{W}_{(y_n)}^{(j)}(f)_N = 2|S_{y_n}{}^{(j)}(f)_N|^2/(2N+1)T_0 \tag{4.3}$$

with an ensemble density

$$\mathcal{W}_{(y_n)}(f) \equiv \lim_{N\to\infty} \frac{1}{2N+1} \boldsymbol{E}\left\{\frac{2|S_{y_n}{}^{(j)}(f)_N|^2}{T_0}\right\} \tag{4.4a}$$

[cf. Eq. (3.43)]. As before, if there are steady and periodic components in addition, at the "line" frequencies $f = mf_a$ ($m = 0, 1, 2, \ldots$) Eq. (4.4a) holds for the continuous part of the spectrum and is interpreted as

$$\left[\lim_{N\to\infty} \frac{1}{(2N+1)^2T_0{}^2} \boldsymbol{E}\{2|S_{y_n}{}^{(j)}(f)_N|^2\}\right] \delta(f - mf_a) \tag{4.4b}$$

at these discrete frequencies.

Starting now with the autocorrelation coefficient of the truncated series y_n, $(-N,N)$, we obtain† in the limit

$$R_{y_n}{}^{(j)}(kT_0) \equiv \lim_{N\to\infty} \frac{1}{2N+1} \sum_{-N}^{N} y_n{}^{(j)} y_{n+k}^{(j)}$$

$$= \lim_{N\to\infty} \sum_{-N}^{N} \iint_{-f_0/2}^{f_0/2} \frac{S_{y_n}{}^{(j)}(f)_N S_{y_n}{}^{(j)}(f')_N}{2N+1} \times \exp\,[i\omega nT_0 + i\omega'(n+k)T_0]\, df\, df' \tag{4.5}$$

Taking the ensemble average of both sides of Eq. (4.5) and using the fact that y_n is wide-sense stationary, we have‡

$$\overline{R_{y_n}{}^{(j)}(k)} = M_y(kT_0) = M_y(k) \tag{4.6}$$

which becomes

$$M_y(k) = \lim_{N\to\infty} \frac{1}{2N+1} \iint_{-f_0/2}^{f_0/2} df\, df'\, \overline{S_{y_n}{}^{(j)}(f)_N S_{y_n}{}^{(j)}(f')_N} e^{i\omega' kT_0} \sum_{-\infty}^{\infty} e^{i(\omega+\omega')nT_0} \tag{4.7}$$

(We can extend the limits on the summation, since $y_n{}^{(j)}$ in the truncated series vanishes for $|n| > N$.) At this point, we use a result of Prob. 4.2, namely,

$$\sum_{-\infty}^{\infty} e^{i(\omega\mp\omega')nT_0} = \frac{1}{T_0}\,\delta(f' \mp f) \qquad -\frac{f_0}{2} < f, f' < \frac{f_0}{2} \tag{4.8}$$

† We assume that $R_y{}^{(j)}(t)$ exists (all t) and is continuous at $t = 0$ and hence that $R_y{}^{(j)}(kT_0)$ is also defined for all k.

‡ With an obvious abbreviation, here and subsequently.

for f, f' in the primary interval, to get at once

$$M_y(k) = \lim_{N\to\infty} \frac{1}{2} \int_{-f_0/2}^{f_0/2} \frac{2\overline{S_{y_n}{}^{j}(f)_N{}^2}}{(2N+1)T_0} e^{-i\omega kT_0}\, df$$
$$= \int_0^{f_0/2} \mathcal{W}_{(y_n)}(f) \cos \omega kT_0 \, df \tag{4.9}$$

where we interpret $\mathcal{W}_{(y_n)}$ according to Eq. (4.4a) or (4.4b) (see Secs. 3.2-2, 3.2-4). From Eq. (4.9), the mean intensity is

$$\overline{y_n{}^2} = \overline{y(t_n)^2} = M_y(0) = \int_0^{f_0/2} \mathcal{W}_{(y_n)}(f)\, df \tag{4.10}$$

and if we multiply both sides of Eq. (4.9) by $e^{-i\omega kT_0}$ and sum over all k, we get at once with the help of Eq. (4.8)

$$\mathcal{W}_{(y_n)}(f) = 2T_0 \sum_{-\infty}^{\infty} M_y(k) e^{-i\omega kT_0} = 2T_0 \sum_{k=0}^{\infty} \epsilon_k M_y(k) \cos \omega kT_0 \tag{4.11}$$

which exhibits the periodic nature of the spectral density as well.† The relations (4.9) and (4.11) are the analogues, for the discrete case, of the ensemble form of the W-K theorem (3.42).

For an ergodic process, y_n is also ergodic, and so $M_y(kT_0) = R_y{}^{(j)}(k)$, $\overline{y_n{}^2} = \langle y_n{}^{(j)2}\rangle \equiv \lim_{N\to\infty} (2N+1)^{-1} \sum_{-N}^{N} y_n{}^{(j)2}$ (probability 1). Note that, for a *series*, $\int_0^{\infty} \mathcal{W}_{y_n}(f)\, df \to \infty$ but that $\overline{y_n{}^2}$, $\langle y_n{}^{(j)2}\rangle$ are still finite [cf. Eq. (4.10)]. If there are no periodicities and $\bar{y}_n$ vanishes, then $M_y(k) = K_y(k)$, where $K_y(\pm\infty)$ vanishes also. In a similar way, letting $k \to \pm\infty$ in Eq. (4.9) establishes the presence or absence of possible deterministic components, just as in the continuous cases. We observe, finally, that an individual spectral density $\mathcal{W}_{y_n}{}^{(j)}(f)$ can be defined in the limit $N \to \infty$, with suitable restrictions on the process y and the series y_n derived from it, corresponding to a similar result already considered for the continuous case (Sec. 3.2). The individual W-K theorem is then given by Eqs. (4.9) and (4.11), with $\mathcal{W}_{y_n}(f)$ and $M_y(k)$ replaced by $\mathcal{W}_{y_n}{}^{(j)}(f)$, $R_{y_n}{}^{(j)}(k)$, respectively. When y_n is an ergodic series, these two sets of relations are equivalent (probability 1).

Let us illustrate Eqs. (4.9), (4.11) with a simple example. Suppose that the sampling interval T_0 and the bandwidth of the process y are both sufficiently large for us to regard the sampled values y_n as effectively

† For period intervals other than $(-f_0/2, f_0/2)$, we must recall that $\sum_{-\infty}^{\infty} e^{i(\omega\pm\omega')nT_0} = T_0{}^{-1} \sum_{-\infty}^{\infty} \delta(f' \pm f + m/T_0)$ (cf. Prob. 4.2).

uncorrelated, with $\bar{y}_n = \bar{y} \neq 0$, $M_y(0) = \overline{y_n^2} = \overline{y^2}$, and $M_y(k) = \bar{y}_n^2 = \bar{y}^2$ ($k \neq 0$). The spectrum of Eq. (4.11) becomes, in the primary interval ($-f_0/2 < f < f_0/2$),

$$\mathcal{W}_{(y_n)}(f) = 2T_0\overline{y^2} + 2T_0\bar{y}^2 \sum_{-\infty}^{\infty}{}' e^{-i\omega k T_0} = 2T_0(\overline{y^2} - \bar{y}^2) + 2\bar{y}^2\delta(f - 0) \qquad (4.12)$$

where the prime on the summation indicates that the term $k = 0$ is omitted; Eq. (4.8) is then used to give us the second relation. This result is easily checked by applying Eq. (4.9) to it, yielding $M_y(k)$, with the above properties, as it should.

Generalizations of the preceding are easily made. Let us consider the *filtered series*, where now the filter consists of a sequence of discrete weights h_m, subject to

$$\begin{aligned} h_m &= 0 \qquad m < 0 \\ \sum_0^\infty |h_m| &< \infty \\ \text{with} \qquad Y(i\omega) &= \sum_{m=0}^{\infty} h_m e^{-im\omega T_0} \end{aligned} \qquad (4.13)$$

in direct analogy with the linear, time-invariant, and stable filters associated with continuous processes [Secs. 2.2-3 and 2.2-5(2)]. Thus, if y_m is the input series and z_n [$= z(t_n)$] is the corresponding output ensemble, we have for the analogue of the convolution integral [Eq. (2.51)]

$$z_n = \sum_{-\infty}^{\infty} h_m y_{n-m} = \sum_{-\infty}^{\infty} h_{n-m} y_m \qquad (4.14)$$

The second-moment function becomes here [cf. Eq. (3.101), Prob. 4.3]

$$M_z(n) = \sum_{l=-\infty}^{\infty} \sum_{m=-\infty}^{\infty} h_l M_y(n + l - m) h_m \qquad (4.15)$$

and the ensemble spectral density is, in turn,

$$\mathcal{W}_{(z_n)}(f) = |Y(i\omega)|^2 \mathcal{W}_{(y_n)}(f) \qquad -f_0/2 < f < f_0/2 \qquad (4.16)$$

corresponding to Eq. (3.104) for continuous processes.

Cross correlation with series is also possible. For a pair of sampled waves y_n, z_n, we get, analogous to Eqs. (3.149), the W-K relations

$$\begin{aligned} M_{yz}(k) &= \frac{1}{2}\int_{-f_0/2}^{f_0/2} \mathcal{W}_{(yz)_n}(f) e^{i\omega k T_0}\, df = M_{zy}(-k) \\ \mathcal{W}_{(yz)_n}(f) &= 2T_0 \sum_{-\infty}^{\infty} M_{yz}(k) e^{-i\omega k T_0} = \mathcal{W}_{(zy)_n}(-f)^* \end{aligned} \qquad (4.17)$$

with
$$\mathcal{W}_{(yz)_n}(f) = \lim_{N\to\infty} \frac{2\overline{S_{y_n}{}^{(j)}(f)_N S_{z_n}{}^{(j)}(f)_N^*}}{(2N+1)T_0} \tag{4.17a}$$

and the corresponding generalization of Eq. (4.4*b*) for any d-c and periodic terms (cf. Prob. 4.4). Still other generalizations, e.g., for narrowband waves, etc., can be made in the same way on replacing continuous processes and operations by discrete series in the appropriate places.

PROBLEMS

4.1 (*a*) Discuss the derivation of Eqs. (4.9), (4.11) when the sampling frequency is commensurable with periodic components in the original mixed process $y(t)$.

(*b*) Derive the *individual* W-K theorem [Eqs. (3.62)] for the discrete series y_n, and discuss the conditions (Chap. 3, Ref. 6) for the existence of the spectral density.

4.2 Obtain the following developments:

$$\sum_{-\infty}^{\infty} e^{2\pi i n(x-x')T_0} = \sum_{n=0}^{\infty} \epsilon_n \cos 2\pi n(x-x')T_0 = T_0^{-1}\sum_{-\infty}^{\infty} \delta(x'-x+m/T_0) \tag{1}$$

HINT: Observe that $\delta(z-na)$ $(n = 0, \pm 1, \pm 2, \ldots)$ is *periodic*, with period a; hence, develop $\delta(z-na)$ in a Fourier series.

$$\sum_{-\infty}^{\infty} \epsilon_n(-1)^n \cos 2\pi n(x'-x)T_0 = \frac{1}{T_0}\sum_{-\infty}^{\infty} \delta\left(x'-x-\frac{2m+1}{2T_0}\right) \tag{2}$$

$$\sum_{-\infty}^{\infty} \epsilon_n \cos 2\pi n y T_0 \cos 2\pi(x'-x)T_0$$
$$= \frac{1}{2T_0}\sum_{-\infty}^{\infty}\left[\delta\left(x'-x+y+\frac{m}{T_0}\right)+\delta\left(x'-x-y+\frac{m}{T_0}\right)\right] \tag{3}$$

$$\sum_{-\infty}^{\infty} \epsilon_n \cos 4\pi n(x'-x)T_0 = \frac{1}{2T_0}\sum_{-\infty}^{\infty}\delta\left(x-x'-\frac{m}{2T_0}\right) \tag{4}$$

$$2\sum_{-\infty}^{\infty} \cos (2n+1)2\pi(x'-x)T_0$$
$$= \frac{1}{2T_0}\sum_{-\infty}^{\infty}\left[\delta\left(x-x'-\frac{m}{T_0}\right)-\delta\left(x-x'-\frac{2m+1}{2T_0}\right)\right] \tag{5}$$

4.3 Show, for the filtered series z_n [Eq. (4.14)], that the second-moment function is

$$M_z(n) = \sum_{l=0}^{\infty}\sum_{m=0}^{\infty} h_l M_y(n+l-m)h_m \tag{1}$$

with the spectral intensity

$$\mathcal{W}_{(z_n)}(f) = |Y(i\omega)|^2 \mathcal{W}_{(y_n)}(f) \tag{2}$$

[cf. Eqs. (4.16), (4.13)].

4.4 Obtain the generalization of Eqs. (4.15), (4.16) for a pair of sampled processes y_n and z_n, namely,

$$M_{(yz)}(n) = \sum_{l=0}^{\infty} \sum_{m=0}^{\infty} h_l(y) M_x(n + l - m) h_m(z) \tag{1}$$

$$\mathcal{W}_{(yz)_n}(f) = Y_y(i\omega) Y_z(i\omega)^* \mathcal{W}_{(x_n)}(f) \tag{2}$$

where y, z stem from an ensemble of common sources, x, and are derived from this source after periodic sampling by passage through (discrete) filters (y), (z) [cf. Eqs. (4.13)].

4.2 *Discrete Sampling (Interpolation)*

It is often necessary, as well as convenient, to analyze the waveforms of message, signal, and noise that occur in communication problems in terms of sets of sampled values at discrete time (and, sometimes, discrete frequency) intervals. The essential motivation of such a sampling process here, unlike the generation of series discussed in Sec. 4.1, is to preserve the information-bearing elements of the original wave. Thus, in terms of a suitably selected set of these sampled values, we would like to be able to reconstruct, uniquely, the waveform of the original disturbance, not only at the sampling instants, but at all intermediate times as well. In addition to the sampled values, then, we shall also need a suitable set of elementary time functions, or sampling "pulses." When these are combined with the sampled values, the original wave is everywhere restored, by analogy very much like the familiar Fourier-series development of a periodic wave, where the sine and cosine play the role of the elementary time functions, and their coefficients, that of the sampled values. The sampling process, by which the sampled wave is at all instants again uniquely specified, is called *interpolation*. Procedures of this type have been examined from a

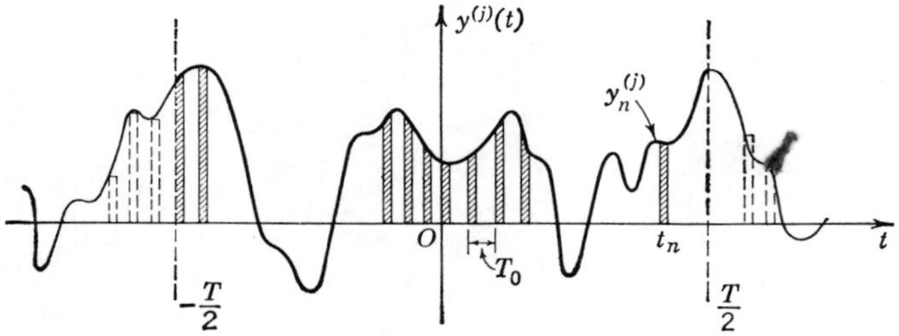

FIG. 4.2. Pulsed sampling of a continuous waveform in the time domain; truncated pulse train in $(-T/2, T/2)$.

purely mathematical viewpoint by Whittaker[1] and by Levinson,[2] while applications to communication theory have been made by Nyquist,[3] Gabor,[4] Shannon,[5] Oswald,[6] and Ville,[7] with generalizations by Kohlenberg,[8] Woodward,[9] Goldman,[10] and others.[11,11a] In this section some of the principal features of this sampling theory, and the conditions under which it can be applied, are reviewed and extended to include ensembles of random waves, as well as the individual waveforms assumed in most of the earlier treatments.

4.2-1 The Sampled Wave. We distinguish two situations wherein such interpolation may be carried out: (1) the case where the ensemble $y(t)$ is strictly *band-limited* in the frequency interval $(-B,B)$ (cf. Figs. 4.2, 4.3*a*), so that each member persists in some fashion for all time $(-\infty < t < \infty)$ with a finite total intensity; and, conversely, (2), the case in which the waveform is strictly *time-limited* (cf. Fig. 4-3*b*), i.e., is different from zero in at most a finite number of intervals within a single finite interval, so that its (amplitude and intensity) spectra are distributed in some fashion over all frequencies.

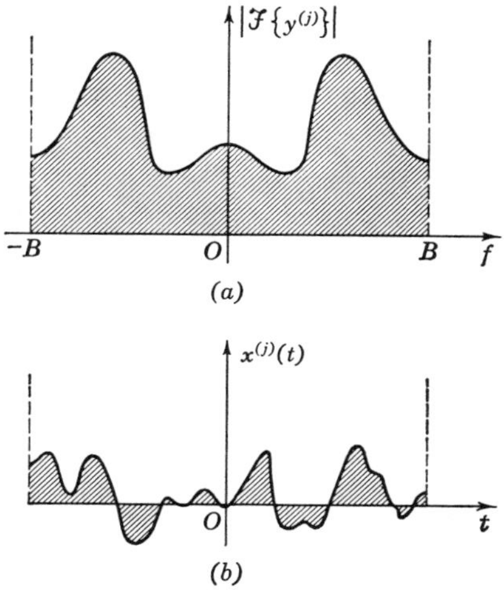

FIG. 4.3. (*a*) Modulus of the amplitude spectrum of a strictly band-limited wave. (*b*) A strictly time-limited wave $x^{(j)}(t)$.

Before we extend the treatment to random waves, however, let us begin with case 1 above and consider only a single member $y^{(j)}(t)$ of the ensemble. We require that the waveform be band-limited, with a definite *amplitude* spectral density, which exists, at worst, in the delta-function sense† [cf. (2.45)]. Then, if $u(t - t_n)$ is a typical sampling pulse or wave, applied at t_n $(-\infty < n < \infty)$, we obtain the sampled wave

$$z^{(j)}(t) = \boldsymbol{T}_S\{y^{(j)}\} = \int_{-\infty}^{\infty} y^{(j)}(t_0)u(t - t_0) \sum_{-\infty}^{\infty} \delta(t_0 - t_n)\, dt_0$$

$$= \sum_{-\infty}^{\infty} y^{(j)}(t_n)u(t - t_n) \tag{4.18}$$

Let us further require that u, $y^{(j)}$, $z^{(j)}$ possess Fourier transforms, so that

$$S^{(j)}_{(u,y,z)}(f) = \int_{-\infty}^{\infty} (u,y^{(j)},z^{(j)})e^{-i\omega t}\, dt \qquad \omega = 2\pi f$$
$$(u,y^{(j)},z^{(j)}) = \int_{-\infty}^{\infty} S^{(j)}_{(u,y,z)}(f)e^{i\omega t}\, df \tag{4.19}$$

respectively. For simplicity, we impose the added condition that the steady component $\langle y^{(j)}\rangle$ vanish, and then sample $y^{(j)}$ periodically, at intervals

† We cannot, of course, use the subsequent argument for a member of a band-limited wide-sense *stationary* random process in the steady state, since its amplitude spectral density does *not* exist [cf. Sec. 2.2-5(4) and remarks in Sec. 3.2-1].

T_0 sec apart, e.g., at times $t_n = nT_0$, where $1/T_0$ is incommensurable with any periodic components that may appear in $y^{(j)}(t)$. [The reason for this restriction will be apparent from the discussion following Eq. (4.24); it is to avoid possible divergence of the resulting series.] With the Fourier expansion of $\delta(t - nT_0)$ (cf. Prob. 4.2), we can then write Eq. (4.18) as

$$z^{(j)}(t) = \frac{1}{T_0} \sum_{-\infty}^{\infty} \int_{-\infty}^{\infty} u(t - t_0) y^{(j)}(t_0) e^{2\pi i n t_0/T_0} \, dt_0 \tag{4.20}$$

Since

$$\begin{aligned} \int_{-\infty}^{\infty} u(t - t_0) e^{-i\omega t} \, dt &= e^{-i\omega t_0} S_u(f) \\ \int_{-\infty}^{\infty} y^{(j)}(t_0) e^{2\pi i n t_0/T_0 - i\omega t_0} \, dt_0 &= S_y^{(j)}(f - n/T_0) \end{aligned} \tag{4.21}$$

we may take the Fourier transform of both members of Eq. (4.20) to obtain finally

$$S_z^{(j)}(f) = \frac{1}{T_0} S_u(f) \sum_{-\infty}^{\infty} S_y^{(j)}\left(f - \frac{n}{T_0}\right) \tag{4.22}$$

which is the frequency form of Eq. (4.18).

We next choose a particular sampling function $u_1(t)$ and sampling points nT_0, *so that the sampled wave is the same as the original* (for all t), that is, $z^{(j)}(t) = y^{(j)}(t)$ in Eq. (4.18), or $S_z^{(j)}(f)$ replaced by $S_y^{(j)}(f)$ in Eq. (4.22), remembering now that $y^{(j)}$ is strictly band-limited in the interval $(-B,B)$. This identity depends, of course, on the location of the sampling points. We see at once that, for any arbitrary $y^{(j)}$ (fulfilling the conditions above), only one choice of $S_{u_1}(f)$ and T_0 is possible: first, if $T_0 > 1/2B$, no single $S_u(f)$ can account for the overlapping of the successive shifted spectra $S_y^{(j)}(f - n/T_0)$, and therefore the smallest number of sampling points by which $y^{(j)}$ is specified for all t is achieved by taking† $T_0 = 1/2B$, with $S_{u_1}(f) = 0$ $(|f| > B)$. Since $S_y^{(j)}(f - n/T_0)$ is the Fourier transform of an arbitrary function, it must remain unchanged in shape and scale by $S_{u_1}(f)$, with the result that specifically $S_{u_1}(f) = T_0$ $(-B < f < B)$. From Eqs. (4.19), the explicit form of the sampling pulses follows immediately, viz.,

$$u_1(t) = \mathfrak{F}^{-1}\{S_{u_1}\} = \int_{-B}^{B} T_0 e^{i\omega t} \, df = \frac{\sin 2\pi Bt}{2\pi Bt} \qquad T_0 = \frac{1}{2B} \tag{4.23}$$

The sampled wave of Eq. (4.18) itself becomes

$$y^{(j)}(t) = \sum_{-\infty}^{\infty} y^{(j)}\left(\frac{n}{2B}\right) u_1\left(t - \frac{n}{2B}\right) = \sum_{-\infty}^{\infty} y^{(j)}\left(\frac{n}{2B}\right) \frac{\sin \pi(2Bt - n)}{\pi(2Bt - n)}$$
$$\langle y^{(j)} \rangle = 0 \tag{4.24}$$

† If $T_0 < 1/2B$, that is, if we sample too rapidly, the choice of $S_{u_1}(f)$ is not unique: these spectra do not overlap or touch, so that it is possible to use sampling functions with arbitrary distributions in the regions outside $(-B,B)$, etc.

which represents $y^{(j)}(t)$ uniquely at $t = n/2B$ and at all points $(-\infty < t < \infty)$ in between. The requirement that the frequency of samples $(1/T_0)$ be incommensurable with periodic components contained in $y^{(j)}(t)$ is illustrated by the example: suppose that $T_0^{-1} = 2B = 2f_c$, f_c the frequency of a periodic component. Then $y^{(j)}(n/2B)$ for all n, alternates in sign and the resulting series diverges for almost all t, for example, $t = (2k + 1)/4B$ $(k = 0, \pm 1, \pm 2, \ldots)$. If $y^{(j)}$ contains a steady component, the latter's sampled values are obviously unaffected by our choice of T_0, remaining constant for all n and T_0, so that (4.24) converges harmonically. The same argument can also be applied to the more general set of sampling functions $u(t - t_n)$ in Eq. (4.18). Finally, if a steady component $\langle y^{(j)} \rangle$ is actually present in $y^{(j)}(t)$, we see, by sampling the *fluctuation* $y^{(j)} - \langle y^{(j)} \rangle$ and inserting this into Eq. (4.24), that the interpolation formula (4.24) becomes, instead,

$$y^{(j)}(t) = \langle y^{(j)} \rangle + \sum_{-\infty}^{\infty} [y^{(j)}(n/2B) - \langle y^{(j)} \rangle] u_1(t - n/2B) \qquad (4.24a)$$

Equation (4.24) is *the (first-order) sampling or interpolation formula in the time domain*, as used by Shannon[5] and others[9–11] in information-theory applications. Note that, if $t_k = k/2B$ $(k \neq n)$, then $u_1(t_k - t_n)$ vanishes: the values at the sampling points are algebraically independent (as distinct from wide-sense statistical independence) in the sense that one and only one $y^{(j)}(t)$ can be drawn through these sample points when the various $y^{(j)}(n/2B)$ are given.† Furthermore, changing one or more of the sampled values $y^{(j)}(n/2B)$ does not alter the rest, unlike the Fourier-series development, for example, where a similar change in waveform at such points would require a modification of *all* the Fourier coefficients. Thus, it is seen from the above that the minimum number of sampled values completely specifying $y^{(j)}(t)$ occurs at the rate $2B$/sec. At a slower rate, the reproduction is not unique (except at the t_n), while, at a faster rate, not only has one superfluous values, as well as a more complicated form of sampling pulse, but one loses the algebraic independence of the sampled values besides.

Alternatively, when the wave in question is strictly *time-limited*, in the interval $(-T/2,T/2)$ (cf. Fig. 4.3*b*), there is a corresponding (first-order) *sampling or interpolation formula in the frequency domain.*‡ Here, the amplitude spectrum can be said to exist for each member $x^{(j)}(t)$ of a random, as well as a deterministic, ensemble, provided, of course, that the energy in $(-T/2,T/2)$ is finite, so that analogous to Eq. (4.24) we may write, with

† A case where algebraic independence does not apply follows from Eq. (4.24) for a set of sampling times such that $u_1(t - t_k) \neq 0$ $(k \neq n)$. Then a set of values a_n [in place of the $y^{(j)}(n/2B)$], with $u_1(t - nT_0)$ as before, still defines the function $y^{(j)}(t)$ with the correct bandwidth limitations, but $y^{(j)}(t_n)$ does not in general equal a_n, nor does a different set of a_n necessarily yield a different $y^{(j)}(t)$.

‡ See, for example, Woodward,[9] chap. 2, eq. (28); Goldman,[10] sec. 2.2.

$S_x^{(j)}(f) = \mathfrak{F}\{x^{(j)}\},$

$$S_x^{(j)}(f) = \sum_{-\infty}^{\infty} S_x^{(j)}\left(\frac{m}{T}\right) s_1\left(f - \frac{m}{T}\right) = \sum_{-\infty}^{\infty} S_x^{(j)}\left(\frac{m}{T}\right) \frac{\sin \pi(fT - m)}{\pi(fT - m)} \tag{4.25}$$

$x^{(j)}(t)$ follows from Eqs. (4.19), while $s_1(f)$ is the "sampling spectrum," which corresponds in the time domain to $v_1(t) = \mathfrak{F}^{-1}\{s_1\} = T^{-1}$ $(0 < |t| < T/2)$, $v_1(t) = \mathfrak{F}^{-1}\{s_1\} = 0$ $(|t| > T/2)$. The Fourier transform of Eq. (4.25) is simply $x^{(j)}(t)$, which vanishes outside $(-T/2, T/2)$ (cf. Prob. 4.5).

4.2-2 Properties of the Sampling Function $u_1(t)$. The sampling function $u_1(t)$ [Eq. (4.23)] possesses a number of useful properties. We observe first that $u_1(k/2B) = 0$ $(k \neq 0)$ and that $u_1(0) = 1$. Moreover, the sampling function has a Fourier transform, which is easily obtained with the help of

$$\int_{-\infty}^{\infty} \frac{\sin \pi(x - a)}{\pi(x - a)} e^{\pm 2\pi i x y}\, dx = \begin{cases} e^{\pm 2\pi i y a} & |y| < \frac{1}{2} \\ 0 & |y| > \frac{1}{2} \end{cases} \tag{4.26}$$

We find that

$$\mathfrak{F}\left\{u_1\left(t - \frac{n}{2B}\right)\right\} = \begin{cases} \displaystyle\int_{-\infty}^{\infty} u_1\left(t - \frac{n}{2B}\right) e^{-i\omega t}\, dt = \frac{1}{2B} e^{-i\omega n/2B} & -B < f < B \\ 0 & |f| > B \end{cases} \tag{4.27}$$

Applying this to Eq. (4.24) when we take its Fourier transform gives

$$S_y^{(j)}(f) = \mathfrak{F}\{y^{(j)}\}$$
$$= \begin{cases} \displaystyle\sum_{-\infty}^{\infty} y^{(j)}\left(\frac{n}{2B}\right) \mathfrak{F}\left\{u_1\left(t - \frac{n}{2B}\right)\right\} = \frac{1}{2B} \sum_{-\infty}^{\infty} y^{(j)}\left(\frac{n}{2B}\right) e^{-i\omega t_n} & \\ \qquad t_n = \dfrac{n}{2B},\ -B < f < B & \\ 0 \qquad |f| > B & \end{cases} \tag{4.28}$$

From this, it follows at once that the amplitudes of the sampled wave at the sampling points are $2B$ times the amplitudes of the Fourier components at $f = t_n^{-1} = 2B/n$. Note also that $\mathfrak{F}^{-1}\{(2B)^{-1}e^{-i\omega n/2B}\} = u_1(t - n/2B)$, as expected.

In a similar way, we find for the interpolation theorem in the frequency domain [Eq. (4.25)], with the aid of Eqs. (4.26) and the fact that $\mathfrak{F}^{-1}\{s_1(f - m/T)\} = T^{-1}e^{2\pi i m t/T}$ $(-T/2 < t < T/2)$, that

$$x^{(j)}(t) = \mathfrak{F}^{-1}\{S_x^{(j)}\} = \begin{cases} \displaystyle\frac{1}{T} \sum_{-\infty}^{\infty} S_x^{(j)}\left(\frac{m}{T}\right) e^{2\pi i m t/T} & |t| < \dfrac{T}{2} \\ 0 & |t| > \dfrac{T}{2} \end{cases} \tag{4.29}$$

which is the analogue of Eqs. (4.28) in the time domain. Comparison of Eqs. (4.28) and (4.29) suggests that, if $x^{(j)}(t) \doteq y^{(j)}(t)$, we have a set of transformations between the amplitude and frequency components, $y^{(j)}(n/2B)$ and $S_{y^{(j)}}(m/T)$, in the two forms of the sampling theorem. Thus, if $n = 2BT \gg 1$, that is, if T is very large (but finite) and B fixed, $x^{(j)}(t)$ can be set equal to $y^{(j)}(t)$ to a good approximation within the interval $(-T/2,T/2)$, and we can write, respectively, for Eqs. (4.28) and (4.29) with $f = f_m = m/T$ the pair of relations

$$S_{y^{(j)}}\left(\frac{m}{T}\right) \doteq \frac{1}{2B} \sum_{[-BT]}^{[BT]} y^{(j)}\left(\frac{n}{2B}\right) e^{-\pi imn/BT} \qquad \text{band-limited} \qquad (4.30a)$$

$$y^{(j)}\left(\frac{n}{2B}\right) \doteq \frac{1}{T} \sum_{[-BT]}^{[BT]} S_{y^{(j)}}\left(\frac{m}{T}\right) e^{+\pi imn/BT} \qquad \text{time-limited} \qquad (4.30b)$$

These are, of course, not exact, since $x^{(j)}$ is time-limited, while $y^{(j)}$ extends in some fashion over the entire time interval $(-\infty < t < \infty)$, so that $S_{x^{(j)}}(m/T) \doteq S_{y^{(j)}}(m/T)$, at best.† It is not possible to equate the spectra of two waves one of which is strictly band-limited and the other of which is strictly time-limited.‡

Finally, it can be shown that the sampling functions $\sqrt{2B}\, u_1(t - n/2B)$, $T^{-1/2}s_1(f - m/T)$ form (separate) complete orthonormal sets in the respec-

† It is in this sense that the results of Goldman, *op. cit.*, sec. 2.5 and app. VI, are to be considered. In general, if $2BT \gg 1$, that is, if the sample is long compared with its fluctuation time, we can use the interpolation formula (4.24) to represent the wave in the interval $(-T/2,T/2)$ to within an error that is $O(1/T)$, with best agreement (apart from the exact values at $t_n = n/2B$) near the middle of the interval and poorest near the ends. While the region at the ends, in which the sampled wave is most in error, does not decrease as T is made larger (B being fixed) (cf. Prob. 4.7), the relative contribution is $O(1/T)$ and vanishes as $T \to \infty$. For a discussion of sampling errors (which can be quite large), see Stewart[12] and Middleton and Van Meter.[24]

‡ No wave can be different from zero in a single finite time interval and simultaneously possess a band-limited spectral density. In fact, there is here a direct analogue of the Heisenberg uncertainty relation in quantum mechanics, which sets a definite limit to the spread of the wave in time vis-à-vis its corresponding spread in frequency, e.g.,

$$\Delta^2 t\, \Delta^2 f \geq (4\pi)^{-1}$$

where $\Delta^2 t = \langle (t - \langle t\rangle)^2\rangle = \int_{-\infty}^{\infty} y^{(j)}(t)(t - \langle t\rangle)^2 y^{(j)}(t)^*\, dt/E_{y^{(j)}}$ and

$$\langle t\rangle = \int_{-\infty}^{\infty} |t|\, |y^{(j)}(t)|^2\, dt/E_{y^{(j)}}$$

in which $E_{y^{(j)}} \equiv \int_{-\infty}^{\infty} |y^{(j)}(t)|^2\, dt = \int_{-\infty}^{\infty} |S_{y^{(j)}}(f)|^2\, df$. [Similar relations for $\langle (f - \langle f\rangle)^2\rangle$ are obtained by replacing $y^{(j)}(t)$ by its Fourier transform $\mathfrak{F}\{y^{(j)}\}$ in the above.] See Gabor[4,13] and also Woodward[14] for a further discussion of this question.

tive intervals $(-\infty < t < \infty)$, $(-\infty < f < \infty)$. To establish this result, we need the relation (cf. Prob. 4.6*b*)

$$I(a,b) = \int_{-\infty}^{\infty} \frac{\sin \pi(x-a)}{\pi(x-a)} \frac{\sin \pi(x-b)}{\pi(x-b)} dx = \frac{\sin \pi(a-b)}{\pi(a-b)}$$
$$= \delta_a^b \qquad \text{if } a, b \text{ integers} \tag{4.31}$$

from which it at once becomes evident that

$$\begin{aligned} \int_{-\infty}^{\infty} u_1(t - t_m)u_1(t - t_n)\,dt &= \frac{1}{2B}\,\delta_m^n \\ \int_{-\infty}^{\infty} s_1(f - f_m)s_1(f - f_n)\,df &= \frac{1}{T}\,\delta_m^n \end{aligned} \tag{4.32}$$

It is clear that completeness is required if $y^{(j)}$ and $x^{(j)}$ are to be uniquely specified everywhere in the time and frequency domains. Proof of this is left to the reader (cf. Prob. 4.6*d*).

4.2-3 Spectra and Covariance Functions. Let us now extend the sampling theorems (4.24), (4.24*a*) to include stochastic processes, rather than just their individual members. For the moment, we restrict $y(t)$ to be at most a wide-sense stationary mixed process, with zero mean and a covariance function that is continuous at $t = 0$, in the usual fashion. As we shall see below, the interpolation formula is identical with Eq. (4.24), except that now the sampling operation is performed on the ensemble as a whole, rather than on a single member, viz.,

$$y(t) = \sum_{-\infty}^{\infty} y(n/2B)u_1(t - n/2B) \tag{4.33}$$

The equivalence of the left- and right-hand members is now to be understood in the sense that the series converges stochastically, i.e., mean square (cf. Sec. 2.1-1), to $y(t)$. Here B is the bandwidth† $(0,B)$ of the ensemble spectral intensity $\mathcal{W}_y(f)$, which is required to vanish for $|f| > B$; Eq. (4.33) is also valid for mixed processes with periodic components that are incommensurable with the sampling frequency $2B$; $\bar{y}$ (and $\langle y^{(j)} \rangle$), however, need not vanish [cf. remarks following Eq. (4.24)].

We begin by observing that, if $y_N(t) = \sum_{-N}^{N} y(n/2B)u_1(t - n/2B)$ represents an ensemble of truncated waves in the interval $(-T/2,T/2)$, we have to show that $\lim_{N\to\infty} y_N(t) = y(t)$ [Eq. (4.33)] mean square, or, equivalently,

† In the case of a complex process, we have seen that $\mathcal{W}_y(f) \neq \mathcal{W}_y(-f)$, and so the bandwidth $(-B,B)$ of the intensity density is defined to include the negative frequencies. See the remarks of Sec. 3.2-2 and footnote on p. 145, regarding the absence of the factor 2 in the definition of the spectral density.

that $\lim_{N\to\infty} K_{y_N}(t) = K_y(t)$ (all t). To do this, we consider first an individual member $y_N{}^{(j)}(t)$ of the truncated ensemble and examine its autocorrelation function $R_{y_N}{}^{(j)}(t)$ in the limit N (or T) $\to \infty$. Since by assumption the process $y(t)$ is at least wide-sense stationary, we have $\overline{R_{y_N}{}^{(j)}(t)} = K_{y_N}(t)$, and so if we can verify that $\lim_{N\to\infty} \overline{R_{y_N}{}^{(j)}(t)} = K_y(t)$, the validity (mean square) of the interpolation formula (4.33) is proved.

Accordingly, let us divide the truncation interval $(-T/2,T/2)$ into $2N + 1$ intervals $T_0 = 1/2B$ sec long [with a remainder $O(T_0)$]. Observing that $y^{(j)}(n/2B) = 0$ ($|n| > N$) for the *truncated* wave and that $y_N{}^{(j)}(t)$ itself falls off to zero as $|t| > T/2$, we use Eq. (3.165) for the autocorrelation function of a truncated wave to write

$$R_{y_N}{}^{(j)}(t) = \frac{1}{(2N+1)T_0} \sum_{-N}^{N} \sum_{-N}^{N} y^{(j)}\left(\frac{m}{2B}\right) y^{(j)}\left(\frac{n}{2B}\right) \int_{-\infty}^{\infty} u_1\left(t_0 - \frac{m}{2B}\right) \times u_1\left(t_0 + t - \frac{n}{2B}\right) dt_0 + O\left(\frac{1}{N}\right) \tag{4.34}$$

Setting $k = n - m$, remembering that $y^{(j)}(n/2B) = 0$ ($|n| > N$), and using Eqs. (4.32), with $T_0 = (2B)^{-1}$, we get directly, on integration followed by the ensemble average,

$$\overline{R_{y_N}{}^{(j)}(t)} = \sum_{-2N}^{2N} \overline{y^{(j)}\left(\frac{m}{2B}\right) y^{(j)}\left(\frac{m+k}{2B}\right)}\, u_1\left(t - \frac{k}{2B}\right) + O\left(\frac{1}{N}\right) \tag{4.35}$$

Since $y(t)$ is wide-sense stationary, $\overline{y^{(j)}(m/2B)y^{(j)}[(m+k)/2B]}$ is equal to $K_y(k/2B)$ and for a similar reason we have $\overline{R_{y_N}{}^{(j)}(t)} = K_{y_N}(t)$, with the result that, in the limit, Eq. (4.35) becomes

$$\lim_{N\to\infty} \overline{R_{y_N}{}^{(j)}(t)} = \lim_{N\to\infty} K_{y_N}(t) = \sum_{-\infty}^{\infty} K_y(k/2B)u_1(t - k/2B) \tag{4.36}$$

At this point, we recall that $K_y(t)$ possesses a Fourier transform, $\mathcal{W}_y(f)$, which is band-limited in $(0,B)$. *We can therefore apply the sampling theorem* (4.24) *directly to* $K_y(t)$ as a definite, nonrandom function of t, getting at once

$$K_y(t) = \sum_{-\infty}^{\infty} K_y(k/2B)u_1(t - k/2B) \tag{4.37}$$

which specifies $K_y(t)$ uniquely for all t. This supplies the desired connection between the covariance of y and the limit of the covariance of y_N, so that from Eq. (4.37) in Eq. (4.36) we have at once $\lim_{N\to\infty} K_{y_N}(t) = K_y(t)$, which establishes the interpolation formula (4.33), provided that y is wide-sense stationary with zero mean.†

If $\bar{y} \neq 0$, we replace y by $y - \bar{y}$ in both members of Eq. (4.33) to obtain the ensemble version of Eq. (4.24*a*). Thus, when the process $y(t)$ contains a nonvanishing mean value, we apply Eq. (4.33) to $y - \bar{y}$, getting for the second-moment function

$$M_y(t) = \bar{y}^2 + \sum_{-\infty}^{\infty} K_y(k/2B)u_1(t - k/2B) \qquad (= \tfrac{1}{2}\mathfrak{F}^{-1}\{\mathcal{W}_y\}) \tag{4.38}$$

The Wiener-Khintchine theorem (3.78), with Eqs. (4.27), gives, in turn,

$$\mathcal{W}_y(f) = 2\mathfrak{F}\{M_y\} = \begin{cases} 2\bar{y}^2\delta(f - 0) + \dfrac{1}{B}\displaystyle\sum_{0}^{\infty} \epsilon_k K_y\left(\frac{k}{2B}\right)\cos\frac{\omega k}{2B} \\ \text{or} \\ \dfrac{1}{B}\displaystyle\sum_{-\infty}^{\infty} M_y\left(\frac{k}{2B}\right)e^{i\omega k/2B} & 0 \le f < B \\ 0 & |f| > B \end{cases} \tag{4.39}$$

where for the second relation we have used the fact that $M_y(k) = \bar{y}^2 + K_y(k)$ and that $\sum_{-\infty}^{\infty} e^{i\omega k/2B} = 2B\delta(f - 0)$ [cf. Eq. (4.8)]. The pair of relations (4.38), (4.39) are the "sampled" form of the W-K theorem, applied to a continuous process. We note, incidentally, that

$$2\int_{-\infty}^{\infty} K_y(t)\,dt = \mathcal{W}_{y-\bar{y}}(0) = B^{-1}\sum_{-\infty}^{\infty} K_y(k/2B)$$

As an example of some later interest (cf. Secs. 6.4-2, 7.4) let us suppose that it is possible to choose the sample points at the zeros of the covariance function, so that $K_y(k/2B) = 0$ $(k \neq 0)$. What must the spectrum and covariance function be for such a process? From Eqs. (4.38) and (4.39), applied to it, we get directly

$$M_y(t) = \bar{y}^2 + K_y(0)u_1(t) = \bar{y}^2 + (\overline{y^2} - \bar{y}^2)u_1(t)$$
$$\mathcal{W}_y(f) = \begin{cases} 2\bar{y}^2\delta(f - 0) + (\overline{y^2} - \bar{y}^2)/B & 0 \le f < B \\ 0 & |f| > B \end{cases} \tag{4.39a}$$

† An equivalent proof, without the device of truncation, can be obtained by starting with $\left\langle\left[y(t) - \sum_{-\infty}^{\infty} y(n/2B)u_1(t - n/2B)\right]^2\right\rangle_{stat\ av}$, that is, the mean-square condition of stochastic convergence. With the help of Eq. (4.37) and related expressions, one next shows that $\overline{y^2}$ is indeed equal to $K_y(0)$, as derived from the interpolation formula (cf. Prob. 4.6*c*). A similar technique may then be used for $K_y(t)$. Still other proofs of the relation (4.33) for ensembles have been given independently by Bennett[15] and Slepian.

Except for the spectral line at $f = 0$, y represents band-limited white noise of spectral density $W_0 = (\overline{y^2} - \bar{y}^2)/B$ [cf. Sec. 3.2-3(2)].

In a similar way [cf. Sec. 3.2-3(3)], we may write the *individual* sampling formula for almost all members of the random process $y(t)$, viz.,

$$y^{(j)}(t) = \sum_{-\infty}^{\infty} y^{(j)}(n/2B^{(j)})u_1(t - n/2B^{(j)}) \tag{4.40}$$

Here $B^{(j)}$ is the bandwidth of the individual intensity spectrum, which may differ from member to member of the ensemble. Again, if y is ergodic, $B^{(j)} = B$ (for almost all j) and Eqs. (4.38), (4.39) apply for the individual representation where $\mathcal{W}_y$ and $\mathcal{W}_y^{(j)}$, M_y and $R_y^{(j)}$ are interchangeable (probability 1).

4.2-4 Generalizations. When the process y is (strictly) narrowband or has an intensity spectrum of the type sketched in Fig. 4.4† we can still use the sampling formulas (4.33), (4.40) provided that we now regard $B_0 + B$ as the new bandwidth. However, this leads to a superfluity of sampled values for the unique determination of $y(t)$, since now $t_n = n/2(B + B_0)$ and these sampled values are no longer algebraically independent. It is therefore natural to seek a modification of the sampling procedure, which again uses a minimum number of sampled values per second and retains algebraic independence. Except in special cases, as Kohlenberg[8] has shown, the simple sampling process described above [Eq. (4.24)] is not adequate, so that a higher-order sampling process is needed[8] for an exact representation. The second-order interpolation formula that is required takes the form

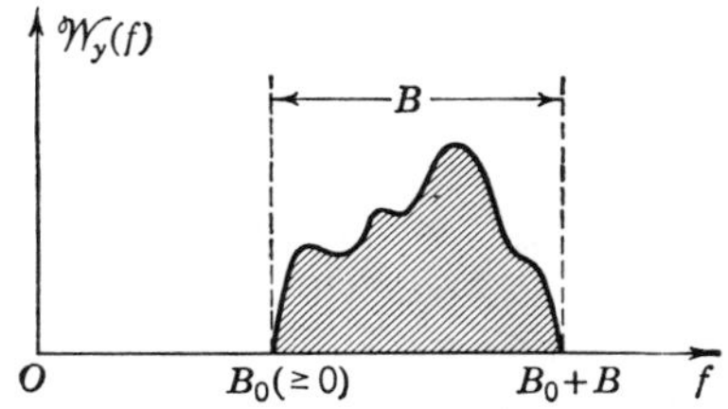

FIG. 4.4. Intensity spectrum of a strictly band-limited process.

$$y(t) = \sum_{-\infty}^{\infty} [y(n/B)u_2(t - n/B) + y(n/B + c)u_2(n/B + c - t)] \tag{4.41}$$

where c is a constant (time) satisfying the conditions that lBc, $(l + 1)Bc \neq 0$, $+1$, etc., and where l is an integer such that $2B_0/B < l < 2B_0/B + 1$ (d-c and periodic terms commensurable with B_0, $B_0 + lB$, and $lB - B_0$ are excluded). The second-order sampling function u_2 is

$$u_2(t) = \frac{1}{B}\int_{B_0}^{lB+B_0} [\cos \omega t + \cot (\pi Blc) \sin \omega t]\, df + \frac{1}{B}\int_{lB-B_0}^{B_0+B} [\cos \omega t + \cot \pi(l + 1)Bc \sin \omega t]\, df \tag{4.41a}$$

† The amplitude spectrum, when it exists, is located at $(-B - B_0, -B_0)$ and $(B_0, B_0 + B)$.

$$u_2(t) = \frac{\cos [2\pi(B_0 + B)t - (l+1)\pi Bc] - \cos [2\pi(lB - B_0)t - (l+1)\pi Bc]}{2\pi Bt \sin (l+1)\pi Bc} + \frac{\cos [2\pi(lB - B_0)t - \pi lBc] - \cos (2\pi B_0 t - \pi lBc)}{2\pi Bt \sin \pi lBc} \qquad (4.41b)$$

with the desired properties that $u_2(0) = 1$, $u_2(n/B) = u_2(n/B \pm c) = 0$ if $n \neq 0$. When B_0/B is an integer, so that $c = (2B)^{-1}$, Eqs. (4.41) reduce to the first-order relation (4.33), at a sample rate of $2B$ values/sec. In any case, the first-order theory gives a good approximation to the exact relation if $B_0 \gg B$, that is, if the ensemble is very narrowband, but, in general, a second-order process must be used for an exact interpolation. Here, the over-all sampling rate is still $2B$ values/sec, but now the sampling times fall into two similar groups, shifted by $|c|$ sec with respect to one another, each at the data rate of B values/sec.

Equivalent results may be obtained with the help of Hilbert transforms† applied to the wave to be sampled, the purpose being to eliminate the difficulty caused by the fact that $|B_0| > 0$. Thus, if we define a *complex* time function $z^{(j)}(t) \equiv y^{(j)}(t) - i\mathfrak{H}\{y^{(j)}\}$, the amplitude spectrum can be shown to be $S_y^{(j)}(f)$ over the positive (or negative) frequencies only;[4] for example, $\mathfrak{F}\{z^{(j)}\} = S_y^{(j)}(f) \neq 0$ in $(B_0, B_0 + B)$, $\mathfrak{F}\{z^{(j)}\} = S_y^{(j)}(f) = 0$ for all other frequencies, while $z^{(j)}(t) = \mathfrak{F}\{S_z^{(j)}\} = \int_{B_0}^{B_0+B} e^{i\omega t} S_y^{(j)}(f)\, df$. Consequently, the problem of spectral overlap, caused by the asymmetry of the image frequencies, is removed and reduced to the simple case of the first-order theory [cf. Eq. (4.24)]. This we may now apply to the complex wave $z^{(j)}(t)$, writing

$$\begin{aligned} z^{(j)}(t) &= \boldsymbol{T}_S\{z^{(j)}\} = \sum_{-\infty}^{\infty} z^{(j)}\left(\frac{n}{B}\right) \frac{e^{i2\pi(B_0+B)(t-n/B)} - e^{i2\pi B_0(t-n/B)}}{2\pi i(Bt - n)} \\ &= y^{(j)}(t) - i\mathfrak{H}\{y^{(j)}\} \end{aligned} \qquad (4.42)$$

From this, we get the sampled form

$$\begin{aligned} y^{(j)}(t) &= \mathrm{Re}\,[z^{(j)}(t)] \\ &= \sum_{-\infty}^{\infty} \left(y^{(j)}\left(\frac{n}{B}\right) \sin \pi B\left(t - \frac{n}{B}\right) \frac{\cos [\pi(2B_0 + B)(t - n/B)]}{\pi(Bt - n)} \right. \\ &\quad \left. + \mathfrak{H}\left\{y^{(j)}\left(\frac{n}{B}\right)\right\} \sin \pi B\left(t - \frac{n}{B}\right) \frac{\sin [\pi(2B_0 + B)(t - n/B)]}{\pi(Bt - n)} \right) \end{aligned} \qquad (4.43)$$

Note that, although the sample points are $1/B$ sec apart, instead of $1/2B$ sec, the minimum number of sampled values per second required for exact

† See Goldman, *op. cit.*, sec. 2.3.

interpolation is still $2B$/sec, since we must have sampled values of both $y^{(j)}$ and $\mathcal{H}\{y^{(j)}\}$. The extension to ensembles is made in a straightforward way, following the argument of Sec. 4.2-3,† while other generalizations, involving the cross-variance function and filtered processes, may be obtained in similar fashion (cf. Probs. 4.8, 4.9).

PROBLEMS

4.5 (*a*) Prove the sampling theorem (4.25) in the frequency domain, and show that $\mathfrak{F}^{-1}\{S_x{}^{(j)}\}$ does indeed vanish everywhere outside $(-T/2,T/2)$.

(*b*) Obtain the interpolation formula for a strictly time-limited wave in the interval (T_1,T_2) when $T_2 - T_1 = T\ (> 0)$, $T_1 \neq -T/2$.

(*c*) Discuss the sampling relation (4.24) for band-limited, purely periodic ensembles.

4.6 (*a*) Show that

$$\int_{-\infty}^{\infty} \frac{\sin \pi(x-a)}{\pi(x-a)} e^{\pm 2\pi i x y}\,dx = \begin{cases} e^{\pm 2\pi i y a} & |y| < \frac{1}{2} \\ 0 & |y| > \frac{1}{2} \end{cases} \tag{1}$$

(*b*) Prove Eq. (4.31), i.e., that

$$\int_{-\infty}^{\infty} \frac{\sin \pi(x-a)}{\pi(x-a)} \frac{\sin \pi(x-b)}{\pi(x-b)}\,dx = \begin{cases} \dfrac{\sin \pi(a-b)}{\pi(a-b)} \\ \delta_a^b \qquad \text{if } a, b \text{ integers} \end{cases} \tag{2}$$

and obtain the generalization

$$\int_{-\infty}^{\infty} \frac{\sin \pi(x-a)}{\pi(x-a)} \frac{\sin \pi(xc-b)}{\pi(xc-b)}\,dx = \begin{cases} \dfrac{\sin [\pi(b-ac)/c]}{\pi(b-ac)} & c \geq 1 \\ \dfrac{\sin [\pi(b-ac)]}{\pi c(b-ac)} & 0 \leq c \leq 1 \end{cases} \tag{3}$$

(*c*) Establish the identity

$$\frac{\sin \pi(x-y)}{x-y} = \sum_{-\infty}^{\infty} \frac{\sin (x-n)}{\pi(x-n)} \frac{\sin \pi(y-n)}{\pi(y-n)} \tag{4}$$

HINT: Observe that $\sin \pi(x-n)/\pi(x-n)$, etc., is band-limited [cf. Eq. (1)].

(*d*) Prove that $\sqrt{2B}\, u_1(t - t_m)$ and $T^{-1/2}s_1(f - f_n)$, where $t_m = m/2B$, $f_n = n/T$ $(m, n = 0, \pm 1, \ldots)$, form a complete (orthonormal) set in each case.

4.7 (*a*) Consider the approximate truncated representation of a stationary random process $y(t)$ in the interval $(-T/2,T/2)$, and show that the error in interpolation near the ends of the interval is larger than at the middle and that the region of significant error near the ends is essentially independent of interval length. Show, with an example, that increasing T diminishes the errors (mean square).

(*b*) Show that, if two waves $x^{(j)}$ and $y^{(j)}$ are band-limited to the same interval $(-B,B)$,

$$\int_{-T/2}^{T/2} x^{(j)}(t)y^{(j)}(t)\,dt \doteq \frac{1}{2B} \sum_{-N}^{N} x^{(j)}\left(\frac{m}{2B}\right) y^{(j)}\left(\frac{m}{2B}\right) \tag{1}$$

† A possible difficulty with Eq. (4.43) in practical applications, *when only sampled values of* $y^{(j)}$ *are available*, is the determination of $\mathcal{H}\{y^{(j)}\}$, that is, of $y^{(j)}(t)$ itself ultimately, which is needed before we can insert the *sampled values* of $\mathcal{H}\{y^{(j)}\}$ in Eq. (4.43).

(c) If $\mathbf{K} = \psi[k(t_k - t_l)]$ is a covariance matrix of a noise process band-limited in $(0,B)$ and such that $k(t_k - t_l) = W_0\psi^{-1}\int_0^\infty |Y_K(i\omega)|^2\, df$, with $\mathbf{K}^{-1}$ defined, obtain the approximate reciprocation of the matrix K, namely,

$$\mathbf{K}^{-1} \doteq \left[\int_{-B}^{B} \frac{e^{i\omega(t_k - t_l)}\, df}{2B\psi|Y_K(i\omega)|^2}\right] \tag{2}$$

4.8 Two entirely random processes, derived from a common source and band-limited by *different* filters to the same spectral interval $(-B,B)$, are sampled according to the interpolation formula (4.33). Show that the cross variance and cross spectral density are given by

$$K_{yz}(t) = \overline{y(t_1)z(t_2)} = \sum_{-\infty}^{\infty} K_{yz}(k/2B)u_1(t - k/2B)$$
$$W_{yz}(f) = \frac{1}{B}\sum_{-\infty}^{\infty} K_{yz}(k/2B)e^{i\omega k/2B} \tag{1}$$

and extend to processes containing finite mean values.

4.9 (a) The sampled ensemble of Sec. 4.2-3 is passed through a linear invariant filter, in the steady state, for example, $z(t) = T_L\{T_S\{y\}\}$. Show that $M_z(t)$ for this sampled, filtered wave is indeed given by $M_z(t) = \int_{-\infty}^{\infty} \rho(t \pm x)M_y(x)\, dx$ [Eq. (3.102)] when $\rho(t)$ is the autocorrelation function of the filter [cf. Eqs. (3.89)].

(b) Repeat for Prob. 4.8 to show that, again, $M_{yz}(t) = \int \rho_{[yz]}(t \mp t_0)M_x(t_0)\, dt_0$ [Eqs. (3.158)].

(c) Show that $2\int_{-\infty}^{\infty} K_z(t)\, dt = B^{-1}\sum_{-\infty}^{\infty} K_y(k/2B)\int_{-\infty}^{\infty} \rho(t_0)$

RANDOM PULSE TRAINS†

4.3 *Random Pulse Trains with Periodic Structure*

This section shifts the discussion of Sec. 4.2 to consider now indefinitely long sequences of pulses produced by periodic sampling of at least partially random processes (which may occur, for example, in such data-handling systems as computers and servomechanisms) and by random modulation of deterministic waves in pulsed time modulation and other types of keyed-carrier communication. In contrast with interpolation methods (Sec. (4.2), we are not concerned here directly with reproducing an original waveform, nor do we require the sampled wave to be strictly band-limited, since now the choice of sampling pulse [cf. $u_1(t)$, Eq. (4.23)] is to some degree arbitrary. However, we are still interested in the steady-state second-moment properties of such waves, e.g., their intensity spectra, covariance functions, etc., when a variety of stochastic effects are introduced into their

† Sections 4.3 to 4.5 are based mainly on earlier work of Middleton.[16]

amplitudes, durations, phases, epochs, and general structure, as these properties are the simplest useful statistics by which system behavior and waveform structure can be described. The examples below illustrate the salient features of the analysis, which for convenience we illustrate with the three categories of (1) nonoverlapping periodic, (2) aperiodic, and (3) random overlapping pulse trains.

4.3-1 Periodic Pulsed Sampling of a Random Wave. Let us suppose that a member of a real wide-sense stationary process $y(t)$ is sampled by a series of periodic pulses $u(t - nT_0)$ whose shape and position are fixed but whose amplitude from pulse to pulse is determined by the magnitude of the original wave at the various sampling times, as shown in Fig. 4.2. In order to obtain the covariance function K_{ys} and intensity spectrum $\mathcal{W}_{ys}$ of the sampled ensemble, we first derive the autocorrelation function of some jth member and use the fact that $K_{ys}(t)$ [or $M_{ys}(t)$] equals $\overline{R_{ys}^{(j)}(t)}$ for a wide-sense stationary process. The Wiener-Khintchine theorem then gives us the desired ensemble spectral density.

To begin with, let us assume that $y(t)$ is an entirely random process [the extension to mixed processes is easily made; cf. Eqs. (3.78)]. Let us consider next a representation $y_N^{(j)}$ truncated in the principal interval $(-T/2,T/2)$, which consists for the moment of exactly $2N$ periods, each T_0 sec long, viz.,

$$y_N^{(j)}(t) = \begin{cases} \boldsymbol{T}_S\{y_T^{(j)}\} = \sum\limits_{-N}^{N} y_n^{(j)} u(t - nT_0) & -T/2 \leq t \leq T/2 \\ 0 & |t| > T/2 \end{cases} \tag{4.44}$$

and where $u(t - nT_0)$ equals $u(t)$ $[nT_0 < t < (n + 1)T_0]$, with unity the maximum amplitude, and is zero elsewhere. The autocorrelation function of the sampled wave $y^{(j)}(t) = \lim\limits_{N\to\infty} y_N^{(j)}(t)$ is then [cf. Eq. (1.113)]

$$\begin{aligned} R_{ys}^{(j)}(t) &= \lim_{T\to\infty} \frac{1}{T} \int_{-\infty}^{\infty} y_N^{(j)}(t_0) y_N^{(j)}(t_0 + t)\, dt_0 \\ &= \lim_{N\to\infty} \left[\frac{1}{(2N+1)T_0} \sum_{m,n=-N}^{N} y_m^{(j)} y_n^{(j)} \right. \\ &\qquad \left. \times \int_{-\infty}^{\infty} u(t_0 - mT_0) u(t_0 + t - nT_0)\, dt_0 + O\left(\frac{1}{N}\right) \right] \end{aligned} \tag{4.45}$$

Introducing the amplitude spectral density of the sampling pulses by

$$S_u(f) = \mathfrak{F}\{u\} = \int_{-\infty}^{\infty} u(t) e^{-i\omega t}\, dt \qquad u(t) = \mathfrak{F}^{-1}\{S_u\} = \int_{-\infty}^{\infty} S_u(f) e^{i\omega t}\, df \tag{4.46}$$

and substituting the second relation in Eq. (4.45), while we remember that

$\int_{-\infty}^{\infty} e^{i(\omega+\omega')t_0}\, dt_0 = \delta(f' + f)$ [cf. Eq. (1.16*a*)], we obtain

$$R_{ys}^{(j)}(t) = \frac{1}{T_0}\int_{-\infty}^{\infty} |S_u(f)|^2 e^{-i\omega t}\left[\lim_{N\to\infty}\frac{1}{2N+1}\sum_{m,n=-N}^{N} y_m^{(j)} y_n^{(j)} e^{i\omega(n-m)T_0}\right] df \tag{4.47}$$

We now set $n - m = k$ and recall that, for the *truncated wave*, the $y_{m,n}^{(j)}$ vanish for all $|m|$, $|n| > N$, so that the double series in Eq. (4.47) can be rewritten $\sum_{-\infty}^{\infty} e^{i\omega kT_0} \sum_{-N}^{N} y_m^{(j)} y_{m+k}^{(j)}$. But, from Eq. (4.5),

$$\lim_{N\to\infty}\frac{1}{2N+1}\sum_{-N}^{N} y_m^{(j)} y_{m+k}^{(j)} = R_y^{(j)}(k)_d \tag{4.48}$$

is simply the autocorrelation coefficient of the discrete random series formed from the sampled amplitudes $y_n^{(j)}$.

At this point, it is convenient to introduce the *autocorrelation function* and *intensity spectrum of a typical sampling pulse*, defined by†

$$\rho_u(t) \equiv \int_{-\infty}^{\infty} |S_u(f)|^2 e^{i\omega t}\, df \qquad w_u(f) \equiv 2|S_u(f)|^2/T_0 \tag{4.49a}$$

$$\rho_u(t) = \frac{T_0}{2}\int_{-\infty}^{\infty} w_u(f) e^{i\omega t}\, df \qquad w_u(f) = \frac{2}{T_0}\int_{-\infty}^{\infty} \rho_u(t) e^{-i\omega t}\, dt \tag{4.49b}$$

Applying Eqs. (4.48), (4.49), with the rearranged series ($n - m = k$), we obtain for the autocorrelation function of the sampled member

$$R_{ys}^{(j)}(t) = \frac{1}{T_0}\sum_{-\infty}^{\infty} R_y^{(j)}(k)_d \rho_u(t - kT_0) \tag{4.50}$$

since $\rho_u(t) = \rho_u(-t)$. Taking the ensemble average of both sides of Eq. (4.50) and remembering that here the covariance function of the sampled wave is $K_{ys}(t) = \overline{R_{ys}^{(j)}(t)}$, we may use the W-K theorem to write finally, for this entirely random process $y_S(t) = \mathbf{T}_S\{y\}$,

$$K_{ys}(t) = \frac{1}{T_0}\sum_{-\infty}^{\infty} K_y(kT_0)\rho_u(t - kT_0) \tag{4.51a}$$

$$\mathcal{W}_{ys}(f) = 2\mathfrak{F}\{K_{ys}\} = w_u(f)\sum_{-\infty}^{\infty} K_y(kT_0) e^{i\omega kT_0} \tag{4.51b}$$

this last with the help of Eqs. (4.49*b*).

† Observe that $\rho_u(t)$ is the direct analogue of the correlation function of a stable linear (passive) filter [Eqs. (3.89)].

The extension to wide-sense stationary mixed processes is immediate. Provided that we do not sample at a rate commensurable with any possible periodic components, we may generalize Eqs. (4.51a) and (4.51b) to

$$M_y(t) = \frac{1}{T_0} \sum_{-\infty}^{\infty} M_y(kT_0)\rho_u(t - kT_0) \tag{4.52a}$$

$$\begin{aligned} \mathcal{W}_y(f) &= w_u(f) \sum_{-\infty}^{\infty} M_y(kT_0)e^{i\omega kT_0} \\ &= \bar{y}^2 T_0^{-1} w_u(f) \sum_{m=0}^{\infty} \delta\left(f - \frac{m}{T_0}\right) + w_u(f) \sum_{-\infty}^{\infty} K_y(kT_0)e^{i\omega kT_0} \end{aligned} \tag{4.52b}$$

with the help of the results of Prob. 4.2.† Observe that, if $\bar{y} \neq 0$, an infinite series of periodic terms is generated by the sampling process, in addition to the continuum associated with $K_y(kT_0)$. The usual equivalence (probability 1) between $R_{ys}^{(j)}$ and M_{ys}, $\mathcal{W}_{ys}^{(j)}$ and $\mathcal{W}_{ys}$ for the ensemble holds if the original process is ergodic.

An example of some interest occurs when the sampling period T_0 is large compared with the fluctuation time of the process $y(t)$. We can then treat successive sampled amplitudes y_n as at least wide-sense independent (apart from any d-c and possible periodic components). Assuming, now, that there are no periodic terms in $y(t)$, we see that under these conditions we have $K_y(kT_0) = 0$ $(k \neq 0)$, $K_y(0) = \overline{y^2} - \bar{y}^2$. Then, Eqs. (4.52$a$) and (4.52$b$) become simply

$$M_{ys}(t) = (\overline{y^2} - \bar{y}^2)T_0^{-1}\rho_u(t) + \bar{y}^2 T_0^{-1} \sum_{-\infty}^{\infty} \rho_u(t - kT_0) \tag{4.53a}$$

$$\mathcal{W}_{ys}(f) = (\overline{y^2} - \bar{y}^2)w_u(f) + \bar{y}^2 T_0^{-1} w_u(f) \sum_{m=0}^{\infty} \delta(f - m/T_0) \tag{4.53b}$$

a result obtained by Lawson and Uhlenbeck.[17] The spectrum, as predicted, consists of a continuous part, whose distribution is determined solely by that of the sampling pulse and whose intensity is given by the mean-square fluctuation of the original process. In addition, there is a discrete spectrum, with components at multiples of the sampling rate $1/T_0$, whose intensity is similarly determined by the sampling pulse. Figure 4.5 shows a typical spectrum and second-moment function M_{ys} when $u(t)$ is a rectangular pulse of duration τ $(< T_0)$. From Eqs. (4.49) we get in this case $\rho_u(t) = \tau - |t|$ $(|t| < \tau)$, $\rho_u(t) = 0$ $(|t| > \tau)$, with $w_u(f) = (2\tau^2/T_0)\{[\sin(\omega\tau/2)]/(\omega\tau/2)\}^2$ [cf. Eqs. (4.49)]. Note that if $\overline{y^2} = \bar{y}^2$, that is, if there is no random element

† From the definition of $\mathcal{W}_y(f)$, only frequencies for which $f > 0$ are used in the right-hand member of Eq. (4.52b).

in the original process, we get the expected intensity spectrum [second term only of Eq. (4.53*b*)] of a periodic pulse train, while if $\bar{y}$ vanishes, all periodic structure is lost and the spectrum is a pure continuum, $\overline{y^2}w_u(f)$. A simple example is the random telegraph signal considered by Rice,[18] where the amplitudes are independent from pulse to pulse and may take the values $\pm a$ with equal probability, for example, $\overline{y^2} = a^2$.

The total intensity of the sampled wave follows at once from Eq. (4.53*a*) on setting $t = 0$. For the example above, we get the expected result $\overline{y_S^2} = \bar{y}^2\tau/T_0$, and if $\bar{y} = 0$, the covariance function is now $M_{ys} = K_{ys}(t) = \overline{y^2}T_0^{-1}\rho_u(t)$, or simply the single triangular curve about $t = 0$ in Fig. 4.5.

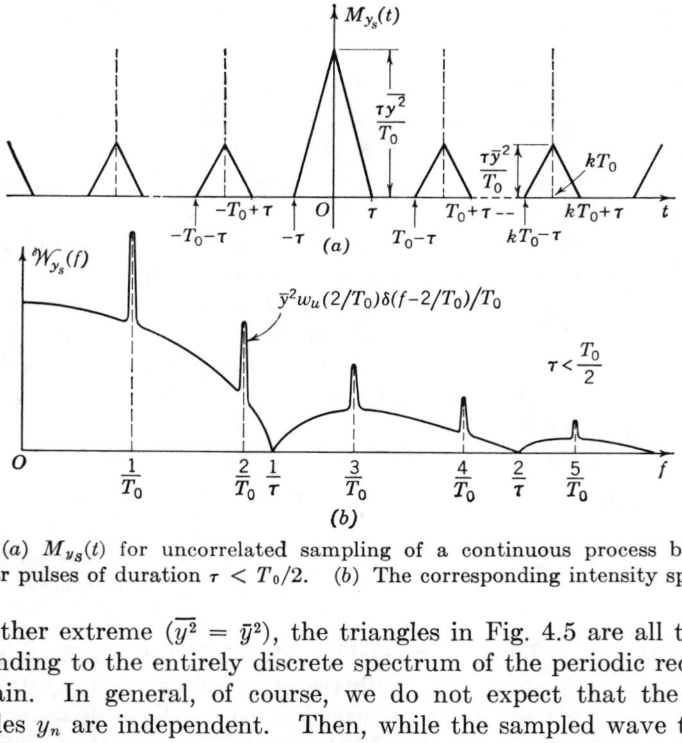

FIG. 4.5. (*a*) $M_{ys}(t)$ for uncorrelated sampling of a continuous process by periodic rectangular pulses of duration $\tau < T_0/2$. (*b*) The corresponding intensity spectrum.

At the other extreme ($\overline{y^2} = \bar{y}^2$), the triangles in Fig. 4.5 are all the same, corresponding to the entirely discrete spectrum of the periodic rectangular pulse train. In general, of course, we do not expect that the sampled amplitudes y_n are independent. Then, while the sampled wave train still may exhibit periodicities at multiples of the sampling rate ($1/T_0$) (if $\bar{y} \neq 0$), the second-moment function $M_{ys}(t)$ no longer has the comparatively simple structure shown in Fig. 4.5*a*. It may be "smeared out" between the period intervals, with a corresponding distortion of the intensity spectrum, which is now the resultant distribution of Eq. (4.52*b*), no longer proportional to the spectrum of the sampling pulses alone.

4.3-2 Pulse-time Modulation. In some communication systems, information is conveyed by periodic pulse trains where the position and sometimes the duration (and amplitude) of the pulses are varied within each period interval. Generation of such pulse trains is known as *pulse-time modulation*, of which we distinguish these main types:

1. *Pulse-position modulation* (PPM), where the pulses are fixed in duration but may vary in position within the period interval (cf. Fig. 4.6*a*)

2. *One-sided pulse-duration modulation* (PDM_1), in which the impulses all start at intervals separated by a period T_0, but whose duration is variable within a specified interval (less than T_0) (cf. Fig. 4.6*b*)

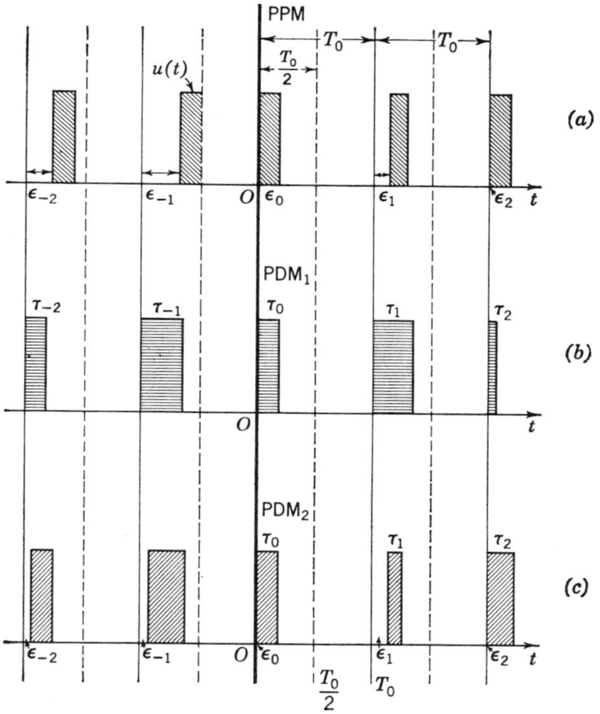

FIG. 4.6. (*a*) Pulse-position modulation (PPM). (*b*) One-sided pulse-duration modulation (PDM_1). (*c*) Two-sided pulse-duration modulation (PDM_2).

3. *Two-sided pulse-duration modulation* (PDM_2), where the pulses have variable durations and leading edges, again such that the period interval is not exceeded (cf. Fig. 4.6*c*)

(In the figure we have taken the amplitudes of the pulses to be constant and have shown some typical modulations before they are impressed upon a carrier. In any case, no overlapping of pulses is permitted, the pulses themselves being confined to a half-period interval, to avoid ambiguity in the message.) In the following discussion, we shall consider a somewhat generalized version of case 1 only, since the analysis is readily extended (cf. Probs. 4.11, 4.12) to include the more involved examples of PDM_1, PDM_2.

Let us therefore examine a PPM produced by a "filtered" periodic sampling of a continuous wide-sense stationary random process, with no inherent deterministic structure except for a possible steady component. As before [cf. Eqs. (4.44)], we start with a member of the truncated ensemble,

$$y_N^{(j)}(t) = \boldsymbol{T}_S\{y_T^{(j)}\} = \sum_{-N}^{N} y_n^{(j)} u(t - nT_0 - \epsilon_n^{(j)}) \tag{4.54}$$

where $u(t)_{\max}$ is unity, $u(t)$ vanishes outside an interval $\tau < T_0$ (or $T_0/2$), and τ is the maximum duration of a typical sampling pulse. Similarly, ϵ is also bounded, so that ϵ is less than $T_0 - \tau$ (or $T_0/2 - \tau$) to prevent overlapping between period intervals. The tighter restriction of strict stationarity must be imposed on the process ϵ if we are to obtain from $\overline{R_{ys}^{(j)}(t)}$ the desired second-moment function $M_{ys}(t)$ of the sampled ensemble in the steady state [this will be apparent from the form of Eqs. (4.57)]. Next, we follow the steps outlined in Eqs. (4.45) to (4.50). With the help of the autocorrelation function of a complex series

$$R_y^{(j)}(k;\omega)_d = \lim_{N\to\infty} \frac{1}{2N+1} \sum_{-N}^{N} a_m^{(j)}(\omega) a_{m+k}^{(j)}(\omega)^* = R_y^{(j)}(\pm k; \mp\omega)_d^* \tag{4.55}$$

where specifically here $a_m^{(j)}(\omega) = y_m^{(j)} e^{-i\omega\epsilon_m^{(j)}}$, we now obtain the autocorrelation function of the sampled wave:

$$R_{ys}^{(j)}(t) = \frac{1}{2} \sum_{-\infty}^{\infty} \int_{-\infty}^{\infty} w_u(f) R_y^{(j)}(k,\omega)_d e^{i\omega(kT_0 - t)}\, df \tag{4.56}$$

Taking the ensemble average of both members of Eq. (4.56) and remembering that y and ϵ are, respectively, wide-sense and strictly stationary, so that $\overline{R_{ys}^{(j)}(t)} = M_{ys}(t)$, we get for the sampled ensemble

$$M_{ys}(t) = \frac{1}{2} \sum_{-\infty}^{\infty} \int_{-\infty}^{\infty} w_u(f) \boldsymbol{E}\{y_1 y_2 e^{i\omega(\epsilon_2 - \epsilon_1)}\}_{t_2 - t_1 = kT_0} e^{i\omega(kT_0 - t)}\, df \tag{4.57a}$$

and

$$\mathcal{W}_{ys}(f) = w_u(f) \sum_{-\infty}^{\infty} \boldsymbol{E}\{y_1 y_2 e^{i\omega(\epsilon_2 - \epsilon_1)}\}_{kT_0} e^{i\omega kT_0} \tag{4.57b}$$

since $\boldsymbol{E}\{\ \}_{kT_0} = \boldsymbol{E}\{\ \}_{-kT_0}$, so that $M_{ys}(t) = M_{ys}(-t)$, as required. The W-K theorem gives us Eq. (4.57b) directly. In contrast with the unjittered sampling of Sec. 4.3-1, the correlation function of the sampling pulses and covariance function of the process y (at kT_0) are "mixed" together because of their mutual frequency dependence.

We note, incidentally, that, as long as the sum of the maximum duration τ of the sampling pulses and the jitter ϵ does not exceed $T_0/2$, the functions of time represented by the integrals in Eq. (4.57*a*) do not overlap in any period interval $[(k - \frac{1}{2})T_0, (k + \frac{1}{2})T_0]$. From this, and, in fact, if the average number of pulses (viz., one) per period interval is not altered, we see that the average intensity of $y_S(t)$ is

$$\overline{y_S^2} = M_{y_S}(0) = \overline{y^2}\rho_u(0)/T_0 \tag{4.58}$$

Moreover, we may obtain the d-c and periodic terms in the sampled ensemble by observing that the deterministic part of $\boldsymbol{E}\{y_1y_2e^{i\omega(\epsilon_2-\epsilon_1)}\}_{kT_0}$ is simply $|\boldsymbol{E}\{ye^{-i\omega\epsilon}\}|^2$, so that, again with the aid of Prob. 4.2, we find that

$$\lim_{t'\to\infty} M_{y_S}(t + t') = \frac{1}{2T_0}\sum_{0}^{\infty} w_u\left(\frac{m}{T_0}\right)|\boldsymbol{E}\{ye^{-2\pi i m\epsilon/T_0}\}|^2 \cos\frac{2\pi mt}{T_0} \tag{4.59}$$

The intensity of the mth component is therefore simply

$$\frac{w_u(m/T_0)|\boldsymbol{E}\{ye^{-2\pi i m\epsilon/T_0}\}|^2}{2T_0}$$

We observe, finally, that if ϵ is strictly determined, that is, if $\epsilon_1 = \epsilon_2 = \epsilon_0$, Eqs. (4.57*a*), (4.57*b*) reduce at once with the help of Eqs. (4.49) to our earlier expressions (4.52).

A simple example occurs when ϵ and y are independent of each other and from sampling interval to sampling interval. Then

$$\boldsymbol{E}\{y_1y_2e^{i\omega(\epsilon_2-\epsilon_1)}\}_{kT_0} = \begin{cases} \overline{y^2} & k = 0 \\ \bar{y}^2|\boldsymbol{E}\{e^{-i\omega\epsilon}\}|^2 & k \neq 0 \end{cases} \tag{4.60}$$

and Eqs. (4.57) reduce to

$$M_{y_S}(t) = \frac{\overline{y^2}\rho_u(t)}{T_0} + \frac{\bar{y}^2}{2}\sum_{-\infty}^{\infty}{}' \int_{-\infty}^{\infty} w_u(f)|\overline{e^{-i\omega\epsilon}}|^2 e^{i\omega(kT_0-t)}\,df \tag{4.61a}$$

and

$$\mathcal{W}_{y_S}(f) = w_u(f)\left[(\overline{y^2} - \bar{y}^2|\overline{e^{-i\omega\epsilon}}|^2) + \bar{y}^2T_0^{-1}|\overline{e^{-i\omega\epsilon}}|^2\sum_{m=0}^{\infty}\delta\left(f - \frac{m}{T_0}\right)\right] \tag{4.61b}$$

Again, if the direct current vanishes, so also does the discrete portion of the spectrum. However, if $\overline{y^2} = \bar{y}^2 = A_0^2$, that is, if we have the simpler constant-amplitude version of PPM shown in Fig. 4.6*a*, we still get a mixed spectrum because of the random epochs. This spectrum reduces specifically to

$$\mathcal{W}_{y_S}(f) = w_u(f)A_0^2\left[1 - |\overline{e^{-i\omega\epsilon}}|^2 + T_0^{-1}|\overline{e^{-i\omega\epsilon}}|^2\sum_{m=0}^{\infty}\delta\left(f - \frac{m}{T_0}\right)\right] \tag{4.61c}$$

a result obtained by Lawson and Uhlenbeck.[17]† The effect of the randomized portion of successive pulses is to spread and distort the original spectrum $w_u(f)$ of a typical sampling pulse and to *decrease* the power available in the discrete components, since $|E\{e^{-i\omega\epsilon}\}| < 1$ usually. A representative spectrum and second-moment function M_{ys} are shown in Fig. 4.7 for the

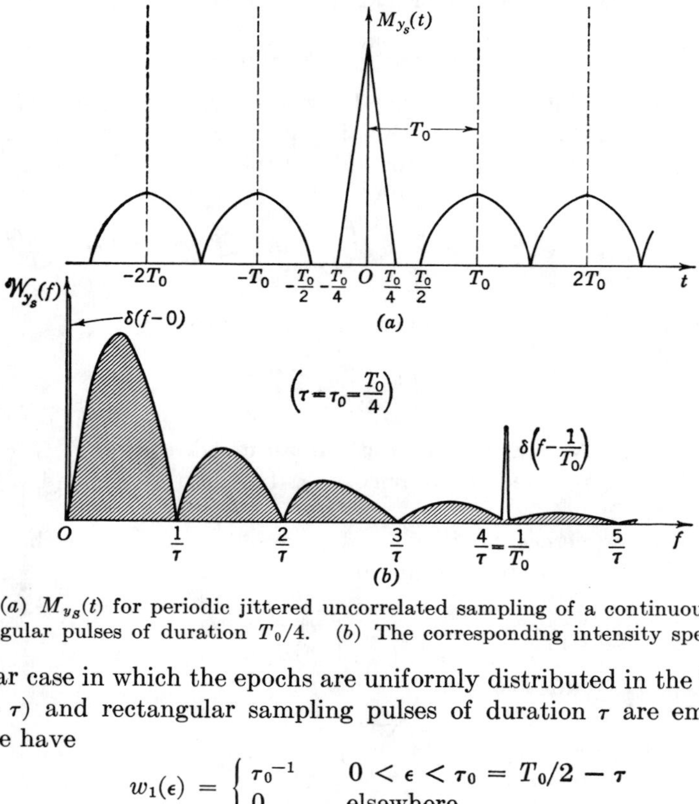

FIG. 4.7. (*a*) $M_{ys}(t)$ for periodic jittered uncorrelated sampling of a continuous process by rectangular pulses of duration $T_0/4$. (*b*) The corresponding intensity spectrum.

particular case in which the epochs are uniformly distributed in the interval $(T_0/2 - \tau)$ and rectangular sampling pulses of duration τ are employed. Thus, we have

$$w_1(\epsilon) = \begin{cases} \tau_0^{-1} & 0 < \epsilon < \tau_0 = T_0/2 - \tau \\ 0 & \text{elsewhere} \end{cases}$$

and
$$|E\{e^{-i\omega\epsilon}\}|^2 = [\sin(\omega\tau_0/2)]^2/(\omega\tau_0/2)^2$$

The intensity spectrum of Eq. (4.61*c*) becomes specifically

$$\mathcal{W}_y(f) = \frac{2A_0{}^2\tau^2}{T_0}\frac{\sin^2 x}{x^2}\left[1 - \frac{\sin^2 x_0}{x_0{}^2} + \frac{1}{T_0}\frac{\sin^2 x_0}{x_0{}^2}\sum_{m=0}^{\infty}\delta\left(f - \frac{m}{T_0}\right)\right]$$

$$x \equiv \frac{\omega\tau}{2},\ x_0 \equiv \frac{\omega\tau_0}{2} \qquad (4.62)$$

$\{M_{ys}(t)$ [Eq. (4.61*a*)] is sketched in Fig. 4.7*a* for $\tau = \tau_0 = T_0/2; \overline{y^2} = \bar{y}^2 =$

† Cf. Middleton,[16] chap. 3, eq. (35). Note that $\overline{e^{-i\omega\epsilon}}$ is the characteristic function of $-\epsilon$.

A_0^2.} Note the characteristic difference of $M_{ys}(t)$ in the first period interval, compared with its value in the others (cf. Fig. 4.5*a*).

More complicated sampling procedures, still with a basic periodic structure, can be treated in the same way. Pulse-duration modulation ($PDM_{1,2}$), periodic or jittered gating of random processes, sampling of narrowband waves, and linear filtering and cross correlation of sampled processes all provide examples of practical interest. A number of specific results, along with some generalizations of the present methods, are illustrated in the problems.

PROBLEMS

4.10 Show that, if a sampled ensemble $y_S(t) = \mathbf{T}_S\{y\}$, where $\mathbf{T}_S$ is the transformation indicated in Eq. (4.54), is passed through a linear filter $\mathbf{T}_L$, the relations (3.102), (3.104) apply to y_S, that is, to $z = \mathbf{T}_L\mathbf{T}_S\{y\}$. What is $u = \mathbf{T}_S\mathbf{T}_L\{y\}$? When are u and z equal? What is $v = \mathbf{T}_S^2\{y\}$ when y is entirely random?

4.11 (*a*) Calculate the intensity spectrum of the *one-sided pulse-duration modulation* (PDM_1) shown in Fig. 4.6*b*. Assume that the pulse durations never exceed $T_0/2$, are uniformly distributed in the interval $(0 < \tau < T_0/2)$, and are independent from pulse to pulse. Hence, show that for a general pulse shape with maximum amplitude A_0

$$\mathcal{W}_{y_S}(f) = \frac{2A_0^2}{T_0}\left[\overline{|S_u(f;\tau)|^2} - |\overline{S_u(f;\tau)}|^2 + |\overline{S_u(f;\tau)}|^2\, T_0^{-1}\sum_{m=0}^{\infty}\delta\left(f - \frac{m}{T_0}\right)\right]$$

$$S_u(f;\tau) = \mathfrak{F}\{u(t;\tau)\} \quad (1)$$

In particular, for rectangular pulses, obtain

$$\mathcal{W}_{y_S}(f) = \frac{A_0^2T_0}{8}\frac{1}{x^2}\left[1 - \frac{\sin^2 x}{x^2} + \frac{\sin^4 x + (x - \cos x \sin x)^2}{T_0}\sum_{m=0}^{\infty}\delta\left(f - \frac{m}{T_0}\right)\right] \quad (2)$$

with $x = T_0/4 = \pi f T_0/2$. Sketch the spectrum, and show that the mean intensity in the continuum is $A_0^2/12$.

(*b*) Obtain $M_{y_S}(t) = \frac{1}{2}\mathfrak{F}^{-1}\{\mathcal{W}_{y_S}\}$ for Eq. (2).

4.12 (*a*) Obtain the spectrum of a general *two-sided pulse-duration modulation* (PDM_2) of the type shown in Fig. 4.6*c*, on the assumption that the pulse duration τ and epoch ϵ are statistically independent from pulse to pulse, with $\tau + \epsilon \le T_0/2$, so that one has specifically

$$\mathcal{W}_{y_S}(f) = \frac{2A_0^2}{T_0}\left[\overline{|S_u(f;\tau)|^2} - |\overline{S_u(f;\tau)e^{-i\omega\epsilon}}|^2 + T_0^{-1}|\overline{S_u(f;\tau)e^{-i\omega\epsilon}}|^2 \sum_{m=0}^{\infty}\delta\left(f - \frac{m}{T_0}\right)\right] \quad (1)$$

where again A_0 is the maximum amplitude of a typical pulse. Next, show that for rectangular pulses whose epochs and durations are independent and uniformly distributed in $(0,T_0/2)$, subject, of course, to the conditions $0 \le \epsilon + \tau \le T_0/2$, Eq. (1) is specifically

$$\mathcal{W}_{y_S}(f) = \frac{A_0^2T_0}{2}\frac{1}{x^2}\Bigg\{[2x^4 - 2x^2 - 8 - (2x^2 - 8)(\cos x + x\sin x)]x^{-4}$$

$$+ T_0^{-1}x^{-4}[2x^2 + 8 + (2x^2 - 8)\cos x - 8x\sin x]\sum_{m=0}^{\infty}\delta\left(f - \frac{m}{T_0}\right)\Bigg\} \quad (2)$$

where $x = T_0/2$, $w_1(\epsilon) = 1/\epsilon_0$, $w_2(\tau) = 1/\tau_0$, $\epsilon_0 = \tau_0 = T_0/2$. Show also that the intensity in the d-c component is $A_0^2/72$.

(*b*) Obtain the second-moment function $M_{y_S}(t)$ for Eq. (2), and sketch the curve.

4.13 A wide-sense stationary random noise process, with a possible steady component, is periodically "gated." If $g(t)$ is the gate "aperture" [taken as a rectangular aperture of constant weight in $(0,\tau)$], show that $M_y(t)$ and the intensity spectrum for this gated noise are given by

$$M_{y_S}(t) = \frac{1}{T_0} M_y(t) \sum_{-\infty}^{\infty} \rho_g(t - kT_0) \tag{1}$$

where $\rho_g(t)$ is the autocorrelation function of the gate $g(t)$ [cf. Eq. (4.49*a*)]. Thus, for a rectangular gate of aperture τ, $\rho_g(t) = \tau - |t|$, $\rho_g(t) = 0$, $(|t| > \tau)$. For the spectrum obtain

$$\mathcal{W}_{y_S}(f) = \frac{1}{2} \sum_{-\infty}^{\infty} w_g(f_0) \frac{\mathcal{W}_y(mf_0 + f) + \mathcal{W}_y(mf_0 - f)}{2} \qquad f_0 = \frac{1}{T_0} \tag{2}$$

where [cf. Eq. (4.49*b*)] $w_g(f)$ is the intensity spectrum of the gate. Sketch the spectral density of Eq. (2), and determine the intensities of the discrete components. [Observe that we can regard the gating process as a simple multiplication, or switching, operation

$$y_S(t) = T_S\{y\} = \sum_{-\infty}^{\infty} g(t - nT_0)y(t) \tag{3}$$

and proceed as in Sec. 4.3-1.] Compare Eqs. (1) and (2) with the results of Sec. 4.3-1.

4.14 Repeat Prob. 4.13, but now with a jittered periodic gate, $g(t - nT_0 - n\epsilon)$ $(\epsilon + \tau < T_0)$, to show that

$$M_{y_S}(t) = \tfrac{1}{2} M_y(t) \sum_{-\infty}^{\infty} \int_{-\infty}^{\infty} \overline{(e^{i\omega(\epsilon_2 - \epsilon_1)})_{kT_0}} w_g(f) e^{i\omega(kT_0 - t)}\, df \tag{1}$$

$$\mathcal{W}_{y_S}(f) = \tfrac{1}{2} \sum_{-\infty}^{\infty} \int_{-\infty}^{\infty} (e^{i\omega'(\epsilon_2 - \epsilon_1)})_{kT_0} w_g(f') \mathcal{W}_y(f' + f) e^{i\omega' kT_0}\, df' \tag{2}$$

Obtain the spectrum if ϵ is uniformly distributed in each period interval $(\epsilon + \tau < T_0)$ and independent from period to period.

4.4 *Aperiodic Nonoverlapping Random Pulse Trains*

The preceding discussion (Sec. 4.3) can now be modified to include situations where the ensemble of nonoverlapping pulses does not have a basic periodic structure. Instead, the termination of one pulse or pulse group provides the beginning of the next, and so on, while their durations are random, as shown in Fig. 4.8. Certain classes of random telegraphing signal,[19] randomly pulse-modulated carrier waves in keyed carrier communication are typical examples.

4.4-1 Spectrum and Covariance Functions. To find the intensity spectrum and covariance function of such waves, we begin, as before, by considering a member of the truncated ensemble

$$y_N^{(j)}(t) = \boldsymbol{T}_S\{y_T^{(j)}\} = \sum_{-N}^{N} y_n^{(j)}u(t - \epsilon_n^{(j)}; \tau_n^{(j)}) \tag{4.63}$$

containing exactly $2N + 1$ pulses in the principal interval $(-T^{(j)}/2, T^{(j)}/2)$, where again $u(t)$ is the characteristic pulse shape, with $u_{max} = 1$, $y_n^{(j)}$ the amplitude of the nth pulse, and τ_n its duration; ϵ_n is an epoch, which marks

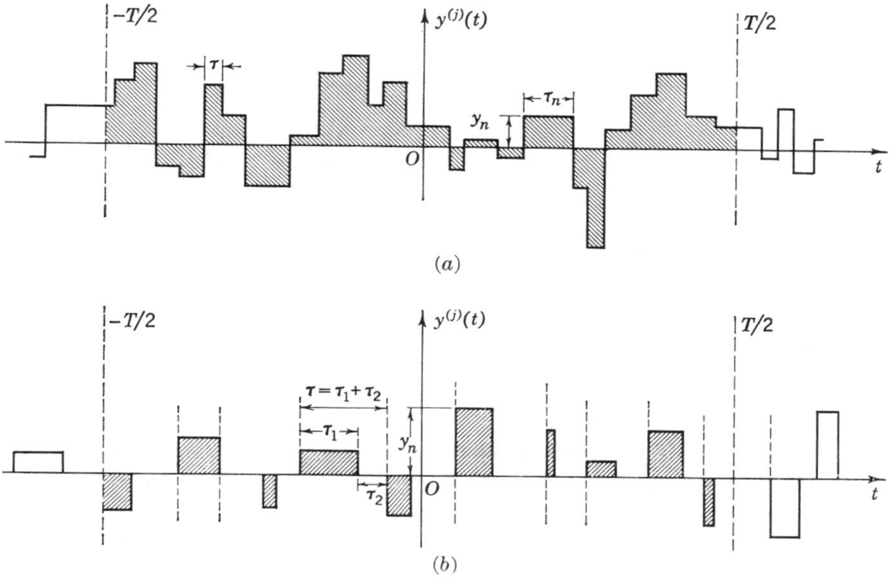

FIG. 4.8. Nonoverlapping pulse trains with random amplitudes and duration.

the start of the nth impulse. Both τ_n and ϵ_n may belong to a random series, as well as y_n. We assume, in addition, strict stationarity for amplitudes y_n.

For a typical pulse we have

$$u(t - \epsilon_n^{(j)}; \tau_n^{(j)}) = \mathfrak{F}^{-1}\{S_u^{(j)}\} = \int_{-\infty}^{\infty} S_u^{(j)}(f;\tau_n^{(j)})e^{i\omega(t-\epsilon_n^{(j)})}\,df$$
$$= \begin{cases} u(t;\tau_n^{(j)}) & \epsilon_n^{(j)} < t < \epsilon_{n+1}^{(j)} \\ 0 & \text{elsewhere} \end{cases} \tag{4.64}$$

The random durations τ_n are related to the random epochs by $\tau_n = \epsilon_{n+1} - \epsilon_n$. Here, τ_n has the probability density

$$w_T(\tau_n) = \int_{-\infty}^{\infty} W_T(\epsilon_n, \epsilon_n + \tau_n)\,d\epsilon_n \tag{4.65a}$$

where W_T is the basic probability (density) for the pair of successive epochs $\epsilon_n, \epsilon_{n+1}$ for all intervals of average duration $(-T/2, T/2)$ during which there are precisely $2N + 1$ random pulses; w_T, as well as W_T, vanishes outside the interval.

The average period T may be expressed as a function of the mean pulse duration $\bar{\tau}$: we note that the expected number of pulses in the truncated ensemble having lengths between τ and $\tau + d\tau$ is $(2N + 1)w_T(\tau)\,d\tau$, so that the contribution of all pulse lengths between τ and $\tau + d\tau$ is then $(2N + 1)\tau w_T(\tau)\,d\tau$. Consequently, the average interval T is

$$T = (2N + 1)\int_0^\infty \tau w_T(\tau)\,d\tau = \bar{\tau}_T(2N + 1) = (2N + 1)\bar{\tau} + O(N^0) \quad (4.65b)$$

where $\bar{\tau}$ is lim $(T \to \infty)$ of $\bar{\tau}_T$. Repeating the steps outlined in Eqs. (4.45) to (4.50),† we obtain for the autocorrelation function of a particular member

$$R_{y_S^{(j)}}(t) = (\bar{\tau})^{-1} \sum_{-\infty}^{\infty} \int_{-\infty}^{\infty} R_{y^{(j)}}(k,\omega)_d e^{-i\omega t}\,df \quad (4.66)$$

analogous to Eq. (4.56), where now the autocorrelation coefficient $R_{y^{(j)}}(k,\omega)_d$ is specifically

$$R_{y^{(j)}}(k,\omega)_d \equiv \lim_{N\to\infty} \frac{1}{2N + 1} \sum_{-N}^{N} y_m{}^{(j)} y_{m+k}^{(j)} S_u{}^{(j)}(f;\tau_m) S_u{}^{(j)}(f;\tau_{m+k})^* e^{i\omega(\epsilon_{m+k}-\epsilon_m)} \quad (4.66a)$$

Taking the ensemble average of both sides of Eq. (4.66) and recalling the assumption of stationarity above for τ, ϵ, y, as before [Eqs. (4.56) et seq.], we can obtain the desired second-moment function $M_{y_S}(t)$, and from it the spectral density, with the help of the W-K theorem (3.78). The result is

$$M_{y_S}(t) = (\bar{\tau})^{-1} \sum_{-\infty}^{\infty} \int_{-\infty}^{\infty} E\{y_1 y_2 S_u(f;\tau_1) S_u(f;\tau_2)^* e^{i\omega(\epsilon_2-\epsilon_1)}\}_k e^{i\omega t}\,df \quad (4.67a)$$

$$\mathcal{W}_{y_S}(f) = 2(\bar{\tau})^{-1} \sum_{-\infty}^{\infty} E\{y_1 y_2 S_u(f;\tau_1) S_u(f;\tau_2)^* e^{i\omega(\epsilon_2-\epsilon_1)}\}_k \quad (4.67b)$$

where $E\{\quad\}_k = E\{\quad\}_{-k}$ and the subscripts 1 and 2 in the statistical averages refer to times t_1, t_2 separated by k successive random pulse lengths, the summation being over all possible differences ($k = 0, \pm 1, \pm 2, \ldots$). Since the pulses do not overlap, one can show (cf. Prob. 4.15) that the mean intensity of the process y_S is

$$\overline{y_S{}^2} = M_{y_S}(0) = 2\int_0^\infty \overline{y^2 |S_u(f;\tau)|^2}\,df/\bar{\tau} = \int_0^\infty \overline{W_u(f,\tau) y^2}\,df \quad (4.68a)$$

† We replace $T^{(j)}$ by T, etc., and assume that $\lim_{T^{(j)}\to\infty} T^{(j)} = \lim_{T\to\infty} T = \lim_{N\to\infty} (2N + 1)\bar{\tau}$ for almost all members of the ensemble.

while the contribution of a possible d-c term is given by†

$$\lim_{t' \to \pm\infty} M_{ys}(t + t') = \bar{y}_s^2 = |\overline{yS_u(0;\tau)}|^2/\bar{\tau}^2 \tag{4.68b}$$

4.4-2 An Example. The simplest case, and the one often of greatest practical interest, arises when y, ϵ, τ are all statistically independent from pulse to pulse. If now we express the random epochs ϵ in terms of pulse duration τ, so that in particular

$$(\epsilon_2 - \epsilon_1)_k = \epsilon_{m+k} - \epsilon_m = \tau_m + \Delta_{m,k}(\tau) \qquad \Delta_{m,k}(\tau) = \begin{cases} \sum\limits_{\lambda=m}^{m+k-1} \tau_\lambda & k \geq 2 \\ 0 & k = 1 \\ -\tau_m & k = 0 \\ -\sum\limits_{m}^{m+k} \tau_\lambda & k \leq -1 \end{cases} \tag{4.69a}$$

and if we let

$$\begin{aligned} \alpha_m(\omega) &\equiv y_m S_u(f;\tau_m) e^{i\omega\tau_m} \\ \beta_{m+k}(\omega) &\equiv y_{m+k} S_u(f;\tau_{m+k}) \\ z(\omega) &\equiv e^{i\omega\tau} \end{aligned} \tag{4.69b}$$

we can sum Eq. (4.67a) directly, observing that the joint average therein factors into a product of averages when $k \neq 0$, since a purely random series for y, τ, ϵ has been assumed. We get explicitly

$$M_{ys}(t)_{k\geq 1} = (\bar{\tau})^{-1} \sum_{k=1}^{\infty} \int_{-\infty}^{\infty} [\overline{\alpha(\omega)}\,\overline{\beta(\omega)^*}\,\overline{z(\omega)}^{k-1}] e^{i\omega t}\, df \tag{4.70a}$$

$$M_{ys}(t)_{k=0} = (\bar{\tau})^{-1} \int_{-\infty}^{\infty} \overline{\alpha(\omega)\beta(\omega)^* e^{-i\omega\tau}} e^{i\omega t}\, df \tag{4.70b}$$

$$M_{ys}(t)_{k\leq -1} = (\bar{\tau})^{-1} \sum_{k=1}^{\infty} \int_{-\infty}^{\infty} [\overline{\alpha(\omega)}^*\,\overline{\beta(\omega)}\,\overline{z(\omega)}^{*k-1}] e^{i\omega t}\, df \tag{4.70c}$$

Excluding now the case $\overline{z(\omega)} = 1$, which for $\tau > 0$ ($\bar{\tau} > 0$) corresponds to a possible spectral line at $f = 0$, we sum the geometric series to obtain at once

$$K_{ys}(t) = (\bar{\tau})^{-1} \int_{-\infty}^{\infty} e^{i\omega t} \{\overline{\alpha\beta^* e^{-i\omega\tau}} + 2\,\mathrm{Re}\,[\bar{\alpha}\overline{\beta^*}/(1 - \bar{z})]\}\, df \tag{4.71a}$$

with the continuous part of the intensity spectrum given by

$$\mathcal{W}_{y_s - \bar{y}_s}(f) = 2(\bar{\tau})^{-1}\{\overline{\alpha\beta^* e^{-i\omega\tau}} + 2\,\mathrm{Re}\,[\bar{\alpha}\overline{\beta^*}/(1 - \bar{z})]\} \tag{4.71b}$$

Adding Eq. (4.68b) to $K_{ys}(t)$ then yields $M_{ys}(t)$, while, for the complete spectrum, Eq. (4.68b) multiplied by $2\delta(f - 0)$ must be added to Eq. (4.71b).

† This is easily established if we apply the above approach to the *first* time moment $\langle y_s^{(j)} \rangle$, with the help of Eqs. (4.64) et seq., to get $\bar{y}_s = \overline{yS_u(0;\tau)}/\bar{\tau}$.

To illustrate with an example of some importance to us later, let us further require that τ and y be independent within the pulse and that the pulses themselves be rectangular (cf. Fig. 4.8*a*), so that $S_u(f;\tau) = (1 - e^{-i\omega\tau})/i\omega$. Then, a little manipulation gives directly [from Eqs. (4.71*b*), (4.68*b*)]

$$\mathcal{W}_{ys}(f) = 4(\overline{y^2} - \bar{y}^2)(\bar{\tau})^{-1}[1 - \text{Re}\ \overline{z(\omega)}]/\omega^2 + 2\bar{y}^2\delta(f - 0) \tag{4.72}$$

Observe that the spectral density of the continuum at $f = 0$ is $2\overline{\tau^2}(\overline{y^2} - \bar{y}^2)/\bar{\tau}$. When the pulse durations obey a distribution law of the *Poisson*† type, for example, when $w(\tau) = \beta e^{-\beta\tau}\ [\beta = (\bar{\tau})^{-1}, \tau \geq 0]$, we get specifically

$$M_{ys}(t) = (\overline{y^2} - \bar{y}^2)e^{-\beta|t|} + \bar{y}^2$$

$$\mathcal{W}_{ys}(f) = \frac{4\beta(\overline{y^2} - \bar{y}^2)}{\beta^2 + \omega^2} + 2\bar{y}^2\delta(f - 0) \tag{4.73}$$

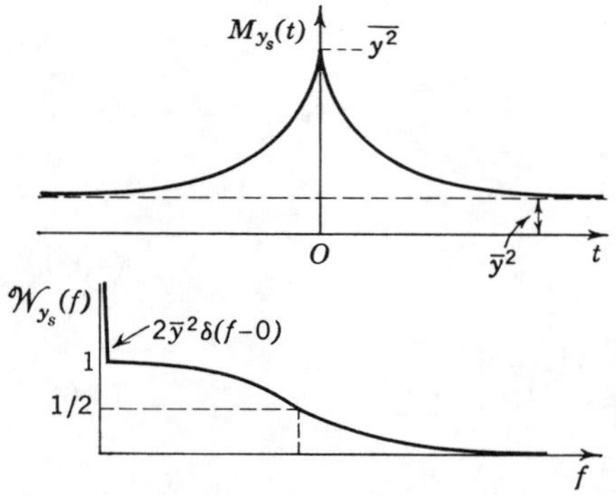

FIG. 4.9. Second-moment function and spectrum of a train of (independent) rectangular pulses of random amplitudes and of durations which obey a Poisson law $w(\tau) = \beta e^{-\beta\tau}$ $(\tau > 0)$.

The spectrum and $M_{ys}(t)$ are sketched in Fig. 4.9. If we neglect the d-c term, spectra of this kind are typical of the current in a resistance and inductance in series when the driving voltage is an (infinitely) wide band of thermal noise (cf. Chap. 11). They are also characteristic of the inputs of automatic tracking radar systems, when $y(t)$ represents the changes in angular velocity with time.[20] Other examples and generalizations of interest are included in the problems below.

PROBLEMS

4.15 Show that the mean intensity of the random pulse train of Eq. (4.63), as $N \to \infty$, is

$$\overline{ys^2} = 2(\bar{\tau})^{-1}\int_0^\infty \overline{|y^2S_u(f;\tau)|^2}\,df$$

and obtain the d-c component [Eq. (4.68*b*)].

4.16 Obtain the intensity spectrum of an ensemble of wave trains where the pulses are rectangular, of fixed duration τ_0, while the spacing $\tau - \tau_0$ between pulses varies randomly and independently from pulse to pulse. Hence, show that

$$\mathcal{W}_{y_s}(f) = \frac{4A_0^2}{\bar{\tau}\omega^2}\sin^2\frac{\omega\tau_0}{2}\left[1 + \text{Re}\ \overline{z(\omega)}\left(\frac{1}{1-\bar{z}} + \frac{1}{1-\bar{z}^*}\right)\right] + A_0^2(\tau_0\bar{\tau})^2\delta(f - 0) \tag{1}$$

4.17 (*a*) Determine the intensity spectrum and second-moment function of a rectangular pulse train where the amplitudes are fixed but the duration of each pulse is

† See Sec. 11.2 and Refs. 16 and 18 at the end of this chapter.

random and independent from pulse to pulse (as well as from gap to gap). Thus, show that

$$\mathfrak{W}_{y_S}(f) = \frac{A_0^2}{2}\,\delta(f-0) + \frac{A_0^2\tau^{-1}}{2\omega^2}\left\{2 - z(\omega) - z(\omega)^* + \mathrm{Re}\,\frac{[1-\overline{z(\omega)}]^2}{1+\overline{z(\omega)}}\right\} \tag{1}$$

(b) If pulse (and gap) duration obey the Poisson law $w(\tau) = (\bar{\tau})^{-1}e^{-\tau/\bar{\tau}}$ $(\tau > 0)$, show from Eq. (1) that

$$\mathfrak{W}_{y_S}(f) = \frac{A_0^2}{2}\,\delta(f-0) + \frac{A_0^2\bar{\tau}}{2}\,\frac{1}{1+(\omega\bar{\tau}/2)^2} \tag{2a}$$

$$M_{y_S}(t) = \frac{A_0^2}{4}\,(1 + e^{-2|t|/\bar{\tau}}) \tag{2b}$$

which, apart from the d-c term, is Rice's result [eq. (2.7-5) of Ref. 18]. Verify this.

4.18 An ideal switch is opened and shut, randomly, and with random "on" and "off" intervals. The gate is applied to a steady-state process $y(t)$ (random, mixed, or deterministic). Show that the spectrum of the gated output z is, in general,

$$\mathfrak{W}_z(f) = \frac{1}{2}\int_{-\infty}^{\infty} \mathfrak{W}_y(f')\,\mathfrak{W}_g(f-f')\,df' \tag{1}$$

where $\mathfrak{W}_g(f)$ is the spectrum of the gate. Show that, for Prob. 4.17b, we get specifically for gated RC noise of intensity ψ_N and time constant $RC = \alpha^{-1}$

$$\mathfrak{W}_z(f) = \psi_N\left[\frac{\alpha}{\alpha^2+\omega^2} + \frac{\alpha + 2/\bar{\tau}}{(\alpha+2/\bar{\tau})^2+\omega^2}\right] \tag{2}$$

4.5 *Overlapping Random Pulses*

In the preceding sections, we have considered ensembles of partially random and deterministic waves obtained by a sampling process $\boldsymbol{T}_S$ whose common descriptive feature has been the nonoverlapping occurrence of the individual impulses that make up the disturbance. Here, we examine a different type of random wave where this restriction is relaxed: the elementary impulses can overlap and, moreover, are assumed to be statistically independent of one another. Such a model is a useful one for describing many types of noise and interference, among which shot noise, impulsive static, precipitation and electron multiplier noise, Barkhausen noise, and a number of statistically similar phenomena are most prominent (cf. Chap. 11). Since for these it is reasonable to assume that the individual impulses are all independent, we may expect a corresponding simplification in the calculation of the intensity spectrum and covariance function, similar to that observed above in the nonoverlapping cases. Figure 4.10 shows a typical impulse noise wave, with the low density (in time) characteristic of many types of "static."

4.5-1 Spectrum and Covariance Functions. Let us proceed as in the previous examples to calculate the spectral intensity and $M_{y_S}(t)$ for the real process $y_S(t)$. We consider, as before, a (long) interval $(-T/2, T/2)$, containing now exactly $N^{(j)}$ impulses, and we write for a particular member of the ensemble of such truncated waves†

$$y_{N^{(j)}}(t) = \sum_{n=0}^{N^{(j)}} y_n{}^{(j)} u(t - \epsilon_n{}^{(j)}; \tau_n{}^{(j)}) \tag{4.74}$$

where $u(t - \epsilon_n{}^{(j)}; \tau_n{}^{(j)})$ vanishes outside an interval $\Delta\tau \ll T$ and has some finite value within $\Delta\tau$ and where $y_0{}^{(j)} = 0$, the case of no pulses in the interval, when $N^{(j)} = 0$. It should be emphasized that here, as distinct from the wave trains discussed in Secs. 4.3, 4.4, the numbering from $n = 0$ to $N^{(j)}$ is purely a matter of convenience and has nothing to do with the order

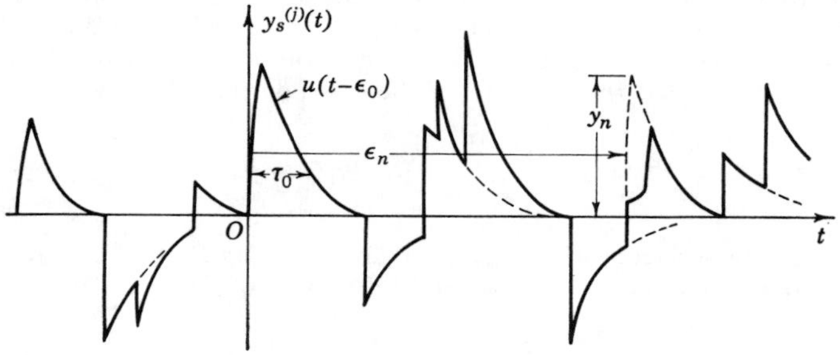

FIG. 4.10. Impulsive "static," consisting of random overlapping pulses. (A very wide filter is assumed here, to pass these pulses without distortion.)

in which the individual effects occur in $(-T/2,T/2)$. Each pulse has the same probability $w_T(\epsilon)$ of occurring in the interval but is unaffected by the occurrence of any other. Here, $y_n{}^{(j)}$, $\tau_n{}^{(j)}$ are again, respectively, the random amplitude and pulse duration, while $\epsilon_n{}^{(j)}$ is the epoch of the nth impulse, measured from some arbitrary time origin (cf. Fig. 4.10); y_n, τ_n, ϵ_n are all independent from pulse to pulse.

Using Eqs. (4.64) (without that restriction on ϵ) and following the steps in Eqs. (4.45) to (4.47), we can write for the autocorrelation function of a typical member of the ensemble in $(-\infty < t < \infty)$

$$R_{ys}{}^{(j)}(t) = \lim_{T\to\infty} \frac{1}{T} \left[\sum_{m,n=0}^{N^{(j)}(T)} y_m{}^{(j)} y_n{}^{(j)} \int_{-\infty}^{\infty} S_u{}^{(j)}(f;\tau_m) \times S_u{}^{(j)}(f;\tau_n)^* e^{-i\omega t + i\omega(\epsilon_n{}^{(j)} - \epsilon_m{}^{(j)})} \, df \right] \tag{4.75}$$

where $N^{(j)}(T)$ is, from our steady-state assumption, a suitably increasing function of the interval T. Now, taking the ensemble average of both

† We neglect the contributions of the elementary impulses near the ends of the interval $(-T/2,T/2)$ which do not fall completely within $(-T/2,T/2)$. This introduces an error $0(\Delta\tau/T)$, which vanishes in the limit $T \to \infty$. The truncation is with respect to the steady-state stationary ensemble in the interval $(-\infty, \infty)$.

sides of Eq. (4.75), including the average over $N^{(j)}$, we note because all pulses are identical in form and are independent [and provided that the epochs are also independent of pulse structure (y,τ)] that the double sum reduces to a single sum of N similar elements when $m = n$ and to a residual double sum of $N(N-1)$ identical terms when $m \neq n$, in the following manner:

$$\overline{R_{ys}{}^{(j)}(t)} = \lim_{T\to\infty} \frac{1}{T}\left[\sum_{n=0}^{\bar{N}} \int_{-\infty}^{\infty} \boldsymbol{E}\{y_m{}^{(j)2}\,|S_u{}^{(j)}(f;\tau_m)|^2\}e^{-i\omega t}\,df\right.$$
$$\left. + \sum_{m,n=0}^{\bar{N}} \int_{-\infty}^{\infty} e^{-i\omega t}\boldsymbol{E}\{y_m{}^{(j)}S_u{}^{(j)}(f;\tau_m)e^{i\omega\epsilon_m}\}\boldsymbol{E}\{y_n{}^{(j)}S_u{}^{(j)}(f;\tau_n)^* e^{-i\omega\epsilon_n}\}\,df\right] \quad (4.76a)$$

Because of the postulated (wide-sense) stationarity of y, τ and the (strict) stationarity of the epochs ϵ, $\overline{R_{ys}{}^{(j)}(t)}$ is equal to the second-moment function $M_{ys}(t)$; $\bar{N}$, of course, is a function of interval length, for example, $\bar{N} = \bar{N}(T)$. We get, therefore, from Eq. (4.76a)

$$M_{ys}(t) = \left(\lim_{T\to\infty}\frac{\bar{N}}{T}\right)\boldsymbol{E}\left\{y^2\int_{-\infty}^{\infty}|S_u(f;\tau)|^2 e^{-i\omega t}\,df\right\}$$
$$+ \lim_{T\to\infty}\left[\frac{\bar{N}(\bar{N}-1)}{T}\int_{-\infty}^{\infty}|\overline{yS_u(f;\tau)}|^2|\overline{e^{i\omega\epsilon}}|_T^2 e^{-i\omega t}\,df\right] \quad (4.76b)$$

where the subscript T reminds us that the average over the epochs is also a function of the truncation interval T.

At this point, we assume that the process is such that

$$\lim_{T\to\infty}\frac{\bar{N}(T)}{T} \equiv \bar{n}_0 \qquad \lim_{T\to\infty}\frac{\bar{N}(\bar{N}-1)}{T^2} = \bar{n}_0{}^2 \quad (4.77)$$

exist and now further require *that the epochs are uniformly distributed in* $(-T/2,T/2)$, so that $w_T(\epsilon) = 1/T$ $(-T/2 < \epsilon < T/2)$. Consequently, we have

$$|\overline{e^{i\omega\epsilon}}|_T^2 = \frac{1}{T^2}\left|\int_{-T/2}^{T/2} e^{i\omega\epsilon}\,d\epsilon\right|^2 = \left[\frac{2\sin(\omega T/2)}{\omega T}\right]^2 = 2\,\frac{1-\cos\omega T}{\omega^2 T^2} \quad (4.78a)$$

Since
$$\lim_{T\to\infty}\frac{2}{T}\int_{-\infty}^{\infty} G(\omega)e^{-i\omega t}\,\frac{1-\cos\omega T}{\omega^2}\,df = G(0) \quad (4.78b)$$

we obtain directly for Eq. (4.76b) the desired result,

$$M_{ys}(t) = \bar{n}_0\boldsymbol{E}\left\{y^2\int_{-\infty}^{\infty}|S_u(f;\tau)|^2 e^{-i\omega t}\,df\right\} + \bar{n}_0{}^2|\boldsymbol{E}\{yS_u(0;\tau)\}|^2 \quad (4.79a)$$

With the help of Eqs. (4.49), and on replacing T_0 by $\bar{\tau}$, we can express Eq. (4.79a) alternatively in terms of a spectral intensity w_u and correlation function ρ_u of a typical impulse, viz.,

$$M_{ys}(t) = \gamma\overline{y^2\rho_u(t;\tau)/\bar{\tau}} + \gamma^2|\overline{yS_u(0,\tau)/\bar{\tau}}|^2 \qquad \gamma \equiv \bar{n}_0\bar{\tau} \tag{4.79b}$$

Here $\gamma = \bar{n}_0\bar{\tau}$ is the "density" of the impulses in time, since $\bar{n}_0$ [Eq. (4.77)] is the expected number of pulses per second, while $\bar{\tau}$ is the average duration of a typical impulse. The larger γ, the greater is the overlapping and the less the resultant noise wave exhibits the time structure of the elementary pulses, while, for small γ (< 1), the reverse is true. Applying the W-K theorem (3.78) to Eqs. (4.79) gives at once the corresponding intensity spectrum,

$$\mathcal{W}_{ys}(f) = \gamma\overline{y^2w_u(f;\tau)} + 2\gamma^2|\overline{yS_u(0;\tau)/\bar{\tau}}|^2\delta(f-0) \tag{4.80}$$

with $w_u(f;\tau) = 2\overline{|y^2S_u(f;\tau)|^2}/\bar{\tau}$ [cf. Eqs. (4.49)]. Equation (4.80) is a generalization of Rice's[18] earlier result, eq. (2.6-4).

4.5-2 Campbell's Theorem. When we set t equal to zero in Eqs. (4.79) and make use of Parseval's theorem, we have an extension of Campbell's theorem,[21–23] namely,

$$M_{ys}(0) = \bar{n}_0\boldsymbol{E}\left\{y^2\int_{-\infty}^{\infty}u(t_0;\tau)^2\,dt_0\right\} + \bar{n}_0^2\boldsymbol{E}^2\left\{y\int_{-\infty}^{\infty}u(t_0;\tau)\,dt_0\right\} = \overline{ys^2} \tag{4.81}$$

As an example, let us suppose that the elementary pulses are rectangular and of duration τ, while τ itself obeys the distribution $w_1(\tau) = \beta e^{-\beta\tau}$ [$\beta = (\bar{\tau})^{-1}$] and the amplitudes y are independent of τ. Then, from the above generalization of Campbell's theorem, we obtain for the total intensity and mean-square fluctuation

$$\overline{ys^2} = \gamma\overline{y^2} + 2\gamma^2\bar{y}^2 \qquad \overline{ys^2} - \bar{y}s^2 = \gamma\overline{y^2} \tag{4.82}$$

From Eqs. (4.79) and (4.80), we easily see that, more generally,

$$M_{ys}(t) = \gamma\overline{y^2}e^{-\beta|t|} + 2\gamma^2\bar{y}^2 \qquad \beta = (\bar{\tau})^{-1} \tag{4.83a}$$

$$\mathcal{W}_{ys}(f) = \frac{4\beta\gamma\overline{y^2}}{\beta^2+\omega^2} + 4\gamma^2\bar{y}^2\delta(f-0) \tag{4.83b}$$

Compare this with the spectrum of Eqs. (4.73) for the nonoverlapping random pulses, when $\bar{y}$ vanishes in both instances: except for the density factor γ, the two spectra are identical, although the time structures of the two ensembles are quite different. This is not too surprising if, on reflection, we remember that the statistical features (as far as second moments are concerned) are the same: both sets of random parameters belong to purely random series, the pulse durations in both cases obey the same distribution law, and because we are dealing with an (auto) intensity spectrum, the differences between the phase structure of the two ensembles are suppressed. Other examples involving $M_{ys}(t)$ and spectra are referred to in the problems below, while, for other generalizations, the reader is referred to Chap. 11, where the problem of the probability distributions of such noise waves is considered.

PROBLEMS

4.19 In the expression (4.74) for $y_N(t)$, let us assume, instead of the uniform distribution of epochs ϵ used in deriving Eqs. (4.79), that the ϵ_n have the following structure, $\epsilon_2 = \epsilon_1 + \lambda_1$, $\epsilon_3 = \epsilon_2 + \lambda_2 = \epsilon_1 + \lambda_1 + \lambda_2$, etc., where $w(\lambda)$ is specifically $\bar{n}_0 e^{-\bar{n}_0\lambda}$ $(\lambda \geq 0)$ and $\bar{n}_0$ is the expected number of pulses, or "events," per second. Taking the pulse shape to be $u(t) = e^{-\alpha t}$ $(t > 0)$, $u(t) = 0$ $(t < 0)$, and with the usual assumptions of independence for y_n, λ_n ($\tau_n \sim \alpha^{-1}$, a constant), show that

$$\bar{y}_S = \frac{\bar{n}_0 \bar{y}}{\alpha} \qquad \overline{y_S{}^2} = \frac{\bar{n}_0 \overline{y^2}}{2\alpha} + \frac{\bar{n}_0 \bar{y}^2 A(\alpha)}{\alpha[1 - A(\alpha)]} \tag{1}$$

where $A(\alpha) = \int_0^\infty w(\lambda) e^{-\alpha\lambda}\, d\lambda$. S. O. Rice,[18] sec. 1.5.

4.20 Instead of the uniform distribution of epochs assumed in the derivation of $M_{y_S}(t)$, $\mathcal{W}_{y_S}(f)$ above [cf. Eqs. (4.79), (4.80)], let us consider a more general distribution $w_T(\epsilon)$ such that $\int_{-T/2}^{T/2} w_T(\epsilon)\, d\epsilon = 1$. Show therefore, for the *truncated* ensemble, that the spectral density is

$$\mathcal{W}_T(f) = 2\bar{n}_0 \frac{|S_u(f)|^2}{T} \left[1 + \bar{n}_0 \left| \int_{-T/2}^{T/2} w_T(\epsilon) e^{-i\omega\epsilon} d \right|^2 \right] \tag{1}$$

provided that the events are independent, and $u(t)$ is now fixed in structure. Here $\bar{n}_0$ is $E\{N^{(j)}T\}$ and $N^{(j)}$ is the number of pulses in the jth member of the truncated ensemble. S. O. Rice,[18] sec. 2.6.

4.21 Show that the generalization of Eqs. (4.79), (4.80) to include the second-moment functions and cross spectral densities of *pairs* of independent overlapping pulse trains, with the same basic statistics governing their amplitudes y and durations τ, but with different individual pulse types, that is, $u_l \neq u_k$ $(l \neq k)$, is

$$M_{lm}(t) = \bar{n}_0 E\left\{ y^2 \int_{-\infty}^{\infty} e^{-i\omega t} S_u{}^{(l)}(f;\tau) S_u{}^{(m)}(f;\tau)^*\, df \right\} + \bar{n}_0{}^2 E\{y S_u{}^{(l)}(0;\tau)\} E\{y S_u{}^{(m)}(0;\tau)^*\}$$
$$(m, l = 1, 2, \ldots, M) \tag{1}$$

and

$$\mathcal{W}_{lm}(f) = 2\bar{n}_0 E\{y^2 S_u{}^{(l)}(f;\tau) S_u{}^{(m)}(f;\tau)^*\} + 2\bar{n}_0{}^2 \overline{y S_u{}^{(l)}(0;\tau)}\, \overline{y S_u{}^{(m)}(0;\tau)}\, \delta(f - 0) \tag{2}$$

Also obtain the following generalization of Campbell's theorem:

$$M_{lm}(0) = \bar{n}_0 E\left\{ y^2 \int_{-\infty}^{\infty} u_l(t_0;\tau) u_m(t_0;\tau)\, dt_0 \right\}$$
$$+ \bar{n}_0{}^2 E\left\{ y \int_{-\infty}^{\infty} u_l(t_0;\tau)\, dt_0 \right\} E\left\{ y \int_{-\infty}^{\infty} u_m(t_0;\tau)\, dt_0 \right\} \tag{3}$$

REFERENCES

1. Whittaker, J. M.: Interpolatory Function Theory, *Cambridge Tract* 33, in "Mathematics and Mathematical Physics," Cambridge, New York, 1935.
2. Levinson, N.: Gap and Density Theorems, *Am. Math. Soc. Colloq. Publ.*, vol. 26, 1940.
3. Nyquist, H.: Certain Topics in Telegraph Transmission Theory, *Trans. AIEE*, **47**: 617 (1928).

4. Gabor, D.: Theory of Communication, *J. IEE (London)*, **93:** 111, 429 (1946), especially p. 429.
5. Shannon, C. E.: Communication in the Presence of Noise, *Proc. IRE*, **37:** 10 (1949).
6. Oswald, J.: *Compt. rend.*, **229:** 21 (1949). In particular, see Signals with Limited Spectra and Their Transformations, *Cables et transmissions*, **4:** 197 (1950); Random Signals with Limited Spectra, *ibid.*, **5:** 158 (1951).
7. Ville, J.: Signaux analytiques à spectre borne, *Cables et transmissions*, **4:** 9 (1950); also, *ibid.*, **7:** 44 (1953).
8. Kohlenberg, A.: Exact Interpolation of Band-limited Functions, *J. Appl. Phys.*, **24:** 1432 (1953).
9. Woodward, P. M.: "Probability and Information Theory, with Applications to Radar," McGraw-Hill, New York, 1953; cf. secs. 2.4, 2.5.
10. Goldman, S.: "Information Theory," Prentice-Hall, Englewood Cliffs, N.J., 1953; cf. chap. II.
11. See the bibliographies of F. H. L. M. Stumpers, e.g., A Bibliography of Information Theory, Communication Theory, Cybernetics, *IRE Trans. on Inform. Theory*, vol. PGIT-2, November, 1953; *ibid.*, **IT-1:** 31–47 (September, 1955). See also P. L. Chessin, A Bibliography on Noise, *ibid.*, **IT-1:** 15–31 (September, 1955).
11a. Balakrishnan, A. V.: A Note on the Sampling Principle for Continuous Signals, *IRE Trans. on Group Inform. Theory*, **IT-3:** 143 (1957).
12. Stewart, R. M.: Statistical Design and Evaluation of Filters for the Restoration of Sampled Data, *Proc. IRE*, **44:** 253 (1956).
13. Gabor, D.: A Summary of Communication Theory, in Willis Jackson (ed.), "Communication Theory," pp. 1-23, Academic Press, New York, 1953, especially pp. 10–12.
14. Woodward, P. M.: Time and Frequency Uncertainty in Waveform Analysis, *Phil. Mag.*, (7)**42:** 883 (1951).
15. Bennett, W. R.: Methods of Solving Noise Problems, *Proc. IRE*, **44:** 609, 627ff. (1956).
16. Middleton, D.: On the Theory of Random Noise. Phenomenological Models. II, *J. Appl. Phys.*, **22:** 1153 (1951).
17. Lawson, J. L., and G. E. Uhlenbeck: "Threshold Signals," chap. 3, eq. (33), Massachusetts Institute of Technology Radiation Laboratory Series, vol. 24, McGraw-Hill, New York, 1950.
18. Rice, S. O.: Mathematical Analysis of Random Noise. I, *Bell System Tech. J.*, **23:** 282 (1944), eq. (2.7-9).
19. Kenrick, G. W.: The Analysis of Irregular Motions with Applications to the Energy Frequency Spectrum of Static and of Telegraph Signals, *Phil. Mag.*, (7)**7:** 176 (1929). See also Ref. 18, sec. 2.7.
20. James, H. M., N. B. Nichols, and R. S. Phillips: "Theory of Servomechanisms," Massachusetts Institute of Technology Radiation Laboratory Series, vol. 25, McGraw-Hill, New York, 1947; cf. sec. 6.12.
21. Campbell, N.: *Proc. Cambridge Phil. Soc.*, **15:** 117, 310 (1909).
22. Rowland, E. N.: The Theory of the Mean Square Variation of a Function Formed by Adding Known Functions with Random Phases and Applications to the Theories of the Shot Effect and of Light, *Proc. Cambridge Phil. Soc.*, **32:** 580 (1936). See also J. M. Whittaker, The Shot Effect for Showers, *ibid.*, **33:** 451 (1937), and A. Khintchine, *Bull. Acad. Sci. U.S.S.R*, sér. math., no. 3, p. 313, 1938.
23. Rivlin, R. S.: An Extension of Campbell's Theorem of Random Fluctuations, *Phil. Mag.*, (7)**36:** 688 (1945). For some other forms, see Ref. 18, sec. 1.5.
24. Middleton, D., and D. Van Meter: Detection and Extraction of Signals in Noise from the Point of View of Statistical Decision Theory, *J. Soc. Ind. Appl. Math.*, **3**(4): 192 (December, 1955), especially p. 241.

CHAPTER 5

SIGNALS AND NOISE IN NONLINEAR SYSTEMS

In Chap. 2 (cf. Sec. 2.3), we sketched the treatment of zero-memory nonlinear devices when certain statistical properties of the resulting output ensembles are desired. Chapters 3 and 4, on the other hand, were devoted to the more special situations of the finite-memory linear transformations that are represented by a wide class of linear filters. In each instance, we were concerned primarily with the second-moment properties of physical interest, e.g., the intensity spectrum, the total average intensity, and so on. From this point of view, then, we shall go on to examine somewhat more fully the important class of zero-memory nonlinear operations $\boldsymbol{T}_g$, discussed in Secs. 2.3-1, 2.3-2, reserving the main account of specific applications to Chaps. 12 to 15 (Part 3).

As mentioned earlier (cf. Secs. 1.6, 2.1-3, and 2.3), the transformations $\boldsymbol{T}_g$ are important because they occur frequently at one stage or another of many existing communication systems. Mixing, modulation, detection, discrimination, etc., are typical of the nonlinear operations which $\boldsymbol{T}_g$ may represent and which act to combine ensembles of signals and (unavoidable) noise in the course of transmission or encoding and to separate them in some fashion in the process of reception or decoding. The characteristic feature of these nonlinear operations is the "scrambling," or mixing, of the various elements of the input ensemble, so that it is impossible by purely linear operations alone to resolve these original elements in the resulting output ensemble.† This, of course, is a direct consequence of the fact that the principle of superposition no longer applies in such cases [Sec. 2.1-3(2)]. Specifically, then, we shall outline in Secs. 5.1, 5.2 a second-moment theory whose chief aim is to determine such quantities as $M_z(t_1,t_2)$, the spectral density $\mathcal{W}_z(f)$, etc., for the output ensemble $z = \boldsymbol{T}_g\{y\}$ when sufficient statistical information about the input process y has been given. From this, in turn, we shall construct in Sec. 5.3 simple criteria of system performance, whose scope, however, is limited, since these criteria make use of second-moment data only. More general methods, utilizing the higher-order statistics of the *output* process, are considered in Chap. 6

† In fact, since most of the nonlinear transformations with which we have to deal are *irreversible* [cf. Secs. 1.2-14 and 2.1-3(2)], neither can we resolve these original elements by nonlinear methods.

and later in Part 4. The present analysis is based on some earlier work of Rice[1] and Middleton[2] and is intended as an introduction to the problems treated more extensively in Parts 3 and 4.

SECOND-MOMENT THEORY

5.1 *Rectification with Zero-memory Devices*

The operations with which we are now concerned are represented by the relation $z = \mathbf{T}_g\{y\}$, in which y and z are, respectively, the input and output ensembles; y and z themselves may be entirely random, deterministic, or mixed processes, of which the last is common in communication applications. A very wide class of input ensembles for such situations can be described by the additive process

$$y = \mathbf{T}_{M^{(N)}}\{x\} = x + cN \qquad x = \mathbf{T}_T\{u\} = S(t) \tag{5.1}$$

or specifically

$$y(t) = S(t) + cN(t) = aA(t;\phi,\phi_A) \cos [\omega_c t + \phi + b\Phi(t;\phi,\phi_F)] + cN(t) \tag{5.1a}$$

where a, b, c are (real) constants whose values we may choose to indicate some special class of input. For example, when the input is noise alone, we set $c = 1$ and $a = 0$, so that $y(t) = N(t)$, as required. Or if y represents an amplitude-modulated wave accompanied by a background noise, we may set $b = 0$ and $c = a = 1$, where now $A(t)$ is slowly varying compared with $\cos (\omega_c t + \phi)$, and so on. The ensemble $S(t)$ is constructed by assigning to the phases ϕ, ϕ_A, and ϕ_F appropriate probability measures; ϕ refers to the phases of the "carrier" (of frequency $f_c = \omega_c/2\pi$) and has the primary interval $(0,2\pi)$, with $(0,2\pi/\omega_c)$ the primary interval in epoch (cf. Sec. 1.3-5), while ϕ_A and ϕ_F refer to phases (or epochs, if a time measure is used) of the amplitude and phase variations $A(t)$ and $\Phi(t)$. In this way, a variety of different signal ensembles can be generated.

For our present discussion, it is convenient to restrict the class of input ensembles somewhat by requiring that $S(t)$ and $N(t)$ be independent and that $A(t)$ and $\Phi(t)$ be slowly varying compared with $\cos \omega_c t$, $\sin \omega_c t$, so that $S(t)$ is narrowband and so that A and Φ can be regarded as modulations in the usual sense [cf. Sec. 2.2-5(6)]. This, then, is the basic model for the modulated sinusoid, accompanied by additive noise if $c \neq 0$ in Eq. (5.1). The modulations themselves may be entirely random or mixed processes, as the examples of Table 5.1 indicate.† Because of their slowly varying nature and the fact that such modulations are usually impressed upon the carrier incoherently, i.e., without a fixed phase relation to this higher-frequency

† The list is representative, but clearly not exhaustive.

Table 5.1. Some Representative Signal Waveforms

$\{u\}$; description ($c = 0$)	$\boldsymbol{T}_T\{u\} = x(t) = S(t)$ Analytical expression	Wave structure of typical member
1. Sinusoid ($a = 1, b = 0$); $A(t) = A_0$	$S(t) = A_0 \cos(\omega_c t + \phi)$	
2. Sinusoidally modulated carrier (amplitude modulation) ($a = 1, b = 0$)	$S(t) = A_0(1 + \lambda \cos \omega_a t) \cos(\omega_c t + \phi)$	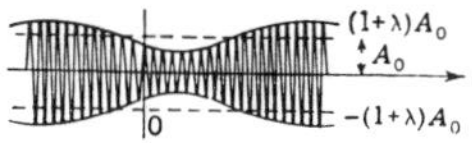
3. Pulse-modulated carrier (simple radar) ($a = 1, b = 0$) ($\tau \ll T_0$)	$S(t) = \left[\sum_k A_0 \Delta_k(t)\right] \cos(\omega_c t + \phi)$ $\Delta_k(t) = \begin{cases} 1 & T_0 k - \frac{\tau}{2} < t < T_0 k + \frac{\tau}{2} \\ 0 & \text{elsewhere} \end{cases}$	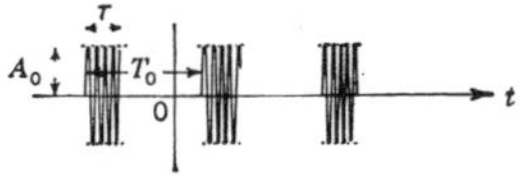
4. Random amplitude modulation of a carrier ($a = 1, b = 0$); $A(t)$ random	$S(t) = \begin{cases} A(t) \cos(\omega_c t + \phi) & A > 0 \\ 0, A < 0 \end{cases}$ ($S = 0$; overmodulation)	
5. Simple sinusoidal angle modulation ($a = 1, b = 1$); $A = A_0$	$S(t) = A_0 \cos(\omega_c t + \phi + \alpha_0 \cos \omega_a t)$	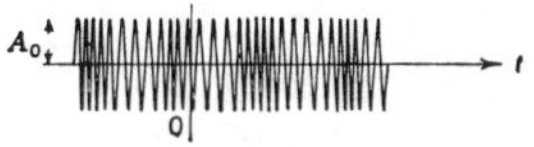
6. Random angle modulation of a carrier ($a = 1, b = 1$); $A = A_0$	$S(t) = A_0 \cos[\omega_c t + \phi + \Phi(t)]$	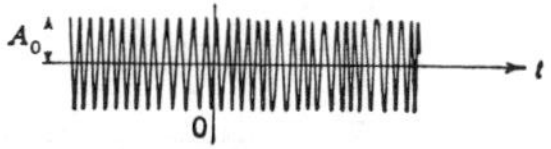
7. Simultaneous amplitude and frequency modulation; general modulations ($a = b = 1$)	$S(t) = \begin{cases} A(t) \cos[\omega_c t + \phi + \Phi(t)] & A > 0 \\ 0 & A < 0 \end{cases}$ ($S = 0$; overmodulation)	

oscillation, $A(t)$ and $\Phi(t)$ can be regarded as essentially independent† of ϕ. If phase Φ and amplitude A are statistically related, we may indicate this by writing $\phi_F = \phi_F(\phi_A)$, etc. More often, ϕ_A and ϕ_F are independent, so that Φ and A are likewise, unless they are shared by some common statistical process. More complicated ensembles are easily constructed: we have only to assign additional random parameters (time-dependent or not), with suitable measures, to $A(t)$ and $\Phi(t)$. For most applications, however, whenever A and Φ are deterministic, it is sufficient to give ϕ_A and ϕ_F the appropriate statistical properties.

In what follows, unless otherwise indicated, we shall assume that the noise process $N(t)$ is ergodic, entirely random, and independent of the signal ensemble $S(t)$ and that the component ensembles of the signal process are at least widely stationary. We shall also take the phase ϕ to be uniformly distributed in the primary interval, although this in no serious way limits the generality of the approach. In fact, the treatment of nonstationary modulations follows readily by obvious extensions of the present structure. Thus, for example, the second-moment function of the additive process y [Eq. (5.1a)] becomes

$$M_y(t_1,t_2) = \boldsymbol{E}\{y(t_1)y(t_2)\} \doteq \frac{a^2}{2}\,\overline{\langle A(t_1;\phi_A)A(t_2;\phi_A)\cos[b\Phi(t_2;\phi_F) - b\Phi(t_1;\phi_F)]} \\ \overline{\times \cos\omega_c(t_2 - t_1)\rangle} + c^2K_N(t_2 - t_1) \qquad (5.1b)$$

from which the ensemble spectral density may be obtained in the usual way with the help of the W-K theorem, provided that the modulations A, Φ are (strictly) stationary. [In fact, Eq. (5.1b) and the spectra derived from it are special cases of the more general relations derived below for the output of a zero-memory nonlinear element, specifically when this nonlinear element reduces to an ideal linear amplifier.] Spectra and covariance functions for many of the signal processes listed in Table 5.1 are considered in Chaps. 12 and 14.

Note, however, that even with stationary subensembles the signal ensemble $S(t)$ itself may not be stationary (cf. Sec. 1.3-7), so that the correlation functions of individual members of the input, and of the rectified process $z(t)$, are not equal to the corresponding ensemble averages. Additional averages over the epochs of the various modulations are also required (cf. Sec. 1.6, Example 3). Also, for the time being, we shall limit the discussion to zero-memory nonlinear elements $\boldsymbol{T}_g$ [such as (ideal) mixers, second detectors, certain types of limiters, and saturated amplifiers] which act only on the instantaneous *amplitudes* of the input.

The performance of nonlinear devices which convert amplitude into frequency, which is done (for example) in frequency and phase modulation or demodulation, is considered later in Chaps. 14, 15. More general

† An exception arises when coherence is maintained, as in certain types of radar, or in coherent measuring systems (cf. Part 4).

theories, not restricted to additive processes, can be constructed in a similar way, but as we might expect, these lead to much more complex expressions for M_z and $\mathcal{W}_z$ and at the same time are not nearly so common in physical situations. For this reason, almost all *input* signal and noise processes considered here will be additive, like Eq. (5.1).

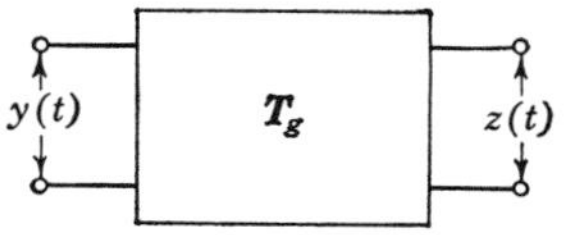

FIG. 5.1. A zero-memory nonlinear system, with input and output ensembles y, z.

5.1-1 The Second-moment Function $M_z(t)$. A typical situation is indicated in Fig. 5.1. Making use of the results of Example 3 of Sec. 1.6 and of Sec. 2.2, we see that under the above conditions the average of the autocorrelation function, $\boldsymbol{E}\{R_z^{(j)}(t)\}$, is equal to the second-moment function $M_z(t)$, and we can therefore write for $z = \boldsymbol{T}\{y\}$

$$M_z(t) = \boldsymbol{E}\{\boldsymbol{T}_g\{y_1\}\boldsymbol{T}_g\{y_2\}\} = \boldsymbol{E}\{g[y(t_1)]g[y(t_2)]\} \tag{5.2}$$

and, from Eq. (5.1),

$$M_z(t) = \overline{g(S_1 + N_1)g(S_2 + N_2)} = \iint\limits_{-\infty}^{\infty} g(y_1)g(y_2)W_2(y_1,y_2;t)\, dy_1\, dy_2 \qquad t = t_2 - t_1 \tag{5.2a}$$

Here $S_1 = S(t_1)$, etc., and M_z is a function of $t = t_2 - t_1$, since all component ensembles of the input signal are also assumed to be stationary, and $\boldsymbol{T}_g$ is invariant (cf. Sec. 2.3).

Our aim at this stage is to develop $M_z(t)$ in as explicit a form as possible, taking advantage of any special features, like the narrowband structure of the component signal and noise processes. Let us accordingly replace $g(\ \ )$ by its contour integral representation [Eq. (2.99), for example] and the probability density by its associated characteristic function, continued analytically [cf. Eqs. (2.95) et seq.], in accordance with our earlier observation that the characteristic-function method usually leads to usable results when the more direct approach proves intractable (cf. Sec. 2.3-2). Then, Eq. (2.97) becomes

$$M_z(t) = \frac{1}{(2\pi)^2}\int_{\mathbf{C}}\int_{\mathbf{C}} f(i\xi_1)f(i\xi_2)F_{y_1,y_2}(i\xi_1,i\xi_2;t)\, d\xi_1\, d\xi_2 \tag{5.3}$$

where we have used the equivalent contour $\mathbf{C}$, extending along the real axis from $-\infty$ to $+\infty$ and indented downward† for the one-sided dynamic path, about a possible singularity at $\xi = 0$. As before, $f(i\xi)$ is the complex Fourier transform of the dynamic path represented by the operation

† For equivalent contours, see Fig. 2.12 and Sec. 2.3-3. The two-sided dynamic path is handled in a similar way by Eq. (2.92b), with the contours $\mathbf{C}_1$, $\mathbf{C}_2$ shown in Fig. 2.12b.

$T_g = g(\ \ )$ (cf. Sec. 2.3-2). Since $S(t)$ and $N(t)$ are additive and independent, the characteristic function F_{y_1,y_2} factors into two parts [Sec. 1.2-10, especially Eq. (1.50)], for example, $F_{y_1,y_2} = F_{(S+N)} = F_S F_N$. Equation (5.3) can then be written

$$M_z(t) = \frac{1}{4\pi^2} \int_C \int_C f(i\xi_1) f(i\xi_2) F_2(i\xi_1, i\xi_2; t)_N F_2(i\xi_1, i\xi_2; t)_S \, d\xi_1 \, d\xi_2 \qquad (5.4)$$

with $(F_2)_S$ and $(F_2)_N$ the characteristic functions of the input signal and background noise, respectively. Equation (5.4) is the starting point for much of our subsequent study of nonlinear systems.

Since the spectrum of the output process is also a quantity of interest, we would like a representation of $M_z(t)$ wherein the time-dependent portion appears simply as a factor of the integrations over ξ_1 and ξ_2, so that the Wiener-Khintchine theorem (3.42) can then be directly applied to give the desired spectral density. For this purpose, let us expand F_S and F_N in suitable series. We begin with F_N and develop it formally† as

$$F_2(i\xi_1, i\xi_2; t)_N = F_1(i\xi_1)_N F_1(i\xi_2)_N \sum_{m,n=0}^{\infty} \frac{a_{m,n}^{(N)}(t)}{m!n!} (i\xi_1)^m (i\xi_2)^n \qquad (5.5)$$

In the limit $t \to \infty$, $a_{m,n}^{(N)}(t)$ vanishes $(m, n > 0)$, since $N(t)$ is entirely random and all "memory" between events at t_1 and t_2 disappears; $a_{0,0}^{(N)}(t) = 1$ for all t, inasmuch as $F_2(0,0;t) = 1$. As $t \to 0$, on the other hand, the $a_{m,n}^{(N)}(t)$ are such that $F_2(i\xi_1, i\xi_2; t)_N$ approaches $F_1(i\xi_1 + i\xi_2)_N$ [cf. Eqs. (1.92)].

For many noise processes, F_N is "diagonal"; that is, $a_{m,n}^{(N)}(t)$ takes the form‡ $a_n{}^{(N)}(t)\delta_n^m$. An important example of this is given by the case of *normal* random noise,§ whose second-order characteristic function is

$$F_2(i\xi_1, i\xi_2; t)_N = \exp \{-[\xi_1^2 + \xi_2^2 + 2k_N(t)\xi_1\xi_2]\psi/2\}$$
$$\psi = \overline{N^2};\ \bar{N} = 0;\ k_N(t) = \overline{N_1 N_2}/\psi \qquad (5.6a)$$

(cf. Prob. 1.8 and Sec. 7.2). Since $F_1(i\xi)_N$ is $e^{-\psi\xi^2/2}$, the expansion (5.5)

† We do not attempt to establish the general conditions for such an expansion here, observing that these can usually be easily determined in specific instances. For the convergence of the integrals in Eq. (5.4) when a series of the type (5.5) is used, the results of E. C. Titchmarsh, "Fourier Integrals," 2d ed., chap. 1, Oxford, New York, 1948, may be applied. In addition, see Levin[18] and also Leipnik,[19] who gives additional illustrations and references to the general technique of using suitable expansions, mainly of $\mathcal{W}_2$, however, rather than its characteristic function.

‡ δ_n^m (or δ_{mm}) is the familiar Kronecker delta, with the properties $\delta_n^m = 1$ $(m = n)$, $\delta_n^m = 0$ $(n \neq m)$.

§ Here, as in preceding chapters, we introduce normal random noise as characteristic of a great many communication processes, and to provide specific examples. For details of the normal process and processes derived from it, see Chaps. 7 to 9; for physical origins, Chaps. 10, 11, also.

reduces at once to

$$F_2(i\xi_1,i\xi_2;t)_N = e^{-\psi\xi_1^2/2}e^{-\psi\xi_2^2/2} \sum_{n=0}^{\infty} \frac{(-1)^n k_N(t)^n \psi^n (i\xi_1)^n (i\xi_2)^n}{n!}$$

Therefore, $$a_n^{(N)}(t) = (-1)^n k_N(t)^n \psi^n \tag{5.6b}$$

We shall encounter other examples like this in Part 3.

A development similar to Eq. (5.5) can be made for the characteristic function of the signal component, with $a_{m,n}^{(N)}(t)$ replaced therein by $a_{m,n}^{(S)}(t)$. But when $S(t)$ is narrowband and represents a modulated sinusoid, for example, Eq. (5.1a), we can take further advantage of this structure to obtain a series that is diagonal, in the above sense, where now $(i\xi_1)^m$, $(i\xi_2)^n$ are replaced by more general functions of ξ_1 and ξ_2. Let us begin by writing F_S as

$$F_2(i\xi_1,i\xi_2;t)_S = \langle \exp \{i\xi_1 aA(t_1,\phi_A) \cos [\omega_c t_1 + \phi + b\Phi(t_1,\phi_F)] + i\xi_2 aA(t_2,\phi_A) \cos [\omega_c t_2 + \phi + b\Phi(t_2,\phi_F)]\}\rangle_{\phi,\phi_F,\phi_A} \tag{5.7}$$

where we have used the fact that A and Φ are effectively independent of ϕ (see the beginning of Sec. 5.1). Next, we employ the relation†

$$e^{i\alpha \cos (\phi+\beta)} = \sum_{m=0}^{\infty} i^m \epsilon_m J_m(\alpha) \cos m(\beta + \phi) \tag{5.8}$$

α, β, ϕ real and ϵ_m the Neumann factor $\epsilon_0 = 1$, $\epsilon_m = 2$ $(m \neq 0)$; J_m is a Bessel function of the first kind and order m. Equation (5.7) then becomes

$$F_S = \sum_{m,n=0}^{\infty} i^{m+n} \epsilon_m \epsilon_n \langle J_m(\alpha_1) J_n(\alpha_2) \langle \cos m(\phi + \beta_1) \cos n(\phi + \beta_2)\rangle_\phi \rangle_{\phi_A,\phi_F} \tag{5.9}$$

with obvious interpretations of α_1, β_1, etc. With ϕ uniformly distributed over an RF cycle and with the help of the relation

$$\int_0^{2\pi} \cos m(\phi + \beta_1) \cos n(\phi + \beta_2) \frac{d\phi}{2\pi} = \delta_m^n \epsilon_m^{-1} \cos m(\beta_2 - \beta_1) \tag{5.10}$$

(cf. Prob. 5.2a) we easily reduce Eq. (5.9) to the following expression:

$$F_S = \sum_{m=0}^{\infty} (-1)^m \epsilon_m \langle J_m(\alpha_1) J_m(\alpha_2) \cos m(\beta_2 - \beta_1)\rangle_{\phi_A,\phi_F} \tag{5.11}$$

In greater detail,

$$\langle J_m(\alpha_1) J_m(\alpha_2) \cos m(\beta_2 - \beta_1)\rangle_{\phi_A,\phi_F} = \langle J_m[a\xi_1 A(t_1,\phi_A)] J_m[a\xi_2 A(t_2,\phi_A)] \times \cos m[\omega_c(t_2 - t_1) + b\Phi(t_2,\phi_F) - b\Phi(t_1,\phi_F)]\rangle_{\phi_A,\phi_F} \tag{5.11a}$$

† For example, G. N. Watson, "Theory of Bessel Functions," sec. 2.22, Cambridge, New York, 1944, or N. W. McLachlan, "Bessel Functions for Engineers," chap. 3, eq. (13), Oxford, New York, 1934.

which reduces still further for the stationary ensembles assumed here to

$$\langle J_m[a\xi_1 A(0,\phi_A)]J_m[a\xi_2 A(t,\phi_A)] \cos m[\omega_c t + b\Phi(t,\phi_F) - b\Phi(0,\phi_F)]\rangle_{\phi_A,\phi_F} \tag{5.11b}$$

(If the modulations are not stationary, F_S depends, of course, on t_1 and t_2, as does M_z eventually.)

Two special cases of particular interest are pure *amplitude modulation* (abbreviated AM) ($a = 1$, $b = 0$) and pure *angle modulation* (abbreviated ∠M) ($a = 1$, $A = A_0$, $b = 1$). Equation (5.11) reduces to

$$F_S\Big|_{\text{AM}} = \sum_{m=0}^{\infty} (-1)^m \epsilon_m \langle J_m[\xi_1 A(0,\phi_A)]J_m[\xi_2 A(t,\phi_A)]\rangle_{\phi_A} \cos m\omega_c t \tag{5.12a}$$

or

$$F_S\Big|_{\angle\text{M}} = \sum_{m=0}^{\infty} (-1)^m \epsilon_m J_m(\xi_1 A_0)J_m(\xi_2 A_0)\langle \cos m[\omega_c t + \Phi(t,\phi_F) - \Phi(0,\phi_F)]\rangle_{\phi_F} \tag{5.12b}$$

For simultaneous amplitude and angle modulation, we must use Eq. (5.11), and if the modulations are independent, the averages over ϕ_A, ϕ_F factor.

We have thus nearly achieved the desired factorization of the characteristic function of the input signal ensemble and can combine the factored forms (5.5) and (5.11) to write $M_z(t)$ as

$$M_z(t) = \sum_{m,n,q=0}^{\infty} \frac{(-1)^m \epsilon_m a_{n,q}^{(N)}(t)}{n!q!} \times \langle H_{m,n,q}(0;\phi_A)H_{m,n,q}(t;\phi_A) \cos m[\omega_c t + b\Phi(t;\phi_F) - b\Phi(0;\phi_F)]\rangle_{\phi_A,\phi_F} \tag{5.13}$$

in which $H_{m,n,q}(t;\phi_A)$ is an amplitude function, specified by

$$H_{m,n,q}(t;\phi_A) \equiv \frac{1}{2\pi}\int_C f(i\xi)\xi^{n+q}F_1(i\xi)_N J_m[a\xi A(t;\phi_A)]\, d\xi \tag{5.14}$$

Note, however, that in this general case the time still appears implicitly. A further development of the Bessel functions is required to put Eq. (5.13) in a form suitable for determining the spectrum explicitly, a point which will be considered in more detail later. In the meantime, observe from Eqs. (5.13), (5.14) that, for the normal noise of Eqs. (5.6), special cases of interest are now:

1. *Noise alone*

$$M_z(t) = \sum_{n=0}^{\infty} (-1)^n \frac{\psi^n k_N(t)^n}{n!} h_{0,n}^2 \tag{5.15a}$$

with
$$h_{0,n} \equiv H_{0,n,0}(0,0) = \frac{1}{2\pi}\int_C f(i\xi)\xi^n e^{-\psi\xi^2/2}\, d\xi \tag{5.15b}$$

2. *Carrier and noise*

$$M_z(t) = \sum_{m,n=0}^{\infty} \frac{\epsilon_m(-1)^{m+n}}{n!} \psi^n k_N(t)^n h_{m,n}^2 \cos m\omega_c t \tag{5.16a}$$

with
$$h_{m,n} \equiv H_{m,n,0}(0,0) = \frac{1}{2\pi} \int_C f(i\xi)\xi^n J_m(A_0\xi) e^{-\psi\xi^2/2}\, d\xi \tag{5.16b}$$

3. *Amplitude modulation and noise*

$$M_z(t) = \sum_{m,n=0}^{\infty} \frac{\epsilon_m(-1)^{m+n}}{n!} \psi^n k_N(t)^n \langle H_{m,n}(t,\phi_A) H_{m,n}(0,\phi_A)\rangle_{\phi_A} \cos m\omega_c t \tag{5.17a}$$

with $H_{m,n} = H_{m,n,0}$, etc. (5.17b)

4. *Angle modulation and noise*

$$M_z(t) = \sum_{m,n=0}^{\infty} \frac{\epsilon_m(-1)^{m+n}}{n!} \psi^n k_N(t)^n h_{m,n}^2 \langle \cos m[\omega_c t + \Phi(t,\phi_F) - \Phi(0,\phi_F)]\rangle_{\phi_F} \tag{5.18}$$

The relations (5.15), (5.16) were obtained originally by Rice.† Finally, we remark again that, if the signal ensemble is not narrowband, Eq. (5.4), in conjunction with Eq. (5.5) or Eqs. (5.6), must then be used, while $(F_2)_S$ is evaluated from the definition $\langle e^{i\xi_1 S_1 + \xi_2 S_2}\rangle_S$. The rectified output (like the input) is a broadband phenomenon.

The mean total intensity $\overline{z^2}$ of the output of the nonlinear device may be obtained from the above simply by setting $t = 0$ in $M_z(t)$ [cf. Eqs. (1.92)]. The contributions of any steady or periodic terms, on the other hand, follow from whatever remains when $\lim_{\tau\to\infty} M_z(t+\tau)$ is computed [cf. Eqs. (1.93)], since $k(\infty)$ vanishes. Thus, we have

$$P_{\text{dc+periodic}} = \lim_{\tau\to\infty} M_z(t+\tau) \qquad P_{\text{total}} = \overline{z^2} = M_z(0) \tag{5.19}$$

Of course, instead of calculating $M_z(t)$ first, we can use the appropriate first-order distribution density of y, applying Eqs. (1.92) and (1.93) to Eq. (5.2a), to write directly

$$\overline{z^2} = \overline{g(S_1+N_1)^2} = \int_{-\infty}^{\infty} g(y)^2 W_1(y)\, dy \left[= \int_0^{\infty} \mathcal{W}_z(f)\, df \right] \tag{5.20a}$$

$$P_{\text{dc+periodic}} = \left[\int_{-\infty}^{\infty} g(y) W_1(y)\, dy\right]^2 \tag{5.20b}$$

As we shall see, the former approach has the advantage of revealing the contributions of the various spectral components generated in the process

† Cf. Rice,[1] pt. IV, especially secs. 4.8, 4.9; see also Middleton,[2] sec. 2.

of rectification, while the latter is often analytically simpler and more convenient if only an estimate of intensity is desired.

5.1-2 The Spectrum. We obtain the intensity spectrum of the output process directly from Eq. (5.13) with the help of the W-K theorem (3.78), since $\mathcal{W}_z(f) = 2F\{M_z\}$. For the special cases of interest, Eqs. (5.15) to (5.18), we can write at once:

1. *Noise alone*

$$\mathcal{W}_z(f) = \sum_{n=0}^{\infty} (-1)^n \frac{\psi^n h_{0,n}^2}{n!} C_{0,n}(f) \tag{5.21}$$

with $$C_{0,n}(f) = 4 \int_0^{\infty} k_N(t)^n \cos \omega t \, dt$$

2. *Carrier and noise*

$$\mathcal{W}_z(f) = \sum_{m,n=0}^{\infty} \frac{(-1)^{m+n}\epsilon_m}{n!} h_{m,n}^2 \psi^n C_{m,n}(f) \tag{5.22}$$

with $$C_{m,n}(f) = 4 \int_0^{\infty} k_N(t)^n \cos \omega t \cos m\omega_c t \, dt$$

3. *Amplitude modulation and noise*

$$\mathcal{W}_z(f) = \sum_{m,n=0}^{\infty} \frac{(-1)^{m+n}\epsilon_m}{n!} \psi^n D_{m,n}(f) \tag{5.23}$$

with

$$D_{m,n}(f) = 4 \int_0^{\infty} k_N(t)^n \langle H_{m,n}(t,\phi_A) H_{m,n}(0,\phi_A)\rangle_{\phi_A} \cos \omega t \cos m\omega_c t \, dt$$

4. *Angle modulation and noise*

$$\mathcal{W}_z(f) = \sum_{m,n=0}^{\infty} \frac{(-1)^{m+n}\epsilon_m}{n!} \psi^n h_{m,n}^2 E_{m,n}(f) \tag{5.24}$$

with

$$E_{m,n}(f) = 4 \int_0^{\infty} k_N(t)^n \cos \omega t \langle \cos m[\omega_c t + \Phi(t,\phi_F) - \Phi(0,\phi_F)]\rangle_{\phi_F} \, dt$$

with the more general case (5.13) handled in a similar fashion.

From these relations, we observe that rectification generates a new spectrum, different from that of the input waves, as a result of the distortion produced in the instantaneous waveforms of the input ensemble. These spectral components, appearing after rectification, are called *modulation*, or *cross, products* and are formed by the intermodulation, or "mixing" together, of the original input components, in amount and intensity determined, of course, by the particular transformation $\mathbf{T}_g$ and by the statistical properties of the input ensemble itself, as Eqs. (5.13) to (5.24) clearly indicate. Of these modulation products, in turn, we can distinguish three principal classes, specified by the fact that they depend† only on the various input frequency terms from which they are produced. To illustrate, let us consider case 2, Eqs. (5.22). Then the three classes of modulation product are:

1. *Noise × noise, or n × n terms.* These are the result of intermodulation of the original noise components with one another, to produce a continuum $\mathcal{W}_z(f)_{n\times n}$. Their contribution is derived from terms of $O(m = 0, n \geq 1)$ in Eqs. (5.22).

2. *Signal × noise, or s × n products.* These are generated by the intermodulation of the signal (here the carrier) terms and its harmonics (f_c, $2f_c$, . . .) with the background noise continuum. The result is also a continuous spectrum, $\mathcal{W}_z(f)_{s\times n}$, derived from the terms of $O(m \geq 1, n \geq 1)$ in Eqs. (5.22).

3. *Signal × signal, or s × s terms.* These are produced by the cross modulation of the signal and its harmonics with one another and are always represented by a discrete, or "line," spectrum, $\mathcal{W}_z(f)_{s\times s}$, described by terms of $O(m \geq 1, n = 0)$ in Eqs. (5.22).

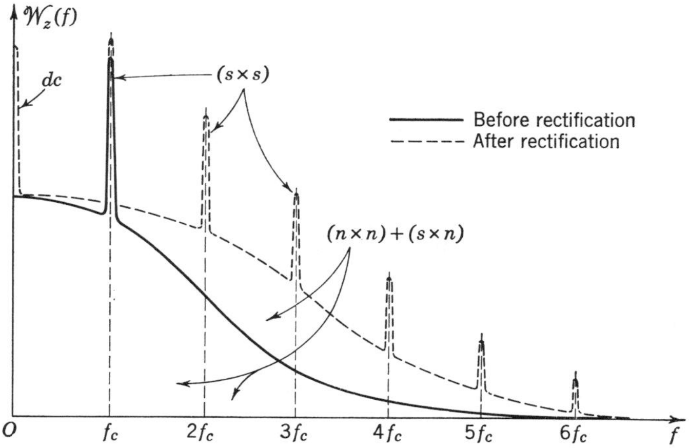

FIG. 5.2. Intensity spectra of a (broadband) carrier and noise before and after rectification. (The continuum after rectification is arbitrarily normalized at $f = 0$ to that before rectification.)

Finally, the component $m = n = 0$ yields the d-c or constant output, if any. Like the $n \times n$, $s \times n$, and $s \times s$ spectra cited above, it, too, is influenced by the input carrier and noise, and, like the $s \times s$ terms, it is represented by a spectral line, at $f = 0$. Figure 5.2 shows a typical broadband spectrum obtained by rectification of a broadband input ensemble

† Spectrally, but not necessarily in intensity.

each member of which consists of an unmodulated carrier and noise. For other examples, see Part 3 and, in particular, Chaps. 12 to 15.

A deterministic modulation ($|m|$) adds further to the $s \times s$ signal products, as well as to the $s \times n$ terms. Another set of signal products is now distinguished: these are the ones for which neither the carrier f_c nor noise components appear spectrally. In a similar way, one gets $s \times m$ terms as well, distributed about the carrier products. The general features of these spectra remain unaltered: intermodulation between components with "discrete," or line, spectra produces further line spectra at the sum and difference frequencies, while $s \times m \times n$ terms appear again as a continuum. If the modulation is itself entirely random, the only contributions to the discrete spectrum are the $s \times s$ components of the rectified carrier. Note, too, that this classification of modulation products, in conjunction with the spectrum $\mathcal{W}_z(f)$ [cf. Eqs. (5.21) et seq.], provides a useful method of determining the contributions of these different classes to the total output. For example, in case 2, Eq. (5.22), we find at once that

$$P_{s\times s} = \int_0^\infty \mathcal{W}_z(f)_{s\times s}\, df = \sum_{m=1}^{\infty} \epsilon_m(-1)^m h_{m,0}^2 \int_0^\infty C_{m,0}(f)\, df \tag{5.25a}$$

while the continuum consists of the $n \times n$ and $s \times n$ cross products

$$P_{n\times n} = \int_0^\infty \mathcal{W}_z(f)_{n\times n}\, df = \sum_{n=1}^{\infty} \frac{(-1)^n}{n!} \psi^n h_{0,n}^2 \int_0^\infty C_{0,n}(f)\, df \tag{5.25b}$$

$$P_{s\times n} = \int_0^\infty \mathcal{W}_z(f)_{s\times n}\, df = \sum_{m=1}^{\infty} \sum_{n=1}^{\infty} \frac{\epsilon_m(-1)^{m+n}}{n!} \psi^n h_{m,n}^2 \int_0^\infty C_{m,n}(f)\, df \tag{5.25c}$$

respectively. The d-c contribution is simply

$$P_{\text{dc}} = \int_0^\infty \mathcal{W}_z(f)_{\text{dc}}\, df = 4 \int_0^\infty h_{0,0}^2 \cos \omega t\, dt = 2h_{0,0}^2\delta(f-0) \tag{5.25d}$$

as expected, with the amplitude $h_{0,0}$. The other cases, (1), (3), and (4), are handled in the same way, while the total intensity P_T may be obtained with the help of Eq. (5.19) or Eq. (5.20a).

5.1-3 Narrowband Input Ensembles. Of particular importance in applications are narrowband mixed input ensembles, where now the component signal and noise processes, $S(t)$ and $N(t)$, each possess narrowband features. This is a common situation in most communication problems, as a glance at Table 5.1 indicates; the transmitted information is distributed in a narrow band of frequencies about a carrier, of much greater frequency than the message bandwidth. Thus, while Eq. (5.4) must be used for broadband inputs and Eqs. (5.7) to (5.18) are particularly suited to the case of narrowband signals in broadband noise, we can modify these relations still further for the completely narrowband input.

Let us postulate once more that the input noise ensemble $N(t)$ is normal, so that Eqs. (5.6) may be used, and let us require in addition that $N(t)$ be narrowband, so that the covariance function of N can be written approximately $K_N(t) \doteq K_0(t)_N \cos [\omega_0 t - \phi_0(t)_N]$, where, specifically, $K_0(t)_N = W_0 \sqrt{\rho_0^2 + \lambda_0^2}$ [Eq. (3.92)† or (3.108*b*)]. Here ω_0 is the (angular) resonant or central frequency of the noise band [cf. Sec. 2.2-5(5)] and $\phi_0(t)_N$ is a possible phase term, characteristic of the original noise source or attributable to the filters through which it may have passed (cf. Secs. 3.3-2, 3.3-4). Using the normalized covariance function $k_0(t) \equiv K_0(t)_N/K_0(0)_N$ and setting $f_0 = f_c$, the carrier frequency, to simplify the subsequent calculations somewhat, we recall for the normal noise assumed here that $a_{n,q}^{(N)}(t)$ in Eq. (5.13) reduces to $(-1)^n[\psi k_0(t)_N \cos (\omega_0 t - \Phi_0)]^n$ and the summation over q is dropped [cf. Eq. (5.6*b*)]. The amplitude functions $H_{m,n}(t;\phi_A)$, $k_0(t)_N$, and the phases $\Phi(t;\phi_F)$, etc., are slowly varying compared with $\cos \omega_0 t$, $\sin \omega_0 t$, $\cos \omega_c t$, etc. Resolving $\cos^n \omega_0 t$ into harmonics of $\omega_0 t$, we next form the sum and difference terms $\cos (p\omega_0 \pm q\omega_c)t$ $(p, q \geq 0)$ and thus reduce $M_z(t)$ to a set of terms consisting of slowly varying components in product with a rapidly varying one of the type $\cos l\omega_0 t$, etc. Accordingly, we use

$$\cos^n \omega_0 t = 2^{-n} \sum_{k=0}^{\frac{n}{2},\frac{n-1}{2}} \epsilon_{n-2k}\, {}_nC_k \cos (n - 2k)\omega_0 t \qquad {}_nC_k = \frac{n!}{(n-k)!k!} \tag{5.26}$$

where the upper limit applies as n is even or odd, respectively. Remembering that $\omega_0 = \omega_c$ here, we collect terms of the type $F_l(t) \cos l\omega_0 t$ and $G_l(t) \sin \omega_0 lt$, where $F_l(f)$ and $G_l(t)$ are slowly varying vis-à-vis $\cos l\omega_0 t$, $\sin l\omega_0 t$, and obtain finally for the second-moment function of the output (cf. Prob. 5.2*b*)‡

$$M_z(t) = \sum_{l=0}^{\infty} M_z(t)_l = \sum_{l=0}^{\infty} [F_l(t) \cos l\omega_0 t + G_l(t) \sin l\omega_0 t] \tag{5.27a}$$

† See Table 3-1 for $K_0(t)_N$ when the noise is shaped by various simple linear high-Q networks.

‡ A less direct, though somewhat simpler, method in the case of normal noise is to replace the time-dependent term in the characteristic function (5.6*a*) by the alternative series $\exp [-k_N(t)\psi\xi_1\xi_2] = \exp [-\xi_1\xi_2 k_0(t) \cos \omega_0 t] = \sum_{m=0}^{\infty} \epsilon_m (-1)^m I_m[\xi_1\xi_2 k_0(t)] \cos m\omega_0 t$, where I_m is a modified Bessel function of the first kind, order m. See Middleton, *op. cit.*, sec. 3, and footnote on p. 252.

where specifically†

$$M_z(t)_l = (-1)^l \Big\langle \sum_{m=0}^{\infty} \{\epsilon_m B_{m,|m-l|}(t,\phi_A) \times \cos [l\omega_0 t + mb\Phi(t,\phi_F) - mb\Phi(0,\phi_F) + |m-l|\phi_0(t)_N] + (1-\delta_l^0)\epsilon_m B_{m,m+l}(t,\phi_A) \cos [l\omega_0 t - mb\Phi(t,\phi_F) + mb\Phi(0,\phi_F) + (m+l)\phi_0(t)_N]\} \Big\rangle \tag{5.27b}$$

and

$$B_{m,|\alpha|}(t,\phi_A) = \sum_{q=0}^{\infty} \frac{k_0(t)_N{}^{2q+|\alpha|}\psi^{2q+|\alpha|}}{q!(q+|\alpha|)!2^{2q+|\alpha|}} H_{m,2q+|\alpha|}(t;\phi_A)H_{m,2q+|\alpha|}(0;\phi_A) \tag{5.27c}$$

with $$|\alpha| = |m \pm l| \tag{5.27d}$$

These amplitude functions are given by Eq. (5.14), with q therein set equal to zero and n replaced by $2q + |\alpha|$. For the normal noise assumed here, $F_1(i\xi)_N$ becomes specifically $e^{-\psi\xi^2/2}$. The intensity spectrum of the rectified input is given by

$$\mathcal{W}_z(f) = 2\mathfrak{F}\left\{\sum_{l=0}^{\infty} M_z(t)_l\right\} = \sum_{l=0}^{\infty} 4\int_0^{\infty} M_z(t)_l \cos \omega t \, dt = \sum_{l=0}^{\infty} \mathcal{W}_z(f)_l \tag{5.28}$$

The essential feature of Eqs. (5.27) is that the coefficients of $\cos l\omega_0 t$, $\sin l\omega_0 t$ in Eq. (5.27a) are slowly varying compared with $\cos l\omega_0 t$, $\sin l\omega_0 t$ and consequently the output spectrum can be resolved into an (infinite) sequence of spectral contributions $\mathcal{W}_z(f)_l$ that are all essentially distinct. Each of the spectra is concentrated about the appropriate harmonic, lf_0, of the input carrier (or noise center frequency), in the manner of Fig. 5.3, where $l \geq 0$ indicates the *spectral zone*, or region, in which $\mathcal{W}_z(f)_l$ predominantly lies. These spectral regions do not noticeably overlap, provided, of course, that both modulation and background noise are comparatively low-frequency (l-f). They may be resolved with the help of appropriate band-pass filters, each of which is wide enough to include an entire zone, but not so wide as to pass noticeable contributions from the other zones as well.[3]

In most communication problems, the spectral zones $l = 0, 1$ are the ones of chief interest. In reception, the former contains the desired message-bearing l-f terms, while, in the course of the modulation or mixing operations required in both transmission and reception, it is the latter ($l = 1$) that is significant. Note, also, that all three classes of modulation product, $n \times n$, $s \times n$, and $s \times s$ components, are present in these zonal spectra. The reader can easily identify each class directly from Eqs. (5.27b), (5.27c),

† Middleton, *op. cit.*, eq. (3.3); the term in $(1 - \delta_l^0)B_{m,m+l}(t)$ was inadvertently omitted in *op. cit*, (3.3). See also Fellows and Middleton.[3] The present relation (5-27b) is a generalization of these earlier results.

once the type of modulation (if any) has been specified. The mean intensity associated with each zone is given by

$$P_l = \int_0^\infty \mathcal{W}_z(f)_l \, df = M_z(0)_l \qquad l \geq 0 \tag{5.29}$$

which, in conjunction with Eqs. (5.27), shows that for normal noise the mean intensity of a zone depends only on the mean intensity of the input waves,† S and N, *and not on how the intensity is distributed* (spectrally) *in the original input.* We shall encounter many examples of this result later, in Part 3, as well as applications of Eqs. (5.27), (5.28).

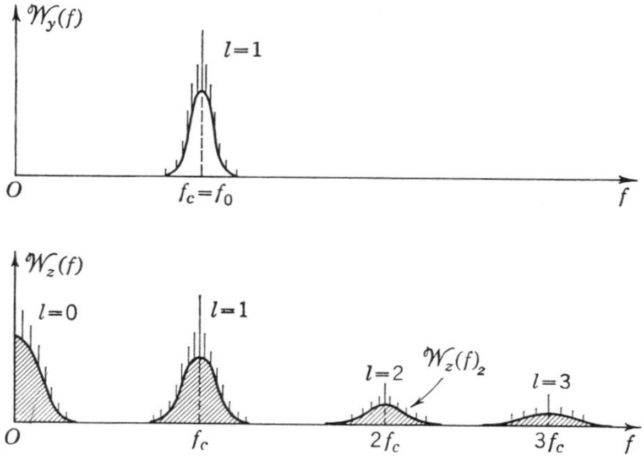

FIG. 5.3. Typical intensity spectra at input and output of a zero-memory nonlinear device. Narrowband input, consisting of a modulated carrier and noise.

Finally, we remark that higher-order statistics, such as the higher second-order moments $\overline{z^m(t_1)z^m(t_2)}$ $(m > 1)$, etc., may be obtained in the same way, while the distribution densities $w_1(z)$, $w_2(z_1,z_2;t)$ of the rectified (but unfiltered) output follow directly from $W_1(y)$, $W_2(y_1,y_2;t)$ of the input, in the usual way [cf. Eq. (1.63) and footnote ‡, page 21], where $z = \boldsymbol{T}_g\{y\}$ and $\boldsymbol{T}_g$ is, as before, a zero-memory nonlinear operation. However, if we are interested in the *filtered* output of this nonlinear device, the analysis becomes more intricate: for the moments one can use an obvious modification of Eq. (2.88*c*), but for the distribution densities explicit results can be found in certain cases only (see Chap. 17).

5.1-4 Envelope and Phase Representations. When the background noise, as well as the signal, is a narrowband process, it is sometimes necessary as well as convenient to obtain corresponding results for the output, second-moment functions, and spectra in terms of the envelope and phase character-

† This is true for other noise statistics, provided that $a_{m,n}^{(N)}(t)$ [cf. Eqs. (5.5), (5.6)] is "diagonal." The magnitude of P_l, of course, depends also on the structure of the rectifier, through $f(i\xi)$ in the amplitude functions $H_{m,n}$, etc. See Rice, *op. cit.*, eqs. (4.9-19) et seq.

istics of the input ensemble. Thus, following the treatment of Sec. 2.2-5(5) [cf. Eq. (2.67)], we can express the input noise ensemble here as $N(t) = N_c(t) \cos(\omega_c t + \phi) + N_s(t) \sin(\omega_c t + \phi)$, where now N_c, N_s are slowly varying vis-à-vis $\cos(\omega_c t + \phi)$, $\sin(\omega_c t + \phi)$. Combining this with the modulated carrier of Eq. (5.1a) gives

$$y(t) = y_c(t) \cos(\omega_c t + \phi) + y_s(t) \sin(\omega_c t + \phi) = E_0(t)_y \cos[\omega_c t + \phi - \phi_0(t)_y] \tag{5.30}$$

where

$$y_c = aA \cos b\Phi + cN_c = E_{0y} \cos \phi_{0y}$$
$$y_s = -aA \sin b\Phi + cN_S = E_{0y} \sin \phi_{0y}$$

$$E_0(t)_y = \sqrt{y_c^2 + y_s^2} = (a^2A^2 + c^2N_c^2 + c^2N_s^2 + 2acAN_c \cos b\Phi - 2acAN_s \sin b\Phi)^{1/2} \tag{5.31}$$
$$\phi_0(t)_y = \tan^{-1} \frac{cN_s - aA \sin b\Phi}{cN_c + aA \cos b\Phi} \pm \pi q \qquad q = 0, 1, 2, \ldots$$

The envelope and phase, E_{0y}, ϕ_{0y}, like y_c, y_s, are also slowly varying compared with $\cos(\omega_c t + \phi)$, etc., and are, of course, statistically dependent on ϕ_A, ϕ_F, A, Φ, etc.

The instantaneous output of our zero-memory nonlinear device $\mathbf{T}_g$ is then given at once by Eqs. (2.103) and consists, as before, of the linear sum of the contributions from the l (≥ 0) spectral zones produced in rectification, viz.,

$$z(t) = \sum_{l=0}^{\infty} B_l(t;\phi_A,\phi_F) \cos l[\omega_c t + \phi - \phi_0(t;\phi_A,\phi_F)_y] \tag{5.32a}$$

where

$$B_l(t;\phi_A,\phi_F) = \frac{\epsilon_l i^l}{2\pi} \int_C f(i\xi) J_l[\xi E_0(t;\phi_A,\phi_F)_y]\, d\xi \tag{5.32b}$$

As a simple example, we let $\mathbf{T}_g$ be the half-wave νth-law device of Eq. (2.85c) (cf. Fig. 2.7a) when the additive process (5.1a) is the input to it. With the aid of Eq. (d) of Prob. 2.16, we readily find that the instantaneous output associated with the lth zone is

$$z(t)_l = \frac{\beta E_{0y}{}^{\nu} \Gamma(\nu + 1) \epsilon_l \cos l(\omega_c t + \phi - \phi_{0y})}{2^{\nu+1} \Gamma[(2 + \nu - l)/2] \Gamma[(2 + \nu + l)/2]} \tag{5.33}$$

The second-moment function now follows immediately from Eq. (5.2) once the joint distribution density of ϕ, E_{0y}, and ϕ_{0y} (or the associated densities of ϕ_A, ϕ_F, A, Φ, etc.) is given. Provided that the modulations A and Φ are (strictly) stationary, the ensemble spectral density for $z(t)_l$ may then be obtained with the assistance of the W-K theorem. The spectrum of the entire output, of course, is simply the sum of the spectra in the individual zones. Here, however, we are usually interested in only one zone, so that when there is no angle modulation ($b = 0$; $c = a = 1$), we get from Eqs. (5.2) and (5.32) in the l-f case ($l = 0$)

Amplitude modulation + noise

$$M_z(t)_0 = \begin{cases} E\{z(t_1)_0 z(t_2)_0\} = \overline{B_0(t_1;\phi_A)B_0(t_2,\phi_A)} \qquad t = t_2 - t_1 \\ \dfrac{1}{4\pi^2} \displaystyle\iint_C f(i\xi_1)f(i\xi_2)\overline{J_0[\xi_1 E_0(t_1;\phi_A)_y]J_0[\xi_2 E_0(t_2;\phi_A)_y]}\, d\xi_1\, d\xi_2 \end{cases} \tag{5.34a}$$

while if angle modulation alone occurs ($A = A_0$; $a = b = c = 1$), we get instead

Angle modulation + noise

$$\begin{aligned} M_z(t)_0 &= \overline{B_0(t_1;\phi_F)B_0(t_2;\phi_F)} \\ &= \frac{1}{4\pi^2}\iint_C f(i\xi_1)f(i\xi_2)\overline{J_0[\xi_1 E_0(t_1;\phi_F)_y]J_0[\xi_2 E_0(t_2;\phi_F)_y]}\, d\xi_1\, d\xi_2 \end{aligned} \tag{5.34b}$$

The statistical averages in either case are to be taken over ϕ_A, ϕ_F, A, Φ as they may appear in E_{0y} [Eqs. (5.31)].

Similarly, for noise alone ($a = b = 0$; $c = 1$) we have

Noise

$$\begin{aligned} M_z(t)_0 &= \overline{B_0(t_1)B_0(t_2)} \\ &= \frac{1}{4\pi^2}\iint_C f(i\xi_1)f(i\xi_2)\overline{J_0(\xi_1 E_{01})J_0(\xi_2 E_{02})}\, d\xi_1\, d\xi_2 \end{aligned} \tag{5.34c}$$

and the ensemble average is over the pair of random variables $E_{01} = E_0(t_1)_y$, $E_{02} = E_0(t_2)_y$, or over $N_{c1,2}$, $N_{s1,2}$, through the transformations $E_{01} = \sqrt{N_{c1}^2 + N_{s1}^2}$, $E_{02} = \sqrt{N_{c2}^2 + N_{s2}^2}$ [Eqs. (5.31)]. Similar results follow at once for the other spectral zones ($l \geq 1$), except that now we need in addition the appropriate distribution densities for ϕ and for ϕ_{0y} in order to obtain $M_z(t)_l$ and the corresponding spectrum. Thus, if ϕ is uniformly distributed in its primary interval, we have for the lth zone

$$M_z(t)_l = \tfrac{1}{2}\langle B_l(t_1;\phi_A,\phi_F)B_l(t_2;\phi_A,\phi_F)\cos l[\omega_c t + \phi_0(t_1;\phi_A,\phi_F)_y - \phi_0(t_2;\phi_A,\phi_F)_y]\rangle_{stat\ av} \tag{5.35}$$

The essential feature, in any case, is that the appropriate second-order distribution densities of envelope and phase, or of the modulations, their parameters, and the noise components N_c, N_s (as well as ϕ), must be given if we are to determine $M_z(t)_l$ and $M_z(t)$ for the rectified output disturbance. Moreover, these statistical properties of the envelope and phase are different from those of $y(t)$ itself. Thus, for example, while $y(t)$ may represent a gaussian process, E_{0y} and ϕ_{0y} will certainly not, by virtue of the nonlinear relations (5.31). For this reason and from the consequent fact that the statistics of envelope and phase are usually much more complicated than for the process y itself, we choose the direct rather than the envelope-phase representation whenever this is possible. In certain situations, however, particularly in the study of angle-modulated ($\angle$M) waves, where the envelope-phase representation is the natural choice, this is not usually possible, and such a representation must then be employed. We shall encounter specific examples presently, in Chaps. 14 and 15 (see also Chap. 8). Note, in any case, that the statistical properties of the process following $\boldsymbol{T}_g$ are no longer even of the same type as those of the input ensemble but are often greatly modified (cf. Chap. 17).

PROBLEMS

5.1 (*a*) Generalize the treatment of Secs. 5.1-1, 5.1-2 to include nonstationary modulations and the cases where modulation and carrier are coherent.

(*b*) Outline a possible theory for $M_z(t)$ in the case of the multiplicative process $z = S(t)N(t)$, where S and N have the properties assumed in Sec. 5.1-1.

5.2 (*a*) Establish Eqs. (5.10), (5.26). What is the form of Eq. (5.13) when in place of a sinusoid a narrow band of noise is used for a carrier?

(*b*) Verify the procedure mentioned in Sec. 5.1-3 leading to Eqs. (5.27*b*), (5.27*c*).

(*c*) Show that, if the carrier frequency f_c and the center of the noise band f_0 are not coincident, one inserts in Eq. (5.27*b*) an additional phase factor $m\omega_d t$ ($\omega_d \equiv \omega_c - \omega_0$) whenever $b\Phi(t;\phi_F)$ appears.

5.2 *Three Examples of Rectification*

With the theory above in mind, let us illustrate it with some examples that exhibit one or more of the features expected in practice, e.g., an accompanying background noise; a possible carrier or signal, which may or may not be modulated; a narrowband structure for the input ensemble, with a corresponding zonal output; the effect of the zero-memory nonlinear element; and finite, as well as infinitely large, samples at the output stage. This last condition simply reflects the fact that for many practical situations we cannot treat the observed output ensemble as if it were of infinite duration but must rather take into account the finite, or limited, time at our disposal for observing and processing data. Apart from their illustrative aspect, the examples below are also of interest to us in their own right, representing situations of more or less common occurrence in reception and measurement.

5.2-1 Half-wave Linear Rectification of Normal Noise.† The purpose of this example is to indicate how the zonal structure of the output spectrum arises in a specific instance and to show how the characteristic-function method (cf. Secs. 2.3-2, 5.1) may be applied to give us quantitative results. Here the input ensemble $y(t)$ consists of normal random noise, for the present purposes adequately described by the characteristic functions (5.6). In addition, the noise is assumed to be ergodic and narrowband, shaped by frequency-sensitive networks in such a way that the phase $\phi_0(t)_y$ in the covariance function $K_y(t)_N = K_0(t)_y \cos[\omega_0 t - \phi_0(t)_y]$ is essentially zero.‡

† See footnote on p. 107.

‡ As occurs, for instance, when white noise is passed through a high-Q LCR filter or filters of this type in cascade (cf. Table 3.1 and also Sec. 3.3-2).

We may now apply these conditions to the general relations (5.13), (5.27*a*) to (5.27*d*), observing that only terms with $m = 0$ therein contribute, to write directly

$$M_z(t) = \sum_{l=0}^{\infty} (-1)^l (2 - \delta_l^0) B_{0,l}(0,0) \cos l\omega_0 t \tag{5.36}$$

where

$$B_{0,l} = \sum_{q=0}^{\infty} \frac{k_0(t)_y{}^{2q+l}\psi^{2q+l}}{q!(q+l)!2^{2q+l}} h_{0,2q+l}^2 \qquad h_{0,2q+l} = \frac{1}{2\pi}\int_C f(i\xi)e^{-\psi\xi^2/2}\xi^{2q+l}\,d\xi \tag{5.37}$$

The amplitude functions $H_{m,n}(t,0)$, etc., reduce to the simpler forms $h_{0,2q+l}$ [cf. Eqs. (5.14), (5.15*b*)]; $k_0(t)_y$ is the normalized covariance function $K_0(t)_y/K_0(0)_y$. From Eq. (5.28), the intensity spectrum of the lth zone becomes

$$\mathfrak{W}_l(f) = \sum_{q=0}^{\infty} \frac{\psi^{2q+l}(2-\delta_l^0)h_{0,2q+l}^2(-1)^l}{q!(q+l)!2^{2q+l}} 4\int_0^\infty k_0(t)_y{}^{2q+l} \cos l\omega_0 t \cos \omega t\, dt \tag{5.38}$$

To be more specific, let us now choose for our nonlinear element the half-wave linear rectifier of Fig. 5.4 which is such that the output ensemble

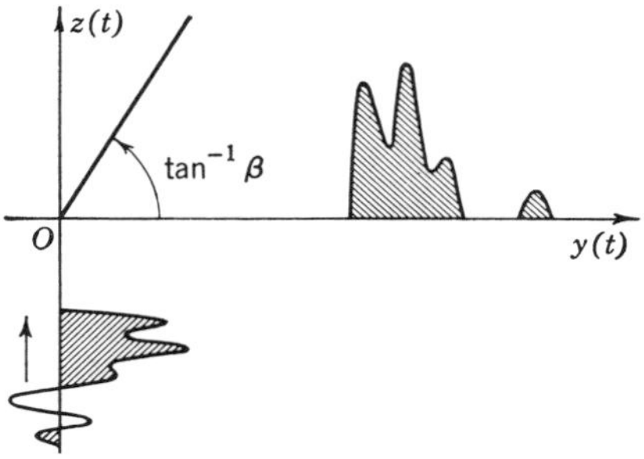

FIG. 5.4. Half-wave linear rectifier characteristic.

is $z = \boldsymbol{T}_g\{y\} = \beta_0 y \ (y > 0)$, $z = \boldsymbol{T}_g\{y\} = 0 \ (y < 0)$ (cf. item 1 of Table 2.2). Then $f(i\xi)$ in Eq. (5.37) becomes $\beta_0/(i\xi)^2$, and with the result of Prob. 2.16*a* we easily show that

$$h_{0,2q+l} = i^{-2q-l}\beta_0{}^2 2^{(2q+l-1)/2}\psi^{(-2q-l+1)/2}\Gamma\left(\frac{1}{2} - q - \frac{l}{2}\right)^{-1} \tag{5.38a}$$

Combining this with Eq. (5.38), we get explicitly for the zonal spectra

$$\mathfrak{W}_l(f) = \frac{\psi\beta_0{}^2}{2}(2-\delta_l^0)\sum_{q=0}^{\infty}\frac{C_{0,l+2q}(f)}{q!(q+l)!\Gamma(\frac{1}{2} - q - l/2)^2} \tag{5.38b}$$

where we write, as before [Eqs. (5.22)],

$$C_{0,l+2q}(f) = 4 \int_0^\infty k_0(t)_y{}^{2q+l} \cos l\omega_0 t \cos \omega t \, dt \tag{5.38c}$$

At this point, let us suppose that the input noise has been passed through a high-Q LCR filter, of the type mentioned above, so that $k_0(t)_y = e^{-\omega_F|t|}$ ($\omega_F = R/2L$), specifically. Inserting this into Eq. (5.38c) and carrying out the indicated integration, we find that

$$C_{0,l+2q}(f) = \frac{2}{(2q+l)\omega_F}\left(\left\{1 + \left[\frac{\omega + l\omega_0}{\omega_F(2q+l)}\right]^2\right\}^{-1} + \left\{1 + \left[\frac{\omega - l\omega_0}{\omega_F(2q+l)}\right]^2\right\}^{-1}\right) \tag{5.39a}$$

and since $\omega_0 > \omega_F$ (the high-Q condition), only the second term of Eq. (5.39a) makes a significant contribution for all $l \geq 1$. We are left with

$$C_{0,l+2q}(f) \doteq \frac{2}{(2q+l)\omega_F}\left\{1 + \left[\frac{\omega - l\omega_0}{\omega_F(2q+l)}\right]^2\right\}^{-1} \tag{5.39b}$$

which clearly reveals the zonal nature of the output spectrum: $C_{0,l+2q}(f)$ is noticeably different from zero only when f is in the immediate spectral vicinity of lf_0 ($l \geq 1$), that is, within several multiples of the spectral width ($\sim\omega_F$) of the input process.

A second important feature of this result is the comparative broadening of the zonal spectra as l increases. This is suggested by the factor $(2q + l)$ in the denominator of $[(\omega - l\omega_0)/\omega_F(2q + l)]^2$ above. Actual computation

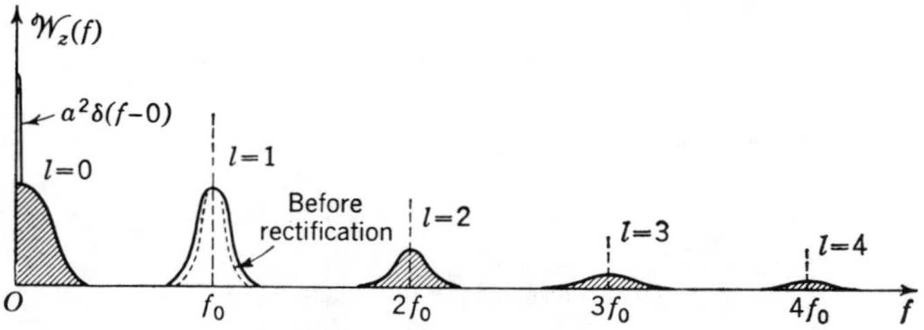

FIG. 5.5. Narrowband noise after (half-wave linear) rectification.

of Eq. (5.38b) does indeed show this effect, which has been confirmed (experimentally) with considerable accuracy by Fellows.[4] A typical spectrum is sketched in Fig. 5.5, along with the d-c component

$$\mathcal{W}_z(f)_0\Big|_{\text{dc}} = 2h_{0,0}^2\delta(f - 0) = \frac{\beta_0{}^2\psi}{\pi^2}\,\delta(f - 0) \tag{5.40}$$

Spectral broadening is, of course, the result of the redistribution of energy among the original noise components, and the generation of new ones, by the nonlinear action of the rectifying element. The result is still a noise process, whose modulation products are of the $n \times n$ type (cf. class 1 of Sec. 5.1-2, page 248). The spectrum remains a continuum, except for the steady component of Eq. (5.40) produced in rectification, but the output process is no longer normal like the input. With the help of Eq. (5.29) in Eq. (5.38*b*), the mean intensity associated with each zone can also be computed, the mean total P_T being given by $M_z(0)$. Specific results are obtained in Chap. 13, where closed expressions, as well as series, are derived for $M_z(t)_l$ and the whole topic of half-wave rectification of normal noise is much more fully treated.

5.2-2 Full-wave Square-law Rectification of a Carrier and Noise. As our second example, let us turn now to a somewhat different and in some ways more general situation, where the input ensemble $y(t)$ consists of a carrier $A_0 \cos(\omega_c t + \phi)$ and (eventually) normal random noise, while the nonlinear element has, instead of the half-wave linear characteristic of Fig. 5.4, a full-wave quadratic response of the kind shown in Fig. 5.6. With noise that can be both broad- and narrowband, the new features of this example are (1) the added presence of a sinusoidal carrier and (2) the different structure of the nonlinear transformation $z = \boldsymbol{T}_g\{y\} = \alpha y^2$ $(-\infty < y < \infty)$, which now makes a direct calculation† of the second-moment function $\overline{z_1 z_2}$ particularly simple, as we shall see below.

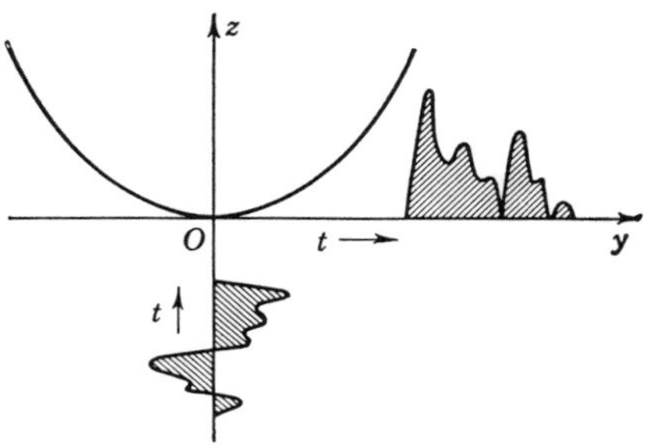

FIG. 5.6. Full-wave square-law characteristic (cf. Table 2.2, item 7, $n = 2$).

The input ensemble is $y(t) = S(t) + N(t) = A_0 \cos(\omega_c t + \phi) + N(t)$, and, as usual, carrier and noise ensembles are assumed independent. A direct calculation gives us

$$M_z(t) = \overline{z_1 z_2} = \alpha^2 \overline{[N_1 + S_1(\phi)]^2 [N_2 + S_2(\phi)]^2} \tag{5.41a}$$

$$M_z(t) = \alpha^2[\overline{N_1^2 N_2^2} + 4\overline{N_1 N_2}\,\overline{S_1 S_2} + \overline{S_1^2 S_2^2} + \overline{S_1^2}\,\overline{N_2^2} + \overline{S_2^2}\,\overline{N_1^2} + (2\overline{N_1^2 N_2}\,\overline{S_2} + 2\overline{N_2^2 N_1}\,\overline{S_1} + \overline{N_1}\,\overline{S_2^2 S_1} + \overline{N_2}\,\overline{S_1^2 S_2})] \tag{5.41b}$$

where the averages over $S(t)$ are now with respect to the random phase ϕ.

To simplify the results, but without sacrificing the principal features of our example, let us further require that ϕ be uniformly distributed, $N(t)$ be ergodic, and, in addition, that $\bar{N} = 0$. Then we can at once make use of the particular structure of $S(t) = A_0 \cos(\omega_c t + \phi)$ to obtain

$$\overline{S_1} = \overline{S_2} = \overline{S_1 S_2^2} = \overline{S_1^2 S_2} = 0 \qquad \overline{S_1 S_2} = \frac{A_0^2}{2} \cos \omega_c t$$

$$\overline{S_1^2 S_2^2} = \frac{A_0^4}{4}\left(1 + \frac{1}{2} \cos 2\omega_c t\right) \tag{5.42}$$

† See Sec. 2.3-2.

so that Eq. (5.41) becomes

$$M_z(t) = \alpha^2 \left[\overline{N_1{}^2N_2{}^2} + 2A_0{}^2\overline{N_1N_2} \cos \omega_c t + \frac{A_0{}^4}{8} \cos 2\omega_c t + \left(\frac{A_0{}^4}{4} + A_0{}^2\overline{N^2} \right) \right] \tag{5.43}$$

For the noise moments, we may use Eq. (1.57a) to write

$$\overline{N_1N_2} = -\frac{\partial^2}{\partial \xi_1\, \partial \xi_2} F_2(i\xi_1,i\xi_2;t)_N \Big|_{\xi_1=\xi_2=0}$$
$$\overline{N_1{}^2N_2{}^2} = \frac{\partial^4}{\partial \xi_1{}^2\, \partial \xi_2{}^2} F_2(i\xi_1,i\xi_2;t)_N \Big|_{\xi_1=\xi_2=0} \tag{5.44}$$

The first expression is simply $K_N(t)$, the covariance function of the input noise (since $\bar{N} = 0$), while the second is the second-moment function of N^2. In the specific instance of the normal process, the characteristic function F_2 of $N(t)$ is given by Eqs. (5.6), so that, carrying out the indicated differentiation in Eqs. (5.44), we obtain

$$\overline{N_1N_2} = \psi k_N(t) \qquad \overline{N_1{}^2N_2{}^2} = \psi^2[1 + 2k^2(t)] \qquad \psi = \overline{N^2};\ k_N(t) = K_N(t)/K_N(0) \tag{5.45}$$

Putting this into Eq. (5.43) and writing $a_0{}^2 = A_0{}^2/2\psi$ then gives us the desired result,

$$M_z(t) = \alpha^2\psi^2[\underset{\text{dc}}{(a_0{}^2 + 1)^2} + \underset{n \times n}{2k_N(t)^2} + \underset{s \times n}{4k_N(t)a_0{}^2 \cos \omega_c t} + \underset{s \times s}{\frac{a_0{}^4}{2} \cos 2\omega_c t}]$$
$$a_0{}^2 \equiv \frac{A_0{}^2}{2\psi} \tag{5.46}$$

Below each term, we have indicated the type of modulation product that is generated in the course of rectification. The quantity $a_0{}^2 \equiv A_0{}^2/2\psi$ is the *input signal-to-noise* (intensity) *ratio*, about whose role in the analysis of system performance (here the nonlinear device of Fig. 5.6) we shall have more to say in later sections. All classes of modulation product are present; note that the d-c term consists of $n \times n$, $s \times n$, and $s \times s$ contributions and is a perfect square, as it is for all inputs.

It is instructive to calculate a spectrum. Let the input noise be broadband and shaped by an RC filter [cf. Eqs. (3.86), (3.90)] so that $K_N(t) = \psi e^{-\omega_F|t|}$ [$\omega_F = (RC)^{-1}$]. Applying the W-K theorem and performing the indicated integrations yields directly

$$\mathcal{W}_z(f) = 2\mathfrak{F}\{M_z\} = 2\alpha^2\psi^2(a_0{}^2 + 1)^2\delta(f - 0)\Big]_{\text{dc}} + \frac{4\alpha^2\psi^2/\omega_F}{1 + (\omega/2\omega_F)^2}\Big]_{n \times n}$$
$$+ \frac{8a_0{}^2\alpha^2\psi^2}{\omega_F} \left\{ \frac{1}{1 + \left(\dfrac{\omega + \omega_c}{\omega_F}\right)^2} + \frac{1}{1 + \left(\dfrac{\omega - \omega_c}{\omega_F}\right)^2} \right\} \Bigg]_{s \times n}$$
$$+ a_0{}^4\alpha^2\psi^2\delta(f - 2f_c)\Big]_{s \times s} \tag{5.47}$$

which is shown in Fig. 5.7. The input spectrum is simply

$$\mathcal{W}_y(f) = 2\mathfrak{F}\{M_y\} = \frac{4\psi/\omega_F}{1 + \omega^2/\omega_F{}^2} + \frac{A_0{}^2}{2}\,\delta(f - f_c) \tag{5.48}$$

(cf. Sec. 3.3-2, 3.3-6). The spread in the spectrum due to rectification is immediately evident: one has only to compare the $n \times n$ contribution of

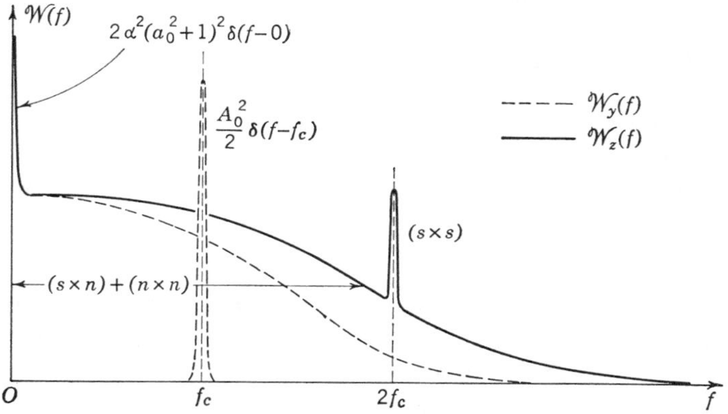

FIG. 5.7. Input and output intensity spectra for a carrier and broadband normal noise after full-wave quadratic rectification. (Output and input densities normalized to unity at $f = 0$.)

$\mathcal{W}_z(f)$ and $\mathcal{W}_y(f)$ [Eqs. (5.47), (5.48)]. Observe that if there is no carrier present at the input, i.e., if $a_0{}^2 = 0$, Eq. (5.47) reduces to

$$\mathcal{W}_z(f)\Big|_{a_0{}^2=0} = 2\alpha^2\psi^2\delta(f - 0) + \frac{4\alpha^2\psi^2/\omega_F}{1 + (\omega/2\omega_F)^2} \tag{5.49}$$

which is also to be compared with $\mathcal{W}_y(f)\Big|_{a_0{}^2=0}$ [cf. Eq. (5.48)]. One characteristic feature of the nonlinear operation $\boldsymbol{T}_g$ is the production of new components, for instance, the d-c term above, as well as the additional $n \times n$ terms that appear in the broadened spectrum.

An interesting modification of the present situation occurs if the input noise is narrowband. Then, as we have noted earlier, its normalized covariance function $k_N(t)$ can be written† $k_0(t) \cos \omega_0 t$. The $n \times n$ and $s \times n$ terms in Eq. (5.46) become $k_0(t)^2 + k_0(t)^2 \cos 2\omega_0 t$, $2a_0{}^2k_0(t)[\cos (\omega_0 - \omega_c)t + \cos (\omega_0 + \omega_c)t]$, respectively, and if we set $\omega_c = \omega_0$, that is, if the input carrier is tuned to the center of the noise band, the $s \times n$ component simplifies further to $2a_0{}^2k_0(t)(1 + \cos 2\omega_0 t)$. Both classes of modulation product are now resolved into slowly and rapidly varying terms, which

† For simplicity, we assume again that the noise is shaped spectrally by a filter for which $\gamma_0(t)$ is zero or a constant (cf. Table 2.1).

appear together in an l-f zone near $f = 0$ and in a narrowband h-f zone, centered about $2f_0$ ($f_0 \gg$ bandwidth of the input noise wave). This is readily illustrated in the case of high-Q LCR noise, where specifically $k_0(t) = e^{-\omega_F|t|}$ (and $\omega_F = R/2L$; cf. Table 3.1). Applying the W-K theorem once more to Eq. (5.46) but with the above modification of $k_N(t)$, we easily find that the intensity spectrum of the rectified process is now

$$\mathcal{W}_z(f) = 2\alpha^2\psi^2(a_0^2+1)^2\delta(f-0)\Big]_{\text{dc}} + a_0^4\alpha^2\psi^2\delta(f-2f_0)\Big]_{s\times s}$$
$$+ \left\{\frac{2\alpha^2\psi^2/\omega_F}{1+\left(\dfrac{\omega}{2\omega_F}\right)^2} + \frac{2\alpha^2\psi^2}{\omega_F}\left[\frac{1}{1+\left(\dfrac{\omega+2\omega_0}{2\omega_F}\right)^2} + \frac{1}{1+\left(\dfrac{\omega-2\omega_0}{2\omega_F}\right)^2}\right]\right\}_{n\times n}$$
$$+ \left\{\frac{4a_0^2\alpha^2\psi^2/\omega_F}{1+\left(\dfrac{\omega}{\omega_F}\right)^2} + \frac{4a_0^2\alpha^2\psi^2}{\omega_F}\left[\frac{1}{1+\left(\dfrac{\omega+2\omega_0}{\omega_F}\right)^2} + \frac{1}{1+\left(\dfrac{\omega-2\omega_0}{\omega_F}\right)^2}\right]\right\}_{s\times n} \tag{5.50}$$

and since $\omega_0 \gg \omega_F$, the contributions of the first term in each of the square brackets may be neglected. Both the l-f and h-f zones ($l = 0$, $l = 2f_0$) contain $s \times n$ and $n \times n$ noise products, as well as an $s \times s$ component. A typical spectrum is shown in Fig. 5.8. Note that there is no noticeable

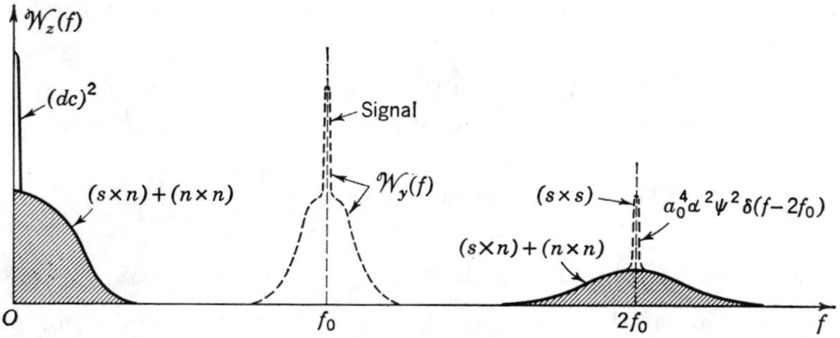

FIG. 5.8. Output spectrum of an on-tune carrier and narrowband normal noise after full-wave square-law rectification.

contribution to the original spectral zone centered at $f = f_0$, a behavior characteristic of all full-wave square-law rectifiers with zero memory when the input ensemble is narrowband.

The total intensity of the various modulation products is easily found from Eq. (5.46). Setting $t = 0$ and letting $t \to \infty$ therein, we have at once

$$M_z(0) = P_\tau = \alpha^2\psi^2[(a_0^2+1)^2 + 2 + 4a_0^2 + a_0^4/2] \tag{5.51a}$$
$$M_z(\infty) = P_{\text{dc}+per} = \alpha^2\psi^2(a_0^2+1)^2\Big]_{\text{dc}} + \frac{a_0^4}{2}\alpha^2\psi^2\Big]_{per} \tag{5.51b}$$
$$P_{\text{continuum}} \equiv M_z(0) - M_z(\infty) = 2\alpha^2\psi^2(1+2a_0^2) = P_{n\times n+s\times n} \tag{5.51c}$$

for the contribution to the total, d-c, and periodic terms and the total spectral continuum, which consists entirely of the $s \times n$ and $n \times n$ modulation products. These results are readily extended to include a modulated carrier, but we reserve a detailed discussion of this to Part 3, our main effort here having been to illustrate with these first two examples the direct and indirect (or characteristic-function) method of handling the zero-memory class of nonlinear problem and to show how the spectrum is modified when random and mixed processes undergo a nonlinear transformation.

5.2-3 A Simple Spectrum Analyzer. Our third example of a nonlinear system, the spectrum analyzer of Fig. 5.9, we shall consider at somewhat

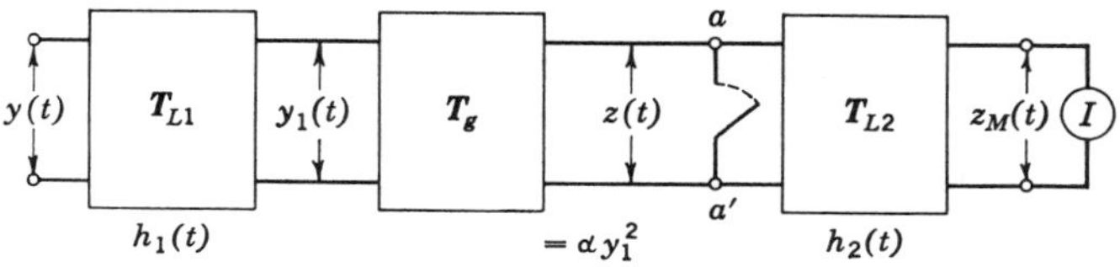

FIG. 5.9. Scheme of a simple spectrum analyzer; T_{L1} is tunable over a frequency range $(f_1 < f_0 < f_2)$.

greater length. The new features are now the explicit role of several linear filters in modifying the performance of the nonlinear element and in particular the effect of a finite observation or processing time on the system's output. As before, a zero-memory nonlinear device is assumed, and, as in Sec. 5.2-2, it is also a full-wave square-law element. That a square-law rectifier should be chosen here follows, of course, from our desire to measure mean power or intensity, specifically the mean intensity of the input wave within certain narrow frequency limits $(f_0 - \Delta f/2, f_0 + \Delta f/2)$, and so obtain an estimate of the ensemble spectral density $\mathcal{W}_y(f)$ for frequencies f within these spectral limits. By examining the output of this spectrum analyzer, we may establish a relationship between what is actually observed physically and what has been analytically defined as the spectral intensity density of the ensemble and in this way impart to the latter a proper operational significance.

We start with an input ensemble $y(t)$ whose intensity spectrum we wish to measure and pass it through a high-Q linear filter T_{L1} called the "aperture" filter, spectrally centered at f_0, to get a new ensemble $y_1(t)$. This in turn enters the nonlinear element T_g, appearing at the output as $z(t)$, which is then smoothed by the linear low-pass filter T_{L2} and is presented finally as $z_M(t)$ on the indicator (cf. Fig. 5.9). The indicator itself may be a simple meter or a tape recorder; in any case, we include in T_{L2} any additional smoothing introduced by the indicator. The input and output ensembles are accordingly related by

$$z_M(t) = T_{L2}T_gT_{L1}\{y\} \qquad \text{with} \quad \begin{cases} z_M(t) = T_{L2}\{z\} \\ z(t) = T_gT_{L1}\{y\} = T_g\{y_1\} \end{cases} \tag{5.52}$$

The following assumptions are now made: (1) $\boldsymbol{T}_{L1}$, $\boldsymbol{T}_g$, $\boldsymbol{T}_{L2}$ are ideal, i.e., free of noise sources, and are suitably isolated [cf. Sec. 2.1-3(1)]; (2) $\boldsymbol{T}_g$ is a zero-memory square-law transformation, that is, $z = \alpha y_1^2\,(-\infty < y_1 < \infty)$; and (3) $y(t)$ is at least a wide-sense stationary process (with $\bar{y}$ and $\langle y\rangle$ zero, for the moment). We can therefore write at once

$$y_1(t) = \boldsymbol{T}_{L1}\{y\} = \int_{-\infty}^{\infty} y(t-\tau)h_1(\tau)\,d\tau \tag{5.53a}$$

$$z(t) = \alpha y_1^2(t) = \alpha \iint\limits_{-\infty}^{\infty} y(t-\tau_1)y(t-\tau_2)h_1(\tau_1)h_1(\tau_2)\,d\tau_1\,d\tau_2 \tag{5.53b}$$

Since $\boldsymbol{T}_{L1}$ is narrowband compared with the process $y(t)$, we can take advantage of this fact and use Eq. (2.63a) to express $h_1(\tau)$ as $2h_{01}(\tau)\cos[\omega_0\tau - \gamma_{01}(\tau)]$. Observing that $h_{01}(\tau)$, $\gamma_{01}(\tau)$ are slowly varying vis-à-vis $\cos\omega_0\tau$, $\sin\omega_0\tau$, we find that $z(t)$ becomes

$$z(t) \doteq 2\alpha \iint\limits_{-\infty}^{\infty} h_{01}(\tau_1)h_{01}(\tau_2)y(t-\tau_1)y(t-\tau_2) \times \cos[\omega_0(\tau_1-\tau_2) + \gamma_{01}(\tau_2) - \gamma_{01}(\tau_1)]\,d\tau_1\,d\tau_2 \tag{5.54}$$

where the term involving $\cos[\omega_0(\tau_1+\tau_2) - \gamma_{01}(\tau_1) - \gamma_{01}(\tau_2)]$ can be discarded, since it makes a negligible contribution to the integral.

To determine $z_M(t)$, the quantity presented at the indicator, it is necessary now to specify the mode of observation. Inasmuch as only finite times are available in practical situations, let us start by applying the rectified wave $z(t)$ to the smoothing filter at, say, time $t = 0$, by opening the short-circuiting switch aa' (cf. Fig. 5.9) and closing it T sec later, cutting off the input to $\boldsymbol{T}_{L2}$ at time $t = T$. The integrating filter $\boldsymbol{T}_{L2}$ thus has a portion of $z(t)$, $(0,T)$ sec long, to operate upon. The output on the indicator can then be read off at any instant $t \geq 0^-$. [We shall see in a moment what value of $t\ (>0)$ to choose for a suitable reading.] The output of $\boldsymbol{T}_{L2}$ accordingly builds up from zero at $t = 0$ to some value at or before $t = T$, after which it falls to zero again as the source is cut off. The ensemble output is specifically

$$z_M(t;T) = \begin{cases} \int_{0-}^{T} h_2(t-\tau)z(\tau)\,d\tau & t \geq 0 \\ 0 & t < 0 \end{cases} \tag{5.55}$$

this last since $h_2(x) = 0$ if $x < 0-$, inasmuch as $\boldsymbol{T}_{L2}$ is taken to be a stable, physically realizable filter [cf. Sec. 2.2-5(3)].

Before considering the behavior with t of a particular member $z_M^{(j)}(t;T)$ of this output ensemble, let us take the ensemble average of both sides of Eq. (5.55), using Eq. (5.54). Let us also simplify the analysis slightly by assuming that, for the high-Q filter $\boldsymbol{T}_{L1}$, $\gamma_{01}(\tau)$ is a constant or zero, as occurs, for instance, if $\boldsymbol{T}_{L1}$ is a series LCR filter or a group of such filters in cascade. The result is

$$\overline{z_M(t;T)} \doteq \left[2\alpha \int_{0-}^{T} h_2(t-\tau)\,d\tau\right] \int_{-\infty}^{\infty} K_y(x)\rho_{01}(x) \cos \omega_0 x\,dx \quad (5.56)$$

where $\rho_{01}(x)$ is the autocorrelation function of $h_{01}(\tau)$, namely,

$$\int_{-\infty}^{\infty} h_{01}(\tau)h_{01}(\tau + x)\,d\tau$$

[cf. Eq. (3.97)].

Since y is broadband compared with $h_{01}(x)$, or, correspondingly, with $\rho_{01}(x)$, the latter is a slowly varying function which to a good first approximation we can safely replace by $\rho_{01}(0)$, the leading term of its Taylor's expansion (cf. Sec. 3.3-5). From the W-K theorem (3.42), it follows at once that $\int_{-\infty}^{\infty} K_y(x) \cos \omega_0 x\,dx = \frac{1}{2}\mathcal{W}_y(f_0)$, and so Eq. (5.56) reduces to

$$\overline{z_M(t;T)} \doteq C(t;T)\mathcal{W}_y(f_0) \qquad C(t;T) \equiv \left[\alpha \int_{0-}^{T} h_2(t-\tau)\,d\tau\right] \rho_{01}(0) \quad (5.57)$$

Thus, *the ensemble average of the smoothing filter's output is directly proportional to the ensemble spectral density at the input filter's center frequency* f_0, the constant of proportionality C being fixed for specified instants t at which readings on the indicator are made. For example, if we let the integrating filter be a simple RC network, so that $h_2(x) = \beta e^{-\beta x}$ [$x > 0$, $\beta = (RC)^{-1}$], $h_2(x) = 0$ ($x < 0$), we easily find that

$$C(t;T) = \begin{cases} 0 & t \leq 0 \\ \alpha(1 - e^{-\beta t})\rho_{01}(0) & 0 < t \leq T \\ \alpha\rho_{01}(0)e^{-\beta t}(e^{\beta T} - 1) & t \geq T \end{cases} \quad (5.58)$$

From this it is clear, incidentally, that for a maximum average response we should set $t = T$ and make βT as large as convenient; $\rho_{01}(0)$ is fixed by the Q of the filter $\boldsymbol{T}_{L1}$, preceding the rectifier. In the case of an LCR filter, $h_{01}(x) = \omega_F e^{-\omega_F t}$ ($t > 0$), where $\omega_F = R/2L$, so that $\rho_{01}(0) = \omega_F/2$. Since for a filter of this type the width Δf_e of the equivalent rectangular filter (cf. Sec. 3.3-2) is $2f_F$ ($= \omega_F/\pi$), $\rho_{01}(0)$ can also be written $(\pi/2)\,\Delta f_e$, while $C(T;T)$ becomes $(\pi\,\Delta f_e/2)\alpha(1 - e^{-\beta T})$.

Finally, by changing f_0 (that is, by changing L or C in the LCR filter, $f_0 = 1/\sqrt{LC}$), one can cover a range of frequencies and thus determine $\mathcal{W}_y(f)$, wholly or in part. This depends, of course, on our ability to keep $\boldsymbol{T}_{L1}$ high-Q,† or at least narrowband, at the same time as f_0 is adjusted from, say, $f_0 = 0$ to a suitably high frequency. In practice, of course, more complicated (usually active) linear filters are used for this purpose.

Having obtained the ensemble average of the indicator readings $z_M(t)$, let us examine a typical member $z_M^{(j)}(t)$ of the ensemble to see how this

† If $\boldsymbol{T}_{L1}$ is not high-Q, that is, narrowband compared with $y(t)$, then from Eq. (5.54) it is evident that $\overline{z_M(t;T)}$ will no longer be proportional to $\mathcal{W}_y(f_0)$.

measured quantity can be related to the desired spectral density. From Eq. (5.54) in Eqs. (5.55), we write [setting $\gamma_{01}(\tau)$ = constant, again]

$$z_M^{(j)}(t) = 2\alpha \int_{0-}^{T} h_2(t-\tau)\, d\tau \iint_{-\infty}^{\infty} h_{01}(\tau_1)h_{01}(\tau_2)y^{(j)}(\tau-\tau_1)y^{(j)}(\tau-\tau_2) \times \cos \omega_0(\tau_2-\tau_1)\, d\tau_1\, d\tau_2 \qquad t > 0 \quad (5.59)$$

The indicator presents a reading which varies slowly with time t about the nonzero value $\overline{z_M(t;T)}$, which is, of course, the quantity we would ideally like to determine, since it is directly proportional to the spectral density at $f = f_0$ [cf. Eqs. (5.57)].

To see how this comes about, let us review the history of the input wave $y^{(j)}(t)$ in the analysis, with the help of Eq. (5.52) and Fig. 5.9. The aperture filter $\boldsymbol{T}_{L1}$ selects a portion of the input in the immediate spectral vicinity of f_0, and, as the new disturbance $y_1^{(j)}(t)$, it then passes through the

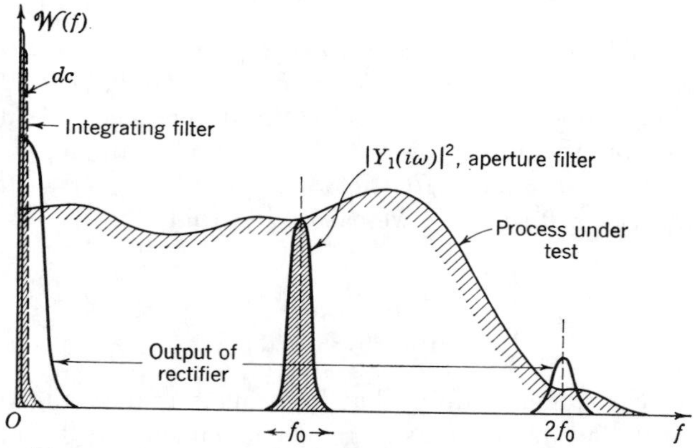

FIG. 5.10. Spectral relations in the analyzer of Fig. 5.9.

square-law rectifier $\boldsymbol{T}_g$. As we have already observed in the preceding example [Sec. 5.2-2, Eq. (5.50)], the output under these circumstances consists of a slowly varying component in the neighborhood of $f = 0$ and a comparatively h-f narrowband disturbance in the vicinity of $2f_0$, characteristic of quadratic devices when the input is spectrally narrow. The final filter $\boldsymbol{T}_{L2}$ then removes this h-f component, and what appears on the indicator is a reading that fluctuates slowly about some nonvanishing value at a mean rate proportional to the bandwidth of this final, integrating filter. The situation is illustrated spectrally in Fig. 5.10.

Our next step is to show how this observed output can be related to the spectral density itself. Let us impose the customary operating conditions that the "aperture" Δf_e of the tunable filter $\boldsymbol{T}_{L1}$ is small vis-à-vis the bandwidth of the wave under measurement and that the final filter is even narrower, with an effective time constant β^{-1} that is also much larger than

Δf_e^{-1}. Furthermore, to simplify our subsequent manipulations somewhat, but without altering the basic operation of the analyzer in any way, let us integrate for a time $t = T$, which is somewhat less than the smoothing filter's time constant β^{-1}, so that $h_2(t - \tau)$ in Eq. (5.59) can be reasonably approximated by the first term of its Taylor's-series development about τ, for example, $h_2(t)\Big|_{t=T}$; T, however, should not be so short that there is no observable reading. Accordingly, subject to the conditions $(\Delta f_e)^{-1} \ll T < \beta^{-1}$, or $T_1 \ll T < T_2$, where T_1, T_2 are the corresponding "memories" of $\boldsymbol{T}_{L1}$ and $\boldsymbol{T}_{L2}$ (cf. Sec. 2.2-3), we can rewrite Eq. (5.59) as

$$z_M^{(j)}(T;T) \doteq 2\alpha T h_2(T) \iint_{-\infty}^{\infty} h_{01}(\tau_1)h_{01}(\tau_2) \cos \omega_0(\tau_2 - \tau_1)\, d\tau_1\, d\tau_2 \times \left[\frac{1}{T}\int_0^T y^{(j)}(\tau - \tau_1)y^{(j)}(\tau - \tau_2)\, d\tau\right] \tag{5.60}$$

We observe now that

$$\begin{aligned}\frac{1}{T}\int_0^T y^{(j)}(\tau - \tau_1)y^{(j)}(\tau - \tau_2)\, d\tau &= \frac{1}{T}\int_{-\tau_1}^{T-\tau_1} y^{(j)}(\lambda)y^{(j)}(\lambda + \tau_1 - \tau_2)\, d\lambda \\ &= \frac{1}{T}\int_0^T y^{(j)}(\lambda)y^{(j)}(\lambda + \tau_1 - \tau_2)\, d\lambda + O\left(\frac{\Delta f_e^{-1}}{T}\right) \\ &\doteq R_y^{(j)}(\tau_1 - \tau_2)\end{aligned} \tag{5.61}$$

since in the second equation only those values of $|\tau_1| \le \Delta f_e^{-1}$ yield a noticeable contribution in the integrations over τ_1, τ_2, and since in the third the interval $(0,T)$ is very long compared with the mean period of $y^{(j)}(t)$. We therefore replace the original time average over τ in Eq. (5.60) by the autocorrelation function of $y^{(j)}$. The result is approximately

$$z_M^{(j)}(T;T) \approx 2\alpha h_2(T)T \int_{-\infty}^{\infty} \rho_{01}(x)R_y^{(j)}(x) \cos \omega_0 x\, dx \tag{5.62}$$

and this, apart from the scale factor, is the direct analogue of (5.56).

To relate this time average to the calculated ensemble property, we must make the additional assumption (always implicitly made in applications) that $y^{(j)}$ belongs to an ergodic process. Then, with probability 1, $R_y^{(j)}(x)$ may be replaced by the covariance $K_y(x)$ (cf. Sec. 1.6), and so the measured value $z_M^{(j)}$ [Eq. (5.62)] is (approximately) equal to the ensemble average $\bar{z}_M$. By Eqs. (5.57), this in turn is proportional to the desired spectral density $\mathcal{W}_y(f_0)$. The actual magnitude of $z^{(j)}(T;T)$ varies from observation to observation, but as long as the analyzer is stable and the input process remains ergodic, these observations will fluctuate about $\bar{z}_M$, differing from it to an extent that is on the average less as the time constant of the smoothing filter is made larger. Under such conditions, a particular observation gives a good measure of the spectral density at f_0. Note, incidentally, that $z_M^{(j)}(T;T)$ is directly proportional to $P_y^{(j)}(f_0 + \Delta f_e, f_0)_T \approx \Delta f_e \mathcal{W}_y(f_0)$ [by Eqs. (5.57) et seq.] and is the average power in $y^{(j)}(t)$, as measured over the interval $(0,T)$ for an "aperture" Δf_e.

How "good" is a good measure of the spectral density, however? Since z_M is a statistical quantity, we may expect that in this regard some statistic of z_M will prove suitable as an estimate. A useful measure is the standard deviation, or rms fluctuation, $\sigma_z = (E\{(z_M^{(j)} - \overline{z_M^{(j)}})^2\})^{1/2}$, which tells us on the average by how much we may expect an observation to differ from the mean. To find $\sigma_z(t;T)$, we need the second moment, $\overline{z_M^2(t;T)}$. From Eqs. (5.55) we write, therefore,

$$\overline{z_M^2(t;T)} = \iint_{0-}^{T} h_2(t-\tau_1)h_2(t-\tau_2)\overline{z(\tau_1)z(\tau_2)}\, d\tau_1\, d\tau_2$$

$$= \iint_{0-}^{t} h_2(\tau_1)h_2(\tau_2)M_z(\tau_2-\tau_1)\, d\tau_1\, d\tau_2 \qquad 0 \le t \le T \tag{5.63}$$

for the stationary input processes assumed here. Introducing the truncated weighting function

$$h_2(\tau)_t = \begin{cases} h_2(\tau) & 0- \le \tau \le t \\ 0 & \tau < 0, \tau > t\ (0 \le t \le T) \end{cases} \tag{5.64}$$

we observe that

$$\overline{z_M^2(t;T)} = \int_{-\infty}^{\infty} M_z(x)\rho_2(x)_t\, dx \tag{5.65}$$

where $\rho_2(x)_t$ is the truncated autocorrelation function of the integrating filter, that is, $\int_{-\infty}^{\infty} h_2(\tau)_t h_2(\tau + x)_t\, d\tau$ [cf. Eqs. (3.89)]. From this, it is at once evident that $\overline{z_M^2}$ requires the second-moment function of the output of the nonlinear element, which for our simple spectrum analyzer is specifically $\alpha^2\overline{y_1(t_1)^2y_1(t_1+x)^2}$.

As an example, we choose as input a broadband stationary *normal* noise. Then, from Eq. (5.46), with $a_0^2 = 0$, we have

$$M_z(t) = \alpha^2\psi_{y_1}^2[1 + 2k_{y_1}^2(x)] \tag{5.66}$$

in which $k_{y_1}(x)$ is the (normalized) covariance function of the normal noise passed by the aperture, which is assumed to be symmetrical so that $\gamma_0(\tau) = 0$ [cf. Eqs. (2.63a), (5.54), filter T_{L1}]. This is a narrowband process which may be expressed approximately as

$$\psi_{y_1}k_{y_1}(x) \doteq K_0(x)_{y_1}\cos\omega_0 x = 2\int_{-\infty}^{\infty} K_y(\tau)\rho_{01}(\tau - x)\cos\omega_0(\tau - x)\, d\tau$$
$$\doteq \rho_{01}(x)\mathcal{W}_y(f_0)\cos\omega_0 x$$

Therefore,

$$\psi_{y_1} \doteq \rho_{01}(0)\mathcal{W}_y(f_0) \tag{5.67}$$

where we have taken advantage of the fact that ρ_{01} is slowly varying compared with K_y and $\cos \omega_0\tau$ and that $\lambda_{0z}(t) \doteq 0$ [cf. Eqs. (3.108b), etc.]. The normalized covariance $k_{y_1}(x)$ accordingly becomes $[\rho_{01}(x)/\rho_{01}(0)] \cos \omega_0 x$. By squaring $k_{y_1}(x)$ and observing that $\rho_2(x)_t$ is essentially constant vis-à-vis $\cos 2\omega_0 x$ because of the long time constant of the smoothing filter, we find that Eqs. (5.65) to (5.67) yield

$$\overline{z_M{}^2(t;T)} = \alpha^2\psi_{y_1}{}^2 \int_{-\infty}^{\infty} \rho_2(x)_t[1 + k_{y_1}{}^2(x)_0]\, dx \qquad 0- < t < T \quad (5.68a)$$

Therefore,

$$\overline{z_M{}^2(t;T)} - \overline{z_M(t;T)}^2 = \alpha^2\psi_{y_1}{}^2 \int_{-\infty}^{\infty} \rho_2(x)_t k_{y_1}{}^2(x)_0\, dx \qquad (5.68b)$$

this last with the help of Eqs. (5.57), (5.58). Equation (5.68b) is the desired variance $\sigma_z{}^2$, defined above.

In practice, a somewhat more convenient measure of the deviation from the mean value is the *relative* rms deviation, or *expected* (rms) *relative error*, defined here by

$$B_M(t;T) \equiv \sigma_z(t;T)/\overline{z_M(t;T)} \qquad (5.69)$$

A specific example of this expected error in the observed spectrum, as a function of observation time t and the data-acquisition period $(0,T)$, is derived in Prob. 5.8b for a high-Q series LCR aperture filter and an RC integrator. Note that, as $T \to \infty$, $B_M(T,T)$ does *not* vanish but approaches a finite, nonzero limit, reflecting the fact that the smoothing filter has a finite passband, so that some fluctuating components are always passed. Additional examples and a further discussion of the statistical errors to be expected in the measurement of random processes are given in Chap. 16.

Apart from the importance of the spectrum analyzer as a practical tool of great utility to the engineer, the example above is a typical illustration of the kind of problem encountered when observation or integration time is finite, i.e., when we must take into account explicitly the finite size of the sample. To interpret the analyzer's results properly, (1) we require effective ergodicity of the input, as reflected in input stability over the entire integration period, and (2) we must know enough about the structure of the input ensemble from which the member in question comes to estimate the error in the measurement, which in the present case requires fourth-order moments of the input if a simple full-wave square-law device $\boldsymbol{T}_g$ is used. Thus, higher-order statistics of the input ensemble are needed in order to give lower-order statistics of the output.

This situation, of course, is not peculiar to the analyzer alone but is quite generally characteristic of all rectifying elements, regardless of whether sample size is finite or infinite. For instance, suppose that $z(\tau)$ in Eqs. (5.55) were the output ensemble of the half-wave linear rectifier discussed in the example of Sec. 5.2-1. Then, to determine $\overline{z_M(t,T)}$, we need $\bar{z} = \beta_0 \int_{-\infty}^{\infty} y_1 W_1(y_1)\, dy_1$, and for $\overline{z_M{}^2(t,T)}$, from Eq. (5.63), we require $M_z(\tau_2 - \tau_1)$

[Eqs. (5.63) et seq.] ($t = \tau_2 - \tau_1$), and so forth, while for the infinite sample we simply omit here the final operation $\boldsymbol{T}_{L2}$. In either case, the complete first- and second-order statistics of the input, as embodied in the probability-density functions $W_1(y_1)$ and $W_2[(y_1)_1,(y_1)_2;t]$, are necessary. From this, finally, we can make the further general observation that, if the structure of a nonlinear system is given and all statistics of the output observed, it is still *not possible uniquely to specify the statistical structure of the input ensemble* without further testing of the observations against the universe of possible inputs. Even then, there is no guarantee of uniqueness, since it is possible that several different input structures may yield identical results. This indeterminacy, of course, is a direct consequence of the irreversibility [cf. Sec. 2.1-3(2)] of a nonlinear transformation. Many examples of similar behavior will be observed later, in Part 3.

PROBLEMS

NOTE: The noise processes in the following problems are assumed to be normal and ergodic, with the first- and second-order densities

$$W_1(N) = \frac{\exp(-N^2/2\psi^2)}{(2\pi\psi)^{1/2}}$$

$$W_2(N_1,N_2;t) = [1 - k_N(t)^2]^{-1/2}(2\pi\psi)^{-1} \exp\left\{-\frac{N_1^2 + N_2^2 - 2N_1N_2k_N(t)}{2\psi[1 - k_N(t)^2]}\right\}$$

$$\text{with } \psi = \overline{N^2};\ \bar{N} = 0;\ K_N(t) = \psi k_N(t)$$

and the corresponding characteristic functions

$$F_1(i\xi) = e^{-\psi\xi^2/2} \qquad F_2(i\xi_1,i\xi_2;t) = e^{-\psi(\xi_1^2+\xi_2^2)/2}e^{-\xi_1\xi_2K_N(t)}$$

(cf. Prob. 1.8). For generalizations, in particular, other laws of rectification, saturation effects, mixed input processes, etc., that are common in communication systems, see Part 3 (text and problems).

5.3 (*a*) Using the characteristic function outlined in Secs. 5.1 and 5.2-1, show for a half-wave linear rectifier $z = \beta y$ ($y > 0$), $z = 0$ ($y < 0$), into which broadband noise is passed, that

$$M_z(t) = \overline{\boldsymbol{T}_g\{y_1\}\boldsymbol{T}_g\{y_2\}} = \frac{\beta^2\psi}{2\pi}\left[k(t)\sin^{-1}k(t) + \pi\frac{k(t)}{2} + \sqrt{1 - k(t)^2}\right] \tag{1}$$

where $k(t)$ is the normalized covariance function of the input noise, that is, $K_y(t) = \psi k(t)$. Hence, observe that

$$(P_z)_{\mathrm{dc}} = \frac{\beta^2\psi}{2\pi} \qquad (P_z)_{n\times n} = \frac{\beta^2\psi}{2} \tag{2}$$

Explain $(P_z)_{n\times n}$.

(*b*) Obtain the spectrum

$$\mathcal{W}_z(f) = \frac{\beta^2\psi}{2\pi}\left[2\delta(f - 0) + \frac{\pi}{2}C_{0,1}(f) + \sum_{n=1}^{\infty}\frac{(\frac{1}{2})_n}{n!(2n - 1)^2}C_{0,n}(f)\right] \tag{3}$$

with $C_{0,n}(f) = 4\int_0^\infty k(t)^n \cos\omega t\, dt$; here $(\frac{1}{2})_n = (\frac{1}{2})(\frac{3}{2}) \cdots (\frac{1}{2} + n - 1)$, $(\frac{1}{2})_0 = 1$. If the noise is shaped by an RC filter, so that $k(t) = e^{-|t|/RC}$, obtain Eq. (3) explicitly, and plot.

(*c*) Do by the direct method. D. Middleton,[2] sec. 7.

5.4 Narrowband noise, with $\phi_0 = 0$ so that $K_y(t) = \psi k_0(t) \cos \omega_0 t$, is rectified by the half-wave linear (zero-memory) device of Prob. 5.3. Show that the covariance functions of the various spectral zones generated in rectification are given by

$$\begin{aligned} M_z(t)_0 &= \frac{\beta^2\psi}{2\pi}\, {}_2F_1\left[-\frac{1}{2},-\frac{1}{2};1;k_0{}^2(t)\right] \\ K_z(t)_{2l} &= \frac{\beta^2\psi}{\pi}\left(-\frac{1}{2}\right)_l^2 \frac{1}{(2l)!}\, k_0(t)^{2l}\, {}_2F_1\left[l-\frac{1}{2}, l-\frac{1}{2}; 2l+1; k_0{}^2(t)\right]\cos 2l\omega_0 t \qquad l \geq 1 \\ K_z(t)_{2l+1} &= 0 \qquad l \geq 0 \end{aligned} \tag{1}$$

and explain why the odd-order zones vanish. Hence, verify that the total intensity in each zone is given by the general relation

$$(P_z)_l = \frac{\beta^2\psi}{2\pi}\, \epsilon_l \left(-\frac{1}{2}\right)_l^2 \Gamma\left(l+\frac{3}{2}\right)^{-2} \qquad l \geq 0 \tag{2}$$

Here, ${}_2F_1$ is a hypergeometric function: ${}_2F_1(\alpha,\beta;\gamma;x) = \sum_{n=0}^{\infty} \frac{(\alpha)_n(\beta)_n}{(\gamma)_n n!} x^n$; see, for example, Sec. A.1.3.

5.5 (*a*) Broadband normal noise with a normalized correlation function $k(t)$ is passed through a *half-wave* quadratic rectifier $z = \boldsymbol{T}_g\{y\} = \alpha y^2$ $(y > 0)$, $z = \boldsymbol{T}_g\{y\} = 0$ $(y < 0)$. Using the characteristic-function approach, show that the second-moment function of the output is

$$M_z(t) = \frac{\alpha^2\psi^2}{2\pi}\left[(1+2k^2)\left(\frac{\pi}{2}+\sin^{-1} k\right) + 3k\sqrt{1-k^2}\right] \tag{1}$$

(*b*) Verify that the intensity spectrum is

$$\mathfrak{W}_z(f) = \frac{\alpha^2\psi^2}{2\pi}\left[\pi\delta(f-0) + 4C_{0,1}(f) + \pi C_{0,2}(f) + 4\sum_{n=1}^{\infty} \frac{(\frac{1}{2})_n}{(2n+1)(2n-1)^2 n!} C_{0,2n+1}(f)\right] \tag{2}$$

and evaluate it specifically for the *RC* input noise of Prob. 5.3*b*.

(*c*) Do by the direct method. D. Middleton,[2] sec. 7.

5.6 Consider the nonlinear operation $z = \boldsymbol{T}_g\{y\} = ay$ $(y \geq 0)$, $z = \boldsymbol{T}_g\{y\} = by$ $(y \leq 0)$. Let $x(t)$ and $y(t)$ be two input (ergodic) normal noise processes to the nonlinear devices $\boldsymbol{T}_g$. Then show that the cross-variance function of the rectified outputs is

$$M_{z_x z_y}(t) = \overline{z(x_1)z(y_2)} = (a^2+b^2)G[K_{xy}(t)] - 2abG[-K_{xy}(t)] \tag{1}$$

where K_{xz} is the cross variance before rectification and

$$G(s) = \frac{1}{2\pi}\left(\sqrt{1-s^2} + \frac{s\pi}{2} + s\tan^{-1}\frac{s}{\sqrt{1-s^2}}\right) \tag{2}$$

[It is assumed, as usual, that $\bar{y} = \bar{x} = 0$ and that x and y come from a common source, so that $K_{xy}(t) \neq 0$ for arbitrary t $(= t_2 - t_1)$.]

5.7 (*a*) A sinusoidally modulated carrier and a narrowband of normal noise, centered about the carrier frequency, i.e., $f_c = f_0$, are passed through a full-wave square-law rectifier $z = \alpha y^2$ $(-\infty < y < \infty)$. If $A_0(t) = A_0[1 + \lambda \cos(\omega_a t + \phi_A)]$ represents the modulation ensemble $(-1 \leq \lambda \leq 1)$ and λ is thus the modulation index, show that the second-moment function of the *l-f output* is

$$M_z(t)_0 = \alpha^2\psi^2 \left\{k_0(t)^2 + 2k_0(t)a_0^2\left(1 + \frac{\lambda^2}{2}\cos\omega_a t\right) + a_0^4\left[\left(1 + \frac{\lambda^2}{2}\right)^2 + 2\lambda^2\cos\omega_a t + \frac{\lambda^4}{8}\cos 2\omega_a t\right]\right\} \quad (1)$$

where $k_0(t)$ is the (normalized) slowly varying part of the covariance function of the accompanying noise and, as before, $a_0^2 \equiv A_0^2/2\psi$ ($\psi = \overline{N^2}$, $\bar{N} = 0$).

(*b*) Determine the spectrum, and sketch a typical result for high-*Q* *LCR* noise.

5.8 (*a*) For the spectrum analyzer of the example of Sec. 5.2-3, show that, if an *RC* smoothing filter is used for $\boldsymbol{T}_{L2}$ (cf. Fig. 5.9), the truncated autocorrelation function of this filter is specifically [$\beta = (RC)^{-1}$]

$$\rho_2(x)_t = \frac{\beta}{2}e^{-\beta|x|}(1 - e^{-2\beta t + 2\beta|x|}) \qquad 0 \leq |x| \leq t\ (\leq T) \quad (1)$$
$$\boldsymbol{T}_{L2} = 0\ (|x| > |t|)$$

(or T, whichever is smaller, that is, $0 \leq t \leq T$).

(*b*) Show, therefore, that the expected rms relative error at $t = T$ becomes in the case where the aperture filter $\boldsymbol{T}_{L2}$ is a high-*Q* *LCR* network, of decay constant ω_F ($\ll \beta$),

$$B_M(T;T) = \frac{[(1 + 2\omega_F/\beta)^{-1} + e^{-2\beta T}(1 - 2\omega_F/\beta)^{-1} - e^{-(\beta+2\omega_F)T}(1 - 2\omega_F^2/\alpha^2)^{-1}]^{1/2}}{1 - e^{-\beta T}} \quad (2)$$

Note that, as $T \to \infty$, $B_M(T;T) \to (1 + 2\omega_F/\beta)^{-1/2}$, which is independent of the duration of the smoothing period $(0,T)$. Explain this result [note the factor $\sqrt{2}$ difference in the numerator between Eq. (2) and the broadband case examined in Sec. 16.1-2 [Eq. (16.23)].

SECOND-MOMENT CRITERIA OF SYSTEM PERFORMANCE

5.3 *Noise Figures and Signal-to-Noise Ratios*

With the ingredients of a second-moment theory at our disposal (cf. Chap. 3 and Secs. 5.1, 5.2), we can now attempt to evaluate system performance where, as before [cf. Sec. 2.1-3(1)], by *system* is understood any combination of elements out of which a four-pole network (with source and output loads) is formed (cf. Fig. 2.2, also). Our first task is to construct suitable criteria for this purpose. We begin by observing that choice of a criterion is in part arbitrary and not uniquely determined by the system itself. Depending on the operational situation, some choices will be better than others, and what is meant by "better" likewise depends on the application and may change accordingly. Nevertheless, a suitable criterion should have the feature of reasonable simplicity and retain at the same time enough of the system's characteristics to yield realistic results. In the present section, we shall consider some simple criteria, based largely on the second-moment statistics of input and output processes of linear and nonlinear systems, as a forerunner of the more sophisticated and powerful approaches discussed subsequently in Chaps. 6 and 18.

Since the ultimate application of these criteria is to an over-all communication system, let us return for the moment to the basic relations (2.20) for the single link, including possible noise sources in various parts of this system. Recalling that u and v are, respectively, the transmitted and received message ensembles, we have

$$\{v\} = \mathbf{T}_R{}^{(N)}\mathbf{T}_M{}^{(N)}\mathbf{T}_T{}^{(N)}\{u\} \tag{5.70}$$

where $\mathbf{T}_R{}^{(N)}$, $\mathbf{T}_M{}^{(N)}$, $\mathbf{T}_T{}^{(N)}$ represent the transformations of the original message ensemble u, produced by the transmitter (or encoder), the effects of the medium, and by the receiver (or decoder), in that order (see Sec. 2.1-3). We illustrate Eq. (5.70) with a typical system involving amplitude modulation in the encoding and decoding processes.†

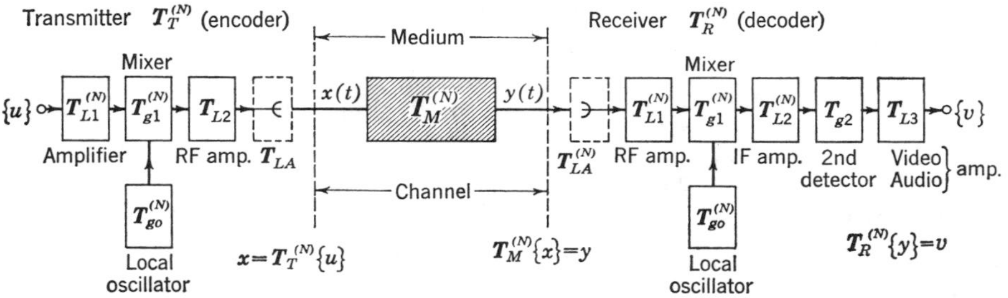

FIG. 5.11. A typical single-link (AM) system for transmission and reception in electronic communication, with a superheterodyne receiver.

A block diagram of the communication link is shown in Fig. 5.11, where the composite operations $\mathbf{T}_T{}^{(N)}$ and $\mathbf{T}_R{}^{(N)}$ have been resolved into their major elements. As before, the $\mathbf{T}_L{}^{(N)}$ are linear passive or active filters for frequency selection and amplification, which may contribute a significant amount of locally generated noise. Similarly, $\mathbf{T}_{g0}{}^{(N)}$, $\mathbf{T}_{g1,2}^{(N)}$ are nonlinear devices, the first of which in this particular example represents a local oscillator, while the $\mathbf{T}_{g1,2}^{(N)}$ are the *mixer* (or first detector) and the second detector, respectively. In most cases, we can treat these as zero-memory nonlinear elements, followed by suitable linear filters (cf. Sec. 2.3). The transformations $\mathbf{T}_{LA}{}^{(N)}$ represent the coupling of transmitter and receiver to the medium; in electromagnetic cases, $\mathbf{T}_{LA}{}^{(N)}$ is the antenna system. In nearly all instances, the locally generated noise in transmitter and receiver is due to the random motion of electrical charges‡ and is a limiting physical property of the circuit components out of which the various networks $\mathbf{T}_L{}^{(N)}$, $\mathbf{T}_g{}^{(N)}$, etc., are built. The noise arising in the medium, or channel, however, may be man-made or natural—the former appears as

† Phase and frequency modulation are considered specifically in Chaps. 14, 15.

‡ The origins and statistical properties of this locally generated noise are considered briefly in Chap. 11.

relatively localized, artificial sources, while the latter is a property of the medium itself.

To estimate system performance, let us first examine the role of the noise sources in the linear elements of $\boldsymbol{T}_T^{(N)}$ and $\boldsymbol{T}_R^{(N)}$ above and then consider the more complex situations imposed by the nonlinear devices present therein.

5.3-1 Linear-network Criteria; Noise Figures (I).† As a linear system, the network $\boldsymbol{T}_L^{(N)}$ simply adds its inherent noise to whatever is presented to it at the input terminals, so that the amplified input ensemble appears, along with this locally generated noise process. Since $\boldsymbol{T}_L^{(N)}$ is linear, input and local noise can be treated independently, the precise structure of the input is not critical, and we may accordingly expect a description of the inherent noise process in terms of its covariance function to be adequate for most purposes.

For the input, then, it is convenient to use a constant voltage source, or *standard signal generator*, which produces a sinusoid of constant maximum (rms) amplitude V_G, in series with the generator's impedance $Z_G = R_G + iX_G$. The network whose inherent noise we wish to observe is then made the load across the generator, in the manner of Fig. 5.12. The mean signal power delivered to the load (when $X_L = -X_G$) is $\overline{I_L{}^2}R_L = \overline{V_G{}^2}R_L(R_L + R_G)^{-2}$, and this is a maximum if $R_G = R_L$, that is, if the generator is matched to the load. Since $\overline{V_G{}^2} = V_G{}^2$ for the sinusoid, we may define an *available signal power* $S_G{}^2$ as $(\overline{I_L{}^2}R_L)_{\max} = V_G{}^2/4R_G$ under these conditions. In addition, since the generator is a physical system, it will produce its own inherent noise ensemble (not shown in the figure), with a mean-square voltage in the frequency interval $(f, f + \Delta f)$ given by $\Delta N_v{}^2 = \int_f^{f+\Delta f} \mathcal{W}_N(f)\, df$, where $\mathcal{W}_N(f)$ is the noise voltage spectral density (volts² per frequency).‡ As we shall see in Chap. 11, this spectrum is effectively constant over the useful range of frequencies, with a density $W_0 = 4k_BT_0R_G$, where k_B is Boltzmann's constant and T_0 is the absolute temperature of the generator's resistance (here at the same temperature as R_L). Thus, $\Delta N_v{}^2$ becomes $W_0\,\Delta f/4R_G = k_BT_0\,\Delta f \equiv \Delta N_G{}^2$,

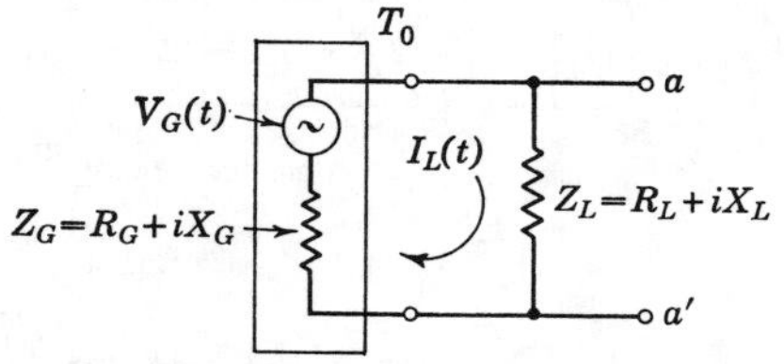

FIG. 5.12. A standard signal generator in series with an impedance load at the same temperature T_0.

† Our brief discussion here of noise figures is based largely on the work of Friis[5] and on extensions by Roberts[6] and others[7] to crystal mixers and local oscillator systems. See Lawson and Uhlenbeck,[9] secs. 5.1–5.5 and chap. 5, for additional references. Davenport and Root,[8] chap. 10, contains further examples.

‡ Often one is interested in the ratio of output signal to noise power—i.e., the ratio of a "gain" to a noise power here. Since the ratio $(P_{out})_{meas}/(P_{in})_{meas}$ for signal or for noise is the same for both signal and noise, even when the network is connected to a mismatched load, it is sufficient to consider available gains only.

and ΔN_G^2 is called the *available* (average) *noise power for the frequency range* $(f, f + \Delta f)$. This is clearly proportional to the spectral density of the noise in the neighborhood of the frequency f and thus depends directly on its second-moment properties.

In a similar way, we may define an *available* (average) *power gain* G for the linear network: this is the ratio of available *output* signal power (where Z_L in Fig. 5.12 is adjusted for maximum output power) to the available *input* signal power, delivered either by the standard signal generator or by a preceding network and observed across aa' in Fig. 5.12, where the network is disconnected at aa' and replaced by the matching load across the generator. Note in Fig. 5.12 that $G = 1$. Thus, we have $G \equiv S_{out}^2/S_{in}^2 = S_{out}^2/S_G^2$, while the *actual* (average) *power gain* G_A is defined as the ratio $(S_{out}^2/S_{in}^2)_{meas}$ of the measured output and input signal process. (In Fig. 5.12, also, $G_A = 1$.) Of course, G and G_A are not equal unless suitable lossless networks matching generator and load are inserted between these latter. *Over-all gain* (available† or actual) of a series 1, 2, . . . , n of cascaded linear networks is simply the product of the respective gains of each network, that is, $G_{\text{over-all}} = G_1G_2 \cdots G_n$.

We are now in a position to construct a second-moment measure of the inherent noise produced by a typical nonideal linear network $\mathbf{T}_L^{(N)}$. Besides the noise and signal from the standard signal generator (cf. above), the noise generated locally in the network results in an *increase* in the ratio $\Delta N_G^2/S_G^2$ of the available input noise-to-signal powers to a new value $\Delta N_0^2/S_0^2$, where ΔN_0^2 is the total output noise power in the band $(f, f + \Delta f)$ centered about the sinusoidal signal.‡ Accordingly, $\Delta N_0^2/S_0^2$ is some multiple F (≥ 1) of the input ratio $\Delta N_G^2/S_G^2$, and F itself exceeds unity, unless the network is noise-free. Thus, the defining relation for F is

$$\frac{\Delta N_0^2}{S_0^2} = F\,\frac{\Delta N_G^2}{S_G^2} \qquad F \geq 1,\ F = F(f) \tag{5.71}$$

and F is called the (available) *noise figure of the network*. F varies with the frequency response of the network and depends on the fact that ΔN_0^2 and ΔN_G^2 are *available* noise powers, measured in the *same* frequency interval $(f, f + \Delta f)$.§

Since the available noise figure F is a measure of the spectral (power) intensity of the noise generated in the network, i.e., the excess noise, let us restate Eq. (5.71) in terms of ΔN_{net}^2, this excess (available) noise power in the range $(f, f + \Delta f)$, that is, $\Delta N_{net}^2 = \int_f^{f+\Delta f} \mathcal{W}_N(f)\,df \doteq \mathcal{W}_N(f)\,\Delta f$. With

† See previous footnote.

‡ Unless otherwise indicated, we shall take Δf to be sufficiently small that variations in $\mathcal{W}_N(f)$, and in the other quantities defined below that are dependent on frequency, can be ignored over the interval $(f, f + \Delta f)$.

§ The further assumption is also made that there is a one-to-one correspondence between the frequencies of the output and input waves. See also previous footnote.

$G\,\Delta N_G^2$ as the amplified generator noise at the network's output, we can write

$$\Delta N_0^2/S_0^2 = (G\,\Delta N_G^2 + \Delta N_{net}^2)/S_0^2 = F\,\Delta N_G^2/S_G^2 \tag{5.72a}$$

from Eq. (5.71), and since $S_0^2 = GS_G^2$, $\Delta N_G^2 = W_0\,\Delta f/4R_G = k_BT_0\,\Delta f$ (from the above), we have

$$\Delta N_0^2 = FGk_BT_0\,\Delta f \qquad \text{with } F - 1 = \Delta N_{net}^2/Gk_BT_0\,\Delta f \tag{5.72b}$$

Other useful relations, easily derived from the above, are

$$\Delta N_0^2/\Delta N_G^2 = FS_0^2/S_G^2 = FG \qquad \Delta N_G^2 = (\Delta N^2)_{in} = k_BT_0\,\Delta f \tag{5.72c}$$

Note, incidentally, that the gain G is a simple ratio of signal powers, while the noise figure F depends as well on a ratio of noise powers.†

Finally, observe that the noise figure is proportional to the spectral intensity of the total noise power output of the network, for if $z(t)$ is the output noise voltage ensemble, letting $\Delta f \to 0$, we obtain from Eq. (5.72b)

$$\mathcal{W}_z(f) = F(f)G(f)k_BT_0 \qquad \mathcal{W}_z(f) = \mathcal{W}_N(f) + \mathcal{W}_G(f) \tag{5.73}$$

where $\mathcal{W}_G(f) = G(f)k_BT_0$ is the spectral intensity of the available noise power from the standard source, at frequency f, *after* passage through the network, and $\mathcal{W}_N(f) = [F(f) - 1]G(f)k_BT_0$ is the corresponding spectral power density of the excess noise voltage produced in the network itself. We have written $G = G(f)$ to emphasize that the (available) power gain, like the noise figure, is also a function of frequency. Specifically, G is equal to $|Y(i\omega)|^2$ for passive linear circuits and for active networks can be written $G = G(f) = G_0|Y(i\omega)|^2$, where G_0 (≥ 0) is a power-amplification ratio which we arbitrarily set to be the value of $G(f)_{\max}$.

A useful modification of F is the *integrated noise figure* F_i, which includes the effects of inherent noise over all frequencies. This is a total power ratio, rather than a power-density ratio like $F(f)$, and is defined as

$$F_i \equiv \frac{N_{out}^2}{N_{0-\text{ideal}}^2} = \frac{\int_0^\infty \mathcal{W}_z(f)\,df}{\int_0^\infty \mathcal{W}_G(f)\,df} = \frac{k_BT_0\int_0^\infty F(f)G(f)\,df}{k_BT_0\int_0^\infty G(f)\,df} \tag{5.74}$$

where the numerator is the (available) noise power integrated over all frequencies and the denominator is the noise power output of an ideal, i.e., noise-free, network in all other respects identical with the actual physical

† The quantity $\Delta N_0^2/\Delta N_G^2 = FG$ is also called the *equivalent noise temperature* or, better, the *noise ratio,* and as such is frequently used to describe the second-moment noise properties of crystal converters and microwave amplifiers (Lawson and Uhlenbeck[9]).

system. Being a ratio of total intensities, F_i is a simpler and less informative statistic than $F(f)$, which is proportional to a spectral density [cf. Eqs. (5.73)].

5.3-2 Linear-network Criteria; Noise Figures (II).† Frequently, the sources of excess noise in electronic devices, such as amplifiers, cathode followers, etc., appear at each of the input circuits of their succeeding stages; i.e., there are sources of noise internal to each stage, and the noise output of any one depends also on all preceding stages. However, if the gain of the over-all system, or of its first stage, is small, its noise sources may make an appreciable contribution to the total noise passed on by this network to the next system. In effect, these noise sources are then located in the input circuit of the second stage of the following network. Let us therefore indicate briefly how the various noise figures of a sequence of suitably isolated linear networks may be determined.

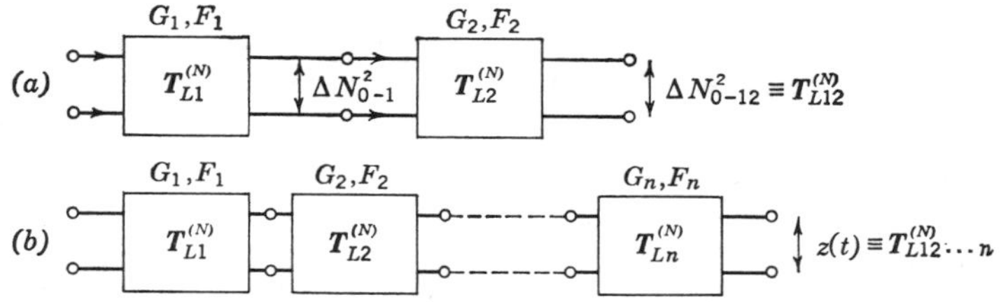

FIG. 5.13. (a) A pair of isolated cascaded linear active networks with gains G_1, G_2 and noise figures F_1, F_2. (b) A set of n isolated cascaded linear active networks with gains $G_1, G_2, \ldots, G_n$ and noise figures $F_1, \ldots, F_n$.

Consider first a pair of cascaded linear networks $\mathbf{T}_{L12}{}^{(N)} = \mathbf{T}_{L2}{}^{(N)}\mathbf{T}_{L1}{}^{(N)}$, for example, two amplifiers in series (and at the same temperature), as indicated in Fig. 5.13*a*. Restricting ourselves for the moment to very narrow bands Δf about the frequency of the signal generator (applied at a,b), we see that the average over-all noise output power ΔN_{0-12}^2 consists of two parts. The first contribution is ΔN_{0-1}^2, from the first stage, amplified by the gain G_2 of the second network (for the moment, taken to be ideal). The second part consists of the *added*, or excess, noise produced by the second stage itself, which from Eq. (5.72*b*) is $(F_2 - 1)G_2k_BT_0\,\Delta f$, so that the total available noise power output is $\Delta N_{0-12}^2 = \Delta N_{0-1}^2\, G_2 + (F_2 - 1)G_2k_BT_0\,\Delta f$. Since, from Eq. (5.72*b*), $F_{12} = \Delta N_{0-12}^2/G_{12}k_BT_0\,\Delta f$, $G_{12} = G_1G_2$, and $\Delta N_{0-1}^2 = F_1G_1k_BT_0\,\Delta f$, we may use these relations to obtain directly

$$\Delta N_{0-12}^2 = \left[F_1(f) + \frac{F_2(f) - 1}{G_1}\right] k_BT_0\,\Delta f\, G_1G_2 \qquad F_{12}(f) = F_1(f) + \frac{F_2(f) - 1}{G_1(f)} \tag{5.75}$$

respectively, for output noise power and the noise figure of the pair of cascaded linear networks in Fig. 5.13*a*.

† For generalizations and modifications of these ideas, see Haus and Adler.[20]

In the same way, we can extend Eqs. (5.75) to three or more cascaded systems. Thus, for the n (isolated) linear networks of Fig. 5.13b, we have directly

$$F_{12\cdots n}(f) = F_1 + \frac{F_2 - 1}{G_1} + \frac{F_3 - 1}{G_1G_2} + \cdots + \frac{F_n - 1}{G_1 \cdots G_{n-1}} \qquad n \geq 2 \tag{5.76}$$

and in terms of spectral intensities this becomes

$$\mathcal{W}_z(f) = \mathcal{W}_z(f)_G + \mathcal{W}_{N_1}(f)G_{2\cdots n}(f) + \mathcal{W}_{N_2}(f)G_{3\cdots n}(f) + \cdots + \mathcal{W}_{N_n}(f) \tag{5.77}$$

where $\mathcal{W}_z(f)_G = G_{1\cdots n}(f)k_BT_0$ is the spectral density of the available noise power from the standard source as the output of the over-all network. $\mathcal{W}_{Nm}(f) = [F_m(f) - 1]k_BT_0G_m(f)$ is the spectral power density of the *excess* noise voltages produced in the mth network ($m = 1, \ldots, n$) [cf. Eqs. (5.73)]. From Eq. (5.76), we observe that if, say, the mth stage has an exceptionally high gain, e.g., if $G_m \gg 1$, *the over-all noise figure* $F_{1\cdots n}(f)$ of the total n ($> m$) stages *is not noticeably affected by the contributions of the succeeding stages* ($m + 1, m + 2, \ldots, n$).† As we shall see presently (cf. Sec. 5.3-3), this often leads to considerable simplification in the analysis of composite systems.

The integrated noise figure for the composite linear network of Fig. 5.13a, b may be obtained directly from the definition in Eq. (5.74), with the help of Eq. (5.76). The result for n cascaded linear networks is

$$F_{i(1\cdots n)} = \frac{k_BT_0 \int_0^\infty F_{1\cdots n}(f)G_{1\cdots n}(f)\,df}{k_BT_0 \int_0^\infty G_{1\cdots n}(f)\,df} \tag{5.78}$$

Now, if the bandwidth of the final (or nth) stage is narrow compared with the bandwidth of any preceding section, Eq. (5.78) can be simplified to resemble Eq. (5.76). Since $F_n(f)$, $G_n(f)$ here alone depend on frequency in the range $(f, f + \Delta f)$, while F_j, G_j ($j = 1, 2, \ldots, n - 1$) are essentially constant over the range, the integrated noise figures of the first $n - 1$ stages are the same as the noise figures $F_1, F_2, \ldots, F_{n-1}(f)$, so that $F_{1\cdots(n-1)}(f)$ is given by Eq. (5.76) ($n \to n - 1$). The integrated noise figure of the nth stage, $F_{i(n)}$, is then obtained from Eq. (5.74). The result is

$$F_{i(1\cdots n)} \doteq F_1 + \frac{F_2 - 1}{G_1} + \cdots + \frac{F_{n-1} - 1}{G_1 \cdots G_{n-2}} + \frac{F_{i(n-1)} - 1}{G_1 \cdots G_{n-1}} \tag{5.79}$$

which, except for the last term, is by Eq. (5.77) proportional to a spectral

† It is assumed that $G_{m+1}(f)$, $G_{m+2}(f)$, etc., are all greater than unity. Even when some G_l ($m < l < n$) are less than unity, as in the case of cathode followers, for example, G_m is usually large enough so that G_mG_j, G_mG_{m+1}, $\ldots$, G_j, etc., are also noticeably greater than unity.

density. Other combinations are clearly possible, depending on the relative bandwidths of the individual stages.

We remark, finally, that the noise figures F and F_i are second-moment properties of the noise (and signal) *ensembles* involved, although the choice of the signal process is more one of convenience than of necessity, since the networks are linear: the noise-to-signal ratios of Eqs. (5.72) by which F is defined can be obtained (at least in principle) by separate observations of signal and noise. Signal and noise do not interact, and hence output signal and noise ensembles can be given specifically and uniquely in terms of their corresponding inputs. By proper circuit design and selection of the network components, e.g., tubes and passive elements, the inherent network noise can often be much reduced, and the amount of inherent noise in the over-all system can then frequently be made much less. A full account of such problems, including the measurement of noise figures, their detailed role in transmitter and receiver design, their determination for traveling-wave tubes and other microwave amplifiers and oscillators, is given elsewhere in a number of standard works and papers.[8–13] For the purpose of this book, however, the essential feature of the noise figure is that it can indicate at what stage and to what extent the various linear elements of $\boldsymbol{T}_R^{(N)}$ and $\boldsymbol{T}_T^{(N)}$ depart from ideal, noise-free operation and where, in the communication link, inherent noise can be ignored and where it must be taken into account. The noise figure is thus a convenient way of specifying the effect of a linear network on the signal-to-noise ratio of a combined process.

5.3-3 Criteria for Nonlinear Systems† (I). Remarks on the Observation Process. As we have just remarked, a useful account of linear systems with respect to input signals and to input and inherent noise can be obtained from their (available) noise figures $F(f)$, F_i and their gains $G(f)$. The choice of input or test signal in these instances is not critical, since the networks are linear. However, as soon as one considers a nonlinear system, this situation changes radically. To see this, let us suppose that there is always an inherent (or injected) noise process accompanying the input signal ensemble. We may expect the output ensemble to contain the original signal and noise processes, but now no longer in their simple input combination. From Secs. 5.1-1 to 5.1-3, we recall that the rectifying properties of a nonlinear element $\boldsymbol{T}_g$ produce cross modulation, or mixing, of the input signal and noise with one another. The resulting output ensemble consists, then, of a mixture of $s \times s$, or signal intermodulation, products and of $n \times n$ and $s \times n$ noise terms which have lost the statistical character of the original input process. Because of this cross modulation, or "scrambling," of signal and noise, it is *not* possible to resolve the output into separate signal and noise components that depend alone on their respective input processes:

† Here and in Sec. 5.3-4, we shall use the term "system" rather generally to mean a single network or a set of networks in parallel or series, where in either case the components may be linear or nonlinear, ideal or noisy, or combinations of both (cf. Fig. 5.11).

the $s \times s$ terms are influenced by the amount (and kind) of noise, and the $s \times s$ and $n \times n$ noise terms are in turn governed to varying degrees by the original signals. For this reason, any description of the behavior of a nonlinear system will depend on the input ensemble and its statistical properties, as well as on the specific nonlinear operation $\boldsymbol{T}_g$ itself. The examples of Sec. 5.2 illustrate this remark.

When the system in question is nonlinear or contains a nonlinear element, we can extend the second-moment theory for linear cases and construct corresponding second-moment criteria here also. This is no longer so easy to do, however, as we must now specify at the output, in terms of the input processes, what is meant by signal, as well as noise, and how these are observed.

Let us begin first with the observation process. We distinguish three main types:

1. *The nondecision case*, where the output of the system is merely the processed data of the input, e.g., a reading on a meter, a pattern on a cathode-ray oscilloscope, a waveform, etc. Here we do not consider, or are not interested in, how the output is to be used.

2. *The automatic decision process.* Here the processed output is passed into some automatic, nonhuman, decision-making system, whose response might be simple "yes" or "no" indications or more complicated readings corresponding to a measurement or an estimation, say, of a waveform. In either case, the decision to accept or reject, i.e., to make a statement or a selection of values, is made automatically when the processed data of the original system are presented to our automatic "observer."

3. *The human observer.* In this case, the automatic system of (2) is replaced by the human operator, who is then called upon to make some type of decision on the basis of the processed data that he senses. For over-all system performance, one should include the observer's psychophysical responses and his operating conditions, as well as the behavior of the linear and nonlinear elements of the processing equipment.

More complicated systems are, of course, possible: we can combine the automatic and the human observer, each of whom may contribute to the final decision output of an over-all system. Our discussion in this book, however, is deliberately confined to cases 1 and 2 and to the inanimate portions of case 3, since an adequate treatment of the human operator is beyond the competence of this work and would require far more space than is at our disposal here. We do not belittle the human observer's importance or the fact that communication systems whose direct function depends at some point on such an observer are certainly incompletely described when his role is omitted.†

† In this connection, see the theoretical and experimental discussion of the human observer in radar operation given by Lawson and Uhlenbeck,[9] chap. 8, for example, and in subsequent experimental studies by Tanner and Swets.[14] For some of the acoustical factors governing the human operation performance, see also Tanner and Swets.

Now, in any observation process, we must make a distinction between *acquisition time* and *processing time*. The former is a measure of sample size, i.e., it is the period during which data are acquired, while the latter represents the interval during which the data are subjected to various linear and nonlinear transformations $\mathbf{T}_L$, $\mathbf{T}_g$, for example, linear filtering, amplification, rectification, etc. Usually, acquisition and processing take place simultaneously, or nearly so, but often there may be a decided lag between the two, as in certain types of computer operation, in message decoding, in signal estimation, etc. Thus, acquisition may be nearly or even entirely completed before processing begins and a final decision recorded. The over-all period in which these two take place we shall call the *observation period*, which is accordingly measured from the moment data enter the system to the time at which the final decision is made. When a human observer is part of the system, we must include his own reaction and processing time as well.

There remains the question of what constitutes a signal at the output of a nonlinear element or system when noise, as well as signal, is present at the input. As we have already noted above, the output ensemble is a complicated, irreducible mixture of the original signal and noise ensembles, and it is not possible to identify any one component or set of components in this output as depending alone on the signal. For this reason, whatever we may choose analytically as an output signal is at best a partial representation. The same is true of the noise, since it, also, is influenced by the presence of the signal. To some extent, then, our choice of "signal" at the output of a nonlinear device, however reasonable or obvious, is always to some degree an *ad hoc* selection which we cannot expect to be fully descriptive of system performance. Typical choices, for second-moment criteria, are total signal intensity, intensity of a single spectral line (if the input process is periodic), the contribution of a number of spectral components, or any other second-moment statistic involving $s \times s$ terms alone. For the noise, one usually takes the total intensity of the $s \times n$ and $n \times n$ components, produced in rectification, or a portion integrated over some comparatively narrow spectral interval, whichever is appropriate to the problem at hand.

One more comment should be made before we attempt to construct second-moment criteria: In all practical situations we have but a finite period $(0,T)$ for observation, i.e., for acquisition and processing. Thus, we may expect this interval $(0,T)$ to influence our operation, and, in fact, from observation to observation we observe that each sample differs somewhat from every other. The ensemble average of our observed values will also be a function of the observation time T, as our example in Sec. 5.2-3 for the simple spectrum analyzer shows. In any case, this statistical fluctuation of the finite sample must ultimately be considered if we are to achieve a realistic appraisal of system performance, and, as such, the finite observation period becomes a central element of the more general theory discussed later in Part 4.

5.3-4 Criteria for Nonlinear Systems (II). The Signal-to-Noise Ratio. With the above remarks in mind, let us now examine a form of second-moment criterion which is particularly common in practice. This is the *signal-to-noise ratio* $(S/N)_{out}$, which is customarily defined as the ratio of an (ensemble) mean or rms amplitude $\bar{S}_{out}$ or $\sqrt{\overline{S^2_{out}}}$ of a suitably representative "signal" S_{out}, computed at a designated point in the system, to the rms (ensemble) mean intensity $\sqrt{\overline{N^2_{out}}}$ of what is considered to be the effective noise background accompanying the "signal" at this point. In its simplest form, an infinite observation period is assumed (that is, $T \to \infty$), although it is not a difficult matter to extend the definition to include ensembles of finite samples ($T < \infty$) as well [cf. Eqs. (5.87) et seq.]. Whether or not one chooses $\bar{S}_{out}$ or $\sqrt{\overline{S^2_{out}}}$ as signal depends on the method of observation in the actual situation. In any case, it is clear that $(S/N)_{out}$ will be some function of the mean power or intensities of the input signal and noise processes so that we can write

$$(S/N)_{out} = G_\infty[(P_S)_{in},(P_N)_{in}] = G_\infty[(P_S/P_N)_{in};(P_N)_{in}] \tag{5.80}$$

and usually $(P_S/P_N)_{in} \equiv (S/N)_{in}{}^2$. With suitable choice of what constitutes an output signal, $(S/N)_{out}$ for these infinite samples can be regarded as a function of $(S/N)_{in}{}^2$ alone, and we may accordingly rewrite Eqs. (5.80) as

$$\left(\frac{S}{N}\right)_{out} \equiv \begin{Bmatrix} \dfrac{\bar{S}_{out}}{\sqrt{\overline{N^2_{out}}}} \\ \text{or} \\ \dfrac{\sqrt{\overline{S^2_{out}}}}{\sqrt{\overline{N^2_{out}}}} \end{Bmatrix} = G_\infty\left[\left(\frac{S}{N}\right)^2_{in}\right] \tag{5.81}$$

Just what function G_∞ will be depends on how output signal and noise are defined, on the linear and nonlinear elements preceding the point at which these output quantities are defined, and on the statistical structure of the input signal and noise processes. [The $(S/N)_{out}$ in Eqs. (5.80), (5.81) are amplitude ratios; one can equally well use intensity ratios $(S/N)^2_{out} = G_\infty{}^2$; the functional dependence of $(S/N)_{out}$ on $(S/N)_{in}$ in either case remains unchanged.]

It is important to remember that $(S/N)_{out}$ is an ensemble value, computed from the ensemble averages of S_{out} and N_{out}. This ratio is never actually observed, of course, either by the automatic or by the human operator. Like our choice of what constitutes signal and noise at the output of a nonlinear device, it is a construction whose chief justification lies in its comparative simplicity and, if reasonable skill is exercised, in its faithfulness to some of the major features of system performance. Our selection of S_{out} must reflect this in some way. For example, if performance is judged on the basis of a particular output value or indicator reading (whether by an automatic or by a human observer), it is the *magnitude*, or amplitude, of this value or reading, in comparison with some standard magnitude or scale, that is significant, so that one naturally chooses here an average value of this magnitude, $\bar{S}_{out}$, to represent the desired output signal. On the other hand, if the decision process is based directly on the intensity, or power, of the output phenomenon, in some way, the natural choice is $\sqrt{\overline{S^2_{out}}}$, rather than $\bar{S}_{out}$, in Eq. (5.81). The former situation we shall refer to as an *amplitude*, or *deflection*, *criterion*, that is, $(S/N)_{out} = \bar{S}_{out}/\sqrt{\overline{N^2}}$, and the latter, as an *intensity*, or *power*, *criterion*, that is, $(S/N)_{out} = (\overline{S^2_{out}}/\overline{N^2})^{1/2}$. Examples of both types have been discussed in the radar case by Lawson and Uhlenbeck[15] and by Van Vleck and Middleton.[16] We shall encounter additional examples later, in Part 3.

Let us illustrate the deflection criterion in more detail, using the results of Sec. 5.2-2. We begin by passing the output of the full-wave square-law rectifier ($\boldsymbol{T}_g$ in this example) through a low-pass smoothing filter $\boldsymbol{T}_{L2}$ in the steady state (i.e., infinite observation time is assumed) and this output,

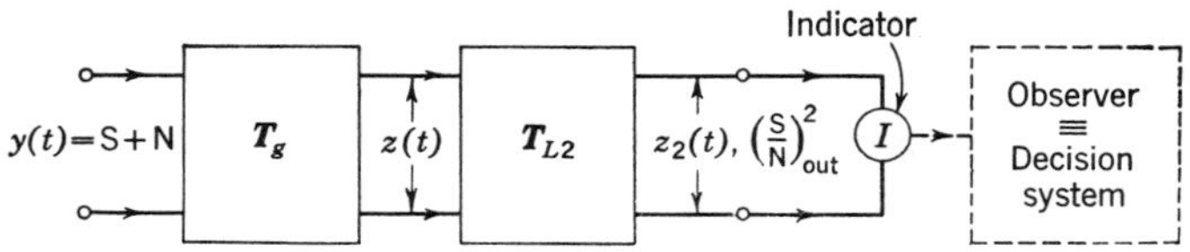

FIG. 5.14. Schema for rectification and smoothing of an input ensemble, consisting of a (sinusoidal) signal and (broadband *RC* normal) noise.

in turn, to some indicator, as sketched in Fig. 5.14. In addition, we assume once more that the input ensemble consists of a broadband normal *RC* noise process (of mean intensity ψ) and an ensemble of sine waves of frequency f_c, amplitude A_0. (Typical spectra before and after rectification by $\boldsymbol{T}_g$ are shown in Fig. 5.7.) It is natural here to choose as our output signal the average d-c increment or deflection due to the $s \times n$ and $s \times s$ modulation products and for the output noise that portion of the spectral continuum [for $(f \geq 0)$] that is passed by $\boldsymbol{T}_{L2}$.† The average increment is therefore $\overline{z_2(t)}_{S+N} - \overline{z_2(t)}_N$, in which $z_2(t) = \int_{-\infty}^{\infty} h_2(t - \tau)z(\tau)\,d\tau$. From Eqs. (5.46), this becomes specifically

$$\bar{S}_{out} = \lambda_\infty \alpha\psi(a_0^2 + 1) - \lambda_\infty \alpha\psi = \lambda_\infty \alpha\psi a_0^2 \tag{5.82}$$

where λ_∞ depends on the smoothing filter and is $\lambda_\infty = \int_{-\infty}^{\infty} h_2(x)\,dx$. The

† We assume that the d-c component after rectification also appears at the output of $\boldsymbol{T}_{L2}$. For our signal choice, we regard the indicator as calibrated to zero at the d-c level produced by an input of noise alone.

mean intensity of the accompanying output noise from $\boldsymbol{T}_{L2}$ is

$$\overline{N^2_{out}} = \int_0^\infty df\,|Y_2(i\omega)|^2 \mathcal{W}_z(f)_{s\times n+n\times n} \tag{5.83}$$

in which Y_2 is the system function, corresponding to $h_2(x)$, of the linear passive smoothing filter and $\mathcal{W}_z(f)_{s\times n+n\times n}$ is the spectrum of the output noise from the nonlinear element $\boldsymbol{T}_g$. The bandwidth of the input noise is proportional to ω_F $(= 2\pi/RC)$, as before.

At this point, let us require $\boldsymbol{T}_{L2}$ to be a narrowband, low-pass rectangular filter, for which $|Y_2(i\omega)| = 1$ $(0 < f < \Delta f_2)$, $|Y_2(i\omega)| = 0$ $(f > \Delta f_2)$, with $\Delta f_2 \ll \omega_F/2\pi$; that is, the spectrum of the input noise is much broader than the response of this filter. Then $\mathcal{W}_z(f)_{n\times n+s\times n}$ is essentially constant in $(0,\Delta f_2)$, and so Eq. (5.83) reduces to $\overline{P^2_{out}} = \Delta f_2\, \mathcal{W}_z(0)_{s\times n+n\times n}$, which from Eq. (5.47) becomes explicitly

$$\overline{N^2_{out}} = 4\,\Delta f_2\, \alpha^2\psi^2\omega_F{}^{-1}[1 + 4a_0{}^2/(1 + \omega_c{}^2/\omega_F{}^2)] \tag{5.84}$$

The desired signal-to-noise ratio (5.81) is now

$$\left(\frac{S}{N}\right)_{out} \equiv \frac{\bar{S}_{out}}{(\overline{N^2_{out}})^{1/2}} = \frac{(\bar{z}_2)_{S+N} - (\bar{z}_2)_N}{[\overline{z_2(t)^2}_{s\times n+n\times n}]^{1/2}} = \frac{(\lambda_\infty{}^2\omega_F/4\,\Delta f_2)^{1/2}a_0{}^2}{[1 + 4a_0{}^2/(1 + \omega_c{}^2/\omega_F{}^2)]^{1/2}} \tag{5.85}$$

where $a_0{}^2$ $(\equiv A_0{}^2/2\psi)$ is the input signal-to-noise mean power ratio $(S/N)_{in}{}^2$.

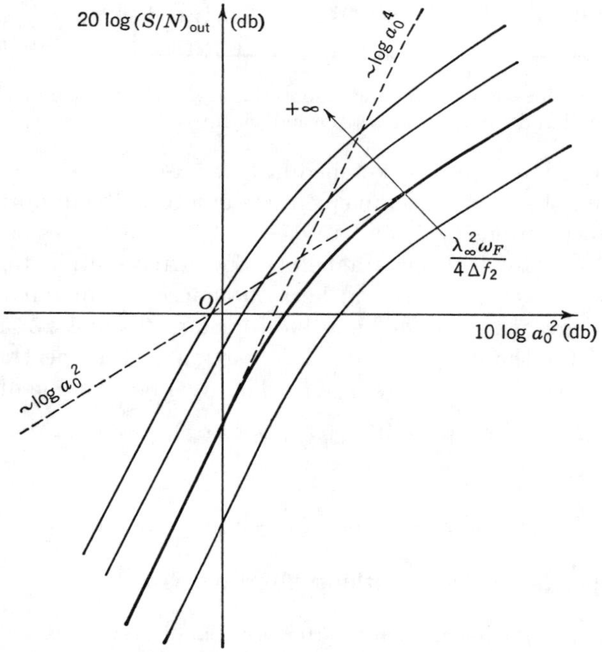

FIG. 5.15. Input and output signal-to-noise ratios for a quadratic rectifier and a low-pass filter; normal noise and sinusoidal input, infinite observation time, and fixed frequency f_c.

Characteristic curves of $(S/N)_{out}$ versus a_0^2 are shown in Fig. 5.15. Since a steady state is assumed for both input and output ensembles, this is equivalent to samples of infinite duration and so each curve in Fig. 5.15 represents a result, averaged over the ensemble, for these infinite observation times. Because of the finite bandwidth of the smoothing filter $\mathbf{T}_{L2}$, an individual member of the ensemble of indicator readings will fluctuate (at a rate determined by Δf_2) about the mean increment $(\bar{z}_2)_{S+N} - (\bar{z}_2)_N$, while this fluctuation, in turn, has a mean intensity given by Eqs. (5.83), (5.84).

Observe that as the *effective* smoothing time ($\sim\omega_F/\Delta f_2$) is made larger, with fixed input, smaller input signal-to-noise ratios yield† the same values of $(S/N)_{out}$. Thus, by increasing the effective smoothing time indefinitely (that is, $\Delta f_2 \to 0$), in principle one could observe as weak an input signal as desired. (In practice, of course, one never has the time, nor can one maintain the necessary stability of the system, to achieve such results.) Certainly, however, the general behavior of $(S/N)_{out}$ versus $(S/N)_{in}$ agrees qualitatively with what we would expect for an individual observation: the signal-to-noise ratio $(S/N)_{out}$ decreases for weak inputs and increases for strong ones.

Limiting behavior at either extreme, here, is

$$\begin{aligned} \left.\left(\frac{S}{N}\right)_{out}\right|_{a_0^2\ll 1} &\doteq \left(\frac{\lambda_\infty{}^2\omega_F}{4\,\Delta f_2}\right)^{1/2} a_0^2 \\ \left.\left(\frac{S}{N}\right)_{out}\right|_{a_0^2\gg 1} &\simeq \sqrt{\frac{\lambda_\infty{}^2\omega_F}{16\,\Delta f_2}\left(1+\frac{\omega_c{}^2}{\omega_F{}^2}\right)}\, a_0 \end{aligned} \tag{5.86}$$

We remark that the dependence on $(S/N)_{in}$ $(= a_0)$ in each instance is not peculiar to quadratic rectifiers and normal noise inputs: in the weak-signal case ($a_0^2 \ll 1$), we have here an example of the phenomenon of *signal suppression*,‡ whereby a weak signal, accompanied by noise through a nonlinear element, is made even weaker at the output, that is, $(S/N)_{out} \sim a_0^2$, rather than a_0 (under certain conditions of observation and when only l-f components of the rectified output are used in the output "signal"). The second relation in Eqs. (5.86) shows the characteristic linear dependence of $(S/N)_{out}$ on a_0 when the input signal is the dominating component.

The generalization of Eq. (5.85) is also easily found for the case of finite observation periods $(0,T)$. Let us suppose for our input ensemble y above that we have again the situation outlined earlier (cf. Fig. 5.9 and Sec. 5.2-3 for the simple spectrum analyzer). The steady-state output $z(t)$ of $\mathbf{T}_g$ is applied at $t = 0$ to the smoothing filter $\mathbf{T}_{L2}$ until $t = T$, at which instant the observation is made. Equation (5.82) becomes, with a slight modification of Eqs. (5.55),

$$\bar{S}_{out} = \int_{0-}^{T+} h_2(T-\tau)[\overline{z(t)}_{S+N} - \overline{z(t)}_N]\,d\tau = \alpha\psi a_0^2\lambda_T \tag{5.87}$$

† A similar effect is observed if ω_c is made larger vis-à-vis ω_F.

‡ For the general theory of this effect, see Sec. 19.4-3.

where $\lambda_T = \int_{0-}^{T+} h_2(T - \tau)\, d\tau = \int_{0-}^{T+} h_2(x)\, dx$, while the mean intensity of the output noise continuum (exclusive of d-c and $s \times s$ terms) is simply Eq. (5.65) with $t = T$ and $M_z(t)$ replaced therein by $K_z(t)_{s\times n+n\times n}$. Equation (5.83) is now

$$\overline{N^2_{out}} = \int_{-\infty}^{\infty} \rho_2(t)_T K_z(t)_{s\times n+n\times n}\, dt = 2\int_{-\infty}^{\infty} |Y_2(i\omega)_T|^2 \mathfrak{W}_z(f)_{s\times n+n\times n}\, df \qquad (5.88)$$

in which $\rho_2(t)_T$ is the autocorrelation function of the filter with the truncated weighting function $h_T(x) = h(x)$ $(0- < x < T+)$, $h_T(x) = 0$ $(x < 0,$ $x > T)$, for example, $\rho_2(t)_T = \int_{-\infty}^{\infty} h_T(x)h_T(x + t)\, dx$, while $Y_T(i\omega) = \mathfrak{F}\{h_T\}$. In the present case, we find from Eqs. (5.46) that $K_z(t)_{s\times n+n\times n} = 2\alpha^2\psi[k_N(t)^2 + 2k_N(t)a_0^2 \cos \omega_c t]$ and $k_N(t) = e^{-\omega_F|t|}$ for the RC noise background. Combining Eqs. (5.87), (5.88) gives us the analogue of Eq. (5.85), viz.,

$$\left(\frac{S}{N}\right)_{out} = \frac{\bar{S}_{out}}{(\overline{N^2_{out}})^{1/2}} = \frac{a_0^2\lambda_T}{\left\{\int_{-\infty}^{\infty} 2\rho_2(\tau)_T[k_N(\tau)^2 + 2a_0^2k_N(t) \cos \omega_c\tau]\, d\tau\right\}^{1/2}} = G_T\left[\left(\frac{S}{N}\right)_{in}^2\right] \qquad (5.89)$$

Equation (5.89) exhibits exactly the same kind of dependence on a_0^2 $[= (S/N)_{in}^2]$ as does the case of infinite observation time above [cf. Eq. (5.85)], which is not surprising, since $\boldsymbol{T}_g$ and our definition of S_{out} and of N_{out} at the output of $\boldsymbol{T}_g$ have not changed. When $T \to \infty$, it is not difficult to show that Eq. (5.89) reduces to our previous expression (5.85).

Other examples of signal-to-noise ratios, involving a variety of different nonlinear transformations $\boldsymbol{T}_g$ encountered in AM and FM receivers, are given in Part 3 (cf. Chaps. 13 and 15 in particular). In all cases, complete second-order statistics (i.e., the appropriate second-order densities W_2) of the input noise process (and sometimes the signal) are at most required, in order to obtain $\overline{N^2_{out}}$. Occasionally, when a power criterion can be used, first-order statistics (W_1) may be sufficient. We remember, of course, that these signal-to-noise ratios are at best incomplete descriptions of system behavior, as any second-moment theory is bound to be, since the output signal is rather arbitrarily defined, and since not all statistical information concerning the input processes is used in the criterion.

5.3-5 Summary Remarks on Second-moment Criteria. Let us return now to the set of transformations represented by Eq. (5.70) and consider again the typical AM communication system† outlined schematically in Fig. 5.11. With the help of the second-moment criteria described in Secs. 5.3-1 to 5.3-4, it is possible to describe the performance of the component

† Although our language is suited to electronic systems primarily, the general features of the discussion are clearly not confined to such systems alone.

operations $T_T^{(N)}$, $T_M^{(N)}$, $T_R^{(N)}$ of transmitter, medium, and receiver, respectively, for a wide class of signal and noise processes.

Reserving the details to Part 3, we may make the following brief observations: First, in the transmitter, or encoder, $T_T^{(N)}$, it is the inherent noise of amplifier $T_{G_1}^{(N)}$ (and possibly the local oscillator $T_{g0}^{(N)}$) (cf. Fig. 5.11) which makes the significant contribution, particularly if the RF amplifier T_{L2} has a large available gain. From Eq. (5.76) and subsequent comments, it is evident that the spectral properties of the locally generated transmitter noise are embodied in the noise figure $F(f)$ of the preceding stages: the RF and antenna (T_{LA}) stages may be regarded, as far as the transmitted wave is concerned, as essentially noise-free. The mixer, or converter, $T_{g1}^{(N)}$ likewise may be regarded as essentially noise-free here. Although the mixer, or converter, $T_{g1}^{(N)}$ is basically a nonlinear device, it may be treated approximately as linear, in so far as the transmitted signal ensemble (and the inherent noise previous to $T_{g1}^{(N)}$) is concerned, under suitable operating conditions (cf. Chap. 12). Actually, with respect to the inherent noise of mixer and LO, the former is divided, some appearing with that of $T_{L1}^{(N)}$ before conversion, some appearing after, while the LO noise itself may be regarded as appearing before the mixing operation.[17]

Thus, in a second-moment theory of the transmitter $T_T^{(N)}$, the performance of the linear stages may be described, as we have seen, by the spectrum of the noise and signal processes (cf. Secs. 5.3-1, 5.3-2), while that of the nonlinear portions can be given in terms of suitably constructed signal-to-noise ratios (Secs. 5.3-3, 5.3-4). However, in most applications, it is the effect of $T_T^{(N)}$ on the original message ensemble $\{u\}$, that is, the encoding, or modulation, that is significant to the rest of the communication process. The inherent noise can usually be disregarded, since it is normally small compared with that generated in the medium $T_M^{(N)}$ and reception process $T_R^{(N)}$.

The medium, or channel, $T_M^{(N)}$, on the other hand, is always noisy, either because of natural processes within the medium itself which generate noise (cf. Chap. 11) or because of man-made disturbances, or both. The inherent noise of the medium usually occurs additively with the transmitted ensemble $x = T^{(N)}\{u\}$, although in some cases, notably scatter propagation and related multipath phenomena, the noise may appear as a multiplicative process (cf. Probs. 11.10, 11.12, 11.13).

The receiver $T_R^{(N)}$ acts essentially like the transmitter $T_T^{(N)}$ with respect to inherent noise and impressed signal, although in reverse: it is the antenna, RF, and first-detector (or converter) as well as LO stages where most of the noise arises (if a heterodyne system is assumed; cf. Fig. 5.11). Part of the IF section also contributes to the inherent background noise.[9] The elements of the second detector T_{g2} and any subsequent linear and nonlinear operations may be regarded as noise-free. This is fortunate, since it means that we can in most cases combine the noise from the medium and from the predetector stages of the receiver with the incoming signal ensemble. Then, with due regard for preceding *linear* operations,† we may treat this combination of signal and noise ensembles as the effective input of an equivalent receiver, consisting of a noise-free linear filter, a second detector $\boldsymbol{T}_{g2}$, and subsequent linear or nonlinear filters.

From Eqs. (5.73) or Eq. (5.76), the intensity spectrum of inherent noise in the effective input is simply $\mathcal{W}_z(f) = F(f)G(f)k_BT_0$, where now $F(f)$ is the noise figure of the successive elements from the antenna through the last significant substage of the IF amplifier. The performance of the equivalent receiver can then be described by the criterion of the signal-to-noise ratio considered in the preceding sections. Note, however, for this, that considerably more statistical information than the spectrum is needed; in fact, the complete second-order statistics, i.e., the second-order density, of the input noise process and of the signal, if it, too, is random, are usually required. This is a characteristic feature of nonlinear systems in general, that a statistical description of the output process, whether it be complete or incomplete (cf. Sec. 1.3-2), requires a correspondingly greater knowledge of the statistical properties of the input process. The examples of Sec. 5.2 and 5.3-4 are typical: to obtain the second-moment function of the output ensemble, higher second-order statistics are needed for a quadratic rectifier, while the complete second-order description of the input noise is necessary in the case of the half-wave linear rectifier. Many further examples will be noted later in Part 3.

A second-moment theory of system behavior has the great advantage of "realizability." It is always possible when the complete second-order statistics (that is, W_2) are given to compute a noise figure and a signal-to-noise ratio and thus obtain some measure of the expected performance. In addition, compared with criteria based on fuller statistical data, it is comparatively simple, analytically and conceptually. However, second-moment theory is incomplete: it can give at best only a partial account of system operation, and it leaves out entirely the actual mechanism of the decision process that occurs whenever an automatic or human observer completes the communication link. Furthermore, the account of performance that it does give cannot be expected, in the more important cases of nonlinear systems, to coincide in a precise way with what is observed experimentally, since it is no longer clear what, exactly, is the observed output signal and what the output noise. These difficulties can be removed, however, when the role of the decision, based on all available information as to the noise and signal processes, is specifically included. The observation problem is then recast into one of describing performance in terms of the probabilities of decision *errors* or, if nondecision systems are used, in terms of output probabilities or probability densities of the final process. A full account of these more general methods is included in Part 4. It is upon

† Again, the LO-mixer section is considered to be linear as far as input noise and signal from the preceding RF stages is concerned.

such an approach, then, and the methods of information theory (considered briefly in Chap. 6) that a statistical communication theory may be expected to have its foundation.

REFERENCES

1. Rice, S. O.: Mathematical Analysis of Random Noise, *Bell System Tech. J.*, **24:** 46 (1945), pt. IV.
2. Middleton, D.: Some General Results in the Theory of Noise through Nonlinear Devices, *Quart. Appl. Math.*, **5:** 445 (1948).
3. Fellows, G. E., and D. Middleton: An Experimental Study of Intensity Spectra after Half-wave Rectification of Signals in Noise, *Proc. IEE (London), Mon.* 160R, January, 1956. See sec. 2.
4. Ref. 3, fig. 9, $p = 0$ for $l = 0$. Other experimental results are given by Fellows, Intensity Spectra after Half-wave Detection of Signals in Noise. II. Experimental Discussion, *Cruft Lab. Tech. Rept.* 218, Feb. 20, 1955.
5. Friis, H. T.: Noise Figures of Radio Receivers, *Proc. IRE,* **33:** 458 (1945).
6. Roberts, S.: Some Considerations Governing Noise Measurements on Crystal Mixers, *Proc. IRE,* **35:** 257 (1947).
7. Twiss, R. Q., and Y. Beers: Minimal Noise Circuits, in G. E. Valley, Jr., and H. Wallman, "Vacuum Tube Amplifiers," chap. 13, Massachusetts Institute of Technology Radiation Laboratory Series, vol. 18, McGraw-Hill, New York, 1948. See also chap. 14.
8. Davenport, W. B., Jr., and W. L. Root: "An Introduction to the Theory of Random Signals and Noise," chap. 10, McGraw-Hill, New York, 1958.
9. Lawson, J. L., and G. E. Uhlenbeck: "Threshold Signals," Massachusetts Institute of Technology Radiation Laboratory Series, vol. 24, McGraw-Hill, New York, 1950, secs. 7.3, 7.4, for example. See also Ref. 6.
10. Valley, G. E., Jr., and H. Wallman: "Vacuum Tube Amplifiers," chaps. 13, 14, Massachusetts Institute of Technology Radiation Laboratory Series, vol. 18, McGraw-Hill, New York, 1948; cf. Ref. 5.
11. Ref. 9, chap. 5.
12. Torrey, H. C., and C. A. Whitmer: "Crystal Rectifiers," Massachusetts Institute of Technology Radiation Laboratory Series, vol. 15, McGraw-Hill, New York, 1948.
13. Van der Ziel, A.: "Noise," chaps. 3, 7, 8, Prentice-Hall, Englewood Cliffs, N.J., 1954.
14. Tanner, W. P., Jr. and J. A. Swets: The Human Use of Information. I, Signal Detection for the Case of Signal Known Exactly; II, Signal Detection for the Case of an Unknown Signal Parameter, Symposium on Information Theory, Sept. 15–17, 1954, *IRE Trans. on Inform. Theory,* **PGIT-4:** 213 (1954).
15. Ref. 9, sec. 7.3.
16. Van Vleck, J. H., and D. Middleton: A Theoretical Comparison of the Visual, Aural, and Meter Reception of Pulsed Signals in the Presence of Noise, *J. Appl. Phys.*, vol. 17, 1946, sec. II(*c*) and also pt. I, sec. I.
17. See, for example, Ref. 9, secs. 5.3 and 5.4.
18. Levin, B. R.: "Theory of Random Processes and Its Application in Radio Engineering," secs. 6A, 7.1, 8.3, 8.4, 8.9, Sovetskoe Radio, Moscow, 1957.
19. Leipnik, R.: Integral Equations, Biorthonormal Expansions, and Noise, *J. Soc. Ind. Appl. Math.*, **7**(1): 6 (March, 1959).
20. Haus, H. A., and R. B. Adler: Optimum Noise Performance of Linear Amplifiers, *Proc. IRE*, **46:** 1517 (1958).

CHAPTER 6

AN INTRODUCTION TO INFORMATION THEORY

As we have indicated more than once in previous chapters (cf. Secs. 1.1, 2.1, 5.3), statistical communication theory is the mathematical theory of certain selected operations on ensembles. Important elements of such a theory, in turn, deal with system design and behavior, optimization, evaluation, and comparison. All these features appear in the case of the single communication link, whose performance is symbolized by the relation [cf. Eq. (2.20)]

$$\{v\} = \boldsymbol{T}_R{}^{(N)}\boldsymbol{T}_M{}^{(N)}\boldsymbol{T}_T{}^{(N)}\{u\} \tag{6.1}$$

Here $\{u\}$ and $\{v\}$ are the original and received message (or decision) ensembles, while the $\boldsymbol{T}_T{}^{(N)}$, $\boldsymbol{T}_R{}^{(N)}$ represent the transformations of modulation (or encoding) and demodulation (or decoding) and $\boldsymbol{T}_M{}^{(N)}$ describes the effect of the medium of propagation (or channel) on the encoded message set $\boldsymbol{T}_T{}^{(N)}\{u\}$. The superscript (N), as before, indicates the possible injection of noise at various stages of the communication process.† Many of the problems encountered in applications can be described within this framework. Thus, for example, in reception attention is directed mainly to the properties of the transformations $\boldsymbol{T}_R{}^{(N)}$, while for encoding (or modulation) the $\boldsymbol{T}_T{}^{(N)}$ are of principal interest. Of course, in general, both $\boldsymbol{T}_R{}^{(N)}$, $\boldsymbol{T}_T{}^{(N)}$ must be considered together, and in all cases the properties of the channel, embodied in $\boldsymbol{T}_M{}^{(N)}$, represent unavoidable constraints.

In this chapter, we shall examine mainly the $\boldsymbol{T}_T{}^{(N)}$, from the point of view of Shannon's information theory,[1] reserving to Part 4 a study of reception problems and system design that depend primarily on the $\boldsymbol{T}_R{}^{(N)}$. The treatment here is deliberately brief and necessarily incomplete, our aim being to present in concise fashion some of the ideas and results of information theory which are useful in our present work and which we shall specifically employ in later sections.‡

† See Sec. 2.1-3 for additional details.

‡ For a more comprehensive account of the theory, the reader is referred to the original papers of Shannon[1] and Fano[2] and to subsequent work of Woodward and Davies[3] and others.[4] See also the book by Woodward.[5] Early attempts to cover the field are represented by the books of Goldman,[6] Bell,[7] and Brillouin.[8] For historical accounts of the development of the theory, see Refs. 4, 7, 8 and the recent book by Cherry.[9] Additional bibliography is available at the end of this chapter and in the references cited above. See, in particular, for a mathematical treatment, the books of Feinstein[38] and Khintchine.[39]

6.1 *Preliminary Remarks*

Let us describe the communication problem above in more detail. The general situation is illustrated in Fig. 6.1, where now $\{x(t)\}$ is the ensemble

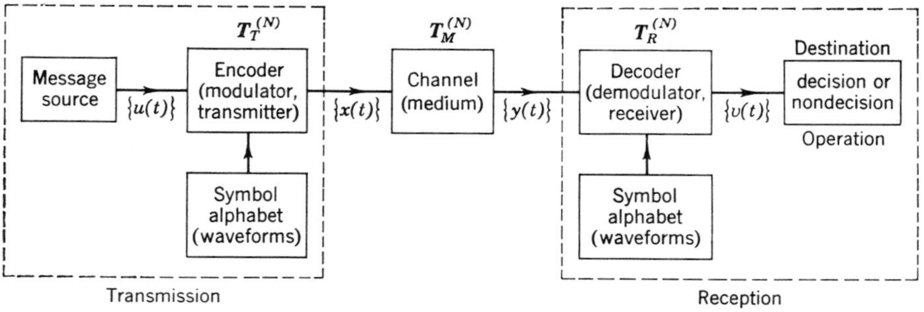

FIG. 6.1. Schematic diagram of a general single-link communication system, where the ultimate outputs are messages or decisions. For details of $T_T^{(N)}$, $T_R^{(N)}$, etc., in a typical case, see Fig. 5.11.

of encoded messages, or transmitted *signals*, and $\{y(t)\}$ is the corresponding ensemble of received signals, after passage through the channel $T_M^{(N)}$ Noise may appear at all three stages of the process.

Two interrelated groups of problems are next to be considered. These are:

1. *Encoding of the original message set* $\{u(t)\}$. This requires first the selection of an appropriate "alphabet" of symbols, or signal waveforms, and then the assignment of a particular sequence, or waveform (e.g., a "code word"), to each message, leading to the choice of transformations $T_T^{(N)}$ and hence to the *transmission of the signal symbols or waveforms* $\{x(t)\}$ in such a way that a "best" use is made of the given (noisy) channel $T_M^{(N)}$. The resulting output is the corrupted signal set $\{y(t)\} = T_M^{(N)}\{x(t)\}$. Which particular alphabet or ensemble of waveforms is used in the encoding process $T_T^{(N)}$ is, of course, strongly governed by these channel considerations, and also by item 2 below.

2. *The reception, or decoding, of the received signal process* $y(t)$, where we must estimate in some "best" fashion, i.e., by suitable choice of $T_R^{(N)}$, just what message was actually sent.

Our selection of appropriate signals (or symbols) and the method of handling them in the communication process depend, as expected, on the criteria of performance that we postulate at the outset. For example, since we are dealing in a general way with the transmission of information, a first problem is to obtain some quantitative measure of this information. A subsequent step is to use the properties of this measure to determine how the system affects the flow of information and to establish, and compare, optimum and suboptimum conditions of operation. Information theory provides an approach of this type, for, in essence, it is a theory of the communication process that sets limits on the rate of transmission (symbols per second) for error-free communication over noiseless and noisy channels.

Thus, in information theory, we are concerned initially with a statistical definition of "information," which in turn is used to define the "informational" capacity of our signal alphabet. In the idealized case of noiseless channels $\boldsymbol{T}_M$, the theory then establishes a lower bound on the average length of encoded messages (e.g., code words) $\{x(t)\}$ for any ergodic process which preserves information. By a proper encoding procedure, this lower bound can be approached as closely as desired if the duration T or number of symbols N in a typical signal sequence becomes indefinitely large in the limit. Extended to noisy channels $\boldsymbol{T}_M^{(N)}$, the theory defines the informational capacity of such channels and shows that it is possible to transmit information through the channel at a rate as close as one wishes to this channel capacity, with arbitrarily small probability of error, again for the limiting cases in which the duration of the encoded messages is made sufficiently long. Reception (e.g., the choice of $\boldsymbol{T}_R^{(N)}$) is postulated to maximize the a posteriori probability $p(x|y)$ of the transmitted signal x, given the received data y. Once $\boldsymbol{T}_T$ is found† for these limiting situations, the receiver is simply the error-free set of inverse operations $\boldsymbol{T}_R = \boldsymbol{T}_T^{-1}\boldsymbol{T}_M^{-1}$, or $\boldsymbol{T}_T^{-1}\{\boldsymbol{T}_M^{(N)}\}^{-1}$, respectively, for the noiseless or noisy channels.

Emphasis here is, first, on suitable measures of information (cf. Secs. 6.2 to 6.4) and, second, on some of the principal results for coding in noiseless and noisy channels (without attention to actual coding procedures). The concern in this chapter with the definition and properties of the information measure may also be justified, apart from the specific problem above of the single-link communication system, by the extensive use made of these concepts in other fields, such as linguistics, biology, psychology, etc. Details of coding procedures, however, are outside the scope of this work.‡

6.2 *Measures of Information: The Discrete Case*

Since the significant properties of message and signal sets in communication theory are statistical (cf. Sec. 1.1), we expect that any definition and any measure implied by the term information in this connection must likewise be based on probability considerations. Accordingly, the measure of information, as used in the technical sense here, is essentially a measure of choice, or uncertainty, regarding distinguishable elements and with respect to some random mechanism; it is not a measure of "meaning" in the usual

† In most cases, it is possible to ignore the noise generated in the encoding process itself.

‡ See, for example, Refs. 1 to 9. For related work in the communication area, consult Stumpers's bibliography[10] and Wiener's work on statistical filtering, prediction, and control.[11]

sense that implies a subjective evaluation. As we shall note presently, there are many possible statistical definitions of the amount of information, each of which may in some degree reflect the element of meaning which gives it significance to a user. Some are more satisfactory than others, and our first problem is to formulate a statistical definition which will withstand the test of utility: an appropriate choice is thus one that is established ultimately by its implications.

6.2-1 Uncertainty, Ignorance, and Information. Let us now make these rather general notions more precise. Our first task is to introduce the statistical mechanism explicitly. We begin by letting $X = \{x_i\}$ $(i = 1, \ldots, n)$ be a set of possible, distinct events (here, messages, symbols, waveforms, etc.), each with a probability of occurrence $p_i(x_i)$, such that $\sum_i^n p_i = 1$. Thus, $p_i(x_i)$ is the a priori, or prior, probability that event x_i may occur. Next, let us construct a function that is a measure of the *uncertainty* as to the occurrence of event x_i; i.e., before x_i occurs, there is an uncertainty as to its happening. How then can we construct such a measure? Clearly, it will depend alone on the a priori probability $p_i(x_i)$ in some fashion, so that we may write

$$\mathcal{U}[p_i(x_i)] = \mathcal{U}(x_i) \equiv \text{prior or initial uncertainty as to occurrence of event } x_i \text{ (before it has happened)} \quad (6.2)$$

When x_i actually occurs, this uncertainty is removed and we can speak of a *gain of information* about the occurrence of x_i, which we shall accordingly define (after Fano[2] and Woodward[3]) as

$$\begin{aligned} \mathcal{I}(x_i|x_i) &\equiv \text{gain of information about } x_i \text{ that is achieved after } x_i \text{ occurs} \\ &= \text{decrease of uncertainty about } x_i,\ \textit{after } x_i \text{ has occurred} \end{aligned}$$

Therefore, (6.3)

$$\mathcal{I}(x_i|x_i) = \mathcal{U}(x_i), \text{ a priori uncertainty in } x_i,\ \textit{before } x_i \text{ has happened}$$

This last follows directly from the notions of information gain and uncertainty: the initial uncertainty as to the possible occurrence of x_i has been removed when x_i occurs and is therefore equal to the information gain after the event. Information, in the above sense, is thus measured in terms of "uncertainty." Since uncertainty is also an expression of ignorance, we can equally well regard this information measure as a measure of ignorance, in the obvious sense that a gain of information represents a corresponding reduction in our lack of knowledge as to the event's occurrence. In any case, our language is one of probability, not meaning; of form, rather than content.

We can readily extend the above to pairs of events. Suppose that now we have two sets of events $X = \{x_i\}$ $(i = 1, \ldots, n)$, $Y = \{y_k\}$ $(k = 1,$

. . . , m) which may be related in some statistical way. For example, x_i may represent an encoded message at the input of a channel $\boldsymbol{T}_M^{(N)}$, while y_k is the output of this channel, because of noise not necessarily identified with x_i. The initial uncertainty as to the joint occurrence of x_i and y_k, like the single event [cf. Eq. (6.2)], may now be written directly as an information gain† [cf. Eqs. (6.3)]:

$$\begin{aligned} \mathcal{U}[p_{ik}(x_i,y_k)] = \mathcal{U}(x_i,y_k) &\equiv \text{gain in information about } x_i \text{ and } y_k, \\ &\quad \text{when both } x_i \text{ and } y_k \text{ are known to} \\ &\quad \text{occur} \\ &= \mathcal{I}(x_i,y_k|x_i,y_k) \end{aligned} \tag{6.4}$$

with an obvious generalization to three or more events.

The more interesting situations from a communication viewpoint arise, however, from the various conditional states that can occur when two or more events take place. Thus, for example, following the discussion above, we may define

$$\mathcal{U}(x_i|y_k) \equiv \textit{posterior} \text{ (or a posteriori) uncertainty about original occurrence of event } x_i, \textit{ after } y_k \text{ has been observed} \tag{6.5a}$$

On the basis of a received signal y_k, this is a measure of the uncertainty as to whether the input signal x_i was actually sent. Here, $\mathcal{U}(x_i|y_k)$ is naturally a function of the conditional probability $p(x_i|y_k)$, which because of the relation (1.45) of inverse probability‡ can be written in a number of equivalent ways:

$$\mathcal{U}(x_i|y_k) = \mathcal{U}[p(x_i|y_k)] = \mathcal{U}[p(x_i,y_k)/p(y_k)] = \mathcal{U}[p(x_i)p(y_k|x_i)/p(y_k)] \tag{6.5b}$$

From Eqs. (6.2) and (6.5a), we next determine the gain in information $\mathcal{I}(x_i|y_k)$ about x_i when the y_k are received. This is simply the net decrease in uncertainty about the occurrence of x_i, from the initial state before y_k is observed to the final state after y_k has been observed. We have

$$\begin{aligned} \mathcal{I}(x_i|y_k) &\equiv \mathcal{U}(x_i) - \mathcal{U}(x_i|y_k) \\ &= (\text{a priori uncertainty about } x_i) - (\text{a posteriori uncertainty} \\ &\quad \text{about } x_i \text{ remaining after } y_k \text{ is observed}) \\ &= (\text{initial ignorance about } x_i) - (\text{final ignorance after } y_k \text{ is} \\ &\quad \text{observed}) \end{aligned} \tag{6.6a}$$

Equivalent interpretations are

$$\begin{aligned} \mathcal{I}(x_i|y_k) &= \text{gain of information about } x_i \text{ on receipt of } y_k \\ &= \text{decrease in uncertainty about } x_i \text{ after } y_k \text{ is given} \end{aligned} \tag{6.6b}$$

† For convenience we shall henceforth use the same symbol, p, for probability, different probabilities being indicated by different arguments, for example, $p_i(x_i) = p(x_i)$, $p(y_k) = p_k(y_k)$, etc., unless there is ambiguity.

‡ See, for example, Woodward, *op. cit.*, sec. 4.2, and Woodward and Davies, *op. cit.*, sec. 2.

We also speak of $\mathcal{I}(x_i|y_k)$ as the information in y_k about x_i, where some sort of process like the above is understood. For instance, if $y_k = x_i$, then $\mathcal{U}(x_i|x_i) = 0$, since there is no longer any posterior uncertainty about x_i when x_i is actually observed, and so from Eq. (6.6*a*) $\mathcal{I}(x_i|y_k) = \mathcal{U}(x_i) = \mathcal{I}(x_i|x_i)$, as it should. On the other hand, if y_k conveys no information about x_i, then $\mathcal{U}(x_i|y_k) = \mathcal{U}(x_i)$ and $\mathcal{I}(x_i|y_k) = 0$: there is no gain in information about x_i possible from y_k under this circumstance. The $\mathcal{I}(x_i|y_k)$ are, of course, functions of the probabilities $p(x_i)$, $p(x_i|y_k)$ [cf. Eqs. (6.2), (6.5*b*)].

Further extensions can be made. Let us suppose that $\mathcal{I}(x_i|y_k)_C$ represents the information in y_k about x_i under a condition or constraint C. Equations (6.6) become now

$$\mathcal{I}(x_i|y_k)_C = \mathcal{U}(x_i)_C - \mathcal{U}(x_i|y_k)_C \equiv \text{information in } y_k \text{ about } x_i \text{ under condition } C \tag{6.7a}$$

Moreover, if condition C is the actual occurrence of an event z_j, which may be statistically related to the occurrence of x_i and y_k, we have

$$\begin{aligned} \mathcal{I}(x_i|y_k)_{z_j} &= \mathcal{U}(x_i)_{z_j} - \mathcal{U}(x_i|y_k)_{z_j} \\ \text{or explicitly} \qquad \mathcal{I}(x_i|y_k)_{z_j} &= \mathcal{U}[p(x_i|z_j)] - \mathcal{U}[p(x_i|y_k,z_j)] \end{aligned} \tag{6.7b}$$

In this way, we can construct a hierarchy of information gains for various contingent events, all of which may contribute to our knowledge as to the occurrence of x_i.

6.2-2 Structure of $\mathcal{U}$. While many choices of $\mathcal{U}$ are, of course, possible, only a comparatively few may be expected to be fruitful in applications. One such form is obtained if we make the reasonable *assumption that information should be additive.* Two axioms concerning additivity are necessary, in order to specify $\mathcal{I}(x_i|x_i)$ and $\mathcal{I}(x_i|y_k)$, and higher-order measures. Once these have been introduced, an explicit mathematical theory can then be developed. Following Woodward and Davies,[3] we employ the two postulates:

Postulate 1. When two independent events (here, messages or signals) occur, the total gain in information concerning their occurrence is the sum of the individual gains when each event is considered separately.

Postulate 2. When two observations representing the same initial event occur, the total gain in information concerning this event is equal to the sum of the gains from each observation. (Here the observer regards the a posteriori probability after the first observation as the a priori probability before the second.)

The first postulate specifies $\mathcal{U}(x_i)$, or, equivalently, $\mathcal{I}(x_i|x_i)$. For let x_i and y_k be the two independent events: Eq. (6.4) shows that the gain in information about the joint events x_i, y_k (when these occur) is $\mathcal{I}(x_i,y_k|x_i,y_k) = \mathcal{U}(x_i,y_k) = \mathcal{U}[p(x_i,y_k)] = \mathcal{U}[p(x_i)p(y_k)]$, since x_i and y_k are independent. Similarly, from Eqs. (6.3), the gains in information about x_i and y_k, separately, are $\mathcal{I}(x_i|x_i) = \mathcal{U}[p(x_i)]$, $\mathcal{I}(y_k|y_k) = \mathcal{U}[p(y_k)]$, so that the first axiom leads to the following functional equation:

$$\mathcal{U}[p(x_i)p(y_k)] = \mathcal{U}[p(x_i)] + \mathcal{U}[p(y_k)] \tag{6.8}$$

The only nontrivial solution for this† is found to be (cf. Prob. 6.1 and Woodward and Davies[3])

$$\mathcal{U}(p) = \lambda \log p = -S_0 \log p \qquad S_0 > 0 \tag{6.9a}$$

We have taken $-\lambda$ to be S_0, a positive constant. Choosing λ to be negative thus makes a gain in information (or reduction in uncertainty) a positive quantity. Here S_0 is a scale factor, which depends on the particular units in which it may be convenient to measure information; S_0 is usually absorbed into the base of the logarithm, so that Eq. (6.9*a*) becomes, for example,

$$\mathcal{U}[p(x)] = -\log_{[\ ]} p(x) \qquad \mathcal{U}[p(x|y)] = -\log_{[\ ]} p(x|y)$$
$$\mathcal{U}[p(x,y)] = -\log_{[\ ]} p(x,y) \tag{6.9b}$$

where the subscript [ ] indicates the choice of base. A base 2 unit is a *binary digit*, or *bit*, common usage in systems which are coded so that symbol elements have but two possible interpretations, e.g., a 1 or 0, etc. Base ϵ (natural units) and base 10 (decimal digits, or Hartleys) are also sometimes used; for mathematical convenience, we shall usually employ natural units. Conversion from one to another is made with the help of the relation

$$(\log_a b) \log_b x = \log_a x$$

To specify $\mathcal{I}(x_i|y_k)$, we now need the second postulate. Letting x_i be the original event and y_k, z_j the two observations referring to x_i, and observing that $\mathcal{I}(x_i|y_k)$ is a function only of the a priori and a posteriori probabilities $p(x_i)$, $p(x_i|y_k)$, we recall [cf. Eq. (6.6*a*)] that $\mathcal{I}(x_i|y_k) = \mathcal{U}[p(x_i),p(x_i|y_k)]$, with a similar relation for $\mathcal{I}(x_i|z_j)$. The second postulate then gives

$$\mathcal{U}[p(x_i),p(x_i|z_j)] = \mathcal{U}[p(x_i),p(x_i|y_k)] + \mathcal{U}[p(x_i|y_k),p(x_i|z_j)] \tag{6.10}$$

The solution of this functional equation can be shown to require the following relations (Prob. 6.1*b*),

$$\begin{aligned}\mathcal{U}[p(x_i),p(x_i|y_k)] &= \mathcal{U}[p(x_i)] - \mathcal{U}[p(x_i|y_k)] \\ &= \mathcal{U}(x_i) - \mathcal{U}(x_i|y_k)\end{aligned} \tag{6.11}$$

which from Eq. (6.9*b*), the results of the first axiom, becomes explicitly

$$\begin{aligned}\mathcal{I}(x_i|y_k) &= -\log p(x_i) - [-\log p(x_i|y_k)] = \log [p(x_i|y_k)/p(x_i)] \\ &= \text{gain in information about } x_i, \text{ on receipt of } y_k\end{aligned} \tag{6.12}$$

At this point, let us introduce special symbols for the prior and posterior

† We impose the added mathematical condition that $\mathcal{U}(p)$ is a continuous function of its argument, differentiable to the first order at least.

uncertainties $\mathcal{U}(x_i)$, $\mathcal{U}(x_i|y_k)$, to emphasize that the logarithmic form is the result of the (reasonable) assumption that information is additive. Thus, we write

$$\mathcal{H}(x_i) \equiv -\log p(x_i) = \text{initial uncertainty about (occurrence of) } x_i \tag{6.13a}$$

$$\mathcal{H}(x_i|y_k) \equiv -\log p(x_i|y_k) = \text{posterior uncertainty about } x_i, \text{ when } y_k \text{ has been given} \tag{6.13b}$$

With these and with the help of Eqs. (1.45) and (6.5b), we have also

$$\mathcal{I}(x_i|x_i) = \mathcal{H}(x_i) = -\log p(x_i) \tag{6.14}$$

$$\mathcal{I}(x_i|y_k) = \mathcal{H}(x_i) - \mathcal{H}(x_i|y_k) = \log [p(x_i|y_k)/p(x_i)] = \log [p(y_k|x_i)/p(y_k)] \tag{6.15a}$$

$$\mathcal{I}(x_i|y_k) = \mathcal{H}(x_i) - \mathcal{H}(x_i|y_k) = \log [p(x_i,y_k)/p(x_i)p(y_k)] \tag{6.15b}$$

Therefore,

$$\mathcal{I}(x_i|y_k) = \mathcal{H}(x_i) + \mathcal{H}(y_k) - \mathcal{H}(x_i,y_k) \qquad \mathcal{H}(x_i,y_k) = -\log p(x_i,y_k) \tag{6.15c}$$

with the interpretations of Eqs. (6.3), (6.6), etc. The extension to include one or more conditions C is directly made from Eqs. (6.7) and (6.15), viz.,

$$\mathcal{I}(x_i|y_k)_{zj} = \mathcal{H}(x_i)_{z_j} - \mathcal{H}(x_i|y_k)_{z_j} = \log [p(x_i|y_k,z_j)/p(x_i|z_j)] \tag{6.16}$$

and so on, for more involved dependencies.

6.2-3 An Example. Let us consider the noisy channel, symbolized by Fig. 6.2, where two possible signals, for example (x_1,x_2), can arise at the

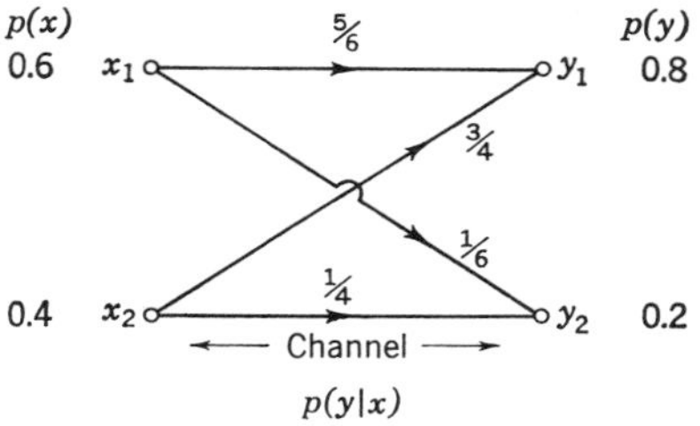

FIG. 6.2. Schematic diagram of a binary transmission through a noisy channel.

channel input, with a corresponding signal set (y_1,y_2) appearing at the output. The prior probabilities $p(x_i)$, $p(y_i|x_k)$ $(i, k = 1, 2)$ are shown in Fig. 6.2, and the $p(y_k)$ follow, of course, as the marginal probabilities

$$p(y_k) = \sum_{i=1}^{2} p(x_i,y_k) \qquad k = 1, 2$$

The various conditional probabilities are obtained from

$$p(x_i|y_k) = p(x_i,y_k)/p(y_k) \qquad p(y_k|x_i) = p(x_i,y_k)/p(x_i) \qquad i, k = 1, 2 \tag{6.17}$$

Results for a specific example, based on the a priori data shown in Fig. 6.2, are summarized in Eqs. (6.19).

From these, it is a simple matter to compute the various information measures associated with this simple system. We find from Eq. (6.14) that the a priori uncertainties of x_1 and x_2 are

$$\begin{aligned} \mathcal{I}(x_1|x_1) &= \mathcal{H}(x_1) = -\log 0.6 = 0.513 \text{ nu, or } 0.714 \text{ bit} \\ \mathcal{I}(x_2|x_2) &= \mathcal{H}(x_2) = -\log 0.4 = 0.913 \text{ nu, or } 1.32 \text{ bits} \end{aligned} \tag{6.18}$$

(1 natural unit = 1.44 bits), while the gain in information about x_1 and x_2, when y_1, y_2, respectively, are observed, becomes from Eqs. (6.15)

$$\begin{aligned} \mathcal{I}(x_1|y_1) &= \log \frac{5/8}{0.6} = \log \frac{5/6}{0.8} = 0.042 \text{ nu, or } 0.061 \text{ bit} \\ \mathcal{I}(x_2|y_1) &= \log \frac{3/8}{0.4} = \log \frac{3/4}{0.8} = -0.056 \text{ nu, or } -0.081 \text{ bit} \\ \mathcal{I}(x_1|y_2) &= \log \frac{1/2}{0.6} = \log \frac{1/6}{0.2} = -0.182 \text{ nu, or } -0.263 \text{ bit} \\ \mathcal{I}(x_2|y_2) &= \log \frac{1/2}{0.4} = \log \frac{1/4}{0.2} = 0.223 \text{ nu, or } 0.322 \text{ bit} \end{aligned} \tag{6.19}$$

Note that here $\mathcal{I}(x_1|y_2)$ and $\mathcal{I}(x_2|y_1)$ are both negative: the posterior probabilities of the x's, when the y's are given, are both less than the prior probabilities of the x's—we are made *less* certain that the x_i occurred after y_k is received ($i \neq k$) than before receipt of the y_k. This is desirable, since we wish ideally to associate a y_1 with an initial x_1 alone, etc., but, of course, the noisy channel prevents our doing this with complete certainty. Moreover, the $\mathcal{I}(x_i|y_i)$ are positive, indicating a gain of information, or reduction of uncertainty, that x_1 or x_2 occurred when y_1 or y_2 is received, again a desirable result.

6.2-4 Properties of $\mathcal{I}(x|x)$, $\mathcal{I}(x|y)$. From the foregoing [e.g., Eqs. (6.14), (6.15) in particular], we can easily obtain a number of important properties of the information measures $\mathcal{I}(x|x)$, $\mathcal{I}(x|y)$ defined above. We obtain first that

(1) $\mathcal{I}(x_i|y_k) = \mathcal{I}(y_k|x_i)$. *The information in* (i.e., gained on receipt of) y_k *about* x_i *is equal to the information in* x_i *about* y_k (6.20*a*)

This follows at once from the equivalent results of Eqs. (6.15). Moreover, we see that, when there is a condition, it is further true that $\mathcal{I}(x_i|y_k)_{z_j} = \mathcal{I}(y_k|x_i)_{z_j}$, since from Eqs. (6.16), (1.45) we can write

$$\begin{aligned} \mathcal{I}(x_i|y_k)_{z_j} &= \log \frac{p(x_i|y_k,z_j)}{p(x_i|z_j)} = \log \frac{p(x_i,y_k,z_j)}{p(y_k,z_j)p(x_i|z_j)} \\ &= \log \left[\frac{p(y_k|x_i,z_j)}{p(y_k|z_j)} \frac{p(x_i,z_j)}{p(z_j)p(x_i|z_j)} \right] \\ &= \log \frac{p(y_k|x_i,z_j)}{p(y_k|z_j)} = \mathcal{I}(y_k|x_i,z_j) = \mathcal{I}(y_k|x_i)_{z_j} \end{aligned} \tag{6.20b}$$

(2) $\mathcal{I}(x_i|y_k) \leq \mathcal{H}(x_i)$. *The information in y_k about x_i cannot exceed the initial uncertainty about x_i* (6.21)

This is easily shown, for Eqs. (6.15) give

$$\mathcal{I}(x_i|y_k) = \mathcal{H}(x_i) - \mathcal{H}(x_i|y_k) \leq \mathcal{H}(x_i) \qquad \text{since} \begin{cases} \mathcal{H}(x_i) \geq 0 \\ \mathcal{H}(x_i|y_k) \geq 0 \end{cases} \tag{6.22}$$

A similar relation holds for the y_k, with the help of Eq. (6.20a), viz.,

$$\mathcal{I}(y_k|x_i) \leq \mathcal{H}(y_k) \tag{6.23}$$

and in conjunction with Eq. (6.15c) it follows that

$$0 \leq \mathcal{H}(x_i),\ \mathcal{H}(y_k) \leq \mathcal{H}(x_i,y_k) \tag{6.24}$$

which is simply a statement of the fact that there is more information in the joint relationship for x_i, y_k than in the measure for either of these alone. (If x_i, y_k are statistically unrelated, then the equality holds.) From Eq. (6.22), it is clear that $\mathcal{I}(x_i|y_k)$ can be less than zero ($i \neq k$), as our example in Sec. 6.2-3 indicated. However, if y_k specifies x_i completely, as in noiseless coding, for example, that is, if $p(x_i|y_k) = 1$, then $\mathcal{H}(x_i|y_k) = -\log p(x_i|y_k)$ vanishes and the equalities of (6.21) and (6.23) apply.

PROBLEMS

6.1 (a) Show that the only nontrivial continuous solution of the functional equation

$$f[p(x)p(y)] = f[p(x)] + f[p(y)]$$

is

$$f(p) = a \log p + b \qquad \text{where } p(x) \neq p(y) \tag{1}$$

(b) Verify that the solution of

$$f[p(x),p(x|z)] = f[p(x),p(x|y)] + f[p(x|y),p(x|z)] \tag{2}$$

requires that

$$f[p(x),p(x|y)] = f[p(x)] - f[p(z|y)] \tag{3}$$

and hence, from (a), that $f(p) = a \log p + b$ (each p), again.

6.2 (a) Compute $\mathcal{H}(x_i)$, $\mathcal{I}(x_i|y_k)$ for the channel outlined in Fig. P 6.2.

$p(x)$	x	y_1	y_2	y_3	y_4
0.2	x_1	0.1	0.05	0.05	0.0
0.3	x_2	0.1	0.0	0.10	0.0
0.3	x_3	0.0	0.1	0.05	0.15
0.2	x_4	0.0	0.2	0.0	0.10

(b) Repeat part a for the channel below:

		$p(x,y)$	
$p(x)$	x	y_1	y_2
0.1	x_1	0.1	0.2
0.5	x_2	0.2	0.1
0.1	x_3	0.2	0.05
0.3	x_4	0.1	0.05

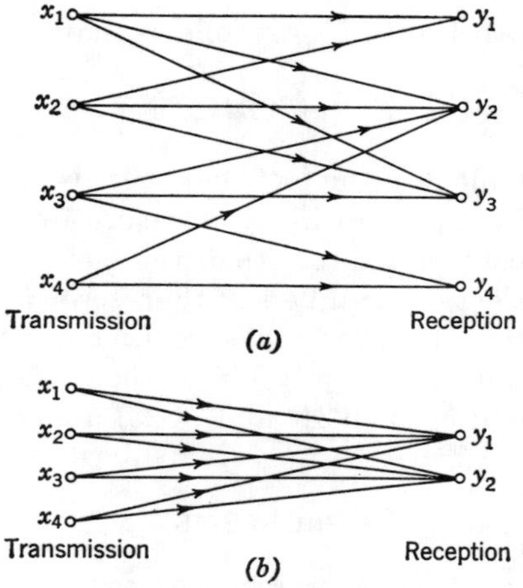

FIG. P 6.2

6.3 *Entropy and Other Average Measures: Discrete Cases (Continued)*

In communication theory, we are interested not merely in system behavior for individual messages or signals but in the totality of situations that may occur. We are therefore concerned with the statistical properties of our random variables, here x, y, and, in particular, in the information measures $\mathcal{H}(x)$, $\mathcal{H}(x|y)$. While a complete description involves the distributions of $\mathcal{H}(x)$, etc., in practice it is usually not possible to deal with these distributions *in toto*, so that various associated averages, or similar statistics, are used instead. Of course, by so doing we necessarily throw away useful knowledge, since an average generally contains less information about the quantity in question than does the distribution from which the average is obtained. However, this is the price that we must often pay for practical utility. Here, we shall consider a number of average measures of information.

6.3-1 Communication Entropy.† Let us obtain first the average uncertainty of the occurrence of an event x_i $(i = 1, \ldots, n)$, out of the totality of possible events $X = (x_1, \ldots, x_n)$. This is the usual statistical average of $\mathcal{H}(x_i)$, with respect to the prior probabilities $p(x_i)$ that govern x_i, namely,

$$\begin{aligned} H(X) \equiv \boldsymbol{E}\{\mathcal{H}(x)\} &= -\sum_{i=1}^{n} p_i(x_i) \log p_i(x_i) \qquad \text{nu} \\ &= \text{average prior uncertainty about occurrence of } x \text{, before any } x_i \text{ occurs} \\ &= \text{average gain in information about occurrence of } x \text{, after any } x_i \text{ has been observed} \\ &= \text{inherent information about } X \text{ in set } X = \{x_i\} \end{aligned} \tag{6.25}$$

$H(X)$ is also called the (communication) *entropy* of the X ensemble, since apart from the scale factor S_0 it is the direct mathematical analogue of the more familiar entropy measure of statistical mechanics.† Like its physical counterpart, $H(X)$ can also be regarded as a measure of the average disorder (or, correspondingly, average order, or "structure") that a given system $X = \{x_i\}$ possesses at a particular instant. From Eq. (6.25), it is clear that, since all $p_i \geq 0$, $p_i \leq 1$,

$$H(X) \geq 0 \qquad [H(X) = 0 \text{ only if some } p_i(x_i) = 1, \text{ all } p_k(x_k) = 0, i \neq k] \tag{6.26}$$

It is also true that

$$H(X) \leq \log n \qquad n \geq 1 \tag{6.27}$$

[or $H(X) \leq \log_2 n$ if H is measured in bits]. One can easily show this with a simple variational technique (Prob. 6.3) or as follows: From Eq. (6.25), write

$$\begin{aligned} H(X) - \log n &= \sum_{i}^{n} p_i(x_i)[\log p_i(x_i)^{-1} + \log n^{-1}] \qquad \sum p_i = 1, \text{ all } p_i \geq 0 \\ &\leq \sum_{i}^{n} p_i(x_i)\{[np_i(x_i)]^{-1} - 1\} = 0 \qquad \text{since } \log x \leq x - 1 \end{aligned} \tag{6.28}$$

The maximum value clearly occurs when the a priori probabilities $p(x_i)$ $(i = 1, \ldots, n)$ are all equal.

† We use the term *communication entropy* (following Fano[2]) to distinguish $H(X)$ from the entropy of a physical system, for example, in thermodynamical equilibrium. In this latter situation, $S_0 = k_B T_0$, where k_B is Boltzmann's constant and T_0 is the absolute temperature. Henceforth, here we shall simply refer to $H(X)$, etc., as an entropy, measured in natural units or in bits.

Analogous to $\mathcal{H}(x_i|y_k)$ [cf. Eq. (6.13*b*)], we can also define a *conditional entropy* for the $\{x_i\}$, given the $\{y_k\}$, according to

$$
\begin{aligned}
H(X|Y) = \boldsymbol{E}_{xy}\{\mathcal{H}(x|y)\} \equiv -\sum_{i,k} p(x_i,y_k) \log p(x_i|y_k) \\
&= \text{average uncertainty in set of events } \{y_k\} \text{ about set } \{x_i\} \\
&= \text{average gain in information from the } \{y_k\} \text{ about set } \{x_i\} \\
&= \text{average information in set } \{y_k\} \text{ about set } \{x_i\}, \text{ etc.}
\end{aligned} \tag{6.29}
$$

From the form of Eq. (6.29), it is at once clear that $H(X|Y) \geq 0$; that is, knowledge of the $\{y_k\}$ always increases our information about X, unless having the y_k in no way influences our knowledge of the x_i, that is, unless $p(x_i|y_k) = 1$ (all y_k). In a similar way, we get for the *joint entropy*

$$
H(X,Y) = \boldsymbol{E}_{xy}\{\mathcal{H}(x,y)\} = -\sum_{i,k} p(x_i,y_k) \log p(x_i,y_k) \geq 0 \tag{6.30}
$$

This and the fact that $p(x_i|y_k) = p(x_i,y_k)/p(y_k)$ allow us to rewrite Eq. (6.29) as

$$
H(X|Y) = -\sum_{i,k} p(x_i,y_k)[\log p(x_i,y_k) - \log p(y_k)] = H(X,Y) - H(Y) \tag{6.31a}
$$

since $\sum_i p(x_i,y_k) = p(y_k)$, etc. It is also clear that these relations are symmetrical, so that one has as well

$$
H(Y|X) = H(X,Y) - H(X) \tag{6.31b}
$$

Noting that $H(X)$, $H(X|Y)$, $H(X,Y)$, etc., are all positive (or zero), we easily obtain the inequalities

$$
0 \leq H(X|Y) \leq H(X) \qquad \text{or} \qquad 0 \leq H(Y|X) \leq H(Y) \tag{6.32a}
$$

[by an argument similar to that used in Eq. (6.28)], from which it follows that

$$
0 \leq H(X,Y) = \begin{Bmatrix} H(X|Y) + H(Y) \\ H(Y|X) + H(X) \end{Bmatrix} \leq H(X) + H(Y) \tag{6.32b}
$$

The entropies of the sets X and Y taken separately are greater than the joint entropy $H(X,Y)$, unless X and Y are independent, for which the equality then holds.

6.3-2 Average Information Gains. We define the average information about $\{x_i\}$ contained in *a particular* y_k or, equivalently, the average gain in information about the occurrence of an x_i in the set $\{x_i\}$ when a y_k is observed by

$$I(X|y_k) = \boldsymbol{E}_x\{\mathcal{I}(x_i|y_k)\} \equiv \sum_i \mathcal{I}(x_i|y_k)p(x_i|y_k)$$
$$= \sum_i p(x_i|y_k) \log [p(x_i|y_k)/p(x_i)] \tag{6.33}$$

As in Eq. (6.28), one can show that

$$I(X|y_k) \geq 0 \tag{6.34}$$

That is, the average information about the x's in a particular y_k is always equal to or greater than zero, unless the x_i are independent of the y_k, in other words, unless $p(x_i|y_k) = p(x_i)$, for which the equality then applies. This is to be compared with the information in a particular y_k about a particular x_i, $\mathcal{I}(x_i|y_k)$ [cf. Eqs. (6.15)], which may be negative, as the example of Sec. 6.2-3 shows.

The average information about the ensemble X, contained in the ensemble Y, may be similarly defined as

$$I(X|Y) = \boldsymbol{E}_y\{I(X|y_k)\} = \boldsymbol{E}_{xy}\{\mathcal{I}(x_i|y_k)\}$$
$$= \sum_k I(X|y_k)p(y_k) = \sum_{i,k} p(y_k)p(x_i|y_k)\mathcal{I}(x_i|y_k)$$
$$= \sum_{i,k} p(x_i,y_k)[\log p(x_i|y_k) - \log p(x_i)] \tag{6.35}$$

$$I(X|Y) = H(X) - H(X|Y) \tag{6.36a}$$

or

$$I(X|Y) = H(Y) - H(Y|X) = I(Y|X) \tag{6.36b}$$

since $\mathcal{I}(x_i|y_k) = \mathcal{I}(y_k|x_i)$ [cf. Eq. (6.20*a*)]. From Eqs. (6.34) and (6.35), it is also evident that

$$I(X|Y) \geq 0 \tag{6.37}$$

Thus, a gain in information is a reduction in entropy, as Eqs. (6.36) indicate. The joint entropy $H(X,Y)$ here is the joint, average information about X and Y, namely,

$$I(X,Y) = \boldsymbol{E}_{xy}\{\mathcal{H}(x,y)\} = H(X,Y) \tag{6.38}$$

from Eq. (6.30), and similarly $I(X|X) \equiv I(X) = H(X)$, $I(Y|Y) \equiv I(Y) = H(Y)$.

As a numerical illustration, let us calculate these entropies for the example discussed in Sec. 6.2-3. From Eqs. (6.25), (6.30), (6.31), (6.33), and (6.35), with Eqs. (6.19), we get directly

$$H(X) = -\sum_i p(x_i) \log p(x_i) = -(0.6 \log 0.6 + 0.4 \log 0.4)$$
$$= 0.674 \text{ nu} = 0.971 \text{ bit}$$

$$H(Y) = -\sum_k p(y_k) \log p(y_k) = -(0.8 \log 0.8 + 0.2 \log 0.2)$$
$$= 0.501 \text{ nu} = 0.721 \text{ bit}$$
$$H(X|Y) = -\sum_{ik} p(x_i,y_k) \log p(x_i|y_k) = -(0.5 \log 5/8 + 0.1 \log 1/2$$
$$+ 0.3 \log 3/8 + 0.1 \log 1/2) \quad (6.39)$$
$$= 0.669 \text{ nu} = 0.964 \text{ bit}$$

Therefore,

$$H(X,Y) = H(X|Y) + H(Y) = 1.170 \text{ nu} = 1.686 \text{ bits}$$
$$H(Y|X) = H(X,Y) - H(X) = 0.496 \text{ nu} = 0.714 \text{ bit}$$
$$I(X|y) = 0.005 \text{ nu} = 0.0072 \text{ bit} = I(Y|X)$$

and

$$I(X|y_k) = \sum_i p(x_i|y_k)\mathcal{I}(x_i|y_k) = \begin{cases} 0.005 \text{ nu} = 0.0072 \text{ bit} & \text{for } y_1 \\ 0.020 \text{ nu} = 0.029 \text{ bit} & \text{for } y_2 \end{cases}$$

6.3-3 Sequences and Sources. So far, we have required that the x_i, y_k represent events, or messages and signals. Let us now examine these messages in more detail, observing that the x_i, etc., consist of sequences of symbols, ordered in time, and of arbitrary duration.

We assume first that the symbols in a typical sequence are discrete in both magnitude and time. Thus, if $X^{(j)} = \{x_i^{(j)}, \ldots, x_n^{(j)}\}$ is a symbol sequence, the subscripts refer to distinct instants, for example, $x_i^{(j)} = x^{(j)}(t_i)$ $(t_1 \leq t_2 \leq \cdots \leq t_n)$, while the $x_i^{(j)}$ $(i = 1, \ldots, n)$ can assume only certain specific values—for example, $X^{(j)} = \{x_i^{(j)}\}$ is one member sequence of the *discrete random series* $X = \{x_i\}$ (cf. Sec. 1.3-3). In addition, each x_i can take on one of l_i distinct values $(1 \leq l_i \leq L, i = 1, \ldots, n)$ so that, letting

$$\mathcal{H}\{\mathbf{x}\} = -\log p(x_1, \ldots, x_n) = -\log p(\mathbf{x}) \quad (6.40)$$

be the prior uncertainty about the sequence $(x_1, \ldots, x_n)$, we can define the average information (or communication entropy) of this set by an obvious generalization of Eq. (6.25), e.g.,

$$H(X) = \boldsymbol{E}_{\mathbf{x}}\{\mathcal{H}(\mathbf{x})\} = -\sum_{l_1=1}^{L} \cdots \sum_{l_n=1}^{L} p(\mathbf{x}) \log p(\mathbf{x}) \quad (6.41)$$

where the sums are over all possible values of each of the symbols x_i in the sequence $X = \{x_i\} = \mathbf{x}$. H and $\mathcal{H}$ are here measured in natural units per sequence (or per n symbols).

Conditional entropies can be defined in the same way. For example, if $Y = \{y_k\}$ $(k = 1, \ldots, m)$ be a discrete random series, representing a received signal sequence for a transmitted sequence $\{x_i\}$, we can set

$$\mathcal{H}(\mathbf{x}|\mathbf{y}) = -\log p(x_1, \ldots, x_n|y_1, \ldots, y_m) = -\log p(\mathbf{x}|\mathbf{y})$$
$$= \text{initial uncertainty of sequence } \mathbf{x}, \text{ given sequence } \mathbf{y} \quad (6.42)$$

and so quite generally we can define an average conditional entropy

$$H(X|Y) = \boldsymbol{E}_{\mathbf{xy}}\{\mathcal{H}(\mathbf{x}|\mathbf{y})\} = -\sum_{l_1=1}^{L} \cdots \sum_{l_n=1}^{L} \sum_{m_1=1}^{M} \cdots \sum_{m_m=1}^{M}$$
$$\times\, p(x_1, \ldots, x_n; y_1, \ldots, y_m) \log p(x_1, \ldots, x_n; y_1, \ldots, y_m) \quad (6.43)$$

where again the sums are over all possible values of the **x** and **y**. Depending on the base, $H(X)$ and $H(X|Y)$ are measured in natural units or bits per sequence of n symbols. Introducing an abbreviated notation, we have in the same fashion, for the joint entropy of X and Y,

$$H(X,Y) = \boldsymbol{E}_{\mathbf{xy}}\{\mathcal{H}(\mathbf{x},\mathbf{y})\} = -\sum_{(\mathbf{x})}\sum_{(\mathbf{y})} p(\mathbf{x},\mathbf{y}) \log p(\mathbf{x},\mathbf{y}) \qquad \text{nu}/(n+m) \text{ symbols} \quad (6.44)$$

and the average information gain [cf. Eqs. (6.35), (6.36)] becomes

$$I(X|Y) = \boldsymbol{E}_{\mathbf{y}}\{I(X|\mathbf{y})\} = \sum_{(\mathbf{x}),(\mathbf{y})} p(\mathbf{x},\mathbf{y}) \log [p(\mathbf{x}|\mathbf{y})/p(\mathbf{x})]$$
$$= H(X) - H(X|Y) = I(Y|X), \text{ etc.} \qquad \text{nu}/n \text{ symbols} \quad (6.45)$$

and so on, for the other measures outlined in Secs. 6.3-1, 6.3-2.

We are in a position now to obtain an average measure of the information generated by a message source. Such sources generate a message or signal ensemble in the form of a discrete random series, of length T (or total number of symbols n), symbol by symbol. Now the $(n + 1)$st symbol in general will depend statistically on the preceding n symbols, so that we expect the conditional entropy of the $(n + 1)$st symbol, when the preceding n are given, to appear in our measure of information generated by the source. This conditional entropy is

$$H(x_{n+1}|\mathbf{x}^{(j)}) = -\sum_{l_{n+1}=1}^{L} p(x_{n+1}|\mathbf{x}^{(j)}) \log p(x_{n+1}|\mathbf{x}^{(j)}) \qquad \text{nu/symbol} \quad (6.46)$$

To define the entropy of the source, about to produce the $(n + 1)$st symbol, we need now the conditional entropy of the $(n + 1)$st symbol, averaged over all possible preceding n states, e.g.,

$$H(x_{n+1}|X) = \boldsymbol{E}_{\mathbf{x}}\{H(x_{n+1}|\mathbf{x})\} = \sum_{(\mathbf{x})} p(\mathbf{x})H(x_{n+1}|\mathbf{x})$$
$$= -\sum_{(\mathbf{x}),(x_{n+1})} p(\mathbf{x},x_{n+1}) \log p(x_{n+1}|\mathbf{x}) \qquad \text{nu/symbol} \quad (6.47)$$

Next, we observe that there are several different classes of information source, according as the random series **x** is nonstationary, stationary, or ergodic (cf. Secs. 1.3-6, 1.6). A sequence of finite duration belongs to a nonstationary random series, while those of infinite extent may belong to the latter two types. Let us assume, therefore, that our source is ergodic and that the series so generated is Markovian. Then the intersymbol correlation extends over only a finite number of symbols (or a finite interval, $t_2 - t_1$) in a typical subsequence whose magnitude depends on the order of the process. Since the process is stationary, the origin of time, or, equivalently, the initial symbol in any subsequence, may be chosen arbitrarily as far as the statistics of the sequence are concerned (cf. Sec. 1.3-6). We can now define the entropy of an ergodic source, or an average measure of the information to be gained from the $(n + 1)$st symbol when the process is Markovian, order n, as the conditional entropy of this $(n + 1)$st symbol, averaged over the preceding n symbols [e.g., Eq. (6.47)], so that

$$H_{\text{source}} = H_s \equiv H(x_{n+1}|X) = - \sum_{(\mathbf{x}),(x_{n+1})} p(\mathbf{x}_{n+1}) \log p(x_{n+1}|\mathbf{x}_n) \qquad (6.48)$$

$\mathbf{x}_n = (x_1, \ldots, x_n)$, $\mathbf{x}_{n+1} = (x_1, \ldots, x_{n+1})$, etc. Here H_s is measured in natural units per symbol and hence is called the *entropy rate of the source* (in these units).

For very long sequences ($n \to \infty$ or $T \to \infty$), it can be shown (cf. Prob. 6.4) that the *entropy rate* H_R associated with sequences of n symbols, generated by an ergodic source, becomes

$$H_R \equiv \lim_{T\to\infty} \frac{H_T}{T} = \lim_{T\to\infty} \frac{\log P^{-1}}{T} \qquad (6.49a)$$

$$H_R \equiv \lim_{n\to\infty} \frac{H_n}{n} = \lim_{T\to\infty} \frac{\log P^{-1}}{n} \qquad (6.49b)$$

where H_n or H_T is the entropy of the *sequence*, for example, $H_n = H(X) = -\boldsymbol{E}\{\log p(\mathbf{x})\}$ [cf. Eq. (6.41)].† Here P is the probability of occurrence of a significant message sequence (i.e., one belonging to the set of total measure 1). Thus we may write approximately,‡ for $T, n \gg 1$,

$$P \doteq \begin{pmatrix} e^{-nH_R} \\ e^{-TH_R} \end{pmatrix} \text{ or } \begin{pmatrix} 2^{-nH_R} \\ 2^{-TH_R} \end{pmatrix} \qquad (6.50a)$$

where again H_R is measured in natural units or bits, respectively, per symbol or per second. We have also, approximately,

$$\begin{aligned} N &\equiv \text{no. of statistically significant sequences} \\ &\doteq e^{TH_R}, \text{ or } e^{nH_R}, \text{ etc.} \end{aligned} \qquad (6.50b)$$

All other sequences from this ergodic source belong to a set of measure zero. For further properties of ergodic sources and sequences, see Shannon[1] (sec.

† Note that the entropies of the source $H(x_{n+1}|X)$ and of the sequence $H(X)$ are defined differently. Also, Eqs. (6.49) are expressed in natural units or bits, per second or per symbol, depending on the base.

‡ The equality holds strictly in the limit, of course; see Shannon,[1] theorem 4.

7) and Prob. 6.4 and, for Markoff processes and chains, see, for example, Blanc-LaPierre and Fortet[12] and Bartlett.[13]

PROBLEMS

6.3 (a) Prove that $H(X) \leq \log n$, by variational methods.

(b) Establish the inequalities corresponding to Eqs. (6.26), (6.32), (6.37) for $H(X)$, $H(X|Y)$, $H(X,Y)$, $I(X|Y)$, etc., when $X = x$, $Y = y$ are sequences from discrete random series.

6.4 Prove for ergodic sources that, as sequences of length n (or T) become indefinitely long, the entropy rate is

$$H_R = \lim_{n \text{ or } T \to \infty} \frac{\log P^{-1}}{n \text{ or } T} \tag{1}$$

where P is the probability of a statistically significant sequence.

C. E. Shannon,[1] theorem 3.

6.4 *Measures of Information: Continuous Cases*

In the discrete cases considered above, the various measures are necessarily expressed in terms of probabilities. It is now required to examine what happens to these measures when the random variables in question are no longer discrete, but may assume a continuum of values in magnitude (and duration). Here we expect that probabilities will be replaced in a suitable way by probability densities, although, as we shall see presently, this is not entirely an elementary operation, since some of the measures may become infinite.

6.4-1 Entropy and Information Gains. Let us begin with the entropy $H(X)$ of $X = \{x_i\}$, where the $\mathbf{x} = (x_1, \ldots, x_n)$ are continuously variable over some region $[x]$, and let us assume, as in Sec. 6.3-3, that the $x_i = x_i(t_i)$ are random variables at various moments in time. The continuous case can now be obtained formally from the discrete by dividing X-space into unit cells, with a "mesh" that becomes progressively finer as the approximation is improved. Thus if $\mathbf{\Delta x}$ $(\equiv \Delta x_1 \, \Delta x_2 \cdots \Delta x_n)$ is the cell size, $p(\mathbf{x})$ becomes essentially $w(\mathbf{x}) \, \mathbf{\Delta x}$, where $w(\mathbf{x})$ is the joint distribution density of the $(x_1, \ldots, x_n)$. Inserting this into the definition (6.41) gives us the following approximate result for the average prior uncertainty of $\mathbf{x}$,

$$H(X) = -E\{\log p(\mathbf{x})\} \doteq -\sum_{[\mathbf{x}]} w(\mathbf{x}) \log [w(\mathbf{x}) \, \mathbf{\Delta x}] \, \mathbf{\Delta x} \tag{6.51a}$$

$$H(X) = -E\{\log p(\mathbf{x})\} \doteq -\int_{[\mathbf{x}]} d\mathbf{x} \, w(\mathbf{x}) \log w(\mathbf{x}) + H_0(\mathbf{x}) \tag{6.51b}$$

where

$$H_0(\mathbf{x}) = -\sum_{[\mathbf{x}]} (\log \mathbf{\Delta x}) w(\mathbf{x}) \, \mathbf{\Delta x} = -\log \mathbf{\Delta x} \tag{6.51c}$$

When pushed to the limit in physical cases, $H_0(\mathbf{x})$ is the *absolute entropy* (of the $\mathbf{x}$) necessarily involving the smallest magnitudes $\mathbf{\Delta x}$ into which a given structure can be resolved and still preserve the essential features of "distinguishability" or "countability" that is implied by the definition of entropy as a measure of order or configuration. In all physical situations this bound on cell size is ultimately imposed by the Heisenberg uncertainty principle, which establishes the limiting magnitudes of the time and energy differences by which different configurations can be distinguished. The physical dimension of a unit cell is here proportional to Planck's constant h ($= 6.61 \times 10^{-27}$ erg-sec). However, because of background noise and finite observation times, this is so much smaller than the energy differences that our usual transmission and reception operations $\boldsymbol{T}_T^{(N)}$, $\boldsymbol{T}_R^{(N)}$ can resolve that the approximation (6.51*b*) is an entirely acceptable substitute for the more precise relation (6.51*a*).† Note, of course, that, if $\boldsymbol{\Delta}\mathbf{x} \to 0$, $H(x)$ becomes infinite, as we would expect from the fact that there are then an infinite number of possible $\mathbf{x}$ values.

In a similar way, we may approximate the conditional entropy $H(X|Y)$ [Eq. (6.43)], where now $Y = \{y_k\}$ and $\mathbf{y} = (y_1, \ldots, y_m)$ is a continuous random series like $\mathbf{x}$ above.

$$H(X|Y) \doteq -\int_{(\mathbf{x})} d\mathbf{x} \int_{(\mathbf{y})} d\mathbf{y}\, w(\mathbf{x},\mathbf{y}) \log w(\mathbf{x}|\mathbf{y}) + H_0(\mathbf{x}) \tag{6.52}$$

The $w(\mathbf{x},\mathbf{y})$, $w(\mathbf{x}|\mathbf{y})$ are, respectively, the joint and conditional d.d.'s of $\mathbf{x}$ and $\mathbf{y}$.

A more convenient representation, which eliminates the dimensional difficulties of the integrals in Eqs. (6.51*b*) and (6.52), combines H_0 with these integrals by using Eq. (6.51*c*). We write, again to a good approximation,

$$H(X) \doteq -\int_{(\mathbf{x})} w(\mathbf{x}) \log [\boldsymbol{\Delta}\mathbf{x}\, w(\mathbf{x})]\, d\mathbf{x} \tag{6.53a}$$

$$H(X|Y) \doteq -\int_{(\mathbf{x})} d\mathbf{x} \int_{(\mathbf{y})} d\mathbf{y}\, w(\mathbf{x},\mathbf{y}) \log [\boldsymbol{\Delta}\mathbf{x}\, w(\mathbf{x}|\mathbf{y})] \tag{6.53b}$$

where, as before, both $H(X)$ and $H(X|Y)$ are measured in natural units per sequence (of n symbols).

We may determine the average information in Y, about X, from Eq. (6.45), either using Eqs. (6.53*a*) and (6.53*b*) or by the same procedure used to obtain $H(X)$ applied directly to Eq. (6.45), viz.,

$$I(X|Y) \doteq \sum_{(\mathbf{x}),(\mathbf{y})} w(\mathbf{x},\mathbf{y})\, \boldsymbol{\Delta}\mathbf{x}\, \boldsymbol{\Delta}\mathbf{y} \log \frac{w(\mathbf{x}|\mathbf{y})\, \boldsymbol{\Delta}\mathbf{x}}{w(\mathbf{x})\, \boldsymbol{\Delta}\mathbf{x}} \tag{6.54a}$$

† An exception occurs in the case of delicate measurements, where the energy differences involved are comparable to those needed to make the observations, i.e., where the properties of the system and the effects of the observer are no longer separable. For an extensive discussion of these limiting situations, see Brillouin,[8] chap. 16. Many additional references are included therein.

which in the limit $\Delta\mathbf{x}$, $\Delta\mathbf{y} \rightarrow 0$ now reduces *exactly* to

$$I(X|Y) = \int_{(\mathbf{x})} d\mathbf{x} \int_{(\mathbf{y})} d\mathbf{y}\, w(\mathbf{x},\mathbf{y}) \log \frac{w(\mathbf{x}|\mathbf{y})}{w(\mathbf{x})} \qquad \text{nu/sequence} \tag{6.54b}$$

$$I(X|Y) = H(X) - H(X|Y) = I(Y|X) \tag{6.54c}$$

by Eqs. (6.53*a*), (6.53*b*). In the same way, we find that

$$\mathcal{I}(\mathbf{x}|\mathbf{y}) = -\log [w(\mathbf{x}|\mathbf{y})/w(\mathbf{x})] = \mathcal{I}(\mathbf{y}|\mathbf{x}) \tag{6.55}$$

exactly. Cell size no longer appears in these expressions for information gain, since they represent the *difference* between two states of ignorance or uncertainty. This is true of all the present measures of a change from an initial to a final information state, unlike the individual entropies $\mathcal{H}(\mathbf{x})$, $H(X)$, etc.

Not only do individual entropies depend on cell size here, but they also depend on the coordinate system in which they are measured. To see this, let a new series $\mathbf{x}'$ be obtained from $\mathbf{x}$ by a set of transformations $\mathbf{x}' = \boldsymbol{T}\{\mathbf{x}\}$, $\mathbf{x} = \boldsymbol{T}^{-1}\{\mathbf{x}'\}$, for which it is assumed that the jacobians $J(\mathbf{x}/\mathbf{x}')$ and $J(\mathbf{x}'/\mathbf{x})$ exist. From Eq. (6.53*a*), it follows that the entropy for $\mathbf{x}'$ is $H'(\mathbf{x}') = -\int_{(\mathbf{x}')} d\mathbf{x}'\, w(\mathbf{x}') \log [w(\mathbf{x}')\, \Delta\mathbf{x}']$. But

$$w(\mathbf{x}') = w(\mathbf{x})|J(\mathbf{x}/\mathbf{x}')| \qquad d\mathbf{x}' = |J(\mathbf{x}'/\mathbf{x})|\, d\mathbf{x} \tag{6.56}$$

so that

$$H'(\mathbf{x}') = -\int_{\mathbf{x}'} d\mathbf{x}'\, w(\mathbf{x})|J(\mathbf{x}/\mathbf{x}')| \log [w(\mathbf{x})|J(\mathbf{x}/\mathbf{x}')|\, \Delta\mathbf{x}'] \tag{6.57a}$$

Transforming back to $\mathbf{x}$-space and noting that the absolute entropies are equal, that is, $\Delta\mathbf{x}' = \Delta\mathbf{x}$, since we assume physical laws to be invariant of the coordinate system, we see at once [inasmuch as $J(\mathbf{x}/\mathbf{x}')J(\mathbf{x}'/\mathbf{x}) = 1$] that

$$H'(\mathbf{x}') = H(\mathbf{x}) - \boldsymbol{E}_{\mathbf{x}}\{\log |J(\mathbf{x}/\mathbf{x}')|\} \qquad \mathbf{x} = \boldsymbol{T}^{-1}\{\mathbf{x}'\} \tag{6.57b}$$

Accordingly, unless the transformations relating $\mathbf{x}$ and $\mathbf{x}'$ are *measure-preserving*, i.e., unless $|J| = 1$, there is an entropy change in passing from one coordinate system to another.

As an example of this behavior, consider the linear relation

$$\mathbf{x}' = \mathbf{Q}\mathbf{x} \qquad \det \mathbf{Q} \neq 0 \tag{6.58}$$

Then we have $J(\mathbf{x}'/\mathbf{x}) = \det \mathbf{Q}$, $J(\mathbf{x}/\mathbf{x}') = \det \mathbf{Q}^{-1} = (\det \mathbf{Q})^{-1}$, so that Eq. (6.57*b*) becomes directly

$$H'(\mathbf{x}') = H(\mathbf{x}) + \log |\det \mathbf{Q}| \tag{6.59}$$

and there is an entropy change. However, if $\mathbf{Q}$ is an orthogonal matrix ($\tilde{\mathbf{Q}}\mathbf{Q} = \mathbf{I}$), then $\det \mathbf{Q} = 1$ and $H'(\mathbf{x}') = H(\mathbf{x})$. A number of other transformations are considered in Prob. 6.5. Note that, since $I(X|Y)$ is a *difference* of two entropies [Eqs. (6.54*b*), (6.54*c*)], it is unaffected by the change in metric† and is still independent of the coordinate system used. If this were not the case, of course, the information measure would lose its general applicability and another, invariant measure would be needed.

6.4-2 Maximum Entropies. So far our results have been expressed in terms of a general distribution density $w(\mathbf{x})$. Let us now ask what particular d.d. maximizes $H(X)$, for $X = \mathbf{x}$, subject to one or more constraints. We begin by requiring that $E\{\mathbf{x}\} = 0$ (to simplify the analysis) and that the elements μ_{ij} of the covariance matrix $E\{\mathbf{x}\tilde{\mathbf{x}}\}$ be given initially.‡ Since $w(\mathbf{x})$ must be everywhere positive (and/or zero) and normalized to unity, a further condition is $\int_{(\mathbf{x})} w(\mathbf{x})\, d\mathbf{x} = 1$. The domain of the $\mathbf{x}$'s is taken to be $(-\infty, \infty)$. Introducing $n(n+1)/2$ Lagrange multipliers γ_{ij} for the different μ_{ij} and one more (λ) for the normalization, we obtain as the variational equation for an extremum of the entropy

$$\delta \left\{ H_0(\mathbf{x}) + \int_{(\mathbf{x})} w(\mathbf{x}) \left[-\log w(\mathbf{x}) + \sum_{ij}^{n} x_i x_j \gamma_{ij} + \lambda \right] d\mathbf{x} \right\} = 0 \tag{6.60}$$

where the variation is carried out with respect to w. The result is

$$w(\mathbf{x})_{\max} = e^{\lambda - 1 + \tilde{\mathbf{x}}\boldsymbol{\gamma}\mathbf{x}} \qquad \boldsymbol{\gamma} = [\gamma_{ij}] \tag{6.61a}$$

This distribution density is *multivariate normal* (cf. Secs. 7.3, 7.4), so that one finds at length that (Prob. 6.6*b*)

$$w(\mathbf{x})_{\max} = \frac{e^{-\frac{1}{2}\tilde{\mathbf{x}}\mathbf{K}_x^{-1}\mathbf{x}}}{(2\pi)^{n/2}(\det \mathbf{K}_x)^{\frac{1}{2}}} \qquad \mathbf{K}_x = \overline{\mathbf{x}\tilde{\mathbf{x}}} = [\mu_{ij}] \tag{6.61b}$$

That $w(\mathbf{x})$ of Eqs. (6.61*a*), (6.61*b*) maximizes $H(X)$ is easily seen if we take the second variation of (6.60), for then we get

$$\frac{-\int_{(\mathbf{x})} (\delta w)^2\, d\mathbf{x}}{w(\mathbf{x})} < 0 \qquad \text{since } w(\mathbf{x}) \geq 0 \tag{6.62}$$

The maximum entropy for Eqs. (6.61) is finally found to be (Prob. 6.6*b*)

$$H_{\max}(X) = \log\,[(2\pi e)^n \det \mathbf{K}_x]^{\frac{1}{2}} + H_0(\mathbf{x}) \tag{6.63}$$

† It is assumed that $\mathbf{x} \to \mathbf{x}'$ and $\mathbf{y} \to \mathbf{y}'$ and that $\mathbf{x}'$, $\mathbf{y}'$ are not functionally related.

‡ Specifically, we require that $\mu_{ii} > 0$ $(i = 1, \ldots, n)$ and $\sum_i \mu_{ii} = \text{constant} < \infty$, where the μ_{ij} are such that $\sum_{ij} \mu_{ij}\lambda_i\lambda_j$ is a positive definite quadratic form. This will ensure that $w(\mathbf{x})$ is nonsingular [cf. Eq. (6.61*b*)].

Although the autovariances μ_{ii} are fixed (and subject to the constraints above; see footnote †, page 310), they are otherwise unspecified, so that, by some additional choice of particular values for the μ_{ii}, we may be able further to maximize $H(X)$. Instead of applying a variational procedure directly, we make use of the fact above [cf. Eq. (6.57*b*)] that a change of coordinate system produces a change in the entropy. Now the $w(\mathbf{x})$ which maximizes $H(X)$ is logarithmically an n-dimensional ellipsoid, so that by a suitable transformation we should be able to deform it into an n-dimensional sphere, which by proper choice of radius will then maximize $H_{\max}$ still further.

Accordingly, let us first use an orthogonal transformation $\mathbf{Q}$ to reduce $\tilde{\mathbf{x}}\mathbf{K}_x^{-1}\mathbf{x}$ to principal-axis form. However, as we have just noted above, such a transformation is measure-preserving and $H_{\max}$ is therefore unchanged; e.g., from Eq. (6.63), we write

$$H(\mathbf{x}) = H'_{\max}(\mathbf{x}') = \log \left[(2\pi e)^{n/2} \prod_{i=1}^{n} \sqrt{\mu_{ii}} \right] + H_0(\mathbf{x}') \tag{6.64}$$

But the product $\prod_i^n \mu_{ii}$, subject to the original constraints $\mu_{ii} > 0$, $\sum_i \mu_{ii} = C$ $(< \infty)$ above, is a maximum provided that all μ_{ii} are equal, i.e., provided that $\mu_{ii} \equiv \sigma^2 = C/n$, so that we have finally

$$H_{\max\,\max}(X) = \log\,(2\pi\sigma^2 e)^{n/2} + H_0(\mathbf{x}) = \log\,[2\pi e C/n(\Delta x)^2]^{n/2} \tag{6.65}$$

where $H_{\max\,\max}$ is still measured in natural units per sequence (of n symbols) and we have set $\mathbf{\Delta x} = (\Delta x)^n$ in the absolute entropy [Eq. (6.51*c*)]. The associated d.d. is now simply

$$w(\mathbf{x})_{\max\,\max} = \prod_{i=1}^{n} \frac{e^{-x_i^2/2\sigma^2}}{\sqrt{2\pi\sigma^2}} = \prod_{i=1}^{n} w(x_i) \tag{6.66}$$

and so the various x_i $(i = 1, \ldots, n)$ are *statistically independent*, with equal variances, as well as normal. (We shall make use of this fact later, in the case of coding for the continuous noisy channel.) The *entropy rate* here becomes [cf. Eq. (6.49*b*)]

$$H_R \equiv \lim_{n\to\infty} \frac{H_{\max\,\max}}{n} = \log \sqrt{\frac{2\pi\sigma^2 e}{(\Delta x)^2}} = \log \sqrt{2\pi e P_N} \qquad \text{nu/symbol} \tag{6.67}$$

where $P_N \equiv \sigma^2/(\Delta x)^2$ is the (normalized) intensity of the series $\{x(t_i)\}$, for example, $P_n = \overline{x_i^2}/(\Delta x)^2$ (for all i).

6.4-3 Sampling, Entropy Loss, and the Destruction of Information. In the preceding sections, we have determined entropies for discrete and continuous random *series*. However, many applications make use of continuous random *processes*. In the case of band-limited processes, these can still be handled by series methods if we employ a suitable sampling procedure, quantizing the process in time and effectively reducing it once more to series form. Sampling is carried out in such a way that the process is completely represented over the infinite interval in question. In the case of a band-limited (and therefore at least wide-sense stationary) random process $x(t)$, the desired representation has been found to be [cf. Sec. 4.2 and Eq. (4.33)]

$$x(t) = \sum_{-\infty}^{\infty} x\left(\frac{n}{2B}\right) \frac{\sin \pi(2Bt - n)}{\pi(2Bt - n)} \qquad -\infty \leq t \leq \infty \tag{6.68}$$

with $E\{x\} = 0$; MSC is assumed here, with B the bandwidth of the intensity spectrum of the process.

Since the sampled values $x(n/2B)$ are a denumerably infinite set, the entropy of such a sequence will not converge but will be infinite also. It is therefore reasonable under such conditions to consider, instead, *entropy rate* as a measure of prior uncertainty, etc., for such processes. Thus, by definition, the entropy rate is here

$$H_R(X) \equiv \lim_{n \to \infty} \frac{H_n(X)}{n} \qquad \text{nu/symbol} \tag{6.69}$$

and this is usually finite, while $H_n(X) \to \infty$. In order to calculate these limiting results, we shall start in the usual way with a truncated process, $x_T(t)$, such that $x_T(t)$ and $x(t)$ coincide (with probability 1) in the limit $T \to \infty$ and $n \to \infty$. Specifically, we let

$$x_T(t) \begin{cases} \equiv \displaystyle\sum_{-BT}^{BT} x\left(\frac{n}{2B}\right) \frac{\sin \pi(2Bt - n)}{\pi(2Bt - n)} \doteq x(t) & -\dfrac{T}{2} < t < \dfrac{T}{2} \\ = 0 & |t| > \dfrac{T}{2} \end{cases} \tag{6.70}$$

Then the $\{x(n/2B)\}$ form a finite sequence of random variables, and our usual methods for calculating entropies can be applied (Sec. 6.4-1).

Now let $x(t)$ be an ergodic band-limited normal noise process, with zero mean. From Eqs. (4.39), it is evident that sample values taken at the times $t_n = n/2B$ are uncorrelated, provided that the intensity spectrum is white in $(0,B)$ (cf. Sec. 3.2-3). Furthermore, since the spectrum (or the covariance function) essentially determines the statistical structure of a normal process (Sec. 7.4), the sample values x_n are statistically independent, as well as uncorrelated. But, from Eq. (6.66), this series (as $n \to \infty$) has just the d.d. which maximizes the entropy for an already normal process (of the same bandwidth), so that it is clear that a band-limited white normal noise process produces the maximum entropy and entropy rate with respect to all other normal and nonnormal processes of the same average intensity.† We note also that the sample values of $x(t)$ are generated at the sampling rate, $2B$ symbols/sec, so that the entropy rate [Eq. (6.67)] becomes alternatively

$$H_R = 2B \log (2\pi e P_N)^{1/2} = B \log P_N + B \log 2\pi e \qquad \text{nu/sec} \quad (6.70a)$$

With arbitrary continuous processes, it is often convenient to measure their "randomness," or "noisiness," not directly by their entropies, but by comparison with a suitable normal noise. For this purpose, we choose a normal process, with uniform spectrum in the same bandwidth as that of the original process and having the same entropy rate H'_R. The average intensity P'_N of this gauss noise is then called the *entropy power* of the arbitrary process in question and from Eq. (6.67) is at once

$$P'_N = \frac{1}{2\pi e} e^{2H'_R} \tag{6.71}$$

Since this normal noise also has the greatest entropy rate for a given intensity, the entropy power of any other process, under the same bandwidth and rate conditions, must be less than (or equal to) its own average intensity.

As we have already seen [cf. Eqs. (6.57*b*), (6.59)], there is an entropy change in going from one coordinate system to another. For this reason, and from the fact that filtering operations are equivalent to transformations of coordinate systems, i.e., represent a mapping from an input signal space to an output one, we expect that filtered processes will also show a change in entropy and entropy rate. In fact, for linear filters $\boldsymbol{T}_L$, this change represents an entropy *loss*.

To see this, consider the linear invariant operation $y(t) = \boldsymbol{T}_L\{x(t)\} = \int_{-\infty}^{\infty} h(t - \tau)x(\tau)\, d\tau$ [cf. Eq. (2.51)], where h is the weighting function of a stable, physically realizable linear network and $x(t)$ is the input process, which is assumed to be ergodic (with $\boldsymbol{E}\{x\} = 0$). We wish to find the new entropy rate for $y(t)$, and so we begin with the truncated process $x_T(t)$ [Eq. (6.70)]. In discrete form, the filtering operation becomes in the time domain $\mathbf{y} = \mathbf{Q}\mathbf{x}$, where $\mathbf{Q} \equiv h(t_m - t_k)\, \Delta t$. Here, however, it is simpler to use a sampling representation in the frequency domain, as then the effect of the linear filter appears as a simple weighting of the frequency components of the input process, and the frequency equivalent of $\boldsymbol{T}_L$ is essentially a diagonal matrix, whose determinant is easily calculated. Equation (6.59) in conjunction with Eq. (6.69) gives the desired change in entropy rate for the steady-state processes.

To use Eq. (6.59), however, we must first show that entropy rate measured in the frequency domain is equal to that measured in the time domain if appropriate sampling in the frequency domain is employed. Equa-

† For geometric interpretations, see Shannon,[1] sec. 21, and Shannon,[14] secs. III–VII.

tion (4.30a), with a change of scale, gives us the desired form of the frequency components of $x_T(t)$ ($n = 2[BT] \gg 1$), for if we let $S'_x(f_m)_T = (2B/T)^{1/2} S_x(f_m)_T$ ($f_m = m/T$), then Eqs. (4.30a), (4.30b) can be expressed as the pair of transformations

$$(\mathbf{S}'_x)_T = \mathbf{A}\mathbf{x}_T \qquad \mathbf{x}_T = \mathbf{A}^{-1}(\mathbf{S}'_x)_T \qquad \mathbf{A} = \left[\frac{e^{-\pi i m k/BT}}{\sqrt{2BT}}\right]$$
$$m, k = \left(-\frac{n}{2}, \ldots, \frac{n}{2}\right) \tag{6.72}$$

and the $n \times n$ matrix $\mathbf{A}$ is essentially unitary† for $2BT \gg 1$ (Prob. 6.8a). This unitary matrix, in turn, can be resolved into a pair of orthogonal matrices associated, respectively, with the real and imaginary parts of $(\mathbf{S}'_x)_T = \mathbf{A}\mathbf{x}_T$, in the manner of Eqs. (6.74), (6.75), so that from Eq. (6.59) it follows that the entropy for this *approximate* representation of the original process $x(t)$ in $(-T/2,T/2)$ is unchanged when measured in the frequency domain, if $(2B/T)^{1/2} S_x(f_m)_T$ is taken as the sample value.‡ It is left for the reader to show that for ergodic processes the entropy rate is also the same when measured in either domain (Prob. 6.8b).

The frequency equivalent of $\mathbf{y} = \mathbf{Q}\mathbf{x}$ in the time domain is

$$(\mathbf{S}'_y)_T = [Y(i\omega_m)\delta_{mk}](\mathbf{S}'_x)_T \tag{6.73}$$

to a good approximation, provided that $n = 2[BT] \gg 1$. Writing $(\mathbf{S}'_y)_T = \mathbf{C}_y + i\mathbf{D}_y$, $(\mathbf{S}'_x)_T = \mathbf{C}_x + i\mathbf{D}_x$, and $\mathbf{Y} = \mathbf{A} + i\mathbf{B}$, and letting $\mathbf{a}_m$ be the 2×2 matrix

$$\mathbf{a}_m = \begin{bmatrix} A_m & -B_m \\ B_m & A_m \end{bmatrix} \tag{6.74}$$

we can reexpress (6.73) as a column matrix of $n = 2[BT]$ elements:

$$\begin{bmatrix} \mathbf{C}_y \\ \mathbf{D}_y \end{bmatrix} = [\mathbf{a}_m \delta_{mk}]\mathbf{S}''_x \qquad \mathbf{S}''_x = \begin{bmatrix} C_{x1} \\ D_{x1} \\ C_{x2} \\ D_{x2} \\ \cdot \\ \cdot \\ \cdot \\ C_{x,[BT]} \\ D_{x,[BT]} \end{bmatrix} \tag{6.75}$$

† Namely, with the property that $\tilde{\mathbf{A}}\mathbf{A}^* = \mathbf{I}$ and $A_{ij} = A^*_{ji}$.

‡ $S'_x(f_m)_T = S_x(f_m)_T(2B/T)^{1/2}$ is, of course, a random variable, considered over the ensemble, and has finite moments for $T < \infty$, unlike the amplitude spectrum of the process in $(-\infty \le t \le \infty)$ (cf. Sec. 3.2). Note that the intensity spectrum of $x(t)$ is here $w_x(f) = \lim_{T\to\infty} E[2|S_x(f)_T|^2/T]$, as required [cf. Eq. (3.47)].

Applying Eq. (6.59) again then gives for the entropy change

$$\Delta H = \log |\det [\mathbf{a}_m \delta_{mk}]| = \log \left[\prod_{m=1}^{[BT]} (A_m{}^2 + B_m{}^2) \right]$$
$$= \log \left[\prod_{m=1}^{[BT]} |Y(i\omega_m)|^2 \right] \tag{6.76}$$

But since $2BT \gg 1$, this can be written approximately

$$\Delta H = T \sum_{m=1}^{[BT]} [\log |Y(i\omega_m)|^2] \Big/ T \doteq T \int_0^B \log |Y(i\omega)|^2 \, df \tag{6.77}$$

and since $n = [2BT]$, the change ΔH_R in entropy rate due to linear filtering becomes finally, in the limit $T \to \infty$, from Eq. (6.69),

$$\Delta H_R = \frac{1}{2B} \int_0^B \log |Y(i\omega)|^2 \, df = H_R(y) - H_R(x) \tag{6.78}$$

From Eq. (6.71), we may also write this in terms of the entropy powers of the input and output processes from the filter, e.g.,

$$(P'_N)_y/(P'_N)_x = e^{2\Delta H_R} \qquad \text{or} \qquad \Delta H_R = \left(\frac{1}{2}\right) \log [(P'_N)_y/(P'_N)_x] \tag{6.78a}$$

For band-limited ergodic processes in spectral intervals other than $(0,B)$, this change in entropy rate is $(2B)^{-1} \int_{(B)} \log |Y(i\omega)|^2 \, df$, where (B) denotes the interval in question. Note that the change in entropy rate is independent of the detailed statistical properties of the input process and depends only on the linear filter used. Further, since $|Y|_{\max}$ is usually less than unity, it follows that ΔH_R represents a decrease in rate (cf. Prob. 6.9). For nonlinear filters, the problem is more involved, since the weighting in the frequency domain is no longer the "diagonal" operation implied by Eq. (6.73): intermodulation products appear as well, and a direct calculation using $\Delta H = -\boldsymbol{E}\{\log |J(\mathbf{x}/\mathbf{x}')|\}$ is probably the most profitable approach.

Just as entropy in physical systems tends to increase in the course of time, the reverse is true of information about an information source: as information about the source is processed, it tends to decrease, becoming more corrupt or noisy until it is eventually destroyed unless additional information is made available.† In fact, whenever information is processed, it decreases unless the transformations are measure-preserving, or nonsingular, and, in any case, it cannot increase once all contact with the source has been eliminated. For a proof of this, see Prob. 6.10.

† Total information in the corrupted message increases, of course, but most of this increase is for information about the accompanying noise, not about the desired message itself.

As an example, consider the communication system of Fig. 6.1. From the results of Prob. 6.10, it is easy to show that the average information in **u**, **x**, **y**, and **v**, respectively, about the message set **u** obeys the following relations,

$$I(\mathbf{u}|\mathbf{v}) \leq I(\mathbf{u}|\mathbf{y}) \leq I(\mathbf{u}|\mathbf{x}) \leq I(\mathbf{u}|\mathbf{u}) \tag{6.79}$$

and since $I(\mathbf{a}|\mathbf{b}) = I(\mathbf{b}|\mathbf{a})$ [cf. Eqs. (6.45), (6.54*c*)], we have also

$$I(\mathbf{x}|\mathbf{v}) \leq I(\mathbf{x}|\mathbf{y}) \leq I(\mathbf{x}|\mathbf{u}) \leq I(\mathbf{u}|\mathbf{u}) \tag{6.80}$$

The inequalities represent destruction of information as we go from source **u** to received message set **v**, where the loss of information is due to system noise and possibly inadequate coding operations. However, for reversible, or nonsingular, coding, we have $I(\mathbf{u}|\mathbf{x}) = I(\mathbf{x}|\mathbf{u}) = I(\mathbf{u}|\mathbf{u})$: no information is destroyed. Also, as we shall see in Sec. 6.5, under certain limiting conditions it is even possible, by transmitting at suitable rates from ergodic sources, for the equalities in Eqs. (6.79), (6.80) to hold throughout: information is not lost. We shall consider these important cases in more detail presently.

PROBLEMS

6.5 (*a*) A band-limited ergodic normal noise process, with zero mean, is passed through a zero-memory square-law device $y = \beta x^2$. Show how the entropy rate of the rectified output may be calculated, and discuss the analytical problems involved.

(*b*) If the noise at both input and output stages is sampled at the same rate and at intervals sufficiently far apart in time to remove all correlation between sampled values, show that the change in entropy rate between the *sampled* processes is

$$\Delta H_R = \log (2\sqrt{2}\,\beta\sigma) \qquad \text{nu/symbol} \tag{1}$$

and calculate the entropy rate for the output as well.

(*c*) Repeat (*b*) for a half-wave linear detector $y = \gamma x$ $(x > 0)$, $y = 0$ $(x \leq 0)$, and show that

$$\Delta H_R = \log (\gamma/2e^{1/4}) \qquad \text{nu/symbol} \tag{2}$$

Discuss the use of Eq. (6.57*b*) for Eqs. (1) and (2).

6.6 (*a*) Show, if $a_i > 0$, $\sum_i^n a_i = C$ $(i = 1, \ldots, n)$, that $\prod_i^n a_i$ is a maximum if $a_i = C/n$ (all i).

(*b*) Verify that the multivariate gauss d.d. (6.61*b*) maximizes the entropy of the process **x**, and carry out the details of the evaluation of $w(\mathbf{x})_{\max}$, $H_{\max}$ [Eqs. (6.61*b*), (6.63)]. Derive the corresponding results when $E\{\mathbf{x}\} = 0$.

6.7 (*a*) Under a *peak*, or *maximum value*, constraint $\int_{X_0} w(\mathbf{x})\,\mathbf{dx} = 1$ [which is also the normalizing condition for $w(x)$] show that the $w(\mathbf{x})$ which yields the greatest average prior uncertainty as to the **x** is (for all **x** in the finite region X_0)

$$w(\mathbf{x}) = \begin{cases} \dfrac{1}{X_0} & \mathbf{x} \in X_0 \\ 0 & \mathbf{x} \notin X_0 \end{cases} \tag{1}$$

and that $H(\mathbf{x})_{\max} = \log X_0 + H_0(\mathbf{x})$. Verify the maximum.

(b) Suppose that $\boldsymbol{E}\{\mathbf{x}\} = [a_i]$ (all $a_i > 0$). Show that the distribution density with the greatest prior uncertainty for $\mathbf{x}$ is

$$w(\mathbf{x})_{\max} = \exp\left(\gamma + \sum_i^n x_i/a_i\right) \qquad \gamma = \log \prod_{i=1}^n a_i(e^{d_i/a_i} - e^{c_i/a_i}) \tag{2}$$

where $c_1 < x_1 < d_1, \ldots, c_n < x_n < d_n$ are the boundaries of X_0 for $\mathbf{x}$.

When X_0 is the semi-infinite region $(0, \infty)$, verify that

$$w(\mathbf{x})_{\max} = \begin{cases} \prod_{i=1}^n a_i^{-x_i/a_i} & x_i > 0 \\ 0 & x_i < 0 \end{cases} \tag{3}$$

with $H_{\max} = \log\left(e^n \prod_i^n a_i\right) + H_0(\mathbf{x})$. From this, show, then, that the distribution of the (time-independent) amplitude A_0 (≥ 0) of a deterministic process, representing the greatest prior uncertainty as to the actual value of A_0, is

(1) A uniform distribution for a peak-power limitation, for $0 < A_0 < a$ ($< \infty$).

(2) The Rayleigh distribution [cf. Eq. (7.99)], $w(A_0) = 2A_0 e^{-A_0^2/\overline{A_0^2}}/\overline{A_0^2}$ ($A_0 \geq 0$), $w(A_0) = 0$ ($A_0 < 0$) for an average power limitation; $\overline{A_0^2}$ = constant; in terms of the signal-to-noise ratio $a_0 = A_0/(2\psi)^{1/2}$, this becomes $w(a_0) = a_0 e^{-a_0^2/2P_0}/P_0$, with $\overline{A_0^2} = 4\psi P$.

6.8 (a) Carry out the details of Eqs. (6.72) to (6.78), and show that $\mathbf{A}$ therein is approximately a unitary matrix.

(b) Verify that for band-limited ergodic processes entropy rate calculated in the time domain is the same as entropy rate determined by sampling in the frequency domain. Discuss the stochastic convergence of the operations involved.

6.9 If $\mathbf{z} = \mathbf{x} + \mathbf{y}$ is the sum of two statistically independent, ergodic random series, show that the following inequalities hold,

$$(P'_N)_x + (P'_N)_y \leq (P'_N)_z \leq P_x + P_y \tag{1}$$

where $(P'_N)_x$, P_x are, respectively, the entropy power and average power (or intensity) of x, etc. C. E. Shannon,[1] app. 6; M. S. Bartlett,[13] p. 215.

6.10 (a) Prove that, if $G(\mathbf{x}) \geq 0$, then

$$\int_{(x)} F(\mathbf{x}) \log G(\mathbf{x})\, d\mathbf{x}$$

is a maximum, subject to the constraint $\int G(\mathbf{x})\, d\mathbf{x} = 1$, provided that $F(\mathbf{x}) \equiv G(\mathbf{x})$ (all values of $\mathbf{x}$).

(b) Use this result (a) to show that, unless $w(\mathbf{u}|\mathbf{y}) \equiv w(\mathbf{u}|\mathbf{x})$ [Eqs. (6.79), (6.80)], information is always lost in the course of the operations implied by the communication system of Fig. 6.1. Note, also, that $w(\mathbf{u}|\mathbf{y}) \equiv w(\mathbf{u}|\mathbf{x})$ is equivalent to $w(\mathbf{x}|\mathbf{y}) \equiv w(\mathbf{x}|\mathbf{u},\mathbf{y})$, that is, although the transformation $y = \boldsymbol{T}_M^{(N)}\{\mathbf{x}\}$ is irreversible, no information about $\mathbf{u}$ is lost in passing from $\mathbf{u}$ to $\mathbf{y}$ if the above identities hold. P. M. Woodward,[5] sec. 3.7.

6.5 *Rates and Channel Capacity*

As we have already observed, information theory is essentially a theory of the communication process that sets limits on the rate of transmission over channels with given characteristics, where the main problem is to match a message source to the channel for best performance in some specified sense. Here "best performance" is taken to mean a maximum *rate* of transmission, consistent with certain standards of accuracy at the receiver and consistent with the constraint imposed by the channel ($\boldsymbol{T}_M^{(N)}$). From this viewpoint, and with the results of the preceding sections, we can now describe several very important theorems governing the transmission of information in a communication system and obtain further insight into the significance of entropy rates. [We shall not, however, consider coding procedures in any detail, nor shall we attempt rigorous proofs of the principal theorems. For these, the reader should consult the references (cf. Shannon,[1] Fano,[2] Feinstein[19]).]

Depending on the source and channel, there are four classes of interest: for discrete message sources, we have (1) the discrete noiseless channel, (2) the discrete channel with noise, and (3) the continuous channel with noise.† For continuous message sources, we shall (4) consider briefly the continuous noisy channel. In all instances, the channel $\boldsymbol{T}_M$ or $\boldsymbol{T}_M^{(N)}$ is assumed to have zero memory; i.e., the instantaneous signal output $y = \boldsymbol{T}_M^{(N)}\{x\}$ depends only on a corresponding instantaneous signal x (finite delays being permitted). Reception also is postulated to be the set of operations $\boldsymbol{T}_R$ which assigns to the received signal y a particular symbol or waveform for which y yields the greatest posterior probability.

6.5-1 Discrete Noiseless Channel. As the term indicates, "rate" measures the rapidity with which information from a source, now encoded as a signal, is being sent through a particular channel. The channel itself is simply a system of constraints, namely, the structure of the alphabet or waveforms that are used, available bandwidth, the physical nature of the propagation medium, the presence of noise, and so on. *Channel capacity C* is defined as the maximum rate at which information from any source can be transmitted, subject to one or more specific constraints. For the discrete channel, the capacity C is given by

$$C = \lim_{T \to \infty} \frac{\log n(T)}{T} \qquad \text{nu/sec} \tag{6.81}$$

where $n(T)$ is the number of possible signals of length T permitted by the channel constraints.‡

Now let us consider the discrete noiseless case. The symbols may be of different duration, and the number of possible sequences is constrained by the message source, as well as by the channel: for example, only one particular language is used, or the number of different symbols is limited, and so on. Since there is no noise here, the transmission operation $\boldsymbol{T}_T$ is revers-

† The terms "discrete" and "continuous" here refer to the allowed message or signal values at any given instant (cf. Secs. 6.2 to 6.4).

‡ This particular form for C follows, of course, from the logarithmic nature of the information measure (cf. Sec. 6.2-2).

ible and nonsingular, which we may also express statistically by the relation $p(\mathbf{x}_i|\mathbf{y}_i) = \delta_{ij}$, where $\mathbf{x}_i$ is the ith input sequence and $\mathbf{y}_i$ the corresponding output for this noiseless channel. How, then, should the source, of entropy rate H_R (natural units or bits per symbol), be matched to a discrete channel of capacity C (natural units or bits per second)? Clearly, a suitable encoding of the message source must be made. That it is possible to match source and channel in this way is the essence of Shannon's first fundamental theorem (theorem 9):

Theorem I. For a message source of entropy rate $H_R(\mathbf{u})$ (units per symbol) *and a noiseless channel of capacity C* (units per second), *it is possible to encode the source's output so as to transmit at an average rate as near $C/H_R(\mathbf{u})$* (symbols per second) *as desired. It is not possible to transmit at a greater rate.*

It is easy to see that $C/H_R(\mathbf{u})$ is the maximum average rate. For observe first that the entropy of a sequence set $\{u\}$ of length T is maximum when all the allowed sequences have equal probability. This maximum value is just $\log n(T)$, where $n(T)$ again is the number of permissible sequences. But the maximum entropy rate of the channel is its capacity C, and since the entropy and entropy rate for these discrete sequences are not changed by the reversible encoding operation $\boldsymbol{T}_T$ inasmuch as $p(\mathbf{u}) = p(\mathbf{x})$, $\mathbf{x} = \boldsymbol{T}_T\{\mathbf{u}\}$, the entropy rate $H_R(\mathbf{u})$ for source and channel must also be the same. Consequently, $H_R(\mathbf{u})$ for the message source cannot exceed C for the channel. (It is assumed that, although there may be finite delays introduced by the encoding process, no infinite time lags between message symbols and signal symbols can occur.)

The first part of the theorem may be proved in a number of ways, one of which is to demonstrate an actual coding procedure for matching source to channel, as was done, independently, by Shannon[1] and Fano.[2] It turns out, as we might expect, that suitable codes assign the shortest sequences of signal symbols to the most probable sequences of message symbols, and in a certain sense, suitable codes are most efficient codes. For details, see Shannon,[1] secs. 9 and 10. A general procedure for the most efficient encoding of finite messages has been given by Huffman.[15]

The *efficiency of coding* here is defined as

$$\eta \equiv H_R(\mathbf{u})/H_R(\mathbf{x})_{\max} \qquad 0 \leq \eta \leq 1 \tag{6.82}$$

where $H_R(\mathbf{x})_{\max}$ is the *alphabet capacity*, or maximum information rate that can be provided by the set of signal symbols, and, as before, $H_R(\mathbf{u})$ is the entropy rate of the original message source. If the coding alphabet consists of M different symbols, then $H_R(\mathbf{x})_{\max}$ is simply $\log M$ (natural units per symbol) [cf. Eq. (6.27)]. In a similar way, the *redundancy* of the message source is defined by

$$r \equiv 1 - \eta = 1 - H_R(\mathbf{u})/H_R(\mathbf{x})_{\max} \qquad 1 \geq r \geq 0 \tag{6.83}$$

English, for example, has a redundancy in the neighborhood of 75 per cent.[1]

Thus, the justification of the concept of entropy rate $H_R(\mathbf{u})$ as the average rate of generating information is given by the theorem above: $H_R(\mathbf{u})$ determines the channel capacity required for most efficient coding. Note that this result is independent of the receiver structure $\boldsymbol{T}_R$, since the channel is assumed to be noiseless (and memoryless). The receiver itself is simply the set of inverse operations $\boldsymbol{T}_T^{-1}\boldsymbol{T}_M^{-1}$.

6.5-2 Discrete Noisy Channel. While the noiseless channel gives us the significant interpretation of entropy and entropy rate of an information source, the situations of greatest practical importance occur when noise is one of the channel constraints. Thus, if $\mathbf{y}_j$ is a particular output sequence for the discrete channel and if $\mathbf{y}_j = \boldsymbol{T}\{\mathbf{x}_i\}$, corresponding to a discrete input signal sequence $\mathbf{x}_i$, we have $p(\mathbf{x}_i|\mathbf{y}_j) \neq 1$ (for all i, j), in contrast to the relation above, $p(\mathbf{x}_i|\mathbf{y}_j) = \delta_{ij}$, for noiseless cases. The noise, of course, has introduced ambiguities, so that the received message no longer bears a one-to-one relation with the (encoded) original. Channel capacity is still defined as a maximum rate of transmission under specified constraints, but now the effects of the noise must be specifically considered.

To see this, let us compute the average rate of transmission of information in sequences from an ergodic source: it is simply the average information gain at the receiver per symbol or per unit time. From Eq. (6.45), we have therefore for the entropy rate† of transmission through the channel

$$R \equiv \lim_{n\to\infty} \frac{I(X|Y)}{n} \text{ or } \lim_{T\to\infty} \frac{I(X|Y)}{T} = \lim_{n \text{ or } T\to\infty} \frac{H(X) - H(X|Y)}{n \text{ or } T} \tag{6.84a}$$

$$R = H_R(X) - H_R(X|Y) \qquad \text{nu/symbol or nu/sec} \tag{6.84b}$$

The first term represents the rate through $\boldsymbol{T}_M$ if there were no noise, while the second measures the rate† at which information about the original signal set is lost because of the actual noisy channel $\boldsymbol{T}_M^{(N)}$. Here $H_R(X|Y)$ is called the *equivocation:* it is the *rate* of uncertainty about X remaining when Y is received [cf. Eqs. (6.29) et seq.]. Channel capacity C now is defined as the maximum possible† rate of transmission, namely, the rate when the source is properly matched to this noisy channel. Accordingly, we have

$$\begin{aligned} C &\equiv \max_{p(X)} \{R\} = \max_{p(X)} \{H_R(X) - H_R(X|Y)\} \\ &= \lim_{T\to\infty} \max_{p(X)} \frac{1}{T} \boldsymbol{E} \left\{ \log \frac{p(\mathbf{x}|\mathbf{y})}{p(\mathbf{x})} \right\} \qquad \text{nu/symbol or nu/sec} \end{aligned} \tag{6.85}$$

where the maximization is over the ensemble of all possible information

† Whenever we talk about rates here and henceforth in this chapter, we mean *average* rates in the sense that the average is taken over the ensemble of allowed transmitted and received signal representations. Since these processes are assumed to be ergodic in the limit, average rate is also the time average of the rate for a representative member of the ensemble (cf. Sec. 1.6).

sources that can be used as inputs to the channel† (Prob. 6.12). Again, matching of message source to channel is to be achieved by a suitable coding procedure.‡ As before, we can define a *coding efficiency* η_c by

$$\eta_c \equiv H_R(\mathbf{u})/C \tag{6.86}$$

Note, however, that, because of the noise, η and η_c are different§ and that for finite sequences there is a nonzero error probability in the reception of the original signal, that is, $p(\mathbf{x}_i|\mathbf{y}_j) > 0$ (all i, j).

The justification of this definition [Eq. (6.85)] of capacity for the discrete noisy channel is embodied in Shannon's[1] second fundamental theorem (his theorem 11):

Theorem II. A discrete noisy channel has the capacity C [cf. Eq. (6.85)] (units per second) *and is fed by a signal source with entropy rate* $H_R(X)$ (units per second).

1. *If* $H_R(X) \leq C$, *there exists at least one coding procedure which permits the output of the message source to be transmitted over the channel with arbitrarily small error rate (or equivocation).*

2. *If* $H_R(X) > C$, *the message source can be encoded so that the equivocation* $H_R(X|Y)$ *is less than* $H_R(X) - C + \epsilon$, *where* ϵ *is arbitrarily small.*

3. *There is no method of encoding which yields an equivocation* $H_R(X|Y) < H_R(X) - C$.

For a proof of this theorem, see Shannon[1] (secs. 13, 14) and McMillan.[18] A more detailed demonstration has been given by Feinstein,[19] who has shown, in addition, the much stronger result that the posterior uncertainty $H(X|Y)$ (or equivocation per sequence), as well as the equivocation rate $H_R(X|Y)$ [$= \lim_{T\to\infty} H(X|Y)/T$, etc.], vanishes absolutely as the sequence length becomes indefinitely great, i.e., as T (or n) $\to \infty$, with proper matching of source to channel. Proper matching, or encoding, in effect creates a sufficient redundancy in the signal so that the noise is unable to obliterate any of the desired information. Individual symbols may be garbled, but over the infinite interval enough will be interpreted correctly to permit an errorless reconstruction of the original message. Theorem II is essentially an existence statement: no practical method of actually achieving the ideal

† It is important to distinguish the order of the operations as $T \to \infty$ [cf. Eq. (6.85)]. We start always with a truncated process, sampled in the interval $(-T/2, T/2)$, maximize (or minimize) over the appropriate distributions of the $2[BT]$ variables $\mathbf{x}$, $\mathbf{y}$, etc., and then pass to the limit of the steady-state ergodic process. This operation is abbreviated by $p(X)$ or $w(X)$, etc., as indicated in Eq. (6.85) and subsequent relations; in detail, we have always lim $T \to \infty$ following maximization.

‡ It has been shown recently by Kelly,[16] with generalizations by Bellman and Kalaba[17] using the methods of dynamic programming, that the concept of transmission rate can still be significant even though no coding is actually used.

§ If there is no noise, $H_R(X|Y) = 0$ and $C = \max_X H_R(X) = H_R(x)_{\max}$ (natural units per second) and Theorem I applies, with $\eta_c = \eta$, also.

coding implied by the theorem has as yet been found, although a number of useful error-correcting codes† have been developed with this ultimate intent. Finally, in this limiting situation of infinite coding time, the receiver is again simply the set of inverse transformations $\boldsymbol{T}_R^{(N)} = \boldsymbol{T}_T^{-1}\boldsymbol{T}_M^{(N)-1}$: the receiver computes the maximum a posteriori probability of the input sequence $\mathbf{x}$ on the basis of the received sequence $\mathbf{y}$, a probability which by Theorem II is always unity for ideal coding.‡

6.5-3 The Continuous Noisy Channel. Here, as before, the message source remains discrete, but now encoding is carried out in terms of continuous waveforms. As we have seen in Sec. 6.4-3, these may be handled by appropriate sampling in the time domain. Furthermore, since the channel is noisy, the effective information content of the signal process still remains finite (for finite sequences), although the signal values are represented by a continuum. Noise and the physical limitations of actual equipment prevent an infinite information rate. Rate is expressed once more by Eqs. (6.84), but, in place of Eq. (6.45), Eqs. (6.54) are used. We have accordingly

$$\begin{aligned} R &\equiv \lim_{T\to\infty} \frac{1}{T} I(X|Y) \\ &= \lim_{T\to\infty} \frac{1}{T} \int_{(\mathbf{x})} \mathbf{dx} \int_{(\mathbf{y})} \mathbf{dy}\, w(\mathbf{x},\mathbf{y}) \log \frac{w(\mathbf{x}|\mathbf{y})}{w(\mathbf{x})} \qquad \text{nu/sec} \end{aligned} \tag{6.87}$$

for the average rate of transmissions of information through the continuous noisy channel, when, as $T \to \infty$, the sequences $\mathbf{x}$, $\mathbf{y}$ become indefinitely long. Thus, in terms of Eqs. (6.84), the average rate for such sequences is

$$R = H_R(X) - H_R(X|Y) \qquad \text{nu/sec} \tag{6.88}$$

where, of course, probability densities replace probabilities in the expressions for the average posterior information in Y about X, for $I(X|Y)$.

The capacity of the channel is once more the maximum average rate with respect to all possible inputs, so that Eq. (6.85) becomes for this continuous case

$$C \equiv \max_{w(X)} \{R\} = \max_{w(X)} \{H_R(X) - H_R(X|Y)\} \tag{6.89a}$$

$$C = \lim_{T\to\infty} \max_{w(\mathbf{x})} \frac{1}{T} \boldsymbol{E}\left\{\log \frac{w(\mathbf{x}|\mathbf{y})}{w(\mathbf{x})}\right\} \qquad \text{nu/sec} \tag{6.89b}$$

Shannon's second fundamental theorem (Theorem II) still applies, but with Eqs. (6.89), rather than Eq. (6.85), as the measure of channel capacity.

† The subject of error-correcting codes is rather extensive. The reader is referred, for example, to the work of Hamming,[20] Gilbert,[21] Laemmel,[22] Reed,[23] Golay,[24] Silverman and Balser,[25] Elias,[26] and Slepian[27] (see the bibliography at the end of this book). For the effects of a feedback channel, see also the papers by Chang.[28] For basic work on convolutional codes, see Wozencraft.[40]

‡ In effect, this is the justification of the existence of the inverse transformations $\boldsymbol{T}_T^{-1}\{\boldsymbol{T}_M^{(N)}\}^{-1}$.

[For a discussion of some of the problems involved in a rigorous treatment of the continuous channel (and continuous sources), the reader is referred to recent work of Kolmogoroff[29] and Khintchine.[39]]

A simple case of considerable interest arises if the signal and noise processes are independent and additive. Then, we have $y = x + n$, where n is the channel noise, and $w(\mathbf{y}|\mathbf{x}) = w_n(\mathbf{y} - \mathbf{x})$, with w_n the d.d. of the noise series $\mathbf{n}$. The rate [Eq. (6.88)] reduces to

$$\begin{aligned} R &= H_R(X) - H_R(X|Y) = H_R(X) + H_R(Y) - H_R(X,Y) \\ &= H_R(Y) - H_R(N) \qquad \text{nu/sec} \end{aligned} \tag{6.90}$$

inasmuch as $H(X,Y) = H(X,N) = H(X) + H(N)$, since X, N are independent. Accordingly, the capacity [Eqs. (6.89)] becomes

$$C = \max_{w(Y)} \{H_R(Y) - H_R(N)\} \qquad \text{nu/sec} \tag{6.91}$$

again because N is independent of X. Maximization over the ensemble Y of received signals is carried out subject to the constraints on the transmitted signals X [since $\mathbf{y}$ is a function of $\mathbf{x}$; see also the definition, Eqs. (6.89)].

If this additive channel noise N is band-limited, white, and gaussian (as well as ergodic), we can use Eq. (6.91) and the results of Secs. 6.4-2, 6.4-3 to find the capacity of the communication system and determine the statistical structure of the signal set needed for matching source to channel, as indicated by Theorem II, when the signal process obeys a constraint of constant average power (or intensity). Now it follows directly from Eq. (6.67) that the maximum entropy rate for the received signals, Y, occurs if the original signal process is also band-limited gaussian noise with a uniform intensity spectrum of the same width as the channel noise. If $P_Y = P_{S+N}$ is the average intensity of this received sum of normal noise waves and P_N is the intensity of the channel noise, Eq. (6.67) applied to Eq. (6.91) gives (in different units)

$$C = \log (2\pi e P_Y)^{1/2} - \log (2\pi e P_N)^{1/2} = \tfrac{1}{2} \log (1 + P_S/P_N) \qquad \text{nu/symbol} \tag{6.92}$$

And since the independent sampled values of the input signal are generated at a rate $2B$ symbols/sec [see Eq. (6.70a) and Secs. 4.2-1 and 6.4-3], we obtain at once the well-known expression for the capacity in this case,

$$C = B \log (1 + P_S/P_N) = B \log (1 + \overline{S^2}/\overline{N^2}) \qquad \text{nu/sec} \tag{6.93}$$

For very long times, for example, $2BT \gg 1$, we have, for the total information transmitted through this noisy channel, approximately

$$I_T(X|Y)_{\max} \doteq TC = BT \log (1 + \overline{S^2}/\overline{N^2}) \qquad \text{nu} \tag{6.94}$$

Equation (6.93) shows explicitly how signal power, noise power, and bandwidth may be interchanged in this limiting situation of continuous transmission with gaussian ergodic signals and noise. For example, suppose that the channel noise is essentially that of the input stages of a physical receiver, which, as we shall see in Chap. 11, is white and gaussian, but with an effectively infinite bandwidth (before filtering). For finite signal intensity ($\overline{S^2} < \infty$), we see then that the capacity C becomes

$$\lim_{B\to\infty} C \doteq \lim_{B\to\infty} B\overline{S^2}/\overline{N^2} = \overline{S^2}/W_{0N} \qquad \overline{S^2} < \infty \tag{6.95}$$

The capacity depends only on the ratio of signal intensity to average noise intensity per unit bandwidth. Equation (6.95) sets the limit on information rate for any system operating against band-limited white gaussian noise subject to average signal-power limitations. Actual systems, of course, do not realize the rate C, since they employ different signal statistics, and since they may operate in situations where the dominant component of the interfering noise is not normal, e.g., impulse noise, static, etc. If the channel noise is not normal, while the signal process remains so, the capacity of the channel can be shown (Prob. 6.13) to be bounded by the inequalities

$$B \log (1 + P_S/P'_N) \leq C \leq B \log (P_N/P'_N + P_S/P'_N) \qquad \text{nu/sec} \tag{6.96}$$

where P'_N is the entropy power of the new channel noise (also of bandwidth B, but not necessarily white and normal). A calculation of the capacities of some typical systems under various signal and noise conditions has been given by a number of investigators.[30,31]

In the case of band-limited, but nonwhite, normal noise processes, with a spectrum $\mathcal{W}_N(f)$, one obtains for the capacity[14]

$$C = \int_0^B \log [1 + \mathcal{W}_S(f)/\mathcal{W}_N(f)]\, df \qquad \text{nu/sec} \tag{6.97a}$$

and the spectral distribution of signal intensity must obey

$$\mathcal{W}_S(f) + \mathcal{W}_N(f) = \begin{cases} A(>0) & 0 < f < B \\ 0 & f > B \end{cases} \tag{6.97b}$$

where A is adjusted to ensure that the total average signal intensity is P_S. The signal, of course, is band-limited and gaussian, but with a nonuniform spectrum, determined, according to Eq. (6.97b), by the spectral distribution of the noise (Prob. 6.15).

We remark, finally, that as in the discrete case (Sec. 6.5-2) the receiver computes the maximum posterior probability of the transmitted signal,† which in this limit of infinite time (i.e., infinite encoding delay) is always unity as long as message source and channel are properly matched: there is then no error (or equivocation) in reception, and $\mathbf{T}_R = \mathbf{T}_T^{-1}\{\mathbf{T}_M{}^{(N)}\}^{-1}$, as before.

† See also the papers by Shannon,[14] Rice,[32] and Fano.[33]

6.5-4 Continuous Noisy Channel with a Continuous Message Source. In all the above situations, we have postulated that the message source is discrete (in value), so that the rate of generation of information, i.e., the entropy of the source [cf. Eq. (6.48) and Sec. 6.4-1] remains finite. However, if the source is continuous, it can assume (at any instant) any one of an effectively infinite number of values† and accordingly requires an infinite number of natural units (or bits) for a precise representation. A channel of infinite capacity would then be needed to effect an exact recovery. Since actual channels have noise (as well as limitations on bandwidth, etc.), their capacity is finite and exact transmission is impossible. In practice, this difficulty is mitigated by the fact that we are interested in performance of a certain finite "quality" only. Exact reproduction is not required; instead, reproduction to within a certain preset tolerance is sufficient. Thus, at the receiver we may choose any one of an (infinite) number of evaluation (or cost) functions to measure the fidelity with which reception is carried out. For example, a cost function of the type $C(x,y) = C_0[y(t) - x(t)]^2$ is a common choice. The *quality* $\mathcal{C}_0$ is then taken to be the expected value of this quantity, e.g., the average cost

$$\mathcal{C}_0 \equiv E_{x,y}\{C(\mathbf{x},\mathbf{y})\} = C_0 \iint_{[\mathbf{x}],[\mathbf{y}]} (\mathbf{y} - \mathbf{x})^2 w(\mathbf{x},\mathbf{y})\, d\mathbf{x}\, d\mathbf{y} \tag{6.98}$$

with C_0 = cost/amplitude² [where, as usual, to apply the statistical average we have replaced $x(t)$, $y(t)$ by an appropriately sampled $\mathbf{x}$, $\mathbf{y}$ in time; for the continuous limit, see the discussion in Chap. 21].

Let us now define the rate R_c of generation of information from a continuous source under a fidelity requirement like the above. From Eq. (6.87), we set [remembering that $w(\mathbf{x}|\mathbf{y}) = w(\mathbf{x},\mathbf{y})/w(\mathbf{y})$]

$$R_c \equiv \min_{w(Y|X)} \{R\} = \lim_{T\to\infty} \min_{w(\mathbf{y}|\mathbf{x})} \frac{1}{T} E\left\{\log \frac{w(\mathbf{x},\mathbf{y})}{w(\mathbf{x})w(\mathbf{y})}\right\} \qquad \text{nu/sec} \tag{6.99}$$

subject to the constraint that the *rate* of average cost remains fixed at $R_0 \equiv \lim_{T\to\infty} \{\mathcal{C}_0/T\}$ (cost per second). Thus, the rate of generating information at the source for a given quality rate R_0 of reproduction is defined as the *maximum* of the rate R, where $w(\mathbf{y}|\mathbf{x})$ is varied [cf. Eq. (6.87) and footnote on page 320]. The justification of this is the following theorem (Shannon,[1] theorem 21):

Theorem III. 1. *If a signal source has the rate R_c for a given quality rate R_0 at the receiver, it is possible to encode the* (continuous) *message source to have the signal rate R_c and transmit the encoded message set over a channel of capacity C with quality rate as near R_c as desired, provided that $R_c \le C$.*

† In physical systems, of course, the apparent continuum becomes discrete once more at quantum levels [cf. remarks following Eqs. (6.51)]. For a discussion of the mathematical problems implied by strictly continuous sources, see Kolmogoroff.[29]

2. *It is not possible to do this if $R_c > C$.*

(The details of the proof are similar to those for the second fundamental theorem; cf. Shannon.[1])

Note that capacity [Eqs. (6.89)] and rate here [Eq. (6.99)] are somewhat similar: depending on the strictness or looseness of the quality rate, one can obtain a signal source rate R_c equal to the channel capacity C *and* transmit over the channel at this rate. Of course, if the demands on fidelity are too loose, R_c will exceed C and transmission at this rate is not possible, while if a too precise signal reproduction is required, R_c may be much smaller than the available channel capacity, as we would expect. We remark also that the variational procedure implied in Eq. (6.99) may be carried out, not only for fixed R_0, but for other constraints, such as constant average intensity of the message source, and so on (Prob. 6.16). Within the limitations of the fidelity requirement, reception is defined as error-free, provided that $R_c \leq C$. The receiver itself is a device which simply assigns to the received process $\mathbf{y}$ one of the set of $\mathbf{x}$'s which fall within the quality limitations. For the square-distance function in Eq. (6.98), this means simply that the receiver decides message $\mathbf{x}$ was sent if $C_0(\mathbf{y} - \mathbf{x})^2 \leq R_0(T)$ for the particular (long) finite sequence, or in the limit $T \to \infty$, if $\lim_{T\to\infty} [C_0(\mathbf{y} - \mathbf{x})^2/T] \leq R_0$ for the particular infinite sequence received. Strictly speaking, of course, there is always a small probability of error as to the actual message sent, even as $T \to \infty$, since R_0 must exceed zero for actual operation of the system.

6.5-5 Discussion. In matching an information source to a channel of specified capacity, a suitable coding procedure must be used. (See also the footnote on page 320.) This in turn requires an infinite delay, i.e., a mapping of a finite message into a signal of infinite duration, if errorless reception is to be possible. In actual practice, of course, infinite times are not available, and so reception is accompanied always by a nonzero error rate: it is not strictly possible in physical systems to transmit without error in a noisy channel over finite time intervals.† Since the error probabilities are now different from zero, we have a decision situation, in which a correct vs. an incorrect judgment must be made at the receiver in terms of some appropriate criterion of values, or costs, and consequently the methods of statistical decision theory may be applied.‡

† In certain ideal cases (involving discrete sources and channels) where some of the input signal symbols are *nonadjacent*, i.e., where the transition probabilities $p(x_i|y_j)$ of input symbol x_i being received as output symbol y_j vanish for some (i,j), it is possible to code messages of finite length into signals of finite length and still transmit at a positive rate *with zero error probability*. The least upper bound of all rates which can be realized with zero error probability is called the *zero error capacity* C_0 of the channel; C_0 is always less than, or at most equal to, the capacity C of the channel (defined in the usual way for signal processes of infinite duration) (Shannon[34]).

‡ The application of statistical decision theory to communication problems, and particularly to those involving reception, is considered in Part 4 (see also Middleton and Van Meter[35]).

That decision and cost assignments now enter follows at once from the fact that the error probabilities here are nonvanishing: a nonzero *error* probability always implies (1) a decision process and (2) a numerical cost or value assignment to the possible decision. The unit of cost is essentially irrelevant, but the *relative* amounts associated with the possible decisions are not.

The postulated receiver (i.e., one that maximizes a posteriori probability) no longer presents the comparatively simple problem that it does in the case of infinite signal durations, where proper encoding yields errorless reception (the limiting extreme of decision theory, when the error probabilities vanish). Now it is *not* true that $\boldsymbol{T}_R{}^{(N)} = \boldsymbol{T}_T{}^{-1}\{\boldsymbol{T}_M{}^{(N)}\}^{-1}$: the receiver, operationally, is not the inverse transformation representing encoding and the effects of the medium, since these are irreversible functions, without inverses. From the receiver's viewpoint, we usually want to minimize the error rate, or, equivalently, the average cost of decision. This is a somewhat arbitrary procedure, as the choice of cost function, $C(\mathbf{x},\mathbf{y})$, is left open to us. For example, we may use a suitable cost function to obtain the maximum posterior probability estimate of the original signal. Or if a quadratic distance function, like Eq. (6.98), is selected instead, we may obtain another form of optimum estimate (still minimizing average cost). In some cases, depending on the noise and signal statistics, two different cost functions will lead to the same form of optimum receiver (cf. Sec. 21.5).

On the other hand, at the transmitter the emphasis from the viewpoint of information theory has been to maximize rate of transmission, subject, of course, to the channel constraints, the finite duration of the encoded message, and (for finite T) as small an error rate as possible compatible with these demands. We can, and often do, of course, limit optimum performance to that of the transmitter or the receiver alone. But clearly an optimized communication system *as a whole* will require the simultaneous adjustment of the encoding and receiver operations: neither can be pushed to an extreme without affecting the other adversely and leading to a deterioration of over-all performance. Since rate is also at a premium, it is natural to assign a cost to operations that yield insufficient or excessive rates vis-à-vis the channel. Then, one way of formulating this more general communication problem is, in essence, to assign this coding cost in such a fashion that when the message source is matched to the channel $\boldsymbol{T}_M{}^{(N)}$ for errorless reception, i.e., as $T \to \infty$, this cost vanishes. At the same time, we choose another cost function for the receiver and determine the receiver structure for which the average cost of decision in reception is least. *A criterion now is to choose the statistics of the input signal set, or alphabet, so as to minimize the sum of the coding and reception costs*, subject, for example, to the usual average or peak power constraints on the input signal ensemble.†

† Note, however, that, as $T \to \infty$, with best coding the coding cost vanishes and moreover so does the cost of wrong decision in reception, since no incorrect decisions are now

Further reduction of the total average cost of communication may often be achieved by selecting specific alphabet or signal waveforms themselves (with the above statistics), e.g., pulses versus cw FM, or FM versus AM, etc. (For additional discussion, see Sec. 23.4.)

In the general case above, performance for the finite interval will depend on the choice of the two cost functions and, of course, on the allowed duration T of the signal sequences. It is evident that there is a rather broad element of choice open to us through these cost functions and the way we may combine them to measure the total cost of communication. It is also true that the procedures of information theory are not necessarily pertinent to such problems as the detection and estimation of weak signals in noise (for finite observation times), where again the nonvanishing error probabilities (and the value judgments they imply) are controlling elements. For these problems, the methods of decision theory provide an appropriate apparatus (cf. Part 4). Finally, we may mention among important problems as yet unresolved by information theory (1) the question of an explicit nonrandomized encoding process for the message ensemble in the case of discrete and continuous noisy channels, where errorless transmission is desired (in the limit $T \to \infty$), (2) the general treatment of the finite signal cases ($T < \infty$), and (3) the problem of channels with nonzero memory.

The present chapter concludes Part 1. In Part 2, we consider the properties of the normal random process and a few of the salient statistical features of channel and system noise. Part 3 is devoted largely to the second-moment theory of amplitude and angle modulation, while Part 4 describes a general theory of reception.

PROBLEMS

6.11 A signal $\mathbf{x}$ of finite duration is sent through a noisy channel. A waveform $\mathbf{y}$ is received, and a decision "yes"—a message is present—or "no"—only noise is present—is then made. The a priori probabilities of noise, and of signal and of noise, are q and p $(= 1 - q)$, respectively. If α and β represent the probabilities of the two kinds of decision error that can occur, i.e., if α is the conditional probability that, if there is no signal, a signal (as well as noise) is decided, with β the conditional error probability for the inverse situation of a signal (and noise), which is called noise alone, show that the information loss that arises when a decision as to presence or absence of the message is made is given[36] by

$$H(\mathbf{x}|\mathbf{y}) = H(\mathbf{x}) - \Big[p(1-\beta)\log\frac{1-\beta}{p(1-\beta)+\alpha q} + p\beta\log\frac{\beta}{p\beta+q(1-\alpha)} + q\alpha\log\frac{\alpha}{p(1-\beta)+q\alpha} + q(1-\alpha)\log\frac{1-\alpha}{p\beta+q(1-\alpha)}\Big] \quad (1)$$

What is $H(\mathbf{x})$?

---

made. Furthermore, from the transmitter's viewpoint it is a matter of indifference which cost function *at the receiver* we may select. All in the limit $T \to \infty$ result in errorless reception (by the second fundamental theorem, Sec. 6.5-2, etc.), and our problem is no longer one within the purlieus of decision theory.

6.12 In the case of the discrete noisy, but memoryless, channel, show that:

(*a*) Independence of the input signal symbols $\mathbf{x}$ in an ergodic sequence is needed if the output sequence $\mathbf{y}$ is to supply maximum information about $\mathbf{x}$, that is,

$$C = \max_{p(\mathbf{x})} I_R(X|Y) \qquad \text{with } p(\mathbf{y}|\mathbf{x})\text{'s fixed} \tag{1}$$

For example, $p(\mathbf{x}) = \prod_i p_i(x_i)$, and I_R is the average rate of information gain ($I_R \equiv I/n$, n = sequence length).

(*b*) To achieve the maximum transmission rate implied by Eq. (1), find the $p(x_i)$'s, given the $p(y_j|x_i)$'s; that is, code the messages in terms of symbol sequences $\mathbf{x}$ in such a way that the x_i's occur with probability $p(x_i)$ and so that the (original) word length is a minimum *and* the error probability is minimized also (for finite sequences). Consider the simple case that X and Y both have the same number of symbol values M, and show that, for this, we get

$$p(y_j) = \frac{Q(y_j)}{\sum_Y Q(y_j)} \qquad \text{each } y_j \tag{2}$$

where the $p(x)$'s are determined from the M equations

$$p(y_j) = \sum_X p(x_i)p(y_j|x_i) \qquad p(y_j|x_i) \text{ known a priori}$$

Here $Q(y)$ is defined by the M equations

$$\sum_Y p(y_j|x_i) \log p(y_j|x_i) = \sum_Y p(y_j|x_i) \log Q(y_j) \tag{3}$$

An extensive treatment of this maximization problem is given by Muroga.[37]

(*c*) For the transition probabilities given below for $M = 2$,

	y_1	y_2
x_1	p	q
x_2	q	p

$$p + q = 1, \; p \neq q$$

show that the capacity of the channel is $C = 1 + p \log p + q \log q$. Thus, to maximize the received information, the x's should be independent and transmitted with certain special frequencies of occurrence.

6.13 Show that the capacity of a channel with bandwidth B containing noise of arbitrary statistics (but from an ergodic process) is bounded by the inequalities

$$B \log (1 + P_S/P'_N) \leq C \leq B \log (P_N/P'_N + P_S/P'_N) \qquad \text{nu/sec} \tag{1}$$

where P'_N is the entropy power of the noise. C. E. Shannon,[1] theorem 18.

6.14 (*a*) Show that the channel capacity C for a signal ensemble and white gaussian noise of average intensity P_N, both band-limited in $(0,B)$, is bounded by

$$C \geq B \log \left(\frac{2}{\pi e} \frac{\hat{P}_S}{P_N} \right) \tag{1}$$

where $\hat{P}_S$ is the *peak*, or maximum, allowed signal intensity.

(*b*) Show that for sufficiently large values of $\hat{P}_S/P_N$

$$C \leq B \log \frac{\frac{2}{\pi e} \hat{P}_S + P_N}{(1+\epsilon)P_N} \qquad \epsilon \to 0 \tag{2}$$

(*c*) Show that as $\hat{P}_S/P_N \to 0$, B fixed as before, we have

$$C \to B \log (1 + \hat{P}_S/P_N) \tag{3}$$

C. E. Shannon,[1] sec. 26.

6.15 (*a*) Show that the capacity of a channel of bandwidth B, in which the noise is gaussian and ergodic, but not spectrally uniform, becomes

$$C = \int_0^B \log [1 + \mathcal{W}_S(f)/\mathcal{W}_N(f)]\, df = B \log (P_N/P_N'' + P_S/P_N'') \tag{1}$$

with

$$P_N'' = \exp \left\{ B^{-1} \int_0^B \log [B\mathcal{W}_N(f)]\, df \right\} \tag{2}$$

the geometric mean noise power, provided that the signal is also normal noise, in the same band, with an intensity spectrum that obeys the condition

$$\mathcal{W}_S(f) + \mathcal{W}_N(f) = \begin{cases} A(>0) & 0 < f < B \\ 0 & f > B \end{cases} \tag{3}$$

Here A is adjusted to maintain fixed the average total intensity P_S of the signal process.[14]

(*b*) What happens to $\mathcal{W}_S(f)$ when $\mathcal{W}_N(f) = 0$ over some finite interval in $(0,B)$? Show that $P_N'' \geq P_N = \int_0^B \mathcal{W}_N(f)\, df$.

6.16 (*a*) If the cost function $C(\mathbf{x},\mathbf{y})$ is the square-distance function $(\mathbf{y} - \mathbf{x})^2$, measuring the discrepancy between the received signal set $\mathbf{y}$ and coded message set $\mathbf{x}$, and if $\mathbf{u}$, the continuous message ensemble, is an ergodic band-limited white gaussian noise process, show that the average rate of transmission R_c, for a quality rate $\lim_{T\to\infty} \mathcal{C}_0/T = \lambda\overline{(y - x)^2} = R_0$ (λ measured in cost per amperes² per second), becomes

$$R_c = B_1 \log (P_M/P_Q) \qquad \text{nu/sec}$$

Here P_M is the average intensity of the message source, of bandwidth B_1, and P_Q is the average intensity $\overline{(y - x)^2} = R_0/\lambda$.

(*b*) Show that R_c for an ergodic source of any statistics and bandwidth B is bounded by

$$B_1 \log (P_M'/P_M) \leq R_c \leq B_1 \log (P_M/P_Q)$$

where P_M' is the entropy power of the message source and $P_Q = R_0/\lambda$, as above.

See Shannon,[1] sec. 29.

REFERENCES

1. Shannon, C. E.: A Mathematical Theory of Communication, *Bell System Tech. J.*, **27**: 379, 623 (1948). See also C. E. Shannon and W. Weaver, "The Mathematical Theory of Communication," University of Illinois Press, Urbana, Ill., 1949.
2. Fano, R.: The Transmission of Information. I, II, *MIT Research Lab. Electronics Tech. Repts.* 65, 1949; 149, 1950.
3. Woodward, P. M., and I. L. Davies: Information Theory and Inverse Probability in Telecommunications, *J. IEE (London)*, **99** (pt. III): 58 (1952).
4. For example, D. Gabor, Theory of Communication, *J. IEE (London)*, **93** (pt. III): 429 (1946), and A Summary of Communication Theory, *Proc. London Symposium*, 1952, in W. Jackson (ed.), "Communication Theory," Academic Press, New York, 1953; E. C. Cherry, A History of the Theory of Information, *Proc. London Symposium*, 1950, pp. 22–43, 167, 168, and *J. IEE (London)*, **98** (pt. III): 383 (1951).
5. Woodward, P. M.: "Probability and Information Theory, with Applications to Radar," McGraw-Hill, New York, 1953.
6. Goldman, S.: "Information Theory," Prentice-Hall, Englewood Cliffs, N.J., 1953.
7. Bell, D. A.: "Information Theory and Its Engineering Applications," Pitman, New York, 1953.
8. Brillouin, L.: "Science and Information Theory," Academic Press, New York, 1956.
9. Cherry, E. C.: "On Human Communication," Technology Press, Cambridge, Mass., and Wiley, New York, 1957.
10. Stumpers, F. L. H. M.: A Bibliography of Information Theory, Communication Theory, Cybernetics, *IRE Trans. on Inform. Theory*, vol. PGIT-2, November, 1953; suppl., *ibid.*, **IT-1**: 31 (September, 1955).
11. Wiener, N.: "Cybernetics," Wiley, New York, 1948, and "Extrapolation, Interpolation, and Smoothing of Stationary Time Series," Wiley, New York, 1950.
12. Blanc-LaPierre, A., and R. Fortet: "Théorie des fonctions aléatoires," chaps. VI, VII, Masson, Paris, 1953.
13. Bartlett, M. S.: "An Introduction to Stochastic Processes," chaps. 2, 3, Cambridge, New York, 1955.
14. Shannon, C. E.: Communication in the Presence of Noise, *Proc. IRE*, **37**: 10 (1949).
15. Huffman, D. A.: A Method for the Construction of Minimum Redundancy Codes, in Willis Jackson (ed.), "Communication Theory," pp. 102–110, Academic Press, New York, 1953.
16. Kelly, J. L., Jr.: A New Interpretation of Information Rate, *Bell System Tech. J.*, **35**: 917 (1956).
17. Bellman, R., and R. E. Kalaba: On the Role of Dynamic Programming in Statistical Communication Theory, *IRE Trans. on Inform. Theory*, **IT-3**: 197 (September, 1957).
18. McMillan, B.: The Basic Theorems of Information Theory, *Ann. Math. Stat.*, **24**: 196 (1952).
19. Feinstein, A.: A New Basic Theorem of Information Theory, *IRE Trans. on Inform. Theory, Symposium*, vol. PGIT-4, September, 1954, sec. III, p. 9.
20. Hamming, R. W.: Error Detecting and Error Correcting Codes, *Bell System Tech. J.*, **29**: 147 (1950).
21. Gilbert, E. N.: A Comparison of Signalling Alphabets, *Bell System Tech. J.*, **31**: 504 (1952).
22. Laemmel, A. E.: Efficiency of Noise Reducing Codes, in W. Jackson (ed.), "Communication Theory," p. 111, Academic Press, New York, 1953.
23. Reed, I. S.: A Class of Multiple-error-correcting Codes and Decoding Scheme, *IRE Trans. on Inform. Theory, Symposium*, **PGIT-4**: 38 (September, 1954).

24. Golay, M. J. E.: Binary Coding, *IRE Trans. on Inform. Theory, Symposium,* **PGIT-4:** 23 (September, 1954).
25. Silverman, R. A., and M. Balser: Coding for Constant-data-rate Systems, *IRE Trans. on Inform. Theory, Symposium,* **PGIT-4:** 50 (September, 1954); *Proc. IRE,* **42:** 1428 (1954), **43:** 728 (1955).
26. Elias, P.: Error-free Coding, *IRE Trans. on Inform. Theory, Symposium,* **PGIT-4:** 29 (September, 1954); Coding for Noisy Channels, *IRE Conv. Record,* pt. 4, p. 37, 1955; Coding for Two Noisy Channels, 3d London Symposium, in E. C. Cherry (ed.), "Information Theory," Academic Press, New York, 1956.
27. Slepian, D.: A Class of Binary Signalling Alphabets, *Bell System Tech. J.,* **35:** 203 (1956).
28. Chang, S. S. L.: Theory of Information Feedback Systems, *IRE Trans. on Inform. Theory, Symposium,* **II-2:** 29 (September, 1956).
29. Kolmogoroff, A. N.: On the Shannon Theory of Information Transmission in the Case of Continuous Signals, *IRE Trans. on Inform. Theory,* **IT-2:** 102 (December, 1956).
30. Oliver, B. M., J. R. Pierce, and C. E. Shannon: The Philosophy of PCM, *Proc. IRE,* **36:** 1324 (1948).
31. Jelonek, Z.: A Comparison of Transmission Systems, in W. Jackson (ed.), "Communication Theory," p. 44, Academic Press, New York, 1953. See Ref. 10 for other pertinent papers.
32. Rice, S. O.: Communication in the Presence of Noise—Probability of Error of Two Encoding Schemes, *Bell System Tech. J.,* **29:** 60 (1950).
33. Fano, R.: Communication in the Presence of Additive Gaussian Noise, in W. Jackson (ed.), "Communication Theory," p. 169, Academic Press, New York, 1953.
34. Shannon, C. E.: The Zero Error Capacity of a Noisy Channel, *IRE Trans. on Inform. Theory, Symposium,* **IT-2**(3): 8 (September, 1956).
35. Middleton, D., and D. Van Meter: Detection and Extraction of Signals in Noise from the Point of View of Statistical Decision Theory, *J. Soc. Ind. Appl. Math.,* **3:** 192 (1955), **4:** 86 (1956).
36. Middleton, D.: Information Loss Attending the Decision Operation in Binary Detection, *J. Appl. Phys.,* **25:** 127 (1954).
37. Muroga, S.: On the Capacity of a Discrete Channel, *J. Phys. Soc. Japan,* **8:** 484 (1953).
38. Feinstein, A.: "Foundation of Information Theory," McGraw-Hill, New York, 1958.
39. Khintchine, A. I.: "Mathematical Foundations of Information Theory," trans. by R. A. Silverman and M. D. Friedman, Dover, New York, 1957.
40. Wozencraft, J. M.: Sequential Decoding for Reliable Communication, *IRE Conv. Record,* pt. 2, pp. 11–25, 1957.

PART 2

RANDOM NOISE PROCESSES

CHAPTER 7

THE NORMAL RANDOM PROCESS: GAUSSIAN VARIATES

A large class of random processes which arise in the various stages of a typical communication link possesses *normal* or *gaussian* statistics. By this we mean that, if $x(t)$ represents an ensemble of noise voltages [implied in certain situations by (N) in $\boldsymbol{T}_R^{(N)}$, etc.; cf. Sec. 2.1-3(1)], then the set of random variables $x_1 = x(t_1)$, $x_2 = x(t_2)$, . . . , $x_n = x(t_n)$ obeys a *normal law*, provided that the d.d. of x_1, . . . , x_n is given by an expression of the type

$$W(x_1, \ldots ,x_n) = [(2\pi)^n \det \mathbf{K}_x]^{-1/2} e^{-1/2 \tilde{\mathbf{x}} \mathbf{K}_x^{-1} \mathbf{x}} \qquad n \geq 1 \tag{7.1}$$

Here $\mathbf{x}$ is a (column) vector whose elements are x_j $(j = 1, \ldots , n)$, $\tilde{\mathbf{x}}$ is the transposed vector, and $\mathbf{K}_x = [K_x(t_j,t_k)] = [\boldsymbol{E}\{x(t_j)x(t_k)\}]$ is the $n \times n$ covariance matrix† of the process $x(t)$, at the times t_j, t_k $(j, k = 1, \ldots , n)$. Normal random variables and processes have already been introduced *ad hoc* in Part 1 (cf. Chaps. 1, 5, 6) for purposes of illustration. However, in view of the importance of such processes in applications, we shall present a more systematic treatment in the present chapter. We begin, therefore, by examining the discrete-parameter situation, where t in $x(t)$ can take only selected discrete values, and then extend the analysis, in Chap. 8, to include the gauss functional, where now t can assume a continuum of values in a finite or infinite interval.

7.1 *The Normal Random Variable*[1]

Before we describe the normal random process, let us consider first the case of a single real gaussian variable x. Here x may be obtained, for example, by sampling a suitable random process $x(t)$ at some fixed instant t_1, so that $x = x_1 = x(t_1)$. We say then that x_1 is a random variable generated from a continuous random process, where the parameter $t = t_1$ itself may take any fixed value in a specified range. In particular, if x has the d.d.‡

$$W(x) = \frac{e^{-(x-\nu)^2/2\sigma_x^2}}{\sqrt{2\pi\sigma_x^2}} \; [= W_1(x_1;t_1)] \tag{7.2}$$

† It is assumed for the moment that $\boldsymbol{E}\{\mathbf{x}\} = \bar{\mathbf{x}} = 0$; for $\boldsymbol{E}\{\mathbf{x}\} \neq 0$, see Sec. 7.3.

‡ In this section, for convenience we drop the subscripts 1, referring to the time $t = t_1$, on x_1, ν_1, etc.

then x is called a *normal*, or *gaussian, random variable.* The associated characteristic function is found at once from Eqs. (1.25), (1.30) to be

$$F_x(i\xi) \equiv E\{e^{i\xi x}\} = \mathfrak{F}\{W(x)\} = e^{i\nu\xi - \sigma_x{}^2\xi^2/2} \tag{7.3}$$

while the distribution of x follows from the definition (1.8) and is

$$D(x) = \int_{-\infty}^{x} W(x)\,dx = \frac{1}{2}\left[1 + \Theta\left(\frac{x-\nu}{\sigma_x\sqrt{2}}\right)\right] \tag{7.4}$$

where $\Theta(x) \equiv (2/\sqrt{\pi})\int_0^x e^{-u^2}\,du$ is called the *error function.*† For large values of the argument, the distribution may be found with the aid of the asymptotic development

$$\Theta(y) \simeq 1 - \frac{e^{-y^2}}{y\sqrt{\pi}}\sum_{m=0}^{\infty}\frac{(-1)^m(2m)!}{2^m m!(2y^2)^m} \tag{7.5}$$

and the fact that $\Theta(y) = -\Theta(-y)$.

Differentiation of the characteristic function (7.3) shows at once that $\nu = \bar{x}$ and $\sigma_x{}^2 = \overline{x^2} - \bar{x}^2$, respectively the mean and variance of x (cf. Secs. 1.2-5, 1.2-6), while the nth moment of x is

$$\overline{x^n} = n!\sum_{k=0}^{\frac{n}{2},\frac{n-1}{2}} \nu^{n-2k}\sigma_x{}^{2k}/2^k k!(n-2k)! \qquad n \geq 0 \tag{7.6}$$

The upper limits on the summation apply according as n is even or odd. From this, it is immediately evident that all odd moments ($n = 1, 3, 5, \ldots$) vanish if the mean ν vanishes, whereas for the even moments ($n = 0, 2, 4, \ldots$) we have the simple relation

$$\overline{x^{2n}} = (2n)!\sigma_x{}^{2n}/2^n n! \qquad \nu = \bar{x} = 0 \tag{7.7}$$

The semi-invariants (or cumulants) of this distribution may be obtained directly from Eq. (7.3) and the definition (1.35), viz.,

$$\log F_x(i\xi) \equiv \sum_{m=1}^{\infty}\frac{(i\xi)^m}{m!}\lambda_m = i\nu\xi - \tfrac{1}{2}\sigma_x{}^2\xi^2 \tag{7.8a}$$

and therefore, $\lambda_1 = \nu \qquad \lambda_2 = \sigma_x{}^2 \qquad \lambda_m = 0 \qquad m \geq 3$ (7.8*b*)

We remark, finally, that, since the characteristic function and its distribution density are uniquely related,‡ we could equally well choose $F_x(i\xi)$ as the defining relation for the normal random variable x. Furthermore,

† This function and related derivatives are tabulated. See, for example, "Tables of the Error Function and Its First Twenty Derivatives," Annals Computation Laboratory, Harvard University Press, Cambridge, Mass., 1950; also see Eq. (A.1.1).

‡ See Cramér,[1] sec. 10.3.

since $F_x(i\xi)$ is here differentiable to all orders at $\xi = 0$, all the moments $\overline{x^n}$ likewise exist and we might use Eq. (7.6), in conjunction with Eqs. (1.31), as still another, alternative definition (cf. Prob. 7.1).

PROBLEM

7.1 Establish Eq. (7.6) for $\overline{x^n}$ in the case of a normal random variable, and conversely show how this result may be used to define a normal random variable. What is $\lim_{\sigma_x \to 0} W(x)$ [Eq. (7.2)]?

7.2 *The Bivariate Normal Distribution*†

Instead of considering a single normal random variable, we can extend the preceding remarks to the more general situation involving a pair of (real) random variables x_1 and x_2. Again, we may suppose that x_1 and x_2 are obtained by observation of a suitable random process $x(t)$ at the two times t_1, t_2. Then x_1 and x_2 are said to be normal or gaussian random variables if they possess the *bivariate* d.d.

$$W(x_1,x_2) = (2\pi\sigma_1\sigma_2\sqrt{1-\rho^2})^{-1} \times \exp\left\{-\left[\left(\frac{x_1-\nu_1}{\sigma_1}\right)^2 + \left(\frac{x_2-\nu_2}{\sigma_2}\right)^2 - 2\rho\left(\frac{x_1-\nu_1}{\sigma_1}\right)\left(\frac{x_2-\nu_2}{\sigma_2}\right)\right] \Big/ 2(1-\rho^2)\right\} \tag{7.9}$$

with the associated distribution

$$D(x_1,x_2) = \int_{-\infty}^{x_1} du \int_{-\infty}^{x_2} dv\, W(u,v) \tag{7.9a}$$

With the help of the transformation $z = (x_2 - \nu_2)/\sigma_2 - \rho(x_1 - \nu_1)/\sigma_1$, we readily find from Eq. (1.40*a*) that the characteristic function for x_1, x_2 is

$$\begin{aligned} F_{x_1,x_2}(i\xi_1,i\xi_2) &\equiv E\{\exp\,(i\xi_1x_1 + i\xi_2x_2)\} \\ &= \exp\,[i\xi_1\nu_1 + i\xi_2\nu_2 - \tfrac{1}{2}(\xi_1^2\sigma_1^2 + \xi_2^2\sigma_2^2 + 2\sigma_1\sigma_2\rho\xi_1\xi_2)] \end{aligned} \tag{7.10}$$

The significance of the parameters ν_1, ν_2, σ_1, σ_2, and ρ is now easily established. We calculate the various first- and second-order moments of x_1 and x_2 by differentiating F_{x_1,x_2} according to Eq. (1.41), setting $\xi_1 = \xi_2 = 0$. The results are

$$\nu_1 = \bar{x}_1 \qquad \nu_2 = \bar{x}_2 \qquad \sigma_1^2 = \overline{x_1^2} - \bar{x}_1^2 \qquad \sigma_2^2 = \overline{x_2^2} - \bar{x}_2^2 \tag{7.11a}$$

and

$$\rho = \overline{(x_1 - \bar{x}_1)(x_2 - \bar{x}_2)}/\sigma_1\sigma_2 \tag{7.11b}$$

so that once again the σ's are the variances of the two random variables, the ν's their mean values, while ρ is the *correlation coefficient*, or normalized covariance, of x_1 and x_2 [cf. Eqs. (1.52)]. From Eqs. (7.11), it is clear that $\sigma_{1,2} \geq 0$, $|\rho| \leq 1$.

† See *ibid.*, sec. 21.12, for example.

If the random variables x_1 and x_2 are obtained, as suggested above, by sampling a random process at two distinct times t_1 and t_2, ρ can be written $\rho_x(t_1,t_2)$, which is now the normalized covariance function of the process $x(t)$. Likewise, the d.d. $W(x_1,x_2)$ becomes the second-order probability density $W_2(x_1,t_1;x_2,t_2)$ associated with $x(t)$ [see Eq. (1.70)]. Frequently, the mean values $\bar{x}_1$ and $\bar{x}_2$ vanish, so that $\sigma_1^2 = \overline{x_1^2}$, $\sigma_2^2 = \overline{x_2^2}$, and $\rho_x(t_1,t_2)\sigma_1\sigma_2 = M_x(t_1,t_2) = \overline{x_1x_2}$ is the second-moment function of the process as well.

When $x(t)$ is at least wide-sense stationary, so that $\rho_x(t_1,t_2) = \rho_x(|t_2 - t_1|)$ and $\sigma_1^2 = \sigma_2^2 \equiv \psi$, we obtain from the W-K theorem (3.42) the equivalent expression

$$\begin{aligned} K_x(t_1 - t_2) &= \psi\rho_x(t_1 - t_2) = \tfrac{1}{2}\mathfrak{F}^{-1}\{\mathcal{W}_x\} \\ &= \int_0^\infty \mathcal{W}_x(f) \cos \omega(t_1 - t_2)\, df\ [= K_x(t_2 - t_1)] \end{aligned} \tag{7.12}$$

for the covariance function, in terms of the spectral density of the process.

When $\bar{x}_1 = \bar{x}_2 = 0$, the distribution density and characteristic function (7.9), (7.10) assume in this stationary case the somewhat simpler forms (cf. Prob. 1.8)

$$W_2(x_1,x_2;t) = \frac{\exp[-(x_1^2 + x_2^2 - 2\rho_x x_1 x_2)/2\psi(1 - \rho_x^2)]}{2\pi\psi\sqrt{1 - \rho_x^2}} \qquad t = t_2 - t_1 \tag{7.13a}$$

and

$$F_2(i\xi_1,i\xi_2;t) = \exp[-\tfrac{1}{2}\psi(\xi_1^2 + \xi_2^2 + 2\rho_x\xi_1\xi_2)] \qquad \psi = \overline{x_1^2} = \overline{x_2^2} \tag{7.13b}$$

In matrix notation, we can write Eqs. (7.13) more compactly as

$$W_2(\mathbf{x};t) = (2\pi)^{-1}(\det \mathbf{K}_x)^{-\frac{1}{2}}e^{-\frac{1}{2}\tilde{\mathbf{x}}\mathbf{K}_x^{-1}\mathbf{x}} \qquad F_2(i\xi;t) = e^{-\frac{1}{2}\tilde{\xi}\mathbf{K}_x\xi} \tag{7.14a}$$

where $\mathbf{x}$ and ξ are column vectors and $\mathbf{K}_x$ is the covariance matrix

$$\mathbf{K}_x = \psi\begin{bmatrix} 1 & \rho_x \\ \rho_x & 1 \end{bmatrix} \qquad \text{with } \det \mathbf{K}_x = \psi^2(1 - \rho_x^2)$$

and therefore (7.14b)

$$\mathbf{K}_x^{-1} = \psi^{-1}(1 - \rho_x^2)^{-1}\begin{bmatrix} 1 & -\rho_x \\ -\rho_x & 1 \end{bmatrix}$$

The cumulant function $\mathcal{K}(i\xi)$ [Eq. (1.60)] is simply $-\frac{1}{2}\tilde{\xi}\mathbf{K}_x\xi$, so that the semi-invariants are

$$\begin{aligned} \lambda_{20}^{(2)} &= \lambda_{02}^{(2)} = \psi \\ \lambda_{11}^{(2)} &= \rho_x\psi = K_x(t) \\ \lambda_{mn}^{(2)} &= 0 \qquad m + n \geq 3 \\ \text{and} \qquad \lambda_{10}^{(2)} &= \lambda_{01}^{(2)} = 0 \end{aligned} \tag{7.15}$$

For the joint moments $\overline{x_1^m x_2^n}$, see Prob. 7.3.

We observe, finally, that if ρ_x vanishes, the joint d.d. (7.13a) factors into a pair of first-order densities, viz.,

$$\lim_{|t|\to\infty} W_2(x_1,x_2;t) = \frac{e^{-x_1^2/2\psi}}{\sqrt{2\pi\psi}}\frac{e^{-x_2^2/2\psi}}{\sqrt{2\pi\psi}} = W_1(x_1)W_1(x_2) \tag{7.16a}$$

with

$$\lim_{|t|\to\infty} F_2(i\xi_1,i\xi_2;t) = e^{-\frac{1}{2}\psi\xi_1^2}e^{-\frac{1}{2}\psi\xi_2^2} = F_1(i\xi_1)F_1(i\xi_2) \tag{7.16b}$$

Under this condition, the normal random variables x_1 and x_2 are *statistically independent*† (cf. Sec. 1.2-10). Conditional d.d.'s, moments, and characteristic functions may be obtained in a similar way (cf. Sec. 1.2-9). A typical conditional density for $x = x_2$, at time t (after) $x = x_1$ at t_1, for a given x_1 is‡

$$W_2(x_1|x_2,t) = (2\pi\psi\sqrt{1-\rho_x^2})^{-1}\, e^{-(x_2-\rho_x x_1)^2/2\psi(1-\rho_x^2)} \tag{7.17a}$$

with the associated c.f.

$$F_2(i\xi_2|x_1) = e^{i\xi_2\rho_x x_1 - \psi(1-\rho_x^2)\xi_2^2/2} = \mathfrak{F}\{W_2(x_1|x_2;t)\} \tag{7.17b}$$

(see Prob. 1.8).

PROBLEMS

7.2 (*a*) Verify that the bivariate density $W_2(x_1,x_2;t)$ has its maximum value at $x_1 = x_2 = 0$. Show also that the maximum of $W_2(x_1,x_2)$ [Eq. (7.9)] occurs at the point (ν_1,ν_2).

(*b*) If the exponent of $W_2(x_1,x_2;t)$ defines an ellipse of constant probability, for example, $\iint_{\text{inside ellipse}} W_2\, dx_1\, dx_2 = F$ $(F < 1)$, show that the area of the annulus between ellipses corresponding to $(-\text{exponent } W_2) = a$ and $a + da$ is $4\pi\psi\sqrt{1-\rho_x^2}\, a\, da$ and hence that $\iint_{\text{outside ellipse}} W_2\, dx_1\, dx_2 = e^{-a^2}$. Verify that this is true also of the general bivariate form (7.9).

(*c*) Show that $\lim_{\rho\to\pm 1} W(x_1,x_2) = W(x_1)\delta(x_2 - x_1)$

7.3 (*a*) Show that the moments of $W_2(x_1,x_2;t)$ [Eq. (7.13*a*)] are [for m, n integers (≥ 0)]

$$\overline{x_1^m x_2^n} = \begin{cases} 0 & m+n \text{ odd} \\ \dfrac{(2\psi)^{(m+n)/2}}{\sqrt{\pi}}\,\Gamma\left(\dfrac{m+n+1}{2}\right)\, {}_2F_1\left(\dfrac{-m}{2},\dfrac{-n}{2};\dfrac{1-m-n}{2};1-\rho_x^2\right) & m, n \text{ even} \\ \dfrac{(2\psi)^{(m+n)/2}}{\sqrt{\pi}}\,\Gamma\left(\dfrac{m+n+1}{2}\right)\,\rho_x\, {}_2F_1\left(\dfrac{1-m}{2},\dfrac{1-n}{2};\dfrac{-1-m-n}{2};1-\rho_x^2\right) & m, n \text{ odd} \end{cases} \tag{1}$$

† Note, incidentally, from Eqs. (7.9), (7.10), that, if $\bar{x}_1 = \bar{x}_2 \neq 0$, x_1 and x_2 are still statistically independent as $|t| \to \infty$ but that, while "memory," or coherence (cf. Sec. 1.4-1), as measured by $\overline{(x_1 - \bar{x}_1)(x_2 - \bar{x}_2)}$, between the random components of the observations at t_1, t_2, vanishes, that between the deterministic components persists unweakened, for example, $\lim_{|t|\to\infty} \overline{x_1x_2} = \bar{x}_1\bar{x}_2 \neq 0$. Thus, it is not necessary (though it is sufficient) that memory, or coherence, between observed values of a process go to zero in the limit to ensure statistical independence of gaussian random variables.

‡ We use the alternative notation of Sec. 1.4-1 here.

or, equivalently, that

$$
\overline{x_1{}^m x_2{}^n} = \begin{cases} \dfrac{(2\psi)^{(m+n)/2}}{\pi}\,\Gamma\left(\dfrac{m+1}{2}\right)\Gamma\left(\dfrac{n+1}{2}\right){}_2F_1\left(-\dfrac{m}{2},-\dfrac{n}{2};\dfrac{1}{2};\rho_x{}^2\right) & m,\ n \text{ both even} \\ \dfrac{2(2\psi)^{(m+n)/2}}{\pi}\,\rho_x\Gamma\left(1+\dfrac{m}{2}\right)\Gamma\left(1+\dfrac{n}{2}\right){}_2F_1\left(\dfrac{1-m}{2},\dfrac{1-n}{2};\dfrac{3}{2};\rho_x{}^2\right) & m,\ n \text{ both odd} \end{cases} \tag{2}
$$

(Here ${}_2F_1$ is a hypergeometric function; see Appendix 1.)

(*b*) Verify that

$$
\overline{x_1{}^m x_2{}^n} = \begin{cases} 0 & m+n \text{ odd} \\ \displaystyle\sum_{\text{all pairs}} \prod \overline{x_i x_j} = \text{Eqs. (1), (2)} & m+n \text{ even} \end{cases}
$$

7.4 Obtain the following integral results, useful in subsequent applications (cf. Chaps. 12, 13):

(*a*)
$$\iint_0^\infty dx_1\,dx_2\,e^{-(x_1{}^2+2ax_1x_2+x_2{}^2)} = \tfrac{1}{2}\phi \csc \phi \qquad \cos\phi = a \qquad 0 \le \phi \le \pi$$

(*b*)
$$\iint_0^\infty x_1x_2\,dx_1\,dx_2\,e^{-(x_1{}^2+2ax_1x_2+x_2{}^2)} = \tfrac{1}{4}\csc^2\phi(1-\phi\cot\phi)$$

(*c*)
$$\iint_0^\infty dx_1\,dx_2\,x_2e^{-(x_1{}^2+2ax_1x_2+x_2{}^2)} = \frac{\sqrt{\pi}}{4}(1+a)^{-1}$$

(*d*)
$$\iint_0^\infty x_1{}^\mu x_2{}^\nu e^{-(x_1{}^2+2ax_1x_2+x_2{}^2)}\,dx_1\,dx_2$$
$$= \frac{1}{4}\sum_{l=0}^{\infty}\frac{(-2a)^l}{l!}\Gamma\left(\frac{\mu+l+1}{2}\right)\Gamma\left(\frac{\nu+l+1}{2}\right) \qquad \text{Re}\,(\mu,\nu) > -1$$

S. O. Rice, *Bell System Tech. J.*, **24:** 46 (1945), sec. 3.5.

7.3 *The Multivariate Normal Distribution*†

The extension to the case of n real normal random variables $x_1, \ldots, x_n$ is now readily made. The generalization of Eq. (7.1) to include the situation $E\{x_j\} \neq 0$ is

$$W(x_1, \ldots, x_n) \equiv W(\mathbf{x}) = \frac{\exp\left[-\tfrac{1}{2}(\tilde{\mathbf{x}}-\bar{\tilde{\mathbf{x}}})\mathbf{K}_x{}^{-1}(\mathbf{x}-\bar{\mathbf{x}})\right]}{(2\pi)^{n/2}(\det \mathbf{K}_x)^{1/2}} \tag{7.18a}$$

$$= (2\pi)^{-n/2}(\det \mathbf{K}_x)^{-1/2} \times \exp\left[-\tfrac{1}{2}\sum_{jk}^{n}(x_j-\bar{x}_j)K_x{}^{jk}(x_k-\bar{x}_k)/\det \mathbf{K}_x\right] \tag{7.18b}$$

where $\mathbf{K}_x$ is the covariance matrix $[E\{(x_j-\bar{x}_j)(x_k-\bar{x}_k)\}] = (\overline{(\mathbf{x}-\bar{\mathbf{x}})(\tilde{\mathbf{x}}-\bar{\tilde{\mathbf{x}}})})$,

† See Cramér,[1] secs. 24.1, 24.2. For singular cases, see also sec. 24.3.

which (except in certain singular cases†) possesses an inverse $\mathbf{K}_x^{-1} = [K_x^{jk}/\det \mathbf{K}_x]$. As before, $\det \mathbf{K}_x$ is the determinant of $\mathbf{K}_x$, while K_x^{jk} is the cofactor‡ of $(K_x)_{jk}$ in $\mathbf{K}_x$. The relations (7.18) define a *multivariate* normal or gaussian distribution density for the n random variables $x_1, \ldots, x_n$; the corresponding multivariate distribution follows from Eq. (1.54).

7.3-1 The Characteristic Function. From Eqs. (7.18), we may apply Eqs. (1.56) in principle to get the desired characteristic function $F_{\mathbf{x}}(i\boldsymbol{\xi}) = F_{x_1,\ldots,x_n}(i\xi_1, \ldots, i\xi_n)$. However, to determine $F_{\mathbf{x}}(i\boldsymbol{\xi})$ explicitly, and for many later applications (cf. Chaps. 17, 20, etc.), we shall need to evaluate the following important integral,

$$I(\boldsymbol{\xi}) \equiv \int_{-\infty}^{\infty} \cdots \int e^{i\tilde{\boldsymbol{\xi}}\mathbf{u} - \frac{1}{2}\tilde{\mathbf{u}}\mathbf{A}\mathbf{u}}\, d\mathbf{u} \qquad d\mathbf{u} = du_1\, du_2 \cdots du_n \tag{7.19}$$

where $\mathbf{A}$ is a real $n \times n$ symmetric matrix, with an inverse $\mathbf{A}^{-1}$, such that $\tilde{\mathbf{u}}\mathbf{A}\mathbf{u}$ is positive definite. Here $\mathbf{u}$ is real, while $\boldsymbol{\xi}$ may have complex components.

We proceed as follows: First we observe that, since $\mathbf{A}$ is a (real) symmetric matrix, we can always find an orthogonal transformation $\mathbf{Q}$ (for example, $\mathbf{Q}\tilde{\mathbf{Q}} = \mathbf{I}$, where $\mathbf{I}$ is the unit matrix $[\delta_{jk}]$) which diagonalizes $\mathbf{A}$ and by means of which the quadratic form $\tilde{\mathbf{u}}\mathbf{A}\mathbf{u}$ can be reduced to the sum of squares.§ Accordingly, if we introduce the transformation $\mathbf{v} = \mathbf{Q}^{-1}\mathbf{u}$, so that $\mathbf{u} = \mathbf{Q}\mathbf{v}$, we see that $\tilde{\mathbf{u}}\mathbf{A}\mathbf{u}$ becomes $\tilde{\mathbf{v}}\mathbf{Q}\mathbf{A}\mathbf{Q}\mathbf{v} = \tilde{\mathbf{v}}\mathbf{Q}^{-1}\mathbf{A}\mathbf{Q}\mathbf{v}$, since $\mathbf{Q}^{-1} = \tilde{\mathbf{Q}}$ here, also. But $\mathbf{Q}$ is chosen to diagonalize $\mathbf{A}$, that is,

$$\mathbf{Q}^{-1}\mathbf{A}\mathbf{Q} = \boldsymbol{\Lambda} = [\lambda_j\delta_{jk}] \tag{7.20a}$$

where the λ_j $(j = 1, \ldots, n)$ are the eigenvalues of $\mathbf{A}$ and δ_{jk} is the Kronecker delta $\delta_{jk} = 0$ $(j \neq k)$, $\delta_{jk} = 1$ $(j = k)$. From this, we have, moreover,

$$\mathbf{A}\mathbf{Q} = \mathbf{Q}\boldsymbol{\Lambda} \tag{7.20b}$$

so that if we let the (column) vectors $\mathbf{f}_j = [f_{(\ )j}]$ be the eigenvectors of $\mathbf{A}$, corresponding to the eigenvalues λ_j, the diagonalizing matrix $\mathbf{Q}$ itself is simply the $n \times n$ matrix formed by the $\mathbf{f}_j$ as columns,¶ for example, $\mathbf{Q} = [\mathbf{f}_1, \mathbf{f}_2, \ldots, \mathbf{f}_n]$. Accordingly, if we take the lth row and jth column of Eq. (7.20b), we may write

$$\sum_k A_{lk} f_{kj} = \lambda_j f_{lj} \qquad j, k = 1, \ldots, n$$

or, in vector form,

$$\mathbf{A}\mathbf{f}_j = \lambda_j \mathbf{f}_j \qquad j = 1, \ldots, n \tag{7.21}$$

† See Cramér,[1] secs. 24.1, 24.2. For singular cases, see also sec. 24.3.

‡ For a discussion of properties of matrices, see, for example, Margenau and Murphy[2] or Frazer, Duncan, and Collar;[3] also see Cramér,[1] chap. 11.

§ Margenau and Murphy, *op. cit.*, secs. 10.16 to 10.19.

¶ We could equally well use row vectors.

The eigenvalues λ_j are first found by solving the secular equation

$$\det (\mathbf{A} - \lambda \mathbf{I}) = 0 \tag{7.22}$$

while the eigenvectors are then obtained from the n linearly independent relations $\mathbf{A}\mathbf{f}_j = \lambda_j \mathbf{f}_j$ [cf. Eqs. (7.21)], subject to the condition of orthonormalization†

$$\sum_k^n Q_{kl} Q_{kj} = \delta_{lj} \qquad \text{or} \qquad \sum_k^n f_{kl} f_{kj} = \delta_{lj} \tag{7.23}$$

Returning to the integral $I(\boldsymbol{\xi})$ [Eq. (7.19)], we see that‡ $\tilde{\boldsymbol{\xi}}\mathbf{u} = \tilde{\boldsymbol{\xi}}\mathbf{Q}\mathbf{v} = \tilde{\boldsymbol{\zeta}}\mathbf{v}$, where $\boldsymbol{\zeta} = \mathbf{Q}^{-1}\boldsymbol{\xi}$. The jacobian of the transformation $\mathbf{v} = \mathbf{Q}^{-1}\mathbf{u}$ is unity, since $\mathbf{Q}$ is orthogonal, and we can write therefore

$$\begin{aligned} I(\boldsymbol{\xi}) &= \int_{-\infty}^{\infty} \cdots \int e^{i\tilde{\boldsymbol{\zeta}}\mathbf{v} - \frac{1}{2}\tilde{\mathbf{v}}\boldsymbol{\Lambda}\mathbf{v}} \, d\mathbf{v} = \prod_{k=1}^{n} \int_{-\infty}^{\infty} e^{i\zeta_k v_k - \frac{1}{2} v_k{}^2 \lambda_k} \, dv_k \\ &= \frac{(2\pi)^{n/2}}{(\det \boldsymbol{\Lambda})^{\frac{1}{2}}} e^{-\frac{1}{2}\tilde{\boldsymbol{\zeta}}\boldsymbol{\Lambda}^{-1}\boldsymbol{\zeta}} \qquad \det \boldsymbol{\Lambda} = \Pi_k \lambda_k \end{aligned} \tag{7.24}$$

But inasmuch as $\det \boldsymbol{\Lambda} = \det \mathbf{A}$, from Eq. (7.22), and since $\mathbf{Q}^{-1} = \tilde{\mathbf{Q}}$, with $\boldsymbol{\Lambda} = \mathbf{Q}^{-1}\mathbf{A}\mathbf{Q}$, it follows that

$$\boldsymbol{\Lambda}^{-1} = \tilde{\mathbf{Q}}\mathbf{A}^{-1}\mathbf{Q}$$

and therefore

$$\tilde{\boldsymbol{\zeta}}\boldsymbol{\Lambda}^{-1}\boldsymbol{\zeta} = \tilde{\boldsymbol{\xi}}\mathbf{Q}\tilde{\mathbf{Q}}\mathbf{A}^{-1}\mathbf{Q}\mathbf{Q}^{-1}\boldsymbol{\xi} = \tilde{\boldsymbol{\xi}}\mathbf{A}^{-1}\boldsymbol{\xi} \tag{7.25}$$

and so we obtain finally the desired result

$$I(\boldsymbol{\xi}) = \int_{(\mathbf{u})} e^{i\tilde{\boldsymbol{\xi}}\mathbf{u} - \frac{1}{2}\tilde{\mathbf{u}}\mathbf{A}\mathbf{u}} \, d\mathbf{u} = \frac{(2\pi)^{n/2}}{(\det \mathbf{A})^{\frac{1}{2}}} e^{-\frac{1}{2}\tilde{\boldsymbol{\xi}}\mathbf{A}^{-1}\boldsymbol{\xi}} \tag{7.26}$$

The characteristic function corresponding to Eqs. (7.18) now follows immediately if we recall that $\det \mathbf{A}^{-1} = (\det \mathbf{A})^{-1}$. We have

$$F_{\mathbf{x}}(i\boldsymbol{\xi}) = \mathfrak{F}\{W(\mathbf{x})\} = \exp\,(i\tilde{\boldsymbol{\xi}}\bar{\mathbf{x}} - \tfrac{1}{2}\tilde{\boldsymbol{\xi}}\mathbf{K}_x\boldsymbol{\xi}) = \exp\left[i\Sigma_k \xi_k \bar{x}_k - \frac{1}{2}\sum_{jk} \xi_j \xi_k (\mathbf{K}_x)_{jk}\right] \tag{7.27}$$

Conversely, applying Eq. (7.26) to $\mathfrak{F}^{-1}\{F_{\mathbf{x}}\}$ gives $W(\mathbf{x})$ [Eqs. (7.18); cf. Secs. 1.2-12, 1.2-13, also]. Observe that by differentiating the c.f. (7.27) with respect to $\boldsymbol{\xi}$ and then setting $\boldsymbol{\xi} = \mathbf{0}$ [cf. Eqs. (1.57)], we find, indeed,

† If $\mathbf{A}$ is *not* symmetrical but has all its eigenvalues distinct, we can still diagonalize it by means of the similarity transformation $\mathbf{Q}$: $\mathbf{Q}^{-1}\mathbf{A}\mathbf{Q} = \boldsymbol{\Lambda}' = [\lambda_j' \delta_{jk}]$. Now, however, since $\mathbf{Q}$ is no longer orthogonal in general, Eq. (7.23) ceases to apply, and the most we can do is to determine the eigenvectors from Eqs. (7.21) to within an arbitrary constant in each component (cf. Margenau and Murphy, *op. cit.*, secs. 10.9, 10.15). The eigenvalues are found as before from Eq. (7.22).

‡ See also Cramér,[1] sec. 11.12.

that for the first- and second-order moments

$$E\{\mathbf{x}\} = \bar{\mathbf{x}} \qquad E\{\mathbf{x}\tilde{\mathbf{x}}\} = E\{[x_jx_k]\} = \mathbf{K}_x + \bar{\mathbf{x}}\bar{\tilde{\mathbf{x}}} \tag{7.27a}$$

Marginal and conditional d.d.'s, with their characteristic functions, may be constructed† from Eqs. (7.18) or (7.27), as before in the case $n = 2$ (Sec. 7.2).

Finally, we remark that if the random variables $x_1, \ldots, x_n$, instead of being real, are complex, e.g.,

$$x_j = \text{Re}\ x_j + i\ \text{Im}\ x_j = u_j + iv_j \qquad j = 1, \ldots, n$$

then the real random variables u_j, v_j together possess a joint $2n$-dimensional multivariate normal distribution, provided that the u_j, v_j $(j = 1, \ldots, n)$ obey the appropriate distribution law (7.18), where now $n \to 2n$ and $\mathbf{x}$ is replaced by the column vector $[\mathbf{u},\mathbf{v}]$.

7.3-2 Moments, Semi-invariants, and Statistical Independence. From Eq. (7.27), we see at once that the cumulant function for the multivariate normal distribution of $x_1, \ldots, x_n$ is $\mathcal{K}(i\boldsymbol{\xi}) = i\tilde{\boldsymbol{\xi}}\bar{\mathbf{x}} - \frac{1}{2}\tilde{\boldsymbol{\xi}}\mathbf{K}_x\boldsymbol{\xi}$, and only the semi-invariants of order $L = 1, 2$ [cf. Eq. (1.58)] are nonvanishing. For the moments, on the other hand, it can be shown (cf. Prob. 7.5*a*), if we write $z_j = x_j - \bar{x}_j$, that

$$\begin{aligned} E\{z_1z_2 \cdots z_{2m}\} &= \sum_{\text{all pairs}} \left(\prod_{j\neq k}^{m} \overline{z_jz_k}\right) \\ &= \sum_{\text{all pairs}} (\overline{z_jz_k}\,\overline{z_lz_p} \cdots \overline{z_qz_s})_{j\neq k,\, l\neq p,\, \text{etc.}} \\ E\{z_1z_2 \cdots z_{2m+1}\} &= 0 \end{aligned} \tag{7.28}$$

Now, the number of such averages over pairs is equal to the number of *different* ways $2m$ different variables $z_1, \ldots, z_{2m}$ can be chosen in pairs, which is‡ $(2m)!/2^m m!$. In the case $m = 2$, for example, there are just three ways into which $\overline{z_1z_2z_3z_4}$ factors into products of covariances, viz.,

$$m = 2: \qquad E\{z_1z_2z_3z_4\} = \overline{z_1z_2}\,\overline{z_3z_4} + \overline{z_2z_3}\,\overline{z_1z_4} + \overline{z_1z_3}\,\overline{z_2z_4} \tag{7.29a}$$

and if $z_2 = z_3$, we get directly

$$E\{z_1z_2^2z_4\} = 2\overline{z_1z_2}\,\overline{z_2z_4} + \overline{z_2^2}\,\overline{z_1z_4} \tag{7.29b}$$

and so on, for higher-order moments and other combinations. Equations (7.28) in conjunction with Eq. (7.27) can also be used as the defining relation for a multivariate normal distribution (cf. Prob. 7.5*b*).

The condition for statistical independence is easily given. If $\mathbf{K}_x$ is diagonal, i.e., if $\mathbf{K}_x = [\sigma_j^2\delta_{jk}]$ when $\sigma_j^2 > 0$, then so is $\mathbf{K}_x^{-1}$ and from Eqs.

† For details, see Cramér,[1] sec. 24.6, and Secs. 1.2-8, 1.2-9, 1.2-12 in the present text.

‡ There are $(2m)!$ permutations, but 2^m interchanges of argument and $m!$ permutations of the factors yield no additional separations into new pairs. See, for instance, a recent paper of Siegert,[4] eq. (3.9).

(7.18) the joint nth-order normal d.d. of the $x_1, \ldots, x_n$ factors into the product of n gaussian first-order densities, viz.,

$$W(x_1, \ldots, x_n) = \prod_{k=1}^{n} \frac{e^{-(x_k-\bar{x}_k)^2/2\sigma_k^2}}{\sqrt{2\pi\sigma_k^2}} = \prod_{k=1}^{n} w_1(x_k) \tag{7.30a}$$

while the c.f. (7.27) reduces at once to

$$F_{\mathbf{x}}(i\boldsymbol{\xi}) = \prod_{k=1}^{n} e^{i\xi_k\bar{x}_k - \frac{1}{2}\sigma_k^2\xi_k^2} = \prod_{k=1}^{n} F_{x_k}(i\xi_k) \tag{7.30b}$$

where $w_1(x_k)$ and $F_{x_k}(i\xi_k)$ are the d.d. and c.f. of a single normal variable (cf. Sec. 7.1). Thus, if either Eq. (7.30*a*) or Eq. (7.30*b*) is true, we say that the normal random variables $x_1, \ldots, x_n$ are statistically independent. Equivalently, statistical independence here occurs if the covariance matrix $\mathbf{K}_x$ is diagonal. This is a necessary and sufficient condition.

PROBLEMS

7.5 (*a*) Show that Eqs. (7.28) follow for the normal random variables $x_1, \ldots, x_n$, whose d.d. is Eqs. (7.18).

(*b*) Starting from Eqs. (7.28) as a postulate, show that the random variables $x_1, \ldots,$ x_n are gaussian with the d.d. (7.18).

7.6 (*a*) Show that, for $G(x_1, \ldots, x_n) = a^2 = (\mathbf{x} - \bar{\mathbf{x}})(\tilde{\mathbf{x}} - \bar{\tilde{\mathbf{x}}})$, the volume contained within the surface G is

$$\int \cdots \int_{G<a^2} dx_1 \cdots dx_n = \frac{\pi^{n/2}a^n}{\Gamma(n/2+1)\sqrt{\det \mathbf{K}_x}} \tag{1}$$

(*b*) Similarly, obtain

$$\int \cdots \int_{G<a^2} x_j x_k \, dx_1 \cdots dx_n = \frac{a^{n+2}\pi^{n/2}(\mathbf{K}_x)_{jk}}{2\Gamma(n/2+2)(\det \mathbf{K}_x)^{3/2}} \tag{2}$$

H. Cramér,[1] sec. 11.12.

7.7 (*a*) Prove that for the $n \times n$ matrix $\mathbf{A}$ and the n-element column matrix $\mathbf{b}$

$$\int_{-\infty}^{\infty} \cdots \int dy_1 \cdots dy_n \, H_1(\tilde{\mathbf{y}}\mathbf{A}\mathbf{y})H_2(\tilde{\mathbf{y}}\mathbf{b})$$
$$= \frac{2}{\Gamma\left(\frac{n-1}{2}\right)} \left(\frac{\pi^{n-1}}{\det \mathbf{A}}\right)^{1/2} \int_{-\infty}^{\infty} du \int_0^{\infty} dv \, v^{n-2} H_1(u^2+v^2) H_2(u\tilde{\mathbf{b}}\mathbf{A}^{-1}\mathbf{b}) \tag{1}$$

State conditions on $\mathbf{A}$, $\mathbf{b}$, H_1, H_2 for convergence.

(*b*) Using Eq. (1), show that

$$\int_{-\infty}^{\infty} \cdots \int dy_1 \cdots dy_n \, e^{-\tilde{\mathbf{y}}\mathbf{A}\mathbf{y}} = (\pi^n/\det \mathbf{A})^{1/2} \tag{2}$$

$$\int_{-\infty}^{\infty} \cdots \int dy_1 \cdots dy_n \, y_j y_k e^{-\tilde{\mathbf{y}}\mathbf{A}\mathbf{y}} = (\pi^n/\det^3 \mathbf{A})^{1/2} \frac{A^{jk}}{2} \tag{3}$$

S. O. Rice, *Bell System Tech. J.*, **24:** 46 (1945), sec. 3.5.

7.4 *The Normal Random Process*

Let us return again to the real random process $x(t)$, which in physical situations may represent, for example, a noise voltage, current, or electromagnetic wave arising in some part of the communication operation $\{v\} = \boldsymbol{T}_R^{(N)}\boldsymbol{T}_M^{(N)}\boldsymbol{T}_T^{(N)}\{u\}$ [cf. Sec. 2.1-3(1)]. We say now that $x(t)$ *is a* (real) *gaussian* or *normal random process if, for every finite set of times*† $(t_1, \ldots, t_n)$ $(n \geq 1)$, *the random variables* (or variates) $x_1\,[= x(t_1)], \ldots, x_n\,[= x(t_n)]$ *possess the multivariate normal frequency function*

$$W_n(x_1,t_1; \ldots ;x_n,t_n) = \frac{1}{(2\pi)^{n/2}(\det \mathbf{K}_x)^{1/2}} e^{-\frac{1}{2}(\tilde{\mathbf{x}}-\bar{\tilde{\mathbf{x}}})\mathbf{K}_x^{-1}(\mathbf{x}-\bar{\mathbf{x}})} \qquad n \geq 1 \tag{7.31}$$

We have rewritten $W(\mathbf{x}) = W_n(x_1,t_1; \ldots ;x_n,t_n)$ to emphasize the dependence of x on t parametrically, or, equivalently, to stress the fact that here the normal variates are obtained (at least in principle) by observations at times $t_1, \ldots, t_n$ on the ensemble constituting the random process $x(t)$. For convenience, these observation times are ordered: $t_1 \leq t_2 \leq \cdots \leq t_n$. Here, also, the matrices of the first- and second-order statistical moments of the process $x(t)$ are specifically

$$\boldsymbol{E}\{\mathbf{x}\} = [\boldsymbol{E}\{x(t_k)\}] = \bar{\mathbf{x}} \qquad k = 1, \ldots, n \tag{7.32a}$$

$$\begin{aligned} \mathbf{K}_x &= \boldsymbol{E}\{(\mathbf{x} - \bar{\mathbf{x}})(\tilde{\mathbf{x}} - \bar{\tilde{\mathbf{x}}})\} = \overline{(\mathbf{x} - \bar{\mathbf{x}})(\tilde{\mathbf{x}} - \bar{\tilde{\mathbf{x}}})} = \overline{\mathbf{x}\tilde{\mathbf{x}}} - \bar{\mathbf{x}}\bar{\tilde{\mathbf{x}}} \\ &= [\overline{x(t_j)x(t_k)} - \overline{x(t_j)}\,\overline{x(t_k)}] \qquad j, k = 1, \ldots, n \end{aligned} \tag{7.32b}$$

The elements of $\mathbf{K}_x$ are obtained from the covariance or the second-moment function according to

$$K_x(t_j,t_k) = M_x(t_j,t_k) - \overline{x(t_j)}\,\overline{x(t_k)} \tag{7.32c}$$

If the process is wide-sense stationary (cf. Sec. 1.3-6), then $K_x(t_j,t_k) = K_x(t_j - t_k) = K_x(t_k - t_j)$, so that Eq. (7.32c) becomes $K_x(|t_j - t_k|) = M_x(|t_j - t_k|) - \overline{x(0)}^2$. But since W_n for the normal process depends on $(t_1, \ldots, t_n)$ only through the covariance function K_x, it follows at once that, if $x(t)$ is wide-sense stationary, it is also strictly stationary.‡ The process itself is then completely described (in the sense of Sec. 1.3-2) by the hierarchy of normal density functions $W_1, W_2, \ldots, W_n, \ldots$, where W_n is given by Eq. (7.31).

† Note that, if we restrict the $(t_1, \ldots, t_n)$ to fixed values in a denumerably infinite set (as $n \to \infty$), we have more precisely a continuous gaussian random *series* (Sec. 1.3-3). For the gaussian process, however, the set $(t_1, \ldots, t_n)$ can take any discrete values in $(-\infty, \infty)$.

‡ For nonnormal statistics, this is, of course, no longer true.

As a case of typical interest in our study of noise in communication systems, let us consider the real stationary normal random voltage process $x(t)$, for which $\overline{x_j^2} = \psi\ (<\infty)\ (j = 1, \ldots, n)$ and $\bar{x}_j = 0$ (all j). Then, from Eqs. (7.31) and (7.27), we have for the joint d.d. and c.f. of $x_1, \ldots, x_n$

$$W_n(x_1,t_1; \ldots ;x_n,t_n) = [(\det \mathbf{k}_x)(2\pi\psi)^n]^{-1}e^{-(\tilde{\mathbf{x}}\mathbf{k}_x^{-1}\mathbf{x})/2\psi} \qquad \bar{\mathbf{x}} = 0,\ n \geq 1 \tag{7.33a}$$

$$F_n(\xi_1,t_1; \ldots ;\xi_n,t_n) = e^{-(\psi/2)\tilde{\xi}\mathbf{k}_x\xi} \tag{7.33b}$$

where $\mathbf{k}_x$ is the normalized covariance matrix $\mathbf{k}_x = \psi^{-1}\overline{\mathbf{x}\tilde{\mathbf{x}}} = [\psi^{-1}K_x(|t_j - t_k|)] = [k_x(|t_j - t_k|)]$. Letting τ represent $t_k - t_j$, we see also that $\mathbf{k}(\tau) = \mathbf{k}(-\tau) = \tilde{\mathbf{k}}(\tau)$. Applying the W-K theorem (3.42) to each element of $\mathbf{k}_x$, we see that the covariance function, and hence the d.d. and c.f., also depends on the ensemble spectral density through

$$k_x(|t_k - t_j|) = \tfrac{1}{2}\psi^{-1}\mathfrak{F}^{-1}\{\mathcal{W}_x(f)\}$$
$$= \psi^{-1}\int_0^\infty \mathcal{W}_x(f) \cos \omega(t_k - t_j)\, df \qquad \omega = 2\pi f \tag{7.34}$$

with $\mathcal{W}_x(f) = 2\psi\mathfrak{F}\{k_x\}$ conversely.

Note that, if the intervals between successive observations can be made sufficiently great (ideally, if $\lim |\tau| \to \infty$), we get the purely random series $(x_1, \ldots ,x_n)$, described by Eqs. (7.30*a*), (7.30*b*), with $\sigma_k^2 = \psi$, $\bar{x}_k = 0$ (all k), since now $\mathbf{k}_x \to \mathbf{I}$, the unit matrix. On the other hand, if $t_n \to t_{n-1} \to \cdots \to t_2 \to t_1$, that is, if all observation times coincide, every element of $\mathbf{k}_x$ is now unity, and therefore $\mathbf{k}_x$ is singular. Here, W_n is still normal but is likewise a singular density function (cf. Cramér,[1] sec. 24.3). But from Eqs. (1.92) applied to the hierarchy (1.89), (1.90) we see at once that

$$W_n(x_1,t_1; \ldots ;x_n,t_n) = W_1(x_1,t_1) \prod_{k=1}^{n-1} \delta(x_{k+1} - x_k) \tag{7.35a}$$

so that Eqs. (7.33) become specifically†

$$W_n = \frac{e^{-x_1^2/2\psi}}{\sqrt{2\pi\psi}} \prod_{k=1}^{n-1} \delta(x_{k+1} - x_k) \qquad F_n = e^{-(\psi/2)(\xi_1+\xi_2+\cdots+\xi_n)^2} \tag{7.35b}$$

As a further illustration, let us suppose that the stationary process $x(t)$, described by Eqs. (7.33*a*), (7.33*b*), is also a band-limited normal white noise process (cf. Fig. 3.11). From Eqs. (3.54*a*), (3.54*b*), we have the spectrum and covariance function

$$\mathcal{W}_x(f) = \begin{cases} W_0 & 0 < f < B \\ 0 & f > B \end{cases}$$
$$K_x(\tau) = \psi \frac{\sin 2\pi B\tau}{2\pi B\tau} \qquad \psi = BW_0 \tag{7.36}$$

† For an alternative approach, see Prob. 7.5*b*.

where B is the bandwidth and W_0 the spectral intensity within the band. The normalized covariance matrix is now simply $\mathbf{k}_x = [\sin 2\pi B(t_j - t_k)/2\pi B(t_j - t_k)]$ (for all $-\infty < t_j, t_k < \infty$). Observe here, however, that, if the observation or sampling times $t_{j,k}$ are restricted to multiples of $1/2B$, $\mathbf{k}$ equals $\mathbf{I}$ again [cf. Eq. (4.39a)] and the $x_1, \ldots, x_n$ are now statistically independent, forming a continuous, purely random sequence whose d.d. and c.f. are

$$W_n = (2\pi\psi)^{-n/2} \exp\left[-\frac{1}{2\psi}\sum_j^n x^2\left(\frac{j}{2B}\right)\right]$$
$$F_n = \exp\left(-\frac{\psi}{2}\tilde{\xi}\xi\right) = \exp\left(-\frac{\psi}{2}\sum_j^n \xi_j^2\right) \tag{7.37}$$

For other choices of $t_{j,k}$ this is, of course, no longer the case. We shall have occasion in subsequent work to make use of Eqs. (7.37).

7.5 *Some Properties of the Normal Process*

In applications, we have frequently to deal not just with a single normal random process, but with various combinations of two or more such processes. Here, we shall examine briefly what happens when linear combinations and linear transformations of gaussian processes are constructed: we shall find as a principal result that the new processes are also gaussian.

7.5-1 Additivity. We begin by considering the two sets of real normal random variables

$$\mathbf{x}_1 = [x_1^{(1)}, x_2^{(1)}, \ldots, x_n^{(1)}] \quad \text{and} \quad \mathbf{x}_2 = [x_1^{(2)}, x_2^{(2)}, \ldots, x_n^{(2)}]$$

obtained† from two real gaussian processes $x_1(t)$, $x_2(t)$. What is the d.d. of the n-component *sum* $\mathbf{X} = a\mathbf{x}_1 + b\mathbf{x}_2$, with a and b for the moment real?

Starting with the c.f. of $\mathbf{X}$, using Eqs. (1.62), (1.64), and noting here, however, that $\mathbf{x}_1$ and $\mathbf{x}_2$ are not necessarily independent [for example, $x_1(t)$ and $x_2(t)$ may have a nonvanishing cross-variance function], we have

$$F_{\mathbf{X}}(i\xi) = \int_{(\mathbf{x}_1,\mathbf{x}_2)} e^{i\tilde{\xi}(a\mathbf{x}_1+b\mathbf{x}_2)} W_{2n}(\mathbf{x}_1,\mathbf{x}_2)\, d\mathbf{x}_1\, d\mathbf{x}_2 \tag{7.38}$$

From Eq. (7.27), the $2n$-dimensional characteristic function of $\mathbf{x}_1$ and $\mathbf{x}_2$ is

$$F_{\mathbf{x}_1,\mathbf{x}_2}(i\zeta) = e^{i\tilde{\zeta}\mathbf{v} - \frac{1}{2}\tilde{\zeta}\mathbf{K}_X\zeta} \tag{7.39}$$

† For convenience, it is assumed that the dimensionality of the vectors $\mathbf{x}_1$ and $\mathbf{x}_2$ is the same (see Prob. 7.8 for the more general case).

where $\boldsymbol{\zeta}$ and $\boldsymbol{\nu}$ are the $2n$-component column vectors

$$\boldsymbol{\zeta} = \begin{bmatrix} \mathbf{n}_1 \\ \mathbf{n}_2 \end{bmatrix} \qquad \tilde{\mathbf{n}}_1 = [\eta_1^{(1)}, \ldots, \eta_n^{(1)}], \text{ etc.} \qquad \boldsymbol{\nu} = \begin{bmatrix} \bar{\mathbf{x}}_1 \\ \bar{\mathbf{x}}_2 \end{bmatrix} \tag{7.40}$$

The $2n \times 2n$ covariance matrix of the joint distribution of $\mathbf{x}_1$ and $\mathbf{x}_2$ is

$$\mathbf{K}_X(t_j;t_k) = \begin{bmatrix} \mathbf{K}_{11}(t_j,t_k) & \mathbf{K}_{12}(t_j,t_k) \\ \mathbf{K}_{21}(t_j,t_k) & \mathbf{K}_{22}(t_j,t_k) \end{bmatrix}$$

with

$$\begin{aligned} \mathbf{K}_{11} &= \overline{\mathbf{x}_1\tilde{\mathbf{x}}_1} - \bar{\mathbf{x}}_1\bar{\tilde{\mathbf{x}}}_1 = [E\{[x^{(1)}(t_j) - \overline{x_j^{(1)}}][x^{(1)}(t_k) - \overline{x_k^{(1)}}]\}] \\ \mathbf{K}_{22} &= \overline{\mathbf{x}_2\tilde{\mathbf{x}}_2} - \bar{\mathbf{x}}_2\bar{\tilde{\mathbf{x}}}_2 = [E\{[x^{(2)}(t_j) - \overline{x_j^{(2)}}][x^{(2)}(t_k) - \overline{x_k^{(2)}}]\}] \\ \mathbf{K}_{12} &= \overline{\mathbf{x}_1\tilde{\mathbf{x}}_2} - \bar{\mathbf{x}}_1\bar{\tilde{\mathbf{x}}}_2 = [E\{[x^{(1)}(t_j) - \overline{x_j^{(1)}}][x^{(2)}(t_k) - \overline{x_k^{(2)}}]\}] \end{aligned} \tag{7.41}$$

As before, $\bar{\mathbf{x}}_1$, $\bar{\mathbf{x}}_2$, $\mathbf{n}_1$, etc., are n-component column vectors, while $\mathbf{K}_{11}$, $\mathbf{K}_{12}$, etc., are the $n \times n$ matrices of the auto- and cross variances of the two processes at times t_j, t_k. Next, we replace $W_{2n}(\mathbf{x}_1,\mathbf{x}_2)$ by $\mathfrak{F}^{-1}\{F_{\mathbf{x}_1,\mathbf{x}_2}\}$ [Eq. (1.56*b*)], rewriting Eq. (7.38) as

$$F_{\mathbf{X}}(i\boldsymbol{\xi}) = \int_{(\boldsymbol{\zeta})} d\boldsymbol{\zeta}\, e^{i\tilde{\boldsymbol{\zeta}}\boldsymbol{\nu} - \frac{1}{2}\tilde{\boldsymbol{\zeta}}\mathbf{K}_X\boldsymbol{\zeta}} \int_{-\infty}^{\infty} \cdots \int e^{i\tilde{\mathbf{x}}_1(a\boldsymbol{\xi}-\mathbf{n}_1)+i\tilde{\mathbf{x}}_2(b\boldsymbol{\xi}-\mathbf{n}_2)} \frac{d\mathbf{x}_1\, d\mathbf{x}_2}{(2\pi)^{2n}} \tag{7.42}$$

But the integrals over $(\mathbf{x}_1,\mathbf{x}_2)$ are simply the multiple delta functions $\boldsymbol{\delta}(\mathbf{n}_1 - a\boldsymbol{\xi})\boldsymbol{\delta}(\mathbf{n}_2 - b\boldsymbol{\xi})$ [cf. Eq. (1.16*b*)], so that if we insert these into Eq. (7.42) and carry out the indicated matrix multiplications, we obtain directly

$$F_{\mathbf{X}}(i\boldsymbol{\xi}) = e^{i\tilde{\boldsymbol{\xi}}(a\bar{\mathbf{x}}_1+b\bar{\mathbf{x}}_2)-\frac{1}{2}\tilde{\boldsymbol{\xi}}\mathbf{K}_X\boldsymbol{\xi}} \tag{7.43a}$$

where

$$\mathbf{K}_X = a^2\mathbf{K}_{11} + ab(\mathbf{K}_{12} + \tilde{\mathbf{K}}_{12}) + b^2\mathbf{K}_{22} \tag{7.43b}$$

Applying the inverse transform (1.56*b*) to $F_{\mathbf{X}}(i\boldsymbol{\xi})$ with the aid of Eq. (7.26) then shows that $\mathbf{X}$ is also normally distributed, with an n-dimensional density $W_n(\mathbf{X})$, where $E\{\mathbf{X}\} = a\bar{\mathbf{x}}_1 + b\bar{\mathbf{x}}_2$ and the covariance $\mathbf{K}_X$ is given by Eq. (7.43*b*).†

The extension to three or more normal random variables is now immediate. We let $\mathbf{X}_1 = a\mathbf{x}_1 + b\mathbf{x}_2$, $\mathbf{X}_2 = c\mathbf{x}_3$, and then apply the preceding argument to $\mathbf{X} \equiv \mathbf{X}_1 + \mathbf{X}_2$, etc. Thus, the (real) linear combination of any number of real gaussian random variables is also a (real) gaussian random variable, and from Sec. 7.4 this same statement can then be made for the normal random processes $x_1(t)$, $x_2(t)$, . . . : if the $x_l(t)$ $(l = 1, 2, \ldots)$ are (real) normal random processes, then so also is $\sum_l a_l x_l(t)$ (with a_l real).‡

† Observe that if $\mathbf{K}_{12} = \mathbf{K}_{21} = \mathbf{0}$, then $\mathbf{x}_1$ and $\mathbf{x}_2$ are statistically independent.

‡ When the dimensionalities of $\mathbf{x}_1$, $\mathbf{x}_2$, etc., are not equal, we may use the generalization of Prob. 7.8. For a, b, . . . complex, the procedure above is repeated for the joint d.d. of the real and imaginary components (cf. Sec. 7.3-1).

7.5-2 Linear Transformations. Let $\mathbf{y} = \mathbf{Qx}$ be a linear transformation such that $\mathbf{Q}^{-1}$ exists. What is the d.d. of $\mathbf{y}$ if $\mathbf{x}$ possesses a multivariate normal distribution? Again, we observe from Eq. (1.62) that the c.f. of $\mathbf{y}$ is

$$F_{\mathbf{y}}(i\boldsymbol{\xi}) = E\{e^{i\tilde{\boldsymbol{\xi}}\mathbf{y}(\mathbf{x})}\} = E\{e^{i\tilde{\boldsymbol{\xi}}\mathbf{Qx}}\} = \int_{(\mathbf{x})} e^{i\tilde{\boldsymbol{\xi}}\mathbf{Qx}} W_n(\mathbf{x})\, d\mathbf{x} \qquad (7.44)$$

With Eqs. (7.18), we have explicitly

$$F_{\mathbf{y}}(i\boldsymbol{\xi}) = e^{-\frac{1}{2}\tilde{\bar{\mathbf{x}}}\mathbf{K}_x^{-1}\bar{\mathbf{x}}} \int_{(\mathbf{x})} e^{i(\tilde{\boldsymbol{\xi}}\mathbf{Q}-i\tilde{\bar{\mathbf{x}}}\mathbf{K}_x^{-1})\mathbf{x}-\frac{1}{2}\tilde{\mathbf{x}}\mathbf{K}_x^{-1}\mathbf{x}} \frac{d\mathbf{x}}{(2\pi)^{n/2}(\det \mathbf{K}_x)^{\frac{1}{2}}} \qquad (7.45)$$

Now with the help of the integral (7.26), where $\boldsymbol{\xi}$ is in this instance replaced by $\tilde{\mathbf{Q}}\boldsymbol{\xi} - i\mathbf{K}_x^{-1}\bar{\mathbf{x}}$, remembering that $\det \mathbf{A}^{-1} = (\det \mathbf{A})^{-1}$, we get directly

$$F_{\mathbf{y}}(i\boldsymbol{\xi}) = e^{i\tilde{\boldsymbol{\xi}}\bar{\mathbf{y}}-\frac{1}{2}\tilde{\boldsymbol{\xi}}\mathbf{K}_y\boldsymbol{\xi}} \qquad (7.46)$$

with $\bar{\mathbf{y}} = \mathbf{Q}\bar{\mathbf{x}}$ and $\mathbf{K}_y = \overline{\mathbf{y}\tilde{\mathbf{y}}} - \bar{\mathbf{y}}\tilde{\bar{\mathbf{y}}} = \mathbf{Q}(\overline{\mathbf{x}\tilde{\mathbf{x}}} - \bar{\mathbf{x}}\tilde{\bar{\mathbf{x}}})\tilde{\mathbf{Q}} = \mathbf{QK}_x\tilde{\mathbf{Q}}$ (7.46*a*)

From Eq. (7.27), we see that $\mathbf{y}$ also possesses a characteristic function associated with the multivariate normal distribution, but now having a mean and variance matrix appropriate to the transformation $\mathbf{y} = \mathbf{Qx}$ [cf. Eq. (7.46*a*)]. Thus, if $\mathbf{x}$ is obtained from a normal process $x(t)$, it follows that $y(t)$, obtained from $\mathbf{y} = \mathbf{Qx}$, is a gaussian process also, for t_j, t_k unrestricted in $(-\infty, \infty)$. Finally, by combining the results here and in Sec. 7.5-1, we observe that any linear combination† of linearly transformed gaussian processes is also a gaussian process.‡ Even when $\mathbf{Q}$ is singular, we may still have a normal distribution for $\mathbf{y}$ but it will then be a *singular* distribution,§ with one or more delta functions in the probability density.

Note, incidentally, that, if the choice of transformation $\mathbf{Q}$ is at our disposal, we can take it to be orthogonal and such that it diagonalizes $\mathbf{K}_x$, since $\mathbf{K}_x$ is symmetrical (cf. Sec. 7.3-1). Accordingly, we write $\mathbf{QK}_x\tilde{\mathbf{Q}} = \tilde{\mathbf{Q}}^{-1}\mathbf{K}_x\tilde{\mathbf{Q}} = \mathbf{\Lambda}_x = [\lambda_j\delta_{jk}]$, where the λ_j are the eigenvalues of $\mathbf{K}_x$. Then the components $y_1, \ldots, y_n$ of $\mathbf{y} = \mathbf{Qx}$ are n *independent* normal random variables with the particularly simple d.d. and c.f.

$$W_n(\mathbf{y}) = \prod_{j=1}^{n} \frac{e^{-[y_j-(\mathbf{Q}\bar{\mathbf{x}})_j]^2/2\lambda_j^2}}{\sqrt{2\pi\lambda_j^2}} \qquad F_{\mathbf{y}}(i\boldsymbol{\xi}) = \prod_{j=1}^{n} e^{i\xi_j(\mathbf{Q}\bar{\mathbf{x}})_j-\lambda_j^2\xi_j^2/2} \qquad (7.47)$$

From this, we see that by a suitable orthogonal transformation it is always possible to combine (linearly) a set of n statistically related normal variates $(x_1, \ldots, x_n)$ in such a way that each element y_k $(k = 1, \ldots, n)$ of the combination is statistically independent of the others. The explicit structure of $\mathbf{Q}$ here depends, of course, on the original covariance matrix $\mathbf{K}_x$ and

† Complex processes are easily included by considering the joint d.d. of the real and imaginary parts.

‡ As we shall see in Sec. 8.1, this applies also for continuous linear transformations of normal processes, as well as for the discrete transformations considered here.

§ See Cramér,[1] sec. 23.3, and Prob. 23.3.

may be found (if the eigenvalues of $\mathbf{K}_x$ are distinct) by solving the simultaneous equations (7.21) for the eigenvectors. However, if $\bar{\mathbf{x}}$ vanishes, only the eigenvalues are needed in Eqs. (7.47).

7.5-3 Examples. To illustrate the general remarks of the preceding Secs. 7.5-1, 7.5-2, let us consider several examples needed in subsequent applications.

(1) The sum of two normal noise ensembles $N_1 + N_2$. This is a very simple case where the additive property of the normal process is evident. Setting $a = b = 1$ in Sec. 7.5-1, we see at once that $N(t) = N_1(t) + N_2(t)$ is also a normal process, with mean $\bar{N} = \bar{N}_1 + \bar{N}_2$ and a covariance function (7.43*b*)

$$K_N(t_1,t_2) = K_{11}(t_1,t_2) + K_{12}(t_1,t_2) + K_{21}(t_1,t_2) + K_{22}(t_1,t_2) \qquad (7.48)$$

with $K_{11} = \boldsymbol{E}\{N_1(t_1)N_1(t_2)\}$, etc.

(2) Additive signal and noise processes $Y = S + N$. Another case of considerable practical importance arises when a signal process $S(t)$ and a normal noise process $N(t)$ are additively and independently combined. If the signal is already a normal noise process, then Example 1 may be invoked at once to show that Y is also. Often, however, S is a deterministic process. Then, for the subensemble $Y_0 = Y(t;\boldsymbol{\theta}_0) = N(t) + S(t;\boldsymbol{\theta}_0)$, where $\boldsymbol{\theta}$ represents the information-bearing statistical parameters of the signal (cf. Secs. 1.3-4, 1.3-5) and $\boldsymbol{\theta}_0$ is a given (nonrandom) set of values of $\boldsymbol{\theta}$, it is evident from substitution into Eqs. (7.18) and the fact that the jacobian of this transformation is unity that

$$w_n(\mathbf{Y}_0) = W_n(\mathbf{Y}_0 - \mathbf{S}_0)_N = \frac{e^{-\frac{1}{2}(\tilde{\mathbf{Y}}_0-\tilde{\mathbf{S}}_0)\mathbf{K}_N^{-1}(\mathbf{Y}_0-\mathbf{S}_0)}}{(2\pi)^{n/2}\sqrt{\det \mathbf{K}_N}} \qquad \bar{N} = 0;\, S_0 = S(t,\boldsymbol{\theta}_0) \qquad (7.49)$$

where $W_n(\ )_N$ is the d.d. of the noise alone. Thus, $Y(t,\boldsymbol{\theta}_0)$ is also a normal process, with the same covariance function as the noise component, but with the new mean value $\bar{\mathbf{Y}}_0 = \mathbf{S}_0$.

When $\boldsymbol{\theta}$ can assume more than one value, however, according to the d.d. $w(\boldsymbol{\theta})$, for example, the complete ensemble Y will no longer in general represent a gaussian process, since now

$$w_n(\mathbf{Y}) = \int_{\boldsymbol{\theta}} w(\boldsymbol{\theta})W_n[\mathbf{Y} - S(\boldsymbol{\theta})]_N\, d\boldsymbol{\theta} \qquad (7.50)$$

and even if $\mathbf{S}(\boldsymbol{\theta})$ is linear in $\boldsymbol{\theta}$, a nonnormal d.d. in $\mathbf{Y}$ usually results. An exception is the case mentioned above, which follows from Eq. (7.50) if $w(\boldsymbol{\theta}) = \boldsymbol{\delta}(\boldsymbol{\theta} - \boldsymbol{\theta}_0)$, an entirely trivial situation for all actual communication processes,† since no information can be conveyed by such a signal alone.

† Of course, for such signals the analysis of system performance in specific instances is by no means trivial if each single (deterministic) signal can be regarded in some respect (i.e., in waveform, intensity, epoch, etc.) as typical of a larger class, including the null signal. It is in this sense that much of the second-moment theory of AM and FM systems described in Part 3 finds its practical justification.

(3) Stationary normal noise. Let us consider the stationary noise process $N(t)$, with covariance function $K_N(t)$, zero mean, and $\overline{N^2} = \psi$, and let us write it in the alternative form†

$$N(t) = N_c(t) \cos \omega_0 t + N_s(t) \sin \omega_0 t \qquad \omega_0 = 2\pi f_0 \tag{7.51}$$

where the "center" frequency f_0 is determined in practice by suitable amplitude and phase detectors or, analytically, by Hilbert transforms [cf. Sec. 2.2-5(6)]. Next, we *postulate*‡ that N_c and N_s are also stationary normal processes, with zero means, and covariance functions given by

$$\overline{N_c(t_j)N_c(t_k)} = \overline{N_s(t_j)N_s(t_k)} \equiv \psi\rho_0(|t_k - t_j|)_N \tag{7.52a}$$
$$\overline{N_c(t_j)N_s(t_k)} = -\overline{N_c(t_k)N_s(t_j)} \equiv \psi\lambda_0(t_k - t_j)_N = -\psi\lambda_0(t_j - t_k)_N \tag{7.52b}$$

Then, by the additivity property cited earlier for gauss processes (cf. Sec. 7.5-1), $N(t)$ is also gaussian (and stationary). Its covariance function $K_N(t)$ can now be written according to Eqs. (3.110) and (7.52) as

$$\begin{aligned} K_N(t_j,t_k) &= \psi[\rho_0(|t_k - t_j|)_N \cos \omega_0(t_k - t_j) + \lambda_0(t_k - t_j)_N \sin \omega_0(t_k - t_j)] \\ &= K_N(t_k,t_j) \end{aligned} \tag{7.53}$$

In terms of the spectral density $\mathcal{W}_N(f)$, we find from Eqs. (3.108*e*) that the covariance functions ρ_0 and λ_0 above can be expressed as

$$\begin{aligned} \rho_0(\tau)_N &= \psi^{-1} \int_0^\infty \mathcal{W}_N(f) \cos (\omega - \omega_0)\tau \, df \\ \lambda_0(\tau)_N &= -\psi^{-1} \int_0^\infty \mathcal{W}_N(f) \sin (\omega - \omega_0)\tau \, df \qquad \tau \equiv t_k - t_j \end{aligned} \tag{7.54}$$

We can also use Eqs. (7.39) to (7.41) to determine the joint $2n$-fold multivariate density for $N_c(t_j)$, $N_s(t_j)$ $(j = 1, \ldots, n)$. Thus, we have for the c.f. of N_c and N_s

$$F_{\mathbf{N}_c,\mathbf{N}_s}(i\zeta) = e^{-\frac{1}{2}\tilde{\zeta}\mathbf{K}_0\zeta} \tag{7.55}$$

where the $2n \times 2n$ matrix $\mathbf{K}_0$ becomes

$$\mathbf{K}_0 \equiv \psi\mathbf{k}_0 = \psi \begin{bmatrix} \boldsymbol{\rho}_0(\tau) & \boldsymbol{\lambda}_0(\tau) \\ \tilde{\boldsymbol{\lambda}}_0(\tau) & \boldsymbol{\rho}_0(\tau) \end{bmatrix} \qquad \tau = t_k - t_j \tag{7.56a}$$

with the $n \times n$ submatrices

$$\begin{aligned} \boldsymbol{\rho}_0(\tau) &= \psi^{-1}[E\{N_c(t_j)N_c(t_k)\}] = \psi^{-1}[E\{N_s(t_j)N_s(t_k)\}] = \tilde{\boldsymbol{\rho}}_0(\pm\tau) \\ \boldsymbol{\lambda}_0(\tau) &= \psi^{-1}[E\{N_c(t_j)N_s(t_k)\}] = -\psi^{-1}[E\{N_c(t_k)N_s(t_j)\}] = \pm\tilde{\boldsymbol{\lambda}}_0(\mp\tau) \end{aligned} \tag{7.56b}$$

† Although Eq. (7.51) is a representation frequently used for narrowband waves, where N_c and N_s are slowly varying compared with $\cos \omega_0 t$, $\sin \omega_0 t$, it is by no means limited to such waves. Broadband disturbances may also be described by Eq. (7.51), but the components N_c and N_s are no longer slowly varying, and, indeed, the concepts of envelope and phase lose their physical interpretation.

‡ As we shall see presently, in Sec. 8.3, this is equivalent to the usual assumption of normal properties for the coefficients in the Fourier-series development of $N(t)$ over the interval $(-T/2,T/2)$ as $T \to \infty$.

The associated d.d. follows from Eq. (7.26) and is

$$W_{2n}(\mathbf{N}_c,\mathbf{N}_s) = \frac{e^{-\frac{1}{2}\tilde{\mathbf{X}}\mathbf{K}_0^{-1}\mathbf{X}}}{(2\pi)^n \sqrt{\det \mathbf{K}_0}} \qquad \mathbf{X} \equiv \begin{bmatrix} \mathbf{N}_c \\ \mathbf{N}_s \end{bmatrix} \tag{7.57}$$

Since $\lambda_0(\tau)$ vanishes, in general, only for $\tau = 0$ (or $|\tau| \to \infty$), $N_c(t)$ and $N_s(t)$ are statistically *dependent* normal processes. We remark again that $N(t)$ [Eq. (7.51)] may be broadband as well as narrowband.

(4) Stationary narrowband normal noise. An important exception to the situation above arises if $N(t)$ is a narrowband noise process. Then N_c, N_s, $\rho_0(\tau)_N$, $\lambda_0(\tau)_N$ are slowly varying vis-à-vis $\cos \omega_0 t$, $\sin \omega_0 t$. Moreover, we can take advantage of this narrowband property to replace $\mathcal{W}_N(f' + f_0)$ by $\mathcal{W}_N(f')$ [cf. Eqs. (3.123) et seq.] after making the substitution $f' = f - f_0$ in Eqs. (7.54). We then replace the lower limit $-f_0$ on the integration by $(-\infty)$ to get approximately

$$\begin{aligned} \rho_0(\tau)_N &\doteq \psi^{-1} \int_{-\infty}^{\infty} \mathcal{W}_N(f') \cos \omega'\tau \, df' \\ \lambda_0(\tau)_N &\doteq -\psi^{-1} \int_{-\infty}^{\infty} \mathcal{W}_N(f') \sin \omega'\tau \, df' = 0 \end{aligned} \tag{7.58}$$

this last provided that $\mathcal{W}_N(f)$ is *symmetrical* (about f_0). [If the spectrum is not symmetrical about its "center" frequency f_0, $\lambda_0(\tau)_N$ will not vanish and may even be appreciable with respect to $\rho_0(\tau)_N$.]

Assuming, then, that the spectrum is symmetrical, we see that $\mathbf{K}_0$ [Eq. (7.56*a*)] becomes approximately for this (stationary) narrowband noise

$$\mathbf{K}_0 \doteq \psi \begin{bmatrix} \boldsymbol{\rho}_0(\tau) & \mathbf{0} \\ \mathbf{0} & \boldsymbol{\rho}_0(\tau) \end{bmatrix} \tag{7.59}$$

and the c.f. and d.d. of $\mathbf{N}_c$, $\mathbf{N}_s$ are then

$$F_{\mathbf{N}_c,\mathbf{N}_s}(i\xi_1,i\xi_2) \doteq e^{-\psi\tilde{\xi}_1\boldsymbol{\rho}_0\xi_1/2}e^{-\psi\tilde{\xi}_2\boldsymbol{\rho}_0\xi_2/2} \tag{7.60a}$$

$$\begin{aligned} W_{2n}(\mathbf{N}_c,\mathbf{N}_s) &= W_n(\mathbf{N}_c)W_n(\mathbf{N}_s) \\ &= [(2\pi\psi)^n \det \boldsymbol{\rho}_0]^{-1} \exp\left(-\frac{\psi}{2}\tilde{\mathbf{N}}_c\boldsymbol{\rho}_0^{-1}\mathbf{N}_c - \frac{\psi}{2}\tilde{\mathbf{N}}_s\boldsymbol{\rho}_0^{-1}\mathbf{N}_s\right) \end{aligned} \tag{7.60b}$$

Thus, $N_c(t)$ and $N_s(t)$ are essentially *independent* normal processes. It should be emphasized however, that Eqs. (7.59), (7.60) are only approximate, although satisfactory, representations in almost all practical applications where spectral symmetry of the narrowband noise process can reasonably be assumed. In any case, the exact expressions for the d.d. and c.f. are given by Eqs. (7.55) and (7.57).

(5) Stationary white normal noise. Although the components N_c, N_s of the narrowband normal noise above are only approximately independent, there is an important limiting situation (apart from the trivial one of zero bandwidth) where they are exactly so. Let us consider again the stationary band-limited white noise of Fig. 3.11 [cf. also Eqs. (7.36)]. Inserting Eqs. (7.36) into Eqs. (7.54) and remembering that $\psi = BW_0$, we find that

$$\rho_0(\tau)_N = \frac{\sin \omega_0\tau + \sin 2\pi(B - f_0)\tau}{2\pi B\tau} \qquad \lambda_0(\tau)_N = \frac{\cos \omega_0\tau - \cos 2\pi(B - f_0)\tau}{2\pi B\tau} \tag{7.61}$$

Now, for unlimited white noise we have $B \to \infty$. Setting $\tau = 0$ and $\tau \neq 0$, respectively, in Eqs. (7.61), and then passing to the limit $B \to \infty$ *afterward*, we find that

$$\begin{aligned} &\lim_{B\to\infty} \rho_0(0)_N = 1 \\ &\lim_{B\to\infty} \rho_0(\tau)_N = 0 \qquad \tau \neq 0 \\ &\lim_{B\to\infty} \lambda_0(\tau)_N = 0 \qquad \text{all } \tau \end{aligned} \tag{7.62}$$

From this, it follows at once that N_c and N_s are truly independent in the limit of a normal stationary white noise process.†

PROBLEM

7.8 If $\mathbf{x}_1$ and $\mathbf{x}_2$ are, respectively, m- and n-component normal random variates ($n \geq m$), with finite means and covariance functions, show that $\mathbf{X} = a\mathbf{x}_1 + b\mathbf{x}_2$ is an n-component vector, with a normal distribution (provided that a, b are real), such that

$$E\{\mathbf{X}\} = a\bar{\mathbf{x}}_1 + b\bar{\mathbf{x}}_2 = [a\bar{x}_j^{(1)} + b\bar{x}_k^{(2)}] \qquad \begin{cases} j = 1, \ldots, m; \ \bar{x}_j^{(1)} = 0 \ (j > m) \\ k = 1, \ldots, n \end{cases} \tag{1}$$

and

$$\begin{aligned} \tilde{\zeta}\mathbf{K}_X\zeta &= a^2\tilde{\xi}_m\mathbf{K}_{11}\xi_m + b^2\tilde{\xi}_n\mathbf{K}_{22}\xi_n + ab[\tilde{\xi}_n\tilde{\mathbf{K}}_{21}\xi_m + \tilde{\xi}_m\mathbf{K}_{12}\xi_n] \\ \zeta &= \begin{bmatrix} \xi_m \\ \xi_n \end{bmatrix} \qquad \tilde{\xi}_m = [\xi_1, \xi_2, \ldots, \xi_m], \text{ etc.} \\ \tilde{\zeta} &= [\zeta_1, \ldots, \zeta_{m+n}] \end{aligned} \tag{2}$$

in the c.f.

$$F_{\mathbf{X}}(i\zeta) = \exp(i\tilde{\xi}_m\bar{\mathbf{x}}_1 + i\tilde{\xi}_n\bar{\mathbf{x}}_2 - \tfrac{1}{2}\tilde{\zeta}\mathbf{K}_X\zeta) \tag{3}$$

7.6 *Classification (Doob's Theorem)*

From the defining relations for a normal process [cf. Eqs. (7.31) et seq.], it is clear that if the covariance function $K_x(t,\tau)$ [or second-moment function $M_x(t,\tau)$ if $E\{x\} \neq 0$] is known, the process is completely described in the sense of Sec. 1.3-2. For stationary processes, knowledge of the ensemble spectral density $\mathcal{W}_x(f)$ serves equally well. Sometimes, particu-

† Of course, in this limit $K_N(t)$ is unbounded and $W_{2n}(\mathbf{N}_c,\mathbf{N}_s) = 0$, since $\psi \to \infty$ [cf. Eq. (7.60*b*)]. Alternatively, using Eqs. (7.36) and taking $\tau\ (= t_k - t_j) = (k - j)/2B$, we see that $\mathbf{K}_N = [K_N(t_k - t_j)] = [W_0B\delta_{jk}] = [(W_0n/2T)\delta_{jk}]$, this last from the fact that for these n points, taken at the zeros of $K_N(t)$, we have $n \doteq 2BT$ ($\gg 1$) (cf. Sec. 4.2-1). Thus, as $B \to \infty$, we have $n \to \infty$ also and each (diagonal) element of $\mathbf{K}_N$ is likewise unbounded. The results (7.62) for the *normalized* covariance, however, are particularly important in many applications involving the detection of signals in white normal noise backgrounds (cf. Sec. 20.2).

larly in the accounts of physical noise mechanisms or in the specification of signal ensembles, it is necessary to know how much statistical information is actually needed to describe the process. Thus, if the process is purely random (an idealized case), only a first-order density is required to give the hierarchy W_n ($n \geq 2$), while if the process is simple Markovian (e.g., of order 1), W_2 is sufficient, and so on, for more complicated statistical structures (cf. Sec. 1.4-5). Now most physical random processes are simple Markovian [or projections of the same (cf. Sec. 1.4-5)], while message and signal processes often belong to a higher-order category, e.g., Markovian of order 2, etc. Considering here only normal processes, characteristic of many physical noise sources (e.g., shot noise, thermal noise, scatter mechanisms, etc.), we suspect from the central role played by the covariance (or spectrum) in such cases that under certain conditions these normal processes may also be Markovian (of order 1). That this is so was shown originally by Doob[5] (see also Wang and Uhlenbeck[6]). Specifically, we have the following theorems:

Theorem I. A one-dimensional stationary normal random process $x(t)$ *will be simple Markovian if* (and only if) *the covariance function has the form*

$$K_x(\tau) = \boldsymbol{E}\{x(t_0)x(t_0+\tau)\} = K_x(0)e^{-\beta|\tau|}$$

or equivalently if (7.63)

$$\mathcal{W}_x(f) = \beta K_x(0)/(\beta^2+\omega^2)$$

For this to be the case, $K_x(\tau)$ above is found to obey the functional equation[5,6]

$$K_x(t_3-t_1) = K_x(t_2-t_1)K_x(t_3-t_2) \tag{7.64}$$

(cf. Prob. 7.9).

Theorem I can also be generalized to an M-dimensional vector process $\mathbf{u}(t) = \hat{\mathbf{i}}_1x^{(1)}(t) + \hat{\mathbf{i}}_2x^{(2)}(t) + \cdots + \hat{\mathbf{i}}_Mx^{(M)}(t)$, where the components of $\mathbf{u}(t)$, for example, the $x^{(k)}(t)$ $(k = 1, \ldots, M)$, are stationary gaussian processes. In this instance, we have now the covariance matrix† [cf. Eqs. (3.143)]

$$\mathbf{K_u}(\tau) = [\boldsymbol{E}\{x^{(k)}(t_0)x^{(l)}(t_0+\tau)\}] = \tilde{\mathbf{K}}_\mathbf{u}(-\tau) \tag{7.65a}$$

or, equivalently, the spectral matrix

$$\boldsymbol{\mathcal{W}}_\mathbf{u}(f) = [\mathcal{W}_{(x^{(k)}x^{(l)})}(f)] = [\mathcal{W}_\mathbf{u}(f) = \tilde{\mathcal{W}}_\mathbf{u}(f)^*] \tag{7.65b}$$

whose elements are the auto- and cross-spectral densities [cf. Eqs. (3.148)]. Then, it can be shown (cf. Prob. 7.10) that the following theorem is true.

Theorem II. A stationary normal random vector process [$\mathbf{u}(t)$ *above*] *is simple Markovian if*

$$\mathbf{K_u}(\tau) = \begin{cases} e^{\mathbf{G}\tau} & \tau > 0 \\ e^{-\tilde{\mathbf{G}}\tau} & \tau < 0 \end{cases} \tag{7.66}$$

† Note that this is *not* the covariance matrix of the distribution density of the various components $x^{(k)}$ of the vector process $\mathbf{u}(t)$, at the times t_0, t, $+t$. This latter matrix is symmetrical [cf. Eqs. (8.22), (8.23)].

Here $\mathbf{G}$ is in general a constant nonsymmetric $M \times M$ matrix ($M \geq 2$), with complex eigenvalues,† and $\mathbf{K_u}$ is now found to satisfy the generalization

$$\mathbf{K_u}(t_3 - t_1) = \mathbf{K_u}(t_2 - t_1)\mathbf{K_u}(t_3 - t_2) \tag{7.67}$$

of the functional equation (7.64).‡ Instead of a single spectrum and covariance function, characteristic of the one-dimensional case, a larger variety of covariance functions and associated spectra are now possible, depending on the different values that $\mathbf{G}$ may take. But since $\mathbf{G}$ is a constant matrix, the various $\mathbf{K_u}(\tau)$ still cannot be chosen arbitrarily. For example (in the one-dimensional situation), a common and convenient choice of covariance function of the output process, when the effects of a number of linear cascaded single-tuned circuits on input normal noise are considered, is the gaussian form $e^{-\omega_b^2 t^2/4}$, corresponding to an (approximate) over-all frequency response of $e^{-\omega^2/\omega_b^2}$. But from Theorem I it is immediately evident that although the output process can be shown to be normal also§ it is not Markovian and $W_2(x_1,x_2;t)$ does not completely describe it: all orders W_n $(n \to \infty)$ are now required. In a similar way for the multidimensional processes, if we have fewer than all the components of the vector process, their covariance function will not arbitrarily satisfy Eq. (7.67). At best, we may have only a *projection* of a Markoff process (cf. Sec. 1.4-5), and W_2 of the right dimensionality is then required. For some systems, the covariances, like the example above, will not take the form (7.66), in any finite number of dimensions, and again the W_n $(n = 1, 2, \ldots)$ are needed for classification.

For thermal noise in linear networks (cf. Chap. 11), which is the analogue of the classical problem of the Brownian movement of coupled harmonic oscillators and which obeys normal statistics, classification is a central problem, so that knowledge of the spectrum or covariance function is always required.¶ This in turn, as we shall see in Chap. 10, must be obtained from the Langevin, or dynamical, equations which govern the potentials and motion of electric charges n the system. On the other hand, the second-moment theory of both linear and nonlinear systems demands only the second-order density (cf. Chap. 5 and Part 3), and the question of spectral shape can be left open: W_2 contains sufficient statistical information to specify the desired spectrum and covariance function of the output process. As soon as we ask, however, for more complicated statistics, such as the distribution of the mean or mean square, adding to this the

† By $e^{\mathbf{G}\tau}$ here we mean $e^{\mathbf{G}\tau} = \mathbf{I} + \tau\mathbf{G} + (\tau^2/2!)\mathbf{G}^2 + \cdots$, etc. (cf. Frazer, Duncan, and Collar, *op. cit.*, sec. 25).

‡ It is assumed that $\mathbf{K_u}(0) = \mathbf{I}$, which may always be achieved by a proper transformation of the components of $\mathbf{u}$ (cf. Sec. 7.5-2; also, Wang and Uhlenbeck,[6] footnote 19).

§ For proof of this, see Sec. 8.1.

¶ This, as we have seen previously, is sufficient to describe a stationary normal random process (cf. Sec. 7.4).

fact that the output process may no longer be gaussian, nth-order densities are needed and classification is helpful once again in assessing the degree of statistical information actually required.

PROBLEMS

7.9 Prove Doob's theorem [Eq. (7.63)] for the one-dimensional stationary normal process $x(t)$.

7.10 Prove the extension of Doob's theorem [Eq. (7.66)] to the M-dimensional stationary normal vector process $\mathbf{u}(t)$, and show that Eq. (7.67) is the only condition on $\mathbf{K_u}$. (HINT: See the approach of Wang and Uhlenbeck, *op. cit.*, app. II.)

7.7 *Unrestricted Random Walk and the Central-limit Theorem*

Although the normal random process has been discussed in preceding sections, we have not as yet attempted to show how such processes occur and why they are so common in applications. That the gauss process, and the processes derived from it by the system transformations $\boldsymbol{T}_L$, $\boldsymbol{T}_g$, $\boldsymbol{T}_R^{(N)}$, etc., are so prevalent is not entirely accidental, since, as we shall see presently, almost any reasonable microscopic random mechanism will exhibit normal properties when viewed at a macroscopic level. For instance, thermal noise currents in a resistor, as observed on a galvanometer, possess gaussian statistics. However, these currents are themselves the superposition of a very large number of individual current impulses, attributable to the thermal motion of electrons in the resistor, none of which necessarily obeys a normal law by itself. Similar statements can also be made about shot noise, Brownian movement, diffusion process, etc. (cf. Chaps. 10, 11). For many such situations, a useful model for the random mechanism is provided by the classical problem of the *random walk*. This is illustrated in Sec. 7.7-1 with a simple example, from which we show how and under what conditions a normal law is approached if we are ultimately to obtain a gauss process (cf. Secs. 7.7-2, 7.7-3).

7.7-1 The Unrestricted Random Walk.† Let us use the following model to introduce the problem of the unrestricted random walk: Consider first the motion (in a three-dimensional space‡) of a particle, which may be a molecule, an electron, or any other suitable vehicle for the physical system in question. Next, we assume that we have an ensemble of such separate systems, each with its individual particle in motion under a set of common

† The literature dealing with the random walk and its applications is voluminous. Among the earliest to consider the problem were Pearson,[7] Kluyver,[8] and Rayleigh.[9] For more modern treatments, including various constraints such as reflecting and absorbing barriers, and applications to runs, renewals, queues, and so on, see Chandrasekhar,[10] Feller,[11] and Bartlett,[12] for example, where many additional references may also be found.

‡ The number of dimensions does not affect the general argument.

conditions. Now let us suppose that the particle in each system starts at time $t = 0$ from the origin of a rectangular coordinate system. At the end of every interval of τ sec, the particle (in the jth system) is displaced instantaneously by an amount $\mathbf{r}_k^{(j)}$ ($k = 1, \ldots, n$), where the kth displacement is a vector with the rectangular components $(x_k^{(j)}, y_k^{(j)}, z_k^{(j)})$. The movement of individual particles is assumed subject to the following conditions:

1. The probability that $\mathbf{r}_k$ lies in the volume $(\mathbf{r}_k, \mathbf{r}_k + \mathbf{dr}_k)$ is governed a priori by $w_k(\mathbf{r}_k)$ $[= w_k(x_k,y_k,z_k)]$, where $w_k \neq w_l$ in general.

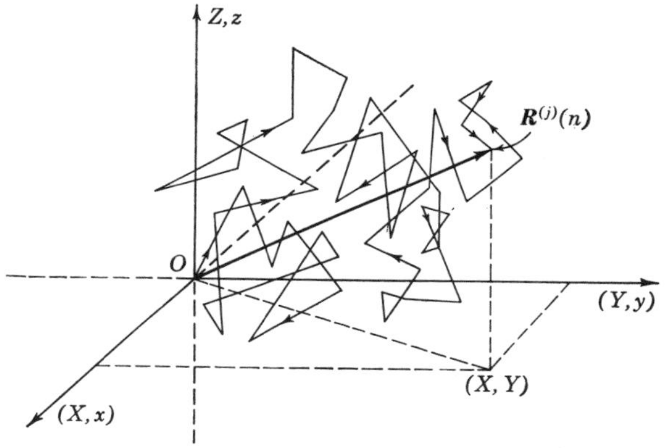

FIG. 7.1. Trajectory of a particle in a three-dimensional unrestricted random walk.

2. The magnitude and direction of each step are statistically independent of those for all other steps, that is, $W_n(\mathbf{r}_1, \ldots, \mathbf{r}_n) = \prod_k^n w_k(\mathbf{r}_k)$ [cf. Eq. (1.59)].

3. The particles encounter no absorbing or reflecting barriers but can move anywhere in space, subject, of course, to (1) and (2) above; this is the simplest class of boundary condition, characteristic of the unrestricted random walk.

The problem in its most elementary form now is to determine the second-order conditional probability† $W_2(0,0|\mathbf{R},n\tau)\,\mathbf{dR}$ that, if a particle started at $\mathbf{R} = 0$ at time $t = 0$, it is located n sec later in the volume element $(\mathbf{R}, \mathbf{R} + \mathbf{dR})$. Here $\mathbf{R}(n) = \hat{\mathbf{i}}_1X(n) + \hat{\mathbf{i}}_2Y(n) + \hat{\mathbf{i}}_3Z(n)$ is a random vector with rectangular components (X,Y,Z), and $\mathbf{dR} = dX\,dY\,dZ$. A typical trajectory for the jth particle is shown in Fig. 7.1. The possible positions

† One can also ask for more complicated statistics, such as the higher-order conditional densities $W_n[0,\ 0;\ \mathbf{R}(1),\ \tau;\ \ldots;\ \mathbf{R}(n-1),\ (n-1)\tau|\mathbf{R}(n),\ n\tau]$, etc., and from these construct the joint densities $W_n[\mathbf{R}(1),\tau_1;\ \ldots;\mathbf{R}(n),\tau n]$ $(n \geq 1)$, with the help of Secs. 1.4-3, 1.4-4. The succession of random vectors $\mathbf{R}(1), \mathbf{R}(2), \ldots, \mathbf{R}(n)$ forms *Markoff chains* if $\mathbf{R}$ is restricted to discrete values through the choice of the $w_k(\mathbf{r}_k)$ (cf. Bartlett, *op. cit.*, sec. 3.2, for example).

of a particle at $t = n\tau$, as viewed over the set of individual systems, are represented by the random variable

$$\mathbf{R}(n) = \sum_{k=1}^{n} \mathbf{r}_k \tag{7.68}$$

and from Eqs. (1.62) et seq. the c.f. of **R** can be written (if we adopt matrix notation)

$$\begin{aligned} F_{\mathbf{R}}(i\boldsymbol{\xi})_n &= \int \cdots \int e^{i\tilde{\boldsymbol{\xi}}\mathbf{R}(\mathbf{r}_1,\ldots,\mathbf{r}_n)} W_n(\mathbf{r}_1, \ldots, \mathbf{r}_n)\, d\mathbf{r}_1 \cdots d\mathbf{r}_n \\ &= \int \cdots \int e^{i\tilde{\boldsymbol{\xi}}\Sigma_k \mathbf{r}_k} W_n(\mathbf{r}_1, \ldots, \mathbf{r}_j)\, d\mathbf{r}_1 \cdots d\mathbf{r}_n \end{aligned} \tag{7.69}$$

where $\tilde{\boldsymbol{\xi}}$ is the row vector (ξ_1,ξ_2,ξ_3) and **R** is a column vector, in the usual way. By assumption 2, the individual displacements are statistically independent, so that Eq. (7.69) can be expressed as

$$F_{\mathbf{R}}(i\boldsymbol{\xi})_n = \prod_{j=k}^{n} F_{\mathbf{r}_k}(i\boldsymbol{\xi}) \tag{7.70a}$$

with

$$F_{\mathbf{r}_k}(i\boldsymbol{\xi}) = \int_{-\infty}^{\infty} \cdots \int \prod_k e^{i\tilde{\boldsymbol{\xi}}\mathbf{r}_k} w_k(\mathbf{r}_k)\, d\mathbf{r}_k$$

Abbreviating $W_2(0,0|\mathbf{R},n\tau)$ by $W_n(\mathbf{R})$, we have in any case for the probability density

$$W_n(\mathbf{R}) = \mathfrak{F}^{-1}\{F_{\mathbf{R}}\} = \iiint_{-\infty}^{\infty} e^{-i\tilde{\boldsymbol{\xi}}\mathbf{R}} F_{\mathbf{R}}(i\boldsymbol{\xi})_n \, d\xi_1\, d\xi_2\, d\xi_3/(2\pi)^3 \tag{7.70b}$$

whether or not the $\mathbf{r}_k$ are independent. The essential feature of the two- or more-dimensional random walk to be observed here, not present in the one-dimensional cases, is that there may be correlation between the components x_k, y_k, z_k of a typical step, since $w_k(r_k)$ may not factor into $w_k^{(1)}(x_k)w_k^{(2)}(y_k)w_k^{(3)}(z_k)$. Only when the steps are independent *and* the separate components are also does this correlation disappear and the components of **R** become statistically independent as a consequence.

Moments of the resultant displacement $\mathbf{R}(n)$ are found in the usual way by differentiating Eq. (7.69) or Eqs. (7.70*a*). For example, the mean of **R** (in matrix form) becomes

$$\overline{\mathbf{R}(n)} = \begin{bmatrix} \bar{X} \\ \bar{Y} \\ \bar{Z} \end{bmatrix} = -i\nabla F_{\mathbf{R}}(i\boldsymbol{\xi})_n \Big|_{\boldsymbol{\xi}=0} = \begin{bmatrix} \sum_k \bar{x}_k \\ \sum_k \bar{y}_k \\ \sum_k \bar{z}_k \end{bmatrix} \tag{7.71}$$

where ∇ is the gradient operator in rectangular coordinates; that is, ∇ is the column vector $[\partial/\partial\xi_1, \partial/\partial\xi_2, \partial/\partial\xi_3]$. For the second moments we get here a 3×3 matrix,

$$\overline{\mathbf{R}\tilde{\mathbf{R}}} = \begin{bmatrix} \overline{X^2} & \overline{XY} & \overline{XZ} \\ \overline{YX} & \overline{Y^2} & \overline{YZ} \\ \overline{ZX} & \overline{ZY} & \overline{Z^2} \end{bmatrix} = -\nabla\tilde{\nabla} F_{\mathbf{R}}(i\boldsymbol{\xi})_n \Big|_{\boldsymbol{\xi}=0} \tag{7.72}$$

The elements of $E\{\mathbf{R}\tilde{\mathbf{R}}\}$ are

$$\overline{X^2} = \sum_k^n \overline{x_k{}^2} + \sum_k^n{}' \overline{x_k x_l} = \sum_k^n (\overline{x_k{}^2} - \bar{x}_k{}^2) + \bar{X}^2 \tag{7.73a}$$

$$\overline{XY} = \sum_k^n \overline{x_k y_k} + \sum_{kl}^n{}' \overline{x_k y_l} = \sum_k^n \overline{(x_k - \bar{x}_k)(y_k - \bar{y}_k)} + \bar{X}\bar{Y}, \text{ etc.} \tag{7.73b}$$

From Eqs. (7.72), (7.73), we observe that the covariance matrix for this random-walk process can be written at once as

$$\mathbf{K}_R = E\{(\mathbf{R} - \bar{\mathbf{R}})(\tilde{\mathbf{R}} - \bar{\tilde{\mathbf{R}}})\} = \overline{\mathbf{R}\tilde{\mathbf{R}}} - \bar{\mathbf{R}}\bar{\tilde{\mathbf{R}}} = \sum_{k=1}^n \mathbf{K}_{\mathbf{r}_k} \tag{7.74}$$

where $\mathbf{K}_{\mathbf{r}_k}$ is the 3×3 covariance matrix of the kth displacement, with typical elements $\overline{x_k{}^2} - \bar{x}_k{}^2$, $\overline{(x_k - \bar{x}_k)(y_k - \bar{y}_k)}$, etc. Higher-order moments may be determined in a similar way.

7.7-2 Random Walk with a Large Number of Steps. The specific structure of the c.f. and d.d. of (the components of) $\mathbf{R}(n)$ will depend, of course, on the number of steps and on the d.d.'s of the individual displacements $\mathbf{r}_k$. However, when the $\mathbf{r}_k$ are statistically independent and the number n of steps becomes large, the d.d. $W_n(\mathbf{R})$ [Eq. (7.70b)] under these conditions approaches a normal law,† whose properties we have considered already in the preceding sections.

To show this, let us start with Eqs. (7.70a) and rewrite them as

$$F_{\mathbf{R}}(i\boldsymbol{\xi})_n = e^{i\tilde{\boldsymbol{\xi}}\bar{\mathbf{R}}} \prod_{k=1}^n [1 + f_k(i\boldsymbol{\xi})] \tag{7.75a}$$

where

$$f_k(i\boldsymbol{\xi}) \equiv \int_{(\mathbf{r}_k)} (e^{i\tilde{\boldsymbol{\xi}}(\mathbf{r}_k - \bar{\mathbf{r}}_k)} - 1) w_k(\mathbf{r}_k)\, d\mathbf{r}_k \qquad \text{since} \int_{(\mathbf{r}_k)} w_k(\mathbf{r}_k)\, d\mathbf{r}_k = 1 \tag{7.75b}$$

We take the logarithm of both members of Eq. (7.75a) to get the cumulant function $\mathcal{K}_R(i\boldsymbol{\xi})$ [cf. Eq. (1.58)], and then we expand $\sum_k^n \log [1 + f_k(i\boldsymbol{\xi})]$ to

† Sufficient conditions are given in Sec. 7.7-3.

write finally, with the help of Eqs. (7.71) et seq.,

$$\mathcal{K}_R(i\boldsymbol{\xi}) = i\tilde{\boldsymbol{\xi}}\bar{\mathbf{R}} - \tfrac{1}{2}\tilde{\boldsymbol{\xi}}\mathbf{K}_R\boldsymbol{\xi} + \sum_{k=1}^{n} \left\{ \frac{-ia_3(\boldsymbol{\xi})_k}{3!} + \left[\frac{a_4(\boldsymbol{\xi})_k}{4!} - \frac{a_2(\boldsymbol{\xi})_k{}^2}{8}\right] + i\left[\frac{a_5(\boldsymbol{\xi})_k}{5!} - \frac{a_2(\boldsymbol{\xi})_k a_3(\boldsymbol{\xi})_k}{2!3!}\right] + \left[\frac{-a_6(\boldsymbol{\xi})_k}{6!} + \frac{a_3(\boldsymbol{\xi})_k{}^2}{2 \times 3!} + \frac{a_2(\boldsymbol{\xi})_k a_4(\boldsymbol{\xi})_k}{2!4!} - \frac{a_2(\boldsymbol{\xi})_k{}^3}{24}\right] + \cdots \right\} \tag{7.76}$$

in which

$$a_m(\boldsymbol{\xi})_k \equiv E\{[\tilde{\boldsymbol{\xi}}(\mathbf{r}_k - \bar{\mathbf{r}}_k)]^m\} \tag{7.76a}$$

At this point, it is convenient to introduce the semi-invariants $\lambda^{(L)}_{l_1l_2l_3}$, $l_1 + l_2 + l_3 = L\ (\geq 1)$ in terms of which the cumulant function may alternatively be written [cf. Eqs. (1.58), (1.58*a*)]. Comparing Eq. (7.76) with the definition of Eqs. (1.58), (1.58*a*), we find (cf. Prob. 7.12) that for the first three orders ($L \leq 3$) these semi-invariants are given by

$$\lambda^{(L)}_{l_1l_2l_3} = \sum_{k=1}^{n} E\{(x_k - \bar{x}_k)^{l_1}(y_k - \bar{y}_k)^{l_2}(z_k - \bar{z}_k)^{l_3}\} \qquad l_1 + l_2 + l_3 = 1,\ 2,\ \text{or}\ 3\ (\text{all}\ l \geq 0) \tag{7.77}$$

This simple relation does not hold, however, for the higher-order semi-invariants ($L \geq 4$), as a glance at the table of Prob. 7.13 will show. [Note that Eq. (7.77) is not restricted to three-dimensional vectors: for q dimensions, we can write analogously

$$\lambda^{(L)}_{l_1,\ldots,l_{(q)}} = \sum_{k=1}^{n} E\{(x_{1k} - \bar{x}_{1k})^{l_1}(x_{2k} - \bar{x}_{2k})^{l_2} \cdots (x_{qk} - \bar{x}_{qk})^{l_q}\} \qquad \sum_{j=1}^{q} l_j = L = 1,\ 2,\ \text{or}\ 3\ (q \geq 1) \tag{7.77a}$$

where as before $L \geq l_j \geq 0$.]

Reexpressing the cumulant function $\mathcal{K}_R$ above in terms of the semi-invariants according to Eqs. (1.58), (1.58*a*), we now retain terms of $O(\boldsymbol{\xi},\boldsymbol{\xi}^2)$ only in the exponent and develop the remainder,

$$\exp\left[\sum_{L=3}^{\infty} \lambda^{(L)}_{l_1l_2l_3}(i\xi_1)^{l_1} \cdots (i\xi_3)^{l_3}/l_1! \cdots l_3!\right]$$

in a series in (ξ_1,ξ_2,ξ_3). The justification of this will appear as soon as we obtain these remainder terms as functions of n, as n grows large. Accordingly, we can write

$$F_{\mathbf{R}}(i\boldsymbol{\xi})_n = e^{-i\tilde{\boldsymbol{\xi}}\bar{\mathbf{R}} - \frac{1}{2}\tilde{\boldsymbol{\xi}}\mathbf{K}_R\boldsymbol{\xi}}[1 + D(\xi_1,\xi_2,\xi_3)] \tag{7.78}$$

where the correction D is explicitly

$$D(\xi_1,\xi_2,\xi_3) = \sum_3 A_3 + \left[\sum_4 A_4 + \frac{1}{2}\left(\sum_3 A_3\right)^2\right] + O_{\mathfrak{F}}(n^{-3/2}) \quad (7.79)$$

with $\Sigma_L A_L = \Sigma_L \lambda^{(L)}_{l_1l_2l_3}(i\xi_1)^{l_1}(i\xi_2)^{l_2}(i\xi_3)^{l_3}/l_1!l_2!l_3!$ ($L = 3, 4, \ldots$). Since $\lambda^{(L)}$ is $O(n)$ (all $L \geq 1$), as is directly evident from Eq. (7.77) for $L = 1, 2, 3$ and from Eq. (7.76) in conjunction with Eq. (1.58) for all L (≥ 1), in general, we have collected terms in Eq. (7.79) in such a way that the first yields a "correction" of $O(n^{-1/2})$ relative to the leading term in the corresponding d.d. ($= \mathfrak{F}^{-1}\{F_{\mathbf{R}}\}$); the second, of $O(n^{-1})$; and so on. The symbol $O_{\mathfrak{F}}(n^{-3/2})$ in Eq. (7.79) indicates that the additional terms in the development are $O(n^{-3/2})$ as $n \to \infty$, when the Fourier transform of Eq. (7.79) is taken. The characteristic function (7.78) is therefore

$$F_{\mathbf{R}}(i\boldsymbol{\xi})_n \doteq e^{i\tilde{\boldsymbol{\xi}}\bar{\mathbf{R}} - \frac{1}{2}\tilde{\boldsymbol{\xi}}K_R\boldsymbol{\xi}} \left\{1 + \sum_{L=3}^{3} A_L + \left[\sum_{L=4}^{4} A_L + \frac{1}{2}\left(\sum_{L=3}^{3} A_L\right)^2\right] + O_{\mathfrak{F}}(n^{-3/2})\right\} \quad (7.80)$$

The d.d. of $\mathbf{R}$ is found to be† (cf. Prob. 7.12)

$$W_n(\mathbf{R}) \simeq \left\{1 + \sum_{L=3}^{3} B_L + \left[\sum_{L=4}^{4} B_L + \frac{1}{2}\left(\sum_{L=3}^{3} B_L\right)^2\right] + O(n^{-3/2})\right\} \times e^{-\frac{1}{2}(\tilde{\mathbf{R}}-\tilde{\bar{\mathbf{R}}})\mathbf{K}_R^{-1}(\mathbf{R}-\bar{\mathbf{R}})}(2\pi)^{-3/2}(\det \mathbf{K}_R)^{-1/2} \quad (7.81)$$

in which

$$B_L \equiv (-1)^L \frac{\lambda^{(L)}_{l_1l_2l_3}}{l_1!l_2!l_3!} \frac{\partial^L}{\partial X^{l_1}\,\partial Y^{l_2}\,\partial Z^{l_3}}$$

is a differential operator (in rectangular coordinates) acting on the multivariate normal density above in Eq. (7.81). The one-dimensional case of Eq. (7.81) is known as *Edgeworth's series*[14] (cf. Prob. 7.11).

Thus, from Eq. (7.81), we see in a formal way that for large numbers of independent steps a normal law for the components of the resultant position vector $\mathbf{R}(n)$ is asymptotically approached, and in the limit $n \to \infty$ we get precisely the form of a normal law, with $\bar{\mathbf{R}} = O(n)$, $\mathbf{K}_R = \mathbf{O}(n)$, and $\det \mathbf{K}_R = O(n^3)$, so that $\mathbf{K}_R^{-1} = \mathbf{O}(n^{-1})$ and $\lim_{n\to\infty} W_n(\mathbf{R}) \to 0$. Since $\bar{\mathbf{R}}$ varies with n, that is, with $n\tau$, the continuous random series that is our present model of the random walk is also nonstationary. However, if we change the scale of observations and introduce now the standardized random variables as components of the random vector $\mathbf{r}$,

$$\mathbf{r} = \hat{\mathbf{i}}_1\left(\frac{X - \bar{X}}{\sqrt{\mu_{11}}}\right) + \hat{\mathbf{i}}_2\left(\frac{Y - \bar{Y}}{\sqrt{\mu_{22}}}\right) + \hat{\mathbf{i}}_3\left(\frac{Z - \bar{Z}}{\sqrt{\mu_{33}}}\right) = \hat{\mathbf{i}}_1 r_1 + \hat{\mathbf{i}}_2 r_2 + \hat{\mathbf{i}}_3 r_3 \quad (7.82)$$

† The asymptotic nature of the density function $W_n(\mathbf{R})$ has been examined, for example, by Cramér;[13] see also Cramér,[1] sec. 17.7, for a discussion of the one-dimensional case and Gnedenko and Kolmogoroff[15] for a general treatment.

where $\mu_{kk} = \boldsymbol{E}\{(X - \bar{X})^2\}, \boldsymbol{E}\{(Y - \bar{Y})^2\}, \boldsymbol{E}\{(Z - \bar{Z})^2\}$ ($k = 1, 2, 3$, respectively), we see that $\bar{\mathbf{r}} = 0$, identically, and $\overline{\mathbf{r}\tilde{\mathbf{r}}} = O(n^0)$, as $n \to \infty$. With the transformations (7.82) we find (cf. Prob. 7.12) that the c.f. and d.d. of the new random variables $\mathbf{r}$ take forms paralleling Eqs. (7.80), (7.81), namely,

$$F_{\mathbf{r}}(i\boldsymbol{\xi})_n = e^{-\frac{1}{2}\tilde{\boldsymbol{\xi}}\boldsymbol{\mu}_r\boldsymbol{\xi}} \left\{1 + \sum_{L=3}^{3} a_L + \left[\sum_{L=4}^{4} a_L + \frac{1}{2}\left(\sum_{L=3}^{3} a_L\right)^2\right] + O_{\mathfrak{F}}(n^{-3/2})\right\} \tag{7.83}$$

where

$$a_L = \Sigma_L \lambda^{(L)}_{l_1l_2l_3}(i\xi_1 \sqrt{\mu_{11}})^{l_1}(i\xi_2 \sqrt{\mu_{22}})^{l_2}(i\xi_3 \sqrt{\mu_{33}})^{l_3}/l_1!l_2!l_3!$$

and

$$W_n(\mathbf{r}) \simeq \left\{1 + \sum_{L=3}^{3} b_L + \left[\sum_{4}^{4} b_L + \frac{1}{2}\left(\sum_{3}^{3} b_L\right)^2\right] + O(n^{-3/2})\right\} e^{-\frac{1}{2}\tilde{\mathbf{r}}\boldsymbol{\mu}_r^{-1}\mathbf{r}} \tag{7.84}$$

in which

$$b_L \equiv (-1)^L \lambda^{(L)}_{l_1l_2l_3}(l_1!l_2!l_3!\mu_{11}{}^{l_1/2}\mu_{22}{}^{l_2/2}\mu_{33}{}^{l_3/2})^{-1} \frac{\partial^L}{\partial r_1{}^{l_1}\,\partial r_2{}^{l_2}\,\partial r_3{}^{l_3}} [= O(n^{1-L/2})] \tag{7.84a}$$

The variance matrix $\boldsymbol{\mu}_r$ is specifically $\boldsymbol{\mu}_r = \boldsymbol{E}\{\mathbf{r}\tilde{\mathbf{r}}\}$, with $\boldsymbol{E}\{\mathbf{r}\} = 0$. In the limit $n \to \infty$, this change of scale (7.82) converts the discrete or continuous *series* represented by the vector random walk $\mathbf{R}(n)$ into a continuous (vector) random *process* $\mathbf{r}(t)$, which is also stationary (since independent of n). In this way, the components of $\mathbf{r}(t)$ become related random processes, descriptive of the large-scale phenomena for which the random-walk model provides a microscopic counterpart. We note here in particular that the normal law achieved in the limit is independent of the detailed statistical laws governing individual displacements. This is an example of a fundamental result in probability theory, known as the central-limit theorem, which is described briefly in Sec. 7.7-3.

7.7-3 The Central-limit Theorem.† As we have already noted in Sec. 7.7-2, under certain conditions the sums of large numbers n of independent random variables approach a normal distribution as $n \to \infty$. Here we shall state a very general form of the central-limit theorem, due to Liapounoff,[17] which gives the sufficient conditions for such behavior. We shall also illustrate its application with a simple example of a two-dimensional random

† Because of its great importance in probability theory, the central-limit theorem has been extensively investigated by many mathematicians, from Laplace, who discovered it, to Tchebycheff, Markoff, and Liapounoff, at the end of the nineteenth century, and, in more recent times, Bernstein, Feller, Gnedenko, Kolmogoroff, Khintchine, Lévy, and many others. Like the problem of the random walk, the literature on the subject is vast; for a comprehensive account, see Gnedenko and Kolmogoroff's recent book[15] and Uspensky's earlier publication,[16] chaps. 14, 15; both works have extensive bibliographies.

walk. Since the statement of the theorem for the two- or more-dimensional process is basically no more complicated than for the one-dimensional case, we shall present it here as follows:[18]

Central-limit Theorem[16–18]

1. Let $x_k^{(1)}, x_k^{(2)}, \ldots, x_k^{(m)}, \ldots, x_k^{(M)}$ $(k = 1, \ldots, n;\ m = 1, \ldots, M)$ be M sets of n *independent* random variables, with $\boldsymbol{E}\{x_k^{(m)}\} < \infty$, $\boldsymbol{E}\{(x_k^{(m)} - \bar{x}_k^{(m)})^2\} < \infty$, and with d.d. $w_k^{(m)}(x_k^{(m)})$, for all m, k.

2. The absolute moments

$$\beta_{k,2+\delta}^{(m)} \equiv \overline{|x_k^{(m)} - \bar{x}_k^{(m)}|^{2+\delta}} = \int_{-\infty}^{\infty} |x_k^{(m)} - \bar{x}_k^{(m)}|^{2+\delta} w_k^{(m)}(x_k^{(m)})\, dx_k^{(m)}$$

exist, for some $\delta > 0$.

Now let us write

$$C_n^{(m)} \equiv \sum_{k=1}^{n} \overline{(x_k^{(m)} - \bar{x}_k^{(m)})^2} \qquad D_n^{(m)} \equiv \sum_{k=1}^{n} \beta_{k,2+\delta}^{(m)} \tag{7.85}$$

3. If

$$\lim_{n\to\infty} D_n^{(m)}/\{C_n^{(m)}\}^{1+\delta/2} \to 0 \qquad \text{all } m = 1, \ldots, M \tag{7.86}$$

then the probability density of the M standardized random variables

$$u_m = \sum_{k=1}^{n} (x_k^{(m)} - \bar{x}_k^{(m)})/\{C_n^{(m)}\}^{1/2} \qquad m = 1, \ldots, M \tag{7.87}$$

is asymptotically normal, with $\boldsymbol{E}\{u_m\} = 0$, $\boldsymbol{E}\{u_m^2\} = 1$, and the (normalized) $M \times M$ covariance matrix $\mathbf{k_u} = \boldsymbol{E}\{u_i u_j\}$ $(i, j = 1, \ldots, M)$.

Thus, we have

$$\lim_{n\to\infty} W_n(u_1, \ldots, u_m) \cong (2\pi)^{-M/2}(\det \mathbf{k}_u)^{-1/2} e^{-\frac{1}{2}\tilde{\mathbf{u}}\mathbf{k}_u^{-1}\mathbf{u}} \tag{7.88}$$

where this limit is approached uniformly. Note that the distribution densities $w_k^{(m)}(x_k^{(m)})$ of the components of u_m can be quite arbitrary, subject to (1) and (2), of course. The condition of independence ensures that the d.d.'s of no one set of random variables dominate the form of the resulting distribution, and the existence of $\beta_{k,2+\delta}^{(m)}$ does the same for any individual variable, while condition 3 is enough to ensure that all higher-order corrections vanish in the limit, as in the three-dimensional random walk above [cf. Eqs. (7.81), (7.84)]. It should be pointed out, of course, that, as these correction terms are $O(n^{-1/2})$ [if the $x_k^{(m)}$'s possess finite (> 0) third moments†], the approach to the normal law is not very rapid and n must be quite large, particularly if we are concerned with the tails of the d.d. $W_n(u_1, \ldots, u_m)$. We remark, finally, that the condition of independence on the $x_k^{(m)}$ can be relaxed in certain circumstances, as Bernstein[18] has shown. Thus, when we have large numbers of *sets* of correlated

† See Gnedenko and Kolmogoroff,[15] par. 40, theorem 1.

variables, where the sets are, however, statistically independent, we may expect a form of the central-limit theorem for the sum of the sets, and so on.

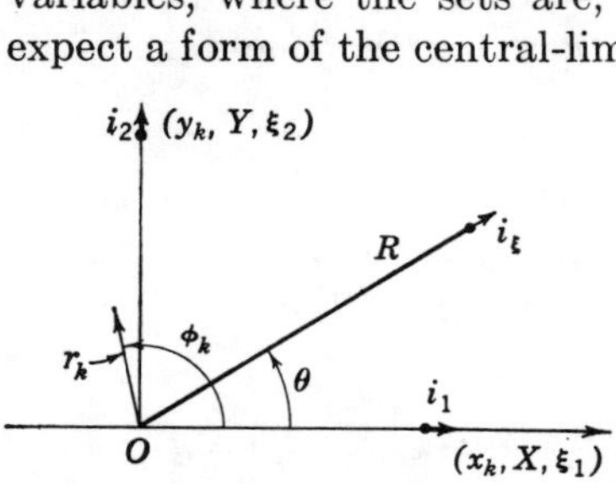

FIG. 7.2. Coordinate systems for $\mathbf{R}$, $\mathbf{r}_k$, and $\mathbf{i}_\xi$.

Let us illustrate the application of the central-limit theorem with the example of a two-dimensional random walk. Here the steps are all taken with fixed lengths a_k but are uniformly distributed in direction; i.e., the angle ϕ_k which each step vector makes with some rectangular coordinate system is uniformly distributed in $(0,2\pi)$ (cf. Fig. 7.2). Then the d.d. of the displacements can conveniently be expressed in terms of the polar coordinates ρ_k, ϕ_k, for example, $x_k = \rho_k \cos \phi_k$, $y_k = \rho_k \sin \phi_k$, and

$$w_k(\mathbf{r}_k) = W_k(\rho_k,\phi_k) = \delta(\rho_k - a_k)\,\frac{1}{2\pi} \qquad 0 \leq \phi_k < 2\pi,\ 0 \leq \rho_k \leq \infty \tag{7.89}$$

where $\mathbf{r}_k = \mathbf{i}_1 x_k + \mathbf{i}_2 y_k$ in the usual way. Now, to obtain the needed ratio (7.86), let us adopt the following device, from which these moments may be computed: We choose $\mathbf{v}_\xi$ as a unit vector in the direction of $\boldsymbol{\xi}$, which we regard as fixed in the (ξ_1,ξ_2) plane, at the angle θ with respect to the ξ_1 and x axes (cf. Fig. 7.2). Consider next the random vector

$$\mathbf{v} = \mathbf{i}_\xi \sum_{k=1}^{n} |\mathbf{r}_k - \bar{\mathbf{r}}_k|^\gamma \qquad \gamma > 0 \tag{7.90}$$

and observe that, since the $\mathbf{r}_k$ are independent, so also are the ρ_k, ϕ_k from step to step [from Eq. (7.89) ρ_k and ϕ_k are also independent (all k)]. Since $\bar{\mathbf{r}}_k = 0$, we obtain from Eq. (7.89) the following c.f. for $\mathbf{v}$:

$$F_{\mathbf{v}}(i\zeta) = \prod_{k=1}^{n} \iint_{-\infty}^{\infty} e^{i\zeta\rho_k{}^\gamma} W_k(\rho_k,\phi_k)\, d\rho_k\, d\phi_k = e^{i\zeta \sum_{k=1}^{n} a_k{}^\gamma} \tag{7.91}$$

Differentiation with respect to ζ gives

$$\beta_{k,\gamma} \equiv \overline{|\mathbf{r}_k - \bar{\mathbf{r}}_k|^\gamma} = -i\,\frac{d}{d\zeta} F_{\mathbf{v}}(i\zeta)\Big|_{\zeta=0} = \sum_{k=1}^{n} a_k{}^\gamma \tag{7.92}$$

so that the ratio (7.86) becomes (for $\gamma = 2,\ 2+\delta$)

$$\lim_{n\to\infty} \frac{D_n{}^{(1)}}{(C_n{}^{(1)})^{1+\delta/2}} = \lim_{n\to\infty} \frac{\sum_{k=1}^{n} a_k{}^{2+\delta}}{\left(\sum_{k=1}^{n} a_k{}^2\right)^{1+\delta/2}} \to 0: \qquad O(n^{-\delta/2}) \tag{7.93}$$

at least, since all a_k are bounded. Condition 3 then holds, and the central-limit theorem can now be applied in the limit to yield a two-dimensional normal density in the *rectangular* components X, Y† of the resultant vector **R**.

From Eq. (7.89) and the fact that $\mathbf{R} = \sum_{k=1}^{n} \mathbf{r}_k$ [cf. Eq. (7.68)], we can equivalently express X, Y in terms of the polar coordinates R ($= |\mathbf{R}|$) and θ, for example,

$$X = R\cos\theta = \sum_k^n \rho_k \cos\phi_k \qquad Y = R\sin\theta = \sum_k^n \rho_k \sin\phi_k \tag{7.94}$$

Next we use Eqs. (7.71) to (7.74) to obtain the first and second moments $\boldsymbol{E}\{X\} = \boldsymbol{E}\{Y\} = 0$, and

$$\begin{aligned}\overline{X^2} &= \sum_k^n a_k^2\,\overline{\cos^2\phi_k} = \frac{1}{2}\sum_k^n a_k^2 \\ \overline{Y^2} &= \sum_k^n a_k^2\,\overline{\sin^2\phi_k} = \frac{1}{2}\sum_k^n a_k^2 \\ \overline{XY} &= \sum_k^n a_k^2\,\overline{\sin\phi_k\cos\phi_k} = 0\end{aligned} \tag{7.95}$$

Since the ϕ_k are independent of ρ_k and uniformly distributed in $(0,2\pi)$, θ is likewise uniformly distributed, independent of R, and so from Eqs. (7.94), (7.95) we have also

$$\overline{R^2} = \sum_{kl} \overline{\rho_k\rho_l\cos(\phi_k - \phi_l)} = \sum_{k=1}^n a_k^2 \tag{7.96}$$

The variance matrix of the random variables X and Y is accordingly

$$\mathbf{K}_R = \overline{\mathbf{R}\tilde{\mathbf{R}}} = \begin{bmatrix} \frac{1}{2}\overline{R^2} & 0 \\ 0 & \frac{1}{2}\overline{R^2}\end{bmatrix} = \begin{bmatrix} \frac{1}{2}\sum_k^n a_k^2 & 0 \\ 0 & \frac{1}{2}\sum_k^n a_k^2\end{bmatrix} \tag{7.97}$$

† Strictly speaking, the central-limit theorem yields a normal density for the *standardized* variables $x = X/(\overline{X^2})^{1/2}$, $y = Y/(\overline{Y^2})^{1/2}$ [cf. Eqs. (7.87), (7.88)]; however, for later reference we shall consider X, Y the variables of interest here and obtain the approximate results $W_n(\mathbf{R}) = W_n(X,Y)$ for large n from the d.d. of x, y.

so that, for large n, the central-limit theorem shows that

$$W_n(x,y) \simeq \frac{e^{-\frac{1}{2}(x^2+y^2)}}{2\pi}$$

and therefore

$$W_n(X,Y) \simeq \frac{e^{-(X^2+Y^2)/\overline{R^2}}}{\pi \overline{R^2}} \tag{7.98}$$

Transforming to polar coordinates and remembering that the jacobian is R, we get from Eqs. (7.94) in Eqs. (7.98) the associated d.d.

$$W_n(R,\theta) \simeq \frac{Re^{-R^2/\overline{R^2}}}{\pi \overline{R^2}} \qquad 0 \le R < \infty, 0 \le \theta < 2\pi \tag{7.99}$$

which was originally derived by Rayleigh.[9] Because of circular symmetry, $W_n(R,\theta)$ does not depend on θ, a result which we might have predicted at once from the isotropy of the individual displacements. From Eq. (7.99), we obtain the marginal densities

$$W_n(R) \simeq \frac{2R}{\overline{R^2}} e^{-R^2/\overline{R^2}} \qquad 0 \le R < \infty$$

$$W_n(\theta) \simeq \frac{1}{2\pi} \qquad 0 \le 0 < 2\pi \tag{7.100}$$

the first of which is one form of the familiar *Rayleigh distribution* (density). We shall encounter it again, with some modifications, in Chap. 9 in dealing with the statistics of the envelope of narrowband normal noise processes. Note, too, that, as stated above, θ is uniformly distributed in the primary interval $(0,2\pi)$.

From the preceding discussion (here and Sec. 7.7-2), then, we may say in summary that, whenever the underlying mechanism of a noise or signal source can be represented by a linear superposition of large numbers of essentially independent random effects, the statistics of the resultant process are normal. Because this is, in fact, true of many of the inherent noise sources encountered in actual systems, the normal process plays a dominant role in the theory of communication.

PROBLEMS

7.11 (*a*) For the one-dimensional random walk, with independent steps, obtain the Edgeworth series

$$W_n(\mathbf{R}) = W_n(X) \simeq \left[1 - \frac{\lambda_3^{(3)}}{3!}\frac{d^3}{dX^3} + \left(\frac{\lambda_4^{(4)}}{4!}\frac{d^4}{dX^4} + \frac{\lambda_3^{(3)2}}{3 \times 4!}\frac{d^6}{dX^6}\right) + \cdots\right] \times \frac{e^{-(X-\bar{X})^2/2\mu_{11}}}{\sqrt{2\pi\mu_{11}}} \tag{1}$$

with $\mu_{11} = \mathbf{E}\{(X - \bar{X})^2\}$ and the c.f.

$$F_X(i\xi) \doteq e^{i\bar{X}\xi - \frac{1}{2}\mu_{11}\xi^2} \left\{1 + \frac{\lambda_3^{(3)}}{3!}(i\xi)^3 + \left[\frac{\lambda_4^{(4)}}{4!}(i\xi)^4 + \frac{\lambda_3^{(3)2}}{72}(i\xi)^6\right] + \left[\frac{\lambda_5^{(5)}}{5!}(i\xi)^5 + \frac{\lambda_3^{(3)}\lambda_4^{(4)}}{144}(i\xi)^7 + \frac{\lambda_3^{(3)3}}{6^4}(i\xi)^9\right] + \cdots\right\} \quad (2)$$

(*b*) Verify that

$$\lambda_1^{(1)} = \sum_k^n \bar{X}_k \qquad \lambda_2^{(2)} = \sum_k^n \overline{(x_k - \bar{x}_k)^2} \qquad \lambda_3^{(3)} = \sum_k^n \overline{(x_k - \bar{x}_k)^3}$$

$$\lambda_4^{(4)} = \sum_k^n [\overline{(x_k - \bar{x}_k)^4} - 3\overline{(x_k - \bar{x})^2}^2] \quad (3)$$

7.12 (*a*) Prove Eq. (7.77) for the semi-invariants $L = 1, 2, 3$ in the three-dimensional random walk of Sec. 7.7-2, and obtain Eq. (7.76).

(*b*) Carry out the details of the work leading to Eq. (7.81) and to Eqs. (7.83), (7.84).

7.13 For the three-dimensional random walk of Sec. 7.7-2, derive the following table for the fourth-order semi-invariants $\lambda^{(4)}_{l_1l_2l_3}$ $(l_1 + l_2 + l_3 = 4)$:

TABLE OF FOURTH-ORDER SEMI-INVARIANTS

$\lambda^{(4)}_{400} = \Sigma_k(\overline{a_k{}^4} - 3\overline{a_k{}^2}^2)$	$\lambda^{(4)}_{040} = \Sigma_k(\overline{b_k{}^4} - 3\overline{b_k{}^2}^2)$
$\lambda^{(4)}_{301} = \Sigma_k(\overline{a_k{}^3c_k} - 3\overline{a_k{}^2}\,\overline{a_kc_k})$	$\lambda^{(4)}_{031} = \Sigma_k(\overline{b_k{}^3c_k} - 3\overline{b_k{}^2}\,\overline{b_kc_k})$
$\lambda^{(4)}_{310} = \Sigma_k(\overline{a_k{}^3b_k} - 3\overline{a_k{}^2}\,\overline{a_kb_k})$	$\lambda^{(4)}_{130} = \Sigma_k(\overline{a_kb_k{}^3} - 3\overline{b_k{}^2}\,\overline{a_kb_k})$
$\lambda^{(4)}_{220} = \Sigma_k(\overline{a_k{}^2b_k{}^2} - 2\overline{a_kb_k}^2 - \overline{a_k{}^2}\,\overline{b_k{}^2})$	$\lambda^{(4)}_{202} = \Sigma_k(\overline{a_k{}^2c_k{}^2} - 2\overline{a_kc_k}^2 - \overline{a_k{}^2}\,\overline{c_k{}^2})$
$\lambda^{(4)}_{211} = \Sigma(\overline{a_k{}^2b_kc_k} - \overline{a_k{}^2}\,\overline{b_kc_k})$	$\lambda^{(4)}_{121} = \Sigma_k(\overline{a_kb_k{}^2c_k} - \overline{b_k{}^2}\,\overline{a_kc_k})$

$\lambda^{(4)}_{004} = \Sigma_k(\overline{c_k{}^4} - 3\overline{c_k{}^2}^2)$
$\lambda^{(4)}_{013} = \Sigma_k(\overline{b_kc_k{}^3} - 3\overline{c_k{}^2}\,\overline{b_kc_k})$
$\lambda^{(4)}_{103} = \Sigma_k(\overline{a_kc_k{}^3} - 3\overline{c_k{}^2}\,\overline{a_kc_k})$
$\lambda^{(4)}_{022} = \Sigma_k(\overline{b_k{}^2c_k{}^2} - 2\overline{b_kc_k}^2 - \overline{b_k{}^2}\,\overline{c_k{}^2})$
$\lambda^{(4)}_{112} = \Sigma_k(\overline{a_kb_kc_k{}^2} - \overline{c_k{}^2}\,\overline{b_ka_k})$

$a_k = x_k - \bar{x}_k,\ b_k = y_k - \bar{y}_k,\ c_k = z_k - \bar{z}_k$

7.14 If the basic probability density $w_k(\mathbf{r}_k) = w_k(x_k,y_k)$ is the same for each step in a two-dimensional random walk and if the $\mathbf{r}_k$ are isotropically distributed so that in polar coordinates one can write

$$w_k(\mathbf{r}_k)\,\mathbf{dr}_k = w(\mathbf{r})\,\mathbf{dr} = \frac{1}{2\pi}G(r)\,dr\,d\phi \qquad r = |\mathbf{r}| \quad (1)$$

show that as the number n of steps becomes large

$$W_n(\mathbf{R}) = W_n(X,Y) \simeq \left[1 + \frac{\overline{R^2}^2}{64n}\left(\frac{\overline{r^4}}{\overline{r^2}^2} - 2\right)(\tilde{\nabla}\nabla)^2 + O(n^{-2})\right](\pi\overline{R^2})^{-1}e^{-(X^2+Y^2)/\overline{R^2}} \quad (2)$$

where $\overline{R^2} = n\overline{r^2}$, and $\overline{r^2} = \int_0^\infty G(r)r^2\,dr$; $\overline{r^4} = \int_0^\infty G(r)r^4\,dr$. Here $\tilde{\nabla}\nabla$ is the Laplacian $(\partial^2/\partial X^2 + \partial^2/\partial Y^2)$; Eq. (2) is a generalization of the example discussed in Sec. 7.7-3, including a correction term.

7.15 Repeat Prob. 7.14, but now for the three-dimensional case. Here we have

$$w(\mathbf{r})\,d\mathbf{r} = \frac{1}{4\pi}\sin\theta\, G(r)\,dr\,d\theta\,d\phi \qquad (0 \le \theta < \pi;\ 0 \le \phi < 2\pi;\ 0 \le r < \infty) \tag{1}$$

Hence, show that

$$W_n(\mathbf{R}) = W_n(X,Y,Z) \simeq \left[1 + \frac{\overline{R^2}^2}{120n}\left(\frac{\overline{r^4}}{\overline{r^2}^2} - \frac{5}{3}\right)(\tilde{\nabla}\nabla)^2 + O(n^{-2})\right]\frac{e^{-3(X^2+Y^2+Z^2)/2\overline{R^2}}}{(\frac{2}{3}\pi\overline{R^2})^{3/2}} \tag{2}$$

where, as before, $\overline{R^2} = n\overline{r^2} = \overline{X^2} + \overline{Y^2} + \overline{Z^2}$ and

$$\tilde{\nabla}\nabla = \frac{\partial^2}{\partial X^2} + \frac{\partial^2}{\partial Y^2} + \frac{\partial^2}{\partial Z^2} \tag{3}$$

REFERENCES

1. Cramér, H.: "Mathematical Methods of Statistics," chap. 17, Princeton University Press, Princeton, N.J., 1946.
2. Margenau, H., and G. Murphy: "The Mathematics of Physics and Chemistry," chap. 10, Van Nostrand, Princeton, N.J., 1943.
3. Frazer, R. A., W. J. Duncan, and A. R. Collar: "Elementary Matrices," Cambridge, New York, 1950.
4. Siegert, A. J. F.: Passage of Stationary Processes through Linear and Non Linear Devices, *IRE Trans. on Inform. Theory*, **PGIT-3**: 4 (March, 1954).
5. Doob, J. L.: The Brownian Movement and Stochastic Equations, *Annals Math. Stat.*, **13**: 351 (1942), **15**: 229 (1944).
6. Wang, M. C., and G. E. Uhlenbeck: On the Theory of the Brownian Motion. II, *Revs. Modern Phys.*, **17**: 323 (1945), sec. 7 and app. II.
7. Pearson, K.: *Nature*, **72**: 294, 342 (1905). See also *Draper's Company Research Mem.*, biometric ser. III, 1906.
8. Kluyver, J. C.: *Proc. Sec. Sci. Koninkl. Akad. Wetenschap. Amsterdam*, **8**: 341 (1906).
9. Rayleigh, Lord: *Phil. Mag.*, (6)**37**: 321 (1919); *Sci. Papers*, **6**: 604 (1920).
10. Chandrasekhar, S.: Stochastic Problems in Physics and Astronomy, *Revs. Modern Phys.*, **15**: 1 (1943), chap. 1.
11. Feller, W.: "An Introduction to Probability Theory and Its Applications," vol. 1, chap. 14, Wiley, New York, 1950.
12. Bartlett, M. S.: "An Introduction to Stochastic Processes," chaps. 2–4, Cambridge, New York, 1955.
13. Cramér, H.: "Random Variables and Probability Distributions," no. 36 in "Cambridge Tracts in Mathematics and Mathematical Physics," Cambridge, New York, 1937; *Skand. Aktuarietidskr.*, **1928**: 13, 141.
14. Edgeworth, F. V.: The Law of Error, *Proc. Cambridge Phil. Soc.*, vol. 36, 1905.
15. Gnedenko, B. V., and A. N. Kolmogoroff: "Limit Distributions for Sums of Independent Random Variables," translated by K. L. Chung, Addison-Wesley, Reading, Mass., 1954.
16. Uspensky, J. V.: "Introduction to Mathematical Probability," chaps. 14, 15, McGraw-Hill, New York, 1937.
17. Liapounoff, A.: Sur une proposition de la théorie des probabilités, *Bull. acad. sci. St. Petersbourg*, (5)**13**(4): 359 (1900); Nouvelle forme du théorème sur la limite de probabilité, *ibid.*, (8)**12**(5): 1 (1901).
18. Bernstein, S.: Sur l'extension du théorème limite du calcul des probabilités aux sommes des quantités dépendantes, *Math. Ann.*, **97**: 1 (1926).

CHAPTER 8

THE NORMAL RANDOM PROCESS: GAUSSIAN FUNCTIONALS

8.1 *Gaussian Functionals*

We have observed in previous chapters that the transformations common to communication theory are usually continuous time-dependent operations (cf. Secs. 2.1, 3.3, 5.1, 5.2, 6.1). The linear passive or active filter, the zero-memory nonlinear device, the various operations of modulation and demodulation used in the transmission and reception of messages and signals are typical. If, now, $x(t)$ and $y(t)$ are, respectively, the input and output ensembles in such cases, they can in most instances be related by an expression of the form†

$$y(t) = \mathbf{T}\{x(t)\} = \int_{a(t)}^{b(t)} A(t,\tau)G[x(\tau)]\,d\tau \tag{8.1}$$

where $A(t,\tau)$ is an appropriate weighting function; $a(t)$, $b(t)$ are usually constants or linear nonrandom functions of the time; and G may be some dynamic characteristic of a linear or nonlinear circuit element or, say, a differential operator, etc. For example, if $y(t)$ is the output process that results from linear time-invariant filtering of an input process $x(t)$, Eq. (8.1) becomes at once (cf. Secs. 2.1-2, 2.1-3)

$$y(t) = \mathbf{T}_{L1}\{x\} = \int_{-\infty}^{\infty} x(\tau)h(t-\tau)\,d\tau \qquad A(t,\tau) = h(t-\tau) \qquad G(x) = x, \text{ etc.} \tag{8.2}$$

with $a \to -\infty$, $b \to +\infty$, and h, as before, the weighting function of the filter. Closing a switch at $t = 0$ to pass $x(t)$ and opening it at $t = T$ to cut it off, for instance, further modifies Eq. (8.1) or (8.2) to

$$y(t) = \mathbf{T}_{L2}\{x\} = \int_0^T x(\tau)h(t-\tau)\,d\tau \tag{8.3}$$

where now $a(t) = 0$, $b(t) = T$.

† This is, of course, a stochastic integral, whose existence in some average sense (e.g., in the mean square) is assumed (cf. Sec. 2.1).

In a similar way, we can use Eq. (8.1) to represent a zero-memory nonlinear operation. Setting G above equal to g, the dynamic path of the element in question, and letting $A(t,\tau) = \delta(t - \tau)$ (with $a \to +\infty$, $b \to -\infty$), we have

$$y(t) = \boldsymbol{T}_g\{x\} = g[x(t)] \tag{8.4}$$

(cf. Sec. 2.3-1). Or if G is the differential operator $G = d/dt$ in the above, we obtain

$$y(t) = \boldsymbol{T}_{L3}\{x\} = \dot{x}(t) \tag{8.5}$$

and so on, for many other operations that form the elements of the general transformations $\boldsymbol{T}_T^{(N)}$ and $\boldsymbol{T}_R^{(N)}$ (cf. Sec. 2.1-3).

Relations of the types (8.1) to (8.5) are *functionals* of x, and, more precisely here, the process y is a *stochastic functional* of the process x. We can have nonlinear as well as linear functionals. Equation (8.4) is a particularly simple example of the former. A more sophisticated example occurs in the case of the filtered output of, say, a full-wave square-law rectifier, where [cf. Eq. (5.52)]

$$y(t) = \boldsymbol{T}_{L2}\boldsymbol{T}_g\{x\} = \beta \int_0^T x(\tau)^2 h(t - \tau)\, d\tau \tag{8.6}$$

Equations (8.2), (8.3), (8.5), on the other hand, are linear functionals. Although no general statistical theory of functionals like Eq. (8.1) is as yet available, a comparatively simple and more or less systematic treatment is possible when $x(t)$ is a normal process and G is a linear operator.

Under these conditions, if $y(t)$ is a gaussian process, then y is also called a gaussian functional (of x). We shall confine our attention in the present chapter to functionals of this class, leaving to Chap. 17 a short account of one or two of the much more difficult nonlinear cases, e.g., Eq. (8.6), where some results of engineering interest are available. Specifically, we seek the various distribution densities W_1, W_2, etc., of the new process $y(t)$ for subsequent application in system analysis and design. The results are presented in a series of examples.

8.1-1 Derivatives of a Normal Process. When all operations on $x(t)$ are linear and $x(t)$ itself is a normal process, we suspect from our earlier results for discrete transformations of a normal process (cf. Sec. 7.5-2) that $y(t)$, as given by Eq. (8.1), will also be normal. As we shall see shortly, this is indeed the case.

As our first step in this direction, therefore, let us find $W_1(y_1,t)$ for the derivative $y = \dot{x}$ of the normal process $x(t)$. We start with the c.f. of $y = \dot{x}$, with $\boldsymbol{E}\{x\} = 0$ for convenience,† and expand in series,‡ getting

† This is no loss of generality, since we can repeat the argument below for $z = x - \bar{x}$ ($\bar{x} \neq 0$), in place of x, without change.

‡ See, for example, eqs. (3.10) et seq. of Siegert's paper.[1]

$$F_y(i\xi) = E\{e^{i\xi(d/dt)x(t)}\} = \sum_{m=0}^{\infty} \frac{(i\xi)^m}{m!} E\left\{\left(\frac{dx}{dt}\right)^m\right\}$$

$$= \sum_{m=0}^{\infty} \frac{(i\xi)^m}{m!} \left[\frac{\partial^m}{\partial t_1 \cdots \partial t_m} E\{x(t_1), \ldots ,x(t_m)\} \Big|_{t_1 = \cdots = t_m = t}\right] \tag{8.7}$$

this last from Sec. 2.1-2, where it is shown that the expectation and limiting operations, including the derivative, commute, provided here that $\partial^m E\{x_1, \ldots ,x_m\}/\partial t_1 \cdots \partial t_m$ exists and is continuous. But since $x(t)$ is a normal process, we may use Eqs. (7.28) to write

$$F_y(i\xi) = \sum_{m=0}^{\infty} \frac{(i\xi)^{2m}}{(2m)!} \sum_{\substack{\text{combinations} \\ \text{of all pairs}}} \prod \left[\frac{\partial^2}{\partial t_j\, \partial t_k} E\{x(t_j)x(t_k)\}\right]_{t_j = t_k = t}$$

$$= \sum_{m=0}^{\infty} \frac{(-1)^m \xi^m}{(2m)!} \frac{(2m)!}{m!\, 2^m} \left[\frac{\partial^2}{\partial t_1\, \partial t_2} K_x(t_1,t_2) \Big|_{t_1 = t_2 = t}\right]^m \tag{8.8}$$

Summing the series gives at once

$$F_y(i\xi) = e^{-\xi^2 K_y(t,t)/2} \qquad \text{with } K_y(t,t) = K_{\dot{x}}(t,t) \equiv \frac{\partial^2}{\partial t_1\, \partial t_2} K_x(t_1,t_2) \Big|_{t_1 = t_2 = t} \tag{8.9}$$

The desired d.d. of y follows immediately (cf. Sec. 7.1) and is

$$W_1(y,t) = \frac{e^{-y^2/2K_{\dot{x}}(t,t)}}{\sqrt{2\pi K_{\dot{x}}(t,t)}} \tag{8.10}$$

Note that because x is a gauss process the condition for the result (8.8) is simply that $\partial^2 K_x/(\partial t_1\, \partial t_2) \Big|_{t_1 = t_2 = t}$ exists and is continuous for all t; the condition on the higher moments is automatically taken care of by Eqs. (7.8).

When x, in addition, is stationary, the variance of y can also be expressed in terms of the spectrum of x [cf. Sec. 3.2-3(4)], according to

$$K_{\dot{x}}(t,t) = K_{\dot{x}}(0) = -\frac{d^2}{dt^2} K_x(t) \Big|_{t=0} = \int_0^{\infty} \omega^2 \mathcal{W}_x(f)\, df \qquad \omega = 2\pi f \tag{8.11}$$

where we observe also that $\dot{x}$ is (at least wide-sense) stationary.

So far we have shown that $y = \dot{x}(t)$ has a normal first-order density $W_1(y,t)$. To verify that $y = \dot{x}$ is also a gaussian *process*, we must show that $W_n(y_1,t_1; \ldots ;y_n,t_n)$ ($n \geq 1$, all $t_1, \ldots ,t_n$) is a multivariate normal density (cf. Sec. 7.4). This can easily be done, for we observe that, since $\dot{x}(t_1)$ and $\dot{x}(t_2)$ are each normal random variates, then together they form a pair of gaussian random variables, where the statistical relation between them is determined solely† by $\boldsymbol{E}\{x(t_1)x(t_2)\} = K_x(t_1,t_2)$, $\boldsymbol{E}\{x\} = 0$ here. The joint d.d. of $\dot{x}_1$ and $\dot{x}_2$ is therefore normal with the covariance matrix

$$\mathbf{K}_{\dot{x}} = [\boldsymbol{E}\{\dot{x}(t_j)\dot{x}(t_k)\}] = \left[\frac{\partial^2}{\partial t_j\,\partial t_k} K_x(t_j,t_k)\right] \qquad j, k = 1, 2 \tag{8.12}$$

provided that each element exists for all t_j, t_k, that is, provided that $K_{\dot{x}}(t_1,t_2)$ is a continuous bounded function of its arguments. The extension to any number of random variables $\dot{x}_1, \ldots, \dot{x}_n$ is immediate. It accordingly follows that, *if $x(t)$ is a gauss process, so also is $\dot{x}(t)$*, provided that $K_{\dot{x}}(t_j,t_k)$ exists (all t_j, t_k; $j, k = 1, \ldots, n$). In a similar way, the argument is extended to higher derivatives $y^{(M)} = d^Mx/dt^M$ $(M \geq 1)$. These are also gauss processes if $x(t)$ is, now provided that $K_y(t_1,t_2) = \partial^{2M}K_x(t_1,t_2)/\partial t_1{}^M\,\partial t_2{}^M$ exists and is continuous (all t_1, t_2), a condition on the spectrum of the original process that is alternatively expressed by $\int_0^\infty \omega^{2M}\mathcal{W}_x(f)\,df < \infty$. We note, finally, that, if x is stationary, so also is $\dot{x}$.

8.1-2 Integrals of a Gauss Process.[1] When y is a linear functional of x, say

$$y(t) = \boldsymbol{T}_L\{x(t)\} = \int_{a(t)}^{b(t)} A(t,\tau)x(\tau)\,d\tau \tag{8.13}$$

and $x(t)$ is a gaussian process [cf. Eqs. (8.2), (8.3)], for example, we expect once more that $y(t)$ is also a normal process. This is readily shown by the approach of Sec. 7.7-1.‡ We expand the c.f. of y as before, using the fact that the expectation and integral operations commute, provided that the integrals so defined exist (cf. Sec. 2.1-2), to get

$$\begin{aligned} F_y(i\xi) &= \sum_{m=0}^{\infty} \frac{(i\xi)^m}{m!} \boldsymbol{E}\left\{\left[\int_a^b A(t,\tau)x(\tau)\,d\tau\right]^m\right\} \\ &= \sum_{m=0}^{\infty} \frac{(i\xi)^m}{m!} \int_{a(t)}^{b(t)} \cdots \int \boldsymbol{E}\{x(\tau_1)\cdots x(\tau_m)\} \prod_{j=1}^{m} A(t,\tau_j)\,d\tau_j \end{aligned} \tag{8.14}$$

Since x is normal, Eqs. (7.28) can be invoked, with the result that

$$\begin{aligned} F_y(i\xi) &= \sum_{m=0}^{\infty} \frac{(-1)^m\xi^{2m}}{(2m)!} \int_{a(t)}^{b(t)} \cdots \int \prod_{j}^{2m} A(t,\tau_j)\,d\tau_j \sum_{\text{all pairs}} \prod \boldsymbol{E}\{x(t_k)x(t_l)\} \\ &= \sum_{m=0}^{\infty} \frac{(-1)^m\xi^{2m}}{(2m)!}\,\frac{(2m)!}{2^m m!}\left[\iint_{a(t)}^{b(t)} A(t,\tau_1)K_x(\tau_1,\tau_2)A(t,\tau_2)\,d\tau_1\,d\tau_2\right]^m \end{aligned} \tag{8.15}$$

† We can also establish this by applying the argument above to $\boldsymbol{E}\{e^{i\xi_1\dot{x}_1+i\xi_2\dot{x}_2}\}$ and using the fact that $x(t)$ is a normal process. Similar remarks apply to the calculation of W_n (cf. Prob. 8.1).

‡ For an alternative method that is also useful in the case of integral functions, see Prob. 8.2 of the present text and sec. 8.4 of Davenport and Root.[2]

Summing the series gives at once

$$F_y(i\xi) = e^{-\xi^2 K_y(t,t)/2} \qquad \text{with } K_y(t,t) = \iint_{a(t)}^{b(t)} A(t,\tau_1)A(t,\tau_2)K_x(\tau_1,\tau_2)\, d\tau_1\, d\tau_2 \tag{8.16}$$

and the corresponding d.d. is

$$W_1(y,t) = \mathfrak{F}^{-1}\{F_y\} = \frac{e^{-y^2/2K_y(t,t)}}{\sqrt{2\pi K_y(t,t)}} \tag{8.17}$$

so that $y(t)$ possesses a first-order normal density function, provided that $K_x(t,t)$ is a bounded continuous function (all t) and is positive definite so that $K_y > 0$ (cf. Sec. 3.2-2). For example, if $x(t)$ is stationary and y represents the output process of a (stable) linear filter in the steady state [e.g., Eqs. (8.2)], we have from Eqs. (8.16), (3.102), (3.104) the following alternative expressions:

$$K_y(t,t) = \iint_{-\infty}^{\infty} h(t-\tau_1)h(t-\tau_2)K_x(\tau_1-\tau_2)\, d\tau_1\, d\tau_2$$
$$= \int_{-\infty}^{\infty} \rho(\tau)K_x(\tau)\, d\tau = K_y(0) \tag{8.18}$$
$$= \int_0^{\infty} |Y(i\omega)|^2 \mathcal{W}_x(f)\, df$$

where ρ is the autocorrelation function of the filter, with weighting and system functions h, Y, respectively. For the filter with the switch [e.g., Eq. (8.3)], the transformation $\boldsymbol{T}_{L2}$ is no longer invariant, so that

$$K_y(t,t) = \iint_0^T h(t-\tau_1)h(t-\tau_2)K_x(\tau_1-\tau_2)\, d\tau_1\, d\tau_2 \tag{8.18a}$$

depends on the time: y is not stationary, although x is.

Higher-order densities follow in a similar way if we observe that for every finite set of times $(t_1, \ldots, t_n)$ the $y(t_1), y(t_2), \ldots, y(t_n)$ together, as well as individually, possess a normal distribution [cf. remarks following Eq. (8.11)]. The characteristic function of $\mathbf{y}$ is

$$F_{\mathbf{y}}(i\boldsymbol{\xi}) = e^{-\frac{1}{2}\tilde{\boldsymbol{\xi}}\mathbf{K}_y\boldsymbol{\xi}} \qquad \text{with } \mathbf{K}_y = [K_y(t_j,t_k)] \qquad j, k = 1, \ldots, n \tag{8.19}$$

and the covariance function that governs the n-fold distribution is specifically

$$K_y(t_j,t_k) = \int_{a(t_j)}^{b(t_j)} d\tau_1 \int_{a(t_k)}^{b(t_k)} d\tau_2\, A(t_j,\tau_1)A(t_k,\tau_2)K_x(\tau_1,\tau_2) \qquad j, k = 1, \ldots, n \tag{8.20}$$

Thus, from Sec. 7.4 and the above, *the integral of a gauss process is also a gauss process* provided that the covariance function of the original process is positive definite and continuous for all (τ_1,τ_2), so that $K_y(t_j,t_j) > 0$ (all

$|t_j| < \infty$). Linear multiple integrals of normal processes can be shown in the same way to be normal also (cf. Prob. 8.4).

8.1-3 Joint Distributions. The results of Secs. 7.7-1, 7.7-2 can be extended readily to include more general situations involving two or more gaussian processes, themselves gaussian functionals. To see this, we observe again that a process which consists of several normal processes is also normal. Thus, we might have the vector process $\mathbf{u}(t) = \mathbf{i}_1 x(t) + \mathbf{i}_2 y(t)$ (cf. Sec. 3.4-1), with x and y normal. The statistical connection between the components is now determined solely by the appropriate cross-variance functions, in this instance by $K_{xy}(t_1,t_2)$ and $K_{yx}(t_1,t_2)$ [all (t_1,t_2)]. Similar remarks apply to a larger number of normal processes. The composite process is also normal, and the statistical information relative to the component processes $x(t)$, $y(t)$, $z(t)$, . . . is contained in the covariance matrix† [cf. Eqs. (3.143), (7.65a)]

$$\mathbf{K}_{xyz\cdots} = \begin{bmatrix} K_x(t_1,t_2) & K_{xy}(t_1,t_2) & K_{xz}(t_1,t_2) & \cdots \\ K_{yx}(t_1,t_2) & K_y(t_1,t_2) & K_{yz}(t_1t_2) & \cdots \\ K_{zx}(t_1,t_2) & K_{zy}(t_1,t_2) & K_z(t_1,t_2) & \cdots \\ \cdots & \cdots & \cdots & \cdots \end{bmatrix} \qquad -\infty < t_1,\, t_2 < \infty \tag{8.21}$$

Usually, we are interested in finding various joint distribution densities of the *components* of these general vector processes, *at selected times*. For example, we might want the d.d. of $x(t_1)$, $y(t_2)$, and $z(t_3)$ or of $x(t_1)$ and $y(t_2) = \dot{x}(t_2)$, and so on. Since these are now normal variates, at the times t_1, t_2, . . . , their joint d.d.'s are easily calculated. Letting $u_1 = x(t_1)$, $u_2 = y(t_2)$, $u_3 = z(t_3)$, $u_4 = \dot{x}(t_2)$, etc., we can write the general expressions for the characteristic function and probability density of the gaussian variates $\mathbf{u} = (u_1,u_2, \ldots ,u_n)$ simply as [cf. Eqs. (7.18) and (7.27)]

$$W_n(u_1, \ldots ,u_n) = \frac{e^{-\frac{1}{2}(\tilde{\mathbf{u}}-\bar{\tilde{\mathbf{u}}})\mathbf{K}_u^{-1}(\mathbf{u}-\bar{\mathbf{u}})}}{(\det \mathbf{K}_u)^{\frac{1}{2}}(2\pi)^{n/2}} \qquad F_{\mathbf{u}}(i\boldsymbol{\xi}) = e^{i\tilde{\boldsymbol{\xi}}\bar{\mathbf{u}}-\frac{1}{2}\tilde{\boldsymbol{\xi}}\mathbf{K}_u\boldsymbol{\xi}} \tag{8.22}$$

Note, however, that the interpretation in terms of the components of the vector $\mathbf{u}$ is *not* the same as for the components of the process $x(t)$ in Eqs. (7.31) et seq.: we have here selected normal variates of more than one normal process. The covariance matrix $\mathbf{K}_u$ reveals this structure at once,

$$\mathbf{K}_u = \begin{bmatrix} \overline{u_1{}^2} & \overline{u_1u_2} & \overline{u_1u_3} & \cdots & \overline{u_1u_n} \\ \overline{u_2u_1} & \overline{u_2{}^2} & \cdots & \cdots & \cdots \\ \cdots & \cdots & \overline{u_3{}^2} & \cdots & \cdots \\ \cdots & \cdots & \cdots & \overline{u_{n-1}^2} & \cdots \\ \overline{u_nu_1} & \cdots & \cdots & \cdots & \overline{u_n{}^2} \end{bmatrix} = [\overline{u_ju_k}] = \tilde{\mathbf{K}}_u \qquad j, k = 1, \ldots , n \tag{8.23}$$

† See footnote referring to Eq. (7.65a). Note that $\mathbf{K}_{xyz\cdots}$ is *not* symmetrical and hence cannot be used as the covariance matrix of a gaussian d.d. [cf. Eqs. (8.22), (8.23)].

where, for instance, $\overline{u_1^2} = \overline{x_1^2}$, $\overline{u_1u_2} = \overline{x_1y_2}$, $\overline{u_1u_4} = \overline{x_1\dot{x}_2}$, and so on. The examples below provide a number of useful results and illustrate the preceding discussion.

(1) $W_2(x_1,t_1;\dot{x}_2,t_2)$. Let us consider the normal process $x(t)$, with $E\{x\} = 0$, such that $K_{\dot{x}} = [\partial^2/(\partial t_1\,\partial t_2)]K_x$ and $K_{x\dot{x}} = (\partial/\partial t_2)K_x$ are continuous and exist for all (t_1,t_2). Our problem now is to find the joint d.d. of $x_1 = x(t_1)$ and the functional $\dot{x}_2 = \dot{x}(t_2)$. Letting $u_1 = x_1$, $u_2 = \dot{x}_2$, we see from Eq. (8.23) and Sec. 7.7-1 that the appropriate covariance matrix is

$$\mathbf{K}_u = \begin{bmatrix} \overline{x_1^2} & \overline{x_1\dot{x}_2} \\ \overline{x_1\dot{x}_2} & \overline{\dot{x}_2^2} \end{bmatrix} = \begin{bmatrix} K_x(t,t_1) & K_{x\dot{x}}(t_1,t_2) \\ K_{x\dot{x}}(t_1,t_2) & K_{\dot{x}}(t_2,t_2) \end{bmatrix} = \tilde{\mathbf{K}}_u \tag{8.24}$$

From Eqs. (8.22), the distribution is normal with this covariance matrix. In the case of a stationary process, Eq. (8.24) reduces to

$$\mathbf{K}_u = \begin{bmatrix} b_0 & b_0\lambda_1(t) \\ b_0\lambda_1(t) & b_2 \end{bmatrix} \tag{8.25a}$$

where we have used the relations in Sec. 3.2-3(4), the W-K theorem (3.42), and various results of Sec. 8.1-1 to write for these matrix elements

$$\begin{aligned} b_0 &= K_x(0) = \int_0^\infty \mathcal{W}_x(f)\,df \qquad b_2 = K_{\dot{x}}(0) = \int_0^\infty \mathcal{W}_x(f)\omega^2\,df \\ b_0\lambda_1(t) &\equiv K_{x\dot{x}}(t) = \frac{d}{dt}K_x(t) = -\int_0^\infty \mathcal{W}_x(f)\omega\sin\omega t\,df \end{aligned} \tag{8.25b}$$

The characteristic function is therefore

$$F_{\mathbf{u}}(i\xi) = \exp(-\tfrac{1}{2}\tilde{\xi}\mathbf{K}_u\xi) = \exp\left[-\frac{b_0\xi_1^2 + b_2\xi_2^2 + 2b_0\lambda_1(t)\xi_1\xi_2}{2}\right] \tag{8.26}$$

To find the corresponding d.d., we compute the inverse matrix $\mathbf{K}_u^{-1}$,

$$\mathbf{K}_u^{-1} = \begin{bmatrix} b_2 & -b_0\lambda_1(t) \\ -b_0\lambda_1(t) & b_0 \end{bmatrix}(\det \mathbf{K}_u)^{-1} \qquad \det \mathbf{K}_u = b_0[b_2 - b_0\lambda_1(t)^2] \tag{8.27}$$

and then apply this to Eqs. (8.22), getting explicitly

$$W_2(\mathbf{u}) = W_2(x_1,\dot{x}_2;t) = [2\pi b_0\sqrt{\rho^2 - \lambda_1^2(t)}]^{-1}\exp\left\{\frac{x_1^2\rho^2 + \dot{x}_2^2 - 2\lambda_1(t)x_1\dot{x}_2}{-2b_0[\rho^2 - \lambda_1^2(t)]}\right\} \tag{8.28}$$

with
$$\rho^2 \equiv b_2/b_0 \tag{8.28a}$$

It is interesting to observe from this that the "velocity" $\dot{x}$ of a normal process is correlated with the "position" x for times other than $t = 0$ and $\lim |t| \to \infty$. Conversely, position and velocity, observed at the *same* instant, are statistically independent.

(2) $W_2(x_1,t_1;\int x_2,t_2)$. Instead of the derivative process $\dot{x}(t)$, let us consider now the single integral process $y = \int xh\,dt = \int_0^T x(\tau)h(t_2 - \tau)\,d\tau$, corresponding to the output, at time t_2, of a linear filter with switch at $t = 0, T$ [cf. Eq. (8.3)] when the input is a normal process. Here we wish to determine the joint d.d. of $u_1 = x(t_1)$ and $u_2 = y(t_2)$. Following the general procedure of the previous example, we find with the help of Sec. 7.7-2 that the covariance matrix in this instance is specifically

$$\mathbf{K}_u = \begin{bmatrix} \overline{x(t_1)^2} & \int_0^T h(t_2 - \tau)K_x(t_1,\tau)\,d\tau \\ \int_0^T h(t_2 - \tau)K_x(t_1,\tau)\,d\tau & \int_0^T\int_0^T h(t_2 - \tau_1)h(t_2 - \tau_2)K_x(\tau_1,\tau_2)\,d\tau_1\,d\tau_2 \end{bmatrix} \tag{8.29}$$

The desired distribution density and characteristic function then follow from Eqs. (8.22), as above. Note that, even if $x(t)$ is a stationary process, $y(t)$ is not.

(3) The FM functional. As our final example here, let us examine the joint distribution of an integral functional at two different instants that has important applications in the theory of frequency modulation, when the modulation is a normal stochastic process. The functional in question is

$$\Psi(t) = \int_{t_0}^{t_0+t} D_0 x(\tau)\,d\tau \tag{8.30}$$

where x is the modulation and D_0 is a scale factor with the dimension angular frequency per volts (rms) if $x(\tau)$ is measured in volts. Thus, $\Psi(t)$ is a random *phase* (cf. Sec. 14.1-2). We seek first $W_2(\Psi_1,t_1;\Psi_2,t_2)$, which we obtain in the usual way from the characteristic function. Here, $u_1 = \int_{t_0}^{t_0+t_1} D_0x(\tau)\,d\tau$, $u_2 = \int_{t_0}^{t_0+t_2} D_0x(\tau)\,d\tau$, and the appropriate covariance matrix follows from Eqs. (8.20) and (8.23), since x is assumed to be normal, with $\boldsymbol{E}\{x\} = 0$. We have specifically

$$\mathbf{K}_u = D_0^2 \begin{bmatrix} \iint_{t_0}^{t_0+t_1} K_x(\tau_1,\tau_2)\,d\tau_1\,d\tau_2 & \int_{t_0}^{t_0+t_1} d\tau_1 \int_{t_0}^{t_0+t_2} d\tau_2\,K_x(\tau_1,\tau_2) \\ \int_{t_0}^{t_0+t_1} d\tau_1 \int_{t_0}^{t_0+t_2} d\tau_2\,K_x(\tau_1,\tau_2) & \iint_{t_0}^{t_0+t_2} K_x(\tau_1,\tau_2)\,d\tau_1\,d\tau_2 \end{bmatrix} \tag{8.31}$$

The off-diagonal elements are, of course, equal, since $K_x(\tau_1,\tau_2) = K_x(\tau_2,\tau_1)$. The characteristic function can be written

$$F_{\mathbf{u}}(i\boldsymbol{\xi}) = F_{\Psi}(i\boldsymbol{\xi}) = e^{-\frac{1}{2}\Phi(t_1,t_2;\boldsymbol{\xi})} \tag{8.32}$$

where the quadratic form Φ is

$$\Phi(t_1,t_2;\xi) = \tilde{\xi}\mathbf{K}_u\xi = D_0{}^2\mathbf{E}\left\{\left[\xi_1\int_{t_0}^{t_0+t_1} x(\tau)\,d\tau + \xi_2\int_{t_0}^{t_0+t_2} x(\tau)\,d\tau\right]^2\right\} \tag{8.33}$$

Applying Eqs. (8.22) for W_2 then gives us the desired distribution density.

We can also use the results above to obtain a number of special statistical averages. For example, to determine the intensity spectrum of a sinusoidal carrier frequency-modulated by a stationary normal noise (cf. Middleton[3] and Sec. 14.1-2), we need an expression of the type

$$\mathbf{E}\{e^{-i\Psi(t_1)+i\Psi(t_2)}\} = F_2(-i,i;t)_\Psi \qquad t = t_2 - t_1 \tag{8.34}$$

From Eq. (8.33), we can write at once for the stationary process assumed here

$$\tilde{\xi}\mathbf{K}_u\xi\Big|_{-\xi_1=\xi_2=1} = D_0{}^2\mathbf{E}\left\{\left[\int_{t_0+t_1}^{t_0+t_2} x(\tau)\,d\tau\right]^2\right\} = D_0{}^2\iint_0^t K_x(\tau_1-\tau_2)\,d\tau_1\,d\tau_2 \tag{8.35}$$

and using the fact that (since K_x is symmetrical in $\tau_1 - \tau_2$)

$$\iint_0^t K_x(\tau_1-\tau_2)\,d\tau_1\,d\tau_2 = 2\int_0^t (t-\tau)K_x(\tau)\,d\tau \tag{8.36}$$

we have the desired average in the following forms,

$$\mathbf{E}\{e^{-i\Psi(t_1)+i\Psi(t_2)}\} = e^{-[\overline{\Psi^2}-\overline{\Psi(t_1)\Psi(t_2)}]} = e^{-K_\Psi(0)+K_\Psi(t)} \tag{8.37a}$$

$$\mathbf{E}\{e^{-i\Psi(t_1)+i\Psi(t_2)}\} = e^{-D_0{}^2\int_0^t (t-\tau)K_x(\tau)\,d\tau} \tag{8.37b}$$

where the first expression (8.37a) makes use of the fact that Ψ also is normal and stationary and the second employs Eq. (8.36). A third form of Eqs. (8.37) is found with the help of the W-K theorem. We replace $K_x(\tau)$ by $\frac{1}{2}\mathfrak{F}^{-1}\{\mathcal{W}_x\}$ [cf. Eqs. (3.42)] and observe that $\int_0^t (t-\tau)\cos\omega\tau\,d\tau = (1-\cos\omega t)/\omega^2$. The result is

$$\mathbf{E}\{\exp\,[-i\Psi(t_1) + i\Psi(t_2)]\} = \exp\left[-D_0{}^2\int_0^\infty \mathcal{W}_x(f)(1-\cos\omega t)\,df/\omega^2\right] \tag{8.38}$$

which is the form more commonly used than Eqs. (8.37) in applications (cf. Sec. 14.1-2).

Note that these results [Eqs. (8.37b), (8.38)] could equally well have been obtained by considering

$$F_1(i\xi)_\Phi = \mathbf{E}\{e^{i\xi\Phi(t)}\} \qquad \Phi(t) = D_0\int_0^t x(\tau)\,d\tau \tag{8.39}$$

after which we specialize Eq. (8.16) in an obvious way to get

$$F_1(i\xi)_\Phi = \exp\left[-\tfrac{1}{2}D_0{}^2\xi \iint_0^t K_x(\tau_1 - \tau_2)\,d\tau_1\,d\tau_2\right]$$
$$= \exp\left[-D_0{}^2\xi^2 \int_0^t (t-\tau)K_x(\tau)\,d\tau\right] \quad (8.40)$$

which is the c.f. for the functional $\Phi(t)$. Setting $\xi = \pm 1$ then gives Eqs. (8.37) or (8.38). Observe from Eqs. (8.37) or (8.38) that

$$\begin{aligned} \boldsymbol{E}\{\cos[\Psi(t_2) - \Psi(t_1)]\} &= \exp\left[-D_0{}^2 \int_0^t (t-\tau)K_x(\tau)\,d\tau\right] \\ \boldsymbol{E}\{\sin[\Psi(t_2) - \Psi(t_1)]\} &= 0 \end{aligned} \quad (8.41)$$

Many other results of a similar nature can be found in the same way (cf. Probs. 14.15 and 14.16 to 14.18). Since the statistical properties of normal processes are essentially determined by the covariance function or by the covariance matrix if more than one process is involved, the problem of finding the d.d.'s of these processes and their linear functionals becomes one of finding the appropriate covariance functions (or spectra, if stationarity can be assumed), typical examples of which for both broad- and narrow-band waves have been considered in Secs. 3.3, 3.4. Combining these results with those of Chap. 7 and the methods of the present section makes possible a complete description of many of the gaussian processes encountered in communication theory.

PROBLEMS

8.1 Show by the argument of Eqs. (8.7) et seq., that is, by calculating $\boldsymbol{E}\{e^{i\xi_1\dot{x}_1+i\xi\dot{x}_2}\}$, $\boldsymbol{E}\{e^{i\xi_1\dot{x}_1+i\xi_2\dot{x}_2+i\xi_3\dot{x}_3}\}$, etc. $(n \geq 1)$, that, if $x(t)$ is a normal process, $\dot{x}(t)$ is also.

8.2 In dealing with a (linear) integral functional of a gauss process $x(t)$, show that the following approach gives results equivalent to those of Sec. 8.1-2:

(*a*) Consider the integral functional $y(t) = \int_a^b A(t,\tau)x(\tau)\,d\tau$. Dividing the interval $[a < \tau < b]$ into n subintervals $\tau_{k-1} < \tau < \tau_k$ $(k = 1, \ldots, n)$, where $\tau_0 = a$ and $\tau_n = b$, replace the functional by the approximating sum $y_n = \sum_{k=1}^{n} A(t,\tau_k)x(\tau_k)\,\Delta\tau_k$ $(\Delta\tau_k = \tau_k - \tau_{k-1})$. Calculate the joint d.d. of these weighted $x(\tau_k)$, and hence obtain in the limit $\Delta\tau_k \to 0$ $(n \to \infty)$, etc., the d.d. and c.f. of the functional $y(t)$ [cf. Eqs. (8.16), (8.17)]. Justify the various steps.

(*b*) Extend the results to the deterministic time-dependent limits $a(t)$, $b(t)$.

HINT: To justify this procedure, show that if $\boldsymbol{E}\{y\}$, $\boldsymbol{E}\{y^2\} < \infty$ and are continuous functions of t, along with the covariance function of x, $\lim_{n\to\infty} \boldsymbol{E}\{y_n\} = \bar{y}$, $\lim_{n\to\infty} \boldsymbol{E}\{y_n{}^2\} = \overline{y^2}$, and y_n converges mean square to y and hence, in probability, to y; also, note that the probability densities of y_n converge to the d.d. of y.

W. B. Davenport, Jr., and W. L. Root,[2] sec. 8.4, Linear Transformations.

8.3 In dealing with a derivative functional $\dot{x}$, $\ddot{x}$, etc., of a gauss process $x(t)$, show that the following approach gives results equivalent to those of Sec. 8.1-1:

(*a*) Consider, for $\dot{x}(t)$, the pair of random variables $x(t+\epsilon)$ and $x(t)$, and determine the d.d. of $[x(t+\epsilon)-x(t)]/\epsilon$; then, passing to the limit $\epsilon \to 0$, obtain $W_1(\dot{x},t)$. Justify the steps, as in Prob. 8.2.

(*b*) Extend to the higher derivatives $\ddot{x}$, $\dddot{x}$, etc.

8.4 Show that linear multiple integrals of a normal process are also normally distributed.

8.5 When $x(t)$ is a stationary gauss process, differentiable mean square to the first order, show that

$$W_1(\dot{x}) = \frac{e^{-\dot{x}^2/2b_2}}{\sqrt{2\pi b_2}} \tag{1}$$

$$W_2(\dot{x}_1,\dot{x}_2;t) = [2\pi b_2\sqrt{1-\rho_2(t)^2}]^{-1}\exp\{-[\dot{x}_1^2+\dot{x}_2^2-2\dot{x}_1\dot{x}_2\rho_2(t)]/2b_2[1-\rho_2(t)^2]\} \tag{2}$$

where $b_2 = -\frac{d^2}{dt^2}K_x(t)\Big|_{t=0}$ and $b_2\rho_2(t) = -\frac{d^2}{dt^2}K_x(t) = -\frac{d^2}{dt^2}\int_0^\infty \mathcal{W}_x(f)\cos\omega t\,df$

8.6 (*a*) If $x(t)$ is differentiable mean square to the second order and is normal, verify that, for $u_1 = x_1$, $u_2 = \ddot{x}_2$, $W_2(x_1,t_1;\ddot{x}_2,t_2)$ is normal, with the covariance matrix

$$\mathbf{K}_u = \begin{bmatrix} K_x(t_1,t_1) & \frac{\partial^2}{\partial t_2^2}K_x(t_1,t_2) \\ \frac{\partial^2}{\partial t_2^2}K_x(t_1,t_2) & \frac{\partial^4}{\partial t_2^4}K_x(t_1,t_4) \end{bmatrix} \tag{1}$$

and show, similarly, that $W_3(x_1,t_1;\dot{x}_2,t_2;\ddot{x}_3,t_3)$ is also normal, with

$$\mathbf{K}_u = \begin{bmatrix} K_x(t_1,t_1) & \frac{\partial^2}{\partial t_2}K_x(t_1,t_2) & \frac{\partial^2}{\partial t_3^2}K_x(t_1,t_3) \\ \cdots & \frac{\partial^2}{\partial t_2^2}K_x(t_2,t_2) & \frac{\partial^3}{\partial t_2\,\partial t_3^2}K_x(t_2,t_3) \\ [(K_u)_{jk} = (K_u)_{kj}] & \cdots & \frac{\partial^4}{\partial t_3^4}K_x(t_3,t_3) \end{bmatrix} \tag{2}$$

(*b*) Hence, if x is stationary, verify that Eqs. (1) and (2) become

$$\mathbf{K}_u = \begin{bmatrix} b_0 & -b_2\rho_2(t) \\ -b_2\rho_2(t) & b_4\rho_4(t) \end{bmatrix} \qquad \mathbf{K}_u = \begin{bmatrix} b_0 & b_0\lambda_1(t) & -b_2\rho_2(t) \\ b_0\lambda_1(t) & b_2 & b_0\lambda_3(t) \\ -b_2\rho_2(t) & b_0\lambda_3(t) & b_4 \end{bmatrix} \tag{3}$$

where

$$b_2\rho_2(t) = -\frac{d^2}{dt^2}K_x = \int_0^\infty \mathcal{W}_x(f)\omega^2\cos\omega t\,df$$

$$b_0\lambda_n(t) \equiv \frac{d^n}{dt^n}K_x(t) = (-1)^{(n+1)/2}\int_0^\infty \omega^n\mathcal{W}_x(f)\sin\omega t\,df \qquad n = 1, 3, 5, \ldots$$

and $b_n = \int_0^\infty \omega^n\mathcal{W}_x(f)\,df$ [$n = 0, 2, 4, \ldots$; cf. Sec. 3.2-3(4). When $t = 0$ in Eq. (3), we have $W_1(x_1,\dot{x}_1,\ddot{x}_1)$, an expression useful in the study of the maxima of a normal wave (cf. Sec. 9.4).

8.7 A joint density needed in the problem of the distribution of zeros of stationary normal processes is $W_2(x_1,\dot{x}_1;x_2,\dot{x}_2;t)$ [cf. Eq. (9.86)]. Show that x_1, $\dot{x}_1$, x_2, $\dot{x}_2$ are normally distributed with the covariance matrix

$$\mathbf{K}_u = \begin{bmatrix} b_0 & 0 & b_0\rho_0(t) & b_1\lambda_1(t) \\ 0 & b_2 & -b_1\lambda_1(t) & b_2\rho_2(t) \\ b_0\rho_0(t) & -b_1\lambda_1(t) & b_0 & 0 \\ b_1\lambda_1(t) & b_2\rho_2(t) & 0 & b_2 \end{bmatrix} \tag{1}$$

(cf. Prob. 8.6 for b_0, b_2, etc.).

8.2 *Orthogonal Expansions of a Random Process*†

For many applications, it is advantageous to be able to develop a stochastic process in a suitable series over some finite (or infinite) interval. We have already encountered examples of this in Chap. 4 in the case of band-limited processes (cf. Sec. 4.2). Here, we shall extend the treatment to the broader problem of the expansion of a general random process (not necessarily band-limited) in an arbitrary interval (a,b). The Fourier-series development and the application to normal processes are particularly important examples in subsequent work (cf. Chap. 9 and Parts 3, 4).

We begin with a second-moment theory.

8.2-1 General Expansions. Let $x(t)$ be a real stochastic process, with finite mean and a bounded continuous second-moment function $M_x(t_1,t_2)$. Let us expand $x(t)$ formally in the general interval (a,b) as follows,

$$x(t) = \sum_m C_m \phi_m(t) \qquad a \le t \le b \tag{8.42}$$

where the $\{\phi_m(t)\}$ are any real, complete orthonormal set appropriate to the interval.‡ Then, the random variables C_m are given by§

$$C_m = \int_a^b x(t)\phi_m(t)\,dt \tag{8.43a}$$

with

$$\int_a^b \phi_m(t)\phi_n(t)\,dt = \delta_{mn} \tag{8.43b}$$

† A detailed mathematical treatment is given by Karhunen;[4] see also Kac and Siegert,[5] Loève's book,[6] sec. 34.5, and additional bibliography in Grenander's paper,[7] as well as the exposition of Davenport and Root[2] (cf. sec. 6.4).

‡ A class of orthonormal functions $\{\phi_k(t)\}$ is *complete*, and square-integrable in an interval (a,b), for example, $\int_a^b |\phi_k|^2\,dt < \infty$, if any function $y(t)$, also square-integrable in this interval, can be represented (limit in the mean, l.i.m.) by a series of the type $y(t) = \Sigma_k A_k\phi_k(t)$ $(a \le t \le b)$, where the A_k are obtained from $A_k = \int_a^b y(t)\phi_k(t)\,dt$. Here l.i.m. signifies specifically $\lim_{K\to\infty} \int_a^b \left| y(t) - \sum_1^K A_k\phi_k(t)\right|^2 dt = 0$.

§ The expansion can also be made in terms of complex orthonormal functions $\{\phi_m\}$, where now the coefficients C_m are also complex. Equations (8.43a), (8.43b) become

$$C_m = \int_a^b x(t)\phi_m(t)^*\,dt \qquad \text{with} \int_a^b \phi_m(t)\phi_n(t)^*\,dt = \delta_{mn}$$

The integral (8.43a) is assumed to exist mean square (cf. Sec. 2.1), of course, and similarly the series (8.42) is postulated to converge mean square to the process $x(t)$, that is,

$$\lim_{M\to\infty} \boldsymbol{E}\left\{\left[x - \sum_{m}^{M} C_m\phi_m(t)\right]^2\right\} = 0 \tag{8.44}$$

for all $a \leq t \leq b$. A sufficient condition for this is that†

$$\sum_m |\phi_m(t)|(\overline{C_m{}^2})^{1/2} < \infty \qquad a \leq t \leq b \tag{8.45}$$

The statistical properties of $x(t)$ are thus embodied in the expansion coefficients C_m. From Eqs. (8.42), (8.43), for example, we find that the first and second moments of these coefficients are

$$\boldsymbol{E}\{C_m\} = \int_a^b \boldsymbol{E}\{x(t)\}\phi_m(t)\,dt \tag{8.46a}$$

$$\boldsymbol{E}\{C_mC_n\} = \int_a^b M_x(t_1,t_2)\phi_m(t_1)\phi_m(t_2)\,dt_1\,dt_2 \tag{8.46b}$$

and since the $\{\phi_m\}$ are an arbitrary set of orthonormal functions on the interval, it is clear that $\boldsymbol{E}\{C_mC_n\}$ does not in general vanish: in this situation, the C_m are therefore not statistically independent. Higher moments of the coefficients may be determined in a similar way from Eq. (8.43a), with a corresponding correlation among coefficients.

In some cases a development more general than Eq. (8.42) may be needed, e.g.,

$$x(t) = \sum_m [a_m\phi_m(t) + b_m\psi_m(t)] \qquad a \leq t \leq b \tag{8.47a}$$

with

$$a_m = \int_a^b x(t)\phi_m(t)\,dt \qquad b_m = \int_a^b x(t)\psi_m(t)\,dt \tag{8.47b}$$

where $\{\phi_m\}$ and $\{\psi_m\}$ are each complete orthonormal sets of functions, which together form a biorthonormal set, with the additional property that

$$\int_a^b \phi_m(t)\psi_n(t)\,dt = 0 \qquad \text{all } m, n \tag{8.48}$$

The expansion (mean square) of the process $x(t)$ in a Fourier series on the interval (a, $b = a + T$) is an important example. Here, we have

$$\{\phi_m\} = \left\{\left(\frac{2}{T}\right)^{1/2} \cos m\omega_0 t\right\} \qquad \{\psi_m\} = \left\{\left(\frac{2}{T}\right)^{1/2} \sin m\omega_0 t\right\} \qquad \omega_0 = \frac{2\pi}{T} \tag{8.49}$$

which certainly obey Eqs. (8.43b) and (8.48). Sufficient conditions for the MSC of the expansion (8.47a) are given by Eq. (8.45) for ϕ_m and a_m, ψ_m and b_m, respectively. Note for an arbitrary process, i.e., here, an arbi-

† See Prob. 2.2 and Bartlett,[8] p. 138.

trary $M_x(t_1,t_2)$, that while the second moments of the random variables a_m, b_n become from Eq. (8.47b)

$$\begin{aligned} \boldsymbol{E}\{a_m a_n\} &= \iint_a^b M_x(t_1,t_2)\phi_m(t_1)\phi_m(t_2)\,dt_1\,dt_2 \\ \boldsymbol{E}\{b_m b_n\} &= \iint_a^b M_x(t_1,t_2)\psi_m(t_1)\psi_n(t_2)\,dt_1\,dt_2 \qquad (8.50) \\ \text{and} \qquad \boldsymbol{E}\{a_m b_n\} &= \iint_a^b M_x(t_1,t_2)\phi_m(t_1)\psi_n(t_2)\,dt_1\,dt_2 \end{aligned}$$

these are not generally zero if $m \neq n$, nor does $\boldsymbol{E}\{a_m b_m\}$ vanish. Consequently the a_n's and b_n's cannot be statistically independent. In fact, as we shall show in detail later (cf. Sec. 8.3), the coefficients in the Fourier-series development of an arbitrary (real) process are statistically independent only (1) if the process is stationary and normal and (2) if the expansion interval (a,b) becomes infinite (i.e., if $b - a \to \infty$).

Although our discussion above has been given in terms of the real process, the extension to the complex process is easily made. We have for a complex process $z(t)$, analogous to Eq. (8.42) for the real case,

$$z(t) = \Sigma_m D_m \phi_m(t) \qquad a \leq t \leq b \qquad (8.51)$$

$$\text{with} \qquad D_m = \int_a^b z(t)\phi_m(t)^*\,dt \qquad \int_a^b \phi_m(t)\phi_n(t)^*\,dt = \delta_{mn} \qquad (8.52)$$

where the $\{\phi_m\}$ are now complex functions, forming a complete orthonormal set in the interval (the asterisk denotes the complex conjugate), and the D_m are complex random variables. Here $M_z(t_1,t_2) = M_z^*(t_2,t_1) = \boldsymbol{E}\{z(t_1)z(t_2)^*\}$, and

$$\boldsymbol{E}\{D_m D_n^*\} = \iint_a^b M_z(t_1,t_2)\phi_m(t_1)\phi_n^*(t_2)\,dt_1\,dt_2 \qquad (8.53)$$

The series (8.51), like Eqs. (8.42) and (8.47a), is assumed to converge mean square to $z(t)$; that is, the appropriate modification of Eq. (8.44) holds, for which Eq. (8.45) is the sufficient condition (with $\overline{C_m{}^2}$ replaced by $\overline{D_m D_m^*}$ therein). An important example is again the Fourier series: here, an orthonormal set is

$$\{\phi_m\} = T^{-1/2}e^{im\omega_0 t} \qquad \omega_0 = 1/T \qquad (a \leq t \leq b = a + T) \qquad (8.54)$$

Once more, it is evident that, for such a choice $\{\phi_m\}$ and an arbitrary process z, we cannot expect statistical independence of the expansion coefficients (cf. Sec. 8.3) for these entirely random processes.

One exception to this general lack of statistical independence is observed for the limiting case of a real stationary white noise process (cf. Sec. 3.2-3(1)].

From Eq. (3.56*b*), we recall that $M_x(t_1,t_2) = K_x(t_2 - t_1) = (W_0/2)\delta(t_2 - t_1)$, so that Eqs. (8.46*b*) and (8.50), with the orthonormal conditions (8.43*b*), (8.48), reduce to

$$\begin{aligned} \boldsymbol{E}\{C_m C_n\} &= \delta_{mn}\frac{W_0}{2} \\ \boldsymbol{E}\{a_m a_n\} = \boldsymbol{E}\{b_m b_n\} &= \delta_{mn}\frac{W_0}{2} \\ \boldsymbol{E}\{a_m b_n\} &= 0 \end{aligned} \tag{8.55}$$

for *any* orthonormal sets of function ϕ_m, ψ_m, etc., in the interval (a,b). The expansion coefficients are then independent in the mean square, and if the white noise process is also gaussian, these coefficients are statistically independent as well, with unit variances and zero means, as implied by Eqs. (8.55). Here, any orthonormal systems of functions (appropriate to the interval, of course) will serve as the series representation of $x(t)$. Similar remarks apply to the complex process above.

8.2-2 Orthogonal Expansion of a Random Process. The expansions above have the frequently undesirable feature that the coefficients are statistically interconnected. Since our choice of the real orthonormal set of expansion functions ϕ_m, ψ_m, etc., is to a large degree open, we expect that it is possible to choose this set in such a way that the second moments of the expansion coefficients are also orthogonal; that is, $\boldsymbol{E}\{C_m C_n\} = \lambda_m \delta_{mn}$ [cf. Eqs. (8.55)], with $\boldsymbol{E}\{C_m\} = 0$, as implied by the former relation. Here, the λ_m are a set of real nonnegative numbers if the process $x(t)$ is real. Writing $C_m = \lambda_m^{1/2} c_m$, we may express the series (8.42) formally now as†

$$x(t) = \Sigma_m \sqrt{\lambda_m}\, c_m \phi_m(t) \qquad a \le t \le b \tag{8.56}$$

and as before the equality is to be understood as MSC of the series to the process $x(t)$ [cf. Eqs. (8.44) et seq.]. The coefficients c_m are given by

$$c_m = (\lambda_m)^{-1/2} \int_a^b x(t)\phi_m(t)\, dt \tag{8.57}$$

and the conditions on $\{\phi_m\}$ and $\{c_m\}$ are now

Orthonormality [cf. Eq. (8.43*b*)]

$$\int_a^b \phi_m(t)\phi_n(t)\, dt = \delta_{mn} \tag{8.58a}$$

Orthogonality and therefore

$$\begin{aligned} \boldsymbol{E}\{c_m c_n\} &= \delta_{mn} \\ \boldsymbol{E}\{c_m\} &= 0 \end{aligned} \tag{8.58b}$$

When Eqs. (8.58*b*) hold, the development (8.56), with Eq. (8.57), is an *orthogonal expansion* of the process $x(t)$ on the interval (a,b), and the $\{\lambda_m\}$, $\{\phi_m\}$ are, respectively, the *eigenvalues* and *eigenfunctions* associated with the expansion.

† For an expansion in a complex orthonormal set [λ_m, $x(t)$ still real], see Prob. 8.8.

These eigenfunctions and eigenvalues are not arbitrarily selected but depend on the covariance function $K_x(t_1,t_2)$ of the process in the following way: Let us for the moment assume that Eqs. (8.56) to (8.58) are satisfied by some $\{\lambda_m\}$ and $\{\phi_m\}$. Then we can write at once

$$\begin{aligned} K_x(t_1,t_2) &= \boldsymbol{E}\{x(t_1)x(t_2)\} = \Sigma_{m,n} \sqrt{\lambda_m\lambda_n}\, \phi_m(t_1)\phi_n(t_2)\boldsymbol{E}\{c_mc_n\} \\ &= \Sigma_m\lambda_m\phi_m(t_1)\phi_m(t_2) \qquad a \le t_{1,2} \le b \end{aligned} \tag{8.59}$$

Multiplying Eq. (8.59) by $\phi_m(t_2)$ and integrating, with the help of Eq. (8.58a), gives directly

$$\int_a^b K_x(t_1,t_2)\phi_m(t_2)\,dt_2 = \lambda_m\phi_m(t_1) \qquad a \le t_1 \le b \tag{8.60}$$

which is the integral equation determining the eigenvalues and eigenfunctions.

To complete the demonstration for the expansion of the process x we must show the MSC of Eq. (8.56). First, however, we shall need *Mercer's theorem.*[9] This states that under certain conditions it is always possible to expand a real function $G(s,\tau)$ in a uniformly and absolutely convergent series on the interval (a,b), namely,

$$G(s,\tau) = \sum_{k=1}^{\infty} \lambda_k\phi_k(s)\phi_k(\tau) \qquad a \le s, \tau \le b \tag{8.61}$$

Sufficient conditions are that:

1. $G(s,\tau)$ is a symmetric real function, for example, $G(s,\tau) = G(\tau,s)$.
2. $G(s,\tau)$ is *positive semidefinite* in (a,b), for example, $G(s,\tau)$ is piecewise continuous, such that

$$\infty > \iint_a^b G(s,\tau)f(s)f(\tau)\,ds\,d\tau \ge 0 \tag{8.62}$$

for any $f(s)$ that are integrable square in (a,b), that is, for $\int_a^b |f(\tau)|^2\,d\tau < \infty$

Here $\{\lambda_k\}$ and $\{\phi_k(s)\}$ are the eigenvalues and (real) eigenfunctions of the integral equation

$$\int_a^b G(s,\tau)\phi_k(\tau)\,d\tau = \lambda_k\phi_k(s) \qquad a \le s \le b \tag{8.63}$$

Moreover, it can be shown† (cf. Prob. 8.9) that, if the (real) kernel $G(s,\tau)$ of this integral equation is positive semidefinite, all the nonzero eigenvalues λ_k are positive real numbers (which, however, need not all be distinct; cf. the example of band-limited white noise discussed in Sec. 8.2-3). In addition,

† For the general treatment of the homogeneous integral equation (8.63), see Courant-Hilbert,[9] chap. 3; Sec. A.2.2 of the present text; and Davenport and Root,[2] app. II.

if $G(s,\tau)$ is *positive definite* [e.g., if Eq. (8.62) applies without the equal sign], the orthonormal set of eigenfunctions $\{\phi_k(s)\}$ is also complete and the eigenvalues have at most a finite degeneracy.† Accordingly, since $K_x(t_1,t_2)$ in Eq. (8.60) is always real and positive definite for the processes assumed here (cf. Sec. 3.2), the nonzero eigenvalues λ_m are positive real numbers and the ϕ_m form a complete set of real eigenfunctions. Mercer's theorem [Eq. (8.61)] then shows that the expansion (8.59) is suitably convergent on (a,b) and the development (8.56) is therefore possible.

The MSC of the series (8.56) is now easily demonstrated. First observe that the orthogonal property (8.58*b*) of the expansion coefficients is satisfied: from Eqs. (8.57) and (8.60), we get

$$\begin{aligned} \boldsymbol{E}\{c_m c_n\} &= (\lambda_m \lambda_n)^{-1/2} \iint\limits_a^b K_x(t_1,t_2)\phi_m(t_1)\phi_n(t_2)\, dt_1\, dt_2 \\ &= \lambda_m \int_a^b \phi_m(t_1)\phi_n(t_1)\, dt_1\, (\lambda_m\lambda_n)^{-1/2} = \delta_{mn} \end{aligned} \tag{8.64}$$

this last from the orthonormal property (8.58*a*) of the eigenfunctions. Next, writing

$$x_M(t) = \sum_{m=1}^{M} \sqrt{\lambda_m}\, c_m \phi_m(t) \qquad a \le t \le b \tag{8.65}$$

we may use Eq. (8.64) to show that

$$\boldsymbol{E}\{x(t)x_M(t)\} = \boldsymbol{E}\{x_M(t)^2\} = \sum_{m=1}^{M} \lambda_m \phi_m(t)^2 \tag{8.66a}$$

Hence, Eq. (8.44) becomes

$$\lim_{M\to\infty} \boldsymbol{E}\{[x(t) - x_M(t)]^2\} = K_x(t,t) - \lim_{M\to\infty} \sum_{m=1}^{M} \lambda_m \phi_m(t)^2 = 0 \tag{8.66b}$$

as required for MSC, by virtue of Mercer's theorem (8.61).

From the above, it is clear that the covariance function, along with the interval of expansion, determines the structure of the eigenfunctions and the magnitude of the eigenvalues through Eq. (8.60). Note, also, that the mean "energy" of the random process over the interval (a,b) is, from Eqs. (8.56), (8.58*a*), (8.58*b*),

$$\begin{aligned} \bar{E} \equiv \boldsymbol{E}\left\{\int_a^b x(t)^2\, dt\right\} &= \int_a^b \sum_{m,n=1} \sqrt{\lambda_m\lambda_n}\, \phi_m(t)\phi_n(t)\boldsymbol{E}\{c_m c_n\}\, dt \\ &= \sum_{m=1}^{\infty} \lambda_m = \int_a^b K_x(t,t)\, dt \end{aligned} \tag{8.67}$$

† See, for example, Youla's paper,[10] app. A, and Courant-Hilbert, *op. cit.*, chap. III, secs. 4 and 5.

which for a (wide-sense) stationary process is simply $\bar{E} = TK_x(0) = T\overline{x^2} = \sum_{m=1}^{\infty} \lambda_m$. While the orthogonal expansions (8.56) et seq. are very useful in theoretical applications, their utility is limited by the fact that, apart from the cases of stationary processes with rational spectra (cf. Sec. 3.3-1), few solutions of Eq. (8.60) are known.† We remark again that since the coefficients of the Fourier-series development in (a,b) are not orthogonal, e.g., since Eq. (8.58*b*) is not satisfied except in the limiting situation of a stationary white noise process‡ (cf. Sec. 8.2-1), we are not able to expand the process $x(t)$ in an *orthogonal* Fourier series on the interval (cf. Prob. 8.10). For complex processes, the procedures above must be somewhat modified (cf. Prob. 8.8).

8.2-3 Applications to the Gauss Process. So far, we have considered the statistical properties of the expansion coefficients through the first and second moments only [cf. Eqs. (8.46), (8.50), (8.53), (8.55), (8.58), and (8.64)]. Higher-order statistics may be handled in the same way, but in general the results are quite complicated. However, for the important special case of a normal process considerable simplification is possible.

Accordingly, let the real process $x(t)$ of Sec. 8.2-2 be gaussian with§ $\boldsymbol{E}\{x\} = 0$, and let us first find the distribution of the expansion coefficients C_m [Eq. (8.43*a*)]. We start in the usual way with the characteristic function for the C_m,

$$F_{\mathbf{C}}(i\zeta) = \boldsymbol{E}\{\exp\,(i\Sigma_m \zeta_m C_m)\} = \boldsymbol{E}\left\{\exp\left[i\int_a^b x(t)\sum_m \phi_m(t)\zeta_m\,dt\right]\right\} \tag{8.68}$$

from Eq. (8.43*a*). But from Sec. 8.1-2 [Eqs. (8.13) et seq.] we may write $y \equiv \int_a^b x(t) \sum_m \phi_m(t)\zeta_m\,dt$, where $a(t) = a$, $b(t) = b$, and $A(t,\tau) = \sum_m \zeta_m \phi_m(t)$, and consider now the c.f. of the gaussian functional y when $\xi = 1$, namely, $\boldsymbol{E}\{e^{i\xi y}\}\Big|_{\xi=1}$. From Eq. (8.16), the result here is at once

$$\begin{aligned} F_{\mathbf{C}}(i\zeta) &= F_y(i) = e^{-\frac{1}{2}K_y(0,0)} \qquad \text{where } K_y(0,0) = \tilde{\boldsymbol{\zeta}}\mathbf{A}\boldsymbol{\zeta} \\ A_{kl} &= \iint_a^b \phi_k(t_1)\phi_l(t_2)K_x(t_1,t_2)\,dt_1\,dt_2 = \boldsymbol{E}\{C_k C_l\} \end{aligned} \tag{8.69}$$

this last according to Eq. (8.46*b*). The distribution density of the $\{C_m\}$ is therefore multivariate gaussian, with the covariance matrix $\mathbf{A} = [A_{kl}] =$

† See Sec. A.2.2 and Prob. 8.12.

‡ Here $\lambda_m = W_0/2$, and the expansion of $x(t)$ is therefore $x(t) = \sqrt{W_0/2}\,\Sigma_m c_m \phi_m(t)$, where $\{\phi_m\}$ can be any complete orthonormal set. Note, as expected, that $\bar{E} = \sum_{m=1}^{\infty} \lambda_m \to \infty$: the process does not possess a finite second moment, although $\boldsymbol{E}\{c_m^2\} = 1$ (all m).

§ If $\boldsymbol{E}\{x\} \neq 0$, then we may use $x' = x - \bar{x}$ in the subsequent argument.

$[\overline{C_k C_l}]$. If $\bar{C}_k \neq 0$, the d.d. is, of course, still normal, but now with means $[\bar{C}_k]$ and the covariance matrix $\mathbf{A} = [\overline{C_k C_l} - \bar{C}_k \bar{C}_l]$ [cf. Eqs. (7.18*a*), (7.18*b*)].

It is evident once more, from Eqs. (8.69), that the expansion coefficients $\{C_m\}$ are not statistically independent if an arbitrary orthonormal set of functions $\{\phi_m\}$ is used for the development of the normal process on the interval (a,b). When the orthogonal expansions (8.56), (8.57), etc., are used, however, the new coefficients $\{c_m\}$ become independent, since, by Eq. (8.64), the new covariance matrix $\mathbf{A}' = [\overline{c_k c_l}]$ becomes the unit matrix $\mathbf{I} = [\delta_{kl}]$. Then, it follows that the expansion coefficients are normally distributed with zero means and unit variances. As mentioned earlier (in Sec. 8.2-2), this is not true of the Fourier-series development on the finite interval (a,b) (cf. Sec. 8.3, also).

Noting that a number of examples for stationary normal processes having different covariance functions are included in Probs. 8.11, 8.12, we close this section with the simple but important case of a stationary normal band-limited white noise process where $\boldsymbol{E}\{x\} = 0$ [cf. Sec. 3.2-3(1)]. From Eqs. (4.31), (4.32), we observe that

$$\{\phi_m(t)\} = \left\{\sqrt{2B}\, u_1\left(t - \frac{m}{2B}\right)\right\} = \left\{\sqrt{2B}\, \frac{\sin 2\pi B(t - m/2B)}{2\pi B(t - m/2B)}\right\} \tag{8.70}$$

form a complete orthonormal set for $(-\infty \leq t \leq \infty)$ and all $(-\infty \leq m \leq \infty)$, while B is, as before, the (ensemble) bandwidth of the noise process. Moreover, the covariance function of the process is [cf. Eqs. (3.54)]

$$K_x(t_1,t_2) = \psi \frac{\sin 2\pi B(t_2 - t_1)}{2\pi B(t_2 - t_1)} \qquad \psi = \overline{x^2} = W_0 B \tag{8.71}$$

so that inserting this relation into the basic integral equation (8.60) for the orthogonal expansion (8.56), along with Eq. (8.70), we easily show, using Eq. (4.31), that the $\{\phi_m\}$ are indeed the eigenfunctions and the $\lambda_m = W_0/2\ (> 0)$ are the eigenvalues of the integral equation. Consequently, the process $x(t)$ has the orthogonal development

$$x(t) = \psi^{-1/2} \sum_{-\infty}^{\infty} c_m \frac{\sin 2\pi B(t - m/2B)}{2\pi B(t - m/2B)} \qquad -\infty \leq t \leq \infty \tag{8.72a}$$

with $$c_m = 2B\psi^{-1/2} \int_{-\infty}^{\infty} dt\, x(t) \frac{\sin 2\pi B(t - m/2B)}{2\pi B(t - m/2B)} \tag{8.72b}$$

One can also readily verify that $K_x(t_1,t_2)$ is positive definite [and from this fact alternatively establish the completeness of the (infinitely) degenerate set of eigenvalues $\lambda_m = W_0/2$, all m]. Mercer's theorem (8.61) then gives us

$$K_x(t_1,t_2) = \psi \frac{\sin 2\pi B(t_2 - t_1)}{2\pi B(t_2 - t_1)} = \frac{W_0}{2} \sum_{-\infty}^{\infty} \phi_m(t_1)\phi_m(t_2) \qquad -\infty \leq t_1,\ t_2 \leq \infty \tag{8.73}$$

from which in turn we have the interesting result

$$\frac{\sin 2\pi B(t_2 - t_1)}{2\pi B(t_2 - t_1)} = \sum_{-\infty}^{\infty} \frac{\sin 2\pi B(t_1 - m/2B)}{2\pi B(t_1 - m/2B)} \frac{\sin 2\pi B(t_2 - m/2B)}{2\pi B(t_2 - m/2B)} \qquad (8.73a)$$

[cf. Prob. 4.6*c*, Eq. (4)]. As expected, the average "energy" $\bar{E}$ [Eq. (8.67)] is infinite, since the interval is indefinitely long and the process stationary.

Finally, we see from Eqs. (8.69) that, since $x(t)$ is a normal process, the expansion coefficients c_m [Eq. (8.72*b*)] are independent and normally distributed with zero means and unit variances. Of course, our results (8.72*a*), (8.72*b*), (8.73) hold for any (real) wide-sense stationary random process (with $E\{x\} = 0$) on the interval $(-\infty, \infty)$, as long as it is band-limited *with a uniform spectrum.* If the spectrum is not white in $(0 < f < B)$, however, then the $\{\phi_m\}$ are not the eigenfunctions of the basic integral equation (8.60), so that an orthogonal development of the random process in $(-\infty, \infty)$ is not possible with the sampling functions $\{\phi_m\}$ [cf. Eq. (8.70)]. As we have already noted, under these conditions the coefficients C_m of the general expansion (8.42), employing the orthogonal set (8.70), are statistically related, even if the process is gaussian. Similar remarks apply if the interval is finite, say $(0,T)$ (cf. Prob. 8.12).

PROBLEMS

8.8 (*a*) Extend the theory of the orthogonal expansion to the complex process $z(t)$, $E\{z\} = 0$, in the general interval $(a \le t \le b)$, and hence show that

$$z(t) = \sum_m \sqrt{\lambda_m}\, d_m \phi_m(t) \qquad a \le t \le b \qquad (1)$$

with
$$d_m = \lambda_m^{-1/2} \int_a^b z(t)\phi_m^*(t)\,dt \qquad \int_a^b \phi_m(t)\phi_n^*(t)\,dt = \delta_{mn} \qquad (2)$$

and
$$E\{d_m d_n^*\} = \delta_{mn} \qquad (3)$$

Here, λ_m is real and positive (when not zero), while ϕ_m, d_m are complex. The complex orthonormal set $\{\phi_m\}$ are the eigenfunctions and λ_m are the eigenvalues of the basic integral equation

$$\int_a^b K_z(t_1,t_2)\phi_m(t_2)\,dt_2 = \lambda_m \phi_m(t_1) \qquad a \le t_1 \le b \qquad (4)$$

with the kernel $K_z(t_1,t_2) = K_z^*(t_2,t_1) = E\{z(t_1)z(t_2)^*\}$. State the conditions for completeness and the existence of solutions to Eq. (4).

(*b*) Prove Mercer's theorem for the complex kernel $G(s,\tau)$; that is, show that

$$G(s,\tau) = \sum_k \lambda_k \phi_k(s)\phi_k(\tau)^* \qquad a \le s, \tau \le b \qquad (5)$$

uniformly and absolutely. Show that sufficient conditions are Hermitian symmetry, that is, $G(s,\tau) = G(\tau,s)^*$, and that $G(s,\tau)$ is positive definite.

(*c*) From (*a*) and (*b*), show that the *real* process $x(t)$ may be developed in terms of the complex eigenfunctions $\{\phi_m\}$ and eigenvalues d_m according to Eqs. (1) to (3), where Eq. (8.60) holds for $\{\phi_m\}$ complex. See Loève,[6] pp. 478, 479.

8.9 If the complex kernel $G(s,\tau)$ is Hermitian symmetric, and if $\iint_a^b |G(s,\tau)|^2\, ds\, d\tau < \infty$, consider the integral equation

$$\int_a^b G(s,\tau)\phi_m(\tau)\, d\tau = \lambda_m \phi_m(s) \qquad a \le s \le b \tag{1}$$

and show that:

(a) If $G(s,\tau)$ is positive semidefinite, all nonzero eigenvalues λ_m are positive real numbers.

(b) If $G(s,\tau)$ is positive definite, the eigenfunctions form a complete orthonormal set.

(c) $G(s,\tau)$ can be expanded (l.i.m.) in the series

$$G(s,\tau) = \underset{K\to\infty}{\text{l.i.m.}} \sum_k^K \lambda_k \phi_k(s)\phi_k(\tau)^* \qquad a \le s,\, \tau \le b \tag{2}$$

(d) If $\phi_m(t)$ and $\phi_{m'}(t)$ are eigenfunctions associated with the eigenvalue λ_m, then $a\phi_m + b\phi_{m'}$ is also an eigenfunction for λ_m; therefore, to each λ_m there corresponds at least one (normalized) eigenfunction.

(e) If $\lambda_m \neq \lambda_n$, then ϕ_m and ϕ_n are orthogonal.

(f) There are at most a countably infinite set of eigenvalues, and, for a constant $C < \infty$, $|\lambda_k|$ is less than C (all k). Courant-Hilbert.[9]

8.10 From the basic integral equation (8.60) for the orthogonal expansion of a real random process, show that it is not possible to obtain a Fourier-series expansion of such a process on any finite interval, where the coefficients are statistically independent. (Exclude the limiting case of a white noise process, assume an entirely random process.)

8.11 An ensemble of stationary white noise voltages, with zero mean, is passed through an RC filter, and the outputs are observed over the interval $(-T/2,T/2)$. If ψ is the mean intensity of the output ensemble and $K_x(\tau - s) = \psi e^{-\alpha|\tau-s|}$ $(\alpha = 1/RC)$ is the covariance function of this output process $x(t)$, show that the expansion of $x(t)$ in $(-T/2,T/2)$ is given by

$$x(t) = \sum_{m=-\infty}^{+\infty} \sqrt{\lambda_m}\, c_m \phi_m(t) \tag{1}$$

where

$$\begin{aligned}
\lambda_m &= \frac{2\psi}{(b_m^2+1)\alpha} & \phi_m &= \left(\frac{T}{2} + \frac{\sin T\alpha b_m}{2\alpha b_m}\right)^{1/2} \cos \alpha b_m t & m &= -1, -2, \ldots, -\infty \\
\lambda_m &= \frac{2\psi}{(b_m^2+1)\alpha} & \phi_m &= \left(\frac{T}{2} - \frac{\sin T\alpha b_m}{2\alpha b_m}\right)^{1/2} \sin \alpha b_m t & m &= 1, 2, \ldots, \infty
\end{aligned} \tag{2}$$

$$\begin{aligned}
c_m &= \frac{\lambda_m^{-1/2}}{\left(\dfrac{T}{2} + \dfrac{\sin T\alpha b_m}{2\alpha b_m}\right)^{1/2}} \int_{-T/2}^{T/2} x(t) \cos \alpha b_m t\, dt & m &\le -1 \ (\lambda_0 = 0) \\
c_m &= \frac{\lambda_m^{-1/2}}{\left(\dfrac{T}{2} - \dfrac{\sin T\alpha b_m}{2\alpha b_m}\right)^{1/2}} \int_{-T/2}^{T/2} x(t) \sin \alpha b_m t\, dt & m &\ge 1 \ (\lambda_0 = 0)
\end{aligned} \tag{3}$$

and $b_m|_{m\ge 1}$ are the solutions of $b \tan (b\alpha T/2) = 1$ and $b_m|_{m\le -1}$ are the solutions of $b \cot (b\alpha T/2) = -1$. Here $\phi_m(t)|_{m\ge 1}$, $\phi(t)|_{m\le -1}$ are each a complete orthonormal set of eigenfunctions, with the corresponding eigenvalues $\lambda_m|_{m\ge 1}$, $\lambda_m|_{m\le -1}$ above, satisfying the integral equation

$$\int_{-T/2}^{T/2} \psi e^{-\alpha|t-s|}\phi_m(t)\, dt = \lambda_m \phi_m(s) \qquad -T/2 \le s \le T/2 \tag{4}$$

W. B. Davenport, Jr., and W. L. Root,[2] sec. 6.4.

8.12 (*a*) If $K_x(t_1,t_2) = \psi \dfrac{\sin 2\pi B(t_2 - t_1)}{2\pi B(t_2 - t_1)}$ $(\psi = \overline{x^2})$ is the covariance function of band-limited white stationary normal noise, with $\bar{x} = 0$, show that the process $x(t)$ has the expansion in $(0,T)$

$$x(t) = \Sigma \sqrt{\lambda_m}\, c_m \phi_m(t) \qquad 0 \le t \le T$$
$$c_m = \lambda_m^{-1/2} \int_0^T x(t)\phi_m(t)\, dt \tag{1}$$

where the eigenvalues are

$$\lambda_m = \psi T\{R_{0m}{}^{(1)}(BT,1)\}^2 \tag{2}$$

and the (normalized) eigenfunctions are

$$\phi_m(t) = \frac{S_{0m}{}^{(1)}[\pi BT,\ 2/T(t-1)]}{\sqrt{N_{0m}}} \qquad 0 \le t \le T \tag{3}$$

where N_{0m} is given by

$$N_{0m} = 2 \sum_n{}' \frac{(d_n{}^m)^2}{2n+1} \tag{4}$$

$R_{ml}{}^{(1)}(c,y)$ and $S_{ml}{}^{(1)}(c,\cos\theta)$ are, respectively, the prolate spheroidal Bessel function and prolate spheroidal Legendre function, as given in Morse and Feshbach[11] and tabulated by Stratton, Morse, Chu, and Hutner,[12] the notation being that of the named citations. Observe that the $\{c_m\}$ are independent and normally distributed with zero mean and unit variance.

(*b*) Show that as $T \to \infty$, the eigenvalues and eigenfunctions [Eqs. (2), (3)] reduce, respectively, to

$$\lambda'_m = \frac{W_0}{2} \qquad \text{and} \qquad \phi'_m(t) = \sqrt{2B}\, \frac{\sin 2\pi B(t - m/2B)}{2\pi B(t - m/2B)} \tag{5}$$

and verify that the sampling functions $\phi'_m(t)$ [Eq. (5)] are *not* solutions of the governing integral equation here:

$$\int_0^T K_x(|t_1 - t_2|)\phi_m(t_2)\, dt_2 = \lambda_m \phi_m(t_1) \qquad 0 \le t_1 \le T \tag{6}$$

D. Slepian, *IRE Trans. on Inform. Theory*, **PGIT-3:** 68 (March, 1954), app. II.

8.3 *Fourier-series Expansions*

In Secs. 8.2-1, 8.2-2, we have observed that a general series representation of a random process on any finite interval (a,b) cannot be an orthogonal expansion,† since in general the expansion coefficients are correlated [cf. Eqs. (8.49) and (8.50)]. Here, we shall show that this is indeed the case for the Fourier-series development and that, unless the interval becomes infinite, statistical independence of these coefficients is not possible.

We begin by considering a wide-sense stationary real *periodic* random

† An exception occurs in the limiting situation of white noise with a uniform spectrum for all frequencies (cf. Sec. 8.2-2).

process $x(t)$ [necessarily on $(-\infty, \infty)$]. Such a process is called periodic if its covariance function $K_x(t)$ has a period T, and consequently the random variables $x(t)$, $x(t+T)$ are equal (probability 1) for any t in $(-\infty, \infty)$ (cf. Prob. 8.13). It is assumed that except for a subensemble of probability zero all the sample functions of the process are periodic. Let us represent this process on the interval $(-\infty, \infty)$ by the Fourier series

$$x(t) = \sum_{m=0}^{\infty} (a_m \cos \omega_m t + b_m \sin \omega_m t) \qquad \omega_m = m\omega_0 = 2\pi m/T \tag{8.74}$$

where the random variables a_m, b_m are given by

$$a_m = \frac{\epsilon_m}{T} \int_0^T x(t) \cos m\omega_0 t \, dt \qquad b_m = \frac{2}{T} \int_0^T x(t) \sin m\omega_0 t \, dt$$
$$\epsilon_0 = 1, \ \epsilon_m = 2 \ (m \geq 1) \tag{8.75}$$

in the usual way and the series converges mean square to the process $x(t)$. Without loss of generality, we assume also that $\boldsymbol{E}\{x\} = 0$, from which it is immediately evident that $\boldsymbol{E}\{a_m\} = \boldsymbol{E}\{b_m\} = 0$ (all m) as well.

Let us now examine the covariances $\boldsymbol{E}\{a_m a_n\}$, etc., of the expansion coefficients. From Eqs. (8.75), we get

$$\boldsymbol{E}\{a_m a_n\} = \frac{\epsilon_m \epsilon_n}{T^2} \boldsymbol{E}\left\{\iint_0^T x(t_1)x(t_2) \cos m\omega_0 t_1 \cos n\omega_0 t_2 \, dt_1 \, dt_2\right\}$$
$$= \frac{\epsilon_m \epsilon_n}{T^2} \iint_0^T K_x(t_2 - t_1) \cos m\omega_0 t_1 \cos n\omega_0 t_2 \, dt_1 \, dt_2 \tag{8.76}$$

But since $K_x(t_2 - t_1)$ is periodic and even, we can write it as

$$K_x(t_2 - t_1) = \sum_{k=0}^{\infty} A_k \epsilon_k \cos k\omega_0 (t_2 - t_1) \qquad A_k = \frac{1}{T} \int_0^T K_x(t) \cos k\omega_0 t \, dt \tag{8.77}$$

and substitute this into Eq. (8.76) to get

$$\boldsymbol{E}\{a_m a_n\} = \frac{\epsilon_m \epsilon_n}{T^2} \sum_{k=0}^{\infty} \epsilon_k A_k \iint_0^T \cos m\omega_0 t_1 \cos n\omega_0 t_2 \cos k\omega_0 (t_2 - t_1) \, dt_1 \, dt_2 \tag{8.78}$$

With the help of the orthogonal relations

$$\frac{1}{2\pi} \int_0^{2\pi} \begin{Bmatrix} \cos lx \cos mx \\ \sin lx \sin mx \end{Bmatrix} dx = \frac{\delta_{lm}}{\epsilon_l} \qquad \frac{1}{2\pi} \int_0^{2\pi} \begin{Bmatrix} \cos lx \sin mx \\ \sin lx \sin mx \end{Bmatrix} dx = 0 \tag{8.79}$$

we see that Eq. (8.78) reduces directly to

$$\boldsymbol{E}\{a_m a_n\} = \sum_{k=0}^{\infty} \epsilon_k A_k \delta_{kn} \delta_{km} = \epsilon_m A_m \delta_{mn} \tag{8.80}$$

For $\boldsymbol{E}\{b_m b_n\}$, we find in a similar way that $\boldsymbol{E}\{b_m b_n\} = \epsilon_m A_m \delta_{mn}$ also, and, for $\boldsymbol{E}\{a_m b_n\}$, we get

$$\boldsymbol{E}\{a_m b_n\} = \frac{\epsilon_m \epsilon_n}{T^2} \sum_{k=0}^{\infty} \epsilon_k A_k \iint_0^T \cos m\omega_0 t_1 \sin n\omega_0 t_2 \cos k\omega_0(t_2 - t_1)\, dt_1\, dt_2 = 0 \tag{8.81}$$

Thus, *for periodic random processes the random variables* (a_m,b_m) *are orthogonal* in the above statistical sense. Conversely, one may show that if the expansion coefficients (a_m,b_m) are orthogonal [according to Eqs. (8.80) and (8.81)] then the process must be periodic (cf. Prob. 8.13). If, in addition, the process is normal, the coefficients are statistically independent and normally distributed with zero means and with the variances

$$\boldsymbol{E}\{a_m a_n\} = \boldsymbol{E}\{b_m b_n\} = \epsilon_m A_m \delta_{mn} \qquad \boldsymbol{E}\{a_m b_n\} = 0 \tag{8.82}$$

The mean total intensity of the process is found from Eqs. (8.74) and (8.80) to (8.82) to be

$$\boldsymbol{E}\{x(t)^2\} = \sum_{m=0}^{\infty} \frac{\overline{a_m{}^2} + \overline{b_m{}^2}}{2} = \sum_{m=0}^{\infty} \epsilon_m A_m = K_x(0) \tag{8.83}$$

The reason why a Fourier-series expansion of a process on a finite interval does not lead to independent coefficients is, basically, that the process is unspecified outside the interval by such an expansion. Any periodic representation, while identical with the original process and its development on the interval, represents a different process outside the interval. The lack of periodicity in the former case, and the unspecified nature of the development outside the interval, are insufficient restrictions on the process for us to be able to ensure the orthogonal nature of the coefficients. As we have already seen in Sec. 8.2-2, for orthogonality we must choose a particular type of expansion, where the orthonormal set involved is the solution of a certain integral equation [cf. Eq. (8.60)]. It is this added second-moment information about the process, contained in the kernel of the basic integral equation, that enables us to construct an orthogonal expansion with the desired orthogonality of the expansion coefficients.

It remains now to show that, when the interval $(0,T)$ becomes infinitely large, the Fourier-series representation (8.74) is orthogonal; i.e., the coefficients (a_m,b_m) obey relations like Eqs. (8.82). This is easily done if we start with Eq. (8.76) and let $u = t_2 - t_1$, writing

$$E\{a_m a_n\} = \frac{\epsilon_m \epsilon_n}{T^2} \int_0^T \cos m\omega_0 t_1 \, dt_1 \int_{-t_1}^{T-t_1} du \, K_x(u) \cos n\omega_0(u + t_1) \quad (8.84)$$

If we set $v = t_1/T$, we see that this becomes

$$\begin{aligned} TE\{a_m a_n\} &= \epsilon_m \epsilon_n \int_0^1 dv \cos 2\pi m v \int_{-vT}^{T(1-v)} K_x(u) \cos [2\pi n(u/T + v)] \, du \\ &= \epsilon_m \epsilon_n \int_0^1 dv \cos 2\pi m v \left[\cos 2\pi n v \int_{-vT}^{T(1-v)} du \, K_x(u) \cos (2\pi n u/T) \right. \\ &\qquad \left. - \sin 2\pi n v \int_{-vT}^{T(1-v)} du \, K_x(u) \sin (2\pi n u/T) \right] \quad (8.85) \end{aligned}$$

For every $v \neq 0$, the inner integrals are, respectively, $\frac{1}{2}\mathcal{W}_x(f_n)$ and 0, as $T \to \infty$, with $f_n \equiv n/T$ [cf. Eqs. (3.42)], where $\mathcal{W}_x(f_n)$ is the spectral density of the process at frequency f_n, and $f_n \to 0$ for n fixed, unless we choose n such that $\lim_{n,T\to\infty} (n/T = f_n) \to f\ (> 0)$. From Eqs. (8.79), we have δ_{mn}/ϵ_n, so that we can write

$$\lim_{T\to\infty} TE\{a_m a_n\} = \lim_{T\to\infty} \frac{\epsilon_m}{2} \mathcal{W}_x(f_n)\delta_{mn} \to \mathcal{W}_x(f_m)\delta_{mn} \qquad m, n > 0 \quad (8.86)$$

or for the normalized expansion coefficients $\hat{a}_m$, $\hat{a}_n$, for example, $\hat{a}_m \equiv a_m/(\overline{a_m^2})^{1/2}$, etc.,

$$\lim_{T\to\infty} E\{\hat{a}_m \hat{a}_n\} = \delta_{mn} \quad (8.86a)$$

In precisely the same way, we easily show that

$$\begin{aligned} &\lim_{T\to\infty} TE\{b_m b_n\} = \mathcal{W}_x(f_m)\delta_{mn} \qquad m, n > 0 \\ &\lim_{T\to\infty} TE\{a_m b_n\} = 0 \qquad \text{all } m, n \geq 0 \end{aligned} \quad (8.87)$$

so that the expansion coefficients of the periodic process become orthogonal as the period interval becomes indefinitely large.

The mean intensity of the process is now, from Eq. (8.74) and the above,

$$\begin{aligned} \overline{x^2} &= \lim_{T\to\infty} \sum_{m,n=0}^{\infty} \left(\overline{a_m^2} \cos^2 \omega_m t + \overline{a_m b_n} \cos \omega_m t \sin \omega_n t + \overline{a_n b_m} \sin \omega_m t \cos \omega_n t \right. \\ &\qquad \left. + \overline{b_n^2} \sin^2 \omega_m t \right) \\ &= \lim_{T\to\infty} \sum_{m=0}^{\infty} (\overline{a_m^2} + \overline{b_m^2}) = \lim_{T\to\infty} \sum_{m=0}^{\infty} \mathcal{W}_x(f_m) \frac{\epsilon_m}{2T} \to \int_0^\infty \mathcal{W}_x(f) \, df \\ &\qquad \lim_{T\to\infty} \frac{1}{T} = df \to 0 \end{aligned} \quad (8.88)$$

as expected. The covariance function may be found in similar fashion,

$$\begin{aligned}
K_x(t) &= \boldsymbol{E}\{x(t_0)x(t_0 + t)\} \\
&= \lim_{T\to\infty} \sum_{m,n} \Big[\overline{a_m{}^2} \cos \omega_m t_0 \cos \omega_m(t_0 + t) + \overline{a_m b_n} \cos \omega_m t_0 \sin \omega_n(t_0 + t) \\
&\qquad + \overline{b_n{}^2} \sin \omega_n t_0 \sin \omega_n(t_0 + t) + \overline{a_n b_m} \sin \omega_n t_0 \cos \omega_m(t_0 + t) \Big] \\
&= \lim_{T\to\infty} \sum_{m=0}^{\infty} (\overline{a_m{}^2} + \overline{b_m{}^2}) \cos \omega_m t \to \int_0^\infty \mathcal{W}_x(f) \cos \omega t \, df
\end{aligned} \tag{8.89}$$

which is consistent with our earlier results (cf. Sec. 3.2) on the W-K theorem Finally, if $x(t)$ is a normal process (with $\boldsymbol{E}\{x\} = 0$), in the limit $T \to \infty$ the expansion coefficients (a_n,b_n) become normal and independent, with zero means and variances $\mathcal{W}_x(f_n)\,\Delta f$ $(\Delta f \equiv 1/T \to df)$.†

Because it is convenient to have an orthogonal Fourier expansion in many applications (cf. Secs. 9.1, 10.4) where the process is normal, we shall often replace the gaussian random process in question by a periodic normal process which is identical with the original one on the fundamental interval $(0,T)$, so that the coefficients have this useful property of orthogonality. Then, allowing $T \to \infty$, we see from the above that this periodic representation coincides (probability 1) with a normal process having the orthogonal representation of the original nonperiodic gaussian process. In this way, the coefficients (a_m,b_m) of the Fourier-series representation (8.74) become identical with the coefficients of the orthogonal expansion of the original process. It is in this sense, then, that the Fourier-series development of a normal random process where orthogonality of the coefficients is postulated, i.e., where periodicity is implied, is to be understood.

PROBLEMS

8.13 If the Fourier expansion coefficients (a_m,b_m) of a random process on a finite interval (a,b) are orthogonal, show that the process must be periodic.

W. L. Root and T. S. Pitcher, *Ann. Math. Stat.*, **26**(2): 313 (1955).

8.14 If a complex process $z(t)$ can be represented mean square in the interval (a,b) by the Fourier series

$$z(t) = \sum_{m=-\infty}^{\infty} d_m e^{im\omega_0 t} \qquad \omega_0 = 2\pi/(b - a) \qquad (a \le t \le b) \tag{1}$$

$$d_m = \frac{1}{b - a} \int_a^b z(t) e^{-im\omega_0 t}\, dt \tag{2}$$

show that:

(a) $\boldsymbol{E}\{d_m d_n^*\} = \delta_{nm} B_m$ $(B_m > 0)$, if the process is periodic.

(b) $\boldsymbol{E}\{d_m d_n^*\} \neq 0$ $(m \neq n)$ and, as T $(= b - a) \to \infty$, $\boldsymbol{E}\{d_m d_n^* T\} \to \mathcal{W}_z(f_n)$, where now $\mathcal{W}_z(f_n)$ is no longer an even function of f_n.

† Strictly speaking, the normalized coefficients $\hat{a}_m$, $\hat{b}_m$ obey a (nonzero) normal law and are independent with zero means and unit variances [cf. Eq. (8.86a)].

(*c*) Repeat Eqs. (8.76) to (8.89) for a real process $x(t)$, represented by

$$x(t) = \sum_{m=-\infty}^{\infty} c_m e^{im\omega_0 t} \tag{3}$$

where c_m is complex, in general. What are sufficient conditions for MSC?

REFERENCES

1. Siegert, A. J. F.: Passage of Stationary Processes through Linear and Nonlinear Devices, *IRE Trans. on Inform. Theory*, **PGIT-3:** 4 (March, 1954).
2. Davenport, W. B., Jr., and W. L. Root: "An Introduction to the Theory of Random Signals and Noise," McGraw-Hill, New York, 1958.
3. Middleton, D.: The Distribution of Energy in Randomly Modulated Waves, *Phil. Mag.*, (7)**62:** 689 (1951). See Chap. 14 for additional references.
4. Karhunen, K.: Zur Spektraltheorie stochastischer Prozesse, *Ann. Acad. Sci. Fennicae*, (A)I, no. 34, 1946; Über lineare Methoden in der Wahrscheinlichkeitsrechnung, *ibid.*, (A)I, no. 37, 1947.
5. Kac, M., and A. J. F. Siegert: An Explicit Representation of a Stationary Gaussian Process, *Ann. Math. Stat.*, **18:** 38 (1947).
6. Loève, M.: "Probability Theory," Van Nostrand, Princeton, N.J., 1955.
7. Grenander, U.: Stochastic Processes and Statistical Inference, *Arkiv Mat.*, vol. 1, no. 3, secs. 1.3, 3.1, April, 1950.
8. Bartlett, M. S.: "An Introduction to Stochastic Processes," Cambridge, New York, 1955.
9. Courant, R., and D. Hilbert: "Mathematische Physik," vol. 1, p. 117, Springer, Berlin, 1931; see also W. V. Lovitt, "Linear Integral Equations," p. 145, Dover, New York, 1950.
10. Youla, D. C.: The Use of the Method of Maximum Likelihood in Estimating Continuous-modulated Intelligence Which Has Been Corrupted by Noise, *IRE Trans. on Inform. Theory*, **PGIT-3:** 90 (March, 1954).
11. Morse, P. M., and H. Feshbach: "Methods of Theoretical Physics," pp. 642, 1503, McGraw-Hill, New York, 1953.
12. Stratton, J. A., P. M. Morse, L. Chu, and Hutner: "Elliptic Cylinder and Spheroidal Wave Functions," pp. 62ff., Wiley, New York, 1941.

CHAPTER 9

PROCESSES DERIVED FROM THE NORMAL

As we have remarked in previous chapters, the largest class of noise processes that originate in the various stages of transmission $\boldsymbol{T}_T^{(N)}$ and reception $\boldsymbol{T}_R^{(N)}$, and even in the medium of propagation $\boldsymbol{T}_M^{(N)}$, is normal. Moreover, as we have observed in Chaps. 7, 8, a normal process that undergoes a linear transformation $\boldsymbol{T}_L$ is still normal. However, since a communication link $\boldsymbol{T}_R^{(N)}\boldsymbol{T}_M^{(N)}\boldsymbol{T}_T^{(N)}$ [cf. Eq. (2.20)] will generally contain nonlinear elements as well as linear ones, we expect that, under a nonlinear transformation $\boldsymbol{T}_g$, these processes will lose their normal character and be transformed into new ensembles, with quite different statistical properties.

For example, normal (i.e., thermal and shot) noise arising in the RF and IF stages of a radar receiver will no longer appear gaussian in the video because of the nonlinear action of the second detector. Similarly, a carrier frequency-modulated by a normal signal process possesses quite different statistics from the original message ensemble because of the highly nonlinear operation of encoding in this instance. Then, too, apart from the effect of an actual nonlinear operation, we may also be interested in statistical features of a normal process which are themselves *not* normal: the envelope and phase of a narrowband noise wave are an important example; others are the average number of zero crossings of filtered thermal noise, their distributions, and the distributions of their maxima; the statistical properties of signal and additive noise ensembles; and so on. Processes of this type, including those which are obtained by nonlinear operations $\boldsymbol{T}_g$, we shall call *normally derived processes* when the original process is itself normal. In this chapter, touching briefly the questions of the zeros, maxima, etc., of such waves, we shall consider primarily first- and second-order statistics of the envelope and of the phase of narrowband, gaussian noise and additive signal processes (needed in our subsequent discussion of FM receivers). Several problems involving the distributions (for example, W_1, W_2, etc.) of normally derived processes—filtered thermal noise after a square-law rectifier [Eq. (8.6)], for example—are examined later, in Chap. 17.

9.1 *Statistical Properties of the Envelope and Phase of Narrowband Normal Noise*†

Here we wish to determine the first- and second-order d.d. of the envelope and phase of a (real) normal narrowband noise process $N(t)$, where $\boldsymbol{E}\{N\} = 0$ and the process is stationary. Let us begin by considering first the narrowband process on a finite interval (a,b), and from Secs. 2.2-6, 7.5-3(3) let us write

$$N(t) = N_c(t) \cos \omega_0 t + N_s(t) \sin \omega_0 t \qquad a \le t \le b \tag{9.1a}$$
$$N(t) = E(t) \cos [\omega_0 t - \theta(t)] \tag{9.1b}$$

where $E(t)$ and $\theta(t)$ are, respectively, the envelope and phase of this narrowband process, obtained from the "cosine" and "sine" components N_c, N_s by the transformations

$$E_c = \sqrt{N_c^2 + N_s^2} \qquad \theta = \tan^{-1} (N_s/N_c) \pm \pi q \qquad q = 0, 1, 2, \ldots \tag{9.2a}$$

with the inverse relations

$$N_c = E \cos \theta \qquad N_s = E \sin \theta \tag{9.2b}$$

Both (N_c,N_s) and (E,θ) are slowly varying functions of time compared with $\cos \omega_0 t$, $\sin \omega_0 t$. As before, ω_0 $(= 2\pi f_0)$ is the central (angular) frequency of the narrowband process and is specified analytically with the help of the Hilbert transform or physically by suitable measurement with phase and envelope detectors (cf. Sec. 2.2-7).

Next, to obtain the statistical properties of N_c and N_s let us expand $N(t)$ [Eq. (9.1a)] in a Fourier series on the interval (a,b), which we shall presently allow to become infinite,

$$N(t) = \sum_{n=0}^{\infty} (a_n \cos \omega_n t + b_n \sin \omega_n t)$$
$$a \le t \le b \; (\omega_n = 2\pi n/T + \omega_0,\; T = b - a) \tag{9.3}$$

where (a_n,b_n) are normally distributed with zero means (since $\bar{N} = 0$) and finite variances. Note that since the interval (a,b) is finite, the coefficients (a_n,b_n) are *not* statistically independent, nor are $(\sin \omega_n t, \cos \omega_n t)$ the eigenfunctions associated with the covariance function $K_N(t_1,t_2)$ of the process (cf. Secs. 8.2, 8.3).‡ Setting $\omega_n = \omega_n' + \omega_0$ in Eq. (9.3) and comparing with

† See also Rice,[1] sec. 3.7.

‡ We could, of course, use any other complete orthonormal set $\{\phi_m\}$ to make the expansion. In fact, we might choose the appropriate set for the kernel K_N, to ensure the statistical independence of the (a_n,b_n) at the outset, i.e., to give an orthogonal expansion of the process on (a,b) (cf. Sec. 8.2-2). The trigonometric functions, however, have the advantage of analytic simplicity and are the natural functions to use when we are deal-

Eq. (9.1a), we see that

$$N_c(t) = \sum_{n=0}^{\infty} (a_n \cos \omega_n' t + b_n \sin \omega_n' t)$$
$$N_s(t) = \sum_{n=0}^{\infty} (-a_n \sin \omega_n' t + b_n \cos \omega_n' t) \tag{9.4}$$
$$\omega_n' = 2\pi n/T,\ T = b - a\ (a \le t \le b)$$

and, as expected, the "cosine" and "sine" components $N_c(t)$, $N_s(t)$ are also normal random processes [cf. Sec. 7.5-3(4)]. If we observe now from Sec. 8.3 that

$$\boldsymbol{E}\{a_m a_n\} = \boldsymbol{E}\{b_m b_n\} \qquad \boldsymbol{E}\{a_m b_n\} = \boldsymbol{E}\{a_n b_m\} = 0 \tag{9.5}$$

and that $\boldsymbol{E}\{a_m\} = \boldsymbol{E}\{b_m\} = 0$, then $\bar{N}_c = \bar{N}_s = 0$. With the help of Eqs. (9.4), we can write the covariance functions of $N_c(t)$ and $N_s(t)$ as

$$\overline{N_c(t_1)N_c(t_2)} = \sum_{m,n} [\overline{a_m a_n} \cos (\omega_n' t_2 - \omega_m' t_1) + \overline{a_m b_n} \sin (\omega_n' t_2 - \omega_m' t_1)] = \overline{N_s(t_1)N_s(t_2)} \tag{9.6a}$$

$$\overline{N_c(t_1)N_s(t_2)} = \sum_{m,n} [-\overline{a_m a_n} \sin (\omega_n' t_2 - \omega_m' t_1) + \overline{a_m b_n} \cos (\omega_n' t_2 - \omega_m' t_1)] \tag{9.6b}$$

$$\overline{N_c(t_2)N_s(t_1)} = \sum_{m,n} [\overline{a_m a_n} \sin (\omega_n' t_2 - \omega_m' t_1) + \overline{a_m b_n} \cos (\omega_n' t_2 - \omega_m' t_1)] \tag{9.6c}$$

for all $(a \le t_1, t_2 \le b)$.

Having determined the covariance functions of N_c and N_s for the finite interval (a,b), let us now pass to the infinite interval $(-\infty < t < \infty)$, where considerable simplification in the expressions (9.6) can be made. In addition to our original assumptions of normality, we assume at this stage that the process $N(t)$, and hence $N_c(t)$, $N_s(t)$, are stationary. From Eq. (8.86), we can accordingly write

$$\lim_{T\to\infty} \overline{a_m a_n} = \lim_{T\to\infty} \frac{1}{T} \delta_{mn} \mathcal{W}_N(f_m) \to \delta_{mn} \mathcal{W}_N(f_m)\, df' \qquad \omega_m \to \omega$$
$$\lim_{T\to\infty} \{T\overline{a_m b_n}\} = 0 \qquad \text{all } m, n \tag{9.7}$$

---

ing with the spectrum of the process. Moreover, since we shall ultimately pass to the infinite interval ($T \to \infty$), we can equally well represent the actual process $N(t)$ on (a,b) by a periodic process $N'(t)$ which coincides with it on this interval. This has the further technical advantage that now the periodic Fourier-series development (over all t) does indeed possess independent coefficients (a_n', b_n') for any period T. In any case, as $T \to \infty$, the process $N'(t)$ and the original process $N(t)$ are identical (probability 1), as are their mean-square Fourier developments, whose coefficients are in both instances independent (cf. Sec. 8.3).

so that Eqs. (9.6a) to (9.6c) become

$$\overline{N_{c1}N_{c2}} = \lim_{T\to\infty} \sum_m \frac{1}{T} \mathcal{W}_N(f_m) \cos \omega_m'(t_2 - t_1)$$
$$\to \int_0^\infty \mathcal{W}_N(f) \cos (\omega - \omega_0)t \, df \equiv \psi\rho_0(t) \tag{9.8a}$$
$$\overline{N_{c1}N_{s2}} = - \lim_{T\to\infty} \sum_m \frac{1}{T} \mathcal{W}_N(f_m) \sin \omega_m'(t_2 - t_1)$$
$$\to - \int_0^\infty \mathcal{W}_N(f) \sin (\omega - \omega_0)t \, df \equiv \psi\lambda_0(t) \tag{9.8b}$$

where we have used the fact that $f' = f - f_0$ and have written $t = t_2 - t_1$, with $1/T = \Delta f' \to df'$ in the limit. Here, also, we have

$$\overline{N_c^2} = \overline{N_s^2} = \int_0^\infty \mathcal{W}_N(f)\, df \equiv \psi \tag{9.9a}$$

and

$$\overline{N_{c1}N_{s2}} = -\overline{N_{c2}N_{s1}} = \psi\lambda_0(t) \qquad \begin{cases} \rho_0(t) = \rho_0(-t);\ \lambda_0(t) = -\lambda_0(-t) \\ \lambda_0(0) = 0;\ \rho_0(0) = 1 \end{cases} \tag{9.9b}$$

In terms of the covariance function of the process, we recall our earlier result [cf. Eq. (7.53)]

$$K_N(t) = \psi[\rho_0(t) \cos \omega_0 t + \lambda_0(t) \sin \omega_0 t] \tag{9.10a}$$

with

$$\rho_0(t) = \psi^{-1} \int_0^\infty \mathcal{W}_N(f) \cos (\omega - \omega_0)t \, df$$
$$\lambda_0(t) = -\psi^{-1} \int_0^\infty \mathcal{W}_N(f) \sin (\omega - \omega_0)t \, df \tag{9.10b}$$

Observe that the postulated normality of (a_n,b_n) in the original expression (9.3) as $T \to \infty$ leads to precisely the conditions on N_c and N_s postulated for Eqs. (7.52). Furthermore, since $N(t)$ is narrowband, and if the spectrum of the process is reasonably symmetrical about f_0, then $\lambda_0(t)$ is essentially zero for all t, as we have already noted in Sec. 7.5-3(4), and $N_c(t)$, $N_s(t)$ are consequently independent normal processes. However, if the spectrum is not symmetrical about f_0, we cannot safely set $\lambda_0(t) = 0$ (all t) and N_c, N_s remain statistically related. In any case, the joint first- and second-order distribution densities for N_c and N_s can be written with the aid of the above and Eq. (7.57) as follows:

$$W_1(N_c,N_s)_N = \frac{e^{-(N_c^2+N_s^2)/2\psi}}{2\pi\psi} \tag{9.11}$$

and

$$W_2(N_{c1},N_{s1};N_{c2},N_{s2};t)_N = \frac{e^{-\frac{1}{2}\tilde{\mathbf{X}}\mathbf{K}_0^{-1}\mathbf{X}}}{(2\pi)^2(\det \mathbf{K}_0)^{\frac{1}{2}}} \tag{9.12}$$

where $N_{c1} = N_c(t_1)$, etc., and

$$\mathbf{X} = \begin{bmatrix} N_{c1} \\ N_{s1} \\ N_{c2} \\ N_{s2} \end{bmatrix} \qquad \mathbf{K}_0 = \psi \begin{bmatrix} 1 & 0 & \rho_0 & \lambda_0 \\ 0 & 1 & -\lambda_0 & \rho_0 \\ \rho_0 & -\lambda_0 & 1 & 0 \\ \lambda_0 & \rho_0 & 0 & 1 \end{bmatrix} \tag{9.12a}$$

From this we get

$$\begin{aligned} \det \mathbf{K}_0 &= \psi^4(1 - \lambda_0^2 - \rho_0^2)^2 = \psi^4(1 - k_0^2)^2 \\ k_0(t)^2 &= \rho_0^2(t) + \lambda_0^2(t) \leq 1 \qquad k_0(t) = k_0(-t) \end{aligned} \tag{9.12b}$$

$$\mathbf{K}_0^{-1} = [\psi(1 - k_0^2)]^{-1} \begin{bmatrix} 1 & 0 & -\rho_0 & -\lambda_0 \\ 0 & 1 & \lambda_0 & -\rho_0 \\ -\rho_0 & \lambda_0 & 1 & 0 \\ -\lambda_0 & -\rho_0 & 0 & 1 \end{bmatrix}$$

$$\doteq \psi^{-1}(1 - \rho_0^2)^{-1} \begin{bmatrix} 1 & 0 & -\rho_0 & 0 \\ 0 & 1 & 0 & -\rho_0 \\ -\rho_0 & 0 & 1 & 0 \\ 0 & -\rho_0 & 0 & 1 \end{bmatrix} \tag{9.12c}$$

if $\lambda_0 \doteq 0$.

With these results and the transformations (9.2), we are now ready to find the corresponding joint distribution densities of the envelope and phase, $w_1(E,\theta)_N$ and $w_2(E_1,E_2;\theta_1,\theta_2;t)_N$, of our stationary narrowband normal process $N(t)$. Specifically, we have

$$w_1(E,\theta)_N = W_1[N_c(E,\theta),N_s(E,\theta)]_N \left| \frac{\partial(N_c,N_s)}{\partial(E,\theta)} \right| \tag{9.13}$$

with a similar relation for $w_2(E_1,E_2;\theta_1,\theta_2;t)_N$ (cf. Sec. 1.2-14). The jacobian in the two cases is E and E_1E_2, respectively, and the regions $(-\infty < N_c, N_s < \infty)$ for N_c, N_s map into $(0 \leq E < \infty)$, $(0 \leq \theta \leq 2\pi)$ for E and θ. In the general situation where $\lambda_0 \neq 0$, we obtain† from Eq. (9.2b) in Eqs. (9.11), (9.12), (9.13), etc.,

$$w_1(E,\theta)_N = \frac{Ee^{-E^2/2\psi}}{2\pi\psi} \qquad 0 \leq E < \infty, 0 \leq \theta \leq 2\pi \tag{9.14}$$

and, finally,

$$\begin{aligned} w_2(E_1,E_2;\theta_1,\theta_2;t)_N = \frac{E_1E_2}{(2\pi\psi)^2(1 - k_0^2)} \exp \biggl\{ -\frac{1}{2\psi(1 - k_0^2)} \\ \times [E_1^2 + E_2^2 - 2\rho_0E_1E_2 \cos(\theta_2 - \theta_1) - 2\lambda_0E_1E_2 \sin(\theta_2 - \theta_1)] \biggr\} \\ 0 \leq E_1, E_2 < \infty \quad (0 \leq \theta_1, \theta_2 \leq 2\pi) \end{aligned} \tag{9.15}$$

with $E_1 = E(t_1)$, etc. The corresponding characteristic functions, however, are not so simple, as Probs. 9.1a and 9.1b indicate. Note, incidentally,

† See also Rice,[1] eqs. (3.7-9) et seq.

that in terms of the envelope and phase of the narrowband noise process the covariance function is here

$$K_N(t) = \overline{N(t_1)N(t_2)} = \tfrac{1}{2}\boldsymbol{E}\{E_1E_2 \cos(\omega_0 t + \theta_2 - \theta_1)\}$$
$$= \psi[\rho_0(t) \cos \omega_0 t + \lambda_0(t) \sin \omega_0 t] \qquad (9.16a)$$
$$K_N(t) = \overline{N(t_1)N(t_2)} = \tfrac{1}{2}\boldsymbol{E}\{E_1E_2 \cos(\omega_0 t + \theta_2 - \theta_1)\}$$
$$= \psi k_0(t) \cos[\omega_0 t - \phi_0(t)] \qquad \phi_0(t) = \tan^{-1}(\lambda_0/\rho_0) \qquad (9.16b)$$

[cf. Eqs. (3.108)], as a direct evaluation of Eq. (9.16*a*) using Eq. (9.15) shows (cf. Prob. 9.2*c*).

9.1-1 Distribution Densities and Moments of the Envelope. Here we are interested in the first- and second-order distribution densities of the envelope E, as well as the associated moments. Specifically, we want the marginal d.d.'s

$$W_1(E)_N = \int_0^{2\pi} w_1(E,\theta)_N \, d\theta$$
$$W_2(E_1,E_2;t)_N = \iint_0^{2\pi} w_1(E_1,E_2;\theta_1,\theta_2;t)_N \, d\theta_1 \, d\theta_2 \qquad (9.17)$$

In the first-order case, we get at once from Eq. (9.14) the *Rayleigh* d.d. (cf. Sec. 7.7-3)

$$W_1(E)_N = \frac{Ee^{-E^2/2\psi}}{\psi} \qquad 0 \le E \le \infty \qquad (9.18)$$

which is shown in Fig. 9.1.

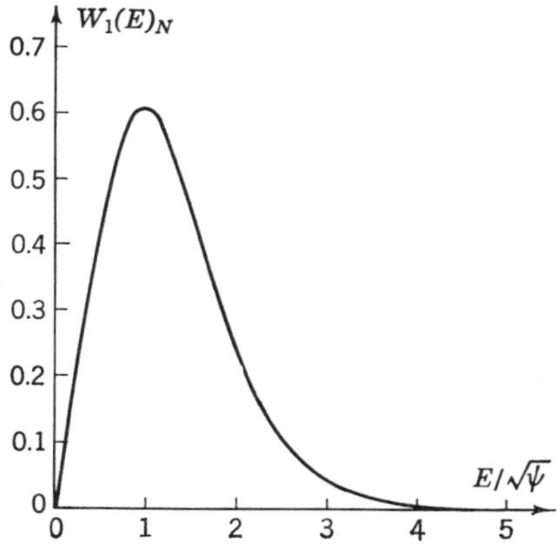

FIG. 9.1. The Rayleigh d.d. for the envelopes of a narrowband normal noise process.

The corresponding d.d. in the second-order case is more involved. From Eq. (9.15), we observe that the exponent is periodic in $(\theta_2 - \theta_1)$ and from Eq. (9.16*b*) that $\rho_0 \cos(\theta_2 - \theta_1) + \lambda_0 \sin(\theta_2 - \theta_1) = k_0 \cos(\theta_2 - \theta_1 - \phi_0)$, with $k_0 = (\rho_0^2 + \lambda_0^2)^{1/2}$, $\phi_0 = \tan^{-1}(\lambda_0/\rho_0)$. Now since $d\theta_1 \, d\theta_2/(2\pi)^2$ is equivalent to $d(\theta_2 - \theta_1)/2\pi$ and the double integration in Eqs. (9.17) can be replaced by a single integration over $(\theta_2 - \theta_1)$, we can use†

$$\int_0^{2\pi} e^{a \cos \psi} \frac{d\psi}{2\pi} = I_0(a) \qquad (9.19)$$

(where I_0 is the modified Bessel function of zero order and first kind) to

† See Sec. A.1.4.

obtain $W_2(E_1,E_2;t)_N$ directly. We find that†

$$W_2(E_1,E_2;t)_N = \frac{E_1E_2}{\psi^2(1-k_0^2)} e^{-(E_1^2+E_2^2)/2\psi(1-k_0^2)} I_0\left[\frac{k_0E_1E_2}{\psi(1-k_0^2)}\right] \qquad 0 \le E_1, E_2 \le \infty \tag{9.20}$$

Note that, as $|t| \to \infty$, k_0 approaches zero, and so

$$W_2(E_1,E_2;t)_N \to W_1(E_1)_N W_1(E_2)_N$$

as required [cf. Eqs. (1.93)], while if $t \to 0$, then k_0^2 becomes unity and we get

$$\lim_{t\to\infty} W_2(E_1,E_2;t)_N = W_1(E_1)\,\delta(E_2 - E_1)$$

where $\lim_{t\to 0} \dfrac{E_2 e^{-(E_2^2+E_1^2k_0^2)/2\psi(1-k_0^2)} I_0[k_0E_1E_2/\psi(1-k_0^2)]}{\psi(1-k_0^2)} = \delta(E_2 - E_1)$

also as expected [cf. Eqs. (1.92)]. The characteristic functions of $W_1(E)_N$, $W_2(E_1,E_2;t)_N$ are more complicated, as indicated by the results of Probs. 9.1*a*, 9.1*c*.

Also of considerable interest in applications are the moments of these distributions. For the first-order case, we obtain from Eq. (9.18)

$$\overline{E^\nu} = \int_0^\infty E^{\nu+1} e^{-E^2/2\psi}\, dE/\psi = (2\psi)^{\nu/2}\Gamma(\nu/2+1) \qquad \nu \ge 0 \tag{9.21}$$

where ν is a (real) number (not necessarily an integer). Note that all moments of the envelope are positive and in particular that

$$\bar{E} = (\pi\psi/2)^{1/2} \qquad \overline{E^2} = 2\psi \tag{9.21a}$$

The joint second moments are found in a similar way. Expanding the Bessel function in Eq. (9.20) and using Eq. (9.21) with appropriate modifications, we get

$$\begin{aligned} \boldsymbol{E}\{E_1^\nu E_2^\eta\} &= \iint_0^\infty E_1^\nu E_2^\eta W_2(E_1,E_2;t)_N\, dE_1\, dE_2 \\ &= (2\psi)^{(\nu+\eta)/2}(1-k_0^2)^{1+(\nu+\eta)/2} \sum_{l=0}^\infty \frac{k_0^{2l}\Gamma(\nu/2+l+1)\Gamma(\eta/2+l+1)}{l!\,l!} \\ &= (2\psi)^{(\nu+\eta)/2}\Gamma(\nu/2+1)\Gamma(\eta/2+1) \begin{cases} (1-k_0^2)^{\frac{\nu+\eta}{2}} \\ \times\, {}_2F_1(\nu/2+1, \eta/2+1; 1; k_0^2) \\ \text{or} \\ {}_2F_1(-\nu/2, -\eta/2; 1; k_0^2) \end{cases} \end{aligned} \tag{9.22}$$

† For example, see Rice,[1] eq. (3.7-13), and Uhlenbeck.[2]

where ${}_2F_1$ is a hypergeometric function (for details, see Sec. A.1.3). In the limiting cases ($|t| \to \infty$, $t \to 0$), we easily see that $\lim_{|t|\to\infty} \overline{E_1{}^\nu E_2{}^\eta} = \overline{E_1{}^\nu}\,\overline{E_2{}^\eta}$ [cf. Eq. (9.21)], and since

$$ {}_2F_1(-\nu/2,-\eta/2;1;1) = \Gamma[1 + (\nu + \eta)/2]/\Gamma(1 + \nu/2) $$

(cf. Appendix 1), it follows that $\lim_{t\to 0} \overline{E_1{}^\nu E_2{}^\eta} = \overline{E^{\nu+\eta}} = \Gamma[1 + (\nu + \eta)/2]$ $(2\psi)^{(\nu+\eta)/2}$, as expected from Eq. (9.21).

Of particular interest are the cases $\nu = \eta = 1, 2$. The former ($\nu = \eta = 1$) gives us the second-moment function of the low-frequency output of a half-wave linear detector when narrowband normal noise $N(t)$ is the input process; e.g., if $x(t)$ is this low-frequency output, we have $x(t) = \boldsymbol{T}_L\boldsymbol{T}_g\{N(t)\}$, $\boldsymbol{T}_g\{N\} = N$ $(N \geq 0)$, $x(t) = \boldsymbol{T}_L\boldsymbol{T}_g\{N(t)\} = 0$ $(N < 0)$, and $\boldsymbol{T}_L$ excludes all spectral zones higher than the zeroth order generated in the course of rectification [see Eq. (2.106)]. Similarly, when $\nu = \eta = 2$, we have the case of a half- or full-wave quadratic rectifier $\boldsymbol{T}_g\{N\} = N^2$ $(N \geq 0)$, $\boldsymbol{T}_g\{N\} = 0$ $(N < 0)$, or $\boldsymbol{T}_g\{N\} = N^2$ (all N). Equation (9.22) then becomes†

$$ M_{x=E}(t) = \overline{E_1E_2} = \frac{\psi\pi}{2}\,{}_2F_1(-\tfrac{1}{2},-\tfrac{1}{2};1;k_0{}^2) = \psi[2E(k_0) - (1 - k_0{}^2)K(k_0)] \tag{9.23} $$

where K and E are, respectively, elliptic integrals of the first and second kind [cf. Appendix 1]. For the square-law case, we get

$$ M_{x=E^2}(t) = \overline{E_1{}^2E_2{}^2} = 4\psi^2(1 + k_0{}^2) \tag{9.24} $$

while for a general νth-law rectifier, preceding the low-pass filter, $M_x(t) = \overline{E_1{}^\nu E_2{}^\nu}$ is given by Eq. (9.22) ($\eta = \nu$).‡ In all cases, the intensity spectra follow on application of the W-K theorems (3.42), (3.78), e.g., $\mathcal{W}_x(f) = 2\mathfrak{F}\{M_x\}$. Some specific examples are considered later, in Chaps. 12, 13.

9.1-2 Moments and Distribution Densities of the Phase. Let us consider now the corresponding problem for the phases of our stationary narrowband normal noise process. The first- and second-order densities are the alternative set of marginal d.d.'s,

$$ \begin{aligned} w_1(\theta)_N &= \int_0^\infty w_1(E,\theta)_N\,dE \\ w_2(\theta_1,\theta_2;t)_N &= \iint_0^\infty w_2(E_1,E_2;\theta_1,\theta_2;t)_N\,dE_1\,dE_2 \end{aligned} \tag{9.25} $$

† A result due originally to Uhlenbeck, *op. cit.*

‡ See Middleton,[3] secs. 6, 7, especially eqs. (6.11), (7.14).

The result in the first-order case is particularly simple,

$$w_1(\theta)_N = \frac{1}{2\pi}\int_0^\infty \frac{E}{\psi} e^{-E^2/2\psi}\,dE = \frac{1}{2\pi} \qquad 0 \le \theta \le 2\pi \tag{9.26}$$

showing, as expected, that the phase is uniformly distributed in the primary interval [which is, of course, arbitrarily chosen, mod (2π); cf. Sec. 7.7-3; see Fig. 9.4].

The second-order density, however, is more complicated. With Eq. (9.16b) in Eq. (9.15) applied to Eqs. (9.25), we may write

$$w_2(\theta_1,\theta_2;t)_N = \iint_0^\infty \frac{E_1E_2}{4\pi^2\psi^2(1-k_0^2)} \exp\left[-\frac{E_1^2+E_2^2-2\beta E_1E_2}{2\psi(1-k_0^2)}\right] dE_1\,dE_2 \tag{9-27}$$

with β given in Eq. (9.32). This can be evaluated with the help of the transformations†

$$E_1 = \sqrt{\psi(1-k_0^2)}\,z^{1/2}e^{\phi/2} \qquad E_2 = \sqrt{\psi(1-k_0^2)}\,z^{1/2}e^{-\phi/2} \tag{9.28}$$

where the domain of E_1, E_2, for example $(0 \le E_1, E_2 \le \infty)$, goes over into $(0 \le z \le \infty)$, $(-\infty \le \phi \le \infty)$ for the new variables (z,ϕ) and the magnitude of the jacobian is found to be $\psi(1-k_0^2)/2$. We get, then, for Eq. (9.27)

$$w_2(\theta_1,\theta_2;t)_N = \frac{1-k_0^2}{4\pi^2}\int_0^\infty ze^{\beta z}\,\frac{1}{2}\int_{-\infty}^\infty e^{-z\cosh\phi}\,d\phi \tag{9.29}$$

and since

$$\frac{1}{2}\int_{-\infty}^\infty e^{-z\cosh\phi}\,d\phi = K_0(z) \qquad z \ge 0 \tag{9.30}$$

[cf. Watson,[4] p. 182, eq. (7)], where K_0 is a modified Bessel function of the second kind, we can write

$$w_2(\theta_1,\theta_2;t)_N = \frac{1-k_0^2}{4\pi^2}\int_0^\infty ze^{\beta z}K_0(z)\,dz \qquad |\beta| \le 1 \tag{9.31}$$

This in turn can be evaluated with the help of Watson,[4] p. 410, eq. (1), to give us finally the desired result (cf. Prob. 9.2a),

$$w_2(\theta_1,\theta_2;t)_N = \frac{1-k_0^2}{4\pi^2}(1-\beta^2)^{-3/2}\left(\beta\sin^{-1}\beta + \frac{\pi\beta}{2} + \sqrt{1-\beta^2}\right)$$

$$-1 \le \beta \le 1,\ \beta \equiv k_0(t)\cos[\theta_2-\theta_1-\phi_0(t)],\ \phi_0(t) = \tan^{-1}\frac{\lambda_0}{\rho_0} \tag{9.32}$$

a relation obtained independently‡ by MacDonald[5] and Bunimovitch.[6]

† For example, Middleton,[3] eqs. (6.8).

‡ Equation (9.32) is a special case of the somewhat more general relation [Middleton,[3] eq. (5.20)], as the reader can readily verify.

Note that $(w_2)_N$ is a function only of the difference $\theta_2 - \theta_1$ of the phases; Fig. 9.2*a*, *b* shows some typical curves of w_2 and the distribution

$$D(\Phi;k_0) = \int_{-\pi}^{\theta_2-\theta_1-\phi_0(t)} w_2\, d(\theta_2 - \theta_1)$$

(cf. Prob. 9.2*b*), as† a function of $\theta_2 - \theta_1$, for various $k_0(t)$ and $\phi_0(t)$. When $|t| \to \infty$, $w_2(\theta_1,\theta_2;t)_N$ becomes $w_1(\theta_1)w_1(\theta_2) = (1/2\pi)^2$ and at the other extreme, $t \to 0$, $w_2(\theta_1,\theta_2;t) \to (1/2\pi)\,\delta(\theta_2 - \theta_1)$ (cf. Prob. 9.3*a*). An alternative expression for the second-order density [Eq. (9.32)] which is useful in certain applications may be obtained by expanding the exponential $e^{\beta E_1 E_2/\psi(1-k_0^2)}$ and applying Eq. (9.21), with minor modifications. The result is the Fourier-series development

$$w_2(\theta_1,\theta_2;t)_N = \frac{1}{(2\pi)^2}\sum_{n=0}^{\infty}\frac{\epsilon_n k_0^n \Gamma(n/2+1)^2}{n!}$$
$$\times\, {}_2F_1\left(\frac{n}{2},\frac{n}{2};n+1;k_0^2\right)\cos n(\theta_2-\theta_1-\phi_0) \quad 0 \le \theta_1,\ \theta_2 \le 2\pi \tag{9.33}$$

The hypergeometric functions ${}_2F_1$ can also be expressed in terms of the complete elliptic integrals $K(k_0)$, $E(k_0)$.

With the help of Eq. (9.33), it is a simple matter to determine the covariance function of the random process

$$y(t) = A_0 \cos\,[\omega_0 t - \theta(t)] \tag{9.34}$$

where A_0 is a constant and $\theta(t)$ is a random phase with the above statistics. Here, for example, $y(t)$ may be thought of as the output (in the spectral zone about the central frequency f_0) of an ideal "super" limiter when a narrowband normal noise process $N(t)$ is applied at the input [cf. Sec. 13.4-2(2)]. We have

$$K_y(t) = \frac{A_0^2}{2}\,E\{\cos\,(\omega_0 t + \theta_2 - \theta_1)\}$$
$$= \frac{A_0^2}{2}\iint_0^{2\pi} w_2(\theta_2,\theta_1;t)_N \cos\,(\omega_0 t + \theta_2 - \theta_1)\, d\theta_1\, d\theta_2 \tag{9.35}$$

From this and Eq. (9.33) it is evident that only the term for $n = 1$ in the series for w_2 contributes to give us directly the equivalent results for the covariance function of y [Eq. (9.34)].

† These were computed for a closely related problem of phase detection, considered by Huggins and Middleton.[7]

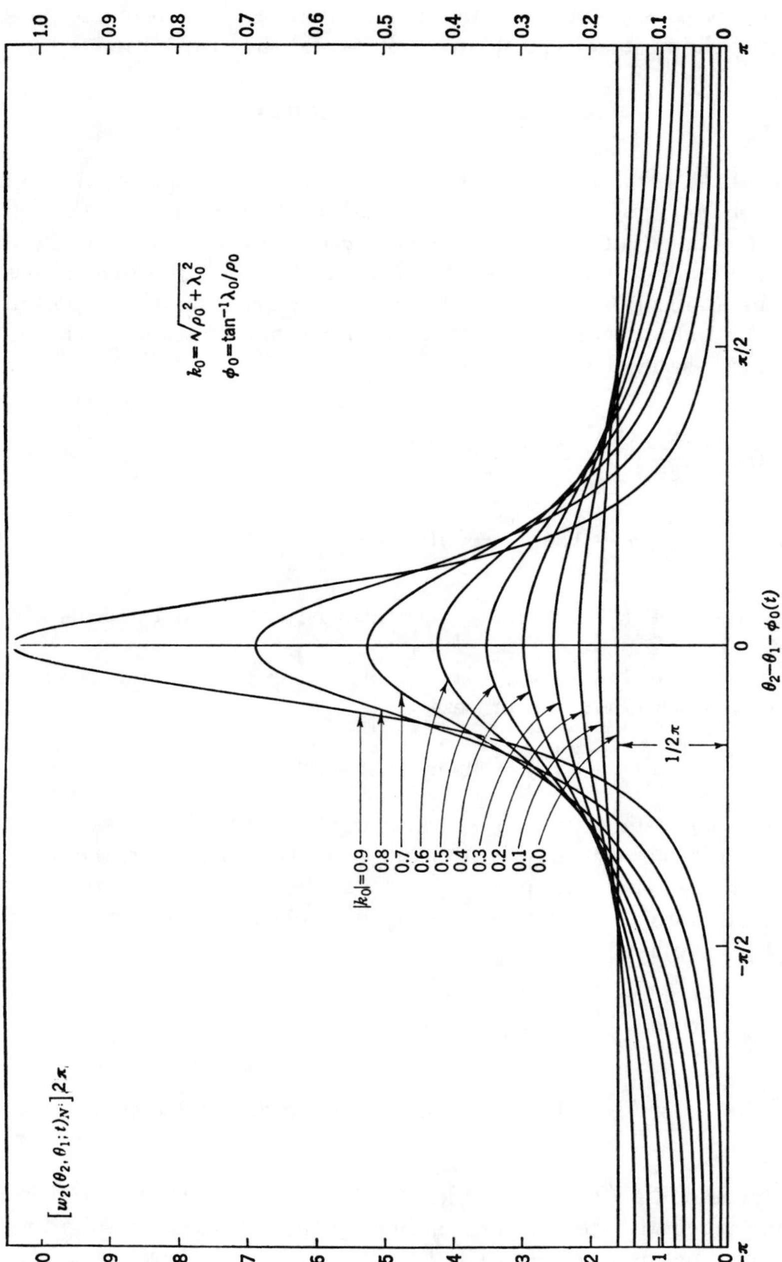

$[w_2(\theta_2, \theta_1; t)_N]\,2\pi$
$k_0 = \sqrt{\rho_0^2 + \lambda_0^2}$
$\phi_0 = \tan^{-1}\lambda_0/\rho_0$
$|k_0| = 0.9$
0.8
0.7
0.6
0.5
0.4
0.3
0.2
0.1
0.0
$1/2\pi$
$\theta_2 - \theta_1 - \phi_0(t)$
$-\pi$
$-\pi/2$
0
$\pi/2$
π
1.0
0.9
0.8
0.7
0.6
0.5
0.4
0.3
0.2
0.1
0

(a)

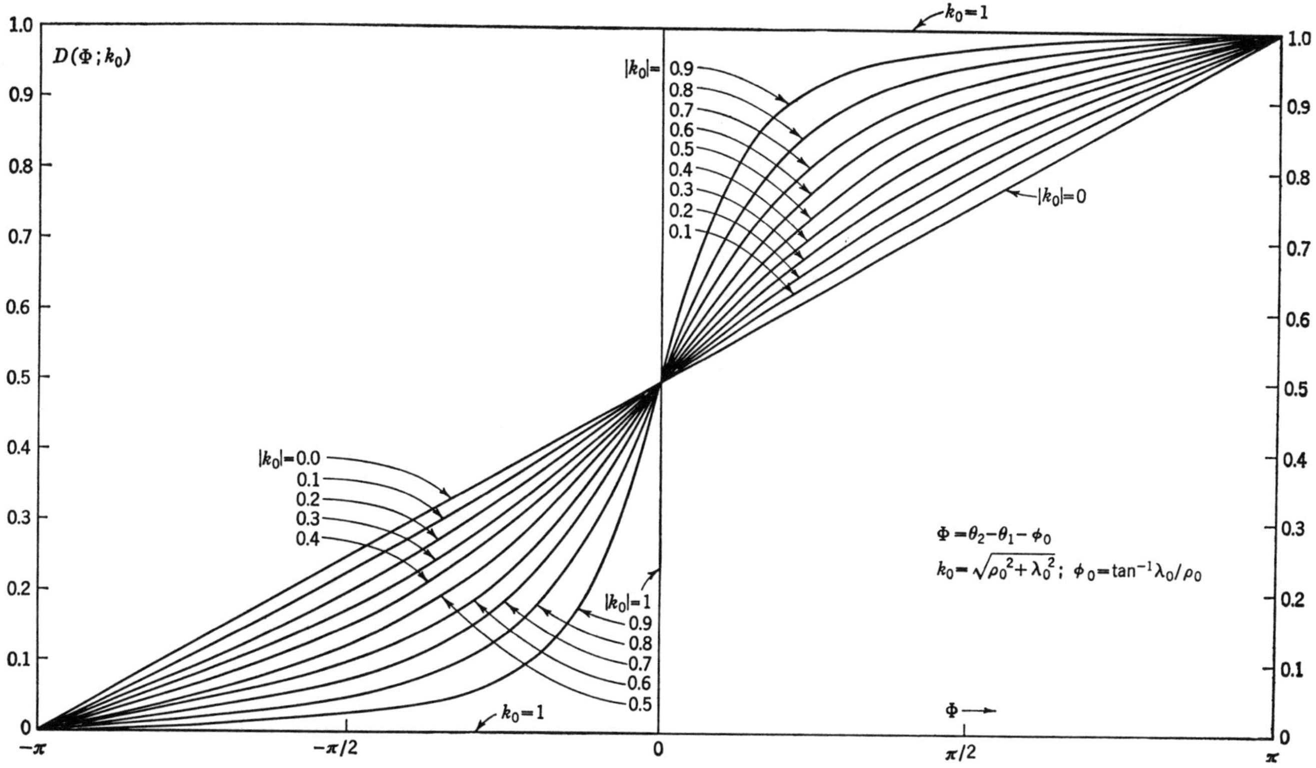

(b)

FIG. 9.2. The second-order probability density and distribution of the phases (θ_1, θ_2) of a narrowband normal process.

$$K_y(t) = \frac{A_0^2\pi}{8} k_0 \, {}_2F_1\left(\frac{1}{2},\frac{1}{2};2;k_0^2\right) \cos(\omega_0 t - \phi_0)$$
$$= \frac{A_0^2}{2} k_0^{-1}[E(k_0) - (1 - k_0^2)K(k_0)] \cos(\omega_0 t - \phi_0)$$
$$= \frac{A_0^2\pi}{8} \sum_{n=0}^{\infty} \left[\frac{(2n)!}{2^{2n}n!^2}\right]^2 \frac{k_0^{2n+1}}{n+1} \cos(\omega_0 t - \phi_0) \tag{9.36}$$

Apart from the constant factors, and in varying degrees of generality, these expressions have been obtained at different times by Van Vleck,[8] Middleton,[9] Davenport,[10] and Price.[11] Spectra of $y(t)$, when the original noise process is band-limited white normal noise, spectrally centered about f_0, have been calculated by Van Vleck[8] and in greater detail by Price,[11] including experimental results in good agreement with theory. Observe from Eq. (9.36) that $K_y(t)$ in these instances depends on both $k_0(t)$ and $\phi_0(t)$, that is, on both the "envelope" and "phase" of the input covariance function $K_N(t)$ [Eq. (9.16*b*)]. Comparing this with the expression (9.23) for the envelope, we remark that $K_y(t)$, for the phase, contains more information about the covariance function (or, equivalently, about the spectrum) of the original noise process $N(t)$ than does the second-moment function $M_x(t)$ for the envelopes alone. The latter is not sufficient for us to deduce $K_N(t)$, while knowledge of the former is; in the case of spectral symmetry about f_0, where $\phi_0 = 0$ (or $m\pi$), either $M_x(t)$ or $K_y(t)$ [cf. Eq. (9.23) or (9.36)] will serve.

Besides processes like Eq. (9.34) generated by ideal bandpass limiters, we can also use these results for the phase densities of Eqs. (9.26), (9.32), (9.33) to calculate the spectrum of the noise output of an idealized *phase detector*, where extreme limiting is employed to remove all amplitude variations due to noise. Here, the final output process is

$$y(t) = \alpha\theta(t) \qquad 0 \le \theta \le 2\pi \; (-\infty < t < \infty) \tag{9.37}$$

and the desired spectrum is $\mathcal{W}_y(f) = 2\mathfrak{F}\{K_y\}$, $K_y(t) = \alpha^2\overline{\theta_1\theta_2}$. To evaluate $\overline{\theta_1\theta_2}$, it is convenient to consider the symmetrical form $\overline{(\theta_2 - \theta_1)^2} = 2(\overline{\theta^2} - \overline{\theta_2\theta_1})$ and apply Eq. (9.32). The result finally is (cf. Prob. 9.4*a*)

$$\overline{\theta_1\theta_2} = \pi^2[1 + 2\gamma - 4\gamma^2 + \tfrac{1}{12}\Omega(k_0)] \tag{9.38}$$

with
$$\gamma \equiv \frac{1}{2\pi}\sin^{-1} k_0$$
$$\Omega(k_0) \equiv \frac{6}{\pi^2}\sum_{n=1}^{\infty}\frac{k_0^{2n}}{n^2} \qquad 0 \le \theta_1,\theta_2 \le 2\pi$$

where $\Omega(1) = 1$, $\Omega(0) = 0$. Thus, $K_\theta(t)$ is a function of ρ_0 and k_0, or k_0 and ϕ_0, etc., and Eq. (9.38) provides a simple form from which the corresponding spectrum may be determined. Other results for phase detectors, FM receivers, and the like are considered in Chap. 15.

Since the original process is gaussian, the spectrum (or covariance function) still determines all W_n for the envelope and phase but these latter are clearly no longer normal, so that we may expect very complicated structures for the higher-order densities. Finally, by comparing Eqs. (9.15), (9.20), and (9.32) or Eq. (9.33) for, say, $\theta_1 = \theta_2$, we observe that the envelope and phase processes are also statistically dependent: although E and θ are in the first order independent, for example, $w_1(E,\theta)_N = W_1(E)_N w_1(\theta)_N$, this is not true for the second- and higher-order distributions, for example, $w_2(E_1,E_2;\theta_1,\theta_2;t)_N \neq W_2(E_1,E_2;t)_N w_2(\theta_1,\theta_2;t)_N$, etc. For these reasons, it is generally simpler in applications to deal directly with the original normal process whenever this is possible.

PROBLEMS

9.1 (*a*) Show that the first-order characteristic function of the envelope and phase of narrowband normal noise is

$$F_1(i\xi,i\eta)_{E,\theta} = F_1(i\xi)_E F_1(i\eta)_\theta \qquad 0 \leq \theta \leq 2\pi \tag{1}$$

where

$$F_1(i\xi)_E = {}_1F_1\left(1;\frac{1}{2};-\frac{\xi^2\psi}{2}\right) + i\xi\left(\frac{\psi\pi}{2}\right)^{1/2} e^{-\xi^2\psi/2} \tag{2}$$

$$F_1(i\eta)_\theta = \frac{e^{2\pi i}-1}{2\pi i} \qquad 0 \leq \theta \leq 2\pi \tag{3}$$

(*b*) Show that the second-order characteristic function for the above process is

$$F_2(i\xi_1,i\xi_2;i\eta_1,i\eta_2;t)_{E,\theta} = \frac{1-k_0^2}{4\pi^2}\sum_{m=0}^{\infty}\epsilon_m k_0{}^m G_m(i\eta_1,i\eta_2;\phi_0) \times \sum_{n=0}^{\infty}\frac{k_0{}^{2n}}{n!(n+m)!}L_{m,n}(i\xi_1;k_0)L_{m,n}(i\xi_2;k_0) \tag{4}$$

where

$$G_m(i\eta_1,i\eta_2;\phi_0) = 2^{-1}\left[\frac{e^{2\pi i(\eta_1-m)}-1}{2\pi i(\eta_1-m)}\,\frac{e^{2\pi i(\eta_2+m)}-1}{2\pi i(\eta_2+m)}e^{-im\phi_0} + \frac{e^{2\pi i(\eta_1+m)}-1}{2\pi i(\eta_1+m)}\,\frac{e^{2\pi i(\eta_2-m)}-1}{2\pi i(\eta_2-m)}e^{im\phi_0}\right] \tag{5}$$

$$L_{m,n}(i\xi;k_0) = \Gamma\left(\frac{m+2n+2}{2}\right){}_1F_1\left[\frac{m+2n+2}{2};\frac{1}{2};-\frac{\xi^2\psi(1-k_0^2)}{2}\right] + i\xi\sqrt{2\psi(1-k_0^2)}\,\Gamma\left(\frac{m+2n+3}{2}\right){}_1F_1\left[\frac{m+2n+3}{2};\frac{3}{2};-\frac{\xi^2\psi(1-k_0^2)}{2}\right] \tag{6}$$

Note from this that E and θ are not independent random processes.

(*c*) Obtain the second-order characteristic function for the envelopes E_1, E_2.

$$F_2(i\xi_1,i\xi_2;t)_E = \sum_{n=0}^{\infty}\frac{k_0{}^{2n}(1-k_0^2)}{4\pi^2(n!)^2}L_{0,n}(i\xi_1;k_0)L_{0,n}(i\xi_2;k_0) \tag{7}$$

(d) Repeat this, but for the phases θ_1, θ_2, to obtain

$$F_2(i\eta_1,i\eta_2;t)_\theta = \frac{1}{4\pi^2}\sum_{m=0}^{\infty}\frac{\epsilon_m k_0{}^m}{m!}G_m(i\eta_1,i\eta_2;\phi_0)\Gamma\left(\frac{m}{2}+1\right)^2 {}_2F_1\left(\frac{m}{2},\frac{m}{2};m+1;k_0{}^2\right) \quad (8)$$

and verify that this gives us $w_2(\theta_1,\theta_2;t)_N$ [Eq. (9.33)].

(e) Show that

$$\begin{aligned} J^{(m)}(\xi_1,\xi_2) &\equiv \iint_{-\pi}^{\pi} e^{i\xi_1\theta_1+i\xi_2\theta_2}\cos m(\theta_1-\theta_2-\phi_0)\,d\theta_1\,d\theta_2 \\ &= \pi^2\left[e^{-im\phi_0}\frac{\sin\pi(\xi_1-m)}{\pi(\xi_1-m)}\frac{\sin\pi(\xi_2+m)}{\pi(\xi_2+m)} + e^{im\phi_0}\frac{\sin\pi(\xi_1+m)}{\pi(\xi_1+m)}\frac{\sin\pi(\xi_2-m)}{\pi(\xi_2-m)}\right] \end{aligned} \quad (9)$$

and obtain

$$\begin{aligned} H^{(m)}_{\mu,\nu} &\equiv \iint_0^{\infty} E_1{}^{\mu}E_2{}^{\nu} I_m\left[\frac{E_1E_2k_0}{\psi(1-k_0{}^2)}\right] e^{-(E_1{}^2+E_2{}^2)/2\psi(1-k_0{}^2)}\frac{dE_1\,dE_2}{4\pi^2\psi^2(1-k_0{}^2)} \\ &= \frac{\psi^{(\nu+\mu)/2-1}2^{(\nu+\mu)/2-3}}{m!\pi^2}k_0{}^m\,\Gamma\left[\frac{m+(\nu+\mu)/2+1}{2}\right]^2 {}_2F_1\left(\frac{m+1}{2}-\frac{\nu+\mu}{2},\right. \\ &\qquad \left.\frac{m+1}{2}-\frac{\nu+\mu}{2};m+1;k_0{}^2\right) \qquad \mathrm{Re}\,(m-\mu) > -1,\ \mathrm{Re}\,(m-\nu) > -1 \end{aligned} \quad (10)$$

Use these results to show that for the envelope E and phase θ of stationary narrowband normal noise,

$$\begin{aligned} \boldsymbol{E}\{E_1E_2e^{i\xi_1\theta_1+i\xi_2\theta_2}\} &= \frac{\psi}{2\pi^2}\sum_{m=0}^{\infty}\frac{\epsilon_m}{m!}k_0{}^m\Gamma\left(\frac{m+3}{2}\right)^2 \\ &\qquad \times {}_2F_1\left(\frac{m-1}{2},\frac{m-1}{2};m+1;k_0{}^2\right)J^{(m)}(\xi_1,\xi_2) \end{aligned} \quad (11)$$

and hence verify that

$$\boldsymbol{E}\{E_1E_2e^{i(\theta_2-\theta_1)}\} = \psi e^{i\phi_0}k_0 \quad (12)$$

Obtain $\boldsymbol{E}\{E_1{}^{\mu}E_2{}^{\nu}e^{i(\nu\theta_2-\mu\theta_1)}\}$ with the help of the above. These results are useful in the study of phase detectors, with and without limiting. (For properties of the various hypergeometric functions ${}_1F_1$, ${}_2F_1$, see Appendix 1.)

9.2 (a) Carry out the details of the evaluation of $w_2(\theta_1,\theta_2;t)_N$ [Eq. (9.32)].

(b) Show that the distribution of $\Phi \equiv \theta_2 - \theta_1 - \phi_0(t)$, from Eq. (9.32), is

$$\begin{aligned} D(\Phi;k_0) &= \int_{-\pi}^{\Phi} w_2(\theta_1;\theta_2;t)_N\,d(\theta_2-\theta_1) \\ &= \frac{1}{2}\left[1+\frac{\Phi}{\pi}+\frac{k_0\sin\Phi}{\pi(1-z_0{}^2\cos^2\Phi)^{1/2}}\cos^{-1}(-k_0\cos\Phi)\right] \qquad -\pi\le\Phi\le\pi \end{aligned} \quad (1)$$

(c) Obtain $K_N(t) = \psi k_0(t)\cos[\omega_0 t - \phi_0(t)]$ for a narrowband (stationary) normal noise process by direct evaluation of

$$K_N(t) = \tfrac{1}{2}\boldsymbol{E}\{E_1E_2\cos(\omega_0 t+\theta_2-\theta_1)\} \quad (2)$$

using the joint d.d. of E and θ (cf. Middleton[3]). HINT: Follow the procedure used in (a).

9.3 (*a*) Show that $\lim_{t\to 0} w_2(\theta_1,\theta_2;t)_N = (1/2\pi)\,\delta(\theta_2 - \theta_1)$ by a direct evaluation.

(*b*) The output of an ideal (e.g., super) limiter and a bandpass filter, centered about mf_0, when narrowband normal noise is the input process, is

$$y_m(t) = A_{0m} \cos m[\omega_0 t - \theta(t)],\ m = \text{odd}; \qquad y = 0,\ m = \text{even} \tag{1}$$

(We assume for simplicity that the bandpass filter passes the mth spectral zone without distortion.) Show that the covariance function for $y_m(t)$ is

$$K_{y_m}(t) = \frac{A_{0m}^2}{2} \frac{k_0(t)^m \Gamma(m/2 + 1)^2}{m!} {}_2F_1\left(\frac{m}{2}, \frac{m}{2}; m + 1; k_0^2\right) \cos m[\omega_0 t - \phi_0(t)] \tag{2}$$

Alternative representations of the hypergeometric functions may be given in terms of the complete elliptic integrals $K(k_0)$, $E(k_0)$ (cf. Sec. A.1.3).

9.4 (*a*) For narrowband normal noise show that

$$E\{|\theta_2 - \theta_1|\} = \pi - 2\sin^{-1} k_0 \qquad 0 \le \theta_1,\ \theta_2 \le 2\pi \tag{1}$$

$$E\{|\theta_2 - \theta_1|^2\} = \frac{\pi^2}{6}[4 - \Omega(k_0)] - 2\pi \sin^{-1} k_0 + 2(\sin^{-1} k_0)^2 \tag{2}$$

$$\Omega(k_0) = \frac{6}{\pi^2} \sum_{n=1}^{\infty} \frac{k_0^{2n}}{n^2} \tag{3}$$

HINT: Use Eq. (9.32), and integrate by parts.

(*b*) Obtain

$$\overline{\theta_1\theta_2} = \pi^2 + 2 \sum_{n=1}^{\infty} \frac{\Gamma(n/2 + 1)^2}{n!} k_0^n {}_1F_1(n/2, n/2; n + 1; k_0^2) \cos n\phi_0 \qquad t \ne 0 \tag{4}$$

noting that, for $t \to 0$, the series does not converge. Hence, use Eq. (2) to obtain Eq. (9.38).

9.5 (*a*) If y is a stationary narrowband normal noise process, show that the first-order d.d. of the product $z = E_1E_2$ of the envelopes at times t_1, t_2 is

$$\left.\begin{aligned} W_1(z;t) &= \frac{z}{\psi(1 - k_0^2)} I_0\left[\frac{k_0 z}{\psi(1 - k_0^2)}\right] K_0\left[\frac{z}{\psi(1 - k_0^2)}\right] && 0 \le z \le \infty \\ &= 0 && z < 0 \end{aligned}\right\} t = t_2 - t_1 \tag{1}$$

where $-1 < k_0 = \pm\sqrt{\rho_0(t)^2 + \lambda_0(t)^2} < 1$ and I_0 and K_0 are, respectively, modified Bessel functions of the first and second kinds.

(*b*) Obtain the moments from Eq. (1) directly,

$$\overline{z^n(t)} = (2\psi)^n \Gamma(n/2 + 1)^2 {}_2F_1(-n/2, -n/2; 1; k_0^2) \tag{2}$$

[cf. Eq. (9.22)].

9.2 *Statistical Properties of Additive Narrowband Signal and Normal Noise Processes*†

Here we shall extend the results of the previous section to include the important case of a narrowband normal noise process which is additively

† The treatment of the present section is based mainly on Middleton,[3] sec. 5; see also Rice,[1] sec. 3.10, and Bunimovitch's book,[12] sec. 17.

combined with a narrowband signal ensemble. This is a very common situation in communication theory: for example, in reception the desired signal process is accompanied at the input stages of a receiver by shot or thermal noise, or external interference, which because of the narrowband filters in these stages becomes effectively a narrowband noise (and signal) process for subsequent operations (cf. Sec. 5.1). The background noise ensemble is usually normal and stationary [cf. Eqs. (9.1) et seq.], while a broad class of signal processes with a variety of statistical properties can be written (cf. Secs. 2.2-6, 2.2-7)

$$\begin{aligned} S(t;\boldsymbol{\phi}) &= \alpha(t;\boldsymbol{\phi}) \cos \omega_c t + \beta(t;\boldsymbol{\phi}) \sin \omega_c t = A(t;\boldsymbol{\phi}) \cos [\omega_c t - \Psi(t;\boldsymbol{\phi})] \\ A(t;\boldsymbol{\phi}) &= \sqrt{\alpha^2 + \beta^2};\ \Psi(t;\boldsymbol{\phi}) = \tan^{-1} (\beta/\alpha) \pm \pi q;\ q = 0, 1, \ldots \end{aligned} \tag{9.39}$$

Here ω_c ($= 2\pi f_c$) is the (angular) carrier or central frequency, which may or may not be equal to ω_0 ($= 2\pi f_0$) for the noise, and $\boldsymbol{\phi}$ represents a set of random parameters, appropriate to the signal ensemble in question. Thus, if $S(t,\boldsymbol{\phi})$ is a set of simple sinusoids, $\boldsymbol{\phi}$ may be so chosen that $A(t;\boldsymbol{\phi}) = A_0$ and $\Psi(t;\boldsymbol{\phi}) = \phi_F$, where ϕ_F has a uniform d.d. in the primary interval; or if an amplitude modulation alone occurs, we may choose $\boldsymbol{\phi}$ so that $\Psi(t;\boldsymbol{\phi}) = \phi_F$, $A(t;\boldsymbol{\phi}) = A(t,\boldsymbol{\phi}_A)$, and so on, depending on the class of signal process under study. (See Secs. 5.1-1 to 5.1-3 for some specific examples.)

Since these signal and noise processes are additive and narrowband, we can write now

$$V(t) = S(t;\boldsymbol{\phi}) + N(t) = A(t;\boldsymbol{\phi}) \cos [\omega_c t - \Psi(t;\boldsymbol{\phi})] + N_c(t) \cos \omega_0 t + N_s(t) \sin \omega_0 t \tag{9.40a}$$

$$\equiv V_c(t) \cos \omega_0 t + V_s(t) \sin \omega_0 t \tag{9.40b}$$

Setting $\omega_d \equiv \omega_c - \omega_0$, we take advantage of the narrowband properties of the mixed process $V(t)$ to express it in envelope and phase form,

$$V(t) = E(t;\boldsymbol{\phi}) \cos [\omega_0 t - \theta(t;\boldsymbol{\phi})] = [N_c + A(t;\boldsymbol{\phi}) \cos \Phi(t;\boldsymbol{\phi})] \cos \omega_0 t + [N_s - A(t;\boldsymbol{\phi}) \sin \Phi(t;\boldsymbol{\phi})] \sin \omega_0 t \tag{9.41}$$

where the envelope and phase are defined in the usual way by†

$$\begin{aligned} E &= [N_c^2 + N_s^2 + A^2 + 2A(N_c \cos \Phi - N_s \sin \Phi)]^{1/2} \\ \theta &= \tan^{-1} \frac{N_s - A \sin \Phi}{N_c + A \cos \Phi} \pm \pi q \end{aligned} \tag{9.42a}$$

and

$$\Phi \equiv \omega_d t - \Psi(t;\boldsymbol{\phi}) \qquad \omega_d \equiv \omega_c - \omega_0 \tag{9.42b}$$

The "cosine" and "sine" components of V are specifically

$$V_c = E \cos \theta = N_c + A \cos \Phi \qquad V_s = E \sin \theta = N_s - A \sin \Phi \tag{9.43}$$

† For convenience we shall omit the arguments of A, Φ except where needed for clarity.

As in Sec. 9.1, we shall limit our discussion to the first- and second-order densities

$$w_1(E,\theta)_{S+N} = \int W_1(E,\theta|\boldsymbol{\phi})_{S+N} w(\boldsymbol{\phi})\, d\boldsymbol{\phi} \tag{9.44a}$$
$$w_2(E_1,E_2;\theta_1,\theta_2;t)_{S+N} = \int W_2(E_1,E_2;\theta_1,\theta_2;t|\boldsymbol{\phi})_{S+N} w(\boldsymbol{\phi})\, d\boldsymbol{\phi} \tag{9.44b}$$

Because of the variety of possible choices of $w(\boldsymbol{\phi})$, we shall confine our attention largely to the conditional d.d. in the integrals above. For the first-order case, we write $\mathbf{X} = \mathbf{V} - \mathbf{S}$, where

$$\mathbf{X} = \begin{bmatrix} N_c \\ N_s \end{bmatrix} = \begin{bmatrix} V_c - A \cos \Phi \\ V_s + A \sin \Phi \end{bmatrix} = \begin{bmatrix} E \cos \theta - A \cos \Phi \\ E \sin \theta + A \sin \Phi \end{bmatrix} \tag{9.45}$$

Inserting this into Eq. (9.11), we see that the jacobian of this transformation $\mathbf{X} \to [E,\theta]$ is E, as before, and that the joint d.d. of (E,θ) becomes

$$W_1(E,\theta|\boldsymbol{\phi})_{S+N} = \frac{E \exp\left(-\{E^2 + A(t)^2 - 2EA(t) \cos [\theta - \Phi(t)]\}/2\psi\right)}{2\pi\psi}$$
$$0 \le E \le \infty,\ 0 \le \theta \le 2\pi \tag{9.46}$$

where, of course, (A,Φ) depend on the random parameters $\boldsymbol{\phi}$.

The second-order density is obtained from Eqs. (9.12) et seq. and from Eqs. (9.40) to (9.43) in a similar way. Here $\mathbf{X} = \mathbf{V} - \mathbf{S}$, with

$$\mathbf{X} = \begin{bmatrix} N_{c1} \\ N_{s1} \\ N_{c2} \\ N_{s2} \end{bmatrix} = \begin{bmatrix} V_{c1} - A_1 \cos \Phi_1 \\ V_{s1} + A_1 \sin \Phi_1 \\ V_{c2} - A_2 \cos \Phi_2 \\ V_{s2} + A_2 \sin \Phi_2 \end{bmatrix} = \begin{bmatrix} E_1 \cos \theta_1 - A_1 \cos \Phi_1 \\ E_1 \sin \theta_1 + A_1 \sin \Phi_1 \\ E_2 \cos \theta_2 - A_2 \cos \Phi_2 \\ E_2 \sin \theta_2 + A_2 \sin \Phi_2 \end{bmatrix} \tag{9.47}$$

and $V_{c1} = V_c(t_1)$, $A_1 = A(t_1,\boldsymbol{\phi})$, etc. The jacobian of this transformation $(N_{c1}, \ldots ,N_{s2}) \to (E_1, \ldots ,\theta_2)$ is E_1E_2 once again. After a little algebra, we find from Eqs. (9.12) [and (9.16)] that the analogue of (9.15) when there is an additive signal becomes

$$\begin{aligned} W_2(E_1,E_2;\theta_1,\theta_2;t|\boldsymbol{\phi})_{S+N} = {} & \frac{E_1E_2}{(2\pi\psi)^2(1-k_0^2)} \exp\left\{-\frac{1}{2\psi(1-k_0^2)}\right. \\ & \times [E_1^2 + E_2^2 - 2k_0E_1E_2 \cos (\theta_2 - \theta_1 - \phi_0) \\ & + A_1^2 + A_2^2 - 2k_0A_1A_2 \cos (\Phi_2 - \Phi_1 + \phi_0) \\ & - 2E_1A_1 \cos (\theta_1 + \Phi_1) - 2E_2A_2 \cos (\theta_2 + \Phi_2) \\ & + 2k_0E_1A_2 \cos (\theta_1 + \Phi_2 + \phi_0) \\ & \left. + 2k_0E_2A_1 \cos (\theta_2 + \Phi_1 + \phi_0)]\right\} \end{aligned} \tag{9.48a}$$
$$\begin{aligned} = {} & w_2(E_1,E_2;\theta_1,\theta_2;t)_N B_2(A_1,A_2;\Phi_1,\Phi_2;t) \\ & \times C_2(A_1,A_2;\Phi_1,\Phi_2;E_1,E_2;\theta_1,\theta_2;t) \end{aligned} \tag{9.48b}$$

where $0 \leq E_1, E_2 \leq \infty$, $0 \leq \theta_1, \theta_2 \leq 2\pi$, and

$$B_2 \equiv \exp \{-[A_1^2 + A_2^2 - 2k_0 A_1 A_2 \cos (\Phi_2 - \Phi_1 + \phi_0)]/2\psi(1 - k_0^2)\} \tag{9.49a}$$

$$C_2 \equiv \exp \{[E_1 A_1 \cos (\theta_1 + \Phi_1) - k_0 E_1 A_2 \cos (\theta_1 + \Phi_2 + \phi_0) + E_2 A_2 \cos (\theta_2 + \Phi_2) - k_0 E_2 A_1 \cos (\theta_2 + \Phi_1 + \phi_0)]/\psi(1 - k_0^2)\} \tag{9.49b}$$

As before [cf. Eqs. (9.12b), (9.16b)], $k_0(t) = \sqrt{\rho_0(t)^2 + \lambda_0(t)^2}$, and $\phi_0(t) = \tan^{-1} (\lambda_0/\rho_0)$, while $\bar{N} = 0$, $\overline{N^2} = \psi$. Observe that, as $|t| \to \infty$, $W_2\Big|_{S+N}$ becomes $W_1(E_1,\theta_1|\boldsymbol{\phi})_{S+N} W_1(E_2,\theta_2|\boldsymbol{\phi})_{S+N}$ and, as $t \to 0$, we get

$$W_1(E_1,\theta_1|\boldsymbol{\phi})_{S+N} = \delta(\theta_2 - \theta_1)\,\delta(E_2 - E_1)$$

as required. It is clear from Eqs. (9.46) and (9.48) that E and θ are not independent random processes, even to the first order, when a signal is present [cf. Eq. (9.14)].

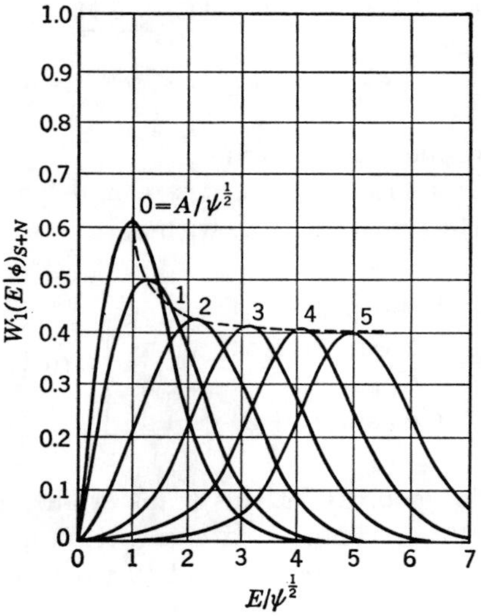

FIG. 9.3. First-order d.d. of the envelope of an additive narrowband signal and normal noise process $A = A(t) =$ constant, that is, t fixed.

9.2-1 Statistics of the Envelope of Signal and Noise. To obtain $W_1(E,\theta|\boldsymbol{\phi})_{S+N}$, we average over θ in Eq. (9.46) with the help of Eq. (9.19). The result is†

$$W_1(E|\boldsymbol{\phi})_{S+N} = \frac{E}{\psi} e^{-[E^2 + A(t)^2]/2\psi} \times I_0\left[\frac{A(t)E}{\psi}\right] \qquad 0 \leq E \leq \infty \tag{9.50}$$

Figure 9.3 illustrates this d.d.‡ for various choices of $A(t)/\psi^{1/2}$. When $EA(t)/\psi \gg 1$, we have the asymptotic result, obtained from the appropriate expansion of I_0 in this instance,

$$W_1(E|\boldsymbol{\phi})_{S+N} \simeq \left[1 + \frac{\psi}{8A(t)E}\right] \times \left[\frac{E}{2\pi A(t)\psi}\right]^{1/2} e^{-[E-A(t)]^2/2\psi} \tag{9.51}$$

which shows that either when E is far out on the tail of the curve or when $A(t)$ is large, so that $E \approx A$, the d.d. behaves like a normal law. The (conditional) characteristic function $F_1(i\xi|\boldsymbol{\phi})_{S+N}$ corresponding to $W_1(E|\boldsymbol{\phi})_{S+N}$ is given in Prob. 9.6b.

† See, for example, Middleton,[3] eq. (5.15), or, in the somewhat less general form $A(t) = A_0$ (constant), cf. Rice,[1] sec. 3.10.

‡ See Rice,[1] p. 109, fig. 7, for curves of the *distribution* $D(E|0)_{S+N} = \int_0^E W_1(E|0)_{S+N}\, dE$.

The νth moments (for the average over E only) are directly found to be

$$\overline{E^\nu(\phi)} = \int_0^\infty \psi^{-1} E^{\nu+1} e^{-(E^2+A^2)/2\psi} I_0(AE/\psi)\, dE$$
$$= (2\psi)^{\nu/2}\Gamma(\nu/2 + 1)\ {}_1F_1[-\nu/2;1;-a_0^2(t)] \quad (9.52)$$

with $a_0^2(t) \equiv A(t)^2/2\psi$, $\nu \geq 0$, where we have used Eqs. (A.1.17), (A.1.49). Some moments for integral values of ν are given in Table 9.1.

TABLE 9.1†

ν	$\overline{E_N^\nu}$, noise alone	Signal alone $A = A(t)$	Signal and noise $\psi = \overline{N^2}$ $a_0 = a_0(t) = A(t)/\sqrt{2\psi}$
1	$\left(\frac{\pi\psi}{2}\right)^{1/2}$	A	$\overline{E}_N\ {}_1F_1(-\tfrac{1}{2};1;-a_0^2)$
2	2ψ	A^2	$2\psi + A^2$
3	$3\psi\left(\frac{\pi\psi}{2}\right)^{1/2}$	A^3	$\overline{E_N^3}\ {}_1F_1(-\tfrac{3}{2};1;-a_0^2)$
4	$8\psi^2$	A^4	$8\psi^2 + 8\psi A^2 + A^4$
5	$15\psi^2\left(\frac{\pi\psi}{2}\right)^{1/2}$	A^5	$\overline{E_N^5}\ {}_1F_1(-\tfrac{5}{2};1;-a_0^2)$
6	$48\psi^3$	A^6	$48\psi^3 + 72\psi^2A^2 + 18\psi A^4 + A^6$
7	$105\psi^3\left(\frac{\pi\psi}{2}\right)^{1/2}$	A^7	$\overline{E_N^7}\ {}_1F_1(-\tfrac{7}{2};1;-a_0^2)$
8	$384\psi^4$	A^8	$384\psi^4 + 768\psi^3A^2 + 288\psi^2A^4 + 32\psi A^6 + A^8$

† The confluent hypergeometric functions may be expressed in terms of exponential and the modified Bessel functions I_0, I_1 (cf. Sec. A.1.2).

The second-order density, like that for noise alone [cf. Eq. (9.20)], is considerably more complex. By suitable rearrangement of the exponential terms in Eqs. (9.48) we get finally

$$W_2(E_1,E_2;t|\phi)_{S+N} = \frac{E_1E_2e^{-(E_1^2+E_2^2)/2\psi(1-k_0^2)}}{\psi^2(1-k_0^2)} B_2(A_1,A_2;\Phi_1,\Phi_2;t) \times$$
$$\sum_{m=0}^\infty \epsilon_m I_m\left(\psi\frac{k_0E_1E_2}{1-k_0^2}\right) I_m\left[\frac{\gamma_{12}E_1}{\psi(1-k_0^2)}\right] I_m\left[\frac{\gamma_{21}E_2}{\psi(1-k_0^2)}\right] \cos m(\Psi_{21} - \Psi_{12} + \phi_0) \quad (9.53)$$

where
$$\gamma_{12} \equiv [A_1^2 + k_0^2A_2^2 - 2k_0A_1A_2\cos(\Phi_2 - \Phi_1 + \phi_0)]^{1/2} \quad (9.54a)$$
$$\Psi_{12} \equiv \tan^{-1}\frac{-A_1\sin\Phi_1 + k_0A_2\sin(\Phi_2+\phi_0)}{A_1\cos\Phi_1 - k_0A_2\cos(\Phi_2+\phi_0)} \quad (9.54b)$$

and γ_{21}, Ψ_{21} are obtained from Eqs. (9.54a), (9.54b) by interchanging subscripts 1 and 2.

Some simplifications are possible in the special case of an ensemble of unmodulated sinusoids, where now $A_1 = A_2 = A_0$, and where $\omega_d = 0$, that is, where $f_c = f_0$ [cf. Eq. (9.42*b*)]. For the first-order density, one gets at once Eq. (9.50), $A(t) = A_0$, typical values of which are shown in Fig. 9.3. For the second-order density, Eq. (9.53) is somewhat simplified, viz.,

$$B_2 = \exp\left[\frac{-A_0^2}{(1 - k_0^2)\psi}(1 - k_0 \cos \phi_0)\right] \tag{9.55a}$$

$$\gamma_{12} = A_0 \sqrt{1 + k_0^2 - 2k_0 \cos \phi_0} = \gamma_{21}$$
$$\Psi_{12} = \tan^{-1} \frac{\sin \phi_F + k_0 \sin (\phi_0 - \phi_F)}{\cos \phi_F - k_0 \cos (\phi_0 - \phi_F)} = \Psi_{21} \tag{9.55b}$$

If we postulate, in addition, that the spectrum of the noise is essentially symmetrical about $f_0 = f_c$, so that $\lambda_0 \doteq 0$ [cf. Eq. (9.8*b*)], then $k_0 = \rho_0$ and $\phi_0 = 0$ and the joint second-order density (9.53) reduces still further to

$$W_2(E_1,E_2;t|\boldsymbol{\phi})_{S+N} = \frac{E_1E_2e^{-(E_1^2+E_2^2)/2\psi(1-\rho_0^2)}}{\psi^2(1-\rho_0^2)} e^{-A_0^2/\psi(1+\rho_0)} \sum_{m=0} \epsilon_m I_m\left[\frac{\rho_0 E_1 E_2}{\psi(1-\rho_0^2)}\right] \times I_m\left[\frac{A_0E_1}{\psi(1+\rho_0)}\right] I_m\left[\frac{A_0E_2}{\psi(1+\rho_0)}\right] \tag{9.56}$$

The second-order moments $\boldsymbol{E}\{E_1{}^\nu E_2{}^\nu\}$ and $\boldsymbol{E}\{E_1{}^\nu E_2{}^\eta\}$ (ν, $\eta \geq 0$) may be found in the usual way, although the analysis is rather intricate. We shall be content here to indicate three main ways of evaluating expressions of the type $\boldsymbol{E}\{E_1{}^\nu E_2{}^\nu\}$. One method is to use the transformation (9.28), with successive applications of the results for the product of two Bessel functions [cf. Watson,[4] p. 148, eq. (2)]; the integral forms of the modified Bessel function of the second kind [cf. Watson,[4] p. 182, eq. (7)], of argument $n - 2k$; and, finally, by Watson's[4] eq. (1) (p. 410). The second method expands $I_0[k_0E_1E_2/\psi(1 - k_0^2)]$ and uses termwise integration with the help of Eq. (A.1.49), while the third approach employs a contour integral representation of the Bessel functions in Eq. (9.53) applied to the addition formula for $I_0[(x^2 + y^2 + 2xy \cos \phi)^{1/2}]$, along with reversal of the order of integration. Some results in the case $\lambda_0 = 0$ are given by Middleton[3] [sec. 6, eqs. (6.9) et seq.]; extension to the general case $\lambda_0 \neq 0$ can be made by inspection.

9.2-2 Statistics of the Phase of Narrowband Signals and Normal Noise.† Let us now consider a corresponding treatment for the phases of narrowband signal and noise ensembles. The first-order density here follows directly from Eq. (9.46) in

$$w_1(\theta|\boldsymbol{\phi})_{S+N} = \int_0^\infty dE\, W_1(E,\theta|\boldsymbol{\phi})_{S+N} \tag{9.57}$$

† See Middleton,[3] sec. 5.

to give us a number of equivalent results,

$$w_1(\theta|\phi)_{S+N} = \frac{1}{2\pi} e^{-a_0{}^2(t)\sin^2\gamma} \left\{ \sqrt{\pi}\, a_0(t)\cos\gamma + {}_1F_1\left[-\frac{1}{2};\frac{1}{2};-a_0(t)^2\cos^2\gamma \right]\right\}$$
$$0 \le \theta \le 2\pi \qquad (9.58a)$$
$$= \frac{e^{-a_0{}^2}}{2\pi}\left[1 + 2\sqrt{\pi}\, a_0 \cos\gamma \, \frac{1+\Theta(a_0\cos\gamma)}{2}\, e^{a_0{}^2\cos^2\gamma}\right] \qquad (9.58b)$$

with $a_0{}^2(t) = A(t)^2/2\psi$, $\gamma = \theta - \Phi(t)$, where we have used Eq. (A.1.28*a*) to obtain these alternative representations. Expanding the exponential $e^{EA\cos\gamma/\psi}$ in Eq. (9.46) with the help of Eq. (A.1-48b) gives us a third form,

$$w_1(\theta|\phi)_{S+N} = \frac{1}{2\pi}\sum_{m=0}^{\infty} \epsilon_m \frac{a_0(t)^m}{m!}\Gamma\left(\frac{m}{2}+1\right) {}_1F_1\left[\frac{m}{2};\, m+1;\, -a_0{}^2(t)\right]$$
$$\times \cos m\gamma(t) \qquad 0 \le \theta \le 2\pi \qquad (9.58c)$$

For strong signals [i.e., whenever $a_0(t)^2 \gg 1$], we can use the asymptotic development of the confluent hypergeometric functions [cf. Eq. (A.1.16)] to write Eq. (9.58*c*) at once as

$a_0{}^2 \gg 1$:

$$w_1(\theta|\phi)_{S+N} \simeq \frac{1}{2\pi}\sum_{m=0}^{\infty} \epsilon_m \cos m\gamma(t)\left[1 - \frac{m^2}{4a_0{}^2} + \frac{m^2(m^2-4)}{32a_0{}^4} + \cdots\right] \qquad (9.59a)$$

$$w_1(\theta|\phi)_{S+N} \simeq \delta\{\theta - [\omega_d t - \Psi(t)]\} \qquad 0 \le \theta \le 2\pi \qquad (9.59b)$$

where we recall that $(0,2\pi)$ is the primary interval for θ and $\gamma = \theta - \omega_d t + \Psi(t)$ [cf. Eq. (9.42*b*)]. At the other extreme of weak signals [i.e., whenever $a_0{}^2(t) \ll 1$], direct expansion of Eqs. (9.58) yields

$a_0{}^2 \ll 1$:

$$w_1(\theta|\phi)_{S+N} \doteq \frac{1}{2\pi}[1 + a_0\sqrt{\pi}\cos\gamma + a_0{}^2\cos 2\gamma + O(a_0{}^3)]$$
$$0 \le \theta \le 2\pi \qquad (9.60)$$

which indicates that now, as expected, the phase is essentially that of the noise process [cf. Eq. (9.26)]. Figure 9.4 shows some typical distribution densities for the phase at some specified instant t, with $a_0(t) = A_0(t)/\sqrt{2\psi}$ as parameter. When the noise is strong compared with the signal, the characteristic uniform distribution (density) prevails, while as the signal intensity increases, the phase of this mixed narrowband process is determined more and more by the phase of the signal process alone [cf. Eq. (9.59*b*)].

The first-order moments $\overline{\cos n\theta}$, $\overline{\theta^\nu}$ are found directly from Eq. (9.58c). For example (considering the average over θ only), we have

$$\begin{aligned} \boldsymbol{E}\{\cos n\theta\} &= \int_0^{2\pi} w_1(\theta|\phi)_{S+N} \cos n\theta \, d\theta \\ &= \frac{a_0(t)^n}{n!} \Gamma\left(\frac{n}{2}+1\right) {}_1F_1\left[\frac{n}{2}; n+1; -a_0^2(t)\right] \cos n[\omega_d t - \Psi(t)] \end{aligned} \tag{9.61}$$

In the same way, we obtain the characteristic function associated with $w_1(\theta|\phi)_{S+N}$ by suitable specialization of the results of Prob. 9.6, e.g.,

$$\begin{aligned} F_1(i\xi|\phi)_\theta = \boldsymbol{E}\{e^{i\xi\theta}\} &= \sum_{m=0}^{\infty} \epsilon_m \frac{a_0(t)^m}{2m!} \Gamma\left(\frac{m}{2}+1\right) {}_1F_1\left(\frac{m}{2}; m+1; -a_0^2\right) \\ &\times \left[\frac{(e^{2\pi i(m+\xi)}-1)e^{-im\Phi}}{2\pi i(m+\xi)} + \frac{(e^{2\pi i(\xi-m)}-1)e^{im\Phi}}{2\pi i(\xi-m)}\right] \end{aligned} \tag{9.62}$$

When $a_0 = 0$, observe that this reduces, as required, to Eq. (3), Prob. 9.1.

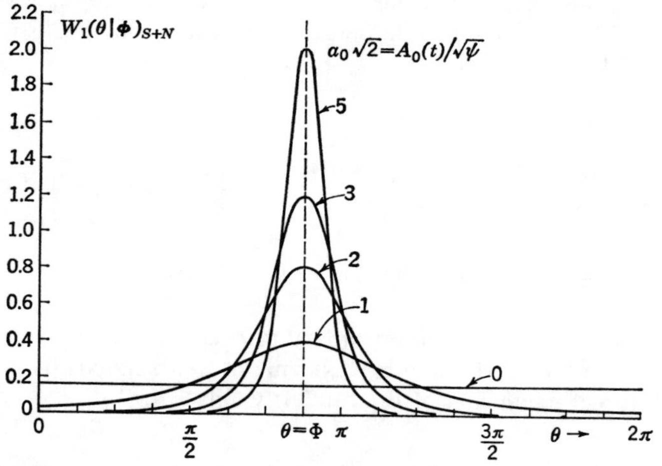

FIG. 9.4. Distribution density of the phase θ of a narrowband signal and noise ensemble at time t, for various signal strengths.

For the second-order densities, we expect much more complicated expressions, even in the simplest case of an unmodulated, on-tune sinusoid, as a glance at Eqs. 9.48 reveals at once. See, for example, Prob. 9.6c. We can use the various methods outlined above for calculating the moments $\boldsymbol{E}\{E_1^\nu E_2^\nu\}$ to obtain these distribution densities, but the forms of the results do not appear to be particularly useful in applications. We shall introduce instead some alternative methods for the particular problems of phase detection and FM reception in Chaps. 14, 15, where interest centers on $\boldsymbol{E}\{\theta_1\theta_2\}$, $\boldsymbol{E}\{\dot\theta_1\dot\theta_2\}$ and their associated spectra.

PROBLEMS

9.6 (*a*) Show that the (conditional) characteristic function for $W_1(E,\theta|\Phi)_{S+N}$ [Eq. (9.46)] is

$$F_1(i\xi,i\eta|\Phi)_{S+N} = e^{-a_0^2} \sum_{m=0}^{\infty} \epsilon_m G_m(i\eta) \sum_{n=0}^{\infty} \frac{a_0^{m+2n}}{n!(n+m)!} \left[\Gamma\left(\frac{m}{2}+n+1\right) \times \right.$$

$$\left. {}_1F_1\left(\frac{m}{2}+n+1;\frac{1}{2};-\frac{\xi^2\psi^2}{2}\right) + i\xi\sqrt{2\psi}\,\Gamma\left(\frac{m}{2}+n+\frac{3}{2}\right) {}_1F_1\left(\frac{m}{2}+\frac{3}{2}+n;\frac{3}{2};-\frac{\xi^2\psi^2}{2}\right)\right] \quad (1)$$

where $$G_m(i\eta) = \frac{1}{2}\frac{e^{2\pi i(\eta+m)}-1}{2\pi i(\eta+m)}e^{-im\Phi} + \frac{1}{2}\frac{e^{2\pi i(\eta-m)}-1}{2\pi i(\eta-m)}e^{im\Phi} \quad (2)$$

Obtain the alternative form

$$F_1(i\xi,i\eta|\Phi)_{S+N} = \sum_{m=0}^{\infty} \frac{\epsilon_m a_0^m}{m!} G_m(i\eta) \sum_{n=0}^{\infty} \frac{(i\xi\sqrt{2\psi})^n}{n!} \Gamma\left(\frac{m+n}{2}+1\right)$$

$$\times {}_1F_1\left(\frac{m-n}{2};m+1;-a_0^2\right) \quad (3)$$

(*b*) Show also that the first-order characteristic function of the envelope of a narrowband signal and (normal) noise process is

$$F_1(i\xi|\Phi)_{S+N} = e^{-a_0^2} \sum_{n=0}^{\infty} \frac{a_0^{2n}}{n!} \left[{}_1F_1\left(n+1;\frac{1}{2};-\frac{\xi^2\psi}{2}\right)\right.$$

$$\left. + \frac{i\xi\sqrt{2\pi}}{n!}\Gamma\left(n+\frac{3}{2}\right) {}_1F_1\left(\frac{n}{2},\frac{3}{2};-\frac{\xi^2\psi}{2}\right)\right] \quad (4)$$

or

$$F_1(i\xi|\Phi)_{S+N} = \sum_{n=0}^{\infty} \frac{[i\xi(2\psi)]^{\frac{1}{2}n}}{n!} \Gamma\left(\frac{n}{2}+1\right) {}_1F_1\left(-\frac{n}{2};1;-a_0^2\right) \quad (5)$$

(*c*) Show, for an additive cw signal of peak amplitude A_0 in a stationary narrowband normal noise process with a symmetrical spectrum, in tune with the center of the noise band, that the second-order d.d. of the phase is

$$w_2(\theta_1,\theta_2;t|\Phi)_{S+N} = \frac{1-k_0^2}{(2\pi)^2} e^{-2a_0^2k_0/(1+k_0)} \sum_{lmn=0}^{\infty} \epsilon_l\epsilon_m\epsilon_n \frac{a_0^{l+m}k_0^n}{l!m!} \left(\frac{1-k_0}{1+k_0}\right)^{l+m} A_{lm}^{(n)}$$

$$\times \cos l\theta_1 \cos m\theta_2 \cos n(\theta_2-\theta_1) \quad (6)$$

where $$A_{lm}^{(n)} \equiv \sum_{q=0}^{\infty} \frac{k_0^{2q}}{q!(q+n)!} \Gamma\left(\frac{l+n}{2}+q+1\right)\Gamma\left(\frac{m+n}{2}+q+1\right)$$

$$\times {}_1F_1\left(\frac{l-n}{2}-q;l+1;-\alpha_0^2\right) {}_1F_1\left(\frac{m-n}{2}-q;m+1;-\alpha_0^2\right) \quad (7)$$

with $$\alpha_0^2 \equiv a_0^2\left(\frac{1-k_0}{1+k_0}\right) \qquad k_0 = k_0(t) \quad (8)$$

Verify the result in the limiting situation of no signal and when $t \to \infty$.

(*d*) When the signal is strong, show that Eq. (6) becomes

$$w_2(\theta_1,\theta_2;t|\phi)_{S+N} \simeq \frac{1-k_0^2}{(2\pi)^{3/2}} \frac{e^{-2k_0^2a_0^2}}{(2a_0k_0)^{1/2}z} \left(\sum_{l=0}^{\infty} \epsilon_l z^l \cos l\theta_1 \sum \epsilon_n z^n \cos m\theta_2 \right) \delta(\theta_2 - \theta_1) \quad (9)$$

with

$$z \equiv \left(\frac{1-k_0}{1+k_0}\right)^{1/2} \qquad \alpha_0^2 \gg 1 \quad (10)$$

(*e*) Obtain the following moments:

$$\overline{\theta_1\theta_2}\Big|_{S+N} = (1-k_0^2)e^{-2a_0^2k_0/(1+k_0)} \left[\pi^2 A_{00}^{(0)} + \sum_{n=1}^{\infty} \left(2\pi^2 + \frac{1}{2n^2}\right) A_{nn}^{(n)} \frac{\alpha_0^{2n}k_0^n}{n!^2} \right.$$

$$\left. + 2 \sum_{\substack{l,m=0 \\ n=1}}^{\substack{\infty \\ (l,m)\neq n}} \frac{n^2 k_0^n \alpha_0^{l+m} A_{lm}^{(n)}}{(l^2 - n^2)(m^2 - n^2) l! m!} \right] \quad (11)$$

At $t = 0$ show that

$$\overline{\theta^2}\Big|_{S+N} = \overline{\theta^2}\Big|_N + 4 \sum_{l=1}^{\infty} \frac{a_0^l}{l!l^2} \Gamma\left(\frac{l}{2}+1\right) {}_1F_1\left(\frac{l}{2}; l+1; -a_0^2\right) \simeq 2\pi^2 + O(a_0^{-2}) \quad (12)$$

9.7 (*a*) Obtain the first-order probability density of the envelope and its first derivative, of narrowband normal noise, to show that

$$W_1(E,\dot{E}) = \frac{E}{(2\pi\mu_{33})^{1/2}} \exp\left(-\mu_{33}\dot{E}^2/2|\mu|\right) \exp\left\{-E^2[\mu_{11}\mu_{33} - (\mu_{14})^2]/2\mu_{33}|\mu|\right\} \quad (1)$$

where $|\mu| = (b_0b_2 - b_1^2)^2 \qquad \mu_{11} = b_2|\mu|^{1/2} \qquad \mu_{33} = b_0|\mu|^{1/2} \qquad \mu_{14} = b_1|\mu|^{1/2}$ (2)

and the b_n are the frequency moments

$$b_n = \int_0^{\infty} \mathcal{W}_N(f)(\omega - \omega_0)^n\, df \qquad \omega = 2\pi f,\ b_0 = \psi \quad (3)$$

If the spectrum is symmetrical (about f_0), we get

$$W_1(E,\dot{E}) = \frac{E}{b_0\sqrt{2\pi b_2}} e^{-E^2/2b_0 - \dot{E}^2/2b_2} \quad (4)$$

(*b*) Repeat (*a*), but now for an unmodulated carrier added to the normal noise, where the noise spectrum is symmetrical and centered about the carrier frequency, to show that

$$W_1(E,\dot{E})_{S+N} = \frac{E}{b_0\sqrt{2\pi b_2}} I_0\left(\frac{AE}{b_0}\right) e^{-(E^2+A^2)/2b_0 - \dot{E}^2/2b_2} \quad (5)$$

D. Middleton,[23] sec. 5; cf. D. K. C. MacDonald.[5]

9.8 Obtain the joint (conditional) first-order density of the envelope and phase, and their first time derivatives, of an (amplitude) modulated carrier which is off-tune ($\omega_d \neq 0$), and a narrowband normal noise process (whose spectrum is essentially symmetrical about f_0). Hence, show that

$$W_1(E,\dot{E};\theta,\dot{\theta};|\phi)_{S+N} = \frac{E^2}{4\pi b_0 b_2} \exp\left[-\frac{E^2 + A^2 - 2EA\cos(\theta+\phi_0)}{2b_0}\right] \exp\left(-\frac{1}{2b_2} \times \right.$$

$$\left. \{\dot{E}^2 + \dot{\theta}^2E^2 + \dot{A}^2 + \omega_d^2A^2 + [2E\dot{\theta}\sin(\theta+\psi_0) - 2\dot{E}\cos(\theta+\psi_0)](\dot{A}^2 + \omega_d^2A^2)^{1/2}\}\right) \quad (1)$$

where the angles ψ_0, ϕ_0 are defined by

$$\cos\phi_0 = \cos\omega_d t$$

and therefore

$$\begin{aligned}\omega_d t &= \phi_0 \\ \cos\psi_0 &= (\dot{A}\cos\omega_d t - \omega_d A\sin\omega_d t)/(\dot{A}^2 + \omega_d^2 A^2)^{1/2} \\ \sin\psi_0 &= (\dot{A}\sin\omega_d t + \omega_d A\cos\omega_d t)/(\dot{A}^2 + \omega_d^2 A^2)^{1/2}\end{aligned} \tag{2}$$

and $A = A(t)$, etc. When the carrier is on-tune, i.e., when $\omega_d = 0$, this reduces to

$$W_1(E,\dot{E};\theta,\dot{\theta}|\phi)_{S+N} = \frac{E^2}{4\pi^2 b_0 b_2}\exp\left\{-\frac{E^2 + A^2 - 2EA\cos\theta}{2b_0} - [\dot{E}^2 + \dot{A}^2 + \dot{\theta}^2 E^2 + (2E\dot{\theta}\sin\theta - 2\dot{E}\cos\theta)\dot{A}]/2b_2\right\} \tag{3}$$

D. Middleton.[23]

9.9 (*a*) For the noise and signal of Prob. 9.8, show that

$$\bar{\dot{\theta}} = (b_1/b_0)e^{-a_0^2} \tag{1}$$

(*b*) Assuming now that the spectrum of the narrowband noise is symmetrical about f_0 (for example, $b_1 = 0$), show that

$$W_1(\dot{\theta})_{S+N} = \frac{1}{2}\left(\frac{b_0}{b_2}\right)^{1/2}\left(1 + \frac{b_0\dot{\theta}^2}{b_2}\right)^{-3/2}\exp -a_0^2\,{}_1F_1\left(\frac{3}{2};1;\frac{a_0^2}{1 + b_0\dot{\theta}^2/b_2}\right) \tag{2}$$

[For a representation in terms of modified Bessel functions, see Eq. (A.1.31*c*) for ${}_1F_1(\frac{3}{2};1;x)$.] Hence, when $a_0^2 \gg 1$, e.g., the strong-signal case, observe that

$$W_1(\dot{\theta})_{S+N} \simeq \frac{A_0}{(2\pi b_2)^{1/2}}e^{-A_0^2\dot{\theta}^2/2b_2} \qquad \frac{b_0\dot{\theta}^2}{b_2} \ll 1 \tag{3}$$

Observe, finally, that for noise alone one gets for the present case of the symmetrical spectrum

$$W_1(\dot{\theta})_N = \tfrac{1}{2}(b_0/b_2)^{1/2}(1 + b_0\dot{\theta}^2/b_2)^{-3/2} \tag{4}$$

(*c*) Verify that

$$\overline{|\dot{\theta}|}_{S+N} = (b_2/b_0)^{1/2}e^{-a_0^2/2}I_0(a_0^2/2) \qquad \overline{\dot{\theta}^2}\Big|_{S+N} \to \infty \tag{5}$$

S. O. Rice,[21] sec. 5.

9.3 *Statistics of Signals and Broadband Normal Noise Processes*

In the preceding two sections, we have considered some of the more important statistical properties of mixed signal and normal narrowband ensembles. Similar results may be obtained in the less special situation of broadband processes. Here, we shall once more limit ourselves to first- and second-order statistics, where the mixed ensemble may be represented by Eqs. (9.40) if narrowband or simply by $V(t) = S(t) + N(t)$ if either or both the component processes have a broadband structure. The noise is assumed to be stationary, normal, and additive with S, such that $\boldsymbol{E}\{N\} = 0$. Writing $S = S(t;\boldsymbol{\theta},\phi)$, where $\boldsymbol{\theta}$ is a set of descriptive parameters which may or may not be random variables (cf. Sec. 1.3-5) and ϕ is a random phase, we see that the first-order conditional density of V is

$$W_1(V|\boldsymbol{\theta})_{S+N} = \mathfrak{F}^{-1}\{F_1(i\xi)_N F_1(i\xi|\boldsymbol{\theta})_S\} \tag{9.63a}$$

since S and N are assumed to be statistically independent. This may be written alternatively as

$$W_1(V|\boldsymbol{\theta})_{S+N} = \int_\phi W_1[V - S(t;\boldsymbol{\theta},\phi)]_N w(\phi)\, d\phi \tag{9.63b}$$

in which $W_1(\ )_N$ is the d.d. of the noise alone. Note that, for the unconditional density of V, we may apply the methods of Sec. 1.3-8 to average over the descriptive parameters $\boldsymbol{\theta}$.

As an example, let us consider the familiar case of a narrowband signal process and broadband normal noise. For the former we may write quite generally $S(t,\boldsymbol{\theta},\phi) = A(t,\boldsymbol{\phi}_A) \cos [\omega_c t - \Psi(t,\boldsymbol{\phi}_F) - \phi]$, with $\boldsymbol{\theta} = (\boldsymbol{\phi}_A,\boldsymbol{\phi}_F)$ [cf. Eq. (9.39) or (5.1*a*)], so that, if we assume in addition that the phases ϕ are uniformly distributed in $(0,2\pi)$, we obtain†

$$F_1(i\xi|\boldsymbol{\theta})_S = \int_{-\infty}^{\infty} w(\phi) e^{i\xi A \cos(\omega_c t - \Psi - \phi)}\, d\phi \qquad w(\phi) = \frac{1}{2\pi}\ (0 \le \phi \le 2\pi) \tag{9.64a}$$

$$= J_0(A\xi) \tag{9.64b}$$

(cf. Prob. 1.11). For the noise, we have as usual

$$F_1(i\xi)_N = e^{-\xi^2\psi/2} \qquad \psi = \boldsymbol{E}\{N^2\} \tag{9.65a}$$

where the corresponding d.d. is

$$W_1(N)_N = (2\pi\psi)^{-1/2} e^{-N^2/2\psi} \tag{9.65b}$$

The result from either Eq. (9.63*a*) or Eq. (9.63*b*) is therefore‡

$$W_1(V|\boldsymbol{\theta})_{S+N} = \mathfrak{F}^{-1}\{J_0(A\xi)e^{-\xi^2\psi/2}\} \tag{9.66a}$$

or

$$W_1(V|\boldsymbol{\theta})_{S+N} = \int_0^{2\pi} \exp\left\{-\frac{[V - A\cos(\omega_c t - \Psi - \phi)]^2}{2\psi}\right\} \frac{d\phi}{2\pi\sqrt{2\pi\psi}} \tag{9.66b}$$

Now, by expanding the Bessel function or the trigonometric term, respectively, in Eqs. (9.66*a*), (9.66*b*), we get the equivalent expressions§

$$W_1(V|\boldsymbol{\theta})_{S+N} = \frac{1}{(2\pi\psi)^{1/2}} \sum_{n=0}^{\infty} \frac{(-V^2/2\psi)^n}{n!}\, {}_1F_1\left(n + \frac{1}{2};\, 1;\, -a_0^2\right) \tag{9.67a}$$

or

$$W_1(V|\boldsymbol{\theta})_{S+N} = \psi^{-1/2} \sum_{n=0}^{\infty} \frac{a_0^{2n}}{2^n (n!)^2}\, \phi^{(2n)}\left(\frac{V}{\psi^{1/2}}\right) \qquad \text{with } a_0^2 \equiv \frac{A(t,\boldsymbol{\phi}_A)^2}{2\psi} \tag{9.67b}$$

† We abbreviate $A(t,\boldsymbol{\phi}_A)$ by A, etc.

‡ See Rice,[1] sec. 3.10.

§ Similar results were obtained independently by Rice[21] (cf. sec. 1), who gives the distribution, as well as the distribution density in the case of the pure sinusoidal signal.

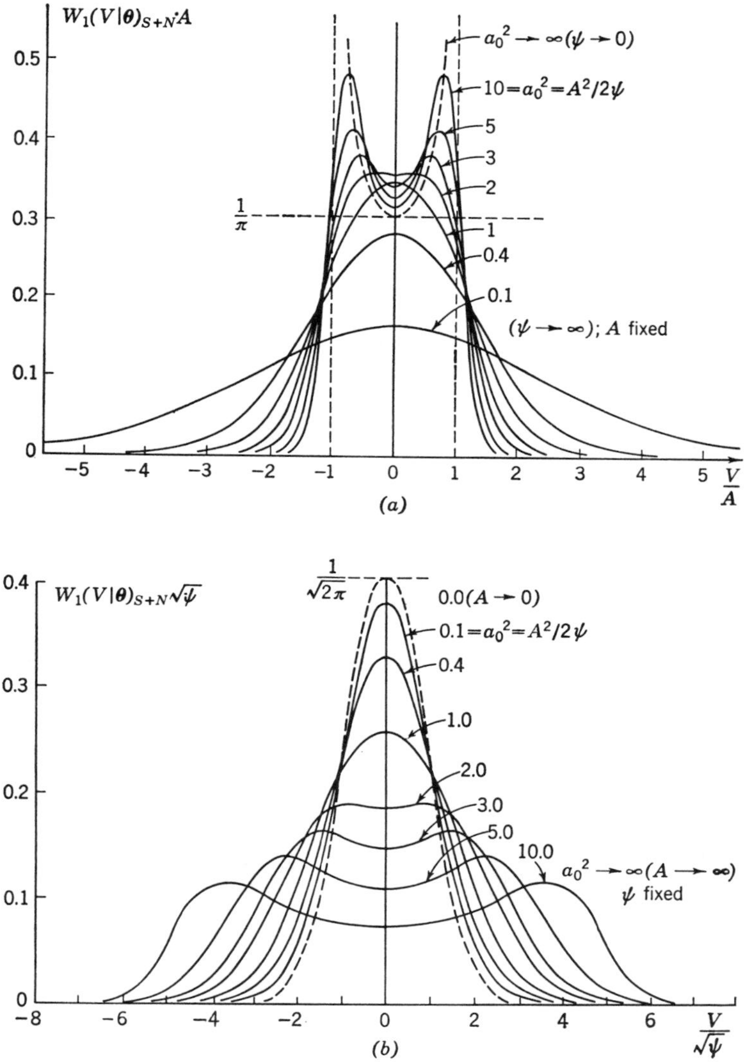

FIG. 9.5. (a) First-order (conditional) distribution density of a (modulated) sine wave and normal noise, for various levels of noise. (b) The same as (a) but for various levels of signal.

and $\phi^{(2n)}$ is a derivative of the error function, for example,

$$\phi^{(k)} \equiv \frac{d^k}{dx^k} \frac{e^{-x^2/2}}{\sqrt{2\pi}}$$

(cf. Appendix 1). The first result is useful when $V/\sqrt{2\psi}$ is reasonably small, while the second is convenient for large $V/\sqrt{2\psi}$ and small a_0. When a_0 is at all large, the asymptotic expression for the confluent hypergeometric function ${}_1F_1$ can be used (cf. Appendix 1). Figure 9.5 shows some typical distribution densities.†

The various (conditional) moments of $V = A \cos (\omega_c t - \Psi) + N$ may be found by differentiating the characteristic function $J_0(A\xi)e^{-\xi^2\psi/2}$ [cf. Eq. (9.66a)] or directly from the definition $E\{V^k\}$, with the help of Eqs. (9.67a), (9.67b). We get

$$\overline{V^k} = [2\pi(2\pi\psi)^{1/2}]^{-1} \int_{-\infty}^{\infty} V^k \, dV \int_0^{2\pi} \exp \left\{ - \frac{[V - A \cos (\omega_c t - \Psi - \phi)]^2}{2\psi} \right\} d\phi \tag{9.68a}$$

$$\overline{V^k} = \begin{cases} \displaystyle\sum_{q=0}^{k/2} \frac{k!}{(q!)^2(k/2 - q)!} 2^{-k/2-q} A^{2q} \psi^{k/2-q} & k = 0, 2, 4, \ldots \\ 0 & k = 1, 3, 5, \ldots \end{cases} \tag{9.68b}$$

For noise or signal alone, this reduces to

$$\begin{aligned} \overline{V^k} &= \frac{k!}{(k/2)!2^{k/2}} \psi^{k/2} & &\text{noise alone} \\ \overline{V^k} &= \frac{k!}{(k/2)!^2 2^k} A^k & &\text{signal alone} \end{aligned} \tag{9.69}$$

TABLE 9.2

k	Noise alone $V = N$	Signal alone $V = S$	Signal and noise $V = N + A \cos (\omega_c t - \Psi)$
1	0	0	0
2	ψ	$\frac{A^2}{2}$	$\psi + \frac{A^2}{2}$
3	0	0	0
4	$3\psi^2$	$\frac{3A^4}{8}$	$3\psi^2 + 3\psi A^2 + \frac{3A^4}{8}$
5	0	0	0
6	$15\psi^3$	$\frac{5A^6}{16}$	$15\psi^3 + \frac{45\psi^2 A^2}{2} + \frac{45\psi A^4}{8} + \frac{5A^6}{16}$
7	0	0	0
8	$105\psi^4$	$\frac{35A^8}{128}$	$105\psi^4 + 210\psi^3 A^2 + \frac{315\psi^2 A^4}{4} + \frac{35A^6\psi}{4} + \frac{35A^8}{128}$

Table 9.2 gives the first eight (*conditional*) moments.

Higher-order densities and moments may be obtained in the same way, although the results rapidly become very intricate. For example, the second-order conditional density, like Eq. (9.63b), may be written here

$$W_2(V_1,t_1;V_2,t_2|\boldsymbol{\theta})_{S+N} = \int_\phi w(\phi) W_2[V_1 - S(t_1;\boldsymbol{\theta},\phi),\ V_2 - S(t_2;\boldsymbol{\theta},\phi)]_N \, d\phi \tag{9.70}$$

† See Prob. 14.13 (p. 628).

where $W_2(\ \ )_N$ is the normal d.d. [Eq. (7.13a)], now with N_1 ($= x_1$) replaced by $V_1 - S(t_1;\boldsymbol{\theta},\phi)$, etc. A somewhat more convenient form for manipulation can perhaps be found if we use instead the characteristic functions $(F_2)_N$, $(F_2)_S$ in the alternative expression [cf. Eq. (9.63a)]

$$W_2(V_1,t_1;V_2;t_2|\boldsymbol{\theta})_{S+N} = \mathfrak{F}^{-1}\{F_2(i\xi_1,i\xi_2;t)_N F_2(i\xi_1,t_1;i\xi_2;t_2|\boldsymbol{\theta})_S\} \tag{9.71}$$

Note, incidentally, that V is not usually a stationary process unless the modulation is particularly simple (cf. Sec. 1.3-7), nor is $(F_2)_S$ represented in general by elementary functions (cf. Sec. 1.3-8). However, for the broadband noise and narrowband signal processes [cf. Eqs. (9.39), (5.1a)], we can use Eq. (5.11) to write

$$(F_2)_S = \sum_{m=0}^{\infty} \epsilon_m(-1)^m J_m(A_1\xi)J_m(A_2\xi) \cos m(\omega_c t + \Psi_2 - \Psi_1) \qquad t = t_2 - t_1 \tag{9.72}$$

where $\Psi_2 = \Psi(t_2)$, $A_1 = A(t_1)$, etc. (cf. Prob. 1.12). Since $(F_2)_N$ is $\exp\{(-\psi/2)[\xi_1^2 + \xi_2^2 + 2\xi_1\xi_2 k(t)]\}$ for the normal noise assumed here, we find finally that Eq. (9.71) becomes

$$(W_2)_{S+N} = \sum_{m=0}^{\infty}\sum_{n=0}^{\infty} \frac{\epsilon_m(-1)^{m+n}}{n!}[k(t)\psi]^n H_{m,n}(V_1;\boldsymbol{\theta})H_{m,n}(V_2;\boldsymbol{\theta}) \times \cos m(\omega_c t + \Psi_2 - \Psi_1) \tag{9.73a}$$

with

$$H_{m,n}(V;\boldsymbol{\theta}) = \int_C \xi^n e^{-\psi\xi^2/2 + i\xi V} J_m(i\xi A)\, d\xi/2\pi \tag{9.73b}$$

But Eq. (9.73b) is just the amplitude function (5.14), which may be evaluated by the methods of Sec. 2.3-2 (see also Prob. 2.15 and Appendix 1). Since the additive process $V = S + N$ is *not* normal, we may expect the resulting statistics to be rather complicated, as our discussion above indicates. For applications of these results in the second-moment theory of linear and nonlinear systems, see Chap. 5 and Part 3.

Other distribution problems involving the envelopes and phases of narrowband normal processes, with and without signals, and through nonlinear as well as linear devices, have been considered by Bussgang,[13] Barrett and Lampard,[14] Zadeh,[15] Price,[16] and others.[17–21]

PROBLEM

9.10 Show as $a_0^2 \to \infty$ that Eqs. (9.66) become

$$W_1(V|\boldsymbol{\theta})_{S+N} \simeq \frac{1}{\pi A}\sum_{m=0}^{\infty}\left[\frac{(\frac{1}{2})_m{}^2}{m!}\, a_0^{-2m}\, {}_2F_1\left(m+\frac{1}{2}, m+\frac{1}{2}; \frac{1}{2}; \frac{V^2}{A^2}\right) + O(a_0^{-2m-2})\right] \qquad |V| \le |A| \qquad a_0^2 = \frac{A^2}{2\psi} \tag{1}$$

$$W_1(V|\boldsymbol{\theta})_{S+N} = 0 \qquad \text{elsewhere}$$

and hence that

$$\lim_{\psi \to 0} W_1(V|\theta)_{S+N} = \begin{cases} \dfrac{1}{\pi\sqrt{A^2 - V^2}} & |V| \le |A| \\ 0 & |V| > |A| \end{cases} \qquad (2)$$

9.4 *Zero Crossings and Extrema of a Random Process*

Although a random process $y(t)$ is said to be completely described if the hierarchy of distribution densities $W_1, W_2, \ldots, W_n, \ldots$, as $n \to \infty$, is given (cf. Sec. 1.3-2), there are other important statistical properties associated with the process that are not immediately implied by this hierarchy. The moments and distribution of the "zeros," or "zero crossings," of $y(t)$ in an interval $(a \le t \le b)$, the moments and distribution of the interval between successive zeros, the expected number and distribution of extrema in the interval, and so on, are typical examples. In this section, based mainly on the work of Rice,[21,22] Middleton,[23] Kohlenberg,[24] and Steinberg et al.,[25] only a few results for the moments are examined, when the original process is normal or normally derived (in the sense of Sec. 9.1). The distribution problems, however, remain largely unsolved as yet.

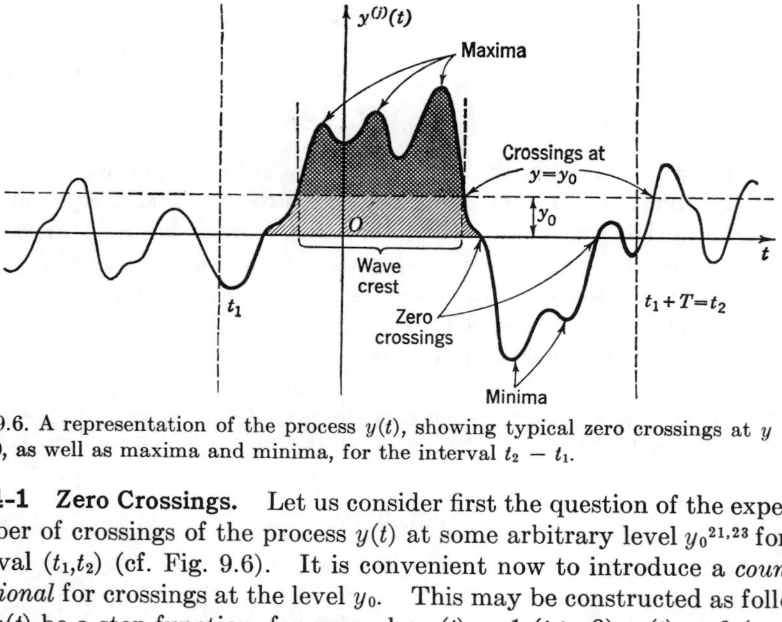

FIG. 9.6. A representation of the process $y(t)$, showing typical zero crossings at $y = y_0$, $y = 0$, as well as maxima and minima, for the interval $t_2 - t_1$.

9.4-1 Zero Crossings. Let us consider first the question of the expected number of crossings of the process $y(t)$ at some arbitrary level y_0[21,23] for the interval (t_1,t_2) (cf. Fig. 9.6). It is convenient now to introduce a *counting functional* for crossings at the level y_0. This may be constructed as follows: Let $u(t)$ be a step function, for example, $u(t) = 1$ $(t > 0)$, $u(t) = 0$ $(t < 0)$. Then, consider $u[y^{(j)}(t) - y_0]$. Differentiating this step function gives

$$\frac{du}{dt} = \dot{y}^{(j)}(t)\,\delta(y^{(j)} - y_0) \qquad \dot{y} = \frac{dy}{dt} \qquad (9.74)$$

which vanishes except when $y^{(j)} = y_0$, at which point one gets a spike (of unit area) directed positively or negatively, depending on whether the slope of $y^{(j)}(t)$ at y_0 is upward or downward. The desired counting functional for this jth member of the ensemble is then simply the number of crossings per second (at a time t) and, including both upward and downward passages, may be written†

$$n_0^{(j)}(y_0;t) = |\dot{y}^{(j)}(t)|\ \delta(y^{(j)} - y_0) \qquad -\infty < y^{(j)} < \infty \tag{9.75a}$$

The total number of crossings in an interval $t_2 - t_1$ is therefore

$$N_0^{(j)}(y_0;\, t_2 - t_1) = \int_{t_1}^{t_2} n_0^{(j)}(y_0;\tau)\, d\tau = \int_{t_1}^{t_2} |\dot{y}^{(j)}(\tau)|\ \delta[y^{(j)}(\tau) - y_0]\, d\tau \tag{9.76}$$

Setting $y_0 = 0$ then gives the number of zeros in the interval.†

Various moments of $n_0(y_0;t)$ and $N_0(y_0;\, t_2 - t_1)$ can now be obtained. We observe first that the process $y(t)$ must be differentiable (mean square) to the first order, at least. Then we see that[22–25]

$$\begin{aligned} \boldsymbol{E}\{n_0(y_0;t)\} &= \boldsymbol{E}\{|\dot{y}(t)|\ \delta(y - y_0)\} \\ \overline{N_0(y_0;T)} \equiv \boldsymbol{E}\{N_0(y_0;\, t_2 - t_1)\} &= \int_{t_1}^{t_2} \boldsymbol{E}\{n_0(y_0;\tau)\}\, d\tau \end{aligned} \tag{9.77}$$

The average number of crossings per second, defined over the interval $(t_2 - t_1)$, is simply $\boldsymbol{E}\{N_0(y_0;\, t_2 - t_1)\}/(t_2 - t_1)$ and is different from $\bar{n}_0$ $(t_1 \le t \le t_2)$ unless the process is stationary. In more detail, Eqs. (9.77) can be written

$$\bar{n}_0 = \iint_{-\infty}^{\infty} |\dot{y}|\ \delta(y - y_0)\ W_1(y,\dot{y};t)\, dy\, d\dot{y} = \int_{-\infty}^{\infty} |\dot{y}| W_1(y_0,\dot{y};t)\, d\dot{y} \tag{9.78}$$

Similarly, for the second moments $\overline{n_0{}^2}$, $\overline{N_0{}^2}$ one finds directly that

$$\overline{n_0{}^2} = \iint_{-\infty}^{\infty} |\dot{y}_1|\ |\dot{y}_2| W_2(y_0,\dot{y}_1,\tau_1;y_0,\dot{y}_2,\tau_2)\, d\dot{y}_1\, d\dot{y}_2 \tag{9.79a}$$

† If, on the other hand, we are interested in the number of *peaks*, or wave crests, above a certain level y_0 (see the shaded areas in Fig. 9.6; note that the peak, or wave crest, is not identified with a maximum here), we count the number of traversals with positive slope, so that Eq. (9.75a) becomes now

$$n_0^{(j)}(y_0;t)_+ = \dot{y}^{(j)}(t)\ \delta(y^{(j)} - y_0) \qquad 0 \le \dot{y}^{(j)} \le \infty \tag{9.75b}$$

A similar expression for the number of peaks per second *below* y_0 is

$$n_{0,-}^{(j)} = \dot{y}^{(j)}\ \delta(y^{(j)} - y_0) \qquad -\infty < \dot{y}^{(j)} < 0 \tag{9.75c}$$

and the total number of peaks in $t_2 - t_1$ is given by the appropriate $n_{0,+}^{(j)}$, $n_{0,-}^{(j)}$ in Eq. (9.76).

and

$$\overline{N_0^2} = \iint_{t_1}^{t_2} \overline{n_0^2(y_0;\tau_1,\tau_2)}\, d\tau_1\, d\tau_2 \qquad E\{n_0(y_0;\tau_1)n_0(y_0;\tau_2)\} \equiv \overline{n_0^2(y_0;\tau_1,\tau_2)} \tag{9.79b}$$

Analogous expressions for the higher moments follow in the same way, but, because of the presence of $|\dot{y}|$ in the integrand, the actual integrations in most instances rapidly become intractable. Note that the process $y(t)$ may be mixed, as well as entirely random: for example, $y(t)$ may represent the sum of a signal and noise process. In any event, it is the joint d.d. of y and $\dot{y}$, at one or more instants $(\tau_1,\tau_2, \ldots)$, that is needed for $\overline{n_0^m}$ and $\overline{N_0^m}$.

We may illustrate this with a few examples for $\bar{n}_0$ and $\overline{n_0^2}$, when $y(t)$ is stationary, normal, and differentiable (mean square) to the first order, with $E\{y\} = E\{\dot{y}\} = 0$ and covariance function $K_y(t)$. From Eq. (8.28), we have

$$W_1(y,\dot{y};t) = \frac{e^{-y^2/2b_0-\dot{y}^2/2b_2}}{2\pi\sqrt{b_0b_2}} = W_1(y,\dot{y}) \tag{9.80}$$

so that the average number of crossings per second of the level y_0, at any time t, is from Eq. (9.78)

$$\bar{n}_0 = 2\int_0^\infty \frac{\dot{y}e^{-\dot{y}^2/2b_2-y_0^2/2b_0}}{2\pi\sqrt{b_0b_2}}\, d\dot{y} = \frac{1}{\pi}\rho_y e^{-y_0^2/2b_0} \tag{9.81}$$

Here, ρ_y is an (angular) frequency, defined as the ratio of two frequency moments [cf. Sec. 3.2-3(4), Eqs. (3.65), (3.66)], according to

$$\rho_y^2 \equiv \frac{b_2}{b_0} = \frac{\int_0^\infty \mathcal{W}_y(f)\omega^2\, df}{\int_0^\infty \mathcal{W}_y(f)\, df} = \frac{K_{\dot{y}}(0)}{K_y(0)} = \frac{-\ddot{K}_y(0)}{K_y(0)} \qquad \omega = 2\pi f \tag{9.82}$$

Observe that, because of the stationarity of the process, $\bar{N}_0$ is equal to $(t_2 - t_1)\bar{n}_0$. Figure 9.8 shows a typical set of experimental values of $\bar{N}_0$ when only the zero crossings in the upward direction are considered (see White[28]).

As a second example, let us calculate the expected number of crossings per second, at a level E_0, of the envelope $E(t)$ when $y(t) = E\cos(\omega_0 t - \theta)$ is a narrowband normal process with the properties[21–23,†] mentioned in the above paragraphs. Then, replacing y by E, from Eq. (9.78) we can write

$$\overline{n_0(E_0)} = \int_{-\infty}^{\infty} |\dot{E}|W_1(E_0,\dot{E})\, d\dot{E}$$

† For additional results, involving the crossings of θ, E, and $\dot{\theta}$, see Rice,[21] sec. 6.

and, using the result (1) of Prob. 9.7 for $W_1(E_0,\dot{E})$, we obtain

$$\begin{aligned}\overline{n_0(E_0)} &= 2(|\mu|/\mu_{33}\sqrt{2\pi\mu_{33}})\, E_0 \exp\{-E_0^2[\mu_{11}\mu_{33} - (\mu_{14})^2]/2\mu_{33}|\mu|\} \\ &\doteq \rho_E E_0(2/\pi b_0)^{1/2} \exp\left(-\frac{E_0^2}{2b_0}\right) \end{aligned} \tag{9.83}$$

this last if the spectrum is symmetrical about f_0, so that $\mu_{14} = 0$. Here, ρ_E is given by

$$\rho_E^2 \equiv \left(\frac{b_2}{b_0}\right)_E = b_0^{-1}\int_0^\infty \mathcal{W}_y(f)(\omega - \omega_0)^2\, df = \frac{-\ddot{K}_{0y}(0)}{K_{0y}(0)} = \frac{-\ddot{\rho}_{0y}(0)}{\rho_{0y}(0)} \tag{9.84}$$

[cf. Eqs. (3.108)], with $b_0 = \overline{y^2}$ in the usual way. Setting $\rho_0^2 \equiv \omega_0^2$ and using the fact that $\mathcal{W}_y(f' + f_0) \doteq \mathcal{W}_y(f') = \mathcal{W}_y(-f')$ for the narrowband process, with the additional assumption of symmetry about f_0 (or $f' = 0$), we see that Eq. (9.82) gives

$$\begin{aligned}\rho_y^2 &= b_0^{-1}\int_{-f_0}^\infty \mathcal{W}_y(f' + f_0)(\omega' + \omega_0)^2\, df' \doteq b_0^{-1}\int_0^\infty \mathcal{W}_y(f)(\omega - \omega_0)^2\, df + \omega_0^2 \\ &= \rho_E^2 + \rho_0^2 \end{aligned} \tag{9.85}$$

As expected, ρ_y is the "centroid" (angular) frequency about $f = 0$, while ρ_E is the centroid about $f' = 0$ (or $f = f_0$) (cf. Fig. 9.7). The relation (9.85) is analogous to the corresponding situation in mechanics.[26]

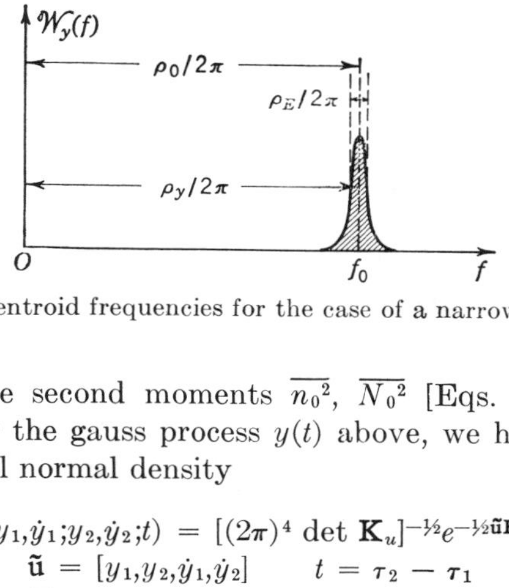

FIG. 9.7. The various centroid frequencies for the case of a narrowband process [cf. Eqs. (9.82), (9.84), (9.85)].

Evaluation of the second moments $\overline{n_0^2}$, $\overline{N_0^2}$ [Eqs. (9.79a), (9.79b)] is more difficult. For the gauss process $y(t)$ above, we have from Prob. 8.7 the four-dimensional normal density

$$W_2(y_1,\dot{y}_1;y_2,\dot{y}_2;t) = [(2\pi)^4 \det \mathbf{K}_u]^{-1/2} e^{-\frac{1}{2}\tilde{\mathbf{u}}\mathbf{K}_u^{-1}\mathbf{u}}$$

with $\quad \tilde{\mathbf{u}} = [y_1, y_2, \dot{y}_1, \dot{y}_2] \qquad t = \tau_2 - \tau_1$

and

$$\mathbf{K}_\mathbf{u} = \begin{bmatrix} b_0 & K_y(t) & 0 & \dot{K}_y(t) \\ K_y(t) & b_0 & -\dot{K}_y(t) & 0 \\ 0 & -\dot{K}_y(t) & b_0\rho_y^2 & -\ddot{K}_y(t) \\ \dot{K}_y(t) & 0 & -\ddot{K}_y(t) & b_0\rho_y^2 \end{bmatrix} \tag{9.86}$$

Restricting our attention to zero crossings, i.e., $y = y_0 = 0$, we see that

W_2 above reduces to a considerably simpler form, so that we may write for $\overline{n^2}$ now[23]

$$\overline{n_0{}^2} = \iint_{-\infty}^{\infty} \frac{d\dot{y}_1\, d\dot{y}_2}{(2\pi)^2(\det \mathbf{K}_u)^{1/2}} |\dot{y}_1|\, |\dot{y}_2| \exp\left[-\frac{1}{2|\mu|}(\dot{y}_1{}^2\mu_{33} + \dot{y}_2{}^2\mu_{34} + 2\dot{y}_1\dot{y}_2\mu_{34})\right]$$

$$|\mu| = \det \mathbf{K}_u \qquad (9.87)$$

where μ_{jk} is the cofactor of $(K_{\mathbf{u}})_{jk}$ in Eq. (9.86). Using Prob. 7.4*b*, we obtain finally the results of Steinberg et al.[25] (cf. Prob. 9.14 and Rice,[21] pp. 70–73):

$$\overline{n_0{}^2} = \frac{1}{2\pi}\rho_y(\tau_2 - \tau_1) + \frac{1}{\pi^2}\frac{(\mu_{33}{}^2 - \mu_{34}{}^2)^{1/2}}{b_0{}^3(1 - k_y{}^2)^{3/2}} \times \left[1 + \frac{\mu_{34}}{(\mu_{33}{}^2 - \mu_{34}{}^2)^{1/2}} \tan^{-1}\frac{\mu_{34}}{(\mu_{33}{}^2 - \mu_{34}{}^2)^{1/2}}\right]$$

$$k_y = k_y(\tau_2 - \tau_1) = \frac{K_y(\tau_2 - \tau_1)}{K_y(0)} \qquad (9.88)$$

For the mean-square number of zeros in the interval $T = t_2 - t_1$, we use Eq. (9.88) in Eq. (9.79*b*), observing that $\overline{n_0{}^2}$ is even in $\tau_2 - \tau_1$. The result is

$$\overline{N_0{}^2(0;T)} = \iint_{0-}^{T+} \overline{n_0{}^2(0, \tau_2 - \tau_1)}\, d\tau_1\, d\tau_2 = 2\int_{0-}^{T+} (T - \tau)\overline{n_0{}^2(0;\tau)}\, d\tau \qquad (9.89)$$

where we write $(0-,T+)$ because of the symmetrical nature of our delta functions [cf. Eq. (1.14) and Sec. 1.2-4]. The evaluation of $\overline{N_0{}^2}$ must usually be done by numerical methods.† This has been carried through and verified experimentally by White,[28] who has considered the cases involving both broadband (low-Q) and narrowband (high-Q) LRC normal noise. Figure 9.8*b* shows some typical values of $\overline{N_0{}^2(0,T)}$, for the positive zero crossings, in the low-Q case. White has also studied experimentally the evaluation of $P(n_0,t)_y$, the probability that the noise process $y(t)$ has n_0 zeros in the interval $(0,t)$, a typical result for which is shown in Fig. 9.8*c*, where again only the positive zero crossings are considered and $n_0 = 2$ ($t = T$ here). (Results for $n_0 = 0, 1, 3, 4, 7$ are included in White.[28])

A number of other results of interest in connection with the zero-crossing problem are included in the Problems. However, even in the case of normal processes, it is apparent that great analytical difficulties lie in the path of more complicated statistics such as the higher moments and the distributions of n_0, N_0, etc.

9.4-2 Extrema.‡ Another statistical problem of some interest in applications is the calculation of the number of extrema, as distinct from "peaks," or wave crests (cf. Fig. 9.6), of a random process $y(t)$. For example, we

† Results for $\overline{N_0(y_{01};T)N_0(y_{02};T)}$ $(y_{01}, y_{02} \neq 0)$ have been calculated by Miller and Freund.[27]

‡ See also Rice,[22] secs. 3.6, 3.8.

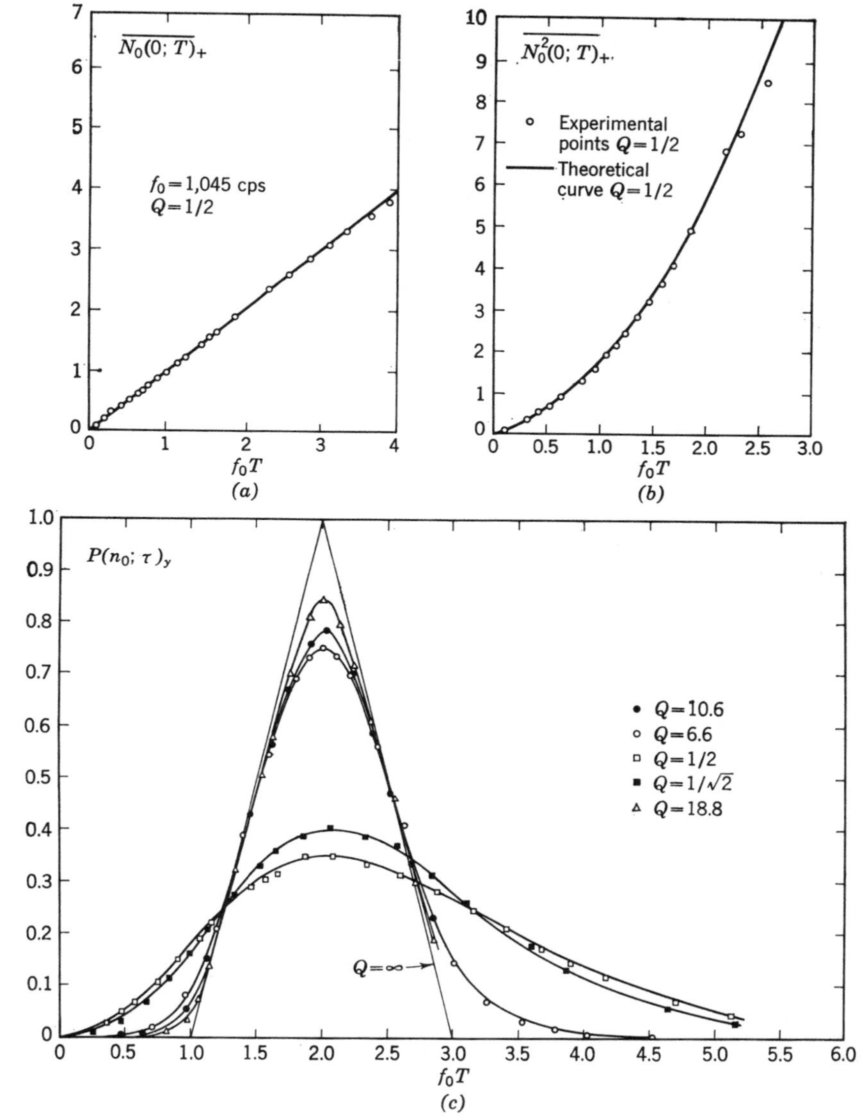

FIG. 9.8. (a) Average number of positive zero crossings in $(0,T)$ for low-Q normal LRC noise. (b) Mean-square number of positive zero crossings in $(0,T)$ for (a). (c) Probability of two zeros in interval $(0,T)$ for normal LRC noise, various Q's. [*After G. M. White, An Experimental System for Studying the Zeros of Noise, Cruft Lab. Tech. Rept.* 261, *May* 15, 1957, *and Experimental Determination of the Zero-crossing Distribution* $W(N;T)$, *Cruft Lab. Tech. Rept.* 265, *September,* 1957.]

may need the average number of maxima (exceeding a given level $y = y_0$) during an interval $(t_2 - t_1)$, or it may be necessary to find the expected number of such extrema per second, and so on.

As in the zero-crossing cases above, we can also construct a counting functional here. We start with the pair of unit step functions $u(\dot{y} - v_0)$, $u(y - y_0)$ and observe that, if $\dot{y}$ and y together exceed a "velocity" v_0 and the level y_0, respectively, then the product $u(\dot{y} - v_0)u(y - y_0)$ is unity, while if $\dot{y} < v_0$ and/or $y < y_0$, $u(\dot{y} - v_0)u(y - y_0)$ vanishes. Differentiating this product, we get for a particular member of the ensemble at time t

$$\frac{d}{dt}[u(\dot{y}^{(j)} - v_0)u(y^{(j)} - y_0)] \\ = \ddot{y}^{(j)}\,\delta(\dot{y}^{(j)} - v_0)u(y^{(j)} - y_0) + \dot{y}^{(j)}\,\delta(y^{(j)} - y_0)u(\dot{y}^{(j)} - v_0) \quad (9.90)$$

The first term is the desired counting functional, since $\delta(\dot{y}^{(j)} - v_0)$ gives a spike (at time t) whenever the slope of $y^{(j)}$ is equal to v_0, the sign of which is determined by the "acceleration" $\ddot{y}^{(j)}$. The step function $u(y^{(j)} - y_0)$ ensures that the counting operation is carried out for all $y^{(j)} \geq y_0$. [In a similar way, the second term of Eq. (9.90) is the counting functional for all crossings of the level $y^{(j)} = y_0$, when the slope of $y^{(j)}$ exceeds v_0.] The number of *extrema* per second (at time t) for $y^{(j)} > y_0$ is found by setting $\dot{y}^{(j)} = v_0 = 0$, for example,

$$n_e^{(j)}(y_0,0;t) = (\pm)\ddot{y}^{(j)}\,\delta(\dot{y}^{(j)} - 0)u(y^{(j)} - y_0) \qquad y^{(j)} \geq y_0 \quad (9.91)$$

The plus and minus signs apply as $\ddot{y}$ is positive or negative, since we are concerned here only with the number of such effects. Similarly, the total number of extrema, for $y^{(j)} > y_0$, in the interval $(t_2 - t_1)$ is

$$N_e^{(j)}(y_0, 0; t_2 - t_1) = \int_{t_1}^{t_2} n_e^{(j)}(y_0,0;\tau)\,d\tau \quad (9.92)$$

For *maxima* (when $y^{(j)} > y_0$), we choose only those $y^{(j)}_{\max}$ with $\ddot{y}^{(j)} < 0$, while, for *minima*, the choice is made when $\ddot{y}^{(j)} > 0$.

Moments of n_e and N_e may be calculated in the usual way. For the expected number of maxima per second (at time t) of y above the level y_0, Eq. (9.91) becomes (for all $\ddot{y} < 0$)

$$\overline{n_e(y_0,0;t)_+} = -\int_{-\infty}^{\infty} dy \int_{-\infty}^{\infty} d\dot{y} \int_{-\infty}^{0} d\ddot{y}\;\ddot{y}\,\delta(\dot{y} - 0)u(y - y_0)W_1(y,\dot{y},\ddot{y};t) \\ = -\int_{y_0}^{\infty} dy \int_{-\infty}^{0} \ddot{y}W_1(y,0,\ddot{y};t)\,d\ddot{y} \quad (9.93a)$$

while, for *minima* above $y = y_0$, we get

$$\overline{n_e(y_0,0;t)_-} = \int_{y_0}^{\infty} dy \int_0^{\infty} \ddot{y}W_1(y,0,\ddot{y};t)\,d\ddot{y} \quad (9.93b)$$

A related expression is

$$W_{\max}(y;t)\,dy\,dt = -dy\,dt\int_{-\infty}^{0} \ddot{y}W_1(y,0,\ddot{y};t)\,d\ddot{y}$$
$$= \text{probability that } y \text{ has maximum in range } (y,\ y+dy) \text{ in interval } (t,\ t+dt) \qquad (9.93c)$$

Higher moments may be found in the same way: in place of $W_1(y,\dot{y},\ddot{y};t)$, one needs the corresponding d.d.'s W_2, W_3, etc., in the triple of random variables y, $\dot{y}$, $\ddot{y}$. Note that $y(t)$ must be differentiable (mean square) to the second order.

As an example, let us calculate the expected number of maxima per second for a stationary normal random process $y(t)$ which is differentiable (mean square) to order 2 and for which $E\{y\} = 0$, a problem originally solved by Rice.[22] Here, we set $y_0 = -\infty$, since all maxima in the range of possible values of y are to be counted. If $\tilde{\mathbf{u}} = [y,\dot{y},\ddot{y}]$ is a normal vector process, then $[y,\dot{y},\ddot{y}]$ have a joint normal d.d. with covariance matrix

$$\mathbf{K_u} = \begin{bmatrix} b_0 & 0 & -b_2 \\ 0 & b_2 & 0 \\ -b_2 & 0 & b_4 \end{bmatrix} \qquad b_{2n} = \int_0^\infty \mathcal{W}_y(f)\omega^{2n}\,df \qquad (9.94)$$

[Eq. (3.66*b*)]. The desired d.d. $W_1(y,0,\ddot{y};t)$ is accordingly†

$$W_1(y,0,\ddot{y}) = (2\pi)^{-3/2}(\det \mathbf{K_u})^{-1/2}\exp\left[-\frac{1}{2|\mu|}(\mu_{11}y^2 + \mu_{33}\ddot{y}^2 + 2\mu_{13}y\ddot{y})\right] \qquad (9.95a)$$

where $|\mu|$ and the cofactors μ_{11}, μ_{33}, μ_{13} of $\overline{u_iu_j}$ in $\mathbf{K_u}$ are, in terms of the spectral moments,

$$|\mu| = \det \mathbf{K_u} = b_2(b_0b_4 - b_2^2) \qquad \mu_{11} = b_2b_4 \qquad \mu_{13} = b_2^2 \qquad \mu_{33} = b_0b_2 \qquad (9.95b)$$

Inserting this into Eq. (9.93*a*) and integrating first over y in the range $(-\infty < y < \infty)$, we obtain directly

$$\overline{n_e(-\infty,0)_+} = \frac{-\mu_{11}^{1/2}}{2\pi}\int_{-\infty}^{0} d\ddot{y}\;\ddot{y}e^{-\ddot{y}^2(\mu_{33}-\mu_{13}^2/\mu_{11})/2|\mu|}$$
$$= \frac{1}{2\pi}\sqrt{\frac{b_4}{b_2}} = \frac{1}{2\pi}\left[\frac{\int_0^\infty \omega^4\mathcal{W}_y(f)\,df}{\int_0^\infty \omega^2\mathcal{W}_y(f)\,df}\right]^{1/2} = \frac{\overline{\ddot{y}^2}}{\overline{\dot{y}^2}} \qquad (9.96)$$

For a bandpass filter with uniform response in $(f_2 - f_1) > 0$, Eq. (9.96)

† W_1 is independent of t since $y(t)$ is stationary. We remark, also, that the result (9.96) may be obtained at once with obvious modifications from Eqs. (9.81) and (9.82), if we consider $\dot{y}(t)$ now as the basic process.

becomes

$$\overline{n_e(-\infty;0)_+} = \left(\frac{3}{5}\frac{f_2^5 - f_1^5}{f_2^3 - f_1^3}\right)^{1/2} \tag{9.97}$$

and if $f_1 = 0$,

$$\overline{(n_e)_+} = \sqrt{3/5}\, f_2$$

As we might expect physically, the expected number of maxima (per second) is proportional to the bandwidth of the process; i.e., the wider the band, the greater the number of oscillations of the waves in any given interval.

The probability (density) that y has a maximum in the range $(y, y + dy)$ at any time t is also easily found (for the normal process of our example) with the help of Eq. (9.95*a*) in Eq. (9.93*c*). The result is[22]

$$\begin{aligned} W_{\max}(y) &= -\int_{-\infty}^{0} \ddot{y} W_1(y,0,\ddot{y})\, d\ddot{y} \\ &= (2\pi)^{-3/2}\mu_{33}^{-1}\Big(|\mu|^{1/2} e^{-\mu_{11}y^2/2|\mu|} + y\mu_{13}(\pi/2\mu_{33})^{1/2} e^{-y^2/2b_0} \\ &\qquad \times \{1 + \Theta[\mu_{13}y/(2\mu_{33}|\mu|)^{1/2}]\}\Big) \end{aligned} \tag{9.98}$$

Figure 9.9 shows a typical curve for $W_{\max}(y)$ when $y(t)$ is the output of the rectangular low-pass filter ($f_1 = 0$) of Eq. (9.97). For results on the maxima of the envelope E of a narrowband normal process, we refer the reader once more to Rice[22] (sec. 3.8).

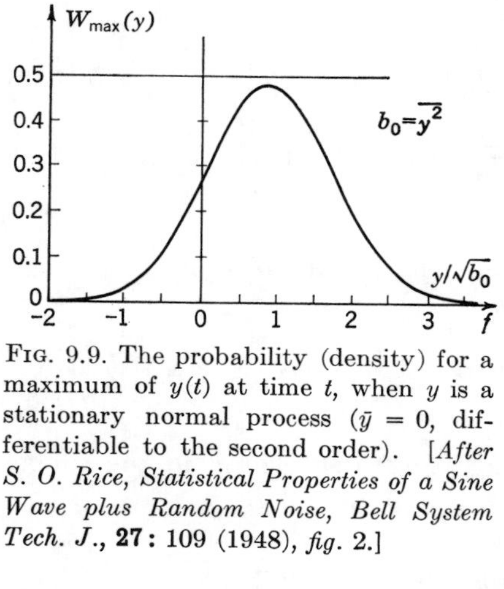

FIG. 9.9. The probability (density) for a maximum of $y(t)$ at time t, when y is a stationary normal process ($\bar{y} = 0$, differentiable to the second order). [*After S. O. Rice, Statistical Properties of a Sine Wave plus Random Noise, Bell System Tech. J.*, **27**: 109 (1948), *fig.* 2.]

9.4-3 Remarks. Here, we have considered only a few situations involving the zeros, maxima, etc., of a random process. Some additional results of interest are included in the problems following. We remark again that the distributions of zeros, intervals between successive zeros and maxima, and so on, present great technical difficulties, a detailed account of which lies outside the aim and scope of this book. Closely related, also, are *recurrence* and *first-passage time* problems, where one asks for the statistics of the interval between a certain state of a system and the instant at which the system first returns to that state or to some other one. For example, a capacitor, initially at zero charge, is charged by a random voltage: what is the average value of the time at which the capacitor *first* reaches a charge Q? Other situations of this type, of interest in communication theory, can also be suggested, e.g., the effects of noise on "zero counters" for determining frequency, on the choice of thresholds for trigger systems,[23] and on the operation of more general threshold devices. For some pertinent analysis, the reader is referred to recent work of Darling[29] and

Siegert,[30,31] to Helstrom,[32] and to Bartlett[33] and McFadden[34] for some other applications.

PROBLEMS

9.11 Show that for $y(t) = N(t) + S(t)$, where $S(t)$ is deterministic and $N(t)$ is normal, with $\bar{N} = 0$, the average number of crossings per second of y_0 at time t is

$$\overline{n_0(y_0;t)} = \frac{\rho_y}{\pi} e^{-(y_0-S)^2/2b_0} \left(e^{-\dot{S}^2/2b_0\rho_y^2} + \dot{S} \sqrt{\frac{\pi}{2b_0\rho_y^2}}\, \Theta \left[\frac{\dot{S}}{(2b_0\rho_y^2)^{1/2}} \right] \right) \tag{1}$$

$$= \frac{\rho_y}{\pi} e^{-(y_0-S)^2/2b_0} \, {}_1F_1 \left(-\frac{1}{2}; \frac{1}{2}; -\frac{\dot{S}^2}{2\rho_y^2 b_0} \right) \tag{2}$$

where $\rho_y = (b_2/b_0)^{1/2}$ [cf. Eq. (9.82)]. D. Middleton,[23] sec. II-3.

9.12 A sine-wave signal $S = A_0 \cos \omega_0 t$ and stationary normal noise $N(t)$, with $\bar{N} = 0$, are combined additively. Show that in a period $(0,T)$, where $T = kT_0$ and T_0 $(= 2\pi/\omega_0)$ is the period of the sinusoid, the average number of zeros is given by

$$\overline{N_0(0;T)} = \frac{kT_0}{\pi} \rho_y \sum_{n=0}^{\infty} \frac{(-\frac{1}{2})_n}{(n!)^2} (-\alpha^2 a_0^2)^n \, {}_1F_1 \left(\frac{1}{2}; n+1; -a_0^2 \right) \tag{1}$$

$$= \frac{kT_0}{\pi} \rho_y \sum_{m=0}^{\infty} \frac{(\frac{1}{2})_m}{(m!)^2} (-a_0^2)^m \, {}_1F_1 \left(-\frac{1}{2}; m+1; -\alpha^2 a_0^2 \right) \tag{2}$$

where $\alpha = \omega_0/\rho_y \qquad a_0^2 = A_0^2/2b_0 \qquad \rho_y^2 = b_2/b_0$ (3)

as before. Note that, when $\alpha = 1$, $\overline{N_0(0;T)}_{S+N} = \overline{N_0(0;T)}_N$ and, in fact, Eqs. (1) and (2) hold for all strengths of the sinusoidal signal assumed here.

9.13 (*a*) If $W(n_0;\tau)_y$ is the d.d. for stationary normal noise $y(t)$ having n_0 zeros in $(0,\tau)$, show that the probability density $W_0(\tau)_y$ for two *consecutive* zeros separated by τ sec obeys

$$W_0(\tau)_y = -\pi \left[\frac{K_y(0)}{-\ddot{K}_y(0)} \right]^{1/2} \frac{d^2W(n_0;\tau)_y}{d\tau^2} \tag{1}$$

A. Kohlenberg.[24]

(*b*) If the noise process in (*a*) has a covariance function of the form

$$K_y(t) = K_0(t)_y \cos \omega_0 t$$

that is, if the process is narrowband with a *symmetrical* spectrum, show that d.d. $W_0(\tau)_y$ for the interval τ between two successive zero crossings is approximately

$$W_0(\tau)_y \doteq \frac{A}{2} \left[1 + A^2 \left(\tau - \frac{\pi}{\omega_0} \right)^2 \right]^{-3/2} \qquad A = \left[\frac{K_0(0)_y \omega_0^4}{-\ddot{K}_0(0)_y \pi^2} \right]^{1/2} \tag{2}$$

[Note that $\overline{(\tau - \pi/\omega_0)^2} \to \infty$, while $\overline{|\tau - \pi/\omega_0|} < \infty$.] S. O. Rice,[22] sec. 3.4.

(*c*) Show in general that the d.d. of the spacing τ between consecutive zeros can be expressed as

$$W_0(\tau)_y = w_0(\tau) + \sum_{n=1}^{\infty} \frac{(-1)^n}{n!} \int_0^{\tau} \cdots \int w_n(t_1,t_2,\ldots,t_n)\, dt_1 \cdots dt_n \tag{3}$$

where $w_n(t_1, \ldots, t_n)\, dt_1 \cdots dt_n$ is the joint probability of the processes having zeros in $dt_1 \cdots dt_n$. S. O. Rice,[22] sec. 3.4.

9.14 (*a*) Show that the probability (density) $W_y(0+,t_1;0-;t_2)$ that y will pass through a zero at t_1 with positive slope and through zero at t_2 ($> t_1$) with negative slope is

$$W_y(0+,t_1;0-,t_2) = -\int_0^\infty d\dot{y}_1 \int_{-\infty}^0 d\dot{y}_2\, \dot{y}_1\dot{y}_2 W_2(0,\dot{y}_1,t_1;0,\dot{y}_2,t_2) \tag{1}$$

and, specifically for the normal process $y(t)$ above (cf. Sec. 9.4-1), obtain

$$W_y(0+,0-;\tau) = \frac{1}{2\pi}\left(\frac{b_0}{b_2}\right)^{1/2} (\mu_{22}{}^2 - \mu_{23}{}^2)^{1/2}[K_y(0)^2 - K_y(\tau)^2]^{-3/2}[1 + H\cot(-H)] \tag{2}$$

where $\tau = t_2 - t_1$ and $(y_1,\dot{y}_1,\dot{y}_2,y_2)$ may be regarded as the components of a normal vector process $\mathbf{u}$, in that order, for which

$$\mathbf{K_u} = [\overline{u_iu_j}] = \begin{bmatrix} b_0 & 0 & \dot{K}_y(\tau) & K_y(\tau) \\ 0 & b_2 & -\ddot{K}_y(\tau) & -\dot{K}_y(\tau) \\ \dot{K}_y(\tau) & -\ddot{K}_y(\tau) & b_2 & 0 \\ K_y(\tau) & -K_y(\tau) & 0 & b_0 \end{bmatrix} \qquad \mathbf{K_u}^{-1} = \left[\frac{\mu_{ij}}{|\mu|}\right] \tag{3}$$

and $H \equiv (\mu_{23}\mu_{22}{}^2 - \mu_{23}{}^2)^{-1/2}$.

(*b*) Show that the probability (density) that y passes through zero at time τ ($= t_2 - t_1$), when y passes through zero at $\tau = 0$ ($t = t_1$), is

$$W_y(0,0;\tau) = \frac{1}{\pi}\sqrt{\frac{b_0}{b_2}}\frac{\mu_{23}}{H}[K_y(0)^2 - K_y{}^2(\tau)]^{-3/2}(1 + H\tan^{-1}H) \qquad |\tan^{-1}H| \leq \frac{\pi}{2} \tag{4}$$

S. O. Rice,[22] sec. 3.4.

REFERENCES

1. Rice, S. O.: Mathematical Analysis of Random Noise. II, *Bell System Tech. J.*, **24:** 46 (1945).
2. Uhlenbeck, G. E.: Theory of Random Process, *MIT Radiation Lab. Rept.* 454, Oct. 15, 1943.
3. Middleton, D.: Some General Results in the Theory of Noise through Non Linear Devices, *Quart. Appl. Math.*, **5:** 445 (1948).
4. Watson, G. N.: "Theory of Bessel Functions," 2d ed., Cambridge, New York, 1944.
5. MacDonald, D. K. C.: Some Statistical Properties of Random Noise, *Proc. Cambridge Phil. Soc.*, **45:** 368 (1949).
6. Bunimovitch, V. I.: Fluctuating Processes as Oscillations with Random Amplitudes and Phases, *Zhur. Tekh. Fiz.*, **19:** 1231 (1949).
7. Huggins, W. H., and D. Middleton: A Comparison of the Phase and Amplitude Principles in Signal Detection, *Proc. Natl. Electronics Conf.*, vol. 11, p. 304, October, 1955.
8. Van Vleck, J. H.: The Spectrum of Clipped Noise, *Harvard RRL Rept.* 51, July 21, 1943. See also J. L. Lawson and G. E. Uhlenbeck, "Threshold Signals," pp. 56–59, Massachusetts Institute of Technology Radiation Laboratory Series, vol. 24, McGraw-Hill, New York, 1950.
9. Ref. 3, eq. (7-15), with $l = 1$, $\nu = 0$; see also Sec. 13.4-2 in this book.
10. Davenport, W. B., Jr.: Signal-to-Noise Ratios in Band-pass Limiters, *J. Appl. Phys.*, **24:** 720 (1953).
11. Price, R.: A Note on the Envelope and Phase-modulated Components of Narrow-band Gaussian Noise, *IRE Trans. on Inform. Theory*, **IT-1**(2): 9 (September, 1955).
12. Bunimovitch, V. I.: "Fluctuation Processes in Radio Receivers," N. A. Shorin (ed.), Moscow, 1951.

13. Bussgang, J. J.: Cross Correlation Functions of Amplitude Distorted Gaussian Signals, *MIT Research Lab. Electronics Tech. Rept.* 216, March, 1952.
14. Barrett, J. F., and D. G. Lampard: An Expansion for Some Second-order Probability Distributions and Its Application to Noise Problems, *IRE Trans. on Inform. Theory*, **IT-1:** 10 (1955).
15. Zadeh, L. A.: On the Representation of Non Linear Operators, *IRE WESCON Conv. Record*, pt. 2, p. 105, 1957.
16. Price, R.: A Useful Theorem for Nonlinear Devices Having Gaussian Inputs, *IRE Trans. on Inform. Theory*, **IT-4:** 69 (1958).
17. Brown, J. L., Jr.: On a Cross-correlation Property for Stationary Random Processes, *IRE Trans. on Inform. Theory*, **IT-3:** 28 (1957).
18. Leipnik, R.: The Effect of Instantaneous Nonlinear Devices on Cross-correlation, *IRE Trans. on Inform. Theory*, **IT-4:** 73 (1958).
19. Miller, K. S., R. I. Bernstein, and L. E. Blumenson: Generalized Rayleigh Processes, *Quart. Appl. Math.*, **16:** 151 (1958).
20. Campbell, L. L.: On the Use of Hermite Expansions in Noise Problems, *J. Soc. Ind. Appl. Math.*, **5:** 244 (1957).
21. Rice, S. O.: Statistical Properties of a Sine Wave plus Random Noise, *Bell System Tech. J.*, **27:** 109 (1948).
22. Ref. 1, secs. 3.3, 3.4, 3.6, 3.8.
23. Middleton, D.: Spurious Signals Caused by Noise in Triggered Circuits, *J. Appl. Phys.*, **19:** 817 (1948). Some additional references to the "zeros" problem are given in Refs. 24, 28.
24. Kohlenberg, A.: Notes on the Zero Distribution of Gaussian Noise, *MIT Lincoln Lab. Tech. Memo.* 44, Oct. 16, 1953.
25. Steinberg, H., P. M. Schultheiss, C. A. Wogrin, and F. Zweig: Short-time Frequency Measurement of Narrow-band Random Signals by Means of a Zero-counting Process, *J. Appl. Phys.*, **26:** 195 (1955).
26. Page, L.: "Introduction to Theoretical Physics," 2d ed., Van Nostrand, Princeton, N.J., 1935, secs. 38, 40, for example.
27. Miller, I., and J. E. Freund: Some Results on the Analysis of Random Signals by Means of a Cut-counting Process, *J. Appl. Phys.*, **27:** 1290 (1956).
28. White, G. M.: An Experimental System for Studying the Zeros of Noise, *Cruft Lab. Tech. Rept.* 261, May 15, 1957, and Experimental Determination of the Zero-crossing Distribution $W(N;T)$, *Cruft Lab. Tech. Rept.* 265, September, **1957**. See also Zeros of Gaussian Noise, *J. Appl. Phys.*, **29:** 722 (1958).
29. Darling, D. A., and A. J. F. Siegert: A Systematic Approach to a Class of Problems in the Theory of Noise and Other Random Phenomena, *RAND Papers* P-730, p. 738, Sept. 1, Sept. 10, 1955. Other references are included. See also *IRE Trans. on Inform. Theory*, **IT-3:** 32–37 (March, 1957).
30. Siegert, A. J. F.: Passage of Stationary Processes through Linear and Non Linear Devices, *IRE Trans. on Inform. Theory*, **PGIT-3:** 4 (March, 1954), pp. 20–25 in particular. Also, On the First Passage Time Probability Function, *Phys. Rev.*, **81:** 617 (1951).
31. Siegert, A. J. F.: A Systematic Approach to a Class of Problems in the Theory of Noise and Other Random Phenomena. Part II, Examples, *IRE Trans. on Inform. Theory*, **IT-3:** 38 (March, 1957).
32. Helstrom, C. W.: The Distribution of the Number of Crossings of a Gaussian Stochastic Processes, *IRE Trans. on Inform. Theory*, **IT-3:** 232 (December, 1957).
33. Bartlett, M. S.: "An Introduction to Stochastic Processes," chap. 3, Cambridge, New York, 1955.
34. McFadden, J. A.: The Axis-crossing Intervals of Random Functions. I, II, *IRE Trans. on Inform. Theory*, **IT-2:** 146 (1956), **IT-4:** 14 (1958).

CHAPTER 10

THE EQUATIONS OF LANGEVIN, FOKKER-PLANCK, AND BOLTZMANN

Instead of determining the distribution densities $W_1, W_2, \ldots, W_n, \ldots$ for our random process directly by postulating certain statistical properties at the outset, as was done in Chaps. 7 to 9 for normal and related processes, we can often attain the same results from a physically more fundamental viewpoint by studying the *differential equations* which govern the process itself. Our postulational approach above yields specific distribution densities W_1, W_2, etc., more or less directly and is therefore better suited to problems of noise or signals and noise in nonlinear systems (cf. Chaps. 5, 12 to 15). On the other hand, an approach based on the differential equations of the physical process is, perhaps, a more natural starting point for the study of noise in linear systems (and for the mathematically identical theory of the Brownian motion).[1,2] We shall assume in what follows that the random processes to be discussed are all stationary Markoff processes (cf. Sec. 1.4). These restrictions, however, do not seriously limit the applications and under a number of additional conditions (to be stated presently) lead to the gaussian random processes which play such a central role in statistical communication theory. For a comprehensive discussion of stochastic differential equations in their own right, we refer the reader to any of the standard works.[3,4]

10.1 *Formulation in Terms of a Stochastic Differential Equation: The Langevin Equation*

Our present approach starts explicitly from an appropriate differential equation. As an example, let us consider the current $I^{(j)}(t)$ in the simple linear network of Fig. 10.1a, where the applied emf is a thermal noise voltage $V^{(j)}(t)$ arising in the resistance. Now, let us suppose that we have an ensemble of such systems, identical in L and R and operating under identical conditions. Then, the circuit equations for each $V^{(j)}$ and $I^{(j)}$ may be represented as an ensemble of differential equations, viz.,†

$$L\frac{dI}{dt} + RI = V(t) \tag{10.1}$$

† Mathematically, the derivative of the stochastic variable I in Eq. (10.1) is to be interpreted in the usual mean-square sense (cf. Sec. 2.1).

The $V^{(j)}(t)$ are random voltage waves with the ensemble spectral density† $\mathcal{W}_v(f) = W_0$.

An analogous expression is encountered in the theory of the Brownian movement when we consider a free (heavy) particle of mass m moving with velocity $\mathbf{v}$ in a viscous medium.[1,2] The corresponding ensemble of equations of motion is

$$m \frac{d\mathbf{v}}{dt} + \gamma \mathbf{v} = \mathbf{K}(t) \tag{10.2}$$

where $\mathbf{K}(t)$ is a random force acting on the heavy particle, owing to its collision with the (much) lighter particles of the supporting medium, and γ is a constant, representing the effect of the (mean) dynamical friction

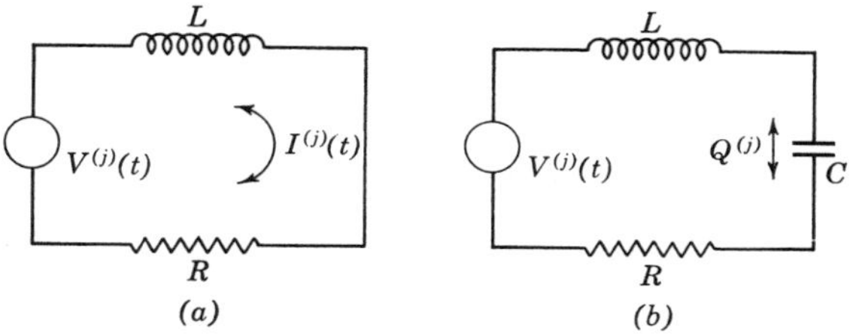

FIG. 10.1. (a) Series circuit for a resistance and inductance, with a thermal noise voltage. (b) Series LCR circuit driven by a thermal noise voltage $V^{(j)}(t)$.

which opposes the motion of the particle through the medium. Equation (10.2) is known as *Langevin's equation*.[6] Comparing Eqs. (10.1) and (10.2), we see that both can be expressed in terms of general first-order equations of the Langevin type

$$\frac{dy}{dt} + \beta y = F(t) \tag{10.3}$$

when β and $F(t)$ are appropriately defined, so that dy/dt exists (mean square) (cf. Sec. 2.1).

Similarly, we may have second-order differential equations of the Langevin type. A more general example is provided by the movement of charge Q in an ensemble of circuits containing a resistance R, inductance L, and capacitance C in series (cf. Fig. 10.1*b*) or by particles in Brownian movement subject to harmonic restoring forces. The fundamental differential equations are now

$$L \frac{d^2Q}{dt^2} + R \frac{dQ}{dt} + \frac{Q}{C} = V(t) \tag{10.4a}$$

and

$$m \frac{d^2\mathbf{r}}{dt^2} + \gamma \frac{d\mathbf{r}}{dt} + \mathcal{K}\mathbf{r} = \mathbf{K}(t) \tag{10.4b}$$

† As shown in Secs. 11.1-2, 11.1-3, in a state of thermal equilibrium the spectral density W_0 is equal to $4k_BT_0R$, where k_B is Boltzmann's constant (ergs per degree) and T_0 represents the absolute temperature of the resistance.

respectively, and the generic Langevin equation is of the type

$$\frac{d^2y}{dt^2} + \beta \frac{dy}{dt} + \omega_0{}^2 y = F(t) \tag{10.5}$$

with appropriate interpretations of β, $\omega_0{}^2$, and $F(t)$, where now both dy/dt and d^2y/dt^2 exist (mean square) (cf. Sec. 2.1). One can go on in this fashion to consider higher-order differential equations and more than one dependent variable. Nonlinear equations like

$$\frac{dy}{dt} + G(y) = F(t)H(y) \qquad \text{or} \qquad \frac{d^2y}{dt^2} + g(y)\left(\frac{dy}{dt}\right)^2 + h(y) = F(t)J(y), \text{ etc.} \tag{10.6}$$

can also be included in the discussion; in each instance, one has the corresponding generalized versions of the Langevin type.

The stochastic differential equations given above [Eqs. (10.1) to (10.6)], however, are fundamentally different from the differential equations that govern individual members of the ensemble. The distinction arises because $y(t)$ is now a stochastic variable, representing an ensemble of variables $y^{(j)}(t)$, since the $F^{(j)}(t)$ are random functions whose properties are statistical ones only. Accordingly, "solving" stochastic differential equations of the above types is ultimately significant only in an ensemble sense: the "solutions" that one can obtain are expressed as probability distributions (or densities) of the random variable $y(t)$. These probability densities depend on the original set of differential equations, their initial conditions, and the statistical structure of the driving, or "force," term $F(t)$. Because the random processes here are assumed to be stationary, one can say at once that the boundary conditions imply a steady-state solution: the systems in question have been operating long enough (ideally, for an infinite time) for all transient effects to disappear. On the other hand, for nonstationary systems the "initial" time is not arbitrary but is determined by the instant at which the ensemble of (similarly prepared) systems is set into movement; transients must then be included and may be quite significant.

The formal method of solving stochastic differential equations like Eq. (10.3) can now be described more precisely; i.e., we seek the probability density of $y(t)$, governing the probability that y lies in the range $(y, y + dy)$ at a time t, measured from some arbitrary origin $t = 0$, if at this time y has the value y_0. The desired result is therefore a second-order *conditional*† probability density $W_2(y_0|y;t)$. This is the direct consequence of the initial assumption that the random process is Markovian.‡ We

† See Sec. 1.4-1.

‡ If the process is not Markovian, one requires a solution in terms of some higher-order conditional distribution W_3, W_4, etc., and, instead of a single independent variable t representing the time difference between observations, one has two or more such time differences (t_2, t_3, etc.) as independent variables.

may expect that the statistical properties of y obeying Eq. (10.3) are completely described by $W_2(y_0|y;t)$, for $\lim_{t\to 0} W_2(y_0|y;t) = \delta(y - y_0)$, which clearly satisfies the initial condition that $y = y_0$ with certainty, i.e., with probability 1. For second-order equations of the Langevin type [Eqs. (10.4), (10.5)], the desired solution is of the form $W_2(y_0,\dot{y}_0|y,\dot{y};t)$, where the usual abbreviation for the (total) time derivative is introduced: $dy/dt \equiv \dot{y}$, $dy/dt\Big|_{y=y_0} \equiv \dot{y}_0$, etc. There are now two initial conditions, of course. At time $t = 0$, $y = y_0$, and $\dot{y} = \dot{y}_0$, and our solution $W_2(y_0,\dot{y}_0|y,\dot{y};t)$ accordingly must reduce† in the limit $t \to 0$ to $\delta(y - y_0)\,\delta(\dot{y} - \dot{y}_0)$. In general, the higher-order Langevin equations require solutions of the form

$$W_2(y_{10},y_{20}, \ldots |y_1,y_2, \ldots ;t)$$

where, for example, y_1 may represent position; $y_2 = \dot{y}$, a velocity; $y_3 = z$, another coordinate; and so on. The initial conditions in any case are always such that

$$\lim_{t\to 0} W_2(y_{10},y_{20}, \ldots |y_1,y_2, \ldots ;t) = \delta(y_1 - y_{10})\,\delta(y_2 - y_{20}) \cdots \tag{10.7}$$

Some specific instances of this kind are examined in Chap. 11 when the circuit equations of a linear network of n meshes with thermal noise sources are discussed.

To solve the general equation of Langevin therefore requires that we shift our attention from the differential equation governing the behavior of a particular member $y^{(j)}(t)$ of the ensemble to the ensemble $y(t)$ itself and the associated ensemble of differential equations. The conditional probability density $W_2(y_{10},y_{20}, \ldots |y_1,y_2, \ldots ;t)$ is now the desired solution, rather than any one $y^{(j)}(t)$. How is this transition from a physical formulation in terms of an arbitrary member of the ensemble to the statistical solution to be accomplished? It is instructive to begin first by considering a *discrete* model of the continuous process as an example of the random walk (cf. Chap. 7). In this form, $W_2(\ \ |\ \ )$ is now a probability, rather than a probability density. As we have seen from Sec. 1.4, W_2 must satisfy the Smoluchowski (or Chapman-Kolmogoroff) equation. With its help, the basic *difference* equation for the resulting random series can next be derived. Passing in suitable fashion to the limit of infinitesimal times and displacements, one obtains finally a partial differential equation‡ which is satisfied by W_2, and W_2 again becomes in the limit a probability density. We must now relate this differential equation in W_2 to the original Langevin equa-

† For the stationary processes assumed here, $\overline{y_1\dot{y}_1} = \frac{1}{2}(d/dt_1)\overline{y_1^2} = 0$, and similarly for the higher moments, so that y and $\dot{y}$ are statistically independent at all $t = t_1$.

‡ In its general form, the equation of Smoluchowski (which is really a form of Boltzmann's equation; cf. Sec. 10.3-3) is an integrodifferential relation, which reduces to a differential equation only if in small times one has correspondingly small displacements. For a discussion, see Secs. 10.3, 10.4.

tion, which is the more fundamental expression and on which, for example, the theory of noise in linear systems and the mathematically identical theory of the Brownian motion are based. This connection is achieved through the fundamental assumptions on the statistical properties of the driving term $F(t)$, which appears originally in the Langevin equation [cf. (10.3), (10.5), (10.6)] and which, through the Langevin equation, therefore controls the statistical properties of the random variable y whose distribution we seek. Note that W_2 in the limit $t \to \infty$ [namely, $\lim_{t\to\infty} W_2(y_0|y;t) = W_1(y)$] depends on the original Langevin equation and hence on the *postulated* properties of the random force term $F(t)$, whereas $\lim_{t\to 0} W_2$ must always be given by Eq. (10.7), quite independently of the particular physical model.

10.2 *Some Examples Leading to a Diffusion Equation*

A discrete model of the random process to which $y^{(j)}$ belongs may be constructed by treating $y^{(j)}$ as the sum of a very large number of independent random effects. Specifically, we have $y^{(j)}(t) = \sum_k^N x_k^{(j)}$, where N is a very large number, from the assumption that $y^{(j)}$ is the cumulative effect of many elementary contributions. Accordingly, $y^{(j)} = \sum_k^N x_k^{(j)}$ can be described by a random walk (cf. Sec. 7.7-1), where for the moment the precise distributions of the x_k are left unspecified, save that $\overline{y^{(j)}} = \sum_k^N \overline{x_k^{(j)}}$ approaches a finite nonzero or zero value as time unfolds (i.e., as N increases; see the remarks in Sec. 7.7-1). The principal advantage of the discrete approach, at least in the building of a model, is that many of the conceptual difficulties inherent in a continuous treatment are avoided, and from the general physical point of view the discrete phenomenological model is also often closer to reality. Whatever the $y^{(j)}(t)$ may represent—the displacement of an electron beam on an oscilloscope screen, the path of a particle in Brownian motion, the temperature of a gas, etc.—the precise nomenclature is not too important. It is sufficient to emphasize the basically discrete character of the physical process, and hence of the preliminary model, without inquiring too profoundly into the exact mechanism. For the problems of interest to us here, after results are obtained for the discrete case it is then usually possible to replace the discrete model by a continuous one, derived from it by appropriate limiting procedures.

Accordingly, let us again employ the concept of the random walk to obtain a discrete version of the random process. For the moment, a simple one-dimensional case is considered, in which the "particle" moves along the

y axis in such a way that in each step it goes a distance $k\Delta$ to the right or to the left, each step lasting a time τ; the transition from one position to the next is made instantaneously. The displacement in either direction depends in general on the location $n\Delta$ at any given instant $s\tau$. The conditional probability that the particle is at the position $y = n\Delta$ after a time $s\tau$ ($s = 0, 1, 2, \ldots$) if initially it was at the point $m\Delta$ is therefore $P_2(m\Delta,0|n\Delta;s\tau)$, which is conveniently abbreviated with the simpler notation $P_2(m|n;s)$. Because we have assumed a stationary Markoff process, P_2 must also satisfy the Smoluchowski (or Chapman-Kolmogoroff) equation of "continuity of probability" [Eq. (1.103b)]. However, instead of the continuous expression

$$W_2(y_0|y;t) = \int W_2(y_0|z;t_0)W_2(z|y, t - t_0)\, dz \qquad 0 \le t_0 \le t \tag{10.8}$$

for the probability density, the probability P_2 obeys the discrete analogue of Eq. (10.8), namely,

$$P_2(m|n; s + 1) = \sum_k P_2(m|k;s)P_2(k|n;1) \tag{10.9}$$

where for the moment the total number of allowed positions is restricted, to ensure stationarity.† The time interval changes by the unit τ, instead of varying continuously from 0 to t; the coordinates y_0, z, and y also vary discretely, by units of length Δ, so that in place of y_0, z, y one has $m\Delta$, $k\Delta$, and $n\Delta$, respectively. Finally, the integral is replaced by a summation over all allowed positions of the particle.[7,8] To emphasize the fact that $P_2(k|n;1)$ is the fundamental probability for the particular random series under examination, we rewrite Eq. (10.9), replacing $P_2(k|n;1)$ by $\Omega(k,n)$; Ω is a *transition probability*, determined by the underlying physical mechanism of the process. We have therefore

$$P_2(m|n; s + 1) = \sum_k P_2(m|k;s)\Omega(k,n) \tag{10.10a}$$

with the further condition that

$$\sum_n \Omega(k,n) = 1 \tag{10.10b}$$

which is simply an expression of the fact that the total probability that there is a transition, allowing for all possible transitions, must be unity.‡

10.2-1 Random Walk of a Free Particle. Here the particle can move freely by unit steps in either direction along the y axis, so that the probability p of a displacement to the right and the probability q ($= 1 - p$) of a

† For a development of the theory with the help of matrix methods, see Bartlett,[5] sec. 2.2.

‡ For the present, it is assumed that the particle cannot remain at a given position for more than τ sec; i.e., a transition to another point must occur at the end of the τ interval. See Prob. 10.3.

step to the left are equal. The transition probability $\Omega(k,n)$ is then

$$\Omega(k,n) = p\delta_n^{k+1} + q\delta_n^{k-1} = \tfrac{1}{2}(\delta_n^{k+1} + \delta_n^{k-1}) \qquad (10.11)$$

and n therefore locates the position arrived at after a unit displacement from the previous position; δ_j^l is the Kronecker delta, with the properties that $\delta_j^l = 0$ $(j \neq l)$, $\delta_j^l = 1$ $(j = l)$. It is clear that Eq. (10.11) satisfies the condition (10.10*b*) that $\sum_k \Omega(k,n)$ give a total transition probability of unity. Figure 10.2 shows a typical free random walk: in (*a*), the duration τ of each step is comparatively long and each step Δ is large, while, in (*b*), the motion is so fine-grained as to appear continuous.

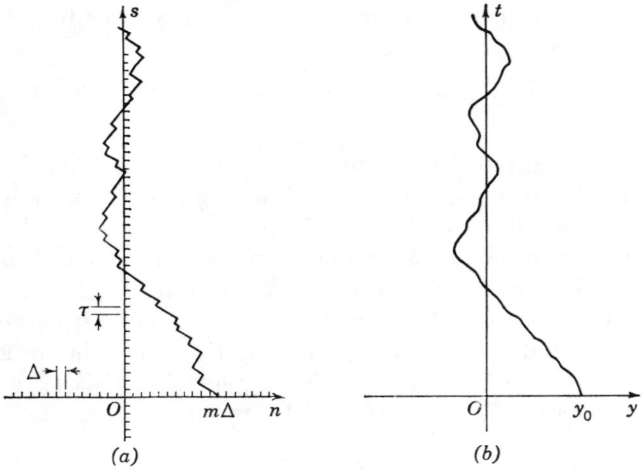

FIG. 10.2. One-dimensional random walk of a free particle.

If now we substitute Eq. (10.11) into the Smoluchowski equation (10.10*a*), we get at once the desired *difference* equation for the conditional probability P_2, namely,

$$P_2(m|n; s+1) = \tfrac{1}{2}P_2(m|n-1; s) + \tfrac{1}{2}P_2(m|n+1; s) \qquad (10.12)$$

with the initial condition $P_2(m|n;0) = \delta_n^m$. The solution of this difference equation is easily shown to be[7] the binomial distribution

$$P_2(m|n;s) = \frac{2^{-s}s!}{[(s+l)/2]![(s-l)/2]!} \quad l \equiv |n-m| \quad (l + s = \text{even}) \qquad (10.13)$$

To clarify the passage to a differential equation, it is helpful to rewrite Eq. (10.12) in greater detail:

$$P_2[m\Delta|n\Delta; (s+1)\tau] = \tfrac{1}{2}P_2[m\Delta|(n-1)\Delta; s\tau] + \tfrac{1}{2}P_2[m\Delta|(n+1)\Delta; s\tau] \qquad (10.14)$$

Subtracting $P_2(m\Delta|n\Delta;s\tau)$ from both members and dividing by τ, one has

$$\frac{P_2[m\Delta|n\Delta;(s+1)\tau] - P_2(m\Delta|n\Delta;s\tau)}{\tau}$$
$$= \frac{\Delta^2}{2\tau}\{P_2[m\Delta|(n+1)\Delta;s\tau] - 2P_2(m\Delta|n\Delta;s\tau) + P_2[m\Delta|(n-1)\Delta;s\tau]\}\Delta^{-2} \tag{10.15}$$

Formally, in the limits $\tau \to 0$, $\Delta \to 0$, where $m\Delta \to y_0$, $n\Delta \to y$, $s\tau \to t$, $P_2 \to W_2$, this becomes a one-dimensional example of the well-known diffusion equation, familiar in the theory of heat, namely,

$$\frac{\partial W_2}{\partial t} = D\frac{\partial^2 W_2}{\partial y^2} \tag{10.16}$$

where $W_2 = W_2(y_0|y;t)$ and $D \equiv \lim_{\Delta,\tau\to 0} \Delta^2/\tau$ are assumed to exist. The probability $P_2(m|n;s)$ has gone over into the probability *density* $W_2(y_0|y;t)$, and the initial condition $P_2(m|n;0) = \delta_n^m$ has become

$$\lim_{\Delta,\tau\to 0} P_2(m\Delta|n\Delta;0) = \lim_{t\to 0} W_2(y_0|y;t) = \delta(y - y_0) \tag{10.16a}$$

as expected.

In physical problems, the *diffusion coefficient* D must be finite, so that one finds in the limits $\Delta, \tau \to 0$ that the velocity Δ/τ of our particle becomes infinite. This is not surprising, since in the discrete case the particle was required to change position instantaneously. Requiring D to be finite is a reflection of the fact that for the continuous examples such instantaneous transitions must now occur gradually, with finite velocity: the particle no longer spends a finite time τ in each position but moves more or less smoothly from point to point. From the macroscopic point of view, when the random series is so fine-grained as to appear as a continuous random process, Eq. (10.16) is the one naturally chosen to describe the statistical properties of the process.

10.2-2 Random Walk with a Harmonic Restoring Force. Here the particle is no longer allowed to move freely but is subject to an elastic restoring force. Such a force is directly proportional to the distance of the particle from some force center (here, at $m = 0$). Therefore, if the particle is at the point $k\Delta$, the probability of moving a step Δ to the right or left is given by

$$p = \frac{1}{2}\left(1 - \frac{k}{K}\right) \quad \text{with } q = 1 - p = \frac{1}{2}\left(1 + \frac{k}{K}\right) \quad (-K \le k \le K) \tag{10.17}$$

K is (for the moment) a fixed positive integer; the possible positions of the particle are accordingly restricted to the interval $(-K\Delta \le k\Delta \le K\Delta)$.

Figure 10.3, like Fig. 10.2, shows a typical history of the particle under these conditions. Unlike the free particle (*a*), which may wander more widely about the average position (here, $m = 0$), the bound particle stays much closer to the center of force at $m = 0$, and extended excursions away from this point are much rarer. [Observe, also, that now the probability governing a step to the left or right depends on the previous history of the

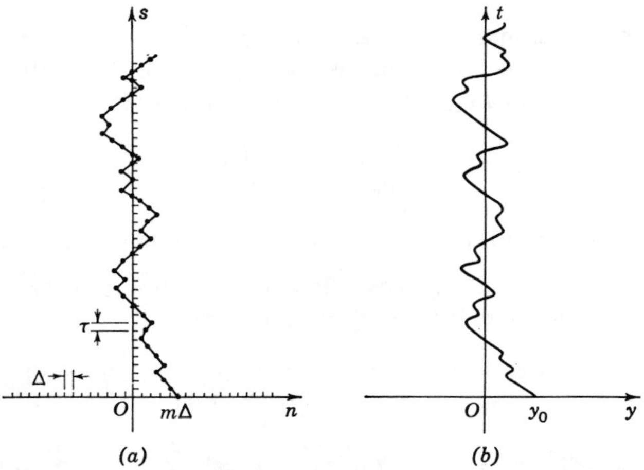

FIG. 10.3. One-dimensional random walk of a harmonically bound particle.

particle—through k in Eq. (10.17)—which was not the case for the free particle; cf. Eq. (10.11).] The basic transition probability here is

$$\Omega(k,n) = p(k)\delta_n^{k+1} + q(k)\delta_n^{k-1} = \frac{1}{2}\left(1 - \frac{k}{K}\right)\delta_n^{k+1} + \frac{1}{2}\left(1 + \frac{k}{K}\right)\delta_n^{k-1} \tag{10.18}$$

and the corresponding difference equation obtained from Eq. (10.10*a*) is

$$P_2(m|n; s+1) = \frac{K-n+1}{2K} P_2(m|n-1; s) + \frac{K+n+1}{2K} P_2(m|n+1; s) \tag{10.19}$$

with the initial condition $P_2(m|n;0) = \delta_n^m$ as before.

The solution of Eq. (10.19) is considerably more difficult to achieve than in the free-particle case [Eq. (10.13)] (see Kac[7] and Table 10.1 on page 460). In the limit $\Delta \to 0$, $\tau \to 0$, $K \to \infty$, we have formally $s\tau \to t$, $m\Delta \to y_0$, $n\Delta \to y$, in such a way that $\lim\limits_{\Delta,\tau\to 0} \Delta^2/2\tau \to D$, $1/K\tau \to \gamma$, and $P_2(m\Delta|n\Delta;s\tau) \to W_2(y_0|y;t)$ and the difference equation obtained from Eq. (10.19) reduces to the more general type of diffusion equation,

$$\frac{\partial W_2}{\partial t} = D\frac{\partial^2 W_2}{\partial y^2} + \gamma y \frac{\partial W_2}{\partial y} + \gamma W_2 = D\frac{\partial^2 W_2}{\partial y^2} + \gamma \frac{\partial}{\partial y}(yW_2) \tag{10.20}$$

The boundary condition $W_2(y_0|y;0)$ is again given by Eq. (10.16*a*).

The examples above show how appropriate models of the discrete random walk can be used to give, in the formal limit, differential equations governing the probability density W_2. Corresponding results for a particle in a constant field of force, and for a particle in a general force field, can be obtained in the same way (cf. Probs. 10.1, 10.2). The results are summarized in Table 10.1. The fourth relation in Table 10.1 was obtained by Smoluchowski[9] in his studies of the Brownian motion and is a special case of the still more general equation of the diffusion type derived by Fokker[10] and Planck[11] (cf. Sec. 10.3). As mentioned earlier (Sec. 10.1), that we obtain a diffusion equation here from our discrete models is a direct result of the assumption that the "particle" can alter its position within a time τ by at most a single step Δ, so that n can change by 0, ± 1 only (see Prob. 10.3). For the continuous cases, this assumption is expressed somewhat less precisely by the remark that in small intervals of time the spatial coordinates can change by small amounts only. In the next section, we shall see how this statement applies specifically to the equation of Fokker-Planck.

PROBLEMS

10.1 (*a*) Obtain the following difference equation for the one-dimensional motion of a particle in a constant force field, where the probability of a displacement to the right is $p = \frac{1}{2} - \beta\Delta$ ($\beta\Delta < \frac{1}{2}$), namely,

$$P_2(m|n; s+1) = (\tfrac{1}{2} - \beta\Delta)P_2(m|n-1; s) + (\tfrac{1}{2} + \beta\Delta)P_2(m|n+1; s) \qquad (1)$$

Hence, show that in the limits τ, $\Delta \to 0$ one has the corresponding diffusion equation

$$\frac{\partial W_2}{\partial t} = D\frac{\partial^2 W_2}{\partial y^2} + 4\beta D\frac{\partial W_2}{\partial y} \qquad P_2 \to W_2(y_0|y;t) \qquad (2)$$

(*b*) Show that

$$P(m|n;s) = \begin{cases} \dfrac{s!}{[(l+s)/2]![(s-l)/2]!}\left(\dfrac{1}{2} - \beta\Delta\right)^{(s+l)/2}\left(\dfrac{1}{2} + \beta\Delta\right)^{(s-l)/2} & s \geq 0 \\ 0 & \text{elsewhere} \end{cases} \qquad l = |n-m| \qquad (3)$$

is the solution of Eq. (1) for the random walk of a particle in a constant field of force.

10.2 If the particle of Prob. 10.1 is subject to an arbitrary force field, obtain the difference equation

$$P_2(m|n; s+1) = \frac{\Delta - \tau F(n-1)}{2\Delta}P_2(m|n-1; s) + \frac{\Delta + \tau F(n+1)}{2\Delta}P_2(m|n+1; s) \qquad (1)$$

[Observe that $F(k\Delta)$ has the dimensions of a velocity, and consequently $\lim_{\tau,\Delta\to 0} F(k\Delta)/\tau$ is an acceleration, or a force per unit mass.] Show in the limit Δ, $\tau \to 0$ that Eq. (1) becomes

$$\frac{\partial W_2}{\partial t} = D\frac{\partial^2 W_2}{\partial y^2} + \frac{\partial}{\partial y}[F(y)W_2] \qquad (2)$$

which is also in general a partial differential equation.

10.3 (*a*) Obtain the corresponding difference equation for P_2, for the models 1 to 3, Table 10.1, when there is a nonzero probability of a zero-step transition, as well as for unit steps $\pm\Delta$.

(*b*) Show for the models above [cf. Eqs. (10.12), (10.19)] that the same results are obtained in the limit of the continuous process when the particle can make a zero displacement, as well as steps of $\pm\Delta$, on any one transition.

A comprehensive discussion of discrete-random-walk models, with and without absorbing and reflecting barriers, may be found, for example, in Feller[8] (chaps. 14–16); see also Bartlett[5] and the supplementary references at the end of the volume.

10.3 *The Equation of Fokker-Planck and Its Relation to the Langevin Equation*

It can be observed from the examples of the preceding paragraphs that the conditional probability density $W_2(y_0|y;t)$, which completely describes the stationary Markoff process to which the $y^{(j)}(t)$ belong, obeys a partial differential equation of the diffusion type (see Table 10.1). A yet more general class of equation with the same basic assumptions can be constructed, which is needed in our subsequent discussion of thermal noise in a general linear network.

10.3-1 The Fokker-Planck Equation. We consider first a one-dimensional process and start with Smoluchowski's equation in its continuous form (1.103*b*). Since we are concerned with small changes t in time, the Smoluchowski relation (1.103*b*) may be written

$$W_2(y_0|y;\, t + \Delta t) = \int_{-\infty}^{\infty} W_2(y_0|z;t)W_2(z|y;\Delta t)\, dz \qquad 0 \le t \le t + \Delta t \quad (10.21)$$

This is the so-called *forward* equation, defining the random process† subject, of course, to the initial condition $W_2(y_0|y;0) = \delta(y - y_0)$ and our assumption of Markoff properties.

Let us now consider the integral

$$I \equiv \int_{-\infty}^{\infty} Q(y)\, \frac{\partial W_2(y_0|y;t)}{\partial t}\, dy \quad (10.22)$$

where $Q(y)$ is so chosen that it (and all its derivatives) vanish rapidly enough at $\pm\infty$ for I to converge satisfactorily. Formally replacing the differential coefficient $\partial/\partial t$ by the limit of a difference quotient allows us to write Eq. (10.22) with the help of Eq. (10.21) as

$$\begin{aligned} I &= \lim_{\Delta t \to 0} \int_{-\infty}^{\infty} dy\, Q(y)\, \frac{W_2(y_0|y;\, t+\Delta t) - W_2(y_0|y;t)}{\Delta t} \\ &= \lim_{\Delta t \to 0} \frac{1}{\Delta t}\left[\int_{-\infty}^{\infty} dy\, Q(y) \int_{-\infty}^{\infty} W_2(y_0|z;t)W_2(z|y;\Delta t)\, dz \right. \\ &\qquad \left. - \int_{-\infty}^{\infty} dz\, Q(z)W_2(y_0|z;t)\right] \end{aligned} \quad (10.23)$$

† For a discussion of the role of "forward" and "backward" equations for discrete stochastic processes, see Feller,[8] chap. 17, and in particular sec. 9. For the original derivations, A. Kolmogoroff, *Math. Ann.*, **104:** 415 (1931), should be consulted.

Interchanging the order of integration and expanding $Q(y)$ in a Taylor's series about the point z then gives for the double integral in Eq. (10.23)

$$\int_{-\infty}^{\infty} dz\, W_2(y_0|z;t) \sum_{n=0}^{\infty} \frac{Q^{(n)}(z)}{n!} \int_{-\infty}^{\infty} (y - z)^n W_2(z|y;\Delta t)\, dy$$

where $Q^{(n)}(z) \equiv d^nQ(z)/dz^n$.

Next, a set of moments $a_n(z;\Delta t)$ is defined by

$$a_n(z;\Delta t) \equiv \int_{-\infty}^{\infty} (y - z)^n W_2(z|y;\Delta t)\, dy \qquad (10.24)$$

where the $a_n(z;\Delta t)$ are specifically the moments of the change in the spatial coordinates (about the point z) corresponding to a small change in time Δt. [Note that Eq. (10.24) also includes the more general situations for which this change in position is not small in the short interval Δt.] It is clear from Eq. (10.24) that $a_0(z;\Delta t)$ is unity. We can now write the integral I [Eq. (10.23)] as

$$I = \sum_{n=1}^{\infty} \frac{1}{n!} \int_{-\infty}^{\infty} dz\, W_2(y_0|z;t) Q^{(n)}(z) A_n(z) \qquad (10.25a)$$

defining a new set of moments $A_n(z)$ by

$$\begin{aligned} A_n(z) &\equiv \lim_{\Delta t \to 0} \frac{a_n(z;\Delta t)}{\Delta t} \\ &= \lim_{\Delta t \to 0} \left[(\Delta t)^{-1} \int_{-\infty}^{\infty} (y - z)^n W_2(z|y;\Delta t)\, dy \right] \qquad n \geq 1 \qquad (10.25b) \end{aligned}$$

It is assumed for the present discussion that all $A_n(z)$ exist and are at most finite quantities, so that the series (10.25a) converges. If Eq. (10.25a) is now integrated by parts, and if we use the assumption that $Q^{(n)}(z)$ $(n \geq 0)$ vanishes sufficiently rapidly at $z = \pm\infty$ for us to set $Q(z)A_1(z)W_2\Big|_{-\infty}^{\infty}$, etc., equal to zero, we find on changing the variable of integration from z to y in the above that

$$\int_{-\infty}^{\infty} Q(y) \left\{ \frac{\partial W_2}{\partial t} - \sum_{n=1}^{\infty} \frac{(-1)^n}{n!} \frac{\partial^n}{\partial y^n} [A_n(y) W_2] \right\} dy = 0 \qquad (10.26)$$

Since $Q(y)$ is otherwise an arbitrary function (except for the above conditions on it and its derivatives at $z = \pm\infty$), the expression in the braces must vanish, leaving finally

$$\frac{\partial W_2(y_0|y;t)}{\partial t} = \sum_{n=1}^{\infty} \frac{(-1)^n}{n!} \frac{\partial^n}{\partial y^n} [A_n(y) W_2(y_0|y;t)] \qquad (10.27)$$

where as before W_2 satisfies the initial condition $W_2(y_0|y;0) = \delta(y - y_0)$. This, then, is a series form of the original Smoluchowski equation.

When the first and second moments a_1, a_2 [Eq. (10.24)] are $O(\Delta t)$, so that the limit (10.25*b*) exists, and when the third and higher moments a_n $(n \geq 3)$ vanish faster than $O(\Delta t)$ as $\Delta t \to 0$, all terms in Eq. (10.27) for $n \geq 3$ become zero and we are left with the one-dimensional *equation of Fokker-Planck*,[10,11]

$$\frac{\partial W_2}{\partial t} = -\frac{\partial}{\partial y}[A_1(y)W_2] + \frac{1}{2}\frac{\partial^2}{\partial y^2}[A_2(y)W_2] \tag{10.28}$$

The assumption that all the higher moments $a_n(z,\Delta t)$ $(n \geq 3)$ approach zero faster than $O(\Delta t)$ as $\Delta t \to 0$ is the analytical expression of the condition previously mentioned, that for random processes of this type the spatial coordinates can change in small intervals of time only by essentially equally small amounts. Thus, if y represents the displacement of a particle in Brownian motion, for example, this assumption means that in a very short time Δt the position of the particle can have altered only by a correspondingly small amount Δy. Higher-order effects $[O(\Delta y^2)]$ are ignorable in the limit. For situations where the random process describes thermal noise, for instance (cf. Chap. 11), the assumption that $A_1(y)$, $A_2(y)$ are finite, while $A_n(y)$ $(n \geq 3)$ vanishes, can be proved, as can also be done in the mathematically similar case of the Brownian movement, under rather nonrestrictive assumptions.

10.3-2 The Moments $A_n(y)$. There remains now to show how the Langevin equation can be applied to the general partial differential equation (10.27) for W_2 in order to determine the moments $A_n(y)$ [Eq. (10.25*b*)]. The appropriate Langevin equation here is Eq. (10.3), viz.,

$$\dot{y} + \beta y = F(t) \tag{10.3}$$

To obtain $A_n(y)$, we need, therefore, to express the moments of the change in the spatial coordinate in terms of the short interval of time Δt. If first we note that initially y has the value z and at a time Δt later $y = z + \Delta z$, where z is for the present regarded as a fixed quantity, then $dy = d(\Delta z)$ and therefore

$$\begin{aligned} A_n(y) &= \lim_{\Delta t \to 0} \left[(\Delta t)^{-1} \int_{-\infty}^{\infty} (y - z)^n W_2(z|y;\Delta t)\, dy \right] \\ &= \lim_{\Delta t \to 0} \frac{1}{\Delta t} \int_{-\infty}^{\infty} (\Delta z)^n W_2(z|z + \Delta z;\, \Delta t)\, d(\Delta z) \end{aligned} \tag{10.29a}$$

Since z is essentially arbitrary, $W_2(z|z + \Delta z;\, \Delta t)$ is equal to the first-order probability density $W_1(\Delta z;\Delta t)$, and the integral (10.29*a*) is simply the mean value of the displacement Δz, namely,

$$\overline{(\Delta z)^n} = \int_{-\infty}^{\infty} (\Delta z)^n W_1(\Delta z;\Delta t)\, d(\Delta z) = \overline{(\Delta y)^n} \tag{10.29b}$$

on a simple change of notation z back again to y. The moments $A_n(y)$ reduce finally to

$$A_n(y) = \lim_{\Delta t \to 0} \frac{\overline{(\Delta y)^n}}{\Delta t} = \lim_{\Delta t \to 0} \frac{1}{\Delta t} \int_{-\infty}^{\infty} (\Delta y)^n W_1(\Delta y; \Delta t)\, d(\Delta y) \qquad n \geq 1 \quad (10.30)$$

For a typical member $y^{(j)}(t)$ of the process, we next integrate the fundamental differential equation (10.3), between the times t and $t + \Delta t$, for which there is the corresponding change of $y^{(j)}$ to $y^{(j)} + \Delta y^{(j)}$,

$$\int_{y^{(j)}}^{y^{(j)}+\Delta y^{(j)}} dy = -\beta \int_t^{t+\Delta t} y^{(j)}\, dt + \int_t^{t+\Delta t} F^{(j)}(t)\, dt \qquad (10.31a)$$

or

$$\Delta y^{(j)} = -\beta y^{(j)}\, \Delta t + O(\Delta y^{(j)}\, \Delta t) + F^{(j)}(t)\, \Delta t + O(\Delta t^2) \qquad (10.31b)$$

Taking the ensemble averages over the incremental changes in position in both members of Eq. (10.31b), we get

$$\overline{\Delta y} = -\beta y\, \Delta t + \overline{O(\Delta y\, \Delta t)} + \overline{F^{(j)}(t)}\, \Delta t + O(\Delta t^2) \qquad (10.31c)$$

from which it follows that

$$A_1(y) = \lim_{\Delta t \to 0} \frac{\overline{\Delta y}}{\Delta t} = -\beta y + \overline{F^{(j)}(t)} \qquad (10.32)$$

The second moment $A_2(y)$ is obtained in the same way with the help of the basic Langevin equation. If the *product* of a pair of the relations (10.31) is formed and the ensemble average performed as before, one has

$$\overline{(\Delta y^{(j)})^2} = \beta^2 y^2 (\Delta t)^2 - 2\beta y \overline{F^{(j)}(t)} (\Delta t)^2 + \iint_t^{t+dt} \overline{F^{(j)}(t_1) F^{(j)}(t_2)}\, dt_1\, dt_2 + \overline{O[(\Delta t)^2\, (\Delta y)^2]} + \overline{O[\Delta y\, (\Delta t)^3]}$$

which in the limit $\Delta t \to 0$ gives for the second moment

$$A_2(y) = \lim_{\Delta t \to 0} \frac{\overline{(\Delta y)^2}}{\Delta t} = \lim_{\Delta t \to 0} \left[\frac{1}{\Delta t} \iint_t^{t+\Delta t} \overline{F^{(j)}(t_1) F^{(j)}(t_2)}\, dt_1\, dt_2 \right] \qquad (10.33)$$

$A_2(y)$ is observed to be independent of y. This is due to the fact that the basic Langevin equation is *linear*.

For the third and higher moments, one finds similarly

$$A_n(y) = \lim_{\Delta t \to 0} \frac{\overline{(\Delta y)^n}}{\Delta t} = \lim_{\Delta t \to 0} \left[\frac{1}{\Delta t} \int \cdots \int_t^{t+\Delta t} \overline{F^{(j)}(t_1) F^{(j)}(t_2) \cdots F^{(j)}(t_n)}\, dt_1 \cdots dt_n \right] \qquad n \geq 3 \quad (10.34)$$

provided, of course, that the limit exists. If the multiple integral is $O(\Delta t)$ at the least, a finite, nonvanishing value for $A_n(y)$ can be expected. If, on the other hand, the integral is $O[(\Delta t)^m]$ $(m \geq 2)$, all the higher moments drop out and the one-dimensional equation (10.28) of Fokker-Planck is the result, where the first and second moments $A_1(y)$ and $A_2(y)$ are given specifically by Eqs. (10.32) and (10.33).

The Fokker-Planck equation in two or more dimensions is easily derived in a similar way. Assuming as before that the random process under consideration is Markovian (now in two dimensions), we start with the corresponding (two-dimensional) Smoluchowski equation, analogous to Eq. (10.21), namely,

$$W_2(x_0, y_0|x, y; t + \Delta t) = \iint\limits_{-\infty}^{\infty} W_2(x_0,y_0|u,v;t) W_2(u,v|x,y;\Delta t)\, du\, dv \qquad 0 \leq t \leq t + \Delta t \quad (10.35)$$

and consider the integral [cf. Eq. (10.22)]

$$J \equiv \iint\limits_{-\infty}^{\infty} Q(x,y) \frac{\partial}{\partial t} W_2(x_0,y_0|x,y;t)\, dx\, dy \quad (10.36)$$

in which Q is now an arbitrary function of x and y, which with its derivatives vanishes sufficiently rapidly as $x, y \to \pm\infty$ to secure convergence of the integral and its series development. The derivative $\partial W_2/\partial t$ is next replaced by its equivalent as the limit of a difference quotient, $Q(x,y)$ is expanded in a *double* Taylor's series about the point (u,v), and with an appropriate integration by parts and change of notation one obtains finally

$$\frac{\partial W_2}{\partial t} = \sum_{m+n>0} \frac{(-1)^{m+n}}{m!n!} \frac{\partial^{m+n}}{\partial x^m\, \partial y^n} [A_{m,n}(x,y) W_2] \quad (10.37)$$

where $A_{m,n}$ is the joint m, nth moment, analogous to Eq. (10.25*b*), which, by the identical argument used to derive $A_n(y)$ in terms of the elementary displacement Δy [cf. Eq. (10.30)], one finds with a change of notation $(u \to x,\ v \to y)$ to be

$$A_{m,n}(x,y) = \lim_{\Delta t \to 0} \frac{\overline{(\Delta x)^m (\Delta y)^n}}{\Delta t} = \lim_{\Delta t \to 0} \frac{1}{\Delta t} \iint\limits_{-\infty}^{\infty} (\Delta x)^m (\Delta y)^n W_1(\Delta x, \Delta y; \Delta t)\, d(\Delta x)\, d(\Delta y) \quad m + n \geq 1 \quad (10.38)$$

Again, if only small displacements Δx, Δy can occur in short times Δt,

Eq. (10.37) reduces to a Fokker-Planck equation,

$$\frac{\partial W_2}{\partial t} = -\frac{\partial}{\partial x}(A_{10}W_2) - \frac{\partial}{\partial y}(A_{01}W_2) + \frac{1}{2!}\left[\frac{\partial^2}{\partial x^2}(A_{20}W_2) + 2\frac{\partial^2}{\partial x\,\partial y}(A_{11}W_2) + \frac{\partial^2}{\partial y^2}(A_{02}W_2)\right] \tag{10.39}$$

For the nondegenerate two-dimensional process [where at least a pair of moments $A_{m,n}$ $(m, n \neq 0)$ exist], we expect a second-order Langevin equation, from which the explicit form of the joint moments $A_{m,n}$ may be found. As an illustration, consider the general type, Eq. (10.5), with a linear "force" term, viz.,

$$\ddot{y} + \beta\dot{y} + \omega_0^2 y = F(t) \tag{10.5}$$

characteristic, for example, of an *LCR* series circuit driven by, say, a thermal noise voltage. It should be pointed out that now $y(t)$ alone does not belong to a Markoff process, but to a *projection* of a Markoff process [Eq. (1.105)] (cf. Sec. 1.4-5). From the theorem of Doob (Sec. 7.6), however, one expects that y and $\dot{y}$ together are Markovian. In terms of y and $\dot{y}$, Eq. (10.5) can be expressed as a pair of first-order relations,

$$\frac{d\dot{y}}{dt} + \beta\dot{y} + \omega_0^2 y = F(t) \qquad \frac{dy}{dt} = \dot{y} \tag{10.5a}$$

and from these two equations the moments $A_{m,n}$ of the Fokker-Planck equation (10.39) are obtained. Setting $x = \dot{y}$, $y = y$, one finds from the Langevin equation and Eq. (10.38), on taking the ensemble average of both members of Eq. (10.5*a*) as before [cf. Eqs. (10.31*c*) et seq.],

$$A_{10}(\dot{y},y) = \lim_{\Delta t\to 0}\frac{\overline{\Delta\dot{y}}}{\Delta t} = -\beta\dot{y} - \omega_0^2 y + \overline{F^{(j)}(t)} \tag{10.40a}$$

$$A_{01}(\dot{y},y) = \lim_{\Delta t\to 0}\frac{\overline{\Delta y}}{\Delta t} = \dot{y} \tag{10.40b}$$

$$A_{20}(\dot{y},y) = \lim_{\Delta t\to 0}\frac{\overline{(\Delta\dot{y})^2}}{\Delta t} = \lim_{\Delta t\to 0}\left[\frac{1}{\Delta t}\iint_t^{t+\Delta t}\overline{F^{(j)}(t_1)F^{(j)}(t_2)}\,dt_1\,dt_2\right] \tag{10.40c}$$

$$A_{11}(\dot{y},y) = \lim_{\Delta t\to 0}\frac{\overline{(\Delta\dot{y})(\Delta y)}}{\Delta t} = 0 \qquad A_{02}(\dot{y},y) = \lim_{\Delta t\to 0}\frac{\overline{(\Delta y)^2}}{\Delta t} = 0 \tag{10.40d}$$

The higher moments $A_{m,n}(\dot{y},y) = \lim_{\Delta t\to 0}[\overline{(\Delta\dot{y})^m(\Delta y)^n/\Delta t}]$ $(m + n \geq 3)$ are derived similarly and are found to vanish if $A_{m,n}$ is $O(\Delta t^k)$ $(k > m + n)$. For the extension to the general s-dimensional case, see Prob. 10.7.

10.3-3 The Equations of Boltzmann and Smoluchowski. We must remember that the equations of Fokker-Planck (10.28), (10.39) are obtained only if the third and higher-order moments of the mean displacement vs. time, for example, $A_n(y)$, $A_{m,n}(x,y)$, etc., vanish, i.e., only if for infinitesimal changes in time the spatial coordinates likewise alter by infinitesimal amounts of the same or higher order. When this is not so and comparatively large deviations in position follow upon small changes in time, these higher moments do not drop out in the limit $\Delta t \rightarrow 0$ and one is left with the general series development [Eqs. (10.27), (10.37)] of the Smoluchowski equation, which, as we shall note below, is simply one form of the integro-differential equation of Boltzmann[12] and which is always the ultimate starting point in the formulation of stochastic problems of this type.

The connection between the Smoluchowski and Boltzmann equations is readily established as follows: Let us start with a one-dimensional form of Boltzmann's equation,[13] namely,

$$\frac{\partial W_2(y_0|y;t)}{\partial t} = \int W_2(y_0|z;t)A(z,y)\,dz - W_2(y_0|y;t)\int A(y,z)\,dz \tag{10.41}$$

where $A(z,y)\,dz$ is the probability per unit time that, in the language of the random walk, particles in a region dz about a position z undergo transitions to a region dy about a new position y, analogous to the transition probability $\Omega(k,n)$ [cf. Eqs. (10.10)]. In a similar fashion, $A(y,z)\,dz$ represents the probability per unit time that particles in dy about position y move into the region dz about the location z. Equation (10.41) is simply a statement of the fact that the rate of change of population in this coordinate space (y,z) is the difference between the rate of departures from the element dz and the rate of arrivals.

From the form of Eq. (10.41), we suspect that the Smoluchowski equation (10.8) or (10.21) is essentially an integral form of the Boltzmann relation (10.41). To show this, let us follow the procedure used in Sec. 10.3-1 to derive the equation of Fokker-Planck (10.28). Multiplying both sides of Eq. (10.41) by $Q(y)$, as before, and carrying out the steps corresponding to Eqs. (10.22) to (10.27) in these earlier relations, we obtain as the series form of the Boltzmann equation

$$\frac{\partial W_2(y_0|y;t)}{\partial t} = \sum_{n=1}^{\infty} \frac{(-1)^n}{n!}\frac{\partial^n}{\partial y^n}[B_n(y)W_2(y_0|y;t)] \tag{10.42}$$

Now, with y fixed and equal to $z + \Delta z$, we have

$$B_n(y) \equiv \int (y - z)^n A(z,y)\,dy = \int (\Delta z)^n A(z, z + \Delta z)\,d(\Delta z) \tag{10.43}$$

But since z is fixed in this expression [cf. Eqs. (10.29*a*), (10.29*b*)], the probability per unit time, A, is by definition simply

$$\begin{aligned} A(z,y) = A(z, z + \Delta z) &\equiv \lim_{\Delta t \to 0} [W_2(z|z + \Delta z; \Delta t)/\Delta t] \\ &= \lim_{\Delta t \to 0} [W_1(\Delta z;\Delta t)/\Delta t] \end{aligned} \tag{10.44}$$

so that

$$B_n(y) = A_n(y) \tag{10.45}$$

[Eqs. (10.29*a*), (10.29*b*)] and thus the expanded form of Boltzmann's equation (10.42) is the same as the series (10.27) obtained above from the time derivative of the Smoluchowski equation. Solutions of the Smoluchowski equation are also solutions of the corresponding Boltzmann equation, and vice versa (cf. Prob. 10.9). Note that, if moments higher than the second are negligible in Eq. (10.42), we obtain once more the equation of Fokker-Planck, as derived here alternatively from Eq. (10.41).†

10.4 *The Gaussian Random Process*

It is easily seen from Eqs. (10.32), (10.33), and (10.34) that the moments $A_n(y)$ depend not only on the particular form of the basic Langevin equation for the process to which the $y^{(j)}(t)$ belong, but on the statistical properties of the "driving" terms $F^{(j)}(t)$ as well. We must therefore examine the conditions to be imposed on $F(t)$ if we wish y to have corresponding statistical properties and, specifically here, so that $y(t)$ will be a normal or gaussian random process (cf. Sec. 7.4).

10.4-1 Assumptions on the Force Term $F(t)$. First of all, it is clearly no restriction to require that

$$\overline{F(t)} = 0 \tag{10.46}$$

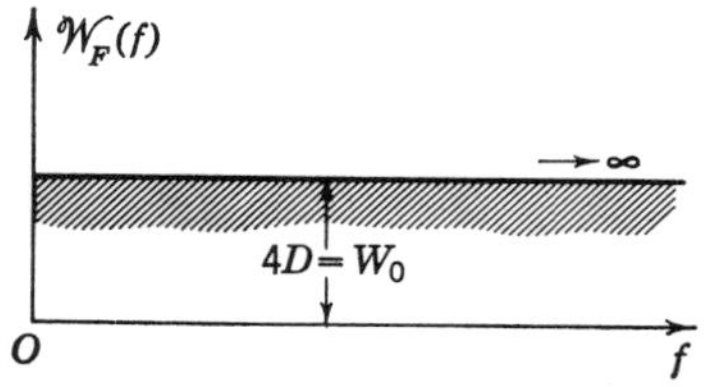

FIG. 10.4. Spectrum of the force term $F(t)$.

The assumptions on the higher moments $\overline{F(t_1)F(t_2)}$, $\overline{F(t_1)F(t_2)F(t_3)}$, etc., are more critical. The simplest useful conditions here [both in the theory of thermal noise (Chap. 11) and in the corresponding theory of the Brownian movement] require that $F(t)$ be at least wide-sense stationary and that the mean intensity spectrum of the process $F(t)$ be uniform for all frequencies, as shown in Fig. 10.4, with spectral density $W_0 = 4D$. The covariance function for $F(t)$ reduces then to

$$\overline{F(t_1)F(t_2)} = K_F(t_2 - t_1) = \int_0^\infty 4D \cos \omega(t_2 - t_1)\, df = 2D\, \delta(t_2 - t_1) \tag{10.47}$$

The reason for assuming a spectrum of the type shown in Fig. 10.4 is the fact that many of the primary noise sources (to be considered later in more detail) produce spectra which are essentially uniform up to frequencies that are well beyond the usual range of interest, so that the concept of a purely random primary process gives significant results, provided that in the final stages of the analysis one deals with but a small portion of the

† For a discussion of the approximative nature of the Fokker-Planck equation vis-à-vis an exact treatment based on Boltzmann's (or, equivalently, Smoluchowski's) equation, see, for example, recent work of Storer and Keilson[14] and also Probs. 10.9, 10.10.

idealized spectrum. This ensures that for a finite D ($\neq 0$) a finite amount of power is stored or dissipated in the system as far as the end result is concerned. In physical systems, of course, the mean total power is always finite—one never actually has a purely random process.†

It is not enough, however, to give the first and second moments of $F(t)$ if a solution of the corresponding Fokker-Planck equation is required. We must assume, in addition, knowledge of *all* the higher moments of all orders; these are needed to show that $A_n(y)$ ($n \geq 3$) drop out as $\Delta t \to 0$. Gaussian or normal statistics for $F(t)$ are the type of main importance in physical problems. We therefore assume, as the additional condition on $F(t)$ besides Eqs. (10.46) and (10.47), that $F(t)$ is a normal random process. This may be expressed in two equivalent ways. The first uses a Fourier series in the usual way (cf. Sec. 8.3) to represent the process $F(t)$ throughout the interval $(0,T)$, namely,

$$F(t) = \sum_{n=1}^{\infty} (a_n \cos \omega_n t + b_n \sin \omega_n t) \qquad \omega_n = 2\pi n/T \tag{10.48}$$

and the Fourier coefficients (a_n,b_n) are *postulated* to be independent‡ and normally distributed with the variance $\sigma_n^2 = 4D/T$, so that the probability density for the coefficients is

$$W(a_1, \ldots ,a_n, \ldots ;b_1, \ldots ,b_n, \ldots) = \prod_{n=1}^{\infty} \frac{1}{2\pi\sigma_n^2} e^{-(a_n^2+b_n^2)/2\sigma_n^2} \tag{10.49}$$

Applying Eq. (10.49), we easily see that

$$\bar{a}_n = \bar{b}_n = 0 \qquad \overline{a_n a_p} = \overline{b_n b_p} = \delta_n^p 4D/T \qquad \overline{a_n b_p} = 0 \tag{10.50}$$

The second method assumes, instead, the following properties[15] on $F(t)$ [cf. Eqs. (7.28)],

$$\begin{aligned} \overline{F(t_1)F(t_2) \cdots F(t_{2m+1})} &= 0 \\ \overline{F(t_1)F(t_2) \cdots F(t_{2m})} &= \sum_{\text{all pairs}} [\overline{F(t_i)F(t_j)}\ \overline{F(t_k)F(t_l)} \cdots \overline{F(t_p)F(t_q)}] \end{aligned} \tag{10.51}$$

where the sum is to be taken over all the different ways one can divide $2m$ points in time $t_1, \ldots , t_{2m}$ into m pairs, namely, $(2m)!/2^m m!$. It is not

† For spectral distributions of the driving term which are not uniform, the present approach cannot give us the d.d. of the process. Since we do not then in general have a Markoff process, unless the Langevin equation is of the appropriate order and the spectrum is rational, our procedure breaks down: we are dealing instead with a *projection* of a Markoff process (cf. Sec. 1.4-5). Our statistical information is incomplete, and consequently we cannot find the appropriate Fokker-Planck equation without additional information on the complete Langevin equation.

‡ This means that $F(t)$ is a periodic process on the interval $(0,T)$. See the discussion in Sec. 8.3, Eqs. (8.81) et seq.

difficult to show (cf. Prob. 10.4) that these two sets of assumptions on $F(t)$ are equivalent and therefore lead to identical results. The process y, of course, will also be normal, since the Langevin equation is *linear* and the Fourier coefficients are postulated to be gaussian (cf. Sec. 7.3).

10.4-2 Two Examples. Returning to the random wave $y(t)$, we consider as our first example the one-dimensional Langevin equation given by Eq. (10.3). We now use the gaussian properties of the driving term to determine the moments $A_n(y)$. Here, it is more convenient to use the alternative relations (10.51). With the help of Eq. (10.46), the first moment $A_1(y)$ [Eq. (10.32)] is immediately

$$A_1(y) = -\beta y \tag{10.52}$$

Equation (10.47) allows us to write the second moment $A_2(y)$ [Eq. (10.33)] as

$$\begin{aligned} A_2(y) &= \lim_{\Delta t \to 0} \left[\frac{1}{\Delta t} \iint_{t=t_1}^{t+\Delta t = t_2} 2D\, \delta(t_2 - t_1)\, dt_1\, dt_2 \right] \\ &= 2D \lim_{\Delta t \to 0} \frac{1}{\Delta t} \int_t^{t+\Delta t} dt = 2D \end{aligned} \tag{10.53}$$

The higher moments follow at once when Eqs. (10.51) are used and are

$$A_n(y) = 0 \qquad n = 3, 5, \ldots \tag{10.54a}$$

$$\begin{aligned} A_n(y) &= \lim_{\Delta t \to 0} \left\{ \frac{n!(\Delta t)^{-1}}{(n/2)!2^{n/2}} \left[\iint_t^{t+\Delta t} \overline{F(t_1)F(t_2)}\, dt_1\, dt_2 \right]^{n/2} \right\} \\ &= \lim_{\Delta t \to 0} \left[\frac{n!D}{(n/2)!} (D\,\Delta t)^{n/2-1} \right] = 0 \qquad n = 4, 6, \ldots \end{aligned} \tag{10.54b}$$

Accordingly, for the gaussian random process the third and higher moments all vanish, and so the Smoluchowski (or Boltzmann) equation reduces in the limit to the Fokker-Planck equation (10.20) of the harmonically bound particle ($\gamma = \beta$; $D = D$),

$$\frac{\partial W_2}{\partial t} = D \frac{\partial^2 W_2}{\partial y^2} + \beta \frac{\partial}{\partial y}(yW_2) \qquad W_2 = W_2(y_0|y;t) \tag{10.55}$$

with the associated Langevin equation $\dot{y} + \beta y = F(t)$.

Similarly as our second example, from Eqs. (10.40) we find for the two-dimensional random process, in y and $\dot{y}$, obeying the Langevin equation $\ddot{y} + \beta\dot{y} + \omega_0^2 y = F(t)$, that

$$\begin{aligned} A_{10}(\dot{y},y) &= -\beta\dot{y} - \omega_0^2 y \qquad & A_{01}(\dot{y},y) &= \dot{y} \\ A_{11}(\dot{y},y) &= A_{02}(\dot{y},y) = 0 \qquad & A_{20}(\dot{y},y) &= 2D \end{aligned} \tag{10.56}$$

provided that $F(t)$ is a gaussian random process. All the higher moments $A_{m,n}(\dot{y},y)$ $(m + n \geq 3)$ vanish [cf. Eq. (10.54b)], and the result is again an equation of the Fokker-Planck type, viz.,

$$\frac{\partial W_2}{\partial t} = \frac{\partial}{\partial \dot{y}} [(\beta\dot{y} + \omega_0{}^2 y)W_2] - \frac{\partial}{\partial y}(\dot{y}W_2) + D\frac{\partial^2 W_2}{\partial \dot{y}^2} \qquad W_2 = W_2(y_0,\dot{y}_0|y,\dot{y};t) \tag{10.57}$$

with the initial condition $W_2 = \delta(y - y_0)\,\delta(\dot{y} - \dot{y}_0)$ as $t \to 0$.

In much the same way, a Fokker-Planck equation also follows for the s-dimensional random process (of which $y_1, y_2, \ldots, y_s$ are the components). As before, the second-order moments are constants (cf. Prob. 10.7) since the Langevin equations from which they are derived are all *linear*. It is this last condition that ensures that the conditional probability density W_2 determines a gaussian Markoff process. One can also have a Fokker-Planck equation for a Markoff process where the force term is nonlinear, i.e., where the Langevin equation is of the type $\ddot{y} + \beta\dot{y} + K(y) = F(t)$, with $K(y) = y^2$, for example, but the statistical properties of y, $\dot{y}$ are then no longer normal.† An illustration of this in the case of the distribution of a rectified noise voltage after filtering is given in Sec. 2.1-3.

In the next chapter, a number of specific linear examples will be discussed of the effects of thermal noise in some simple electrical networks containing resistances, inductances, and capacitances. The appropriate Fokker-Planck equations are the direct result of the assumptions on $F(t)$. Note, however, that $F(t)$ need not be normal: it is sufficient only that the higher moments of the change in the spatial coordinates vanish at least $O(\Delta t)$ as $\Delta t \to 0$. Finally, it should be emphasized that the *physical* justification for the assumed gaussian properties in the theories of the Brownian motion and thermal noise stems from the Maxwell-Boltzmann law of velocity distribution[16] for the particle in question, namely,

$$\lim_{t\to\infty} W_2(\mathbf{v}_0|\mathbf{v};t) = W_1(\mathbf{v}) = (m/2\pi k_B T_0)^{3/2} e^{-m(\mathbf{v}\cdot\mathbf{v})/2k_B T_0} \tag{10.58}$$

where T_0 is the absolute temperature, k_B is Boltzmann's constant, and m is the mass of the particle. The Maxwell-Boltzmann law is always postulated here, not derived, just as the gaussian properties of the driving term $F(t)$ are always assumed and not derived.

10.4-3 A Solution of the One-dimensional Fokker-Planck Equation. There remains now to determine the solution of Eq. (10.55) for the conditional probability $W_2(y_0|y;t)$, subject to the initial condition $\lim_{t\to 0} W_2(y_0|y;t) = \delta(y - y_0)$. In terms of the Brownian-motion model, y represents the velocities of an ensemble of harmonically bound particles, while, in the electrical-circuit cases, y may be the currents in a set of series LR circuits excited by a thermal noise source (cf. Fig. 10.1a).

† The Langevin equation $\ddot{y} + \beta\dot{y} + K(y) = F(t)$ has been considered by Kramers [*Physica*, **7**: 284 (1940)] in the problem of particles escaping over a potential barrier. See also Chandrasekhar's article,[1] chap. II, sec. 4(vi), and chap. III, sec. 7.

We note that to solve Eq. (10.55) [or Eq. (10.20)] it is convenient to introduce first the Fourier transform of W_2, namely,

$$f_2(y_0|\xi_2;t) \equiv \mathfrak{F}\{W_2(y_0|y;t)\} \equiv \int_{-\infty}^{\infty} e^{iy\xi_2}W_2(y_0|y;t)\,dy \tag{10.59}$$

with $W_2(y_0|y;t) = \mathfrak{F}^{-1}\{f_2\}$. One finds immediately for $\partial W_2/\partial t$ and $\partial^2 W_2/\partial y^2$ appearing in Eq. (10.55) that

$$\frac{\partial W_2}{\partial t} = \mathfrak{F}^{-1}\left\{\frac{\partial f_2}{\partial t}\right\} \qquad \text{and} \qquad \frac{\partial^2 W_2}{\partial y^2} = \mathfrak{F}^{-1}\{-f_2\xi_2{}^2\}$$

and for $(\partial/\partial y)(yW_2)$ an integration by parts gives, after a little algebra, $\mathfrak{F}^{-1}\{-\xi_2(\partial f_2/\partial \xi_2)\}$, where it is assumed that $f_2(y_0|\pm\infty;t)$ goes to zero sufficiently rapidly. Applying these results to Eq. (10.55), we find that the Fourier transform of

$$\frac{\partial W_2}{\partial t} = D\,\frac{\partial^2 W_2}{\partial y^2} + \beta\,\frac{\partial}{\partial y}(yW_2)$$

is

$$\frac{\partial f_2}{\partial t} + \beta\xi_2\,\frac{\partial f_2}{\partial \xi_2} = -D\xi_2{}^2 f_2 \tag{10.60}$$

where by this artifice the second-order equation has been reduced to two first-order ones.

Equation (10.60) is a special case of the standard first-order partial differential equation

$$P(x,y;u)\,\frac{\partial u}{\partial x} + Q(x,y;u)\,\frac{\partial u}{\partial y} = R(x,y;u) \tag{10.61}$$

where $u = u(x,y)$ and P, Q, and R may be functions of both the dependent and independent variables.[17] The system of associated characteristic equations is

$$dx/P = dy/Q = du/R \tag{10.62}$$

Independent solutions of Eq. (10.61) are $\phi(x,y;u) = a$, $\psi(x,y;u) = b$ (a, b constants), and therefore $g(\phi,\psi) = 0$ is the general solution of Eq. (10.61), where g is some arbitrary function to be established by the specific structure of Eq. (10.61) and the boundary conditions. From Eq. (10.60) the characteristic equations are accordingly $\phi(t,\xi_2;f_2) = \xi_2 e^{-\beta t}$ and $\psi(t,\xi_2;f_2) = f_2 e^{\xi_2{}^2 D/2\beta}$. For the general solution $g(\phi,\psi) = 0$, we can write equivalently $\psi = H(\phi)$, and H is to be determined from the boundary conditions. This gives

$$f_2(y_0|\xi_2;t) = e^{-\xi_2{}^2D/2\beta}H(\xi_2 e^{-\beta t}) \tag{10.63}$$

and if we now apply the boundary condition $\lim_{t\to 0} W_2 = \delta(y - y_0)$ to Eq. (10.59), we find that

$$H(\xi_2)_{t=0} = e^{i\xi_2 y_0 + D\xi_2{}^2/2\beta} \tag{10.64a}$$

Table 10.1. Selected One-Dimensional Random-Walk Models†

Continuous case: Force term	Continuous case: Fokker-Planck and Langevin equations	Continuous case: Solution	Model of the random walk‡	Discrete case: Force term	Discrete case: Difference equation	Discrete case: Solution
0	$\frac{\partial W_2}{\partial t} = D \frac{\partial^2 W_2}{\partial y^2}$ $\dot{y} = F(t)$	$W_2(y_0 \mid y;t) = \frac{e^{-(y-\hat{y})^2/2\sigma^2}}{\sqrt{2\pi\sigma^2}}$ $\sigma^2 = 2Dt,\ \hat{y} = y_0$	1. Free particle	0	$P_2(m \mid n; s+1) = \frac{1}{2}P_2(m \mid n-1; s) + \frac{1}{2}P_2(m \mid n+1; s)$	$\frac{s!2^{-s}}{\left(\frac{s+\lvert n-m\rvert}{2}\right)!\left(\frac{s-\lvert n-m\rvert}{2}\right)!}$
$4\beta D$	$\frac{\partial W_2}{\partial t} = D \frac{\partial^2 W_2}{\partial y^2} + 4\beta D \frac{\partial W_2}{\partial y}$ $\dot{y} + 4\beta D = F(t)$	$W_2 = \frac{e^{-(y-\hat{y})^2/2\sigma^2}}{\sqrt{2\pi\sigma^2}}$	2. Particle in a field of constant force	$\frac{2\beta\Delta^2}{\tau}$	$P_2 = (\frac{1}{2} - \beta\Delta)P_2(m \mid n-1; s) + (\frac{1}{2} + \beta\Delta)P_2(m \mid n+1; s)$	$\frac{s!(\frac{1}{2} - \beta\Delta)^{s+l/2}(\frac{1}{2} + \beta\Delta)^{s-l/2}}{(s + \lvert n-m\rvert/2)!(s - \lvert n-m\rvert/2)!}$ (Prob. 10.1)
βy	$\frac{\partial W_2}{\partial t} = D \frac{\partial^2 W_2}{\partial y^2} + \beta \frac{\partial}{\partial y}(yW_2)$ $\dot{y} + \beta y = F(t)$	$W_2 = \frac{e^{-(y-\hat{y})^2/2\sigma^2}}{\sqrt{2\pi\sigma^2}}$ $\sigma^2 = \frac{D}{\beta}(1 - e^{-2\beta t}),$ $\hat{y} = y_0 e^{-\beta t}$	3. Particle harmonically bound to a force center (linear attractive force)	$\frac{k\Delta}{K\tau}$	$P_2 = \frac{K-n+1}{2K}P_2(m \mid n-1; s) + \frac{K+n+1}{2K}P_2(m \mid n+1; s)$	$\frac{(-1)^{K-m}}{2^{2K}} \sum_{j=-K}^{K} \left(\frac{j}{K}\right)^s \times C_{K+j}^{(-m)} C_{K+n}^{(j)}$ §
$F_0(y)$	$\frac{\partial W_2}{\partial t} = D \frac{\partial^2 W_2}{\partial y^2} + \frac{\partial}{\partial y}(F_0 W_2)$ $\dot{y} + F_0(y) = F(t)$	Depends on F_0; not known in the nonlinear cases $F_0(y) = y^2, y^3, \ldots$, etc.	4. Particle in a general field of force	$F_0(k\Delta)$	$P_2 = \frac{\Delta - \tau F_0(n-1)}{2\Delta} \times P_2(m \mid n-1; s) + \frac{\Delta + \tau F_0(n+1)}{2\Delta} \times P_2(m \mid n+1; s)$	Depends on F_0; for nonlinear cases see Anchonov, Pontryagin, Witt, (Soviet) *J. Exp. and Theo. Physics*, **3**, 165 (1933).

† 1. Initial conditions: $P_2(m \mid n;0) = \delta_n^m;\ \lim_{t \to 0} W_2(y_0 \mid y;t) = \delta(y - y_0)$.

2. Assumptions: Process is Markovian; $F(t)$, the driving, or force, term in the Langevin equation, belongs to a purely (stationary) normal random process: $\overline{F(t)} = 0$, with $\overline{F(t_1)F(t_2)} = 2D\,\delta(t_2 - t_1)$.

3. In the limit $\tau, \Delta \to 0$: $s\tau \to t$; $m\Delta \to y_0$; $n\Delta \to y$; $1/K\tau \to \beta$; $\Delta^2/2\tau \to D$ (finite).

‡ Small changes in position for small changes in time in limit $\Delta t \to 0$.

§ M. Kac, Random Walk and the Theory of the Brownian Motion, *Am. Math. Monthly*, **14**: 369 (1947)

But since ξ_2 can have any value, we can replace ξ_2 by $\xi_2 e^{-\beta t}$ in Eq. (10.64*a*) to obtain the explicit form of $H(\xi_2 e^{-\beta t})$, namely,

$$H(\xi_2 e^{-\beta t}) = \exp\,(i\xi_2 y_0 e^{-\beta t} + D\xi_2^2 e^{-2\beta t}/2\beta) \tag{10.64b}$$

The transform (10.60) of the conditional probability density W_2 is therefore finally

$$f_2(y_0|\xi_2;t) = e^{i\hat{y}\xi_2 - \sigma^2\xi_2^2/2} \qquad \text{where } \hat{y} \equiv y_0 e^{-\beta t} \text{ and } \sigma^2 \equiv \frac{D}{\beta}(1 - e^{-2\beta t}) \tag{10.65}$$

Taking the Fourier transform gives us finally the well-known† normal form for W_2,

$$\begin{aligned} W_2(y_0|y;t) &= (2\pi\sigma^2)^{-1/2} \exp\,[-(y - \hat{y})^2/2\sigma^2] \\ &= \{2\pi D[1 - \exp\,(-2\beta t)]/\beta\}^{-1/2} \\ &\quad \times \exp\,\{-\beta[y - y_0 \exp\,(-\beta t)]^2/2D[1 - \exp\,(-2\beta t)]\} \end{aligned} \tag{10.66}$$

This is the solution corresponding in the continuous case to Kac's result[7] 3, Table 10.1, for the discrete random walk of the harmonically bound particle. The first-order probability density follows directly from Eq. (10.66) when $t \to \infty$; one has specifically the normal frequency function

$$W_1(y) = \lim_{t\to\infty} W_2(y_0|y;t) = \frac{1}{\sqrt{2\pi D/\beta}}\, e^{-\beta y^2/2D} \tag{10.67}$$

If we write $\rho(t) = e^{-\beta t}$ and $\sigma_\infty^2 = D/\beta$, then $\sigma^2 = \sigma_\infty^2(1 - \rho^2)$; $\hat{y} = y_0\rho$, and we have finally the familiar bivariate normal distribution density [Eq. (7.13*a*)]

$$W_2(y_0,y;t) = \frac{1}{2\pi\sigma_\infty^2\sqrt{1 - \rho^2}}\, e^{-(y_0^2 + y^2 - 2y_0 y\rho)/2\sigma_\infty^2(1-\rho^2)} \tag{10.68}$$

Since the random process has been assumed to be simple Markovian, $W_2(y_0,y;t)$ or $W_2(y_0|y;t)$ above completely describes it (cf. Sec. 1.4-3) and the original problem of solving the Langevin equation $\dot{y} + \beta y = F(t)$ (with the present simple boundary conditions) when $F(t)$ belongs to a purely random normal process has been resolved.

10.4-4 Moments. Various moments of interest may be obtained directly from Eqs. (10.64) and (10.65). For example, the mean value of the displacement $\bar{y}$ is clearly zero, while the covariance function is $K_y(t) = \overline{y_1 y_2} = \sigma_\infty^2(t) = (D/\beta)e^{-\beta t}$. The intensity spectrum of y is therefore, from the W-K theorem,

$$\mathcal{W}_y(f) = 2\mathfrak{F}\{K_y\} = \frac{4D}{\beta}\int_0^\infty e^{-\beta t}\cos \omega t\, dt = \frac{4D}{\omega^2 + \beta^2} \tag{10.69}$$

† Equation (10.66) was first derived by Lord Rayleigh [*Sci. Papers*, **3**: 473 (1920)] for the Brownian motion of a heavy particle.

which is illustrated in Fig. 10.5. [Compare Eq. (10.69) with the single-tuned spectrum in Table 3.1, with Eqs. (4.73) for random nonoverlapping pulse trains, and, finally, with Eq. (4.83b) for overlapping pulses of random duration.]

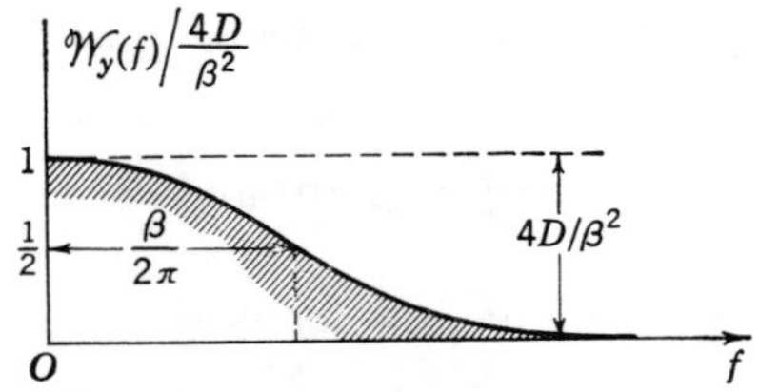

FIG. 10.5. Intensity spectrum of $y(t)$, when $y(t)$ obeys the Langevin equation $\dot{y} + \beta y = F(t)$ and $F(t)$ is a purely random normal process.

We observe that, to find the covariance function, spectrum, mean, and mean-square values of $y(t)$, it is not necessary, of course, to derive the defining distribution W_2 first. These quantities may be calculated *directly* from the appropriate Langevin equation. To see this in the case of the spectrum, multiply the Langevin equation (10.3) by itself, to get, with the help of Eq. (10.47),

$$\int_0^\infty (\omega^2 + \beta^2)\mathcal{W}_y(f) \cos \omega t \, df = 2D \, \delta(t - 0) \tag{10.70}$$

since $\beta\overline{\dot{y}_1 y_2} = -\beta\overline{y_1\dot{y}_2}$. Next, take the Fourier transform of both members of Eq. (10.70). The result is the expression (10.69). Higher moments are found in a similar way from the Langevin equation and the assumed statistical properties of the driving term $F(t)$.

A short table (cf. Table 10.1) lists the solutions for a few simple models of the random walk and their continuous analogues. The treatment of higher-dimensional random processes parallels the analysis of the one-dimensional systems examined above; solutions of these more complicated Fokker-Planck equations are included in the problems following.

We remark finally that even in those cases where the transition probabilities for the displacements in the discrete random walk depend on the previous history, i.e., on the position of the particle at any given instant [cf. (3) in Table 10.1], the limiting distribution for very large numbers of steps is still normal; a linear force field is at most assumed. It is not, however, always possible here to apply the central-limit theorem directly to this model since the individual displacements are not independent.

PROBLEMS

10.4 Show that the following alternative postulates are equivalent: on the interval $(0,T)$, (a) the driving term, i.e., the process $F(t)$, represented as a Fourier series in the interval $(0,T)$ whose coefficients a_n and b_n are independent and normally disturbed with zero means and variances $\sigma_n^2 = 4D/T = \overline{a_n^2} = \overline{b_n^2}$; (b) $F(t)$ such that

$$\begin{aligned} \overline{F(t_1) \cdots F(t_{2m+1})} &= 0 & m \geq 0 \\ \overline{F(t_1) \cdots F(t_{2m})} &= \sum_{\text{all pairs}} [\overline{F(t_i)F(t_j)} \cdots \overline{F(t_k)F(t_l)}] & m \geq 2 \end{aligned} \tag{1}$$

and

$$\overline{F(t_1)F(t_2)} = 2D\,\delta(t_2 - t_1)$$

PROCEDURE: Start with Eqs. (1), and show that a_n and b_n are independent and normally distributed with zero means and standard deviations $\sigma_n = \sqrt{4D/T}$. Thus, the Fourier coefficients a_n, b_n are

$$a_n = \frac{2}{T}\int_0^T F(t)\cos\omega_n t\,dt \qquad b_n = \frac{2}{T}\int_0^T F(t)\sin\omega_n t\,dt \tag{2}$$

Next, show that

$$\begin{aligned} \overline{a_n^{2m}} &= \frac{(2m)!}{2^m m!}\left(\frac{4D}{T}\right)^m = \overline{b_n^{2m}} \qquad \overline{a_n^{2m+1}} = \overline{b_n^{2m+1}} = 0 \\ \overline{a_n^p b_m^q} &= \begin{cases} 0 & p \text{ (and/or } q) \text{ odd} \\ \overline{a_n^p}\;\overline{b_m^q} & p \text{ and } q \text{ even} \end{cases} \end{aligned} \tag{3}$$

From knowledge of the moments $\overline{a_n^m}$, etc., determine the characteristic functions $\overline{e^{i\xi a_n}}$, $\overline{e^{i\xi b_n}}$, and hence show that typical probability densities for a_n, b_n are

$$W(a_n) = \left(\frac{T}{8\pi D}\right)^{1/2} e^{-a_n^2 T/8D}, \text{ etc.} \tag{4}$$

Finally, obtain $W(a_1, \ldots, a_n, \ldots; b_1, \ldots, b_n, \ldots)$ explicitly, and compare with the first postulate [Eq. (10.49)]. M. C. Wang and G. E. Uhlenbeck,[2] appendix, note III.

10.5 If $y(t)$ obeys the Langevin equation

$$\ddot{y} + \beta\dot{y} + \omega_0^2 y = F(t) \tag{1}$$

where $F(t)$ is a purely random, normally distributed process with the explicit properties given in Sec. 10.4, obtain the following moments:

$$\bar{\dot{y}} = \frac{\dot{y}_0 e^{-\frac{1}{2}\beta t}}{\omega_1}\left(\omega_1\cos\omega_1 t - \frac{\beta}{2}\sin\omega_1 t\right) - \frac{\omega_0^2}{\omega_1} y_0 e^{-\frac{1}{2}\beta t}\sin\omega_1 t \tag{2}$$

$$\bar{y} = \frac{\dot{y}_0 e^{-\frac{1}{2}\beta t}}{\omega_1}\sin\omega_1 t + \frac{y_0}{\omega_1} e^{-\frac{1}{2}\beta t}\left(\omega_1\cos\omega_1 t + \frac{\beta}{2}\sin\omega_1 t\right) \tag{3}$$

$$\overline{(\dot{y} - \bar{\dot{y}})^2} = \frac{D}{\beta}\left[1 - \frac{1}{\omega_1^2} e^{-\beta t}(\omega_1^2 + \tfrac{1}{2}\beta^2\sin^2\omega_1 t - \beta\omega_1\sin\omega_1 t\cos\omega_1 t)\right] \tag{4}$$

$$\omega_0^2\overline{(y - \bar{y})^2} = \frac{D}{\beta}\left[1 - \frac{1}{\omega_1^2} e^{-\beta t}(\omega_1^2 + \tfrac{1}{2}\beta^2\sin^2\omega_1 t + \beta\omega_1\sin\omega_1 t\cos\omega_1 t)\right] \tag{5}$$

$$\omega_0\overline{(\dot{y} - \bar{\dot{y}})(y - \bar{y})} = \frac{D\omega_0}{\omega_1^2} e^{-\beta t}\sin^2\omega_1 t \tag{6}$$

where y_0 and $\dot{y}_0$ are the initial values (at $t = 0$) of y and $\dot{y}$ and $\omega_1^2 = \omega_0^2 - \beta^2/4$. These results are written for the *underdamped* case ($\omega_1 > 0$). For the *aperiodic* and *overdamped* case, one simply sets $\omega_1 \to 0$ and $\omega_1 \to i\omega_1$, respectively, in Eqs. (2) to (6).

S. Chandrasekhar,[1] chap. II, sec. 3. M. C. Wang and G. E. Uhlenbeck,[2] sec. 10.

10.6 (*a*) For the random wave $y(t)$ of Prob. 10.5, show also that

$$\overline{y_0 y} = \frac{D}{\beta\omega_0^2} e^{-\beta t/2} \left(\cos \omega_1 t + \frac{\beta}{2\omega_1} \sin \omega_1 t\right) \tag{1}$$

$$\overline{\dot{y}_0 \dot{y}} = \frac{D}{\beta} e^{-\beta t/2} \left(\cos \omega_1 t - \frac{\beta}{2\omega_1} \sin \omega_1 t\right) \tag{2}$$

$$\overline{\dot{y}_0 y} = \frac{D}{\beta\omega_1} e^{-\beta t/2} \sin \omega_1 t \tag{3}$$

(*b*) Show for y and $\dot{y}$ of Prob. 10.5 that for small values of t,

$$\overline{(\dot{y} - \bar{\dot{y}})^2} \doteq 2Dt \tag{4}$$

$$\overline{(y - \bar{y})^2} \doteq \tfrac{2}{3}Dt^3 \tag{5}$$

$$\overline{(y - \bar{y})(\dot{y} - \bar{\dot{y}})} = Dt^2 \tag{6}$$

(underdamped case).

10.7 (*a*) Derive the general s-dimensional Fokker-Planck equation for

$$W_2(y_{01}, \ldots, y_{0s}|y_1, \ldots, y_s;t)$$

$$\frac{\partial W_2}{\partial t} = -\sum_{k=1}^{s} \frac{\partial}{\partial y_k}(B_k W_2) + \frac{1}{2!}\sum_{kl}^{s} \frac{\partial^2}{\partial y_k\, \partial y_l}(C_{kl} W_2) \tag{1}$$

where

$$B_k = B_k(y_1, \ldots, y_s) = \lim_{\Delta t \to 0} [(\Delta y_1)^{\lambda(1)} \cdots (\Delta y_s)^{\lambda(s)}/\Delta t]$$

$$\lambda(1) + \lambda(2) + \cdots + \lambda(s) = 1, \text{ etc.} = \text{integers} \geq 0$$

$$C_{kl} = C_{kl}(y_1, \ldots, y_s) = \lim_{\Delta t \to 0} [(\Delta y_1)^{\lambda(1)} \cdots (\Delta y_k)^{\lambda(k)} \cdots (\Delta y_l)^{\lambda(l)} \cdots (\Delta y_s)^{\lambda(s)}]/\Delta t$$

$$\sum_{j=1}^{s} \lambda(j) = 2$$

(*b*) Show that the solution of the s-dimensional Fokker-Planck equation (1), when $B_k = -\beta_k y_k$ and $C_{kl} = \sigma_{kl}$ are constants, viz.,

$$\frac{\partial W_2}{\partial t} = \sum_{k}^{s} \beta_k \frac{\partial}{\partial y_k}(y_k W_2) + \frac{1}{2}\sum_{k}^{s}\sum_{l}^{s} \sigma_{kl} \frac{\partial^2 W_2}{\partial y_k\, \partial y_l} \qquad W_2 = W_2(y_{01}, \ldots, y_{0s}|y_1, \ldots, y_s;t) \tag{2}$$

where the y_k ($k = 1, \ldots, s$) belong to a stationary Markoff process, is

$$W_2 = (2\pi)^{s/2}|\mu|^{-1/2} \exp\left[-\frac{1}{2|\mu|}\sum_{k}^{s}\sum_{l}^{s} (y_k - \nu_k)(y_l - \nu_l)\mu^{kl}\right] \tag{3}$$

with

$$\boldsymbol{\mu} = \|\mu_{kl}\| = \|\overline{(y_k - \nu_k)(y_l - \nu_l)}\|$$

$$\nu_k \equiv \bar{y}_k = y_{0k}e^{-\beta_k t}$$

$$\mu_{kl}(t) = (1 - e^{-(\beta_l + \beta_k)t}) \frac{\sigma_{kl}}{\beta_k + \beta_l} \qquad t \geq 0$$

$$|\mu| = \det \boldsymbol{\mu} \qquad \mu^{kl} = \text{cofactor of } \mu_{kl}$$

Observe that the solution (3) is formally the s-dimensional multivariate normal probability density considered in Sec. 7.3. W_2 here is also seen to be the solution to the s-dimensional random walk of an aggregate of "particles" subject to harmonic restoring forces. [For application to s-coupled linear (LCR) networks excited by thermal noise voltages, see Sec. 11.1-6.] M. C. Wang and G. E. Uhlenbeck,[2] appendix, note iv.

10.8 Obtain and solve the Fokker-Planck equation below [cf. Eq. (10.57)] for the charge in a series LCR circuit driven by a thermal noise voltage of spectral density $4k_BT_0R$:

$$\frac{\partial W_2}{\partial t} = \frac{\partial}{\partial I}\left[\left(\frac{RI}{L} + \omega_0{}^2Q\right)W_2\right] - \frac{\partial}{\partial Q}(IW_2) + \frac{k_BT_0R}{L^2}\frac{\partial^2 W_2}{\partial I^2}$$

$$W_2 = W_2(Q_0,I_0|Q,I;t);\ I = \frac{dQ}{dt} = \dot{Q};\ \omega_0{}^2 = \frac{1}{LC} \qquad (1)$$

HINT: Transform I and Q to a new set of independent variables with the help of

$$\begin{aligned} y_1 &= I + aQ \qquad y_2 = I + bQ \\ a &= \frac{1}{2}\frac{R}{L} + i\omega_1 \\ b &= \frac{1}{2}\frac{R}{L} - i\omega_1 \\ \omega_1 &= \sqrt{\frac{1}{LC} - \frac{R^2}{4L^2}} \end{aligned} \qquad (2)$$

and obtain the more symmetrical form of Eq. (1),

$$\frac{\partial W_2}{\partial t} = b\frac{\partial}{\partial y_1}(y_1W_2) + a\frac{\partial}{\partial y_2}(y_2W) + \frac{k_BT_0R}{L^2}\left(\frac{\partial}{\partial y_1} + \frac{\partial}{\partial y_2}\right)^2 W_2 \qquad (3)$$

Next, use the results of Prob. 10.7.

10.9 Show that solutions of the Boltzmann equation (10.41) are solutions of the corresponding Smoluchowski equation. HINT: Use Fourier transforms.

10.10 (*a*) Solve the Boltzmann equation

$$\frac{\partial W_2}{\partial t} = A_0 \int W_2(y_0|z;t)\,e^{-\beta(y-\gamma z)^2}\,dz - W_2(y_0|y;t)\tau^{-1} \qquad (1)$$

where $\tau = A_0{}^{-1}(\beta/\pi)^{1/2}$. Show that the solution is, for the usual initial condition $\lim_{t\to 0} W_2 = \delta(y - y_0)$ $(t \geq 0)$,

$$W_2(y_0|y;t) = \left[\delta(y - y_0) + \sum_{n=0}^{\infty} \frac{1}{n!}\left(\frac{t}{\tau}\right)^n \left(\frac{\beta}{\pi\Delta_n}\right)^{1/2} e^{-\beta(y-\gamma^n y_0)^2/\Delta_n}\right] e^{-t/\tau} \qquad (2)$$

with $\Delta_n = (1 - \gamma^{2n})/(1 - \gamma^2)$.

(*b*) Letting $D = 1/2\beta\tau$ and $\eta = (1 - \gamma)/\tau$, show that

$$\frac{d\bar{y}}{dt} = -\eta\bar{y} \qquad \frac{d^2(\overline{y^2})}{dt^2} = D - 2\eta\overline{y^2} \qquad (3)$$

(*c*) Compare the results of (*a*), (*b*) with the solution of the corresponding Fokker-Planck equation,

$$\frac{\partial W_2}{\partial t} = \tfrac{1}{2}D\frac{\partial^2}{\partial y^2}W_2 + \eta\frac{\partial}{\partial y}(yW_2) \qquad \text{[Ref. 14]} \qquad (4)$$

REFERENCES

1. Chandrasekhar, S.: Stochastic Processes in Physics and Astronomy, *Revs. Modern Phys.*, **15**: 1 (1943).
2. Wang, M. C., and G. E. Uhlenbeck: On the Theory of the Brownian Motion. II, *Revs. Modern Phys.*, **17**: 323 (1945), in particular, secs. 1, 9, 10.

3. Doob, J. L.: "Stochastic Processes," Wiley, New York, 1953.
4. Blanc-Lapierre, A., and R. Fortet: "Théorie des fonctions aléatoires," chap. 7, Masson, Paris, 1953.
5. Bartlett, M. S.: "An Introduction to Stochastic Processes," chaps. 2–4, 6, Cambridge, New York, 1956.
6. Langevin, P.: *Compt. rend.*, **146**: 530 (1908). See also F. Zernicke, "Handbuch der Physik," vol. 3, p. 456, Berlin, 1928, and, more recently, Ref. 1, chap. II.
7. Kac, M.: Random Walk and the Theory of the Brownian Motion, *Am. Math. Monthly*, **14**: 369 (1947). See also Ref. 2, sec. 5.
8. Feller, W.: "An Introduction to Probability Theory and Its Applications," chap. 15, Wiley, New York, 1950. See also N. Arley, "Stochastic Processes and Cosmic Radiation," Wiley, New York, 1949.
9. Smoluchowski, M. V.: *Physik. Z.*, **17**: 447, 585 (1916).
10. Fokker, A. D.: *Ann. Physik*, **43**: 812 (1914).
11. Planck, M.: *Sitzber. kgl. preuss. Akad. Wiss.*, p. 324, 1917. See also Ref. 1, references to chaps. I and II.
12. Boltzmann, L.: "Vorlesungen uber Gasentheorie," vol. 1, chaps. II, III.
13. Seitz, F.: "The Modern Theory of Solids," chap. IV, sec. 3.1, McGraw-Hill, New York, 1940.
14. Storer, J. E., and J. Keilson: On Brownian Motion, Boltzmann's Equation, and the Fokker-Planck Equation, *Quart. Appl. Math.*, **10**: 244 (1952).
15. Wiener, N.: Generalized Harmonic Analysis, *Acta Math.*, **55**: 117 (1930). See also Ref. 2, sec. 9*a*.
16. Ref. 1, chap. II, sec. 2. See also Ref. 2, sec. 9*a*, and in this connection G. E. Uhlenbeck and L. S. Ornstein, On the Theory of the Brownian Motion, *Phys. Rev.*, **36**: 823 (1930).
17. Madelung, E.: "Die mathematischen Hilfsmittel des Physikers," pp. 189ff., Dover, New York, 1943.

CHAPTER 11

THERMAL, SHOT, AND IMPULSE NOISE

In this chapter, we shall describe briefly some of the noise processes which are encountered in electrical and electronic systems. Comparatively little attention is given here to the details of the physical mechanism, as our primary interest is in the statistical description of the macroscopic process, e.g., in the spectra, covariance functions, first-, second-, and higher-order distribution densities, and so on. Since most of the important noise mechanisms in practice are found to generate normal processes (an important exception is *impulse* noise, also discussed below), a second-moment theory is usually sufficient (cf. Secs. 7.4ff.) and the main effort reduces to that of finding the spectrum or the associated covariance function (cf. Sec. 3.2). Once either of these has been found, the desired d.d.'s, W_1, $W_2, \ldots, W_n$, follow in the usual way from the results of Sec. 7.4, etc. A few examples are included below for the cases of thermal and shot noise. However, if the underlying physical mechanism does not lead to a normal process, a second-moment theory is not enough. All higher-order moments play an important role, so that much more involved relations result for even the first-order probability density, while the expressions for W_2, W_3, etc., rapidly become intractable, as the material of Sec. 9.1 indicates. For this reason, not much can usually be done for these nonnormal cases beyond examining the spectrum and perhaps determining the first-order and second-order densities W_1, W_2.

With this in mind, we shall consider first in Sec. 11.1 some of the statistical properties of thermal noise, in Sec. 11.2 those of impulse noise, and in Sec. 11.3 a few results for shot noise. For a much more complete account of the physical origins of the random processes in question, including quantum-statistical treatments, the reader should consult the references† at the end of the chapter.

11.1 *Thermal Noise*

Thermal noise is the result of the random motion of the free electrons in a conductor. These electrons are in continuous thermal agitation in the

† See, in particular, the book by Van der Ziel[1] and also the papers of Chandrasekhar,[2] Wang and Uhlenbeck,[3] and Moyal.[4]

conductor, which is assumed to be at some absolute temperature T_0. Because of this movement, a random current $I(t)$ exists, and if a sufficiently sensitive ammeter were placed across the conductor's terminals, a fluctuating reading would be observed. In a similar way, a voltmeter would show a fluctuating voltage $V(t)$. Schematically, this conductor can be represented as a noiseless resistance in series with a voltage source, with a spectral intensity proportional to that of the current generated therein. (We can equally well represent the conductor as a *current* source in parallel with a noiseless resistance.) The noiseless resistance in either instance has the same numerical measure as the actual, noisy one. Since $V(t) = I(t)R$, the spectral intensities are (in the steady, or equilibrium, state) related by $\mathcal{W}_V(f) = R^2\mathcal{W}_I(f)$. Figure 11.1*a*, *b* shows two equivalent representations of a physical resistance at some equilibrium temperature T_0.

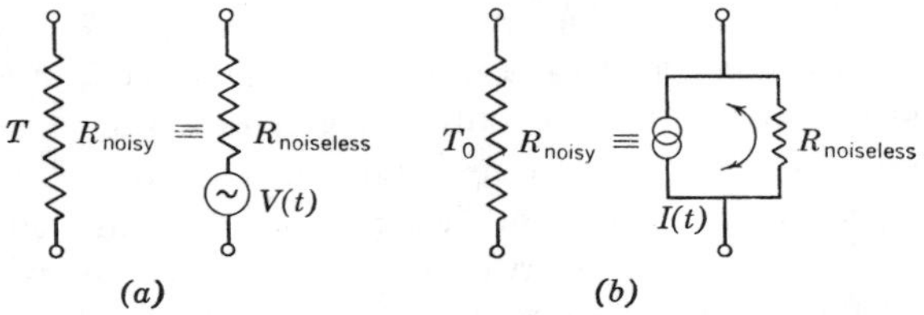

FIG. 11.1. Equivalent representations of a resistance (at temperature T_0).

Our first task now is to determine the spectral intensities $\mathcal{W}_V(f)$, $\mathcal{W}_I(f)$ and from them the other second-moment statistic, e.g., the mean total power $\overline{V^2}/R$ (or $\overline{I^2}R$). This can be done in two ways: (1) directly on the basis of a kinetic-theory model of the conduction process; (2) by appropriate thermodynamical arguments, which do not depend on the detailed mechanism of conduction. The former has the advantage of explicitness; the latter is more abstract, while avoiding some of the technical difficulties inherent in any detailed model of the conduction process. Finally, after determining the spectrum, we shall show that to a satisfactory approximation thermal noise can be regarded as a normal random process, which may under certain conditions also be Markovian (cf. Sec. 1.4).

11.1-1 The Intensity Spectrum (Kinetic Derivation).† Let us begin by calculating in a rather elementary way the spectral distribution of the thermal noise current in a metallic conductor. The voltage spectrum then follows at once from the relation above: $\mathcal{W}_V(f) = R^2\mathcal{W}_I(f)$.

We postulate the following simple model of a conductor, of length L and cross section A, in thermal equilibrium at temperature T_0: A certain number n_0 (per cubic centimeter) of "free" electrons move with velocities $\mathbf{v}$ randomly through the lattice of the metal and collide at random with the lattice centers. We assume first that the collisions of the electrons with

† See, for example, the original papers of Bernamont,[5] Bakker and Heller,[7] and Spenke.[8] A concise account is also given in Lawson and Uhlenbeck,[9] sec. 4.3.

the lattice centers are perfectly elastic—there is no exchange of energy with the lattice structure. Consequently, all electrons have essentially the same speed v ($\equiv |\mathbf{v}|$), but since they may alter their directions of motion after impact, their velocities are likewise altered. It is further assumed that in the equilibrium state the scattering of these electrons is isotropic—there are no preferred directions of movement. Restricting ourselves to classical statistics, we may expect that the electrons are also independent.† Each particle will travel a certain distance l between successive collisions, with time $\tau = l/v$ between collisions. In addition, the free-path length l is assumed to vary randomly and independently from collision to collision, so that l's have a d.d. $w(l) = \lambda^{-1}e^{-l/\lambda}$ $(l > 0)$, where λ is the *mean free path* for the conductor in question.‡ Consequently, the times between collisions are distributed according to

$$w(\tau) = (\bar{\tau})^{-1}e^{-\tau/\bar{\tau}} \qquad \tau > 0,\ \bar{\tau} = \lambda/v \tag{11.1}$$

(We assume, also, that λ, and therefore $\bar{\tau}$ in this model, are constants; in general, λ is taken to be independent of v as well.)

If now we follow a single electron in the course of time, we find that the current in the conductor due to this one electron consists of consecutive pulses of amplitude $ev_xL^{-1} = ev(\cos\phi)/L$ and duration τ, where ϕ is the angle between the direction of motion and the axis of the conductor. Because of changes in direction after each collision, these pulses vary

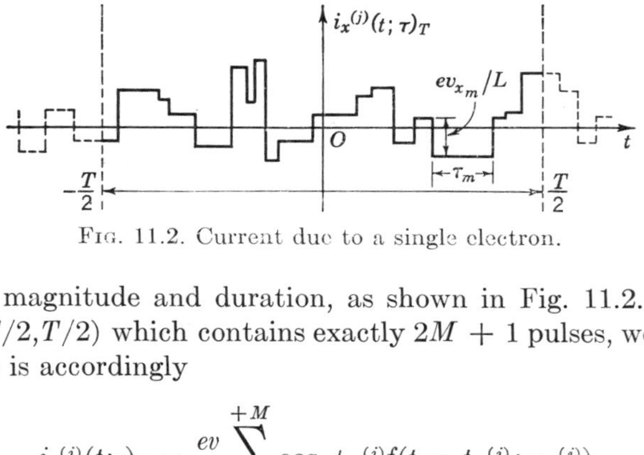

FIG. 11.2. Current due to a single electron.

randomly in magnitude and duration, as shown in Fig. 11.2. Taking an interval $(-T/2, T/2)$ which contains exactly $2M + 1$ pulses, we see that the current wave is accordingly

$$i_x^{(j)}(t;\tau)_T = \frac{ev}{L}\sum_{-M}^{+M} \cos\phi_m^{(j)} f(t - t_m^{(j)};\ \tau_m^{(j)}) \tag{11.2}$$

† As Prob. 11.1*b* indicates, Fermi-Dirac statistics yield essentially the same results for the spectrum. See also Prob. 11.1*a*, where the classical Maxwell-Boltzmann velocity distribution of the electrons is used in place of our more elementary assumption of constant speeds.

‡ See H. A. Lorentz, "The Theory of Electrons," Teubner Verlagsgesellschaft, Leipzig, 1909.

[cf. Eq. (4.63]. Here, the $f(t - t_m; \tau_m)$ are unity† when $t_m < t < t_{m+1} = t_m + \tau_m$ and vanish elsewhere. Next, to determine the covariance function of the process $i_x(t;\tau)$ when $T \to \infty$, we observe that Eq. (11.2) is a special case of an ensemble of nonoverlapping consecutive pulses of random duration, whose second-moment properties have been considered earlier in Sec. 4.4. Since successive collisions are independent, $\phi^{(j)}$ belongs to a purely random process ($\bar{\phi} = 0$), and therefore $\boldsymbol{E}\{\cos \phi_m \cos \phi_n\} = \frac{1}{3}\delta_{mn}$, because of the assumed isotropy of the scattering. The individual pulses are rectangular, with durations τ that obey Eq. (11.1), so that, setting $\beta \equiv v/\lambda = \bar{\tau}^{-1}$, we see that Eqs. (4.73) apply immediately, with $\overline{y^2} = e^2v^2/3L^2$ and $\bar{y} = 0$, this last since there is, of course, no steady drift of electrons in the equilibrium state. The autovariance of the individual current waves is therefore, from Eqs. (4.73),

$$K_{i_x}(t) = \frac{e^2v^2}{3L^2\lambda} e^{-|t|/\bar{\tau}} \tag{11.3}$$

To obtain the spectrum of the total current I, we have only to sum over all electrons N which can contribute to the fluctuations and then apply the W-K theorem (3.42). Making use of the fact that the mean total kinetic energy of these electrons is $\bar{E} = \frac{1}{2}mNv^2$, where m is the mass of a typical electron, and that the equipartition theorem[10] for classical (Maxwell-Boltzmann) statistics allows us to write $\bar{E} = 3k_BT_0N/2$ (since there are three translational degrees of freedom for each electron and k_B is Boltzmann's constant), we get the current spectrum

$$\mathcal{W}_I(f) = (4Ne^2\bar{\tau}k_BT_0/mL^2)(1 + \omega^2\bar{\tau}^2)^{-1} \tag{11.4}$$

With $\lambda = 10^{-6}$ cm‡ and $T_0 = 300°$K, we easily find from $\bar{\tau} = \lambda/v = \lambda(m/3k_BT_0)^{1/2}$ that $\bar{\tau}$ is $O(10^{-13}$ sec), so that the spectrum begins to depart from the uniform response when f is $O(10^{13}$ cps). This order of magnitude for the spectral falling off and, in fact, the shape of the spectrum at these high frequencies are, however, not correct, because the assumptions of isotropic scattering and elastic collisions are no longer adequate here.

To obtain the voltage spectrum $\mathcal{W}_V(f)$, we must first obtain the resistance R from our kinetic model. Confining our attention to frequencies such that $\omega^2\bar{\tau}^2 \ll 1$, that is, where $\mathcal{W}_I(f)$ is constant, we may expect that R also is frequency-insensitive under these conditions. Letting R_0 be the d-c resistance, we can obtain $R \doteq R_0$, $\omega^2\bar{\tau}^2 \ll 1$ by observing first that, when a small d-c field is applied to a conductor, the free electrons gain energy from the field and because of their (elastic) collisions with the lattice a new equilibrium state§ is established whereby the electrons have a net mean velocity in the (opposite) direction to the field. This net drift velocity is

† The rectangular pulse shape is a result of the constant speed v between collisions.

‡ This value is about 10^2 greater than the interatomic separations; for comment, see Seitz,[11] chap. IV.

§ As distinct from a purely thermal equilibrium.

proportional to $\bar{\tau}$ and therefore produces a net current. Collisions with the lattice accordingly yield a finite d-c conductivity σ_0, which for Maxwell-Boltzmann statistics can be shown[11] to have the value $\sigma_0 = e^2\bar{\tau}N/mV_0$, where V_0 $(= AL)$ is the volume of the piece of conductor in question. Since $R_0 \equiv L/A\sigma_0 = L^2/V_0\sigma_0$, we have from this and Eq. (11.4)

$$\mathfrak{W}_V(f) \doteq R_0{}^2\mathfrak{W}_I(f) = 4k_BT_0R_0 \qquad R_0 = mL^2/e^2\bar{\tau}N,\ f < 10^{12} \text{ cps} \tag{11.5}$$

which is the well-known result derived by Nyquist[12] from thermodynamical considerations (cf. Sec. 11.1-2) and confirmed experimentally many times.[13,14] Since, in communication problems, we shall rarely have to deal with frequencies in excess of 10^{12} cps, Eq. (11.5) and its generalizations (discussed in Sec. 11.1-3) are satisfactory for almost all applications.

11.1-2 The Intensity Spectrum (Thermodynamical Argument). With the help of a general thermodynamical argument, which avoids the difficulties inherent in a detailed model, it is also possible to derive the l-f spectrum $\mathfrak{W}_V(f) = 4k_BT_0R_0$ and to obtain in addition an expression for the h-f behavior of the thermal noise voltage. This was done originally by Nyquist,[12] the principal features of whose argument are outlined in the following paragraphs.

First, let us assume a pair of resistances R_{01} and R_{02}, independent of frequency, in equilibrium at temperature T_0, and coupled together to form a closed circuit in the manner of Fig. 11.3*a*. There is a thermal noise

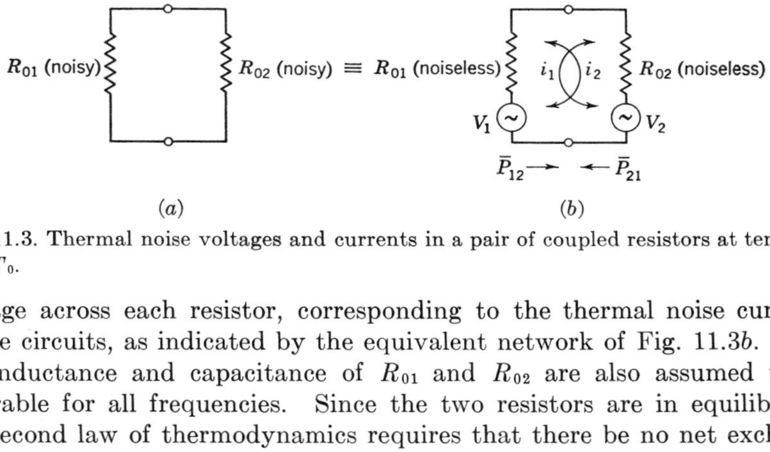

FIG. 11.3. Thermal noise voltages and currents in a pair of coupled resistors at temperature T_0.

voltage across each resistor, corresponding to the thermal noise currents in the circuits, as indicated by the equivalent network of Fig. 11.3*b*. The self-inductance and capacitance of R_{01} and R_{02} are also assumed to be ignorable for all frequencies. Since the two resistors are in equilibrium, the second law of thermodynamics requires that there be no net exchange of energy between them. Consequently, if we consider an arbitrary frequency range $(f, f + \Delta f)$, the power transferred on the average from R_{01} to R_{02} must be the same as that conveyed to R_{01} from R_{02}, *independent* of the spectral location of the interval $(f, f + \Delta f)$.† Therefore, the mean power $\bar{P}_{12}$ for the range Δf transferred from resistor 1 to resistor 2 must equal

† This is clearly true, since placing between the resistors a suitable filter which passes only the components in the range $(f, f + \Delta f)$ ensures that currents of these frequencies are the only ones which exchange power between the two elements.

$\bar{P}_{21}$, the power in this range Δf transferred back from resistor 2 to resistor 1. (In obedience to the second law of thermodynamics, there is, of course, no *net* current or voltage in the system.) Moreover, if v_1, v_2 and i_1, i_2 are, respectively, the voltage and currents for the frequency range Δf, then for the circuit of Fig. 11.3 we have

$$i_1 = v_1/(R_{01} + R_{02}) \qquad i_2 = v_2/(R_{01} + R_{02}) \qquad \text{and} \qquad \begin{aligned} \bar{P}_{12} &= \overline{i_1{}^2} R_{02} \\ \bar{P}_{21} &= \overline{i_2{}^2} R_{01} \end{aligned} \tag{11.6}$$

Substituting for i_1 and i_2 and using the facts that $\bar{P}_{12} = \bar{P}_{21}$ and that $\overline{v_{1,2}^2}$ can be written in terms of the voltage spectra as $\mathcal{W}_{V_1}(f)$, $\mathcal{W}_{V_2}(f)\,\Delta f$, respectively, we find directly that

$$\mathcal{W}_{V_1}(f) R_{02} = \mathcal{W}_{V_2}(f) R_{01} \tag{11.7}$$

Setting $R_{01} = R_{02}$ in Eq. (11.7) shows that the spectral intensity must be independent of the particular nature of R_{01} and R_{02} and of the detailed mechanism of conduction; the voltage spectrum is therefore a *universal* function of the resistance, temperature, and frequency [cf. Eq. (11.5)].

This universal function $\mathcal{W}_V(f)$ can be determined as follows: We couple two equal resistances R_0 together, with the help of an ideal lossless transmission line of characteristic impedance $Z_c = R_0$. The line is assumed to be matched and reflectionless at either end, and all power emitted by one resistor is absorbed by the other. Now, in the equilibrium state a certain mean amount of electromagnetic energy $U_{\Delta f}$ is stored in the line for the frequency interval $(f, f + \Delta f)$. This mean energy is $U_{\Delta f} = \Delta k(\bar{U}_H + \bar{U}_E)$, where $\bar{U}_H$ and $\bar{U}_E$ are, respectively, the mean magnetic and mean electromagnetic energies per mode and $\Delta k = 2L\,\Delta f/v$ is the number of modes in the frequency range $(f, f + \Delta f)$ for a line L cm long and a velocity v of propagation. The equipartition theorem[10] now allows us to assign energy of amount $k_B T_0/2$ per degree of freedom at equilibrium. Because our one-dimensional line has two degrees of freedom (one magnetic and one electric), the mean energy per mode is

$$\bar{U}_H + \bar{U}_E = k_B T_0 \qquad \text{and therefore} \qquad U_{\Delta f} = \Delta k(\bar{U}_H + \bar{U}_E) = 2k_B T_0 L\,\Delta f/v \tag{11.8}$$

[Equation (11.8) is the one-dimensional form of the Rayleigh-Jeans law of black-body radiation.[11]] Finally, since the system is in equilibrium, one half the energy goes from resistor 1 to resistor 2, and vice versa, and consequently the mean energy per second (i.e., the *mean power*) into resistor 2 from resistor 1 is simply $\frac{1}{2}U_{\Delta f}/(L/v) = \Delta f\, k_B T_0 = \bar{P}_{12}$, with a similar expression for $\bar{P}_{21}$. From Eqs. (11.6) we have therefore $\bar{P}_{12} = k_B T_0\,\Delta f = \mathcal{W}_{V_1}(f) R_{02}\,\Delta f/(R_{01} + R_{02})^2$, so that if $R_{01} = R_{02} = R_0$, we obtain at once the desired result of Nyquist, namely,

$$\mathcal{W}_V(f) = 4k_B T_0 R_0 \tag{11.9}$$

The result (11.9) is, strictly speaking, valid only for the l-f end of the spectrum. If we were to use this relation for all frequencies, the *power spectrum* $\mathcal{W}_V(f)/R_0$ $(= 4k_BT_0)$ would then be independent of frequency in this model ($R = R_0$ has been *assumed* to be frequency-insensitive everywhere) and one would have an infinite total power. The difficulty is readily overcome, for as Nyquist indicated in his original paper,[12] the equipartition value k_BT_0 for the energy per mode of the one-dimensional transmission line must be replaced by Planck's expression

$$hf(e^{hf/k_BT_0} - 1)^{-1}$$

giving, in place of Eq. (11.9), the relation

$$\mathcal{W}_V(f) = \frac{4R_0hf}{e^{hf/k_BT_0} - 1} \tag{11.10}$$

The spectrum departs noticeably from its constant l-f value at about $hf_0 = k_BT_0$, or at $f_0 = k_BT_0/h = 2.1 \times 10^{10}T_0$ cps, where f_0 may be considered the critical frequency [analogous to $f_c = \bar{\tau}^{-1}$ in Eq. (11.4)] at which this deviation becomes significant. The mean power $R_0^{-1}\int_0^\infty \mathcal{W}_V(f)\,df$ is now finite, as required, with the value $2\pi^2k_B^2T_0^2/3h$. (For additional details, see Van der Ziel,[1] sec. 2.1.)

11.1-3 Generalizations. Nyquist's result (11.9) is not limited to purely resistive elements in an equilibrium state but can be directly extended to general (passive) linear systems. To see this, let us suppose that such a system includes a general resistance $R(f)$, which may also be a function of frequency. The derivation proceeds in the same way as above, except that we replace the resistor R_{02} (cf. Fig. 11.3) with a general impedance† $Z(f) = R(f) + iX(f)$, also at the same temperature T_0 as the other resistance R_{01}, as shown in Fig. 11.4. With $Z(f)$ is associated a fluctuating noise voltage $V(t)$. Again applying the second law of thermodynamics and equating the power flowing from one section to the other of the network, we find as in Eq. (11.7) that

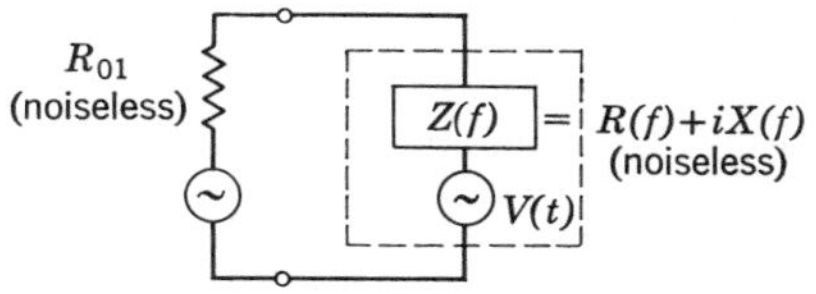

FIG. 11.4. A resistance and a general impedance in series, at temperature T_0.

$$\mathcal{W}_{V_1}(f)R(f) = \mathcal{W}_V(f)R_{01} \tag{11.11}$$

But since $\mathcal{W}_{V_1}(f) = 4k_BT_0R_{01}$ in the simplified model [and $\mathcal{W}_{V_1}(f) = 4R_0hf/(e^{hf/k_BT_0} - 1)$ if one takes account of the proper equipartition value

† To ensure a proper match, an ideal adjustable filter is used to pass only the frequencies in the interval $(f, f + \Delta f)$. This filter is then matched to the line and load for this frequency interval. By suitable readjustment of the filter the entire frequency range $(f = 0, \infty)$ can in principle be covered.

of the energy per mode in the transmission line], we have finally

$$\begin{aligned} \mathcal{W}_V(f) &= 4k_BT_0R(f) && f < 10^{12} \\ \mathcal{W}_V(f) &= 4R(f)hf/(e^{hf/k_BT_0} - 1) && \text{all } f \geq 0 \end{aligned} \tag{11.12}$$

or

$$\mathcal{W}_V(f) \equiv 4k_BT_0R'(f) \qquad \text{with } R'(f) \equiv R(f)(hf/k_BT_0)(e^{hf/k_BT_0} - 1)^{-1} \tag{11.12a}$$

An extensive experimental verification of the first of these relations was carried out by Williams,[15] who showed that $\mathcal{W}_V(f)$ did *not* change as the reactive parts of the network were varied, thus confirming the general notion that *it is the resistance alone which is to be regarded as the source of the fluctuating thermal electromotive force* (emf).

From our previous discussion, we see that, if we confine our attention to frequencies below 10^{12} cps, we can replace $R'(f)$ [Eq. (11.12*a*)] by the spectrum of the thermal noise in resistance R_{02}, that is, $\mathcal{W}_V(f) = 4k_BT_0R_0$. For almost all practical purposes, it is now quite accurate to regard this as a valid expression for *all* frequencies. The covariance function for this thermal noise process is therefore, by the W-K theorem,

$$K_V(t) = \boldsymbol{E}\{V_1V_2\} = \int_0^\infty \mathcal{W}_V(f) \cos \omega t \, df = 2k_BT_0R_0 \, \delta(t - 0) \tag{11.13}$$

where $\boldsymbol{E}\{V\} = 0$ and $t = t_2 - t_1$ as before. *Provided* that the higher moments $\overline{V_1^kV_2^l}$ $(k + l \geq 3)$ reduce to $\overline{V_1^k}\,\overline{V_2^l}$, for all $t > 0$, one has with this approximation a *purely random process* (cf. Sec. 1.4-2). [Note that this is identical with the conditions (10.51) on the random-force term $F(t)$ in the preceding discussion of the gaussian random process; cf. Sec. 10.4.] It is not necessary to require that the voltage fluctuations be due to a random emf: one can equally well say that they arise from a fluctuating current source (cf. Fig. 11.1*b*) whose spectral density and covariance function (subject to the above idealization) are equivalently

$$\mathcal{W}_I(f) = \frac{4k_BT_0}{R_0} \qquad \text{and} \qquad K_I(t) = \boldsymbol{E}\{I_1I_2\} = \frac{2k_BT_0}{R_0}\,\delta(t - 0) \qquad \bar{I} = 0 \tag{11.14}$$

The mean power associated with the thermal fluctuations when this approximation is made is infinite, as we saw previously a behavior characteristic of all purely random processes, which, of course, are physically unrealizable [cf. Sec. 3.2-3(1)*c*].

The generalization equations (11.12) permit us to treat a general impedance $Z(f) = R(f) + iX(f)$ in a similar way. Analogous to the case of the simple resistance above, we have, for the mutually equivalent voltage and current sources associated with the resistive part $R(f)$ of a complex (linear passive) network, the following spectra and covariance functions:†

$$\mathcal{W}_V(f) = 4k_BT_0R(f)$$
$$\text{and} \qquad K_V(t) = \boldsymbol{E}\{V_1V_2\} = 4k_BT_0 \int_0^\infty R(f) \cos \omega t \, df \qquad (11.15)$$

Similarly, if $G(f)$ is the *conductance* for the general admittance $Y(i\omega) = Z(i\omega)^{-1}$ in which the noise arises, we find that

$$\mathcal{W}_I(f) = 4k_BT_0G(f) = \frac{4k_BT_0R(f)}{R(f)^2 + X(f)^2} = \frac{4k_BT_0R(f)}{|Z(i\omega)|^2} \qquad (11.16a)$$

$$\text{and} \qquad K_I(t) = \boldsymbol{E}\{I_1I_2\} = 4k_BT_0 \int_0^\infty R(f)|Y(i\omega)|^2 \cos \omega t \, df \qquad (11.16b)$$

And as shown earlier [cf. Eq. (3.104)], one has the expected relations $\mathcal{W}_I(f) = |Y(i\omega)|^2 \mathcal{W}_V(f) = |Z(i\omega)|^{-2} \mathcal{W}_V(f)$ between the current and voltage spectra, as comparison of Eqs. (11.15), (11.16) indicates. Note, however, that the noise voltage process V_N, or current process I_N [associated with the impedance $Z(t_1)$], and $V(t_2)$ are statistically related ($t_2 - t_1 \neq 0$); that is, $R(f)$ is not generally a constant in Eq. (11.15). The mean total power $\bar{P}$ dissipated in the resistive elements of this network is simply

$$\bar{P} \equiv \int_0^\infty \mathcal{W}_I(f)R(f) \, df = 4k_BT_0 \int_0^\infty R(f)^2|Y(i\omega)|^2 \, df$$
$$= \int_0^\infty \mathcal{W}_V(f)|Y(i\omega)|^2 \, df \qquad (11.17)$$

from Eq. (11.15). For all practical networks, $\bar{P}$ is finite, unlike the simpler case above [Eq. (11.13)]. Here, of course, finite power is due to the selective properties of the circuit, and not to the *actual* high-frequency behavior of the resistive elements, since the "low-frequency" approximation is still assumed.

11.1-4 The Equipartition Theorem. Nyquist's proof of Eqs. (11.12), (11.12*a*) assumed that the equipartition law, which requires the mean electric and magnetic energy stored in the system to be $\frac{1}{2}k_BT_0$, respectively, holds even for a (linear) impedance whose resistive part is not independent of frequency. With the assumption that the thermal emf is a random process with the properties (11.13), it is possible to prove for all (stable) linear passive networks that the equipartition theorem remains valid. To see this, we first present a simple example.

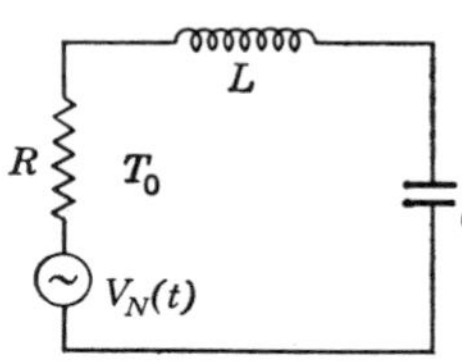

FIG. 11.5. A series *LCR* circuit excited by thermal fluctuations.

(1) The series (*LRC*) network. Let us consider the series circuit of Fig. 11.5. Here, the idealized form of Nyquist's theo n is assumed, and R ($= R_0$) is therefore a constan for all frequencies, so that for the thermal emf the second-moment statistics of Eqs. (11.9) and (11.13) apply. Because

† The practical "low-frequency" approximation $R(f) \doteq R'(f)$ [cf. Eq. (11.12*a*)] is again made; in effect, we replace the exact expressions $R'(f)$ by $R(f)$ for all frequencies.

of this thermal driving voltage, a fluctuating charge $Q(t)$ will flow in the circuit, and a current $I(t)$ [$= \dot{Q}(t)$] will exist there, also. As usual, we shall assume a steady-state condition for an ensemble of such systems (with identical elements) equivalent statistically to the statement that voltage, charge, and current are all at least wide-sense stationary processes. The appropriate Langevin equation (cf. Sec. 10.1) for the random charge Q is easily seen to be (see type 3, Table 2.1, with $CV_0 = Q$, etc.)

$$L\ddot{Q} + R\dot{Q} + \frac{Q}{C} = V_N(t) \tag{11.18}$$

The spectrum of the random charge is found by the methods of Sec. 3.3 to be

$$\mathcal{W}_Q(f) = \frac{4k_BT_0R\omega^{-2}}{R^2 + (\omega L - 1/\omega C)^2} = \frac{4k_BT_0R}{\omega^2R^2 + L^2(\omega_0{}^2 - \omega^2)^2} \qquad \omega_0{}^2 = (LC)^{-1} \tag{11.19}$$

and the associated covariance function becomes, with the help of the W-K theorem [cf. type 3, Table 3.1],

$$K_Q(t) = L^2Ck_BT_0e^{-\omega_F|t|}[\cos \omega_1 t + (\omega_F/\omega_1) \sin \omega_1|t|] \qquad \omega_1 = \sqrt{(\omega_0{}^2 - \omega_F{}^2)} > 0 \tag{11.20}$$

In a similar way, we find the modified Langevin equation for the *current* I ($= \dot{Q}$) to be

$$L\ddot{I} + R\dot{I} + IC^{-1} = \dot{V}_N(t) \tag{11.21}$$

(cf. type 4, Table 2.1, with $V_0 = IR$), and from Eqs. (11.16) the associated spectrum is

$$\mathcal{W}_I(z) = 4k_BT_0R[R^2 + (\omega L - 1/\omega C)^2]^{-1} \tag{11.22}$$

since $Z_I(i\omega) = R + i(\omega L - 1/\omega C)$, as usual. The covariance function here is obtained with the help of type 4, Table 3.1, and is specifically

$$K_I(t) = k_BT_0Le^{-\omega_F|t|}[\cos \omega_1 t - (\omega_F/\omega_1) \sin \omega_1|t|] \tag{11.23}$$

where again $\omega_1 = \sqrt{(\omega_0{}^2 - \omega_F{}^2)} > 0$ in the underdamped case, while, for critical damping and overdamping, $\omega_1 = 0$ and is replaced by $i\omega_1$, respectively. The mean power dissipated in this circuit is easily calculated from Eqs. (11.17) and (11.22); one finds that

$$\bar{P} = 4k_BT_0R^2 \int_0^\infty \frac{df}{R^2 + (L\omega - 1/\omega C)^2} = \frac{k_BT_0R}{L} = 2k_BT_0\omega_F \tag{11.24}$$

To prove the equipartition theorem, one notes that here there are *two* modes in which energy can be stored in the system: the mean magnetic energy $\bar{U}_H$ ($= \frac{1}{2}L\overline{I^2}$) represents the average energy stored in the inductance, while the mean electric energy $\bar{U}_E$ ($= \overline{Q^2}/2C$) is the average energy stored in the capacitance C. Specifically, these quantities are

$$\bar{U}_H = \tfrac{1}{2}L\overline{I^2} = 2k_BT_0RL\int_0^\infty \frac{df}{R^2 + (L\omega - 1/\omega C)^2} = \frac{k_BT_0}{2} \quad (11.25)$$

$$\bar{U}_E = \frac{\overline{Q^2}}{2C} = \frac{2k_BT_0R}{C}\int_0^\infty \frac{df}{\omega^2[R^2 + (L\omega - 1/\omega C)^2]} = \frac{k_BT_0}{2} \quad (11.26)$$

as direct integration shows, demonstrating the equipartition theorem for this particular network. Note, again, that the currents and charges cannot be purely random processes since their spectra are not uniform for all frequencies.

(2) The equipartition theorem for general linear passive networks. So far, we have proved the equipartition theorem only for the simple, single-mesh circuit of the preceding example. This result, however, can be readily generalized for an arbitrary system consisting of n-coupled linear, passive networks, in equilibrium at some temperature T_0, provided that with each resistance is associated, as before, a thermal emf obeying the idealized Nyquist law $\mathcal{W}_V(f) = 4k_BT_0R_0$, for all frequencies. Here we shall briefly outline the proof for the general network of Fig. 2.2. (Details may be found in Wang and Uhlenbeck.[3])

Letting the $n \times n$ matrix $\mathbf{B}$ represent the mesh operator for the n-coupled circuits, we have

$$\mathbf{B} = \left[L_{kl}\frac{d^2}{dt^2} + R_{kl}\frac{d}{dt} + S_{kl}\right] \qquad k, l = 1, \ldots, n \quad (11.27)$$

where L_{kl}, R_{kl}, and S_{kl} $(= C_{kl}^{-1})$ are the *mesh parameters*, namely, the various inductances, resistances, and *elastances*. The matrices $\mathbf{L}$, $\mathbf{R}$, $\mathbf{S}$ are all symmetric.† As a consequence of the thermal emf $V_{kl}(t)$, one has also the mesh charges $Q_{kl}(t)$ and mesh currents $I_{kl}(t)$; the thermal $V_{kl}(t)$ are associated with the resistances R_{kl}. The circuit equations relating the mesh *charges* and *voltages* can be written for the equilibrium (steady-state) condition in the following compact form: Here $V_k(t) = \sum_l V_{kl}(t)$, since the effective emfs for the kth mesh are $\sum_k V_{kl}$ (Chap. 4, Ref. 9, sec. 4). Thus, we have‡ as the Langevin equation for these circuits

$$\sum_{l=1} (L_{kl}\ddot{Q}_l + R_{kl}\dot{Q}_l + S_{kl}Q_l) = \sum_{l=1} V_{kl}(t) \qquad k = 1, \ldots, n \quad (11.28)$$

† It is not necessary that R_{kl} $(k \neq l)$ be positive; R_{kl} is negative when, for the resistance in common to the kth and lth meshes, the positive directions chosen for the *currents* I_{kl} are opposite to each other; L_{kl} and S_{kl} can also be negative $(k \neq l)$.

‡ If we are interested instead in the mesh *currents* $I_l(t)$, the differential operator $\mathbf{B}$ [Eq. (11.27)] is replaced by the *differential-integral* operator $\mathbf{A} = \|L_{kl}(d/dt) + R_{kl} + S_{kl}\int(\ )\,dt\|$ [cf. Eqs. (2.27a)].

The $V_{kk}(t)$ $(k = 1, \ldots, n)$ are the thermal emfs in any part of the resistances of the kth mesh which are *not* in common with any other mesh. The $V_{kl}(t)$ $(k \neq l)$, on the other hand, are the thermal noise voltages in the resistances R_{kl}, where we have chosen the convention of positive directions for the currents. Accordingly, if in R_{kl} these positive directions are in the same direction, then $V_{kl} = V_{lk}$, whereas if they are opposed, then $V_{kl} = -V_{lk}$. The various V_{kl} are, as before, assumed to have a vanishing mean value and joint second moments which obey Eq. (11.13). Specifically, we write

$$\begin{aligned} \overline{V_{kl}(t)} &= 0 \\ \overline{V_{kl}(t_1)V_{lk}(t_2)} &= 2R_{kl}k_BT_0\,\delta(t_2 - t_1) \\ \overline{V_{kl}(t_1)V_{kl}(t_2)} &= 2|R_{kl}|k_BT_0\,\delta(t_2 - t_1) \\ \overline{V_{kl}(t_1)V_{mn}(t_2)} &= 0 \qquad k, l, m, n \text{ different} \end{aligned} \tag{11.29}$$

To prove the equipartition theorem, we now outline a method employed by Wang and Uhlenbeck.[3] First, we note that in terms of the inductance and elastic matrix elements L_{kl}, S_{kl} the electric and magnetic energies of the n-coupled system are

$$U_E = \frac{1}{2}\sum_{k,l} S_{kl}Q_kQ_l \qquad \text{and} \qquad U_H = \frac{1}{2}\sum_{k,l} L_{kl}\dot{Q}_k\dot{Q}_l \tag{11.30}$$

To establish the equipartition theorem, it is now required that we show that†

$$\boldsymbol{E}\left\{Q_k\frac{\partial U_E}{\partial Q_k}\right\} = \boldsymbol{E}\left\{\dot{Q}_k\frac{\partial U_H}{\partial \dot{Q}_k}\right\} = k_BT_0 \qquad k = 1, \ldots, n \tag{11.31}$$

From this, it is evident that the statistical averages

$$\overline{Q_k(t_1)Q_l(t_2)} \qquad \text{and} \qquad \overline{\dot{Q}_k(t_1)\dot{Q}_l(t_2)}$$

are needed later, in which we shall set t $(= t_2 - t_1)$ equal to zero. These elements of the covariance matrix of the mesh charges and currents may be readily determined by the procedure used in Sec. 10.4-4 [cf. Eq. (10.70)]. We first multiply the Langevin (or circuit) equations (11.28) by themselves considered at a later time t $(= t_2 - t_1)$ and take the ensemble average of the product, which is then expressed in terms of the various auto- and cross-spectral densities (cf. Sec. 3.4) of the mesh charges. Introducing the *impedance matrix* $\mathbf{Z}_Q(i\omega)$ for these charges, namely,

$$\mathbf{Z}_Q(i\omega) = (i\omega)^2\mathbf{L} + (i\omega)\mathbf{R} + \mathbf{S} = [-\omega^2L_{kl} + i\omega R_{kl} + S_{kl}] \tag{11.32}$$

and using Eqs. (11.29), we find that

$$\boldsymbol{E}\{(\mathbf{BQ})_1(\widetilde{\mathbf{BQ}})_2\} = \boldsymbol{E}\{\mathbf{V}_1\tilde{\mathbf{V}}_2\} \tag{11.33a}$$

becomes

$$\int_{-\infty}^{\infty} \mathbf{Z}_Q(i\omega)\mathbf{W}_Q(\omega)\mathbf{Z}_Q(-i\omega)e^{-i\omega t}\,df = 4k_BT_0R_{kl}\,\delta(t_2 - t_1) \tag{11.33b}$$

† See, for example, Tolman,[10] chap. 4, sec. 25.

where now $\mathcal{W}_Q(\omega)$ is the $n \times n$ matrix whose elements are the auto- and cross-spectral densities $\mathcal{W}_{kl}(\omega)_Q$ of the mesh charges. The Fourier transform of Eqs. (11.33) accordingly gives as a typical k, lth element

$$\begin{aligned} [\mathcal{W}(\omega)_Q]_{kl} &= 4k_BT_0[\mathbf{Z}_Q(i\omega)^{-1}\mathbf{R}\mathbf{Z}_Q(-i\omega)^{-1}]_{kl}\,\delta_l^k \\ \text{or} \qquad [\mathcal{W}(\omega)_Q]_{kl} &= 4k_BT_0[\mathbf{Y}_Q(i\omega)\mathbf{R}\mathbf{Y}_Q(-i\omega)]_{kl}\delta_l^k \end{aligned} \tag{11.34}$$

where $\mathbf{Y}_Q = \mathbf{Z}_Q^{-1}$ is the *admittance matrix* for the charges in the n-coupled system.

Since the charge and current impedances are related by

$$\mathbf{Z}_I(i\omega) = (i\omega)^{-1}\mathbf{Z}_Q(i\omega) = (i\omega)\mathbf{L} + \mathbf{R} + \frac{1}{i\omega}\mathbf{S} = \left[i\omega L_{kl} + R_{kl} - \frac{iS_{kl}}{\omega}\right] \tag{11.35}$$

similarly to Eqs. (11.34), we find for the mesh currents the following cross-spectral densities:

$$\begin{aligned} [\mathcal{W}_I(\omega)]_{kl} &= 4k_BT_0[\mathbf{Z}_I(i\omega)^{-1}\mathbf{R}\mathbf{Z}_I(-i\omega)^{-1}]_{kl}\delta_l^k \\ &= 4\omega^2k_BT_0\delta_l^k[\mathbf{Z}_Q(i\omega)^{-1}\mathbf{R}\mathbf{Z}_Q^{-1}(-i\omega)]_{kl} \end{aligned} \tag{11.36}$$

The generalized Wiener-Khintchine theorem (cf. Sec. 3.4) applied to each element of Eqs. (11.34) and (11.35) then gives the various covariance functions of Q and $\dot{Q}$ $(= I)$.

For $E\{Q_k\,\partial U_E/\partial Q_k\}$, etc., this gives, at $t = 0$,

$$E\left\{Q_k\frac{\partial U_E}{\partial Q_k}\right\} = \frac{k_BT_0\delta_l^k}{\pi}\int_{-\infty}^{\infty}[\mathbf{S}\mathbf{Z}_Q(i\omega)^{-1}\mathbf{R}\mathbf{Z}_Q(-i\omega)^{-1}]_{kl}\,d\omega \tag{11.37}$$

The integral is readily evaluated if we note first from Eq. (11.32) that

$$\begin{aligned} \mathbf{Z}_Q(i\omega) - \mathbf{Z}_Q(-i\omega) &= 2i\omega\mathbf{R} \\ \text{and therefore} \qquad \mathbf{S}\mathbf{Z}_Q^{-1}\mathbf{R}(\mathbf{Z}_Q^{-1})^* &= -\mathbf{S}\mathbf{Z}_Q^{-1}/i\omega \end{aligned}$$

Substituting into Eq. (11.37), where now we take the principal value of the integral, allows us to write

$$E\left\{Q_k\frac{\partial U_E}{\partial Q_k}\right\} = -\frac{k_BT_0}{\pi i}\int_{\mathbf{C}}[\mathbf{S}\mathbf{Z}_Q^{-1}(i\omega)]_{kk}\frac{d\omega}{\omega} \tag{11.38}$$

where $\mathbf{C}$ is the contour of Fig. 2.6. Since the determinant of $\mathbf{Z}_Q(i\omega)$ has no zeros in the lower half plane, inasmuch as the network is in general *dissipative*,[16] the elements of $\mathbf{Z}_Q(p)^{-1}$, with p as a complex variable, have only poles in the left half plane (cf. Fig. 2.6). For the contour $\mathbf{C}$, the only contribution to the integral now is $-\pi i$ times the residue at $\omega = 0$. Since $\mathbf{Z}_Q^{-1}(0) = \mathbf{S}^{-1}$, a constant from Eq. (11.32), we have at once

$$E\left\{Q_k\frac{\partial U_E}{\partial Q_k}\right\} = -\frac{k_BT_0}{\pi i}(-\pi i) = k_BT_0$$

as required. That $E\{\dot{Q}_k\,\partial U_H/\partial\dot{Q}_k\}$ equals k_BT_0 can be shown in similar fashion (cf. Prob. 11.3).

11.1-5 Nonequilibrium Conditions. We expect that the generalized Nyquist formula $\mathcal{W}_V(f) = 4k_BT_0R(f)$ [Eq. (11.15)] can also be derived for a general (linear passive) network by associating with each constant resistance a fluctuating emf with a constant spectrum of the type (11.9). Then, with the help of the preceding analysis, we can calculate the combined effect of all these thermal voltages, as seen across the two poles of the network (cf. Fig. 2.2), and this must be equal to a composite random voltage whose spectral distribution is just $4k_BT_0R(f)$, where $R(f)$ is the effective resistance of the general network. This has been experimentally verified by Williams[17] as part of an extensive series of experiments with combinations of resistive and reactive elements at the same and at different temperatures.

That each resistor R_0 can be considered to have its own thermal emf of spectral intensity $4k_BT_0R_0$, regardless of its orientation in a complex network, is of considerable importance because it enables us to predict what will occur when different resistances in the network *are no longer at the same temperature.* Just as in our previous discussion of the thermal noise spectrum, where equilibrium conditions were assumed, it seems plausible here still to associate with each constant resistance R_l a random thermal voltage $V_l(t)$, even though the system is not now in equilibrium. As before, the mean value of $V_l(t)$ is assumed to vanish, and the covariance function and spectrum are given by expressions of the types (11.9) and (11.13), viz.,

$$K_{V_l}(t) = \overline{V_l(t_1)V_l(t_2)} = 2k_BT_lR_l\,\delta(t_2 - t_1) \qquad \mathcal{W}_{V_l}(f) = 4k_BT_lR_l \quad (11.39)$$

where now T_l is the temperature of the lth resistor. Ornstein's investigation[18] of the way in which cooling a resistance affected the sensitivity of a galvanometer first led to this assumption, while the complete experimental demonstration was provided by Williams[17] at a somewhat later date.†

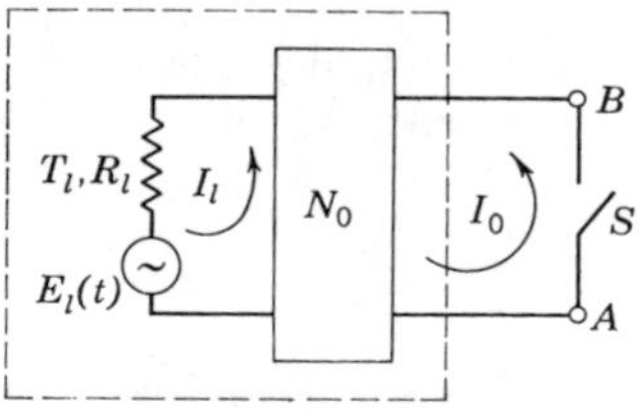

FIG. 11.6. A general linear passive network, with a resistance R_l at temperature T_l, driven by a voltage source $E_l(t)$.

On the basis of this assumption, the spectral density $\mathcal{W}_V(f)$ of the effective thermal voltage fluctuations across any two terminals A, B of a general network, whose various elements are at *different* temperatures, may be calculated in the following way:[19]

Consider any linear passive network, where R_l is a typical resistive element, at temperature T_l. A generator $E_l(t)$ is then connected in series with R_l, as shown in Fig. 11.6. The arbitrary terminals A, B are provided with a switch S, so that they may be

† As Lawson and Uhlenbeck have indicated (*op. cit.*, sec. 4.4), that each resistance in a network whose components are *not* in thermodynamical equilibrium can have associated with it a thermal emf with the properties (11.39) must be an *assumption*, since no general statistical proof can be given in this case.

short-circuited. We first assume that the generator delivers a sinusoidal voltage $E_l(t) = E_{0l}e^{i\omega t}$ and that the switch S is closed. The (complex) current I_0 from A to B is then found from the pair of relations

$$\begin{aligned} E_{0l} &= Z_{11}I_l + Z_{12}I_0 \\ 0 &= Z_{21}I_l + Z_{22}I_0 \end{aligned} \tag{11.40}$$

where Z_{11} and Z_{22} are the open-circuited impedances of $N_0 + R_l$ and $Z_{21} = Z_{12}$ (since we are dealing with a linear passive system), where Z_{12}, Z_{21} are the transfer impedances of N_0. Consequently, we can write

$$I_0 = \frac{E_{0l}}{Z_{11} - Z_{12}{}^2/Z_{22}} = \frac{E_{0l}}{Z_{Al}} \tag{11.41}$$

in which Z_{Al} $(= Z_{11}{}^2 - Z_{12}{}^2/Z_{22})$ is the transfer impedance from the source E_l to the short circuit at AB for a general frequency f $(= \omega/2\pi)$. Replacing the sine-wave source now by an equivalent thermal source $V_l(t)$, whose mean-square voltage output in the range $(f, f + \Delta f)$ is $\mathcal{W}_V(f)\,\Delta f = 4k_BT_lR_l\,\Delta f$, one finds from Eq. (11.41) that the mean-square current through S in this interval $(f, f + \Delta f)$, attributable to the noise source $E_l(t) = V_l(t)$, is

$$\mathcal{W}_{I_l}(f)\,\Delta f = 4k_B\,\Delta f\,\frac{T_lR_l}{|Z_{Al}|^2} \tag{11.42}$$

Similar expressions follow for every other resistance in the complex network, and since the system is *linear* and the noise sources are independent, the spectral distribution of the total mean-square current fluctuations is the sum of terms like Eq. (11.42), namely,

$$\mathcal{W}_I(f) = 4k_B \sum_l \frac{T_lR_l}{|Z_{Al}(f)|^2} \tag{11.43}$$

To obtain the voltage spectrum $\mathcal{W}_V(f)$ for the potential drop $V(t)$ across AB, we observe that by Thévenin's theorem[20] the open-circuit voltage is equal to the short-circuit current (already found above) divided by the impedance looking in at the terminals A, B with all voltage sources replaced by short circuits. Since $\mathcal{W}_I(f)|Z(i\omega)|^2 = \mathcal{W}_V(f)$, we obtain at once from Eq. (11.43) the desired result

$$\mathcal{W}_V(f) = 4k_B|Z(i\omega)|^2 \sum_l \frac{T_lR_l}{|Z_{Al}(i\omega)|^2} \tag{11.44}$$

The real part of $Z(i\omega)$ is found to be

$$R(f) = \operatorname{Re} Z(i\omega) = |Z(i\omega)|^2 \sum_l \frac{R_l}{|Z_{Al}(i\omega)|^2} \tag{11.44a}$$

When the temperatures T_l are all equal (to T_0), one has the simpler generalization of Nyquist [Eqs. (11.12)] for the equilibrium state:

$$\mathcal{W}_V(f) = 4k_BT_0R(f) = 4k_BT_0|Z(i\omega)|^2 \sum_l R_l/|Z_{Al}(i\omega)|^2 \qquad (11.45)$$

We remark again that these results have received experimental confirmation in a variety of ways.

11.1-6 Probability Distributions for Thermal Noise. In addition to the mean value, the spectrum, and the variance function, it is also possible to find the probability densities W_n for thermal noise in a (passive) linear network if certain reasonable assumptions are made. In fact, all orders of W_n can then be obtained, and thus the random processes to which the noise currents and charges belong can be completely described (cf. Sec. 1.3-2).

To see how this may be achieved, let us first recall that for frequencies below some critical value $O(10^{13}$ cps) the spectra of the thermal noise currents and voltages associated with individual resistances in the network are uniform. Next, we again make the approximation [cf. Eqs. (11.14)] that this uniformity extends to all frequencies and observe, moreover, that in the equilibrium state ($t \to \infty$) the first-order distribution density $W_1(I)$ of the thermal currents I in the network must obey a Maxwell-Boltzmann distribution of the type

$$W_1(I) = \frac{e^{-I^2/2\overline{I^2}}}{\sqrt{2\pi\overline{I^2}}} = Be^{-A\mathbf{v}\cdot\mathbf{v}} \qquad (11.46)$$

where $A = (2\overline{v^2})^{-1}$ and $B = (2\pi\overline{v^2})^{-1/2}$ from the kinetic-theory model discussed in Sec. 11.1-1, and where, of course, $\overline{I^2}$ depends on the specific circuit in question. Now, the necessary and sufficient condition that the thermal currents in the network obey equilibrium distribution densities of the form (11.46) is that *the thermal emfs* arising in each resistance *belong to a purely random gaussian process.*[21] From Sec. 10.4, this means that the thermal emfs associated with the resistances obey the following statistics,

$$\begin{aligned} \overline{V(t_1)_N V(t_2)_N \cdots V(t_{2m+1})_N} &= 0 \\ \overline{V(t_1)_N V(t_2)_N \cdots V(t_{2m})_N} &= \sum_{\text{all pairs}} [\overline{V(t_i)_N V(t_j)_N} \cdots \overline{V(t_p)_N V(t_q)_N}] \end{aligned} \qquad (11.47)$$

where the sum is to be taken over the $(2m)!/2^m m!$ different ways in which the $2m$ time points $t_1, \ldots, t_{2m}$ can be combined in pairs; as before $\overline{V(t_i)V(t_j)} = 2k_BT_0R_0\,\delta(t_j - t_i)$ [cf. Eq. (11.13)]. [We can also use the alternative Fourier-series representation, which is entirely equivalent to Eqs. (11.47) in the limit of infinite sample durations (cf. Secs. 8.3 and 10.4).]

To determine $W_1(I)$, $W_2(I_1,I_2;t)$, etc., for the currents in the given network, we must solve the appropriate Fokker-Planck equation, which in turn is to be established specifically with the help of the given *Langevin*, or *circuit*, equations, according to the method outlined in Sec. 10.3-2. With the assumptions above on $V(t)_N$, this is always possible (Sec. 10.3-2) provided that the Langevin equation is *linear*.† Doob's theorem[22] (Sec. 7.6) may next be applied to make sure that we are dealing with a Markoff process in the proper number of variables, and not with incomplete information in the form of a *projection* of a Markoff process (Sec. 1.4-5). Thus, for example, for a simple circuit containing a resistance and inductance in series, the Langevin equation is of the type $\dot{y} + \beta y = F(t)$, where $F(t)$ belongs to a purely random gaussian process, and Doob's theorem then tells us that y alone is the variable to be considered. On the other hand, in a series circuit containing inductance, capacitance, and resistance, where the Langevin equation is described by $\ddot{y} + \beta\dot{y} + \omega_0^2 y = F(t)$, y ($\sim$ charge) or $\dot{y}$ ($\sim$ current) alone does not contain all the information needed for a complete description. The Fokker-Planck equation must be expressed in terms of *both* y and $\dot{y}$.

Example 1. The series *LCR* network. Let us consider the network shown in Fig. 11.5. The appropriate Langevin equation is

$$L\ddot{Q} + R\dot{Q} + \frac{Q}{C} = V(t)_N \tag{11.48}$$

According to Doob's theorem, there are now *two* random variables Q and $\dot{Q} = I$, the charge and current, which are needed to describe the performance of the system, so that Q and I jointly belong to a two-dimensional Markoff process. The Fokker-Planck equation follows from Prob. 10.8. We see that the conditional d.d. $W_2(Q_0,I_0|Q,I;t)$ obeys

$$\frac{\partial W_2}{\partial t} = \frac{\partial}{\partial I}\left[\left(\frac{RI}{L} + \frac{Q}{LC}\right) W_2\right] - \frac{\partial}{\partial Q}(IW_2) + \frac{k_B T_0 R}{L^2}\frac{\partial^2 W_2}{\partial Q^2} \tag{11.49}$$

subject now to the more general initial condition $W_2(Q_0,I_0|Q,I;0) = \delta(Q - Q_0)\,\delta(I - I_0)$. With the help of the results of Prob. 10.7, the explicit solution is found to be the two-dimensional normal frequency function

$$W_2(Q_0,I_0|Q,I;t) = \exp\left\{-\frac{(Z_1 - \bar{Z}_1)^2/\sigma_1^2 + (Z_2 - \bar{Z}_2)^2/\sigma_2^2 - 2(Z_1 - \bar{Z}_1)(Z_2 - \bar{Z}_2)\rho(t)/\sigma_1\sigma_2}{2[1 - \rho(t)^2]}\right\} \times [2\pi\sigma_1\sigma_2\sqrt{1 - \rho(t)^2}]^{-1} \tag{11.50}$$

where‡

$$Z_1 = I + aQ \qquad Z_2 = I + bQ \tag{11.51a}$$

† If it were not, of course, the thermal current would not then belong to a normal random process, even though $V(t)_N$ follows the gaussian structure implied by Eqs. (11.47).

‡ We replace t by $|t|$ since we have assumed a stationary process.

and

$$a = \omega_F + i\omega_1 \qquad b = \omega_F - i\omega_1 \qquad \omega_1 = \sqrt{\omega_0^2 - \omega_F^2} \qquad \omega_F = R/2L \qquad \omega_0^2 = 1/LC$$

$$\bar{Z}_1 = (I_0 + aQ_0)e^{-b|t|} \qquad \bar{Z}_2 = (I_0 + bQ_0)e^{-a|t|} \tag{11.51b}$$

$$\sigma_1^2 = \overline{(Z_1 - \bar{Z}_1)^2} = \frac{k_B T_0 R}{bL^2}(1 - e^{-2b|t|})$$
$$\sigma_2^2 = \overline{(Z_2 - \bar{Z}_2)^2} = \frac{k_B T_0 R}{aL^2}(1 - e^{-2a|t|}) \tag{11.51c}$$

$$\rho(t) \equiv \frac{\overline{(Z_1 - \bar{Z}_1)(Z_2 - \bar{Z}_2)}}{\sigma_1\sigma_2} = \frac{2}{R}\sqrt{\frac{L}{C}} \frac{1 - e^{-2\omega_F|t|}}{\sqrt{(1 - e^{-2a|t|})(1 - e^{-2b|t|})}} \tag{11.51d}$$

As before, $\omega_1 > 0$ applies for the underdamped case; for critical damping, $\omega_1 = 0$, or $R = \frac{1}{2}(L/C)^{\frac{1}{2}}$, while, for overdamping, ω_1 is replaced by $i\omega_1$ in the above.

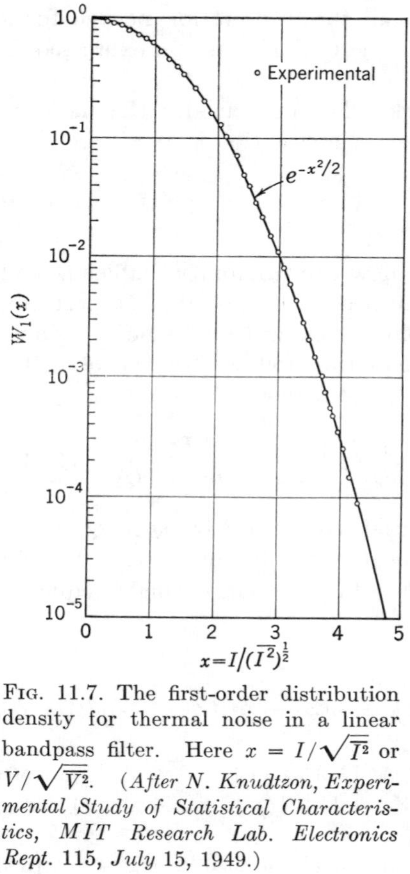

FIG. 11.7. The first-order distribution density for thermal noise in a linear bandpass filter. Here $x = I/\sqrt{\overline{I^2}}$ or $V/\sqrt{\overline{V^2}}$. (*After N. Knudtzon, Experimental Study of Statistical Characteristics, MIT Research Lab. Electronics Rept.* 115, *July* 15, 1949.)

The Maxwell-Boltzmann distribution is easily established when $t \to \infty$. We have

$$\lim_{t\to\infty} W_2(I_0,Q_0|I,Q;t) = W_1(I,Q) = \frac{1}{2\pi k_B T_0}\sqrt{\frac{L}{C}}\, e^{-(Q^2/2k_BT_0C + LI^2/2k_BT_0)} \tag{11.52}$$

As is immediately apparent from Eq. (11.52), there is no correlation at any given instant between the random charge and its derivative, the current. We note too that $\overline{I^2} = k_B T_0/L$ and $\overline{Q^2} = k_B T_0 C$, again in agreement with the equipartition values [Eqs. (11.25), (11.26)]. The four-dimensional second-order probability density

$$W_2(I_1,Q_1;I_2,Q_2;t)$$

which now completely describes the Markoff process to which I and Q jointly belong, follows immediately from Eqs. (11.50) and (11.52) in the usual way. We can clearly go on in this fashion to consider more complicated systems. Note that, although the thermal emfs are (by our simplified theory) purely random gaussian processes, the random charges and currents are not. They are, however, still gaussian, since the basic Langevin equations [cf. Eq. (11.48)] are linear.

The first-order distribution density $W_1(I)$ has been very accurately verified experimentally by Knudtzon[23] for a number of different linear networks. Figure 11.7 shows $W_1(I) = \int_{-\infty}^{\infty} W_1(I,Q)\, dQ$ for a bandpass filter, similar to the circuit of Fig. 11.5.

Example 2. Thermal noise in general meshes.† The analysis of the preceding paragraphs may now be extended to the general case of the arbitrary linear (passive) network of n meshes, considered previously in Sec. 11.1-4(2) (cf. Fig. 2.2). The Langevin, or circuit, equations are given as before by Eq. (11.28), and the thermal emfs not only obey the conditions (11.29) in the first and second moments but, in line with the preceding argument, are assumed to belong as well to a purely random gaussian process, for which Eqs. (11.47) apply in each instance. Doob's theorem (Sec. 7.6) shows then that the system of $2n$ random charges and currents Q_k, $\dot{Q}_k$ $(k = 1, \ldots, n)$ form a $2n$-dimensional *Markoff process.* The corresponding equation of Fokker-Planck for the conditional probability density $W_2 = W_2(Q_{10}, \ldots, Q_{n0}, I_{10}, \ldots, I_{n0}|Q_1, \ldots, Q_n, I_1, \ldots, I_n; t)$ can therefore be written (cf. Prob. 10.7)

$$\frac{\partial W_2}{\partial t} = - \sum_{k=1}^{2n} \frac{\partial}{\partial z_k} (B_k W_2) + \frac{1}{2} \sum_{k=1}^{2n} \sum_{l=1}^{2n} \frac{\partial^2}{\partial z_k\, \partial z_l} (C_{kl} W_2) \tag{11.53}$$

where $z_k = Q_k$ $(k = 1, \ldots, n)$ and $z_{k+n} = \dot{Q}_k = I_k$ $(k = 1, \ldots, n)$. As usual here, the boundary conditions on W_2 are of the form $\lim_{t \to \infty} W_2 = \delta(z_1 - z_{10})\, \delta(z_2 - z_{20}) \cdots \delta(z_{2n} - z_{2n0})$.

If we now let $\mathfrak{B}$ and $\mathbf{z}$ be column vectors, such that

$$\mathfrak{B} = \begin{vmatrix} B_1 \\ \cdot \\ \cdot \\ \cdot \\ B_k \\ \cdot \\ \cdot \\ \cdot \\ B_{2n} \end{vmatrix} \qquad \mathbf{z} = \begin{vmatrix} Q_1 \\ \cdot \\ \cdot \\ \cdot \\ Q_n \\ \dot{Q}_1 \\ \cdot \\ \cdot \\ \cdot \\ \dot{Q}_n \end{vmatrix}$$

one may show from the circuit equations (11.28) and the conditions (11.29) on the thermal voltages that the coefficients B_k in Eq. (11.53) are related by the transformation

$$\mathfrak{B} = \boldsymbol{\alpha}\mathbf{z} \qquad \text{or} \qquad B_k = \sum_j^{2n} \alpha_{kj} z_j \tag{11.54}$$

† The treatment is essentially that of Wang and Uhlenbeck, *op. cit.*, sec. 11.

where specifically $\boldsymbol{\alpha}$ is the $2n \times 2n$ matrix

$$\alpha = \begin{bmatrix} \mathbf{0} & \mathbf{I} \\ -\mathbf{L}^{-1}\mathbf{S} & -\mathbf{L}^{-1}\mathbf{R} \end{bmatrix} \tag{11.55}$$

In a similar way, one gets for the matrix $\mathbf{C}$ of the second-order coefficients C_{kl} the relation

$$\mathbf{C} = \begin{bmatrix} \mathbf{0} & \mathbf{0} \\ \mathbf{0} & 2k_BT_0\mathbf{L}^{-1}\mathbf{R}\mathbf{L}^{-1} \end{bmatrix} \tag{11.56}$$

Now, to find the fundamental solution of the general Fokker-Planck equation (11.53), it is helpful first to introduce the linear transformation $\mathbf{y} = \mathbf{cz}$, where $\mathbf{c}$ is the $2n \times 2n$ matrix which diagonalizes $\boldsymbol{\alpha}$. Therefore, we can write $\mathbf{c}\boldsymbol{\alpha} = \boldsymbol{\lambda}\mathbf{c}$, and $\boldsymbol{\lambda}$ is the diagonal matrix whose elements are $[\lambda_j\delta_{jk}]$, where λ_j $(j = 1, \ldots, 2n)$ can be shown to be the eigenvalues, or $2n$ roots, of the secular equation

$$\det\,[L_{kl}\lambda^2 + R_{kl}\lambda + S_{kl}] = 0 \tag{11.57}$$

For the stable linear passive networks considered here, the $2n$ roots of this equation must have negative real parts (cf. Sec. 2.2), and therefore Re λ_j must also be less than zero. Now, since $\mathbf{y} = \mathbf{cz}$, $\boldsymbol{\mathcal{B}} = \boldsymbol{\alpha}\mathbf{z}$, we can write $\mathbf{z} = \mathbf{c}^{-1}\mathbf{y}$ and $\boldsymbol{\mathcal{B}} = \mathbf{c}^{-1}\boldsymbol{\lambda}\mathbf{y}$. Substituting these results into the Fokker-Planck equation (11.53), after setting $\beta = -\lambda_k$, and writing $\sigma_{kl} = (\mathbf{cC\tilde{c}})_{kl}$ gives us finally

$$\frac{\partial W_2}{\partial t} = \sum_{k=1}^{2n} \beta_k \frac{\partial}{\partial y_k}(y_kW_2) + \frac{1}{2}\sum_{k,l=1}^{2n} \sigma_{kl}\frac{\partial^2 W_2}{\partial y_k\,\partial y_l} \tag{11.58}$$

We now note that Eq. (11.58) has precisely the form discussed in Prob. (10.7*b*), where also the solution in the n-dimensional case is given. With an immediate extension to $2n$ dimensions, this enables us to write for the fundamental solution of Eq. (11.58)

$$W_2(y_{10}, \ldots, y_{2n,0}|y_1, \ldots, y_{2n};t) = [(2\pi)^n \det \boldsymbol{\mu}]^{-1/2} \exp - [-\tfrac{1}{2}(\tilde{\mathbf{y}} - \tilde{\boldsymbol{\nu}})\boldsymbol{\mu}^{-1}(\mathbf{y} - \boldsymbol{\nu})] \tag{11.59}$$

where the averages and variances are, respectively,†

$$\boldsymbol{\nu} = [\nu_k] = [\bar{y}_k] = [y_{k0}e^{-\beta_k|t|}] \tag{11.60a}$$

$$\boldsymbol{\mu} = [\mu_{kl}] = [\overline{(y_k - \bar{y}_k)(y_l - \bar{y}_l)}] = \frac{\sigma_{kl}}{\beta_k + \beta_l}(1 - e^{-(\beta_k+\beta_l)|t|}) \tag{11.60b}$$

As before, the y_{k0} $(k = 1, \ldots, 2n)$ are the initial values of y_k.

In terms of the original random variables z_k $(k = 1, \ldots, 2n)$, we may use Eqs. (11.54) et seq. to transform back, obtaining finally the following

† We have replaced t by $|t|$ here since the process in question is assumed to be stationary (in the wide or strict sense).

matrix relation between $\bar{z}_k$ and its initial value z_{k0}:

$$\bar{\mathbf{z}} = e^{\boldsymbol{\alpha} t}\mathbf{z}_0 \tag{11.61}$$

In a similar way, the matrix of the variances $\overline{(z_k - \bar{z}_k)(z_l - \bar{z}_l)}$ may be written

$$\boldsymbol{\mu}^{(z)} = [\overline{(z_k - \bar{z}_k)(z_l - \bar{z}_l)}] = \mathbf{c}^{-1}\boldsymbol{\mu}\tilde{\mathbf{c}}^{-1} \tag{11.62}$$

The conditional probability density $W_2(z_{10}, \ldots ,z_{2n,0}|z_1, \ldots ,z_{2n};t)$ becomes in the limit $t \to \infty$ the Maxwell-Boltzmann distribution

$$W_1(z_1, \ldots ,z_{2n}) = [(2\pi)^n \det \boldsymbol{\mu}^{(z)}(\infty)]^{-1/2} \exp [-\tfrac{1}{2}\tilde{\mathbf{z}}\boldsymbol{\mu}^{(z)}(\infty)\mathbf{z}] \tag{11.63}$$

where all $\bar{z}_k$ must vanish and $\boldsymbol{\mu}^{(z)}(\infty)$ must reduce to a matrix which at most has block structure [i.e., only the matrices which form the diagonal elements of $\boldsymbol{\mu}^{(z)}(\infty)$ are different from zero]. Specifically, it is found that now

$$\boldsymbol{\mu}^{(z)}(\infty) = \begin{bmatrix} k_B T_0 \mathbf{S}^{-1} & \mathbf{0} \\ \mathbf{0} & k_B T_0 \mathbf{L}^{-1} \end{bmatrix} \tag{11.64}$$

The general second-order density $W_2(z_1, \ldots ,z_{2n};z_1', \ldots ,z_{2n}';t)$ is now obtained at once from Eqs. (11.59) and (11.63), where z_k' represents z_k at a time t later, after transformation back to the original variables by means of the relation $\mathbf{y} = \mathbf{cz}$.

A corresponding theory for additional mixtures of signal and thermal noise processes in linear networks can be constructed by a direct application of the results above to those of Sec. 7.5-3(2). Note again that, unless the signal process consists of a single member function [cf. Eq. (7.49)], the resulting process for the additive mixture is no longer normal.

11.1-7 Comments. In the previous sections, the spectral and general statistical properties of the thermal noise waves associated with the resistive parts of conducting elements have been described in some detail. A complete statistical theory, which holds for all frequencies, has not yet been constructed.† However, if one is willing to restrict oneself to the spectral region of almost exclusive practical interest, below $O(10^{12}$ cps), where quantum effects can be disregarded, a satisfactory theory is possible which not only accounts for the spectral distribution of thermal noise voltages and currents but is also able to provide a complete statistical description‡ of these random processes as well.

The difficulties with the various thermal noise models become apparent at the h-f end of the spectrum, where a more profound investigation based on a detailed quantum-mechanical picture of the solid state is needed. From the viewpoint of kinetic theory, one must start with the correct Boltzmann equation, including Fermi-Dirac statistics in the mechanism

† For second-order statistics, the generalized Nyquist relations (cf. Sec. 11.1-3) have been further extended to include other types of general (linear) dissipative systems. See, for example, Callen and Welton[24] and associated work.[25]

‡ In the sense of Sec. 1.3-2.

of the electronic movement and interactions with the lattice of the conductor. If a statistical approach similar to Nyquist's is followed, one can enclose the conductor in a black body in equilibrium and, with the proper boundary conditions between conductor and radiation field, establish the specific form of the resistance $R'(f)$ for which the equipartition theorem is accordingly shown to hold. However, in place of the macroscopic boundary conditions usual in the Maxwell theory, the correct microscopic conditions incorporating the required Fermi-Dirac statistics are to be used, taking account also of the detailed surface, or "skin," structure of the conductor. Both the kinetic and statistical methods should lead to identical results, but the latter is undoubtedly the simpler. Note also that the statistical properties of the thermal noise are expected to depend on the *shape* and on the *material* of the conducting element.

The approximate theory of thermal noise in linear systems can also be carried over directly to the study of the Brownian motion, since (as mentioned earlier; cf. Chap. 10) the two theories are mathematically identical. This is the case, in talking of the Brownian movement, provided that we refer to the motion of *heavy* particles through a medium of agitated *light* particles. The Langevin equations then have the same structure as above, and from them similar Fokker-Planck equations follow. However, as we have seen earlier, the approximate electrical (or equivalent Brownian motion) model described here is not correct in the limiting case of light particles moving in clouds of comparably light or even very heavy particles. The Langevin equations are no longer appropriate for this physical situation, and a more refined dynamical structure must be considered. Furthermore, the Fokker-Planck equation is then not adequate to describe the diffusion process, and one must return to the original Smoluchowski [cf. Eq. (10.21)] or Boltzmann relation [cf. Eq. (10.41)] from which it was originally obtained.

The simple model of the random walk (cf. Sec. 7.7), based on the independence of individual displacements, and so useful in the theory of the Brownian motion[2] and thermal noise,[3] can no longer be applied, since the mean free time between collisions is no longer independent of the velocity. Suitable additional averages over the velocity distribution must now be included. Consequently, the normal random processes characteristic of the motion of heavy particles in a medium of particles of lesser mass and appearing as the usual limiting forms of the basic random-walk model are modified, losing their gaussian character. These difficulties are quite analogous to the problems that arise when a more comprehensive (that is, h-f) theory of the thermal noise is attempted.

PROBLEMS

11.1 (*a*) If we generalize the kinetic model of Sec. 11.1-1 to include the effects of a velocity distribution, show that, for classical statistics, the spectrum of the thermal noise current becomes

$$\mathcal{W}_I(f) = \frac{4k_BT_0}{R_0} \frac{\displaystyle\int_0^\infty \frac{\lambda v^3}{1+\omega^2\lambda^2/v^2} e^{-mv^2/2k_BT_0}\, dv}{\displaystyle\int_0^\infty \lambda v^3 e^{-mv^2/2k_BT_0}\, dv} \qquad v^2 = v_x^2 + v_y^2 + v_z^2 \tag{1}$$

and, if λ is taken independent of velocity, that

$$\mathcal{W}_I(f) = \frac{4k_BT_0}{R_0}[1 - \alpha^2 - \alpha^4 e^{\alpha^2} Ei(-\alpha^2)] \qquad Ei(-x) \equiv -\int_x^\infty e^{-y}\frac{dy}{y}$$
$$\alpha \equiv \omega\lambda\sqrt{\frac{2k_BT_0}{m}} \tag{2}$$

(*b*) Repeat for Fermi-Dirac statistics, and verify that approximately

$$\mathcal{W}_I(f) \doteq \frac{4k_BT_0}{R_0}\left(1 + \frac{\omega^2\lambda'^2}{v^2}\right)^{-1} \tag{3}$$

where λ' is the mean free path of electrons with velocity $v' = [(-2k_BT_0/m)A]^{1/2}$ and $A = -n^2/2mk_BT_0(3N/2\pi v)^{2/3} \doteq -E_m(k_BT_0)$. Bakker and Heller.[7]

11.2 (*a*) Prove the equipartition theorem for a series RL circuit, and show that

$$\mathcal{W}_I(f) = \frac{4k_BT_0R}{R^2+\omega^2L^2} \qquad K_I(t) = \frac{k_BT_0}{L}e^{-R|t|/L} \tag{1}$$

and obtain similarly, for the random charges in the circuit,

$$\mathcal{W}_Q(f) = \frac{4k_BT_0R}{\omega^2(R^2+\omega^2L^2)} \tag{2}$$

What is $K_Q(t)$? Show that $P = \overline{I^2}R = k_BT_0R/L$.

(*b*) Also establish the equipartition theorem for the single RC circuit, verifying that

$$\mathcal{W}_Q(f) = 4k_BT_0RC^2[1+(\omega RC)^2]^{-1} \qquad K_Q(t) = k_BT_0Ce^{-|t|/RC} \tag{3}$$

and show that

$$\mathcal{W}_I(f) = \omega^2\mathcal{W}_Q(f) = 4k_BT_0R\omega^2C^2[1+(\omega RC)^2]^{-1} \tag{4}$$

What are $K_I(t)$, P?

11.3 In the general system of n-coupled networks described in Sec. 11.1-4, prove that

$$E\{\dot{Q}_k\, \partial U_H/\partial \dot{Q}_k\} = k_BT_0 \tag{1}$$

thus completing the proof of the equipartition theorem in this instance.

11.4 Obtain, for a linear system of n-coupled harmonic oscillators (n-coupled series LCR circuits), the *mean power* dissipated in the system, namely,

$$\bar{P} = \sum_{k,l}^{n} \overline{\dot{Q}_k\dot{Q}_l R_{kl}} = \overline{\tilde{\dot{\mathbf{Q}}}\mathbf{R}\dot{\mathbf{Q}}} \tag{1}$$

11.5 Show for the general network of Fig. 2.2 that

$$R(f) = |Z(f)|^2 \Sigma R_l/|Z_{Al}(f)|^2 \tag{1}$$

and hence show that, when the resistances are at different temperatures T_m, the mean power $\bar{P}$ is given by

$$\bar{P} = 4k_B\int_0^\infty |Z(f)|^2 \sum_{l,m} \frac{R_lR_mT_m}{|Z_{Al}(f)Z_{Am}(f)|^2}\,df = 4k_B\int_0^\infty \sum_l \frac{R_lT_l}{|Z_{kl}(f)|^2}\,df \tag{2}$$

11.6 For the RL circuit of Prob. 11.2, show that the Fokker-Planck equation is specifically

$$\frac{\partial W_2}{\partial t} = \frac{k_B T_0 R}{L^2} \frac{\partial^2 W_2}{\partial I^2} + \frac{R}{L} \frac{\partial (W_2 I)}{\partial I} \qquad W_2 = W_2(I_0|I;t) \tag{1}$$

and hence show that specifically

$$W_2(I_0|I;t) = \left(\frac{L}{2\pi k_B T_0}\right)^{1/2} \left(1 - \exp\frac{-2k_B|t|R}{L}\right)^{-1/2} \times \exp\left\{-\frac{[I - I_0 \exp(-R|t|/L)]^2}{(2k_B T_0/L)[1 - \exp(-2R|t|/L)]}\right\} \tag{2}$$

Obtain $W_1(I)$ and, finally, $W_2(I_1,I_2;t)$. What are the associated characteristic functions?

11.7 For the RC circuit of Prob. 11.2b, obtain the Fokker-Planck equation

$$\frac{\partial W_2}{\partial t} = \frac{k_B T_0}{R} \frac{\partial^2 W_2}{\partial Q^2} + \frac{1}{RC} \frac{\partial (W_2 Q)}{\partial Q} \tag{1}$$

and show that this has the solution

$$W_2(Q_0|Q,t) = \left[2\pi k_B T_0 C \left(1 - \exp\frac{-2|t|}{RC}\right)\right]^{-1/2} \exp\left\{-\frac{[Q - Q_0 \exp(-|t|/RC)]^2}{2k_B T_0 C[1 - \exp(-2|t|/RC)]}\right\} \tag{2}$$

Obtain $W_1(Q)$ and $W_2(Q_1,Q_2;t)$. What are the associated characteristic functions here?

11.2 *Impulse Noise*

As we have mentioned in a number of previous chapters, the noise inherent in transmitting and receiving systems is for the most part due to thermal effects in both the passive and active elements of the system (e.g., the thermal noise, Sec. 11.1, and shot noise; cf. Sec. 11.3). Additional noise, of course, may enter a communication link through the medium of propagation. One common source is "static," or "interference," natural or man-made, where now the noise may have a noticeably different statistical character from the normal processes to which inherent system noise belongs for the most part (before rectification). The salient feature of this interference is its impulsive nature. These disturbances consist of sequences of pulses, of varying duration and intensity, and with a more or less random occurrence in time of the individual impulses. Some types of interference (usually man-made) appear as trains of nonoverlapping pulses, such as occur in pulse-time modulation (cf. Secs. 4.3-2, 4.4-1), for example, while others are the result of the random superposition of individual effects, e.g., atmospheric and solar static, due to thunder storms, sun spots, and the like. Here, we shall consider the latter type only,† although the present

† The material of this section is based for the most part on Middleton;[26] see also Rice[27] and Bunimovitch.[28] Additional references besides those cited in this chapter are given in Middleton. For a detailed mathematical treatment, see Blanc-Lapierre and Fortet.[29]

approach can be generalized to the analytically more complicated situations that arise when the nonoverlapping character of the elementary impulses is taken into account.† Our principal object, as before, is to determine the various probability densities W_1, W_2, . . . (cf. Sec. 1.3-2) which describe the simple (overlapping) impulsive noise process.

11.2-1 A General Model. Let us begin by constructing the hierarchy of distribution densities W_n $(n \geq 1)$ for a rather general random process $X(t;\mathbf{a})$ defined over the interval $(t_0, t_0 + T)$ when $\mathbf{a}$ represents a set of (time-invariant) random parameters the conditions on whose distribution we leave open for the moment except to state that they must be such that the hierarchy W_n $(n \geq 1)$ exists. The process $X(t;\mathbf{a})$, in addition, is assumed to be the (linear or nonlinear) resultant of exactly N "events" in the interval $(t_0, t_0 + T)$, where $N \geq 0$, and where each subensemble of N events may consist of an infinite number of representations. By "event," we designate here any elementary effect out of which the composite process $X(t;\mathbf{a})$ is formed, like the individual impulses of an atmospheric or a solar static, for example. If now we set

$P_1(N)_T$ = probability of exactly N events in interval $(t_0, t_0 + T)$ (11.65a)

$W_n(X_1,t_1; \ldots ;X_n,t_n|N)_T\, dX_1 \cdots dX_n$ = joint conditional probability that, if there are exactly N events in $(t_0, t_0 + T)$, X_1 lies in the range $(X_1, X_1 + dX_1)$ at a time t_1 $(t_0 \leq t_1 \leq T)$, X_2 falls in the interval $(X_2, X_2 + dX_2)$ at time t_2 $(t_1 \leq t_2 \leq T)$, . . . , etc. (11.65b)

where $X_1 = X(t_1;\mathbf{a})$, etc., we can write formally for the joint probability of $(X_1, \ldots ,X_n)$ in the ranges $(X_1, X_1 + dX_1)$, . . . , $(X_n, X_n + dX_n)$

$$W_n(X_1,t_1; \ldots ;X_n,t_n)_T\, dX_1 \cdots dX_n = \sum_{N=0}^{\infty} P_1(N)_T W_n(X_1,t_1; \ldots ;X_n,t_n|N)_T\, dX_1 \cdots dX_n \quad (11.66)$$

As $n = 1, 2, \ldots, \infty$, the $(W_n)_T$ are the desired distribution densities for our process $X(t;\mathbf{a})$ $(t_0 \leq t \leq T + t_0)$ considered at the various instants $(t_1,t_2, \ldots ,t_n)$ in the interval. We can see this by considering the Fourier transforms of $(W_n)_T$.

Equation (11.66) can be expressed in a fashion more suited to subsequent use. Thus, letting $F_n(i\xi_1,t_1; \ldots ;i\xi_n,t_n|N)_T$ be the (conditional) characteristic function corresponding to $W_n(X_1,t_1; \ldots ;X_n,t_n|N)_T$ in Eq. (11.65b),

† For details, see Middleton, *op. cit.*, pt. I, sec. 2. Spectra and second-moment functions in this case have been considered in Secs. 4.3, 4.4 above.

we can write directly (cf. Sec. 1.2-13)

$$F_n(i\xi_1,t_1; \ldots ;i\xi_n,t_n|N)_T \equiv E\left\{\exp\left[i\sum_{l=1}\xi_l X_N(t_l;\mathbf{a})\right]\right\} \tag{11.67a}$$

$$F_n(i\xi_1,t_1; \ldots ;i\xi_n,t_n|N)_T = \int_{(\mathbf{a})} w(\mathbf{a})\exp\left[i\sum_l \xi_l X_N(t_l;\mathbf{a})\right] d\mathbf{a} \tag{11.67b}$$

and the integration is to be performed over all allowed values of the random parameters. (The subscript N on X_N above reminds us that only the subensembles containing representations with exactly N events are to be considered.) We note from Eq. (11.67b) that when $N = 0$, that is, when there are no events in $(t_0, t_0 + T)$, F_n is unity, and the conditional probability density [Eq. (11.65b)] becomes

$$W_n(X_1,t_1; \ldots ;X_n,t_n|0)_T \equiv \mathfrak{F}^{-1}\{(F_n)_{N=0}\} = \prod_{l=1}^{n} \delta(X_l - 0) \tag{11.68}$$

Depending on the mechanism by which the elementary events are generated, i.e., on the form of $P_1(N)_T$ [Eq. (11.65a)], there may or may not be a finite probability (> 0) of value zero for the process. As we shall see presently, one distinguishing characteristic of impulse noise is just this finite probability of zero magnitude, which corresponds to the fact that there may be finite (> 0) intervals during which no impulse appears, unlike thermal or shot noise. The explicit structure of the hierarchy $W_1, W_2, \ldots$ will depend, of course, on the manner in which the resultants $X_N(t;\mathbf{a})$ are constructed. If the combination of individual impulses is nonlinear, or if individual events are statistically dependent, the result is very involved. However, for a process $X(t;\mathbf{a})$ which is produced by a linear superposition of elementary events, where these, in turn, are statistically independent, considerable simplification occurs. Fortunately, this proves to be the case of greatest general applicability, as the summary in Table 11.1 suggests.

We turn, therefore, to the important class of noise processes where these conditions of independence and linearity are obeyed.

11.2-2 Poisson Noise. In this model, the process $X(t;\mathbf{a})$ is assumed to be the result of the linear superposition of independent impulses, so that for the subensemble $X_N(t;\mathbf{a})$ consisting of exactly N such events in the basic interval $(t_0, t_0 + T)$ we have specifically

$$X_N(t;\mathbf{a}) = \sum_{m=0,1}^{N} a_m u(t - t_m; \tau_m)\cos\phi_m \qquad t_0 < t < t_0 + T \tag{11.69}$$

where† a_m $(-\infty < a < \infty)$, τ_m (≥ 0), ϕ_m $(0 \leq \phi_m \leq 2\pi)$ are the amplitude, duration, and "phase"‡ of the mth impulse; t_m is its epoch; and u itself is

† When $N = 0$, we use $m = 0$; otherwise, $m = 1$ is taken for the lower limit on the summation.

‡ This phase factor, $\cos\phi_m$, represents a change of scale or sign in the amplitude, according to the structure of the particular model [cf. Eq. (11.2)].

the basic waveform† (cf. Fig. 11.8). Since the impulses are statistically independent, the random parameters $\mathbf{a} = (a, \tau, \phi, t)$ are also, although amplitude and duration, phase and epoch, etc., may be related within any one impulse. Now, if the times of occurrence of individual pulses are uniformly as well as independently distributed in $(t_0, t_0 + T)$, this further condition, in addition to our other assumptions above for a, τ, ϕ, and t, is sufficient to define a Poisson distribution for the subensembles of N (≥ 0) events.[30] The basic probability (11.65a) is thus specifically[31] (cf. Prob. 1.4)

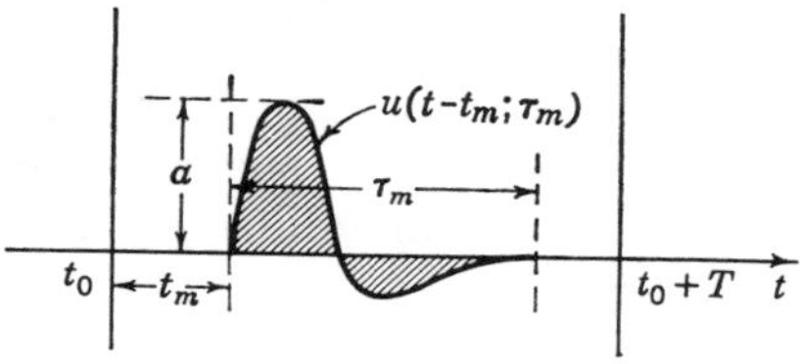

FIG. 11.8. An elementary impulse.

$$P_1(N)_T = \bar{N}_T{}^N e^{-\bar{N}_T}/N! \qquad N \geq 0 \tag{11.70}$$

where $\bar{N}_T$ is the average number of events in $(t_0, t_0 + T)$. Here, we may write $\bar{N}_T = \bar{n}_T T$, with $\bar{n}_T$ the average number of events per second.

We observe next that, on taking Fourier transforms of both sides of Eq. (11.66), the characteristic function of W_n can be written [Eqs. (11.67), (11.69), (11.70)] with the help of

$$F_n(i\xi_1,t_1; \ldots ;i\xi_n,t_n)_T = \sum_{N=0}^{\infty} P_1(N)_T F_n(i\xi_1,t_1; \ldots ;i\xi_n,t_n|N)_T$$
$$= \sum_{N=0}^{\infty} \frac{(\bar{n}_T T)^N}{N!} \exp(-\bar{n}_T T) \left\{ \int_0^\infty d\tau \int_0^\infty da \right.$$
$$\left. \times \int_0^{2\pi} d\phi\, w(a,\tau,\phi) \int_{t_0}^{t_0+T} \frac{d\lambda}{T} \exp\left[ia \cos\phi \sum_l^n \xi_l u(t_l - \lambda; \tau)\right]\right\}^N \tag{11.71a}$$

where we have used the fact that the epochs t_m [cf. Eq. (11.69)] are uniformly distributed in the interval. Summing the series then gives

$$(F_n)_T = \exp\left(\bar{n}_T \int_0^\infty d\tau \int_{-\infty}^\infty da \int_0^{2\pi} d\phi\, w(a,\tau,\phi)\right.$$
$$\left. \times \int_{t_0}^{t_0+T} \left\{\exp\left[ia\cos\phi \sum_l \xi_l u(t_l - \lambda; \tau)\right] - 1\right\} d\lambda\right) \tag{11.71b}$$

The desired nth-order probability density follows as the n-fold Fourier transform $(W_n)_T = \mathfrak{F}^{-1}\{(F_n)_T\}$.

Some simplification of Eq. (11.71b) is possible if we use the general properties of the individual pulses $u(t)$. First, if we postulate that outside the fundamental interval $(t_0, t_0 + T)$ there is no disturbance, we may extend

† All pulses are assumed to have the same waveform here; see Middleton, *op. cit.*, sec. 2, for generalizations.

the limits of integration over λ to $\pm\infty$. A small error is introduced by ignoring the relatively rare pulses which fall within a mean pulse length $\bar{\tau}$ of the ends of the interval, but in most practical applications, where T is much larger than $\bar{\tau}$, this effect is ignorable. In the limit $T \to \infty$, the error vanishes $O(1/T)$. Next, we require pulse shape u to be such that the integration over λ is convergent for all T. This is ensured in physical cases by the fact that individual pulses contain finite energy. Making the following substitutions,

$$z = \beta(t_1 - \lambda) \qquad z_l = (t_l - t_1)\beta \qquad l = 1, 2, \ldots, n \tag{11.72}$$

where $\beta \equiv (\bar{\tau})^{-1}$ is an effective bandwidth of an elementary impulse, and writing $v(z_l;\tau) = u[(z_l + z)/\beta;\tau]$, we can express Eq. (11.71b) finally as

$$F_n(i\xi_1,t_1; \ldots ;i\xi_n,t_n)_T = \exp\left((\bar{n}_T/\beta) \int_0^\infty d\tau \int_{-\infty}^{\infty} da \int_0^{2\pi} d\phi\, w(a,\tau,\phi) \int_{-\infty}^{\infty} dz \times \left\{\exp\left[ia \cos\phi \sum_l \xi_l v(z_l;\tau)\right] - 1\right\}\right) \tag{11.73a}$$

with

$$W_n(X_1,t_1; \ldots ;X_n,t_n)_T = \mathfrak{F}^{-1}\{F_n\} \tag{11.73b}$$

where $v(z_l;\tau)$ vanishes identically outside $(t_0,\ t_0 + T)$.

For those applications where we can safely assume a stationary process (cf. comments at the end of Sec. 1.3-7), we set $T \to \infty$ in the above and observe that Eqs. (11.73a), (11.73b) still apply, now with

$$\begin{aligned} \lim_{T\to\infty} \frac{\bar{n}_T}{\beta} &= \frac{\bar{n}_0}{\beta} = \bar{n}_0\bar{\tau} \\ &\equiv \gamma \\ &= \text{(av no. impulses/sec) (mean duration of typical impulse)} \end{aligned} \tag{11.74}$$

Thus, γ represents the "density" of pulses in a unit time interval. From Eqs. (11.73) it is clear that the magnitude of γ plays a critical role in determining the statistical structure of Poisson noise. For small γ, that is, comparatively little overlapping of pulses, the waveform of the individual pulses is a dominant factor. The distribution density of the pulse parameters, $w(a,\tau,\phi)$, is similarly influential. With little overlap, there will be appreciable gaps (in time) between successive pulses, and accordingly small or zero amplitudes for X are the most probable, with only occasional bursts that are comparable in intensity to that of a single pulse. For large γ, on the other hand, it follows directly from the central-limit theorem (cf. Sec. 7.7-3) that X belongs to a normal process, regardless of the specific form that $w(a,\tau,\phi)$ and pulse shape u may take. These quantities still affect the nature of the various first and second moments by which the gauss process is completely determined (cf. Sec. 7.3-2), but they no longer influence the general type of statistics.

For all γ (≥ 0), regardless of pulse density, we readily obtain the various moments of X by direct differentiation of Eq. (11.73a). Thus, for the second-moment function, we get (when $T \to \infty$)

$$M_X(t) = \frac{-\partial^2}{\partial\xi_1\,\partial\xi_2} F_n \Big|_{\xi_1=\xi_2=0} = \beta\gamma \left\langle a^2 \cos^2 \phi \int_{-\infty}^{\infty} u(t_1,\tau)u(t_1 + t, \tau)\, dt_1 \right\rangle_{stat\ av}$$
$$+ \beta^2\gamma^2 \left\langle a \cos \phi \int_{-\infty}^{\infty} u(t_1,\tau)\, dt_1 \right\rangle^2_{stat\ av} \qquad t = t_2 - t_1 \quad (11.75)$$

[cf. Eqs. (4.79*a*), (4.79*b*)]. The corresponding intensity spectrum follows directly from the W-K theorem [Eqs. (3.42), (3.78)]. We have

$$\mathcal{W}_X(f) = 2\mathfrak{F}\{M_X\} = \gamma\langle a^2 \cos^2 \phi\, w_u(f,\tau)\rangle_{stat\ av}$$
$$+ 2\gamma^2\beta^2 \left\langle a \cos \phi \int_{-\infty}^{\infty} u(t_1,\tau)\, dt_1 \right\rangle^2_{stat\ av} \delta(f - 0) \quad (11.76a)$$
$$\mathcal{W}_X(f) = 2\gamma\beta\langle a^2 \cos^2 \phi\, |S_u(f,\tau)|^2\rangle_{stat\ av}$$
$$+ 2\gamma^2\beta^2|\langle a \cos \phi S_u(f,\tau)\rangle|^2_{stat\ av}\, \delta(f - 0) \quad (11.76b)$$

where the spectral intensity density of the basic impulse u is $w_u(f,\tau) \equiv 2|S_u(f,\tau)|^2/\bar{\tau}$ [cf. Eqs. (4.49)] and $S_u(f,\tau)$ is the amplitude spectral density $S_u(f,\tau) = \int_{-\infty}^{\infty} u(t,\tau)e^{-i\omega t}\, df$ ($\omega = 2\pi f$) as usual [cf. Eqs. (2.43)]. Higher moments follow in the same way. Thus, writing $\Phi_n(i\xi_1,t_1; \ldots ;i\xi_n,t_n)$ for the coefficient of γ ($= \bar{n}_0/\beta$) in Eq. (11.73*a*) when $T \to \infty$, and noting that $\Phi_n(0,t_1;0,t_2; \ldots ;0,t_n)$ vanishes, we can reexpress the characteristic function F_n as

$$F_n(i\xi_1,t_1; \ldots ;i\xi_n,t_n) = 1 + \sum_{k=1}^{\infty} \frac{\gamma^k}{k!} \Phi_n(i\xi_1,t_1; \ldots ;i\xi_n,t_n)^k \quad (11.77)$$

The moments of order k are simply the coefficients of $(i\xi_1)^{k_1}(i\xi_2)^{k_2} \cdots (i\xi_n)^{k_n}/(k_1)! \cdots (k_n)!$, $k_1 + k_2 + \cdots + k_n = k$ (≥ 1) [cf. Eqs. (1.57)] in the expression for F_n above. The particular form of the characteristic functions (11.73) lends itself directly to a representation in terms of semi-invariants [cf. Eq. (1.58)]. We have accordingly (again for $T \to \infty$)

$$\lambda^{(L)}_{l_1 \cdots l_n} = \left\langle (a \cos \phi)^L \int_{-\infty}^{\infty} v(z_1 + z; \tau)^{l_1} v(z_2 + z; \tau)^{l_2} \cdots v(z_n + z; \tau)^{l_n}\, dz \right\rangle \quad (11.78)$$

where $l_1 + l_2 + \cdots + l_n = L$ (≥ 1) (all l_1, $l_2 \geq 0$, etc.) and $z_l + z = (t_l - \lambda)\beta$ [cf. Eq. (11.72)]. Note, also, that, except for the factor γ, Φ_n is the cumulant function of this multivariate process [cf. Eq. (1.58*a*)]. When $\gamma \to \infty$, we have a normal process with the second-moment function (11.75) (cf. Sec. 7.7-3).

11.2-3 Example and Discussion. Let us illustrate the foregoing with a simple example. For this, we assume a stationary train of overlapping *rectangular* impulses, where $w(\phi) = \delta(\phi - 0)$ and the distributions over duration τ and amplitude a may take any meaningful value. From Eq.

(11.73a), we get for the first-order characteristic function of $X(t;a)$ (as $T \to \infty$)

$$F_1(i\xi) = \exp(-\gamma) \exp\left[\gamma \int_{-\infty}^{\infty} da \int_0^{\infty} d\tau\, w(a,\tau)\, \beta\tau \exp(ia\xi)\right]$$
$$= \exp[-\gamma + \gamma F_a(i\xi)] \quad (11.79)$$

The corresponding first-order probability density is

$$W_1(X_1) = e^{-\gamma}\left\{\delta(X_1 - 0) + \sum_{k=1}^{\infty} \frac{\gamma^k}{k!}\left[\int_{-\infty}^{\infty} F_a(i\xi)^k e^{-i\xi X_1} \frac{d\xi}{2\pi}\right]\right\} \quad (11.80)$$

which exhibits the characteristic "spike" at $X_1 = 0$, for a finite (> 0) probability of zero magnitude of the random variable X_1. Figure 11.9 shows a typical density when the amplitude a is normally distributed with mean $\bar{a}$ and variance σ and the pulse density γ is small.

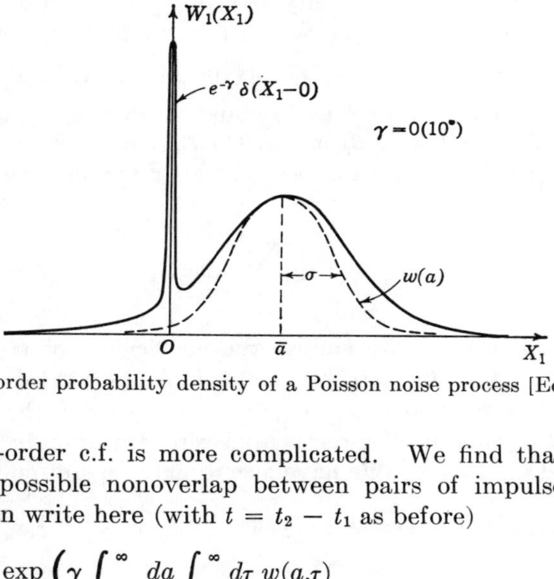

FIG. 11.9. First-order probability density of a Poisson noise process [Eq. (11.80)], small impulse density.

The second-order c.f. is more complicated. We find that now, taking into account possible nonoverlap between pairs of impulses, as well as overlap, we can write here (with $t = t_2 - t_1$ as before)

$$F_2(i\xi_1,i\xi_2;t) = \exp\Big(\gamma \int_{-\infty}^{\infty} da \int_{|t|}^{\infty} d\tau\, w(a,\tau)$$
$$\times \int_0^{\beta(\tau-|t|)} \{\exp[ia(\xi_1 + \xi_2)] - 1\}\, dz + \gamma \int_{-\infty}^{\infty} da \int_0^{\infty} d\tau\, w(a,\tau)$$
$$\times \int_0^{\beta\tau} [\exp(ia\xi_1) - 1 + \exp(ia\xi_2) - 1]\, dz\Big) \quad (11.81a)$$

$$F_2(i\xi_1,i\xi_2;t) = F_1(i\xi_1)F_1(i\xi_2) \exp\Big\{-\gamma\rho(t)$$
$$+ \gamma\rho(t) \int_{-\infty}^{\infty} w(a) \exp[ia(\xi_1 + \xi_2)]\, da\Big\} \quad (11.81b)$$

where $\rho(t)$ is given by

$$\rho(t) = \beta \int_{|t|}^{\infty} (\tau - |t|) w(\tau)\, d\tau \quad (11.81c)$$

The corresponding second-order probability density becomes

$$W_2(X_1;X_2;t) = e^{-\gamma\rho(t)} \sum_{k=1}^{\infty} \frac{\gamma^k \rho^k}{k!} \left[\iint_{-\infty}^{\infty} F_1(i\xi_1)F_1(i\xi_2)F_a(i\xi_1 + i\xi_2)^k e^{-i\xi_1 X_1 - i\xi_2 X_2} \frac{d\xi_1\, d\xi_2}{4\pi^2} \right] \tag{11.82}$$

and again the characteristic singularity appears for zero magnitudes, attributable to the gaps in time that distinguish these random waves from, say, a thermal noise process.

The example above and the results of Sec. 11.2-2 are by no means the most general models of the Poisson noise process: nonstationary sources ($|T| < \infty$), a nonuniform distribution of pulse epochs t_m, more than one pulse shape, and so on, can be included as well (cf. Middleton[26]). Narrow-band impulsive noise, where the individual pulses possess an envelope and phase structure, can be treated in the same way, although the resulting expressions are correspondingly more involved.

A second-moment theory of impulse noise in nonlinear devices is also possible, along the lines described in Chap. 5. Several cases of interest may be distinguished: (1) the low-density situation [$\gamma < O(10^0)$], where a direct development of the characteristic function in powers of γ [cf. Eq. (11.77)] is the natural step whereby the various contour integrals appearing in the relation for the second-moment function may be evaluated; and (2) the high-density case ($\gamma \gg 1$), where one has a *nearly normal process*, i.e., a process described by a generalized Edgeworth series of the forms (7.81), (7.84). Again, a series development of the characteristic function can be used [cf. Eqs. (7.80), (7.83)], with correction term $O(\gamma^{-1/2}$ or $\gamma^{-1})$ in the resulting second-moment function. Rectification of impulse noise by a zero-memory detector (under both these conditions) has been examined by Mullen and Middleton.[32] As expected, the analysis is much more complicated than in the corresponding case of normal noise.

We conclude this section with a brief list (see Table 11.1) of some of the more common physical noise processes which may be said to belong to the Poisson model. The essential feature, of course, of all such mechanisms is the statistical independence of the elementary impulses. If this condition can be reasonably assumed in practice, we may apply the procedure above to the statistical description of the process in question. If this is not possible, we must return to an approach similar to that outlined in Sec. 11.2-1, where, however, an actual calculation of distribution densities is usually very difficult. Fortunately, independence is usually not too critical an assumption—it is quite acceptable in the theory of thermal noise, for example, and as we shall see in the next section, for a common condition of shot noise, also. Moreover, pulse density γ is often very large, so that the limiting normal form of Poisson noise may be used, with a correspondingly great simplification in the analysis. Table 11.1 cites a few† of the Poisson noise processes that appear in communication situations, while Sec. 11.3 discusses shot noise in more detail.

TABLE 11.1

Poisson noise	Magnitude of γ	Character of distribution
1. Impulsive noise	$O(0\text{–}10)$	Depends on individual pulse shape and pulse statistics; strong dependence on magnitude of γ
a. Static; ignition noise; solar interference	$O(10^{-1})$	
b. Underwater sound; reflections from randomly oriented objects moving relative to observer (clutter)	$O(10^{-1})$	
c. Speech (random "burst" model)†	$O(10^{-1})$	
2. Nearly normal noise	$O(10)\text{–}O(10^4)$	Normal distribution with one or more correction terms; these are of order $\gamma^{-1/2}$ or γ^{-1}, depending on whether or not the third moments exist ($\neq 0$); noticeable to weak dependence on magnitude of γ
a. Heavy atmospheric static (ionosphere)		
b. Precipitation noise		
c. Clutter, sea waves, etc.		
d. Underwater sound (heavy overlapping of impulses for randomly moving obstacles) (clutter)		
e. Solar static (ionosphere)		
3. Normal noise	$O(10^4\text{–}\infty)$	Normal distribution; ignorable correction terms (γ enters only as a scale factor for the probability densities, whose *form* now does not depend on γ)
a. Shot noise (tube and crystal)		
b. Photomultiplier noise		
c. Thermal noise		
d. Clutter (scattering from water droplets); ground clutter		

† W. B. Davenport, An Experimental Study of Speech Wave Probability Distributions, *J. Acoust. Soc. Am.*, **24:** 390 (1952).

11.3 *Temperature-limited Shot Noise*

Shot noise is the name usually given to the noise that arises in vacuum tubes and crystals because of the random emission and motion of electrons in these active elements. Noise of this type appears as a randomly fluctuating component of the output current and along with thermal noise is an important factor inhibiting the performance of transmitting and receiving systems. Depending on the device in question, and on its operating conditions, shot noise will have a variety of different statistical properties. For

† A much more comprehensive summary is provided in *Tech. Rept.* 189 (cf. Mullen and Middleton[32]). For other generalizations of the model, see also Middleton, *op. cit.*

example, a tube may be operated as a *temperature-limited diode*, where the essential factor limiting the anode current is the temperature of the cathode (or emitter) element. As we shall see below (Sec. 11.3-1), in this condition the output current belongs effectively to a normal random process. On the other hand, the active element more commonly may be run under a condition of *space-charge limiting*, where now the output current is limited by the cloud of electrons in the cathode-anode region, rather than by a restricted supply of emitted electrons. Now the current may no longer belong to a normal process, nor does it even have the same spectral and other second-moment properties. Further modifications of the physical model occur in multigrid and microwave tubes: apart from space-charge effects, intervening grids may capture some of the emitted electrons, producing various forms of *partition noise*, while the presence of applied magnetic (as well as the usual electric) fields—as in the magnetron or traveling-wave tube, for example—greatly complicates the description of space-charge and output current. Since our primary purpose here is to illustrate the general methods described in previous chapters (cf. Secs. 7.3, 7.4, 7.6, 11.2, etc.), rather than to pursue in detail the question of noise generation in physical elements, we shall not consider these more involved situations, but only the comparatively simple case of the temperature-limited diode. For this, a fairly complete statistical (as well as physical) theory is available. For details in the more general case, the reader may consult the standard references,† although comparatively little is available, as yet, concerning the statistical description of these more complicated processes.

11.3-1 Probability Densities. Before we examine particular waveforms of the shot noise current induced in the anode of a temperature-limited diode, let us first obtain general expressions for the various probability densities W_n $(n \geq 1)$ describing this noise process. This we can readily do with the help of Sec. 11.2-2 and a few reasonable assumptions as to the underlying statistical mechanism.

We observe, to begin with, that the induced current at the anode is the cumulative effect of a (large) number of individual current pulses, $i_e(t)$, each of which results from the emission of a single electron from the cathode and from its subsequent motion to the anode under the applied electric field. Now, it is reasonable to assume that these individual emissions occur at random times and independently of one another.‡ Furthermore, because of the comparatively high potential V_0 in temperature-limited operation, individual electrons do not noticeably influence each other in transit, which is another way of saying that space-charge effects are ignor-

† See, for example, the books of Valley and Wallman,[34] Spangenberg,[35] Lawson and Uhlenbeck,[36] Harman,[37] Van der Ziel,[38] and Smullin and Haus.[50] See also Davenport and Root,[39] chap. VII.

‡ The emitting surface is assumed to be pure and homogeneous, so that *flicker effect*[40] can be ignored.

able here. The induced anode current can therefore be regarded as the *linear* superposition of the individual current pulses $i_e(t - t_m)$, where the epochs (i.e., initial times; cf. Fig. 11.8) t_m are randomly and independently distributed in any interval $(t_0, t_0 + T)$. We assume, in addition, that the diode is operating in a stable steady state, so that these initial times are also uniformly distributed. A typical subensemble of current waves, consisting of exactly N (≥ 0) pulses in the interval $(t_0, t_0 + T)$,† is accordingly written [cf. Eq. (11.69)]

$$I(t;\boldsymbol{\theta}) = \sum_{m=0,1}^{N} i_e(t - t_m; \boldsymbol{\theta}) \qquad t_0 < t,\, t_m < t_0 + T \tag{11.83}$$

where $\boldsymbol{\theta}$ is a set of possible random parameters that may enter in the detailed description of waveform, e.g., the Maxwell-Boltzmann distribution of emission velocities, etc.

The assumption that the individual pulses i_e are independent,‡ with a uniform distribution of emission times, is sufficient to define a Poisson distribution for the number of pulses N in the subensemble (11.83); that is, $P_1(N)_T$ is given by Eq. (11.70). The results of Sec. 11.2-2 may now be applied directly to give all orders of probability density W_n. Letting $T \to \infty$ and using Eqs. (11.73) in this limit, with the help of Eq. (11.74) we obtain for the n random variables $I_1 = I(t_1), \ldots, I_n = I(t_n)$ representing the induced anode current at the various times $t_1, t_2, \ldots, t_n$ the characteristic function $(n \geq 1)$

$$F_n(i\xi_1;i\xi_2,t_2; \ldots ;i\xi_n,t_n) = \exp\left(\gamma\beta \int_{(\boldsymbol{\theta})} w(\boldsymbol{\theta})\, d\boldsymbol{\theta} \int_{-\infty}^{\infty} \left\{\exp\left[i \sum_l \xi_l i_e(t_l + \lambda; \boldsymbol{\theta})\right] - 1\right\} d\lambda\right) \tag{11.84}$$

where t_1 is set equal to zero, since the process is stationary, and $\beta^{-1} = \bar{\tau}$ is the mean duration of a typical impulse. The corresponding probability density is

$$W_n(I_1;I_2,t_2; \ldots ;I_n,t_n) = \int \cdots \int_{-\infty}^{\infty} \exp\left(-i \sum_l \xi_l I_l\right) \times \left[\exp\left(\gamma\beta \int_{(\boldsymbol{\theta})} w(\boldsymbol{\theta})\, d\boldsymbol{\theta} \times \int_{-\infty}^{\infty} \left\{\exp\left[i \sum_l \xi_l i_e(t_l + \lambda; \boldsymbol{\theta})\right] - 1\right\} d\lambda\right)\right] \frac{d\xi_1 \cdots d\xi_n}{(2\pi)^n} \tag{11.85}$$

Temperature-limited shot noise is thus one form of impulsive Poisson noise (Sec. 11.2-2).

† End effects are ignorable here as long as T is very much larger than the transit time τ of an individual electron.

‡ This implies that the $\boldsymbol{\theta}$ are likewise independent from pulse to pulse.

In temperature-limited operation, we can usually neglect the initial velocities of the emitted electrons because of the large applied potential and correspondingly great transit velocities. Thus, β^{-1} is equal to τ, the transit time. If, further, we assume the diode to consist of parallel plane anode and cathode surfaces and neglect edge effects, we can omit the random parameters $\boldsymbol{\theta}$ entirely in the above. For example, the first-order density becomes now, from Eq. (11.85),

$$W_1(I) = \int_{-\infty}^{\infty} \exp(-i\xi I) \left[\exp\left(\gamma\beta \int_{-\infty}^{\infty} \{\exp[i\xi i_e(\lambda)] - 1\}\, d\lambda \right) \right] \frac{d\xi}{2\pi} \qquad \beta = \tau^{-1} \tag{11.86}$$

a result obtained originally by Rice.[27] The second-order density is similarly

$$W_2(I_1,I_2;t) = \iint_{-\infty}^{\infty} \exp(-i\xi_1 I_1 - i\xi_2 I_2) \times \left[\exp\left(\gamma\beta \int_{-\infty}^{\infty} \{\exp[i\xi_1 i_e(\lambda) + i\xi_2 i_e(\lambda + t)] - 1\}\, d\lambda \right) \right] \frac{d\xi_1\, d\xi_2}{(2\pi)^2} \tag{11.87}$$

and so on, for the higher-order cases ($n \geq 3$).

When the pulse density γ is very large (and this is the usual condition in practice), the distribution densities (11.85) to (11.87) all approach an appropriate normal form, as we have already noted in Sec. 11.2-2. To see how the normal structure is approached, let us apply Eq. (7.80) to Eqs. (11.86), (11.87) specifically. For the characteristic function [e.g., the expressions in the large brackets of the integrands of Eqs. (11.86), (11.87)], we obtain

$$F_1(i\xi) \doteq \exp\left(i\xi\lambda_1^{(1)} - \tfrac{1}{2}\lambda_2^{(2)}\xi^2\right) \left\{ 1 + \lambda_3^{(3)} \frac{(i\xi)^3}{3!} + \left[\frac{\lambda_4^{(4)}}{4!} (i\xi)^4 + \frac{\lambda_3^{(3)2}}{72} (i\xi)^6 \right] + \cdots \right\} \tag{11.88}$$

$$\begin{aligned} F_2(i\xi_1,i\xi_2;t) &\doteq \exp\left[i\xi_1\lambda_{10}^{(2)} + i\xi_2\lambda_{01}^{(2)} - \tfrac{1}{2}(\lambda_{20}^{(2)}\xi_1^2 + 2\lambda_{11}^{(2)}\xi_1\xi_2 + \lambda_{02}^{(2)}\xi_2^2)\right] \\ &\times \left[1 + \frac{\lambda_{30}^{(3)}(i\xi_1)^3 + 3\lambda_{21}^{(3)}(i\xi_1)^2(i\xi_2) + 3\lambda_{12}^{(3)}(i\xi_1)(i\xi_2)^2 + \lambda_{03}^{(3)}(i\xi_2)^3}{3!} + \cdots \right] \end{aligned} \tag{11.89}$$

where the λ's are the semi-invariants [cf. Eq. (1.58)] from Eq. (11.78), here specifically

$$\lambda_1^{(1)} = \gamma\beta \int_{-\infty}^{\infty} i_e(x)\, dx = \lambda_{10}^{(2)} = \lambda_{01}^{(2)} \qquad \beta = \tau^{-1} \tag{11.90a}$$

$$\lambda_2^{(2)} = \gamma\beta \int_{-\infty}^{\infty} i_e(x)^2\, dx = \lambda_{20}^{(2)} = \lambda_{02}^{(2)} \tag{11.90b}$$

and

$$\lambda_{11}^{(2)} = \gamma\beta \int_{-\infty}^{\infty} i_e(x) i_e(x + t)\, dx = \lambda_{11}^{(2)}(t) \tag{11.90c}$$

Higher-order semi-invariants for the first- and second-order distributions may be obtained from

$$\lambda_{l_1}^{(l_1)} = \gamma\beta \int_{-\infty}^{\infty} i_e(x)^{l_1}\,dx \qquad \lambda_{l_1l_2}^{(L)} = \int_{-\infty}^{\infty} i_e(x)^{l_1} i_e(x+t)^{l_2}\,dx \quad (11.90d)$$

where $L = l_1 + l_2 \geq 3$ [cf. Eq. (11.78) again].

The semi-invariants are related to the various moments of the distributions in the usual way (cf. Prob. 1.2), the most important cases of which are

$$\lambda_1^{(1)} = \bar{I} \qquad \lambda_2 = \overline{I^2} - \bar{I}^2 \equiv \sigma^2 \qquad \lambda_{11}^{(2)}(t) = E\{(I_1 - \bar{I}_1)(I_2 - \bar{I}_2)\} \qquad t = t_2 - t_1 \quad (11.91)$$

If we use these in Eqs. (11.88), (11.89), taking the Fourier transform, we find that [cf. Eq. (7.81)] the corresponding first- and second-order probability densities are given in the high-density cases ($\gamma \gg 1$) by the Edgeworth series

$$W_1(I) \simeq \frac{1}{\sigma}\,\phi(x) + \frac{1}{\sigma}\left\{\frac{-\lambda_3^{(3)}}{3!\sigma^3}\,\phi^{(3)} + \left[\frac{\lambda_4^{(4)}}{4!\sigma^4}\,\phi^{(4)}(x) + \frac{\lambda_3^{(3)2}}{72\sigma^6}\,\phi^{(6)}(x)\right] - \cdots\right\} \quad (11.92)$$

where $x \equiv (I - \bar{I})/\sigma$, and

$$W_2(I_1,I_2;t) \simeq \left(1 - \sum_{L=3}^{3} \frac{\lambda_{l_1l_2}^{(3)}}{l_1!l_2!}\,\frac{\partial^L}{\partial x_1^{l_1}\,\partial x_2^{l_2}} + \cdots\right) \times \exp\left[-\frac{x_1^2 + x_2^2 - 2x_1x_2\lambda_{11}^{(2)}\sigma^{-2}}{2(1 - \lambda_{11}^{(2)}/\sigma^2)}\right] \quad (11.93)$$

with $x_1 \equiv (I_1 - \bar{I})/\sigma$, etc.

To see precisely when these higher-order terms may be neglected and a purely normal process results, as well as to determine the spectrum and other moments of the process, we must now obtain the explicit waveform of a typical induced current impulse $i_e(t)$

11.3-2 Waveform of the Induced Current. We begin by assuming once more the specific case of a parallel-plane diode, with a cathode-anode spacing of d (centimeters) and a (steady) potential drop of V_0 volts. Because of the large value of V_0 in temperature-limited operation, all electrons emitted from the cathode achieve such high velocities that a negligible fraction can contribute to a space charge, with the result that the space-charge density, between anode and cathode, vanishes. The potential $V(x)$ at any point in the tube (but away from the edges of the plates) is accordingly governed by Laplace's equation,† $\nabla^2 V(x) = 0$, or $d^2V/dx^2 = 0$, which has the simple solution $V(x) = V_0x/d$, if cathode and anode are located at $x = 0$, $x = d$, respectively. The electric field intensity is then $\mathbf{E} = \mathbf{i}_x\,dV/dx = \mathbf{i}_xV_0/d$, where $\mathbf{i}_x$ is a unit vector in the x direction. The force on a typical

† See the original paper of Langmuir[41] for a full account of the spatial current calculations in plane and cylindrical diodes.

electron is then $\mathbf{F} = -e\mathbf{E}$ (e = electronic charge = 4.8×10^{-10} esu), so that the differential equation of motion is

$$\mathbf{F} = m\frac{d^2x}{dt^2} = -e\mathbf{E} = \frac{eV_0}{d} \tag{11.94}$$

If v_0 is the initial velocity of the electron (at $x = 0$, $t = 0$), we find the integrated equations of motion to be

$$\dot{x} = v(t) = (eV_0/md)t + v_0 \qquad x(t) = (eV_0/2md)t^2 + v_0t \tag{11.95}$$

From these, the transit time τ of the electron is easily found ($x = d$, $t = \tau$), namely,

$$\tau = \eta_0 d \qquad \text{where } \eta_0 \equiv -v_0m/eV_0 + \sqrt{2m/eV_0 + (v_0m/eV_0)^2} \tag{11.96}$$

Since the initial velocity v_0 is usually negligible here,† these relations simplify to

$$\tau \doteq (2m/eV_0)^{1/2}d \qquad v(t) \doteq (eV_0/md)t = (2d/\tau)(t/\tau)$$
$$x(t) \doteq (eV_0/2md)t^2 = d(t/\tau)^2 \tag{11.97}$$

with $0 \le t \le \tau$, of course.

The waveform of the current pulse induced in the anode by the motion of the electron toward it can now be obtained from the time derivative of the charge induced in the anode by this moving electron. We let Q be the induced charge and observe that the potential energy U gained by the electron in moving a distance x through the potential difference $V(x)$ is just $U = eV(x) = eV_0x/d$. This, in turn, is equal to the work, V_0Q, that must be done to induce Q at the anode when the anode is at potential V_0. Equating and solving for Q gives $Q = ex/d$, so that the induced current pulse becomes

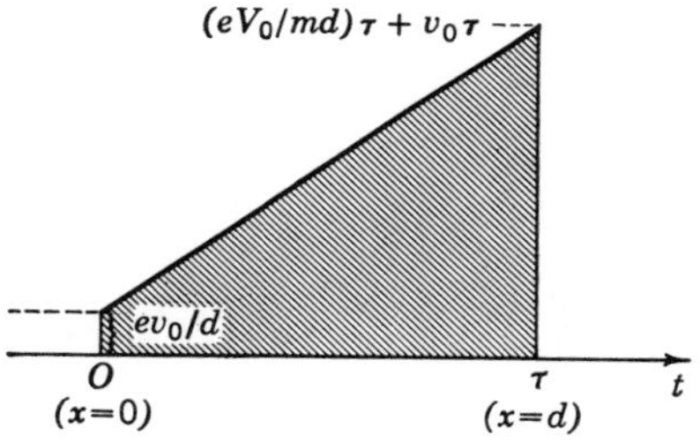

FIG. 11.10. A typical current pulse induced in the anode of a temperature-limited (plane) diode.

$$i_e(t) = \frac{dQ}{dt} = \frac{e\dot{x}}{d} = \begin{cases} \left(\dfrac{e}{d}\right)^2 \dfrac{V_0}{m}t + \dfrac{e}{d}v_0 & 0 \le t \le \tau \\ 0 & \text{otherwise} \end{cases} \tag{11.98a}$$

For negligible initial velocities, this becomes

$$i_e(t) \begin{cases} \doteq (e/d)^2(V_0/m)t = (2e/\tau)(t/\tau) & 0 \le t \le \tau \\ = 0 & \text{otherwise} \end{cases} \tag{11.98b}$$

A typical impulse is shown in Fig. 11.10.

† 3000°K is equivalent to about 0.25 ev, while V_0 is O(100).

11.3-3 Moments and Spectra. The various moments of the shot noise current, $I(t)$, in the anode of our planar diode can now be found from the characteristic functions (11.88), (11.89) by differentiation [cf. Eqs. (1.31)] or from the semi-invariants (11.90), when Eq. (11.98a) or (11.98b) is used for $i_e(x)$. We find that (neglecting initial velocities) the average diode current is

$$\bar{I} = \lambda_1^{(1)} = \gamma\beta e = \bar{n}_0 e \tag{11.99}$$

where $\bar{n}_0$ is the average number of electron emissions (or arrivals) per second, since $\gamma = \bar{n}_0/\beta$ [cf. Eq. (11.74)]. The mean-square current and the variance of I follows from Eqs. (11.90b) and (11.91). We get for the variance

$$\sigma^2 \equiv \overline{I^2} - \bar{I}^2 = \lambda_2^{(1)} = \frac{4}{3\tau}\gamma\beta e^2 = \tfrac{4}{3}\bar{n}_0\frac{e^2}{\tau} = \frac{4}{3}e\bar{I} \tag{11.100a}$$

while $\overline{I^2}$ becomes

$$\overline{I^2} = \bar{n}_0\frac{e^2}{\tau}\left(\frac{4}{3} + \bar{n}_0\tau\right) \tag{11.100b}$$

The covariance function of the diode noise process is simply Eq. (11.90c) [or the third relation in Eqs. (11.91)]. With Eq. (11.98b), this reduces specifically to

$$K_{I-\bar{I}}(t) \equiv \mathbf{E}\{(I_1 - \bar{I})(I_2 - \bar{I})\} = \begin{cases} \lambda_{11}^{(2)}(t) \\ \dfrac{\gamma\beta}{\tau^2}\displaystyle\int_0^{\tau-|t|}\left(\frac{2e}{\tau}\right)^2 x(x+|t|)\,dx & |t| \le \tau \\ 0 & \text{otherwise} \end{cases} \tag{11.101a}$$

$$K_{I-\bar{I}}(t) \equiv \mathbf{E}\{(I_1 - \bar{I})(I_2 - \bar{I})\} = \begin{cases} \dfrac{4}{3}\dfrac{e}{\tau}\bar{I}\left(1 - \dfrac{3}{2}\dfrac{|t|}{\tau} + \dfrac{1}{2}\dfrac{|t|^3}{\tau^3}\right) & |t| < \tau \\ 0 & \text{otherwise} \end{cases} \tag{11.101b}$$

The spectrum of the fluctuation $I - \bar{I}$ is obtained by applying the W-K theorem [Eqs. (3.42)] to Eqs. (11.101). The result is[42]

$$\mathcal{W}_{I-\bar{I}}(f) = 2\mathfrak{F}\{K_{I-\bar{I}}\} = e\bar{I}\frac{8}{(\omega\tau)^4}[(\omega\tau)^2 + 2(1 - \cos\omega\tau - \omega\tau\sin\omega\tau)] \qquad \omega = 2\pi f \tag{11.102}$$

which is shown in Fig. 11.11. Note that the l-f spectrum, where $(\omega\tau)^2 \ll 1$, reduces to

$$\mathcal{W}_{I-\bar{I}}(f) = 2e\bar{I} \qquad f \ll 1/\tau \tag{11.102a}$$

which is Schottky's original result,[43] based on the more elementary model of current impulses of the form $i_e(t) = e\,\delta(t - 0)$. In fact, applying this expression in Eq. (11.90c) gives

$$K_{I-\bar{I}}(t) \doteq \gamma\beta e^2\,\delta(t - 0) \qquad \mathcal{W}_{I-\bar{I}}(f) \simeq 2\gamma\beta e^2 = 2e\bar{I} \tag{11.102b}$$

which is, of course, not valid for frequencies $O(\tau^{-1})$ or higher; Eq. (11.102b) implies essentially an infinite transit velocity for the electrons, or, equivalently, a zero transit time.† Note that while this simple current model,

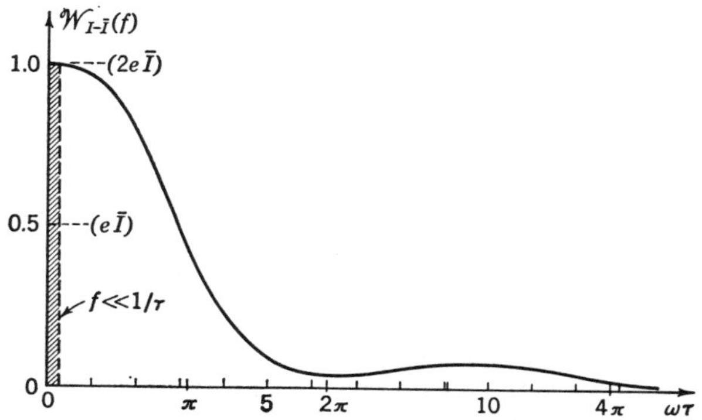

FIG. 11.11. The intensity spectrum of the random component, $I - \bar{I}$, of the shot noise current in a temperature-limited planar diode.

$i_e(t) = e\,\delta(t - 0)$, yields a correct average value for the shot noise current, e.g.,

$$\bar{I} = \gamma\beta \int_{-\infty}^{\infty} e\,\delta(x - 0)\,dx = \bar{n}_0 e$$

the higher moments become infinite again because of this assumption of zero transit time. The complete spectrum of the shot noise current at the anode, including the steady component, follows at once from Eqs. (11.99), (11.102) and

$$\mathcal{W}_I(f) = 2\mathfrak{F}\{\bar{I}^2 + K_{I-\bar{I}}(t)\} = 2\bar{I}^2\,\delta(f - 0) + \mathcal{W}_{I-\bar{I}}(f) \quad (11.103)$$

Figure 11.11 is simply modified by adding a spike, $2\bar{I}^2\,\delta(f - 0)$, at $f = 0$.

There remains for us to establish quantitatively how large the pulse density γ must be for the shot noise process to obey gaussian statistics. Let us do this, in part, by examining the correction terms in the Edgeworth series (11.92) when a typical operating condition of, say, an anode-cathode potential of $V_0 = 200$ volts, a mean current of 20 ma, and a spacing of 2 cm is assumed. Ignoring initial emission velocities, we find from Eqs. (11.97) that here the transit time τ is 2.4×10^{-9} sec (corresponding to an upper spectral limit of about 400 Mc/sec), so that the density factor γ is 4.7×10^8, with $\gamma\beta = \bar{n}_0 = \bar{I}/e = 1.2 \times 10^{17}$ arrivals per second, on the average. Inserting these values into Eq. (11.92), with the help of Eqs. (11.90a), (11.90b), (11.90d) we see that the maximum value of the first correction term is $(\sqrt{3}/\sqrt{\gamma})\phi^{(3)}_{\max}$, where $|\phi^{(3)}|_{\max}$ is approximately 0.55. The correction is therefore clearly negligible compared with the first term, so that we

† Setting $\tau \to 0$ in Eq. (11.102) also yields Eqs. (11.102a), (11.102b) for all $f > 0$.

may safely say that shot noise in the temperature-limited cases is well represented by a normal process.†

The fact that we are able to obtain a complete statistical description of the shot noise process in the present case is, of course, due to our assumption of independence of the effects of the individual electrons. This assumption is quite reasonable, since space-charge effects are essentially nonexistent in temperature-limited operation. However, as soon as the space charge becomes at all appreciable (at lower operating voltages than in the temperature-limited case), electron motion from cathode to anode is influenced by that of other electrons and the induced current pulses at the anode lose their independent character. The potential distribution in the conduction region is also no longer linear but takes a rather complicated form, with a potential minimum that is now at some point between cathode and anode. For these reasons, reflected mainly in the statistically dependent nature of the induced current pulses, it has not proved possible‡ as yet to obtain the various probability densities describing space-charge-limited shot noise processes. A second-moment theory, including results for the current spectrum, has, however, been developed by North,[44] Schottky and Spenke,[45] Rack,[42] and others.[46] (See also Lawson and Uhlenbeck,[36] secs. 4.7, 4.8, as well as Van der Ziel,[38] chap. 6, for a more complete account and additional references.) We remark, finally, that the random character of both thermal and shot noise stems from the common feature of a random movement of discrete charges. Randomness in the former case arises because of the thermal agitation of the conducting electrons within the material in question, while in the latter it is the random emission from the cathode surface of the active element that produces the fluctuating current at the anode. Similar observations apply for multigrid tubes and other, more complicated devices, which are outside the scope of the present study.

PROBLEMS

11.8 Obtain the various nth-order characteristic functions and distribution densities of temperature-limited shot noise for a finite interval $(t_0, t_0 + T)$, and examine the role of the end effects as $T \to \tau$, the transit time.

11.9 Investigate the approach to the normal process as the impulse density in temperature-limited shot noise becomes increasingly great. Assume a typical impulse of the form (11.98*b*), and consider $W_2, W_3, \ldots, W_n$. Anode-cathode spacing is 2 cm, and $V_0 = 150$ volts, with $\bar{I} = 80$ ma.

Another class of noise process, related to the impulse types discussed in Secs. 11.2, 11.3, is the so-called "clutter," or scatter, noise, appearing as the return from large

† Strictly speaking, we must also verify the gaussian nature of W_n $(n \geq 2)$, but from Eq. (7.81) and our results above it is not too difficult to show that higher-order terms in the generalized Edgeworth series can also be discarded.

‡ Although individual electrons are not independent, the emitting cathode can often be divided into a large number of independently emitting areas, so that γ is very large and the resulting noise is gaussian. A second-moment theory then completes the statistical description. (See reference 22 in Ref. 32.)

numbers of more or less independent scatterers when an incident wave illuminates them. When the number of independent scatterers is large, the scatter return usually belongs to a normal process, provided that the original source is a sinusoid.[47-49] Thus, if $x^{(i)}(t)$ is the incident sine wave, the resulting noise disturbance is the narrowband return

$$z^{(j)}(t)' = \mathbf{T}_M^{(N)}\{x^{(i)}(t)\} = z_c'(t) \cos \omega_0 t + z_s'(t) \sin \omega_0 t$$

where ω_0 $(= 2\pi f_0)$ is the frequency of the sinusoid. In the following problems, the basic scattering mechanism is assumed to be stationary.

11.10 (a) Show that, when a receiver is moving relative to the assembly of scatterers, the return intercepted by it is, at fixed range,

$$z(t) = z_c(t) \cos \omega_c t + z_s(t) \sin \omega_c t \qquad \omega_c = \omega_0 + \omega_d \tag{1}$$

where ω_d is the (angular) Doppler frequency produced by the relative motion.

(b) By considering z_c, z_s as the components of a two-dimensional random walk, show that the first-order density of z_c, z_s approaches

$$W_1(z_c,z_s) = \frac{e^{-(z_c^2+z_s^2)/2\psi_z}}{2\pi\psi_z} \tag{2}$$

where $\overline{z^2} = \overline{z_c^2} = \overline{z_s^2} = \psi_z$ and $\bar{z}_c = \bar{z}_s = \bar{z} = 0$ if there is no specular reflection or coherent scattering. Thus, show that the first-order d.d. of the intensity $I_z(t) = z_c^2(t) + z_s^2(t)$ is given by

$$W_1(I_z) \simeq (\bar{I}_z)^{-1}e^{-I_z/\bar{I}_z} \qquad I_z > 0;\ \bar{I}_z = \overline{z_c^2} + \overline{z_s^2} = 2\psi_z \tag{3}$$

(c) Discuss the assumptions here by which the central-limit theorem is applied to the scatter model in question. Note that the results above are for the scatter returns received at the same instant at a receiver, i.e., in radar applications, at fixed range.

11.11 (a) Obtain the second-order d.d. of the intensity $I_z(t)$ in Prob. 11.10, and hence show specifically that

$$W_2(I_{z1},I_{z2};t) \simeq [\bar{I}_z^2(1 - k_0^2)]^{-1}e^{-(I_{z1}+I_{z2})/\bar{I}_z(1-k_0^2)}I_0\left[\frac{2k_0\sqrt{I_{z1}I_{z2}}}{(1 - k_0^2)\bar{I}_z}\right] \qquad 0 \le I_{z1}, I_{z2} \le \infty \tag{1}$$

where $$k_0^2(t) = (\overline{I_{z1}I_{z2}} - \bar{I}_z^2)/(\overline{I_z^2} - \bar{I}_z^2) \tag{2}$$

with $\overline{I_z^2} = 2\bar{I}_z^2$.

(b) In terms of the spectrum of the fixed-range return of $z(t)$, show that

$$\rho_0(t)_z + i\lambda_0(t)_z = \psi_z^{-1}\int_{-\infty}^{\infty} \mathcal{W}_z(f)e^{i(\omega-\omega_c)t}\,df \tag{3}$$

where $k_0(t) = |\rho_0(t)_z + i\lambda_0(t)_z|$ and $\rho_0(t)_z = \overline{z_{c1}z_{c2}} = \overline{z_{s1}z_{s2}}$; $\lambda_0(t)_z = \overline{z_{c1}z_{s2}} = -\overline{z_{c2}z_{s1}}$.

Lawson and Uhlenbeck.[47]

11.12 (a) If $X(t) = X_c(t) \cos \omega_0 t + X_s(t) \sin \omega_0 t = E_X(t) \cos [\omega_0 t - \Phi_X(t)]$ is a narrowband process incident upon an assembly of scatterers, show that the complex random fixed-range return intercepted by a receiver in relative motion with the scatterers is proportional to

$$\begin{aligned} Z(t) = \mathbf{T}_M^{(N)}\{X(t)\} &\doteq (X_c z_c + X_s z_c \cos \phi_d - X_s z_s \sin \phi_d) \cos \omega_c t \\ &\quad + (X_c z_s + X_s z_s \cos \phi_d + X_s z_c \sin \phi_d) \sin \omega_c t \qquad (1a) \\ &= Z_c(t) \cos \omega_c t + Z_s(t) \sin \omega_c t \qquad (1b) \end{aligned}$$

where $\phi_d = \pi\omega_c/2\omega_0 = \pi(1 + \omega_d/\omega_0)/2$ $(\omega_c, \omega_0 \gg \omega_d)$ in all practical cases. Here, $z(t)$ is the normal random noise return when the scatterers are illuminated by a sinusoidal process. The form of Eq. (1a) shows that the components of the incident wave X and basic scatter noise z are multiplied together in various combinations, from which the term "multiplicative" process for $Z(t)$ arises.

(b) If $E_z = \sqrt{z_c^2 + z_s^2}$, $E_Z = \sqrt{Z_c^2 + Z_s^2}$, $\phi_z = \tan^{-1}(z_s/z_c)$, $\Phi_Z = \tan^{-1}(Z_s/Z_c)$ are, respectively, the envelopes and phases of $z(t)$, $Z(t)$ above (and in Probs. 11.10, 11.11), show that

$$E_Z(t) = E_z(t)E_{X_0}(t) \qquad E_{X_0} = \sqrt{X_c^2 + X_s^2 + 2X_cX_s \cos \phi_d} \tag{2}$$

where X_c, X_s are the components of the incident complex signal process, and that

$$\Phi_Z = \tan^{-1} \frac{z_s/z_c + (X_s/X_c)(z_c \sin \phi_d + z_s \cos \phi_d)/z_c}{1 + (z_s/z_c)(X_s/X_c)(z_c \cos \phi_d - z_s \sin \phi_d)/z_s} \tag{3}$$

Hence, in the usual case where $\omega_d/\omega_0 \ll 1$, obtain

$$E_Z(t) = E_z(t)E_X(t) \qquad \Phi_Z(t) = \phi_z(t) + \Phi_X(t) \tag{4}$$

with the instantaneous complex scatter return at fixed range,

$$Z(t) = E_z(t)E_X(t) \cos [(\omega_0 + \omega_d)t - \Phi_X(t) - \phi_z(t)] \tag{5}$$

11.13 (*a*) Show that the *conditional* d.d.'s of $Z(t)$ are normal (for Probs. 11.10 to 11.12), and obtain W_1, W_2 explicitly, with the help of the transformations

$$\mathbf{Z} = \begin{bmatrix} Z_c \\ Z_s \end{bmatrix} = \begin{bmatrix} z_c & z_c \cos \phi_d - z_s \sin \phi_d \\ z_s & z_s \cos \phi_d + z_c \sin \phi_d \end{bmatrix} \begin{bmatrix} X_c \\ X_s \end{bmatrix} \tag{1}$$

$$\mathbf{z} = \begin{bmatrix} z_c \\ z_s \end{bmatrix} = \begin{bmatrix} X_c + X_s \cos \phi_d & -X_s \sin \phi_d \\ X_s \sin \phi_d & X_c + X_s \cos \phi_d \end{bmatrix}^{-1} \begin{bmatrix} Z_c \\ Z_s \end{bmatrix} \tag{2}$$

Obtain these relations also.

(*b*) Show that the covariance function of the complex fixed-range scatter return is

$$K_Z(t_1,t_2) = \tfrac{1}{2} \operatorname{Re} [\overline{(Z_{c1} - iZ_{s1})(Z_{c1} + iZ_{s2})e^{-i\omega_c t}}] \qquad t = t_2 - t_1 \tag{3}$$

with

$$\begin{aligned} \overline{(Z_{c1} - iZ_{s1})(Z_{c2} + iZ_{s2})} = {} & 2\{\rho_{0z}\overline{X_{c1}X_{c2}} + [\rho_{0z} \cos (\phi_2 - \phi_1) - \lambda_{0z} \sin (\phi_2 - \phi_1)]\overline{X_{s1}X_{s2}} \\ & + \overline{X_{s1}X_{c2}}(\rho_{0z} \cos \phi_1 + \lambda_{0z} \sin \phi_1) + \overline{X_{c1}X_{s2}}(\rho_{0z} \cos \phi_2 - \lambda_{1z} \sin \phi_2)\} \\ & + 2i\{\lambda_{1z}\overline{X_{c1}X_{c2}} + [\rho_{0z} \sin(\phi_2 - \phi_1) + \lambda_{1z} \cos (\phi_2 - \phi_1)]\overline{X_{s1}X_{s2}} \\ & + \overline{X_{s1}X_{c2}}(-\rho_{0z} \sin \phi_1 + \lambda_{0z} \cos \phi_1) + \overline{X_{c1}X_{s2}}(\rho_{0z} \sin \phi_2 + \lambda_{0z} \cos \phi_2)\} \end{aligned} \tag{4}$$

where $\rho_{0z} = \rho_{0z}(t)$, etc., and $\phi_1 = \phi(t_1)$, $X_{c1} = X_c(t_1)$, etc.

11.14 An airplane flying over a homogeneous surface receives a certain clutter return (without a specular component) if the receiving (and transmitting) antennas are sufficiently oriented away from vertical incidence. A pulsed continuous wave is emitted. Show that the (normalized) covariance functions ρ_0, λ_0 of the gaussian return (*at fixed range*) are given by

$$\rho_0(t) + i\lambda_0(t) = \frac{1}{A} \iint\limits_{[A]} r\, dr\, d\theta\, e^{i \cos \Psi(r,\alpha;t)} \qquad t = t_2 - t_1 \tag{1}$$

where

$$\Psi = \frac{2\omega_0}{c} \frac{rd \cos \alpha - \frac{1}{2}d^2}{\sqrt{r^2 + z^2}} \qquad \alpha = \alpha(t) \tag{2}$$

and A is the illuminated ground surface, d is the distance traveled by the airplane between pulses, along a horizontal course, at altitude z, and r is the distance from the spot directly beneath the airplane at the initial time t_1 (that is, T_0 sec before, when T_0 is the pulse-repetition time). The angle between a typical scatterer and the line of flight (along the x axis) is α. Show also for constant velocity that

$$\Psi = 2\omega_0 vt/c \tag{3}$$

What is the video spectrum of the clutter return if the pulses are rectangular?

T. S. George.[49]

REFERENCES

1. Van der Ziel, A.: "Noise," Prentice-Hall, Englewood Cliffs, N.J., 1954.
2. Chandrasekhar, S.: Stochastic Problems in Physics and Astronomy, *Revs. Modern Phys.*, **15**: 1 (1943).
3. Wang, M. C., and G. E. Uhlenbeck: On the Theory of the Brownian Motion. II, *Revs. Modern Phys.*, **17**: 323 (1945).
4. Moyal, J. E.: Stochastic Processes and Statistical Physics, *J. Research Roy. Stat. Soc.*, (3)**11**: 150 (1949).
5. Bernamont, J.: Fluctuations de potentiel aux bornes d'un conducteur métallique de faible volume parcouru par un courant, *Ann. phys.*, **7**: 71 (1937).
6. Bell, D. A.: A Theory of Fluctuation Noise, *J. IEE (London)*, **82**: 522 (1938).
7. Bakker, C. J., and G. Heller: On the Brownian Motion in Electrical Resistances, *Physica*, **6**: 262 (1939).
8. Spenke, E.: *Wiss. Veröffentl. Siemens-Werken*, **18**: 54 (1939).
9. Lawson, J. L., and G. E. Uhlenbeck: "Threshold Signals," Massachusetts Institute of Technology Radiation Laboratory Series, vol. 24, McGraw-Hill, New York, 1950.
10. Tolman, R.: "The Principles of Statistical Mechanics," chap. 4, Oxford, New York, 1938. See also R. B. Lindsay, "Introduction to Physical Statistics," chap. V, sec. 5, Wiley, New York, 1941.
11. Seitz, F.: "The Modern Theory of Solids," secs. 31, 32 and eq. 5, p. 190, McGraw-Hill, New York, 1940.
12. Nyquist, H.: Thermal Agitation of Electric Charge in Conductors, *Phys. Rev.*, **32**: 110 (1928).
13. Johnson, J. B.: Thermal Agitation of Electricity in Conductors, *Phys. Rev.*, **32**: 97 (1928).
14. Moullin, E. B., and H. D. M. Ellis: The Spontaneous Background Noise in Amplifiers Due to Thermal Agitation and Shot Effects, *J. IEE (London)*, **74**: 331 (1934). See also Ref. 1, chaps. 2, 3, and Ref. 9, sec. 4.5.
15. Williams, F. C.: Thermal Fluctuations in Complex Networks, *J. IEE (London)*, **81**: 751 (1937).
16. Guillemin, E. A.: Communication Networks, vol. II, chap. V, sec. 2, Wiley, New York, 1935.
17. Williams, F. C.: Fluctuation Noise in Vacuum Tubes Which Are Not Temperature Limited, *J. IEE (London)*, **78**: 326 (1936); Fluctuation Voltage in Diodes and in Multielectrode Valves, *ibid.*, **79**: 349 (1937), **81**: 751 (1937), secs. 3–6.
18. Ornstein, L. S.: Zur Theorie der Brownschen Bewegung für Systeme worin mehrere Temperaturen Vorkommen, *Z. Physik*, **41**: 848 (1927). L. S. Ornstein, H. C. Burger, J. Taylor, and W. Clarkson, The Brownian Movement of a Galvanometer Coil and the Influence of the Temperature of the Outer Circuit, *Proc. Roy. Soc. (London)*, **115**: 391 (1927).
19. Ref. 15, secs. 2ff.
20. Shea, T. E.: "Transmission Networks and Wave Filters," chap. II, sec. 11, Van Nostrand, Princeton, N.J., 1929.
21. Uhlenbeck, G. E., and L. S. Ornstein: The Theory of the Brownian Motion. I, *Phys. Rev.*, **36**: 823 (1930). Also Ref. 3.
22. Doob, J. L.: The Brownian Movement and Stochastic Equations, *Ann. Math. Stat.*, **13**: 351 (1942); *ibid.*, **15**: 229 (1944). Also Ref. 3, appendix, note II.
23. Knudtzon, N.: Experimental Study of Statistical Characteristics of Filtered Random Noise, *MIT Research Lab. Electronics Rept.* 115, July 15, 1949. For an earlier experimental study, see R. Fürth and D. K. C. MacDonald, Statistical Analysis of Spontaneous Electrical Fluctuations, *Proc. Phys. Soc.*, **59**: 388 (1947).

24. Callen, H. B., and T. A. Welton: Irreversibility and Generalized Noise, *Phys. Rev.*, **83**: 34 (1951).
25. Green, H. F., and H. B. Callen: On the Formalism of Thermodynamic Fluctuation Theory, *Phys. Rev.*, **83**: 1231 (1951). See also the supplementary references at the end of the volume.
26. Middleton, D.: On the Theory of Random Noise. Phenomenological Models. I, II, *J. Appl. Phys.*, **22**: 1143, 1153, 1326 (1951).
27. Rice, S. O.: Mathematical Analysis of Random Noise, *Bell System Tech. J.*, **23**: 282 (1944), **24**: 46 (1945).
28. Bunimovitch, V. I.: Effects of the Fluctuations and Signal Voltages on a Non Linear System, *J. Phys.* (*USSR*), **10**: 35 (1946).
29. Blanc-Lapierre, A., and R. Fortet: "Théorie des fonctions aléatoires," Masson, Paris, 1953.
30. Hurwitz, H., and M. Kac: Statistical Analysis of Certain Types of Random Functions, *Ann. Math. Stat.*, **15**: 173 (1944).
31. Cramér, H.: "Mathematical Methods of Statistics," sec. 16.5, Princeton University Press, Princeton, N.J., 1946.
32. Mullen, J. A., and D. Middleton: The Rectification of Non-gaussian Noise, *Quart. Appl. Math.*, **15**: 395 (1958).
33. Davenport, W. B.: An Experimental Study of Speech Wave Probability Distributions, *J. Acoust. Soc. Am.*, **24**: 390 (1952).
34. Valley, G. E., Jr., and H. Wallman: "Vacuum Tube Amplifiers," Massachusetts Institute of Technology Radiation Laboratory Series, vol. 18, McGraw-Hill, New York, 1948.
35. Spangenberg, K. R.: "Vacuum Tubes," chaps. 8, 16, McGraw-Hill, New York, 1948.
36. Ref. 9, chap. 4, secs. 4.7ff.
37. Harman, W. W.: "Fundamentals of Electronic Motion," chaps. 5, 6, McGraw-Hill, New York, 1953.
38. Ref. 1, chap. 6.
39. Davenport, W. B., Jr., and W. L. Root: "An Introduction to the Theory of Random Signals and Noise," McGraw-Hill, New York, 1958.
40. Schottky, W.: Small-shot Effect and Flicker Effect, *Phys. Rev.*, **28**: 74 (1926).
41. Langmuir, I.: The Effect of Space-charge and Initial Velocities on the Potential Distribution and Thermionic Current between Parallel Plane Electrodes, *Phys. Rev.*, **21**: 419 (1923).
42. Rack, A. J.: Effect of Space Charge and Transit Time on the Shot Noise in Diodes, *Bell System Tech. J.*, **17**: 592 (1938), pt. III.
43. Schottky, W.: Zusammenhänge zwischen korpuskularen und thermischen Schwankungen in Elektronenröhren, *Z. Physik*, **104**: 748 (1937).
44. North, D. O.: Fluctuations in Space-charge-limited Currents at Moderately High Frequencies, Part II, *RCA Rev.*, **4**: 441 (1940).
45. Schottky, W., and E. Spenke: *Wiss. Veröffentl. Siemens-Werken*, **16**(2): 1, 19 (1937).
46. Moullin, E. B.: "Spontaneous Fluctuations of Voltage," chaps. 3, 4, Oxford, New York, 1938.
47. Ref. 9, chap. 6.
48. Kerr, D.: "Propagation of Short Radio Waves," secs. 6.18, 6.19 and app. B, Massachusetts Institute of Technology Radiation Laboratory Series, vol. 13, McGraw-Hill, New York, 1951.
49. George, T. S.: Fluctuations of Ground Clutter Return in Airborne Radar Equipment, *Mon.* 22, Feb. 15, 1952, *Proc. IEE* (*London*), **99** (pt. IV): 92 (1952).
50. Smullin, L. D., and H. A. Haus (eds.): "Noise in Electron Devices," Wiley, New York, 1959.

PART 3

APPLICATIONS TO SPECIAL SYSTEMS

CHAPTER 12

AMPLITUDE MODULATION AND CONVERSION

In Parts 1 and 2, we have outlined some of the salient features of a statistical communication theory for the single-link system

$$\{v\} = \boldsymbol{T}_R^{(N)}\boldsymbol{T}_M^{(N)}\boldsymbol{T}_T^{(N)}\{u\}$$

(cf. Fig. 6.1). Part 1 has been devoted mainly to the statistical properties of the transformations $\boldsymbol{T}_R^{(N)}$, $\boldsymbol{T}_T^{(N)}$, etc., while Part 2 has described in some detail the normal process, and those derived from it, which constitute the principal noise mechanisms encountered in the channel $\boldsymbol{T}_M^{(N)}$ and in the reception operations $\boldsymbol{T}_R^{(N)}$. Part 3 applies these results to a variety of specific problems associated with the transformations $\boldsymbol{T}_R^{(N)}$ and $\boldsymbol{T}_T^{(N)}$, such as amplitude modulation, mixing (or conversion), angle modulation, various demodulation or detection problems, filtering, linear prediction, and the calculation of probability distributions after nonlinear operations. A complete coverage is not attempted here. The intent is to provide pertinent illustrations of general methods described in earlier chapters, as well as specific results of technical interest. For many of the analytical details, the reader should consult the indicated references. The present treatment is based largely on the results of Chap. 5 and, except for the material of Chap. 17, is essentially a second-moment theory; i.e., the aim is to determine spectra and covariance functions following the linear and nonlinear operations embodied in the specific transformations $\boldsymbol{T}_T^{(N)}$, $\boldsymbol{T}_R^{(N)}$, etc. We leave to Part 4 (Chaps. 18 to 23) the more general questions of over-all system optimization and evaluation and turn now to problems involving amplitude modulation.

12.1 *Methods of Modulation*

There are a variety of ways of producing amplitude modulation of a cw carrier. These depend in some fashion upon an ultimately nonlinear operation, which in most practical cases can be represented by suitable analytic expressions. Our first task is to describe these mathematical relations and consider conditions under which they may be applied, before we attempt to determine spectra and covariance functions.

Modulation is usually carried out with the aid of a suitable tube or crystal element. We distinguish three common states of operation: these are (1) class A operation, wherein the input disturbance is confined entirely to the amplifying or transmission portion of the element's dynamic characteristic (cf. Fig. 12.1*a*); (2) class B operation, where a part, up to half the input wave on the average, lies in the cutoff region of the dynamic response (cf. Fig. 12.1*b*); and finally (3) class C operation (Fig. 12.1*c*), in which less than half the incoming disturbance is passed by the device in question. The last two clearly represent nonlinear transformations $\boldsymbol{T}_g$, while the first may be linear, $\boldsymbol{T}_L$, unless the excursions of the input wave fall in the regions of cutoff and saturation (in which instances we have class AB operation). A half-wave linear rectifier is an example of class C or extreme class B ($b_0 = 0$) operation,† whereas an element used for amplification alone is generally run class A.

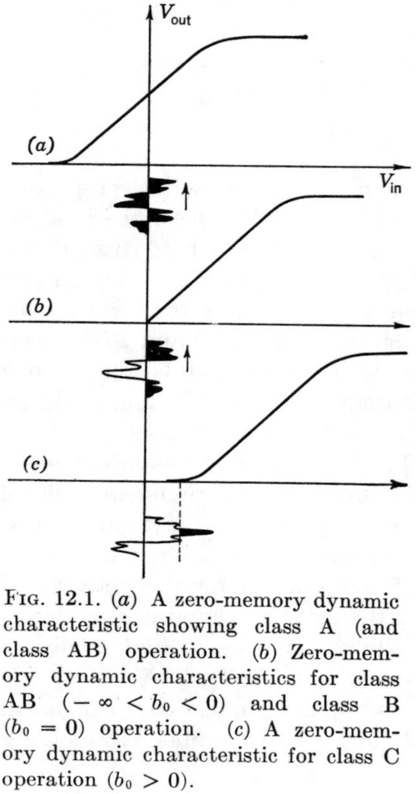

FIG. 12.1. (*a*) A zero-memory dynamic characteristic showing class A (and class AB) operation. (*b*) Zero-memory dynamic characteristics for class AB ($-\infty < b_0 < 0$) and class B ($b_0 = 0$) operation. (*c*) A zero-memory dynamic characteristic for class C operation ($b_0 > 0$).

One method of obtaining amplitude modulation uses a plate-modulated class C amplifier. A carrier from an oscillator of frequency f_c is applied to the grid of, say, a triode in class C operation. The unmodulated plate current then consists of the top halves ($b_0 = 0$) of a sine wave of frequency f_c, of which only the carrier component $A_0' \cos \omega_c t$ is available when a tuned circuit is introduced into the plate network. Next, one applies the modulation (from another source) to the plate circuit, with the result that the amplitude of the sinusoidal oscillations in the tank circuit is changed according to the instantaneous value of the modulating wave. The ensemble of possible tank-circuit outputs in the case of an ideal zero-memory system, where there is no distortion of any kind, is then represented by

$$V(t)_0 = A_0' \cos (\omega_c t + \phi)$$

and

$$V(t)_{mod} = \boldsymbol{T}_g\{V(t)_0\} = A_0[1 + \lambda V_M(t)] \cos (\omega_c t + \phi) \qquad \lambda V_M(t) \geq -1, \text{ all } t \tag{12.1}$$

† See Sec. 2.1-3 for general properties of these transformations and also Fig. 5.11 for a schematic diagram of a typical AM communication system in which these elements appear.

where $V_M(t)$ is the modulation and λ is some constant which depends on tube parameters and the auxiliary circuits.

When there is distortion, Eq. (12.1) is modified in a number of ways. If only the modulation is changed in the course of its transmission to the plate circuit, we then have

$$V_{mod}(t) = \boldsymbol{T}_{g_1}\{\boldsymbol{T}_{g_2}\{V_M(t)\}\} = A_0\{1 + \lambda g[V_M(t)]\} \cos (\omega_c t + \phi) \tag{12.2a}$$

while if there is overmodulation, such that $\lambda V_M < -1$ part of the time, so that the class C amplifier is cut off at those times, we get

$$V_{mod}(t) = \begin{cases} A_0[1 + \lambda V_M(t)] \cos (\omega_c t + \phi) & \lambda V_M \geq -1 \\ 0 & \lambda V_M \leq -1 \end{cases} \tag{12.2b}$$

[For a distortion of the above type and overmodulation, $\lambda V_M \geq -1$ is replaced by $\lambda g(V_M) \geq -1$, etc.] A nonlinear swing of the dynamic characteristic in the process of modulation produces a distortion which modifies Eq. (12.1) further to

$$V_{mod}(t) = A_0\boldsymbol{T}_{g_1}\{\boldsymbol{T}_{g_2}\{1 + \lambda V_M\}\} = A_0 g[1 + \lambda V_M(t)] \cos (\omega_c t + \phi) \tag{12.2c}$$

while if the dynamic path itself is nonlinear, the most general representation of the tank circuit's output is of the type

$$V_{mod}(t) = G[V_M(t)] \cos (\omega_c t + \phi) \tag{12.2d}$$

where G is generally a very complicated function of $V_M(t)$, embodying a finite, nonzero memory. In practice, of course, the less distortion the better, so that, to achieve this, the Q (or quality factor) of the tank circuit must not be so large that the modulation is altered because of frequency selection, nor must the operating point Q' be allowed to wander, with the result that the swing of the dynamic characteristic is no longer uniform. This usually means that the maximum excursion about rest is not too great compared with the magnitude of the unmodulated carrier. Overmodulation is also avoided in this way.

Another method of obtaining an amplitude-modulated carrier is to apply both the carrier and the modulation to the grid of a class C amplifier. The input to the nonlinear device is then

$$V_{\text{grid}}(t) = A'_0 \cos (\omega_c t + \phi) + b_0 + \eta V_M(t) \tag{12.3}$$

where $b_0 + \eta V_M(t)$ represents a modulated "bias," including a steady component and one that is slowly varying compared with the carrier $A'_0 \cos (\omega_c t + \phi)$. The output of the resonant circuit that is coupled to the final stages of the system ($\boldsymbol{T}_T$) consists of the band of frequencies spectrally located about f_c. There is much greater distortion here compared with the plate-modulation system: in general, the output of the tuned circuit cannot be represented by the simple relation (12.1). As we shall see in Secs. 12.3, 12.4, one gets the linear model only when the amplitude A'_0 of the carrier oscillation is large compared with the maximum amplitude of the modulation itself. This is known as the *mixer condition*, which on a basis of average intensity is here $A_0'^2/2 \gg \langle V_M^2 \rangle$.

Frequently, however, instead of a nonlinear element such as the class C (triode) amplifier described above, an oscillator is used, generating a carrier whose amplitude is controlled directly by the modulation. Here, the local oscillation and modulation are combined directly in a single device, with a resulting linearity which is often as good as or better than other systems.

Instead of elements in class C operation, class A may be employed, when the operating point is adjusted to lie in the curved portion of the dynamic characteristic and both carrier and modulation are applied to the grid, so that

$$V_{\text{grid}}(t) = A_0' \cos(\omega_c t + \phi) + V_M(t) \tag{12.4}$$

The instantaneous output (with a resistive load only) is given by

$$V_{\text{no tank}}(t) = \sum_{l=0}^{\infty} \alpha_l V_{\text{grid}}(t)^l \qquad \alpha_l = \frac{d^l V}{l!\, dV_{\text{grid}}^l} \tag{12.5}$$

where the α_l are governed by the tube and circuit parameters. As before, the tuned circuit (of impedance Z_L) selects those components of $V(t)$ which lie in the immediate spectral vicinity of the carrier frequency. In practical operation, the carrier must be strong relative to the modulation, for little distortion. If this is not the case, the distortion in the modulation caused by the curvature of the dynamic path will be severe. In this respect, the performance resembles the preceding grid-modulated class C amplifier.

Other types of nonlinear modulators are copper oxide rectifiers and crystals. Here, as in the small-signal case above, the modulation is achieved through the curvature of the dynamic path, in class A or AB operation, and frequently in class C operation. The copper oxide element is a half-wave square-law device, while the crystal can be considered either as a full-wave or as an essentially class C device, depending on the magnitude of the input and particular back-bias response. A good analytic representation of the current output response of the copper oxide rectifier is

$$I(t) = \begin{cases} \alpha V^2 & V > 0 \\ 0 & V < 0 \end{cases} \tag{12.6}$$

A reasonably satisfactory representation of the crystal response is

$$I(t) = \begin{cases} \beta_1(e^{\alpha_1 V} - 1) & V > 0,\ \alpha_1, \alpha_2, \beta_1, \beta_2 > 0 \\ -\beta_2(e^{-\alpha_2 V} - 1) & V < 0 \end{cases} \tag{12.7}$$

and when there is a negligible back current, $\beta_2 \doteq 0$, or $\alpha_1 \doteq 0$. A comprehensive discussion of these and other methods of amplitude modulation is available in the literature.[1]

12.2 *Covariance Functions and Intensity Spectra*

As we have seen in Sec. 12.1, there are two principal ways of producing an amplitude modulation of an h-f carrier: (1) by using the nonlinear portions of a rectifier's dynamic characteristic, either in small-signal or class C operation; or (2) by producing the desired amplitude changes by a direct modification of the oscillator's output. In practical systems, a considerable degree of linearity can often be obtained, even for large modulations, so that Eq. (12.1) is a valid representation of the output of the modulating system, as long as $\lambda V_M > -1$, at least.

We wish now to determine the spectrum and covariance function of these modulated waves under a variety of conditions. Unless otherwise indicated, it is assumed that modulation and carrier belong to ergodic processes, so that the interchange of time and ensemble averages (with probability 1) may be effected in the usual way (cf. Sec. 1.6 and examples).

We begin, therefore, by considering the covariance function for an ensemble of modulated waves. When $V(t)$ is the simple product of independent† carrier and modulation processes, we have directly

$$K_V(t) = \mathbf{E}\{V(t_1)V(t_2)\} = K_{V_0}(t)K_c(t) \qquad t = t_2 - t_1 \tag{12.8}$$

where, by Eq. (12.1), $V_0(t) = 1 + \lambda V_M(t)$, $\lambda V_M > -1$, $V_0 = 0$, $\lambda V_M < -1$. Thus $K_{V_0}(t)$ becomes in the important case of no overmodulation, for example, $\lambda V_M > -1$ at all times,

$$K_{V_0}(t) = 1 + \lambda^2 K_M(t) \tag{12.9}$$

and for a periodic carrier $K_c(t)$ is equal to $(A_0^2/2) \cos \omega_c t$ [cf. Eq. (3.67)]. The intensity spectrum of the modulated process then follows with a direct application of the Wiener-Khintchine theorem [cf. Eqs. (3.42)], giving

$$\mathcal{W}_V(f) = 2\mathfrak{F}\{K_V\} = \mathcal{W}_C(f) + 4\lambda^2 \int_0^\infty K_c(t)K_M(t) \cos \omega t \, dt \tag{12.10a}$$

With the usual periodic carrier, this becomes

$$\mathcal{W}_V(f) \doteq \frac{A_0^2}{2} \delta(f - f_c) + \lambda^2 A_0^2 \int_0^\infty K_M(t) \cos (\omega_c - \omega)t \, dt \tag{12.10b}$$

where the contribution to the integral from the term $\cos (\omega_c + \omega)t$ has been discarded, since this term oscillates very rapidly over the region for which $K_M(t)$ is significant (see Sec. 3.3-4 for some examples).

The condition that $V_0(t)$ vanishes when $\lambda V_M < -1$ represents a state of *overmodulation* in most cases: when $V_M < -1/\lambda$, the oscillator is cut off

† This corresponds in the time domain to incoherent carrier and modulation for almost all members of the product ensemble.

and there is no output. With deterministic modulations, overmodulation occurs at $(V_M)_{\min} < -1/\lambda$, while, for a stochastic modulation, there is usually no definite limit to $(V_M)_{\min}$. However, if λ is made small enough, modulation becomes again essentially a linear operation and overmodulation occurs an ignorable percentage of the time. For example, if γ is this fraction of the time for which V_M falls below an amplitude $V_0 = -1/\lambda$, then λ and γ are related by

$$\gamma = \int_{-\infty}^{V_0=-1/\lambda} W_1(V_M)\, dV_M \tag{12.11}$$

In the case of normal noise modulation, it is easily shown [cf. Eqs. (7.2) and (7.4)] that

$$\gamma = \tfrac{1}{2}\{1 - \Theta[1/(2\lambda^2\psi)^{1/2}]\} \tag{12.12}$$

where Θ is the familiar error function [cf. Eqs. (7.4) and (A.1.1)]. When $\lambda\sqrt{2\psi} \leq 0.6$, for instance, overmodulation occurs less than 1 per cent of the time, for example, $\gamma \leq 0.01$, and so the restriction on V_0 for $\lambda V_M < -1$ may be safely removed in most cases.

When overmodulation or distortion cannot be ignored, its effects on the spectrum can be determined directly with the help of the appropriate complex Fourier transform of the modulation characteristic (cf. Sec. 2.3-2). Thus, for Eqs. (12.2*b*) $f(i\xi)$ becomes $-\xi^{-2}e^{i\xi}$, and we can write

$$V(t) = -\frac{A_0 \cos(\omega_c t + \phi)}{2\pi} \int_C \frac{d\xi}{\xi^2} e^{i\xi[1+\lambda V_M(t)]} \tag{12.13}$$

In the case of distortion of V_M before modulation, where Eq. (12.2*a*) applies, corresponding to Eq. (12.13) we have

$$V(t) = A_0 \cos(\omega_c t + \phi) + \frac{\lambda A_0}{2\pi} \cos(\omega_c t + \phi) \int_C f(i\xi) e^{iV_M(t)\xi}\, d\xi \tag{12.14}$$

where $f(i\xi)$ is the Fourier transform of $g(V_M)$. The covariance function $K_V(t)$ is then determined from Eq. (12.8), and the spectrum follows in turn by the Wiener-Khintchine theorem (3.42). The method is illustrated by the examples below.

Example 1. Sine wave × carrier.† First, let us consider the simple problem of a sinusoidally modulated carrier, where the modulation is linear while it is on, and where any degree of modulation can occur, so that Eq. (12.1) or Eq. (12.13) is applicable.[2] The covariance function of $V(t)$ [Eq. (12.13)] becomes

$$K_V(t) = \frac{A_0^2}{2} \cos \omega_c t \int_C \int_C \frac{e^{i\xi_1 + i\xi_2}}{4\pi^2} \frac{F_2(\xi_1,\xi_2;t)_S\, d\xi_1\, d\xi_2}{(\xi_1\xi_2)^2} \tag{12.15}$$

where F_2 is the characteristic function of $\lambda V_M(t) = \mu B_0 \cos(\omega_a t + \phi)$.

† The reader may find it convenient to review Chap. 5, in particular Sec. 5.1-1.

This, from Eq. (5.12a), with A therein replaced by μB_0, becomes

$$F_2(\xi_1,\xi_2;t)_S = \sum_{m=0}^{\infty} (-1)^m \epsilon_m J_m(\mu B_0 \xi_1) J_m(\mu B_0 \xi_2) \cos m\omega_a t \qquad (12.16)$$

The double integral in Eq. (12.15) may now be factored into the product of two similar integrals (cf. Sec. 5.1-1) so that the covariance function can be written finally as

$$K_V(t) = \frac{A_0{}^2}{2} \sum_{m=0}^{\infty} \epsilon_m (-1)^m \frac{h_{m,0}^2}{2} [\cos (\omega_c + m\omega_a)t + \cos (\omega_c - m\omega_a)t] \qquad (12.17)$$

in which† [cf. Eq. (5.16b)]

$$\begin{aligned} h_{m,0} &= \frac{-1}{2\pi} \int_C \frac{J_m(\mu B_0 \xi) e^{i\xi}\, d\xi}{\xi^2} \\ &= \frac{-i\mu B_0}{4} \left\{ \frac{{}_2F_1\left[\frac{m-1}{2}, \frac{-m-1}{2}; \frac{1}{2}; (\mu B_0)^{-2}\right]}{\Gamma\left(\frac{3+m}{2}\right)\Gamma\left(\frac{3-m}{2}\right)} \right. \\ &\qquad \left. + \frac{2}{\mu B_0} \frac{{}_2F_1\left[\frac{m}{2}, \frac{-m}{2}; \frac{3}{2}; (\mu B_0)^{-2}\right]}{\Gamma\left(\frac{2+m}{2}\right)\Gamma\left(\frac{2-m}{2}\right)} \right\} \qquad \mu B_0 \geq 1 \qquad (12.18a) \end{aligned}$$

or

$$h_{m,0} = \begin{cases} -1 & m = 0 \\ i\mu B_0/2 & m = 1 \\ 0 & m \geq 2 \end{cases} \qquad 0 \leq \mu B_0 \leq 1 \qquad (12.18b)$$

The corresponding intensity spectrum is

$$\mathcal{W}_V(f) = 2\mathfrak{F}\{K_V\} = \frac{A_0{}^2}{2} \sum_{m=0}^{\infty} \epsilon_m (-1)^m h_{m,0}^2 [\delta(f + mf_a) + \delta(f - mf_a)] \qquad (12.19)$$

The quantity ${}_2F_1$ is a hypergeometric function, some of whose properties are discussed in Sec. A.1.3. As expected [cf. Eqs. (12.18b)], when there is no overmodulation, i.e., when $|\mu B_0| \leq 1$, the simple linear result follows. However, when $|\mu B_0| > 1$, the resulting overmodulation generates additional $s \times s$ components ($m \geq 2$) [cf. Eq. (12.18a)], the extent of the distortion depending on the magnitude of the $h_{m,0}$. The intensity spectrum $\mathcal{W}_V(f)$ [Eq. (12.19)] in either instance is, of course, a line spectrum (cf. the $s \times s$ terms in Fig. 5.2, for example), with components at frequency intervals f_a.

† Here, $h_{m,0}$ is recognized as a special case of the Weber-Schafheitlin integral.[3,4]

The mean total intensity in the modulated carrier is found with the help of Eq. (5.20a), which is modified in this instance to

$$P_\tau = \frac{A_0^2}{2} \int_{-1/\mu}^{\infty} (1 + \mu V_M)^2 W_1(V_M)\, dV_M = \frac{A_0^2}{2\pi} \int_{-Z_0}^{1} \frac{(\mu B_0 Z + 1)^2}{\sqrt{1 - Z^2}}\, dZ$$

$$\text{with } Z_0 = \frac{1}{\mu B_0},\ \mu B_0 \geq 1;\ Z_0 = 1,\ 0 \leq \mu B_0 \leq 1 \qquad (12.20)$$

where we have used the probability density for a sine wave (Prob. 1.10). Integration gives

$$P_\tau = \frac{A_0^2}{2}\left[\left(\frac{1}{2} + \frac{1}{\pi} \sin^{-1} Z_0\right)\left(1 + \frac{\mu^2 B_0^2}{2}\right) + \mu B_0 \left(1 - \frac{\mu B_0 Z_0}{4}\right) \sqrt{1 - Z_0^2}\right] \qquad (12.21)$$

The mean intensity of the residual carrier is [cf. Eq. (5.20b)]

$$P_{f_c} = \frac{A_0^2}{2} h_{0,0}^2 = \frac{A_0^2}{2} \left| \int_{-Z_0}^{1} \frac{\mu B_0 Z + 1}{\pi\sqrt{1 - Z^2}}\, dZ \right|^2$$

$$= \frac{A_0^2}{2} \left[\frac{1}{2} + \frac{1}{\pi} \sin^{-1} Z_0 + \frac{\mu B_0}{\pi} \sqrt{1 - Z_0^2}\right]^2 \qquad (12.22)$$

and the difference between Eqs. (12.21) and (12.22) represents the mean intensity P_c in the sidebands.

The intensity spectrum consists of three spikes, at $f = f_0, f_0 \pm f_a$ when $|\mu B_0| \leq 1$, and an infinite number of higher-order components at $f_0 \pm mf_a$ when $|\mu B_0| > 1$. We note also that, when $|\mu B_0| \gg 1$, only the carrier, the two original sidebands, and the even-order components ($m = 2, 4, 6, \ldots$) are significant in this limiting instance of essentially 50 per cent overmodulation.

Example 2. Noise × carrier. Instead of the deterministic modulation, let us consider the other statistical extreme of a modulation that belongs to a normal random process. The noise $N(t)$ is taken to be a broadband l-f wave. The covariance function of such a modulated carrier is given by Eq. (12.15) provided that we replace $F_2(\xi_1,\xi_2;t)_S$ therein by $F_2(\xi_1,\xi_2;t)_N$ [Eqs. (5.6)], where ψ is now replaced by $\lambda^2\psi$ $(= \lambda^2\overline{N^2})$. Using Eq. (5.6$b$) and following the procedures of Sec. 5.1-1 gives us the desired covariance function of the modulated carrier,

$$K_V(t) = h_{0,0}^2 \frac{A_0^2}{2} \cos \omega_c t + \frac{A_0^2}{2} h_{0,1}^2[-\lambda^2\psi k(t)] \cos \omega_c t$$

$$+ \left\{\frac{A_0^2}{2} \sum_{n=2}^{\infty} \frac{[-\lambda^2\psi k(t)]^n}{n!} h_{0,n}^2 \cos \omega_0 t\right\} \qquad (12.23)$$

where $k(t)$ is the normalized covariance function of the noise. The amplitude functions $h_{0,n}$ are specifically [cf. Eq. (5.15*b*)], with now $f(i\xi) = e^{i\xi}(-\xi^{-2})$,

$$h_{0,n} = \frac{-1}{2\pi}\int_C \xi^{n-2}\,d\xi\, e^{i\xi-\lambda^2\psi\xi^2/2}$$
$$= i^{-n}(\lambda^2\psi)^{(1-n)/2}2^{(n-3)/2}\left[\frac{{}_1F_1\left(\frac{n-1}{2};\frac{1}{2};-\frac{1}{2\lambda^2\psi}\right)}{\Gamma\left(\frac{3-n}{2}\right)} + \frac{\frac{2}{\sqrt{2\lambda^2\psi}}\,{}_1F_1\left(\frac{n}{2};\frac{3}{2};-\frac{1}{2\lambda^2\psi}\right)}{\Gamma\left(\frac{2-n}{2}\right)}\right] \quad (12.24)$$

Here, ${}_1F_1$ is a confluent hypergeometric function (cf. Sec. A.1.2) and the $h_{0,n}$ may alternatively be expressed in terms of derivatives of $e^{-x^2/2}/\sqrt{2\pi}$ (cf. Sec. A.1.1), according to

$$h_{0,0} = [1 + \Theta(x_0^{-1})]/2 - x_0(4\pi)^{-1/2}e^{-1/x_0^2} \qquad x_0 \equiv \sqrt{2\lambda^2\psi} \quad (12.25a)$$
$$h_{0,1} = -i[1 + \Theta(x_0^{-1})]/2 \quad (12.25b)$$
$$h_{0,n} = -(-1)^{n/2}\left(\frac{x_0}{\sqrt{2}}\right)^{1-n}\phi^{(n-2)}\left(\frac{\sqrt{2}}{x_0}\right) \qquad n = 2, 3, 4, \ldots \quad (12.25c)$$

where $\phi^{(n)}(x) = (d^n/dx^n)e^{-x^2/2}/\sqrt{2\pi}$ and Θ is the error function [Eq. (7.4)].

The mean intensity spectrum follows at once from Eq. (12.23) and the Wiener-Khintchine relation (3.42). For a modulating noise with a gaussian spectrum $\mathcal{W}_N(f) = W_0e^{-\omega^2/\omega_b^2}$, the spectrum is

$$\mathcal{W}_V(f) = \frac{A_0^2h_{0,0}^2}{2}\,\delta(f - f_c) + \sqrt{\pi}\,\frac{A_0^2\lambda^2\psi}{\omega_b}\left[-h_{0,1}^2e^{-(\omega-\omega_c)^2/n\omega_b^2} + \sum_{n=2}^{\infty}\frac{\phi^{(n-2)}\,(\sqrt{2}/x_0)^2}{n!n^{1/2}}\,e^{-(\omega-\omega_c)^2/n\omega_b^2}\right] \quad (12.26)$$

which is shown in Fig. 12.2, where the spectra have been normalized for a maximum value unity at $f = f_c$ (a table of normalization factors, by which the indicated ordinate must be *multiplied* to get the true value, is also given in the figure).

Overmodulation gives a spectrum which is more widely spread about f_c, as we would expect, than when such distortion is negligible (i.e., when $\gamma \leq 0.01$, or $\sqrt{2\lambda^2\psi} < 0.6$). The additional harmonics represent (*carrier* $\times$ *noise*) noise products resulting from the amplitude clipping inherent in overmodulation. Note that there is a definite limit to the ultimate spread of this spectrum, which is determined by the fact that overmodulation can take place, at most, 50 per cent of the time. The effect is small even when $x_0 \to \infty$ and is usually ignorable when $x_0 \leq 1.0$, unless we are concerned

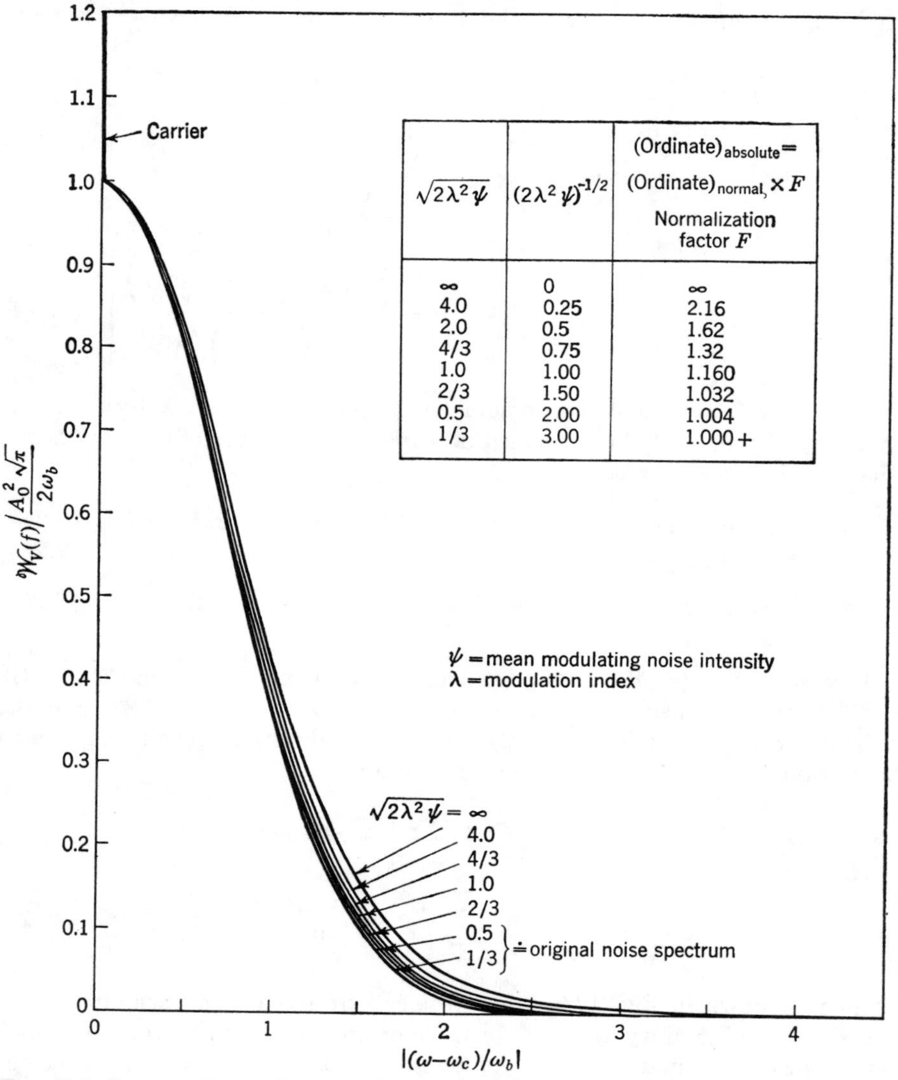

FIG. 12.2. Spectra of a carrier amplitude-modulated by noise with a gaussian spectrum ($f > f_c$).

with the spectrum well away from the half-intensity points. When clipping is not significant, the covariance function reduces effectively to the first two terms of Eq. (12.23) and the corresponding spectrum is given by Eq. (12.26), where the contributions from the sum for $n \geq 2$ are ignored; the amplitude functions (12.25a) are explicitly $h_{0,0} \doteq 1$, $h_{0,1} \doteq -i$, $h_{0,n} \doteq 0$ $(n \geq 2)$.

If we recall that $\lambda N(t)$ is normally distributed with zero mean and standard deviation $\lambda^2\psi$, the mean total and continuum intensities can be determined from

$$P_\tau = K_V(0) = \frac{A_0^2}{2} \int_{-1}^{\infty} (1 + z)^2 W_1(z)\, dz \quad \text{with } W_1(z) = \frac{1}{\sqrt{2\pi\lambda^2\psi}}\, e^{-z^2/2\lambda^2\psi}$$

$$= \frac{A_0^2}{2} \left\{ \frac{1 + x_0^2/2}{2} \left[\Theta(x_0^{-1}) + 1\right] - \frac{x_0^2}{2}\, \phi^{(1)}\left(\frac{\sqrt{2}}{x_0}\right) \right\} \tag{12.27a}$$

and

$$P_{f_c} = K_V(\infty) = \frac{A_0^2}{2} \left| \int_{-1}^{\infty} (1 + z) W_1(z)\, dz \right|^2$$

$$= \frac{A_0^2}{2} \left[\frac{1 + \Theta(1/x_0)}{2} + \frac{x_0}{2\sqrt{\pi}}\, e^{-1/x_0^2} \right]^2 = \frac{A_0^2}{2}\, h_{0,0}^2 \tag{12.27b}$$

and $P_c = P_\tau - P_{f_c}$. The variation of P_τ, P_c, P_{f_c} with the intensity of the modulating noise is indicated in Fig. 12.3. Their limiting conditions are instructive. As the intensity of the modulating noise is increased, a correspondingly greater proportion of the modulated wave's intensity is distributed in (*noise* × *noise*) noise sidebands, generated in the process of modulation and overmodulation. We remark that the intensities of the carrier component and of the continuum are quite independent of the particular spectral distribution of the original noise and depend only on the clipping level at which significant overmodulation occurs, since intensity in a given band of frequencies is proportional to the integrated intensity of the band [cf. the remarks at the end of Sec. 5.1-3, following Eq. (5.29)].

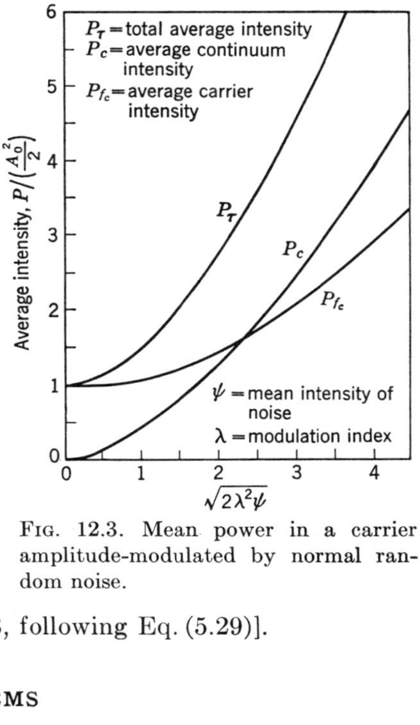

FIG. 12.3. Mean power in a carrier amplitude-modulated by normal random noise.

PROBLEMS

12.1 (*a*) Obtain the covariance function of a carrier very heavily amplitude-modulated by (ergodic) normal noise, that is, $2\lambda^2\psi \to \infty$, thus showing that

$$K_V(t)\Big|_{(2\lambda^2\psi\to\infty)} \cong \frac{A_0^2\lambda^2\psi}{4\pi} \cos \omega_c t \left[1 + \frac{\pi}{2} k(t) + \sum_{n=0}^{\infty} \frac{k(t)^{2n+2}(2n)!}{2^{2n} n!^2 (2n+1)(2n+2)} \right] + O(\sqrt{2\lambda^2\psi}) \tag{1}$$

$$K_V(t)\Big|_{(2\lambda^2\psi\to\infty)} \cong \frac{A_0^2\lambda^2\psi}{4\pi} \cos \omega_c t \left[\sqrt{1 - k^2} + k\left(\frac{\pi}{2} + \sin^{-1} k\right) \right] \qquad k = k(t) \tag{2}$$

(*b*) Verify that, when there is less than 1 per cent overmodulation, the correlation function is

$$K_V(t)\Big|_{(2\lambda^2\psi\to 0)} \doteq \frac{A_0^2}{2}\cos\omega_c t + \frac{A_0^2\lambda^2\psi}{2}k(t)\cos\omega_c t + O[e^{-1/2\lambda^2\psi}k(t)^3] \quad (3)$$

D. Middleton.[2]

12.2 (*a*) If $y = \lambda N(t)$ represents a normal noise modulation and $z = \mu V_S(t)$ a deterministic signal modulation, show that the mean total intensity of a (sinusoidal) carrier modulated by the sum of V_S and N can be written

$$P_T = K_V(0) = \frac{A_0^2}{2}\boldsymbol{E}\{(1+y+z)^2\} \qquad \text{for all } u = y+z \geq -1 \quad (1a)$$

$$P_T = K_V(0) = \frac{A_0^2}{2}\int_{-1}^{\infty} W_1(u)(1+u)^2\,du \quad (1b)$$

where

$$W_1(u) = \int_{-\infty}^{\infty} e^{-iu\xi-\lambda^2\psi\xi^2/2}\frac{d\xi}{2\pi}\int_0^{2\pi} e^{iu\xi V_S(0;\phi)}\frac{d\phi}{2\pi} \quad (2)$$

(*b*) Show that

$$P_{f_c} = \frac{A_0^2}{2}h_{0,0}^2 = \frac{A_0^2}{2}(\boldsymbol{E}\{1+y+z\})^2 = \frac{A_0^2}{2}\left[\int_{-1}^{\infty}(1+u)W_1(u)\,du\right]^2 \quad (3)$$

(*c*) Obtain for a sinusoidal signal $V_S = B_0\cos(\omega_a t + \phi)$ the explicit density $W_1(u)$, in Eq. (2) above. D. Middleton,[2] pp. 348–350.

12.3 (*a*) For a sinusoidal carrier amplitude-modulated by the sum of normal noise and a sinusoidal signal $V_S = B_0\cos(\omega_a t + \phi)$, show that the covariance function of the modulated wave is in general

$$K_V(t) = \frac{A_0^2}{2}\sum_{m=0}^{\infty}\epsilon_m(-1)^m\frac{1}{2}[\cos(\omega_c + m\omega_a t) + \cos(\omega_c - m\omega_a t)] \times\left\{h_{m,0}^2 + \sum_{n=1}^{\infty}(-1)^n\frac{[\lambda^2\psi k(t)]^n}{n!}h_{m,n}^2\right\} \quad (1)$$

where

$$h_{m,n} = \frac{1}{2\pi}\int_C \xi^{n-2}J_m(\mu B_0\xi)e^{-i\xi-\lambda^2\psi\xi^2/2}\,d\xi \quad (2)$$

and λ, μ are, respectively, the modulation factors of noise and signal (cf. Prob. 12.2).

(*b*) Show that, specifically,

$$h_{m,n} = \frac{-i^{m+n}}{2}\left(\frac{x_0^2}{4}\right)^{(1-n)/2}\left(\frac{\mu B_0}{x_0}\right)^m\sum_{q=0}^{\infty}\frac{2^q\,{}_1F_1[(q+m+n-1)/2;\ m+1;\ -\mu^2B_0^2/x_0^2]}{x_0^q q!\Gamma[(3-m-n-q)/2]} \quad (3)$$

or

$$h_{m,n} = \frac{-i^{m+n}}{2}\left(\frac{x_0^2}{4}\right)^{(1-n)/2}\left(\frac{\mu B_0}{x_0}\right)^m\sum_{q=0}^{\infty}\alpha_{mnq}\left(\frac{\mu B_0}{x_0}\right)^{2q} \quad (4a)$$

where

$$\alpha_{mnq} = [q!(q+m)!]^{-1}\left\{\frac{{}_1F_1[(2q+m+n+1)/2;\ \tfrac{1}{2};\ -x_0^{-2}]}{\Gamma[(-2q-m-n)/2]} + \frac{2}{x_0}\frac{{}_1F_1[(2q+m+n+1)/2;\ \tfrac{3}{2};\ -x_0^{-2}]}{\Gamma[(2-2q-m-n)/2]}\right\} \quad (4b)$$

Discuss the behavior of the spectrum as the signal and noise components of the modulation change in intensity and degree of modulation. D. Middleton.[2]

12.4 (*a*) Show that for a carrier modulated by a sine wave and random noise

$$P_\tau = \frac{A_0^2}{2}\left\{\left(\frac{1}{2}+\frac{\lambda^2\psi}{2}+\frac{\mu^2B_0^2}{4}\right)\{1+\Theta[(2\lambda^2\psi)^{-1/2}]\} - k^2\psi\phi^{(1)}[(\lambda^2\psi)^{-1/2}] + \lambda^2\psi\sum_{n=2}^{\infty}\left(\frac{\mu^2B_0^2}{2\lambda^2\psi}\right)^n\frac{1}{n!^2 2^n}\phi^{(2n-3)}[(\lambda^2\psi)^{-1/2}]\right\} \tag{1}$$

(*b*) As in (*a*), obtain

$$P_{f_0} = \frac{A_0^2}{2}\left\{\frac{1+\Theta[(2\lambda^2\psi)^{-1/2}]}{2} - \lambda^2\psi\phi^{(1)}[(\lambda^2\psi)^{-1/2}] + \sqrt{\lambda^2\psi}\sum_{n=1}^{\infty}\left(\frac{\mu^2B_0^2}{2\lambda^2\psi}\right)^n\frac{1}{n!^2 2^n}\phi^{(2n-2)}[(\lambda^2\psi)^{-1/2}]\right\}^2 \tag{2}$$

(*c*) For a general periodic modulation, show that Eq. (1), Prob. 12.3, applies when

$$h_{m,n} = \frac{-1}{2\pi}\int_C \xi^{n-2}B_m(\xi)e^{i\xi-\lambda^2\psi\xi^2/2}\,d\xi \tag{3}$$

with

$$B_m(\xi) = \frac{1}{T_0}\int_0^{T_0} e^{i\xi\mu V(t_0)-im\omega_a t_0}\,dt_0 \tag{4}$$

where f_a $(=\omega_a/2\pi)$ is the fundamental frequency of $V_S(t)$ and T_0 is the period of the modulation.
D. Middleton,[2] pp. 350, 351.

12.3 *Conversion I. Preliminary Remarks*

The theory of rectification by zero-memory devices discussed in Secs. 2.3, 5.1, and 5.2 has shown that any nonlinear system will produce a cross modulation of the components of the wave in question, generating large numbers of terms at sum and difference frequencies, as well as producing a

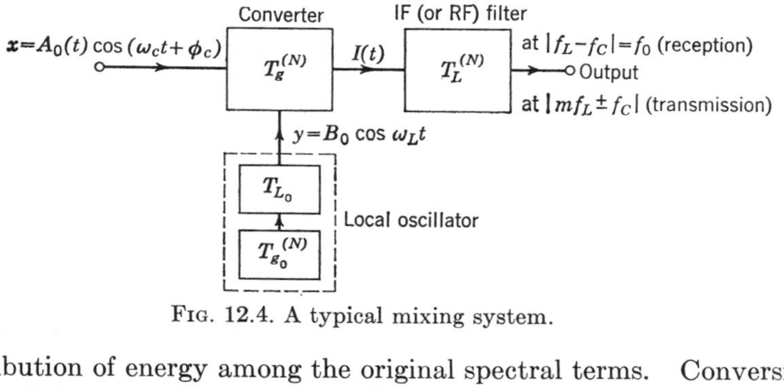

FIG. 12.4. A typical mixing system.

redistribution of energy among the original spectral terms. Conversion, or "mixing," is a special case of rectification where the (modulated) carrier or signal process and a local oscillation are combined in a nonlinear system in such a way that a part of the output consists of the original wave, undistorted by rectification, and shifted in frequency so that ideally no change in the spacing and in the relative phases of the components occurs. Any zero-memory nonlinear device can in principle be used, but some are better than others, since the critical condition on the mixing operation is that the instantaneous signal process be translated *linearly* to some new reference frequency. Figure 12.4 shows a typical mixing scheme.

To see how the mixing operation takes place, let us consider the input to the converter to be the sum of a modulated carrier $x = A_0(t) \cos (\omega_c t + \phi_c)$ and a local oscillation $y = B_0 \cos \omega_L t$, the case, for instance, when a grid-modulated class C amplifier is used as a converter. Let us also assume for the moment a general zero-memory nonlinear dynamic characteristic (cf. Sec. 2.3) and, as usual, ergodic modulations, carriers, and local oscillations. The instantaneous output of the converter (before filtering) can be represented in the usual way by

$$I(t) = \mathbf{T}_g^{(N)}\{x + y\}$$
$$= \frac{1}{2\pi} \int_C f(i\xi) \exp \{i\xi[A_0(t) \cos (\omega_c t + \phi_c) + B_0 \cos \omega_L t]\} \, d\xi \qquad A_0(t) \geq 0 \tag{12.28}$$

The instantaneous output may be expressed in terms of the harmonics of the carrier and local oscillation in the following way, if we use the Fourier development of $e^{ia \cos \theta}$ [cf. Eq. (5.8)]. We obtain for Eq. (12.28)

$$I(t) = \frac{1}{2\pi} \sum_{m=0}^{\infty} \sum_{n=0}^{\infty} \epsilon_m \epsilon_n i^{m+n} \cos m\omega_L t \cos n(\omega_c t + \phi_c)$$
$$\times \int_C f(i\xi) J_m(B_0\xi) J_n[A_0(t)\xi] \, d\xi \tag{12.29}$$

from which it is immediately evident that all orders of modulation products between signal and local oscillation appear. In general, when $A_0(t)$ is comparable to B_0 in mean intensity, terms linear in $A_0(t)$ are not the only significant ones. However, if B_0 is large compared with $A_0(t)$, that is, if we have a strong local oscillation, a considerable simplification is possible.

Writing $f_0 = f_L - f_c$ and introducing the important additional conditions[1] that the modulation be suitably band-limited, that the frequency interval between f_c and f_L be large compared with the modulation's spectral spread, but small compared with both f_0, f_c so that there is no overlapping of spectral zones, we see that

$$A_0(t) \cos (\omega_c t + \phi_c) + B_0 \cos \omega_L t = \sqrt{\alpha^2(t) + \beta^2(t)} \cos (\omega_L t - \phi) \tag{12.30a}$$

where now the envelope $C = \sqrt{\alpha^2 + \beta^2}$ and phase ϕ (cf. Sec. 2.2-6) are specifically

$$C(t)^2 = \alpha^2 + \beta^2 = B_0^2 + A_0^2(t) + 2A_0(t)B_0 \cos (\omega_0 t + \phi_c)$$
$$\phi(t) = \left[\tan^{-1} \frac{A_0(t) \sin (\omega_0 t + \phi_c)}{B_0 + A_0(t) \cos (\omega_0 t + \phi_c)}\right] \pm \pi q \qquad \omega_0 \equiv \omega_L - \omega_c \tag{12.30b}$$

For strong local oscillations, this can be written approximately as

$$C(t) \doteq B_0\left[1 + \frac{A_0(t)}{B_0}\cos(\omega_0 t + \phi_c)\right]\cos\left[\omega_L t - \frac{A_0(t)}{B_0}\sin(\omega_0 t + \phi_c)\right]$$
$$\equiv B_0(t)\cos[\omega_L t - \phi(t)] \tag{12.30c}$$

and in many cases (but not all; cf. Prob. 12.5) ϕ can be set equal to zero. Now, for $B_0 \gg A_0(t)_{\max}$, Eq. (12.29) becomes

$$I(t) \doteq \sum_{m=0}^{\infty} I_m \cos m\omega_L t + A_0(t)\sum_{m=1}^{\infty} g_{cm}\{\cos(m\omega_L t + \omega_c t + \phi_c)$$
$$+ \cos[(m\omega_L - \omega_c)t - \phi_c]\} + A_0(t)g\cos(\omega_c t + \phi_c) \qquad B_0^2 \gg A_0(t)^2_{\max} \tag{12.31}$$

where specifically

$$I_m = \frac{i^m\epsilon_m}{2\pi}\int_C f(i\xi)J_m(B_0\xi)\,d\xi \tag{12.32a}$$

$$g = \frac{1}{2\pi}\int_C i\xi f(i\xi)J_0(B_0\xi)\,d\xi \tag{12.32b}$$

$$g_{cm} = \frac{i_m\epsilon_m}{4\pi}\int_C i\xi f(i\xi)J_m(B_0\xi)\,d\xi \tag{12.32c}$$

As an example, let us consider a converter with a half-wave νth-law dynamic response [cf. Eq. (2.101a)]. Then Eq. (12.29) becomes

$$I(t) = \sum_{m=0}^{\infty}\sum_{n=0}^{\infty} \cos m\omega_L t \cos n(\omega_c t + \phi_c)\left\{\beta\,\frac{\epsilon_m\epsilon_n i^{m+n}}{2\pi}\int_C \frac{J_m(B_0\xi)J_n[A_0(t)\xi]}{(i\xi)^{\nu+1}}\,d\xi\right\} \tag{12.33}$$

and the integral in Eq. (12.33) may be found from Eq. (A.1.56) to be

$$B_{m,n}(t) = \frac{\epsilon_m\epsilon_n i^{m+n}}{2\pi}\,\beta\int_C \frac{J_m(B_0\xi)J_n[A_0(t)\xi]}{(i\xi)^{\nu+1}}\,d\xi\Bigg|_{B_0 \geq A_0(t)_{\max}}$$
$$= \frac{\beta\epsilon_m\epsilon_n B_0^{\nu+1}}{2^{\nu+1}n!}\left[\frac{A_0(t)}{B_0}\right]^n$$
$$\times\frac{{}_2F_1[(m+n-\nu)/2,\,(-m+n-\nu)/2;\,n+1;\,A_0(t)^2/B_0^2]}{\Gamma[(2+\nu-m-n)/2]\Gamma[(2+\nu+m-n)/2]} \tag{12.34}$$

This shows that, for a linear translation of the modulated wave to a new carrier frequency, the intensity of the local oscillator, $B_0^2/2$, must be much larger than $A_0(t)^2_{\max}$, which is our previously quoted "mixer" condition. A further condition involving the curvature of the dynamic characteristic also enters (Prob. 12.5), for if the amplifying portion of the dynamic response is not linear over the region of effective excursion of the "modulation" $A_0(t)\cos(\omega_0 t + \phi_c)$, which "rides" on the local oscillation ($B_0 \cos \omega_L t$), according to Eq. (12.30c) $A_0(t)$ is distorted. We therefore also require this amplitude deviation to fall in the essentially linear portions of the characteristic about the point $V(t) = B_0$. A typical spectrum of the converter's output under these conditions is sketched in Fig. 12.5.

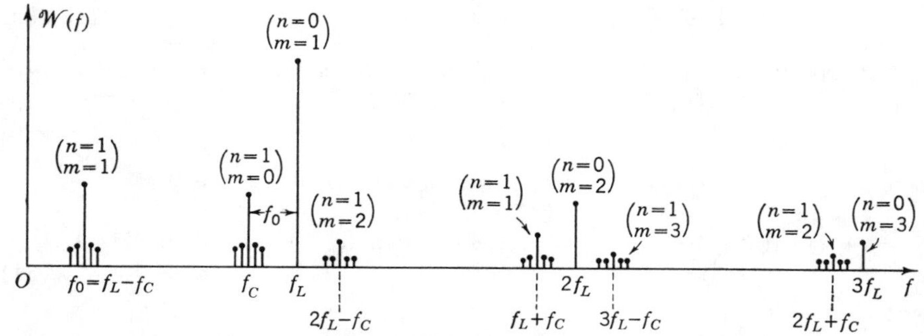

FIG. 12.5. Spectrum of a periodically modulated carrier and strong local oscillation after conversion (but before IF or RF filtering), showing typical modulation products.

Aims of the Present Analysis. So far, we have assumed the mixer conditions in the simple cases where noise is absent. In actual practice, however, we cannot always ignore the noise accompanying the signal and local oscillation and inherent in the mixer itself. Furthermore, the well-known requirements of a strong local oscillation vis-à-vis the signal, and a linear characteristic, do not tell us specifically what departures from these ideal operating conditions may be expected where (1) the local oscillation is *not* overwhelmingly more intense than $A_0(t)$; (2) when there is noticeable noise; and (3) when the dynamic characteristic possesses a finite curvature. Accordingly, here, and in Sec. 12.4, we attempt a short account of (1) to (3), involving noise as well as sinusoidal signals. For more comprehensive treatments and the physical details of converters, their conversion gains and losses, noise temperatures, etc., the reader may consult the literature.[5–8]

The quantities of greatest statistical utility in estimating the departure from ideal operation are, as before, the covariance function and spectrum of the mixer's output. To describe the rectifier more fully as a converter, however, a number of additional parameters are needed. These are (1) the *conversion conductances*, for large- and small-signal operation, (2) the mean power absorbed by the system, and (3) the conversion losses and equivalent noise temperatures. We shall consider only the first two here.

(1) Small-signal and conversion conductances. The small-signal conductance is defined here as the slope of the dynamic characteristic at the instantaneous operating point $y = B_0 \cos \omega_L t$. In the usual operation of the nonlinear device $\boldsymbol{T_g}$ as a mixer, we can represent the instantaneous output current $I(t)$ approximately by

$$I(t) \doteq I\Big|_y + \frac{\partial I}{\partial V}\Big|_y \Delta V(t) \tag{12.35}$$

where $\Delta V = A_0(t) \cos(\omega_c t + \phi_c)$ and $y = B_0 \cos \omega_L t$. The coefficient of ΔV in Eq. (12.35) is called the (instantaneous) *small-signal* conductance,

$$g(t) \equiv \left.\frac{\partial I}{\partial V}\right|_{y=y(t)} \tag{12.36}$$

A direct evaluation of Eq. (12.36), i.e.,

$$g(t) = \frac{1}{2\pi} \int_C i\xi f(i\xi) e^{i\xi B_0 \cos \omega_L t}\, d\xi \tag{12.36a}$$

shows that

$$g(t) = g + 2 \sum_{m=1}^{\infty} g_{cm} \cos m\omega_L t \tag{12.37}$$

where g and g_{cm} are given by Eqs. (12.32b), (12.32c). We note next from Eq. (12.37) that g is the *average* small-signal conductance, while the g_{cm} are often called the (average) *conversion conductances, for mixing* at the mth harmonic of the local oscillator's frequency. The g_{cm} are here defined only for small-signal operation. We remark, therefore, that, in this approximation, the presence of mixer, local oscillator, and signal noise does not change the specific character of the transconductances: Eqs. (12.35) and (12.36) still apply, except that in Eq. (12.35) ΔV becomes now $A_0(t) \cos(\omega_c t + \phi_c) + N(t)$, where $N(t)$ represents the combined effects of system noise.

(2) Large-signal conductances. The instantaneous conductance presented to a signal of large amplitude is defined by

$$G(t) \equiv I(t)/V(t) = g[V(t)]/V(t) \tag{12.38}$$

The average large-signal conductance $\langle G \rangle$ (or $\bar{G}$) is the steady component of $G(t)$, which is

$$\langle G \rangle \equiv \bar{G} = \int_{-\infty}^{\infty} \frac{g(V)}{V} w_1(V)_S\, dV = \int_C f'(i\xi) F_1(\xi) \frac{d\xi}{2\pi} \tag{12.39a}$$

where $f'(i\xi)$ is the (complex) Fourier transform of $g(V)/V$ and $F_1(\xi)_S$ is the characteristic function, corresponding to the probability density $w_1(V)_S$ of the signal. For the strong local oscillations assumed here, $V = B_0 \cos \omega_L t$, so that $w_1(V)_S$ is simply the probability density of a sine wave of amplitude B_0, for which the associated characteristic function is $J_0(B_0\xi)$ (cf. Prob. 1.11); therefore, we get

$$\bar{G} = \frac{1}{\pi} \int_{-B_0}^{B_0} \frac{g(V)}{V\sqrt{B_0^2 - V^2}}\, dV = \frac{1}{2\pi} \int_C J_0(B_0\xi) f'(i\xi)\, d\xi \tag{12.39b}$$

For a half-wave linear conversion characteristic, for example, the mean large-signal conductance becomes at once $\beta/2$, where β is the dynamic transconductance of this linear rectifier [cf. Eq. (2.101a)].

(3) Mean absorbed power. This quantity represents the power absorbed by the mixing device itself and is defined as

$$P_a = \langle I(t_0)V(t_0)\rangle = \overline{IV} \qquad \text{probability 1} \tag{12.40a}$$

$$= \int_{-\infty}^{\infty} Vg(V)w_1(V)_S\,dV = \frac{-i}{2\pi}\int_C f(i\xi)\,\frac{dF_1(\xi)_S}{d\xi}\,d\xi \tag{12.40b}$$

For essentially ideal and noise-free conversion, $V(t)$ equals $B_0 \cos \omega_L t$ and $I(t)$ is given explicitly by Eq. (12.31) in this approximation. The result is

$$P_a = |I_1|\frac{B_0}{2} = \left|\frac{iB_0}{2\pi}\int_C f(i\xi)J_1(B_0\xi)\,d\xi\right| \tag{12.41}$$

Generalizations of $\bar{G}$ and P_a to include the effects of noise and incoming signal as correction terms are derived in straightforward fashion.

PROBLEMS

12.5 Show, starting with Eq. (12.28) and using Eqs. (12.30), that the output of the converter (before filtering) can be alternatively expressed as

$$I(t) = \sum_{m=0}^{\infty} B_m(t)\cos m(\omega_L t - \phi) \tag{1}$$

where the envelopes B_m are specifically

$$B_m(t) = \frac{i^m\epsilon_m}{2\pi}\int_C J_m[\xi C(t)]f(i\xi)\,d\xi \tag{2}$$

and for a half-wave νth-law dynamic characteristic

$$B_m(t) = \frac{\epsilon_m\Gamma(\nu+1)\beta C(t)^\nu}{2^{\nu+1}\Gamma[(2+\nu-m)/2]\Gamma[(2+\nu+m)/2]} \tag{3}$$

Hence, show that, for $\nu \neq 1$, the signal term must not extend appreciably into the regions of finite curvature of the dynamic response for the mixing to be linear in the signal.

12.6 (*a*) Show for a half-wave νth-law characteristic that the small-signal conversion conductances become

$$g = \frac{\beta B_0^{\nu-1}\Gamma(\nu+1)}{2^\nu\Gamma[(\nu+1)/2]^2} \qquad g_{cm} = \frac{\beta B_0^{\nu-1}\epsilon_m\Gamma(\nu+1)}{2^{\nu+1}\Gamma[(1+\nu-m)/2]\Gamma[(1+\nu+m)/2]} \tag{1}$$

(*b*) Verify also that the components of the current at mf_L are

$$I_m = \frac{\beta B_0^{\nu}\epsilon_m}{2^{\nu+1}}\,\frac{\Gamma(\nu+1)}{\Gamma[(2+\nu-m)/2]\Gamma[(2+\nu+m)/2]} \tag{2}$$

12.4 *Conversion II. Second-moment Theory*

Before we examine the general case of the mixing of signals and noise, let us introduce the problem by considering the idealized, i.e., noise-free, example of a modulated carrier and a local-oscillation process alone, so that Eq. (12.28) represents the instantaneous output (before filtering) and Fig. 12.6 shows the spectrum of the input.

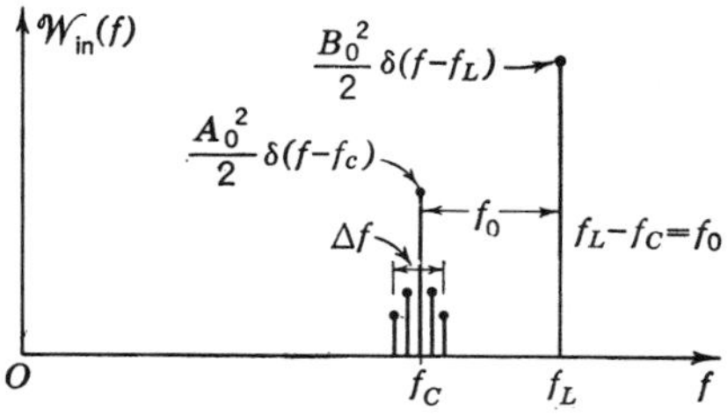

FIG. 12.6. Spectrum of the combined inputs to the converter.

We begin here by putting the following restrictions on the mixing process:

1. That the bandwidth of the modulation is much less than the intermediate frequency $f_0 = |f_L - f_c|$, a condition almost always met in heterodyning problems.

2. That $f_L, f_c \gg f_0$, the usual requirement in mixer operation.

3. That there be no coherence between modulation, carrier, and local oscillation for almost all members of their respective ensembles, that is, f_L, f_c (and hence f_0), and f_{mod} are incommensurable. Statistically, modulation, carrier, and local oscillation processes are independent. The covariance function of the output process $I(t)$ is therefore

$$K_I(t) = \int_C \int_C f(i\xi_1)f(i\xi_2)F_2(\xi_1,\xi_2;t)_S \frac{d\xi_1\, d\xi_2}{(2\pi)^2} \qquad S = A_0(t) + B_0(t) \tag{12.42}$$

and since the various component parts of the input process are independent, the characteristic function F_2 factors into

$$\begin{aligned} F_2(\xi_1,\xi_2;t)_{A_0+B_0} &= F_2(\xi_1,\xi_2;t)_{A_0}F_2(\xi_1,\xi_2;t)_{B_0} \\ &= \sum_{m=0}^{\infty}\sum_{n=0}^{\infty} (-1)^{m+n}\epsilon_m\epsilon_n\langle J_n[\xi_1 A_0(t_1)]J_n[\xi_2 A_0(t_2)]\rangle_{mod} \\ &\qquad \times J_m(B_0\xi_1)J_m(B_0\xi_2)\cos m\omega_L t \cos n\omega_c t \end{aligned} \tag{12.43}$$

where we have used Eq. (5.8) in Eq. (5.7) appropriately modified. The expression for $K_I(t)$ becomes finally

$$K_I(t) = \sum_{m=0}^{\infty}\sum_{n=0}^{\infty} (-1)^{m+n}\epsilon_m\epsilon_n\langle H_{m,n}(t_0')H_{m,n}(t_0'+t)\rangle_{mod} \cos m\omega_L t \cos n\omega_c t \tag{12.44a}$$

in which the amplitude functions $H_{m,n}$ [cf. Eqs. (5.14), (5.17)] are specifically

$$H_{m,n}(t) = \frac{1}{2\pi}\int_C f(i\xi)J_m(B_0\xi)J_n[\xi A_0(t)]\, d\xi \tag{12.44b}$$

(1) IF output. Half-wave νth-law response. The components of interest at the intermediate frequency are obtained when $m = n = 1$ in Eq. (12.44) above. The covariance function of the IF output is therefore

$$K_I(t)_{\mathrm{IF}} = 2\langle H_{1,1}(t_0')H_{1,1}(t_0'+t)\rangle_{mod} \cos \omega_0 t \tag{12.45}$$

for all meaningful values of A_0 and B_0. As an example, we again choose a mixer with a half-wave νth-law characteristic so that, with the help of Eq. (12.34), we get

$$K_I(t)_{\mathrm{IF}} = \beta^2 B_0^{2\nu}(\cos \omega_0 t)\langle A_1 A_2 \, {}_2F_1(1 - \nu/2, -\nu/2; 2; A_1^2/B_0^2) \times {}_2F_1(1 - \nu/2, -\nu/2; 2; A_2^2/B_0^2)\rangle / 2^{2\nu+1}\Gamma(\nu/2)^2\Gamma(\nu/2 + 1)^2$$
$$A_1 = A_0(t_0'),\ A_2 = A_0(t_0' + t) \quad (12.46)$$

For a half-wave quadratic response ($\nu = 2$) this is particularly simple,

$$K_I(t)_{\mathrm{IF}}\Big|_{\nu=2} = \frac{\beta^2 B_0^4}{32}\langle A_1 A_2\rangle \cos \omega_0 t \quad (12.47a)$$

which is linear in the modulation for *all* signal intensities, an exceptional case, since for all values of ν the mixer condition $\langle A^2\rangle \ll B_0^2$ must be satisfied for linear conversion. The linear case ($\nu = 1$) illustrates this,

$$K_I(t)_{\mathrm{IF}}\Big|_{\nu=1} \doteq \frac{B_0^2\beta^2}{8\pi^2}\left[\langle A_1 A_2\rangle - \frac{\langle A_1^3 A_2^3\rangle}{4B_0^4} + \cdots\right]\cos \omega_0 t \quad (12.47b)$$

and if the mixer condition is obeyed, the correction term [and others not shown in Eq. (12.47*b*)] becomes quite negligible. The IF spectrum therefore contains only the original components of the modulation (with their original phase relations preserved), and the carrier is shifted to the new frequency f_0.

(2) Harmonic generation. When $A_0(t)$ and B_0 are of comparable intensity, all orders of cross-modulation products are significant, according to Eq. (12.44*a*). But when the mixer condition is obeyed, one gets instead Eq. (12.31) for the instantaneous output and only the terms at mf_L and $mf_L \pm f_c$ are significant. The covariance function and spectrum of the output process, before filtering, are then (for periodic modulation of fundamental frequency $\omega_a/2\pi$)

$$K_I(t) \doteq \sum_{m=0}^{\infty} \frac{I_m^2}{2}\cos m\omega_L t + g\left\langle\frac{A_1 A_2}{2}\right\rangle \cos \omega_c t + \langle A_1 A_2\rangle \sum_{m=1}^{\infty} g_{cm}^2 \frac{1}{2}\cos (m\omega_L t \pm \omega_c t) \quad (12.48a)$$

and therefore

$$\mathcal{W}_I(f) \doteq \sum_{m=0}^{\infty} \frac{I_m^2}{2}\delta(f - mf_L) + \frac{g^2}{2}\sum_{k}^{\infty} a_k^2\,\delta[f - (f_c \pm kf_a)] + \sum_{m=1}^{\infty}\frac{g_{cm}^2}{2}\sum_{k=1}^{\infty} a_k^2\,\delta[f - (mf_L \pm f_c \pm kf_a)] \quad (12.48b)$$

a typical example of which is shown in Fig. 12.5. (Note that the above also applies for random modulations obeying the mixer condition.)

(3) **Noisy mixing.** In many cases, it is not possible to ignore the noise accompanying the signal and local oscillations. We need now to examine the effects of noise on the output modulation products. For this purpose, we have for the input process

$$V(t) = A_0(t) \cos(\omega_c t + \phi_c) + B_0 \cos \omega_L t + N(t) \qquad (12.49)$$

whose spectrum is sketched in Fig. 12.7, and where now $N(t)$ represents the sum of three independent† noise voltage processes:

1. $N_B(t)$, the narrow band of noise accompanying the local oscillation (in Fig. 12.4 a suitable filter between converter and local oscillator is assumed).

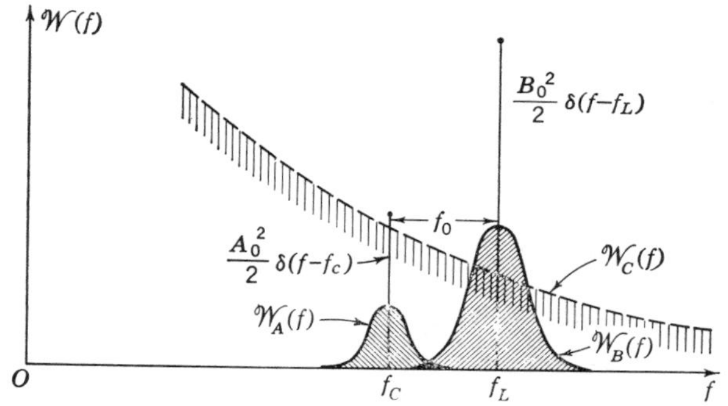

FIG. 12.7. Spectra of the input when there is accompanying noise.

2. $N_A(t)$, the narrowband noise accompanying the RF signal.
3. $N_C(t)$, the noise inherent in the converter itself, which in the analysis below is assumed to arise as a kind of shot effect (cf. Chap. 11), effectively accompanying the input signals before rectification takes place.‡ This inherent noise must be treated as broadband, unlike N_A and N_B, which are spectrally narrow, owing to previous frequency selection.

The autocorrelation function of the converter's output [Eq. (12.42)] now becomes

$$K_I(t) = \int_C \int_C f(i\xi_1)f(i\xi_2)F_2(\xi_1,\xi_2;t)_S F_2(\xi_1,\xi_2;t)_N \, d\xi_1 \, d\xi_2 \qquad (12.50)$$

where $(F_2)_S$ is given by Eq. (12.43). We assume the noise processes to have the usual normal statistics, so that, from their independence,

$$\begin{aligned} F_2(\xi_1,\xi_2;t)_N &= F_2(\xi_1,\xi_2;t)_A F_2(\xi_1,\xi_2;t)_B F_2(\xi_1,\xi_2;t)_C \\ &= \exp\left[-\tfrac{1}{2}\Psi(\xi_1^2 + \xi_2^2) - \Psi_A(t)\xi_1\xi_2 - \Psi_B(t)\xi_1\xi_2 - \Psi_C(t)\xi_1\xi_2\right] \\ &\qquad \text{with } \Psi = \Psi_A + \Psi_B + \Psi_C \qquad (12.51) \end{aligned}$$

† N_A, N_B, N_C are certainly independent processes, as they arise from different sources.

‡ For crystal mixers, some of the inherent noise appears before the nonlinear operation and the rest, additively, afterward. In this case, $N_C(t)$ refers to that part of the inherent noise appearing before mixing.

where the $\Psi_{A,B,C}(t)$ are the covariance functions of the separate noise terms, and $\Psi_A = \overline{N_A{}^2}$, $\Psi_B = \overline{N_B{}^2}$, $\Psi_C = \overline{N_C{}^2}$. With the help of Eq. (12.45), $K_I(t)$ may next be written as

$$K_I(t) = \sum_{m=0}^{\infty}\sum_{n=0}^{\infty} (-1)^{m+n}\epsilon_m\epsilon_n \cos m\omega_L t \cos n\omega_c t$$
$$\times \left\langle \frac{1}{4\pi^2} \int_C f(i\xi_1) J_n[A_0(t)\xi_1] J_m(B_0\xi_1) \exp(-\Psi\xi_1{}^2/2)\right.$$
$$\times \int_C f(i\xi_2) J_n[A_0(t_2)\xi_2]\, J_m(B_0\xi_2) \exp(-\Psi\xi_2{}^2/2)$$
$$\left.\times \exp\{-[\Psi_A(t) + \Psi_B(t) + \Psi_C(t)]\xi_1\xi_2\}\, d\xi_1\, d\xi_2 \right\rangle_{mod} \quad (12.52)$$

Taking advantage of the narrowband character of $N_A(t)$ and $N_B(t)$ allows us to write $\Psi_A(t) + \Psi_B(t)$ as

$$\Psi_0(t)_A \cos \omega_c t + \Psi_0(t)_B \cos \omega_L t = \mu(t) \cos [\omega_L t - \Phi(t)] \quad (12.53a)$$

in which

$$\mu(t) = [\Psi_0(t)_A{}^2 + \Psi_0(t)_B{}^2 + 2\Psi_0(t)_A\Psi_0(t)_B \cos \omega_0 t]^{1/2} = \mu(t;\omega_0)$$
$$\omega_L - \omega_c = \omega_0 \quad (12.53b)$$
$$\Phi(t) = \tan^{-1} \{\Psi_0(t)_A \sin \omega_0 t/[\Psi_0(t)_B + \Psi_0(t)_A \cos \omega_0 t]\} \pm q\pi \cdots = \Phi(t;\omega_0) \quad (12.53c)$$

Consequently, making use of Eq. (5.8) to expand

$$\exp\{-[\Psi_A(t) + \Psi_B(t)]\xi_1\xi_2\}$$

we get finally for the complete covariance function

$$K_I(t) = \sum_{c,m,n,q,s}^{\infty} \frac{\epsilon_m\epsilon_n\epsilon_q(-1)^{c+m+n+q}}{c!q!(q+s)!2^{q+2s}} \Psi_C(t)^c \mu(t;\omega_0)^{q+2s}$$
$$\times \langle H_{cmnqs}(t_0')H_{cmnqs}(t_0'+t)\rangle \cos m\omega_L t \cos n\omega_c t \cos q[\omega_L t - \Phi(t;\omega_0)] \quad (12.54)$$

in which the amplitude functions $H_{cmnqs}(t)$ are specifically

$$H_{cmnqs}(t) = \frac{1}{2\pi} \int_C f(i\xi)\, \xi^{c+q+2s} e^{-\Psi\xi^2/2} J_m(B_0\xi) J_n[A_0(t)\xi]\, d\xi \quad (12.54a)$$

Our next task is to rearrange the sums in Eq. (12.54) in such a way that the zonal structure of the covariance function is revealed and, in particular, the terms contributing to the spectrum in the IF band. To select from the triple cosine product [and from $\mu(t;\omega_0)$] those terms which yield $\cos \omega_0 t$ is a very tedious process in general. However, by introducing the mixer conditions

$$B_0{}^2 \gg 2\overline{A_0(t)^2} \qquad B_0{}^2 \gg 2\Psi_A,\ 2\Psi_B,\ 2\Psi_C \quad (12.55)$$

we can greatly reduce the labor.

(4) Half-wave linear mixer. We illustrate with a half-wave linear characteristic ($\nu = 1$); there the amplitude functions are

$$H_{cmnqs}(t) = -\frac{\beta}{2\pi}\sum_{l=0}^{\infty}\frac{(-1)^l}{l!(l+n)}\left[\frac{A_0(t)}{2}\right]^{2l+n}\int_C \xi^{\gamma}e^{-\Psi\xi^2/2}J_m(B_0\xi)\,d\xi \qquad \gamma = c+q+2s$$

$$\simeq \beta\Psi^{1/2-c/2-q/2-s}2^{-3/2+c/2+q/2+s}i^{-c-m-n-q-2s} \times [p_A(t)/p_B]^{n/2}p_B^{1/2-c/2-q/2-s}(n!)^{-1}$$

$$\times\left\{\frac{{}_2F_1[(m+n+\gamma-1)/2,\ (-m+n+\gamma-1)/2;\ n+1;\ p_A(t)/p_B]}{\Gamma[(3-m-n-\gamma)/2]\Gamma[(3+m-n-\gamma)/2]} + p_B^{-1}\frac{{}_2F_1[(m+n+\gamma+1)/2,\ (-m+n+\gamma+1)/2;\ n+1;\ p_A(t)/p_B] + O(p_B^{-2})}{\Gamma[(1-m-n-\gamma)/2]\Gamma[(1+m-n-\gamma)/2]}\right\} \tag{12.56}$$

with $\quad p_A(t) \equiv A_0(t)^2/2\Psi \qquad p_B \equiv B_0^2/2\Psi \qquad 0 \le p_A(t)^2 \ll 1$

$\quad p_A \equiv A_0^2(t)/2\Psi \qquad \eta^2 \equiv p_A/p_B$

For a first-order theory, it is sufficient to use only terms of order $[p_A(t)/p_B]^0$, but to see how the ideal mixer operation is approached, a second-order development including all contributions through order $p_A(t)/p_B$ is needed. The significant modulation products are found in a straightforward way. We examine first the factors

$$(p_A/p_B)^{n/2}p_B^{1/2-c/2-q/2-s}\ {}_2F_1 \qquad (p_A/p_B)^{n/2}p_B^{-1/2-c/2-q/2-s}\ {}_2F_1$$

in Eq. (12.56) and select all terms of order $(p_A/p_B)^{0,1}$, remembering that $\eta^2 = p_A/p_B \ll 1$ here. After these values of c, q, n, s have been chosen, we next determine from $\mu(t)^{q+2s} \times$ (trigonometric factors) those values of m which yield $\cos \omega_0 t$.

The covariance function of the mixer's output before filtering and in the spectral vicinity of f_0 is found in the first-order theory to be

$$K(t)\Big|_{O(\eta^0)} \simeq \frac{\beta^2}{\pi^2}\left\{\left[\frac{A_1A_2}{2} + \Psi_0(t)_A\right]\cos \omega_0 t + \Psi_C(t)\left[\frac{\pi^2}{4} + 2\sum_{m=0}^{\infty}\frac{\cos (2m+1)\omega_L t}{(2m+1)^2}\right]\right\} \tag{12.57}$$

where $\Psi_0(t)$ is the slowly varying part of $\Psi(t)$ in the narrowband cases. The first term alone represents the shifted signal and noise, while the second is simply the sum of that part of the inherent noise passed in mixing and the $s \times n$ cross terms produced in the conversion of this noise and the local oscillation. Observe that no contributions from the local oscillator, either carrier or noise, are present in a first-order theory, as expected when the mixer condition is satisfied. For a second-order theory, however, terms of

$O(\eta^2)$ must be included, and here we may expect contributions involving the local-oscillator noise and its cross products with N_A and N_C. We obtain finally for the covariance function of the correction terms

$$K(t)\Big|_{O(\eta^2)} \simeq \frac{\eta^2\beta^2}{2\pi^2}\frac{\overline{A_1A_2}}{\overline{A_1^2}}\cos\omega_0 t \times \left\{\Psi_0(t)_B - \Psi_0(t)_A - \Psi - \frac{\overline{A_1A_2}}{4} + \frac{[\Psi_0(t)_B - 2\Psi]\Psi_0(t)_A}{\sqrt{\overline{A_1A_2}}}\right\}$$
$$+ \frac{\eta^2\beta^2}{2\pi\omega_c}\frac{\Psi_C(t)}{\overline{A_1^2}}\sum_{k=-\infty}^{\infty}\delta(t - kT_L)\left\{\Psi_C(t) - 2\left(\Psi + \frac{\overline{A_1A_2}}{2}\right)\right.$$
$$\left. + [\overline{A_1A_2} + 2\Psi_0(t)_B]\cos\omega_c t + 2\Psi_0(t)_B\cos\omega_L t\right\} \qquad T_L = \frac{2\pi}{\omega_L} \tag{12.58}$$

The complete covariance function before filtering is therefore the sum of Eqs. (12.57) and (12.58) in the present approximation.

Since the inherent noise is broadband and consequently is the disturbance produced in the mixing process, the output following conversion will be noticeably modified by any subsequent narrowband filter (cf. Sec. 3.3-3). The IF stage of a radio or radar receiver provides an important example. The covariance function of this output wave is, from Eq. (3.102),

$$K_{IF}(t) = \int_{-\infty}^{\infty} K(t')_{\text{input}}\rho_{IF}(t' + t)\,dt' \tag{12.59}$$

where ρ_{IF} is the correlation function of the IF filter, $K(t)_{\text{input}}$ is here simply the sum of Eqs. (12.57) and (12.58), and Eq. (12.59) is itself the Fourier transform of the more familiar relation

$$\mathcal{W}_{IF}(f) = |Y_{IF}(i\omega)|^2\mathcal{W}_{\text{in}}(f)$$

[cf. Eq. (3.104)]. Taking advantage of the spectral narrowness of $N_C(t)$, we can write (at least for the first-order terms)

$$\Psi_C(t) \doteq \frac{W_C}{2}\,\delta(t - 0) \tag{12.60}$$

where W_C is the spectral intensity in the neighborhood of f_0. The mean intensity of the IF filter's output is then

$$P_{(\nu=1)} \simeq \frac{\beta^2}{\pi^2}\int_{-\infty}^{\infty} dt'\rho_{IF}(t')\cos\omega_0 t'[K_A(t') + \Psi_0(t')_A] + \beta^2\frac{W_C}{4}\rho_{IF}(0)$$
$$+ \frac{\eta^2\beta^2 K_A(t')}{4\pi\omega_c K_A(0)^2}\rho_{IF}(0)(2\Psi_B - 2\Psi_A - \Psi_C)$$
$$- \frac{\eta^2\beta^2}{2\pi^2}\left[\Psi_C + 2\Psi_A + \frac{K_A(0)}{2} + \frac{\Psi_A(2\Psi_A + \Psi_B + 2\Psi_C)}{2K_A(0)}\right] \tag{12.61}$$

in which all terms involving $\Psi_C(kT_L)$ ($|k| > 0$) have been omitted because of the rapid decay of $\Psi_C(t)$. Here also, $K_A(t) = \frac{1}{2}\overline{A_0(t_0)A_0(t_0 + t)}$.

For an estimate of the magnitude of the correction term, let us assume a wide IF filter, so that $\rho_{IF}(t) = \delta(t - 0)$ may be used in the integrand of Eq. (12.61). For finite results, this IF filter, though wide, must pass only a limited amount of power. For the terms in Eq. (12.61) involving η^2, we let $\rho_{IF}(t) = (\omega_F/\sqrt{\pi})e^{-t^2\omega_F^2}$ [so that $\lim_{\omega_F \to \infty} \rho_{IF}(t) = \delta(t - 0)$], where ω_F is proportional to the half-width of the filter. Then, the mean intensity [Eq. (12.61)] becomes

$$P_{(\nu=1)} \cong \frac{\beta^2}{\pi^2}[K_A(0) + \Psi_A + 2\pi^2\Psi'_C] - \frac{\eta^2\beta^2}{\pi^2}\left[\Psi_A + \frac{\Psi_C}{2} + \frac{K_A(0)}{4} + \frac{\Psi_A(2\Psi_A + \Psi_B + 2\Psi_C)}{4K_A(0)}\right] \quad (12.62)$$

where the contribution of the term with $\rho_{IF}(0)$ as factor is negligible, since $\omega_c \gg \omega_F$.† Note that the correction effects subtract from the limiting case (when $\eta \to 0$). For small values of η^2, this correction term may be negligible: in some instances, values of $\eta^2 = -13$ db have proved an upper bound for essentially "linear" mixer performance, while in other cases much smaller values may be required. In threshold reception, when $K_A(0) \approx \Psi_A + 2\pi^2\Psi'_C$, the leading term of Eq. (12.62) shows that Ψ'_C should be at least 13 db ($\doteq 2\pi^2$) or more below Ψ_A. Furthermore, even if this is so, Ψ_C itself may be so large that values of η^2 considerably below -13 db are needed to make $\frac{\eta^2\beta^2}{\pi^2}\frac{7\Psi_A + \Psi_B + 4\Psi_C}{4}$ negligible [$K_A(0) \approx \Psi_A$]. On the other hand, if the strength of the incoming signal is the controlling factor, if, for example, $K_A(0) \gg \Psi'_C, \Psi_A, \Psi_C, \Psi_B$, the correction term depends on $[\eta^2 K_A(0)/4\pi^2]\beta^2$.

(5) Crystal mixers.[9] Many mixing schemes using vacuum tubes can be constructed to give half-wave linear rectification characteristics ($\nu = 1$). When crystals are used, however, the dynamic response may take quite different forms.[8] For tubes, it is possible to adjust the rectifying element (with suitable bias) so that instantaneous input and output are related by the familiar half-wave νth-law $I = \beta V^\nu$ ($V > 0$), $I = 0$ ($V < 0$). For crystals, on the other hand, a number of different responses may be appropriate, depending on the magnitude of the input voltage. A half-wave exponential response is often suitable for comparatively small voltages, while for moderately large inputs the half-wave νth law can be used, bias being negligible in both cases. For very intense excitation, however, the negative back current cannot be ignored, and one has the general

† Here Ψ'_C ($= W_c\omega_F/2\sqrt{\pi}$) is that part of the inherent mixer noise that is passed by a filter having the gaussian response assumed above, on the assumption that the noise spectrum is flat over the frequency range $f_F - f_0 < f < f_0 + f_F$. The factor 2 in $2\beta^2\Psi'_C$ is attributable to the cross products between local oscillator and background noise. The actual factor depends, of course, on the dynamic response of the mixer, but for most practical cases, when $0 < \nu < 2$, we expect it to be larger than 1 and less than 4 or 5.

characteristic

$$I(t) = \begin{cases} \beta_1(e^{\alpha_1[V(t)-b_0]} - 1) & V > b_0 \\ -\beta_2(e^{-\alpha_2[V(t)-b_0]} - 1) & V < b_0 \ (\alpha_1, \beta_1, \ldots > 0) \end{cases} \qquad (12.63)$$

where b_0 is the bias; Eq. (12.63) is sufficiently general to be used for weak or very strong signals, as well as moderate inputs.

PROBLEMS

12.7 (*a*) Show that

$$\sum_{l=0}^{\infty} \frac{x^l}{l!(l+n)!\Gamma(\alpha' - l)\Gamma(\beta' - l)} = [n!\Gamma(\alpha')\Gamma(\beta')]^{-1} {}_2F_1(1 - \alpha', 1 - \beta' : n + 1; x) \qquad (1)$$

(*b*) Hence, obtain Eq. (12.56) for $H_{cmnqs}(t)$ from Eq. (12.54*a*) in the case of the half-wave linear mixer response.

12.8 (*a*) Show that for an unbiased, quadratic characteristic ($\nu = 2$) the covariance function of the mixer's output (before filtering and for the contribution in the vicinity of the intermediate frequency) is to the first order

$$K_{(\mathrm{IF})}(t) \simeq \beta^2 \frac{B_0{}^2}{4} \cos \omega_0 t \left[K_A(t) + \Psi_0(t)_A + \frac{\Psi_C(t)}{\pi^2} \sum_{m=0}^{\infty} \frac{\epsilon_m \cos m\omega_L t}{(4m^2 - 1)^2} \right] \qquad (1)$$

(*b*) Obtain the correction term $O(\eta^2)$.

REFERENCES

1. Terman, F. E.: "Radio Engineers' Handbook," chap. 7, pp. 531–553, McGraw-Hill, New York, 1943. For a discussion of crystal characteristics, see H. C. Torrey and C. A. Whitmer, "Crystal Rectifiers," Massachusetts Institute of Technology Radiation Laboratory Series, vol. 15, McGraw-Hill, New York, 1948.
2. Middleton, D.: On the Distribution of Energy in Noise and Signal-modulated Waves. I. Amplitude Modulation, *Quart. Appl. Math.*, **9**: 337 (1952).
3. Watson, G. N.: "Theory of Bessel Functions," 2d ed., sec. 13.4, Cambridge, New York, 1944.
4. Middleton, D.: Some General Results in the Theory of Noise through Non-linear Devices, *Quart. Appl. Math.*, **5**: 445 (1948); for details of the integration, see eqs. 4-19, 4-20, 4-22.
5. Herold, E. W., and L. Malter: Some Aspects of Radio Reception at Ultra-high Frequencies, *Proc. IRE*, **31**: 575 (1943).
6. Peterson, L. C., and F. B. Llewellyn: The Performance and Measurement of Mixers in Terms of Linear Network Theory, *Proc. IRE*, **33**: 458 (1945).
7. Torrey, H. C., and C. A. Whitmer: "Crystal Rectifiers," Massachusetts Institute of Technology Radiation Laboratory Series, vol. 15, McGraw-Hill, New York, 1948.
8. Strum, P. D.: Some Aspects of Mixer Crystal Performance, *Proc. IRE*, **41**: 875 (1953). Additional references are included here.
9. For detailed accounts of mixing systems using crystals, see R. V. Pound, "Microwave Mixers," vol. 16, and S. N. Van Voorhis, "Microwave Receivers," vol. 23, Massachusetts Institute of Technology Radiation Laboratory Series, McGraw-Hill, New York, 1948.

CHAPTER 13

RECTIFICATION OF AMPLITUDE-MODULATED WAVES: SECOND-MOMENT THEORY

The preceding chapter has considered the second-moment theory of amplitude modulation. Here, however, we shift our attention to rectification problems centered in the nonlinear elements common to most receiving systems and measuring devices. Instead of amplitude modulation, we consider the demodulation, or detection, of such waves when they are accompanied by additive (and independent) normal noise. As before, and unless otherwise indicated, we assume that modulation, carrier, and noise processes are ergodic and that the nonlinear elements in question are zero-memory invariant operators $\boldsymbol{T}_g$ (cf. Secs. 2.1, 2.3). The accompanying noise is also assumed to belong to a gaussian random process (cf. Sec. 7.4) whose first moments vanish and whose spectrum is shaped by the linear filter $\boldsymbol{T}_L$ preceding the nonlinear element. The general methods of Chap. 5 are used here to determine various second-moment properties,† e.g., the covariance function, second-moment function, intensity spectrum, mean intensity, etc., of the output process. While specific results are obtained for a variety of examples of engineering interest, no attempt at a complete coverage is made. The present treatment is purposely concise; for many of the analytical details, the reader is referred to the original papers.‡

13.1 *Rectification of Broad- and Narrowband Noise Processes*

Before the more general discussion of the rectification of signals in noise is attempted, it is instructive to consider the rectification, or detection, of noise alone. In detection problems, the input is usually narrowband, and one is chiefly interested in the l-f portion of the output,§ unlike the cases

† The reader may find it profitable to review the material of Secs. 5.1-1 to 5.1-3 before continuing with the present chapter. Section 2.3 is also recommended, as is Appendix 1 for many of the special functions and integrals that appear in the subsequent analysis.

‡ See the general analyses of Rice[1] and Middleton[2] and the papers by Fränz,[3] Bennett,[4] North,[5] Middleton,[6] and Burgess.[7] The larger part of the material of this chapter, however, is based on Middleton[2] and extensions of these earlier results. Further references will be found in subsequent sections.

§ If the disturbance is broadband, a resolution into distinct spectral zones is, of course, impossible. The whole spectrum after rectification must be considered.

of modulation and mixing treated in the preceding sections. It is assumed, as before (cf. Sec. 2.3), that reactive loading of the nonlinear elements is negligible, so that one has a (stable) one-valued dynamic characteristic relating the instantaneous input and output disturbances. The present discussion is divided into two parts, depending on whether or not the input process is spectrally broad or narrow; a half-wave νth-law rectifier (Fig. 13.1) is also assumed in both instances.

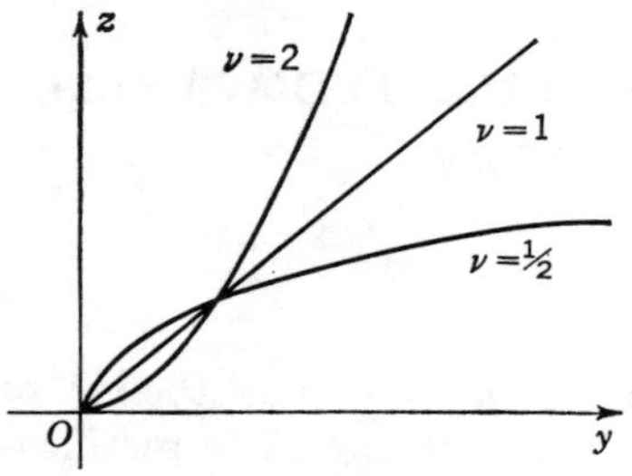

FIG. 13.1. Typical dynamic characteristics of a half-wave νth-law rectifying element with zero memory.

13.1-1 Broadband Noise. The dynamic characteristic relating the input and output processes is

$$z(t) = \begin{cases} \boldsymbol{T}_g\{y\} = \beta y(t)^\nu & y > 0 \\ 0 & y < 0 \ (\nu \geq 0) \end{cases} \tag{13.1}$$

where β is an appropriate scale factor and no bias or saturation effects are present (cf. Fig. 12.1, for example). From Eqs. (2.88) and (7.13a), the second-moment function of the output is specifically†

$$M_z(t) = \boldsymbol{E}\{z_1z_2\} = \frac{\beta^2\psi^{-1}}{2\pi}(1-k^2)^{-\frac{1}{2}} \iint_0^\infty y_1^\nu y_2^\nu e^{-(y_1^2+y_2^2-2y_1y_2k)/2\psi(1-k^2)}\,dy_1\,dy_2 \tag{13.2}$$

where as usual $\psi = \overline{y_1^2} = \overline{y_2^2}$ and $k = k(t)$ is the normalized covariance function of a broadband input noise process [cf. Eq. (5.6a)]. Next, we introduce the normalizations $x_1 = y_1/\psi^{\frac{1}{2}}$, $x_2 = y_2/\psi^{\frac{1}{2}}$, transform to polar coordinates $x_1 = \rho\cos\theta$, $x_2 = \rho\sin\theta$, with the further substitution $\phi = 2\theta$, and carry out the integrations over ρ and ϕ, in that order, with the help of the relations (in which $|k| \leq 1$)

$$\begin{aligned} \sin^\nu\phi(1-k\sin\phi)^{-\nu-1} &= \sum_{n=0}^\infty k^n(\sin\phi)^{\nu+n}\Gamma(\nu+n+1)/n!\Gamma(\nu+1) \\ \int_0^\pi \sin^{\nu+n}\phi\,d\phi &= \sqrt{\pi}\,\Gamma[(\nu+n+1)/2]/\Gamma[(\nu+n)/2+1] \end{aligned} \tag{13.3}$$

Using Eq. (A.1.35) after summing the series, we get finally the desired result,

$$\begin{aligned} M_z(t) = \beta^2\frac{(2\psi)^\nu}{4\pi}\Bigg[\Gamma\left(\frac{\nu+1}{2}\right)^2 {}_2F_1\left(-\frac{\nu}{2},-\frac{\nu}{2};\frac{1}{2};k^2\right) \\ + 2k\Gamma\left(\frac{\nu}{2}+1\right)^2 {}_2F_1\left(\frac{1-\nu}{2},\frac{1-\nu}{2};\frac{3}{2};k^2\right)\Bigg] \end{aligned} \tag{13.4a}$$

† We can equally well use the characteristic-function approach [Eqs. (5.15a), (5.15b)], where now $f(i\xi) = \beta\Gamma(\nu+1)/(i\xi)^{\nu+1}$, from Eq. (2.101$a$).

or

$$M_z(t) = \beta^2 \frac{(2\psi)^\nu}{4\pi} \left[\sum_{n=0}^{\infty} \frac{2^{2n}(-\nu/2)_n{}^2}{(2n)!} \Gamma\left(\frac{\nu+1}{2}\right)^2 k^{2n} \right.$$
$$\left. + \sum_{n=0}^{\infty} \frac{2^{2n+1}}{(2n+1)!} \left(\frac{1-\nu}{2}\right)_n^2 \Gamma\left(\frac{\nu}{2}+1\right)^2 k^{2n+1} \right] \qquad (13.4b)$$

Here ${}_2F_1$ is a hypergeometric function (cf. Sec. A.1.3). The expressions for the half-wave linear and quadratic detectors follow immediately (cf. Probs. 5.3, 5.5, also):

$$M_z(t)_{\nu=1} = \frac{\beta^2\psi}{2\pi} \left[\frac{\pi k}{2} + {}_2F_1\left(-\frac{1}{2}, -\frac{1}{2}; \frac{1}{2}; k^2\right) \right]$$
$$= \frac{\beta^2\psi}{2\pi} \left(k \sin^{-1} k + \frac{\pi k}{2} + \sqrt{1-k^2} \right) \qquad (13.5)$$

and

$$M_z(t)_{\nu=2} = \frac{\beta^2\psi^2}{2\pi} \left[\frac{\pi}{2}(1+2k^2) + 4k\, {}_2F_1\left(-\frac{1}{2}, -\frac{1}{2}; \frac{3}{2}; k^2\right) \right]$$
$$= \frac{\beta^2\psi^2}{2\pi} \left[(1+2k^2)\left(\frac{\pi}{2} + \sin^{-1} k\right) + 3k\sqrt{1-k^2} \right] \qquad (13.6)$$

When ν is integral, the hypergeometric function can always be given in terms of k, $\sqrt{1-k^2}$, and $\sin^{-1} k$, with the help of the appropriate recurrence relations (see Sec. A.1.3).

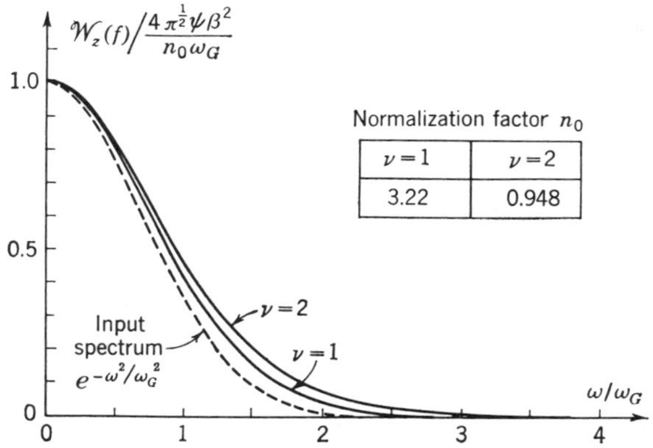

FIG. 13.2. Intensity spectra of broadband noise after rectification by half-wave zero-memory νth-law devices. (Normalized to unity at $f = 0$.)

The intensity spectrum corresponding to $M_z(t)$ is easily obtained with the help of the theorem of Wiener and Khintchine. Specific results for the half-wave linear and quadratic rectifiers, corresponding to Eqs. (13.5) and (13.6), are given in Probs. 5.3, 5.5. Figure 13.2 shows some typical spectra when the input spectrum has the form $\mathcal{W}_y(f) = W_0 e^{-\omega^2/\omega_F^2}$, or correspondingly $k(t) = e^{-\omega_F^2 t^2/4}$ We observe, as expected, that clipping of the input process due to rectification results in a spread in the intensity distribution of the output, or equivalently a "narrowing" of the second-moment function $M_z(t)$.

The mean intensity after detection is most easily obtained from Eqs. (5.20) (with $S_1 = 0$ therein). The total, d-c, and continuum powers are therefore

$$P_\tau = \frac{\beta^2(2\psi)^\nu}{2\sqrt{\pi}}\,\Gamma\left(\nu + \frac{1}{2}\right) \qquad P_{\mathrm{dc}} = \frac{\beta^2\psi^\nu 2^\nu}{4\pi}\,\Gamma\left(\frac{\nu+1}{2}\right)^2 \qquad P_c = P_\tau - P_{\mathrm{dc}} \tag{13.7}$$

curves of which are shown in Fig. 13.3 as a function of the law (ν) of the detector. The interesting limiting cases $\nu = 0$, $\nu \to \infty$ are considered in Prob. 13.4.

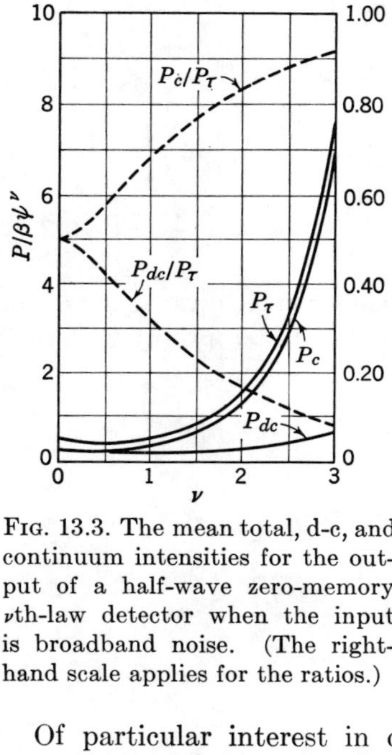

FIG. 13.3. The mean total, d-c, and continuum intensities for the output of a half-wave zero-memory νth-law detector when the input is broadband noise. (The right-hand scale applies for the ratios.)

13.1-2 Narrowband Noise. As shown earlier in Secs. 2.3-2 and 5.1-3, the output process after rectification of a narrowband input consists of spectral zones, located about harmonics of the central frequency f_0. The spectra associated with these harmonic zones are distorted from their original shape: here, an infinite number of $n \times n$ noise components are generated, redistributing the original intensity into new spectral terms. The second-moment function of the output may be determined in a straightforward way from the expansion of $k(t)^n = k_0(t)^n \cos^n \omega_0 t$ in Eqs. (13.4) and subsequent collection of all terms of order $\cos l\omega_0 t$ ($l \geq 0$), as indicated in Sec. 5.1-3. The results are the covariance (or second-moment) functions of the respective spectral zones ($l = 0, 1, 2, \ldots$) produced in rectification (cf. Prob. 13.1).

Of particular interest in detection is the l-f contribution ($l = 0$), for which the second-moment function becomes[2]

$$M_z(t)_0 = \frac{\beta^2\psi^\nu 2^{\nu-2}}{\pi}\,\Gamma\left(\frac{\nu+1}{2}\right)^2 {}_2F_1\left(-\frac{\nu}{2}, -\frac{\nu}{2}; 1; k_0^2\right) \qquad k_0 = k_0(t) \tag{13.8}$$

For the half-wave linear detector, this reduces to

$$M_z(t)_0\Big|_{\nu=1} = \frac{\beta^2\psi}{2\pi}\,{}_2F_1\left(-\frac{1}{2}, -\frac{1}{2}; 1; k_0^2\right) = 2\,\frac{\beta^2\psi}{\pi^2}\left[\,E(k_0) - \frac{1}{2}(1 - k_0^2)K(k_0)\right] \tag{13.9}$$

where K and E are, respectively, complete elliptic integrals of the first and second kinds and of modulus k_0 (see Sec. A.1.3). For the half-wave quadratic detector, on the other hand, the expression is much simpler,

$$M_z(t)_0 \bigg|_{\nu=2} = \frac{\beta^2\psi^2}{4}(1 + k_0^2) \tag{13.10}$$

and in fact, for all even values of ν, ${}_2F_1$ terminates to give us the polynomial

$$M_2(t)_0 \bigg|_{\nu=2n} = \left[\frac{\beta\psi^n(2n)!}{2^{n+1}n!}\right]^2 \sum_{k=0}^{n} {}_nC_k{}^2 k_0(t)^{2(n-k)} \qquad {}_nC_k = \frac{n!}{(n-k)!k!} \tag{13.11}$$

In a number of cases, these second-moment functions have been verified experimentally, with excellent agreement with the theory. Johnson[8] has examined the case $\nu = 1$; his results are shown in Fig. 13.4 for shot noise

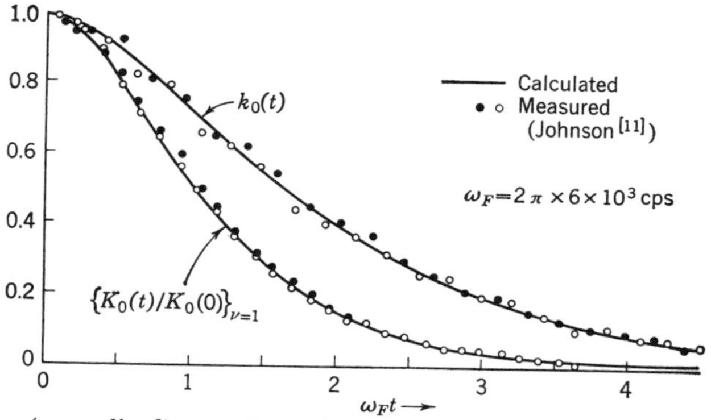

FIG. 13.4. The (normalized) covariance functions for (1) the l-f output of a half-wave linear detector and (2) the slowly varying part, $k_0(t)$, of the input covariance functions.

from a temperature-limited diode which has been passed through two cascaded single-tuned filters, each of (angular) bandwidth $2\omega_F$. (The steady component in the output has been removed by filtering.) The normalized covariance function here is $k(t) = k_0(t) \cos \omega_0 t = (1 + \omega_F|t|)e^{-\omega_F|t|} \cos \omega_0 t$. Note the more rapid decay of the covariance function of the output, corresponding to the spectral spread caused by the nonlinear operation (13.1).

The spectra associated with Eq. (13.8) follow in the usual way.† We get, for $\nu = 1, 2,$

$$\mathcal{W}_z(f)_0 \bigg|_{\nu=1} = \frac{\beta^2\psi}{2\pi}\left\{2\delta(f-0) + \sum_{n=1}^{\infty}\left[\frac{(\tfrac{1}{2})_n}{n!(2n-1)}\right]^2 C_{0,2n}^{(0)}(f)\right\} \tag{13.12}$$

$$\mathcal{W}_z(f)_0 \bigg|_{\nu=2} = \frac{\beta^2\psi^2}{4}[2\delta(f-0) + C_{0,2}^{(0)}(f)] \tag{13.13}$$

where
$$C_{0,m}^{(0)}(f) = 4\int_0^\infty k_0(t)^m \cos \omega t\, dt \qquad \omega = 2\pi f \tag{13.14}$$

† An extensive set of spectra for these cases has been obtained experimentally by Fellows[9] and is in excellent agreement with the theory.

The mean intensities associated with the total l-f disturbance, its direct current and continuum, are also readily found [cf. Eq. (5.29)]. One has

$$(P_\tau)_0 = \frac{\beta^2\psi^\nu 2^{\nu-2}}{\pi}\,\Gamma(\nu+1)\Gamma\left(\frac{\nu+1}{2}\right)^2 \Big/ \Gamma\left(\frac{\nu}{2}+1\right)^2$$

$$P_{dc} = \frac{\beta^2\psi^\nu 2^\nu}{4\pi}\,\Gamma\left(\frac{\nu+1}{2}\right)^2 \qquad P_c = P_\tau - P_{dc} \tag{13.15}$$

which are illustrated in Fig. 13.5 as functions of the detector's law.

Results for the narrowband cases above may also be obtained if we carry through the analysis in terms of the envelope E and phase θ of the input process. From the results of Sec. 5.1-4 and specifically Eq. (5.34c), we find for the l-f zone ($l = 0$), when the half-wave νth-law element is used, that

$$M_z(t)_0 = \gamma^2\overline{E_1{}^\nu E_2{}^\nu} = \gamma^2 \iint\limits_0^\infty E_1{}^{\nu+1}E_2{}^{\nu+1} I_0\left[\frac{k_0E_1E_2}{\psi(1-k_0{}^2)}\right] \times e^{-(E_1{}^2+E_2{}^2)/2\psi(1-k_0{}^2)}\,dE_1\,dE_2$$

$$= \gamma^2(2\psi)^\nu\Gamma(\nu/2+1)^2\,{}_2F_1(-\nu/2,-\nu/2;1;k_0{}^2) \tag{13.16}$$

where we have used Eq. (9.22). The scale factor γ here is related to β in Eqs. (13.1) by

$$\gamma = \frac{\beta}{2\pi}\int_0^\pi \sin^\nu\theta\,d\theta = \frac{\beta}{2\sqrt{\pi}}\,\frac{\Gamma[(\nu+1)/2]}{\Gamma(\nu/2+1)} \tag{13.17}$$

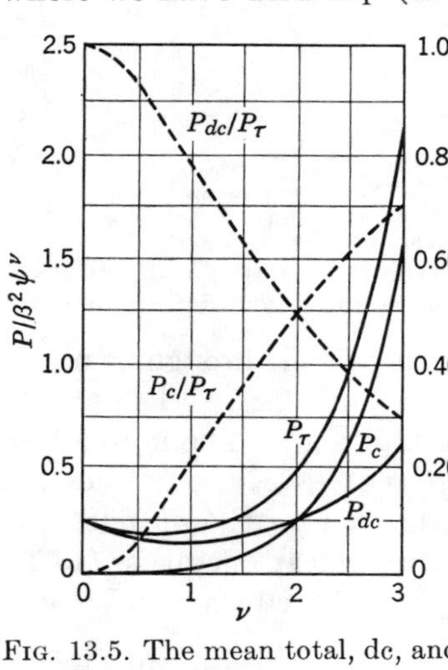

FIG. 13.5. The mean total, dc, and continuum intensities associated with the l-f output when the input noise process is narrowband. (The right-hand scale applies for the ratios.)

This follows from the fact that we are dealing with the envelopes E rather than the instantaneous amplitudes $V(t)$ of the incoming process. The output z after rectification (but before filtering) consists of the positive halves of the oscillations of βV^ν. But the area under the loops of z is proportional to the area under βE^ν in the same ratio as the area under one loop of $\sin^\nu\theta$ is to an area of unit height and duration 2π, so that as far as the l-f portion of the output is concerned, the loops of z are "smeared" together into a disturbance which varies as γE^ν, where now γ is given by Eq. (13.17). Accordingly, we observe that Eqs. (13.16) and (13.8) are identical, as expected.

Finally, for additional calculations of spectra, power, dependence on the law of the rectifier, and so on, we refer the reader to the publications cited on page 597 and, in particular, to Refs. 2, 6, 8, and 9. See also the supplementary references at the end of the book.

PROBLEMS

13.1 (*a*) In the case of narrowband normal noise which has been passed through a half-wave (zero-memory) νth-law device, show that the second-moment functions associated with the spectral zones centered about $(0,f_0,2f_0, \ldots)$ are, respectively, for the even- and odd-order zones, where the input spectrum is symmetrical,

$$M_z(t)_{2l} = \beta^2 \frac{\psi^\nu 2^{\nu-2}}{\pi} \epsilon_l \left(\frac{-\nu}{2}\right)^2 \frac{\Gamma[(\nu+1)/2]^2}{(2l)!} k_0^{2l} \, {}_2F_1\left(l - \frac{\nu}{2}, l - \frac{\nu}{2}; 2l+1; k_0^2\right) \cos 2l\omega_0 t \quad (1)$$

$$K_z(t)_{2l+1} = \beta^2 \frac{\psi^\nu 2^{\nu-1}}{\pi} \left(\frac{1-\nu}{2}\right)_l^2 \frac{\Gamma(\nu/2+1)^2}{(2l+1)!} k_0^{2l+1} \times {}_2F_1\left(l + \frac{1-\nu}{2}, l + \frac{1-\nu}{2}; 2l+2; k_0^2\right) \cos(2l+1)\omega_0 t \quad (2)$$

Observe that when ν is even only those zones for which l is even and equal to or less than ν are produced, along with all odd harmonic regions (and vice versa when ν is odd). For the general case one simply replaces $\omega_0 t$ by $\omega_0 t - \phi_0(t)$.

(*b*) Show that the mean intensities of the various spectral regions are given by

$$(P_T)_{2l} = \frac{\beta^2(2\psi)^\nu}{4\pi} \epsilon_l \left(-\frac{\nu}{2}\right)_l^2 \frac{\Gamma[(\nu+1)/2]^2}{\Gamma(\nu/2+l+1)^2} \Gamma(\nu+1)$$
$$= P_c \qquad l > 0 \quad (3)$$

$$(P_T)_{2l+1} = \frac{\beta^2(2\psi)^\nu}{2\pi} \left(\frac{1-\nu}{2}\right)_l^2 \frac{\Gamma(\nu/2+1)^2}{\Gamma(\nu/2+l+3/2)^2} \Gamma(\nu+1)$$
$$= P_c \qquad l \geq 0 \quad (4)$$

13.2 The *small-signal* dynamic path can be represented by a series of the form

$$z = \sum_{k=0}^{\infty} \left.\frac{\partial^k z}{\partial y^k}\right|_Q y^k = \alpha_0 + \alpha_1 y + \alpha_2 y^2 + \cdots + \alpha_k y^k \quad (1)$$

$$\alpha_k \equiv \left.\frac{\partial^k z}{\partial y^k}\right|_Q \quad (2)$$

measured at operating point Q.

(*a*) Show that, when normal random noise (with no d-c component) is rectified by a device of this type, then

$$M_z(t) = \sum_{k,j}^{\infty} \alpha_j \alpha_k \left.\frac{\partial^{j+k}}{\partial \xi_1^j \, \partial \xi_2^k} F_2(\xi_1,\xi_2;t)_N \right|_{\xi_1=\xi_2=0} = \begin{cases} \displaystyle\sum_{j=0}^{\infty} \sum_{k=0}^{\infty} \alpha_j \alpha_k \psi^{(j+k)/2} Q_{j,k}[k(t)] & j+k = \text{even} \\ 0 & j+k = \text{odd} \end{cases} \quad (3)$$

where

$$Q_{j,k} = \frac{2k(t)}{\pi} 2^{(j+k)/2} \Gamma\left(\frac{j}{2}+1\right) \Gamma\left(\frac{k}{2}+1\right) {}_2F_1\left[\frac{1-j}{2}, \frac{1-k}{2}; \frac{3}{2}; k(t)^2\right] \qquad j, k \text{ odd} \quad (4a)$$

$$Q_{j,k} = \frac{2^{(j+k)/2}}{\pi} \Gamma\left(\frac{j+1}{2}\right) \Gamma\left(\frac{k+1}{2}\right) {}_2F_1\left[-\frac{j}{2}, -\frac{k}{2}; \frac{1}{2}; k(t)^2\right] \qquad j, k \text{ even} \quad (4b)$$

(b) Verify that

$$P_T = \sum_{j=0}^{\infty}\sum_{k=0}^{\infty} \alpha_j\alpha_k \left(\frac{\psi}{2}\right)^{(j+k)/2} \frac{(j+k)!}{[(j+k)/2]!}$$
$$P_{dc} = \left[\sum_{j=0}^{\infty} \alpha_j \left(\frac{\psi}{2}\right)^{j/2} \frac{j!}{(j/2)!}\right]^2 \qquad j = 0, 2, \ldots \tag{5}$$

In practice, only the first few terms of Eqs. (3) are significant since $\alpha_j \to 0$ quite quickly if the excursion of the input is confined on the average to a relatively small portion of the curved path. D. Middleton,[2] sec. 8.

13.3 (a) When the noise is narrowband, show that for Prob. 13.2 one obtains for the correlation function corresponding to the spectral zones (0,1,2, . . .)

$$M_z(t)_{2l} = \frac{\epsilon_l k_0^{2l} \cos 2l\omega_0 t}{(2l)!} \sum_{j,k=0}^{\infty} \alpha_j\alpha_k (2\psi)^{(j+k)/2}(-j/2)_l(-k/2)_l\Gamma[(j+1)/2]\Gamma[(k+1)/2]$$
$$\times {}_2F_1(l - j/2, l - k/2; 2l+1; k_0^2) \qquad j, k \text{ even} \tag{1}$$

$$K_z(t)_{2l+1} = \frac{2k_0^{2l+1} \cos (2l+1)\omega_0 t}{(2l+1)!} \sum_{j,k=0}^{\infty} \alpha_j\alpha_k$$
$$\times (2\psi)^{(j+k)/2}[(1-j)/2]_l[(1-k)/2]_l\Gamma(j/2+1)\Gamma(l/2+1)$$
$$\times {}_2F_1[l + (1-j)/2, l + (1-k)/2; 2l+2; k_0^2] \qquad j, k \text{ odd} \tag{2}$$

(b) Verify that

$$(P_T)_{2l} = \epsilon_l \sum_{j,k=0}^{\infty} \alpha_j\alpha_k (2\psi)^{(j+k)/2} \frac{\Gamma[(j+1)/2]\Gamma[(k+1)/2]\Gamma[1+(k+j)/2]}{\Gamma(1+k/2)\Gamma(1+j/2)}$$
$$\times \frac{(-j/2)_l(-k/2)_l}{(1+j/2)_l(1+k/2)_l} = P_c \qquad l > 0 \tag{3}$$

$$(P_T)_{2l+1} = 2 \sum_{j,k=0}^{\infty} \alpha_j\alpha_k (2\psi)^{(j+k)/2} \frac{\Gamma(j/2+1)\Gamma(k/2+1)\Gamma[1+(j+k)/2]}{\Gamma[1+(k+1)/2]\Gamma[1+(j+1)/2]}$$
$$\times \frac{[(1-j)/2]_l[(1-k)/2]_l}{(3/2+j/2)_l(3/2+k/2)_l} = P_c \tag{4}$$

[For unsymmetrical spectra replace $\omega_0 t$ by $\omega_0 t - \phi_0(t)$.]

13.4 (a) Extreme values of ν ($\to 0, \to \infty$) provide some interesting results. When $\nu = 0$, that is, for "superclipping" by a half-wave detector, one gets

$$M_z(t)\Big|_{\nu=0} = \beta^2 \left[\frac{1}{4} + \frac{1}{2\pi}\sin^{-1} k(t)\right] \tag{1}$$

This is also the second-moment function of a sequence of rectangular pulses and spaces with random durations, in the broadband case. Verify that the power content of this wave lies half in the steady component and half in the continuum of the $n \times n$ noise products.

(b) Show now that, as $\nu \to \infty$, one gets for broadband noise through the half-wave νth-law device

$$M_z(t)\Big|_{\nu\to\infty} \simeq \frac{\beta^2\psi^\nu e^{\nu \log \nu - \nu - 1}}{8\pi^2} [\cosh (k\nu) + (e\nu)^{-1} \sinh (k\nu)] \qquad k = k(t) \tag{2}$$

The continuum intensity is therefore

$$P_c \simeq \frac{\beta^2\psi^\nu e^{\nu \log \nu - \nu - 1}}{8\pi^2} [\cosh \nu - 1 + (e\nu)^{-1} \sinh \nu] \tag{3}$$

(c) In the case of narrowband noise, one finds for the limit $\nu \to 0$ that the l-f region consists of the d-c term alone: all the continuum power is associated with the higher spectral zones ($l \geq 1$). Show that

$$M_z(t)_{2l}\bigg|_{\nu=0} = 0 \qquad l > 0 \tag{4}$$

$$K_z(t)_{2l+1}\bigg|_{\nu=0} = \frac{\beta^2}{2\pi} (\tfrac{1}{2})_l^2 \frac{k_0(t)^{2l+1}}{(2l+1)!} \, {}_2F_1(l + \tfrac{1}{2}, l + \tfrac{1}{2}; 2l + 2; k_0^2) \cos (2l + 1)\omega_0 t \tag{5}$$

Verify, on the other hand, that when $\nu \to \infty$ we get

$$M_z(t)_0\bigg|_{\nu\to\infty} \simeq \frac{\beta^2\psi^\nu}{8\pi^2} e^{\nu \log \nu - \nu - 1} I_0(k_0\nu) \tag{6}$$

[For unsymmetrical spectra replace $\omega_0 t$ by $\omega_0 t - \phi_0(t)$.]

13.5 (a) Show in the narrowband case above that, in the limit $\nu \to \infty$,

$$M_z(t)_{2l} \simeq \frac{\epsilon_{2l}\beta^2\psi^\nu}{8\pi^2} e^{\nu \log \nu - \nu - 1} I_{2l}(k_0\nu) \cos 2l\omega_0 t \tag{1}$$

$$K_z(t)_{2l+1} \simeq \frac{\beta^2\psi^\nu}{8\pi^2} e^{\nu \log \nu - \nu} I_{2l+1}(k_0\nu) \cos (2l + 1)\omega_0 t \tag{2}$$

(b) From Prob. 13.1b, obtain, in the limit $\nu \to \infty$,

$$(P_T)_l \simeq \frac{\beta^2\psi^\nu 2^\nu \epsilon_l}{(2\pi)^{3/2}\nu^{1/2}} e^{\nu \log \nu - \nu - 1} \frac{[1 - 2l/(\nu + 1)]^{2l-1}}{[1 + (2l + 1)/(\nu + 2)]^{2l+2}} \tag{3}$$

[For unsymmetrical spectra replace $\omega_0 t$ by $\omega_0 t - \phi_0(t)$.]

13.2 *Rectification of Noise and a Sinusoidal Carrier*[1,2]

In the previous section, the rectification of normal random noise alone was considered. The treatment is now extended to the more complicated cases of a sinusoidal carrier and noise. Here, a simple sinusoidal process is assumed, additively combined with the noise, and as in Sec. 13.1 a zero-memory half-wave νth-law rectifier is once again employed. Our interest centers mainly on the second-moment function $M_z(t)$ of the output process and, to a lesser extent, on the spectrum. As before, we distinguish between inputs which are spectrally broad and those which have an envelope structure. As before, the present treatment is based on the general analysis of Secs. 5.1-1 to 5.1-3, appropriately specialized to this simple signal structure. A discussion of more complicated signals (e.g., modulated carriers) is reserved for Sec. 13.3.

13.2-1 Carrier and Broadband Noise. The second-moment function of the output of a zero-memory nonlinear element when the input consists of the sum $y(t)$ of normal random noise $N(t)$ and a sinusoidal carrier process $A_0 \cos(\omega_0 t + \phi)$ may be written down at once from Eqs. (5.16). We have[1,2]

$$M_z(t) = \sum_{m=0}^{\infty} \sum_{n=0}^{\infty} \frac{\epsilon_m(-1)^{n+m}}{n!} \psi^n k(t)^n h_{m,n}^2 \cos m\omega_0 t$$

$$\omega_0 = 2\pi f_0,\ \psi = \overline{N^2};\ \overline{N} = 0 \qquad (13.18)$$

where the amplitude functions $h_{m,n}$ are

$$h_{m,n} = \frac{1}{2\pi} \int_C \xi^n f(i\xi) J_m(A_0\xi) e^{-\psi\xi^2/2}\, d\xi \qquad (13.19)$$

For a half-wave νth-law detector, $f(i\xi) = \beta\Gamma(\nu + 1)/(i\xi)^{\nu+1}$ from Eq. (2.101a), and with the help of equations in Appendix 1 we obtain specifically†

$$h_{m,n} = \frac{\beta\Gamma(\nu + 1) i^{-(m+n)}}{2m!} \left(\frac{\psi}{2}\right)^{(\nu-n)/2} a_0{}^m \frac{{}_1F_1\left(\frac{m + n - \nu}{2}; m + 1; -a_0{}^2\right)}{\Gamma[(2 + \nu - m - n)/2]} \qquad (13.20)$$

where ${}_1F_1$ is a confluent hypergeometric function [cf. Eq. (A.1.14)]. Here, $a_0{}^2$ is the carrier-to-noise (or signal-to-noise) intensity ratio,

$$a_0{}^2 = \frac{A_0{}^2}{2\psi} = \frac{\text{mean carrier intensity at input}}{\text{mean noise intensity at input}} \equiv \left[\frac{S}{N}\right]_{in}^2 \qquad (13.21)$$

The spectrum follows in the usual way on application of the Wiener-Khintchine theorem to Eq. (13.18) [cf. Eq. (5.22)].

We observe that the output consists in part of a d-c term,

$$M_z(t)_{dc} = h_{0,0}^2 = \frac{\beta^2\Gamma(\nu + 1)^2\psi^\nu}{2^{\nu+2}\Gamma(1 + \nu/2)^2}\, {}_1F_1{}^2\left(-\frac{\nu}{2};1;-a_0{}^2\right)$$

$$= \beta^2 \frac{2^{\nu-2}}{\pi} \Gamma\left(\frac{\nu + 1}{2}\right)^2 \psi^\nu\, {}_1F_1{}^2\left(-\frac{\nu}{2};1;-a_0{}^2\right) \qquad (13.22)$$

in agreement with Eqs. (13.7) and (13.15) when $a_0{}^2 = 0$ in the above.‡ The portion of the output with a continuous spectrum results from the cross modulation of noise and signal components. It consists of two portions, an $n \times n$ and an $s \times n$ continuum (cf. Sec. 5.1-2), for which the respective contributions to $M_z(t)$ occur for the terms $m = 0, n \geq 1$ and $m \geq 1, n \geq 1$ in the double summation of Eq. (13.18). There is also the set of discrete $s \times s$ components, attributable to $m \geq 1, n = 0$ in Eq. (13.18). The mean

† When ν is integral, note that $h_{m,n}$ vanishes for all $m + n$ such that $m + n \geq 2 + \nu$. Tables of the confluent hypergeometric function are available (Middleton and Johnson[10]) by which values of ${}_1F_1$ in Eq. (13.20) may be obtained.

‡ From the duplication formula of the gamma function, it follows at once that $2^{2\nu}\Gamma[(\nu + 1)/2]^2/\pi = \Gamma(\nu + 1)^2/\Gamma(\nu/2 + 1)^2$ [cf. Eq. (A.1.45)].

total intensity of the rectified wave may be found on setting $t = 0$ in Eq. (13.18), but a simpler† result is obtained with the help of Eqs. (5.19) and (5.20) and the expression (9.46) for the distribution density of the additive process $y = N + A_0 \cos(\omega_0 t + \phi)$. We find finally that

$$P_\tau = \beta^2 \int_0^\infty y^{2\nu} W_1(y)\, dy = \frac{\beta^2 \psi^\nu \Gamma(2\nu + 1)}{2^{\nu+1}\Gamma(\nu + 1)}\, {}_1F_1(-\nu;1;-a_0^2)$$
$$= \frac{\beta^2 \psi^\nu 2^{\nu-1}}{\sqrt{\pi}}\, \Gamma(\nu + \tfrac{1}{2})\, {}_1F_1(-\nu;1;-a_0^2) \qquad (13.23)$$

which is also in agreement with Eqs. (13.7) when there is no carrier ($a_0 = 0$). Note that increasing ν (> 1) produces a noticeable rise in the relative output intensities ($P_\tau/\beta^2\psi$), etc. For the specific cases of the half-wave linear ($\nu = 1$) and quadratic ($\nu = 2$) detectors, Eqs. (13.22) and (13.23) assume the comparatively simple forms

$$\nu = 1: \qquad P_\tau = \frac{\beta^2 \psi}{2}(1 + a_0^2) \qquad (13.24a)$$

$$P_{dc} = \frac{\beta^2 \psi}{2\pi}\, {}_1F_1^2\left(-\frac{1}{2};1;-a_0^2\right)$$
$$= \frac{\beta^2}{2\pi} e^{-a_0^2}\left[(1 + a_0^2) I_0\left(\frac{a_0^2}{2}\right) + a_0^2 I_1\left(\frac{a_0^2}{2}\right)\right]^2 \qquad (13.24b)$$

and

$$\nu = 2: \qquad P_\tau = \beta^2 \psi^2 \left(\frac{3}{2} + 3a_0^2 + \frac{3a_0^4}{4}\right) \qquad (13.25a)$$

$$P_{dc} = \frac{\beta^2 \psi^2}{4}(1 + a_0^2)^2 \qquad (13.25b)$$

(1) Weak signals in noise. When the input signal is weak compared with the accompanying noise, we may express $M_z(t)$ [Eq. (13.18)] by the series

$$M_z(t) = \frac{\beta^2 (2\psi)^\nu}{4\pi}[G_0(t) + a_0^2 G_1(t) + a_0^4 G_2(t) + \cdots]$$
$$= M_z(t)_{dc} + K_z(t)_{n\times n} + K_z(t)_{s\times n} + K_z(t)_{s\times s} \qquad (13.26)$$

The functions $G_0, G_1, \ldots$ consist of the $n \times n$, $s \times n$, d-c, and $s \times s$ modulation products generated in rectification. For the half-wave νth-law detector considered here, we easily find that $G_0(t)_{dc} = \Gamma[(\nu + 1)/2]^2$, $G_1(t)_{dc} = \nu\Gamma[(\nu + 1)/2]^2$, $G_0(t)_{s\times n} = G_0(t)_{s\times s} = 0$, etc., with more complicated expressions for the higher-order terms $G_1, G_2, \ldots$.

The numerically significant term in Eq. (13.26) is $G_0(t)$, which is the second-moment function (13.4a) when noise alone is present, and the various G_1, G_2, etc., are the perturbations of this leading term. The discrete $s \times s$ contributions are small and the effect of the $s \times n$ terms is also slight if

† This is simpler only because we are dealing here with a *half-wave* device; the method breaks down for biased rectifiers and when there is saturation.

$a_0^4 \ll 1$, as one would expect. A similar effect is observed in P_τ, P_{dc}:

$$P_\tau \doteq \frac{(2\psi)^\nu \beta^2}{2\sqrt{\pi}} \Gamma\left(\nu + \frac{1}{2}\right)\left[1 + \nu a_0^2 + \frac{\nu(\nu-1)}{4} a_0^4 + \cdots\right] \quad (13.27a)$$

$$P_{dc} \doteq \frac{(2\psi)^\nu \beta^2}{4\pi} \Gamma\left(\frac{\nu+1}{2}\right)^2\left[1 + \nu a_0^2 + \frac{\nu(3\nu-2)}{8} a_0^4 + \cdots\right] \quad (13.27b)$$

When $\nu \to \infty$, these expressions may be found from the results of Prob. 13.4, applied to each term with suitable modifications. A typical spectrum for this broadband noise and signal combination is sketched in Fig. 5.2.

(2) Strong signals in noise. Instead of a signal which is weak compared with the (fixed) noise background, let us consider the reverse situation, in which the signal is now very strong. With the help of the asymptotic series (A.1.16) for the hypergeometric functions [cf. Eq. (13.20)] which appear in the second-moment function of the output, we obtain directly

$$M_z(t)_{dc} \simeq \frac{\beta^2(2\psi)^\nu}{4\pi} \left\{\frac{\Gamma[(\nu+1)/2]}{\Gamma(\nu/2+1)}\right\}^2 a_0^{2\nu}\left[1 + \frac{\nu^2}{2a_0^2} + \frac{\nu^2}{8a_0^4}(\nu^2 - 2\nu + 2) + \cdots\right] \qquad a_0^2 \gg 1 \quad (13.28a)$$

for the d-c component, and for the other terms in the rectified wave we get similarly

$$M_z(t)_{n\times n} \simeq \frac{\beta^2(2\psi)^\nu}{4\pi} \left\{\frac{\Gamma(\nu/2+1)}{\Gamma[(\nu+1)/2]}\right\}^2 a_0^{2\nu}\left\{\frac{2k(t)}{a_0^2} + \frac{k(t)}{a_0^4}\left[(1-\nu)^2 + \frac{\nu^2 k(t)}{2}\frac{\Gamma[(\nu+1)/2]^4}{\Gamma(\nu/2+1)^4}\right] + \cdots\right\} \quad (13.28b)$$

$$M_z(t)_{s\times n} \simeq \beta^2\Gamma(\nu+1)^2\psi^\nu 2^{-\nu} a_0^{2\nu} \times \left\{\frac{k(t)}{a_0^2}\sum_{m=1}^{\infty}\frac{\cos m\omega_0 t}{\Gamma[(1+\nu+m)/2]^2\Gamma[(1+\nu-m)/2]^2} \times\left[1 - \frac{m^2-(1-\nu)^2}{2a_0^2} + \cdots\right] + \frac{k(t)^2}{a_0^4} \times \sum_{m=1}^{\infty}\frac{\cos m\omega_0 t}{\Gamma[(\nu+m)/2]^2\Gamma[(\nu-m)/2]^2}(1+\cdots) + \cdots\right\} \quad (13.28c)$$

and

$$M_z(t)_{s\times s} \simeq \beta^2\Gamma(\nu+1)^2\psi^\nu 2^{-1-\nu} a_0^{2\nu} \times \left\{\sum_{m=1}^{\infty}\frac{\cos m\omega_0 t}{\Gamma[(2+\nu-m)/2]^2\Gamma[(2+\nu+m)/2]^2} \times\left[1 - \frac{m^2-\nu^2}{4a_0^2} + \frac{(m^2-\nu^2)[m^2-(\nu-1)^2]}{32a_0^4}\cdots\right]\right\} \quad (13.28d)$$

where the last two expressions may be summed when ν is integral. As before, the spectrum is given by the cosine Fourier transform (of the sum) of Eqs. (13.28*a*) to (13.28*d*).

We observe that now, as expected, it is the carrier and its harmonics (after rectification) which are chiefly significant. The contribution of the noise, moreover, is *suppressed* relative to the signal, as we can see from Eqs. (13.28). The $s \times n$ and $n \times n$ components are $O(a_0^{2\nu-2})$; the d-c and $s \times s$ terms are $O(a_0^{2\nu})$. This suppression effect may be explained by noting that when the carrier is strong the noise at the input effectively "rides" on the carrier, well above the level at which clipping occurs. Only for a small fraction of the time, at the intervals near the "zeros" of the carrier, are carrier and noise together rectified, and, correspondingly, only here are the $s \times n$ and $n \times n$ noise products generated. As $a_0 \to \infty$, we observe further that if $\nu > 1$ the suppression of the noise is only relative: the actual $s \times n$ and $n \times n$ terms are increased, the discrete contributions growing at a more rapid rate. However, if $\nu < 1$, the suppression of the noise is absolute as $a_0 \to \infty$: the d-c and $s \times s$ components alone remain in the limit. For $\nu = 1$, there is a finite residuum of noise whose intensity is independent of the input signal-to-noise ratio a_0. The precise analytical structure of this suppression effect depends, of course, on (1) the form of the signal, (2) the statistics of the noise, and (3) the law of the detector.

A similar behavior is noted in the expressions for the intensity of the rectified wave. From Eqs. (13.23) and (13.22), we obtain, for $a_0^2 \gg 1$,

$$P_\tau \simeq \frac{2^{\nu-1}\Gamma(\nu + \frac{1}{2})\psi^\nu}{\sqrt{\pi}\,\Gamma(\nu + 1)} a_0^{2\nu}\left[1 + \frac{\nu^2}{a_0^2} + \frac{\nu^2(\nu - 1)^2}{2a_0^4} + \cdots\right]$$
$$P_{dc} \simeq \frac{2^{\nu-2}}{\pi}\beta^2 \frac{\Gamma[(\nu + 1)/2]^2}{\Gamma(\nu/2 + 1)^2} a_0^{2\nu}\psi^\nu\left[1 + \frac{\nu^2}{2a_0^2} + \frac{\nu^2(\nu^2 - 2\nu + 2)}{8a_0^4} + \cdots\right] \tag{13.29}$$

and since $P_c = P_\tau - P_{dc}$, we see, moreover, that $P_c = O(a_0^{2\nu})$. It is also instructive to observe the limiting behavior for different ranges of ν: P_τ, P_{dc}, and P_c fall off as ν is decreased below $\nu = 1$ (for a large but constant a_0), while the reverse is true when $\nu > 1$. This is attributable to the fact that, for $\nu < 1$, the curvature of the detector's dynamic characteristic *decreases* the higher amplitudes relative to the smaller ones—i.e., the fluctuations in the wave are relatively diminished—while, if $\nu > 1$, these larger amplitudes are relatively enhanced, on the average. At $\nu = 1$, all parts of the wave passed in rectification are weighted equally.

(3) The signal-to-noise ratio. As we have noted in Sec. 5.3-4, the signal-to-noise ratio is a useful measure of the nonlinear device's performance. Here, we specifically wish to relate the output signal-to-noise ratio to the corresponding input ratio (a_0) and to show how this relationship depends on (1) a_0, (2) the detector's law, and (3) the response of a linear filter at the input (i.e., preceding the nonlinear element), through which signal and noise are passed before rectification.† The output signal is again defined to be the increment in the *amplitude*‡ of the steady component after detection, due to the presence of the signal. The effect of the interfering noise, which is represented by the $s \times n$ and $n \times n$ noise products produced in rectification (without the d-c contributions from these products, however), appears in the fluctuating background which accompanies the signal after smoothing by a low-pass filter of effective width Δf. This is the *amplitude*, or *deflection, criterion*, discussed in Sec. 5.3-4. Thus, we have

$$\begin{aligned} \bar{S}_{out} &\equiv \bar{z}\Big|_{a_0>0} - \bar{z}\Big|_{a_0=0} \\ \overline{N^2_{out}} &\equiv \int_0^{\Delta f} \mathcal{W}_z(f)_{\text{noise}}\,df \doteq \Delta f\,\mathcal{W}_z(0)_{\text{noise}} = 4\,\Delta f \int_0^\infty K_z(t)\,dt \end{aligned} \tag{13.30}$$

This last is an expression of the fact that the smoothing time is long compared with the mean fluctuation time of the input disturbance. From Eq. (13.22), we get specifically for the half-wave νth-law detector ($\nu \geq 0$)

$$\left(\frac{S}{N}\right)_{out} = \frac{\bar{S}_{out}}{(\overline{N^2_{out}})^{1/2}} = \frac{\beta 2^{\nu/2-1}\Gamma[(\nu+1)/2]\psi^{\nu/2}[{}_1F_1(-\nu/2;1;-a_0^2) - 1]}{\sqrt{\pi}\left[\Delta f \int_0^\infty K_z(t)\,dt\right]^{1/2}}$$
$$K_z(t) = K_z(t)_{n\times n} + K_z(t)_{s\times n} \tag{13.31}$$

For weak and strong signals, the results of Secs. 1 and 2 above gives us at once

$a_0^4 \ll 1$:

$$\left(\frac{S}{N}\right)_{out} \doteq \frac{\nu\Gamma[(\nu+1)/2]a_0^2[1 - (1 - \nu/2)a_0^2 \cdots]}{4\sqrt{\Delta f}\left\{\int_0^\infty [G_0(t)_{n\times n} + a_0^2 G_1(t)_{n\times n+s\times n} + a_0^4 G_2(t)_{n\times n+s\times n} + \cdots]\,dt\right\}^{1/2}} \tag{13.31a}$$

$a_0^4 \gg 1$:

$$\left(\frac{S}{N}\right)_{out} \simeq \frac{\Gamma[(\nu+1)/2]^2}{2\sqrt{\Delta f}\,\Gamma(\nu/2+1)^2}\,\frac{a_0^{\nu}(1 + \nu^2/4a_0^2 \cdots) - 1}{a_0^{\nu-1}\left\{\int_0^\infty [L_1(t) + a_0^{-2}L_2(t) + \cdots]\,dt\right\}^{1/2}} \tag{13.31b}$$

Here, $L_1(t)$, $L_2(t)$ are the coefficients of a_0^{-2}, a_0^{-4}, etc., in Eqs. (13.28*b*), (13.28*c*) after the factor $\beta^2[(2\psi)^\nu/4\pi]\Gamma(\nu/2+1)^2/\Gamma[(\nu+1)/2]^2 a_0^{2\nu}$ has been removed.

† In the discussion of the S/N ratio and its calculation, effective smoothing times long compared with the mean period of the fluctuating background are postulated. Here, infinite observation times are assumed, and time averages are replaced by equivalent ensemble averages.

‡ One always observed an *amplitude* against a fluctuating background; the units in which these amplitudes are measured, e.g., volts$^\nu$, power$^{\nu/2}$, etc., do not, of course, alter the structure of the detection criterion.

(4) Signal suppression. Figure 13.6 sketches the variation of the output signal-to-noise ratio as a function of the corresponding input ratio. Several important features of the detector's behavior are now apparent.

(*a*) *When the input signal is weak relative to the noise*, then $(S/N)_{out} \sim (S/N)_{in}^2$ for all types of (unsaturated) half-wave detectors. In other words, all half-wave νth-law detectors behave like a *quadratic* rectifier. This

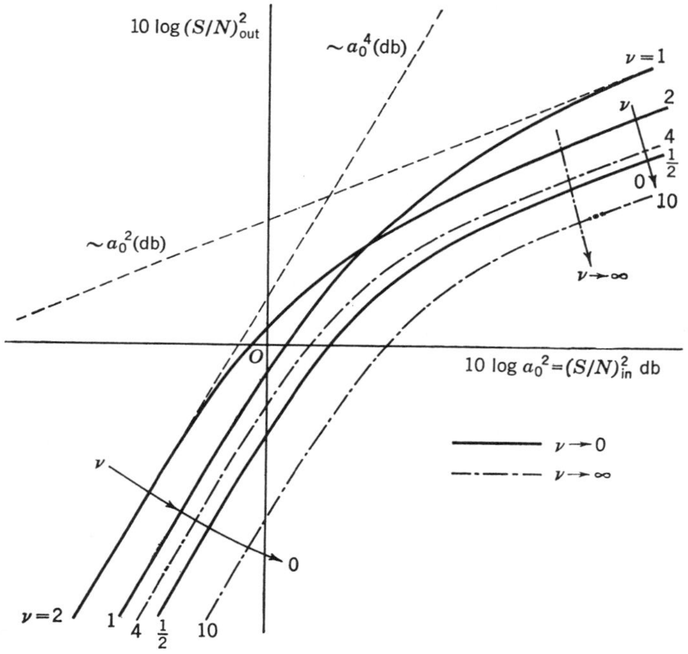

FIG. 13.6. The input and output signal-to-noise ratios (on a decibel scale) for *broadband noise* and a sinusoidal carrier after half-wave νth-law rectification. (The relationships are qualitative only.)

effect, whereby the output ratio is made smaller than if it were simply proportional to the input ratio, is called *signal suppression:* the desired signal is suppressed relative to the accompanying noise.† The generality of this result leads us to suspect that the characteristic quadratic behavior for weak signals may be true for more complicated systems and more general signals. This is indeed the case, as we shall show later in Chap. 19, where the detection problem is more fully formulated.

(*b*) *When the signal is strong compared with the noise*, on the other hand, one has effectively a *linear* system for $(S/N)_{out} \sim (S/N)_{in}$, as shown by Eq. (13.31*b*) for all half-wave νth-law systems. The dependence of the

† See the discussion in Sec. 5.3-4 and Refs. 16, 17 of Chap. 5.

signal alone on ν is given by

$$\bar{S}_{out} \simeq \frac{\beta 2^{\nu/2-1}}{\sqrt{\pi}} \frac{\Gamma[(\nu+1)/2]}{\Gamma(\nu/2+1)} \left[a_0{}^{\nu/2}\psi^{\nu/2}\left(1 + \frac{\nu^2}{4a_0{}^2} + \cdots\right) - 1 \right] \qquad \nu \geq 0 \tag{13.32}$$

and from this one sees that now it is the noise that is suppressed relative to the signal [since $(a_0{}^2\psi)^{\nu/2} = (A_0{}^2/2)^{\nu/2}$ is independent of ψ]. However, the actual signal strength is governed by ν, unlike the weak-signal case, where $\bar{S}_{out} \sim a_0{}^2$, no matter what ν may be.

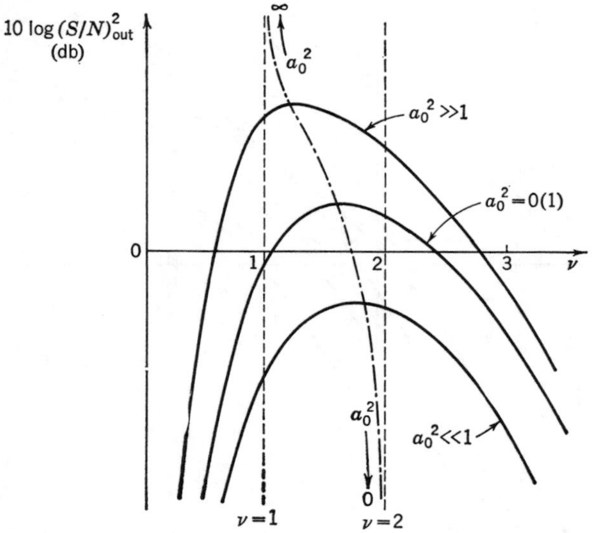

FIG. 13.7. The output signal-to-noise ratio as a function of the detector's law. (Half-wave rectification of a sine wave and *broadband noise;* the relationships are qualitative only.)

(*c*) *The effect of the law of the detector* can be deduced from the preceding analyses. We observe first that, when $\nu = 0$, $(S/N)_{out}$ is identically zero for *all* signal strengths ($a_0 > 0$), since no d-c *increment* is then possible. The noise, however, remains. Next, one can show that as ν becomes larger, for $a_0{}^2 \gg 1$, $(S/N)_{out} \simeq (a_0/\sqrt{\Delta f}\,\nu^{3/2}) \Big/ \left[\int_0^\infty L_0(t)\,dt\right]^{1/2} \to 0$ (for $\nu \to \infty$), where specifically

$$L_0(t) = k(t)\left(1 + 16\pi^2 \sum_{m=0}^{\infty} \cos m\omega_0 t\right) = k(t)\left[1 + 8\pi^2 + 8\pi^2 T_0 \times \sum_{k=-\infty}^{\infty} \delta(t - kT_0)\right] \tag{13.33}$$

Consequently, there is a value of ν for which $(S/N)_{out}$ is a maximum. This value turns out to be $\nu = 2$ for weak signals and $\nu = 1$ for strong signals, in agreement with what we would expect from the more fundamental detection theory (cf. Chap. 19). For intermediate signal strengths, ν_{opt} lies between 1 and 2; the typical behavior of $(S/N)_{out}$ with ν is sketched in Fig. 13.7.

13.2-2 Carrier and Narrowband Noise. The second-moment function of the l-f output when narrowband noise accompanies the carrier through the half-wave νth-law detector is found at once from Eqs. (5.27). Letting $l = 0$, $\phi_A = 0$, $\Phi = 0$, etc., therein, we get for this unmodulated carrier

$$M_z(t)_0 = \sum_{m=0}^{\infty} \epsilon_m \sum_{q=0}^{\infty} \frac{[\psi k_0(t)]^{m+2q} h^2_{m,m+2q}}{2^{m+2q} q!(q+m)!} \tag{13.34}$$

where $\psi k_0(t) = K_0(t)$ is the slowly varying part of the covariance function of the input noise and $h_{m,m+2q}$ is given by (13.20) with $n = 2q + m$ now. Only this l-f output appears, of course, since we remove the higher-order contributions in the usual way with a suitable low-pass filter.

The d-c term is just Eq. (13.22),† while the portions of the l-f continuum produced by the $n \times n$ and $s \times n$ modulation products are explicitly

$$K_z(t)_0 \bigg|_{n\times n} = \frac{\beta^2 \Gamma(\nu+1)^2}{4} \left(\frac{\psi}{2}\right)^{\nu} \sum_{q=1}^{\infty} \frac{k_0(t)^{2q}\ {}_1F_1{}^2(q - \nu/2; 1; -a_0{}^2)}{q!^2 \Gamma[(2+\nu)/2 - q]^2} \tag{13.35a}$$

$$K_z(t)_0 \bigg|_{s\times n} = \frac{\beta^2 \Gamma(\nu+1)^2}{2} \left(\frac{\psi}{2}\right)^{\nu} \sum_{m=1}^{\infty} \sum_{q=0}^{\infty} \frac{k_0(t)^{m+2q} a_0{}^{2m}\ {}_1F_1{}^2(m+q-\nu/2; m+1; -a_0{}^2)}{q!(q+m)!m!^2 \Gamma[(2+\nu)/2 - m - q]^2} \tag{13.35b}$$

Here, there are no discrete $s \times s$ contributions other than those appearing in the steady component. Note that for even powers ($\nu = 2, 4, \ldots$) these series terminate, but not for odd or nonintegral values of ν. In particular, one finds for the *half-wave quadratic detector* (cf. Table 2.1)

$$M_z(t)_0 \bigg|_{\nu=2} = \frac{\psi^2\beta^2}{4} [(1 + a_0{}^2)^2_{dc} + k_0(t)^2_{n\times n} + 2a_0{}^2 k_0(t)_{s\times n}] \tag{13.36}$$

Typical spectra for $\nu = 1, 2$ and various input signal-to-noise ratios are shown in Figs. 13.8, 13.9 for both experimental and theoretical cases, after Fellows.[9]

The simplest way of obtaining the total average intensity associated with the zone ($l = 0$) is to use the envelope representation $z_0 = \gamma E^{\nu}$, where

† See Fellows, *op. cit.*, fig. 11, for a comparison of measured and calculated values of P_{dc}, $\nu = 1$.

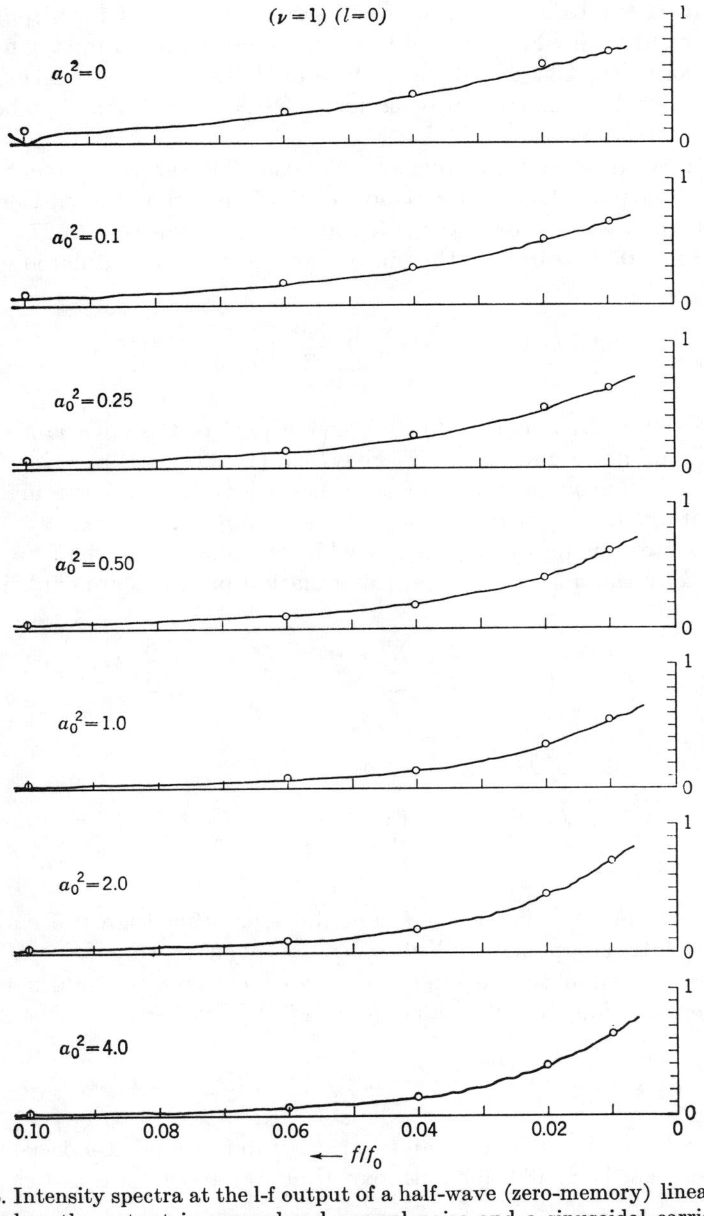

FIG. 13.8. Intensity spectra at the l-f output of a half-wave (zero-memory) linear rectifier ($\nu = 1$) when the output is narrowband normal noise and a sinusoidal carrier. (The circles represent calculated values; the solid lines, experimental data.)

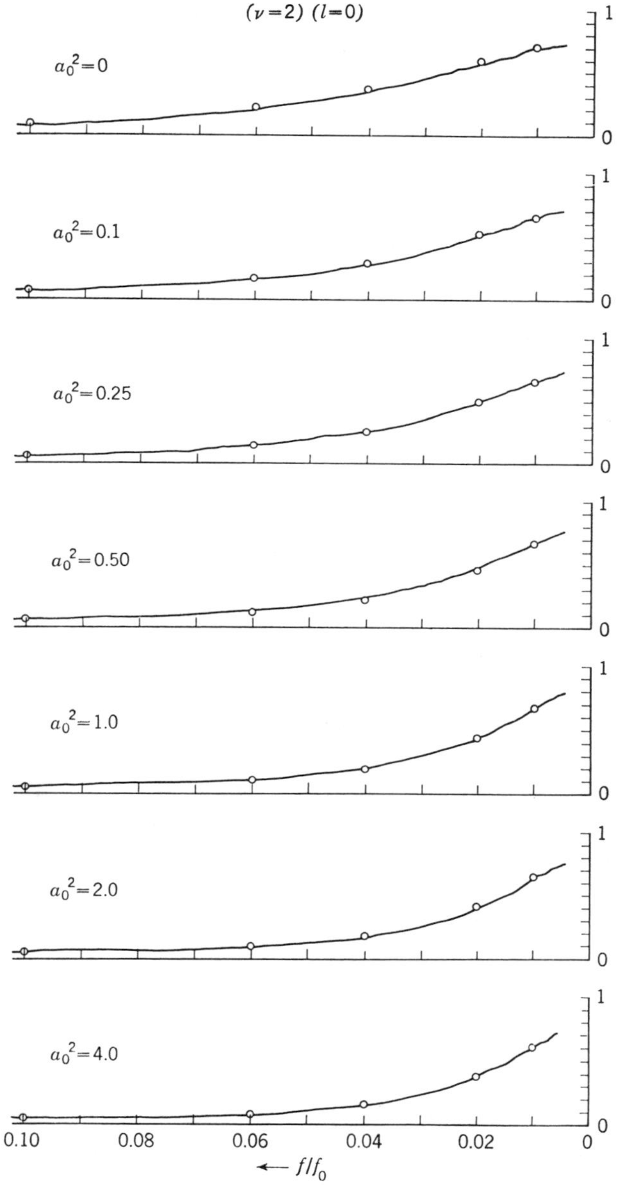

FIG. 13.9. The same as Fig. 13.8 for a half-wave (zero-memory) quadratic rectifier ($\nu = 2$). (The circles represent calculated values; the solid lines, experimental data.)

γ is given by Eq. (13.17). We have with the help of Eq. (9.50) that

$$(P_\tau)_0 = \overline{z_0^2} = \boldsymbol{E}\{\gamma^2 E^{2\nu}\} = \gamma^2 e^{-a_0^2} \int_0^\infty E^{2\nu+1} I_0(Ea_0 \sqrt{2/\psi})\, e^{-E^2/2\psi}\, dE/\psi$$
$$= \beta^2 \frac{\Gamma[(\nu+1)/2]^4 (2\psi)^\nu 2^{2\nu}}{4\pi^2 \Gamma(\nu+1)}\, {}_1F_1(-\nu;1;-a_0^2) \tag{13.37}$$

The average intensity of the continuum is consequently†

$$(P_c)_0 = (P_\tau)_0 - P_{\text{dc}} = \frac{(2\psi)^\nu \beta^2 \Gamma[(\nu+1)/2]^2}{4\pi}$$
$$\times \left\{ \frac{\Gamma[(\nu+1)/2]^2 2^{2\nu}}{\pi \Gamma(\nu+1)}\, {}_1F_1(-\nu;1;-a_0^2) - {}_1F_1{}^2\left(-\frac{\nu}{2};1;-a_0^2\right) \right\} \tag{13.38}$$

Note that the ratio of the average total intensity in the l-f zone to the average total intensity in the rectified wave (all zones) is independent of the input carrier-to-noise ratio, viz.,

$$(P_\tau)_0/P_\tau = 2^{\nu-1}\Gamma[(\nu+1)/2]^3/\pi\Gamma(\nu/2+1)\Gamma(\nu+\tfrac{1}{2}) \tag{13.39}$$

As expected for $\nu \to \infty$, progressively more of the available power lies in the spectral zones about ω_0, $2\omega_0$, . . . , as the nonlinear characteristic is made more sensitive to the larger fluctuations. Specifically, it is easily shown that

$$\lim_{\nu\to\infty} [(P_\tau)_0/P_\tau] = \lim_{\nu\to\infty} (2\pi\nu)^{-\frac{1}{2}} \tag{13.39a}$$

For $\nu = 0$, there is no l-f fluctuation, but only a d-c term, of intensity $\beta^2/4$ (from the fact that the clipped wave has an amplitude β and is on but half the time). All continuum energy is associated with the higher zones generated in rectification and is excluded here by the low-pass filter following the detector.

(1) Weak signals in noise. As in Sec. 13.2-1, we can express the second-moment function (13.34) as a power series in the input signal-to-noise ratio a_0^2,

$$M_z(t)_0 = \frac{\beta^2\Gamma(\nu+1)^2}{4}\left(\frac{\psi}{2}\right)^\nu [G_0^{(0)}(t) + a_0^2 G_1^{(0)}(t) + a_0^4 G_2^{(0)}(t) + \cdots]$$
$$= M_z(t)_{\text{dc}} + K_z(t)_0\Big|_{n\times n} + K_z(t)_0\Big|_{s\times n} \tag{13.40}$$

where $G_0^{(0)}(t) = G_0^{(0)}(t)_{\text{dc}} + G^{(0)}(t)_{n\times n} + G^{(0)}(t)_{s\times n}$, etc. (Note that $G_k^{(0)} \neq G_k$ both in structure and in the normalizing factor.) Specifically, one finds that $G_0^{(0)}(t)_{\text{dc}}$, $G_1^{(0)}(t)_{\text{dc}}$, etc., are $2^{2\nu}/\pi\Gamma(\nu+1)^2$ times $G_0(t)_{\text{dc}}$, etc., for the broadband case. For $G_0^{(0)}(t)_{n\times n}$, one has

$$\Gamma(\nu/2+1)^{-2}[{}_2F_1(-\nu/2,-\nu/2;1;k_0^2) - 1]$$

† See Fellows, *op. cit.*, fig. 3 for calculated value of $(P_c)_0$ $(0 < \nu < 3)$, fig. 4 for P_{dc}, and fig. 11.

etc., while $G_0^{(0)}(t)_{s\times n}$ vanishes and $G_1^{(0)}(t)_{s\times n}$ is

$$2k_0\Gamma(\nu/2)^{-2}[{}_2F_1(1-\nu/2,\ 1-\nu/2;\ 2;\ k_0^2)]$$

and so on. When ν is zero, there is only a d-c output; no l-f fluctuations occur. Explicit results for $\nu = 1, 2$ are included as well in Probs. 13.6, 13.7. Note also that the total average intensity becomes

$$\nu = 1: \qquad (P_\tau)_0 = \frac{2\beta^2\psi}{\pi^2}(1 + a_0^2) \tag{13.41a}$$

$$\nu = 2: \qquad (P_\tau)_0 = \frac{\psi^2}{2}\left(1 + 2a_0^2 + \frac{a_0^4}{2}\right) \tag{13.41b}$$

[cf. Eqs. (13.24*a*), (13.25*a*)].

(2) Strong signals in noise. Again, the contribution to the steady component in the output [as represented by Eq. (13.28*a*)] is unchanged. However, the second-moment functions for the $n \times n$ and $s \times n$ products appearing in the l-f continuum now become ($a_0^2 \gg 1$)

$$K_z(t)_0\Big|_{n\times n} \simeq \frac{\beta^2\Gamma(\nu+1)^2}{4}\left(\frac{\psi}{2}\right)^\nu a_0^{2\nu}\frac{k_0^2}{\Gamma(\nu/2)^4}\left\{\frac{1}{a_0^4} + \frac{2(1-\nu/2)^2}{a_0^6} + \frac{1}{a_0^8}\left[1 + \left(1-\frac{\nu}{2}\right)^4 + \left(1-\frac{\nu}{2}\right)^2\left(2-\frac{\nu}{2}\right)^2 + \frac{\Gamma(\nu/2)^4k_0^2}{4\Gamma(\nu/2-1)^4}\right] + \cdots\right\} \tag{13.42a}$$

$$K_z(t)_0\Big|_{s\times n} \simeq \frac{\beta^2\Gamma(\nu+1)^2}{4}\left(\frac{\psi}{2}\right)^\nu a_0^{2\nu}\frac{2k_0}{\Gamma(\nu/2)^2\Gamma(\nu/2+1)^2} \times \left(\frac{1}{a_0^2} + \frac{1}{a_0^4}\left[\nu\left(\frac{\nu}{2}-1\right) + \frac{k_0\Gamma(\nu/2)^2}{2\Gamma(\nu/2-1)^2}\right] + \frac{1}{a_0^6}\left\{\frac{\nu}{2}(\nu-2)\left(1-\frac{\nu}{2}\right)^2 + k_0\frac{\Gamma(\nu/2)^2(\nu/2)(\nu/2-2)}{\Gamma(\nu/2-1)^2} + k_0^2\left[\frac{\Gamma(\nu/2+1)^2}{2\Gamma(\nu/2-1)^2} + \frac{\Gamma(\nu/2)^2}{6\Gamma(\nu/2-2)^2}\right]\right\} + \cdots\right) \tag{13.42b}$$

Observe that for $\nu = 0$ there is still no contribution to the continuum.

Cases of particular interest occur when $\nu = 1, 2$: for the half-wave linear detector one finds that

$(\nu = 1)\ a_0^2 \gg 1$:

$$M_z(t)_0\Big|_{dc} \simeq \frac{\beta^2\psi}{\pi^2}\left(2a_0^2 + 1 + \frac{1}{4a_0^2} + \frac{1}{8a_0^4} + \cdots\right) \tag{13.43a}$$

$$K_z(t)_0\Big|_{n\times n} \simeq \frac{\beta^2\psi}{\pi^2}k_0^2\left[\frac{1}{8a_0^2} + \frac{1}{16a_0^4} + \frac{(26+k_0^2)k_0^2}{128a_0^6} + \cdots\right] \tag{13.43b}$$

$$K_z(t)_0\Big|_{s\times n} \simeq \frac{\beta^2\psi}{\pi^2}k_0\left[1 - \frac{4-k_0}{8a_0^2} - \frac{1}{16a_0^4}(2 + 3k_0 - 2k_0^2) + \cdots\right] \tag{13.43c}$$

while for the half-wave quadratic device the correlation function is given by Eq. (13.36), for *all* a_0 (≥ 0).

As a_0 is increased, the d-c component becomes progressively more significant relative to the $n \times n$ and $s \times n$ contributions, in the ratios $a_0^{2\nu}$: $a_0^{2\nu-4}$: $a_0^{2\nu-2}$, respectively. As in the broadband situation above, the components of the noise are suppressed relative to the steady components, but, unlike the former, the $n \times n$ terms drop out $O(a_0^{-2})$ faster than the $s \times n$ terms, since most of the $n \times n$ energy lies in the higher zones produced in rectification. We observe that, for $\nu > 1$, this suppression is relative, while, for $\nu < 1$, it is absolute. At the transition point $\nu = 1$, there is again a finite $s \times n$ noise contribution, which is ultimately independent of the input signal-to-noise ratio as $a_0 \to \infty$. The mean intensities associated with the zone ($l = 0$) are asymptotically

$$(P_r)_0 \simeq \beta^2 \frac{(2\psi)^\nu \Gamma[(\nu+1)/2]^2}{4\pi\Gamma(\nu/2+1)^2} a_0^{2\nu} \left[1 + \frac{\nu^2}{a_0^2} + \frac{\nu^2(\nu-1)^2}{2a_0^4} + \cdots\right] \tag{13.44a}$$

$$(P_c)_0 \simeq \beta^2 \frac{(2\psi)^\nu \Gamma[(\nu+1)/2]^2}{4\pi\Gamma(\nu/2+1)^2} a_0^{2\nu} \left[\frac{\nu^2}{2a_0^2} + \frac{\nu^2(3\nu^2-6\nu+2)}{8a_0^4} + \cdots\right] \tag{13.44b}$$

showing now that the l-f continuum power is $O(a_0^{2\nu-2})$, unlike the broadband case [cf. Eqs. (13.29)].

(3) The signal-to-noise ratio and signal suppression. We employ again the usual deflection criterion of Secs. 5.3-4 and 13.2-1(3), with a postrectification smoothing filter of effective spectral width Δf. Our definition of output signal remains unchanged [cf. Eqs. (13.30)], but the accompanying background fluctuations stem from the noise in the l-f zone only. In terms of the input signal-to-noise ratio, $(S/N)_{out}$ is once again given by Eq. (13.31), with $K_z(t)$ replaced by $K_z(t)_0$ [Eqs. (13.35*a*), (13.35*b*)]. For weak and strong signals, we write explicitly [cf. Eq. (13.40)]

$(a_0^4 \ll 1)$:

$$\left(\frac{S}{N}\right)_{out} \doteq \frac{2^{\nu-2}\Gamma[(\nu+1)/2]\nu a_0^2[1-(1-\nu/2)a_0^2+\cdots]}{\sqrt{\pi\,\Delta f}\,\Gamma(\nu+1)\left\{\int_0^\infty [G_0^{(0)}(t)_{n\times n} + a_0^2 G_1^{(0)}(t)_{n\times n+s\times n} + \cdots]\,dt\right\}^{1/2}} \tag{13.45a}$$

$(a_0^4 \gg 1)$:

$$\left(\frac{S}{N}\right)_{out} \simeq \frac{2^{\nu-1}\Gamma[(\nu+1)/2][a_0^\nu(1+\nu^2/4a_0^2+\cdots)-1]}{\Gamma(\nu+1)\Gamma(\nu/2+1)\sqrt{\pi\,\Delta f}\,a_0^{\nu-1}\left\{\int_0^\infty dt\,[L_1^{(0)}(t) + a_0^{-2}L_2^{(0)}(t) + \cdots]\right\}^{1/2}} \tag{13.45b}$$

Here, $L_1^{(0)}(t)$ is the coefficient of a_0^{-2} in Eq. (13.42*b*) after the factor $\beta^2\Gamma(\nu+1)^2(\psi/2)^\nu a_0^{2\nu}/4$ has been removed, while $L_2^{(0)}(t)$ is the coefficient of a_0^{-4} in Eqs. (13.42*a*), (13.42*b*), etc. In contrast to the broadband cases, observe that only the $s \times n$ noise terms are primarily evident in the fluctuating background. Figure 13.8 also suggests the general behavior of $(S/N)_{out}$ as a function of $a_0^2 = (S/N)_{in}^2$ for these narrowband waves. The same phenomenon of signal suppression when the input is weak and suppression of the noise when the signal is strong is observed. The behavior of $(S/N)_{out}$ as a function of the detector's law is of similar quality. There exists for each value of $(S/N)_{in}$ a value of ν for which $(S/N)_{out}$ is a maximum (cf. Fig. 13.7). Extensive calculations of average intensities, covariance functions, spectra, and experimental results in excellent agreement with theory have been obtained by Johnson[11] for the case $\nu = 1$ and by Fellows[9] not only for $\nu = 1$ but for $\nu = \frac{1}{2}$ and $\nu = 2$ as well.

Finally, observe that we can alternatively attack the narrowband case in terms of the statistical properties of the envelope, where again $z_0 = \gamma E^\nu$, in the present half-wave cases. The second-moment function of the output is $M_z(t)_0 = \boldsymbol{E}\{\gamma^2 E_1{}^\nu E_2{}^\nu\}$. For noise alone, we easily obtain Eq. (13.16), but for a general signal we must use Eq. (9.53) in this average, and the results are usually very complicated. We conclude, therefore, that, when a choice is available between using the statistics of the instantaneous amplitude (which are normal, as far as the noise is concerned) and using those of envelope and phase (which are *not* normal), the former is to be preferred for its generally simpler expressions.

PROBLEMS

13.6 (*a*) Show that the average power in the discrete spectrum is given by

$$P_{s\times s} = 2\sum_{m=1}^{\infty} h_{m,0}^2 = \frac{\beta^2\Gamma(\nu+1)^2}{2}\left(\frac{\psi}{2}\right)^\nu \sum_{m=1}^{\infty} \frac{a_0{}^{2m}\,{}_1F_1{}^2[(m-\nu)/2;\, m+1;\, -a_0{}^2]}{m!^2\Gamma[(2+\nu-m)/2]^2} \tag{1}$$

and obtain the asymptotic form when $a_0 \to \infty$.

(*b*) Show that, for $\nu = 1$ and $\nu = 2$,

$$P_{s\times s} \simeq \frac{\beta^2\psi}{\pi^2}\left[a_0{}^2\left(\frac{\pi^2-4}{2}\right) - \frac{2}{\pi} + \frac{1}{4\pi a_0{}^2}\cdots\right] \qquad \nu = 1 \tag{2}$$

$$P_{s\times s} \simeq \frac{\beta^2\psi^2}{2}\left[a_0{}^4 - \frac{32a_0{}^2}{\pi^2}\sum_{m=0}^{\infty}\frac{1}{(4m^2-1)(2m+1)(2m+3)} + \frac{8}{\pi^2}\sum_{m=1}^{\infty}\frac{m(m+1)}{(4m^2-1)(2m+1)(2m+3)} + \cdots\right] \qquad \nu = 2 \tag{3}$$

13.7 For weak carriers in broadband noise, show that the second-moment function (13.26) takes the following explicit forms when $\nu = 0, 1, 2$:

$$M_z(t)_{\nu=0} \doteq \beta^2\left\{\left(\frac{1}{4} + \frac{1}{2\pi}\sin^{-1}k\right) + \frac{a_0{}^2}{2\pi}\frac{\cos\omega_0 t - k}{\sqrt{1-k^2}} + \frac{a_0{}^4}{16\pi} \times\left[-6\sin^{-1}k + \frac{k(1+\cos 2\omega_0 t - 2\cos\omega_0 t)}{(1-k^2)^{3/2}} - \frac{4(1-k^2)\cos\omega_0 t}{(1-k^2)^{3/2}}\right] + \cdots\right\} \tag{1}$$

$$M_z(t)_{\nu=1} \doteq \frac{\beta^2\psi}{2\pi}\left\{\left[k\left(\frac{\pi}{2} + \sin^{-1} k\right) + \sqrt{1-k^2}\right] + a_0^2\left[\sqrt{1-k^2} + (\sin^{-1} k)\cos\omega_0 t + \frac{\pi}{2}\cos\omega_0 t\right] + \cdots\right\} \tag{2}$$

$$M_z(t)_{\nu=2} \doteq \frac{\beta^2\psi}{\pi}\left\{\left(\frac{1+2k^2}{2}\sin^{-1} k + \frac{\pi k^2}{2} + \frac{\pi}{4} + \frac{3k}{2}\sqrt{1-k^2}\right) + a_0^2\left[\frac{\pi}{2} + \sin^{-1} k + k\sqrt{1-k^2} + 2\cos\omega_0 t\left(k\sin^{-1} k + \sqrt{1-k^2} + \frac{\pi k}{2}\right)\right] + \cdots\right\} \tag{3}$$

Examine the behavior as $t \to 0$ and $t \to \infty$, and compare with the exact results of Eqs. (13.22), (13.23), etc., in these instances.

13.8 A carrier of frequency f_c off-tune from a narrowband of noise centered about f_0 is passed into a zero-memory nonlinear device. Verify that Eqs. (13.43) are now modified, in the case $\nu = 1$, to

$$K_z(t)_0\Big|_{n\times n} \simeq k_0^2\frac{\beta^2\psi}{\pi^2}\left(\frac{1}{8p} + \frac{1}{16p^2} + \cdots\right) \tag{1}$$

$$K_z(t)_0\Big|_{s\times n} \simeq \frac{\beta^2\psi}{\pi^2}\left(\cos\omega_D t - \frac{4\cos\omega_D t - k_0\cos 2\omega_0 t}{8p} - \frac{1}{32p^2}\{[4 - k_0(t)^2]\cos\omega_D t + 6k_0\cos 2\omega_D t - 3k_0^2\cos 3\omega_D t\} + \cdots\right) \tag{2}$$

where $p \equiv a_0^2|Y(i\omega_c)|^2$ and $\omega_D = \omega_c - \omega_0$.

13.9 (*a*) Extend the results of Prob. 13.2 for small-signal detection when the input now consists of the sum of signal and noise. Thus, if $s_1 = S(t_1)/\psi^{1/2}$, $s_2 = S(t_2)/\psi^{1/2}$ are normalized signal values at the times t_1, t_2, show that the second-moment function of the output wave is

$$M_z(t) = \sum_{j=0}^{\infty}\sum_{k=0}^{\infty} \alpha_j\alpha_k\psi^{(j+k)/2}\langle I_{j,k}(t_2 - t_1)\rangle_S \tag{1}$$

where

$$I_{j,k}(t) = \iint_{-\infty}^{\infty} (x + s_1)^j(y + s_2)^k W_2(x,y;t)\,dx\,dy$$

$$= j!k!2^{-(j+k)/2}\sum_{q_1,q_2=0}^{\infty}\frac{2^{(q_1+q_2)/2}s_1^{q_1}s_2^{q_2}}{q_1!q_2![(j-q_1)/2]![(k-q_2)/2]!} \times {}_2F_1\left[\frac{q_1-j}{2},\frac{q_2-k}{2};\frac{1}{2};k(t)^2\right] \qquad j - q_1,\ k - q_2 \text{ even} \tag{2a}$$

$$I_{j,k}(t) = 2k(t)j!k!2^{-(j+k)/2}\sum_{q_1,q_2=0}^{\infty}\frac{2^{(q_1+q_2)/2}s_1^{q_1}s_2^{q_2}}{q_2!q_2![(j-q_1-1)/2]![(k-q_2-1)/2]!} \times {}_2F_1\left[\frac{1-j+q_1}{2},\frac{1-k+q_2}{2};\frac{3}{2};k(t)^2\right] \qquad j - q_1,\ k - q_2 \text{ odd} \tag{2b}$$

$$I_{j,k}(t) = 0 \qquad j + k - (q_1 + q_2) \geq 0 \text{ odd} \tag{2c}$$

(*b*) In the case of the general quadratic response ($j, k \leq 2$), obtain specifically

$$M_z(t) = \alpha_0^2 + \sqrt{\psi}\,\alpha_0\alpha_1\langle s_1 + s_2\rangle + \psi\alpha_0\alpha_2\langle 2 + s_1^2 + s_2^2\rangle + \psi\alpha_1^2(k + \langle s_1s_2\rangle) + \psi^{3/2}\alpha_1\alpha_2\langle(s + s_1)(s_1s_2 + 2k + 1)\rangle + \psi^2\alpha_2^2\langle 1 + 2k^2 + s_1^2 + s_2^2 + 4ks_1s_2 + s_1^2s_2^2\rangle \tag{3}$$

(*c*) If $S(t) = A_0(t')\cos\omega_0 t$, show that the covariance function of the l-f part exclusive of the steady component is

$$M_z(t)_{LF} = \psi^2\alpha_2^2[k_0(t)^2 + k_0(t)\langle(A_{0_1}A_{0_2})_{LF}\rangle + 4\langle(A_{0_1}{}^2A_{0_2}{}^2)_{LF}\rangle] \tag{4}$$

where the subscript LF indicates that the steady component has been removed.

13.3 *Detection of a Sinusoidally Modulated Carrier in Noise*

Only the cases of an unmodulated carrier have been considered so far. This is the simplest type of signal, where the "message" conveyed to the receiver consists solely of the presence or absence of this carrier in its accompanying noise. The theory may be extended to include more general signal structures, in the manner of Sec. 5.1. Here we shall examine† the simplest class of deterministic modulation process, e.g., an ensemble of sinusoidally amplitude-modulated carriers accompanied by an additive narrowband noise centered about the carrier at f_0.

If now we extend Eq. (13.21) and write for the time-dependent case $p(t) = a_0^2g(t)^2$, where a_0 is given as before and $g(t)$ is the time response of the *filtered* input to the detector, we obtain for the second-moment function of the output process in the l-f zone following rectification

$$M_z(t)_0 = K_z(t)_0\Big|_{n\times n} + K_z(t)_0\Big|_{s\times n} + M_z(t)_0\Big|_{s\times s+\mathrm{dc}} \tag{13.46}$$

Letting $p_1 = p(t_0)$ and $p_2 = p(t_0 + t)$, we find that this becomes specifically, from Eqs. (13.34) and (13.35),

$$K_z(t)_0\Big|_{n\times n} = \frac{\beta^2\Gamma(\nu+1)^2}{4}\left(\frac{\psi}{2}\right)^\nu \times \sum_{q=1}^{\infty} \frac{k_0(t)^{2q}\langle {}_1F_1(q-\nu/2;1;-p_1)\,{}_1F_1(q-\nu/2;1;-p_2)\rangle}{(q!)^2\Gamma[(2+\nu)/2-q]^2} \tag{13.47a}$$

$$K_z(t)_0\Big|_{s\times n} = \frac{\beta^2\Gamma(\nu+1)^2}{4}\left(\frac{\psi}{2}\right)^\nu \sum_{m=1}^{\infty}\sum_{q=0}^{\infty} \frac{2k_0(t)^{m+2q}\langle(p_1p_2)^{m/2}\,{}_1F_1(m+q-\nu/2;m+1;-p_1) \times {}_1F_1(m+q-\nu/2;m+1;-p_2)\rangle}{q!(q+m)!(m!)^2\Gamma[(2+\nu)/2-m-q]^2} \tag{13.47b}$$

and

$$M_z(t)_0\Big|_{s\times s+\mathrm{dc}} = \frac{\beta^2\Gamma(\nu+1)^2}{4\Gamma(\nu/2+1)^2}\left(\frac{\psi}{2}\right)^\nu \left\langle {}_1F_1\left(-\frac{\nu}{2};1;-p_1\right)\,{}_1F_1\left(-\frac{\nu}{2};1;-p_2\right)\right\rangle \tag{13.47c}$$

Next, let us take advantage of the periodic nature of the modulation to

† The present section largely follows the analysis of Middleton[12] and the experimental results of R. A. Johnson[11] and of Fubini and D. C. Johnson[13] for the half-wave linear detector. See also Fellows, *op. cit.*, in the case of the half-wave νth-law devices. Other pertinent material is found in the papers of Burgess,[7] Ragazzini,[14] and Weinberg and Kraft.[15]

expand Eq. (13.47c) in a Fourier series,† obtaining for the d-c term and the harmonics of the modulation, generated in detection,

$$M_z(t)_{s\times s+\mathrm{dc}} = S_0^2 + \sum_{l=1}^{\infty} \frac{|S_l|^2}{2} \cos l\omega_a t \tag{13.48a}$$

where

$$|S_l| = \beta \left(\frac{\psi}{2}\right)^{\nu/2} \frac{\epsilon_l \Gamma(\nu+1)}{2\Gamma(\nu/2+1)} \left| \frac{1}{2\pi} \int_0^{2\pi} {}_1F_1\left[-\frac{\nu}{2};1;-p(\theta)\right] e^{-il\theta}\, d\theta \right| \qquad l \geq 0 \tag{13.48b}$$

This is quite general.

To relate the input *before* filtering to the input to the nonlinear element (i.e., to the input *after* the predetection filters) we must use Eq. (2.51). For a sinusoidally modulated carrier as input, such that $A_0(t,t') = A_0(1 + \lambda' \cos \omega_a t') \cos (\omega_c t + \phi)$, we find that $A_0(t,t') = V_{in\ nonlin}(t,t') = A_0\{1 + \lambda' |Y_F(i\omega_a)_0| \cos [\omega_a t' + \phi_0(\omega_a)]\} \cos (\omega_c t + \phi')$ $(\omega_c = \omega_0)$, in which $Y_F(i\omega_a)_0 = |Y_F(\omega_a)_0| e^{i\phi_0(\omega_a)}$. Since carrier and modulation are assumed to be independent processes, we may rewrite this as

$$A_0(t',t) = A_0(1 + \lambda \cos \omega_a t') \cos (\omega_c t + \phi) \tag{13.49a}$$

where

$$\lambda = \lambda' |Y_F(i\omega_a)_0| \qquad \phi = \phi' - \omega_c \frac{\phi_0(\omega_a)}{\omega_a} \tag{13.49b}$$

Therefore, we see that, for a sinusoidal signal, $p(\theta)$ becomes

$$p(\theta) = a_0^2 (1 + \lambda \cos \theta)^2 \qquad \theta = \omega_a t \tag{13.49c}$$

corresponding to the original input $A_0'(t,t') = A_0(1 + \lambda' \cos \omega_a t') \cos (\omega_c t + \phi')$. Here, λ is the effective *modulation index* and is restricted to the range $(0 \leq |\lambda| \leq 1)$ so that no overmodulation occurs (cf. Sec. 12.1, etc.); ω_a is the angular frequency of the modulating wave.

With the help of the relation (Prob. 13.10a)

$$\frac{1}{2\pi} \int_0^{2\pi} (1 + \lambda \cos \theta)^\alpha \cos l\theta\, d\theta = \frac{(-\alpha)_l (-\lambda)^l}{2^l l!} \times {}_2F_1\left(\frac{l-\alpha}{2}, \frac{l-\alpha+1}{2}; l+1; \lambda^2\right) \qquad l = 0, 1, \ldots \tag{13.50}$$

Eq. (13.48b) may be evaluated to give the general expression

$$|S_l| = \beta \left(\frac{\psi}{2}\right)^{\nu/2} \frac{\epsilon_l \Gamma(\nu+1)}{2\Gamma(\nu/2+1)} \left| \frac{(-\lambda)^l}{2^l l!} \sum_{n=0}^{\infty} \frac{(-\nu/2)_n a_0^{2n}}{n!} \times {}_1F_1\left(n - \frac{\nu}{2}; n+1; -a_0^2\right) \sum_{m=0}^{n} \frac{(-1)^m (-2m)_l}{m!(n-m)!} \times {}_2F_1\left(\frac{l-2m}{2}, \frac{l-2m+1}{2}; l+1; \lambda^2\right) \right| \tag{13.51}$$

† We may replace $e^{-il\theta}$ by $\cos l\theta$ below, for the particular choice of zero phase at $t = 0$ for $p(t)$.

In practice, it is natural to select as the detected signal the component $l = 1$, that is, the first harmonic of the modulation. For small values of a_0, a direct expansion of Eqs. (13.47) is useful, while for strong signals it can be shown that (cf. Prob. 13.10c)

$$|S_l| \simeq \beta \left(\frac{\psi}{2}\right)^{\nu/2} \frac{\epsilon_l \Gamma(\nu + 1) a_0{}^\nu}{2\Gamma(\nu/2 + 1)^2} \left| \frac{(-\lambda)^l}{2^l l!} \left[(-\nu)_l \, {}_2F_1\left(\frac{l-\nu}{2}, \frac{l-\nu+1}{2}; l+1; \lambda^2\right) + \frac{\nu^2(2-\nu)_l}{4a_0{}^2} \, {}_2F_1\left(\frac{l-\nu}{2}+1; \frac{l-\nu+3}{2}; l+1; \lambda^2\right) + \frac{\nu^2(\nu+2)^2(4-\nu)_l}{32a_0{}^4} \, {}_2F_1\left(\frac{l-\nu}{2}+2; \frac{l-\nu+5}{2}; l+1; \lambda^2\right) + \cdots \right] \right| \tag{13.52}$$

provided that $a_0{}^2(1 - \lambda)^2 \gg 1$, so that the asymptotic development of ${}_1F_1$ may be safely applied over the whole of a modulation cycle. No simple expressions are available for the noise terms in the general case. However, for weak signals one can expand the hypergeometric functions in a series in $p_1 = p(\theta)$, $p_2 = p(\theta + \phi)$, using the results of Prob. 13.11a to evaluate terms of the type $\langle p(\theta)^{a_1} p(\theta + \phi)^{a_2} \rangle$. Clearly, this procedure breaks down if $a_0{}^2$ is at all large. At the other extreme of $a_0{}^2 \gg 1$, one can use the asymptotic expression for the hypergeometric function to write the noise terms

$$K_z(t)_0 \Big|_{\text{noise}} \simeq \beta^2 \frac{\Gamma(\nu + 1)^2}{4} \left(\frac{\psi}{2}\right)^\nu \times \sum_{m=0}^{\infty} \sum_{q=0}^{\infty} \frac{\epsilon_m (1 - \delta_m^0 \delta_q^0) k_0(t)^{m+2q}}{\Gamma[(2+\nu)/2 - q]^2 \Gamma[(2+\nu)/2 - q - m]^2 q!(q+m)!} \times \left\langle (p_1 p_2)^{(\nu - m)/2 - q} \left[1 + \frac{(m + q - \nu/2)(q - \nu/2)}{1!p_1} + \cdots \right] \times \left[1 + \frac{(m + q - \nu/2)(q - \nu/2)}{1!p_2} + \cdots \right] \right\rangle \tag{13.53}$$

and then apply the result of Prob. 13.11b. An example is given below (Sec. 13.3-2), in the important special case of the half-wave linear detector ($\nu = 1$). The spectrum of the l-f output fellows in the usual way from the Wiener-Khintchine theorem [cf. Sec. 3.2-4(2)].

(1) The quadratic detector ($\nu = 2$). Before we discuss the linear detector, it will be helpful to illustrate the analysis with the simpler example of the half-wave quadratic rectifier. The input signal to the detector is, from Eq. (13.49a),

$$A_0(t') = \sqrt{2}\, s_F(t') = s_F \sqrt{2}\,(1 + \lambda \cos \omega_a t') \qquad a_0{}^2 = s_F{}^2/\psi \tag{13.54}$$

and the mean intensity of the modulated wave is $P_S = s_F{}^2(1 + \lambda^2/2)$, with $\lambda^2 s_F{}^2/2$ attributable to the modulation; s_F itself is the rms amplitude of the unmodulated carrier. The second-moment function of the detector's output is, from Eqs. (13.46), (13.47),†

$$M_z(t)_0 = \frac{\beta^2\psi^2}{4}\{(1 + \langle p_1 + p_2\rangle + \langle p_1p_2\rangle_{s\times s+\mathrm{dc}} + k_0(t)^2_{n\times n} + [2\langle\sqrt{p_1p_2}\rangle k_0(t)]_{s\times n}\} \tag{13.55a}$$

For the sinusoidal modulation of Eq. (13.54), this reduces to

$$R_0(t) = \frac{\beta^2\psi^2}{4}\left(\left\{\left[1 + a_0{}^2\left(1 + \frac{\lambda^2}{2}\right)\right]^2 + a_0{}^4\left(2\lambda^2 \cos \omega_a t + \frac{\lambda^4}{8}\cos 2\omega_a t\right)\right\}_{s\times s+\mathrm{dc}} + k_0(t)^2_{n\times n} + \left[2a_0{}^2\left(1 + \frac{\lambda^2}{2}\cos \omega_a t\right)k_0(t)\right]_{s\times n}\right) \tag{13.55b}$$

By choice, the output signal is the first harmonic of the modulation, and its intensity is

$$(P_S)_{out} = \frac{|S_1|^2}{2} = \frac{\beta^2\lambda^2 s_F{}^4}{2} \equiv s_a{}^2 \tag{13.56}$$

As before, we assume a linear, postdetection filter, located about f_a, of fixed width Δf much smaller than the effective spectral width Δf_N of the input noise. The effective intensity of the noise accompanying the output signal is therefore

$$N_a{}^2 = [\mathcal{W}_z(f)_{\mathrm{noise}}\,\Delta f]_{f=0} \tag{13.57}$$

since the spectrum is essentially constant for all $f \sim f_a \ll \Delta f_N$. Furthermore, if we next assume that $\Delta f_N \gg f_a$, then $k_0(t)$ falls off much more rapidly than $\cos \omega_a t$ and we may replace $\cos \omega_a t$, as well as $\cos \omega t$, by unity in the Wiener-Khintchine relation. Note that then $\lambda \approx \lambda'$, as $|Y(i\omega)_0| \approx 1$ [cf. Eq. (13.49*b*)]. From Eq. (13.55*b*) we obtain finally the output signal-to-noise ratio‡

$$\left(\frac{S}{N}\right)_a = \frac{\lambda a_0{}^2}{\sqrt{2\,\Delta f}}\left\{\left[\int_0^\infty k_0(t)^2\,dt\right]_{n\times n} + \left[(2 + \lambda^2)a_0{}^2\int_0^\infty k_0(t)\,dt\right]_{s\times n}\right\}^{-1/2} \tag{13.58}$$

a result which holds equally well for the full- and for the half-wave quadratic detector [cf. Eq. (13.36)]. Note the effect of *signal suppression* when $a_0{}^2 < 1$: $(S/N)_{out} \doteq O(a_0{}^2)$. For strong signals, one gets the predicted linear dependence of $(S/N)_{out}$ on a_0, as now the $s \times n$ noise fluctuations become dominant.

We observe from Eq. (13.58) that the integrals therein depend on the spectral shape of the network preceding the detector. This dependence may be expected to differ among the $n \times n$ and $s \times n$ terms by a not entirely ignorable amount, and such behavior may be explained by the

† This is the generalization of Eq. (13.36) to include modulation.

‡ When $\Delta f_N \sim f_a$, the unmodified $s \times n$ term in Eq. (13.55*b*) must be used and the effect of the predetection filter on the signal must also be specifically taken into account.

following argument: The spectral contribution near $f = 0$ is produced by beats between closely adjacent components. Hence, the intensity of modulation products of this kind, when generated at high frequencies—i.e., out on the "wings" of the spectrum—depends on the spectral density in their neighborhood. Thus, the spectral distribution of the IF output determines in varying degrees the spectral ordinate near $f = 0$ after rectification, as this ordinate may be considered the sum, weighted according to the law of the detector, of the beats generated in the process of detection. Choosing as representative the three filter responses of (1) to (3) following Eq. (3.99*b*), we may write Eq. (13.58) specifically as

$$\left(\frac{S}{N}\right)_a = \lambda a_0^2 \left(\frac{\omega_F}{2\,\Delta f}\right)^{1/2} \left\{ \begin{pmatrix} 1 \\ \frac{1}{\sqrt{2}} \\ \frac{1}{2} \end{pmatrix}_{n\times n} + a_0^2(2+\lambda^2) \begin{pmatrix} 1 \\ 1 \\ 1 \end{pmatrix}_{s\times n} \right\}^{-1/2} \quad \begin{matrix} \text{rectangular } (R) \\ \text{gaussian } (G) \\ \text{"optical" } (F) \end{matrix} \tag{13.59}$$

where the parentheses apply downward in the order indicated. The filter parameters are related by normalizing on the basis of equal power outputs ($n_F^2 = \psi$) and the same maximum spectral intensity at $f = f_c$, so that [cf. Eq. (3.100)]

$$\omega_F = \frac{\omega_G}{\sqrt{\pi}} = \frac{2}{\pi}\,\omega_R$$

The output signal-to-noise ratio [Eq. (13.59)] is shown in Fig. 13.10, where we observe again the signal suppression characteristic of weak signals. When $a_0 \to \infty$, the ratio of Eq. (13.59) also simplifies to

$$\left(\frac{S}{N}\right)_a \simeq \frac{\lambda a_0}{\sqrt{2}\,(2+\lambda^2)^{1/2}} \left(\frac{\omega_F}{\Delta f\,\Gamma^{(1)}}\right)^{1/2} \tag{13.60}$$

where $\Gamma^{(1)} = \omega_F \int_0^\infty k_0(t)\,dt = \pi/2, \sqrt{\pi}, 1$, in the order of the filters given above. We observe, as expected, that the effect of predetector filter shape on $(S/N)_a$ is noticeable for the weak signals only, as then the $s \times n$ noise terms play a secondary role with respect to the $n \times n$ contributions. For very weak signals, this effect is O(1–2) db at most, for $\lambda = 1.0$. (From Sec. 13.4-3, it is noted that for $S_{out} \sim |S_1|$ the curves of Fig. 13.10 apply as well for *full-wave* quadratic rectification.)

(2) The linear detector ($\nu = 1$). Here, as in the quadratic case, we assume primarily resistive loads, so that the dynamic path has zero memory and (for $\nu = 1$) is linear over the conducting region, representing in effect an ideal "envelope tracer." The analysis, however, no longer leads to simple forms, as none of the series (13.47) terminates. For the threshold cases, it is nevertheless possible to construct useful approximations to the exact expressions.

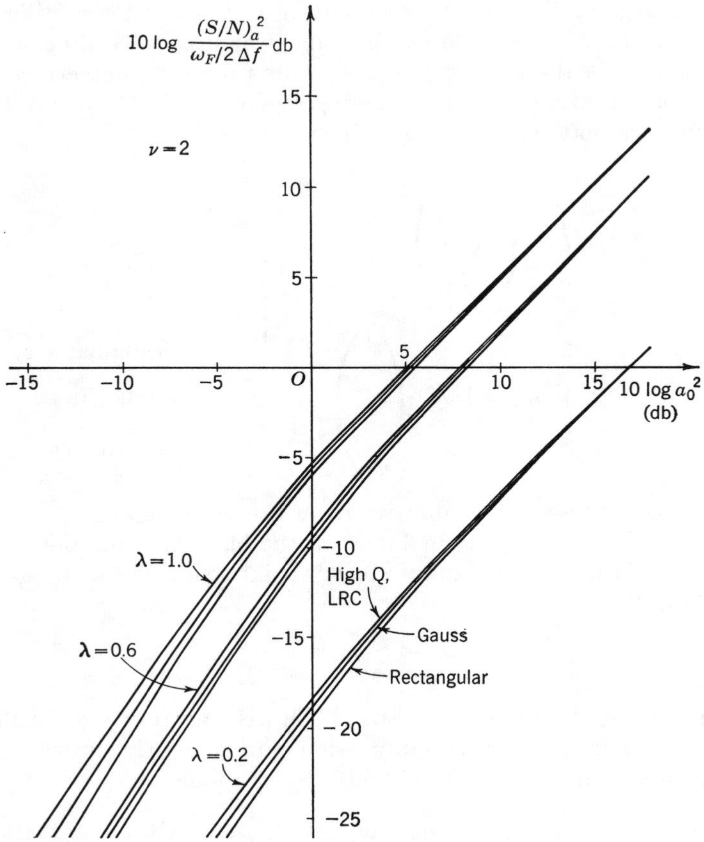

FIG. 13.10. $(S/N)_{out}$ following half-wave *quadratic* rectification for a sinusoidally modulated carrier in narrowband noise through (1) a rectangular, (2) a gauss, (3) a single-tuned predetector filter.

Let us consider first the signal at the output and begin by replacing the hypergeometric function in Eq. (13.48*b*) by a three-term fit† over the range $0 \leq a_0^2(1 + \lambda)^2 \leq 10$, for $\nu = 1$, $l = 1$. The rms amplitude of the output signal, corresponding to Eq. (13.56), is then

$$s_a = \frac{|S_1|}{\sqrt{2}} \doteq \frac{\beta\psi^{1/2}}{\sqrt{4\pi}}\left[1 - 0.2289a_0\left(1 + \frac{\lambda^2}{4}\right)\right] \qquad a_0^2(1 + \lambda)^2 \leq 10 \qquad (13.61)$$

† For a detailed discussion of the approximation and the errors involved, see Middleton,[12] sec. IV, and ref. 10 in that paper.

On the other hand, for strong input signals we may use Eq. (13.52) to get

$$s_a \simeq \beta\lambda \frac{a_0\psi^{1/2}}{\pi}\left[1 - \frac{1}{4a_0^2}\,{}_2F_1\left(1,\frac{3}{2};2;\lambda^2\right) - \frac{3}{32a_0^4}\,{}_2F_1\left(2,\frac{5}{2};2;\lambda^2\right) + \cdots\right] \qquad a_0^2(1-\lambda)^2 \gg 1 \quad (13.62)$$

The noise contributions are more involved. For weak signals, we again approximate the hypergeometric functions in Eqs. (13.47), keeping now only the leading terms. We have finally for the covariance function[12]

$$\begin{aligned} K_z(t)_0\Big|_{n\times n} &\doteq \frac{\beta^2\psi}{8\pi}\,\sigma_1(a_0^2,\lambda;1)k_0(t)^2 \\ K_z(t)_0\Big|_{s\times n} &\doteq \frac{\beta^2\psi}{4\pi}\,\sigma_2(a_0^2,\lambda;1)k_0(t) \end{aligned} \qquad (13.63)$$

where specifically

$$\begin{aligned} \sigma_1(a_0^2,\lambda;1) = 1 &- a_0^2(1+\lambda^2/2) - 0.334a_0^3(1+3\lambda^2/2) \\ &+ a_0^4(1+3\lambda^2+3\lambda^4/8)/4 - 0.167a_0^5(1+5\lambda^2+15\lambda^4/8) \\ &\quad + 0.0279a_0^6(1+15\lambda^2/2+45\lambda^4/8+5\lambda^6/16) \qquad (13.64a) \end{aligned}$$

$$\begin{aligned} \sigma_2(a_0^2,\lambda;1) = 2a_0^2[&(1+\lambda^2/2) - a_0^2(1+3\lambda^2+3\lambda^4/8)/2 \\ &+ 0.131a_0^3(1+5\lambda^2+15\lambda^4/8) \\ &+ a_0^4(1+15\lambda^2/2+45\lambda^4/8+5\lambda^6/16) \\ &- 0.0327a_0^5(1+21\lambda^2/2+105\lambda^4/8+35\lambda^6/16) \\ &+ 4.29\times10^{-3}a_0^6(1+14\lambda^2/16+105\lambda^4/4+35\lambda^6/4 \\ &\qquad + 35\lambda^8/128)] \qquad (13.64b) \end{aligned}$$

Figures 13.11, 13.12 show how σ_1 and σ_2 vary with a_0 and λ. Figure 13.13 illustrates the total l-f power $\sigma_1 + \sigma_2$, with the experimental data of R. A.

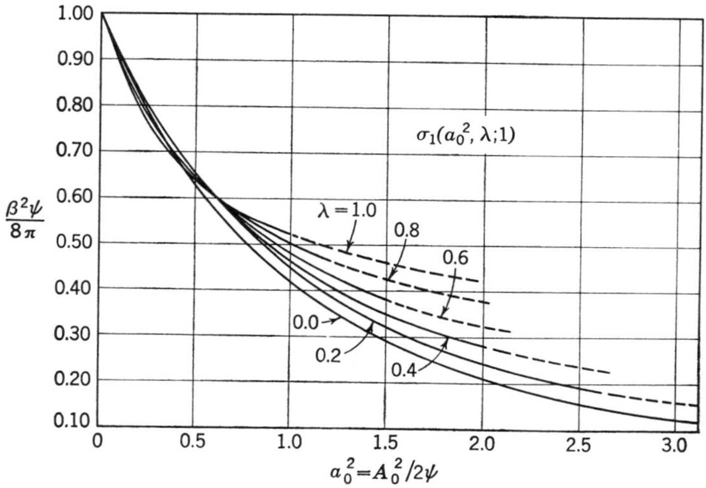

FIG. 13.11. Mean l-f output $n \times n$ noise intensity, linear rectifier, sinusoidal modulation.

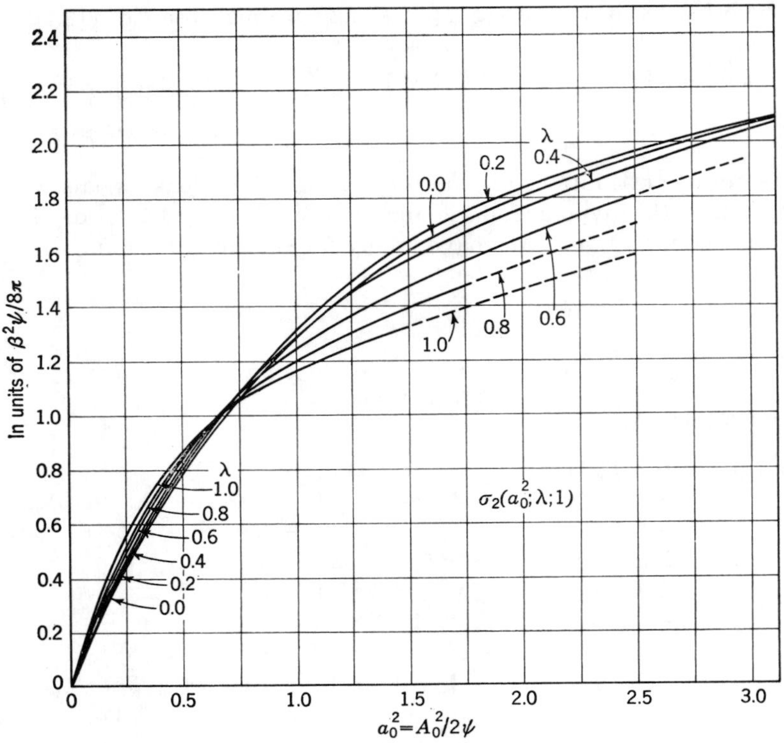

FIG. 13.12. Mean l-f output $s \times n$ noise intensity, linear rectifier, sinusoidal modulation.

Johnson.[11] For strong signals, the covariance function of the l-f noise is modified here (cf. Prob. 3.11b) to

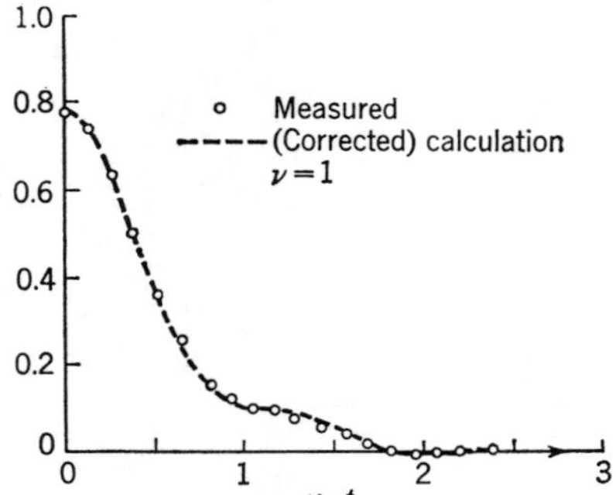

FIG. 13.13. Low-frequency noise intensity after rectification of a sinusoidally modulated carrier and noise by a linear detector.

$$K_z(t)_0 \simeq \frac{\beta^2\psi}{\pi^2}\left[k_0(t) + \frac{k_0{}^2(t)}{4a_0{}^2(1-\lambda^2)}\frac{2(1 - a_\lambda{}^2\cos\phi)}{1 - 2a_\lambda{}^2\cos\phi + a_\lambda{}^4} + O(a_0{}^4)\right] \tag{13.65}$$

with $a_\lambda = (1 - \sqrt{1-\lambda^2})/\lambda$, $\phi = \omega_a t$.

The $n \times n$ products are $O(a_0{}^{-2})$, unlike the $s \times n$ terms, which are $O(a_0{}^0)$. In fact, the $n \times n$ components are always suppressed relative to the $s \times n$ terms as $a_0 \to \infty$. For $\nu = 1$, this suppression (cf. Fig. 13.11) is largest for the lesser degrees of modulation, while the $s \times n$ products increase with increasing a_0. In this instance, the increase is least for the larger modulation indices. Physically, this behavior is explained by the

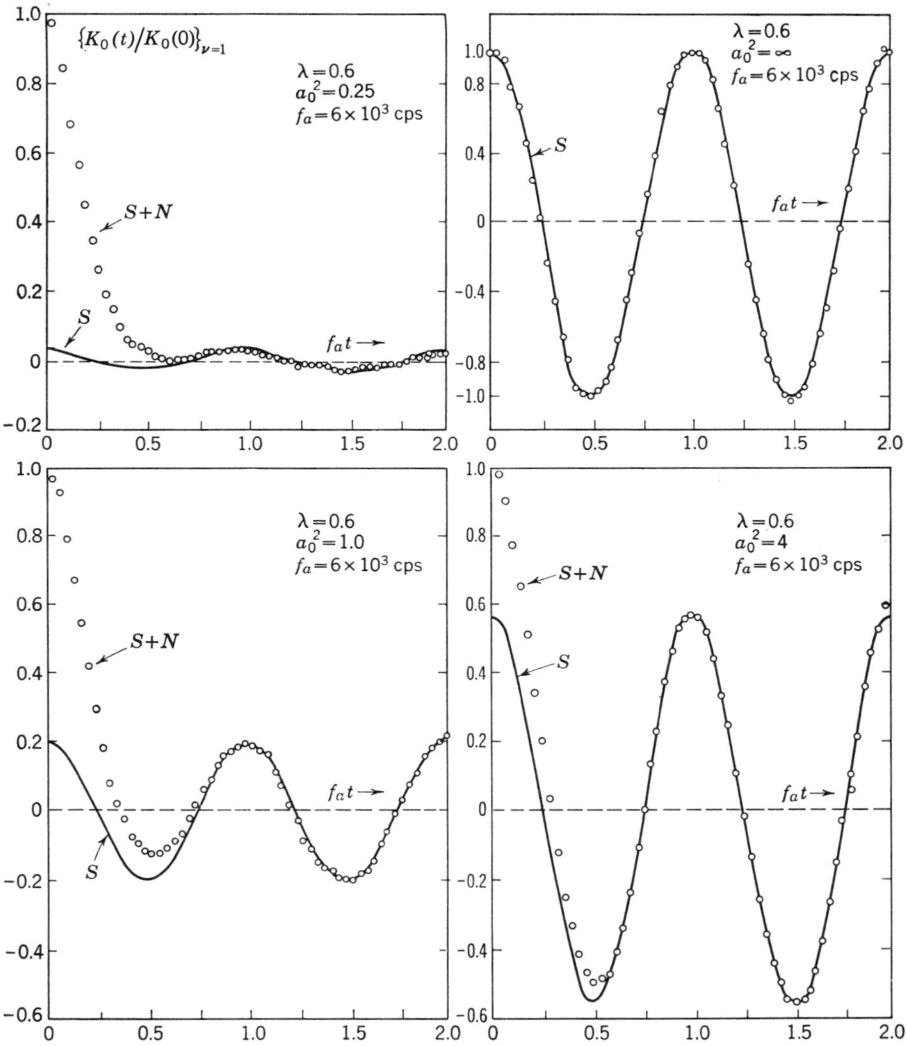

FIG. 13.14. Normalized covariance functions of the l-f output of a sinusoidally modulated carrier after a half-wave linear rectifier ($\lambda = 0.6$; $f_a = 6.10^3$ cps).

fact that, when the signal is large, the noise "rides" on top of it, so that for small modulation indices (λ) both positive and negative noise peaks (relative to $\bar{N} = 0$) are passed by the detector without much clipping on the average due to cutoff. Greater percentages of modulation mean less noise passed near the bottom of a modulation cycle; hence, the decrease on the average in $s \times n$ noise for $\lambda \approx 1$. When $a_0 \to \infty$, however, there is not much difference between noise outputs with $\lambda \approx 1$ or $\lambda \approx 0$, as Eq. (13.65) indicates, as relatively little noise (per modulation cycle) is blocked off in either extreme. Some examples of the output covariance function are shown in Fig. 13.14, contrasting theory and experiment. In general, close agreement with theory has been established in all cases where comparisons were possible.[9,11]

The signal-to-noise ratio [cf. Eq. (13.59)] now takes the following two forms, respectively, for weak and for strong carriers,†

$$\left(\frac{S}{N}\right)_a \doteq \lambda a_0^2 \left(\frac{\omega_F}{2\,\Delta f}\right)^{1/2} \left[1 - 0.2289 a_0 \left(1 + \frac{\lambda^2}{4}\right)\right] \left\{ \begin{pmatrix} 1 \\ \dfrac{1}{\sqrt{2}} \\ \dfrac{1}{2} \end{pmatrix} \sigma_1(a_0^2,\lambda;1) + \begin{pmatrix} 1 \\ 1 \\ 1 \end{pmatrix} \sigma_2(a_0^2,\lambda;1) \right\}^{-1/2} \qquad a_0^2(1+\lambda)^2 \le 4 \tag{13.66}$$

$$\left(\frac{S}{N}\right)_a \simeq \frac{\lambda a_0}{\sqrt{2}} \left(\frac{\omega_F}{2\,\Delta f \Gamma^{(1)}}\right)^{1/2} \qquad a_0^2(1-\lambda)^2 \gg 1 \tag{13.67}$$

where additional terms may be added from Eqs. (13.62) and (13.65).

Figure 13.15 shows $(S/N)_a$ for the case $\nu = 1$, with the three IF filters of Fig. 13.10. Signal suppression is again observed when $a_0 \le 1$, and the linear behavior of $(S/N)_a$, characteristic of strong signals, appears once more as $a_0 \to \infty$. Taking the ratio of Eq. (13.60) to Eq. (13.67) yields

$$\frac{(S/N)_a\Big|_{\nu=2}}{(S/N)_a\Big|_{\nu=1}} \simeq \left(1 + \frac{\lambda^2}{2}\right)^{-1/2} \tag{13.68}$$

which is evident from a comparison of Figs. 13.10 and 13.15: the output ratio for the quadratic detector is always reduced relative to that for the linear rectifier because of the greater production of $s \times n$ noise terms in the former, regardless of the shape of the predetection filter. Note again (cf. Sec. 13.4-3) that these relations apply equally well for the *full-wave* linear rectifier, provided that we consider as our output signal a single harmonic (here $l = 1$) of the modulation.

PROBLEMS

13.10 (*a*) Show that, if l is an integer and $|\lambda| < 1$,

$$\frac{1}{2\pi}\int_0^{2\pi} (1 + \lambda \cos\theta)^\alpha \cos l\theta \, d\theta = \frac{(-\alpha)_l(-\lambda)^l}{2^l l!} \, {}_2F_1\left(\frac{l-\alpha}{2}, \frac{l-\alpha+1}{2}; l+1; \lambda^2\right) \tag{1}$$

What conditions on α must be satisfied if $|\lambda| < 1$, $|\lambda| = 1$?

† The order reads downward in the parentheses above with respect to the various predetection filters of Eq. (13.59).

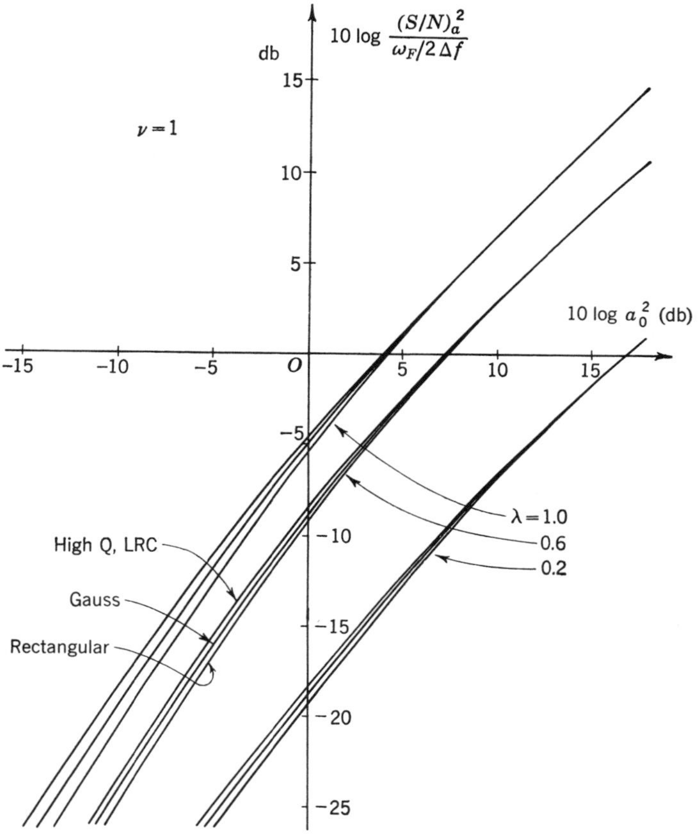

FIG. 13.15. $(S/N)_{out}$ for a sinusoidally modulated carrier in noise through a half-wave *linear* detector, for the three IF filters of Fig. 13.10.

(*b*) By expanding ${}_1F_1[\alpha;\beta;-p(\theta)]$ in a power series about $p(0) = a_0^2$ and with the help of Eq. (1), obtain Eq. (13.51) for $|S_l|$ $(l \geq 0)$.

(*c*) Obtain Eq. (13.52) from the asymptotic expansion of ${}_1F_1[\alpha;\beta;-p(\theta)]$.

13.11 (*a*) Show that

$$I_{a_1,a_2}(\phi) \equiv \int_0^{2\pi} (1 + \lambda \cos \theta)^{a_1}[1 + \lambda \cos (\theta + \phi)]^{a_2} \frac{d\theta}{2\pi} \qquad 0 \leq |\lambda| \leq 1 \tag{1}$$

$$= \sum_{k=0}^{\infty} \epsilon_k \frac{(-a_1)_k(-a_2)_k(-\lambda)^{2k} \cos k\phi}{2^{2k}(k!)^2}$$

$$\times\, {}_2F_1\left(\frac{k-a_1}{2}, \frac{k-a_1+1}{2}; k+1; \lambda^2\right) {}_2F_1\left(\frac{k-a_2}{2}, \frac{k-a_2+1}{2}; k+1; \lambda^2\right) \tag{2}$$

(*b*) Show that

$$I_{-1,-1}(\phi) = \frac{1}{2\pi}\int_0^{2\pi} \{(1 + \lambda \cos \theta)[1 + \lambda \cos (\theta + \phi)]\}^{-1}\, d\theta \tag{3}$$

$$= \frac{1}{1-\lambda^2}\left[\frac{2(1 - a_\lambda{}^2 \cos \phi)}{1 - 2a_\lambda{}^2 \cos \phi + a_\lambda{}^4} - 1\right] \tag{4}$$

where $a_\lambda = (1 - \sqrt{1-\lambda^2})/4$, $|\lambda| \leq 1$.

HINT: Use the identity

$$ {}_2F_1\left[\frac{k+1}{2}, \frac{k+2}{2}; k+1; (\cosh x)^{-2}\right] = e^{-kx}(2\cosh x)^k \frac{\cosh x}{\sinh x} \tag{5}$$

What is $I_{-1,-1}(\phi)_{\lambda=1}$?

13.12 (*a*) For the detection problems discussed in Secs. 13.3(1), 13.3(2), but extended now to the half-wave νth-law detector, show that the limiting forms of the l-f background fluctuations for weak and strong carriers are, respectively,

$$a_0 \to 0: \qquad N_a{}^2(\nu) \doteq \frac{\beta^2\Gamma(\nu+1)^2\,\Delta f\,W_0{}^\nu\Gamma^{(0)}\omega_F{}^{\nu-1}\Gamma^{(2)}}{2^\nu\Gamma(\nu/2)^2} \qquad n \times n \tag{1}$$

$$a_0 \to \infty: \qquad N_a{}^2(\nu) \simeq \frac{\beta^2\Gamma(\nu+1)^2\,\Delta f\,W_0\Gamma^{(0)}\Gamma^{(1)}s_F{}^{2\nu-2}}{2^{\nu-1}\Gamma(\nu/2)^2\Gamma(\nu/2+1)^2} \qquad s \times n \tag{2}$$

where

$$\Gamma^{(j)} = \omega_F \int_0^\infty k_0(t)^j\,dt \qquad j \geq 1$$
$$\Gamma^{(0)} = \psi/W_0\omega_F \qquad j = 0 \tag{3}$$

W_0 is the maximum spectral intensity of the input noise, and the other symbols are defined in Sec. 13.3.

(*b*) Sketch $N_a{}^2(\nu)$ as a function of $a_0{}^2$ for $\nu > 1$, $\nu = 1$, $\nu < 1$, and explain the behavior.

13.4 *Other Rectification Problems*

Our attention so far has been confined to the more usual cases of simple (zero-memory) half-wave rectification. Occasionally, however, "clipping" (in amplitude) of the input wave is introduced in the nonlinear system either

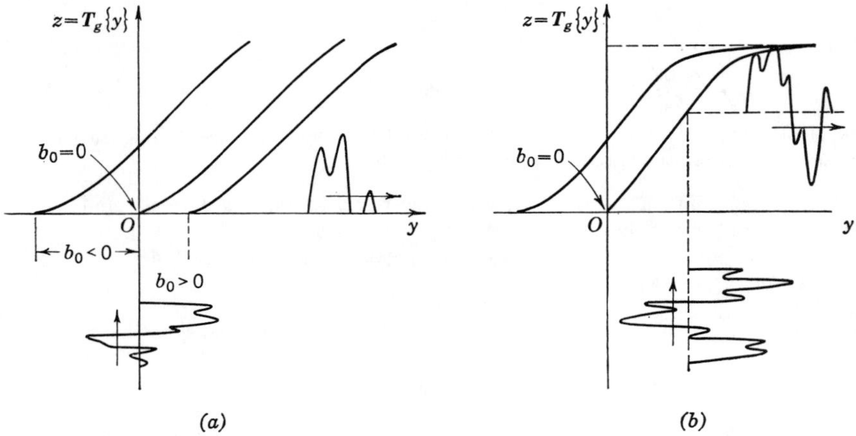

Fig. 13.16. (*a*) Zero-memory dynamic characteristic, illustrating amplitude clipping due to *bias*. (*b*) Zero-memory dynamic path, illustrating limiting due to *saturation*.

(1) because too negative a cathode or grid voltage with respect to the plate is allowed, resulting in a bias,† or (2) because of an inability to supply sufficient current, for the larger applied plate-to-cathode potentials, which

† We define *bias* here as any class C or AB operation (cf. Fig. 12.1) other than $b_0 = 0$.

results in *saturation*, or *limiting* of the input waves' instantaneous amplitude (cf. Fig. 13.16*a*, *b*). Sometimes this limiting is deliberately introduced, as in FM receivers (Chap. 15), in order to eliminate unwanted amplitude variations. In other instances, it may be due to unavoidable overloading of the system. A direct and simple generalization of the usual transform techniques in the half-wave cases provides us with the desired results here.[6,16] As before, clipping spreads the spectrum, whether because of bias or saturation, though the extent of these effects is naturally governed by the relation between the dynamic path and the excursions in amplitude of the incoming wave. Some of the principal results are given in Secs. 13.4-1, 13.4-2.

13.4-1 Bias. (1) Noise alone. As our first example let us consider the case of normal noise alone. From Eqs. (5.15*a*), (5.15*b*), the second-moment function of $z = T_g\{y\}$ is

$$M_z(t) = \sum_{n=0}^{\infty} \frac{\psi^n k(t)^n}{n!} h_{0,n}^2 \qquad \text{with } h_{0,n} = \frac{1}{2\pi} \int_C \xi^n f(i\xi) e^{-\psi^2/2}\, d\xi \tag{13.69a}$$

The associated intensity spectrum [cf. Eqs. (5.21)] is

$$\begin{aligned} \mathcal{W}_z(f) &= \sum_{n=0}^{\infty} (-1)^n \psi^n C_{0,n}(f) h_{0,n}^2 / n! \\ C_{0,n}(t) &= 4 \int_0^\infty k(t)^n \cos \omega t\, dt \qquad \omega = 2\pi f \end{aligned} \tag{13.69b}$$

These results apply equally well for broad- and narrowband noise processes. However, in the narrowband cases here, it is the individual spectral zones following rectification that are of chief interest, in particular the zones $l = 0, 1$ (cf. Sec. 5.1-3). Equations (5.27*b*), (5.27*c*) reduce now to

$$M_z(t)_l = (-1)^l (2 - \delta_l^0) \cos \omega_0 t \sum_{q=0}^{\infty} \frac{\psi^{2q+l} k_0(t)^{2q+l}}{2^{2q+l} q!(q+l)!} h_{0,2q+l}^2 \tag{13.70}$$

and for the l-f and "carrier" regions one gets specifically

$$\begin{aligned} M_z(t)_0 &= \sum_{q=0}^{\infty} \frac{[\psi k_0(t)]^q h_{0,2q}^2}{2^{2q} q!^2} \\ M_z(t)_1 &= -\cos \omega_0 t \sum_{q=0}^{\infty} \frac{[\psi k_0(t)]^{2q+1} h_{0,2q+1}^2}{2^{2q} q!(q+1)!} \end{aligned} \tag{13.71}$$

As before, $\overline{N^2} = \psi$, and $k_0(t)$ is the normalized covariance function for the slowly varying part of the input process.

The effect of the bias appears in the amplitude functions through $f(i\xi)$. For the biased νth-law detector, where specifically (cf. Table 2.1)

$$z = T_g\{y\} = \begin{cases} \beta(y - b_0)^\nu & y > b_0 \\ 0 & y < b_0 \end{cases}$$

we get

$$f(i\xi) = \frac{\beta\Gamma(\nu + 1)e^{-ib_0\xi}}{(i\xi)^{\nu+1}} \tag{13.72}$$

and the $h_{0,n}$ become with the help of Prob. 2.16c

$$h_{0,n} = \frac{\beta\Gamma(\nu + 1)i^{-n}}{2}\left(\frac{\psi}{2}\right)^{(\nu-n)/2}\left\{\frac{{}_1F_1[-(n - \nu)/2; \tfrac{1}{2}; -b^2/2]}{\Gamma[(2 + \nu - n)/2]} - \sqrt{2}\,b\,\frac{{}_1F_1[(n - \nu + 1)/2; \tfrac{3}{2}; -b^2/2]}{\Gamma[(1 + \nu - n)/2]}\right\} \qquad b = \frac{b_0}{\psi^{1/2}} \tag{13.73}$$

When ν is a (positive) integer, the hypergeometric functions are expressible in terms of $\phi^{(k)}(x) = (d^k/dx^k)(e^{-x^2/2}/\sqrt{2\pi})$, the derivatives of the error function (cf. Sec. A.1.1) according to Eqs. (A.1.21a), (A.1.21b). Figures 13.17a, b show some typical intensity spectra when normal noise alone is passed through a biased linear (zero-memory) rectifier ($\nu = 1$). The characteristic spectral spread is observed, with an accompanying drop in total intensity, attributable to the new modulation products generated by the nonlinear element. A large number of similar cases are also illustrated in Middleton.[6]

(2) **Carrier and noise.** Here we may use Eqs. (13.18), (13.19) directly, the only modification appearing in the explicit form of the amplitude functions (13.20). In the case of the biased νth-law device (13.72), these become (cf. Middleton,[2] sec. 4)

$$h_{m,n} = \frac{\beta\Gamma(\nu + 1)}{2\pi}\int_C \xi^n e^{-ib_0\xi-\psi\xi^2/2}\,\frac{J_m(A_0\xi)\,d\xi}{(i\xi)^{\nu+1}} \tag{13.74a}$$

$$= \frac{\beta\Gamma(\nu + 1)i^{-(m+n)}}{2n!}\left(\frac{\psi}{2}\right)^{(\nu-n)/2} a_0^m \times \sum_{k=0}^{\infty}\frac{(-1)^k(\sqrt{2}\,b)^k\,{}_1F_1[(k + m + n - \nu)/2; m + 1; -a_0^2]}{k!\Gamma[(2 + \nu - m - k - n)/2]} \tag{13.74b}$$

or

$$h_{m,n} = \frac{\beta\Gamma(\nu + 1)i^{-m-n}}{2}\left(\frac{\psi}{2}\right)^{(\nu-n)/2} a_0^m \sum_{k=0}^{\infty}\alpha_{mnk}a_0^k \qquad b \equiv \frac{b_0}{\psi^{1/2}};\ a_0^2 \equiv \frac{A_0^2}{2\psi} \tag{13.74c}$$

where

$$\alpha_{mnk} = \frac{1}{k!(k + m)!}\left\{\frac{{}_1F_1[(2k + m + n - \nu)/2; \tfrac{1}{2}; -b^2/2]}{\Gamma[(2 + \nu - 2k - m - n)/2]} - \frac{\sqrt{2}\,b\,{}_1F_1[(2k + m + n - \nu + 1)/2; \tfrac{3}{2}; -b^2/2]}{\Gamma[(1 + \nu - 2k - m - n)/2]}\right\} \tag{13.74d}$$

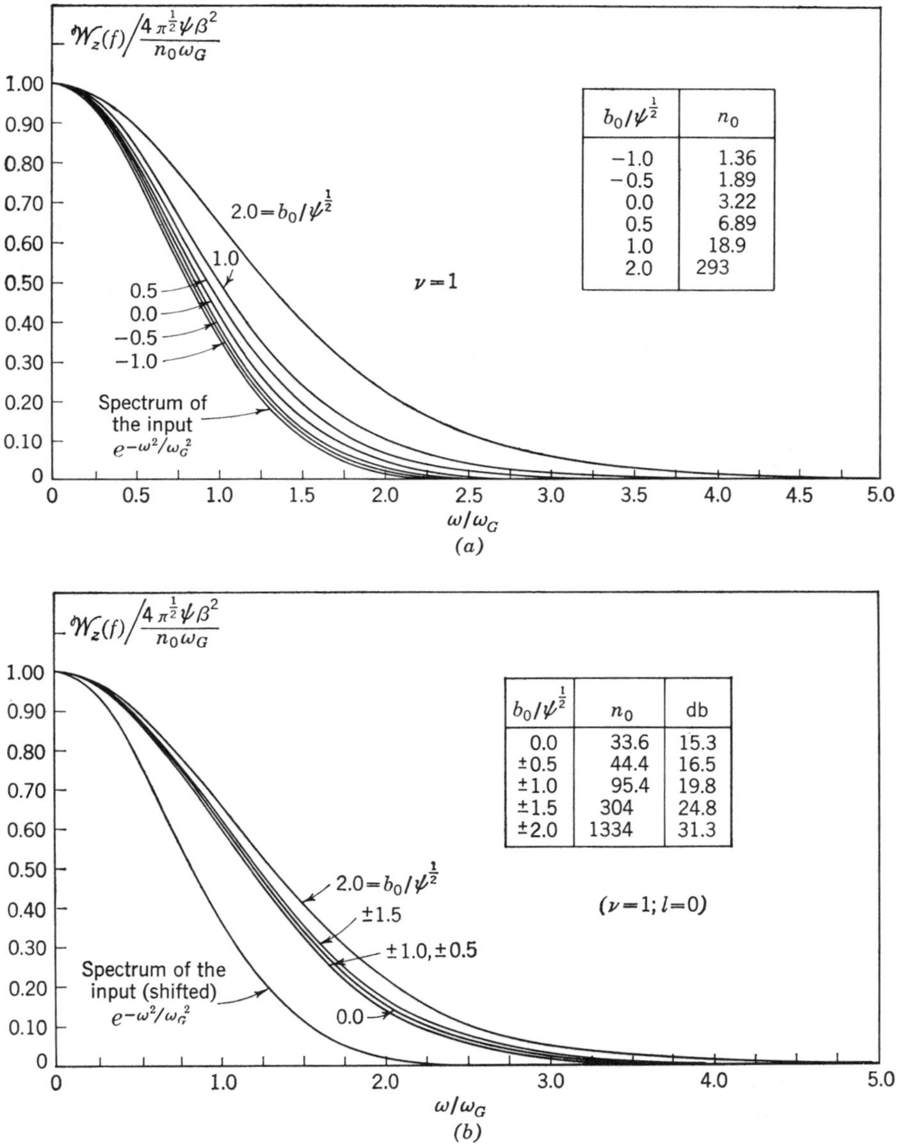

FIG. 13.17. (a) Normalized output spectra for broadband noise, from a biased linear rectifier ($\nu = 1$). (b) Normalized output spectra in the l-f zone ($l = 0$) following narrowband noise into a biased linear rectifier ($\nu = 1$).

Equation (13.74*b*) is particularly useful when b is reasonably small, while Eqs. (13.74*c*), (13.74*d*) are convenient when a_0^2 is small. For large values of the input signal-to-noise ratio (a_0^2), the asymptotic series for ${}_1F_1$ may be used [cf. Eq. (A.1.16) and Prob. 13.13]. We observe that, when $b \to -\infty$, the only rectification stems from the curvature of the dynamic path ($\nu \neq 1$), while, for $b \to \infty$, $h_{m,n}$ vanishes, as expected, since the device is then completely cut off.

Modifications similar to those in (1) for the narrowband inputs (carrier on-tune with the noise) follow directly from Eqs. (5.27), etc., for all spectral zones ($l \geq 0$). Equation (13.34) may be used here for the l-f cases, on replacing n by $m + 2q$ in Eqs. (13.74). The spectrum is obtained in the usual way from the Wiener-Khintchine theorem, and the same phenomenon of spectral broadening due to clipping is observed once again. The various modulation products described in Sec. 5.1-2 arise here also, and with essentially the same effect, although the degree of bias necessarily modifies their absolute contribution. The basic dependence on signal-to-noise ratio is similar to that indicated in Secs. 13.2, 13.3. Note that considerable analytic simplification results for the "even" detector laws ($\nu = 0, 2, 4, \ldots$), as then the sum over k [Eq. 13.74*b*)] terminates after a few terms. A number of other special cases are considered in Probs. 13.14, 13.15.

13.4-2 Saturation. (1) Noise, $S + N$, and modulated carriers. Here, we assume half-wave operation as before, but now with the restriction that the more positive amplitudes of the input are limited in some fashion (cf. Fig. 13.16*b*). A simple modification of the biased cases enables us to account for this effect in a direct way. We easily see that for a half-wave *saturated* νth-law detector

$$z = \begin{cases} \beta y^\nu & 0 < y < y_0 \\ \beta y_0^\nu & y > y_0 \\ 0 & y < 0 \end{cases} \tag{13.75a}$$

and that the transform of this (cf. type 4, Table 2.1) is

$$f(i\xi) = \beta\Gamma(\nu + 1)(1 - e^{-iy_0\xi})/(i\xi)^{\nu+1} \tag{13.75b}$$

The expressions derived earlier for the second-moment function and spectrum are altered only through the explicit changes in the amplitude functions $h_{0,n}$, $h_{m,n}$, etc. For example, when noise alone is the input to the device of Eqs. (13.75*a*), we get

$$h_{0,n} = \frac{\beta\Gamma(\nu + 1)i^{-n}}{2}\left(\frac{\psi}{2}\right)^{(\nu-n)/2}\left\{\frac{1 - {}_1F_1[(n - \nu)/2; \tfrac{1}{2}; -y_0^2/2\psi]}{\Gamma[(2 + \nu - n)/2]} + \sqrt{\frac{2}{\psi}}\, y_0 \frac{{}_1F_1[(n - \nu + 1)/2; \tfrac{3}{2}; -y_0^2/2\psi]}{\Gamma[(1 + \nu - n)/2]}\right\} \tag{13.76}$$

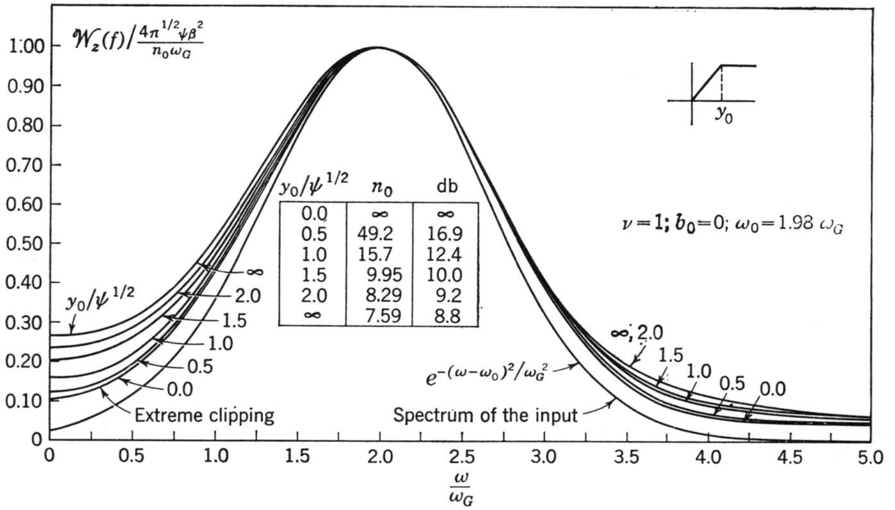

FIG. 13.18. Normalized intensity spectra of normal noise after half-wave linear rectification and saturation.

Figure 13.18 shows typical spectra for a half-wave linear device with saturation; again, the spectrum is spread relative to the unclipped case. We note, however, that saturation under these circumstances actually reduces this spectral broadening somewhat in the neighborhood of the spectral maximum. Extreme limiting ($y_0 \to 0$) has not only the least intensity of output ($\to 0$) but also the least spread, except for frequencies very far out on the "tails" of the spectrum, where the relatively larger numbers of h-f components make themselves felt.

When a carrier, as well as noise, is present (with both broad- and narrow-band spectra), for this nonlinear element we may use the amplitude functions of Eq. (13.74*d*), replacing b by $y_0/\psi^{1/2}$ and ${}_1F_1(—;\tfrac{1}{2};—)$ by $1 - {}_1F_1(—;\tfrac{1}{2};—)$ therein, and in Eq. (13.74*b*) by starting the summation at $k = 1$. A similar modification may be made directly for the noise and carrier of Sec. 13.2. Finally, both bias and saturation effects may be handled simultaneously, with appropriate additional modifications of the amplitude functions (cf. Prob. 13.15, for example).

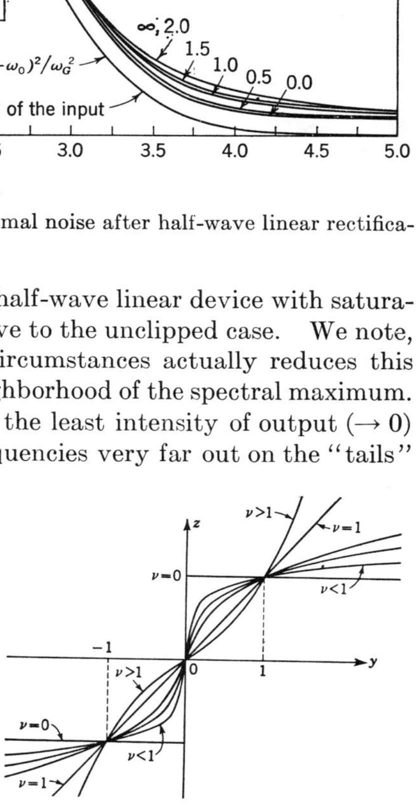

FIG. 13.19. Dynamic response of a band-pass symmetrical limiter (or antisymmetrical amplifier with distortion). ($\nu = 0$: ideal limiter.)

(2) Symmetrical limiters. As a final topic, let us examine the effects of an antisymmetrical rectifier, or *symmetrical limiter*, when $\nu < 1$. The dynamic characteristic is

$$z = \boldsymbol{T}_g\{y\} = g_o(y) = \begin{cases} \beta y^\nu & y > 0 \\ -\beta|y|^\nu & y < 0;\ \nu \geq 0 \end{cases} \tag{13.77}$$

as sketched in Fig. 13.19. Here, we properly speak of limiting action only when $\nu < 1$. When $\nu = 1$, the system is effectively an ideal linear amplifier, while, for $\nu > 1$, it becomes in effect an amplifier with distortion. From the results of Prob. 2.12, we see that here $g_e(y)$ vanishes and only the odd characteristic $g_o(y)$ remains, so that the output in this instance of an antisymmetric dynamic response is

$$z = \frac{i}{\pi} \int_C f(i\xi) \sin \xi y \, d\xi \tag{13.78}$$

where $f(i\xi) = \beta\Gamma(\nu + 1)/(i\xi)^{\nu+1}$, as before, and the second-moment function for the sum of an input signal and normal noise process becomes for signals and noise possessing symmetrical distributions (cf. Prob. 13.17)

$$M_z(t)_{S+N} = \frac{1}{2\pi^2} \iint\limits_C f(i\xi_1)f(i\xi_2)[F_2(\xi_1,\xi_2;t)_{S+N} - F_2(\xi_1, -\xi_2;t)_{S+N}] \, d\xi_1 \, d\xi_2 \tag{13.79}$$

We find, developing $M_z(t)_{S+N}$ as a double series in (m,n), following Eq. (13.18), that the terms for which $m + n$ (in the broadband case) is *odd* remain, while the even sums drop out. The covariance function here is

Broadband

$$K_z(t)_{S+N} = -4 \sum_{m+n=1,3,4,\ldots}^{\infty} \frac{\epsilon_m h_{m,n}^2}{n!} [\psi k(t)]^n \cos m\omega_0 t \tag{13.80}$$

with $h_{m,n}$ given by Eqs. (13.19), (13.20) for the νth-law response above. For noise alone, this reduces to

$$K_z(t)_N = -4 \sum_{n=0}^{\infty} \frac{[\psi k(t)]^{2n+1}}{(2n+1)!} h_{0,2n+1}^2 \tag{13.81a}$$

$$K_z(t)_N = \frac{\psi^\nu}{2^{\nu-1}} k(t)\beta^2 \frac{\Gamma(\nu+1)^2}{\Gamma[(\nu+1)/2]^2} \, {}_2F_1\left[\frac{1-\nu}{2};\frac{1-\nu}{2};\frac{3}{2};k(t)^2\right] \tag{13.81b}$$

which reduces still further to $\psi k(t)\beta^2$ when $\nu = 1$, as then one simply has an ideal linear amplifier. The total intensity of the process [cf. Eqs. (13.81)] is easily found to be $\beta^2(2\psi)^\nu\Gamma(\nu + \frac{1}{2})/\sqrt{\pi}$.

The *narrowband cases* are handled in a similar way. From Eqs. (5.27) applied to Eq. (13.79), we find that only the *odd* zones appear in the output, as expected, and the covariance functions of these regions are specifically†

$$K_z(t)_{2l+1}\Big|_{S+N} = -4\cos(2l+1)\omega_0 t \sum_{m=0}^{\infty} \epsilon_m \sum_{q=0}^{\infty} \frac{[\psi k_0(t)]^{2q}}{2^{2q}q!}$$
$$\times \left[\frac{(\psi k_0)^{|m-2l-1|}h^2_{m,|m-2l-1|+2q}}{2^{|m-2l-1|}(q+|m-2l-1|)!} + \frac{(\psi k_0)^{m+2l+1}h^2_{m,m+2q+2l+1}}{2^{m+2l+1}(q+m+2l+1)!}\right] \qquad l = 0, 1, 2, \ldots \quad (13.82)$$

and the $h_{m,2q+l'}$ $(l' = |m \pm 2l \pm 1|)$ are obtained from Eqs. (13.19), (13.20) in the usual way. The first harmonic zone gives specifically

$$K_z(t)_1\Big|_{S+N} = \Gamma(\nu+1)^2\beta^2\left(\frac{\psi}{2}\right)^{\nu}\cos\omega_0 t \sum_{m=0}^{\infty}\frac{\epsilon_m a_0{}^{2m}}{(m!)^2}\sum_{q=0}^{\infty}\frac{k_0(t)^{2q}}{q!}$$
$$\times\left\{\frac{k_0(t)^{|m-1|}\,{}_1F_1{}^2[(m+|m-1|+2q-\nu)/2;\,m+1;\,-a_0{}^2]}{(q+|m-1|)!\Gamma^2[(2+\nu-m-|m-1|-2q)/2]} + \frac{k_0(t)^{m+1}\,{}_1F_1{}^2[(1-\nu)/2+m+q;\,m+1;\,-a_0{}^2]}{(q+m+1)!\Gamma^2[(1+\nu)/2-m-q]}\right\} \quad (13.83)$$

For noise alone, Eqs. (13.82) and (13.83) reduce readily to

$$K_z(t)_{2l+1}\Big|_N = 2\beta^2\frac{(2\psi)^{\nu}}{\pi}\left(\frac{1-\nu}{2}\right)_l^2\frac{\Gamma(\nu/2+1)^2}{(2l+1)!}k_0{}^{2l+1}$$
$$\times\,{}_2F_1(l+\tfrac{1}{2}-\nu/2,\,l+\tfrac{1}{2}-\nu/2;\,2l+2;\,k_0{}^2)\cos(2l+1)\omega_0 t \quad (13.84)$$

and

$$K_z(t)_1\Big|_N = 2\beta^2\frac{(2\psi)^{\nu}}{\pi}\Gamma\left(\frac{\nu}{2}+1\right)^2 k_0(t)\cos\omega_0 t\,{}_2F_1\left(\frac{1-\nu}{2},\frac{1-\nu}{2};2;k_0{}^2\right) \quad (13.85)$$

The mean intensity of the noise in the first zone is here

$$K_z(0)_1\Big|_N = \frac{2^{1-\nu}\psi^{\nu}\beta^2\Gamma(\nu+1)^3}{\Gamma[(\nu+1)/2]^2\Gamma[(\nu+3)/2]^2} \quad (13.85a)$$

We may now compute the signal-to-noise ratios, as in Sec. 13.2-2. For *broadband waves* and unmodulated carriers, we define the output signal to be the *first* harmonic $(m = 1, n = 0)$ of the carrier after rectification, since there is no d-c term, owing to the asymmetry of the dynamic path. The signal intensity is therefore

$$S^2_{out} = -8h^2_{1,0} = \frac{2\beta^2\Gamma(\nu+1)^2}{\Gamma[(1+\nu)/2]^2}\left(\frac{\psi}{2}\right)^{\nu}a_0{}^2\,{}_1F_1{}^2\left(\frac{1-\nu}{2};2;-a_0{}^2\right) \quad (13.86)$$

† The noise spectrum is also assumed to be symmetrical about f_0; that is, $\phi_K(t) = 0$ in Eq. (5.27*b*). If, however, the noise is not spectrally symmetrical, we can generalize these results directly with the help of Eqs. (5.27).

The background noise accompanying the signal follows from Eq. (13.80), for all components of order $m + n \geq 1, 3, \ldots$ $(n \geq 1)$. A narrow filter, of effective width Δf, centered about f_0, is used to reduce the accompanying fluctuation. The background noise intensity $\overline{N^2}$ is then found from Eq. (13.80), as in Eqs. (13.30). For weak signals, the output signal-to-noise (amplitude) ratio becomes

$$\left(\frac{S}{N}\right)_{out} \doteq \frac{a_0/(2\,\Delta f^{1/2})}{\left\{\int_0^\infty k(t)\,{}_2F_1\left[\frac{1-\nu}{2}, \frac{1-\nu}{2}; \frac{3}{2}; k(t)^2\right] \cos \omega_0 t\,dt\right\}^{1/2}} + O(a_0^2) \quad (13.87)$$

the important feature of which is the absence of the signal suppression [cf. Sec. 13.2-1(4)] which occurs in the usual half-wave detection, e.g., as $(S/N)_{out} \sim a_0^2$. This different behavior is attributable to the relative increase in the $s \times s$ contribution *at the carrier frequency* f_0 over the contribution to the d-c terms (representing the signal increment) in the half-wave cases.

When the signal is intense relative to the noise, Eq. (13.86) becomes

$$S_{out}^2 \simeq \frac{2\beta^2\Gamma(\nu+1)^2}{\Gamma[(\nu+1)/2]^2\Gamma[(\nu+3)/2]^2}\left(\frac{A_0}{2}\right)^{2\nu} + O(\psi^\nu a_0^{2\nu-2}) \quad (13.88)$$

showing that the output signal strength is now independent of the noise, as in the half-wave cases [cf. Eq. (13.28*d*)]. The signal-to-noise (amplitude) ratio here is

$$\left(\frac{S}{N}\right)_{out} \simeq \frac{a_0\Gamma[(\nu+1)/2]/(2\sqrt{\Delta f})}{\left[\int_0^\infty k(t)\cos\omega_0 t\,dt\right]^{1/2}} \quad (13.89)$$

and now the characteristic linear relation between $(S/N)_{out}$ and a_0 is observed [cf. Eq. (13.31*b*)]. Again, it is the $s \times n$ noise terms which predominate in the interference.

A similar analysis for the *narrowband* cases, when attention is focused on the first spectral zone, shows that the signal contribution [$(m = 1, q = 0)$ and $k_0(t)^{(0)}$] in Eq. (13.83) is just once more Eq. (13.86). The noise terms follow for all other values of m, q, where $k_0(t)$ appears to the first or higher power in the covariance function. If now we use once again a narrowband filter, broad enough, however, to pass the entire first zone without attenuation, but not so wide as to admit any of the higher zones, we see that $\overline{N_{ou}^2}$ follows directly from Eq. (13.85) (with $t = 0$ and the signal term excluded). For weak signals, the analogue of Eq. (13.87) reduces to

$$\left(\frac{S}{N}\right)_{out} \doteq a_0 \left\{\frac{\Gamma[(\nu+3)/2]}{\Gamma(\nu+1)^{1/2}} + O(a_0^2)\right\} \quad (13.90)$$

and when the signal is strong, the leading terms for the noise occur for

$(m = q = 0)$ and $(m = 2, q = 0)$. We obtain

$$\overline{N^2} \simeq 2\beta^2\Gamma(\nu + 1)^2 \left(\frac{\psi}{2}\right)^\nu a_0^{2\nu-2} \left[\Gamma\left(\frac{\nu+1}{2}\right)^{-4} + \Gamma\left(\frac{\nu-1}{2}\right)^{-2} \Gamma\left(\frac{\nu+3}{2}\right)^{-2}\right] \tag{13.90a}$$

and so

$$\left(\frac{S}{N}\right)_{out} \simeq \left(\frac{2a_0^2}{\nu^2+1}\right)^{1/2} [1 + O(a_0^2)] \tag{13.91}$$

Note that, in both instances above, $(S/N)_{out} = (S/N)_{in} = a_0$ when $\nu = 1$, as it should for linear amplification. Again, we see that here there is no

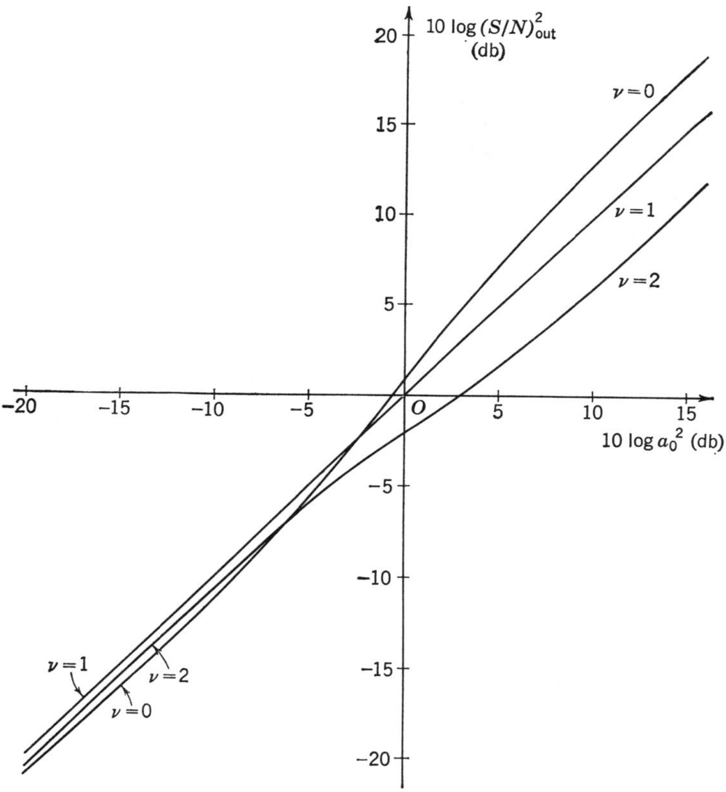

FIG. 13.20. The output signal-to-noise ratio for symmetrical bandpass limiters, when carrier and narrowband noise constitute the input. (Antisymmetrical "rooters.")

signal suppression, as there would be if we had only a half-wave device, since now the $s \times s$ contribution is increased relative to the background noise *at the carrier frequency zone.* Figure 13.20 shows how the output signal-to-noise ratio depends on a_0 for various limiter laws ($\nu \geq 0$). For the ideal limiter ($\nu = 0$), we see that

$$\left(\frac{S}{N}\right)_{out} \doteq \frac{\sqrt{\pi}\, a_0}{2} \qquad a_0^2 \ll 1 \tag{13.92a}$$

while, for strong signals, we get

$$\left(\frac{S}{N}\right)_{out} \simeq \sqrt{2}\, a_0 \qquad a_0^2 \gg 1 \tag{13.92b}$$

a result in agreement with Tucker[17] and Davenport,[16] showing that for very intense signals the output intensity ratio (divided by a_0^2) is increased by about 4.1 db† over the weak-signal case. For ν greater than unity, the difference between these ratios for the strong- and weak-signal cases is progressively reduced [cf. Eq. (13.91)]; the scale of $(S/N)_{out}$ varies as ν^{-1}, $\nu \gg 1$.

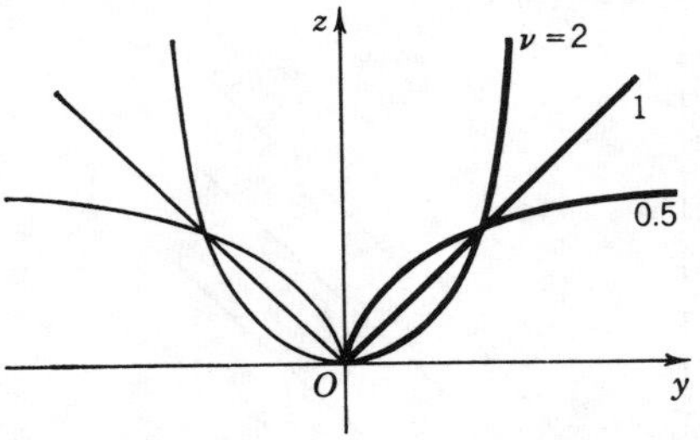

FIG. 13.21. Typical full-wave symmetrical (zero-memory) dynamic characteristics ($\nu > 0$). (Heavy curves ≡ half-wave responses.)

13.4-3 Full-wave Rectification. While most applications of nonlinear elements in communication theory require half-wave (or biased) nonlinear devices, in some operations, such as certain types of limiting, power rectification, saturated amplifiers, and small- and large-signal detection, symmetrical full-wave rectification may occur or may be used. Typical dynamic characteristics are shown in Fig. 13.21.

The analysis of these symmetrical systems, whose dynamic responses are given by

$$z = \boldsymbol{T}_g\{y\} = \beta|y|^{\nu} \qquad -\infty < y < \infty;\ \nu > 0 \tag{13.93}$$

is readily carried out as an extension of the half-wave theory of Secs. 13.2, 13.3. Here, the calculation of $M_z(t)$ for the output process follows the treatment of Stone and Middleton.[18] Thus, for the symmetrical full-wave devices to be discussed in this section, we make use of the results of Prob. 2.12 to write for the output process of Eq. (13.93)

$$z = \frac{1}{\pi} \int_C f(i\xi) \cos \xi y \, d\xi \tag{13.94}$$

where again $f(i\xi)$ is the Fourier transform of the corresponding *half-wave* response. [For the antisymmetrical, or limiter, responses, see Sec. 13.4-2(2).]

When noise alone enters such a system, the second-moment function of

† $(2 \div \pi/4)$ in decibels = 4.1 db.

the output is† readily found to be

$$M_z(t) = \frac{1}{2\pi^2} \iint_C f(i\xi_1)f(i\xi_2)[F_2(\xi_1,\xi_2;t)_N + F_2(\xi_1,-\xi_2;t)_N]\, d\xi_1\, d\xi_2 \quad (13.95)$$

while, for (independent) input signal and noise processes, we have only to replace $(F_2)_N$ by $(F_2)_N(F_2)_S$ [cf. Eq. (5.4)] and proceed with the analysis as in Secs. 13.2, 13.3. Some modifications of the earlier results may be expected to occur, however, as we shall observe directly with the following examples.

(1) Broadband noise; signal and noise. With (broadband) gaussian noise alone, the characteristic function is given by Eq. (5.6a), and so Eq. (13.95) becomes

$$M_z(t)_N = \frac{1}{\pi^2} \iint_C d\xi_1\, d\xi_2\, f(i\xi_1)f(i\xi_2)e^{-\psi(\xi_1^2+\xi_2^2)/2} \cosh [\psi\xi_1\xi_2 k(t)] \quad (13.96)$$

which in terms of the amplitude functions (5.37), (5.38a) reduces to

$$M_z(t)_N = 4 \sum_{n=0}^{\infty} \frac{[\psi k(t)]^{2n}}{(2n)!} h_{0,2n}^2 \quad (13.97a)$$

$$M_z(t)_N = \beta^2 \Gamma\left(\frac{\nu+1}{2}\right)^2 \frac{(2\psi)^\nu}{\pi}\, {}_2F_1\left[-\frac{\nu}{2}, -\frac{\nu}{2}; \frac{1}{2}; k(t)^2\right] \quad (13.97b)$$

Comparison with Eqs. (13.4) for the half-wave case shows that the "odd" terms here are missing, as we might expect from the symmetry of the response.

The addition of a signal is effected in the usual way (Sec. 13.2). Here, we use for the characteristic function of a simple sinusoidal signal Eqs. (5.12), suitably modified; Eq. (13.95) accordingly becomes

$$M_z(t)_{S+N} = \left\{\begin{matrix} 2 \sum\limits_{m,n=0} [(-1)^{m+n} + 1] \\ \text{or} \\ 4 \sum\limits_{m+n=0,2,4} \end{matrix}\right\} \frac{\epsilon_m h_{m,n}^2}{n!} [\psi k(t)]^n \cos m\omega_0 t \quad (13.98)$$

where now the amplitude functions $h_{m,n}$ are specifically given by Eq. (13.20). For more complicated signals (e.g., periodic modulation of a carrier), the results of Sec. 13.3 may be applied directly.

(2) Narrowband noise; signal and noise. Again we have a simple modification of the analysis of Sec. 13.2-2. When the input involves only

† Equation (13.95) applies only for noise (and signals) having symmetrical amplitude distributions, so that $F_2(-\xi_1,\xi_2;t) = F_2(\xi_1,-\xi_2;t)$, etc. For a more general situation, see Prob. 13.17.

narrowband noise, Eq. (13.95) is modified to exhibit a zonal structure and the second-moment function of a particular zone is

$$M_z(t)_{2l} = \beta^2 \frac{(2\psi)^\nu}{\pi} \Gamma\left(\frac{\nu+1}{2}\right)^2 \epsilon_{2l} \left(-\frac{\nu}{2}\right)_l^2 \frac{k_0(t)^{2l}}{(2l)!} \times {}_2F_1(l - \nu/2, l - \nu/2; 2l + 1; k_0^2) \cos 2l\omega_0 t \quad (13.99)$$

[cf. Prob. 13.1, Eq. (1)]. Only the even zones are present again: the odd harmonics drop out because of the symmetry of the dynamic response. The output has as its fundamental twice the (central) frequency f_0 of the input disturbance. As usual, the spectral zone of chief interest occurs for $l = 0$, and one can write for the corresponding l-f term

$$M_z(t)_0 = \beta^2 \frac{(2\psi)^\nu}{\pi} \Gamma\left(\frac{\nu+1}{2}\right)^2 {}_2F_1\left[-\frac{\nu}{2}, -\frac{\nu}{2}; 1; k_0^2(t)\right] \quad (13.100)$$

which, except for the factor 4, is identical with the half-wave expression (13.8).

When a sinusoidal signal is added to the narrowband input noise, the extension of Eq. (13.95) is made directly with the help of the resolution of $M_z(t)$ into spectral zones, following Eqs. (5.27). We have finally [again for symmetrical noise spectra, $\phi_K = 0$ in Eq. (5.27b)]

$$K_z(t)_{2l} = 4 \cos l\omega_0 t \sum_{m,q=0}^{\infty} \frac{\epsilon_m (\psi k_0)^{2q}}{q! 2^{2q}} \left[\frac{h^2_{m,|m-2l|+2q} (\psi k_0)^{|m-2l|}}{(q + |m - 2l|)! 2^{|m-2l|}} + (1 - \delta^0_{2l}) \frac{h^2_{m,m+2l+2q} (\psi k_0)^{m+2l}}{(q + m + 2l)! 2^{m+2l}} \right] \quad (13.101)$$

and $h_{m,|m\pm 2l|+2q}$ is given by Eq. (13.20).

The l-f zone becomes specifically

$$M_z(t)_0 = \beta^2 \frac{(2\psi)^\nu}{\pi} \Gamma\left(\frac{\nu+1}{2}\right)^2 \sum_{m,q=0}^{\infty} \epsilon_m a_0^{2m} k_0(t)^{m+2q} \times \left(-\frac{\nu}{2}\right)_{m+q}^2 \frac{{}_1F_1^2(m + q - \nu/2; m + 1; -a_0^2)}{(m!)^2 q! (m + q)!} \quad (13.102)$$

Observe that, if the carrier is modulated, Eqs. (13.101), (13.102) can be extended at once, with the help of Eq. (13.34), to give Eqs. (13.46), (13.47) for a periodic modulation, with only a change of scale: the half-wave expressions are multiplied by a factor 4. This factor 4 follows in the l-f case because we are doubling the instantaneous envelope of the narrow band in the process of full-wave rectification.

(3) **Signal-to-noise ratios.** We consider first only the simplest situation of an unmodulated carrier, as in Sec. 13.2. As before, the output signal

is taken to be the d-c increment [Eqs. (13.30)], which becomes here from Eq. (13.98) for the *broadband cases*

$$(\bar{S})_{out} = \frac{(2\psi)^{\nu/2}}{\sqrt{\pi}}\,\beta\Gamma\left(\frac{\nu+1}{2}\right)\left[{}_1F_1\left(-\frac{\nu}{2};1;-a_0^2\right)-1\right] \tag{13.103}$$

The mean-square fluctuation follows from Eq. (13.98) when the term $m + n = 0$ is excluded and when Eqs. (13.30) are used. This l-f fluctuation, due to the finite width of the smoothing filter in the output, is specifically

$$\begin{aligned}\overline{N^2} = \Delta f\, \mathcal{W}_z(0) &= 4\,\Delta f\,\beta^2\Gamma\left(\frac{\nu}{2}+1\right)^2\Gamma\left(\frac{\nu+1}{2}\right)^2\frac{(2\psi)^\nu}{\pi} \\ &\times \sum_{\substack{m+n=2,4,\ldots\\ n>0}}^{\infty} \epsilon_m 2^n a_0^{2m}\int_0^\infty k(t)^n \cos m\omega_0 t\,dt \\ &\times \frac{{}_1F_1[(m+n-\nu)/2;\,m+1;\,-a_0^2]^2}{(m!)^2 n!\,\Gamma[(\nu+2-m-n)/2]^2}\end{aligned} \tag{13.104}$$

The signal-to-noise ratio is formed in the usual way [cf. Eq. (13.31)]. Note that, as the law of the detector increases, the output ratio is reduced because the background fluctuations are enhanced owing to the heavier weighting of the more positive noise peaks compared with the (bounded) sinusoidal signal. Nevertheless, for any ν (> 0), as $a_0 \to \infty$, so does $(S/N)_{out} \to \infty$ as expected: the d-c increment increases more rapidly than the noise because of the coherence of the $s \times s$ terms vis-à-vis the incoherent $s \times n$ noise of the background. We remark, finally, that no simple numerical relation exists between the half- and the full-wave situations, unlike the case of the l-f output in narrowband rectification [Eqs. (13.100) and (13.102)].

The signal-to-noise ratio for *narrowband inputs* reduces to the cases already examined in Secs. 13.2-2, 13.3(2), even for more general modulations. However, for a sinusoidal or periodic signal process, we note a difference vis-à-vis the corresponding half-wave system. This difference depends on how we define and observe the output signal. For example, if we take the output signal to be a d-c increment, there is then a 3-db improvement in $(S/N)_{out}$ for the full-wave system. This occurs because the *d-c amplitude* is doubled [cf. Eq. (13.103)], while the accompanying noise fluctuation, being incoherent, is increased in the average by $\sqrt{2}$ only. On the other hand, if the output signal is taken to be a (nonzero) *harmonic* ($l \geq 1$) *of the input signal's repetition frequency*, the rms value of this harmonic is increased by $\sqrt{2}$ only [Eq. (13.48*a*)], for example, S_0 versus $|S_l|$, as is also the noise in full-wave operation, so that there is now no difference in the levels of $(S/N)_{out}$ between half- and full-wave systems. The possible improvement of 3 db in performance of the full-wave device when increments are observed must, of course, be weighed against the requirement of additional circuit elements and the difficulty with which local d-c drifts can be controlled.

13.4-4 A Method of Residues. We have seen from the preceding sections that a central problem is the calculation of the amplitude functions $h_{0,n}$, $h_{m,n}$, $H_{m,n}$, etc., which appear in the expressions for the second-moment function of the rectified process. If there is modulation, the direct treatment is almost always very complicated because of the additional averages [cf. Eq. (5.27*b*)]. For biased and saturated characteristics, further complexities may also arise, due to the structure of the dynamic path. Under these conditions, then, we are often faced with a difficult technical problem when a specific result is required. The difficulties can to a large extent be resolved, or at least greatly lessened, with the help of the following analytic device, which attempts to represent in useful form the many series and transcendental functions that appear.

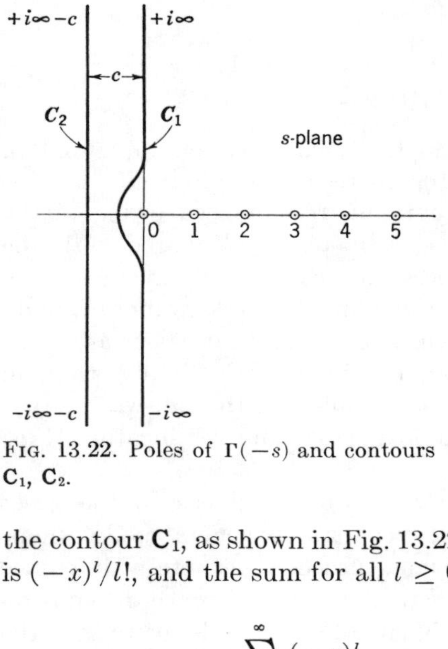

FIG. 13.22. Poles of $\Gamma(-s)$ and contours $\mathbf{C}_1$, $\mathbf{C}_2$.

The essence of the procedure is to express series and functions as integral representations of the Mellin-Barnes[19] type, involving the gamma function $\Gamma(-s)$. The method is based on the property of the gamma function $\Gamma(-s)$ that it has only simple poles at $s = 0, 1, 2, \ldots, l, \ldots$, of residue $(-1)^l/l!$, and that

$$\int_{\mathbf{C}} \Gamma(-s) \frac{ds}{2\pi i}$$

converges everywhere to the right of the contour $\mathbf{C}_1$, as shown in Fig. 13.22. Thus, the residue of $\Gamma(-s)x^s$ at $s = l$ is $(-x)^l/l!$, and the sum for all $l \geq 0$ can be written at once as

$$\sum_{l=0}^{\infty} \frac{(-x)^l}{l!} = e^{-x} = \int_{-\infty i}^{\infty i} \Gamma(-s) x^s \frac{ds}{2\pi i} \tag{13.105}$$

where it is now understood that $(-\infty i, \infty i)$ represents the contour $\mathbf{C}_1$, including the pole at the origin. A more sophisticated example is provided by the Bessel function $J_\nu(ax)$,

$$J_\nu(ax) = \sum_{l=0}^{\infty} \frac{(ax/2)^{2l+\nu}}{\Gamma(\nu + l + 1)l!} (-1)^l = \int_{-\infty i - c}^{\infty i - c} \frac{\Gamma(-s)}{\Gamma(\nu + s + 1)} \left(\frac{ax}{2}\right)^{2s+\nu} \frac{ds}{2\pi i} \tag{13.106}$$

where now c (> 0) is chosen to ensure convergence along the contour. For $\operatorname{Re} \nu > 0$, c may be set equal to zero.

Still more involved expressions may be deduced by a slight modification of the technique. Consider, for example, the integral representation

$$J_\nu(az)e^{-q^2z^2} = \frac{(a/2)^\nu}{\Gamma(\nu+1)} \int_{-\infty i - c}^{\infty i - c} \Gamma(-s)\, {}_1F_1\left(-s; \nu+1; -\frac{a^2}{4q^2}\right) z^{2s+\nu} q^{2s} \frac{ds}{2\pi i} \tag{13.107a}$$

$$= \left(\frac{az}{2}\right)^\nu \sum_{l=0}^{\infty} \frac{(izq)^{2l}}{l!\Gamma(\nu+1)}\, {}_1F_1\left(-l; \nu+1; -\frac{a^2}{4q^2}\right) \qquad \nu \neq -1, -2, \ldots \tag{13.107b}$$

To obtain this, we observe first that $\Gamma(m-s)(qx)^{2s}$ has residues of value $(-1)^l(qx)^{2m+2l}/l!$ at the poles $s = m + l$ ($l = 0, 1, \ldots$). The sum of these residues is therefore $(qx)^{2m}e^{-q^2x^2}$, and so expanding the Bessel function and taking the mth term allows us to write

$$\frac{(az/2)^\nu(aiz/2q)^{2m}e^{-q^2z^2}}{m!\Gamma(\nu+m+1)} = \frac{(az/2)^\nu(ai/2q)^{2m}}{m!\Gamma(\nu+m+1)} \int_{-\infty i - c}^{\infty i - c} \Gamma(m-s)(qz)^{2s} \frac{ds}{2\pi i} \tag{13.108}$$

Summing over m with the aid of $\Gamma(m-s) = \Gamma(-s)(-s)_m$, we have

$$\left(\frac{az}{2}\right)^\nu \sum_{m=0}^{\infty} \frac{\Gamma(m-s)(ai/2q)^m}{m!\Gamma(m+\nu+1)} = \left(\frac{az}{2}\right)^\nu \frac{\Gamma(-s)}{\Gamma(\nu+1)}\, {}_1F_1\left(-s; \nu+1; -\frac{a^2}{4q^2}\right) \tag{13.109}$$

which leads to Eq. (13.107a). In this manner, the well-known result of Kummer[19] is also easily established, viz.,

$$J_\nu(z)e^{-iz} = (z/2)^\nu\, {}_1F_1(\nu + \tfrac{1}{2}; 2\nu+1; 2iz)/\Gamma(\nu+1) \tag{13.110}$$

An extension of this, which is of some use in biased rectification, may be obtained in similar fashion. We have

$$J_\nu(az)e^{ibz} = \frac{(az/2)^\nu}{\Gamma(\nu+1)} \int_{-\infty i - c}^{\infty i - c} \Gamma(-s)\, {}_2F_1\left(-\frac{s}{2}, \frac{1-s}{2}; \nu+1; \frac{a^2}{b^2}\right) (ibz)^s \frac{ds}{2\pi i} \tag{13.111}$$

Evaluating Eq. (13.111), for instance, leads to

$$J_\nu(az)e^{ibz} = \frac{(az/2)^\nu}{\Gamma(\nu+1)} \sum_{l=0}^{\infty} \frac{(-ibz)^l}{l!}\, {}_2F_1\left(-\frac{l}{2}, \frac{1-l}{2}; \nu+1; \frac{a^2}{b^2}\right) \tag{13.112}$$

which is one generalization of Kummer's relation (13.110). It should be emphasized that the *function* ${}_2F_1$ in the integrand of Eq. (13.111) is convergent for all finite a^2/b^2, $|\arg (a^2/b^2)| < \pi - \delta$, while the usual hypergeometric *series* in general is not, unless $|a^2/b^2| < 1$. Here, however, the question of convergence does not arise, since the series (13.112) terminates.

To see how these integral representations may simplify the analytical operations, let us consider the case of a general amplitude modulation followed by rectification. The amplitude functions $H_{m,n}(t)$ [Eqs. (5.16*b*), (5.17*b*)], with A_0 replaced by $A_0(t)$, contain a Bessel function $J_m[A_0(t)\xi]$, and the average over the epochs of the modulation now requires the evaluation of terms of the type $\overline{J_m(A_1\xi_1)J_m(A_2\xi_2)}$, or $\langle J_m(A_1\xi_1)J_m(A_2\xi_2)\rangle$. The average is further complicated by the fact that $A(t)$ may vanish for some values of t, as in most cases overmodulation leads to zero amplitudes† (cf. Sec. 12.1). With the help of Eq. (13.106), the average over the Bessel functions is reduced to that over the product $\langle A(t_1)^{2s+m}A(t_2)^{2t+m}\rangle$, and, if necessary, to account for the effects of overmodulation, we can use the familiar representation of Eq. (12.13) to write

$$\langle A_0(t_1)^{2s+m}A_0(t_2)^{2t+m}\rangle = \frac{\Gamma(2s+m+1)\Gamma(2t+m+1)}{4\pi^2} \times \int_C \int_C \frac{\langle e^{i\lambda_1 A_0(t_1)+i\lambda_2 A_0(t_2)}\rangle}{(i\lambda_1)^{2s+m+1}(i\lambda_2)^{2t+m+1}}\, d\lambda_1\, d\lambda_2 \qquad (13.113)$$

where $\langle e^{i\lambda_1 A_1+i\lambda_2 A_2}\rangle$ is the characteristic function of the modulation and the integrations over λ_1 and λ_2 are carried out before those over s and t, etc. Similar techniques may be used for angle modulation (Chap. 14), saturated and biased rectification, and where modulated carriers undergo general nonlinear operation, such as arises, for instance, when an amplitude-modulated carrier is passed through an FM receiver, and vice versa. An example follows presently.

Frequently, in the course of evaluating contour integrals like Eqs. (13.107), (13.108), etc., double as well as single poles are encountered. As an example, let us evaluate

$$L = \frac{1}{2\pi i}\int_{-\infty i}^{\infty i} \Gamma(-s)^2 F(s)\, ds \qquad (13.114)$$

where $F(s)$ contains no poles or other singularities in the right-half complex plane Re $s \geq 0$. The poles of the integrand are now double, e.g., at $s = l = 0, 1, 2, \ldots$, in $\Gamma(-s)^2$. From the Laurent expansion, the residues at $s = l$ are given by

$$R_l = \left\{\frac{d}{ds}[(-s+l)^2\Gamma(-s)^2F(s)]\right\}_{s=l\geq 0} \qquad (13.115)$$

† This is true of most electronic transmission systems, where the amplitude of an oscillator is varied by the modulation; an exception is the case of amplitude perturbations caused by ionospheric changes in the course of propagation. Here, a typical input is $V_{in} = N_{in} + M(t)S(t)$, where $M(t)$ is the modulation of the original signal $S(t)$, which now may assume negative, as well as positive, values, unlike $A_0(t)$ above.

Carrying out the indicated differentiations and limiting operations in Eq. (13.115), we get finally

$$L = \sum_{l=0}^{\infty} \frac{1}{(l!)^2} [F'(l) - 2\psi(l+1)F(l)] \tag{13.116}$$

where $\psi(z)$ is the logarithmic derivative of the Γ function:[20]

$$\psi(z) = \frac{d}{dz} \log \Gamma(z) = \frac{\Gamma'(z)}{\Gamma(z)} = \sum_{n=1}^{\infty} \left(\frac{1}{n} - \frac{1}{z+n-1} \right)$$

$$\text{and} \quad \psi(1) = -\gamma \qquad \psi(m+1) = 1 + \frac{1}{2} + \frac{1}{3} + \cdots + \frac{1}{m} - \gamma \tag{13.117}$$

$$\text{with} \quad \gamma = \lim_{n\to\infty} \left(1 + \frac{1}{2} + \frac{1}{3} + \cdots + \frac{1}{n} - \log n \right) = 0.5772 \cdots$$

(= Euler-Mascheroni constant)

We remark, finally, that sometimes these integral representations may be used to derive asymptotic forms and series. If the integrand contains one or more poles in the *left*-half plane, so that Re $s < 0$, we can displace the contour to the left, provided that (1) the integral along the new contour exists and the contributions from the infinite arcs vanish and (2) the residue for Re $s < 0$ does indeed constitute the main contribution vis-à-vis the integral along the original, unshifted contour. Other applications of this residue method are mentioned in Chap. 15.

(1) An example. A problem of importance in the theory of noise measurements and in communication theory arises when a carrier amplitude-modulated by normal random noise $V_N(t)$ is demodulated by an AM receiver. As we have seen (Chap. 6), signal processes of this type are of particular interest from the information-theory point of view. The problem is outlined in some generality below (1) to include an uncorrelated background noise in addition to the signal, (2) to take into account possible overmodulation, and (3) to illustrate specifically the analytical techniques described above. Linear heterodyning and predetection filtering are assumed as usual, so that we may write for the input to the second detector of our AM receiver, $\boldsymbol{T}_R^{(N)}$,

$$V(t) = \begin{Bmatrix} A_0[1 + \lambda N(t)] \cos \omega_0 t + N_b(t) & \lambda N > -1 \\ 0 + N_b(t) & \lambda N < -1 \end{Bmatrix}$$
$$= G(t) \cos \omega_0 t + N_b(t) \tag{13.118}$$

The IF (as well as the RF) filter stages merely modify the spectrum of N and N_b in the usual way, while the carrier amplitude A_0 is unchanged, except for possible gain factors.

From Eqs. (5.13), (5.14), (5.27), the second-moment function of the l-f output of the second detector is

$$M_z(t)_0 = \sum_{m,q=0} \epsilon_m \psi_b{}^\alpha \frac{2^{-\alpha} k_0(t)_b{}^\alpha}{q!(q+m)!} \overline{H_{m,\alpha}(t_1) H_{m,\alpha}(t_2)} \qquad \alpha = m + 2q \tag{13.119}$$

where $k_0(t)_b$ is the slowly varying part of the normalized covariance function of $N_b(t)$, and $\overline{N_b{}^2} = \psi_b$. The statistical average in Eq. (13.119) becomes

$$\mathfrak{M}^{(0)} \equiv \overline{H_{m,\alpha}(t_1) H_{m,\alpha}(t_2)} = \frac{1}{4\pi^2} \iint_{\mathrm{C}} f(i\xi_1) f(i\xi_2) (\xi_1 \xi_2)^\alpha e^{-\psi_b(\xi_1{}^2 + \xi_2{}^2)/2} \times \overline{J_m(G_1\xi_1) J_m(G_2\xi_2)}\, d\xi_1\, d\xi_2 \tag{13.120}$$

With the help of Eq. (13.107a), we may represent the Bessel functions as a complex power of G_1 and G_2, over which the required average may be taken with the aid of

$$G(t)^\mu = \frac{A_0{}^\mu \Gamma(\mu+1)}{2\pi} \int_{\mathrm{C}} e^{i\xi[1+\lambda N(t)]} \frac{d\xi}{(i\xi)^{\mu+1}} \tag{13.121}$$

This average becomes

$$\overline{G_1{}^{2s+m} G_2{}^{2t+m}} = \frac{A_0{}^{2s+2t+2m}}{4\pi^2} \Gamma(m+2s+1)\Gamma(m+2t+1) \times \iint_{\mathrm{C}} \frac{\overline{e^{i\zeta_1(1+\lambda N_1)+i\zeta_2(1+\lambda N_2)}}}{(i\zeta_1)^{2s+m+1}(i\zeta_2)^{2t+m+1}}\, d\zeta_1\, d\zeta_2 \tag{13.122a}$$

$$= \frac{A_0{}^{2m+2s+2t}\Gamma(m+2s+1)\Gamma(m+2t+1)}{i^{2s+2t+2m+2}} \times \sum_{n=0}^{\infty} \frac{(-1)^n [\lambda^2 K_N(t)]^n}{n!} h^{(1)}_{0,n-2s-m+1} h^{(1)}_{0,n-2t-m+1} \tag{13.122b}$$

where we have used†

$$h_{0,\mu}(b) \equiv \frac{1}{2\pi} \int_{\mathrm{C}} z^{\mu-2} e^{ibz - q^2 z^2}\, dz = \frac{i^{2-\mu}}{2q^{\mu-1}} \left\{ \frac{{}_1F_1[(\mu-1)/2; \tfrac{1}{2}; -b^2/4q^2]}{\Gamma[(3-\mu)/2]} + \frac{b}{q} \frac{{}_1F_1(\mu/2; \tfrac{3}{2}; -b^2/4q^2)}{\Gamma[(2-\mu)/2]} \right\} \tag{13.123}$$

Here, $q^2 = \lambda\psi_N/2$, and $b = 1$. The hypergeometric functions are recognized as parabolic cylinder functions[21] (of complex order) in Eq. (13.123). We may write finally

$$\mathfrak{M}^{(0)} = \sum_{n=0}^{\infty} \frac{(-1)^n}{n!} [\lambda^2 K_N(t)]^n \mathcal{H}^2_{0,m,n,q} \tag{13.124a}$$

† See Eq. (A.1.55) and Prob. 2.15.

where the amplitude functions $\mathcal{H}$ are

$$\mathcal{H}_{0,m,n,q} = \int_{-\infty i-c}^{\infty i-c} \frac{\Gamma(-s)\Gamma(2s+m+1)A_0^{m+2s}}{\Gamma(m+s+1)i^{2s+m+1}2^{2s+m}} h^{(1)}_{0,n-2s-m+1} \frac{ds}{2\pi i} \times \int_C \frac{f(i\xi)\,\xi^{\alpha+m+2s}}{2\pi} e^{-\psi_b\xi^2/2}\,d\xi \quad (13.124b)$$

and where c (> 0) is so chosen as to include the pole at $s = 0$ and $\alpha = m + 2q$.

When a half-wave νth-law detector is used, we find that Eq. (13.124b) reduces to

$$\mathcal{H}_{0,m,n,q} = \frac{\beta\Gamma(\nu+1)}{2i^{1+\alpha}} 2^{(\alpha-\nu)/2}\psi_b^{(\nu-\alpha)/2} \times \int_{-\infty i-c}^{\infty i-c} \frac{(-a_0^2)^{s+m/2}\Gamma(-s)\Gamma(2s+m+1)}{\Gamma[1-s-(\alpha+m-\nu)/2]\Gamma(s+m+1)} h^{(1)}_{0,n-2s-m+1} \frac{ds}{2\pi i} \quad (13.125)$$

where $a_0^2 = A_0^2/2\psi_b$. This result, when substituted into Eq. (13.124a), and that in turn into Eq. (13.119), gives us finally the desired second-moment function of the detected wave. For the strong carrier cases, alternative expressions are readily obtained with the aid of Prob. 13.18, Eqs. (2), (4).

Let us illustrate the above in more detail with the example of a half-wave quadratic rectifier.

First, we note from Eq. (13.125) that only for $m + q \leq 1$ is there any contribution to the integral; when $m = q = 0$, there are two simple poles at $s = 0$, 1, while for $m = 1$, $q = 0$, $m = 0$, $q = 1$, there is but a single pole at $s = 0$. The amplitude functions are easily found to be

$$\mathcal{H}_{0,0,n,0} = -i\beta\psi_b h_{0,n+1} + 2a_0^2 h_{0,n-1}/2 \quad (13.126a)$$

$$\mathcal{H}_{0,1,n,0} = -i\beta(\psi_b a_0^2/2)^{1/2} h_{0,n} \quad \text{and} \quad \mathcal{H}_{0,0,n,1} = i\beta h_{0,n+1} \quad (13.126b)$$

The second-moment function (13.119) becomes

$$M_z(t)_0 = \frac{\beta^2\psi_b^2}{4}\left(\sum_{n=0}^{\infty} \frac{(-1)^{n+1}}{n!}[\lambda K_N(t)]^n\{h^2_{0,n+1}[1 + k_0^2(t)_b] + a_0^2[2k_0(t)_b h^2_{0,n} + 4h_{0,n+1}h_{0,n-1}] + a_0^4(4h^2_{0,n-1})\}\right) \quad (13.127)$$

The terms of $O(a_0^0)$ represent cross products between the modulation and the background noise, while, for $n = 0$ here, we have simply the intermodulation of $N_b(t)$. The remaining contributions are the result of the cross modulation between carrier, noise, and modulation, which arise in the nonlinear operation of the rectifier. For no background noise, we get

$$M_z(t)_0\Big|_{\psi_b=0} = \frac{\beta^2 A_0^4}{4}\sum_{n=0}^{\infty}(-1)^{n-1}h^2_{0,n-1}\frac{[\lambda^2 K_N(t)]^n}{n!} \quad (13.128)$$

while, for strong signals ($a_0^2 \gg 1$), the leading term [$O(a_0^4)$] of Eq. (13.127) illustrates the behavior.

The amplitude functions $h_{0,l}$ [Eq. (13.123)] are identical with those described in Eq. (12.24). Using the latter, we can easily obtain results for the extreme cases of weak modulation ($\lambda_N 2\psi \ll 1$) and of strong modulation ($\lambda_N 2\psi \to \infty$). When $\lambda^2\psi_N < 0.18$ (Sec. 12.2), overmodulation occurs less than 1 per cent of the time, and only for $n = 0$, $n = 1$ are there significant contributions in this approximation. We find that $h_{0,0} \doteq 1$, $h_{0,1} \doteq i$, $h_{0,-1} = -i(2 + \lambda^2\psi_N)/4$, and so Eq. (13.127) becomes

$$M_z(t)_0\Big|_{\lambda^2\psi_N \ll 1} \doteq \frac{\beta^2\psi_b}{4}\{1 + k_0(t)_b{}^2 - a_0{}^2[2k_0(t)_b + 2 + \lambda^2\psi_N] + a_0{}^4(1 + \lambda^2\psi_N)\} + O[(\lambda^2\psi_N)^2] \quad (13.129)$$

For strong modulations, the leading terms in Eq. (13.127) are provided by $h_{0,n-1}^2$. The result is

$$M_z(t)_0\Big|_{\lambda^2\psi_N} \simeq \left(\frac{\beta A_0{}^2\lambda^2\psi_N}{8}\right)^2 \sum_{n=0}^{\infty} \frac{k_N(t)^n 2^n}{n!\Gamma(2 - n/2)^2} + O[(\lambda^2\psi_N)^{3/2}] \qquad k_N = k_N(t)$$

$$\simeq \left(\frac{\beta A_0{}^2k^2\psi_N}{8}\right)^2 \left[1 + 2k_N{}^2 + \frac{2}{\pi}(1 + 2k_N{}^2)\sin^{-1} k_N + \frac{6}{\pi} k_N\sqrt{1 - k_N{}^2}\right] + \cdots \quad (13.130)$$

As we can see from Eq. (13.127), the spectrum is spread from the contributions of the additional components generated by overmodulation and by the half-wave rectification.

The applicability of the theory for nonlinear systems outlined in a general way in Chap. 5, and in some detail in the preceding section of this chapter, is based on the fact that the nonlinear devices in question are zero-memory elements—i.e., there is no hysteresis or feedback whereby the output at any instant depends on its past history. This has meant effectively resistive loads, and no frequency sensitivity as far as the nonlinear action is concerned. Frequency-selective elements following the rectifier are sufficiently isolated from the nonlinear device that they do not influence its operation. Fortunately, this is the situation for most practical systems, or can be made so without seriously altering their performance. However, when the nonlinear characteristic has a nonzero memory, the analysis must be modified and analytical difficulties now arise, which present serious technical problems. Instances of this behavior occur in the *RC* loaded diode rectifier and in nonlinear elements used as measuring devices. For a discussion of the latter, see the papers of Bennett,[4] Johnson,[11] Burgess,[22] and Aiken[26] and, for the former, the work of Wiener,[23] Ikehara,[24] and Deutsch.[25]

PROBLEMS

13.13 (*a*) When the carrier is large relative to the noise, show that

$$h_{m,n} \simeq \frac{\beta\Gamma(\nu+1)(i)^{-m-n}2^{n-\nu-1}A_0^{\nu-n}}{2\Gamma[(2+\nu-n+m)/2]}\left\{\frac{1}{\Gamma[(2+\nu-m-n)/2]} - \frac{2b_0}{A_0}\frac{\Gamma[(2+\nu-n+m)/2]}{\Gamma[(1+\nu+m-n)/2]\Gamma[(1+\nu-m-n)/2]} + \cdots\right\} \qquad a_0^4 \gg 1 \quad (1)$$

(*b*) Show that

$$h_{m,n}\Big|_{b_0\to\infty} \simeq 0 \quad (2)$$

and that

$$h_{m,n}\Big|_{b_0\to\infty} \simeq \frac{\beta\Gamma(\nu+1)i^{-m-n}\sqrt{\pi}\,2^{n-\nu}|b_0|^{-m-n+\nu}A_0^m}{m!\Gamma[(1+\nu-m-n)/2]\Gamma[(2+\nu-m-n)/2]}\left[1+\frac{\psi}{2b_0^2}\left(1-\frac{a_0^2}{m+1}\right)(m+n-\nu+1)(m+n-\nu)+\cdots\right] \quad (3)$$

Thus, for the biased linear rectifier in this limit

$$h_{0,0} = \beta|b_0| \qquad h_{0,1} = -i\beta \qquad h_{1,0} = -i\beta A_0/2 \qquad h_{m,n} \to 0,\ m+n > 1 \quad (4)$$

D. Middleton,[2] sec. 4.

13.14 When the noise accompanying the signal vanishes, for biased νth-law rectification, show that the amplitude functions become

$$h_{m,0} = \frac{\beta\Gamma(\nu+1)i^{-m}\sqrt{\pi}\,|b_0|^2(A_0/|b_0|)^m}{2^\nu\Gamma[(2+\nu-m)/2]\Gamma[(1+\nu-m)/2]m!}\,{}_2F_1\left(m-\nu;\,m-\nu+1;\,m+1;\,\frac{A_0^2}{b_0^2}\right) \qquad b_0 < 0,\ |b_0| \geq A_0 \quad (1)$$

and

$$h_{m,0} = \frac{\beta\Gamma(\nu+1)i^{-m}A_0^\nu}{2^{\nu+1}}\left\{\frac{{}_2F_1[(m-\nu)/2,\,(-m-\nu)/2;\,\tfrac{1}{2};\,b_0^2/A_0^2]}{\Gamma[(2+\nu+m)/2]\Gamma[(2+\nu-m)/2]} - \frac{2b_0}{A_0}\frac{{}_2F_1[(m-\nu+1)/2,\,(-m-\nu+1)/2;\,\tfrac{3}{2};\,b_0^2/A_0^2]}{\Gamma[(1+\nu+m)/2]\Gamma[(1+\nu-m)/2]}\right\} \qquad 0 \leq |b_0| < A_0 \quad (2)$$

Show also that, for half-wave detection ($b_0 = 0$), here

$$h_{m,0} = \frac{\beta\Gamma(\nu+1)i^{-m}A_0^\nu/2^{\nu+1}}{\Gamma[(2+\nu+m)/2]\Gamma[(2+\nu-m)/2]} \quad (3)$$

D. Middleton,[2] sec. 4.

13.15 Show that the second-moment function for the output process after *extreme limiting*, due to bias and saturation, i.e., for $C - b_0 = \epsilon \ll 1$ ($C > b_0$; $\epsilon > 0$), becomes for *any* (zero-memory) dynamic path

$$M_z(t) \doteq \epsilon^2\lambda^2\left\{\left[\frac{1-\Theta(b/\sqrt{2})}{2}\right]^2 + \sum_{n=1}^{\infty}\frac{\phi^{(n-1)}(b)^2k(t)^n}{n!}\right\} \quad (1)$$

where λ is a suitable constant, depending on the mean intensity of the input noise and on the specific law of the limiter's response, and $b = b_0/\sqrt{\psi}$ ($\psi = \overline{N^2}$).

D. Middleton,[6] sec. VI.

13.16 A normal noise process is passed through a full-wave rectifier, or limiter, with the antisymmetrical characteristic response

$$z = c\Theta[y/(2\psi)^{1/2}] \qquad c > 0;\ \psi = \overline{y^2} \quad (1)$$

(a) Show that the covariance function of the resulting output process is

$$K_z(t) = \frac{c^2k(t)}{\pi}\,{}_2F_1\left[\frac{1}{2},\frac{1}{2};\frac{3}{2};\frac{k^2(t)}{4}\right] = \frac{2c^2}{\pi}\sin^{-1}\frac{k(t)}{2} \tag{2}$$

and if the input is RC noise, for which $k(t) = e^{-|t|\alpha}$ $[\alpha = (RC)^{-1}]$, verify that the intensity spectrum of the output is

$$\mathcal{W}_z(f) = \frac{8c^2}{\pi\alpha}\sum_{m=0}^{\infty}\frac{(\frac{1}{2})_m}{(2m+1)^2m!2^{2m+1}}\left[1+\frac{\omega^2}{(2m+1)^2\alpha^2}\right]^{-1} \tag{3}$$

(b) Obtain corresponding results for the antisymmetrical νth-law characteristic

$$z = \boldsymbol{T}_g\{y\} = \begin{cases} c|y|^\nu & y > 0 \\ -c|y|^\nu & y < 0 \end{cases} \tag{4}$$

when $\nu = 0, 1$, namely,

$$K_z(t)\Big|_{\nu=0} = \frac{2c^2}{\pi\alpha}\sin^{-1}k(t) \qquad K_z(t)\Big|_{\nu=1} = \psi c^2k(t) \tag{5}$$

and, correspondingly, the spectra

$$\mathcal{W}_z(f)\Big|_{\nu=0} = \frac{8c^2}{\pi\alpha}\sum_{m=0}^{\infty}\frac{(\frac{1}{2})_m}{(2m+1)^2m!}\left[\frac{1+\omega^2}{(2m+1)^2\alpha^2}\right]^{-1} \tag{6}$$

$$\mathcal{W}_z(f)\Big|_{\nu=1} = \frac{8c^2}{\pi\alpha}\frac{\pi/2}{1+\omega^2/\alpha^2} \tag{7}$$

Compare Eqs. (3), (6), and (7).

13.17 Show that for a general full-wave dynamic characteristic (not necessarily symmetrical or antisymmetrical) the second-moment function of the output when the statistics of the input noise are not symmetrical becomes

$$\begin{aligned} M_z(t) = \frac{1}{4\pi^2}\int_C\int_C & f_e(i\xi_1)f_e(i\xi_2)[F_2(\xi_1,\xi_2;t)_N + F_2(-\xi_1,-\xi_2;t)_N \\ & + F_2(\xi_1,-\xi_2;t)_N + F_2(-\xi_1,\xi_2;t)_N]\,d\xi_1\,d\xi_2 \\ + \frac{1}{4\pi^2}\int_C\int_C & f_o(i\xi_1)f_o(i\xi_2)[F_2(\xi_1,\xi_2;t)_N + F_2(-\xi_1,-\xi_2;t)_N \\ & - F_2(\xi_1,-\xi_2;t)_N - F_2(-\xi_1,\xi_2;t)_N]\,d\xi_1\,d\xi_2 \end{aligned}$$

where f_e, f_o refer to the symmetrical and antisymmetrical portions of the dynamic response (cf. Prob. 2.12).

13.18 With the help of the appropriate relations in the text (Sec. 13.4-4), show that

$$\begin{aligned} I^{(1)} &= \int_C z^{\mu-1}e^{ibz-q^2z^2}J_\nu(az)\,dz \\ &= \frac{\pi(a/2)^\nu i^{1-\mu-\nu}}{q^{\mu+\nu}}\int_{-\infty i-c(\nu)}^{\infty i-c(\nu)}\frac{ds}{2\pi i}\frac{\Gamma(-s)(a/2iq)^{2s}}{\Gamma(\nu+s+1)}\left\{\frac{{}_1F_1[(\mu+\nu)/2+s;\,\frac{1}{2};\,-b^2/4q^2]}{\Gamma[1-s-(\mu+\nu)/2]}\right. \\ &\qquad\left.+\frac{b}{q}\frac{{}_1F_1[(\mu+\nu+1)/2+s;\,\frac{3}{2};\,-b^2/4q^2]}{\Gamma[1-s-(\nu+\mu+1)/2]}\right\} \end{aligned} \tag{1}$$

with an alternative expression

$$\begin{aligned} I^{(1)} = \left(\frac{a}{2}\right)^\nu\left(\frac{2}{b}\right)^{\nu+\mu}\frac{2\pi^{3/2}i^{1-\mu-\nu}}{\Gamma(\nu+1)} \\ \times\int_{-\infty i-c}^{\infty i-c}\frac{\Gamma(-s)\,{}_1F_1(-s;\,v+1;\,-a^2/4q^2)(2q/ib)^{2s}}{\Gamma[1-s-(\nu+\mu)/2]\Gamma[1-s-(\nu+\mu+1)/2]}\frac{ds}{2\pi i} \end{aligned} \tag{2}$$

which is useful for large a and large b. For small b and large a, obtain the useful third relation

$$I^{(1)} = \frac{\pi}{\Gamma(\nu+1)}\left(\frac{a}{2q}\right)^{\nu}\frac{i^{1-\mu-\nu}}{q^{\mu}} \times \int_{-\infty i - c(\nu)}^{\infty i - c(\nu)} \frac{i^{-2s}\Gamma(-s)\ {}_1F_1[(\mu+\nu+s)/2;\ \nu+1;\ -a^2/4q^2]}{\Gamma[1-(\nu+\mu+s)/2]}\left(\frac{b}{q}\right)^{s}\frac{ds}{2\pi i} \tag{3}$$

Still another expression, particularly useful when $b < q$, $a < b$, is

$$I^{(1)} = \left(\frac{a}{2}\right)^{\nu} q^{-\mu-\nu}\frac{\pi i^{1-\mu-\nu}}{\Gamma(\nu+1)} \times \int_{-\infty i - c}^{\infty i - c} \frac{\Gamma(-s)\ {}_2F_1[-s/2,\ (1-s)/2;\ \nu+1;\ a^2/b^2]}{\Gamma[1-(\mu+\nu+s)/2]}\left(\frac{b}{q}\right)^{s}\frac{ds}{2\pi i} \tag{4}$$

The integral $I^{(1)}$ is recognized as the basic type arising in the theory of the biased νth-law detector when an additive carrier and noise process appears at the input.

REFERENCES

1. Rice, S. O.: Mathematical Analysis of Random Noise, *Bell System Tech. J.*, **24:** 46 (1945), pt. IV.
2. Middleton, D.: Some General Results in the Theory of Noise through Nonlinear Devices, *Quart. Appl. Math.*, **4:** 445 (1948).
3. Fränz, K.: Die Übertragung von Rauschspannung uber den linearen Gleichrichter, *Hoch. u. Elek.*, **57:** 146 (1947); Beitrage zur Berechnung der Verhältnisses von Signalspannung zu Rauschspannung am Ausgang von Empfänger, *Elek. Nachr.-Tech.*, **17:** 215 (1940), **19:** 285 (1942).
4. Bennett, W. R.: Response of a Linear Rectifier to Signal and Noise, *Bell System Tech. J.*, **23:** 97 (1944); *J. Acoust. Soc. Amer.*, **15:** 164 (1944).
5. North, D. O.: The Modification of Noise by Certain Nonlinear Devices, IRE Winter Technical Meeting, New York, Jan. 28, 1944.
6. Middleton, D.: The Response of Biased, Saturated Linear and Quadratic Rectifiers to Random Noise, *J. Appl. Phys.*, **17:** 778 (1946).
7. Burgess, R. E.: The Rectification and Observation of Signals in the Presence of Noise, *Phil. Mag.*, **42:** 475 (1951).
8. Johnson, R. A.: Rectification of Signals and Random Noise by a Linear Detector, dissertation, Harvard, June, 1952; see also *Cruft Lab. Tech. Rept.* 144, March, 1952, and 145, May, 1952. Also, R. A. Johnson and D. Middleton, Measurement of Correlation Functions of Modulated Carriers and Noise Following a Nonlinear Device, in W. Jackson (ed.), "Communication Theory," Butterworth Scientific Publications, London, 1953.
9. Fellows, G. E., and D. Middleton: An Experimental Study of Intensity Spectra after Half-wave Rectification of Signals in Noise, *J. IEE (London), Mon.* 160R, January, 1956. See also G. E. Fellows, Experimental Study of Intensity Spectra after Half-wave Detection of Signals in Noise, dissertation, Harvard, January, 1955, and *Cruft Lab. Tech. Rept.* 217, 218, February, 1955.
10. Middleton, D., and V. Johnson: A Tabulation of Selected Confluent Hypergeometric Functions, *Cruft Lab. Rept.* 140, January, 1952.
11. Johnson, R. A., and D. Middleton: "Measurement of Correlation Functions of Modulated Carrier and Noise Following a Nonlinear Device," in W. Jackson (ed.), "Communication Theory," p. 195, Academic Press, New York, 1953. See also Ref. 9.

12. Middleton, D.: Rectification of a Sinusoidally Modulated Carrier in the Presence of Noise, *Proc. IRE*, **36**: 1467 (1948).
13. Fubini, E. G., and D. C. Johnson: Signal-to-noise Ratio in AM Modulated Receivers, *Proc. IRE*, **36**: 146 (1948).
14. Ragazzini, J. R.: The Effect of Fluctuation Voltages on the Linear Detector, *Proc. IRE*, **30**: 277 (1942).
15. Weinberg, L., and L. G. Kraft: Measurement of Detector Output Spectra by Correlation Methods, *Proc. IRE*, **41**: 1157 (1953).
16. Davenport, W. B., Jr.: Signal-to-noise Ratios in Band-pass Limiters, *J. Appl. Phys.*, **24**: 720 (1953).
17. Tucker, G. D.: Linear Rectifiers and Limiters, *Wireless Engr.* (*London*), **29**: 128 (1952).
18. N. Stone and D. Middleton: Full-wave Detection of Signals in Noise, *Cruft Lab. Tech. Rept.* 182, Aug. 1, 1953.
19. Mellin, R. H.: *Math. Ann.*, **48**: 305 (1910); E. W. Barnes, *Trans. Cambridge Phil. Soc.*, **20**: 270 (1908); and specifically G. N. Watson, "Theory of Bessel Functions," 2d ed., secs. 6.5, 6.51, Macmillan, New York, 1945.
20. Whittaker, E. T., and G. N. Watson: "Modern Analysis," chaps. 7, 8, Macmillan, New York, 1944. See also E. Jahnke and F. Emde, "Tables of Functions," Dover, New York, 1945, for numerical values and other relations. Note, however, that $\psi_{JE}(z)$ is related to $\psi(z + 1)$ here by $\psi_{JE}(z) = \psi(z + 1)$.
21. Magnus, W., and F. Oberhettinger: "Special Function of Mathematical Physics," p. 91, Chelsea, New York, 1948.
22. Burgess, R. E.: The Measurement of Fluctuation Noise by Diode and Anode Bend Voltmeters, *Proc. Phys. Soc.* (*London*), **B64**: 508 (1951); see also Ref. 7.
23. Wiener, N.: Response of a Nonlinear Device to Noise, *MIT Radiation Lab. Rept.* 129, 1942.
24. Ikehara, S.: A Method of Wiener in a Nonlinear Circuit, *MIT Research Lab. Electronics Tech. Rept.* 217, Dec. 10, 1951.
25. Deutsch, R.: On a Method of Wiener for Noise through Nonlinear Devices, *IRE Conv. Record*, pt. 4, pp. 186–192, March, 1955.
26. Aiken, C. B.: Theory of the Diode Voltmeter, *Proc. IRE*, **26**: 859 (1938).

CHAPTER 14

PHASE AND FREQUENCY MODULATION

Two common and important methods of mapping messages into signals are amplitude and angle modulation. In Chaps. 5 and 12 we have considered some aspects of the former; here, we shall examine the latter, and, as before (cf. Chaps. 5, 12), mainly in terms of a second-moment theory, where we shall be concerned for the most part with covariance functions and intensity spectra of angle-modulated waves. The modulations themselves fall into three categories: (1) certain classes of deterministic processes, e.g., periodic waves (in the steady state); (2) random modulations, such as (stationary) broadband normal noise processes; (3) modulations that consist of additive mixtures of (1) and (2). The results of the analysis are applicable in a wide variety of problems:† in communication systems, in physical measurements, and in the theory of certain types of microwave tubes. We shall not, however, discuss such applications here, as they are treated extensively elsewhere.‡

In angle modulation, it is the instantaneous phase of a carrier that is changed. Thus, if we set

$$y(t) = \boldsymbol{T}_g\{\Psi(t)\} = A_0 \cos \Psi(t) \qquad \text{with } \Psi(t) \equiv \omega_c t - \Phi(t) \tag{14.1}$$

$y(t)$ represents such a signal process, and it is the instantaneous phase $\Psi(t)$ or $\Phi(t)$ that contains the encoded message. Two cases are distinguished, depending on how the original message set is introduced into this phase term: (1) *phase modulation*, where Ψ or Φ is directly proportional to the message waveform, and (2) *frequency modulation*, where it is $\dot{\Psi}$ or $\dot{\Phi}$ that exhibits this proportionality.

Let us consider first *phase modulation*. One scheme§ whereby this is realized is to combine after a phase shift of 90° an unmodulated cw carrier with a portion of the original carrier which is modulated by the desired message waveform. The combined waves are then passed through a limiter, where with heavy limiting essentially all amplitude variations are removed and the resulting output is an angle-modulated carrier of the form (14.1).

† The material of this chapter is based largely on the work of Middleton.[1–3]

‡ See Gottschalk,[7] La Plante,[8] the additional references therein, and the bibliography at the end of this chapter.

§ For modifications and other methods, see F. E. Terman, "Radio Engineers' Handbook," sec. 7, pars. 16–17, McGraw-Hill, New York, 1943.

For *frequency modulation*, a similar procedure may be used, but now $V_{\text{mod}}(t)$ is replaced by $\int^t V_{mod}(t')\,dt' + C$, where C depends on the instant at which modulation was initiated. In either case, we can write accordingly

$$\begin{aligned} V_{out}(t) &= A_0 \cos\,[\omega_c t - \Phi(t)] \\ \Phi(t)_{\phi\mathrm{M}} &= D_\phi V_{mod}(t) \\ \Phi(t)_{\mathrm{FM}} &= D_F \int^t V_{mod}(t')\,dt' + C \end{aligned} \tag{14.2}$$

Here D_ϕ and D_F have, respectively, the units of radians per volt and angular frequency per volt. Figure 14.1 shows a typical angle-modulated

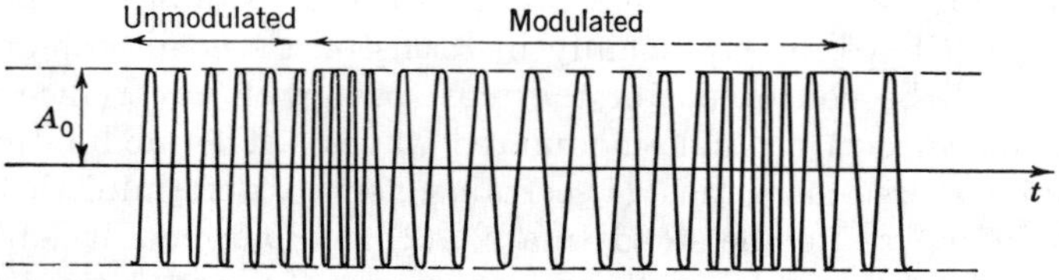

FIG. 14.1. Phase (or frequency) modulation of a carrier by a sine wave.

cw carrier wave. Other schemes of modulation employ an oscillator directly, where the frequency (and hence phase) is varied by changing the effective capacitance or inductance of the resonant system, and where a limiter at the output serves to remove the unwanted amplitude variations. Without limiting, both amplitude and angle modulation may occur.

We note that, as in amplitude modulation (cf. Secs. 12.1, 12.2), certain restrictions govern the spectral character of both the carrier and the wave which is to be impressed upon it as modulation. For example, in both phase and frequency modulation if we are properly to speak of a "modulation" as such, rather than of a multiplication or mixing operation, and if Eqs. (14.2) are to be a valid representation of the output wave, the maximum angular frequency deviation $|\dot{\Phi}|_{\max}$ *and* the maximum *rate* ($\sim|\ddot{\Phi}|_{\max}$) at which this deviation is carried out must be small compared with the carrier frequency.[4] For stochastic modulations one replaces the maximum by the usual mean-square condition, in the form of rms values. These conditions are also sometimes referred to as the "quasi-stationary" state of operation,[5] meaning that the system is completely capable of following through stationary states the variable frequency of the modulation. However, unlike simple amplitude modulation, the process of phase or frequency modulation is highly nonlinear [cf. Eqs. (14.2)]. A large variety of modulation products are generated among the original components of the *modulating* wave, although, as we shall see presently, the number of significant sidebands of the carrier depends markedly on frequency deviation and on the rate at which this deviation is accomplished.

14.1 *Covariance Functions and Spectra*

In this section, we shall consider the second-moment theory of (1) periodic, (2) random, and (3) mixed modulating processes for both phase and frequency modulation. A discussion of the important limiting cases is, however, reserved to Sec. 14.2.

14.1-1 Periodic Modulations. Let us write for a typical member of the modulated carrier ensemble

$$V^{(j)}(t) = \boldsymbol{T}_g\{\Phi^{(j)}(t)\} = A_0 \operatorname{Re} e^{i\omega_c t - i\Phi^{(j)}(t)} \tag{14.3}$$

and develop the exponential in a Fourier series, so that the amplitude spectrum [cf. Eqs. (2.43)] becomes

$$V^{(j)}(t) = A_0 \operatorname{Re} \left(e^{i\omega_c t} \sum_{-\infty}^{\infty} a_m{}^{(j)} e^{im\omega_a t} \right) \tag{14.4a}$$

where the (complex) amplitudes $a_m{}^{(j)}$ are given by

$$a_m{}^{(j)} = T_0{}^{-1} \int_0^{T_0} e^{-i\Phi^{(j)}(t_0) - im\omega_a t_0}\, dt_0 \qquad T_0 = 2\pi/\omega_a \tag{14.4b}$$

We illustrate this with the simple example of a sinusoidal modulation $V_{mod}^{(j)}(t) = B_0 \cos(\omega_a t + \phi^{(j)})$, where $\phi^{(j)}$ is set equal to zero for convenience in subsequent work. The waveform of the modulated carrier is easily obtained with the help of Eq. (A.1.48*a*).[6] For *phase modulation*, we have specifically

$$\begin{aligned} V^{(j)}(t)_{\phi M} &= A_0 \operatorname{Re} e^{i\omega_c t - i\mu_{\phi M} \cos \omega_a t} \\ &= A_0 \operatorname{Re} \left[\sum_{-\infty}^{\infty} (-i)^m J_m(\mu_{\phi M}) e^{im\omega_a t + i\omega_c t} \right] \end{aligned} \tag{14.5}$$

where $a_m{}^{(j)} = (-i)^m J_m(\mu_{\phi M})$ and $\mu_{\phi M}$ ($\equiv D_\phi B_0$) is the *modulation index* for (sinusoidal) phase modulation.

A similar expression is readily derived for *frequency modulation*, viz.,

$$\begin{aligned} V^{(j)}(t)_{FM} &= A_0 \operatorname{Re} e^{i\omega_c t - i\mu_{FM} \sin \omega_a t} \\ &= A_0 \operatorname{Re} \left[\sum_{-\infty}^{\infty} (-1)^m J_m(\mu_{FM})\, e^{im\omega_a t + i\omega_c t} \right] \end{aligned} \tag{14.6}$$

where now one finds $a_m{}^{(j)} = (-1)^m J_m(\mu_{FM})$ and μ_{FM} ($\equiv D_F B_0/\omega_a$) is the appropriate modulation index for (sinusoidal) frequency modulation. The modulation conditions are here

$$\begin{array}{ll} \textit{Phase modulation} & \omega_a D_\phi B_0 \ll \omega_c;\ \omega_a \ll \omega_c \\ \textit{Frequency modulation} & D_F B_0 \ll \omega_c;\ \omega_a \ll \omega_c \end{array} \tag{14.7}$$

The components generated in the course of modulation are spaced f_a ($= \omega_a/2\pi$) cps apart, and since $J_m(\mu)$ falls off rapidly to zero for all $m \sim O(\mu)$ or greater, the effective spectral spread is essentially equal to the frequency deviation $\omega_a D_\phi B_0/2\pi$ for phase modulation and to $D_F B_0/2\pi$ for frequency modulation, as indicated in Fig. 14.2.

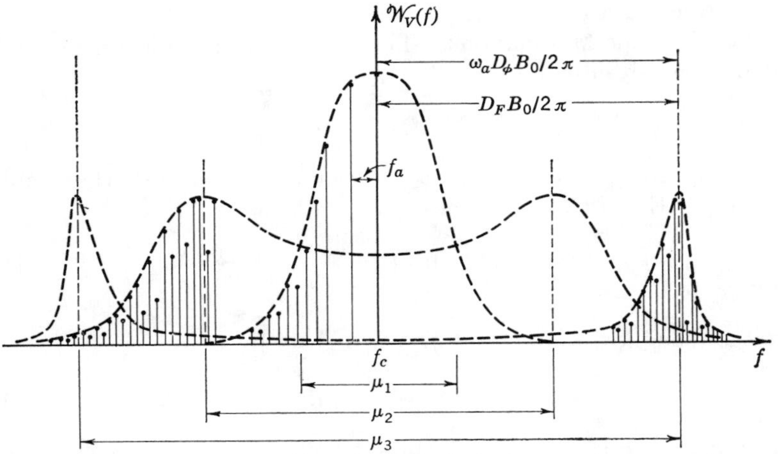

FIG. 14.2. Intensity spectra of angle modulation by a sine wave, showing the envelopes of the discrete spectra, for small (μ_1), intermediate (μ_2), and large (μ_3) modulation indices and fixed rate of angular deviation.

The covariance function of the ensemble of modulated waves, of which Eq. (14.3) is a typical member, is found in the usual way by an appropriate statistical average over the random parameters of the process. If, however, we restrict ourselves to ergodic modulations, the covariance function is now easily obtained (probability 1) from a suitable time average, namely, the autocorrelation function, whenever the modulation is deterministic (as well as ergodic) so that we can take advantage of the specific time structure of an individual wave (cf. Sec. 1.6). This autocorrelation function of $V^{(j)}(t)$ [cf. Eq. (14.4a)] is found specifically from Eq. (3.23), e.g.,

$$R_{V^{(j)}}(t) = \frac{A_0^2}{2} \operatorname{Re} \{\exp(-i\omega_c t)\langle \exp [i\Phi^{(j)}(t_2) - i\Phi^{(j)}(t_1)]\rangle\} \qquad t = t_2 - t_1 \tag{14.8}$$

$$= K_V(t) \qquad \text{probability 1} \tag{14.8a}$$

for ergodic modulations. Since $R_{V^{(j)}}(t)$ is necessarily even in $t = t_2 - t_1$, we can develop it in a cosine Fourier series [cf. Eq. (3.11)] of the form

$$R_{V^{(j)}}(t) = \frac{A_0^2}{2} \cos \omega_c t \sum_{m=0}^{\infty} \epsilon_m B_m^{(j)2} \cos m\omega_a t \tag{14.9a}$$

where now

$$B_m^{(j)2} = T_0^{-1} \int_0^{T_0} \cos m\omega_a t \left\{ T_0^{-1} \int_0^{T_0} \exp \left[i\Phi^{(j)}(t_0') - i\Phi^{(j)}(t_0' + t) \right] dt_0' \right\} dt$$
$$= a_m^{(j)} a_m^{(j)*} \qquad (14.9b)$$

For the sinusoidal example above, we let $\Phi^{(j)}(t) = \mu_{\phi M} \cos \omega_a t,\ \mu_{FM} \sin \omega_a t$, respectively, to obtain from Eqs. (14.8), (14.9) the covariance functions†

$$K_V(t) = \frac{A_0^2}{2} \sum_{m=0}^{\infty} \frac{\epsilon_m}{2} J_m^2(\mu)[\cos (\omega_c + m\omega_a)t + \cos (\omega_c - m\omega_a)t] \qquad (14.10a)$$

$$= \frac{A_0^2}{2} J_0\left(2\mu \sin \frac{\omega_a t}{2}\right) \cos \omega_c t \qquad (14.10b)$$

where $\mu = \mu_{\phi M}$ or μ_{FM} for phase or frequency modulation by the sinusoidal process $B_0 \cos (\omega_a t + \phi)$ [ϕ uniformly distributed in the primary interval $(0,2\pi)$; cf. Sec. 1.3-7]. Total intensity in either case is the same,

$$K_V(0) = \frac{A_0^2}{2} \sum_{m=0}^{\infty} \epsilon_m J_m^2(\mu) = \frac{A_0^2}{2} \qquad (14.11)$$

showing that the total intensity of the wave is unchanged by angle modulation, although the intensity in the original carrier is now redistributed among the various components $f_c \pm mf_a$ produced in modulation. The carrier itself is just another component, which intensitywise contributes but little to the total if μ is at all large. The intensity spectrum follows from the W-K theorem [Eqs. (3.42)]. We have directly

$$\mathcal{W}_V(f) = \frac{A_0^2}{2} \sum_{m=0}^{\infty} \frac{\epsilon_m}{2} J_m^2(\mu) \{ \delta[f - (f_c + mf_a)] + \delta[f - (f_c - mf_a)] \} \qquad (14.12)$$

(cf. Fig. 14.2). With large μ, when the frequency deviation is held fixed, the significant spectral lines are numerous and crowded together, while, for small μ, there are few effective components, and these are more widely separated. On the other hand, if we require the modulation frequency to be fixed, the effective components now remain spectrally equidistant, while the deviation is increased (parametrically) as μ is made larger. The spectra

† With the help of the Fourier series (A.1.48a) and the relation

$$J_0(\sqrt{a^2 + b^2 \pm 2ab \cos \psi}) = \sum_{m=0}^{\infty} \epsilon_m (\mp 1)^m J_m(a) J_m(b) \cos m\psi$$

(cf. G. N. Watson, "Theory of Bessel Functions," chap. 11, Cambridge, New York, 1948).

of Fig. 14.2 then become progressively broadened, spreading out on either side of the carrier component. Additional examples of periodic modulations, other than sine waves, are provided in Probs. 14.1 to 14.4.

14.1-2 Random Modulations. When the modulation is a stationary l-f noise process $V_N(t)$, for example, representing a complex message set, or the intrinsic modulations produced by a shot-noise mechanism in certain types of microwave oscillators,† the covariance function of the modulated carrier $V(t) = \boldsymbol{T}_g\{\Phi(t)\} = A_0 \cos [\omega_c t - \Phi(t)]$ can be obtained from Eq. (3.81),

$$K_V(t) = \frac{A_0^2}{2} \operatorname{Re} (e^{-i\omega_c t} \boldsymbol{E}\{e^{i\Phi_N(t_0+t)-i\Phi_N(t_0)}\})$$

$$= \frac{A_0^2}{2} \operatorname{Re} [e^{-i\omega_c t} F_2(-1,1;t)_{\Phi_N}] \tag{14.13a}$$

or

$$K_V(t) = \frac{A_0^2}{2} \operatorname{Re} (e^{-i\omega_c t} \boldsymbol{E}\{e^{i\Psi_N(t,t_0)}\})$$

$$= \frac{A_0^2}{2} \operatorname{Re} [e^{-i\omega_c t} F_1(1;t)_{\Psi_N}] \tag{14.13b}$$

where the first average is recognized as the characteristic function of Φ_N, with $\xi_1 = -1$, $\xi_2 = 1$, and the second average is the characteristic function of $\Psi_N \equiv \Phi_N(t_0 + t) - \Phi_N(t_0)$, with $\xi = 1$.

While Eq. (14.13) holds for quite general statistics, we consider here only the tractable case (fortunately of chief interest in applications) when $V_N(t)$ and hence $\Phi_N(t)$ represent normal random processes.[1] Applying the results of Sec. 8.1-3(3), where now $\Phi(t)_{\phi M} = D_\phi V_N(t)$ is a stationary gauss process and $(\Psi_N)_{FM} = \int_{t_0}^{t_0+t} D_F V_N(t')\, dt'$ is a gauss functional,‡ we obtain

$$F_2(-1,1;t)_{\Phi_N} = F_1(1;t)_{\Psi_N} = e^{-D_0^2\Omega(t)} \tag{14.14a}$$

where

$$\Omega(t) = (\overline{\Phi_N^2} - \overline{\Phi_{N1}\Phi_{N2}})/D_0^2 \tag{14.14b}$$

Here D_0^2 ($= D_\phi^2$ or D_F^2) is obtained from the relations (14.2), viz.,

$$\overline{\Phi_{\phi M}^2} = D_\phi^2 \overline{V_N^2} \qquad \overline{\Phi_{FM}^2} = D_F^2 \overline{V_N^2} \tag{14.14c}$$

so that D_ϕ has the dimensions of angle per (rms) volt and D_F those of *angular* frequency (or radians) per (rms) volt, if V_N is measured in volts. From Eqs. (8.37), (8.38), it follows at once that

$$\Omega(t)_{FM} = \int_0^\infty \mathcal{W}_N(f)(1 - \cos \omega t)\, df/\omega^2 = \int_0^{|t|} (|t| - z) K_V(z)\, dz \tag{14.15a}$$

$$\Omega(t)_{\phi M} = \int_0^\infty \mathcal{W}_N(f)(1 - \cos \omega t)\, df = K_V(0) - K_V(t) \tag{14.15b}$$

† For some recent work in this connection, see Gottschalk, *op. cit.*, La Plante, *op. cit.*, Middleton,[9] and Stewart.[10]

‡ Since $V_N(t')$ is stationary, we can set $t_0 = 0$ here.

in which $\mathcal{W}_N(f)$ is the intensity spectrum and $K_V(t)$ is the covariance function of this modulating noise process, with $K_V(0) = \overline{V_N^2}$ as usual. The covariance functions and intensity spectra for normal modulations are thus

$$K_V(t) = \frac{A_0^2}{2} e^{-D_0^2\Omega(t)} \cos \omega_c t \tag{14.16a}$$

$$\mathcal{W}_V(f) \doteq A_0^2 \int_0^\infty e^{-D_0^2\Omega(t)} \cos (\omega_c - \omega)t \, dt \tag{14.16b}$$

where the term in $\cos (\omega_c + \omega)t$ in the integrand of Eq. (14.16*b*) has been neglected, as it yields an ignorable contribution to the spectrum of the modulated carrier.

Since $\Omega(0)$ vanishes, the mean total power is $A_0^2/2$, as expected, inasmuch as the modulation is entirely in the phase term. The intensity of the carrier component is

$$P_{f_c} = K_V(\infty) = \frac{A_0^2}{2} e^{-D_0^2\Omega(\infty)} \tag{14.17a}$$

and that of the continuum becomes

$$P_c = K_V(0) - K_V(\infty) = \frac{A_0^2}{2} (1 - e^{-D_0^2\Omega(\infty)}) \tag{14.17b}$$

For phase modulation [cf. Eq. (14.15*b*)], it is clear that $\Omega(\infty)_{\phi M} = \overline{V_N^2} > 0$, so that there is always a discrete carrier component, although it may represent a trivial fraction of the total intensity if $D_0^2\overline{V_N^2}$ is at all large.

The situation for frequency modulation is less simple. Taking the limit $t \to \infty$ in Eq. (14.15*a*) and integrating by parts, we get

$$\lim_{t\to\infty} \Omega(t) = \int_0^\infty \mathcal{W}_N(f) \, df/\omega^2$$

and if we expand the integrand† about $f = 0$, we can write

$$\lim_{t\to\infty} \Omega(t) = \int_0^\infty \mathcal{W}_N(f) \, df/\omega^2 = O\left(\frac{C_0}{\omega} - C_1 \log \omega - C_2 + \cdots\right)_{\omega\to 0+} \tag{14.18}$$

distinguishing three cases:

1. $C_0 > 0$; therefore, $\Omega(\infty)_{\text{FM}} \to +\infty$ as $O(1/\omega)_{\omega\to 0+}$.
2. $C_0 = 0$, $C_1 > 0$; $\Omega(\infty)_{\text{FM}} \to +\infty$, $O(-\log \omega)_{\omega\to 0+}$.
3. $C_0 = C_1 = 0$, $C_2 \neq 0$, etc.; $\Omega(\infty)_{\text{FM}} \to$ finite constant (> 0).

Thus, if $\int_0^\infty \mathcal{W}_N(f) \, df/\omega^2$ is divergent, or, in more detail, if $\mathcal{W}_N(0+)$ is finite [i.e., if $C_0 > 0$, or if $\mathcal{W}_N(0+)$ vanishes no more rapidly than $O(\omega)$ as $\omega \to 0+$], $\Omega(\infty)_{\text{FM}}$ is positively infinite and so, by Eq. (14.17*a*), *all the*

† It is assumed that $\mathcal{W}_N(f)$ can be represented by a series of the form $C_0 + C_1\omega + C_2\omega^2 + \cdots$.

unmodulated carrier intensity is now distributed among the sideband continuum produced in the course of modulation (see the example in Sec. 14.1-3). On the other hand, if $\int_0^\infty \mathcal{W}_N(f)\, df/\omega^2$ is convergent, i.e., if the modulating noise contains no spectral contribution at or near $f = 0$, so that $\mathcal{W}_N(f)$ vanishes with sufficient rapidity as $f \to 0$, then $\Omega(\infty)_{\text{FM}}$ may be a finite (positive) constant and the spectrum $\mathcal{W}_V(f)$ contains a discrete carrier term (cf. Prob. 14.4). Note also that frequency modulation may be interpreted here as an instance of phase modulation by a new noise wave, with an intensity spectrum modified according to $\mathcal{W}_{\phi\text{M}}(f) = \mathcal{W}_{\text{FM}}(f)/\omega^2$ [cf. Eqs. (14.15)].

The quantity $\Omega(t)$ depends, of course, on the spectral distribution of the input noise $\mathcal{W}_N(f)$. When this noise is shaped by a realizable filter [cf. Sec. 2.2-5(3)], $\Omega(t)$ may be expressed in series of the type

$$\begin{aligned} \Omega(t)_{\text{FM}} &= a_2t^2 + a_3|t|^3 + a_4t^4 + a_5|t|^5 + \cdots \\ \Omega(t)_{\phi\text{M}} &= b_0 + b_1|t| + b_2|t|^2 + \cdots \end{aligned} \qquad a_2, b_0 \neq 0 \qquad (14.19a)$$

where, however, the coefficients $a_2, a_3, \ldots, b_0, b_1, \ldots$, etc., are determined only when $\mathcal{W}_N(f)$ is given. A general Taylor's series is not possible, since $\Omega(t)$ possesses only a finite number of finite derivatives at $t = 0$. On the other hand, when nonrealizable filters (e.g., gauss or rectangular responses) determine the spectrum $\mathcal{W}_N(f)$, a Taylor's development of $\Omega(t)$ about $t = 0$ is possible and the series (14.19*a*) now take the form

$$\Omega(t)_{\text{FM}} = a_2't^2 + a_4't^4 + \cdots \qquad \Omega(t)_{\phi\text{M}} = b_0' + b_2't^2 + b_4't^4 + \cdots \qquad (14.19b)$$

where the coefficients may be established from Eqs. (14.15) by differentiation to all orders. Note, as expected, that $\Omega(t)$ is always an even function of t; for some examples, see Sec. 14.1-3 and Probs. 14.1 to 14.3, 14.5.

14.1-3 Examples of Phase Modulation and Frequency Modulation by Stationary Normal Noise. As an illustration of the foregoing remarks, let us suppose that the modulating noise has been shaped by an *RC* filter, so that Eqs. (3.55) apply,

$$\mathcal{W}_N(f) = \frac{4\psi/\Delta\omega_N}{1 + \omega^2/\Delta\omega_N{}^2} \qquad K_N(t) = \psi e^{-\Delta\omega_N|t|} \qquad K_N(0) \equiv \psi = \overline{V_N{}^2} \quad (14.20)$$

and let us first examine the case of *phase modulation.*

From Eq. (14.15*b*), we get at once

$$\Omega(\zeta)_{\phi\text{M}} = \psi(1 - e^{-|\zeta|}) \qquad \zeta = \Delta\omega_N t \qquad (14.21)$$

which is expressible as a series of the type (14.19*a*), since the *RC* network is realizable. The covariance function and spectrum of the modulated carrier thus become explicitly

$$K_V(\zeta)_{\phi\text{M}} = \frac{A_0{}^2}{2} e^{-\mu_\phi{}^2(1-e^{-|\zeta|})} \cos \frac{\omega_c\zeta}{\Delta\omega_N} \qquad \mu_\phi{}^2 = D_\phi{}^2\psi \qquad (14.22a)$$

and therefore

$$\mathcal{W}_V(f)_{\phi\mathrm{M}} = \frac{A_0^2 e^{-\mu_\phi^2}}{\Delta\omega_N} \sum_{n=0}^{\infty} \frac{(\mu_\phi^2)^n}{n!} \int_0^\infty e^{-n\zeta} \cos \beta\zeta \, d\zeta \qquad \beta \equiv \frac{\omega - \omega_c}{\Delta\omega_N}$$

$$= \frac{A_0^2}{\Delta\omega_N} e^{-\mu_\phi^2} \sum_{n=0}^{\infty} \frac{(\mu_\phi^2)^n}{n!} \frac{n}{n^2 + \beta^2}$$

$$= \frac{A_0^2 e^{-\mu_\phi^2}}{\Delta\omega_N} \left[\pi\delta(\beta - 0) + \sum_{n=1}^{\infty} \frac{(\mu_\phi^2)^n}{(n-1)!(n^2 + \beta^2)} \right] \qquad (14.22b)$$

which is shown in Fig. 14.3 for various values of μ_ϕ^2. Here, μ_ϕ is the modulation index for phase modulation by the normal noise process of average

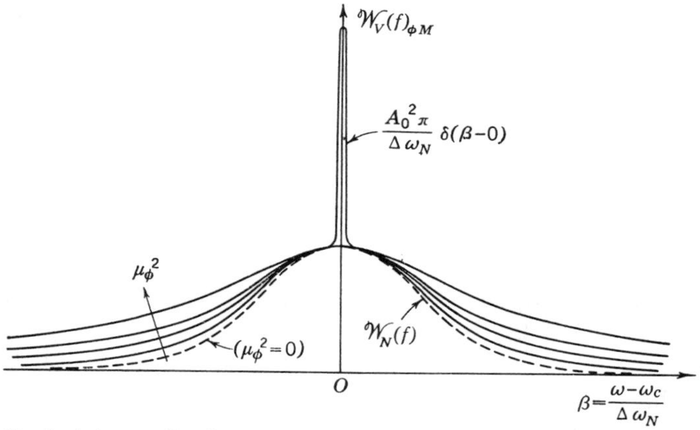

FIG. 14.3. Typical (normalized) spectra of a carrier *phase-modulated* by *RC* noise, with fixed bandwidth, for various rms angular deviations $(\overline{\Phi_N^2})^{1/2}$. (Not numerically precise.)

intensity ψ; μ_ϕ is thus an rms angular deviation [cf. Eq. (14.22*a*)]. Note the discrete carrier term, and observe that, as the rms phase deviation is increased, the spectrum (here normalized to unity at $f = f_c$, excluding the carrier component) is spread relative to that of the modulation $V_N(t)$. At the other extreme of $\mu_\phi \to 0$, the (normalized) spectrum $\mathcal{W}_V(f)_{\phi\mathrm{M}}$ never becomes narrower than $\mathcal{W}_N(f)$, although in the limit only the "spike" at the carrier frequency is significant.

In the more common cases of *frequency modulation*, we readily find by contour integration for the *RC* modulating noise assumed here that[3]

$$\Omega(\zeta)_{\mathrm{FM}} = \frac{\overline{V_N^2}}{\Delta\omega_N^2} \left[\frac{2}{\pi} \int_0^\infty \frac{1 - \cos \zeta}{\zeta^2(1 + \zeta^2)} \, d\zeta \right] = \frac{\psi(e^{-|\zeta|} + |\zeta| - 1)}{\Delta\omega_N^2} \qquad (14.23)$$

which is another example leading to series of the form (14.19*a*). The covariance function and spectrum (14.16) now become

$$K_V(\zeta)_{FM} = \frac{A_0^2}{2}\, e^{-\mu_F^2(e^{-|\zeta|}+|\zeta|-1)} \cos \frac{\omega_c \zeta}{\Delta\omega_N} \qquad \mu_F^2 \begin{cases} \equiv \dfrac{\overline{\dot{\Phi}_N^2}}{\Delta\omega_N^2} \\ = \dfrac{D_F^2 \psi}{\Delta\omega_N^2} \end{cases} \tag{14.24}$$

and

$$\mathcal{W}_V(f)_{FM} = \frac{A_0^2}{\Delta\omega_N} \int_0^\infty \cos \beta\zeta e^{-\mu_F^2(e^{-|\zeta|}+|\zeta|-1)}\, d\zeta$$

$$= \frac{A_0^2}{\Delta\omega_N} e^{\mu_F^2} \sum_{n=0}^{\infty} \frac{(-\mu_F^2)^n(\mu_F^2 + n)}{n![\beta^2 + (\mu_F^2 + n)^2]} \qquad \beta \equiv \frac{\omega - \omega_c}{\Delta\omega_N} \tag{14.25a}$$

or

$$\mathcal{W}_V(f)_{FM} = A_0^2(\overline{\dot{\Phi}_N^2})^{-\frac{1}{2}} \mu_F e^{\mu_F^2} \sum_{n=0}^{\infty} \frac{(-\mu_F^2)^n(\mu_F^2 + n)}{[(\beta')^2 + (\mu_F^2 + n)^2]n!} \qquad \beta' \equiv \frac{\omega - \omega_c}{(\overline{\dot{\Phi}^2})^{\frac{1}{2}}} \tag{14.25b}$$

curves of which are shown in Fig. 14.4a for different values of μ_F when the bandwidth $\Delta\omega_N$ (measured between half-power points) of the modulating noise is held constant and in Fig. 14.4b when the rms angular frequency deviation is fixed instead.

For $\mu_F \ll 1$, the spectra become very narrow and sharp about β, $\beta' = 0$, while, for $\mu_F \gg 1$, these spectra approach the expected limiting form in both instances (cf. Sec. 14.2), but on quite different absolute scales. Thus, when the deviation is increased, the spectrum spreads indefinitely as $\mu_N \to \infty$, whereas, for $\overline{\dot{\Phi}_N^2}$ fixed, the spectral broadening is limited to a (gaussian) distribution of constant width. Also, unlike phase modulation with noise shaped by this filter, all the original carrier intensity now appears in the sideband continuum produced in modulation.

We may now explain the various spectral structures obtained above on more physical grounds, if we observe first that frequency modulation or phase modulation is a highly nonlinear operation, in so far as the components of the modulating wave are concerned. (The conditions of a proper modulation ensure that there may be present only one component of the carrier—i.e., a portion of the carrier itself—and that the carrier and components of the modulation are combined to form sidebands in the usual way.) Referring to Fig. 14.5, we illustrate several conditions of a modulating noise.

We observe first in general that only $c \times c$ (carrier $\times$ carrier) and $c \times n$ (carrier $\times$ noise) terms remain after the very nonlinear operation of modulation. The $n \times n$ terms and other harmonics of the carrier are removed by appropriate filters since they lie in spectral zones outside that of the carrier region. The spectrum of the modulated carrier in general has the

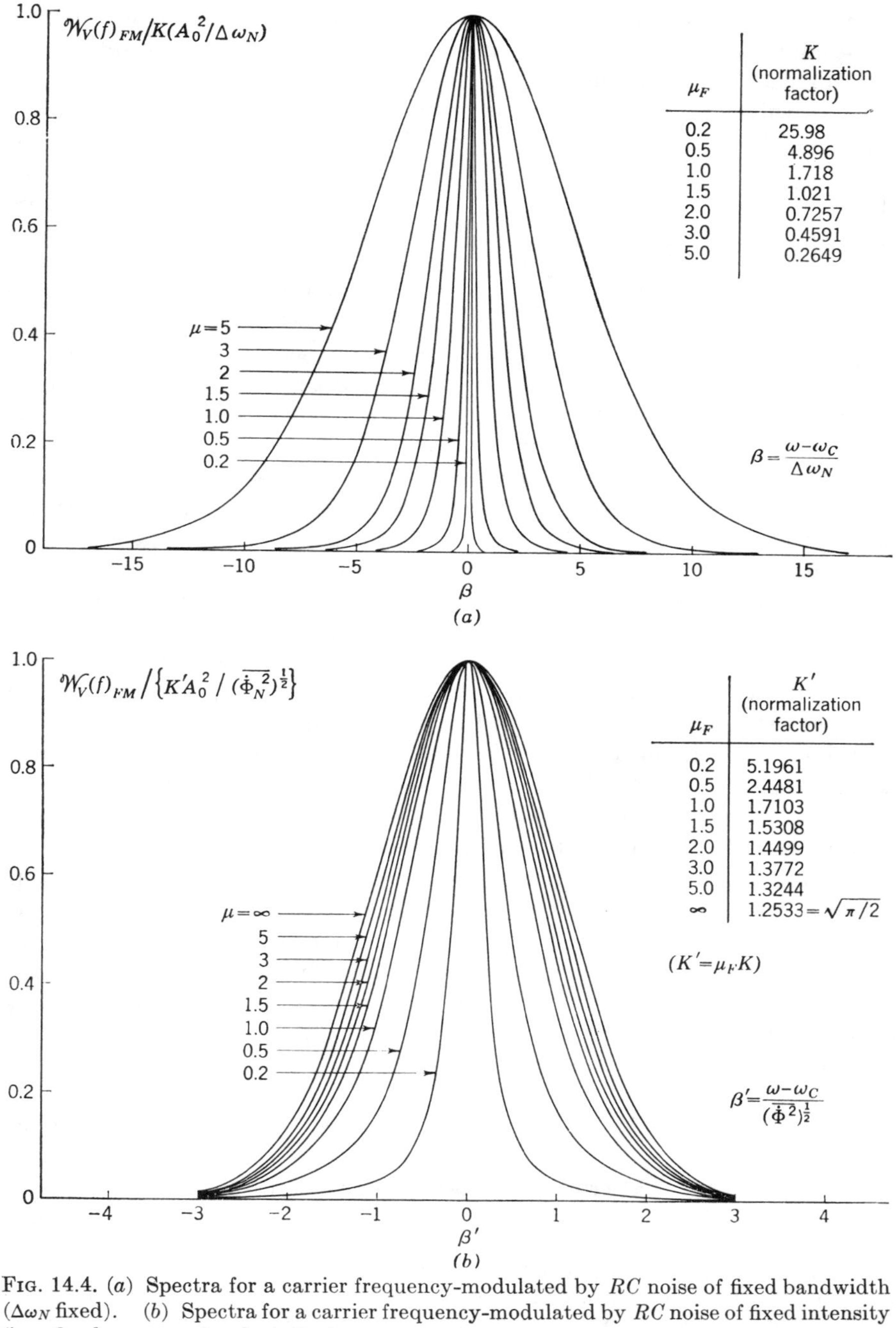

FIG. 14.4. (*a*) Spectra for a carrier frequency-modulated by *RC* noise of fixed bandwidth ($\Delta\omega_N$ fixed). (*b*) Spectra for a carrier frequency-modulated by *RC* noise of fixed intensity (i.e., fixed rms angular deviation).

form shown in Fig. 14.3, with a finite discrete carrier component, for both phase and frequency modulation. Thus, from Fig. 14.5 we expect with a modulation (*a*) that there is a nonvanishing carrier. Now, as the spectrum is widened [(*b*), (*c*), etc.], the number of $c \times n$ components forming the sideband continuum increases, not only for sum and difference terms near $f_c \pm 0$, but at all spectral points. This produces a broader and more intense spectrum than does (*a*).

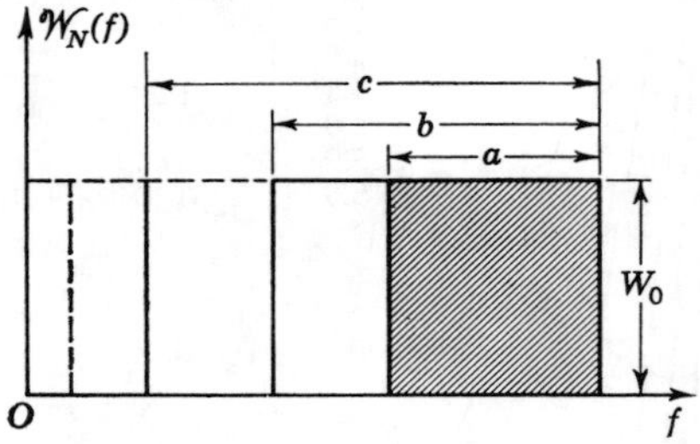

FIG. 14.5. Different bandwidths of the spectrum of a modulating noise process.

Since the total mean power in the wave necessarily remains unchanged for this pure angle modulation, the increase in sideband power as the band is widened can take place only at the expense of the carrier component, which is therefore reduced. The vanishing of the discrete term as $\mathcal{W}(f)_{FM}$ is broadened in the above fashion depends on the shape of the spectrum near $f = 0$ [e.g., Eq. (14.18)]. Accordingly, it is the *low-frequency* terms of the modulating wave which are significant in determining the presence or absence of a discrete carrier term. For the noise of the ϕM example, the spectrum $\mathcal{W}(f)_N$ vanishes sufficiently rapidly near $f = 0$ to yield a residual carrier term, while, in the FM case above [Eqs. (14.23) et seq.], the carrier is completely absorbed. If $\mathcal{W}(f)_N$ should behave $O(\omega^2)$ as $f \to 0$, or should have the distributions of Fig. 14.5*a* to *c*, a carrier spike would remain.†

14.1-4 Modulation Indices. In the case of frequency modulation by a sine wave, the concept of modulation index is quite straightforward and natural. However, when we attempt to extend the concept to more complicated waveforms, difficulties arise as to precisely what constitutes the rate at which the prescribed (rms) frequency deviation is carried out, as this rate is a rather involved function of the modulating waveform. We may, for example, define the *mean frequency* of deviation as the average number of "zero crossings" per second (cf. Sec. 9.4-1) and choose this as our measure. Another closely related possibility is provided by the *centroid frequency* $\rho/2\pi$ [cf. Eqs. (9.82), (9.84)], defined in terms of the intensity spectrum according to

$$\rho^2 \equiv \int_0^\infty \omega^2 \mathcal{W}_N(f)_{FM}\, df \Big/ \int_0^\infty \mathcal{W}_N(f)_{FM}\, df = \overline{\dot{V}_{FM}{}^2}/\overline{V_{FM}{}^2} \qquad \omega = 2\pi f \quad (14.26)$$

while several somewhat cruder choices are the median spectral frequency or simply the bandwidth of the equivalent rectangular spectrum [Eq. (3.99*a*)]

† A similar effect is noted for periodic modulations of finite bandwidth, where, of course, the resulting sidebands necessarily have a discrete, rather than a continuous, structure. The argument above also applies equally well to phase modulation, with the help of the relation $\mathcal{W}_{\phi M}(f) = \mathcal{W}_{FM}(f)/\omega^2$.

of the modulating noise. The *mean frequency* ($\sim\overline{|\dot{V}|}$), while more accurately a measure of what is actually taking place, is not easy to obtain analytically and frequently has no finite measure, while the centroid frequency, though useful, often does not exist either, as in the case of a nondifferentiable random process such as thermal noise shaped by an *RC* filter.

Compromise measures, such as the median frequency, bandwidth, or combinations of these quantities, can be constructed for specific modulations, but a single general measure based on the intensity spectrum alone does not appear adequate. For example, with the modulation (*a*) of Fig. 14.5, the central or the median frequency of the band provides a much better indication of the average rate of swing than does the bandwidth, while, for the modulation (*c*), either measure may serve. Since most modulations may be classed as broadband (cf. Sec. 3.3-2), we adopt here the bandwidth ($\Delta\omega/2\pi$) of the modulation itself as the measure of the rate of frequency deviation. This is done for random disturbances, and as an alternative to the choice of fundamental frequency for periodic modulations. We remember, however, that in narrowband cases it is the central frequency, which may be many multiples of the bandwidth, that is a more descriptive measure. Accordingly, for frequency modulation we define the *modulation index* as

$$\mu_{\mathrm{FM}} \equiv D_F \langle V_{\mathrm{FM}}{}^2 \rangle^{1/2} / \Delta\omega \tag{14.27a}$$

which for noise modulations becomes

$$\mu_F = D_F (\overline{V_N{}^2})^{1/2} / \Delta\omega_N \tag{14.27b}$$

For narrow (finite) bands, we can also use an *effective* modulation index, represented by

$$\mu_e \equiv \mu_{\mathrm{FM}} / (2\pi f_c' / \Delta\omega) \tag{14.27c}$$

where $f_c'/\Delta f$ is the ratio of central frequency to the effective bandwidth of the modulating wave.

It is clear from the somewhat arbitrary nature of these definitions that factors 2, π, etc., that may appear because of normalization, spectral shape, definition of bandwidth, etc., are not too critical. Note, also, that with periodic modulations it may be more convenient for analytical purposes to use as the index $\mu_{\mathrm{FM}} = D_F B_0 / \omega_a$, etc., expressed in terms of *peak* amplitudes and the fundamental angular frequency. Thus, for the sinusoid of Sec. 14.1-1, $\mu'_{\mathrm{FM}} = \sqrt{2}\,\mu_{\mathrm{FM}}$ (rms) of Eq. (14.27*a*). Similarly, for phase modulation we have

$$\mu_{\phi\mathrm{M}} = D_\phi \langle V_{\phi\mathrm{M}}{}^2 \rangle^{1/2} \qquad \mu_{\phi(\mathrm{noise})} = D_\phi \psi^{1/2} \tag{14.27d}$$

and for periodic waves $\mu_{\phi\mathrm{M}}$ may also be given in terms of peak amplitudes, e.g., for the sinusoid $\mu'_{\phi\mathrm{M}} = \sqrt{2}\,\mu_{\phi\mathrm{M}}$ (rms) of Eq. (14.27*d*).

14.1-5 Mixed Angle Modulation. The analysis of the preceding section is easily extended to cases where the modulation consists of the sum of a periodic and a stationary random process, e.g.,

$$V_{mod}(t) = V_N(t) + V_S(t) \qquad \text{with} \begin{cases} \dot{\Phi}(t)_{\text{FM}} = D_F(V_N + V_S) \\ \Phi(t)_{\phi\text{M}} = D_\phi(V_N + V_S) \end{cases} \tag{14.28}$$

as before [cf. Eqs. (14.2)]. If the modulations are independent, a simple extension of Eqs. (14.13) gives for the covariance function of the modulated carrier

$$K_V(t) = \frac{A_0{}^2}{2} \operatorname{Re} \left(e^{-i\omega_c t} \boldsymbol{E}\{e^{-i\Phi(t_1)_{S+N} + i\Phi(t_2)_{S+N}}\}\right) \qquad t = t_2 - t_1 \tag{14.29a}$$

$$= \frac{A_0{}^2}{2} \operatorname{Re} \left[e^{-i\omega_c t} F_2(-1,1;t)_{\Phi_N} F_2(-1,1;t)_{\Phi_S}\right] \tag{14.29b}$$

where the characteristic function for the periodic signal is obtained from Eq. (5.7), viz.,

$$F_2(-1,1;t)_{\Phi_S} = \frac{1}{T_0} \int_0^{T_0} e^{-i\Phi(t_0)_S + i\Phi(t_0+t)_S}\, dt_0 \tag{14.30}$$

or equivalently from Eqs. (14.8), (14.9).

To illustrate with a specific example, let us find the spectrum when the modulation consists of a sine wave $B_0 \cos \omega_a t$ and broadband normal noise shaped by the RC filter of Eq. (14.20). For both phase and frequency modulation, $F_2(-1,1;t)_{\Phi_S} = \sum_0^\infty \epsilon_m J_m{}^2(\mu) \cos m\omega_a t$ [cf. Eq. (5.12b)], while $\Omega(\zeta)$ follows at once from Eqs. (14.21), (14.23). The covariance functions of the modulated carrier become

$$K_V(\zeta)_{\phi\text{M}} = \frac{A_0{}^2}{2} \sum_{m=0}^\infty \epsilon_m J_m{}^2(\mu_{\phi\text{M}}) e^{-\mu_\phi{}^2(1-e^{-|\zeta|})} \cos \frac{\omega_c \zeta}{\Delta\omega_N} \cos \frac{m\omega_a \zeta}{\Delta\omega_N} \tag{14.31}$$

$$K_V(\zeta)_{\text{FM}} = \frac{A_0{}^2}{2} \sum_{m=0}^\infty \epsilon_m J_m{}^2(\mu_{\text{FM}}) e^{-\mu_F{}^2(e^{-|\zeta|}+|\zeta|-1)} \cos \frac{\omega_c \zeta}{\Delta\omega_N} \cos \frac{m\omega_a \zeta}{\Delta\omega_N} \tag{14.32}$$

showing that the nonlinear process of modulation generates a continuum which contains $(n \times n) \times c$ terms (for $m = 0$) and $(s \times n) \times c$ terms $(m \geq 1)$ in both instances, with an additional discrete spectrum for phase modulation composed of $c \times c$ and $s \times c$ components. The average intensity in this discrete spectrum is

$$P_{f_c} = |K_V(\infty)_{\phi\text{M}}| = \frac{A_0{}^2}{2} e^{-\overline{\Phi_N{}^2}} \tag{14.33}$$

as before [cf. Eq. (14.17a)].

The spectral distribution corresponding to Eq. (14.31) becomes

$$\mathcal{W}_V(f)_{\phi M} = \frac{A_0^2 e^{-\mu_\phi^2}}{\Delta\omega_N} \sum_{m=0}^{\infty} \frac{\epsilon_m}{2} J_m^2(\mu_{\phi M}) \sum_{n=0}^{\infty} \frac{(\mu_\phi^2)^n}{n!} \left(\frac{n}{n^2 + \beta_{+m}^2} + \frac{n}{n^2 + \beta_{-m}^2} \right)$$

$$\text{with } \beta_{\pm m} = \frac{\omega - \omega_c \mp m\omega_a}{\Delta\omega_N} \qquad (14.34)$$

and is sketched in Fig. 14.6. Note the discrete, as well as the continuous,

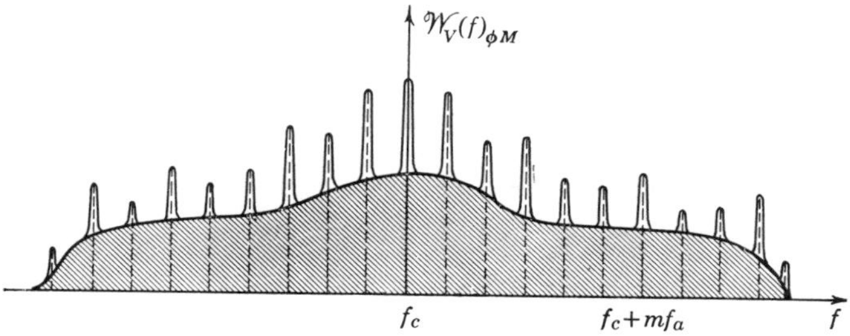

FIG. 14.6. Intensity spectrum of a carrier phase-modulated by a sine wave and normal noise through an RC filter.

character of this spectrum, at $f = f_c \pm mf_a$. Similarly, for frequency modulation by this composite modulation, one finds from Eq. (14.32) that

$$\mathcal{W}_V(f)_{FM} = \frac{A_0^2}{\Delta\omega_N} e^{\mu_F^2} \sum_{m=0}^{\infty} \frac{\epsilon_m}{2} J_m^2(\mu_{FM}) \sum_{n=0}^{\infty} \frac{(-\mu_F^2)^n(\mu_F^2 + n)}{n!}$$
$$\times \left[\frac{1}{(\mu_F^2 + n)^2 + \beta_{+m}^2} + \frac{1}{(\mu_F^2 + n)^2 + \beta_{-m}^2} \right] \qquad (14.35)$$

At this point, it is convenient to extend the definition of modulation index discussed in Sec. 14.1-4 to include the added signal (or noise). Let us begin by defining an effective bandwidth $\Delta\omega_{S+N}$ for the sum of a periodic signal and noise,

$$\Delta\omega_{S+N}^2 \equiv \frac{\Delta\omega_N^2 + p\,\Delta\omega_S^2}{1 + p} = \frac{\Delta\omega_N^2(1 + p\gamma^2)}{1 + p} \qquad \gamma \equiv \frac{\Delta\omega_S}{\Delta\omega_N} \qquad (14.36)$$

where $p \equiv \overline{V_S^2}/\overline{V_N^2} = \mu_{FM}^2/\mu_F^2$, and the modulation index for the signal is $\mu_{FM} = (\overline{V_S^2})^{1/2}/\Delta\omega_S$ here [cf. Eqs. (14.27)]. For this composite phase or frequency modulation, we accordingly define the indices

$$(\mu_{S+N}^2)_{FM} \equiv \frac{\overline{\dot\Phi_S^2} + \overline{\dot\Phi_N^2}}{\Delta\omega_{S+N}^2} = \frac{(1 + p)^2\mu_F^2}{1 + p\gamma^2}$$
$$= \frac{(1 + p)(\mu_{FM}^2\gamma^2 + \mu_F^2)}{1 + p\gamma^2} \qquad (14.37a)$$
$$(\mu_{S+N}^2)_{\phi M} \equiv \mu_\phi^2(1 + p) = \mu_{\phi M}^2(1 + p)/p \qquad (14.37b)$$

Here, $\mu_{FM} \equiv (\overline{\dot{\Phi}_S{}^2})^{1/2}/\Delta\omega_S$, and $\mu_F \equiv (\overline{\dot{\Phi}_N{}^2})^{1/2}/\Delta\omega_N$. Figure 14.7*a*, *b* shows a variety of FM spectra, based on Eq. (14.35) for different modulation ratios *p*. Observe that there is no discrete spectrum (as long as $\overline{\dot{\Phi}_N{}^2} > 0$); only $(s \times n) \times c$ and $(n \times n) \times c$ components are produced in this instance.

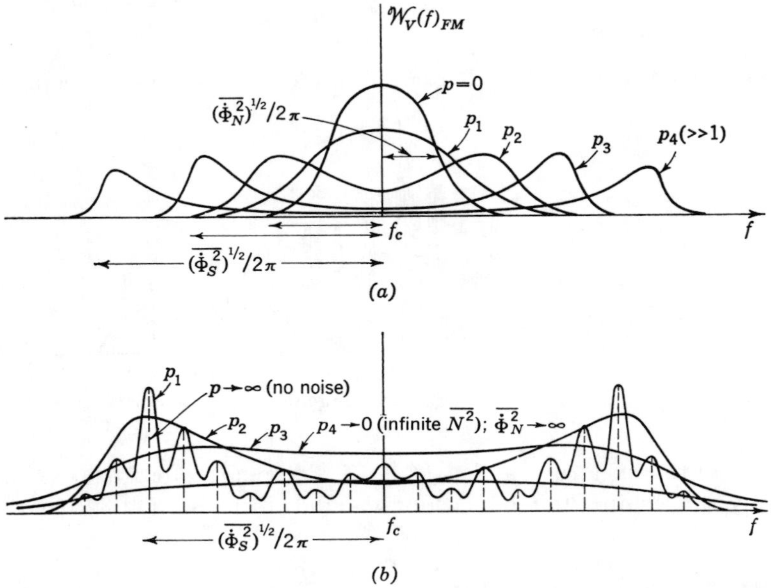

FIG. 14.7. (*a*) Intensity spectra of a carrier frequency-modulated by a sinusoid and normal noise of fixed intensity (μ_N and $\Delta\omega_S$ fixed). (*b*) Intensity spectra of a carrier frequency-modulated by a sinusoid of constant strength and normal noise (μ_S fixed, $\Delta\omega_N$ fixed).

Clearly, many different spectral shapes are possible, depending on our choice of p, μ_{FM}, μ_F, ω_a, $\overline{\dot{\Phi}_N{}^2}$. See also Fig. 9.5 and Prob. 14.13.

PROBLEMS

14.1 A carrier is frequency-modulated by a square wave of period T_0 $(= 2\pi/\omega_a)$ (cf. Fig. P 14.1). Show that the modulated wave is

$$V^{(i)}(t) = A_0 \operatorname{Re} \left\{ e^{i\omega_c t} \sum_{-\infty}^{\infty} \frac{i\mu_{FM}}{\pi(m^2 - \mu_{FM}{}^2)} [1 - e^{-\pi i \mu_{FM}}(-1)^m] e^{im\omega_a t} \right\} \tag{1}$$

Obtain the covariance function

$$K_V(t) = \frac{A_0{}^2\mu_F{}^2}{\pi^2} \sum_{m=0}^{\infty} \frac{\sin^2(\pi\mu_{FM}/2)\delta^m_{\text{even}} + \cos^2(\pi\mu_{FM}/2)\delta^m_{\text{odd}}}{(m^2 - \mu_{FM}{}^2)^2} \times [\cos(\omega_c + m\omega_a)t + \cos(\omega_c - m\omega_a)t] \tag{2}$$

where $\delta^m_{\text{odd}} = 1$ (m odd), $\delta^m_{\text{odd}} = 0$ (m even), etc., and $\mu_{FM} = D_F B_0/\omega_a$.

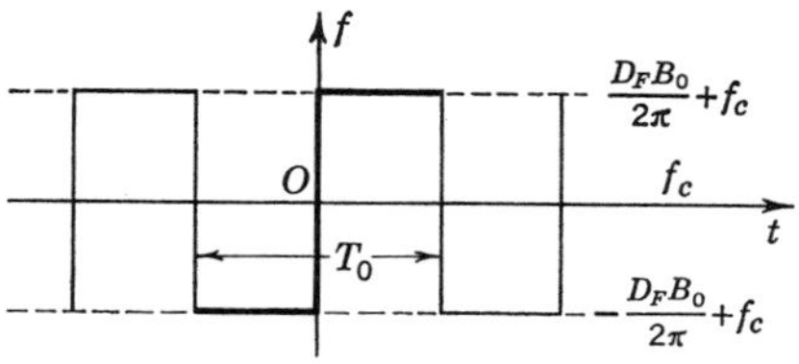

FIG. P 14.1. Square-wave frequency modulation.

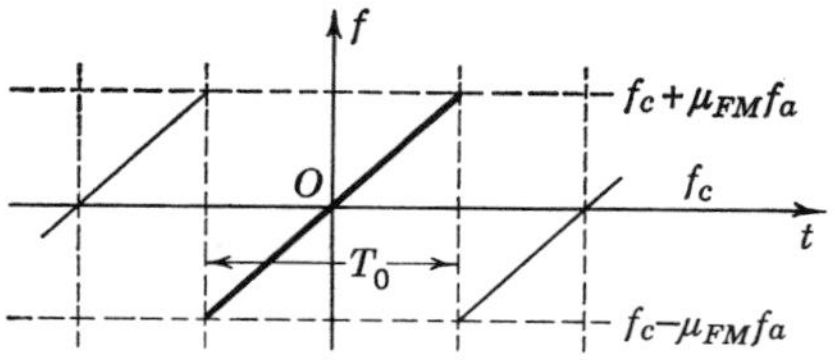

FIG. P 14.2. Saw-tooth frequency sweep.

14.2 Instead of the square wave of Prob. 14.1, a linear frequency variation of the type shown in Fig. P 14.2 is used. Show that the instantaneous modulated wave $V^{(i)}(t)$ becomes

$$V^{(i)}(t) = A_0 \operatorname{Re}\left(e^{i\omega_c t} \sum_{m=-\infty}^{\infty} a_m e^{im\omega_a t}\right) \tag{1}$$

where $\mu_F = D_F B_0/\omega_a$ and

$$a_m = \frac{e^{\pi i m^2/2\mu_{FM}}}{2\sqrt{\mu_{FM}}}\left[C\left(\frac{m+\mu_{FM}}{\sqrt{\mu_{FM}}}\right) - C\left(\frac{m-\mu_{FM}}{\sqrt{\mu_{FM}}}\right) - iS\left(\frac{m+\mu_{FM}}{\sqrt{\mu_{FM}}}\right) + iS\left(\frac{m-\mu_{FM}}{\sqrt{\mu_{FM}}}\right)\right] \tag{2}$$

in which C and S are Fresnel integrals, defined by

$$C(x) \equiv \int_0^x \cos\frac{\pi}{2}t^2\,dt \qquad S(x) \equiv \int_0^x \sin\frac{\pi}{2}t^2\,dt \tag{3}$$

With the help of Cornu's spiral, discuss the behavior of the amplitude and intensity spectrum, with particular attention to large modulation indices and fixed deviations. Note that the intensity of the mth component is

$$P_m = a_m a_m^* \frac{A_0^2}{2} \tag{4}$$

14.3 In this case, Fig. P 14.3 shows the time structure of the frequency modulation, which is now a symmetrical linear sweep. Equation (1) of Prob. 14.2 again applies, where we must show that

$$2\sqrt{2\mu_{FM}}a_m = 2\operatorname{Re}\{e^{\pi i(\mu_{FM}+m)^2/4\mu_{FM}}[C(a) + C(b) - iS(a) - iS(b)]\} \tag{1}$$

with

$$\mu_{FM} = \frac{D_F B_0}{\omega_a} \qquad a = \sqrt{2\mu_{FM}} - \frac{\mu_{FM}+m}{\sqrt{2\mu_{FM}}} \qquad b = \frac{\mu_{FM}+m}{\sqrt{2\mu_{FM}}} \tag{2}$$

and C and S the Fresnel integrals of Eqs. (3), Prob. 14.2. As in the above problem, discuss the behavior of $|a_m|$ with large modulation indices.

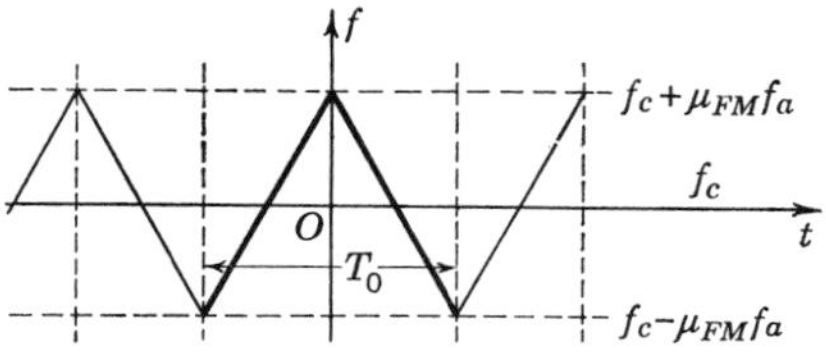

FIG. P 14.3. Symmetrical linear frequency sweep.

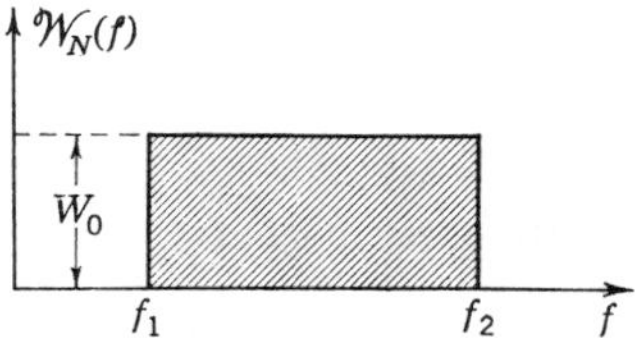

FIG. P 14.4. Intensity spectrum of frequency-modulating noise.

14.4 A carrier is frequency-modulated by gaussian noise having the intensity spectrum of Fig. P 14.4. Show that

$$\Omega(t)_{\text{FM}} = \frac{W_0}{2\pi}\left[\frac{\omega_2 - \omega_1}{\omega_1\omega_2} + \frac{\cos \omega_2 t}{\omega_2} - \frac{\cos \omega_1 t}{\omega_1} + tSi(\omega_2 t) - tSi(\omega_1 t)\right] \tag{1}$$

where
$$Si(x) \equiv \int_0^x \frac{\sin y}{y}\, dy \tag{2}$$

Show also that the intensity in the carrier after modulation is

$$P_{f_c} = \frac{A_0^2}{2}\, e^{-\overline{\dot{\Phi}_N^2}/\omega_1\omega_2} \tag{3}$$

Defining Γ as an effective modulation index, in terms of the centroid frequency $\rho/2\pi$ [Eq. (14.26)], show that for $\Gamma \gg 1$ the spectrum becomes

$$\mathcal{W}_V(f)_{\text{FM}} \simeq \frac{\pi A_0^2}{(\overline{\dot{\Phi}_N^2})^{1/2}}\left[\phi^{(0)}\left(\frac{\omega - \omega_c}{\rho\Gamma}\right) + \frac{1}{24\Gamma^2}\phi^{(4)}\left(\frac{\omega - \omega_c}{\rho\Gamma}\right) + O(\Gamma^{-4})\right] \tag{4}$$

where $\phi^{(n)}(x)$ is given by Eq. (A.1.3).

14.5 A carrier is frequency-modulated by a sinusoid, reflected from a moving target, and is then received at a time τ later at the transmitter. If this received wave is suitably combined with the original (undelayed) modulation and observed by an ideal phase detector (with heavy limiting), the resulting l-f output is found to be

$$V(t) = A_0 \cos\left[\omega_d(t - \tau) - \omega_c\tau + \phi(t - \tau) - \phi(t)\right] \tag{1}$$

where ω_c is the original carrier's (angular) frequency and ω_d is the Doppler shift of the moving target (with respect to the transmitter and receiver). Here,

$$\phi(t) = B \cos \omega_a t \qquad B \equiv \text{FM modulation index} \tag{2}$$

(*a*) Show that the covariance function of the output V is

$$\begin{aligned} K_V(t) &= \frac{A_0^2}{2}\operatorname{Re}\left[F_\phi(-1,1;t)e^{i\omega_d t}\right] \qquad \omega_a B \ll \omega_d \qquad \omega_a \ll \omega_d \\ &= \frac{A_0^2}{2}\cos \omega_d t\left[\sum_{m=0}^{\infty} \epsilon_m J_m^2\left(2B \sin\frac{\omega_a\tau}{2}\right)\cos m\omega_a t\right] \end{aligned} \tag{3}$$

Obtain the spectrum, and sketch for small and large modulation indices and delays.

(*b*) Repeat (*a*), now for a stationary normal noise modulation

$$\phi(t) = \int_{t_0}^{t} D_0 N(t')\, dt' \tag{4}$$

and show specifically that

$$K_V(t) = \frac{A_0^2}{2}\left\{\exp\left[-2D_0^2\int_0^\infty \mathcal{W}_N(f;\tau)(1 - \cos \omega t)\frac{df}{\omega^2}\right]\right\}\cos \omega_d t \tag{5}$$

where
$$\mathcal{W}_N(f;\tau) = \mathcal{W}_N(f)(1 - \cos \omega\tau) \tag{6}$$

and $\mathcal{W}_N(f)$ is the intensity of the modulation $N(t)$.

14.2 *Limiting Forms*

While the analysis of the preceding section is valid for all values of modulation index (and bandwidth, spectra, etc.) that do not violate the modulation conditions, these general relations are not particularly useful under conditions of high or low modulation index. However, at such extremes, as we shall see below, it is possible to recast the results for the covariance function and spectrum into more convenient forms, which lend themselves much more readily to computation. In the following, we shall confine our attention chiefly to random modulations, although the general conclusions apply equally well for the deterministic cases.

We distinguish two principal conditions of operation: (1) low modulation indices ($\mu^2 \ll 1$), and (2) high modulation indices ($\mu^2 \gg 1$). Let us start with the former.

14.2-1 Low Modulation Indices.[11] In the case of *phase modulation* by a stationary random process, we may express the ensemble of modulated carriers quite generally as

$$V(t)_{\phi M} = A_0 \operatorname{Re}\left(e^{i\omega_c t - i\mu_\phi \psi^{-1/2} V_N(t)}\right) \tag{14.38}$$

where $\mu_\phi = D_\phi \psi^{1/2}$, $\psi = \overline{V_N^2}$. If we assume further that $V_N(t)$ represents a process for which the mean-square expansion of the exponential in Eq. (14.38) is convergent, and that (without much loss of generality) the odd moments of $V_N(t)$ vanish, we can write, for suitably small indices μ_ϕ,

$$V(t)_{\phi M} = A_0 \left[\cos \omega_c t - \mu_\phi \psi^{-1/2} V_N(t) \sin \omega_c t - \frac{\mu_\phi^2}{2} \psi^{-1} V_N(t)^2 \cos \omega_c t + \cdots \right] \tag{14.39}$$

The covariance function of this becomes directly

$$K_V(t)_{\phi M} = \frac{A_0^2}{2} \cos \omega_c t \left[\left(1 - \frac{\mu_\phi^2}{2} - \frac{\mu_\phi^4}{24} \psi^{-2} \overline{V_N^4} + \cdots \right) + \psi^{-1} \mu_\phi^2 K_N(t) + \frac{\mu_\phi^4}{12} \psi^{-2} \left(3\overline{V_{N1}^2 V_{N2}^2} - 2\overline{V_{N1} V_{N2}^3}\right) + \cdots \right] \tag{14.40}$$

Thus, *independent of the particular statistics of the modulation*, we obtain a result which is identical $O(\mu_\phi^2)$ with what we have already found in the case of a suitably weak *amplitude* modulation by the same process, now with an AM modulation index $\lambda = \mu_\phi \psi^{1/2}$ [cf. Sec. 12.1 and Eq. (14.40)]. For normal noise modulations, Eq. (14.40) reduces specifically to

$$K_V(t)_{\phi M} = \frac{A_0^2}{2} \cos \omega_c t \left\{\left(1 - \frac{\mu_\phi^2}{2} - \frac{\mu_\phi^4}{8} + \cdots \right) + \mu_\phi^2 k_N(t) + \frac{\mu_\phi^4}{4} [1 + 2k_N^2(t)] + \cdots \right\} \tag{14.40a}$$

where as usual $k_N(t)$ is the normalized covariance function of the modulation $V_N(t)$. The corresponding intensity spectrum of the modulated carrier is undistinguishable $O(\mu_\phi^2)$ from that of amplitude modulation by the same process [cf. Eq. (12.10*b*)]. In general, when the index is sufficiently small, knowledge of the second-order moments (e.g., covariance function, spectrum) of the phase modulation is sufficient to give us the corresponding second-order moments for the phase-modulated process.

Low-index *frequency modulation*, however, presents a more involved picture, particularly if we wish to estimate spectral behavior well away from the carrier frequency. For the covariance function $K_V(t)$ in this instance, we can use Eqs. (14.13), where $(\Psi_N)_{FM}$ is the linear functional $\int_0^t D_F V_N(t')\,dt'$ of the stationary random modulating process $V_N(t')$.† Letting $a \equiv \psi^{-1/2}\,\Delta\omega_N$, with $\mu_F = D_F\psi^{1/2}/\Delta\omega_N$ [cf. Eq. (14.24)], we postulate once more that $K_V(t)$ can be expanded mean square, to give

$$K_V(t)_{FM} = \frac{A_0^2}{2}\cos\omega_c t\left[1 - \frac{\mu_F^2 a^2}{2}\iint_0^{|t|} K_N(|x_1 - x_2|)\,dx_1\,dx_2 + \frac{\mu_F^4 a^4}{4!}\iint_0^{|t|}\overline{V_N(x_1)\cdots V_N(x_4)}\,dx_1\cdots dx_4 + \cdots\right] \tag{14.41}$$

The term $O(\mu_F^2)$ may also be represented by the alternative spectral form in Eq. (14.15*a*). When $\mathcal{W}_N(f)$ vanishes with sufficient rapidity near $f = 0$, we may expect each term in the series above to be finite for all t, including $\lim |t| \to \infty$. However, if $\mathcal{W}_N(f)$ approaches a constant value at $f = 0$, for example, convergence of the various integrals fails as $\lim |t| \to \infty$, which reflects the fact that a discrete carrier component remains in the spectrum of the modulated process.

To determine the spectral distribution of the modulated wave, even for low indices the complete second-order statistics of the modulation V_N [embodied in $W_2(V_{N1},V_{N2};t)$] must generally be available, at least to enable us to decide when higher-order approximations should be retained or neglected. Let us illustrate this with the important example of a normal process. We assume now that $V_N(t)$ is once more gaussian, with $\overline{V_N} = 0$. Then, $K_V(t)$ is given by Eqs. (14.15*a*), (14.16*a*), and, according to whether $L \equiv \lim_{t\to\infty}\Omega(t) = \int_0^\infty \mathcal{W}_N(f)\,df/\omega^2$ is finite or not, two different sets of expressions for the spectrum $\mathcal{W}_V(f)$ can be obtained when μ_F is small.

Let us consider the case where $L < \infty$. Here, it is appropriate to separate the exponent (14.15*a*) into two terms and to expand the second (containing the carrier) in a power series. We find that

$L < \infty$:

$$K_V(t)_{FM} = \frac{A_0^2}{2}e^{-a^2\mu_F^2 L}\cos\omega_c t\left[1 + a^2\mu_F^2\int_0^\infty \mathcal{W}_N(f)\cos\omega t\,df/\omega^2 + \cdots\right] \tag{14.42a}$$

and the corresponding intensity spectrum becomes[11,12]

$L < \infty$:

$$\mathcal{W}_V(f)_{FM} \doteq \frac{A_0^2}{2}e^{-a^2\mu_F^2 L}\left[\delta(f - f_c) + \frac{a^2\mu_F^2}{2}\frac{\mathcal{W}_N(f_c - f)}{(f_c - f)^2} + O(\mu_F^4)\right] \tag{14.42b}$$

† All of whose odd moments again are assumed to vanish, for convenience.

where the term $\mathcal{W}_N(f_c + f)/(f_c + f)^2$ has been discarded as quite negligible. The resulting spectrum exhibits the expected carrier spike [cf. Eqs. (14.18) et seq.], while the continuum is that of the spectrum of the modulating noise, now centered about f_c and divided by $(f_c - f)^2$, provided that $|f_c - f|$ represents frequencies less than the bandwidth of the modulation. For frequencies out on the tails of the spectrum, however, the higher-order terms (in μ_F^2) must be used. With terms $O(\mu_F^4)$ included the spectral continuum is distributed about the carrier, f_c, *without any spectral gaps*, even though the spectrum of the modulating noise may vanish for all frequencies below some lower limit f_1 (> 0) (cf. Fig. 14.5). This continuum is the result of the many intermodulation products that are generated in the highly nonlinear operation of frequency modulation to which terms of all orders in μ_F^2 contribute. Our second case occurs for $L \to \infty$, that is, when $\int_0^\infty \mathcal{W}_N(f)\, df/\omega^2$ diverges, as happens, for example, when $\mathcal{W}_N(f)$ approaches a constant value (> 0) in the neighborhood of $f = 0$ [cf. Eq. (14.20)]. Thus, we cannot separate the exponent $\Omega(t)$ [Eq. (14.15*a*)] into the two terms above but must seek another division. This is achieved by rewriting Eq. (14.16*a*), with the help of the time-domain form (14.15*a*). We have[11]

$$K_V(t)_{FM} = \frac{A_0^2}{2} \cos \omega_c t \left\{ \exp \left[-\mu_F^2 a^2 \int_0^{|t|} (|t| - z) K_N(z)\, dz \right] \right\}$$

$$a^2 = \psi^{-1} \Delta\omega_N^2 \qquad (14.43a)$$

$$= \frac{A_0^2}{2} \cos \omega_c t \left(\exp \left\{ \mu_F^2 a^2 \left[-|t| \int_0^\infty K_N(z)\, dz + |t| \int_{|t|}^\infty K_N(z)\, dz + \int_0^{|t|} z K_N(z)\, dz \right] \right\} \right) \qquad (14.43b)$$

Retaining the first exponential and expanding the rest in a power series then gives us the desired development in this instance,

$L \to \infty$:

$$K_V(t)_{FM} = \frac{A_0^2}{2} \cos \omega_c t \left\{ e^{-\mu_F^2 a^2 \mathcal{W}_N(0)|t|/4} \times \left[1 + \mu_F^2 a^2 \int_0^\infty K_N(z) H(z,t)\, dz + O(\mu_F^4) \right] \right\} \qquad (14.44)$$

where $H(z,t) = z$ ($z < |t|$), $H(z,t) = |t|$ ($z > |t|$). The corresponding intensity spectrum is found to have the canonical form† (cf. Prob. 14.6)

$L \to \infty$:

$$\mathcal{W}_V(f)_{FM} \doteq \frac{\gamma A_0^2}{\gamma^2 + (\omega - \omega_c)^2} + \frac{\gamma^2 A_0^2}{\gamma^2 + (\omega - \omega_c)^2} \times \int_{-\infty}^{\infty} dx \frac{[\gamma^2 - (\omega - \omega_c)^2][\mathcal{W}_N(0) - \mathcal{W}_N(x)] - 2(\omega - \omega_c)(2\pi x - \omega + \omega_c)\mathcal{W}_N(x)}{[(2\pi x - \omega + \omega_c)^2 + \gamma^2]\mathcal{W}_N(0)} + O(\mu_F^6) \qquad (14.45)$$

† The first term of the relation (14.45) was obtained originally by Stewart,[13] for a specific modulation spectrum.

in which $\gamma \equiv \mu_F{}^2 a^2 \mathcal{W}_N(0)/4 = \bar{\Phi}_N{}^2 \psi^{-1} \mathcal{W}_N(0)/4 = D_F{}^2 \mathcal{W}_N(0)/4$ (14.45a)

and we have retained the terms in $f - f_c$ alone, since those in $f + f_c$ are quite negligible in comparison ($f \geq 0$). Note that the leading term of the resulting spectrum is independent of the spectral distribution of the modulating noise. However, this leading term, as in Eq. (14.42b), is a satisfactory representation only for frequencies less than the bandwidth $\Delta\omega_N/2\pi$ of the modulating noise. The higher-order terms become significant when we are concerned with the tails of the spectrum $\mathcal{W}_V(f)_{\text{FM}}$. In fact, it can be shown that the expansion (14.45) is not well suited for these higher frequencies, so that we must return to Eq. (14.43b), which is exact, if we are to obtain a suitable development of $\mathcal{W}_V(f)_{\text{FM}}$ on the spectral tails. The leading term of such a development is†

$L \to \infty$:

$$\mathcal{W}_V(f)_{\text{FM}} \simeq \frac{\gamma \mathcal{W}_N(f - f_c)}{4(\omega - \omega_c)^2 \mathcal{W}_N(0)} = \frac{D_F{}^2 \mathcal{W}_N(f - f_c)}{16(\omega - \omega_c)^2} \qquad \frac{\omega - \omega_c}{\Delta\omega_N} \gg 1 \tag{14.46}$$

This is not too surprising, since from Eq. (14.42b) we might expect the spectral intensity well away from the carrier to be $O[\mathcal{W}_N(f - f_c)/(\omega - \omega_c)^2]$, on the basis of the observation that the behavior of the modulated spectrum at sufficiently high frequencies should not depend very noticeably on whether or not the modulating spectrum extends to zero at the lower frequency limit. The results of the exact analysis bear this out.

From the above, then, we may conclude, for *low-index frequency-modulation by a normal noise process*, that, when the spectrum of the modulation extends to zero frequency [i.e., when $\mathcal{W}_N(0) \geq 0$ and $L \to \infty$], the spectrum of the modulated process has a distribution which is similar in shape to the frequency response of a (high-Q) single-tuned circuit (cf. type 4, Table 3.1), regardless of the specific spectral form of the modulation, for frequencies closer to the carrier than the bandwidth of the modulation. At frequencies farther from the carrier than this, the spectrum has a form which depends more and more on the spectral distribution of the modulation, as the tails of the spectrum are approached. In particular, if $\mathcal{W}_N(f)$ has tails of $O(f^{-m})$, then the spectrum of the modulated carrier has tails of $O(f^{-m-2})$, as Eq. (14.46) indicates. On the other hand, if $\mathcal{W}_N(f)$ vanishes at some nonzero lower limit ($f_1 > 0$) or is such that $L = \int_0^\infty \mathcal{W}_N(f)\, df/\omega^2$ is otherwise convergent, we see from Eq. (14.42b) that the spectral distribution of the modulated carrier is everywhere determined by the spectrum of the modulating noise, and it is only on the spectral tails that the two types of modulation (distinguished by $L < \infty$ and $L \to \infty$) exhibit the same form of dependence on the original modulation [cf. Eqs. (14.42b), (14.46)]. In any case, for an adequate second-moment theory of low-index frequency

† For details, see Mullen and Middleton.[11]

modulation, complete second-order statistics are required, as in amplitude modulation whenever overmodulation or other general nonlinearities occur. This is, of course, a direct consequence of the highly nonlinear character of angle modulation, and of frequency modulation in particular.

14.2-2 High Modulation Indices. When the modulation index (μ_ϕ or μ_F) is large, we may expect results that are quite different from the low-index case, since the former represents a relatively rapid frequency sweep, while the latter corresponds to one that is comparatively slow. If the modulating process is suitably (mean-square) differentiable, we can obtain general relations for the covariance functions and spectra, quite independent of the explicit statistics of the modulation. Even if the modulation is (mean-square) differentiable only once (for phase modulation) or not at all (for frequency modulation), it is still possible to obtain the limiting spectral distribution in a canonical form, as we shall demonstrate below.

(1) Phase modulation. We begin again by considering first the case of *phase modulation* by a stationary (not necessarily normal) random process [cf. Eq. (14.38)]. The covariance function of the modulated carrier is quite generally given by Eqs. (14.13*a*), (14.13*b*), where in particular

$$\begin{aligned} \Psi_N(t,t_0) &= \mu_\phi \psi^{1/2}[V_N(t_0+t) - V_N(t_0)] \\ K_V(t)_{\phi\mathrm{M}} &= \frac{A_0{}^2}{2} \operatorname{Re} \left(e^{\mp i\omega_c t} \boldsymbol{E}\{e^{i\Psi_N(t,t_0)_{\phi\mathrm{M}}}\}\right) \end{aligned} \tag{14.47}$$

Our approach when μ_ϕ is large is to recognize that this means, in effect, slow variations of $V_N(t)$ or $\Psi_{\phi\mathrm{M}}$, so that the first few terms in the (mean-square) expansion of $\Psi_{\phi\mathrm{M}}$ about $t = 0$ (with $t_0 = 0$, for the stationary processes assumed here) should ultimately yield the principal contributions to the spectrum. Accordingly, we have

$$\Psi_N(t,0) = \mu_\phi \psi^{-1/2} \left[t\dot{V}_N(0) + \frac{t^2}{2} \ddot{V}_N(0) + \frac{t^3}{3!} \dddot{V}_N(0) + \cdots \right] \tag{14.48}$$

where it is assumed that $\overline{\dot{V}_N{}^2}$, $\overline{\ddot{V}_N(0)^2}$, $\overline{\dot{V}_N(0)\ddot{V}_N(0)}$, etc., exist for mean-square convergence; i.e., the modulating process is differentiable to some finite (> 0) order at least (cf. Secs. 3.2-3, 3.2-4). Inserting Eq. (14.48) in Eq. (14.47), we may write

$$K_V(t)_{\phi\mathrm{M}} = \frac{A_0{}^2}{2} \operatorname{Re} \left[e^{\mp i\omega_c t} F_1(\xi_1,\xi_2,\xi_3, \ldots)_{\dot{V},\ddot{V},\dddot{V},\ldots} \Big|_{\xi_1 = \mu_\phi \psi^{-1/2} t,\ \xi_2 = \mu_\phi \psi^{-1/2} t^2/2,\ \ldots} \right] \tag{14.49}$$

in which F_1 is the joint first-order characteristic function of the random variables $\dot{V}$, $\ddot{V}$, etc.

Our next step is to expand Eq. (14.49), keeping only the term $i\xi_1\dot{V}_N = i\mu_\phi\psi^{-1/2}t\dot{V}_N(0)$ in the exponent of the statistical average (14.47), (14.49)

We have

$$K_V(t)_{\phi M} \doteq \frac{A_0^2}{2} \operatorname{Re} [e^{\mp i\omega_c t}(E\{e^{i\xi_1 \dot{V}_N}\} + i\xi_2 E\{\ddot{V}_N e^{i\xi_1 \dot{V}_N}\} + i\xi_3 E\{\dddot{V}_N e^{i\xi_1 \dot{V}_N}\} + \cdots)] \quad (14.50)$$

Splitting these averages into real and imaginary parts allows us to write

$$E\{e^{i\xi_1 \dot{V}_N}\} = \operatorname{Re}_1(t) + i \operatorname{Im}_1(t) = \int_{-\infty}^{\infty} [\cos(\mu_\phi \psi^{-1/2} \dot{V}_N t) + i \sin(\mu_\phi \psi^{-1/2} \dot{V}_N t)] w_1(\dot{V}_N)\, d\dot{V}_N \quad (14.51a)$$

$$E\{\ddot{V}_N e^{i\xi_2 \dot{V}_N}\} = \operatorname{Re}_2(t) + i \operatorname{Im}_2(t) = \iint_{-\infty}^{\infty} [\cos(\mu_\phi \psi^{-1/2} \dot{V}_N t) + i \sin(\mu_\phi \psi^{-1/2} \dot{V}_N t)] w_1(\dot{V}_N, \ddot{V}_N) \ddot{V}_N\, d\dot{V}_N\, d\ddot{V}_N \quad (14.51b)$$

and so on, so that the covariance function (14.50) can be expressed as

$$K_V(t)_{\phi M} \doteq \frac{A_0^2}{2} \Big\{ [\operatorname{Re}_1(t) \cos \omega_c t \pm \operatorname{Im}_1(t) \sin \omega_c t] + \frac{\mu_\phi \psi^{-1/2}}{2!} t^2 [-\operatorname{Im}_2(t) \cos \omega_c t \pm \operatorname{Re}_2(t) \sin \omega_c t] + \mu_\phi \psi^{-1/2} \frac{t^3}{3!} [-\operatorname{Im}_3(t) \cos \omega_c t \pm \operatorname{Re}_3(t) \sin \omega_c t] + O(t^4) \Big\} \quad (14.52)$$

We observe that, since $K_V(t)_{\phi M}$ must be an *even* function of t, the right-hand member of Eq. (14.52) can contain nonvanishing terms in t only if these are even. This means that the coefficient of $\mu_\phi \psi^{-1/2} t^2/2!$ vanishes, since $\operatorname{Im}_2(t)$ is odd in t, while $\operatorname{Re}_2(t)$ is even. In other words, we have $\overline{\ddot{V}_N} = 0$, from Eq. (14.51b).

The intensity spectrum is obtained in the usual way with the help of the W-K theorem (3.42) applied to Eq. (14.52). For explicit results, we shall need integral relations of the form

$$\int_0^\infty dt \cos \omega t \cos \omega_c t \int_{-\infty}^{\infty} (\cos b\dot{V}_N t) w_1(\dot{V}_N)\, d\dot{V}_N = \frac{1}{8} \frac{2\pi}{b} \left[w_1(x) \Big|_{x=(\omega-\omega_c)/b} + w_1(x) \Big|_{x=-[(\omega-\omega_c)/b]} \right] \quad (14.53a)$$

$$\int_0^\infty dt \cos \omega t \sin \omega_c t \int_{-\infty}^{\infty} (\sin b\dot{V}_N t) w_1(\dot{V}_N)\, d\dot{V}_N = \frac{1}{8} \frac{2\pi}{b} \left[w_1(x) \Big|_{x=-[(\omega-\omega_c)/b]} - w_1(x) \Big|_{x=(\omega-\omega_c)/b} \right] \quad (14.53b)$$

where $b \equiv \mu_\phi \psi^{-1/2}$ and $x = \dot{V}_N$ in w_1, for the first term of Eq. (14.52). The second term, as was indicated above, is zero, so that for the first *correction* term (in t^3) we have integrals of the type

$$\iiint_{-\infty}^{\infty} dt\, t^3 \cos \omega t \begin{Bmatrix} \cos b\dot{V}_N t \sin \omega_c t \\ \sin b\dot{V}_N t \cos \omega_c t \end{Bmatrix} \ddot{V}_N w_1(\dot{V}_N, \ddot{V}_N)\, d\dot{V}_N\, d\ddot{V}_N \tag{14.54a}$$

which can be evaluated with the aid of

$$\int_{-\infty}^{\infty} t^3 \sin 2\pi\beta t\, dt = (2\pi)^{-3}\delta^{(3)}(\beta - 0) \tag{14.54b}$$

where $\delta^{(3)}$ is the third derivative of the operator $\delta(\beta - 0)$. Applying this to Eq. (14.54a), integrating over t, after some algebra we obtain for the first correction term of the intensity spectrum

$$\frac{A_0^2}{6b^3} \int_{-\infty}^{\infty} d\ddot{V}_N\, \ddot{V}_N \frac{\partial^3}{\partial x^3} w_1(x, \ddot{V}_N) \Big|_{x = \mp[(\omega - \omega_c)/b]} \qquad x = \dot{V}_N \tag{14.55}$$

The intensity spectrum of the high-index phase-modulated carrier becomes finally [cf. Eqs. (14.53a), (14.53b), and (14.55)]

$$\mathcal{W}_V(f)_{\phi M} \Big|_{\mu_\phi \gg 1} \simeq \Bigg[\frac{A_0^2 \pi \sqrt{\psi}}{\mu_\phi} w_1(x) + \frac{A_0^2 \pi \psi^{3/2}}{6\mu_\phi^3} \int_{-\infty}^{\infty} \ddot{V}_N\, d\ddot{V}_N \frac{\partial^3}{\partial x^3} w_1(x, \ddot{V}_N) + \cdots \Bigg]_{x = \mp[(\omega - \omega_c)/\mu_\phi]\sqrt{\psi} = \dot{V}_N} \tag{14.56}$$

where the upper sign on $x = \mp(\omega - \omega_c)\psi^{1/2}/\mu_\phi$ refers to the argument $\omega_c t - \Phi$ and the lower to $-\omega_c t - \Phi$ in Eq. (14.47); $w_1(x)$ is the first-order d.d. of the *derivative*, $\dot{V}_N$, of the modulation; $w_1(x, \ddot{V}_N)$ is the joint first-order d.d. of $\dot{V}_N = x$, $\ddot{V}_N$, etc. Equation (14.56) requires a (mean-square) differentiable modulation process, at least through the first order, if the leading term is to give us an indication of spectral behavior. The higher-order terms become increasingly important out on the tails of the spectrum.

(2) Frequency modulation. Since phase modulation by $V_N(t)$ is equivalent to frequency modulation by $\dot{V}_N(t)$, we can at once apply the results of (1) to the important case where $V_N(t)$ is now taken to be the frequency modulation, simply by replacing $\dot{V}_N$ therein by V_N, $\ddot{V}_N$ by $\dot{V}_N$, etc. The covariance function (14.50) becomes directly

$$K_V(t)_{FM} \doteq \frac{A_0^2}{2} \operatorname{Re} \left[e^{\mp i\omega_c t}(\boldsymbol{E}\{e^{i\xi_1 V_N}\} + i\xi_2 \boldsymbol{E}\{\dot{V}_N e^{i\xi_1 V_N}\} + \cdots)\right] \tag{14.57}$$

with $\xi_1 = \mu_F a t$, $a = \psi^{-1/2}\, \Delta\omega_N$ as before, $\xi_2 = \mu_F a t^2/2!$, etc., and similar modifications of Eqs. (14.51) et seq. The limiting form of the intensity spectrum is now [from Eq. (14.56)]

$$\mathcal{W}_V(f)_{FM} \Big|_{\mu_F \gg 1} \simeq \Bigg[\frac{A_0^2 \pi}{\mu_F a} w_1(y) + \frac{A_0^2 \pi}{6\mu_F^3 a^3} \int_{-\infty}^{\infty} \dot{V}_N\, d\dot{V}_N \frac{\partial^3}{\partial y^3} w_1(y, \dot{V}_N) + \cdots \Bigg]_{y = \mp[(\omega - \omega_c)/\mu_F a] = V_N} \tag{14.58}$$

Here, w_1 is the first-order d.d. of $y = V_N$, while $w_1(y,\dot{V}_N)$ is the joint first-order d.d. of $y = V_N$ and $\dot{V}_N$. Even if the modulation is not differentiable once (mean square), the leading terms of Eqs. (14.57), (14.58) still represent the limiting forms as the modulation index $\mu_F \to \infty$.

The first terms of Eqs. (14.50), (14.56), (14.57), and (14.58) embody what we may call the *principle of adiabatic frequency sweeps.*† This states that, *in angle modulation of a carrier by stationary processes, the intensity density of the resulting modulated carrier is proportional to the first-order probability density of the equivalent frequency-modulating process,*‡ *for sufficiently slow frequency deviations or sweeps about the equilibrium carrier frequency* (i.e., for sufficiently high modulation indices, μ_ϕ, $\mu_F \to \infty$). Observe that only first-order distribution densities are needed. Moreover, the analysis is equally applicable to deterministic modulations and to mixed modulating processes, consisting of additive combinations of random and periodic components. A number of specific results are available in Probs. 14.8 to 14.14. Finally, from Eqs. (14.56), (14.58) it is evident that one can have *unsymmetrical* spectra (about f_c) if $w_1(x)$, $w_1(y)$ are themselves unsymmetrical about $\bar{\dot{V}}_N = 0$ or $\bar{V}_N = 0$; these spectra are generated without simultaneous amplitude modulation (cf. Sec. 14.3-1). It should be noted, however, that when μ_ϕ or μ_F becomes large one can always obtain useful expressions for the spectrum and covariance function if the complete second-order statistics of the modulating process are once more available. For example, in the gaussian case [Eqs. (14.14*a*) et seq.] we can use Eqs. (14.19*a*) or Eqs. (14.19*b*) directly as a development of the exponent of the covariance function (14.16*a*). Retaining the first term in $|t|$ only and expanding the rest in series then gives us the desired approximation for $K_V(t)$, after which the spectrum follows as before by the W-K theorem.[1]

The examples of Sec. 14.1-3 provide convenient illustrations of the foregoing remarks. From Eqs. (14.21) and (14.23) we get for the RC modulating noise assumed therein

$$\begin{aligned} \Omega(\zeta)_{\phi M} &= \psi(|\zeta| - \zeta^2/2! + |\zeta|^3/3! - \cdots) \\ \Omega(\zeta)_{FM} &= \psi(\zeta^2/2! - |\zeta|^3/3! + \zeta^4/4! - \cdots) \end{aligned} \tag{14.59}$$

with $\zeta = t\,\Delta\omega_N$, from which in Eqs. (14.22*a*), (14.24), following the procedure above we can write

$$K_V(\zeta)_{\phi M} \doteq \frac{A_0{}^2}{2} \cos\frac{\omega_c\zeta}{\Delta\omega_N}\, e^{-\mu_\phi{}^2|\zeta|}\left(1 + \frac{\zeta^2\mu_\phi{}^2}{2!} - \frac{|\zeta|^3\mu_\phi{}^2}{3!}\cdots\right) \qquad \mu_\phi{}^2 = \psi D_\phi{}^2 \tag{14.60a}$$

† This principle, in less general form, has been used earlier by Blachman[14] and by Middleton[3,15] for both periodic and random modulations.

‡ Here $\dot{V}_N(t)$ is used for phase modulation, $V_N(t)$ for frequency modulation.

$$K_V(\zeta)_{FM} \doteq \frac{A_0^2}{2} \cos \frac{\omega_c \zeta}{\Delta\omega_N} e^{-\mu_F^2\zeta^2/2} \left[1 + \frac{|\zeta|^3 \mu_F^2}{3!} - \frac{\zeta^4 \mu_F^2}{3!} \cdots + \frac{\mu_F^2}{2!} \left(\frac{\zeta^6}{3!^2} + \cdots \right) + \cdots \right] \qquad \mu_F^2 = \frac{\psi D_F^2}{\Delta\omega_N^2} \tag{14.60b}$$

With the help of Eq. (A.1.59*a*), we find the limiting intensity spectra in these high-index cases to be

$$\mathcal{W}_V(f)_{\phi M}\Big|_{\mu_\phi^2 \gg 1} \simeq \frac{A_0^2}{\Delta\omega_N} \left[\frac{\mu_\phi^2}{\mu_\phi^4 + \beta^2} + \frac{\mu_\phi^2}{2} \frac{2\mu_\phi^2(\mu_\phi^4 - 3\beta^2)}{(\mu_\phi^4 + \beta^2)^3} + O(\mu_\phi^{-6}) \right] \qquad \beta = \frac{\omega - \omega_c}{\Delta\omega_N} \tag{14.61a}$$

$$\mathcal{W}_V(f)_{FM}\Big|_{\mu_F^2 \gg 1} \simeq \frac{A_0^2 \pi}{\mu_F \, \Delta\omega_N} \left\{ \phi^{(0)}\left(\frac{\beta}{\mu_F}\right) + \frac{1}{3\mu_F} {}_1F_1\left(2; \tfrac{1}{2}; \frac{-\beta^2}{2\mu_F^2}\right) - \frac{1}{24\mu_F^2} \left[\phi^{(4)}\left(\frac{\beta}{\mu_F}\right) + \tfrac{1}{3}\phi^{(6)}\left(\frac{\beta}{\mu_F}\right) \right] + O(\mu_F^{-3}) \right\} \tag{14.61b}$$

where ${}_1F_1$ is a confluent hypergeometric function [Sec. A.1.2 and Eq. (A.1.29*b*)] and $\phi^{(m)}(x)$ is the mth derivative of $e^{-x^2/2}/(2\pi)^{1/2}$ (cf. Sec. A.1.1). Note that we cannot obtain Eq. (14.61*a*) from (14.56), since the phase modulation is a normal *RC* noise process, which is not differentiable (mean square) even to the first order [cf. Sec. 3.2-3(4)]. The first term of Eq. (14.61*b*) for frequency modulation, however, does reveal the expected gaussian frequency distribution, as predicted by Eq. (14.58) from the principle of adiabatic frequency sweeps.† On the other hand, the higher-order terms in Eq. (14.58) cannot be used, since $V_N(t)$ is nondifferentiable: Eq. (14.61*b*) exhibits the appropriate development. Additional results are included in the problems at the end of this section.

PROBLEMS

14.6 Obtain the relation (14.45) for low-index frequency modulation with a normal noise modulation, when $L = \int_0^\infty \mathcal{W}_N(f)\, df/\omega^2 \to \infty$.

14.7 Derive the result (14.46) for the conditions of Prob. 14.6, for frequencies on the tails of the spectrum.

14.8 (*a*) In the case of high-modulation-index angle modulation of a carrier by a sinusoid $B_0 \cos \omega_a t$, show that the covariance function and spectrum are to the first approximation given by

$$K_V(t) \doteq \frac{A_0^2}{2} J_0(\mu\omega_a t) \cos \omega_c t \tag{1}$$

$$\mathcal{W}_V(f) \begin{cases} \simeq \dfrac{A_0^2}{\omega_a} \left[\mu^2 - \left(\dfrac{\omega_c - \omega}{\omega_a} \right)^2 \right]^{-1/2} & \mu > \left| \dfrac{\omega_c - \omega}{\omega_a} \right| \\ = \infty & \mu = \left| \dfrac{\omega_c - \omega}{\omega_a} \right| \\ = 0 & \mu < \left| \dfrac{\omega_c - \omega}{\omega_a} \right| \end{cases} \tag{2}$$

† Since $w_1(y) = (2\pi \overline{y^2})^{-1/2} e^{-y^2/2\overline{y^2}}$ and $\overline{y^2} = \psi$, $y = \psi^{1/2}(\omega - \omega_c)/\mu_F \,\Delta\omega_N$ [Eq. (14.58)], we see that the leading terms of Eqs. (14.58), (14.61*b*) are identical.

The limiting spectrum is shown in Fig. P 14.8; for frequency modulation, we have $\mu_{FM} = D_F B_0/\omega_a$; for phase modulation, $\mu_{\phi M} = D_\phi B_0$.

(*b*) Using Eq. (14.58) with the results of Prob. 14.6, show that the limiting spectrum above can be written

$$\mathcal{W}_V(f) \simeq \frac{A_0^2}{D_F B_0}\left(1 - \frac{y^2}{B_0^2}\right)^{-\frac{1}{2}}\left[1 - \frac{1}{2\mu_{FM}^2}\frac{1 + 4y^2/B_0^2}{(1 - y^2/B_0^2)^3} + O(\mu_{FM}^{-4})\right] \tag{3}$$

where $y/B_0 = (\omega - \omega_c)/\omega_a\mu_{FM} = (\omega - \omega_c)/D_F B_0$.

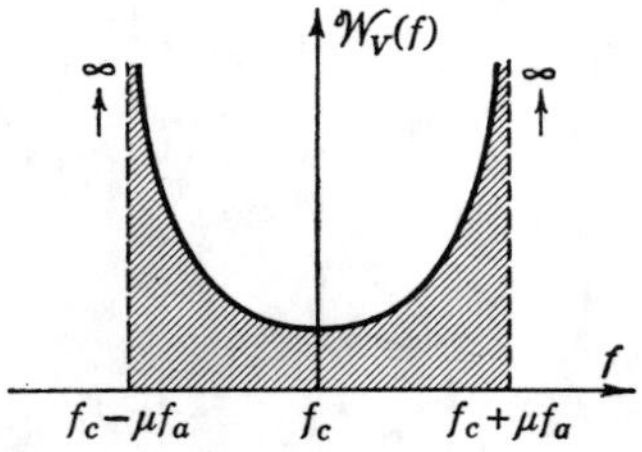

FIG. P 14.8. Limiting spectrum of a carrier angle-modulated by a sinusoidal process.

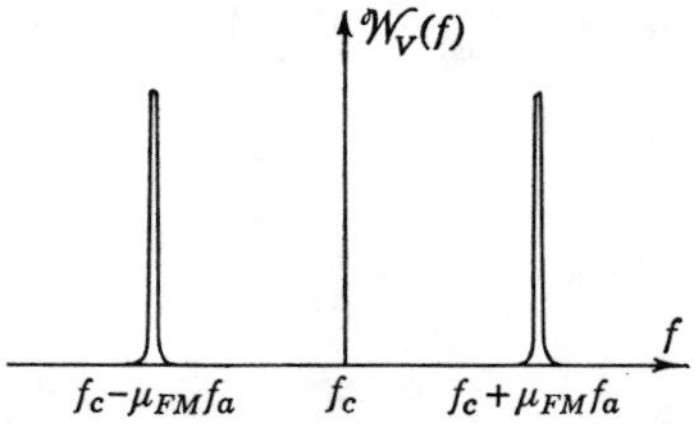

FIG. P 14.9. Limiting form of intensity spectrum of carrier frequency-modulated by a square wave; fixed deviation.

14.9 Show that the covariance function and spectrum of a carrier frequency-modulated by the square wave of Fig. P 14.1 have the limiting forms

$$\begin{aligned} K_V(t) &\doteq \frac{A_0^2}{2}\cos\omega_a\mu_{FM}t\cos\omega_c t \\ \mathcal{W}_V(f) &\simeq \frac{A_0^2}{4}\{\delta[f - (f_c - \mu_{FM}f_a)] + \delta[f - (f_c + \mu_{FM}f_a)]\} \end{aligned} \tag{1}$$

when the modulation index becomes indefinitely large (cf. Fig. P 14.9).

14.10 (*a*) Show that, for the frequency modulation of Fig. P 14.2, the spectrum and covariance functions have the limiting forms ($\mu_{FM} \to \infty$)

$$\begin{aligned} K_V(t) &\doteq \frac{A_0^2}{2}\frac{\sin\mu_{FM}\omega_a t}{\mu_{FM}\omega_a t}\cos\omega_c t \\ \mathcal{W}_V(f) &\simeq \begin{cases} \dfrac{A_0^2\pi}{2\omega_a\mu_{FM}} & 0 < |f - f_c| < \mu_{FM}f_a \\ 0 & |f - f_c| > \mu_{FM}f_a \end{cases} \end{aligned} \tag{1}$$

(cf. Fig. P 14.10); explain and discuss.

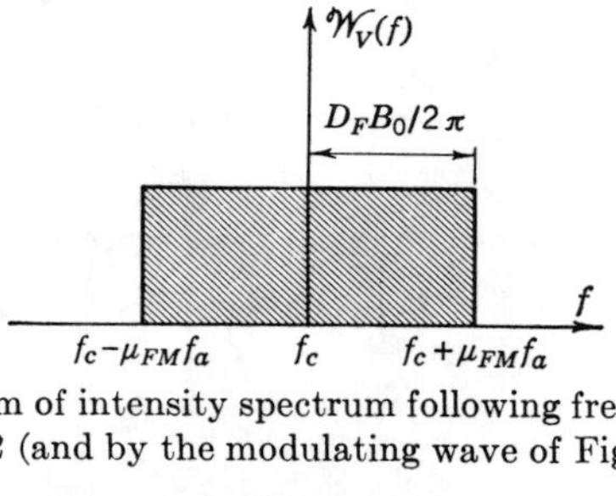

FIG. P 14.10. Limiting form of intensity spectrum following frequency modulation by the linear sweep of Fig. P 14.2 (and by the modulating wave of Fig. P 14.3); fixed deviation.

(*b*) Verify that the saw-tooth modulation of Fig. P 14.3 also yields Eqs. (1) for the limiting covariance function and spectrum. Explain the equivalence.

14.11 A carrier is frequency-modulated by a periodic wave of the type $B_0t^2\omega_a{}^2/2\pi$ $(0 < t < T_0/2)$, $-B_0t^2\omega_a{}^2/2\pi$ $(-T_0/2 < t < 0)$ (cf. Fig. P 14.11*a*). Verify that the

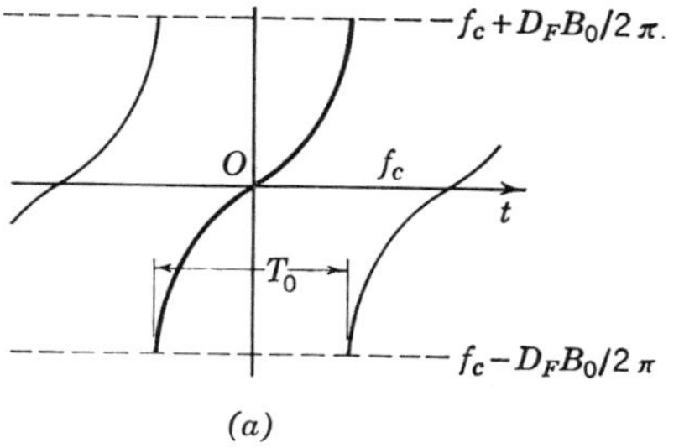

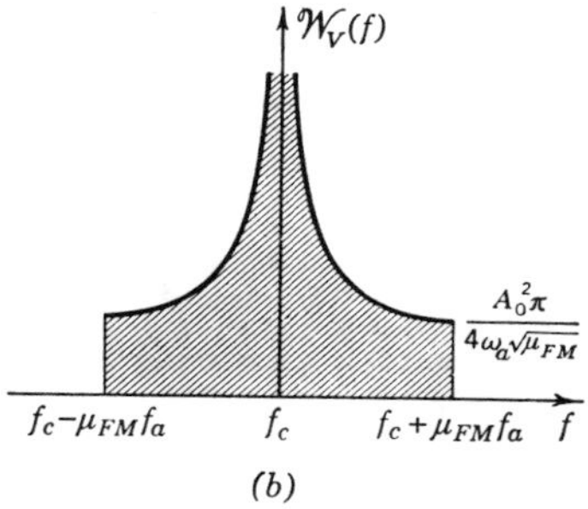

FIG. P 14.11. (*a*) Frequency-modulation wave. (*b*) Limiting intensity spectrum.

limiting forms (as $\mu_{FM} \to \infty$) of covariance function and spectrum are (cf. Fig. P 14.11*b*)

$$K_V(t) \doteq \frac{A_0{}^2}{2}\left[\int_0^1 \cos\,(\mu_{FM}\omega_a t x^2)\,dx\right]\cos\,\omega_c t$$

$$= \frac{A_0{}^2}{2}\left(\frac{\pi}{2\omega_a\mu_{FM}t}\right)^{1/2} C\left(\sqrt{\frac{\mu_{FM}\omega_a t\pi}{2}}\right) \tag{1}$$

with C a Fresnel integral, and

$$\mathcal{W}_V(f)\begin{cases} \simeq \dfrac{A_0{}^2\pi}{4\omega_a\sqrt{\mu_{FM}}}\left|\dfrac{\omega-\omega_c}{\omega_a}\right|^{-1/2} & |f-f_c| < \mu_{FM}f_a \\ = 0 & \text{elsewhere}\end{cases} \tag{2}$$

14.12 Generalize the result of Prob. 14.11 to include the cases where the instantaneous angular frequency $\dot{\Phi}(t) = D_FB_0\alpha|t|^\nu$ $(0 < t < T_0/2)$, $\dot{\Phi} = -D_FB_0\alpha|t|^\nu$ $(-T_0/2 < t < 0)$, and hence show that

$$K_V(t) \doteq \frac{A_0{}^2}{2}\left[\int_0^1 \cos\,(\omega_a t\mu_{FM}|x|^\nu)\,dx\right]\cos\,\omega_c t \qquad \nu \geq 0 \tag{1}$$

with the limiting spectral distribution, as $\mu_F \to \infty$,

$$\mathcal{W}_V(f)\begin{cases} \simeq \dfrac{A_0{}^2}{4\nu}(f_a\mu_{FM})^{-1/\nu}|f-f_c|^{1/\nu-1} & 0 < |f-f_c| < f_a\mu_{FM} \\ = 0 & \text{elsewhere}\end{cases} \tag{2}$$

For $\nu = 1/2$, obtain $\mathcal{W}_V(f) \simeq (A_0{}^2/2)(f_a\mu_{FM})^{-2}|f-f_c|$ (cf. Fig. P 14.12), and verify that for $\nu = 0, 1, 2$ we obtain, respectively, the results of Probs. 14.9 to 14.11.

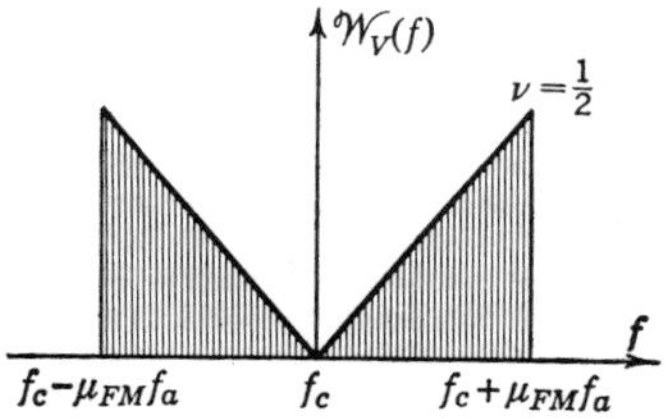

FIG. P 14.12. Limiting spectral intensity when $\nu = 1/2$.

14.13 A sinusoidal process V_S of peak amplitude B_0 is added to an l-f stationary normal noise process V_N, in the manner of Sec. 14.1-5. For high modulation indices μ_{FM}, $\mu_F \gg 1$, show that the leading terms for the covariance function and for the intensity spectrum are, for frequency modulation,

$$K_V(t)_{FM} \doteq \frac{A_0^2}{2} e^{-\mu_F^2 t^2 \Delta\omega_N^2/2} J_0(\mu_{FM} \sqrt{2}\, \omega_a t) \cos \omega_c t \tag{1}$$

and
$$\mathcal{W}_V(f)_{FM} \simeq \frac{A_0^2 \pi}{\mu_F \Delta\omega_N} \sum_{n=0}^{\infty} \frac{p^n}{2^n (n!)^2} \phi^{(2n)}\left(\frac{\beta}{\mu_F}\right) \tag{2a}$$

or
$$\mathcal{W}_V(f)_{FM} \simeq \frac{A_0^2 \sqrt{\pi/2}}{\mu_F \Delta\omega_N} \sum_{n=0}^{\infty} \frac{(-1)^n}{n!} \left(\frac{\beta}{\mu_F \sqrt{2}}\right)^{2n} {}_1F_1(n + \tfrac{1}{2}; 1; -p) \tag{2b}$$

where $\beta = (\omega - \omega_c)/\Delta\omega_N$ and $p = \overline{V_S^2}/\overline{V_N^2} = \mu_{FM}^2/\mu_F^2$. [Note that $\mathcal{W}_V(f)_{FM}$ is proportional to the probability density of the sum of a sinusoid and normal noise, when μ_F, μ_{FM} (together) approach infinity; see Fig. 9.5*a*, *b*.]

14.14 Extend the analysis of Sec. 14.2-2 to the high-modulation-index case of frequency modulation by the sum of a random and periodic process, $y_S + y_N$, and hence show that the limiting spectrum becomes

$$\begin{aligned}\mathcal{W}_V(f)_{FM} \simeq{} & \frac{A_0^2 \pi}{D_F} \left[\int_{-\infty}^{\infty} w_1(y - x)_N w_1(x)_S \, dx \right]_{y = \mp[(\omega - \omega_c)/D_F]} \\ & + \frac{A_0^2 \pi}{6 D_F^3} \left[\iint_{-\infty}^{\infty} \ddot{y}_S \, d\ddot{y}_S \frac{\partial^3}{\partial y_S^3} w_1(y_S, y_N; \ddot{y}_S) \Big|_{y_S = \mp[(\omega - \omega_c)/D_F] - y_N} dy_N \right. \\ & \left. + \iint_{-\infty}^{\infty} \ddot{y}_N \, d\ddot{y}_N \frac{\partial^3}{\partial y_S^3} w_1(y_S, y_N; \ddot{y}_N) \Big|_{y_S = \mp[(\omega - \omega_c)/D_F] - y_N} dy_N \right] + \cdots \end{aligned} \tag{1}$$

Obtain the corresponding expression for high-index phase modulation in this case.

14.3 *Generalizations*

Sometimes in applications we have to deal not with simple angle modulations alone, but with mixtures of amplitude and angle modulation, which may be produced in a variety of ways: from the source of modulation itself, from the effects of a filter, or from a combination of such mechanisms. For example, in the case of cw magnetrons, angle and amplitude modulation are found to occur simultaneously:[7,9] the output of a linear filter which is spectrally narrower than the deviations of an angle-modulated input exhibits a similar kind of dual modulation. In this section, we shall examine a few of the features of these modulations, where the periodic or stationary random processes of the preceding sections are used for modulation.

14.3-1 Simultaneous Amplitude and Angle Modulation. By an obvious modification of Eqs. (14.2) and (12.2*c*), the modulated carrier process can be represented by

$$V(t) = \text{Re}\,[A_0(t) e^{i\omega_c t - i\Phi(t)}] \tag{14.62}$$

where $A_0(t) = A_0[1 + \lambda V_M(t + \tau)]$ if we restrict ourselves to weak amplitude modulations.† Here, $\Phi(t)$ is equal to $D_\phi V_M(t)$, or $\dot{\Phi}(t) = D_F V_M(t)$, while τ in $\lambda V_M(t + \tau)$ is a possible lag ($\tau > 0$) or lead ($\tau < 0$) of the angle over the amplitude effects. The modulations are correlated, since they involve the common source $V_M(t)$, and for this reason may be expected to yield a variety of spectral distributions, which in general are *unsymmetrical.*

(1) **Periodic modulations.** When $V_M(t)$ is a periodic process, we may expand $e^{-i\Phi}$ and $V_M(t)$ in Fourier series,

$$e^{-i\Phi(t)} = \sum_{-\infty}^{\infty} a_m e^{im\omega_a t} \qquad V_M(t - \tau) = \sum_{-\infty}^{\infty} b_n e^{in\omega_a(t-\tau)} \qquad \omega_a = \frac{2\pi}{T_0} \tag{14.63}$$

where

$$a_m = \frac{1}{T_0}\int_0^{T_0} e^{-i\Phi(t_0)-im\omega_a t_0}\,dt_0 \qquad \text{and} \qquad b_n = \frac{1}{T_0}\int_0^{T_0} V_M(t_0)e^{-in\omega_a t_0}\,dt_0 \tag{14.64}$$

The modulated carrier is therefore

$$V(t) = A_0 \operatorname{Re}\left[e^{i\omega_c t} \sum_{m=-\infty}^{\infty} a_m \left(e^{im\omega_a t} + \lambda \sum_{n=-\infty}^{\infty} b_n e^{in\omega_a t - in\omega_a \tau}\right)\right] \tag{14.65}$$

For the example of a sine wave $V_M = B_0 \cos \omega_a t$, this becomes at once, from Sec. 14.1-1,

$$a_m = \frac{1 + (-1)^m}{2} J_M(\mu_{FM}), \text{ or } (-i)^m J_M(\mu_{\phi M}) \qquad b_n = B_0 \delta_n{}^1 \cos n\omega_a t \tag{14.66}$$

The intensity spectrum is asymmetrical (cf. Fig. 14.8), the more so the larger the value of λ. The asymmetry itself may be toward the low or high side of the carrier frequency f_c, depending on τ and on the magnitude of the modulation index.

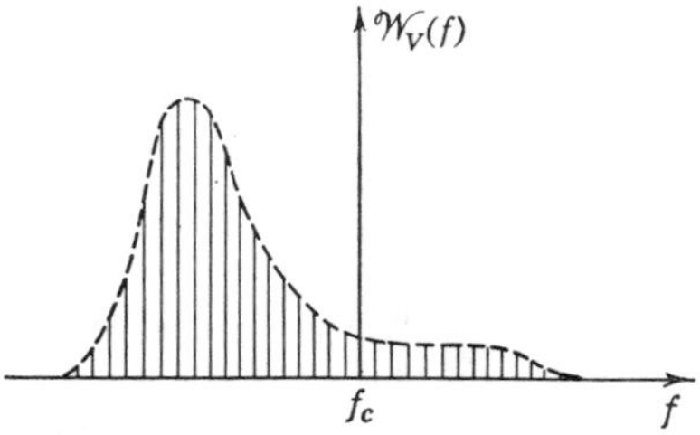

FIG. 14.8. A typical intensity spectrum of a carrier simultaneously amplitude- and angle-modulated by a sine wave.

(2) **Random modulations.** If $V_M(t)$ is an entirely random process, we obtain from Eq. (3.81) for the covariance function of the modulated carrier

$$K_V(t) = \tfrac{1}{2}A_0{}^2 \operatorname{Re}\left(e^{-i\omega_c t}\boldsymbol{E}\{a_1 a_2 e^{-i\Phi_1+i\Phi_2}\}\right) \tag{14.67a}$$

$$= \tfrac{1}{2}A_0{}^2 \operatorname{Re}\left\{e^{-i\omega_c t}\left[-\frac{\partial^2}{\partial\xi_1\,\partial\xi_2}\left(F_2 e^{i\xi_1+i\xi_2}\right)\right]_{\xi_1=\xi_2=0}\right\} \tag{14.67b}$$

where F_2 is now the characteristic function

$$F_2(\xi_1,\xi_2;-1,1;t) = \boldsymbol{E}\{e^{i\lambda V_{M1}\xi_1+i\lambda V_{M2}\xi_2-i\Phi_1+i\Phi_2}\} \tag{14.67c}$$

from the fact that $a(t)e^{-i\Phi(t)} = (1 + \lambda V_M)e^{-i\Phi}$ [Eq. (14.62)].

† Cases of overmodulation are readily included with the help of Eq. (12.13), etc.

When $V_M(t)$ $[= V_N(t)]$ belongs to a gaussian random process, we may show that[2] the characteristic function becomes explicitly

$$F_2(\xi_1,\xi_2;-1,1;t) = \exp\left[-\frac{\lambda^2\psi(\xi_1^2+\xi_2^2)}{2} - \lambda^2\xi_1\xi_2K_N(t)\right]\exp\left[-D_0^2\Omega(t)\right] \times \exp\{\lambda D_0[\xi_1\Lambda^{(+)}(t) + \xi_2\Lambda^{(-)}(t)]\} \quad (14.68)$$

where $\overline{\lambda V_N^2} = \lambda\psi \leq 0.18$ for ignorable overmodulation [cf. Eqs. (12.12) et seq.], $K_N(t)$ is the covariance function of the modulation, and $\Omega(t)$ is given by Eqs. (14.15) as before. Here, the $\Lambda^{(\pm)}(t)$ are the cross-variance functions between the amplitude and phase modulations, viz.,

$$\Lambda^{(\pm)}(t) = \overline{V_{\binom{N_1}{N_2}}(\Phi_1 - \Phi_2)}/D_0 \equiv (\pm)\Lambda(\pm t,\tau) \quad (14.69a)$$

with $V_{N1} = V_N(t_1)$, etc. Specifically, one can show that[2]

$$\Lambda^{(\pm)}(t)_{\mathrm{FM}} = (\mp)\int_0^\infty \mathcal{W}_N(f)[\sin\omega\tau - \sin\omega(\tau \mp t)]\,df/\omega \quad (14.69b)$$

$$\Lambda^{(\pm)}(t)_{\phi\mathrm{M}} = (\pm)\int_0^\infty \mathcal{W}_N(f)[\cos\omega\tau - \cos\omega(\tau \mp t)]\,df \quad (14.69c)$$

in which $\mathcal{W}_N(f)$ is the intensity spectrum of the noise modulation V_N. Carrying out the indicated differentiation in Eq. (14.67b), we find now that

$$K_V(t) = \frac{A_0^2}{2}\{[1 + k^2K_N(t) - \lambda^2D_0^2\Lambda^{(+)}(t)\Lambda^{(-)}(t)]\cos\omega_c t - \lambda D_0[\Lambda^{(+)}(t) + \Lambda^{(-)}(t)]\sin\omega_c t\}e^{-D_0^2\Omega(t)} \quad (14.70)$$

We observe first that the total intensity of the process is influenced, as expected, by the amplitude modulation alone.

$$P_\tau = K_V(0) = \frac{A_0^2}{2}(1 + \lambda^2\overline{V_M^2})$$

Part of this intensity is distributed in the sideband continuum, and part remains in the carrier itself. How much actually remains in the discrete component depends on the spectrum of the modulating noise (cf. Middleton[2]).

The intensity spectrum consists of the following intermodulation terms: AM $\times$ FM and (AM $\times$ FM) $\times$ FM representing the quantities in the brackets [Eq. (14.70)], each multiplied by the FM effect, $e^{-D_c^2\Omega(t)}$. The significant feature of Eq. (14.70) is the antisymmetry of the cross-variance expression $[\Lambda^{(+)}(t) + \Lambda^{(-)}(t)]e^{-D_0^2\Omega(t)}$, which yields a positive spectral contribution for frequencies on one side of the carrier (f_c) and a negative contribution for frequencies on the other side, just which side depending on τ and on whether or not frequency modulation or phase modulation by the noise is indicated. The net result of the combination of symmetrical and antisymmetrical terms [in Eq. (14.70)] is an *antisymmetrical* power spectrum, in the manner of Fig. 14.9. Since the intensity spectra of pure amplitude modulation and pure frequency modulation are always symmetrical† (about $f_c \gg \Delta f_{mod}$), an antisymmetrical power spectrum can only be produced by *correlated*, simultaneous amplitude and angle modulations. When there

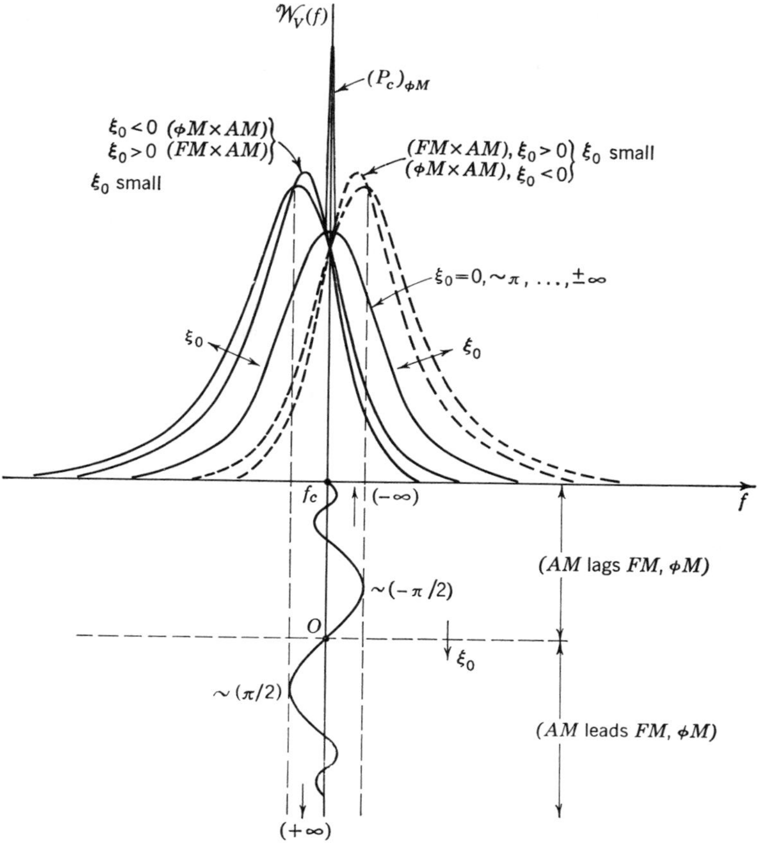

FIG. 14.9. Sketches of typical intensity spectra of a carrier amplitude- and angle-modulated by correlated *RC* noise modulations, for various delays $\xi_0 = \tau\,\Delta\omega_{Nj}$ (cf. Prob. 14.15).

is no correlation between modulations—when, for example, different sources are used—the cross terms vanish and Eq. (14.70) becomes simply

No correlation

$$K_V(t) = \frac{A_0^2}{2}[1 + \lambda^2 K_N(t)]e^{-D_0^2\Omega(t)} \cos \omega_c t \tag{14.71}$$

† An exception occurs for high modulation indexes when $w_1(x)$ is not symmetrical [cf. remarks following Eq. (14.58)].

as expected. A similar effect is also observed in the correlated case if the delay τ is made very large and provided that $\mathfrak{W}_N(f)$ behaves suitably near $f = 0$; $\Lambda(\pm t, \infty)$ accordingly vanishes.

14.3-2 The Steady-state Effects of a Narrowband Linear Filter. As we remarked earlier (Sec. 14.3-1), simultaneous amplitude and angle modulation can occur when a frequency- (or phase-) modulated wave is passed through a narrowband linear filter with a suitable system function $Y_F(i\omega)$. An additional amplitude modulation may be introduced by the finite slope (with frequency) of this system function, as the instantaneous frequency is displaced with respect to it. The resulting modulations are necessarily correlated and are combined in such a way that it is not possible to distinguish the two effects without the help of a suitable AM and FM receiver.

(1) Periodic modulations. We assume first a cosinusoidal frequency modulation where $V_{in}(t)$ is given by Eq. (14.6) (with $\sin \omega_a t$ replaced by $\cos \omega_a t$) and the output process is $V_0(t) = \boldsymbol{T}_L\{V_{in}\} = \int_{-\infty}^{\infty} d\tau\, V_{in}(t - \tau)h_F(\tau)$ in the usual way [cf. Eq. (2.16)]. For a narrowband filter, the weighting function $h_F(\tau)$ may be written $h_F(\tau) = 2h_0(\tau)_F \cos [\omega_0\tau - \gamma_0(\tau)_F]$ [cf. Eqs. (2.63*a*)] so that the steady-state instantaneous output becomes

$$V_{out}(t) = A_0 \operatorname{Re} \left[e^{i\omega_0 t} \int_{-\infty}^{\infty} h_0(\tau)_F e^{i\mu_F \cos \omega_a(t-\tau) + i\omega_D(t-\tau) - i\gamma_0(\tau)_F} \, d\tau \right]$$
$$\omega_D = \omega_c - \omega_0 \qquad (14.72)$$

Expanding the trigonometric exponential and using the fact that the system function $Y_0(i\omega)_F$ is the Fourier transform of the weighting function $h_0(t)_F e^{i\gamma_0(t)_F}$ [cf. Eqs. (2.63*b*)] permits us to write finally

$$V_0(t) = \frac{A_0}{2} \sum_{m=0}^{\infty} \operatorname{Re} \{ i^m \epsilon_m J_m(\mu_F) [Y_0(im\omega_a + i\omega_D) e^{i(\omega_0 + m\omega_a + \omega_D)t} + Y_0(im\omega_a - i\omega_D)^* e^{i(\omega_0 - m\omega_a + \omega_D)t}] \} \qquad (14.73)$$

In general, the intensity spectrum is *unsymmetrical*, depending on the degree of detuning ω_D, as is apparent at once from Eq. (14.73). Furthermore, the total intensity of the wave may be noticeably reduced if ω_D is at all large compared with the spectral width of the filter. The induced amplitude modulation disappears, as we expect, when $\omega_D \to 0$ *and* when the filter is made sufficiently wide to contain the frequency deviations of the input wave within the uniform portion of the passband. Similar expressions are easily derived for phase-modulated inputs.

(2) Random modulations. Let us extend the above remarks to the more general situation of a randomly angle-modulated wave. From Eq. (3.119), the covariance function of the filtered wave is

$$K_{V_0}(t) = \boldsymbol{E}\{\boldsymbol{T}_L\{V_{in}(t_1)\}\boldsymbol{T}_L\{V_{in}(t_2)\}\} = \tfrac{1}{2} [\operatorname{Re} e^{-i\omega_0 t} \mathcal{K}_{\alpha,\beta}(t)] \qquad (14.74a)$$

where now

$$\mathcal{K}_{\alpha,\beta}(t) = A_0{}^2 e^{-i\omega_D t} \int_{-\infty}^{\infty} \rho_F(\tau)\, {}_0F_2(-1, 1; t \mp \tau)_{\Phi_N} e^{\pm i\omega_D \tau}\, d\tau \qquad \omega_D = \omega_c - \omega_0 \tag{14.74b}$$

in which ρ_F is the autocorrelation function of the filter, referred to f_0 as "zero" frequency, and F_2 is the characteristic function of the random phase $\Phi_N(t)$. [We assume that $\gamma_0(t)_F$ is a constant for the class of filters used here.] The associated spectral density is simply

$$\mathcal{W}_{V_0}(f) = |Y(i\omega)_F|^2 \mathcal{W}_V(f) = \frac{A_0{}^2}{2} |Y(i\omega)_F|^2 \operatorname{Re}\left[\int_{-\infty}^{\infty} e^{i(\omega_0-\omega)t} F_2(-1,1;t)_{\Phi_N}\, dt\right] \tag{14.75}$$

The instantaneous phase $\Phi_N(t)$ is related to the modulating noise, according to Eqs. (14.2).

When $V_N(t)$ and therefore $\Phi_N(t)$ belong to a normal random process, F_2 is given explicitly by Eqs. (14.14a) and (14.15). As an example, let us assume that the modulating noise obeys Eqs. (14.20) and that the narrow filter through which the modulated wave is passed is a high-Q single-tuned circuit for which $\rho_F(t)_0 = (\Delta\omega_F/2\pi)e^{-\Delta\omega_F|t|/2\pi}$, where $\Delta\omega_F$ is the angular frequency width of the equivalent rectangular filter [cf. Eq. (3.99b)]. For frequency modulation under these conditions, we find that here [cf. Eq. (14.23) in Eq. (14.74b)]

$$K_{V_0}(\zeta)_{\mathrm{FM}} = \frac{A_0{}^2 a}{4} \operatorname{Re}\left[\exp\left(-\frac{i\omega_0\zeta}{\Delta\omega_N}\right)\int_{-\infty}^{\infty} \exp\{-|z-\zeta|a - \mu_F{}^2[\exp(-|z|) + |z| - 1] - ibz\}\, dz\right] \tag{14.76}$$

with $a \equiv \Delta\omega_F/(\pi\,\Delta\omega_N)$, $b \equiv \omega_D/\Delta\omega_N$, and $\zeta = t\,\Delta\omega_N$. The intensity spectrum may be obtained from Eq. (14.76), but a simpler procedure is to use Eq. (14.75), after $\mathcal{W}_V(f)$ has been determined with the help of Eqs. (14.23) to (14.25). The total average intensity after filtering follows at once on setting $\zeta = 0$ in Eq. (14.76) in this example or in Eq. (14.75) for the general input. Limiting expressions for large μ_N, large a, b, etc., are given in Prob. 14.16, while spectra in specific instances have been calculated in Middleton.[3] The same procedures apply equally well when the input involves simultaneous amplitude and angle modulations.

PROBLEMS

14.15 A carrier is simultaneously amplitude- and angle-modulated by normal RC noise $V_N(t)$, where there is a delay τ of the amplitude modulation with respect to the angle modulation. For weak amplitude modulation, show that the covariance function of the modulated carrier becomes

$$K_V(\zeta)_{\phi\mathrm{M}} \doteq \frac{A_0{}^2}{2}\left\{[1 + \lambda^2\psi e^{-|\zeta|} + \lambda^2\psi\mu_\phi{}^2(e^{-|\zeta_0|} - e^{-|\zeta_0+\zeta|})(e^{-|\zeta|} - e^{|\zeta_0-\zeta|})]\cos\frac{\omega_c\zeta}{\Delta\omega_N} + (\lambda^2\psi)^{1/2}\mu_\phi(e^{-|\zeta_0-\zeta|} - e^{-|\zeta_0+\zeta|})\sin\frac{\omega_c\zeta}{\Delta\omega_N}\right\} e^{-\mu_\phi{}^2(1-e^{-|\zeta|})} \tag{1}$$

where $\zeta_0 = \tau \Delta\omega_N$, $\zeta = t \Delta\omega_N$, $\psi = \overline{V_N^2}$, λ is the AM index, and $\mu_\phi = D_\phi \psi^{1/2}$ is the ϕM index. In the case of frequency modulation, show also that

$$K_V(\zeta)_{FM} \doteq \frac{A_0^2}{2} \Big\{ [1 + \lambda^2 \psi e^{-|\zeta|} + \lambda^2 \mu_F^2 (e^{-|\zeta_0|} - 1 - \{e^{-|\zeta_0+\zeta|} - 1\} \operatorname{sgn}[\zeta_0 + \zeta])(e^{-|\zeta_0|} - 1$$

$$- \{e^{-|\zeta_0-\zeta|} - 1\} \operatorname{sgn}[\zeta_0 - \zeta]) \cos \frac{\omega_c \zeta}{\Delta\omega_N} + (\lambda^2\psi)^{1/2} \mu_F (\{e^{-|\zeta_0+\zeta|} - 1\} \operatorname{sgn}[\zeta_0 + \zeta]$$

$$- \{e^{-|\zeta_0-\zeta|} - 1\} \operatorname{sgn}[\zeta_0 - \zeta]) \sin \frac{\omega_c \zeta}{\Delta\omega_N} \Big\} e^{-\mu_F^2 (e^{-|\zeta|} + |\zeta| - 1)} \quad (2)$$

Typical spectra are shown in Fig. 14.9. Verify that the intensity in the carrier after modulation is

$$(P_{f_c})_{\phi M} = \frac{A_0^2}{2} e^{-\mu_\phi^2} (1 + \lambda^2 \psi \mu_\phi^2 e^{-2\tau \Delta\omega_N}) \qquad (P_{f_c})_{FM} = 0 \quad (3)$$

14.16 In the case of a carrier frequency-modulated by RC normal noise through a high-Q single-tuned circuit, where Eq. (14.76) gives the covariance function of the filter's output, show that $K_V(0)_{FM}$ has the following limiting forms,[3] and discuss the physical implications:

(a) $\mu_F \to \infty$ ($\overline{\dot{\Phi}_N^2} \to \infty$; $\Delta\omega_N$, a, b fixed)

$$K_V(0) \simeq \frac{A_0^2}{2} \sqrt{\frac{\pi}{2}} \frac{a}{\mu_F} [1 + O(\mu_F^{-1})] \quad (1)$$

(b) $a \to \infty$ ($\mu_F < \infty$; $\Delta\omega_N \to \infty$, b fixed)

$$K_V(0) \simeq \frac{A_0^2}{2} \left\{ \frac{1}{1 + b^2/a^2} - \left(\frac{\mu_F}{a}\right)^2 \frac{1 - 3b^2/a^4}{(1 + b^2/a^2)^3} + O\left[\left(\frac{\mu_F}{a}\right)^4\right] \right\} \quad (2)$$

(c) $\mu_F \gg 1$, $a/\mu_F \to \infty$, or $a/\mu_F \to 0$

$$K_V(0) \simeq \frac{A_0^2}{2} \frac{a}{\mu_F} \sqrt{\frac{\pi}{2}} e^{a^2/2\mu_F^2} \left[1 - \Theta\left(\frac{a}{\mu_F \sqrt{2}}\right) + O\left(\frac{1}{\mu_F}\right)\right] \quad (3)$$

14.17 Develop the theory of a sinusoidal carrier ensemble that is simultaneously amplitude- and angle-modulated by a stationary noise process (when there may be delays, τ, between the modulations), including the effects of a linear narrowband filter through which the modulated process is passed. Show, in particular, that the covariance function of this output process is given by

$$K_V(t) = \tfrac{1}{2} A_0^2 \operatorname{Re} \left\{ e^{-i(\omega_0 + \omega_D)t} \int_{-\infty}^{\infty} \rho_F(\tau)_0 \left[- \frac{\partial^2}{\partial \xi_1 \, \partial \xi_2} (F_2 e^{i\xi_1 + i\xi_2}) \right]_{\xi_1 = \xi_2 = 0} e^{\pm i\omega_D \tau} \, d\tau \right\} \quad (14.36a)$$

with $\omega_c = \omega_0 + \omega_D$, and that the characteristic function F_2 is given by Eqs. (14.67c), (14.68) for weak amplitude modulation. Discuss the spectrum of the output, and use the results of Prob. 14.18 to extend the theory to the case of general amplitude modulation, where overmodulation occurs.

14.18 Show, in the general case of different modulating spectra and possible overmodulation in amplitude, that when there is coherence between the AM and ∠M mechanisms the covariance function of the modulated wave can be written

$$K_V(t) = \frac{A_0^2}{2} e^{-D_0^2 \Omega(t)} \left\{ \sum_{n=0}^{\infty} \frac{(-\lambda^2 \psi_A)^n}{n!} k_0(t)_A{}^n \right.$$

$$\left. \sum_{jl=0}^{\infty} \frac{[-\lambda D_0 (\psi_A \psi_F)^{1/2}]^{j+l}}{j!\,l!} \lambda^{(+)}(t)^j \lambda^{(-)}(t)^l \operatorname{Re} (e^{-i\omega_c t} h_{0,n+j} h_{0,n+l}) \right\} \quad (1)$$

for gaussian noise; $\psi_A = \overline{V_{AM}^2}$; $\psi_F = \overline{V_{\phi M}^2}_{\text{ or FM}}$; and $\lambda^{(\pm)}(t) \equiv \Lambda^{(\pm)}(t)/\sqrt{\psi_A\psi_F}$, with $k_0(t)_A$ the normalized covariance function of the AM noise. The amplitude functions $h_{0,n+q}$ are here

$$h_{0,n+q} = \frac{1}{2\pi} \int_C e^{i\xi - \lambda^2\psi_A\xi^2/2} \xi^{q+n-2} \, d\xi \tag{2}$$

The relation (1) may be used for both phase and frequency modulation.

D. Middleton.[2]

REFERENCES

1. Middleton, D.: The Distribution of Energy in Randomly Modulated Waves, *Phil. Mag.*, (7)**42:** 689 (1951).
2. Middleton, D.: On the Distribution of Energy in Noise and Signal Modulated Waves. II. Simultaneous Amplitude and Angle-modulation, *Quart. Appl. Math.*, **10:** 35 (1952).
3. Middleton, D.: Random Frequency Modulated Waves, *Johns Hopkins Univ. Radiation Lab. Tech. Rept.* AF-9, 1955, and The Effects of a Linear, Narrow-band Filter on a Carrier Frequency-modulated by Normal Noise, *Johns Hopkins Univ. Radiation Lab. Tech. Rept.* AF-24, November, 1955.
4. Van der Pol, B.: The Fundamental Principles of Frequency Modulation, *J. IEE (London)*, **93** (pt. III): 153 (1946); for a more refined approximation, involving amplitude as well as angle changes, see sec. 7 of this reference.
5. Carson, J. R., and T. C. Fry: Variable Frequency Electric Circuit Theory with Application to the Theory of Frequency Modulation, *Bell System Tech. J.*, **16:** 513 (1937).
6. Van der Pol, B.: Frequency Modulation, *Proc. IRE*, **18:** 1194 (1930).
7. Gottschalk, W. M.: Direct Detection of Noise in CW Magnetrons, *Trans. IRE*, **PGED-1**(4): 91 (1954).
8. La Plante, R. A.: Development of a Low-noise, X-band CW Klystron Power Oscillator, *Trans. IRE*, **PGED-1**(4): 99 (1954).
9. Middleton, D.: Theory of Phenomenological Models and Direct Measurements of the Fluctuating Output of CW Magnetrons, *Trans. IRE*, **PGED-1**(4): 56 (1954). For some further modifications of the results, a more complete treatment is available in *Rept.* R-5, Nov. 1, 1954, and Feb. 3, 1956, Raytheon Manufacturing Co., Waltham, Mass., Research Division.
10. Stewart, J. L.: Theory of Frequency Modulation Noise in Tubes Employing Phase Focusing, *J. Appl. Phys.*, **26:** 409 (1955).
11. Mullen, J. A., and D. Middleton: Limiting Forms of FM Noise Spectra, *Proc. IRE*, **45:** 874 (1957).
12. Medhurst, R. G.: RF Bandwidth of Frequency-division Multiplex Systems Using Frequency Modulation, *Proc. IRE*, **44:** 189 (1956).
13. Stewart, J. L.: The Power Spectrum of a Carrier Frequency Modulated by Gaussian Noise, *Proc. IRE*, **42:** 1539 (1954).
14. Blachman, N.: The Demodulation of a Frequency-modulated Carrier and Random Noise by a Discriminator, *J. Appl. Phys.*, **20:** 976 (1949), eqs. (5) et seq. For a more detailed treatment, see Limiting Frequency-modulation Spectra, *Inform. and Control*, **1:** 26 (1957).
15. Middleton, D.: On the Distribution of Energy in Noise- and Signal-modulated Waves, *Cruft Lab. Tech. Rept.* 99, Mar. 1, 1950.

CHAPTER 15

DETECTION OF FREQUENCY-MODULATED WAVES; SECOND-MOMENT THEORY

In Chap. 14 we have examined the spectrum and covariance function of sinusoidal carriers angle-modulated by various random and deterministic processes. Here we shall consider the "inverse" problem that arises when such angle-modulated waves are demodulated, or detected by an appropriate receiver, $\boldsymbol{T}_R^{(N)}$. As in Chap. 13 for the AM case, we shall outline a second-moment theory for the received disturbances, where the principal results are the spectrum and covariance function of the final output process and, as an estimate of receiver performance, signal-to-noise ratios based upon these quantities. The signal processes, however, are deterministic,† while the noise is entirely random, appears additively with the signal at the input of the first nonlinear elements of the receiving system, and represents random current and voltage fluctuations arising from thermal and shot effects in the initial stages of the receiver. Stationary normal noise‡ is assumed as before, and the signal itself (for the most part) is taken to be a frequency-modulated cw carrier.§ With simple modifications (essentially in the intensity and spectral distribution of the input processes), the analysis also applies directly to situations where external noise ($\boldsymbol{T}_M^{(N)}$) is the principal interference, as long as it is still normal and additive with the input signal. Only a few of the principal results can be given here; for a

† In certain limiting cases (of weak or strong carriers, no limiting, etc.), a second-moment theory of the received process is feasible for random signal modulations.

‡ Impulse noise (cf. Sec. 11.2) is not treated here. In the common situation of isolated impulsive bursts, receiver performance can be analyzed by conventional methods,[1] but when the pulse density γ [cf. Eq. (11.74)] is $O(10^0)$ or larger, a statistical treatment is necessary, along the lines described in the sections following. However, the analysis itself is far more difficult, since the background noise is not normal. On the other hand, when pulse density is large [$\gamma \geq O(10)$], the noise is nearly gaussian, becoming normal as $\gamma \to \infty$, so that the present approach can be applied, with one or two correction terms in the normal distributions [cf. Eqs. (11.92), (11.93), for example]. When γ is extremely large, we have, of course, the normal case again for all practical purposes.

§ The results here for frequency modulation apply also for phase modulation, when the two are related by $\mathcal{W}_{\phi M}(f) = \mathcal{W}_{FM}(f)/\omega^2$. For a discussion of phase detectors and a comparison with AM detectors from the point of view of signal-to-noise ratios, see Huggins and Middleton[2] and E. Gott.[3]

more complete treatment the reader should consult the appropriate references.†

15.1 *The FM Receiver*

Figure 15.1 shows a typical FM receiver $T_R^{(N)}$. Except for the limiter and discriminator elements, the structure is essentially similar to the heterodyne AM receiver described earlier in Sec. 5.3.

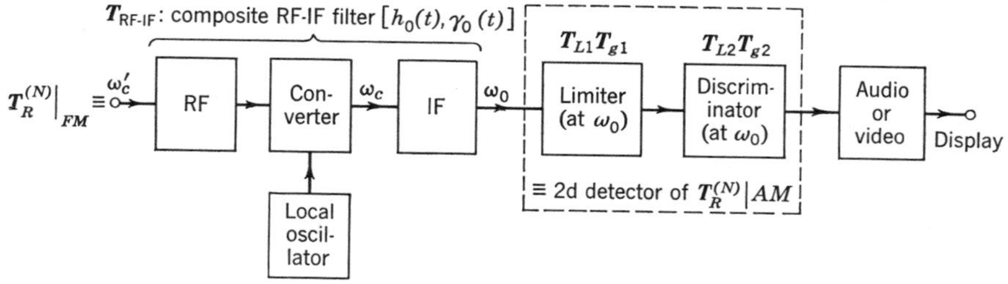

FIG. 15.1. Block diagram of a typical FM receiver $(T_R^{(N)})_{FM}$.

15.1-1 RF and IF Receiver Stages.

If we let

$$V_{in}(t) = \mathrm{Re}\,\{[\alpha_0(t) - i\beta_0(t)]e^{i\omega_c' t}\} \tag{15.1}$$

represent an input narrowband process to the receiver, we may write the output of the IF stage as

$$V_0(t) = \mathrm{Re}\,\{[\alpha(t) - i\beta(t)]e^{i\omega_0 t}\} \tag{15.2}$$

where in the usual way (cf. Sec. 2.2-6)

$$\alpha(t) - i\beta(t) = \int_{-\infty}^{\infty} h_0(\tau)[\alpha_0(t-\tau) - i\beta_0(t-\tau)]e^{i\omega_d(t-\tau)-i\gamma_0(\tau)}\,d\tau$$

$$\omega_d = \omega_c - \omega_0 \tag{15.3}$$

and f_c, f_0 are, respectively, the frequencies of the *shifted* carrier and of the center of the IF band. Here, $h_0(\tau)$ and $\gamma_0(\tau)$ are the slowly varying parts of the weighting function [cf. Eq. (2.63*a*)] of the *composite* RF-IF strip, on the assumption that conversion is linear as far as the input process is concerned (cf. Sec. 12.4): the output of the RF stages is simply shifted to the new IF frequency, without distortion of the envelope and phase portions

† In this connection, we mention the papers of Carson and Fry[4] and Crosby[5] and the later investigations of Blachman,[6] Rice,[7] Stumpers,[8] Wang,[9] Middleton,[10,11] and Fuller.[12] The earlier efforts were largely concerned with the case of carriers strong compared with the background noise, while the more difficult problem of threshold signals was examined theoretically somewhat later (and independently) by Blachman, Stumpers, Wang, and Middleton. Extensive experimental results for the intensity and the spectra of the demodulated noise and signal processes have been obtained by Fuller.[12] The discussion in the present chapter is based mainly on the general analysis of Middleton, with experimental data from Fuller.

of the wave. The IF output is also a narrowband disturbance which can be represented in terms of an envelope and phase according to

$$V_0(t) = E(t) \cos [\omega_0 t - \theta(t)] \tag{15.4}$$

where E and θ are slowly varying functions of the time compared with $\cos \omega_0 t$, etc. Through Eq. (15.3), the signal component of the IF output is easily related to the incoming signal, which is the real part of the purely angle-modulated carrier $(\alpha_0 - i\beta_0)e^{i\omega_c t} = B_0 e^{-i\Phi(t)+i\omega_c t}$. We write for this output signal process

$$\alpha_s - i\beta_s = B_0 \int_{-\infty}^{+\infty} h_0(\tau) e^{-i\Phi(t-\tau)+i\omega_d(t-\tau)-i\gamma_0(\tau)}\, d\tau \tag{15.5a}$$

$$\alpha_s - i\beta_s \equiv A_0(t) e^{-i\Psi(t)-i\omega_d t} \tag{15.5b}$$

in which $A_0(t)$ and $\Psi(t) + \omega_d t$ are the new envelope and phase, respectively. Note that, unless the composite linear filter preceding the limiter is sufficiently wide, the original frequency-modulated wave is now converted into one containing a certain amount of amplitude modulation, as well as some distortion of the original angle Φ [cf. Eqs. (15.5*a*), (15.5*b*)].

If the interference accompanying the signal is system (e.g., shot and thermal) noise, then it may be considered to have a spectral shape determined by this composite RF-IF filter, since such noise arises in the RF, mixer, and first stages of the IF section alone (cf. Secs. 5.3-2, 5.3-3). Moreover, as we have observed earlier, system noise in the linear portions of a network is normal, so that to a good approximation it is reasonable to regard the output of the IF stages now as the sum of a narrowband gaussian noise and deterministic FM signal process. Alternatively, Eq. (15.4) becomes

$$\begin{aligned} V_{\text{IF}\,0}(t) &= A_0(t) \cos [\omega_0 t - \Psi(t) - \omega_d t] + N(t) \\ &= \alpha_s(t) \cos \omega_0 t + \beta_s(t) \sin \omega_0 t + N(t) \end{aligned} \tag{15.6}$$

Since the noise is also narrowband, we can write $N(t) = N_c \cos \omega_0 t + N_s \sin \omega_0 t$ [cf. Eq. (9.1*a*)] and Eq. (15.6) as

$$\begin{aligned} V_{\text{IF}\,0}(t) &= [\alpha_s(t) + N_c] \cos \omega_0 t + [N_s + \beta_s(t)] \sin \omega_0 t \\ &= E(t) \cos [\omega_0 t - \theta(t)] \end{aligned} \tag{15.7a}$$

with

$$E(t) = \sqrt{(N_c + \alpha_s)^2 + (N_s + \beta_s)^2} \qquad \theta(t) = \tan^{-1} \frac{N_s + \beta_s}{N_c + \alpha_s} \pm \pi q \tag{15.7b}$$

and, for the signal components,

$$\alpha_s(t) = A_0(t) \cos [\omega_d t + \Psi(t)] \qquad \text{and} \qquad \beta_s(t) = A_0(t) \sin [\omega_d t + \Psi(t)] \tag{15.7c}$$

15.1-2 The Limiter. After the IF follows the limiter, which is usually a device with symmetrical dynamic characteristics similar to those shown in Fig. 13.19. In actual practice, the clipping of the input wave is never abrupt, but if the rms amplitude of the input is large compared with the clipping level, the idealization of Fig. 15.2 is a good model of actual situations. Since we shall ultimately be concerned only with the effects of the limiter on the envelope, we may use the half-wave characteristic (solid

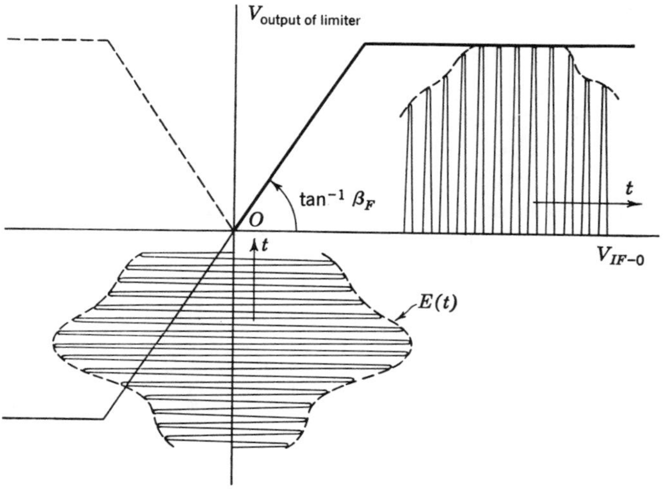

FIG. 15.2. Symmetrical ideal limiter (solid line).

line, Fig. 15.2), doubling it to account for both the positive and negative portions of the input wave. Thus,

$$\boldsymbol{T}_{g1}\{V_{\text{IF }0}\} = \begin{cases} \beta_F|V| & 0 < |V| < R_0 \\ \beta_F R_0 & |V| > R_0 \end{cases} \tag{15.8a}$$

for which the transform is (cf. type 4, Table 2.1)

$$f_L(i\xi) = 2\beta_F \frac{1 - e^{-i\xi R_0}}{(i\xi)^2} \tag{15.8b}$$

The output of the limiter (before filtering) is easily found from the relations (2.103*a*), (2.103*b*) to be

$$\begin{aligned} V_{0L}(t) &= \boldsymbol{T}_{g1}\{V_{\text{IF }0}\} = \frac{1}{2\pi}\int_C f_L(i\xi)e^{i\xi E\cos(\omega_0 t - \theta)}\,d\xi \\ &= \sum_{n=0}^{\infty} B_n(E)\cos n(\omega_0 t - \theta) \end{aligned} \tag{15.9}$$

with $$B_n(E) = \frac{i^n\epsilon_n}{2\pi}\int_C f_L(i\xi)J_n(E\xi)\,d\xi$$

At this point, we introduce the important condition that the passband of the limiter be wide enough to transmit the first spectral zone (i.e., about f_0) without attenuation, but not so wide as to include contributions from any of the other zones. Consequently, with the further, not unreasonable condition that the limiter's frequency response† be uniform over the spectral region of this first zone, we can now write for the input to the discriminator stage (cf. Fig. 15.1)

$$V_{id}(t) = \boldsymbol{T}_{L1}\boldsymbol{T}_{g1}\{V_{\mathrm{IF}\,0}\} = \boldsymbol{T}_L\{V_{0L}\} = B_1(E)\cos(\omega_0 t - \theta) \quad (15.10)$$

It should be emphasized that effective limiting requires that the frequency response of the limiter be sufficiently broad, as otherwise the higher-frequency components representing the sharp cutoff of the wave in time (cf. Fig. 15.2) are not passed. This "ceiling effect" of heavy limiting is then partly or wholly destroyed and the limiting itself rendered ineffective.

15.1-3 The Discriminator. In actual practice, a typical discriminator is the equivalent of a pair of off-tune narrowband filters followed by a

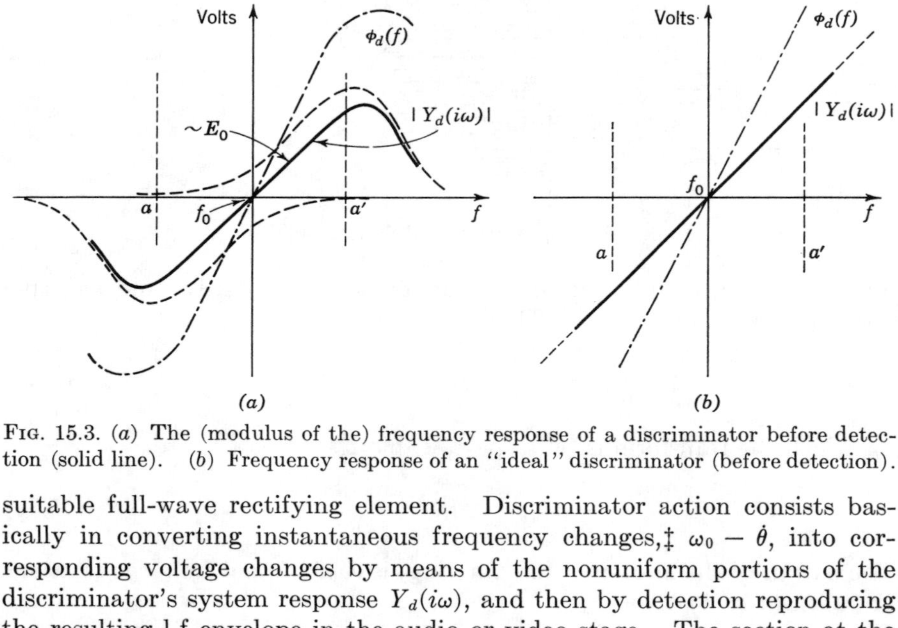

FIG. 15.3. (a) The (modulus of the) frequency response of a discriminator before detection (solid line). (b) Frequency response of an "ideal" discriminator (before detection).

suitable full-wave rectifying element. Discriminator action consists basically in converting instantaneous frequency changes,‡ $\omega_0 - \dot{\theta}$, into corresponding voltage changes by means of the nonuniform portions of the discriminator's system response $Y_d(i\omega)$, and then by detection reproducing the resulting l-f envelope in the audio or video stage. The section at the IF frequency is sketched in Fig. 15.3*a*.

The actual demodulating elements in the FM receiver are too complex to yield satisfactorily to a direct treatment in all details, but an acceptable model of the essential circuit elements may be constructed if we assume that the physical discriminator is replaced by the "ideal" one of Fig. 15.3*b*, which responds everywhere linearly with frequency (the "quasi-stationary" assumption of Carson and Fry[4]). Accordingly, the output voltage (or

† Here and subsequently, we assume that the limiter consists of a zero-memory nonlinear element $\boldsymbol{T}_{g1}$ followed by a linear filter $\boldsymbol{T}_{L1}$.

‡ For comments on the *amplitude* response of the discriminator, see Sec. 15.1-4.

current) is directly proportional to the instantaneous difference frequency between the wave and the central or resonant frequency of the IF, limiter, and discriminator bands (all tuned to the same frequency and assumed to be symmetrical). The passband of the limiter is taken wide enough to pass the IF portion of the limited wave without distortion due to frequency selection. In practice, of course, the real discriminator saturates for sufficiently large excursions about f_0, but this can be made quite unimportant if the rms frequency excursions of the input are kept well within the linear region (aa', Fig. 15.3a) and if only an insignificant portion of the input wave's intensity is redistributed into higher harmonics caused by the spectral "spreading" of the limiter.

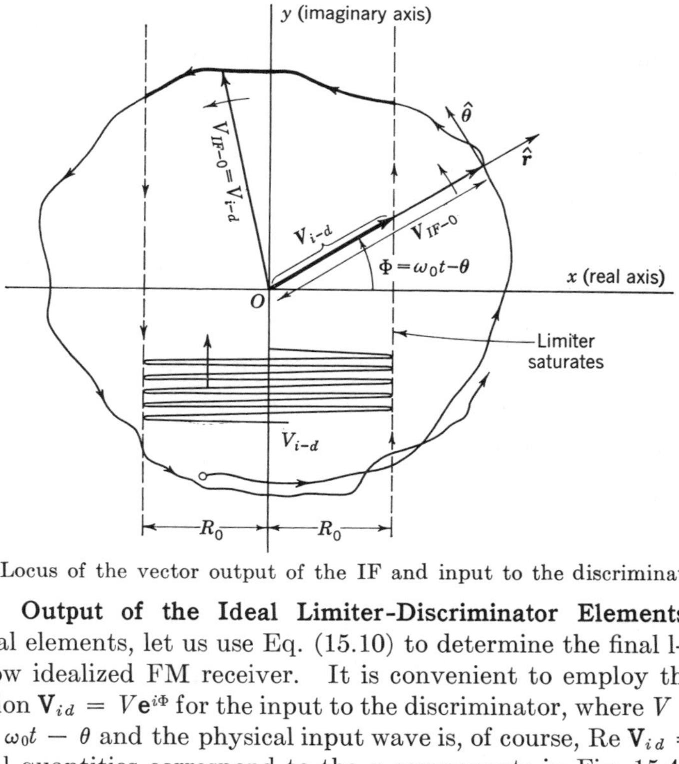

FIG. 15.4. Locus of the vector output of the IF and input to the discriminator stages.

15.1-4 Output of the Ideal Limiter-Discriminator Elements. With these ideal elements, let us use Eq. (15.10) to determine the final l-f output of the now idealized FM receiver. It is convenient to employ the vector formulation $\mathbf{V}_{id} = V\mathbf{e}^{i\Phi}$ for the input to the discriminator, where $V = B_1(E)$ and $\Phi = \omega_0 t - \theta$ and the physical input wave is, of course, Re $\mathbf{V}_{id} = V_{id}(t)$. Here, real quantities correspond to the x components in Fig. 15.4 and the imaginary quantities to the y components. Then we can write

$$\begin{aligned} \dot{\mathbf{V}}_{id} &= \dot{V}(\cos\Phi + i\sin\Phi) + V\dot{\Phi}(-\sin\Phi + i\cos\Phi) \\ &= \mathbf{r}\dot{V} + \boldsymbol{\theta}V\dot{\Phi} \end{aligned} \tag{15.11}$$

since the polar unit vectors ($\mathbf{r}$,$\boldsymbol{\theta}$) are related to the rectangular ones by the transformation ($\mathbf{r} = \mathbf{e}^{i\Phi}$, $\boldsymbol{\theta} = i\mathbf{e}^{i\Phi}$). Thus, $\dot{V}$ is the time rate of change of the modulus of this vector input, and $V\dot{\Phi}$ is proportional to the time derivative of the phase, i.e., to the instantaneous frequency. We emphasize, however, that this discriminator output is *not* equal to the instantaneous frequency alone of the input but varies linearly with the instantaneous amplitude of the input, as well. Thus, physically, the factor V in the l-f output $V\dot{\Phi}$ arises because of this amplitude response of the discriminator, and although the instantaneous frequency is $\dot{\Phi}$, the (l-f) discriminator output is $V\dot{\Phi}$.

The output $E_0(t)$ of the idealized discriminator, before detection, is precisely the second term in Eq. (15.11), viz.,

$$E_0'(t) = \kappa V\dot{\Phi} = \kappa B_1(E)(\omega_0 - \dot{\theta}) \tag{15.12}$$

where κ is a circuit constant with the dimension seconds. The h-f term, represented by the coefficient of ω_0 in Eq. (15.12), is not passed by the audio (or video) filter after detection, and so the final discriminator output is†

$$E_0(t) = \kappa B_1(E)\dot{\theta} = \frac{\kappa\dot{\theta}i}{\pi}\int_C J_1(E\xi)f_L(i\xi)\,d\xi \tag{15.13}$$

from Eq. (15.9), subject to the approximations mentioned above. For the particular limiter [Eqs. (15.8)] chosen here, we readily find that $E_0(t)$ can be written specifically

$$E_0(t) = \kappa\beta_F\dot{\theta}E \qquad 0 \le E \le R_0 \tag{15.14a}$$

$$E_0(t) = \kappa\beta_F\dot{\theta}\begin{cases}(4/\pi)R_0\,{}_2F_1(-\tfrac{1}{2},\tfrac{1}{2};\tfrac{3}{2};R_0^2/E^2) \text{ or} \\ (2/\pi)E\{(R_0/E)[1-(R_0/E)^2]^{1/2} \\ \qquad + \sin^{-1}(R_0/E)\}\end{cases} \qquad R_0 \le E \tag{15.14b}$$

The cases of principal interest arise (1) *when there is no limiting* ($R_0 \to \infty$), so that Eq. (15.14a) applies for all values of the envelope E, and (2) *when the limiting is very heavy* $[R_0/(2\overline{N^2})^{1/2} \to 0]$, so that effectively the output wave is independent of amplitude variations:

1. *No limiting* ($R_0 \to \infty$)

$$E_0(t) = \kappa\beta_F E\dot{\theta} \qquad 0 \le E \tag{15.15a}$$

2. *Heavy limiting* $[R_0/(2\overline{N^2})^{1/2} \to 0]$

$$E_0(t) \doteq \frac{4\kappa\beta_F}{\pi} R_0\dot{\theta} \qquad E > R_0 \text{ always} \tag{15.15b}$$

The situation of arbitrary limiting, where Eqs. (15.14) apply in their entirety, is considered in Sec. 15.3.

15.2 *The Mean and Mean-square Output of an FM Receiver*

Here we shall consider $\bar{E}_0$ and $\overline{E_0^2}$ following the discriminator and a wide video filter which passes this l-f output without attenuation. Specifically, we write

† The polarity of E_0 versus E_0' has been reversed, for convenience.

$$\overline{E_0(t)} = \kappa\overline{B_1(E)\dot{\theta}} = \kappa \int_0^\infty B_1(E)\, dE \int_{-\infty}^{\infty} \dot{\theta} W_1(E,\dot{\theta};t)\, d\dot{\theta} \tag{15.16a}$$

$$\overline{E_0(t)^2} = \kappa^2\overline{B_1(E)^2\dot{\theta}^2} = \kappa^2 \int_0^\infty B_1(E)^2\, dE \int_{-\infty}^{\infty} \dot{\theta}^2 W_1(E,\dot{\theta};t)\, d\dot{\theta} \tag{15.16b}$$

where only the statistical average over the accompanying noise is indicated. With a deterministic signal, we must, of course, take an additional average to get $\langle\overline{E_0(t)}\rangle$, etc.

We first obtain the conditional marginal d.d. $W_1(E,\dot{\theta};t)_{S+N}$ from the conditional distribution density $W_1(E,\dot{E};\theta,\dot{\theta};t)_{S+N}$. From Prob. 9.8, this is specifically

$$W_1(E,\dot{E};\theta,\dot{\theta};t)_{S+N} = \frac{E^2}{4\pi^2 b_0 b_2}\left\{\exp \frac{-[E^2 + A_0(t)^2 - 2A_0(t)E\cos(\theta - \chi_1)]}{2b_0}\right\}$$
$$\exp\frac{-\{\dot{E}^2 + E^2\dot{\theta}^2 + \dot{\alpha}_s^2 + \dot{\beta}_s^2 + 2(\dot{\alpha}_s^2 + \dot{\beta}_s^2)^{1/2} \times [E\dot{\theta}\sin(\theta - \chi_2) - \dot{E}\cos(\theta - \chi_2)]\}}{2b_2} \tag{15.17}$$

where $\chi_1 = \tan^{-1}(\beta_s/\alpha_s) = \omega_d t + \Psi(t)$ [Eq. (15.5b)] and $\chi_2 = \tan^{-1}(\dot{\beta}_s/\dot{\alpha}_s)$, with $b_n \equiv \int_0^\infty (\omega - \omega_0)^n \mathcal{W}_N(f)\, df$, in which $\mathcal{W}_N(f)$ is the spectrum of the noise leaving the IF. Thus, $b_0 = \overline{N^2}$ [Eqs. (15.6) et seq.].† Now, integration over $\dot{E}$ and θ gives at once

$$W_1(E,\dot{\theta};t)_{S+N} = \frac{E^2}{b_0\sqrt{2\pi b_2}} \exp\left\{\frac{-[E^2 + A_0^2(t)]}{2b_0} - \frac{\dot{\theta}^2E^2 + \dot{\alpha}_s^2 + \dot{\beta}_s^2}{2b_2}\right\}$$
$$\times \int_0^{2\pi} \frac{d\theta}{2\pi} \exp\left[\frac{(\dot{\alpha}_s^2 + \dot{\beta}_s^2)\cos^2(\theta - \chi_2)}{2b_2} + \frac{A_0(t)E\cos(\theta - \chi_1)}{b_0} - \frac{\sqrt{\dot{\alpha}_s^2 + \dot{\beta}_s^2}\, E\dot{\theta}\sin(\theta - \chi_2)}{b_2}\right] \tag{15.18}$$

Applying this to Eqs. (15.16) and integrating over θ last, with the observation that

$$\sin(\chi_1 - \chi_2) = \frac{-\dot{\phi}A_0}{\sqrt{A_0^2\dot{\phi}^2 + \dot{A}_0^2}} \qquad \cos^2(\chi_2 - \chi_1) = \frac{\dot{A}_0^2}{\dot{A}_0^2 + A_0^2\dot{\phi}^2}$$
$$\dot{\alpha}_s^2 + \dot{\beta}_s^2 = \dot{A}_0^2 + A_0^2\dot{\phi}^2 \quad \text{and} \quad \dot{\phi} \equiv \omega_d + \dot{\Psi}(t) \qquad A_0 = A_0(t),\ \text{etc.} \tag{15.19}$$

we normalize according to $r = R/(2b_0)^{1/2}$, $a_0 \equiv a_0(t) = A_0(t)/(2b_0)^{1/2}$, $r_0 \equiv R_0/(2b_0)^{1/2}$ and let $\rho_E{}^2 \equiv b_2/b_0$ [cf. Eq. (9.84)]. We obtain finally the general expressions

† The b_0 ($= \overline{N^2}$) here and in subsequent sections is not to be confused with the bias b_0 used in Chaps. 12, 13.

$$\overline{E_0(t)} = 2\kappa\beta_F\sqrt{2b_0}a_0\dot{\phi}(t)e^{-a_0^2}\int_0^\infty B_1(r)I_1(2a_0r)e^{-r^2}\,dr \qquad a_0 = a_0(t) \tag{15.20}$$

$$\overline{E_0(t)^2} = 2\kappa^2\beta_F{}^2b_0a_0{}^2e^{-a_0^2}\left\{\int_0^\infty \frac{B_1(r)^2}{r}\,e^{-r^2}[(\dot{a}_0{}^2 + a_0{}^2\dot{\phi}^2)I_0(2a_0r) + (a_0{}^2\dot{\phi}^2 - \dot{a}_0{}^2)I_2(2a_0r)]\,dr + \rho_E{}^2\int_0^\infty B_1(r)^2e^{-r^2}I_0(2a_0r)\,\frac{dr}{r}\right\} \tag{15.21}$$

where I_0, I_1, etc., are modified Bessel functions of the first kind and the appropriate $B_1(r)$ is obtained directly from Eq. (15.13) in conjunction with Eq. (15.8*b*), for example, $B_1(r) = (\sqrt{2b_0})^{-1}B_1(E)$, with $E = r\sqrt{2b_0}$. When the IF and preceding filters are sufficiently broad and flat, so as to cause no incidental amplitude modulation, e.g., when $h_0(\tau) \doteq \delta(\tau - 0)$, then we have $A_0(t) = A_0\cos[\omega_d t + \Phi(t)]$ and $\Phi(t) = \int_{t_0}^t D_0(t')\,dt'$. Consequently, we can write $a_0(t) = a_0$; $\dot{a}_0{}^2 + a_0{}^2\dot{\phi}^2$ reduces to $a_0{}^2\dot{\phi}^2$, with $\dot{\phi} = \omega_d + \dot{\Phi}(t) = \omega_d + D_0(t)$, where $D_0(t)$ is proportional to the original modulation.[8] An important feature of Eq. (15.20) is the *linear* dependence of the output process on the modulation (that is, $\bar{E}_0 \sim \dot{\phi}$). This is a direct consequence, of course, of the ideal linear discriminator characteristic assumed in the present analysis.

15.2-1 The Mean Output. Although a closed expression for $\bar{E}_0$ [Eq. (15.20)] is not generally possible, in the two limiting cases of interest [cf. Eqs. (15.15)], namely, for no limiting and extreme limiting, we can obtain simple results. In what follows, we restrict ourselves to IF outputs which contain frequency modulation only. Then, for the limiter of Sec. 15.1-2 we find that

$$\begin{aligned}\overline{E_0(t)}\Big|_{r_0\to\infty} &= 2\kappa\beta_F\sqrt{2b_0}\,\dot{\phi}(t)a_0e^{-a_0^2}\int_0^\infty rI_1(2a_0r)e^{-r^2}\,dr \\ &= \kappa\beta_F\left(\frac{\pi b_0}{2}\right)^{1/2}\dot{\phi}(t)a_0{}^2e^{-a_0^2/2}\left[I_0\left(\frac{a_0{}^2}{2}\right) + I_1\left(\frac{a_0{}^2}{2}\right)\right] \\ &= \kappa\beta_F\dot{\phi}\left(\frac{\pi b_0}{2}\right)^{1/2}a_0{}^2\,{}_1F_1\left(\frac{1}{2};2;-a_0{}^2\right)\end{aligned} \tag{15.22}$$

When the input carrier-to-noise ratio is large or small, we can write

$$\overline{E_0(t)}\Big|_{r_0\to\infty,\,a_0^2\ll1} \doteq \left(\frac{\pi b_0}{2}\right)^{1/2}\kappa\beta_F\dot{\phi}(t)a_0{}^2\left(1 - \frac{a_0{}^2}{4} + \frac{a_0{}^4}{16} + \cdots\right) \tag{15.23a}$$

$$\overline{E_0(t)}\Big|_{r_0\to\infty,\,a_0^2\gg1} \simeq \kappa\beta_F\dot{\phi}(t)\sqrt{2b_0}\,a_0(1 - 1/4a_0{}^2 - 3/32a_0{}^4 - \cdots) \tag{15.23b}$$

Note that for weak carriers the mean output (at arbitrary t) is proportional to the *square* of the input ratio a_0. This is another example of *signal suppression* [cf. Secs. 13.2-1(4) and Eq. (19.59)]. At the other extreme of strong carriers, on the other hand, the noise is suppressed rather than the signal, as we see from Eq. (15.23*b*), where $\bar{E}_0 \simeq \kappa\beta_F\dot{\phi}(t)A_0$.

With a state of very heavy limiting, where $r_0^2 \ll 1$, we observe that the termwise expansion of the integrand of Eq. (15.20) leads finally to

$$\overline{E_0(t)}\Big|_{r_0 \to 0} \doteq \frac{4}{\pi} \beta_{FK} \dot{\phi} r_0 (2b_0)^{1/2} (1 - e^{-a_0^2}) \tag{15.24}$$

For weak and strong carriers, respectively, this becomes

$$\overline{E_0(t)}\Big|_{r_0 \to 0, a_0^2 \ll 1} \doteq \frac{4}{\pi} \beta_{FK} \dot{\phi} r_0 (2b_0)^{1/2} a_0^2 \left(1 - \frac{a_0^2}{2} + \cdots \right) \tag{15.25a}$$

$$\overline{E_0(t)}\Big|_{r_0 \to 0, a_0^2 \gg 1} \simeq 4\beta_{FK} \dot{\phi}(t) R_0/\pi \tag{15.25b}$$

Equations (15.22) and (15.24) apply also for finite limiting ($r_0 \neq \infty$) provided that $a_0^2 \ll r_0^2$ or $a_0^2 \gg r_0^2$, respectively—i.e., provided either that

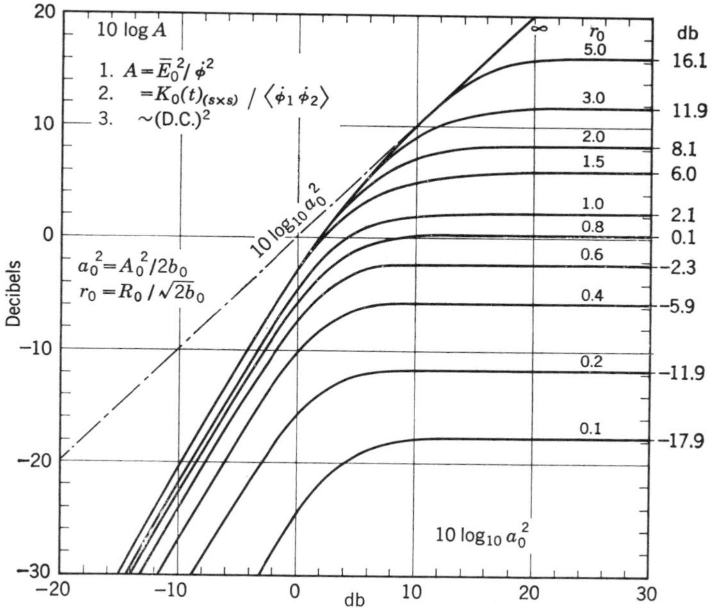

FIG. 15.5. (1) Square of the mean l-f output. (2) Covariance function of the l-f output. (3) Square of the d-c output.

there is essentially no clipping of the carrier and noise because the limiting level is rarely exceeded or that this limiting threshold is exceeded most of the time. Note again the characteristic suppression of the signal ($\sim \dot{\phi}$) in the weak-carrier cases [cf. Eq. (15.25a)]. A number of curves of $(\bar{E}_0/\dot{\phi})^2$ showing the effects of limiting for different input carrier-to-noise ratios are included in Fig. 15.5.

15.2-2 The Mean-square Output. In a similar way, and with assumptions identical to those in Sec. 15.2-1, we find that for no limiting Eq. (15.21) reduces to

$$\overline{E_0(t)^2}\Big|_{r_0\to\infty} = 2\kappa^2\beta_F^2 b_0\left[a_0^2\phi^2(t)\left(1-\frac{1-e^{-a_0^2}}{2a_0^2}\right)+\frac{\rho_E^2}{2}\right] \qquad (15.26)$$

For weak or strong carriers, this becomes alternatively

$$\overline{E_0(t)^2}\Big|_{r_0\to\infty,a_0^2\ll 1} \doteq b_0\kappa^2\beta_F^2\left[\phi^2 a_0^2\left(1+\frac{a_0^2}{2!}-\frac{a_0^4}{3!}\cdots\right)+\rho_E^2\right] \qquad (15.27a)$$

$$\overline{E_0(t)^2}\Big|_{r_0\to\infty,a_0^2\gg 1} \simeq 2b_0\kappa^2\beta_F^2 a_0^2\left[\phi^2\left(1-\frac{1}{2a_0^2}\right)+\frac{\rho_E^2}{2a_0^2}\right] \qquad (15.27b)$$

The suppression effects mentioned above for $\bar{E}_0$ are also seen to be in operation here.

At the other extreme of very heavy limiting, we have

$$\overline{E_0^2}\Big|_{r_0^2\ll 1} \doteq \frac{16}{\pi^2}\kappa^2\beta_F^2 r_0^2 b_0 e^{-a_0^2}\left\{(a_0^2\phi^2+\rho_E^2)[\gamma_M - Ei(-r_0^2)] + \sum_{m=0}^{\infty}\frac{\phi^2 a_0^{2m+4}(2m+3)+\rho_E^2 a_0^{2m+2}(m+2)}{(m+1)(m+2)!}\right\} \qquad (15.28)$$

where[10] $\gamma_M = \pi^2/12 - 0.3444 = 0.2731$ and $Ei(-x)$ is the exponential integral $Ei(-x) \equiv -\int_x^\infty e^{-y}\,dy/y$. When the carrier is weak compared with the noise, Eq. (15.28) simplifies to

$$\overline{E_0^2}\Big|_{r_0^2\ll 1;\, a_0^2\ll 1} \doteq \frac{16}{\pi^2}\kappa^2\beta_F^2 b_0 r_0^2\{\lambda_0\rho_E^2 + a_0^2[\rho_E^2(1-\lambda_0)+\lambda_0\phi^2] + O(a_0^4)\} \qquad (15.29a)$$

where $\lambda_0 \equiv \gamma_M - Ei(-r_0^2)$.

For strong carriers, one can show that[10]

$$\overline{E_0^2}\Big|_{r_0^2\ll 1;a_0^2\gg 1} \simeq \frac{16}{\pi^2}\kappa^2\beta_F^2 r_0^2 b_0\left[\phi^2\left(2+\frac{1!}{a_0^2}+\frac{2!}{a_0^4}+\cdots\right) + \frac{\rho_E^2(1+1!/a_0^2+2!/a_0^4+\cdots)}{a_0^2}\right] \qquad (15.29b)$$

where now the suppression of the noise by the carrier is observed, that is, $2\phi^2$ versus ρ_E^2/a_0^2, as $a_0^2 \to \infty$. Curves of $\langle\overline{E_0^2}\rangle$ are shown in Fig. 15.6 for various limiting levels and a particular ratio of $\langle\phi^2\rangle$ to ρ_E^2; good experimental agreement with this has been obtained by Fuller[12] (dissertation, figs. 5.5 to 5.8). If $a_0^2 \gg r_0^2$, or if $a_0^2 \ll r_0^2$, we observe the effects of heavy or no limiting, respectively [cf. Eqs. (15.28), (15.26)]. When there is no modulation ($\phi = 0$), $\bar{E}_0$ vanishes as expected and $\overline{E_0^2}$ becomes progressively smaller as the carrier is increased relative to the limit level R_0. From Eq. (15.21),

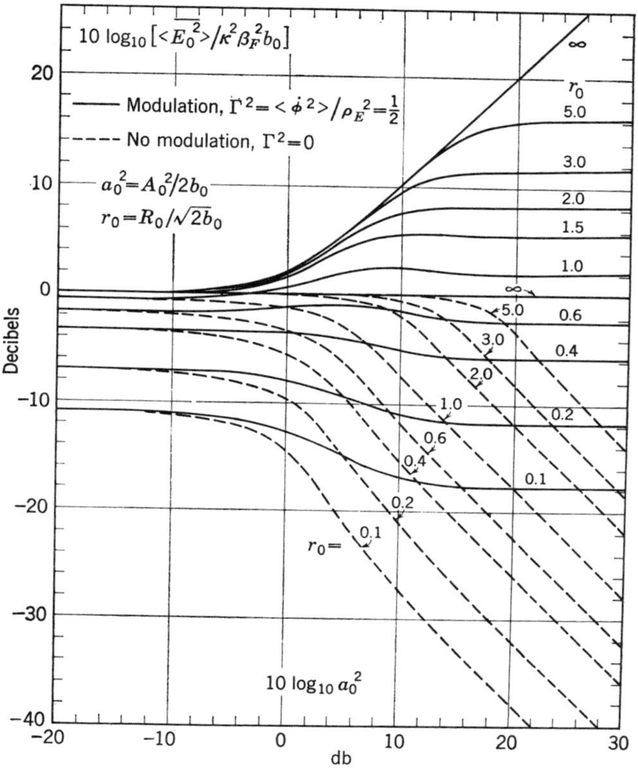

FIG. 15.6. Mean intensity of the l-f output of an ideal discriminator, with limiting, and with and without modulation; $\langle\dot\phi^2\rangle/\rho_E^2 = \frac{1}{2}$ or 0.

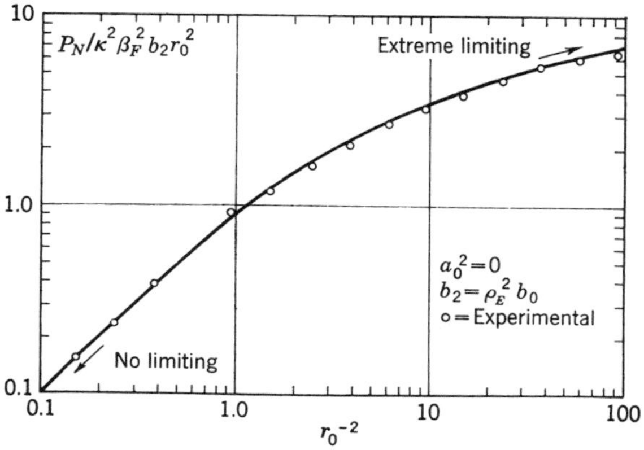

FIG. 15.7. Mean intensity of the l-f noise output of the ideal limit discriminator, as a function of limit level (no carrier).

we accordingly see that the mean intensity of the noise appearing in the discriminator's output becomes (for a nonvanishing carrier)

$$P_{N+C} = \kappa^2\beta_F{}^2 b_0 e^{-a_0{}^2}\rho_E{}^2 \int_0^\infty B_1(r)^2 e^{-r^2} I_0(2a_0 r)\,\frac{dr}{r} \tag{15.30}$$

as indicated by the dashed curves of Fig. 15.6. If now the carrier is removed, Eq. (15.21) reduces to

$$P_N = \kappa^2\beta_F{}^2 b_0 \rho_E{}^2 \int_0^\infty B_1(r)^2 e^{-r^2}\, dr/r \tag{15.31}$$

which is illustrated in Fig. 15.7 (as a function of r_0^{-2}) for all limiting levels and the limiter of Eqs. (15.8*a*), (15.8*b*); the experimental data were obtained by Fuller.[12] Note that, when $r_0 \to \infty$ (for practical purposes, $r_0 \doteq 3.0$), there is no further increase in the available output noise level.

PROBLEMS

15.1 If there is no modulation and no limiting, show that the mean output and mean-square output of an ideal discriminator are proportional to

$$\bar{E}_0 \sim \overline{\cos\alpha} \qquad \overline{E_0{}^2} \sim \overline{\cos^2\alpha}$$

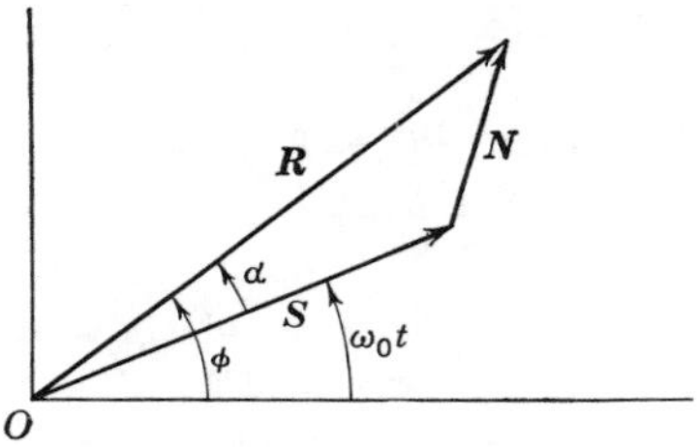

FIG. P 15.1. Vector diagram of input to a discriminator (no limiting).

(cf. Fig. P 15.1), where

$$\overline{\cos\alpha} = \frac{\sqrt{\pi}}{2}\, a_0 e^{-a_0{}^2/2}\left[I_0\left(\frac{a_0{}^2}{2}\right) + I_1\left(\frac{a_0{}^2}{2}\right)\right]$$

$$\overline{\cos^2\alpha} = 1 - \frac{1 - e^{-a_0{}^2}}{2a_0{}^2}$$

HINT: Obtain $w(\phi) = \int_0^\infty R e^{-(R^2+A_0{}^2-2RA_0\cos\phi)/2b_0}(dR/2\pi b_0)$, and find $\cos\alpha = f(\phi)$.

N. M. Blachman.[6]

15.2 When only noise enters the ideal FM receiver of the text, show that the first-order *distribution density* of $E_0(t)$ becomes

$$W_1(E_0) = (\kappa\beta_F\sqrt{2\pi b_2})^{-1} e^{-E_0{}^2/2\kappa^2\beta_F{}^2 b_2} \qquad \text{no limiting} \tag{1}$$

$$W_1(E_0) = (2\rho_E a)^{-1}\left(1 + \frac{E_0{}^2}{a^2\rho_E{}^2}\right)^{-3/2} \qquad a = \frac{4}{\pi}\kappa\beta_F R_0 \qquad \text{heavy limiting, } R_0 \to 0 \tag{2}$$

Verify that $\overline{E_0{}^2} \to \infty$ $(R_0 > 0)$, $\overline{E_0{}^2} \to 0$ $(R_0 \to 0)$.

15.3 Show, in the case of noise alone as input to the ideal limiter-discriminator combination of the text, that the mean intensity [Eq. (15.31)] of the (l-f) output of the discriminator can be expressed for *all* limit levels as

$$P_N = \kappa^2\beta_F{}^2\rho_E{}^2 b_0 \left[1 - e^{-r_0{}^2} - \frac{16}{\pi^2} r_0{}^2 \sum_{l=0}^{\infty} \frac{(-1)^l}{l!} C_l r_0{}^{2l} Ei(-r_0{}^2) \right.$$
$$\left. - \frac{16}{\pi^2} e^{-r_0{}^2} \sum_{l=1}^{\infty} C_l \sum_{n=1}^{l} \frac{r_0{}^{2n}}{(-l)_n} \right] \tag{1}$$

where

$$C_l = \sum_{m=0}^{l} \frac{(-\frac{1}{2})_{l-m}(-\frac{1}{2})_m(\frac{1}{2})_{l-m}(\frac{1}{2})_m}{(\frac{3}{2})_{l-m}(\frac{3}{2})_m(l-m)!m!} \tag{2}$$

H. W. Fuller,[12] app. 2-2.

15.3 *The Reception of Narrowband Frequency Modulation*

In order to analyze the performance of our (idealized) FM receiver, we must first distinguish two types of frequency modulation for the input signal process, depending on the relative widths of the RF-IF and l-f pass-bands. These are (1) *narrowband frequency modulation*, for which the maximum (or rms) deviation in the (angular) frequency of the carrier is comparable to or less than the highest significant modulation frequency, and (2) *broadband frequency modulation*, which is characterized by a maximum change in the carrier frequency that may be much greater than the highest modulating component. A correspondingly wide RF-IF band is then required for the transmission of the signal. To determine the l-f noise power after detection for narrowband frequency modulation (1), we may consider the entire l-f spectrum. Thus, for a given carrier strength, spectral shape is relatively unimportant—only the *area* under the spectral curve is required. For broadband frequency modulation (2), however, the intensity of the noise accompanying the signal in the final stage of the receiver must now be obtained from a portion of the l-f spectrum passed by the video (or audio) filter. Spectral shape is therefore a significant factor, as we shall see presently in Sec. 15.4. Analytically, the former case demands only first-order densities of E, $\dot{\theta}$ [cf. Eqs. (15.18), (15.20), (15.21)], while the latter requires an appropriate second-order distribution (cf. Sec. 15.4). Similar remarks apply, of course, for narrowband frequency modulation whenever the signal or one of its components is observed in a band narrower than the l-f noise spectrum following discrimination.

Before considering specifically the performance of a receiver for narrowband frequency modulation, let us first establish the particular structure of the signal-to-noise criteria to be used here and in the broadband cases.

15.3-1 Signal-to-Noise Ratios. Here we shall need a suitable modification of the signal-to-noise ratio already introduced in the treatment of AM receiving systems (Secs. 5.3, 13.2 to 13.4). To illustrate, let us assume as before a periodic signal process. Then $\overline{E_0(t)}_{per}$ represents the signal portion of the output from the discriminator, where specifically

$$\overline{E_0(t)}_{per} = \sum_{-\infty}^{\infty} b_k e^{i\omega_a kt} \qquad b_k = \frac{1}{T_0}\int_0^{T_0} \overline{E_0(t_0)}_{per} e^{-ik\omega_a t_0}\,dt_0 \qquad T_0 = 2\pi/\omega_a \tag{15.32}$$

since $\overline{E_0(t)}_{per}$ is the periodic part of $\overline{E_0(t)}$ [cf. Eq. (15.20)], including a d-c term, if any. The signal power P_S is simply $P_S = \sum_{-\infty}^{\infty} b_k b_k^* = \langle \overline{E_0(t)}^2_{per} \rangle$. The mean intensity associated with the nondeterministic, or fluctuating, part of the output is similarly found to be

$$P_{s\times n} = \langle \overline{E_0(t)^2} \rangle - \langle \overline{E_0(t)}^2_{per} \rangle \tag{15.33}$$

since $\langle \overline{E_0(t)^2} \rangle$ represents the mean total intensity of the output, including both signal and noise. The d-c contribution is

$$E_0(t)_{dc} = \langle \overline{E_0(t)} \rangle = \langle \overline{E_0(t)}_{per} \rangle = b_0 = \frac{1}{T_0}\int_0^{T_0} \overline{E_0(t_0)}_{per}\,dt_0 \tag{15.34}$$

Taking P_S to be a measure of signal intensity, or as an indication of the presence or absence of a signal, as in the "deflection" criterion (Sec. 5.3-4), we construct the following signal-to-noise ratio for the output when *narrow-band frequency modulation* is employed:

$$\left[\frac{S}{N}\right]^2_{\mathrm{I}} \equiv \frac{P_S - (P_S)_{dc}}{P_{s\times n}} = \frac{\langle \overline{E_0(t)}^2_{per} \rangle - \langle \overline{E_0(t)}_{per} \rangle^2}{\langle \overline{E_0(t)^2} \rangle - \langle \overline{E_0(t)}^2_{per} \rangle} \tag{15.35}$$

For *broadband frequency modulation*, however, only part of the output noise accompanies the signal, so that the corresponding noise intensity here is now

$$P'_{s\times n} = \int_0^\infty |Y_V(i\omega)|^2 \mathcal{W}_0(f)\,df \tag{15.36}$$

where $\mathcal{W}_0(f)$ is the intensity spectrum of the random part of $E_0(t)$, the l-f output of the discriminator *before* video or audio filtering and now with the d-c and discrete terms removed. Y_V is the system function of this final video or audio filter. The deflection criterion here becomes

$$\left[\frac{S}{N}\right]^2_{\mathrm{II}} = \frac{P_S - (P_S)_{dc}}{P'_{s\times n}} = \frac{\langle \overline{E_0(t)}^2_{per} \rangle - \langle \overline{E_0(t)}_{per} \rangle^2}{\int_0^\infty |Y_V(i\omega)|^2 \mathcal{W}_0(f)\,df} \tag{15.37}$$

When only one component of $\overline{E_0(t)}_{per}$ is taken as the signal output, we obtain still another form of $[S/N]^2_{\mathrm{I}}$ and $[S/N]^2_{\mathrm{II}}$:

$$\left[\frac{S}{N}\right]^2_{\mathrm{III}} \equiv \frac{(P_S)_k}{P'_{s\times n}}, \text{ narrowband FM} \qquad \left[\frac{S}{N}\right]^2_{\mathrm{III}} \equiv \frac{(P_S)_k}{P'_{s\times n}}, \text{ broadband FM} \tag{15.38}$$

Here, $(P_S)_k = b_k b_k^* + b_{-k} b_{-k}^* = \epsilon_k b_k b_k^*$ [cf. Eqs. (15.32) et seq.] is the intensity of the kth component, and $P'_{s \times n}$ is given by Eq. (15.36), where now Y_V is to be interpreted as a filter centered about this kth component and passing the noise in the immediate vicinity of it.†

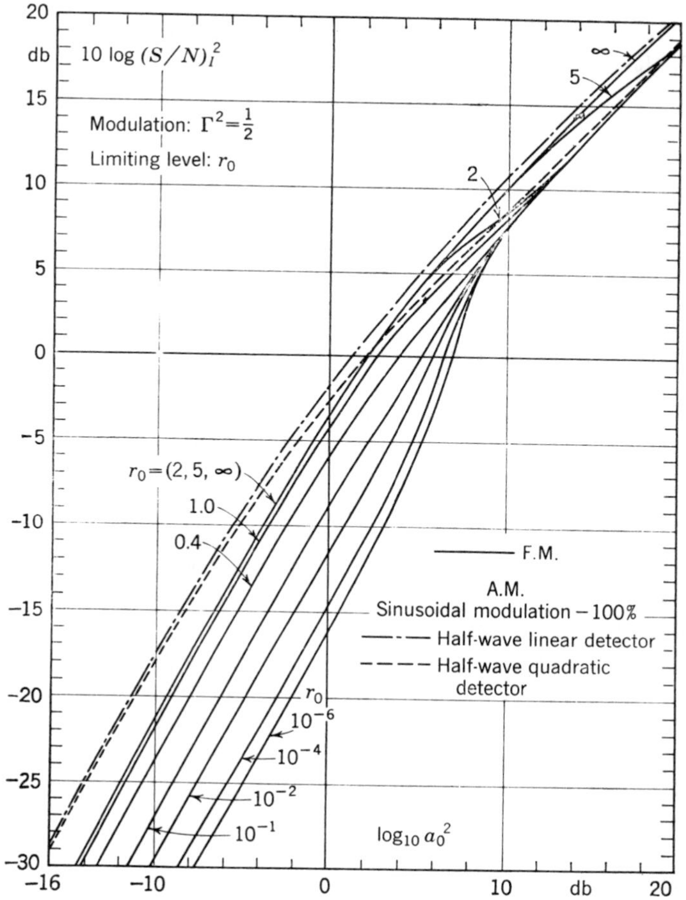

FIG. 15.8. Output signal-to-noise ratios for narrowband frequency modulation as a function of the input (S/N) ratio a_0 for heavy to no limiting (gaussian composite RF-IF filter).

15.3-2 Signal-to-Noise Ratios (Narrowband Frequency Modulation). We restrict ourselves again to a flat passband in the prelimiter stages of the receiver, so that if FM signals are present, we may use the results of Secs. 15.2-1, 15.2-2 to determine the output signal-to-noise ratio for narrowband frequency modulation. Next, let us introduce the *modulation factor* Γ in the definition

$$\Gamma^2 \equiv \langle \dot{\phi}^2 \rangle / \rho_E{}^2 \tag{15.39}$$

† Middleton,[10] sec. 2.

where ϕ is the modulation and ρ_E is the centroid (angular) frequency of the accompanying noise [cf. Eq. (9.84)]. We then use Eqs. (15.20) to (15.31) in Eq. (15.35) to obtain the curves of output signal-to-noise ratio shown in Fig. 15.8.

A number of special results follow. When there is effectively no limiting, we can write in the case $\langle\phi\rangle = 0$, that is, for an unmodulated carrier or one for which the average value of the modulation vanishes,

(1) $r_0^2 \to \infty$, or $r_0^2 \gg a_0^2$

$$\left[\frac{S}{N}\right]_{\mathrm{I}}^2 \doteq \frac{\pi a_0^4\Gamma^2(1 + a_0^2/2 + \cdots)}{2(1 + a_0^2\Gamma^2\{1 + [(1-\pi)/2]a_0^2 + \cdots\})} \qquad a_0^2 \ll 1 \qquad (15.40)$$

and, at the other extreme,

$$\left[\frac{S}{N}\right]_{\mathrm{I}}^2 \simeq 2a_0^2\Gamma^2\left(1 - \frac{1}{2a_0^2} - \cdots\right) \qquad a_0^2 \gg 1 \qquad (15.41)$$

The characteristic suppression of the signal is again observed when the carrier is weak, while for strong carriers the linear behavior in a_0 (and in ϕ) is noted: the input noise is now suppressed. Thus in the threshold case, if Γ^2 is small (< 1), the output *signal* ($\sim \phi$) still appears linearly, but its *amplitude* is proportional to a_0^2 rather than to a_0.

When the spectral character of the interfering noise is determined principally by the filters preceding the nonlinear elements, Γ cannot be much larger than unity if amplitude modulation due to the nonuniform response of these filters' system functions is to be avoided. For example, with a rectangular IF of width Δf_{I} and a sinusoidal frequency modulation of peak amplitude B_0, we find that $\Gamma^2 = 3(D_0B_0)^2/(8\pi^2\,\Delta f_{\mathrm{I}}^2)$. Since, for pure frequency modulation, the angular frequency deviation D_0B_0 should not exceed $2\pi\,\Delta f_{\mathrm{I}}$, then $\Gamma_{\max} = (3/2)^{1/2}$. However, if the principal source of noise is spectrally narrow compared with the filters preceding the detection elements, then ρ_E ($\sim$ bandwidth of this noise) is much less than Δf_{I} and consequently Γ^2 is much larger than unity. Under these circumstances, if $a_0^2\Gamma^2 \gg 1$ [cf. Eq. (15.40)] although a_0^2 is small, we see that $[S/N]_{\mathrm{I}}^2 \doteq \pi a_0^2/2$, and the effects of the signal suppression are removed.† In contrast to this, for heavy limiting the signal-to-noise ratios become

(2) $r_0^2 \ll 1$, or $r_0^2 \ll a_0^2$

$$\left[\frac{S}{N}\right]_{\mathrm{I}}^2 \doteq \frac{2\Gamma^2 a_0^4(1 - \cdots)}{\lambda_0(1 + a_0^2\Gamma^2) + a_0^2(1 - \lambda_0^2)} \qquad a_0^2 \ll 1,\ \lambda_0 \equiv \gamma_M - Ei(-r_0^2) \qquad (15.42)$$

† Of course, the bandwidths of the predetection filters are limited by the receiver noise (among other factors), so that we cannot increase bandwidths indefinitely without soon reaching the point where the inherent noise of the system itself once more becomes the predominant factor. The condition $a_0^2\Gamma^2 \gg 1$ is therefore no longer satisfied and signal suppression (of the carrier *and* signal) again occurs.

and for strong carriers

$$\left[\frac{S}{N}\right]_{\mathrm{I}}^2 \simeq \frac{2a_0^2\Gamma^2}{1+\Gamma^2} \qquad a_0^2 \gg 1 \tag{15.43}$$

The suppression of the signal is again evident in the threshold case $a_0^2 \ll 1$. If now we can allow $a_0^2\Gamma^2 \gg 1$, without introducing extraneous amplitude modulation, as mentioned above, the suppression of the signal is removed as before and with strong carriers the output ratio is essentially independent of Γ (where $\Gamma^2 \gg 1$), unlike the unlimited case (15.41).

Besides rectangular IF (and RF) passbands, filters with gaussian responses of the type $|Y(i\omega)| = e^{-(\omega-\omega_0)^2/2\omega_b^2}$ are also considered.[10] For these, $\rho_E = \omega_b/\sqrt{2}$, and with simple sinusoidal frequency modulation we obtain $\Gamma^2 = D_0^2B_0^2/2\rho_E^2 = D_0^2B_0^2/\omega_b^2$. Here, we choose $\Gamma^2 \leq \frac{1}{2}$, so that the maximum frequency deviation of this signal is $0.424(\Delta f)_{1/2} = 0.399\,\Delta f_E$, where $(\Delta f)_{1/2}$ and Δf_E are, respectively, the widths of the gaussian filter between half-power points and of the equivalent rectangular filter (cf. Sec. 3.3-2), viz.,

$$f_b = \frac{\omega_b}{2\pi} = \frac{\Delta f_E}{\sqrt{\pi}} \qquad (\Delta f)_{1/2} = \Delta f_E \sqrt{\frac{\log 16}{\pi}} = f_b \sqrt{\log 16} \tag{15.44}$$

This is small enough for us to disregard the extraneous amplitude modulation that occurs because of the nonuniformity of $Y(i\omega)$ in the vicinity of the (shifted) carrier frequency f_0.†

Curves of $[S/N]_{\mathrm{I}}^2$ for all a_0^2 and various limiting levels are shown in Fig. 15.8, based on Eqs. (15.22) et seq. and for gaussian responses of the composite RF-IF filter. For weak carriers ($a_0^2 \ll 1$), the corresponding curves for the rectangular filter (not shown in the figure) lie about 4.8 db above those for the gauss case (since $\Gamma^2_{\max\,rect} = \frac{3}{2}$, while here $\Gamma^2_{\max\,\mathrm{gauss}} = \frac{1}{2}$) [cf. Eq. (15.42)]. For strong carriers, on the other hand, the superiority of the former over the latter falls to 2.5 db $\{$since $[\Gamma^2/(1+\Gamma^2)]_{rect} = \frac{3}{5}$, $[\Gamma^2/(1+\Gamma^2)]_{\mathrm{gauss}} = \frac{1}{3}\}$ [cf. Eq. (15.43)].

When the input carrier is weak relative to the noise, the mean total intensity of the output is essentially independent of carrier strength, the carrier being suppressed by the noise. The output accordingly depends heavily on the degree of limiting, as Fig. 15.8 indicates, being considerably reduced for heavy clipping because less of the original wave is passed to the discriminator, and much of that appears as $s \times n$ noise after limiting. As

† The maximum change in amplitude produced by the excursion of the carrier about $f = f_0$ is $\Delta \equiv |Y(0)| - |Y[i(\omega_0 \pm D_0B_0)]| = 1 - e^{-D_0^2B_0^2/2\omega_b^2}$, and, for $\Gamma^2 \leq \frac{1}{2}$, this becomes $\Delta \leq (1 - e^{-1/4}) = 0.22$, or 22 per cent or less. With more complex modulations, where B_0 is the peak amplitude, this incidental amplitude modulation is even less (cf. Secs. 14.3-1, 14.3-2). Our subsequent approximations, for both narrow- and broadband frequency modulation, where gaussian filter responses are assumed, consist in replacing the more precise expressions (15.20), (15.21) with the subsequent relations (15.22) et seq., which assume no AM effects at all; for $\Gamma^2 \leq \frac{1}{2}$, this is not a very serious modification, leading to results in $[S/N]_{\mathrm{I}}^2$ accurate to within 1 db or less.

the carrier is increased, however, the contribution due to the modulation (e.g., the desired signal) becomes significant and approaches a limit proportional to the clipping level r_0. The purely noise contribution is now suppressed [cf. Eqs. (15.27*b*), (15.29*b*)].

The character of the signal-to-noise ratios follows from the discussion above. A number of interesting conclusions can be drawn at once. First, limiting in narrowband frequency modulation always decreases the sensitivity—i.e., raises the input carrier level necessary to obtain a given value of the output ratio $[S/N]_{\mathrm{I}}^2$. This is especially noticeable in the threshold cases [where $(S/N)_{\mathrm{I}} \leq 3$ db]. Furthermore, for strong carriers ($a_0^2 > 10$ db) the output ratio $(S/N)_{\mathrm{I}}$ is essentially independent of limiting, provided that the clipping level is less than the peak carrier amplitude. Here, the weak noise is suppressed, and the signal, represented by the "zero crossings" of the clipped wave, is affected only slightly. The deleterious effect of limiting arises because of the additional $s \times n$ and $n \times n$ noise products generated in the operation, which effectively increase the amount of noise accompanying the signal at the output of the discriminator.

15.3-3 Narrowband Frequency Modulation vs. Amplitude Modulation. It is instructive to compare the reception of AM and narrowband FM signals on the basis of our second-moment theory, e.g., by a comparison of signal-to-noise ratios in the two cases. For the corresponding case in AM reception, we need the results of Sec. 13.3. Simple sinusoidal modulations are assumed as before, and both linear and quadratic rectifiers are possible, so that, when appropriate modifications are made for the audio or video response, $(S/N)_{\mathrm{I}}$ [cf. Secs. 13.3(1), 13.3(2) and Eqs. (13.58 to 13.60), (13.66), (13.67)] becomes

$$\left[\frac{S}{N}\right]_{\mathrm{I}\,quad}^2 = \left[\frac{S}{N}\right]_a^2 = \frac{\lambda^2 a_0^4(2 + \lambda^2/8)}{[(\frac{1}{2})^{\frac{1}{2}} \text{ or } 1] + a_0^2(2 + \lambda^2)} \qquad \lambda \leq 1,\ a_0 \geq 0 \qquad (15.45)$$

Here $1/\sqrt{2}$ or 1 in the denominator applies for the now composite gauss or rectangular RF-IF filters, respectively, λ is the (AM) modulation index (cf. Secs. 12.1, 12.2), and a quadratic second detector is used. For a half-wave linear second detector in our AM receiver, we obtain two expressions, appropriate for weak or strong carriers, respectively,

$$\left[\frac{S}{N}\right]_{\mathrm{I}\,lin}^2 \doteq \frac{a_0^4\lambda^2(2 + \lambda^2/8)}{(16/\pi - 4) + (4/\pi - \frac{1}{2})(2 + \lambda^2)a_0^2 + \cdots} \qquad a_0^2 < \frac{1}{2},\ |\lambda| < 1 \qquad (15.46a)$$

$$\left[\frac{S}{N}\right]_{\mathrm{I}\,lin}^2 \simeq a_0^2\lambda^2 \qquad a_0^2 > 10 \qquad (15.46b)$$

Choosing the maximum index ($\lambda = 1$) for amplitude modulation, we have included $[S/N]_{\mathrm{I}\,lin,quad}^2$ among the curves of Fig. 15.8. This gives the desired comparison between amplitude modulation and narrowband frequency modulation (of comparable carrier level) for the same conditions of operation; that is, Γ^2 is also taken to have the maximum values permitted above, and the predetection stages are identical in both instances. Only the types of modulation and, correspondingly, the demodulating elements are different.

For the gaussian RF-IF, narrowband frequency modulation in this idealized situation is less effective than amplitude modulation ($\lambda = 1$) by about 3.0 db for either type of second detector in threshold operation. For strong carriers and no limiting ($r_0 \to \infty$), the FM and AM receivers (linear second detector) yield essentially the same performance with respect to $(S/N)_I$ as the criterion. On the other hand, when limiting is heavy, there is considerable degradation of the behavior of the FM receiver, as we have already remarked. Only when the carrier level becomes sufficiently large does the receiver approach AM performance (with a square-law second detector). If, however, a rectangular RF-IF filter is used, so that $\Gamma^2_{max} = 3/2$, a slight improvement over amplitude modulation for weak carriers may be expected: with no limiting, the most we can expect is about 1.8 db ($\doteq 4.8 - 3.0$) ($a_0{}^2 \ll 1$) (cf. Fig. 15.8), rising to about 2.5 db over amplitude modulation with a linear detector when $a_0{}^2 \gg 1$. When $r_0 \doteq 0.40$, amplitude modulation and frequency modulation have the same signal-to-noise ratios ($a_0{}^2 \ll 1$), and for heavier limiting the performance of the FM receiver again falls off rapidly for weak signals. In general, we conclude on the basis of our second-moment criteria that for normal random noise ("white" over the composite RF-IF band) there is theoretically no significant superiority of narrowband FM reception over AM reception with large modulation indices and, in addition, that there is a serious degradation of threshold performance in the FM receiver if heavy limiting is used. For smaller indices, performance deteriorates still further. Experience has in general borne this out in applications, even though, as noted previously (cf. Sec. 5.3-5), system comparisons based on signal-to-noise ratios are not by themselves sufficient for a complete description of system performance.

15.4 *Covariance Function and Spectrum of the Output of the FM Receiver (General Theory)*

In this section, we shall outline the general theory on which is based the calculation of the video (or audio) intensity spectrum at the output of the discriminator, including the effects of arbitrary limiting and input carrier-to-noise levels. For additional details, see Middleton.[10]

As usual, in a second-moment theory the principal problem is to determine the covariance function of the l-f output here, $E_0(t)$ [Eq. (15.13)], from which the intensity spectrum follows in the usual way on application of the W-K theorem. We write therefore

$$K_0(t) = \langle\overline{E_0(t_0)E_0(t_0 + t)}\rangle = \kappa^2\langle\overline{B_1(E_1)B_1(E_2)\dot{\theta}_1\dot{\theta}_2}\rangle \tag{15.47}$$

We observe first of all from Eq. (15.7*b*) that

$$\dot{\theta} = [(N_c + \alpha_s)(\dot{N}_s + \dot{\beta}_s) - (N_s + \beta_s)(\dot{N}_c + \dot{\alpha}_s)]/E^2 \qquad (15.48)$$

Combining this with the expressions for E [Eqs. (15.7*b*), (15.13)], we may express the output $E_0(t)$ in terms of the various normal components $(N_c,\dot{N}_c,N_s,\dot{N}_s)$ of the accompanying noise:

$$E_0(t) = i\kappa\{(N_c + \alpha_s)(\dot{N}_s + \dot{\beta}_s) - (N_s + \beta_s)(\dot{N}_c + \dot{\alpha}_s)\} \times \int_C J_1(E\xi)f_L(i\xi)\, d\xi/\pi E^2 \qquad (15.49)$$

In order to determine the covariance function of $E_0(t)$, let us attempt to obtain first the conditional moment $K_0(t)_N = \overline{E_0(t_0)E_0(t_0 + t)}$. For this purpose, we need the eightfold distribution density of the normal random variables $(N_{c1},N_{c2}, \ldots ,\dot{N}_{s1},\dot{N}_{s2})$; actually, it is more convenient to deal with the corresponding characteristic function. We find that the desired characteristic function is†

$$F_2(z_1, \ldots ,z_8;t) = \exp\left[-\frac{b_0}{2}(z_1^2 + z_2^2 + z_3^2 + z_4^2) - \frac{b_2}{2}(z_5^2 + z_6^2 + z_7^2 + z_8^2)\right] \times \exp[-\Phi_0(t)(z_1z_2 + z_3z_4) - \Phi_1(t)(z_1z_6 - z_2z_5 - z_3z_8 - z_4z_7) + \Phi_2(t)(z_5z_6 + z_7z_8)] \qquad (15.50)$$

where [cf. Eqs. (3.66)]

$$b_n \equiv \int_0^\infty \mathcal{W}(f)_N(\omega - \omega_0)^n\, df \qquad \Phi_n(t) \equiv \frac{\partial^n}{\partial t^n}\int_0^\infty \mathcal{W}(f)_N \cos(\omega - \omega_0)t\, df$$

$$b_{2n} = (-1)^n\Phi_{2n}(0) \qquad k_n(t) \equiv \frac{\Phi_n(t)}{b_0} \qquad (15.50a)$$

and $b_0 = \overline{N^2}$, $\Phi_0(t)/b_0 = k_0(t)$, the normalized covariance function of the input noise, whose narrowband intensity spectrum $\mathcal{W}(f)_N$ is symmetrical about the frequency f_0.

The covariance function $K_0(t)_N$ $(= \overline{E_{01}E_{02}})$ can now be written

$$K_0(t)_N = \kappa^2(I_1 - I_2 - I_3 + I_4) \qquad (15.51a)$$

where

$$I_j = \int_{-\infty}^\infty dN_{c1} \cdots \int_{-\infty}^\infty \frac{d\dot{N}_{s2}}{(2\pi)^8}\int_{-\infty}^\infty dz_1 \cdots \int_{-\infty}^\infty dz_8\, F_2(z_1, \ldots ,z_8;t)G_j(N_{c1}, \ldots ,\dot{N}_{s2}) \times e^{-i(z_1N_{c1}+\cdots+z_8\dot{N}_{s2})}\int_C\int_C \frac{i^2J_1(E_1\xi_1)J_1(E_2\xi_2)f_L(i\xi_1)f_L(i\xi_2)\, d\xi_1\, d\xi_2}{\pi E_1^2 \times \pi E_2^2}$$

$$j = 1, \ldots , 4 \qquad (15.51b)$$

† It is assumed that the intensity spectrum of the noise is symmetrical about f_0 here.

with
$$G_1 = (N_{c1} + \alpha_{s1})(\dot{N}_{s1} + \dot{\beta}_{s1})(N_{c2} + \alpha_{s2})(\dot{N}_{s2} + \dot{\beta}_{s2}) \tag{15.51c}$$
$$G_2 = (N_{c1} + \alpha_{s1})(\dot{N}_{s1} + \dot{\beta}_{s1})(N_{s2} + \beta_{s2})(\dot{N}_{c2} + \dot{\alpha}_{s2}) \tag{15.51d}$$
$$G_3 = (N_{s1} + \beta_{s1})(\dot{N}_{c1} + \dot{\alpha}_{s1})(N_{c2} + \alpha_{s2})(\dot{N}_{s2} + \dot{\beta}_{s2}) \tag{15.51e}$$
$$G_4 = (N_{s1} + \beta_{s1})(\dot{N}_{c1} + \dot{\alpha}_{s1})(N_{s2} + \beta_{s2})(\dot{N}_{c2} + \dot{\alpha}_{s2}) \tag{15.51f}$$

Our next step is to perform the integrations over $N_{c1}, \ldots, \dot{N}_{s2}$, for which purpose we examine the integrals

$$K_{1,2} \equiv \int_{-\infty}^{\infty} dx \int_{-\infty}^{\infty} dy \begin{pmatrix} x \\ y \end{pmatrix} \frac{J_1(\xi \sqrt{x^2 + y^2})}{x^2 + y^2} e^{-izx - iz'y} \tag{15.52}$$

which we can reduce to a single integral with the help of a transformation to polar coordinates:

$$K_{1,2} = -\frac{2\pi i \begin{pmatrix} z \\ z' \end{pmatrix}}{Z} \int_0^{\infty} J_1(\lambda\xi) J_1(Z\lambda)\, d\lambda \qquad Z = \sqrt{z^2 + z'^2} \tag{15.53}$$

Now, letting $x = \alpha_{s1} + V_{c1}$, $y = \beta_{s1} + V_{s1}$, etc., and using Eq. (15.53), we can reduce the eighteenfold integrals (15.51b) for I_j to a set of sixteenfold integrals:

$$\begin{aligned} I_1 = \int_{-\infty}^{\infty} d\dot{N}_{c1} \cdots \int_{-\infty}^{\infty} d\dot{N}_{s2} (2\pi)^{-8} \int_{-\infty}^{\infty} dz_1 \cdots \int_{-\infty}^{\infty} dz_8 \\ \times \{\exp[-i(z_5 \dot{N}_{c1} + \cdots + z_8 \dot{N}_{s2})]\} F_1(z_1, \ldots, z_8; t)(\dot{N}_{s1} + \dot{\beta}_{s1})(\dot{N}_{s2} + \dot{\beta}_{s2}) \\ \times \frac{(-2\pi i)^2 z_1 z_2 \exp[i(z_1\alpha_{s1} + z_2\alpha_{s2} + z_3\beta_{s1} + z_4\beta_{s2})]}{\sqrt{z_1^2 + z_3^2}\sqrt{z_2^2 + z_4^2}} \\ \times \iiint_0^{\infty} \int_C J_1(\lambda_1\xi_1)\, d\lambda_1\, d\lambda_2\, J_1(\lambda_2\xi_2) J_1(\lambda_1 \sqrt{z_1^2 + z_3^2}) J_1(\lambda_2 \sqrt{z_2^2 + z_4^2}) \\ \times \frac{f_L(i\xi_1) f_L(i\xi_2)\, d\xi_1\, d\xi_2}{(\pi i)^2} \end{aligned} \tag{15.54}$$

The integrals $I_2, \ldots, I_4$ [Eqs. (15.51)] are found in the same way if we replace $z_1z_2(\dot{N}_{s1} + \dot{\beta}_{s1})(\dot{N}_{s2} + \dot{\beta}_{s2})$ by $z_1z_4(\dot{N}_{s1} + \dot{\beta}_{s1})(\dot{N}_{c2} + \dot{\alpha}_{s2})$, $z_2z_3(\dot{N}_{c1} + \dot{\alpha}_{s1})(\dot{N}_{s2} + \dot{\beta}_{s2})$, and $z_3z_4(\dot{N}_{c1} + \dot{\alpha}_{s1})(\dot{N}_{c2} + \dot{\alpha}_{s2})$, successively.

The integrations over $\dot{N}_{c1} \cdots \dot{N}_{s2}$ are effected with the help of

$$\frac{1}{2\pi} \int_{-\infty}^{\infty} u^n\, du \int_{-\infty}^{\infty} e^{-iuz} g(z)\, dz = (-i)^n \left[\frac{\partial^n g}{\partial z^n}\right]_{z=0} \tag{15.55}$$

where it is assumed that $g^{(n)}(\pm\infty)$ vanishes satisfactorily. The sixteenfold integrals, like Eq. (15.54), can now be reduced to eightfold integrals, which when combined according to Eq. (15.51a) give us the following expression for the output covariance function:

$$K_0(t)_N = \frac{\kappa^2}{4\pi^2} \int_{-\infty}^{\infty} dz_1 \cdots \int_{-\infty}^{\infty} dz_4 [\Phi_2(t)(z_1z_2 + z_3z_4) + \Phi_1(t)^2(z_1z_4 - z_2z_4)^2$$
$$- i\Phi_1(t)(z_1z_4 - z_2z_3)(\alpha_{s1}z_3 + \dot{\alpha}_{s2}z_4 - \dot{\beta}_{s1}z_1 - \dot{\beta}_{s2}z_2)$$
$$- (\dot{\beta}_{s1}\dot{\beta}_{s2}z_1z_2 + \dot{\alpha}_{s1}\dot{\alpha}_{s2}z_3z_4 - \dot{\alpha}_{s2}\dot{\beta}_{s1}z_1z_4 - \dot{\alpha}_{s1}\dot{\beta}_{s2}z_2z_3)]$$
$$\times (z_1^2 + z_3^2)^{-1/2}(z_2^2 + z_4^2)^{-1/2}$$
$$\times \exp\left[-\frac{b_0^2}{2}(z_1^2 + z_2^2 + z_3^2 + z_4^2) - \Phi_0(t)(z_1z_2 + z_3z_4) \right.$$
$$\left. + i(\alpha_{s1}z_1 + \alpha_{s2}z_2 + \beta_{s1}z_3 + \beta_{s2}z_4) \right]$$
$$\times \iint_0^\infty d\lambda_1\, d\lambda_2 \int_C \int_C d\xi_1\, d\xi_2\, J_1(\lambda_1\xi_1)J_1(\lambda_2\xi_2)J_1(\lambda_1 \sqrt{z_1^2 + z_3^2})$$
$$\times \frac{J_1(\lambda_2 \sqrt{z_2^2 + z_4^2}) f_L(i\xi_1) f_L(i\xi_2)}{(i\pi)^2} \qquad (15.56)$$

At this point, we make the obvious transformation to polar coordinates, $z_1 = \rho_1 \cos\theta_1$, $z_3 = \rho_1 \sin\theta_1$, etc., with the additional assumption that no significant incidental amplitude modulation of the pure FM signal is introduced by the IF and preceding filters.† This last enables us to write

$$\phi(t) = \omega_d t + \Psi(t) \qquad \dot{\phi} = \omega_d + \dot{\Psi} = \omega_d + \int_{t_0}^{t} D(t')\, dt'$$
$$\text{with} \begin{cases} \dot{\alpha}_s^2 + \dot{\beta}_s^2 = \dot{\phi}^2 A_0^2 \\ \alpha_s^2 + \beta_s^2 = A_0^2 \end{cases} \qquad (15.57)$$

and $\phi_1 = \phi(t_0)$, $\phi_2 = \phi(t_0 + t)$, etc. Carrying out the integrations over θ_1, θ_2, expanding the resulting modified Bessel functions, and collecting terms gives finally†

$$K_0(t)_N = \sum_{k=0}^{\infty} \sum_{m=0}^{\infty} k_0(t)^{k+2m}$$
$$\left\{ -\Phi_2(t) \frac{\epsilon_k}{2} [\mathcal{H}^2_{k,2m,k+1} \cos (k+1)(\phi_2 - \phi_1) + \mathcal{H}^2_{k,2m,|k-1|} \cos (k-1)(\phi_2 - \phi_1)] \right.$$
$$+ \Phi_1(t)^2 \frac{\epsilon_k}{2} [\mathcal{H}^2_{k,2m+1,k} \cos k(\phi_2 - \phi_1) - \tfrac{1}{2}\mathcal{H}^2_{k,2m+1,k+2} \cos (k+2)(\phi_2 - \phi_1) - \tfrac{1}{2}\mathcal{H}^2_{k,2m+1,|k-2|} \cos (k-2)(\phi_2 - \phi_1)]$$
$$+ (2b_0a_0^2)^{1/2} \frac{\dot{\phi}_1 + \dot{\phi}_2}{2} \Phi_1(t)[(1 - \delta_0^k)\mathcal{H}_{k,2m,k+1}(\mathcal{H}_{k,2m+1,k+2} - \mathcal{H}_{k,2m+1,k}) \sin (k+1)(\phi_2 - \phi_1)$$
$$- \mathcal{H}_{k,2m,|k-1|}(\mathcal{H}_{k,2m+1,k} - \mathcal{H}_{k,2m+1,|k-2|}) \sin |k-1|(\phi_2 - \phi_1)]$$
$$+ 2a_0^2 b_0 \dot{\phi}_1 \dot{\phi}_2 \cos k(\phi_2 - \phi_1)$$
$$\left. [\delta_0^k \mathcal{H}^2_{k,2m,k+1} + \tfrac{1}{2}(\mathcal{H}_{k,2m,k+1} - \mathcal{H}_{k,2m,|k-1|})^2] \right\} \qquad (15.58)$$

† The effects of incidental amplitude modulation can be included in straightforward fashion with the aid of Eqs. (15.19), but this necessitates a recalculation of terms in Eq. (15.56) when the integrations over θ_1, θ_2 are performed. For the details of the evaluation for pure FM signals, see Middleton,[11] sec. 2 and app. III.

The amplitude functions $\mathcal{H}$ are now given by †

$$\mathcal{H}_{k,2m+n,q}(a_0,r_0) = \frac{b_0^{(2m+k)/2}}{2^{(2m+k)/2}\sqrt{m!(m+k)!}}\frac{\kappa}{\pi}\int_C f_L(i\xi)\,d\xi \times \int_0^\infty J_1(\lambda\xi)\,d\lambda \int_0^\infty \rho^{\mu+1}J_1(\lambda\rho)J_q(A_0\rho)e^{-\rho^2 b_0/2}\,d\rho$$
$$\mu = 2m + n + k = 1, 3, 5, \ldots \quad (15.59)$$

which for the limiter of Eq. (15.8*b*) becomes

$$\mathcal{H}_{k,2m+n,q} = \frac{\kappa\beta_F a_0^q(2/b_0)^{n/2}}{q!\sqrt{\pi}\sqrt{m!(m+k)!}}\int_{-\infty i-c}^{\infty i-c}\frac{\Gamma(-s)\Gamma(1-s)r_0^{2s+1}\Gamma(\nu_0+s)}{(s+\frac{1}{2})\Gamma(s+1)\Gamma(\frac{3}{2}-s)} \times {}_1F_1(\nu_0+s;\,q+1;\,-a_0^2)\frac{ds}{2\pi i} \quad (15.60)$$

where now $\nu_0 = (2m + n + k + q + 1)/2 = 1, 2, 3, \ldots$.

The unconditional covariance function $K_{E_0}(t)$ is obtained after the average over the modulation in Eq. (15.58) has been performed. Then the desired intensity spectrum follows directly from the Fourier transform of $K_{E_0}(t)$, in the usual way. We note from Eq. (15.58) that, as in the rectification of amplitude-modulated waves (Chaps. 5, 13), the l-f output consists of the three classes of modulation products. The first is (1) $n \times n$ noise, which is represented by the terms in the first two brackets of Eq. (15.58) when $q = |k-1|, k, |k-2| = 0$. The remaining terms of these first and second brackets and all those of the third and fourth are (2) $s \times n$ noise components, except in the fourth, when $m = k = 0$, in which case we have (3) the $s \times s$ terms appearing in the output wave. Note from the above that, when there is no noise,

No noise $$K_0(t)_N \to K_0(t)_{s\times s} = A_0^2\langle\phi_1\phi_2\rangle\mathcal{H}_{001}^2 \quad (15.61)$$

while, for noise alone, we obtain simply the terms in the first two brackets of Eq. (15.58).

15.5 *Special Cases for Broadband Frequency Modulation*‡

Here we shall examine a number of important limiting situations. In each instance the general result (15.58), in conjunction with Eq. (15.60), forms the starting point of our treatment. We begin by considering the signal component in the l-f output of the FM receiver for various degrees of limiting and in the presence of noise. The case of noise alone is then discussed, and, finally, some special examples involving an unmodulated carrier complete our account of FM receiver performance.

† For footnote see opposite page.

‡ The results in this section correct a number of misprints, minor errors, and faulty calculations in Ref. 11, pt. II, as a comparison of Eqs. (15.62), (15.65), (15.81*b*) with the corresponding expressions therein [cf. eqs. (2.8*b*), (A.2.2), as well as (A.1.15) for $\mathcal{H}$ (and Prob. 15.4 here)] shows. See also Fuller and Middleton,[12] pt. I, chaps. 2 and 3.

15.5-1 Signal Output; Arbitrary Limiting. The signal term is obtained from Eq. (15.58) on setting $m = k = 0$ therein. We find with the help of Sec. 13.4-4 applied to Eq. (15.60) (cf. Prob. 15.4*a*) that Eq. (15.61) becomes explicitly

$$K_0(t)_{s\times s} = \frac{16}{\pi^2}\beta^2\kappa^2 R_0^2\langle\phi_1\phi_2\rangle a_0^4\left(-\frac{1-e^{-a_0^2}}{a_0^2}\right.$$
$$+\sum_{l=1}^{\infty}\frac{(-\tfrac{1}{2})_l(-r_0^2)^l}{l!^3}\frac{2l}{2l+1}\left\{{}_1F_1(l+1;2;-a_0^2)\right.$$
$$\times\left[\tfrac{3}{2}\psi(l+1)-\tfrac{1}{2}\psi\left(\frac{3}{2}-l\right)-\log r_0-\frac{1}{2l(2l+1)}\right]$$
$$\left.\left.-\frac{1}{2}\sum_{n=0}^{\infty}\frac{(l+1)_n\psi(l+n+1)(-a_0^2)^n}{(2)_n n!}\right\}\right)^2 \qquad (15.62)$$

where ψ is the logarithmic derivative of the gamma function [cf. Eq. (13.117)]. An alternative expression for $K_0(t)_{s\times s}$ above may be obtained directly from Eq. (15.20), viz.,

$$K_0(t)_{s\times s} = 4\kappa^2\beta_F^2 a_0^2\langle\phi_1\phi_2\rangle\left[e^{-a_0^2}\int_0^\infty B_1(r)I_1(2a_0r)e^{-r^2}\,dr\right]^2 \qquad (15.63)$$

The results are shown in Fig. 15.5, in units of $\langle\phi_1\phi_2\rangle$. Observe once more that, because of our assumption of an ideal discriminator (Sec. 15.1-3), the *signal* ϕ appears undistorted, a situation certainly not obeyed when there is incidental amplitude modulation from the IF or preceding stages or when the discriminator's characteristic is nonlinear.

Various limiting forms of $K_0(t)_{s\times s}$ follow directly from Sec. 15.2-1. We note, moreover, that for arbitrary limiting, such that $R_0 < A_0$, Eq. (15.62) becomes (cf. Prob. 15.5*a*)

$$K_0(t)_{s\times s}\Big|_{a_0\to\infty} \simeq \kappa^2\beta_F^2\langle\dot\phi_1\dot\phi_2\rangle\frac{16}{\pi^2}R_0^2\,{}_2F_1{}^2\left(-\frac{1}{2},\frac{1}{2};\frac{3}{2};\frac{R_0^2}{A_0^2}\right)$$
$$\simeq\frac{4}{\pi^2}\kappa^2\beta_F^2\langle\phi_1\phi_2\rangle R_0^2\left[\left(1-\frac{R_0^2}{A_0^2}\right)^{1/2}+\frac{A_0}{R_0}\sin^{-1}\frac{R_0}{A_0}\right]^2$$
$$R_0 < A_0 \qquad (15.64)$$

Here the noise is suppressed by the carrier. At the other extreme of weak carriers, Eq. (15.62) reduces to

$$K_0(t)_{s\times s}\Big|_{a_0\to 0} \doteq \left(\frac{4\beta_F\kappa R_0}{\pi}\right)^2\langle\phi_1\phi_2\rangle a_0^4\left\{-1+\sum_{m=1}^{\infty}\frac{(-\tfrac{1}{2})_m(-r_0^2)^m}{(m!)^3}\frac{2m}{2m+1}\right.$$
$$\left.\times\left[\psi(m+1)-\tfrac{1}{2}\psi\left(\frac{3}{2}-m\right)-\log r_0-\frac{1}{2m(2m+1)}\right]\right\}^2 \qquad (15.65)$$

We see once more the characteristic effects of *signal suppression;* that is, $S(t)^2_{out}$ is proportional to a_0^4, rather than to a_0^2, in this threshold situation. This dependence on the amplitude of the carrier is identical with that of amplitude modulation in threshold reception (see Chap. 13) and, as will be demonstrated more fully in Chap. 19, is always present whenever reception is incoherent. Finally, the direct current and mean intensity associated with the $s \times s$ terms follow directly from Eq. (15.20).

15.5-2 Covariance Functions and Spectra with Strong Carriers and Moderate to Heavy Limiting. In this case our general expression (15.58) simplifies considerably if we assume that the noise level is small compared with the peak carrier amplitude and that there is always some limiting, i.e., that $R_0 < A_0$. Retaining terms in b_0^0, b_0 only as $a_0 \to \infty$, we find that the (conditional) covariance function of the final l-f output process reduces to

$$\begin{aligned} K_0(t)_N \simeq [-\mathcal{H}^2_{001}\Phi_2(t) \cos(\phi_2 - \phi_1) \\ + \tfrac{1}{2}A_0\mathcal{H}_{001}\Phi_1(t)(\dot\phi_1 + \dot\phi_2)(\mathcal{H}_{012} - \mathcal{H}_{010}) \sin(\phi_2 - \phi_1) \\ + a_0^2\Phi_0(t)(\dot\phi_1\dot\phi_2)(\mathcal{H}_{102} - \mathcal{H}_{100})^2] + A_0^2\dot\phi_1\dot\phi_2\mathcal{H}^2_{001} \end{aligned} \quad (15.66)$$

where the amplitude functions $\mathcal{H}$ are obtained from Eq. (15.60) when the asymptotic development of the hypergeometric function ${}_1F_1$ is used [cf. Eq. (A.1.16)] and only the leading terms of the resulting series are retained.† We have the following relations,

$$\mathcal{H}_{012} - \mathcal{H}_{010} = \frac{2}{A_0}\mathcal{H}_{001} \qquad \mathcal{H}_{102} - \mathcal{H}_{100} = \frac{\sqrt{2b_0}}{A_0}\mathcal{H}_{001} \quad (15.67a)$$

where now (cf. Prob. 15.5a) $\mathcal{H}_{001}$ is specifically

$$\begin{aligned} \mathcal{H}_{001} &\simeq \frac{\kappa\beta_F R_0}{\sqrt{\pi}\,A_0}\frac{1}{2\pi i}\int_{-\infty i-c}^{\infty i-c}\frac{\Gamma(-s)(R_0/A_0)^{2s}}{(s+\frac{1}{2})\Gamma(\frac{3}{2}-s)}\,ds \\ &= \frac{2\kappa\beta_F}{\pi}\left(\frac{R_0}{A_0}\sqrt{1-\frac{R_0^2}{A_0^2}} + \sin^{-1}\frac{R_0}{A_0}\right) \qquad R_0 < A_0 \end{aligned} \quad (15.67b)$$

Taking the time average over the phases of the modulation, we find that the covariance function in these circumstances is accordingly

$$\begin{aligned} K_0(t) \simeq [-\Phi_2(t)\langle\cos(\phi_2 - \phi_1)\rangle + \Phi_1(t)\langle(\dot\phi_1 + \dot\phi_2)\sin(\phi_2 - \phi_1)\rangle \\ + \Phi_0(t)\langle\dot\phi_1\dot\phi_2\cos(\phi_2 - \phi_1)\rangle]\mathcal{H}^2_{001} + A_0^2\langle\dot\phi_1\dot\phi_2\rangle\mathcal{H}^2_{001} \\ a_0 \to \infty,\ R_0 < A_0 \end{aligned} \quad (15.68)$$

† Equivalently, we can derive this relation from Eq. (15.59) when $r_0 \to 0$, for all input carrier-to-noise ratios a_0 (≥ 0). Thus, it can be shown (cf. Prob. 15.4b) that

$$\lim_{r_0 \to 0} \mathcal{H}_{k,2m+n,q} = \frac{4}{\pi} r_0\beta_F\kappa\left(\frac{2}{b_0}\right)^{n/2}\frac{a_0^q\Gamma(\nu_0)}{q![m!(m+k)!]^{1/2}}\,{}_2F_1(\nu_0;\, q+1;\, -a_0^2) \quad (15.66a)$$

The asymptotic value of ${}_1F_1$ [Eq. (A.1.16)] then gives us the desired relations for strong carriers.

The terms in the brackets represent the significant $s \times n$ noise components. Again, we observe the effects of signal suppression here: the signal contribution $A_0^2\langle\phi_1\phi_2\rangle\mathcal{H}_{001}^2$ is the dominant factor, while the accompanying noise represents a perturbation of this signal, independent of carrier strength since the $n \times n$ terms are $O(a_0^{-2})$ and are absolutely suppressed as $a_0 \to \infty$.

The intensity of the noise components is

$$K_0(0)_{\text{noise}} \simeq [-\Phi_2(0) + \Phi_0(0)\langle\phi^2\rangle]\mathcal{H}_{001}^2 = (b_2 + b_0\langle\phi^2\rangle)\mathcal{H}_{001}^2 \quad (15.69)$$

Note that it depends on the strength of the modulation, as well as on the clipping level and on the intensity of the (RF-IF) noise output. Finally, the only steady or periodic components in the video output arise from the modulation itself and from the degree of possible detuning ($\omega_d \neq 0$), since $K_0(\infty)_{s\times s} = \lim_{\tau\to\infty} [\omega_d + \dot{\Psi}(t)]\mathcal{H}_{001}^2[\omega_d + \Psi(t+\tau)] \neq 0$ for arbitrary values of t.

The spectrum of the fluctuating output (apart from signal) is of considerable interest. Starting with Eq. (15.68), we may alternatively express $K_0(t)_{\text{noise}}$ as[11]

$$K_0(t)_{\text{noise}} \simeq \mathcal{H}_{001}^2 \left\{\left(-\frac{\partial}{\partial t} + \frac{\partial}{\partial x}\right)\left(\frac{\partial}{\partial t} + \frac{\partial}{\partial y}\right)\Phi_0(t) \times \int_0^{x_0} \cos\left[\omega_d t + \int_x^y D_0(t')\,dt'\right]\frac{dx}{x_0}\right\} \quad (15.70)$$

where $x_0\ (= 2\pi/\omega_a)$ is the period of the modulation here and $(\partial^n/\partial t^n)\Phi_0 = \Phi_n(t)$ [cf. Eq. (15.50a)]. Using the integral form of $\Phi_n(t)$ and the W-K theorem [Eqs. (3.42), (3.78), etc.; cf. Prob. 15.6a], we obtain finally for the ntensity spectrum of the l-f noise output

$$\mathcal{W}_0(f)_N \simeq \left(\frac{4\kappa\beta_F R_0}{\pi A_0}\right)^2 {}_2F_1{}^2\left(-\frac{1}{2},\frac{1}{2};\frac{3}{2};\frac{R_0^2}{A_0^2}\right) \times \sum_{n=0}^{\infty} A_n\omega^2[\mathcal{W}_I(\omega_0 + n\omega_a + \omega_d + \omega) + \mathcal{W}_I(\omega_0 + n\omega_a + \omega_d - \omega)] \quad (15.71)$$

where A_n is the amplitude of the nth term in the Fourier-series development of $\langle\cos(\phi_2 - \phi_1)\rangle$ and $\mathcal{W}_I(f)$ is the spectrum of the input noise. For the degenerate case of our on-tune carrier ($\omega_d = 0$) and no modulation, this reduces to the simpler result

$$\mathcal{W}_0(f)_N \simeq \frac{32\kappa^2\beta_F{}^2}{\pi^2}\left(\frac{R_0}{A_0}\right)^2 {}_1F_1{}^2\left(-\frac{1}{2},\frac{1}{2};\frac{3}{2};\frac{R_0^2}{A_0^2}\right)\omega^2\mathcal{W}_I(\omega_0 + \omega) \qquad R_0 < A_0,\ A_0 \to \infty \quad (15.71a)$$

Thus, we see that, when the *carrier* is strong and there is limiting at or below the maximum carrier level, the output noise spectrum vanishes at $f = 0$ and is small in the vicinity of zero frequency. This accounts for the great improvement in the signal vs. the noise background when *broadband frequency modulation* (maximum modulation frequency small compared with the maximum frequency deviation of the carrier) is used at *large* carrier levels, with *sufficient limiting* ($R_0 < A_0$). This last is important, since it is limiting which removes the amplitude variations (almost completely, if $R_0 \ll A_0$, and to a very large degree if $R_0 < A_0$, since $A_0^2 \gg 2\overline{N^2}$). The intensity of the perturbing noise is redistributed by the limiter-discriminator action into spectra of the type (15.71), as illustrated in Fig. 15.9. The factor ω^2 represents the effect of the discriminator on its constant amplitude input: the noise appears only as a small, random displacement of the zeros of the clipped wave $R_0\dot{\theta}$. We note also that these $s \times n$ noise products are the result of the intermodulation of the signal (e.g., carrier and modulation) with the noise. The output noise spectrum is thus a superposition of the displaced, original noise spectrum $\mathcal{W}_I(f)$, weighted by the intensity of the components of the modulation and by ω^2. The degree of limiting changes the scale and not the shape of the output noise spectrum under the above conditions as long as the peak carrier amplitude exceeds the limit level R_0. Note from Eq. (15.71) that the intensity of the noise decreases $O(a_0^{-2})$. Thus, with heavy limiting an even larger reduction of the interfering noise is achieved absolutely, provided that a suitable filter is used to pass only the l-f terms, containing the signal ($\sim\phi$) (cf. shaded portion in Fig. 15.9). Note that the ratio of the total signal to total l-f noise intensity (before the final video or audio filter) is given by

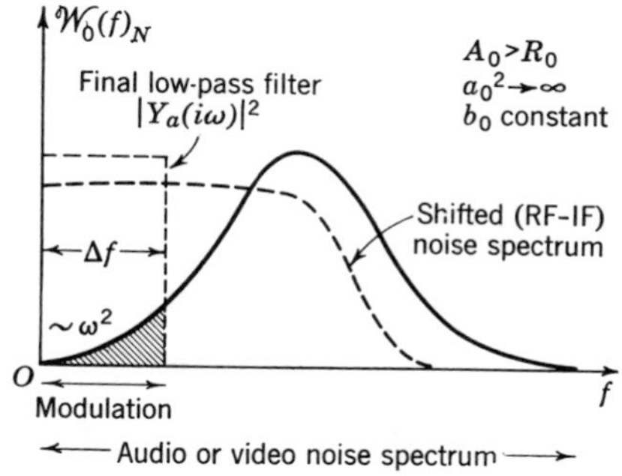

FIG. 15.9. A typical video (or audio) noise spectrum for broadband FM reception of a frequency-modulated carrier strong compared with the noise and with moderate to heavy *limiting* ($R_0 < A_0$).

$$\left.\frac{P_{s\times s}}{P_{s\times n}}\right|_{a_0\to\infty, R_0<A_0} \simeq \frac{2a_0^2\langle\phi^2\rangle}{\rho_E^2\langle\dot{\phi}^2\rangle} \tag{15.72}$$

15.5-3 Covariance Functions and Spectra with Strong Carriers; Little or No Limiting. For this situation, we have $A_0 < R_0$, $A_0 \gg \sqrt{2b_0}$, and to the first approximation we get again, with the help of Sec. 13.4-4 applied to Eq. (15.60), the amplitude functions for no limiting, which are (cf. Prob. 15.4*b*)

$$\lim_{r_0\to\infty} \mathcal{K}(a_0;r_0) = \frac{\kappa\beta_F}{2}\left(\frac{2}{b_0}\right)^{n/2} \frac{a_0^q}{q!}\frac{\Gamma(\nu_0 - \frac{1}{2})}{[m!(m+k)!]^{1/2}}\, {}_1F_1\left(\nu_0 - \frac{1}{2}; q+1; -a_0^2\right)$$
$$\nu_0 = \frac{2m+n+q+k+1}{2} \tag{15.73}$$

[which is to be compared with $\lim_{r_0 \to 0} \mathcal{H}(a_0,r_0)$ in Eq. (15.66a)]. With strong carriers, this becomes

$$\mathcal{H}_{k,2m+n,q}\Big|_{a_0\to\infty} \simeq \frac{\kappa\beta_F}{2}\left(\frac{2}{b_0}\right)^{n/2} \frac{a_0^{-2m-k-n}\Gamma[(2m+k+n+q)/2]}{m!(m+k)!\Gamma[(q-2m-k-n+2)/2]} \tag{15.74}$$

so that the covariance function (15.58) reduces in this instance to

$$K_0(t)\Big|_{a_0\to\infty} \simeq \kappa^2\beta_F^2[-\Phi_2(t)\langle\cos(\phi_2-\phi_1)\rangle + 2b_0a_0^2\langle\dot\phi_1\dot\phi_2\rangle] \qquad R_0 > A_0 \tag{15.75}$$

For a tuned unmodulated carrier, this reduces still further to

$$K_0(t)\Big|_{a_0\to\infty, r_0>a_0} \simeq \beta_F^2\kappa^2 b_0\rho_E^2 \tag{15.75a}$$

There is only one set of significant $s \times n$ noise terms, unlike the limited case above [Eq. (15.68)]; the signal contribution predominates, as expected, and increases relative to the noise, which has a fixed level of intensity.

The spectrum follows directly with the help of the W-K theorem; we find that now for the output noise (Prob. 15.6b)

$$\mathcal{W}_0(f)_N \simeq \kappa^2\beta_F^2 \sum_{n=0}^{\infty} A_n[(n\omega_a+\omega_d+\omega)^2\mathcal{W}_I(\omega_0+n\omega_a+\omega_d+\omega) + (n\omega_a+\omega_d-\omega)^2\mathcal{W}_I(\omega_0+n\omega_a+\omega_d-\omega)] \qquad a_0\to\infty\ (A_0<R_0) \tag{15.76}$$

which in contrast to Eq. (15.71) does not in general have ω^2 as a common factor: the spectral intensity at and near zero frequency does not vanish but has rather the type of distribution shown in Fig. 15.10. Broadband frequency modulation under these conditions is accordingly much less satisfactory than when moderate to heavy limiting is used [cf. Eq. (15.71) and Fig. 15.9]. The $s \times n$ noise terms here do not fall off near $f = 0$ but remain significant. The total noise intensity, likewise, does not vanish as $a_0 \to \infty$ (b_0 fixed), in contrast to the situations in Sec. 15.5-2 for heavy limiting [cf. Eq. (15.71)]. Both these results are attributable to the lack of limiting of $E\dot\theta$, so that amplitude, as well as phase variations, is present. Comparison of the *total* video signal and noise intensities in Eq. (15.75)

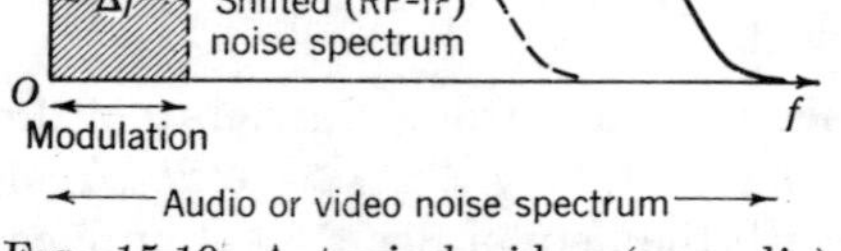

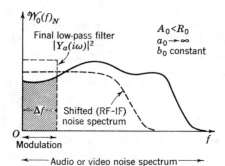

FIG. 15.10. A typical video (or audio) noise spectrum for broadband FM reception of a modulated carrier strong compared with the noise, but with *no limiting* ($R_0 > A_0$).

(before the final filter) shows that now

$$\left.\frac{P_{s\times s}}{P_{s\times n}}\right|_{a_0\to\infty,\ R_0>A_0} \simeq \frac{2a_0^2\langle\dot\phi^2\rangle}{\rho_E^2} \tag{15.77}$$

[cf. Eq. (15.72)]. Note, however, that an exception to the spectral distribution of Eq. (15.76) is the degenerate case of a tuned unmodulated carrier, where Eq. (15.76) reduces to

$$\mathcal{W}_0(f)_N \simeq 2\kappa^2\beta_F^2\omega^2\mathcal{W}_I(\omega_0 \pm \omega) \tag{15.78}$$

similar to Eq. (15.71*a*) in frequency dependence. Further discussion of spectral behavior in the strong-carrier case may be found in Sec. 15.5-6 and in Blachman[6] and Middleton.[11]

15.5-4 Threshold Signals. Before going on to consider the more specialized examples of noise alone (Sec. 15.5-5) and noise and an unmodulated carrier (Sec. 15.5-6), let us briefly outline the analysis for the case of weak, or threshold, signals. Here, the carrier-to-noise intensity ratio a_0^2 is unity or less, and we have the usual problem of determining receiver performance in this limiting situation. We expect that the covariance function will be dominated by the contributions due to noise alone, while the effects of the signal will appear as a perturbation. This can be shown from Eqs. (15.58) to (15.60) by expanding in powers of a_0^2; the results for arbitrary limiting levels are very involved (because of the structure of the amplitude function $\mathcal{H}$ in this instance). However, general developments have been obtained [Middleton,[11] eqs. (2.48) et seq.] for the two extremes of no limiting ($r_0 \to \infty$) and "superlimiting" ($r_0 \to 0$). These take the form

$$r_0 \to \infty: \quad K_0(t) = \frac{\pi}{4}\beta_F^2\kappa^2 b_0\{L_0(t) + a_0^2L_1(t) + a_0^4[L_2(t) + 2\langle\dot\phi_1\dot\phi_2\rangle] + O(a_0^6)\} \tag{15.79a}$$

and

$$r_0 \to 0: \quad K_0(t) = \frac{16\beta_F^2\kappa^2 r_0^2 b_0}{\pi^2}\{M_0(t) + a_0^2M_1(t) + a_0^4[M_2(t) + 2\langle\dot\phi_1\dot\phi_2\rangle] + O(a_0^6)\} \tag{15.79b}$$

where $L_n(t)$, $M_n(t)$ ($n = 0, 1, \ldots$) are complicated functions† of the limit level r_0 and the modulation ϕ and represent the $n \times n$ and $s \times n$ noise products produced in the course of demodulation. The important feature of this result (15.79*a*), (15.79*b*) is again the presence of signal suppression: the desired output signal ($\sim\dot\phi$) appears now $O(a_0^4)$, instead of $O(a_0^2)$ [see Eqs. (15.23*a*), (15.40), (15.42), (15.45), and in particular (15.65), for the signal term alone, with general limiting]. Weak signals are made weaker after detection in the presence of noise (or of another, stronger signal, for that matter). The general theory of this effect is discussed later, in Sec. 19.4.

† See Middleton,[11] pt. I, eqs. (2.49*b*), (2.49*c*).

15.5-5 Noise Alone; Arbitrary Limiting.[11] When normal noise alone is the input to the nonlinear elements of the FM receiver, we can obtain the covariance function and spectrum of the output $E_0(t)$ once more by a suitable reduction of Eq. (15.58). Here we set $\phi_1 = \phi_2 = 0 = a_0$, so that the covariance function of this l-f output becomes simply

$$K_0(t)_N = \sum_{m=0}^{\infty} [-\mathcal{H}_{1,2m,0}^2\Phi_2(t)k_0(t)^{2m+1} + \tfrac{1}{2}\mathcal{H}_{0,2m+1,0}^2\Phi_1(t)^2k_0(t)^{2m} - \tfrac{1}{2}\mathcal{H}_{2,2m+1,0}^2\Phi_1(t)^2k_0(t)^{2m+2}] \quad (15.80)$$

The mean intensity $K_0(0)_N$ has already been given by Eq. (15.31) and is shown in Fig. 15.7 as a function of the normalized limit level r_0. The amplitude functions $\mathcal{H}$ are found from Eq. (15.60) and the results of Prob. 15.4 to be

$$\mathcal{H}_{1,2m,0} = \frac{4\beta_F\kappa r_0}{\pi}(m+1)^{-1/2}G_m \qquad \mathcal{H}_{0,2m+1,0} = \frac{4\beta_F\mathcal{K}r_0}{\pi}\left(\frac{2}{b_0}\right)^{1/2}G_m$$

$$\mathcal{H}_{2,2m+1,0} = \frac{4\beta_F\mathcal{K}r_0}{\pi}\left(\frac{m+1}{m+2}\right)^{1/2}\left(\frac{2}{b_0}\right)^{1/2}G_{m+1} \quad (15.81a)$$

where
$$G_m(r_0) = -1 + \sum_{l=1}^{\infty}\frac{(-r_0^2)^l(-\frac{1}{2})_l(m+1)_l}{(l!)^3}\frac{2l}{2l+1}$$
$$\times\left[\tfrac{3}{2}\psi(l+1) - \tfrac{1}{2}\psi\left(\frac{3}{2}-l\right) - \tfrac{1}{2}\psi(m+l+1) - \frac{1}{2l(2l+1)} - \log r_0\right] \quad (15.81b)$$

Substitution of Eqs. (15.81) into Eq. (15.80) enables us to write the covariance function more compactly as

$$K_0(t)_N = \frac{16\beta_F{}^2\kappa^2r_0{}^2b_0}{\pi^2}[k_1(t)^2 - k_2(t)k_0(t)]\sum_{m=0}^{\infty}\frac{G_m(r_0)^2}{m+1}k_0(t)^{2m} \quad (15.82)$$

from which the corresponding spectrum follows directly, for example, $\mathcal{W}(f)_N = 2\mathfrak{F}\{K_0(t)_N\}$ [cf. Eqs. (3.42)]. Figure 15.11 shows how $G_m(r_0)$ varies with order m, when the limiting level r_0 increases from "super-clipping" ($r_0 \doteq 0$) to essentially no limiting ($r_0 \to \infty$).

Let us assume again (cf. Sec. 15.3-2) that the intensity spectrum of the input noise to the limiter-discriminator combination has the narrowband gaussian distribution

$$\mathcal{W}_I(f) = W_0e^{-(\omega-\omega_0)^2/\omega_b^2} \qquad \omega_b = 2\pi f_b \quad (15.83)$$

so that $b_0 = W_0\omega_b/(2\sqrt{\pi})$ and $K_I(t) = \frac{1}{2}\mathfrak{F}^{-1}\{\mathcal{W}_I\} = b_0e^{-\omega_b^2t^2/4}$. Applying this to the various relations in Eq. (15.50a), we find that (cf. Prob. 15.8)

$$k_1(t)^2 - k_2(t)k_0(t) = \frac{\omega_b^2}{2}e^{-x^2/2} \qquad x \equiv \omega_bt \quad (15.84)$$

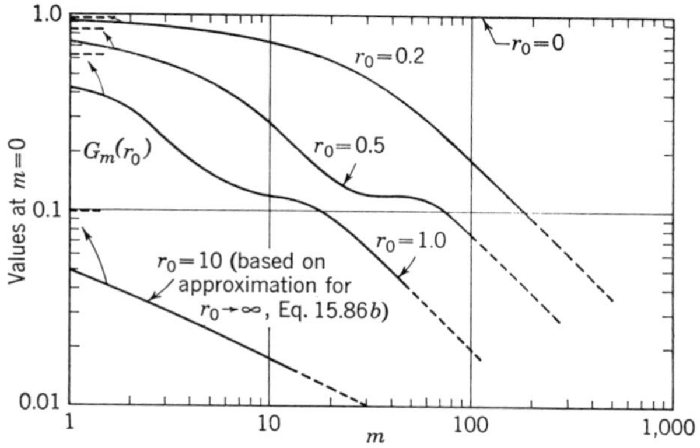

FIG. 15.11. The function $G_m(r_0)$ (for noise alone).

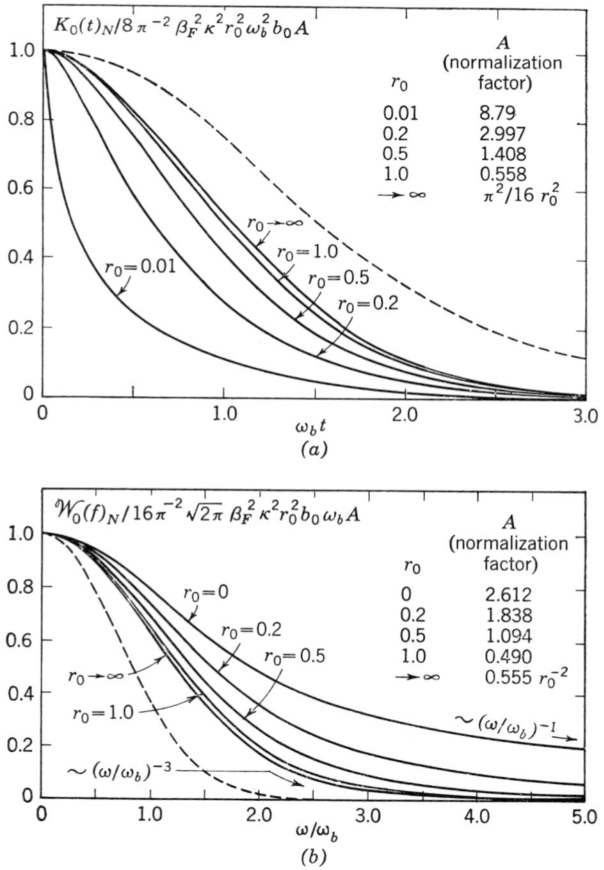

FIG. 15.12. (a) (Normalized) output covariance functions at various limit levels for input noise alone, shaped by a gaussian (composite) RF-IF filter (dashed line). (b) (Normalized) output spectra for the cases shown in (a).

The covariance function and spectrum, as obtained from Eqs. (15.82) et seq., in this instance become specifically

$$K_0(t)_N = \frac{8\beta_F{}^2\kappa^2\omega_b{}^2b_0r_0{}^2}{\pi^2} e^{-\omega_b{}^2t^2/2} \sum_{m=0}^{\infty} \frac{G_m(r_0)^2}{m+1} e^{-m\omega_b{}^2t^2/2} \qquad (15.85a)$$

$$\mathcal{W}_0(f)_N = \frac{16\sqrt{2}\,\beta_F{}^2\kappa^2r_0{}^2b_0\omega_b}{\pi^{3/2}} \sum_{m=0}^{\infty} \frac{G_m(r_0)^2}{(m+1)^{3/2}} e^{-\omega^2/2(m+1)\omega_b{}^2} \qquad (15.85b)$$

Typical examples are shown in Fig. 15.12*a*, *b* for Eqs. (15.85*a*), (15.85*b*), respectively.

The extreme cases of heavy and no limiting are now easily obtained from Eq. (15.82). We note first from Eq. (15.81*b*) that, for very heavy limiting,

$$r_0 \to 0: \qquad G_m(r_0) \to -1 \qquad (15.86a)$$

while, for no limiting, we find on comparing Eq. (15.82) with Eq. (15.68) for little or no limiting that

$$r_0 \to \infty: \qquad G_m(r_0) \to -r_0{}^{-1} \frac{\pi^{3/2}(\frac{1}{2})_m}{4\sqrt{2}\,m!} \qquad (15.86b)$$

The covariance functions and spectra for these two cases follow at once from Eqs. (15.86) in Eq. (15.82). We can write in general[11]

$$r_0 \to 0: \quad K_0(t)_N \doteq \frac{16}{\pi^2}\beta_F{}^2\kappa^2r_0{}^2b_0 \begin{cases} [k_1{}^2(t) - k_0(t)k_2(t)]\ {}_2F_1[1,1;2;k_0(t)^2] \\ -\dfrac{k_1{}^2(t) - k_0(t)k_2(t)}{k_0(t)^2} \log\,[1 - k_0(t)^2] \end{cases} \qquad (15.87a)$$

and

$$r_0 \to \infty: \quad K_0(t)_N \simeq \frac{\pi}{4}\kappa^2\beta_F{}^2b_0[k_1{}^2(t) - k_0(t)k_2(t)]\ {}_2F_1\left[\frac{1}{2},\frac{1}{2};2;k_0(t)^2\right]$$

$$\text{or} \qquad \simeq \kappa^2\beta_F{}^2b_0[k_1{}^2(t) - k_0(t)k_2(t)][E(k_0) - (1-k_0{}^2)K(k_0{}^2)]/k_0{}^2 \qquad (15.87b)$$

where K and E are, respectively, complete elliptic integrals of the first and second kind [cf. Eq. (A.1.38*a*)], while the spectrum is determined from the W-K theorem in the usual way.

With the gaussian (RF-IF) spectrum [Eq. (15.83)] for the input noise, these relations (15.87*a*), (15.87*b*) are specifically

$$r_0 \to 0: \qquad K_0(t)_N \doteq \frac{8}{\pi^2}\beta_F{}^2\kappa^2r_0{}^2\omega_b{}^2b_0e^{-\omega_b{}^2t^2/2}\ {}_2F_1(1,1;2;e^{-\omega_b{}^2t^2/2}) \qquad (15.88a)$$

$$\mathcal{W}_0(f)_N \simeq \frac{32}{\pi\sqrt{2\pi}}\beta_F{}^2\kappa^2\omega_bb_0 \sum_{m=0}^{\infty} \frac{e^{-\omega^2/2(m+1)\omega_b{}^2}}{(m+1)^{3/2}} \qquad (15.88b)$$

and

$$r_0 \to \infty: \qquad K_0(t)_N \simeq \frac{\pi}{8} \beta_F{}^2 \kappa^2 \omega_b{}^2 b_0 e^{-\omega_b{}^2 t^2/2} \, {}_2F_1\left(\frac{1}{2},\frac{1}{2};2;e^{-\omega_b{}^2 t^2/2}\right) \tag{15.89a}$$

$$\mathcal{W}_0(f)_N \doteq \left(\frac{\pi}{2}\right)^{3/2} \beta_F{}^2 \kappa^2 \omega_b b_0 \sum_{m=0}^{\infty} \frac{(\frac{1}{2})_m{}^2}{(m!)^2 (m+1)^{3/2}} e^{-\omega^2/2(m+1)\omega_b{}^2} \tag{15.89b}$$

both sets of which are illustrated in Fig. 15.12*a*, *b*. The spectral behavior out on the tails of the spectrum is found to be

$$r_0 \to 0: \qquad \mathcal{W}_0(f)_N \to \frac{32}{\pi} \beta_F{}^2 \kappa^2 r_0{}^2 \omega_b b_0 \left(\frac{\omega}{\omega_b}\right)^{-1} \tag{15.90a}$$

$$r_0 \to \infty: \qquad \mathcal{W}_0(f)_N \to \frac{\pi}{2} \beta_F{}^2 \kappa^2 \omega_b b_0 \left(\frac{\omega}{\omega_b}\right)^{-3} \qquad \frac{\omega}{\omega_b} \to \infty \tag{15.90b}$$

In all cases, we observe the characteristic spectral spreading due to the nonlinear elements in the receiver. Even when there is no limiting ($r_0 \to \infty$), the output spectrum is broader than that of the input, as a consequence of discrimination. Heavy limiting ($r_0 \to 0$) redistributes most of the noise intensity into the tails of the spectrum, which from Eq. (15.90*a*) fall off slowly $O(\omega^{-1})$, in contrast to the unlimited case, where the spectrum is $O(\omega^{-3})$ for frequencies well away from $f = 0$. Similar behavior is observed for other types of (RF-IF) filter response (e.g., rectangular, single-tuned, double stagger-tuned, and double synchronous-tuned filters), detailed results for which have been obtained by Fuller.[12] Figure 15.13 shows a typical set of experimental spectra[10] at various limit levels.† For considerable additional data, see also Chap. 5, Ref. 12.

15.5-6 Noise and an Unmodulated Carrier. As our final example, let us consider briefly the situation where the input carrier process is unmodulated and on tune ($\omega_d = 0$), so that $\phi_1 = \phi_2 = 0$, etc., in the general relation (15.58). The covariance function is now given by the double series

$$K_0(t)_{N+C} = \sum_{k=0}^{\infty} \sum_{m=0}^{\infty} k_0(t)^{k+2m} \frac{\epsilon_k}{2} [-\Phi_2(t)(\mathcal{H}^2_{k,2m,k+1} + \mathcal{H}^2_{k,2m,|k-1|}) + \Phi_1(t)^2(\mathcal{H}^2_{k,2m+1,k} - \tfrac{1}{2}\mathcal{H}^2_{k,2m+1,k+2} - \tfrac{1}{2}\mathcal{H}^2_{k,2m+1,|k-2|})] \tag{15.91}$$

where the amplitude functions are found as before from Eq. (15.60) and Prob. 15.4. Explicit results for the spectrum are available in the extreme cases of no limiting ($r_0 \to \infty$, or $r_0/a_0 > 1$, $a_0 \to \infty$) and superlimiting ($r_0 \to 0$, or $r_0/a_0 < 1$, $a_0 \to \infty$), when the input noise has the gauss-shaped spectrum of Eq. (15.83) and the amplitude functions are given by Eqs.

† The relative attenuation (Rel. Att.) on the scale represents an adjustment to give close to full-scale reading in each instance.

(15.73) and (15.66*a*), respectively.† Output spectra for all input carrier-to-noise levels have been computed and are shown in Figs. 15.14 and 15.15*a*, *b*. In addition, extensive measurements of spectra in these extreme

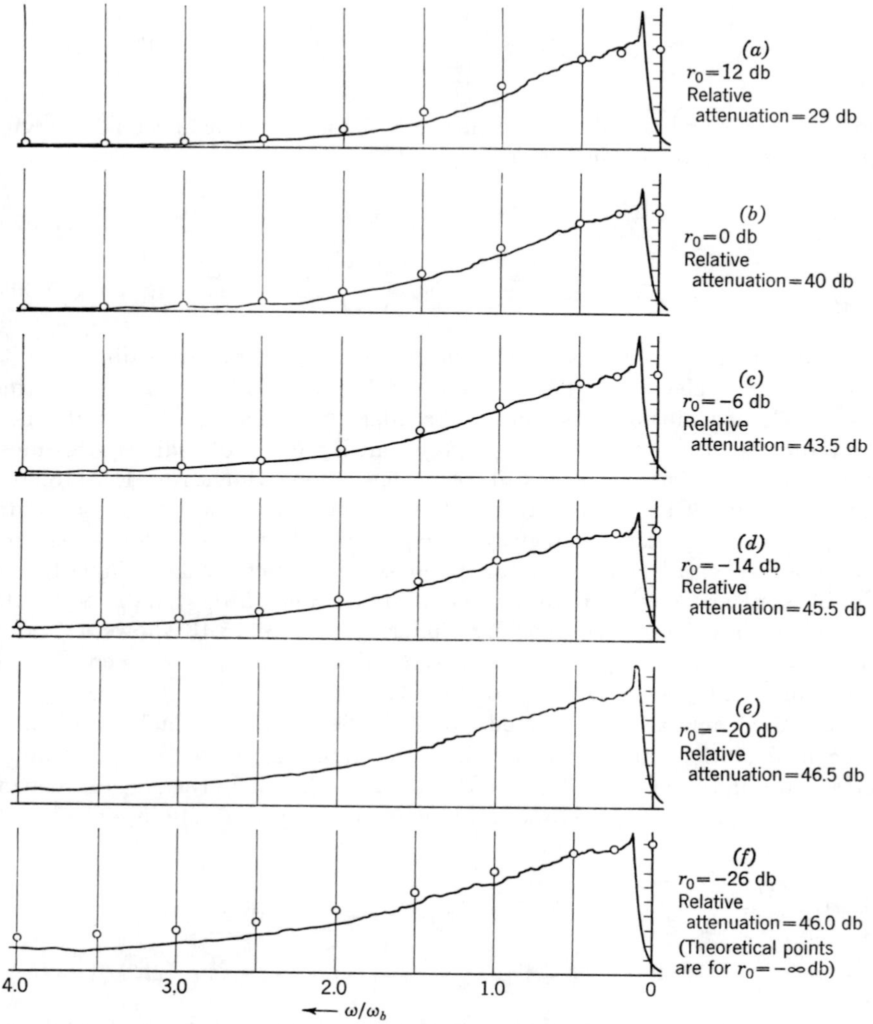

FIG. 15.13. Output spectra for various limit levels, noise alone ($a_0^2 \to -\infty$ db) and a gauss-shaped (composite) RF-IF filter. Theoretical points are circled.

cases have been carried out by Fuller[12] (figs. 5.12 to 5.17), with results in good agreement with the theory.

Limiting spectral distributions for strong carriers may also be obtained from Eq. (15.71*a*) and from Eq. (15.78) in conjunction with Eq. (15.83).

† See Middleton,[11] pt. I, eq. (3.13).

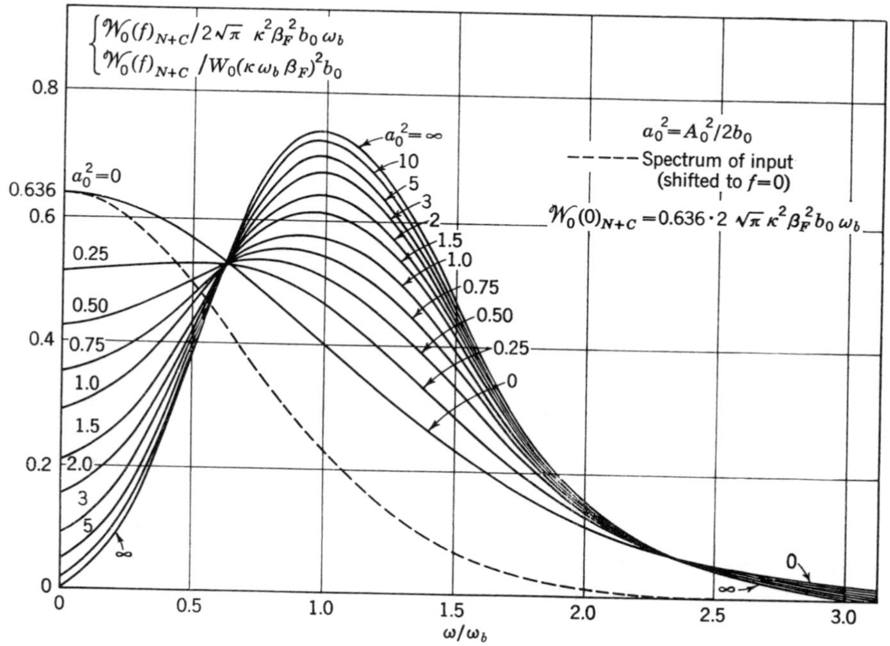

FIG. 15-14. Low-frequency output noise spectrum following discrimination for normal noise and a tuned unmodulated carrier. No limiting; gauss-shaped noise spectrum.

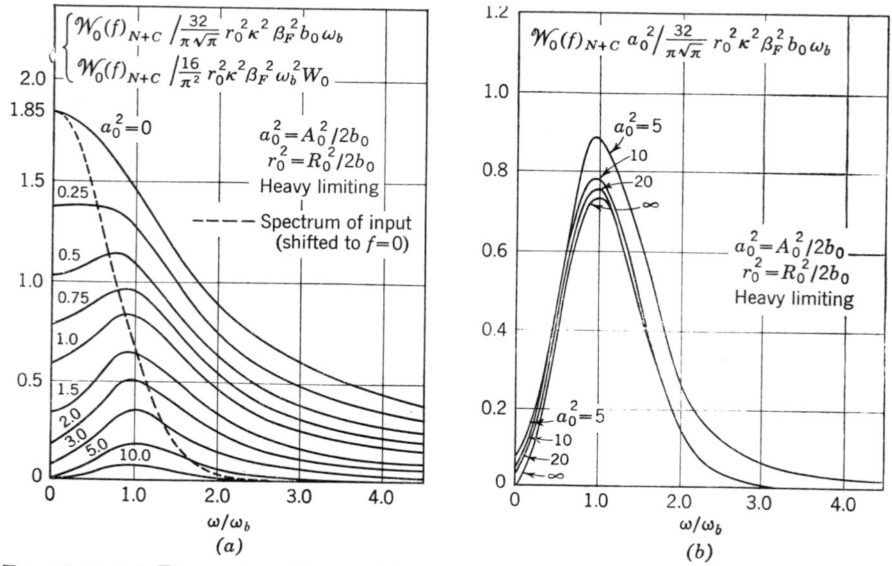

FIG. 15.15. (a) The same as Fig. 15.14, but with extreme limiting. (b) The same as (a), but for strong carriers. (Note change in scale.)

For these latter, we have specifically

$r_0 \to \infty$, or $r_0^2/a_0^2 > 1$; $a_0^2 \to \infty$:

$$K_0(t)_{N+C} \simeq \tfrac{1}{4}\beta_F{}^2\kappa^2\omega_b{}^2 b_0(2 - \omega_b{}^2 t^2)e^{-\omega_b{}^2 t^2/4} \tag{15.92a}$$

and

$$\mathcal{W}_0(f)_{N+C} \doteq 4\sqrt{\pi}\,\beta_F{}^2\kappa^2\omega_b b_0 \left(\frac{\omega}{\omega_b}\right)^2 e^{-\omega^2/\omega_b{}^2} \tag{15.92b}$$

$r_0 \to 0$, or $r_0^2/a_0^2 < 1$, $a_0^2 \to \infty$:

$$K_0(t)_{N+C} \simeq \frac{4}{\pi^2}\,\beta_F{}^2\kappa^2\omega_b{}^2 b_0\,\frac{r_0^2}{a_0^2}\,{}_2F_1{}^2\left(-\frac{1}{2},\frac{1}{2};\frac{3}{2};\frac{r_0^2}{a_0^2}\right)(2 - \omega_b{}^2t^2)e^{-\omega_b{}^2t^2/4} \tag{15.93a}$$

$$\mathcal{W}_0(f)_{N+C} \doteq \frac{64}{\pi\sqrt{\pi}}\,\beta_F{}^2\kappa^2\omega_b b_0\,\frac{r_0^2}{a_0^2}\,{}_2F_1{}^2\left(-\frac{1}{2},\frac{1}{2};\frac{3}{2};\frac{r_0^2}{a_0^2}\right)\left(\frac{\omega}{\omega_b}\right)^2 e^{-(\omega/\omega_b)^2} \tag{15.93b}$$

Notice once more the absolute attenuation of the output as $a_0 \to \infty$ (fixed limit level). A similar investigation of the off-tune carrier ($\omega_d \neq 0$) has been carried out,[11,12] but the results are too involved to be considered here; for further details the reader should consult the literature.[6,11,12]

15.5-7 Signal-to-Noise Ratios (Broadband Frequency Modulation). For effective reception of broadband frequency modulation in normal noise, the second-moment theory above indicates that two conditions must now be met: (1) as in AM reception (cf. Chap. 13), the carrier must be strong compared with the input noise background, and (2) heavy limiting must be used for best results. This can be seen at once from the expressions for the output signal-to-noise ratios when either extreme limiting or no limiting is used.

Here, the output signal is taken to be all the $s \times s$ signal components† [there is no distortion of the original modulation because of the ideal nature of the (RF-IF) filter, limiter, and discriminator], while the interfering noise is represented by that portion of the l-f output of the discriminator which is passed by the final video or audio filter. The appropriate signal-to-noise ratio is accordingly given by Eq. (15.37). For convenience, the final audio or video filter is assumed to be rectangular [that is, $|Y_a(i\omega)|^2$], with an effective width Δf (cf. Figs. 15.9, 15.10).‡

From the results of the preceding sections, it is now a simple matter to construct the desired signal-to-noise ratios $[S/N]^2_{\text{II}}$ for the limiting cases of weak and strong carriers, with heavy or no limiting. With heavy limiting we find from Eqs. (15.65), (15.79*b*) for the signal and from Eqs. (15.79*b*), (15.87*a*) for the noise that the threshold expression for $[S/N]^2_{\text{II}}$ [Eq. (15.37)] becomes

$$\left[\frac{S}{N}\right]^2_{\text{II}}\Bigg|_{a_0^2 \ll 1, r_0 \doteq 0} \doteq \frac{a_0{}^4\langle\dot{\phi}^2\rangle}{2\displaystyle\int_0^{\Delta f} df \int_0^\infty [(k_1{}^2 - k_0k_2)\,{}_2F_1(1,1;2;k_0{}^2) + O(a_0{}^2)]\cos\omega t\,dt} \tag{15.94a}$$

† Except the d-c term, which is zero for balanced discriminator and symmetrical limiters, when the carrier is on-tune.

‡ The extension to more general filters is obvious.

while, at the other extreme of strong signals, we get from Eqs. (15.67*b*), (15.68) for the signal and from Eq. (15.71) for the noise the following relation:

$$\left[\frac{S}{N}\right]^2_{\text{II}} \Bigg|_{a_0^2 \gg 1, r_0 \doteq 0} \simeq \frac{b_0 a_0^2 \langle \phi^2 \rangle}{\int_0^{\Delta f} \omega^2 \, df \sum_{n=0}^{\infty} A_n[\mathcal{W}_{\text{I}}(\omega_0 + n\omega_a + \omega) + \mathcal{W}_{\text{I}}(\omega_0 + n\omega_a - \omega)]} \tag{15.94b}$$

On the other hand, when there is no limiting, we may use Eq. (15.79*a*) for the threshold signal and Eqs. (15.79*a*), (15.87*b*) for the accompanying noise, to write

$$\left[\frac{S}{N}\right]^2_{\text{II}} \Bigg|_{a_0^2 \ll 1, r_0 \to \infty} \doteq \frac{a_0^4 \langle \phi^2 \rangle}{2 \int_0^{\Delta f} df \int_0^{\infty} [(k_1^2 - k_0 k_2)\, {}_2F_1(\tfrac{1}{2},\tfrac{1}{2};2;k_0^2) + O(a_0^2)] \cos \omega t \, dt} \tag{15.95a}$$

The strong-carrier case may be determined from Eq. (15.75) for the signal and from Eq. (15.76) for the noise. We get

$$\left[\frac{S}{N}\right]^2_{\text{II}} \Bigg|_{a_0^2 \gg 1, r_0 \to \infty} \simeq \frac{2 b_0 a_0^2 \langle \phi^2 \rangle}{\int_0^{\Delta f} df \sum_{n=0}^{\infty} A_n[(n\omega_a + \omega)^2 \mathcal{W}_{\text{I}}(\omega_0 + n\omega_a + \omega) + (n\omega_a - \omega)^2 \mathcal{W}_{\text{I}}(\omega_0 + n\omega_a - \omega)]} \tag{15.95b}$$

(In the above, it is assumed that the intensity of the input noise remains constant.)

When the carrier is strong, it is clear from Eqs. (15.94*b*), (15.95*b*) that narrowing the final low-pass filter, i.e., decreasing Δf, will improve the signal-to-noise ratio in both cases. However, it is the further decrease of the spectral distribution with ω^2 as $f \to 0$ (cf. Fig. 15.9 with Fig. 15.10), and hence of the total noise, that gives broadband frequency modulation *with heavy limiting* its superiority† over broadband frequency modulation with little or no limiting. Broadband frequency modulation, in any case, is always as good as if not better† than narrowband frequency modulation at large carrier levels, as is apparent if we compare Fig. 15.8 with Fig. 15.16. Note again, in contrast to the broadband situation, that limiting is actually

† Of course, with respect to the particular criterion of performance, here an output signal-to-noise ratio.

deleterious in narrowband frequency modulation, since the effective background noise contribution is now increased, rather than decreased, relative to the desired signal. On the other hand, threshold performance [Eqs. (15.94*a*), (15.95*a*)] in every instance exhibits the expected signal suppression, just as in AM reception; for example, $(S/N)^2 \sim a_0^4$. Here, frequency

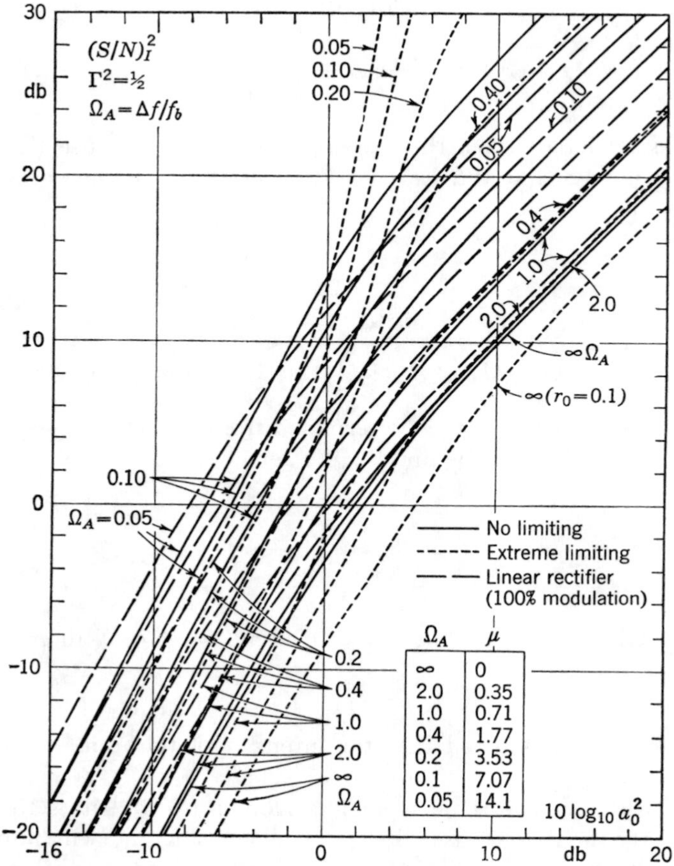

FIG. 15.16. Output signal-to-noise ratios for broadband frequency modulation after audio filtering. [Sinusoidal modulation, constant RF-IF widths ($\sim f_b$), and different final filter widths, Δf. Here $\mu = \Delta f/f_b$; for f_b, see Eq. (15.83).]

modulation of either variety is not noticeably better than amplitude modulation under comparable circumstances, and, in fact, amplitude modulation may be slightly superior, as the curves of Figs. 15.8, 15.16 indicate, for $\Gamma^2 = \langle \dot{\phi}^2 \rangle / \rho_E = \frac{1}{2}$ with gaussian (RF-IF) filter responses. [For the rectangular (RF-IF), Γ^2 equals $\frac{3}{2}$, and broadband frequency modulation shows a slight improvement (about 2 db) over amplitude modulation in this case;

cf. remarks following Eq. (15.46b).] For further details, including the results of Fig. 15.16, see Middleton,[10] sec. 5.†

Finally, as Fuller's extensive experimental investigation[12] has shown, good to excellent agreement with theory‡ is obtained whenever conditions approximating the idealizations of limiter and discriminator are reproduced in the actual receiver. Fortunately, as Fuller's work indicates, with careful design serviceable approximations to the idealized structure can be realized in practical cases, which in turn yield a performance close to the theoretically predicted one. In this connection, we stress again the importance of the limiter's frequency response: the heavier the limiting, the wider this (uniform) response should be, so that the effects of limiting are not destroyed by frequency selection. The h-f components of the clipped input to the discriminator are those which, in the time domain, give the wave its clipped appearance. Too narrow a filter accordingly removes these components and in effect destroys the action of the limiter. In a similar way, the frequency response of the discriminator should be of comparable width, so that demodulation can take place without distortion and without unnecessary degradation of the signal by the noise.

PROBLEMS

15.4 (a) Show that $\mathcal{H}_{k,2m+n,q}$ [Eq. (15.60)] becomes

$$\mathcal{H}_{k,2m+n,q} = \frac{4}{\pi}\beta_F \frac{\kappa r_0 a_0^q}{q![m!(m+k)!]^{1/2}} \left(\frac{2}{b_0}\right)^{n/2} \Gamma(\nu_0) \Bigg(-{}_1F_1(\nu_0; q+1; -a_0^2)$$
$$+ \sum_{l=1}^{\infty} \frac{(\nu_0)_l(-\tfrac{1}{2})_l(-r_0)^{2l}}{(l!)^3} \frac{2l}{2l+1} \left\{ \left[\tfrac{3}{2}\psi(l+1) - \tfrac{1}{2}\psi(\tfrac{3}{2} - l) - \frac{1}{2l(2l+1)} - \log r_0 \right] \right.$$
$$\left. \times\, {}_1F_1(\nu_0 + l; q+1; -a_0^2) - \frac{1}{2}\sum_{j=0}^{\infty} \frac{(\nu_0+l)_j\psi(\nu_0+l+j)(-a_0^2)^j}{j!(q+1)_j} \right\} \Bigg) \qquad (1)$$

where ψ is the logarithmic derivative of the gamma function.

(b) Obtain from this

$$\lim_{r_0\to 0} \mathcal{H}_{k,2m+n,q} \doteq \frac{4}{\pi}\beta_F\kappa r_0 \left(\frac{2}{b_0}\right)^{n/2} \frac{a_0^q}{q!} \frac{\Gamma(\nu_0)}{[m!(m+k)!]^{1/2}}\, {}_1F_1(\nu_0; q+1; -a_0^2) \qquad (2)$$

and
$$\lim_{r_0\to\infty} \mathcal{H}_{k,2m+n,q} = \frac{\beta_F\kappa}{2}\left(\frac{2}{b_0}\right)^{n/2} \frac{a_0^q}{q!} \frac{\Gamma(\nu_0 - \tfrac{1}{2})}{[m!(m+k)!]^{1/2}}\, {}_1F_1\left(\nu_0 - \frac{1}{2}; q+1; -a_0^2\right) \qquad (3)$$

† The preceding comments apply, of course, for normal noise interference. If the background disturbance is low-density impulse noise, it is well known that broadband frequency modulation with strong limiting also gives better performance than amplitude modulation under similar conditions.

‡ These results may also be extended to more complicated input waves—e.g., carriers simultaneously amplitude- and angle-modulated by noise (cf. Sec. 14.3-1). Examples of this have been discussed, for instance, by Middleton[13] and Gottschalk and Mullen[14] in the measurement of the noise output of cw magnetrons, where no limiting was used.

15.5 (*a*) Show for the ideal limiter of the text that

$$\mathcal{H}_{001}\Big|_{a_0\to\infty} \simeq \frac{4\kappa\beta_F}{\pi}\left(\frac{R_0}{A_0}\right) {}_2F_1\left(-\frac{1}{2},\frac{1}{2};\frac{3}{2};\frac{R_0^2}{A_0^2}\right) \qquad R_0 < A_0 \tag{1}$$

(*b*) For the symmetrical νth-law limiters of Sec. 13.4, show that the amplitude functions (15.59) now become

$$\mathcal{H}_{k,2m+n,q} = -\beta_F\kappa\left(\frac{2}{b_0}\right)^{(n-\nu+1)/2} a_0^q \frac{\Gamma[(\nu+2)/2]\Gamma[(1-\nu)/2]\Gamma(\nu+1)}{\Gamma(\nu/2)\Gamma[(\nu+3)/2]\Gamma(1-\nu/2)^2}$$
$$\frac{\Gamma[(q+2m+n+k-1-\nu)/2]}{[q!^2n!(m+k)!]^{1/2}} {}_1F_1\left(\frac{q+2m+n+k-\nu+1}{2};q+1;-a_0^2\right)$$
$$0 \le \nu < 1 \tag{2}$$

15.6 (*a*) Obtain the result of Eq. (15.71) for the l-f spectral output of the discriminator assumed in the text, before final video or audio filtering, in the case of moderate to heavy limiting and strong carriers.

(*b*) Repeat the above, obtaining now Eq. (15.76), but with no limiting and with strong carriers.

15.7 (*a*) Show that the first zonal output of the limiter used in the text has the covariance function for large carriers:

$$K_L(t)\Big|_{a_0\to\infty} \simeq \frac{8\beta_F^2R_0^2}{\pi^2}\left[1 - \frac{1}{2a_0^2} + \frac{k_0(t)}{2a_0^4} + \cdots\right]\cos\omega_0 t \tag{1}$$

(*b*) In the weak-carrier case, obtain, for the above,

$$K_L(t)\Big|_{a_0^2\ll 1} \doteq \frac{8\beta_F^2R_0^2}{\pi^2}\left[k_0(t)\,{}_2F_1\left(\frac{1}{2},\frac{1}{2};2;k_0^2\right) + a_0^2(1-k_0)\,{}_2F_1\left(\frac{1}{2},\frac{3}{2};2;k_0^2\right) + O(a_0^4)\right]\cos\omega_0 t \tag{2a}$$

$$\doteq \frac{8\beta_F^2R_0^2}{\pi^2}\left\{\frac{4}{\pi k_0}[E(k_0) - (1-k_0^2)K(k_0)] + a_0^2\frac{1-k_0}{k_0^2}[K(k_0) - E(k_0)] + O(a_0^4)\right\}\cos\omega_0 t \tag{2b}$$

where K and E are complete elliptic integrals of the first and second kind.

H. W. Fuller.[12]

15.8 Obtain the relations shown in Table 15.1 for these different composite RF-IF filter characteristics.

15.9 A cw carrier, amplitude-modulated by stationary normal noise $V_N(t)$, is passed through the ideal FM receiver discussed in this chapter. This carrier may be off-tune by an amount f_d.

(*a*) If the input to the limiter-discriminator combination is

$$V(t) = A_0[1 + \lambda V_N(t)]\cos\omega_c t = g(V_N)\cos\omega_c t \tag{1}$$

show that the covariance function of the audio (or video) output is

$$K_0(t) = \frac{16}{\pi^2}\kappa^2\beta_F^2\omega_d^2 \iint_0^\infty \frac{\sin R_0\xi_1 \sin R_0\xi_2}{\xi_1^2\xi_2^2} \boldsymbol{E}\{J_1(\xi_1 g_1)J_1(\xi_2 g_2)\}\, d\xi_1\, d\xi_2 \tag{2}$$

where $g_1 = g(t_1)$, etc.

(*b*) In the case of *no limiting*, obtain

$$K_0(t) = \kappa^2\beta_F^2\omega_d^2\boldsymbol{E}\{g_1g_2\} \tag{3}$$

and use Eqs. (12.23) et seq. to obtain $\boldsymbol{E}\{g_1g_2\}$ explicitly for a linear modulation, except for the effects of overmodulation.

TABLE 15.1

Composite RF-IF filter characteristic	Analytic form, $\mathcal{W}_N(f)$	b_0	$k_0(t)$	$k_1(t)$	$k_2(t)$	$k_1(t)^2 - k_0(t)k_2(t)$
A. Gauss-shaped	$\mathcal{W}_0 e^{-(\omega-\omega_0)^2/\omega_b^2}$	$\frac{\mathcal{W}_0\omega_b}{2\sqrt{\pi}}$	$e^{-x^2/4}$, $x = \omega_b t$	$-\frac{\omega_b x}{2} e^{-x^2/4}$	$\frac{\omega_b^2}{2}\left(\frac{x^2}{2} - 1\right) e^{-x^2/4}$	$\frac{\omega_b^2}{2} e^{-x^2/2}$
B. Rectangular	$\mathcal{W}_0$, $\omega_0 - \omega_b < \omega < \omega_0 + \omega_b$ 0, $\omega_0 - \omega_b > \omega$ $\omega_0 + \omega_b < \omega$	$\frac{\mathcal{W}_0\omega_b}{\pi}$	$\frac{\sin x}{x}$	$-\omega_b\left(\frac{\sin x}{x^2} - \frac{\cos x}{x}\right)$	$\omega_b^2\left(2\frac{\sin x}{x^3} - 2\frac{\cos x}{x^2} - \frac{\sin x}{x}\right)$	$\omega_b^2 \frac{x^2 - \sin^2 x}{x^4}$
C. Single-tuned	$\frac{\mathcal{W}_0}{1 + (\omega - \omega_0)^2/\omega_b^2}$	$\frac{\mathcal{W}_0\omega_b}{2}$	$e^{-\lvert x\rvert}$	$\omega_b e^x$, $x < 0$ $-\omega_b e^{-x}$, $x > 0$ 0, $x = 0$	$\omega_b^2 e^{-\lvert x\rvert}$, $x \neq 0$ $-2\omega_b^2\delta(x - 0)$, $x = 0$	0, $x \neq 0$ $2\omega_b^2\delta(x - 0)$, $x = 0$
D. Stagger-tuned double	$\frac{\mathcal{W}_0}{1 + (\omega - \omega_0)^4/\omega_b^4}$	$\frac{\mathcal{W}_0\omega_b}{2\sqrt{2}}$	$\sqrt{2}\, e^{-\lvert x\rvert/\sqrt{2}} \cos\left(\frac{\lvert x\rvert}{\sqrt{2}} - \frac{\pi}{4}\right)$	$-\sqrt{2}\,\omega_b e^{-\lvert x\rvert/\sqrt{2}} \sin\frac{\lvert x\rvert}{\sqrt{2}}$	$-\sqrt{2}\,\omega_b^2 e^{-\lvert x\rvert/\sqrt{2}} \cos\left(\frac{\lvert x\rvert}{\sqrt{2}} + \frac{\pi}{2}\right)$	$\omega_b^2 e^{-\sqrt{2}\lvert x\rvert}$
E. Synchronous-tuned double	$\frac{\mathcal{W}_0}{[1 + (\omega - \omega_0)^2/\omega_b^2]^2}$	$\frac{\mathcal{W}_0\omega_b}{4}$	$e^{-\lvert x\rvert}(1 + \lvert x\rvert)$	$-\omega_b x e^{-\lvert x\rvert}$	$-\omega_b^2 e^{-\lvert x\rvert}(1 + \lvert x\rvert)$	$\omega_b^2 e^{-2\lvert x\rvert}$

(*c*) For *extreme limiting*, show that

$$K_0(t) \doteq \frac{16}{\pi^2} R_0^2 \kappa^2 \beta_F^2 \omega_d^2 \qquad R_0 \ll \sqrt{\overline{g^2}} \tag{4}$$

That is, only the d-c output, attributable to detuning, is nonvanishing here; there is no l-f continuum.

REFERENCES

1. Smith, D. B., and W. E. Bradley: The Theory of Impulse Noise in Ideal Frequency-modulation Receivers, *Proc. IRE*, **34**: 743 (1946).
2. Huggins, W. H., and D. Middleton: The Phase and Amplitude Principles in Signal Detection, National Electronics Conference, Chicago, 1955, vol. 11, p. 304.
3. Gott, E.: A Study of the Phase Principle of Signal Detection, *Johns Hopkins Univ. Radiation Lab. Tech. Rept.* AF-70, July, 1959.
4. Carson, J. R., and T. C. Fry: Variable Frequency Electric Circuit Theory with Application to the Theory of Frequency-modulation, *Bell System Tech. J.*, **16**: 513 (1937).
5. Crosby, M. C.: Frequency-modulation Noise Characteristics, *Proc. IRE*, **25**: 472 (1937).
6. Blachman, N. M.: The Demodulation of a Frequency-modulated Carrier and Random Noise by a Limiter and Discriminator, *J. Appl. Phys.*, **20**: 38, 976 (1949); also *Cruft Lab. Tech. Rept.* 31, March, 1948.
7. Rice, S. O.: Statistical Properties of a Sine-wave and Random Noise, *Bell System Tech. J.*, **27**: 109 (1948), secs. 7, 8.
8. Stumpers, F. L. H. M.: Theory of Frequency-modulation Noise, *Proc. IRE*, **36**: 1080 (1948).
9. Wang, M. C.: chap. 13 in J. L. Lawson and G. E. Uhlenbeck, "Threshold Signals," Massachusetts Institute of Technology Radiation Laboratory Series, vol. 24, McGraw-Hill, New York, 1950.
10. Middleton, D.: On Theoretical Signal-to-Noise Ratios in FM Receivers: A Comparison with Amplitude Modulation, *J. Appl. Phys.*, **20**: 334 (1949).
11. Middleton, D.: The Spectrum of Frequency-modulated Waves after Reception in Random Noise. I, II, *Quart. Appl. Math.*, **7**: 129 (1949), **8**: 59 (1950).
12. Fuller, H. W.: Experimental Study of Signals and Noise in a Frequency-modulation Receiver, doctoral dissertation, Harvard University, May, 1956. See also H. W. Fuller and D. Middleton, Signals and Noise in an FM Receiver. I. Theoretical Discussion, *Cruft Lab. Tech. Rept.* 242, February, 1957; and H. W. Fuller, Signals and Noise in an FM Receiver. II. Experimental Discussion, *Cruft Lab. Tech. Rept.* 243, February, 1957.
13. Middleton, D.: Theory of Phenomenological Models and Direct Measurements of the Fluctuating Output of CW Magnetrons, Symposium on Fluctuation Phenomena in Microwave Sources, *IRE Trans. on Electron Devices*, **ED-1**: 56 (December, 1954); also, *Rept.* R-5(A), November, 1954, Raytheon Manufacturing Co., Research Division, Waltham, Mass., for a more complete account.
14. Gottschalk, W. M.: Direct Detection Measurements of Noise in CW Magnetrons, Symposium on Fluctuation Phenomena in Microwave Sources, *IRE Trans. on Electron Devices*, **ED-1**: 91 (December, 1954); W. M. Gottschalk and J. A. Mullen, *Rept.* R-5(B), Raytheon Manufacturing Co., Research Division, Waltham, Mass.

CHAPTER 16

LINEAR MEASUREMENTS, PREDICTION, AND OPTIMUM FILTERING

In Chaps. 2 to 4, we have considered linear systems and some of their effects on the statistical character of the input processes to them. Here, we shall return again to an examination of linear systems, but now chiefly from the viewpoint of optimized performance, where the sampling (or observation) time is for the most part restricted to finite intervals $(0,T)$ and the systems in question are limited to the simplest class of noninvariant† filters, e.g., linear invariant filters with switches at $t = 0$, T [cf. Sec. 2.1-3(2) and Fig. 5.9].

Among the problems considered in this chapter are (1) the role of a linear filter in finite-time measurements;[1] (2) prediction and filtering with realizable (and nonrealizable†) linear filters (the Wiener[2]-Kolmogoroff[3] theory), with some generalizations;[4,5] and (3) the maximization of signal-to-noise ratios by linear systems, including the concept of the "matched" filter.[6,7] The analysis is confined to linear systems and to the first and second moments of the various input and output disturbances, including deterministic as well as entirely random signal and noise processes, additively combined for optimum and suboptimum conditions of operation. The treatment here is introductory in two respects: not only have these topics been extensively explored in recent years, so that a considerable body of detailed literature is now available, but they may also be treated from the more general viewpoint of statistical decision theory, as we shall observe later in Part 4, where it is recognized that these problems are all special cases of linear estimation, in a more general formulation which includes nonlinear systems as well. The present analysis therefore makes no attempt at completeness, since it is preliminary to the more systematic discussion in subsequent chapters (cf. Chaps. 18, 20, 21, 23, and in particular Chap. 21, Secs. 21.3, 21.4). A comprehensive account of the various communication applications of linear filtering, linear prediction, and the like, is available in a number of recent publications.[8,9]

16.1 *Linear Finite-time Measurements*

Up to this point, we have generally assumed infinite smoothing periods for our processed data (for an exception, see Sec. 5.2-3). In practice, such

† For a discussion of realizability, invariance, etc., see Sec. 2.2-5(3).

periods are, of course, never infinite, although as we have remarked earlier (cf. Sec. 1.3-6) they may be effectively treated as such if the interval in question is sufficiently large vis-à-vis the fluctuation time of the process under investigation. However, in many applications, this is neither a possible nor a safe assumption, so that we must take specifically into account the role of the finite sample or finite data-processing and data-acquisition periods. This is particularly important, for example, in measurements which involve a random component, where the component may represent either the desired quantity to be estimated or an undesired, accompanying noise background. Accordingly, the purpose of the present section is to describe briefly some of the statistical errors introduced by finite averaging and observation intervals when linear measurements $\boldsymbol{T}_L$ on random functions are undertaken. The discussion for the most part is limited to the first and second moments of the sample mean.† A more general approach using nonlinear operations is briefly described in Chap. 21.

16.1-1 Statistical Errors. Let us consider first the errors that occur when a finite sample is used to obtain a linear finite-time measurement, or estimate, on an interval $(t_1, t_1 + T)$, of various statistical properties of a random process $Z(t)$, when these properties are defined over the ensemble on the infinite interval $(-\infty < t < \infty)$. We are concerned here for the most part with time averages only, since practical measuring devices operate on only a single or, at most, on a relatively few member functions of the ensemble for a finite period.

(1) Continuous sampling. A typical situation is shown in Fig. 16.1, where now $Z(t) = Z[y(t)]$ is a process derived from another, originally stationary process $y(t)$ (by some previous circuit operation), such that Z is at least wide-sense stationary (cf. Sec. 1.3-6), for the time being free of any deterministic components, and possesses a positive definite covariance function. Now, an essential property of measuring systems is an ability to smooth or weight the input data in a manner appropriate to the type of measurement desired. Thus, if $h(t)$ is the effective weighting function introduced by our (postulated) linear measuring device, a measurement $M_{Z^{(j)}}(T) = \boldsymbol{T}_L\{Z^j\}$ made at time $t = T$ on a particular member $Z^{(j)}$ of the Z ensemble, after $Z^{(j)}$ has been introduced at time $t = 0$, can be expressed in the usual way as the convolution

$$M_{Z^{(j)}}(T) = \boldsymbol{T}_L\{Z^{(j)}\} = \int_{0-}^{T+} h(u) Z^{(j)}(T - u)\, du \qquad h(u) = 0,\ u < 0- \tag{16.1}$$

where $(0,T)$ is the observation or acquisition interval on which the value $M_{Z^{(j)}}(T)$ is based. $Z^{(j)}(t)$, of course, may itself be the result of some non-

† Based on the treatment of Davenport, Johnson, and Middleton;[1] for a rigorous account of the mathematical questions involved, see Grenander and Rosenblatt.[10]

linear operation on an original input† $y^{(j)}(t)$. For example, to find the mean intensity $\langle y^{(j)2}\rangle$, we first put $y^{(j)}$ through a suitable squaring device $Z^{(j)}(t) = \alpha y^{(j)}(t)^2$ and then into an integrating circuit (cf. Fig. 16.1), or if we wish a point on the correlation curve of $y^{(j)}(t)$ (cf. Figs. 13.13, 13.14), we set $Z^{(j)}(t) = y^{(j)}(t)y^{(j)}(t - \tau)$ and average, etc. The operation (16.1) may be performed by a very simple time-varying linear network, consisting of a simple switch followed by a passive linear circuit, turned on at $t = 0$ and read off at $t = T$.

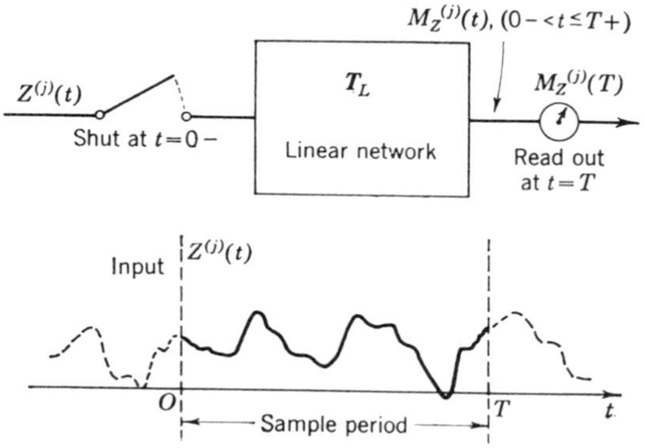

FIG. 16.1. Finite-time linear measurement with a stationary input.

While $M_{Z^{(j)}}(T)$ is a definite number (for a particular $Z^{(j)}$), $M_{Z^{(j)}}(T)$ will vary from representation to representation, fluctuating about the ensemble mean $\overline{M_Z(T)}$, with a variance $\overline{M_Z(T)^2} - \overline{M_Z(T)}^2$. Let us consider first the relationship between the sample mean value $\overline{M_Z(T)}$ and a particular measurement $M_{Z^{(j)}}(T)$. A natural definition of the *error* in a particular measurement of the mean value $\overline{M_Z(T)}$ is simply

$$\epsilon_M^{(j)} \equiv M_{Z^{(j)}}(T) - \overline{M_Z(T)} \tag{16.2}$$

from which it is at once clear that the average error $\overline{\epsilon_M^{(j)}}$ is zero. Similarly, the mean-square fluctuation in $M(T)$ is given by the sample variance $\sigma_Z^2 = \overline{M_Z(T)^2} - \overline{M_Z(T)}^2$.

To evaluate the performance of the linear measuring system [as represented by the operation $M_Z = \mathbf{T}_L\{Z\}$ in Eq. (16.1)], let us now introduce several simple second-moment criteria. These are the *accuracy* $A_M^{(j)}$ of an actual observation and the (*rms*) *relative error* B_M, respectively given by

$$A_M^{(j)} \equiv \epsilon_M^{(j)}/(\overline{\epsilon_M^2})^{1/2} \qquad B_M \equiv \sqrt{\overline{\epsilon_M^2}}/\overline{M(T)} \tag{16.3}$$

† In this sense, our measuring system may be thought of more generally as a nonlinear device in that it relates an original input $y^{(j)}(t)$, by means of a linear measurement $M^{(j)}(T)$ on $Z^{(j)}(y^{(j)})$, to some average nonlinear property ($\sim Z$) of the process $y(t)$ which we may ultimately wish to estimate. We shall not in what follows, however, be particularly concerned with the original process y but rather with linear operations on the derived process Z.

this last in all situations where $\overline{M(T)} \neq 0$. The former measures the error (16.2) in units of the sample variance for individual observations. The latter is a measure of the relative spread in observed values over the ensemble of possible observations. It is with this latter that we are primarily concerned, here and in Secs. 16.1-2, 16.1-3.

In terms of the input process Z, we easily see that

$$\overline{M_Z(T)} = \int_{0-}^{T+} h(u)\overline{Z(T-u)}\,du = \bar{Z}\lambda \qquad \lambda \equiv \int_{0-}^{T+} h(u)\,du\ (>0) \tag{16.4}$$

since Z is at least wide-sense stationary, and

$$\overline{M_Z(T)^2} = \iint_{0-}^{T+} h(u)\overline{Z(T-u)Z(T-v)}h(v)\,du\,dv$$
$$= \iint_{0-}^{T+} h(u)K_Z(|u-v|)h(v)\,du\,dv \tag{16.5}$$

If Z is, in addition, ergodic, we can replace the covariance function K_Z by the autocorrelation function $R_Z(|u-v|)$ in the usual way (cf. Sec. 1.6-1). Now, defining a weighting function of the linear filter above with switches at $t = 0-,\ T+$, for example,

$$h_T(u)\begin{cases} \equiv h(u) & 0- < t < T+ \\ = 0 & t < 0-,\ t > T+ \end{cases}$$

and therefore

$$\lambda = \int_{-\infty}^{\infty} h_T(u)\,du \tag{16.6}$$

we may put Eq. (16.5) into the simpler form

$$\overline{M_Z(T)^2} = \int_{-\infty}^{\infty} \rho_T(x)K_Z(x)\,dx \qquad \text{where } \rho_T(x) = \int_{-\infty}^{\infty} h_T(u)h_T(u+x)\,du \tag{16.7}$$

is the *truncated* autocorrelation function of the original linear filter (Sec. 3.3). Note that $\rho_T(x)$ vanishes identically for all $|x| > T$ and that $\int_{-\infty}^{\infty} \rho_T(x)\,dx = \lambda^2$. The variance $\overline{\epsilon_M{}^2}$ then becomes, from Eqs. (16.4), (16.7),

$$\overline{\epsilon_M{}^2} = \overline{M_Z{}^2} - \bar{M}_Z{}^2 = \int_{-\infty}^{\infty} \rho_T(x)[K_Z(x) - \bar{Z}^2]\,dx \tag{16.8}$$

so that the (rms) relative error [Eqs. (16.3)] is now

$$B_M = \left\{\int_{-\infty}^{\infty} \rho_T(x)[K_Z(x) - \bar{Z}^2]\,dx \Big/ \bar{Z}^2\lambda^2\right\}^{1/2}$$
$$= \left[(\bar{Z}\lambda)^{-2}\int_{-\infty}^{\infty} \rho_T(x)K_Z(x)\,dx - 1\right]^{1/2} \geq 0 \tag{16.9a}$$

With the help of the Fourier transform $Y_T(i\omega)$ of $h_T(u)$ and the W-K theorem (3.42), we obtain the alternative expression for B_M in the frequency domain:

$$B_M = \left[\int_0^{\infty} |Y_T(i\omega)|^2 \mathcal{W}_{Z-\bar{Z}}(f)\,df \Big/ |Y_T(0)|^2\bar{Z}^2\right]^{1/2} \geq 0 \tag{16.9b}$$

(2) An ideal integrator. A weighting function of considerable theoretical interest is the one for which $h_T(u)$ is a constant (> 0) in $(0-,T+)$, such that $\lambda = 1$, that is,

$$h_T(u) = \begin{cases} \frac{1}{T} & 0 < u < T \\ 0 & \text{elsewhere} \end{cases} \tag{16.10a}$$

with

$$\rho_T(x) = \begin{cases} \frac{1}{T}\left(1 - \frac{|x|}{T}\right) & |x| < T \\ 0 & |x| > T \end{cases} \tag{16.10b}$$

and

$$Y_T(i\omega) = \frac{1 - e^{-i\omega T}}{i\omega}$$

as is easily seen from Eqs. (16.6), (16.7). Here, $\bar{M}$ is simply $\bar{Z}$, and, for stationary Z, $\overline{M^2}$ [Eq. (16.7)] becomes

$$\overline{M^2} = \frac{2}{T}\int_0^T \left(1 - \frac{x}{T}\right) K_Z(x)\, dx \tag{16.11}$$

with a corresponding expression for B_M and $\overline{\epsilon_M{}^2}$. The spectral counterpart of Eq. (16.9b) is

$$B_M = \left[\int_0^\infty \frac{\sin (\omega T/2)^2}{(\omega T/2)^2} \frac{\mathcal{W}_{Z-\bar{Z}}(f)\, df}{\bar{Z}^2}\right]^{1/2} \geq 0 \tag{16.12}$$

We observe that to calculate the (rms) relative error, involving the second moment of the measurement $M^{(j)}(T)$, the covariance function of Z must be known–or calculated from the statistics of the original process $y(t)$. This usually requires *greater* knowledge of the statistics of y than can be obtained from the measurement of Z. The situation is very similar to that encountered in the measurement of most physical quantities: more statistical information concerning the input process is required than is obtainable by the corresponding observations of the output process. A typical case is that of noise through a general zero-memory nonlinear device: to determine, say, a second-order statistic like the spectrum of the output, we need the entire second-order distribution density of the input process; for more complicated output statistics, still higher-order input densities are required, and so on (cf. Chap. 5).

(3) Limiting forms. When the acquisition period $(0,T)$ becomes sufficiently short $(T \to 0)$ or long $(T \to \infty)$, the expressions above for the rms relative error reduce to a number of useful relations. For the ideal integrator [Eqs. (16.10)], where $\lambda = 1$, we find that

$$\lim T \to 0: \quad B_M \doteq \frac{[K_Z(0) - \bar{Z}^2]^{1/2}}{\bar{Z}} = \left[\frac{\int_0^\infty \mathcal{W}_{Z-\bar{Z}}(f)\, df}{\bar{Z}^2}\right]^{1/2} = \frac{\sigma_Z}{\bar{Z}} \tag{16.13a}$$

$$\lim T \to \infty: \quad B_M \simeq \left\{\frac{2}{T\bar{Z}}\int_0^\infty [K_Z(x) - \bar{Z}^2]\, dx\right\}^{1/2} = \frac{\mathcal{W}_{Z-Z}(0+)^{1/2}}{\bar{Z}\sqrt{2T}} \tag{16.13b}$$

where for Eq. (16.13b) Z is assumed free of periodic terms. If Z does contain periodicities, so that we can write $Z(t) = N(t) + S(t)$, where S is periodic (and uncorrelated with N), one can easily show that

$$(\overline{M^2} - \bar{M}^2)_{Z=S+N} = (\overline{M^2} - \bar{M}^2)_N + \sum_{n=1}^{\infty} \frac{c_n{}^2}{2} \frac{1 - \cos n\omega_0 T}{(n\omega_0 T)^2} + \frac{c_0{}^2}{4} \quad (16.14)$$

so that, as $T \to \infty$, B_M is inversely proportional to $T^{1/2}$ for the noise but includes a constant term $O(T^0)$ for the d-c component of the signal, if any. Otherwise, we may use Eq. (16.13b) for $Z = S + N$ ($c_0 = 0$). The important feature of Eq. (16.13b) is that as smoothing time T becomes longer the rms relative error (apart from a possible d-c term) becomes vanishingly small $O(T^{-1/2})$.

With the more general weighting [Eqs. (16.6)], we observe that, as $T \to 0$, the effective filter response $Y_T(i\omega)$ is broad compared with the spectral density $\mathcal{W}_{Z-\bar{Z}}(f)$, so that $|Y_T(i\omega)|^2 \approx |Y_T(0)|^2$ over the range of the principal nonzero values of the spectrum, or, equivalently, $\rho_T(x) \to \lambda^2$ and is significant only in the immediate neighborhood of $x = 0$, vis-à-vis $K_Z - \bar{Z}^2$. Equations (16.9) then become

$$(\lim T \to 0): \qquad B_M \doteq \sigma_Z/\bar{Z} = \left[\int_0^\infty \mathcal{W}_{Z-\bar{Z}}(f)\, df\right]^{1/2} \Big/ \bar{Z} \quad (16.15a)$$

which is identical with the result (16.13a) for the ideal integrator [Eqs. (16.10)].

At the other extreme of long smoothing periods, the effective bandwidth of the averaging filter is assumed very narrow compared with the effective spectral width of $\mathcal{W}_{Z-\bar{Z}}(f)$, so that $\mathcal{W}_{Z-\bar{Z}}(f) \doteq \mathcal{W}_{Z-\bar{Z}}(0+)$, as above [cf. Eq. (16.13b)]. Now, as $T \to \infty$, $h_T(u) \to h(u)$ and therefore $Y_T(i\omega) \simeq Y(i\omega)$, for which Eqs. (16.9$a$), (16.9$b$) become

$$\lim T \to \infty: \quad B_M \simeq \mathcal{W}_{Z-\bar{Z}}(0+)^{1/2} \left[\int_0^\infty |Y(i\omega)|^2\, df\right]^{1/2} \Big/ \bar{Z}|Y(0)| \quad (16.15b)$$

Defining an *effective bandwidth* B_e by

$$B_e \equiv \int_0^\infty |Y(i\omega)|^2\, df \Big/ |Y(0)|^2 \quad (16.16)$$

we have finally

$$B_M \simeq [\mathcal{W}_{Z-\bar{Z}}(0+) B_e]^{1/2}/\bar{Z} \quad (16.17)$$

for the value of the rms relative error as the integration time becomes very long, provided that $0 < B_e \ll B_Z$, where B_Z is the bandwidth of Z. The important feature of this relation is that for *fixed* filter characteristics (as opposed to the "adjustable" nature of the ideal integrator), the (rms) relative error approaches a *nonvanishing constant value in the limit.*† The fact that a nonvanishing limit for this error estimate is obtained follows, of course, from the finite (> 0) aperture of the smoothing filter: rapid fluctuations are removed, but the very slow ones remain, corresponding to spectral energy of the input wave within the passband B_e of the filter. A particular example of this is shown in Fig. 16.2 (see also Sec. 5.2-3).

(4) Discrete sampling. While it is usually more efficient to use continuous (or analogue) sampling, sometimes it is necessary to employ discrete (or digital) sampling. We may calculate $\bar{M}_Z$, etc., in the latter case, following the procedures above for the former. For example, it can be shown that[11] a (time-) average value $M_{Z^{(j)}}$ of $Z^{(j)}(t)$, analogous to that produced by the ideal filter [Eqs. (16.10)], but now computed by sampling $Z^{(j)}(t)$ periodically (with period T_0), is

$$M_{Z^{(j)}} \to M_{Z^{(j)}}(n) = \frac{1}{n} \sum_{k=1}^{n} Z^{(j)}(kT_0) \tag{16.18}$$

Similarly, the first two moments of the distribution of $M_{Z^{(j)}}(n)$ can be found as in Secs. 16.1-1(1), 16.1-1(2), viz.,

$$\overline{M_Z(n)} = \frac{1}{n} \sum_{k=1}^{n} \overline{Z(kT_0)} = \bar{Z} \tag{16.19a}$$

$$\overline{\epsilon_{M_n}{}^2} = \overline{M_Z(n)^2} - \overline{M_Z(n)}^2 = \frac{\sigma_Z{}^2}{n} + \frac{2}{n^2} \sum_{k=1}^{n-1} (n-k)[K_Z(kT_0) - \bar{Z}^2] \tag{16.19b}$$

with $\sigma_Z{}^2 = K_Z(0) - \bar{Z}^2$. In the limits $n \to \infty$, $T_0 \to dt$, $kT_0 \to x$ (T fixed), we get formally the results above for the ideal integrator. Observe that the mean-square error or fluctuation is always *larger* for a finite number of sample elements in $(0,T)$ than for continuous sampling in the same interval. This is not surprising, since discrete sampling does not use all the available data about $Z(t)$ in the period $(0,T)$. We also observe that, if Z is entirely random and if $T_0 \gg B_Z{}^{-1}$, $K_Z(kT_0) - \bar{Z}^2$ is essentially zero and we get the familiar expression for the variance of the sum of n independent samples, namely, $\overline{\epsilon_{M_n}{}^2} \doteq \sigma_Z{}^2/n$. If Z contains a periodic component, with period commensurable with the sampling interval T_0, then $K_Z(kT_0) - \bar{Z}^2 \neq 0$ as $k \to \infty$ and so $\overline{\epsilon_{M_n}{}^2}$ is effectively independent of n and can be of the order of $\sigma_Z{}^2$.

† We remark, however, that an approach to the ideal integrator using fixed filter characteristics can be made if, for instance, the filtered data are passed through a recording meter; a "$1/T$" average can then be carried out on the recorded wave. "Second averaging" of this kind is also naturally done by the human observer up to the point where his memory fails.

16.1-2 An Example: Mean Intensity of a Random Wave. Let us illustrate the preceding discussion with an example of considerable practical interest. Specifically, we wish to measure the mean intensity of a stationary normal random process $y(t)$. Here, it is assumed that the process $y(t)$ is comparatively broadband and spectrally shaped by an *RC* filter, so that the covariance function is [cf. Eqs. (3.55)]

$$K_y(t) = K_y(0)e^{-\omega_F|t|} \qquad \omega_F = (R_F C_F)^{-1} \tag{16.20}$$

For our measuring device, let us also use a simple *RC* filter. For the measurement itself, we square the original wave and pass the result $Z^{(j)} = (y^{(j)})^2$ into this *RC* smoothing filter (cf. Fig. 16.1). The mean intensity in y is given simply by Eq. (16.4),

$$\overline{M_Z(T)} = \bar{Z}\lambda = \overline{y^2}(1 - e^{-\alpha T}) \qquad \alpha = 1/RC \tag{16.21a}$$

(cf. Table 16.1), and from Eq. (16.1) a measured value is simply

$$M_Z{}^{(j)}(T) = \boldsymbol{T}_L\{Z^{(j)}(T)\} = \alpha e^{-\alpha T} \int_0^T e^{\alpha u} Z^{(j)}(u)\, du \tag{16.21b}$$

Thus, by choosing a sufficiently long integration time $\alpha T(\gg 1)$, the average of the measured value (and the measured value itself) approaches the ensemble average (or infinite-time average, if the input is ergodic).

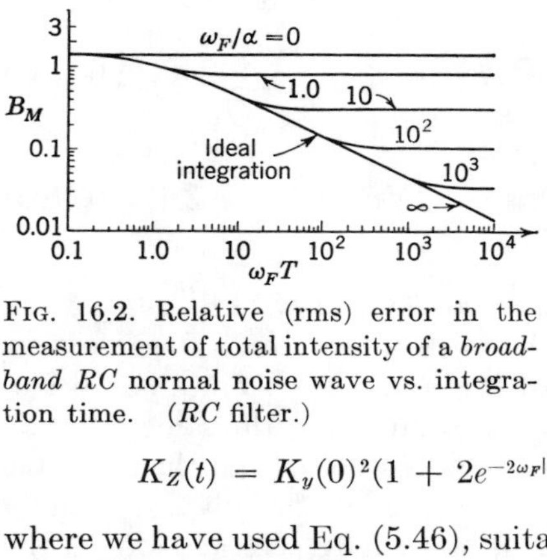

FIG. 16.2. Relative (rms) error in the measurement of total intensity of a *broadband RC* normal noise wave vs. integration time. (*RC* filter.)

The quantity of greatest interest, however, is the (rms) relative error [Eqs. (16.9)], which gives us an estimate of the expected spread in the observed values, $M(T)$, with respect to the "true"—i.e., the ensemble—average $\overline{M(T)}$. For this, we need

$$K_Z(t) = K_y(0)^2(1 + 2e^{-2\omega_F|t|}) \qquad K_Z(t) - \bar{Z}^2 = 2\bar{Z}^2 e^{-2\omega_F|t|} \tag{16.22}$$

where we have used Eq. (5.46), suitably modified for the case of normal noise $y(t)$, alone. We obtain directly

$$B_M = \frac{[(\tfrac{1}{2} + \omega_F/\alpha)^{-1} + e^{-2\alpha T}(\tfrac{1}{2} - \omega_F/\alpha)^{-1} - e^{-(\alpha + 2\omega_F)T}(\tfrac{1}{2} - \omega_F{}^2/\alpha^2)^{-1}]^{1/2}}{1 - e^{-\alpha T}} \tag{16.23}$$

[cf. Prob. 5.8, Eq. (2)]. Clearly, this measure of error depends only on the ratio of sample length T to the relevant (reciprocal) bandwidths of the input wave and averaging filter; B_M is shown in Fig. 16.2 as a function of the dimensionless integration time $\omega_F T$, with ω_F/α the ratio of bandwidth of the input wave to that of the integrating filter as parameter.

If we use the *ideal integrator* ($\lambda = 1$) [Eqs. (16.1), (16.10)], we obtain specifically

$$\frac{B_M = (2\omega_F T - 1 + e^{-2\omega_F T})^{1/2}/\omega_F T}{M_Z(T) = \bar{Z} = \overline{y^2}} \tag{16.24}$$

and, for long integration times,

$$B_M \bigg|_{T\to\infty} \simeq \sqrt{\frac{1}{\pi f_F T}} \qquad \omega_F = 2\pi f_F \tag{16.25}$$

To get some idea of the actual smoothing times involved, suppose that $f_F = 10^3$ cps and we wish $B_M \leq 1$ per cent. From Eq. (16.25), it follows at once that $T \geq 3$ sec. However, if the wave in question is spectrally very narrow, say $f_F = 1$ cps, we find that $T \doteq 50$ min, a smoothing period often beyond the interval during which stationarity of input and output can be maintained. If we are interested in, say, spectral intensities in bands 10^{-2} cps wide, with this same accuracy, T becomes even more astronomical [$T = O(400 \text{ days})$]! From this, we can begin to see how critical requirements of accuracy may be and how severely the available smoothing time T may limit it, depending on the bandwidth of the wave in question. For many kinds of measurements, the period during which stationarity can be maintained sets an upper bound to T and a lower one to the magnitude of B_M; for others (e.g., in certain radar and communication problems) T may be much shorter than the period of stationarity.

Equation (16.17) is convenient for comparing the values of the rms relative error when different low-pass integrating filters are used, in the limit of long observation times and very narrow filter bandwidths. For the problem examined above, we have $\mathcal{W}_{Z-\bar{Z}}(0+) = 4\bar{Z}^2/\omega_F$, from the transform of $K_{Z-\bar{Z}}(t)$ [Eq. (16.22)]. Inserting this into Eq. (16.17) gives

$$B_M = 2(B_e/\omega_F)^{1/2} \tag{16.26}$$

with B_e for various filters indicated in Table 16.1. This limiting expression is sufficiently accurate for design use in cases where the relative rms error is required to be O(10 per cent) or less. From Eqs. (16.25), (16.26), note again that the relative error can be made as small as we wish with ideal integration (Fig. 16.2) but approaches a finite value when fixed-constant smoothing systems [i.e., with finite (> 0) bandwidths] are used.

The preceding approach is, of course, not confined alone to square-law elements and measurements of mean intensity. As we remarked at the beginning of Sec. 16.1-1, the input $Z^{(j)}(t)$ to the final linear smoothing filter (cf. Fig. 16.1), which gives us our linear estimate $M^{(j)}(T)$ of $Z^{(j)}(t)$, may be derived in quite nonlinear fashion from some original source $y(t)$. $Z^{(j)}(t)$ itself may appear in the calculation of signal-to-noise ratios (cf. Sec. 5.3-4) or may represent the output of a correlator, a spectrum analyzer (cf. Sec. 5.2-3), and, in fact, any general network. A few generalizations of this type are given in the problems (cf. Prob. 16.1), where some nonlinear

TABLE 16.1. EFFECTIVE BANDWIDTH B_e, FINITE-TIME WEIGHTING, AND SYSTEM FUNCTIONS OF SEVERAL TYPES OF SMOOTHING FILTERS T_L

Smoothing filter	$h(t) \begin{cases} = 0, & t < 0- \\ \neq 0, & t > 0- \end{cases}$	λ	B_e	$Y_T(i\omega)$	$\rho_T(x),\ \|x\| \leq T$
1. Ideal integrator	$\frac{1}{T},\ 0 < t < T$	1	$\frac{1}{2T}$	$\frac{1 - e^{-i\omega T}}{i\omega}$	$\frac{1}{T}\left(1 - \frac{\|x\|}{T}\right)$
2. *RC* filter	$\alpha e^{-\alpha t}$	$1 - e^{-\alpha T}$	$\frac{\alpha}{4}$	$\frac{\alpha(1 - e^{-(\alpha + i\omega)T})}{\alpha + i\omega}$	$\frac{\alpha}{2} e^{-\alpha\|x\|}(1 - e^{-2\alpha T + 2\alpha\|x\|})$
3. Critically damped meter	$\mu^2 t e^{-\mu t}$	$1 - e^{-\mu T}(1 + \mu T)$	$\frac{\mu}{2}$	$\frac{\mu^2\{1 - e^{-T(\mu + i\omega)}[1 + T(\mu + i\omega)]\}}{(\mu + i\omega)^2}$	$\frac{\mu e^{-\mu\|x\|}}{2}\left(\frac{1}{2} + \mu\|x\|\right)$ $+ \frac{1}{2}\mu e^{-2\mu T + \mu\|x\|}[(\mu x)^2 - \frac{1}{2} - \mu T - \mu^2 T^2]$

elements other than the simple square-law device assumed above are employed.

We remark, finally, that the *interpretation* of the measurement $M^{(j)}(T)$ as some definite function of the statistics of the original process $y(t)$ naturally depends on the set of transformations $\boldsymbol{T}_L$, $\boldsymbol{T}_g$, etc., relating $y(t)$ and $Z(t)$ *and* on the general statistical character of $y(t)$. The latter, however, is often not available to the observer, who has only his measurements, and perhaps some knowledge of the transformations relating y and Z, to guide him. Under these conditions, it is then very difficult to obtain indications of the statistical properties of y: only by extensive measurements, using essentially a trial-and-error approach with different postulated models of the original input $y(t)$, can the observer expect to deduce the structure of this process. For these reasons, he is usually content with a few of the lower-order statistics, e.g., spectrum, covariance, and possibly some third- and fourth-order moments. There is as yet no satisfactory general approach to the problem of determining the nature of an original process from observations after known transformation of this process. A similar statement can be made for the related situation in which the structure of the input is given and the observer is asked to determine an unknown combination of linear and nonlinear transformations $\boldsymbol{T}$ which yields him a known set of final observations.

16.1-3 Optimum Linear Finite-time Measurements; Noise Alone. When the covariance function of the input $Z(t)$ to a linear smoothing filter is known, it is possible to choose this filter so as to minimize either the mean-square error $\overline{[M_Z^{(j)}(T) - \overline{M_Z(t)}]^2}$ or the relative mean-square error $B_M{}^2$ with respect to the "true" value $\overline{M_Z(t)}$, as distinguished from the estimate $M_Z^{(j)}(T)$ obtained on any one observation. We begin with the latter problem here, although, as we shall see presently, the two are essentially the same (provided that $\bar{M}_Z \neq 0$) and lead to similar results: both are examples of optimum linear estimation, minimizing a mean-square error.

Let us first consider the case where the input $Z(t)$ is a stationary, entirely random process. We require initially that optimization be subject to the constraint of Eqs. (16.6), i.e., that

$$\int_{-\infty}^{\infty} h_T(u)_{op}\, du = \int_{0-}^{T+} h_T(u)_{op}\, du = \lambda > 0 \qquad \lambda < \infty \tag{16.27}$$

where λ is some specified finite nonvanishing constant and h_T is zero outside $(0-,T+)$. Our task is to find the weighting function $(h_T)_{op}$ of the optimum smoothing filter so as to minimize $B_M{}^2$ [Eqs. (16.9)]:

$$B_M{}^2 = \frac{\overline{[M_Z^{(j)}(T) - \overline{M_Z(T)}]^2}}{\overline{M_Z(T)}^2} = \frac{\iint\limits_{-\infty}^{\infty} K_{Z-\bar{Z}}(|u-v|) h_T(u) h_T(v)\, du\, dv}{\bar{Z}^2 \iint\limits_{-\infty}^{\infty} h_T(u) h_T(v)\, du\, dv} \tag{16.28}$$

To do this, let us compute the variation of $B_M{}^2$ with respect to h_T, or, equivalently, the variation of the numerator of $B_M{}^2$, with the denominator as a constraint. Taking advantage of the symmetry of the kernel $K_{Z-\bar{Z}}$, we get for an extremum

$$\delta B_M{}^2 = 2\int_{0-}^{T+} \delta h_T(v)\,dv \int_{0-}^{T+} [K_{Z-\bar{Z}}(|u-v|) - \gamma\bar{Z}^2]h_T(u)\,du = 0 \tag{16.29}$$

from which the basic integral equation is

$$\int_{0-}^{T+} K_{Z-\bar{Z}}(|u-v|)h_T(u)_{op}\,du = \gamma\bar{Z}^2\lambda \equiv \Gamma \qquad 0- < v < T+ \tag{16.30}$$

Here, γ (or Γ) is an as yet undetermined (constant) multiplier, which may be found from Eq. (16.27) once $h_T(u)_{op}$ has been determined. Finally, we remark that when $K_{Z-\bar{Z}}$ is a rational kernel, i.e., when its Fourier transform is a rational function in $(i\omega)$, the integral equation may be solved by the procedures of Appendix 2.

Before we establish the condition that $h_T(t)_{op}$, when it exists, represents a solution for which the relative mean-square error $B_M{}^2$ is a minimum, let us determine the constant γ (or Γ). This is easily done if we multiply both members of Eq. (16.30) by $h_T(v)_{op}$ and integrate, using Eq. (16.27), to get

$$\iint_{-\infty}^{\infty} K_{Z-\bar{Z}}(|u-v|)h_T(u)_{op}h_T(v)_{op}\,du\,dv = \Gamma\lambda = \gamma\lambda^2\bar{Z}^2 = (B_M{}^2)_{ext}\lambda^2\bar{Z}^2 \tag{16.31}$$

this last from Eq. (16.28), also with the help of condition (16.27). Consequently, we have

$$\gamma = (B_M{}^2)_{ext} > 0 \tag{16.32}$$

With this, we can readily obtain the condition that $h_T(t)_{op}$ actually represents a minimum of $B_M{}^2$ [Eq. (16.28)]. Taking the second variation of Eq. (16.28) with respect to h_T, we get with the help of Eq. (16.30)

$$\delta^2(B_M{}^2) = 2\iint_0^T \delta h_T(u)[K_{Z-\bar{Z}}(|u-v|) - \bar{Z}^2\gamma]\delta h_T(v)\,du\,dv \tag{16.33a}$$

Using Eqs. (16.11), (16.12), (16.32) and the fact that $K_{Z-\bar{Z}}$ is symmetric *and* positive definite on $(0,T)$ [provided that δh_T is square integrable on $(0,T)$], we may rewrite Eq. (16.33a) as

$$\begin{aligned} \delta^2(B_M{}^2) &= \eta^2 \iint_0^T [K_{Z-\bar{Z}}(|u-v|) - \bar{Z}^2\gamma]\,du\,dv > 0 \\ \text{or} \qquad \delta^2(B_M{}^2) &= \eta^2[(B_M{}^2)_{\text{ideal}} - (B_M{}^2)_{ext}] > 0 \end{aligned} \tag{16.33b}$$

as the condition for a minimum, where η^2 (> 0) is some undetermined constant, depending on the particular variation δh_T chosen. For all the situations we shall usually encounter, Eqs. (16.33) apply, and the optimum filter indeed gives the desired minimum (cf. Prob. 16.2*c*). The example below is typical.

(1) An example. Let us consider again the example of Sec. 16.1-2 for the measurement of the intensity of a broadband stationary normal *RC* noise $y(t)$, with $(RC)^{-1} = \omega_F$. From Eqs. (16.22), we write for the kernel of the integral equation (16.30)

$$K_{Z-\bar{Z}}(|u - v|) = 2\bar{Z}^2 e^{-2\omega_F|u-v|} \tag{16.34}$$

Using Sec. A.2-3 (cf. Prob. 16.3*a*), we get, after a little manipulation,

$$h_T(t)_{op} = \begin{cases} \dfrac{\omega_F \Gamma}{2\bar{Z}^2} + \dfrac{\Gamma}{4\bar{Z}^2}[\delta(t-0) + \delta(t-T)] & 0- < t < T+ \\ 0 & \text{elsewhere} \end{cases} \tag{16.35a}$$

$$\begin{aligned} Y_T(i\omega)_{op} &= \int_{0-}^{T+} h_T(t) e^{-i\omega t}\, dt \\ &= \frac{\Gamma}{4\bar{Z}^2}\left[1 + e^{-i\omega T} + \frac{2\omega_F(1 - e^{-i\omega T})}{i\omega}\right] \end{aligned} \tag{16.35b}$$

From Eqs. (16.27) and (16.28), we find that

$$\Gamma = 2\lambda\bar{Z}^2(1 + \omega_F T)^{-1} \qquad (B^2)_{ext} = 2/(1 + \omega_F T) \tag{16.36}$$

so that Eq. (16.35*a*) becomes finally

$$h_T(t)_{op} = \begin{cases} \dfrac{\lambda\omega_F}{1 + \omega_F T} + \dfrac{\lambda}{2(1 + \omega_F T)}[\delta(t-0) + \delta(t-T)] & 0- < t < T+ \\ 0 & \text{elsewhere} \end{cases} \tag{16.37}$$

Since we do not know the epoch of the random wave (y or Z) relative to, say, the beginning of the observation (at $t = 0$), the solution that the best weighting of the input in the least mean-square-error sense above is a uniform one in the interval $(0,T)$ is not surprising. From a physical point of view, one consequently expects that all points within the interval are to be given the same emphasis. The delta functions at $t = 0$, T may be explained also on the basis of "equal weighting": at $t = 0$, the first *observed* value is completely uncorrelated with any other value of the wave. As seen by the filter, this point then belongs to a purely random process (cf. Sec. 1.4-2), whereas the actual wave observed in $(0,T)$ (and existing for all t, but unobserved outside this interval) must belong to a simple or higher-order

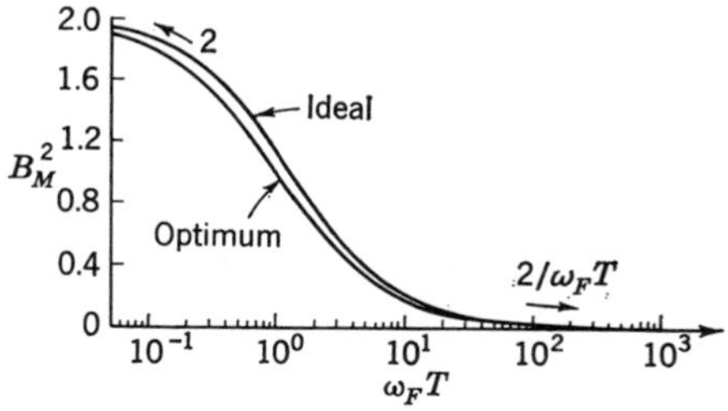

FIG. 16.3. Comparison of (relative) mean-square error for ideal and optimum filtering of a squared (stationary) *RC* broadband normal noise wave.

Markoff process, since $K_{Z-\bar{Z}}$ is finite. The delta-function weighting at $t = 0$ is needed to compensate for the filter's "interpretation" of the initial value. Since the input is stationary, this effect is symmetrical in time, accounting for the singularity in $h_T(t)_{op}$ at $t = T$ also.

The relative importance of these end points can be estimated by comparing the optimum and ideal filters for $B_M{}^2$. From Eqs. (16.24) and (16.36), we have

$$(B_M{}^2)_{ext} = \frac{2}{1 + \omega_F T} \qquad (B_M{}^2)_{\text{ideal}} = \frac{e^{-2\omega_F T} + 2\omega_F T - 1}{(\omega_F T)^2} \tag{16.38}$$

so that, as $\omega_F T \to 0$ and ∞, $(B_M{}^2)_{ext} \to (B_M)_{\text{ideal}}$, with the largest differences for $\omega_F T = O(10°)$, as shown in Fig. 16.3. Except at these limits, $(B_M{}^2)_{ext}$ is always less than $(B_M{}^2)_{\text{ideal}}$, and $h_T(t)_{op}$ does indeed yield a minimum mean-square relative error $(B_M{}^2)_{ext} = (B_M{}^2)_{\min}$, as expected. For most applications, however, the difference between the optimum and ideal systems may not be great enough to justify the additional complexity of the former (at $t = 0, T$), so that ideal integration may be used for essentially optimum performance.

(2) Extensions. A number of related results, generalizations, and special cases are easily obtained from the preceding analysis.

For example, let us suppose that, instead of a finite integration period $(0-,T+)$, the input $Z^{(j)}(t)$ has been applied steadily since $t = -\infty$ to the smoothing filter $h(t)$. Then, with the usual condition of physical realizability, that $h(t) = 0$ $(t < 0-)$, what now is the structure of the optimum filter that minimizes the relative mean-square error for measured values observed at $t = T$? We still postulate a constraint like that of Eq. (16.27) but leave open for the moment the conditions on λ that $\lambda \neq 0$, $\lambda < \infty$, imposed in the finite-sample case. The measured values of $Z(t)$ are given by Eq. (16.1), modified to

$$M^{(j)}(T) = \boldsymbol{T}_L\{Z^{(j)}(T)\} = \int_{0-}^{\infty} h(u) Z^{(j)}(T - u)\, du \tag{16.39}$$

Repeating the steps of Eqs. (16.28) to (16.30), we get formally, for the basic integral equation determining $h(t)_{op}$,

$$\int_{0-}^{\infty} K_{Z-\bar{Z}}(|u - v|) h(u)_{op}\, du = \gamma \bar{Z}^2 \lambda \qquad 0- < v < \infty \tag{16.40}$$

If λ is required to be greater than zero, then a nontrivial solution $(\neq 0)$ for $h(t)_{op}$ is found (for rational kernels) when $\lambda \to \infty$. Similarly, one has the trivial solution $h(t)_{op} = 0$ if $0 \leq \lambda < \infty$; see, for instance, Eqs. (16.37) in the example above. This is to be expected, since the "best" filter here is always one that essentially provides uniform weighting over the smoothing interval (apart from a bounded end effect at $t = T$; cf. Prob. 16.2a), so that condition (16.27) implies for $(0 \leq \lambda < \infty)$ an $h(t) \to O(1/\text{interval}) \to 0$ as the interval becomes indefinitely large.†

The present treatment may also be extended to include nonstationary input processes $Z(t)$, as well as the stationary ones postulated above. For specified filters [cf. Eq. (16.1)], Eqs. (16.4) and (16.5) become

$$\overline{M_Z(T)} = \int_{0-}^{T+} h(u)\overline{Z(T-u)}\,du \qquad \overline{M_Z(T)^2} = \iint_{0-}^{T+} h(u)K_Z(u,v)h(v)\,du\,dv \tag{16.41}$$

and the relative mean-square error is given by

$$B_M{}^2 = [\overline{M_Z(T)^2} - \overline{M_Z(T)}^2]/\overline{M_Z(T)}^2 \tag{16.42}$$

with the usual constraint in Eq. (16.4). The optimum linear filter here follows as the solution, if any, of

$$\int_{0-}^{T+} [K_Z(v,u) - (\gamma + 1)\overline{Z(T-u)}\,\overline{Z(T-v)}]h_T(u)_{op}\,du = 0 \qquad 0- < v < T+ \tag{16.43}$$

For the usual kernels K_Z, which are real, symmetric, and positive definite in $(0-,T+)$, however, one can show that there are no nonzero solutions to Eq. (16.43) (cf. Prob. 16.2*d*). This does not mean that there is no operator $h(t)$ which will minimize the relative mean-square error in this general nonstationary situation, only that this operation cannot be a linear time-invariant‡ one [cf. Eq. (16.1)]. We expect, rather, that a time-varying nonlinear operator is needed (cf. Sec. 21.4-1).

We remark finally that filters for minimizing the mean-square error $\overline{\epsilon_M{}^2} = \overline{[M_Z{}^{(j)}(T) - \overline{M_Z(T)}]^2}$ are identical with those for minimizing the relative mean-square error, subject to the constraint of Eq. (16.27), provided that $\bar{M}_Z \neq 0$. This is easily seen by a straightforward variational calculation (cf. Prob. 16.2*b*). Of course, when $\bar{M}_Z$ vanishes, we cannot use B_M as a criterion of performance. However, we can still minimize the mean-square error by proper choice of linear filters above, subject again to Eq. (16.27). We get for the basic integral equation here simply

$$\int_{0-}^{T+} K_Z(v,u)h_T(u)_{op}\,du = \gamma \qquad 0- < v < T+,\ \bar{M}_Z = 0 \tag{16.44}$$

where $\gamma = \overline{\epsilon_M{}^2}/\lambda$ $(0 < \lambda < \infty)$ [cf. Eq. (16.31)].

Note that now, unlike Eqs. (16.42) and (16.43) for $B_M{}^2$, we can obtain nontrivial solutions when the input process is not stationary.§ To see this,

† $h(t)$ may have a nonzero "end effect" in the neighborhood of $t = T$, particularly if the kernel $K_{Z-\bar{Z}}$ corresponds to a rational spectrum $\mathcal{W}_{Z-\bar{Z}}$ of higher order than the *RC* case [cf. Eq. (A.2.3)].

‡ Apart from simple switches at $t = 0, T$.

§ As long as K_Z is (real) symmetrical, and positive definite so that $\gamma > 0$, or, equivalently, as long as $\overline{\epsilon_M{}^2} > 0$.

let us separate $h_T(u)_{op}$ into a continuous part $h_T(u)_C$ and one containing only delta functions (and possibly their derivatives) at $t = 0, T$; for example, we write

$$h_T(u)_{op} = h_T(u)_C + \sum_{k=0}^{M} [A_k\delta^{(k)}(t - 0) + C_k\delta^{(k)}(t - T)] \tag{16.45}$$

where M (≥ 0) depends on the order of the kernel [see Eq. (A.2.3)]. Next, we apply Mercer's theorem (8.61) to (16.44); expanding $h_T(u)_C$ in the appropriate orthonormal set on $(0,T)$ obtained from

$$\int_0^T K_Z(v,u)\phi_k(u)\,du = \lambda_k\phi_k(v) \qquad 0 \leq v \leq T \tag{16.46a}$$

so that

$$h_T(u)_C = \sum_{k=1}^{\infty} b_k\phi_k(u) \qquad b_k = \int_0^T h_T(u)_C\phi_k(u)\,du \tag{16.46b}$$

we find that Eq. (16.44) becomes

$$\sum_k^{\infty} \int_0^T \lambda_k\phi_k(u)\phi_k(v)h_T(u)_C\,du + G(v) = \gamma \tag{16.47a}$$

or

$$\sum_k^{\infty} \lambda_k b_k\phi_k(v) + G(v) = \gamma \qquad 0 \leq v \leq T \tag{16.47b}$$

Here, $G(v)$ is the function of v that results when the integrations over u with the second term of $h_T(u)_{op}$ [cf. Eq. (16.45)] are performed in Eq. (16.44). Expanding γ and $G(v)$ on $0, T$ in the same orthonormal set, we get

$$\sum_{k=1}^{\infty} (\lambda_k b_k + c_k)\phi_k(v) = \sum_{k=1}^{\infty} d_k\phi_k(v) \tag{16.48}$$

or

$$b_k = (d_k - c_k)/\lambda_k$$

where

$$c_k = \int_0^T G(v)\phi_k(v)\,dv \qquad d_k = \int_0^T \gamma\phi_k(v)\,dv \tag{16.48a}$$

In general, $d_k - c_k \neq 0$, for some k, so that b_k is nonvanishing (for some k), and consequently $h_T(u)_C$ is specified [cf. Eq. (16.46b)]. To determine the arbitrary constants A_k, C_k in $G(v)$, we substitute the solution (16.46b), with b_k as given by Eq. (16.48), back into the original integral equation and treat the resulting expression as an identity [see Prob. 16.3 in the case of the RC kernel, Eq. (16.34), for example]. Finally, if the symmetrical kernel $K_Z(v,u)$ can be written in the form $A(u)A(v)H(|u - v|)$, typical, for example, of the random process due to certain types of ground-clutter return (cf. Probs. 11.10 to 11.14 and Sec. 20.4-6), we can use the results of Sec. A.2.3 for the stationary case to obtain a solution for the inhomogeneous integral equation (16.44) directly.

The preceding discussion, which has considered random processes with no deterministic components (other than the possible steady element

$\bar{Z} \neq 0$), may also be extended to include the cases where an estimate $M^{(j)}(T)_{S+N}$ is made on a wave consisting of a mixed process and we wish to determine the relative mean-square error, etc. Thus, $M^{(j)}(T)$ may consist of a signal and a noise background, when the signal structure is deterministic, with various random parameters whose values we may wish to measure. Another possibility is that the signal, like the noise, is another entirely random process, or that it is made up of both random and deterministic components. These situations may be handled with appropriate extensions of the procedures above. A number of results are given in Prob. 16.4. For further discussion of this topic, see Secs. 21.1, 21.3, and 21.4 and the references at the end of the chapter.

PROBLEMS

16.1 Stationary narrowband normal noise $y(t)$ is passed through a zero-memory νth-law rectifier. Show that the mean-square fluctuation in the l-f output process $Z(t)$ is, after smoothing for a period $(0,T)$,

$$\overline{\epsilon_M(T)^2} = 2^{\nu-2}\psi_y{}^\nu\beta^2\Gamma\left(\frac{\nu+1}{2}\right)^2\pi^{-1}\left\{\int_{-\infty}^{\infty}\rho_T(x)\ {}_2F_1\left[-\frac{\nu}{2},-\frac{\nu}{2};1;k_0{}^2(x)\right]dx - \lambda^2\right\} \quad (1)$$

where ρ_T is the truncated autocorrelation function of the smoothing filter in $(0,T)$. Hence, if the narrowband noise before rectification is such that the envelope of its covariance function corresponds to the high-Q single-tuned response $k_0(t) = e^{-\beta|t|}$, obtain the relative rms error

$$B_M = \sum_{n=1}^{\infty}\frac{(-\nu/2)_n{}^2}{(n!)^2}G_n(\beta T) \quad (2)$$

Show also that for

1. Ideal integrator

$$G_n(\beta T) = \frac{1}{n\beta T}\left(1+\frac{e^{-2n\beta T}-1}{2n\beta T}\right) \quad (3)$$

2. RC integrator

$$G_n(\beta T) = \frac{1-e^{-(2n\beta/\alpha+1)T\alpha}}{2n\beta/\alpha+1} - \frac{e^{-2T\alpha}-e^{-(2n\beta/\alpha+1)T\alpha}}{2n\beta/\alpha-1} \qquad \alpha = (RC)^{-1} \quad (4)$$

Sketch the behavior of B_M with βT, β/α as parameter.

16.2 (*a*) Show for general kernels (continuous everywhere) that, if the smoothing time T is long compared with the correlation time T_c of the input noise (considered in Sec. 16.1-3), the optimum filter for least rms relative error B_M is effectively the ideal filter [e.g., constant weighting $\sim 1/T$ in $(0,T)$], even though the precise form of the kernel may not be known. Verify the result for the filter [Eqs. (16.37)] (see Davenport, Johnson, and Middleton,[1] page 388).

(*b*) Show that when $\overline{M_Z(T)} \neq 0$ the linear filters minimizing mean-square error in the measurement $M_Z(T)$ are the same as those minimizing the relative rms error B_M, subject to the usual constraint [Eq. (16.27)].

(*c*) Prove that Eqs. (16.33) represent the necessary and sufficient condition for a minimum.

(*d*) Verify for the nonstationary cases [Eq. (16.42)] that there are no nonzero solutions to the integral equation (16.43) that are linear and time-invariant (apart from switches at $t = 0, T$). HINT: Use Mercer's theorem (8.61).

16.3 (*a*) For the *RC* kernel [Eq. (16.34)], carry out the solution of the basic integral equation (16.30) by the method of orthogonal functions, using the approach of Eqs. (16.45) to (16.48). Show that the complete orthonormal set $\{\phi_j\}$ here is specifically [cf. Eq. (16.46*a*)]

$$\phi_j(t) = [T/2 - (\sin 2\gamma_j T)/2\gamma_j]^{-\frac{1}{2}} \sin \gamma_j t \qquad j \geq 1 \tag{1}$$

where $\gamma_j = 2\omega_F q_j$ and the q_j are the positive roots of

$$\tan 2\omega_F T q_j = -2q_j/(1 - q_j^2) \tag{2}$$

(*b*) Show, therefore, that, alternative to Eqs. (16.37), one has

$$h(t)_{op} = \gamma \sum_{j=1}^{\infty} \frac{(1 + q_j^2)\omega_F}{2\bar{Z}^2} \frac{\sin 2\omega_F q_j t}{[(T/2 - \sin 4\omega_F q_j T)/4\omega_F q_j]^{\frac{1}{2}}} + \frac{\lambda\omega_F}{2(1 + \omega_F T)} [\delta(t - 0) + \delta(t - T)] \qquad 0- < t < T+ \tag{3}$$

and that $h(t)_{op} = 0$ for t outside the interval $(0-, T+)$; verify that $\gamma = B_M{}^2$.

16.4 (*a*) The process to be measured now consists not only of an entirely random noise $Z_N(t)$ but of a signal with a deterministic component $Z_S(t;\boldsymbol{\theta})$ and a random component $Z_{S_N}(t)$. If $Z(t) = Z_N + Z_S + Z_{S_N}$ is stationary, with $\bar{Z}_N \neq 0$ at least, show that the minimum relative mean-square error $(B_M{}^2)_{S+N}$ follows for that linear filter whose weighting function is the solution of

$$\int_{0-}^{T+} \{K_{Z-\bar{Z}}(u - v)_{S+N} - \overline{\gamma Z_S(T - u, \boldsymbol{\theta})[Z_S(T - v; \boldsymbol{\theta})} + \bar{Z}_{S_N} + \bar{Z}_N]\} h_T(u)_{op}\, du = \gamma\lambda(\bar{Z}_{S_N} + \bar{Z}_N)[\overline{Z_S(T - v; \boldsymbol{\theta})} + \bar{Z}_{S_N} + \bar{Z}_N] \qquad 0- < v < T+ \tag{1}$$

Verify that $\gamma = [(B_M{}^2)_{S+N}]_{\min}$.

(*b*) The input $Z(t)$ to our linear measuring device consists of the sum of a sinusoidal signal of frequency f_0 (and random phase) and a stationary broadband *RC* noise ($Z_{S_N} = 0$).

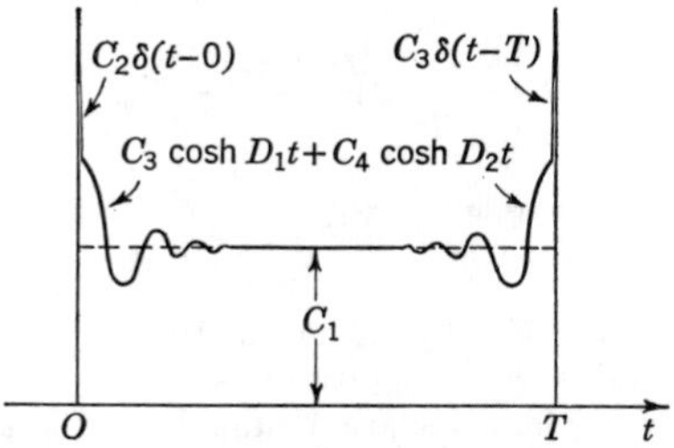

FIG. P 16.4. Weighting function of optimum linear smoothing filter for measuring intensity of a sine wave and broadband noise.

We wish a measure of the intensity of this wave in $(0,T)$. Show that the optimum filter has the weighting function

$$h(t)_{op} = \begin{cases} C_1 + C_2[\delta(t - 0) + \delta(t - T)] + C_3 \cosh D_1 t + C_4 \cosh D_2 t & 0- < t < T \\ 0 & t < 0-,\ t > T+ \end{cases} \tag{2}$$

Find the constants C_1, C_2, C_3, C_4, D_1, D_2, and calculate the system function $Y(i\omega)_{op}$. Figure P 16.4 shows a typical response. HINT: Use the continuity properties of the kernel, and follow the procedure of Prob. 16.3*a*.

16.2 *Optimum Linear Prediction and Filtering*

In the preceding sections, we have seen that by a proper choice of linear filter $\boldsymbol{T}_L$ it is possible with random processes possessing a suitable covariance function to minimize the mean-square or relative mean-square error in the estimation of the sample mean $\overline{M_Z(T)}$. Closely related problems are those of estimating the magnitude of the waveform (or some parameter of the waveform such as its average value, intensity, etc.) of a representative member of the process in question, for times t_λ both inside and outside the sample interval $(0,T)$. The former is often called *interpolation*, or smoothing, and the latter *extrapolation*, or *prediction* (if the time t_λ for which the estimate is desired lies in the "future" of the sample, if, for example, $t_\lambda > T$; cf. Fig. 16.4). The estimate, whether the result of smoothing,

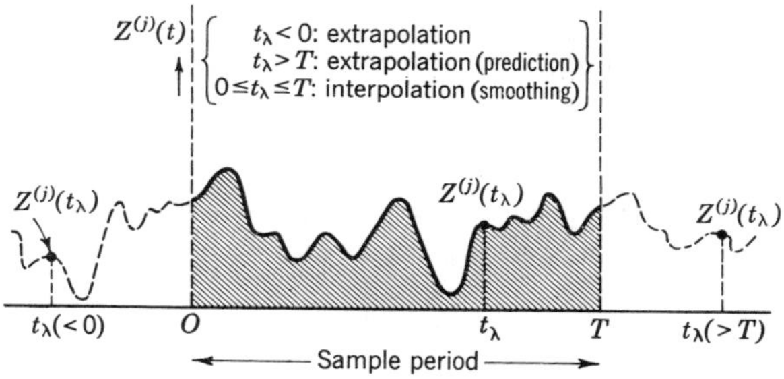

FIG. 16.4. Interpolation and extrapolation times t_λ for a typical sample of the process $Z(t)$.

prediction, or extrapolation, is, of course, based in some fashion on the observed waveform in the sample period $(0,T)$.

Here we seek, for the most part, time-invariant† linear realizable filters which will optimize the estimation process in the sense of minimizing a mean-square error. Other criteria are possible and often desirable (as we shall note in Sec. 21.2), but this one has the added advantage of analytical simplicity. Equally important where linear devices are required, the choice of a mean-square-error criterion is also sufficient‡ to give us a solution in terms of a linear system. Prediction and interpolation are but two of a variety of operations which we can perform with linear systems under this criterion: integration (smoothing or estimation of waveform at $t_\lambda = T$; cf. Sec. 16.1-3), differentiation (at various t_λ), extrapolation, in general, are other important examples. Here, we continue the discussion of optimum linear filtering begun in Sec. 16.1-3, extending it specifically to include cases of interpolation and extrapolation.

† Except, possibly, for simple switches at $t = 0$, T.

‡ In particular for normal processes. This criterion is, however, not always necessary: for certain classes of process (e.g., the normal process), one can find linear estimators that are simultaneously optimum, not only with respect to mean-square error, but for other criteria as well (e.g., maximum likelihood) (cf. Secs. 21.2 and 22.3-2).

16.2-1 Formulation. We consider now in more detail the class of linear-estimation problems described above. Specifically, we wish to modify in a linear fashion an original signal process† $S(t;\boldsymbol{\theta})$ in order to produce some desired output signal waveform $S_d(t;\boldsymbol{\theta})$. This output may represent a predicted value $S_d(t;\boldsymbol{\theta}) = S(t + t_\lambda, \boldsymbol{\theta})$ $(t_\lambda > 0)$, or it may represent a smoothed wave $(t_\lambda < 0)$ or one that is differentiated (say at $t_\lambda = 0$), etc. In any case, an output waveform is produced that is the result of some linear operation on the past and present of the original input.

An essential feature of these problems is the presence of noise $N(t)$, which makes it impossible for us to obtain the desired modified input signal exactly. We can only approximate it. Thus, a certain ensemble of errors

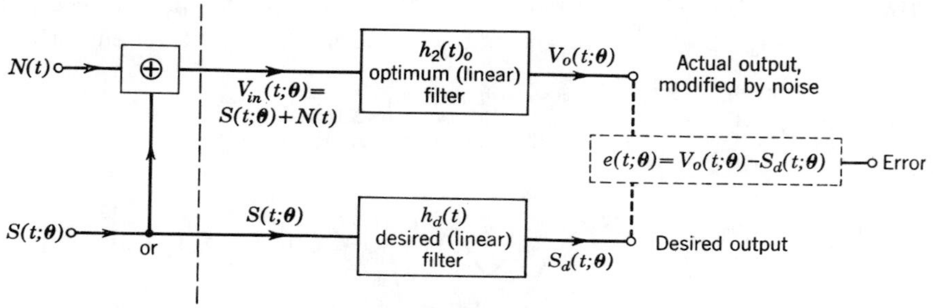

FIG. 16.5. Actual and desired outputs when the input wave consists of a pure signal or a signal contaminated by noise, $V_{in}(t,\boldsymbol{\theta})$.

between the actual output process $V_o(t;\boldsymbol{\theta})$ and the ideal, or desired, one $S_d(t;\boldsymbol{\theta})$ may be defined by

$$e(t;\boldsymbol{\theta}) \equiv V_o(t;\boldsymbol{\theta}) - S_d(t;\boldsymbol{\theta}) \tag{16.49}$$

If $V_{in}(t;\boldsymbol{\theta}) = S(t;\boldsymbol{\theta}) + N(t)$ represents the input process to the actual modifying filter $\boldsymbol{T}_L$, and if $h_\lambda(t)_0$ is the weighting function of this linear filter, then at time t the instantaneous output process is

$$V_o(t;\boldsymbol{\theta}) = \boldsymbol{T}_L\{V_{in}\}_0 = \int_{-\infty}^{t+} V_{in}(t - \tau; \boldsymbol{\theta}) h_\lambda(\tau)_0 \, d\tau \tag{16.50}$$

Letting $h_d(t)$ be the weighting function of the linear filter which is to produce the desired uncontaminated modification $S_d(t;\boldsymbol{\theta})$ of the original $S(t;\boldsymbol{\theta})$, we can also write

$$S_d(t;\boldsymbol{\theta}) = \boldsymbol{T}_L\{S\}_d = \int_{-\infty}^{t+} S(t - \tau; \boldsymbol{\theta}) h_d(\tau) \, d\tau \tag{16.51}$$

so that in terms of the various possible inputs the error, Eq. (16.49), becomes

$$e(t;\boldsymbol{\theta}) = \boldsymbol{T}_L\{V_{in}\}_0 - \boldsymbol{T}_L\{S\}_d = \int_{-\infty}^{t+} \{h_\lambda(\tau)_0[S(t - \tau; \boldsymbol{\theta}) + N(t - \tau)] - h_d(\tau) S(t - \tau; \boldsymbol{\theta})\} \, d\tau \tag{16.52}$$

The situation is illustrated in Fig. 16.5.

† Where $\boldsymbol{\theta}$, as before, represents a set of statistical parameters (cf. Secs. 1.3-5, 1.3-7) if S is deterministic.

Since $h_d(t)$ is given, i.e., since the form of the desired output is specified a priori, there remains to choose $h_\lambda(t)_0$ so that the error, Eq. (16.52), becomes as small as possible. The error cannot at all times be made to vanish, except in trivial cases, nor can the average error $\bar{e} = \bar{V}_o - \bar{S}_d$, unless both $\bar{V}_o$ and $\bar{S}_d$ are zero. However, if we require that the mean error $\bar{e}$ take some specified (positive or negative) value e_0, we may determine that linear filter $h_\lambda(t)_0$ which minimizes the spread in error about the mean. Taking the variance $\overline{e^2} - \bar{e}^2$ as the usual measure of spread, we may express the optimization in the usual way as a variational problem,

$$\delta(\overline{e^2} - \bar{e}^2 + \gamma\bar{e}) = 0 \tag{16.53}$$

where the variation is with respect to h_0 and γ is as yet an undetermined multiplier. For a minimum, we have in addition the condition $\delta^2(\overline{e^2} - \bar{e}^2 + \gamma\bar{e}) > 0$.

We may now use Eq. (16.52) to write the variance directly in terms of the covariance functions of the input or, as is frequently the case, in terms of their correlation functions on the additional assumption that the purely random components of the input wave are ergodic. We obtain

$$\begin{aligned}\overline{e^2} - \bar{e}^2 = \iint_{-\infty}^{t+} \{ &[h_\lambda(\tau_1)_0 - h_d(\tau_1)][h_\lambda(\tau_2)_0 - h_d(\tau_2)]K_{S-\bar{S}}(\tau_1,\tau_2;t) \\ &+ h_\lambda(\tau_1)_0 h_\lambda(\tau_2)_0[K_{N-\bar{N}}(\tau_1,\tau_2;t) + K_{N-\bar{N},S-\bar{S}}(\tau_1,\tau_2;t) \\ &+ K_{S-\bar{S},N-\bar{N}}(\tau_1,\tau_2;t)] - h_d(\tau_1)h_\lambda(\tau_2)_0 K_{S-\bar{S},N-\bar{N}}(\tau_1,\tau_2;t) \\ &- h_\lambda(\tau_1)_0 h_d(\tau_2)K_{N-\bar{N},S-\bar{S}}(\tau_1,\tau_2;t)\}\, d\tau_1\, d\tau_2 \end{aligned} \tag{16.54}$$

where the autovariance functions are specifically

$$K_{S-\bar{S}} = \boldsymbol{E}_{S,\boldsymbol{\theta}}\{S(t-\tau_1;\boldsymbol{\theta})S(t-\tau_2;\boldsymbol{\theta}) - \overline{S(t-\tau_1;\boldsymbol{\theta})}\,\overline{S(t-\tau_2;\boldsymbol{\theta})}\} \tag{16.55a}$$

$$K_{N-\bar{N}} = \boldsymbol{E}\{N(t-\tau_1)N(t-\tau_2) - \overline{N(t-\tau_1)}\,\overline{N(t-\tau_2)}\} \tag{16.55b}$$

while the cross variances are

$$K_{S-\bar{S},N-\bar{N}} = \boldsymbol{E}_{S,\boldsymbol{\theta},N}\{[S(t-\tau_1;\boldsymbol{\theta}) - \overline{S(t-\tau_1;\boldsymbol{\theta})}][N(t-\tau_2) - \overline{N(t-\tau_2)}]\} \tag{16.55c}$$

$$K_{N-\bar{N},S-\bar{S}} = \boldsymbol{E}_{S,\boldsymbol{\theta},N}\{[N(t-\tau_1) - \overline{N(t-\tau_1)}][S(t-\tau_2;\boldsymbol{\theta}) - \overline{S(t-\tau_2;\boldsymbol{\theta})}]\} \tag{16.55d}$$

(The statistical averages are over the random parameters $\boldsymbol{\theta}$, if any, as well as the processes N, S, etc.). The expression (16.54), with $\bar{e} \neq 0$ [cf. Eq. (16.52)], is the general relation, special cases of which we shall consider in Sec. 16.2-2.

16.2-2 Noise Signals in Noise Backgrounds; The Theory of Wiener[2]† and Kolmogoroff.[3] In this theory, the signal $S(t)$ is assumed to have sta-

† See also Wiener's original report (with the same title), No. DIC-6037, sec. D_2, Massachusetts Institute of Technology, Feb. 1, 1942. See Wiener,[2] preface and footnote, p. 59.

tistical properties only; i.e., it is an entirely random process. A background or interfering noise $N(t)$, also entirely random, accompanies this signal additively and may be statistically related to it. It is further assumed that data acquisition and processing take place only over the past of the input processes, which are themselves postulated to be wide-sense stationary with positive definite covariance functions. Moreover, for the most part optimum linear filters are sought which are not only realizable but time-invariant as well (for exceptions, see Secs. 16.2-3, 16.2-4). Thus, the general expressions (16.50), (16.51) become here

$$V_o(t) = \int_{-\infty}^{\infty} V_{in}(t-\tau)h_\lambda(\tau)_0\,d\tau \qquad S_d(t) = \int_{-\infty}^{\infty} S(t-\tau)h_d(\tau)\,d\tau \quad (16.56)$$

and from Eq. (16.52) it is also evident that $\bar{e} = 0$. Accordingly, the quantity to be minimized is given by Eq. (16.54), with appropriate modifications of the covariance functions (16.55), namely, $K_{S-\bar{S}} = K_S$ ($\bar{S} = 0$), $K_{N-\bar{N}} = K_N$ ($\bar{N} = 0$), etc.

Let us now carry out the variation of $\overline{e^2}$ with respect to the weighting function h_0, setting the former equal to zero for an extremum. Using the fact that the covariance functions are symmetrical in their arguments, we obtain

$$\begin{aligned}\delta\overline{e^2} = 2\int_{-\infty}^{\infty} \delta h_\lambda(\tau_1)_0\,d\tau_1 \int_{-\infty}^{\infty} d\tau_2\{[h_\lambda(\tau_2)_0 - h_d(\tau_2)]K_S(\tau_1-\tau_2)\\ + [h_\lambda(\tau_2)_0 - h_d(\tau_2)]K_{SN}(\tau_1-\tau_2)\\ + h_\lambda(\tau_2)_0\,K_{NS}(\tau_1-\tau_2) + h_\lambda(\tau_2)_0K_N(\tau_1-\tau_2)\} = 0 \quad (16.57)\end{aligned}$$

and so the condition determining h_0 is†

$$\int_{-\infty}^{\infty} K_{S+N}(t-u)h_\lambda(u)_0\,du = \int_{-\infty}^{\infty} h_d(u)K_{SV_{in}}(t-u)\,du \equiv G(t) \quad (16.58a)$$

where

$$K_{S+N}(t) = K_S(t) + K_{SN}(t) + K_{NS}(t) + K_N(t) \qquad K_{SV_{in}}(t) = K_S(t) + K_{SN}(t) \quad (16.58b)$$

since $V_{in} = S + N$. More compactly, the integral equation (16.58a) can be written‡

$$\int_{-\infty}^{\infty} K_{S+N}(t-u)h_\lambda(u)_0\,du = G(t) \qquad 0- < t < \infty \quad (16.59)$$

where $G(t)$ is defined over the *past*, only, of the input, in accordance with the conditions above. The lower limit on the integral in Eq. (16.59), in conjunction with the definition of $G(t)$ in the right member, for $(0- < t < \infty)$,

† The function $h_\lambda(u)_0$ depends, of course, on the parameters of the desired filtering operation $h_d(u)$, for example, on the time t_λ for which an estimate is to be computed and on the nature of the operation $h_d(u)$ itself.

‡ If both signal and noise are, in addition, assumed to be ergodic, we may replace the covariance functions by their appropriate correlation functions (probability 1), in the usual way (cf. Sec. 1.6).

expresses the requirement that $h_\lambda(t)_0$ represent a *realizable* linear (and, for the most part, time-invariant) filter. Note that the second variation of Eq. (16.59) gives

$$\delta^2(\overline{e^2}) = \iint_{-\infty}^{\infty} \delta h_0(t) K_{S+N}(t-u) \delta h_0(u) \, dt \, du > 0$$

for the processes assumed here (since K_{S+N} is positive definite), so that Eq. (16.57) does indeed yield a minimum, as expected. It can also be shown that Eq. (16.59) is the necessary and sufficient condition for the minimum (cf. Prob. 16.5).

The solution of the integral equation† (16.59) when the kernel K_{S+N} is rational may be carried out by the methods of Appendix 2 (in the limit $T \to \infty$). We shall not go into the details here. However, it is instructive to point out a few of the considerations that are needed for physically realizable solutions. We observe first that, as in all equations of the type (16.58*a*) and, for that matter, like Eqs. (16.30), (16.40), (16.43), (16.44), the result of the integration is equal to some as yet unspecified function $H(t)$ outside the interval in question. Here, $H(t)$ is generally different from zero ($t < 0-$) and is such that $H(t) = 0$ ($t > 0-$). The requirement of realizability, expressed mathematically in the operation of finding $H(t)$, establishes a certain condition on the covariance function $K_{S+N}(t)$, or, equivalently, on the intensity spectrum $\mathcal{W}_{S+N}(f)$. Specifically, the condition on the spectrum $\mathcal{W}_{S+N}(f)$ is that it be a rational fraction in (the complex variable) p and that it can be factored into two parts, in the complex p plane, viz.,

$$\mathcal{W}_{S+N}(p/2\pi i) = \Phi(p)\Phi(-p) \tag{16.60}$$

such that $\Phi(p)$ has poles *and* zeros only in the left-half p plane, so that consequently $\Phi(-p)$ has them only in the right half plane. In other words, $\mathcal{W}_{S+N}(p/2\pi i)$ is of the form

$$\Phi(p)\Phi(-p) = \frac{\prod_{k=1}^{m} (c_k^+ + p) \prod_{k=1}^{m} (c_k^- - p)}{\prod_{k=1}^{n} (d_k^+ + p) \prod_{k=1}^{n} (d_k^- - p)} \tag{16.60a}$$

where c_k^+, d_k^+ and c_k^-, d_k^- are, respectively, the zeros and poles of $\mathcal{W}_{S+N}$ in the left- and right-half complex planes. In the time domain, the condition (16.60) takes the equivalent form

$$K_{S+N}(t) = \int_{-\infty}^{\infty} \phi_2(x)\phi_1(t-x)\,dx \qquad \phi_1(t) = 0,\ t < 0-;\ \phi_2(t) = 0,\ t > 0- \tag{16.60b}$$

† This is a Wiener-Hopf integral equation of the first kind (cf. Wiener,[2] app. C, footnote, sec. 2).

where, in fact, $\Phi(p)$ is the Fourier transform of ϕ_1 and $\Phi(-p)$ of ϕ_2. Our ability to factor $\mathcal{W}_{S+N}(p/2\pi i)$, thus determining $H(t)$ and permitting the solution of the integral equation (16.59) by the usual Fourier-transform methods (cf. Sec. A.2.3), depends on whether or not $\mathcal{W}_{S+N}(f)$ satisfies the Paley-Wiener[13] criterion

$$J \equiv \int_{-\infty}^{\infty} \frac{|\log \mathcal{W}_{S+N}(f)|}{1 + \omega^2} \, df < \infty \qquad \omega = 2\pi f \tag{16.60c}$$

From the above, it is clear that not any covariance function $K_{S+N}(t)$ or spectrum $\mathcal{W}_{S+N}(f)$ will give a physically realizable solution to Eq. (16.59) for the minimization of mean-square error, but only those which obey Eq. (16.60c). For example, if the intensity spectrum has the form $(a^2 + \omega^2)^{-1}$, characteristic of RC noise [cf. Eqs. (3.55)], then the integral J above is convergent; $\mathcal{W}_{S+N}(f)$ can indeed be factored, namely, $\Phi(p) = (a + p)^{-1}$ $(a > 0)$, and $\mathcal{W}_{S+N}(p/2\pi i) = (a + p)^{-1}(a - p)^{-1}$. On the other hand, if the approach to zero of $\mathcal{W}_{S+N}$ as $f \to \infty$ is too rapid, as in the case for the spectrum $\mathcal{W}_{S+N}(f) \sim e^{-\omega^2}$, for instance, then J is infinite and factorization is not possible. This in turn means that there is no physically realizable (time-invariant) linear filter which will minimize the mean-square error.

The criterion (16.60c) also determines whether or not a wide-sense stationary random process is deterministic. For it can be shown† that if, and only if, the integral J converges [cf. Eq. (16.60c)] the random process in question is nondeterministic. If it does not converge, the process is deterministic and hence (by definition of a deterministic process; cf. Sec. 1.3-4) any future value can be predicted exactly by only a linear (invariant) operation on the past of the process.

To illustrate this, let us consider a simple example.‡ We take first the real process $S(t) = A_0 \cos(\omega t + \phi)$ $(\omega = 2\pi f)$, of fixed amplitude, where ω (or f) and ϕ are independent random variables, with ϕ uniformly distributed in the primary interval $(0 \leq \phi < 2\pi)$, so that $S(t)$ is stationary (cf. Sec. 1.3-6). We find directly that $K_S(t) = (A_0{}^2/2)\langle \cos \omega t \rangle_f = (A_0{}^2/2) \int_{-\infty}^{\infty} \cos \omega t w(f) \, df$. Let us require further that $w(f)$ is an even but arbitrary function of the random variable f. Since $K_S(t) = \frac{1}{2} \int_{-\infty}^{\infty} \cos \omega t \mathcal{W}_S(f) \, df$ by the W-K theorem (3.42), comparing the two expressions for $K_S(t)$ shows that $\mathcal{W}_S(f) = A_0{}^2 w(f)$. Therefore, by choosing $w(f)$ to satisfy the condition (16.60c), we ensure the nondeterminancy of the process $S(t)$. On the other hand, by taking $w(f)$ such that the integral J does not converge, we obtain an exactly predictable deterministic process.§

† Wiener,[2] chap. 3, sec. 2.4; also J. L. Doob, "Stochastic Processes," chap. XII, sec. 5, theorem (5.1), Wiley, New York, 1950.

‡ Due to Davenport and Root,[12] sec. 11.3.

§ We remark that in the former case $S(t)$ can also be predicted exactly, on the basis of its past, but not by a linear operation.

Great simplification occurs, on the other hand, if we relax the condition on $h_\lambda(u)_0$ for realizability, and correspondingly, on $\mathcal{W}_{S+N}(f)$ above, the integral equation (16.59) becomes the much more elementary relation

$$\int_{-\infty}^{\infty} h_\lambda(u)_0 K_{S+N}(t-u)\,du = G(t) \qquad -\infty < t < \infty \tag{16.61}$$

This is formally solved at once by taking the Fourier transform of both sides. The result is

$$Y_\lambda(p)_0 = \frac{Y_d(p)[\mathcal{W}_S(p/2\pi i) + \mathcal{W}_{NS}(p/2\pi i)]}{\mathcal{W}_{S+N}(p/2\pi i)} \tag{16.62a}$$

where $Y_\lambda(p)_{0,d}$ are the transfer functions [cf. Eq. (2.56)] of the optimum and desired filters, respectively. The weighting function corresponding to $Y(p)_0$ is

$$h_\lambda(t)_0 = \int_{-\infty i}^{\infty i} \frac{e^{pt} Y_d(p)[\mathcal{W}_S(p/2\pi i) + \mathcal{W}_{NS}(p/2\pi i)]\,dp}{2\pi i \mathcal{W}_{S+N}(p/2\pi i)} \tag{16.62b}$$

Now, it is *not* true in general that $h_\lambda(t)_0$ vanishes for $t < 0-$, and so $h_\lambda(t)_0$ is not physically realizable. As we can see from the preceding discussions, it is the condition of realizability that produces the complexities of this theory. For example, the solution to Eq. (16.61) operates on the future of the received wave, as well as on its past. However, this weighting function can be approximated by realizable elements although, of course, the resulting operation is not strictly an optimum one.[14]

In the above, the integral equation (16.59) has been obtained for the weighting function of the optimum filter by minimizing the mean-square error $\overline{e^2}$ in the time domain [cf. Eqs. (16.54), (16.57), with $\bar{e} = 0$ here]. An equivalent alternative result can be derived for the system function $Y_\lambda(i\omega)_0$ of this optimum filter if we carry out the variation of $\overline{e^2}$ with (16.54) expressed now in the frequency domain with the aid of the W-K theorem (3.42). Thus, remembering that $Y_d(i\omega)$ is $\mathfrak{F}\{h_d(t)\}$ and $Y_\lambda(i\omega)_0$ is $\mathfrak{F}\{h_\lambda(t)_0\}$ [cf. Eqs. (2.53)] and that the input processes are stationary, with $\bar{N} = \bar{S} = 0$, we take the variation of $\overline{e^2}$ with respect to $Y_0\,[\equiv Y_\lambda(i\omega)_0]$ and Y_0^*, to obtain

$$\begin{aligned}\delta\overline{e^2} = \tfrac{1}{2}\int_{-\infty}^{\infty} (\delta Y_0^*\{Y_\lambda(i\omega)_0\mathcal{W}_{S+N}(f) - Y_d(i\omega)[\mathcal{W}_S(f) + \mathcal{W}_{NS}(f)]\}\\ + \delta Y_0\{Y_\lambda(i\omega)_0^*\mathcal{W}_{S+N}(f) - Y_d(i\omega)^*[\mathcal{W}_S(f) + \mathcal{W}_{SN}(f)]\})\,df\end{aligned} \tag{16.63}$$

where† as before $\mathcal{W}_{S+N}(f) = \mathcal{W}_S(f) + \mathcal{W}_{SN}(f) + \mathcal{W}_{NS}(f) + \mathcal{W}_N(f)$.

We recall now that, for physical realizability of the filter, the spectrum $\mathcal{W}_{S+N}(f)$ must satisfy the condition of factorability [cf. Eqs. (16.60), (16.60*a*), (16.60*c*)]. Inserting Eq. (16.60) with $f \to p/2\pi i$, we evaluate the resulting integrals in Eqs. (16.62), after splitting the integrand into two

† We need ultimately consider only the first integral, as the integrand of the second is simply the complex conjugate of the first, if we remember that $\mathcal{W}_{SN}(f) = \mathcal{W}_{NS}(f)^*$ and that $\mathcal{W}_{SN}(f) + \mathcal{W}_{NS}(f)$ is real [cf. Eq. (3.149*a*)].

parts with poles only in the right-half or left-half p plane. We obtain finally for $\delta\overline{e^2} = 0$ the desired result

$$Y_\lambda(p)_0 = \left[Y_d(p) \frac{\mathcal{W}_S(p/2\pi i) + \mathcal{W}_{NS}(p/2\pi i)}{\Phi(-p)} \right]_+ \frac{1}{\Phi(p)} \tag{16.64}$$

Here, $[\quad]_+$ indicates that this expression has poles in the left-half p plane only. Because of this, and the fact that $\Phi(p)$ has *zeros* only in this region [cf. Eq. (16.60*a*)], the weighting function [cf. Eq. (2.56)]

$$h_\lambda(t)_0 = \int_{-\infty i}^{\infty i} Y_\lambda(p)_0 e^{pt} \frac{dp}{2\pi i} \tag{16.65}$$

does indeed vanish for $t < 0-$, as required by our originally imposed condition of physical realizability. Thus, to obtain the weighting function of the optimum filter, we first find the factors of $\mathcal{W}_{S+N}(p/2\pi i)$, according to Eq. (16.60), and then determine the terms $[\quad]_+$ (with poles only in the left-half plane). Equation (16.65), with Eq. (16.64), then gives us $h_\lambda(t)_0$. Finding the terms $[\quad]_+$ may also be achieved analytically by evaluating

$$Y_\lambda(p)_0 = \frac{1}{\Phi(p)} \int_0^\infty e^{-pt}\, dt \int_{-\infty i}^{\infty i} Y_d(p) \frac{\mathcal{W}_S(p/2\pi i) + \mathcal{W}_{NS}(p/2\pi i)}{\Phi(-p)} \frac{dp}{2\pi i} \tag{16.66}$$

directly, but it is usually simpler to use Eq. (16.64).

Comparison of Eqs. (16.64), (16.65) with Eqs. (16.62*a*), (16.62*b*) shows at once the modification that physical realizability imposes on the spectrum $\mathcal{W}_{S+N}$, namely, Eq. (16.60): the weighting function in the latter case is nonvanishing for *all* $(-\infty < t < \infty)$, in general, unlike $h_\lambda(t)_0$ determined according to Eqs. (16.64), (16.65). Moreover, even if there is no background noise, the optimum filter is not usually the same as the desired response, as the example of Sec. 16.2-3(1) shows. This is also true, of course, when there is an accompanying noise, and, in fact, the optimum filters here are also quite different from the desired filter, particularly if the noise is at all intense relative to the "signal" (cf. Prob. 16.6*c*).

The minimum mean-square error may now be obtained from Eqs. (16.58*a*) in the appropriate modification of Eq. (16.54) for $\overline{e^2}$ (with $\bar{e} = 0$ here). We easily find that in terms of the auto- and cross-correlation functions of the optimum and desired filters, for example, $\rho_\lambda(\tau)_o$, $\rho(\tau)_d$, $\rho_\lambda(\tau)_{od}$, we can write

$$\overline{e^2} = \int_{-\infty}^{\infty} \rho(\tau)_o K_{S+N}(\tau)\, d\tau - 2 \int_{-\infty}^{\infty} [K_S(\tau) + K_{SN}(\tau)] \rho_\lambda(\tau)_{od}\, d\tau + \int_{-\infty}^{\infty} \rho(\tau)_d K_S(\tau)\, d\tau \tag{16.67}$$

where we have taken advantage of the fact that $K_{NS}(\tau) = K_{SN}(-\tau)$, $\rho_\lambda(\tau)_{do} = \rho_\lambda(-\tau)_{od}$, $\rho_\lambda(\tau)_o = \rho_\lambda(-\tau)_o$, etc. The filter's correlation functions

ρ are specifically here (cf. Sec. 3.4-2)

$$\begin{aligned} \rho_\lambda(\tau)_o &\equiv \int_{-\infty}^{\infty} h_\lambda(u)_0 h_\lambda(u+\tau)_0\, du \\ \rho(\tau)_d &\equiv \int_{-\infty}^{\infty} h_d(u) h_d(u+\tau)\, du \\ \rho_\lambda(\tau)_{od} &\equiv \int_{-\infty}^{\infty} h_\lambda(u)_0 h_d(u+\tau)\, du \end{aligned} \tag{16.68}$$

Next, we multiply both sides of the integral equation (16.59) by $h_\lambda(t)_0$ and integrate for *all* t, remembering again that $h_\lambda(t)_0 = 0$ $(t < 0-)$. The result becomes (for the *optimum* filter h_0)

$$\int_{-\infty}^{\infty} \rho_\lambda(\tau)_o K_{S+N}(\tau)\, d\tau = \int_{-\infty}^{\infty} \rho_\lambda(\tau)_{od}[K_S(\tau) + K_{SN}(\tau)]\, d\tau \tag{16.69}$$

which, when inserted into Eq. (16.67) gives us two equivalent expressions for the minimum mean-square error:

$$\overline{e^2_{\min}} = \int_{-\infty}^{\infty} \rho(\tau)_d K_S(\tau)\, d\tau - \int_{-\infty}^{\infty} [K_S(\tau) + K_{SN}(\tau)]\rho_\lambda(\tau)_{od}\, d\tau \tag{16.70a}$$

or

$$\overline{e^2_{\min}} = \int_{-\infty}^{\infty} \rho(\tau)_d K_S(\tau)\, d\tau - \int_{-\infty}^{\infty} \rho_\lambda(\tau)_o K_{S+N}(\tau)\, d\tau \qquad (\geq 0) \tag{16.70b}$$

We can also express the minimum mean-square error in terms of the associated spectra and the system functions of the various filters, if we recall (cf. Sec. 3.4-2) that

$$\begin{aligned} \rho_\lambda(\tau)_o &= \int_{-\infty}^{\infty} |Y_\lambda(i\omega)_o|^2 e^{i\omega\tau}\, df \\ \rho(\tau)_d &= \int_{-\infty}^{\infty} |Y_d(i\omega)|^2 e^{i\omega\tau}\, df \qquad \omega = 2\pi f \\ \rho_\lambda(\tau)_{od} &= \int_{-\infty}^{\infty} Y_\lambda(i\omega)_o Y_d(i\omega)^* e^{-i\omega\tau}\, df \end{aligned} \tag{16.71}$$

Applying these to Eqs. (16.70*a*), (16.70*b*) gives alternatively

$$\overline{e^2_{\min}} = \tfrac{1}{2} \int_{-\infty}^{\infty} |Y_d(i\omega)|^2 \mathcal{W}_S(f)\, df - \tfrac{1}{2} \int_{-\infty}^{\infty} [\mathcal{W}_S(f) + \mathcal{W}_{SN}(f)] Y_\lambda(i\omega)_o Y_d(i\omega)^*\, df \tag{16.72a}$$

or

$$\overline{e^2_{\min}} = \tfrac{1}{2} \int_{-\infty}^{\infty} |Y_d(i\omega)|^2 \mathcal{W}_S(f)\, df - \tfrac{1}{2} \int_{-\infty}^{\infty} \mathcal{W}_{S+N}(f) |Y_\lambda(i\omega)_o|^2\, df \tag{16.72b}$$

Since each of the integrals in Eqs. (16.70*b*), (16.72*b*), for example, represents a mean total output intensity for its respective filter, we may say that $\overline{e^2}$ is the difference between the average output powers (or intensities) of the desired filter when the signal is uncontaminated by noise and the average output power of the optimum filter in the actual case where the signal is accompanied by a background noise. The latter is always equal to or less than the former (since $\overline{e^2_{\min}} \geq 0$), and usually is different from zero, because of the interference.

Extensive application of these ideas to practical problems involving optimized linear smoothing and prediction has been carried out during the last decade.[4,5,7–9,15,16] This approach has also been applied to servomechanisms,[8,9,17,18] and a number of generalizations to take account of finite observation periods, nonstationary inputs, and more sophisticated signal structures have recently extended the theory to a wider class of problems,[4,5,19–22] a brief reference to which is given in Sec. 16.2-4.

16.2-3 Some Examples. We begin with several cases of optimum linear prediction (extrapolation) and smoothing (interpolation) by realizable time-invariant filters. As in Sec. 16.2-2, the signal and noise processes are here additive, entirely random, and wide-sense stationary, and the various filtering operations make use of the entire past $(-\infty, t)$ of the available waveforms. Additional examples are given in the problems at the end of Sec. 16.2-4.

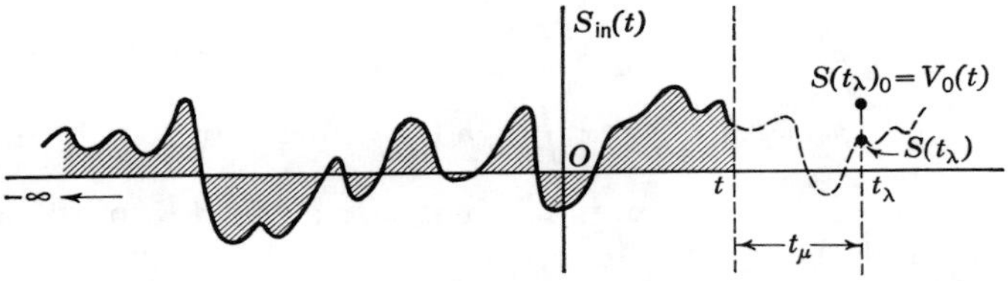

FIG. 16.6. Prediction of a random waveform at $t_\lambda > t$ (t_μ = lead time).

(1) **Optimum linear prediction** $(t_\lambda > t)$. In this case, we set the desired signal $S_d(t)$ equal to the actual signal at some time in the observer's future $(t_\lambda > t)$, in the manner of Fig. 16.6, where now t represents the observer's present. From Eqs. (16.56), we have accordingly

$$S_d(t) = S(t_\lambda) = \int_{-\infty}^{\infty} S(t - \tau)h_d(\tau)\, d\tau \tag{16.73}$$

and also, from Eqs. (16.56), the optimum prediction $S(t_\lambda)_0$ is simply

$$V_o(t) \equiv S(t_\lambda)_0 = \int_{-\infty}^{\infty} V_{in}(t - \tau)h_\lambda(\tau)_0\, d\tau \tag{16.74}$$

where $V_{in} = S + N$ as usual. The desired weighting function h_d, which yields the true predicted value $S(t_\lambda)$, is clearly

$$h_d(\tau) = \delta(\tau + t_\lambda - t) = \delta(\tau + t_\mu) \qquad t_\mu = t_\lambda - t\ (> 0) \tag{16.75a}$$

with system function

$$Y(i\omega)_d = e^{i\omega t_\mu} \tag{16.75b}$$

where t_μ is the *lead time* of the predicted value over the actual value $V_o(t) = S(t_\lambda)_0$ available at the observer's present $(t = t)$ as a result of optimum filtering on the entire past of the input. Inserting Eq. (16.75a) into the basic integral equation (16.59) gives us at once the relation

$$\int_{0-}^{\infty} K_{S+N}(t-u)h_\lambda(u)_0\,du = \int_{-\infty}^{\infty} h_d(\tau)K_{SV_{in}}(t-\tau)\,d\tau = K_{SV_{in}}(t_\lambda)$$
$$= K_{SV_{in}}(t+t_\mu) \qquad 0- < t,\, t_\mu < \infty \tag{16.76}$$

from which $h_\lambda(u)_0$ may be determined.

For the rational kernels assumed here, we may extend the results of Secs. A.2.3, A.2.4 to obtain explicit solutions, or we may equally well use the frequency form of solution (16.64) or (16.66) to obtain the system function $Y_\lambda(i\omega)_0$ and from this the corresponding weighting function $h(t)_0 = \mathfrak{F}^{-1}\{Y_0\}$. For instance, consider the case of an RC noise signal in additive independent white noise of spectral density W_{0N}. Then, $K_S(t) = \psi_S e^{-\alpha|t|}$, with $\mathcal{W}_S(f) = W_{0S}[1 + (\omega/\alpha)^2]^{-1}$ [cf. Eqs. (3.55)], and $K_N(t) = (W_{0N}/2)\delta(t-0)$ [cf. Eqs. (3.56)], so that Eq. (16.76) becomes

$$\int_{0-}^{\infty} [K_S(t-u) + K_N(t-u)]h_\lambda(u)_0\,du = K_{SV_{in}}(t_\lambda) = K_S(t+t_\mu) \tag{16.77a}$$

or specifically

$$\int_{0-}^{\infty} \left[\psi_S e^{-\alpha|t-u|} + \frac{W_{0N}}{2}\,\delta(t-u)\right] h_\lambda(u)_0\,du = \psi_S e^{-\alpha(t+t_\mu)} \qquad 0- < t,\, t_\mu < \infty \tag{16.77b}$$

where ψ_S and α are, respectively, the mean intensity and shape factor $[\alpha = (RC)^{-1}]$ of the noise "signal." From Secs. A.2.3, A.2.4, we obtain directly the weighting function of the optimum predictor,

$$h_\lambda(t)_0 = A_\mu e^{-\alpha\sqrt{1+\gamma_0^2}\,t} \qquad t > 0 \tag{16.78a}$$

where
$$A_\mu \equiv \frac{\alpha\gamma_0^2 e^{-\alpha t_\mu}}{1+\sqrt{1+\gamma_0^2}} \qquad \gamma_0^2 \equiv \frac{W_{0S}}{W_{0N}} = \frac{4\psi_S}{\alpha W_{0N}} \tag{16.78b}$$

The corresponding system function is

$$Y_\lambda(i\omega)_0 = \mathfrak{F}\{h_\lambda(t)_0\} = \frac{A_\mu\alpha^{-1}}{\sqrt{1+\gamma_0^2} + i\omega/\alpha} \tag{16.79}$$

Thus, the best predictor for RC noise in white noise is an RC filter, with the parameters above [Eqs. (16.78*b*)].

The minimum mean-square error follows from Eqs. (16.70) or Eqs. (16.72). Using the latter, we get for this example

$$\overline{e^2_{\min}} = \psi_S\left[1 - \frac{\gamma_0^2 e^{-2\alpha t_\mu}}{(1+\sqrt{1+\gamma_0^2})^2}\right] \qquad t_\mu > 0 \tag{16.80}$$

Note that when the noise background has a small spectral density compared with that of the RC noise signal, e.g., when $\gamma_0 \gg 1$, Eq. (16.80) reduces to

$$\overline{e^2_{\min}}\Big|_{\gamma\gg1} = \psi_S\left[1 - \left(1 - \frac{2}{\gamma_0}\right)e^{-2\alpha t_\mu}\right] \to \psi_S(1 - e^{-2\alpha t_\mu})\Big|_{\gamma_0\to\infty} \tag{16.80a}$$

For intense interference ($\gamma_0 \to 0$), the minimum mean-square error approaches its maximum value, ψ_S.

A special case of the above is *pure prediction*, where there is no accompanying noise, only the noise "signal." Then, Eq. (16.76) becomes

Pure prediction

$$\int_{0-}^{\infty} K_S(t-u)h_\lambda(u)_0\,du = K_S(t_\lambda) = K_S(t+t_\mu) \qquad 0- < t, t_\mu < \infty \tag{16.81}$$

This may now be solved for rational kernels directly by the methods of Secs. A.2.3, A.2.4. Considering once again the specific example above of an RC noise signal, we find from Eq. (A.2.48) that

$$h_\lambda(u)_0 = \begin{cases} (2\psi_S\alpha)^{-1}[\alpha^2 K_S(t+t_\mu) - \ddot{K}_S(t+t_\mu) \\ \qquad + (-\dot{K}_S + \alpha K_S)_{t_\mu}\delta(t-0)] & 0- < t < \infty \\ 0 & t < 0- \end{cases} \tag{16.82a}$$

which reduces directly to

$$h_\lambda(u)_0 = e^{-\alpha t_\mu}\delta(t-0) \qquad Y_\lambda(i\omega)_0 = \mathfrak{F}\{h_0\} = e^{-\alpha t_\mu} \tag{16.82b}$$

The optimum predicting filter here is a simple attenuator, of magnitude $e^{-\alpha t_\mu}$ for all frequencies (with $t_\mu \geq 0$). With Eqs. (16.82b) in Eqs. (16.72), the minimum mean-square error is readily found to be

$$\overline{e^2_{\min}} = \psi_S(1 - e^{-2\alpha t_\mu}) \tag{16.83}$$

which is, as expected, the limiting case of $\overline{e^2_{\min}}$ [Eq. (16.80)], when the accompanying noise becomes vanishingly small. Note that, if the lead time t_μ also vanishes, $\overline{e^2_{\min}}$ is zero, also to be expected, since then the output is identical with the signal input (for all $t < 0+$).

(2) Optimum linear interpolation ($-\infty < t_\lambda \leq t$). Here the instant t_λ at which we wish an estimate of the true signal input, $S(t_\lambda)$, lies at or within the observation period ($-\infty < t_\lambda \leq t$). We distinguish two cases: (1) simple filtering, $t_\lambda = t$, that is, $t_\mu = t_\lambda - t = 0$, for which Eqs. (16.75$a$), (16.75$b$) reduce to $h_d(\tau) = \delta(\tau - 0)$, $Y_d(i\omega) = 1$, respectively; (2) a general interpolation, where $t_\mu = t_\lambda - t < 0$ (cf. Fig. 16.4).

For the former, the optimum linear (time-invariant) realizable filter may be found directly from Eq. (16.76), with $t_\mu = 0$ therein. Thus, in the case of the RC noise signal above, in a white noise background, this filter structure follows at once from Eqs. (16.78), (16.79), viz.,

Simple filtering: $t_\mu = 0$

$$h_\lambda(t)_0 = \begin{cases} \dfrac{\gamma_0{}^2\alpha}{1+\sqrt{1+\gamma_0{}^2}}\, e^{-\alpha\sqrt{1+\gamma_0{}^2}\,t} & t > 0- \\ 0 & t < 0 \end{cases} \tag{16.84a}$$

$$Y_\lambda(i\omega)_0 = \frac{\gamma_0{}^2(1+\sqrt{1+\gamma_0{}^2})^{-1}}{\sqrt{1+\gamma_0{}^2} + i\omega/\alpha} \tag{16.84b}$$

and is again an RC filter with suitable gain and time constant. The minimum mean-square error from Eq. (16.80) is simply

$$\overline{e_{\min}^2} = \psi_S[1 - \gamma_0{}^2/(1 + \sqrt{1 + \gamma_0{}^2})^2] \tag{16.85}$$

which approaches zero as the condition of *pure smoothing* (e.g., no background noise, $\gamma_0 \to \infty$) is reached.

However, when t_μ is negative, i.e., when t_μ represents a *lag time*, rather than a lead time, it is no longer possible to find a linear *time-invariant* realizable filter to minimize the mean-square error $e(t)$. Approaches to minimization using suitable long-lag filters as approximations[12,13,18] to the linear nonrealizable filters of Eqs. (16.62a), (16.62b) offer one way of handling this problem. Another is to relax the constraint of time invariance while reimposing that of realizability. In fact, it is this last which the solution of the basic integral equation (16.76) suggests in the present case.

To see this, let us first observe that now the weighting and system functions of the desired operation are [cf. Eqs. (16.73), (16.75)]

$$h_d(\tau) = \delta(\tau - |t_\mu|) \qquad Y(i\omega)_d = e^{-i\omega|t_\mu|} \qquad t_\mu < 0 \tag{16.86}$$

so that the basic integral equation (16.76) is modified to

$$\begin{aligned} \int_{0-}^{\infty} K_{S+N}(t - u)h_\lambda(u)_0 \, du &= \int_{-\infty}^{\infty} h_d(\tau)K_{SV_{in}}(t - \tau)\, d\tau \\ &= K_{SV_{in}}(t - |t_\mu|) \qquad 0- < t < \infty \end{aligned} \tag{16.87}$$

For rational kernels, the solutions have the forms indicated in Secs. A.2.3, A.2.4. The essential feature here is that while $h_\lambda(t)_0$ vanishes for $t < 0-$, as it should to satisfy the realizability condition, this weighting function contains terms of the type $Be^{-b|t-|t_\mu||}$, which imply a time-varying element, in order to account for the behavior at $0- < t < |t_\mu|$. The time-varying element may be a simple switch at $t = |t_\mu|$ and an active network for the preceding interval $0- < t < |t_\mu|$. In any case, it is not possible to minimize the mean-square error with finite lag (i.e., interpolation) filters and still require, besides linearity, both realizability and time invariance: we can keep the latter constraint and obtain the results (16.62), which are not realizable, or we can keep the former, if we do not insist on invariance.

Let us illustrate these observations, using again the RC noise signal and independent noise background of our first example [Sec. 16.2-3(1)]. The basic integral equation (16.87) reduces to Eq. (16.77b) with t_μ now replaced by $-|t_\mu|$. Applying Eqs. (A.2.107), (A.2.112) here, we find that the optimum realizable weighting function is

$$h_\lambda(t)_0 = \begin{cases} h(t,t_\mu)_0 = \dfrac{\alpha\gamma_0{}^2}{b}\left(e^{-\alpha b|t-|t_\mu||} - 2\,\dfrac{1-b}{1+b}\,e^{-\alpha b(t+|t_\mu|)}\right) & t > 0- \\ 0 & t < 0- \end{cases} \tag{16.88a}$$

with $b = \sqrt{1 + \gamma_0^2}$. The corresponding system function follows from $Y_\lambda(i\omega)_0 = \mathfrak{F}\{h_0\}$, viz.,†

$$Y_\lambda(i\omega)_0 = \frac{\alpha\gamma_0^2}{b}\left(\frac{e^{-i\omega|t_\mu|}}{\alpha b + i\omega} + \frac{e^{-i\omega|t_\mu|} - e^{-\alpha b|t_\mu|}}{\alpha b - i\omega} - 2\,\frac{1-b}{1+b}\,\frac{e^{-\alpha b|t_\mu|}}{\alpha b + i\omega}\right) \quad (16.88b)$$

The calculation of $\overline{e^2_{\min}}$ is left as an exercise. Observe now the time-varying nature of this filter, as represented by the term $e^{-\alpha b|t-|t_\mu||}$ in Eqs. (16.88*a*).

When the background noise can be ignored, e.g., when $\gamma_0, b \to \infty$, $h_\lambda(t)_0$ reduces to the simple delay network

$$h_\lambda(t)_0 = \delta(t - |t_\mu|) \qquad \text{or} \qquad Y_\lambda(i\omega)_0 = e^{-i\omega|t_\mu|} = Y(i\omega)_d \qquad t_\mu < 0 \quad (16.89)$$

this last from Eqs. (16.86). Accordingly, in this case of pure smoothing, when $t_\mu < 0$ the optimum filter is identical with the desired filter, both of which are now time-invariant and realizable. It is also at once evident that $\overline{e^2_{\min}}$ vanishes here. A background noise, however, removes the time invariance, as we have seen above, and leads to minimum mean-square smoothing errors greater than zero.

16.2-4 Extensions. The linear theory of optimum prediction and smoothing described in Sec. 16.2-2 is not restricted to stationary, entirely stochastic input signals and semi-infinite observation intervals. All these conditions, to varying degrees, can be removed if we are willing to permit time-varying systems. Realizability is still required for obvious practical reasons, however, and it is this constraint that presents the principal analytical difficulties, as the examples of Sec. 16.2-3 and the results of Appendix 2 demonstrate. A few of the more common generalizations are briefly summarized below; for a detailed treatment, the reader is referred to the literature.‡

(1) Differentiation and other desired (linear) operations. The present approach is, of course, not limited alone to extrapolation or interpolation. We may wish, for example, for a best predicted, or smoothed, value of the derivative dS/dt of the input signal process§ at t_λ. Then, Eq. (16.73) gives

$$h_d(t) = \delta'(t + t_\mu) \qquad Y_d(i\omega) = i\omega e^{i\omega t_\mu} \qquad t_\mu \gtrless 0 \quad (16.90)$$

and the basic integral equation (16.59) now becomes

$$\int_{0-}^{\infty} K_{S+N}(t-u)h_\lambda(u)_0\,du = -\frac{d}{d\tau}K_{SV_{in}}(\tau)\bigg|_{\tau=t+t_\mu} \qquad 0- < t < \infty \quad (16.91)$$

For rational kernels, the solution of this is once more obtained by the methods of Sec. A.2.3, or in some instances we can use the spectral approach

† Here, as above in Eqs. (16.78), (16.82), $t = 0$ is the "present" of the weighting function so that $t_\lambda = t_\mu$ (< 0).

‡ See, for example, the book by Laning and Battin,[9] chaps. 6–8, and the papers of Zadeh and Ragazzini,[4] Davis,[5] Booton,[19] and Bendat.[20]

§ Stationarity is still assumed, as is the entirely nondeterministic nature of the signal and noise processes.

[Eqs. (16.63) to (16.66)] above. The signal process must, however, be differentiable mean square at least once. Higher-order derivatives may be handled in the same way. Time-invariant optimum filters can be expected for the prediction cases, but again, unless we relax the constraint of realizability, the optimum interpolation of dS/dt ($t_\lambda < 0$) requires a time-varying system.

(2) **Finite samples.**[4,5] Instead of operating on the entire past of the input, we frequently have only a limited period of data on which to compute the desired estimate. The theory of Sec. 16.2-2 is easily extended to take this into account: from Eq. (16.54), we find that the basic integral equation (16.59) becomes here

$$\int_{0-}^{T+} K_{S+N}(t-u)h_\lambda(u)_0\,du = \int_{-\infty}^{\infty} K_{S+N}(t-u)h_\lambda(u;T)_0\,du = G(t) \qquad 0- < t < T+ \quad (16.92a)$$

where $h_\lambda(u;T)_0 = 0$ outside $(0-,T+)$. Thus, for prediction ($t_\mu > 0$) or smoothing ($t_\mu \leq 0$), we get from Eqs. (16.75) in Eq. (16.58a), as before,

$$\int_{0-}^{T+} K_{S+N}(t-u)h_\lambda(u)_0\,du = K_{SV_{in}}(t+t_\mu) \qquad 0- < t < T+ \quad (16.92b)$$

and the solution is obtained once more with the help of the results of Appendix 2. See Secs. 21.4-1 and 21.4-2 for some specific examples.

(3) **Nonstationary inputs.**[19,5] So far, we have postulated that the noise "signal," and the accompanying background noise, are at least wide-sense stationary. However, this is not always a valid assumption in practice, and their nonstationarity must then be taken specifically into account. Again, by suitable modification of Eq. (16.54), including finite sample size, we find that the fundamental integral equation for the optimum filter in this case is generalized to

$$\int_{0-}^{T+} K_{S+N}(\tau-t,\tau-u)h_\lambda(u,\tau)_0\,du = G(t,\tau) \qquad 0-- < t < T+ \quad (16.93a)$$

which for prediction and smoothing becomes

$$\int_{0-}^{T+} K_{S+N}(t_\lambda-t,t_\lambda-u)h_\lambda(u,t_\lambda)_0\,du = K_{SV_{in}}(t_\lambda-t,t_\lambda+t_\mu) \qquad 0- < t < T+ \quad (16.93b)$$

In most cases, the optimum filter here is a time-varying system. Solutions to Eqs. (16.93a), (16.93b) now, however, cannot be obtained directly by the methods of Appendix 2 [unless the kernel has the form $K_{S+N}(t,\tau) = F(t)F(\tau)H(|t-\tau|)$]; instead, we can use† Mercer's theorem (8.61) to reduce Eqs. (16.93) to a set of homogeneous integral equations, in the manner of

† This is the approach employed by Davis in his original extension of the prediction problem. For details, see Davis, *op. cit.*

Eqs. (16.45) to (16.48), the solutions of which then enable us to construct the optimum weighting function h_0.

Still other modifications are possible. The signal can contain besides the noise term a deterministic component, in the form of a polynomial in time of known degree and (invariant) random coefficients, $\boldsymbol{\theta}$; for example, $S(t,\boldsymbol{\theta}) = S_N(t) + \theta_0 + \theta_1 t + \theta_2 t^2 + \cdots + \theta_q t^q$ (cf. Sec. 21.4-3). A less general case of this occurs when the $\theta_0, \ldots, \theta_q$ are all known a priori[4,5,21] (cf. Prob. 16.8). More involved operations for the desired filter h_d can also be introduced, along with general nonstationarity and finite sampling, nor need signal and noise be additively combined. We shall return briefly to these questions in Chap. 21, where optimum prediction and filtering are discussed from the more general standpoint of decision theory, and where nonlinear, as well as linear, systems may be used.

PROBLEMS

16.5 Prove that Eq. (16.59) represents both a necessary and sufficient condition for a minimum mean-square error for the particular input processes assumed in Sec. 16.2-1.

16.6 (*a*) A stationary entirely random signal process is additively combined with a stationary white noise background of spectral density W_{0N}. The noise signal has the intensity spectrum

$$\mathcal{W}_S(f) = W_{0S}/[1 + (a\omega)^4] \qquad \omega = 2\pi f \ (a > 0) \tag{1}$$

If the noise and signal are statistically unrelated, using the approach in the frequency domain [cf. Eqs. (16.63) to (16.66)], show that the system function of the optimum linear (realizable) filter for differentiating the signal at $t_\lambda = t$ (that is, $t_\mu = 0$) is

$$Y(i\omega)_0 = \gamma_0^2(1 + \sqrt{\gamma_0^2 + 1})^{-1}[1 + (1 + \gamma_0^2)^{1/4}]^{-1} \times \frac{(\sqrt[4]{1 + \gamma_0^2} - 1)i\omega - \sqrt{2}}{(ai\omega)^2 + (ai\omega)\sqrt{2}\sqrt[4]{1 + \gamma_0^2} + \sqrt{1 + \gamma_0^2}} \tag{2}$$

with γ_0 given by Eq. (16.78*b*). What is the weighting function?

Show that, as $\gamma_0 \to \infty$ (that is, vanishing noise), $Y(i\omega)_o \to Y_d(i\omega)$, while for intense noise ($\gamma_0 \to 0$) one gets

$$Y(i\omega)_0 \simeq -\frac{\gamma_0^2\sqrt{2}}{4}[(ai\omega)^2 + \sqrt{2}\, ai\omega + 1]^{-1} \qquad \gamma_0 \to 0 \tag{3}$$

which is quite different from $Y(i\omega)_0 = i\omega$ essentially for signal alone.

(*b*) Obtain the minimum mean-square error for Eq. (2), viz.,

$$\overline{e^2_{\min}} = \frac{W_{0S}\sqrt{2}}{8a}\left[1 - \gamma_0^2\frac{\sqrt{1 + \gamma_0^2} + 3 - 2\sqrt[4]{1 + \gamma_0^2}}{(1 + \sqrt{1 + \gamma_0^2})^2(1 + \sqrt[4]{1 + \gamma_0^2})^2}\right] \tag{4}$$

Note that, as $\gamma_0 \to \infty$ (that is, pure differentiation), $\overline{e^2_{\min}} \to 0$, while for intense noise backgrounds ($\gamma_0 \to 0$) there is a residual error $W_{0S}\sqrt{2}/8a$. Show that

$$\begin{aligned} \overline{e^2_{\min}}\Big|_{\gamma_0 \to 0} &\doteq \frac{W_{0S}\sqrt{2}}{8a}\left(1 - \frac{\gamma_0^2}{8} + \frac{3\gamma_0^4}{32} - \cdots\right) \\ \overline{e^2_{\min}}\Big|_{\gamma_0 \to \infty} &\simeq \frac{W_{0S}\sqrt{2}}{2a}\left(\frac{1}{\sqrt{\gamma_0}} - \frac{1}{\gamma_0} + \cdots\right) \end{aligned} \tag{5}$$

(c) For high noise levels, show that the optimum filter has the approximate structure

$$Y(i\omega)_0 \doteq \gamma_0^2 [\gamma_0^2 + G(i\omega)]^{-1} \left[\frac{Y_d(i\omega) w_s(f)}{\gamma_0^2 + G(-i\omega)} \right]_+ \qquad \gamma_0 \to 0 \tag{6}$$

where $G(i\omega)$ is that part of $w_s(f)$ with poles and zeros in the left half plane only, i.e.,

$$G(p) = \int_0^\infty e^{-pt}\, dt \int_{-\infty}^{\infty} w_s(f) e^{i\omega t}\, df = [w_s(p/2\pi i)]_+ \tag{7}$$

with $w_s(f) = \gamma_0^{-2} \mathcal{W}_S(f)$.

16.7 (a) Show that the optimum linear realizable filter *for interpolation* $(0- < t_\lambda < T+)$ of a stationary *RC* noise signal in (additive) stationary white noise, based only on data in the sampling interval $(0,T)$, is

$$h_\lambda(t)_0 = \begin{cases} \dfrac{\gamma_0^2 \alpha}{\sqrt{1+\gamma_0^2}} \left[e^{-\alpha b |t - |t_\mu||} + \dfrac{2}{W_{0N}} \dfrac{1-b}{1+b} (A_1^{(+)} e^{\alpha b (t-T)} + A_1^{(-)} e^{-\alpha b t}) \right] & 0- < t < T+ \\ 0 & \text{elsewhere} \end{cases} \tag{1}$$

where $b = \sqrt{1 + \psi_0^2}$ $(0- < |t_\mu| = T - t_\lambda < T+)$ and

$$A_1^{(+)} = \alpha \gamma_0^2 \frac{(1+b) e^{\alpha b |t_\mu|} - (1-b) e^{-\alpha b |t_\mu|}}{(1-b)^2 e^{-\alpha b T} - (1+b)^2 e^{\alpha b T}} \tag{2}$$

$$A_1^{(-)} = \alpha \gamma_0^2 \frac{(1+b) e^{\alpha b (T - |t_\mu|)} - (1-b) e^{\alpha b (|t_\mu| - T)}}{(1-b)^2 e^{-\alpha b T} - (1-b)^2 e^{\alpha b T}} \tag{3}$$

(b) Obtain the system function $Y_\lambda(i\omega)_0$, and then obtain $\overline{e^2_{\min}}$.

(c) Verify that, for *pure interpolation* (i.e., no noise background), we get $h_\lambda(t)_0 = \delta(t - |t_\mu|)$ $(0- < |t_\mu| < T+)$, so that $h_\lambda(t)_0 = h_d(t)$ and therefore $\overline{e^2_{\min}} = 0$ in this limiting case.

16.8 If the input signal process contains a completely specified nonrandom component, of the form $S_D(t) = \theta_0 + \theta_1 t + \theta_2 t^2 + \cdots + \theta_q t^q$, as well as a stationary, entirely random component, show that the optimum linear realizable filter for minimum mean-square error in the actual vs. the desired output (such as prediction, differentiation, or interpolation) must satisfy the integral equation

$$\int_{0-}^{T+} K_{S+N}(t-u) h_\lambda(u)_0\, du = \gamma_0 + \gamma_1 t + \cdots + \gamma_q t^q + \int_{-\infty}^{\infty} h_d(u) K_{SV_{in}}(t-u)\, du \qquad 0- < t < T+ \tag{1}$$

Here, $\gamma_1, \ldots, \gamma_q$ are $q+1$ Lagrange multipliers, which are determined by the $q+1$ constraints implied by

$$S_D(t)_{out} = \int_{0-}^{T+} h(\tau)_0 S_D(t-\tau)\, d\tau \tag{2}$$

where $S_D(t)$ is given, above, in terms of the known parameters $\theta_0, \ldots, \theta_q$. HINT: Expand $S_D(t-\tau)$ about $\tau = 0$, and observe that

$$S_D(t-\tau) = \sum_{k=0}^{q} \frac{(-1)^k}{k!} \eta_k S_D^{(k)}(t) \qquad \eta_k \equiv \int_{0-}^{T+} \tau^k h_\lambda(\tau)_0\, d\tau \tag{3}$$

The η_k are the $q+1$ constraints to which the calculation of $\overline{e^2_{\min}}$ is subject.

L. A. Zadeh and J. R. Ragazzini,[4] sec. III.

16.9 (*a*) Calculate $\overline{e^2_{\min}}$ for Prob. 16.8, and determine the $\gamma_0, \ldots, \gamma_q$.

(*b*) If in Prob. 16.8 the random part of the signal vanishes and $S_D(t) = \theta_0 + \theta_1 t$, $K_N(t) = \psi_N e^{-\alpha|t|}$, show that the optimum linear predictor has the form

$$h_\lambda(t)_0 = \begin{cases} a_0 + a_1 t + c_1\delta(t - 0) + c_2\delta(t - T) & 0- < t < T+ \\ 0 & \text{elsewhere} \end{cases} \tag{1}$$

and find a_0, a_1, c_1, c_2.

(*c*) Instead of the *RC* noise of (*b*), we have now a narrowband single-tuned noise, with covariance function $K_N(t) \doteq \psi_N e^{-\alpha|t|} \cos \omega_0 t$. Show that the optimum weighting function here is of the form

$$h_\lambda(t)_0 = \begin{cases} b_0 + b_1 t + d_1 e^{at} + d_2 e^{-at} + e_1\delta(t - 0) + e_2\delta(t - T) & 0- < t < T+ \\ 0 & \text{elsewhere} \end{cases} \tag{2}$$

$a^2 = \omega_0^2 + \alpha^2$. Find the constants $b_0, \ldots, e_2$, and calculate $\overline{e^2_{\min}}$.

16.10 (*a*) Repeat Prob. 16.7, but now for prediction ($t_\lambda > T$) and extrapolation ($t_\lambda < 0$).

(*b*) Do Prob. 16.7 for interpolation ($0- < t_\mu < T+$), with an *LRC* noise signal process. HINT: Use Sec. A.2.4, Example 3.

16.11 (*a*) Use Mercer's theorem (8.61) to show that the solution of the prediction and interpolation equation for the case of the stationary noise signal and background noise processes of Sec. 16.2, e.g.,

$$\int_{0-}^{T+} K_{S+N}(t - u) h_\lambda(u)_0 \, du = K_{SV_{in}}(t + t_\mu) \qquad t_\mu \gtreqless 0 \ (0- < t < T+) \tag{1}$$

can be expressed as

$$h_\lambda(t)_0 = \sum_{k=1}^{\infty} \frac{\phi_k(t)}{\lambda_k} \int_0^T K_{SV_{in}}(u + t_\mu)\phi_k(u) \, du \tag{2}$$

(*b*) Extend this approach to the case of nonstationary signals and noise. R. C. Davis.[5]

16.3 *Maximization of Signal-to-Noise Ratios—Matched Filters*

In Chap. 5 (cf. Sec. 5.3-4) and in Chaps. 13 to 15, we have seen that the signal-to-noise ratio is a useful quantity for evaluating system performance. Although, as pointed out earlier, it gives only a partial indication of how the system in question operates, it has the advantage of being simple to calculate, at least vis-à-vis some of the more sophisticated (and realistic) criteria of the general theory (cf. Chaps. 19, 21). Since the signal-to-noise ratio may be taken as one measure of system behavior, and since systems with small $[S/N]^2_{in}$ (for given $[S/N]^2_{out}$) are usually (but not always)† better than those with larger $[S/N]^2_{in}$ (for the same $[S/N]^2_{out}$), it is natural to try further to improve performance by maximizing the signal-to-noise ratio. This may be accomplished in several ways: (1) for a given input signal process, we may attempt to adjust the various linear and nonlinear operations intervening between input and final output, or (2) with a given receiver we may attempt

† This depends on the criterion employed and on the signal and noise processes (cf. Sec. 23.4).

the enhancement of $(S/N)_{out}$ by selecting a suitable input signal process.† Here, we are concerned with (1), and in this chapter only with those *linear* filtering operations by which the output signal-to-noise ratio may be maximized, when the input signal process is specified in some definite sense.

16.3-1 Maximization of $(S/N)_{out}$. The input signal and the noise in receiving systems are, of course, mixed together by the nonlinear elements of the receiver in such a way that it is not strictly possible to separate signal uniquely from noise at the output (cf. Sec. 5.3-4). However, it is still possible to define a signal output in some reasonable sense, so that an output signal-to-noise ratio can be constructed which is a monotonic function of the input ratio $[S/N]_{in}^2$ (for example, the deflection and power criteria of Sec. 5.3-4). Here, $[S/N]_{in}^2$ is the signal-to-noise ratio at a point just preceding the nonlinear element (i.e., the second detector if an AM receiver is used), but following one or more effectively linear filters. Furthermore, in threshold reception this output (amplitude) ratio is given by‡

$$(S/N)_{out} \doteq A[S/N]_{in}^2 \qquad [S/N]_{in}^2 < 1 \tag{16.94}$$

where A is some constant of proportionality. Thus, inasmuch as $(S/N)_{in}$ appears to no power higher than the second in Eqs. (16.94) for threshold detection, we are able to maximize $(S/N)_{out}$ with *linear* filters. In fact, in these weak-signal cases maximization of $(S/N)_{out}$ is essentially the same problem we have already considered in Sec. 16.1-3, for the case of linear measurements with least (relative) mean-square error. The former involves a signal-to-noise ratio, the latter a "noise-to-signal" ratio; in the former we expect a maximum, for the latter a minimum. Actually, maximization of $(S/N)_{out}$ is really an estimation problem, since it is basically a question of measurement: its applicability to detection problems stems from relations (16.94) and the not unreasonable *assumption* that, if $(S/N)_{out}$ is maximized, so also will be the performance of the system in detecting the signal. Of course, for linear systems these complications do not arise: a threshold condition is not necessary, since $(S/N)_{out} = (S/N)_{in}$ for all signal strengths. Accordingly, it is sufficient here, for optimizing $(S/N)_{out}$, to consider the maximization of $[S/N]_{in}^2$. This is the problem to which we address ourselves below. In other words, observing that S and N in $[S/N]_{in}^2$ are themselves the result of some linear filtering operation, $\boldsymbol{T}_L$, on an original input signal $S_{in}(t)$ and noise $N_{in}(t)$, we see that our problem is essentially that of finding that linear filter $\boldsymbol{T}_L = (\boldsymbol{T}_L)_{op}$ which maximizes $[S/N]_{in}^2$.

Let us begin by requiring $S_{in}(t)$ to be a known function of time and the noise $N_{in}(t)$ to be a wide-sense stationary process, with $\bar{N}_{in} = 0$. We wish

† The latter problem is an example of an encoding technique which we shall examine from the viewpoint of the more general theory discussed in Part 4 (cf. Sec. 23.2).

‡ This is the well-known phenomenon of signal suppression [cf. Secs. 13.2-1(4), 15.2-1, for example]; for a general theory of this effect in the case of detection, see Sec. 19.4.

to maximize

$$\left[\frac{S}{N}\right]_{in}^2 \equiv \frac{S(t)^2}{\overline{N(t)^2}} \tag{16.95}$$

at some fixed time $t = \epsilon_0$. The filter $\boldsymbol{T}_L$ which gives us $S(t)$, $N(t)$, respectively, when $S(t)_{in}$, $N(t)_{in}$ are inputs, acts for a finite period $(0,T)$, so that, for $t = \epsilon_0$ $(0- < \epsilon_0 < T+)$, we have

$$S(\epsilon_0) = \boldsymbol{T}_L\{S_{in}\} = \int_{0-}^{T+} h(u)S_{in}(\epsilon_0 - u)\,du \tag{16.96a}$$

$$N(\epsilon_0) = \boldsymbol{T}_L\{N_{in}\} = \int_{0-}^{T+} h(u)N_{in}(\epsilon_0 - u)\,du \tag{16.96b}$$

and

$$\overline{N(\epsilon_0)^2} = \iint_{0-}^{T+} h(u)K_{N_{in}}(|t-u|)h(t)\,du\,dt > 0 \tag{16.97}$$

Now, let us suppose that $[S/N]_{in\,\max}^2$ is some fixed (finite) value γ_0^{-1} Then, Eq. (16.95) can be written, for $t = \epsilon_0$,

$$\overline{N(\epsilon_0)^2} - \gamma_0 S(\epsilon_0) \geq 0 \tag{16.98}$$

where the equality holds only for the optimum filter. Taking the variation of this relation with respect to $h(u)$, using the symmetry property $K_N(|u - v|)$, and setting the result equal to zero, we get in the usual way [cf. Eqs. (16.28) to (16.30)] the condition†

$$\int_{0-}^{T+} K_{N_{in}}(|t-u|)h_T(u)_{op}\,du = \gamma_0 S_{in}(\epsilon_0 - t) \qquad 0- < t,\, \epsilon_0 < T+ \tag{16.99}$$

which determines the optimum linear filter that maximizes $[S/N]_{in}^2$ [the proof that Eq. (16.99) is a necessary and sufficient condition for a maximum is left to the reader; cf. Prob. 16.12]. This filter is, of course, a time-varying network (with switch at $t = T$), as well as a realizable one ($h_T = 0$, $t < 0-$).

Except for the time dependence of the right-hand member, this is the same sort of relation we encountered previously in our linear-measurement problems [cf. Eq. (16.30)]. It is also identical in form with the basic integral equation for linear interpolation (or extrapolation) [Eq. (16.92b)], when the sample is finite $(0,T)$, and is solved, for rational kernels, by the methods mentioned earlier (cf. Secs. 16.2-3, 16.2-4; see also Sec. A.2.3). We easily find the constant γ_0 by multiplying Eq. (16.99) by $h_T(v)_{op}$ and integrating, remembering that $h_T(t)_{op}$ is proportional to γ_0. We obtain directly

† The result (16.99) was obtained originally by Zadeh and Ragazzini,[7] generalizing some earlier work of North,[23] Van Vleck and Middleton,[6] Dwork,[24] and others.[25] We shall encounter this relation again in our treatment of simple-alternative detection problems by decision-theory methods (cf. Sec. 20.2-4).

$$\gamma_0 = \frac{\iint_{0-}^{T+} h_T(t)_{op} K_N(|t-u|) h_T(u)_{op}\, du\, dt}{\int_{0-}^{T+} S_{in}(\epsilon - t) h_T(t)_{op}\, dt} = \frac{\overline{N_{out}^2}}{S_{out}} \tag{16.100}$$

Finally, since $(S/N)_{out} = A[S/N]_{in}^2$ in threshold detection [cf. Eqs. (16.94)] and $[S/N]_{out}^2 = [S/N]_{in}^2$ for linear systems, by maximizing $[S/N]_{in}^2$ we have made $(S/N)_{out}$ a maximum as well.

16.3-2 Examples. Matched Filters. The linear filter $h_T(t)_{op}$, which maximizes $(S/N)_{in}^2$, and hence $(S/N)_{out}$ above, is called a *matched*† *filter*.‡ As we shall make use of generalizations of this type of filter later, in the statistical theory of reception (Part 4), we shall refer to the matched filters here as matched (S/N) filters, i.e., linear filters that are optimum in the sense of maximizing a signal-to-noise ratio, in order to distinguish them from filters matched on the basis of other criteria. Filters of this class are properly associated with problems of signal estimation, where quadratic cost functions are usually assumed (cf. Sec. 21.3).

(1) White noise background. As our first example, let us suppose that the noise $N_{in}(t)$ has a uniform spectrum, for all frequencies, with density W_0. Then, $K_{N_{in}}(|u-v|) = (W_0/2)\delta(u-v)$, and Eq. (16.99) is solved at once, with

$$h_T(t)_{op} = \begin{cases} \dfrac{2\gamma_0}{W_0} S_{in}(\epsilon_0 - t) & 0 \le t \le T \\ 0 & \text{elsewhere} \end{cases} \tag{16.101}$$

The matched filter is accordingly proportional to the time reverse of the input signal (starting at $t = \epsilon_0$). Its system function is obtained from $Y_T(i\omega)_{op} = \mathfrak{F}\{h_T\}$, namely,

$$Y_T(i\omega)_{op} = \frac{2\gamma_0 e^{-i\omega\epsilon_0}}{W_0} \int_{-\infty}^{\infty} S_{in}(\tau)_T e^{i\omega\tau}\, d\tau = \frac{2\gamma_0 e^{-i\omega\epsilon_0}}{W_0} S_{in}(f)_T^* \tag{16.102}$$

where $S_{in}(f)_T$ is the amplitude spectral density of the truncated input signal $S_{in}(t)_T = S_{in}(t)$ $(0 \le t \le T)$, $S_{in}(t)_T = 0$ (elsewhere). Thus, in the frequency domain, we see that the matched filter (for white noise backgrounds) is one whose system function is (proportional to) the complex conjugate of the amplitude spectrum of the truncated input signal. Figure 16.7*a* shows $|Y_T(i\omega)_{op}|^2$ for a signal S_{in} consisting of a long train of periodic rectangular pulses in white noise.

† This term was originally employed by Van Vleck and Middleton, *op. cit.*, in the more restricted sense of maximizing $(S/N)_{out}$ for signals in *white* noise backgrounds (cf. Examples 1 and 3 following).

‡ Matched filters are always postulated to be linear in the present treatment.

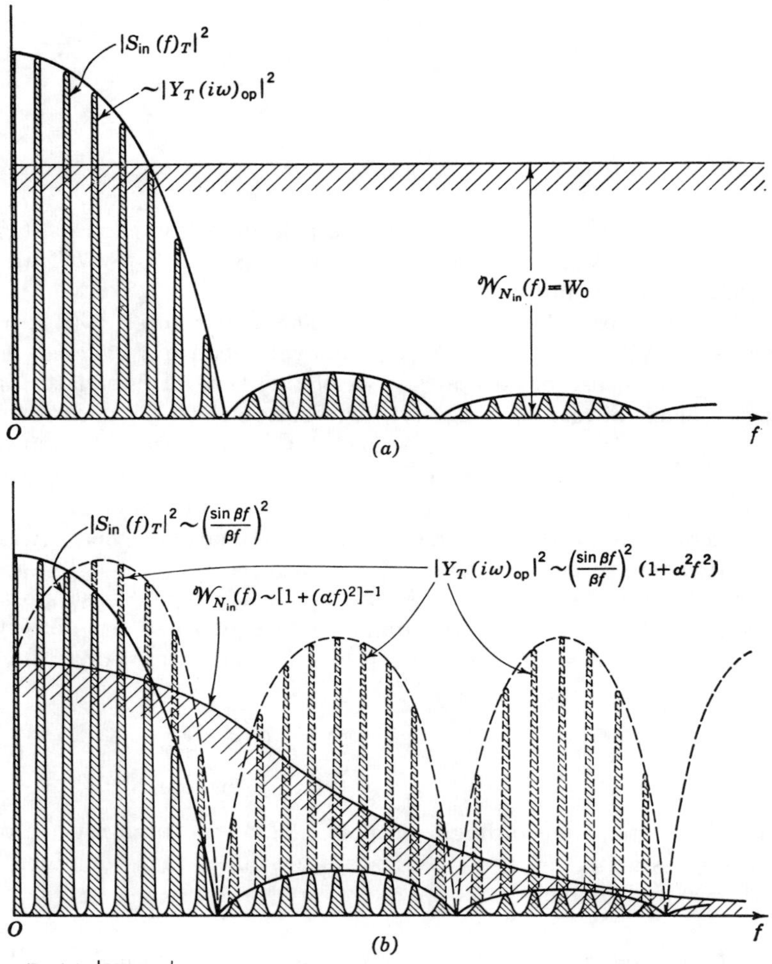

FIG. 16.7. (a) $|Y_T(i\omega)|^2$ of matched filters maximizing signal-to-noise ratio for a train of periodically spaced rectangular pulses in white noise. (b) The same as (a), but now for RC noise backgrounds.

(2) ***RC* noise background.** In place of the white noise above, let us assume, instead, that the accompanying disturbance is RC noise, with $K_{N_{in}}(t) = \psi e^{-\alpha|t|}$ $[\alpha = (RC)^{-1}]$ in the usual way [cf. Eqs. (3.55)]. The basic integral equation (16.99) now becomes

$$\psi \int_{0-}^{T+} e^{-\alpha|t-u|} h_T(u)_{op}\, du = \gamma_0 S_{in}(\epsilon_0 - t) \qquad 0- < t < T+ \qquad (16.103)$$

The weighting function of the optimum filter is readily found with the help of Eq. (A.2.48) to be

$$h_T(t)_{op} = \begin{cases} \gamma_0(2\psi\alpha)^{-1}\{[\alpha^2 S_{in}(\epsilon_0 - t) - \ddot{S}_{in}(\epsilon_0 - t)] \\ \quad + (\dot{S}_{in} + \alpha S_{in})_{\epsilon_0 - T}\delta(t - T) \\ \quad + (-\dot{S}_{in} + \alpha S_{in})_{\epsilon_0}\delta(t - 0)\} & 0- < t,\ \epsilon_0 < T+ \\ 0 & t < 0-,\ t > T+ \end{cases} \tag{16.104}$$

and the system function follows in the usual way as the Fourier transform of $h_T(t)_{op}$. Figure 16.7*b* illustrates $|Y_T(i\omega)_{op}|^2$ for the pulse train of Fig. 16.7*a*, but now with *RC* noise backgrounds. (For cases involving modulated carriers and narrowband noise, see Probs. 16.13 and 16.14.)

(3) Infinite samples. The linear filters of the two preceding examples are time-varying and realizable systems. Let us now consider the somewhat more idealized situation where the sample time is allowed to become infinite, so that the filter uses the entire past of the input signal and noise. Here $T \to \infty$, and Eq. (16.99) becomes

$$\int_{0-}^{\infty} K_{N_{in}}(t - u)h_T(u)_{op}\, du = \gamma_0 S_{in}(\epsilon_0 - t) \qquad 0- < t,\ \epsilon_0 < \infty \tag{16.105}$$

where again the solution is obtained by the methods of Sec. A.2.3 when the kernel is rational. For our examples above, we get at once

White noise

$$\begin{aligned} h_\infty(t)_{op} &= \frac{2\gamma_0}{W_0} S_{in}(\epsilon_0 - t) \qquad 0- < t \\ Y_T(i\omega)_{op} &= \frac{2\gamma_0 e^{-i\omega\epsilon_0}}{W_0} S_{in}(f)_\infty^* \end{aligned} \tag{16.106a}$$

this last when $S_{in}(f)_\infty$ exists,† and

RC noise

$$h_\infty(t)_{op} = \begin{cases} \gamma(2\psi\alpha)^{-1}\{[\alpha^2 S_{in}(\epsilon_0 - t) - \ddot{S}_{in}(\epsilon_0 - t)] \\ \qquad + (-\dot{S}_{in} + \alpha S_{in})_{\epsilon_0}\delta(t - 0)\} & 0- < t \\ 0 & t < 0 \end{cases} \tag{16.106b}$$

Both these filters are physically realizable and are now time-invariant as well.

If we relax the constraint of realizability, so that $h_\infty(t)_{op}$ operates on the entire future, as well as the past, of the input waves, we see that Eq. (16.99) is modified to

$$\int_{-\infty}^{\infty} K_{N_{in}}(t - u)h_\infty(u)\, du = \gamma_0 S_{in}(\epsilon_0 - t) \qquad -\infty < t,\ \epsilon_0 < \infty \tag{16.107}$$

analogous to Eq. (16.61) in this respect. Simple Fourier transformation

† We recall that a system can have a weighting function [and a corresponding transfer function $Y_\infty(p)_{op}$] even though its system function is not defined [cf. Sec. 2.2-5(2)].

of both sides then gives

$$Y_\infty(i\omega) = 2\gamma_0 \frac{S_{in}(f) \overset{*}{\infty} e^{-i\omega\epsilon_0}}{\mathcal{W}_{N_{in}}(\omega/2\pi)} \qquad h_\infty(t) = \mathfrak{F}^{-1}\{Y_\infty\} \tag{16.108a}$$

For rational kernels, $\mathcal{W}_{N_{in}}(\omega/2\pi) = W_0|Y_N(i\omega)|^2$ so that Eq. (16.108*a*) takes the more familiar form

$$Y_\infty(i\omega) = \frac{2\gamma_0}{W_0} \frac{S_{in}(f) \overset{*}{\infty} e^{-i\omega\epsilon_0}}{|Y_N(i\omega)|^2} \tag{16.108b}$$

Here, $Y_N(i\omega)$ is the system function of a linear (stable, realizable) filter through which originally white noise of spectral density W_0 has been passed (cf. Sec. 3.3-2). The optimizing filter is, of course, not realizable. However, it can be approximated as closely as desired if sufficient delays are allowed. A situation where Eqs. (16.108) can be used for quite accurate results occurs, for example, in pulsed radar when the duration of the pulse train is long compared with the pulse repetition period.[6] Figure 16.7 shows some typical limiting cases.

16.3-3 Summary Remarks. Because the criteria of performance in the problems above do not require more than second-order statistical information (e.g., spectra and covariance functions) *and* because of the initially imposed constraint of linearity, these solutions in turn all have the common feature of linearity: the desired optimum filters are linear, usually time-invariant, and, for the most part, realizable. This does not mean, however, that we could not do better with an appropriate *nonlinear* system. In fact, as we shall see in Chap. 21, if we remove the condition of linearity, these same criteria (e.g., minimum mean-square error, maximum signal-to-noise ratio, etc.) generally imply nonlinear operations. An example of this occurs in the incoherent estimation of signal amplitude (or intensity) when the signal is accompanied by a normal noise (cf. Secs. 21.3-2, 21.3-3).

Our discussion above is by no means complete. For example, the important question of optimum linear feedback (e.g., servo) systems, where the ideas of prediction and smoothing are particularly relevant, has not been considered. Moreover, as treated above, the problem of maximization of signal-to-noise ratios has not touched upon such generalizations of conventional systems as correlation detectors,[26–29] where the present techniques are certainly appropriate. However, because of limitations of space, these topics are beyond the range of our study here. We refer the reader to the literature[8,9] and to Chaps. 19 to 21, where some of these questions are reconsidered from a more general statistical viewpoint.

PROBLEMS

16.12 Show that Eq. (16.99) is a necessary and sufficient condition for maximizing the signal-to-noise ratio $[S/N]^2_{in}$ [Eq. (16.95)].

16.13 If the input noise to the linear filter which is to minimize $[S/N]_{in}^2$ is narrowband *LRC* noise, show that for finite sampling times $(0,T)$ the optimum filter is

$$\begin{aligned} h_T(t)_{op} \doteq {} & 2(W_0^{-1}\omega_0^{-4})[(\omega_0^4 S_{in} + 2\omega_0^2 \ddot{S}_{in} + S_{in}^{(\mathrm{iv})})_{\epsilon_0 - t} \\ & + \delta(t - T)(-\dddot{S}_{in} + \omega_0^2 \dot{S}_{in})_{\epsilon_0 - T} + \delta'(t - T)(-\ddot{S}_{in} + \omega_0^2 S_{in})_{\epsilon_0 - T} \\ & - \delta(t - 0)(\dddot{S}_{in} - \omega_0^2 \dot{S}_{in})_{\epsilon_0} + \delta'(t - 0)(\ddot{S}_{in} - \omega_0^2 S_{in})_{\epsilon_0}] \\ & \qquad 0- < t < T+ \qquad (\omega_F \doteq 0) \qquad (1) \end{aligned}$$

where $\psi_N = W_0\omega_0^2/8\omega_F$, $\omega_F = R/2L$, $\omega_0 = (LC)^{-1}$, $Q \equiv \omega_0/2\omega_F \gg 1$.

16.14 (*a*) Calculate $[S/N]_{in\ \max}^2$ for the filter of Prob. 16.13.

(*b*) Obtain $h_T(t)_{op}$ above (Prob. 16.13) when S_{in} is a carrier (at f_0) modulated by a train of periodically spaced rectangular pulses.

16.15 (*a*) Calculate $h_\infty(t)$ in the nonrealizable case with a general *LRC* noise background, and sketch $|Y_\infty(i\omega)|^2$ for the high-Q condition.

(*b*) A general signal is observed in an additive mixture of system noise N_S and clutter N_C. Show that the condition for which $[S/N]_{in}^2$ [Eq. (16.95)] is a maximum is

$$\int_{0-}^{T+} [K_C(t,u) + K_{N_S}(|t - u|)]h_T(u)_{op}\, du = \gamma_0 S_{in}(\epsilon_0 - t) \qquad 0- < t,\ \epsilon_0 < T+ \qquad (1)$$

If $K_C(t,u) = F_{in}(t)F_{in}(u)k_c(t - u)$, $S_{in}(t)$ is a periodic pulse train at radio frequencies, and the system noise is white, obtain $h_T(u)_{op}$.

REFERENCES

1. Davenport, W. B., Jr., R. A. Johnson, and D. Middleton: Statistical Errors in Measurements on Random Time Functions, *J. Appl. Phys.*, **23:** 377 (1952).
2. Wiener, N.: "Extrapolation, Interpolation, and Smoothing of Stationary Time Series," Technology Press, Cambridge, Mass.; Wiley, New York, 1949.
3. Kolmogoroff, A. N.: Interpolation und Extrapolation, *Bull. Acad. Sci. U.S.S.R.*, sér. math., **5:** 3–14 (1944).
4. Zadeh, L. A., and J. R. Ragazzini: An Extension of Wiener's Theory of Prediction, *J. Appl. Phys.*, **21:** 645 (1950).
5. Davis, R. C.: On the Theory of Prediction of Nonstationary Stochastic Processes, *J. Appl. Phys.*, **23:** 1047 (1952).
6. Van Vleck, J. H., and D. Middleton: A Theoretical Comparison of the Visual, Aural, and Meter Reception of Pulsed Signals in the Presence of Noise, *J. Appl. Phys.*, **17:** 940 (1946).
7. Zadeh, L. A., and J. R. Ragazzini: Optimum Filters for the Detection of Signals in Noise, *Proc. IRE*, **40:** 1223 (1952).
8. Truxal, J. G.: "Automatic Feedback Control System Synthesis," chap. 8, McGraw-Hill, New York, 1955.
9. Laning, J. H., Jr., and R. H. Battin: "Random Processes in Automatic Control," chaps. 6–8, McGraw-Hill, New York, 1956.
10. Grenander, U., and M. Rosenblatt: "Statistical Analysis of Stationary Time Series," Wiley, New York, 1957.
11. Costas, J. R.: Periodic Sampling of Stationary Time Series, *MIT Research Lab. Electronics Rept.* 156, May, 1950.
12. Davenport, W. B., Jr., and W. L. Root: "An Introduction to the Theory of Random Signals and Noise," McGraw-Hill, New York, 1957, secs. 11.2, 11.3, for example. See also Ref. 2, app. C(2).

13. Paley, R. E. A. C., and N. Wiener: Fourier Transforms in the Complex Domain, *Am. Math. Soc. Colloq. Publ.*, **10**: 16–19 (1934), theorem XII.
14. Bode, H. W., and C. E. Shannon: A Simplified Derivation of Linear Least-square Smoothing and Prediction Theory, *Proc. IRE*, **38**: 417 (1950).
15. Lee, Y. W., and C. A. Stutt: Statistical Prediction of Noise, *MIT Research Lab. Electronics Tech. Rept.* 129, 1949. See also Y. W. Lee, Communication Applications of Correlation Analysis, Symposium on Applications of Autocorrelation Analysis to Physical Problems (Office of Naval Research), Woods Hole, June, 1949.
16. Stutt, C. A.: Experimental Study of Optimum Filters, *MIT Research Lab. Electronics Tech. Rept.* 182, May, 1951.
17. James, H. M., N. B. Nichols, and R. S. Phillips: "Theory of Servomechanisms," chap. 7, pp. 369–370, Massachusetts Institute of Technology Radiation Laboratory Series, vol. 25, McGraw-Hill, New York, 1947.
18. Tsien, H. S.: "Engineering Cybernetics," sec. 16.5, McGraw-Hill, New York, 1954.
19. Booton, R. C.: An Optimization Theory for Time-varying Linear Systems with Nonstationary Statistical Inputs, *Proc. IRE*, **40**: 977 (1952).
20. Bendat, J.: A General Theory of Linear Prediction and Filtering, *J. Soc. Ind. Appl. Math.*, **4**: 131 (1956).
21. Phillips, R. S., and P. R. Weiss: Theoretical Calculations on Best Smoothing of Position Data for Gunnery Predictions, *MIT Radiation Lab. Rept.* 532, February, 1944.
22. Blum, M.: Generalization of the Class of Nonrandom Inputs of the Zadeh-Ragazzini Prediction Model, *IRE Trans. on Inform. Theory*, **IT-2**: 76 (June, 1956).
23. North, D. O.: Analysis of Factors Which Determine Signal-Noise Discrimination in Pulsed Carrier Systems, *RCA Tech. Rept.* PTR-6C, June, 1943.
24. Dwork, B. M.: Detection of a Pulse Superimposed on Fluctuation Noise, *Proc. IRE*, **38**: 771 (1950).
25. George, T. S.: Fluctuation of Ground Clutter Return in Airborne Radar Equipment, *Mon.* 22, Feb. 15, 1952, *Proc. IEE (London)*, **99** (pt. IV): 92 (1952).
26. Lee, Y. W., T. P. Cheatham, and J. B. Wiesner: Application of Correlation Analysis to the Detection of Periodic Signals in Noise, *Proc. IRE*, **38**: 1165 (1950).
27. Fano, R. M.: Signal-to-Noise Ratios in Correlation Detectors, *MIT Research Lab. Electronics Tech. Rept.* 186, February, 1951.
28. Faran, J. J., and R. Hills: Correlators for Signal Reception, *Harvard Univ. Acoustics Research Lab. Tech. Mem.* 27, September, 1952; see also The Application of Correlation Techniques to Acoustic Receiving Systems, *Harvard Univ. Acoustics Research Lab. Tech. Mem.* 28, November, 1952.
29. Green, P. E., Jr.: The Output Signal-to-Noise Ratio of Correlation Detectors, *IRE Trans. on Inform. Theory*, **IT-3**: 10 (March, 1957).

CHAPTER 17

SOME DISTRIBUTION PROBLEMS

So far, we have been concerned principally with a second-moment theory of the input and output processes characteristic of many common communication systems (cf. Chaps. 5, 12 to 15). Here, we shall extend the treatment to consider a few of the important, and analytically much more involved, situations where higher-order statistical information is required. One such problem arises in the case of the hierarchy of distribution densities $W_1, W_2, \ldots, W_n$, etc. (cf. Sec. 1.3-2), for the output process following a nonlinear operation $\boldsymbol{T}_g$, like rectification, clipping, etc. Another occurs in the calculation of the distributions of various linear and nonlinear functions and functionals of the input process. As we shall see in Chaps. 20, 21, and 23, this is typical of optimum and suboptimum systems for the detection and extraction of signals in noise. A central problem here, for instance, requires the evaluation of integrals of the form

$$\int_{-\infty}^{\infty} \frac{e^{-i\xi x}}{\det (\mathbf{I} - i\xi\mathbf{G})} \frac{d\xi}{2\pi} \qquad \int_{-\infty}^{\infty} \frac{e^{-i\xi x + g(i\xi)}}{[\det (\mathbf{I} - i\xi\mathbf{G})]^{1/2}} \frac{d\xi}{2\pi} \tag{17.1}$$

Other examples can also be found: the problem of the zero-crossing distributions and "first-passage times" (mentioned earlier, in Sec. 9.4-1), the higher-order statistics of linear and nonlinear systems with randomly time-varying elements, and the general nature of the output of linear and nonlinear systems with nonnormal inputs are representative. All these are related in the sense that they require for their solution a common and rather specialized body of technical apparatus.

The aim of the present chapter, accordingly, is to describe some of the techniques now available for handling the problems mentioned above. We begin in Sec. 17.1 with some preliminary results on the reduction of the general determinant $\det (\mathbf{I} + \gamma\mathbf{G})$ needed in the evaluation of integrals like Eqs. (17.1). With the help of Sec. 17.1, we next examine in Sec. 17.2 some specific distribution densities and quadratic functionals of interest in detection and extraction theory (cf. Chaps. 20, 21). Finally, in Sec. 17.3 and the associated exercises, we consider the question of the probability densities $W_1, W_2, \ldots$ after a variety of nonlinear operations followed by linear filtering. Our discussion concludes with a brief mention of several extensions and alternative approaches developed for these and related problems. Although space prevents anything like a comprehensive treatment here, some idea of the scope and limitations of these methods is indicated by the text and accompanying exercises. For further details and additional examples, the reader may consult the references at the end of the chapter.†

17.1 *Preliminary Results: Reduction of det* ($\mathbf{I} + \gamma\mathbf{G}$)

As we shall see later (cf. Secs. 19.3-1, 20.3), integrals of the type (17.1) occur frequently in the evaluation of optimum system performance. A principal technical problem in such cases is the reduction of quantities like det ($\mathbf{I} + \gamma\mathbf{G}$) to a form more convenient for analysis and, ultimately, for computation. Here, we shall consider two methods for achieving the desired reduction. The first requires the eigenvalues of a certain homogeneous integral equation and is exact for all γ. The second, which yields approximate expressions in many instances, avoids the calculation of eigenvalues employed in the first method and is particularly suited to problems of threshold reception (when γ is small) and the evaluation of moments (where γ is set equal to zero eventually).

17.1-1 The Eigenvalue Method.‡ We assume initially for the various constituents of det ($\mathbf{I} + \gamma\mathbf{G}$) that:

1. γ is in general a complex quantity ($0 \leq |\gamma| \leq \infty$).
2. $\mathbf{I}$ is the unit matrix, as before (cf. Sec. 7.3-1).
3. $\mathbf{G}$ is an $n \times n$ matrix all of whose eigenvalues are distinct and all of whose elements are real quantities. ($\mathbf{G}$, however, is not necessarily symmetrical.)

From (3), we can always find an $n \times n$ matrix $\mathbf{Q}$ which diagonalizes $\mathbf{G}$ by means of the similarity transformation‡

$$\mathbf{Q}^{-1}\mathbf{G}\mathbf{Q} = \mathbf{\Lambda} = [\lambda_j^{(n)}\delta_{jk}] \qquad \text{or} \qquad \mathbf{G}\mathbf{Q} = \mathbf{Q}\mathbf{\Lambda} \tag{17.2}$$

where the $\lambda_j^{(n)}$ ($j = 1, \ldots, n$) are the n distinct *eigenvalues* of $\mathbf{G}$ [and δ_{jk} is the usual Kronecker delta $\delta_{jj} = 1$, $\delta_{jk} = 0$ ($j \neq k$)]. The *eigenvectors* $\mathbf{f}_j$ (formed from the rows or columns of $\mathbf{Q}$) are found from the n linearly independent relations

$$\mathbf{G}\mathbf{f}_j = \lambda_j^{(n)}\mathbf{f}_j \qquad j = 1, \ldots, n \tag{17.3a}$$

† For the most part, Sec. 17.1 follows the work of Kac and Siegert[1] (Sec. 17.1-1) and Middleton[2] (Sec. 17.1-2). Section 17.2 is based on Middleton, while Sec. 17.3 and its associated problems borrow heavily from the investigations of Kac and Siegert, Emerson,[3] Meyer and Middleton,[4] Arthur,[5] and Lampard.[6] Also, in Sec. 17.3 reference is made to recent work of Rosenbloom,[7] Heilfron,[8] Trautman,[9] and Darling[10] and Siegert,[11,12] among others, which extends the present treatment.

‡ For further details, see, for example, Kac and Siegert, *op. cit.*, Middleton, *op. cit.*, and H. Margenau and G. Murphy, "The Mathematics of Physics and Chemistry," secs. 10.9, 10.14, 10.15, Van Nostrand, Princeton, N.J., 1943. See also above, Sec. 7.3-1, Eqs. (7.20a) to (7.23), and footnote referring to these equations.

subject to the orthonormality condition

$$\tilde{\mathbf{f}}_j\mathbf{f}_k = \delta_{jk} \qquad j, k = 1, \ldots, n \tag{17.3b}$$

after the eigenvalues have been determined from the secular equation $\det(\mathbf{G} - \lambda^{(n)}\mathbf{I}) = 0$. In any case, for matrices satisfying (3) above, we can use Eq. (17.2) and the fact that $\det \mathbf{AB} = \det \mathbf{A} \cdot \det \mathbf{B}$ to write

$$\det(\mathbf{I} + \gamma\mathbf{G}) = \det(\mathbf{Q}^{-1}\mathbf{I}\mathbf{Q} + \gamma\mathbf{Q}^{-1}\mathbf{G}\mathbf{Q}) = \det(\mathbf{I} + \gamma\mathbf{\Lambda})$$
$$= \prod_{j=1}^{n} (1 + \gamma\lambda_j^{(n)}) \tag{17.4}$$

This is the factored, or reduced, form of the nth-degree polynomial in γ represented by $\det(\mathbf{I} + \gamma\mathbf{G})$. Note that, when $\mathbf{G}$ is symmetrical (as well as real†), $\mathbf{Q}$ can be an orthogonal matrix and we can relax the constraint above on $\mathbf{G}$ that all its eigenvalues be distinct. If $\mathbf{G}$ is not symmetrical, we must reimpose this constraint.‡ However, for the physical processes considered here this is not a serious restriction.

In many of our subsequent applications, continuous (or analogue) sampling is ultimately required. Here, in passing from the continuous series to the continuous process the intervals between sampled values are allowed to become arbitrarily close, while the total number (n) of sampled values becomes infinite. We distinguish two situations: (1) the case where the data interval $(0,T)$ remains finite, and (2) where $T \to \infty$. In either instance, we postulate now that $\lim_{n\to\infty} \mathbf{G}$ goes over into $G(t_1,t_2)$ for all (t_1,t_2) in $(0,T)$, where G now has suitable continuity and convergence properties. The determinant (17.4) becomes

$$\mathfrak{D}_T(\gamma) \equiv \lim_{n\to\infty} \det(\mathbf{I} + \gamma\mathbf{G}) = \prod_{j=1}^{\infty} (1 + \gamma\lambda_j) \tag{17.5}$$

where $\mathfrak{D}_T$ is called a *Fredholm determinant.*§ Since the eigenvalues of $\mathbf{G}$ are all distinct, we may use the Hilbert theory of integral equations to write λ_j as the appropriate limiting form of the eigenvalues $\lambda_j^{(n)}$: more precisely, we have

$$\lambda_j = \lim_{n\to\infty} \left(\frac{T}{n}\lambda_j^{(n)}\right) \qquad T < \infty \tag{17.5a}$$

† If $\mathbf{G}$ is complex, the argument proceeds as above, except that now $\mathbf{Q}$ and $\mathbf{f}$ have complex elements and Eq. (17.3b) becomes $\tilde{\mathbf{f}}_j^* \mathbf{f}_k = \delta_{jk}$, etc. Then, if $\mathbf{G}$ is Hermitian (e.g., if $G_{ij} = G_{ji}^*$), $\mathbf{Q}$ can be a unitary matrix (cf. Margenau and Murphy, *op. cit.*)

‡ Even when the eigenvalues of $\mathbf{G}$ are not distinct *and* $\mathbf{G}$ is not symmetrical, it is still possible to write $\prod_j^n (1 + \gamma\lambda_j^{(n)})$ for $\det(\mathbf{I} + \gamma\mathbf{G})$, but now there no longer exists a matrix $\mathbf{Q}$ (and hence a set of eigenvectors $\mathbf{f}_j$) which can be used to diagonalize $\mathbf{G}$, since $\mathbf{G}$ cannot then be put into completely diagonal form.

§ See, for example, Lovitt,[18] sec. 17, p. 32.

[cf. Eqs. (17.6a), (17.6b)]. The Fredholm determinant $\mathfrak{D}_T(\gamma)$ is absolutely convergent† for all $0 \leq (t_1,t_2) \leq T$, provided that $|\gamma G(t_1,t_2)| \leq M_G$, where M_G is the maximum value of $|\gamma G|$ in the interval $(0,T)$. Similar remarks apply for the infinite interval when $T \to \infty$. Equation (17.5) accordingly represents the factored, or reduced, form of $\mathfrak{D}_T(\gamma)$ for the continuous case, analogous to the discrete situation above [Eq. (17.4)].

The integral equation from which the λ_j are found is, of course, the limit of the set of n simultaneous equations (17.3a) in the discrete case. The matrix $\mathbf{G}$ goes over into the kernel $G(t_1,t_2)$, the eigenvectors $\mathbf{f}_j$ become the eigenfunctions $\phi_j(t)$, and in place of the sum (17.3a) one gets an integral instead. Thus, multiplying both sides of Eq. (17.3a) by T/n, setting $t_k = kT/n$, $\Delta t = T/n$, $f_{jk} = \phi_j(t_k)$, etc. $(0 \leq t_k,\, t_l \leq T)$, so that we have

$$\sum_{k=1}^{n} \frac{T}{n} G(t_l,t_k)\phi_j(t_k) = \frac{T}{n} \lambda_j^{(n)} \phi_j(t_l) \qquad 0 \leq t_l \leq T \qquad j = 1, \ldots, n \tag{17.6a}$$

with the help of Eq. (17.5a) we obtain formally in the limit $n \to \infty$ the homogeneous integral equation

$$\int_0^T G(t,u)\phi_j(u)\,du = \lambda_j\phi_j(t) \qquad 0 \leq t \leq T\ (j = 1, \ldots, n) \tag{17.6b}$$

A sufficient condition that the eigenvalues λ_j be discrete and that the $\{\phi_j\}$ form a complete orthonormal set[13] is that the (real) kernel $G(t,u)$ be symmetric and positive definite on $(0 \leq t,\ u \leq T)$. Usually, $G(t,u)$ is positive semidefinite and such that at most there may be only a finite number of negative (real) eigenvalues; all other eigenvalues are (real) and positive. If the λ_j remain distinct (as well as discrete), Eq. (17.6b) holds as well for nonsymmetric kernels (we remember that symmetry is a sufficient condition, not a necessary one: there are nonsymmetric kernels where the $\{\phi_j\}$ form a complete orthonormal set). We shall encounter a few illustrations of this in Sec. 17.3. The orthonormality condition (17.3b) in either case becomes

$$\int_0^T \phi_j(u)\phi_k(u)\,du = \delta_{jk} \qquad j, k = 1, \ldots, \infty \tag{17.7}$$

where the $\{\phi_j\}$ form a complete orthonormal set (of weight 1) on $(0,T)$. Another useful, sufficient condition that the $\{\phi_j\}$ form a complete set is that the kernel $G(t,u)$ be the Fourier transform of a spectral density,[14] i.e., that $G(t,u) = G(|t-u|) = \mathfrak{F}^{-1}\{\mathfrak{W}(f)\}$ (cf. Prob. 17.3). For proofs of these statements, the reader may consult the appropriate references (cf. Lovitt,[18] Courant and Hilbert,[13] and Youla[14]). Note, incidentally, that when, as is frequently the case in applications, an unsymmetric kernel has the form

† *Ibid.*

$G(t,u) = A(u)K(|t-u|)$, where $A(u) \geq 0$ (and $\neq 0$, for some u), we can symmetrize it by writing

$$g(t,u) \equiv A(t)^{1/2}A(u)^{-1/2}G(t,u) = A(t)^{1/2}K(|t-u|)A(u)^{1/2} = g(u,t) \quad (17.8a)$$

and forming a new set of eigenfunctions

$$\psi_j(t) = A(t)^{1/2}\phi_j(t) \quad (17.8b)$$

The integral equation (17.6b) becomes now

$$\int_0^T g(t,u)\psi_j(u)\,du = \lambda_j\psi_j(t) \qquad 0 \leq t \leq T \quad (17.9)$$

The $\{\psi_j\}$ form a complete orthonormal set on $(0,T)$ with weight $A(t)^{-1}$; for example, Eq. (17.7) is specifically

$$\int_0^T A(u)^{-1}\psi_j(u)\psi_k(u)\,du = \delta_{jk} \quad (17.10)$$

provided, of course, that $g(t,u)$ is positive definite $\left[\text{and that } \int_0^T |A(u)|\,du < \infty\right]$. Kernels of the type $G(t,u) = A(u)K(|t-u|)$ are called *polar kernels*.[15] Eigenvalues and eigenfunctions for a number of different kernels $G(t,u)$ are listed in Table 17.1 with reference to several problems considered later (cf. Probs. 17.5, 17.6), in which they occur. A general method of solution for homogeneous integral equations like Eqs. (17.6b), (17.9) when the kernel is rational is outlined briefly in Sec. A.2.2.

17.1-2 The Trace Method.† Our second method of reducing det $(\mathbf{I} + \gamma\mathbf{G})$ to a more manageable form depends on the *trace* of the matrix $\mathbf{G}$ and its higher powers. To see how this comes about, let us start with the following development of the determinant as a polynomial[18] in γ:

$$\det(\mathbf{I} + \gamma\mathbf{G}) = \sum_{k=0}^{n} \frac{\gamma^k}{k!} D_k^{(n)} \qquad D_k^{(n)} \equiv \sum_{l_1 \cdots l_k}^{n} \begin{vmatrix} G_{l_1l_1} & G_{l_1l_2} & \cdots & G_{l_1l_k} \\ G_{l_2l_1} & G_{l_2l_2} & \cdots & \cdots \\ \cdots & \cdots & \cdots & \cdots \\ G_{l_kl_1} & \cdots & \cdots & G_{l_kl_k} \end{vmatrix}$$

$$1 \leq k \leq n \quad (17.11)$$

Evaluating the determinants $D_k^{(n)}$ shows that $D_k^{(n)}$ is a function of the traces of $\mathbf{G}, \mathbf{G}^2, \ldots, \mathbf{G}^k$, namely,

$$D_k^{(n)} = D_k^{(n)}(\text{tr }\mathbf{G}, \ldots, \text{tr }\mathbf{G}^k)$$

One finds that

$$D_0^{(n)} = 1 \qquad D_1^{(n)} = \text{tr }\mathbf{G} \qquad D_2^{(n)} = \text{tr}^2\,\mathbf{G} - \text{tr }\mathbf{G}^2$$
$$D_3^{(n)} = \text{tr}^3\,\mathbf{G} - 3\,\text{tr }\mathbf{G}\cdot\text{tr }\mathbf{G}^2 + 2\,\text{tr }\mathbf{G}^3$$
$$D_4^{(n)} = \text{tr}^4\,\mathbf{G} - 6\,\text{tr}^2\,\mathbf{G}\cdot\text{tr }\mathbf{G}^2 + 3\,\text{tr}^2\,\mathbf{G}^2 + 8\,\text{tr }\mathbf{G}\cdot\text{tr }\mathbf{G}^3 - 6\,\text{tr }\mathbf{G}^4, \text{ etc.} \quad (17.12)$$

† See Middleton, *op. cit.*, app. 2.

TABLE 17.1. $\int_0^T A(u)K(|t-u|)\phi_j(u)\,du = \lambda_j\phi_j(t)$

Model $\alpha, \beta, a^2 \geq 0; B > 0$	$A(u)$	$K(\|t-u\|)$	Eigenvalues	Eigenfunctions
1. $\beta = 0, T < \infty$ (Kac and Siegert†); distribution density of $\int_0^T x(t)^2\,dt$ (x = gauss process; $\bar{x} = 0$; RC noise; high-Q LRC noise)	a^2	$Be^{-\alpha\|t-u\|}$	$\lambda_j = \dfrac{2Ba^2}{\alpha(1+q_j^2)}$ q_j = positive roots of $\tan \alpha T q_j = \dfrac{-2q_j}{1-q_j^2}$	$A_j e^{-i\gamma_j(T-t)} + B_j e^{i\gamma_j(T-t)}$ $\gamma_j = \alpha q_j = \alpha\sqrt{\dfrac{2Ba^2}{\alpha\lambda_j} - 1}$
2. $0 < T < \infty$ (Price‡) *a*. Markoff scatter $A(u) = (ae^{-\beta u})^2$ (optimum detection of RC or high-Q LRC normal noise "signal" in white noise) *b*. RC or high-Q LRC noise into a square-law detector, followed by an RC low-pass filter $A(u)$; distributions of $\int_0^T A(u)I(t-a)^2\,du$	$a^2e^{-2\beta u}$	$Be^{-\alpha\|t-u\|}$	q_j = positive roots of $J_{\nu+1}(e^{-\beta T}q_j)N_\nu(q_j)$ $= N_{\nu+1}(e^{-\beta T}q_j)J_{\nu-1}(q_j)$ $= \sqrt{\dfrac{2\alpha Ba^2}{\lambda_j\beta^2}} \qquad \nu = \dfrac{\alpha}{\beta}$	$B_j\left[\dfrac{-N_{\nu-1}(q_j)}{J_{\nu-1}(q_j)} J_\nu(q_je^{-\beta t}) + N_\nu(q_je^{-\beta t})\right]$
3. $T \to \infty$ (Juncosa,§ Kac and Siegert†) *a*. As in (2*a*) *b*. As in (2*b*)	$a^2e^{-2\beta u}$	$Be^{-\alpha\|t-u\|}$	q_j = positive roots of $J_{\nu-1}(q_j) = 0$ $= \sqrt{\dfrac{2\alpha Ba^2}{\lambda_j\beta^2}} \qquad \nu = \dfrac{\alpha}{\beta}$	$A_jJ_\nu(q_je^{-\beta t})$

† M. Kac and A. J. F. Siegert, On the Theory of Noise in Radio Receivers with Square-law Detectors, *J. Appl. Phys.*, **18**: 383 (1947).

‡ R. Price, Optimum Detection of Random Signals in Noise, with Application to Scatter-multipath Communication. I, *IRE Trans. on Inform. Theory*, **IT-2**: 125 (December, 1956).

§ M. L. Juncosa, An Integral Equation Related to Bessel Functions, *Duke Math. J.*, **12**: 465 (1945).

From Eqs. (17.11) and (17.12), we can now, in fact, establish the following identity, which is basic to the trace method for reducing det $(\mathbf{I} + \gamma\mathbf{G})$ to more manageable forms. The identity in question is specifically (cf. Middleton,[2] appendix 2)

$$\exp\left[\sum_{m=1}^{\infty} \frac{(-1)^{m+1}}{m} \gamma^m \operatorname{tr} \mathbf{G}^m\right] \equiv \det(\mathbf{I} + \gamma\mathbf{G}) \qquad (17.13)$$

which holds (as we shall see presently) whenever the exponential series converges. Proof of Eq. (17.13) can be given in several ways. A direct method uses the determinantal expansion (17.11): we develop both members of Eq. (17.13) as a power series in γ, compare coefficients of γ^k, and observe for all $k > n$ that the coefficients of γ^k are identically zero. For $k \leq n$, we obtain simply the determinantal expansion (17.11), with $D_k^{(n)}$ [Eqs. (17.12)] the coefficient† of γ^k.

However, this direct approach does not readily reveal the condition of convergence for the exponential series. A more satisfactory demonstration of Eq. (17.13), which at the same time establishes the desired interval of convergence (in γ), starts with the eigenvalue representation (17.4) and makes use of the well-known result (cf. Prob. 17.4*b*) for any matrix $\mathbf{G}$ that

$$\sum_{j=1}^{n} \lambda_j^{(n)m} = \operatorname{tr} \mathbf{G}^m \qquad m \geq 0 \qquad (17.14)$$

Assuming again that $\mathbf{G}$ has only distinct eigenvalues‡ (cf. Sec. 17.1-1, items 1 to 3), we can write from Eq. (17.4)

$$\det(\mathbf{I} + \gamma\mathbf{G}) = \exp\left[\log \prod_{j=1}^{n} (1 + \gamma\lambda_j^{(n)})\right] = \exp\left[\sum_{j=1}^{n} \log(1 + \gamma\lambda_j^{(n)})\right] \qquad (17.15)$$

Now, let the n distinct eigenvalues $\lambda_j^{(n)}$ $(j = 1, \ldots, n)$ be arranged in descending order of magnitude with $|\lambda_1^{(n)}|$ the magnitude of the largest, for example, $|\lambda_1^{(n)}| > |\lambda_2^{(n)}| > \cdots > |\lambda_j^{(n)}| > \cdots > |\lambda_n^{(n)}|$. Then the logarithm can be expanded in an absolutely convergent series for all $\lambda_j^{(n)}$, provided that $|\gamma\lambda_1^{(n)}| < 1$, to give us

$$\sum_{j=1}^{n} \log(1 + \gamma\lambda_j^{(n)}) = \sum_{j=1}^{n}\sum_{m=1}^{\infty} \frac{(-1)^{m+1}}{m} (\gamma\lambda_j^{(n)})^m$$
$$= \sum_{m=1}^{\infty} \frac{(-1)^{m+1}}{m} \gamma^m \left(\sum_{j=1}^{n} \lambda_j^{(n)m}\right) \qquad |\gamma\lambda_1^{(n)}| < 1 \qquad (17.16)$$

† This is considerably simpler, in practice, than a direct evaluation of $D_k^{(n)}$ in Eq. (17.11).

‡ This is not a necessary restriction, but it simplifies the proof somewhat; see Prob. 17.4*c*.

where the interchange of summations is permitted, since the m series is absolutely convergent, $|\gamma\lambda_1^{(n)}| < 1$. But the summation over j is just tr $\mathbf{G}^m$ [cf. Eq. (17.14)], and so the identity (17.13), with the domain of convergence in γ, is established.

The region of convergence in the complex γ plane is determined solely by the *largest* eigenvalue of $\mathbf{G}$ (in absolute magnitude) and is a circle of radius $\rho < |\lambda_1^{(n)}|^{-1}$. The right-hand member of Eq. (17.13) is, of course, an entire function of γ, defined for all $|\gamma| < \infty$, and represents, in effect, the analytic continuation of the left-hand member outside the circle of convergence $|\lambda_1^{(n)}|^{-1}$. The convergence condition on the exponential suggests that this form may be particularly useful in analytic operations where the principal contributions occur for values of γ within the region of convergence. We shall see a number of examples of this in the next section, when we attempt to evaluate integrals of the type (17.1). Note, finally, that starting with a relation (17.14) we can also establish the determinantal expansion (17.12), or given Eq. (17.13), we can equally well prove the validity of Eq. (17.14).

In the continuous cases, when $n \to \infty$ (T fixed), we make use of Eq. (17.5a) again, multiplying both sides of Eq. (17.14) by $(T/n)^m$ and passing to the limit, to get formally

$$\sum_{j=1}^{\infty} \lambda_j^m = \lim_{n\to\infty} \left[\left(\frac{T}{n}\right)^m \operatorname{tr} \mathbf{G}^m\right] = \lim_{n\to\infty} \left[\sum_{l_1 \cdots l_m}^{n} G_{l_1l_2}G_{l_2l_3} \cdots G_{l_ml_1}\left(\frac{T}{n}\right)^m\right]$$
$$= \int_0^T \cdots \int_0^T G(t_1,t_2)G(t_2,t_3) \cdots G(t_m,t_1)\, dt_1 \cdots dt_m \equiv B_m^{(T)} \qquad m \geq 1 \quad (17.17)$$

[As before, it is assumed that $G(t,\tau)$ is quadratically integrable on $(0,T)$.] The $B_m^{(T)}$ are the *iterated kernels* of the basic integral equation (17.6b). Thus, for $m = 1, 2$, we get specifically

$$B_1^{(T)} = \int_0^T G(t,t)\, dt = \sum_{j=1}^{\infty} \lambda_j \tag{17.18a}$$

$$B_2^{(T)} = \iint_0^T G(t_1,t_2)G(t_2,t_1)\, dt_1\, dt_2 = \sum_{j=1}^{\infty} \lambda_j^2, \text{ etc.} \tag{17.18b}$$

(where G need not be symmetrical†). The fundamental identity (17.13) now becomes, with the help of Eq. (17.5) and the above,

$$\mathfrak{D}_T(\gamma) = \prod_{j=1}^{\infty} (1 + \gamma\lambda_j) = \exp\left[\sum_{m=1}^{\infty} \frac{(-1)^{m+1}}{m} \gamma^m B_m^{(T)}\right] \tag{17.19}$$

† Lovitt, *op. cit.*, pp. 32, 33.

The continuous analogue of Eqs. (17.11) is simply the power series in γ,

$$\mathfrak{D}_T(\gamma) = \sum_{k=0}^{\infty} \frac{\gamma^k}{k!} D_k^{(\infty)}(B_1^{(T)}, \ldots, B_k^{(T)}) \qquad (17.19a)$$

where $D_k^{(\infty)}$ is given by Eqs. (17.12), with the various traces replaced by the appropriate iterated kernels $B_k^{(T)}$ [cf. Eq. (17.17)]. Observe that this relation and the first equation of Eq. (17.19) are in fact valid for all γ, since the Fredholm determinant is absolutely and permanently convergent for the kernels G assumed here.†

The practical importance of the trace method is that frequently it permits an evaluation of the Fredholm determinant, and integrals depending upon it, without having to solve the associated homogeneous integral equation. In place of the eigenvalues λ_j and eigenfunctions $\{\phi_j\}$, we have instead to calculate the iterated kernels of Eq. (17.17) and then use the fundamental identity (17.19). For all γ within the circle of convergence, this is exact; for γ outside, we are usually led to asymptotic expressions. As we shall see in the next section, the chief utility of this approach occurs in those situations where the principal contributions occur for γ within the interval of convergence and where only the first few iterated kernels are then significant. The method, in any case, is particularly well suited to the evaluation of the various integrals that arise in threshold-detection theory (cf. Sec. 20.3) and in a variety of estimation problems (cf. Secs. 21.3-2, 21.3-3). The eigenvalue approach, on the other hand, is exact (as far as the Fredholm determinant is concerned) but frequently requires extensive calculations since the set of eigenvalues $\{\lambda_j\}$ may often form a but slowly converging series.‡ Moreover, subsequent operations [like the integrations (17.1) in the discrete case, for example] for the most part cannot be carried out without making use of the approximations inherent in the trace method itself, so that from a practical viewpoint there is then no ultimate advantage of the latter approach over the former. Similar remarks apply as well for the discrete situation: the discrete form (17.13) of the identity (17.19) is usually more convenient than the eigenvalue representation (17.4).

PROBLEMS

17.1 (*a*) Show that, when the eigenvalues of **G** are not distinct and $G_{ij} \neq G_{ji}$, it is still possible to write $\det(\mathbf{I} + \gamma\mathbf{G}) = \prod_{j=1}^{n} (1 + \gamma\lambda_j^{(n)})$.

† When G is symmetrical, we can establish these relations alternatively with the help of Mercer's theorem [cf. Eq. (8.61)].

‡ We mention that under certain conditions and for certain specific kernels G the Fredholm determinant [Eqs. (17.5), (17.19), (17.19*a*)] can also be expressed in closed form (Siegert[12]) [cf. Eq. (17.43*a*)].

(*b*) Obtain the condition for the absolute convergence of the Fredholm determinant $\mathfrak{D}_T$ on $(0 \leq t_1, t_2 \leq T)$. Discuss the limit $T \to \infty$.

(*c*) Show that, if the eigenvalues $\lambda_j = \lim_{n\to\infty} \left(\frac{T}{n} \lambda_j^{(n)}\right)$ of the integral equation (17.6*b*) are distinct, the kernel G can be nonsymmetric.

17.2 (*a*) Show with the help of Bessel's inequality

$$\sum_{j=1}^{m} \lambda_j^2 \leq \iint_0^T G^2(t,u)\, dt\, du < \infty \tag{1}$$

that the eigenvalues of symmetric, quadratically integrable kernels G are discrete.

(*b*) Extend this to nonsymmetric kernels G with at most a finite number of negative eigenvalues. What conditions on G must be satisfied?

(*c*) Discuss the necessity and sufficiency of these conditions.

17.3 Prove that a sufficient condition that the eigenfunctions $\{\phi_j\}$ of the homogeneous integral equation (17.6*b*) form a complete orthonormal set is that the kernel is the Fourier transform of a spectral density. D. C. Youla.[14]

17.4 (*a*) Establish the relation (17.11).

(*b*) Prove Eq. (17.14) for any (real) matrix **G**.

(*c*) Show that the basic identity (17.13) holds when **G** has multiple eigenvalues and that the condition of convergence for the exponential series is (again)

$$|\lambda_{1,0}\gamma| < 1$$

where $\lambda_{1,0}$ is the largest eigenvalue (apart from its multiplicity).

17.5 (*a*) Obtain the eigenvalues and eigenfunctions of

$$\int_0^T a^2 B e^{-\alpha|t-u|} \phi_j(u)\, du = \lambda_j \phi_j(t) \qquad 0 \leq t \leq T \tag{1}$$

(cf. Table 17.1, item 1).

(*b*) Show that for large orders

$$\lambda_j \doteq 2Ba^2Tg/[\pi^2(j-1)^2 + g(4+g)] \qquad j \gg 1,\ g \equiv \alpha T \tag{2}$$

17.6 (*a*) Obtain the eigenvalues and eigenfunctions of

$$\int_0^T a^2 B e^{-2\beta u - \alpha|t-u|} \phi_j(u)\, du = \lambda_j \phi_j(t) \qquad 0 \leq t \leq T \tag{1}$$

(cf. Table 17.1, item 2).

(*b*) Repeat, for

$$B \int_0^\infty (e^{-\beta u})^2 e^{-\alpha|t-u|} \phi_j(u)\, du = \lambda_j \phi_j(t) \qquad 0 \leq t \leq \infty \tag{2}$$

(cf. Table 17.1, item 3).

(*c*) For (*b*), when $\alpha/\beta = \nu = \frac{1}{2}, \frac{3}{2}$, verify that for all orders $(j \geq 1)$

$$\begin{aligned} \nu &= \tfrac{1}{2} \qquad \lambda_j = \frac{8\alpha B a^2}{(2j-1)^2\pi^2\beta} \\ \nu &= \tfrac{3}{2} \qquad \lambda_j = \frac{4\alpha B a^2}{j^2\pi^2\beta} \end{aligned} \tag{3}$$

17.7 (*a*) Obtain the following iterated kernels on $(0,T)$ for an *RC* noise process with covariance function $K(t) = \psi e^{-\alpha_0|t|}$, $\alpha_0 = (RC)^{-1}$:

$$B_1^{(T)} = T \qquad B_2^{(T)} = T^2 \frac{2\mu + e^{-2\mu} - 1}{2\mu^2} \tag{1}$$

$$B_3^{(T)} = \frac{3T^3}{2} \frac{(\mu - 1) + (\mu + 1)e^{-2\mu}}{\mu^3} \tag{2}$$

$$B_4^{(T)} = T^4 \left(\frac{2e^{-2\mu}}{\mu^2} + \frac{10e^{-2\mu} + 5}{2\mu^3} + \frac{28e^{-2\mu} + e^{-4\mu} - 29}{8\mu^4} \right) \tag{3}$$

$$B_5^{(T)} = 5T^5 \left(\frac{e^{-2\mu}}{3\mu^2} + \frac{3e^{-2\mu}}{2\mu^3} + \frac{20e^{-2\mu} + e^{-4\mu} + 7}{8\mu^4} + \frac{12e^{-2\mu} + e^{-4\mu} - 13}{8\mu^5} \right) \tag{4}$$

where $\mu \equiv \alpha_0 T$.

(*b*) Verify that for *LRC* noise, of covariance function $K(t) = \psi e^{-\alpha_1|t|}[\cos \omega_1 t + (\alpha_1/\omega_1) \sin \omega_1|t|]$ $[\omega_1 = \sqrt{\omega_0^2 - \alpha_1^2}, \omega_0^2 = (LC)^{-1}, \alpha_1 = R/2L]$, the iterated kernels in the high-*Q* case $(\omega_0/2\alpha_1 \gg 1)$ become

$$B_m^{(T)} \Big|_{LRC} \doteq 2^{1-m} B_m^{(T)} \Big|_{RC} \tag{5}$$

where μ $(= \alpha_0 T)$ is replaced by $\mu_{LRC} \equiv \alpha_1 T$ in the right-hand member of Eq. (5).

(*c*) Show that for large μ, that is, for $\alpha_0 T \gg 1$,

$$B_m^{(T)} \Big|_{RC} \simeq \frac{(2T)^m (\tfrac{1}{2})_{m-1}}{2(m-1)!} \mu_{RC}{}^{1-m} \qquad m \geq 1$$

while for small μ one gets

$$B_m^{(T)} \Big|_{RC} \doteq T^m \left\{ 1 - \frac{m\mu}{3} + \frac{\mu^3}{2!} \left[\frac{17m}{60} + \frac{m(m-3)}{9} \right] + O(\mu^3) \right\}_{RC} \qquad m \geq 2 \tag{6}$$

Use Eq. (5) to obtain the corresponding expressions for $B_m^{(T)} \Big|_{LRC}$ in the high-*Q* case.

17.2 *Distribution Densities and Functionals*†

Let us suppose that a random variable x is obtained from a random process $V(t)$ by a transformation of the type

$$x = \boldsymbol{T}_g\{V\} = \log G_n(\mathbf{V}) \tag{17.20}$$

where $\mathbf{V} = [V(t_1), \ldots, V(t_n)]$ is a column vector formed from the sequence (at the times $t_1 \leq t_2 \leq \cdots \leq t_n$) of n sampled values of $V(t)$ in an interval $(t_0, t_0 + T)$ and G_n is some real nonnegative function of V. Given G_n and the joint d.d. $W_n(V_1, \ldots, V_n)$, we ask now for the (first-order) density $W_1(x)$ of x. A closely related problem arises in the limit $n \to \infty$ (T fixed): we require here the d.d. of the *functional* $x(t) = \lim_{n\to\infty} \log G_n(\mathbf{V}) = \log G_T[V(t)]$. In principle, the desired distribution densities in either instance may be obtained from (1.28*b*), e.g.,

$$W_1(x;n) = \mathfrak{F}^{-1}\{F_1(i\xi;n)_x\} \tag{17.21a}$$

† See, for example, Appendix 2 and Middleton, *op. cit.*, secs. III, IV.

with the characteristic functions

$$F_1(i\xi;n)_x = \mathbf{E}_{\mathbf{V}}\{e^{i\xi \log G_n(\mathbf{V})}\} \qquad F_1(i\xi;\infty,T)_x \equiv \lim_{n\to\infty} F_1(i\xi;n)_x \tag{17.21b}$$

where $W_1(x;T) \equiv \lim_{n\to\infty} W_1(x;n)$, whenever these limits exist.

Both classes of problem occur frequently in noise theory and in the analysis of optimum and suboptimum communication systems, where we are ultimately interested in obtaining various statistics (like the moments and the distribution densities themselves) of the output process after selected nonlinear operations. In the present section, we shall confine our attention to the important cases where the original process V is normal and the transformation $\boldsymbol{T}_g$ [Eq. (17.20)] is quadratic. The evaluation of expressions like Eqs. (17.21) is illustrated with several examples of use to us in subsequent applications (cf. Sec. 20.3). Additional results are included in the problems.

17.2-1 Evaluation of Integrals. Here we consider three examples that are representative of the type of problem mentioned above.

Example 1. A generalized χ^2 distribution (noise alone). Let us determine the first-order distribution density of x [Eq. (17.20)], where specifically the transformation $\boldsymbol{T}_g\{V\}$ is

$$x = \log G_n(\mathbf{V}) = A_n + \tilde{\mathbf{v}}\mathbf{J}\mathbf{v} \qquad \mathbf{v} = \mathbf{V}/\psi^{1/2} \tag{17.22}$$

and $\mathbf{V}$ is obtained from a not necessarily stationary normal random noise process with zero mean and intensity ψ. The $n \times n$ matrix $\mathbf{J}$ is real and symmetrical, such that the quadratic form in Eq. (17.22) is positive definite. With the help of Eqs. (7.18) for $W_n(\mathbf{V})$, we can write directly for the characteristic function (17.21b)

$$F_1(i\xi;n)_x = \int_{-\infty}^{\infty} \cdots \int W_n(\mathbf{V}) e^{i\xi(A_n + \tilde{\mathbf{v}}\mathbf{J}\mathbf{v})}\, d\mathbf{V}$$

$$= e^{i\xi A_n} \int_{-\infty}^{\infty} \cdots \int \frac{e^{-\frac{1}{2}\tilde{\mathbf{v}}(\mathbf{k}_N^{-1} - 2i\xi\mathbf{J})\mathbf{v}}}{(2\pi)^{n/2}(\det \mathbf{k}_N)^{1/2}}\, d\mathbf{v} \tag{17.23}$$

where $\mathbf{k}_N$ $(= \psi^{-1}\mathbf{K}_N)$ is the normalized covariance matrix of the gaussian process V and $\mathbf{v} = \mathbf{V}/\psi^{1/2}$ [cf. Eq. (17.22)]. Using Eq. (7.26), we easily evaluate Eq. (17.23), obtaining the desired c.f.

$$F_1(i\xi;n)_x = e^{i\xi A_n}[\det(\mathbf{I} - 2i\xi\mathbf{k}_N\mathbf{J})]^{-1/2} \tag{17.24a}$$

(in which $\mathbf{k}_N\mathbf{J} \neq \mathbf{J}\mathbf{k}_N$ generally, although $\mathbf{k}_N$ and $\mathbf{J}$ are themselves symmetric). Alternatively, if $\lambda_j^{(n)}$ are the n eigenvalues of $\mathbf{G} \equiv \mathbf{k}_N\mathbf{J}$, we can write from Eq. (17.4), setting $\gamma = -2i\xi$,

$$F_1(i\xi;n)_x = e^{i\xi A_n} \prod_{j=1}^{n} (1 - 2i\xi\lambda_j^{(n)})^{-1/2} \tag{17.24b}$$

The corresponding probability density $W_1(x;n)$ is the Fourier transform [Eq. (17.21*a*)] of Eq. (17.24*a*) or Eq. (17.24*b*). Comparing Eq. (17.24*b*) with the usual expression $\boldsymbol{E}\{e^{i\xi x^2}\} = (1 - 2i\xi)^{-n/2}$ [cf. Cramér,[19] eq. (18.1-2)], we call $W_1(x)$ a *generalized* χ^2 *distribution density* (with n degrees of freedom). Note that, in the degenerate case of independent samples, where $\lambda_j^{(n)} = \lambda_0$ $(1 \leq j \leq n)$, we obtain the familiar χ^2 form [for the random variable $(x - A_n)\lambda_0^{-1}$], with the c.f. $F_1 = e^{i\xi A_n}(1 - 2i\xi\lambda_0)^{-n/2}$.

When $n = 1, 2$, $W_1(x;n)$ can be obtained in closed form, but for $n \geq 3$ exact expressions are not generally possible† (except in the degenerate situation $\mathbf{k}_N\mathbf{J} = \lambda_0\mathbf{I}$ and for special values of n). We must therefore resort to various approximate methods to obtain the desired distribution density. However, when the eigenvalues of $\mathbf{G} \equiv \mathbf{k}_N\mathbf{J}$ are distinct,‡ it is possible to obtain asymptotic forms§ of $W_1(x;n)$ directly with the aid of the basic identity (17.13). Letting $\gamma = -2i\xi$ therein, we can write at once

$$[\det(\mathbf{I} - 2i\xi\mathbf{k}_N\mathbf{J})]^{-1/2} = \exp\left[\sum_{m=1}^{\infty} \frac{2^{m-1}}{m}(i\xi)^m \operatorname{tr}(\mathbf{k}_N\mathbf{J})^m\right] \tag{17.25}$$

where the equality holds for all $|\xi|$ less than the reciprocal of the magnitude of the largest eigenvalue of $\mathbf{G}$ [cf. Eq. (17.16)]. Applying Eq. (17.25) to Eq. (17.24*a*) and using Eq. (17.21*a*), we have the asymptotic relation

$$W_1(x;n)_N \simeq \int_{-\infty}^{\infty} \frac{d\xi}{2\pi} \exp\{-i\xi[x - (A_n + \operatorname{tr}\mathbf{k}_N\mathbf{J})] - \xi^2 \operatorname{tr}(\mathbf{k}_N\mathbf{J})^2\} \times \exp\left[\sum_{m=3}^{\infty} \frac{2^{m-1}}{m}(i\xi)^m \operatorname{tr}(\mathbf{k}_N\mathbf{J})^m\right] \tag{17.26}$$

Retaining terms $O(\xi,\xi^2)$ only in the exponent, we obtain the normal d.d. [cf. Eqs. (7.2), (7.3)]

$$\begin{aligned} W_1(x;n)_N &\simeq \frac{e^{-(x-C_n^{(N)})^2/2D_n^{(N)}}}{\sqrt{2\pi D_n^{(N)}}} \\ C_n^{(N)} &\equiv A_n + \operatorname{tr}\mathbf{k}_N\mathbf{J} = A_n + \sum_j^n \lambda_j^{(n)} \\ D_n^{(N)} &\equiv 2\operatorname{tr}(\mathbf{k}_N\mathbf{J})^2 = 2\sum_j^n \lambda_j^{(n)2} \end{aligned} \tag{17.27}$$

† Numerical approximations can be obtained as indicated by Slepian.[19a]

‡ This is, in effect, another condition on $\mathbf{J}$, one which is not at all restrictive in actual applications.

§ The asymptotic character of expressions like (17.26), (17.27) may be established by the methods described, for example, by Cramér.[20] See also Gnedenko and Kolmogoroff,[21] secs. 45–47, and Chap. 7, Refs. 17, 18.

where higher-order terms [$m \geq 3$ in Eq. (17.26)] are for the moment assumed to be negligible. The same result is, of course, found from Eq. (17.24*b*) on expanding $\prod_j^n (1 - 2i\xi\lambda_j^{(n)})^{-1/2}$ as an exponential series and keeping only terms $O(\xi,\xi^2)$ in the exponent, as above; Eq. (17.14) gives us the eigenvalue representation of the mean $C_n^{(N)}$ and variance $D_n^{(N)}$ in Eq. (17.27).

The higher-order elements [$m \geq 3$ in Eq. (17.26)] yield a succession of correction terms, forming a *generalized Edgeworth series* (cf. Sec. 7.7-2). Following the procedure of Eqs. (7.76) et seq. for the one-dimensional case, we obtain specifically (cf. Prob. 17.9)

$$W_1(x;n)_N \simeq \{D_n^{(N)}\}^{-1/2}\left[\phi^{(0)}(z) - \frac{1}{3!}E_3^{(N)}\phi^{(3)}(z) + \frac{1}{4!}E_4^{(N)}\phi^{(4)}(z) + \tfrac{1}{72}E_3^{(N)2}\phi^{(6)}(z) + \cdots\right] \tag{17.28a}$$

with
$$z \equiv \frac{x - C_n^{(N)}}{(D_n^{(N)})^{1/2}} \qquad \phi^{(l)}(z) \equiv \frac{d^l}{dz^l}\frac{e^{-z^2/2}}{\sqrt{2\pi}}$$
and
$$E_{l(\geq 3)}^{(N)} \equiv \frac{2^{l-1}(l-1)!\,\mathrm{tr}\,(\mathbf{k}_N\mathbf{J})^l}{D_n^{(N)l/2}} \tag{17.28b}$$

The extent to which these higher-order terms can be ignored depends on the magnitudes of the $E_l^{(N)}$. In the degenerate case of independent samples, $\mathbf{k}_N\mathbf{J} = \lambda_0\mathbf{I}$, we have $\mathrm{tr}\,(\mathbf{k}_N\mathbf{J})^l = n\lambda_0^l$, so that

$$E_l^{(N)} = 2^{l/2-1}(l-1)!n^{1-l/2} \qquad (l \geq 3)$$

and the series (17.28*a*) is the usual Edgeworth series, with the first correction term $O(n^{-1/2})$ vis-à-vis the leading term. Here, however, we do not usually have independent sample values since the intervals between the various sampling instants t_k, t_{k+1}, etc., are not sufficiently large for us to treat $V(t_k)$ and $V(t_{k+1})$ as uncorrelated. The result is that $E_l^{(N)}$ is $O(n^{-\epsilon l})$ $(0 < \epsilon \leq 1/2)$, where ϵ may be quite small. Values of ϵ near zero imply, in effect, intervals $(t_0, t_0 + T)$ that are comparable to or less than the correlation time† T_c of the process $V(t)$. The correction terms in Eq. (17.28*a*) are no longer insignificant, and, in fact, the asymptotic development above fails: the higher-order terms must be retained in the exponent of Eq. (17.26), or, equivalently, the exact expressions, Eqs. (17.24*a*), (17.24*b*), for the integrand must be used for all ξ. In subsequent applications, unless otherwise indicated, we shall require that the interval $(t_0, t_0 + T)$ be sufficiently great, and n sufficiently large, that the number of effectively independent components of $\mathbf{V}$ is also large. Just how large T versus T_c and n versus unity

† The *correlation time* T_c of an (entirely) random process is defined as that interval $t = t_2 - t_1$ between any pair of observations for which the covariance function has fallen to some specified fraction η of its value at $t = 0$, for example, to $1/e$th, or to 1 per cent, etc.—that is, $K(T_c) = \eta K(0)$.

must be is determined from E_3, E_4, etc. [cf. Eqs. (17.28*b*)], and depends, of course, on $\mathbf{J}$ and on the specific covariance function $\mathbf{K}_N$ of the noise. Then the correction terms in Eqs. (17.28) are negligible and the desired d.d. [Eqs. (17.27)] is asymptotically normal, with mean $C_n^{(N)}$ and variance $D_n^{(N)}$ [Eqs. (17.27)]. This is, as expected, a special case of the central-limit theorem (cf. Sec. 7.7-3), where the original distribution is itself normal and is characteristic of a large-sample theory.

When $n \to \infty$ and T is fixed, we have the continuous version of the above, where x now becomes the quadratic functional $x_T(t) = A_T + \Phi_T[V(t)]$, provided, of course that $\lim_{n\to\infty} A_n \to A_T$ exists and $\lim_{n\to\infty} \tilde{\mathbf{v}}\mathbf{J}\mathbf{v} \equiv \lim_{n\to\infty} \Phi_n \to \Phi_T$ exists. Assuming that these limits do exist, then, we find from Eq. (17.5) applied to Eqs. (17.24*a*), (17.24*b*) that the characteristic function and the corresponding distribution density [Eqs. (17.21)] become

$$F_1(i\xi;\infty,T)_x = e^{i\xi A_T}\mathfrak{D}_T(-2i\xi)^{-\frac{1}{2}} \qquad W_1(x_T;T)_N = \int_{-\infty}^{\infty} \frac{e^{-i\xi(x_T-A_T)}}{\mathfrak{D}_T(-2i\xi)^{\frac{1}{2}}}\frac{d\xi}{2\pi} \tag{17.29}$$

where $\mathfrak{D}_T$ is the Fredholm determinant $\prod_{j=1}^{\infty} (1 - 2i\xi\lambda_j)$ whose eigenvalues λ_j are found from the basic integral equation (17.6*b*) in which the kernel $G(t,u)$ is the (continuous) limit of the l, kth element of $\mathbf{k}_N\mathbf{J}$ [cf. Eq. (17.6*a*)].

Again, it is not possible to evaluate $W_1(x_T;T)$ in closed form, except in certain degenerate cases, but for our subsequent applications we can readily obtain large-sample asymptotic results analogous to Eqs. (17.27), (17.28) for the discrete case. Proceeding as before [cf. Eqs. (17.25) et seq.], using now the continuous version [Eq. (17.19)] of the fundamental identity (17.13), we get directly

$$W_1(x_T;T)_N \simeq \frac{e^{-(x_T-C_T^{(N)})^2/2D_T^{(N)}}}{\sqrt{2\pi D_T^{(N)}}}$$

$$C_T^{(N)} = A_T + B_1^{(T)} = A_T + \sum_{j=1}^{\infty} \lambda_j \tag{17.30}$$

$$D_T^{(N)} = 2B_2^{(T)} = 2\sum_{j=1}^{\infty} \lambda_j^2$$

where $B_1^{(T)}$, $B_2^{(T)}$ are the first and second iterated kernels of the integral equation (17.6*b*) [cf. Eqs. (17.18) and (17.17)]. The generalized Edgeworth series corresponding to Eqs. (17.28) is obtained on replacing E_l therein by

$$E_l^{(T)} \equiv 2^{l-1}(l-1)!B_l^{(T)}/(2B_2^{(T)})^{l/2} \tag{17.30a}$$

and A_n by A_T, $C_n^{(N)}$ by $C_T^{(N)}$, $D_n^{(N)}$ by $2B_2^{(T)}$, etc. The extent to which the asymptotic normal d.d. [Eqs. (17.30)] is an adequate representation of the exact distribution density [Eqs. (17.29)] depends once more on the relation between the correlation time T_c of the original process $V(t)$ and the sample size T: if $T \gg T_c$, we can safely use the normal form (17.30), while if $T \sim T_c$ or less, even the generalized Edgeworth series breaks down and we are left with the usually intractable though exact expression (17.29). Fortunately, in most of our system design problems threshold performance is the critical condition of interest. Since this, in turn, is described by a large-sample theory,† where $T \gg T_c$ (cf. Sec. 19.4-3 for signal detection, Secs. 21.3-2, 21.3-3 for signal extraction), we may expect normal forms like Eqs. (17.27), (17.30) in most instances.

Example 2. A generalized χ^2 distribution (signal and noise). An extension of the preceding example, and a result of some later importance, is obtained when the original random process $V(t)$ in Eq. (17.22) is the sum of a deterministic signal and independent noise process, e.g., when $V(t) = S(t;\boldsymbol{\theta}) + N(t)$. Once more, we assume that $N(t)$ is normal, with $\bar{N} = 0$, $\overline{N^2} = \psi$, but not necessarily stationary.

To determine the first-order d.d. of x, as given by Eq. (17.22), we proceed as before by calculating the characteristic functions (17.21b), where now, however, the average over $\mathbf{V}$ is with respect to the d.d. of $S + N$ and S for the moment is regarded as fixed. Then, with $\mathbf{S} = [S(t_1), \ldots, S(t_n)]$, a column vector like $\mathbf{V}$ in Eq. (17.20), we have

$$F_1(i\xi;n)_x = \int_{-\infty}^{\infty} \cdots \int F_n(\mathbf{V}|\mathbf{S}) e^{i\xi(A_n+\tilde{\mathbf{V}}\mathbf{J}\mathbf{V})}\, d\mathbf{V} \tag{17.31a}$$

Here $F_n(\mathbf{V}|\mathbf{S})$ is simply the d.d. of $\mathbf{N} = \mathbf{V} - \mathbf{S}$, which for a normal process is, from Eq. (7.49),

$$F_n(\mathbf{V}|\mathbf{S}) = W_n(\mathbf{V} - \mathbf{S})_N = \frac{e^{-\frac{1}{2}(\tilde{\mathbf{v}}-a_0\tilde{\mathbf{s}})\mathbf{k}_N^{-1}(\mathbf{v}-a_0\mathbf{s})}}{(2\pi)^{n/2}\sqrt{\det \mathbf{k}_N}} \tag{17.31b}$$

where we have used the convenient normalizations

$$a_0 = A_0/\sqrt{2\psi} \qquad \sqrt{\psi}\, a_0\mathbf{s} = \mathbf{S} \qquad \mathbf{v}\sqrt{\psi} = \mathbf{V} \tag{17.31c}$$

The characteristic function (17.31a) now becomes

$$F_1(i\xi;n)_x = e^{i\xi A_n - \frac{1}{2}a_0^2\tilde{\mathbf{s}}\mathbf{k}_N^{-1}\mathbf{s}} \int_{-\infty}^{\infty} \cdots \int \frac{e^{a_0\tilde{\mathbf{s}}\mathbf{k}_N^{-1}\mathbf{v} - \frac{1}{2}\tilde{\mathbf{v}}(\mathbf{k}_N^{-1} - 2i\xi\mathbf{J})\mathbf{v}}}{(2\pi)^{n/2}(\det \mathbf{k}_N)^{\frac{1}{2}}}\, d\mathbf{v} \tag{17.32a}$$

which with the help of Eq. (7.26) reduces at once to

$$F_1(i\xi;n)_x = \frac{\exp\{i\xi A_n - \frac{1}{2}a_0^2\tilde{\mathbf{s}}\mathbf{k}_N^{-1}[\mathbf{I} - (\mathbf{I} - 2i\xi\mathbf{k}_N\mathbf{J})^{-1}]\mathbf{s}\}}{[\det(\mathbf{I} - 2i\xi\mathbf{k}_N\mathbf{J})]^{\frac{1}{2}}} \tag{17.32b}$$

This is to be compared with the case of noise alone [Eqs. (17.24a), (17.24b)], to which it clearly reduces when $S = 0$ (or $a_0 = 0$).

† Whenever the probability of decision errors is required to be small, the useful case in practice.

Although Eq. (17.32*b*) is exact, the corresponding probability density [Eq. (17.21*a*)] is intractable except for $n = 1, 2$ and the degenerate case where $\mathbf{G} \equiv \mathbf{k}_N\mathbf{J} = \lambda_0\mathbf{I}$ (cf. Prob. 17.9*c*). However, the asymptotic d.d. and associated generalized Edgeworth series, characteristic of the large-sample theory, can be derived directly by the methods of Example 1. Thus, using Eq. (17.25) and expanding Eq. (17.32*b*) about $\xi = 0$, for example,

$$\mathbf{I} - (\mathbf{I} - 2i\xi\mathbf{G})^{-1} = -2i\xi\mathbf{G} - (2i\xi)^2\mathbf{G}^2 - \sum_{m=3}^{\infty} 2^m(i\xi)^m\mathbf{G}^m \qquad \mathbf{G} \equiv \mathbf{k}_N\mathbf{J}$$

we have as the extension of Eq. (17.26), now for signal and noise,

$$W_1(x;n)_{S+N} \simeq \int_{-\infty}^{\infty} \frac{d\xi}{2\pi} \exp\left[-i\xi(x - C_n^{(S+N)}) - \tfrac{1}{2}\xi^2 D_n^{(S+N)}\right] \times \exp\left[\sum_{m=3}^{\infty} (i\xi)^m 2^{m-1}\left(\frac{1}{m}\operatorname{tr}\mathbf{G}^m + 2a_0^2\tilde{\mathbf{s}}\mathbf{J}\mathbf{G}^{m-1}\mathbf{s}\right)\right] \tag{17.33a}$$

where

$$C_n^{(S+N)} \equiv A_n + \operatorname{tr}\mathbf{G} + a_0^2\tilde{\mathbf{s}}\mathbf{J}\mathbf{s} \qquad D_n^{(S+N)} = 2(\operatorname{tr}\mathbf{G}^2 + 2a_0^2\tilde{\mathbf{s}}\mathbf{J}\mathbf{G}\mathbf{s}) \tag{17.33b}$$

Retaining terms $O(\xi,\xi^2)$ in the exponent only, we integrate to get the generalized Edgeworth series [cf. Prob. 17.9*b* and Eq. (17.28*a*)]

$$W_1(x;n)_{S+N} \simeq (D_n^{(S+N)})^{-\frac{1}{2}}\left\{\phi^{(0)}(u) - \frac{1}{3!}E_3^{(S+N)}\phi^{(3)}(u) + \left[\frac{1}{4!}E_4^{(S+N)}\phi^{(4)}(u) + \tfrac{1}{72}E_3^{(S+N)2}\phi^{(6)}(u)\right] + \cdots\right\} \tag{17.34a}$$

in which specifically

$$u \equiv \frac{x - C_n^{(S+N)}}{(D_n^{(S+N)})^{\frac{1}{2}}} \qquad E_{l(\geq 3)}^{(S+N)} \equiv \frac{2^{l-1}(l-1)!}{(D_n^{(S+N)})^{l/2}}(\operatorname{tr}\mathbf{G}^l + 2la_0^2\tilde{\mathbf{s}}\mathbf{J}\mathbf{G}^{l-1}\mathbf{s}) \tag{17.34b}$$

Again, for a sufficiently large number of independent sampled values, we can neglect the higher-order terms ($l \geq 3$) in the series (17.34*a*) and obtain once more the limiting normal distribution density for x, similar to Eqs. (17.27) but now with different mean and variance [Eq. (17.33*b*)].

Similar results are obtained for the quadratic functional $x_T = A_T + \Phi_T[V(t)]$, where it is postulated once more that the various functions $\lim_{n\to\infty} \tilde{\mathbf{s}}\mathbf{J}\mathbf{s}$, $\tilde{\mathbf{s}}\mathbf{J}\mathbf{G}^{l-1}\mathbf{s}$ are defined and that $\lim_{n\to\infty} A_n \to A_T$ exists as well. The characteristic function of the random functional x_T now becomes

$$F_1(i\xi;\infty,T)_x = e^{i\xi A_T}\mathfrak{D}_T(-2i\xi)^{-\frac{1}{2}} \lim_{n\to\infty} \exp\left\{-\tfrac{1}{2}a_0^2\tilde{\mathbf{s}}\mathbf{k}_N^{-1}[\mathbf{I} - (\mathbf{I} - 2i\xi\mathbf{G})^{-1}]\mathbf{s}\right\} \tag{17.35}$$

with $W_1(x_T;T)_{S+N} = \mathfrak{F}^{-1}\{F_1\}$ [Eq. (17.35)]. Although exact expressions for $W_1(x_T;T)$ are not known in the general case, whenever the correlation time T_c of the original noise process $N(t)$ is short compared with the sample time T, we get from Eqs. (17.34*b*), on replacing trace $\mathbf{G}^l$ and the various quadratic forms by the appropriate iterated kernels $B_m^{(T)}$ [Eq. (17.17)] and functionals, the expected limiting normal d.d. for continuous sampling, viz.,

$$\begin{aligned} W_1(x_T;T)_{S+N} &\simeq \frac{e^{-(x_T - C_T^{(S+N)})^2/2D_T^{(S+N)}}}{\sqrt{2\pi D_T^{(S+N)}}} \\ C_T^{(S+N)} &\equiv A_T + B_1^{(T)} + a_0^2\Phi_T(s,s)_J \\ D_T^{(S+N)} &\equiv 2B_2^{(T)} + 4a_0^2\Phi_T(s,s)_{JG} \end{aligned} \tag{17.36}$$

where now $\Phi_T(s,s)_J \equiv \lim_{n\to\infty} \tilde{\mathbf{s}}\mathbf{J}\mathbf{s}$ and $\Phi_T(s,s)_{JG} \equiv \lim_{n\to\infty} \tilde{\mathbf{s}}\mathbf{J}\mathbf{G}\mathbf{s}$. The corresponding generalized Edgeworth series is obtained from Eq. (17.34*a*) if we replace $D_n^{(S+N)}$ by $D_T^{(S+N)}$, etc., and $E_l^{(S+N)}$ by

$$E_{l(\geq 3)}^{(T)} = 2^{l-1}(l-1)!(D_T^{(S+N)})^{-l/2}(B_l^{(T)} + 2la_0^2 \lim_{n\to\infty} \tilde{\mathbf{s}}\mathbf{J}\mathbf{G}^{l-1}\mathbf{s}) \tag{17.37}$$

We shall encounter some examples of Eqs. (17.36), as well as Eqs. (17.30), later in the discussion of binary detection theory (cf. Chap. 20).

Example 3. An exact distribution density. As our final example, let us consider a case where an exact expression can be obtained for the probability density of x, regardless of sample size T vis-à-vis correlation time T_c. Again, the original process is assumed to be normal, and the transformation (17.22) is now modified to the quadratic form

$$x = \log G_n(\mathbf{V}) = \log G_{2n}(\mathbf{V}_c,\mathbf{V}_s) = A_n + \tilde{\mathbf{v}}_c\mathbf{J}\mathbf{v}_c + \tilde{\mathbf{v}}_s\mathbf{J}\mathbf{v}_s \tag{17.38}$$

where $V(t)$ is narrowband, such that $V(t) = V_c(t)\cos\omega_0 t + V_s(t)\sin\omega_0 t$ [cf. Secs. 7.5-3(3), 7.5-3(4)], and the spectrum $\mathcal{W}_V(f)$ is *symmetrical* (about $f = f_0$), so that $\lambda_0(t) = 0$ (all t) [cf. Eqs. (7.58)]. (We employ the usual normalizations $\mathbf{v}_c = \mathbf{V}_c/\psi^{1/2}$, $\mathbf{v}_s = \mathbf{V}_s/\psi^{1/2}$.) The characteristic function of x is now

$$\begin{aligned} F_1(i\xi,n)_x &= E_{\mathbf{V}_c,\mathbf{V}_s}\{e^{i\xi\log G_{2n}(\mathbf{V}_c,\mathbf{V}_s)}\} \\ &= \int\cdots\int_{-\infty}^{\infty} e^{i\xi(A_n + \tilde{\mathbf{v}}_c\mathbf{J}\mathbf{v}_c + \tilde{\mathbf{v}}_s\mathbf{J}\mathbf{v}_s)} \frac{e^{-\frac{1}{2}\tilde{\mathbf{v}}_c\boldsymbol{\varrho}_0^{-1}\mathbf{v}_c - \frac{1}{2}\tilde{\mathbf{v}}_s\boldsymbol{\varrho}_0^{-1}\mathbf{v}_s}}{(2\pi)^n \det \boldsymbol{\varrho}_0}\, d\mathbf{v}_c\, d\mathbf{v}_s \end{aligned} \tag{17.39a}$$

where we have used Eqs. (7.56), (7.57) for the d.d. of $(\mathbf{v}_c,\mathbf{v}_s)$. This becomes (on dropping subscripts)

$$F_1(i\xi,n)_x = e^{i\xi A_n}\left[\int\cdots\int_{-\infty}^{\infty} \frac{e^{-\frac{1}{2}\tilde{\mathbf{v}}(\boldsymbol{\varrho}_0^{-1} - 2i\xi\mathbf{J})\mathbf{v}}}{(2\pi)^{n/2}(\det \boldsymbol{\varrho}_0)^{1/2}}\, d\mathbf{v}\right]^2$$

or

$$F_1(i\xi,n)_x = e^{i\xi A_n}[\det(\mathbf{I} - 2i\xi\boldsymbol{\varrho}_0\mathbf{J})]^{-1} = e^{i\xi A_n}\prod_{j=1}^{n}(1 - 2i\xi\lambda_j^{(n)})^{-1} \tag{17.39b}$$

with the help of Eqs. (7.26) and (17.4); the $\lambda_j^{(n)}$, as before, are the eigenvalues of the $n \times n$ matrix $\mathbf{G} = \varrho_0\mathbf{J}$.

On the assumption that $\mathbf{J}$ is such that all eigenvalues of $\mathbf{G}$ are distinct, it is a simple matter to obtain an exact result for the probability density

$$W_1(x;n) = \int_{-\infty}^{\infty} e^{-i\xi(x-A_n)} \prod_{j=1}^{n} (1 - 2i\xi\lambda_j^{(n)})^{-1} \frac{d\xi}{2\pi} \tag{17.40}$$

Observing that all poles of the integrand are simple (at $\xi = 1/2i\lambda_j^{(n)}$), we find[1] by a straightforward application of the calculus of residues that

$$W_1(x;n) = \sum_{l=1}^{n(+)} \frac{e^{-(x-A_n)/\lambda_l^{(n)}}}{\lambda_l^{(n)} \prod_k^{n}{}' (1 - \lambda_k^{(n)}/\lambda_l^{(n)})} \qquad x > A_n \tag{17.41a}$$

$$W_1(x;n) = \sum_{m=1}^{n(-)} \frac{e^{-(x-A_n)/\lambda_m^{(n)}}}{\lambda_m^{(n)} \prod_k^{n}{}' (1 - \lambda_k^{(n)}/\lambda_m^{(n)})} \qquad x < A_n \tag{17.41b}$$

Here, $n^{(+)}$, $n^{(-)}$ are, respectively, the number of positive and negative eigenvalues of $\mathbf{G} = \mathbf{k}_0\mathbf{J}$ such that $n^{(+)} + n^{(-)} = n$; the summations are over these positive or negative eigenvalues, while the prime on the product indicates that the factor for $k = l$ or $k = m$ is omitted.

With continuous sampling, Eqs. (17.39) become

$$F_1(i\xi;\infty,T) = e^{i\xi A_T}\mathfrak{D}_T(-2i\xi)^{-1} = e^{i\xi A_T} \prod_{j=1}^{\infty} (1 - 2i\xi\lambda_j)^{-1} \tag{17.42}$$

and Eqs. (17.41a), (17.41b) still apply, now for the d.d. of the quadratic functional

$$x_T = A_T + \Phi_T(v_c,v_c)_J + \Phi_T(v_s,v_s)_J \tag{17.43}$$

if the $\lambda_{l,m}^{(n)}$ are replaced by $\lambda_{l,m}$, where these eigenvalues are now obtained from the basic integral equation (17.6b) with the kernel $G(t,u)$ as the continuous limit of the l, kth element of $\mathbf{G}$, T fixed. (A finite number of negative eigenvalues at most are assumed, while $n^+ \to \infty$.)

Finally, in certain special cases, as Siegert has shown,[12] it is possible to express the Fredholm determinant $\mathfrak{D}_T(-2i\xi)$ in closed form. For example, if $G(t,u)$ is the RC noise kernel $Be^{-\alpha|t-u|}$ (cf. Table 17.1), it can be shown that, for all ξ [Siegert,[12] eqs. (29) to (36)],

$$\mathfrak{D}_T(-2i\xi) = e^{-\alpha T}\left\{\cosh\,[\alpha T b(i\xi)] + b(i\xi)^{-1}\left(1 - \frac{2i\xi B}{\alpha}\right)\sinh\,[\alpha T b(i\xi)]\right\}$$

$$b(i\xi) = 1 - \frac{4iB\xi}{\alpha} \tag{17.43a}$$

When $T \gg T_c$, so that a large-sample theory is meaningful, we can of course use the asymptotic approach outlined in the previous two examples to obtain the limiting normal d.d. of x. This is given by Eqs. (17.27), (17.30), (17.36) for both the discrete and continuous cases, except that now the means and variances are modified to

$$C_n^{(N)} \to C_{2n}^{(N)} = A_n + 2 \operatorname{tr} \boldsymbol{\varrho}_0\mathbf{J} \qquad D_n^{(N)} \to D_{2n}^{(N)} = 4 \operatorname{tr} (\boldsymbol{\varrho}_0\mathbf{J})^2 \tag{17.44a}$$
$$C_T^{(N)} \to C_{2T}^{(N)} = A_T + 2B_1^{(T)} \qquad D_T^{(N)} \to D_{2T}^{(N)} = 4B_2^{(T)} \tag{17.44b}$$

as a direct application of the basic identities (17.13), (17.19) shows. The expressions for $E_l^{(N)}$, $E_l^{(T)}$ [Eqs. (17.28b), (17.37)] follow at once when $\operatorname{tr} \mathbf{G}^m$ is replaced by $2 \operatorname{tr} \mathbf{G}^m$ $(m \geq 2)$ in numerator and denominator.

17.2-2 Moments and Semi-invariants. Although exact expressions for $W_1(x;n)$, $W_1(x;T)$ above are not generally possible, we can calculate precisely any desired moment of these distributions, since the characteristic function is available in each instance. Thus, from Eq. (1.31b) applied to Eq. (17.24a), for example, we have

$$\overline{x_N{}^k} = (-i)^k \frac{d^k}{d\xi^k} F_1(i\xi;n)_x \bigg|_{\xi=0} = (-i)^k \frac{d^k}{d\xi^k} \{e^{i\xi A_n}[\det (\mathbf{I} - 2i\xi\mathbf{G})]^{-\frac{1}{2}}\} \bigg|_{\xi=0} \tag{17.45}$$

for the kth moment of the quadratic functional x [Eq. (17.22)] when the original process $V(t)$ is normal noise and $\mathbf{G} = \mathbf{k}_N\mathbf{J}$. The basic identity (17.13) is particularly useful here: with the help of Eq. (17.25), we easily obtain the first two moments $\bar{x}_N$, $\overline{x_N{}^2}$ from Eq. (17.45), viz.,

$$\begin{aligned} \bar{x}_N &= -i \frac{d}{d\xi} (\exp \{i\xi(A_n + \operatorname{tr} \mathbf{G}) - \xi^2 \operatorname{tr} \mathbf{G}^2 + O[(i\xi)^3]\}) \bigg|_{\xi=0} \\ &= A_n + \operatorname{tr} \mathbf{G} = C_n^{(N)} \end{aligned} \tag{17.46a}$$
$$\begin{aligned} \overline{x_N{}^2} &= -\frac{d^2}{d\xi^2} (\exp \{i\xi(A_n + \operatorname{tr} \mathbf{G}) - \xi^2 \operatorname{tr} \mathbf{G}^2 + O[(i\xi)^3]\}) \bigg|_{\xi=0} \\ &= (A_n + \operatorname{tr} \mathbf{G})^2 + 2 \operatorname{tr} \mathbf{G}^2 = C_n^{(N)2} + D_n^{(N)} \end{aligned} \tag{17.46b}$$

Higher-order moments are obtained in the same way.

When the original process is the sum of signal and noise (as in Example 2 above), we may apply Eq. (17.45) to the characteristic function (17.32b), again using the basic identity $\mathbf{G} \equiv \mathbf{k}_N\mathbf{J}$ (17.13), with Eq. (17.25). Straightforward differentiation gives

$$\bar{x}_{S+N} = A_n + \operatorname{tr} \mathbf{G} + a_0{}^2\tilde{\mathbf{s}}\mathbf{J}\mathbf{s} = C_n{}^{(S+N)} \tag{17.47a}$$
$$\begin{aligned} \overline{x^2_{S+N}} &= (A_n + \operatorname{tr} \mathbf{G} + a_0{}^2\tilde{\mathbf{s}}\mathbf{J}\mathbf{s})^2 + 2 \operatorname{tr} \mathbf{G}^2 + 4a_0{}^2\tilde{\mathbf{s}}\mathbf{J}\mathbf{G}\mathbf{s} \\ &= C_n{}^{(S+N)2} + D_n{}^{(S+N)} \end{aligned} \tag{17.47b}$$

with the higher-order moments obtained in a similar fashion.

In the case of continuous sampling, $\operatorname{tr} \mathbf{G}$, $\operatorname{tr} \mathbf{G}^2$, etc., and the quadratic forms $\tilde{\mathbf{s}}\mathbf{J}\mathbf{s}$ are replaced by the appropriate iterated kernels and quadratic functionals. We have, then, directly from Eqs. (17.46), (17.47),

$$\overline{(x_T)_N} = A_T + B_1^{(T)} \qquad \overline{(x_T)_N^2} = (A_T + B_1^{(T)})^2 + 2B_2^{(T)} \tag{17.48a}$$

$$\begin{aligned} \overline{(x_T)_{S+N}} &= A_T + B_1^{(T)} + a_0^2\Phi(s,s)_J \\ \overline{(x_T)^2_{S+N}} &= [A_T + B_1^{(T)} + a_0^2\Phi_T(s,s)_J]^2 + 2B_2^{(T)} + 4a_0^2\Phi_T(s,s)_{JG} \end{aligned} \tag{17.48b}$$

The basic identities (17.13), (17.19), besides being particularly useful in the calculation of moments, also yield the semi-invariants L_m of the distribution by inspection. Thus, comparing Eq. (1.35) with $\log F_1(i\xi;n)_x$ [cf. Eqs. (17.24a), (17.25)], we get for noise alone†

$$L_1^{(N)} = A_n + \text{tr } \mathbf{G} \qquad L^{(N)}_{m(\geq 2)} = 2^{m-1}(m-1)! \text{ tr } \mathbf{G}^m \tag{17.49}$$

so that the $E_l^{(N)}$ in the Edgeworth series (17.28a) are equivalently

$$(L_2^{(N)})^{-l/2}L_l^{(N)}$$

For the additive signal and noise in Example 2, the semi-invariants are

$$\begin{aligned} L_1^{(S+N)} &= A_n + \text{tr } \mathbf{G} + a_0^2\tilde{\mathbf{s}}\mathbf{J}\mathbf{s} \\ L^{(S+N)}_{m(\geq 2)} &= 2^{m-1}(m-1)!(\text{tr } \mathbf{G}^m + 2ma_0^2\tilde{\mathbf{s}}\mathbf{J}\mathbf{G}^{m-1}\mathbf{s}) \end{aligned} \tag{17.50}$$

with $E_l^{(S+N)} = (L_2^{(S+N)})^{-l/2}L_l^{(S+N)}$. Corresponding expressions for continuous sampling follow as before on replacing trace $\mathbf{G}^m$ by the iterated kernel $B_m^{(T)}$, etc. Additional examples are given in the problems at the end of the section and in Chap. 20.

17.2-3 Probability Distribution of the Spectral Density of a Normal Process. In Sec. 3.2, we have examined the first moment of the spectral intensity density‡ $\mathcal{W}_y(f)_T \equiv 2|S_y(f)_T|^2/T$ of the truncated process y_T when $\mathcal{W}_y(f)_T$ is considered over the ensemble of representations $\mathcal{W}_y^{(j)}(f)_T$ (all j) [cf. Eq. (3.32)]. Here, as a further illustration of the techniques of the preceding sections, we shall determine the probability density of this random variable $\mathcal{W}_y(f)_T \equiv \mathcal{W}_T$ when the truncated process y_T is obtained from an originally stationary normal process $y(t)$ with zero mean ($\bar{y} = 0$). We shall also, incidentally, verify the statement in the earlier argument (cf. pages 140 and 141) that $\lim_{T\to\infty} \mathcal{W}_y^{(j)}(f)_T$, where $\mathcal{W}_y^{(j)}(f)_T$ is defined according to Eq. (3.32), does not exist (for almost all j) by showing that the random variable $\mathcal{W}_\infty = \lim_{T\to\infty} \mathcal{W}_y(f)_T$ has a nonvanishing variance.

† In place of tr $\mathbf{G}^m$, we can, of course, use $\sum_{j=1}^{n} \lambda_j^{(n)m}$ [Eq. (17.14)].

‡ In Sec. 17.2-3 we have modified the notation somewhat vis-à-vis Sec. 3.2-2; specifically $\mathcal{W}_y(f)_T$ here is the random variable, i.e., the *unaveraged* spectral intensity density, while in Eqs. (3.37) et seq. $\mathcal{W}_y(f)_T$ refers to the first moment of this variable. Comparisons with Chap. 3 are accordingly made without ambiguity if this is borne in mind.

Let us begin by defining a pair of random variables u_T, v_T according to

$$u_T - iv_T \equiv \left(\frac{2}{T}\right)^{1/2} \int_{-T/2}^{T/2} y(t)e^{-i\omega t}\,dt \qquad \omega = 2\pi f \tag{17.51a}$$

$$= \left(\frac{2}{T}\right)^{1/2} S_y(f)_T \tag{17.51b}$$

this last from Eq. (3.31), where $S_y(f)_T$ is the *amplitude spectral density* of the truncated process $y_T = y(t)$ $(-T/2 \le t \le T/2)$, $y_T = 0$ outside this interval.† In terms of (u_T,v_T), the random variable $\mathcal{W}_T$ that is the spectral intensity density is now simply‡

$$\mathcal{W}_T = |u_T - iv_T|^2 = u_T^2 + v_T^2 = \frac{2}{T}|S_y(f)_T|^2 \equiv \mathcal{W}_y(f)_T \tag{17.52}$$

Since $y(t)$ is normal, so also are u_T and v_T normal *functionals* (cf. Sec. 8.1-3), with a d.d. which from Eqs. (8.21) to (8.23) is easily found to be (cf. Prob. 17.12*a*)

$$W_1(u_T,v_T) = \frac{e^{-\frac{1}{2}\tilde{\mathbf{X}}\mathbf{K}_X^{-1}\mathbf{X}}}{2\pi\sqrt{\det \mathbf{K}_X}} \tag{17.53a}$$

with
$$\mathbf{X} = \begin{bmatrix} u_T \\ v_T \end{bmatrix} \qquad \mathbf{K}_X = \begin{bmatrix} A_T(\omega) & B_T(\omega) \\ B_T(\omega) & C_T(\omega) \end{bmatrix} \tag{17.53b}$$

where specifically

$$\begin{aligned}
A_T(\omega) &= \frac{2}{T}\iint_{-T/2}^{T/2} \cos\omega\tau_1\, K_y(\tau_1,\tau_2)\cos\omega\tau_2\, d\tau_1\, d\tau_2 \\
C_T(\omega) &= \frac{2}{T}\iint_{-T/2}^{T/2} \sin\omega\tau_1\, K_y(\tau_1,\tau_2)\sin\omega\tau_2\, d\tau_1\, d\tau_2 \\
B_T(\omega) &= \frac{1}{T}\iint_{-T/2}^{T/2} \sin\omega(\tau_1+\tau_2)\, K_y(\tau_1,\tau_2)\, d\tau_1\, d\tau_2
\end{aligned} \tag{17.54}$$

Here, K_y is the covariance function of the original process $y(t)$, and since $y(t)$ is also assumed to be stationary, we may replace $K_y(\tau_1,\tau_2)$ by $K_y(|\tau_1 - \tau_2|)$ in the usual way.§

† The integral (17.51*a*) is understood in the usual mean-square sense [cf. Eqs. (2.12) et seq.]. We can also equally well define $y_T(t)$ on the interval $(0,T)$.

‡ See footnote ‡ on p. 743.

§ Note that it is not necessary for the success of our analysis that $y(t)$ be stationary; however, $\mathcal{W}_T$ defined according to Eq. (17.52), with Eqs. (17.51), is different from a spectral intensity density [Eq. (17.52)] where y is truncated on the interval $(0,T)$, say. The *form* [cf. Eqs. (17.58)] of our ultimate result remains the same, but the parameters [cf. Eqs. (17.54)] of the distribution are, of course, different, depending on the interval chosen.

From Eq. (17.52), we observe next that $\mathcal{W}_T$ is the result of a quadratic transformation of the type (17.22), where specifically we have $A_n = 0$ and $\mathbf{J} = \mathbf{I}$, that is,

$$\mathcal{W}_T = \boldsymbol{T}_g\{\mathbf{X}\} = \tilde{\mathbf{X}}\mathbf{J}\mathbf{X} = \tilde{\mathbf{X}}\mathbf{X} \tag{17.55}$$

with $n = 2$. The characteristic function of $\mathcal{W}_T$ follows at once from Eqs. (17.23), (17.24), with $\mathbf{v}$ replaced by $\mathbf{X}$ and $\mathbf{k}_N$ by $\mathbf{K}_X$, so that $\mathbf{G} \equiv \mathbf{k}_N\mathbf{J} = \mathbf{K}_X$ here. The result is

$$F_1(i\xi;2)_{\mathcal{W}_T} = [\det(\mathbf{I} - 2i\xi\mathbf{K}_X)]^{-1/2} = [1 - 2ia_T(\omega)\xi + (i\xi)^2 b_T(\omega)]^{-1/2} \tag{17.56a}$$

where we have expanded the 2×2 determinant with the help of Eq. (17.53b), setting

$$a_T(\omega) \equiv A_T + C_T = \frac{2}{T} \iint\limits_{-T/2}^{T/2} \cos\omega(\tau_2 - \tau_1)\, K_y(\tau_1,\tau_2)\, d\tau_1\, d\tau_2 = \overline{\mathcal{W}_y(f)_T} \tag{17.56b}$$

[cf. Eq. (3.40) and footnote on page 142]

$$b_T(\omega) \equiv 4(A_TC_T - B_T{}^2) = \frac{4}{T^2} \int_{-T/2}^{T/2} \cdots \int K_y(\tau_1,\tau_2)K_y(t_1,t_2)$$
$$\times \{[\cos\omega(\tau_1 + \tau_2) + \cos\omega(\tau_1 - \tau_2)][\cos\omega(t_1 - t_2) - \cos\omega(t_1 + t_2)]$$
$$- \sin\omega(\tau_1 + \tau_2)\sin\omega(t_1 + t_2)\}\, d\tau_1 \cdots dt_2 \tag{17.56c}$$

The random variable $\mathcal{W}_T$ accordingly possesses a generalized χ^2 distribution with $n = 2$ degrees of freedom. Moreover, the probability density corresponding to the c.f. (17.56a) can be obtained exactly, as we shall see directly.

Setting $p = i\xi\sqrt{b_T}$, we can express the d.d. of $\mathcal{W}_T$ (for all $f > 0$†) as

$$W_1(\mathcal{W}_T)_{f>0} = \mathfrak{F}^{-1}\{F_1(i\xi;2)_{\mathcal{W}_T}\} = \frac{1}{\sqrt{b_T}} \int_{-\infty i}^{\infty i} \frac{e^{-p\mathcal{W}_T/\sqrt{b_T}}}{\sqrt{1 - 2\mu_T p + p^2}} \frac{dp}{2\pi i} \qquad \mu_T \equiv \frac{a_T}{\sqrt{b_T}} \tag{17.57}$$

The denominator can be factored; for example, $\sqrt{1 - 2\mu_T p + p^2} = \sqrt{(p - p_1)(p - p_2)}$, where $p_{1,2} = \mu_T \pm \sqrt{\mu_T{}^2 - 1}$ are the two roots of $1 - 2\mu_T p + p^2 = 0$. Moreover, one can show that a_T, $b_T > 0$, that is, $\mu_T > 1$, and that $a_T{}^2 - b_T > 0$ (cf. Prob. 17.12c), so that the roots $p_{1,2}$ are also positive. Integration‡ then gives (cf. Prob. 17.12c) for Eq. (17.57)

† When $f = 0$, one can show that $b_T = 0$ (cf. Prob. 17.12d).

‡ See, for example, No. 555 in G. A. Campbell and R. M. Foster, "Fourier Integrals," Van Nostrand, Princeton, N.J., 1948, where $\rho = -p_1$, $\sigma = -p_2$, and $\sqrt{1 - 2\mu_T p + p^2} = \sqrt{(p + \rho)(p + \sigma)}$.

the desired result,

$$W_1(\mathcal{W}_T)_{f>0} = \begin{cases} \dfrac{e^{-\mu_T \mathcal{W}_T/\sqrt{b_T}}}{\sqrt{b_T}} I_0(\sqrt{\mu_T^2 - 1}\, \mathcal{W}_T/\sqrt{b_T}) & \mathcal{W}_T > 0 \\ 0 & \mathcal{W}_T < 0 \end{cases} \tag{17.58}$$

where the frequency f $(= \omega/2\pi)$ and sample size T are parameters of the distribution [for $f = 0$, see Prob. 17.12, Eqs. (4)]. Figure 17.1 shows some

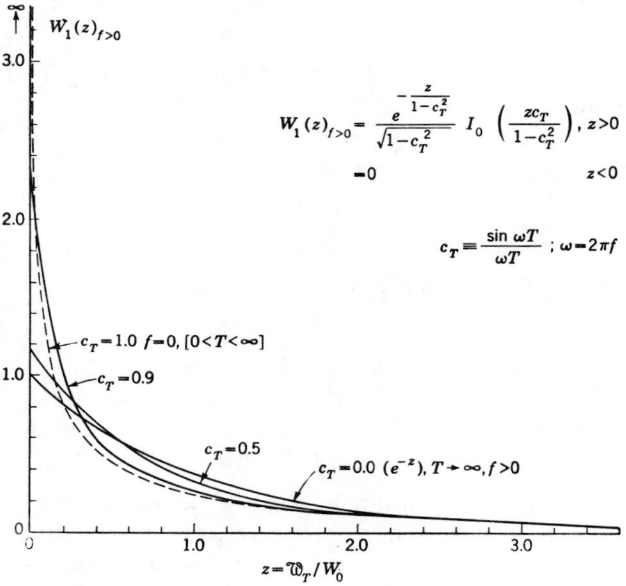

FIG. 17.1. The probability density of the spectral density $\mathcal{W}_T$ $(f > 0)$ for an originally stationary white normal random process.

typical curves of $W_1(\mathcal{W}_T)$ in the case where $y(t)$ is a purely random process with spectral density W_0 for all frequencies. Note in any case that $\mathcal{W}_T$ is not normally distributed even though $y_T(t)$ is gaussian.

For stationary processes, it can be shown when $T \to \infty$ that (cf. Prob. 17.12b)

$$\lim_{T\to\infty} b_T(\omega) = \lim_{T\to\infty} a_T^2(\omega) = \lim_{T\to\infty} [\overline{\mathcal{W}_y(f)_T}]^2 = \overline{\mathcal{W}_y(f)_\infty}^2 = \mathcal{W}_y(f)^2 \tag{17.59}$$

this last from Eqs. (3.40), (3.41), where $\mathcal{W}_y(f)$ is the spectral density of the original process $y(t)$. Then, $\lim_{T\to\infty} \mu_T = 1$, and Eqs. (17.58) reduce to the simple expression

$$W_1(\mathcal{W}_\infty)_{f>0} = \begin{cases} \dfrac{e^{-\mathcal{W}_\infty/\overline{\mathcal{W}}_\infty}}{\overline{\mathcal{W}}_\infty} & \mathcal{W}_\infty > 0 \\ 0 & \mathcal{W}_\infty < 0 \end{cases} \tag{17.60}$$

[cf. Eqs. (3.35*b*)]. This is the usual χ^2 distribution, now for $\mathcal{W}_\infty = u_\infty{}^2 + v_\infty{}^2$, with two degrees of freedom [cf. Eq. (17.52)]. [When $f = 0$, however, see Prob. 17.12, Eqs. (5).]

From Eqs. (17.58) and (17.60), it is a simple matter to determine the d.d. of the modulus $Z_T = |u_T - iv_T|$ of $u_T - iv_T$, for example, of $Z_T = (2/T)^{1/2}|S_y(f)_T| = \sqrt{\mathcal{W}_T}$. We have directly (cf. footnote, Sec. 1.2-14)

$$W_1(Z_T)_{f>0} = \begin{cases} W_1[\mathcal{W}_T(Z_T)] \left| \dfrac{d\mathcal{W}_T}{dZ_T} \right| = \\ \dfrac{2Z_T e^{-\mu_T Z_T{}^2/b_T{}^{1/2}}}{\sqrt{b_T}} I_0\left(\dfrac{\sqrt{\mu_T{}^2 - 1}}{b_T{}^{1/2}} Z_T{}^2\right) & Z_T > 0 \\ 0 & Z_T < 0 \end{cases} \tag{17.61a}$$

from Eqs. (17.58), and from Eqs. (17.60), with $\lim\limits_{T\to\infty} Z_T = \mathcal{W}_\infty{}^{1/2}$, we get

$$W_1(Z_\infty)_{f>0} = \begin{cases} \dfrac{2Z_\infty e^{-Z_\infty{}^2/\overline{Z_\infty{}^2}}}{\overline{Z_\infty{}^2}} & Z_\infty \geq 0 \\ 0 & Z_\infty < 0;\ \overline{Z_\infty{}^2} = \overline{\mathcal{W}}_\infty \end{cases} \tag{17.61b}$$

which is a simple Rayleigh d.d. [cf. Eqs. (7.100)].

The various moments of $\mathcal{W}_T$ may be found by differentiating the characteristic function (17.56*a*) [cf. Eq. (17.45)]. Thus, we easily obtain for the mean and mean square

$$\overline{\mathcal{W}}_T = a_T(\omega) \qquad \overline{\mathcal{W}_T{}^2} = 3a_T{}^2(\omega) - b_T(\omega) \tag{17.62}$$

Accordingly, the variance of $\mathcal{W}_T$ is

$$\overline{\mathcal{W}_T{}^2} - \overline{\mathcal{W}}_T{}^2 = 2a_T{}^2(\omega) - b_T(\omega) \qquad (> 0) \tag{17.63a}$$

which, in the limit $T \to \infty$, becomes, with the help of Eq. (17.59), simply

$$\overline{\mathcal{W}_\infty{}^2} - \overline{\mathcal{W}}_\infty{}^2 = \lim_{T\to\infty} a_T{}^2(\omega) = \mathcal{W}_y(f)^2\ (> 0) \qquad \text{for some } f > 0 \tag{17.63b}$$

Thus, for a gaussian random process (truncated or not) the variance of $\mathcal{W}_T$ is greater than zero, at least for some values of frequency f, a result which remains true in the limit $T \to \infty$. Then, by the argument of pages 140 and 141, it follows that $\lim\limits_{T\to\infty} \mathcal{W}_y{}^{(j)}(f)_T$ [cf. Eq. (3.32)] does *not* have a single given value for almost all members of the ensemble, i.e., for almost all j.

An important special case of our general result (17.58) arises when the original process $y(t)$ is a stationary normal "white" noise, e.g., a purely random process, with a uniform spectral intensity W_0 (> 0) for all frequencies. Then, by Eq. (3.56*b*), $K_y(\tau_1,\tau_2) = (W_0/2)\delta(\tau_2 - \tau_1)$; the elements of the matrix $\mathbf{K}_X$ [Eqs. (17.53*b*)] are explicitly [cf. Eqs. (17.54)]

$$\begin{aligned} A_T(\omega) &= \frac{W_0}{2}\left(1 + \frac{\sin 2\omega T}{2\omega T}\right) \\ C_T(\omega) &= \frac{W_0}{2}\left(1 - \frac{\sin 2\omega T}{2\omega T}\right) \\ B_T(\omega) &= W_0 \frac{1 - \cos 2\omega T}{4\omega T} \end{aligned} \tag{17.64}$$

so that from Eqs. (17.56*b*), (17.56*c*) we get

$$a_T(\omega) = W_0 \qquad b_T(\omega) = W_0{}^2\left[1 - \frac{\sin^2 \omega T}{(\omega T)^2}\right] \tag{17.65}$$

Letting

$$c_T \equiv \frac{\sin \omega T}{\omega T} \tag{17.66}$$

and using Eqs. (17.65) in Eqs. (17.58), we obtain finally for the d.d. of $\mathcal{W}_T$ here

$$W_1(\mathcal{W}_T)_{\text{white},f>0} = \begin{cases} \dfrac{\exp\left[-\dfrac{\mathcal{W}_T}{\mathcal{W}_0}\dfrac{1}{(1 - c_T{}^2)}\right]}{W_0(1 - c_T{}^2)^{1/2}} I_0\left[\dfrac{\mathcal{W}_T}{W_0}\dfrac{c_T}{(1 - c_T{}^2)}\right] & \mathcal{W}_T > 0 \\ 0 & \mathcal{W}_T < 0 \end{cases} \tag{17.67}$$

In the limit of indefinitely long samples, this reduces to Eqs. (17.60), e.g.,

$$W_1(\mathcal{W}_\infty)_{\text{white},f>0} = \begin{cases} \dfrac{e^{-\mathcal{W}_\infty/W_0}}{W_0} & \mathcal{W}_\infty > 0 \\ 0 & \mathcal{W}_\infty < 0 \end{cases} \tag{17.68}$$

A number of curves of $W_1(\mathcal{W}_T)_{\text{white}}$ are shown in Fig. 17.1 for various values of the parameter c_T, or, equivalently, of the parameter product ωT. The first and second moments of $\mathcal{W}_T$ in this instance are easily obtained from Eqs. (17.62), (17.63), with the aid of Eqs. (17.65). We have

$$\overline{\mathcal{W}_T} = W_0 \qquad \overline{\mathcal{W}_T{}^2} = W_0{}^2\left[2 + \frac{\sin^2 \omega T}{(\omega T)^2}\right] \tag{17.69a}$$

with the variance

$$\overline{\mathcal{W}_T{}^2} - \overline{\mathcal{W}_T}^2 = W_0{}^2\left[1 + \frac{\sin^2 \omega T}{(\omega T)^2}\right] > 0 \tag{17.69b}$$

When $T \to \infty$, we get simply W_0, $2W_0{}^2$, $W_0{}^2$ for Eqs. (17.69*a*), (17.69*b*), respectively, as may be readily verified by a direct calculation, using Eqs. (17.68). An extension of these results [Eqs. (17.56*a*) to (17.69)] for additive (independent) signal and noise processes is given in Probs. 17.13, 17.14.

Although the original process, whose spectrum we are considering here, is normal, the d.d. of the spectrum is not, even in the limit $T \to \infty$. For finite T, we are not surprised at this, since Eq. (17.55) is a nonlinear transformation, but when $T \to \infty$, at first glance we might expect the usual passage to a normal process, as in the examples of Sec. 17.2-1. However, a more careful inspection shows that, although $T \to \infty$, the effective number of independent sample values does not: in fact, we have a χ^2 distribution for $\mathcal{W}_\infty$, with $n = 2$ degrees of freedom only.

The limiting results (17.60), (17.61*b*), (17.68) are nevertheless not unexpected from the viewpoint of a large-sample theory, as the following interpretation in terms of a random walk (Secs. 7.7-1, 7.7-2) suggests. For when T is sufficiently large, if we split the interval $(-T/2, T/2)$ into a large number of long subintervals, then the u_T, v_T [cf. Eq. (17.51*a*)] for these subintervals can be regarded as the effectively independent components of a set of two-dimensional vectors. Equation (17.52) thus represents the square of the over-all vector length for the vector resultant of a random walk with a large number of independent steps. Applying the results of Sec. 7.7-3 then gives at once the Rayleigh d.d. [Eqs. (17.61*b*)], which in turn goes over into Eqs. (17.60) with the transformation $\mathcal{W}_\infty = Z_\infty^2$. This, of course, is an example of the central-limit theorem (Sec. 7.7-3) and is not restricted to originally normal processes. The same argument may be used for a nongaussian process $y(t)$ (under the conditions 1 to 3, Sec. 7.7-3) to give the Rayleigh d.d. for Z_∞, or, correspondingly, the χ^2 distribution for $\mathcal{W}_\infty$ $(n = 2)$, with the spectral intensity $\mathcal{W}_y(f)$ for the mean of $\mathcal{W}_\infty$ as parameter. If $y(t)$ contains a d-c or otherwise deterministic component, the Rayleigh d.d. is modified to a conditional d.d. of the form (9.50), as the results [Eqs. (3), (4)] of Prob. 17.14*b* for an additive and independent signal and noise indicate, in the special situation of an originally normal noise process. The significant parameter, in any case, is the ratio of the spectral densities $\mathcal{W}_S(f)/\mathcal{W}_N(f)$, if S is ergodic and N is stationary.

PROBLEMS

17.8 If $x = \log \Lambda_n(\mathbf{V})$ and $\Lambda_n(\mathbf{V}) \equiv \mu[\langle F_n(\mathbf{V}|\mathbf{S})\rangle_S / F_n(\mathbf{V}|0)]$ is a generalized likelihood ratio [cf. Eq. (19.14)], show that

$$Q_n(x) = \mu e^{-x} P_n(x) \tag{1}$$

where Q_n, P_n are, respectively, the probability densities of x with respect to the weightings $F_n(\mathbf{V}|0)$ and $\langle F_n(\mathbf{V}|\mathbf{S})\rangle_S$. Discuss the case where the c.f.'s of $P_n(x)$ and $Q_n(x)$ have singularities in the upper-half ξ plane, as well as in the lower. HINT: Start with the identity

$$\int_\Gamma \left[\frac{\mu\langle F_n(\mathbf{V}|\mathbf{S})\rangle_S}{F_n(\mathbf{V}|0)}\right]^{i\xi} F(\mathbf{V}|0)\, d\mathbf{V} = \mu \int_\Gamma \left[\frac{\mu\langle F_n(\mathbf{V}|\mathbf{S})\rangle_S}{F_n(\mathbf{V}|0)}\right]^{i\xi-1} \langle F_n(\mathbf{V}|\mathbf{S})\rangle_S\, d\mathbf{V} \tag{2}$$

17.9 (*a*) Carry out the details leading to Eqs. (17.28*a*), (17.28*b*) for an original normal noise process alone.

(*b*) Repeat, for Eqs. (17.34*a*), (17.34*b*), in the case of a deterministic signal process and additive normal noise.

(*c*) Show, in the degenerate case $\mathbf{k}_N\mathbf{J} \equiv \mathbf{G} = \lambda_0 \mathbf{I}$, that

$$F_1(i\xi;n)_x = e^{i\xi A_n + i\xi\lambda_0 B/(1-2i\xi\lambda_0)}(1 - 2i\xi\lambda_0)^{-n/2} \tag{1}$$

and hence that

$$W_1(x;n)_{S+N} = \begin{cases} \dfrac{e^{-B/2}}{2\lambda_0}\left(\dfrac{B}{2}\right)^{1-n/2}\left(\dfrac{x-A_n}{2\lambda_0}\right)^{(n-2)/4} e^{-[(x-A_n)/2\lambda_0]} I_{n/2-1}\left(\sqrt{2B}\sqrt{\dfrac{x-A_n}{2\lambda_0}}\right) & x > A_n \\ 0 & x < A_n \end{cases} \quad (2)$$

where

$$B = a_0^2\tilde{\mathbf{s}}\mathbf{k}_N^{-1}\mathbf{s} \quad (3)$$

Verify that, when $B = 0$,

$$W_1(x;n)_N = \lambda_0^{-1}2^{-n/2}\Gamma\left(\frac{n}{2}\right)^{-1}\left(\frac{x-A_n}{\lambda_0}\right)^{n/2-1} e^{-[(x-A_n)/2\lambda_0]} \qquad x > A_n \quad (4)$$

17.10 (*a*) If $V(t)$ is a normal random process with $\bar{V} = 0$, show that $x = -\log W_n(\mathbf{V})$, with $\mathbf{V} = [V(t_1), \ldots, V(t_n)]$, has a χ^2 distribution (about $x = B_n$) with n degrees of freedom, e.g.,

$$W_1(x;n) = \begin{cases} \dfrac{(x - \log B_n)^{n/2-1}e^{-x+\log B_n}}{\Gamma(n/2)} & x > \log B_n \\ 0 & x < \log B_n \end{cases} \quad (1)$$

where

$$B_n = (2\pi)^{n/2}(\det \mathbf{K}_N)^{1/2} \quad (2)$$

(*b*) Show that

$$\bar{x} = \log(e^{n/2}B_n) \qquad \overline{x^2} = \log^2 B_n + \log(B_n{}^n e^{n(n/2+1)/2}) \quad (3)$$

and hence show that

$$\sigma_x{}^2 \equiv \overline{x^2} - \bar{x}^2 = n/2 \quad (4)$$

independent of B_n.

17.11 If $\mathbf{z} = [z_1, \ldots, z_n]$ is normal, with covariance matrix $\mathbf{F}$ $(= \tilde{\mathbf{F}})$ and mean $\bar{\mathbf{z}} \neq 0$, show that the characteristic function of

$$x = C_0 + \tilde{\mathbf{C}}_1\mathbf{z} + \tfrac{1}{2}\tilde{\mathbf{z}}\mathbf{C}_2\mathbf{z} \quad (1)$$

is

$$F_1(i\xi)_x = (\exp\{i\xi[C_0 + \tfrac{1}{2}\tilde{\bar{\mathbf{z}}}(\boldsymbol{\Delta}^{-1} + \mathbf{F}^{-1}\boldsymbol{\Delta}^{-1}\mathbf{F})\mathbf{C}_1] - \tfrac{1}{2}\xi^2\tilde{\mathbf{C}}_1\boldsymbol{\Delta}^{-1}\mathbf{F}\mathbf{C}_1 - \tfrac{1}{2}\tilde{\bar{\mathbf{z}}}\mathbf{F}^{-1}(\mathbf{I} - \boldsymbol{\Delta}^{-1})\bar{\mathbf{z}}\}) \det \boldsymbol{\Delta}^{-1} \quad (2)$$

where

$$\boldsymbol{\Delta} = \boldsymbol{\Delta}(i\xi) = \mathbf{I} - i\xi\mathbf{F}\mathbf{C}_2 \quad (2a)$$

Show that this can be written equivalently

$$F_1(i\xi)_x = e^{i\xi C_0}\prod_{j=1}^{n}\frac{\exp\{[i\xi(b_j + \tfrac{1}{2}\lambda_j a_j) - \tfrac{1}{2}\xi^2 c_j]/(1 - i\xi\lambda_j)\}}{\sqrt{1 - i\xi\lambda_j}} \quad (3)$$

where

$$a_j \equiv \bar{z}_j\sum_k^n \bar{z}_k(\mathbf{F}^{-1})_{jk} \qquad b_j \equiv \bar{z}_jC_j \qquad c_j \equiv C_j\sum_k^n C_kF_{jk} \quad (3a)$$

and the λ_j are the eigenvalues of $\mathbf{FC}_2$.

17.12 (*a*) Obtain the joint normal distribution density [Eqs. (17.53)] of the gaussian functionals u_T, v_T when $\bar{u}_T = \bar{v}_T = 0$.

(*b*) Show that, for an originally stationary process $N(t)$, A_T, B_T, C_T in Eqs. (17.54) become

$$\lim_{T\to\infty} A_T(\omega) = \lim_{T\to\infty} \frac{1}{T} \iint_{-T/2}^{T/2} K_N(\tau_1 - \tau_2) \cos \omega(\tau_2 - \tau_1)\, d\tau_1\, d\tau_2$$
$$= \tfrac{1}{2}\mathcal{W}_N(f) = \lim_{T\to\infty} C_T(\omega) \tag{1}$$

and

$$\lim_{T\to\infty} B_T(\omega) \to O(T^{-1}) \tag{2}$$

Hence, show that $\lim_{T\to\infty} a_T(\omega)^2 = \lim_{T\to\infty} b_T(\omega) = \mathcal{W}_N(f)^2$ [cf. Eq. (17.59)].

(c) For finite intervals, show also that

$$a_T,\, b_T > 0 \qquad a_T^2 - b_T > 0 \tag{3}$$

[cf. Eqs. (17.56)].

(d) Show that, when $f = 0$ $(0 < T < \infty)$, the d.d. [Eqs. (17.58)] of the intensity spectrum $\mathcal{W}_T$ reduces to

$$W_1(\mathcal{W}_T)_{f=0} = \begin{cases} \dfrac{1}{\sqrt{2\pi}} [a_T(0)\mathcal{W}_T]^{-\frac{1}{2}} e^{-\mathcal{W}_T/2a_T(0)} & \mathcal{W}_T > 0 \\ 0 & \mathcal{W}_T < 0 \end{cases} \tag{4}$$

so that

$$W_1(\mathcal{W}_\infty)_{f=0} = \begin{cases} [2\pi\mathcal{W}_y(0)\mathcal{W}_\infty]^{-\frac{1}{2}} e^{-\mathcal{W}_\infty/2\mathcal{W}_y(0)} & \mathcal{W}_\infty > 0 \\ 0 & \mathcal{W}_\infty < 0 \end{cases} \tag{5}$$

(e) Evaluate

$$\int_{-\infty}^{\infty} \frac{e^{-i\xi z}}{\sqrt{1 - 2\alpha z + z^2}} \frac{dz}{2\pi} \qquad \int_{-\infty}^{\infty} (z - z_1)^{-\mu}(z - z_2)^{-\nu} \frac{dz}{2\pi} \qquad \operatorname{Re} \mu, \nu \geq \frac{1}{2} \tag{6}$$

17.13 (a) If a deterministic signal S and normal noise N ($\bar{N} = 0$) are additively combined, show that the (conditional) characteristic function of the spectral intensity density $(\mathcal{W}_T)_{S+N}$ of the process $V = S + N$ on $(-T/2, T/2)$ is

$$F_1(i\xi;2|S)_{S+N} = \frac{e^{-\frac{1}{2}\tilde{\bar{\mathbf{X}}}\mathbf{K}^{-1}_X[\mathbf{I}-(\mathbf{I}-2i\xi\mathbf{K}_X)^{-1}]\bar{\mathbf{X}}}}{\sqrt{\det (\mathbf{I} - 2i\xi\mathbf{K}_X)}} \tag{1}$$

where $\mathbf{K}_X$ is given by Eqs. (17.53b) and now $\tilde{\bar{\mathbf{X}}} = [\bar{u}_T, \bar{v}_T] \neq \mathbf{0}$.

(b) Setting

$$A_T(\omega)_S \equiv \frac{2}{T} \iint_{-T/2}^{T/2} \cos \omega\tau_1 \cos \omega\tau_2\, S(\tau_1)S(\tau_2)\, d\tau_1\, d\tau_2 = \overline{u_T}^2$$
$$C_T(\omega)_S \equiv \frac{2}{T} \iint_{-T/2}^{T/2} \sin \omega\tau_1 \sin \omega\tau_2\, S(\tau_1)S(\tau_2)\, d\tau_1\, d\tau_2 = \overline{v_T}^2 \tag{2}$$
$$B_T(\omega)_S \equiv \frac{2}{T} \iint_{-T/2}^{T/2} \cos \omega\tau_1 \sin \omega\tau_2\, S(\tau_1)S(\tau_2)\, d\tau_1\, d\tau_2 = \overline{u_T}\,\overline{v_T}$$

reduce Eq. (1) to the equivalent result

$$F_1(i\xi;2|S)_{S+N} = \frac{e^{(i\xi\alpha + 2\xi^2\beta)/(1-2ia\xi-b\xi^2)}}{\sqrt{1 - 2ia\xi - b\xi^2}} \tag{3}$$

where $a = a_T$, $b = b_T$ [cf. Eqs. (17.56)], and

$$\alpha_T \equiv \alpha_T(\omega)_S = \frac{2}{T} \iint_{-T/2}^{T/2} S(\tau_1)S(\tau_2) \cos \omega(\tau_2 - \tau_1)\, d\tau_1\, d\tau_2 = \overline{u_T^2} + \overline{v_T^2} \tag{4}$$

$$\beta_T \equiv \beta_T(\omega)_{SN} = A_T(\omega)_S C_T(\omega)_N + C_T(\omega)_S A_T(\omega)_N + 2B_T(\omega)_N B_T(\omega)_S$$

in which $A_T(\omega)_N$, etc., are given by Eqs. (17.54) for the noise alone.

(c) Show that the first and second moments of the spectrum $(\mathcal{W}_T)_{S+N}$ are

$$\overline{(\mathcal{W}_T)_{S+N}} = \overline{\alpha_T(\omega)_S} + a_T(\omega) = \overline{\mathcal{W}_T(f)_S} + \overline{\mathcal{W}_T(f)_N} \tag{5}$$

$$\overline{(\mathcal{W}_T)^2_{S+N}} = 3a_T{}^2 - b_T + \overline{\alpha_T{}^2} + 6a_T\bar{\alpha}_T - 4\bar{\beta}_T = \overline{(\mathcal{W}_T)_N{}^2} + \overline{(\mathcal{W}_T)_S{}^2} + 6\overline{(\mathcal{W}_T)_N}\,\overline{(\mathcal{W}_T)_S} - 4\bar{\beta}_T \tag{6}$$

Hence, show that the variance $(\sigma_T{}^2)_{S+N} = [\overline{(\mathcal{W}_T)^2} - \overline{(\mathcal{W}_T)}^2]_{S+N}$ is given by

$$(\sigma_T{}^2)_{S+N} = (\sigma_T{}^2)_S + (\sigma_T{}^2)_N + 4[\overline{(\mathcal{W}_T)_S}\,\overline{(\mathcal{W}_T)_N} - \overline{\beta_T(\omega)}] \tag{7}$$

In the limit $T \to \infty$, obtain for ergodic S (and N)

$$(\sigma_\infty{}^2)_{S+N} = (\sigma_\infty{}^2)_S + \mathcal{W}_N(f)^2 + 2\mathcal{W}_S(f)\mathcal{W}_N(f) \tag{8}$$

[Note that $(\sigma_\infty{}^2)_S$ does not generally reduce to $\mathcal{W}_S(f)^2$; cf. Eq. (17.59) for the noise.]

17.14 (a) Obtain the conditional distribution density of $(\mathcal{W}_T)_{S+N}$ $(f > 0)$,

$$W_1(\mathcal{W}_T|S)_{S+N} = \begin{cases} \dfrac{e^{-B-(\mu+\sqrt{\mu^2-1})Z}}{\sqrt{b}} \displaystyle\sum_{m,n=0}^{\infty} \frac{(GZ)^m(HZ)^n}{m!(m+n)!n!} \\ \qquad \times\, {}_1F_1\left(n + \frac{1}{2};\, m + n + 1;\, 2\sqrt{\mu^2 - 1}\, Z\right) & Z > 0 \\ 0 & Z < 0 \end{cases} \tag{1}$$

where $a = a_T$, $b = b_T$, $\mu = \mu_T$ (in the text) and α, β are given by Prob. 17.13, Eqs. (4), with

$$Z = \frac{\mathcal{W}_T}{\sqrt{b}} \qquad B = \frac{2\beta}{b} \qquad A' = \frac{\alpha b - 4a\beta}{b^{3/2}}$$
$$G = \frac{B - p_1A'}{p_1 - p_2} \qquad H = \frac{B - p_2A'}{p_2 - p_1} \qquad p_{1,2} = -\mu \mp \sqrt{\mu^2 - 1} \tag{2}$$

(b) In the limit $T \to \infty$, with an ergodic signal process S, obtain for $f > 0$

$$W_1(\mathcal{W}_\infty|S)_{S+N} = \begin{cases} \dfrac{e^{-\gamma-Z}}{\mathcal{W}_N(f)} I_0(2\sqrt{\gamma Z}) & Z > 0 \\ 0 & Z < 0 \end{cases} \tag{3}$$

where $\quad Z = \mathcal{W}_\infty(f)/\mathcal{W}_N(f) \qquad \gamma = \gamma(f) \equiv \mathcal{W}_S(f)/\mathcal{W}_N(f) \quad (4)$

[This is just the form of the generalized Rayleigh d.d. for the envelope of narrowband signal and noise; cf. Eq. (9.50).]

(c) For $f = 0$ $(0 < T)$, show that

$$W_1(\mathcal{W}_T|S)_{S+N}\Big|_{f=0} = \int_{-\infty}^{\infty} \frac{e^{-i\xi\mathcal{W}_T+i\xi\alpha/(1-2ia\xi)}}{\sqrt{1 - 2ia\xi}} \frac{d\xi}{2\pi} \tag{5}$$

which is just Eqs. (2), Prob. 17.9c, with $A_n = 0$, $B = \alpha/a$, λ, a, and $n = 1$ therein.

17.3 *Distribution Densities after Nonlinear Operations and Filtering*

In Secs. 17.1, 17.2, we have described several methods of finding the probability densities of a nonlinear (specifically, a quadratic) function or functional of an originally normal random process. Let us now apply these results to several problems of interest in system design and analysis:

1. The first-order d.d. of the output process following square-law rectification and video (i.e., postdetection) filtering, when the input is a stationary normal noise.[1,3,4]

Other related problems of interest discussed in the literature, but reserved as exercises in our present treatment, are:

2. The extension of (1) to the nth-order d.d. of the rectified filtered output.[4]

3. The first-order probability density of the output of an (idealized) FM discriminator (cf. Sec. 15.1-3), when the discriminator's nonlinear elements

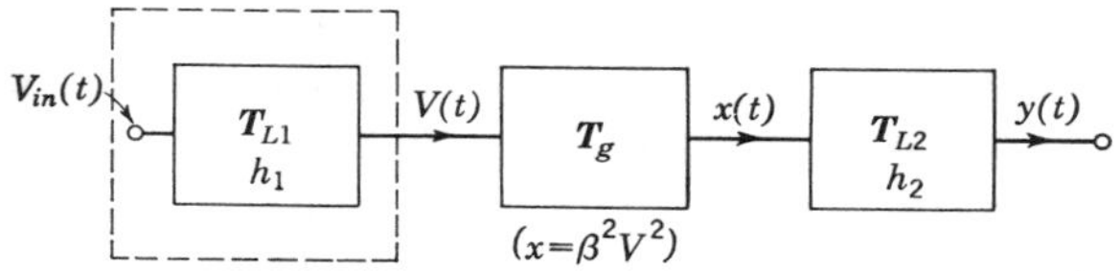

FIG. 17.2. A system with a zero-memory quadratic rectifier and filter.

are quadratic and when the input is normal noise, with and without a subsequent low-pass linear filter.[5]

4. The first-order d.d. for the filtered output of a zero-memory multiplier whose inputs are two statistically related stationary normal processes.[6]

The first problem illustrates the techniques of Secs. 17.1-1, 17.1-2, while the second may be handled by an extension of the analysis to include vector processes (cf. Probs. 17.15 to 17.17). The third example is essentially that of determining the d.d. of the difference of two statistically dependent, nonnormal processes (cf. Prob. 17.18), while the fourth considers their product, in either case with or without a subsequent linear filtering operation (cf. Prob. 17.19). References to additional examples and further generalizations are given in Sec. 17.3-3.

17.3-1 First-order Distribution Densities Following a Quadratic Rectifier and Linear Filter.† The system in question is shown in Fig. 17.2 Here, a stationary normal input noise process $V_{in}(t)$ (with $\bar{V}_{in} = 0$) is first passed through a stable linear (invariant) filter $\boldsymbol{T}_{L1}$, whose output $V(t)$ $(= \boldsymbol{T}_{L1}\{V_{in}\})$ is also a stationary normal process, now the input to a zero-memory (full-wave) square-law device:

$$x(t) = \boldsymbol{T}_g\{V\} = \beta^2 V(t)^2 \qquad V(t) = \int_{-\infty}^{\infty} V_{in}(t-\tau)h_1(\tau)\,d\tau \quad (17.70)$$

† This problem was originally solved by Fränz[1a] and later and independently by Kac and Siegert[1] (see also Emerson[3] and Meyer and Middleton[2]).

The output $x(t)$ of the rectifier is next passed through another linear (usually low-pass) filter $\mathbf{T}_{L2}$ to give finally

$$y_T(t) = \mathbf{T}_{L2}\{x\} = \int_{0-}^{T+} h_2(\tau)x(t-\tau)\,d\tau \tag{17.71}$$

where smoothing is for the time being carried out only over T sec of the past of the input x. The rectified outputs x, y are, of course, no longer normal.

Our problem now is to find the first-order d.d. $W_1(y)$ of the nongaussian functional $y_T(t)$. We distinguish two cases: (1) *narrowband inputs* $V(t) = V_C(t)\cos\omega_0 t + V_S(t)\sin\omega_0 t$, for which

$$x(t) = \beta^2 V^2 = \frac{\beta^2}{2}\left[(V_C{}^2 + V_S{}^2) + (V_C{}^2 - V_S{}^2)\cos 2\omega_0 t + 2V_C V_S \sin 2\omega_0 t\right] \tag{17.72}$$

and (2) *broadband inputs*, for which Eq. (17.70) may be applied directly. We begin first with the former, as exact expressions for the probability density $W_1(y_T)$ can be obtained here for all T when the spectrum of the original normal process $V(t)$ is symmetrical about its center frequency f_0, as will be assumed henceforth in the narrowband case.

Case 1. Narrowband inputs. Here, the filtered output [Eq. (17.71)] in the narrowband case [Eq. (17.72)] is well approximated by

$$y_T(t) \doteq \frac{\beta^2}{2}\int_{0-}^{T+} [V_C(t-\tau)^2 + V_S(t-\tau)^2]h_2(\tau)\,d\tau \tag{17.73}$$

in the usual situation of a postdetection low-pass filter $\mathbf{T}_{L2}$, whose effective input is thus $x(t) = (\beta^2/2)[V_C(t)^2 + V_S(t)^2]$.

The desired first-order d.d. of y_T can now be found directly from the results of Sec. 17.2-1, Example 3. It is convenient first to consider the discrete analogue of Eq. (17.73), e.g.,

$$y_n(t) = \frac{\beta^2}{2}\sum_{j=1}^{n} [V_C(t-\tau_j)^2 + V_S(t-\tau_j)^2]h_2(\tau_j)\,\Delta_n\tau \tag{17.74}$$

where ultimately $\lim_{n\to\infty} \Delta_n\tau \to d\tau$ and the sum may be replaced by the integral (17.73). Setting

$$\mathbf{J} = \mathbf{J}_1 = \left[\frac{\psi\beta^2}{2} h_2(\tau_j)\,\Delta_n\tau\,\delta_{ij}\right] \qquad \psi = \overline{V^2}\ (= \overline{V_C{}^2} = \overline{V_S{}^2}) \tag{17.75}$$

in Eq. (17.38), with $A_n = 0$, we see that x therein is just $y_n(t)$ above. Equations (17.39) accordingly give at once the characteristic function for the first-order d.d. of $y_n(t)$, namely,

$$F_1(i\xi;n)_{y_n} = [\det(\mathbf{I} - 2i\xi\boldsymbol{\varrho}_0\mathbf{J}_1)]^{-1} = \prod_{j=1}^{n} (1 - 2i\xi\lambda_j^{(n)})^{-1} \quad (17.76)$$

where $\boldsymbol{\varrho}_0 = (\mathbf{K})_V\psi^{-1}$ is the normalized covariance matrix of each of the slowly varying components V_C, V_S of the noise input $V(t)$ to the quadratic rectifier [cf. Eqs. (7.56), (7.57)] and the $\lambda_j^{(n)}$ are the (distinct, i.e., nonmultiple) eigenvalues of the $n \times n$ matrix $\mathbf{G} \equiv \boldsymbol{\varrho}_0\mathbf{J}_1$ (cf. Sec. 17.1-1). The associated distribution density for $y_n(t)$ is then given by Eqs. (17.41*a*), (17.41*b*), $A_n = 0$.

Passing to the limit ($n \to \infty$) of continuous sampling in $(0-,T+)$, we may use Eq. (17.42) to write for the characteristic function of $y_T(t)$

Narrowband

$$F_1(i\xi;\infty;T)_{y_T} = \mathfrak{D}_T(-2i\xi)^{-1} = \prod_{j=1}^{\infty} (1 - 2i\xi\lambda_j)^{-1} \quad (17.77)$$

where now the eigenvalues λ_j are found from the basic integral equation (17.6*b*), which is here specifically†

$$\int_{0-}^{T+} \frac{\beta^2}{2} h_2(u)K_V(|t-u|)_0\phi_j(u)\,du = \lambda_j\phi_j(t) \qquad 0- < t < T+ \quad (17.78)$$

with $K_V(\tau)_0 = \psi\rho(\tau)_0$.

The corresponding first-order d.d. $W_1(y_T)$ of the functional y_T is again given by Eqs. (17.41*a*), (17.41*b*), with $\lambda_j^{(n)}$ replaced by λ_j, etc., and with $A_n = 0$. When the postrectification filtering is over the entire past of the input (e.g., when $T \to \infty$), Eq. (17.78) becomes

$$\int_{0-}^{\infty} \frac{\beta^2}{2} h_2(u)K_V(|t-u|)_0\phi_j(u)\,du = \lambda_j\phi_j(t) \qquad 0- < t < \infty \quad (17.78a)$$

with corresponding modifications of F_1 and W_1.

Note that, because of the stationarity of the process, the first-order c.f. and d.d. of $y_T(t)$ are invariant of t. Observe also that there may be a finite number of negative eigenvalues associated with the integral equation (17.78), which itself must always be homogeneous, depending on the structure of the kernel and, in particular here, on the postrectification filter $\mathbf{T}_{L2}$. Thus, for example, in the case of an RC filter $\mathbf{T}_{L2}$, $h_2(t) = \mu e^{-\mu t}$ $(t > 0)$, $\mu = (RC)^{-1}$ (≥ 0, all t). Only positive eigenvalues can occur [for the covariance functions $K_V(\tau)_0$ usually assumed in practice]. This corresponds physically to the fact that the final output $y_T(t)$ of such a filter is always

† It is assumed that the elements of $\mathbf{G} = \boldsymbol{\varrho}_0\mathbf{J}$ possess the necessary continuity properties, so that $\psi\boldsymbol{\varrho}_0 \to K_V(|t-u|)_0$ and the kernel $G(t,u) = (\beta^2/2)h_2(u)K_V(|t-u|)_0$ is positive definite on $(0,T)$; the $\{\phi_j\}$ form a complete orthonormal set, subject to Eq. (17.7); see also Eq. (17.3*b*) in the discrete case.

positive [since $\boldsymbol{T}_g$ is also quadratic; cf. Eq. (17.70)]. On the other hand, when $h_2(t)$ can be negative for some $t\ (> 0)$, there are a finite number of negative eigenvalues: the output $y_T(t)$ can be less than zero occasionally. This occurs when a CR filter $\boldsymbol{T}_{L2}$ is used, with the output observed across the resistance, as then $h_2(t) = \delta(t - 0) - \mu e^{-\mu t}\ (t > 0-)$. Note in the RC case that Eq. (17.78) is a homogeneous integral equation, and for the CR filter it is still homogeneous.†

Finally, when T and the reciprocal of the bandwidth of the final filter h_1 are both large compared with the correlation time of the original normal process, we obtain a normal d.d. of the form (17.30), where the mean and variance are given by Eqs. (17.44*b*), which here are specifically, in terms of the appropriate eigenvalues and iterated kernels [cf. Eq. (17.17)],

$$C_{2T} = 2 \sum_{j=1}^{\infty} \lambda_j = \beta^2 \int_{0-}^{T+} h_2(t_1) K_V(|t_1 - t_1|)_0 \, dt_1$$
$$= \beta^2 \psi \int_{0-}^{T+} h_2(t_1) \, dt_1 \tag{17.79a}$$
$$D_{2T} = 4 \sum_{j=1}^{\infty} \lambda_j^2 = \beta^4 \iint_{0-}^{T+} h_2(t_1) K_V(|t_1 - t_2|)_0{}^2 h_2(t_2) \, dt_1 \, dt_2 \tag{17.79b}$$

Case 2. Broadband inputs. Here Eq. (17.70) applies, in conjunction with Eq. (17.71), so that, in place of Eq. (17.73), we have

$$y_T(t) = \beta^2 \int_{0-}^{T+} V(t - \tau)^2 h_2(\tau) \, d\tau \tag{17.80}$$

the discrete analogue of which is now [cf. Eq. (17.74)]

$$y_n(t) = \beta^2 \sum_{j=1}^{n} V(t - \tau_j)^2 h_2(\tau_j) \, \Delta_n \tau \tag{17.81}$$

Letting

$$\mathbf{J} = \mathbf{J}_2 = [\psi \beta^2 h_2(\tau_j) \, \Delta_n \tau \, \delta_{ij}] \tag{17.82}$$

we may use the results of Sec. 17.2-1, Example 1, as in Case 1, to obtain for the c.f. in this instance [Eqs. (17.24*a*), (17.24*b*), with $A_n = 0$]

$$F_1(i\xi;n)_{y_n} = [\det(\mathbf{I} - 2i\xi \mathbf{k}_V \mathbf{J}_2)]^{-1/2} = \prod_{j=1}^{n} (1 - 2i\xi \lambda_j^{(n)})^{-1/2} \tag{17.83}$$

where $\mathbf{k}_V\ (= \psi^{-1} \mathbf{K}_V)$ is the normalized covariance matrix of $V(t)$ and the $\lambda_j^{(n)}$ are now the eigenvalues of $\mathbf{k}_V \mathbf{J}_2$.

Passing to the limit $n \to \infty$ (T fixed), following the argument of Eqs. (17.5) et seq. and (17.27) et seq., we obtain for the first-order c.f. of the functional $y_T(t)$ [Eq. (17.80)]

† In particular, Eq. (17.78) remains a homogeneous Fredholm integral equation of the second kind [cf. Morse and Feshbach,[15] eq. (8.1.17), p. 905].

Broadband

$$F_1(i\xi;\infty,T)_{y_T} = \mathfrak{D}_T(-2i\xi)^{-1/2} = \prod_{j=1}^{\infty} (1 - 2i\xi\lambda_j)^{-1/2} \tag{17.84}$$

where now the eigenvalues λ_j are obtained on solution of the integral equation†

$$\int_{0-}^{T+} \beta^2 h_2(u) K_V(t-u)\phi_j(u)\, du = \lambda_j \phi_j(t) \qquad 0- < t < T+ \tag{17.85}$$

The corresponding d.d., $W_1(y_T)$, however, cannot be found in a useful form unless the effective "memory" or smoothing time of the postdetection filter and the sample size T are both large compared with the correlation time T_C of the original noise.[19a] In this latter instance, we have the asymptotically normal situation described earlier [cf. Eqs. (17.30), (17.30*a*)], which is most easily obtained with the help of the basic identities (17.13), (17.25). Thus, for the dominant normal term in $W_1(y_T)$, it is not necessary to solve the integral equation (17.85), since the required mean and variance of y_T are more readily obtained from the first and second iterated kernels[3] [Eqs. (17.17), (17.30)]:

$$C_T^{(N)} = \sum_{j=1}^{\infty} \lambda_j = \beta^2 \int_{0-}^{T+} h_2(t) K_V(t,t)\, dt = \psi\beta^2 \int_{0-}^{T+} h_2(t)\, dt \tag{17.86a}$$

$$D_T^{(N)} = 2\sum_{j=1}^{\infty} \lambda_j^{\,2} = 2\beta^4 \iint_{0-}^{T+} h_2(t_1) K_V(|t_1 - t_2|)^2 h_2(t_2)\, dt_1\, dt_2 \tag{17.86b}$$

The first-order density $W_1(y_T)$ is itself, apart from correction terms in the generalized Edgeworth series, simply the normal form

$$W_1(y_T) \simeq \frac{e^{-(y_T - C_T^{(N)})^2/2D_T^{(N)}}}{(2\pi D_T^{(N)})^{1/2}} \tag{17.87}$$

The case of infinite sample size follows directly on setting $T \to \infty$ in Eqs. (17.80) to (17.86).

17.3-2 Further Examples. Let us illustrate the preceding discussion with two simple examples.

Example 1. Here, we assume that $V(t)$ is a narrowband normal noise spectrally shaped by a high-Q LRC filter, so that $K_V(\tau)_0 \doteq \psi e^{-\alpha|\tau|}$. The filter following the square-law rectifier is a realizable RC filter with weighting function $h_2(t) = \gamma e^{-\gamma t}$ $(t > 0)$, $h_2(t) = 0$ $(t < 0)$, and, unless otherwise indicated, smoothing is over the entire past of the input $x(t)$ to this filter, so

† Note that Eq. (17.85) is generally *not* the same as Eq. (17.78), as $K_V(|t-u|)$ is usually different from $K_V(|t-u|)_0$; hence, eigenvalues and eigenfunctions are likewise different from those of Case 1.

that $T \to \infty$. From Eq. (17.78a), the basic integral equation is

$$\frac{\psi\beta^2\gamma}{2}\int_0^\infty e^{-\gamma u-\alpha|t-u|}\phi_j(u)\,du = \lambda_j\phi_j(t) \qquad 0 \le t \tag{17.88}$$

the eigenvalues of which are found[1] from

$$\lambda_j = 4\alpha\psi\beta^2/\gamma q_j^2 \qquad \text{with } J_{2\alpha/\gamma-1}(q_j) = 0\ (q_j > 0) \tag{17.89}$$

where the q_j are the positive roots of the second equation in Eqs. (17.89) (cf. Table 17.1, item 3). The d.d. of y_∞ [Eq. (17.73), $T \to \infty$] corresponding to the c.f. [Eq. (17.77)] is, from Eq. (17.41a),

$$W_1(y_\infty) = \begin{cases} \displaystyle\sum_{l=1}^{\infty} \frac{e^{-y_\infty/\lambda_l}}{\lambda_l} \prod_{k=1}^{\infty}{}' \left(1 - \frac{\lambda_k}{\lambda_l}\right)^{-1} & y_\infty > 0 \\ 0 & y_\infty < 0 \end{cases} \tag{17.90}$$

(Since the kernel is everywhere positive here, there are no negative eigenvalues.)

When $2\alpha/\gamma = \frac{3}{2}$, the eigenvalues become particularly simple functions of j. Since $J_{1/2}(x) = (2/\pi)^{1/2}(\sin x)/\sqrt{x}$, setting $x = q_j$ gives $q_j = \pi j$ ($j = 1, 2, \ldots$), and from Eqs. (17.89) we get directly

$$\lambda_j = \frac{3\psi\beta^2}{\pi^2 j^2} \qquad j = 1, 2, \ldots \tag{17.91}$$

Equation (17.90) becomes†

$$W_1(y_\infty) = \frac{\pi^2}{3\psi\beta^2}\sum_{l=1}^{\infty} l^2 e^{-\pi^2 l^2 y_\infty/3\psi\beta^2} \prod_{k=1}^{\infty}{}' \left(1 - \frac{l^2}{k^2}\right)^{-1} \tag{17.92a}$$

$$= \frac{2\pi^2}{3\psi\beta^2}\sum_{l=1}^{\infty} (-1)^{l+1} l^2 e^{-\pi^2 l^2 y_\infty/3\psi\beta^2} \tag{17.92b}$$

inasmuch as $\prod_{k=1}^{\infty}{}' (1 - l^2/k^2) = -\frac{1}{2}\cos \pi l$. Figure 17.3 shows the first-order density [Eqs. (17.92)] for $2\alpha/\gamma = \frac{3}{2}$, following Jastram,[22] with good agreement between experiment and theory.

† Equations (17.92a), (17.92b) can also be expressed in terms of the derivative of the theta function: $\theta(r;0) = 1 - 2r^2 + 4r^4 \cdots$; for example,

$$W_1(y_\infty) = -\frac{\pi^2}{3\psi\beta^2} e^{\pi^2 y_\infty/3\psi\beta^2}\theta'(e^{-\pi^2 y_\infty/3\psi\beta^2};0)$$

(cf. Kac and Siegert, *op. cit.*, sec. VII, 1).

When T is finite, the basic integral equation (17.78) becomes here

$$\frac{\psi\beta^2\gamma}{2}\int_0^T e^{-\gamma u-\alpha|t-u|}\,\phi_j(u)\,du = \lambda_j\phi_j(t) \qquad 0 \le t \le T \qquad (17.93)$$

and the eigenvalues are again found from $\lambda_j = 4\alpha\psi\beta^2/\gamma q_j^2$ [cf. Eqs. (17.89)], where now, however, the q_j are the positive roots of[16]

$$J_{2\alpha/\gamma+1}(e^{-\gamma T/2}q_j)N_{2\alpha/\gamma-1}(q_j) = J_{2\alpha/\gamma-1}(q_j)N_{2\alpha/\gamma+1}(e^{-\gamma T/2}q_j) \qquad (17.94)$$

(cf. Table 17.1, item 2). The first-order density is again given by Eqs. (17.90), with the modified eigenvalues λ_j above [cf. Eq. (17.94)], and with a behavior similar in general form to that shown in Fig. 17.3 for the case $T \to \infty$.

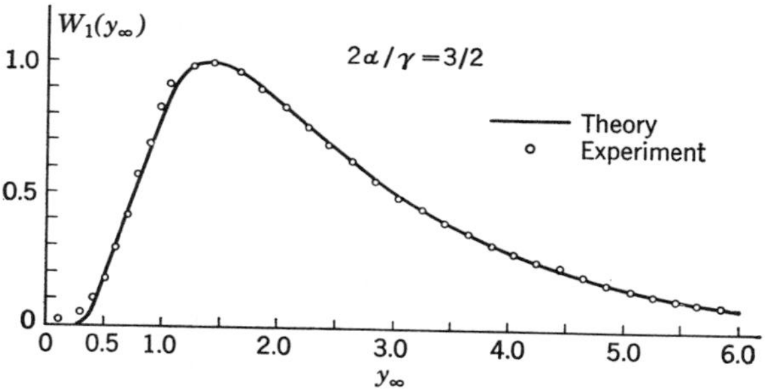

FIG. 17.3. First probability density of y_∞ [Eq. (17.73), $T \to \infty$] after square-law detection of a narrowband normal process, followed by a low-pass RC filter. The ordinate is normalized to a maximum value of unity. (*After P. Jastram, The Effect of Nonlinearity and Frequency Distortions on the Amplitude Distribution for Stationary Random Processes, doctoral dissertation, University of Michigan, Ann Arbor, Mich.,* 1947.)

When the postdetection filter h_2 responds to all the *low-frequency* terms in the rectified output $x(t)$—the so-called case of "no video" filter, where the RF terms ($\sim f_0$) are, however, excluded—we find that Eq. (17.78*a*) has only one eigenvalue and one eigenfunction. This follows since $h_2(\tau)$ may be set equal to $\delta(\tau - 0)$ here, so that we get directly from Eq. (17.78*a*)

$$\lambda_1 = \frac{\beta^2\psi}{2} \qquad \phi_1(t) = \phi_1(0)\rho_v(t)_0 \qquad (17.95)$$

with $\rho_v(t)_0 = K_V(t)_0/\psi$ in the usual way. The orthonormality condition (17.7) ($T \to \infty$) gives us $\phi_1(0)$ [and hence $\phi_1(t)$] explicitly in terms of $\rho_v(t)_0$, for example, $\int_0^\infty \phi_1{}^2(t)\,dt = \int_0^\infty \phi_1(0)^2\rho_v{}^2(t)_0\,dt = 1$. In the present example, $\rho_v(t)_0 = e^{-\alpha|t|}$, so that $\phi_1(0) = \sqrt{2\alpha}$. From Eq. (17.77), the c.f. of y_∞ now is simply

$$F_1(i\xi;\infty;\infty)_{y_\infty} = (1 - i\xi\beta^2\psi)^{-1} \qquad (17.96a)$$

with the resulting first-order d.d.

$$W_1(y_\infty)_\infty = \begin{cases} \displaystyle\int_{-\infty}^{\infty} \frac{e^{-i\xi y_\infty}}{1 - i\xi\beta^2\psi} \frac{d\xi}{2\pi} = (\psi\beta^2)^{-1} e^{-y_\infty/\psi\beta^2} & y_\infty > 0 \\ 0 & y_\infty < 0 \end{cases} \tag{17.96b}$$

which is, as expected, a simple χ^2 d.d. with two degrees of freedom. The normal character of the input is, of course, destroyed by the subsequent square-law rectification $\boldsymbol{T}_g$. Note that in the broadband case where the postdetection filter passes the h-f components of $V(t)$ [cf. Eq. (17.72)], as well as the l-f terms, Eq. (17.85) is the appropriate integral equation (since the entire input is now considered). Here, $K_V(t) = K_V(t)_0 \cos[\omega_0 t - \phi_0(t)]$, and, for the noise assumed above, Eq. (17.85) reduces to Eq. (17.78*a*) except for an additional factor 2; that is, $\beta^2/2$ in the latter is replaced by β^2, so that Eqs. (17.95), (17.96*a*), (17.96*b*) still apply, with this modification. Moreover, when this postrectification filter is wide compared with the bandwidth of the original input noise, but not infinitely so, it is possible to determine the departures from the exponential d.d. [Eqs. (17.96*b*)] to varying degrees of approximation (cf. Prob. 17.18).

At the other extreme is the case of a low-pass postdetection filter spectrally narrow (i.e., with a long time constant) compared with the original noise $V(t)$. Here, there are large numbers of eigenvalues, however, such that the various iterated kernels appearing in Eq. (17.30*a*) lead to small correction terms in the resulting generalized Edgeworth series. This is expected from the fact that the time constant of the filter is very much larger than the correlation time T_c of the input noise (with $T \to \infty$ in addition), so that an asymptotically normal d.d. characteristic of a large-sample theory results. Accordingly, the d.d. of y_T (T for the moment large vis-à-vis T_c but finite, for example, $\alpha T \gg 1$, $\alpha \gg \gamma$) is given by the normal form (17.30), where specifically for the input noise and RC filter of the present example [Eqs. (17.88) et seq.] we find from Eqs. (17.79*a*), (17.79*b*) that

$$C_{2T}{}^{(N)} = \beta^2\psi(1 - e^{-\gamma T}) \qquad D_{2T}{}^{(N)} = \frac{2\beta^4\psi^2\gamma^2}{\gamma - \alpha}\left(\frac{1 - e^{-(\gamma+\alpha)T}}{\gamma + \alpha} - \frac{1 - e^{-2\gamma T}}{2\gamma}\right) \tag{17.97a}$$

or, as $T \to \infty$, that

$$C_\infty{}^{(N)} = \beta^2\psi \qquad D_\infty{}^{(N)} = \frac{\beta^4\psi^2\gamma}{\alpha + \gamma} \doteq \beta^4\psi^2\left[\frac{\gamma}{\alpha} - \left(\frac{\gamma}{\alpha}\right)^2 + \cdots\right] \qquad \frac{\gamma}{\alpha} \ll 1 \tag{17.97b}$$

Thus, the first-order distribution density of this narrowband rectified and filtered output $y_\infty(t)$ is normal, with mean $C_\infty{}^{(N)}$ and variance $D_\infty{}^{(N)}$, as sketched in Fig. 17.4. (Note again that, strictly speaking, the d.d. for y_∞ must vanish here for all $y_\infty < 0$, a condition well approximated here when $\alpha \gg \gamma$.) For additional examples and results, including the modification of $W_1(y_\infty)$ that occurs when a signal accompanies the input, see Emerson[3] and Probs. 17.18, 17.19.

Example 2. In Sec. 17.2-3, we derived the first-order d.d. of the spectral density of a normal random process for finite $(-T/2,T/2)$ and infinite $(-\infty,\infty)$ intervals. As our second, and final, example, let us now outline the solution of the corresponding problem for the *integral* of the spectral density, or the average intensity of the process in question. Thus, if $\mathcal{W}_V(f)_T \equiv (2/T)|S_V(f)_T|^2$ is the random variable representing the spectral intensity density of the stationary and ergodic normal process $V(t)$ on the interval $(-T/2,T/2)$, our problem here is to determine the first-order d.d. of the nongaussian functional†

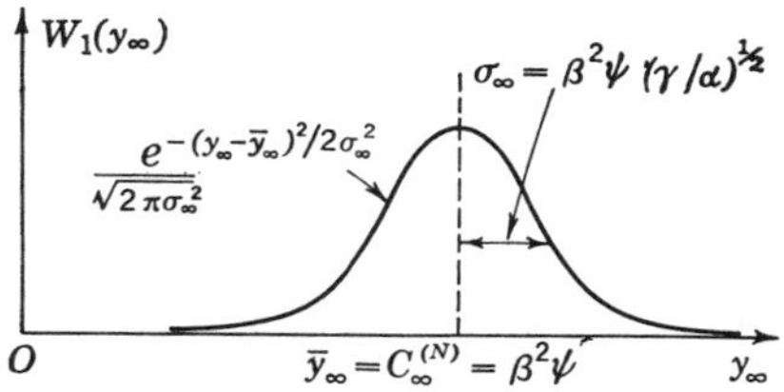

FIG. 17.4. The first-order d.d. of $y_\infty(t)$ after square-law rectification for an RC video filter spectrally much narrower than the original noise input.

$$P_T \equiv \int_0^\infty \mathcal{W}_V(f)_T\, df = \frac{1}{T}\int_{-\infty}^\infty |S_V(f)_T|^2\, df \tag{17.98a}$$

From the definition (3.31) for the amplitude spectral density $S_V(f)_T$ of V on $(-T/2,T/2)$, we easily see that the random variable P_T may be alternatively expressed as†

$$P_T = \frac{1}{T}\int_{-T/2}^{T/2} V(\tau)^2\, d\tau = \frac{1}{T}\int_0^T V(\tau)^2\, d\tau = \frac{1}{T}\int_0^T V(t-\tau)^2\, d\tau \tag{17.98b}$$

these last for the stationary processes assumed here.

The analysis of Case 2 (Sec. 17.3-1) can be applied at once if we set $h_2(\tau) = 1/T$ $(0 \le \tau \le T)$, $h_2(\tau) = 0$ elsewhere, and let $\beta^2 = 1$. Then, Eqs. (17.98*b*) and (17.80) are identical, and we may proceed as Eqs. (17.81) et seq. indicate. The governing integral equation (17.85) becomes now the homogeneous relation

$$\int_0^T K_V(|t-u|)\phi_j(u)\, du = \lambda_j T\phi_j(t) \qquad 0 \le t \le T \tag{17.99}$$

whose solution may be found by the methods described in Appendix 2. A simple case arises when $V(t)$ is an RC normal noise process. Then $K_V(\tau) = \psi e^{-\alpha_0|\tau|}$ $[\alpha_0 = (RC)^{-1}]$, and it can be shown that the eigenvalues are determined by

$$\lambda_j = \frac{2\psi}{\alpha_0 T(1+q_j^2)} \tag{17.100}$$

where q_j are the (positive) roots of $\tan \alpha_0 T q_j = -2q_j/(1-q_j^2)$ (cf. Table 17.1, item 1, Kac and Siegert,[1] and Probs. 16.3*a*, 16.3*b*).

† The stochastic integrals in Eqs. (17.98*a*), (17.98*b*) are interpreted in the usual mean-square sense [cf. Eqs. (2.12) et seq.].

For broadband processes $V(t)$, Eq. (17.84) gives the desired characteristic function of the first-order d.d. An explicit form of the latter, however, is not achievable, except in the large-sample case where T is large compared with the correlation time T_c of the process.[19a] Then by the methods of Example 1, Sec. 17.2-1, we get the asymptotically normal density

$$W_1(P_T) \simeq e^{-(P_T - C_T^{(N)})^2/2D_T^{(N)}} / \sqrt{2\pi D_T^{(N)}} \tag{17.101}$$

where from Eqs. (17.86*a*), (17.86*b*) for the specific *RC* noise assumed above

$$C_T^{(N)} = \psi \qquad D_T^{(N)} = \psi^2 \frac{2\mu_0 + e^{-2\mu_0} - 1}{\mu_0{}^2} \doteq \frac{2\psi^2}{\mu_0} \qquad \mu_0 = \alpha_0 T \gg 1 \tag{17.101a}$$

Note that, for sufficiently long times ($T \to \infty$),

$$W_1(P_\infty) \simeq \delta(P_\infty - \psi) \tag{17.102}$$

as expected. The random variable that is the average intensity [Eqs. (17.98*a*), (17.98*b*)] has the value $\psi = \overline{V^2}$ (probability 1), in contrast to the spectral intensity *density* $\mathcal{W}_V(f)_\infty$ itself, which has the d.d. given by Eqs. (17.60), with a nonvanishing variance† [cf. Eq. (17.63*b*)]. (The consequences of this for the interpretation and calculation of spectral densities has already been commented upon; cf. pages 140 and 141 and Sec. 3.2, in particular.)

Finally, it should be noted that, if $V(t)$ is narrowband (with a symmetrical spectrum), the average intensity P_T is well approximated by

$$P_T \doteq \frac{1}{2T} \int_0^T [V_C(t-\tau)^2 + V_S(t-\tau)^2]\, d\tau \tag{17.103}$$

[cf. Eq. (17.73)]. Consequently, setting $\beta^2 = 1$ and $h_2 = 1/T$ $(0 < \tau < T)$, $h_2 = 0$ elsewhere, once more and applying Eqs. (17.77) to (17.79), we may use Eqs. (17.90) for the d.d. of P_T if we replace y_∞ by P_T therein and obtain the eigenvalues now from Eq. (17.78) modified to

$$\int_{0-}^{T+} K_V(|t-u|)_0 \phi_j(u)\, du = 2T\lambda_j \phi_j(t) \qquad 0 \le t \le T \tag{17.104}$$

In the case of narrowband *LRC* noise, for which $K_V(\tau)_0 = \psi e^{-\alpha|\tau|}$, $\alpha = R/2L$, the eigenvalues here are just one-half those of the broadband *RC* noise above [cf. Eq. (17.100)]. For large T $(\gg \alpha^{-1})$, we again get a normal distribution like Eq. (17.101), with the same mean $C_{2T}^{(N)} = \psi = C_T^{(N)}$, but with a variance

$$D_{2T}^{(N)} = \psi^2 \frac{2\mu + e^{-2\mu} - 1}{2\mu^2} \approx \frac{\psi^2}{\mu} \qquad \mu = \alpha T \gg 1 \tag{17.105}$$

† This latter case corresponds to the situation of no postrectification filtering, cited as a special case of Example 1; compare Eqs. (17.96*b*) with Eqs. (17.60) here.

where $\alpha \neq \alpha_0$ in general. Once more, as $T \to \infty$, P_∞ equals ψ (probability 1), as in Eq. (17.102).

17.3-3 Concluding Remarks. In the preceding discussion, we have observed that, if the linear filter $\boldsymbol{T}_{L2}$ following the square-law rectifier in Fig. 17.2 is spectrally narrow, the final output process is now normal, with first- (and higher-) order d.d.'s that depend on the sample size T, as well as on the nature of the nonlinear element $\boldsymbol{T}_g$ and the statistics of the original input process. Similar remarks also apply in the more general situation of nongaussian inputs and nonlinear elements other than the simple quadratic devices considered explicitly here, although no detailed analytic theory is yet available in such situations.

Thus, in more detail, if the postrectification filter $\boldsymbol{T}_{L2}$ is spectrally narrow (and sample size T is sufficiently large vis-à-vis the correlation time T_c of the normal input), the final output process $y_T(t)$, or $y_\infty(t)$, is again (asymptotically) normal, but with a different mean and variance from those of the original input. [This has been demonstrated here only for the first-order d.d., but can also be shown to be true for the higher-order densities W_n $(n \geq 2)$.] However, for originally nongaussian inputs (such as impulse noise, for example), even under these conditions the output processes usually remain nonnormal, although, of course, modified by both the nonlinear element and the subsequent linear filter. Some of the general properties (e.g., the Markovian nature) of such output processes have been considered elsewhere.[23] For our present purposes, perhaps the most significant general result of the preceding analysis is the approach to normality that occurs whenever (1) the original input is gaussian and (2) the effective sample size is sufficiently large.† In practice, this means that, if the linear filter $\boldsymbol{T}_L$ between two (zero-memory) nonlinear elements $\boldsymbol{T}_{g_1}$, $\boldsymbol{T}_{g_2}$ is sufficiently narrow, the output of this filter can be regarded as once again a normal process, so that the standard techniques of earlier chapters may be used once more: for a normal input to the second, zero-memory nonlinear device $\boldsymbol{T}_{g_2}$. Note, incidentally, at the other extreme, that, if this intervening filter $\boldsymbol{T}_L$ is spectrally very broad, the effect of these two nonlinear elements $\boldsymbol{T}_{g_1}$, $\boldsymbol{T}_{g_2}$ is equivalent to the single composite operation $\boldsymbol{T}_g = \boldsymbol{T}_{g_2}\boldsymbol{T}_{g_1}$.

Finally, we remark that the methods and examples of the preceding sections represent only a portion of the available analysis for problems of this type: many specific results, including those for additive signal and noise processes, are summarized in the exercises following. Moreover, techniques equivalent to and in some instances more general than the trace and eigenvalue approaches discussed here are also available, for these and related

† We have shown this explicitly here for zero-memory quadratic devices only, but the argument cited in Sec. 17.2-1, Example 1, may be applied in the same way for more general zero-memory nonlinear elements. See, for example, the discussion at the end of Sec. 19.4-3.

problems, the details of which, however, are beyond the scope of the present treatment.†

PROBLEMS

17.15 A stationary narrowband normal noise process $V(t)$ is accompanied by an additive signal $S(t) = M(t) \cos \omega_0 t$ at the center frequency of the noise. The corresponding l-f output following full-wave square-law rectification is $x_{LF}(t) = (\beta^2/2)[V_C{}^2(t) + V_S{}^2(t) + 2V_C(t)M(t) + M(t)]$ (cf. Fig. 17.2).

(*a*) Show that the *n*th-order characteristic function of $x_{LF}(t)$, for example, of the set of random variables $x_{LF}(t_1 + \tau_1)$, $x_{LF}(t_1 + \tau_k)$, $x_{LF}(t_1 + \tau_n)$, is given by

$$F_n(i\xi_1, \ldots, i\xi_n)_{x_{LF}} = \prod_{j=1}^{n} \frac{\exp\left\{i\left[\sum_{j=1}^{n} M(t + \tau_l)\xi_l f_{j(l)}\right]^2 \Big/ (1 - i\lambda_j)\right\}}{1 - i\lambda_j} \tag{1}$$

where the eigenvalues $\lambda_j = \lambda_j(\xi_1, \ldots, \xi_n)$ and eigenfunctions $f_{j(l)}$ are obtained from

$$\frac{\beta^2}{2}\psi \sum_{l=1}^{n} \xi_l(\rho_0)_{lm} f_{j(k)} = \lambda_j f_{j(m)} \tag{2}$$

and the orthonormality condition $\sum_{l=1}^{n} \xi_l f_{k(l)} f_{j(l)} = \delta_{kj}$, with $\psi(\rho_0)_{lm} = K_V(|t_l - t_m|)_0$, $\psi = \overline{V^2}$.

(*b*) If the noise $V(t)$ is broadband and $S(t)$ again represents the additive signal, show that the *n*th-order c.f. of the output $x(t) = \beta^2[V^2(t) + S^2(t) + 2V(t)S(t)]$ is

$$F_n(i\xi_1, \ldots, i\xi_n)_x = \prod_{j=1}^{n} \frac{\exp\left\{i\left[\sum_{l=1}^{n} S_l(t + \tau_l)\xi_l f_{j(l)}\right]^2 \Big/ (1 - i\lambda_j)\right\}}{(1 - i\lambda_j)^{1/2}} \tag{3}$$

where Eq. (2) still applies for the eigenfunctions and eigenvalues provided that $\beta^2/2$ is replaced by β^2 and ρ_0 by $\rho = K_V/\psi$ in the usual way.

M. A. Meyer and D. Middleton,[4] sec. 3.

17.16 The noise of Prob. 17.15, without the signal, is after square-law rectification passed through a linear filter of weighting function h_2, yielding an output $y_T(t) = \int_{0-}^{T+} h_2(\tau)x(t - \tau)\,d\tau$ [cf. Eq. (17.80) and Fig. 17.2].

(*a*) Show that the *n*th-order c.f. of the vector process $y_T(t_1 + \tau_1), \ldots, y_T(t_1 + \tau_n)$ is

$$F_n(i\xi)_{y_T} = \prod_{j=1}^{\infty} [1 - i\lambda_j(\xi)]^{-1} \qquad \text{narrowband noise} \tag{1a}$$

† A concise review of these methods and some of their generalizations is given by Rosenbloom, Heilfron, and Trautman;[9] see also Meyer and Middleton,[4] Rosenbloom,[7] Heilfron,[8] Darling and Siegert,[10] and Siegert,[11,12] and the supplementary bibliography at the end of the volume. Still other related material may be found in Chap. 9, Refs. 20 to 22.

$$F_n(i\xi)_{y_T} = \prod_{j=1}^{\infty} [1 - i\lambda_j(\xi)]^{-1/2} \qquad \text{broadband noise} \tag{1b}$$

with ξ the set of variables $(\xi_1, \ldots, \xi_n)$, where the eigenvalues $\lambda_j(\xi)$ and eigenfunctions $\phi_j(t)$ are determined from the integral equations (with $\tau_0 = 0$)

$$\psi \begin{Bmatrix} \frac{\beta^2}{2} \\ \beta^2 \end{Bmatrix} \int_{0-}^{T+} \begin{Bmatrix} \rho(|t-u|)_0 \\ \rho(t-u) \end{Bmatrix} \left[\sum_{k=1}^{n} \xi_k h_2(u - \tau_{k-1}) \right] \phi_j(u)\, du = \lambda_j(\xi)\phi_j(t) \qquad 0- < t < T+ \tag{2}$$

(The upper and lower lines apply, respectively, for narrow- and broadband noise inputs.)

(*b*) Hence, show that for $n = 1$ (the first-order case), Eqs. (1*a*) and (2) reduce to the results [Eqs. (17.77), (17.78)] of Kac and Siegert[1] and that for $n = 2$ we get for the basic integral equations

$$\psi \begin{Bmatrix} \frac{\beta^2}{2} \\ \beta^2 \end{Bmatrix} \int_{0-}^{T+} [\xi_1 h_2(u) + \xi_2 h_2(u-\tau)] \begin{Bmatrix} \rho(|t-u|)_0 \\ \rho(|t-u|) \end{Bmatrix} \phi_j(u) = \lambda_j(\xi_1,\xi_2)\phi_j(t) \qquad 0- < t < T+ \tag{3}$$

(*c*) Using the results of Prob. 17.15, extend Eqs. (1*a*), (1*b*) to obtain the *n*th-order c.f.'s:

Narrowband signal and noise

$$F_n(i\xi)_{y_T}\Big|_{S+N} = \prod_{j=1}^{\infty} \frac{\exp\left(i \left\{ \int_{0-}^{T+} M(t+u) \left[\sum_{l=1}^{n} \xi_l h_2(u - \tau_{l-1}) \right] \phi_j(u)\, du \right\}^2 \Big/ (1 - i\lambda_j) \right)}{1 - i\lambda_j} \tag{4}$$

For broadband signal and noise, derive Eq. (4), with M replaced by S and $(1 - i\lambda_j)^{-1}$ by $(1 - i\lambda_j)^{-1/2}$ in the denominator [cf. Prob. 17.15, Eq. (3)]. Besides being the solution of Eq. (2), the eigenfunctions obey the usual orthonormality condition $\int_{0-}^{T+} \phi_j(u)\phi_l(u)\, du = \delta_{jl}$. From Eq. (4), it is clear that in dealing with a signal, as well as noise, by this method we must know the eigenfunctions in addition to the eigenvalues.

M. A. Meyer and D. Middleton,[4] sec. 4.

17.17 (*a*) Discuss the evaluation of the probability densities $W_2, W_3, \ldots, W_n$ for noise alone and the signal and noise of Prob. 17.16.

(*b*) Letting $\lambda_j \equiv \lambda_j'(\beta^2/2)\psi\xi_1 = \xi_1\lambda_j''$, obtain the first-order density $(W_1)_{S+N}$ for narrowband noise and signal, viz.,

$$W_1(y_T)_{S+N} = \int_{-\infty}^{\infty} e^{-i\xi_1 y_T} \left(\prod_{j=1}^{\infty} \frac{e^{i\xi_1{}^2 A_j(t)^2/(1-i\lambda_j''\xi_1)}}{1 - i\lambda_j''\xi_1} \right) \frac{d\xi_1}{2\pi} \tag{1}$$

and evaluate. Here, $A_j(t) \equiv \int_{0-}^{T+} M(t+u)h_2(u)\phi_j(u)\, du$.

17.18 Figure P 17.18 shows an idealized frequency-discriminator system, where T_{L1}, T_{L2} are two narrowband linear filters antisymmetrically detuned from center frequency f_0. Here, $V_{in}(t)$ is a stationary broadband noise, so that $V_1(t)$, $V_2(t)$ are likewise stationary narrowband noise processes. The nonlinear elements are identical (full-wave) square-law zero-memory devices, whose outputs x_1, x_2 are subtracted to form $x(t) = x_1(t) - x_2(t)$; this in turn may be smoothed by a low-pass linear filter T_{L3} [which, however, in some cases can be broad enough to pass all the l-f components of $x(t)$].

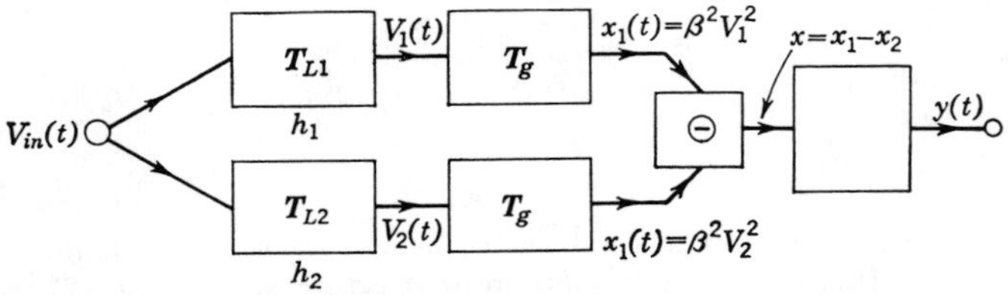

Fig. P 17.18. An idealized discriminator system.

(a) Show that the first-order c.f. for $y(t)$, in the steady state, is

$$F_1(i\xi)_{y_\infty} = \prod_{j=1}^{\infty} (1 - 2i\xi\lambda_j)^{-1/2} \tag{1}$$

where the λ_j are determined from

$$\int_{-\infty}^{\infty} G(u,t)\phi_j(u)\,du = \lambda_j\phi_j(t) \qquad -\infty < t < \infty \tag{2}$$

in which $$G(u,t) = \beta^2 \int_{-\infty}^{\infty} h_3(\tau)[h_1(u-\tau)h_1(t-\tau) - h_2(u-\tau)h_2(t-\tau)]\,d\tau \tag{3}$$

(b) When T_{L3} is spectrally very wide, show that Eq. (2) has only two eigenvalues,

$$\lambda_{1,2} = \tfrac{1}{2}(\rho_{F1} - \rho_{F2}) \pm \tfrac{1}{2}[(\rho_{F1} + \rho_{F2})^2 - 4\rho_{F12}^2]^{1/2} \qquad 0 \le \rho_F \le 1 \tag{4}$$

where ρ_{F1}, ρ_{F2} are the envelopes of the covariance functions of $V_1(t)$ and $V_2(t)$, normalized to unity, and ρ_{12} is the corresponding cross-variance function between $V_1(t)$ and $V_2(t)$, all at $\tau = t_2 - t_1 = 0$. Accordingly, show that λ_1 is positive, while λ_2 is negative, so that the c.f. [Eq. (1)] becomes

$$F_1(i\xi)_{y_\infty} = [(1 - 2i\xi\lambda_1)(1 + 2i\xi|\lambda_2|)]^{-1/2} \tag{5}$$

Show, therefore, that

$$W_1(y_\infty) = \frac{1}{2\pi} (\lambda_1|\lambda_2|)^{-1/2} e^{y_\infty(\lambda_1 - |\lambda_2|)/4\lambda_1|\lambda_2|} K_0\left(|y_\infty| \frac{\lambda_1 + |\lambda_2|}{4\lambda_1|\lambda_2|}\right) \qquad -\infty < y_\infty < \infty \tag{6}$$

(c) When the low-pass filter T_{L3} is spectrally narrow, verify that y_∞ has an asymptotically normal d.d. with mean and variance given by

$$\bar{y}_\infty \simeq [\rho_{F1}(0) - \rho_{F2}(0)] \int_{-\infty}^{\infty} h_3(\tau)\,d\tau \tag{7}$$

$$\overline{y_\infty^2} - \bar{y}_\infty^2 \simeq 2 \int_{-\infty}^{\infty} h_3(\tau_1)\,d\tau_1 \int_{-\infty}^{\infty} d\tau_2\, h_3(\tau_2)[\rho_{F1}(\tau_2-\tau_1)^2 + \rho_{F2}(\tau_2-\tau_1)^2 - \rho_{12}(\tau_2-\tau_1)^2 - \rho_{21}(\tau_2-\tau_1)^2] \tag{8}$$

(d) By evaluating $E_3^{(\infty)}$ [Eq. (17.30a)] here, show how the normal law of (c) is approached.

G. R. Arthur.[5]

17.19 Two statistically related stationary normal noise processes $V_1(t)$, $V_2(t)$, with zero means, are passed into the multiplier-filter system of Fig. P 17.19. The multiplier

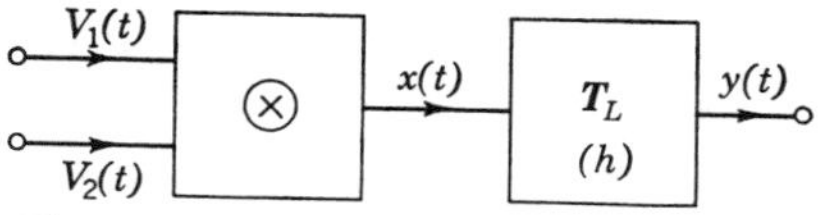

Fig. P 17.19. Multiplier-filter system.

has zero memory and is ideal, and the linear filter T_L is stable and time-invariant, with weighting function h, so that in the steady state

$$x(t) = \gamma V_1(t) V_2(t) \tag{1a}$$

$$y_\infty(t) = \gamma \int_{-\infty}^{\infty} V_1(t-\tau) V_2(t-\tau) h(\tau)\, d\tau \tag{1b}$$

(*a*) Show that the characteristic function of the first-order density of the nongaussian functional $y(t)$ is

$$F_1(i\xi)_y = \prod_{-\infty}^{\infty}{}' (1 - 2i\xi\lambda_j)^{-1/2} \tag{2}$$

where the prime indicates that the term for $j = 0$ is omitted and the eigenvalues λ_j are determined from the pair of simultaneous integral equations

$$\left.\begin{aligned} &\gamma \int_{-\infty}^{\infty} h(u) K_{11}(t-u) \phi_j^{(1)}(u)\, du \\ &\qquad - \gamma \int_{-\infty}^{\infty} h(u) K_{12}(t-u) \phi_j^{(2)}(u)\, du = \lambda_j \phi_j^{(1)}(t) \\ &\gamma \int_{-\infty}^{\infty} h(u) K_{21}(t-u) \phi_j^{(1)}(u)\, du \\ &\qquad - \gamma \int_{-\infty}^{\infty} h(u) K_{22}(t-u) \phi_j^{(2)}(u)\, du = \lambda_j \phi_j^{(2)}(t) \end{aligned}\right\} \quad -\infty < t < \infty \tag{3}$$

Here,

$$\begin{aligned} K_{11}(\tau) &= \tfrac{1}{4}[\psi_{11}(\tau) + \psi_{12}(\tau) + \psi_{21}(\tau) + \psi_{22}(\tau)] \\ K_{12}(\tau) &= \tfrac{1}{4}[\psi_{11}(\tau) - \psi_{12}(\tau) + \psi_{21}(\tau) - \psi_{22}(\tau)] \\ K_{21}(\tau) &= \tfrac{1}{4}[\psi_{11}(\tau) + \psi_{12}(\tau) - \psi_{21}(\tau) - \psi_{22}(\tau)] \\ K_{22}(\tau) &= \tfrac{1}{4}[\psi_{11}(\tau) - \psi_{12}(\tau) - \psi_{21}(\tau) + \psi_{22}(\tau)] \end{aligned} \tag{4}$$

where $\psi_{ij}(\tau) = \boldsymbol{E}\{V_i(t)V_j(t+\tau)\}$ $(i, j = 1, 2)$.

(*b*) In the case of no postmultiplier filter [that is, $h(\tau) = \delta(\tau - 0)$], show that

$$F_1(i\xi)_{y_\infty} = [(1 - 2i\xi\lambda_1)(1 - 2i\xi\lambda_2)]^{-1/2} \tag{5}$$

where $\lambda_{1,2} = \gamma(\psi_{12} \pm \sqrt{\psi_{11}\psi_{22}})/2$, with $\psi_{12} = \psi_{12}(0)$, etc. Verify that one root is positive and the other negative, so that

$$\lambda_1 = \gamma(\psi_{12} + \sqrt{\psi_{11}\psi_{22}})/2 > 0 \qquad |\lambda_2| = |\gamma(-\psi_{12} + \sqrt{\psi_{11}\psi_{22}})/2| > 0 \tag{6}$$

Hence, obtain the first-order d.d. in this case:

$$W_1(y_\infty) = \frac{1}{\pi} \gamma^{-1} (\psi_{11}\psi_{22} - \psi_{12}^2)^{-1/2} e^{y_\infty \psi_{12}/\gamma(\psi_{12}\psi_{22} - \psi_{12}^2)} K_0 \left[\frac{|y_\infty| \sqrt{\psi_{11}\psi_{22}}}{\gamma(\psi_{11}\psi_{22} - \psi_{12}^2)} \right] \quad -\infty < y_\infty < \infty \tag{7}$$

(*c*) If V_1 is white normal noise of spectral density W_0 and V_2 is originally the same white noise after passage through an RC filter of time constant $\alpha^{-1} = R_1C_1$, and if the postmultiplier filter is also an RC element with time constant $\beta^{-1} = R_2C_2$, show that the eigenvalues of Eqs. (3) are the positive *and* negative roots of

$$J_{2\alpha/\beta-2}\left[2\sqrt{\frac{\alpha}{\gamma_{0j}}\left(\frac{2\alpha}{\beta}-1\right)}\right] = 0 \qquad \gamma_{0j} \equiv \frac{4\lambda_j}{W_0\gamma} \tag{8}$$

Hence, in the special case $\alpha/\beta = 5/4$, show that (with $\psi = W_0\alpha/4$)

$$W_1(y_\infty) = \begin{cases} \dfrac{\pi^2}{6\psi\gamma} \displaystyle\sum_{m=1}^{\infty} (-1)^{m+1} m^2 e^{-\pi^2 m^2 y_\infty/12\gamma\psi_0} & y_\infty > 0 \\ 0 & y_\infty < 0 \end{cases} \tag{9}$$

(*d*) Show that the first four semi-invariants of $y_\infty(t)$ for (*c*) (general α, β) are

$$L_1^{(N)} = \psi_0\gamma \qquad L_2^{(N)} = \psi_0^2\gamma^2\frac{\beta}{\alpha} \qquad L_3^{(N)} = 8\psi_0^3\gamma^3\frac{\beta}{\alpha}\frac{\beta}{2\alpha\beta}$$
$$L_4^{(N)} = 6\psi_0^4\gamma^4\frac{\beta}{\alpha}\frac{\beta}{\alpha\beta}\frac{\beta}{2\alpha\beta}\frac{10\alpha+\beta}{\alpha} \tag{10}$$

Accordingly, obtain the generalized Edgeworth series for large α/β, in terms of the normalized variable $q = [y_\infty(t) - L_1^{(N)}]/(L_2^{(N)})^{1/2}$, namely,

$$W_1(z) \simeq \phi^{(0)}(z) + A_3\phi^{(3)}(z) + [A_4\phi^{(4)}(z) + A_6\phi^{(6)}(z)] + \cdots \tag{11}$$

where specifically $\phi^{(k)}(x) = (d^k/dx^k)(e^{-x^2/2}/\sqrt{2\pi})$ and

$$A_3 = \frac{-4/3(\alpha/\beta)^{1/2}}{2\alpha/\beta+1} \qquad A_4 = \frac{5}{2}\frac{\alpha/\beta + 1/10}{(2\alpha/\beta+1)(\alpha/\beta+1)} \tag{11a}$$

D. G. Lampard.[6]

REFERENCES

1. Kac, M., and A. J. F. Siegert: On the Theory of Noise in Radio Receivers with Square-law Detectors, *J. Appl. Phys.*, **18:** 383 (1947).
1a. Fränz, K.: Die Amplituden von Gerauschpannung, *Elek. Nachr.-Tech.*, **19:** 166 (1942).
2. Middleton, D.: On the Detection of Stochastic Signals in Additive Normal Noise. I, *IRE Trans. on Inform. Theory*, **IT-3:** 86 (June, 1957); cf. apps. 1, 2, 3. Typographical corrections, **IT-3:** 256 (December, 1957).
3. Emerson, R. C.: First Probability Densities for Receivers with Square-law Detectors, *J. Appl. Phys.*, **24:** 1168 (1953).
4. Meyer, M. A., and D. Middleton: On the Distribution of Signals and Noise after Rectification and Filtering, *J. Appl. Phys.*, **25:** 1037 (1954).
5. Arthur, G. R.: The Statistical Properties of the Output of a Frequency Sensitive Device, *J. Appl. Phys.*, **25:** 1185 (1954).
6. Lampard, D. G.: The Probability Distribution for the Filtered Output of a Multiplier Whose Inputs Are Correlated, Stationary, Gaussian Time-series, *IRE Trans. on Inform. Theory*, **IT-2:** 4 (March, 1956).
7. Rosenbloom, A.: Analysis of Randomly Time-varying Linear Systems, *Univ. Calif. (Los Angeles) Dept. Eng. Rept.* 54-95, October, 1954. See also Analysis of Linear Systems with Randomly Time-varying Parameters, Polytechnic Institute of Brooklyn, *Proc. Symposium on Inform. Networks*, **8:** 145 (April, 1954); and Ref. 9.

8. Heilfron, J.: On the Response of Linear Systems to Non-gaussian Noise, *Univ. Calif. (Los Angeles) Dept. Eng. Rept.* 54-82, September, 1954. See also On the Response of a Certain Class of Systems to Random Inputs, *IRE Trans. on Inform. Theory*, **IT-1**: 59 (March, 1955).
9. Rosenbloom, A., J. Heilfron, and D. L. Trautman: Analysis of Linear Systems with Randomly Varying Inputs and Parameters, *IRE Conv. Record*, March, 1955, pt. 4, p. 106.
10. Darling, D. A., and A. J. F. Siegert: A Systematic Approach to a Class of Problems in the Theory of Noise and Other Random Phenomena. Part I, *IRE Trans. on Inform. Theory*, **IT-3**: 32 (March, 1957).
11. Siegert, A. J. F.: Passage of Stationary Processes through Linear and Non Linear Devices, *IRE Trans. on Inform. Theory*, **PGIT-3**: 4 (March, 1954).
12. Siegert, A. J. F.: A Systematic Approach to a Class of Problems in the Theory of Noise and Other Random Phenomena. Part II, Examples, *IRE Trans. on Inform. Theory*, **IT-3**: 38 (March, 1957).
13. Courant, R., and D. Hilbert: "Methods of Mathematical Physics," vol. I, chap. III, arts. 4, 5, Interscience, New York, 1953; also pp. 50, 122, 160.
14. Youla, D. C.: The Use of the Method of Maximum Likelihood in Estimating Continuous-modulated Intelligence Which Has Been Corrupted by Noise, *IRE Trans. on Inform. Theory*, **PGIT-3**: 90 (March, 1954), app. A.
15. Morse, P. M., and H. Feshbach: "Methods of Theoretical Physics," vol. I, p. 905, McGraw-Hill, New York, 1953.
16. Price, R.: Optimum Detection of Random Signals in Noise, with Application to Scatter-multipath Communication. I, *IRE Trans. on Inform. Theory*, **IT-2**: 125 (December, 1956).
17. Juncosa, M. L.: An Integral Equation Related to Bessel Functions, *Duke Math. J.*, **12**: 465 (1945).
18. Lovitt, W. V.: "Linear Integral Equations," sec. 15, p. 24, Dover, New York, 1950.
19. Cramér, H.: "Mathematical Methods of Statistics," Princeton University Press, Princeton, N.J., 1946, sec. 18.1, for example.
19a. Slepian, D.: Fluctuations of Random Noise Power, *Bell System Tech. J.*, **37**: 163 (1958), secs. IV, V especially.
20. Cramér, H.: Random Variables and Probability Distributions, *Cambridge Tracts in Math.* 36, 1937.
21. Gnedenko, B. V., and A. N. Kolmogoroff: "Limit Distributions for Sums of Independent Random Variables," Addison-Wesley, Reading, Mass., 1954.
22. Jastram, P.: The Effect of Nonlinearity and Frequency Distortions on the Amplitude Distribution for Stationary Random Processes, doctoral dissertation, University of Michigan, Ann Arbor, Mich., 1947.
23. Heilfron, J.: On the Response of a Certain Class of Systems to Random Inputs, *IRE Trans. on Inform. Theory*, **IT-1**: 59 (March, 1955).

PART 4

A STATISTICAL THEORY OF RECEPTION

CHAPTER 18

RECEPTION AS A DECISION PROBLEM

In preceding chapters, we have considered various elements of the single-link communication process, $\{v\} = \boldsymbol{T}_R{}^{(N)}\boldsymbol{T}_M{}^{(N)}\boldsymbol{T}_T{}^{(N)}\{u\}$ [cf. Eq. (2.20)], where, as before, u, v are the original and received message ensembles, or decisions stemming from them, and $\boldsymbol{T}_T{}^{(N)}$, $\boldsymbol{T}_R{}^{(N)}$, etc., are the operations of transmission, reception, etc., that take place at different stages of the process. This earlier discussion has dealt with the effects of noise and the role of individual system elements, or groups of elements, on system performance and for the most part has been based on the results of the appropriate second-moment theory (cf. Chaps. 5, 12 to 16), e.g., spectra, covariance functions, and signal-to-noise ratios. Here and in the remaining chapters, however, we shall adopt a broader viewpoint, where the aim now is to replace this limited and rather piecemeal approach with a more general and more powerful treatment, utilizing all available information and giving a more realistic account of performance in actual situations. Accordingly, in a very broad sense the purpose of this and succeeding chapters is to outline and illustrate recent developments in statistical communication theory which provide such a systematic and unified approach.

The minimum requirements of a theory of this nature are threefold; they embody (1) methods for determining the explicit structure of optimal, or "best," systems for the problem at hand; (2) procedures for evaluating the performance of such optimum systems; and (3) techniques of system comparison, whereby different systems, optimum and suboptimum, may be quantitatively compared. In addition, an important practical requirement for system design is the interpretation of the analytic results in terms of ordered combinations of physically realizable network elements, e.g., time-varying linear filters, zero-memory nonlinear devices, and the like. We must remember, too, that the general conditions of operation must take into account the finite times available for data acquisition and processing. The approach followed here in constructing a theory of this type uses the methods of statistical decision theory as developed in recent years, principally by Wald[1,2] and others,[3] and is concerned specifically with the wide class of practical communication situations where definite decisions of one kind or another are required. While we shall be concerned almost entirely with reception problems, i.e., with the properties of $\boldsymbol{T}_R{}^{(N)}$, when $\boldsymbol{T}_T{}^{(N)}$ as well as

$\boldsymbol{T}_M^{(N)}$ is given, the approach is not limited to those alone but is capable of a direct extension to the more general cases, where both $\boldsymbol{T}_R^{(N)}$ and $\boldsymbol{T}_T^{(N)}$ are considered for simultaneous adjustment, as several of the examples in Chap. 23 indicate. The emphasis on reception here simply reflects the pattern in which this aspect of communication theory has developed in the last few years (for further comments, see Secs. 23.4, 23.5).

As we have mentioned a number of times before, the uses of statistical concepts in communication theory have evolved rapidly since the early 1940s as witnessed, for example, in the case of reception by the great variety of contributions in various ways devoted to questions of optimum detection and extraction of signals in noise (cf. the references at the end of this chapter). These contributions, in turn, may properly be included within the unifying structure of decision theory, as the analysis in succeeding chapters will show. Here, of course, the emphasis is on applications to communication theory, and by way of discussion and illustration a considerable portion of the material of Chaps. 18 to 23 is based on the original presentations of Middleton[4] and Van Meter.[5] Part 4 of the book accordingly consists of essentially three main parts: Chap. 18 provides a general introduction, with Chaps. 19, 20 devoted to the detection of signals in noise; Chaps. 21, 22 examine problems of signal estimation or measurement; and Chap. 23 concludes with a brief discussion of special, related topics, such as game theory, as well as generalizations of some of the results of the preceding chapters. In the final sections, the principal accomplishments and difficulties of the decision-theory approach to date are surveyed, and a number of promising future applications are briefly indicated.

18.1 *Introduction*

The reception of signals in noise presents problems of critical importance in the theory of communication, since noise to varying degrees always obscures the desired signal or message. Because the observation period during which the signal may be recovered is necessarily limited, and because of the inherently statistical character of signal and interference, information is lost and recovery incomplete. Of course, reception of signals under such conditions can usually be carried out in a variety of ways, but very few of these possess optimum properties. The reception problem here may then be loosely described as the task of finding "best," or optimum, systems—i.e., optimum choices of $\boldsymbol{T}_R^{(N)}$, for given $\boldsymbol{T}_M^{(N)}$, $\boldsymbol{T}_T^{(N)}$—in order to mitigate the deleterious effects of the accompanying noise.

A reception problem in communication theory arises when it is necessary to make a specified type of judgment, or decision, about a signal, based on observations of the signal after it has become contaminated by noise. The judgment itself may concern the presence or absence of one or more signals, as in *detection*, or it may be an estimate of one or more of the information-bearing features of a signal (e.g., amplitude, frequency, waveform, etc.), as in *extraction*. It is possible to conceive also of other judgments for which a receiving system can be designed, a fact which has not so far been utilized. In any case, a receiving system, as the term is used here, processes the observed data in conjunction with the available a priori information to produce an output quantity which is the required judgment. Reception systems of this nature should be distinguished from other types which have been described elsewhere (cf. Woodward[6]). The latter systems are not designed to yield the required judgment but only certain a posteriori information on the basis of which an observer can come to a decision himself. Such detection systems, therefore, assist the observer but leave actual decisions to his discretion. The detection processes considered here for the most part, however, furnish indications of the type "yes"—a signal is present— or "no"—only noise occurs—while a nondecision system might present the a posteriori probabilities of signal and no signal at its output, leaving the option of a definite decision as to the choice between them to the observer himself.[6] Similarly, a decision system for extraction provides an actual estimate of the signal or its descriptive parameters at the output, whereas a nondecision system might be designed simply to maximize signal-to-noise ratio (cf. Sec. 16.3).

As previous chapters have indicated, communication theory is concerned largely with operations (for example, $\boldsymbol{T}_R^{(N)}$, $\boldsymbol{T}_T^{(N)}$, etc.) on ensembles, where the effective techniques are basically statistical. Thus, communication theory deals with random processes and time series and their pertinent statistical properties. Strategically, reception ($\boldsymbol{T}_R^{(N)}$) is distinguished by the fact that the receiver has only a limited knowledge of the input signal (necessarily so, if any information is to be conveyed) and little or no control over it. In other words, the judgment required about the input must be a *statistical inference*.[4,5,7] This suggests the application of statistical decision theory, which in fact provides a very general method for the synthesis of optimum procedures for statistical inference, and hence for the design of optimum reception systems. The latter problem has been referred to as a central problem of reception, and it should be at least intuitively evident that decision theory is likely to be eminently well suited to the problem. One should expect it to lead in an elegant and general fashion to known solutions, and also to open access to new ones. That this is actually so will, it is hoped, become clear from the results of the succeeding chapters.†

In order to attack a problem by statistical decision theory, certain information must be available beforehand. One should know, more particularly, the statistics of the noise and, if possible, also the statistics of the signal. The less one can assume known concerning these, the more difficult is the solution in general. A clear statement should be available concerning the number of alternatives among which the decision must be made. In

† For an evaluation, see Sec. 23.3.

addition to this, another datum is required for the problem, namely, a criterion of excellence by which the performance of a reception system can be rated and with respect to which the optimization can be carried out. Once the criterion is selected, the optimum system is in principle determined.

It is important, however, to understand the strength and the limitations of this theoretical approach. In practice, performance specifications are rarely explicit enough to fix the optimum system uniquely. More often, the designer must himself supply a definition of "best," for the situation at hand, which is sufficiently precise to determine a unique system and which at the same time accurately reflects the given design constraints. Clearly, this is a creative phase of design beyond the reach of any specified theory, where insight and experience are prerequisites of success.

In the present theory, however, analysis still plays an important part (1) by proposing general methods of system evaluation, (2) by exhibiting the properties of the resulting optimum systems, and (3) by comparing these with others and determining the extent to which performance is degraded by system simplification. By such comparative study, a repertoire of optimum designs and characteristics becomes available to the designer, allowing a more adequate appraisal of the advantages and disadvantages of any particular criterion. Of equal importance, also, are the evaluation of actual, nonideal systems and comparison of these with the theoretical optimum for the same purpose and with respect to the same criterion. In this way, more definite estimates of the costs and gains of improvements can be obtained, and in this respect, also, decision theory supplies a framework upon which specific systems can be tested and compared,[4,5] a framework for the most part inclusive of earlier analyses.

Optimum design depends on the nature of the signal and noise statistics, and it is usually assumed that some knowledge of these is available. However, in practice reliable estimates of these distributions are not easy to obtain, especially estimates of the a priori signal probabilities. Another function of the analytical approach, therefore, is to suggest methods whereby optimum performance may be defined and determined in situations where the signal probabilities are unknown, or only partially known, and to calculate the improvement in performance which might be expected if full knowledge of these were available to the designer. In this way, the cost of obtaining missing data can be included in system planning, and the penalties of such ignorance assessed.

With these introductory comments, let us now formulate the problems of optimum and suboptimum reception as decision processes.

18.2 *Signal Detection and Extraction*

We begin the discussion with a general account of the two main reception problems, those of detection and extraction of signals in noise. Here we introduce some terminology, taken partly from the field of statistics, partly from communication engineering, and review the problems in these terms. We shall also point out some considerations which must be kept in mind concerning the given data of the problems. Later, in Secs. 18.3 to 18.5, we generalize the reception problem, state it in mathematical language, and outline the nature of its solutions.†

18.2-1 Detection. The problem of the detection of a signal in noise is equivalent to one which, in statistical terminology, is called the problem of *testing hypotheses:* here, the hypothesis that noise alone is present is to be tested, on the basis of some received data, against the hypothesis (or hypotheses) that a signal (or one of several possible signals) is present.

Detection problems can be classified in a number of ways: by the number of possible signals which need to be distinguished, by the nature of the hypotheses, by the nature of the data and their processing, and by the characteristics of the signal and noise statistics. These will now be described in greater detail.

(1) The number of signal classes to be distinguished. This is equal to the number of hypotheses to be tested but does not depend on their nature. A *binary detection system* can make but two decisions, corresponding to two hypotheses, while a *multiple alternative detection system*[11] makes more than two decisions. For the time being, we deal only with the binary detection problem (the multiple cases are discussed briefly in Sec. 23.1).

(2) The nature of the hypotheses. A signal is a desired system input during the interval available for observation of the mixture of signal and noise. Noise (stationary or nonstationary) is an undesired input, considered to enter the system independently‡ of the signal and to affect each observation according to an appropriate scheme whereby the two are combined.§ The class of all possible (desired) system inputs is called the *signal class* and is conveniently represented as an abstract space (*signal space*) in which each point corresponds to an individual signal.

A hypothesis which asserts the presence of a single nonrandom signal at the input is termed a simple *hypothesis.* A *class* (or *composite*) *hypothesis,* on the other hand, asserts the presence at the input of an unspecified member of a specified subclass of signals; i.e., it reads "some member of subclass k (it does not matter which member) is present at the input." Such a sub-

† As an introduction to the methods of statistical inference, see, for example, Hoel[8] and the more advanced treatments of Kendall[9] and Cramér.[10] The reader unfamiliar with these ideas will find this helpful background material in the subsequent discussion.

‡ In most applications.

§ There can be noiselike signals also, but these are not to be confused with the noise background. It is frequently convenient to speak of "noise alone" at the input, and this is to be interpreted as "no signal of any kind" present. In physical systems, which do not, of course, use ideal (i.e., noise-free) elements, noise may be introduced at various points in the system, so that care must be taken in accounting for the manner in which signal and noise are combined (cf. Secs. 5.3-3, 5.3-4).

class is called a *hypothesis class*. Hypothesis classes may or may not overlap (cf. Fig. 18.1*e*).

Usually, one hypothesis in detection asserts the presence of noise alone (or the absence of any signal) and is termed the *null hypothesis*. In binary detection, the other hypothesis is called the *alternative*. If the alternative is a class hypothesis and the class includes all nonzero signals involved in the problem, it is termed a *one-sided alternative*. It is a *simple alternative* if there is but one nonzero signal in the entire signal space (which signal must therefore contain nonrandom parameters only). Figure 18.1 illustrates some typical situations. In each case, the class of all possible system inputs

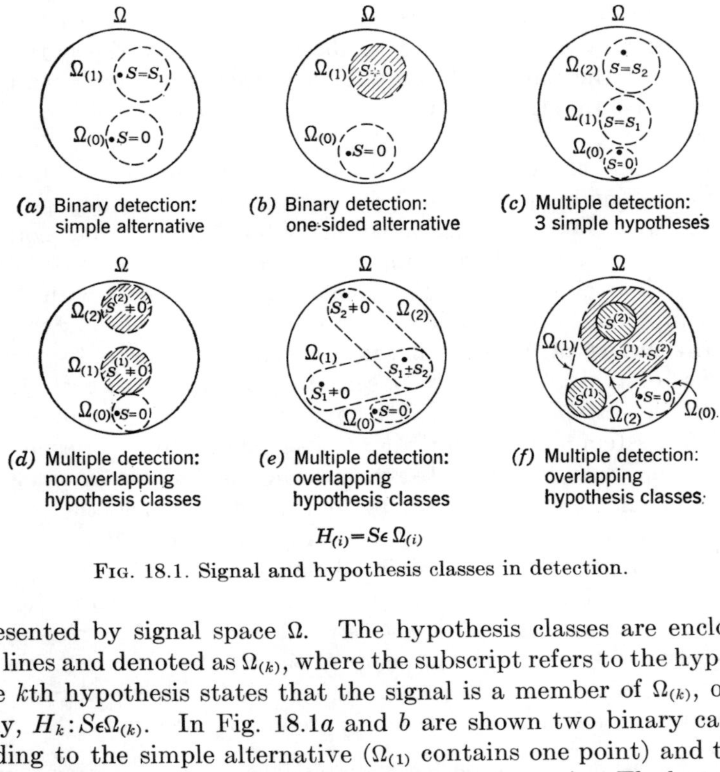

FIG. 18.1. Signal and hypothesis classes in detection.

is represented by signal space Ω. The hypothesis classes are enclosed by dashed lines and denoted as $\Omega_{(k)}$, where the subscript refers to the hypothesis; i.e., the kth hypothesis states that the signal is a member of $\Omega_{(k)}$, or, symbolically, $H_k: S\epsilon\Omega_{(k)}$. In Fig. 18.1*a* and *b* are shown two binary cases corresponding to the simple alternative ($\Omega_{(1)}$ contains one point) and the one-sided alternative ($\Omega_{(1)}$ contains all nonzero system inputs). The latter would occur, for example, if all signals in a binary detection problem were the same except for a random amplitude scale factor, governed by, say, a gaussian distribution. Figure 18.1*c* and *d* show multiple hypothesis situations where the hypothesis classes do not overlap, while Fig. 18.1*e* and *f* represent situations where overlapping can occur, in (*e*) with single-point classes and in (*f*) when the classes are one-sided or composite. Many different combinations can be constructed, depending on the actual problems at hand. In the present treatment, we shall confine our attention to the nonoverlapping cases, although the general approach is in no way restricted by our so doing.

(3) The nature of the data and their processing. The observations made on the mixture of signal and noise during the observation period may consist of a discrete set of values (*discrete or digital sampling*) or may include a continuum of values throughout the interval (*continuous*, or *analogue, sampling*)[7] (cf. Fig. 18.2). Whether one procedure or the other is used is a datum of the problem. In radar, for example, detection may (to a first approximation) be based on a discrete set of successive observations, while, in certain communication cases, a continuous-wave signal may be sampled continuously.

Similarly, it is a datum of the problem whether or not the observation interval, i.e., the interval over which the reception system can store the

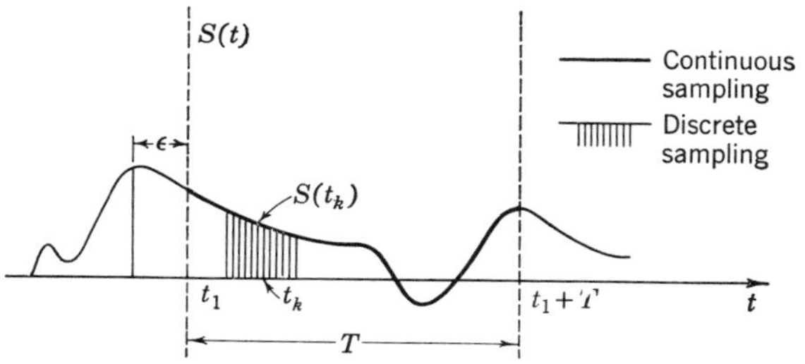

FIG. 18.2. A signal waveform, showing discrete and continuous sampling, the epoch ϵ, and the observation period.

data for analysis, is fixed or variable. In the latter case, one can consider *sequential detection*. A sequential test proceeds in steps, deciding at each stage either to terminate the test or to postpone termination and repeat the test with additional data. In applications of decision theory, it turns out that the analysis divides conveniently at the choice between the sequential and nonsequential. The theory of each type is complete in a certain sense, and additional restrictions on the tests may not be imposed without compromising this completeness. It is, of course, true that, since the class of sequential tests includes nonsequential tests as a special subclass, a higher grade of performance may be expected, on the average, under the wider assumption (see references at end of Sec. 20.4-8).

(4) The signal and noise statistics. The nature of these quantities is clearly of central importance, as it is upon them that specific calculations of performance depend. In general, individual sample values *cannot* be treated as statistically independent, and this inherent correlation between the sample values over the observation period, in both the continuous and discrete cases, is an essential feature of the problem.

The signal itself may be described in quite general terms involving both random and deterministic parameters. Thus, we write $S(t) = S(t,\epsilon;a_0,\boldsymbol{\theta})$

(cf. Sec. 1.3-5), where ϵ is an *epoch*, or time interval, measured between some selected point in the "history" of the signal S and, say, the beginning of the observation period $(t_1, t_1 + T)$, relating the observer's to the signal's time scale, as indicated in Fig. 18.2; a_0^2 is a scale factor, measuring (relative) intensity of the signal with respect to the noise background; and $\boldsymbol{\theta}$ denotes all other descriptive parameters, such as pulse duration, period, etc., that may be needed to specify the signal; S itself gives the "shape," or functional *form*, of the wave in time.

No restriction is placed on the signal other than that it have finite energy in the observation interval; it may be entirely random, partly random (e.g., a "square wave" with random durations), or entirely causal or deterministic [e.g., a sinusoid, or a more complex structure which is nevertheless uniquely specified by $S(t)$]. Signals for which the epoch ϵ assumes a fixed value are said to be *coherent*, while if ϵ is a random variable, such signals are called *incoherent.*† Coherent signals may have random parameters and thus belong to subclasses of Ω containing more than one member. Coherent signals corresponding to subclasses containing but a single member will be called *completely coherent.* From these remarks, it is clear that an incoherent signal cannot belong to such an elementary class. The description of the noise is necessarily statistical, and here we distinguish between noise belonging to *stationary* and *nonstationary* processes.[12,13] Generalizations of the noise structure to include partially deterministic waves offer no conceptual difficulties.

18.2-2 Types of Extraction. We use the term *extraction* here to describe a reception process which calls for an estimate of the signal itself or one or more of its descriptive parameters.

Signal extraction, like detection, is a problem which in other areas has received considerable attention from statisticians and has been known under the name of *parameter estimation.* A certain terminology has become traditional in the field, which we shall mention presently. We can classify extraction problems under three headings: the nature of the estimate, the nature of the data processing, and the statistics of signal and noise. Much of what can be said under these headings has already been mentioned above. A few more comments may be useful.

Information about the signal may be available in either of two forms: it may be given as an elementary random process in time, defined by the usual hierarchy of multidimensional distribution functions,[14] or it may be a known function of time, containing one or more random parameters with specified distributions. In the latter case, the random parameters may be time-independent, or, more generally, they may be themselves random processes (e.g., a noise-modulated sine wave). Clearly there is, as in detection, a wide variety of possible situations. They may be conveniently classified as follows:

† For a fuller discussion of coherent and incoherent reception, see Sec. 19.4-3.

(1) The nature of the estimate. A *point estimate*† is a decision that the signal or one or more of its parameters have a definite value. An *interval estimate*† is a decision that such a value lies within a certain interval with a given probability. Among point estimates, it is useful to make a further distinction between *one-dimensional* and *multidimensional estimates.* An illustration of the former is the estimate of an amplitude scale factor constant throughout the interval, while an estimate of the signal itself throughout the observation period is an example of the latter.

(2) The nature of the data processing. When the value of a time-varying quantity $V(t)$ at a particular instant is being estimated, the relationship between the time t_λ for which the estimate is valid and the times at which data are collected becomes important (cf. Fig. 18.3). If t_λ coincides with one of the sampling instants, the estimation process is termed *simple estimation*, or *simple extraction*. If, on the other hand, t_λ does not coincide with

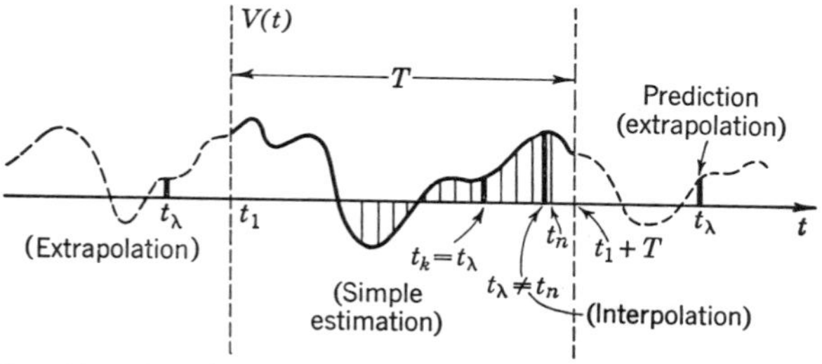

FIG. 18.3. Simple estimation, interpolation (smoothing), and extrapolation (prediction).

any sampling instant, the process is called *interpolation*, or *smoothing*, when t_λ lies within the observation interval $(t_1, t_1 + T)$ and *extrapolation*, or *prediction*, when t_λ lies outside $(t_1, t_1 + T)$. Systems of these types may estimate the value of the signal itself or alternatively that of a time-varying signal parameter or some functional of the signal, such as its derivative or integral.

Frequently, a requirement of linearity may be imposed on the optimum system (which is otherwise almost always nonlinear), so that its operations may be performed by a linear network, or sometimes certain specific classes of nonlinearity may be allowed. An important question, then, is the extent to which performance is degraded by such constraints.

(3) The signal and noise statistics. As in detection, the finite sample upon which the estimate is based may be discrete or continuous, correlated or uncorrelated, and the random processes stationary or nonstationary, ergodic or nonergodic. In a similar way, we may speak of *coherent* and *incoherent extraction* according to whether the signal's epoch ϵ is known exactly or is a random variable. The signals themselves may be *structurally determinate*, i.e., the functions S have definite analytic forms; or they may be

† See Cramér, *op. cit.*, for a further discussion of conventional applications.

structurally indeterminate, when the S are described only in terms of a probability distribution. A sinusoid is a simple example of the former, while a random function is typical of the latter. The case where the signal is known completely does not arise in extraction.

18.2-3 Other Reception Problems. Reception itself may require a combination of detection and extraction operations. Extraction presupposes the presence of a signal at the input, and sometimes this cannot be assumed. We may then perform detection and extraction simultaneously and judge the acceptability of the estimate according to the outcome of the detection process. The procedure is schematically illustrated in Fig. 18.4.

In our reception problems here, the system designer has no control over the input signal: $\boldsymbol{T}_T^{(N)}$ is specified a priori. The present definition of the problem states that each possible signal is prescribed, together with its probability of occurrence, and the designer cannot change these data. However, a different strategic situation confronts the designer of a system for

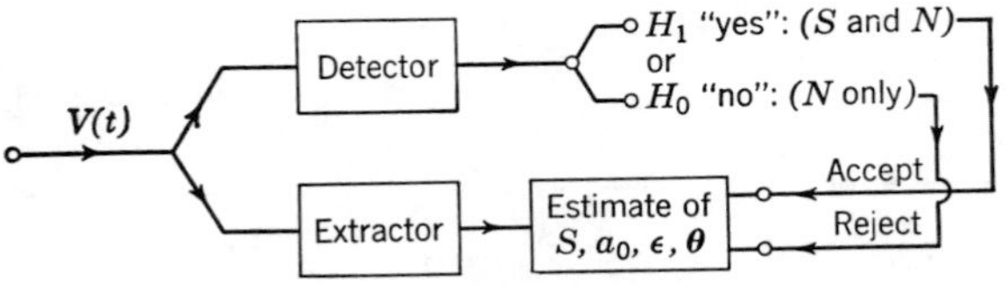

FIG. 18.4. Reception involving signal detection and extraction.

transmitting messages from point to point through a noisy channel, since he is then permitted to control the way in which he matches the signal to the channel. The encoding process ($\boldsymbol{T}_T^{(N)}$) is accordingly concerned with finding what class of signal is most effective against channel noise ($\boldsymbol{T}_M^{(N)}$) and how best to represent messages by such signals (cf. Secs. 6.1, 6.5-5). It is not directly a reception problem, except in the more general situation mentioned earlier (cf. also Sec. 23.2), where simultaneous adjustment of the transmission and reception operations $\boldsymbol{T}_T^{(N)}$, $\boldsymbol{T}_R^{(N)}$ is allowed. Decoding, of course, is a special form of reception in which the nature of the signals and their distributions are intimately related to the noise characteristics. Moreover, for the finite samples and finite delays available in practice, this is always a nontrivial problem, since it is impossible in physical cases† to receive messages from a noisy channel without error.

18.3 *The Reception Situation in General Terms*

Let us now consider the main elements of a general reception problem. We have pointed out earlier that the reception problem can be formulated

† Strictly speaking, there is always some noise, although in certain limiting situations this may be a very small effect and hence to an excellent approximation ignorable vis-à-vis the signal.

as a decision problem and that consequently certain information must be available concerning the statistics of signal and noise. We have also indicated that some assumptions are necessary concerning the nature of the data and of the sampling interval. Finally, we must prescribe a criterion of excellence by which to select an optimum system and must specify the set of alternatives among the decisions to be made.

In our present formulation, we shall make certain assumptions concerning these elements. For definiteness, these assumptions will not be the most general, but they will be sufficiently unrestrictive to exhibit the generality of the approach. Later, in Sec. 18.3-3, we shall discuss the reasoning by which some of these restrictions are removed.

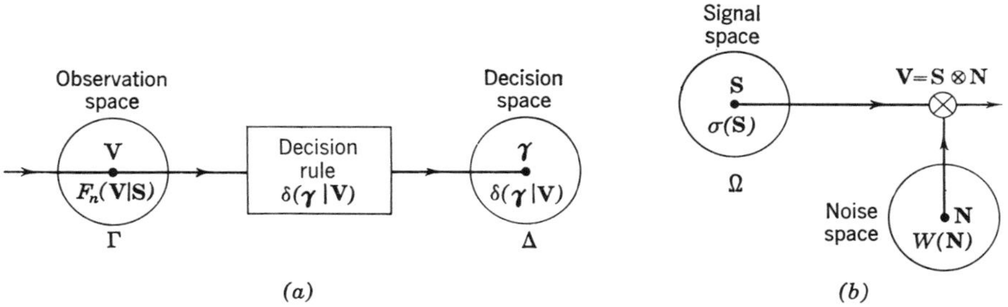

FIG. 18.5. The reception situation. (a) Observation and decision space; (b) signal space and noise space.

18.3-1 Assumptions. Concerning the statistics of signal and noise, we shall assume that both are known a priori. We shall further assume that the data are sampled discretely and that the sampling interval is of a fixed and finite length T during which n data points can be acquired (cf. Fig. 18.3). We make no special assumption concerning the criterion of excellence, but we do assume, for the sake of simplicity, that the decision to be made by the system is to select among a finite number m of alternatives. Figure 18.5*a*, *b* illustrates the problem. A decision $\boldsymbol{\gamma}$ is to be made about a signal $\mathbf{S}$, based on data $\mathbf{V}$, in accordance with a *decision rule* $\delta(\boldsymbol{\gamma}|\mathbf{V})$, as shown at Fig. 18.5*a*. Here, $\boldsymbol{\gamma} = (\gamma_1, \ldots, \gamma_m)$, $\mathbf{S} = (S_1, \ldots, S_n)$, $\mathbf{V} = (V_1, \ldots, V_n)$ are vectors, and the subscripts on the components of $\mathbf{S}$ and $\mathbf{V}$ are ordered in time so that $S_k = S(t_k)$, $V_k = V(t_k)$, etc., with $0 \le t_1 \le t_2 \le \cdots \le t_k \le \cdots \le t_n \le T$. Thus, the components of $\mathbf{V}$ form the a posteriori data of the sample upon which some decision γ_j is to be made.

Each of the quantities $\mathbf{S}$, $\mathbf{N}$, $\mathbf{V}$, $\boldsymbol{\gamma}$ can be represented by a point in an abstract space of the appropriate dimensionality, and the occurrence of particular values is governed in each instance by an appropriate probability-density function. These are multidimensional density functions, which are to be considered discrete or continuous depending on the discrete or continuous nature of the spaces and of corresponding dimensionality.

Here, we introduce $\sigma(\mathbf{S})$, $W(\mathbf{N})$, and $F_n(\mathbf{V}|\mathbf{S})$, respectively, as the probability-density functions for signal, for noise, and for the data **V** when **S** is given. As mentioned earlier, the possible signals **S** may be represented as points in a space Ω over which the a priori distribution $\sigma(\mathbf{S})$ is defined. Information about the signal and its distribution may be available in either of two forms: it may be given directly as an elementary random process, i.e., the distribution $\sigma(\mathbf{S})$ is immediately available, as a datum of the problem (stationary and gaussian, etc.), or, as is more common, the signal **S** is a known function of one or more random parameters $\boldsymbol{\theta} = (\theta_1, \ldots, \theta_j)$, and it is the distribution† $\sigma(\boldsymbol{\theta})$ of these parameters that is given rather than $\sigma(\mathbf{S})$ itself. In fact, the reception problem may require decisions about the parameters instead of the signals.

We can also raise the question of what to do when $\sigma(\mathbf{S})$ is not known beforehand (or perhaps only partially known), contrary to our assumption here. Such situations are in fact encountered in practice, where it is considered risky or otherwise unreasonable to assume that complete knowledge of $\sigma(\mathbf{S})$ is available. This question is a difficult one, and a considerable portion of decision theory is devoted to providing a reasonable answer to it. It is taken up later (cf. Sec. 19.2-3), and it is shown there that, even in this case, the above formulation of the reception problem can be retained in its essentials.

We can further compound the complexity of the reception problem and inquire into what to do when not only $\sigma(\mathbf{S})$, but also $W(\mathbf{N})$, the distribution density of the noise, is partially or completely unknown a priori. In such a case, specification of **S** is not enough to determine $F_n(\mathbf{V}|\mathbf{S})$. In statistical terminology, $F_n(\mathbf{V}|\mathbf{S})$ is then said to belong to a *nonparametric* family. Error probabilities, associated with possible incorrect decisions, cannot then be computed directly and the system can no longer be evaluated in such terms, so that the question of optimization is reopened. Nonparametric inference from the general point of view of decision theory has been discussed by several investigators[15–17] but will not be considered further here.

The discussion of possible generalizations, which has so far dealt with the decision space and with the statistics of signal and noise, can also be extended to one other topic, i.e., the method of data acquisition. In the line of reasoning which led to the above formulation of the reception problem, we assumed for convenience that the data were sampled discretely and that the sampling interval was fixed and finite. Actually, neither of these assumptions is strictly necessary. The sampling process can be continuous, and we shall in fact discuss cases of this type in Chaps. 19 to 23.

We observe also that the length of the sampling interval need not be kept fixed. In fact, the idea of a variable sampling interval leads to the notion of the *sequential* decision. A reception system which is based on sequential principles proceeds in steps, deciding, after each sample of data has been

† See footnote § on p. 790.

recorded, whether or not to come to a conclusion or whether to extend the sampling interval and to take another reading. The class of sequential reception systems is broad and contains the nonsequential type discussed so far as a subclass. A reference to sequential detectors is given in Sec. 20.4-8.

18.3-2 The Decision Rule. We begin by observing that the decision rule is represented as a probability. This may seem somewhat surprising. A reception system operating according to such a decision rule would not function like a conventional receiver, which generates a certain and definite output γ from a given set of inputs $\mathbf{V}$. Rather, it would contain a battery of chance mechanisms of a well-specified character. A given set of inputs would actuate the corresponding mechanism, which in turn would generate one of the m possible outputs γ with a certain probability, each mechanism in general with a different probability.

Arrangements such as this will probably appear quite artificial but they are necessary concepts, at least in principle, for it can be shown that devices with chance mechanisms as their outputs can be superior in performance, under certain circumstances, to the conventional ones.

Accordingly, $\delta(\gamma|\mathbf{V})$ is the conditional probability of deciding γ when $\mathbf{V}$ is given. More specifically, since the space Δ is here assumed to contain a finite number of decisions γ, the decision rule $\delta(\gamma|\mathbf{V})$ assigns a probability between (or equal to) 0 and 1 to each decision γ_j ($j = 1, \ldots, m$), the distribution depending on $\mathbf{V}$. In most cases of practical interest, δ is either 0 or 1 for each $\mathbf{V}$ and γ in this case and is called a *nonrandomized decision rule.* The opposite case, a *randomized decision rule,* is not excluded from this general formulation, although, as we shall see in subsequent applications, the decision rules reduce to the nonrandom case for all the systems treated here.

We note now that the *key feature of the decision situation is that* $\delta(\gamma|\mathbf{V})$ *is a rule for making the decision* γ *from a posteriori data* $\mathbf{V}$ *alone*, i.e., without knowledge of, or dependence upon, the particular $\mathbf{S}$ that results in the data $\mathbf{V}$. The a priori knowledge of the signal class and signal distribution, of course, is built into the optimum-decision rule, but the probability of deciding γ, given $\mathbf{V}$, is independent of the particular $\mathbf{S}$; that is, γ is algebraically independent of $\mathbf{S}$, although statistically dependent upon it. This may be expressed as

$$\delta(\gamma|\mathbf{V}) = \delta(\gamma|\mathbf{V},\mathbf{S}) \tag{18.1}$$

which states that the probability (density) of deciding γ, given $\mathbf{V}$, is the same as the probability of γ, given both $\mathbf{V}$ and $\mathbf{S}$. Thus, *the decision rule* $\delta(\gamma|\mathbf{V})$ *is the mathematical embodiment of the physical system used to process the data and yield decisions.*

Both fixed and sequential procedures are included in this formulation, and in both cases we deal with terminal decisions. We remark also that the Wald theory of sequential tests[18] introduces a further degree of freedom over the fixed-sample cases through the adoption of a second cost function, the "cost of experimentation"[18,19] (cf. references in Sec. 20.4-8). In the general theory, we are free to limit the class of decision rules, in advance, to either of the above types without compromising the completeness of the theory of either type.

18.3-3 The Decision Problem. In order to give definite structure to the decision problem, we must prescribe a criterion of excellence, in addition to the a priori probabilities $\sigma(\mathbf{S})$ and $W(\mathbf{N})$. By this we mean the following: The decisions which are to be made by the reception system must be based on the given data **V**, which, because of their contamination with noise, constitute only incomplete clues to the underlying signal **S**. Therefore, whatever the decision rule $\delta(\boldsymbol{\gamma}|\mathbf{V})$ which is finally adopted, the decisions to which it leads cannot always be correct (except possibly in the unrealizable limit $T \to \infty$). Thus, it is clear that whenever there is a nonzero probability of error some sort of value judgment is implied; in fact, the former always implies (1) a decision process and (2) a numerical cost assignment of some kind to the possible decisions. The units in which such a cost, or value, is measured are essentially irrelevant, but the relative amounts associated with the possible decisions are not.

In order to determine the decision problem, a *loss* $\mathfrak{F}(\mathbf{S},\boldsymbol{\gamma})$ is assigned to each combination of decision $\boldsymbol{\gamma}$ and signal **S** (the latter selecting a particular distribution function of **V**), in accordance with some prior judgment of the relative importance of the various correct and incorrect decisions. Each decision rule may then be rated by adopting an *evaluation function* (for example, the mathematical expectation of loss), which takes into consideration both the probabilities of correct and incorrect decisions and the losses associated with them. There are, of course, many ways of assigning loss, and hence many different risk functions. One example, which has been very common in statistics and in communication theory, is the squared-error loss. This type of loss is used in extraction problems in which the decision to be rendered is an estimation of a signal after it has been contaminated with noise. In this case, the loss is taken to be proportional to the square of the error in this estimation. Other examples are discussed in Chaps. 20 to 22.

We may now state the reception problem in the following general terms:

Given the family of distribution functions $F_n(\mathbf{V}|\mathbf{S})$, the a priori signal probability distribution $\sigma(\mathbf{S})$, the class of possible decisions, and the loss (and evaluation) function(s) $\mathfrak{F}$ [and $\mathcal{E}(\mathfrak{F})$], the problem is to determine the best rule $\delta(\boldsymbol{\gamma}|\mathbf{V})$ for using the data to make decisions.

In arriving at this statement, we have introduced a number of somewhat restrictive assumptions. We now give a brief heuristic discussion of what can be done to remove them. To begin with, the statement of the reception problem in these terms is actually more general than the argument which led up to it, a fact which requires some comment. A quick review of that argument shows, on the one hand, that the restriction of the decision $\boldsymbol{\gamma}$ to a finite number m of alternatives $\boldsymbol{\gamma} = (\gamma_1, \gamma_2, \ldots, \gamma_m)$ is irrelevant and that a denumerably infinite number may equally well be used. In fact, the extension to a continuum of possible alternatives is a mere matter of reinterpretation. The decision rule $\delta(\boldsymbol{\gamma}|\mathbf{V})$ which was introduced above as a discrete probability distribution must in this case be interpreted as a probability-density function; that is, $\delta(\boldsymbol{\gamma}|\mathbf{V})\, \mathbf{d}\boldsymbol{\gamma}$ is the probability that $\boldsymbol{\gamma}$ lies between $\boldsymbol{\gamma}$ and $\boldsymbol{\gamma} + \mathbf{d}\boldsymbol{\gamma}$, given $\mathbf{V}$. To represent a nonrandomized decision rule in this case, we interpret $\delta(\boldsymbol{\gamma}|\mathbf{V})$ as a Dirac δ function [see Eq. (18.15), for example]. Usually, the family of distribution functions is not given directly and must be found from a given noise distribution $W(\mathbf{N})$ and the mode of combining signal and noise.

18.3-4 The Generic Similarity of Detection and Extraction. Figure 18.5 emphasizes that decision rules are essentially transformations that map observation space into decision space. In detection, each point of observation space Γ (or $\mathbf{V}$) is mapped into the various points constituting the space Δ of terminal decisions. For example, in binary detection that is the same as dividing Γ into two regions, one corresponding to "no signal" and the other to "signal and noise," and carrying out the operation of decision in one step, since only a single alternative in involved. The binary detection problem is then the problem of how best to make this division. Similarly, in extraction each point of Γ is mapped into a point of the space Δ of terminal decisions, which in this instance has the same structure as the signal space Ω. If the dimensionality of Δ is smaller than that of Γ (as is usually the case in estimating a signal parameter), the transformation is "irreversible"; i.e., many points of Γ go into a single point of Δ. In this way, extraction may also be thought of as a division of Γ into regions, so that, basically, detection and extraction have this common and generic feature and are thus not ultimately different operations. It is merely necessary to group the points of Δ corresponding to $\mathbf{S} \neq 0$ into a single class labeled "signal and noise" to transform an extractor into a detector. Conversely, detection systems may be regarded as extractors followed by a threshold device which separates, say, $\mathbf{S} = 0$ from $\mathbf{S} \neq 0$. However, *a system optimized for the one function may not necessarily be optimized for the other*, and it is in this sense, therefore, that we consider detection and extraction as separate problems for analysis.

18.4 *System Evaluation*

In this section, we shall apply the concepts discussed above to a description of the problem of evaluating system performance, including that of both optimum and suboptimum types. It is necessary first to establish some reasonable method of evaluation, after which a number of criteria of excellence may be postulated, with respect to which optimization may then be specifically defined.

18.4-1 Evaluation Functions. As mentioned in Sec. 18.3-3, $\mathfrak{F}(\mathbf{S},\gamma)$ is a *generalized loss function*, adopted in advance of any optimization procedure, which assigns a *loss*, or *cost*, to every combination of system input and decision (system output) in a way that may or may not depend on the system's operation. Actual evaluation of system performance is now made as mentioned earlier, provided that we adopt an evaluation function $\mathcal{E}(\mathfrak{F})$ which takes into account all possible modes of system behavior and their relative frequencies of occurrence and assigns an over-all loss rating to each system or decision rule. One obvious choice of $\mathcal{E}$ is the *mathematical expectation*, or *average value*, of $\mathfrak{F}$, and it is on this reasonable but arbitrary choice that the present theory is based for the most part.†

At this point, it is convenient to define two different loss ratings for a system, one of which is used to rate performance when the signal input is fixed and the other to take account of a priori signal probabilities. For a given $\mathbf{S}$, we have first:

The conditional loss rating‡ $\mathcal{L}(\mathbf{S},\delta)$ *of* δ *is defined as the conditional expectation of loss:*

$$\mathcal{L}(\mathbf{S},\delta) = E_{\mathbf{V}|\mathbf{S}}\{\mathfrak{F}[\mathbf{S},\gamma(\mathbf{V})]\} = \int_\Gamma d\mathbf{V} \int_\Delta d\gamma\, \mathfrak{F}(\mathbf{S},\gamma) F_n(\mathbf{V}|\mathbf{S})\delta(\gamma|\mathbf{V}) \quad (18.2)$$

By this notation, we include discrete as well as continuous spaces Δ; for the former, the integral over Δ is to be interpreted as a sum and $\delta(\gamma|\mathbf{V})$ as a probability, rather than as a probability density. (See the remarks at the end of Sec. 18.3-3.)

Actually, as will be seen in Sec. 18.4-4, the conditional loss rating is most significant when the a priori probability $\sigma(\mathbf{S})$ is unknown. However, when $\sigma(\mathbf{S})$ is known, we use this information to rate the system by averaging the loss over both the sample and the signal distributions:

The average loss rating $\mathcal{L}(\sigma,\delta)$ *of* δ *is defined as the* (*unconditional*) *expectation of loss when the signal distribution is* $\sigma(\mathbf{S})$:

$$\mathcal{L}(\sigma,\delta) = E_{\mathbf{V},\mathbf{S}}\{\mathfrak{F}(\mathbf{S},\gamma)\} = \int_\Omega d\mathbf{S} \int_\Gamma d\mathbf{V} \int_\Delta d\gamma\, \mathfrak{F}(\mathbf{S},\gamma)\sigma(\mathbf{S})F_n(\mathbf{V}|\mathbf{S})\delta(\gamma|\mathbf{V}) \quad (18.3)$$

Some remarks are appropriate concerning the loss function $\mathfrak{F}$. In statistical literature, $\mathfrak{F}$ is usually a function which assigns to each combination of signal and decision a certain loss, or *cost*, which is independent of δ:

$$\mathfrak{F}_1 = C(\mathbf{S},\gamma) \quad (18.4)$$

† Other linear or nonlinear operations for $\mathcal{E}$ are possible and should not be overlooked in subsequent generalizations (see the comments in Sec. 18.5-4).

‡ This quantity is called the *a priori risk* in Wald's terminology.[1]

In the present analysis, we restrict our discussion chiefly to systems whose performance is rated according to simple loss functions† of this nature. There exists a substantial body of theory for this case, and certain very general statements can be made about optimum systems derived under this restriction (cf. Wald's complete class theorem, admissibility,[1] etc.; see Sec. 18.5).

We point out, however, that a more general type of loss function can be constructed. In fact, one such function is suggested by information theory. For, if we let

$$\mathfrak{F}_2 = -\log p(\mathbf{S}|\boldsymbol{\gamma}) \tag{18.5}$$

where $p(\mathbf{S}|\boldsymbol{\gamma})$ is the a posteriori probability of $\mathbf{S}$ given $\boldsymbol{\gamma}$, the average loss rating [Eq. (18.3)] becomes the well-known equivocation of information theory[6,20] (cf. Sec. 6.5-2). This loss function can be interpreted as a measure of the "uncertainty" (or "surprisal") about $\mathbf{S}$ when $\boldsymbol{\gamma}$ is known[6,21] (cf. Sec. 6.2-1). It is an example of a more general type than the simple cost function [Eq. (18.4)]. For, unlike $C(\mathbf{S},\boldsymbol{\gamma})$, which depends on $\mathbf{S}$ and $\boldsymbol{\gamma}$ alone, Eq. (18.5) depends also on the decision rule in use and cannot be preassigned independently of δ. Loss functions like Eq. (18.5) are more difficult to deal with, and some of the general statements (cf. Sec. 18.5) which can be derived for Eq. (18.4) clearly do not hold true for Eq. (18.5). In Chap. 22, it is shown that close connections may exist between results based on the two types of loss function.

The conditional and average loss ratings of δ may now be written, from Eqs. (18.2) to (18.5), as

Conditional risk

$$r(\mathbf{S},\delta) = \int_\Gamma d\mathbf{V}\, F_n(\mathbf{V}|\mathbf{S}) \int_\Delta d\boldsymbol{\gamma}\, C(\mathbf{S},\boldsymbol{\gamma})\delta(\boldsymbol{\gamma}|\mathbf{V}) \tag{18.6}$$

Average risk

$$R(\sigma,\delta) = E\{r(\mathbf{S},\delta)\} = \int_\Omega r(\mathbf{S},\delta)\sigma(\mathbf{S})\, d\mathbf{S} \tag{18.7a}$$

or

$$R(\sigma,\delta) = \int_\Omega \sigma(\mathbf{S})\, d\mathbf{S} \int_\Gamma d\mathbf{V}\, F_n(\mathbf{V}|\mathbf{S}) \int_\Delta d\boldsymbol{\gamma}\, C(\mathbf{S},\boldsymbol{\gamma})\delta(\boldsymbol{\gamma}|\mathbf{V}) \tag{18.7b}$$

and

Conditional information loss

$$h(\mathbf{S},\delta) = -\int_\Gamma d\mathbf{V}\, F_n(\mathbf{V}|\mathbf{S}) \int_\Delta d\boldsymbol{\gamma}\, [\log p(\mathbf{S}|\boldsymbol{\gamma})]\delta(\boldsymbol{\gamma}|\mathbf{V}) \tag{18.8}$$

Average information loss

$$H(\sigma,\delta) = E\{h(\mathbf{S},\delta)\} = \int_\Omega h(\mathbf{S},\delta)\sigma(\mathbf{S})\, d\mathbf{S} \tag{18.9a}$$

or

$$H(\sigma,\delta) = -\int_\Omega \sigma(\mathbf{S})\, d\mathbf{S} \int_\Gamma d\mathbf{V}\, F_n(\mathbf{V}|\mathbf{S}) \int_\Delta d\boldsymbol{\gamma}\, [\log p(\mathbf{S}|\boldsymbol{\gamma})]\delta(\boldsymbol{\gamma}|\mathbf{V}) \tag{18.9b}$$

† We shall use the term *risk*, henceforth, as synonymous with this simple cost, or loss.

The last of these is the well-known "equivocation" of information theory† (cf. Secs. 6.5-2 and 6.5-3).

As we have already mentioned in Sec. 18.1, **S**, when deterministic, is a function of a set of random parameters‡ $\boldsymbol{\theta}$, and frequently it is the parameters $\boldsymbol{\theta}$ about which decisions are to be made, rather than about **S** itself (see, for example, Sec. 21.2). Similar to Eqs. (18.6) and (18.7), the conditional and average risks for this situation may be expressed as§

$$r(\boldsymbol{\theta},\delta) = \int_\Gamma \mathbf{dV}\, F_n[\mathbf{V}|\mathbf{S}(\boldsymbol{\theta})] \int_\Delta \mathbf{d}\boldsymbol{\gamma}\, C(\boldsymbol{\theta},\boldsymbol{\gamma})\delta(\boldsymbol{\gamma}|\mathbf{V}) \tag{18.10}$$

and

$$R(\sigma,\delta)_\theta = \int_{\Omega\theta} r(\boldsymbol{\theta},\delta)\sigma(\boldsymbol{\theta})\, \mathbf{d}\boldsymbol{\theta} \tag{18.11}$$

Here, of course, $r(\boldsymbol{\theta},\delta)$ and $R(\sigma,\delta)_\theta$ are not necessarily the same as $r(\mathbf{S},\delta)$, $R(\sigma,\delta)$ above, nor is the form of $\boldsymbol{\gamma}$ either. Notice that the cost function $C(\boldsymbol{\theta},\boldsymbol{\gamma})$ is usually a different function of $\boldsymbol{\theta}$ from $C[\mathbf{S}(\boldsymbol{\theta}),\boldsymbol{\gamma}]$, also. Considerable freedom of choice as to the particular conditional and average risks is thus frequently available to the system analyst, although the appropriate choice is often dictated by the problem in question. Finally, observe that $r(\mathbf{S},\delta)$ and $R(\sigma,\delta)$ for decisions about $\mathbf{S} = \mathbf{S}(\boldsymbol{\theta})$ are still given by Eqs. (18.6), (18.7), where $\sigma(\mathbf{S})\,\mathbf{dS}$ is replaced by its equivalent $\sigma(\boldsymbol{\theta})\,\mathbf{d}\boldsymbol{\theta}$ in Eqs. (18.7*a*), (18.7*b*), with a corresponding change from Ω-space (for **S**) to Ω_θ-space (for $\boldsymbol{\theta}$) according to the transformations implied by $\mathbf{S} = \mathbf{S}(\boldsymbol{\theta})$. Similar remarks apply for the conditional and average information losses, Eqs. (18.8), (18.9), as well.

18.4-2 System Comparisons and Error Probabilities. The expressions (18.2) and (18.3) for the loss ratings can be put into another and often more revealing form, which exhibits directly the role of the error probabilities associated with the various possible decisions. Let $p(\boldsymbol{\gamma}|\mathbf{S})$ be the conditional probability¶ that the system in question makes a decision $\boldsymbol{\gamma}$ when the signal is **S** and a decision rule $\delta(\boldsymbol{\gamma}|\mathbf{V})$ is adopted, so that

$$p(\boldsymbol{\gamma}|\mathbf{S}) = \int_\Gamma F_n(\mathbf{V}|\mathbf{S})\delta(\boldsymbol{\gamma}|\mathbf{V})\, \mathbf{dV} \tag{18.12}$$

Comparison with the conditional risk (18.6) shows that the latter may be written

$$r(\mathbf{S},\delta) = \int_\Delta \mathbf{d}\boldsymbol{\gamma}\, p(\boldsymbol{\gamma}|\mathbf{S})C(\mathbf{S},\boldsymbol{\gamma}) \tag{18.13}$$

† Note that, when **S** can assume a continuum of values (as is usually the case), we must replace the probability $p(\mathbf{S}|\boldsymbol{\gamma})$ in Eq. (18.5) by the corresponding probability *density* $w(\mathbf{S}|\boldsymbol{\gamma})$ and include in Eqs. (18.8) and (18.9) the absolute entropy H_0 (cf. Sec. 6.4-1). See, for example, Eqs. (21.87) et seq.

‡ For simplicity, these are assumed to be time-invariant here; the generalization to include time variations $\boldsymbol{\theta} = \boldsymbol{\theta}(t)$ is straightforward.

§ Here and henceforth, unless otherwise indicated, we adopt the notational convention that the principal argument of a function distinguishes that function from other functions: thus, $\sigma(\mathbf{S}) \neq \sigma(\boldsymbol{\theta})$, $p(\mathbf{V}) \neq p(\boldsymbol{\theta})$, etc.; however, $\sigma[\mathbf{S}(\boldsymbol{\theta})] = \sigma(\mathbf{S})$, and so on.

¶ Or probability density, when $\boldsymbol{\gamma}$ represents a continuum of decisions (as in extraction; cf. Chap. 21).

which is simply the sum of the costs associated with all possible decisions for the given **S**, weighted according to their probability of occurrence. In a similar way, we can obtain the probability (density†) of the decisions $\boldsymbol{\gamma}$ by averaging Eq. (18.12) with respect to **S**, for example,

$$p(\boldsymbol{\gamma}) = \langle p(\boldsymbol{\gamma}|\mathbf{S})\rangle_S = \int_\Omega \sigma(\mathbf{S})\, d\mathbf{S} \int_\Gamma d\mathbf{V}\, F_n(\mathbf{V}|\mathbf{S})\delta(\boldsymbol{\gamma}|\mathbf{V}) \qquad (18.14)$$

Since we shall be concerned in what follows almost exclusively with nonrandomized decision rules, particularly in applications, we see that $\delta(\boldsymbol{\gamma}|\mathbf{V})$ may be expressed as

$$\delta(\boldsymbol{\gamma}|\mathbf{V}) = \delta[\boldsymbol{\gamma} - \boldsymbol{\gamma}_\sigma(\mathbf{V})] \qquad (18.15)$$

where the δ of the right-hand member is now the Dirac δ function. Here it is essential to distinguish between the decision $\boldsymbol{\gamma}$ and the functional operation $\boldsymbol{\gamma}_\sigma(\mathbf{V})$ performed on the data by the system. The subscript σ reminds us that this operation depends in general on signal statistics. With Eq. (18.15), the probability (density) of decision $\boldsymbol{\gamma}$ on condition **S**, which may represent correct or incorrect decisions, can be written

$$p(\boldsymbol{\gamma}|\mathbf{S}) = \int_\Gamma F_n(\mathbf{V}|\mathbf{S})\delta[\boldsymbol{\gamma} - \boldsymbol{\gamma}_\delta(\mathbf{V})]\, d\mathbf{V} \qquad (18.16a)$$

$$= \int_{-\infty}^{\infty} \cdots \int e^{i\boldsymbol{\xi}\boldsymbol{\gamma}} \frac{d\boldsymbol{\xi}}{(2\pi)^m} \int_\Gamma F_n(\mathbf{V}|\mathbf{S}) e^{-i\boldsymbol{\xi}\boldsymbol{\gamma}_\delta(\mathbf{V})}\, d\mathbf{V} \qquad (18.16b)$$

(cf. Sec. 17.2-1). This reveals the explicit system operation, and Eq. (18.16*b*) in particular provides a direct way of calculating $p(\boldsymbol{\gamma}|\mathbf{S})$ for any system once its system structure $\boldsymbol{\gamma}_\delta(\mathbf{V})$ is known. As for Eq. (18.14), we can also obtain the probability density of $\boldsymbol{\gamma}$ itself by averaging $p(\boldsymbol{\gamma}|\mathbf{S})$ [Eqs. (18.16)] over **S**.

Comparison of explicit decision systems now follows directly. For example, this may be done by determining which has the smallest average loss rating $\mathfrak{L}(\sigma,\delta)$, which, in terms of *average* risk [Eqs. (18.7)], involves the comparison of $R(\sigma,\delta_1)$ and $R(\sigma,\delta_2)$ for two systems with system functions $\boldsymbol{\gamma}_\delta(\mathbf{V})_1$ and $\boldsymbol{\gamma}_\delta(\mathbf{V})_2$ [cf. Eq. (18.15)]. In a similar fashion, one can compare also $H(\sigma,\delta_1)$ and $H(\sigma,\delta_2)$ [cf. Eqs. (18.9)]. Note that not only optimum but suboptimum systems may be so handled, once $\boldsymbol{\gamma}_\delta(\mathbf{V})$ is specified, so that now one has a possible quantitative method of deciding in practical situations between "good," "bad," "fair," "best," etc., where the comparisons are consistently made within a common criterion and where the available information can be incorporated in ways appropriate to each system under study. Examples of both optimum and nonoptimum systems are provided in Chaps. 20 and 21. We emphasize that this consistent framework for system comparison is one of the most important practical features of the theory, along with its ability to indicate the explicit structure of optimum and suboptimum systems, embodied in the decision rule $\delta(\boldsymbol{\gamma}|\mathbf{V})$.

† Here and subsequently we use the pointed brackets (⟨ ⟩) to indicate a statistical average rather than a time average (cf. Sec. 1.5).

18.4-3 Optimization: Bayes Systems. In Sec. 18.4-1, we have seen how average and conditional loss ratings may be assigned to any system, once the evaluation and cost functions have been selected. We now define what we mean by an optimum decision system. We state a definition first for the case where complete knowledge of the a priori signal probabilities $\sigma(\mathbf{S})$ is assumed and in which, from what has been said above, evaluation from the point of view of the *average* loss $R(\sigma,\delta)$ is most appropriate. Consider, then, that *one system is "better" than another if its average loss rating is smaller for the same application (and criterion) and that the "best," or optimum, system is the one with the smallest average loss rating.* (The preassigned costs, of course, are the same.) We call this optimum system a *Bayes system:*

A Bayes system obeys a Bayes decision rule δ^*, where δ^* is a decision rule whose average loss rating $\mathfrak{L}$ is smallest for a given a priori distribution σ.

For the risk and information criteria of Eqs. (18.7) and (18.9), this becomes

$$R^* = \min_{\delta} R(\sigma,\delta) = R(\sigma,\delta^*) \qquad \text{Bayes risk} \tag{18.17a}$$

and

$$H^* = \min_{\delta} H(\sigma,\delta) = H(\sigma,\delta^*) \qquad \text{Bayes equivocation} \tag{18.17b}$$

The former minimizes the average risk (or cost), while the latter minimizes the equivocation. Bayes decision rules (for the given $\mathfrak{F}$) form a *Bayes class,* each member of which corresponds to a different a priori distribution† $\sigma(\mathbf{S})$.

18.4-4 Optimization: Minimax Systems. When the a priori signal probabilities are not known or are only incompletely given, definition of the optimum system is still open. A possible criterion for optimization in such cases is provided by the Minimax decision rule δ_M^*, or Bayes rule associated with conditional risk $r(\mathbf{S},\delta)$. As indicated by our notation, there is one conditional risk figure attached to each possible signal $\mathbf{S}$. In general, these risks will be different for different signals, and there will be a maximum among them, say $r(\mathbf{S},\delta)_{\max}$. The Minimax rule is, roughly speaking, the decision rule which reduces this maximum as far as possible. More precisely:

The Minimax decision rule δ_M^* is the rule for which the maximum conditional loss rating $\mathfrak{L}(\mathbf{S},\delta)_{\max}$, as the signal $\mathbf{S}$ ranges over all possible values, is not greater than the maximum conditional loss rating of any other decision rule δ.

Thus, in terms of conditional risk r, or conditional information loss h, we may write

$$\begin{aligned} \max_{\mathbf{S}} r(\mathbf{S},\delta_M^*) &= \max_{\mathbf{S}} \min_{\delta} r(\mathbf{S},\delta) \le \max_{\mathbf{S}} r(\mathbf{S},\delta) \\ \max_{\mathbf{S}} h(S,\delta_M^*) &= \max_{\mathbf{S}} \min_{\delta} h(\mathbf{S},\delta) \le \max_{\mathbf{S}} h(\mathbf{S},\delta) \end{aligned} \tag{18.18}$$

† Of course, it is possible that different $\sigma(\mathbf{S})$ may lead to identical decision rules, but aside from this possible ambiguity we observe that a Bayes criterion is entirely appropriate when $\sigma(\mathbf{S})$ is known, since it makes full use of all available information.

Wald[1] has shown† under certain rather broad conditions (see Secs. 18.5-2, 18.5-3) that $\max_{\mathbf{S}} \min_{\delta} r(\mathbf{S},\delta) = \min_{\delta} \max_{\mathbf{S}} r(\mathbf{S},\delta)$ for the risk (i.e., simple cost) formulation, from which the significance of the term "Minimax" becomes apparent. (Whether or not a corresponding result holds for the information-loss formulation remains to be established.)

We may also express the Minimax decision process in terms of the resulting average risk. From Sec. 18.5-3, theorems 1, 4, 5, 9, we have the equivalent Minimax formulation

$$R_M^*(\sigma_0,\delta_M^*) = \max_{\sigma} R^*(\sigma,\delta^*) = \max_{\sigma} \min_{\delta} R(\sigma,\delta) = \min_{\delta} \max_{\sigma} R(\sigma,\delta) \qquad \text{Minimax average risk‡} \tag{18.19}$$

this last from Eq. (18.17*a*) and Sec. 18.5-2, definition 7*a*. Thus, the Minimax average risk is the largest of all the Bayes risks, considered over the class

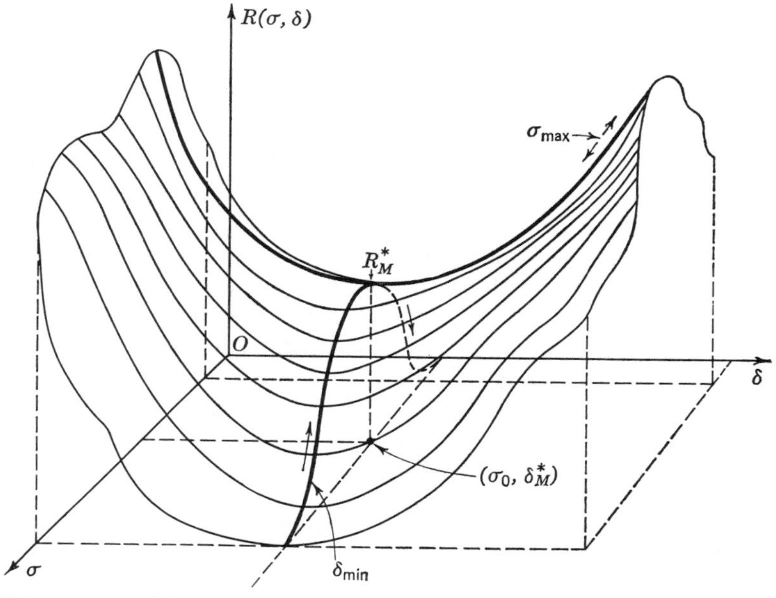

FIG. 18.6. Average risk as a function of decision rule and a priori signal distribution, showing a Minimax saddle point.

of a priori signal distribution $\{\sigma(\mathbf{S})\}$. The distribution σ_0 ($= \sigma_M^*$) for which this occurs is called the *least favorable distribution*, and accordingly the Minimax decision rule δ_M^* [obtained by adjusting the Bayes rule δ^* as σ is varied;

† The Minimax theorem was first introduced and proved by Von Neumann, in an early paper on the theory of games.[22] For a further account, see Von Neumann and Morgenstern,[23] sec. 17.6, p. 154.

‡ This is the average risk associated with the Minimax decision rule.

cf. Eq. (18.19)] is one which gives us the least favorable, or "worst," of all Bayes—i.e., "best"—systems. Geometrically, the Minimax situation of $\sigma \to \sigma_0$, $\delta \to \delta_M^*$, $R(\sigma,\delta) \to R_M^*(\sigma_0,\delta_M^*)$ is represented by a saddle point of the average-risk surface over the (σ,δ) plane, as Fig. 18.6 indicates. The existence of σ_0, δ_M^* and this saddle point follows from the appropriate theorems (cf. Sec. 18.5-3).

The Minimax decision rule has been the subject of much study and also of some adverse criticism. It has been argued that it is often too conservative to be very useful. However, it is also true that there are situations in which the Minimax rule is unquestionably an excellent choice. Figure 18.7*a*

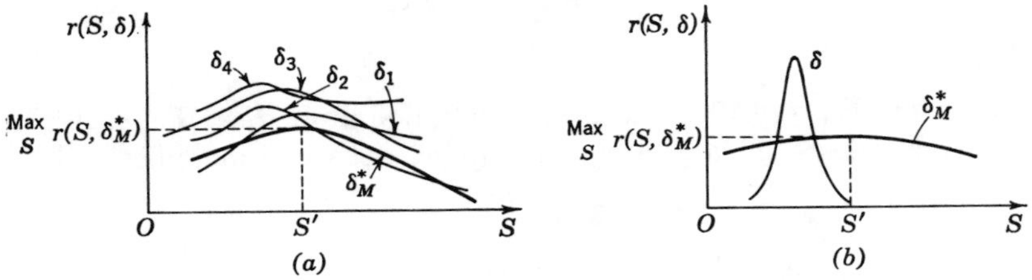

FIG. 18.7. (*a*) An acceptable Minimax situation. (*b*) A Minimax situation that is possibly too conservative.

illustrates these remarks. Here we have presented the case where the maximum conditional loss rating of all other decision rules $\delta_1, \delta_2, \ldots$ exceeds that for δ_M^* and where even most of the minimum loss ratings are also noticeably larger than the corresponding minimum for δ_M^*. Sometimes, however, we may have the situation shown in Fig. 18.7*b*, where δ_M^* leads to excessive loss ratings, except for a comparatively narrow range of values of **S**. In the latter case, δ_M^* is perhaps too conservative, and a more acceptable decision rule might be sought.† The Minimax procedure does, at any rate, have the advantage of guarding against the worst case, but also may be too cautious for the more probable states of the input to the system. When the costs are preassigned and immutable, the possible conservatism of Minimax

† These Minimax risk curves have a single distinct maximum. The least favorable a priori distribution σ_0 in this case consequently concentrates all of its probability mass at the signal value corresponding to the maximum conditional risk (a δ-function distribution for continuous signal space), since by definition σ_0 must maximize Bayes (average) risk. Existence of a least favorable distribution is here ensured by our assumptions *A* to *D*, which correspond to Wald's assumptions 3.1 to 3.7 (see Secs. 18.5-2, 18.5-3). Roughly speaking, the Minimax conditional risk must equal its maximum value for all signals to which the least favorable distribution assigns a nonzero a priori probability (see Wald's theorems 3.10, 3.11). Thus, a Minimax conditional risk curve with two distinct and equal maxima could have a corresponding σ_0 with probability concentrated at either of the two maxima or distributed between them, while if one maximum were larger than the other, the mass would have to be concentrated only at the larger, etc. Or again, if σ_0 were nonzero over a finite interval, the corresponding Minimax conditional risk would be constant over this range (but might take on other, smaller values outside).

cannot be avoided without choosing another criterion,† but in many cases where the actual values of the preassigned costs are left open to an a posteriori adjustment it may be possible by a more judicious cost assignment to modify δ_M^* more along the lines of Fig. 18.7*a*, where the "tails" of $\max_S r(\mathbf{S},\delta_M^*)$ are comparable to those of $r(\mathbf{S},\delta)$ (all $\mathbf{S}$), and thus eliminate, at least in part, the conservative nature of the decision process.‡

The Bayes decision rule makes the fullest use of a priori probabilities (when these are known) and in a sense assumes the most favorable system outcome. The Minimax decision rule, on the other hand, makes no use at all of these a priori probabilities (for the good reason that they are not available to the observer) and in the same sense assumes the worst case [cf. Eq. (18.19) and Fig. 18.6]. In practical cases, an important problem is to find δ_M^*. No general, simple procedure is available, although δ_M^* always exists in the risk formulation. From the definitions of δ^* and δ_M^*, however, it can be shown that a *Bayes decision rule whose conditional loss rating* $\mathfrak{L}$ *is the same for all signals is a Minimax rule* (see Sec. 18.5-3 and theorem 7). Thus, if we can find a δ^* for which this is the case, we have also determined a least favorable a priori distribution $\sigma_M^*(S)$ for this δ^* $(= \delta_M^*)$, which follows from the above, i.e.,

$$r(\mathbf{S},\delta^*)\ [\text{or } h(\mathbf{S},\delta^*)] = \text{constant} \qquad \text{all } \mathbf{S} \tag{18.20}$$

Note that a non-Bayes rule whose conditional loss rating is constant for all $\mathbf{S}$ is not necessarily Minimax. When Eq. (18.20) holds, it furnishes a useful method for finding δ_M^* and δ_M^*, as we shall see presently in Sec. 19.2-3.

18.5 *A Summary of Basic Definitions and Principal Theorems*

We conclude this chapter now with a short summary of some of the principal results, theorems, etc., that give decision theory its general scope and power in our present application to communication problems. For proofs and further discussion, the reader is referred to the appropriate sections of Wald[1] and other pertinent references.

18.5-1 Some General Properties of Optimum Decision Rules. The practical utility of an optimum procedure lies to a considerable extent in its uniqueness: there are not a number of different reception systems with the same optimal properties. For the unique optimum, the problem of choosing the simplest (or least expensive) from the point of view of design is automatically resolved. For this reason and for the central one of optimization itself, it is important to know the properties of optimum decision rules and when they may be expected to apply in physical systems. We state now two of the main results on which subsequent applications are based:

† For example, the "Minimax regret" criterion or the Hurwicz criterion, etc.[24]

‡ Hodges and Lehmann[25] have discussed some intermediate situations where $\sigma(\mathbf{S})$ is partially known on the basis of previous experience. Also, see Sec. 18.5.

(1) **Admissible decision rules.** We note, first, that the conditional loss rating of a decision rule depends, of course, on the particular signal present at the input. One decision rule may have a smaller rating than another for some signals and a larger one for others. If the conditional loss rating of δ_1 never exceeds that of δ_2 for *any* value of **S**, and is actually less than that of δ_2 for at least one **S**, then δ_1 is said to be *uniformly better* than δ_2. This leads, accordingly, to the notion of *admissibility:*

A decision rule is admissible if no uniformly better one exists.

Observe that with this definition an admissible rule is not necessarily uniformly better than any other; other rules can have smaller ratings at particular **S**'s (cf. Fig. 18.7*a*). However, they cannot be better for *all* **S**.

It follows, then, that, *if a Bayes or Minimax rule is unique, it is admissible.* The converse is not true, since an admissible rule is not necessarily Bayes or Minimax. Accordingly, no system that does not minimize the average risk (loss rating) can be uniformly better than a Bayes system [for the same $\sigma(\mathbf{S})$], and no system that does not minimize the maximum conditional risk (loss rating) can be uniformly better than a Minimax system. Admissibility is an important additional optimum property of unique Bayes and Minimax decision systems.

(2) **The complete class theorem.** This is Wald's fundamental theorem[1] [theorem (3.20)] concerning complete classes of decision rules. We say first that a class D of decision rules is *complete* if, for any δ not in D, we can find a δ^* in D such that δ^* is uniformly better than δ. If D contains no subclass which is complete, D is a *minimal complete class.* Wald has shown that *for the simple loss functions* [Eqs. (18.4), (18.6), (18.7)] *the class of all admissible Bayes decision rules is a minimal complete class,* under a set of conditions which are certainly satisfied for most if not all physical situations (cf. Secs. 18.5-2, 18.5-3). For the same set of conditions, any Minimax decision rule can be shown to be a Bayes rule with respect to a certain *least favorable a priori* distribution $\sigma_M^*(\mathbf{S})$, and the existence of $\sigma_M^*(\mathbf{S})$, as well as of the Bayes and Minimax rules themselves, is assured. The complete class theorem thus establishes an optimum property of the Bayes class as a whole. Note that no complete class theorem has as yet been demonstrated for the information loss ratings of Eqs. (18.8), (18.9), nor have the general conditions for the existence of Bayes and Minimax rules for such measures been established. However, some results on the characterization of Bayes tests with this measure, for detection, are given in Chap. 22.

18.5-2 Definitions.† It is assumed that a decision γ is to be made about a signal **S**, based on observations **V** whose occurrence is governed by the conditional distribution-density function $F_n(\mathbf{V}|\mathbf{S})$. The decision rule $\delta(\gamma|\mathbf{V})$ is the probability (density) that γ will be decided when the observation is **V**, regardless of **S**.

† Wald's book,[1] referred to in the following as SDF. Wald's paper[2] is recommended as an introduction to the subject.

Risk theory is based on the following definitions:

1. It is assumed that a *cost* $C(\mathbf{S},\boldsymbol{\gamma})$ is preassigned to every possible combination of signal $\mathbf{S}$ and decision $\boldsymbol{\gamma}$ in the problem.
2. The *conditional risk* $r(\mathbf{S},\delta)$ of using a decision rule δ is the expected value of the cost when the signal is $\mathbf{S}$:

$$r(\mathbf{S},\delta) = \int_\Gamma \int_\Delta C(\mathbf{S},\boldsymbol{\gamma})\delta(\boldsymbol{\gamma}|\mathbf{V})F_n(\mathbf{V}|\mathbf{S})\, d\mathbf{V}\, d\boldsymbol{\gamma}$$

3. The *average risk* $R(\sigma,\delta)$ of using δ is the expected value of $r(\mathbf{S},\delta)$ in view of the a priori probability (density) $\sigma(\mathbf{S})$:

$$R(\sigma,\delta) = \int_\Omega r(\mathbf{S},\delta)\sigma(\mathbf{S})\, d\mathbf{S}$$

4. A *Minimax* decision rule δ_M^* is one whose maximum conditional risk is not greater than the maximum conditional risk of any other δ:

$$\max_{\mathbf{S}} r(\mathbf{S},\delta_M^*) \leq \max_{\mathbf{S}} r(\mathbf{S},\delta) \qquad \text{for all } \delta$$

5. A Bayes decision rule δ^* is one whose average risk is smallest for a given a priori distribution $\sigma(S)$:

$$R(\sigma,\delta^*) = \min_{\delta} R(\sigma,\delta) \qquad \text{for all } \delta$$

6. A decision rule δ_1 is *uniformly better* than a decision rule δ_2 if the conditional risk of δ_1 does not exceed that of δ_2 for any value of $\mathbf{S}$ and is actually less than that of δ_2 for some particular $\mathbf{S}$.
7. A decision rule is *admissible* if no uniformly better one exists.
   a. An admissible rule is *not* necessarily uniformly better than any other; i.e., other rules can have smaller risks at a particular value of $\mathbf{S}$. The point is that they cannot be better at *all* values of $\mathbf{S}$.
   b. An admissible rule need not be Minimax. Clearly, δ_M^* could have a larger risk than δ at some values of $\mathbf{S}$ and still have a smaller maximum risk.
8. A class D of decision rules is complete if for any δ not in D we can find a δ'^* in D such that δ'^* is uniformly better than δ. If D contains no subclass which is complete, it is a *minimal complete* class.

18.5-3 Principal Theorems. We assume that the following conditions are fulfilled:

A. $F_n(\mathbf{V}|\mathbf{S})$ is continuous in $\mathbf{S}$.

B. $C(\mathbf{S},\boldsymbol{\gamma})$ is bounded in $\mathbf{S}$ and $\boldsymbol{\gamma}$.

C. The class of decision rules considered is restricted to either (1) nonsequential rules or (2) sequential rules.

D. $\mathbf{S}$ and $\boldsymbol{\gamma}$ are restricted to finite closed domains.

These conditions are more restrictive in some cases than those imposed by Wald but are sufficient for our purposes. Specifically, Wald's assumptions[1]

(3.1), (3.2), (3.3) are covered by conditions A and B, (3.5) and (3.6) by condition C, and (3.4) and (3.7) by condition D.

Under these assumptions the following theorems hold:

1. The decision problem viewed as a zero-sum two-person game is *strictly determined:*

$$\max_{\sigma} \min_{\delta} R(\sigma,\delta) = \min_{\delta} \max_{\sigma} R(\sigma,\delta) \qquad \text{(SDF Th. 3.4)}$$

2. For any a priori $\sigma(\mathbf{S})$, there exists a Bayes decision rule δ^* relative to $\sigma(\mathbf{S})$ (SDF Th. 3.5).

3. A Minimax decision rule exists (SDF Th. 3.7).

4. A *least favorable* a priori distribution $\sigma_0(\mathbf{S})$ exists:

$$\min_{\delta} R(\sigma_0,\delta) = \max_{\sigma} \min_{\delta} R(\sigma,\delta) \qquad \text{(SDF Th. 3.14)}$$

5. Any Minimax decision rule is Bayes relative to a least favorable a priori distribution (SDF Th. 3.9).

6. The class of all Bayes decision rules is complete relative to the class of all decision rules for which the conditional risk is a bounded function of $\mathbf{S}$ (SDF Th. 3.20). (Kiefer[26] shows that the restriction of the set of decision rules to those for which the conditional risk is a bounded function of $\mathbf{S}$ is unnecessary. He also shows that the class of all admissible decision functions is minimal complete. See, in addition, Wald's remarks following Th. 3.20 in SDF.)

The facts below follow from the definitions of Sec. 18.5-2:

7. A Bayes decision rule δ_M^* whose conditional risk is constant is a Minimax decision rule. [This follows from definitions 4, 5, and 7. For suppose δ_M^* were not Minimax. Then there would exist a δ' with smaller maximum risk and smaller average risk with respect to $\sigma_0(\mathbf{S}) = \sigma_M^*(\mathbf{S})$. This contradicts the definition of δ_M.]

8. If a Bayes decision rule is unique, it is admissible. [For suppose δ^* were Bayes with respect to $\sigma(\mathbf{S})$ and not admissible. Then a uniformly better δ' would exist; that is, $r(S,\delta') \leq r(S,\overline{\delta^*})$ for all $\mathbf{S}$, with equality for some $\mathbf{S}$. But this implies that the average risk of δ' with respect to $\sigma(\mathbf{S})$ is less than that of δ^* with respect to $\sigma(\mathbf{S})$. This contradicts the definition of δ^*.]

9. A Minimax decision rule has a smaller *maximum* average risk than any other. [This follows from the fact that the average risk cannot exceed the maximum conditional risk. Of course, for some particular $\sigma(\mathbf{S})$ another test might have smaller average risk than the Minimax with the same $\sigma(\mathbf{S})$.]

Finally, it is of considerable importance in practice to be able to avoid randomized decision rules. We have quoted one theorem due to Hodges and Lehmann[15] on this point. Others may be found in some work of Dvoretzky, Wald, and Wolfowitz.[27]

18.5-4 Remarks. From our discussion, it is clear that a priori probabilities play a critical part in the formulation and application of decision theory. When such probabilities do not exist, such a theory is inapplicable. When they do exist, however, *or when they can be conceived of as existing,* even if not known or available to the observer, then the theory can be used with various methods (Minimax among them) to provide the needed distributions. In this sense, the search for a priori distributions is often one of the chief problems to be faced in practical situations; in any case, the role of a priori information cannot be shrugged off or avoided.

The problem of cost assignment also must be carefully examined, since it provides the important link to the actual situation and significance in the larger world of events. In this way, the connection between, say, the design of an optimum (or near-optimum) system and the operational aspects of the original problem is made with a theory of values, which seeks some over-all *raison d'être* for cost assignments in the particular case. Similar remarks apply for other loss functions, and a task of subsequent investigation is to seek out other meaningful criteria (besides $\mathfrak{F}_1$ and $\mathfrak{F}_2$) and establish (if this be possible) optimal properties such as admissibility and the complete class theorem for them also.

Another problem of general importance is to discover the "invariants" of various classes of detection and extraction systems, for example, the general type of weak-signal behavior in additive noise backgrounds, or the influence of such signal parameters as epoch, waveform, etc., on optimum performance, and so on. Some of these have already been obtained (cf. Sec. 19.4); others will have to be deduced from the solutions of many individual problems, as yet unsolved in specific instances. Finally, it is evident that still other types of optimization than Bayes and Minimax are possible: one can take as a criterion minimum average risk *with constraints,* say, on higher moments of the risk function, for example, or other evaluation functions, like $\mathfrak{F}_2$ [Eq. (18.5)]. However, an analytical theory for uniqueness, admissibility, etc., comparable to Wald's for the simple cost function remains to be developed in such instances.

We turn now to the principal subject of the remaining chapters: to apply in detail the method of statistical decision theory to problems of communication, in particular the central problems of signal detection and extraction in noise.

REFERENCES

1. Wald, A.: "Statistical Decision Functions," Wiley, New York, 1950.
2. Wald, A.: Basic Ideas of a General Theory of Statistical Decision Rules, *Proc. Intern. Cong. Math.,* 1950, p. 231; an introductory account.
3. Blackwell, D., and M. A. Girshick: "Theory of Games and Statistical Decisions," Wiley, New York, 1954. An extensive bibliography of the mathematical literature is given here and in Ref. 1.
4. Middleton, D., and D. Van Meter: Detection and Extraction of Signals in Noise from the Point of View of Statistical Decision Theory, *J. Soc. Ind. Appl. Math.,* **3:** 192 (1955), **4:** 86 (1956).

5. Van Meter, D., and D. Middleton: Modern Statistical Approaches to Reception in Communication Theory, Symposium on Information Theory, *IRE Trans. on Inform. Theory,* **PGIT-4:** 119 (September, 1954).
6. Woodward, P. M.: "Probability and Information Theory with Applications to Radar," chaps. 4, 5, McGraw-Hill, New York, 1953.
7. Middleton, D.: Statistical Theory of Signal Detection, Symposium on Statistical Methods in Communication Engineering, Berkeley, Calif., Aug. 17, 1953, *IRE Trans. on Inform. Theory,* **PGIT-3:** 26 (March, 1954).
8. Hoel, P. G.: "Introduction to Mathematical Statistics," chap. 11, Wiley, New York, 1947.
9. Kendall, M. G.: "The Advanced Theory of Statistics," 5th ed., vol. II, Griffin, London, 1952.
10. Cramér, H.: "Mathematical Methods of Statistics," Princeton University Press, Princeton, N.J., 1946.
11. Middleton, D., and D. Van Meter: On Optimum Multiple Alternative Detection of Signals in Noise, *IRE Trans. on Inform. Theory,* **IT-1:** 1 (September, 1955).
12. Doob, J. L.: "Stochastic Processes," chaps. 10, 11, Wiley, New York, 1953.
13. Blanc-Lapierre, A., and R. Fortet: "Théorie des fonctions aléatoires," Masson, Paris, 1953.
14. Wang, M. C., and G. E. Uhlenbeck: On the Theory of the Brownian Motion. II, *Revs. Modern Phys.,* **17:** 323 (1945).
15. Hodges, J. L., and E. L. Lehmann: Some Problems in Minimax Point Estimation, *Am. Math. Stat.,* **21:** 182 (1950).
16. Hoeffding, W.: Optimum Non-parametric Tests, *Proc. 2d Berkeley Symposium on Mathematical Statistics and Probability,* 1950, p. 83.
17. Lehmann, E. L., and C. Stein: On the Theory of Some Nonparametric Hypotheses, *Am. Math. Stat.,* **20:** 28 (1949).
18. Wald, A.: "Sequential Analysis," Wiley, New York, 1947.
19. Bussgang, J., and D. Middleton: Optimum Sequential Detection of Signals in Noise, *IRE Trans. on Inform. Theory,* **IT-1:** 5 (1955).
20. Shannon, C. E.: Mathematical Theory of Communication, *Bell System Tech. J.,* **27:** 379, 623 (1948).
21. Samson, E. W.: Fundamental Natural Concepts of Information Theory, *AFCRC Rept.* E5079, sec. 14, October, 1951.
22. Von Neumann, J.: Zur Theorie der Gesellschaftsspiele, *Math. Ann.,* **100:** 295 (1928).
23. Von Neumann, J., and O. Morgenstern: "Theory of Games and Economic Behavior," 3d ed., Princeton University Press, Princeton, N.J., 1953.
24. Radner, R., and J. Marschak: Note on Some Proposed Decision Criteria, in R. M. Thrall, C. H. Coombs, and R. L. Davis (eds.), "Decision Processes," sec. V, Wiley, New York, 1954.
25. Hodges, J. L., and E. L. Lehmann: The Use of Previous Experience in Reaching Statistical Decisions, *Ann. Math. Stat.,* **23:** 396 (1952).
26. Kiefer, J.: On Wald's Complete Class Theorem, *Ann. Math. Stat.,* **24:** 70 (1953).
27. Dvoretzky, A., A. Wald, and J. Wolfowitz: Elimination of Randomization in Certain Statistical Decision Procedures and Zero-sum Two Person Games, *Ann. Math. Stat.,* **22:** 1 (1951).

CHAPTER 19

BINARY DETECTION SYSTEMS MINIMIZING AVERAGE RISK. GENERAL THEORY

Here, we shall outline the general theory of single-alternative detection systems $\boldsymbol{T}_R^{(N)} = (\boldsymbol{T}_R^{(N)})_{det}$, for the common and important cases where the data acquisition period (or, as we shall somewhat more loosely call it, the observation period) is fixed at the outset.† Since the decisions treated here have only two possible outcomes, we call them *binary decisions*, and the corresponding detection process, *binary detection*, in order to distinguish them from the multiple-alternative situations examined later (cf. Sec. 23.3-1).

19.1 *Formulation*

Binary detection problems have been extensively studied, since about 1942, from various points of view, although the explicit formulation in terms of statistical decision theory is comparatively recent.[1–3]‡ The principal objectives of the following sections are to obtain (1) a formulation of the binary detection problem itself and (2), by so doing, to indicate how these different viewpoints can not only be reestablished by the decision theoretical approach but extended to include situations of general practical significance. Specifically, we first derive the optimum Bayes system and point out that several other detection systems considered previously are special cases of this. Considerable detail is devoted to optimum detection for the particularly important case of a weak signal embedded in additive noise. It is shown here, for instance, that the effects like *signal suppression* [cf. Secs. 13.2-1(4), 15.5-4] are fundamental and unavoidable in the detection

† Variable observation periods are referred to in Sec. 20.4-8.

‡ For the earlier studies, based for the most part on a second-moment theory (e.g., signal-to-noise ratios, etc.), see, for example, the papers of North,[4] Van Vleck and Middleton,[5] Dwork,[6] George,[7] Zadeh,[8,9] and Ragazzini.[8] The development of a more complete statistical approach, leading up to and in some instances coinciding with certain aspects of the present theory, is described in the work of Siegert,[10] Marcum,[11] Singleton,[12] Kaplan and McFall,[13] Hance,[14] Schwartz,[15] Davies,[16] Slattery,[17] Reich and Swerling,[18] Davis,[19,20] Middleton,[21,22] Stone,[23] and Benner and Drenick[24] and in the study of Peterson, Birdsall, and Fox.[25] A more complete list is available in Middleton and Van Meter[2] and in the references at the end of this and succeeding chapters.

of weak signals.† (In the next chapter, we shall also see that there are a variety of ways in which these optimum systems may be realized by combination of physical elements such as invariant and time-varying linear filters, zero-memory nonlinear devices, ideal integrators, and the like.)

19.1-1 The Average Risk. First we use $\mathfrak{F}_1$ [Eq. (18.4)] as our loss function and determine optimum systems of the Bayes class [cf. Eq. (18.17*a*)], which, as we have seen, are defined by minimizing the average risk

$$R(\sigma,\delta) = \int_{\Omega} d\mathbf{S}\, \sigma(\mathbf{S}) \int_{\Gamma} d\mathbf{V}\, F_n(\mathbf{V}|\mathbf{S}) \int_{\Delta} d\boldsymbol{\gamma}\, C(\mathbf{S},\boldsymbol{\gamma})\delta(\boldsymbol{\gamma}|\mathbf{V}) \tag{19.1}$$

We recall that in binary detection we test the hypothesis H_0 that noise alone is present against the alternative H_1 of a signal and noise, so that there are but two points $\gamma = (\gamma_0,\gamma_1)$, respectively, in decision space Δ. For the moment, allowing the possibility that the decision rule δ may be randomized, we let $\delta(\gamma_0|\mathbf{V})$ and $\delta(\gamma_1|\mathbf{V})$ be the probabilities‡ that γ_1 and γ_0 are decided, given $\mathbf{V}$. Since definite, terminal decisions are postulated here, some decision is always made and therefore

$$\delta(\gamma_0|\mathbf{V}) + \delta(\gamma_1|\mathbf{V}) = 1 \tag{19.2}$$

Denoting by $\mathbf{S}$ the input signal that may occur during the observation interval, we may express the two hypotheses concisely as $H_0:\mathbf{S}\epsilon\Omega_0$ and $H_1:\mathbf{S}\epsilon\Omega_1$, where Ω_0 and Ω_1 are the appropriate nonoverlapping hypothesis classes, as discussed in Sec. 18.2-1(2). In binary detection, the null class Ω_0 usually contains only one member, corresponding to no signal.§ The signal class Ω_1 may consist of one or more nonzero signals. It is now convenient to describe the occurrence of signals within the nonoverlapping classes Ω_0, Ω_1 by density functions $w_0(\mathbf{S})$, $w_1(\mathbf{S})$, normalized over the corresponding spaces, e.g.,

$$\int_{\Omega_0} w_0(\mathbf{S})\, d\mathbf{S} = 1 \qquad \int_{\Omega_1} w_1(\mathbf{S})\, d\mathbf{S} = 1 \tag{19.3}$$

If q and p $(=1-q)$ are, respectively, the a priori probabilities that some one signal from Ω_0 and Ω_1 will occur, the a priori probability distribution $\sigma(\mathbf{S})$ over the total signal space $\Omega = \Omega_0 + \Omega_1$ becomes

$$\sigma(\mathbf{S}) = qw_0(\mathbf{S}) + pw_1(\mathbf{S}) = q\delta(\mathbf{S} - 0) + pw_1(\mathbf{S}) \tag{19.4}$$

this last when there is but one (zero) signal in class Ω_0. Equation (19.4)

† Except in certain rather special situations (cf. Sec. 19.4-3).

‡ Since the number of alternatives is finite and discrete, the decision rule is represented by a probability (cf. the remarks following the statement of the general reception problem in Sec. 18.3-3).

§ Another important binary problem arises when Ω_0 contains $\mathbf{S}_0$ and noise, while Ω_1 contains $\mathbf{S}_1 + \mathbf{N}$ ($\mathbf{S}_0$, $\mathbf{S}_1 \neq 0$); the treatment is identical with the more usual problem discussed here, except for obvious modifications in the signal and noise distributions. See, for example, Sec. 20.4-5.

represents the *one-sided alternative*,† mentioned in Sec. 18.2-1(2), while if there is only a single signal in class Ω_1 as well, Eq. (19.4) becomes $\sigma(\mathbf{S}) = q\delta(\mathbf{S} - 0) + p\delta(\mathbf{S} - \mathbf{S}_1)$ $(\mathbf{S}_1 \neq 0)$ and we have an example of the *simple alternative* situation. In both cases, $\int \sigma(\mathbf{S})\, d\mathbf{S} = 1$, by definition of p, q, and w.

The next step in the application of the risk theory is to assign a set of costs $C(\mathbf{S},\gamma) = \mathfrak{F}_1$ [cf. Eq. (18.4)] to each possible combination of signal input and decision. We illustrate the discussion with the assumption of one-sided alternatives and uniform costs, although the method is not restricted by such choice. Accordingly, we have

$$\begin{aligned} C(\mathbf{S} \in \Omega_0; \gamma_0) &= C_{1-\alpha} \qquad & C(\mathbf{S} \in \Omega_1; \gamma_0) &= C_\beta \\ C(\mathbf{S} \in \Omega_0; \gamma_1) &= C_\alpha \qquad & C(\mathbf{S} \in \Omega_1; \gamma_1) &= C_{1-\beta} \end{aligned} \tag{19.5}$$

where $C_{1-\alpha}$, $C_{1-\beta}$ are the costs associated with correct decisions, or "success," while C_α, C_β are associated with the possible incorrect decisions, or "failure." Clearly, it is necessary from the definition of success and failure that we require $C_\alpha > C_{1-\alpha}$, $C_\beta > C_{1-\beta}$. We can also postulate that $C_{1-\alpha} \geq 0$, $C_{1-\beta} \geq 0$, that is, that there is no net gain, or "profit," from a correct decision. The best we can expect here (if we are at liberty to adjust the C's) is that success may cost us nothing: $C_{1-\alpha} = C_{1-\beta} = 0$. However, note that, in any case, the cost assignments are made with respect to the signal class, and not with respect to any one member of the class.

Specifying that $F_n(\mathbf{V}|\mathbf{S})$ is continuous in $\mathbf{S}$, that fixed-sample tests only are considered (cf. Sec. 20.4-8 for reference to variable-sample cases), and that the assumptions needed for the validity of risk theory (cf. Sec. 18.5-3) are satisfied, we see that the integration over decision space Δ in Eq. (19.1) is replaced by an appropriate summation, so that with Eqs. (19.5) the average risk is now

$$\begin{aligned} R(\sigma,\delta) = \int_\Gamma d\mathbf{V}\ \{ &\delta(\gamma_0|\mathbf{V})[qC_{1-\alpha}F_n(\mathbf{V}|0) + pC_\beta\langle F_n(\mathbf{V}|\mathbf{S})\rangle_S] \\ &+ \delta(\gamma_1|\mathbf{V})[pC_{1-\beta}\langle F_n(\mathbf{V}|\mathbf{S})\rangle_S + qC_\alpha F_n(\mathbf{V}|0)]\} \end{aligned} \tag{19.6}$$

where $p + q = 1$, and, from Eq. (19.4),

$$\int_{\Omega_1} \sigma(\mathbf{S})F_n(\mathbf{V}|\mathbf{S})\, d\mathbf{S} = p\langle F_n(\mathbf{V}|\mathbf{S})\rangle_S = p\int_S w_1(\mathbf{S})F_n(\mathbf{V}|\mathbf{S})\, d\mathbf{S} \tag{19.7a}$$

$$\int_{\Omega_0} \sigma(\mathbf{S})F_n(\mathbf{V}|\mathbf{S})\, d\mathbf{S} = qF_n(\mathbf{V}|0) = q\int_S w_0(\mathbf{S})F_n(\mathbf{V}|\mathbf{S})\, d\mathbf{S} \tag{19.7b}$$

When $\mathbf{S} = \mathbf{S}(\boldsymbol{\theta})$, these become alternatively

$$p\langle F_n(\mathbf{V}|\mathbf{S})\rangle_S = p\int_\theta w_1(\boldsymbol{\theta})F_n[\mathbf{V}|\mathbf{S}(\boldsymbol{\theta})]\, d\boldsymbol{\theta} \tag{19.7c}$$

$$qF_n(\mathbf{V}|0) = q\int_\theta w_0(\boldsymbol{\theta})F_n[\mathbf{V}|\mathbf{S}(\boldsymbol{\theta})]\, d\boldsymbol{\theta} \tag{19.7d}$$

† This corresponds to a composite hypothesis H_1.

There are two possible classes of error here: the type I error of calling noise alone a signal and noise, and the type II error of deciding noise only, when a signal as well as noise is actually present. The class conditional probabilities of these two types of error (obtained by averaging over the respective hypothesis classes) are defined by

$$\alpha \equiv \int_\Gamma F_n(\mathbf{V}|0)\delta(\gamma_1|\mathbf{V})\, d\mathbf{V} \qquad \beta \equiv \int_\Gamma \langle F_n(\mathbf{V}|\mathbf{S})\rangle_S \delta(\gamma_0|\mathbf{V})\, d\mathbf{V} \tag{19.8}$$

so that with the help of Eqs. (19.2), (19.7), (19.8) we may write the average risk [Eq. (19.6)] more compactly as

$$R = qC_{1-\alpha} + pC_{1-\beta} + q\alpha(C_\alpha - C_{1-\alpha}) + p\beta(C_\beta - C_{1-\beta}) \tag{19.9a}$$

$$= \mathcal{R}_0 + q\alpha(C_\alpha - C_{1-\alpha}) + p\beta(C_\beta - C_{1-\beta}) \tag{19.9b}$$

where

$$\mathcal{R}_0 = qC_{1-\alpha} + pC_{1-\beta} \tag{19.9c}$$

From Eq. (18.6), in a similar way, the *conditional risk* becomes

$$r(\mathbf{S}) = \begin{cases} (1-\alpha')C_{1-\alpha} + \alpha' C_\alpha & \mathbf{S} = 0 \\ [1-\beta'(\mathbf{S})]C_{1-\beta} + \beta'(\mathbf{S})C_\beta & \mathbf{S} \neq 0 \end{cases} \tag{19.10}$$

where the *simple conditional error probabilities* α' and β' are distinguished from the class conditional error probabilities α and β [Eqs. (19.8)] according to

$$\begin{aligned} \alpha' &\equiv \int_\Gamma F_n(\mathbf{V}|0)\delta(\gamma_1|\mathbf{V})\, d\mathbf{V} \ (= \alpha) \\ \beta'(\mathbf{S}) &= \int_\Gamma F_n(\mathbf{V}|\mathbf{S})\delta(\gamma_0|\mathbf{V})\, d\mathbf{V} \ (\neq \beta) \end{aligned} \tag{19.11}$$

19.1-2 Optimum Detection. The Bayes decision rule δ^* (cf. Eq. 18.17a) can now be found from Eq. (19.6). First let us rewrite Eq. (19.6) to include the constraint of Eq. (19.2). We get directly

$$\begin{aligned} R(\sigma,\delta) = qC_\alpha + pC_{1-\beta} + \int_\Gamma \delta(\gamma_0|\mathbf{V})[p(C_\beta - C_{1-\beta})\langle F_n(\mathbf{V}|\mathbf{S})\rangle_S \\ - q(C_\alpha - C_{1-\alpha})F_n(\mathbf{V}|0)]\, d\mathbf{V} \end{aligned} \tag{19.12}$$

Our problem then is to choose $\delta(\gamma_0|\mathbf{V})$, and hence $\delta(\gamma_1|\mathbf{V})$, in such a way as to minimize the average risk. Since both δ's must be positive and (equal to or) less than 1, it is clear that the integral in Eq. (19.12) is least if for each $\mathbf{V}\epsilon\Gamma$ we choose the δ's in the following way:

When $p(C_\beta - C_{1-\beta})\langle F_n(\mathbf{V}|\mathbf{S})\rangle_S \geq q(C_\alpha - C_{1-\alpha})F_n(\mathbf{V}|0)$, *choose*

$$\begin{aligned} \delta^*(\gamma_1|\mathbf{V}) &= 1 \qquad \text{i.e., } \textit{decide signal and noise} \\ \delta^*(\gamma_0|\mathbf{V}) &= 0 \end{aligned} \tag{19.13a}$$

When $p(C_\beta - C_{1-\beta})\langle F_n(\mathbf{V}|\mathbf{S})\rangle_S < q(C_\alpha - C_{1-\alpha})F_n(\mathbf{V}|0)$, *choose*

$$\begin{aligned} \delta^*(\gamma_1|\mathbf{V}) &= 0 \\ \delta^*(\gamma_0|\mathbf{V}) &= 1 \qquad \text{i.e., } \textit{decide noise only} \end{aligned} \tag{19.13b}$$

Note that the Bayes decision rule turns out to be nonrandomized (Sec. 18.3-2), in spite of the fact that the derivations (19.12) et seq. allowed for the possibility of a randomized optimum rule.

At this point, it is convenient to introduce the *generalized likelihood ratio*,†

$$\Lambda_n = \frac{p\langle F_n(\mathbf{V}|\mathbf{S})\rangle_S}{qF_n(\mathbf{V}|0)} \ (\geqq 0) \tag{19.14}$$

in terms of which the Bayes decision rule (for nonoverlapping hypothesis classes) in the binary case [Eqs. (19.13*a*), (19.13*b*)] may be stated more compactly as

$$\begin{array}{l} \textit{Decide } \gamma_1 \textit{ when } \Lambda_n \geq \mathcal{K} \\ \textit{Decide } \gamma_0 \textit{ when } \Lambda_n < \mathcal{K} \end{array} \qquad \mathcal{K} = \frac{C_\alpha - C_{1-\alpha}}{C_\beta - C_{1-\beta}} > 0 \tag{19.15}$$

where $\mathcal{K}$ is called the *threshold* and depends only on the (preassigned) costs. [The subscript n reminds us that Λ_n and F $(= F_n)$ are functions of the sample size or observation period.]

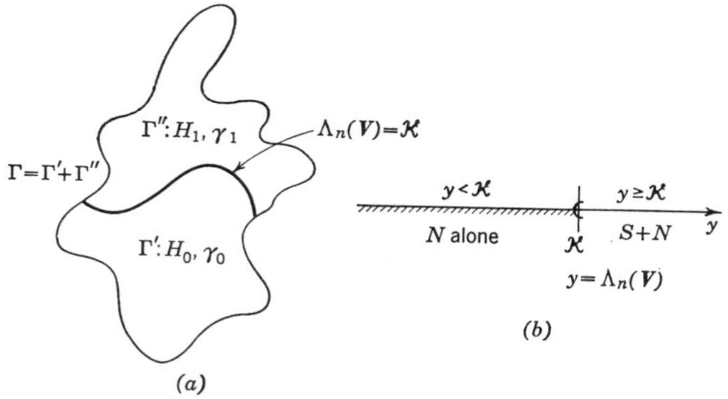

FIG. 19.1. (*a*) Decision regions for **V** in binary (fixed-sample) detection. (*b*) The same for $y = \Lambda_n(\mathbf{V})$.

Thus, the Bayes rule essentially amounts to a division of observation space Γ into two regions separated by the **V** satisfying the equation

$$\Lambda_n(\mathbf{V}) = \mathcal{K} \tag{19.16}$$

as suggested in Fig. 19.1. The *acceptance region* Γ' [that is, for the hypothesis H_0: $(\mathbf{S}\epsilon\Omega_0)$] contains all **V** for which $\Lambda_n(\mathbf{V}) < \mathcal{K}$, and the corresponding *rejection region* Γ'' [that is, the acceptance region for the alternative hypothesis H_1: $(\mathbf{S}\epsilon\Omega_1)$] similarly contains those **V** for which $\Lambda_n(\mathbf{V}) \geq \mathcal{K}$. Upon receiving the data **V**, the Bayes system then decides whether the data belong to Γ' or Γ''. It does this by computing the generalized likelihood ratio

† We note that Λ_n is more general than the classical likelihood ratio $F_n(\mathbf{V}|\mathbf{S})/F_n(\mathbf{V}|0)$, that is, the ratio of the conditional probability densities of **V** with and without **S** fixed. The generalized likelihood ratio (19.14) reduces to this form when the a priori probabilities p and q are equal and the signal space contains but one point, corresponding to the very special case of a completely deterministic signal.

(19.14) for the particular **V** received in the fixed observation period and then by comparing $\Lambda_n(\mathbf{V})$ with the fixed threshold $\mathcal{K}$, established by the preassigned costs. In order to calculate Λ_n, the system must "know" the a priori probabilities p, q, $w_1(\mathbf{S})$, as well as $F_n(\mathbf{V}|\mathbf{S})$; that is, the designer's knowledge of these must be built into the computer which constitutes the optimum system.† In this way, *all* (available) a priori information is used in the decision process, up to the point at which the definite decision "yes, signal and noise" or "no, only noise" is made. As soon as γ_0 or γ_1 is decided, some information is inevitably lost because of noise and the finite observation period; the amount of this loss is calculated for some of the examples in Chaps. 20 to 23.

In general, the *optimum detector* $(\boldsymbol{T}_R^{(N)})_{opt}$ *is a computer which processes the received data* **V** *in a nonlinear fashion.* Its precise form depends on the statistics of the background noise and signal structure, as well as on the a priori probabilities, although in certain interesting cases (chiefly with normal statistics) only linear operations on **V** are required. Thus, Λ_n (or any monotonic function‡ of Λ_n) embodies the explicit structure of the optimum binary detection system, that is, $\Lambda_n(\mathbf{V}) = \boldsymbol{T}_R^{(N)}\{\mathbf{V}\}_{opt}$. As mentioned in the preceding chapter, one of our principal objectives in system analysis and design is to interpret the analytical operations implied by this generalized likelihood ratio in terms of physically realizable system elements, including ordered sequences of linear invariant and noninvariant filters, zero-memory nonlinear devices, and the like. Frequently, the general computer Λ_n (or $\log \Lambda_n$)‡ can be reduced to the standard types of direct or heterodyne receiver, with suitable modifications of the input and final output stages as required by the signal structure and noise statistics. In more involved cases, it may be necessary to add further data processing (in the form of a simple analogue or digital computer) to this receiver output before decisions are made. Specific examples of both types are discussed presently in Chap. 20.

19.1-3 Some Further Properties of the Bayes Detection Rule. The likelihood-ratio test [Eqs. (19.15)] may also be interpreted in a somewhat different way. Noting that $F_n(\mathbf{V}|\mathbf{S}) = w(\mathbf{V})w(\mathbf{S}|\mathbf{V})/\sigma(\mathbf{S})$, where $w(\mathbf{V})$ is the (total) probability (density) of **V**'s occurrence and $w(\mathbf{S}|\mathbf{V})$ is the a posteriori (conditional) probability (density) of **S** given **V**, we may write for the coefficients of $\delta(\gamma_0|\mathbf{V})$ and $\delta(\gamma_1|\mathbf{V})$ in Eq. (19.6)

$$A_0(\mathbf{V},\sigma) = w(\mathbf{V}) \left[\int_{\Omega_0} C_{1-\alpha} w(\mathbf{S}|\mathbf{V}) \, d\mathbf{S} + \int_{\Omega_1} C_\beta w(\mathbf{S}|\mathbf{V}) \, d\mathbf{S} \right] \quad (19.17a)$$

$$A_1(\mathbf{V},\sigma) = w(\mathbf{V}) \left[\int_{\Omega_0} C_\alpha w(\mathbf{S}|\mathbf{V}) \, d\mathbf{S} + \int_{\Omega_1} C_{1-\beta} w(\mathbf{S}|\mathbf{V}) \, d\mathbf{S} \right] \quad (19.17b)$$

† For the cases in which p, q, $w_1(\mathbf{S})$, etc., may not be available to the observer, see the discussion in Secs. 19.2-3 and 20.4-8.

‡ For various reasons, which will be mentioned later, it is convenient to take $\log \Lambda_n$, rather than Λ_n, for the structure of the optimum detector.

For given $\mathbf{V}$, the integrals in Eqs. (19.17) are the conditional expected costs of the decisions γ_0, γ_1; that is, they are the *a posteriori risks* associated with the decisions γ_0 and γ_1. Thus, the Bayes decision rule [Eqs. (19.13) or Eqs. (19.15)] may be stated alternatively: *Make the decision for which the a posteriori risk is least.*

The second property is of interest for its connections with information theory. Note that the logarithm of the likelihood ratio [Eq. (19.14)] is proportional to the measure (cf. Sec. 6.2-1) of the difference between the uncertainties about the states H_0 and H_1 when $\mathbf{V}$ is known; viz., the optimum receiver (for binary detection) $\boldsymbol{T}_R^{(N)}\{\mathbf{V}\}_{opt}$ is alternatively

$$\log \Lambda_n(\mathbf{V}) = [-\log qF_n(\mathbf{V}|0)] - [-\log p\langle F_n(\mathbf{V}|\mathbf{S})\rangle_S] \tag{19.18}$$

We see from this that the decision rule [Eq. (19.15)] is equivalent to deciding in favor of H_1 when the uncertainty in H_1 is less than the uncertainty about H_0 by an amount $\log \mathcal{K}$.

Finally, we remark that the complete class theorem (Sec. 18.5-1) for the risk formulation assures us that we have an optimum test and, further, that all such tests based on the likelihood ratio [Eq. (19.14)] are Bayes tests (see Sec. 18.5-1). The Bayes risk R^* and the average risk R [Eqs. (19.9)] for general (not necessarily optimum) systems become, respectively,

$$R^* = \mathcal{R}_0 + p(C_\beta - C_{1-\beta})\left(\frac{\mathcal{K}}{\mu}\,\alpha^* + \beta^*\right) \tag{19.19a}$$

$$\mu \equiv \frac{p}{q}$$

$$R = \mathcal{R}_0 + p(C_\beta - C_{1-\beta})\left(\frac{\mathcal{K}'}{\mu}\,\alpha + \beta\right) \tag{19.19b}$$

with $\mathcal{K}$ given by Eqs. (19.15), while $\mathcal{R}_0 = qC_{1-\alpha} + pC_{1-\beta}$ in either instance [cf. Eq. (19.9*c*)]. The threshold $\mathcal{K}'$ for the nonideal case† may or may not be equal to $\mathcal{K}$. Note, incidentally, from Eq. (19.19*a*) that if we differentiate $(R^{(*)} - \mathcal{R}_0)/p(C_\beta - C_{1-\beta})$ we have at once

$$d\beta^*/d\alpha^* = -\mathcal{K}/\mu \qquad d\beta/d\alpha = -\mathcal{K}'/\mu \tag{19.19c}$$

relations which are of some use in describing a receiver's performance characteristics (cf. Sec. 20.4-3).

19.2 *Special Optimum Detection Systems*

As we have seen in Sec. 19.1-2, where a criterion of minimum average cost, or risk, is chosen, the optimum (or Bayes) receiver for binary detection is the likelihood system‡

$$\boldsymbol{T}_R^{(N)}\{\mathbf{V}\}_{det\ opt} = \log \Lambda_n(\mathbf{V}) = \log \mu + \log [\langle F_n(\mathbf{V}|\mathbf{S})\rangle_S / F_n(\mathbf{V}|0)] \tag{19.20}$$

† One always has a definite threshold when a definite decision is required.

‡ We consider only discrete (or digital) sampling for the time being [cf. Sec. 19.4-2 for continuous (or analogue) processing].

Other criteria, however, may also lead to likelihood systems and as such can be interpreted as special cases of the Bayes tests derived above. Three (fixed-sample) systems of this kind are considered in the following sections, namely, the Neyman-Pearson, Ideal, and Minimax receivers. The first two of these turn out to be Bayes with special cost assumptions, while the third, as we have already noted in Sec. 18.4, is an extremum of the Bayes class, still with the general cost assignments of Eqs. (19.5), however.

19.2-1 The Neyman-Pearson Detection System. Preliminary to the more general detection situation involving a priori probabilities and one-sided alternatives [cf. Sec. 18.2-1(2)], let us briefly describe the classical Neyman-Pearson test of a hypothesis[26,27] against a *simple* alternative. The line of reasoning which is usually employed, however, disguises the explicit use of the cost concept or, for that matter, of a priori probabilities. Rather, it involves the simple (conditional) probabilities of erroneous decisions. One begins by introducing two (nonrandomized) decision functions $\delta(\gamma_0|\mathbf{V})$ and $\delta(\gamma_1|\mathbf{V})$ of, for the moment, an unspecified nature. The simple conditional probabilities of error α and $\beta'(\mathbf{S})$ are then defined according to Eqs. (19.11). The Neyman-Pearson test itself *is then defined as that procedure which minimizes* β' *for fixed* α *and* $\mathbf{S}$. As we shall see presently [Eqs. (19.23*a*) et seq.], the test which is optimum in this sense is a likelihood-ratio test with a threshold $\mathcal{K}$ depending on α. Fixing α and minimizing β' is therefore equivalent to assuming a certain cost ratio depending on the type I error probability α. In its usual form, the Neyman-Pearson test is appropriate for simple alternatives only. Thus, when applied to detection problems, it can be used only for the case where Ω_1 consists of a single element;[11,14,15,18] i.e., only one possible signal $\mathbf{S} = \mathbf{S}_1$ other than zero can occur, and $w_1(\mathbf{S}) = \delta(\mathbf{S} - \mathbf{S}_1)$.

Extensions of this classical test, however, to include the more general situation (more than one element in Ω_1 and a priori probabilities) encountered in detection problems have been made.[22] To achieve the desired extension from the point of view of decision theory, we modify the classical procedure now so as to minimize the *total* type II error probability $p\beta$ [$= p\langle\beta'(\mathbf{S})\rangle_S$] [cf. Eqs. (19.21), (19.22)] with the total type I error probability $q\alpha$ fixed;† i.e., we calculate

$$\mathcal{R}^*_{NP} \equiv \min_{\delta} \, (p\beta + \lambda q\alpha) = p\beta^*_{NP} + \lambda q\alpha \tag{19.21}$$

where λ is an as yet undetermined multiplier. Minimization, of course, is with respect to the decision rule δ, subject to the usual constraint [Eq. (19.2)]

† We also have the reverse Neyman-Pearson system, where β is fixed and α is minimized, for example, $\mathcal{R}_{(RNP)} \equiv \min_{\delta} \, (\lambda p\beta + q\alpha)$. This is appropriate in those cases where the type II error probability is to be kept fixed and the "false alarm" rate ($\sim\alpha$) is to be minimized.

of a definite decision. From Eqs. (19.8), we write explicitly

$$\begin{aligned}\mathcal{R}^*_{NP} &= \min_{\delta} \left[p \int_\Gamma \langle F_n(\mathbf{V}|\mathbf{S})\rangle_S \delta(\gamma_0|\mathbf{V})\, d\mathbf{V} + \lambda q \int_\Gamma F_n(\mathbf{V}|0)\delta(\gamma_1|\mathbf{V})\, d\mathbf{V} \right] \\ &= \min_{\delta} \left\{ \int_\Gamma d\mathbf{V}\; \delta(\gamma_0|\mathbf{V})[p\langle F_n(\mathbf{V}|\mathbf{S})\rangle_S - \lambda q F_n(\mathbf{V}|0)] \right\} + \lambda q \end{aligned} \tag{19.22}$$

and for this to be a minimum it is clear that we must choose the δ's such that when

$p\langle F_n(\mathbf{V}|\mathbf{S})\rangle_S \geq \lambda q F_n(\mathbf{V}|0)$, *we set* $\delta^*(\gamma_1|\mathbf{V}) = 1$, $\delta^*(\gamma_0|\mathbf{V}) = 0$ *and decide signal and noise*

or when (19.23)

$p\langle F_n(\mathbf{V}|\mathbf{S})\rangle_S < \lambda q F_n(\mathbf{V}|0)$, *we set* $\delta^*(\gamma_0|\mathbf{V}) = 1$, $\delta^*(\gamma_1|\mathbf{V}) = 0$ *and decide noise alone*

Thus, we have

$$\boldsymbol{T}^{(N)}\{\mathbf{V}\}_{NP} = \Lambda_n = \frac{p\langle F_n(\mathbf{V}|\mathbf{S})\rangle_S}{qF_n(\mathbf{V}|0)} \begin{cases} \geq \lambda & \text{for } \gamma_1 \\ < \lambda & \text{for } \gamma_0 \end{cases} \tag{19.23a}$$

which establishes the likelihood nature of the detection system [cf. Eq. (19.14)].

Comparison with Eqs. (19.15) et seq. shows that the undetermined multiplier λ here plays the role of threshold $\mathcal{K}$ and $\mathcal{R}^*_{NP}$ is (except for a scale factor) the corresponding Bayes risk for this threshold $\lambda \equiv \mathcal{K}_{NP}$. The decision regions Γ', Γ'' for $\mathbf{V}$ are pictured in Fig. 19.1. However, λ is not arbitrary but is determined by the constraint of a preassigned value of the conditional type I error probability α, for example,

$$\alpha_{NP} = \int_\Gamma F_n(\mathbf{V}|0)\delta^*(\gamma_1|\mathbf{V})\, d\mathbf{V} = \alpha_{NP}(\lambda = \mathcal{K}_{NP}) \tag{19.24}$$

from Eqs. (19.8) and the nature of the optimum decision rule [Eqs. (19.23)].†

We can also write Eqs. (19.23a) in the more classical form, either by an obvious modification or from a computation of $\min_\delta (\beta + \lambda\alpha)$ by precisely the same sort of argument given above, viz.,

$$\Lambda_n \equiv \frac{\langle F_n(\mathbf{V}|\mathbf{S})\rangle_S}{F_n(\mathbf{V}|0)} = \frac{\lambda}{\mu} \equiv \mathcal{K}'_{NP} \tag{19.25}$$

where now $\mathcal{K}'_{NP}$ is a new threshold or significance level, into which have been absorbed the a priori probabilities (p,q) and the cost ratio λ. Thus, in the subclass of Neyman-Pearson tests the significance level $\mathcal{K}'_{NP}$ is set by choosing α_{NP}, or, equivalently, α_{NP} is specified for a predetermined level $\mathcal{K}'_{NP}$. In either instance, it is clear from the preceding remarks that such a formulation *implies* a specific set of a priori probabilities (p,q) and a cost ratio $\mathcal{K}_{NP}$ if we are to apply this optimum detection procedure to physical situations.

† See Eqs. (19.33) for the explicit dependence on $\lambda = \mathcal{K}_{NP}$.

19.2-2 The Ideal Observer Detection System. Another way of designing a fixed-sample one-sided alternative test is to require that the total probability of error $q\alpha + p\beta$ be minimized, instead of just $p\beta$ as above. An observer who makes a decision in this way is called an *Ideal Observer*,[10,22] and the corresponding receiver an Ideal Observer detection system. As in the Neyman-Pearson case, this may be set up as a variational problem and shown to yield a likelihood-ratio test with $\mathcal{K} = \mathcal{K}_I = 1$. Specifically, we want

$$\mathcal{R}_I^* \equiv \min_\delta (q\alpha + p\beta) = q\alpha_I^* + p\beta_I^* \tag{19.26a}$$

where now α and β are jointly minimized in the sum by proper choice of the decision rule. Using Eqs. (19.8) again, we can write Eq. (19.26a) as

$$\mathcal{R}_I^* = \min_\delta \left\{ \int_\Gamma \delta(\gamma_0|\mathbf{V})[p\langle F_n(\mathbf{V}|\mathbf{S})\rangle_S - qF_n(\mathbf{V}|0)]\, \mathbf{dV} \right\} + q \tag{19.26b}$$

from which it follows at once that the decision procedure for the Ideal Observer is

Decide signal and noise when $\Lambda_n \geq 1$, that is, set

$$\delta^*(\gamma_1|\mathbf{V}) = 1 \qquad \delta^*(\gamma_0|\mathbf{V}) = 0$$

or (19.27a)

Decide noise alone when $\Lambda_n < 1$, that is, set

$$\delta^*(\gamma_0|\mathbf{V}) = 1 \qquad \delta^*(\gamma_1|\mathbf{V}) = 0$$

Accordingly, the Ideal Observer system $\boldsymbol{T}^{(N)}\{\mathbf{V}\}_I$ is a Bayes detector with threshold $\mathcal{K}_I$ of unity. The fact that both the Neyman-Pearson and Ideal Observer systems yield likelihood-ratio tests of the type (19.15) follows from the optimum performance they require. Since they *are* likelihood-ratio tests, they belong to the Bayes risk class and accordingly share the general optimum properties possessed by that class, including uniqueness and admissibility (cf. Sec. 18.5-1). Note, finally, that both these special classes of optimum detection system employ nonrandomized decision rules.

19.2-3 The Minimax Detection Rule. There is yet another solution to the detection problem which, like the two just discussed, is optimum in a certain sense and which leads to a likelihood-ratio test. This is the Minimax detection rule.

When the a priori signal probabilities $\sigma(\mathbf{S})$ are unknown, the Minimax criterion discussed in Sec. 18.4-4 provides one possible definition of optimum system performance. As we have seen, a Minimax system for binary detection (and the disjoint hypothesis classes of the present chapter) can be regarded as a likelihood-ratio system for some least favorable distribution

$\sigma = \sigma_0 = \sigma_M^*$. Once this distribution is found, the Bayes system is completely determined. Now, in order to find it, we may take advantage of the fact that a likelihood-ratio system with the same conditional risks for all signals is Minimax, in consequence of the definitions of Bayes and Minimax systems (Sec. 18.4-4). The procedure is briefly illustrated below:

First, we require the conditional probabilities α', $\beta' = \beta'(\mathbf{S})$ [cf. Eqs. (19.11)] to be the result of a Bayes decision rule, which here means a likelihood-ratio test, with σ as yet unspecified. Then, as different σ are tried, different α', β' result (for the same threshold $\mathcal{K}$, which depends only on the preassigned costs). The conditional risks [Eqs. (19.10)] for $\mathbf{S} = 0$ and for $\mathbf{S} \neq 0$ (all $\mathbf{S}$) will vary. As one increases, the other must decrease, in consequence of the admissibility and uniqueness of the particular Bayes rule corresponding to our choice of σ. If, then, there exists a σ for which the conditional risks [Eqs. (19.10)] are equal† for all $\mathbf{S}$, we have the required Minimax rule δ_M^* and associated least favorable a priori distribution σ_M^*. We remember, moreover, that, if the equation between conditional risks has no solution, this does *not* mean that a Minimax rule or least favorable distribution does not exist, only that other methods must be discovered for determining them.

As an example, suppose we have the *simple* alternative detection problem, when p and q are unknown, $w_1(\mathbf{S}) = \delta(\mathbf{S} - \mathbf{S}_1)$. From Eq. (19.14), we have for the Bayes test

$$\Lambda_n = \frac{pF_n(\mathbf{V}|\mathbf{S}_1)}{qF_n(\mathbf{V}|0)} \gtrless \mathcal{K} \tag{19.27b}$$

where α' $(= \alpha)$, β' $(= \beta)$ are the corresponding error probabilities [cf. Eqs. (19.11)]. The α, β are functions of p and q since the decision rule $\delta = \delta_M^*$ depends on p and q through Λ_n. Thus, as p and q are varied, α, β also are changed as the boundary between the critical and acceptance region varies. Equating the conditional risks [Eqs. (19.10)] in order to determine the least favorable $p = p_M^*$, $q = q_M^*$ $(= 1 - p_M^*)$, we have

$$[1 - \alpha(p_M^*,q_M^*)]C_{1-\alpha} + \alpha(p_M^*,q_M^*)C_\alpha = [1 - \beta(p_M^*,q_M^*)]C_{1-\beta} + \beta(p_M^*,q_M^*)C_\beta \tag{19.28}$$

provided that a solution exists.

Alternatively, the least favorable a priori distribution σ_M^*, and therefore the Minimax decision rule, may be found in principle from the basic definitions (see Sec. 18.5-3, theorems 1 to 5). That is, $\sigma_0 = \sigma_M^*$ is the a priori distribution that maximizes the Bayes risk (cf. Fig. 18.6). Since every Bayes decision rule is associated with a specific a priori distribution, however, the Bayes rule changes as this distribution is varied for maximum risk. As a result, this method of finding the extremum may be technically difficult

† There can be no system for which both conditional risks together can be less than this, since this is a Bayes test.

to carry through. It is applicable, however, when the previous method (based on uniform conditional risk) fails.

In the case of the one-sided alternative where $w_1(\mathbf{S})$ is known but again p and q are not, the same procedure may be tried when now Eq. (19.14) is used in place of Eq. (19.27*b*). Finally, if $w_1(\mathbf{S})$ is unspecified, or if neither p, q, nor $w_1(\mathbf{S})$ is given [i.e., if $\sigma(\mathbf{S})$ is completely unavailable to the observer], Eq. (19.28) with Eq. (19.14) still applies when a solution exists, although the task of finding σ_M^* may be excessively formidable. In any case, an explicit evaluation of $(\alpha')_M^*$ and $(\beta')_M^*$ from Eqs. (19.11), when $\delta = \delta_M^*$ therein, may be carried out by methods outlined in Sec. 19.3-1 and illustrated in succeeding sections.

We observe that the Minimax error probabilities $(\alpha)_M^*$, $(\beta)_M^*$ are fixed quantities, independent of the actual a priori probabilities p, q, $w_1(\mathbf{S})$ chosen by nature. The average Minimax risk R_M^* is given directly by writing R_M^* for R^* in Eqs. (19.19) and replacing p, q, etc., and α^*, β^* by p_M^*, q_M^*, . . . and α_M^*, β_M^* therein. The difference $(R_M^* - R^* \geq 0)$ between the Bayes (σ known) and Minimax average risk (σ unknown) is thus one useful measure of the price we must pay for our ignorance of nature's strategy (i.e., here nature's choice of p, q, etc.).

Finally, we remark that, even when σ is unavailable to the observer, other classes of Bayes systems can be selected as optimum receivers if it is not appropriate to use a Minimax decision rule. For example, when σ is unknown, we may adopt as our second criterion† the choice of σ $(= \sigma_U)$ for which there is the greatest average uncertainty about $\mathbf{S}$ or the parameters $\boldsymbol{\theta}$ of $\mathbf{S}$; that is, we choose that σ_U for which $-\boldsymbol{E}\{\log \sigma(\mathbf{S})\}$ is a maximum, subject to one or more constraints that may be indicated by the problem at hand. A Bayes test is still carried out according to Eqs. (19.13) or Eqs. (19.15), but now with $\sigma = \sigma_U$. Other criteria for σ may be appropriate‡ and may be handled in the same way, yielding a different set of Bayes systems, the critical element, as always, being the appositeness of our choice of criterion vis-à-vis the physical world.

19.3 *Evaluation of Performance*

A second requirement of the theory (cf. Sec. 18.1), comparable in importance to the determination and interpretation of structure, is the evaluation of system performance for both optimum and nonoptimum systems. This may be achieved in a number of ways, all essentially based on the calculation of the appropriate (conditional) error probabilities α, β, etc. [Eqs. (19.8)], since these are required in the expressions for the average and Bayes risks [cf. Eqs. (19.19)] if we are to obtain a comparison of performance

† The first is, of course, still the Bayes condition $\min_{\delta} R(\sigma,\delta)$.

‡ For example, one may use maximum likelihood estimates of the signal parameters in question (cf. Sec. 20.4-8).

with respect to cost. Moreover, the error probabilities are also functions of the input signal and noise parameters, so that comparisons of performance can also be made in terms of such parameters as well, under a variety of conditions. We shall consider both these aspects in the following three sections.

19.3-1 Error Probabilities; Optimum Systems. Our first problem is to provide some way of determining the various error probabilities α, α', β, β', α^*, . . . that occur for Bayes, non-Bayes, and Minimax detection systems.

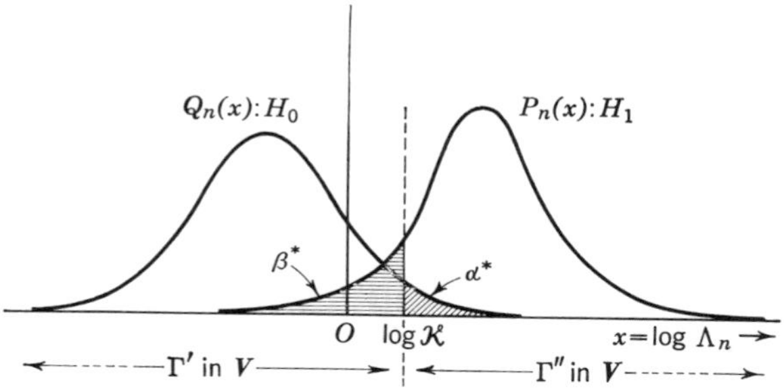

FIG. 19.2. Probability densities and decision regions for $x = \log \Lambda_n$.

We begin with the Bayes class and with $\delta^*(\gamma_0|\mathbf{V})$, $\delta^*(\gamma_1|\mathbf{V})$ determined according to Eqs. (19.13). With the help of the transformation $x = \boldsymbol{T}\{\mathbf{V}\} = \log \Lambda_n(\mathbf{V})$ and Secs. 1.2-14 and 17.2, we can write for the corresponding conditional class probabilities of the type I and II errors

$$\alpha^* = \int_{\log \mathcal{K}}^{\infty} dx \int_{\Gamma} F_n(\mathbf{V}|0)\,\delta[x - \log \Lambda_n(\mathbf{V})]\,\mathbf{dV} = \int_{\log \mathcal{K}}^{\infty} Q_n(x)\,dx \qquad (19.29a)$$

$$\beta^* = \int_{-\infty}^{\log \mathcal{K}} dx \int_{\Gamma} \langle F_n(\mathbf{V}|\mathbf{S})\rangle_S\,\delta[x - \log \Lambda_n(\mathbf{V})]\,\mathbf{dV} = \int_{-\infty}^{\log \mathcal{K}} P_n(x)\,dx \qquad (19.29b)$$

where the second relation in each case makes clear the definition of Q_n and P_n, and where Λ_n is given by Eq. (19.14). The mapping from $\mathbf{V}$- to x-space by means of the transformation $x = \log \Lambda_n$ is quite arbitrary; any monotonic function of $\Lambda_n(\mathbf{V})$ could be used. The logarithm, however, turns out to be particularly convenient when it becomes necessary to obtain α^* and β^* (or α, α', . . .) explicitly. [For other reasons–cf. Eq. (19.18)–the logarithm is also suggested.] We see that the δ function picks out that region Γ'' in $\mathbf{V}$-space for which $\delta(\gamma_1|\mathbf{V}) = 1$ in the case of α^*, with a corresponding interpretation for β^* when $\delta(\gamma_0|\mathbf{V}) = 1$. Figure 19.2 shows the two regions in x-space when x is the random variable $\log \Lambda_n$.

From the fact that $F_n(\mathbf{V}|0)$, $\langle F_n(\mathbf{V}|\mathbf{S})\rangle_S$ are probability densities and that $x = \log \Lambda_n(\mathbf{V})$ is also a random variable when considered over the ensemble of possible values $\mathbf{V}$, it follows that $Q_n(x)$, $P_n(x)$ are the probability densities of x with respect to the distributions appropriate to the hypotheses H_0 and H_1, respectively (cf. Sec. 17.2). To see this, observe that the characteristic functions under these two conditions are

$$F_1(i\xi)_Q = \mathbf{E}_{\mathbf{V}|H_0}\{e^{i\xi \log \Lambda_n(\mathbf{V})}\} = \int_\Gamma e^{i\xi \log \Lambda_n(\mathbf{V})} F_n(\mathbf{V}|0)\, d\mathbf{V} \tag{19.30a}$$

$$F_1(i\xi)_P = \mathbf{E}_{\mathbf{V}|H_1}\{e^{i\xi \log \Lambda_n(\mathbf{V})}\} = \int_\Gamma e^{i\xi \log \Lambda_n(\mathbf{V})} \langle F_n(\mathbf{V}|\mathbf{S})\rangle_S\, d\mathbf{V} \tag{19.30b}$$

from which it follows that

$$\begin{aligned} Q_n(x) &= \mathfrak{F}^{-1}\{F_1(i\xi)_Q\} = \int_{-\infty}^{\infty} e^{-i\xi x} \frac{d\xi}{2\pi} \int_\Gamma e^{i\xi \log \Lambda_n(\mathbf{V})} F_n(\mathbf{V}|0)\, d\mathbf{V} \\ &= \int_\Gamma F_n(\mathbf{V}|0)\,\delta[x - \log \Lambda_n(\mathbf{V})]\, d\mathbf{V} \end{aligned} \tag{19.31}$$

[cf. Eq. (19.29a)], from the integral form of the delta function. In a similar fashion, we have

$$P_n(x) = \mathfrak{F}^{-1}\{F_1(i\xi)_P\} = \int_\Gamma \langle F_n(\mathbf{V}|\mathbf{S})\rangle_S \delta[x - \log \Lambda_n(\mathbf{V})]\, d\mathbf{V} \tag{19.32}$$

As was shown in Prob. 17.8, P_n and Q_n are related by

$$P_n(x) = \frac{1}{\mu} e^x Q_n(x) \tag{19.32a}$$

which is sometimes useful in the explicit evaluation of error probabilities (cf. Sec. 20.4-1). Note also that $(\alpha')^*$ and $(\beta')^*$ [cf. Eqs. (19.10), (19.11) for Bayes tests] follow at once from Eqs. (19.29a), (19.29b) if we omit the average $\langle \quad \rangle_S$.

While Eqs. (19.29) hold for all (fixed-sample) Bayes systems, it is instructive to apply Eqs. (19.31), (19.32) to the Bayes subclass of Neyman-Pearson and Ideal Observer detectors. The error probabilities for each type are given specifically, with the help of Eqs. (19.23), (19.25), and (19.27), by

$$\alpha_{NP} = \int_{\log \mathcal{K}_{NP}}^{\infty} Q_n(x)\, dx \qquad \beta^*_{NP} = \int_{-\infty}^{\log \mathcal{K}_{NP}} P_n(x)\, dx \tag{19.33}$$

$$\alpha^*_I = \int_0^\infty Q_n(x)\, dx \qquad \beta^*_I = \int_{-\infty}^{0} P_n(x)\, dx \tag{19.34}$$

and the corresponding Bayes risks are calculated from Eqs. (19.21), (19.26), viz.,

$$R^*_{NP} = C_0\mathcal{R}^*_{NP} = C_0(p\beta^*_{NP} + \mathcal{K}_{NP} q\alpha_{NP}) \tag{19.35a}$$

$$R^*_I = C_0\mathcal{R}^*_I = C_0(q\alpha^*_I + p\beta^*_I) \qquad \mathcal{K}_I = 1 \tag{19.35b}$$

where C_0 is an appropriate cost scale factor for the problem in question. The first relation in Eqs. (19.33) establishes the explicit connection between the preset type I error probability and threshold characteristic of the Neyman-Pearson system; from this, $\mathcal{K}_{NP}$ is determined, giving β^*_{NP} according to the second relation. Finally, we observe that α^*, β^* are also functions of the a priori probabilities (p,q), and the parameters of signal and noise, through the structure $\log \Lambda_n$ of the optimum detector itself [cf. Eqs. (19.29)].

19.3-2 Error Probabilities; Suboptimum Systems. The approach of Sec. 19.3-1 is in no way restricted to optimum systems. For example, in the case of an actual, nonideal detection system, with threshold $\mathcal{K}'$ (implying at least a cost *ratio*), and with a structure represented by $G_n(\mathbf{V})$ [$\neq \Lambda_n(\mathbf{V})$ usually], the conditional probabilities of the type I and II errors become

$$\alpha = \int_{\log \mathcal{K}'}^{\infty} dy \int_{\Gamma} F_n(\mathbf{V}|0)\,\delta[y - \log G_n(\mathbf{V})]\,\mathbf{dV} = \int_{\log \mathcal{K}'}^{\infty} q_n(y)\,dy \tag{19.36a}$$

$$\beta = \int_{-\infty}^{\log \mathcal{K}'} dy \int_{\Gamma} \langle F_n(\mathbf{V}|\mathbf{S})\rangle_S\,\delta[y - \log G_n(\mathbf{V})]\,\mathbf{dV} = \int_{-\infty}^{\log \mathcal{K}'} p_n(y)\,dy \tag{19.36b}$$

and β' is again found by omitting the average $\langle\ \rangle_S$; $p_n(y)$ and $q_n(y)$ are now the distribution densities of $y = \log G_n(\mathbf{V})$, with respect to H_0, H_1. Their characteristic functions are specifically

$$F_1(i\xi)_{q_n} = E_{\mathbf{V}|H_0}\{e^{i\xi \log G_n(\mathbf{V})}\} = \int_{\Gamma} e^{i\xi \log G_n(\mathbf{V})} F_n(\mathbf{V}|0)\,\mathbf{dV} \tag{19.37a}$$

$$F_1(i\xi)_{p_n} = E_{\mathbf{V}|H_1}\{e^{i\xi \log G_n(\mathbf{V})}\} = \int_{\Gamma} e^{i\xi \log G_n(\mathbf{V})} \langle F_n(\mathbf{V}|\mathbf{S})\rangle_S\,\mathbf{dV} \tag{19.37b}$$

(Figure 19.2 applies here also, if we replace $\mathcal{K}$ by $\mathcal{K}'$, α^* by α, Q_n by q_n, etc.) The same sort of procedure for evaluating α, β is employed here as for α^*, β^* above; for details, see Sec. 17.2 and the examples in Chap. 20. With Eqs. (19.36), we are able to determine the average risk R [cf. Eqs. (19.9), (19.19*b*)] and so compare the performance of the nonoptimum system with that of the corresponding Bayes detector, as outlined below.

19.3-3 Decision Curves and System Comparisons. The relations (19.29) with the Bayes risk R^*, (19.19), and (19.36) with the average risk R enable us to compare the performance of actual and optimum systems for the same purpose and, of course, for the same input waves and noise statistics.

We note first that the error probabilities α^*, β^* and α, β, which appear in R^* and in R, are functions of a_0, the *input signal-to-noise* [(rms) amplitude] *ratio*† defined according to $a_0 = (\langle S^2\rangle / \langle V_N^2\rangle)^{1/2}$. Curves of average risk as a function of a_0 (or of any other pertinent signal parameter) are called *decision curves*, and it is in terms of these that specific system comparisons may be made. Figure 19.3 illustrates a typical situation, involving an ideal and an actual detection system for the same purpose. Thus, if we choose the same thresholds ($\mathcal{K} = \mathcal{K}'$) and assign the same costs [Eqs. (19.5)] to each possible decision, then the average risk R for all a_0 will exceed the corresponding Bayes risk R^*, as indicated.

Next, let us define a *minimum detectable signal*[1,2] *as that input signal-to-*

† Frequently, a_0 is random over the signal class, so that the appropriate ratio is $\bar{a}_0$, or $\overline{a_0^2}$, etc., depending on the system [cf. Eqs. (19.50), (19.56) and examples in Chap. 20].

noise ratio $(a_0)_{min}$ *which yields an average risk* R_0 *which is some specified fraction of the maximum average risk*, i.e., *the* a_0 *for which* $R_0 = \eta R_{max}$ $(0 < \eta < 1)$. (R_{max}, in physical situations at least, occurs for $a_0 = 0$.) System comparison can now be carried out in a variety of ways, of which the following are examples (cf. Fig. 19.3):

1. $(a_0)^*_{min\,\eta}$ versus $(a_0)_{min\,\eta}$ (in general, $R_0 \neq R_0^*$ for the same η)
2. $(a_0)^*_{min\,\eta_1}$ versus $(a_0)_{min\,\eta_2}$ (for $R_0^* = R_0''$, which determines η_1, η_2)
3. R_0^* versus R' [for the same $(a_0)^*_{min\,\eta}$]

Another decision curve, also useful for comparison, is the *betting curve*, introduced originally by Siegert,[10] which relates the probability $W_1(a_0;n)$ of a correct decision[21,22] to the input signal-to-noise ratio a_0. This is defined by

$$W_1(a_0;n) = 1 - (\alpha q + \beta p) \tag{19.38}$$

For optimum systems such as the Neyman-Pearson and Ideal Observer, we may replace α and β by the appropriate α^* and β^* [Eqs. (19.29)], since

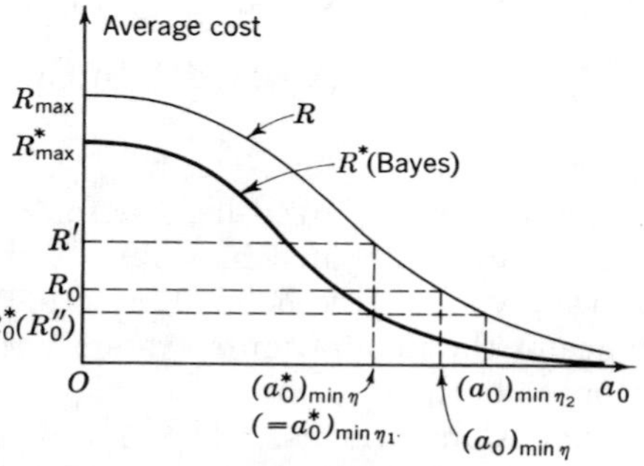

FIG. 19.3. Typical situations of comparison, showing average and Bayes risks and minimum detectable signals.

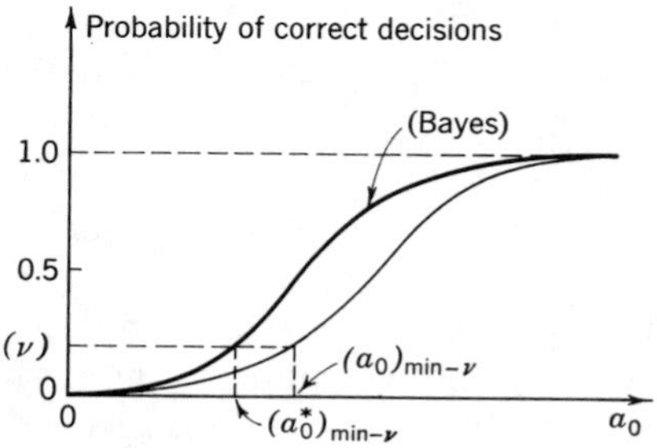

FIG. 19.4. Betting curves and associated mininum detectable signals.

these systems were shown in Secs. 19.2-1, 19.2-2 to be Bayes with suitable assumptions on the cost ratio $\mathcal{K}$. The conditional probabilities may be calculated as before [cf. Eqs. (19.21) and (19.36)] and may be used in a similar way for system comparison, with the minimum detectable signal defined now in terms of an input which leads to a given percentage ν of *successful decisions* at the output (cf. Fig. 19.4). Each point on the betting curve, considered as a function of α and β, corresponds to a point on the risk curve, which is also a function of these α, β. Thus, comparisons in terms of "success" are equivalent to those on a risk basis (except for the scale which sets the absolute cost).

An additional description and comparison of system performance, often of considerable interest, are given by the behavior of the minimum detectable signal as a function of the acquisition, or integration, time T ($\sim n$, the sample size). This relationship is found from the set of average risk curves

[Eqs. (19.9)] (cf. Fig. 19.3) or the betting curves [Eq. (19.38)], as n assumes all allowed values, as sketched in Fig. 19.5. The examples considered in Sec. 20.4 provide some further illustrations.

Still another variant of the general average risk curve is given by the probability of successfully deciding that a signal is present (i.e., the alternative hypothesis H_1), as a function of input signal-to-noise ratio or other significant system parameter. Thus, one may use the conditional probability $P(S|S) = 1 - \beta^*$, or the total probability $P(S) = pP(S|S) = p(1 - \beta^*)$, versus a_0, T (or n), etc., for optimum systems. Here, usually, α is fixed, so that a Neyman-Pearson system is essentially employed (cf. Sec. 19.2-1). For suboptimum systems, one has similarly $P(S|S) = p(1 - \beta)$, $P(S) = p(1 - \beta)$, versus a_0, etc., for the desired performance-curve and system comparisons.

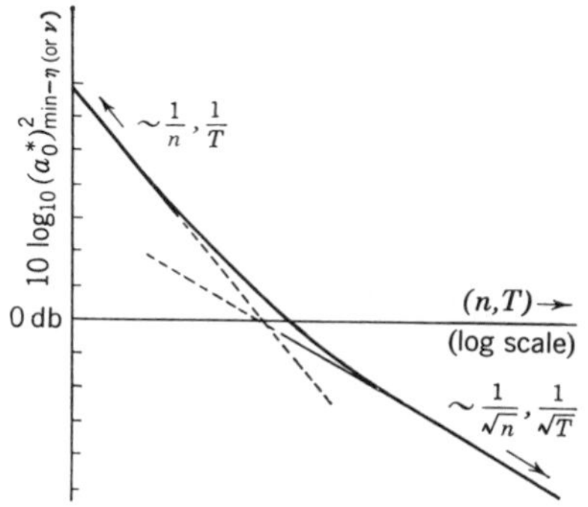

FIG. 19.5. Typical behavior of minimum detectable signal with data-acquisition time (for incoherent reception, cf. Sec. 19.4-3).

It is also sometimes convenient to use a *normalized decision curve* for defining minimum detectable signals and making system comparisons.[1,21,22] One way of doing this is to observe first that since $\alpha, \beta \rightarrow 0$ as $a_0 \rightarrow \infty$ [or as $T, n \rightarrow \infty$, a_0 fixed (>0)], inasmuch as the effective background noise vanishes (under these conditions vis-à-vis the desired signal), Eqs. (19.19a), (19.19b) become $R(\sigma,\delta) \rightarrow \mathcal{R}_0 = qC_{1-\alpha} + pC_{1-\beta}$. Furthermore, as $a_0 \rightarrow 0$, we have $R(\sigma,\delta)_{a_0=0} = qC_{1-\alpha} + pC_\beta$, $\mathcal{K}/\mu > 1$, or $qC_\alpha + pC_{1-\beta}$, $\mathcal{K}/\mu < 1$, and so we may define a normalized (Bayes) risk curve by

$$\mathcal{R}(\sigma,\delta) \equiv \frac{R(\sigma,\delta) - \mathcal{R}_0}{R(\sigma,\delta)_{a_0=0} - \mathcal{R}_0} = \begin{cases} \alpha \dfrac{\mathcal{K}}{\mu} + \beta & \dfrac{\mathcal{K}}{\mu} \geq 1 \\ \alpha + \dfrac{\mu}{\mathcal{K}} \beta & \dfrac{\mathcal{K}}{\mu} \leq 1 \end{cases} \qquad 0 \leq \mathcal{R} \leq 1 \quad (19.39)$$

Figure 20.8 is typical for threshold detection systems. However, since normalization is an arbitrary procedure, there is no unique or compelling general reason for using it. In any case, for system comparison care must be taken that common criteria be used under identical conditions. Often this may mean that comparisons should be made on the basis of the unnormalized or absolute risk curves instead, as normalization may sometimes disguise or diminish significant differences.

We remark, finally, that in the construction, operation, and evaluation of these binary detection processes we have assumed that the signal parameters, *or their average values,* if they are originally random, are known or preset beforehand from a decision curve and that they are then inserted into Λ_n [Eq. (19.14)] so that the scale of Λ_n can be fixed and an actual test [Eqs. (19.15)] carried out with the same parameter values. However, it may be that these "true" parameters, i.e., the values actually occurring when a signal is present or average values appropriate to the signal class in question, are not given beforehand, in which case the test of Eqs. (19.15) can still be carried out, but we are unable to specify the error probabilities α, β uniquely and so cannot determine the Bayes or average risk uniquely. We note some examples of this in the case of the Bayes sequential detectors referred to in Sec. 20.4-8. For the most part, however, it is not unrealistic to assume at least a knowledge of the required moments of the signal parameters or by some such process as Minimax, etc. (cf. Sec. 19.2-3), to define a class of Bayes receivers for the problem at hand which guards against least favorable situations in some operationally meaningful sense. Unless otherwise indicated, we shall assume henceforth that the appropriate statistics of the signal (and noise) parameters are specified and used.

19.4 *Structure; Threshold Detection*

In the preceding sections, we have discussed several detection procedures which are optimum under certain circumstances and have shown that all are basically variants on the Bayes decision rule and, hence, on the likelihood-ratio test. Physical systems realizing these procedures must therefore be receivers which execute likelihood-ratio tests. We have mentioned earlier that the processing of the data needed for the tests will in general be nonlinear. In this section, we shall explore some general properties of these data processes for cases of signal detection which are of central importance in practice. These are mainly the cases in which signal and noise are additive. In this respect, the discussion contained in the next two sections is somewhat more restricted than the preceding. We remark, however, that the discussion is also capable of direct generalization to include nonadditive communication situations, as some of the references in Chap. 20 indicate. We shall consider both discrete and continuous data-sampling techniques, and we shall emphasize the threshold condition of receiver performance, since this is almost always the critical state where optimum or near-optimum design is essential for effective operation.

19.4-1 Discrete Sampling.† We begin by limiting the scope of our argument to (conditionally) additive and independent noise, i.e., noise such that the received data $\mathbf{V}$ are given by

$$\mathbf{V} = \mathbf{S} + \mathbf{N}$$

or, in continuous form, by (19.40)

$$V(t) = S(t) + N(t)$$

Then, $F_n(\mathbf{V}|0) = W_n(\mathbf{V})_N$, and $F_n(\mathbf{V}|\mathbf{S}) = W_n(\mathbf{V} - \mathbf{S})_N$, where W_N is the distribution density of the accompanying noise. As we have seen from Eqs. (19.14) et seq. (and Sec. 19.3-1), the optimum binary detector is a computer which here has the structure

$$\boldsymbol{T}_R^{(N)}\{\mathbf{V}\}_{opt\,det} = \log \Lambda_n = \log \mu + \log \langle W_n(\mathbf{V} - \mathbf{S})_N\rangle_S - \log W_n(\mathbf{V})_N \tag{19.41}$$

where $\mu = p/q$.

Since we are concerned mainly with the problems of *threshold*, or weak-signal, detection, where the intensity of the accompanying noise is comparable to, or even much greater than, that of the signal (if present), we can consider the added contribution of this threshold signal as a perturbation of the background noise. This suggests the development of $\log \Lambda_n$ in a suitable power series in S, where the higher-order terms (in S) are progressively less significant. Accordingly, let us examine in more detail the manner in which the received data $\mathbf{V}$ are processed. We start by expanding $W_n(\mathbf{V})$ formally in a power series‡ about $\mathbf{V} = 0$,

$$W_n(\mathbf{V})_N = A^{(0)}[1 + \tilde{\mathbf{V}}\mathbf{a}^{(1)} + (\tilde{\mathbf{V}}\mathbf{a}^{(2)})^2 + (\tilde{\mathbf{V}}\mathbf{a}^{(3)})^3 + \cdots] \tag{19.42}$$

where $\mathbf{a}^{(m)}$ is a vertical array of components $a_j^{(m)}$ $(j = 1, \ldots, n)$ which, when juxtaposed, form a symmetrical m-index quantity

$$\underbrace{a_j^{(m)}a_k^{(m)} \cdots}_{m} = \frac{(A^{(0)})^{-1}}{m!}\left(\frac{\partial}{\partial V_j}\frac{\partial}{\partial V_k}\cdots\right)_m W_n(\mathbf{V})_N\bigg|_{V=0}$$

$$A^{(0)} = W_n(O)_N \tag{19.43}$$

Because of Eq. (19.41), it is convenient to write Eq. (19.42) in the form

$$W_n(\mathbf{V}) = \exp(B^{(0)}) \exp[\tilde{\mathbf{V}}\mathbf{b}^{(1)} + (\tilde{\mathbf{V}}\mathbf{b}^{(2)})^2 + (\tilde{\mathbf{V}}\mathbf{b}^{(3)})^3 + \cdots] \tag{19.44a}$$

† The present treatment generally follows the approach used by Middleton in some earlier studies.[21,2]

‡ $W_n(\mathbf{V})_N$, of course, is a random variable when considered over the data ensemble $\mathbf{V}$, so that termwise mean-square convergence of the series is required if we are to write a relation like Eq. (19.42) (cf. Sec. 2.1). This puts certain conditions on the nature of W_N, for example, the existence of the second moments of the various terms in the expansion, with respect to the null hypothesis H_0, that is, with respect to $W_n(\mathbf{V})_N$ itself, over the entire data space $(-\infty < \mathbf{V} < \infty)$. In the analysis, we shall assume that series like (19.42) are stochastically convergent, a condition which is easily verified for all the gauss or gauss-derived processes encountered in physical applications, as the material of Chap. 17 indicates.

in which $B^{(0)} = \log A^{(0)}$ and

$$\begin{aligned} b_j^{(1)} &= a_j^{(1)} \\ b_j^{(2)}b_k^{(2)} &= a_j^{(2)}a_k^{(2)} - \tfrac{1}{2}a_j^{(1)}a_k^{(1)} \\ b_j^{(3)}b_k^{(3)}b_l^{(3)} &= a_j^{(3)}a_k^{(3)}a_l^{(3)} - \tfrac{1}{3}a_j^{(2)}a_l^{(2)}a_k^{(1)} - \tfrac{1}{3}a_k^{(2)}a_l^{(2)}a_j^{(1)} \\ &\qquad + \tfrac{1}{3}a_j^{(1)}a_k^{(1)}a_l^{(1)}, \text{ etc.} \end{aligned} \tag{19.44b}$$

These are obtained by comparing the two developments (19.42) and (19.44). For additive signal and noise, we merely replace $\mathbf{V}$ by $\mathbf{V} - \mathbf{S}$ in the above, again with the assumption of stochastic convergence, now of the series expansion of $W_n(\mathbf{V} - \mathbf{S})_N$ with respect to H_0.

Regrouping into terms containing $\mathbf{V}$ alone, and those with $\mathbf{S}$, $\mathbf{S}^2$, etc., we find after taking the logarithm according to Eqs. (19.41), (19.44) that the more detailed structure of the optimum detector becomes finally (cf. Prob. 19.1)

$$\begin{aligned} \log \Lambda_n = \log \mu &+ [-\langle \tilde{\mathbf{S}}\mathbf{b}^{(1)}\rangle - 2(\tilde{\mathbf{V}}\mathbf{b}^{(2)})\langle \tilde{\mathbf{S}}\mathbf{b}^{(2)}\rangle - O(\mathbf{S};\mathbf{V}^2,\mathbf{V}^3, \ldots)] \\ &+ [\langle(\tilde{\mathbf{S}}\mathbf{b}^{(2)})^2\rangle + \tfrac{1}{2}\langle(\tilde{\mathbf{S}}\mathbf{b}^{(1)})^2\rangle - \tfrac{1}{2}\langle\tilde{\mathbf{S}}\mathbf{b}^{(1)}\rangle^2 \\ &+ 2(\tilde{\mathbf{V}}\mathbf{b}^{(2)})^2\langle(\tilde{\mathbf{S}}\mathbf{b}^{(2)})^2\rangle - 2(\tilde{\mathbf{V}}\mathbf{b}^{(2)})^2\langle\tilde{\mathbf{S}}\mathbf{b}^{(2)}\rangle^2 \\ &+ 2\langle(\tilde{\mathbf{S}}\mathbf{b}^{(1)})(\tilde{\mathbf{V}}\mathbf{b}^{(2)})(\tilde{\mathbf{S}}\mathbf{b}^{(2)})\rangle - 2\langle\tilde{\mathbf{S}}\mathbf{b}^{(1)}\rangle(\tilde{\mathbf{V}}\mathbf{b}^{(2)})\langle\tilde{\mathbf{S}}\mathbf{b}^{(2)}\rangle \\ &+ O(\mathbf{S}^2;\mathbf{V}, \ldots)] + O(\mathbf{S}^3) \end{aligned} \tag{19.45}$$

on omitting all terms in $\mathbf{b}^{(3)}$, $\mathbf{b}^{(4)}$, etc. Here, $O(\mathbf{S}^2;\mathbf{V}^3, \ldots)$ signifies that the next term involves $\mathbf{S}$ to the second order and $\mathbf{V}$ to the third and higher orders, etc. [Note that both the first and second (and higher) orders of S generally involve all $\mathbf{b}^{(m)}$.]

For the noise characteristic of so many physical situations,[28–30] we may write specifically the *normal* density [cf. Eq. (7.31)]

$$W_n(\mathbf{V})_N = (2\pi)^{-n/2}(\det \mathbf{K})^{-1/2} \exp\left[-\tfrac{1}{2}(\tilde{\mathbf{V}} - \tilde{\bar{\mathbf{V}}}_N)\mathbf{K}^{-1}(\mathbf{V} - \bar{\mathbf{V}}_N)\right] \tag{19.46}$$

where $\mathbf{K}$ is the covariance matrix $\|\overline{V_iV_j} - \bar{V}_{Ni}\bar{V}_{Nj}\|$ and $\bar{\mathbf{V}}_N = \bar{\mathbf{V}}$. Comparison with Eq. (19.44a) gives directly

$$\begin{aligned} B^{(0)} &= \tfrac{1}{2}\tilde{\bar{\mathbf{V}}}_N\mathbf{K}^{-1}\bar{\mathbf{V}}_N - \tfrac{1}{2}\log[(2\pi)^n \det \mathbf{K}] \\ \mathbf{b}^{(1)} &= \mathbf{K}^{-1}\bar{\mathbf{V}}_N \\ \mathbf{b}^{(2)}\tilde{\mathbf{b}}^{(2)} &= -\tfrac{1}{2}\mathbf{K}^{-1} \\ \mathbf{b}^{(m)} &= 0 \qquad m \geqq 3 \end{aligned} \tag{19.47}$$

The structure of the optimum detector [Eq. (19.45)] accordingly becomes

$$\begin{aligned} \log \Lambda_n = \log \mu &+ (\tilde{\mathbf{V}}\mathbf{K}^{-1}\langle\mathbf{S}\rangle - \tilde{\bar{\mathbf{V}}}_N\mathbf{K}^{-1}\langle\mathbf{S}\rangle) \\ &+ [-\tfrac{1}{2}\langle\tilde{\mathbf{S}}\mathbf{K}^{-1}\mathbf{S}\rangle + \tfrac{1}{2}\langle(\tilde{\mathbf{S}}\mathbf{K}^{-1}\bar{\mathbf{V}}_N)^2\rangle - \tfrac{1}{2}\langle\tilde{\mathbf{S}}\mathbf{K}^{-1}\bar{\mathbf{V}}_N\rangle^2 \\ &- \langle(\tilde{\mathbf{S}}\mathbf{K}^{-1}\bar{\mathbf{V}}_N)(\tilde{\mathbf{V}}\mathbf{K}^{-1}\mathbf{S})\rangle + \langle\tilde{\mathbf{S}}\mathbf{K}^{-1}\bar{\mathbf{V}}_N\rangle\langle\tilde{\mathbf{V}}\mathbf{K}^{-1}\mathbf{S}\rangle \\ &+ \tfrac{1}{2}\langle(\tilde{\mathbf{V}}\mathbf{K}^{-1}\mathbf{S})^2\rangle - \tfrac{1}{2}\langle\tilde{\mathbf{V}}\mathbf{K}^{-1}\mathbf{S}\rangle^2] + O(S^3) \end{aligned} \tag{19.48}$$

which is exact through $O(S^2)$. For another example, see Prob. 19.2.

The general detection computer, as can be seen from the formula (19.45), usually requires nonlinear, as well as linear, operations on its input $\mathbf{V}$. Since we are concerned here with threshold operation, we expect terms of higher order in the signal to become less significant: it is only the terms $O(S)$ and $O(S^2)$ that are of primary importance in describing the weak-signal structure of the system. This effect can be more readily assessed if a normalization is introduced. We first normalize the noise by measuring it in terms of multiples of its rms value $\psi^{1/2}$. For the signal, we proceed by defining a "mean amplitude" A_0 over the sampling interval $(t_0, t_0 + T)$ according to

$$A_0^2/2 = (1/T) \int_{t_0}^{t_0+T} S^2(t)\, dt \tag{19.49a}$$

and normalize the signal to S by putting $a_0 s = S/\psi^{1/2}$, where $a_0 \equiv A_0/\sqrt{2\psi}$. Thus, we write

$$V(t) = \sqrt{\psi}\, v(t) \qquad S(t) = \sqrt{\psi}\, a_0 s(t) \tag{19.49b}$$

Here, a_0 is recognized as an input signal-to-noise ratio where for the moment we consider that the signal is "on" during this acquisition period $(t_0, t_0 + T)$ (cf. remarks in Sec. 19.4-3). If this notation is now introduced into Eq. (19.45), the various terms can be ordered by ascending powers of a_0. We find more particularly that†

$$\begin{aligned}\log \Lambda_n = \log \mu &+ \bar{a}_0[\langle -A_n^{(1)}(s)\rangle - 2\langle A_n^{(2)}(v,s)\rangle + O(s;v^2, \ldots)] \\ &+ \overline{a_0^2}[\langle A_n^{(2)}(s,s)\rangle + \tfrac{1}{2}\langle A_n^{(1)}(s)^2\rangle - \tfrac{1}{2}\langle A_n^{(1)}(s)\rangle^2(\bar{a}_0^2/\overline{a_0^2}) \\ &+ 2\langle A_n^{(1)}(s)A_n^{(2)}(v,s)\rangle - 2\langle A_n^{(1)}(s)\rangle\langle A_n^{(2)}(v,s)\rangle(\bar{a}_0^2/\overline{a_0^2}) \\ &+ 2\langle A_n^{(2)}(v,s)^2\rangle - 2\langle A_n^{(2)}(v,s)\rangle^2(\bar{a}_0^2/\overline{a_0^2}) + O(s^2;v^3, \ldots)] \\ &+ O(\overline{a_0^3})\end{aligned} \tag{19.50}$$

where we have put

$$\begin{aligned}A_n^{(1)}(s) &= \sqrt{\psi}\, \tilde{\mathbf{s}}\mathbf{b}^{(1)} \\ A_n^{(2)}(v,s) &= \psi(\tilde{\mathbf{v}}\mathbf{b}^{(2)})(\tilde{\mathbf{s}}\mathbf{b}^{(2)}) = \psi\tilde{\mathbf{v}}\mathbf{b}^{(2)}\tilde{\mathbf{b}}^{(2)}\mathbf{s} \\ A_n^{(2)}(s,s) &= \psi(\tilde{\mathbf{s}}\mathbf{b}^{(2)})^2\end{aligned} \tag{19.50a}$$

The expression $A_n^{(2)}(v,s)$ can be interpreted as an averaged, general cross correlation of the signal with the observed data, and $A_n^{(2)}(s,s)$ as the corresponding averaged, generalized autocorrelation of the signal with itself. Carrying the explicit representation of Eqs. (19.50*a*) still further, we obtain the averaged higher-order cross- and autocorrelations, such as $A_n^{(3)}(v^2,s)$, $A_n^{(4)}(v^3,s)$, etc. Because of the fact that the signal is weak, on the average, in the observation period $(t_1, t_1 + T)$, most of these higher-order contributions can be neglected or replaced by simple approximations, the precise

† In the expansion (19.50), we assume that the amplitude A_0 of the signal is independent of the other signal parameters. This is usually a reasonable assumption, except in certain radar problems, where a target may be located anywhere in a large range interval. Then both the amplitude and the phase delay of the return depend on the same (random) parameter, as we shall see in the example of Sec. 20.4-6. In such instances, we replace $\bar{a}_0\langle A_N^{(1)}(s)\rangle$ by $\langle a_0 A_n^{(1)}(s)\rangle$, etc., in the general development (cf. Prob. 20.2).

nature of which, of course, depends on the statistics of the noise and the way in which it appears with the signal. We shall return to this point again in Sec. 19.4-3 before considering some specific examples.

19.4-2 Continuous Sampling. When continuous data processing is used in place of discrete sampling methods, it is reasonable to expect that the general methods of statistical inference (here, testing statistical hypotheses) can still be applied, provided that the appropriate limits exist as the number of sample points n becomes infinite in the fixed observation interval $(0,T)$. Then, quadratic forms like $Q_n = \tilde{\mathbf{z}}\mathbf{M}^{-1}\mathbf{y}$ become quadratic functionals, $Q_T = \int_{0-}^{T+} z(t)X(t)\,dt < \infty$, and the general likelihood ratio $\Lambda_n(\mathbf{V})$ is transformed into the corresponding *likelihood ratio functional* $\lim_{n\to\infty} \Lambda_n(\mathbf{V}) = \Lambda_T[V(t)]$, with $0 < \Lambda_T < \infty$. Statistical tests are carried out as before [cf. Eqs. (19.15), (19.23), (19.28)], with Λ_n now replaced by Λ_T. The associated error probabilities and Bayes risks are likewise determined as the limits of the corresponding quantities in the discrete case; i.e., quadratic forms are replaced by quadratic functionals, and so on. When this occurs we have what is called the *regular*† case, which fortunately is the physically significant situation in applications, as we shall note presently. However, under certain conditions the limits of quadratic forms as $n \to \infty$ may not exist, and similarly the likelihood-ratio functional Λ_T itself may become zero or infinite. Then we have the *singular* case† where decisions may be made with probability unity on the basis of arbitrarily small samples $(T > 0)$, unlike the practical situation where the probabilities of incorrect decisions are always greater than zero (for finite samples, $T < \infty$). Here we shall merely outline some of the steps in the passage to the limit and the conditions leading to both the regular and singular cases, referring the reader to the appropriate references‡ for the analytical details (cf. Prob. 19.4).

Let us begin with the various quadratic forms that appear in the threshold expansion§ of the likelihood ratio for signals and a general additive noise process (not necessarily normal) [cf. Eq. (19.45)]. As we have seen above [cf. Eq. (19.50)], the optimum threshold detector structure is represented by a number of such quadratic forms, for the most part of the type

$$Q_n = \tilde{\mathbf{z}}\mathbf{M}^{-1}\mathbf{y} \ (\geq 0) \tag{19.51}$$

where $\mathbf{y}$ and $\mathbf{z}$ are column vectors and $\mathbf{M}$ (and hence $\mathbf{M}^{-1}$) is a symmetrical nonsingular $n \times n$ matrix. The desired limit $n \to \infty$ (T fixed) of Q_n for continuous sampling may be found formally as follows:¶

† Following Grenander[31] (see especially secs. 4.2, 4.4, 4.7).

‡ A rigorous treatment is given by Grenander, *ibid.*, chap. 4, and the examples in secs. 4.4, 4.7. See also Davenport and Root,[33] pp. 340–343.

§ For a discussion of the higher-order terms, see Sec. 19.4-3.

¶ See Reich and Swerling[18] for an alternative method and also Prob. 19.3.

Starting with $\tilde{\mathbf{z}}\mathbf{M}^{-1}\mathbf{y}$, we first write

$$\mathbf{x} \equiv \mathbf{M}^{-1}\mathbf{y}$$

and therefore

$$Q_n = \tilde{\mathbf{z}}\mathbf{x} = \tilde{\mathbf{x}}\mathbf{z} \tag{19.52}$$

We assume that $y(t)$ at least and a finite number (>0) of its derivatives are bounded and continuous functions of t and that $z(t)$ is a bounded continuous function of t (with the additional requirement in many instances that $\dot{z}$, $\ddot{z}$, etc., exist at $t = 0, T$). (If y or z, or both, are stochastic processes, these same properties are assumed mean square.) The sum $\tilde{\mathbf{x}}\mathbf{z} = \sum_i^n x(t_i)z(t_i)$ can be represented by a Stieltjes integral

$$\int_{0-}^{T+} z(t)\, d\Gamma_n(t) = \sum_i^n x(t_i)z(t_i) = Q_n \tag{19.53}$$

where $\Gamma_n(t)$ is a step function with jump $x(t_i)$ at t_i and $\Gamma_n(t_i+) - \Gamma_n(t_i-) = x(t_i)$ $(i = 1, \ldots, n)$. Now, if $\lim_{n\to\infty} d\Gamma_n(t) = d\Gamma(t)$ exists, and consequently if $d\Gamma(t) = X(t)\,dt$ in the Riemannian sense, where at most $X(t)$ contains a finite number of delta functions and its derivatives (at $t = 0, T$ here), we have formally for Eq. (19.53)

$$\begin{aligned} \lim_{n\to\infty} Q_n \equiv Q_T &= \int_{0-}^{T+} z(t)\, d\Gamma(t) = \int_{0-}^{T+} z(t)X(t)\,dt \\ &= \int_{-\infty}^{\infty} z(t)X_T(t)\,dt \end{aligned} \tag{19.54}$$

where $X_T = X(t)$ $(0- < t < T+)$ and $X_T = 0$ outside this region. Here, X_T is the solution of the continuous limit of $\mathbf{y} = \mathbf{Mx}$ [cf. Eqs. (19.52)], e.g., of the inhomogeneous Fredholm integral equation (of the first kind)

$$\int_{0-}^{T+} M(t,u)X_T(u)\,du = y(t) \qquad 0- < t < T+ \tag{19.55}$$

with $M(t,u)$ $[= M(u,t)]$ the limiting form of a general element of the matrix M_{kl} (with $0- \leq t, u \leq T+$). For stochastic processes $z(t)$, or $X_T(t)$, $y(t)$, etc., Eqs. (19.54), (19.55) are, of course, to be understood in their mean-square sense (cf. Sec. 2.1-2).

In applications, the main task [apart from solving Eq. (19.55)] is to show that $\lim_{n\to\infty} d\Gamma_n(t) = X_T(t)\,dt$, with $X_T(t)$ a unique and finite solution (except for a finite number of delta functions and their derivatives). Such solutions require a suitably rapid and continuous approach of the matrix elements in a typical mesh of adjacent elements to a definite limiting value. For matrices of the form $M_{kl} = e^{-\alpha\delta|k-l|}$ $(t_k = k\delta$, etc.), for example, a direct evaluation of $\mathbf{x} = \mathbf{M}^{-1}\mathbf{y}$ with the expansion of y_{k+1} about t_k, in the limits $\delta \to 0$, $u \to \infty$, can be shown[18] to give Q_T, as obtained alternatively from

Eqs. (19.54), (19.55) [cf. Prob. 19.3 and Eq. (A.2.48)]. In fact, if we use the orthogonal expansion techniques of Sec. 8.2 and the results of Prob. 19.3b, a unique solution X_T to Eq. (19.55) (of integrable square on 0, T) can be shown to occur if (and only if) the kernel $M(t,u)$ is positive definite and the series $\sum_{m=1}^{\infty} b_m^2/\lambda_m^2$ converges, where the λ_m are the eigenvalues of the related homogeneous Fredholm integral equation (of the second kind):

$$\int_0^T M(t,u)\phi_m(u)\,du = \lambda_m\phi_m(t) \qquad 0 \le t \le T \tag{19.55a}$$

Because of the positive definiteness of $M(t,u)$, the eigenfunctions $\{\phi_m\}$ form a complete orthonormal set, thus ensuring the uniqueness of X_T.†

With the above in mind, let us now apply Eqs. (19.53), (19.54) to the threshold development [Eq. (19.50)] of the optimum detector characteristic to obtain instead of log $\Lambda_n(\mathbf{V})$ the likelihood ratio functional

$$\begin{aligned}\lim_{n\to\infty} \log \Lambda_n &= \log \Lambda_T[V(t)] \\ &= \log \mu + \bar{a}_0[\langle -A^{(1)}(s)\rangle - 2\langle A^{(2)}(v,s)\rangle + O(s;v^2, \ldots)] \\ &\quad + \overline{a_0^2}[\langle A^{(2)}(s,s)\rangle + \tfrac{1}{2}\langle A^{(1)}(s)^2\rangle - \tfrac{1}{2}\langle A^{(1)}(s)\rangle^2(\bar{a}_0^2/\overline{a_0^2}) \\ &\quad + 2\langle A^{(1)}(s)A^{(2)}(v,s)\rangle - 2\langle A^{(1)}(s)\rangle\langle A^{(2)}(v,s)\rangle(\bar{a}_0^2/\overline{a_0^2}) \\ &\quad + 2\langle A^{(2)}(v,s)^2\rangle - 2\langle A^{(2)}(v,s)\rangle^2(\bar{a}_0^2/\overline{a_0^2}) \\ &\quad + O(s^2;v^3, \ldots)] + O(\overline{a_0^3})\end{aligned} \tag{19.56}$$

where specifically from Eqs. (19.50a) we have

$$A^{(1)}(s) = \psi^{1/2}\int_{0-}^{T+} s(t-\epsilon;\boldsymbol{\theta})\,d\Gamma^{(1)}(t) \tag{19.57a}$$

where $\Gamma^{(1)}(t+) - \Gamma^{(1)}(t-) = b^{(1)}(t)$ [cf. Eqs. (19.53) et seq.] and $A^{(1)}(s)$ is assumed to exist if $d\Gamma^{(1)}(t) = X_T^{(1)}(t)\,dt$ and $X_T^{(1)}$ can be regarded as the solution of an equation like (19.55). Similarly, we have

$$\begin{aligned} A^{(2)}(v,s) &= \psi\int_{0-}^{T+} v(t)X_T(t;\epsilon,\boldsymbol{\theta})\,dt \\ A^{(2)}(s,s) &= \psi\int_{0-}^{T+} s(t-\epsilon,\boldsymbol{\theta})X_T(t;\epsilon,\boldsymbol{\theta})\,dt\end{aligned} \tag{19.57b}$$

Here, $b^{(1)}(t) = \lim_{n\to\infty} \mathbf{b}^{(1)}$ and $M(t,u) = \lim_{n\to\infty} (\mathbf{b}^{(2)}\tilde{\mathbf{b}}^{(2)})^{-1}$, with $y(t)$ equal to the normalized signal $s(t-\epsilon;\boldsymbol{\theta})$ and $X(t,\epsilon;\boldsymbol{\theta})$ the solution of Eq. (19.55) in this instance. As before (cf. Sec. 18.3-1), $\boldsymbol{\theta}$ represents all random parameters of the signal set not previously indicated explicitly, i.e., here all parameters except a_0 and the epoch ϵ, over which the statistical average is to be performed according to Eq. (19.56).

As we have already remarked, the likelihood ratio $\Lambda_n(\mathbf{V})$ in the limit $n \to \infty$ may converge (in probability) to a zero or infinite value, leading to the singular case where correct decisions can theoretically be made with

† See Sec. 8.2, references therein, Probs. 8-8, 8-9, and also Ref. 33, sec. A.2-2.

arbitrarily small samples ($T > 0$). This, of course, contradicts our experience. Such behavior is basically a result of inappropriate assumptions in the mathematical model concerning the precision to which the received data can be processed, the a priori information available to the observer about the background noise process, and in certain instances (e.g., stochastic signals) the signal process as well. It is accordingly important to be able to show in practical applications that our analytical models of actual detection situations do indeed yield the regular case in accordance with experience, where the likelihood ratio converges in probability to a likelihood functional Λ_T which is (almost) never zero or infinite for finite samples ($T < \infty$), and where, consequently, the probabilities of decision error are nonvanishing.

Since the threshold structure of the likelihood ratio is governed by quadratic forms [cf. Eq. (19.50)], it is sufficient to consider the resulting quadratic functionals of the terms in Eq. (19.56). We distinguish two main situations: (1) deterministic signals and noise, and (2) stochastic signals in noise. For the former, when the noise is gaussian and the signal completely deterministic, it can be shown† that if X_T is of integrable square on $(0,T)$ and is a unique solution of Eq. (19.55) then it can be used in Eq. (19.56) to give the logarithm of the likelihood-ratio functional, which is accordingly seen to be bounded ($-\infty < \log \Lambda_T < \infty$). Conversely, if Λ_T is bounded, the various X_T's in the quadratic functionals representing Λ_T are quadratically integrable unique solutions of integral equations of the type (19.55). The necessary and sufficient conditions for this are the positive definiteness of the kernel $M(t,u)$ and the quadratic integrability of X_T. These are conditions‡ which are always satisfied when the input noise is filtered white noise, and the signal function [$y(t)$ here] is itself quadratically integrable, with a suitable number of bounded derivatives at $t = 0$, T, the number depending on the precise nature of the kernel (cf. Sec. A.2.3). Similarly, one can apply the same argument in the threshold case when the signal is a deterministic random process, even when the noise process is not gaussian: the test is regular when the appropriate integral equations possess unique solutions.§

For purely stochastic signals, the details of the approach must be somewhat modified. In the case of gaussian signals, regularity can be established for most practical situations (cf. Prob. 19.4). Thus, for a colored noise signal in a white noise background vs. white noise alone, it is necessary and sufficient that the covariance of the signal process be positive definite on the observation interval; a more complicated condition is invoked

† See Grenander,[31] sec. 4.6, and also Ref. 33, p. 343.

‡ The former ensures the completeness of the set of eigenfunctions in the related homogeneous integral equation (19.55a), which, in conjunction with the latter, ensures the uniqueness of the solution of the corresponding inhomogeneous integral equation (19.55).

§ For an example of the singular case with completely deterministic signals and normal noise, see Grenander,[31] sec. 4.4, and Davenport and Root,[33] pp. 340–343.

for colored noise backgrounds (cf. Prob. 19.4). However, if the noise and signal processes here have the *same* covariance functions (except for a possible difference in scale), the singular case results.† Such a model is, of course, not a suitable representation of the actual state of affairs. The difficulty in the singular case arises, first, because too rigid demands are made on the observer's a priori knowledge of the noise spectra—it is assumed that he knows spectral behavior *everywhere*, even for infinite frequencies ($f \to \infty$)—and, second, because it is supposed that he can process the received data to an arbitrary degree of refinement in going from discrete to continuous sampling procedures.‡

In the regular case, on the other hand, these assumptions can be safely treated as the kind of approximation which one ordinarily makes in setting up a model of a physical event: for example, the assumed spectra, while not strictly identical with the true distributions, are close enough where it counts, e.g., in the spectral regions, where most of the signal and noise energy is located. Moreover, the structure of the detection system here, i.e., the various realizable linear filters and nonlinear devices needed to carry out the indicated data processing (e.g., the integrations, rectifications, etc.), can be approximated with existing equipment, theoretically to any desired degree of accuracy if enough equipment and care are exercised. Finally, we remark that for all the detection models described subsequently (cf. Chap. 20), with either deterministic or stochastic signal processes, regularity is seen to apply, as a consequence of our choice of covariance functions and signal structures, so that these models can be safely regarded as valid representations of the corresponding physical situations.

19.4-3 General Remarks on Optimum Threshold Detection. Several important practical conclusions can be drawn from the preceding developments in their normalized forms [Eqs. (19.50), (19.56)]. Apart from the general threshold condition of operation, these conclusions depend on the statistical character of the epoch ϵ, which [cf. Fig. 18.2 and Sec. 18.2-1(4)] relates signal waveform to the observer's time scale. We distinguish three cases: (1) ϵ is known a priori; i.e., there is no *sample uncertainty;* (2) ϵ is uniformly distributed in a given interval; and (3) ϵ has some more general distribution over the interval. These last represent states of sample uncertainty, of course, since the observer does not know definitely how his received data sample is related to (any) specified reference time point of the possible signal. The first case is sometimes called *coherent*, or *synchronous*, detection, while the latter are termed *incoherent*. When we use the term coherent here, sample certainty is always understood, while incoherent, or asynchronous, always implies sample uncertainty, even though the signal

† See Grenander,[31] sec. 4.7, and Prob. 19.4.

‡ A condition for determining when the singular case arises for stationary normal noise signals in normal noise is that the ratio of the spectra ($S + N$ versus N) be different from unity as $f \to \infty$ (see Slepian[34]).

process itself may be "coherent" in the engineering sense of maintaining a definite phase structure with time, the so-called coherent pulsed Doppler or cw radar systems, for example.

In the first case, i.e., for *coherent* detection (e.g., with *sample certainty*), apart from random parameters the signal is a known function of time $s(t)$ with known epoch ϵ throughout the sampling interval. For such a signal, $\langle S \rangle$ will not in general vanish,[21] and Eq. (19.45) will therefore usually contain a term proportional to $\bar{a}_0$. Since $\bar{a}_0 \sim \sqrt{\overline{a_0^2}}$, we in fact have directly the important result for both discrete and continuous sampling that

> *Optimum coherent (binary) threshold detection of a weak signal in additive noise depends essentially on the square root of the input signal-to-noise intensity ratio $\overline{a_0^2}$, and the detector itself requires an averaged, generalized cross correlation of the received data with the signal* (19.58)

The averaged, generalized cross correlations $A^{(2)}(v,s)$, $A^{(3)}(v^2,s)$, . . . are the operations which must be performed by the ideal detection computer. Higher-order terms, such as $A^{(3)}(v^2,s)$, $A^{(4)}(v^3,s)$, . . . can often be neglected so that the single averaged, weighted cross correlation $A^{(2)}(v,s)$ is a good approximation to the operation required of the detection computer. The constant terms $\log \mu$, $A^{(1)}(s)$, etc., in Eqs. (19.50), (19.56) are "biases" which must be retained in the computation for threshold comparison [cf. Eqs. (19.15)].

The second type of signal, which is more often encountered in practice, involves *sample uncertainty*. This incoherent signal is again assumed to be a known function of time $s(t - \epsilon; \boldsymbol{\theta})$ with random epoch ϵ. For a wide class of problems, this epoch is uniformly distributed† over some specified interval (T_0), with density $1/T_0$, so that $\langle S \rangle$ vanishes. The principal contribution to $\log \Lambda - \log \mu$ is then proportional to $\overline{a_0^2}$. We may say accordingly that

> *Optimum incoherent (binary) threshold detection of a signal in additive noise depends essentially on the first power of the input signal-to-noise intensity ratio $\overline{a_0^2}$, and the detector itself requires an averaged, generalized autocorrelation of the received data with itself*‡ (19.59)

Moreover, this contribution is made up of terms of the form $(vB^{(1)}v)$, $(v^2B^{(2)}v)$, etc., where the quantities $B^{(1)}$, $B^{(2)}$ depend alone on the noise and the various autocorrelation functions of the signals. As before, higher-order terms in a_0 are usually negligible or replaceable by suitable averages,

† This corresponds to the greatest average uncertainty in ϵ over T_0, for a maximum value limitation (cf. Prob. 6.7*a*). Here T_0 is usually large vis-à-vis the mean period of the signal.

‡ True in most practical cases, when $A^{(1)} = 0$, cf. (19.50), (19.56).

so that the optimum detector computes the simple averaged, weighted autocorrelation $(vB^{(1)}v)$. Incoherence, for example, occurs when the "fine structure," or phases, of the carrier in a modulated wave is unknown or when the signal's epoch ϵ in broadband cases has a uniform distribution, etc.

Other possibilities, e.g., (3) above, arise when ϵ is not uniformly distributed or where both coherent and incoherent operation can occur because of the nature of the signal. These we term *semicoherent*, or *semisynchronous*, detection, and now both the contributions in $\sqrt{\overline{a_0{}^2}}$ and $\overline{a_0{}^2}$ must be included in threshold operation. Sample uncertainty for a signal with a d-c component, for example, provides a simple illustration of this mixed situation, for while the a-c part is observed incoherently (ϵ uniformly distributed), the d-c, or steady, component by definition is always coherent with respect to the observer, so that terms linear in $\bar{a}_0$ (for example, $\sim \sqrt{\overline{a_0{}^2}}$) enter as well.

The result (19.59) is a general statement of what in communication theory is often called the *law of signal suppression*, whereby in incoherent operation a weak signal (in additive noise) at the receiver's input is made even weaker at the output; input performance is proportional to $\sqrt{\overline{a_0{}^2}}$, while at the output it is proportional to $\overline{a_0{}^2}$. In this latter case, detection depends primarily on the amount of signal vs. noise *energy* received during the observation period, while for coherent detection, where output and input both are $\sim \sqrt{\overline{a_0{}^2}}$, it is the *waveform of the signal*, in addition to its energy content, which is critical. In the former case, fine-structure phase information is unavailable, as indicated by the dependence on the various signal autocorrelation functions, whereas, for the latter, the detailed time structure, or *amplitude* spectrum, of the signal, with its attendant phase information, is made use of. Earlier examples of signal suppression [cf. Secs. 13.2-1(4), 15.5-4, for example], based on calculations of the output signal-to-noise ratio, have been shown for particular receivers (AM and FM) and gaussian noise. Our expression (19.59) is the generalization of these earlier results,† for all optimum threshold detection systems operating in additive noise, in the following respects: (1) no particular system is postulated beforehand; (2) all available information is used optimally; and (3) the complete signal and noise structures are included. Note, finally, that in threshold cases it is the distribution of the background noise which dominates receiver structure, while for strong signals the signal statistics are the controlling factor, with the noise as a small perturbation.

We should point out that in our preceding discussion we have assumed that the signal is on (if present) during the entire observation period. However, when this is not the case, it is easy to see that no improvement in performance can be expected for data acquisition during intervals longer

† This includes correlation detectors also[27–29] for both coherent and incoherent reception.

than that of the signal duration. This is evident from the above, since *a detector is in essence an energy-measuring device, which discriminates between the two states of signal and noise, and noise alone, on the basis of the difference in energy which these two states represent.* Consequently, longer observation periods merely add more noise energy, without a corresponding gain in signal energy. The maximum effective observation time T, therefore, is simply equal to the over-all duration of the signal. The precise fashion in which the optimum detector processes the received data $V(t)$ during this interval, of course, is embodied in the general detector structure [Eqs. (19.50), (19.56)]. Just how this can be accomplished by the various classes of system components in common use, e.g., linear and nonlinear filters, etc., is illustrated by the specific examples considered in Chap. 20.

Finally, there remains the question of the higher-order terms in the representation of the optimum threshold detectors. As we have noted, these involve averaged higher-order correlations between signal and received data. Instead of expressions quadratic in $V(t)$ or $S(t)$, we have quantities quadratic in V^2 or S^2, and so on. These can be replaced by appropriate averages, with respect to H_0 (noise alone) in the present threshold situation, which appear then as additional bias terms. The resulting nearly optimum system is very close to the true one but becomes less so, of course, for stronger signals, where the higher-order structure terms become significant. However, for most applications it is only at threshold that one worries about optimum performance, for although the optimum (or approximate near-optimum) threshold system is no longer optimum for stronger signals, its performance is usually still quite satisfactory, on an absolute scale, so that by designing the system for best threshold operation we almost always ensure adequate performance at the higher levels.

For acceptably small probabilities of decision error, weak signals require long integration times, while strong ones need much less. Thus, *threshold detection is in most cases a large-sample theory with asymptotically normal statistics* [which follows whenever the signal process possesses incoherent structure in $(t_1, t_1 + T)$, on application of the central-limit theorem]. It is this fact which makes possible the explicit analytic treatment of practical receiver problems. A general quantitative estimate of the convergence properties of our expanded detector structures [Eqs. (19.50), (19.56)] does not appear practicable, but in specific cases it is always possible to obtain useful estimates of the way in which the nearly optimum system approaches its limiting threshold performance, as we shall see from a number of the examples considered later.

PROBLEMS

19.1 (*a*) Carry out the steps from Eqs. (19.41) to (19.45), obtaining $\log \Lambda_n$ as indicated.

(*b*) Obtain all terms through $O(S^4)$ in Eq. (19.45).

19.2 For the case of the n-fold Rayleigh distribution with independent samples $\mathbf{E}$ of the envelopes of narrowband normal noise, and signal and noise,

$$
\begin{aligned}
W_n(\mathbf{E})_N &= \prod_{l=1}^{n} \psi^{-1} E_l e^{-E_l^2/2\psi} \\
\langle W_n(\mathbf{E})_{S+N} \rangle_S &= \left\langle \prod_{l=1}^{n} \psi^{-1} E_l e^{-E_l^2/2\psi - S_l^2/2\psi} I_0(S_l E_l/\psi) \right\rangle_S \qquad E_l \geqq 0
\end{aligned} \tag{1}
$$

obtain the following structure of the optimum threshold detector:

$$
\begin{aligned}
\log \Lambda_n = \log \mu &+ \left(-\frac{1}{2\psi} \langle \tilde{\mathbf{S}}\mathbf{S} \rangle \right) + \left[\frac{1}{4\psi^2} \langle (\tilde{\mathbf{S}}\mathbf{S})^2 \rangle - \frac{1}{2\psi^2} \langle \tilde{\mathbf{S}}\mathbf{S} \rangle^2 \right] + O(\mathbf{S}^6) \\
&+ \operatorname{tr} \frac{\langle (\tilde{\mathbf{S}}\mathbf{E})^2 \rangle}{4\psi^2} + \frac{\langle \operatorname{tr} (\tilde{\mathbf{S}}\mathbf{E})^2 \rangle}{16\psi^4} - \frac{1}{4\psi^3} \langle \operatorname{tr} (\tilde{\mathbf{S}}\mathbf{E})^2 \cdot \tilde{\mathbf{S}}\mathbf{S} \rangle - \frac{1}{32\psi^4} \langle \operatorname{tr} (\tilde{\mathbf{S}}\mathbf{E})^2 \rangle \cdot \operatorname{tr} \langle (\tilde{\mathbf{S}}\mathbf{E})^2 \rangle \\
&+ \frac{1}{8\psi^3} \langle \operatorname{tr} (\tilde{\mathbf{S}}\mathbf{E})^2 \rangle \langle \tilde{\mathbf{S}}\mathbf{S} \rangle + O(\mathbf{S}^6, \mathbf{E}^2, \mathbf{E}^4, \ldots)
\end{aligned} \tag{2}
$$

19.3 (*a*) Show, by a direct evaluation of $\mathbf{x} = \mathbf{M}^{-1}\mathbf{y}$ with expansion of $y_{k+1} = y(t_{k+1})$ about t_k in the limits $n \to \infty$, $\delta \to 0$ ($n\delta = T$ fixed), that for $M_{ij} = e^{-\alpha\delta|i-j|}$

$$
Q_T = \lim_{n\to\infty} Q_n = \lim_{n\to\infty} \tilde{\mathbf{z}}\mathbf{x} = \int_{0-}^{T+} z(t) X(t)\, dt \tag{1}
$$

with $X(t)$ the solution of the integral equation

$$
\int_{0-}^{T+} e^{-\alpha|t-u|} X(u)\, du = y(t) \qquad 0- < t < T+ \tag{2}
$$

(*b*) If $\{\phi_m\}$ is a complete orthonormal set satisfying the integral equation

$$
\int_0^T M(t,u) \phi_m(u)\, du = \lambda_m \phi_m(t) \qquad 0 \le t \le T \tag{3}
$$

where $M(t,u) = M(u,t)$ is positive definite on $(0,T)$, show that the solution X_T of the integral equation (19.55) is unique and integrable square on $(0-,T+)$ if and only if

$$
y(t) = \lim_{N\to\infty} \sum_{m=1}^{N} y_m \phi_m(t) \tag{4}
$$

where $y_m = \int_0^T y(t)\phi_m(t)\, dt$, and if the series

$$
\sum_{m=1}^{\infty} \frac{y_m^2}{\lambda_m^2} < \infty \tag{5}
$$

(*c*) Show that the solution X_T is (mean square)

$$
X_T(t) = \sum_{m=1}^{\infty} \frac{y_m}{\lambda_m} \phi_m(t) \qquad 0- < t < T+ \tag{6}
$$

19.4 Given a normal background noise $N(t)$ with vanishing mean but not necessarily stationary, whose covariance function $K_N(t,u)$ is continuous and positive definite on

$(0,T)$, and given an independent signal process $S(t)$ which is also normal with zero mean and covariance function $K_S(t,u)$, likewise continuous and positive definite on $(0,T)$, obtain the necessary and sufficient conditions that a statistical test of the hypothesis $H_1{:}S + N$ versus $H_0{:}N$ be regular, and show specifically for colored noise signals in white noise backgrounds that such a test is regular, while if $K_S = \gamma K_N$ (γ a positive constant), one has instead the extreme singular case. For stationary processes, show that the spectrum of $S(t)$ when $N(t)$ is spectrally white must vanish as $f^{-1-\epsilon}$ ($\epsilon > 0$) as $f \to \infty$ in order that the test be regular. HINT: Expand $N(t)$ in a complete orthonormal set $\{\phi_m\}$ on $(0,T)$ with respect to the kernel $K_N(t,u)$ and $S(t)$ in a similarly complete orthonormal set $\{\psi_m\}$ with respect to $K_S(t,u)$ (cf. Sec. 8.2-2). Let λ_{Nj}, λ_{Sj} be the respective eigenvalues. Consider the first n components of these expansions, and show that the logarithm of the likelihood ratio by which the test is carried out becomes, for the first n coefficients $\mathbf{V} = (V_1, \ldots, V_n)$ of the expansion of $V(t)$ in the set $\{\phi_m\}$,

$$\log \Lambda_n = \log \mu - \tfrac{1}{2} \log \det (\mathbf{I} + \mathbf{G}_{SN}) + \tfrac{1}{2}\tilde{\mathbf{V}}\mathbf{\Lambda}_N^{-1}[\mathbf{I} - (\mathbf{I} + \mathbf{G}_{SN})^{-1}]\mathbf{V} \tag{1}$$

where $V(t) = N(t)$ for H_0 and $V(t) = N(t) + S(t)$ for H_1. Here, $\mathbf{G}_{SN} = \tilde{\mathbf{G}}_{SN}$ is given by

$$\mathbf{G}_{SN} = (\mathbf{b}\mathbf{\Lambda}_S\tilde{\mathbf{b}})\mathbf{\Lambda}_N^{-1} = \left[\sum_{l=1}^{n} b_{il}\lambda_{Sl}b_{jl}/\lambda_{Nj}\right] \tag{2}$$

in which

$$\mathbf{\Lambda}_S = [\lambda_{Sj}\delta_{ij}] \qquad \mathbf{\Lambda}_N = [\lambda_{Nj}\delta_{ij}] \qquad \lambda_{Sj} = \sigma_{Sj}^2 \qquad \lambda_{Nj} = \sigma_{Nj}^2 \tag{3}$$

$$(\mathbf{b})_{ml} = \int_0^T \phi_m(t)\psi_l(t)\,dt \tag{4}$$

Regularity now is established if the bias, $\log \mu - \tfrac{1}{2} \log \det (\mathbf{I} + \mathbf{G}_{SN})$, and $E_{\mathbf{V}|H_0}$, $E_{\mathbf{V}|H_1}$ of the data terms in Eq. (1) remain finite as $n \to \infty$. First note that $\lim_{n\to\infty} \det (\mathbf{I} + \mathbf{G}_{SN}) = \prod_{j=1}^{\infty} (1 + \lambda_{jG}) = \mathfrak{D}_G(1)$, the Fredholm determinant, where the λ_{jG} are the eigenvalues of $G(t,u)_{SN}$, which is the continuous form of the matrix elements $G(t_k,t_l)_{SN}$. If $G(t,u)$ is positive definite (and bounded) on $(0,T)$, then $\mathfrak{D}_G(1)$ exists. Show convergence in probability to H_0, H_1 for the data terms, and hence show that the resulting likelihood functional is bounded.

If the background noise is stationary and white, then $\lambda_{Nj} = W_0/2$ (all $j \geq 1$) and we can set $\phi_j(t) = \psi_j(t)$ (cf. footnote ‡, page 386). Then, $b_{ml} = \delta_{ml}$ and $(\mathbf{G}_{SN})_j = (2/W_0)\delta_{ij}\sigma_{Sj}^2$ (> 0), with $\lim_{n\to\infty} \det (\mathbf{I} + \mathbf{G}_{SN}) = \prod_{j=1}^{\infty} \left(1 + \frac{2}{W_0}\lambda_{Sj}\right) < \infty$, since $\sum_{j=1}^{\infty} \lambda_{Sj} = \int_0^T K_S(t,t)\,dt = \psi_S T < \infty$ for colored noise whose mean intensity is finite. [In the stationary case, this clearly means that $\mathcal{W}_S(f) \to 0$, $O(f^{-1-\epsilon})$, as $f \to \infty$.]

In the singular case $K_S = \gamma K_N$ ($\gamma > 0$), set $\psi_m = \phi_m$, and show that $(\mathbf{G}_{SN})_{ij} = \gamma^2\delta_{ij}$ and hence that $\lim_{n\to\infty} \det (\mathbf{I} + \mathbf{G}_{SN}) = \lim_{n\to\infty} (1 + \gamma^2)^n \to \infty$. Show then that $\lim_{n\to\infty} \log \Lambda_n \to -\infty$ in probability if H_0 is valid and $\lim_{n\to\infty} \log \Lambda_n \to +\infty$ in probability if H_1 is valid. (See Sec. 19.4-2 above, Grenander,[31] sec. 4.7, and Grenander and Rosenblatt.[32])

REFERENCES

1. Van Meter, D., and D. Middleton: Modern Statistical Approaches to Reception in Communication Theory, Symposium on Information Theory, *IRE Trans. on Inform. Theory*, **PGIT-4:** 119 (September, 1954).

2. Middleton, D., and D. Van Meter: Detection and Extraction of Signals in Noise from the Point of View of Statistical Decision Theory, *J. Soc. Ind. Appl. Math.*, **3:** 192 (1955), **4:** 86 (1956).
3. Bussgang, J., and D. Middleton: Optimum Sequential Detection of Signals in Noise, *IRE Trans. on Inform. Theory*, **IT-1:** 5 (December, 1955).
4. North, D. O.: Analysis of the Factors Which Determine Signal-to-Noise Discrimination in Radar, *RCA Tech. Rept.* PTR-6C, June, 1943.
5. Van Vleck, J. H., and D. Middleton: A Theoretical Comparison of the Visual, Aural, and Meter Reception of Pulsed Signals in the Presence of Noise, *J. Appl. Phys.*, **17:** 940 (1946).
6. Dwork, B. M.: Detection of a Pulse Superimposed on Fluctuation Noise, *Proc. IRE.*, **38:** 771 (1950).
7. George, T. S.: Fluctuations of Ground Clutter Return in Airborne Radar Equipment, *J. IEE (London)*, **99** (pt. IV): 92 (1952).
8. Zadeh, L. A., and J. R. Ragazzini: Optimum Filters for the Detection of Signals in Noise, *Proc. IRE*, **40:** 1223 (1952).
9. Zadeh, L. A.: Optimum Non Linear Filters for the Extraction and Detection of Signals, *J. Appl. Phys.*, **24:** 396 (1953), and *IRE Conv. Record*, pt. 8, p. 57, 1953.
10. Siegert, A. J. F.: Preface and sec. 7.5 in J. L. Lawson and G. E. Uhlenbeck, "Threshold Signals," Massachusetts Institute of Technology Radiation Laboratory Series, vol. 24, McGraw-Hill, New York, 1950.
11. Marcum, J. I.: A Statistical Theory of Target Detection by Pulsed Radar, *RAND Corp. Repts.* RM-753, R-113, 1948; RM-15061, 1947.
12. Singleton, H. E.: Theory of Non Linear Transducers, *MIT Research Lab. Electronics Tech. Rept.* 160, August, 1950.
13. Kaplan, S. M., and R. W. McFall: The Statistical Properties of Noise Applied to Radar Range Performance, *Proc. IRE*, **39:** 56 (1951).
14. Hance, H.: The Optimization and Analysis of Systems for the Detection of Pulsed Signals in Random Noise, doctoral dissertation, Massachusetts Institute of Technology, January, 1951.
15. Schwartz, M.: A Statistical Approach to the Automatic Search Problem, doctoral dissertation, Harvard University, June, 1951. See also A Coincidence Procedure for Signal Detection, *IRE Trans. on Inform. Theory*, **IT-2:** 135 (December, 1956).
16. Davies, I. L.: On Determining the Presence of Signals in Noise, *Proc. IEE (London)*, **99** (pt. III): 45 (1952).
17. Slattery, T. G.: The Detection of a Sine Wave in the Presence of Noise by the Use of a Non Linear Filter, *Proc. IRE*, **40:** 1232 (1952).
18. Reich, E., and P. Swerling, Jr.: Detection of a Sine Wave in Gaussian Noise, *J. Appl. Phys.*, **24:** 289 (1953)
19. Davis, R. C.: On the Detection of Sure Signals in Noise, *J. Appl. Phys.*, **25:** 76 (1953).
20. Davis, R. C.: The Detectability of Random Signals in the Presence of Noise, Symposium on Statistical Methods in Communication Engineering, *IRE Trans. on Inform. Theory*, **PGIT-3:** 52 (March, 1954).
21. Middleton, D.: Statistical Theory of Signal Detection, Symposium on Statistical Methods in Communication Engineering, *IRE Trans. on Inform. Theory*, **PGIT-3:** 26 (March, 1954).
22. Middleton, D.: Statistical Criteria for the Detection of Pulsed Carriers in Noise. I, II, *J. Appl. Phys.*, **24:** 371, 379 (1953); also *J. Appl. Phys.*, **25:** 127 (1954).
23. Stone, W. M.: On the Statistical Theory of Detection of a Randomly Modulated Carrier, *J. Appl. Phys.*, **24:** 935 (1953).
24. Benner, A. H., and R. F. Drenick: On the Problem of Optimum Detection of Pulsed Signals in Noise, *RCA Rev.*, **16:** 463 (1955).

25. Peterson, W. W., T. G. Birdsall, and W. C. Fox: The Theory of Signal Detectability, Symposium on Information Theory, *IRE Trans. on Inform. Theory,* **PGIT-4:** 171 (1954).

26. Neyman, J., and E. S. Pearson: On the Problem of the Most Efficient Tests of Statistical Hypotheses, *Trans. Roy. Soc. (London),* **A231:** 289 (1933); see also J. Neyman, Basic Ideas and Some Recent Results of the Theory of Testing Statistical Hypotheses, *J. Research Roy. Stat. Soc.,* **105:** 292 (1942).

27. For a detailed discussion, see A. M. Mood, "Introduction to the Theory of Statistics," chaps. 12, 15, McGraw-Hill, New York, 1950, and H. Cramér, "Mathematical Methods of Statistics," chaps. 30–37, Princeton University Press, Princeton, N.J., 1946.

28. Middleton, D.: On the Theory of Random Noise. Phenomenological Models. I, II, *J. Appl. Phys.,* **22:** 1143, 1153 (1951).

29. Rice, S. O.: Mathematical Analysis of Random Noise, *Bell System Tech. J.,* **23:** 282 (1944), **24:** 46 (1945).

30. Mullen, J. A., and D. Middleton: The Rectification of Nearly Gaussian Noise, *Cruft Lab. Tech. Rept.* 189, June, 1954, and also The Rectification of Non Gaussian Noise, *Quart. Appl. Math,* **15:** 395 (1958).

31. Grenander, U.: Stochastic Processes and Statistical Inference, *Arkiv Math.,* **1:** 195 (1950).

32. Grenander, U., and M. Rosenblatt: "Statistical Analysis of Stationary Time Series," Wiley, New York, 1957.

33. Davenport, W. B., Jr., and W. L. Root: "An Introduction to the Theory of Random Signals and Noise," McGraw-Hill, New York, 1958.

34. Slepian, D.: Some Comments on the Detection of Gaussian Signals in Gaussian Noise, *IRE Trans. on Inform. Theory,* **IT-4:** 65 (June, 1958).

CHAPTER 20

BINARY DETECTION SYSTEMS MINIMIZING AVERAGE RISK. EXAMPLES†

In Chap. 19, we have outlined a general theory of binary detection systems which minimize average risk or cost. We have also indicated how nonideal systems for similar purposes may be compared with the theoretical optimum. Here, our principal aim is to show how these rather general methods of system analysis and comparison can be applied in a variety of actual detection situations. The present chapter, therefore, is largely devoted to specific examples of varying degrees of realism, which illustrate the techniques of solution and which provide some useful results for problems of general communication interest.

As we noted earlier, the goal of a general theory of system design and analysis (like that described in Chap. 18, for instance) is threefold: it seeks (1) to obtain the explicit structure of the optimum (and near-optimum) system for the task at hand, including a resolution of the system into an ordered sequence of physically realizable elements; (2) to evaluate the performance of such systems quantitatively; and (3) to provide a comparison between the theoretical optimum and the actual suboptimum systems with which we are almost always forced to work in practice. Thus, item 1 guides the designer in the practical approach to optimality, item 2 provides an upper limit on effective performance, toward which his practice strives, and item 3 enables him to estimate how far from this limit an actual system may be and on the basis of this whether or not further improvement is worth the effort. In the examples following, we attempt to illustrate each of these features for binary detection situations. A similar philosophy also governs our treatment of signal extraction, or estimation, as described in Chap. 21. For the general background in both instances, we refer the reader again to Chap. 18.

The present chapter discusses *fixed-sample systems* only, where the observation and processing intervals available to the observer are regarded as fixed. A treatment of *variable-sample*, or *sequential*, systems, where these

† The following presentation is based for the most part on the author's published and unpublished work. For additional references, see also Middleton and Van Meter,[1] Bussgang and Middleton,[2] and Chap. 19, Refs. 1 to 25, as well as the references at the end of this chapter.

periods are adjusted in accordance with the actual data received, is described in Refs. 38 to 40. We distinguish two main methods of processing the received data: (1) discrete, or "digital," sampling techniques and (2) continuous, or "analogue," sampling [cf. Sec. 18.2-1(3)]. The discrete formulation is naturally considered first, since it is the starting point for the limiting situations of continuous sampling. Unless otherwise indicated, we shall assume here (cf. Sec. 19.4-1) that the received signal and noise processes are additive and independent, that the signal is deterministic, and that the noise is normal, with vanishing mean. However, the noise process may be nonstationary. Both coherent and incoherent† reception are also considered, and following the general analysis of Secs. 20.1 to 20.3 may be found additional examples illustrating the method (cf. Sec. 20.4).

20.1 *Threshold Structure I. Discrete Sampling*

Let us begin without any essential loss of generality by considering a normal noise of vanishing mean ($\bar{V}_N = 0$). The appropriate nth-order distribution density is given by Eq. (19.46), which, for the additive and independent signal and noise processes assumed throughout, enables us to write the optimum detector structure [Eq. (19.41)] specifically as

$$\begin{aligned} \boldsymbol{T}_R{}^{(N)}\{V\} = \log \Lambda_n(\mathbf{V}) &= \log [p\langle W_n(\mathbf{V} - \mathbf{S})_N\rangle_S / qW_n(\mathbf{V})_N] \\ &= \log \mu + \log [\langle \exp (-\tfrac{1}{2}a_0{}^2\tilde{\mathbf{s}}\mathbf{k}_N{}^{-1}\mathbf{s} \\ &\qquad + a_0\tilde{\mathbf{v}}\mathbf{k}_N{}^{-1}\mathbf{s})\rangle_{S=a_0,\epsilon,\boldsymbol{\theta}}] \end{aligned} \tag{20.1}$$

Here, we have used the normalizations (19.49); as before, the average $\langle\ \ \rangle_S$ is over all pertinent parameters of the signal process, including amplitude ($\sim a_0$) and epoch ϵ. Setting

$$\Phi_n(v,s) \equiv \tilde{\mathbf{v}}\mathbf{k}_N{}^{-1}\mathbf{s} \qquad \Phi_n(s,s) = \tilde{\mathbf{s}}\mathbf{k}_N{}^{-1}\mathbf{s} \tag{20.2a}$$

and abbreviating with $\Phi_n(v,s) = \Phi_v$, $\Phi_n(s,s) = \Phi_s$, we note first from Eqs. (19.50a) that‡

$$A_s{}^{(1)} = \sqrt{\psi}\,\tilde{\mathbf{s}}\mathbf{b}^{(1)} = 0 \qquad -2A_n{}^{(2)}(v,s) = \Phi_v \qquad -2A_n{}^{(2)}(s,s) = \Phi_s \tag{20.2b}$$

Next, we follow the development of Eq. (19.50), which gives us at once the desired expression for the optimum threshold detector,§ namely,

$$\begin{aligned} \boldsymbol{T}_R{}^{(N)}\{V\} = \log \Lambda_n = \log \mu + \bar{a}_0\langle\Phi_v(v,s)\rangle + \tfrac{1}{2}\overline{a_0{}^2}[\langle\Phi_v(v,s)^2\rangle \\ - (\bar{a}_0{}^2/\overline{a_0{}^2})\langle\Phi_v(v,s)\rangle^2 - \langle\Phi_s(s,s)\rangle] + O(\overline{a_0{}^3}) \end{aligned} \tag{20.3}$$

† See the remarks in Sec. 19.4-3 on coherency, sample certainty, etc.

‡ The notation is that of Sec. 19.4.

§ The pointed brackets $\langle\ \ \rangle$ indicate the statistical average over all random components ϵ, $\boldsymbol{\theta}$ of $s(t - \epsilon, \boldsymbol{\theta})$ in the usual way. In general, the pointed brackets and the vinculum are to be interpreted here as ensemble averages over any and all random parameters that may be present.

Note that we assume here that the amplitude (or scale) of the signal ($\sim a_0$) is statistically independent of the other parameters of the signal; later (cf. Probs. 20.2, 20.3 and the example of Sec. 20.4-6), we shall consider the case where a_0 is statistically related to the other signal parameters.

While it is sufficient for threshold systems, as far as the actual operations on the received data $\mathbf{V}$ are concerned, to retain terms through $O(a_0^2)$ only, to evaluate system performance it is often necessary to include the contributions† of terms $O(a_0^3, a_0^4)$ in $\log \Lambda_n$ as well, as we shall see presently (Sec. 20.3-1). However, these higher-order terms can be replaced by their appropriate average values, here with respect to the hypothesis H_0 (noise alone) for threshold operation, as noted earlier, in Sec. 19.4-3. Following the procedure leading to Eq. (19.50) and carrying out the indicated averages $E_{\mathbf{V}|H_0}\{\ \ \}$ with respect to $\mathbf{V}$ under the hypothesis H_0, we get finally, for these higher-order terms,‡

$$O(a_0^3):\quad E_{\mathbf{V}|H_0}\left\{\overline{a_0^3}\left(\frac{\langle\Phi_v^3\rangle}{3!} - \frac{\langle\Phi_v\Phi_s\rangle}{2!}\right) - \frac{\bar{a}_0\overline{a_0^2}}{2}\left(\langle\Phi_v^3\rangle - \langle\Phi_v\rangle\langle\Phi_s\rangle\right) + \tfrac{1}{3}\overline{a_0^3}\langle\Phi_v\rangle^3\right\} = 0 \qquad (20.4a)$$

where each term $O(a_0^3)$ vanishes separately. Setting

$$\bar{\mathbf{G}} \equiv \mathbf{k}_N^{-1}\overline{\mathbf{s}\tilde{\mathbf{s}}}\mathbf{k}_N^{-1} \qquad \Phi_{\bar{s}} = \tilde{\bar{\mathbf{s}}}\mathbf{k}_N^{-1}\bar{\mathbf{s}}$$

we have, for the next higher-order terms,

$$\begin{aligned} O(a_0^4):\quad & E_{\mathbf{V}|H_0}\left\{\overline{a_0^4}\left\langle\frac{\Phi_v^4}{4!} - \tfrac{1}{4}\Phi_v^2\Phi_s + \tfrac{1}{8}\Phi_s^2\right\rangle - \bar{a}_0\overline{a_0^3}\left\langle\Phi_v\frac{\Phi_s^3}{3!}\right.\right. \\ & \left.\left. - \frac{\Phi_v^2\Phi_s}{2!}\right\rangle - \frac{\overline{a_0^2}^2}{8}\langle\Phi_v^2 - \Phi_s\rangle^2 + \frac{\bar{a}_0^2\overline{a_0^2}}{2}\langle\Phi_v\rangle^2\langle\Phi_v^2 - \Phi_s\rangle - \tfrac{1}{4}\bar{a}_0^4\langle\Phi_v\rangle^4\right\} \\ & = \overline{a_0^4}\times 0 - \frac{\bar{a}_0\overline{a_0^3}}{2}\left[\langle(\tilde{\mathbf{s}}\mathbf{k}_N^{-1}\bar{\mathbf{s}})\Phi_s\rangle - \langle\tilde{\bar{\mathbf{s}}}\mathbf{G}\mathbf{s}\rangle\right] \\ & \quad - \frac{\overline{a_0^2}^2}{8}\left[2\operatorname{tr}(\mathbf{k}_N\bar{\mathbf{G}})^2 + \operatorname{tr}^2\mathbf{k}_N\bar{\mathbf{G}} - 2\langle\Phi_s\rangle\operatorname{tr}\mathbf{k}_N\bar{\mathbf{G}} + \langle\Phi_s\rangle^2\right] \\ & \quad + \bar{a}_0^2\overline{a_0^2}\langle(\tilde{\mathbf{s}}\mathbf{k}_N^{-1}\bar{\mathbf{s}})^2\rangle - \tfrac{3}{4}\bar{a}_0^4\langle\Phi_{\bar{s}}\rangle^2 \end{aligned} \qquad (20.4b)$$

Here $\langle\ \ \rangle$ as before indicates the average over the parameters of $\mathbf{s}$ (as does $\bar{\mathbf{s}}$).

The structure of the optimum threshold detector given by Eq. (20.3), with these higher-order terms [Eqs. (20.4a), (20.4b)], can now be expressed

† For incoherent reception, cf. Sec. 20.3-2; for coherent reception, on the other hand, terms $O(a_0^2)$ are sufficient (cf. Sec. 20.3-1).

‡ See Prob. 20.1.

more compactly as

$$\log \Lambda_n(\mathbf{V}) = [\log \mu + \mathcal{B}_n^{(2)} + \mathcal{B}_n^{(4)} + O(a_0^6)] + \bar{a}_0\langle\Phi_v\rangle + \frac{\overline{a_0^2}}{2}\left(\langle\Phi_v^2\rangle - \frac{\bar{a}_0^2}{\overline{a_0^2}}\langle\Phi_v\rangle^2\right) \quad (20.5)$$

where explicitly

$$\mathcal{B}_n^{(2)} \equiv -\frac{\overline{a_0^2}}{2}\langle\Phi_s\rangle \qquad \mathcal{B}_n^{(4)} \equiv \text{Eq. (20.4}b) \quad (20.5a)$$

Along with $\log \mu = \log (p/q)$, $\mathcal{B}_n^{(2)}$ and $\mathcal{B}_n^{(4)}$ are *bias* terms, $O(a_0^2)$, $O(a_0^4)$, respectively. Note that this threshold system involves both linear and nonlinear (i.e., quadratic) operations on the received data in general.

20.1-1 Coherent Detection. Let us consider now the special case of *coherent detection*, where there is sample certainty, i.e., where the epoch ϵ_0 is known to the observer (cf. Sec. 19.4-3). Then $\langle s\rangle_\epsilon = s(t - \epsilon_0, \boldsymbol{\theta})$ does not vanish, and consequently $\langle\Phi_v\rangle = \tilde{\mathbf{v}}\mathbf{k}_N^{-1}\bar{\mathbf{s}} = \langle\langle\tilde{\mathbf{v}}\mathbf{k}_N^{-1}\mathbf{s}\rangle_\epsilon\rangle_{\boldsymbol{\theta}} \neq 0$ also, so that Eq. (20.5) gives the structure of the optimum threshold system in this instance, where quadratic operations on the received data are included as well as linear ones.

However, a more convenient and simpler form may be derived if we retain only the linear term in $\mathbf{V}$ and replace those $O(V^2)$ by the appropriate statistical average $E_{\mathbf{V}|H_0}\{\quad\}$, which becomes then an additional contribution to the bias, $O(a_0^2)$. Thus, we have in effect a *linear* system, as expected from the general result (19.58). This system now takes the specific form

$$\mathbf{T}_R^{(N)}\{V\} = \log \Lambda_n(\mathbf{V}) \doteq \left(\log \mu + \frac{\overline{a_0^2}}{2} B_n^{(2)}\right) + \bar{a}_0\tilde{\mathbf{v}}\mathbf{k}_N^{-1}\bar{\mathbf{s}} \quad (20.6)$$

where

$$\begin{aligned} B_n^{(2)} &= E_{\mathbf{V}|H_0}\{\tilde{\mathbf{v}}\mathbf{k}_N^{-1}[\overline{\mathbf{s}\tilde{\mathbf{s}}} - (\bar{a}_0^2/\overline{a_0^2})\bar{\mathbf{s}}\bar{\tilde{\mathbf{s}}}]\mathbf{k}_N^{-1}\mathbf{v}\} - \overline{\tilde{\mathbf{s}}\mathbf{k}_N^{-1}\mathbf{s}} \\ &= \operatorname{tr}\{[\overline{\mathbf{s}\tilde{\mathbf{s}}} - (\bar{a}_0^2/\overline{a_0^2})\bar{\mathbf{s}}\bar{\tilde{\mathbf{s}}}]\mathbf{k}_N^{-1}\} - \overline{\tilde{\mathbf{s}}\mathbf{k}_N^{-1}\mathbf{s}} \\ &= (-\bar{a}_0^2/\overline{a_0^2})\bar{\tilde{\mathbf{s}}}\mathbf{k}_N^{-1}\bar{\mathbf{s}} \end{aligned} \quad (20.7)$$

this last with the help of Eq. (17.46a) and the identity $\operatorname{tr} \bar{\mathbf{s}}\bar{\tilde{\mathbf{s}}}\mathbf{k}_N^{-1} = \operatorname{tr} \mathbf{k}_N^{-1}\bar{\mathbf{s}}\bar{\tilde{\mathbf{s}}} = \bar{\tilde{\mathbf{s}}}\mathbf{k}_N^{-1}\bar{\mathbf{s}}$ (cf. Prob. 20.2).

The threshold structure [Eq. (20.6)] can be interpreted in a number of ways. As we have already seen from Eq. (19.58), the operations leading to the number $\bar{a}_0\tilde{\mathbf{v}}\mathbf{k}_N^{-1}\bar{\mathbf{s}}$ can be represented as a discrete cross correlation, without delay, of the received data $\mathbf{v}$ with the known weighted averaged signal $\bar{\mathbf{s}}_0 = \bar{a}_0\mathbf{k}_N^{-1}\bar{\mathbf{s}}$, for example, $\bar{a}_0\tilde{\mathbf{v}}\mathbf{k}_N^{-1}\bar{\mathbf{s}} = \tilde{\mathbf{v}}\bar{\mathbf{s}}_0 = \sum_{j=1}^{n} v_j\bar{s}_{0j} = \sum_{j=1}^{n} v(t_j)\overline{s_0(t_j)}$. This representation is not unique, however, since we can also regard $\bar{a}_0\tilde{\mathbf{v}}\mathbf{k}_N^{-1}\bar{\mathbf{s}}$ as $a_0\bar{\tilde{\mathbf{s}}}\mathbf{k}_N^{-1}\mathbf{v} \equiv \bar{a}_0\bar{\tilde{\mathbf{s}}}\mathbf{v}_F = a_0 \sum_{j=1}^{n} \overline{s(t_j)}v_F(t_j)$, where $\mathbf{v}_F$ is the *filtered* data

$$\mathbf{v}_F \equiv \mathbf{Q}_c\mathbf{v} \qquad \text{with } \mathbf{Q}_c = \mathbf{k}_N^{-1} = \tilde{\mathbf{Q}}_c \quad (20.8)$$

Again, we have a discrete cross correlation, without delay, but now between the averaged, but unweighted, signal $\bar{a}_0\bar{\mathbf{s}}$ and the filtered data $\mathbf{v}_F$. Both these possibilities are shown schematically in Fig. 20.1, which includes the bias and terminal decision process, where Eq. (20.6) is employed according to the binary-test procedure† of Eqs. (19.15) et seq.

Here, $\mathbf{Q}_c$ is the discrete analogue of a linear filter, which may or may not be required to be time-invariant. In fact, we can often impose the further constraint of time invariance characteristic of the simplest types of linear

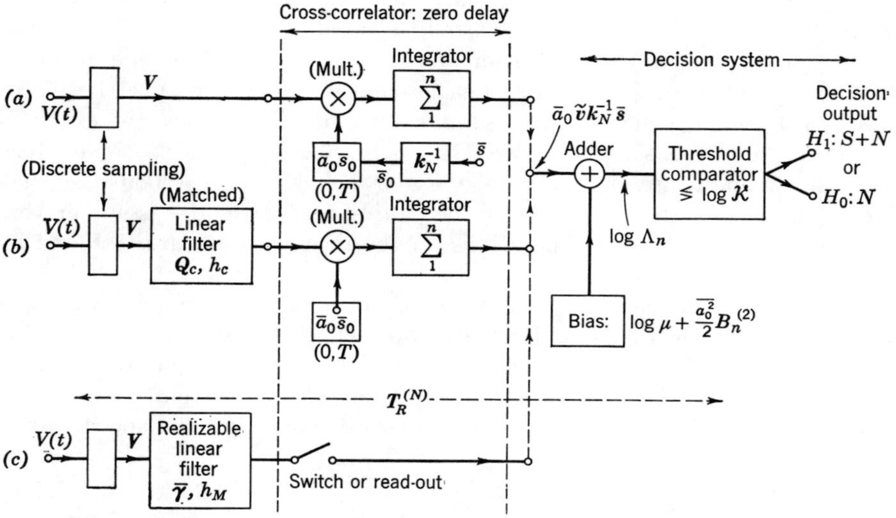

FIG. 20.1. Two equivalent interpretations of optimum threshold structure for binary coherent detection of a signal in normal noise.

filters (cf. Sec. 2.2) without altering Eq. (20.6) in any way. Thus, writing $\mathbf{Q}_c = [h_c(t_i,t_j)\ \Delta t] = \tilde{\mathbf{Q}}_c$ for Eq. (20.8), where $\Delta t = t_{l+1} - t_l$ is the interval between consecutive sampling times, we may require that

(1) $\mathbf{Q}_c$ is *linear* [this follows directly from Eq. (20.8)] (20.9*a*)

(2) $\mathbf{Q}_c = [h_c(|t_i - t_j|)\ \Delta t]$; that is, $\mathbf{Q}_c = \mathbf{k}_N^{-1}$ is *time-invariant*, provided that the background noise is stationary (20.9*b*)

We observe, however, that since $\mathbf{Q}_c$ is given in terms of a covariance function and is hence symmetric, if condition (2) is imposed, it cannot represent a realizable filter, where $h_c(t_i - t_j) = 0$ (for all $t_i < t_j$). Even if this condition is relaxed, $\mathbf{Q}_c$ cannot be constructed even as a physically realizable, *time-varying* linear network since $\tilde{\mathbf{v}}\mathbf{k}_N^{-1}\mathbf{s}$ is not symmetric in $\mathbf{v}$.

With the above interpretations, the operations on the received data are

† See footnote on p. 805.

given by

$$\bar{a}_0\tilde{\mathbf{v}}\mathbf{k}_N{}^{-1}\bar{\mathbf{s}} = \bar{a}_0\tilde{\bar{\mathbf{s}}}\mathbf{v}_F = \bar{a}_0 \sum_{j=1}^{n} \overline{s(t_j)} \sum_{i=1}^{n} v(t_i)h_c(|t_j - t_i|)\,\Delta t \tag{20.10}$$

and the finiteness of the sample is further ensured by the fact that one of the inputs, $\bar{\mathbf{s}}$, vanishes outside the integration period $(0,T)$ (cf. Fig. 20.1). For further comments on the role of the filter $\mathbf{Q}_c$, see Sec. 20.2-4.

Finally, there is a third possible interpretation of the operation (20.6) which does make use of a realizable invariant linear filter. To see this, let $\boldsymbol{\gamma}$ be a column vector, such that $\mathbf{K}_N{}^{-1}\bar{\mathbf{s}} = \bar{\boldsymbol{\gamma}}$; then $\tilde{\mathbf{v}}\mathbf{k}_N{}^{-1}\bar{\mathbf{s}}$ in Eq. (20.6) becomes simply $\psi\tilde{\mathbf{v}}\bar{\boldsymbol{\gamma}} = \psi \sum_{i=1}^{n} v(t_i)\bar{\gamma}_i$. Now, let

$$\bar{\boldsymbol{\gamma}} \equiv [h_M(T - t_i;\, T)\,\Delta t] \tag{20.10a}$$

where h_M is the weighting function of a time-invariant realizable linear filter, with a read-out at time T $(= n\,\Delta t)$ (the additional T in h_M is to remind us that its structure also depends on the observation interval). Then, Eq. (20.6) becomes

$$\tilde{\mathbf{v}}\mathbf{k}_N{}^{-1}\bar{\mathbf{s}} = \psi \sum_{i=1}^{n} v(t_i)h_M(T - t_i;\, T)\,\Delta t \tag{20.10b}$$

so that $\tilde{\mathbf{v}}\mathbf{k}_N{}^{-1}\bar{\mathbf{s}}$ can indeed be represented as the output at time T of this (discrete) linear realizable filter, when the sampled data $\mathbf{v}$ constitute the input (cf. Fig. 20.1c). From Eqs. (20.8), (20.9), it is clear that the filter $\mathbf{Q}_c$ is related to $[(h_M)_i\,\Delta t]$ by

$$\bar{\boldsymbol{\gamma}} = \psi^{-1}\mathbf{Q}_c\bar{\mathbf{s}} \quad \text{or} \quad h_M(T - t_i;\, T) = \psi^{-1} \sum_{j=1}^{n} h_c(t_i - t_j)\langle s(t_j - \epsilon;\, \boldsymbol{\theta})\rangle \tag{20.10c}$$

where, of course, $\bar{\boldsymbol{\gamma}}$ is found from the defining relation $\mathbf{K}_N\bar{\boldsymbol{\gamma}} = \bar{\mathbf{s}}$ above [see Eqs. (20.41) to (20.43d) for continuous sampling].

20.1-2 Incoherent Detection I. General Signals. In almost all applications, the *sample certainty* needed for coherent operation does not occur: the most that is known (to the observer) about the epoch ϵ is that it may take several values or, more commonly, a continuum of values, subject in either instance to some specified probability distribution, so that the detection process is *incoherent* in the sense of Sec. 19.4-3. Then, depending on the d.d. of ϵ, $\langle s\rangle_\epsilon$ may or may not vanish. In the latter case, the structure of the optimum threshold system demands both linear and quadratic operations on the received data, as Eq. (20.5) indicates.

However, it is frequently possible to regard ϵ as uniformly distributed†

† This is the condition for which there is the greatest average uncertainty as to the actual value of ϵ (≥ 0), subject to a maximum value constraint (cf. Prob. 6.7 and the examples of Secs. 20.4-3, 20.4-5, 20.4-6). Such a condition represents an additional subcriterion restricting the class of optimum systems here.

over a sufficiently long interval,† so that $\langle s \rangle_\epsilon$ is zero and consequently $\langle \Phi_v \rangle = 0$ also [cf. Eq. (20.5)]. This discrete threshold system is now entirely nonlinear, with the specific structure

$$T_R^{(N)}\{V\} = \log \Lambda_n(\mathbf{V}) \doteq (\log \mu + \mathfrak{B}_n^{(2)} + \mathfrak{B}_n^{(4)}) + \frac{\overline{a_0^2}}{2}\,\tilde{\mathbf{v}}\bar{\mathbf{G}}\mathbf{v} = \{\text{bias}\} + \frac{\overline{a_0^2}}{2}\,\Psi_n(v) \tag{20.11a}$$

from Eq. (20.5), where

$$\bar{\mathbf{G}} = \mathbf{k}_N{}^{-1}\overline{\mathbf{s}\tilde{\mathbf{s}}}\mathbf{k}_N{}^{-1} \qquad \text{and} \qquad \Psi_n(\mathbf{v}) \equiv \tilde{\mathbf{v}}\bar{\mathbf{G}}\mathbf{v} \tag{20.11b}$$

with

$$\mathfrak{B}_n^{(2)} = -\frac{\overline{a_0^2}}{2}\,\overline{\tilde{\mathbf{s}}\mathbf{k}_N{}^{-1}\mathbf{s}} = -\frac{\overline{a_0^2}}{2}\,\text{tr}\;\mathbf{k}_N\bar{\mathbf{G}} \qquad \mathfrak{B}_n^{(4)} = -\frac{\overline{a_0^2}^2}{4}\,\text{tr}\;(\mathbf{k}_N\bar{\mathbf{G}})^2 \tag{20.11c}$$

since by Prob. 20.2 tr $\mathbf{k}_N\bar{\mathbf{G}} = \langle \Phi_s \rangle$.

As in the coherent case above, we can interpret the threshold structure of Eq. (20.11*a*) in a variety of ways. First we observe that, instead of the cross correlation characteristic of linear systems [Eqs. (19.58), (20.6)], we have now a *generalized* (*averaged*) *autocorrelation* (with weight $\bar{\mathbf{G}}$) [cf. Eq. (19.59)] and *signal suppression;* that is, log $\Lambda_n(\mathbf{V})$ depends on the *square* ($\sim\overline{a_0^2}$) of the input signal-to-noise (amplitude) ratio, rather than on the first power ($\sim\bar{a}_0$ or $\sqrt{\overline{a_0^2}}$). Next, we can, in fact, interpret $\tilde{\mathbf{v}}\bar{\mathbf{G}}\mathbf{v}$ in Eqs. (20.11) as a sequence of elements consisting of a time-varying linear filter and a zero-memory square-law rectifier, followed by a time-varying linear filter or simple integrator. As we shall see below, this prerectifier time-varying linear filter may be resolved into many different forms. Here the principal ones we shall consider are (1) a linear invariant filter, with a time-varying switch, and (2) a parallel combination of time-varying gains and linear invariant filters. In special cases, still other structures are possible [cf. Eqs. (20.69*c*), (20.69*d*)]. To see this, let us begin with case 1 and introduce a (real) linear discrete filter $\mathbf{Q}$ [$\neq \mathbf{Q}_c$, Eq. (20.8)], such that

$$\mathbf{v}_F = \mathbf{Q}\mathbf{v} \tag{20.12}$$

Then we have

$$\Psi_n(\mathbf{v}) \equiv \tilde{\mathbf{v}}\bar{\mathbf{G}}\mathbf{v} = \tilde{\mathbf{v}}_F\tilde{\mathbf{Q}}^{-1}\bar{\mathbf{G}}\mathbf{Q}^{-1}\mathbf{v}_F > 0 \tag{20.13}$$

where $\mathbf{v}_F$ is the output of this filter. Note not only that the (real) matrix $\bar{\mathbf{G}}$ is symmetric [cf. Eq. (20.11*b*)] but also that it is such as to make the quadratic form Ψ_n positive definite, since for arbitrary (non-null) inputs there is always some output energy from the system.

† Compared with the duration of the signal or with its mean fluctuation time, whichever is smaller.

Our next step is to select $\mathbf{Q}$ in Eq. (20.12) so that $\bar{\mathbf{G}}$ is diagonalized and, in fact, is transformed into the unit matrix $\mathbf{I}$. Because of the symmetry of $\bar{\mathbf{G}}$ and the positive definiteness of Ψ_n, this, too, can always be accomplished with the help of either a suitable *congruent* transformation or a product of a suitable diagonal and *orthogonal* transformation.† We consider first diagonalization with a congruent transformation, which, as we shall see presently, leads to the invariant filter and time-varying switch of case 1. We have, accordingly,

$$\tilde{\mathbf{Q}}^{-1}\bar{\mathbf{G}}\mathbf{Q}^{-1} = \mathbf{I} \qquad \text{or} \qquad \bar{\mathbf{G}} = \tilde{\mathbf{Q}}\mathbf{Q} \tag{20.14}$$

Accordingly, Ψ_n is reduced to an unweighted sum of squares and from Eq. (20.14) becomes directly

$$\Psi_n = \tilde{\mathbf{v}}_F\mathbf{v}_F = \sum_{j=1}^{n} (v_F)_j{}^2 \tag{20.15}$$

which is just a zero-memory square-law operation on $(v_F)_j$ followed by a simple "integration" over the interval $(0,T)$.

So far, $\mathbf{Q}$ is required to be linear and such that Eq. (20.14) holds. Again because of the symmetry and positive definiteness of $\bar{\mathbf{G}}$, an additional condition of physical realizability can now be imposed upon this discrete filter $\mathbf{Q}$ without altering the value of the quadratic form Ψ_n. Specifically, we may say that

(1) $\mathbf{Q}$ is *linear* [this follows at once from Eq. (20.12)] (20.16*a*)

(2) $Q_{ij} = 0,\ t_i < t_j$ (or $i < j$); $\mathbf{Q}$ is *physically realizable*, in the sense of Sec. 2.2-5(3). Thus, $\mathbf{Q}$ acts only on the past of the input. This follows at once from the fact that it is possible to find a set of congruent transforma-

† The argument may be outlined as follows: We remark first that to obtain a congruent transformation which reduces $\bar{\mathbf{G}}$ to the unit matrix $\mathbf{I}$ we must show that the eigenvalues of $\bar{\mathbf{G}}$ are positive. Now, since $\bar{\mathbf{G}} = \tilde{\bar{\mathbf{G}}}$, we can always find an *orthogonal* transformation $\mathbf{Q}_0$ ($\neq \mathbf{Q}$) to reduce $\tilde{\mathbf{v}}\bar{\mathbf{G}}\mathbf{v}$ to diagonal form. Thus, if $\mathbf{y} = \mathbf{Q}_0\mathbf{v}$, then $\Psi_n = \tilde{\mathbf{y}}\mathbf{\Lambda}\mathbf{y}$, where $\tilde{\mathbf{Q}}_0{}^{-1}\bar{\mathbf{G}}\mathbf{Q}_0{}^{-1} = \mathbf{Q}_0\bar{\mathbf{G}}\mathbf{Q}_0{}^{-1} = \mathbf{\Lambda} = [\lambda_j d_{ij}]$ since $\mathbf{Q}_0 = \tilde{\mathbf{Q}}_0{}^{-1}$. But, inasmuch as $\mathbf{v}$ (and $\mathbf{y}$) are arbitrary, and $\Psi_n > 0$, therefore $\tilde{\mathbf{y}}\mathbf{\Lambda}\mathbf{y} > 0$ and all eigenvalues λ_j of $\bar{\mathbf{G}}$ are likewise positive. [This is easily seen by setting $y_i = 1$ and all other elements zero in y; then, $\tilde{\mathbf{y}}\mathbf{\Lambda}\mathbf{y} = \lambda_i\ (> 0)$, etc., for $i = 1, \ldots, n$.] Next, since $\lambda_j > 0$ (all j), and again since $\bar{\mathbf{G}} = \tilde{\bar{\mathbf{G}}}$, a congruent transformation $\mathbf{Q}$ can now be obtained which will reduce $\bar{\mathbf{G}}$ to the unit matrix, according to Eq. (20.14) (for the proof and further details, see, for example, Margenau and Murphy,[3] secs. 10.16, 10.17). Note that generally $\mathbf{Q} \neq \mathbf{Q}_0$; $\mathbf{Q}$ is not unique, while $\mathbf{Q}_0$ is. Moreover, unless $\bar{\mathbf{G}} = \mathbf{I}$, it is not possible to impose orthogonality on $\mathbf{Q}$ *and* reduce $\mathbf{\Lambda}$ to $\mathbf{I}$ at the same time. A second transformation is required, namely, $\mathbf{y}' = \mathbf{Q}'\mathbf{y}$, where $\mathbf{Q}' = [\lambda_j{}^{1/2}\delta_{ij}]$, so that one has finally $\mathbf{y}' = \mathbf{Q}_0'\mathbf{v}$, with $\mathbf{Q}_0' = \mathbf{Q}'\mathbf{Q}_0$. Note that $\mathbf{Q}_0' = \tilde{\mathbf{Q}}_0'$ if $(\mathbf{Q}_0)_{ij} = (\lambda_j/\lambda_i)^{1/2}(\mathbf{Q}_0)_{ji}$, as assumed for the example, footnote p. 844.

tions (e.g., a set of $\mathbf{Q}$'s) diagonalizing $\bar{\mathbf{G}}$ such that for each transformation $Q_{ij} = 0$ $(i < j)$ (cf. Margenau and Murphy, sec. 10.16)† (20.16*b*)

(3) $\mathbf{Q} = [H(t_i,t_j)\,\Delta t] = [h(t_i - t_j,\, t_i)\,\Delta t]$, all t_i, t_j; that is, $\mathbf{Q}$ here is *time-varying*. [Note that $H(t_i,t_j) \neq H(t_j,t_i)$, since $\mathbf{Q} \neq \tilde{\mathbf{Q}}$.] (20.16*c*)

The general relation determining $\mathbf{Q}$ [and hence $H(t_i,t_j) = h(t_i - t_j,\, t_i)$] is in any case the set (20.14) of simultaneous *nonlinear* equations for the Q_{ij}, namely,

$$\bar{\mathbf{G}} = \tilde{\mathbf{Q}}\mathbf{Q} \tag{20.14}$$

Writing

$$\bar{\mathbf{G}} \equiv [\rho_G(t_i,t_j)\,\Delta t] = [\rho_G(t_j,t_i)\,\Delta t] = \boldsymbol{\varrho}_G\,\Delta t \tag{20.17}$$

we observe further that $\boldsymbol{\varrho}_G$ vanishes for all t_i, t_j outside $(0,T)$, that is, for i, $j > n$ or ≤ 1, since only operations on data acquired in $(0,T)$ are to be used in the decision process. Equation (20.14) reduces explicitly to

$$\rho_G(t_i,t_j) = \begin{cases} \sum_{l=1}^{n} H(t_l,t_i)H(t_l,t_j)\,\Delta t = \sum_{l=1}^{n} h(t_l - t_i,\, t_l)h(t_l - t_j,\, t_l)\,\Delta t & 1 \leq i, j \leq n \\ 0 & \text{elsewhere} \end{cases} \tag{20.18a}$$

To obtain h we next observe that this (realizable) time-varying filter is equivalent to the (realizable) *time-invariant* filter and a time-varying switch of case 1; that is, we can replace $h(t_l - t_i,\, t_l)$ by $h(t_l - t_i)$, etc., in Eq. (20.18*a*), *provided* that we impose in addition to realizability [that is, $h(t_l - t_i) = 0$, $t_i > t_l$] the further condition that $h(t_l - t_i) = 0$, all $t_i < 0$, as well. This is equivalent to $h(z_i) = 0$, $z_i > t_l$ (> 0), $i = 1, \ldots, n$, and is required by the original condition on ρ_G that ρ_G vanish outside $(i,j = 1, \ldots, n)$ or $(0,T)$. Accordingly, we can rewrite Eq. (20.18*a*) as

$$\rho_G(t_i,t_j) = \begin{cases} \sum_{l=1}^{n} h(t_l - t_i)h(t_l - t_j)\,\Delta t & 1 \leq i, j \leq n \\ 0 & \text{elsewhere} \end{cases} \tag{20.18b}$$

† Note that $\mathbf{Q}^{-1}$ here is equivalent to $\mathbf{Q} \equiv \mathbf{Q}_{MM}$ ($MM \equiv$ Margenau and Murphy, *op. cit.*, sec. 10.16), so that the realizability condition $Q_{ij} = 0$ $(i < j)$ here corresponds to the fact that $(\mathbf{Q}_{MM})_{ij} = 0$ $(i > j)$ since the elements of an inverse matrix are formed from the transpose of the *adjoint* matrix, divided by the determinant of the original matrix. Observe that $\mathbf{Q}_0' = \mathbf{Q}'\mathbf{Q}_0$, $\mathbf{Q}' \equiv [\lambda_i^{1/2}\delta_{ij}]$ (cf. preceding footnote) is also a congruent transformation which reduces $\bar{\mathbf{G}}$ to $\mathbf{I}$, as does $\mathbf{Q}$ above. But $(\mathbf{Q}_0')_{ji} \neq 0$ and $(\mathbf{Q}_0')_{ij} \neq 0$ $(i < j)$, so that $\mathbf{Q}_0'$ is not acceptable if we require physical realizability in the sense of Sec. 2.2-5(3).

Thus, under conditions 1 to 3, and subject to Eq. (20.14), Eq. (20.15) becomes

$$\Psi_n = \tilde{\mathbf{v}}\bar{\mathbf{G}}\mathbf{v} = \sum_{j=1}^{n} \left[\sum_{i=1}^{\infty \text{ or } j} h(t_j - t_i)v(t_i)\,\Delta t \right]^2 \tag{20.19a}$$

$$= \sum_{j=1}^{n} \left[\sum_{i=1}^{j} h(t_i)v(t_j - t_i)\,\Delta t \right]^2 \tag{20.19b}$$

which reveals the time-varying switch ($j = 1, \ldots, n$). Finally, finite sample size is ensured by a suitable switch, or read-out, at the end of the observation period $(0,T)$. We remark that without the additional feature of a time-varying switch associated with the prerectifier filter h it is not possible to write Eq. (20.18b) with $\rho_G = 0$ ($i, j > n, < 1$). In fact, we

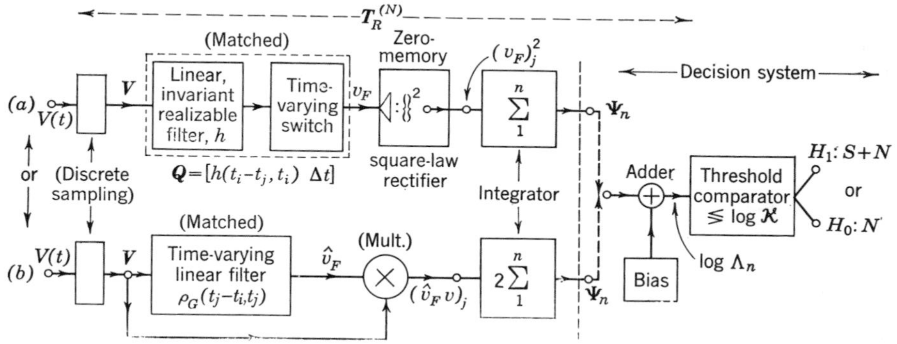

FIG. 20.2. Two equivalent interpretations of optimum threshold structure for binary incoherent detection of a signal in noise.

cannot have this (or *any*) time-invariant filter h alone (which would be equivalent to replacing $i = 1$ by $i = -\infty$ in the lower limit of the summation) and still satisfy the condition that ρ_G vanish outside $1 \le i, j \le n$, that is, maintain a finite sample size.† The structure of this optimum threshold system is indicated in Fig. 20.2*a*, where the time-varying switch "scans" the instantaneous output of the invariant filter $h(t_i - t_j)$, passing the result to the zero-memory quadratic detector, as indicated. Note that

† This is simply illustrated by the following example: Consider the 2×2 matrix for time-invariant $\mathbf{Q} = [h(t_i - t_j)\,\Delta t]$, $1 \le i, j \le 2$, namely,

$$\mathbf{Q} = \begin{bmatrix} x & y \\ 0 & x \end{bmatrix} \qquad \text{where now } \bar{\mathbf{G}} = \begin{bmatrix} a & c \\ c & b \end{bmatrix}$$

Then, $\tilde{\mathbf{Q}}\mathbf{Q} \neq \bar{\mathbf{G}}$ in general; we get $a = x^2 + y^2$, $b = x^2$, $c = xy$, and with a, b, c given there are too many relations for x and y to satisfy. The time-varying switch condition, however, provides the needed third variable in $\mathbf{Q}$.

Eqs. (20.19) can also be regarded as a simple autocorrelation of the filtered wave $\mathbf{v}_F$ with itself, subject to zero delay† [cf. Eq. (20.15)].

The interpretation above is not, however, unique: still another equivalent representation of system structure may be made in terms of a linear (discrete) time-varying filter and multiplier with zero delay. For this, we observe that $\bar{\mathbf{G}}$ [Eq. (20.17)] can be regarded as the weighting function $(\cdot\,\Delta t)$ of a time-varying linear filter, since $\bar{\mathbf{G}}$ is symmetrical and since $\rho_G(t_i,t_j) \equiv \hat{h}(t_j - t_i, t_j) = A(t_i - t_j, t_i) = B(t_i, t_i - t_j)$, etc. Thus, if $\hat{\mathbf{v}}_F$ is the output of this filter, we have

$$\hat{\mathbf{v}}_F = \bar{\mathbf{G}}\mathbf{v} = \sum_{i=1}^{j} \hat{h}(t_j - t_i, t_j)v(t_i)\,\Delta t \tag{20.20}$$

† A further class of possible representations for the time-varying filter preceding square-law rectification [implied by Eq. (20-13)], namely, case 2 mentioned above, follows when we choose a different congruent transformation $\mathbf{Q}_0' = \mathbf{Q}'\mathbf{Q}_0$, based on the orthogonal transformation $\mathbf{Q}_0$ of the footnote following Eq. (20.16*b*). Here $\mathbf{Q}_0' = [H'(t_i,t_j)\,\Delta t]$ is symmetric, unlike the $\mathbf{Q}$ of Eq. (20.16*a*), so that $H'(t_i,t_j) = H'(t_j,t_i) \neq H(t_i,t_j)$ of Eq. (20.16*c*). The basic relation (20.14) is now modified to

$$\bar{\mathbf{G}} = \tilde{\mathbf{Q}}_0'\mathbf{Q}_0' \tag{1}$$

and the analogue of Eqs. (20.18*a*) is accordingly

$$\rho_G(t_i,t_j) = \begin{cases} \displaystyle\sum_{l=1}^{n} H'(t_l,t_i)H'(t_l,t_j)\,\Delta t & 1 \le i, j \le n \\ 0 & \text{elsewhere} \end{cases} \tag{2}$$

The solution of this for H' may be effected if we write

$$H'(t_l,t_i) = \sum_{m=1}^{n} a_m(t_l)g_m(t_i) \tag{3}$$

such that a_m, g_m vanish for all l, $i > n$, < 1 and the a_m $(= g_m)$ are orthogonal on i, j, $m = 1, \ldots, n$, etc., with respect to ρ_G in the manner of Eqs. (17.3*a*), (17.3*b*), for example. Then we can write $g_m(t_i) = h_m(T - t_i)$, where h_m is realizable and the a_m are time-varying gains, so that we have the structure of case 2 mentioned earlier. Since H' is symmetrical, the time-varying gains and invariant filters are interchangeable. (For details, see footnote ‡ on page 857 for continuous sampling.) Note that $H'(t_i,t_j) = h'(t_i - t_j, t_i)$, etc., is not realizable in the usual sense of Sec. 2.2-5(3), since $h' \neq 0$, $t_j > t_i$, in view of the symmetry of $\mathbf{Q}_0'$.

The procedure sketched here is somewhat similar to the methods for solving the nonlinear (integral) equations analogous to case 2 (in the continuous case) employed independently by Kailath,[41] who, in addition, independently obtained representations analogous to case 1 in the text, using an invariant filter and time-varying switch. Finally, we remark that unless otherwise indicated we shall in our discussion of structure for incoherent reception refer to the realizable cases only, e.g., case 1 [Eqs. (20.16) et seq.].

and therefore

$$\Psi_n = 2\tilde{\hat{\mathbf{v}}}_F{}^{-1}\mathbf{v} = 2\tilde{\mathbf{v}}\hat{\mathbf{v}}_F = 2\sum_{j=1}^{n}(\hat{v}_F)_j v_j = 2\sum_{j=1}^{n} v_j \sum_{i=1}^{j} v_i \hat{h}(t_j - t_i, t_j)\,\Delta t \qquad (20.21)$$

We note also that $\hat{h}$ does not have to vanish $(i, j > n)$; however, physical realizability again requires $\hat{h}(t_j - t_i, t_j) = 0$ for $t_i > t_j$ (see remarks on page 858). A switch, or readout, at $t = T$ following the "integration" gives Ψ_n. The structure of this system is shown in Fig. 20.2*b*, including threshold comparison and decision output. The essential features of incoherent reception are again emphasized: the optimum threshold system is nonlinear, with signal suppression [cf. Eq. (19.59)], in contrast to the corresponding situation for coherent systems, which are essentially linear [cf. Eq. (19.58)]. We remark again that if $\langle s \rangle_\epsilon \neq 0$, although ϵ is not precisely known to the observer, the linear, as well as the quadratic, operation on the received data is required [cf. Eq. (20.5)]. Structurally, the resulting optimum system is an appropriate linear combination of the systems illustrated in Figs. 20.1, 20.2.

Finally, we observe that unlike the situation in coherent detection, it is *not generally possible to employ a linear invariant and realizable filter* h_M *defined*, analogous to Eq. (20.10*a*), *by*

$$\bar{\gamma} = [h_M(T - t_i; T)\,\Delta t] \qquad \mathbf{K}_N{}^{-1}\bar{\mathbf{s}} = \bar{\gamma} \qquad (20.21a)$$

This is easily seen by examining $\bar{\mathbf{G}}$ [Eqs. (20.11*b*)]; we can write from Eqs. (20.21*a*)

$$\tilde{\mathbf{v}}\bar{\mathbf{G}}\mathbf{v} = \psi^2\tilde{\mathbf{v}}\overline{\gamma\tilde{\gamma}}\mathbf{v} = \psi^2\langle(\tilde{\gamma}\mathbf{v})^2\rangle \neq \psi^2(\tilde{\bar{\gamma}}\mathbf{v})^2 \qquad (20.21b)$$

so that we cannot simply regard $\tilde{\mathbf{v}}\bar{\mathbf{G}}\mathbf{v}$ as the square of the output of the filter h_M when $\mathbf{v}$ is the input. This is possible only when the signal class has but one member, i.e., is entirely determinate, in contradiction, however, to the original assumption of incoherent observation, which gives epoch ϵ at least more than one possible value. An approximate exception occurs in the case of narrowband signals, if amplitude and RF epoch are the only distributed parameters [cf. Eq. (20.31*a*) and Prob. 20.6].

20.1-3 Incoherent Detection II. Narrowband Signals. Although Eq. (20.11*a*) is adequate for incoherent threshold reception (in normal noise), regardless of the structure of the possible signal input and regardless of the particular statistics of ϵ, when the signals are narrowband *and* ϵ is uniformly distributed we can take advantage of the narrowband character of the input signal to use an alternative development in place of Eq. (20.3). This is obtained from the exact relation (19.41) by explicitly carrying out the average over the epoch ϵ before $\log \Lambda_n$ is expanded in a power series in the input signal-to-noise ratio a_0. Such a procedure often leads to comparatively simple closed forms for detector structure, as will be seen presently.

We begin, accordingly, with Eq. (20.1) and observe that a narrowband deterministic signal process $s(t - \epsilon, \boldsymbol{\theta})$ to a considerable degree of generality can be represented by

$$s(t - \epsilon_c, \boldsymbol{\theta}) = F_s(t;\boldsymbol{\theta}_1) \cos [\omega_s(t - \epsilon_c) - \phi_s(t;\boldsymbol{\theta}_2)] \tag{20.22}$$

Here $\boldsymbol{\theta}$ is the set of signal parameters ω_s, $\boldsymbol{\theta}_1$, $\boldsymbol{\theta}_2$, and ϵ_0 is replaced by ϵ_c, the epoch over a cycle of the carrier frequency f_s $(= \omega_s/2\pi)$; F_s, ϕ_s are, respectively, slowly varying envelope and phase functions, essentially independent of ϵ_c, while $\boldsymbol{\theta}_1$, $\boldsymbol{\theta}_2$ represent all other pertinent statistical parameters.

The quadratic forms Φ_s, Φ_v [cf. Eqs. (20.2*a*)] may now be obtained explicitly. Inserting Eq. (20.22) into Eqs. (20.2*b*), we find that

$$\begin{aligned}\Phi_s = \tilde{\mathbf{s}}\mathbf{k}_N^{-1}\mathbf{s} \doteq \frac{1}{2} \sum_{ij} (\mathbf{k}_N^{-1})_{ij} F_s(t_i;\boldsymbol{\theta}_1) F_s(t_j;\boldsymbol{\theta}_1) \\ \times \{\cos [\omega_s(t_i - t_j) - \phi_s(t_i;\boldsymbol{\theta}_2) + \phi_s(t_j;\boldsymbol{\theta}_2)] \\ + \cos [\omega_s(t_i + t_j) - \phi_s(t_i,\boldsymbol{\theta}_2) - \phi_s(t_j;\boldsymbol{\theta}_2)]\}\end{aligned} \tag{20.23}$$

Abbreviating $F_s(t_i;\boldsymbol{\theta}_1)$ by F_{si}, $\phi_s(t_i;\boldsymbol{\theta}_2)$ by ϕ_{si}, etc., we define a pair of column vectors by

$$\mathbf{a} \equiv [F_{si} \cos (\omega_s t_i - \phi_{si})] \qquad \mathbf{b} \equiv [F_{si} \sin (\omega_s t_i - \phi_{si})] \tag{20.24a}$$

so that Eq. (20.22) can be alternatively written

$$s(t_i - \epsilon_c, \boldsymbol{\theta}) = (\mathbf{a})_i \cos \omega_s\epsilon_c + (\mathbf{b})_i \sin \omega_s\epsilon_c \tag{20.24b}$$

Then we can express Φ_s [Eq. (20.23)] more compactly as

$$\begin{aligned}\Phi_s = \tfrac{1}{2}(\tilde{\mathbf{a}}\mathbf{k}_N^{-1}\mathbf{a} + \tilde{\mathbf{b}}\mathbf{k}_N^{-1}\mathbf{b}) + \tfrac{1}{2}(\tilde{\mathbf{a}}\mathbf{k}_N^{-1}\mathbf{a} - \tilde{\mathbf{b}}\mathbf{k}_N^{-1}\mathbf{b}) \cos 2\omega_s\epsilon_c \\ + \tilde{\mathbf{a}}\mathbf{k}_N^{-1}\mathbf{b} \sin 2\omega_s\epsilon_c\end{aligned} \tag{20.25}$$

But these last two sets of terms are just the sum terms in the cosine product [cf. Eq. (20.23)], which contribute negligibly to the total (when summed over i, j),† because of the rapid oscillations of this portion of the summand.‡ Consequently, Φ_s reduces to the still more compact expression

$$\Phi_s \doteq \tfrac{1}{2}(\tilde{\mathbf{a}}\mathbf{k}_N^{-1}\mathbf{a} + \tilde{\mathbf{b}}\mathbf{k}_N^{-1}\mathbf{b}) \equiv B_n(\boldsymbol{\theta}) \tag{20.26}$$

The significant feature here is that Φ_s is effectively independent of ϵ_c, an approximation dependent on the narrowband character of the signal process.

The quantity Φ_v may be handled in a similar way, although it will prove

† It is assumed that the sampling points are not commensurable with the period of the carrier.

‡ The narrowband character of the signal process is sufficient to ensure this, even if the noise process is broadband. This result may be demonstrated, for example, quite directly with the help of the continuous forms of Sec. 20.2-3.

dependent on ϵ_c. Using Eqs. (20.24*a*), (20.24*b*) in Eqs. (20.2*a*), we get

$$\Phi_v = \tilde{\mathbf{v}}\mathbf{k}_N^{-1}\mathbf{s} = \sum_{ij} v_i(\mathbf{k}_N^{-1})_{ij}F_{sj}\cos[\omega_s(t_j - \epsilon_c) - \phi_{sj}]$$

$$= (\tilde{\mathbf{v}}\mathbf{k}_N^{-1}\mathbf{a})\cos\omega_s\epsilon_c + (\tilde{\mathbf{v}}\mathbf{k}_N^{-1}\mathbf{b})\sin\omega_s\epsilon_c \tag{20.27a}$$

$$\Phi_v = \tilde{\mathbf{v}}\mathbf{k}_N^{-1}\mathbf{s} = \gamma(\mathbf{v};\boldsymbol{\theta})\cos[\omega_s\epsilon_c - \eta(\mathbf{v};\boldsymbol{\theta})] \tag{20.27b}$$

where now†

$$\gamma = [(\tilde{\mathbf{v}}\mathbf{k}_N^{-1}\mathbf{a})^2 + (\tilde{\mathbf{v}}\mathbf{k}_N^{-1}\mathbf{b})^2]^{1/2} \qquad \eta = \tan^{-1}(\tilde{\mathbf{v}}\mathbf{k}_N^{-1}\mathbf{b}/\tilde{\mathbf{v}}\mathbf{k}_N^{-1}\mathbf{a}) \tag{20.28}$$

Note that γ^2 can be written

$$\gamma^2 = \tilde{\mathbf{v}}(\mathbf{k}_N^{-1}\mathbf{a}\tilde{\mathbf{a}}\mathbf{k}_N^{-1} + \mathbf{k}_N^{-1}\mathbf{b}\tilde{\mathbf{b}}\mathbf{k}_N^{-1})\mathbf{v} \equiv \tilde{\mathbf{v}}\mathbf{G}_0\mathbf{v} \tag{20.29}$$

with

$$\mathbf{G}_0 = \mathbf{k}_N^{-1}(\mathbf{a}\tilde{\mathbf{a}} + \mathbf{b}\tilde{\mathbf{b}})\mathbf{k}_N^{-1} \equiv \mathbf{G}_0(\boldsymbol{\theta}) \doteq 2\mathbf{k}_N^{-1}\langle\mathbf{s}\tilde{\mathbf{s}}\rangle_{\epsilon_c}\mathbf{k}_N^{-1} = 2\mathbf{G}(\boldsymbol{\theta}) \tag{20.29a}$$

[cf. Eqs. (20.11*b*)], so that $\mathbf{a}\tilde{\mathbf{a}} + \mathbf{b}\tilde{\mathbf{b}} \doteq 2\mathbf{s}\tilde{\mathbf{s}}$. Here we write $\mathbf{G}_0 = \mathbf{G}_0(\boldsymbol{\theta})$ to remind ourselves that $\mathbf{G}_0$ (and $\mathbf{G}$) depend on the statistical parameters of the signal, as well as on its structure and on the background noise.

To carry out the average over ϵ_c in Eq. (20.1), we need Eq. (A.1.48*b*), observing that for the uniform d.d. of ϵ_c over an RF cycle we get a modified Bessel function:

$$\int_{\epsilon_c} w(\epsilon_c)e^{\gamma\cos(\omega_0\epsilon_0 - \eta)}\,d\epsilon_c = \int_0^{2\pi}\frac{e^{\gamma\cos(x-\eta)}}{2\pi}\,dx = I_0(\gamma) \tag{20.30}$$

The result for Eq. (20.1) is finally

$$T_R^{(N)}\{V\} = \log\Lambda_n(\mathbf{V}) = \log\mu + \log\langle e^{-a_0{}^2B_n(\boldsymbol{\theta})/2}I_0[a_0\sqrt{\tilde{\mathbf{v}}\mathbf{G}_0(\boldsymbol{\theta})\mathbf{v}}]\rangle_{a_0,\boldsymbol{\theta}} \tag{20.31}$$

with $B_n(\boldsymbol{\theta})$ given specifically by Eq. (20.26). This expression is still valid for all signal strengths, and in the special case where only single values of a_0 and $\boldsymbol{\theta}$ can occur, for example, $w(\boldsymbol{\theta}) = \delta(\boldsymbol{\theta} - \boldsymbol{\theta}_0)$, etc., reduces at once to the exact‡ relation

$$\log\Lambda_n(\mathbf{V}) = \left[\log\mu - \frac{a_0{}^2}{2}B_n(\boldsymbol{\theta}_0)\right] + \log I_0[a_0\sqrt{\tilde{\mathbf{v}}\mathbf{G}_0(\boldsymbol{\theta}_0)\mathbf{v}}] \tag{20.31a}$$

However, for the general situation represented by Eq. (20.31) it is not possible to obtain results in closed form, like Eq. (20.31*a*). We must develop $\log\Lambda_n$ once again as a power series in the signal-to-noise ratio a_0 to achieve the explicit structure of the optimum threshold system here.

† Here, and unless otherwise indicated, $\boldsymbol{\theta}$ represents all signal parameters other than those exhibited explicitly, e.g., other than ϵ_c, ω_s in Eq. (20.27*b*).

‡ "Exact," consistent with the narrowband approximation inherent in Eqs. (20.26) and (20.29*a*).

The result is finally

$$T_R^{(N)}\{V\} = \log \Lambda_n(\mathbf{V}) \doteq (\log \mu + \mathcal{B}_{0n}^{(2)} + \mathcal{B}_{0n}^{(4)}) + \frac{\overline{a_0^2}}{4} \tilde{\mathbf{v}}\overline{\mathbf{G}_0(\boldsymbol{\theta})}\mathbf{v} \quad (20.32)$$

where the bias terms are†

$$\mathcal{B}_{0n}^{(2)} = -\frac{\overline{a_0^2}}{2} \langle B_n(\boldsymbol{\theta})\rangle \doteq \mathcal{B}_n^{(2)} = -\frac{\overline{a_0^2}}{2} \operatorname{tr} \mathbf{k}_N\overline{\mathbf{G}} \quad (20.33a)$$

$$\mathcal{B}_{0n}^{(4)} \doteq \mathcal{B}_n^{(4)} = -\frac{\overline{a_0^2}^2}{4} \operatorname{tr} (\mathbf{k}_N\overline{\mathbf{G}})^2 \doteq -\frac{\overline{a_0^2}^2}{16} \operatorname{tr} (\mathbf{k}_N\overline{\mathbf{G}}_0)^2 \quad (20.33b)$$

consistent with the narrowband approximations [Eqs. (20.26), (20.29a)], namely, $B_n \doteq \Phi_s$, $\mathbf{G}_0 \doteq 2\mathbf{G}$, with $\Psi_{0n} \equiv \tilde{\mathbf{v}}\overline{\mathbf{G}}_0\mathbf{v}$. Equation (20.32) is, as expected, identical with the precise threshold relation (20.11a) to within this approximation, subject to the large-sample (i.e., long-observation-time) condition required of threshold operation (cf. Prob 20.3).

The structure of the optimum threshold system of Eq. (20.32) is similar to that outlined in Sec. 20.1-2, except that $\overline{\mathbf{G}}$ is now replaced by $\overline{\mathbf{G}}_0$, altering the particular elements but not their general purpose and order: Fig. 20.2 still applies, with this modification. As we shall remark in more detail presently (Sec. 20.2-4), the linear prerectification filters $\mathbf{Q}$ belong to a general class of matched filters (cf. Sec. 16.3-2), with the important property of minimizing the effects of the background noise vis-à-vis the desired signal. Finally, we remind ourselves that the log I_0 structure exhibited in Eqs. (20.31), (20.31a) and the expansion (20.32) all depend on (1) the additive normal character of the noise, (2) threshold operation, (3) the uniform distribution of epoch ϵ_c over a typical RF cycle, and (4) the narrowband character of the desired signal, in contrast with the more general relations (20.4), (20.5) where (3) and (4) are not necessarily imposed.

PROBLEMS

20.1 (a) Obtain the following statistical averages ($\mathbf{V}$ normal with $\bar{\mathbf{V}} = 0$):

$$E_{\mathbf{V}|H_0}\{(\tilde{\mathbf{v}}\mathbf{k}_N^{-1}\mathbf{a})(\tilde{\mathbf{v}}\mathbf{G}\mathbf{v})\} = 0 \quad (1)$$

$$E_{\mathbf{V}|H_0}\{(\tilde{\mathbf{v}}\mathbf{k}_N^{-1}\mathbf{a})(\tilde{\mathbf{v}}\mathbf{b})\} = \tfrac{1}{2}(\tilde{\mathbf{a}}\mathbf{b} + \tilde{\mathbf{b}}\mathbf{a}) \quad (2)$$

$$E_{\mathbf{V}|H_0}\{(\tilde{\mathbf{v}}\mathbf{k}_N^{-1}\mathbf{a})(\tilde{\mathbf{v}}\mathbf{k}_N^{-1}\mathbf{c})\} = 3(\tilde{\mathbf{a}}\mathbf{k}_N^{-1}\mathbf{c})(\tilde{\mathbf{c}}\mathbf{k}_N^{-1}\mathbf{c}) \quad (3)$$

HINT: Obtain first the characteristic functions

$$E_{\mathbf{V}|H_0}\{\exp\,(i\xi_1\tilde{\mathbf{v}}\mathbf{k}_N^{-1}\mathbf{a} + i\xi_2\tilde{\mathbf{v}}\mathbf{G}\mathbf{v})\} = \frac{\exp\,[-\tfrac{1}{2}\xi_1^2\tilde{\mathbf{a}}\mathbf{k}_N^{-1}(\mathbf{I} - 2i\xi_2\mathbf{k}_N\mathbf{G})^{-1}\mathbf{a}]}{\{\det\,(\mathbf{I} - 2i\xi_2\mathbf{k}_N\mathbf{G})\}^{1/2}} \quad (4)$$

$$E_{\mathbf{V}|H_0}\{\exp\,(i\xi_1\tilde{\mathbf{v}}\mathbf{k}_N^{-1}\mathbf{a} + i\xi_2\tilde{\mathbf{v}}\mathbf{b})\} = \exp\,[-\tfrac{1}{2}\xi_1^2\tilde{\mathbf{a}}\mathbf{k}_N^{-1}\mathbf{a} - \xi_2^2\tilde{\mathbf{b}}\mathbf{k}_N^{-1}\mathbf{b} - \xi_1\xi_2(\tilde{\mathbf{a}}\mathbf{b} + \tilde{\mathbf{b}}\mathbf{a})] \quad (5)$$

where the $\mathbf{v}$ are normally distributed ($\bar{\mathbf{v}} = 0$), with covariance $\mathbf{k}_N$. Differentiation for the appropriate averages, with the help of Eq. (17.25), then gives Eqs. (1) to (3).

(b) With the results of (a), obtain Eqs. (20.4a), (20.4b).

† See Prob. 20.3 for the still more general situation where a_0 depends on other signal parameters as well.

20.2 When signal strength ($\sim a_0$) is also dependent on one or more of the other signal parameters, show that the threshold structure of the optimum detector of Eq. (20.5) becomes

$$\log \Lambda_n(\mathbf{V}) = [\log \mu + \mathcal{B}_n^{(2)} + \mathcal{B}_n^{(4)} + O(a_0^6)] + \langle a_0 \Phi_v \rangle + \tfrac{1}{2}[\langle (a_0\Phi_v)^2 \rangle - \langle a_0 \Phi_v \rangle^2] \tag{1}$$

with $\Phi_s = \tilde{\mathbf{s}}\mathbf{k}_N^{-1}\mathbf{s}$, $\Phi_v = \tilde{\mathbf{v}}\mathbf{k}_N^{-1}\mathbf{s}$, where specifically

$$\mathcal{B}_n^{(2)} = -\tfrac{1}{2}\langle a_0^2 \Phi_s \rangle \qquad \text{bias } O(a_0^3) = 0 \tag{2}$$

$$\begin{aligned} \mathcal{B}_n^{(4)} = \boldsymbol{E}_{\mathbf{V}|H_0} \Big\{ & \Big\langle a_0^4 \Big(\frac{\Phi_v^4}{4!} - \tfrac{1}{4}\Phi_v^2\Phi_s + \tfrac{1}{8}\Phi_s^2 \Big) \Big\rangle - \langle a_0 \Phi_v \rangle \Big\langle a_0^3 \Big(\frac{\Phi_v^3}{3!} - \frac{\Phi_v \Phi_s}{2!} \Big) \Big\rangle \\ & - \frac{1}{8} \langle a_0^2(\Phi_v^2 - \Phi_s) \rangle^2 + \frac{1}{2} \langle a_0 \Phi_v \rangle^2 \langle a_0^2(\Phi_v^2 - \Phi_s) \rangle - \frac{1}{4} \langle a_0 \Phi_v \rangle^4 \Big\} \end{aligned} \tag{3a}$$

$$\begin{aligned} = \{0 & - \tfrac{1}{2}[\langle (\overline{a_0\tilde{\mathbf{s}}}\mathbf{k}_N^{-1}\mathbf{s}a_0)(a_0^2\tilde{\mathbf{s}}\mathbf{k}_N^{-1}\mathbf{s}) \rangle - \langle \overline{a_0\tilde{\mathbf{s}}}\mathbf{s}a_0^3\tilde{\mathbf{s}}\mathbf{k}_N^{-1}\mathbf{s} \rangle] \\ & - \tfrac{1}{8}[2 \operatorname{tr} (\mathbf{k}_N\overline{\mathbf{G}'})^2 + \operatorname{tr}^2 (\mathbf{k}_N\overline{\mathbf{G}'}) - 2 \operatorname{tr} \mathbf{k}_N\overline{\mathbf{G}'}\langle a_0^2\Phi_s \rangle + \langle a_0^2 \Phi_s \rangle^2] \\ & + \overline{a_0\tilde{\mathbf{s}}}\,\overline{\mathbf{G}'}\,\overline{a_0\tilde{\mathbf{s}}} - \tfrac{3}{4}(\overline{a_0\tilde{\mathbf{s}}}\mathbf{k}_N^{-1}\overline{a_0\mathbf{s}})^2 \} \end{aligned} \tag{3b}$$

and $\bar{\mathbf{G}}' \equiv \mathbf{k}_N^{-1}\overline{\mathbf{a}_0^2\mathbf{s}\tilde{\mathbf{s}}}\mathbf{k}_N^{-1}$. Prove also that

$$\operatorname{tr} \overline{a_0^2\mathbf{s}\tilde{\mathbf{s}}}\mathbf{k}_N^{-1} = \overline{a_0^2\tilde{\mathbf{s}}\mathbf{k}_N^{-1}\mathbf{s}} = \langle a_0^2 \Phi_s \rangle \tag{4a}$$

$$\operatorname{tr} (\mathbf{k}_N\overline{a_0^2\mathbf{G}})^2 = \langle a_0^2\tilde{\mathbf{s}}\overline{a_0^2\mathbf{G}}\mathbf{s} \rangle \tag{4b}$$

20.3 (*a*) When signal amplitude a_0 is dependent on one or more of the other signal parameters, show that the threshold detector structure of Eq. (20.32) for narrowband signals becomes

$$\boldsymbol{T}_R\{V\} = \log \Lambda_n(\mathbf{V}) \doteq \log \mu - \tfrac{1}{2}\langle a_0^2 B_n \rangle + \mathcal{B}_{0n}^{(4)} + \tfrac{1}{4}\tilde{\mathbf{v}}\langle a_0^2 \mathbf{G}_0(\boldsymbol{\theta}) \rangle \mathbf{v} \tag{1}$$

where B_n is given by Eq. (20.26), $\mathbf{G}_0$ by Eq. (20.29), and

$$\begin{aligned} \mathcal{B}_{0n}^{(4)} = \tfrac{1}{8}[-\langle a_0^2 B_n \rangle^2 + \langle a_0^4 B_n^2 \rangle + \boldsymbol{E}_{\mathbf{V}|H_0}\{ & -\tilde{\mathbf{v}}\langle a_0^4 B_n \mathbf{G}_0 \rangle \mathbf{v} + \langle a_0^2 B_n \rangle \tilde{\mathbf{v}} \langle a_0^2 \mathbf{G}_0 \rangle \mathbf{v} \\ & - \tfrac{1}{4}(\tilde{\mathbf{v}}\langle a_0^2 \mathbf{G}_0 \rangle \mathbf{v})^2 + \tfrac{1}{8}\langle a_0^4 (\tilde{\mathbf{v}}\mathbf{G}_0\mathbf{v})^2 \rangle \}] \doteq \mathcal{B}_n^{(4)} \end{aligned} \tag{2}$$

The various averages $\boldsymbol{E}_{\mathbf{V}|H_0}$ may be obtained directly with the help of Eqs. (17.46*a*), (17.46*b*):

$$\boldsymbol{E}_{\mathbf{V}|H_0}\{\tilde{\mathbf{v}}\mathbf{G}_0\mathbf{v}\} = \operatorname{tr} \mathbf{k}_N\mathbf{G}_0 \qquad \boldsymbol{E}_{\mathbf{V}|H_0}\{(\tilde{\mathbf{v}}\mathbf{G}_0\mathbf{v})^2\} = 2 \operatorname{tr} (\mathbf{k}_N\mathbf{G}_0)^2 + \operatorname{tr}^2 \mathbf{k}_N\mathbf{G}_0 \tag{3}$$

(*b*) Obtain Eq. (20.32) for optimum threshold detection of narrowband signals, and verify Eqs. (20.33*a*), (20.33*b*) for the postulated narrowband and large-sample conditions.

20.2 *Threshold Structure II. Continuous Sampling*

In many applications, it is often more convenient to process the received data continuously, rather than to sample at discrete instants. This has the advantage of including all information concerning the signal, which in the discrete cases may be lost. However, a greater advantage is that analogue systems are often simpler than their corresponding discrete counterparts. Here we shall consider continuous optimum threshold systems as the natural limit of the discrete cases examined earlier, in Sec. 20.1, where we specifically employ the procedures of Sec. 19.4-2 in going from the discrete to the continuous state, and where we shall assume that the signal and noise processes are such as to permit passage to this limiting state, which itself always represents the regular, rather than the singular, case.

Let us begin with the general threshold system of Eq. (20.5) and remember that for continuous sampling the number n of sampling instants is made indefinitely large, while the acquisition period $(0,T)$ is held fixed. Thus, the optimum structure $\boldsymbol{T}_R^{(N)}\{V\} = \log \Lambda_n(\mathbf{V})$ is now replaced in the limit $n \to \infty$ by the functional†

$$\begin{aligned} \boldsymbol{T}_R^{(N)}\{V(t)\} &= \log \Lambda_T[V(t)] = \lim_{n\to\infty} \log \Lambda_n(\mathbf{V}) \\ &= [\log \mu + \mathcal{B}_T^{(2)} + \mathcal{B}_T^{(4)} + O(a_0^6)] \\ &\qquad + \bar{a}_0\langle(\Phi_T)_v\rangle + \frac{\overline{a_0^2}}{2}\left[\langle(\Phi_T)_v^2\rangle - \frac{\bar{a}_0^2}{\overline{a_0^2}}\langle(\Phi_T)_v\rangle^2\right] \end{aligned} \tag{20.34}$$

where

$$(\Phi_T)_v = \Phi_T(v,s) = \lim_{n\to\infty} \Phi_n(v,s) = \lim_{n\to\infty} \tilde{\mathbf{v}}\mathbf{k}_N^{-1}\mathbf{s} \qquad \text{with } v(t) = V(t)/\sqrt{\psi} \tag{20.35a}$$

$$(\Phi_T)_s = \Phi_T(s,s) = \lim_{n\to\infty} \Phi_n(s,s) = \lim_{n\to\infty} \tilde{\mathbf{s}}\mathbf{k}_N^{-1}\mathbf{s} \tag{20.35b}$$

and

$$\mathcal{B}_T^{(2)} = \lim_{n\to\infty} \mathcal{B}_n^{(2)} \qquad \mathcal{B}_T^{(4)} = \lim_{n\to\infty} \mathcal{B}_n^{(4)} \tag{20.36}$$

whenever these limits exist. As before, the $\langle\ \rangle$ indicate the appropriate statistical averages over signal parameters [cf. Eqs. (20.1) to (20.3)].

Application of Eqs. (19.50) through (19.57) to Eqs. (20.1) through (20.2*a*), (20.2*b*) in the present gaussian case gives, for the quadratic functionals $(\Phi_T)_v$, $(\Phi_T)_s$,

$$\begin{aligned} \Phi_T(v,s) &= \psi \int_{0-}^{T+} v(t)X(t;\epsilon,\boldsymbol{\theta})\,dt \\ \Phi_T(s,s) &= \psi \int_{0-}^{T+} s(t-\epsilon;\boldsymbol{\theta})X(t;\epsilon,\boldsymbol{\theta})\,dt \end{aligned} \tag{20.37}$$

where X is the solution of the following inhomogeneous Fredholm integral equation of the first kind:‡

$$\int_{0-}^{T+} K_N(t,u)X(u;\epsilon,\boldsymbol{\theta})\,du = s(t-\epsilon;\boldsymbol{\theta}) \qquad 0- < t < T+ \tag{20.38}$$

In a similar fashion, the bias term $\mathcal{B}_T^{(2)}$ is found to be

$$\mathcal{B}_T^{(2)} = -\frac{\overline{a_0^2}}{2}\langle\Phi_T(s,s)\rangle = -\frac{\overline{a_0^2}}{2}\psi\int_{0-}^{T+}\langle s(t-\epsilon;\boldsymbol{\theta})X(t;\epsilon,\boldsymbol{\theta})\rangle\,dt \tag{20.39a}$$

For the higher-order term, $O(a_0^4)$, we use the approach described presently

† For the still more general situation where $a_0 = a_0(\boldsymbol{\theta})$, see Prob. 20.4.

‡ See Sec. A.2.3 for methods of solution and specific results for rational covariance functions K_N. Note here that X is equal to $-2X$ of Eqs. (19.57*b*).

in Sec. 20.2-2 to achieve finally, as the continuous limit of Eq. (20.4b),

$$
\begin{aligned}
\mathcal{B}_T{}^{(4)} = & -\frac{\bar{a}_0\overline{a_0{}^3}}{2}\Bigg[\langle\Phi_T(s,s)\Phi_T(s,\bar{s})\rangle \\
& - \psi^2 \iint_{0-}^{T+} \langle s(t-\epsilon;\boldsymbol{\theta})\rangle\langle s(u-\epsilon;\boldsymbol{\theta})X(t;\epsilon,\boldsymbol{\theta})X(u;\epsilon,\boldsymbol{\theta})\rangle\,dt\,du\Bigg] \\
& - \frac{\overline{a_0{}^2}^2}{4}\psi^2 \iint_{0-}^{T+} \langle X(t;\epsilon,\boldsymbol{\theta})X(u;\epsilon,\boldsymbol{\theta})\rangle\langle s(t-\epsilon;\boldsymbol{\theta})s(u-\epsilon;\boldsymbol{\theta})\rangle\,dt\,du \\
& + \bar{a}_0{}^2\overline{a_0{}^2}\langle\Phi_T(s,\bar{s})^2\rangle - \tfrac{3}{4}\bar{a}_0{}^4\langle\Phi_T(\bar{s},\bar{s})\rangle^2 \qquad (20.39b)
\end{aligned}
$$

where

$$
\begin{aligned}
\Phi_T(s,\bar{s}) &= \psi \int_{0-}^{T+} \langle s(t-\epsilon,\boldsymbol{\theta})\rangle X(t;\epsilon,\boldsymbol{\theta})\,dt \\
\Phi_T(\bar{s},\bar{s}) &= \psi \int_{0-}^{T+} \langle s(t-\epsilon,\boldsymbol{\theta})\rangle\langle X(t;\epsilon,\boldsymbol{\theta})\rangle\,dt \qquad (20.39c)
\end{aligned}
$$

and X is the solution of Eq. (20.38), as before.†

As succeeding sections show in detail, the interpretation of detector structure $\log \Lambda_T$ is essentially identical with that already described above when discrete data sampling is employed, the only important distinction being the substitution of continuous elements and operations for the corresponding discrete ones. Again, when $\langle s\rangle$ does not vanish, the optimum threshold system may require both linear and square-law operations on the data $V(t)$ in $(0,T)$, according to Eq. (20.34).

20.2-1 Coherent Detection. Here, there is sample certainty, so that $\langle s\rangle_\epsilon$ does not vanish and the results of Sec. 20.1-1 may now be appropriately modified in the limit of continuous sampling. Again, the optimum threshold structure is primarily linear in the received data. With the help of Eqs. (20.37) to (20.39c) we readily find that Eq. (20.34) reduces to the continuous version of Eqs. (20.6), (20.7), where terms $O(\overline{a_0{}^2})$ in V are replaced by their average values, $\boldsymbol{E}_{\mathbf{V}|H_0}$, namely,

$$
T_R{}^{(N)}\{V(t)\} = \log \Lambda_T[V(t)] \doteq \left(\log\mu + \frac{\overline{a_0{}^2}}{2}B_T{}^{(2)}\right) + \bar{a}_0\psi\int_{0-}^{T+} v(t)\langle X(t;\epsilon,\boldsymbol{\theta})\rangle\,dt \qquad (20.40a)
$$

† By multiplying Eq. (20.38) by X and integrating over $(0-,T+)$, note that

$$
\int_{0-}^{T+} s(t-\epsilon,\boldsymbol{\theta})X(t;\epsilon,\boldsymbol{\theta})\,dt = \iint_{0-}^{T+} K_N(t,u)X(t;\epsilon,\boldsymbol{\theta})X(u;\epsilon,\boldsymbol{\theta})\,dt\,du = \Phi_T(s,s)\psi^{-1} \qquad (20.39d)
$$

We have used this relation to simplify $\mathcal{B}_T{}^{(4)}$ above.

where now specifically [cf. Eq. (20.7)]

$$B_T^{(2)} = -(\bar{a}_0^2/\overline{a_0^2})\psi \int_{0-}^{T+} \langle s(t-\epsilon;\theta)\rangle\langle X(t;\epsilon,\boldsymbol{\theta})\rangle\, dt \tag{20.40b}$$

The interpretation of detector structure here precisely parallels the discrete cases treated earlier [cf. Eqs. (20.8) to (20.10) and Fig. 20.1*a*, *b*], where now, of course, discrete sampling is omitted and the summation $\sum_1^n$ is replaced by the integration $\int_{0-}^{T+} (\quad)\, dt$ in the cross correlator. Thus, a direct interpretation of the operations $\langle(\Phi_T)_v\rangle$, on $v(t)$, in Eq. (20.40*a*) is simply the cross correlation (with zero delay) of the received data $v(t)$ with the weighting function $\langle X(t,\epsilon,\boldsymbol{\theta})\rangle$ followed by integration over the interval $(0-,T+)$.

An alternative interpretation in terms of the continuous analogue of the discrete linear filter $\mathbf{Q}_c$ [Eq. (20.8)] is also easily achieved. Using the relation† $\mathbf{Q}_c = [h_c(|t_i - t_j|)\,\Delta t] = \mathbf{k}_N^{-1}$ [Eqs. (20.8), (20.9*a*)] and following the procedure of Eqs. (19.51) to (19.55), we obtain first

$$\langle X(t;\epsilon,\boldsymbol{\theta})\rangle = \begin{cases} \int_{-\infty}^{\infty} h_c(|t-u|)\langle s_T(u-\epsilon,\boldsymbol{\theta})\rangle\, du & s_T = s(t-\epsilon,\boldsymbol{\theta}),\ 0- < t < T+ \\ 0 & \text{elsewhere} \end{cases} \tag{20.41a}$$

Inserting this in $\langle(\Phi_T)_v\rangle$ in Eq. (20.40*a*) and writing

$$X_T = \begin{cases} X & 0- < t < T+ \\ 0 & \text{elsewhere} \end{cases} \tag{20.41b}$$

we have

$$\langle(\Phi_T)_v\rangle = \psi \iint_{0-}^{T+} v(t)h_c(|t-u|)\langle s_T(u-\epsilon;\boldsymbol{\theta})\rangle\, dt\, du$$

$$= \psi \int_{-\infty}^{\infty} \langle s_T(u-\epsilon;\boldsymbol{\theta})\rangle \left[\int_0^T v(t)h_c(|u-t|)\, dt\right] du \tag{20.42a}$$

$$\langle(\Phi_T)_v\rangle = \psi \int_{0-}^{T+} \langle s_T(u-\epsilon;\boldsymbol{\theta})\rangle\, v_F(u)\, du \tag{20.42b}$$

where

$$v_F(t) = \int_0^T v(\tau)h_c(|t-\tau|)\, d\tau \tag{20.42c}$$

is the continuous analogue of Eq. (20.8) and Eq. (20.42*a*) that of Eq. (20.10). Here h_c represents a "matched" filter, in the sense of an optimum decision system (cf. Sec. 20.2-4). This optimum system again employs cross correlation with zero delay (i.e., simple multiplication), now of the filtered data $v_F(t)$, with $\langle s_T\rangle$. The finiteness of the integration period $(0,T)$ is ensured by switches at $t = 0, T$ (cf. Fig. 20.1*b*). Here $v_F(t)$ is the output of a linear

† This noise is assumed stationary; if it is not, then we may write $\mathbf{Q}_c = [H(t_i,t_j)\,\Delta t] = [h(t_j - t_i, t_j)\,\Delta t]$, etc., and the subsequent interpretation must be given in terms of a linear *time-varying* filter. Note, moreover, from its definition here that h_c has the dimensions $\sec^{-2}$ and that h_c is also not realizable [cf. Eqs. (20.9) and comments].

time-invariant filter with weighting function h_c [cf. Eqs. (20.9a), (20.9b)] when $v(t)$ is the input. Note, however, that this filter is not physically realizable, since $h_c(|t - \tau|) \neq 0$ $(t - \tau < 0)$.

Finally, to obtain the weighting function h_c of the matched filter, we must solve Eqs. (20.41a), once X has been found from Eq. (20.38). Since both $\langle X_T \rangle$ and $\langle s_T \rangle$ are defined on $(-\infty, \infty)$, Fourier transformation of Eqs. (20.41a) gives formally

$$h_c(|t|) = \int_{-\infty i}^{\infty i} \frac{S_{\langle X \rangle}(p)_T}{S_{\langle S \rangle}(p)_T} e^{pt} \frac{dp}{2\pi i} \tag{20.43}$$

where $S_{\langle X \rangle}(p)_T = \mathfrak{F}\{\langle X_T \rangle\}$, $S_{\langle S \rangle}(p)_T = \mathfrak{F}\{\langle s_T \rangle\}$ are the Fourier transforms of $\langle X_T \rangle$, $\langle s_T \rangle$, respectively. If the noise process is spectrally white, then $K_N(t - u) = (W_0/2)\delta(t - u)$ [Eq. (3.56b)] and from Eq. (20.38) we get

$$X_T(t;\epsilon,\boldsymbol{\theta}) = 2s(t - \epsilon, \boldsymbol{\theta})/W_0 \tag{20.43a}$$

Taking the average over the random parameters and inserting into Eq. (20.43), we find, not unexpectedly, that

$$h_c(|t|) = \frac{2}{W_0} \delta(t - 0) \tag{20.43b}$$

For colored noise backgrounds (when K_N is a rational kernel), we may use the methods of Sec. A.2.3 to obtain $h_c(|t|)$ (see Prob. 20.8).

In a similar way, the threshold structure of Eq. (20.40a) is found by straightforward application of the solution of Eq. (20.38). For white noise, we may use Eq. (20.43a) in Eq. (20.40a), first factoring out the ψ's in a_0 and $v(t)$ before passing to the limit ψ $(= BW_0) \to \infty$ (that is, $B \to \infty$), to get directly

$$\log \Lambda_T \doteq \left[\log \mu - \frac{\bar{A}_0^2}{2W_0} \int_0^T \langle s(t - \epsilon, \boldsymbol{\theta}) \rangle^2 \, dt\right] + \frac{\bar{A}_0 \sqrt{2}}{W_0} \int_0^T V(t) \langle s(t - \epsilon, \boldsymbol{\theta}) \rangle \, dt \tag{20.43c}$$

Observe, finally, that for both colored and white noise backgrounds we may also employ a realizable invariant linear filter h_M, in place of h_c [Eqs. (20.41a)]. From Eqs. (20.10a), (20.10c), we have simply

$$\langle X(t;\epsilon,\boldsymbol{\theta}) \rangle \equiv h_M(T - t; T) \qquad 0- < t < T+ \tag{20.43d}$$

and the data read-out occurs at $t = T$; h_M and h_c are related by Eqs. (20.41a), (20.43). This filter, h_M, is closely related to that used for maximizing signal-to-noise ratio [cf. Eq. (16.99)].

20.2-2 Incoherent Detection I. General Signals. Here we must consider the limiting situation of the discrete threshold system described in Sec. 20.1-2. It is again assumed that epoch ϵ is distributed so that $\langle s \rangle_\epsilon$

vanishes; Eqs. (20.11a) to (20.11c) may then serve as starting points for the continuous theory.

We begin with the quadratic functional $\langle \Psi_T[V(t)] \rangle = \lim_{n\to\infty} \langle \Psi_n(\mathbf{v}) \rangle = \lim_{n\to\infty} \tilde{\mathbf{v}}\bar{\mathbf{G}}\mathbf{v}$ [cf. Eqs. (20.11a), (20.11b)]. Setting

$$\boldsymbol{\gamma} \equiv \psi^{-1}\mathbf{k}_N{}^{-1}\mathbf{s}$$

and therefore

$$\mathbf{G} = \mathbf{k}_N{}^{-1}\mathbf{s}\tilde{\mathbf{s}}\mathbf{k}_N{}^{-1} = \psi^2\boldsymbol{\gamma}\tilde{\boldsymbol{\gamma}} \tag{20.44a}$$

we have

$$\mathbf{s} = \mathbf{K}_N\boldsymbol{\gamma} \tag{20.44b}$$

Following the procedure of Eqs. (19.51) to (19.55) in the limit $n \to \infty$, T fixed, we obtain for the continuous version of Eq. (20.44b) in $(0- < t < T+)$

$$s(t-\epsilon;\,\boldsymbol{\theta}) = \int_{0-}^{T+} K_N(t,u)\,d\Gamma(u;\epsilon,\boldsymbol{\theta}) = \int_{-\infty}^{\infty} K_N(t,u)X_T(u;\epsilon,\boldsymbol{\theta})\,du \tag{20.45}$$

which is just the basic integral equation (20.38), where $\boldsymbol{\gamma} \to d\Gamma(u;\epsilon,\boldsymbol{\theta}) \equiv X_T(u;\epsilon,\boldsymbol{\theta})\,du$ and Eqs. (20.41b) are used to replace X by X_T. We may now express Ψ_T in terms of X_T by a similar procedure applied to Eqs. (20.44a), (20.44b) in Ψ_n:

$$\begin{aligned}\langle \Psi_T \rangle &= \psi^2 \lim_{n\to\infty} \langle (\tilde{\mathbf{v}}\boldsymbol{\gamma})^2 \rangle = \psi^2 \left\langle \left[\int_{0-}^{T+} v(t)\,d\Gamma(t;\epsilon,\boldsymbol{\theta}) \right]^2 \right\rangle \\ &= \psi^2 \left\langle \left[\int_{-\infty}^{\infty} v(t)X_T(t;\epsilon,\boldsymbol{\theta})\,dt \right]^2 \right\rangle\end{aligned} \tag{20.46a}$$

Note that if we write

$$D_T(t,u;\epsilon,\boldsymbol{\theta}) \begin{cases} \equiv X_T(t;\epsilon,\boldsymbol{\theta})X_T(u;\epsilon,\boldsymbol{\theta}) & 0- < t,\ u < T+ \\ = 0 & \text{elsewhere} \end{cases} \tag{20.46b}$$

we can also describe $\langle \Psi_T \rangle$ alternatively as

$$\langle \Psi_T \rangle = \psi^2 \iint_{-\infty}^{\infty} v(t)\langle D_T(t,u;\epsilon,\boldsymbol{\theta})\rangle v(u)\,dt\,du \tag{20.46c}$$

The bias may be found in the same way. $\mathcal{B}_T{}^{(2)}$ is given by Eq. (20.39a) and $\mathcal{B}_T{}^{(4)}$ by Eq. (20.39b), where only the terms $O(\overline{a_0{}^2}^2)$ remain, since $\langle s \rangle_\epsilon$ is assumed to vanish here. The trace terms in Eq. (20.39b) go over into their continuous counterparts by an argument similar to that used above for $\langle \Psi_T \rangle$. Thus, we have

$$\mathcal{B}_T{}^{(2)} = -\frac{\overline{a_0{}^2}}{2} \int_{-\infty}^{\infty} \langle s(t-\epsilon;\,\boldsymbol{\theta})X_T(t;\epsilon,\boldsymbol{\theta})\rangle\,dt \tag{20.47a}$$

equivalent to Eq. (20.39a), and

$$\mathcal{B}_T{}^{(4)} = -\frac{\overline{a_0{}^2}^2}{4}\psi^2 \iint_{-\infty}^{\infty} \langle D_T(t,u;\epsilon,\boldsymbol{\theta})\rangle\langle s(t-\epsilon;\,\boldsymbol{\theta})s(u-\epsilon;\,\boldsymbol{\theta})\rangle\,dt\,du \tag{20.47b}$$

The structure of the optimum threshold detector is finally

$$T_R^{(N)}\{V(t)\} = \log \Lambda_T \doteq (\log \mu + \mathcal{B}_T^{(2)} + \mathcal{B}_T^{(4)}) + \frac{\overline{a_0^2}}{2}\psi^2 \iint_{-\infty}^{\infty} v(t)\langle D_T(t,u;\epsilon,\boldsymbol{\theta})\rangle v(u)\,dt\,du \tag{20.48}$$

There remains the question of interpreting the operations on $v(t)$ in terms of physically realizable system elements. This is easily done with the aid of Eqs. (20.12) and (20.16), as $n \to \infty$ (and $\Delta t \to dt$). Thus, the (discrete) linear realizable time-varying filter represented by $\mathbf{Q}$ in $\mathbf{v}_F = \mathbf{Qv}$ [cf. Eq. (20.12)] goes over into the corresponding continuous version, which like the discrete case above [cf. Eqs. (20.18), (20.19)] may also be interpreted as an *invariant*, realizable filter and a time-varying switch. We have explicitly

$$v_F(t;T) = \int_0^t v(\tau)h(t-\tau|T)\,d\tau = \int_0^\infty v(\tau)h(t-\tau|T)\,d\tau \tag{20.49}$$

where we have set $h(t) \equiv h(t|T)$ to remind ourselves that the weighting function of this prerectification linear filter depends on the integration period $(0,T)$. Here h, in combination with the time-varying switch, is, of course, the continuous analogue of $\mathbf{Q}$ and is itself the weighting function of a *linear* [Eq. (20.16*a*)], *time-invariant* [Eqs. (20.18*b*), $h(t,\tau|T) = h(t-\tau|T)$], and *physically realizable* filter [Eq. (20.16*b*), $h = 0,\ t < \tau$]. Equations (20.18*a*) then enable us to write in the limit

$$\rho_G(t,u)_T = \begin{cases} \int_{0-}^{T+} H(x,t)H(x,u)\,dx & 0- < t,\ u < T+ \\ 0 & \text{elsewhere} \end{cases} \tag{20.50a}$$

which for the invariant filter here reduces to the analogue of Eqs. (20.18*b*), namely,

$$\rho_G(t,u)_T = \begin{cases} \int_{0-}^{T+} h(x-t|T)h(x-u|T)\,dx & 0- < t,\ u < T+ \\ 0 & \text{elsewhere} \end{cases} \tag{20.50b}$$

Again, the time-varying switch, embodied in the further condition $h(x - t|T) = 0,\ t < 0$, etc., is necessary [in addition to realizability, for example, $h(x-t|T) = 0,\ t > x\ (x > 0)$, etc.] to ensure that ρ_G vanishes outside $(0-,T+)$ as required for finite data samples [cf. also the remarks following Eqs. (20.18*a*)]. Of course, the operations here are equivalent to those of an averaged, generalized autocorrelation of the received data with itself on $(0-,T+)$ [cf. Eq. (20.46*c*) and Eq. (19.59), as well as the remarks in Sec. 20.1-2].

The weighting function† $h(t|T)$ of this linear invariant filter (with switch) is now found by taking the iterated Fourier transform‡ of the nonlinear integral equation (20.50*b*). The result is

$$P_G(i\omega)_T \equiv \iint_{-\infty}^{\infty} \rho_G(t,u)_T e^{-i\omega(t-u)}\, dt\, du = \int_0^T |Y(i\omega;x)|^2\, dx > 0 \qquad (20.51a)$$

since the time-varying switch must be included; where

$$Y(i\omega;x) \equiv \int_{-\infty}^{\infty} \hat{h}(z)e^{-i\omega z}\, dz = \int_{-0}^{x} h(z|T)e^{i\omega z}\, dz \qquad (20.51b)$$

$$\hat{h}(x-t) = h(x-t|T); \qquad = 0,\ t > x \text{ (realizability)}$$
$$= 0,\ t < 0 \qquad \text{(switch condition)} \qquad (20.51c)$$

this last since $|Y(i\omega;x)|$ is real and positive. An infinite number of such optimum linear filters are seen to be available, since only the modulus of the system function $Y(i\omega;x)$ is determined by the defining relation (20.50*b*). This is a consequence of the condition of sample uncertainty, so that a priori phase information about the input is unavailable to the observer. We cannot, therefore, expect to specify this linear, invariant (and realizable) prerectifier filter any more precisely than above. This unavoidable lack of uniqueness, however, is an advantage to the system designer, since it provides him with an additional degree of freedom: he is at liberty to select those phase responses, $\phi(i\omega;x)$, in $Y(i\omega;x) = |Y(i\omega;x)|e^{i\phi(i\omega;x)}$ which are most convenient for the system at hand. Finally, observe that a unique predetection filter of the type of Eqs. (20.10*a*) or (20.21*a*) is not generally possible here, except under very special conditions [cf. Prob. 20.6 and Eqs. (20.21*a*), (20.21*b*) and comments].

The relation (20.50*b*) also permits us to write the quadratic functional equations (20.46*a*), (20.46*c*) as

$$\langle \Psi_T \rangle \equiv \lim_{n\to\infty} \Psi_n = \iint_{-\infty}^{\infty} v(t)\rho_G(t,u)_T v(u)\, dt\, du \qquad (20.52)$$

Comparing this with Eqs. (20.46*b*) shows at once that

$$\rho_G(t,u)_T = \psi^2\langle D_T(t,u;\epsilon,\boldsymbol{\theta})\rangle = \psi^2\langle X_T(t;\epsilon,\boldsymbol{\theta})X_T(u;\epsilon,\boldsymbol{\theta})\rangle \qquad (20.53)$$

Accordingly, with Eqs. (20.49), (20.50*b*) we obtain from Eqs. (20.46*c*),

† Note that h now has the dimensions $\sec^{-3/2}$ since ρ_G is $O(\sec^{-2})$. The dimensionality of h can easily be adjusted to the usual dependence ($\sec^{-1}$) on multiplication by $T^{1/2}$: $h''(t|T) \equiv T^{1/2}h(t|T)$. In the case of white noise backgrounds, however, see Eqs. (20.56) to (20.58*c*).

‡ For a similar procedure in the case of entirely random signals in normal noise, see Middleton,[4] app. V.

(20.52) the equivalent relations

$$\langle \Psi_T \rangle = \int_{0-}^{T+} dx \iint_0^{\infty \text{ or } x} v(t)h(x - t|T)v(u)h(x - u|T)\,dt\,du \tag{20.54a}$$

$$= \int_{0-}^{T+} v_F(t;T)^2\,dt \tag{20.54b}$$

Thus, it is clear from Eq. (20.54*b*) that one possible structure of the optimum threshold detector is essentially that shown in Fig. 20.2*a* with discrete operations replaced by continuous ones: a linear time-varying filter (or, equivalently, a linear invariant filter h followed by a time-varying switch) whose output $v_F(t;T)$ is passed through a quadratic rectifier is then followed by simple integration† $\int_0^T (\quad)\,dt$ in $(0,T)$, at the end of which the desired number $\langle \Psi_T \rangle$ is read out and passed in turn to the decision system for a decision according to the usual procedure [cf. Eq. (19.15)]. This result [Eq. (20.54*b*)] also emphasizes again the fact (cf. Sec. 19.4-3) that these optimum incoherent systems are basically *energy*-sensitive devices which distinguish between the states of signal and no signal on an energy basis, where signal and noise are suitably processed to minimize the effects of the noise, while enhancing those of the signal. The filter $h(t|T)$ is one type of *matched* filter, about which we shall have more to say in Sec. 20.2-4.‡

† A "simple" integrator is proportional to the ideal integrator of Eqs. (16.10*a*), where $\lambda = T$ in the former and $\lambda = 1$ in the latter instance. Whether a factor T appears specifically in the postdetection integration or, equivalently, a factor $T^{1/2}$ is included in the predetection filter's weighting function is purely a matter of design convenience, since the quadratic forms remain unchanged. We shall employ both representations: simple integration for optimum systems and ideal integration for suboptimum ones [cf. Sec. 20.3-2(1)]. Finally, the integration itself may be achieved in practice with a suitable delay line [of appropriate weighting in $(0,T)$], where the desired output is obtained at the tap-off point representing T sec delay. Thus, at any instant an output read off at this point is just the integrated output of the rectifier, over an interval T sec into the past. In this way, the "start" and "finish" of any one particular acquisition period $(0,T)$ is automatically determined.

‡ The continuous analogue $H'(t,u)$ of the nonrealizable class of filters based on the symmetric congruent transformation $\mathbf{Q}_0'$ and discussed in the footnote on p. 844 is found in similar fashion: The nonlinear integral equation governing H' is specifically [cf. Eqs. 20.50*a*)]

$$\rho_G(t,u) = \begin{cases} \int_{0-}^{T+} H'(x,t)H'(x,u)\,dx & 0- < t,\ u < T+ \\ 0 & \text{elsewhere} \end{cases} \tag{1}$$

A formal solution may be achieved as follows: Let

$$H'(x,t) = \sum_m a_m(x)g_m(t) \tag{2}$$

where the as yet unspecified a_m, g_m are each zero outside $(0-,T+)$ to ensure the vanishing

Finally, an alternative representation of the optimum threshold structure is the combination of time-varying linear filter and multiplier shown in Fig. 20.2*b*, again with discrete operations replaced by their corresponding continuous analogues. The weighting function of this time-varying linear filter is just $\rho_G(t,u)_T = \hat{\rho}_G(t - u, t)_T = \hat{\rho}_G(u - t, u)_T$, etc., of Eqs. (20.50), (20.53), which is determined as soon as X_T has been found from the fundamental integral equation (20.45). Accordingly, we have still another version of Ψ_T here. Thus, since $\iint_0^T A(x,y)\,dx\,dy = 2\int_0^T dx \int_0^x A(x,y)\,dy$, if $A(x,y) = A(y,x)$, and since $\rho_G(t,u)_T = \rho_G(u,t)_T$ [cf. Eq. (20.52)], we can write Eq. (20.52) as

$$\langle \Psi_T \rangle = 2\int_{0-}^{T+} v(t) \int_{0-}^{t} v(u)\hat{\rho}_G(t - u, t)_T\,du = 2\int_{0-}^{T+} v(t)\hat{v}_F(t;T)\,dt \qquad (20.55)$$

where $\hat{v}_F$ is the output (at time t) of a physically realizable† linear time-varying filter, viz.,

$$\hat{v}_F(t;T) = \int_{0-}^{t} v(u)\hat{\rho}_G(t - u, t)_T\,du \qquad (20.55a)$$

Unlike the linear invariant filters $h(t|T)$ of the first representation, of which there are potentially an infinite number of possible choices, the linear time-varying filter in this instance is uniquely specified by ρ_G [Eq. (20.53)].

---

of ρ_G as required in (1). Since ρ_G is symmetric and positive definite, apply next to ρ_G Mercer's theorem (8.61), extended to include possible delta-function singularities at $0-$, $T+$. Choosing $a_m(x) = \lambda_m^{1/4}\phi_m(x)$ $(0- < x < T+$ and zero elsewhere), with λ_m, ϕ_m respectively the eigenvalues and eigenfunctions of the associated integral equation (of kernel ρ_G), we readily find that $g_m(t) = a_m(x)$, with the desired result for (2) that

$$H'(x,t) = \begin{cases} \sum_m \lambda_m^{1/4}\phi_m(x)\lambda_m^{1/4}\phi_m(t) = H'(t,x) & 0- < x,\ t < T+ \\ 0 & \text{elsewhere} \end{cases} \qquad (3)$$

Writing $\phi_m(t) = h_m(T - t)$, with h_m a realizable invariant filter, we can accordingly interpret h' as a parallel combination of time-varying gains in series with invariant realizable filters (see also Kailath[41] for a similar, independent approach).

Note that still another interpretation of $\langle \Psi_T \rangle$ is possible if we use a generalization of Mercer's theorem directly on ρ_G. We get

$$\langle \Psi_T \rangle = \sum_m \lambda_m \left[\int_{0-}^{T+} v(t)\phi_m(t)\,dt\right]^2 \qquad (4)$$

with $\phi_m(t) = h_m(T - t)$. In this form we can interpret $\langle \Psi_T \rangle$ as an (infinite) set of linear, realizable, invariant filters h_m, each with read-out at $t = T+$, whose outputs are simultaneously squared, weighted according to λ_m, and then added to give the desired number $\langle \Psi_T \rangle$.

† Since we may set $\rho_G(t,u)_T = \hat{\rho}_G(t - u, t)_T = 0$ for all $t - u < 0$ without altering the value of Ψ_T.

The system designer has no other alternative if such a filter is to be used. For this reason, and because of the possibly simpler character of the time-invariant element h, the first representation may be preferred.

For examples of both types, when the noise is spectrally white [i.e., when Eq. (20.43a) applies], we follow the procedure leading to Eq. (20.43c), first factoring out ψ before passing to the limit $\psi \to \infty$. Thus, defining with the help of Eq. (20.43a),

$$\rho_w \equiv \lim_{\psi \to \infty} \{\rho_G \psi^{-2}\} = \frac{4}{W_0{}^2} \langle s(t - \epsilon; \boldsymbol{\theta}) s(u - \epsilon, \boldsymbol{\theta})\rangle \qquad (20.56a)$$

from Eq. (20.53), we obtain the weighting function of the matched invariant filter from Eqs. (20.50), now modified to

$$\rho_w(t,u)_T = \begin{cases} \int_0^T h_w(x - t|T) h_w(x - u|T)\, dx & 0 \le t,\, u \le T \\ 0 & \text{elsewhere} \end{cases} \qquad (20.56b)$$

[We remember, of course, that the time-varying switch is needed in conjunction with this filter to ensure that ρ_w vanishes outside $(0-,T+)$.] The modulus of the corresponding system function is specified by Eqs. (20.51), which become here

$$\left(\int_0^T |Y_w(i\omega;x)|^2\, dx \right)^{1/2} = \frac{2}{W_0} \left[\iint_0^T \langle s(t - \epsilon; \boldsymbol{\theta}) s(u - \epsilon, \boldsymbol{\theta})\rangle \cos \omega(t - u)\, dt\, du \right]^{1/2} \qquad (20.57)$$

The operations on the received data $V(t)$, as given by Eq. (20.48), reduce directly to

$$\frac{\overline{a_0{}^2}}{2} \langle \Psi_T \rangle = \frac{\overline{A_0{}^2}}{W_0{}^2} \left\langle \left[\int_0^T V(t) s(t - \epsilon, \boldsymbol{\theta})\, dt \right]^2 \right\rangle \qquad (20.58a)$$

where the bias terms $\mathcal{B}_T{}^{(2)}$, $\mathcal{B}_T{}^{(4)}$ [Eqs. (20.39a), (20.39b)] are easily found to be

$$\mathcal{B}_T{}^{(2)} = -\frac{\overline{A_0{}^2}}{2W_0} \int_0^T \langle s(t - \epsilon, \boldsymbol{\theta})^2 \rangle\, dt \qquad (20.58b)$$

$$\mathcal{B}_T{}^{(4)} = -\frac{\overline{A_0{}^2}^2}{4W_0{}^2} \iint_0^T \langle s(t - \epsilon, \boldsymbol{\theta}) s(u - \epsilon, \boldsymbol{\theta})\rangle^2\, dt\, du \qquad (20.58c)$$

For the alternative representation using the time-varying filter and multiplier, we have ρ_w [Eq. (20.56a)] as weighting function, with appropriate modifications of Eqs. (20.55), (20.55a) for the white noise background. In the more general situations of nonuniform noise spectra, we must follow Eqs. (20.44) to (20.55), where the solution of the basic integral equation (20.45) may be achieved by the methods of Appendix 2 in the case of rational kernels.

20.2-3 Incoherent Detection II. Narrowband Signals. These same procedures may now be applied specifically to the important special class of detection problems involving narrowband signal processes, discussed in Sec. 20.1-3 for discrete data sampling. As in Sec. 20.1-3, epoch is assumed to be uniformly distributed over an RF cycle,† and the (normalized) signal process is given by Eq. (20.22).

To obtain the corresponding continuous versions of Eqs. (20.31) to (20.33), let us follow the steps suggested by Eqs. (20.44*a*) to (20.47) and begin by defining the vectors

$$\boldsymbol{\gamma}_1 \equiv \psi^{-1}\mathbf{k}_N^{-1}\mathbf{a} \qquad \boldsymbol{\gamma}_2 \equiv \psi^{-1}\mathbf{k}_N^{-1}\mathbf{b} \tag{20.59a}$$

so that

$$\mathbf{a} = \mathbf{K}_N\boldsymbol{\gamma}_1 \qquad \mathbf{b} = \mathbf{K}_N\boldsymbol{\gamma}_2 \tag{20.59b}$$

We can then write $\mathbf{G}_0(\boldsymbol{\theta})$ [Eq. (20.29*a*)] as

$$\mathbf{G}_0(\boldsymbol{\theta}) \doteq \mathbf{k}_N^{-1}(\mathbf{a}\tilde{\mathbf{a}} + \mathbf{b}\tilde{\mathbf{b}})\mathbf{k}_N^{-1} = \psi^2(\boldsymbol{\gamma}_1\tilde{\boldsymbol{\gamma}}_1 + \boldsymbol{\gamma}_2\tilde{\boldsymbol{\gamma}}_2) \tag{20.60a}$$

while $B_n(\boldsymbol{\theta})$ [Eq. (20.26)] becomes

$$B_n(\boldsymbol{\theta}) = \frac{1}{2}(\tilde{\mathbf{a}}\mathbf{k}_N^{-1}\mathbf{a} + \tilde{\mathbf{b}}\mathbf{k}_N^{-1}\mathbf{b}) = \frac{\psi}{2}(\tilde{\mathbf{a}}\boldsymbol{\gamma}_1 + \tilde{\mathbf{b}}\boldsymbol{\gamma}_2) \tag{20.60b}$$

Passing now to the limit $n \to \infty$ (T fixed) [cf. Eqs. (19.51) to (19.55)], we have

$$\boldsymbol{\gamma}_1 \to d\Gamma_1(u;\epsilon,\boldsymbol{\theta}) \equiv X_T(u;\epsilon,\boldsymbol{\theta})_0\,du \qquad \boldsymbol{\gamma}_2 \to d\Gamma_2(u;\epsilon,\boldsymbol{\theta}) \equiv Y_T(u;\epsilon,\boldsymbol{\theta})_0\,du \tag{20.61a}$$

where X_{0T}, Y_{0T} are the solutions of the basic inhomogeneous integral equations obtained in the limit from Eqs. (20.59*b*), viz.,

$$\begin{aligned} a(t;\epsilon,\boldsymbol{\theta}) &= \int_{0-}^{T+} K_N(t,u)\,d\Gamma_1(u;\epsilon,\boldsymbol{\theta}) = \int_{-\infty}^{\infty} K_N(t,u)X_T(u;\epsilon,\boldsymbol{\theta})_0\,du \\ b(t;\epsilon,\boldsymbol{\theta}) &= \int_{0-}^{T+} K_N(t,u)\,d\Gamma_2(u;\epsilon,\boldsymbol{\theta}) = \int_{-\infty}^{\infty} K_N(t,u)Y_T(u;\epsilon,\boldsymbol{\theta})_0\,du \\ & \qquad 0- < t < T+ \end{aligned} \tag{20.61b}$$

and X_{0T}, Y_{0T} vanish outside $(0-,T+)$ as before [cf. Eqs. (20.41*b*)].‡ Finally, observe that multiplying Eqs. (20.61*b*) by $\cos\omega_s\epsilon$, adding and using Eq. (20.24*b*), on comparison with Eq. (20.38), gives us the following relations,

$$X_T = X_{0T}\cos\omega_0\epsilon + Y_{0T}\sin\omega_0\epsilon \tag{20.61c}$$

which we shall need subsequently.

In a similar fashion, we obtain with the help of Eq. (20.60*a*) the functional

† Cf. footnote on p. 839.

‡ In the present case, $a(t;\epsilon,\boldsymbol{\theta})$ and $b(t;\epsilon,\boldsymbol{\theta})$, and hence X_{0T}, Y_{0T} are independent of ϵ, but we have kept the more general forms (20.61*b*) for later applications (cf. Sec. 20.4-6). The same is also true of $G_0(\epsilon,\boldsymbol{\theta})$, $B_n(\epsilon,\boldsymbol{\theta})$ here, namely, $\mathbf{G}_0(\epsilon,\boldsymbol{\theta}) = \mathbf{G}_0(\boldsymbol{\theta})$, etc.

$$\Psi_{0T} \equiv \lim_{n\to\infty} \tilde{\mathbf{v}}\mathbf{G}_0(\boldsymbol{\theta})\mathbf{v} = \psi^2 \lim_{n\to\infty} [(\tilde{\mathbf{v}}\boldsymbol{\gamma}_1)^2 + (\tilde{\mathbf{v}}\boldsymbol{\gamma}_2)^2]$$
$$= \psi^2 \left\{ \left[\int_{0-}^{T+} v(t)\, d\Gamma_1(t;\epsilon,\boldsymbol{\theta}) \right]^2 + \left[\int_{0-}^{T+} v(t)\, d\Gamma_2(t;\epsilon,\boldsymbol{\theta}) \right]^2 \right\}$$
$$= \psi^2 \left\{ \left[\int_{-\infty}^{\infty} v(t) X_T(t;\epsilon,\boldsymbol{\theta})_0\, dt \right]^2 + \left[\int_{-\infty}^{\infty} v(t) Y_T(t;\epsilon,\boldsymbol{\theta})_0\, dt \right]^2 \right\} \quad (20.62a)$$

Analogous to Eq. (20.46b), let us define

$$D_T(t,u;\epsilon,\boldsymbol{\theta})_0 \equiv X_T(t;\epsilon,\boldsymbol{\theta})_0 X_T(u;\epsilon,\boldsymbol{\theta})_0 + Y_T(t;\epsilon,\boldsymbol{\theta})_0 Y_T(u;\epsilon,\boldsymbol{\theta})_0 \qquad 0- < t,\, u < T+ \quad (20.62b)$$

with $D_T = 0$ elsewhere. Then we can express Ψ_{0T} alternatively as

$$\Psi_{0T} = \psi^2 \iint_{-\infty}^{\infty} v(t) D_T(t,u;\epsilon,\boldsymbol{\theta})_0 v(u)\, dt\, du \quad (20.62c)$$

[cf. Eq. (20.46c)]. The continuous form of $B_n(\boldsymbol{\theta})$ likewise becomes

$$B_T(\epsilon,\boldsymbol{\theta})_0 \equiv \lim_{n\to\infty} B_n(\epsilon,\boldsymbol{\theta}) = \frac{\psi}{2} \lim_{n\to\infty} (\tilde{\mathbf{a}}\boldsymbol{\gamma}_1 + \tilde{\mathbf{b}}\boldsymbol{\gamma}_2)$$
$$= \frac{\psi}{2} \left[\int_{0-}^{T+} a(t;\epsilon,\boldsymbol{\theta})\, d\Gamma_1(t;\epsilon,\boldsymbol{\theta}) + \int_{0-}^{T+} b(t;\epsilon,\boldsymbol{\theta})\, d\Gamma_2(t;\epsilon,\boldsymbol{\theta}) \right]$$
$$= \frac{\psi}{2} \int_{-\infty}^{\infty} [a(t;\epsilon,\boldsymbol{\theta}) X_T(t;\epsilon,\boldsymbol{\theta})_0 + b(t;\epsilon,\boldsymbol{\theta}) Y_T(t;\epsilon,\boldsymbol{\theta})_0]\, dt \quad (20.63a)$$

An alternative result is found on multiplying each relation in Eqs. (20.61b), respectively, by X_{0T} and Y_{0T}, adding, and integrating over t, using Eq. (20.62b), viz.,

$$B_T(\epsilon,\boldsymbol{\theta}) = \frac{\psi}{2} \iint_{-\infty}^{\infty} K_N(t,u) D_T(t,u;\epsilon,\boldsymbol{\theta})_0\, dt\, du \quad (20.63b)$$

[cf. Eq. (20.39d)].

The continuous version of the discrete detector structure [Eq. (20.31)] can now be written

$$\boldsymbol{T}_R^{(N)}\{V(t)\} = \log \Lambda_T = \log \mu + \log \langle e^{-a_0^2 B_T/2} I_0(a_0 \Psi_{0T}^{1/2}) \rangle_{a_0,\boldsymbol{\theta}} \quad (20.64)$$

and in the cases where only single values of the parameters can occur, for example, $w(\boldsymbol{\theta}) = \delta(\boldsymbol{\theta} - \boldsymbol{\theta}_0)$, etc., this reduces directly to

$$\boldsymbol{T}_R^{(N)}\{V(t)\} = \log \Lambda_T = \log \mu - \frac{a_0^2}{2} B_T(\boldsymbol{\theta}_0) + \log \left[I_0(a_0 \Psi_{0T}^{1/2}) \Big|_{\boldsymbol{\theta}=\boldsymbol{\theta}_0} \right] \quad (20.64a)$$

which holds for all input signal levels as long as the narrowband approximations (20.26), (20.29a) apply.

As an example, consider the instance of a white noise background, so that analogous to Eq. (20.43a) we get for solutions of the basic integral equations (20.61b)

$$X_{0T} = 2a(t;\epsilon,\boldsymbol{\theta})/W_0 \qquad Y_{0T} = 2b(t;\epsilon,\boldsymbol{\theta})/W_0 \qquad 0 \leq t \leq T \quad (20.65)$$

Then $a_0\Psi_{0T}^{1/2}$ and $B_T a_0^2/2$ in Eqs. (20.64), (20.64a) become [after we have factored out the ψ's in $v(t) = V(t)/\sqrt{\psi}$, $a_0 = A_0/\sqrt{2\psi}$, etc., since $\psi = BW_0 \to \infty$ here]

$$a_0\Psi_{0T}^{1/2} = \left\{\frac{2A_0^2}{W_0^2} \iint\limits_0^T [a(t;\epsilon,\boldsymbol{\theta})a(u;\epsilon,\boldsymbol{\theta}) + b(t;\epsilon,\boldsymbol{\theta})b(u;\epsilon,\boldsymbol{\theta})]V(t)V(u)\,dt\,du\right\}^{1/2}$$

$$= \frac{A_0}{W_0}\left\{2 \iint\limits_0^T F_s(t;\boldsymbol{\theta}_1)F_s(u;\boldsymbol{\theta}) \cos\,[\omega_s(t-u) - \phi_s(t;\boldsymbol{\theta}_2) + \phi_s(u;\boldsymbol{\theta}_2)]V(t)V(u)\,dt\,du\right\}^{1/2} \quad (20.66a)$$

with the help of Eq. (20.25), and

$$\frac{a_0^2 B_T}{2} = \frac{A_0^2}{4W_0}\int_0^T [a(t;\epsilon,\boldsymbol{\theta})^2 + b(t;\epsilon,\boldsymbol{\theta})^2]\,dt \quad (20.66b)$$

$$= \frac{A_0^2}{4W_0}\int_0^T F_s(t;\boldsymbol{\theta}_1)^2\,dt \quad (20.66c)$$

either from Eq. (20.63a) or from Eq. (20.63b). Thus, Eqs. (20.66a), (20.66b) in Eq. (20.64) or Eq. (20.64a) yield the corresponding detector structure.

When, as is usually the case, it is not possible to carry out the averages over the signal parameters in closed form in the general expression (20.64), we make once again the threshold development in powers of a_0, so that the continuous version of Eq. (20.32) becomes now the quadratic functional

$$\boldsymbol{T}_R^{(N)}\{V\} = \log \Lambda_T \doteq (\log \mu + \mathcal{B}_{0T}^{(2)} + \mathcal{B}_{0T}^{(4)}) + \frac{\overline{a_0^2}}{4}\psi^2 \iint\limits_{-\infty}^{\infty} v(t)\langle D_T(t,u;\epsilon,\boldsymbol{\theta})_0\rangle v(u)\,dt\,du \quad (20.67)$$

the last term being $\langle\Psi_{0T}\rangle\overline{a_0^2}/4$ [Eqs. (20.62a), (20.62c)]. The bias terms $\mathcal{B}_{0T}^{(2)}$, $\mathcal{B}_{0T}^{(4)}$ are found as before [cf. Eqs. (20.47)], now with $\mathcal{B}_{0T}^{(2)} = \lim_{n\to\infty} \mathcal{B}_{0n}^{(2)}$, $\mathcal{B}_{0T}^{(4)} = \lim_{n\to\infty} \mathcal{B}_{0n}^{(4)}$, in Eqs. (20.33$a$), (20.33$b$). We get specifically

$$\mathcal{B}_{0T}{}^{(2)} = -\frac{\overline{a_0{}^2}\psi}{4}\iint_{-\infty}^{\infty} K_N(t,u)\langle D_T(t,u;\epsilon,\boldsymbol{\theta})_0\rangle\, dt\, du \doteq \mathcal{B}_T{}^{(2)} \tag{20.67a}$$

$$\mathcal{B}_{0T}{}^{(4)} = -\frac{\overline{a_0{}^2}^2\psi^2}{16}\iint_{-\infty}^{\infty} \langle D_T(t,u;\epsilon,\boldsymbol{\theta})_0\rangle\langle F_s(t;\boldsymbol{\theta}_1)F_s(u;\boldsymbol{\theta}_1)\cos[\omega_s(t-u) - \phi_s(t;\boldsymbol{\theta}_2) + \phi_s(u;\boldsymbol{\theta}_2)]\rangle\, dt\, du \doteq \mathcal{B}_T{}^{(4)} \tag{20.67b}$$

For white noise backgrounds, the optimum threshold structure [Eq. (20.67)] becomes

$$\log \Lambda_T \Big|_{\text{white}} \doteq (\log \mu + \mathcal{B}_{0T}{}^{(2)} + \mathcal{B}_{0T}{}^{(4)}) + \frac{\overline{A_0{}^2}}{2W_0{}^2}\iint_0^T V(t)\, V(u) \times \langle F_s(t;\boldsymbol{\theta}_1)F_s(u;\boldsymbol{\theta}_1)\cos[\omega_s(t-u) - \phi_s(t;\boldsymbol{\theta}_2) + \phi_s(u;\boldsymbol{\theta}_2)]\rangle\, dt\, du \tag{20.67c}$$

where the bias terms [Eqs. (20.67*a*), (20.67*b*)] are now

$$\mathcal{B}_{0T}{}^{(2)}\Big|_{\text{white}} = -\frac{\overline{A_0{}^2}}{4W_0}\int_0^T \langle F_s(t;\boldsymbol{\theta}_1)^2\rangle\, dt \doteq \mathcal{B}_T{}^{(2)}\Big|_{\text{white}} \tag{20.67d}$$

$$\mathcal{B}_{0T}{}^{(4)}\Big|_{\text{white}} \doteq -\frac{\overline{A_0{}^2}^2}{16W_0{}^2}\iint_0^T \langle F_s(t;\boldsymbol{\theta}_1)F_s(u;\boldsymbol{\theta}_1)\cos[\omega_s(t-u) - \phi_s(t;\boldsymbol{\theta}_2) + \phi_s(u;\boldsymbol{\theta}_2)]\rangle^2\, dt\, du \doteq \mathcal{B}_T{}^{(4)}\Big|_{\text{white}} \tag{20.67e}$$

Note that the operation on $V(t)$, $V(u)$ in Eq. (20.67*c*) can be expressed alternatively as

$$\frac{\overline{A_0{}^2}}{2W_0{}^2}\left\langle\left|\int_0^T V(t)F_s(t;\boldsymbol{\theta}_1)e^{i\omega_s t - i\phi_s(t;\boldsymbol{\theta}_2)}\, dt\right|^2\right\rangle \tag{20.67f}$$

while from Eq. (20.61*c*), with ϵ replaced by ϵ_c in the sine and cosine terms [cf. Eq. (20.24*b*)], it is easily seen that

$$\langle D(t,u;\epsilon,\epsilon_c,\boldsymbol{\theta})\rangle_{\epsilon_c} = \tfrac{1}{2}D(t,u;\epsilon,\boldsymbol{\theta})_0 \tag{20.67g}$$

Other results, including the dependence of amplitude on one or more of the other signal parameters, may be found in Prob. 20.5.

The interpretation here of threshold detector structure in terms of realizable elements is just that given in the previous section, where now, of course, the character of the individual elements is modified to take specifically into account the narrowband nature of the input signal set. Figure 20.2 still applies, with discrete operations replaced by their continuous analogues: $\langle\Psi_{0T}\rangle$ again embodies the successive operations of linear invariant realizable filtering and time-varying switch, zero-memory quadratic rectification, and simple integration or, equivalently, a linear time-varying filter, followed by a simple multiplier and integrator (cf. Fig. 20.2*b*). The expressions corresponding to ρ_G and $|Y(i\omega)_T|$ [Eqs. (20.50), (20.51), (20.53)] become here, with the help of Eq. (20.62b),

$$\rho_{G_0}(t,u)_T \begin{cases} = \int_{0-}^{T+} h_0(x-t|T)h_0(x-u|T)\,dx & 0- < t,\, u < T+ \\ = 0 & \text{elsewhere} \\ \equiv \psi^2\langle D_T(t,u;\epsilon,\boldsymbol{\theta})_0\rangle \end{cases} \tag{20.68}$$

and

$$\left(\int_0^T |Y_0(i\omega;x)|^2\,dx\right)^{1/2} = \left[\iint_{0-}^{T+} \rho_{G_0}(t,u)_T \cos\omega(t-u)\,dt\,du\right]^{1/2} > 0 \tag{20.69}$$

For the modulus $|Y_0(i\omega)_T|$, there are as before an infinite number of possible linear invariant realizable *matched filters* (with weighting function h_0) that can precede the (time-varying) switch and quadratic rectifier (Fig. 20.2a). The alternative of a single linear time-varying filter, of weighting function $\rho_{G_0} = \rho_{G_0}(t-u,\,t)_T$, is indicated in Fig. 20.2b. In either case, $\langle\Psi_{0T}\rangle$ is given by Eqs. (20.54) or (20.55), with ρ_G and h replaced by ρ_{G_0}, h_0, etc. Note that, for white noise backgrounds, Eqs. (20.68), (20.69) are modified to

$$\rho_{0\,\text{white}} \equiv \lim_{\psi\to\infty} \rho_{G_0}\psi^2 = \frac{4}{W_0{}^2}\langle F_s(t;\boldsymbol{\theta}_1)F_s(u;\boldsymbol{\theta}_1)\cos[\omega_s(t-u) - \phi_s(t;\boldsymbol{\theta}_2) + \phi_s(u;\boldsymbol{\theta}_2)]\rangle \tag{20.69a}$$

and

$$\left(\int_0^T |Y_0(i\omega;x)|^2_{\text{white}}\,dx\right)^{1/2} = 2W_0{}^{-1}\left\{\iint_0^T \langle F_s(t;\boldsymbol{\theta}_1)F_s(u;\boldsymbol{\theta}_1) \times \cos[\omega_s(t-u) - \phi_s(t;\boldsymbol{\theta}_2) + \phi_s(u;\boldsymbol{\theta}_2)]\rangle\,dt\,du\right\}^{1/2}$$

or (20.69b)

$$\left(\int_0^T |Y_0(i\omega;x)|^2_{\text{white}}\,dx\right)^{1/2} = 2W_0{}^{-1}\left\langle\left(\left|\int_0^T F_s(t;\boldsymbol{\theta}_1)e^{i(\omega_s+\omega)t-i\phi_s(t;\boldsymbol{\theta}_2)}\,dt\right|^2 + \left|\int_0^T F_s(t;\boldsymbol{\theta}_1)e^{i(\omega_s-\omega)t-i\phi_s(t;\boldsymbol{\theta}_2)}\,dt\right|^2\right)\right\rangle^{1/2}$$

respectively.

When amplitude a_0 alone is random, so that $\boldsymbol{\theta}$ is equal to $\boldsymbol{\theta}_0$ in Eq. (20.64), it is then possible to use a pair of matched filters, of the type employed earlier to maximize signal-to-noise ratio (cf. Sec. 16.3-1), viz.,

$$h_{M_1}(T-t;\,T) = X_{0T}(t;\boldsymbol{\theta}_0) \qquad 0- < t < T+ \tag{20.69c}$$

$$h_{M_2}(T-t;\,T) = Y_{0T}(t;\boldsymbol{\theta}_0) \tag{20.69d}$$

These matched filters are unique and are related to the set of matched filters defined above by Eqs. (20.68) through (20.69b), when we recall that

$$\rho_{G_0}(t,u) = \psi D_T(t,u;\boldsymbol{\theta}_0)_0 = X_{0T}(t;\boldsymbol{\theta}_0)X_{0T}(u;\boldsymbol{\theta}_0) + Y_{0T}(t;\boldsymbol{\theta}_0)Y_{0T}(u;\boldsymbol{\theta}_0)$$

For this system, there is no postdetection integration. The read-outs of h_{M_1} and h_{M_2} should be made exactly at $t = T$ and the result simply squared and added to give the desired value of ψ_{0T} [cf. Eqs. (20.21a), (20.21b) in the discrete case and Prob. 20.6 with Fig. P 20.6].

20.2-4 Bayes Matched Filters. In the preceding analysis of threshold detection, we have already introduced the notion of the *matched* linear *filter*, which, in connection with subsequent nonlinear and linear elements, optimizes receiver performance in the sense of minimum average risk, or cost, of decision. We can describe the concept of the matched filter more precisely as follows, with the observation that various types of "matched" (linear) filters are possible, according to the criterion and the nature of the decision, coherent or incoherent reception, etc. Thus, for the latter we define:

1. A *Bayes matched filter* of the first kind is the *linear time-varying* (but not necessarily realizable) *filter*, preceding subsequent zero-memory quadratic and finite-memory linear operations, which yields ultimately a minimum average cost of decision.

2. A *Bayes matched filter* of the second kind is the *linear time-varying* (realizable) *filter*, again preceding subsequent zero-memory (other than quadratic) nonlinear and finite-memory linear operations, which also leads to minimum average risk of decision.

The discrete linear filters of Eqs. (20.12), (20.13), (20.14), (20.16) and the continuous linear filters of Eqs. (20.49), (20.50a), (20.50b), (20.51) are examples of Bayes matched filters of the first kind for incoherent threshold detection, while the time-varying linear filters of Eqs. (20.20), (20.55a) are examples of the second kind (cf. Figs. 20.1b, 20.2b). In a similar way, corresponding classes of matched filters can also be found for estimation or signal extraction processes (cf. Sec. 21.3). Bayes matched filters for coherent (threshold) reception h_c, h_M [cf. Eqs. (20.9), (20.10), (20.41) to (20.43d)] are defined analogously, but without the subsequent nonlinear operations.

For both detection and extraction (cf. Sec. 21.3), the Bayes matched filter may be interpreted essentially as a linear operation on the received data which enhances the energy of the desired signal vis-à-vis that of the undesired noise background, so that on the average optimum decisions are made. Thus, in threshold detection (and in certain types of estimation) from a spectral viewpoint the matched filter emphasizes the signal components in those spectral regions where the interfering noise is weak and rejects both signal and noise in those portions of the spectrum where the latter predominates. Alternatively, from a geometric viewpoint, the matched filter represents a scale transformation and rotation of the (hyper-) ellipsoid of received energy $\{\sim\tilde{\mathbf{v}}\mathbf{G}\mathbf{v}$ or $\tilde{\mathbf{v}}\bar{\mathbf{G}}\mathbf{v}$, etc., with $\mathbf{G} = \mathbf{k}_N{}^{-1}\mathbf{s}\tilde{\mathbf{s}}\mathbf{k}_N{}^{-1}$ [cf. Eqs. (20.11b)]; $\sim\langle\Psi_T\rangle$ [Eq. (20.52)], etc., for continuous sampling$\}$ which reduce it to a (hyper-) sphere such that in detection the state $H_1\!:S + N$ is most easily distinguished from the alternative state $H_0\!:N$ (or, in extraction, the true signal from the contaminated signal) for minimum average cost of decision. In these circumstances, a matched filter is a transformation which in effect diagonalizes the matrix $\mathbf{G}$ in the energy ellipsoid, such that the energy sphere for signal and noise "overlaps" as little as possible that for noise alone, consistent with the cost constraints of the decision process. Just how this is accomplished, of course, depends on a variety of conditions: the statistical character of the noise and signal processes, their manner of combination, the mode of observation (coherent, incoherent, etc.), the specific criterion of optimality (within the Bayes theory), and the type of decision. However, apart from these particularities, for threshold situations it is the ultimately optimum nature of the decision process, qualitative in detection and quantitative in extraction, that defines the Bayes matched filter.

Let us compare the Bayes matched filter (1) with the S/N matched filter defined earlier, in Sec. 16.3. We first observe that in certain cases these two classes of matched filter are structurally identical: for example, in the coherent detection of a completely deterministic signal in additive normal noise in the first instance and in the maximization of output signal-to-noise ratio, of this same signal in additive noise, in the second instance. The optimum, or "matched," filter in both cases is determined by the basic integral equation (16.99) or Eqs. (20.38), (20.45), where h_{op} and X are now equivalent. However, this structural identity is not an inherent property of the matched filter concept but is rather the result of (1) the particular signal and noise statistics and (2) the fact that, with the normal process in the Bayes case and the S/N formulation in the other, the respective optimizations are both based on quadratic forms involving the signal and noise energies, so that enhancement of signal vs. noise in either case, albeit for quite different criteria, lead to the same matched filters.† Note that, for the S/N matched filter, there is no restriction to normality for the noise, as there is for the Bayes matched filter here. On the other hand, the latter is the result of an actual decision process minimizing average cost, where all pertinent information concerning signal and noise is utilized optimally, while the former is based on the more limited and incomplete second-moment criterion of signal-to-noise ratio.

For these reasons, the Bayes matched filter is much the more general concept, subsuming the S/N matched filter, just as second-moment operations are subsumed by those making full use of the statistically more complete information contained in the nth-order distributions ($n \geq 1$). Also for these reasons, the structure of the Bayes matched filter may differ considerably from those based on the more elementary criterion of signal-to-noise ratio. For example, in the incoherent threshold detection of signals in additive normal noise, there are an infinite number of possible Bayes matched filters [cf. Eqs. (20.51), (20.69)], whose structures are quite different from that of the S/N matched filter employed in the earlier effort to optimize signal detection by maximizing a suitably defined signal-to-noise

† See Eqs. (20.10a), (20.10b), (20.21a), (20.21b), comments in both instances, and Prob. 20.6.

ratio at the detector's output. As we have already noted (Sec. 16.3-2), the latter operation is really associated with an estimation process, not with detection directly. We shall encounter some examples of this latter type in Chap. 21.

PROBLEMS

20.4 (*a*) For optimum threshold detection of a general deterministic signal in additive normal noise, obtain Eq. (20.39*b*) as the continuous limit of Eq. (20.4*b*).

(*b*) In some applications, such as radar, where signal strength depends on range and angle, the (normalized) amplitude a_0 becomes $a_0(\epsilon,\boldsymbol{\theta})$; that is, signal amplitude depends on one or more of the other descriptive, random parameters of the received signal process. Show that the general optimum threshold structure [Eqs. (20.35), (20.39*a*), (20.39*b*)] (for additive normal noise) becomes explicitly

$$\boldsymbol{T}_R^{(N)}\{V\} = \log \Lambda_T \doteq [\log \mu + \mathcal{B}_T^{(2)} + \mathcal{B}_T^{(4)} + O(a_0^6)] + \langle a_0(\Phi_T)_v\rangle + \tfrac{1}{2}[\langle a_0(\Phi_T)_v\rangle^2 - \langle a_0(\Phi_T)_v\rangle^2] \qquad (1)$$

where $(\Phi_T)_v$, $(\Phi_T)_s$ are given by Eqs. (20.37) and now the bias terms take the form

$$\mathcal{B}_T^{(2)} = -\frac{1}{2}\langle a_0^2\Phi_T(s,s)\rangle = -\frac{\psi}{2}\int_{0-}^{T+} \langle s(t-\epsilon;\boldsymbol{\theta})a_0^2X(t;\epsilon,\boldsymbol{\theta})\rangle\, dt$$
$$= -\frac{1}{2}\iint_{-\infty}^{\infty} \langle a_0^2D_T(t,u;\epsilon,\boldsymbol{\theta})\rangle K_N(t,u)\, dt\, du \qquad (2)$$

$$\mathcal{B}_T^{(4)} = -\frac{1}{2}\left[\langle a_0^3\Phi_T(s,s)\Phi_T(s,a_0s)\rangle - \psi^2\left\langle a_0^3 \iint_{-\infty}^{\infty} \langle a_0s(t-\epsilon;\boldsymbol{\theta})\rangle D_T(t,u;\epsilon,\boldsymbol{\theta})s(u-\epsilon,\boldsymbol{\theta})\, dt\, du\right\rangle\right]$$
$$- \frac{\psi^2}{4}\left[\iint_{-\infty}^{\infty} \langle a_0^2D_T(t,u;\epsilon,\boldsymbol{\theta})\rangle\langle a_0^2s(t-\epsilon;\boldsymbol{\theta})s(u-\epsilon;\boldsymbol{\theta})\rangle\, dt\, du\right]$$
$$+ \langle a_0^2\Phi_T(s,\overline{a_0s})^2\rangle - \tfrac{3}{4}\langle\Phi_T(\overline{a_0s},\overline{a_0s})^2\rangle \qquad (3)$$

where

$$\Phi_T(s,\overline{a_0s}) = \psi\int_{0-}^{T+} \langle a_0s(t-\epsilon,\boldsymbol{\theta})\rangle X(t;\epsilon,\boldsymbol{\theta})\, dt \qquad (4a)$$

$$\Phi_T(\overline{a_0s},\overline{a_0s}) = \psi\int_{0-}^{T+} \langle a_0s(t-\epsilon,\boldsymbol{\theta})\rangle\langle a_0X(t,\epsilon,\boldsymbol{\theta})\rangle\, dt \qquad (4b)$$

and D_T is given by Eqs. (20.46*b*).

Thus, for *coherent* threshold detection, observe that Eq. (20.40*a*) becomes

$$\log \Lambda_T \doteq (\log \mu + \tfrac{1}{2}\langle a_0^2B_T^{(2)}\rangle) + \psi\int_{0-}^{T+} v(t)\langle a_0X(t;\epsilon,\boldsymbol{\theta})\rangle\, dt \qquad (5a)$$

with

$$B_T^{(2)} = -(\overline{a_0^2})^{-1}\psi\int_{0-}^{T+} \langle a_0^2s(t-\epsilon,\boldsymbol{\theta})X(t;\epsilon,\boldsymbol{\theta})\rangle\, dt \qquad (5b)$$

For an interpretation of structure using a matched filter, we may still apply Eqs. (20.41*a*) to (20.43*b*), however, with Eqs. (20.41*a*) now modified to

$$\langle a_0X(t;\epsilon,\boldsymbol{\theta})\rangle = \int_{-\infty}^{\infty} h_c(|t-u|)\langle a_0s_T(u-\epsilon;\boldsymbol{\theta})\rangle\, du \qquad \text{etc.} \qquad (5c)$$

Similarly, note that for *incoherent* threshold detection, where $\langle s\rangle_\epsilon = 0$ *and* a_0 *does not depend on* ϵ, Eq. (20.48) becomes

$$\log \Lambda_T \doteq (\log \mu + \mathcal{B}_T^{(2)} + \mathcal{B}_T^{(4)}) + \frac{\overline{a_0^2}}{2} \iint_{-\infty}^{\infty} v(t)\rho_G(t,u)_T v(u)\, dt\, du \tag{6a}$$

now with
$$\rho_G(t,u)_T \equiv \psi^2 \langle a_0^2 D_T(t,u;\epsilon,\theta)\rangle / \overline{a_0^2} \tag{6b}$$

[cf. Eq. (20.53)]. The bias terms $\mathcal{B}_T^{(2)}$, $\mathcal{B}_T^{(4)}$ are given by Eqs. (2), (3), where specifically for Eq. (3) we get

$$\mathcal{B}_T^{(4)} = -\frac{\psi^{-2}\overline{a_0^2}}{4} \iint_{-\infty}^{\infty} \rho_G(t,u)_T \langle a_0^2 s(t-\epsilon;\, \theta) s(u-\epsilon;\, \theta)\rangle\, dt\, du \tag{7}$$

The relations (20.49) to (20.58) also apply here with appropriate modifications. For example, we insert $a_0^2/\overline{a_0^2}$ as the coefficient of Φ_T, D_T, etc., before taking the indicated averages; thus, $\langle \Psi_T\rangle$ is replaced by $\langle a_0^2 \Psi_T/\overline{a_0^2}\rangle$, $\langle D_T\rangle$ by $\langle a_0^2 D_T/\overline{a_0^2}\rangle$, and the weighting function of the matched filter is still determined by Eqs. (20.50), (20.51). The various interpretations of system structure [Eq. (6a)] likewise are unaltered. For the white noise case examined in Eqs. (20.56) to (20.58), replace $a_0^2/\overline{a_0^2}$ by $A_0^2/\overline{A_0^2}$ therein.

20.5 (*a*) Carry out the details of the evaluation of Eqs. (20.67*a*) to (20.67*c*) for these narrowband signals.

(*b*) If all signal parameters are known, except RF phase (which is uniformly distributed), and if the accompanying normal noise is white, show for the narrowband signal process of Sec. 20.2-3 that the optimum detector's structure for all signal strengths is given by Eq. (20.64*a*), which now becomes

$$\boldsymbol{T}_R^{(N)}\{V\} = \log \Lambda_T = \log \mu - \frac{A_0^2}{4W_0}\int_0^T F_s(t;\theta_1)^2\, dt + \log I_0 \left[\frac{A_0}{W_0}\left|\int_0^T F_s(t;\theta_1) V(t) e^{i\omega_s t - i\phi_s(t,\theta_2)}\, dt\right|\right] \tag{1}$$

Show also that the system function of the Bayes matched filter for this system is obtained from

$$\left(\int_0^T |Y_0(i\omega;x\, dx\right)^{1/2} = 2W_0^{-1}\left[\left|\int_0^T F_s(t,\theta_1) e^{i(\omega_s+\omega)t - i\phi_s(t;\theta_2)}\, dt\right|^2 + \left|\int_0^T F_s(t;\theta_1) e^{i(\omega_s-\omega)t - i\phi_s(t;\theta_2)}\, dt\right|^2\right]^{1/2} \tag{2}$$

Give a block diagram of the detection and decision systems.

(*c*) For the narrowband signal processes of Sec. 20.2-3, show that, when the signal amplitude depends statistically on other signal parameters, for example, $a_0 = a_0(\epsilon,\theta)$ (cf. Prob. 20.4*b*), the optimum threshold structure of Eq. (20.67) becomes

$$\boldsymbol{T}_R^{(N)}\{V\} = \log \Lambda_T \doteq (\log \mu + \mathcal{B}_{0T}^{(2)} + \mathcal{B}_{0T}^{(4)}) + \frac{\overline{a_0^2}}{4} \iint_{-\infty}^{\infty} v(t)\rho_{G_0}(t,u)_T v(u)\, dt\, du \tag{3}$$

where now
$$\rho_{G_0}(t,u)_T \equiv \psi^2 \langle a_0^2 D_T(t,u;\epsilon,\theta)_0\rangle / \overline{a_0^2} \tag{4}$$

[cf. Eqs. (20.62*b*), (20.68)] and the bias terms are given by appropriate modifications of Eqs. (20.39*a*), (20.39*b*), viz.,

$$\mathcal{B}_{0T}^{(2)} = -\frac{\psi}{2}\int_{-\infty}^{\infty} \langle a_0^2 s(t-\epsilon;\, \theta) X_T(t,\epsilon,\theta)\rangle\, dt \doteq \mathcal{B}_T^{(2)} \tag{5}$$

$$\mathcal{B}_{0T}^{(4)} \doteq \frac{-\overline{a_0^2}}{16\psi^2} \iint_{-\infty}^{\infty} \rho_{G0}(t,u)_T \langle a_0^2 s(t-\epsilon;\theta)s(u-\epsilon;\theta)\rangle\, dt\, du \doteq \mathcal{B}_T^{(4)} \tag{6}$$

Again, the results of Eqs. (20.67) to (20.69) apply with suitable modifications: we insert $a_0^2/\overline{a_0^2}$ before the averaging process is carried out over D_T, ψ_{0T}, etc.

(*d*) Verify that the operations on $v(t)$, $v(u)$ in Eq. (3) can be written alternatively in the white noise case as

$$\left\langle A_0^2 \left| \int_0^T V(t)F_s(t;\theta_1)e^{i\omega_s t - i\phi_s(t;\theta_2)}\, dt \right|^2 \right\rangle \Big/ 2W_0^2 \tag{7}$$

where the bias terms [Eqs. (20.67*d*), (20.67*e*)] become specifically

$$\mathcal{B}_{0T}^{(2)} = -\frac{1}{4W_0}\int_0^T \langle A_0^2 F_s(t;\theta_1)\rangle\, dt \tag{8}$$

$$\mathcal{B}_{0T}^{(4)} = -\frac{1}{16W_0^2}\iint_0^T \langle A_0^2 F_s(t;\theta_1)F_s(u;\theta_1) \times \cos\,[\omega_s(t-u) - \phi_s(t;\theta_2) + \phi_s(u;\theta_2)]\rangle^2\, dt\, du \tag{9}$$

Extend Eqs. (20.69*b*) to get for the modulus of the system function of the Bayes matched filter

$$\left(\int_0^T |Y_0(i\omega;x)|^2\, dx\right)^{1/2} = 2W_0^{-1}\left[(\overline{a_0^2})^{-1}\left\langle \left|\int_0^T a_0 F_s(t;\theta_1)e^{i(\omega_s+\omega)t - i\phi_s(t;\theta_2)}\, dt\right|^2 + \left|\int_0^T a_0 F_s(t;\theta_1)e^{i(\omega_s-\omega)t - i\phi_s(t;\theta_2)}\, dt\right|^2\right\rangle\right]^{1/2} \tag{10}$$

20.6 In the case of narrowband signals and incoherent observation with all signal parameters specified except amplitude and RF epoch, which is assumed uniformly distributed over an RF cycle, show for continuous sampling that two realizable time-invariant matched filters, of weighting functions

$$h_{M1}(T-t;\,T) \equiv X_T(t;\theta_0) \qquad h_{M2}(T-t;\,T) \equiv Y_T(t;\theta_0) \qquad 0- < t < T+ \tag{1}$$

can be found [cf. Eqs. (20.61)] to give us Ψ_{0T} [Eq. (20.62*a*)] in threshold detection, with the help of a pair of identical zero-memory square-law rectifiers, in the manner of Fig. P 20.6. (The switch, or read-out, can be before or after the quadratic rectifier, since the latter has zero memory.) Note that here there is no postdetector integration, unlike that shown in Fig. 20.2*a*, *b*. In practice, however, it is usually simpler to use the alternative system of Fig. 20.2*a*, and its continuous analogue, since it may be very difficult to

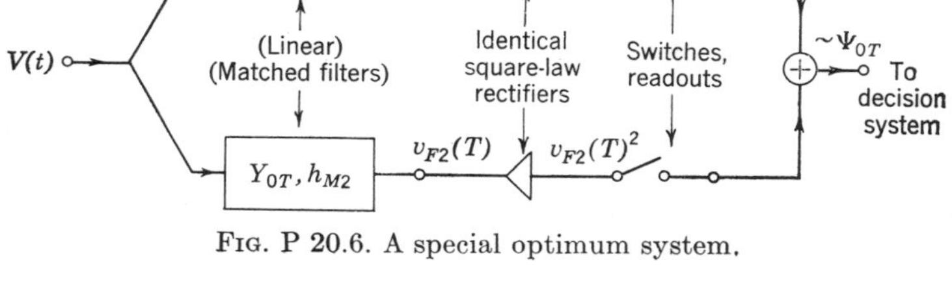

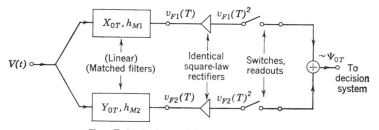

FIG. P 20.6. A special optimum system.

obtain a read-out *exactly* at $t = T$. Because the outputs of $h_{M1,M2}$ are narrowband [cf. Eqs. (20.61*b*) and (1) above], they oscillate rapidly and unless great care is taken to read out exactly at $t = T$ (or at $T \pm mT_0$, T_0 = RF period, m small), where the output is a maximum, considerable error in the result may then occur. For this reason, the alternative approach of the text may be preferred, as well as for the greater latitude it offers the system designer, who is then restricted only to a specific modulus of the transfer function [cf. Eqs. (20.51*a*) et seq.]. On the other hand, the invariant nature of the h_M, vis-à-vis the time-varying switch required in the Bayes matched filter of the first kind, is a point in favor of the former. Note, in the present problem, that there is but one set of matched filters [Eq. (1)].

20.3 *System Evaluation: Error Probabilities and Average Risk*

In the previous two sections, we have obtained the threshold structure of optimum binary detection systems when deterministic signals and normal noise are present at the input. Both coherent and incoherent modes of observation (cf. Sec. 19.4-3) are assumed, and the signal processes may be broad- or narrowband, the latter characteristic of most communication waveforms. Here we shall use the results of these preceding sections, in conjunction with portions of Chap. 17, to determine the error probabilities and minimum average (i.e., Bayes) risks associated with these optimum detection systems (Sec. 19.3-1). Specifically, we need to apply the methods and results of Secs. 17.2-1, 17.2-2, along with those of Secs. 20.1, 20.2, to the general relations of Secs. 19.3-1, 19.3-2, under the conditions on the signal and noise processes described at the beginning of the present chapter. A number of detailed examples may also be found in Sec. 20.4.

20.3-1 Optimum Threshold Systems. The Bayes risk, or minimum average cost, associated with the decision process of binary detection is given by Eq. (19.19*a*), viz.,

$$R^* = \mathcal{R}_0 + p(C_\beta - C_{1-\beta})\left(\frac{\mathcal{K}}{\mu}\alpha^* + \beta^*\right) \qquad \mathcal{R}_0 = qC_{1-\alpha} + pC_{1-\beta} \tag{20.70}$$

where the type I and II conditional error probabilities α^*, β^* are given by Eqs. (19.29*a*), (19.29*b*), with Eqs. (19.30*a*) to (19.32). Thus, we have

$$\alpha^* = \int_{\log \mathcal{K}}^{\infty} Q_n(x)\,dx \qquad \beta^* = \int_{-\infty}^{\log \mathcal{K}} P_n(x)\,dx \tag{20.71}$$

with $x \equiv \boldsymbol{T}_R^{(N)}\{V\} = \log \Lambda_n(V)$; Q_n, P_n are, respectively, the d.d.'s of x with respect to the null hypothesis $H_0{:}N$ and the alternative $H_1{:}S + N$. The *threshold* $\mathcal{K}$ is a cost ratio; for example, $\mathcal{K} = (C_\alpha - C_{1-\alpha})/(C_\beta - C_{1-\beta})$ [cf. Eq. (19.15)], whose value depends on the costs C_α, $C_{1-\alpha}$, etc., preassigned to the outcomes of the four possible correct and incorrect decisions [cf. Eqs. (19.5)].

Accordingly, our first task is to determine the d.d.'s $P_n(x)$, $Q_n(x)$ for the various optimum threshold systems $x = \log \Lambda_n$ which we have derived in

Secs. 20.1, 20.2. Applying these results to the quadratures of Eqs. (20.71) then gives the desired error probabilities α^*, β^* and the associated Bayes risk [Eq. (20.70)]. As before, the natural starting point is the discrete system; results for the corresponding continuous systems follow in the appropriate limit $n \to \infty$, $T < \infty$, as described in Secs. 19.4-2, 20.2.

(1) Coherent reception. Let us begin with the optimum threshold system, where coherent† observation is employed, so that Eq. (20.6) applies, viz.,

$$x = \boldsymbol{T}_R^{(N)}\{V\} = \log \Lambda_n(\mathbf{V}) \doteq \left(\log \mu - \frac{\bar{a}_0^2}{2}\Phi_{\bar{s}}\right) + \bar{a}_0\tilde{\mathbf{v}}\mathbf{k}_N^{-1}\bar{\mathbf{s}} \tag{20.72}$$

where $\Phi_{\bar{s}} = \tilde{\bar{\mathbf{s}}}\mathbf{k}_N^{-1}\bar{\mathbf{s}}$ [cf. Eqs. (20.5*a*), (20.7)] and $(\bar{\mathbf{s}})_j \neq 0$ (one or more $j = 1, \ldots, n$). To obtain $Q_n(x)$, $P_n(x)$ according to Eqs. (19.31), (19.32), we first find the characteristic functions of x [Eqs. (19.30*a*), (19.30*b*)], with the help of Eq. (7.26). Setting $A_n \equiv \log \mu - (\bar{a}_0^2/2)\Phi_{\bar{s}}$, we have

$$F_1(i\xi)_Q = \boldsymbol{E}_{\mathbf{V}|H_0}\{e^{i\xi x(\mathbf{V})}\} = e^{i\xi A_n}\int_{-\infty}^{\infty}\cdots\int e^{i\xi\bar{a}_0\tilde{\mathbf{v}}\mathbf{k}_N^{-1}\bar{\mathbf{s}} - \frac{1}{2}\tilde{\mathbf{v}}\mathbf{k}_N^{-1}\mathbf{v}} \times \frac{d\mathbf{v}}{(2\pi)^{n/2}(\det \mathbf{k}_N)^{1/2}} \tag{20.73a}$$

$$F_1(i\xi)_Q = \boldsymbol{E}_{\mathbf{V}|H_0}\{e^{i\xi x(\mathbf{V})}\} = e^{i\xi A_n - (\bar{a}_0^2/2)\Phi_{\bar{s}}\xi^2} \tag{20.73b}$$

Similarly, from Eqs. (7.26), (17.31*b*), and (19.30*b*) we get, for the additive noise and signal processes assumed here,

$$\begin{aligned} F_1(i\xi)_P &= \boldsymbol{E}_{\mathbf{V}|H_1}\{\exp[i\xi x(\mathbf{V})]\} \\ &= \exp(i\xi A_n)\left\langle \exp\left(-\frac{1}{2}a_0^2\tilde{\mathbf{s}}\mathbf{k}_N^{-1}\mathbf{s}\right)\right. \\ &\qquad \int_{-\infty}^{\infty}\cdots\int \exp\left[i\tilde{\mathbf{v}}\mathbf{k}_N^{-1}(\bar{a}_0\xi\bar{\mathbf{s}} - ia_0\mathbf{s}) - \frac{1}{2}\tilde{\mathbf{v}}\mathbf{k}_N^{-1}\mathbf{v}\right] \\ &\qquad \left.\frac{d\mathbf{v}}{(2\pi)^{n/2}(\det \mathbf{k}_N)^{1/2}}\right\rangle_S \end{aligned} \tag{20.74a}$$

$$F_1(i\xi)_P = F_1(i\xi)_Q \,\langle \exp(i\xi a_0\bar{a}_0\tilde{\bar{\mathbf{s}}}\mathbf{k}_N^{-1}\mathbf{s})\rangle_{S(\equiv a_0,\,\boldsymbol{\theta})} \tag{20.74b}$$

In the present weak-signal condition of operation we can write the indicated average over $\mathbf{s}$ in Eq. (20.74*b*) to the order of approximation of $\log \Lambda_n$ in a_0,

$$\boldsymbol{E}_S\{e^{i\xi a_0\bar{a}_0\tilde{\bar{\mathbf{s}}}\mathbf{k}_N^{-1}\mathbf{s}}\} \doteq 1 + i\xi\bar{a}_0^2\Phi_{\bar{s}} + O[a_0^4;(i\xi)^2] \doteq e^{i\xi\bar{a}_0^2\Phi_{\bar{s}}} \tag{20.75}$$

so that the c.f. $F_1(i\xi)_P$ becomes finally

$$F_1(i\xi)_P \doteq e^{i\xi(\log\mu + \bar{a}_0^2\Phi_{\bar{s}}/2) - \frac{1}{2}\bar{a}_0^2\Phi_{\bar{s}}\xi^2} \tag{20.76}$$

† "Coherence" here implies that $w(\epsilon) = \delta(\epsilon - \epsilon_0)$, that is, epoch is known to the observer, so that $\langle s\rangle_\epsilon = s(t - \epsilon_0; \boldsymbol{\theta}) \neq 0$.

The corresponding d.d.'s of x are given by

$$Q_n(x) = \mathfrak{F}^{-1}\{F_1(i\xi)_Q\} = (2\pi\bar{a}_0{}^2\Phi_{\bar{s}})^{-1/2} \times \exp\{-[x - (\log\mu - \bar{a}_0{}^2\Phi_{\bar{s}}/2)]^2/2\bar{a}_0{}^2\Phi_{\bar{s}}\} \quad (20.77a)$$

$$P_n(x) = \mathfrak{F}^{-1}\{F_1(i\xi)_P\} \doteq (2\pi\bar{a}_0{}^2\Phi_{\bar{s}})^{-1/2} \times \exp\{-[x - (\log\mu + \bar{a}_0{}^2\Phi_{\bar{s}}/2)]^2/2\bar{a}_0{}^2\Phi_{\bar{s}}\} \quad (20.77b)$$

The desired error probabilities follow from Eqs. (20.71). With the help of the error function

$$\Theta(z) = \frac{2}{\sqrt{\pi}} \int_0^z e^{-y^2}\, dy \quad (20.78)$$

[cf. Eqs. (A.1.1) et seq.], these become specifically

$$\begin{Bmatrix} \alpha^* \\ \beta^* \end{Bmatrix} \cong \tfrac{1}{2}\left\{1 - \Theta\left[\frac{\bar{a}_0\Phi_{\bar{s}}^{1/2}}{2\sqrt{2}} \pm \frac{\log(\mathcal{K}/\mu)}{\sqrt{2}\,\bar{a}_0\Phi_{\bar{s}}^{1/2}}\right]\right\} \quad (20.79)$$

With continuous sampling, Eq. (20.79) still applies, if now $\Phi_{\bar{s}}$ is replaced by

$$\lim_{n\to\infty} \Phi_{\bar{s}} = (\Phi_T)_{\bar{s}} \equiv \Phi_T(\bar{s},\bar{s}) = \psi \int_{0-}^{T+} \langle s(t - \epsilon_0;\, \boldsymbol{\theta})\rangle\langle X_T(t;\epsilon_0,\boldsymbol{\theta})\rangle\, dt \quad (20.79a)$$

[cf. Eqs. (20.37) to (20.39*c*)]. For white noise backgrounds, this becomes [cf. Eq. (20.43*c*)]

$$\bar{a}_0{}^2(\Phi_T)_{\bar{s}} = \bar{A}_0{}^2W_0{}^{-1}\int_0^T \langle s(t - \epsilon_0;\, \boldsymbol{\theta})\rangle^2\, dt \qquad (\neq 0) \quad (20.79b)$$

The significant feature of Eq. (20.79) is the dependence of the error probabilities on the product $\bar{a}_0\Phi_{\bar{s}}^{1/2}$ [or $\bar{a}_0(\Phi_T)_{\bar{s}}^{1/2}$ for coherent observation]. This dependence, in turn, permits us to relate in a simple way the minimum detectable signal and the required observation time or sample size for threshold operation. Thus, we note that $\Phi_{\bar{s}}$ depends on sample size like $ng_1(n,s)$, where g_1 is such that $\lim_{n\to\infty} g(n,s) = g_0(s)$, dependent on signal waveform only; similarly, $(\Phi_T)_{\bar{s}} = Tg_2(T,s)$, with $\lim_{T\to\infty} g_2 \to g_0(s)_\infty$, etc. Thus, for a preset Bayes risk (i.e., preset values of α^*, β^*) [Eq. (20.70)], *the* (average, threshold) *minimum detectable signal-to-noise* (power) *ratio* $(\overline{a_0{}^2})^*_{\min} \sim \bar{a}_0{}^2$ (cf. Sec. 19.3-3) *is inversely proportional to observation time*, e.g.,

$$(\overline{a_0{}^2})^*_{\min} \sim 1/n \text{ or } 1/T \qquad \textit{coherent reception} \quad (20.80)$$

As we shall observe later (Sec. 20.3-4), this behavior is characteristic of all binary optimum threshold systems with coherent observation, independent of the specific statistical structure of the interfering noise, as long as the latter is additive, and as long as a high accuracy of decision is required, so that large-sample (e.g., normal) statistics ultimately govern† the decision

† Subject to the conditions obeyed by the central-limit theorem (Sec. 7.7-3).

process (cf. Sec. 19.4-3). Finally, note from Eq. (20.79) as $\bar{a}_0 \to \infty$, with $\Phi_{\bar{s}}$ fixed, or vice versa, that α^*, β^* both vanish, as expected: for sufficiently strong signals or sufficiently long observation periods (during which the signal is, of course, nonvanishing), it becomes ideally possible to determine the presence or absence of a signal with zero probability of error.

(2) Incoherent reception. Let us consider next the more common situation of incoherent reception where $\langle s \rangle_\epsilon$ vanishes. The optimum threshold system (for discrete sampling) is given by Eqs. (20.11), viz.,

$$x = \boldsymbol{T}_R^{(N)}\{V\} \doteq (\log \mu + \mathcal{B}_n^{(2)} + \mathcal{B}_n^{(4)}) + \frac{\overline{a_0^2}}{2}\,\tilde{\mathbf{v}}\bar{\mathbf{G}}\mathbf{v} \qquad \bar{\mathbf{G}} = \mathbf{k}_N^{-1}\overline{\mathbf{s}\tilde{\mathbf{s}}}\mathbf{k}_N^{-1} \tag{20.81}$$

The d.d.'s of x [Eqs. (19.31), (19.32)] are found, as before, by first determining the associated characteristic functions (19.30a). For Eq. (19.30a), we can write explicitly

$$F_1(i\xi)_Q = \boldsymbol{E}_{V|H_0}\{e^{i\xi\left[A_n + \frac{\overline{a_0^2}}{2}\tilde{\mathbf{v}}\bar{\mathbf{G}}\mathbf{v}\right]}\} \qquad A_n \equiv \log \mu + \mathcal{B}_n^{(2)} + \mathcal{B}_n^{(4)}$$

$$= e^{i\xi A_n} \int_{-\infty}^{\infty} \cdots \int e^{-\frac{1}{2}\tilde{\mathbf{v}}\mathbf{k}_N^{-1}(\mathbf{I} - i\xi\overline{a_0^2}\mathbf{k}_N\bar{\mathbf{G}})\mathbf{v}} \frac{d\mathbf{v}}{(2\pi)^{n/2}(\det \mathbf{k}_N)^{\frac{1}{2}}} \tag{20.82}$$

This is just Eq. (17.23) with $\mathbf{J} = \overline{a_0^2}\bar{\mathbf{G}}/2$ here. Applying Eqs. (17.24a), (17.24b), and (17.27) to Eq. (20.82), we can write at once

$$F_1(i\xi)_Q = e^{i\xi A_n}[\det(\mathbf{I} - i\xi\overline{a_0^2}\mathbf{k}_N\bar{\mathbf{G}})]^{-\frac{1}{2}} \tag{20.83}$$

$$Q_n(x) \simeq (2\pi D_n^{(N)})^{-\frac{1}{2}} e^{-(x - C_n^{(N)})^2/2D_n^{(N)}} \tag{20.84}$$

where specifically

$$C_n^{(N)} = \log \mu + \mathcal{B}_n^{(2)} + \mathcal{B}_n^{(4)} + \frac{\overline{a_0^2}}{2} \operatorname{tr} \mathbf{k}_N\bar{\mathbf{G}}$$
$$D_n^{(N)} = \frac{\overline{a_0^2}^2}{2} \operatorname{tr} (\mathbf{k}_N\bar{\mathbf{G}})^2 \tag{20.84a}$$

Equation (20.83) is exact [for the x of Eq. (20.81)], while Eq. (20.84) is the result of the large-sample condition implied by threshold reception [cf. Eqs. (17.25), (17.26), etc.].† The type I error probability α^* [Eq. (20.71)] is now easily found to be

$$\alpha^* \simeq \frac{1}{2}\left\{1 - \Theta\left[\frac{\log \mathcal{K} - C_n^{(N)}}{(2D_n^{(N)})^{\frac{1}{2}}}\right]\right\} \tag{20.85}$$

Finding the type II error probability β^* is somewhat more involved. We begin, as usual, by considering the alternative c.f. of x [Eq. (19.30b)],

† For further terms in the generalized Edgeworth series, of which Eq. (20.84) represents the leading term and significant contribution, see Eqs. (17.28a), (17.28b).

which is now specifically

$$F_1(i\xi)_P = \boldsymbol{E}_{\mathbf{V}|H_1}\left\{\exp\left[i\xi\left(A_n + \frac{\overline{a_0^2}}{2}\,\tilde{\mathbf{v}}\bar{\mathbf{G}}\mathbf{v}\right)\right]\right\}$$

$$= \exp(i\xi A_n)\Big\langle \exp(-\tfrac{1}{2}a_0^2\tilde{\mathbf{s}}\mathbf{k}_N^{-1}\mathbf{s}) \int \cdots \int_{-\infty}^{\infty} \exp[a_0\tilde{\mathbf{s}}\mathbf{k}_N^{-1}\mathbf{v} - \tfrac{1}{2}\tilde{\mathbf{v}}\mathbf{k}_N^{-1}(\mathbf{I} - i\xi\overline{a_0^2}\mathbf{k}_N\bar{\mathbf{G}})\mathbf{v}] \frac{d\mathbf{v}}{(2\pi)^{n/2}(\det \mathbf{k}_N)^{\frac{1}{2}}}\Big\rangle_S \quad (20.86)$$

Applying Eq. (17.32b), we can write this as

$$F_1(i\xi)_P = \exp(i\xi A_n)\,\frac{\langle \exp\{-\tfrac{1}{2}a_0^2\tilde{\mathbf{s}}\mathbf{k}_N^{-1}[\mathbf{I} - (\mathbf{I} - i\xi\overline{a_0^2}\mathbf{k}_N\bar{\mathbf{G}})^{-1}]\mathbf{s}\}\rangle_S}{[\det(\mathbf{I} - i\xi\overline{a_0^2}\mathbf{k}_N\bar{\mathbf{G}})]^{\frac{1}{2}}} \quad (20.87a)$$

Our next step is to expand the exponent, retaining terms through $O(a_0^4)$ only, in keeping with the threshold approximation governing the system structure of Eq. (20.81). The exponent within $\langle\ \rangle_S$ in Eq. (20.87a) becomes

$$M(a_0,s,i\xi) = i\xi a_0^2\overline{a_0^2}\tilde{\mathbf{s}}\bar{\mathbf{G}}\mathbf{s}/2 + O(a_0^6,\xi^2)$$

so that carrying out the indicated average over the parameters of S yields

$$\langle e^M\rangle_S \doteq 1 + i\xi\,\frac{\overline{a_0^2}^2}{2}\langle\tilde{\mathbf{s}}\bar{\mathbf{G}}\mathbf{s}\rangle \doteq e^{(i\xi\overline{a_0^2}^2/2)\langle\Phi_G\rangle} \qquad \Phi_G \equiv \tilde{\mathbf{s}}\bar{\mathbf{G}}\mathbf{s} \quad (20.87b)$$

Expanding the determinant in Eq. (20.87a) according to Eq. (17.25), with the help of the basic identity (17.13), then gives for the principal contribution to $F_1(i\xi)_P$ the normal form

$$F_1(i\xi)_P \doteq e^{i\xi C_n^{(S+N)} - \frac{1}{2}D_n^{(S+N)}\xi^2} \quad (20.87c)$$

with

$$C_n^{(S+N)} = A_n + \frac{\overline{a_0^2}}{2}\,\mathrm{tr}\,\mathbf{k}_N\bar{\mathbf{G}} + \frac{\overline{a_0^2}^2}{2}\langle\Phi_G\rangle$$
$$D_n^{(S+N)} = \frac{\overline{a_0^2}^2}{2}\,\mathrm{tr}\,(\mathbf{k}_N\bar{\mathbf{G}})^2 = D_n^{(N)} \quad (20.87d)$$

The corresponding d.d. is therefore

$$P_n(x) \simeq \{2\pi D_n^{(N)}\}^{-\frac{1}{2}} e^{-(x - C_n^{(S+N)})^2/2D_n^{(N)}} \quad (20.88)$$

and from Eqs. (20.71) the type II error probability is accordingly

$$\beta^* \cong \frac{1}{2}\left\{1 - \Theta\left[\frac{-\log\mathcal{K} + C_n^{(S+N)}}{(2D_n^{(N)})^{\frac{1}{2}}}\right]\right\} \quad (20.89)$$

We can further simplify the arguments of the error functions in Eqs. (20.85), (20.89) by noting that (cf. Prob. 20.2)

$$\operatorname{tr} \mathbf{k}_N\overline{\mathbf{G}} = \operatorname{tr} \overline{\mathbf{s}\tilde{\mathbf{s}}}\mathbf{k}_N^{-1} = \langle \tilde{\mathbf{s}}\mathbf{k}_N^{-1}\mathbf{s}\rangle \equiv \langle \Phi_s \rangle \tag{20.90a}$$

$$\operatorname{tr} (\mathbf{k}_N\overline{\mathbf{G}})^2 = \langle \tilde{\mathbf{s}}\overline{\mathbf{G}}\mathbf{s}\rangle \equiv \langle \Phi_G \rangle > 0 \tag{20.90b}$$

Therefore,

$$\mathcal{B}_n^{(2)} = -\frac{\overline{a_0{}^2}}{2} \operatorname{tr} \mathbf{k}_N\overline{\mathbf{G}} \tag{20.90c}$$

[cf. Eqs. (20.11c)]. Combining these results with Eqs. (20.11c) for $\mathcal{B}_n^{(4)}$ in Eqs. (20.85) and (20.87d), we obtain directly†

$$\begin{Bmatrix} \alpha^* \\ \beta^* \end{Bmatrix} \cong \frac{1}{2}\left\{1 - \Theta\left[\frac{\overline{a_0{}^2}\langle\Phi_G\rangle^{1/2}}{4} \pm \frac{\log (\mathcal{K}/\mu)}{\overline{a_0{}^2}\langle\Phi_G\rangle^{1/2}}\right]\right\} \tag{20.91}$$

which is also valid for continuous data sampling provided that we replace $\langle \Phi_G \rangle$ by

$$\langle \Phi_T \rangle_G \equiv \lim_{n\to\infty} \langle \Phi_G \rangle = \psi^2 \lim_{n\to\infty} \langle \tilde{\boldsymbol{\gamma}}\overline{\mathbf{s}\tilde{\mathbf{s}}}\boldsymbol{\gamma}\rangle$$

$$= \psi^2 \iint\limits_{-\infty}^{\infty} \langle D_T(t,u;\epsilon,\boldsymbol{\theta})\rangle\langle s(t - \epsilon, \boldsymbol{\theta})s(u - \epsilon, \boldsymbol{\theta})\rangle \, dt \, du \tag{20.91a}$$

[cf. Eqs. (20.44a) et seq.]. With white noise backgrounds, Eq. (20.91a) is replaced instead by

$$\overline{a_0{}^2}^2\langle \Phi_T \rangle_G = \frac{\overline{A_0{}^2}^2}{W_0{}^2} \iint\limits_0^T \langle s(t - \epsilon, \boldsymbol{\theta})s(u - \epsilon, \boldsymbol{\theta})\rangle^2 \, dt \, du \tag{20.91b}$$

The salient feature of Eqs. (20.91) is now the dependence of the error probabilities α^*, β^* on the product $\overline{a_0{}^2}\langle\Phi_G\rangle^{1/2}$ in this weak-signal theory. By an argument similar to that leading to Eq. (20.80) in the coherent case, we find that *the* (average, threshold) *minimum detectable signal-to-noise* (power) *ratio* $(\overline{a_0{}^2})^*_{\min}$ (Sec. 19.3-3) *is inversely proportional to the square root of the observation time*, e.g.,

$$(\overline{a_0{}^2})^*_{\min} \sim 1/\sqrt{n} \text{ or } \sim 1/T^{1/2} \qquad \textit{incoherent reception} \tag{20.92}$$

This behavior is characteristic of all binary threshold systems for incoherent reception such that $\langle s \rangle_\epsilon = 0$, regardless of the specific statistical nature of the noise, as long as the signal and noise processes are additive and as long as large-sample (e.g., normal) statistics (for x) apply, i.e., as long as the threshold situation is in force. Again, we observe that α^*, β^* approach

† Note that if we set $\Phi_G = 2\Phi_{0G}$, the argument of Θ in Eq. (20.91) has the standard factors $2\sqrt{2}$ and $\sqrt{2}$, respectively, in the denominators of the first and second terms [cf. Eqs. (20.79), (20.120), (20.131)].

zero, as required, when either $\overline{a_0^2}$ or $\langle\Phi_G\rangle$ and $\langle\Phi_T\rangle_G$ become indefinitely large: indefinitely strong signals or long observation times (with signal present) lead to vanishing probabilities of decision error.†

Finally, observe that we obtain Eq. (20.91) once again for the error probabilities when the signal process is narrowband and we employ the alternative representation [Eqs. (20.32), (20.33)]. Specifically, since $B_n \doteq \langle\Phi_s\rangle$, $\mathbf{G}_0 \doteq 2\mathbf{G}$ [cf. Eqs. (20.26), (20.29a)], we can rewrite Eq. (20.91) as

$$\begin{Bmatrix}\alpha^*\\ \beta^*\end{Bmatrix} \simeq \frac{1}{2}\left\{1 - \Theta\left[\frac{\overline{a_0^2}\langle\Phi_{0G}\rangle^{1/2}}{2\sqrt{2}} \pm \frac{\log(\mathcal{K}/\mu)}{\sqrt{2}\,\overline{a_0^2}\langle\Phi_{0G}\rangle^{1/2}}\right]\right\} \tag{20.93}$$

where now for discrete and continuous sampling we have set

$$\langle\Phi_{0G}\rangle \equiv \tfrac{1}{2}\langle\Phi_G\rangle \doteq \tfrac{1}{8}\,\mathrm{tr}\,(\mathbf{k}_N\bar{\mathbf{G}}_0)^2 \tag{20.93a}$$

$$\langle\Phi_{0T}\rangle_G = \lim_{n\to\infty} \tfrac{1}{2}\langle\Phi_G\rangle \doteq \frac{\psi^2}{8}\iint_{-\infty}^{\infty} \langle D_T(t,u;\epsilon,\boldsymbol{\theta})_0\rangle\langle F_s(t;\boldsymbol{\theta}_1)F_s(u;\boldsymbol{\theta}_1) \times \cos\left[\omega_s(t-u) - \phi_s(t;\boldsymbol{\theta}_2) + \phi_s(u;\boldsymbol{\theta}_2)\right]\rangle\,dt\,du \tag{20.93b}$$

With white noise backgrounds, Eq. (20.93b) becomes

$$\overline{a_0^2}^2\langle\Phi_{0T}\rangle_G \doteq \frac{\overline{A_0^2}^2}{8W_0^2}\iint_0^T \langle F_s(t;\boldsymbol{\theta}_1)F_s(u;\boldsymbol{\theta}_1) \times \cos\left[\omega_s(t-u) - \phi_s(t;\boldsymbol{\theta}_2) + \phi_s(u;\boldsymbol{\theta}_2)\right]\rangle^2\,dt\,du \tag{20.93c}$$

For extensions of Eqs. (20.91), (20.93), including the important case where signal amplitude depends on one or more of the other signal parameters, see Prob. 20.7*b*.

20.3-2 A Suboptimum System. In almost all applications, the optimum system is a theoretical limit, never strictly realizable in practice, because of system complexity, cost, size, etc., or other external constraints. It is therefore important to know how well suboptimum systems under these limitations, but chosen to approximate the essential operations indicated by the optimal system, approach the desired limiting behavior. The methods of decision theory are, of course, applicable to such nonoptimum systems. By computing the average risk R [Eq. (19.19b)], in such cases we can then compare the given (suboptimum) system with the optimum one for the same purpose (cf. Secs. 19.3-2, 19.3-3). In this way, we can obtain a definite measure of the extent to which actual performance is degraded.

† This would, however, not appear from our results for α^*, β^* if we omitted the fourth-order term $\mathcal{B}_n^{(4)}$ in the bias, as Eq. (20.90c) in Eqs. (20.84a) shows. For this reason, we must always include terms through $O(a_0^4)$ in calculating the appropriate first and second moments C_n, D_n of the limiting normal d.d. for $x = \log \Lambda_n$, as we remarked at the beginning of Sec. 20.1.

Here we shall illustrate the general approach with a brief outline of the analysis for a class of nonideal binary threshold detection systems, to show how such comparisons may be made.

(1) Remarks on structure. As our example, let us consider incoherent threshold reception, with $\langle s \rangle_\epsilon = 0$, corresponding to the situation examined in Secs. 20.1-2, 20.2-2, but where instead of the optimum system we choose for our class of suboptimum systems the general square-law operation

$$y = \boldsymbol{T}_R^{(N)}\{V\}_{subopt} = \log G_n(\mathbf{V}) = \Gamma_2 + \frac{\overline{A_0{}^2}}{4} \tilde{\mathbf{V}}\mathbf{G}_2\mathbf{V} \tag{20.94}$$

analogous to Eq. (20.81), with $\mathbf{G}_2$ symmetrical and positive definite. There is little point in introducing a more involved structure, since it has been shown above that the quadratic system with appropriate weighting is optimum [cf. Eq. (20.81)]. Here Γ_2 is the bias, and $\mathbf{G}_2$ is the new weighting, which is generally not equal to $\psi^{-2}\bar{\mathbf{G}}$ of optimum receivers. The bias Γ_2 need not be set equal to the optimum bias, since we may frequently wish to select some other value of Γ_2 to help compensate for the fact that $\mathbf{G}_2$ is different from $\psi^{-2}\bar{\mathbf{G}}$; $\mathbf{G}_2$ itself is specified by the prescribed system structure, although not every $\mathbf{G}_2$ or bias Γ_2 guarantees satisfactory performance, as we shall observe presently.

At this point, we can parallel the argument of Eqs. (20.12) to (20.21) for structural interpretation, but now essentially in reverse. Thus, either of the following two equivalent classes of preset suboptimum systems specifies the same $\mathbf{G}_2$ [Eq. (20.94)]: (1) a time-varying linear filter, zero-memory square-law rectifier, and a weighted linear integrator; (2) a time-varying linear filter, followed by a multiplier and similar integrator, analogous to the systems diagramed in Fig. 20.2*a*, *b*. Here, however, the linear predetector filter is no longer matched, and the final low-pass device may not be the simple integrator employed in the optimum systems.

Let us consider (1) first, with discrete sampling, and let us construct the various operations in sequence. We have accordingly for the output of the linear prerectifier filter

$$\mathbf{V}_{2F} = \mathbf{Q}_2\mathbf{V} \qquad \text{or} \qquad V_{2F}(t_j) = \sum_{i=1}^{\infty} h_2(t_j - t_i)V(t_i)\,\Delta t \tag{20.95}$$

Here $[(\mathbf{Q}_2)_{ij}] = [h_2(t_i - t_j)\,\Delta t]$ is the combination of a (real) linear discrete filter, which is postulated to be invariant and physically realizable [cf. Eqs. (20.16)], and a time-varying switch.† Next, for the output of the zero-memory quadratic element at a time t_j we have

$$(\tilde{\mathbf{V}}_{2F}\mathbf{V}_{2F})_j = (\tilde{\mathbf{V}}\tilde{\mathbf{Q}}_2\mathbf{Q}_2\mathbf{V})_j = V_{2F}(t_j)^2 \tag{20.96}$$

† See also the discussion following Eqs. (20.19).

Now, for the output of the final postrectifier smoothing element we can write

$$\sum_{j=1}^{n} \frac{a_j}{n} V_{2F}(t_j)^2 \equiv (\tilde{\mathbf{V}}_{2F}'^{*})^{*}\mathbf{V}_{2F}' \tag{20.97}$$

where the asterisk denotes the complex conjugate and where specifically the column vector $\mathbf{V}_{2F}'$ is expressed as

$$V_{2F}'(t_i) \equiv \sum_{j=1}^{n} A_{ij}V_{2F}(t_j) \qquad \text{or} \qquad \mathbf{V}_{2F}' = \mathbf{A}\mathbf{V}_{2F} \qquad \text{with } A_{ij} \equiv \left(\frac{a_j}{n}\right)^{1/2} \delta_{ij} \tag{20.97a}$$

The matrix $\mathbf{A}$ is diagonal, such that $a_j = H(T - t_j)$, a dimensionless weighting factor, which may be negative for some t_j, so that A may have imaginary elements. Since $n\,\Delta t = T$, we see that $a_j/n = H(T - t_j)\,\Delta t/T$. Consequently, if we set

$$h_I(T - t_j) \equiv H(T - t_j)/T \tag{20.97b}$$

where h_I is the weighting function† of the final integrating filter, we can rewrite $\mathbf{V}_{2F}'$ as

$$\mathbf{V}_{2F}' = [h_I(T - t_j)^{1/2}V_{2F}(t_j)\,\Delta t^{1/2}] \tag{20.97c}$$

Combining Eqs. (20.95) to (20.97c), we obtain for the final output of this system (apart from some scale factors)

$$\begin{aligned}(\tilde{\mathbf{V}}_{2F}')\mathbf{V}_{2F}' &= \tilde{\mathbf{V}}_{2F}\tilde{\mathbf{A}}\mathbf{A}\mathbf{V}_{2F} = \tilde{\mathbf{V}}\tilde{\mathbf{Q}}_2\tilde{\mathbf{A}}\mathbf{A}\mathbf{Q}_2\mathbf{V} \\ &= \sum_{j=1}^{n} h_I(T - t_j)\left[\sum_{i=1}^{\infty \text{ or } j} h_2(t_j - t_i)V(t_i)\,\Delta t\right]^2 \Delta t \end{aligned} \tag{20.98}$$

(We remark that the weighting functions h_2, h_I here have the customary dimensions $\sec^{-1}$.)

Our choice of $\mathbf{Q}_2$ (and $\mathbf{A}$) determines $\mathbf{G}_2$ ($= \tilde{\mathbf{G}}_2$) in Eq. (20.94). As the analysis above indicates, $\mathbf{G}_2$ can be written

$$\mathbf{G}_2 = \gamma_0\tilde{\mathbf{Q}}_2\tilde{\mathbf{A}}\mathbf{A}\mathbf{Q}_2 = [\rho_{2G}(t_i,t_j)\,\Delta t]$$

or

$$\tilde{\mathbf{A}}^{-1}\tilde{\mathbf{Q}}_2^{-1}\gamma_0^{-1}\mathbf{G}_2\mathbf{Q}_2^{-1}\mathbf{A}^{-1} = \mathbf{I} \tag{20.99}$$

Applying this to Eq. (20.94) gives at once, for the present suboptimum system,

$$y = \Gamma_2 + \frac{\overline{A_0^2}}{4}\tilde{\mathbf{V}}\mathbf{G}_2\mathbf{V} = \Gamma_2 + \frac{\overline{A_0^2}}{4}\gamma_0(\tilde{\mathbf{V}}_{2F}')\mathbf{V}_{2F}' \tag{20.100}$$

[cf. Eq. (20.98)]. The factor γ_0 in Eqs. (20.99), (20.100) is equal to ψ^{-2} for discrete (and continuous) sampling, as long as $\psi < \infty$; γ_0 is introduced to provide a proper scaling of the (dimensionless) output y in terms of the

† It is not necessary that the a_j in Eq. (20.97a) be positive; negative a_j correspond to the fact that h_I may be negative for some t.

background noise level and to effect the correct reduction to the optimum case. The weighting $\mathbf{G}_2$ ($= \boldsymbol{\varrho}_{2G}\,\Delta t$) is determined by h_2, h_I here, in contrast to the optimum case, where $\psi^{-2}\bar{\mathbf{G}}$ [analogous to $\mathbf{G}_1$; cf. Prob. 20.8, Eq. (2)] *determines* a set of matched filters and is itself specified by the input signal and the spectrum of the accompanying noise process [cf. Eq. (20.81)]. From Eqs. (20.98), (20.99), it follows directly that†

$$(\boldsymbol{\varrho}_{2G})_{ij} = \begin{cases} \gamma_0 \sum_{l=1}^{n} h_I(T - t_l)h_2(t_l - t_i)h_2(t_l - t_j)\,\Delta t = (\boldsymbol{\varrho}_{2G})_{ji} & 1 \leq i, j \leq n \\ 0 & \text{elsewhere} \end{cases} \tag{20.101}$$

Finally, the system y [Eqs. (20.94), (20.100)] is also equivalent to the combination of time-varying linear filter, multiplier, and general integrator, where now we have

$$\begin{aligned} \gamma_0\tilde{\mathbf{V}}'_{2F}\mathbf{V}'_{2F} &= 2\sum_{j=1}^{n} h_I(T - t_j)V_{2F}(t_j)V''_{2F}(t_j)\,\Delta t \\ V''_{2F}(t_j) &= \sum_{j=1}^{j} \hat{\rho}_2(t_j - t_i, t_j)V(t_i)\,\Delta t \qquad \gamma_0 = \psi^{-2} \end{aligned} \tag{20.102a}$$

in which

$$\rho_2(t_i,t_j) = \hat{\rho}_2(t_j - t_i, t_j) = \gamma_0 \sum_{l=1}^{n} h_2(t_l - t_i)h_2(t_l - t_j)\,\Delta t \tag{20.102b}$$

is the weighting function of this time-varying filter. Note that, for the ideal integrator of the optimum systems,‡ $h_I(T - x) = 1/T$ in the above, while if this system approaches the optimum, i.e., if $\mathbf{G}_2 \to \psi^{-2}\bar{\mathbf{G}}$, so that $a_i \to 1$, $h_2 \to h$, etc., our results above reduce, as required, to the corresponding expressions in Sec. 20.2-2.

When continuous sampling is used, Eqs. (20.95), (20.100), (20.101), etc., go over formally into§

$$\begin{aligned} V_{2F}(t) &= \int_{0-}^{\infty} h_2(t - \tau)V(\tau)\,d\tau \\ y_T &= \Gamma_2 + \frac{\overline{A_0^2}}{4} \iint_{-\infty}^{\infty} V(t)\rho_{2G}(t,u)_T V(u)\,dt\,du \qquad (20.103a) \\ &= \Gamma_2 + \frac{\overline{A_0^2}}{4}\gamma_0 \int_{0-}^{T+} h_I(T - t)V_{2F}(t)^2\,dt \qquad (20.103b) \end{aligned}$$

† See Eqs. (20.18), (20.19), and associated comments. We remark again that h_2 cannot be used alone; a time-varying switch is also needed to ensure the finite sample, e.g., the vanishing of ρ_{2G} outside $(0,T)$.

‡ See the footnote referring to Eqs. (20.54).

§ Again, $\gamma_0 = \psi^{-2}$ ($\psi < \infty$), while, for white noise backgrounds and continuous sampling, $\gamma_0 = 4T^2/W_0^2$.

where now

$$\rho_{2G}(t,u)_T = \begin{cases} \gamma_0 \int_{0-}^{T+} h_I(T - x)h_2(x - t)h_2(x - u)\, dx & 0 \le t,\, u \le T \\ 0 & \text{elsewhere} \end{cases} \tag{20.103c}$$

With the above in mind, we can make several general statements about system performance:

(*a*) *No improvement in performance of an optimum* (threshold) *receiver can be gained by using a postdetection filter other than the simple integrator.*† (By improved performance here is meant smaller average risk.) Any other filter will decrease the effectiveness of the system. This follows at once from the fact that the optimum system is Bayes and (as a decision rule) belongs to a minimal complete class (Sec. 18.5-1).

(*b*) *With a suboptimum receiver, postrectification filtering other than simple integration*‡ *can give improved performance.* However, from the definition of the optimum or Bayes system as one that minimizes average risk, it is clear that this smaller average risk R can never be less than the Bayes risk R^*. By adjusting h_I we can alter R, so that some h_I will lead to systems with smaller average risk than when simple (or ideal) integration is employed ($h_I \sim 1/T$). To prove this, we must determine R under these various possibilities, but from physical considerations it is evident that other than a uniform weighting in postdetection integration can enhance the signal vis-à-vis the noise if the predetection filter h_2 is not matched‡ (cf. Sec. 20.2-4).

(2) **Error probabilities and acceptable systems.** The average risk for binary suboptimum detection systems is given by Eq. (19.19*b*),

$$R = \mathcal{R}_0 + p(C_\beta - C_{1-\beta})\left(\alpha \frac{\mathcal{K}}{\mu} + \beta\right) \tag{20.104}$$

and the associated error probabilities α, β, like the Bayes error probabilities α^*, β^*, are obtained from Eqs. (19.36*a*), (19.36*b*),

$$\alpha = \int_{\log \mathcal{K}}^{\infty} q_n(y)\, dy \qquad \beta = \int_{-\infty}^{\log \mathcal{K}} p_n(y)\, dy \tag{20.105}$$

where now y is specifically given by Eq. (20.94). The $q_n(y)$, $p_n(y)$ are the d.d.'s of y with respect to the hypotheses $H_0{:}N$ and $H_1{:}S + N$ [cf. Eqs. (19.36), (19.37)]. The preassigned costs of decision are taken to be the same as for the optimum system, since it is assumed that the application in

† See the footnote referring to Eqs. (20.54).

‡ (*a*) and (*b*) are generalizations of earlier theorems[5–7] of Van Vleck and Middleton, who derived analogous results on the basis of calculations of signal-to-noise ratios at the output of a radar receiver. Here, however, the decision process is explicitly taken into consideration, and all available information in the received wave, as well as a priori data, is used. Also, the matched filters referred to here are *Bayes* matched filters.

either instance is the same; accordingly, the threshold $\mathcal{K}$ is still given by Eq. (19.15).

The distribution densities q_n, p_n, their characteristic functions, and, finally, the error probabilities α, β may all be found for threshold performance by the procedures illustrated in Sec. 20.3-1(2). The results are listed below; the details are left to the reader. It is found that through $O(a_0^4)$

$$F_1(i\xi)_{q_n} = \mathbf{E}_{\mathbf{V}|H_0}\{e^{i\xi y}\} \doteq e^{i\xi c_n^{(N)} - \frac{1}{2} d_n^{(N)} \xi^2} \tag{20.106a}$$
$$F_1(i\xi)_{p_n} = \mathbf{E}_{\mathbf{V}|H_1}\{e^{i\xi y}\} \doteq e^{i\xi c_n^{(S+N)} - \frac{1}{2} d_n^{(S+N)} \xi^2} \tag{20.106b}$$

With $\Phi_{G2} \equiv \tilde{\mathbf{s}}\mathbf{G}_2\mathbf{s}$, we have

$$c_n^{(N)} = \Gamma_2 + \frac{\overline{a_0^2}}{2}\Phi_1 \qquad d_n^{(N)} = \frac{\overline{a_0^2}^2}{2}\Phi_2 = d_n^{(S+N)} \tag{20.107a}$$
$$c_n^{(S+N)} = \Gamma_2 + \frac{\overline{a_0^2}}{2}\Phi_1 + \frac{\overline{a_0^2}^2}{2}\Phi_3 \tag{20.107b}$$

where

$$\Phi_1 \equiv \psi^2 \operatorname{tr} \mathbf{k}_N\mathbf{G}_2 \qquad \Phi_2 \equiv \psi^4 \operatorname{tr} (\mathbf{k}_N\mathbf{G}_2)^2$$
$$\Phi_3 \equiv \psi^2\langle\Phi_{G2}\rangle = \psi^2\langle\tilde{\mathbf{s}}\mathbf{G}_2\mathbf{s}\rangle \tag{20.107c}$$

The d.d.'s q_n, p_n are (asymptotically) normal, with means $c_n^{(N)}$, $c_n^{(S+N)}$ and the same variance $d_n^{(N)}$, so that the error probabilities for threshold operation become

$$\alpha \simeq \frac{1}{2}\left[1 - \Theta\left(\frac{-\Gamma_2 - \overline{a_0^2}\Phi_1/2 + \log \mathcal{K}}{\overline{a_0^2}\Phi_2^{\frac{1}{2}}}\right)\right] \tag{20.108a}$$

$$\beta \simeq \frac{1}{2}\left[1 - \Theta\left(\frac{\Gamma_2 + \overline{a_0^2}\Phi_1/2 + \overline{a_0^2}^2\Phi_3/2 - \log \mathcal{K}}{\overline{a_0^2}\Phi_2^{\frac{1}{2}}}\right)\right] \tag{20.108b}$$

With continuous sampling, the quantities Φ_1, Φ_2, Φ_3 are given by

$$\Phi_{1T} = \psi \iint_{-\infty}^{\infty} K_N(t,u)\rho_{2G}(t,u)_T \, dt \, du \tag{20.109a}$$

$$\Phi_{2T} = \psi^2 \int \cdots \int_{-\infty}^{\infty} K_N(t_1,u_1)K_N(t_2,u_2)\rho_{2G}(t_2,u_1)_T\rho_{2G}(t_1,u_2)_T \, dt_1 \cdots du_2 \tag{20.109b}$$

$$\Phi_{3T} = \psi^2 \iint_{-\infty}^{\infty} \langle s(t;\epsilon,\boldsymbol{\theta})\rho_{2G}(t,u)_T s(u;\epsilon,\boldsymbol{\theta})\rangle \, dt \, du \tag{20.109c}$$

For white noise, these reduce, respectively, to

$$\overline{a_0^2}\Phi_{1T} = \frac{\overline{A_0^2}T^2}{W_0} \iint_{0-}^{T+} h_I(T - x)h_2(x - t)^2 \, dx \, dt \tag{20.110a}$$

$$\overline{a_0^2}^2\Phi_{2T} = \frac{\overline{A_0^2}^2 T^4}{W_0^2} \iint_{0-}^{T+} dt_1\, dt_2 \left[\int_{0-}^{T+} h_I(T - x)h_2(x - t_1)h_2(x - t_2)\, dx\right]^2 \quad (20.110b)$$

$$\overline{a_0^2}^2\Phi_{3T} = \frac{\overline{A_0^2}^2 T^2}{W_0^2} \iint_{0-}^{T+} \left[\int_{0-}^{T+} h_I(T - x)h_2(x - t)h_2(x - u)\, dx\right] \langle s(t;\epsilon,\boldsymbol{\theta})s(u;\epsilon,\boldsymbol{\theta})\rangle\, dt\, du \quad (20.110c)$$

Observe for the optimum system, i.e., when $\psi^2\mathbf{G}_2 \to \bar{\mathbf{G}}$ in the above (with $h_I \to 1/T$), that $\Phi_1 \to \langle\Phi(s,s)\rangle$, Φ_2, $\Phi_3 \to \langle\Phi_G\rangle$ [cf. Eq. (20.91a)], and if $\Gamma_2 =$ optimum bias, then $\alpha \to \alpha^*$, $\beta \to \beta^*$ [Eq. (20.91)].

Let us now define an *acceptable system* as one for which the type I and II error probabilities vanish in both the limiting situations: $T \to \infty$, $\overline{a_0^2}$ fixed, and $\overline{a_0^2} \to \infty$, T fixed, as is the case for optimum systems [cf. Eqs. (20.91), (20.93)]. Accordingly, not every suboptimum system of the form (20.94) is necessarily acceptable, as we can see from Eqs. (20.108), (20.109), (20.110): this depends on our choice of bias Γ_2 and pre- and postrectification filters h_2, h_I. For example, let us be guided by the corresponding optimum system above and choose the bias to have the form

$$\Gamma_2 = \log \mu - \frac{\overline{a_0^2}}{2}\lambda_2 - \frac{\overline{a_0^2}^2}{4}\lambda_4 \qquad \lambda_2 \geq 0,\ \lambda_4 \geq 0 \quad (20.111)$$

where λ_2, λ_4 may or may not be functions of T. Then, Eqs. (20.108) become

$$\alpha \simeq \frac{1}{2}\left(1 - \Theta\left\{\left[\frac{\overline{a_0^2}\lambda_4}{4} + \frac{\lambda_2 - \Phi_1}{2} + \frac{\log(\mathcal{K}/\mu)}{\overline{a_0^2}}\right]\Phi_2^{-1/2}\right\}\right) \quad (20.112a)$$

$$\beta \simeq \frac{1}{2}\left(1 - \Theta\left\{\left[\frac{\overline{a_0^2}}{4}(2\Phi_3 - \lambda_4) - \frac{\lambda_2 - \Phi_1}{2} - \frac{\log(\mathcal{K}/\mu)}{\overline{a_0^2}}\right]\Phi_2^{-1/2}\right\}\right) \quad (20.112b)$$

Now unless $2\Phi_3 > \lambda_4$, then $\beta \to 1$ as $\overline{a_0^2} \to \infty$, or if $(\Phi_1 - \lambda_2)\Phi_2^{-1/2} > 0$ as $T \to \infty$, then $\alpha \to 1$ also, and so on; the complete set of conditions depends, of course, on the noise and signal structure as well. For a specific example, see Secs. 20.4-1, 20.4-2, where a given suboptimum system is compared with the corresponding optimum one.

20.3-3 Information Loss in Binary Detection. Since a definite decision is required in all these detection procedures, there is always an uncertainty as to the true state of affairs when the decision is made (unless α, $\beta \to 0$, of course). Now the average information as to the presence or absence of a signal $\mathbf{S}$ in noise that is actually conveyed by the decision "noise alone," or "signal and noise" when a particular data sample $\mathbf{V}$ is available, is given by Eqs. (6.36), viz.,

$$I(\mathbf{S}|\mathbf{V}) = H(\mathbf{S}) - H(\mathbf{S}|\mathbf{V}) \quad (20.113)$$

where $H(\mathbf{S})$ is the communication entropy of $\mathbf{S}$ or the inherent information about $\mathbf{S}$ in the ensemble $\{\mathbf{S}\}$ [cf. Eq. (6.25)] and $H(\mathbf{S}|\mathbf{V})$ is the conditional entropy of $\mathbf{S}$, given $\mathbf{V}$, or average uncertainty in the data $\mathbf{V}$, concerning the signal set $\{\mathbf{S}\}$ [cf. Eq. (6.29)]. Thus, $I(\mathbf{S}|\mathbf{V})$ represents the actual amount of (average) information about S gained by the observer after a decision is made on the basis of the received data $\mathbf{V}$, while $H(\mathbf{S}|\mathbf{V})$ measures the (average) information lost, vis-à-vis the initial state $H(\mathbf{S})$, as a consequence of the definite decision.

The inherent information $H(\mathbf{S})$ is found at once to be

$$H(\mathbf{S}) = -p \log p - q \log q \tag{20.114}$$

where once more p, q $(= 1 - p)$ are the a priori probabilities of signal and no signal in a data sample $\mathbf{V}$. [Note that $H(\mathbf{S})_{\max} = 1$ bit per indication or decision, if 2 is the base of the logarithm.]

The average uncertainty $H(\mathbf{S}|\mathbf{V})$, per decision, or the *equivocation* [if decisions are made at a certain rate; cf. Eq. (6.84*b*)], is easily found as follows:[8] Let x_1 be some number associated with the null signal state $\mathbf{S} = 0$ and x_2 another number associated with the alternative state signal and noise $(\mathbf{S} \neq 0)$. Similarly, let y_1 be associated with the received data $\mathbf{V}$ when a "no, noise alone" decision is made, and y_2 correspondingly when "yes, signal and noise" is decided. From Eq. (6.29) and the relation $p(x_i|y_k) = p(x_i,y_k)/p(y_k)$, we have at once

$$H(\mathbf{S}|\mathbf{V}) = -\sum_{i,k}^{2} p(x_i,y_k) \log [p(x_i,y_k)/p(y_k)] \tag{20.115a}$$

Remembering that the two types of conditional error can occur in any binary decision, with the probabilities α, β, we obtain

$$\begin{aligned} p(x_1,y_1) &= q(1-\alpha) \qquad & p(x_2,y_2) &= p(1-\beta) \\ p(x_1,y_2) &= q\alpha & p(x_2,y_1) &= p\beta \end{aligned} \tag{20.115b}$$

With the marginal probabilities $p(y_k) = \sum_{i}^{2} p(x_i,y_k)$, for example, $p(y_1) = p\beta + q(1 - \alpha)$, $p(y_2) = q\alpha + p(1 - \beta)$, the desired average loss of information vis-à-vis the initial state is specifically (cf. Prob. 6.11 also)

$$\begin{aligned} H(\mathbf{S}|\mathbf{V}) = p(1-\beta) \log \frac{1-\beta}{p(1-\beta) + q\alpha} + p\beta \log \frac{\beta}{q(1-\alpha) + p\beta} \\ + q\alpha \log \frac{\alpha}{p(1-\beta) + q\alpha} + q(1-\alpha) \log \frac{1-\alpha}{q(1-\alpha) + p\beta} \end{aligned} \tag{20.116}$$

This relation applies, of course, for optimum systems as well, where α and β are replaced by α^*, β^*, respectively [cf. Eqs. (20.79), (20.91), (20.93)]. However, it is not necessarily true that $H(\mathbf{S}|\mathbf{V})$ is least for optimum systems using the present simple cost criterion [cf. Eqs. (19.5)]. Thus, Bayes systems, where the average risk is minimized, do not usually yield a minimum equivocation $H(\mathbf{S}|\mathbf{V})$, as the analysis of Sec. 22.2-2 shows, although for certain classes of signal and noise statistics and conditions of operations it is possible to minimize both equivocation and average risk simultaneously. In this regard, see also Sec. 22.2-1 and the associated example where binary detectors that are optimum from the point of view of minimum equivocation, rather than average risk, are briefly considered.

Finally, it is easily shown (cf. Prob. 20.7*c*) that $H(\mathbf{S}|\mathbf{V})$ is a *maximum* when $\alpha + \beta = 1$ and is in fact equal to $H(\mathbf{S})$, as expected, since with completely incorrect decisions the average information as to the true state of affairs vanishes upon detection, that is, $I(\mathbf{S}|\mathbf{V}) = 0$. One does better to guess on the basis of the a priori probabilities p, q. At the other extreme of $\alpha = \beta = 0$, $I(\mathbf{S}|\mathbf{V}) = H(\mathbf{S})$, since error-free decisions produce no loss in (average) information regarding the true presence or absence of a signal.

20.3-4 General Remarks. Let us briefly summarize the principal results of the preceding sections. It is convenient to distinguish between the limiting situations of weak-signal, or threshold, detection and of strong-signal operation. The character of the former is dominated by the statistical properties of the noise, where the signal is essentially a perturbation of the noise process. The latter situation, on the other hand, is effectively governed by the statistics of the signal process, with the noise now acting as a perturbation. The threshold case, of course, is by far the more significant, since a detection system that approaches optimality for weak signals is almost always adequate in strong-signal operation, although it is usually no longer optimum at these higher signal levels.

Technically, our chief problem here is to determine the distribution densities of the (logarithm of the) likelihood ratio $\Lambda_n(\mathbf{V})$ with respect to the null and alternative hypotheses H_0, H_1. For given received data $\mathbf{V}$, $\log \Lambda_n(\mathbf{V})$ is a *number* which is to be compared with some preset threshold $\log \mathcal{K}$ (cf. Sec. 19.1-2); $\log \Lambda_n$ also embodies the structure of the optimum system which computes this number, on which the binary decision operation rests. The d.d.'s of $\log \Lambda_n(\mathbf{V})$ in turn enable us to calculate the error probabilities α^*, β^* and the Bayes risk R^* associated with the decision process. A parallel procedure is required for suboptimum systems, the essential difference here being that $\log \Lambda_n(\mathbf{V})$ is now replaced by a different relation, $\log G_n(\mathbf{V})$ $(\Lambda_n \neq G_n)$.

With this in mind, we observe again (cf. Sec. 19.4-3) that with high accuracy of decision in threshold systems, for example, when α^* (or α) and β^* (or β) are small (say, $q\alpha^{(*)} + p\beta^{(*)} < 0.1$), we have always a *large-sample* theory. For additive and independent normal noise processes, the desired d.d.'s of $x = \log \Lambda_n(\mathbf{V})$ are themselves asymptotically normal, although x is not a linear function of $\mathbf{V}$. Even if the accompanying noise is *not* normal, e.g., if $F_n(\mathbf{V}|\mathbf{S})$, $F_n(\mathbf{V})$ are not gaussian, the normal character of $\log \Lambda_n(\mathbf{V})$ is preserved in all those cases (which are the ones of main physical interest) where the central-limit theorem can be applied (cf. Sec. 7.7-3 and Prob. 20.9). Physically, this gaussian behavior is a general reflection [through the quadratic form in **V** into which $\log \Lambda_n$ is resolved in threshold operation; cf. Eq. (19.59)] of the fact that it is essentially the *energy* in the received wave, and the energy in its two possible components, signal and noise, on which the decision process is based. Similar remarks apply for acceptable suboptimum systems, whose threshold character is linear and/or quadratic in the received data **V** [cf. Eq. (20.94)].

It is this asymptotically normal behavior for either optimum or suboptimum systems which fortunately allows us to develop the threshold theory explicitly. A general strong-signal theory, however, is not yet available, essentially because of the following technical difficulties: (1) the *small samples* (i.e., short observation times) implied here do not yield gaussian statistics for $\log \Lambda_n(\mathbf{V})$, and (2) the signal statistics, which are usually quite non-gaussian, dominate the statistical character of $\log \Lambda_n$. In addition, although the likelihood ratio is valid for all signal levels, the signal class usually contains more than one member (cf. Sec. 18.2-1), and it is this condition, appearing as an average over the signal set, which further complicates the analysis for strong signals.

Characteristic of optimum detection systems (and suitably designed suboptimum ones with a final ideal filter $h_I = 1/T$) is the inverse proportionality of the minimum detectable signal $(a_0{}^2)^*_{\min}$ (Sec. 19.3-3) to the observation time ($\sim T$ or n) for coherent reception, and to the square root of the observation time for entirely incoherent reception, when threshold conditions apply. For strong-signal operation, on the other hand, $(a_0{}^2)^*_{\min}$ depends on the particular signal waveform as well, and it is not necessarily true that $(a_0{}^2)^*_{\min} \sim T^{-1}$ any longer (cf. Secs. 20.4-1, 20.4-2). Various behaviors of $(a_0{}^2)^*_{\min}$ with T (or n) are sketched in Fig. 20.3, where typical curves for suboptimum systems have been included, and typical strong-signal (or small-sample) performance is also indicated. We shall encounter specific examples in Sec. 20.4.

The structure of the optimum threshold systems for coherent and incoherent detection can also be represented in various equivalent ways. In fact, as we have already observed [cf. Eqs. (19.58), (19.59)], these threshold systems are examples, respectively, of generalized cross- and autocorrelation receivers, where the weighting in the correlation process depends on both signal and noise. With sample uncertainty such that $\langle s \rangle_\epsilon$ vanishes, one possible structure using physically realizable and time-varying elements consists of a (linear) Bayes matched filter (cf. Sec. 20.2-4), followed by a zero-memory square-law rectifier and ideal integrator over the observation interval $(0,T)$ (cf. Fig. 20.2*a*). Another alternative structure is a sequence consisting of a linear time-varying matched filter and simple multiplier, again followed by an ideal integration on $(0,T)$, as in Fig. 20.2*b*. Under much more restricted conditions, a third type of matched filter may be employed which does not require postdetection integration [cf. Eqs. (20.10a) et seq., Eqs. (20.69), Fig. P 20.6]. These basic structures apply for both discrete and continuous data processing, as Secs. 20.1, 20.2 indicate. Finally, we observe that these optimum threshold systems are, in effect, generalizations of various correlation detectors considered in earlier investigations,[9–13] whose structure and system performance are based essentially

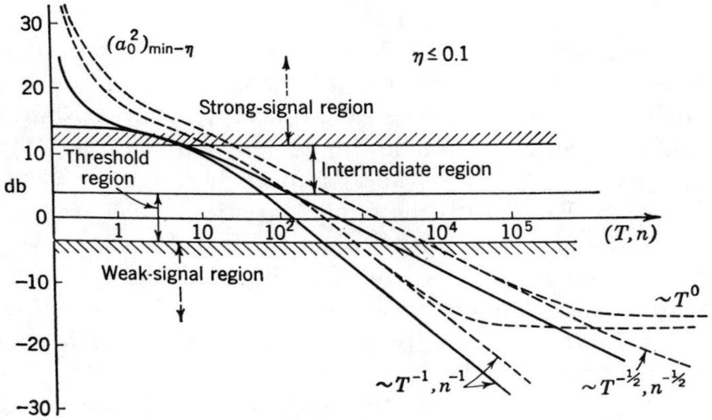

FIG. 20.3. Minimum detectable signal as a function of sample size (observation time); various (qualitative) optimum and suboptimum systems. Solid lines, optimum systems; broken lines, suboptimum systems.

on a signal-to-ratio criterion (cf. Sec. 5.3-4). Here, however, full use is made of signal and noise statistics, the difficulties inherent in defining a signal (in noise) after a nonlinear operation are avoided, and an optimized decision procedure that explicitly represents the detection process is specifically introduced as part of system design and evaluation. In the following section, we shall illustrate these general remarks with additional examples of interest in communication theory.

PROBLEMS

20.7 (a) Extend the results of Sec. 20.3-1 to the case $a_0 = a_0(\boldsymbol{\theta})$. Hence, show that for optimum *coherent* threshold operation, the conditional error probabilities [Eq. (20.79)] become now

$$\begin{Bmatrix} \alpha^* \\ \beta^* \end{Bmatrix} \simeq \frac{1}{2} \left\{ 1 - \Theta \left[\frac{\Phi_{\overline{a_0 s}}^{1/2}}{2\sqrt{2}} \pm \frac{\log (\mathcal{K}/\mu)}{\sqrt{2}\, \Phi_{\overline{a_0 s}}^{1/2}} \right] \right\} \tag{1}$$

where $\bar{a}_0 \Phi_{\bar{s}}$ of the original analysis goes over into $\Phi_{\overline{a_0 s}} \equiv \overline{a_0 \tilde{\mathbf{s}}} \mathbf{k}_N^{-1} \overline{a_0 \mathbf{s}}$, and where the optimum detector structure [Eq. (20.72)] in the discrete case is now

$$\log \Lambda_n \doteq (\log \mu + \tfrac{1}{2} \Phi_{\overline{a_0 s}}) + \tilde{\mathbf{v}} \mathbf{k}_N^{-1} \overline{a_0 \mathbf{s}} \tag{2}$$

For continuous sampling, show that $\Phi_{\overline{a_0 s}}$ is replaced by

$$(\Phi_T)_{\overline{a_0 s}} = \psi \int_{0-}^{T+} \langle a_0 s(t - \epsilon; \theta)\rangle \langle a_0 X_T(t;\epsilon,\theta)\rangle \, dt \tag{3a}$$

$$(\Phi_T)_{\overline{a_0 s}}\Big|_{\text{white}} = W_0^{-1} \int_0^T \langle A_0 s(t - \epsilon; \theta)\rangle \, dt \tag{3b}$$

and

$$\lim_{n\to\infty} \tilde{\mathbf{v}}\mathbf{k}_N^{-1}\overline{a_0 \mathbf{s}} = \psi \int_{0-}^{T+} v(t) \langle a_0 X_T(t;\epsilon,\theta)\rangle \, dt \tag{3c}$$

(*b*) Repeat (*a*) for optimum *incoherent* threshold detection (with $\langle s\rangle_\epsilon = 0$) to obtain as the extension of Eq. (20.91) the error probabilities

$$\begin{Bmatrix} \alpha^* \\ \beta^* \end{Bmatrix} \simeq \frac{1}{2}\left\{1 - \Theta\left[\frac{\overline{a_0^2}\langle\Phi_G(a_0)\rangle^{1/2}}{4} \pm \frac{\log(\mathcal{K}/\mu)}{\overline{a_0^2}\langle\Phi_G(a_0)\rangle^{1/2}}\right]\right\} \tag{4}$$

where now $\langle\Phi_G(a_0)\rangle = \langle a_0^2 \tilde{\mathbf{s}} \overline{a_0^2 \mathbf{G}} \mathbf{s}\rangle / \overline{a_0^2}^2$, and where the optimum detector structure [Eq. (20.81)] becomes

$$\log \Lambda_n \doteq (\log \mu + \mathcal{B}_n^{(2)} + \mathcal{B}_n^{(4)}) + \frac{\tilde{\mathbf{v}}}{2} \overline{a_0^2 \mathbf{G}} \mathbf{v} \tag{5}$$

(cf. Prob. 20.3). Obtain the corresponding continuous relations for $\langle\Phi_G(a_0)\rangle$,

$$\langle\Phi_T(a_0)_G\rangle = \overline{a_0^2}^{-2} \psi^2 \iint_{-\infty}^{\infty} \langle a_0^2 D_T(t,u;\epsilon,\theta)\rangle \langle a_0^2 s(t - \epsilon; \theta) s(u - \epsilon; \theta)\rangle \, dt \, du \tag{6a}$$

which for white noise backgrounds reduces to

$$\overline{a_0^2}^2 \langle\Phi_T(a_0)_G\rangle = \frac{1}{W_0^2} \iint_0^T \langle A_0^2 s(t - \epsilon, \theta) s(u - \epsilon; \theta)\rangle^2 \, dt \, du \tag{6b}$$

Obtain the modifications appropriate to narrowband signal processes.

(*c*) Show that $H(\mathbf{S}|\mathbf{V})_{\max}$ [cf. Eq. (20.116)] occurs for $\alpha + \beta = 1$.

20.8 A linear suboptimum system for coherent detection is suggested by Eq. (20.6), viz.,

$$z = \Gamma_1 + \tilde{\mathbf{V}}\mathbf{G}_1\bar{\mathbf{S}} = \Gamma_1 + \frac{\bar{A}_0}{\sqrt{2}} \tilde{\mathbf{V}}\mathbf{G}_1\bar{\mathbf{s}} \tag{1}$$

where $\mathbf{G}_1$ is positive definite and $\bar{A}_0\bar{\mathbf{s}}/\sqrt{2} = \bar{\mathbf{S}}$. This suboptimum system is assumed to have the following structure, as shown in Fig. P 20.8: (1) an ideal multiplier, which forms the instantaneous product of the (sampled) incoming wave $\mathbf{V}$ with $\mathbf{S}'$, a suitably weighted

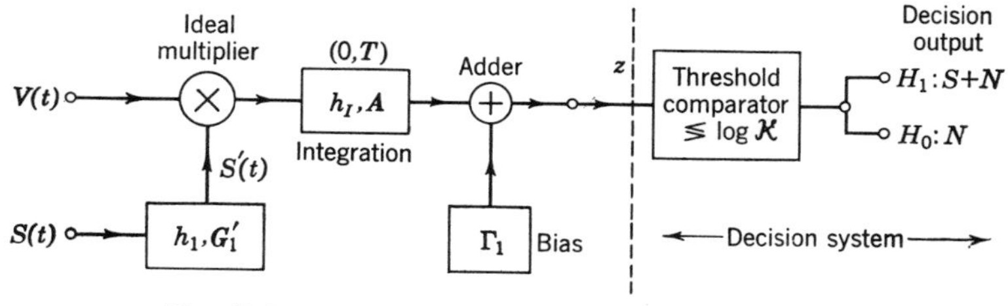

FIG. P 20.8. A linear suboptimum decision system.

signal waveform, and (2) following this multiplier, an integrating filter, whose output is observed at T sec after the start of an observation.

Following an argument similar to that outlined in Sec. 20.3-2, we set $\mathbf{A} = [h_I(T - t_i)\,\Delta t\,\delta_{ij}]$ again to be the filtering operation of this final (discrete) linear filter. If $h_1(t_i - t_j)\,\Delta t \equiv (\mathbf{G}')_{ij}$ is also a linear realizable filter, such that $\mathbf{S}' = \mathbf{G}_1\bar{\mathbf{s}}$, that is, $\mathbf{S}'$ is obtained on passing $\bar{\mathbf{s}}$ through h_1, we can write† $S_i' = \sum_{-\infty\,(\text{or }1)}^{\infty} h_1(t_i - t_j)\bar{s}(t_j)\,\Delta t$ [with $h_1(t_i - t_j) = 0$ for $j > i$; cf. Eqs. (20.16), etc.]. Setting

$$\mathbf{G}_1 \equiv \gamma_0^{1/2}\mathbf{A}\mathbf{G}_1' \tag{2}$$

where $\gamma_0^{1/2} = \psi^{-1}$ as before for discrete sampling [cf. Eq. (20.99)], we can express Eq. (1) equivalently as

$$z = \Gamma_1 + \frac{\bar{A}_0}{\sqrt{2}}\,\tilde{\mathbf{V}}\gamma_0^{1/2}\mathbf{A}\mathbf{G}_1'\bar{\mathbf{s}} = \Gamma_1 + \gamma_0^{1/2}\tilde{\mathbf{V}}\mathbf{A}\mathbf{S}' \tag{3a}$$

$$= \Gamma_1 + \psi^{-1}\sum_{i=1}^{n} V(t_i)S'(t_i)h_I(T - t_i)\,\Delta t \tag{3b}$$

the second term in either instance being the integrated output of the ideal multiplier when V_i and S_i' (the filtered $\bar{\mathbf{S}}$) are its two inputs.

When continuous sampling is used, we get† [cf. Eqs. (20.103)]

$$\begin{aligned} z_T &= \Gamma_1 + \gamma_0^{1/2}\int_{0-}^{T+} V(t)S'(t)h_I(T - t)\,dt \\ &= \Gamma_1 + \gamma_0^{1/2}\int_{0-}^{T+} V(t)\left[\int_{-\infty\,(\text{or }0)}^{\infty} \overline{S(u)}h_1(t - u)\,du\right] h_I(T - t)\,dt \end{aligned} \tag{4}$$

In the case of white noise backgrounds, $\gamma_0^{1/2} = \psi^{-1}$ is replaced by $2T/W_0$ [cf. footnote referring to Eqs. (20.103)].

(*a*) For additive (and independent) normal noise of vanishing mean, show that in the threshold case the error probabilities α, β for the systems of Eqs. (1), (3*b*), (4) become

$$\alpha = \frac{1}{2}\left\{1 - \Theta\left[\frac{\log \mathcal{K} - \Gamma_1}{(2\bar{a}_0{}^2\Phi_2)^{1/2}}\right]\right\} \qquad \beta \simeq \frac{1}{2}\left\{1 - \Theta\left[\frac{-\log \mathcal{K} + \Gamma_1 + \bar{a}_0{}^2\Phi_1}{(2\bar{a}_0{}^2\Phi_2)^{1/2}}\right]\right\} \tag{5}$$

where $\Phi_1 \equiv \psi\tilde{\bar{\mathbf{s}}}\mathbf{G}_1\bar{\mathbf{s}}$, $\Phi_2 \equiv \psi^2\tilde{\bar{\mathbf{s}}}\mathbf{G}_1\mathbf{k}_N\mathbf{G}_1\bar{\mathbf{s}}$, and

$$F_1(i\xi)_{q_n} = e^{i\xi\Gamma_1 - \frac{1}{2}\bar{a}_0{}^2\xi^2\Phi_2} \qquad F_1(i\xi)_{p_n} \doteq e^{i\xi\bar{a}_0{}^2\Phi_1}F_1(i\xi)_{q_n} \tag{6a}$$

$$q_n(z) = \frac{e^{-(z-\Gamma_1)^2/2\bar{a}_0{}^2\Phi_2}}{(2\pi\bar{a}_0{}^2\Phi_2)^{1/2}} \qquad p_n(z) \simeq \frac{e^{-(z-\Gamma_1-\bar{a}_0{}^2\Phi_1)^2/2\bar{a}_0{}^2\Phi_2}}{(2\pi\bar{a}_0{}^2\Phi_2)^{1/2}} \tag{6b}$$

Verify that these results hold for continuous sampling, where now†

$$\bar{a}_0{}^2\Phi_1 \to \gamma_0^{1/2}\int_{0-}^{T+} \overline{S(t)}\left[\int_{-\infty\,(\text{or }0)}^{\infty} \overline{S(u)}h_1(t - u)\,du\right] h_I(T - t)\,dt \tag{7a}$$

$$\bar{a}_0{}^2\Phi_2 \to \gamma_0 \iint_{0+}^{T+} S'(t)h_I(T - t)K_N(t,u)S'(u)h_I(T - u)\,dt\,du \tag{7b}$$

† The lower limit may be extended to $-\infty$, since $\bar{S}$ vanishes for all times outside $(0-, T+)$.

with $S'(t) = \int_{-\infty\ (\text{or } 0)}^{\infty} \overline{S(\tau)}h_1(t - \tau)\,d\tau$, etc. Observe for white noise backgrounds that Eq. (7b) becomes

$$\bar{a}_0{}^2\Phi_2\Big|_{\text{white}} = \frac{2T^2}{W_0}\int_{0-}^{T+} S'(t)^2 h_I(T - t)^2\,dt \tag{7c}$$

since $\gamma_0 = 4T^2/W_0{}^2$.

(b) Show that the results in (a) reduce, as expected, to the earlier relations (cf. Sec. 20.2-1) for optimum systems when $\mathbf{G}_1 \to \mathbf{K}_N{}^{-1}$. For continuous sampling, verify that $h_I = 1/T$ $(0 \le t \le T)$ and h_1 becomes the *matched* filter $\gamma_0{}^{-\frac{1}{2}}h_c$ of Eqs. (20.41a), (20.43) with $\bar{a}_0{}^2\Phi_1 \to \bar{a}_0{}^2\Phi_2 \to \bar{a}_0{}^2\langle(\Phi_T)_S\rangle$. Hence, for white noise backgrounds, obtain also $h_1(t) = \delta(t - 0)$.

(c) When $\Gamma_1 = \log \mu - \bar{a}_0{}^2\lambda_2$ $(\lambda_2 > 0)$, discuss the dependence of α, β on sample size n (and T). Repeat for the minimum detectable signal. Examine the strong-signal case when everything is known about the signal except whether it is present or absent (see Sec. 20.4-2).

20.9 (a) Show for wide classes of input noise processes, *not necessarily normal*, that, under conditions of high-precision optimum threshold detection where the noise is additive and independent of the signal, $x = \log \Lambda_n(\mathbf{V})$ is (asymptotically) normally distributed, with nonvanishing mean and finite variance. Obtain the associated error probabilities and Bayes risk, and establish sufficient conditions for the limiting distributions. [HINT: Start with the general development of Sec. 19.4-1, and follow the procedures outlined in the discussion of the random walk and the central-limit theorem (Secs. 7.7-2, 7.7-3); obtain the correction terms to show how the normal law is approached.]

(b) Extend (a) to suboptimum threshold systems of the type of Eq. (20.94), Sec. 20.3-2, and Prob. 20.8, Eq. (1), above.

20.4 *Examples*

In Secs. 20.1 to 20.3, we have outlined a general approach to optimum and suboptimum binary detection which includes the three main features of a design theory: structure, evaluation, and comparison. Here we shall conclude our treatment of fixed-sample systems with a number of specific examples. Starting with the simplest models and continuing to the more realistic and complex cases, we shall show how the general approach of Chap. 19 and the preceding sections can be applied to a wide variety of communication problems. In due course, we shall discuss a number of new results as well as reestablish some already well-known ones. The present treatment by no means exhausts the possible applications: still other examples of interest are summarized in the problems, and additional results may also be found in the references. Unless otherwise indicated, the conditions and assumptions on the signal and noise processes mentioned at the beginning of the chapter are still in force here: the signal processes are deterministic and independently additive to the background noise, which is normal, but not necessarily stationary.

20.4-1 Coherent Detection (Optimum Simple-alternative Cases). We consider first the simplest of all detection situations: everything is known about the signal (as well as the noise), except whether or not it is present in some specified observation period $(0,T)$.† This is rarely a practical case, except under controlled laboratory conditions. However, it is almost the only system for which an exact analysis is possible for *all* input signal levels and for this reason is important in providing a limiting indication of strong-signal behavior.

Since the signal structure is completely determined, for example, $w(\boldsymbol{\theta}) = \delta(\boldsymbol{\theta} - \boldsymbol{\theta}_0)$, $w_1(\epsilon) = \delta(\epsilon - \epsilon_0)$, etc., each signal parameter can assume but a single value, so that $S(t - \epsilon, \boldsymbol{\theta}) = S(t - \epsilon_0, \boldsymbol{\theta}_0)$, the signal class has but one member, and detection is binary simple-alternative‡ [cf. Sec. 18.2-1(2)]. The optimum system [Eq. (20.1)] reduces directly to the *exact* expression

$$x = \boldsymbol{T}_R\{V\} = \log \Lambda_n(\mathbf{V}) = \left(\log \mu - \frac{a_0{}^2}{2}\Phi_s\right) + a_0\Phi_v \qquad \text{all } a_0 \geq 0 \quad (20.117a)$$

with $\Phi_s \equiv \tilde{\mathbf{s}}\mathbf{k}_N{}^{-1}\mathbf{s}$, $\Phi_v \equiv \tilde{\mathbf{v}}\mathbf{k}_N{}^{-1}\mathbf{s}$ ($\mathbf{s} \neq 0$ as before) [cf. Eqs. (20.2a), (20.2b)], with a_0 the normalized amplitude of the signal [cf. Eqs. (19.49)]. Figures 20.1a to c also show typical structures here, with $\bar{a}_0$, $\bar{s}$ replaced by a_0, s, etc., where a_0 can assume all (positive) values. With continuous sampling in $(0,T)$, Eq. (20.117a) becomes simply§

$$x = \log \Lambda_T(V) = \left[\log \mu - \frac{a_0{}^2}{2}(\Phi_T)_s\right] + a_0(\Phi_T)_v \qquad (20.117b)$$

where from Eqs. (20.37) we write directly

$$\begin{aligned}(\Phi_T)_s &= \psi \int_{0-}^{T+} s(t - \epsilon_0, \boldsymbol{\theta}_0) X_T(t;\epsilon_0,\boldsymbol{\theta}_0)\, dt \\ (\Phi_T)_v &= \psi \int_{0-}^{T+} v(t) X_T(t;\epsilon_0,\boldsymbol{\theta}_0)\, dt\end{aligned} \qquad (20.117c)$$

with X_T the solution of the basic integral equation (20.38), viz.,

$$\int_{-\infty}^{\infty} K_N(t,u) X_T(u;\epsilon_0,\boldsymbol{\theta}_0)\, du = s(t - \epsilon_0, \boldsymbol{\theta}_0) \qquad 0- < t < T+ \quad (20.117d)$$

(see Appendix 2 for solutions in the case of rational kernels). When the accompanying noise is white, so that $K_N(t,u) = (W_0/2)\delta(t - u)$ and $X_T =$

† See, for example, Middleton[14] and Middleton and Van Meter,[1] sec. 3.9 and footnote.

‡ When $w(\boldsymbol{\theta}) \neq \delta(\boldsymbol{\theta} - \boldsymbol{\theta}_0)$, etc., the signal class contains more than one member and we have the more common binary *single*—or one-sided (i.e., composite)—*alternative* models of most practical situations.

§ For an interpretation in terms of a matched filter (Sec. 20.2-4), multiplier, and integrator, see Eqs. (20.41a), (20.43) and Prob. 20.4b. Note also that an alternative representation may be given in terms of still another matched filter (Fig. 20.1c) $X_T(t,\epsilon_0,\theta_0) \equiv h_M(T - t;\, T)$ [cf. Eq. (20.43d) and discussion following]. This particular filter also maximizes signal-to-noise ratio [cf. Sec. 16.3 and Eq. (16.99)].

$(2/W_0)s(t - \epsilon_0, \boldsymbol{\theta}_0)$, Eq. (20.117$b$) reduces at once to

$$\log \Lambda_T(V)\Big|_{\text{white}} = \left[\log \mu - \frac{A_0^2}{2W_0}\int_0^T s(t - \epsilon_0, \boldsymbol{\theta}_0)^2\, dt\right] + \frac{A_0\sqrt{2}}{W_0}\int_0^T V(t)s(t - \epsilon_0, \boldsymbol{\theta}_0)\, dt \quad (20.118)$$

[cf. Eq. (20.43c)], which again exhibits the characteristic unweighted cross correlation in $(0,T)$ of signal with received data $V(t)$ [cf. Eqs. (19.58), (20.37)].

Inasmuch as Eq. (20.117a) is exact (all $a_0 \geq 0$) and linear in $\mathbf{V}$, the calculations of characteristic functions, d.d.'s, and error probabilities are essentially given by Eq. (20.79), now without approximation. It is easily seen that†

$$F_1(i\xi)_Q = \exp\left[-\tfrac{1}{2}a_0^2\xi^2\Phi_s + i\xi\left(\log \mu - \frac{a_0^2}{2}\Phi_s\right)\right] \quad (20.119a)$$
$$F_1(i\xi)_P = [\exp(i\xi a_0^2\Phi_s)]F_1(i\xi)_Q$$

$$\begin{Bmatrix} Q_n(x) \\ P_n(x) \end{Bmatrix} = \frac{e^{-(x-\log \mu \pm a_0^2\Phi_s/2)^2/2a_0^2\Phi_s}}{(2\pi a_0^2\Phi_s)^{1/2}} \quad (20.119b)$$

The desired error probabilities α^*, β^* are given by Eqs. (20.71) for all a_0 (≥ 0):

$$\begin{Bmatrix} \alpha^* \\ \beta^* \end{Bmatrix} = \frac{1}{2}\left\{1 - \Theta\left[\frac{a_0\Phi_s^{1/2}}{2\sqrt{2}} \pm \frac{\log(\mathcal{K}/\mu)}{\sqrt{2}\, a_0\Phi_s^{1/2}}\right]\right\} \quad (20.120)$$

and the Bayes risk R^* follows at once from Eq. (19.19a). For continuous sampling, $\Phi_s = \tilde{\mathbf{s}}\mathbf{k}_N^{-1}\mathbf{s}$ is replaced in Eqs. (20.119), (20.120) by $(\Phi_T)_s$ [Eqs. (20.117c)] as before.

Figure 20.4a shows the type II conditional error probability β^* as a function of the type I probability α^*, for various $a_0\Phi_s^{1/2}$ and $\mathcal{K}/\mu$, while Fig. 20.4b gives the Bayes risk for various values of $\mathcal{K}/\mu$. Specifying any two of the four quantities α^*, β^*, $a_0^2\Phi_s$, and $\mathcal{K}/\mu$ determines the other two, and, for $a_0\Phi_s^{1/2}$ fixed, the values of α^*, β^* are interchanged if $\mathcal{K}/\mu$ is replaced by $\mu/\mathcal{K}$. We may also calculate the *normalized* Bayes risk $\mathcal{R}^*$ from Eqs. (19.39). Typical curves are shown in Fig. 20.8, where $\overline{a_0^2}\sqrt{n}$ therein is replaced by $a_0\Phi_s^{1/2}$ here. Since $\Phi_s \sim n$ (or T), for large n or T, it is clear from Eq. (20.120) that, as $T \to \infty$, α^*, $\beta^* \to 0$ in all cases, as does the Bayes risk R^*. Similarly, for fixed n or T, α^*, $\beta^* \to 0$, etc., as $a_0 \to \infty$; this, too, is the expected behavior for errorless decisions. Observe that again α^*, β^* are universal functions of $a_0^2\Phi_s$ [cf. Eq. (20.120)]. Note, finally, that with white noise backgrounds we have

$$a_0^2(\Phi_T)_s\Big|_{\text{white}} = \frac{A_0^2}{W_0}\int_0^T s(t - \epsilon_0, \boldsymbol{\theta}_0)^2\, dt = \frac{2E_s}{W_0} \quad (20.121)$$

† Note that given Q_n, for example, it is a simple matter here to use the result (1) of Prob. 17.8 to obtain P_n, and vice versa.

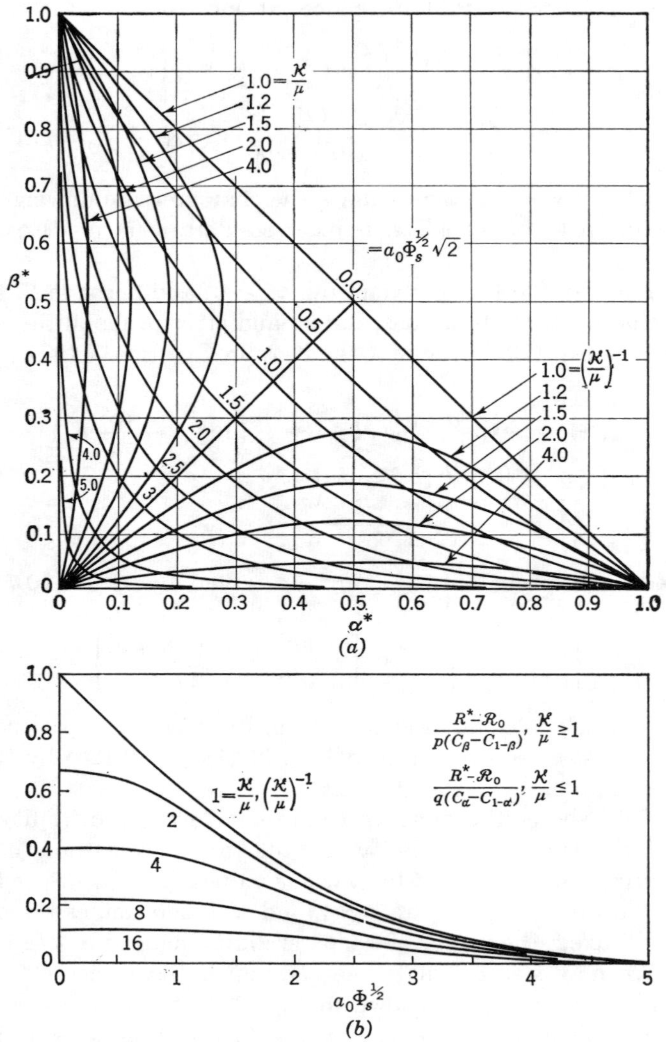

FIG. 20.4. (*a*) β^* versus α^* for binary coherent simple-alternative detection (normal noise). (*b*) normalized Bayes risk for (*a*); $\mathcal{K}/\mu \geqq 1$.

where E_s is proportional to the total *energy* in the signal received in $(0,T)$. Thus, we obtain the familiar result[15] that optimum performance in this very simple situation depends only on the ratio of signal energy to the noise power density W_0 (for a specified threshold $\mathcal{K}$ and given μ).

The behavior of the minimum detectable signal $[\sim(a_0{}^2)^*_{\min}$; cf. Eq. (20.80)] is instructive. Since α^*, β^* are universal functions of $a_0\Phi_s{}^{1/2}$ and since for threshold signals and high accuracy of decision $\Phi_s \sim T$, we have once more

TABLE 20.1. $a_0^2\Phi_s$ FOR OPTIMUM COHERENT SIMPLE-ALTERNATIVE DETECTION: CONTINUOUS SAMPLING IN $(0,T)$

Signal and noise	Signal, s	Noise, $K_N(t)$	X_T $(0-,T+)$	$a_0^2(\Phi_T)_s = \frac{A_0^2}{2}\int_{0-}^{T+} s(t-\epsilon_0,\,\theta_0)X_T(t;\epsilon_0,\theta_0)\,dt$
1. D-c signal in white noise	$\sqrt{2}$	$\frac{W_0}{2}\,\delta(t-0)$	$\frac{2}{W_0}\sqrt{2}$	$2\,\frac{A_0^2T}{W_0} = 2\sigma_e^2\lambda,\ \lambda = bT,\ \sigma_e^2 = \frac{A_0^2}{W_0b}$
2. Sine wave in white noise ($\epsilon_0 = 0$)	$\sqrt{2}\cos\omega_st$	$\frac{W_0}{2}\,\delta(t-0)$	$\frac{2\sqrt{2}}{W_0}\cos\omega_st$	$2\sigma_e^2\lambda\left(1+\frac{\sin 2\lambda}{2\lambda}\right),\ \lambda = \omega_sT,\ \sigma_e^2 = \frac{A_0^2}{2W_0\omega_s}$
3. Sine wave in RC noise ($\epsilon_0 = 0$)	$\sqrt{2}\cos\omega_st$	$\psi e^{-\alpha_0\lvert t\rvert}$	Eq. (A.2.48) $\alpha \to \alpha_0,\ \epsilon = 0$ $G = \sqrt{2}\cos\omega_st$	$a_0^2\,\frac{\lambda}{2\gamma}(1+\gamma^2)\left[1+\frac{\sin 2\lambda}{2\lambda}\left(1-\frac{2\gamma^2}{1+\gamma^2}\right)+\frac{2(1+\cos^2\lambda)}{\lambda(1+\gamma^2)}\,\gamma\right]$ $\lambda = \omega_sT,\ \gamma = \frac{\omega_s}{\alpha_0}$

the expected performance $(a_0^2)^*_{\min} \sim T^{-1}$ (or n^{-1}). For strong signals, however, we no longer have a large-sample theory and consequently $(a_0^2)^*_{\min}$ is not simply proportional to T^{-1} (or n^{-1}) any more. The precise behavior now depends on signal waveform and on the covariance function of the accompanying noise. In Table 20.1, some specific $a_0^2(\Phi_T)_s$ are listed for simple signals in white and colored noise, while Fig. 20.5 shows $(a_0^2)^*_{\min}$ as a function of observation time T for continuous sampling, when $\mathcal{K}/\mu = 1$ and $R^* - \mathcal{R}_0$ is required to be a fraction η_0 $(= 0.5, 0.1)$ of $R^*_{\max} - \mathcal{R}_0$. For white noise backgrounds, $(a_0^2)^*_{\min}$ is replaced by $(\sigma_e^2)^*_{\min}$, where $\sigma_e^2 \equiv A_0^2/W_0 b$, $A_0^2/2W_0\omega_s$, respectively, for the d-c and sinusoidal signals $s(t - \epsilon_0, \boldsymbol{\theta}_0) = \sqrt{2}$,

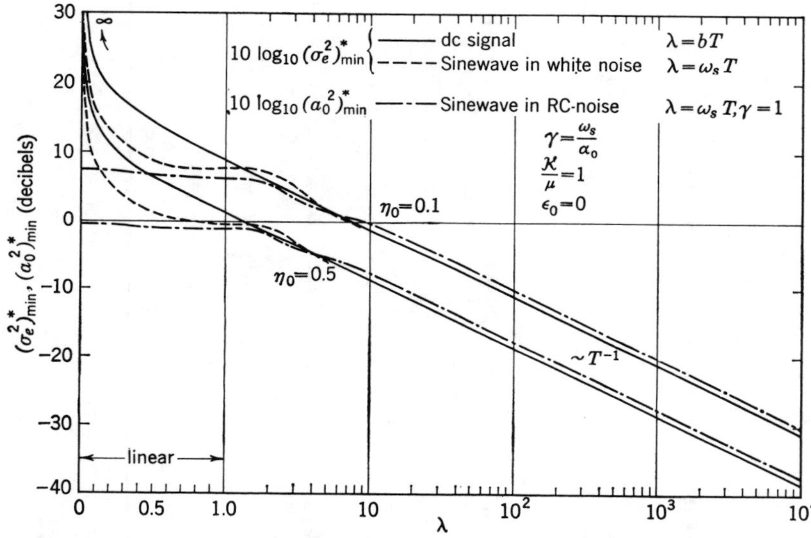

FIG. 20.5. Minimum detectable signals for optimum coherent simple-alternative detection in $(0,T)$.

$\sqrt{2} \cos \omega_s t$ chosen here; b is some convenient (angular) frequency, for scaling purposes, so that A_0^2 is suitably normalized. The quantity λ $(= bT, \omega T_s)$ is a normalized observation time, and $\gamma = \omega_s/\alpha_0$ in the third example, where $\alpha_0 = (RC)^{-1}$ for this background noise. Note, finally, from the curves of normalized Bayes risk $\mathcal{R}^*$ here (cf. Fig. 20.8) that $(a_0^2)^*_{\min}$ [and $(\sigma_e^2)^*_{\min}$] are smallest when $\mathcal{K}/\mu = 1$. Thus, when $\mu = 1$, the Ideal Observer (Sec. 19.2-2) who takes $\mathcal{K} = 1$ not only minimizes the Bayes risk but also obtains the smallest minimum detectable signal [see, however, the remarks in Sec. 20.4-3(2)].

The characteristic optimum weak-signal behavior with observation time T is again observed: $(a_0^2)^*_{\min}$ or $(\sigma_e^2)^*_{\min}$ is proportional to T^{-1}. Note, however, as mentioned earlier (Sec. 20.3-4), that for small samples the minimum detectable signal becomes infinite as $T \to 0$ for d-c and sine-wave signals in white noise while remaining finite for colored noise backgrounds.† Thus, the spectrum of the noise, as well as the signal waveform, influences this limiting behavior, as we might expect. Of course, the greater the accuracy of decision (i.e., the smaller η_0), the larger is $(a_0^2)^*_{\min}$ required to be, as Fig. 20.5 indicates for $\eta_0 = 0.5, 0.1$. The general nature of these results is in no way altered for more involved signal waveforms and more general spectra.

20.4-2 Coherent Detection (Suboptimum Simple-alternative Cases). Instead of the optimum system of Sec. 20.4-1, we may use a similar, but suboptimum, receiver whose general structure embodies the essential operations of Eq. (20.117*a*) but whose elements are not necessarily identical with those of the optimum system. This is typical of practical situations, where it is usually not worth the effort and cost to attempt an exact duplication of a theoretically "best" system if a reasonable approximation can be achieved.

Here we are guided by Eq. (20.117*a*). We choose the linear system of Prob. 20.8, Eqs. (1), (4), for this approximate representation. Thus, for discrete and continuous sampling we have, for the simple-alternative cases examined in this section,

$$z = \boldsymbol{T}_R\{\mathbf{V}\} = \Gamma_1 + \tilde{\mathbf{V}}\mathbf{G}_1\mathbf{S} = \Gamma_1 + \frac{A_0}{\sqrt{2}}\tilde{\mathbf{V}}\gamma_0^{1/2}\mathbf{A}\mathbf{G}_1'\mathbf{S} \tag{20.121a}$$

$$= \Gamma_1 + \gamma_0^{1/2}\int_{0-}^{T+} V(t)\left[\int_0^\infty S(u)h_1(t-u)\,du\right]h_I(T-t)\,dt \tag{20.121b}$$

where as before $\gamma_0 = \psi^{-2}$, or $4T^2/W_0^2$ (for white noise). Replacing $\bar{a}_0$ by a_0, $\bar{\mathbf{s}}$ by $\mathbf{s}$, etc., in Prob. 20.8, Eqs. (5) to (7*c*), we may write at once the desired error probabilities,

$$\alpha = \frac{1}{2}\left\{1 - \Theta\left[\frac{\log \mathcal{K} - \Gamma_1}{(2a_0^2\Phi_2)^{1/2}}\right]\right\} \qquad \beta = \frac{1}{2}\left\{1 - \Theta\left[\frac{\Gamma_1 - \log \mathcal{K} + a_0^2\Phi_1}{(2a_0^2\Phi_2)^{1/2}}\right]\right\} \tag{20.122}$$

in which $\Phi_1 = \psi\tilde{\mathbf{s}}\mathbf{G}_1\mathbf{s}$, $\Phi_2 = \psi^2\tilde{\mathbf{s}}\mathbf{G}_1\mathbf{k}_N\mathbf{G}_1\mathbf{s}$, now with Eqs. (7*a*), (7*b*) of Prob. 20.8 in the continuous case.

If we set the bias Γ_1 equal to that of the optimum system here, for example, $\Gamma_1 = \Gamma_0 = \log \mu - (a_0^2/2)\Phi_s$ [cf. Eqs. (20.117*a*), (20.117*b*) with Φ_s replaced by $(\Phi_T)_s$ for continuous sampling], these error probabilities become (exactly)

$$\alpha = \frac{1}{2}\left\{1 - \Theta\left[\frac{a_0^2\Phi_s}{2\sqrt{2}\,a_0\Phi_2^{1/2}} + \frac{\log(\mathcal{K}/\mu)}{\sqrt{2}\,a_0\Phi_2^{1/2}}\right]\right\} \tag{20.123a}$$

$$\beta = \frac{1}{2}\left\{1 - \Theta\left[\frac{a_0^2(2\Phi_1 - \Phi_s)}{2\sqrt{2}\,a_0\Phi_2^{1/2}} - \frac{\log(\mathcal{K}/\mu)}{\sqrt{2}\,a_0\Phi_2^{1/2}}\right]\right\} \tag{20.123b}$$

† The $(A_0^2)^*_{\min}$ for the different signals (Table 20.1) can be compared if we set $W_0b = W_{0\omega_s} = \psi$ and use Fig. 20.5.

TABLE 20.2. $a_0^2\Phi_1$, $a_0^2\Phi_2$ FOR SUBOPTIMUM COHERENT SIMPLE-ALTERNATIVE DETECTION; CONTINUOUS SAMPLING IN $(0,T)$

Signal and noise	$h_1(t)$ $(t > 0-)$	$h_I(T-t)$ $(0 \le t \le T)$	$a_0^2\Phi_1$	$a_0^2\Phi_2$	$\frac{a_0^2\Phi_1}{(2a_0^2\Phi_2)^{1/2}}$ $(T \to \infty)$
1. D-c signal in white noise	a. $\delta(t-0)$	$be^{-b(T-t)}$	$2\sigma_e^2\lambda(1-e^{-\lambda})$	$\sigma_e^2\lambda^2(1-e^{-2\lambda})$	$\sigma_e\sqrt{2}$
$s(t) = \sqrt{2}$	b. be^{-bt}	T^{-1} (ideal)	$2\sigma_e^2(\lambda + e^{-\lambda} - 1)$	$2\sigma_e^2(\lambda - \frac{5}{2} + \frac{1}{2}e^{-2\lambda} + 2e^{-\lambda})$	$\sigma_e\sqrt{\lambda}$
$\lambda = bT$ $\sigma_e^2 = \frac{A_0^2}{W_0 b}$	c. be^{-bt}	$be^{-b(T-t)}$	$2\sigma_e^2\lambda(1 - e^{-\lambda} - \lambda e^{-\lambda})$	$\sigma_e^2\lambda^2(1 + 3e^{-2\lambda} - 4e^{-\lambda} + 2\lambda e^{-2\lambda})$	$\sigma_e\sqrt{2}$
2. Sine wave in white noise $s(t) = \sqrt{2}\cos\omega_s t$ $\sigma_e^2 = \frac{A_0^2}{2W_0\omega_s}$ $\lambda = \omega_s T$ $\frac{\omega_s}{b} \equiv \gamma$	$\delta(t-0)$	$be^{-b(T-t)}$	$2\lambda\sigma_e^2\left(1 - e^{-\lambda/\gamma} + \frac{2\gamma\sin 2\gamma + \cos 2\gamma - e^{-\lambda/\gamma}}{1+4\gamma^2}\right)$	$\sigma_e^2\lambda^2\left(1 - e^{-2\lambda/\gamma} + \frac{\cos 2\lambda + \gamma\sin 2\lambda - e^{-2\lambda/\gamma}}{1+\gamma^2}\right)$	$\sigma_e\sqrt{2}$ $(\gamma^2 \gg 1)$

which are to be compared with Eq. (20.120) for α^*, β^* in the corresponding situation of the optimum system. Instead of a single "parameter" $a_0{}^2\Phi_s$, there are now three: $a_0{}^2\Phi_s$, $a_0{}^2\Phi_1$, $a_0{}^2\Phi_2$, and consequently in threshold operation α, β are no longer universal functions of $a_0{}^2\Phi_s$, and hence of T. The average risk and minimum detectable signal are accordingly more complicated functions of signal and noise structure. Table 20.2 gives specifically $a_0{}^2\Phi_1$, $a_0{}^2\Phi_2$ for several different signals and noise backgrounds and a number of simple linear filters; $a_0{}^2(\Phi_T)_s$ in corresponding cases may be found from Table 20.1.

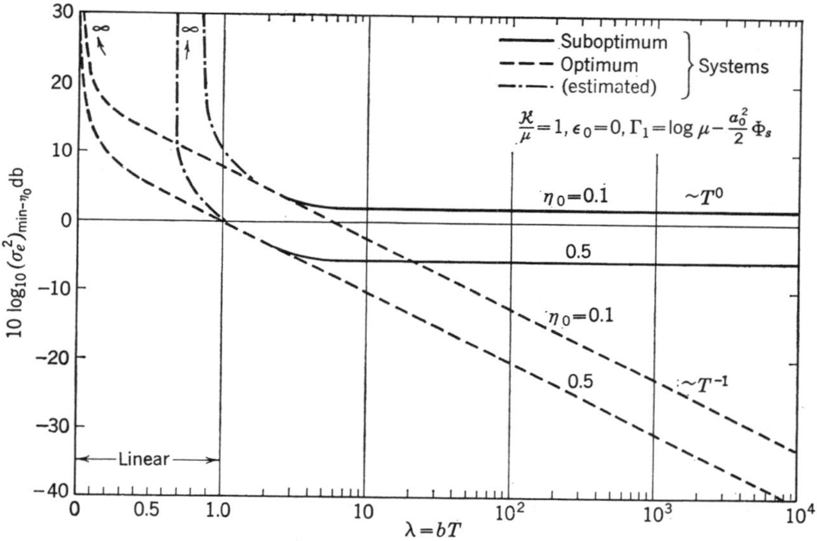

FIG. 20.6. Minimum detectable d-c signals for suboptimum coherent simple-alternative detection in $(0,T)$.

The average risk R is calculated as before with the help of Eq. (19.19b), yielding curves similar to those in Fig. 20.4b, except that the average risk does not usually vanish as $T \to \infty$ (a_0 fixed) unless h_I is an ideal filter; the minimum detectable signal is likewise found as above, from Sec. 19.3-3. Figure 20.6 shows $(\sigma_e{}^2)_{\min\,\eta_0}$ [$\mathcal{K}/\mu = 1$, $R - \mathcal{R}_0 = \eta_0(R - \mathcal{R}_0)_{\max}$, $\epsilon_0 = 0$ as in Fig. 20.5] for the simplest of signals: a d-c, or steady, wave, here in a white noise background when various combinations of pre- and postmultiplication filters h_1, h_I are used. The general behavior is characteristic of more complicated signal waveforms and colored noise interference: when long observation periods are available, $(\sigma_e{}^2)_{\min}$ decreases $\sim T^{-1}$ with an ideal final smoothing filter but approaches a fixed value, independent of T when an ordinary lumped-constant filter is used here. This occurs because in the former instance the final bandwidth becomes progressively narrower with increasing T, while in the latter the bandwidth is fixed, so that there is a limit to the degree of smoothing obtainable. The performance† is similar to that encountered earlier in the measurement problems of Sec. 16.1-2. Finally, with short observation times $(\sigma_e^2)_{\min}$ generally approaches an infinite value well before $(\sigma_e^2)^*_{\min}$, as $T \to 0$, with only a small range of values of T during which $(\sigma_e^2)_{\min}$ is proportional to T^{-1}. All this is entirely what we would predict: small samples require strong signals, and for suboptimum systems it is usually the case that below a certain sample size it is not possible to obtain a signal strong enough to achieve the desired accuracy of decision (here $\eta_0 = 0.5$, or 0.1 in Figs. 20.5, 20.6).

20.4-3 Optimum Incoherent Detection (A Simple Radar Problem). As our next example, let us consider the simple "A-scope" radar problem of determining the presence or absence of a target at some fixed range, without scanning, when there is sample uncertainty, and when the usual pulsed carrier is employed to interrogate the target.‡ Let us further simplify the analysis by adopting a "one-point" theory: the received wave $V(t)$ is sampled periodically during a fixed interval $(0,T)$ at the pulse repetition frequency of the modulated carrier, with one sampled value per period. If a target is present, it is assumed that this sampling time occurs at the maximum of the returned pulse, whose amplitude A_0, moreover, remains unchanged in $(0,T)$. In addition, it is assumed that successive cw pulse trains are generated by turning an oscillator on and off, so that there is no RF phase coherence from pulse train to pulse train. Clearly, reception is incoherent, and if the subcriterion of maximum average uncertainty of RF phase (subject to the fixed maximum 2π) is chosen, then RF phase is uniformly distributed over a typical RF cycle. Reception is further constrained by fixed RF and IF receiver elements (cf. Fig. 5.11), including a given antenna pattern, which here simply modifies the strength of the returned signal since there is no scanning in this particular situation. In addition, the pulse repetition period T_0 is taken large enough so that sampled values of the background noise are statistically independent; that is, T_0 is large compared with the correlation time T_c of the interference, where T_c is essentially the reciprocal of the bandwidth of the composite RF-IF strip. The noise itself may be the inherent system noise which arises primarily in these elements (cf. Secs. 5.3-1, 5.3-2), or it may be due to external sources, strong compared with system noise.

Our problem now is to determine the optimum detector structure and to

† Note that, in spite of the particular form of the signal of Fig. 20.6, cases $1a$, $1b$ in Table 20.2 are not identical; moreover, different performances are also expected for the more general signal waveforms (sine waves and others).

‡ This problem has received considerable theoretical attention in recent years; see, for example, the work of Siegert,[16] Marcum,[17] Hanse,[18] Middleton,[19] Peterson, Birdsall, and Fox,[20] Benner and Drenick[21] and the related studies of Woodward,[15] Drukey,[22] Reich and Swerling,[23] Kaplan and McFall,[24] Harrington,[25] and Siebert.[25a,25b] Additional references are given in these papers and in the supplementary references at the end of the volume.

evaluate this system's performance, subject to the above assumptions and constraints. This means here optimum (second) detection and video processing of the data before a binary decision, and calculation of the error probabilities and Bayes risk of the resulting system, when discrete sampling is employed. While the results of the analysis have limited applicability in most practical cases, they are useful as a general indication of the methods involved and the type of system that is encountered in radar detection problems. We shall return to these problems from a broader viewpoint in Sec. 20.4-6; in any case, note that we consider here only the detection features of a radar system, not the measurement aspects: range (and/or velocity) estimation is not involved in the present model.

(1) **Structure.** Let us begin with the structure of the optimum system. Since successive pulse trains are here observed incoherently and independently, as far as RF signal phase and the background noise are concerned, it is convenient to use the envelope-phase form of the narrowband output $\mathbf{V}$ from the composite RF-IF stages of the receiver. Thus, $\mathbf{V}$ is replaced by $(\mathbf{E},\boldsymbol{\phi})$, where $V(t_j) = E(t_j) \cos [\omega_0 t - \phi(t_j)]$ $[j = 1, \ldots, n, \text{ in } (0,T)]$, and E_j, ϕ_j are, respectively, the envelope and phase of V_j at time t_j (cf. Secs. 9.1, 9.2). Now, because of the way in which signals are generated and received, it is not possible to use the detailed RF phase information. We must therefore average over all possible RF phase states, as well as the signal parameter, to write the optimum structure $x = \log \Lambda_n(\mathbf{V})$ [Eq. (20.1)] alternatively as

$$x = \boldsymbol{T}_R\{\mathbf{V}\} = \log \Lambda_n(\mathbf{E}|\mathbf{S}) = \log \mu + \log \langle F_n(\mathbf{E},\boldsymbol{\phi}|\mathbf{S})\rangle_{\phi,S} - \log \langle F_n(\mathbf{E},\boldsymbol{\phi}_0|0)\rangle_{\phi_0} \tag{20.124a}$$

$$= \log \mu + \log \langle W_n(\mathbf{E}|\mathbf{S})\rangle_S - \log W_n(\mathbf{E}|0) \tag{20.124b}$$

where $\boldsymbol{\phi}$, $\boldsymbol{\phi}_0$ represent the phases of the narrowband wave V with and without a signal.

Since sampled values are here assumed to be independent, the RF phases uniformly distributed, and since the signal amplitude A_0 does not vary from pulse to pulse, we may use Eqs. (9.18), (9.50) or Eqs. (1) of Prob. 19.2 to express the joint d.d.'s of the envelope above as the Rayleigh densities

$$W_n(\mathbf{E}|\mathbf{0}) = \prod_{j=1}^{n} \psi^{-1} E_j e^{-E_j^2/2\psi} \qquad E_j \geq 0 \tag{20.125a}$$

$$W_n(\mathbf{E}|\mathbf{S}) = \prod_{j=1}^{n} \psi^{-1} E_j e^{-E_j^2/2\psi - A_0^2/2\psi} I_0(A_0 E_j/\psi) \qquad E_j \geq 0 \tag{20.125b}$$

with $\psi = \overline{N^2}$, as before. With the normalizations

$$\mathcal{E}_j \equiv E_j/\sqrt{2\psi} \qquad a_0^2 \equiv A_0^2/2\psi \tag{20.126}$$

we find that the optimum system [Eq. (20.124*b*)] reduces directly to

$$x = \log \Lambda_n(\mathbf{E}|\mathbf{S}) = \log \mu + \log \left\langle e^{-na_0^2} \prod_{j=1}^{n} I_0(2a_0\mathcal{E}_j) \right\rangle_{a_0} \qquad (20.127a)$$

If the amplitude ($\sim a_0$) is known to the observer, or arbitrarily set by him, this becomes the familiar result[19]

$$x = \log \Lambda_n(\mathbf{E}|\mathbf{S}) = \log \mu - na_0^2 + \sum_{j=1}^{n} \log I_0(2a_0\mathcal{E}_j) \qquad (20.127b)$$

In either case, the optimum threshold structure is found to be†

$$x = \log \Lambda_n(\mathbf{E}|\mathbf{S}) \doteq (\log \mu - n\overline{a_0^2} - n\overline{a_0^2}^2/2) + \overline{a_0^2} \sum_{j=1}^{n} \mathcal{E}_j^2 \qquad (20.128)$$

with $\overline{a_0^2}$ replaced by a_0^2 for Eq. (20.127*b*).

The operations are indicated in Fig. 20.7. Note again the characteristic square-law behavior of incoherent reception: the individual envelopes are

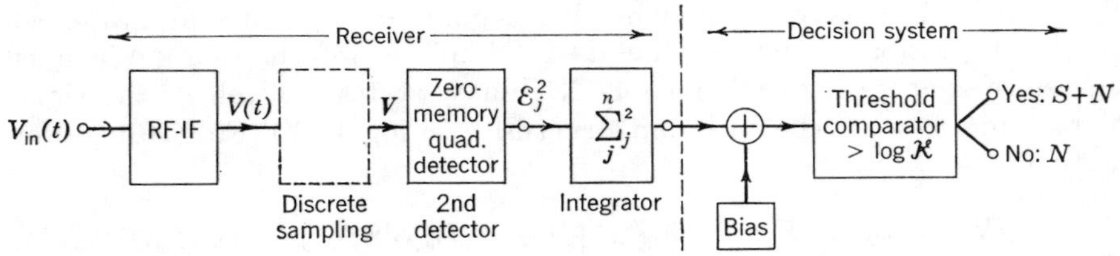

FIG. 20.7. A simple optimum threshold radar detection system (discrete sampling).

squared, then added (i.e., "integrated"), and the resulting number compared with some preset threshold $\log \mathcal{K}$ in the usual way. When a_0 is known, in place of the zero-memory square-law second detector a $\log I_0$ device is employed, according to Eq. (20.127*b*). It is then a simple matter to obtain the approximate strong-signal structure of the optimum system from Eq. (20.127*b*). Using the asymptotic expansion of the Bessel function, we get

$$\log \Lambda_n(\mathbf{E}|\mathbf{S}) \simeq \log \mu - na_0^2 + 2a_0 \sum_{j=1}^{n} \mathcal{E}_j \qquad a_0^2 \gg 1 \qquad (20.129)$$

Instead of the $\log I_0$ rectifier for the receiver's second detector, a linear envelope tracer (or half-wave linear rectifier) is now required. The half-wave linear rectifier actually used in practical systems is an excellent approximation to the ideal characteristic for all signal strengths, since for weak

† Bias terms $O(a_0^4)$ must be included as before (cf. Sec. 20.3-1); these are obtained from $E_{\mathcal{E}_j|H_0}\{O(\mathcal{E}_j^4)\}$ in the expansion of the second term of Eq. (20.127*a*) in the usual way [cf. Eqs. (20.4*a*), (20.4*b*), et seq.].

signals it, too, exhibits the general quadratic, or square-law, behavior demanded by Eq. (20.128) for optimum threshold operation [cf. Sec. 13.2-2(3)].

Finally, it is instructive to compare Eqs. (20.127*a*), (20.127*b*) with Eqs. (20.31), (20.31*a*) under similar conditions: for the latter, we have approximately

$$\log \Lambda_n(\mathbf{V}) \approx \log \mu + \log \left\langle e^{-na_0^2/2} I_0 \left(a_0 \sqrt{\sum_j^n \mathcal{E}_j^2} \right) \right\rangle_{a_0} \tag{20.130}$$

in contrast with Eq. (20.127*a*), where the envelopes $\mathcal{E}_j$ appear singly in a product of Bessel functions. The difference lies essentially in the fact that Eq. (20.130) represents a case where the RF phase relations are maintained from pulse to pulse, although the epoch of the pulse trains is not known at the receiver, while Eq. (20.127*a*) represents a situation of random RF phase from pulse to pulse. The former is an example of the so-called "coherent" radar, where the transmitted (and received) signals are produced by pulsing an oscillator at fixed intervals, the oscillator itself running continuously (and stably). The latter [cf. Eq. (20.127*a*)] is the result of turning the oscillator on and off, so that RF phase coherence is lost, and as expected, the performance of such systems is inferior to the former, whenever the former can be applied. We shall return to this problem again in Sec. 20.4-6.

(2) Error probabilities and Bayes risk. The error probabilities associated with optimum threshold detection are now easily determined from Eq. (20.128) and the methods of Sec. 19.3-1 (cf. Prob. 20.7). We obtain finally

$$\begin{Bmatrix} \alpha^* \\ \beta^* \end{Bmatrix} \simeq \frac{1}{2} \left\{ 1 - \Theta \left[\frac{\overline{a_0^2} \sqrt{n}}{2\sqrt{2}} \pm \frac{\log (\mathcal{K}/\mu)}{\sqrt{2}\, \overline{a_0^2} \sqrt{n}} \right] \right\} \tag{20.131}$$

[cf. Eq. (20.120)]. Figure 20.4*a* summarizes these relations for $\alpha^* + \beta^* < 1$, provided that we replace $a_0\Phi_s^{1/2}$ by $\overline{a_0^2}\sqrt{n}$. For a given $\overline{a_0^2}\sqrt{n}$ the values of α^*, β^* are interchanged on reciprocating $\mathcal{K}/\mu$, and, again, specifying any two of the four quantities α^*, β^*, $a_0^2\sqrt{n}$, and $\mathcal{K}/\mu$ determines the other two. Figure 20.4*b* may also be used for the Bayes risk here [Eq. (20.131) in Eq. (19.19*a*)], with $a_0\Phi_s$ replaced by $\overline{a_0^2}\sqrt{n}$.

Sometimes it is convenient to use the normalized Bayes risk $\mathcal{R}^*(\sigma,\delta^*)$ [Eqs. (19.39)]. Figure 20.8 shows some typical cases, for various threshold $\mathcal{K}$ and a priori signal probabilities $\mu = p/q$. Because of the symmetry of Eq. (20.131) and the normalized Bayes risk $\mathcal{R}^*(\sigma,\delta^*)$, $\mathcal{R}^*$ for given $\overline{a_0^2}\sqrt{n}$ is the same for $\mathcal{K}/\mu$ and $\mu/\mathcal{K}$ when these ratios have the same value. [Note that the Ideal Observer (Sec. 19.2-2), who sets $\mathcal{K} = 1$, minimizes the (normalized) Bayes risk $\mathcal{R}^*$ when $\mu = 1$ (i.e., when $p = q = \frac{1}{2}$).] However, the most favorable values of $\mathcal{R}^*$ occur when $\mathcal{K} = \mu$, not when $\mathcal{K} = 1$ alone. This is true also of the *minimum detectable signal*, $(\overline{a_0^2})^*_{\min\,\eta_0}$, which for these threshold cases is seen from Eq. (20.131) to be once more inversely proportional to the square root of the sampling size, or observation time [cf. Eq. (20.92)]. Some typical weak-signal values of $(a_0{}^2)^*_{\min\,\eta_0}$ for the simple radar receiver of the present problem are shown in Fig. 20.9 when $\mathcal{K}/\mu = 1$ and $\eta_0 \equiv \mathcal{R}^*/\mathcal{R}^*_{\max} = 0.5,\ 0.1$. On the other hand, for some results of a strong-signal theory here, see Prob. 20.11.

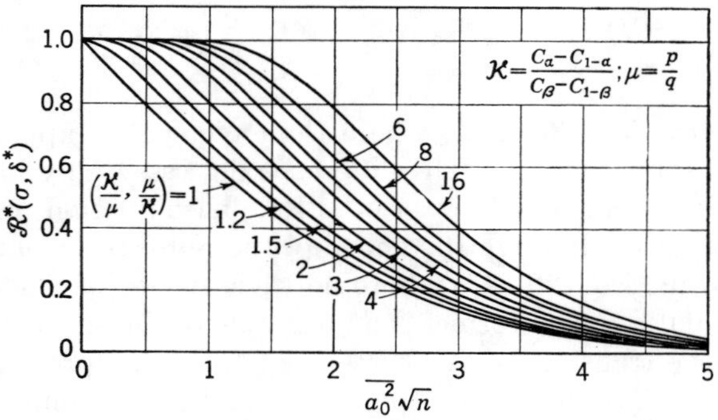

FIG. 20.8. Normalized Bayes risk for optimum incoherent threshold detection; discrete sampling, pulsed carrier, and Rayleigh statistics.

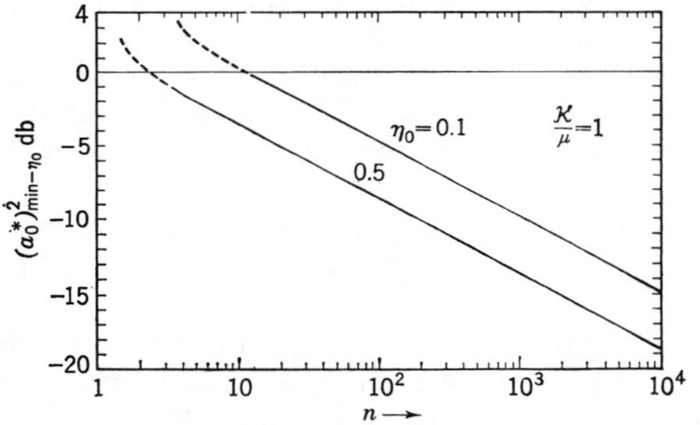

FIG. 20.9. Minimum detectable signals for the system of Fig. 20.8 (normalized Bayes risk).

Finally, we remark that some situations may require an unnormalized average risk R^*, or $R^* - \mathcal{R}_0$, for defining $(\overline{a_0{}^2})^*_{\min\,\eta_0}$. This occurs, for instance, when $C_\alpha - C_{1-\alpha}$ and $C_\beta - C_{1-\beta}$ are fixed *and* the minimum detectable signal is defined as the smallest $\overline{a_0{}^2}$ for which the absolute (average) risk does not exceed a specified fraction of $C_\alpha - C_{1-\alpha}$ or $C_\beta - C_{1-\beta}$, regardless of the a priori probabilities p, q.† Since the cost assumptions determine the detailed performance of the optimum system, we must be careful to decide whether or not normalized (average) risk is appropriate. Thus, if

† A fixed fraction of $C_\alpha - C_{1-\alpha}$ or $C_\beta - C_{1-\beta}$ does not correspond to a given fraction of the maximum average risk $R^*_{\max}$ for all p, q, since $R^*_{\max}$ depends on p, q.

there are practical considerations which fix the cost in advance *without relation to the performance they imply*, then there is but one (optimum) threshold for the job and there is therefore no reason to compare this with other (optimum) systems having different thresholds. On the other hand, if the conditions of the problem permit some latitude in the choice of cost assignments, it is then meaningful to compare the performance that results from different cost assumptions within the permitted range of values and to select those particular costs which lead to the smallest values of Bayes risk and $(\overline{a_0^2})^*_{\min}$.†

(3) Neyman-Pearson and Ideal observers. We have already noted in Secs. 19.2-1, 19.2-2 that the Neyman-Pearson observer, who fixes α $(= \alpha_{NP})$ and minimizes $p\beta$ $(= p\beta^*_{NP})$, and the Ideal Observer, who minimizes the sum $\alpha q + \beta p$ $(= q\alpha^*_I + p\beta^*_I)$, are both special cases of the Bayes observer, who minimizes the average risk. The former makes a certain type of cost assignment, in effect choosing a particular threshold $\mathcal{K}_{NP}$ (which depends on α_{NP}), while the latter chooses his costs so that the ratio equals unity.

Let us now consider the simple radar problem of this section in terms of each of these special examples of a Bayes threshold receiver, where attention is focused on the error probabilities directly, rather than on the average risk. Of course, the one implies the other, and as we have indicated earlier (Sec. 19.2), the Bayes formulation is not only the more general but the more fundamental concept as well.

The conditional error probabilities α, β are, respectively, related to two useful parameters of a radar system: the *false alarm time* T_α and the *false dismissal* (or false rejection) *time* T_β. The former is a measure of the average time interval between successive false alarms, i.e., incorrect decisions that a target is present, while the latter indicates the average interval between successive incorrect decisions that a target is absent—i.e., a target is really present but is called "noise alone." For threshold operation, we can write approximately for normal noise‡

$$T_\alpha \doteq T_0/(\alpha\,\Delta f\,\Delta\tau) \qquad T_\beta \doteq T_0/(\beta\,\Delta f\,\Delta\tau) \tag{20.132}$$

where Δf is the equivalent rectangular bandwidth of the composite RF-IF stages of the receiver and $\Delta\tau$ is the width of the (time) gate about the center of a typical pulse; $\Delta\tau$ is essentially the duration of the pulse, say, between half-intensity points. Similarly, α and β are also called *false alarm* and *false dismissal probabilities*. Thus, the larger α (or β), the smaller the intervals, on the average, between incorrect decisions, as we would expect, while small α (or β) ensures a comparatively long period between erroneous decisions. Values of $\alpha = O(10^{-3}\text{–}10^{-13})$ cover the useful range in the Neyman-Pearson case, the former corresponding to rather short false alarm times O(seconds);

† For further discussion, see the general comments of Secs. 20.4-8 and Sec. 23.4 and Van Meter and Middleton,[26] sec. 3.3.

‡ See Middleton,[19] sec. 7B, pt. II, for details of the derivation.

the latter may run to a period of days. Note finally that since α and β, and the α^*, β^* of optimum systems, are not independently chosen when the observation time ($\sim n$) is fixed, T_α and T_β are likewise functionally related, so that fixing one determines the other.

In place of the average-risk curves of Figs. 20.4*b*, 20.8 [cf. Eqs. (19.19)], we can equally well employ a *betting curve* [Eq. (19.38)] to show how the probability of correct decisions varies with signal strength and sample size for these particular choices of threshold $\mathcal{K}$, for example, $\mathcal{K}_{NP}(\alpha_{NP})$, or $\mathcal{K}_I = 1$. Moreover, it is convenient to use a normalized betting curve, $\mathcal{P}_1$, with the properties that as $\bar{a}_0 \to \infty$, $\mathcal{P}_1 \to 1$ and as $\bar{a}_0 \to 0$, $\mathcal{P}_1 \to 0$. These curves are accordingly defined, respectively, for the Neyman-Pearson and Ideal observers by[19]

$$\mathcal{P}_1(a_0;n)_{NP} = \{W_1(a_0,n)_{NP} - [q + \alpha_{NP}(p - q)]\}(1 - q - \alpha_{NP}p)^{-1} \tag{20.133a}$$

$$\mathcal{P}_1(a_0,n)_I = [W_1(a_0,n)_I - (p \text{ or } q)][1 - (p \text{ or } q)]^{-1} \tag{20.133b}$$

where (p or q) means that the larger of the two is to be used and $W_1(a_0,n)_{NP} = 1 - (q\alpha_{NP} + p\beta^*_{NP})$, $W_1(a_0,n)_I = 1 - (q\alpha^*_I + p\beta^*_I)$, from Eq. (19.38) and Sec. 19.2. Specifically, from Eqs. (19.33), (19.34), the error probabilities are

$$\alpha_{NP} = \int_{\log \mathcal{K}_{NP}}^{+\infty} Q_n(x)\,dx \qquad \beta^*_{NP} = \int_{-\infty}^{\log \mathcal{K}_{NP}} P_n(x)\,dx \tag{20.134}$$

with α^*_I, β^*_I obtained from Eqs. (20.134) on setting $\mathcal{K}_{NP} = 1$ therein. As before, $Q_n(x)$, $P_n(x)$ are the d.d.'s of the optimum receiver characteristic with respect to the hypothesis $H_0\!:N$ and $H_1\!:S + N$, given by Eq. (20.128) in the threshold case. Equation (20.131) in turn yields the desired α^*, β^* here, so that we can write directly with its help

$$\log \mathcal{K}_{NP} \simeq \sqrt{2}\,\overline{a_0^2}\,\sqrt{n}\,\Theta^{-1}(1 - 2\alpha_{NP}) - \frac{(\overline{a_0^2}\,\sqrt{n})^2}{2} + \log \mu > 0$$
$$\beta^*_{NP} \simeq \frac{1}{2}\left\{1 - \Theta\left[\frac{\overline{a_0^2}\,\sqrt{n}}{2\sqrt{2}} - \frac{\log(\mathcal{K}_{NP}/\mu)}{\sqrt{2}\,\overline{a_0^2}\,\sqrt{n}}\right]\right\} \tag{20.135a}$$

$$\text{and} \qquad \begin{Bmatrix}\alpha^*_I\\ \beta^*_I\end{Bmatrix} \simeq \frac{1}{2}\left[1 - \Theta\left(\frac{\overline{a_0^2}\,\sqrt{n}}{2\sqrt{2}} \mp \frac{\log \mu}{\sqrt{2}\,\overline{a_0^2}\,\sqrt{n}}\right)\right] \tag{20.135b}$$

Observe from Eqs. (19.35*a*), (19.35*b*) that the corresponding Bayes risks can be expressed as

$$R^*_{NP} = C_0[1 - (W_1)_{NP} + \alpha_{NP}q(\mathcal{K}_{NP} - 1)] \qquad R^*_I = C_0[1 - (W_1)_I] \tag{20.136}$$

with somewhat more involved relations when W_1 is replaced by its normalized values [Eqs. (20.133*a*), (20.133*b*)].

Figures 20.10, 20.11 show normalized betting curves for the Neyman-Pearson and Ideal observers, respectively, while Fig. 20.12 gives the minimum detectable signal as a function of observation time ($\sim n$), for various

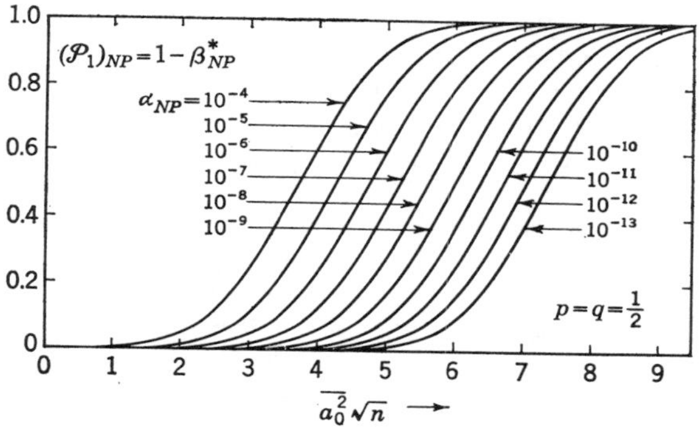

FIG. 20.10. Normalized betting curves $(\mathcal{P}_1)_{NP}$ [Eq. (20.133b)] for the Neyman-Pearson observer; incoherent threshold detection and Rayleigh statistics.

α_{NP}, at the 90 per cent level of correct decisions on these normalized curves, i.e., for $\mathcal{P}_1 = 0.90$. Only a single such curve is possible, of course, for the Ideal Observer, where α_I, β_I are adjusted simultaneously. [That $(a_0^2)^*_{\min I}$ lies slightly above $(a_0^2)^*_{\min NP}$ is due to the nature of the normalizations (20.133) and the fact that different thresholds are implied: $\mathcal{K}_{NP} \neq \mathcal{K}_I = 1$, for the α^* and β^* that occur here; one should not infer that the Neyman-Pearson observer with $\alpha_{NP} = 10^{-2}$ is necessarily better than the Ideal Observer with $\alpha^*_I = \beta^*_I = 0.05$, even though $\beta^*_{NP} \doteq 0.10$ under these conditions.] The strong-signal behavior of $(a_0^2)^*_{\min}$ in Fig. 20.12 has been calculated with the help of Prob. 20.11 and applies only for the cases where a_0 is known a priori or set at some desired level (at which the receiver is to perform optimally).†

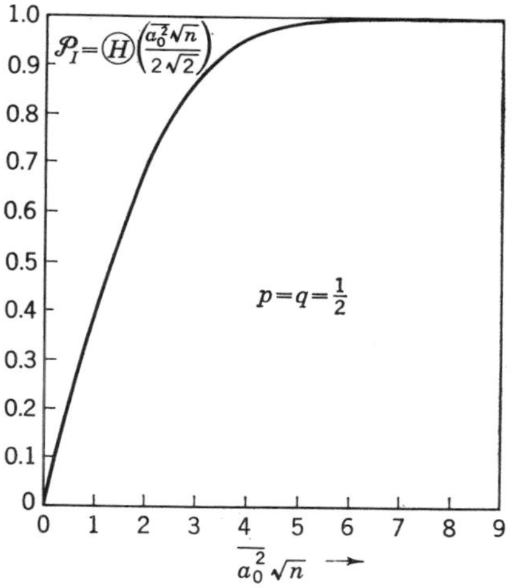

FIG. 20.11. Same as Fig. 20.10 but for the Ideal Observer.

The betting curve is by no means the only convenient way of representing system behavior. Another quantity of some utility in radar problems is the probability $P(S)$ of correctly deciding the presence of a target. This

† Performance is optimal only at this preset level, and not at other levels. See Middleton,[19] sec. 7E, pt. II, for details.

is simply

$$P(S) = pP(S|S) = p(1 - \beta) \qquad (20.136a)$$

If a Neyman-Pearson observer is used, we get $P(S)_{opt\,NP} = p(1 - \beta^*_{NP}) \doteq p \times (\mathcal{P}_1)_{NP}$ (cf. Fig. 20.10), where Eqs. (20.135*a*) apply in the threshold case with $\beta^*_{NP} = \beta^*_{NP}(\alpha_{NP},\mu,\overline{a_0{}^2}\sqrt{n})$ in any case. Observe once more, however, that whenever the criterion chosen involves α and β linearly, as do all the above, we have essentially a criterion based on a calculation of average cost, or risk, with the implied choice of a threshold ratio $\mathcal{K}$. Which form is

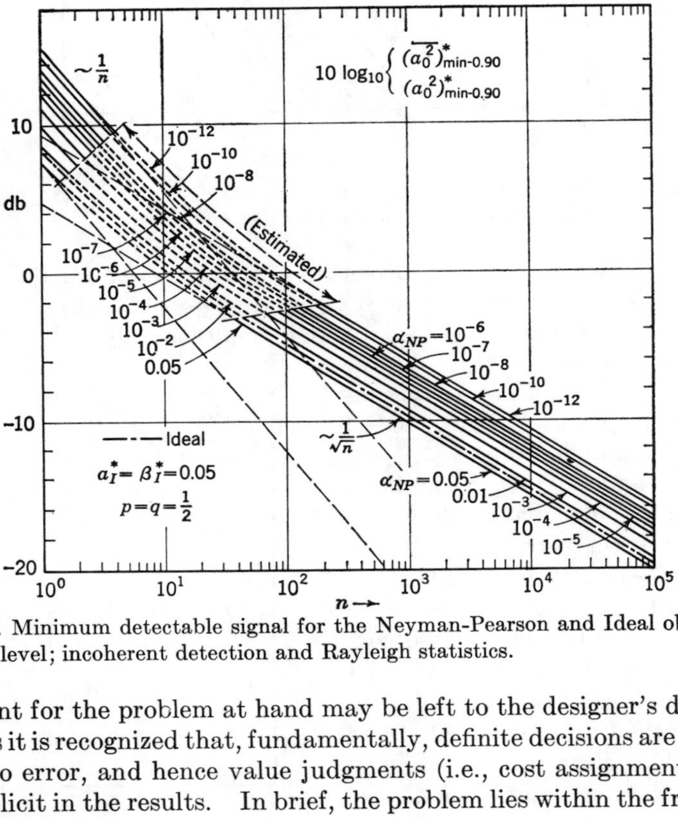

FIG. 20.12. Minimum detectable signal for the Neyman-Pearson and Ideal observers at $\mathcal{P}_1 = 0.90$ level; incoherent detection and Rayleigh statistics.

convenient for the problem at hand may be left to the designer's discretion, as long as it is recognized that, fundamentally, definite decisions are required, subject to error, and hence value judgments (i.e., cost assignments) are at least implicit in the results. In brief, the problem lies within the framework of decision theory.

20.4-4 A Simple Suboptimum Radar System (Incoherent Detection). Optimum or near-optimum systems are not always selected in practice, as we have noted previously. Although cost, circuit simplicity, bulk, and the like, are usually the reasons for the choice of a suboptimum system, not every suboptimum system will necessarily come close to the ideal, or even suboptimum systems based upon the ideal, as the example below will indicate.

Let us consider the following very simple detection system[24] for the radar problem considered in Sec. 20.4-3: Here, n independent observations $E_1, E_2, \ldots, E_n$ are made of the envelope of the pulsed carrier, and a signal is decided when *at least one* of the E_j ($j = 1, \ldots, n$) exceeds a certain threshold value $\mathcal{K}_c$. Otherwise, "noise alone" is the decision. In addition, it is assumed that signal amplitude is known, or preset, at the receiver. A system of this type we shall call an *envelope-threshold* detection system, and our problem here is to determine how its average performance compares with that of the simplified optimum systems of Sec. 20.4-3, in particular, the Neyman-Pearson system.

Our first step is to find the type I and II error probabilities α, β. The d.d. of a single envelope observation E_j is from Eqs. (9.50), (20.125), (20.126)

$$W_1(E_j|S) = \frac{E_j}{\psi} e^{-a_0^2 - E_j^2/2\psi} I_0\left(a_0 E_j \sqrt{\frac{2}{\psi}}\right) \qquad E_j > 0 \tag{20.137}$$

The probability, denoted by $P_1(\gamma_1|a_0)$, that an E_j will exceed $\mathcal{K}_c$ is thus given by

$$P_1(\gamma_1|a_0) = \int_{\mathcal{K}_c}^{\infty} dx \int_0^{\infty} W_1(E_j|S)\delta(x - E_j)\, dE_j = \int_{\mathcal{K}_c}^{\infty} W_1(x|S)\, dx \tag{20.138}$$

The probability, denoted by $P_n(\gamma_1|a_0)$, that *at least one* of the n independently observed E_j's will exceed $\mathcal{K}_c$ is the same as 1 minus the probability that no E_j will exceed $\mathcal{K}_c$:

$$P_n(\gamma_1|a_0) = 1 - [1 - P_1(\gamma_1|a_0)]^n \tag{20.139}$$

But

$$1 - P_1(\gamma_1|a_0) = \int_0^{\mathcal{K}_c} W_1(x|S)\, dx \equiv I(a_0;\mathcal{K}_c) \tag{20.140}$$

so that the probability of deciding "signal" when only noise is present ($a_0 = 0$), or the false alarm probability α, becomes

$$\alpha = 1 - [I(0;\mathcal{K}_c)]^n \tag{20.141a}$$

The false dismissal probability β, or the probability that no E_j will exceed $\mathcal{K}_c$ when signal is present ($a_0 \neq 0$), is then

$$\beta = [I(a_0;\mathcal{K}_c)]^n \tag{20.141b}$$

We note that

$$\alpha \rightarrow \begin{cases} 1 & \text{as } n \rightarrow \infty \text{ or } \mathcal{K}_c \rightarrow 0 \\ 0 & \text{as } n \rightarrow 0 \text{ or } \mathcal{K}_c \rightarrow \infty \end{cases} \qquad \beta \rightarrow \begin{cases} 0 & \text{as } n \rightarrow \infty \text{ or } \mathcal{K}_c \rightarrow 0 \\ 1 & \text{as } n \rightarrow 0 \text{ or } \mathcal{K}_c \rightarrow \infty \end{cases} \tag{20.141c}$$

with $\beta \to 0$, $a_0 \to \infty$; $\beta \to 1 - \alpha$, $a_0 \to 0$. The integral $I(a_0;\mathcal{K}_c)$ is found to be†

$$I(a_0;\mathcal{K}_c) = e^{-a_0^2-\lambda^2} \sum_{m=1}^{\infty} \left(\frac{\lambda}{a_0}\right)^m I_m(2a_0\lambda) \tag{20.142a}$$

$$I(0;\mathcal{K}_c) = 1 - e^{-\lambda^2} \qquad \lambda \equiv \frac{\mathcal{K}_c}{\sqrt{2\psi}} \tag{20.142b}$$

Figure 20.13 shows β, the false dismissal probability, as a function of the signal-to-noise (power) ratio a_0^2 in decibels when $\alpha = 10^{-4}$ for various sample sizes $n = 10, 10^2, 10^3$. The curves for the envelope-threshold system are calculated by using Eq. (20.141*a*) to find the threshold $\mathcal{K}_c$ for which $\alpha = 10^{-4}$ and then Eq. (20.141*b*) to find β for this threshold. Curves showing how β^*_{NP} depends on a_0^2 for the optimum likelihood-ratio detector are also shown in Fig. 20.13 for comparison. These are determined from Eqs. (20.135*a*) with α_{NP} ($= \alpha$) fixed at 10^{-4}. The great superiority of the Neyman-Pearson system is indicated by the signal level required for a given $\beta = \beta^*_{NP}$. For $n = 100$ and $\beta = \beta^*_{NP} = 0.1$, for example, the advantage of this optimum system over the envelope threshold detector here is 10.2 db.

To compare the systems on an average risk basis, we should assume that the costs of type I and II errors are the same for each. The fact that α_{NP} is fixed at 10^{-4} sets the threshold $\mathcal{K}_{NP}$ ($\neq \mathcal{K}_c$, in general) and hence the ratio of costs for the optimum system with any fixed a_0 [cf. Eqs. (20.135*a*)]. For example, the average risk of either system for $p = q = \frac{1}{2}$, $C_{1-\alpha} = C_{1-\beta} = 0$ may be written $R = C_\beta(\beta + \alpha\mathcal{K}_{NP})$, where $\mathcal{K}_{NP}$ is the cost ratio fixed by the *optimum* system. Now, as a_0 is increased, α ($= \alpha_{NP}$) remaining fixed at 10^{-4}, Eq. (20.131) shows that $\mathcal{K}_{NP}$ is increased at first and then decreases, approaching zero as $a_0 \to \infty$. [Of course, Eq. (20.131) is not valid for large a_0, but this qualitative behavior of $\mathcal{K}$ does in fact occur.]

† This is a result due to W. R. Bennett [see Rice,[45] eq. (3.10-17)]. For purposes of calculation, the Bessel functions in Eq. (20.142*a*) are expanded to obtain

$$I(a_0;\mathcal{K}_c) = e^{-a_0^2} \sum_{m=0}^{\infty} B_m a_0^{2m}$$

where B_m depends only on λ and is given by

$$B_m = e^{-\lambda^2} \sum_{k=1}^{\infty} \frac{\lambda^{2m+2k}}{m!(m+k)!}$$

The B_m's are conveniently calculated from the recursion relation

$$B_0 = (1 - \alpha)^{1/n} \doteq 1 - \frac{\alpha}{n} \qquad \text{for small } \alpha$$

$$B_m = \frac{B_{m-1}}{m} - \frac{\lambda^{2m}e^{-\lambda^2}}{m!m!}$$

Thus, it turns out that $\alpha\mathcal{K}_{NP}$ in the average risk R above is completely negligible when the envelope-threshold detector is considered and moreover causes only a slight shift of the curves for the Neyman-Pearson detector from the relative positions shown in Fig. 20.13. For this reason, we may consider the ordinate in Fig. 20.13 as very nearly the *unnormalized average risk per unit of* C_β when the costs of type I and type II errors are related in such a way that $\alpha = 10^{-4}$ for the Bayes system.

With this interpretation of the curves in Fig. 20.13, we may also compare the minimum detectable signals of the two systems on an *unnormalized* risk basis. If the minimum detectable signal is taken as that for which the risk does not exceed 10 per cent of C_β, we observe that, for $n = 10$, 100, or

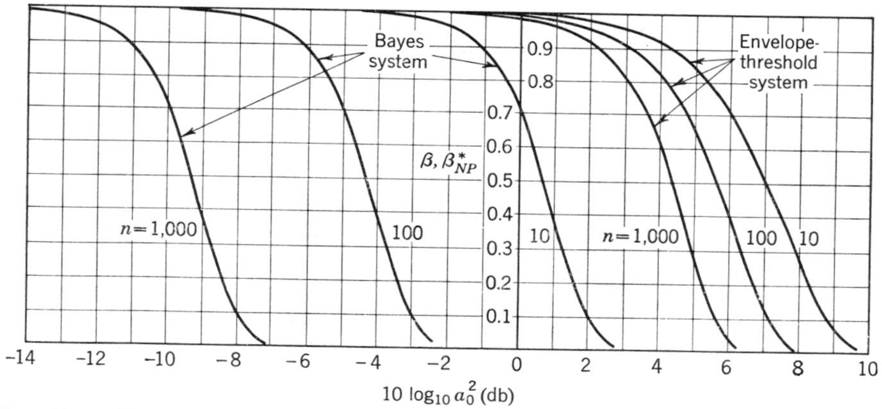

FIG. 20.13. Type II error probabilities β, β^*_{NP} of an envelope-threshold detector and a Neyman-Pearson system (Rayleigh statistics; threshold reception).

1,000, the optimum system has a 6.8-, 10.2-, or 13.6-db advantage, respectively, over the envelope-threshold system. While the minimum detectable signal $(a_0{}^2)^*_{\min\ 0.9}$ is $O(n^{-1/2})$ for the optimum system, the corresponding figure $(a_0{}^2)_{\min\ 0.9}$ is $O(n^{-1/4})$ for the envelope-threshold receiver. The unusually great superiority of the optimum over the envelope-threshold system for weak-signal reception is due primarily to the fact that for weak signals the latter system badly mishandles the data needed for correct decisions. With weak signals, threshold crossings are due primarily to the background noise; the contributions of an additive signal are essentially ignorable. Only when the signal is strong vis-à-vis the noise do we expect an effective distinction between the two states $H_0{:}N$ and $H_1{:}S + N$ to be made by such a system, as then the probability of a threshold crossing at a high level $O(a_0)$ is small for noise alone and large for signal and noise. Note, however, that, while the Bayes system here is much better than this particular suboptimum receiver, this does not, of course, mean that the Bayes receiver will always be an order of magnitude better than other suboptimum systems for this purpose. Those which follow closely the same type and sequence of operations indicated for the optimum (cf. Fig. 20.7) will be within 1 db or so of the optimum [from the point of view of $(a_0^2)_{\min}$]. We say, then, not that the Bayes system here is so good but, rather, that this particular suboptimum one is so poor. (For further remarks on system comparisons, see also Sec. 23.4).

20.4-5 Optimum Incoherent Detection (A Communication Problem). As our next and more involved example of an optimum receiving system, let us consider the following communication problem: It is desired to transmit messages through a noisy channel in finite time. We assume that a suitable (i.e., satisfactory, but not necessarily optimal) encoding of the message ensemble into a signal set has already been carried out and that the signal alphabet (i.e., set of elementary signal waveforms) has been chosen. The encoded messages consist of long sequences of these elementary signal waveforms, each of which lasts T sec, and each of which is accompanied by noise which can be regarded as essentially stationary during these periods [$(0,T)$ sec]. A telegraph signal is one example; radio transmission by a frequency-shift-keyed (FSK) carrier is another; additional applications are readily brought to mind (DFSK, or double FSK, etc.). Reception of a message, correspondingly, is a series of detection operations, resulting in a series of decisions based on the data received in each elementary interval $(0,T)$. Under these conditions, it is sufficient to consider system structure and behavior in a typical period $(0,T)$.

Here, we shall specialize the signal alphabet to just two types of elementary signal waveforms $S^{(1)}(t)$ and $S^{(2)}(t)$, whose precise structure we shall leave open initially. Although still binary detection, our problem is a generalization of the preceding examples, in that now we are required to distinguish optimally between signals of type 1, $S^{(1)}(t)$, in noise and signals of type 2, $S^{(2)}(t)$, during each interval $(0,T)$. Thus, instead of testing the hypotheses $H_1{:}S + N$ of signal and noise versus $H_0{:}N$ of noise alone, we test $H_1{:}S^{(1)} + N$ versus $H_2{:}S^{(2)} + N$. As before, the original background noise is assumed to be normal, additive, and independent of the various signal processes. Furthermore, characteristic of most communication processes, reception is taken to be incoherent, such that both $\langle S^{(1)}(t-\epsilon)\rangle_\epsilon$ and $\langle S^{(2)}(t-\epsilon)\rangle_\epsilon = 0$ here (cf. Secs. 20.1-2, 20.1-3). Proceeding first somewhat generally with threshold systems, we shall specialize our results to simple FSK transmission in white noise, when there may be slow fading of the desired signals† (cf. Fig. 20.14).

(1) Structure. The extension of the theory of Secs. 20.1-2, 20.1-3 to this binary case involving two signals is easily made: we have only to replace the alternative of noise alone by that of a second signal in Eqs.

† For more rapid fading and other situations involving multipath phenomena, see, for example, Price,[27] Turin,[28,28a] Price and Green,[29] Pierce,[42] and Brennan,[43] these last two for an account of diversity reception.

(19.14), (20.1) to write for the optimum detector now

$$x = T_R\{V\} = \log \Lambda_n(\mathbf{V})_{21} = \log [p_2\langle F(\mathbf{V}|\mathbf{S}^{(2)})\rangle_{S^{(2)}}/p_1\langle F(\mathbf{V}|\mathbf{S}^{(1)})\rangle_{S^{(1)}}] \quad (20.143a)$$

For the signal and noise processes assumed here, this becomes specifically

$$\log \Lambda_n(\mathbf{V})_{21} = \log \mu_{21} + \log \langle \exp(-\tfrac{1}{2}a_{02}^2\tilde{\mathbf{s}}_2\mathbf{k}_N^{-1}\mathbf{s}_2 + a_{02}\tilde{\mathbf{v}}\mathbf{k}_N^{-1}\mathbf{s}_2)\rangle_{a_0,\epsilon,\boldsymbol{\theta}} - \log \langle \exp(-\tfrac{1}{2}a_{01}^2\tilde{\mathbf{s}}_1\mathbf{k}_N^{-1}\mathbf{s}_1 + a_{01}\tilde{\mathbf{v}}\mathbf{k}_N^{-1}\mathbf{s}_1)\rangle_{a_0,\epsilon,\boldsymbol{\theta}} \quad (20.143b)$$

with p_1 $(= 1 - p_2)$ the a priori probability of a sample $\mathbf{V}$ containing a signal of type $S^{(1)}$ in noise, etc.

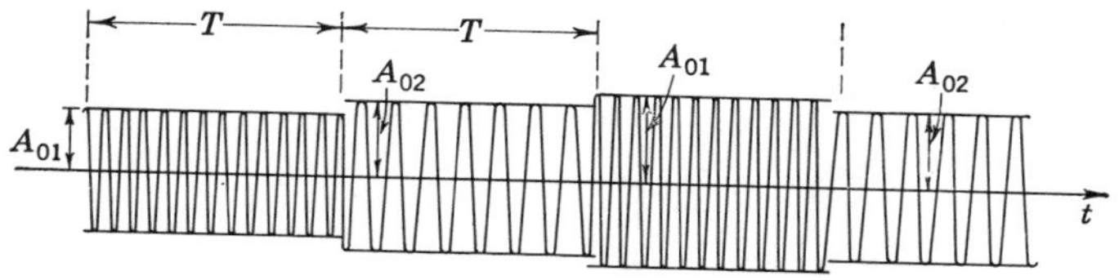

FIG. 20.14. Frequency-shift-keyed signals, with varying amplitudes due to slow fading.

For threshold operation, Eqs. (20.11) may be applied directly for each term in Eq. (20.143b) containing a signal, to give us

$$x = \log \Lambda_n(\mathbf{V})_{21} \doteq \log \mu_{21} + \Gamma_{21} + \tfrac{1}{2}\tilde{\mathbf{v}}\bar{\mathbf{H}}_{21}\mathbf{v} \quad (20.144a)$$

where

$$\Gamma_{21} \equiv -\tfrac{1}{2}(\overline{a_{02}^2}\langle\Phi_s^{(2)}\rangle - \overline{a_{01}^2}\langle\Phi_s^{(1)}\rangle) - \tfrac{1}{4}(\overline{a_{02}^2}^2\langle\Phi_G^{(2)}\rangle - \overline{a_{01}^2}^2\langle\Phi_G^{(1)}\rangle) \quad (20.144b)$$

with $\bar{G}_{1,2} \equiv (\mathbf{k}_N^{-1}\overline{\mathbf{s}\tilde{\mathbf{s}}}\mathbf{k}_N^{-1})_{1,2}$, $\Phi_s^{(1,2)} \equiv (\tilde{\mathbf{s}}\mathbf{k}_N^{-1}\mathbf{s})_{1,2}$, and $\Phi_G^{(1,2)} \equiv (\tilde{\mathbf{s}}\bar{\mathbf{G}}\mathbf{s})_{1,2}$, and where now

$$\bar{\mathbf{H}}_{21} \equiv \overline{a_{02}^2}\,\bar{G}_2 - \overline{a_{01}^2}\,\bar{G}_1 = \tilde{\bar{\mathbf{H}}}_{21} \quad (20.144c)$$

since $\bar{\mathbf{G}} = \tilde{\bar{\mathbf{G}}}$, etc. [cf. Eqs. (20.11b), (20.11c), (20.90a) to (20.90c), etc.]. Again, the characteristic quadratic form of the optimum threshold system is exhibited.

When the signals are narrowband, we can equally well use the alternative results of Sec. 20.1-3 to write Eq. (20.143b) exactly† as

$$\log \Lambda_n(\mathbf{V})_{21} = \log \mu_{21} + \log \langle e^{-a_{02}^2B_n^{(2)}(\boldsymbol{\theta})/2}I_0[a_{02}\sqrt{\tilde{\mathbf{v}}\mathbf{G}_0^{(2)}(\boldsymbol{\theta})\mathbf{v}}]\rangle_{a_0,\boldsymbol{\theta}} - \log \langle e^{-a_{01}^2B_n^{(1)}(\boldsymbol{\theta})/2}I_0[a_{01}\sqrt{\tilde{\mathbf{v}}\mathbf{G}_0^{(1)}(\boldsymbol{\theta})\mathbf{v}}]\rangle_{a_0,\boldsymbol{\theta}} \quad (20.145a)$$

which reduces at once to

$$\log \Lambda_n(\mathbf{V})_{21} = \log \mu_{21} - \left[\frac{a_{02}^2}{2}B_n^{(2)}(\boldsymbol{\theta}_0) - \frac{a_{01}^2}{2}B_n^{(1)}(\boldsymbol{\theta}_0)\right] + \log I_0[a_{02}\sqrt{\tilde{\mathbf{v}}\mathbf{G}_0^{(2)}(\boldsymbol{\theta}_0)\mathbf{v}}] - \log I_0[a_{01}\sqrt{\tilde{\mathbf{v}}\mathbf{G}_0^{(1)}(\boldsymbol{\theta}_0)\mathbf{v}}] \quad (20.145b)$$

† Consistent with the narrowband approximation [cf. Eqs. (20.23) to (20.26), etc.].

when only single values of a_0, $\boldsymbol{\theta}$ can occur [B_n and $\mathbf{G}_0$ are given by Eqs. (20.26), (20.29a)]. When, as is usually the case, a_0, $\boldsymbol{\theta}$ are distributed over several values or a continuum, so that Eq. (20.145a) applies, we may again make the familiar threshold development in powers of a_{02}, a_{01} [cf. Eqs. (20.32), (20.33), and (20.144)] to obtain explicit results. Finally, the expressions above are unchanged in form for continuous data processing in $(0,T)$, since the results of Secs. 20.2-2, 20.2-3 may be applied directly for each component signal term [for narrowband signals, see Eqs. (20.67) in particular].

The structure of this optimum threshold system [Eq. (20.144a)] is a simple combination of the optimum threshold systems for the individual signals $S^{(1)}$ and $S^{(2)}$ described earlier (cf. Secs. 20.1-2, 20.2-2): the input wave is received by identical (broadband) antennas, the output of each

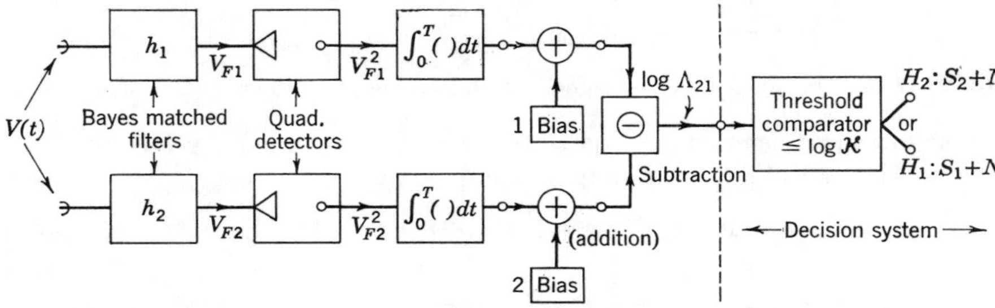

FIG. 20.15. Optimum threshold detection system for determining which of two signals is present in $(0,T)$ for continuous sampling (also, for all signal levels in the case of slow Rayleigh fading).

of which is passed in turn into an appropriate linear (time-varying) Bayes matched filter, zero-memory square-law rectifier, and ideal integrator. The two outputs are then combined [i.e., subtracted, according to Eq. (20.144a)] and the result compared with the threshold log $\mathcal{K}$ for a definite decision: $S^{(1)} + N$ or $S^{(2)} + N$, in the manner of Fig. 20.15. An equivalent system using time-varying filters and simple multipliers (cf. Fig. 20.2) may also be employed here; and in the narrowband case, if $\boldsymbol{\theta} = \boldsymbol{\theta}_0$, so that only a_0 (and ϵ) are random, we may use the alternative structure of Fig. P 20.6 (with a pair of matched filters for each signal), where now postdetection integration is not needed [cf. Prob. 20.6 and discussion following Eqs. (20.21a)]. For details of the discrete and continuous structures, see Sec. 20.1-2, Eqs. (20.12) et seq., Sec. 20.2-2, Eqs. (20.49), (20.58). The Bayes matched filter, square-law device, and integrator embodied in $\tilde{\mathbf{v}}\overline{\mathbf{H}}_{21}\mathbf{v}$ [Eq. (20.144a)] are here just a simple additive combination of the corresponding elements for the individual signals, while individual Bayes matched filters are themselves found by the procedures described in Secs. 20.1-2, 20.2-2. Note that, whereas these Bayes filters are different (since the two signals usually do not have the same waveform, frequency, etc.), the subsequent nonlinear elements and integrating filters should be identical.

A case where the quadratic structure of Eq. (20.144*a*) applies for *all* signal levels arises in the narrowband situation when there is slow fading of the signals. Here, the signal *amplitudes* are found to follow a Rayleigh distribution law† [cf. Eq. (20.125*a*), with E_j replaced by A_{01}, A_{02}, respectively]. Because of the slowness of the fading process vis-à-vis the basic intervals $(0,T)$, we may regard the amplitudes during any period $(0,T)$ as constant, although over longer periods there will be large variations about the steady, nonfading value. Then we find that the optimum system [Eq. (20.145*a*)] becomes

$$\log \Lambda_n(\mathbf{V})_{21} = \log \mu_{21} + \langle \log P_{01}^{-1}(1 + B_n^{(1)}P_{01})\rangle_\theta - \langle \log P_{02}^{-1}(1 + B_n^{(2)}P_{02})\rangle_\theta + \tfrac{1}{2}\tilde{\mathbf{v}}\langle \mathbf{H}_{21}^{(R)}\rangle_\theta \mathbf{v} \tag{20.146a}$$

where now we write

$$\mathbf{H}_{21}^{(R)} = \left[\frac{P_{02}\mathbf{G}_0^{(2)}(\boldsymbol{\theta})}{1 + B_n^{(2)}(\boldsymbol{\theta})P_{02}} - \frac{P_{01}\mathbf{G}_0^{(1)}(\boldsymbol{\theta})}{1 + B_n^{(1)}(\boldsymbol{\theta})P_{01}}\right] \tag{20.146b}$$

with a corresponding expression for continuous sampling obtained as above (Secs. 20.2-2, 20.2-3) for the individual signals.

(2) Error probabilities and Bayes risk. For threshold operation, these are readily obtained from the results of Sec. 20.3-1(2). Here, however, instead of α^*, let us write $\beta_2^{(1)*}$ for the error probability of deciding $S^{(2)}$ when actually $S^{(1)}$ occurs. Similarly, β^* of the earlier analysis becomes $\beta_1^{(2)*}$, the error probability of incorrectly deciding that $S^{(1)}$ occurs when really $S^{(2)}$ is present. In the same way, it is convenient to modify the notation for the preassigned costs: $C_\alpha \to C_2^{(1)}$, $C_\beta \to C_1^{(2)}$, $C_{1-\alpha} \to C_1^{(1)}$, $C_{1-\beta} = C_2^{(2)}$; the superscript notation refers to the true situation, while the subscripts refer to the situation decided upon (cf. Sec. 23.1). The threshold $\mathcal{K}$ is accordingly $\mathcal{K} = (C_2^{(1)} - C_1^{(1)})/(C_1^{(2)} - C_2^{(2)})$. Then the characteristic functions of the d.d.'s of x [Eq. (20.144*a*)] with respect to the hypotheses $H_1\!:S^{(1)} + N$ and $H_2\!:S^{(2)} + N$ are found with the help of Eqs. (20.86), (20.87), (20.144) to be

$$F_1(i\xi)_{H_1,H_2} = \boldsymbol{E}_{\mathbf{V}|H_1,H_2}\{\exp i\xi(\log \mu_{21} + \Gamma_{21} + \tfrac{1}{2}\tilde{\mathbf{v}}\overline{\mathbf{H}}_{21}\mathbf{v})\} \doteq \exp\left(i\xi C_n^{(1,2)} - \tfrac{1}{2}\xi^2 D_n^{(1,2)}\right) \tag{20.147a}$$

where (with $\Phi_G^{(12)} \equiv \tilde{\mathbf{s}}_1\overline{\mathbf{G}}_2\mathbf{s}_1$, $\Phi_G^{(21)} \equiv \tilde{\mathbf{s}}_2\overline{\mathbf{G}}_1\mathbf{s}_2$, and $\langle\Phi_G^{(12)}\rangle = \langle\Phi_G^{(21)}\rangle$)

$$C_n^{(1)} = \log \mu_{21} - \tfrac{1}{4}\overline{a_0^2}^2\Psi_H \qquad C_n^{(2)} = \log \mu_{21} + \frac{\overline{a_0^2}^2}{4}\Psi_H$$

$$D_n^{(1)} = D_n^{(2)} = \tfrac{1}{2}\overline{a_0^2}^2\Psi_H \tag{20.147b}$$

† This is also true of the case where the distribution of amplitudes is not known to the observer, who then chooses a Rayleigh distribution, i.e., adopts the subcriterion of the greatest average uncertainty in amplitude, subject to a fixed average power limitation $(\overline{a_0^2} = 2P_0)$ (cf. Prob. 6.7).

and

$$\Psi_H \equiv (\overline{a_{02}^2}^2\langle\Phi_G^{(2)}\rangle - 2\overline{a_{01}^2}\,\overline{a_{02}^2}\langle\Phi_G^{(12)}\rangle + \overline{a_{01}^2}^2\langle\Phi_G^{(1)}\rangle)/\overline{a_0^2}^2 \qquad (20.147c)$$

[Here $\overline{a_0^2}$ (> 0) is some convenient normalizing square of an amplitude, say $\overline{a_0^2} = \overline{a_{01}^2}$ or $\overline{a_{02}^2}$, etc.] The d.d.'s are consequently normal, with the above means and variances [cf. Eqs. (20.147*b*)], and the desired error probabilities are similar to the results of Eq. (20.91), viz.,

$$\begin{Bmatrix}\beta_2^{(1)*} \\ \beta_1^{(2)*}\end{Bmatrix} \simeq \frac{1}{2}\left\{1 - \Theta\left[\frac{\overline{a_0^2}\Psi_H^{1/2}}{4} \pm \frac{\log(\mathcal{K}/\mu)}{\overline{a_0^2}\Psi_H^{1/2}}\right]\right\} \qquad (20.148a)$$

while the Bayes risk here is now specifically

$$R^* = \mathcal{R}_0 + p_2(C_1^{(2)} - C_2^{(2)})\left(\beta_2^{(1)*}\frac{\mathcal{K}}{\mu_{21}} + \beta_1^{(2)*}\right) \qquad (20.148b)$$

with $\mathcal{R}_0 = p_1C_1^{(1)} + p_2C_2^{(2)}$. Observe again the characteristic dependence of $(\overline{a_0^2})^*_{\min}$ on $n^{-1/2}$ (or $T^{-1/2}$) [cf. Eqs. (20.92) et seq.].

With continuous sampling, Ψ_H takes the following forms [cf. Eq. (20.91*a*), Prob. 20.7*b*], for $\psi < \infty$,

$$\begin{aligned}\Psi_H = \psi^2\overline{A_0^2}^{-2}\iint_{-\infty}^{\infty} [\overline{A_{02}^2}^2\langle D_{2T}(t,u;\epsilon,\boldsymbol{\theta})\rangle\langle s_2(t-\epsilon;\boldsymbol{\theta})s_2(u-\epsilon;\boldsymbol{\theta})\rangle \\ - 2\overline{A_{01}^2}\,\overline{A_{02}^2}\langle D_{1T}(t,u;\epsilon,\boldsymbol{\theta})\rangle\langle s_1(t-\epsilon;\boldsymbol{\theta})s_2(u-\epsilon;\boldsymbol{\theta})\rangle \\ + \overline{A_{01}^2}^2\langle D_{1T}(t,u;\epsilon,\boldsymbol{\theta})\rangle\langle s_1(t-\epsilon;\boldsymbol{\theta})s_1(u-\epsilon;\boldsymbol{\theta})\rangle]\,dt\,du \qquad (20.149a)\end{aligned}$$

where D_{1T}, D_{2T} are given by Eqs. (20.46*b*), with X_{1T}, X_{2T}, etc., obtained from the basic integral equation (20.45) when s_1, s_2 are, respectively, inserted therein ($\overline{a_{01}^2} = \overline{A_{01}^2}/2\psi$, etc.). For white noise ($\psi \to \infty$), we get [cf. Eq. (20.91*b*)] the not surprising result

$$\begin{aligned}\overline{a_0^2}^2\Psi_H = W_0^{-2}\iint_0^T [\overline{A_{02}^2}\langle s_2(t-\epsilon;\boldsymbol{\theta})s_2(u-\epsilon;\boldsymbol{\theta})\rangle \\ - \overline{A_{01}^2}\langle s_1(t-\epsilon;\boldsymbol{\theta})s_1(u-\epsilon;\boldsymbol{\theta})\rangle]^2\,dt\,du \qquad (20.149b)\end{aligned}$$

These relations, as well as Eqs. (20.148), hold for both broad- and narrowband signals. For the latter, we may in the narrowband approximation replace $D_{2T}\langle s_2s_2\rangle$, etc., in Eq. (20.149*a*) by

$$\tfrac{1}{4}(D_{2T})_0\langle F_{s_2}(t)F_{s_2}(u)\cos[\omega_{s_2}(t-u) - \phi_{s_2}(t) + \phi_{s_2}(u)]\rangle$$

etc. [cf. Eqs. (20.93*a*), (20.93*b*)], with a corresponding modification of Eq. (20.149*b*) [cf. Eq. (20.93*c*) and Sec. 20.3-2]. Finally, these results also apply in the case of slow Rayleigh fading [cf. Eqs. (20.146)], however, only for weak signals on the average ($\overline{a_0^2}^2 = 4P_0^2 < 1$). Curves of $\beta_1^{(2)*}$ versus $\beta_2^{(1)*}$ and Bayes risk for this general two-signal system are given (in the threshold case, of course) by Fig. 20.4a, b, respectively [cf. Eq. (20.120) vs. Eq. (20.93)], if we replace β^* by $\beta_1^{(2)*}$, α^* by $\beta_2^{(1)*}$, and $a_0\Phi_s^{1/2}\sqrt{2}$ (in Fig. 20.4a) by $\overline{a_0^2}\Psi_H^{1/2}$ and by $\overline{a_0^2}\Psi_H^{1/2}/\sqrt{2}$ in Fig. 20.4b. [See also the comments following Eq. (20.121).] Neyman-Pearson or Ideal Observer operation is also possible: for the former, we fix $\beta_2^{(1)*}$, thus determining $\mathcal{K}$, with $\beta_1^{(2)*} \to (\beta_1^{(2)*})_{NP}$; for the latter, $\mathcal{K}$ is unity. Figures 20.8 to 20.12 may be applied here, too, with obvious adjustments of the ordinates and parameters. Observe, in any case, that these results hold for general signals and normal noise processes.

(3) **FSK communication in white noise.** As a special case of the analysis above, let us consider the problem of the threshold reception of FSK signals (Fig. 20.14), where there may or may not be slow Rayleigh fading. Here the two signals are sinusoidal wave trains of constant amplitude $A_{01} = A_{02}$ (or $\overline{A_{01}^2} = \overline{A_{02}^2}$), such that $F_{s1} = F_{s2} = \sqrt{2}$, and $\phi_{s1} = \omega_1 t = -\phi_{s2}$; that is, $\omega_s - \omega_1$ is the carrier (angular) frequency (for $S^{(1)}$), $\omega_s + \omega_1$ is the other frequency (for $S^{(2)}$) [cf. Eq. (20.22)], while ω_s is the "center" frequency of the signal system. Then, with continuous sampling in white noise the optimum detector [Eq. (20.144a)] is easily found, with the help of Eq. (20.67) for a single signal, to be

$$x = \log \Lambda_T(V) \doteq \log \mu_{21} + (\Gamma_{21})_T + \frac{\overline{A_0^2}}{W_0^2} \iint_0^T V(t)V(u) \times [\cos(\omega_s + \omega_1)(t - u) - \cos(\omega_s - \omega_1)(t - u)]\, dt\, du \quad (20.150)$$

where the bias is specifically [from Eqs. (20.144), (20.67d), (20.67e)]

$$(\Gamma_{21})_T = -\frac{1}{2}\left(\frac{\overline{A_0^2}T}{2W_0}\right)^2 \left\{\left[\frac{\sin(\omega_s + \omega_1)T}{(\omega_s + \omega_1)T}\right]^2 - \left[\frac{\sin(\omega_s - \omega_1)T}{(\omega_s - \omega_1)T}\right]^2\right\} \quad (20.150a)$$

The moduli of the system functions of the time-invariant portion of the Bayes matched filters preceding the quadratic rectifiers in Fig. 20.15 are readily found from Eqs. (20.69b) for these values of F_s and ϕ_s. We get

$$\left(\int_0^T |Y_{01}(i\omega;x)|^2\, dx\right)^{1/2} = 2^{5/2} W_0^{-1} T^{-1/2} \left|\frac{\sin(\omega - \omega_1)(T/2)}{\omega - \omega_1}\right| \quad (20.150b)$$

with a similar expression for $|Y_{02}(i\omega)_T|$.

The quantity $\overline{a_0^2}^2\Psi_H$ [Eq. (20.149b)] likewise reduces to

$$\overline{a_0^2}^2\Psi_H = (4W_0^2)^{-1} \iint_0^T \{\overline{A_{02}^2}\langle F_{s2}(t;\boldsymbol{\theta}_1)F_{s2}(u;\boldsymbol{\theta}_1) \cos[\omega_{s2}(t - u) - \phi_{s2}(t;\boldsymbol{\theta}_2) + \phi_{s2}(u;\boldsymbol{\theta}_2)]\rangle - \overline{A_{01}^2}\langle F_{s1}(t;\boldsymbol{\theta}_1)F_{s1}(u;\boldsymbol{\theta}_1) \cos[\omega_{s1}(t - u) - \phi_{s1}(t;\boldsymbol{\theta}_2) + \phi_{s2}(u;\boldsymbol{\theta}_2)]\rangle\}^2\, dt\, du \quad (20.151a)$$

$$= \frac{\overline{A_0{}^2}^2}{W_0{}^2} \iint_0^T [\cos (\omega_s + \omega_1)(t - u) - \cos (\omega_s - \omega_1)(t - u)]^2 \, dt \, du \tag{20.151b}$$

$$= \left(\frac{\overline{A_0{}^2}T}{W_0}\right)^2 \left\{1 + \left[\frac{\sin (\omega_s + \omega_1)T}{(\omega_s + \omega_1)T}\right]^2 + \left[\frac{\sin (\omega_s - \omega_1)T}{(\omega_s - \omega_1)T}\right]^2 - \left(\frac{\sin \omega_s T}{\omega_s T}\right)^2 - \left(\frac{\sin \omega_1 T}{\omega_1 T}\right)^2\right\} \doteq \left(\frac{\overline{A_0{}^2}T}{W_0}\right)^2 \tag{20.151c}$$

for the large samples (i.e., large T) required for threshold detection.

The curves of Fig. 20.4*a*, *b* may be applied once more if we replace $a_0\Phi_s{}^{1/2}\sqrt{2}$ by $\overline{a_0{}^2}\Psi_H{}^{1/2}$ in the former and $a_0\Phi_s{}^{1/2}$ by $\overline{a_0{}^2}\Psi_H{}^{1/2}/\sqrt{2}$ in the latter. Figure 20.4*a* then gives $\beta_1{}^{(2)*}$ versus $\beta_2{}^{(1)*}$, while Fig. 20.4*b* shows the Bayes risk [Eq. (20.148*b*)] for this particular Ψ_H [Eqs. (20.151*a*) et seq.]. Similarly, replacing $\overline{a_0{}^2}\sqrt{n}$ by $\overline{a_0{}^2}\Psi_H{}^{1/2}/\sqrt{2}$ in Fig. 20.8 gives curves of the normalized Bayes risk in the present instance. Fixing $\beta_2{}^{(1)*} = (\beta_2{}^{(1)})_{NP}$, we can equally well treat this system now as a Neyman-Pearson observer, $\mathcal{K} \to \mathcal{K}_{NP}$, with appropriate modifications of Eqs. (20.134), (20.135*a*). Results for an Ideal Observer may also be obtained on setting $\mathcal{K} = 1$. The curves of Figs. 20.10 to 20.12 then apply here, too, with the indicated changes in ordinates and parameters. Finally, we note that it is possible to compare suboptimum FSK systems with the optimum above, specifically, by using the approach of Sec. 20.3-2 and choosing the appropriate analogue of Eq. (20.144*a*) [cf. Eq. (20.94)] for each signal.

20.4-6 A General Radar Detection Problem. As one of our final examples of a detailed application of decision theory to detection, let us consider the following radar problem, where we considerably broaden the model assumed earlier in Sec. 20.4-3. Specifically, we are concerned now with the design and evaluation of an optimum threshold receiver for the detection of (single) targets in a noise background, when the signal waveforms are given, the noise is originally normal, and certain components of the receiver, such as the antenna, are specified. The radar here is operating solely in its search phase, so that only its detection features are examined; the system is not simultaneously required to perform any estimations of target position, velocity, etc. However, other, rather general conditions of operation are assumed: the target may be moving, with an unknown velocity, at an unknown position in an area large compared with an antenna beam width,† and apart from system noise N_S there may be the additional interference of external noise N_E such as ground or sea clutter.

Under the above circumstances, reception is incoherent (i.e., there is

† Or, more precisely, in a volume large compared with the beam volume. We shall restrict the model somewhat by considering two-dimensional patterns only, although the approach can be generalized in straightforward fashion.

sample uncertainty; cf. Sec. 19.4-3). The optimum threshold system here has already been determined in a general way (cf. Secs. 20.2-2, 20.2-3). It consists of a suitable Bayes matched filter, zero-memory square-law rectifier, and ideal integrator followed by a threshold comparator, in the manner of Figs. 20.2*a* and 20.16. However, unlike most of our earlier examples, we observe that now the amplitude A_0 of a received signal depends on target location in the area scanned; a given target returns a stronger signal near the receiver than far away from it, of course, so that the magnitude of A_0 is governed not only by target size but by position as well. We expect system structure, in particular the Bayes matched filter, to be influenced accordingly and hence that the extensions of the analysis of Secs. 20.1 to 20.3 given in Probs. 20.2 to 20.5 will be needed here, where the principal task is to incorporate the various features of our particular model into the framework of the general theory.

Figure 20.16 shows the communication situation, with the optimum threshold receivers in some detail: the interrogations $\{u\}$ are encoded as a set of signals $S = \boldsymbol{T}_T\{u\}$, which go out through the medium $\{\boldsymbol{T}_M^{(N)}\}$ to a possible target (also considered to be a part of the medium). A portion of the reflected signals from the target are then intercepted by the receiver, $\boldsymbol{T}_R^{(N)}$, which yields finally a set of decisions $\{v\}$ as to the presence or absence of this target. Noise is injected at two places: in the medium, e.g., as clutter or other interference, and as system noise in the receiver, essentially in the RF stage or matched filter, which being linear may be regarded as a sequence consisting of a noise source and a noise-free element (cf. Sec. 5.3-5). In the case of clutter, the external noise is also influenced by the signal waveform, as indicated by the dashed line in Fig. 20.16, for $N_E = S \times C$, and is, in fact, a multiplicative rather than an additive process (cf. Probs. 11.10 to 11.14 and Chap. 11, Refs. 47 to 49).

(1) The model. Let us begin by considering the salient features of the signal and noise processes involved. These may be briefly summarized under the headings of signal structure, scanning procedure, noise and system parameters.

Signal structure. A general narrowband signal process [cf. Eq. (20.22)] is usually employed. Frequently, this consists of a periodically pulsed carrier, where the pulsing may or may not be incoherent [see discussion following Eq. (20.130)]. Since an area large compared with a beam width is under scrutiny, the antenna responses of transmitter and receiver will modulate the emitted wave and that returned from a desired or undesired (e.g., clutter) target. In addition to this, there is the variation with time of the desired target's return σ_s, in the direction of the receiver, due to target scintillation, etc. If $S(t;\boldsymbol{\theta}) = A_0 f_s(t;\boldsymbol{\theta}_1) \cos [\omega_0 t - \psi_s(t;\boldsymbol{\theta}_2)]$ is the narrowband signal process generated at the transmitter, *before* passing through the transmitter antenna, then we can write finally for the received signal at A,

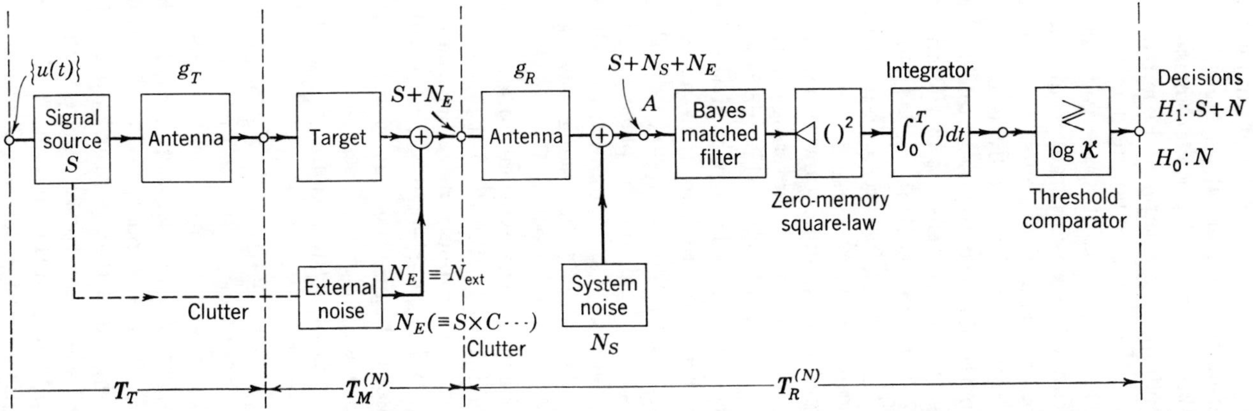

FIG. 20.16. An optimum radar receiver for binary threshold detection and continuous sampling with transmitter, single target, and external noise source.

Fig. 20.16, when there is a target,

$$S_R(t;\epsilon,\boldsymbol{\theta}) = A_0\sigma_s\left(t - \frac{\epsilon}{2}\right) g_T(t - \epsilon)g_R(t)f_s(t - \epsilon;\, \boldsymbol{\theta}_1) \times \cos\,[\omega_s(t - \epsilon) - \psi_s(t - \epsilon;\, \boldsymbol{\theta}_2) - \psi_g(t;\epsilon)_{TR}] \quad (20.152a)$$

where $\epsilon = 2R/c$, with R the range and c the velocity of propagation. Here, $\omega_s = \omega_0 + \omega_{ds}$, where ω_{ds} is the (angular) Doppler frequency of the moving target with respect to transmitter and receiver (assumed in this instance to be located at the same point), and g_T, g_R, $(\psi_g)_{TR}$ are the amplitude and phase modulations of the original signal process, produced in the course of scanning by the antennas at transmitter and receiver. These effects, of course, depend on the geometry of the problem and on the physical structure of the equipment involved.

Next, let us express the epoch ϵ as $\epsilon = \epsilon' + \epsilon_0$, where ϵ_0 is the epoch over a cycle of the carrier frequency f_0 (or f_s);† ϵ' is accordingly measured in multiples of RF cycles and embodies the uncertainty as to target position in range and angle (i.e., azimuth). Because of the assumed narrowband nature of the original signal process, we can therefore write the received signal [Eq. (20.152a)] in normalized form [cf. Eq. (20.22)] as

$$s(t;\epsilon,\boldsymbol{\theta}) = F_s(t;\epsilon',\boldsymbol{\theta}_1') \cos\,[\omega_s(t - \epsilon_0) - \phi_s(t,\epsilon';\boldsymbol{\theta}_2')] \quad (20.152b)$$

[cf. Eq. (20.22)], where $F_s \equiv \sigma_s g_T g_R f_s$ and $\phi_s \equiv \psi_s - \psi_g$, and we have explicitly noted the dependence on target position by ϵ', for example, $\boldsymbol{\theta}_1 \equiv (\epsilon',\boldsymbol{\theta}_1')$, $\boldsymbol{\theta}_2 = (\epsilon',\boldsymbol{\theta}_2')$, etc., with $\boldsymbol{\theta}_1'$, $\boldsymbol{\theta}_2'$ any other possibly random parameters of the signal. Here, F_s, ϕ_s are general, slowly varying functions of time. Figure 20.17, on the other hand, shows a typical signal return from a target (smaller than the beam width) when a pulsed carrier is used with scanning and the side-lobe pattern of the composite transmitter-receiver antenna responses is not symmetrical. Here, T is the width (in time) of the scanning beam, for example, $T = \Delta\theta/\dot{\theta}_s$, where $\Delta\theta$ is the corresponding angular width and $\dot{\theta}_s$ is the (constant) scanning rate. The actual magnitude of T (and $\Delta\theta$) depends on the antenna responses: T starts and stops when the effective return from a target vanishes or is some very small fraction of the main-lobe return.

Scanning. The scanning procedure is indicated in Fig. 20.18. The area in which it is desired to determine the presence or absence of a target is here large (in angle) compared with the effective beam width $\Delta\theta$ and for convenience is taken to be a single multiple of $\Delta\theta_1$, that is, $\theta_2 - \theta_1 = M\,\Delta\theta$; in depth, the sector is also long compared with target size, $\epsilon_2 - \epsilon_1 \gg \epsilon_T$, and in the cases where pulsed carriers are used may consist of several pulse

† Note in Eq. (20.152a) that $\epsilon_0\omega_s = \epsilon_0\omega_0(1 + \omega_{ds}/\omega_0) \doteq \epsilon_0\omega_0$, since $\omega_{ds}/\omega_0 \ll 1$ in all practical cases.

repetition periods T_0 although for other reasons (e.g., range determination) T_0 may be chosen greater† than $\epsilon_2 - \epsilon_1$.

The decision process, which is the final output of the system (cf. Fig. 20.16), may be handled in a variety of ways. For example, at the end of each acquisition period T, a decision may be obtained so that a sequence of

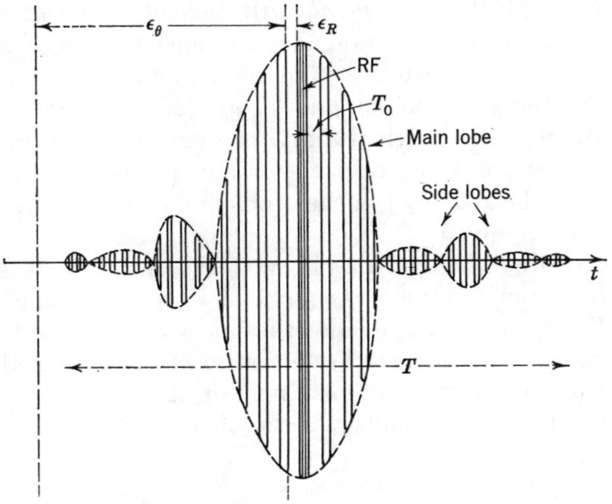

FIG. 20.17. Returned signal from a target (pulsed carrier with scanning).

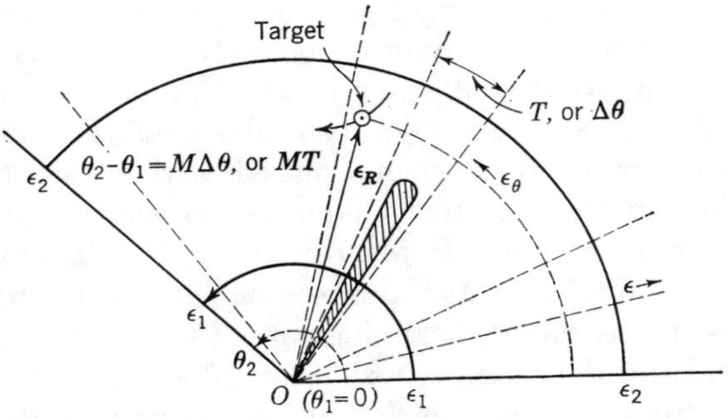

FIG. 20.18. A sector ($\epsilon_2 - \epsilon_1$, $\theta_2 - \theta_1$) scanned for presence or absence of a target.

decisions, covering the sector, is recorded. On the other hand, we may generate a "continuum" of decisions by using the delay-line read-off for the integrator in the optimum threshold system above [cf. footnote referring to Eq. (20.54*b*)], with, in effect, a continuously moving average over T sec of the past and a consequent continuous output of binary decisions, the

† Because of the rapid fall-off of target return with distance ($\sim\epsilon^{-2}$), we can often set $\epsilon_2 \to \infty$.

particular procedure being determined in part by the ultimate uses of those decisions.

Noise. The accompanying noise consists at least of the inherent noise of the receiver, which for practical purposes can be regarded as arising at A, Fig. 20.16, before the nonlinear element of the system (as discussed in Sec. 5.3-5). This noise process is essentially stationary, as well as normal, and in many cases may be considered to have a white spectrum. When there is an external noise component, nonstationary processes can occur. Ground and sea clutter, for example, exhibit a characteristic nonstationarity and are, moreover, dependent on the waveform of the illumination from the transmitter. Typical results for the covariance functions of these multiplicative processes have been described in Probs. 11.10 to 11.14, in the common situations where the basic scattering mechanism leads to normal statistics. The covariance function of the interfering noise $N(t)$ can accordingly be expressed as $K_N(t,u) = K_{N_E}(t,u) + K_{N_S}(|t-u|)$, since $N(t) = N_E(t) + N_S(t)$ and N_E, N_S are statistically independent. The precise form of $K_{N_E}(t,u)$ in the case of clutter depends on the signal waveforms used (cf. Probs. 11.12, 11.13). This is true of distributed clutter return, as well as the return from fixed ranges, although the explicit forms of the covariance functions are, of course, different. The former depend on both the horizontal and vertical beam patterns of the antennas involved, on the scanning procedure, and on the distribution of scatterers over the surface illuminated by the transmitter and subtended by the receiver. While a detailed calculation of the covariance functions under these conditions is outside the immediate scope of this example, we can, however, carry through the general features of the present analysis on the assumption that $K_{N_E}(t,u)$ is either known or obtainable.

Signal parameters. As mentioned at the beginning of this section, the amplitude of the target return depends on range, that is, $A_0 = A_0(\epsilon')$. More specifically, if a pulsed carrier is used (cf. Fig. 20.17), we can further anatomize the epoch ϵ', writing $\epsilon' = \epsilon_R + \epsilon_\theta$, where $0 \leq \epsilon_R \leq T_0$, the pulse repetition period, and ϵ_θ is measured in units of T_0, this last since scanning rate is much smaller than the PRF (pulse repetition frequency $= T_0^{-1}$) in most instances. Thus, ϵ_R locates the target in range (on a time scale), while ϵ_θ locates it similarly in angle (cf. Figs. 20.17, 20.18). The (normalized) amplitude a_0 becomes, then, $a_0(\epsilon_R)$, and two types of statistical variations of a_0 (or A_0) are now distinguished: a distribution over the magnitude of a_0 (at any given range), representing targets of different possible strength, and the dependence on range, ϵ_R, which is also a random variable as far as the receiver is concerned. We write accordingly $A_0 = \alpha_0/\epsilon_R^2$, where α_0 is independent of ϵ_R and equal to a target return at some standard range; α_0 has some distribution over the possible target returns.

Another signal parameter of interest is the carrier frequency f_s of the target return. Because of the unknown velocity of the target vis-à-vis the receiver (and transmitter), the Doppler frequency f_{d_s} of the target is also a random variable here, subject to some distribution law. However, it is assumed here that ground velocity is known, so that the Doppler of the ground clutter return, if it appears, shall in $K_{N_E}(t,u)$ appear as a simple parameter with a definite value. Finally, since the distributions of target amplitude, position, and Doppler are not generally available to the observer, he may choose to optimize his decision process with respect to some subcriterion governing these distributions (cf. Sec. 19.2). A Minimax procedure is one possibility (Sec. 19.2-3), and maximization of the average uncertainty $\langle -\log w\rangle$ for these parameters subject to various constraints is another (see Prob. 6.7). In this way, the indicated statistical averages may be obtained, and the explicit form of the resulting optimum system is then given.

(2) **Structure and average performance.** Figure 20.16 shows a typical optimum threshold receiver (under the usual condition of a uniform distribution of RF phase; cf. Sec. 20.1-3). The specific form of the Bayes matched filter depends, of course, on the signals used, the spectral character of the (normal) noise interference, and the distributions of the various signal (and noise) parameters, as mentioned above. The antenna responses, the acquisition time T—determined here essentially by the antenna patterns and the scanning procedure—target scintillation, etc., also play an important part in specifying the precise nature of this linear prerectifier filter.

The threshold structure of the detector is given by Prob. 20.5, Eqs. (3) to (6), and Prob. 20.7, Eq. (5), while the modulus of the matched filter's system function follows directly from Eq. (20.69), when the basic integral equations (20.61b) have been solved to give us ρ_{G_0} [Prob. 20.5, Eq. (4)]. Here, $a(t;\epsilon,\boldsymbol{\theta}) = F_s(t;\epsilon',\boldsymbol{\theta}_1') \cos [\omega_s t - \phi_s(t;\epsilon',\boldsymbol{\theta}_2')]$,

$$b(t;\epsilon,\boldsymbol{\theta}) = F_s(t;\epsilon',\boldsymbol{\theta}_1') \sin [\omega_s t - \phi_s(t;\epsilon',\boldsymbol{\theta}_2')]$$

[cf. Eq. (20.152b)], and $K_N(t,u) = K_{N_E}(t,u) + (W_0/2)\delta(t-u)$ for a background consisting of system noise and clutter. When the background is white noise alone, we may use Eqs. (7) to (10) of Prob. 20.5 specifically for the structure.

The error probabilities and Bayes risk follow directly from Eqs. (20.91a), (20.91b) in Eq. (20.91), or (20.93a), (20.93b) in Eq. (20.93), with the modifications of Probs. 20.5, 20.7 imposed by the dependence of amplitude a_0 on range. Replacing $D_T(t,u;\epsilon,\boldsymbol{\theta})$ by $a_0^2 D_T(t,u;\epsilon,\boldsymbol{\theta})/\overline{a_0^2}$, we have for $(\Phi_T)_G$ [Eq. (20.91a)] the somewhat more general expression [cf. Prob. 20.7, Eqs. (4), (6)]

$$\overline{a_0^2}\langle\Phi_T\rangle_G = \psi^2 \iint_{-\infty}^{\infty} \langle a_0^2 D_T(t,u;\epsilon,\boldsymbol{\theta})\rangle\langle a_0^2 s(t-\epsilon,\ \boldsymbol{\theta}) s(u-\epsilon;\ \boldsymbol{\theta})\rangle \, dt\, du \qquad (20.153a)$$

which for white noise alone is modified to

$$\overline{a_0^2}^2\langle\Phi_T\rangle_G = W_0^{-2}\iint_0^T \langle A_0^2(\epsilon')s(t-\epsilon,\boldsymbol{\theta})s(u-\epsilon,\boldsymbol{\theta})\rangle^2\,dt\,du \tag{20.153b}$$

with s in either instance given by Eq. (20.152b) and $\epsilon = \epsilon' + \epsilon_0$. Equivalent expressions in the explicit narrowband form may be obtained by a corresponding extension of Eqs. (20.93b), (20.93c). The curves of Figs. 20.4a, b, 20.8 to 20.12 once again may be applied here, with obvious changes of abscissa and parameters, to give us the type I and II error probabilities, the associated Bayes risk, and minimum detectable signal (cf. Secs. 20.4-1, 20.4-3) for threshold operation.

When the narrowband signal consists of a pulse-modulated carrier, we can represent $\langle a_0^2 s(t-\epsilon,\boldsymbol{\theta})s(u-\epsilon,\boldsymbol{\theta})\rangle$ explicitly. Let us assume, in addition to the usually adopted criterion of a uniform distribution of RF epoch (Sec. 20.1-3), that the Doppler frequency of the target is uniformly distributed† between the angular frequency limits $\omega_0 \pm \Omega$.

Now, the pulse-modulated carrier s can be written here as

$$s(t-\epsilon;\boldsymbol{\theta}) = \left\{\sum_{m=1}^{M} f_s[t-(m-1)T_0-\epsilon_R]G(t;\epsilon_\theta)\right\}\cos\left[\omega_s(t-\epsilon_0)-\phi_s(t;\epsilon_\theta)\right] \tag{20.154}$$

with $f_s(t) = 0$ except for $0 < t < \tau$, where τ is the width (suitably defined) of a typical pulse form f_s; $G(t;\epsilon_\theta) = \sigma_s(t-\epsilon_\theta/2)g_T(t-\epsilon_\theta)g_R(t)$; and $\phi_s = \psi_g(t,\epsilon_\theta)_{TR}$ [cf. Eq. (20.152a)]. Because the sweep rate, though comparable to that of target scintillation, is slow compared with the pulse repetition rate, which in turn is slowly varying with respect to the carrier oscillations, we can neglect the effects of ϵ_0 in f_s, G, and ϕ_s and ϵ_R in G and ϕ_s, as indicated. The average $I(t,u) \equiv \langle a_0^2 s(t-\epsilon;\boldsymbol{\theta})s(u-\epsilon;\boldsymbol{\theta})\rangle$ takes different forms, depending on whether or not RF coherence is maintained from pulse to pulse. Where RF phase coherence is maintained, corresponding physically to a synchronous on-off switching of a continuously running oscillator, we find, after carrying out the average over the Doppler frequency ω_{ds} $(=\omega_s - \omega_0)$ with the help of

$$\int_{\omega_0-\Omega}^{\omega_0+\Omega}\cos(\omega_s\tau-\Phi)\frac{d\omega_s}{2\Omega} = \frac{\sin\Omega\tau}{\Omega\tau}\cos(\omega_0\tau-\Phi) \tag{20.154a}$$

and the average over the uniformly distributed‡ ϵ_0, that

† Again, this is the distribution yielding the greatest average uncertainty in ω_{ds}, subject to a maximum value constraint $|\omega_{ds}|_{\max} = \Omega$ (cf. Prob. 6.7).

‡ See footnote on p. 919 referring to Eq. (20.152a).

$$I(t,u)_{coh} = \frac{1}{2}\frac{\sin \Omega(t-u)}{\Omega(t-u)} \left\langle \sum_{m,n=1}^{M} a_0{}^2(\epsilon_R) f_s[t-(m-1)T_0-\epsilon_R] \right.$$
$$\times f_s[u-(n-1)T_0-\epsilon_R]$$
$$\left. \times G(t;\epsilon_\theta)G(u;\epsilon_\theta)\cos[\omega_0(t-u)-\phi_s(t;\epsilon_\theta)+\phi_s(u;\epsilon_\theta)] \right\rangle \quad (20.155)$$

With incoherent pulse generation, such as occurs when the oscillator itself is turned on and off at the repetition frequency, we must modify Eq. (20.154), writing ϵ_{0m}, ϵ_{0n} in place of ϵ_0, where now the ϵ_{0m}, ϵ_{0n} are not only uniformly distributed over an RF cycle but are independent from pulse to pulse ($m \neq n$). The result is that $I(t,u)_{incoh}$ is given by Eq. (20.155), but now with a single series over m in place of the double series in Eq. (20.155).† A similar sort of calculation can be carried out for $\langle a_0{}^2 D_T(t,u;\epsilon,\boldsymbol{\theta})\rangle$, as soon as the basic integral equations (20.61*b*) have been solved, to give us Eqs. (20.153) for pulsed carriers. With white noise interference, in fact, Eq. (20.153*b*) becomes $\overline{a_0{}^2}^2\langle\Phi_G\rangle_G = W_0{}^{-2}\iint_0^T I(t,u)^2\,dt\,du$, with the appropriate $I(t,u)$ for coherent and incoherent pulse trains.

Finally, we emphasize that the present system is optimum for detection only; it is not constrained to carry out simultaneously the estimation operations implied in a tracking process, although the choice of a pulsed carrier signal makes such range estimation possible. Because it is thus unconstrained, it is theoretically superior to mixed systems, also operating on the same a priori information, which require simultaneously (range or velocity) estimation as well as detection. The latter, however, with good design are often close to the performance levels of the former, and their versatility frequently more than outweighs their inferiority in the purely detection, or search, phase of operation. For this reason, one does not necessarily choose the former unless the particular application demands detection only. Observe, in either instance, that the effective integration time T is determined by the beam width and rate of scan (for targets smaller than a beam width), since longer intervals can add only noise, without the desired signal. How the received data are handled within this interval depends, of course, on the signal process itself, as well as on the noise: for pulsed signals, for example, we expect the weighting function of the Bayes filter to vanish for those epochs during which the target return is zero, because of the pulsed character of the transmitter output. With continuous signals, on the other hand, we expect essentially nonvanishing weighting functions, and so on. For remarks on the choice of an optimum signal process for detection, see Sec. 23.2; observe also that the evaluation and comparison of sub-

† This is one result in the continuous version of the "one-point" theory outlined earlier, in Sec. 20.4-3 [cf. Eqs. (20.127), (20.130)].

optimum systems here may be carried out along the lines indicated in Sec. 20.3-2.

20.4-7 Stochastic Signals in Normal Noise.† As a final example of optimum detection with fixed observation times, let us consider the situation where the signal, like the background noise, is itself a normal process. Signals of this type occur in scatter-path communication, when originally deterministic signals are converted by the medium $T_M^{(N)}$ into (narrowband) random disturbances; FSK communication under scatter conditions is typical. Another example arises in radio spectroscopy, where the signal source is represented by one or more spectral lines, which because of collision broadening and similar atomic effects have a definite width, while the radiation process by which these lines are generated may be regarded as normal from a macroscopic viewpoint. A similar phenomenon occurs in radio astronomy. Here the signal sources are radio stars or gas clouds, whose signals are in effect normal noise, with a spectrum of variable intensity, the former producing broadband processes and the latter narrowband ones. Finally, for certain classes of systems the signals themselves may possess such complex time structures that a normal process is a reasonable statistical approximation of the actual signal ensemble.

(1) Optimum structure. Signal and noise are assumed additive and independent with vanishing mean as before, and as before we begin first with discrete sampling. The optimum structure is found from Eq. (20.1), where now for the normal processes assumed here we can write

$$\langle W_n(\mathbf{V} - \mathbf{S})\rangle_S = W_n(\mathbf{V})_{S+N} = \frac{\exp\left[-\frac{1}{2}\tilde{\mathbf{v}}\mathbf{k}_N^{-1}(\mathbf{I} + a_0^2\mathbf{D})^{-1}\mathbf{v}\right]}{(2\pi)^{n/2}[\det \mathbf{K}_N \cdot \det(\mathbf{I} + a_0^2\mathbf{D})]^{1/2}} \qquad (20.156)$$

and $W_n(\mathbf{V})_N$ follows when $a_0 = 0$ in Eq. (20.156). Here, also, we have

$$\mathbf{D} \equiv \mathbf{k}_S\mathbf{k}_N^{-1} \qquad a_0^2 \equiv \psi_S/\psi_N \qquad \mathbf{k}_N = \psi_N^{-1}\mathbf{K}_N \qquad \mathbf{k}_S = \psi_S^{-1}\mathbf{K}_S$$
$$\mathbf{v} = \mathbf{V}/\psi_N^{1/2} \qquad (20.156a)$$

where ψ_S, ψ_N are, respectively, the mean intensities of the signal and noise, so that a_0^2 is again an (input) signal-to-noise (power) ratio. The optimum detector structure is specifically

$$x = T_R\{V\} = \log \Lambda_n(\mathbf{V}) = \log \mu + \log [W_n(\mathbf{V})_{S+N}/W_n(\mathbf{V})_N]$$
$$= \Gamma_0 + \tfrac{1}{2}\Psi_n(\mathbf{v},a_0^2) \qquad (20.157)$$

The bias Γ_0 and operations on $\mathbf{v}$ are given by

$$\Gamma_0 = \log \mu - \tfrac{1}{2} \log \det (\mathbf{I} + a_0^2\mathbf{D}) \qquad \Psi_n(\mathbf{v},a_0^2) = \tilde{\mathbf{v}}\mathbf{F}\mathbf{v} = \tilde{\mathbf{V}}\psi_N^{-1}\mathbf{F}\mathbf{V}\ (> 0) \qquad (20.157a)$$

$$\psi_N^{-1}\mathbf{F} = \mathbf{K}_N^{-1}[\mathbf{I} - (\mathbf{I} + a_0^2\mathbf{D})^{-1}] = \mathbf{K}_N^{-1} - (\mathbf{K}_N + \mathbf{K}_S)^{-1} = \psi_N^{-1}\tilde{\mathbf{F}} \qquad (20.157b)$$

† For further details, see Middleton;[4] specific applications to scatter-multipath communication are considered by Price,[27] Turin,[28] and Price and Green.[29]

Since reception is necessarily incoherent here, and since both the signal and noise processes are normal, the optimum structure involves at most a generalized (averaged) autocorrelation of the received data $\mathbf{V}$ with itself [cf. Eqs. (19.59)] at *all* signal levels ($a_0 > 0$); the optimum receiver is essentially an energy detector (cf. Sec. 19.4-3).

The structure of this optimum receiver is readily seen to consist of the familiar linear Bayes matched filter of the first kind† followed by a zero-memory square-law rectifier and ideal integrator (cf. Fig. 20.2*a*) or, equivalently, a linear time-varying filter, multiplier, and integrator (cf. Fig. 20.2*b*). To show this, we apply the analysis of Sec. 20.1-2 with some slight modifications. Again, we let $\mathbf{Q}_F = [h_F(t_i - t_j)\,\Delta t]$ be a (discrete) linear filter, such that $\mathbf{Q}_F\mathbf{V} = \mathbf{V}_F$. Since $\mathbf{F}$ is symmetric and positive definite, it follows from Eqs. (20.14) to (20.18) that $\mathbf{Q}_F$ is time-varying and physically realizable as well as linear, such that $\Psi_n(\mathbf{v},a_0^2)$ is diagonalized as before. Thus, we can write

$$\tilde{\mathbf{V}}\psi_N^{-1}\mathbf{F}\mathbf{V} = \tilde{\mathbf{V}}_F(\tilde{\mathbf{Q}}_F^{-1}\psi_N^{-1}\mathbf{F}\mathbf{Q}_F^{-1})\mathbf{V}_F = \tilde{\mathbf{V}}_F(\mathbf{I}\,\Delta t)\mathbf{V}_F = \sum_{i=1}^{n} (V_F)_i^2\,\Delta t \tag{20.158a}$$

where we have chosen $\mathbf{Q}_F$ such that $\tilde{\mathbf{Q}}_F^{-1}\psi_N^{-1}\mathbf{F}\mathbf{Q}_F^{-1} = \mathbf{I}\,\Delta t$, or $\tilde{\mathbf{Q}}_F\mathbf{Q}_F = \psi_N^{-1}\mathbf{F}\,(\Delta t)^{-1}$. Now setting

$$\psi_N^{-1}\mathbf{F} = \boldsymbol{\rho}_F\,(\Delta t)^2 \tag{20.158b}$$

[cf. Eq. (20.17)], we have again the familiar condition for ρ_F [cf. Eqs. (20.18*b*)]

$$\tilde{\mathbf{Q}}_F\mathbf{Q}_F = \boldsymbol{\rho}_F\,\Delta t \qquad \text{or} \qquad \sum_{k=1}^{n} h_F(t_k - t_i)h_F(t_k - t_j)\,\Delta t = \rho_F(t_i,t_j) = \rho_F(t_j,t_i) \tag{20.158c}$$

with $(\rho_F)_{ij} = 0$ elsewhere as before, and $h_F = 0$, t_i, $t_j < 0$. Thus, the analogue of Eqs. (20.19) is now

$$\Psi_n(\mathbf{v},a_0^2) = \tilde{\mathbf{v}}\mathbf{F}\mathbf{v} = \sum_{i=1}^{n} \Delta t\left[\sum_{j=1}^{\infty\ (\text{or } i)} h_F(t_i - t_j)V(t_j)\,\Delta t\right]^2 \tag{20.158d}$$

The bias Γ_0 [Eq. (20.157*a*)] may alternatively be represented as

$$\Gamma_0 = \log\mu + \sum_{m=1}^{\infty} \frac{(-1)^m}{2m}\,a_0^{2m}\,\mathrm{tr}\,\mathbf{D}^m \tag{20.159}$$

from Eq. (17.13) in Eq. (20.157*a*), provided that $a_0^2|[\lambda_1]_D| < 1$, where λ_1 is the largest eigenvalue of $\mathbf{D}$.

† For example, an invariant realizable filter with a time-varying switch.

When continuous sampling is used, we find that Eq. (20.157b) goes over into a pair of integral equations, from which $\rho_F(t,u)_T$ may be obtained. Rewriting Eq. (20.157b) as $(\mathbf{K}_S + \mathbf{K}_N)\psi_N^{-1}\mathbf{F} = \mathbf{K}_S\mathbf{K}_N^{-1}$, and setting

$$\mathbf{K}_S\mathbf{K}_N^{-1} = \mathbf{L}\,\Delta t \qquad \text{or} \qquad \mathbf{K}_S = \mathbf{L}\mathbf{K}_N\,\Delta t \tag{20.160a}$$

we see that $(\mathbf{K}_S + \mathbf{K}_N)\psi_N^{-1}\mathbf{F}\,(\Delta t)^{-1} = (\mathbf{K}_S + \mathbf{K}_N)\boldsymbol{\rho}_F\,\Delta t = \mathbf{L}$, so that formally (cf. Sec. 19.4-2) we get

$$\int_{0-}^{T+} [K_S(t,u) + K_N(t,u)]\rho_F(u,\tau)_T\,du = L_T(t,\tau;a_0^2) \quad 0 \le t,\,\tau \le T \tag{20.160b}$$

and from Eq. (20.160a)

$$\int_{0-}^{T+} L_T(t,u;a_0^2)K_N(u,\tau)\,du = K_S(t,\tau) \qquad 0 \le t,\,\tau \le T \tag{20.160c}$$

where ρ_F and L (for u but not t) vanish outside $(0 \le \tau \le T)$. Solving for L enables us to find ρ_F from Eq. (20.160b), whenever solutions exist. In the case of *white noise backgrounds*, $K_N(t,u) = (W_0/2)\delta(t-u)$, so that $L_T = 2K_S(t,\tau)/W_0$ $(0 \le t,\,\tau \le T)$, and hence Eq. (20.160b) reduces to the simpler relation

$$\int_{0-}^{T+} \left[K_S(t,u) + \frac{W_0}{2}\,\delta(t-u)\right]\rho_F(u,\tau)_T\,du = \frac{2}{W_0}\,K_S(t,\tau) \qquad 0 \le t,\,\tau \le T \tag{20.161}$$

[See Appendix 2, Examples 3, 4, in the stationary cases where Eq. (20.161) can be put in the form of Eq. (A.2.76), with the help of Eq. (A.2.12).] Once ρ_F has been found, we obtain the set of possible† Bayes matched filters h_F from Eqs. (20.51), e.g.,

$$\int_0^T |Y_F(i\omega;x)|^2\,dx = \left[\iint_{0-}^{T+} \rho_F(t,u)_T \cos\omega(t-u)\,dt\,du\right] \tag{20.162}$$

Observe that since $\rho_F(t,u)_T$ is symmetrical and positive definite, with the further assumption, valid for rational noise spectra, that ρ_F contains no singularities (at $t = 0,\,T$), we can apply Mercer's theorem [Eq. (8.61)] to write

$$\rho_F(t,u)_T = \sum_{m=1}^{\infty} \lambda_m\phi_m(t)\phi_m(u) \qquad 0 \le t,\,u \le T \tag{20.162a}$$

where the $\{\phi_m\}$ are a complete orthonormal set obtained from the homogeneous integral equation

$$\int_0^T \rho_F(t,u)_T\phi_m(u)\,du = \lambda_m\phi_m(t) \qquad 0 \le t \le T \tag{20.162b}$$

† Note that unlike certain special cases involving deterministic signal processes [e.g., where ρ_G can be *factored;* cf. Eq. (20.69c) and discussion] it is not possible here to obtain matched filters of the type $h_M(T-t)$ (cf. Prob. 20.6).

and the eigenvalues λ_m are all real and positive. Then Eq. (20.162) can be alternatively expressed as

$$\int_0^T |Y_F|(i\omega;x)|^2 \, dx = \left[\sum_{m=1}^{\infty} \lambda_m |A_m(i\omega)|^2 \right] > 0 \qquad A_m(i\omega) = \int_0^T \phi_m(t) e^{-i\omega t} \, dt \tag{20.162c}$$

where in effect the $A_m(i\omega)$ are the Fourier "components" of $h_F(t)$.

The continuous version of the detector structure [Eqs. (20.157), (20.158*d*), (20.159)] is†

$$\log \Lambda_T[V(t)] = \log \mu - \frac{1}{2} \log \mathfrak{D}_T(a_0^2) + \frac{1}{2} \iint_{-\infty}^{\infty} V(t) \rho_F(t,u)_T V(u) \, dt \, du \tag{20.163}$$

where $\mathfrak{D}_T(a_0^2)$ is the Fredholm determinant $\lim_{n\to\infty} \det(\mathbf{I} + a_0^2 \mathbf{D}) = \prod_{j=1}^{\infty} \{1 + a_0^2[\lambda_j^{(T)}]_D\}$, in which the $[\lambda_j^{(T)}]_D$ are the eigenvalues of $\mathbf{D} = \mathbf{k}_S \mathbf{k}_N^{-1}$ in the limit $n \to \infty$ [Eqs. (17.5) et seq.]. For threshold reception, where $a_0^2|[\lambda_j^{(T)}]_D| < 1$, we may alternatively express the bias Γ_0 in terms of the iterated kernels [cf. Eq. (17.17)]

$$B_m^{(T)} = \int_0^T \cdots \int D(t_1,t_2) D(t_2,t_3) \cdots D(t_m,t_1) \, dt_1 \cdots dt_m \qquad m \geq 1 \tag{20.164a}$$

writing for Eq. (20.159) now

$$\Gamma_0 = \log \mu + \sum_{m=1}^{\infty} \frac{(-1)^m}{2m} a_0^{2m} B_m^{(T)} \tag{20.164b}$$

Note, also, as in Sec. 20.2-2, that here the quadratic functional in Eq. (20.163) becomes, with the help of Eq. (20.158*d*),

$$\begin{aligned} \Psi_T[V(t),a_0^2] \equiv \lim_{n\to\infty} \tilde{\mathbf{v}}\mathbf{F}\mathbf{v} &= \int_0^T v_F(t)^2 \, dt \\ &= \int_0^T \left[\int_0^t v(x) h_F(t-x) \, dx \right]^2 dt \end{aligned} \tag{20.165}$$

which is again the characteristic structure consisting of a (time-varying) Bayes matched filter of the first kind, square-law rectifier, and integrator (or time-varying linear filter and multiplier, etc.).

Finally, with white noise backgrounds, ρ_F is obtained from the single

† We deal here always with the regular case (cf. Sec. 19.4-2 and Prob. 19.4).

integral equation (20.161), and the bias becomes now

$$\Gamma_0 = \log \mu - \frac{1}{2} \log \prod_{j=1}^{\infty} \left(1 + \frac{2\psi_s}{W_0} \lambda_j^{(T)}\right) \qquad (20.166a)$$

where the $\lambda_j^{(T)}$ are the eigenvalues of

$$\int_0^T k_S(t,u) \phi_j(u)\, du = \lambda_j^{(T)} \phi_j(t) \qquad 0 \leq t \leq T \qquad (20.166b)$$

The alternative expression (20.164*b*) becomes here

$$\Gamma_0 = \log \mu + \sum_{m=1}^{\infty} \frac{2^{m-1}(-1)^m}{m} \sigma_e^{2m} \lambda^m (T^{-m} B_m^{(T)}) \qquad \lambda \equiv \omega_F T \qquad (20.166c)$$

where now the $B_m^{(T)}$ are the iterated kernels of Eq. (20.164*a*), with $D(t_i,t_j)$ replaced by $k_S(t_i,t_j)$ therein, and $\sigma_e^2 \equiv \psi_s/W_0\omega_F$, with ω_F an angular frequency measure of the spectral width of the signal process (when it possesses a spectrum). Thus, σ_e^2 is an effective input signal-to-noise power ratio, which is related to the ratio of the total input signal and noise energy $\sigma_0^2 \equiv \psi_s T/W_0$ accepted in $(0,T)$ by $\sigma_e^2\lambda = \sigma_0^2$.

(2) Error probabilities and Bayes risk. The error probabilities may be calculated in the usual fashion from Eqs. (19.29) to (19.31). We obtain first the characteristic functions for Eq. (20.157) from appropriate modification of Eq. (20.82), using Eq. (20.156) for $W(\mathbf{V})_{S+N}$ and $W(\mathbf{V})_N$:

$$F_1(i\xi)_Q = \boldsymbol{E}_{\mathbf{V}|H_0}\{e^{i\xi \log \Lambda_n}\} = e^{i\xi\Gamma_0}[\det (\mathbf{I} - i\xi \mathbf{k}_N \mathbf{F})]^{-\frac{1}{2}} \qquad (20.167a)$$

$$F_1(i\xi)_P = \boldsymbol{E}_{\mathbf{V}|H_1}\{e^{i\xi \log \Lambda_v}\} = e^{i\xi\Gamma_0}\{\det [\mathbf{I} - i\xi(a_0^2\mathbf{k}_S + \mathbf{k}_N)\mathbf{F}]\}^{-\frac{1}{2}} \qquad (20.167b)$$

The corresponding probability densities $Q_n(x)$, $P_n(x)$ of $x = \log \Lambda_n$ [Eqs. (20.157)] are next found from Eqs. (19.31), (19.32), and the error probabilities α^*, β^* then follow upon another quadrature. Here it is convenient to write

$$\mathbf{K}_N \mathbf{F} \psi_N^{-1} = \mathbf{I} - (\mathbf{I} + a_0^2 \mathbf{D})^{-1} \equiv \mathbf{G}_N \qquad (a_0^2 \mathbf{k}_S + \mathbf{k}_N)\mathbf{F} = a_0^2 \mathbf{D} \equiv \mathbf{G}_{S+N} \qquad (20.168)$$

However, it is not generally possible to determine the α^*, β^* explicitly, except in the important threshold cases, where we can use the generalized Edgeworth expansions described in Sec. 17.2-1, for instance.† The desired error probabilities are finally found to be[30] [cf. Eq. (20.91)]

$$\begin{Bmatrix} \alpha^* \\ \beta^* \end{Bmatrix} \simeq \frac{1}{2} \left\{1 - \Theta \left[\frac{a_0^2 \sqrt{\operatorname{tr} \mathbf{D}^2}}{4} \pm \frac{\log (\mathcal{K}/\mu)}{a_0^2 \sqrt{\operatorname{tr} \mathbf{D}^2}}\right]\right\} \qquad (20.169)$$

† See Sec. 17.1 for a discussion of the trace and eigenvalue methods of representing these determinantal forms and Sec. 17.2 for their use in the evaluation of averages like Eqs. (20.167*a*), (20.167*b*).

The Bayes risk is then given by Eq. (20.169) in Eq. (20.70).

When continuous sampling is employed, we get for Eq. (20.167a) in the limit

$$F_1(i\xi)_Q = e^{i\xi\Gamma_0}\mathfrak{D}_T(-i\xi)_N^{-\frac{1}{2}} \qquad F_1(i\xi)_P = e^{i\xi\Gamma_0}\mathfrak{D}_T(-i\xi)_{S+N}^{-\frac{1}{2}} \tag{20.170a}$$

where $\mathfrak{D}_T(-i\xi)_N$, $\mathfrak{D}_T(-i\xi)_{S+N}$ are the Fredholm determinants

$$\lim_{n\to\infty} \det\,(\mathbf{I} - i\xi\mathbf{G})$$

with $\mathbf{G} = \mathbf{G}_N, \mathbf{G}_{S+N}$, respectively. Thus, we have

$$\mathfrak{D}_T(-i\xi)_{N,S+N} = \prod_{j=1}^{\infty} \{1 - i\xi[\lambda_j^{(T)}]_{N,S+N}\} \tag{20.170b}$$

where $\lambda_j^{(T)}$ are the eigenvalues of

$$\int_0^T G(t,u)_{N,S+N}\phi_j(u)\,du = [\lambda_j^{(T)}]_{N,S+N}\phi_j(t) \qquad 0 \le t \le T \tag{20.170c}$$

The various d.d.'s of $\log \Lambda_T[V(t)]$ now follow as the Fourier transforms of Eqs. (20.170a), while the error probabilities in the threshold case are still given by Eq. (20.169), when $a_0^4 \operatorname{tr} \mathbf{D}^2$ is replaced in the limit by

$$a_0^4\Phi_D \equiv \lim_{n\to\infty} a_0^4 \operatorname{tr} \mathbf{D}^2 = \iint_0^T L_T\,(t,u;a_0^2)^2\,dt\,du \tag{20.171}$$

this last since $a_0^2\mathbf{D} = \mathbf{K}_S\mathbf{K}_N^{-1} = \mathbf{L}\,\Delta t$, from Eq. (20.160$a$); L_T itself is obtained from Eq. (20.160c). For white noise backgrounds, we have

$$a_0^4\Phi_D\Big|_{\text{white}} = \frac{4}{W_0^2}\iint_0^T K_S(t,u)^2\,dt\,du \tag{20.171a}$$

Note that for average performance we need calculate only L_T from Eq. (20.160c), while both L_T and ρ_T are required if we are to determine the explicit structure of the system (for all $a_0 > 0$). Again, Figs. 20.4, 20.8, 20.10, 20.11 apply here for Eqs. (20.169), (20.171), with appropriate modification of ordinates and parameters. Thus, in Fig. 20.4a we replace $a_0\Phi_s^{\frac{1}{2}}\sqrt{2}$ by $a_0^2\sqrt{\operatorname{tr}\mathbf{D}^2}$, or $a_0^2\Phi_D^{\frac{1}{2}}$, while in Fig. 20.4$b$ for Bayes risk we substitute $(a_0^2/\sqrt{2})\sqrt{\operatorname{tr}\mathbf{D}^2}$, or $a_0^2\Phi_D^{\frac{1}{2}}/\sqrt{2}$, for $a_0\Phi_s^{\frac{1}{2}}$, with similar adjustments for the other figures. Curves of normalized Bayes risk [cf. Eqs. (19.39)] are also readily determined. In the specific instances of broadband RC and narrowband LRC noise signals in white noise backgrounds (with continuous sampling), we find[4] that the arguments of the error functions in Eq. (20.169) are specifically

$$\left[\frac{\sigma_e^2\sqrt{\lambda}}{2} \pm \frac{\log(\mathcal{K}/\mu)}{2\sigma_e^2\sqrt{\lambda}}\right]_{RC} \qquad \left[\frac{\sigma_e^2\sqrt{\lambda}}{2\sqrt{2}} \pm \frac{\log(\mathcal{K}/\mu)}{\sqrt{2}\,\sigma_e^2\sqrt{\lambda}}\right]_{LRC} \tag{20.172}$$

where $\sigma_e^2 = \psi_s/W_0\omega_F$ and $\lambda = \omega_F T$, with $(\omega_F)_{RC} = (RC)^{-1}$, $(\omega_F)_{LRC} = R/2L$, $(\sigma_e^2)_{LRC} = \sqrt{2}\,(\sigma_e^2)_{RC}$, and where λ is reasonably large, O(5). Figure 20.19 shows the normalized Bayes risks $\mathcal{R}^*$ for threshold reception in these two instances, for several values of $\mathcal{K}/\mu$. From this, various minimum detectable signals $(\sigma_e^2)^*_{\min}$ can be established at once. In any case, we observe again the characteristic threshold dependence of $(\sigma_e^2)^*_{\min}$ on integration time, for example, $(\sigma_e^2)^*_{\min} \sim T^{-1/2}$ (cf. Fig. 20.9).

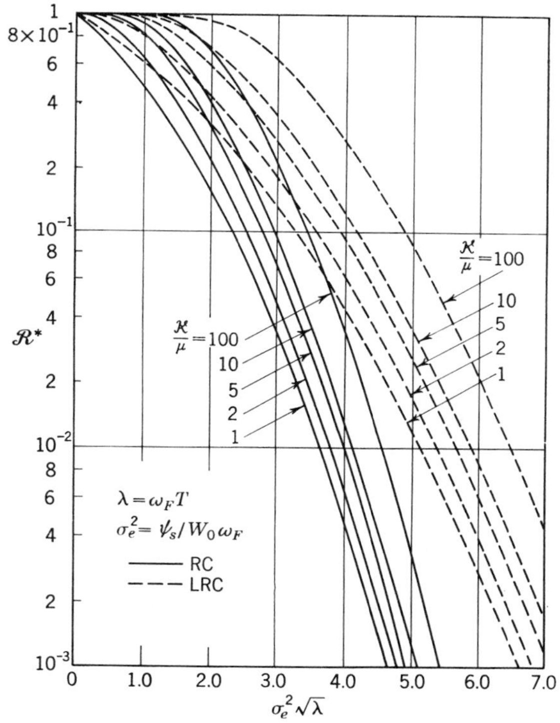

FIG. 20.19. Normalized Bayes risk for RC and narrowband LRC noise signals in white noise backgrounds; threshold reception and continuous sampling.

20.4-8 Remarks. The preceding sections have included a number of examples, of varying degrees of generality and realism, which illustrate specific ways in which the methods of decision theory may be applied to detection problems. Let us consider briefly some of the principal results and questions that these examples have revealed.

First, there is the problem of the a priori probabilities, not only p, q ($\mu = p/q$), but the various d.d.'s governing signal amplitude, phase, frequency, etc. (cf. Sec. 19.1-1). It has been assumed, usually, that these data are available at the receiver. Frequently, however, the observer is partially or totally ignorant of these probabilities and d.d.'s, so that it becomes necessary for him to adopt some further subcriterion by which he can justify a choice of distribution for them.† Two such criteria are the Minimax and maximum-average-uncertainty principles, described in Sec. 19.2-3. A third is the *maximum-likelihood* principle, closely related to the classical idea of maximum-likelihood estimation discussed later, in Sec. 21.1-2. Here, when we are ignorant of the precise distribution governing the signal parameters $\boldsymbol{\theta}$, we select those $\boldsymbol{\theta} = \hat{\boldsymbol{\theta}}$ which make the observation $\mathbf{V}$ most "likely"; i.e., on receiving a given $\mathbf{V}$, we choose $\boldsymbol{\theta} = \hat{\boldsymbol{\theta}}$ so as to maximize the "likelihood" $W_n(\mathbf{V}|\boldsymbol{\theta})$, where $W_n(\mathbf{V}|\boldsymbol{\theta})$ is the conditional d.d. of $\mathbf{V}$, given $\boldsymbol{\theta}$. In effect, we write the unknown d.d. $w(\boldsymbol{\theta})$ of $\boldsymbol{\theta}$ as $w_1(\boldsymbol{\theta}) = \delta(\boldsymbol{\theta} - \hat{\boldsymbol{\theta}})$, where $\hat{\boldsymbol{\theta}}$ is determined from the extremal conditions

$$\frac{\partial}{\partial \boldsymbol{\theta}} [\log W_n(\mathbf{V}|\boldsymbol{\theta})]_{\boldsymbol{\theta}=\hat{\boldsymbol{\theta}}} = 0 \tag{20.173}$$

for those $\hat{\boldsymbol{\theta}}$ which render $W_n(\mathbf{V}|\hat{\boldsymbol{\theta}})$ an (absolute) maximum.‡ Then the generalized likelihood ratio $\Lambda_n(\mathbf{V})$ [Eq. (19.14)], which embodies the optimum receiver and by which the binary test of the hypotheses $H_0{:}N$ versus $H_1{:}S + N$ is carried out in detection, becomes (in logarithmic form)

$$\log \Lambda_n(\mathbf{V}) = \log \mu + \log \frac{\langle F_n(\mathbf{V}|\mathbf{S})\rangle_S}{F(\mathbf{V}|\mathbf{0})} = \log \mu + \log \frac{F_n[\mathbf{V}|\mathbf{S}(\hat{\boldsymbol{\theta}})]}{F(\mathbf{V}|\mathbf{0})} \tag{20.174}$$

We still have a Bayes system here with the general properties summarized in Sec. 18.5. However, the maximum-likelihood principle, like the Minimax or, for that matter, any other criterion involving a parameter distribution which is selected out of partial or total ignorance, may not be acceptable in all situations. For if we design a likelihood receiver using Eq. (20.174) when actually $w_1(\boldsymbol{\theta}) \neq \delta(\boldsymbol{\theta} - \hat{\boldsymbol{\theta}})$, we obtain a Bayes system that lies somewhere between the Bayes system where the actual $w_1(\boldsymbol{\theta})$ is known and the Minimax case, where the unknown $w_1(\boldsymbol{\theta})$ is replaced by a d.d. which maximizes the Bayes risk. The resulting Bayes risk for $\boldsymbol{\theta} = \hat{\boldsymbol{\theta}}$ may be too great, while another choice of $\boldsymbol{\theta} = \hat{\boldsymbol{\theta}}' \neq \hat{\boldsymbol{\theta}}$ may bring it within acceptable limits. Thus, a maximum-likelihood criterion does not necessarily lead to a minimum Bayes or even small Bayes risk. This is, of course, one of the penalties of ignorance: the resulting expected cost of decision may be too high. A reassignment of costs, if permissible, offers one way out of the difficulty; the acquisition of the needed data, if possible, provides another. What is to be done in a given case depends ultimately on the larger situation of which the particular decision process is a part. We shall comment again on this question of the a priori probabilities in Sec. 23.4.

On the other hand, different subcriteria for the a priori probabilities sometimes lead to the same results. An important illustration of this

† It is assumed here that the observer is always required to make a decision.

‡ For details of the theory, see Cramér,[30] chap. 33 and associated bibliography, and Sec. 21.1-2 below.

occurs when p (and q, or $\mu \equiv p/q$) is unknown at the receiver and it is reasonable to assign zero costs for correct decisions ($C_{1-\alpha} = C_{1-\beta} = 0$) and the same cost for incorrect ones, for example, $C_\alpha = C_\beta$. Then we have an example of the Ideal Observer ($\mathcal{K} = 1$; cf. Sec. 19.2-2), with unknown p and q. If a Minimax criterion is next imposed, we may readily apply Eq. (19.28) to the relations in Secs. 20.3, 20.4 for the error probabilities in order to find the least favorable $p = p_M^*$, $q_M^* = 1 - p_M^*$ in the Minimax sense. For coherent systems, with Eq. (20.79) in (19.28) and $\mathcal{K} = 1$, it is at once evident under these conditions that

$$\log \mu_M^* = 0 \qquad \text{or} \qquad p_M^* = q_M^* = \tfrac{1}{2} \qquad \mathcal{K} = 1 \tag{20.175}$$

The a priori probabilities here are equal. With incoherent reception [Eq. (20.91) in Eq. (19.28), $\mathcal{K} = 1$], this is still true, as it is for all the specific optimum detection systems considered in this section, since these are simply special examples of the general results of Sec. 20.3.

Instead of the Minimax criterion, let us now adopt that of the greatest average uncertainty, subject to a maximum value constraint. Thus, maximizing $-p \log p - (1 - p) \log (1 - p)$ with $p_{\max} = 1$ gives $p^* = q^* = \tfrac{1}{2}$ [cf. Prob. 6.7, Eqs. (1)], and we observe from Eqs. (20.175) that these are identical to the Minimax values. Accordingly, an optimum threshold system (with the above cost assignments) where $p = q = \tfrac{1}{2}$ can also be regarded as a Bayes system with respect to either of these criteria. This is an example of the invariance of a system with respect to a set of criteria. Such invariance is clearly a valuable system property, for it broadens the applicability of a given optimum design to several classes of situations. The curve for $\mathcal{K} = \mu = 1$ in Fig. 20.4*b* shows the Bayes risk in these Minimax and maximum-uncertainty cases and, with proper adjustment of abscissa, for nearly all the threshold examples of this section too. From this and the curves for other values of μ, $\mathcal{K} = 1$, one can obtain a quantitative measure of the price one must pay for ignorance of these particular a priori probabilities.

Finally, we observe that this procedure is not confined to the situation $\mathcal{K} = 1$ alone. However, if $\mathcal{K} \neq 1$, as one can readily see from Eqs. (19.28), (20.79), (20.91), p_M^* is no longer $\tfrac{1}{2}$, so that $p_M^* \neq p^*$, which is always $\tfrac{1}{2}$, since it is determined independently of the initial cost assignments. The two criteria do not now yield identical results, although, for these particular a priori probabilities, system *structure* (apart from the bias term $\log \mu$) remains invariant. Even system structure, however, generally loses its invariance with respect to Minimax or maximum-uncertainty criteria when these are applied to the distributions of signal parameters themselves, as is evident from the forms of Eq. (19.28) and the particular results [Eqs. (20.79), (20.91)] of the preceding sections.

The procedures for determining the structure and analyzing the performance of optimum systems in binary detection are quickly summarized.

The generalized likelihood ratio $\Lambda_n(\mathbf{V})$, or, more conveniently, its logarithm $\log \Lambda_n(\mathbf{V})$, embodies the required operations $\boldsymbol{T}_R$ on the received data V. In threshold cases, using the methods of Secs. 20.1, 20.2, we can interpret $\boldsymbol{T}_R\{V\} = \log \Lambda_n(\mathbf{V})$ as a sequence of operations employing realizable linear and nonlinear elements. For the common situation of additive normal noise, optimum threshold structure can always be represented as a Bayes matched filter [of the first or second kind (cf. Sec. 20.3-4)], followed by a zero-memory square-law rectifier and a final ideal integrator. In certain special cases (when the only random signal parameter is its amplitude, and approximately for narrowband signals, when in addition RF epoch is randomly distributed), a third type of matched filter, identical with that used to maximize signal-to-noise ratio (cf. Secs. 16.3-1, 16.3-2), can be employed. Now, however, while square-law rectification is still used, no final smoothing is required: the desired number to be compared with $\log \mathcal{K}$ for an optimum decision is read off directly following rectification.† This third type of matched filter can also be used when a maximum-likelihood principle (cf. above) is invoked to handle the unknown signal parameters.‡ However, for purely stochastic signals (cf. Sec. 20.4-7), such filters are not possible, and we must have recourse to the general Bayes matched filters of the first and second kind (e.g., with invariant and time-varying elements) described above (Secs. 20.1-1, 20.1-2, 20.1-4). Finally, we observe that most practical detection systems structurally are represented by a generalized (averaged) autocorrelation of the received data with themselves [cf. Eq. (19.59)], since sample certainty (Sec. 19.4-3) is not available to the observer, at least initially. After a number of observations, during which the needed epoch data are acquired, it may then be possible to switch to the appropriate "coherent" system for further decisions. Both the coherent and incoherent observation processes have been considered in preceding sections (cf. Secs. 20.1-1, 20.1-2, 20.2-1, 20.2-2).

Finally, we recall that the applications of decision theory to detection in the present study have considered fixed observation periods only. However, it is often convenient and, indeed, in certain circumstances often an advantage (in the sense that smaller samples on the average, or smaller average risks, become possible) to relax the constraint of fixed sample size n, T and allow it to vary from observation to observation, so that decisions are made *sequentially*.§ The detection process above must now be modified to take into account the random nature of the sample size (or observation period). The theory of sequential testing, again due primarily to Wald,[32] was developed during the last decade, and the proof of the optimum nature of certain classes of such tests was found during that period by Wald and

† See the comments following Prob. 20.6, Eqs. (1), and the discussion in Sec. 20.3-4.

‡ For an example, see Davenport and Root,[31] pp. 347–350.

§ Still another possible procedure is to sample the data at random, with fixed over-all observation period. For an analysis of this in a radar example, see Middleton.[44]

Wolfowitz.[33] Since then a considerable number of applications of sequential methods to communication problems have been undertaken.[34–40] The reader is referred in particular to the explicit analysis of Refs. 38 to 40, where a variety of examples of optimum sequential detectors, minimizing average risk, are considered. Reference 38 also examines cases of continuous sequential sampling and contains, as well, a discussion of truncated tests and the distributions of observation time.

PROBLEMS

20.10 (*a*) Obtain the various elements of Table 20.2.

(*b*) Obtain $a_0^2\Phi_1$ and $a_0^2\Phi_2$ for a sine wave in *RC* noise.

20.11 (*a*) Show for the simple radar example of Sec. 20.4-3 that the optimum detector structure *for strong signals* is given by

$$x = \log \Lambda_n \simeq \log \mu - n\overline{a_0^2} + 2\bar{a}_0 \sum_{j=1}^{n} \varepsilon_j \tag{1}$$

and obtain for the d.d.'s of x

$$Q_n(x) \simeq (2\pi\sigma_Q^2)^{-1/2} e^{-(y-\bar{y}_Q)^2/2\sigma_Q^2} \tag{2a}$$

$$P_n(x) \simeq (2\pi\sigma_P^2)^{-1/2} e^{-(y-\bar{y}_P)^2/2\sigma_P^2} \tag{2b}$$

where

$$y = (x - \log \mu + n\overline{a_0^2})/\bar{a}_0 n(2/\psi)^{1/2} \qquad \sigma_Q^2 = \frac{\psi}{n}(2 - \pi/2) \qquad \sigma_P^2 = \psi/n \tag{2c}$$

$$\bar{y}_Q = \sqrt{\pi\psi/2} \qquad \bar{y}_P = a_0\sqrt{2\psi}$$

(*b*) From this, determine the error probabilities, and show that $(\overline{a_0^2})^*_{\min}$ is now $O(n^{-1})$, in contrast to the weak-signal behavior $(\overline{a_0^2})^*_{\min} = O(n^{-1/2})$. Explain the difference.

D. Middleton.[19]

20.12 (*a*) Show for the FSK signals of Sec. 20.4-5 when reception is *coherent* (i.e., fixed epoch $\epsilon = \epsilon_0$) and everything is known about each signal, except which is actually present during a typical signal period $(0,T)$, that the optimum receiver (for discrete sampling) is

$$x = \log \Lambda_{21}(\mathbf{V}) = \log \mu_{21} - \tfrac{1}{2}(a_{02}^2\tilde{\mathbf{s}}_2\mathbf{k}_N^{-1}\mathbf{s}_2 - a_{01}^2\tilde{\mathbf{s}}_1\mathbf{k}_N^{-1}\mathbf{s}_1) + \tilde{\mathbf{v}}\mathbf{k}_N^{-1}\mathbf{s}_{21}$$

$$\mathbf{s}_{21} \equiv a_{02}\mathbf{s}_2 - a_{01}\mathbf{s}_1 \tag{1}$$

for *all signal levels,* where $s_2, s_1 \neq 0$. Optimum detector structure for each signal is shown in Fig. 20.1; the optimum system here consists of a pair of such receivers (into each of which the data **V** is passed), whose outputs are subtracted before the resulting number is passed to the threshold comparator for a decision. When continuous sampling is used, we modify the individual receiver elements in accordance with Sec. 20.2-1. The basic structure of each receiver contains the usual matched filter (of which there are several possible types), with or without a subsequent integration (see Sec. 20.2-1).

(*b*) Show for the more general situation of signals with random parameters, but still coherent reception $\langle \mathbf{s}_1 \rangle, \langle \mathbf{s}_2 \rangle_\epsilon \neq 0$, that Eq. (1) holds for threshold operation, provided that $a_{01}, \mathbf{s}_1, a_{02}, \mathbf{s}_2$ are replaced by $\bar{a}_{01}, \bar{\mathbf{s}}_1, \bar{a}_{02}, \bar{\mathbf{s}}_2$, with $\mathbf{s}_{21} \to \bar{\mathbf{s}}_{21} = \bar{a}_{02}\bar{\mathbf{s}}_2 - \bar{a}_{01}\mathbf{s}_1$.

(*c*) Show that the d.d.'s of $x = \log \Lambda_{21}(\mathbf{V})$ in (*a*) are given by

$$P_{1,2}(x) = (2\pi\Phi_{s_{21}})^{-1/2} \exp\{-[x - (\log \mu_{21} \mp \tfrac{1}{2}\Phi_{s_{21}})]^2/2\Phi_{s_{21}}\} \tag{2}$$

and hence obtain the type I and II error probabilities

$$\begin{Bmatrix}\alpha^* \\ \beta^*\end{Bmatrix} = \frac{1}{2}\left\{1 - \Theta\left[\frac{\Phi_{s_{21}}{}^{1/2}}{2\sqrt{2}} \pm \frac{\log(\mathcal{K}/\mu)}{\sqrt{2}\,\Phi_{s_{21}}{}^{1/2}}\right]\right\} \qquad \text{all } a_{01}, a_{02} \geq 0 \tag{3}$$

with

$$\Phi_{s_{21}} \equiv (a_{02}\tilde{\mathbf{s}}_2 - a_{01}\tilde{\mathbf{s}}_1)\mathbf{k}_N{}^{-1}(a_{02}\mathbf{s}_2 - a_{01}\mathbf{s}_1) \tag{4}$$

(*d*) Verify that the error probabilities in (*b*) are still given by Eq. (3), for threshold reception only, provided that $\Phi_{s_{21}}$ is now replaced by $\Phi_{\bar{s}_{21}}$ (a_{01}, $\mathbf{s}_1$, etc., in $\Phi_{s_{21}}$ replaced by $\bar{a}_{01}$, $\bar{\mathbf{s}}_1$, etc.). Figures 20.4*a* and *b* apply here with obvious modifications of the ordinates.

(*e*) When continuous sampling is used, show that

$$\Phi_{s_{21}} = \psi \int_{0-}^{T+} [a_{02}{}^2 s_2(t - \epsilon_0)X_2(t;\epsilon_0) - 2a_{01}a_{02}s_2(t - \epsilon_0)X_1(t;\epsilon_0) + a_{01}{}^2 s_1(t - \epsilon_0)X_1(t;\epsilon_0)]\,dt \tag{5}$$

where X_1, X_2 are the solutions of the integral equation (20.38) with $s = s_1, s_2$, respectively. [For (*b*), verify that s_2, X_2, etc., are replaced by $\langle s_2\rangle$, $\langle X_2\rangle$, etc., and a_{01} by $\bar{a}_{01}$, and so on.] With white noise backgrounds, obtain

$$\Phi_{s_{21}}\Big|_{\text{white}} = \frac{1}{W_0}\int_0^T [A_{02}s_2(t - \epsilon_0) - A_{01}s_1(t - \epsilon_0)]^2\,dt \tag{6}$$

20.13 (*a*) When the noise signal of Sec. 20.4-7 is narrowband and has a symmetrical spectrum [and when the background noise has the covariance matrix $\mathbf{K}_N = \boldsymbol{\varrho}\psi_N$ ($\boldsymbol{\lambda}_0 = 0$)], show that the optimum structure can be given alternatively by

$$y = \log \Lambda_n(\mathbf{V}_c,\mathbf{V}_s) = \log\mu - \log\det(\mathbf{I} + \mathbf{K}_{S_0}\mathbf{K}_{N_0}{}^{-1}) + \tfrac{1}{2}\tilde{\mathbf{V}}_c\mathbf{F}_0\mathbf{V}_c + \tfrac{1}{2}\tilde{\mathbf{V}}_s\mathbf{F}_0\mathbf{V}_s \tag{1}$$

where $\mathbf{F}_0 = \mathbf{K}_{N_0}{}^{-1} - (\mathbf{K}_{S_0} + \mathbf{K}_{N_0})^{-1}$, $\mathbf{V}_c$, $\mathbf{V}_s$ are the "cosine" and "sine" terms in the relation $V(t) = V_c(t)\cos\omega_0 t + V_s(t)\sin\omega_0 t$ [cf. Sec. 7.5-3(4)]. For continuous sampling, with white noise backgrounds, obtain for Eq. (1)

$$\log \Lambda_T(V_c,V_s) = \Gamma_0' + \tfrac{1}{2}[\Phi_T(V_c) + \Phi_T(V_s)] \tag{2}$$

where

$$\Gamma_0' = \log\mu + \sum_{m=1}^{\infty} \frac{\sigma_0{}^{2m}}{m}(-1)^m\{T^{-m}B_m'\}$$

with B_m' the mth iterated kernel [Eq. (20.164*a*)] for $\mathbf{D}' = \mathbf{k}_{S_0}\mathbf{k}_{N_0}{}^{-1}$ and where $\Phi_T(V_c)$, $\Phi_T(V_s)$ are given by Eq. (20.163), with V replaced by V_c, V_s, respectively, therein.

(*b*) Obtain the characteristic functions of y,

$$F_y(i\xi)_Q = e^{i\xi\Gamma_0'}[\det(\mathbf{I} - i\xi\mathbf{K}_{N_0}\mathbf{F}_0)]^{-1} \tag{3}$$

$$F_y(i\xi)_P = e^{i\xi\Gamma_0'}\{\det[\mathbf{I} - i\xi\psi_N(a_0{}^2\mathbf{k}_{S_0} + \mathbf{k}_{N_0})\mathbf{F}_0]\}^{-1} \tag{4}$$

where $\Gamma_0' = \log\mu - \log\det(\mathbf{I} + \mathbf{K}_{S_0}\mathbf{K}_{N_0}{}^{-1})$, and show that the type I and II error probabilities are

$$\alpha^* = \sum_{k=1}^{n+} e^{-(\log\mathcal{K} - \Gamma_0')/\lambda_{kN}'} \prod_{\substack{j=1\\ j\neq k}}^{n+} (1 - \lambda_j'/\lambda_k')_N{}^{-1} \qquad \log\mathcal{K} - \Gamma_0' \geq 0 \tag{5a}$$

$$\alpha^* = \sum_{k=1}^{n+}\prod_{\substack{j=1\\ j\neq k}}^{n+} (1 - \lambda_j'/\lambda_k')_N{}^{-1} \qquad \log\mathcal{K} - \Gamma_0' < 0 \tag{5b}$$

and

$$\beta^* = \begin{cases} \sum_{k=1}^{n+} (1 - e^{-(\log \mathcal{K} - \Gamma_0')/\lambda'_{k,S+N}}) \prod_{\substack{j=1 \\ j \neq k}}^{n+} (1 - \lambda_j'/\lambda_k')_{S+N}^{-1} & \log \mathcal{K} - \Gamma_0' \geq 0 \quad (6a) \\ 0 & \log \mathcal{K} - \Gamma_0' < 0 \quad (6b) \end{cases}$$

Here $n+$ is the number of positive eigenvalues, with $(\lambda_k')_N$, $(\lambda_k')_{S+N}$ themselves the eigenvalues of $\mathbf{G}_N' = \mathbf{I} - (\mathbf{I} + a_0^2\mathbf{D}')^{-1}$, $\mathbf{G}_{S+N}' = a_0^2\mathbf{D}'$, $\mathbf{D}' = \mathbf{k}_{S_0}\mathbf{k}_{N_0}{}^{-1}$.

D. Middleton,[4] sec. VII.

20.14 (*a*) For the general normal noise signal in normal noise of Sec. 20.4-7, the received wave is sampled at intervals sufficiently far apart in time to ensure the independence of the n sampled values in $(0,T)$. Show that the optimum detector structure in this case becomes

$$x = \log \Lambda_n(\mathbf{V}) = \log \mu - \frac{n}{2} \log (1 + a_0^2) + \frac{1}{2} \frac{a_0^2}{1 + a_0^2} \tilde{\mathbf{v}}\mathbf{v} \tag{1}$$

and, writing $B_0 = a_0^2/(1 + a_0^2)$, obtain the d.d.'s of x,

$$Q_n(x) = B_0^{-n/2}\Gamma(n/2)^{-1}(x - \Gamma_0)^{n/2-1}e^{-(x-\Gamma_0)/B_0} \qquad x > \Gamma_0 \tag{2a}$$
$$P_n(x) = a_0^{-n/2}\Gamma(n/2)^{-1}(x - \Gamma_0)^{n/2-1}e^{-(x-\Gamma_0)/a_0} \qquad x > \Gamma_0 \tag{2b}$$

where $P_n(x) = Q_n(x) = 0$ if $x < \Gamma_0$, with $\Gamma_0 \equiv \log \mu - (n/2) \log (1 + a_0^2)$.

(*b*) Verify that the error probabilities are precisely

$$\alpha^* = 1 - I_c\left(\frac{\log \mathcal{K} - \Gamma_0}{B_0}; \frac{n}{2}\right) \qquad \beta^* = I_c\left(\frac{\log \mathcal{K} - \Gamma_0}{a_0^2}; \frac{n}{2}\right) \tag{3}$$

in which $I_c(\lambda;\gamma) \equiv \Gamma(\gamma)^{-1} \int_0^\lambda e^{-y}y^{\gamma-1}\,dy$ is the incomplete gamma function. In the threshold cases, show also that, approximately,

$$\begin{Bmatrix} \alpha^* \\ \beta^* \end{Bmatrix} \simeq \frac{1}{2}\left\{1 - \Theta\left[\frac{a_0^2\sqrt{n}}{4} \pm \frac{\log(\mathcal{K}/\mu)}{a_0^2\sqrt{n}}\right]\right\} \tag{4}$$

with the characteristic dependence $(a_0^2)^*_{\min} \sim n^{-1/2}$ for the minimum detectable signal.

D. Middleton,[4] sec. VI.

REFERENCES

1. Middleton, D., and D. Van Meter: Detection and Extraction of Signals in Noise from the Point of View of Statistical Decision Theory, *J. Soc. Ind. Appl. Math.*, **3:** 192 (1955).
2. Bussgang, J. J., and D. Middleton: Optimum Sequential Detection of Signals in Noise, *IRE Trans. on Inform. Theory*, **IT-1:** 5 (December, 1955).
3. Margenau, H., and G. M. Murphy: "The Mathematics of Physics and Chemistry," Van Nostrand, Princeton, N.J., 1943.
4. Middleton, D.: On the Detection of Stochastic Signals in Additive Normal Noise. I, *IRE Trans. on Inform. Theory*, **IT-3:** 86 (June, 1957).
5. Van Vleck, J. H., and D. Middleton: A Theoretical Comparison of the Visual, Aural, and Meter Reception of Pulsed Signals in the Presence of Noise, *J. Appl. Phys.*, **17:** 940 (1946).
6. Lawson, J. L., and G. E. Uhlenbeck: "Threshold Signals," sec. 8.7, p. 218, Massachusetts Institute of Technology Radiation Laboratory Series, vol. 24, McGraw-Hill, New York, 1950.

7. Middleton, D.: The Effect of a Video Filter on the Detection of Pulsed Signals in Noise, *J. Appl. Phys.*, **21**: 734 (1950).
8. Middleton, D.: Information Loss Attending the Decision Operation in Detection, *J. Appl. Phys.*, **25**: 127 (1954).
9. Lee, Y. W., T. P. Cheatham, Jr., and J. B. Wiesner: Application of Correlation Analysis to the Detection of Periodic Signals in Noise, *Proc. IRE*, **38**: 1165 (1950).
10. Fano, R. M.: Signal-to-Noise Ratios in Correlation Detectors, *MIT Research Lab. Electronics Tech. Rept.* 186, February, 1951.
11. Davenport, W. B., Jr.: Correlator Errors Due to Finite Observation Integrals, *MIT Research Lab. Electronics Tech. Rept.* 191, March, 1951.
12. Faran, J. J., and R. Hills: Correlators for Signal Reception, *Harvard Univ. Acoustics Research Lab. Tech. Mem.* 27, September, 1952; see also J. J. Faran and R. Hills, The Application of Correlation Techniques to Acoustic Receiving Systems, *Harvard Univ. Acoustics Research Lab. Tech. Mem.* 28, November, 1952.
13. Green, P. E., Jr.: The Output Signal-to-Noise Ratio of Correlation Detectors, *IRE Trans. on Inform. Theory*, **IT-3**: 10 (March, 1957).
14. Middleton, D.: Statistical Theory of Signal Detection, Symposium on Statistical Methods in Communication Engineering, *IRE Trans. on Inform. Theory*, **PGIT-3**: 26 (March, 1954).
15. Woodward, P. M.: "Probability and Information Theory, with Applications to Radar," McGraw-Hill, New York, 1953, secs. 4.3, 5.2, for example.
16. Siegert, A. J. F.: sec. 7.5 in J. L. Lawson and G. E. Uhlenbeck, "Threshold Signals," Massachusetts Institute of Technology Radiation Laboratory Series, vol. 24, McGraw-Hill, New York, 1950.
17. Marcum, J.: A Statistical Theory of Target Detection by Pulsed Radar, *RAND Corp. Repts.* RM-753, R-113, 1948; also, *RAND Corp. Rept.* RM-15061, 1947.
18. Hanse, H.: The Optimization and Analysis of Systems for the Detection of Pulsed Signals in Random Noise, doctoral dissertation, Massachusetts Institute of Technology, January, 1951.
19. Middleton, D.: Statistical Criteria for the Detection of Pulsed Carriers in Noise. I, II, *J. Appl. Phys.*, **24**: 371, 379 (1953).
20. Peterson, W. W., T. G. Birdsall, and W. C. Fox: The Theory of Signal Detectability, *IRE Trans. on Inform. Theory*, **PGIT-4**: 171 (1954).
21. Benner, A. H., and R. F. Drenick: On the Problem of Optimum Detection of Pulsed Signals in Noise, *RCA Rev.*, **16**: 461 (1955).
22. Drukey, D. L.: Optimization Techniques for Detecting Pulse Signals in Noise, *Proc. IRE*, **40**: 217 (1952).
23. Reich, E., and P. Swerling, Jr.: Detection of a Sine Wave in Gaussian Noise, *J. Appl. Phys.*, **24**: 289 (1953).
24. Kaplan, S. M., and R. W. McFall: The Statistical Properties of Noise Applied to Radar Range Performance, *Proc. IRE*, **39**: 56 (1951).
25. Harrington, J. V.: An Analysis of the Detection of Repeated Signals in Noise by Binary Integration, *IRE Trans. on Inform. Theory*, **IT-1**: 1 (March, 1955).
25a. Siebert, W. McC.: A Radar Detection Philosophy, *IRE Trans. on Inform. Theory, Symposium*, **IT-2**: 204 (September, 1954).
25b. Siebert, W. McC.: Some Applications of Detection Theory to Radar, *IRE Natl. Conv. Record*, pt. 4, 5, March, 1958.
26. Van Meter, D., and D. Middleton: Optimum Decision Systems for the Reception of Signals in Noise. Part I, *Cruft Lab. Tech. Rept.* 215, Oct. 5, 1955.
27. Price, R.: Optimum Detection of Random Signals in Noise with Application to Scatter Multipath Communication. I, *IRE Trans. on Inform. Theory*, **IT-2**(4): 125 (1956); see also The Detection of Signals Perturbed by Scatter and Noise, *IRE Trans. on Inform. Theory*, **PGIT-4**: 163 (September, 1954).

28. Turin, G. L.: Communication through Noise Random-multipath Channels, *IRE Conv. Record*, pt. 4, p. 154, 1956.

28a. Turin, G. L.: Error Probabilities for Binary Symmetric Ideal Reception through Nonselective Slow Fading and Noise, *Proc. IRE*, **46:** 1603 (1958).

29. Price, R., and P. E. Green, Jr.: A Communication Technique for Multipath Channels, *Proc. IRE*, **46:** 555 (1958).

30. Cramér, H.: "Mathematical Methods of Statistics," Princeton University Press, Princeton, N.J., 1946.

31. Davenport, W. B., Jr., and W. L. Root: "An Introduction to the Theory of Random Signals and Noise," McGraw-Hill, New York, 1958.

32. Wald, A.: "Sequential Analysis," Wiley, New York, 1947.

33. Wald, A., and J. Wolfowitz: Optimum Character of the Sequential Probability Ratio Test, *Ann. Math. Stat.*, **19:** 326 (1948).

34. Schwartz, M.: A Statistical Approach to the Automatic Search Problem, doctoral dissertation, Harvard University, January, 1951.

35. Harrington, J. V.: An Analysis of Detection of Repeated Signals in Noise by Binary Integration, *IRE Trans. on Inform. Theory*, **IT-1:** 1 (March, 1955).

36. Reed, I. S.: Analysis of Signal Detection by Sequential Observer, *MIT Lincoln Lab. Tech. Rept.* 20, Mar. 12, 1953.

37. Ref. 20, pp. 179–182.

38. Bussgang, J. J., and D. Middleton: Optimum Sequential Detection of Signals in Noise, *IRE Trans. on Inform. Theory*, **IT-1:** 5 (December, 1955); also, J. J. Bussgang, Sequential Detection of Signals in Noise, doctoral dissertation, Harvard University, 1955.

39. Blasbalg, H.: The Relationship of Sequential Filter Theory to Information Theory and Its Application to the Detection of Signals in Noise by Bernoulli Trials, *IRE Trans. on Inform. Theory*, **IT-3:** 122 (June, 1957); also, H. Blasbalg, Theory of Sequential Filtering and Its Application to Signal Detection and Classification, doctoral dissertation, Johns Hopkins University, 1955.

40. Blasbalg, H.: The Sequential Detection of a Sine-wave Carrier of Arbitrary Duty Ratio in Gaussian Noise, *IRE Trans. on Inform. Theory*, **IT-3:** 248 (December, 1957).

41. Kailath, T.: Sampling Models for Linear Time-variant Filters, *MIT Research Lab. Electronics Tech. Rept.* 352, May, 1959.

42. Pierce, J. N.: Theoretical Diversity Improvement in Frequency-shift Keying, *Proc. IRE*, **46:** 903 (1958).

43. Brennan, D. G.: Linear Diversity Combining Techniques, *Proc. IRE*, **47:** 1075 (1959).

44. Middleton, D.: A Comparison of Random and Periodic Data Sampling for the Detection of Signals in Noise, *IRE Trans. on Inform. Theory*, **IT-5:** 234 (May, 1959) (Special Supplement, International Symposium on Circuit and Information Theory, 1959).

45. Rice, S. O.: Mathematical Analysis of Random Noise, *Bell System Tech. J.*, **24:** 46 (1945), pt. III.

CHAPTER 21

EXTRACTION SYSTEMS MINIMIZING AVERAGE RISK; SIGNAL ANALYSIS

In the preceding chapters, we have considered binary detection systems, where the central problem is to determine the presence or absence of a signal in noise. Here, on the other hand, we are concerned with a somewhat more general problem: that of determining the explicit structure or the descriptive parameters of a signal in noise when these are unknown to the observer. Typical examples of communication interest are the measurement of range or velocity of a moving target by a radar; the calculation of signal amplitude, frequency, delay, and other information-bearing features of the signal process in radio telegraphy; and the extraction of waveform data in the analysis and identification of signal structures in various communication operations. In all such cases, we have to deal with a *measurement situation:* our decisions now are not the simple "yes" or "no" of binary detection but are instead specific numbers, associated with the signal process in question. Furthermore, in most practical cases we do not have "pure" data upon which to operate, but data corrupted by noise. Again, since observation time is limited, i.e., since the available samples from which we obtain our measurements are finite, we can expect only imperfect estimates of the desired quantities. Some will be better than others, depending on our methods for reducing the effects of the interfering noise. The principal aim of the present chapter, accordingly, is to find and examine systems $\boldsymbol{T}_R$ which have this desirable property of yielding optimum estimates.

A receiving system $\boldsymbol{T}_R$ which measures the waveform or one or more of the parameters of an incoming signal ensemble we call a *signal extraction system* and the process itself *signal extraction.* As we noticed in Sec. 18.2-2, signal extraction is a form of statistical estimation (e.g., point estimation) analogous to signal detection as a form of hypothesis testing. Both, of course, fall within the domain and techniques of decision theory (cf. Sec. 18.2). Defining *signal analysis* to include methods of estimation as well as representation, we see, then, that signal extraction plays an important part in this theory. Here, as in detection, the goals of an adequate theory remain unchanged: to obtain optimum systems for estimation, to interpret the resulting structures $\boldsymbol{T}_R$ in terms of specific, realizable elements, and to evaluate and compare the performance of optimum and suboptimum systems. We shall start with some of the ideas and results of classical estimation theory in Sec. 21.1 and then in Sec. 21.2 shall indicate briefly how these can be incorporated within the framework of decision theory. These general methods are then applied in Secs. 21.3 to 21.5 to various problems of signal analysis. Because of the variety of possible situations and criteria, our treatment† is necessarily less compact and less comprehensive than the earlier treatment of detection. Additional examples may be found in the Problems and References and in the Supplementary Bibliography at the end of the volume.

21.1 *Some Results of Classical Estimation Theory*

We begin by giving some definitions and, for the most part without proof,‡ some results of classical estimation theory which are needed in subsequent applications. A complete account is not attempted here although estimation techniques employing both discrete and continuous sampling procedures are considered.

Accordingly, we assume once more (cf. Sec. 18.3-1) that the vector $\mathbf{V} = [V(t_1), \ldots, V(t_n)]$ represents a discrete data sample observed in the interval $(0,T)$, where now it is known that $\mathbf{V} = \mathbf{S} \oplus \mathbf{N}$, that is, that $\mathbf{V}$ is some combination of signal and noise vectors $\mathbf{S} = [S(t_1), \ldots, S(t_n)]$ and $\mathbf{N} = [N(t_1), \ldots, N(t_n)]$. The signal $S = S(t;\boldsymbol{\theta})$, of known waveform, contains one or more unknown parameters $\boldsymbol{\theta}$, which it is our task to estimate on the basis of samples $\mathbf{V}$ observed in $(0,T)$. We further postulate that the parameters of the signal, when considered over the signal ensemble, possess a probability density $\sigma(\boldsymbol{\theta})$, and that the noise process likewise is governed by a known distribution. The sample size is fixed and finite; that is, $1 \leq n < \infty$ or $0 \leq T < \infty$. For purposes of illustration, we restrict ourselves momentarily to a single unknown parameter $\boldsymbol{\theta} = \theta$; generalizations to two or more parameters, and to the case of waveform estimation itself, will be made presently (cf. Sec. 21.1-2 and, in particular, Sec. 21.2). Finally, we remark that the techniques to be discussed here and subsequently are forms of *point estimation*, in contrast to methods involving *interval estimation* or confidence intervals.§

21.1-1 Estimates, Estimators, and the Cramér-Rao Inequality. Let θ be the parameter, unknown to the observer, which he desires to estimate in some fashion from a finite data sample $\mathbf{V} = \mathbf{S}(\theta) \oplus \mathbf{N}$. We distinguish

† The present material, except where otherwise indicated, is an extension of the earlier study by Middleton and Van Meter[1] (cf. sec. 4, in particular).

‡ For proofs and further details, see the standard literature, especially Cramér,[2] chaps. 32, 33. For other developments, in particular, applications to time-series analysis, see, for example, Bartlett,[3] chaps. 8, 9, and bibliography.

§ See, for example, Mood,[4] chap. 11, and Cramér,[2] chap. 34.

now two classes of estimates: (1) the *conditional* (point) *estimates* $\gamma_\theta(\mathbf{V})$, of $\boldsymbol{\theta}$, based on the given samples $\mathbf{V}$, and (2) the *unconditional* (point) *estimates* $\gamma_\sigma(\mathbf{V})$, likewise based on the observed data $\mathbf{V}$, but now averaged over all possible values of θ. Thus, the unconditional estimate takes into account the statistical variation of θ over the signal ensemble and is thus an estimate representing an average value of γ_θ and involves $\sigma(\theta)$, where $\sigma(\theta)$ is the a priori distribution density of θ. Note, however, that unlike the detection situation $H_0{:}N$ versus $H_1{:}S \oplus N$, where $\sigma(\theta) = q\delta(S - 0) + pw(\theta)$ [cf. Eq. (19.4)], we have here $\sigma(\theta) = w(\theta)$ $(p = 1)$, since the conditions H_1: $V = S \oplus N$ are postulated: we always deal with representations containing a signal.

At this point, it is necessary to distinguish between an *estimate* and an *estimator:* the former is simply a value of γ_θ or γ_σ, corresponding to a particular data sample $\mathbf{V} = \mathbf{V}'$, while the latter is the operation (decision rule) which produces estimates (decisions) from data samples. We shall often find it convenient, however, to use the term estimator to designate the random variable whose sample values are estimates. Thus, an estimate is some definite number or statistic, based on the particular $\mathbf{V}'$, while the estimator is a random variable, whose statistical properties are derived from those of the signal and noise, in accordance with the relation $\mathbf{V} = \mathbf{S} \oplus \mathbf{N}$ and the particular form that γ_θ (or γ_σ) may take. Observe, moreover, that the analytical form of γ_θ or γ_σ embodies the structure of the estimation system itself: the operations upon $\mathbf{V}$ implied by $\gamma_\theta(\mathbf{V})$ or $\gamma_\sigma(\mathbf{V})$ represent the functions which a physical system is required to carry out in order to obtain the desired estimate when a particular $\mathbf{V}$ is the system input. We shall return to this question of system structure presently, in Secs. 21.1-3, 21.2ff. However, first let us consider some useful properties of estimates and estimators.

(1) Biased and unbiased estimators. An *unbiased estimator* is one whose average value is the same as the "true" or the expected value of the quantity being estimated. For conditional unbiased estimators, therefore, we have

$$E_{\mathbf{V}|\boldsymbol{\theta}}\{\gamma_\theta(\mathbf{V})\} = \int_\Gamma W_n(\mathbf{V}|\theta)\gamma_\theta(\mathbf{V})\,d\mathbf{V} = \theta \tag{21.1}$$

while for unconditional unbiased estimators we obtain

$$E_{\mathbf{V}}\{\gamma_\sigma(\mathbf{V})\} = \int_\Gamma W_n(\mathbf{V})\gamma_\sigma(\mathbf{V})\,d\mathbf{V} = \bar{\theta} \quad \left[= \int_{\Omega_\theta} \theta\sigma(\theta)\,d\theta\right] \tag{21.2}$$

from which we have alternatively

$$\begin{aligned} E_{\mathbf{V},\theta}\{\gamma_\theta(\mathbf{V})\} &= \int_{\Omega_\theta} d\theta \int_\Gamma W_n(\mathbf{V}|\theta)\sigma(\theta)\gamma_\theta(\mathbf{V})\,d\mathbf{V} \\ &= \int_{\Omega_\theta}\int_\Gamma W_n(\mathbf{V},\theta)\gamma_\theta(\mathbf{V})\,d\mathbf{V}\,d\theta = \bar{\theta} \end{aligned} \tag{21.3}$$

Here, as before, $W_n(\mathbf{V}|\theta)$ is the conditional d.d. of $\mathbf{V}$, given θ, while $W_n(\mathbf{V},\theta)$ is the *joint* d.d. of $\mathbf{V}$ and θ.

Biased estimators, on the other hand, do not possess this desirable feature; their expected values contain an additional function $b(\theta)$ of the parameter in question. Accordingly, for biased estimators we have

$$E_{\mathbf{V}|\theta}\{\gamma_\theta(\mathbf{V})\} = \theta + b(\theta) \qquad E_{\mathbf{V}}\{\gamma_\sigma(\mathbf{V})\} = E_\theta\{\theta\} + E_\theta\{b(\theta)\} = \bar{\theta} + \overline{b(\theta)} \tag{21.4}$$

respectively, for the conditional and unconditional estimators.

(2) The Cramér-Rao inequality. An important result which establishes a lower bound on the variance of point estimators is given by the so-called Cramér-Rao inequality,† which for unbiased conditional estimators of the single parameter θ takes the form

$$E_{\mathbf{V}|\theta}\{[\theta - \gamma_\theta(\mathbf{V})]^2\} \geq \frac{1}{\displaystyle\int_\Gamma \left[\frac{\partial \log W_n(\mathbf{V}|\theta)}{\partial\theta}\right]^2 W_n(\mathbf{V}|\theta)\, \mathbf{dV}} \tag{21.5a}$$

With bias, this becomes

$$E_{\mathbf{V}|\theta}\{[\theta - \gamma_\theta(\mathbf{V})]^2\} \geq \frac{\left[1 + \dfrac{db(\theta)}{d\theta}\right]^2}{\displaystyle\int_\Gamma \left[\frac{\partial \log W_n(\mathbf{V}|\theta)}{\partial\theta}\right]^2 W_n(\mathbf{V}|\theta)\, \mathbf{dV}} \tag{21.5b}$$

Equivalent forms of the denominator are

$$E_{\mathbf{V}|\theta}\left\{\left[\frac{\partial \log W_n(\mathbf{V}|\theta)}{\partial\theta}\right]^2\right\} = \int_\Gamma \left(\frac{\partial W_n}{\partial\theta}\right)^2 W_n(\mathbf{V}|\theta)^{-1}\, \mathbf{dV}$$
$$= -E_{\mathbf{V}|\theta}\left\{\frac{\partial^2}{\partial\theta^2}\log W_n(\mathbf{V}|\theta)\right\} \tag{21.5c}$$

this last provided that $W_n(\mathbf{V}|\theta)$ is such that the operations of differentiation with respect to θ and integration over $\mathbf{V}$ can be interchanged. The equality in Eqs. (21.5*a*), (21.5*b*) holds under rather special conditions on the conditional d.d. $W_n(\mathbf{V}|\theta)$ (cf. Cramér, sec. 32.3). The right-hand member of Eqs. (21.5*a*), (21.5*b*) represents a minimum variance of the estimators obtained from all possible conditional estimators γ_θ.

Similar relations may be obtained for the unconditional estimators γ_σ (cf. Prob. 21.1*a*). We have

$$E_{\mathbf{V},\theta}\{[\theta - \gamma_\sigma(\mathbf{V})]^2\} \geq \frac{E_\theta\left\{\left[1 + \dfrac{db(\theta)}{d\theta}\right]^2\right\}}{\displaystyle\int_{\Omega_\theta} d\theta \int_\Gamma \left\{\frac{\partial \log[\sigma(\theta) W_n(\mathbf{V}|\theta)]}{\partial\theta}\right\}^2 W_n(\mathbf{V}|\theta)\sigma(\theta)\, \mathbf{dV}} \tag{21.6}$$

in the case of bias, while for unbiased estimators the numerator is unity; in terms of the joint d.d. of $\mathbf{V}$ and θ, the denominator becomes

† See Cramér[5] and Rao[6]; see also Savage,[7] p. 238, and the discussions in Cramér,[2] sec. 32.3. Some further applications, including sequential procedures, have been examined by Hodges and Lehmann.[8]

$$E_{\mathbf{V},\theta}\left\{\left[\frac{\partial \log W_n(\mathbf{V},\theta)}{\partial\theta}\right]^2\right\} = \int_{\Omega_\theta} d\theta \int_\Gamma d\mathbf{V}\left[\frac{\partial W_n(\mathbf{V},\theta)}{\partial\theta}\right]^2 W_n(\mathbf{V},\theta)^{-1}$$
$$= -E_{\mathbf{V},\theta}\left\{\frac{\partial^2}{\partial\theta^2}\log W_n(\mathbf{V},\theta)\right\} \tag{21.6a}$$

[cf. Eq. (21.5*c*)].

(3) **Sufficiency, efficiency, and consistency.**† If the conditional d.d. of $\mathbf{V}$, given θ, can be written $W_n(\mathbf{V}|\theta) = f_n(\mathbf{V})g(\theta,\gamma_\theta)$, where f_n is independent of θ, γ_θ, then $\gamma_\theta\,[= \gamma_\theta(\mathbf{V})]$ is a *(conditionally) sufficient estimator* of θ. Similarly if $\int_{\Omega_\theta} \sigma(\theta)W_n(\mathbf{V}|\theta)\,d\theta = f_n(\mathbf{V})G[\gamma_\sigma(\mathbf{V})]$, then $\gamma_\sigma(\mathbf{V})$ is an *(unconditionally) sufficient estimator* of θ. With the aid of Eqs. (21.5), (21.6), *efficient estimators* can also be defined. Writing

$$0 \le e(\theta) \equiv E_{\mathbf{V}|\theta}\{(\theta - \gamma_\theta)^2\}_{\min}/E_{\mathbf{V}|\theta}\{(\theta - \gamma_\theta)^2\} \le 1 \tag{21.7a}$$

we call $e(\theta)$ the *conditional efficiency* of the estimators $\gamma_\theta(\mathbf{V})$, where $E_{\mathbf{V}|\theta}\{(\theta - \gamma_\theta)^2\}_{\min}$ is given by the right-hand members of Eqs. (21.5*a*), (21.5*b*). When $e(\theta) = 1$, then γ_θ is termed a (conditionally) *efficient estimator* of θ for samples of size n. In the same way, writing

$$0 \le e_\sigma \equiv E_{\mathbf{V},\theta}\{(\theta - \gamma_\sigma)^2\}_{\min}/E_{\mathbf{V},\theta}\{(\theta - \gamma_\sigma)^2\} \le 1 \tag{21.7b}$$

we call e_σ the *unconditional efficiency* of the estimators $\gamma_\sigma(\mathbf{V})$, where now $E_{\mathbf{V},\theta}\{(\theta - \gamma_\sigma)^2\}_{\min}$ is similarly represented by the right-hand member of Eq. (21.6). When $e_\sigma = 1$, γ_σ is an (unconditionally) *efficient estimator* of θ. When efficient estimators exist, they can be found by the method of *maximum likelihood* (Sec. 21.1-2).

Finally, if $e(\theta)$, e_σ become unity as sample size n (or T) becomes indefinitely large, then γ_θ, γ_σ are, respectively, *asymptotically* (conditional and unconditional) *efficient estimators;* such estimators exist under rather general conditions.‡ Furthermore, if γ_θ, γ_σ converge in probability to θ and $\bar{\theta}$, respectively, as n (or T) $\to \infty$, then γ_θ and γ_σ are (conditionally and unconditionally) *consistent* estimators of θ. Note from the above that an efficient estimator is always also sufficient. It can be shown that,§ if $\lim_{n\to\infty} e(\theta) = e(\theta)_\infty$ and $\lim_{n\to\infty} e_\sigma = (e_\sigma)_\infty$ are the asymptotic efficiencies of γ_θ, γ_σ as $n \to \infty$, then one can write

$$\begin{aligned} 0 \le e(\theta)_\infty &= K_\theta^2 \Big/ \lim_{n\to\infty}\left(E_{\mathbf{V}|\theta}\left\{\left[\frac{\partial \log W_n(\mathbf{V}|\theta)}{\partial\theta}\right]^2\right\}\right) \le 1 \\ 0 \le (e_\sigma)_\infty &= K_\sigma^2 \Big/ \lim_{n\to\infty}\left(E_{\mathbf{V},\theta}\left\{\left\{\frac{\partial \log [\sigma(\theta)W_n(\mathbf{V}|\theta)]}{\partial\theta}\right\}^2\right\}\right) \le 1 \end{aligned} \tag{21.8}$$

where K_θ^2, K_σ^2 are (positive) constants.

(4) **Generalizations.**§ Only a single unknown parameter has been

† See Cramér,[2] sec. 32.4.
‡ *Ibid.*, sec. 33.3.
§ See *ibid.*, secs. 32.6–32.8.

assumed in the above. However, the treatment can be extended to include the frequent cases where two or more unknown parameters $\boldsymbol{\theta} = (\theta_1,\theta_2, \ldots ,\theta_M)$ are involved. Instead of $\gamma_\theta(\mathbf{V})$, $\gamma_\sigma(\mathbf{V})$, one has sets of estimators: $\boldsymbol{\gamma}_{\boldsymbol{\theta}}(\mathbf{V}) = [\gamma_{\theta_1}(\mathbf{V}),\gamma_{\theta_2}(\mathbf{V}), \ldots ,\gamma_{\theta_M}(\mathbf{V})]$, $\boldsymbol{\gamma}_\sigma(\mathbf{V}) = [\gamma_{1\sigma}(\mathbf{V}),\gamma_{2\sigma}(\mathbf{V}), \ldots ,\gamma_{M\sigma}(\mathbf{V})]$. Various *joint* estimators are also possible: $\gamma_{\theta_1,\theta_2}(\mathbf{V})$, $\gamma_{\theta_1,\theta_2,\theta_3}(\mathbf{V})$, etc., with corresponding expressions for the unconditional estimators $\boldsymbol{\gamma}_\sigma$. Nor are the $\boldsymbol{\theta}$ restricted to parameter values alone. Thus, for example, if waveform $\mathbf{S}$ of a signal is being measured on the basis of n sampled values $\mathbf{V} = [V(t_1), \ldots , V(t_n)]$ in a period $(0,T)$, where the waveform at each of the n sampling instants t_k is desired, we have the vector estimator $\gamma_S(\mathbf{V}) = [\gamma_{S_1}(\mathbf{V}), \gamma_{S_2}(\mathbf{V}), \ldots ,\gamma_{S_n}(\mathbf{V})]$ of $\mathbf{S}$, where an estimate of each S_k is made on the basis of the entire set of n data values $\mathbf{V}$. In general, we expect that $\gamma_{\theta_j}(\mathbf{V}) \neq \gamma_{\theta_k}(\mathbf{V})$ $(j \neq k)$. Examples and other extensions are considered in subsequent sections. The notion of biased and unbiased estimators may also be generalized in an obvious way (cf. Secs. 21.2 to 21.4),

$$E_{\mathbf{V}|\boldsymbol{\theta}}\{\boldsymbol{\gamma}_{\boldsymbol{\theta}}(\mathbf{V})\} = \boldsymbol{\theta} + \mathbf{b}(\boldsymbol{\theta}) \tag{21.9}$$

$$E_{\mathbf{V},\boldsymbol{\theta}}\{\boldsymbol{\gamma}_\sigma(\mathbf{V})\} = E_{\boldsymbol{\theta}}\{\boldsymbol{\theta}\} + E_{\boldsymbol{\theta}}\{\mathbf{b}(\boldsymbol{\theta})\} \tag{21.10}$$

with $\mathbf{b}(\boldsymbol{\theta}) = [b_1(\boldsymbol{\theta}),b_2(\boldsymbol{\theta}), \ldots]$; in the unbiased cases, $\mathbf{b}(\boldsymbol{\theta}) = 0$. The extension of the Cramér-Rao inequalities (21.5), (21.6) is more involved,† but the concepts of sufficiency, efficiency, and consistency may still be applied, once more with obvious generalizations of the one-parameter case.

We observe again that, since the estimators γ_θ, γ_σ and their vector extensions $\boldsymbol{\gamma}_{\boldsymbol{\theta}}$, $\boldsymbol{\gamma}_\sigma$ are random variables, they may be expected to possess distributions with respect to the hypothesis $H_1{:}V = S \oplus N$ here. These may be calculated in principle from their corresponding characteristic functions

$$F_{\gamma_\theta}(i\xi) = E_{\mathbf{V}|\theta}\{e^{i\xi\gamma_\theta(\mathbf{V})}\} \qquad F_{\gamma_\sigma}(i\xi) = E_{\mathbf{V}}\{e^{i\xi\gamma_\sigma(\mathbf{V})}\} \tag{21.11a}$$

with the joint averages

$$F_{\boldsymbol{\gamma}_{\boldsymbol{\theta}}}(i\boldsymbol{\xi}) = E_{\mathbf{V}|\boldsymbol{\theta}}\{e^{i\tilde{\boldsymbol{\xi}}\boldsymbol{\gamma}_{\boldsymbol{\theta}}(\mathbf{V})}\} \qquad F_{\boldsymbol{\gamma}_\sigma}(i\boldsymbol{\xi}) = E_{\mathbf{V}}\{e^{i\tilde{\boldsymbol{\xi}}\boldsymbol{\gamma}_\sigma(\mathbf{V})}\} \tag{21.11b}$$

for the vector estimators $\boldsymbol{\gamma}_\theta$, $\boldsymbol{\gamma}_\sigma$, where the averages are carried out with respect to the weightings $W_n(\mathbf{V}|\boldsymbol{\theta})$ and $W_n(\mathbf{V})$ for $\mathbf{V} = \mathbf{S} \oplus \mathbf{N}$, respectively. In place of γ_θ or γ_σ, etc., it may be convenient as in detection (cf. Sec. 19.1-2) to use their logarithms, so that $\log \gamma_\theta$, $\log \gamma_\sigma$ may be called the (conditional and unconditional) *logarithmic estimators* of θ, with corresponding vector forms $\log \boldsymbol{\gamma}_{\boldsymbol{\theta}} = [\log \gamma_{\theta_1}, \log \gamma_{\theta_2}, \ldots , \log \gamma_{\theta_M}]$, $\log \boldsymbol{\gamma}_\sigma = [\log \gamma_{1\sigma}, \log \gamma_{2\sigma}, \ldots , \log \gamma_{M\sigma}]$. The moments and variances of these estimators, when they exist, may be obtained in the usual way from the appropriate derivatives of their characteristic functions (cf. Sec. 1.2-6).

Finally, as in detection, the question of regularity in the limit of continuous sampling again arises: in special circumstances, the functionals $\lim_{n\to\infty} \gamma_\theta(\mathbf{V}) = \gamma_\theta[V(t)]_T$, $\lim_{n\to\infty} \gamma_\sigma(\mathbf{V}) = \gamma_\sigma[V(t)]_T$ may not converge in proba-

† See *ibid.*, secs. 32.6–32.8.

bility to finite values. However, in physical problems the natural constraints on signal and noise generally ensure the existence of these limits, in much the same way as for the likelihood-ratio functionals encountered in detection.† The regular case is assumed in our subsequent discussion and for the examples presented here may be verified along the lines suggested by Grenander.[9] An interpretation of estimators from an information viewpoint is also briefly considered in Chap. 22.

21.1-2 Maximum Likelihood Estimation. While there are many possible estimators for signal parameters $\boldsymbol{\theta}$, waveform $\mathbf{S}$, and the like, with respect to any reasonable criterion most of them do not possess optimal properties, so that it is natural to select those that do, if possible. Just what constitutes an optimum estimator depends, of course, on the criterion of optimality, as well as on the quantity being estimated, the statistics of the background noise, and so on. One possible method of obtaining an estimator with optimal properties is the classical procedure of *maximum likelihood*,‡ due to R. A. Fisher.[10] We shall see in Sec. 21.2-1 how this procedure is incorporated in decision theory; in the meantime, let us briefly describe the method.

We begin by defining the (*conditional*) *likelihood function* (CLF) $l_n(\mathbf{V}|\theta)$ of a data set $\mathbf{V}$, containing a parameter§ θ, as the d.d. of $\mathbf{V}$, given θ, that is, as $l_n(\mathbf{V}|\theta) \equiv W_n(\mathbf{V}|\theta)$. Similarly, we may define the (*unconditional*) *likelihood function* (ULF) as $L_n(\mathbf{V},\theta) \equiv \sigma(\theta)W_n(\mathbf{V}|\theta)$; this is just the joint d.d. of $\mathbf{V}$ and θ. The method of maximum likelihood consists in choosing an estimator $\hat{\gamma}_\theta(\mathbf{V})$ or $\hat{\gamma}_\sigma(\mathbf{V})$ for given $\mathbf{V}$ which maximizes l_n or L_n. Since $\log l_n$, $\log L_n$ have extrema at the same values of θ as do l_n and L_n, let us use the more convenient logarithmic form to write the *likelihood equations*

$$\frac{\partial}{\partial\theta}[\log l_n(\mathbf{V}|\theta)]\Big|_{\theta=\hat{\gamma}_\theta} \equiv \frac{\partial}{\partial\theta}[\log W_n(\mathbf{V}|\theta)]\Big|_{\theta=\hat{\gamma}_\theta} = 0 \tag{21.12a}$$

$$\frac{\partial}{\partial\theta}[\log L_n(\mathbf{V},\theta)]\Big|_{\theta=\hat{\gamma}_\sigma} \equiv \frac{\partial}{\partial\theta}\{\log[\sigma(\theta)W_n(\mathbf{V}|\theta)]\}\Big|_{\theta=\hat{\gamma}_\sigma} = 0 \tag{21.12b}$$

The nonconstant solutions of Eqs. (21.12), when they exist, then give us the *conditional maximum likelihood estimators* (CMLE) and *unconditional maximum likelihood estimators* (UMLE) of θ. It can be shown¶ that:

1. If $\hat{\gamma}_\theta(\mathbf{V})$ is a (conditionally) efficient estimator of θ, then $\theta = \hat{\gamma}_\theta$ is a unique solution of Eq. (21.12*a*).

2. If $\hat{\gamma}_\theta(\mathbf{V})$ is a (conditionally) sufficient estimator of θ, then any (nonconstant) solution of Eq. (21.12*a*) is a function of $\hat{\gamma}_\theta(\mathbf{V})$, for example, $\hat{\gamma}_\theta = f(\hat{\gamma}_\theta)$.

† See Sec. 19.4-2 and, for estimation, Grenander.[9]

‡ See Cramér,[2] chap. 33.

§ More precisely the unknown parameter is assumed to take some particular value θ.

¶ Cramér,[2] sec. 33.2, and Prob. 21.1*b*.

Similarly, for unconditional estimators, it can be established that:

1. If $\hat{\gamma}_\sigma(\mathbf{V})$ is an (unconditionally) efficient estimator, then $\theta = \hat{\gamma}_\sigma$ is a unique solution of Eq. (21.12*b*).

2. If $\hat{\gamma}_\sigma(\mathbf{V})$ is an (unconditionally) sufficient estimator, then any solution of Eq. (21.12*b*) is a function of $\hat{\gamma}_\sigma$.

These expressions are readily generalized to more than one parameter $\boldsymbol{\theta} = (\theta_1, \ldots, \theta_M)$; Eqs. (21.12) become now a set of likelihood equations

$$\frac{\partial}{\partial \boldsymbol{\theta}} [\log l_n(\mathbf{V}|\boldsymbol{\theta})] \bigg|_{\boldsymbol{\theta}=\hat{\boldsymbol{\gamma}}_{\boldsymbol{\theta}}} = 0 \tag{21.13a}$$

$$\frac{\partial}{\partial \boldsymbol{\theta}} [\log L_n(\mathbf{V},\boldsymbol{\theta})] \bigg|_{\boldsymbol{\theta}=\hat{\boldsymbol{\gamma}}_\sigma} = 0 \tag{21.13b}$$

whose simultaneous (nonconstant) solutions yield the M components $[\hat{\gamma}_{\theta},\hat{\gamma}_{\theta_2}, \ldots, \hat{\gamma}_{\theta_M}]$ of $\hat{\boldsymbol{\gamma}}_{\boldsymbol{\theta}}$, with similar results for $\hat{\boldsymbol{\gamma}}_\sigma$. Observe that even when the parameters $\boldsymbol{\theta}$ are statistically independent, i.e., when $\sigma(\boldsymbol{\theta}) = w_1(\theta_1)w_2(\theta_2) \cdots w_M(\theta_M)$, the solutions of Eqs. (21.13) are not necessarily algebraically independent, while each component $\hat{\gamma}_{\theta_j}$ depends in general on the entire data sample $\mathbf{V}$. The former occurs because l_n, L_n are not usually *linear* functions of $\boldsymbol{\theta}$ (an important exception to be considered presently is the case of signal amplitude, where $\theta = A_0$), while the latter arises because of the correlated nature of the sampled data (that is, $\overline{V_iV_j} \neq \bar{V}_i\bar{V}_j$).

When the samples become very large, and in fact when $n \to \infty$ $(T \to \infty)$, it can also be shown† that, under rather general conditions the CMLE $\hat{\gamma}_\theta$ is asymptotically efficient, that is, $\lim_{n\to\infty} \hat{\gamma}_\theta(\mathbf{V}) \to \theta$ (probability 1), and, correspondingly, that the UMLE $\hat{\gamma}_\sigma$ is likewise asymptotically efficient, that is, $\lim_{n\to\infty} \hat{\gamma}_\sigma(\mathbf{V}) \to \bar{\theta}$ (probability 1). Furthermore, the maximum likelihood estimators $\hat{\gamma}_\theta$, $\hat{\gamma}_\sigma$ are themselves asymptotically normally distributed, with mean values $\boldsymbol{E}_{\mathbf{V}|\theta}\{\hat{\gamma}_\theta(\mathbf{V})\} \sim \theta$, $\boldsymbol{E}\,\{\hat{\gamma}_\sigma(\mathbf{V})\} \sim \bar{\theta}$ and variances $\sigma_\theta{}^2$, $\sigma_\sigma{}^2$, for example,

$$\begin{aligned} w(\hat{\gamma}_\theta) &\sim (2\pi\sigma_\theta{}^2)^{-\frac{1}{2}}e^{-(\hat{\gamma}_\theta-\theta)^2/2\sigma_\theta{}^2} \\ w(\hat{\gamma}_\sigma) &\sim (2\pi\sigma_\sigma{}^2)^{-\frac{1}{2}}e^{-(\hat{\gamma}_\sigma-\bar{\theta})^2/2\sigma_\sigma{}^2} \end{aligned} \tag{21.14}$$

where, specifically,‡

† Cramér,[2] sec. 33.3; here obvious extensions to correlated samples $\mathbf{V}$ must be made.

‡ Conditions on the d.d.'s $W_n(\mathbf{V}|\theta)$ and $W_n(\mathbf{V},\theta)$ that must be satisfied in order that the normal distributions [Eqs. (21.14)] occur in large-sample theory are basically those which permit application of the central-limit theorem (cf. Sec. 7.7-3). Here, for example, it is required (1) that integration with respect to $\mathbf{V}$ and differentiation with respect to θ (so that $\boldsymbol{E}\{\log l_n\} = 0$, etc.) can be interchanged; (2) that $(\partial^2/\partial\theta^2) \log l_n$, $(\partial^2/\partial\theta^2) \log L_n$ have finite means and variances; (3) that $n^{-1}(\partial^3/\partial\theta^3)l_n$, $n^{-1}(\partial^3/\partial\theta^3)L_n$ remain bounded as $n \to \infty$ $(T \to \infty)$ for all possible values of θ; and (4) that $\partial l_n/\partial\theta \big|_{\theta=\hat{\gamma}_\theta} = 0$, $\partial L_n/\partial\theta \big|_{\theta=\hat{\gamma}_\sigma} = 0$, respectively. Essentially, one must have a large number of "independent" samples here, so that, as $T \to \infty$, $\overline{V_iV_j} \to \bar{V}_i\bar{V}_j$ (etc., for the higher moments) for a sufficiently large number of such terms. See Cramér,[2] sec. 33.3, and Mood,[4] sec. 10.8.

$$\begin{aligned}\sigma_\theta{}^2 &\sim \left[-E_{\mathbf{V}|\theta} \left\{ \frac{\partial^2}{\partial \theta^2} \log l_n(\mathbf{V}|\theta) \right\} \right]^{-1} \\ \sigma_\sigma{}^2 &\sim \left[-E_{\mathbf{V},\theta} \left\{ \frac{\partial^2}{\partial \theta^2} \log L_n(\mathbf{V},\theta) \right\} \right]^{-1}\end{aligned} \tag{21.14a}$$

[In the limit $n \to \infty$ $(T \to \infty)$, the $\sim$ is replaced by the equalities.] Note specifically from Eqs. (21.14*a*) in Eqs. (21.5*a*), (21.5*b*) that for unbiased cases maximum likelihood estimators are asymptotically efficient, as mentioned above. The results (21.14) are readily extended to the important case of two or more parameters $\boldsymbol{\theta} = (\theta_1, \ldots, \theta_M)$. The maximum likelihood estimators $\hat{\boldsymbol{\gamma}}_{\boldsymbol{\theta}}$, $\hat{\boldsymbol{\gamma}}_{\boldsymbol{\sigma}}$, obtained from Eqs. (21.13), are (as $n \to \infty$, $T \to \infty$) asymptotically distributed in a multivariate normal distribution with means $E_{\mathbf{V}|\theta}\{\hat{\boldsymbol{\gamma}}_{\boldsymbol{\theta}}(\mathbf{V})\} \simeq \boldsymbol{\theta}$, $E_{\mathbf{V}}\{\boldsymbol{\gamma}_\sigma(\mathbf{V})\} \simeq \bar{\boldsymbol{\theta}}$ and covariance matrices

$$(\mathbf{K}_\theta)_{ij} \simeq \left[-E_{\mathbf{V}|\boldsymbol{\theta}} \left\{ \frac{\partial^2}{\partial \theta_i \, \partial \theta_j} \log l_n(\mathbf{V}|\boldsymbol{\theta}) \right\} \right]^{-1} \qquad i, j = 1, \ldots, M \tag{21.15a}$$

$$(\mathbf{K}_\sigma)_{ij} \simeq \left[-E_{\mathbf{V},\boldsymbol{\theta}} \left\{ \frac{\partial^2}{\partial \theta_i \, \partial \theta_j} \log L_n(\mathbf{V},\boldsymbol{\theta}) \right\} \right]^{-1} \tag{21.15b}$$

under conditions essentially the same as those for the single-parameter case. As we shall note presently (cf. Sec. 21.2-5), the asymptotically normal character of the distribution of a wide class of optimum estimators, not only maximum likelihood cases, also occurs for threshold estimation of signals and their parameters in additive noise backgrounds, where the noise process itself need not be normal. Again, this is basically a result of the fact that in threshold operation one has sufficiently large numbers of "independent" sample values, so that the central-limit theorem can be applied.

Finally, for continuous sampling in $(0,T)$ we obtain the *maximum likelihood functionals* $\lim_{n \to \infty} \hat{\gamma}_\theta(\mathbf{V}) = \hat{\gamma}_\theta[V(t)]_T$, $\lim_{n \to \infty} \hat{\gamma}_\sigma(\mathbf{V}) = \hat{\gamma}_\sigma[V(t)]_T$ (probability 1) whenever these limits exist (in the regular case). Thus, in the communication situations considered here, where correlated sampling commonly occurs, the analogue of classical estimation theory based on n independent observations of a random variable is an estimation theory based on n independent realizations (or member functions) of the process $V(t) = S(t) + N(t)$ in a finite observation period $(0,T)$. As Grenander has shown,[9] the estimator obtained by maximizing the simultaneous likelihood functionals of the n realizations then has the asymptotic properties of efficiency and normality (as $n \to \infty$) corresponding to the above in the discrete cases. Moreover, the maximum likelihood functional based on only one realization in $(0,T)$ has these same desirable asymptotic characteristics as $T \to \infty$, provided that the process is ergodic and possesses certain generalized Markoff properties.†

† See, in particular, Grenander,[9] secs. 3.9, 10.11, 10.12, 10.14.

21.1-3 Three Examples of Signal Extraction by Maximum Likelihood. We shall illustrate the foregoing results with three examples of communication interest. In all instances, signal and noise are assumed to be additive, and it is known that a deterministic or entirely random signal is actually present. Maximum likelihood methods are then to be used to estimate one or more of the unknown signal parameters or, in the case of stochastic signals, the signal waveform itself.

(1) Coherent threshold extraction of signal amplitude in additive noise. This situation arises whenever it is necessary to establish input signal levels or otherwise make decisions based on an empirical knowledge of the input signal-to-noise ratio a_0. Let us assume first that everything is known about the signal except its (normalized) amplitude a_0 [cf. Eqs. (19.49)] and that the a priori distribution density of possible amplitudes a_0 is $w_1(a_0)$ $(0 \leq a_0)$. Similarly, the additive noise process has a known d.d. $W_n(\mathbf{V})_N$ of the form of Eqs. (19.44). We seek now the conditional and unconditional maximum likelihood estimators $\hat{\gamma}_\theta = \hat{\gamma}_{a_0}$, $\hat{\gamma}_\sigma \equiv \hat{a}_0$ of $\theta = a_0$ based on the data process $\mathbf{V} = \mathbf{S} + \mathbf{N}$. Normalizing $\mathbf{V}$ $(= \mathbf{v}\psi^{1/2})$ and $\mathbf{S}$ $(= a_0\psi^{1/2}\mathbf{s})$ as before [cf. Eqs. (19.49)], we expand $W_n(\mathbf{V} - \mathbf{S}|a_0)_N \equiv l_n(\mathbf{V}|\theta)$ and $w_1(a_0)W_n(\mathbf{V} - \mathbf{S}|a_0)_N \equiv L_n(\mathbf{V},\theta)$ in powers of a_0, as in threshold detection [cf. Eq. (19.45)], to obtain for the logarithms of l_n, L_n in the corresponding weak-signal cases

$$\begin{aligned} \log l_n = \log W_n(\mathbf{V} - \mathbf{S}|a_0)_N &= B^{(0)} + [\psi^{1/2}\tilde{\mathbf{v}}\mathbf{b}^{(1)} + \psi\tilde{\mathbf{v}}(\mathbf{b}^{(2)}\tilde{\mathbf{b}}^{(2)})\mathbf{v} + \cdots] \\ &\quad - a_0[\psi^{1/2}\tilde{\mathbf{s}}\mathbf{b}^{(1)} + 2\psi\tilde{\mathbf{v}}(\mathbf{b}^{(2)}\tilde{\mathbf{b}}^{(2)})\mathbf{s} + O(v^2)] \\ &\quad + a_0^2[\psi\tilde{\mathbf{s}}(\mathbf{b}^{(2)}\tilde{\mathbf{b}}^{(2)})\mathbf{s} + O(v)] + O(a_0^3) \qquad (21.16a) \end{aligned}$$

$$\log L_n = \log l_n + \log w_1(a_0) \qquad (21.16b)$$

Abbreviating $\mathbf{B}^{(2)} \equiv \mathbf{b}^{(2)}\tilde{\mathbf{b}}^{(2)}$, let us first apply Eqs. (21.13) to Eq. (21.16*a*) in order to determine the CMLE $\hat{\gamma}_{a_0}$. This is easily done to give the CML threshold estimator

$$T_R^{(N)}\{\mathbf{V}\} \equiv \hat{\gamma}_{a_0}(\mathbf{V}) \doteq (\psi^{1/2}\tilde{\mathbf{s}}\mathbf{b}^{(1)} + 2\psi\tilde{\mathbf{v}}\mathbf{B}^{(2)}\mathbf{s} + \cdots)/2\psi\tilde{\mathbf{s}}\mathbf{B}^{(2)}\mathbf{s} \qquad (21.17)$$

Now substituting Eq. (21.17) back into Eq. (21.16*a*), we see that $W_n(\mathbf{V} - \mathbf{S}|a_0)_N$ can be written in this approximation,

$$\begin{aligned} W_n(\mathbf{V} - \mathbf{S}|\hat{\gamma}_{a_0})_N &\doteq \exp(-2\psi\hat{\gamma}_{a_0}a_0\tilde{\mathbf{s}}\mathbf{B}^{(2)}\mathbf{s} + \psi a_0^2\tilde{\mathbf{s}}\mathbf{B}^{(2)}\mathbf{s}) \\ &\quad \times \exp(B^{(0)} + \psi^{1/2}\tilde{\mathbf{v}}\mathbf{b}^{(1)} + \psi\tilde{\mathbf{v}}\mathbf{B}^{(2)}\mathbf{v}) \qquad (21.18) \end{aligned}$$

Since this d.d. factors into one term involving the estimator $\hat{\gamma}_{a_0}$, the parameter a_0, and another term independent of a_0, the estimator† $\hat{\gamma}_{a_0}$ is (conditionally) *sufficient* for threshold signals, when terms $O(v,v^2, \ldots)$ can

† Recall that $\hat{\gamma}_{a_0}$ regarded as a function of an ensemble of values $\mathbf{V}$ is here the (conditional) maximum likelihood *estimator* of a_0, while, for a particular $\mathbf{V} = \mathbf{V}'$, $\hat{\gamma}_{a_0}$ becomes the corresponding *estimate* of a_0, based on these particular values $\mathbf{V}'$.

safely be ignored in Eq. (21.16a)† [cf. Sec. 21.1-1(3)]. Moreover, since $\mathbf{V} = \mathbf{S} + \mathbf{N}$ is equal to $\mathbf{v} = a_0\mathbf{s} + \mathbf{n}$ in normalized form (with $\mathbf{n} = \mathbf{N}/\psi^{1/2}$), we find from Eqs. (21.1), (21.17) that

$$E_{\mathbf{V}|a_0}\{\hat{\gamma}_{a_0}\} \doteq a_0 + (2\psi\tilde{\mathbf{s}}\mathbf{B}^{(2)}\bar{\mathbf{n}} + \psi^{1/2}\tilde{\mathbf{s}}\mathbf{b}^{(1)})/2\psi\tilde{\mathbf{s}}\mathbf{B}^{(2)}\mathbf{s} \tag{21.19}$$

so that $b(a_0) =$ constant. However, with a background noise whose mean value is zero, we have $\mathbf{b}^{(1)} = \bar{\mathbf{n}} = 0$ and consequently $\langle\hat{\gamma}_{a_0}\rangle_{\mathbf{v}} = a_0$, so that now $\hat{\gamma}_{a_0}$ is *unbiased* in the threshold case [cf. Eq. (21.1)]. Thus, the unbiased CMLE here is, from Eq. (21.17), simply

$$\hat{\gamma}_{a_0}(\mathbf{V}) \doteq \tilde{\mathbf{v}}\mathbf{B}^{(2)}\mathbf{s}/\tilde{\mathbf{s}}\mathbf{B}^{(2)}\mathbf{s} \tag{21.20}$$

This unbiased CMLE is also (asymptotically) *efficient*,† for example, $\lim_{n,T\to\infty} e(a_0) = 1$ [cf. Eq. (21.7a)], with the asymptotically normal d.d. [Eqs. (21.14)], where $\bar{\hat{\gamma}}_{a_0} = a_0$ and $\sigma_{a_0}{}^2 \simeq -(2\psi\tilde{\mathbf{s}}\mathbf{B}^{(2)}\mathbf{s})$.

Now let us consider the unconditional maximum likelihood estimator $\hat{\gamma}_\sigma(\mathbf{V}) \equiv \hat{a}_0$ obtained as a nontrivial solution of Eq. (21.13b), with Eqs. (21.16) in the threshold case, viz.,

$$\left[\frac{w_1'(a_0)}{w_1(a_0)} - (\psi^{1/2}\tilde{\mathbf{s}}\mathbf{b}^{(1)} + 2\psi\tilde{\mathbf{v}}\mathbf{B}^{(2)}\mathbf{s} + \cdots) + 2a_0(\psi\tilde{\mathbf{s}}\mathbf{B}^{(2)}\mathbf{s} + \cdots)\right]_{a_0=\hat{a}_0} = 0 \tag{21.21}$$

An explicit solution depends, of course, on the a priori d.d. $w_1(a_0)$. When $w_1(a_0)$ is unknown to the observer, he may employ an additional (sub-) criterion of optimality, choosing for w_1 that d.d. which maximizes the average uncertainty in a_0 (cf. Prob. 6.7). The two cases of chief interest here occur for peak and average power constraints for which $w_1(a_0)$ becomes, respectively,

Peak power:

$$(a_0)^2_{\max} = P_M \qquad w_1(a_0) = \frac{1}{P_M{}^{1/2}} \qquad 0 < a_0 < P_M{}^{1/2} \tag{21.22a}$$

Average power:

$$\overline{a_0{}^2} = 2P_0 \qquad w_1(a_0) = \frac{a_0 e^{-a_0{}^2/2P_0}}{P_0} \qquad 0 \leq a_0 \tag{21.22b}$$

When $\bar{N} = 0$ (for example, when $\mathbf{b}^{(1)} = 0$), we get the following threshold UMLE's from Eqs. (21.22) in Eq. (21.21),‡

† More precisely, since Eqs. (21.17), (21.18) are valid only in the weak-signal cases where sufficiently large samples ensure that the higher-order contributions $O(v,v^2, \ldots)$ can be neglected, $\hat{\gamma}_{a_0}$ is *asymptotically sufficient*, just as $\hat{\gamma}_{a_0}(\mathbf{V}')$ is an asymptotically sufficient estimate. Unless otherwise indicated, we shall assume $W_n(\mathbf{V})_N$ to be such that our estimators possess these large-sample properties.

‡ We exclude the singular points $a_0 = 0$, $a_0 = P_M{}^{1/2}$ in $w_1'(a_0)$ for the uniform distribution, as they clearly do not yield solutions depending on the received data $\mathbf{V}$.

$$\hat{a}_0\Big|_{P_M} \doteq \tilde{\mathbf{v}}\mathbf{B}^{(2)}\mathbf{s}/\tilde{\mathbf{s}}\mathbf{B}^{(2)}\mathbf{s} \tag{21.23a}$$

$$\hat{a}_0\Big|_{P_0} \doteq \frac{\psi P_0\tilde{\mathbf{v}}\mathbf{B}^{(2)}\mathbf{s} \pm [(\psi P_0\tilde{\mathbf{v}}\mathbf{B}^{(2)}\mathbf{s})^2 - 2\psi\tilde{\mathbf{s}}\mathbf{B}^{(2)}\mathbf{s}P_0^2 + P_0]^{1/2}}{2\psi P_0\tilde{\mathbf{s}}\mathbf{B}^{(2)}\mathbf{s} - 1} \doteq \frac{\tilde{\mathbf{v}}\mathbf{B}^{(2)}\mathbf{s}}{\tilde{\mathbf{s}}\mathbf{B}^{(2)}\mathbf{s}} \tag{21.23b}$$

this last in consequence of the threshold condition $P_0^2 \ll 1$ and the fact that $\overline{a_0^2}\tilde{\mathbf{s}}\mathbf{B}^{(2)}\mathbf{s} > \overline{a_0^2}$ and $\overline{a_0^2}\tilde{\mathbf{v}}\mathbf{B}^{(2)}\mathbf{s} > \overline{a_0^2}$, on the average, if any precision is to be obtained. Note here that, for both the uniform† and Rayleigh distribution densities of the normalized amplitude a_0, the unconditional threshold estimators $\hat{a}_0\Big|_{P_M}$, $\hat{a}_0\Big|_{P_0}$ are equal to the conditional threshold estimator $\hat{\gamma}_{a_0}$ [cf. Eq. (21.20)]. Like the latter, they are unbiased, for example, $\bar{\hat{a}}_0 = \bar{a}_0$ [cf. Eq. (21.2)], (asymptotically) efficient [cf. Eq. (21.7b)], with the same normal distributions [cf. Eqs. (21.14)], in which $\hat{\gamma}_\sigma \equiv \bar{\hat{a}}_0 = \bar{a}_0$, $\sigma_\sigma^2 \simeq -(2\psi\tilde{\mathbf{s}}\mathbf{B}^{(2)}\mathbf{s}) = \sigma_{a_0}^2$ above. Note that the maximum likelihood estimators of the *unnormalized* amplitude A_0 may be obtained directly from these results [cf. Eqs. (21.17) et seq.] if we observe that since $A_0 \equiv \sqrt{2\psi}\, a_0$ [Eqs. (19.49)] the corresponding CML and UML estimators are

$$\hat{\gamma}_{A_0}(\mathbf{V}) = \sqrt{2\psi}\,\hat{\gamma}_{a_0}(\mathbf{V}) \qquad \hat{\gamma}_\sigma(\mathbf{V})_{A_0} = \sqrt{2\psi}\,\hat{a}_0 = \hat{A}_0 \tag{21.24a}$$

The d.d.'s of A_0 likewise are given by the "unnormalized" forms of Eqs. (21.22a), (21.22b), that is,

$$w_1(A_0) = \begin{cases} (A_0)_{\max}^{-1} & 0 \le A_0 \le (A_0)_{\max} \\ \dfrac{2A_0 e^{-A_0^2/\overline{A_0^2}}}{\overline{A_0^2}} & 0 \le A_0 < \infty \end{cases} \tag{21.24b}$$

A simple example occurs when the background noise is normal with zero mean. Then $\mathbf{B}^{(2)} = -\frac{1}{2}\psi^{-1}\mathbf{k}_N^{-1}$ [Eqs. (19.47)], $\mathbf{b}^{(1)} = 0$, and the CMLE [Eq. (21.20)] is exact for all signal strengths a_0 and sample sizes n. The corresponding UMLE [Eq. (21.23a)] for a uniform distribution of amplitudes

† Uniform over a finite domain (and zero elsewhere). However, a common approximation in some applications is $P_M \to \infty$: signals of all peak amplitudes may occur, so that the domain of $w_1(a_0)$ is essentially infinite. Actually, from the strict point of view, all domains associated with deterministic signals are assumed finite, as required by the condition D (Sec. 18.5-3), on which the complete-class theorem is based. For the approximation of an "infinite" domain in the case of nonuniform distributions, the intent is that the domain be taken sufficiently large so that the contributions of the "tails" of the distribution are negligible. Correspondingly, for uniform distributions [cf. Eqs. (21.22a)] which are also of particular importance in the discussion of Minimax systems with additive signals and noise (cf. Sec. 21.2-2), it is assumed that such d.d.'s are constant over a large finite domain and are zero outside, so that there is a small but finite probability of each signal value within the domain and a finite, though large, variance.

is likewise given exactly by Eq. (21.20), that is,

$$T_R^{(N)}\{\mathbf{V}\} = \hat{\gamma}_{a_0}(\mathbf{V}) = \tilde{\mathbf{v}}\mathbf{k}_N^{-1}\mathbf{s}/\tilde{\mathbf{s}}\mathbf{k}_N^{-1}\mathbf{s} = \hat{\gamma}_\sigma(\mathbf{V}) \tag{21.25}$$

These estimators are unbiased, efficient, and hence sufficient, with normal distributions [Eqs. (21.14)] for *all* sample sizes $n \geq 1$, and with means and variances that are, respectively, a_0, $\bar{a}_0$ $[= P_M^{1/2}/2 = (a_0)_{\max}/2]$ and $\sigma_{a_0}{}^2 = \tilde{\mathbf{s}}\mathbf{k}_N^{-1}\mathbf{s} = \sigma_\sigma{}^2$. On the other hand, although the first relation in Eq. (21.23*b*) is now exact† for the Rayleigh distribution of amplitudes of Eqs. (21.22*b*), only in the threshold and strong-signal cases [where the second relation of Eq. (21.23*b*) applies] is the resulting UMLE effectively equal to that for the uniform amplitude distribution [Eqs. (21.24)]. For these extremes, $\hat{a}_0\Big|_{P_0}$ is likewise approximately normally distributed with means $\bar{a}_0$ and variances $\tilde{\mathbf{s}}\mathbf{k}_N^{-1}\mathbf{s}$, as well as being unbiased and efficient. The exact d.d. of $\hat{a}_0\Big|_{P_0}$, however, is quite involved. Observe, also, that the estimators of the unnormalized amplitude A_0 [e.g., Eq. (21.25) in Eqs. (21.24)] have corresponding properties.

When continuous sampling is employed, the corresponding functional forms of these maximum likelihood estimators may be obtained by an appropriate passage to the limit, when this limit exists. We use the methods of Sec. 19.4-2, replacing the various quadratic forms in Eqs. (21.17), (21.20), (21.21), (21.23), (21.25) by quadratic functionals in the usual way.‡ Thus, for the normal process above, the CMLE and UMLE with uniform amplitude d.d. [Eq. (21.25)] become the functionals

$$T_R^{(N)}\{\mathbf{V}\} = \hat{\gamma}_{a_0}[V(t)] = \Phi_T(v,s)/\Phi_T(s,s) = \hat{a}_0(\mathbf{V}) \tag{21.26}$$

where $\Phi_T(v,s)$, $\Phi_T(s,s)$ are given by Eqs. (20.37), with $\epsilon = \epsilon_0$, $\boldsymbol{\theta} = \boldsymbol{\theta}_0$ specified parameter values. For white noise backgrounds, $\hat{\gamma}_{a_0} = \hat{a}_0 \to 0$, since $\psi \to \infty$; however, from Eqs. (21.24) and (21.26) we get, for the *unnormalized* amplitude A_0, the CML and UML estimators

$$\hat{\gamma}_{A_0}[V(t)]\Big|_{\text{white}} = \frac{\sqrt{2}\int_0^T V(t)s(t-\epsilon_0;\boldsymbol{\theta}_0)\,dt}{\int_0^T s(t-\epsilon_0;\boldsymbol{\theta}_0)^2\,dt} = \hat{A}_0[V(t)]\Big|_{\text{white}} \tag{21.26a}$$

which have the dimensions of an amplitude, as expected.

For continuous sampling, generally, we obtain definite values of $\hat{\gamma}_{a_0}$, $\hat{a}_0$ (or $\hat{A}_0$, etc.), although $\lim_{n\to\infty} B^{(0)} \to \infty$ in l_n, L_n [Eqs. (19.44), (21.16*a*)],

† For normal noise, where $\mathbf{B}^{(2)} = -(2\psi)^{-1}\mathbf{k}_N^{-1}$, it is clear that only the negative sign before the square root in Eq. (21.23*b*) appears.

‡ The regular case is assumed (cf. end of Sec. 21.1-1).

since basically the maximum likelihood procedure involves the *relative* frequencies of occurrence of a signal with some a_0 (or A_0) against those of other values, $\hat{\gamma}_{a_0}$ of $\hat{a}_0$, etc., being most likely among the possible choices. A similar situation arises in detection, and for the same reasons. In fact, the normalization $B^{(0)}$ [cf. Eqs. (19.47) in the gauss case] could perfectly well be omitted in both instances right at the start. The distribution densities of these functional estimators still approach the normal frequency functions of Eqs. (21.14), as $n \to \infty$ and $T \to \infty$, where now the means and variances are replaced by their corresponding limiting forms. For the estimators of Eq. (21.26), the sample size T can have any value, and the d.d.'s are given *exactly* by Eqs. (21.14) with $\theta = a_0$, $\sigma_\theta^2 = \Phi_T(s,s)$, $\bar{\theta} = (a_0)_{\max}/2$, $\sigma_\sigma^2 = \Phi_T(s,s)$.

Structurally, the CMLE and UMLE of the gaussian case [Eqs. (21.25), (21.26)], for example, may consist of a matched filter of the type of Eq. (20.10) [or Eq. (20.43*d*) for continuous sampling] with read-out at time $t = T$. In fact, this filter is identical (except for scale) with the filters used to maximize signal-to-noise ratio under similar conditions of coherent observation [cf. Eq. (16.99)]. This is not surprising, since the maximum likelihood estimators [Eq. (21.26)] of a_0 are, as we shall see presently (cf. Sec. 21.3), optimum estimators of the input signal-to-noise ratio a_0 with respect to various criteria, which in effect maximizes signal vis-à-vis noise. These estimators may also be expressed in terms of the unrealizable linear filters of Eqs. (20.8 to 20.10), (20.42) and Fig. 20.1.

(2) Incoherent extraction of a stochastic signal in normal noise. In contrast with the preceding example, the signal $S(t)$ is now an entirely random normal process, so that reception is necessarily incoherent. The background noise is also normal and additive, and both signal and noise are assumed to be independent, although not necessarily stationary, with zero means and covariance functions K_S, K_N. Our problem here is to extract by maximum likelihood methods the actual waveforms of the signal process in an interval $(0,T)$ from the received process $V = S + N$ in this interval.

With discrete sampling, our task accordingly is to determine the n components $(\mathbf{S})_j = S(t_j)$ $(j = 1, \ldots, n)$ obtained at the sampling instants t_j in $(0,T)$. Thus $\boldsymbol{\theta}$ of Eqs. (21.13) is the n-component vector $\boldsymbol{\theta} \equiv \mathbf{S}$, and our receiver $T_R\{\mathbf{V}\}$ is chosen to be the unconditional maximum likelihood estimator $\hat{\gamma}_\sigma(\mathbf{V}) = \hat{\mathbf{S}}(\mathbf{V})$ of each of these n components. In general, the UML estimate of *each* $S(t_j)$ depends on *all* the data $\mathbf{V}$ in $(0,T)$. To see how this comes about, let us observe from Eqs. (19.47) that the logarithm of the unconditional likelihood function of $\mathbf{V} = \mathbf{S} + \mathbf{N}$ can be written

$$\log L_n(\mathbf{V},\mathbf{S}) = B^{(0)}_{S,N} - \tfrac{1}{2}(\tilde{\mathbf{V}} - \tilde{\mathbf{S}})\mathbf{K}_N^{-1}(\mathbf{V} - \mathbf{S}) - \tfrac{1}{2}\tilde{\mathbf{S}}\mathbf{K}_S^{-1}\mathbf{S} \quad (21.27)$$

where $B^{(0)}_{S,N} = -\tfrac{1}{2} \log [(2\pi)^{2n} \det \mathbf{K}_S \det \mathbf{K}_N]$. To find the UML estimator $\hat{\mathbf{S}}$ of $\mathbf{S}$ in $(0,T)$, we next differentiate Eq. (21.27) according to Eq.

(21.13*b*), with respect to each component of **S** in turn, and equate the results to zero, obtaining n equations for the n components of $\hat{\gamma}_\sigma = \hat{\mathbf{S}}$. These become

$$\hat{\mathbf{S}} = \mathbf{K}_S\mathbf{K}_N^{-1}(\mathbf{V} - \hat{\mathbf{S}}) \tag{21.28a}$$

Solving for $\hat{\mathbf{S}}$, we introduce a (positive definite) matrix $\mathbf{Q}_E$, which may be expressed in terms of the covariance matrices $\mathbf{K}_S$, $\mathbf{K}_N$ in several equivalent ways:

$$\mathbf{K}_S = \mathbf{Q}_E(\mathbf{K}_S + \mathbf{K}_N) \quad \text{or} \quad \mathbf{Q}_E = (\mathbf{K}_N^{-1} + \mathbf{K}_S^{-1})^{-1}\mathbf{K}_N^{-1} = \mathbf{K}_S(\mathbf{K}_S + \mathbf{K}_N)^{-1} \neq \tilde{\mathbf{Q}}_E \tag{21.28b}$$

$$\mathbf{K}_S = \mathbf{Q}_E(\mathbf{K}_S + \mathbf{K}_N) \quad \text{or} \quad \mathbf{Q}_E = (\mathbf{I} + \mathbf{K}_N\mathbf{K}_S^{-1})^{-1} = [(\mathbf{I} + \mathbf{K}_S\mathbf{K}_N^{-1})(\mathbf{K}_S\mathbf{K}_N^{-1})^{-1}]^{-1} \tag{21.28c}$$

Then the desired UMLE† of **S** is

$$T_R\{\mathbf{V}\} = \hat{\mathbf{S}} = \mathbf{Q}_E\mathbf{V} \tag{21.29}$$

It is easily shown (Prob. 21.2) that $\hat{\mathbf{S}}$ is unbiased (for example, $E\{\hat{\mathbf{S}}\} = \bar{\mathbf{S}} = 0$ here), efficient, with a covariance matrix $\mathbf{K}_{\hat{S}} = \mathbf{K}_S\tilde{\mathbf{Q}}_E$ and a multivariate normal d.d.

$$W_n(\hat{\mathbf{S}}) = [(2\pi)^n \det \mathbf{K}_S\tilde{\mathbf{Q}}_E]^{1/2} e^{-1/2\hat{\mathbf{S}}(\mathbf{K}_S\tilde{\mathbf{Q}}_E)^{-1}\mathbf{S}} \tag{21.30}$$

When continuous sampling is used, we may parallel in part the analysis of Secs. 20.1-2, 20.4-7, writing

$$\tilde{\mathbf{Q}}_E \equiv \boldsymbol{\varrho}_E \, \Delta t \qquad \boldsymbol{\varrho}_E \neq \tilde{\boldsymbol{\varrho}}_E \tag{21.31}$$

where now $\boldsymbol{\varrho}_E = [\rho_E(t_i,t_j)] \equiv [h_E(t_i - t_j, t_i)\,\Delta t]$, the weighting function of a (discrete) time-varying linear filter [cf. Eq. (20.20)]. This weighting function $\rho_E = h_E$ is the solution of the integral equation corresponding to the matrix equations (21.28), i.e.,

$$\int_{0-}^{T+} [K_S(t,u) + K_N(t,u)]\rho_E(u,\tau)_T \, du = K_S(t,\tau) \qquad 0 \leq t, \tau \leq T \tag{21.32}$$

where as usual $\rho_E = 0$ outside $(0,T)$. Then, with Eqs. (21.31) in the limiting form of Eq. (21.29), we find directly that the UML estimator for $S(t)$ at any $t = t_\lambda$ in $(0,T)$ is‡

$$\hat{S}(t_\lambda) = \int_{0-}^{T+} V(\tau)\rho_E(t_\lambda,\tau)_T \, d\tau = \int_{0-}^{T+} V(\tau)h_E(t_\lambda - \tau, t_\lambda) \, d\tau \qquad 0 \leq t_\lambda \leq T \tag{21.33}$$

† The CMLE of **S** is the not surprising result $\hat{\gamma}_\mathbf{S} = \mathbf{V}$ [cf. $\log l_n(\mathbf{V}|\mathbf{S}) = B_N^{(0)} - \frac{1}{2}(\tilde{\mathbf{V}} - \tilde{\mathbf{S}})\mathbf{K}_N^{-1}(\mathbf{V} - \mathbf{S})$ in Eq. (21.13*b*)]. Note that $\hat{\mathbf{S}} = \hat{\gamma}_\mathbf{S}$ if $\mathbf{K}_S^{-1} = \mathbf{0}$, the null matrix, as then $\mathbf{Q}_E = \mathbf{I}$. This has significance for Minimax extractors of **S** (cf. Secs. 21.4-1, 21.4-2).

‡ See Middleton and Van Meter,[1] pp. 95, 96, and Preston.[11]

The integral equation (21.32) may be solved by the methods of Sec. A.2.3 when the kernel $K_S + K_N$ is rational. Explicit results for RC and LRC noise signals in white noise backgrounds are given in Example 4 of Sec. A.2.4, Eqs. (A.2.110), (A.2.111), and (A.2.115), in conjunction with Eq. (A.2.12); see also Eqs. (A.2.103) to (A.2.106) and Eqs. (A.2.12), (A.2.92*b*).

Since the estimation process here is one of smoothing or interpolation, based on the data $V(t) = S(t) + N(t)$ received in the finite interval $(0,T)$ where the interpolation time t_λ likewise falls within the sample period, we may expect these results to coincide with those of the Wiener-Kolmogoroff theory discussed in Secs. 16.2-2, 16.2-3, when this UML estimator $\hat{S}(t_\lambda)$ has the same structure. A comparison of the above with the earlier treatment shows this to be indeed the case: the UML estimator here is also optimum from the standpoint of minimizing the mean-square error, in the sense of Eqs. (16.54) to (16.58). In particular, we note that now because of finite sample size† even if $t_\lambda = T$ (that is, if $t_\mu = 0$, the "now," or "present," of the observation) a linear realizable time-invariant filter is not possible: the optimum linear filter is necessarily time-varying, as our analysis above indicates [cf. Sec. 16.2-3(2)]. Finally, note that the linear structure of the UMLE [Eqs. (21.29), (21.33)] here is due to the gaussian character of both the signal and noise processes.

(3) Transcendental estimation. The comparatively simple nature of the maximum likelihood estimators in the preceding two examples is attributable primarily to the fact that the quantity estimated, i.e., amplitude, or wave shape, appears *linearly* in the signal waveform (and, of course, that signal and noise are additively combined). Sometimes, however, it is desired to estimate a parameter which appears as the argument of some transcendental operation, i.e., of a cosine, exponential, Bessel function, etc. Maximum likelihood methods may still be applied [cf. Eqs. (21.12), (21.13)], but much more involved expressions for estimator structure then occur. In fact, the CML and UML estimators are now obtained as the solutions of certain transcendental equations (hence the term "transcendental estimation"), whose precise nature depends on the way in which the parameter or parameters appear in the signal waveform. Except in very simple cases and in certain limiting situations (e.g., weak-signal large-sample theory), as the example following indicates, it is not possible to reduce these results to explicit form. For this reason maximum likelihood estimation does not generally provide so useful an approach to design and the evaluation of performance as procedures that yield explicit estimators (e.g., the quadratic cost function of Sec. 21.2-2). However, in the weak-signal case approximate relations can sometimes be found, so that maximum likelihood remains a possible tool in such instances.

Let us illustrate these remarks with the problem of UML estimation

† In this sense, Eq. (21.33) is a generalization of Sec. 16.2-3(2) [cf. Sec. 16.2-4(2) also].

of a parameter θ of a narrowband signal† such as carrier frequency, wave shape, or a time delay which appears transcendentally in the waveform $s(t;\epsilon_0;\theta) = F_s(t;\theta) \cos[\omega_s(t - \epsilon_0) - \phi_s(t;\theta)]$. This signal is observed incoherently (i.e., with sample uncertainty; cf. Sec. 19.4-3) on an interval $(0,T)$ in normal random noise, where ϵ_0 is an RF epoch, uniformly distributed over an RF cycle (cf. Sec. 20.1-3). All other signal parameters, e.g., amplitude, etc., are assumed known for the moment. Thus, we can write the unconditional likelihood function as

$$L_n(\mathbf{V},\theta) = \sigma(\theta) W_n(\mathbf{V}|\theta) = \langle F_n(\mathbf{V}|\mathbf{S}(\theta,\boldsymbol{\theta}'))\rangle_{\boldsymbol{\theta}'} \\ = \langle W_n[\mathbf{V} - \mathbf{S}(\theta,\boldsymbol{\theta}')]_N\rangle_{\boldsymbol{\theta}'} \qquad (21.34a)$$

where $\boldsymbol{\theta}'$ represents all other distributed parameters of the signal, here epoch ϵ_0 only. Since the likelihood *function* L_n is equal to the likelihood *ratio* Λ_n multiplied by $\mu^{-1}W_n(\mathbf{V})_N$, we may use Eq. (20.31*a*) to obtain directly

$$\log L_n(\mathbf{V},\theta) = \log W_n(\mathbf{V})_N - \frac{a_0^2}{2} B_n(\theta) + \log I_0[a_0 \sqrt{\tilde{\mathbf{v}}\mathbf{G}_0(\theta)\mathbf{v}}] \qquad (21.34b)$$

where $\mathbf{G}_0$, B_n are given by Eqs. (20.29*a*), (20.26), with $B_n \doteq 2 \operatorname{tr} \mathbf{k}_N\mathbf{G} = \operatorname{tr} \mathbf{k}_N\mathbf{G}_0$, alternatively [cf. Eq. (20.33*a*)]. The UMLE $\hat{\gamma}_\sigma = \hat{\theta}$ is accordingly found from Eq. (21.34*b*) in Eq. (21.12*b*), viz., as the solution of the transcendental equation

$$\frac{\sigma'(\hat{\theta})}{\sigma(\hat{\theta})} + \frac{a_0}{2}\left[\frac{I_1(a_0\Psi_{0n}^{1/2})}{I_0(a_0\Psi_{0n}^{1/2})} \tilde{\mathbf{v}} \frac{\partial \mathbf{G}_0}{\partial\theta} \mathbf{v}\Psi_{0n}^{-1/2} - a_0 \frac{\partial B_n}{\partial\theta}\right]_{\theta=\hat{\theta}} = 0 \qquad (21.35)$$

with $\Psi_{0n} = \tilde{\mathbf{v}}\mathbf{G}_0\mathbf{v}$. In the limit of continuous data sampling, we see as before that $\lim_{n\to\infty} \Psi_{0n} \to \Psi_{0T}$ [Eqs. (20.62)], $\lim_{n\to\infty} B_n \to B_T$ [Eqs. (20.63)], and that‡

$$\lim_{n\to\infty} \tilde{\mathbf{v}} \frac{\partial \mathbf{G}_0}{\partial\theta} \mathbf{v} \equiv \frac{\partial \Psi_{0T}}{\partial\theta} = \psi^2 \iint_{-\infty}^{\infty} v(t) \frac{\partial D_T(t,u;\theta)_0}{\partial\theta} v(u)\, dt\, du \qquad (21.36a)$$

$$\lim_{n\to\infty} \frac{\partial B_n}{\partial\theta} \equiv \frac{\partial B_T}{\partial\theta} = \frac{\psi}{2} \iint_{-\infty}^{\infty} K_N(t,u) \frac{\partial D_T(t,u;\theta)_0}{\partial\theta}\, dt\, du \qquad (21.36b)$$

With white noise backgrounds of spectral intensity W_0, these are modified to

† For further details and additional examples, see a report by Middleton.[12] Other and somewhat earlier results of interest using maximum likelihood methods in radar and communication situations are given in the papers of Slepian,[13] Youla,[13a] Swerling,[14] and Davis[15] and in the report of Bennion.[16]

‡ See the discussion of Sec. 19.4-2.

$$a_0^2 \frac{\partial \Psi_{0T}}{\partial \theta}\bigg|_{\text{white}} = \frac{2A_0^2}{W_0^2}\frac{\partial}{\partial \theta}\left\{\left[\int_0^T V(t)a(t;\theta)\,dt\right]^2 + \left[\int_0^T V(t)b(t;\theta)\,dt\right]^2\right\} \tag{21.37a}$$

$$a_0^2 \frac{\partial B_n}{\partial \theta}\bigg|_{\text{white}} = \frac{A_0^2}{2W_0}\int_0^T \frac{\partial}{\partial \theta}[F_s(t;\theta)^2]\,dt \tag{21.37b}$$

again where a, b are the continuous versions of Eqs. (20.24a) and F_s is the envelope of the (normalized) narrowband signal s [cf. Eq. (20.22)]. When

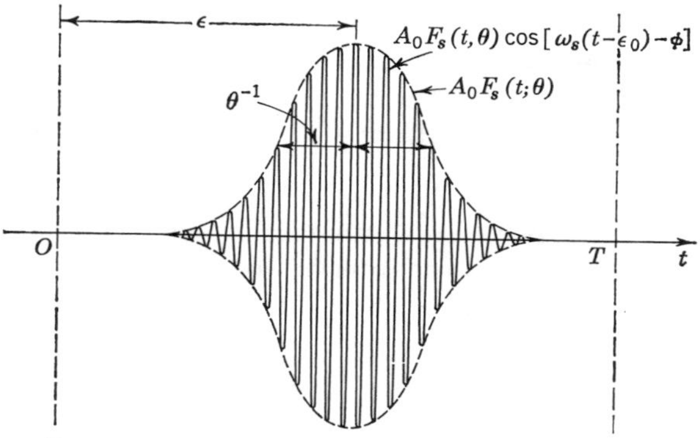

FIG. 21.1. A gaussian RF pulse observed in $(0,T)$.

there are other distributed parameters $\boldsymbol{\theta}'$ (besides amplitude and the RF epoch ϵ_0), Eq. (21.35) becomes the more general relation

$$\left.\frac{\sigma'(\theta)}{\sigma(\theta)} + \frac{\left\langle e^{-a_0^2B_n/2}\left[\frac{a_0}{2}I_1(a_0\Psi_{0n}^{1/2})\left(\tilde{\mathbf{v}}\frac{\partial \mathbf{G}}{\partial \theta}\mathbf{v}\right)\Psi_{0n}^{-1/2} - \frac{a_0^2}{2}\frac{\partial B_n}{\partial \theta}I_0(a_0\Psi_{0n}^{1/2})\right]\right\rangle_{a_0,\boldsymbol{\theta}'}}{\langle I_0(a_0\Psi_{0n}^{1/2})e^{-a_0^2B_n/2}\rangle_{a_0,\boldsymbol{\theta}'}}\right]_{\theta=\hat{\theta}} = 0 \tag{21.38}$$

This expression, however, is usually too difficult to handle, except possibly in limiting cases of strong or weak signals.

To see how approximate UML estimators $\hat{\theta}$ may be obtained from Eq. (21.35) under certain conditions, let us consider the weak-signal situation of a signal consisting of a train of RF gaussian pulses where the original background noise is white and continuous sampling is used in the course of incoherent observation on the interval $(0,T)$. Specifically, it is desired to estimate the "shape factor" θ of a typical pulse, when one pulse is observed in $(0,T)$ at a known (envelope) position ϵ entirely within the interval (cf. Fig. 21.1). In general, the weak-signal version of Eq. (21.35) becomes

$$\frac{\sigma'(\hat{\theta})}{\sigma(\hat{\theta})} + \frac{a_0^2}{2}\left(\tfrac{1}{2}\tilde{\mathbf{v}}\frac{\partial \mathbf{G}}{\partial \theta}\mathbf{v} - \frac{\partial B_n}{\partial \theta}\right)_{\theta=\hat{\theta}} = 0 \qquad a_0^4 \ll 1 \tag{21.39a}$$

which for continuous sampling in white noise reduces to

$$\frac{\sigma'(\hat{\theta})}{\sigma(\hat{\theta})} + \frac{A_0^2}{2W_0^2} \iint_0^T V(t)V(u) \cos \omega_s(t-u) \frac{\partial}{\partial \theta} [F_s(t;\theta)F_s(u;\theta)]_{\theta=\hat{\theta}}\, dt\, du - \frac{A_0^2}{4W_0} \int_0^T \frac{\partial}{\partial \theta} F_s^2(t,\theta) \bigg|_{\theta=\hat{\theta}} dt = 0 \quad (21.39b)$$

with $(A_0^2T/2W_0)^2 \ll 1$ for weak signals.

For the present problem, $F_s(t;\theta)$ is equal to $e^{-(t-\epsilon)^2\theta}$. Putting Eq. (21.39*b*) in the more convenient form

$$\frac{\sigma'(\theta)}{\sigma(\theta)} + \frac{\partial}{\partial \theta} \left[\frac{A_0^2}{2W_0^2} \left| \int_0^T V(t)F_s(t;\theta)e^{i\omega_s t}\, dt \right|^2 - \frac{A_0^2}{4W_0} \int_0^T F_s^2(t;\theta)\, dt \right]_{\theta=\hat{\theta}} = 0 \quad (21.39c)$$

let us expand $V(t)e^{i\omega_s t}$ about $t = \epsilon$ and employ the condition that the pulse is well away from the ends of the interval $(0,T)$, so that $T\theta^{1/2} \gg 1$, $\epsilon\theta^{1/2} \gg 1$. Then we can replace $(0,T)$ by $(-\infty,\infty)$ in the integrations, to obtain

$$\frac{\sigma'(\hat{\theta})}{\sigma(\hat{\theta})} + \frac{\partial}{\partial \theta} \left(\frac{A_0^2\pi}{2W_0^2\theta} \left| A^{(0)} + \frac{A^{(2)}}{2\theta} + \frac{3A^{(4)}}{8\theta^2} + \cdots \right|^2 - \frac{A_0^2\sqrt{\pi}}{4W_0^{1/2}\theta^{1/2}} \right)_{\theta=\hat{\theta}} = 0 \quad (21.40a)$$

where the $A^{(k)}$ are the coefficients $A^{(k)}(\epsilon) = d^k[V(x)e^{i\omega_s x}]/dx^k|_{x=\epsilon}$ in the expansion about $t = \epsilon$. A further simplification may be introduced by choosing $\epsilon = 2\pi m/\omega_s$, which is always approximately the case if ω_s is large, as we can always find an integer m for which $\epsilon' = 2\pi m/\omega_s$ is very close to the actual ϵ. The $A^{(k)}$ are then *real*, so that Eq. (21.40*a*) reduces finally to

$$\frac{\sigma'(\hat{\theta})}{\sigma(\hat{\theta})} - \frac{A_0^2\pi}{2W_0^2} \left(\frac{A^{(0)2}}{\hat{\theta}^2} + \frac{2A^{(0)}A^{(2)}}{\hat{\theta}^3} + \cdots \right) + \frac{A_0^2\sqrt{\pi}}{8W_0\sqrt{2}} \hat{\theta}^{-3/2} = 0 \quad (21.40b)$$

There remains the question of what $\sigma(\hat{\theta})$ to use: if $\sigma(\theta)$ is known a priori, then Eq. (21.40*b*) is to be solved for $\hat{\theta}$ to the desired degree of approximation. However, if $\sigma(\theta)$ is unavailable to the observer, then, as indicated earlier, we can introduce some suitable subcriterion governing the choice of σ. For example, for maximum average uncertainty in θ subject to a peak-value constraint, $\sigma(\hat{\theta})$ is uniform, and $\sigma'(\hat{\theta})$ vanishes,† so that to a first approximation $\hat{\theta}$ is found from Eq. (21.40*b*) to be

$$\hat{\theta} \doteq \hat{\theta}_1 = 32\pi V(\epsilon)^4/W_0^2 \quad (21.41)$$

Second- and higher-order approximations follow as the roots of third- and higher odd-order polynomials in $\hat{\theta}$.

We remark finally that the notion of bias (or its lack), sufficiency, efficiency, etc., may still be invoked, but again they are not easy to establish

† See footnote referring to Eq. (21.23*b*).

specifically, except in limiting circumstances, because of the implicit nature of the results for these estimators. Moreover, there appears to be no reasonably simple, general way of instrumenting the operation on V embodied in such expressions as Eqs. (21.35) to (21.38), short of a large number of separate subsystems to approximate the required differentiations, in contrast to the procedures available in threshold-detection theory. For these reasons, it is almost always more convenient to carry out parameter estimation in the so-called "transcendental" cases with respect to other criteria, as we shall see in Sec. 21.2, where these signal-extraction techniques are incorporated into the more general framework of decision theory.

PROBLEMS

21.1 (*a*) Establish Eq. (21.5*c*), and extend the Cramér-Rao inequality for conditional estimators to obtain Eq. (21.6). (HINT: Use Cramér,[2] pages 480, 481.)

(*b*) Establish the results (1), (2) following Eqs. (21.12*a*), (21.12*b*) for conditionally and unconditionally efficient and sufficient estimators. (HINT: See Cramér,[2] pages 499, 500.)

21.2 In the case of a normal stochastic signal accompanied by an additive and independent normal noise, show that the UML estimator $\hat{\mathbf{S}} = \mathbf{Q}_E\mathbf{V}$ [Eq. (21.29)] is unbiased, efficient, with the multivariate normal d.d. of Eq. (21.30). Show also that $\mathbf{Q}_E$ is positive definite.

21.3 A narrowband deterministic signal of unknown carrier frequency ω_0 is observed in $(0,T)$, accompanied by additive normal noise. Show that the relations determining the CML and UML estimators $\hat{\theta} = \hat{\omega}_0$ of $\theta = \omega_0$ are found from

$$\left[a_0 \frac{\partial \Phi_T(v,s)}{\partial \theta} - \frac{a_0^2}{2} \frac{\partial \Phi_T(s,s)}{\partial \theta}\right]_{\theta=\hat{\theta}} = 0$$
$$\left[a_0 \frac{\partial \Phi_T(v,s)}{\partial \theta} - \frac{a_0^2}{2} \frac{\partial \Phi_T(s,s)}{\partial \theta} + \frac{\sigma'(\theta)}{\sigma(\theta)}\right]_{\theta=\hat{\theta}} = 0 \qquad (1)$$

respectively, where $\Phi_T(v,s)$, $\Phi_T(s,s)$ are given by Eqs. (20.37). Hence, when a sinusoidal signal $A_0 \sin \omega_0 t$ is observed in white noise of spectral density W_0, show that the mean-square error $\overline{\Delta\omega^2} = \overline{(\hat{\omega}_0 - \omega_0)^2}$ in the CML estimate of (angular) frequency obeys the relation

$$(\Delta T)^2\, \overline{\Delta\omega^2}\, \sigma_0^2 \approx 1 \qquad (2)$$

where $\sigma_0^2 = A_0^2T/2W_0$ is the ratio of signal energy in $(0,T)$ to the noise power per unit bandwidth and $\Delta T = T$, provided that $\Delta T\, \bar{\omega}_0 \ll 1$, $\Delta T \omega_0 \ll 1$, that is, provided that the interval is sufficiently short. Discuss the case of longer intervals (including UML estimates), and outline a method for extending the time-frequency "uncertainty" relation (2) in this instance.

21.2 *Decision Theory Formulation*

So far, we have said nothing about the cost, or value, judgments that can be associated with the decisions required of a signal extraction system. As we remarked earlier, these decisions are now numbers or magnitudes, since signal extraction is basically an estimation or measurement process.

In turn, the fact that we have a definite decision implies some sort of cost assignment, so that we may expect that estimation systems, like detection systems, may be naturally incorporated within the framework of decision theory on choice of an appropriate cost function. That this is the case will be demonstrated here and in succeeding sections, where, for example, we shall show that classical estimation methods, like maximum likelihood under rather broad conditions (cf. Sec. 21.1-2), may be regarded as decision systems (minimizing average risk) with respect to certain cost assignments. Optimum systems here are therefore Bayes systems (cf. Sec. 18.4-3), as in the analogous detection theory, although the cost functions themselves and the resulting optimal structures are now quite different. An important feature of estimation from the decision-theory viewpoint is the variety of possible cost functions and corresponding optimum systems that can be generated. The present section is accordingly devoted to signal extraction (or estimation) from this standpoint and, specifically, to a discussion of some cost functions of practical interest, the associated Bayes systems, and their various properties, such as weak-signal structure, the distribution of estimators, and the probabilities of correct decisions.

Extraction in the theory of signal reception is the counterpart of parameter estimation in statistics. Here the signal parameters $\boldsymbol{\theta} = (\theta_1,\theta_2, \ldots ,\theta_M)$ are the parameters of the distribution density $F_n[\mathbf{V}|\mathbf{S}(\boldsymbol{\theta})]$ governing the occurrence of the received data $\mathbf{V}$. Frequently, the signal parameters $\boldsymbol{\theta}$ are taken to be the signal components $\mathbf{S}$ themselves, as in the case of stochastic signals and waveform estimation generally. Point estimation, or the direct estimation of each component of $\boldsymbol{\theta}$ or $\mathbf{S}$, is considered here, rather than estimation by confidence intervals. As before (cf. Sec. 18.3), we let $\boldsymbol{\gamma}$ represent the decision to be made about $\boldsymbol{\theta}$, where equivalently $\boldsymbol{\gamma}$ ($= \boldsymbol{\gamma}_{\boldsymbol{\theta}}$ or $\boldsymbol{\gamma}_{\sigma}$) is the required estimator of $\boldsymbol{\theta}$ [cf. Sec. 21.1-1(4)]. Observe also that, when $\boldsymbol{\gamma}$ is to be an estimate of $\boldsymbol{\theta}$, the parameter and decision spaces Ω_θ, Δ have the same structure, i.e., a one-to-one mapping exists between the two. We assume that each space contains a continuum of points and is a finite, closed region, which may, however, be taken large enough to be essentially infinite for practical purposes.† Similarly, when waveform estimation is desired, we set $\boldsymbol{\theta} = \mathbf{S}$ and the signal and decision spaces Ω, Δ of Fig. 18.5*a*, *b* likewise have the same structure. The cost functions $F_1 = C(\boldsymbol{\theta},\boldsymbol{\gamma})$, $F_1 = C(\mathbf{S},\boldsymbol{\gamma})$ [cf. Eq. (18.4)], to be used in the risk analysis, are as in detection (cf. Sec. 19.1) preassigned in accordance with the external constraints of the problem and are critical in determining the specific structure of the resulting system.

An important theorem, due to Hodges and Lehmann,[17] enables us to avoid randomized decision rules (cf. Sec. 18.3-2) in most applications. This theorem states:

† See the footnote referring to Eq. (21.23*b*); see also Sec. 18.5-3.

If Δ is the real line and if $C(\boldsymbol{\theta},\boldsymbol{\gamma})$ [or $C(\mathbf{S},\boldsymbol{\gamma})$] is a convex function† *of $\boldsymbol{\gamma}$ for every $\boldsymbol{\theta}$ (or $\mathbf{S}$), then for any decision rule δ there exists a nonrandomized decision rule whose average risk is not greater than that of δ for all $\boldsymbol{\theta}$ (or $\mathbf{S}$) in Ω_θ (or Ω).*

This applies to one-dimensional $\boldsymbol{\theta}$ $(= \theta_1)$, or $\mathbf{S}$ $(= S_1)$, and $\boldsymbol{\gamma}$ $(= \gamma_1)$ but can usually be extended to include multidimensional vectors.[18] Thus, when the decision rule is nonrandomized, we have from Eq. (18.15)

$$\delta(\boldsymbol{\gamma}|\mathbf{V}) = \delta[\boldsymbol{\gamma} - \boldsymbol{\gamma}_\sigma(\mathbf{V})] \tag{21.42}$$

where $\boldsymbol{\gamma}_\sigma(\mathbf{V}) = \boldsymbol{T}_R\{\mathbf{V}\}$ is the functional operation performed on the data $\mathbf{V}$ by the system $\boldsymbol{T}_R$ and $\boldsymbol{\gamma}$ is, for particular $\mathbf{V}$, the estimate produced by the system. Since we are interested primarily in Bayes systems, the unconditional estimator $\boldsymbol{\gamma}_\sigma(\mathbf{V})$, rather than the conditional estimator $\boldsymbol{\gamma}_\theta(\mathbf{V})$, is the quantity to be optimized, by minimization of the average risk. Accordingly, for waveform extraction, $\boldsymbol{\theta} = \mathbf{S}$, and $\boldsymbol{\gamma}(\mathbf{V})$ is an estimator of $\mathbf{S}$, based on $\mathbf{V}$ and on the a priori distribution $\sigma(\mathbf{S})$, while, for parameter extraction, the estimator $\boldsymbol{\gamma}_\sigma(\mathbf{V})$ is based instead‡ on $\mathbf{V}$ and $\sigma(\boldsymbol{\theta})$.

For waveform estimation, in contrast to that for time-independent parameters, the instants at which estimates are desired provide a convenient subdivision of the extraction process into three principal types of procedure. Let us write for the waveform value under estimation $S_\lambda = S(t_\lambda)$, where λ may assume a discrete (and, later, a continuous) set of values. Depending on how t_λ is chosen with respect to the times $(t_1,t_2, \ldots ,t_n)$ in $(0,T)$ at which the received data $V(t)$ are sampled (to give $\mathbf{V}$), we distinguish:

1. *Smoothing or interpolation.* Here t_λ for all λ lies within the observation period $(0,T)$, but it does not coincide with any of the times t_k $(k = 1, \ldots, n)$ of the data and signal elements $(V_1, \ldots ,V_k, \ldots ,V_n)$, $(S_1, \ldots ,S_k, \ldots ,S_n)$ on which the estimation of $\mathbf{S}$ is based. [λ may take a single value or many values in $(0,T)$; usually, however, only a single S_λ is to be determined, based on $\mathbf{V}$ and $\mathbf{S}$.]

2. *Simple or coincident estimation.* Here t_λ is one (or more) of the times t_k at which data are obtained and for which a priori information concerning $\mathbf{S}$ is also given. Thus, t_λ again lies in the observation interval $(0,T)$, and extraction consists in obtaining suitable estimates of $\mathbf{S}$ based on $\mathbf{V}$.

3. *Prediction or extrapolation.* In this case t_λ lies outside the data interval $(0,T)$, where no samples are taken, and we are asked now to make an estimate of $\mathbf{S}$ on the basis of $\mathbf{S}$ and $\mathbf{V}$ in $(0,T)$ (see Fig. 18-3).

From Eqs. (18.11) and (18.15), we accordingly write for the average risk

$$R(\sigma,\delta)_S = \int_\Omega \sigma(\mathbf{S},S_\lambda)\, dS_\lambda\, \mathbf{dS} \int_\Gamma C(\mathbf{S},S_\lambda;\boldsymbol{\gamma}_\sigma) F_n(\mathbf{V}|\mathbf{S})\, \mathbf{dV} \tag{21.43}$$

† A real-valued function $g(x)$ is convex in an interval (a,b) if for any x and y in (a,b) and any number $0 < \rho < 1$ one has $\rho g(x) + (1 - \rho)g(y) \geq g[\rho x + (1 - \rho)y]$.

‡ Again, we employ the convention that (unless otherwise indicated) functions of the same form and different arguments are different; that is, $\sigma(\boldsymbol{\theta}) \neq \sigma(\mathbf{S})$ unless $\boldsymbol{\theta} = \mathbf{S}$; $w(\mathbf{V}) \neq w(\mathbf{S}|\mathbf{V})$, etc. [cf. footnote referring to Eq. (18.10)].

where $F_n(\mathbf{V}|\mathbf{S})$ does not contain S_λ if $t_\lambda \neq t_k$ ($k = 1, \ldots, n$) or if $t_\lambda < 0$, $t_\lambda > T$, as in prediction.† Note that in the case of deterministic signals, since $\mathbf{S}$ is usually a function of one or more statistical parameters $\boldsymbol{\theta}$, we can make use of the fact that $\sigma(\mathbf{S},S_\lambda)\,\mathbf{dS}\,dS_\lambda$ when integrated over all Ω-space is equivalent to $\sigma(\boldsymbol{\theta})\,\mathbf{d\boldsymbol{\theta}}$, similarly integrated over all Ω_θ-space [Ω and Ω_θ mapping into each other as a result of the transformation $\mathbf{S} = \mathbf{S}(\boldsymbol{\theta})$]. Then Eq. (21.43) can be put into the alternative and more convenient form

$$R(\sigma,\delta)_S = \int_{\Omega_\theta} \sigma(\boldsymbol{\theta})\,\mathbf{d\boldsymbol{\theta}} \int_\Gamma C[\mathbf{S}(\boldsymbol{\theta}),S_\lambda(\boldsymbol{\theta});\boldsymbol{\gamma}_\sigma]F_n[\mathbf{V}|\mathbf{S}(\boldsymbol{\theta})]\,\mathbf{dV} \tag{21.43a}$$

Similarly, the average risk associated with the parameter estimation of $\boldsymbol{\theta}$ is

$$R(\sigma,\delta)_\theta = \int_{\Omega_\theta} \mathbf{d\boldsymbol{\theta}}\,\sigma(\boldsymbol{\theta}) \int_\Gamma C(\boldsymbol{\theta},\boldsymbol{\gamma}_\sigma)W_n(\mathbf{V}|\boldsymbol{\theta})\,\mathbf{d\boldsymbol{\theta}} \tag{21.44}$$

where $W_n(\mathbf{V}|\boldsymbol{\theta}) = \langle F_n[\mathbf{V}|\mathbf{S}(\boldsymbol{\theta},\boldsymbol{\theta}')]\rangle_{\theta'}$ and where $\boldsymbol{\theta}'$ represents all other parameters, apart from $\boldsymbol{\theta}$ [cf. Eq. (21.34a)]. Of course, if there are no other parameters $\boldsymbol{\theta}'$, then $W_n(\mathbf{V}|\boldsymbol{\theta}) = F_n[\mathbf{V}|\mathbf{S}(\boldsymbol{\theta})]$. Observe here that the $\boldsymbol{\gamma}_\sigma$ for $\boldsymbol{\theta}$ and for $\mathbf{S}$, S_λ are different estimators in general, while the average risks R_θ, R_S likewise assume different values. As in detection, Bayes systems are obtained by the choice of decision rule, here the estimator $\boldsymbol{\gamma}_\sigma$ [cf. Eq. (21.42)], which minimizes the average risk. Examples for special cost functions are considered in Secs. 21.2-1 to 21.2-3.

21.2-1 Bayes Extraction with a Simple Cost Function. Let us now determine the optimum extraction procedure for minimizing the average risk [Eq. (21.43) or (21.44)] when a simple, or "constant," cost function is chosen. By a *simple*, or *constant, cost function* is meant one for which the cost of correct estimates is set equal to C_C, while the costs of incorrect estimates all have the same value C_E ($>C_C$), regardless of how much in error they may be.

Assuming simple, or coincident, estimation ($t_\lambda = t_k$) of the signal parameters $\boldsymbol{\theta} = (\theta_1, \ldots, \theta_M)$ (cf. procedure 2 above) for the moment, as well as discrete parameter and data spaces, so that $\boldsymbol{\theta}$ and $\mathbf{V}$ can take only discrete values, we may write the constant cost function specifically as

$$C(\boldsymbol{\theta},\boldsymbol{\gamma}_\sigma) = \sum_{k=1}^{M} [C_E - (C_E - C_C)\,\delta_{\gamma_k\theta_k}] \tag{21.45}$$

where $\delta_{\gamma_k\theta_k}$ is the Kronecker delta ($\delta = 1$, $\gamma_k = \theta_k$; $\delta = 0$, $\gamma_k \neq \theta_k$). Thus, we are penalized an amount C_E for each parameter ($k = 1, \ldots, M$) when our estimate γ_k is incorrect, i.e., when $\gamma_k \neq \theta_k$, and, correspondingly, are assessed the smaller amount C_C when a correct decision ($\gamma_k = \theta_k$) is made. From Eqs. (18.7b) and (21.45), we can write the discrete analogue of the

† When $t_\lambda = t_k$, the notation is pleonastic: S_λ is absorbed into $\mathbf{S}$; this is illustrated in subsequent examples.

average risk [Eq. (21.44)] here as

$$R(\sigma,\delta)_\theta = MC_E - (C_E - C_C) \sum_{k=1}^{M} \sum_{\Omega_\theta} \sum_{\Gamma} \sum_{\Delta} \delta_{\gamma_k \theta_k}\, p_\sigma(\boldsymbol{\theta}) p_n(\mathbf{V}|\boldsymbol{\theta}) \delta(\boldsymbol{\gamma}|\mathbf{V}) \tag{21.46}$$

where p_σ and p_n are, respectively, the a priori *probabilities* of $\boldsymbol{\theta}$ and the conditional probability of $\mathbf{V}$, given $\boldsymbol{\theta}$. The symbols $\sum_{\Omega_\theta}$, etc., indicate summation over all L_k allowed values of each θ_k $(k = 1, \ldots, M)$, etc.; thus,

$$\sum_{\Omega_\theta} \equiv \sum_{l_1=1}^{L_1} \cdots \sum_{l_M=1}^{L_M} \qquad \sum_{\Gamma} \equiv \sum_{l_1=1}^{N_1} \cdots \sum_{l_n=1}^{N_n} \qquad \text{etc.}$$

Next, let us define

$$\mathfrak{D}_k(\mathbf{V},p_\sigma,\delta) \equiv \sum_{\Delta} \sum_{\Omega_\theta} \delta_{\gamma_k \theta_k}\, p_n(\mathbf{V}|\boldsymbol{\theta}) p_\sigma(\boldsymbol{\theta}) \delta(\boldsymbol{\gamma}|\mathbf{V}) \tag{21.47a}$$

$$= \sum_{\Delta_k} p_n(\mathbf{V}|\gamma_k) p_\sigma(\gamma_k) \delta(\gamma_k|\mathbf{V}) \tag{21.47b}$$

in which $p_n(\mathbf{V}|\gamma_k)$, $p_\sigma(\gamma_k)$ are the *marginal* probabilities

$$p_n(\mathbf{V}|\theta_k) = \sum_{\Omega_\theta - \Omega_{\theta_k}} p_n(\mathbf{V}|\boldsymbol{\theta}) \qquad p_\sigma(\theta_k) = \sum_{\Omega_\theta - \Omega_{\theta_k}} p_\sigma(\boldsymbol{\theta}) \tag{21.48a}$$

The decision rule $\delta(\gamma_k|\mathbf{V})$ is similarly the marginal probability

$$\delta(\gamma_k|\mathbf{V}) = \sum_{\Delta - \Delta_k} \delta(\boldsymbol{\gamma}|\mathbf{V}) \tag{21.48b}$$

$\left(\text{In our present notation, } \sum_{\Delta} = \sum_{\Delta_k} \cdot \sum_{\Delta - \Delta_k}, \text{etc.}\right)$ The average risk [Eq. (21.46)] can now be expressed more compactly as

$$R(\sigma,\delta)_\theta = MC_E - (C_E - C_C) \sum_{\Gamma} \sum_{k=1}^{M} \mathfrak{D}_k(\mathbf{V},p_\sigma,\delta) \tag{21.49}$$

Optimization is next achieved by a suitable choice of the M decision rules $\delta(\gamma_k|\mathbf{V})$ $(k = 1, \ldots, M)$. It is clear that R_θ is smallest when each $\mathfrak{D}_k$ is largest, since $\mathfrak{D}_k \geq 0$ $[1 \geq p_n, p_\sigma, \delta(\gamma_k|\mathbf{V}) \geq 0$ also$]$. Accordingly, we select $\delta(\gamma_k|\mathbf{V})$ to *maximize* $\mathfrak{D}_k$, and this is accomplished if we set

$$\delta(\gamma_k|\mathbf{V}) = \delta_{\gamma_k \hat{\gamma}_k} = 0 \qquad \gamma_k \neq \hat{\gamma}_k(\mathbf{V}) \qquad k = 1, \ldots, M \tag{21.50}$$

where $\hat{\gamma}_k$ $(= \hat{\theta}_k)$ is the *unconditional* maximum likelihood estimator of θ_k, defined by

$$p_\sigma(\hat{\theta}_k) p_n(\mathbf{V}|\hat{\theta}_k) \geq p_\sigma(\theta_k) p_n(\mathbf{V}|\theta_k) \qquad \text{all } \theta_k \text{ in } \Omega_{\theta_k} \tag{21.50a}$$

This UMLE, $\hat{\theta}_k$, may be obtained from Eq. (21.12*b*), where, of course, L_n is now $L(\mathbf{V},\theta_k) = p_\sigma(\theta_k)p_n(\mathbf{V}|\theta_k)$, with probability densities replaced by the appropriate probabilities.† The Bayes risk becomes

$$R_\theta^* \equiv \min_\delta R(\sigma,\delta)_\theta = MC_E - (C_E - C_C) \sum_{k=1}^{M} \sum_\Gamma p_\sigma(\hat{\theta}_k)p_n(\mathbf{V}|\hat{\theta}_k) \qquad (21.51)$$

Note that, because of the particular structure of the cost function [Eq. (21.45)], the Bayes estimators of $\theta_1, \ldots, \theta_M$ here are the individual UML estimators $\hat{\gamma}_k = \hat{\theta}_k$ of each θ_k. The parameters may be statistically related, as $p_n(\mathbf{V}|\boldsymbol{\theta})$ indicates; each $p_\sigma(\theta_k)$, $p_n(\mathbf{V}|\theta_k)$ embodies such interrelationships, but the UMLE's are determined independently from Eq. (21.12*b*), as indicated above.

When the data, parameter, and decision spaces Γ, Ω_θ, Δ are continuous instead of discrete, the continuous analogue of the discrete constant cost function [Eq. (21.45)] becomes

$$C(\boldsymbol{\theta},\boldsymbol{\gamma}_\sigma) = \sum_{k=1}^{M} [C_E A_k' - (C_E - C_C)\delta(\gamma_k - \theta_k)] \qquad (21.52)$$

where the A_k' are (positive) constants, with the dimensions of the delta functions (that is, $|\theta_k|^{-1}$) chosen so that the average risk for each θ_k is also positive (or zero).‡ This can be expressed more compactly as

$$C(\boldsymbol{\theta},\boldsymbol{\gamma}_\sigma) = C_0 \sum_{k=1}^{M} [A_k - \delta(\gamma_k - \theta_k)] \qquad C_0 \equiv C_E - C_C,\ A_k \equiv \frac{C_E}{C_E - C_C} A_k' \qquad (21.52a)$$

Paralleling Eqs. (21.46) to (21.49), we get for the average risk

$$R(\sigma,\delta)_\theta = C_0 \sum_{k=1}^{M} \left[A_k - \int_\Gamma \mathfrak{D}_k(\mathbf{V};\sigma,\delta)\, d\mathbf{V} \right] \qquad (21.53a)$$

with

$$\mathfrak{D}_k(\mathbf{V};\sigma,\delta) = \int_\Delta \sigma(\gamma_k) W_n(\mathbf{V}|\gamma_k)\delta(\gamma_k|\mathbf{V})\, d\gamma_k \qquad (21.53b)$$

[cf. Eq. (21.47*b*)], where now probabilities have been replaced by the corresponding probability densities σ, W_n, δ.

Optimization again is achieved by selecting the decision rules $\delta(\gamma_k|\mathbf{V})$ so as to maximize $\mathfrak{D}_k$. The analogue of Eq. (21.50) for this purpose is

$$\delta(\gamma_k|\mathbf{V}) = \delta[\gamma_k - \hat{\gamma}_k(\mathbf{V})] \qquad k = 1, \ldots, M \qquad (21.54)$$

† Note that generally $p_n(\mathbf{V}|\theta_k) \neq p_n(\mathbf{V}|\theta_l)$ $(k \neq l$, etc.).

‡ The choice of the A_k', while to an extent arbitrary, is closely related to the fineness with which our extraction system can distinguish between observed magnitudes (cf. Sec. 21.2-3).

where $\hat{\gamma}_k(\mathbf{V}) = \hat{\theta}_k(\mathbf{V})$ is the UML estimator of θ_k, defined by

$$\sigma(\hat{\theta}_k)W_n(\mathbf{V}|\hat{\theta}_k) \geq \sigma(\theta_k)W_n(\mathbf{V}|\theta_k) \qquad \text{all } \theta_k \text{ in } \Omega_{\theta_k} \tag{21.54a}$$

The required UMLE's, θ_k, are once more obtained from Eq. (21.12b), with $L_n = \sigma(\theta_k)W_n(\mathbf{V}|\theta_k)$ now. The corresponding Bayes risk for this constant cost function [Eqs. (21.52), (21.54)] is

$$R^*(\sigma,\delta)_\theta = C_0 \sum_{k=1}^{M} \left[A_k - \int_\Gamma \sigma(\hat{\theta}_k)W_n(\mathbf{V}|\hat{\theta}_k)\, d\mathbf{V}\right] \tag{21.55}$$

where we remember, of course, that $\hat{\theta}_k$ is a function of $\mathbf{V}$.

For waveform estimation, the constant cost function [Eq. (21.52a)] becomes, for the coincident estimation $t_\lambda = t_k$ $(k = 1, \ldots, n)$ employed here,

$$C(\mathbf{S},S_\lambda;\boldsymbol{\gamma}_\sigma) = C(\mathbf{S},\boldsymbol{\gamma}_\sigma) = C_0\left[A_n - \sum_{k=1}^{n} \delta(\gamma_k - S_k)\right] \tag{21.56}$$

in which it is assumed that $A_k = A'_n/n$ for each point t_k at which estimates of S_k are desired. Following the argument of Eqs. (21.53) to (21.55), we see that the Bayes estimators of $S_1, \ldots, S_n$ are again the various unconditional maximum likelihood estimators $\hat{\gamma}_k(\mathbf{V}) = \hat{S}_k(\mathbf{V})$, while the Bayes risk [Eq. (21.55)] is modified to†

$$R^*(\sigma,\delta)_S = C_0\left[A_n - \sum_{k=1}^{n} \int_\Gamma \sigma(\hat{S}_k)F_n(\mathbf{V}|\hat{S}_k)\, d\mathbf{V}\right] (> 0) \tag{21.57}$$

Other types of "constant" cost function can be constructed (cf. Sec. 21.2-3), but they do not appear particularly well suited to application, their chief defect being an excessive strictness with regard to accuracy and hence too great an average risk.

In the above instances, the optimum estimators are all UML estimators. The estimates themselves are the parameter or signal values which maximize the joint probabilities (for discrete magnitudes) or probability densities of θ_k (or S_k) and $\mathbf{V}$, regarded as functions of θ_k (or S_k) with $\mathbf{V}$ fixed. For each $\mathbf{V}$, these estimates differ. If all signals S_k or parameters θ_k are a priori equally likely, these estimates are equivalent to the corresponding *conditional* likelihood estimates. However, regardless of the precise form of $\sigma(\theta_k)$ or $\sigma(S_k)$, these Bayes estimates are the parameters (or signals) most likely a posteriori, i.e., for given $\mathbf{V}$. In general, *we can regard unconditional maximum likelihood estimators as Bayes estimators relative to an appropriate constant, or simple, cost function.* Since these optimum estimators are all UMLE's, classical maximum likelihood theory as extended to the uncon-

† Note that in Eq. (21.57) $F_n(\mathbf{V}|S_k) \neq F_n(\mathbf{V}|\mathbf{S})$; for example, $F_n(\mathbf{V}|S_k) = \int \sigma(\mathbf{S})F_n(\mathbf{V}|\mathbf{S})\, d\mathbf{S}'$, where $\mathbf{S}' = S_1, \ldots, S_n$ with S_k omitted.

ditional cases in Sec. 21.1-2 may be applied in detail here. We can accordingly interpret the UMLE's of amplitude a_0 and shape factor θ considered in the examples of Secs. 21.1-3(1), 21.1-3(3) as optimum decision systems which minimize average risk with respect to a constant cost function of the type (21.52). In a similar way, the estimator of Eq. (21.29) for the example of Sec. 21.1-3(2) can be shown to be Bayes relative to the cost function of Eq. (21.56) (cf. Sec. 21.4-1). Finally, observe that the extension of Eqs. (21.56), (21.57) to the interpolation and extrapolation of waveforms ($t_\lambda \neq t_k$) as distinct from simple estimation ($t_\lambda = t_k$) may be carried out in the same way (cf. Prob. 21.4 and Sec. 21.4-2).

21.2-2 Bayes Extraction with a Quadratic Cost Function. We begin again with the case of parameter estimation. Now, however, the cost function to be used in the estimation of the M signal parameters $\boldsymbol{\theta} = (\theta_1, \ldots, \theta_M)$ is the quadratic cost function

$$C(\boldsymbol{\theta},\boldsymbol{\gamma}_\sigma) = C_0(\tilde{\boldsymbol{\theta}} - \tilde{\boldsymbol{\gamma}}_\sigma)(\boldsymbol{\theta} - \boldsymbol{\gamma}_\sigma) = C_0\|\boldsymbol{\theta} - \boldsymbol{\gamma}_\sigma(\mathbf{V})\|^2 = C_0 \sum_{k=1}^{M} (\theta_k - \gamma_k)^2 \tag{21.58}$$

where, of course, $C_0 > 0$. Here C itself is convex so that the inconvenience of considering both randomized and nonrandomized decision rules may be avoided. This cost function is the square of the "distance" between the true value and the estimate. The chief virtues of this "squared-error," or quadratic, cost function for extraction are, first, that it is convenient mathematically; second, that it takes reasonable account of the fact that, usually, large errors are more serious than small ones; and third, that under certain conditions (described below in Sec. 21.2-3) it is also an optimum or Bayes estimator for a wider class of cost functions than the quadratic.[31] Moreover, as we shall see presently, it also leads in certain cases to the earlier extraction procedures[19,20] based on least-mean-squared-error criteria for linear systems, which we wish to include from the more general viewpoint of decision theory.

For Bayes extractors, we minimize the average risk R as before. The conditional risk [Eq. (18.6)] for Eq. (21.58) and the decision rule $\delta = \delta[\boldsymbol{\gamma} - \boldsymbol{\gamma}_\sigma(\mathbf{V})]$ become

$$r(\boldsymbol{\theta},\boldsymbol{\gamma}_\sigma) = C_0 \int_\Gamma \|\boldsymbol{\theta} - \boldsymbol{\gamma}_\sigma(\mathbf{V})_\theta\|^2 W_n(\mathbf{V}|\boldsymbol{\theta})\, d\mathbf{V} \tag{21.59}$$

where care is taken to distinguish between the estimator $\boldsymbol{\gamma}_\sigma(\mathbf{V})$, defined for all possible $\mathbf{V}$, and the estimate $\boldsymbol{\gamma}_\sigma(\mathbf{V}')$, for a particular $\mathbf{V} = \mathbf{V}'$. The average risk follows at once from Eqs. (18.7) and is here

$$R(\sigma,\delta)_\theta = C_0 E_{\mathbf{V},\theta}\{\|\boldsymbol{\theta} - \boldsymbol{\gamma}_\sigma(\mathbf{V})_\theta\|^2\} = C_0 \int_\Gamma d\mathbf{V}\, w_n(\mathbf{V}) \times \int_{\Omega_\theta} \|\boldsymbol{\theta} - \boldsymbol{\gamma}_\sigma(\mathbf{V})_\theta\|^2 w_M(\boldsymbol{\theta}|\mathbf{V})\, d\boldsymbol{\theta} \tag{21.60a}$$

where we have written the probability densities

$$w_n(\mathbf{V}) = \int_{\Omega_\theta} \sigma(\boldsymbol{\theta}) W_n(\mathbf{V}|\boldsymbol{\theta})\, d\boldsymbol{\theta} \qquad w_M(\boldsymbol{\theta}|\mathbf{V}) = \sigma(\boldsymbol{\theta}) W_n(\mathbf{V}|\boldsymbol{\theta})/w_n(\mathbf{V}) \tag{21.60b}$$

When the cost function is differentiable (with respect to the estimator), we obtain the following general condition for an extremum of R (actually a minimum for these concave cost functions; cf. Prob. 21.5):

$$\delta R = \int_\Gamma d\mathbf{V} \int_{\Omega_\theta} \sigma(\boldsymbol{\theta}) W_n(\mathbf{V}|\boldsymbol{\theta}) \frac{\partial C(\boldsymbol{\theta},\boldsymbol{\gamma})}{\partial \boldsymbol{\gamma}}\, d\boldsymbol{\theta}\, \delta\boldsymbol{\gamma} = 0 \tag{21.61a}$$

or

$$\int_{\Omega_\theta} \sigma(\boldsymbol{\theta}) W_n(\mathbf{V}|\boldsymbol{\theta}) \left.\frac{\partial C(\boldsymbol{\theta},\boldsymbol{\gamma})}{\partial \boldsymbol{\gamma}}\right|_{\boldsymbol{\gamma}=\boldsymbol{\gamma}^*} d\boldsymbol{\theta} = 0 \tag{21.61b}$$

With the quadratic cost function (21.58), this becomes simply

$$\begin{aligned}\boldsymbol{\gamma}_\sigma^*(\mathbf{V})_\theta &= \int_{\Omega_\theta} \boldsymbol{\theta} w_M(\boldsymbol{\theta}|\mathbf{V})\, d\boldsymbol{\theta} = \frac{\int_{\Omega_\theta} \boldsymbol{\theta}\sigma(\boldsymbol{\theta}) W_n(\mathbf{V}|\boldsymbol{\theta})\, d\boldsymbol{\theta}}{\int_{\Omega_\theta} \sigma(\boldsymbol{\theta}) W_n(\mathbf{V}|\boldsymbol{\theta})\, d\boldsymbol{\theta}} \\ &= \frac{\int_{\Omega_\theta} \boldsymbol{\theta} w(\mathbf{V},\boldsymbol{\theta})\, d\boldsymbol{\theta}}{\int_{\Omega_\theta} w(\mathbf{V},\boldsymbol{\theta})\, d\boldsymbol{\theta}}\end{aligned} \tag{21.62}$$

in which $w(\mathbf{V},\boldsymbol{\theta})$ is the joint d.d. of $\mathbf{V}$ and $\boldsymbol{\theta}$. Thus, *the Bayes estimator for the squared-error cost function is the conditional expectation*[17] *of* $\boldsymbol{\theta}$, *given* $\mathbf{V}$. Since both members of Eq. (21.62) are vectors, Eq. (21.62) represents M equations between the M components of $\boldsymbol{\gamma}_\sigma^*$ and $\boldsymbol{\theta}$. Accordingly, the kth component of the estimator is

$$\begin{aligned}&[\boldsymbol{\gamma}_\sigma^*(V_1, \ldots, V_n)]_k \\ &\quad = \int \cdots \int \theta_k w_M(\theta_1, \ldots, \theta_M | V_1, \ldots, V_n)\, d\theta_1 \cdots d\theta_M \\ &\qquad\qquad k = 1, \ldots, M\end{aligned} \tag{21.62a}$$

The same result follows, of course, if instead of Eq. (21.58) we use for each component $C = (\theta_k - \gamma_k)^2$, that is, if the Bayes estimate of a single component of $\boldsymbol{\theta}$ is required, rather than the more general vector estimate. This is a consequence of the quadratic nature of the preassigned cost function and clearly does not hold in general. Note again that $W_n(\mathbf{V}|\boldsymbol{\theta}) = \langle F_n[\mathbf{V}|S(\boldsymbol{\theta},\boldsymbol{\theta}')]\rangle_{\theta'}$, and if $\mathbf{S}$ is a function of $\boldsymbol{\theta}$ only, then $W_n(\mathbf{V}|\boldsymbol{\theta}) = F_n[\mathbf{V}|\mathbf{S}(\boldsymbol{\theta})]$.

The Bayes risk is specifically, from Eq. (21.62) in Eq. (20.60a),

$$\begin{aligned}R^*(\sigma,\delta)_\theta &= C_0 E_{\mathbf{V},\boldsymbol{\theta}}\{\|\boldsymbol{\theta} - \boldsymbol{\gamma}_\sigma^*(\mathbf{V})_\theta\|^2\} \\ &= C_0 E_{\boldsymbol{\theta}}\{\tilde{\boldsymbol{\theta}}\boldsymbol{\theta}\} - 2C_0 E_{\mathbf{V},\boldsymbol{\theta}}\{\tilde{\boldsymbol{\theta}}\boldsymbol{\gamma}_\sigma^*\} + C_0 E_{\mathbf{V}}\{\tilde{\boldsymbol{\gamma}}_\sigma^*\boldsymbol{\gamma}_\sigma^*\}\end{aligned} \tag{21.63}$$

since $\boldsymbol{\gamma}_\sigma^*$ is a function of $\mathbf{V}$ only, while $\boldsymbol{\theta}$ is governed solely by the d.d. $\sigma(\boldsymbol{\theta})$. Note, however, that since V is a function of $\boldsymbol{\theta}$, inasmuch as $V = S(\boldsymbol{\theta}) + N$ here, the cross term in Eq. (21.63) does *not* generally factor into the product of the individual averages $E_{\mathbf{V}}\{\boldsymbol{\gamma}_\sigma^*\} \times E_{\boldsymbol{\theta}}\{\tilde{\boldsymbol{\theta}}\}$.

In the case of waveform estimation, the preceding treatment is easily modified: for simple estimation ($t_\lambda = t_k$), instead of Eq. (21.58) one has now the cost function

$$C(\mathbf{S},\boldsymbol{\gamma}_\sigma) = C_0(\tilde{\mathbf{S}} - \tilde{\boldsymbol{\gamma}}_\sigma)(\mathbf{S} - \boldsymbol{\gamma}_\sigma) \tag{21.64}$$

where $\boldsymbol{\gamma}_\sigma = \boldsymbol{\gamma}_\sigma(\mathbf{V})$ is the estimator of $\mathbf{S}$ itself, rather than of $\boldsymbol{\theta}$ in $\mathbf{S} = \mathbf{S}(\boldsymbol{\theta})$. Replacing $\boldsymbol{\theta}$ by $\mathbf{S}$ and $W_n(\mathbf{V}|\boldsymbol{\theta})$ by $F_n(\mathbf{V}|\mathbf{S})$, etc., in Eqs. (21.59) to (21.62), we obtain the corresponding average risk, optimum estimators, and Bayes risk

$$R(\sigma,\delta)_S = C_0 \int_\Omega \int_\Gamma \sigma(\mathbf{S}) \|\mathbf{S} - \boldsymbol{\gamma}_\sigma(\mathbf{V})\|^2 F_n(\mathbf{V}|\mathbf{S})\, d\mathbf{S}\, d\mathbf{V} \tag{21.65}$$

$$\boldsymbol{\gamma}_\sigma^*(\mathbf{V})_S = \frac{\int_\Omega \mathbf{S}\sigma(\mathbf{S})F_n(\mathbf{V}|\mathbf{S})\, d\mathbf{S}}{\int_\Omega \sigma(\mathbf{S})F_n(\mathbf{V}|\mathbf{S})\, d\mathbf{S}} = \frac{\int_\Omega \mathbf{S}W_n(\mathbf{V},\mathbf{S})\, d\mathbf{S}}{\int_\Omega W_n(\mathbf{V},\mathbf{S})\, d\mathbf{S}} \tag{21.66}$$

and

$$R^*(\sigma,\delta)_S = C_0 E_{\mathbf{V},\mathbf{S}}\{\|\mathbf{S} - \boldsymbol{\gamma}_\sigma^*(\mathbf{V})_S\|^2\} \tag{21.67}$$

Like Eq. (21.43*a*), useful alternative forms of Eqs. (21.65) to (21.67) in the case of deterministic signal processes are obtained on replacing $\mathbf{S}$ by $\mathbf{S}(\boldsymbol{\theta})$ explicitly and $\sigma(\mathbf{S})$ by $\sigma(\boldsymbol{\theta})$, where now the integration is over all Ω_θ-space. Thus Eqs. (21.66), (21.67) are modified to

$$\boldsymbol{\gamma}_\sigma^*(\mathbf{V})_S = \frac{\int_{\Omega\theta} \mathbf{S}(\boldsymbol{\theta})\sigma(\boldsymbol{\theta})F_n[\mathbf{V}|\mathbf{S}(\boldsymbol{\theta})]\, d\boldsymbol{\theta}}{\int_{\Omega\theta} \sigma(\boldsymbol{\theta})F_n[\mathbf{V}|\mathbf{S}(\boldsymbol{\theta})]\, d\boldsymbol{\theta}} \tag{21.66a}$$

$$R^*(\sigma,\delta)_S = C_0 E_{\mathbf{V},\boldsymbol{\theta}}\{\|\mathbf{S}(\boldsymbol{\theta}) - \boldsymbol{\gamma}_\sigma^*(\mathbf{V})_S\|^2\} \tag{21.67a}$$

Even when $\bar{\mathbf{S}}$ or $\bar{\boldsymbol{\gamma}}_\sigma^*$ vanishes, the Bayes risk [Eqs. (21.67), (21.67*a*)] still contains the cross term, viz.,

$$R(\sigma,\delta)_S^* = C_0 E_{\mathbf{S}\text{ or }\boldsymbol{\theta}}\{\tilde{\mathbf{S}}\mathbf{S}\} - 2C_0 E_{\mathbf{V},\mathbf{S}\text{ or }\boldsymbol{\theta}}\{\tilde{\mathbf{S}}\boldsymbol{\gamma}_\sigma^*\} + C_0 E_V\{\tilde{\boldsymbol{\gamma}}_\sigma^*\boldsymbol{\gamma}_\sigma^*\} \tag{21.68}$$

again since V (in $\boldsymbol{\gamma}_\sigma^*$) is a function of S or $\boldsymbol{\theta}$.

Further properties (Minimax). Some further properties of the Bayes estimator with this cost function are easily shown. For example, in the case of simple estimation if a fixed signal $\mathbf{S}_1$ is applied at the input, so that $\sigma(\mathbf{S}) = \delta(\mathbf{S} - \mathbf{S}_1)$, the Bayes estimator is $\boldsymbol{\gamma}_\sigma^*(\mathbf{V})_S = \mathbf{S}_1$ and is therefore *conditionally unbiased*, i.e.,

$$\int_\Gamma \boldsymbol{\gamma}_\sigma^*(\mathbf{V})_S F_n(\mathbf{V}|\mathbf{S})\, d\mathbf{V} = \mathbf{S}_1 \tag{21.69}$$

The conditional risk for any unbiased estimator $\boldsymbol{\gamma}_\sigma(\mathbf{V})_S$ is actually its conditional variance, indicated by var $\boldsymbol{\gamma}_\sigma(\mathbf{V})_S$. The average risk [Eq. (21.65)] may be written

$$R(\sigma,\boldsymbol{\gamma}_\sigma) = C_0 \int_\Omega [\text{var}\ \boldsymbol{\gamma}_\sigma(\mathbf{V})_S]\sigma(\mathbf{S})\, d\mathbf{S} \tag{21.70}$$

Since $R(\sigma,\gamma_\sigma)$ is least for the Bayes estimator, we see from Eq. (21.70) that *the Bayes estimator for the quadratic cost function has the smallest average variance among all unbiased estimators.* Similar remarks apply for the parameter estimators $\gamma_\sigma(\mathbf{V})_\theta$ above.

When signal and noise are additive and independent and waveform estimation is required, several more important properties of Bayes estimators with quadratic cost functions and simple estimation can be demonstrated. The first of these is the so-called *translation property*,[21,22] which appears when the a priori signal distribution is uniform. The Bayes estimator of $\mathbf{S}$ is now given by Eq. (21.66), where $F_n(\mathbf{V}|\mathbf{S})$ is replaced by $W_n(\mathbf{V} - \mathbf{S})_N$. With $\sigma(\mathbf{S})$ uniform [cf. footnote referring to Eq. (21.23*b*)], this estimator becomes

$$\gamma_\sigma^*(\mathbf{V})_S = \frac{\int \cdots \int_{-\infty}^{\infty} \mathbf{S} W_n(\mathbf{V}-\mathbf{S})_N \, d\mathbf{S}}{\int \cdots \int_{-\infty}^{\infty} W_n(\mathbf{V}-\mathbf{S})_N \, d\mathbf{S}} \tag{21.71}$$

Now, letting $\boldsymbol{\lambda}$ be an arbitrary fixed vector and introducing a new variable $\mathbf{U}$ such that $\mathbf{U} = \mathbf{S} + \boldsymbol{\lambda}$, we see that Eq. (21.71) gives at once

$$\gamma_\sigma^*(\mathbf{V} \pm \boldsymbol{\lambda})_S = \gamma_\sigma^*(\mathbf{V})_S \pm \boldsymbol{\lambda} \tag{21.72}$$

For example, if a fixed signal $\mathbf{S} = \mathbf{V} - \mathbf{N}$ is applied to a system designed to be Bayes with respect to a uniform a priori distribution $\sigma(\mathbf{S})$, then, if the signal is changed by an amount $\boldsymbol{\lambda}$, the system output is altered by the same amount.

The second property is a Minimax one. We recall from Sec. 18.4-4 that, if a Bayes system designed for a certain signal distribution $\sigma_0(\mathbf{S})$ has a conditional risk that is independent of $\mathbf{S}$, then it is a Minimax system and $\sigma_0(\mathbf{S})$ is called the least favorable distribution. The above Bayes system $\gamma_\sigma^*(\mathbf{V})_S$ has also the conditional risk

$$r(\mathbf{S},\gamma^*) = C_0 \int_\Gamma \|\mathbf{S} - \gamma^*(\mathbf{V})_S\|^2 W_n(\mathbf{V} - \mathbf{S})_N \, d\mathbf{V} \tag{21.73}$$

Letting the domain be infinite for each component and introducing new variables $\mathbf{X} = \mathbf{V} - \mathbf{S}$, we see that Eq. (21.73) becomes

$$r(\mathbf{S},\gamma^*) = C_0 \int \cdots \int_{-\infty}^{\infty} \|\mathbf{S} - \gamma^*(\mathbf{S} + \mathbf{X})_S\|^2 W_n(\mathbf{X})_N \, d\mathbf{X} \tag{21.74}$$

but since the Bayes system γ^* here has the translation property [cf. Eq. (21.72)], we have $\gamma^*(\mathbf{S} + \mathbf{X})_S = \gamma^*(\mathbf{X})_S + \mathbf{S}$. When this is used in Eq. (21.74), the conditional risk becomes independent of $\mathbf{S}$, showing, therefore, that, *when signal and noise are additive and independent, the least favorable* *a priori distribution* $\sigma_0(\mathbf{S})$ *is a uniform one and the Bayes system* $\boldsymbol{\gamma}_\sigma^*$ *for this distribution is Minimax* $(= \boldsymbol{\gamma}_M^*)$.

Observe that, with deterministic signals, the signal parameters (except for amplitude $\sim a_0$) do not appear *linearly* in $\mathbf{S}$, so that the translation property above does not hold any more for the Bayes estimator $\boldsymbol{\gamma}_\sigma^*(\mathbf{V})_\theta$ even if $\sigma(\boldsymbol{\theta})$ is uniform. Moreover, this Bayes system is no longer Minimax, while the least favorable distribution $\sigma_0(\mathbf{S})$ is not uniform. The exception to this occurs for the signal amplitude, because of its linear relation vis-à-vis the accompanying noise (cf. Secs. 21.3-1, 21.3-2).

In cases where signal and noise belong to ergodic processes, so that statistical averages may be replaced by appropriate time averages (except for a set of functions of measure zero), the average risk [Eq. (21.65)] can be written

$$R(\sigma,\delta) = C_0 \lim_{T\to\infty} \frac{1}{T} \int_0^T \{S(t) - \gamma_S[V(t)]\}^2 \, dt \tag{21.75}$$

The usual treatments of extraction have used the minimum value of this as an optimum criterion, just as does the risk formulation when the cost function is of the squared-error type [Eq. (21.64)].

Sometimes the added constraint of linearity is imposed upon the estimator $\boldsymbol{\gamma}(\mathbf{V})$, so that the estimate $\boldsymbol{\gamma}$ is required to be the output of a linear (usually physically realizable) filter with $V(t)$ as its input. Then it is not generally true that optimum extractors under this constraint are Bayes, since from Eq. (21.66) the Bayes extractor is usually a nonlinear operator upon $V(t)$, even for the simple cost assignment of Sec. 21.2-1. In fact, as we can see from the above, $F_n(\mathbf{V}|\mathbf{S})$ must have rather special properties if Bayes and linearity are to be concomitant features of the optimum extractor. We remark, however, that some of the notions of risk theory are useful for such restricted classes of decision rules, even though the main theorems do not apply.[23] That is, if we agree that only the class of linear estimators is to be considered, we may speak of the one with the smallest average risk, the one for which the maximum conditional risk is smallest, the one with the property that no other in the class is uniformly better, etc., settling the questions of existence and uniqueness in specific situations by construction.

Finally, recall again that these estimators $\boldsymbol{\gamma}_\sigma(\mathbf{V})_S$, $\boldsymbol{\gamma}_\sigma(\mathbf{V})_\theta$ embody the actual structure of our receiving and data processing systems, for example, $\boldsymbol{T}_R\{\mathbf{V}\} = \boldsymbol{\gamma}_\sigma(\mathbf{V})_S$ or $\boldsymbol{\gamma}_\sigma(\mathbf{V})_\theta$. Optimal systems from the risk point of view have now been defined with respect to at least two classes of cost function, the simple and the quadratic† [cf. Eqs. (21.45) et seq., (21.58) et seq.]. Some other more or less useful cost functions are considered in the next section, while the questions of structure and the distributions of the estimators themselves are examined in more detail in Secs. 21.2-5, 21.3, 21.4.

† We remark that the Bayes estimators for quadratic cost functions also sometimes possess optimal properties with respect to a wider class of cost functions (cf. Sec. 21.2-3).

21.2-3 Other Cost Functions.† The number of possible cost functions, while theoretically infinite, is in practice limited by the dual requirements of reasonableness and computability: on the one hand, the cost of "errors" must depend on the magnitude of the "error," in some fashion such that correct decisions are penalized (if at all) less than incorrect decisions. On the other hand, the cost function should be such that optimum systems (in the sense of minimum average risk, for example) can be found, at least approximately, from such conditions as Eqs. (21.61*a*), (21.61*b*) and their corresponding Bayes risks in turn determined. Perhaps the most natural choice of cost function is the symmetric "distance" function

$$C(\boldsymbol{\theta},\boldsymbol{\gamma}_\sigma) = C(|\boldsymbol{\theta} - \boldsymbol{\gamma}_\sigma|) \qquad \text{or} \qquad C(\mathbf{S},\boldsymbol{\gamma}_\sigma) = C(|\mathbf{S} - \boldsymbol{\gamma}_\sigma|) \tag{21.76}$$

where correct decisions cost a fixed amount $C^{(0)}$ (≥ 0) and incorrect decisions a greater amount, so that, if C possesses a Taylor expansion, Eq. (21.76) can be alternatively represented by series of the type

$$C(\boldsymbol{\theta},\boldsymbol{\gamma}_\sigma) = C_{\theta^{(0)}} + \sum_{ij} (C_{\theta^{(2)}})_{ij}(\theta_i - \gamma_{\sigma_i})(\theta_j - \gamma_{\sigma_j}) + \sum_{ijkl} (C_{\theta^{(4)}})_{ijkl}(\theta_i - \gamma_{\sigma_i}) \cdots (\theta_l - \gamma_{\sigma_l}) + \cdots \tag{21.76a}$$

with an analogous development for $C(\mathbf{S},\boldsymbol{\gamma}_\sigma)$ in terms of $C_{S^{(0)}}$, $C^{(2)}_{S_{ij}}$, $(S_i - \gamma_{\sigma_i})$, etc.

Let us consider for the M parameters $\boldsymbol{\theta}$ a number of M-dimensional cost functions of the type (21.76) [which, however, are not necessarily developable in a Taylor series like Eq. (21.76*a*)]. The quadratic cost function of Eq. (21.58) is one example which we have already examined in some detail (cf. Sec. 21.2-2). Another is the *exponential cost function*

$$C_2(\boldsymbol{\theta},\boldsymbol{\gamma}_\sigma) = C_0\{1 - \exp[-\tfrac{1}{2}(\tilde{\boldsymbol{\theta}} - \tilde{\boldsymbol{\gamma}}_\sigma)\mathbf{n}^{-1}(\boldsymbol{\theta} - \boldsymbol{\gamma}_\sigma)]\} \tag{21.77}$$

where $\mathbf{n}$ is the diagonal matrix $[\eta_k\delta_{k_j}]$ and the η_k are scale factors with the dimensions of the respective parameters θ_k. The η_k are directly related to the fineness with which the estimator (or receiver $\boldsymbol{T}_R\{\mathbf{V}\} = \boldsymbol{\gamma}_\sigma$) is able to distinguish between correct and incorrect values. According to Eq. (21.77), the receiver is penalized comparatively little if the estimates $\boldsymbol{\gamma}_\sigma(\mathbf{V})$ are close to the actual values $\boldsymbol{\theta}$ and almost the maximum amount C_0 when one or more of these estimates depart noticeably from the correct values. Unlike the expandable cost functions of Eqs. (21.58), (21.77), we may cite two additional examples involving the nondevelopable "rectangular" cost functions

$$C_3(\boldsymbol{\theta},\boldsymbol{\gamma}_\sigma) = C_0\left(1 - \sum_{k=1}^{M} \Delta\left|\frac{\theta_k - \gamma_k}{\eta_k}\right|\right) \tag{21.78}$$

† See the report by Ashby.[24]

and
$$C_4(\boldsymbol{\theta},\boldsymbol{\gamma}_\sigma) = C_0\left(1 - \prod_{k=1}^{M} \Delta \left|\frac{\theta_k - \gamma_k}{\eta_k}\right|\right) \tag{21.79}$$

where $\Delta|x| = 1$ if $|x| < 1$ and $\Delta|x| = 0$ when $|x| > 1$. The former assesses the observer an amount $C_0 l/M$ if l (≥ 1) estimates out of the M fall outside the tolerance limit $\theta_i \pm \eta$, while the latter invokes the maximum penalty C_0 if any *one* estimate lies outside these limits. For many applications, the latter is too strict. It produces too high an average risk (for given η), rapidly approaching the maximum C_0 as $M \to \infty$.

Although, strictly speaking, the derivatives of the rectangular cost functions C_3, C_4 [Eqs. (21.78), (21.79)] do not exist, we can still obtain the solution of the extremum condition (21.61*b*), here for the Bayes estimators $\boldsymbol{\gamma}^*$, in terms of delta functions. Thus, for C_3 we write

$$\frac{\partial}{\partial \boldsymbol{\gamma}} C_0\left(1 - \sum_{k=1}^{M} \Delta \left|\frac{\theta_k - \gamma_k}{\eta_k}\right|\right) = C_0[\delta(\theta_k - \gamma_k + \eta_k) - \delta(\theta_k - \gamma_k - \eta_k)] \qquad k = 1, \ldots, M \tag{21.80}$$

Substituting this into Eq. (21.61*b*) yields the set of M equations

$$\sigma(\gamma_k^* - \eta_k) W_n(\mathbf{V}|\gamma_k^* - \eta_k) = \sigma(\gamma_k^* + \eta_k) W_n(\mathbf{V}|\gamma_k^* + \eta_k) \tag{21.81a}$$

The solutions $\boldsymbol{\gamma}^*$ of these relations are accordingly the optimum estimators of $\boldsymbol{\theta}$. In terms of the joint d.d. $w_k(\mathbf{V},\theta_k)$ of $\mathbf{V}$ and θ_k, we have the equivalent relations

$$w_k(\mathbf{V}, \gamma_k^* - \eta_k) = w_k(\mathbf{V}, \gamma_k^* + \eta_k) \qquad k = 1, \ldots, M \tag{21.81b}$$

Note that, when η_k is small, we may develop both members of Eq. (21.81*b*), after first taking their logarithms, to get approximately (for each $k = 1, \ldots, M$)

$$\log w_k(\mathbf{V},\gamma_k^*) - \eta_k \left.\frac{\partial \log w_k}{\partial \theta_k}\right|_{\theta_k = \gamma_k^*} = \log w_k(\mathbf{V},\gamma_k^*) + \eta_k \left.\frac{\partial \log w_k}{\partial \theta_k}\right|_{\theta_k = \gamma_k^*}$$
$$\therefore \left.\frac{\partial \log w_k(\mathbf{V},\theta_k)}{\partial \theta_k}\right|_{\theta_k = \gamma_k^*} = 0 \tag{21.82}$$

But this is precisely the relation specifying the unconditional maximum likelihood estimator of θ_k [cf. Eq. (21.12*b*)] or, equivalently, obeying the condition (21.54*a*), since $w_k(\mathbf{V},\theta_k) = \sigma(\theta_k) W_n(\mathbf{V}|\theta_k)$. Thus, $\boldsymbol{\gamma}^*$ is Bayes with respect to the continuous version (21.52*a*) of the constant cost function discussed earlier, in Sec. 21.2-1, which in turn is the limiting form as $\eta_k \to 0$ of the rectangular cost function of Eq. (21.78).† The (vector) Bayes estimator $\boldsymbol{\gamma}_\sigma^*$ is again the UML estimator $\hat{\boldsymbol{\gamma}}_\sigma$, each component of which is determined from Eq. (21.82), when one is able to distinguish between

† After appropriate adjustment of the constants C_0, etc., in the two cases.

magnitudes arbitrarily close together in value (that is, $\eta_k \to 0$). While such distinctions cannot be pushed in physical cases to this theoretical limit, often η_k can still be made so small that Eq. (21.82) is quite acceptable in practice.

Similar results may be derived for the stricter rectangular cost function (21.79) (cf. Prob. 21.6), which in the limit $\mathbf{n} \to 0$ leads to the unconditional maximum likelihood estimator $\hat{\boldsymbol{\gamma}}_\sigma(\mathbf{V}) = \boldsymbol{\gamma}_\sigma^*(\mathbf{V})$, defined now by†

$$\sigma(\hat{\boldsymbol{\gamma}}_\sigma) W_n(\mathbf{V}|\hat{\boldsymbol{\gamma}}_\sigma) \geq \sigma(\boldsymbol{\theta}) W_n(\mathbf{V}|\boldsymbol{\theta}) \qquad \text{all } \boldsymbol{\theta} \text{ in } \Omega_\theta \tag{21.83}$$

or, equivalently, determined by the set of joint likelihood equations (21.13*b*). The constant cost function in this instance is given by

$$C = C_0[A_M - \delta(\boldsymbol{\gamma}_\sigma - \boldsymbol{\theta})] \tag{21.84}$$

[cf. Eq. (21.52*a*)], with a corresponding Bayes risk

$$R^*(\sigma,\delta)_\theta = C_0 \left[A_M - \int_\Gamma \sigma(\hat{\boldsymbol{\gamma}}_\sigma) W_n(\mathbf{V}|\hat{\boldsymbol{\gamma}}_\sigma)\, d\mathbf{V} \right] > 0 \tag{21.85}$$

A theorem[24] for determining Bayes estimators $\boldsymbol{\gamma}_\sigma^*$ can sometimes be applied when the cost function is a symmetric differentiable distance function of the type of Eq. (21.76). The theorem states:

Theorem I. If the joint d.d. $w(\mathbf{V},\boldsymbol{\theta})$ *can be factored into the form* $w(\mathbf{V},\boldsymbol{\theta}) = f_1(\mathbf{V})f_2[\boldsymbol{\theta} - \mathbf{g}(\mathbf{V})]$, *where* f_1 *and* g *are any functions of* $\mathbf{V}$, *and if* $f_2[\boldsymbol{\theta} - \mathbf{g}(\mathbf{V})] = f_2[\mathbf{g}(\mathbf{V}) - \boldsymbol{\theta}]$ and is unimodal about $\boldsymbol{\theta} = \mathbf{g}(\mathbf{V})$, *where the ranges of* $\boldsymbol{\theta}$ *are* $(-\infty, \infty)$, *then, for differentiable‡ cost criteria [Eq. (21.76)], the Bayes estimator is*

$$\boldsymbol{\gamma}_\sigma^*(\mathbf{V}) = \mathbf{g}(\mathbf{V}) \tag{21.86}$$

To establish this result, observe that Eq. (21.61*b*) for determining $\boldsymbol{\gamma}_\sigma^*$ is equivalent to

$$\int_{\Omega_\theta} f_1(\mathbf{V})f_2[\boldsymbol{\theta} - g(\mathbf{V})] \frac{\partial C(\boldsymbol{\theta} - \boldsymbol{\gamma}_\sigma^*)}{\partial \boldsymbol{\theta}}\, d\boldsymbol{\theta} = 0 \tag{21.86a}$$

Since $C(\boldsymbol{\theta} - \boldsymbol{\gamma}_\sigma^*)$ is even, its derivatives are odd. With $\mathbf{z} = \boldsymbol{\theta} - \mathbf{g}(\mathbf{V})$, and after the substitution $\mathbf{g}(\mathbf{V}) = \boldsymbol{\gamma}_\sigma^*$, Eq. (21.86) is now equivalent to

$$\int \cdots \int_{-\infty}^{\infty} f_2(\mathbf{z}) \frac{\partial C(\mathbf{z})}{\partial \mathbf{z}}\, d\mathbf{z} = \mathbf{0} \tag{21.86b}$$

which is identically satisfied, since the integrand is odd. This establishes the result of Eq. (21.86), where we note that Eq. (21.86) yields a minimum

† This is the extraction procedure discussed originally in Middleton and Van Meter,[1] sec. 4.1.

‡ The theorem is easily extended to include the rectangular cost functions [Eqs. (21.78), (21.79)] in the limit of arbitrarily small η_k, that is, the constant cost functions [Eq. (21.52*a*), (21.84)].

average risk because of the convex nature of the cost function and the unimodality of f_2.

A second theorem† is also sometimes useful when we attempt to obtain Bayes estimation and Bayes risks for cost functions other than the quadratic:

Theorem II. For all cost functions of the type $C(\theta - \gamma_\sigma)$ *where*

$$C(\theta - \gamma_\sigma) = C(\gamma_\sigma - \theta) \geq 0$$
$$C[(\theta - \gamma_\sigma)_1] < C[(\theta - \gamma_\sigma)_2] \qquad \text{if } |(\theta - \gamma_\sigma)_1| < |(\theta - \gamma_\sigma)_2|$$

the Bayes estimator $\boldsymbol{\gamma}_\sigma^*$ $[= (\gamma_\sigma^*)_{QCF}]$ *for a quadratic cost function is also optimum, that is, minimizes the average risk, for these other cost functions, provided the conditional d.d. of the parameter* (θ), *given* $\mathbf{V}$, that is, $w(\theta|\mathbf{V})$, *is unimodal and symmetric about the mode*‡ [of $w(\theta|\mathbf{V})$].

Some representative cost functions of practical interest are [with $\gamma_\sigma = \gamma^* \equiv (\gamma_\sigma^*)_{QCF}$ here, for optimality]

$$C_1(\theta - \gamma_\sigma^*) = |\theta - \gamma_\sigma^*| \tag{21.87a}$$

$$C_2(\theta - \gamma_\sigma^*) = \begin{cases} 0 & |\theta - \gamma_\sigma^*| < A \\ 1 & |\theta - \gamma_\sigma^*| > A \ (> 0) \end{cases} \tag{21.87b}$$

$$C_3(\theta - \gamma_\sigma^*) = \begin{cases} 0 & |\theta - \gamma_\sigma^*| < A \\ \dfrac{|\theta - \gamma_\sigma^*| - |A|}{|B| - |A|} & |A| \leq |\theta - \gamma_\sigma^*| \leq |B| \\ 1 & |B| < |\theta - \gamma_\sigma^*| \end{cases} \tag{21.87c}$$

Note, again, that $\gamma_\sigma^* = (\gamma_\sigma^*)_{QCF}$ is not generally linear in $\mathbf{V}$ and that, of course, the Bayes risks§ $E_{\mathbf{V},\theta}\{C_1\}$, etc., are not the same as for the quadratic cost function [cf. Eq. (21.63)]. An example where this theorem applies is indicated in Sec. 21.3-1.

Finally, the results of this section may be equally well applied to the important cases of waveform estimation. For simple estimation ($t_\lambda = t_k$, $k = 1, \ldots, n$), one simply replaces $\boldsymbol{\theta}$ by $\mathbf{S}$ and $W_n(\mathbf{V}|\boldsymbol{\theta})$ by $F_n(\mathbf{V}|\mathbf{S})$, etc., as before. For interpolation and extrapolation, similar results are readily found, with an appropriate modification of the cost function $C(\mathbf{S};\boldsymbol{\gamma}_\sigma)$ to $C(\mathbf{S},S_\lambda;\boldsymbol{\gamma}_\sigma)$ (cf. the beginning of Sec. 21.2). While the cost functions of the distance type lead generally to the simplest results for $\boldsymbol{\gamma}_\sigma^*$ and R^*, we are in no way restricted to their use. The informational cost function $\mathfrak{F}_2 = -\log p(\mathbf{S}|\boldsymbol{\gamma})$ is one example of a more complicated type, where the cost assignment depends not only on $\mathbf{S}$ and $\boldsymbol{\gamma}$ but on the decision rule δ as well [cf. Eqs. (18.5) et seq. and Secs. 21.2-4, 22.2].

21.2-4 Information Loss in Extraction. Instead of cost functions like Eq. (21.76) which are independent of the decision rule δ, we may use a measure of information loss [cf. Eq. (18.5)] to assess the performance of an

† This is a modified version of Sherman's results.[31]

‡ Or, equivalently, that $w(\theta - \gamma_\sigma^*|\mathbf{V})$ is unimodal and symmetric about $\theta = \gamma_\sigma^*$.

§ To calculate these other Bayes risks we may profitably employ the Laplace-transform techniques of Sec. 2.3-2 and carry out the required averages over the resulting exponential terms, as in the calculation of output covariance and spectra (cf. Chap. 13 also).

extraction system in a way analogous to that used for detection systems (cf. Sec. 20.3-1). The information about the signal waveform **S** contained in an observation **V**, or the equivocation at the extractor's input, is given by

$$H_V = -\int_\Omega \int_\Gamma w_n(\mathbf{V},\mathbf{S}) \log w_n(\mathbf{S}|\mathbf{V})\, d\mathbf{S}\, d\mathbf{V} + H_0(\mathbf{S}) \qquad (21.88a)$$

where $H_0(\mathbf{S})$ is the absolute entropy associated with the observation of **S** (cf. Secs. 6.4-1 and 18.4). The equivocation at the output of an extractor providing an estimate $\boldsymbol{\gamma}(\mathbf{V})$ is given by Eqs. (18.9) for a nonrandomized decision rule:

$$H(\sigma,\delta) = -\int_\Omega \int_\Gamma w_n(\mathbf{V},\mathbf{S}) \log \{w_n[\mathbf{S}|\boldsymbol{\gamma}(\mathbf{V})]\}\, d\mathbf{S}\, d\mathbf{V} + H_0(\mathbf{S}) \qquad (21.88b)$$

The difference between these is the (average) information about **S** lost in the system operation of transforming from **V** to $\boldsymbol{\gamma} = T_R\{\mathbf{V}\}$, namely,

$$H_V - H(\sigma,\delta) = -\int_\Omega \int_\Gamma w_n(\mathbf{V},\mathbf{S}) \log \frac{w_n(\mathbf{S}|\mathbf{V})}{w_n[\mathbf{S}|\boldsymbol{\gamma}(\mathbf{V})]}\, d\mathbf{S}\, d\mathbf{V} \qquad (21.89)$$

which is, of course, independent of the absolute entropy H_0.

When the extractor $\boldsymbol{\gamma}(\mathbf{V})$ defines a one-to-one reversible point transformation from **V** to $\boldsymbol{\gamma}$, as it does, for example, in the case of Bayes estimation with gaussian signal and noise (cf. Sec. 21.4-1), then the information lost by the system is zero. That is, under these conditions, knowledge of $\boldsymbol{\gamma}$ is as good as knowledge of **V** on the average, with respect to whatever clue either furnishes as to the signal present at the input. The fact that the information loss is zero follows from Eq. (21.89) if we note that

$$w_n(\mathbf{S}|\boldsymbol{\gamma}) = \sigma(\mathbf{S}) \frac{w_n(\boldsymbol{\gamma}|\mathbf{S})}{w_n(\boldsymbol{\gamma})} \qquad (21.90a)$$

and

$$w_n(\boldsymbol{\gamma}|\mathbf{S}) = F_n(\mathbf{V}|\mathbf{S})/|J(\boldsymbol{\gamma}/\mathbf{V})| \qquad (21.90b)$$

$$w_n(\boldsymbol{\gamma}) = w_n(\mathbf{V})/|J(\boldsymbol{\gamma}/\mathbf{V})| \qquad (21.90c)$$

where $|J(\boldsymbol{\gamma}/\mathbf{V})|$ is the (absolute) value of the Jacobian $\|\partial(\gamma_1, \ldots, \gamma_n)/\partial(V_1, \ldots, V_n)\|$. We thus have

$$w_n(\mathbf{S}|\boldsymbol{\gamma}) = \sigma(\mathbf{S}) \frac{F_n(\mathbf{V}|\mathbf{S})}{w_n(\mathbf{V})} = w_n(\mathbf{S}|\mathbf{V}) \qquad (21.91)$$

which makes the input and output equivocations [Eqs. (21.88a) and (21.88b)] of the extractor equal,† or the average information loss zero.

† Note that there is an entropy loss in the extractor equal to

$$-\int_\Gamma w_n(\mathbf{V}) \log \frac{w_n(\mathbf{V})}{w_n(\boldsymbol{\gamma})}\, d\mathbf{V} = -\int_\Gamma w_n(\mathbf{V}) \log \left| J\left(\frac{\boldsymbol{\gamma}}{\mathbf{V}}\right)\right| d\mathbf{V}$$

which, for example, becomes $-\log \det \mathbf{Q}_E$ for the linear transformation $\boldsymbol{\gamma} = \mathbf{Q}_E\mathbf{V}$ encountered in Sec. 21.1-3. This, however, is not information loss in the sense used here, where Eqs. (21.87), (21.88) give the information loss about the signal **S**, given **V**, while the entropy loss above is a measure of the information loss in the posterior data **V** itself, occasioned by the change in metric (cf. Sec. 6.4-1). This shows the importance of specifying carefully which quantity we wish information about.

When the extractor $\gamma(\mathbf{V})$ defines a nonreversible transformation from $\mathbf{V}$ to γ (that is, when knowledge of γ does not imply complete knowledge of $\mathbf{V}$), information is lost unless $\gamma(\mathbf{V})$ happens to be a sufficient statistic [i.e., unless $\gamma(\mathbf{V})$ is such that it contains all relative information concerning the quantity estimated; see Secs. 21.1-1(3) and 22.1-1]. The Bayes estimator of Sec. 21.3-1 for the amplitude of a coherent signal in gaussian noise, for example, transforms n-dimensional data $\mathbf{V}$ into one-dimensional estimates $a^*(\mathbf{V})$, but since it is a sufficient statistic, in this particular instance no information is lost (cf. Sec. 21.4-1 also).

21.2-5 Structure, System Comparisons, and Distributions. As we have already remarked at the beginning of this chapter, the estimators $\boldsymbol{T}_R\{\mathbf{V}\} = \gamma(\mathbf{V})$ embody the explicit operations that a receiving system $\boldsymbol{T}_R$ must carry out on the received data V in order to provide the observer with his desired estimate. For Bayes systems, actual estimator structure, as in detection,

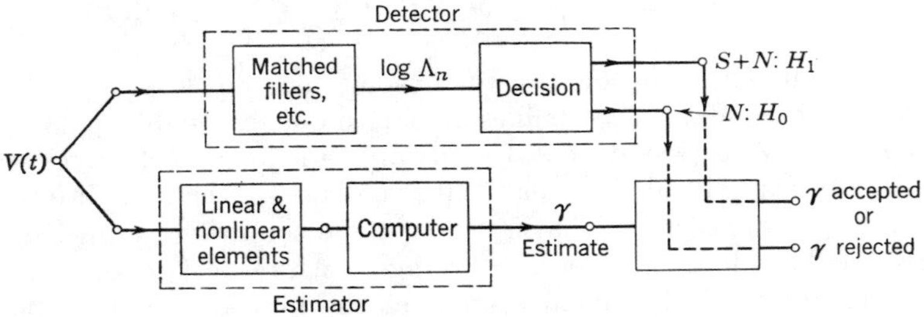

FIG. 21.2. Estimation, with detection.

depends (1) on the statistical character of both the signal and the interfering noise, (2) on the sample size, and (3) here, particularly, on the choice of cost function. Also as in detection, the resolution of a Bayes estimator into physical components, e.g., matched filters, nonlinear elements, etc., is not necessarily unique, especially if the observation process involves sample uncertainty. We shall encounter several specific examples in Secs. 21.3, 21.4. The block diagram of Fig. 21.2 shows a typical scheme for optimal or suboptimal estimation. Since we do not always know that a signal is actually present in many cases, a detection system is operated in concert with the estimator. The estimate from the latter is then accepted or rejected (subject, of course, to the usual type I and II error probabilities) according as the detection system decides that a signal is present or absent.

It is also possible, as in optimum threshold detection, to present a rather general theory for the analogous situation of optimum threshold estimation, by a suitable expansion in powers of the signal-to-noise ratio (cf. Secs. 19.4-1, 19.4-3). Let us illustrate with the following formal development for the quadratic cost function (21.64) and simple estimation. Here $\gamma_k^*(\mathbf{V})_S$ is the kth component of the corresponding Bayes estimator of waveform $\mathbf{S}$ [cf. Eqs. (21.66), (21.66a)], and we desire its weak-signal structure when signal and noise are additive and independent. For the moment, the signal is assumed to be deterministic, with random parameters $\boldsymbol{\theta}$, so that Eq. (21.66*a*) is the appropriate expression for the Bayes estimator here. Observing that now $F_n(\mathbf{V}|\mathbf{S}) = W_n[\mathbf{V} - \mathbf{S}(\boldsymbol{\theta})]_N$, we can write Eq. (21.66*a*) with the help of the normalizations (19.49) as

$$\gamma_k^*(\mathbf{V})_S = \langle a_0 \sqrt{\psi}\, s_k W_n(\mathbf{V} - a_0 \sqrt{\psi}\, \mathbf{s})_N \rangle_{\boldsymbol{\theta}} / \langle W_n(\mathbf{V} - a_0 \sqrt{\psi}\, \mathbf{s})_N \rangle_{\boldsymbol{\theta}} \quad (21.92a)$$

Following the procedures outlined in Sec. 19.4-1 for threshold detection, we obtain

$$\gamma_k^*(\mathbf{V})_S \doteq \langle a_0 \sqrt{\psi}\, s_k \rangle_{\boldsymbol{\theta}} - \langle a_0{}^2 s_k \psi [\tilde{\mathbf{s}}\mathbf{b}^{(1)} + 2\tilde{\mathbf{v}}\mathbf{b}^{(2)}\tilde{\mathbf{b}}^{(2)}\mathbf{s} + O(\mathbf{V}^2,\mathbf{V}^3) \cdots] \rangle_{\boldsymbol{\theta}}$$
$$+ \langle a_0 \psi^{1/2} s_k \rangle_{\boldsymbol{\theta}} \langle a_0 \sqrt{\psi}\, s_k [\tilde{\mathbf{s}}\mathbf{b}^{(1)} + 2\tilde{\mathbf{v}}\mathbf{b}^{(2)}\tilde{\mathbf{b}}^{(2)}\mathbf{s} + O(\mathbf{V}^2,\mathbf{V}^3)] \rangle_{\boldsymbol{\theta}} + O(a_0{}^3) \quad (21.92b)$$

We observe that with sample uncertainty, such that $\langle s_k \rangle_\epsilon$ vanishes (each k), the Bayes threshold estimator depends primarily on the first power of the (mean) input signal-to-noise *intensity* ratio ($\sim\overline{a_0{}^2}$), rather than on its square root, again as in detection. However, the leading contributions to this estimator involve both (averaged) generalized cross correlations and autocorrelations of the received data with (an appropriate average of) the signal. Thus, the dependence on the received data is both linear and nonlinear, like the analogous case in detection, where an (averaged) generalized autocorrelation of the received wave with itself is required [cf. Eq. (19.59)]. Note that with sample certainty ($\langle s \rangle_\epsilon \neq 0$) there is an additional contribution, independent of the received data $\mathbf{V}$, which is simply the expected value of the component s_k, as Eq. (21.92*b*) indicates. This general behavior is attributable to the choice of cost function, which in conjunction with the particular statistical character of the noise and signal is so influential in determining the character of the Bayes estimator. For instance, the simple, or constant, cost function of Sec. 21.2-1, which yields UML estimators, gives a different threshold development. This is easily seen by substituting Eq. (21.16*b*) in Eq. (21.13*b*), after averaging over the relevant parameters of S. The result from which the appropriate Bayes estimator of waveform is to be found is now

$$\frac{\partial}{\partial S_k} [\log \sigma(S_k) F_n(\mathbf{V}|S_k)] = \frac{\partial}{\partial S_k} [\log \sigma(S_k) + \log \langle W_n(\mathbf{V} - \mathbf{S})_N \rangle_{\mathbf{S}'}] = 0 \quad (21.93a)$$

where† $\mathbf{S}'$ represents all components of $\mathbf{S}$ except S_k. We get specifically

$$\frac{\sigma'(\hat{S}_k)}{\sigma(\hat{S}_k)} - [b_k{}^{(1)} + 2\sqrt{\psi}\,(\tilde{\mathbf{v}}\mathbf{b}^{(2)}\tilde{\mathbf{b}}^{(2)})_k + \cdots]$$
$$+ \left(\frac{d}{dS_k} \langle \tilde{\mathbf{S}}\mathbf{b}^{(2)}\tilde{\mathbf{b}}^{(2)}\mathbf{S} \rangle_{\mathbf{S}'} + \cdots \right)_{S_k = \hat{S}_k} = 0 \quad (21.93b)$$

† Remember that $F_n(\mathbf{V}|\mathbf{S}_k) = \int \sigma(\mathbf{S}')F_n(\mathbf{V}|\mathbf{S})\, d\mathbf{S}' \equiv \langle F_n(\mathbf{V}|\mathbf{S})\rangle_{\mathbf{S}'}$ [cf. footnote referring to Eq. (21.57)].

which is to be solved for the desired threshold Bayes (UML) estimator $\gamma_k^* = \hat{S}_k$. Clearly, $\hat{S}_k$ is generally *not* the same as γ_k^* obtained from Eqs. (21.92). That different optimum estimator structures occur for different cost functions is rather to be expected. The more unusual case is that for which the *same* Bayes estimator applies, for *different* cost functions. An example of this sort of structural invariance is observed in Sec. 21.4-1, for the optimum estimation of the waveforms of normal stochastic signals.

In the same way, we may obtain results similar to Eqs. (21.92), (21.93) for signal-parameter estimation, replacing $F_n(\mathbf{V}|\mathbf{S})$ by $W_n(\mathbf{V}|\boldsymbol{\theta})$, etc. However, since, with the exception of amplitude ($\sim a_0$), signal parameters such as frequency, epoch, etc., do not appear linearly in $\mathbf{S}(\boldsymbol{\theta})$, explicit developments like Eqs. (21.92), (21.93) are not possible without our knowing the actual form of the distributions. Finally, again as in detection, it is possible to interpret the various operations on the received data V, required by γ_σ, in terms of the familiar realizable elements of detection theory, e.g., matched linear filters, zero-memory nonlinear devices, and simple integrators (e.g., transversal filters) [cf. Secs. 21.1-3(1), 21.1-3(3)]. In addition, a computer may be required following these elements (cf. Fig. 21.2 and Sec. 21.3-2).

Suboptimum estimation systems may be evaluated within the framework of decision theory in the same fashion as the optimum systems considered above, by computing their average risks. System comparison is then possible in a consistent way: Bayes systems have the smallest average risk, and one system is considered better than another (for the same purpose) if its average risk is less. Thus, if $\boldsymbol{\gamma}(\mathbf{V})_S$ is a suboptimum estimator of waveform S and $\boldsymbol{\gamma}_\sigma^*(\mathbf{V})_S$ is the corresponding Bayes extractor, we may measure the superiority of the latter over the former by comparing the average risks $R^*(\sigma,\delta)$, $R(\sigma,\delta)$. For quadratic cost functions, these become simply

$$C_0 E_{\mathbf{V},} \{\|\mathbf{S} - \boldsymbol{\gamma}_\sigma^*(\mathbf{V})_S\|^2\} < C_0 E_{\mathbf{V},\mathbf{S}}\{\|\mathbf{S} - \boldsymbol{\gamma}(\mathbf{V})_S\|^2\} \tag{21.94}$$

and one can then determine the increase in sample size or input signal-to-noise ratio that a nonoptimum extractor requires to yield the same minimum average risk. Similar comparisons and "trade-offs" between system parameters can be computed in this way, in principle for any suitable cost function, whenever the corresponding optimum estimator is known. Of course, the same procedures apply equally well to comparisons between suboptimum systems themselves.

The estimators themselves, whether optimal or not, are random variables when considered over the ensemble of possible inputs, as we have remarked earlier, in Sec. 21.1-1. Their distribution (densities) may be calculated with the help of Eqs. (21.11*a*), (21.11*b*), and from the associated characteristic functions the various moments of these estimators may be directly obtained. Specific examples may be found in Secs. 21.3, 21.4. Maximum likelihood estimators, of both signal parameters and signal waveforms, are usually asymptotically normally distributed [cf. Eqs. (21.14), (21.15)]; the threshold Bayes estimators of waveform are likewise normal, with quadratic cost functions, but the Bayes estimators for other than the quadratic and simple cost functions are not generally gaussian, even in the large-sample cases characteristic of threshold reception. The situation is more complicated than that of threshold detection, where (in the physically significant cases) one usually has normal statistics for the logarithm of the likelihood ratio, as long as high accuracies of decision are required. This is again because of the nature of the quantity estimated—i.e., how it appears in signal structure—and because of the latitude in the choice of cost functions. For these reasons, a formal theory as comprehensive for extraction as for detection does not as yet exist.

PROBLEMS

21.4 Extend the theory for the simple cost function (21.56) to the cases of interpolation ($0 < t_\lambda < T$) and extrapolation ($t_\lambda < 0$, $t_\lambda > T$).

21.5 Show that the relation (21.61*b*) determining the Bayes estimator with a quadratic cost function actually yields a minimum average risk.

21.6 Carry out the details of the derivation of Eqs. (21.83), (21.85) for the simple cost function (21.84).

21.7 Extend the results of Sec. 21.2-5 to include interpolation and extrapolation.

21.3 *Estimation of Amplitude (Deterministic Signals and Normal Noise)*

A fairly complete theory for amplitude estimation in normal noise can be constructed when signal and noise are additive and independent, and when the quadratic, or simple, cost functions are chosen as measures of the risk. The following three examples illustrate the calculation of Bayes extractors and their associated Bayes risks: (1) coherent extraction of signal amplitude in normal noise with a quadratic cost function and normal d.d. of amplitudes; (2) incoherent estimation of amplitude, again in normal noise with a quadratic cost function; and (3) similar to (2), but with a simple cost function, resulting in an appropriate UML estimator.

21.3-1 Coherent Estimation of Signal Amplitude (Quadratic Cost Function). We examine first the estimation of the amplitude factor a_0 of a deterministic signal $S(t - \epsilon_0, \boldsymbol{\theta}_0)$ which has a *normal* distribution of amplitudes but is otherwise completely specified at the receiver:

$$\sigma(\boldsymbol{\theta}) = \sigma(a_0) = (2\pi\sigma^2)^{-\frac{1}{2}} e^{-(a_0-\bar{a}_0)^2/2\sigma^2} \qquad \sigma^2 = \overline{a_0^2} - \bar{a}_0^2 \tag{21.95}$$

Again we assume that signal and noise are additive and normalized [cf. Eqs. (19.49)],

$$\mathbf{S} = a_0\psi^{\frac{1}{2}}\mathbf{s} \qquad \mathbf{V} = \mathbf{S} + \mathbf{N} = \psi^{\frac{1}{2}}\mathbf{v}$$

The quadratic cost function here is $C_0(\gamma_\sigma - a_0)^2$, where γ_σ is the estimator, so that from Eq. (21.62) or (21.71) the Bayes estimator $\gamma_\sigma^* = a_0^*$ becomes

$$T_{R^{(N)}}(V) = a_0^*(\mathbf{V}) = \int_{-\infty}^{\infty} a_0 w(\mathbf{V},a_0)\, da_0 \Big/ \int_{-\infty}^{\infty} w(\mathbf{V},a_0)\, da_0 \quad (21.96)$$

with the joint distribution density $w(\mathbf{V},a_0)$ here specifically equal to

$$\frac{\exp\{-\tfrac{1}{2}[(\tilde{\mathbf{v}} - a_0\tilde{\mathbf{s}})\mathbf{k}_N^{-1}(\mathbf{v} - a_0\mathbf{s}) + (a_0 - \bar{a}_0)^2/\sigma^2]\}}{(2\pi)^{n/2}(\det \mathbf{k}_N)^{1/2}(2\pi\sigma^2)^{1/2}} \qquad \bar{N} = 0 \quad (21.96a)$$

and with $\mathbf{k}_N$, as before, the (normalized) variance matrix of the normal noise process for sample size n. Noting the identity

$$a_0^* = \frac{\int_{-\infty}^{\infty} a_0 e^{-A(a_0-a_0^*)^2+B(\mathbf{v})}\, da_0}{\int_{-\infty}^{\infty} e^{-A(a_0-a_0^*)^2+B(\mathbf{v})}\, da_0} \quad (21.96b)$$

we can evaluate Eq. (21.96) very simply by completing the square in Eq. (21.96a). The result is the desired Bayes estimator, or optimum receiver,

$$T_{R^{(N)}}(V) = a_0^*(\mathbf{V}) = \frac{\Phi_n(v,s) + \bar{a}_0/\sigma^2}{\Phi_n(s,s) + 1/\sigma^2} \quad (21.97)$$

where $\Phi_n(v,s) = \tilde{\mathbf{v}}\mathbf{k}_N^{-1}\mathbf{s} \equiv \Phi_v \qquad \Phi_n(s,s) = \tilde{\mathbf{s}}\mathbf{k}_N^{-1}\mathbf{s} \equiv \Phi_s$

[cf. Eqs. (20.2a), (20.2b)]. Because of sample certainty and the normality of the noise, this estimator is *linear* in the received data V.

We observe first that as $\sigma^2 \to 0$, that is, as the a priori signal density becomes a delta function at $a_0 = \hat{a}_0$, the Bayes estimator reduces to $a_0^* = \bar{a}_0 = a_0$ as expected. When $\sigma^2 \to \infty$ [so that the a priori distribution density becomes uniform; cf. footnote referring to Eqs. (21.23)], the Bayes estimate becomes equal to $a_0^* = \hat{\gamma}_{a_0} = \hat{a}_0$ [Eq. (21.25)], agreeing with the maximum (conditional and unconditional) likelihood estimates, as it should [for $\bar{N} = 0$; cf. the example of Sec. 21.1-3(1)].

Estimation of a_0 with continuous sampling gives $\Phi_n(v,s) \to \Phi_T(v,s)$, etc., as before [cf. Eqs. (20.35a), (20.35b), (20.37)], and the governing integral equation is Eq. (20.38) (see Sec. 19.4-2 for remarks on the passage to the limit). Thus, Eq. (21.97) becomes, for continuous sampling, the functional

$$a_0^*[V(t)] = \frac{\Phi_T(v,s) + \bar{a}_0/\sigma^2}{\Phi_T(s,s) + \sigma^{-2}} \quad (21.97a)$$

Since $\sigma^2 = (2\psi)^{-1}\sigma_A{}^2$, $\sigma_A{}^2 = \overline{A_0{}^2} - \bar{A}_0{}^2$, and $A_0^*(\mathbf{V}) \equiv \sqrt{2\psi}\, a_0^*(\mathbf{V})$ [cf. Eqs. (21.24)], we easily find for white noise backgrounds ($\psi = BW_0$, $\lim B \to \infty$) that Eq. (21.97a) reduces to

$$A_0^*[V(t)]\Big|_{\text{white}} = \frac{\sqrt{2}\left[(1/W_0)\int_0^T V(t)s(t - \epsilon_0, \boldsymbol{\theta}_0)\, dt + \bar{A}_0/\sqrt{2}\,\sigma_A{}^2\right]}{(1/W_0)\int_0^T s(t - \epsilon_0, \boldsymbol{\theta}_0)^2\, dt + \sigma_A{}^{-2}} \quad (21.97b)$$

The structure of these estimators takes the form indicated in Fig. 21.2. Here, for $\Phi_n(v,s)$ or $\Phi_T(v,s)$ one has essentially the same elements, e.g., matched filters and ideal integrators used in the coherent detection of such signals (cf. Secs. 20.1-1, 20.2-1, and Fig. 20.1), followed, however, by a simple computer which adds $\bar{a}_0/\sigma^2$ and then divides by $\Phi_T + \sigma^{-2}$ according to Eqs. (21.97), (21.97a).

Because of the normal character of the noise, the normal distribution of amplitudes, and the manner in which the signal is observed, the Bayes estimator of amplitude is here a linear function or functional of the received data $V(t)$. The distribution density of this estimator is accordingly normal. To determine the explicit form of the d.d., let us begin first with the characteristic function

$$F_1(i\xi)_{a_0^*} = \boldsymbol{E}_{\mathbf{V}|H_1}\{e^{i\xi a_0^*(\mathbf{V})}\} = \int_\Gamma e^{i\xi(A\tilde{\mathbf{v}}\mathbf{k}_N^{-1}\mathbf{s}+B)}w_n(\mathbf{V})\, \mathbf{dV} \tag{21.98}$$

where $w_n(\mathbf{V})$ is found from Eq. (21.96a) to be specifically

$$w_n(\mathbf{V}) = [(\sigma^2\Phi_s + 1)(2\pi)^n(\det \mathbf{k}_N)]^{-\frac{1}{2}} \exp\,(c_0 - \bar{a}_0{}^2/2\sigma^2) \\ \exp\,[-\tfrac{1}{2}\tilde{\mathbf{v}}(\mathbf{k}_N{}^{-1} - 2c_2\mathbf{G})\mathbf{v} + c_1\tilde{\mathbf{v}}\mathbf{k}^{-1}\mathbf{s}] \tag{21.99}$$

with $\mathbf{G} = \mathbf{k}_N{}^{-1}\mathbf{s}\tilde{\mathbf{s}}\mathbf{k}_N{}^{-1}$ [cf. Eqs. (20.11b)] and

$$c_0 = \bar{a}_0{}^2/(2\sigma^4\Phi_s + 2\sigma^2) \qquad c_1 = \bar{a}_0/(1 + \sigma^2\Phi_s) \qquad c_2 = 2\Phi_s + 2/\sigma^2 \tag{21.99a}$$

and $c_0 > 0$, c_1, $c_2 \geq 0$. Carrying out the indicated operations† in Eq. (21.98) using Eq. (21.99), we get finally

$$F_1(i\xi)_{a_0^*} = e^{i\xi\bar{a}_0 - \frac{1}{2}\sigma^4\Phi_s\xi^2/(1+\sigma^2\Phi_s)} \tag{21.100}$$

so that a_0^* is normally distributed with mean $\bar{a}_0$ and variance $\sigma^4\Phi_s/(1 + \sigma^2\Phi_s)$. Observe that, when $\sigma^2 \to 0$, the Bayes estimator a_0^* equals $\bar{a}_0 = a_0$, as expected, since the only value of amplitude is $\bar{a}_0$ itself, that is, $w_1(a_0^*) = \delta(a_0^* - \bar{a}_0)$. On the other hand, if σ^2 is sufficiently great, $\sigma^2\Phi_s \gg 1$ and then a_0^* is normally distributed with a variance that is independent of waveform and equal to the variance of a_0 itself. This is reasonable, since now the large spread in the possible values of signal amplitude, apart from the effects of the background noise, dominates the distribution of $a_0^*(\mathbf{V})$.

We may use Eq. (21.100) directly to obtain the first and second moments of the Bayes estimator. These are

$$\overline{a_0^*} = \bar{a}_0 \qquad \overline{a_0^{*2}} = \bar{a}_0{}^2 + \frac{\sigma^4\Phi_s}{1 + \sigma^2\Phi_s} \tag{21.101}$$

For the Bayes risk [Eq. (21.63)], we need the d.d. of $a_0^* - a_0$, which like that of a_0^* is also normal, since a_0^* is linear in V [cf. Eq. (21.97)] and a_0 is

† An alternative and simpler method is to compute the c.f. of a_0^* directly from

$$\boldsymbol{E}_{\mathbf{V},a_0|H_1}\{e^{i\xi a_0^*}\}$$

with respect to the joint d.d. $w(\mathbf{V},a_0)$ [Eq. (21.96a)].

normally distributed. Using the fact that

$$\boldsymbol{E}_{\mathbf{V}|H_1}\{\Phi_v\} = \bar{a}_0\Phi_s \qquad \boldsymbol{E}_{\mathbf{V}|H_1}\{\Phi_v{}^2\} = \overline{a_0{}^2}\Phi_s{}^2 + \Phi_s$$

we find after a little manipulation that

$$\boldsymbol{E}_{\mathbf{V},a_0|H_1}\{(a_0^* - a_0)^2\} = \frac{\sigma^2}{\sigma^2\Phi_s + 1} \tag{21.102a}$$

The d.d. of $a_0^* - a_0$ is normal, specifically with zero mean (since $\overline{a_0^*} = \bar{a}_0$) and variance $\sigma^2/(\sigma^2\Phi_s + 1)$ above. The c.f. of $a_0^* - a_0$ is accordingly

$$F_{a_0^*-a_0}(i\xi) = e^{-\xi^2\sigma^2/2(\sigma^2\Phi_s+1)} \tag{21.102b}$$

The Bayes risk itself is proportional to Eq. (21.102a), or to $-d^2F/d\xi^2\Big|_{\xi=0}$, from Eq. (21.102$b$), so that we can write it directly as

$$R^*(\sigma,\delta)_{a_0} = C_0(\overline{a_0^{*2}} - 2\overline{a_0a_0^*} + \overline{a_0{}^2}) = \frac{C_0\sigma^2}{1 + \sigma^2\Phi_s} \tag{21.102c}$$

As expected when $\sigma^2 \to 0$, the average risk vanishes, since extraction is exact on the average, $a_0^* = \bar{a}_0 = a_0$, while for $\sigma^2 \to \infty$ there is necessarily always a finite average cost $C_0\Phi_n(s,s)^{-1}$. The Bayes risk, of course, depends on the structure of the signal and on the spectrum of the accompanying noise. This example is of some practical interest in the case where signals subject to fading are received in noise, when $\bar{a}_0$ is large and σ^2 reasonably small, so that a_0 is essentially always greater than zero. Here the noise arises primarily in the receiver (and hence is unaffected by the propagation mechanism producing the fading). Observe, moreover, that a_0^* [Eq. (21.97) or (21.97a)] with $\sigma^2 \to \infty$ is also a Minimax estimate, as can be seen from the translation property [Eq. (21.72)], which becomes here $a_0^*(v + bs) = a_0^*(v) + bs$, with b a scalar quantity. One then uses an argument precisely parallel to that given above [Eqs. (21.73) et seq.], to show that the conditional risk is independent of a_0. The Minimax average risk is specifically, for discrete sampling,

$$R_M^* = C_0\Phi_n(s,s)^{-1} \qquad (> R^*) \tag{21.102d}$$

The Bayes risk for continuous sampling is given by Eq. (21.102c), with $\Phi_s = \Phi_T(s,s)$ as before. With white noise backgrounds we have $R^* = \lim_{\psi\to\infty} 2\psi C_0\overline{(a_0^* - a_0)^2}$, since $A_0^* = \sqrt{2\psi}\, a_0^*(\mathbf{V})$, etc. The result is

$$R^*(\sigma,\delta)_{A_0^*}\Big|_{\text{white}} = C_0\sigma_A{}^2(1 + 2\sigma_A{}^2E_S/\overline{A_0{}^2}W_0)^{-1} \tag{21.103a}$$

where $E_S = (\overline{A_0{}^2}/2)\int_0^T s(t - \epsilon_0, \boldsymbol{\theta}_0)^2\, dt$ is the average energy in the signal, showing that for white noise backgrounds the Bayes risk here is independent of signal waveform and depends only on the first and second moments of the amplitude A_0, the energy of the signal, and the spectral density of the noise.

The Minimax average risk follows directly from Eq. (21.102*d*) and is

$$R_M^*\Big|_{\text{white}} = C_0\overline{A_0{}^2}W_0/2E_S \tag{21.103b}$$

The Bayes estimator of amplitude, Eq. (21.97) or Eq. (21.97*a*) for continuous sampling, is also optimum for a wide class of cost functions other than the quadratic postulated here. For we readily see from Eq. (21.102*b*) that the conditional d.d. of $a_0 - a_0^*$, given **V** [that is, $a_0^* = a_0^*(\mathbf{V})$], is unimodal and symmetrical about the mode, here $a_0^*(\mathbf{V})$, since the d.d. of $a_0 - a_0^*$ is gaussian. The conditions of Theorem II (Sec. 21.2-3) are accordingly fulfilled, and consequently a_0^* [Eqs. (21.97), (21.97*a*)] minimizes the average risk for the cost functions of Eqs. (21.87*a*) to (21.87*c*) as well. Note that, when the d.d. of a_0 is uniform, that is, $\sigma \to \infty$, the conditional d.d. of $a_0 - a_0^*$ reduces to

$$w(a_0 - a_0^*|\mathbf{V}) = \sqrt{\frac{\Phi_s}{2\pi}}\exp\left[-\tfrac{1}{2}\Phi_s\left(a_0 - \frac{\Phi_v}{\Phi_s}\right)^2\right] \tag{21.103c}$$

and from Eq. (21.25) the resulting CML and UML estimators $\hat{a}_0 = a_0^*$ here are similarly Bayes with respect to this wider class of cost functions.

The analysis is not confined to optimum systems but is equally applicable to suboptimum estimators as well. Thus, instead of Eqs. (21.97), (21.97*a*) under the otherwise similar conditions of the above examples we may choose as a nonoptimum receiver for amplitude extraction

$$\gamma(\mathbf{V})_{a_0} = A_1\tilde{\mathbf{v}}\mathbf{G}_1\mathbf{s} + B_1 \qquad A_1,\ B_1 > 0$$

where $\tilde{\mathbf{v}}\mathbf{G}_1\mathbf{s}$ has the properties of the linear system described in Prob. 20.8. We may obtain the average risk, moments, and d.d. of this suboptimum estimator by the methods illustrated above [Eqs. (21.15)]. Comparisons may then be made with γ_σ^* on the basis of the respective average risks.

21.3-2 Incoherent Estimation of Signal Amplitude (Quadratic Cost Function). As our second example, let us consider the previous problem again, where now the deterministic signal is required to be narrowband and reception involves sample uncertainty, here with a uniform d.d. of RF epochs ϵ_c. All other signal parameters are assumed known, except the amplitudes ($\sim a_0$), which are random, with distributions unknown to the observer, unlike the example above.

As in detection (cf. Secs. 19.2-3, 20.4-8), we shall adopt the subcriterion of selecting an amplitude distribution which maximizes the average uncertainty, subject to (1) a *maximum value constraint* $(a_0)_{\max} = P_M{}^{1/2}$ or to (2) an *average intensity constraint* $\overline{a_0{}^2} = 2P_0$. As shown in Prob. 6.7 and Eqs. (21.22*a*), (21.22*b*), the former leads to a uniform d.d. for a_0 (> 0), with $\overline{a_0{}^m} = P_M{}^{m/2}/(m+1)$, while the latter yields a Rayleigh d.d., with $\overline{a_0{}^m} = (2P_0)^{m/2}\Gamma(m/2 + 1)$. For the class of signals considered here we find from

Eqs. (20.22) to (20.30) that the joint and conditional d.d.'s of $\mathbf{V}$ and a_0 are

$$\begin{aligned} w_n(\mathbf{V},a_0) &= \sigma(a_0)W_n(\mathbf{V}|a_0) \\ &= \sigma(a_0)f_0(\mathbf{V})I_0(a_0\sqrt{\tilde{\mathbf{v}}\mathbf{G}_0\mathbf{v}})\,e^{-a_0^2B_n/2} \end{aligned} \qquad (21.104)$$

with $f_0(\mathbf{V}) = (2\pi\psi)^{-n/2}(\det \mathbf{k}_N)^{-\frac{1}{2}}e^{-\frac{1}{2}\tilde{\mathbf{v}}\mathbf{k}_N^{-1}\mathbf{v}}$ and $\sigma(a_0)$ given specifically by Eqs. (21.22*a*), (21.22*b*). From Eqs. (20.26), (20.29*a*), we have, respectively, $B_n \equiv (\tilde{\mathbf{a}}\mathbf{k}_N^{-1}\mathbf{a} + \tilde{\mathbf{b}}\mathbf{k}_N^{-1}\mathbf{b})/2$ and $\mathbf{G}_0 = \mathbf{k}_N^{-1}(\mathbf{a}\tilde{\mathbf{a}} + \mathbf{b}\tilde{\mathbf{b}})\mathbf{k}_N^{-1}$. Inserting Eq. (21.104) into Eq. (21.96) then gives the desired Bayes estimator of (normalized) amplitude, with quadratic cost function, when reception is incoherent with epoch ϵ_c uniformly distributed over an RF cycle. We write accordingly

$$\boldsymbol{T}_R^{(N)}(V) = a_0^*(\mathbf{V}) = \frac{\int_{-\infty}^{\infty} a_0\sigma(a_0)I_0(a_0\Psi_{0n}^{\frac{1}{2}})e^{-a_0^2B_n/2}\,da_0}{\int_{-\infty}^{\infty} \sigma(a_0)I_0(a_0\Psi_{0n}^{\frac{1}{2}})e^{-a_0^2B_n/2}\,da_0} \qquad \Psi_{0n} \equiv \tilde{\mathbf{v}}\mathbf{G}_0\mathbf{v} \qquad (21.105)$$

Let us consider as our first case where $\sigma(a_0)$ is uniform $[0 < a_0 < (a_0)_{\max}$; cf. Eq. (21.22*a*)]. Then Eq. (21.105) becomes

$$a_0^*(\mathbf{V})_{P_M} = \frac{\int_0^{P_M^{\frac{1}{2}}} a_0I_0(a_0\Psi_{0n}^{\frac{1}{2}})e^{-a_0^2B_n/2}\,da_0}{\int_0^{P_M^{\frac{1}{2}}} I_0(a_0\Psi_{0n}^{\frac{1}{2}})e^{-a_0^2B_n/2}\,da_0} \qquad (21.106)$$

For weak signals, we may use a threshold development similar to Eq. (21.92*b*), obtained by expanding $W_n(\mathbf{V}|a_0)$ [that is, the coefficient of $\sigma(a_0)$ in Eq. (21.104)] in powers of a_0, averaging over a_0 in both numerator and denominator, and then developing the resulting fraction. Quite generally, for arbitrary $\sigma(a_0)$ and Eq. (21.104) we get the threshold estimator

$$a_0^*(\mathbf{V})_{P_M} \doteq [\bar{a}_0 + \tfrac{1}{2}(\bar{a}_0\overline{a_0^2} - \overline{a_0^3})B_n + O(\overline{a_0^5})] + (\overline{a_0^3} - \bar{a}_0\overline{a_0^2})\Psi_{0n}/4 + O(\overline{a_0^5},v^4) \qquad (21.107)$$

This becomes for the uniform d.d. of Eq. (21.22*a*), with

$$\overline{a_0^m} = P_M^{m/2}/(m+1)$$

$$a_0^*(\mathbf{V})_{P_M} = \frac{1}{2}(3\overline{a_{0u}^2})^{\frac{1}{2}}\left(1 - \frac{\overline{a_{0u}^2}}{4}B_n + \cdots\right) + \frac{\sqrt{3}}{16}(\overline{a_{0u}^2})^{\frac{3}{2}}\Psi_{0n} + O(\overline{a_0^5},v^4) \qquad (21.108)$$

where $\overline{a_{0u}^2} = P_M/3$. With strong signals as an upper limit, on the other hand, we may set $P_M \to \infty$ in Eq. (21.106) without seriously altering the result. Evaluation of the integrals then follows directly with the help of Eqs. (A.1.49), (A.1.19*a*), (A.1.31*b*), giving us

$$a_0^*(\mathbf{V})_{P_M\to\infty} = \left(\frac{\pi B_n}{2}\right)^{-1/2} \begin{cases} {}_1F_1\left(\frac{1}{2};1;-\frac{\Psi_{0n}}{2B_n}\right)^{-1} \\ \text{or} \\ e^{\Psi_{0n}/4B_n} I_0\left(\frac{\Psi_{0n}}{4B_n}\right)^{-1} \end{cases} \tag{21.109}$$

which is essentially $\Psi_{0n}^{1/2}/B_n$ as long as $\Psi_{0n} \gg B_n$. Figure 21.3 shows Eqs. (21.109) as a function of Ψ_{0n}, where $x = \Psi_{0n}/4B_n$; note that in the weak-signal case a_0^* depends linearly on Ψ_{0n} [cf. Eq. (21.108)].

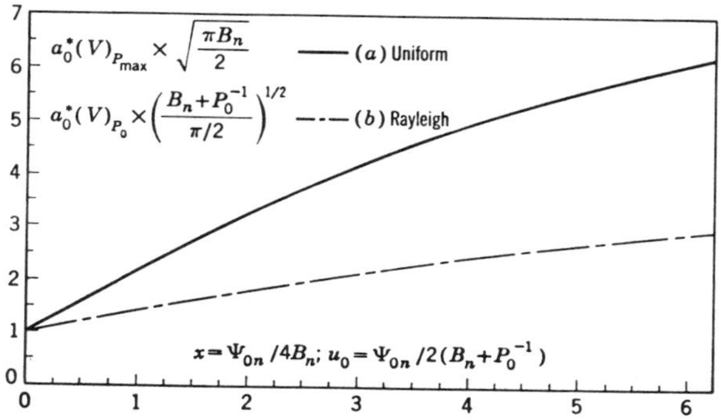

FIG. 21.3. Bayes estimators of amplitude; incoherent reception, quadratic cost function, for (a) peak-value constants and $P_M \to \infty$, (b) mean-square value constraint, all signals.

Similar results are obtained for the Rayleigh d.d. of amplitudes of Eq. (21.22b), corresponding to the maximum power constraint $\overline{a_{0R}^2} = 2P_0$. The Bayes estimator [Eq. (21.105)] now becomes

$$a_0^*(\mathbf{V})_{P_0} = \frac{\int_0^\infty a_0^2 I_0(a_0\Psi_{0n}^{1/2}) e^{-a_0^2(B_n+P_0^{-1})/2}\, da_0}{\int_0^\infty a_0 I_0(a_0\Psi_{0n}^{1/2}) e^{-a_0^2(B_n+P_0^{-1})/2}\, da_0} \tag{21.110}$$

With weak signals, we can use Eq. (21.107) directly, where for the Rayleigh d.d. $\overline{a_{0R}^m} = (2P_0)^{m/2}\Gamma(m/2+1)$, or we can evaluate Eq. (21.110) and apply the condition $P_0^2 \ll 1$. The threshold estimator is found to be

$$a_0^*(\mathbf{V})_{P_0} \doteq \frac{\sqrt{\pi}}{2} (\overline{a_{0R}^2})^{1/2} \left(1 - \frac{\overline{a_{0R}^2}}{4} B_n + \cdots\right) + \frac{\sqrt{\pi}}{16} (\overline{a_{0R}^2})^{3/2} \Psi_{0n} + O(a_0^5, v^4) \tag{21.111}$$

which is seen to be nearly the same as $a_0^*(\mathbf{V})_{P_M}$ for weak signals when $(\overline{a_0^2})_u = (\overline{a_0^2})_R$, that is, when the mean power in the signal is the same under both constraints.

The general expression for $a_0^*(\mathbf{V})_{P_0}$ is found once more, with the help of

Eqs. (A.1.49) and (A.1.31a), to be specifically for all signal strengths

$$a_0^*(\mathbf{V})_{P_0} = \frac{\sqrt{\pi}}{\sqrt{2}}(P_0^{-1} + B_n)^{-\frac{1}{2}} \begin{cases} {}_1F_1(-\frac{1}{2};1;-u_0) \\ \text{or} \\ e^{-u_0/2}[(1 + u_0)I_0(u_0/2) + u_0 I_1(u_0/2)] \end{cases} \tag{21.112}$$

where $u_0 \equiv \Psi_{0n}/(2B_n + 2P_0^{-1})$ (cf. Fig. 21.3). For strong signals ($P_0 B_n \gg 1$), this becomes

$$a_0^*(\mathbf{V})_{P_0} \simeq \Psi_{0n}{}^{\frac{1}{2}}/B_n \tag{21.112a}$$

as in the case of the uniform d.d. above when $\Psi_{0n} \gg B_n$ [cf. Eqs. (21.109)].

When continuous sampling in $(0,T)$ is employed, we may replace Ψ_{0n}, B_n by Ψ_{0T}, B_T in the usual way [cf. Sec. 20.2-3 and Eqs. (20.62c), (20.63a), (20.63b)] without changing the form of the estimators above. With white noise backgrounds, our estimators become $A_0^*(\mathbf{V})_u \equiv \lim_{\psi\to\infty} \sqrt{2\psi}\, a_0^*(\mathbf{V})_{P_M}$ and $A_0^*(\mathbf{V})_R \equiv \lim_{\psi\to\infty} \sqrt{2\psi}\, a_0^*(\mathbf{V})_{P_0}$ [cf. Eqs. (21.24)]. Specifically, we can write for Eqs. (21.108), (21.111) in the threshold cases

$$A_0^*(\mathbf{V}) \doteq \left(\begin{matrix}\sqrt{3}\\ \sqrt{\pi}\end{matrix}\right)\frac{\overline{A_0^2}^{\frac{1}{2}}}{2}\left(1 - \frac{\overline{A_0^2}}{4}B_T^{(w)} + \cdots\right) + \left(\begin{matrix}\sqrt{3}\\ \sqrt{\pi}\end{matrix}\right)\frac{\overline{A_0^2}^{\frac{3}{2}}}{16}\Psi_{0T}^{(w)} + \cdots \tag{21.113}$$

where the upper line applies for Eq. (21.108) and the lower for Eq. (21.111), where similarly $\overline{A_0^2} = \overline{A_{0u}^2}$ or $\overline{A_{0R}^2}$, and where

$$B_T^{(w)} = \frac{1}{2W_0}\int_0^T F_s(t,\boldsymbol{\theta}_1)^2\,dt \tag{21.113a}$$

$$\Psi_{0T}^{(w)} = \frac{2}{W_0^2}\left|\int_0^T V(t)F_s(t;\boldsymbol{\theta}_1)e^{i\omega_s t - i\phi_s(t,\boldsymbol{\theta}_2)}\,dt\right|^2 \tag{21.113b}$$

from Eqs. (20.66b), (20.66c), (20.67). In the case $P_M \to \infty$, Eqs. (21.109) reduce here to

$$A_0^*(\mathbf{V})_u = \left(B_T^{(w)}\frac{\pi}{2}\right)^{-\frac{1}{2}} {}_1F_1\left(\frac{1}{2};1;-\frac{\Psi_{0T}^{(w)}}{2B_T^{(w)}}\right)^{-1} \tag{21.114}$$

with $A_0^* \simeq \sqrt{\Psi_{0T}^{(w)}}/B_T^{(w)}$ when $\Psi_{0T}^{(w)} \gg B_T^{(w)}$. For all signal strengths, on the other hand, Eqs. (21.112) become

$$A_0^*(\mathbf{V})_R = \sqrt{\pi/2}\,(B_T^{(w)} + 2/\overline{A_{0R}^2})^{-\frac{1}{2}}\,{}_1F_1(-\tfrac{1}{2};1;-u_0^{(w)}) \tag{21.115}$$

where now $u_0^{(w)} = \Psi_{0T}^{(w)}/2(B_T^{(w)} + 2/\overline{A_{0R}^2})$. With obvious changes of scale, Fig. 21.3 can also be used for Eqs. (21.114), (21.115).

The structure of these Bayes receivers $\boldsymbol{T}_R^{(N)}\{V\}$ is easily deduced from the relations (21.108), (21.111) and (21.109), (21.112). For threshold signals, one has the same realizable elements as the corresponding Bayes binary detector, e.g., a Bayes matched filter (of the first or second kind; cf. Sec. 20.2-4), a zero-memory square-law rectifier, and an ideal integrator (linear transversal filter; cf. Fig. 20.2*a*, *b* and Secs. 20.1-3, 20.2-3). The resulting output is the quantity Ψ_{0n} (or Ψ_{0T}). This, in turn, is passed into an elementary computer, which simply adds the appropriate constant terms in Eqs. (21.108), (21.111) and combines them with Ψ_{0n}, suitably scaled (i.e., multiplied by $\overline{a_{0u}^2}\sqrt{3}/4$ or $\overline{a_{0R}^2}\sqrt{\pi}/4$ here) to give us the desired optimum *threshold* estimator of amplitude. For general input signal levels, one still retains the data-processing elements of the Bayes binary threshold detector, but the output Ψ_{0n} (or Ψ_{0T}) now is subject to a more involved computation, as the argument of the hypergeometric functions above. Detection may

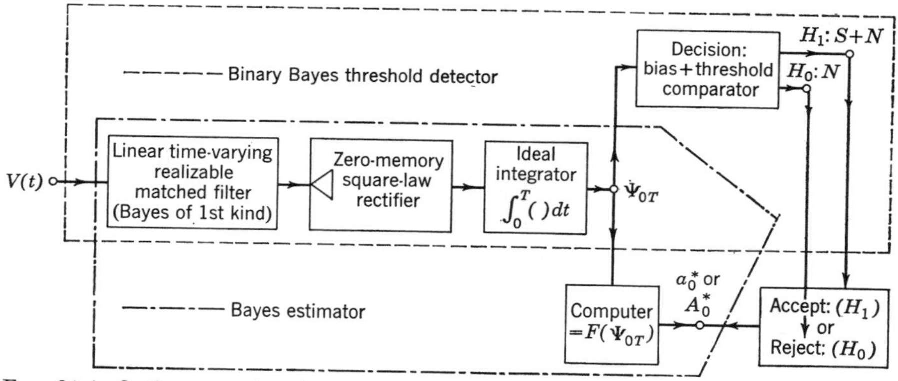

FIG. 21.4. Optimum estimation system for signal amplitude (narrowband signals in normal noise, incoherent observation, quadratic cost functions, continuous sampling, and general signals).

be carried out simultaneously, as indicated in Figs. 21.2 and 21.4, to determine whether or not the signal whose amplitude is being estimated is actually present. Of course, the threshold detector, while satisfactory for detection of the stronger signals, is no longer optimum in this regard, although its output† Ψ_{0n} is still required here for optimum *estimation*.

There remains the calculation of the Bayes risk for these optimum estimators. Here we shall consider the threshold cases only; the results for strong and intermediate signals are more difficult to obtain, although they may be achieved by a combination of analytic and numerical methods. Since the Bayes threshold estimators [Eqs. (21.108), (21.111)] have the form $a_0^*(\mathbf{V}) = B_0 + B_2\tilde{\mathbf{v}}\mathbf{G}_0\mathbf{v}$ here, we can apply the techniques of Sec. 17.2-1 directly to obtain first the characteristic function of $a_0^*(\mathbf{V})$. From this, in turn, the desired first and second moments of a_0^*, which are required in the expression (21.63) for the Bayes risk with quadratic cost function, are

† Before a decision.

easily found. We have

$$F_1(i\xi)_{a_0^*} = E_{\mathbf{V}|H_1}\{e^{i\xi(B_0+B_2\tilde{\mathbf{v}}\mathbf{G}_0\mathbf{v})}\}$$
$$= e^{i\xi B_0}\int_\Gamma e^{i\xi B_2\tilde{\mathbf{v}}\mathbf{G}_0\mathbf{v}}\langle W_n(\mathbf{V} - a_0\sqrt{\psi}\,\mathbf{s})_N\rangle_{\epsilon,a_0}\,d\mathbf{V} \quad (21.116)$$

where W_n is found from Eq. (20.86). Setting $\mathbf{H} = B_2\mathbf{k}_N\mathbf{G}_0$ and letting $\mathbf{\Omega}(i\xi) \equiv \mathbf{k}_N^{-1}[\mathbf{I} - (\mathbf{I} - 2i\xi\mathbf{H})^{-1}]$, we use Eqs. (17.32) to write Eq. (21.116) as

$$F_1(i\xi)_{a_0^*} = \frac{e^{i\xi B_0}\langle e^{-\frac{1}{2}a_0^2\tilde{\mathbf{s}}\mathbf{\Omega}(i\xi)\mathbf{s}}\rangle_{a_0,\epsilon}}{[\det(\mathbf{I} - 2i\xi\mathbf{H})]^{\frac{1}{2}}} \quad (21.117)$$

Next, we employ Eq. (17.25), developing $\mathbf{\Omega}$ and the determinant in an exponential power series in $i\xi$ through $O(\xi^2)$. Differentiation according to Eq. (1.31b) then gives us exactly

$$\overline{a_0^*} = B_0 + B_2 \operatorname{tr} \mathbf{k}_N\mathbf{G}_0 + B_2\langle a_0^2\tilde{\mathbf{s}}\mathbf{G}_0\mathbf{s}\rangle_{a_0,\epsilon} \quad (21.118a)$$
$$\overline{a_0^{*2}} = 2B_2^2 \operatorname{tr} (\mathbf{k}_N\mathbf{G}_0)^2 + 4B_2^2\langle a_0^2\tilde{\mathbf{s}}\mathbf{k}_N^{-1}(\mathbf{k}_N\mathbf{G}_0)^2\mathbf{s}\rangle_{a_0,\epsilon} + \overline{a_0^*}^2 \quad (21.118b)$$

where it is easily shown with the help of Eq. (20.24b) that $\langle\tilde{\mathbf{s}}\mathbf{G}_0\mathbf{s}\rangle_\epsilon = \frac{1}{2}[(\tilde{\mathbf{a}}\mathbf{k}_N^{-1}\mathbf{a})^2 + 2(\tilde{\mathbf{a}}\mathbf{k}_N^{-1}\mathbf{b})^2 + (\tilde{\mathbf{b}}\mathbf{k}_N^{-1}\mathbf{b})^2] \equiv \Psi_{0G}$ [cf. Eqs. (20.29a), (20.90b)]. However, since $a_0^*(\mathbf{V})$ here is $O(a_0^3)$ at most [$B_0 = O(a_0,a_0^3)$, $B_2 = O(a_0^3)$; cf. Eq. (21.111)], the first and second moments [Eqs. (21.118)] become approximately

$$\overline{a_0^*} \doteq B_0 + B_2 \operatorname{tr} \mathbf{k}_N\mathbf{G}_0 \qquad \overline{a_0^{*2}} \doteq 2B_2^2 \operatorname{tr} (\mathbf{k}_N\mathbf{G}_0)^2 + \overline{a_0^*}^2 \quad (21.118c)$$

The cross moment $\overline{a_0^*a_0}$ needed for the Bayes risk with quadratic cost function is readily found by carrying out the average directly (with the results of the footnote on page 992 and the fact that $v = a_0s + n$) to be

$$\overline{a_0^*a_0} = \bar{a}_0B_0 + B_2(\overline{a_0^3}\Psi_{0G} + 2\bar{a}_0B_n) \quad (21.118d)$$

The Bayes risk with QCF is here

$$R^* = C_0(\overline{a_0^{*2}} - 2\overline{a_0^*a_0} + \overline{a_0^2}) \quad (21.119)$$

which can then be calculated from Eqs. (21.118c) and (21.118d) when in addition we recall from Eqs. (20.26), (20.29a), (20.90a), and (20.90b) that $\operatorname{tr} \mathbf{k}_N\mathbf{G}_0 = 2B_n$ and $\operatorname{tr} (\mathbf{k}_N\mathbf{G}_0)^2 = 2\Psi_{0G} = 8\Phi_{0G}$ [cf. Eq. (20.93a)].

Specifically, for the uniform and Rayleigh amplitude distributions above [Eqs. (21.22a), (21.22b)] we have from Eqs. (21.108) and (21.111) under the present weak-signal condition

$$(1)\quad B_{0u} = \tfrac{1}{2}(3\overline{a_{0u}^2})^{\frac{1}{2}}\left(1 - \frac{\overline{a_{0u}^2}}{4}B_n + \cdots\right) \qquad B_{2u} = \frac{\sqrt{3}}{16}\,\overline{a_{0u}^2}^{\frac{3}{2}} \quad (21.120a)$$

$$(2)\quad B_{0R} = \tfrac{1}{2}(\pi\overline{a_{0R}^2})^{\frac{1}{2}}\left(1 - \frac{\overline{a_{0R}^2}}{4}B_n + \cdots\right) \qquad B_{2R} = \frac{\sqrt{\pi}}{16}\,\overline{a_{0R}^2}^{\frac{3}{2}} \quad (21.120b)$$

Combining these with Eqs. (21.118c) in Eqs. (21.119) after some manipulation, being careful to retain terms no higher than $O(a_0^4)$ in conformity with the weak-signal approximation for the estimator, we obtain the Bayes risk (QCF)

$$R_u^* = C_0\overline{(a_0^* - a_0)_u{}^2} \doteq C_0 \frac{\overline{a_{0u}{}^2}}{4}\left[1 - \frac{3}{4} + \overline{a_{0u}{}^2} \times 0 + O(a_0{}^4)\right] \tag{21.120c}$$

with a similar form for $R_R^* = C_0\overline{(a_0^* - a_0)_R{}^2}$, where $\overline{a_{0u}{}^2}$ is replaced by $\overline{a_{0R}{}^2}\cdot\left\{\left(1-\frac{\pi}{4}\right)+\ldots\right\}$. Note that the terms $O(\overline{a_0{}^4})$ are missing in R_u^* and R_R^*.

With continuous sampling, B_n and Ψ_{0n} are replaced as before by B_T and Ψ_{0T}, and for white noise backgrounds ($\psi \to \infty$) we may employ the modifications indicated in Eq. (21.113). Similarly, for the above moments and the Bayes risk [Eq. (21.119)], tr $\mathbf{k}_N\mathbf{G}_0$ and tr $(\mathbf{k}_N\mathbf{G}_0)^2$ are replaced by expressions proportional to Eqs. (20.67a) and (20.67b), respectively, with Eqs. (20.67d) and (20.67e) when the noise is white. Suboptimum systems are handled in the same way. However, instead of the optimum system $a_0^*(v)$, we might employ here a simpler version,

$$\gamma(V)_{a_0} = B_{s0} + B_{s2}\tilde{\mathbf{v}}\mathbf{G}_2\mathbf{v} \qquad B_{s0}, B_{s2} > 0 \tag{21.120d}$$

where $\tilde{\mathbf{v}}\mathbf{G}_2\mathbf{v}$ has properties similar to the nonlinear system discussed in Sec. 20.3-2. Comparison of the average risk for Eq. (21.120d) with Eq. (21.120c) then gives us a measure of the extent by which the suboptimum system departs from the expected performance of the limiting (and usually unachievable) optimum.

In the strong-signal situation, where Eqs. (21.109) and (21.112a) apply, respectively, for the uniform and Rayleigh d.d.'s of amplitude, a direct calculation of the various moments in Eq. (21.119) shows that the cross moment $\overline{a_0^* a_0}$ is not negligible and that $\overline{a_{0u}a_{0u}^*} \simeq \overline{a_{0u}{}^2} + (2\pi B_n)^{-1/2}$.

$$\therefore R_u^* = C_0\overline{(a_0^* - a_0)_u{}^2} \simeq C_0/B_n \tag{21.120e}$$

with an identical expression for R_R^* on replacing $\overline{a_{0u}{}^2}$ by $\overline{a_{0R}{}^2}$ once more. Since $(\tilde{\mathbf{a}}\mathbf{k}_N{}^{-1}\mathbf{a})(\tilde{\mathbf{b}}\mathbf{k}_N{}^{-1}\mathbf{b}) \geq (\tilde{\mathbf{a}}\mathbf{k}_N{}^{-1}\mathbf{b})^2$, as can be established readily by the Schwartz inequality,† it follows that $1 \leq \Psi_{0G}/B_n{}^2 < 2$. With white noise backgrounds one finds, for example, that $\Psi_{0G}/B_n{}^2 \doteq 1$, so that Eq. (21.120e) becomes

$$R_w^* \simeq C_0\overline{A_0{}^2}/2\sigma_{in}{}^2 \tag{21.120f}$$

where $\sigma_{in}{}^2 \equiv \overline{A_0{}^2}\int_0^T F_s{}^2\,dt/4W_0$ is the input signal-to-noise "energy" ratio and we have again used the narrowband condition to simplify Ψ_{0G}. The

† Let $\mathbf{a}_F = \mathbf{Q}\mathbf{a}$, $\mathbf{b}_F = \mathbf{Q}\mathbf{b}$, where $\tilde{\mathbf{Q}}^{-1}\mathbf{k}_N{}^{-1}\mathbf{Q}^{-1} = \mathbf{I}$; then from the Schwartz inequality, $(\tilde{\mathbf{a}}_F\mathbf{a}_F)(\tilde{\mathbf{b}}_F\mathbf{b}_F) \geq (\tilde{\mathbf{a}}_F\mathbf{b}_F)^2$, and the relation above follows directly.

Bayes estimator itself is in this instance

$$A_0^*(V) \simeq \frac{\sqrt{2\psi}\,\Psi_{0T}{}^{1/2}}{B_T}\bigg|_w \simeq 2^{3/2}\left\{\iint_0^T F_s(t)F_s(u)\cos\left[\omega_s(t-u)-\phi_s(t)+\phi_s(u)\right] V(t)V(u)\,dt\,du\right\}^{1/2} \div \int_0^T F_s{}^2(t)\,dt \quad (21.120g)$$

with a structure like those described earlier [cf. comments following Eq. (21.115)].

Moreover one can show in the weak-signal (large-sample) cases that the d.d. of the Bayes estimators [Eqs. (21.107), (21.111)] is asymptotically normal, with a mean $\overline{a_0^*}$ and variance $\overline{a_0^{*2}} - \overline{a_0^*}^2$ given by Eqs. (21.118c). Note, however, that as sample size T becomes larger, the Bayes risk approaches a zero limiting value, not depending on $\overline{a_0{}^2}$, as does the relative error [Eqs. (16.3)], where $\overline{M(T)}$ is here defined to be $\overline{a_0^*} - B_0$, and $\overline{\epsilon_M{}^2} = \overline{a_0^{*2}} - \overline{a_2^*}^2$. This is in agreement with the vanishing of relative error with increasing sample size [cf. Eq. (16.25)] when a parameter is being (conditionally) estimated, where in effect the parameter has one value. With $T < \infty$, it is the fact of many possible values, as represented by $\sigma(a_0)$ above, which in the present (unconditional) estimation procedure leads to limiting forms of estimator and Bayes risk that do not vanish with increasing sample size. In essence, this reflects the inescapable uncertainty of the estimation procedure, which specifically and realistically incorporates the observer's "ignorance" as to the parameter value through the mechanism of the a priori d.d. when $T < \infty$.

Finally, it is readily seen from the fact that $w(a_0|\mathbf{V})$ is not symmetrical or even unimodal in the threshold cases that a_0^* here is not optimum for other cost functions (Sec. 21.2-3, Theorem II). However, unimodality and symmetry about the mode, which is a_0^*, can be shown to occur approximately in the strong-signal situation [Eqs. (21.109), (21.112a)], so that $a_0^*(\approx \Psi_{0n}{}^{1/2}/B_n)$ is now Bayes with respect to a broader class of cost functions than the quadratic alone [cf. Eqs. (21.87)].

21.3-3 Incoherent Estimation of Signal Amplitude (Simple Cost Function). Let us repeat the analysis of the preceding section, but now for the simple cost function of Eq. (21.52a), in place of the quadratic cost function of Eq. (21.58). From Eq. (21.54), we know that the Bayes estimator of amplitude here is the unconditional maximum likelihood estimator $\hat{a}_0^*(\mathbf{V})$, which is obtained by applying Eqs. (21.13) to the particular likelihood function $L(\mathbf{V},a_0) = \sigma(a_0)\langle W_n(\mathbf{V} - a_0\sqrt{\psi}\,s)_N\rangle_\epsilon = w_n(\mathbf{V},a_0)$ [Eq. (21.104)]. Carrying out the indicated operations, we have

$$\frac{\partial}{\partial a_0}\log L(\mathbf{V},a_0)\bigg|_{a_0=\hat{a}_0^*} = \left[\frac{\sigma'(a_0)}{\sigma(a_0)} + \Psi_{0n}{}^{1/2}\frac{I_1(a_0\Psi_{0n}{}^{1/2})}{I_0(a_0\Psi_{0n}{}^{1/2})} - B_n a_0\right]_{a_0=\hat{a}_0^*} = 0 \quad (21.121)$$

as the relation from which the desired UML estimator is obtained. With the uniform and Rayleigh d.d. [Eqs. (21.22)] of amplitude, we can write Eq. (21.121) specifically as

1. *Uniform*

$$\lambda \frac{I_1(\lambda \hat{z}^*)}{I_0(\lambda \hat{z}^*)} = \hat{z}^* \qquad \lambda \equiv \left(\frac{\Psi_{0n}}{B_n}\right)^{1/2}, \ \hat{z}^* \equiv \hat{a}_0^* \sqrt{B_n} \tag{21.122a}$$

2. *Rayleigh*

$$\frac{1}{\hat{z}^*} - (1 + \eta)\hat{z}^* + \lambda \frac{I_1(\lambda \hat{z}^*)}{I_0(\lambda \hat{z}^*)} = 0 \qquad \eta \equiv (P_0 B_n)^{-1} \tag{21.122b}$$

The solutions of Eqs. (21.122*a*), (21.122*b*) are the desired Bayes estimators $\hat{z}^*$ (or a_0^*).

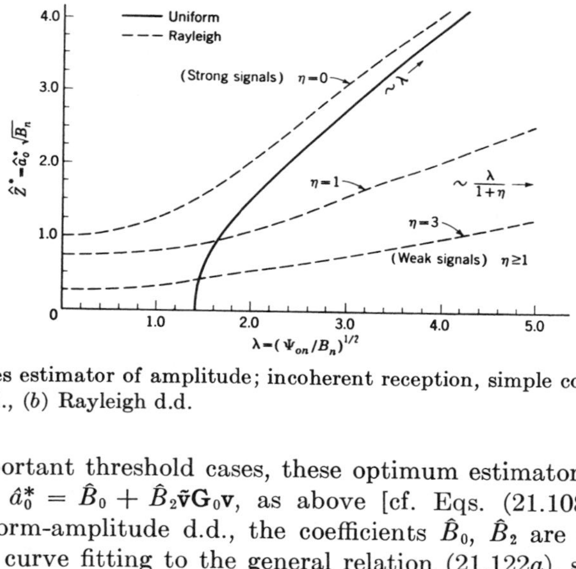

FIG. 21.5. Bayes estimator of amplitude; incoherent reception, simple cost function, for (*a*) uniform d.d., (*b*) Rayleigh d.d.

In the important threshold cases, these optimum estimators can be put in the form $\hat{a}_0^* = \hat{B}_0 + \hat{B}_2 \tilde{\mathbf{v}} \mathbf{G}_0 \mathbf{v}$, as above [cf. Eqs. (21.108), (21.111)]. For the uniform-amplitude d.d., the coefficients $\hat{B}_0$, $\hat{B}_2$ are conveniently obtained by curve fitting to the general relation (21.122*a*), shown in Fig. 21.5. With the Rayleigh d.d., one can easily find the threshold development of this kind directly from the weak-signal expansion of Eq. (21.122*b*). The results in each instance are

1. *Uniform*

$$\hat{z}^* \doteq \hat{b}_{0u} + \hat{b}_{2u}\lambda^2 \qquad \text{or} \qquad \hat{a}_{0u}^*(\mathbf{V}) \doteq \left(-\frac{0.80 \cdot 2}{B_n^{1/2}}\right) + \frac{0.80\Psi_{0n}}{B_n^{3/2}} \tag{21.123a}$$

2. *Rayleigh*

$$\hat{z}^* \doteq \hat{b}_{0R} + \hat{b}_{2R}\lambda^2 \qquad \text{or}$$

$$\hat{a}_{0R}^*(\mathbf{V}) \doteq \frac{1}{\sqrt{1+\eta}}[B_n^{-1/2} + B_n^{-3/2}\,\Psi_{0n}/4(1+\eta)], \quad \text{all } \eta > 0 \tag{21.123b}$$

For strong signals, on the other hand, where $\lambda \gg 1$ on the average,† one has

1. *Uniform*

$$\hat{z}^* \simeq \lambda - \frac{1}{2\lambda} + O\left(\frac{-1}{\lambda^3}\right) \qquad \text{or} \qquad \hat{a}^*_{0u}(\mathbf{V}) \simeq \frac{\Psi_{0n}{}^{1/2}}{B_n} \tag{21.124a}$$

2. *Rayleigh*

$$\hat{z}^* \simeq \frac{\lambda}{1+\eta} + O\left(\frac{1+\eta}{\lambda^3}\right) \quad \text{or} \quad \hat{a}^*_{0R}(\mathbf{V}) \simeq \frac{\Psi_{0n}{}^{1/2}}{B_n} \qquad P_0 B_n \to \infty \tag{21.124b}$$

Figure 21.5 shows $\hat{z}^*$ as a function of the received data ($\sim\lambda$) for both (1), (2) above; note that for the uniform d.d. of amplitudes, $\hat{z}^*$ and λ are uniquely related, while for the Rayleigh d.d. various relationships are possible, depending on the mean signal power P_0. The curve for $\eta = 1$ corresponds to a comparatively weak-signal state, while that for $\eta = 0$ exhibits $\hat{z}$ in the limit of very strong signals, on the average. Observe that, as we would expect, the strong-signal estimators are essentially independent of whether or not the amplitudes are uniformly or Rayleigh distributed, while with weak signals the character of the amplitude distribution becomes significant. In fact, with a uniform d.d. we must reject all λ's less than $\sqrt{2}$, so that, when $\sigma(a_0)$ is unknown to the observer and when λ is actually found to be less than $\sqrt{2}$, it is clear that the subcriterion leading to the d.d. (1), employing a peak-value constraint, is not then appropriate. The criterion 2, leading to the Rayleigh d.d., however, may be acceptable.

The structure of these Bayes receivers is sketched once more in Fig. 21.4. The only change is in the computer, which now performs its operation on Ψ_{0n} according to Eqs. (21.122), rather than Eqs. (21.109) or (21.112). Again, with continuous sampling, and when the background noise is white, we may employ the usual modifications mentioned above; for example, Ψ_{0n} is replaced by Ψ_{0T}, B_n by B_T [cf. Eqs. (20.62), (20.63), et seq.].

Since these Bayes estimators are UML estimators also, we can apply Eqs. (21.14) at once for threshold reception to obtain the expected distributions of $\hat{a}^*_{0u}$, $\hat{a}^*_{0R}$. These are accordingly normal, with means $\overline{\hat{a}^*_{0u}}$, $\overline{\hat{a}^*_{0R}}$ and variances $\sigma_u{}^2$, $\sigma_R{}^2$, respectively, in the large-sample threshold theory, where the variances may be found as in Eqs. (21.118) above.

The Bayes risk R^* for $\hat{a}^*_{0u}$, $\hat{a}^*_{0R}$ is given by Eq. (21.57). For weak signals,

† Note that $\lambda \gg 1$ on the average implies strong signals: since $\mathbf{v} = a_0\mathbf{s} + \mathbf{n}$, we have

$$\begin{aligned}\bar{\Psi}_{0n} = E_{\mathbf{V}|H_1}\{\tilde{\mathbf{v}}\mathbf{G}_0\mathbf{v}\} &= \overline{a_0{}^2}\langle\tilde{\mathbf{s}}\mathbf{G}_0\mathbf{s}\rangle_\epsilon + \langle\tilde{\mathbf{n}}\mathbf{G}_0\mathbf{n}\rangle + 2\bar{a}_0\langle\tilde{\mathbf{n}}\mathbf{G}_0\mathbf{s}\rangle_{\epsilon,n} \\ &= \overline{a_0{}^2}\langle\tilde{\mathbf{s}}\mathbf{G}_0\mathbf{s}\rangle_\epsilon + 2B_n \gg B_n\end{aligned}$$

(cf. $\overline{a_0{}^2} \gg 1$) (we have used $\langle\tilde{\mathbf{n}}\mathbf{G}_0\mathbf{n}\rangle = \tilde{\mathbf{a}}\mathbf{k}_N{}^{-1}\mathbf{a} + \tilde{\mathbf{b}}\mathbf{k}_N{}^{-1}\mathbf{b} = 2B_n = \operatorname{tr} \mathbf{k}_N\mathbf{G}_0$ with $\mathbf{n}\tilde{\mathbf{n}} = \mathbf{k}_N$).

we can obtain R^* explicitly using an appropriate expansion of the integrand, which takes specifically into account the threshold behavior of the estimator. With a Rayleigh distribution of amplitudes, we apply Eq. (21.123*b*) to Eq. (21.57), viz.,

$$R_R^* = C_0\left[A_1 - \int_\Gamma \sigma(\hat{a}_{0R}^*)w_n(\mathbf{V}|\hat{a}_{0R}^*)\, d\mathbf{V}\right] \tag{21.125a}$$

and expand σ and w_n, retaining terms $O(P_0{}^0,P_0)$ only. The integral in Eq. (21.125*a*) becomes here [cf. Eq. (21.104)]

$$\hat{I}_R^* = \int_\Gamma \sigma(\hat{a}_{0R}^*)\langle W_n(\mathbf{V} - \hat{a}_{0R}^*\sqrt{\psi}\,\mathbf{s})_N\rangle_\epsilon\, d\mathbf{V}$$

$$\doteq (eP_0)^{-1/2}\left[\left(1 + \frac{B_nP_0}{4} + \cdots\right) + \frac{P_0}{8} E_{\mathbf{V}|H_0}\{\tilde{\mathbf{v}}\mathbf{G}_0\mathbf{v}\} + O(P_0{}^2)\right] \tag{21.125b}$$

$$= (eP_0)^{-1/2}\left[1 + \frac{B_nP_0}{4} + \cdots + \frac{P_0}{8}\operatorname{tr}\mathbf{k}_N\mathbf{G}_0 + O(P_0{}^2)\right] \tag{21.125c}$$

from Eqs. (3) of Prob. 20.3. Setting A_1 equal to the constant term in $\hat{I}_R^*$ as a matter of convenience, since the choice of A_1 is arbitrary, we have finally for the Bayes risk in the Rayleigh case

$$R_R^* \doteq -\frac{C_0}{4}\sqrt{\frac{P_0}{e}}\operatorname{tr}\mathbf{k}_N\mathbf{G}_0 - O(P_0{}^{3/2}) = -\frac{C_0}{2}\sqrt{\frac{P_0}{e}}B_n - O(P_0{}^{3/2}) \tag{21.126}$$

[cf. Eqs. (20.33)]. As required, the Bayes risk decreases (i.e., becomes more negative) with increasing signal strength (larger P_0) and longer observation periods (larger values of $\operatorname{tr}\mathbf{k}_N\mathbf{G}_0$). With white noise backgrounds and continuous sampling, we have alternatively for $\hat{A}_{0R}^*$ [cf. Eq. (21.113)]

$$R_R^*\Big|_{\text{white}} \doteq -\frac{C_0}{4W_0}\frac{\overline{A_0{}^2}^{1/2}}{\sqrt{e}}\int_0^T F_s(t,\boldsymbol{\theta}_1)^2\,dt - O(\overline{A_0{}^2}^{3/2}) \tag{21.126a}$$

A similar calculation in the case of the uniform amplitude distribution, where Eq. (21.123*a*) gives the desired weak-signal estimator $\hat{a}_{0u}^*$, leads to

$$R_u^* = -0.255\sqrt{P_M}\,C_0\operatorname{tr}\mathbf{k}_N\mathbf{G}_0\ldots = -0.51C_0\sqrt{P_M}\,B_n + (\text{higher-order terms}) \tag{21.127}$$

The decrease of R_u^* with increasing sample size is accounted for by the higher-order terms [not given explicitly in Eq. (21.127)]. The average risk for suboptimum systems, like that of Eq. (21.120*c*), may be found from Eqs. (21.53) in similar fashion.

Finally, note that, since the simple cost function obeys the condition of Theorem II (Sec. 21.2-3), the estimators of Eqs. (21.124*a*), (21.124*b*), being equivalent to the strong-signal forms with QCF, are optimal vis-à-vis other cost functions. However, as in Sec. 21.3-2, this generality does not apply in the threshold cases.

21.4 *Waveform Estimation*[32] *(Stochastic Signals)*

When the signal ensemble is represented by an entirely random process, the methods outlined in Sec. 21.3 for amplitude estimation may be extended to the estimation of signal waveform. With normal signal and background noise processes, the resulting Bayes estimators are, in fact, often simpler than their deterministic counterparts. Here we shall consider specifically the following three problems, all but the last involving purely stochastic signal processes: (1) *simple estimation*† ($t_\lambda = t_k$, $0 \leq t_k \leq T$) of signal waveform in additive normal noise, with respect to the quadratic and the simple cost functions (21.64), (21.56); (2) *smoothing and prediction* ($t_\lambda \neq t_k$) of signal waveform, with respect to a quadratic cost function, including Minimax cases; (3) the *Minimax smoothing and prediction* of a certain class of deterministic signals. In (1) and (2), both signal and noise are postulated to be normal processes, with vanishing means and covariance functions $K_S(t_1,t_2)$, $K_N(t_1,t_2)$ additively and independently combined; in (3), the signal is represented by a polynomial in time.

21.4-1 Normal Noise Signals in Normal Noise (Quadratic Cost Function). Let us begin by choosing a quadratic cost function $C(\mathbf{S},\gamma) = \|\mathbf{S} - \boldsymbol{\gamma}_\sigma\|^2$ [cf. Eq. (21.64)] and applying Eq. (21.66) to obtain first the Bayes estimator $\boldsymbol{T}_R^{(N)}(V) = \boldsymbol{\gamma}_\sigma^*$ of waveform $\mathbf{S}$ in $(0,T)$, when simple estimation procedures are employed, i.e., when $t_\lambda = t_k$ $(k = 1, \ldots, n)$. Here, specifically, we have

$$\sigma(\mathbf{S}) = [(2\pi)^n \det \mathbf{K}_S]^{-1/2} e^{-1/2 \tilde{\mathbf{S}} \mathbf{K}_S^{-1} \mathbf{S}} \tag{21.128a}$$

and

$$F_n(\mathbf{V}|\mathbf{S}) = W_n(\mathbf{V} - \mathbf{S})_N = [(2\pi)^n \det \mathbf{K}_N]^{-1/2} e^{-1/2 (\tilde{\mathbf{V}} - \tilde{\mathbf{S}}) \mathbf{K}_N^{-1} (\mathbf{V} - \mathbf{S})} \tag{21.128b}$$

Applying these to Eq. (21.66), we can write for the desired optimum estimator (of each component S_k of $\mathbf{S}$)

$$\boldsymbol{\gamma}_\sigma^*(\mathbf{V})_S = \frac{\int_{\Omega_S} \mathbf{S} \exp\left[-\frac{1}{2}\tilde{\mathbf{S}}(\mathbf{K}_S^{-1} + \mathbf{K}_N^{-1})\mathbf{S} + \tilde{\mathbf{V}}\mathbf{K}_N^{-1}\mathbf{S}\right] d\mathbf{S}}{\int_{\Omega_S} \exp\left[-\frac{1}{2}\tilde{\mathbf{S}}(\mathbf{K}_S^{-1} + \mathbf{K}_N^{-1})\mathbf{S} + \tilde{\mathbf{V}}\mathbf{K}_N^{-1}\mathbf{S}\right] d\mathbf{S}} \tag{21.129}$$

where Ω_S is the region $(-\infty, \infty)$ for each S_k. To evaluate Eq. (21.129), let us consider the conditional characteristic function

$$F_n(i\xi|\mathbf{V})_S \equiv \int_{\Omega_S} e^{i\tilde{\xi}\mathbf{S}} \sigma(\mathbf{S}) F_n(\mathbf{V}|\mathbf{S})\, d\mathbf{S} \tag{21.130}$$

Then Eq. (21.129) becomes alternatively

$$\boldsymbol{\gamma}_\sigma^*(\mathbf{V})_S = -i \frac{d}{d\xi} \log F_n(i\xi|\mathbf{V})_S \Big|_{\xi=0} \tag{21.131}$$

† See Sec. 21.2; see also Fig. 18.3.

Writing $\mathbf{M} \equiv \mathbf{K}_S^{-1} + \mathbf{K}_N^{-1}$, we obtain from Eqs. (21.128) in Eq. (21.130) with the help of Eq. (7.26)

$$F_n(i\boldsymbol{\xi}|\mathbf{V})_S = [(2\pi)^n \det \mathbf{K}_S\mathbf{K}_N \det \mathbf{M}]^{-1/2} \exp(\tfrac{1}{2}\tilde{\mathbf{V}}\mathbf{K}_N^{-1}\mathbf{M}^{-1}\mathbf{K}_N^{-1}\mathbf{V}) \\ \exp(-\tfrac{1}{2}\tilde{\boldsymbol{\xi}}\mathbf{M}^{-1}\boldsymbol{\xi} + i\tilde{\boldsymbol{\xi}}\mathbf{M}^{-1}\mathbf{K}_N^{-1}\mathbf{V}) \quad (21.132)$$

Applying this to Eq. (21.131), we find directly that

$$T_R^{(N)}\{V\} = \gamma_\sigma^*(\mathbf{V})_S = \mathbf{S}^* = \mathbf{M}^{-1}\mathbf{K}_N^{-1}\mathbf{V} = \mathbf{Q}_E\mathbf{V} \quad (21.133)$$

where $\mathbf{Q}_E$ is $\mathbf{M}^{-1}\mathbf{K}_N^{-1}$, or the equivalent expressions (21.28).

Thus, the Bayes estimator of $\mathbf{S}$ with respect to the quadratic cost function (21.64) is given by Eq. (21.133). As we see at once on comparison with Eq. (21.29), it is also the UML estimator of $\mathbf{S}$, when $\sigma(\mathbf{S})$ is given by Eq. (21.128*a*), e.g.,

$$\mathbf{S}^* = \hat{\mathbf{S}} = \mathbf{Q}_E\mathbf{V} \qquad \mathbf{M}^{-1} = \mathbf{Q}_E\mathbf{K}_N \quad (21.134)$$

Moreover, from Eqs. (21.83) to (21.85) we know that $\hat{\mathbf{S}}$ can be interpreted as a Bayes estimator with respect to the "strict" simple cost function (21.84), so that, in effect, we can write $\hat{\mathbf{S}} \equiv \hat{\mathbf{S}}^* = \mathbf{S}^*$ of Eq. (21.134). Finally, one can also show that, in terms of the *less strict* simple cost function (21.56), S_k^* is equal to $(\mathbf{Q}_E\mathbf{V})_k$ (all $k = 1, \ldots, n$; cf. Prob. 21.8), and consequently $\mathbf{S}^* = \hat{\mathbf{S}}^* = \mathbf{Q}_E\mathbf{V}$ here as well. This is another example of a system which is optimum with respect to two or more different criteria, involving in this case three different cost functions. That such an invariance can occur is clearly quite strongly dependent on the statistical structure of the noise and signal processes, as well as on the choice of cost function.

Before determining the Bayes risk, let us first find the characteristic functions of $\mathbf{S}$, of the estimator $\mathbf{S}^*$, and of $\mathbf{S}^* - \mathbf{S}$. Since $\mathbf{S}$ and $\mathbf{S}^*(\mathbf{V})$ are both normal with zero means, the former with variance $\mathbf{K}_S$ and the latter with $\mathbf{K}_{S^*} = E_{\mathbf{V}|H_1}\{\mathbf{S}^*\tilde{\mathbf{S}}^*\} = \mathbf{Q}_E E_{\mathbf{V}|H_1}\{\mathbf{V}\tilde{\mathbf{V}}\}\tilde{\mathbf{Q}}_E = \mathbf{K}_S\tilde{\mathbf{Q}}_E$ [cf. Eq. (21.28*c*)], their characteristic functions are

$$F_n(i\boldsymbol{\xi})_S = E_S\{e^{i\tilde{\boldsymbol{\xi}}\mathbf{S}}\} = e^{-\frac{1}{2}\tilde{\boldsymbol{\xi}}\mathbf{K}_S\boldsymbol{\xi}} \quad (21.135)$$

$$F_n(i\boldsymbol{\xi})_{S^*} = E_{\mathbf{V}|H_1}\{e^{i\tilde{\boldsymbol{\xi}}\mathbf{S}^*}\} = e^{-\frac{1}{2}\tilde{\boldsymbol{\xi}}\mathbf{K}_S\tilde{\mathbf{Q}}_E\boldsymbol{\xi}} \quad (21.136)$$

For the c.f. of $\mathbf{S}^* - \mathbf{S}$, we must remember that $\mathbf{S}$ and $\mathbf{S}^*$ are *not* independent: $\mathbf{S}^* = \mathbf{Q}_E\mathbf{V} = \mathbf{Q}_E(\mathbf{S} + \mathbf{N})$, but since both $\mathbf{S}$ and $\mathbf{N}$ or $\mathbf{S}$ and $\mathbf{V}$ are normally distributed, we again expect a normal d.d. for $\mathbf{S}^* - \mathbf{S}$. Specifically, the means of $\mathbf{S}^* - \mathbf{S}$ vanish, and the covariance matrix is

$$\mathbf{K}_{S^*-S} \equiv E_{S,N}\{(\mathbf{S}^* - \mathbf{S})(\tilde{\mathbf{S}}^* - \tilde{\mathbf{S}})\} = \mathbf{K}_S(\mathbf{I} - \tilde{\mathbf{Q}}_E) \quad (21.137)$$

where we have used $E_{\mathbf{V}|H_1}\{\mathbf{S}^*\tilde{\mathbf{S}}^*\} = \mathbf{K}_S\tilde{\mathbf{Q}}_E$ and Eq. (21.28*c*). Again [cf. Eq. (21.30)] the gaussian character of these distributions follows from the linear nature of the Bayes estimator $\mathbf{S}^*$ ($= \hat{\mathbf{S}}^*$, etc.) and the normal properties of both the original signal and noise processes. Note also that the d.d.

of a single estimate S_k^* is normal, with zero mean and variance $(\mathbf{M})_{kk}^{-1}$, since $F_1(i\xi_k)_{S_k^*} = F_n(0,0, \ldots ,i\xi_k, \ldots ,0)_{S^*}$, etc. [cf. Eqs. (1.42)].

The Bayes risk for $\hat{\mathbf{S}}^*$ in the case of the quadratic cost function is now easily found from Eq. (21.137). Let us consider the Bayes risk associated with the single estimation of waveform S_k at time $t_\lambda = t_k$, where t_k $(k = 1, \ldots , n)$ is any one of the instants at which the data $\mathbf{V}$ is acquired. From Eqs. (21.67), (21.134), (21.137), we have for the Bayes risk here

$$R_k^* \equiv C_0 E_{\mathbf{V},S_k}\{(S_k - S_k^*)^2\} = C_0(\mathbf{K}_S - \mathbf{K}_S\tilde{\mathbf{Q}}_E)_{kk} \tag{21.138}$$

The Bayes risk associated with the simple estimation of, say, m values of the waveform, for example, $S_{k_1}, \ldots , S_{k_m}$ $(1 \leq k_1, \ldots , k_l, \ldots , k_m \leq n)$, where k_l, etc., is any one data-sampling instant in the interval $(0,T)$, becomes

$$R_{[m]}^* = \sum_{l=1}^{m} R_{k_l}^* = C_0 \sum_{l=1}^{m} \Big[(\mathbf{K}_S)_{k_lk_l} - \sum_{j=1}^{n} (\mathbf{K}_S)_{k_lj}(\tilde{\mathbf{Q}}_E)_{jk_l} \Big] \tag{21.139a}$$

and if $m = n$, that is, if all points $(1 \leq k \leq n)$ are considered, we have†

$$R_{[n]}^* = C_0 \operatorname{tr} (\mathbf{K}_S - \mathbf{K}_S\tilde{\mathbf{Q}}_E) \tag{21.139b}$$

With continuous instead of discrete sampling on $(0,T)$, we may apply Eq. (21.31) directly. Thus, again we may write

$$\mathbf{Q}_E \equiv \boldsymbol{\rho}_E \, \Delta t \qquad \boldsymbol{\rho}_E \to \rho_E(t,\tau)_T \neq \boldsymbol{\delta}_E \tag{21.140a}$$

where for continuous data processing ρ_E is now the solution of the inhomogeneous integral equation [cf. Eq. (21.32)]

$$\int_{0-}^{T+} [K_S(t,u) + K_N(t,u)]\rho_E(u,\tau)_T \, du = K_S(t,\tau) \qquad 0 \leq t, \tau \leq T \tag{21.140b}$$

where $\rho_E = 0$ for u outside $(0,T)$. The Bayes estimator S^* goes over into [cf. Eq. (21.33)]

$$S^*(t_\lambda) = \int_{0-}^{T+} \rho_E(t_\lambda,u)_T V(u) \, du \; [= \hat{S}(t_\lambda^*)] \qquad 0 \leq t_\lambda \leq T \tag{21.140c}$$

and, as discussed earlier, $\rho_E = h_E(t_\lambda - u, t_\lambda)_T$ can be interpreted as the weighting function of a time-varying linear realizable filter. When the spectra of the noise and signal processes are rational, we may use the methods of Appendix 2 to obtain ρ_E specifically (see also Sec. 16.2-3).

The Bayes risk for the estimate of waveform at any one instant t_λ $(0 \leq t_\lambda \leq T)$ follows at once from Eq. (21.138), where we write $\mathbf{K}_S - \mathbf{K}_S\tilde{\mathbf{Q}}_E \equiv \mathbf{J}$. Then Eq. (21.138) becomes

$$R_\lambda^* = C_0 \Big[K_S(t_\lambda,t_\lambda) - \int_{0-}^{T+} \rho_E(t_\lambda,u)_T K_S(u,t_\lambda) \, du \Big] \equiv C_0 J_T(t_\lambda,t_\lambda) \tag{21.141}$$

† Equations (4.49), (4.50) in Middleton and Van Meter[1] are incorrect.

The Bayes risk for m points t_{λ_l} ($l = 1, \ldots, m$) follows similarly from Eq. (21.139a). We have

$$R^*_{\lambda[m]} = C_0 \sum_{l=1}^{m} J_T(t_{\lambda_l},t_{\lambda_l}) \qquad 0 \le t_{\lambda_l} \le T \tag{21.142}$$

with a corresponding result for $R^*_{\lambda[n]}$ [cf. Eq. (21.139b)].

Note that, as $n \to \infty$, the Bayes risk $R^*_{\lambda[n]}$ for the values of S at the n sampling points t_λ ($l = 1, \ldots, n$) also becomes infinite, since the Bayes risk for each sampling instant t_λ in the interval $(0,T)$ is itself finite and there are now an infinite number of such points in this interval. However, by modifying the definition of $R^*_{\lambda[n]}$ so as to define instead the Bayes risk $\mathcal{R}^*_{\lambda[n]}$ *per sampling point*, by

$$\mathcal{R}^*_{\lambda[n]} \equiv R^*_{\lambda[n_1]}/n \tag{21.143a}$$

we obtain in the limit

$$\mathcal{R}^*_T \equiv \lim_{n\to\infty} \mathcal{R}^*_{\lambda[n]} = \frac{C_0}{T}\int_0^T J_T(u,u)\,du \tag{21.143b}$$

a finite quantity which may be used as a measure of the expected performance of the Bayes estimator $S^*(t_\lambda)$ [Eq. (21.140c)].

Finally, we observe here that since the Minimax estimator is Bayes for a uniform a priori signal distribution (cf. Sec. 21.2-2) we can obtain the specific Minimax extractor on setting $\mathbf{K}_S{}^{-1} = \mathbf{0}$, the null matrix. Then, $\mathbf{Q}_E \to \mathbf{Q}_M = \mathbf{I}$ [cf. Eq. (21.28c)], and consequently

$$\mathbf{S}^* \to \mathbf{S}^*_M = \hat{\mathbf{S}}^*_M = \mathbf{V} \tag{21.144}$$

i.e., the Minimax estimates are just the sampled values themselves. The Minimax average risks R^*_{Mk}, $R^*_{M[n]}$ [Eqs. (21.139)] are readily computed from Eq. (21.138) with the help of Eq. (21.144). They are, respectively, ψ $(= \overline{N^2})$ and $n\psi$. Since $\mathbf{S}^*$ for quadratic cost functions is also equal to the UMLE $\hat{\mathbf{S}}^*$, these remarks apply equally well for the corresponding simple cost functions and the associated Minimax estimators.

21.4-2 Smoothing and Prediction (Gaussian Signal and Quadratic Cost Functions). Let us now extend the analysis of Secs. 21.1-3(2), 21.4-1 to the cases of extrapolation and interpolation, where the time t_λ at which an estimate is desired does *not* coincide with any of the times at which data are collected and sampled and for which $\mathbf{S}$ is available, that is, $t_\lambda \neq t_k$ (cf. Secs. 18.2-2 and 21.2-1).

Quite generally, the average risk is given by Eq. (21.43). For a *quadratic cost function* (21.64), the Bayes estimator S^*_λ of signal waveform is found by minimizing the average risk with respect to γ as before:

$$S^*_\lambda(\mathbf{V}) = \frac{\int_{\Omega_S} S_\lambda \sigma(\mathbf{S},S_\lambda)F_n(\mathbf{V}|\mathbf{S})\,\mathbf{dS}\,dS_\lambda}{\int_{\Omega_S} \sigma(\mathbf{S},S_\lambda)F_n(\mathbf{V}|\mathbf{S})\,\mathbf{dS}\,dS_\lambda} \tag{21.145}$$

We observe at once that if $\mathbf{S}$ and S_λ are independent, so that $\sigma(\mathbf{S},S_\lambda) = \sigma_1(\mathbf{S})\sigma_2(S_\lambda)$, Eq. (21.145) reduces to $S_\lambda^* = \bar{S}_\lambda$; that is, the optimum estimate does not depend on $\mathbf{V}$ [in $(0,T)$] unless the signal is correlated. This bears directly on the Minimax estimator for this case. Now it is easily verified (cf. Prob. 21.10) that, when the a priori distribution $\sigma(\mathbf{S},S_\lambda)$ is uniform, the Bayes estimator [Eq. (21.145)] has the translation property $S_\lambda^*(\mathbf{V} + \mathbf{u}) = S_\lambda^*(\mathbf{V}) + u_\lambda$ and a conditional risk that is independent of S_λ. The uniform distribution is thus the one for which Eq. (21.145) becomes the Minimax estimator. We conclude, accordingly, that in order to obtain the Minimax estimator when the signal is stochastic (and normal) we must discard the only information upon which an estimate can reasonably be based, namely, the correlation inherent in the signal. Therefore, in these cases the Minimax estimate presents no advantage over an estimate based solely on the a priori probability of S_λ.

Let us now consider as a particular example the signal and noise of Sec. 21.4-1. Again both signal and noise are gaussian, independent, and additively combined. The Bayes estimator [Eq. (21.145)] may now be obtained as follows: We observe first that

$$\sigma(\mathbf{S},S_\lambda) \sim \exp\left[-\tfrac{1}{2}(\tilde{\mathbf{S}}\mathbf{K}_S^{-1}\mathbf{S} + 2\mathbf{S}_\lambda\tilde{\mathbf{k}}_\lambda\mathbf{S} + S_\lambda{}^2k_{\lambda\lambda})\right] \tag{21.146}$$

where the column vector $\mathbf{k}_\lambda$ has the components $k_{\lambda_j} = \mathbf{K}_S{}^{-1}(t_\lambda,t_j)$, that is, the k_{λ_j} are certain elements of an inverse signal variance matrix which is identical with the $n \times n$ matrix $\mathbf{K}_S{}^{-1}$ [cf. Eq. (21.146)] except that it has an extra row and column, corresponding to the time t_λ. The joint distribution of $\mathbf{V}$, $\mathbf{S}$, and S_λ needed in Eq. (21.145), for example, $\sigma(\mathbf{S},\mathbf{S}_\lambda)W_n(\mathbf{V} - \mathbf{S})_N$, is then proportional to

$$\exp\left\{-\tfrac{1}{2}[\tilde{\mathbf{S}}(\mathbf{K}_N{}^{-1} + \mathbf{K}_S{}^{-1})\mathbf{S} - 2\tilde{\mathbf{S}}\mathbf{K}_N{}^{-1}(\mathbf{V} - \mathbf{K}_N\mathbf{k}_\lambda S_\lambda) + S_\lambda{}^2k_{\lambda\lambda}]\right\} \tag{21.146a}$$

where terms involving $\mathbf{V}$ alone are not indicated, since they cancel identical terms in denominator (or numerator) [cf. Eq. (21.145)]. The integration over $\mathbf{S}$ is carried out as before (cf. the example of Sec. 21.4-1), with $\mathbf{M} = \mathbf{K}_S{}^{-1} + \mathbf{K}_N{}^{-1}$ and the substitution $\mathbf{K}_N{}^{-1} = (\mathbf{K}_N{}^{-1} + \mathbf{K}_S{}^{-1})\mathbf{Q}_E = \mathbf{M}\mathbf{Q}_E$, which may also be written $\mathbf{K}_S = \mathbf{Q}_E(\mathbf{K}_S + \mathbf{K}_N)$ [cf. Eqs. (21.28b), (21.28c)]. The result of this integration is independent of S_λ and cancels a corresponding contribution from the denominator (numerator). We are accordingly left with

$$S_\lambda^*(\mathbf{V}) = \frac{\int_{-\infty}^{\infty} S_\lambda e^{-\frac{1}{2}(C_1S_\lambda{}^2 + 2\tilde{\mathbf{k}}_\lambda\mathbf{D}\mathbf{V}S_\lambda)}\,dS_\lambda}{\int_{-\infty}^{\infty} e^{-\frac{1}{2}(C_1S_\lambda{}^2 + 2\tilde{\mathbf{k}}_\lambda\mathbf{D}\mathbf{V}S_\lambda)}\,dS_\lambda} \tag{21.147}$$

where $$C_1 = k_{\lambda\lambda}{}^2 - \tilde{\mathbf{k}}_\lambda\mathbf{M}^{-1}\mathbf{k}_\lambda \qquad \mathbf{D} = \mathbf{K}_N\mathbf{Q}_E\mathbf{M}\mathbf{Q}_E = \mathbf{Q}_E \tag{21.147a}$$

Equation (21.147) is readily integrated to give the desired Bayes estimator,

$$\boldsymbol{T}_{R^{(N)}}\{V\} = S_\lambda^*(\mathbf{V}) = -\tilde{\mathbf{k}}_\lambda\mathbf{Q}_E\mathbf{V}/C_1 \tag{21.148}$$

where C_1 is a scale factor depending on the correlation matrices of the signal and noise.

The essential character of this estimator is contained in the linear operator $\tilde{\mathbf{k}}_\lambda \mathbf{Q}_E$, which determines the optimum system for processing the data $\mathbf{V}$. It is convenient to introduce the (row) vector

$$\tilde{\mathbf{q}}_\lambda \equiv -k_{\lambda\lambda}{}^{-1}\tilde{\mathbf{k}}_\lambda \mathbf{Q}_E \tag{21.148a}$$

as the optimum system operator. We then observe from Eqs. (21.28) that the optimum system is determined by the following relation:

$$\tilde{\mathbf{q}}_\lambda(\mathbf{K}_S + \mathbf{K}_N) = -k_{\lambda\lambda}{}^{-1}\tilde{\mathbf{k}}_\lambda \mathbf{K}_S \tag{21.149}$$

Recalling that the components of $\tilde{\mathbf{k}}_\lambda$ are certain elements of a matrix like $\mathbf{K}_S{}^{-1}$, with an extra row and column, we observe finally

$$(\tilde{\mathbf{k}}_\lambda \mathbf{K}_S)_j = -k_{\lambda\lambda} K_S(t_\lambda,t_j) \tag{21.150}$$

With Eq. (21.150), Eq. (21.149) becomes specifically

$$K_S(t_\lambda,t_j) = K_S(t_j - t_\lambda,\, t_j) = \sum_{l=1}^{n} q_\lambda(t_l)[K_S(t_j,t_l) + K_N(t_j,t_l)] \tag{21.151}$$

from which the optimum operator q_λ is accordingly found.

The distribution density of the optimum estimator S_λ^* here is normal, as expected, since $S_\lambda^* = a_\lambda \tilde{\mathbf{q}}_\lambda \mathbf{V}$ is linear in $\mathbf{V}$ [cf. Eq. (21.148)]. The mean and variance of S_λ^* are specifically

$$\begin{aligned} E_{\mathbf{V}|H_1}\{S_\lambda^*\} &= 0 \\ E_{\mathbf{V}|H_1}\{(S_\lambda^*)^2\} &= a_\lambda{}^2 E_{\mathbf{V}|H_1}\{\tilde{\mathbf{V}}\mathbf{q}_\lambda\tilde{\mathbf{q}}_\lambda\mathbf{V}\} \\ &= a_\lambda{}^2\tilde{\mathbf{q}}_\lambda(\mathbf{K}_S + \mathbf{K}_N)\mathbf{q}_\lambda \\ \text{or}\qquad E_{\mathbf{V}|H_1}\{(S_\lambda^*)\} &= a_\lambda{}^2 k_{\lambda\lambda}{}^{-2}\tilde{\mathbf{k}}_\lambda\mathbf{K}_S\mathbf{Q}_E\mathbf{k}_\lambda \end{aligned} \tag{21.152}$$

where $a_\lambda \equiv k_{\lambda\lambda}/C_1$. Since the d.d. of S_λ is also normal, with zero mean and variance $\overline{S_\lambda{}^2} = K_S(0) \equiv \psi_S$, we may obtain the Bayes risk directly from this fact [Eqs. (21.152)] and the definition (21.67):

$$\begin{aligned} R_\lambda^* &= C_0 E_{\mathbf{V},S_\lambda|H_1}\{(S_\lambda - S_\lambda^*)^2\} = C_0(\overline{S_\lambda{}^2} - 2\overline{S_\lambda S_\lambda^*} + \overline{S_\lambda^{*2}}) \\ &= \psi_S - 2a_\lambda\tilde{\mathbf{q}}_\lambda(\mathbf{K}_S)_\lambda + a_\lambda{}^2\tilde{\mathbf{q}}_\lambda(\mathbf{K}_S + \mathbf{K}_N)\mathbf{q}_\lambda \end{aligned} \tag{21.153}$$

with $(\mathbf{K}_S)_\lambda = [K_S(t_i,t_\lambda)]$.

When continuous sampling is used, we let

$$\mathbf{q}_\lambda \equiv [h_\lambda(T - t_j)\,\Delta t] \tag{21.154}$$

where h_λ is the weighting function of a time-varying realizable linear filter, with read-out at time $t = T$. Then the Bayes estimator (21.148) becomes formally† a continuous operation

$$S_\lambda^* \to S_T^*(t_\lambda) = (k_{\lambda\lambda}/C_1) \int_{0-}^{T+} h_\lambda(T - u)V(u)\, du \tag{21.155}$$

where h_λ is found from the inhomogeneous integral equation corresponding to Eq. (21.151), viz.,

$$\int_{0-}^{T+} h_\lambda(T - u)[K_S(t,u) + K_N(t,u)]\, du = K_S(t - t_\lambda, t_\lambda) \qquad 0- < t \leq T+ \tag{21.156}$$

The structure of the Bayes estimator here is simply a suitable time-varying linear filter, with a read-out at the end of the smoothing period $(0,T)$ which yields the desired interpolations or extrapolation at t_λ. In a similar way, the Bayes risk with quadratic cost function (21.153) is found to be

$$R_\lambda^* = \psi_S - 2a_\lambda \int_{0-}^{T+} h_\lambda(T - \tau)K_S(\tau,t_\lambda)\, d\tau + a_\lambda^2 \iint_{0-}^{T+} h_\lambda(T - \tau)[K_S(\tau,u) + K_N(\tau,u)]h_\lambda(T - u)\, d\tau\, du \tag{21.157}$$

Equations (21.155) to (21.157) hold for both smoothing and prediction, i.e., for t_λ either inside or outside the data interval $(0,T)$ (cf. Prob. 21.10). We remark that these results are just those given by the Wiener theory of prediction and smoothing,[19] as extended by Zadeh and Ragazzini[20] (and also by Secs. 16.2-2, 16.2-3). In the Wiener theory, however, the assumption of gaussian statistics is not made, while on the other hand the estimator is constrained to be linear (in $\mathbf{V}$). The above Bayes solution, which is not constrained to be linear, turns out to be so because of the normal statistics, and thus the best linear estimator, for both smoothing and prediction, is the best least-mean-squared-error estimator‡ when the statistics are gaussian.[25] In general, Eq. (21.145) holds for any statistics, with the advantage that it applies for nonstationary as well as stationary processes; the difference in treatment appears only in the interpretation of the results derived from Eq. (21.145), and not in their derivation; e.g., if the noise is stationary, $K_N(t_j,t_l) = K_N(|t_j - t_l|)$ in Eqs. (21.151) et seq. Finally, observe again for Minimax estimation, where $\sigma(\mathbf{S},S_\lambda)$ is uniform, that $\mathbf{K}_S^{-1} = 0$ and therefore $\mathbf{Q}_E \to \mathbf{I}$, so that the Minimax estimator becomes, from Eq. (21.148),

$$S_M = -\tilde{\mathbf{k}}_\lambda \mathbf{V}/C_1 \tag{21.157a}$$

and the Minimax average risk may be computed from $\overline{(S_\lambda - S_M)^2}$, now with the help of Eq. (21.157*a*).

21.4-3 Minimax Smoothing and Prediction of Deterministic Signals (Quadratic Cost Functions). We conclude Sec. 21.4 with a brief treatment

† See the remarks at the end of Sec. 21.1-1 on passage to the continuous limit.

‡ For an extension of Theorem II (Sec. 21.2-3) to include prediction, see Sherman.[31]

of the linear estimation of signal waveform when the signal has a particular type of time structure. Specifically, we consider signals that are functions of, say, $q + 1$ parameters $\boldsymbol{\theta} = (\theta_0, \ldots, \theta_q)$, whose occurrence is governed by the $(q + 1)$st-order density function $\sigma(\boldsymbol{\theta})$. We assume, in addition, that each signal is a *linear* function of these parameters, i.e., if $\mathbf{S} = (S_1, \ldots, S_n)$, etc., then

$$\mathbf{S} = \mathbf{T}_S\boldsymbol{\theta} \tag{21.158}$$

where $\mathbf{T}_S$ is an $n \times (q + 1)$ matrix of coefficients. The intention here is to include those cases of prediction in which the signal is a polynomial of order q in the time, e.g.,

$$S(t) = \theta_0 + \theta_1 t + \theta_2 t^2 + \cdots + \theta_q t^q \tag{21.158a}$$

as considered, for example, by Phillips and Weiss,[26] Zadeh and Ragazzini,[20] and Drenick.[27] A typical element of the matrix $\mathbf{T}_S$ is then

$$(T_S)_{jk} = t_j^k \qquad j = 1, \ldots, n; k = 0, 1, \ldots, q \tag{21.158b}$$

Let it be desired to estimate another linear function of $\boldsymbol{\theta}$, say $\mathbf{d} = (d_1, \ldots, d_m)$, $d_j = d(t_j)$, etc.,

$$\mathbf{d} = \mathbf{T}_d\boldsymbol{\theta} \tag{21.159}$$

where $\mathbf{T}_d$ is an $m \times (q + 1)$ matrix of coefficients. In particular, if m is larger than n and $d_k = S(t_k)$, the problem is one of smoothing when $k < n$ and prediction when $k > n$.

Denote any estimate of $\mathbf{d}$ by $\mathbf{A}(\mathbf{V})$, where $\mathbf{A}$ is an m-dimensional column vector, each element of which is a function of $\mathbf{V} = (V_1, \ldots, V_n)$, $V_j = V(t_j)$, etc. Specifically, for the *squared-error cost function* the conditional risk of $\mathbf{A}$ is

$$r(\boldsymbol{\theta},\mathbf{A}) = C_0 \int_\Gamma \|\mathbf{d}(\boldsymbol{\theta}) - \mathbf{A}(\mathbf{V})\|^2 W_n(\mathbf{V} - \mathbf{T}_S\boldsymbol{\theta})_N \, d\mathbf{V} \tag{21.160}$$

when signal $\mathbf{S}$ and noise $\mathbf{N}$ are additive and independent.† The average risk is obtained as usual by averaging Eq. (21.160) with respect to the a priori distribution $\sigma(\boldsymbol{\theta})$,

$$R(\sigma,\mathbf{A}) = C_0 \int_{\Omega\theta} \int_\Gamma \|\mathbf{T}_d\boldsymbol{\theta} - \mathbf{A}(\mathbf{V})\|^2 \sigma(\boldsymbol{\theta}) W_n(\mathbf{V} - \mathbf{T}_S\boldsymbol{\theta})_N \, d\mathbf{V}\, d\boldsymbol{\theta} \tag{21.161}$$

The Bayes estimator of $\mathbf{d}$, denoted by $\mathbf{d}^*(\mathbf{V})$ $[= T_R^{(N)}(V)]$, is found as before by minimizing Eq. (21.161) with respect to the estimator $\mathbf{A}(\mathbf{V})$,

$$\mathbf{d}^*(\mathbf{V}) = \frac{\int_{\Omega\theta} \mathbf{T}_d\boldsymbol{\theta}\sigma(\boldsymbol{\theta}) W_n(\mathbf{V} - \mathbf{T}_S\boldsymbol{\theta})_N \, d\boldsymbol{\theta}}{\int_{\Omega\theta} \sigma(\boldsymbol{\theta}) W_n(\mathbf{V} - \mathbf{T}_S\boldsymbol{\theta})_N \, d\boldsymbol{\theta}} \tag{21.162}$$

We note immediately that, when $\sigma(\boldsymbol{\theta})$ is uniform, the Bayes estimator

† Here we assume also that $\bar{\mathbf{N}} = 0$.

has the following translation property,

$$\mathbf{d}^*(\mathbf{V} + \mathbf{T}_S\boldsymbol{\theta}') = \mathbf{d}^*(\mathbf{V}) + \mathbf{T}_d\boldsymbol{\theta}' \tag{21.163}$$

and that when this is used in Eq. (21.160) the conditional risk becomes constant for all $\boldsymbol{\theta}$. Thus the *Minimax estimator* (cf. Sec. 21.2-2) is

$$\mathbf{d}_M^*(\mathbf{V}) = \frac{\int_{-\infty}^{\infty} \cdots \int \mathbf{T}_d\boldsymbol{\theta} W_n(\mathbf{V} - \mathbf{T}_S\boldsymbol{\theta})_N \, d\boldsymbol{\theta}}{\int_{-\infty}^{\infty} \cdots \int W_n(\mathbf{V} - \mathbf{T}_S\boldsymbol{\theta})_N \, d\boldsymbol{\theta}} \tag{21.164}$$

When the noise is gaussian, this may be integrated as follows: Define the symmetrical $(q + 1) \times (q + 1)$ matrix $\mathbf{D}$ as

$$\mathbf{D} \equiv \tilde{\mathbf{T}}_S \mathbf{K}_N^{-1} \mathbf{T}_S \tag{21.165}$$

so that the density function W_n may be expressed as

$$W_n(\mathbf{V} - \mathbf{T}_S\mathbf{Q})_N \sim \exp\left[-\tfrac{1}{2}(\boldsymbol{\theta} - \mathbf{D}^{-1}\tilde{\mathbf{T}}_S\mathbf{K}_N^{-1}\mathbf{V})_{trans}\mathbf{D}(\boldsymbol{\theta} - \mathbf{D}^{-1}\tilde{\mathbf{T}}_S\mathbf{K}_N^{-1}\mathbf{V})\right] \tag{21.166}$$

where terms not involving $\boldsymbol{\theta}$ have been omitted since they cancel corresponding terms in the denominator. Then the substitution $\mathbf{RU} = \boldsymbol{\theta} - \mathbf{D}^{-1}\tilde{\mathbf{T}}_S\mathbf{K}_N^{-1}\mathbf{V}$, where $\mathbf{R}$ diagonalizes $\mathbf{D}$, that is, $\tilde{\mathbf{R}}\mathbf{D}\mathbf{R} = \boldsymbol{\Lambda}$, yields the result

$$\mathbf{d}^*(\mathbf{V}) = \mathfrak{L}_E\mathbf{V} \tag{21.167}$$

where $\mathfrak{L}_E$ is an $m \times n$ matrix, with

$$\mathbf{T}_d\mathbf{D}^{-1}\tilde{\mathbf{T}}_S = \mathfrak{L}_E\mathbf{K}_N \tag{21.167a}$$

This last equation defines the Minimax estimator $\mathfrak{L}_E$ in terms of the correlation matrix of the (gaussian) noise $\mathbf{K}_N$ and the matrices $\mathbf{T}_S$ and $\mathbf{T}_d$ by means of which the signal and the quantity whose estimate is desired are expressed according to Eqs. (21.158) and (21.159). Note that the estimator [Eq. (21.167)] has the translation property [Eq. (21.72)], as it should, and that it is linear in the received data $\mathbf{V}$, as expected from the normal character of the noise and the linear dependence of $\mathbf{S}$ on the parameters $\boldsymbol{\theta}$.

The d.d. of $\mathbf{d}^*(\mathbf{V})$ is multivariate gaussian, order m, with means and variances

$$\begin{aligned} E_{\mathbf{V}|H_1}\{\mathbf{d}^*(\mathbf{V})\} &= \mathbf{T}_d\bar{\boldsymbol{\theta}} \\ E_{\mathbf{V}|H_1}\{\mathbf{d}^*(\mathbf{V})\tilde{\mathbf{d}}^*(\mathbf{V})\} - \mathbf{T}_d\bar{\boldsymbol{\theta}}\bar{\tilde{\boldsymbol{\theta}}}\tilde{\mathbf{T}}_d &= \mathfrak{L}_E(\mathbf{T}_S\overline{\boldsymbol{\theta}\tilde{\boldsymbol{\theta}}}\tilde{\mathbf{T}}_S + \mathbf{K}_N)\tilde{\mathfrak{L}}_E - \mathbf{T}_d\bar{\boldsymbol{\theta}}\bar{\tilde{\boldsymbol{\theta}}}\tilde{\mathbf{T}}_d \end{aligned} \tag{21.168}$$

so that the Bayes risk [Eq. (21.161)] with quadratic cost function becomes

$$\begin{aligned} R^*(\sigma,\boldsymbol{\delta}^*) &= C_0 E_{\mathbf{V}|H_1}\{\|\mathbf{T}_d\boldsymbol{\theta} - \mathfrak{L}_E\mathbf{V}\|^2\} \\ &= \mathbf{T}_d\overline{\boldsymbol{\theta}\tilde{\boldsymbol{\theta}}}\tilde{\mathbf{T}}_d + \mathfrak{L}_E(\mathbf{T}_S\overline{\boldsymbol{\theta}\tilde{\boldsymbol{\theta}}}\mathbf{T}_S + \mathbf{K}_N)\tilde{\mathfrak{L}}_E - 2\mathfrak{L}_E\mathbf{T}_S\overline{\boldsymbol{\theta}\tilde{\boldsymbol{\theta}}}\tilde{\mathbf{T}}_d \end{aligned} \tag{21.169}$$

In order to obtain the solution for smoothing or prediction of $S(t)$ when the sampling is continuous within $(0,T)$, we first define $\mathbf{T}_S$ and $\mathbf{T}_d$ according to Eq. (21.158*a*). Then, if t_λ denotes the time at which a signal estimate is desired, whether inside $(0,T)$ (smoothing) or outside $(0,T)$ (prediction), Eq. (21.167) gives for continuous sampling within $(0,T)$

$$S^*(t_\lambda) = \int_0^T L_\lambda(\tau)_E V(\tau)\,d\tau \tag{21.170}$$

where

$$\mathfrak{L}_E \equiv [L_\lambda(t_j)_E\,\Delta t] = [h_\lambda(T - t_j)_L\,\Delta t] \tag{21.170a}$$

(See remarks in Sec. 19.4-2 on similar passage to the continuous limit.) To obtain the corresponding representation of Eq. (21.167*a*), we introduce a new $(q + 1) \times n$ matrix $\mathbf{X}$ such that

$$\mathbf{X} = \mathbf{D}^{-1}\tilde{\mathbf{T}}_S \qquad \text{or} \qquad \mathbf{T}_S\mathbf{X} = \mathbf{K}_N \tag{21.171}$$

With this, the continuous form of Eq. (21.167*a*) becomes

$$\sum_{k=0}^{q} t_\lambda^k X_k(t_0) = \int_0^T h_\lambda(T - \tau)_L K_N(\tau - t_0)\,d\tau \qquad 0 \le t_0 \le T \tag{21.172}$$

where the coefficients $X_k(t_0)$ are determined from the continuous form of Eq. (21.171),

$$\sum_{k=0}^{q} t^k X_k(t_0) = K_N(t - t_0) \qquad 0 \le t_0 \le T \tag{21.173}$$

Here, t_0 is an arbitrary instant inside $(0,T)$ which may be taken at any convenient value.

With $t_0 = t_\lambda$, the results (21.170), (21.172) agree in form with those given by Phillips and Weiss[26] and Zadeh and Ragazzini[20] for smoothing. Equation (21.173) suggests, however, that the solution above is Minimax only if the correlation function of the noise can be represented as a polynomial of order q in time. The result of Phillips and Weiss is derived without assumptions as to the statistics, but with a constraint of linearity on the system. The result above assumes gaussian noise and does not constrain the system.

The Minimax characterization of Eq. (21.167) is valid, of course, only for gaussian noise. For different statistics, the Minimax extractor is not generally linear. It is of interest, therefore, to determine the extent to which the linear Minimax predictor falls short of the optimum nonlinear one. Drenick[27] has treated some aspects of this problem. His treatment is based on an extension of a theorem due to Girshick and Savage[22] and expresses the Minimax predictor of any order q [the order of the polynomial of Eq. (21.158*a*)] as a linear predictor of order q plus a correction term. The correction term depends upon a hierarchy of lower-order linear predictors, each of which is required to have a translation property identical with Eq. (21.163). Drenick gives an application of the theory showing the magnitude of the correction term for weakly non-gaussian noise.

PROBLEMS

21.8 Show that the Bayes estimator $\mathbf{S} = \hat{\mathbf{S}} = \mathbf{Q}_E\mathbf{V}$ [Eq. (21.134)] also holds for the simple cost function of Eq. (21.56), as well as for its stricter version [Eq. (21.84)], $\theta_k = S_k$.

21.9 (*a*) Show that for the strict version [Eq. (21.84)] of the simple cost function the corresponding Bayes risk [Eq. (21.85)] becomes, for the signals of Sec. 21.4-1,

$$R^* = C_0\{A_n - [(2\pi)^n \det \mathbf{K}_S \det (\mathbf{I} - \tilde{\mathbf{Q}}_E)]^{-\frac{1}{2}}\} \tag{1}$$

(*b*) Repeat (*a*) for the less strict cost function [Eq. (21.56)].

21.10 In order to show that a uniform d.d. of $\mathbf{S}$ and S_λ is a Minimax d.d., verify that, when the a priori d.d. $\sigma(\mathbf{S},S_\lambda)$ is uniform, the Bayes estimator of S_λ (with quadratic cost function) possesses the translation property $S_\lambda^*(\mathbf{V} + \mathbf{u}) = S_\lambda^*(\mathbf{V}) + u_\lambda$ and has a conditional risk that is independent of S_λ.

21.11 Show that the results of Sec. 21.4-2 reduce to those of Sec. 21.4-1 for simple estimation (i.e., for $t_\lambda = t_k$, $k = 1, \ldots, n$), and verify for continuous sampling.

21.5 *Remarks*

While estimation theory is necessarily less "compact" than the corresponding theory for detection, since the system designer has more degrees of freedom at his disposal, in particular the choice of cost function, and frequently the quantity to be estimated, it is still possible to make certain rather general observations about the results obtained in Secs. 21.1 to 21.4. For example, weak-signal or threshold estimators usually provide asymptotically sufficient estimates, and the d.d.'s of these estimators are themselves usually normal. This is true for maximum likelihood estimators of amplitude and waveform, which are Bayes relative to simple cost functions (cf. Secs. 21.1-2, 21.2-1), and also for corresponding systems based on quadratic cost functions (cf. Sec. 21.4). Strong-signal estimators of these quantities also exhibit a similar invariance: regardless of whether or not a simple or a quadratic cost function is chosen, the Bayes system has the same structure in this limiting situation [although the Bayes *risks* may (of course) be different; cf. Secs. 21.3-2, 21.3-3)]. Finally, the maximum likelihood approach (Sec. 21.1-2) is found to have a broader interpretation from the viewpoint of decision theory, as Secs. 21.2-1, 21.2-2, 21.4 have indicated.

Let us briefly review some of the implications of this last statement. Further insight into the significance of maximum likelihood estimation may be gained from Eq. (18.12) in the case of waveform estimation if we set $\boldsymbol{\gamma} = \mathbf{S}$ therein, according to Eq. (21.42). Then we have

$$p(\boldsymbol{\gamma}|\mathbf{S}) = p(\boldsymbol{\gamma} = \mathbf{S}|\mathbf{S}) = \int_\Gamma F_n(\mathbf{V}|\mathbf{S})\,\delta[\mathbf{S} - \boldsymbol{\gamma}_S(\mathbf{V})]\,d\mathbf{V} \tag{21.174}$$

This is the probability of a correct decision when the signal is $\mathbf{S}$. It is clearly greatest when for each $\mathbf{V}$ we choose $\boldsymbol{\gamma}_S(\mathbf{V})$ equal to the maximum conditional likelihood estimate $\hat{\gamma}_S$. In other words, the *maximum (conditional) likelihood estimator maximizes the probability of a correct decision,* without regard to incorrect decisions or their cost.

By taking the average of both sides of Eq. (21.174) with respect to the a priori signal distribution $\sigma(\mathbf{S})$, we obtain the average probability of a correct estimate,

$$\int_{\Omega_S} p(\mathbf{S}|\mathbf{S})\sigma(\mathbf{S})\, d\mathbf{S} = \int_\Gamma \int_{\Omega_S} \sigma(\mathbf{S}) F_n(\mathbf{V}|\mathbf{S})\delta[\mathbf{S} - \boldsymbol{\gamma}_\sigma(\mathbf{V})]\, d\mathbf{S}\, d\mathbf{V} \tag{21.175}$$

where $p(\mathbf{S}|\mathbf{S})$ represents $p(\boldsymbol{\gamma} = \mathbf{S}|\mathbf{S})$ of Eq. (21.174). By the same reasoning as for Eq. (21.174), we see that Eq. (21.175) is largest when, for each $\mathbf{V}$, $\boldsymbol{\gamma}_\sigma(\mathbf{V})$ is chosen as the particular value of $\mathbf{S}$ that makes the unconditional likelihood function (or a posteriori probability of $\mathbf{S}$, given $\mathbf{V}$) a maximum. Thus, *the maximum* unconditional *likelihood estimator* $\hat{\mathbf{S}}$ *maximizes the* average *probability of a correct decision* when all possible signals are taken into account, and again without particular regard for incorrect decisions and their costs.

Now, since the cost of a correct decision is always less (by definition) than that of any other, the maximum likelihood estimator, because it effectively assigns the greatest probabilities to the least costs (i.e., the closest estimates have the greatest probabilities), may be expected to minimize the average risk for cost functions other than the simple one of Sec. 21.2-2, provided that certain symmetries are present in the joint density $p(\mathbf{S},\mathbf{V}) = \sigma(\mathbf{S})F_n(\mathbf{V}|\mathbf{S})$ and in the cost function itself. Specifically, in the case of the squared-error cost function (21.58), it is found that,[28] if *the joint distribution of* $\mathbf{V}$ *and* $\mathbf{S}$ *is symmetrical about the unconditional maximum likelihood estimate* $\hat{\mathbf{S}}$ *for every* $\mathbf{V}$ (remember that $\hat{\mathbf{S}}$ depends on $\mathbf{V}$), *then* $\hat{\mathbf{S}}$ *is a Bayes solution with respect to this cost function.*†

A receiver which presents the a posteriori probability $p(\mathbf{S}|\mathbf{V})$ as a function of $\mathbf{S}$ at its output (such as those of Woodward and Davies[30]) may also be operated as a decision system by taking the maximum value of the output as the decision or estimate. From the above, we see that such a receiver maximizes the probability of a correct decision for each $\mathbf{S}$, and also maximizes the average probability of a correct decision [cf. Eq. (21.175)] when the various $\mathbf{S}$ appear at random. The receiver's limitation is that it ignores the possibly different relative importance of the various system errors, i.e., discrepancies between $\boldsymbol{\gamma}$ and $\mathbf{S}$. This is equivalent to saying that it is a Bayes extractor with the simple cost function of Sec. 21.2-1. On the other hand, as we have just seen, under certain symmetry conditions the Bayes extractor for the squared-error cost function is also the (unconditional)

† Wald[29] uses a similar approach to show that the maximum conditional likelihood estimate $\hat{\gamma}_S$ (our notation) is Minimax when the cost function depends only on the error, signal and noise are additive (in our terminology), the signal is one-dimensional ($\mathbf{S} = S$), and the a priori signal distribution is uniform (over a bounded interval).

maximum likelihood estimator $\hat{\mathbf{S}}$, so that from this point of view the maximum likelihood receiver may be considered as one which penalizes incorrect decisions according to $C(\mathbf{S},\boldsymbol{\gamma}_\sigma) = C_0\|\mathbf{S} - \boldsymbol{\gamma}_\sigma\|^2$ and simultaneously maximizes the average probability of a correct decision. It is evident, therefore, that under certain conditions *an extraction system may minimize average risk for more than one cost function*, just as in the case of detection (cf. Sec. 21.4-1), where a given system may be optimum for many cost assignments provided that the threshold $\mathcal{K}$ is held fixed. In fact, as we have demonstrated in Sec. 21.4-1, it is possible for the same system, $\mathbf{S}^* = \hat{\mathbf{S}} = \hat{\mathbf{S}}^*$, to be optimum with respect to three different cost functions, where, for example, the signal, as well as the noise, is a normal process.† Some further properties of maximum likelihood estimators are considered in Chap. 22 (cf. Sec. 22.3).

Other system invariances are possible: under certain conditions, this may occur for the uniform d.d. of signal amplitude or waveform. For quadratic cost functions, this leads on the one hand to the Minimax class of Bayes extractors. On the other hand, when the d.d. of amplitude or waveform is unknown to the observer, he may invoke the subcriterion of maximum average uncertainty, subject to a maximum value constraint to obtain the same optimum system, since the d.d. of amplitude or waveform is again uniform. Observe, finally, that the conditional and unconditional maximum likelihood estimators are identical for uniform distributions.

REFERENCES

1. Middleton, D., and D. Van Meter: Detection and Extraction of Signals in Noise from the Point of View of Statistical Decision Theory, *J. Soc. Ind. Appl. Math.*, **3:** 192 (1955), **4:** 86 (1956).
2. Cramér, H.: "Mathematical Methods of Statistics," Princeton University Press, Princeton, N.J., 1946.
3. Bartlett, M. S.: "An Introduction to Stochastic Processes," Cambridge, New York, 1955.
4. Mood, A. M.: "Introduction to the Theory of Statistics," McGraw-Hill, New York, 1950.
5. Cramér, H.: A Contribution to the Theory of Statistical Estimation, *Skand. Aktuarietidskr.*, **29:** 85 (1946).
6. Rao, C. R.: Information and the Accuracy Attainable in the Estimation of Statistical Parameters, *Bull. Calcutta Math. Soc.*, **37**(3): 89 (1945).
7. Savage, L. J.: "The Foundations of Statistics," Wiley, New York, 1954.
8. Hodges, J. L., Jr., and E. L. Lehmann: Some Applications of the Cramér-Rao Inequality, *Proc. 2d Berkeley Symposium on Mathematical Statistics and Probability*, July–August, 1950.
9. Grenander, U.: Stochastic Processes and Statistical Inference, *Arkiv Mat.*, **1:** 195 (1950).
10. Fisher, R. A.: On an Absolute Criterion for Fitting Frequency Curves, *Messenger of*

† Theorem II (Sec. 21.2-3) states a similar result, now, however, without specifying the particular nature of the signal and noise processes, but with respect to the Bayes estimator derived from a quadratic cost function.

Math., **41**: 155 (1912); see also refs. 89, 96, 103, 104 in Ref. 2, pp. 564, 565, and also ref. 76.

11. Preston, G. W.: The Equivalence of Optimum Transducers and Sufficient and Most Efficient Statistics, *J. Appl. Phys.*, **24**: 841 (1953).
12. Middleton, D.: Signal Analysis. I. Estimation of Signal Parameters and Waveform Structure in Noise Backgrounds, *Johns Hopkins Univ. Radiation Lab. Tech. Rept.* AF-50, June, 1958.
13. Slepian, D.: Estimation of Signal Parameters in the Presence of Noise, Symposium on Statistical Methods in Communication Engineering, *IRE Trans. on Inform. Theory*, **PGIT-3**: 68 (March, 1954).

13a. Youla, D. C.: The Use of the Method of Maximum Likelihood in Estimating Continuous Modulated Intelligence Which Has Been Corrupted by Noise, *IRE Trans. on Inform. Theory*, **PGIT-3**: 90 (March, 1954).

14. Swerling, P.: Maximum Angular Accuracy of a Pulsed Search Radar, *Proc. IRE*, **44**: 1146 (1956).
15. Davis, R. C.: Optimum vs. Correlation Methods in Tracking Random Signals in Background Noise, *Quart. Appl. Math.*, **15**: 123 (1957).
16. Bennion, D. R.: On the Statistical Estimation of Radar Signal Parameters, *Stanford Electronic Labs. Tech. Rept.* 361-2, May 7, 1956; see also the bibliography.
17. Hodges, J. L., and E. L. Lehmann: Some Problems in Minimax Point Estimation, *Ann. Math. Stat.*, **21**: 182 (1950).
18. Blackwell, D., and M. A. Girshick: "Theory of Games and Statistical Decisions," Wiley, New York, 1954.
19. Wiener, N.: "The Extrapolation, Interpolation, and Smoothing of Stationary Time Series," Wiley, New York, 1949.
20. Zadeh, L. A., and J. R. Ragazzini: An Extension of Wiener's Theory of Prediction, *J. Appl. Phys.*, **21**: 645 (1950).
21. Pitman, E. J. G.: The Estimation of Location and Scale Parameters of a Continuous Population of Any Given Form, *Biometrika*, **30**: 391 (1939).
22. Girshick, M. A., and L. J. Savage: Bayes and Minimax Estimates for Quadratic Loss Functions, *Proc. 2d Berkeley Symposium on Mathematical Statistics and Probability*, 1950, p. 53.
23. Zadeh, L. A.: General Filters for Separation of Signal and Noise, *Symposium on Inform. Networks*, April, 1954, p. 3.
24. Ashby, N.: On the Extraction of Noise-like Signals from a Noisy Background from the Risk Point of View, *Air Force Cambridge Research Center Tech. Rept.* AFCRC-TR-56-123, December, 1956.
25. Singleton, H. E.: Theory of Non Linear Transducers, *MIT Research Lab. Electronics Tech. Rept.* 160, August, 1950.
26. Phillips, R. C., and P. R. Weiss: Theoretical Calculation on Best Smoothing of Position Data for Gunnery Prediction, *MIT Radiation Lab. Rept.* 532, 1944.
27. Drenick, R. F.: A Nonlinear Prediction Theory, *IRE Trans. on Inform. Theory, Symposium*, **PGIT-4**: 146 (1954).
28. Van Meter, D.: Optimum Decision Systems for the Reception of Signals in Noise, sec. 4.2*b*, doctoral dissertation, Harvard University, 1955; also, *Cruft Lab. Tech. Repts.* 215, 216, October, 1955.
29. Wald, A.: Contributions to the Theory of Statistical Estimation and Testing Hypotheses, *Ann. Math. Stat.*, **10**: 299 (1939), theorems 5, 6.
30. Woodward, P. M., and I. L. Davies: Information Theory and Inverse Probability in Telecommunications, *Proc. IEE (London)*, **99** (pt. III): 37 (1952).
31. Sherman, S.: Non-mean-square Error Criteria, *IRE Trans. on Inform. Theory*, **IT-4**: 125 (1958).
32. Middleton, D.: A Note on the Estimation of Signal Waveform, *IRE Trans. on Inform. Theory*, **IT-5**: 86 (1959).

CHAPTER 22

INFORMATION MEASURES IN RECEPTION

In this chapter, we shall examine in a little more detail the use of information measures as criteria of system performance, within the general framework of decision theory.† Again, we are concerned specifically with the reception processes ($\boldsymbol{T}_R^{(N)}$) of detection and extraction, and while the emphasis is on systems which are optimal in the sense of minimizing information loss, we consider also cases which are not optimal in this respect, but which may be on the basis of other criteria, e.g., the simple cost or risk measures of Sec. 18.4.

As the discussion in Sec. 18.4 indicates, the general formulation of reception problems from the point of view of information loss is the same as the risk formulation, except that the cost function $C(\mathbf{S},\boldsymbol{\gamma})$ of the latter is replaced by the uncertainty measure $-\log p(\mathbf{S}|\boldsymbol{\gamma})$ [cf. Eq. (18.5)]. An important problem here is to determine under what conditions, if at all, receiving systems that are optimum from the viewpoint of minimum average risk (or cost) are also optimum in the simultaneous minimization of, say, equivocation, and vice versa. Because of the different natures of these two criteria, we do not expect that systems generally can be optimized with respect to both criteria at once. However, in special cases this may be possible and, in fact, is actually achieved, as the example of Sec. 22.2-2 indicates.

We begin the discussion by considering first the Fisher and Shannon information measures, in connection with the notion of a sufficient statistic (cf. Sec. 22.1). Then, with this as background, in Sec. 22.2 we show how equivocation (the analogue of average risk) can also be used to evaluate the performance of binary detection systems $(\boldsymbol{T}_R^{(N)})_{det}$ (optimum or not). These procedures are then extended to signal extraction, for example, $(\boldsymbol{T}_R^{(N)})_{ext}$ in Sec. 22.3. The chapter concludes with a brief summary of some of the important properties of the maximum likelihood extractor (cf. Sec. 21.1) with respect to information-loss criteria, as well as to the simple risk theory.

22.1 *Information and Sufficiency*

Preliminary to our treatment of reception systems that use information loss as a criterion of performance (instead of risk or simple cost), let us

† The material of the present chapter is based for the most part on Middleton[1] and Van Meter,[2] sec. 5.

first examine and compare the Fisher and Shannon measures of information. To facilitate this, we begin with the somewhat older notion of sufficient statistics.

22.1-1 Sufficient Statistics. The classical concept of sufficiency, due to Fisher,† is based on the view that the parameters governing the distribution of a random variable exert a "causal" influence on the value of the variate, which is more or less obscured by the "randomness" of the distribution. That is, the value of the variate furnishes some clue, however imperfect, to the value of the parameter "behind" the randomness. An indication of the extent to which the influence of the signal **S** "comes through" the randomness and is reflected in the data **V** is provided by the variation of the probability $p(\mathbf{S},\mathbf{V})$ with **S** when **V** is fixed, or of the d.d. $w(\mathbf{S},\mathbf{V})$, where **S** and **V** can assume a continuum of values. This variation constitutes the unconditional likelihood function (see Sec. 21.1-2); its value for a particular **S** and **V** is thought of as the *likelihood* that **S** was responsible for **V**.

We recall that a *statistic* of a probability $p(\mathbf{V}|\mathbf{S})$, or d.d. $w(\mathbf{V}|\mathbf{S})$, is any quantity $\gamma(\mathbf{V})$ formed by the transformation of the variate **V** (sometimes the transformation itself is referred to as the statistic[3]). An estimator of **S**, for example, is a statistic. A statistic $\gamma(\mathbf{V})$ is said to be *sufficient*, roughly speaking, if knowledge of γ is as good as knowledge of **V** itself, as far as determining the **S** "responsible" for both is concerned. Now, from Eq. (18.1) we have

$$p(\gamma|\mathbf{V}) = p(\gamma|\mathbf{V},\mathbf{S}) \tag{22.1}$$

With this, the *likelihood function* (cf. Sec. 21.1-2) may be arranged to read‡

$$p(\mathbf{S},\mathbf{V}) = p(\mathbf{S},\gamma)p(\mathbf{V}|\gamma,\mathbf{S})/p(\gamma|\mathbf{V}) \tag{22.2}$$

Now let us assume that all these functions are known and that we are given a value of γ in ignorance of the **V** and **S** that produced it. Under what conditions will $\gamma(\mathbf{V})$ tell us as much about **S** as would **V** directly if **V** were known? The only kind of knowledge about **S** we get by knowing **V** is the dependence of $p(\mathbf{V},\mathbf{S})$ on **S** with **V** fixed. If γ alone enables us to reproduce this variation, then γ is a *sufficient statistic* for the distribution. In Eq. (22.2), the value of $p(\gamma|\mathbf{V})$ is unknown but independent of **S**. The quantity $p(\mathbf{V}|\gamma,\mathbf{S})$, on the other hand, does depend on **S**, but in an unknown way, since **V** is unknown. Clearly, γ can be a sufficient statistic only if $p(\mathbf{V}|\gamma,\mathbf{S})$ is independent of **S**, that is, if

$$p(\mathbf{V}|\gamma,\mathbf{S}) = p(\mathbf{V}|\gamma) \qquad \text{all } \mathbf{S} \text{ and } \mathbf{V} \tag{22.3}$$

† For a discussion and listing of the original contributions by Fisher and others to this subject, see Cramér[3] and Fisher's collected works.[4] A recent treatment has been given by Halmös and Savage.[5]

‡ The convention that functions with different arguments are different functions (unless otherwise noted) is in force here [cf. footnote referring to Eq. (18.10)].

With Eq. (22.3), Eq. (22.1) is equivalent to

$$p(\mathbf{S}|\mathbf{V}) = p(\mathbf{S}|\boldsymbol{\gamma}) \qquad \text{all } \mathbf{S}, \mathbf{V} \tag{22.4}$$

Either relation (22.3) or (22.4) may be taken as the definition of sufficiency. When $\boldsymbol{\gamma}$ is a sufficient statistic, specification of $\mathbf{V}$ in addition to $\boldsymbol{\gamma}$ does not in any way improve our knowledge of $\mathbf{S}$. The relation (22.2) shows that only distributions that can be factored into a product of two terms, one involving $\boldsymbol{\gamma}$ and $\mathbf{S}$ alone and the other $\mathbf{V}$ and $\boldsymbol{\gamma}$ alone, admit a sufficient statistic. The factorization condition was introduced by Fisher; the definition above, however, differs somewhat from the classical one in that $\mathbf{S}$ is considered as a random variable.†

22.1-2 Information Measures. Closely related to sufficiency is the idea, also due to Fisher,[6]‡ of associating with an observation a numerical measure of the information it contains about the distribution parameter (or parameters), for example, θ or $\boldsymbol{\theta}$, in $S(\boldsymbol{\theta})$ (cf. Sec. 21.1-1). Fisher originally introduced the quantity

$$I_j = \int_{\Gamma_j} \left[\frac{\partial \log p(V_j|\theta)}{\partial \theta} \right]^2 p(V_j|\theta)\, dV_j \tag{22.5}$$

as a measure of the information about the parameter θ per member V_j of the sample (the V_j are here assumed uncorrelated), or, as he called it, the "intrinsic accuracy" of the distribution $p(V_j|\theta)$ with respect to the parameter θ. Here, θ is a single one-dimensional parameter, the same for all V_j. In the terminology of reception theory, with the V_j identified as observations ($\sim \boldsymbol{T}_R{}^{(N)}$) on a mixture of signal and noise, the case considered corresponds to discrete independent sampling when the signal is either a d-c wave (that is, $\theta = S_0 =$ constant throughout the observation interval) or a signal whose functional dependence on a single parameter θ (epoch, for example) is known. For the normal distribution with $\theta = \bar{V}_j$ (for example, for a d-c signal mixed additively with gaussian noise whose mean is zero), this I_j $(= I)$ is the reciprocal of the variance. For any statistic $\boldsymbol{\gamma}(\mathbf{V})$ formed from n independent samples, it can be shown[1,4-6] (cf. Prob. 22.1) that:

1. $\boldsymbol{\gamma}$ cannot contain more than nI units of information about θ.
2. $\boldsymbol{\gamma}$ can contain nI units if and only if $\boldsymbol{\gamma}$ is a sufficient statistic.
3. If $\boldsymbol{\gamma}$ is unbiased ($\bar{\boldsymbol{\gamma}} = \theta$), then the variance of $\boldsymbol{\gamma}$ cannot exceed $(nI)^{-1}$.

Shannon's concept[10] of the information transmitted by a noisy channel in communication problems (cf. Secs. 6.4, 6.5) can also be used to define a measure of the information about $\mathbf{S}$ contained in $\mathbf{V}$, as we have already

† Similar remarks clearly apply when distributions are replaced by probability densities: the various p's in Eqs. (22.1) to (22.4) are replaced by the appropriate probability densities $w(\boldsymbol{\gamma}|\mathbf{V})$, $w(\mathbf{S}|\mathbf{V})$, etc.

‡ For an account of the properties of Fisher's information measure, see Pitman[7] and Hodges and Lehmann.[8] Doob[9] discusses another information measure with similar properties.

shown in Sec. 6.3-1. Thus, we may regard **V** as the output of a noisy channel with signal **S** applied at the input, for example, $\mathbf{V} = T_{M^{(N)}}\{\mathbf{S}\}$. The uncertainty about **S** when **V** is known is defined according to Eq. (6.5*a*) as

$$\mathcal{U}(\mathbf{S}|\mathbf{V}) = -\log p(\mathbf{S}|\mathbf{V}) \tag{22.6}$$

where the base of the logarithm determines the particular unit of measure. The uncertainty about **S** before **V** is known is just $-\log p(\mathbf{S})$. Thus, knowledge of **V** reduces the uncertainty by an amount

$$\mathcal{I}(\mathbf{S}|\mathbf{V}) = \log \frac{p(\mathbf{S}|\mathbf{V})}{p(\mathbf{S})} = \log \frac{p(\mathbf{V}|\mathbf{S})}{p(\mathbf{V})} \tag{22.7}$$

[cf. Eqs. (6.6)]. This is the information about **S** contained in the sample **V**.

Similarly, the information about **S** contained in a statistic $\gamma(\mathbf{V})$ is

$$\mathcal{I}(\mathbf{S}|\gamma) = \log \frac{p(\mathbf{S}|\gamma)}{p(\mathbf{S})} = \log \frac{p(\gamma|\mathbf{S})}{p(\gamma)} \tag{22.8}$$

The difference between these [cf. Eq. (6.6*a*)], or

$$-\Delta\mathcal{I}(\mathbf{S}|\mathbf{V},\gamma) \equiv \mathcal{I}(\mathbf{S}|\mathbf{V}) - \mathcal{I}(\mathbf{S}|\gamma) = -\log [p(\mathbf{S}|\gamma)/p(\mathbf{S}|\mathbf{V})] \tag{22.9}$$

is thus the *loss* of information about **S** attending formation of the statistic $\gamma = \gamma(\mathbf{V})$ from **V**. Again, if **S**, **V**, γ possess a continuum of values, the probabilities $p(\mathbf{S})$, $p(\mathbf{V})$, $p(\mathbf{S}|\mathbf{V})$, etc., are replaced by the corresponding probability densities $w(\mathbf{S})$, etc., in Eqs. (22.7) to (22.9), as described in Sec. 6.4-1. Henceforth here, unless otherwise indicated, we shall assume that **S**, **V**, etc., are continuous random variables.

With these last remarks in mind, let us compute the *average* information loss ΔH attending formation of the statistic γ from **V**. This is simply the average of $-\Delta\mathcal{I}$ with respect to all γ, **V**, and **S**,

$$\Delta H = -\overline{\Delta\mathcal{I}} = -\int_\Gamma \int_\Omega \int_\Delta w(\mathbf{V},\mathbf{S},\gamma) \log \frac{w(\mathbf{S}|\gamma)}{w(\mathbf{S}|\mathbf{V})}\, d\mathbf{V}\, d\mathbf{S}\, d\gamma \tag{22.10}$$

For a nonrandomized decision rule for forming γ from **V**, this may be expressed as

$$\Delta H = \int_\Gamma d\mathbf{V}\, w(\mathbf{V}) \left[\int_\Omega w(\mathbf{S}|\mathbf{V}) \log w(\mathbf{S}|\mathbf{V})\, d\mathbf{S} - \int_\Omega w(\mathbf{S}|\mathbf{V}) \log w(\mathbf{S}|\gamma)\, d\mathbf{S} \right] \tag{22.11a}$$

or

$$\Delta H = \int_\Gamma d\mathbf{V}\, w(\mathbf{V}) \int_\Omega w(\mathbf{S}|\mathbf{V}) \log w(\mathbf{S}|\mathbf{V})\, d\mathbf{S} - H(\sigma,\delta) \tag{22.11b}$$

where $H(\sigma,\delta)$ is the *equivocation* [cf. Eqs. (6.84) et seq. and Sec. 6.5-3]. Note that the statistic γ is contained only in the last integral, e.g., in $H(\sigma,\delta)$ [cf. Eq. (18.9*b*)]. When this integral is maximized by variation of the function $w(\mathbf{S}|\gamma)$ subject only to the constraint $\int_\Omega w(\mathbf{S}|\gamma)\, d\mathbf{S} = 1$, the result is the condition (22.4), which makes Eqs. (22.11) identically zero. Thus, *the average information loss is zero if and only if* γ *is a sufficient statistic.*

In order to compare the Fisher and Shannon measures of information, it must be noticed first that the ideas behind the two definitions are somewhat different. We observe that Fisher's measure† [Eq. (22.5)] is the statistical average over Γ_j of

$$\frac{1}{w(V_j|\theta)^2}\left[\frac{\partial w(V_j|\theta)}{\partial \theta}\right]^2 \tag{22.12}$$

so that this may be regarded as the information about θ contained in a particular V_j, and is the appropriate quantity for comparison with the Shannon definition [Eq. (22.7)]. Clearly, the measure (22.12) is actually a measure of the sensitivity of V_j to small *changes* in θ, since, if $\partial w(V_j|\theta)/\partial\theta$ vanishes for some V_j and θ, the information about that θ contained in V_j is zero. The Shannon measure (22.7), on the other hand, is more an absolute measure of statistical dependence, reducing to zero for a $\mathbf{V}$ and $\mathbf{S}$ such that $w(\mathbf{V}|\mathbf{S}) = w(\mathbf{V})$.

The proper Shannon measure for comparison with Eq. (22.5), therefore, is the average of the *difference* of the information in $\mathbf{V}$ about $\mathbf{S}$ and $\mathbf{S} + \mathbf{\Delta S}$,

$$\Delta I(\mathbf{V}|\mathbf{S}) \equiv -\int_\Gamma w(\mathbf{V}|\mathbf{S}) \log \frac{w(\mathbf{V}|\mathbf{S}+\mathbf{\Delta S})}{w(\mathbf{V}|\mathbf{S})}\, d\mathbf{V} \tag{22.13}$$

For small ΔS_j, we have approximately

$$w(\mathbf{V}|\mathbf{S}+\mathbf{\Delta S}) \doteq w(\mathbf{V}|\mathbf{S}) + \widetilde{\boldsymbol{\partial}\mathbf{w}}\, \mathbf{\Delta S} \tag{22.14}$$

where $\boldsymbol{\partial}\mathbf{w}$ is a column vector with components

$$\partial w_j \equiv \frac{\partial w(\mathbf{V}|\mathbf{S})}{\partial S_j} \tag{22.15}$$

Using this in Eq. (22.13) with the approximation

$$\log\left(1 + \frac{1}{w}\widetilde{\boldsymbol{\partial}\mathbf{w}}\,\mathbf{\Delta S}\right) \doteq \frac{1}{w}\widetilde{\boldsymbol{\partial}\mathbf{w}}\,\mathbf{\Delta S} - \frac{1}{2w^2}\widetilde{\mathbf{\Delta S}}\,\boldsymbol{\partial}\mathbf{w}\,\widetilde{\boldsymbol{\partial}\mathbf{w}}\,\mathbf{\Delta S} \tag{22.16}$$

we find that‡

$$\Delta I(\mathbf{V}|\mathbf{S}) = \tfrac{1}{2}\widetilde{\mathbf{\Delta S}}\,\mathbf{B}\,\mathbf{\Delta S} \tag{22.17}$$

where $\mathbf{B}$ is the Fisher information matrix with coefficients

$$B_{ij} = \int_\Gamma \frac{\partial \log w}{\partial S_i}\frac{\partial \log w}{\partial S_j}\, w(\mathbf{V}|\mathbf{S})\, d\mathbf{V} \qquad w = w(\mathbf{V}|\mathbf{S}) \tag{22.18}$$

† Recall that we are dealing now with continuous random variables.

‡ Kullback and Leibler[11] give the result of Eq. (22.17). See also Bartlett[12] for comments on the relation between the Fisher and Shannon measures of information.

When the samples are independent and **S** is a one-dimensional parameter, **B** becomes a diagonal matrix with its leading elements equal to the quantity (22.5), originally introduced by Fisher.

Finally, in their discussion of binary hypothesis testing, Kullback and Leibler also use the logarithm of the (conditional) likelihood ratio $\Lambda_n(\mathbf{V})$ as a measure of the information contained in an observation for discrimination between the two hypotheses, showing that its average value [Eq. (22.13), with one-dimensional $\mathbf{S} \to \theta$ and arbitrary $\Delta\theta$] is positive, semidefinite, and invariant under a sufficient transformation, i.e., a transformation $\gamma(\mathbf{V})$ that leads to a sufficient statistic. (See the comments in Sec. 19.1-3 on the interpretation of $\log \Lambda_n$ in the average-risk theory of optimum binary detection.)

PROBLEMS

22.1 For any statistic $\gamma(\mathbf{V})$ formed from n independent samples (from the same process), show that:

(*a*) γ cannot contain more than nI units of information about the parameter θ.

(*b*) γ can contain nI units if and only if γ is a sufficient statistic.

(*c*) If γ is unbiased ($\bar{\gamma} = \theta$), then the variance of γ cannot exceed $(nI)^{-1}$.

Here I is given by Eq. (22.5).

22.2 Carry out the details of the demonstration to show that the average information loss ΔH is zero if and only if γ is a sufficient statistic.

22.2 *Information-loss Criterion for Detection*

When a definite decision is made under conditions of uncertainty, information is lost concerning the quantity or state about which the decision is taken. In other words, there is a nonzero equivocation [in natural units (or bits) per decision] associated with the decision process. Just how great this equivocation may be will depend on the signal, the background noise, and the system in question. Here we shall examine the information loss of binary detection systems $(\boldsymbol{T}_R{}^{(N)})_{det}$ measured by the equivocation for each decision.

22.2-1 Equivocation of Binary Detectors. As described in Sec. 18.4-3, the general formulation of reception problems from the point of view of information loss is the same as the risk formulation except that the cost function $C(\mathbf{S},\gamma)$ of the latter is replaced by the uncertainty measure $-\log p(\mathbf{S}|\gamma)$. To specialize the general expressions (18.9) for binary detection, we assume first that the (nonrandomized) decision rule divides observation space Γ into two regions, here denoted as Γ_0 and Γ_1, and that the decision γ_0 (noise alone) is made when $V \in \Gamma_0$ and the decision γ_1 (signal and noise) when $V \in \Gamma_1$. Thus, we write

$$\delta(\gamma_i|V \in \Gamma_j) = \delta_{ij} \qquad i,j = 0,1 \tag{22.19}$$

where $\delta_{ij} = 0$ $(i \neq j)$ or 1 $(i = j)$. With this, the equivocation (18.9) becomes

$$H(\sigma,\delta) = -\int_\Omega \sigma(\mathbf{S})[p(\gamma_0|\mathbf{S}) \log p(\mathbf{S}|\gamma_0) + p(\gamma_1|\mathbf{S}) \log p(\mathbf{S}|\gamma_1)]\, d\mathbf{S} \qquad (22.20)$$

The equivocation in natural units per decision for any given binary detection system may be calculated from Eq. (22.20) and used to judge its performance[13] (cf. Prob. 6.11). As an example, let us suppose for the moment that either $\mathbf{S} = \mathbf{S}_0 = 0$ or $\mathbf{S} = \mathbf{S}_1 \neq 0$ can be present at the input with a priori probabilities $\sigma(\mathbf{S}_0) = q$ and $\sigma(\mathbf{S}_1) = p$ (for example, the simple-alternative case). Then $\sigma(\mathbf{S})$ becomes $\sigma(\mathbf{S}_0)\delta(\mathbf{S} - \mathbf{S}_0) + \sigma(\mathbf{S}_1)\delta(\mathbf{S} - \mathbf{S}_1)$, and Eq. (22.20) gives

$$H(\sigma,\delta) = -\sum_{i,j} \sigma(\mathbf{S}_i)p(\gamma_j|\mathbf{S}_i) \log p(\mathbf{S}_i|\gamma_j) \qquad i, j = 0, 1 \qquad (22.21)$$

The probabilities $p(\gamma_i|\mathbf{S}_j)$ and $p(\mathbf{S}_i|\gamma_j)$ may readily be expressed in terms of the conditional error probabilities α' $(= \alpha)$ and $\beta'(\mathbf{S}_1)$ $(= \beta)$, introduced in Sec. 19.1-1, as follows:

$$p(\gamma_0|\mathbf{S}_0) = 1 - \alpha' \qquad p(\gamma_0|\mathbf{S}_1) = \beta'$$
$$p(\gamma_1|\mathbf{S}_0) = \alpha' \qquad p(\gamma_1|\mathbf{S}_1) = 1 - \beta' \qquad (22.22a)$$

$$p(\mathbf{S}_0|\gamma_0) = \frac{q(1-\alpha')}{q(1-\alpha') + p\beta'} \qquad p(\mathbf{S}_0|\gamma_1) = \frac{q\alpha'}{q\alpha' + p(1-\beta')}$$
$$p(\mathbf{S}_1|\gamma_0) = \frac{p\beta'}{q(1-\alpha') + p\beta'} \qquad p(\mathbf{S}_1|\gamma_1) = \frac{p(1-\beta')}{q\alpha' + p(1-\beta')} \qquad (22.22b)$$

However, if we have, not a simple-alternative situation like the above, where the conditional probabilities α', β' are equal to the class conditional probabilities α, β [cf. Eqs. (19.8)], but rather the more general one-sided alternative situation of Sec. 19.1-1, we must use Eqs. (22.22) in Eq. (22.20), where $\sigma(\mathbf{S}) = q\delta(S - 0) + pw_1(\mathbf{S})_1$ [cf. Eq. (19.4)]. The calculations are now considerably more involved, as Eqs. (22.22) in Eq. (22.20) clearly indicate.

Figure 22.1 is calculated for the simple-alternative example of an incoherent threshold detector with fixed RF-IF in the case of a pulsed radar (cf. Sec. 20.4-3) and shows how the equivocation of Bayes tests depends on the cost ratio $\mathcal{K}$ and on $a_0^2 \sqrt{n}$, for $\mu = 1$ (and Rayleigh statistics). These curves are calculated [with the help of Eq. (20.131) and Fig. 20.8] for α $(= \alpha')$ and β $(= \beta')$ by evaluating Eqs. (22.22), followed by substitution into Eq. (22.21). We see that the Ideal Observer ($\mathcal{K} = 1$) loses the *least* information for a given signal amplitude and integration time. As the cost ratio $\mathcal{K}$ is varied for fixed $a_0^2 \sqrt{n}$, the minimum at $\mathcal{K} = 1$ appears broad. In fact, the information loss for any fixed $a_0^2 \sqrt{n}$ does not vary here by more than 0.2 bit/decision as $\mathcal{K}$ is changed from $1/16$ to 16.

These curves may also be used to define a minimum detectable signal (for threshold detection) in the same way as the Bayes risk curves are used (see Sec. 19.3-3). Accordingly, if 0.2 bit is taken as the largest allowable loss, the minimum value of $a_0^2 \sqrt{n}$ is 3.75 for $\mathcal{K} = 1$ and 4.25 for $\mathcal{K} = 16$, $1/16$. For fixed sample size n, this amounts to a difference of only 0.54 db in the amplitude of the minimum detectable signal. Correspondingly, for fixed signal amplitudes, the change in integration time is only $\sqrt{4.25/3.75} = 1.06$, or 6 per cent. For very small signals, the equivocation approaches 1 bit, so that the system does no better than one who guesses on the basis of the a priori probabilities. For strong signals or long integration times, on the other hand, the equivocation approaches zero, as expected, corresponding to the increasing certainty of a correct decision. For a discussion of information loss in binary detection, see Sec. 20.3-3.

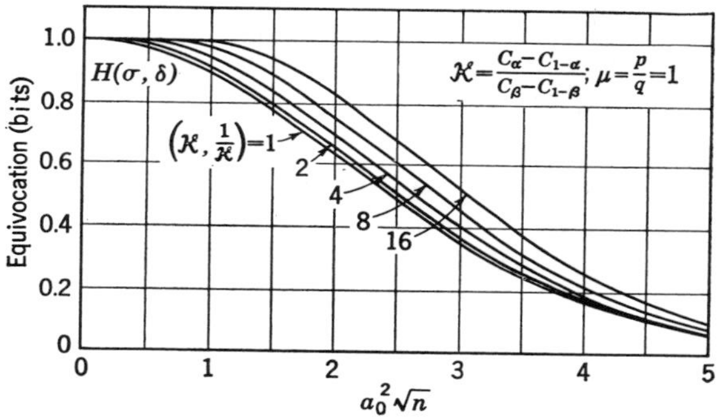

FIG. 22.1. Loss of information with Bayes detectors; binary threshold incoherent detection of a pulsed carrier (Rayleigh statistics); $\mu = 1$.

22.2-2 Detectors That Minimize Equivocation. In the above, we have seen how equivocation may be used to evaluate *arbitrary* detection systems. We turn now to the problem of characterizing the detection system that is "best" from the information-loss point of view, i.e., the system that minimizes equivocation. Since the value of $p(\mathbf{S}|\gamma)$ in the information-loss formulation of reception problems depends on the decision rule in use, while the value of $C(\mathbf{S},\gamma)$ does not, the decision rule that minimizes information loss is more difficult to find than the one that minimizes risk. (Cf. Sec. 20.3-3, where now probabilities are replaced by probability densities in the appropriate places.)

Let us return to the general expression for equivocation [Eq. (22.20)] and seek to minimize it by choice of the boundary between regions Γ_0 and Γ_1, without specific assumptions as yet about the prior signal distribution $\sigma(\mathbf{S})$, save that the "alphabet," or waveforms, $\mathbf{S}$ are specified in some fashion. The probability density function $w(\mathbf{S}|\gamma_i)$ and probability $p(\gamma_i|\mathbf{S})$ may be written

$$w(\mathbf{S}|\gamma_i) = \int_{\Gamma_i} w(\mathbf{S}|\mathbf{V}) w(\mathbf{V}|\gamma_i)\, d\mathbf{V} \tag{22.23}$$

$$p(\gamma_i|\mathbf{S}) = \int_{\Gamma_i} F_n(\mathbf{V}|\mathbf{S})\, d\mathbf{V} \tag{22.24}$$

Here, $w(\mathbf{V}|\gamma_i)$ is the probability (density) that $\mathbf{V}$ was responsible for the decision γ_i. Since the decision rule is nonrandomized, every $\mathbf{V}$ in Γ_i leads to the decision γ_i with the probability 1, and the $\mathbf{V}$'s outside of Γ_i cannot lead to γ_i at all. Thus, $w(\mathbf{V}|\gamma_i)$ is constant throughout Γ_i, equal in fact to the reciprocal of the "volume" of Γ_i, since it must be properly normalized. Accordingly, we have

$$w(\mathbf{V}|\gamma_i) = 1/\Gamma_i \tag{22.25}$$

$$w(\mathbf{S}|\gamma_i) = (1/\Gamma_i) \int_{\Gamma_i} w(\mathbf{S}|\mathbf{V})\, \mathbf{dV} \tag{22.26}$$

where we have let Γ_i stand here for the volume of the region as well as for the domain of $\mathbf{V}$ included within the volume.

Now denoting by $\mathbf{V}'$ a point on the boundary between Γ_0 and Γ_1 and letting Γ_0 be increased to $\Gamma_0 + d\Gamma_0$ by a change of $\mathbf{V}'$ to $\mathbf{V}' + \mathbf{dV}'$, we find for the derivatives involved in the minimization

$$\frac{\partial}{\partial \mathbf{V}'} p(\gamma_0|\mathbf{S}) = -\frac{\partial}{\partial \mathbf{V}'} p(\gamma_1|\mathbf{S}) = F_n(\mathbf{V}'|\mathbf{S}) \tag{22.27}$$

$$\frac{\partial}{\partial \mathbf{V}'} \log w(\mathbf{S}|\gamma_0) = \frac{w(\mathbf{S}|\mathbf{V}')}{\Gamma_0 w(\mathbf{S}|\gamma_0)} - \frac{1}{\Gamma_0} \tag{22.28}$$

$$\frac{\partial}{\partial \mathbf{V}} \log w(\mathbf{S}|\gamma_1) = \frac{-w(\mathbf{S}|\mathbf{V}')}{\Gamma_1 w(\mathbf{S}|\gamma_1)} + \frac{1}{\Gamma_1} \tag{22.29}$$

Here, the derivatives are to be interpreted as follows,

$$\frac{\partial}{\partial \mathbf{V}} f(\mathbf{V}) = \lim_{\Delta\mathbf{V}\to 0} \frac{f(\mathbf{V} + \Delta\mathbf{V}) - f(\mathbf{V})}{\Delta\mathbf{V}} \tag{22.30}$$

and it is assumed that a unique limit exists for all $\Delta\mathbf{V}$. The derivative of the bracketed terms in the integrand of Eq. (22.20) then becomes

$$\frac{w(\mathbf{S}|\mathbf{V}')}{\sigma(\mathbf{S})} \left[\frac{p(\gamma_0)}{\Gamma_0} - \frac{p(\gamma_1)}{\Gamma_1} \right] + \frac{p(\gamma_1|\mathbf{S})}{\Gamma_1} - \frac{p(\gamma_0|\mathbf{S})}{\Gamma_0} + F_n(\mathbf{V}'|\mathbf{S}) \log \frac{w(\mathbf{S}|\gamma_0)}{w(\mathbf{S}|\gamma_1)} \tag{22.31}$$

where we have used the relation

$$w(\mathbf{S}|\gamma_i) = \frac{\sigma(\mathbf{S})}{p(\gamma_i)} p(\gamma_i|\mathbf{S}) \tag{22.32}$$

In these expressions, $p(\gamma_i)$ is the (total) probability of making decision γ_i, given by

$$p(\gamma_i) = \int_\Omega p(\gamma_i|\mathbf{S})\sigma(\mathbf{S})\, \mathbf{dS} \tag{22.33}$$

The integration over Ω indicated in Eq. (22.20) causes the first four terms in expression (22.31) to cancel, leaving finally

$$\frac{\partial H(\sigma,\delta)}{\partial \mathbf{V}'} = -\int_\Omega \sigma(\mathbf{S}) F_n(\mathbf{V}'|\mathbf{S}) \log \frac{w(\mathbf{S}|\gamma_0)}{w(\mathbf{S}|\gamma_1)}\, \mathbf{dS} = 0 \tag{22.34}$$

This is the condition for minimum (or maximum) equivocation, i.e., the boundary between Γ_0 and Γ_1 must be such that this relation is satisfied for an extremum in $H(\sigma,\delta)$.

The results for the one-sided- and simple-alternative tests are obtained from Eq. (22.34) by suitable specialization of $\sigma(\mathbf{S})$. For the one-sided alternative we take as before $\sigma(\mathbf{S}) = q\delta(\mathbf{S} - 0) + pw_1(\mathbf{S})$, which yields [with $F_n(\mathbf{V}'|\mathbf{S})$ now replaced by $W_n(\mathbf{V}' - \mathbf{S})_N$ in the case of additive noise]

$$\frac{p\langle W_n(\mathbf{V}' - \mathbf{S})_N\rangle_{S|\gamma}}{qW(\mathbf{V}')_N} = -\log\frac{p(\mathbf{S} = 0|\gamma_0)}{p(\mathbf{S} = 0|\gamma_1)} \tag{22.35}$$

where $$\langle W_n(\mathbf{V}' - \mathbf{S})_N\rangle_{S|\gamma} = \int_\Omega W_n(\mathbf{V}' - \mathbf{S})_N w_1(\mathbf{S}) \log\frac{w(\mathbf{S}|\gamma_0)}{w(\mathbf{S}|\gamma_1)}\, d\mathbf{S} \tag{22.35a}$$

Equations (22.35), (22.35*a*) show that the optimum (or extremal) division of observation space is achieved here also by a generalized likelihood-ratio test, in which $W_n(\mathbf{V}' - \mathbf{S})_N$ now is averaged with respect to a distribution $w_1(\mathbf{S}) \log [w(\mathbf{S}|\gamma_0)/w(\mathbf{S}|\gamma_1)]$, which itself depends on the optimum (or extremal) division.

We can illustrate the preceding explicitly in the case of simple-alternative detection. We have $\sigma(\mathbf{S}) = q\delta(\mathbf{S} - 0) + p\delta(\mathbf{S} - \mathbf{S}_1)$, so that Eq. (22.34) becomes†

$$\frac{pW_n(\mathbf{V}' - \mathbf{S}_1)_N}{qW_n(\mathbf{V}')_N} = \mathcal{K}_H \tag{22.36}$$

where $$\mathcal{K}_H = \frac{\log z_0}{\log z_1} \tag{22.37}$$

and $$z_0 = \frac{p(\mathbf{S} = 0|\gamma_0)}{p(\mathbf{S} = 0|\gamma_1)} \tag{22.38a}$$

with $$z_1 = \frac{p(\mathbf{S}_1|\gamma_0)}{p(\mathbf{S}_1|\gamma_1)} \tag{22.38b}$$

The values of $\mathbf{V}'$ satisfying Eq. (22.36) define the extremum boundary between Γ_0 and Γ_1; that is, if "noise alone" is decided whenever $\mathbf{V}$ falls within Γ_0 and "signal and noise" when $\mathbf{V}$ falls within Γ_1, the average information loss $H(\sigma,\delta)$ is a maximum or a minimum for this division.

Note that Eq. (22.36) defines a likelihood-ratio test of the same type as the Bayes test for the corresponding problem in the risk formulation. Thus, tests that minimize average information loss (when they exist) belong to the Bayes class and are equivalent to minimum average risk tests with special cost assumptions. It is therefore possible for a system to be optimum simultaneously from the standpoints of both average risk and average information loss, even though the two criteria may have quite different significance in practical applications, e.g., high cost decisions may correspond

† Siebert and Lerner[14] in a recent memorandum independently obtained a similar result.

to situations of low information content, and vice versa (cf. discussion in Sec. 23.4).

To show that there are solutions of Eq. (22.36) that minimize average information loss, we use the appropriate expression in Eqs. (22.22*b*), with $\mathbf{S}_0 = 0$, to express z_0 and z_1 of Eqs. (22.38*a*), (22.38*b*) in terms of p, q, α $(= \alpha')$, and β $(= \beta')$ for the simple-alternative example above. With $\mu \equiv p/q$ as before [cf. Eqs. (19.19)], the result is

$$z_0 = \mu \frac{1 - \alpha}{\alpha} \frac{1 + \alpha/\mu - \beta}{1 - \alpha + \beta\mu} \tag{22.39a}$$

$$z_1 = \frac{1}{\mu} \frac{1 - \beta}{\beta} \frac{1 - \alpha + \beta\mu}{1 + \alpha/\mu - \beta} \tag{22.39b}$$

It may readily be shown that for $\alpha + \beta < 1$ (the case of ordinary interest) z_0 and z_1 are both greater than unity, so that the likelihood threshold $\mathcal{K}_H$

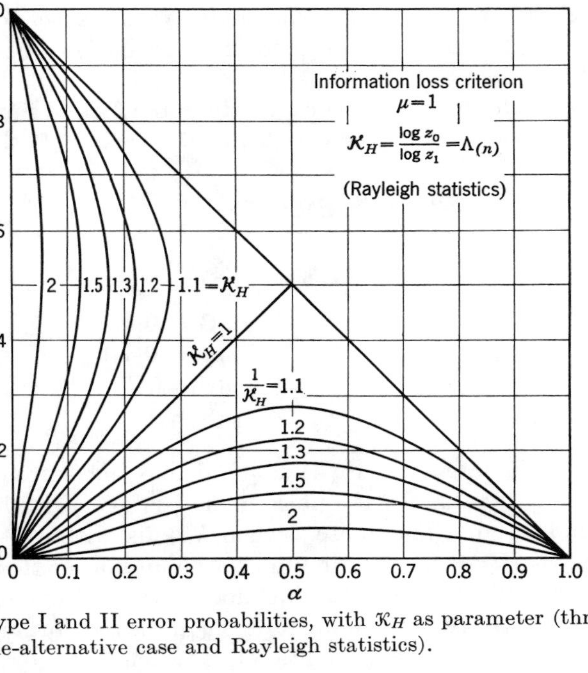

FIG. 22.2. The type I and II error probabilities, with $\mathcal{K}_H$ as parameter (threshold detection in the simple-alternative case and Rayleigh statistics).

is always positive. (We note that interchange of α and β and of p and q interchanges z_0 and z_1, thus inverting $\mathcal{K}_H$.)

The universal curve of Fig. 22.2 shows the relation between α, β, and $\mathcal{K}_H$ for $\mu = 1$, exhibiting this symmetry. We see for this example that the existence of a test with minimum average information loss depends on whether the statistics of the problem admit a likelihood-ratio test with α and β related to the threshold $\mathcal{K}_H$, as shown in Fig. 22.2. If, for instance,

Fig. 22.2 is superimposed on Fig. 20.4*a*, which gives the characteristics for the simple-alternative binary detection problem of Sec. 20.4-3 (e.g., for Rayleigh statistics), we observe that there are no combinations of α, β, and $\mathcal{K}$ (for $\mu = 1$) that fit both at once, except for those along the line $\alpha = \beta$, $\mathcal{K} = \mathcal{K}_H = 1$. Figure 22.1 shows that the average information loss (e.g., the equivocation) is indeed a minimum for this value of $\mathcal{K}$, so that for Rayleigh statistics, when $\mu = 1$, the Ideal Observer (who takes $\mathcal{K} = 1$) minimizes average information loss and average risk simultaneously.

On the other hand, we note that since the Neyman-Pearson observer for this simple-alternative case does not generally minimize average information loss, as the example of Fig. 22.1 shows ($\mathcal{K} \neq 1$), and yet is optimum in a risk sense when fixed false alarm time (that is, fixed α) is important, it is not necessarily desirable for a system to minimize average information loss. Detailed control of decision error may be more important (cf. Sec. 23.4), depending on the particular application.

Thus, although tests that minimize average information loss are likelihood-ratio tests and therefore form a subclass of the Bayes tests, they exist under much less general conditions than do the minimum average risk tests. That is, the latter exist for any given cost ratio and μ, while the former exist only for certain cost ratios and μ's, depending on the statistics. Determination of the broad conditions under which the information-loss extremum exists and, moreover, is a minimum awaits further investigation.

PROBLEMS

22.3 (*a*) Show that for the simple-alternative case discussed above [Eqs. (22.36) et seq.] z_0, $z_1 > 1$ for $\alpha + \beta < 1$ and hence that $\mathcal{K}_H > 0$, that is, that a solution to Eq. (22.36) is possible.

(*b*) Carry out the details of the extremizing of Eq. (22.20), i.e., Eqs. (22.27) to (22.35).

22.4 Obtain the condition that the equivocation be (*a*) a minimum or (*b*) a maximum.

22.3 *Information-loss Criterion for Extraction*

Just as we have shown above how the equivocation $H(\sigma,\delta)$ can be calculated for both optimum and nonoptimum binary detectors, so also can we extend the analysis to the calculation of this average information loss when the system is used for extraction. A brief outline of the argument and a summary of various salient features of the maximum likelihood extractor (in this respect, as well as from the viewpoint of average risk) conclude the present chapter.

22.3-1 Average Information Loss and Its Extrema in Extraction. To specialize Eq. (18.3) for $H(\sigma,\delta)$ in extraction, we first assume that the decision rule is nonrandomized and use Eq. (18.5) to obtain

$$H(\sigma,\delta) = -\int_\Gamma d\mathbf{V}\, w(\mathbf{V}) \int_\Omega w(\mathbf{S}|\mathbf{V}) \log w[\mathbf{S}|\gamma(\mathbf{V})]\, d\mathbf{S} + H_0 \quad (22.40)$$

This expression may be minimized by choosing γ to minimize the second integral for arbitrary fixed $\mathbf{V}$. As before, $w(\mathbf{S}|\gamma)$ may be expressed as

$$w(\mathbf{S}|\gamma) = \int_{\Gamma_\gamma} w(\mathbf{S}|\mathbf{V})w(\mathbf{V}|\gamma)\, d\mathbf{V} \tag{22.41}$$

where Γ_γ denotes the domain of all $\mathbf{V}$'s that lead to the decision γ. By the argument used previously [cf. Eq. (22.25)], $w(\mathbf{V}|\gamma)$ is constant over the region Γ_γ and zero outside, so that

$$w(\mathbf{V}|\gamma) = \frac{1}{N(\gamma)} \qquad N(\gamma) = \int_{\Gamma_\gamma} d\mathbf{V} \tag{22.42}$$

Thus, Eq. (22.41) becomes

$$w(\mathbf{S}|\gamma) = \frac{M_S(\gamma)}{N(\gamma)} \qquad M_S(\gamma) = \int_{\Gamma_\gamma} w(\mathbf{S}|\mathbf{V})\, d\mathbf{V} \tag{22.43}$$

Differentiating Eq. (22.40) with respect to γ, we obtain the following condition for an information-loss extremum,

$$\int_\Omega w(\mathbf{S}|\mathbf{V}) \left[\frac{M'_S(\gamma)}{M_S(\gamma)} - \frac{N'(\gamma)}{N(\gamma)}\right] d\mathbf{S} = 0 \tag{22.44}$$

where the primes denote differentiation with respect to γ. In view of the definitions of Eqs. (22.42) and (22.43), this states the requirements to be fulfilled by Γ_γ. The optimum (or extremal) rule for obtaining γ from $\mathbf{V}$ must be such that it produces Γ_γ's with the properties implied by Eq. (22.44).

We note immediately that Eq. (22.44) is satisfied if γ is a sufficient statistic, i.e., if $w(\mathbf{S}|\mathbf{V}) = w(\mathbf{S}|\gamma)$, $\mathbf{V} \in \Gamma_\gamma$, for in that case we have

$$M_S(\gamma) = w(\mathbf{S}|\gamma)N(\gamma) \tag{22.45}$$

and the quantity in the brackets in Eq. (22.44) becomes

$$\frac{1}{w(\mathbf{S}|\mathbf{V})} \frac{\partial}{\partial \gamma} w(\mathbf{S}|\gamma) \tag{22.46}$$

which satisfies Eq. (22.44) identically. The sufficiency condition results alternatively, if Eq. (22.40) is minimized directly with respect to unconstrained variation of the density function $w(\mathbf{S}|\gamma)$ (see Sec. 22.2-2). When the distribution $F_n(\mathbf{V}|\mathbf{S})$ does not admit a sufficient statistic, however, Eq. (22.44) gives the condition for an extremum.

Equations (22.42) and (22.43) may also be written somewhat more conveniently as follows:

$$N(\gamma) = \int_\Gamma \delta[\gamma - \gamma_\sigma(\mathbf{V})]\, d\mathbf{V} \tag{22.47}$$

$$M_S(\gamma) = \int_\Gamma w(\mathbf{S}|\mathbf{V})\delta[\gamma - \gamma_\sigma(\mathbf{V})]\, d\mathbf{V} \tag{22.48}$$

22.3-2 Minimax and Minimum Equivocation Extraction. Under certain conditions, it is possible for optimum estimators based on one criterion to be optimum simultaneously for another. We have already seen examples of this for detection in Sec. 22.2-2 and for extraction in Sec. 21.4. Here we shall show that in some instances an extractor which minimizes equivocation is also Minimax with respect to a quadratic cost function (cf. Sec. 21.2-2).

We begin by postulating additive signal and noise, *uniform* a priori signal probabilities [see the footnote following Eq. (21.23*b*) in the first example of Sec. 21.1-3], and a quadratic cost function for this class of extractors. Then the a posteriori probability density $w(\mathbf{S}|\mathbf{V})$ is given by

$$w(\mathbf{S}|\mathbf{V}) = \frac{W_n(\mathbf{V} - \mathbf{S})_N}{\int_{-\infty}^{\infty} W_n(\mathbf{V} - \mathbf{S})_N \, d\mathbf{S}} \tag{22.49}$$

so that

$$w(\mathbf{S}|\mathbf{U} + \boldsymbol{\lambda}) = w(\mathbf{S} - \boldsymbol{\lambda}|\mathbf{U}) \tag{22.50}$$

for arbitrary $\boldsymbol{\lambda}$. We recall now from Sec. 21.2-2 that the optimum system† under these conditions is Minimax from the risk point of view and has the translation property $\gamma_\sigma(\mathbf{V} + \boldsymbol{\lambda}) = \gamma_\sigma(\mathbf{V}) + \boldsymbol{\lambda}$.

When γ is replaced by $\gamma + \boldsymbol{\lambda}$ in Eq. (22.47), the translation property and the change of variable $\mathbf{U} = \mathbf{V} - \boldsymbol{\lambda}$ show that $N(\gamma)$ is a constant, or

$$N'(\gamma) = 0 \tag{22.51}$$

Here, it is assumed that the signal domain Γ is infinite. Similarly, the same sequence of operations on Eq. (22.48) and use of Eq. (22.50) show that $M_S(\gamma + \boldsymbol{\lambda}) = M_{S-\lambda}(\gamma)$ for arbitrary $\boldsymbol{\lambda}$. Consequently,

$$M'_S(\gamma) = -\frac{\partial M_S(\gamma)}{\partial \mathbf{S}} \tag{22.52}$$

so that the condition for an extremum of average information loss [Eq. (22.44)] now becomes

$$\int \cdots \int_{-\infty}^{\infty} w(\mathbf{S}|\mathbf{V}) \frac{1}{M_S(\gamma)} \frac{\partial M_S(\gamma)}{\partial \mathbf{S}} \, d\mathbf{S} = 0 \tag{22.53}$$

and $\gamma = \gamma_E^*$, obtained from Eq. (22.53), is that estimator for which $H(\sigma,\delta)$ is an extremum.

For Eq. (22.53) to be satisfied, an additional symmetry condition on $w(\mathbf{S}|\mathbf{V})$ is needed, in effect, a further condition on the statistics of the

† Note, however, that, when $\mathbf{S} = \mathbf{S}(\boldsymbol{\theta})$ and $\mathbf{S}$ is not *linear* in the random parameters $\boldsymbol{\theta}$, a uniform distribution $w(\boldsymbol{\theta})$ does not lead to a Minimax extractor, nor does the latter possess the translation property above. On the other hand, a case of considerable practical importance where $\boldsymbol{\theta}$ may be uniform *and* $\mathbf{S}(\boldsymbol{\theta})$ linear in $\boldsymbol{\theta}$ occurs for $\boldsymbol{\theta}$ = signal amplitude, as the example of Sec. 21.3-1 indicates, but in general $\mathbf{S}(\boldsymbol{\theta})$ is not a linear function of $\boldsymbol{\theta}$.

signal alphabet (or waveform). For example, if the noise distribution is symmetrical in $\mathbf{S}$ about the maximum unconditional likelihood value $\hat{\mathbf{S}}(\mathbf{V})$ for fixed $\mathbf{V}$, that is, if

$$W_n[\mathbf{V} - (\hat{\mathbf{S}} + \boldsymbol{\lambda})]_N = W_n[\mathbf{V} - (\hat{\mathbf{S}} - \boldsymbol{\lambda})]_N \tag{22.54}$$

for arbitrary $\boldsymbol{\lambda}$, then from Eq. (22.49) it follows that

$$w(\hat{\mathbf{S}} + \boldsymbol{\lambda}|\mathbf{V}) = w(\hat{\mathbf{S}} - \boldsymbol{\lambda}|\mathbf{V}) \tag{22.55}$$

and from Eq. (22.48) that

$$M_{\hat{S}+\lambda}(\boldsymbol{\gamma}) = M_{\hat{S}-\lambda}(\boldsymbol{\gamma}) \tag{22.56}$$

When the substitution $\mathbf{S} = \hat{\mathbf{S}} + \boldsymbol{\lambda}$ is made in Eq. (22.53), and the integration ranges $(-\infty,0)$ and $(0,\infty)$ are considered separately, the relations (22.55) and (22.56) çause the condition (22.53) to be satisfied identically. In addition, it can be shown that the estimator $\boldsymbol{\gamma}_E^*$ for which Eq. (22.53) holds also makes the equivocation a minimum (cf. Prob. 22.5*b*). We conclude, therefore, that *when signal and noise are additive, when the distribution* $W_n(\mathbf{V} - \mathbf{S})_N$ *is symmetrical,* as required by Eq. (22.54), *and when the a priori signal distribution is uniform, then the extractor with Minimax average risk* (for the squared-error cost function) *also minimizes average information loss.*

22.3-3 Remarks on Maximum Likelihood Extractors. With the results of Sec. 21.1 and the above in mind, we can summarize a number of useful general properties of the maximum likelihood estimators as they are used in statistical communication theory.

1. First, as we have just shown, the symmetry condition (22.54) makes the maximum likelihood and Minimax extractors coincide, provided the following conditions hold:
   a. Additive signal and noise.
   b. Uniform a priori signal (or signal parameter) distributions.
   c. Linear dependence of $\mathbf{S}$ on signal parameters $\boldsymbol{\theta}$.
   d. $W_n(\mathbf{V} - \mathbf{S})_N$ symmetric about the unconditional maximum likelihood estimate $\hat{\mathbf{S}}(\mathbf{V})$.
2. In addition, if the (additive) noise belongs to a normal process (not necessarily stationary), the maximum likelihood extractor is a sufficient statistic [Eqs. (21.12*a*) et seq.], which indeed maximizes average information loss, as we know from Eqs. (22.11), (22.45), (22.46).
3. From the relations of Sec. 21.3, we remark finally that the maximum likelihood extractor (or estimator) is also optimum in the following senses:
   a. It minimizes average risk for the simple cost function (Sec. 21.3).
   b. It minimizes average risk for the squared-error cost function where $w(\mathbf{V},\mathbf{S})$ is symmetrical about $\hat{\mathbf{S}}$ for every $\mathbf{V}$ (Sec. 21.3).
   c. From Sec. 22.3-2, it is evident that the maximum likelihood extractor

also minimizes equivocation where the conditions *a* to *d* of (1) above are satisfied, with a quadratic cost function.

We note, also, that under certain conditions, where the additive accompanying noise is normal, the optimum extractor in the circumstances above are also *linear* operators on the received data (cf. Secs. 21.3, 21.4). In general, however, as in detection, optimum extraction is a nonlinear process, whose nonlinear character is quite independent of the criterion chosen [unless, of course, a constraint of linearity is imposed upon the system, for example, as in the Wiener-Kolmogoroff theory of prediction (Secs. 16.2, 21.4)].

PROBLEM

22.5 (*a*) Carry out the details of the steps of Eqs. (22.51) to (22.56), and show that Eqs. (22.55), (22.56) indeed satisfy the extremum condition (22.53) identically.

(*b*) Prove that $\gamma = \gamma_E^*$, obtained from Eq. (22.53), makes the equivocation a minimum.

REFERENCES

1. Middleton, D., and D. Van Meter: Detection and Extraction of Signals in Noise from the Point of View of Statistical Decision Theory, Part II, *J. Soc. Ind. Appl. Math.*, **4:** 86 (1956).
2. Van Meter, D.: Optimum Decision Systems for the Reception of Signals in Noise, Part II, *Cruft Lab. Tech. Rept.* 216, October, 1955.
3. Cramér, H.: "Mathematical Methods of Statistics," p. 564, Princeton University Press, Princeton, N.J., 1946.
4. Fisher, R. A.: "Contributions to Mathematical Statistics," Wiley, New York, 1950.
5. Halmös, P. R., and L. J. Savage: Applications of the Radon-Nikodym Theorem to the Theory of Sufficient Statistics, *Am. Math. Stat.*, **20:** 225 (1949).
6. Fisher, R. A.: Theory of Statistical Estimation, *Proc. Cambridge Phil. Soc.*, **22:** 700 (1925).
7. Pitman, E. J. G.: Sufficient Statistics and Intrinsic Accuracy, *Proc. Cambridge Phil. Soc.*, **32:** 567 (1936).
8. Hodges, J. L., Jr., and E. L. Lehmann: Some Applications of the Cramér-Rao Inequality, *Proc. 2d Berkeley Symposium on Mathematical Statistics and Probability*, 1950, p. 13.
9. Doob, J. L.: Statistical Estimation, *Trans. Am. Math. Soc.*, **39:** 410 (1936).
10. Shannon, C. E.: Mathematical Theory of Communication, *Bell System Tech. J.*, **27:** 379, 623 (1948).
11. Kullback, S., and R. A. Leibler: On Information and Sufficiency, *Am. Math. Stat.*, **22:** 79 (1951).
12. Bartlett, M. S.: The Statistical Approach to the Analysis of Time Series, *Proc. London Symposium on Inform. Theory*, September, 1950, p. 81.
13. Middleton, D.: Information Loss Attending the Decision Operation in Detection, *J. Appl. Phys.*, **25**(L): 127 (1954).
14. Siebert, W. M., and R. M. Lerner: Detection of Signals in Noise with Minimum Information Loss, *MIT Lincoln Lab. Group Rept.* 32-8, Apr. 1, 1954.

CHAPTER 23

GENERALIZATIONS AND EXTENSIONS

In Chaps. 19 and 20, we have outlined a theory of optimum binary detection where the final output of our optimum receiver is a definite decision as to the presence or absence of a signal. Following this, in Chap. 21 we have applied decision-theory methods to related problems of signal parameter and waveform estimation, when it is known that a representation of a signal process is present in noise and a quantitative estimate of one or more of the information-bearing features of the signal process is desired.

For the detection situation, so far, only binary cases have been considered, where at most two hypothesis states are in question: signal and noise vs. noise (cf. Sec. 19.1), or a signal of one class and noise vs. a signal of another class and noise (cf. Sec. 20.4-5). Here we shall begin by extending the analysis to multiple-alternative situations consisting of an arbitrary number of (disjoint) hypothesis states, including the case of decision rejection (cf. Sec. 23.1-2). In a corresponding way, the extension of the estimation theory outlined earlier, in Chap. 21, now involves signal extraction using multiple data sources, where the quantity to be estimated is generated simultaneously in some fashion by each source, and we are to combine these data so as to obtain best estimates for the particular cost functions chosen (Sec. 23.1-3).

Our next generalization of earlier results (cf. Sec. 23.2) is described by "cost coding," which is the process of joint optimization of both the transmission and reception operations by suitable choice of signal waveforms in order to minimize the Bayes risk of decision. The procedure is illustrated with a few examples, mainly for the detection process. The relations of these and the preceding methods of signal detection and extraction to game theory are then briefly touched upon in Sec. 23.3. Section 23.4 gives a critique of decision theory as it is applied to the general communication problems discussed in Part 4. Our study concludes in Sec. 23.5 with a brief description of some important, as yet unsolved problems suggested by this earlier work.

23.1 *Multiple-alternative Detection and Estimation*†

As we have noted above, the detection and estimation processes considered in earlier chapters are capable of generalization within the framework

† This section is based in part on Middleton and Van Meter.[1]

of decision theory. We illustrate this by considering first an important extension of the binary detection process.

23.1-1 Detection. Although the binary systems described in Chaps. 19, 20 are quite common in practice, it is frequently necessary to consider situations involving more than two alternatives: instead of having to distinguish one signal (and noise) out of a given class of signals from noise alone, we may be required to determine which one of several possible signals is present. For example, in communication applications a variety of different signal waveforms may be used as a signal alphabet in the course of transmission, and the receiver is asked to determine which particular waveform is actually sent in any given signal period $(0,T)$. A second example arises in radar, where it is desired to distinguish between several targets that may or may not appear simultaneously during an observation interval. In fact, whenever we are required to discriminate between more than two hypothesis classes, we have an example of multiple-alternative detection. Here we shall extend the binary theory of Chaps. 19, 20 to those multiple-alternative situations where only a single signal can appear (with noise) on any one observation—i.e., to the case of disjoint, or nonoverlapping, hypothesis classes. The more involved situations of joint, or overlapping, hypothesis classes are not considered here.

(1) Formulation. Let us begin with a brief formulation of the decision model. The criterion of excellence, as before (cf. Chap. 19), is the minimization of average risk, or cost. The resulting system (i.e., the indicated operations on the received data) that achieves this is the corresponding optimum detection system. Our procedure, accordingly, is first to construct the average risk function [Eq. (18.7*b*)] and then to minimize it by a suitable choice of decision rule (e.g., system structure). From Eq. (18.7*b*), the average risk can be written

$$R(\sigma,\delta) = \int_\Omega \sigma(\mathbf{S})\, \mathbf{dS} \int_\Gamma F_n(\mathbf{V}|\mathbf{S})\, \mathbf{dV} \int_\Delta \mathbf{d}\boldsymbol{\gamma}\, C(\mathbf{S},\boldsymbol{\gamma})\delta(\boldsymbol{\gamma}|\mathbf{V}) \tag{23.1}$$

where now specifically:

1. $\sigma(\mathbf{S})$ is the a priori probability (density) governing all possible signals $\mathbf{S}$, where explicitly

$$\sigma(\mathbf{S}) = \sum_{k=0}^{M} p_k w_k(\mathbf{S})\delta^{(k)}(\mathbf{S}) \qquad \sum_{k=0}^{M} p_k = 1 \tag{23.2}$$

with p_k $(k = 0, \ldots, M)$ the a priori probabilities that a signal of class (or type) k is present on any one observation. Here, w_k is the d.d. of $\mathbf{S}^{(k)}$ [or of the parameters† $\boldsymbol{\theta}$ of $\mathbf{S}^{(k)}(\boldsymbol{\theta})$], where $\mathbf{S}^{(k)}$ represents the signals of

† In such cases we replace $\sigma(\mathbf{S})$ by $\sigma(\boldsymbol{\theta}) = \sum_{k=0}^{M} p_k w_k(\boldsymbol{\theta})\delta^{(k)}(\mathbf{S})$ and Ω by Ω_θ for the region of integration [cf. Eqs. (18.10), (18.11)].

class k. The class $k = 0$ is a class of possible null signals, or "noise alone," and $\delta^{(k)}(\mathbf{S}) = 1$, when $\mathbf{S} = \mathbf{S}^{(k)}$, with $\delta^{(k)}(\mathbf{S}) = 0$, $\mathbf{S} \neq \mathbf{S}^{(k)}$.

2. $\mathbf{S}^{(k)} = [S^{(k)}(t_1), \ldots, S^{(k)}(t_n)]$ is the kth signal vector, whose components are the sampled values of $S^{(k)}(t)$ at times t_m $(m = 1, \ldots, n)$ in the observation period $(0,T)$. Similarly, we have $\mathbf{V} = [V_1, \ldots, V_n]$, the received-data vector, with $V_m = V(t_m)$, etc.

3. $F_n(\mathbf{V}|\mathbf{S})$, $F_n[\mathbf{V}|\mathbf{S}(\theta)]$ are the conditional d.d.'s of $\mathbf{V}$, given $\mathbf{S}$.

4. $\boldsymbol{\gamma} = [\gamma_0, \ldots, \gamma_M]$ is a set of $M + 1$ decisions. γ_j is a decision that a signal of class j is present vs. all other possibilities.

As before, $C(\mathbf{S},\boldsymbol{\gamma})$ is a cost function [cf. Eq. (18.4)] which assigns to the various $\mathbf{S}$, for one or more possible decisions γ_j $(j = 0, \ldots, M)$ about the $\mathbf{S}$, some appropriate preassigned constant costs. These costs are assigned to signal classes: no cost distinction between different signals of the same class or type is made. Also, as before (cf. Sec. 18.3-2), $\delta(\boldsymbol{\gamma}|\mathbf{V})$ is a decision rule, or probability.† The decisions $\boldsymbol{\gamma}$ are governed by the further condition

$$\sum_{j=0}^{M} \delta(\gamma_j|\mathbf{V}) = 1 \tag{23.3}$$

which is simply a statement of the fact that a definite decision must be made.

Constant costs are next assigned to the possible outcomes, according to our usual procedure [cf. Eqs. (19.5) in the binary case]. Here we set $C_j^{(k)}$ = cost of deciding that a signal of class j is present when actually a signal of class k occurs.‡ Thus, if $j \neq k$, $C_j^{(k)}$ is the cost of an incorrect decision, while if $j = k$, $C_j^{(j)}$ represents the cost of a correct decision. In all cases, we have

$$C_j^{(k)} \Big|_{j \neq k} > C_j^{(j)} \tag{23.4}$$

since by definition an "error" must be more "expensive" than a correct choice.

Let us consider the costs $C_j^{(k)}$ in more detail. For "successful," or correct, decisions we have specifically

$C(\mathbf{S}^{(0)};\gamma_0) = C_0^{(0)}$; noise alone is correctly detected
$C(\mathbf{S}^{(k)};\gamma_k) = C_k^{(k)}$; a signal of class k is correctly detected $(k = 1, \ldots, M)$ (23.5a)

The costs preassigned to "failures," or incorrect decisions, are represented by

$C(\mathbf{S}^{(0)};\gamma_j) = C_j^{(0)}$; a signal of class j $(j = 1, \ldots, M)$ is incorrectly decided when noise alone occurs

† With $0 \leq \delta(\gamma_j|\mathbf{V}) \leq 1$ $(j = 0, \ldots, M)$ in the case of detection; for estimation, $\delta(\boldsymbol{\gamma}|\mathbf{V})$ is a probability density (Secs. 18.3-2, 18.3-3).

‡ We adopt the convention that the superscript on the cost $C_j^{(k)}$ refers to the true, or actual, state, while the subscript refers to the decision made.

$C(\mathbf{S}^{(k)};\gamma_j) = C_j^{(k)}$; a signal of class j ($j \neq k$; $j = 0, \ldots, M$; $k = 1, \ldots, M$) is incorrectly detected when a signal of class k occurs (23.5b)

Setting $p_0 \equiv q$, we readily obtain the average risk [Eq. (23.1)] after integrating over decision space Δ. The result is

$$R(\sigma,\delta) = \int_\Gamma \Bigg(\Bigg[C_0^{(0)}\delta(\gamma_0|V) + \sum_{k=1}^{M} C_k^{(0)}\delta(\gamma_k|\mathbf{V}) \Bigg] qF_n(\mathbf{V}|0) + \Bigg\{ \sum_{k=1}^{M} p_k \Bigg[C_k^{(k)}\delta(\gamma_k|\mathbf{V}) + \sum_{j=0}^{M}{}' C_j^{(k)}\delta(\gamma_j|\mathbf{V}) \Bigg] \langle F_n(\mathbf{V}|\mathbf{S}^{(k)})\rangle_k \Bigg\} \Bigg) \, d\mathbf{V} \tag{23.6}$$

subject to Eqs. (23.3), (23.4), where $\langle \quad \rangle_k$ denotes the statistical average over $\mathbf{S}^{(k)}$ (or over the random parameters of $\mathbf{S}^{(k)}$), and the prime on the summation signifies that $j \neq k$. Note that, when $M = 1$ (the binary case), Eq. (23.6) reduces at once to Eq. (19.6), with obvious changes of notation.

(2) **Minimization of the average risk.** At this point, it is convenient to rearrange Eq. (23.6) with the help of Eq. (23.3), by collecting coefficients of $\delta(\gamma_k|\mathbf{V})$. First, let us introduce the expressions

$$\begin{aligned} &\lambda_l^{(0)} \equiv C_l^{(0)} - C_0^{(0)} && l = 1, \ldots, M \\ &\lambda_k^{(k)} \equiv C_k^{(k)} - C_0^{(k)} && \\ &\lambda_k^{(l)} \equiv C_k^{(l)} - C_0^{(l)} && l \neq k \ (k, l = 1, \ldots, M) \\ &\lambda_k^{(l)} \gtreqless 0 && k \neq l \quad [\text{with } \lambda_k^{(0)} > 0 \ (k > 0),\ \lambda_k^{(k)} < 0 \ (k \neq 0)] \end{aligned} \tag{23.7}$$

and write

$$A_k(\mathbf{V}) \equiv \lambda_k^{(0)} + \sum_{l=1}^{M} \lambda_k^{(l)}\Lambda_l(\mathbf{V}) \tag{23.8}$$

where, as before [cf. Eq. (19.14)], the $\Lambda_l(\mathbf{V})$ are generalized likelihood ratios

$$\Lambda_l(\mathbf{V}) \equiv p_l\langle F_n(\mathbf{V}|\mathbf{S}^{(l)})\rangle_l / qF_n(\mathbf{V}|0) \tag{23.9}$$

For additive signals and noise, this reduces to the simpler relation

$$\Lambda_l(V) = p_l\langle W_n(\mathbf{V} - \mathbf{S}^{(l)})_N\rangle_l / qW_n(\mathbf{V})_N \tag{23.9a}$$

[cf. Eq. (19.41)], where, as before, $W_n(\mathbf{V})_N$ is the joint nth-order d.d. of the background noise.

After some algebra, we find that the average risk [Eq. (23.6)] may now be rewritten

$$R_M(\sigma,\delta) = \mathcal{R}_{0M} + \mathcal{R}_M \tag{23.10a}$$

$$\mathcal{R}_{0M} \equiv qC_0^{(0)} + \sum_{k=1}^{M} p_k C_0^{(k)} \ (> 0) \tag{23.10b}$$

$$\mathcal{R}_M \equiv \int_\Gamma \left[\sum_{k=1}^{M} \delta(\gamma_k|\mathbf{V}) A_k(\mathbf{V}) \right] qF_n(\mathbf{V}|0)\, d\mathbf{V} \tag{23.10c}$$

Here $\mathcal{R}_{0M}$ is simply the expected cost of calling every signal (including noise alone) "noise," while $\mathcal{R}_M$ is the portion of the average risk which can be adjusted by choice of the decision rules. Again, the precise form of the system is embodied in the likelihood ratios, while the process of detection is determined by our choice of the δ's and the corresponding regions in data space Γ which are nonoverlapping, since the various hypotheses here are mutually exclusive.

Now, in order to optimize the detection operation we minimize the average risk by proper selection of the $\delta(\gamma_k|\mathbf{V})$. In essence, this is the problem of finding the boundaries of the critical regions for the $\Lambda_k(\mathbf{V})$. The argument for minimization is readily given. Since $qF_n(\mathbf{V}|0) \geq 0$ everywhere in Γ, the average risk [Eq. (23.10c)] is least where for each value of $\mathbf{V}$ we choose δ to minimize $\sum \delta_k A_k$. The procedure, accordingly, is to examine all A's for the given $\mathbf{V}$, selecting the one (A_k) that is algebraically least† *and then for this same* $\mathbf{V}$ *choosing* $\delta(\gamma_k|\mathbf{V}) = 1$, $\delta(\gamma_j|\mathbf{V}) = 0$ (all $j \neq k$). We repeat this for all $\mathbf{V}$ in Γ, to obtain finally a set of conditions on the A_k's ($k = 1, \ldots, M$). Observe from the form of $\mathcal{R}_M$, where the δ's appear linearly, and from the method of minimization itself, that δ is automatically a nonrandomized decision rule, that is, $\delta = 1$ or 0 only [cf. Eqs. (19.13) et seq.]. Since the $A_k(\mathbf{V})$ may contain negative parts [cf. Eqs. (23.7)], we find that, to minimize the average risk of Eq. (23.10a) and make a decision γ_k (signal of type k present in noise) on the basis of received data $\mathbf{V}$, the explicit conditions on the A_k are that the data $\mathbf{V}$ satisfy the following linear *inequalities*,

$$A_k(\mathbf{V}) \leq A_j(\mathbf{V}) \quad \text{or} \quad \lambda_k^{(0)} + \sum_{l=1}^{M} \lambda_k^{(l)} \Lambda_l(\mathbf{V}) \leq \lambda_j^{(0)} + \sum_{l=1}^{M} \lambda_j^{(l)} \Lambda_l(\mathbf{V})$$

and (23.11a)

$$A_k(\mathbf{V}) \leq 0 \quad \text{or} \quad \lambda_k^{(0)} + \sum_{l=1}^{M} \lambda_k^{(l)} \Lambda_l(\mathbf{V}) \leq 0$$

$$\text{all } j \neq k \ (j = 1, \ldots, M)$$

for each decision γ_k ($k = 1, \ldots, M$) in turn, with

† Subject (for the moment) to the assumption that for this given $\mathbf{V}$ there actually exists an $A_k(\mathbf{V})$ which is algebraically less than all other $A_j(\mathbf{V})$ ($j \neq k$).

$$\delta(\gamma_k|\mathbf{V}) = 1 \qquad \delta(\gamma_j|\mathbf{V}) = 0 \qquad \text{all } j \neq k, j = 0, \ldots, M \tag{23.11b}$$

For the remaining case of noise alone ($k = 0$), we have the conditions

$$\Lambda_k(\mathbf{V}) \geq 0 \qquad \text{or} \qquad \lambda_k^{(0)} + \sum_{l=1}^{M} \lambda_k^{(l)}\Lambda_l(\mathbf{V}) \geq 0 \qquad \text{each } k = 1, \ldots, M \tag{23.12a}$$

and
$$\delta(\gamma_0|\mathbf{V}) = 1 \qquad \delta(\gamma_j|\mathbf{V}) = 0 \qquad j = 1, \ldots, M \tag{23.12b}$$

If now we regard the Λ's as independent variables, we can at once give a direct geometric interpretation of the mutually exclusive sets of conditions (23.11), (23.12). Writing (L_k) as the value of the quantity $\lambda_k^{(0)} + \sum_{l=1}^{m} \lambda_k^{(l)}\Lambda_l(\mathbf{V})$ and L_k for the hypersurface† $(L_k) = 0$, we observe that, in conjunction with the hypersurfaces forming the boundaries of the first "2^{M+1}-tant,"‡ the *equalities* in the conditions (23.11) or (23.12) give the boundaries of a *closed* region within which lie all values of $\Lambda_k(\mathbf{V})$ associated with the decision γ_k. Each closed region is distinct from every other, and the M planar hypersurfaces which form its boundaries are then from Eqs. (23.11), (23.12) specified by

$$L_k = 0 \qquad L_k - L_j = 0 \qquad \text{all } j = 1, \ldots M \ (j \neq k)$$
$$(\text{for each } \gamma_k, k = 1, \ldots, M) \tag{23.13a}$$
$$L_k = 0 \qquad \text{all } k = 1, \ldots, M \ (\text{for } \gamma_0) \tag{23.13b}$$

Solving the M linear equations (23.11), (23.12), or Eqs. (23.13*a*), (23.13*b*), one can show in straightforward fashion that the various (distinct) M planar hypersurfaces determining the boundaries of each region all intersect at a point $\mathbf{K} = (K_0^{(1)}, \ldots, K_0^{(M)})$, where now the K's represent a set of M thresholds $\Lambda_k(\mathbf{V}') = K_0^{(k)}$ $(k = 1, \ldots, M)$, in which the $\mathbf{V}'$ are all values of $\mathbf{V}$ satisfying this relation. These thresholds depend explicitly only on the preassigned costs, i.e., only on the λ's of Eqs. (23.11), etc., and the requirement (23.4) ensures that the point $\mathbf{K}$ lies in the first 2^{M+1}-tant, i.e., all $K^{(k)} \geq 0$.

A simpler variant of Eqs. (23.11), also of practical interest, arises when the problem becomes that of testing for the presence of one signal of class k in noise against (any one of the) other possible nonzero signals in *noise.*

†For $M \geq 4$, $L_k = 0$ is a plane hypersurface (in Λ_1-, . . . , Λ_M-space); for $M = 1$, 2, $L_k = 0$ represents a straight line in two dimensions (of Λ_1-, Λ_2-space), while, for $M = 3$, $L_k = 0$ represents a plane surface (in Λ_1-, Λ_2-, Λ_3-space).

‡ For example, if $M = 1$, the first "2^2-tant" = first "quadrant"; for $M = 2$, the first "2^3-tant" = first "octant," etc. Since likelihood ratios Λ_k can never be negative, values of $\Lambda_k(\mathbf{V})$ must always lie in the first "2^{M+1}-tant."

The case of noise alone is here eliminated.† Under these circumstances, the costs $C_0^{(l)}$, $C_0^{(0)}$, $C_l^{(0)}$ drop out, and the λ's of Eqs. (23.7) et seq. are simply $\lambda_k^{(l)} = C_k^{(l)}$ $(k, l = 1, \ldots, M) \geq 0$. Consequently, if $C_k^{(l)} \geq 0$, $C_l^{(l)} \geq 0$, the $A_k(\mathbf{V})$ can now never be negative. Minimization of the average risk then gives only the first set of inequalities in Eqs. (23.11a), with $\lambda_k^{(0)} = \lambda_j^{(0)} = 0$, so that we may write, for the decision γ_k, the modified conditions

$$\sum_{l=1}^{M} \Lambda_l(\mathbf{V}) C_k^{(l)} \leq \sum_{l=1}^{M} \Lambda_l(\mathbf{V}) C_j^{(l)} \qquad j \neq k;\ \text{all } j = 1, \ldots, M \quad (23.14)$$

repeated for each $k = 1, \ldots, M$ in turn. The point $\mathbf{K}$ is zero now, and all decision regions (for the Λ's) have their apexes at the origin. The bounding hyperplanes all intersect at $\mathbf{K} = \mathbf{0}$, and the equations of the boundaries of the kth region are simply

$$L_k - L_j = 0 \qquad \text{all } j = 1, \ldots, M\ (j \neq k) \quad (\text{for each } \gamma_k,\ k = 1, \ldots, M) \quad (23.15)$$

To summarize, we see that the optimum $(M + 1)$-ary (or M-ary) detector consists of a computer which evaluates the $\Lambda_k(\mathbf{V})$ for a given set of data $\mathbf{V}$ over the observation period $(0,T)$, computes the various $A_k(\mathbf{V})$, and then inserts the results into the inequalities (23.11), (23.12), or (23.14), finally making the decision $\gamma_{(k)}$ associated with the one set of inequalities that is satisfied.‡ One possibility is sketched in Fig. 23.1, where a succession of intermediate binary decisions is employed to yield ultimately γ_k.

(**3**) **Examples.** (*a*) *Binary detection.* The simplest and most familiar case of Eqs. (23.11), (23.12) arises when we have to distinguish $\mathbf{S}^{(1)} + \mathbf{N}$ versus $\mathbf{N}$ alone, so that $M = 1$. With the help of Eqs. (23.7) in Eqs. (23.11), (23.12), we easily find that we

$$\left.\begin{aligned} &\textit{Decide } \gamma_0\text{: } \mathbf{N}, \text{ if } \Lambda_1(\mathbf{V}) < \\ &\textit{Decide } \gamma_1\text{: } \mathbf{S}^{(1)} + \mathbf{N}, \text{ if } \Lambda_1(\mathbf{V}) > \end{aligned}\right\} K_0^{(1)} \quad (23.16a)$$

$$K_0^{(1)} \equiv -\frac{\lambda_1^{(0)}}{\lambda_1^{(1)}} = \frac{C_1^{(0)} - C_0^{(0)}}{C_0^{(1)} - C_1^{(1)}} = \frac{C_\alpha - C_{1-\alpha}}{C_\beta - C_{1-\beta}} \quad (23.16b)$$

with the threshold $K_0^{(1)}$ a function of the costs only. [C_α, $C_{1-\alpha}$, etc., are expressed in the earlier notation of Eqs. (19.5).]

The somewhat more general binary problem of distinguishing $\mathbf{S}^{(a)} + \mathbf{N}$ against $\mathbf{S}^{(b)} + \mathbf{N}$, with a, b any single integers in the range $(1 \leq a, b \leq M)$

† This is equivalent to setting $\delta(\gamma_0|\mathbf{V}) = 0 = 1 - \sum_{k=1}^{M} \delta(\gamma_k|\mathbf{V})$ [cf. Eq. (23.3)] in Eq. (23.6) and proceeding as above. $qF_n(\mathbf{V}|0)$ in Λ_l is only a normalizing factor here.

‡ Of course, this is not a unique way of setting up an actual computing scheme.

$(a \neq b,\ M = 2)$, is readily treated. From Eq. (23.14), we write for the decision process

$$\begin{aligned} &\textit{Decide } \gamma_a\text{: } \mathbf{S}^{(a)} + \mathbf{N}, \text{ if } \Lambda_a(\mathbf{V}) > \Lambda_b(\mathbf{V})K_a{}^{(b)} \\ &\textit{Decide } \gamma_b\text{: } \mathbf{S}^{(b)} + \mathbf{N}, \text{ if } \Lambda_a(\mathbf{V}) < \Lambda_b(\mathbf{V})K_a{}^{(b)} \\ &\text{where} \qquad K_a{}^{(b)} = \frac{C_a{}^{(b)} - C_b{}^{(b)}}{C_b{}^{(a)} - C_a{}^{(a)}} > 0 \end{aligned} \tag{23.17}$$

which can also be expressed alternatively in terms of a single likelihood ratio

$$\Lambda_a{}^{(b)}(\mathbf{V}) \equiv p_b\langle F_n(\mathbf{V}|\mathbf{S}^{(b)})\rangle_b / p_a\langle F_n(\mathbf{V}|\mathbf{S}^{(a)})\rangle_a \tag{23.18}$$

or its reciprocal. Typical decision regions are shown in Fig. 23.2 (or in Fig. 23.3 if we replace Λ_1 by Λ_a, Λ_2 by Λ_b and set the point $\mathbf{K} = 0$).

(*b*) *Ternary detection.* In this second example, we assume that noise alone is one of the three possible alternatives, so that $M = 2$. Then, from Eqs. (23.11), (23.12), the decision process is found at once (with $k = 1, 2;\ l = 1, 2$). The two thresholds $K_0{}^{(1)}$, $K_0{}^{(2)}$ are

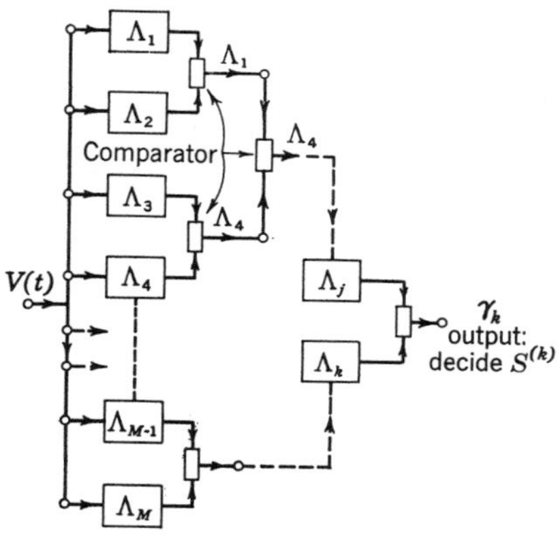

FIG. 23.1. Schematic diagram of the decision process γ_k, for Eqs. (23.11), (23.12), or (23.14).

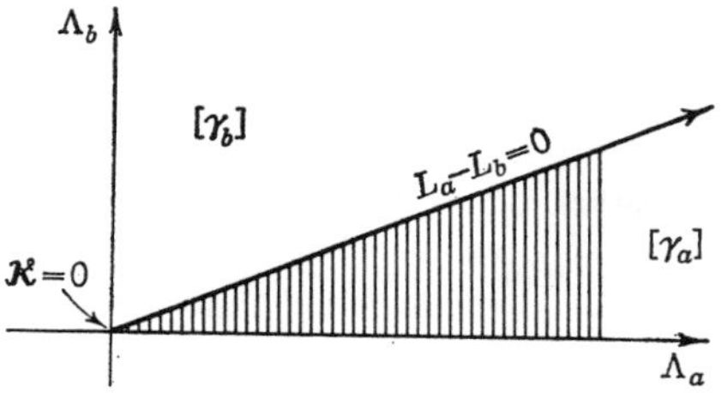

FIG. 23.2. Region of decision for the binary case $(M = 2)$: $\mathbf{S}^{(a)} + \mathbf{N}$ versus $\mathbf{S}^{(b)} + \mathbf{N}$.

$$\begin{aligned} K_0{}^{(1)} &= \frac{\lambda_2{}^{(0)}\lambda_1{}^{(2)} - \lambda_1{}^{(0)}\lambda_2{}^{(2)}}{\Delta} \\ K_0{}^{(2)} &= \frac{\lambda_1{}^{(0)}\lambda_2{}^{(1)} - \lambda_1{}^{(1)}\lambda_2{}^{(0)}}{\Delta} \\ \Delta &= \lambda_1{}^{(1)}\lambda_2{}^{(2)} - \lambda_1{}^{(2)}\lambda_2{}^{(1)} \end{aligned} \tag{23.19}$$

and we decide

$$\begin{aligned} &\gamma_0\text{:}\mathbf{N}, \text{ when } \lambda_1{}^{(0)} + \lambda_1{}^{(1)}\Lambda_1 + \lambda_1{}^{(2)}\Lambda_2 > 0 \\ &\qquad\qquad \lambda_2{}^{(0)} + \lambda_2{}^{(1)}\Lambda_1 + \lambda_2{}^{(2)}\Lambda_2 > 0 \end{aligned} \tag{23.20a}$$

$$\begin{aligned} &\gamma_1\text{:}\mathbf{S}^{(1)} + \mathbf{N}, \text{ when } \lambda_1{}^{(0)} + \lambda_1{}^{(1)}\Lambda_1 + \lambda_1{}^{(2)}\Lambda_2 \\ &\qquad\qquad < \lambda_2{}^{(0)} + \lambda_2{}^{(1)}\Lambda_1 + \lambda_2{}^{(2)}\Lambda_2 \\ &\qquad \lambda_1{}^{(0)} + \lambda_1{}^{(1)}\Lambda_1 + \lambda_1{}^{(2)}\Lambda_2 < 0 \end{aligned} \tag{23.20b}$$

$$\gamma_2:\mathbf{S}^{(2)} + \mathbf{N}, \text{ when } \lambda_2^{(0)} + \lambda_2^{(1)}\Lambda_1 + \lambda_2^{(2)}\Lambda_2 < \lambda_1^{(0)} + \lambda_1^{(1)}\Lambda_1 + \lambda_1^{(2)}\Lambda_2$$
$$\lambda_2^{(0)} + \lambda_2^{(1)}\Lambda_1 + \lambda_2^{(2)}\Lambda_2 < 0 \qquad (23.20c)$$

The boundaries of the various decision regions are easily determined from Eqs. (23.13) and the Λ_1, Λ_2 axes bounding the first quadrant. A typical case is illustrated in Fig. 23.3.

The decision process is particularly simple when the case of noise alone is

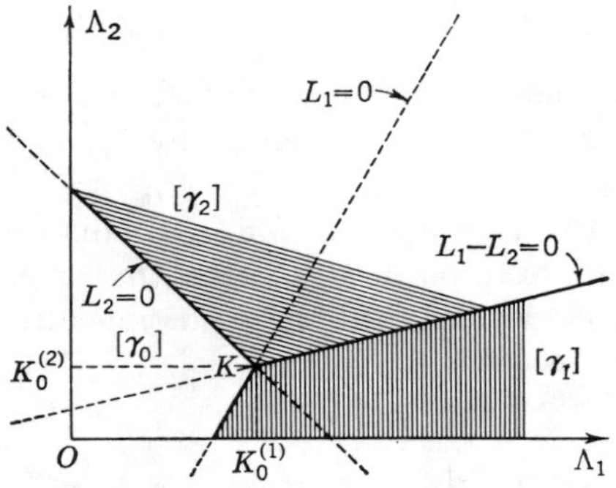

FIG. 23.3. Regions of decision for ternary detection ($M = 2$): $\mathbf{N}$ versus $\mathbf{S}^{(1)} + \mathbf{N}$ versus $\mathbf{S}^{(2)} + \mathbf{N}$; $\lambda_1^{(2)} > 0$; $\lambda_2^{(1)} < 0$.

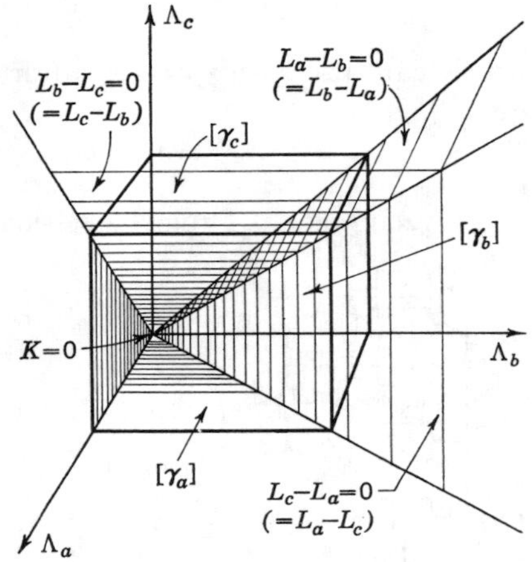

FIG. 23.4. Regions of decision for ternary detection ($M = 3$): $\mathbf{S}^{(a)} + \mathbf{N}$ versus $\mathbf{S}^{(b)} + \mathbf{N}$ versus $\mathbf{S}^{(c)} + \mathbf{N}$; constant cost C_0 of failure, zero cost of success.

removed and one of three possible combinations of signal and noise can now occur, for example, $\mathbf{S}^{(a)} + \mathbf{N}$, $\mathbf{S}^{(b)} + \mathbf{N}$, $\mathbf{S}^{(c)} + \mathbf{N}$ ($a \neq b \neq c$; $1 \leq a, b, c \leq M$; $M = 3$). From Eq. (23.14), we find the decision process to be

Decide $\gamma_a:\mathbf{S}^{(a)} + \mathbf{N}$, when $\Lambda_a > \Lambda_b K_{ab}{}^{(b)} + \Lambda_c K_{ab}{}^{(c)}$

$$\Lambda_a > \Lambda_b K_{ac}{}^{(b)} + \Lambda_c K_{ac}{}^{(c)} \qquad (23.21a)$$

Decide $\gamma_b:\mathbf{S}^{(b)} + \mathbf{N}$, according to Eq. (23.21a), replacing a by b and b by a therein (23.21b)

Decide $\gamma_c:\mathbf{S}^{(c)} + \mathbf{N}$, according to Eq. (23.21a), letting $a \to c$, $b \to a$, $c \to b$ therein (23.21c)

where the thresholds $K_{ab}{}^{(b)}$, etc., are specifically

$$K_{ab}{}^{(b)} \equiv \frac{C_a{}^{(b)} - C_b{}^{(b)}}{C_b{}^{(a)} - C_a{}^{(a)}} \qquad K_{ab}{}^{(c)} \equiv \frac{C_a{}^{(c)} - C_b{}^{(c)}}{C_b{}^{(a)} - C_a{}^{(a)}}$$
$$K_{ac}{}^{(b)} \equiv \frac{C_a{}^{(b)} - C_c{}^{(b)}}{C_c{}^{(a)} - C_a{}^{(a)}} \qquad K_{ac}{}^{(c)} \equiv \frac{C_a{}^{(c)} - C_c{}^{(c)}}{C_c{}^{(a)} - C_a{}^{(a)}} \qquad (23.22)$$

[The thresholds for γ_b, γ_c are found by interchanging a, b, c according to Eqs. (23.21b), (23.21c), respectively.] The bounding surfaces follow from Eq. (23.15), and the planes defining the first *octant* (in Λ_a-, Λ_b-, Λ_c-space) are shown for a typical case [in (*c*) below] in Fig. 23.4. A variety of different regions are clearly possible in these ternary cases, depending on our choice of the preassigned costs, subject to Eq. (23.4), of course.

(*c*) *Simple* $(M+1)$*-ary detection.* A particular case of considerable interest and simplicity [as far as the structure of the detector, e.g., Eqs. (23.11*a*), (23.11*b*), or (23.14), is concerned] occurs when we assign the *same* constant costs $C_0\ (>0)$ to all types of "failure" and zero costs to all types of "success." Then Eqs. (23.7) become $\lambda_k^{(0)} = C_0$, $\lambda_k^{(l)} = 0\ (l \neq k)$, $\lambda_k^{(k)} = -C_0$, and the decision process [Eqs. (23.11)] is governed simply by the inequalities:

$$\text{Decide } \gamma_k, \text{ if } \begin{aligned} \Lambda_k(\mathbf{V}) &\geq \Lambda_j(\mathbf{V}) \qquad \text{all } j \neq k\ (j = 1, \ldots, M) \\ \Lambda_k(\mathbf{V}) &\geq 1 \end{aligned} \tag{23.23}$$

for each k $(= 1, \ldots, M)$ in turn. For noise alone, Eq. (23.12*a*) is simply

$$\text{Decide } \gamma_0, \text{ if } \Lambda_k(\mathbf{V}) \leq 1 \qquad \text{all } k = 1, \ldots, M \tag{23.24}$$

The boundaries of the decision regions are obtained as before from Eqs. (23.13*a*), (23.13*b*) in terms of the equalities above. For M-ary detection, where the case "noise alone" is removed, the first line of Eq. (23.23) provides the desired inequalities, and the boundaries of the decision regions again follow from the indicated equalities [see Eq. (23.15) and Fig. 23.4].

The actual structure of these optimum multiple-alternative systems depends in each case on the explicit form of $\langle F_n(\mathbf{V}|\mathbf{S}^{(k)})\rangle_k$, as in the binary theory. Also as in the binary theory, we may resolve the threshold structure for additive signals and (normal) noise into a sequence of realizable linear and nonlinear elements: for example, when there is sample uncertainty, into the Bayes matched filters, zero-memory nonlinear rectifiers, and ideal integrators (cf. Secs. 20.1-2, 20.1-3, 20.2-2, 20.2-3). This yields $\log \Lambda_l$ $(l = 1, \ldots, M)$, from which in turn the Λ_l may be found to be combined according to Eqs. (23.11), (23.12), or (23.14) for an ultimate decision γ_k (cf. Fig. 23.1). Similar remarks apply in the instances of sample certainty with the structure appropriate to these cases (cf. Secs. 20.1-1, 20.2-1). The logarithmic form of the characteristic may also be used to advantage here in the special situation of the example in (*c*) [and, of course, for the binary cases, the example in (*a*)]. Instead of Eqs. (23.23), (23.24), we can write for this particular cost assignment

$$\text{Decide } \gamma_k, \text{ if } \begin{aligned} \log \Lambda_k(\mathbf{V}) &\geq \log \Lambda_j(V) \\ &\qquad \text{all } j \neq k\ (j = 1, \ldots, M) \\ \log \Lambda_k(\mathbf{V}) &\geq 0 \end{aligned} \tag{23.25}$$

for each k $(= 1, \ldots, M)$ in turn. For noise alone, we have

$$\text{Decide } \gamma_0, \text{ if } \log \Lambda_k(\mathbf{V}) \leq 0 \qquad \text{all } k = 1, \ldots, M \tag{23.26}$$

The boundaries of the decision regions follow from Eqs. (23.13), subject now to the additional logarithmic transformation. When the alternative "noise alone" is eliminated, Eq. (23.25) becomes

$$\textit{Decide } \gamma_k, \text{ if } \log \Lambda_k(\mathbf{V}) \geq \log \Lambda_j(\mathbf{V}) \quad \text{all } j \neq k \; (j = 1, \ldots, M) \tag{23.27}$$

The optimum system consists of a sequence of operations, $\log \Lambda_l(\mathbf{V})$, as in Fig. 23.1, whose ultimate output is once more a decision γ_k and whose structure (each l) is in threshold cases given by developments of the type of Eq. (19.45), or Eq. (20.3) for normal noise backgrounds, with the associated interpretations of Secs. 20.1, 20.2.

(*d*) *Error probabilities, average risk, and system evaluation.* In order to evaluate the performance of these optimum decision systems, it is necessary to determine their Bayes risks, and for this in turn we need the error probabilities associated with the various possible decisions. Similarly, to evaluate the performance of suboptimum systems and compare them with the corresponding optimum cases, we must also determine the appropriate error probabilities. This is a conceptually straightforward generalization of Sec. 19.3 for the earlier binary theory, although, as we shall note presently, there are certain technical problems here not present in the simpler case, problems which make explicit calculations considerably more difficult.

Let us now consider the various conditional probabilities of error. We define

$$\alpha_k^{(0)} \equiv \int_\Gamma \delta(\gamma_k|\mathbf{V}) F_n(\mathbf{V}|\mathbf{S}^{(0)})\, d\mathbf{V}$$

$\equiv$ conditional probability of calling a null signal any one member of kth signal class ($k = 1, \ldots, M$) (23.28)

$$\langle\beta_j^{(k)}\rangle_k \equiv \int_\Gamma \delta(\gamma_j|\mathbf{V}) \langle F_n(\mathbf{V}|\mathbf{S}^{(k)})\rangle_k\, d\mathbf{V}$$

$\equiv$ conditional probability of calling any one member of the kth signal class a member of jth signal class (in noise) ($j \neq k; j = 0, \ldots, M; k = 1, \ldots, M$) (23.29)

The conditional probability of correctly deciding that any one signal member of class k ($k = 0, \ldots, M$) is present in noise is

$$\langle\eta_k^{(k)}\rangle_k \equiv \int_\Gamma \delta(\gamma_k|\mathbf{V}) \langle F_n(\mathbf{V}|\mathbf{S}^{(k)})\rangle_k\, d\mathbf{V} \tag{23.30}$$

$$\langle\eta_k^{(k)}\rangle_k = 1 - \sum_{j=0}^{M}{}' \langle\beta_j^{(k)}\rangle_k \tag{23.30a}$$

In the situation where the case "noise alone" is removed, we have only $\langle\beta_j^{(k)}\rangle_k$ ($j \neq k; j, k = 1, \ldots, M$).

For optimum systems, where Eqs. (23.11), (23.12), (23.14) apply, it is convenient to make a change of variable and consider some monotonic function of the Λ's as our new independent variables, since it is in terms of the Λ's that the (optimum) decision regions for $\gamma_0, \ldots, \gamma_M$ are explicitly given. As before, we let $x_k = \log \Lambda_k$ $(k = 1, \ldots, M)$, so that Eqs. (23.28), (23.29) may now be written

$$\alpha_k^{(0)*} = \int_{[x_1]} \cdots \int_{[x_k]} \cdots \int_{[x_M]} Q_n(x_1, \ldots, x_M)\, dx_1 \cdots dx_M \tag{23.31}$$

$$\langle \beta_j^{(k)*} \rangle_k = \int_{[x_1]} \cdots \int_{[x_j]} \cdots \int_{[x_M]} P_n^{(k)}(x_1, \ldots, x_M)\, dx_1 \cdots dx_M$$

$$P_n^{(0)} = Q_n \tag{23.32}$$

where Q_n, $P_n^{(k)}$ are the *joint probability densities* for the random variables $\log \Lambda_1, \ldots, \log \Lambda_M$ in the first instance with respect to the null hypothesis H_0 and in the second with respect to the alternatives H_k. Here $[x_k]$ signifies the decision region for x_k (some typical cases are illustrated in Figs. 23.2, 23.3 for the transformation $x_k' = \Lambda_k$, rather than $x_k = \log \Lambda_k$). These probability densities may be written in terms of the original **V** as [cf. Eqs. (19.29a), (19.29b)]

$$Q_n(x_1, \ldots, x_M) = \int_\Gamma F_n(\mathbf{V}|0) \prod_{j=1}^{M} \delta(x_j - \log \Lambda_j)\, d\mathbf{V} \tag{23.33}$$

$$P_n^{(k)}(x_1, \ldots, x_M) = \int_\Gamma \langle F_n(\mathbf{V}|\mathbf{S}^{(k)}) \rangle_k \prod_{j=1}^{M} \delta(x_j - \log \Lambda_j)\, d\mathbf{V} \tag{23.34}$$

The corresponding characteristic functions are

$$F_M(i\boldsymbol{\xi})_Q = \int_\Gamma e^{i\tilde{\boldsymbol{\xi}}\mathbf{x}} F_n(\mathbf{V}|0)\, d\mathbf{V} \tag{23.35}$$

$$F_M^{(k)}(i\boldsymbol{\xi})_P = \int_\Gamma e^{i\tilde{\boldsymbol{\xi}}\mathbf{x}} \langle F_n(\mathbf{V}|\mathbf{S}^{(k)}) \rangle_k\, d\mathbf{V} \tag{23.36}$$

[cf. Eqs. (19.30a), (19.30b)]. A specific example with additive gaussian noise is considered in Prob. 23.3.

With Eqs. (23.28) to (23.32), we can now write the average risk [Eq. (23.6) or (23.10)], minimized or not, as

$$R_M(\sigma,\delta) = \mathcal{R}_{0M} + q \sum_{1}^{M} \lambda_k^{(0)} \alpha_k^{(0)(*)} + \sideset{}{'}\sum_{\substack{k,l=0;\ k\geq 1,\ l\geq 0 \\ (k\neq l)}}^{M} p_k \lambda_l^{(k)} \langle \beta_l^{(k)(*)} \rangle_k + \sum_{k=1}^{M} p_k \lambda_k^{(k)} \langle \eta_k^{(k)(*)} \rangle_k \tag{23.37}$$

which for the case of the excluded null signal simply omits the terms in $\lambda_k^{(0)}$ and sets $\mathcal{R}_{0M} = 0$, with $\lambda_k^{(l)}$ $(k, l \geq 1)$ equal to the costs $C_k^{(l)} \geq 0$.

The expressions (23.31), (23.32) for the error probabilities appearing in Eq. (23.35) have particularly simple limits for $[x_k]$ when we make the $(C_0,0)$ cost assumptions of the example in (c) above. We easily find that now (the primes indicate terms $j = k$ or l omitted in the products)

$$\alpha_k^{(0)*} = \int_0^\infty dx_k \left(\prod_{j=1}^{M}{}' \int_{-\infty}^{x_k} dx_j \right) Q_n(x_1, \ldots ,x_M) \qquad k = 1, \ldots , M \tag{23.38}$$

$$\langle \beta_j^{(k)*} \rangle_k = \int_0^\infty dx_j \left(\prod_{l=1}^{M}{}' \int_{-\infty}^{x_j} dx_l \right) P_n^{(k)}(x_1, \ldots ,x_M) \qquad j \neq k; j, k \geq 1 \tag{23.39}$$

and
$$\langle \beta_0^{(k)*} \rangle_k = \int_{-\infty}^{0} dx_1 \cdots \int_{-\infty}^{0} dx_M \, P_n^{(k)}(x_1, \ldots ,x_M) \tag{23.40}$$

When the noise class is omitted, Eqs. (23.38), (23.40) do not apply and the lower limit on the first integral of Eq. (23.39) becomes $-\infty$ instead of 0. For the general cost assumptions [Eqs. (23.7)], we must use Eqs. (23.13) or (23.15) (all k) to specify the boundaries on Λ_k and hence on $\log \Lambda_k$. General results for the ternary case [the example in (*b*) above] follow at once from Eqs. (23.19), (23.20), or (23.21) with the aid of Figs. 23.3, 23.4.

For suboptimum systems, instead of $x_l = \log \Lambda_l(\mathbf{V})$ we have $y_l = \log G_l(\mathbf{V})$, where $G_l(\mathbf{V})$ represents the lth component of the actual suboptimum system in use (cf. Sec. 19.3-2). Then, analogous to Eqs. (19.36*a*), (19.36*b*), etc., Eqs. (23.31), (23.32) are modified to

$$\alpha_k^{(0)} = \int \cdots \int_{[\mathbf{y}]} q_n(y_1, \ldots ,y_M) \, dy_1 \cdots dy_M \neq \alpha_k^{(0)*} \tag{23.41a}$$

$$\langle \beta_j^{(k)} \rangle_k = \int \cdots \int_{[\mathbf{y}]} p_n^{(k)}(y_1, \ldots ,y_M) \, dy_1 \cdots dy_M \neq \langle \beta_j^{(k)*} \rangle \tag{23.41b}$$

where q_n, $p_n^{(k)}$ are now the joint d.d.'s of the random variables $\log G_l(\mathbf{V})$ $(l = 1, \ldots , M)$ and $[\mathbf{y}]$ denotes the decision regions for the $y_1, \ldots , y_M$, which may or may not be the same as $[\mathbf{x}]$ in Eqs. (23.31), (23.32). These probability densities are expressed in terms of the data process $\mathbf{V}$, analogous to Eqs. (23.33), (23.34), as

$$q_n(y_1, \ldots ,y_M) = \int_\Gamma F_n(\mathbf{V}|0) \prod_{j=1}^{M} \delta[x_j - \log G_j(\mathbf{V})] \, d\mathbf{V} \tag{23.42a}$$

$$p_n^{(k)}(y_1, \ldots ,y_M) = \int_\Gamma \langle F_n(\mathbf{V}|\mathbf{S}^{(k)}) \rangle_k \prod_{j=1}^{M} \delta[x_j - \log G_j(\mathbf{V})] \, d\mathbf{V} \tag{23.42b}$$

with the associated characteristic functions [cf. Eqs. (19.37*a*), (19.37*b*)]

$$F_M(i\xi)_q = \int_\Gamma e^{i\xi\mathbf{y}} F_n(\mathbf{V}|0) \, d\mathbf{V} \qquad F_M^{(k)}(i\xi)_p = \int_\Gamma e^{i\tilde{\xi}\mathbf{y}} \langle F_n(\mathbf{V}|\mathbf{S}^{(k)}) \rangle_k \, d\mathbf{V} \tag{23.43}$$

System comparisons are then made on the basis of the respective Bayes and average risks, by an obvious extension of the binary methods of Sec. 19.3-3 to these multialternative cases. However, even when the background noise is normal, additive, and independent, it is difficult to evaluate these error probabilities, since they are, in effect, M-dimensional error functions, whose arguments depend on the successive variables of integration: the technical problem is analogous to that of evaluating the volume cut off from a hyper-ellipsoid by a series of hyperplanes. One special example of some interest, where the multiple integrals "factor" conveniently, is considered in Prob. 23.3. Generally, however, our expressions are not so tractable, so that, while structure can be obtained, the error probabilities require a more formidable computational program.

23.1-2 Detection with Decision Rejection. In the usual procedure described above, if a data sample $\mathbf{V}'$ is such that the inequalities (23.11) or (23.12) are satisfied for a particular k, that is, if the data "point" $\mathbf{V}'$ falls in the decision region associated with γ_k, we make the decision that a signal of class k is present. However, it may be that the point $\mathbf{V}'$ falls too close to the boundaries of the acceptance region for γ_k, that is, for the given cost assignments the probability of decision error is too large, so that the resulting decision is actually incorrect. Because of the background noise, the point $\mathbf{V}'$, which really belongs in the region for γ_j $(j \neq k)$, say, has landed in γ_k. Of course, we can realign the boundaries to diminish this effect—i.e., decrease the probabilities of error—by a readjustment of the various costs involved, when these are at our disposal. But frequently we do not have this option: not only are the costs preassigned (in any case), but these preassigned values are inflexible for other reasons. In such circumstances, we may then decide to reject all excessively doubtful decisions. Thus, we introduce a *new set of decisions*, viz., decisions to reject the decisions that particular signals are present. The purpose of this rejection procedure is to guard further against the penalties of wrong decisions, by substituting "blanks" or other indications of rejection in the noticeably doubtful cases. Accordingly, a typical rejection of a signal of class k might be represented by $S_R^{(k)}$, where both the signal rejected and the fact of rejection are available at the output of the receiver. Rejection procedures have application in coding for communication and computation† whenever it is advantageous to take additional precautions against decision errors. The expense of incorrect decisions is, of course, reduced at the lesser expense‡ of a lower positive decision rate.

We can illustrate this schematically with the simple diagrams of Fig. 23.5*a*, *b*. These show the data space Γ divided into various regions. For the nonrejection case, there are $M + 1$ distinct, nonoverlapping zones, corresponding to the decisions to accept noise or $S^{(k)}$ in noise, for example, N, $S^{(1)} + N$, $S^{(2)} + N$, . . . , $S^{(M)} + N$, while in the rejection case $M + 1$ additional zones have been introduced which represent the regions where

† See, for example, a recent paper of Chow.[2]

‡ This depends on the cost assignments of rejection and acceptance; for meaningful operation, it is clear that rejection is to be assigned a smaller cost than a decision error.

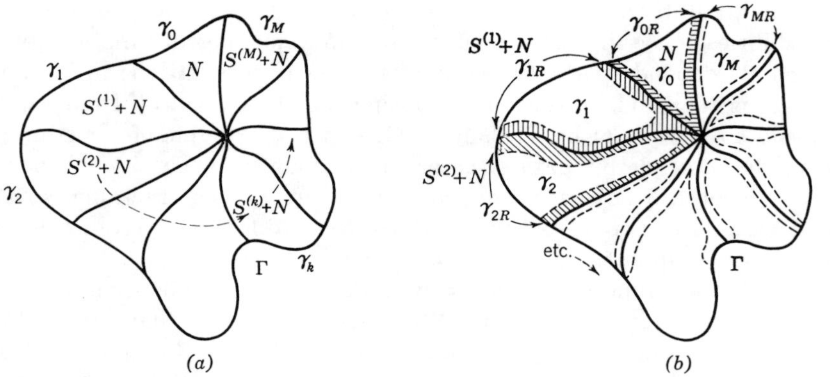

FIG. 23.5. (*a*) Data space Γ, showing decision regions without rejection. (*b*) Same as (*a*), but with rejection regions.

acceptance of a particular $S^{(k)}$ in noise (or noise alone) is excessively doubtful. Thus, if a particular data point $\mathbf{V}'$ falls too close to an original acceptance boundary (solid lines in Fig. 23.5*a*, *b*), our decision is to reject the original decision, whether or not it is correct or false. With this in mind, we can readily extend the analysis of Sec. 23.1-1 to include the additional $M + 1$ decisions with rejection.

(1) Optimum $(M + 1)$-ary decisions with rejection. We begin first with the cost assignments. Let us write

$C_k^{(l)}$ = cost of *incorrectly* deciding $S^{(k)}$ is present when $S^{(l)}$ actually occurs and of accepting the decision [$l \neq k$; l, $k = 0, \ldots, M$ ($l = 0$, $k = 0$ are "noise alone" states)]

$C_l^{(l)}$ = cost of *correctly* deciding a signal of class l is present when such a signal actually occurs and of accepting the decision ($l = 0, \ldots, M$)

C_{kR}^{l} = cost of deciding $S^{(k)}$ is present when $S^{(l)}$ really occurs and of *rejecting* this decision (whether correct or incorrect)

A meaningful interpretation of these costs, and of the subsequent decision procedures, requires that

$$C_l^{(l)} < C_{lR}^{(l)} \leq C_{kR}^{(l)} < C_{k(\neq l)}^{(l)} \qquad l, k = 0, \ldots, M \tag{23.44}$$

i.e., the cost of a correct decision is less than the cost of a rejection, which in turn is less than the cost of an error.

Modifying Eqs. (23.7) to

$$\lambda_k^{(l)} = C_k^{(l)} - C_0^{(l)} \qquad \lambda_{kR}^{(l)} = C_{kR}^{(l)} - C_0^{(l)} \qquad l, k = 0, \ldots, M \tag{23.45a}$$

and setting

$$\gamma_k' = \begin{cases} \gamma_k & k = 1, \ldots, M \\ \gamma_{(k-M)R} & k = M + 1, \ldots, 2M + 1 \end{cases} \tag{23.45b}$$

where $\gamma_{(M+1)R}$ = decision to reject the choice of "noise alone," we extend the condition (23.3) to

$$\sum_{k=0}^{M} \delta(\gamma_k|\mathbf{V}) + \sum_{M+1}^{2M+1} \delta(\gamma_{(k-M)R}|\mathbf{V}) = 1 \tag{23.45c}$$

Then, following the steps leading to Eq. (23.10c), we obtain the rearranged version of the average risk,

$$R(\sigma,\delta)_{2M+1} = (\mathcal{R}_0)_{2M+1} + \int_\Gamma \left[\sum_{k=1}^{2M+1} \delta(\gamma_k'|\mathbf{V}) A_k(\mathbf{V}) \right] qF_n(\mathbf{V}|0)\, d\mathbf{V} \tag{23.46}$$

where now

$$\begin{aligned} A_k(\mathbf{V}) &= \lambda_k^{(0)} + \sum_{l=1}^{M} \lambda_k^{(l)} \Lambda_l(\mathbf{V}) && k = 1, \ldots, M \\ &= \lambda_{(k-M)R}^{(0)} + \sum_{l=M+1}^{2M+1} \lambda_{(k-M)R}^{(l-M)} \Lambda_{(l-M)}(\mathbf{V}) && k = M+1, \ldots, 2M+1 \end{aligned} \tag{23.46a}$$

The generalized likelihood ratios Λ_l, $\Lambda_{(l-M)}$ are given as before by Eqs. (23.9), (23.9a), with $q + \sum_{j=1}^{M} p_j = 1$ [cf. Eq. (23.2)].

For optimum detection, we minimize the average risk [Eq. (23.46)] again by suitable choices of the decision rules, as in the nonrejection case above. The result is the following set of decision procedures:

Decide γ_k' $(k = 1, \ldots, 2M+1)$, if

$$\begin{gathered} \lambda_k^{(0)} + \sum_{l=1}^{2M+1} \lambda_k^{(l)} \Lambda_l(\mathbf{V}) \le \lambda_j^{(0)} + \sum_{l=1}^{2M+1} \lambda_j^{(l)} \Lambda_l(\mathbf{V}) \\ \lambda_k^{(0)} + \sum_{l=1}^{2M+1} \lambda_k^{(l)} \Lambda_l(\mathbf{V}) \le 0 \\ \text{all } j \ne k \; (j = 1, \ldots, 2M+1) \end{gathered} \tag{23.47a}$$

Then set $\delta(\gamma_k'|\mathbf{V}) = 1$ and $\delta(\gamma_j'|\mathbf{V}) = 0$ $(j \ne k)$. For noise alone,

$$\begin{gathered} \textit{Decide } \gamma_0', \text{ if } \lambda_k^{(0)} + \sum_{l=1}^{2M+1} \lambda_k^{(l)} \Lambda_l(\mathbf{V}) \ge 0 \\ \text{all } k = 1, \ldots, 2M+1 \end{gathered} \tag{23.47b}$$

Similarly, $\delta(\gamma_0'|\mathbf{V}) = 1$, $\delta(\gamma_k'|\mathbf{V}) = 0$ $(k \ge 1)$. Here we adopt the convention that

$$\Lambda_l \Big|_{l>M} = \Lambda_{(l-M)} \qquad \lambda_k^{(l)} \Big|_{k \text{ and/or } l>M} = \lambda_{(k-M)R}^{(l) \text{ or } (l-M)} \tag{23.47c}$$

Accordingly, all decisions for $k \geq M + 1$ represent *rejections* of the decisions $k = 1, \ldots, M, 0$, respectively. As in the nonrejection case, the boundaries of the decision regions are obtained from the equalities in Eqs. (23.47*a*), (23.47*b*).

(2) Optimum M-ary decisions with rejection. Here we remove the null signal class: a signal is always present in noise. From Eq. (23.14), by an obvious extension of the analysis we can write the decision procedure as

$$\textit{Decide } \gamma_k', \text{ if } \sum_{l=1}^{2M} \hat{\lambda}_k^{(l)} \Lambda_l(\mathbf{V}) \leq \sum_{l=1}^{2M} \hat{\lambda}_j^{(l)} \Lambda_l(\mathbf{V}) \qquad j \neq k, \text{ all } j = 1, \ldots, 2M \tag{23.48}$$

where now

$$\begin{aligned}
\hat{\lambda}_k^{(l)} &= C_k^{(l)} && l, k = 1, \ldots, M \\
&= C_k^{(k)} && k = 1, \ldots, M \\
&= C^{(l)}_{(k-M)R} && k = M + 1, \ldots, 2M \\
&= C^{(l-M)}_{(k-M)R} && l, k = M + 1, \ldots, 2M \\
&= C^{(l-M)}_{kR} && l \neq k \neq M, \text{ with } C^{(l)}_{(k-M)R} = C^{(l-M)}_{kR} = C^{(l-M)}_{(k-M)R}
\end{aligned} \tag{23.48a}$$

The likelihood ratios Λ_l still obey Eqs. (23.9), (23.9*a*), except that $q \; (> 0)$ is arbitrary, and the condition (23.2) becomes $\sum_1^M p_j = 1$. The convention of Eqs. (23.47*c*) also applies, while $A_k = \sum_{l=1}^{2M} \hat{\lambda}_k^{(l)} \Lambda_l(\mathbf{V})$ [cf. Eq. (23.8)]. The boundaries of the various decision regions are determined by the equalities in Eq. (23.48).

(3) A simple cost assignment. Let us consider as an example the situation where the null signal class is omitted, so that Eq. (23.48) applies. We set C_R = cost of rejection, C_E = cost of an error, and zero the cost of a correct decision that is accepted. Accordingly, Eq. (23.48*a*) becomes

$$\begin{aligned}
\hat{\lambda}_k^{(k)} &= 0 \\
\hat{\lambda}_k^{(l)} &= C_E && l \neq k, l = 1, \ldots, 2M; k = 1, \ldots, M \\
\lambda_k^{(l)} &= C_R && k = M + 1, \ldots, 2M, \text{ with } 0 < C_R < C_E
\end{aligned} \tag{23.49}$$

The decision rules [Eq. (23.48)] reduce explicitly to

Acceptance

Decide that some $S^{(k)}$ $(k = 1, \ldots, M)$ is present and accept the decision, if

$$\Lambda_j \leq \Lambda_k \qquad \text{for some } k \text{ (each } j = 1, \ldots, M; j \neq k)$$

$$0 \leq \frac{C_R - C_E}{C_E} \sum_{l=1}^{M} \Lambda_l + \Lambda_k \qquad \text{for this } k \tag{23.50a}$$

Rejection

Reject a signal $S^{(k)}$ $(k = 1, \ldots, M)$, if

$$0 \geq \frac{C_R - C_E}{C_E} \sum_{l=1}^{M} \Lambda_l + \Lambda_k \tag{23.50b}$$

for the k above for which $\Lambda_j \leq \Lambda_k$ applies $(j \neq k)$. Once more, the equalities in Eqs. (23.50) establish the boundaries of the various decision regions.

(4) Remarks. As in the nonrejection situations of Sec. 23.1-1, the elements of the system are the likelihood ratios Λ_l $(l = 1, \ldots, M)$ as shown in Fig. 23.1, albeit they are combined in a somewhat different fashion because of the rejection operations. Threshold signal structure once again involves a generalized (averaged) cross- or autocorrelation of the received data with itself, in the manner of Eqs. (19.58), (19.59), and if the background noise is normal, we may use the results of Chap. 20 to obtain the specific structures of the Λ_l in such cases. The error probabilities and Bayes risk may be found in principle from Eqs. (23.31), (23.32), although explicit calculations are usually very involved because of the nature of the integrands and the limits. For the simple cost assignments of Secs. 23.1-1(3), 23.1-2(3), when the a priori probabilities q, p_j are not known to the observer, the corresponding Minimax system can be shown to have the same structure as the above, where now $q = p_j = 1/(M + 1)$ (all j): the p_j's are "uniformly" distributed.

23.1-3 Multiple Estimation. Here we shall very briefly outline an extension of the estimation theory described in Chap. 21, where now we wish to examine the role of auxiliary data in determining "best" or Bayes estimates of an associated quantity. For example, in radar applications, given position and velocity data, how should the latter be used to improve the Bayes estimate of the former, i.e., how should we use both sets of data to obtain estimates of position, say, and how much better are they than the more restricted Bayes estimates of position based on position data alone? Similar questions can be asked of velocity estimates. Another example is provided by the communication situation of (spatial or frequency) diversity reception, where the same message is encoded and sent over several different channels and received by a corresponding number of (spatially or frequency-wise) distinct receivers. Here one may wish to extract certain information-bearing features of the waveform structure,†

† Frequently, we may wish simply to detect the presence or absence of two different signals, e.g., "mark" or "space." We then have a *binary detection* situation, where now the decisions depend on the various data $V_1(t), \ldots, V_M(t)$ (for an M channel system). The likelihood ratio $\Lambda_{nM} \equiv p\langle F_n(\mathbf{V}_1, \mathbf{V}_2, \ldots, \mathbf{V}_M|\mathbf{S})\rangle_S / qF_n(\mathbf{V}_1, \ldots, V_M|0)$, instead of the single-channel expression $p\langle F_n(\mathbf{V}_1|\mathbf{S})\rangle_S / qF_n(\mathbf{V}_1|0)$, becomes the fundamental quantity embodying the system structure. The explicit form of Λ_{nM} indicates the precise way in which the various data sets $\mathbf{V}_1, \ldots, \mathbf{V}_M$ are to be combined and processed for optimum detection (cf. Prob. 23.4).

e.g., amplitude, frequency, etc. Then, the problem is, as in the radar case above, to combine these data, i.e., the various received waves, so as to obtain a best (or, in actual practice, a near-best) estimate of the desired quantities.

Let us formulate a typical, rather general problem. There are M signals, $\mathbf{S}^{(1)}, \ldots, \mathbf{S}^{(M)}$, each of which is sampled n times in an interval $(0,T)$, for example, $\mathbf{S}^{(j)} = [S_1^{(j)}, \ldots, S_n^{(j)}]$, in the usual way [cf. Eqs. (23.2) et seq.]. With these M signals, there occur M noise waves, $N_1, \ldots, N_M$, so that the received data are $V_j = S_j \oplus N_j$ (all j). Corresponding to these M signal vectors are the M data vectors $\mathbf{V}_1, \ldots, \mathbf{V}_M$, with $V_j = [V_{j1}, V_{j2}, \ldots, V_{jn}]$. Next, let us assign M cost functions $C_{(j)}$ $(j = 1, \ldots, M)$, one, respectively, to each (simple) estimator $\boldsymbol{\gamma}_j$ of signal waveform S_j.† Note that $\boldsymbol{\gamma}_{jM}$ is in general a function of all M sets of data, for example, $\boldsymbol{\gamma}_{jM} = \boldsymbol{\gamma}_j(\mathbf{V}_1, \ldots, \mathbf{V}_M)$. Finally, with each estimate and estimator‡ there is an average risk $R_j^{(M)}(\sigma,\delta)$ such that the total average risk $R(\sigma,\delta) = \sum_j R_j^{(M)}$. Clearly, there is an element of arbitrariness associated with $R_M(\sigma,\delta)$, since it depends on the choice of each cost function $C_{(j)}$. Different signals may have different cost functions assigned to them.

The average risk for the jth signal is an obvious extension of Eq. (18.7b),

$$R_j^{(M)}(\sigma,\delta) = R_j^{(M)}(\sigma,\gamma_{jM}) = \int_{\Omega_j} \sigma_j(\mathbf{S}_j)\, d\mathbf{S}_j \times \int_{\Gamma_M} F_n(\mathbf{V}_1, \ldots, \mathbf{V}_M|\mathbf{S}_j) C_{(j)}[\mathbf{S}_j|\boldsymbol{\gamma}_j(\mathbf{V}_1, \ldots, \mathbf{V}_M)]\, \mathbf{dV}_1 \cdots \mathbf{dV}_M \tag{23.51}$$

where $F_n(\mathbf{V}_1, \ldots, \mathbf{V}_M|\mathbf{S}_j)$ is the marginal *joint* conditional d.d. of $\mathbf{V}_1, \ldots, \mathbf{V}_M$, given $\mathbf{S}_j$. For the quadratic cost function [cf. Eq. (21.64)], we have specifically

$$R_j^{(M)} = \int_{\Omega_j} \sigma_j(\mathbf{S}_j)\, \mathbf{dS}_j \int_{\Gamma_M} \|\mathbf{S}_j - \boldsymbol{\gamma}_j(\mathbf{V}_1, \ldots, \mathbf{V}_M)\|^2 \times F_n(\mathbf{V}_1, \ldots, \mathbf{V}_M|\mathbf{S}_j)\, \mathbf{dV}_1 \cdots \mathbf{dV}_M \tag{23.52}$$

Note that irrespective of whether the $\mathbf{V}_1, \ldots, \mathbf{V}_M$ are statistically independent or not, if the cost function $C_{(j)}$ depends only on $\mathbf{V}_j$, then $R_j^{(M)}$ [Eq. (23.51)] reduces to our previous result [Eq. (21.65)] for a single, effective data set $\mathbf{V}_j$. It is the choice of cost functions which "scrambles" the various sets of data. The Bayes estimator for Eq. (23.52) is easily found as before

† If, instead of waveform estimation, we are concerned with estimating L signal parameters $\boldsymbol{\theta}$, that is, $S_j = S_j(\theta_1, \ldots, \theta_L)$ $(\boldsymbol{\theta} = \theta_1, \ldots, \theta_L)$, then we may assign in a similar way L cost functions $C_{(k)}$ $(k = 1, \ldots, L)$, one to each estimator γ_{kM} of θ_k $(k = 1, \ldots, L)$. γ_{kM}, however, may still depend on the M signals and M noise waves, for example, $\gamma_{kM} = \gamma_k(V_1, \ldots, V_M)$.

‡ See remarks in Sec. 21.4-1.

[cf. Eqs. (21.61), (21.62), (21.66)] to be

$$\boldsymbol{\gamma}_{jM}^{*} \equiv \boldsymbol{\gamma}_{j}^{*}(\mathbf{V}_1, \ldots, \mathbf{V}_M) = \frac{\int_{\Gamma} \mathbf{S}_j \sigma_j(\mathbf{S}_j) F_n(\mathbf{V}_1, \ldots, \mathbf{V}_M|\mathbf{S}_j)\, d\mathbf{S}_j}{\int_{\Omega_j} \sigma_j(\mathbf{S}_j) F_n(\mathbf{V}_1, \ldots, \mathbf{V}_M|\mathbf{S}_j)\, d\mathbf{S}_j} \tag{23.53}$$

In the same way, we can obtain the Bayes estimators for various extensions of the simple cost functions [Eqs. (21.45), (21.52), (21.78), (21.79)].

When the signal parameters $\boldsymbol{\theta} = [\theta_1, \ldots, \theta_L]$ are the quantities of interest, Eq. (23.51) is modified to

$$R_{(k)}{}^{(L)} = \int_{\Omega_{\theta_k}} \sigma_k(\theta_k)\, d\theta_k \int_{\Omega_M} W_n(\mathbf{V}_1, \ldots, \mathbf{V}_M|\theta_k) \\ C_{(k)}[\theta_k|\gamma_k(\mathbf{V}_1, \ldots, \mathbf{V}_M)]\, d\mathbf{V}_1 \cdots d\mathbf{V}_M \tag{23.54}$$

where $W_n(\mathbf{V}_1, \ldots, \mathbf{V}_M|\theta) = \langle F_n[\mathbf{V}_1, \ldots, \mathbf{V}_M|S_1(\boldsymbol{\theta}',\theta_k), \ldots, S_M(\boldsymbol{\theta}',\theta_k)]\rangle_{\boldsymbol{\theta}'}$ [cf. Eq. (21.44)], with corresponding expressions for Eqs. (23.52), (23.53) in the case of the quadratic cost functions. Finally, we can show (cf. Prob. 23.7) that the Bayes risk $R_j^{*(M)}$ for the multiple data sets is always equal to or smaller than the Bayes risk R_j^*, based on $\mathbf{V}_j$ alone, i.e.,

$$R_j^* - R_j^{*(M)} \geq 0 \tag{23.55}$$

This is, of course, to be expected, since the Bayes estimator $\boldsymbol{\gamma}_{jM}^*$ (for any cost function) takes advantage of the possible statistical interdependence of the $\mathbf{V}_j$ $(j = 1, \ldots, M)$ with respect to the desired signal waveform (or parameter) and the partial (or possibly total) *independence* of the $\mathbf{N}_j$ with respect to one another. The effect is essentially that of an effectively larger statistical sample than is available from, say, $\mathbf{V}_j$ alone. The equality in Eq. (23.55) applies when $\mathbf{V}_j$ is statistically independent of the other data elements $\mathbf{V}_k$ $(k \neq j)$, since then none of the information in the $\mathbf{V}_k$ $(k \neq j)$ is relevant to $\mathbf{S}_j$ in $\mathbf{V}_j$ or is of any assistance to us in the estimation of $\mathbf{S}_j$ (or its parameters). Some results for amplitude estimation are given in Prob. 23.6.

PROBLEMS

23.1 (*a*) Carry out the details leading from Eq. (23.6) to Eq. (23.10*c*).

(*b*) Obtain Eqs. (23.38) to (23.40) for the simple cost assignment.

(*c*) Establish Eqs. (23.50*a*), (23.50*b*) for the simple cost assignment, when there is decision rejection.

23.2 Show that in the multiple-alternative detection case of Sec. 23.1-1 the actual amount of information conveyed, on the average, is

$$H_T = -H(\sigma,\delta) - q \log q - \sum_{i=1}^{M} p_i \log p_i \tag{1}$$

Verify that the equivocation $H(\sigma,\delta)$ is

$$H(\sigma,\delta) = -\sum_{i=0}^{M}\sum_{j=0}^{M} P(\gamma_i,S_j)\log\,[P(\gamma_i,S_j)/P(\gamma_i)] \tag{2}$$

where

$$P(\gamma_i,S_j) = \begin{cases} p_j\beta_i^{(j)} & j \neq 0 \\ \text{or} & \\ q\alpha_i^{(0)} & j = 0 \end{cases} \tag{3}$$

and

$$P(\gamma_i) = \begin{cases} q\alpha_i^{(0)} + p_i\left(1 - \sum\limits_{j\neq i}\beta_k^{(i)}\right) + \sum\limits_{\substack{j\neq i\\ j\neq 0}} p_{ji}^{(j)} & i \neq 0 \\ \text{or} & \\ \sum\limits_{j\neq 0} p_j\beta_0^{(j)} + q\left(1 - \sum\limits_{k\neq 0}\alpha_k^{(0)}\right) & i = 0 \end{cases} \tag{4}$$

23.3 Consider the multiple-alternative detection situation in which the null signal, or any one of M arbitrary nonzero signals (and noise), identical in structure but differing in amplitude, may be present at the input. Detection is assumed to be *coherent*, the signals and noise are additive, each signal set contains only one member, and the costs of correct and incorrect decisions are taken to be zero and unity, respectively. The noise itself is assumed gaussian, with a known variance matrix. The kth hypothesis class contains but one signal, of amplitude a_{0k}, and the amplitudes are ordered: $0 < a_{01} < a_{02} \cdots < a_{0k} < \cdots a_{0M}$.

(*a*) Obtain the characteristic functions and d.d.'s for $\mathbf{y} = \mathbf{x} - (\log \mu_j + a_{0j}^2\Phi_{jj}/2)$:

$$F_M(i\boldsymbol{\xi})_Q = e^{-\frac{1}{2}\tilde{\boldsymbol{\xi}}\mathbf{S}_M\boldsymbol{\xi}} \tag{1}$$

$$F_M^{(k)}(i\boldsymbol{\xi})_P = e^{-i\tilde{\boldsymbol{\xi}}(\mathbf{S}_M)_k - \frac{1}{2}\tilde{\boldsymbol{\xi}}\mathbf{S}_M\boldsymbol{\xi}} \tag{2}$$

$$Q_n(\mathbf{y}) = (2\pi)^{-M/2}(\det \mathbf{S}_M)^{-\frac{1}{2}}e^{-\frac{1}{2}\tilde{\mathbf{y}}\mathbf{S}_M^{-1}\mathbf{y}} \tag{3}$$

$$P_n^{(k)}(\mathbf{y}) = (2\pi)^{-M/2}(\det \mathbf{S}_M)^{-\frac{1}{2}}e^{-\frac{1}{2}[\tilde{\mathbf{y}}-(\tilde{\mathbf{S}}_M)_k]\mathbf{S}_M^{-1}[\mathbf{y}-(\mathbf{S}_M)_k]} \tag{4}$$

where $\mathbf{S}_M = \|a_{0j}a_{0k}\Phi_{jk}\|$, $\Phi_{jk} = \tilde{\mathbf{s}}_j\mathbf{k}_N^{-1}\mathbf{s}_k$, and $\mu_j = p_j/q$.

(*b*) Show that the error probabilities for the optimum system here are

$$\begin{aligned}\alpha_j^{(0)*} &= \int_{-A_j}^{\infty} dy_k\left(\prod_{l=1}^{M}{}' \int_{-\infty}^{y_j+A_j-A_l} dy\right) Q_n(\mathbf{y}) \\ &= a_{0j}^{-1}(2\pi\Phi)^{-\frac{1}{2}}\int_{(C_{j-}\text{ or } -A_j)}^{C_{j+}} e^{-y^2/2a_{0j}^2\Phi}\,dy \qquad j = 1, \ldots, M\end{aligned} \tag{5}$$

$$\begin{aligned}\langle\beta_0^{(k)*}\rangle &= \int_{-\infty}^{-A_1} dy_1 \cdots \int_{-\infty}^{-A_M} dy_M\, P_n^{(k)}(\mathbf{y}) \\ &= a_{0k}^{-1}(2\pi\Phi)^{-\frac{1}{2}}\int_{-\infty}^{C_{0+}} e^{-(y-a_{0k}^2\Phi)^2/2a_{0k}^2\Phi}\,dy \qquad k = 1, \ldots, M\end{aligned} \tag{6}$$

$$\begin{aligned}\langle\beta_j^{(k)*}\rangle &= \int_{-A_j}^{\infty} dy_j\left(\prod_{l=1}^{M}{}' \int_{-\infty}^{y_j+A_j-A_l} dy\right) P_n^{(k)}(\mathbf{y}) \\ &= a_{0j}^{-1}(2\pi)^{-\frac{1}{2}}\int_{(C_{j-}\text{ or } -A_j)}^{C_{j+}} e^{-(y-a_{0j}a_{0k}\Phi)^2/2a_{0j}^2\Phi}\,dy \qquad j, k = 1, \ldots, M;\, j \neq k\end{aligned} \tag{7}$$

where $A_j = \log \mu_j - a_0^2\Phi_{jj}/2$; $\Phi_{jj} \equiv \Phi$; $C_{j-} \equiv \max\limits_{l} C_{jl}$ $(l < j)$, $C_{j+} \equiv \min\limits_{l} C_{jl}$ $(l > j)$, with $C_{jl} \equiv \frac{1}{2}a_{0j}(a_{0j} + a_{0l})\Phi + a_{0j}/(a_{0l} - a_{0j}) - \log\,(\mu_j/\mu_l)$.

(c) Assume next that the amplitudes are uniformly spaced, so that $a_{0j} = j\Delta$ with $\Delta = (\Delta A_0)/(2\psi_N)^{1/2}$, and let $\mu_j = 1$ (all $j = 1, \ldots, M$). Show that the Bayes risk for these simple cost assignments becomes

$$R_M^* = \frac{M}{M+1}\left[1 - \Theta\left(\frac{d}{2}\right)\right] \qquad d \equiv \left[\frac{\Phi_n(s,s)}{2}\right]^{1/2} \Delta \tag{8}$$

(cf. Fig. P 23.3). Verify that this result also holds for the M-ary case as well. Discuss the behavior of the Bayes risk as M increases.

D. Middleton and D. Van Meter,[1] pp. 6–9.

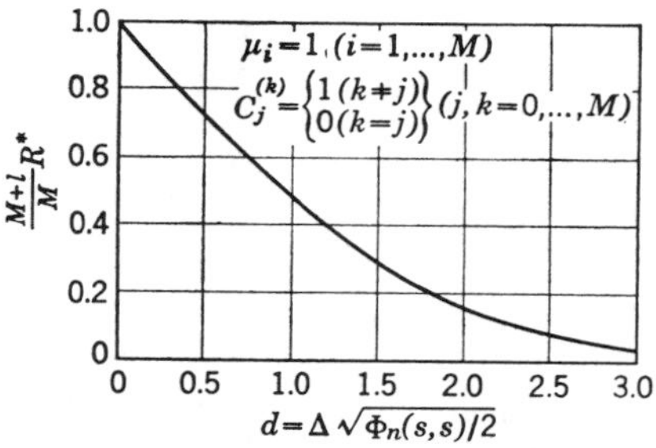

FIG. P 23.3. Bayes risk for the coherent multiple-alternative detection of M signals in gaussian noise; simple cost assignments.

23.4 (a) A signal is present additively with noise in M channels, or there are only the various noises in these M channels. Reception is coherent, and the signal class has only one member per channel. If S_M is the version of the signal in the mth channel, and if the noises are all independent normal processes, show that the optimum detector now uses the data $\mathbf{V}_1, \ldots, \mathbf{V}_M$ from the M channels according to

$$\log \Lambda_n(\mathbf{V}_1, \ldots, \mathbf{V}_M) = \log \mu - \tfrac{1}{2} \sum_{m=1}^{M} a_{0m}{}^2 \Phi_m(s) + \sum_{m=1}^{M} a_{0m} \Phi_m(v,s) \gtrless \log \mathcal{K} \tag{1}$$

where

$$\Phi_m(s) \equiv \tilde{\mathbf{s}}_m(\mathbf{k}_N{}^{-1})_m \mathbf{s}_m \qquad \Phi_m(v,s) \equiv \tilde{\mathbf{v}}_m(\mathbf{k}_N{}^{-1})_m \mathbf{s}_m \tag{1a}$$

in which $\mathbf{s}_m$, $\mathbf{v}_m$ are, respectively, the (normalized) signal and data vectors in the mth channel and $(\mathbf{k}_N)_m = (\mathbf{K}_N)_m/\psi_m$ is the (normalized) covariance matrix of the noise in this channel. As before, $\mathcal{K}$ is a threshold [cf. Eqs. (19.15)], and $a_{0m}{}^2 = A_{0m}{}^2/2\psi_m$, with A_{0m} the amplitude of $S_m(t)$. Thus, one simply takes the outputs $\log \Lambda_n(\mathbf{V}_j) - (1/M) \log \mu + \tfrac{1}{2} a_{0j}{}^2 \Phi_j(s) = a_{0j}\Phi_j(v,s)$, which are the weighted cross correlations [cf. Eq. (1a)] of s_m with v_m, and combines them additively to get Eq. (1), which is then compared with the threshold in the usual way.

(b) Show, next, that the error probabilities of this Bayes system are for all signal levels

$$\begin{Bmatrix} \alpha^* \\ \beta^* \end{Bmatrix} = \frac{1}{2}\left(1 - \Theta\left\{\frac{[\Sigma a_{0m}{}^2 \Phi_m(s)]^{1/2}}{2\sqrt{2}} \pm \frac{\log (\mathcal{K}/\mu)}{\sqrt{2}\,[\Sigma a_{0m}{}^2 \Phi_m(s)]^{1/2}}\right\}\right) \tag{2}$$

In the limiting situation where $s_m = s$, $a_{0m} = a_0$, and $(\mathbf{k})_m = \mathbf{k}_N$ (all m), we accordingly have

$$\begin{Bmatrix} \alpha^* \\ \beta^* \end{Bmatrix} = \frac{1}{2}\left\{1 - \Theta\left[\frac{a_0 \Phi_s{}^{1/2} \sqrt{M}}{2\sqrt{2}} \pm \frac{\log (\mathcal{K}/\mu)}{\sqrt{2}\, a_0 \Phi_s{}^{1/2} \sqrt{M}}\right]\right\} \tag{3}$$

showing that in effect we have increased the "size" of our statistical sample: $a_0\Phi_s{}^{1/2}$ of the single-channel case is made larger by the factor $\sqrt{M}$. Thus, the minimum detectable signal (on an rms basis) is decreased by $\sqrt{M}$.

23.5 (a) Repeat Prob. 23.4, but now for incoherent reception (where $\langle s \rangle_\epsilon = 0$), and where s_m, s_n are essentially statistically independent (like the noises) from channel to channel. Hence, show that the optimum detector structure in the threshold case is

$$\log \Lambda_n(\mathbf{V}_1, \ldots, \mathbf{V}_M) \doteq \log \mu + \sum_{m=1}^{M} (B_{nm}{}^{(2)} + B_{nm}{}^{(4)}) + \tfrac{1}{2} \sum_{m=1}^{M} \overline{a_{0m}{}^2 \tilde{\mathbf{v}}_m \bar{\mathbf{G}}_m \mathbf{v}_m} \tag{1}$$

with $\bar{\mathbf{G}}_m = (\mathbf{k}_N^{-1})_m \overline{\mathbf{s}_m \tilde{\mathbf{s}}_m} (\mathbf{k}_N^{-1})_m$ [cf. Eqs. (20.11b)].

(b) Obtain the Bayes error probabilities

$$\begin{Bmatrix} \alpha^* \\ \beta^* \end{Bmatrix} \simeq \frac{1}{2} \left\{ 1 - \Theta \left[\frac{(\Sigma \overline{a_{0m}^2}^2 \langle \Phi_{Gm} \rangle)^{1/2}}{4} \pm \frac{\log (\mathcal{K}/\mu)}{(\Sigma \overline{a_{0m}^2}^2 \langle \Phi_{Gm} \rangle)^{1/2}} \right] \right\} \tag{2}$$

where $\langle \Phi_{Gm} \rangle = \langle \tilde{\mathbf{s}}_m \bar{\mathbf{G}}_m \mathbf{s} \rangle$ [cf. Eqs. (20.90), (20.91)]. Hence, in the special case $s_m = s$, $a_{0m} = a_0$, $(\mathbf{k}_N)_m = \mathbf{k}_N$, etc., we get

$$\begin{Bmatrix} \alpha^* \\ \beta^* \end{Bmatrix} \simeq \frac{1}{2} \left\{ 1 - \Theta \left[\frac{\overline{a_0^2} \langle \Phi_G \rangle^{1/2} \sqrt{M}}{4} \pm \frac{\log (\mathcal{K}/\mu)}{\overline{a_0^2} \langle \Phi_G \rangle^{1/2} \sqrt{M}} \right] \right\} \tag{3}$$

so that the improvement (on an rms basis) of the minimum detectable signal with M independent channels (all containing the same signal) over the single channel case is $\sqrt[4]{M}$ for incoherent threshold reception.

23.6 Given the two sets of data $V_1 = a_0 S_1 + N_1$, $V_2 = a_0 S_2 + N_2$, where N_1, N_2 are normal processes which may be statistically related, if a_0 is normally distributed according to Eq. (21.95), show that with a quadratic cost function the Bayes estimator of a_0 is (for sampled data)

$$a_0^*(\mathbf{V}_1, \mathbf{V}_2) = \frac{\bar{a}_0/\sigma^2 + \tilde{\mathbf{v}}_1 \mathbf{k}_1^{-1} \mathbf{s}_1 + \tilde{\mathbf{v}}_1 \mathbf{G}_1 \mathbf{s}_1 + \tilde{\mathbf{v}}_2 \mathbf{G}_2 \mathbf{s}_2 - \tilde{\mathbf{v}}_1 \mathbf{G}_3 \mathbf{s}_2 - \tilde{\mathbf{v}}_2 \tilde{\mathbf{G}}_3 \mathbf{s}_1}{1/\sigma^2 + \tilde{\mathbf{s}}_1 \mathbf{k}_1^{-1} \mathbf{s}_1 + \tilde{\mathbf{s}}_1 \mathbf{G}_4 \mathbf{s}_1 - 2 \tilde{\mathbf{s}}_1 \mathbf{G}_3 \mathbf{s}_2 + \tilde{\mathbf{s}}_2 \mathbf{G}_5 \mathbf{s}_1} \tag{1}$$

where

$$\begin{aligned} \mathbf{G}_1 &= \tilde{\mathbf{k}}_{12} \mathbf{k}_1^{-1} (\mathbf{k}_2 - \mathbf{k}_{21} \mathbf{k}_1^{-1} \mathbf{k}_{12})^{-1} \mathbf{k}_1^{-1} \mathbf{k}_{12} \qquad \eta^2 = \psi_2/\psi_1 \\ \mathbf{G}_2 &= \eta^{-2} (\mathbf{k}_2 - \mathbf{k}_{21} \mathbf{k}_1^{-1} \mathbf{k}_{12})^{-1} \\ \mathbf{G}_3 &= \tilde{\mathbf{k}}_{12} \mathbf{k}_1^{-1} (\mathbf{k}_2 - \mathbf{k}_{21} \mathbf{k}_1^{-1} \mathbf{k}_{12})^{-1} \\ \mathbf{G}_4 &= \mathbf{k}_{21} \mathbf{k}_1^{-1} (\mathbf{k}_2 - \mathbf{k}_{21} \mathbf{k}_1^{-1} \mathbf{k}_{12})^{-1} \mathbf{k}_1^{-1} \mathbf{k}_{12} \\ \mathbf{G}_5 &= \eta^{-2} (\mathbf{k}_2 - \mathbf{k}_{21} \mathbf{k}_1^{-1} \mathbf{k}_{12})^{-1} \end{aligned} \tag{2}$$

and the **k**'s are normalized auto- and cross-variance matrices between N_1 and N_2. (Compare these results with those of Sec. 21.3-1.)

23.7 (a) Show that for multiple-estimation procedures the Bayes risk $R_j^{*(M)}$ is always equal to or less than R_j^* using only a single data set $\mathbf{V}_j$ [cf. Eq. (23.55)].

(b) Extend the treatment of Sec. 23.1-3 to the simple cost function.

23.2 *Cost Coding:† Joint Optimization of Transmission and Reception by Choice of Signal Waveform*

In the preceding discussion, the problem of major concern has been the optimization of the reception process, when signals of some preselected class are observed in noise. We recall from Eq. (2.20) that the single-link communication system considered here can be described by $\{v\} = T_R^{(N)} T_M^{(N)} T_T^{(N)} \{u\}$, where the $T^{(N)}\{\ \}$ represent the transformations imposed upon the signal message set $\{u\}$ to yield the final received message or decisions $\{v\}$. So far, we have considered the optimization of the reception process $T_T^{(N)}$ alone, when the operations of the channel, $T_M^{(N)}$, and of the encoding, $T_T^{(N)}$, are assumed known a priori at the receiver.

† Here "cost coding" refers to that part of the encoding process which selects signal alphabets for the encoded messages, i.e., after mapping from message to signal space (cf. Sec. 6.1).

From the decision-theory viewpoint, optimization in reception has been achieved under these conditions by minimizing the average risk, or cost, associated with the ensemble of possible decisions at the receiver, when $T_M^{(N)}$ and $T_T^{(N)}$ are specified. Now we shall extend the theory to include a *simultaneous* adjustment of $T_R^{(N)}$ and $T_T^{(N)}$, in order further to minimize the average risk, whenever this is possible. In effect, we seek the subclass of Bayes decision rules δ^* which yield a *minimum* Bayes risk. Since δ^* is already specified for a given signal class (with respect to this class) by our choice of $T_R^{(N)}$, and since $T_M^{(N)}$ remains unchanged, further optimization can be accomplished only by selection of $T_T^{(N)}$—that is, by choice of signal waveform (or alphabet) at the transmitter. Implied in this operation are one or more constraints: at the least we must require that the total or average signal energy received in the observation interval $(0,T)$ be finite. Other possible constraints are band or peak limitations on signal waveform. Here we shall assume only that the signal energy is fixed and finite in the data-acquisition period.

In order to minimize the Bayes risk, either for detection or for extraction, we must first determine the Bayes risk itself and then obtain the required signal subclasses for further optimization by some appropriate variational procedure whenever this is possible. Let us begin with a short summary of some of the principal results of binary detection (cf. Chap. 20) and of signal extraction (cf. Chap. 21). From these explicit forms, we can then (usually) see directly which quantity must be adjusted for the additional optimization.† We consider binary detection in normal noise first, separating the results into the two detection classes $S + N$ versus N and $S_2 + N$ versus $S_1 + N$. From Chap. 20, we can write at once the following sets of error probabilities:

I. *Coherent simple-alternative detection* ($S + N$ versus N), Eq. (20.120)

$$\begin{Bmatrix} \alpha^* \\ \beta^* \end{Bmatrix} = \frac{1}{2}\left\{1 - \Theta\left[\frac{a_0\Phi_s^{1/2}}{2\sqrt{2}} \pm \frac{\log(\mathcal{K}/\mu)}{\sqrt{2}\,a_0\Phi_s^{1/2}}\right]\right\} \qquad \text{all } a_0 \geq 0 \tag{23.56}$$

II. *Coherent threshold single- (composite-) alternative detection* ($S + N$ versus N), Eq. (20.79)

$$\begin{Bmatrix} \alpha^* \\ \beta^* \end{Bmatrix} \simeq \frac{1}{2}\left\{1 - \Theta\left[\frac{\bar{a}_0\Phi_s^{1/2}}{2\sqrt{2}} \pm \frac{\log(\mathcal{K}/\mu)}{\sqrt{2}\,\bar{a}_0\Phi_s^{1/2}}\right]\right\} \tag{23.57}$$

III. *Incoherent threshold detection* ($S + N$ versus N), Eq. (20.91)

$$\begin{Bmatrix} \alpha^* \\ \beta^* \end{Bmatrix} \simeq \frac{1}{2}\left\{1 - \Theta\left[\frac{\overline{a_0^2}\langle\Phi_G\rangle^{1/2}}{4} \pm \frac{\log(\mathcal{K}/\mu)}{\overline{a_0^2}\langle\Phi_G\rangle^{1/2}}\right]\right\} \tag{23.58}$$

† Some problems of this type for both detection and extraction have been considered by Turin[3,4] and Helstrom.[5]

IV. *Coherent simple-alternative detection* ($S_2 + N$ versus $S_1 + N$), Prob. 20.12*c*, Eq. (3)

$$\begin{Bmatrix} \alpha^* \\ \beta^* \end{Bmatrix} = \frac{1}{2}\left\{1 - \Theta\left[\frac{\Phi_{s_{21}}{}^{1/2}}{2\sqrt{2}} \pm \frac{\log(\mathcal{K}/\mu)}{\sqrt{2}\,\Phi_{s_{21}}{}^{1/2}}\right]\right\} \qquad \text{all } a_{01}, a_{12} \geq 0 \quad (23.59)$$

V. *Coherent threshold single-alternative detection* ($S_2 + N$ versus $S_1 + N$), Prob. 20.12*d*

$$\begin{Bmatrix} \alpha^* \\ \beta^* \end{Bmatrix} \simeq \frac{1}{2}\left\{1 - \Theta\left[\frac{\Phi_{\bar{s}_{21}}{}^{1/2}}{2\sqrt{2}} \pm \frac{\log(\mathcal{K}/\mu)}{\sqrt{2}\,\Phi_{\bar{s}_{21}}{}^{1/2}}\right]\right\} \qquad (23.60)$$

VI. *Incoherent threshold detection* ($S_2 + N$ versus $S_1 + N$), Eq. (20.148*a*)

$$\begin{Bmatrix} \beta_2{}^{(1)*} \\ \beta_1{}^{(2)*} \end{Bmatrix} \simeq \frac{1}{2}\left\{1 - \Theta\left[\frac{\overline{a_0{}^2}\Psi_H{}^{1/2}}{4} \pm \frac{\log(\mathcal{K}/\mu)}{\overline{a_0{}^2}\Psi_H{}^{1/2}}\right]\right\} \qquad (23.61)$$

The Bayes risk R^* in all these cases is given by Eq. (19.19*a*). Here we shall consider specifically only those situations where $\mathcal{K}/\mu = 1$, so that R^* is effectively proportional to α^* ($= \beta^*$) and it is Φ_s, $\Phi_{\bar{s}}$, $\langle\Phi_G\rangle$, $\Phi_{s_{21}}$, $\Phi_{\bar{s}_{21}}$, and Ψ_H, respectively, which we seek to *maximize* in order to minimize the Bayes risk. Similarly, let us mention by way of illustration two simple signal-extraction problems, where the Bayes risk is found to be specifically:

VII. *Coherent estimation of signal amplitude* (quadratic cost function, gauss d.d. of amplitudes), Eq. (21.102*c*)

$$R^*(\sigma,\delta^*)_{a_0} = \frac{C_0\sigma^2}{1 + \sigma^2\Phi_s} \qquad (23.62)$$

VIII. *Incoherent* (*threshold*) *estimation of signal amplitude* (simple cost function; Rayleigh d.d. of amplitudes), Eq. (21.126)

$$R^*(\sigma,\delta^*)_R \doteq -\frac{C_0}{2}\left(\frac{P_0}{e}\right)^{1/2} B_n - O(P_0{}^{3/2}) \qquad (23.62a)$$

Here again, when possible, maximization of Φ_s or B_n achieves the desired minimization of Bayes risk.

Accordingly, our task now in each instance is to maximize Φ_s, $\Phi_{\bar{s}}$, etc., in the above, subject to the necessary constraint† of fixed signal energy in the finite observation period $(0,T)$. From Chap. 20, we have specifically for both colored and white noise backgrounds:‡

† Without such a constraint, it would be possible to make α^*, β^* vanish with any signal of sufficient intensity vis-à-vis the noise background.

‡ Observe a slight, obvious change in notation: Eq. (23.63) vs. Eqs. (20.117*c*), etc.

I. Φ_s, Eqs. (20.117c)

$$\Phi_s = \psi \int_{0-}^{T+} s(t - \epsilon_0, \boldsymbol{\theta}_0) X_T(t;\epsilon_0,\boldsymbol{\theta}_0)\, dt \tag{23.63}$$

$$a_0{}^2\Phi_s \Big|_{\text{white}} = \frac{A_0{}^2}{W_0} \int_0^T s(t - \epsilon_0, \boldsymbol{\theta}_0)^2\, dt = \frac{2E_s}{W_0} \qquad \text{Eq. (20.121)} \tag{23.63a}$$

II. $\Phi_{\bar{s}}$, Eq. (20.79a)

$$\Phi_{\bar{s}} = \psi \int_{0-}^{T+} \langle s(t - \epsilon_0, \boldsymbol{\theta})\rangle \langle X_T(t;\epsilon_0,\boldsymbol{\theta})\rangle\, dt \qquad \bar{s} \neq 0 \tag{23.64}$$

$$\bar{a}_0{}^2\Phi_s \Big|_{\text{white}} = \frac{\bar{A}_0{}^2}{W_0} \int_0^T \langle s(t - \epsilon_0, \boldsymbol{\theta})\rangle^2\, dt \qquad \text{Eq. (20.79b)} \tag{23.64a}$$

III. $\langle\Phi_G\rangle$, Eq. (20.91a)

$$\langle\Phi_G\rangle = \psi^2 \iint_{-\infty}^{\infty} \langle D_T(t,u;\epsilon,\boldsymbol{\theta})\rangle \langle s(t - \epsilon, \boldsymbol{\theta}) s(u - \epsilon, \boldsymbol{\theta})\rangle\, dt\, du \tag{23.65}$$

$$\overline{a_0{}^2}^2 \langle\Phi_G\rangle \Big|_{\text{white}} = \frac{\overline{A_0{}^2}^2}{W_0{}^2} \iint_0^T \langle s(t - \epsilon, \boldsymbol{\theta}) s(u - \epsilon, \boldsymbol{\theta})\rangle^2\, dt\, du \qquad \text{Eq. (20.91b)} \tag{23.65a}$$

IV. $\Phi_{s_{21}}$, Prob. 20.12, Eq. (5)

$$\begin{aligned}\Phi_{s_{21}} = \psi \int_{0-}^{T+} [a_{02}{}^2 s_2(t - \epsilon_0, \boldsymbol{\theta}_{02}) X_2(t;\epsilon_0,\boldsymbol{\theta}_{02}) - 2a_{01}a_{02} s_2(t - \epsilon_0, \boldsymbol{\theta}_{02}) X_1(t;\epsilon_0,\theta_{01}) \\ + a_{01}{}^2 s_1(t - \epsilon_0, \boldsymbol{\theta}_{01}) X_1(t;\epsilon_0,\boldsymbol{\theta}_{01})]\, dt\end{aligned} \tag{23.66}$$

$$\Phi_{s_{21}} \Big|_{\text{white}} = \frac{1}{W_0} \int_0^T [A_{02} s_2(t - \epsilon_0, \boldsymbol{\theta}_{02}) - A_{01} s_1(t - \epsilon_0, \boldsymbol{\theta}_{01})]^2\, dt \qquad \text{Eq. (6)} \tag{23.66a}$$

V. $\Phi_{\bar{s}_{21}}$: Same as Eqs. (23.66), (23.66a), with a_{02}, s_2, X_2, etc., replaced by $\bar{a}_{02}$, $\bar{s}_2$, $\bar{X}_2$, etc.

VI. Ψ_H, Eq. (20.149a)

$$\begin{aligned}\Psi_H = \psi^2 \overline{A_0{}^2}^{-2} \iint_{-\infty}^{\infty} [\overline{A_{02}{}^2} \langle D_{2T}(t,u;\epsilon,\boldsymbol{\theta})\rangle \langle s_2(t - \epsilon, \boldsymbol{\theta}) s_2(u - \epsilon, \boldsymbol{\theta})\rangle \\ - 2\overline{A_{01}{}^2}\,\overline{A_{02}{}^2} \langle D_{1T}(t,u;\epsilon,\boldsymbol{\theta})\rangle \langle s_1(t - \epsilon, \boldsymbol{\theta}) s_2(u - \epsilon, \boldsymbol{\theta})\rangle \\ + \overline{A_{01}{}^2}^2 \langle D_{1T}(t,u;\epsilon,\boldsymbol{\theta})\rangle \langle s_1(t - \epsilon, \boldsymbol{\theta}) s_1(u - \epsilon, \boldsymbol{\theta})\rangle]\, dt\, du\end{aligned} \tag{23.67}$$

$$\begin{aligned}\overline{a_0{}^2}^2 \Psi_H \Big|_{\text{white}} = W_0{}^{-2} \iint_0^T [\overline{A_{02}{}^2} \langle s_2(t - \epsilon, \boldsymbol{\theta}) s_2(u - \epsilon, \boldsymbol{\theta})\rangle \\ - \overline{A_{01}{}^2} \langle s_1(t - \epsilon, \boldsymbol{\theta}) s_1(u - \epsilon, \boldsymbol{\theta})\rangle]^2\, dt\, du \qquad \text{Eq. (20.149b)}\end{aligned} \tag{23.67a}$$

VII. B_T, Eqs. (20.26), (20.39*a*)

$$B_T = \psi \int_{0-}^{T+} \langle s(t - \epsilon, \boldsymbol{\theta}) X_T(t,\epsilon,\boldsymbol{\theta}) \rangle \, dt \tag{23.68}$$

Both X_T and D_T follow from the basic integral equation (20.38) and the definition of Eqs. (20.46*b*), viz.,

$$\int_{0-}^{T+} K(t,u) X_T(u;\epsilon,\boldsymbol{\theta}) \, du = s(t - \epsilon, \boldsymbol{\theta}) \qquad 0- < t < T+ \tag{23.69a}$$

$$\begin{aligned} D_T(t,u;\epsilon,\boldsymbol{\theta}) &= X_T(t;\epsilon,\boldsymbol{\theta}) X_T(u;\epsilon,\boldsymbol{\theta}) \qquad && 0- < t,\, u < T+ \\ &= 0 && t,\, u \text{ outside } (0-,T+) \end{aligned} \tag{23.69b}$$

With these relations, we are now ready to examine the maximization of Φ_s, etc.

23.2-1 Single Signals in Noise (Detection and Extraction). The variational principle we shall invoke here and in much of the subsequent analysis takes the form

$$\max_s \bar{J} \text{ or } \delta \bar{J} = \boldsymbol{E}\{\delta J\} = 0 \qquad \delta^2 J < 0$$

where

$$J \equiv a_0{}^2 \Phi_s - \eta \frac{A_0{}^2}{2} \int_0^T s(t - \epsilon, \boldsymbol{\theta})^2 \, dt \tag{23.70}$$

Here η is a Lagrange multiplier, and the second term of J is simply the signal energy in $(0,T)$.

1. *max* Φ_s. We begin with the simplest case, I [Eq. (23.63)]. Remembering that ϵ and $\boldsymbol{\theta}$, as well as A_0, are specified, we have

$$J = \frac{A_0{}^2}{2} \int_{0-}^{T+} s(t - \epsilon_0, \boldsymbol{\theta}_0) X_T(t;\epsilon_0,\boldsymbol{\theta}_0) \, dt - \eta \frac{A_0{}^2}{2} \int_0^T s(t - \epsilon_0, \boldsymbol{\theta}_0)^2 \, dt \tag{23.71}$$

Taking the variation with respect to s [and hence X_T, by virtue of Eq. (23.69*a*)] we obtain

$$\delta J = \int_{0-}^{T+} \delta X_T(t;\epsilon_0,\boldsymbol{\theta}_0) \left\{ \int_{0-}^{T+} K_N(t,u)[-2\eta s(u - \epsilon_0, \boldsymbol{\theta}_0) + X_T(u;\epsilon_0,\boldsymbol{\theta}_0)] \, du \right\} dt = 0 \tag{23.72}$$

so that for arbitrary δX_T the extremum condition is simply

$$\int_0^T K_N(t,u) s(u - \epsilon_0, \boldsymbol{\theta}_0) \, du = \frac{1}{2\eta} s(t - \epsilon_0, \boldsymbol{\theta}_0) \qquad 0 \leq t \leq T \tag{23.73}$$

where we have used Eq. (23.69*a*) once more in the second term of Eq. (23.72).

Equation (23.73) is now recognized as a homogeneous Fredholm integral equation [of the second kind; cf. Eq. (A.2.1)], which has solutions when $2\eta = \lambda_j^{-1}$ $(j = 1, \ldots)$, and s becomes s_j, the jth member of a complete orthonormal† set $\{s_j\}$, *provided* that $s(t - \epsilon_0, \boldsymbol{\theta}_0) = s_j(t)$ for the general kernel $K_N(t,u)$, or $s_j(t - \epsilon_0)$ for the stationary noise process‡ $K_N(|t - u|)$.

The condition (23.73) is accordingly for nonstationary and stationary noise processes

$$\int_0^T K_N(t,u)s_j(u)\,du = \lambda_j s_j(t) \qquad (23.73a)$$

$$\int_0^T K_N(t-u)s_j(u-\epsilon_0)\,du = \lambda_j s_j(t-\epsilon_0) \qquad 0 \le t \le T \qquad (23.73b)$$

and the desired signals that extremize Φ_s are any members of the orthonormal set§ $\{s_j\}$. Comparing Eq. (23.73*a*) and the basic integral equation (23.69*a*), we have the important relations that

$$0 \le t \le T \qquad X_T(t;\epsilon_0,\boldsymbol{\theta}_0) = \begin{cases} X_T(t) = s_j(t)/\lambda_j \\ \text{or} \\ X_T(t;\epsilon_0) = s_j(t-\epsilon_0)/\lambda_j \end{cases} \qquad (23.74)$$

The condition for a maximum follows from Eq. (23.72). We have

$$\delta^2 J = \int_0^T [\delta s_j(t)]^2\,dt - \lambda_j^{-1} \iint_0^T \delta s_j(t) K_N(t,u)\,\delta s_j(u)\,dt\,du < 0 \qquad (23.75)$$

which is satisfied for colored noise backgrounds, as we shall note presently from other considerations. Observe, also, that the energy in the signal is invariant of the choice of a particular $s_j(t)$: $E_0 \equiv A_0{}^2/2 \int_0^T s_j(t)^2\,dt = A_0{}^2/2$, since the s_j are orthonormal.

Let us now evaluate $(\Phi_s)_{\text{extremum}}$. From Eqs. (23.63) and (23.74), we get directly

$$(\Phi_s)_{\text{extremum}} = \frac{\psi}{\lambda_j} \int_0^T s_j(t)^2\,dt = \frac{\psi}{\lambda_j}\ (>0) \qquad (23.76)$$

For rational kernels $K_N(|t-u|)$, it can be shown that λ_j *decreases* as j increases. Thus, consider, for example, *RC* noise backgrounds, where $K_N = \psi e^{-\alpha|t-u|}$. From item 1, Table 17.1, the eigenvalues λ_j are determined by $\lambda_j = 2\psi/\alpha(1 + q_j{}^2)$, with q_j the positive roots of $\tan \alpha T q_j = -2q_j/(1 - q_j{}^2)$. For large j, λ_j is $O(j^{-2})$ [cf. Prob. 17.5, Eq. (2)]. Consequently, $(\Phi_s)_{\text{extremum}}$ can be made indefinitely large simply by choosing an eigenfunction $s_j(t)$ of sufficiently high order, showing, incidentally, that

† $K_N(t,u)$ is assumed to be positive definite, as well as symmetrical, so that the λ_j are all positive and distinct.

‡ The eigenvalues λ_j and eigenfunctions are determined solely by the kernel $K_N(t,u)$ and the limits $(0,T)$. Since neither K_N nor the limits depend on $\boldsymbol{\theta}_0$, so also must the s_j be independent of $\boldsymbol{\theta}_0$. Similarly, s_j cannot depend on ϵ_0, unless the noise is stationary, as then $K_N(|t-u|) = K_N[|t - \epsilon_0 - (u - \epsilon_0)|]$.

§ Note that these signals also form the orthonormal set in terms of which the background noise process may be represented MS (cf. Secs. 8.2-2 and 8.2-3).

Eq. (23.73) maximizes Φ_s. Physically, these results are easily explained: the higher the order j, the higher the center frequency of $s_j(t)$, so that in effect we have a signal of fixed energy concentrated in a comparatively narrow zone which progressively moves out into spectral regions where the background noise is falling off to zero. It is assumed, of course, that our receiver employs the proper Bayes matched filter [cf. Eqs. (20.41*a*), (20.43*d*)], based on the particular $s_j(t)$ selected as above. Practically, we are limited in both bandwidth and frequency, so that while j may be large, it is not infinite. The extent by which α^* and β^* are decreased in comparison with the error probabilities for arbitrary signal waveforms may be calculated directly from Eq. (23.76) in Eq. (23.56).

There remains the important case of white noise backgrounds. Here $K_N(t,u) = W_0\delta(t-u)/2$, so that $\lambda_j = W_0/2$ (all j), and *any* $s(t-\epsilon_0;\boldsymbol{\theta}_0)$ will serve, as long as $(A_0^2/2)\int_0^T s(t-\epsilon_0;\boldsymbol{\theta}_0)^2\,dt$ remains constant: all signals of the same energy are equally effective. There is no maximum (or minimum), and $\delta^2 J = 0$. This result might have been anticipated at once from Eq. (23.63*a*) and is a typical example of the invariance of system performance with respect to waveform under the rather narrow conditions of the present problem.

2. *max* $\Phi_{\bar{s}}$. Let us now turn to the next more complicated situation of II [Eq. (23.64)], where s and X_T in $\Phi_{\bar{s}}$ are to be averaged over the possible random parameters $\boldsymbol{\theta}$ (we recall that ϵ_0 is specified or adjusted, but not random). We suspect that the results above will apply here also, with an optimum choice of $s = s_j(t-\epsilon_0)$ or $s_j(t)$ again, independent of the parameters $\boldsymbol{\theta}$. That s_j must be independent of $\boldsymbol{\theta}$ (and sometimes of ϵ_0) can be shown from the following argument: We first write

$$J = \frac{A_0^2}{2}\int_{0-}^{T+} X_T(t;\epsilon_0,\boldsymbol{\theta}_1)s(t-\epsilon_0,\boldsymbol{\theta}_2)\,dt - \eta\,\frac{A_0^2}{2}\int_0^T s(t-\epsilon_0,\boldsymbol{\theta}_1)^2\,dt \tag{23.77}$$

For an extremum, we require that $\delta \mathbf{E}_{\boldsymbol{\theta}_1,\boldsymbol{\theta}_2}\{J\} = \mathbf{E}_{\boldsymbol{\theta}_1,\boldsymbol{\theta}_2}\{\delta J\} = 0$. Carrying out the indicated operations as before, we get finally

$$\delta J = \frac{A_0^2}{2}\int_{0-}^{T+} \delta X_T(t;\epsilon_0,\boldsymbol{\theta}_2)\left[s(t-\epsilon_0,\boldsymbol{\theta}_1) - \eta\int_0^T K_N(t,u)s(u-\epsilon_0,\boldsymbol{\theta}_2)\,du\right]dt + \frac{A_0^2}{2}\int_{0-}^{T+} \delta X_T(t;\epsilon_0,\boldsymbol{\theta}_1)s(t-\epsilon_0,\boldsymbol{\theta}_2)\,dt = 0 \tag{23.78}$$

But unless s is independent of $\boldsymbol{\theta}$, there is no solution. With s independent of $\boldsymbol{\theta}$, we get Eq. (23.73) once more, and Eqs. (23.74) to (23.76) apply again, with the same physical interpretation.

3. *max* $\langle\Phi_G\rangle$. Here a straightforward application of variational techniques leads to a condition for a minimum of $\langle\Phi_G\rangle$ [Eq. (23.65)], rather than a maximum. However, for stationary noise backgrounds we can make

$\langle\Phi_G\rangle$ arbitrarily large, as was the case for Φ_s, $\Phi_{\bar{s}}$ above, by again using as the maximizing waveform $s(t - \epsilon, \boldsymbol{\theta}) = s_j(t - \epsilon)$ on $(0,T)$, where s_j is determined from Eq. (23.73) and ϵ is now random. With the help of Eq. (23.74), we obtain

$$\langle\Phi_G\rangle_{\max} = \frac{\psi^2}{\lambda_j^2} \iint_0^T \rho_s(t,u)_j{}^2 \, dt \, du \tag{23.79}$$

where

$$\rho_s(t,u)_j \equiv \langle s_j(t - \epsilon) s_j(u - \epsilon)\rangle_\epsilon \tag{23.79a}$$

For white noise, Eq. (23.65*a*) becomes $\langle\Phi_G\rangle\Big|_{\text{white}} = \dfrac{\overline{A_0{}^2}^2}{W_0{}^2} \displaystyle\iint_0^T \rho_s(t,u)^2 \, dt \, du$ for *any* $s(t - \epsilon)$ which obeys the constraint of constant energy. As in the coherent cases, a suitable matched filter is employed, based on the particular $s_j(t - \epsilon)$ chosen at the transmitter. The decrease in α^* and β^* with j here is similar to that in the coherent cases.

4. *Extraction* [Eqs. (23.62), (23.63)]. We can apply the preceding results at once to the two amplitude-estimation examples VII, VIII (page 1048). Again, Φ_s is maximized according to Eq. (23.73) and is specifically ψ/λ_j [cf. Eq. (23.76)], so that the minimized Bayes risk [Eq. (23.62)] of the first example is

$$\min_s R^*_{a_0} = C_0\sigma^2/(1 + \sigma^2\psi/\lambda_j) \tag{23.80}$$

which approaches zero as $\lambda_j \to 0$. No improvement can be obtained in the white noise case: all signals having the same energy are equally effective.

The Bayes risk [Eq. (23.63)] of the second extraction example (VIII) is made smaller (i.e., more negative) with $s(t - \epsilon, \boldsymbol{\theta}) = s_j(t - \epsilon)$ as in (3) above when the noise process is stationary. The result is

$$\min_s R^*_R = -\frac{C_0}{2}\left(\frac{P_0}{e}\right)^{1/2} \frac{\psi}{\lambda_j} \tag{23.81}$$

again, with all signals of constant energy equally effective when the background noise is white [cf. Eq. (21.126*a*)].

23.2-2 Detection of S_2 versus S_1 in Noise. When we have to distinguish between two signals in the binary detection process, that is, $S_2 + N$ versus $S_1 + N$, a number of new results may be expected. For instance, although the noise may be white, it is possible to choose S_2, say, given S_1, so as to reduce the Bayes error probabilities still further. If the noise is colored, not only can we obtain an additional optimization by choice of the waveform for S_2, but this can be further enhanced, in principle, by choosing S_1 for optimum Bayes performance in noise according to Sec. 23.2-1. Just what waveforms are to be chosen depends ultimately on the noise and on the mode of observation.

1. *max* $\Phi_{s_{21}}$. We consider first the case of two entirely specified signals S_1 and S_2, each of which is observed coherently in normal noise, so that our detector tests† $S_2 + N$ against $S_1 + N$. Let S_1 be selected arbitrarily for the moment. We ask now for the waveform of S_2 (in terms of S_1) so that the Bayes risk—here the error probabilities [Eq. (23.59)]—is minimized, subject to the usual constraints of fixed signal energies for S_2 (and S_1). Alternatively, we are to maximize $\Phi_{s_{21}}$ [Eq. (23.66)], subject to these constraints.

Let us begin by rewriting $\Phi_{s_{21}}$ [cf. Prob. 20.12, Eq. (4)] as

$$\Phi_{s_{21}} = \psi \int_{0-}^{T+} [S_2(t - \epsilon_0, \boldsymbol{\theta}_0) - S_1(t - \epsilon_0, \boldsymbol{\theta}_0)]X_{21}(t;\epsilon_0,\boldsymbol{\theta}_0)\, dt \qquad (23.82)$$

where‡ $\quad S_1 = a_{01}s_1 \qquad S_2 = a_{02}s_2 \qquad X_{21} = a_{02}X_2 - a_{01}X_1 \qquad (23.82a)$

with

$$S_2(t - \epsilon_0, \boldsymbol{\theta}_0) - S_1(t - \epsilon_0, \boldsymbol{\theta}_0) = \int_{0-}^{T+} K_N(t,u)[X_2(u;\epsilon_{01},\boldsymbol{\theta}_0)a_{02} - X_1(u;\epsilon_0,\boldsymbol{\theta}_0)a_{01}]\, du \qquad (23.82b)$$

and X_1, X_2 are the solutions of Eq. (23.69*a*) when $s = s_1, s_2$, respectively.

Next, we seek the variation of§

$$J = \psi \int_{0-}^{T+} [S_2(t) - S_1(t)]X_{21}(t)\, dt + \lambda\psi \int_0^T S_2(t)^2\, dt \qquad (23.83)$$

with respect to S_2. From Eqs. (23.82*b*) and (23.83), we see that

$$\delta S_2 = \int_{0-}^{T+} K_N(t,u)\, \delta X_2(u)\, du \qquad (23.84a)$$

and

$$\delta J = \psi \int_{0-}^{T+} [\delta S_2 X_{21} + (S_2 - S_1)\, \delta X_{21}]\, dt + 2\lambda\psi \int_0^T S_2\, \delta S_2\, dt \qquad (23.84b)$$

The result for δJ is

$$\psi \int_{0-}^{T+} \delta X_{21}(t) \left\{ \int_{0-}^{T+} K_N(t,u)[X_{21}(u) + 2\lambda S_2(u)]\, du + S_2(t) - S_1(t) \right\} dt$$

which when set equal to zero yields the condition

$$\int_{0-}^{T+} [\lambda K_N(t,u) + \delta(t - u)]S_2(u)\, du = S_1(t) \qquad 0- < t < T+ \qquad (23.85)$$

with the help of Eq. (23.82*b*).¶ The second variation similarly gives

$$\delta^2 J = -\psi \iint_{0-}^{T+} \delta X_{21}(t)K_N(t,u)\, \delta X_{21}(u)\, dt\, du - \lambda^{-1}\psi \int_0^T (\delta X_{21})^2\, dt \qquad (23.86)$$

† This is an example of ΦSK, or "phase-shift keying."

‡ Note that here and subsequently S_1, S_2 are normalized [cf. Eqs. (19.49*a*), (19.49*b*)].

§ We drop the explicit notation for ϵ_0, $\boldsymbol{\theta}_0$ for the moment.

¶ Note also that $X_{21} = -\lambda S_2$ from Eqs. (23.85) and (23.82*b*).

This may or may not be negative, depending on λ. Equation (23.85) is the desired relation between S_2 and S_1 for extremizing $\Phi_{s_{21}}$. We may solve Eq. (23.85) by the methods of Appendix 2 when the kernel is rational [cf. Eqs. (A.2.62) and (A.2.75) for a solution in the RC case].

To illustrate the general result [Eq. (23.85)], let us consider again the simple case of white noise, so that $K_N(t,u) = (W_0/2)\delta(t-u)$. This gives directly

$$S_2(t) = \left(1 + \frac{\lambda W_0}{2}\right)^{-1} S_1(t) \qquad 0 \le t \le T \tag{23.87}$$

and λ is determined from

$$E_2 = \psi \int_0^T S_2(t)^2\,dt = \frac{A_{02}^2}{2}\int_0^T s_2(t)^2\,dt = \psi\left(1+\frac{\lambda W_0}{2}\right)^{-2}\int_0^T S_1(t)^2\,dt = \left(1+\frac{\lambda W_0}{2}\right)^{-2} E_1 \tag{23.88}$$

Setting $\gamma \equiv E_2/E_1\ (>0)$, we find that

$$\lambda = \frac{2}{W_0}(\pm\gamma^{1/2} - 1) \tag{23.89a}$$

so that Eq. (23.87) becomes

$$S_2(t) = \pm\sqrt{\gamma}\,S_1(t) \qquad \text{or} \qquad s_2(t-\epsilon_0, \boldsymbol{\theta}_0) = \pm\sqrt{\gamma}\,\frac{a_{01}}{a_{02}}\,s_1(t-\epsilon_0,\boldsymbol{\theta}_0) \tag{23.89b}$$

Substituting this back into Eq. (23.82), we get

$$\Phi_{s_{21}} = 2\,\frac{(\pm\sqrt{\gamma}-1)^2}{W_0}\,E_1 \tag{23.90}$$

which shows that for $+\sqrt{\gamma}$ we have a *minimum* and for $-\sqrt{\gamma}$ we obtain the desired *maximum*. This latter is quite reasonable: for optimum coherent performance (in white noise), the second waveform should be proportional to the *negative* of the first, the factor of proportionality being the ratio of the signal energies. Note when these signal energies are equal ($\gamma = 1$) that $S_2(t) = S_1(t)$ yields the minimum $\Phi_{s_{21}} = 0$, as expected, and, with $S_2(t) = -S_1(t)$, we get the maximum $8E_1/W_0$. Equation (23.59) gives the minimum Bayes error probabilities as before. Finally, observe that, if S_1 is zero, Eq. (23.85) reduces once more to Eq. (23.73), where s_j is independent of $\boldsymbol{\theta}_0$ and possibly ϵ_0.

In the preceding we have assumed that the energies of S_1 and S_2 in $(0,T)$ are individually constrained, that is, E_1, E_2 are separately fixed. However, if we relax this constraint and require only that the *total* energy, $E_1 + E_2$, be specified, we can then ask how this total energy should be divided between the two signals in order further to minimize the Bayes risk. This is easily done if we set $E_2 = \gamma E_1$ [cf. Eq. (23.89*a*)] and observe that $E_2 = C - E_1$

(C = constant), so that $E_1 = C/(\gamma + 1)$. Substitution into Eq. (23.90) then gives directly†

$$(\Phi_{s_{21}})_{\max} = \frac{2C}{W_0}\left(1 + \frac{2\sqrt{\gamma}}{\gamma + 1}\right) \tag{23.90a}$$

which is further maximized for $\gamma = 1$, or $E_1 = E_2 = C/2$. Accordingly, equal signal energies, for a constraint of fixed total energy, yield the desired optimization here.

2. *max* $\Phi_{\bar{s}_{21}}$. This is essentially the same as (1) above, but now in order to achieve the extremum [Eq. (23.85)] $s_{1,2}(t - \epsilon_0, \boldsymbol{\theta})$ must be independent of $\boldsymbol{\theta}$ at least [the argument parallels that in (2), Sec. 23.2-1]. With this condition, Eqs. (23.85) et seq. apply here.

3. *max* Ψ_H. To maximize Ψ_H, we observe directly from Eq. (23.67) that, since $\rho_{s_2}(t,u) \equiv \overline{s_2(t - \epsilon, \boldsymbol{\theta})s_2(u - \epsilon, \boldsymbol{\theta})}$, etc., are all positive definite, we should require that

$$\iint_{0-}^{T+} \rho_{s_2}(t,u)\rho_{1T}(t,u)\, dt\, du = 0 \tag{23.91}$$

i.e., that the covariance functions of s_2 and of the weighting function X_1 associated with s_1 through Eq. (23.69a) be orthogonal on the square $(0- < t, u < T+)$. For white noise, we see from Eq. (23.67a) that this implies

$$\iint_{0}^{T} \rho_{s_2}(t,u)\rho_{s_1}(t,u)\, dt\, du = 0 \tag{23.92}$$

The covariance functions of s_2 and s_1 are orthogonal over $(0 \le t, u \le T)$. Specifically, we have from Eq. (23.67a)

$$\overline{a_0^{2}}^{2}\, \Psi_H \Big|_{\text{white}} = \frac{1}{W_0^{2}} \iint_{0}^{T} [\overline{A_{02}^{2}}^{2}\, \rho_{s_2}(t,u)^2 + \overline{A_{01}^{2}}^{2}\, \rho_{s_1}(t,u)^2]\, dt\, du > 0 \tag{23.93}$$

while, if we choose $\rho_{s2} = \rho_{s1}$, Eq. (23.67a) gives $\overline{a_0^{2}}^{2}\, \Psi_H \Big|_{\text{white}} = 0$, for an obvious minimum. These conditions are clearly consistent with the energy constraint $\int_0^T \rho_{s_2}(t,t)\, dt$ = a constant and $\int_0^T \rho_{s_1}(t,t)\, dt$ = another constant.

Finally, as in case 1 above, we have assumed that the energy in each signal is separately fixed at the outset. However, if we require only that the total energy be fixed, for example, $E_1 + E_2 = C$, a constant, we can easily investigate the additional question of how to divide the total available energy between the two signals, as well as choose their waveforms properly, for further maximization of Ψ_H (and hence minimization of the Bayes risk

† We recall, of course, that $\Phi_{\max}$ produces minimum Bayes risk [cf. the remarks following Eq. (23.61)].

here). Let us illustrate with the case of white noise backgrounds, so that we can write Eq. (23.93) now as

$$\overline{a_0^{2}}^{2}\Psi_H\Big|_{\text{white}} = \alpha_1 E_1^2 + \alpha_2 E_2^2 \tag{23.94a}$$

with

$$E_1 + E_2 = C \qquad (> 0) \tag{23.94b}$$

and α_1, $\alpha_2 > 0$. Taking the variation of Eq. (23.94*a*) with Eq. (23.94*b*) as constraint, we easily obtain $E_1 = C(1 + \alpha_1/\alpha_2)^{-1}$, $E_2 = C\alpha_1/\alpha_2(1 + \alpha_1/\alpha_2)$ and hence

$$\overline{a_0^{2}}^{2}\Psi_H\Big|_{\text{white}} = \frac{\alpha_1 C^2}{1 + \alpha_1/\alpha_2} \tag{23.95}$$

From this it is clear that Eq. (23.95) is a *minimum* if $\alpha_1 = \alpha_2$, of magnitude $C^2\alpha_1/2$. For a *maximum*, on the other hand, we see by inspection that in Eq. (23.94*a*) we should set the coefficient of the smaller of α_1, α_2 equal to zero. Thus, if $\alpha_1 > \alpha_2$, then we get for the maximum $E_1 = C$, $E_2 = 0$, with $\overline{a_0^{2}}^{2}\Psi_H\Big|_{\text{white,max}} = \alpha_1 C^2$. The difference between the maximum and minimum states is $\sqrt{2} = 1.5$ db. This is to be compared with the coherent situation in case 1 above, where a maximum (in Φ_s) is obtained when the two signal energies are equal (in white noise backgrounds). Theoretically, then, incoherent reception should be "on-off," with the total energy in one signal, rather than have two signals with some nonzero proportioning of energy between them. Actual practice, however, chooses the minimum condition ($\alpha_1 = \alpha_2$), i.e., signals of equal energy, because the relatively small loss of about 1.5 db vis-à-vis the "on-off" case is more than compensated for by the greater reliability in distinguishing one signal from the other, when there is fading or other phenomena to which there is an absolute threshold sensitivity.

23.3 *Relations to Game Theory*

Before we conclude our study with a critique of the decision-theory approach, let us briefly mention its connection with game theory. Recent applications of linear and dynamic programming methods to communication problems have been made by Bellman,[6–9] Kalaba,[6,9,10] and Juncosa.[10] For details, see Refs. 11 to 14.

23.3-1 Game Theory. The ideas underlying the present view of reception are closely related to those of game theory, as developed by Von Neumann[15] and others.[16,17] In fact, Wald[18] was led to the fundamental result embodied in his complete class theorem by considering the decision problem as a zero-sum two-person game between Nature and the Observer. Some familiarity with the terminology used in the analysis of such strategic situations is accordingly helpful to our understanding of the basic approach to reception outlined here, and for extending its application to new problems.[19,20]

In its simplest form, a zero-sum two-person game involves two players, each with a set of possible "moves," or courses of action, denoted by $\{a_i\}$ and $\{b_j\}$, respectively, and a *payoff function* $A(a_i,b_j) \geqq 0$. Players 1 and 2 select their moves a_i and b_j, each without knowledge of the other's choice, and thereupon receive payments of $A(a_i,b_j)$ and $-A(a_i,b_j)$, respectively. Player 1 clearly wants to maximize A, while Player 2 seeks to minimize A. Game theory is concerned with finding optimal ways of playing such games.

As an illustration, suppose each player has two possible moves (a_1, a_2 for Player 1 and b_1, b_2 for Player 2) and that the schedule of payments is represented by the following *payoff matrix:*

$$\mathbf{A} = [A(a_i,b_j)] = \begin{bmatrix} 6 & 5 \\ 3 & 4 \end{bmatrix} = \begin{bmatrix} A(a_1,b_1) & A(a_1,b_2) \\ A(a_2,b_1) & A(a_2,b_2) \end{bmatrix} \tag{23.96}$$

Here we may consider that Player 1 moves by choosing a row and Player 2 by choosing a column, the final payoff being indicated by the matrix element at the intersection of the row and column. It is evident that Player 1 can gain at least 5 by the right choice (row 1) regardless of what Player 2 does and also that Player 2 can avoid losing more than 5 (by choice of column 2) regardless of what Player 1 does. If each player is ignorant of the other's choice, there is good justification for them to play in this way. The choices of row 1 by Player 1 and column 2 by Player 2 are called Minimax strategies, since clearly the principle involved is that Player 1 plays to maximize his minimum gain, while Player 2 plays to minimize his maximum loss.

In this example, we have

$$\max_{a_1} \min_{a_2} A(a_1,a_2) = \min_{a_2} \max_{a_1} A(a_1,a_2) = v \tag{23.97}$$

with $v = 5$. When Eq. (23.97) holds, the game is *strictly determined* and is said to have the value v. Under this condition, with both players using their Minimax strategies, neither player would be able to do better, even if his opponent's strategy became known to him. These strategies, therefore, may be defined as optimal for two players, each in complete ignorance of his opponent's choice.

Suppose, on the other hand, that the payoff matrix is

$$\mathbf{A} = \begin{bmatrix} 2 & 5 \\ 4 & 3 \end{bmatrix} \tag{23.98}$$

The Minimax strategies of Players 1 and 2 here are the choices of row 2 and column 1, respectively. The game is not strictly determined, however, because Eq. (23.97) is not satisfied. As a result, if Player 2 finds out that

Player 1 is using Minimax strategy, Player 2 can improve his situation by abandoning his Minimax choice (column 1) and choosing column 2 instead. It then becomes important for Player 1 to make his choice in a way that cannot be foreseen exactly by Player 2. He may do this with the aid of a chance mechanism that selects row 1 with probability P_1 and row 2 with probability P_2. Choice of the probabilities $\mathbf{P} = (P_1,P_2)$ is called a *mixed* strategy, to distinguish it from the *pure* strategies or simple nonrandom choices discussed above. Of course, Player 1 must assume in general that Player 2 will use a mixed strategy $\mathbf{Q} = (Q_1,Q_2)$ as well, and accordingly the Minimax (mixed) strategy of Player 1 here is to choose $\mathbf{P}$ to maximize his minimum expectation of gain in view of the choices of $\mathbf{Q}$ open to Player 2. Similar remarks hold for the Minimax strategy of Player 2. The condition for strict determinateness when mixed strategies are allowed becomes

$$\max_{\mathbf{P}} \min_{\mathbf{Q}} \bar{A}(\mathbf{P},\mathbf{Q}) = \min_{\mathbf{Q}} \max_{\mathbf{P}} \bar{A}(\mathbf{P},\mathbf{Q}) = v \tag{23.99}$$

where

$$\bar{A}(\mathbf{P},\mathbf{Q}) = \Sigma_i\Sigma_j P_i A(a_i,b_j) Q_j \tag{23.100}$$

and is termed the *outcome function* of the game. Again, whenever Eq. (23.99) holds for a game, neither player can improve his expectation by finding out his opponent's strategy if both are using their optimal (Minimax) strategies. Moreover, although a game is not always strictly determined when pure strategies alone are available to the players, as illustrated above, Von Neumann[15] has shown that such games are *always* strictly determined when mixed strategies are allowed, provided that the number of alternatives is finite. It may be verified, for example, that the optimal strategies for the game of Eq. (23.98) are $\mathbf{P} = (\frac{1}{4},\frac{3}{4})$ and $\mathbf{Q} = (\frac{1}{2},\frac{1}{2})$ with $v = 3\frac{1}{2}$.

In the decision theory of reception, the two "players" are Nature, supplying the noise and possibly a signal, and the Observer, or receiver, in the form of a suitable terminal decision process. The payoff function is the conditional risk $r(\mathbf{S},\delta)$, and the outcome function is the average risk $R(\sigma,\delta)$. The pure strategies of Nature and the Observer are choices of signal $\mathbf{S}$ and decision rule δ, respectively, while the mixed strategies are the choice of an a priori distribution $\sigma(\mathbf{S})$ over the class Ω of possible signal inputs by Nature and the choice of a distribution over the class $\mathfrak{D}$ of possible decision rules by the observer. In this connection, note that, since the decision rule δ is itself a probability distribution, the choice of a distribution over all δ is equivalent to a simple choice of δ. This way of formulating the game turns out to be convenient for showing that under certain conditions (see Wald[18]) the relation of strict determinateness analogous to Eq. (23.99), namely,

$$\min_\delta \max_\sigma R(\sigma,\delta) = \max_\sigma \min_\delta R(\sigma,\delta) \tag{23.101}$$

with

$$R(\sigma,\delta) = \int_\Omega r(\mathbf{S},\delta)\sigma(\mathbf{S})\, d\mathbf{S} \tag{23.102}$$

holds, not only when the spaces Ω and $\mathfrak{D}$ of pure strategies are finite, but also when they contain countably or even continuously many elements. Thus, when Nature's mixed strategy $\sigma(\mathbf{S})$ is unknown to the Observer, an optimal strategy for the latter exists under these conditions, requiring in essence that he proceed as if Nature were using the least favorable strategy as far as he is concerned. If, on the other hand, the Observer knows Nature's $\sigma(\mathbf{S})$, there is usually no question then of his choice of optimum strategy, since Bayes procedures are usually unique, as indicated, for example, by the specific Bayes detectors and extractors considered in Chaps. 19 to 21.

23.4 *A Critique of the Decision-theory Approach*

In the preceding chapters, we have shown how the methods of statistical decision theory can be applied to such central communication problems as signal detection (Chaps. 19, 20, Sec. 23.1), signal extraction (Chap. 21), and system optimization by choice of waveform at the transmitter (Sec. 23.2). We conclude our treatment here with a summary of the achievements and difficulties of the method. Only decision systems have been examined, i.e., systems whose output is a "yes," a "no" (in detection), or a magnitude (in estimation). These exist, of course, whenever it is reasonable to speak of error probabilities in connection with system performance: since "error" implies a value judgment or distinction between categories (quantitatively measured by a cost function), and since this in turn implies a decision or selection, we have symbolically "error" $\leftrightarrow$ "value" $\leftrightarrow$ "decision," in any appropriate order.

Let us begin with some of the achievements of the decision-theoretical approach in communication theory. First, we observe that the method is quite general, not only in the way that it uses all available information, but also in the ways it provides of incorporating the observer's "ignorance," as well—reflected, for example, in the random nature of signal parameters, or in the nonzero or nonunity values of signal a priori probabilities. The approach also provides us with the three principal ingredients of an effective theory: (1) a representation of optimum system structure; (2) an evaluation of the expected performance of this optimum system; and (3) a quantitative method for the consistent comparison of actual suboptimum systems with the theoretical optimum for the same purpose. As we have demonstrated in preceding chapters, an explicit threshold theory is possible, giving us (1), (2), and (3) in detail.

Apart from specific systems and their evaluation, decision-theory methods also offer us a much broader and potentially more realistic technique of system design than earlier methods based on lower-order statistics, e.g., the spectra, correlation functions, second-order distribution densities, etc., characteristic of analyses based on calculations of signal-to-noise ratio, for instance. These latter, though useful because they are readily calculable, are incomplete, not only statistically, but also in formulation: the essential decision features of detection or estimation are omitted, and the full implications of the underlying assumptions are not always evident. With a decision-theory formulation, on the other hand, the assumptions and constraints of any particular problem are brought out into the open by its very formulation as a statistical decision process. The analyst and system designer is forced at the outset to a careful statement of his problem, after which the various results (1) to (3) above follow according to the indicated operations and the available a priori information. For example, one avoids the suboptimization that may result in the attempt to answer two questions at once with a single set of operations on the received data, such as a simultaneous detection of a radar signal and estimation of its range. For optimization of each operation, mutually incompatible functions are usually required. With a single operation, optimization for one is achieved at the expense of the other; acceptable performance for both is a compromise, with neither operation an optimum.

The theory also provides a definite method for handling new situations, where current experience and intuition may be inadequate. A systematic approach to system design is another feature which permits us not only to obtain the explicit operations on the data but to interpret these operations as ordered sequences of realizable physical elements, available to the practicing engineer (cf. Secs. 20.1-1 to 20.2-4). Perhaps the least exploited feature so far is the comparison of systems. We remark in this connection that actual systems are never strictly equivalent to the corresponding optima derived from theory. However, the theoretical optimum not only provides us with a limit on expected performance which the actual systems can approach but also acts as a guide to the designer. His real, approximate system should embody the significant features of the optimum structure. Similarly, suboptimum systems can be compared, not only with the optimum, but with one another. With the growing improvement of components—in size, cost, operating ranges, etc.—the designer may expect to find it progressively easier to approximate the ideal structures, and as more analytical experience is gained, still greater realism can be incorporated into the theoretical models (compare, for instance, the examples discussed in Secs. 20.4-1 to 20.4-7).

However, one should not gain the impression that decision-theory methods are without limitations or that they can always be applied for equally effective results. The very generality of the approach puts demands on the model which are never met fully, and sometimes only slightly, in practice. There are three main areas where a decision-theory model presents difficulties: (1) in the apparent arbitrariness of the cost assignments; (2) in the usual inadequacy of the a priori information; and (3) in the selection of the criterion of optimality itself.

Let us begin with (1). We observe first that it is not possible to assign costs accurately from an objective viewpoint. As soon as we attempt this, subjective elements creep in, and we are forced to admit that our particular problem does not possess unique costs of decision. How, then, can we apply the model to actual situations, if there is no one set of cost assignments solely appropriate to our problem? Actually, this lack of objectivity (or uniqueness) in cost is not a difficulty inherent in our communication applications alone, but is basic to all areas where we wish in some sense to match a model to the real world. The subjectivity of the cost assignments here simply reflects the *unavoidable uncertainty* regarding costs that is the price we must pay for an inevitably incomplete knowledge of the world around us, and *as such should be accepted as part of the model.* In fact, from this viewpoint, the lack of uniqueness is not a defect but is rather a mark of realism: no subjectivity, no contact with reality, and hence no ultimate *practical* significance. Moreover, an important mitigating feature here is the invariance of optimum system structure in detection with respect to cost assignments [as embodied in the threshold $\mathcal{K}$; cf. Eq. (19.15)]: no matter what costs are selected for the possible outcomes, the mode of processing the data remains unchanged. This invariance is, of course, usually vital in practical situations. It would be almost always prohibitively expensive to have to redesign the system for each new cost assignment. Similar statements can be made about extraction systems, although now many more classes of optimal systems are possible, depending on the choice of cost function. Thus, to answer the question of what cost assignments to use, we can accept as unique those given to us from "higher authority" (to whom the subjective element is now transferred!). Or when we ourselves must constitute the higher authority, we simply make the selection which in our best judgment at the time is most suitable, recognizing that this is precisely what we in effect do in all decision situations. In practice, this is where the methods of operations research play an important part in providing the needed contact with "reality."

The question of the inadequacy of the a priori information may be considered as basically similar to our inability to fix costs uniquely for a given situation. Here we regard a priori information (embodied as the *functional form* of probability distributions, the known signal parameters, specific noise spectra, and the like) as a measure of our subjective awareness of the facts of the detection or extraction situation in question.[21] Similarly, a lack of a priori data (which may appear as a signal parameter regarded as a random variable, or as an unknown probability or d.d., etc.) reflects our subjective awareness of ignorance in respect to these quantities. Just as cost assignment is ultimately subjective, so also is the assignment of probabilities, probability densities, and the like, to our model a subjective judgment, helped along with whatever facts we may possess from previous experience.

Lack of a priori information can be overcome in a number of more or less satisfactory ways.† A conceptually elegant approach, but one almost impossible to achieve in practice, is to go out and acquire the needed data and then build and operate the system accordingly. A less arduous and more realistic approach applies some additional extremal principle, subject perhaps to one or more constraints: Minimax, or a maximum average uncertainty principle, maximum likelihood, and so on. A disadvantage is that usually different extremal principles lead to different results; so once again we are confronted with the task of having to make a subjective judgment between possible classes of systems for the purpose at hand. Frequently, circumstances (i.e., the environment in which our particular decision process is embedded) will narrow down the choice of possibilities and sometimes, in special instances, eliminate all but one. In any case, the more uncertain our a priori data, the greater the expected cost of operation—we cannot avoid paying for ignorance. It should be pointed out, however, that lack of a priori data may make the design of a Bayes system, for example, practically very difficult. Moreover, the additional extremal principles invoked to remedy this shortage have their own drawbacks which limit the scope of their applicability; Minimax tends to be too conservative, for example. All this should be borne in mind when optimal (and near-optimal) systems are to be designed and evaluated.

The choice of criterion itself is arbitrary, although here our selection is always a compromise between realism and mathematical simplicity: we attempt to construct a criterion which fits a broad class of situations and at the same time permits us to produce some definite numerical conclusions. Decision theory is a good example of this. The notion of cost, or risk, is a natural and quite general quantitative interpretation of more subtle value judgments, an interpretation which, as employed in its present form, also yields quantitative results. Minimization of average cost, or risk, is, of course, just one possible criterion of performance. Other more elaborate ones can be constructed: minimization of the cost variance, subject to fixed average risk, for example, might be more appropriate in some situations. The point is that, in general, different criteria lead to different results, so that again it is our subjective choice, now of criteria, which inescapably influences our decisions and the way in which they are arrived at. Our special results, accordingly, are not applicable in all situations, although sometimes systems exist which are invariant with respect to several criteria simultaneously (cf. Secs. 22.2-2, 22.3-3).

Finally, a practical limitation on the utility of decision-theory methods is analytical complexity. This probably limits a detailed application of these methods to single-link systems, with or without feedback. Over-all optimization of many-link systems in this fashion appears to be intractable, so that other techniques must be employed, such as linear[10] and dynamic

† Or unsatisfactory, depending on one's viewpoint.

programming,[6,9] which, however, may profitably use the results of the decision-theory approach to provide the details of individual link performance needed in the analysis of the over-all system. Another technical limitation, although not usually a serious one, is the difficulty in dealing with the strong-signal situation. As we have seen in Chaps. 19 to 22, the important threshold cases can be handled, while a corresponding treatment for strong signals is usually intractable. However, a system designed for optimum or near-optimum performance at threshold signal levels will usually be quite satisfactory at the higher ones.

We can summarize the preceding comments. The decision-theory approach provides a unified, though somewhat restricted, quantitative theory of system optimality, design, evaluation, and comparison. These methods are limited and qualified by the observer's judgment, the a priori information available to him, and the degree of realism (and hence analytical complexity) he wishes to incorporate into his models. The search for optimality should not be pushed too far: the principal role of an optimum system (which can never be strictly realized in practice) is to provide a limiting measure of performance and a guide to system structure for the approximations actually used in applications.

Optimality is also qualified by its lack of uniqueness in general: different criteria lead to different results.

A point that should not be overlooked in the discussion of optimality is the significance of different degrees of improvement that may be indicated over corresponding suboptimum systems. Some such systems may be "close to" and others "far from" the theoretical limiting performance. Just what this means depends of course on the larger question of the system's role in the communication environment. For example, a difference of a few decibels in certain situations may be quite ignorable or not worth the great increase in structural complexity needed for essentially optimum performance. On the other hand, a few decibels can also mean a significant improvement in over-all performance. This is particularly true in systems where low error rates are required, with a high rate of decision. Then, for certain types of background (e.g., normal noise or processes derived from it, Rayleigh noise, etc.), an increase or decrease of 2 or 3 db in the input signal-to-noise (power) ratio can produce a corresponding increase or decrease of several orders of magnitude in the error probabilities of decision, which may in turn become quite significant for conditions involving high decision rates. The reason for this large shift in the magnitude of the error probabilities is the initial requirement of very small error probabilities per decision: in effect, we are operating well out on the "tail" of the curve for the d.d. in question (i.e., the d.d. of $\log \Lambda_n$ in detection, for instance), so that comparatively small changes in the variance (as measured by changes in the signal-to-noise ratio) produce large changes in the type I and II error probabilities (cf. Fig. 19.2). A few decibels may also represent a significant change, even when a high decision rate is not required. For example, in a radar 3 db can imply an expansion or contraction of effective threshold range by 20 per cent. Experience so far has indicated that "well-designed" systems are usually within a few (say 2 to 4) decibels of the theoretical limit for the given noise and signal conditions, while "poorly designed" systems may be an order of magnitude (or more) away from optimality† (cf. Sec. 20.4-4). Other factors may, of course, intervene in determining whether or not our actual system can be regarded as "close to" optimality.

So far, we have considered some of the achievements and attendant difficulties of the decision-theory approach. It is also instructive to consider the extent by which decision- and information-theory methods both differ and overlap. Here information theory is again defined in the strict sense, as a theory of information measures and coding (cf. Sec. 6.1). The difference between the two is partly a matter of emphasis. Thus, in information theory it is the maximization of average rate of message transmission by suitable encoding, subject to the constraints of the channel (i.e., the noise, and the fidelity requirements at the receiver; cf. Sec. 6.5), that is the principal aim. The significance of the message set and the outcomes of particular transmissions are of secondary importance. On the other hand, in applications of decision theory to communication problems it is precisely these latter factors which are of chief concern: here the reception process (and transmission, also, through proper selection of signal alphabets; cf. Sec. 23.2) is governed by the value judgments associated with the possible decisions which represent the final output from the communication link. The former (i.e., informational) approach is naturally suited to the many situations where one wishes to minimize the cost of utilizing the channel, as in commercial telephone operation, for example. The latter (or decision-theory method) arises where channel considerations are secondary (except, of course, as they influence the physics of the communication process itself). This is the case with many military situations, where it is the significance of particular messages and the decisions consequent upon them that are of principal concern. For example, in a radar we normally seek to minimize the cost of decision error, not necessarily to maximize the rate of information transmission. That these two viewpoints frequently lead to different results is clear from the discussion in Sec. 6.5-5. The reason is that each criterion emphasizes different aspects of performance—the evaluation of average risk is improved by lowering the probabilities of decision errors (for fixed costs), whereas the criterion of information loss favors a system in which the *posterior* signal probabilities before and after the system operation are nearly equal. A small error rate is not necessarily accompanied by a small information loss, although this may occur under certain conditions.‡

† In effect, this provides a quantitative definition of good or poor design.

‡ See Middleton and Van Meter,[1] sec. 6.2.

When the role of the finite sample is specifically taken into account, i.e., when the error probabilities in reception are necessarily nonvanishing, the information- and decision-theory approaches may overlap to varying extents in both method and viewpoint. The recovery and reception of encoded messages are not now simply a matter of applying the inverse operation to the encoding process, e.g., of setting $\boldsymbol{T}_R = \boldsymbol{T}_T^{-1}$. Reception is no longer directly related in this fashion to the transmission operation but requires separate attention and adjustment. As noted earlier, finite error probabilities in reception imply a decision process, and it is precisely here that the techniques of decision theory are applied (cf. Chap. 18, and Chaps. 19 to 21 for detection and extraction). Moreover, further system optimization may frequently be achieved by proper selection of the signal alphabets at the transmitter, i.e. by cost coding, or optimizing that part of the encoding process which relates the coded message set to the ensemble of signal waveforms to be used in the physical process of transmission and reception (cf. Sec. 23.2). Here the criterion is minimization of the probabilities of decision error at the receiver, which is also, of course, a minimization of average risk (with special cost assignments).†

We remark, finally, that with certain classes of interference, increasing transmitter power may achieve improved reception more simply than involved, error-correcting codes. For as noted above in the case of normal noise backgrounds, a few decibels increase in signal-to-noise ratio at the receiver in low-error-probability systems can mean a further decrease of several orders of magnitude in the error probabilities. This in turn is comparable to the changes realized by most error-correcting systems. Similar remarks apply for the improvements achieved with cost coding or proper choice of signal waveform at the transmitter. There are, of course, many situations where coding techniques are essential to improved or even effective communication: whenever it is not possible to obtain a 2- or 3-db increase or more in effective signal power. This may occur in many computer applications, whenever system size is strictly limited, in communication systems operating against impulsive or other highly non-gaussian interference, and whenever the effective signal suffers large changes in level, for example, because of fading. We observe, in conclusion, that the information- and decision-theory approaches are also connected through the selection of cost functions: in place of the usual constant cost or difference cost assignments of detection and estimation, we can use a cost function based on a measure of uncertainty, or on an equivocation when we are concerned with information rate (cf. Chap. 22). Thus, the two approaches, though different in scope and emphasis, complement each other, providing powerful

† Some examples in detection and estimation are considered in Sec. 23.2. In coding situations, with such cost-coding techniques one can often obtain considerable improvement in systems of constant information rate by using suitable binary signal alphabets (Reiger[23]) and further improvement by using alphabets other than binary (Reiger[22]).

tools for a comprehensive attack on the basic problems of system optimization, design, and evaluation.

23.5 *Some Future Problems*

As indicated in the Preface and as illustrated by the preceding text itself, this study is introductory. By no means have all the important problems been indicated or considered, not even those which lie within the scope of present theory and practice. Much still lies ahead. We conclude, accordingly, with a short summary of some of the general problems, suggested by the above, which remain for further investigation:

1. *System invariants.* The central question here is to discover what features of system performance remain invariant of system structure and operation, and vice versa. For example, it is known that, in the simplest class of coherent binary detectors operating against white normal noise, optimum performance (as measured by the Bayes risk) is independent of signal waveform and depends only on the ratio of received signal energy in the observation interval to the noise power per cycle (cf. Sec. 20.4-1). A more general example is provided in signal extraction: under certain conditions, a given system may be optimum with respect to several criteria (cf. Sec. 21.3). Other problems involving invariance concern the choice of cost function (in detection; cf. Sec. 22.2), as well as waveform, background noise statistics, etc., and the parameters needed to specify system performance.

2. *Stability of criteria.* Here we wish to determine to what extent a change in criterion results in a change in the corresponding optimum system. An associated question is that of finding a criterion with respect to which a *given* system is optimum, and how "far" from other criteria this criterion may be. In this way, one seeks a quantitative correspondence between criteria, based on optimum performance.

3. *Search for criteria.* This is the ever-present problem of creating new criteria which have the polar properties of realism and sufficient simplicity for application. This may involve simple combinations of existing criteria, as in decision and information theory, or the generation of new ones.

4. *"Trade-offs."* The problem is to determine how changes in the values of system parameters compensate or interact with one another, i.e., how change in cost assignments, signal parameters, and system structure interact. This can be done for given optimum and suboptimum systems when the error probabilities and average risks have first been obtained. An important result is the extent to which performance is influenced by the change in specific parameters. Having such relationships and a certain degree of flexibility in parameter values, it is possible to achieve at least a near-optimization in many practical situations.

5. *Nonparametric methods.* Here we seek to modify and adapt the decision-theory approach for system design and evaluation to include nonparametric or so-called "distribution-free" methods. The principal aim is to avoid in the initial formalism the use of specific (and possibly unknown) distributions and distribution densities.

6. *Mathematical theory of the information cost function.* The problem now is to extend the original theory of Wald for the constant cost function $\mathfrak{F}_1$ [Eq. (18.4)] to the more involved cost functions of the type $\mathfrak{F}_2$ [Eq. (18.5)], used in Chap. 22, which are not independent of the decision rule. Here one seeks theorems corresponding to admissibility, Minimax, and the complete class theorems of the simpler theory (cf. Sec. 18.5-1), and specifically the more restricted conditions on the signal and noise processes for which these are valid. In brief, a rigorous mathematical theory is required.

7. *"Learning" systems.* Another extension of the single-decision cases is to those situations involving sequences of decisions where existing a priori data at any given stage are used in reception to acquire further information for ultimate optimized processing and decision. Dynamic programming methods should prove valuable here.[9]

8. *System synthesis.* This is the extension of system analysis provided by decision theory to the synthesis of the various components and system elements indicated for optimum and near-optimum performance. Synthesis here is a direct analogue of the synthesis problem in classical network theory.

Other, somewhat more specific problems of interest are:

9. *Multiple-extraction theory.* The specific applications and extensions of the analysis described in Sec. 23.1-3 to various problems, such as diversity reception and optimal data processing.

10. *Background noise stability.* A study of the modifications required in system design for both detection and extraction, when the background noise level is not known precisely or is varying randomly during a typical observation period.

11. *Strong-signal detection theory.* The development of a general approach comparable to that for threshold cases (cf. Secs. 19.3, 19.4).

12. *Nonnormal noise.* Extension of the specific results for originally normal noise backgrounds to cases where one has nearly normal or impulsive noise or other strongly non-gaussian interference.

13. *Feedback systems.* Applications of decision theory to single-link communication systems with one feedback channel, and extensions to decision sequences. An area of particular importance is that of control systems, which for optimality, as in communication applications, are generally nonlinear.

14. *Sequential estimation.* Signal extraction using sequential methods, paralleling the existing treatments for sequential detection (cf. end of Sec. 20.4-8).

Many other interesting problems will be suggested to the reader by the results of Parts 3 and 4. For example, much remains to be done with the specific results of system comparisons, in particular the evaluation of actual suboptimum systems. Cost-coding techniques need further development, while more intimate connections of decision theory with existing coding methods might be sought. Finally, with the growing application of more powerful analytic methods, we may expect and obtain an increasing realism in our communication models. At the same time, a more subtle science and technology may provide us with simple ways of realizing these models: many and more complex operations may be carried out with less effort and greater precision. Thus, perhaps, we may say that, although the beginnings have been made, the best is still before us.

REFERENCES

1. Middleton, D., and D. Van Meter: On Optimum Multiple-alternative Detection of Signals in Noise, *IRE Trans. on Inform. Theory,* **IT-1:** 1 (1955).
2. Chow, C. K.: An Optimum Character Recognition System Using Decision Functions, *IRE Trans. on Electronic Computers,* **EC-6:** 247 (1957).
3. Turin, G.: Communication through Noisy, Random-multipath Channels, *MIT Lincoln Lab. Tech. Rept.* 116, May, 1956. See also Error Probabilities for Binary Symmetric Ideal Reception through Nonselective Slow Fading and Noise, *Proc. IRE,* **46:** 1603 (1958).
4. Turin, G.: On the Estimation in the Presence of Noise of the Impulse Response of a Random Linear Filter, *IRE Trans. on Inform. Theory,* **IT-3:** 5 (1957).
5. Helstrom, C. W.: The Resolution of Signals in White Gaussian Noise, *Proc. IRE,* **43:** 1111 (1955).
6. Bellman, R., and R. E. Kalaba: On the Role of Dynamic Programming in Statistical Communication Theory, *IRE Trans. on Inform. Theory,* **IT-3:** 197 (1957).
7. Bellman, R.: The Theory of Dynamic Programming, *Bull. Am. Math. Soc.,* **60:** 503 (1954).
8. Bellman, R.: "Dynamic Programming," Princeton University Press, Princeton, N.J., 1957.
9. Bellman, R., and R. E. Kalaba: On Communication Processes Involving Learning and Random Duration, *IRE Conv. Record,* 1958, pt. 4, p. 16.
10. Kalaba, R. E., and M. L. Juncosa: Optimal Design and Utilization of Communication Networks, *Management Science,* **3:** 33 (1956); *Proc. IRE,* **44:** 1874 (1956).
11. Charnes, A., W. W. Cooper, and A. Henderson: "An Introduction to Linear Programming," Wiley, New York, 1953.
12. Dorfman, R., P. A. Samuelson, and R. M. Solow: "Linear Programming and Economic Analysis," McGraw-Hill, New York, 1957.
13. Eisemann, K.: Linear Programming, *Quart. Appl. Math.,* **14:** 205 (1956).
14. Goode, H. H., and R. E. Machol: "System Engineering—An Introduction to the Design of Large-scale Systems," secs. 24, 25, McGraw-Hill, New York, 1957.
15. von Neumann, J., and O. Morgenstern: "Theory of Games and Economic Behavior," Princeton University Press, Princeton, N.J., 1947.
16. McKinsey, J. C. C.: "Introduction to the Theory of Games," McGraw-Hill, New York, 1952.
17. Luce, D., and H. Raiffa: "Games and Decisions," Wiley, New York, 1957.
18. Wald, A.: "Statistical Decision Functions," Wiley, New York, 1950.

19. Mandelbrot, B.: "Jeux de communication," vol. 2, pts. 1, 2, University of Paris, Institute of Statistics, Paris, 1953.
20. Blachman. N. M.: Communication as a Game, *IRE Wescon*, 1957, pt. 2, p. 61.
21. Savage, L. J.: "The Foundations of Statistics," Wiley, New York, 1954.
22. Reiger, S.: Error Rates in Data Transmission, *Proc. IRE*, **46**: 919 (1958).
23. Reiger, S.: Error Probabilities in Binary Data Transmission Systems in the Presence of Random Noise, *IRE Conv. Record*, 1953, pt. 8, p. 72.

APPENDIX 1

SPECIAL FUNCTIONS AND INTEGRALS†

This appendix contains a short résumé of a number of mathematical results of special interest in noise and communication problems. Additional details, derivations, and results are for the most part given in the references at the end of this appendix; for application here, see Parts 2 to 4.

A.1.1 *The Error Function and Its Derivatives*[1-3]

A function of common occurrence where normal statistics are involved is the *error function*, defined by

$$\Theta(x) \equiv \frac{2}{\sqrt{\pi}} \int_0^x e^{-t^2}\, dt \tag{A.1.1}$$

with the properties

$$\Theta(0) = 0 \qquad \Theta(-x) = -\Theta(x) \qquad \Theta(\infty) = 1 \tag{A.1.2}$$

Related functions of considerable interest are the derivatives

$$\phi^{(k)}(x) \equiv \frac{d^k}{dx^k} \frac{e^{-x^2/2}}{\sqrt{2\pi}} = (-1)^k H_k(x) \frac{e^{-x^2/2}}{\sqrt{2\pi}} \qquad k = 0, 1, 2, \ldots \tag{A.1.3}$$

Specifically, we can write

$$H_k(x) = \sum_{j=0}^{\frac{k}{2}, \frac{k-1}{2}} (-1)^j x^{k-2j}\, {}_kC_{2j} (2j)!/j!2^j \tag{A.1.3a}$$

and here $H_k(x)$ is a Hermitian polynomial of order k.

When $k = 0$, we obtain from the relation

$$\int_0^\infty \cos bz\, e^{-az^2/2}\, dz = \sqrt{\frac{\pi}{2a}}\, e^{-b^2/2a} \tag{A.1.4}$$

by successive differentiations the following integrals:

$$\int_0^\infty z^{2k} \cos bz\, e^{-az^2/2}\, dz = \pi(-1)^k a^{-k-\frac{1}{2}} \phi^{(2k)}(b/\sqrt{a}) \tag{A.1.5}$$

$$k = 0, 1, 2, \ldots$$

$$\int_0^\infty z^{2k-1} \sin bz\, e^{-az^2/2}\, dz = \pi(-1)^k a^{-k} \phi^{(2k-1)}(b/\sqrt{a}) \tag{A.1.6}$$

† See, for example, D. Middleton, Some General Results in the Theory of Noise through Non Linear Devices, *Quart. Appl. Math*, **5**: 445 (1948), app. III.

In this connection, one also has

$$\int_0^\infty \frac{\sin bz}{z} e^{-az^2/2}\, dz = \frac{\pi}{2}\Theta\left(\frac{b}{\sqrt{2a}}\right) = \pi\phi^{(-1)}\left(\frac{b}{\sqrt{a}}\right) \tag{A.1.7}$$

which defines $\phi^{(-1)}(x)$. The functions $\phi^{(k)}$ also obey the recurrence relation

$$\phi^{(k+1)}(x) = -[x\phi^{(k)}(x) + k\phi^{(k-1)}(x)] \qquad k = 0, 1, \ldots \tag{A.1.8}$$

Other integral representations of $\phi^{(k)}(x)$ are also possible,[4] Eq. (A.1.4) for example,

$$\Theta\left(\frac{x}{\sqrt{2}}\right) = 1 - \frac{1}{2\pi i}\int_\Gamma \frac{d\xi}{\xi} e^{\xi - x\sqrt{2}\,\xi^{1/2}} \tag{A.1.9a}$$

$$\phi^{(k-1)}(x) = (-1)^{k-1}\frac{2^{k/2-1}}{2\pi i}\int_\Gamma \frac{d\xi}{\xi^{1-k/2}} e^{\xi - x\sqrt{2}\,\xi^{1/2}}, \; k\geq 1, \tag{A.1.9b}$$

where Γ is the contour $(-\infty, 0+, -\infty)$ (cf. Fig. 2.14) about the branch line $\xi = xe^{\pi i}$.

Differentiation and integration by parts may also be used to obtain the following results:

$$\int_{-\infty}^\infty \Theta(z)e^{-a^2z^2\pm bz}\, dz = \frac{\sqrt{\pi}}{a} e^{b^2/4a^2}\Theta\left[\frac{\pm b}{2a(a^2+1)^{1/2}}\right] \qquad \text{Re } a^2 > 0 \tag{A.1.10a}$$

$$\int_0^x \Theta(z)\, dz = x\Theta(x) - \frac{1}{\sqrt{\pi}}(1 - e^{-x^2}) \tag{A.1.10b}$$

$$\int_{-\infty}^\infty z\Theta(z)e^{-a^2z^2}\, dz = \frac{2}{a^2(a^2+1)^{1/2}} \tag{A.1.10c}$$

$$\int_0^x z^2e^{-z^2}\, dz = -\frac{xe^{-x^2}}{2} + \frac{\sqrt{\pi}}{4}\Theta(x) \tag{A.1.10d}$$

$$\int_{-\infty}^\infty e^{-az^2\pm bz}\, dz = \sqrt{\frac{\pi}{a}}\, e^{b^2/4a} \qquad \text{Re } a > 0 \tag{A.1.10e}$$

$$\int_{-\infty}^\infty ze^{-az^2\pm bz}\, dz = \pm\frac{b}{2a}\sqrt{\frac{\pi}{a}}\, e^{b^2/4a} \qquad \text{Re } a > 0 \tag{A.1.10f}$$

$$\int_{-\infty}^\infty z^2e^{-az^2\pm bz}\, dz = \frac{1}{2a}\sqrt{\frac{\pi}{a}}\, e^{b^2/4a}\left(1 + \frac{b^2}{2a}\right) \qquad \text{Re } a > 0 \tag{A.1.10g}$$

Many other relations of the above general character can be derived in the same way. For example, from Eq. (A.1.7) we get

$$\int_0^\infty e^{-az^2}\frac{(1-\cos bz)\, dz}{z^2} = \frac{\pi\sqrt{a}}{2}\left[\frac{b}{\sqrt{a}}\Theta\left(\frac{b}{2\sqrt{a}}\right) - \frac{2}{\sqrt{\pi}}(1 - e^{-b^2/4a})\right] \tag{A.1.11a}$$

$$\int_0^\infty e^{-z^2}\frac{\sin^2 az}{z^2}\, dz = \pi a\Theta(a) - \sqrt{\pi}(1 - e^{-a^2}) \tag{A.1.11b}$$

Another useful result, obtained by a Taylor's-series expansion, is

$$e^{-(x\pm y)^2/2} = \sqrt{2\pi}\sum_{n=0}^\infty \frac{(\pm x)^n}{n!}\phi^{(n)}(y) = \sqrt{2\pi}\sum_{m=0}^\infty \frac{(\pm y)^m}{m!}\phi^{(m)}(x) \tag{A.1.12a}$$

from which it is easy to show that

$$e^{-(x+a)^2/2} + e^{-(x-a)^2/2} = 2\sqrt{2\pi}\sum_{n=0}^\infty \frac{x^{2n}\phi^{(2n)}(a)}{(2n)!} \tag{A.1.12b}$$

and, in turn, to obtain

$$\Theta\left(\frac{x-a}{\sqrt{2}}\right)+\Theta\left(\frac{x+a}{\sqrt{2}}\right)=4\sum_{n=0}^{\infty}\frac{x^{2n+1}\phi^{(2n)}(a)}{(2n+1)!} \tag{A.1.12c}$$

A function, related to $\Theta(x)$, that appears occasionally in our work is

$$\Psi(x)\equiv\frac{2}{\sqrt{\pi}}\int_0^x e^{t^2}\,dt=-i\Theta(ix)\qquad\text{with } \Theta(x)=-i\Psi(ix) \tag{A.1.13}$$

The function $\Psi(x)$ is tabulated by Jahnke and Emde.[5]

A.1.2 *The Confluent Hypergeometric Function*†

The confluent hypergeometric function ${}_1F_1(\alpha;\beta;\pm x)$ is represented by

$${}_1F_1(\alpha;\beta;\pm x)=1+\frac{\alpha}{\beta}\frac{(\pm x)}{1!}+\frac{\alpha(\alpha+1)}{\beta(\beta+1)}\frac{(\pm x)^2}{2!}+\cdots+\frac{(\alpha)_n}{(\beta)_n}\frac{(\pm x)^n}{n!}+\cdots \tag{A.1.14}$$

$$=\sum_{n=0}^{\infty}\frac{(\alpha)_n}{(\beta)_n}\frac{(\pm x)^n}{n!} \tag{A.1.14a}$$

where $(\alpha)_n\equiv\alpha(\alpha+1)\cdots(\alpha+n-1)$, $(\alpha)_0=1$. This function is analytic and is a solution of Kummer's differential equation

$$x\frac{d^2F}{dx^2}+(\beta-x)\frac{dF}{dx}-\alpha F=0 \tag{A.1.15}$$

A useful asymptotic development is

$${}_1F_1(\alpha;\beta;-x)\simeq\frac{\Gamma(\beta)x^{-\alpha}}{\Gamma(\beta-\alpha)}\sum_{m=0}^{\infty}\frac{(\alpha)_m(\alpha-\beta+1)_m}{m!x^m}\qquad \operatorname{Re} x>0 \tag{A.1.16a}$$

$$\simeq\frac{\Gamma(\beta)x^{-\alpha}}{\Gamma(\beta-\alpha)}\left[1+\frac{(\alpha)(\alpha-\beta+1)}{1!x}+\cdots\right] \tag{A.1.16b}$$

while functions of positive and negative argument are related by Kummer's transformation

$${}_1F_1(\alpha;\beta;x)=e^{x}\,{}_1F_1(\beta-\alpha;\,\beta;\,-x) \tag{A.1.17a}$$

or

$${}_1F_1(\alpha;\beta;-x)=e^{-x}\,{}_1F_1(\beta-\alpha;\,\beta;\,x) \tag{A.1.17b}$$

The confluent hypergeometric function obeys a set of six independent recurrence relations, as indicated in Table A.1. The subscripts on F heading the columns indicate the addition or subtraction of unity in α or β for $F\equiv{}_1F_1(\alpha;\beta;\pm x)$; the rows list the factors by which the quantities heading the columns are to be multiplied, and the sum of each row is zero. The upper sign on $(\mp x)$, $(\pm x)$ refers to ${}_1F_1(\alpha;\beta;+x)$, the lower to ${}_1F_1(\alpha;\beta;-x)$. Among the more useful recurrence relations are

Row 1

$${}_1F_1(\alpha+1;\,\beta;\,-x)=\alpha^{-1}[(\beta-\alpha)\,{}_1F_1(\alpha-1;\,\beta;\,-x)+(2\alpha-\beta-x)\,{}_1F_1(\alpha;\beta;-x)] \tag{A.1.18a}$$

† See, for example, Rice,[6] Middleton,[7] and, for more extensive discussion, Magnus and Oberhettinger,[1] chap. VI, and Whittaker and Watson,[2] chap. 16.

TABLE A.1

	$F_{\alpha+}$	$F_{\alpha-}$	$F_{\beta+}$	$F_{\beta-}$	F
1	α	$\alpha - \beta$	. . . .	. . . . .	$\beta - 2\alpha \mp x$
2	$\alpha\beta$	. . . . .	$\pm x(\beta - \alpha)$	. . . . .	$-\beta(\alpha \pm x)$
3	α	. . . . .	. . . .	$1 - \beta$	$\beta - \alpha - 1$
4	. . .	$-\beta$	$\mp x$	. . . .	β
5	. . .	$\alpha - \beta$	. . . .	$\beta - 1$	$1 - \alpha \mp x$
6	. . .	. . . . .	$\pm x(\beta - \alpha)$	$\beta(\beta - 1)$	$\beta(1 - \beta \mp x)$

Row 6

$$
\begin{aligned}
{}_1F_1(\alpha;\, \beta + 1;\, -x) = x^{-1}(\beta - \alpha)^{-1}[\beta(\beta - 1)\, {}_1F_1(\alpha;\, \beta - 1;\, -x) \\
+ \beta(1 - \beta + x)\, {}_1F_1(\alpha;\beta;-x)] \qquad \beta \neq \alpha \qquad \text{(A.1.18}b\text{)}
\end{aligned}
$$

Others can be obtained by elimination of one of the F's from each of two rows.

For special values of α and β, ${}_1F_1$ can be expressed in terms of other analytic functions. Thus, if α and β are integral, we get

$$
\begin{aligned}
{}_1F_1(m;m;-x) &= e^{-x} & \text{(A.1.19}a\text{)}\\
{}_1F_1(1;2;-x) &= x^{-1}(1 - e^{-x}) & \text{(A.1.19}b\text{)}\\
{}_1F_1(2;1;-x) &= (1 - x)e^{-x} & \text{(A.1.19}c\text{)}\\
{}_1F_1(2;3;-x) &= 2(1 - e^{-x} - xe^{-x})x^{-2} & \text{(A.1.19}d\text{)}\\
{}_1F_1(3;4;-x) &= 3(2 - 2e^{-x} - 2xe^{-x} - x^2e^{-x})x^{-3} & \text{(A.1.19}e\text{)}\\
{}_1F_1(3;2;-x) &= (1 - x/2)e^{-x} & \text{(A.1.19}f\text{)}
\end{aligned}
$$

and so on, for other combinations of α and β.

When $2\alpha = \beta = 1, 2, 3, \ldots$, ${}_1F_1$ is given in terms of modified Bessel functions,[8]

$$
{}_1F_1(\alpha;2\alpha;\pm x) = 2^{2\alpha-1} \frac{\Gamma(2\alpha)}{(\pm x)^{\alpha-1/2}} e^{\pm x/2} I_{\alpha-1/2}\left(\frac{\pm x}{2}\right) \qquad 2\alpha = 1, 2, \ldots \qquad \text{(A.1.20)}
$$

while it can be shown that[7]

$$
{}_1F_1\left(\frac{2k+1}{2};\frac{1}{2};-x\right) = \frac{\sqrt{2\pi}\,(-1)^k 2^k k!}{(2k)!}\,\phi^{(2k)}(\sqrt{2x}) \qquad \text{(A.1.21}a\text{)}
$$

$$
k = 0, 1, 2, \ldots
$$

$$
{}_1F_1\left(\frac{2k+1}{2};\frac{3}{2};-x\right) = \frac{\sqrt{2\pi}\,(-1)^k 2^k k!}{(2k)!}\,(\sqrt{2x})^{-1}\phi^{(2k-1)}(\sqrt{2x}) \qquad \text{(A.1.21}b\text{)}
$$

The Hermitian polynomials $H_k(x)$ become, from Eq. (A.1.3) and Kummer's transformations (A.1.17),

$$
H_{2k}(x) = \frac{(-1)^k(2k)!}{2^k k!}\,{}_1F_1\left(-k;\frac{1}{2};\frac{x^2}{2}\right) \qquad \text{(A.1.22}a\text{)}
$$

$$
H_{2k-1}(x) = x\,\frac{(-1)^{k+1}(2k)!}{2^k k!}\,{}_1F_1\left(1 - k;\,\frac{3}{2};\,\frac{x^2}{2}\right) \qquad \text{(A.1.22}b\text{)}
$$

from which it follows directly that

$$
\begin{aligned}
\phi^{(2k)}(0) &= H_{2k}(0)(2\pi)^{-1/2} = (-1)^k(2k)!/2^k k!(2\pi)^{1/2} & \text{(A.1.23}a\text{)}\\
\phi^{(2k-1)}(0) &= -(2\pi)^{-1/2}H_{2k-1}(0) = 0 & \text{(A.1.23}b\text{)}
\end{aligned}
$$

With the help of Eqs. (A.1.17) and the recurrence relation (A.1.18*a*), a number of other relations follow. We have

$$
\begin{aligned}
{}_1F_1(k+1;\tfrac{1}{2};-x^2) &= e^{-x^2}\,{}_1F_1(-\tfrac{1}{2}-k;\tfrac{1}{2};x^2) &\text{(A.1.24a)}\\
{}_1F_1(k+1;\tfrac{3}{2};-x^2) &= e^{-x^2}\,{}_1F_1(\tfrac{1}{2}-k;\tfrac{3}{2};x^2) &\text{(A.1.24b)}
\end{aligned}
$$

where, from Eqs. (A.1.22) and Kummer's relations (A.1.17) again, we can also write

$$
{}_1F_1(k+\tfrac{1}{2};\tfrac{1}{2};x^2) = e^{x^2}\,{}_1F_1(-k;\tfrac{1}{2};-x^2) = e^{x^2}2^k k!(-1)^k H_{2k}(ix\sqrt{2})/(2k)! \qquad \text{(A.1.25a)}
$$

$$
{}_1F_1(k+\tfrac{1}{2};\tfrac{3}{2};x^2) = e^{x^2}\,{}_1F_1(1-k;\tfrac{3}{2};-x^2) = e^{x^2}2^k k!(-1)^k H_{2k-1}(ix\sqrt{2})/(2k)!ix\sqrt{2} \qquad \text{(A.1.25b)}
$$

Specifically, we have

$$
\begin{aligned}
{}_1F_1(-\tfrac{1}{2};\tfrac{1}{2};x^2) &= e^{x^2} - x\sqrt{\pi}\,\Psi(x) &\text{(A.1.26a)}\\
{}_1F_1(-\tfrac{1}{2};\tfrac{3}{2};x^2) &= \tfrac{1}{2}e^{x^2} + \frac{\sqrt{\pi}}{4x}(1-2x^2)\Psi(x) &\text{(A.1.26b)}\\
{}_1F_1(\tfrac{1}{2};\tfrac{1}{2};x^2) &= e^{x^2} &\text{(A.1.26c)}\\
{}_1F_1(\tfrac{1}{2};\tfrac{3}{2};x^2) &= \frac{\sqrt{\pi}}{2x}\Psi(x) &\text{(A.1.26d)}
\end{aligned}
$$

from which, and Eqs. (A.1.18), we can obtain the higher-order cases.

In a similar way, we obtain

$$
\begin{aligned}
{}_1F_1(k+1;\tfrac{1}{2};x^2) &= e^{x^2}\,{}_1F_1(-\tfrac{1}{2}-k;\tfrac{1}{2};-x^2) &\text{(A.1.27a)}\\
{}_1F_1(k+1;\tfrac{3}{2};x^2) &= e^{x^2}\,{}_1F_1(\tfrac{1}{2}-k;\tfrac{3}{2};-x^2) &\text{(A.1.27b)}
\end{aligned}
$$

and these are easily found with the help of Eqs. (A.1.18) applied to the right-hand members, in conjunction with Eqs. (A.1.21). Special functions are

$$
{}_1F_1(-\tfrac{1}{2};\tfrac{1}{2};-x^2) = e^{-x^2} + \sqrt{\pi}\,x\Theta(x) \qquad {}_1F_1(\tfrac{1}{2};\tfrac{1}{2};-x^2) = e^{-x^2} \qquad \text{(A.1.28a)}
$$

$$
\begin{aligned}
{}_1F_1(-\tfrac{1}{2};\tfrac{3}{2};-x^2) &= \tfrac{1}{2}e^{-x^2} + \frac{\sqrt{\pi}}{4x}(1+2x^2)\Theta(x)\\
{}_1F_1(\tfrac{1}{2};\tfrac{3}{2};-x^2) &= \frac{\sqrt{\pi}}{2x}\Theta(x) &\text{(A.1.28b)}
\end{aligned}
$$

Other useful results, derived from the above, are

$$
{}_1F_1(1;\tfrac{1}{2};-x^2) = 1 - x\sqrt{\pi}\,e^{-x^2}\Psi(x) \qquad {}_1F_1(1;\tfrac{3}{2};-x^2) = \frac{\sqrt{\pi}}{2x}e^{-x^2}\Psi(x) \qquad \text{(A.1.29a)}
$$

$$
\begin{aligned}
{}_1F_1(2;\tfrac{1}{2};-x^2) &= \tfrac{1}{2}\{(3-2x^2)[1 - x\sqrt{\pi}\,\Psi(x)e^{-x^2}] - 1\} &\text{(A.1.29b)}\\
{}_1F_1(2;\tfrac{3}{2};-x^2) &= \tfrac{1}{2}\left[1 + \frac{\sqrt{\pi}}{2x}(1-2x^2)e^{-x^2}\Psi(x)\right] &\text{(A.1.29c)}
\end{aligned}
$$

and, for positive arguments, we get

$$
{}_1F_1(1;\tfrac{1}{2};x^2) = 1 + e^{x^2}\sqrt{\pi}\,x\Theta(x) \qquad {}_1F_1(1;\tfrac{3}{2};x^2) = \frac{\sqrt{\pi}}{2x}e^{x^2}\Theta(x) \qquad \text{(A.1.30a)}
$$

$$
\begin{aligned}
{}_1F_1(2;\tfrac{1}{2};x^2) &= \tfrac{1}{2}\{(3+2x^2)[1 + e^{x^2}x\sqrt{\pi}\,\Theta(x)] - 1\} &\text{(A.1.30b)}\\
{}_1F_1(2;\tfrac{3}{2};x^2) &= \tfrac{1}{2}\left[1 + \frac{\sqrt{\pi}}{2x}(1+2x^2)e^{x^2}\Theta(x)\right] &\text{(A.1.30c)}
\end{aligned}
$$

As noted above, when α is a half integer and β is a (positive) integer, then ${}_1F_1$ can be expressed in terms of modified Bessel functions of the first kind. A few examples are included below,

$$ {}_1F_1(-\tfrac{1}{2};1;-x) = e^{-x/2}[(1+x)I_0(x/2) + xI_1(x/2)] \tag{A.1.31a} $$

$$ {}_1F_1(\tfrac{1}{2};1;-x) = e^{-x/2}I_0(x/2) \tag{A.1.31b} $$

$$ {}_1F_1(\tfrac{3}{2};1;-x) = e^{-x/2}[(1-x)I_0(x/2) + xI_1(x/2)] \tag{A.1.31c} $$

$$ {}_1F_1(\tfrac{1}{2};2;-x) = e^{-x/2}[I_0(x/2) + I_1(x/2)] \tag{A.1.31d} $$

$$ {}_1F_1(\tfrac{3}{2};2;-x) = e^{-x/2}[I_0(x/2) - I_1(x/2)] \tag{A.1.31e} $$

$$ {}_1F_1(\tfrac{3}{2};3;-x) = \frac{4}{x}\, e^{-x/2}I_1(x/2) \tag{A.1.31f} $$

$$ {}_1F_1(\tfrac{3}{2};4;-x) = \frac{4}{x}\, e^{-x/2}[I_0(x/2) + (1 - 4/x)I_1(x/2)] \tag{A.1.31g} $$

with more involved expressions derivable with the aid of recurrence formulas.

Tabulations of confluent hypergeometric functions ${}_1F_1(\alpha;\beta;-x)$ to five significant figures for $\alpha = -\tfrac{1}{2}, \tfrac{1}{2}, \ldots, \beta = 1, 2, \ldots, 10$ ($\alpha \simeq \beta$) have been prepared by Middleton and Johnson[9] for a mesh of 0.25 for ($0 \le x \le 2.0$) and 0.5 for ($2.0 < x < 10.0$); additional values for $x = 20, 30, \ldots, 100$ are included as well. Curves of these functions in the range ($0 \le x \le 9.5$) and representations in terms of the modified Bessel functions I_0, I_1 [cf. Eqs. (A.1.31*a*), etc.] are also given, along with additional references.

A.1.3 *The Gaussian Hypergeometric Function*†

This function is represented by the series

$$ {}_2F_1(\alpha,\beta;\gamma;x) = 1 + \frac{\alpha\beta}{\gamma}\frac{x}{1!} + \frac{\alpha(\alpha+1)\beta(\beta+1)}{\gamma(\gamma+1)2!}\,x^2 + \cdots \tag{A.1.32a} $$

$$ = \sum_{n=0}^{\infty} \frac{(\alpha)_n(\beta)_n}{(\gamma)_n n!}\,x^n \qquad |x| < 1 \tag{A.1.32b} $$

(Convergence also occurs sometimes for $|x| = 1$.) The confluent hypergeometric function discussed above may be obtained as a limit

$$ \lim_{\beta'\to\infty} {}_2F_1(\alpha,\beta';\beta;x/\beta') = {}_1F_1(\alpha;\beta;x) \tag{A.1.33} $$

A useful relation is

$$ {}_2F_1(\alpha,\beta;\gamma;1) = \frac{\Gamma(\gamma)\Gamma(\gamma-\alpha-\beta)}{\Gamma(\gamma-\alpha)\Gamma(\gamma-\beta)} \qquad \begin{array}{l} \text{Re}\ \gamma \neq 0, -1, -2, \ldots; \\ \text{Re}\ (\gamma-\alpha-\beta) \neq 0, -1, -2, \ldots \end{array} \tag{A.1.34} $$

and another is

$$ {}_2F_1(\alpha,\beta;\gamma;x) = (1-x)^{\gamma-\alpha-\beta}\ {}_2F_1(\gamma-\beta,\ \gamma-\alpha;\ \gamma;\ x) \tag{A.1.35} $$

Some special cases that appear in problems involving the distributions of envelope and phase of a normal process are

$$ {}_2F_1(-\tfrac{1}{2},-\tfrac{1}{2};1;x^2) = \frac{4}{\pi}\,E(x) - \frac{2}{\pi}\,(1-x^2)K(x) \tag{A.1.36a} $$

$$ {}_2F_1(\tfrac{1}{2},\tfrac{1}{2};1;x^2) = \frac{2}{\pi}\,K(x) \tag{A.1.36b} $$

where K and E are, respectively, complete elliptic integrals of the first and second kind.

† See, in particular, Magnus and Oberhettinger,[1] chap. 2, and Whittaker and Watson,[2] chap. XIV.

We have also

$$ {}_2F_1(-\tfrac{1}{2},\tfrac{1}{2};1;x^2) = \frac{2}{\pi} E(x) \qquad \text{(A.1.36c)} $$

A recurrence formula from which higher-order functions may be obtained is

$$ \alpha\beta(\gamma - \alpha - \beta + 1)(1 - y)^2 \, {}_2F_1(\alpha + 1, \beta + 1; \gamma; y) = G \, {}_2F_1(\alpha,\beta;\gamma;y) \\ - (\gamma - \alpha - \beta - 1)(\gamma - \alpha)(\gamma - \beta) \, {}_2F_1(\alpha - 1, \beta - 1; \gamma; y) \qquad \text{(A.1.37)} $$

in which

$$ G \equiv [(\gamma - \alpha - \beta)^2 - 1](\gamma - \alpha - \beta) + (1 - y)(\gamma - \alpha - \beta - 1)(\gamma - \beta)(\beta - 1) \\ + (\gamma - \alpha - \beta + 1)(\gamma - \alpha - 1)\alpha $$

Other cases of interest are

$$ {}_2F_1(\tfrac{1}{2},\tfrac{1}{2};2;x^2) = \frac{4}{\pi x^2} [E(x) - (1 - x^2)K(x)] \qquad \text{(A.1.38a)} $$

$$ {}_2F_1(\tfrac{3}{2},\tfrac{3}{2};3;x^2) = \frac{16}{\pi x^4} [(2 - x^2)K(x) - 2E(x)] \qquad \text{(A.1.38b)} $$

For the general set of recurrence relations, see Magnus and Oberhettinger.[1]

Finally, a number of simple functions have their series representations as ${}_2F_1$ functions. For example,

$$ {}_2F_1(\tfrac{1}{2},\tfrac{1}{2};\tfrac{3}{2};x^2) = \frac{\sin^{-1} x}{x} \qquad \text{(A.1.39a)} $$

$$ {}_2F_1(-\tfrac{1}{2},-\tfrac{1}{2};\tfrac{1}{2};x^2) = x \sin^{-1} x + \sqrt{1 - x^2} \qquad \text{(A.1.39b)} $$

$$ {}_2F_1(-\tfrac{1}{2},-\tfrac{1}{2};\tfrac{3}{2};x^2) = \tfrac{3}{4} \sqrt{1 - x^2} + \frac{1 + 2x^2}{4x} \sin^{-1} x \qquad \text{(A.1.39c)} $$

$$ {}_2F_1(-\tfrac{1}{2},\tfrac{1}{2};\tfrac{3}{2};x^2) = \tfrac{1}{2} \left(\sqrt{1 - x^2} + \frac{\sin^{-1} x}{x} \right) \qquad \text{(A.1.39d)} $$

$$ {}_2F_1(\tfrac{3}{2},\tfrac{3}{2};\tfrac{5}{2};x^2) = \frac{3}{x^2} \left[(1 - x^2)^{-\frac{1}{2}} - \frac{\sin^{-1} x}{x} \right] \qquad \text{(A.1.39e)} $$

$$ {}_2F_1(\tfrac{1}{2},\tfrac{1}{2};\tfrac{5}{2};x^2) = \frac{3}{4x^2} \left[(2x^2 - 1) \frac{\sin^{-1} x}{x} + \sqrt{1 - x^2} \right] \qquad \text{(A.1.39f)} $$

and

$$ (1 - x)^{-\alpha} = {}_2F_1(\alpha;\beta;\beta;x) \qquad (1 + x)^{\alpha} = {}_2F_1(-\alpha,\beta;\beta;-x) \qquad \text{(A.1.40a)} $$

$$ \log (1 \pm x) = \pm x \, {}_2F_1(1,1;2;\mp x) \qquad \text{(A.1.40b)} $$

$$ {}_2F_1(n/2 + 1, n/2 + \tfrac{1}{2}; n + 1; x^2) = (2/x^2)^n (1 - x^2)^{-\frac{1}{2}} (1 - \sqrt{1 - x^2})^n \qquad \text{(A.1.40c)} $$

Other expressions may be obtained with the aid of recurrence relations applied to the above; for additional relations, see Magnus and Oberhettinger.[1]

A.1.4 *Auxiliary Relations*

Summation of series plays an indispensable part in the evaluation of many of the definite, contour, and infinite integrals which arise in special applications of the theory (cf. Secs. 2.3, 5.1, 5.2, and Parts 3, 4 for examples). A number of useful relations for this purpose are given below.

We begin with some results based on the gamma function:

$$ \Gamma(n + \tfrac{1}{2}) = (2n)! \sqrt{\pi}/2^{2n} n! \qquad n = 0, 1, 2, \ldots $$
$$ \Gamma(n - \tfrac{1}{2}) = (2n - 2)! \sqrt{\pi}/2^{2n-2}(n - 1)! \qquad n = 1, 2, \ldots \qquad \text{(A.1.41a)} $$

When n is a negative integer, we have

$$ \Gamma(n + \tfrac{1}{2}) = (-1)^n (-n)! \sqrt{\pi}/(-2n)! 2^{2n} \qquad n = 0, -1, -2, \ldots \qquad \text{(A.1.41b)} $$

Related forms, involving $(\alpha)_n$, $(\alpha)_{2n}$, etc., are

$$\Gamma(\alpha + n) = \Gamma(\alpha)(\alpha)_n \qquad \text{(A.1.42a)}$$
Therefore,
$$(\alpha)_n = \Gamma(\alpha + n)/\Gamma(\alpha) \qquad \alpha \neq -n$$
$$\Gamma(\alpha - n) = \Gamma(\alpha)(-1)^n/(1 - \alpha)_n \qquad \text{(A.1.42b)}$$
Therefore,
$$(1 - \alpha)_n = \Gamma(\alpha)(-1)^n/\Gamma(\alpha - n) \qquad \alpha \neq n$$

From these, we can deduce

$$(\alpha + m)_n = (\alpha)_{m+n}/(\alpha)_m \qquad \text{(A.1.43a)}$$
Therefore,
$$(\alpha)_{m+n} = (\alpha)_m(\alpha + m)_n$$
$$(\alpha + m)_n = (\alpha)_n(\alpha + n)_m/(\alpha)_m \qquad \text{(A.1.43b)}$$
$$(-n)_m = \frac{(-1)^m n!}{\Gamma(n + 1 - m)} \qquad \text{(A.1.43c)}$$
$$\Gamma(m + n + 1) = m!(m + 1)_n = n!(n + 1)_m \qquad \text{(A.1.43d)}$$

Special cases of interest are

$$\left(\frac{1}{2}\right)_n = \frac{(2n)!}{2^{2n}n!} \qquad \left(-\frac{1}{2}\right)_n = -\frac{1}{2n - 1}\left(\frac{1}{2}\right)_n \qquad \text{(A.1.44a)}$$
$$(3/2)_n = (2n + 1)(1/2)_n \qquad \text{(A.1.44b)}$$
$$(1)_n = n! \qquad (2)_n = (n + 1)! \qquad (m)_n = (m + n - 1)!/(m - 1)! \qquad \text{(A.1.44c)}$$

With the help of the multiplication theorem for the gamma function[1] ($n = 2$), viz.,

$$\Gamma(x)\Gamma(x + 1/2) = \sqrt{2\pi}\, 2^{1/2 - 2x}\Gamma(2x) \qquad \text{(A.1.45)}$$
or
$$\Gamma(x + 1/2)\Gamma(x + 1) = \sqrt{\pi}\, 2^{-2x}\Gamma(2x + 1)$$

and the above, we find, in addition, that

$$(\alpha)_{2n} = \frac{2^{2n+\alpha-1}}{\sqrt{\pi}\,\Gamma(\alpha)}\Gamma\left(n + \frac{\alpha}{2}\right)\Gamma\left(n + \frac{1}{2} + \frac{\alpha}{2}\right) = 2^{2n}\left(\frac{\alpha}{2}\right)_n\left(\frac{\alpha}{2} + \frac{1}{2}\right)_n \qquad \text{(A.1.46a)}$$

where $\Gamma(\alpha + 2n) = \Gamma(\alpha)(\alpha)_{2n} = \Gamma(\alpha)2^{2n}\left(\frac{\alpha}{2}\right)_n\left(\frac{\alpha + 1}{2}\right)_n$ (A.1.46b)

Besides these special forms, various alternative ways of representing special functions also have their uses in the evaluation of many of the integral expressions encountered in the theory. Examples are†

$$\cos az = \left(\frac{\pi az}{2}\right)^{1/2} J_{-1/2}(az) \qquad \sin az = \left(\frac{\pi az}{2}\right)^{1/2} J_{1/2}(az) \qquad \text{(A.1.47a)}$$
$$e^{-az} = \sqrt{\frac{2az}{\pi}}\, K_{1/2}(az) = \sqrt{\frac{\pi az}{2}}\,[I_{-1/2}(az) - I_{1/2}(az)] \qquad \text{(A.1.47b)}$$
$$e^{az} = \sqrt{\frac{\pi az}{2}}\,[I_{-1/2}(az) + I_{1/2}(az)] \qquad \text{(A.1.47c)}$$

where J, I, K are, respectively, Bessel functions of the first kind, modified first kind, and modified second kind. Some useful series developments are†

$$e^{ia\cos\theta} = \sum_{m=0}^{\infty} i^m \epsilon_m J_m(a)\cos m\theta \qquad \text{(A.1.48a)}$$
$$e^{\pm a\cos\theta} = \sum_{m=0}^{\infty} \epsilon_m(\pm 1)^m I_m(a)\cos m\theta \qquad \text{(A.1.48b)}$$

† See Watson,[8] secs. 3.4, 3.71, and 2.22.

A.1.5 *Some Special Integrals*

The following results are frequently used in Parts 3 and 4. We shall give only the briefest outline of their derivations, if that, being content to collect together here the results themselves, for reference and application elsewhere. (For derivations, see the references.)

Our first set of integrals is based on Hankel's exponential integral (cf. Watson,[8] sec. 13.3)

$$\int_0^\infty J_\nu(az)z^{\mu-1}e^{-b^2z^2}\,dz = \frac{\Gamma[(\nu+\mu)/2]}{2b^\mu\Gamma(\nu+1)}\left(\frac{a}{2b}\right)^\nu {}_1F_1\left(\frac{\nu+\mu}{2};\nu+1;-\frac{a^2}{4b^2}\right)$$

$$\text{Re}\ \mu+\nu>0;\ |\arg b|<\frac{\pi}{4} \qquad \text{(A.1.49)}$$

which is easily established by direct expansion of the Bessel function and termwise integration with the aid of the gamma function. Making use of Eqs. (A.1.47*a*) in Eq. (A.1.49) for the Bessel function, the results (A.1.5), (A.1.6), (A.1.7), and (A.1.21) are readily obtained.[7] Thus, we can write, alternatively to Eqs. (A.1.5) to (A.1.7),

$$\int_0^\infty z^{2k}\cos bz\, e^{-az^2/2}\,dz = \sqrt{\frac{\pi}{2}}\frac{2^{-k}a^{-k-\frac{1}{2}}}{k!}(2k)!\ {}_1F_1\left(k+\frac{1}{2};\frac{1}{2};-\frac{b^2}{2a}\right) \qquad \text{(A.1.50}a\text{)}$$

$$\int_0^\infty z^{2k-1}\sin bz\, e^{-az^2/2}\,dz = \sqrt{\frac{\pi}{2}}\frac{2^{-k}}{k!}a^{-k}(2k)!\frac{b}{\sqrt{a}}\ {}_1F_1\left(k+\frac{1}{2};\frac{3}{2};-\frac{b^2}{2a}\right) \qquad \text{(A.1.50}b\text{)}$$

and Eq. (A.1.49) can be used in the same way to obtain[10]

$$\int_0^\infty z^{2k+1}\cos bz\, e^{-az^2/2}\,dz = 2^k k!a^{-k-1}\ {}_1F_1(k+1;\tfrac{1}{2};-b^2/2a) \qquad \text{(A.1.51}a\text{)}$$

$$\int_0^\infty z^{2k}\sin bz\, e^{-az^2/2}\,dz = 2^k k!a^{-k-1}b\ {}_1F_1(k+1;\tfrac{3}{2};-b^2/2a) \qquad \text{(A.1.51}b\text{)}$$

A related integral is

$$\int_0^\infty z^k e^{-bz-az^2/2}\,dz = a^{-(k+1)/2}2^{(k-1)/2}\left[\Gamma\left(\frac{k+1}{2}\right){}_1F_1\left(\frac{k+1}{2};\frac{1}{2};\frac{b^2}{2a}\right)\right.$$
$$\left.-b\sqrt{\frac{2}{a}}\,\Gamma\left(\frac{k}{2}+1\right){}_1F_1\left(\frac{k}{2}+1;\frac{3}{2};\frac{b^2}{2a}\right)\right] \qquad \text{(A.1.52)}$$

which is easily extended to nonintegral values of k, provided that Re $k > -1$ and $|\arg a| < \pi/2$ in any case. For k integral, we may use the results of Sec. A.1.2 to obtain the ${}_1F_1$ in terms of tabulated functions.

Closely related results are the following contour integrals, which may in a number of instances be obtained from Eq. (A.1.49) by analytic continuation [cf. Prob. 2.16; Rice,[6] eq. (4.10-5); and Middleton[7]]:

$$\int_C e^{-c^2\xi^2}\xi^{2\mu-1}\,d\xi = \frac{\pi i e^{-\mu\pi i}}{c^{2\mu}\Gamma(1-\mu)} = ic^{-2\mu}\Gamma(\mu)e^{-i\pi\mu}\sin\pi\mu \qquad |\arg c|<\frac{\pi}{4} \qquad \text{(A.1.53)}$$

$$\int_C \xi^{\mu-1}J_\nu(a\xi)e^{-q^2\xi^2}\,d\xi = \frac{\pi i^{1-\nu-\mu}(a/2q)^\nu\ {}_1F_1[(\mu+\nu)/2;\nu+1;-a^2/4q^2]}{q^\mu\Gamma(\nu+1)\Gamma[1-(\nu+\mu)/2]}$$

$$|\arg q|<\frac{\pi}{4} \qquad \text{(A.1.54)}$$

$$\int_C \xi^\mu e^{\pm ib\xi-c^2\xi^2}\,d\xi = \frac{\pi i^{-\mu}}{c^{\mu+1}}\left\{{}_1F_1\left(\frac{\mu+1}{2};\frac{1}{2};\frac{-b^2}{4c^2}\right)\Big/\Gamma[(1-\mu)/2]\right.$$
$$\left.\pm\frac{b}{c}\,i\,{}_1F_1\left(\frac{\mu+2}{2};\frac{3}{2};\frac{-b^2}{4c^2}\right)\Big/\Gamma(-\mu/2)\right\} \qquad |\arg c|<\frac{\pi}{4} \qquad \text{(A.1.55)}$$

The contour **C** runs from $-\infty$ to $+\infty$ and is indented downward about the singularity at $\xi = 0$.

Another result of some utility is an extension of the Weber-Schafheitlin integral† to

$$\int_{\mathbf{C}} J_\alpha(a\xi)J_\beta(b\xi)\xi^{-\gamma}\,d\xi = \frac{\pi e^{-\pi i(\alpha+\beta-\gamma)/2}b^\beta}{2^{\gamma-1}a^{\beta-\gamma+1}\Gamma(\beta+1)\Gamma[(1+\gamma-\alpha-\beta)/2]\Gamma[(1+\gamma+\alpha-\beta)/2]}$$
$$\times\, {}_2F_1\left(\frac{\alpha+\beta-\gamma+1}{2}, \frac{\beta-\gamma-\alpha+1}{2}; \beta+1; \frac{b^2}{a^2}\right) \qquad |b| \le |a|,\ \text{Re}\,\gamma > 0 \quad \text{(A.1.56)}$$

A further modification of Eq. (A.1.56) (for the real integral) is‡

$$\int_0^\infty K_\alpha(az)J_\beta(bz)z^{-\gamma}\,dz = \frac{b^\beta\Gamma[(\beta-\gamma+\alpha+1)/2]\Gamma[(\beta-\gamma-\alpha+1)/2]}{2^{\gamma+1}a^{\beta-\gamma+1}\Gamma(\beta+1)}$$
$$\times\, {}_2F_1\left(\frac{\beta-\gamma+\alpha+1}{2}, \frac{\beta-\gamma-\alpha+1}{2}; \beta+1; -\frac{b^2}{a^2}\right) \quad \text{(A.1.57)}$$

for $|b| < |a|$, Re $(\beta+1) > |\text{Re}\,\alpha|$, Re $a > |\text{Im}\,b|$. Many other results of a similar nature may be found in Watson,[8] Magnus and Oberhettinger,[1] and Erdélyi et al.[11]

Other integral relations of use in receiver problems are[10]

$$\int_0^\infty z^m e^{-az}\cos bz\,dz = \frac{m!\cos[(m+1)\tan^{-1}(b/a)]}{(a^2+b^2)^{m/2}} \quad \text{(A.1.58}a\text{)}$$

$$\text{Re}\,a > 0,\ \text{Im}\,b = 0$$

$$\int_0^\infty z^m e^{-az}\sin bz\,dz = \frac{m!\sin[(m+1)\tan^{-1}(b/a)]}{(a^2+b^2)^{m/2}} \quad \text{(A.1.58}b\text{)}$$

with the alternative expressions

$$\int_0^\infty z^{2m}e^{-az}\cos bz\,dz = (-1)^m\frac{d^{2m}}{db^{2m}}\left(\frac{a}{a^2+b^2}\right) \quad \text{(A.1.59}a\text{)}$$
$$\int_0^\infty z^{2m}e^{-az}\sin bz\,dz = (-1)^m\frac{d^{2m}}{db^{2m}}\left(\frac{b}{a^2+b^2}\right) \quad \text{(A.1.59}b\text{)}$$
$$\int_0^\infty z^{2m+1}e^{-az}\cos bz\,dz = (-1)^m\frac{d^{2m+1}}{db^{2m+1}}\left(\frac{b}{a^2+b^2}\right) \quad \text{(A.1.59}c\text{)}$$
$$\int_0^\infty z^{2m+1}e^{-az}\sin bz\,dz = (-1)^{m+1}\frac{d^{2m+1}}{db^{2m+1}}\left(\frac{a}{a^2+b^2}\right) \quad \text{(A.1.59}d\text{)}$$

Related expressions are

$$K_0 \equiv \int_0^\infty \frac{\sin bz}{z}e^{-az}\,dz = \frac{\pi}{2} - \tan^{-1}\frac{a}{b} \qquad 0 \le |K_0| \le \frac{\pi}{2} \quad \text{(A.1.60}a\text{)}$$
$$K_1 \equiv \int_0^\infty \frac{\sin bz\cos cz}{z}e^{-az}\,dz = \frac{\pi}{2} - \frac{1}{2}\tan^{-1}\frac{2ab}{b^2-a^2-c^2} \qquad 0 \le |K_1| \le \frac{\pi}{2} \quad \text{(A.1.60}b\text{)}$$

Miscellaneous results are

$$\int_0^\infty \frac{dx}{(a^2+x^2)^{n+1}} = \frac{(2n)!\pi}{(n!)^2 2^{2n+1}a^{2n+1}} \qquad n \ge 0 \quad \text{(A.1.61)}$$
$$\int_{-\infty}^\infty \left(e^{ia\xi(b^2+x^2)^{-1}} - 1\right)dx = \frac{\pi ia\xi}{b}\,{}_1F_1\left(\frac{1}{2};2;\frac{ia\xi}{b^2}\right) \quad \text{(A.1.62}a\text{)}$$
$$= \frac{\pi ia\xi}{b}e^{ia\xi/2b^2}\left[J_0\left(\frac{a\xi}{b^2}\right) - iJ_1\left(\frac{a\xi}{b^2}\right)\right] \quad \text{(A.1.62}b\text{)}$$
$$\int_0^\infty \frac{e^{-ax^2}}{b^2+x^2}\,dx = \frac{\pi e^{ab^2}}{2b}[1-\Theta(b\sqrt{a})] \qquad \text{Re}\,a > 0,\ \text{Re}\,b \ne 0 \quad \text{(A.1.63)}$$

† See Watson,[8] sec. 13.4, for the real-variable case.
‡ *Ibid.*, sec. 13.45.

$$\int_0^{2\pi} \sin m(\theta + \eta) \sin n(\theta + \gamma) \frac{d\theta}{2\pi} = \frac{1}{\epsilon_m} \delta_{mn} \cos m(\eta - \gamma) \tag{A.1.64a}$$

$$\int_0^{2\pi} \cos m(\theta + \eta) \cos n(\theta + \gamma) \frac{d\theta}{2\pi} = \frac{1}{\epsilon_m} \delta_{mn} \cos m(\eta - \gamma) \tag{A.1.64b}$$

$$\int_0^{2\pi} \cos m(\theta + \eta) \sin n(\theta + \gamma) \frac{d\theta}{d\pi} = \frac{1}{\epsilon_m} \delta_{mn} \sin m(\gamma - \eta) \tag{A.1.64c}$$

$$\int_0^{2\pi} \sin m(\theta + \eta) \cos n(\theta + \gamma) \frac{d\theta}{2\pi} = \frac{1}{\epsilon_m} \delta_{mn} \sin m(\eta - \gamma) \tag{A.1.64d}$$

Many other expressions may be found in the text (cf. Part 3 in particular).

REFERENCES

1. Magnus, W., and F. Oberhettinger: "Special Functions of Mathematical Physics," chap. VI, sec. 4(*e*), Chelsea, New York, 1949.
2. Whittaker, E. T., and G. N. Watson: "Modern Analysis," chap. XVI, Cambridge, New York, 1940.
3. "Tables of the Error Function and Its First Twenty Derivatives," Annals Computation Laboratory, Harvard University Press, Cambridge, Mass., 1950.
4. McLachlin, N. W.: "Complex Variables and Operational Calculus," sec. 5.2, Cambridge, New York, 1939.
5. Jahnke, F., and F. Emde: "Tables of Functions," Stechert ed., p. 32, New York, 1941.
6. Rice, S. O.: Mathematical Analysis of Random Noise, *Bell System Tech. J.*, **24**: 46 (1945), app. 4B.
7. Middleton, D.: Some General Results in the Theory of Noise through Non Linear Devices, *Quart. Appl. Math.*, **5**: 445 (1948), app. III.
8. Watson, G. N.: "Theory of Bessel Functions," sec. 6.5, Cambridge, New York, 1948.
9. Middleton, D., and V. Johnson: A Tabulation of Selected Confluent Hypergeometric Functions, *Cruft Lab. Tech. Rept.* 140, January, 1952. See also Ref. 5, p. 275.
10. Middleton, D.: The Effects of a Linear, Narrow-band Filter on a Carrier Frequency-modulated by Normal Noise, *Johns Hopkins Univ. Radiation Lab. Tech. Rept.* 24, app. I, November, 1955.
11. Erdélyi, A., et al.: "Higher Transcendental Functions," vols. 1–3, McGraw-Hill, New York, 1953, 1955.

APPENDIX 2

SOLUTIONS OF SELECTED INTEGRAL EQUATIONS

In many of the preceding chapters (cf. Chaps. 8, 16, 17, 19 to 23), we have encountered two types of integral equations that occur frequently in system design and analysis. These are (1) *homogeneous (Fredholm) integral equations* (of the second kind), like

$$\int_a^b K(t,u)\phi_j(u)\,du = \lambda_j\phi_j(t) \qquad a \le t \le b \ (j = 1, 2, \ldots) \tag{A.2.1}$$

where the $\{\lambda_j\}$ and $\{\phi_j\}$ are, respectively, the set of eigenvalues and eigenfunctions that comprise the solutions of Eq. (A.2.1) (when these exist); and (2) *inhomogeneous (Fredholm) integral equations* (of the first kind), such as

$$\int_{a-}^{b+} K(t,u)z(u)\,du = G(t - t_0) \qquad a- < t < b+ \tag{A.2.2}$$

where as in Eq. (A.2.1) $K(t,u)$ is a known kernel, G is a specified function on the interval $(a-, b+)$, and $z(u)$, if it exists, is the desired solution.†

We shall give in the following sections a brief outline of methods for solving these Fredholm integral equations when the kernels $K(t,u)$ possess, for the most part, the special properties 1 to 4 listed in Sec. A.2.1. Although a moderate level of generality is maintained, with examples illustrating the technique, a comprehensive treatment is neither intended nor provided. For such questions as the necessity or sufficiency of the various conditions 1 to 4 on the kernel and the existence and uniqueness of solutions, etc., we refer the reader to some of the more extensive accounts cited at the end of this appendix.‡

A.2.1 *Introduction*

Before we consider solutions of the homogeneous and inhomogeneous integral equations (A.2.1), (A.2.2), let us state the general properties which for the most part the kernels $K(t,u)$ are assumed to possess. These properties are characteristic of many of the noise and signal processes encountered in communication problems, as indicated in the body of the text. It is assumed first that:

1. *The kernels $K(t,u)$ are symmetric functions, positive definite on* $(a-, b+)$, unless otherwise indicated.

2. *$K(t,u)$ is proportional (or equal) to the covariance function of an at least wide-sense*

† $z(u)$ is here and subsequently an abbreviation of $z(u,t_0)$.

‡ See in particular Lovitt,[1] chaps. 3 and 5, and Morse and Feshbach,[2] secs. 8.3 and 8.4, for the case of general kernels, not necessarily possessing the properties 1 to 4 listed in Sec. A.2.1. See also Davenport and Root,[3] app. 2, and Laning and Battin,[4] secs. 7.5, 8.4, 8.5, when the kernels of Eqs. (A.2.1), (A.2.2) possess these properties. Still other references are available in the bibliography of Morse and Feshbach.

stationary process, so that $K(t,u) = K(|t-u|) = K(\tau) = K(-\tau)$ $(\tau = t-u)$. Thus, $K(\tau)$ is *continuous* at $\tau = 0$ and for all other τ [cf. Eq. (3.41*a*)].

3. *$K(\tau)$ is a real, rational kernel;* i.e., its Fourier transform by the W-K theorem (3.42) is a spectral intensity density $\mathcal{W}(f) = \mathcal{W}(\omega/2\pi)$, which is a rational fraction in ω^2, for example,

$$\mathcal{W}(f) = C_0 \prod_{n=0}^{M} (c_n^2 + \omega^2) \Big/ \prod_{n=1}^{N} (b_n^2 + \omega^2) \qquad M+1 \leq N \ (C_0 > 0) \tag{A.2.3}$$

where, for $n = 0$, $c_n^2 + \omega^2$ is replaced by unity.

4. *$K(\tau)$ is expressible as the sum of exponential terms,* e.g.,

$$K(\tau) = \sum_{n=1}^{N} a_n e^{-b_n|\tau|} \qquad \text{Re } b_n > 0 \tag{A.2.4}$$

This is equivalent to postulating that all the poles (and zeros) of $\mathcal{W}(f)$ [Eq. (A.2.3)] are simple† and distinct, that is, $c_n \neq b_n$, with Re $c_n > 0$, for if this is the case, then we can resolve Eq. (A.2.3) into partial fractions of the form

$$\mathcal{W}(f) = 4 \sum_{n=1}^{N} \frac{a_n b_n}{\omega^2 + b_n^2} \tag{A.2.5}$$

and, with the help of the W-K relations (3.42), this gives us directly Eq. (A.2.4). The a_n and b_n here are generally functionally related and complex, but since $K(\tau)$ is real and even, these a_n, b_n always occur in complex conjugate pairs, for example, $a_2 = a_1^*$, $a_4 = a_3^*$, and N is also even. The real parts of b_n are likewise always positive, corresponding to the fact that the process for which $K(\tau)$ is the covariance function is real, entirely random, and (usually) of finite mean square.

For both the homogeneous and inhomogeneous integral equations (A.2.1), (A.2.2), the kernels $K(t,u)$ may represent [by (2)] the covariance function of a simple noise process or, as is frequently the case in prediction and filtering problems (cf. Sec. 16.2), the covariance function of an additive combination of noise "signal" and interfering noise processes. To indicate this in the subsequent analysis, we write

$$K(t,u) = K(|t-u|) = AK_S(|t-u|) + BK_N(|t-u|) + C[K_{SN}(t-u) + K_{NS}(t-u)] \tag{A.2.6}$$

where A, B, C (≥ 0) are real dimensionless numbers and $K_{SN}(\tau) = K_{NS}(-\tau)$ [cf. Eq. (3.149*b*)]. Thus, for noise alone, we set $A = C = 0$ in Eq. (A.2.6), or if the noise "signal" and background noise are uncorrelated, we put $C = 0$, $A = B$ $(= 1)$, and so on.

In the case of the inhomogeneous integral equation (A.2.2), it is advantageous to write this equation explicitly, using Eq. (A.2.6), in order to indicate a number of important special cases which we shall consider presently. We have

$$\int_{a-}^{b+} \{AK_S(|t-u|) + BK_N(|t-u|) + C[K_{SN}(t-u) + K_{NS}(t-u)]\} z(u)\, du = G(t-t_0) \qquad a- < t < b+ \tag{A.2.7}$$

† The condition (4) that the poles and zeros of the intensity spectrum are all simple is not really very restrictive, since by a suitable limiting process, $b_k \to b_{k+2}$, for example [with a corresponding modification of the a_n's and b_n's in Eq. (A.2.5)], we can obtain results for the multiple-pole cases as well. The covariance function (A.2.4) is no longer a sum of simple exponentials but now contains terms of the type $|\tau|^k e^{-b_n|\tau|}$ $(N/2 > k \geq 0)$. We shall see an illustration of this in Sec. A.2.4, Example 2.

Special cases are:

1. $A = C = 0;\ B = 1$

$$\int_{a-}^{b+} K_N(t-u)z(u)\,du = G(t-t_0) \qquad a- < t < b+ \tag{A.2.8a}$$

a common relation in threshold detection problems, where it is desired to determine the structure of the linear matched filter preceding the nonlinear element (cf. Sec. 20.2).

2. $A = B = 1;\ C = 0$

$$\int_{a-}^{b+} [K_S(t-u) + K_N(t-u)]z(u)\,du = G(t-t_0) = K_S(t-t_0) \qquad a- < t < b+ \tag{A.2.8b}$$

which arises in the prediction ($t_0 < 0$) and smoothing ($t_0 > 0$) of a noise "signal" in additive independent noise interference [cf. Eq. (16.77*a*)]. A third case of particular importance in applications occurs when the background noise can be regarded as essentially "white," so that we have now a *mixed* kernel and Eqs. (A.2.7), (A.2.8) reduce to:

3. $A = B = 1;\ C = 0$

$$\int_{a-}^{b+} \left[K_S(t-u) + \frac{W_{0N}}{2}\,\delta(t-u)\right]z(u)\,du = G(t-t_0) = K_S(t-t_0), \text{ etc.} \qquad a- < t < b+ \tag{A.2.8c}$$

The function G may also take the form, with $t_0 = 0$,

$$G(t) = \int_{a-}^{b+} Q(t,t')g(t')\,dt' \qquad Q(t,t') = Q(t',t) \tag{A.2.9}$$

where Q, g are given (and hence G is determined). A special case of this, which we shall examine in detail later (cf. Sec. A.2.4, Example 4), is obtained when $Q(t,t') = K_S(t-t')$.

A related integral equation is

$$\int_{a-}^{b+} K(t,u)\rho(u,\tau)\,du = Q(t,\tau) \qquad a- < t,\tau < b+,\ \rho(u,\tau) = \rho(\tau,u) \tag{A.2.10}$$

with $K(t,u)$ given by Eq. (A.2.6). This may be solved by the same methods used for Eq. (A.2.7), or the solution, for example, $\rho(u,\tau)$, may be obtained by inspection from the solution, $z(u)$, of Eq. (A.2.7) if we note that the latter is a linear functional of the former. Thus, if we multiply both members of Eq. (A.2.10) by $g(\tau)$ and integrate over $(a-,b+)$, we find that

$$\int_{a-}^{b+} K(t,u)\left[\int_{a-}^{b+} \rho(u,\tau)g(\tau)\,d\tau\right]du = \int_{a-}^{b+} Q(t,\tau)g(\tau)\,d\tau$$

or

$$\int_{a-}^{b+} K(t,u)z(u)\,du = G(t) \tag{A.2.11}$$

from Eq. (A.2.9), provided that

$$z(u) = \int_{a-}^{b+} \rho(u,\tau)g(\tau)\,d\tau \tag{A.2.12}$$

Knowing ρ and g, we get z, of course, by the quadrature (A.2.12). Conversely, given z and g, *and* provided that $G(t)$ is also given by Eq. (A.2.9), we may obtain ρ by a direct comparison of the solution z with the right-hand member of Eq. (A.2.12). An illustration of this is given in Sec. A.2.4, Example 4.

Finally, for the exponential kernels assumed here [cf. (4) above], we can write the components of Eq. (A.2.6) in more detail as

$$K_S(\tau) = \sum_{j=1}^{S} a_j^{(S)} e^{-b_j^{(S)}|\tau|} \qquad K_N(\tau) = \sum_{j=1}^{Q} a_j^{(N)} e^{-b_j^{(N)}|\tau|} \tag{A.2.13a}$$

$$K_{SN}(\tau) + K_{NS}(\tau) = \sum_{j=1}^{P} \alpha_j e^{-\beta_j|\tau|} \qquad P = S + Q \tag{A.2.13b}$$

with

$$K_{SN}(\tau) = \begin{cases} \sum_{j=1}^{P} \eta_j e^{-\gamma_j \tau} & \tau > 0 \\ \sum_{j=1}^{P} \mu_j e^{\delta_j \tau} & \tau < 0 \end{cases} \qquad K_{NS}(\tau) = \begin{cases} \sum_{j=1}^{P} \mu_j e^{-\delta_j \tau} & \tau > 0 \\ \sum_{j=1}^{P} \eta_j e^{\gamma_j \tau} & \tau < 0 \end{cases} \tag{A.2.13c}$$

so that, when these are combined according to Eq. (A.2.6) in Eq. (A.2.4), we have $N = S + Q + P = 2(S + Q)$, and $a_n = a_j^{(S)}$ $(j = 1, \ldots, S;\ n = 1, \ldots, S)$; $a_n = a_j^{(N)}$ $(j = 1, \ldots, Q;\ n = S + 1, \ldots, S + Q)$; $a_n = \alpha_j$ $(j = 1, \ldots, P;\ n = S + Q + 1, \ldots, S + Q + P)$, etc., for b_n. We can use the decomposition [Eqs. (A.2.13*a*), (A.2.13*b*)] for $K(\tau)$ [Eq. (A.2.4)] in the generalizations of some of the predictions and filtering problems (cf. Sec. 16.2).

A.2.2 *Homogeneous Integral Equations with Rational Kernels*

When the kernel of Eq. (A.2.1) is rational, e.g., when $2\mathfrak{F}\{K\}$ is given by Eq. (A.2.3), and the conditions 1, 2 (but not necessarily condition 4) are satisfied, we can easily show in a formal way that this homogeneous integral equation (A.2.1) is equivalent to a linear differential equation with constant coefficients, whose solutions $\{\phi_j\}$ are just the eigenfunctions of the original integral equation, when the parameter λ takes the selected values λ_j.

To see this, let us use Eqs. (3.42), setting $i\omega = p$ therein, so that, in terms of the intensity spectrum $\mathcal{W}(p/2\pi i)$ [Eq. (A.2.3)], the covariance function $K(\tau)$ is given by

$$K(\tau) = \frac{1}{2}\int_{-\infty i}^{\infty i} e^{p\tau}\mathcal{W}\left(\frac{p}{2\pi i}\right)\frac{dp}{2\pi i} = \frac{1}{2}\int_{-\infty i}^{\infty i} e^{p\tau}\frac{M(p^2)}{N(p^2)}\frac{dp}{2\pi i} \tag{A.2.14}$$

where $M(p^2)$, $N(p^2)$ are, respectively, the numerator and denominator of the corresponding rational spectrum, and where $N(p^2)$ has no real roots. The degree, M, in p^2 of the polynomial $M(p^2)$ is equal to or less than the degree $N - 1$ (in p^2) of the polynomial $N(p^2)$.

Our next step is to substitute Eq. (A.2.14) into Eq. (A.2.1), where for the moment the λ_j are replaced by the as yet unspecified parameter λ and the $\phi_j(t)$ are replaced by the unknown function $\phi(t)$. We have accordingly

$$\int_a^b du\, \phi(u) \int_{-\infty i}^{\infty i} e^{p(t-u)} \frac{M(p^2)}{N(p^2)}\frac{dp}{2\pi i} = \lambda\phi(t) \qquad a \le t \le b \tag{A.2.15}$$

Since differentiating twice with respect to t is equivalent to multiplying the integrand by p^2, we can rewrite Eq. (A.2.15) as

$$M\left(\frac{d^2}{dt^2}\right)\left[\int_a^b \phi(u)\,du \int_{-\infty i}^{\infty i} e^{p(t-u)} \frac{dp}{2\pi i}\right] = \lambda N\left(\frac{d^2}{dt^2}\right)\{\phi(t)\}$$

and inasmuch as

$$\int_{-\infty i}^{\infty i} e^{p(t-u)} \frac{dp}{2\pi i} = \delta(t-u)$$

we obtain the predicted differential equation

$$M\left(\frac{d^2}{dt^2}\right)\{\phi(t)\} = \lambda N\left(\frac{d^2}{dt^2}\right)\{\phi(t)\} \qquad a \le t \le b \tag{A.2.16}$$

This is a homogeneous linear differential equation with constant coefficients, whose solution $\phi(t)$ contains the parameter λ and $2N$ arbitrary constants $C_1, C_2, \ldots, C_{2N}$.

Now, to solve the integral equation (A.2.1) for this class of kernel, we first solve Eq. (A.2.16) and substitute the $\phi(t)$ so obtained back into the integral equation (A.2.1). It is then found† that the integral equation can be satisfied only for a discrete set of values of λ $(= \{\lambda_j\})$ $(j = 1, 2, \ldots)$ and that for each λ_j the constants $C_1, \ldots, C_{2N}$ must obey certain conditions. The λ_j, of course, are the eigenvalues of Eq. (A.2.1). Thus, any $\phi(t)$ which satisfies Eq. (A.2.16) and the conditions on $C_1, \ldots, C_{2N}$ that occur when $\lambda = \lambda_j$ is an eigenfunction associated with λ_j.

This procedure is not, however, the only one at our disposal: another approach seeks an expansion of the kernel $K(t,u)$ in a suitable orthonormal set, by which means it is possible to reduce the problem to that of finding the nonzero roots of an infinite determinant, which, by proper choice of the orthonormal set, takes a suitably tractable form (cf. Morse and Feshbach,[2] sec. 8.4). We note also that if we relax the condition of symmetry on the kernel $K(t,u)$, requiring only that $K(t,u) = A(u)k(t,u)$, where $k(t,u)$ now has a rational Fourier transform, in the manner of Eq. (A.2.3), we can still use the procedure above, obtaining, in place of the linear differential equation with constant coefficients [Eq. (A.2.16)], the more general homogeneous linear differential equation

$$M\left(\frac{d^2}{dt^2}\right)\{\phi(t)A(t)\} = \lambda N\left(\frac{d^2}{dt}\right)\{\phi(t)\} \qquad a \le t \le b \tag{A.2.17}$$

where now the coefficients are no longer constant. An example of the former kind (symmetric rational kernels) appears in Prob. 8.11 and Eq. (16.34); of the latter [cf. Eq. (A.2.17)], in item 2 of Table 17.1, Sec. 17.3, and Refs. 1, 4 of Chap. 17. Finally, if the kernel itself is not rational and does not contain a rational factor as in Eq. (A.2.17) [but is still the covariance function of a stationary random process with a properly defined spectrum, e.g., band-limited white noise; cf. Eqs. (3.54*a*)], one can still find solutions to Eq. (A.2.1) in special cases, as the results of Prob. 8.12 indicate. There is, however, no single, infallible method yielding *explicit* solutions in this general situation (cf. Morse and Feshbach,[2] sec. 8.4).

A.2.3 *Inhomogeneous Equations with Rational Kernels*

In this section, we shall obtain the solution of the integral equation (A.2.2) when the kernel $K(t,u)$ is the sum of exponentials [Eq. (A.2.4)] and when the other conditions 1 to 3 apply as well.

† For a detailed account of the solution, with less restricted kernels, see Lovitt,[1] chap. 3, secs. 22–24, and chap. 5.

It is convenient, first, to reexpress Eq. (A.2.2) in a form more suited to subsequent manipulation. To do this, we let

$$u' = u - a \qquad t_0' = t_0 + a \qquad Z(u') \equiv z(u' + a) \tag{A.2.18}$$

so that Eq. (A.2.2) becomes directly

$$\int_{0-}^{(b-a)+} K(|t' - u'|)Z(u')\,du' = G(t' - t_0') \qquad 0- < t' < (b - a)+ \tag{A.2.19}$$

Letting

$$Z_T(u') = \begin{cases} Z(u') & 0- < u' < T+ \equiv (b - a)+ \\ 0 & \text{elsewhere} \end{cases} \tag{A.2.20}$$

we obtain finally for our original relation (A.2.2) the equivalent expression (dropping primes)

$$\int_{-\infty}^{\infty} K(|t - u|)Z_T(u)\,du = G(t - t_0 - a) \qquad 0- < t < T+ \tag{A.2.21}$$

Since the kernel is continuous everywhere, writing

$$G^{(+)} \equiv G(T - t_0 - a) \qquad G^{(-)} \equiv G(-t_0 - a) \tag{A.2.22}$$

we can define an "extended" representation of G by

$$G_E(t; t_0 + a) = \begin{cases} G^{(+)}e^{-c^{(+)}(t-T)} & t \geq T \\ G(t - t_0 - a) & 0 \leq t \leq T \\ G^{(-)}e^{c^{(-)}t} & t \leq 0 \end{cases} \tag{A.2.23}$$

where the constants $c^{(\pm)}$ are to be selected presently. These exponential factors ensure the proper convergence of G_E at infinity, where, in addition, it is assumed that G is positive (or negative) definite on $(0-, T+)$. The integral equation (A.2.21) can now be rewritten in a form to which the usual Fourier-transform techniques may be applied. We have

$$\int_{-\infty}^{\infty} K(|t - u|)Z_T(u)\,du = G_E(t; t_0 + a) + \begin{cases} \chi^{(+)}(t) & t > T \\ 0 & 0 \leq t \leq T \\ \chi^{(-)}(t) & t < 0 \end{cases} \tag{A.2.24}$$

where $\chi^{(+)}(t)$ is zero for $t \leq T$ and $\chi^{(-)}(t)$ vanishes for $t \geq 0$. From the continuity of the kernel, it is clear that (at $t = 0, T$) $\chi^{(+)}(T) = \chi^{(-)}(0) = 0$. Essentially our task at this point is to use this continuity property, along with the fact that the kernel is the sum of exponentials, to find these as yet unknown functions $\chi^{(\pm)}(t)$. As soon as this is accomplished, Fourier inversion of both members of Eqs. (A.2.24) gives us (the Fourier transform of) $Z_T(u)$ and our problem is formally solved.

Let $S_Z(p)_T$, $X^{(+)}(p)$, $X^{(-)}(p)$, $S_E(p)$ be the Fourier transforms of Z_T, $\chi^{(+)}$, $\chi^{(-)}$, G_E, respectively, where we have set $p = i\omega = 2\pi if$ and where by analytic continuation p will be allowed at the appropriate stage to assume complex values. Then we can write the following sets of Fourier transform pairs,

$$S_Z(p)_T = \mathfrak{F}\{Z_T\} = \int_{0-}^{T+} Z_T(t)e^{-pt}\,dt \qquad Z_T(t) = \mathfrak{F}^{-1}\{S_Z\} = \int_{-\infty i}^{\infty i} S_Z(p)_T e^{pt}\,\frac{dp}{2\pi i} \tag{A.2.25}$$

$$X^{(+)}(p) = \mathfrak{F}\{\chi^{(+)}\} = \int_{T}^{\infty} \chi^{(+)}(t)e^{-pt}\,dt \qquad \chi^{(+)}(t) = \mathfrak{F}^{-1}\{X^{(+)}\} = \int_{-\infty i}^{\infty i} X^{(+)}(p)e^{pt}\,\frac{dp}{2\pi i} \tag{A.2.26}$$

$$X^{(-)}(p) = \mathfrak{F}\{\chi^{(-)}\} = \int_{-\infty}^{0} \chi^{(-)}(t)e^{-pt}\,dt \qquad \chi^{(-)}(t) = \mathfrak{F}^{-1}\{X^{(-)}\} = \int_{-\infty i}^{\infty i} X^{(-)}(p)e^{-pt}\,\frac{dp}{2\pi i} \tag{A.2.27}$$

and, in particular with G_E given by Eqs. (A.2.23), we obtain

$$S_E(p) = \mathfrak{F}\{G_E\} = \int_{0-}^{T+} G(t - t_0 - a)e^{-pt}\,dt + \frac{G^{(+)}e^{-pT}}{p + c^{(+)}} - \frac{G^{(-)}}{p - c^{(-)}} \quad \text{(A.2.28)}$$

Now, by the Wiener-Khintchine theorem [cf. Eqs. (3.42) et seq.] we easily see that

$$\int_{-\infty}^{\infty} K(|t - u|)e^{-pt}\,dt = \tfrac{1}{2}e^{-pu}\mathcal{W}(p/2\pi i) \quad \text{(A.2.29)}$$

where $\mathcal{W}$ is the spectral density corresponding to the covariance function K, so that taking the Fourier transform of both sides of Eqs. (A.2.24), with the help of Eqs. (A.2.25) to (A.2.29), gives after a little manipulation

$$S_Z(p)_T = \frac{2S_E(p)}{\mathcal{W}(p/2\pi i)} + 2\,\frac{X^{(+)}(p) + X^{(-)}(p)}{\mathcal{W}(p/2\pi i)} \quad \text{(A.2.30)}$$

and the desired solution $Z_T(t)$ follows from

$$Z_T(t) = \int_{-\infty i}^{\infty i} S_Z(p)_T e^{pt}\,\frac{dp}{2\pi i} \quad \text{(A.2.31)}$$

[cf. Eqs. (A.2.25)], once the $X^{(\pm)}(p)$ have been determined.

There remains, then, to obtain $X^{(\pm)}(p)$, or, equivalently, $\chi^{(\pm)}(t)$. Again, we shall use the continuity of the kernel, as well as its specific form [Eq. (A.2.4)]. Then, taking the Fourier transform of both members of Eqs. (A.2.24), with Eq. (A.2.4) for $K(|t - u|)$, when $t > T$ we get

$$X^{(+)}(p) = \sum_{n=1}^{N} \frac{e^{-pT}}{b_n + p} \int_{-\infty}^{\infty} a_n e^{b_n(\tau - T)} Z_T(\tau)\,d\tau - \frac{G^{(+)}e^{-pT}}{c^{(+)} + p} \quad \text{(A.2.32)}$$

Letting

$$\gamma_n^{(+)} \equiv a_n e^{-b_nT} \int_{-\infty}^{\infty} e^{b_n\tau} Z_T(\tau)\,d\tau \quad \text{(A.2.33)}$$

where the $\gamma_n^{(+)}$ are a set of constants, gives us for Eq. (A.2.32)

$$X^{(+)}(p) = \sum_{n=1}^{N} \frac{\gamma_n^{(+)}e^{-pT}}{b_n + p} - \frac{G^{(+)}e^{-pT}}{c^{(+)} + p} \quad \text{(A.2.34)}$$

A similar approach when $t < 0$ yields

$$X^{(-)}(p) = \sum_{n=1}^{N} \frac{1}{b_n - p} \int_{-\infty}^{\infty} a_n e^{-b_n\tau} Z_T(\tau)\,d\tau - \frac{G^{(-)}}{c^{(-)} - p} \quad \text{(A.2.35)}$$

so that, paralleling Eq. (A.2.33), we may write

$$\gamma_n^{(-)} \equiv a_n \int_{-\infty}^{\infty} e^{-b_n\tau} Z_T(\tau)\,d\tau \quad \text{(A.2.36)}$$

Equation (A.2.35) accordingly reduces to

$$X^{(-)}(p) = \sum_{n=1}^{N} \frac{\gamma_n^{(-)}}{b_n - p} - \frac{G^{(-)}}{c^{(-)} - p} \tag{A.2.37}$$

At this point, let us use the continuity of the kernel at $t = 0, T$ to reduce the number of as yet undetermined parameters $\gamma_n^{(\pm)}$. Thus, since $\chi^{(+)}(T) = 0$, we find from Eq. (A.2.34) and the second relation in Eqs. (A.2.26) that

$$\chi^{(+)}(T) = \left[\int_{-\infty i}^{\infty i} e^{p(t-T)} \left(\sum_{n=1}^{N} \frac{\gamma_n^{(+)}}{b_n + p} - \frac{G^{(+)}}{c^{(+)} + p} \right) \frac{dp}{2\pi i}\right]_{t=T} = 0$$

or

$$G^{(+)} = \sum_{n=1}^{N} \gamma_n^{(+)} \tag{A.2.38}$$

Setting $c^{(+)} = b_1$ (or any other of the b_n), with the help of Eqs. (A.2.38) we can rewrite Eq. (A.2.34) finally as

$$X^{(+)}(p) = \sum_{n=2}^{N} \gamma_n^{(+)} e^{-pT} \frac{b_1 - b_n}{(b_n + p)(b_1 + p)} \tag{A.2.39}$$

A similar procedure (for the continuity condition at $t = 0$) yields

$$G^{(-)} = \sum_{n=1}^{N} \gamma_n^{(-)} \tag{A.2.40}$$

so that, putting $c^{(-)} = b_1$, we obtain from Eq. (A.2.37) as the analogue of Eq. (A.2.39)

$$X^{(-)}(p) = \sum_{n=2}^{N} \gamma_n^{(-)} \frac{b_1 - b_n}{(b_n - p)(b_1 - p)} \tag{A 2.41}$$

Inserting this, with Eq. (A.2.39), into Eq. (A.2.30), we then have the desired solution of the integral equation (A.2.21), with the $2N - 2$ as yet undetermined constants $\gamma_n^{(\pm)}$ $(n = 2, \ldots, N)$. These constants, however, may always be found by introducing the solution (A.2.30) with Eqs. (A.2.38), (A.2.39), (A.2.41) into the original integral equation (A.2.21) and treating the ensuing relation as an identity.† The $\gamma_n^{(\pm)}$ are usually complex, while $Z_T(t)$ itself is real, vanishing outside $(0-, T+)$.‡

Besides the above, some further properties of the solution Z_T, including restrictions on the function G, are to be noted. Let us begin by writing the various terms of Eq. (A.2.30) in more detail, remembering that the spectrum $\mathcal{W}$ is assumed to be rational, with simple, distinct poles and zeros. Applying Eq. (A.2.3) to Eq. (A.2.30), with Eq.

† It is not necessary to employ the continuity condition at $t = 0, T$ directly, as in Eqs. (A.2.38) to (A.2.41). We can equally well use Eqs. (A.2.34), (A.2.37) in Eq. (A.2.30), determining the now $2N$ unknown constants again by regarding the integral equation as an identity.

‡ These properties sometimes enable us to find the $\gamma_n^{(\pm)}$ more directly without having to insert Eq. (A.2.30) into the integral equation itself.

(A.2.28), we obtain

$$\frac{2S_E(p)}{\mathfrak{W}(p/2\pi i)} = 2C_0^{-1}\int_{0-}^{T+} G(\tau - t_0 - a)\,d\tau\,\frac{e^{-p\tau}\prod_{n=1}^{N}(b_n^2 - p^2)}{\prod_{n=0}^{M}(c_n^2 - p^2)} + [2G^{(+)}e^{-pT}(b_1 - p) + 2G^{(-)}(b_1 + p)]\,\frac{C_0^{-1}\prod_{n=2}^{N}(b_n^2 - p^2)}{\prod_{n=0}^{M}(c_n^2 - p^2)} \tag{A.2.42}$$

Similarly, from Eqs. (A.2.39), (A.2.41) in Eq. (A.2.30) we get

$$\frac{2X^{(\pm)}(p)}{\mathfrak{W}(p/2\pi i)} = 2C_0^{-1}\sum_{n=2}^{N}\gamma_n^{(\pm)}\,\frac{(e^{-pT}\text{ or }1)(b_1 - b_n)(b_1 \mp p)}{b_n \pm p}\,\frac{\prod_{j=2}^{N}(b_j^2 - p^2)}{\prod_{j=0}^{M}(c_j^2 - p^2)} \tag{A.2.43}$$

From Eq. (A.2.42), we observe that the right-hand member is $O(p^{2N-2M-1})$ and that the terms containing $X^{(\pm)}(p)$ are $O(p^{2N-2M-2})$. Now, from the fact that

$$\int_{-\infty i}^{\infty i} p^m e^{p(t-\tau)}\,\frac{dp}{2\pi i} \equiv \delta^{(m)}(t - \tau) = \frac{d^m}{dt^m}\,\delta(t - \tau) = (-1)^m\,\frac{d^m}{d\tau^m}\,\delta(t - \tau) \tag{A.2.44}$$

we have

$$\begin{aligned}\int_{0-}^{T+} G(\tau - t_0')\delta^{(m)}(t - \tau)\,d\tau &= G(-t_0')\delta^{(m-1)}(t - 0) - G(T - t_0')\delta^{(m-1)}(t - T)\\ &\quad + \dot G(-t_0')\delta^{(m-2)}(t - 0) - \dot G(T - t_0')\delta^{(m-2)}(t - T)\\ &\quad + \cdots + G^{(m-1)}(-t_0')\delta(t - 0) - G^{(m-1)}(T - t_0')\delta(t - T)\\ &\quad + G^{(m)}(t - t_0')\Big|_{0\le t\le T}\end{aligned} \tag{A.2.45}$$

Accordingly, it follows that the solution Z_T contains terms involving the first $2N - 2M$ derivatives of G in the interval $(-t_0', T - t_0')$, including the end points of this interval.†

† When the derivative G^m is defined in $(-t_0', T - t_0')$, but $G^{(m)}$ does not exist at $-t_0+$ and at $T - t_0'$, we can take the limit of the solution as $t \to 0$ positively and $t \to T$ negatively. An example of this is given in Eqs. (16.81), (16.82) for pure prediction, when $t_\mu \to 0$ positively: the basic integral equation now has the form

$$\int_{0-}^{\infty} K_S(t - u)h_\lambda(u)_0\,du = K_S(t) \qquad 0- < t < \infty$$

and it is clear that the solution is simply $h_\lambda(t)_0 = \delta(t - 0)$, or from Eqs. (16.82)

$$\lim_{t_\mu \to 0+}\ (2\psi_S)^{-1}(-\dot K_S + \alpha K_S)_{t_\mu}\delta(t - 0)$$

where $\dot K_S = -\alpha\psi_S e^{-\alpha_S t}$ $(t \geq 0+)$. When G is a rational kernel, like K above, this can be demonstrated in general by carrying out the steps indicated in Eqs. (A.2.42), (A.2.43), followed by inversion according to Eq. (A.2.31), etc. The details are left to the reader.

The differentiability of G to the required order is thus a condition for the existence of the solution. The singularities at the ends of the interval $(0-,T+)$ can be of order $2N - 2M - 1$ (≥ 0) but usually are of no higher order than $N - M - 1$, as the contributions of order $2N - 2M - 1, \ldots, N - M$ in $2S_E(p)/\mathcal{W}$ cancel corresponding terms in $2X^{(\pm)}(p)/\mathcal{W}$ (after inversion). Illustrations of this are shown in examples 1, 2 of Sec. A.2.4.

Finally, when $T \to \infty$ [or when the interval in question is $(-\infty, 0\pm)$, say], the integral equation reduces to one of the Wiener-Hopf type.† There are then $N - 1$ undetermined constants $\gamma_n^{(-)}$ [$n = 1, \ldots, N - 1$; cf. Eq. (A.2.41)] as $X^{(+)}(p)$ vanishes, and the requirement on G is still differentiability through order $2N - 2M$, as before. The solution contains possible singularities only at $t = 0$, and the factorization method discussed in Sec. 16.2.2 may alternatively be applied provided that the condition (16.60c) holds. It should be noted, however, that this factorization technique cannot be applied for the finite interval $(0-,T+)$ above, as G is not normally known outside $(0- < t < T+)$. The theories of the semi-infinite and finite regions are distinct. In fact, solutions in the latter case are known, to date, only in the case of the rational kernel.

A.2.4 *Examples*

From the discussion in Sec. A.2.1, it is evident that a rational kernel, where the poles and zeros of the corresponding Fourier transform are distinct and simple, can always be expressed as a sum of simple exponential terms, according to Eq. (A.2.4), with Eqs. (A.2.13a). In this section, we shall illustrate the procedure of Sec. A.2.3 for kernels of this important class with a number of examples of importance to us in the communication problems discussed in the text.

Example 1. The RC Kernel. Let us suppose that stationary white noise of density W_0 is passed through an RC filter [cf. Eqs. (3.55)] and that the covariance function of this process is the kernel $K(t,u)$ of our integral equation (A.2.21), where $b - a \equiv T$ and $t_0' = t_0 + a = \epsilon$. The kernel and its Fourier transform [by the W-K theorem (3.42)] are specifically

$$K(\tau) = \psi e^{-\alpha|\tau|} \qquad \mathcal{W}\left(\frac{p}{2\pi i}\right) = \frac{4\psi\alpha}{\alpha^2 - p^2} \qquad \psi = \frac{W_0\alpha}{4} \tag{A.2.46}$$

Here, from Eqs. (A.2.5) and (A.2.39), (A.2.41), we have $N = 1$; $M = 0$; $a_1 = \psi$; $b_1 = \alpha$; $\gamma_n^{(\pm)} = 0$; $n > 1$; and $\gamma_1^{(\pm)} = G^{(\pm)} = G(T - \epsilon)$, $G(-\epsilon)$ [cf. Eqs. (A.2.38), (A.2.40), with Eqs. (A.2.22)]. From Eqs. (A.2.30), (A.2.31), we can write

$$Z_T(t) = \frac{1}{2\psi\alpha}\left[\int_{0-}^{T+} G(\tau - \epsilon)\,d\tau \int_{-\infty i}^{\infty i} (\alpha^2 - p^2)e^{p(t-\tau)}\frac{dp}{2\pi i} + \int_{-\infty i}^{\infty i} (\alpha^2 - p^2)\left(\frac{G^{(+)}e^{-pT}}{\alpha + p} + \frac{G^{(-)}}{\alpha - p}\right)e^{pt}\frac{dp}{2\pi i}\right] \tag{A.2.47}$$

Using Eqs. (A.2.44), (A.2.45) and carrying out the indicated operations by the usual methods of contour integration, we get finally‡

$$Z_T(t) = \frac{1}{2\psi\alpha}\{[\alpha^2 G(t - \epsilon) - \ddot{G}(t - \epsilon)] + (\dot{G} + \alpha G)_{T-\epsilon}\delta(t - T) + (\alpha G - \dot{G})_{-\epsilon}\delta(t - 0)\} \qquad 0- < t < T+ \tag{A.2.48}$$

† For a discussion, see Morse and Feshbach,[2] pp. 978–985.

‡ This is a slight generalization of a result of Reich and Swerling,[5] arising in certain detection problems (cf. Sec. 20.4-1).

The corresponding Wiener-Hopf solution ($T \to \infty$) is simply

$$Z_T(t) = \begin{cases} \frac{1}{2\psi\alpha}[(\alpha^2 G - \ddot{G})_{t-\epsilon} + (\alpha G - \dot{G})_{-\epsilon}\delta(t-0)] & 0- < t < \infty \\ 0 & t < 0- \end{cases} \tag{A.2.48a}$$

Example 2. The *LRC* Kernel. This case is considerably more involved than the preceding example, inasmuch as the undetermined constants are no longer easily found by inspection but rather by substituting the solution (A.2.30) back into the original integral equation and treating the ensuing relation as an identity. Here, the kernel $K(t,u)$ is now the covariance of the random process obtained by passing stationary white noise through the *LRC* filter of Table 3.1. Thus,

$$K(\tau) = a_1 e^{-b_1|\tau|} + a_2 e^{-b_2|\tau|} = \psi e^{-\omega_F|\tau|}\left(\cos\omega_1\tau + \frac{\omega_F}{\omega_1}\sin\omega_1|\tau|\right) \tag{A.2.49}$$

wherein now $N = 2$, $M = 0$, and

$$a_1 = \frac{-\omega_0{}^4 W_0}{4b_1}\frac{1}{b_1{}^2 - b_2{}^2} \qquad a_2 = \frac{\omega_0{}^4 W_0}{4b_2}\frac{1}{b_1{}^2 - b_2{}^2} \qquad W_0 = \frac{8\psi\omega_F}{\omega_0{}^2} \tag{A.2.49a}$$

and

$$2\mathfrak{F}\{K\} = \mathcal{W}\left(\frac{p}{2\pi i}\right) = W_0 Y(p) Y(-p) = \frac{W_0\omega_0{}^4}{(b_1{}^2 - p^2)(b_2{}^2 - p^2)} \tag{A.2.49b}$$

where, in turn,

$$b_1 = \omega_F - i\omega_1 \qquad b_2 = \omega_F + i\omega_1 \qquad b_1 b_2 = \omega_0{}^2 \qquad b_1{}^2 + b_2{}^2 = 2(\omega_F{}^2 - \omega_1{}^2) \qquad \text{etc.} \tag{A.2.49c}$$

The circuit parameters ω_F, ω_1, ω_0 are specifically

$$\omega_F = R/2L \qquad \omega_0 = 1/\sqrt{LC} \qquad \omega_1 = \sqrt{\omega_0{}^2 - \omega_F{}^2} \tag{A.2.49d}$$

Setting

$$H^{(\pm)}(p) \equiv 2X^{(\pm)}(p)/W_0 Y(p) Y(-p) \tag{A.2.50}$$

we readily find from Eqs. (A.2.39), (A.2.41), (A.2.49*b*), etc., in Eq. (A.2.30) that, on inversion,

$$\mathfrak{F}^{-1}\{H^{(+)}(p)\} = \frac{-4i\omega_1\gamma_2{}^{(+)}}{W_0\omega_0{}^4}[b_1 b_2\delta(t-T) - (b_1+b_2)\delta'(t-T) + \delta''(t-T)] \tag{A.2.51a}$$

$$\mathfrak{F}^{-1}\{H^{(-)}(p)\} = \frac{-4i\omega_1\gamma_2{}^{(-)}}{W_0\omega_0{}^4}[b_1 b_2\delta(t-0) + (b_1+b_2)\delta'(t-0) + \delta''(t-0)] \tag{A.2.51b}$$

We have also

$$\mathfrak{F}^{-1}\left\{\frac{2S_E(p)}{\mathcal{W}}\right\} = \frac{2}{W_0\omega_0{}^4}\left\{\int_{0-}^{T+} G(\tau-\epsilon)[b_1{}^2 b_2{}^2\delta(t-\tau) - (b_1{}^2 + b_2{}^2)\delta^{(2)}(t-\tau) + \delta^{(4)}(t-\tau)]\,d\tau + G(T-\epsilon)(b_1 b_2{}^2\delta - b_2{}^2\delta^{(1)} - b_1\delta^{(2)} + \delta^{(3)})_{t-T} + G(-\epsilon)(b_1 b_2{}^2\delta + b_2{}^2\delta^{(1)} - b_1\delta^{(2)} - \delta^{(3)})_{t-0}\right\} \tag{A.2.52}$$

Integrating Eq. (A.2.52), combining with Eqs. (A.2.51*a*), (A.2.51*b*), and substituting into the original integral equation† give finally

$$\gamma_2{}^{(+)} = -[\dot{G}(T-\epsilon) + b_1 G(T-\epsilon)]/2i\omega_1 \qquad \gamma_2{}^{(-)} = [\dot{G}(-\epsilon) - b_1 G(-\epsilon)]/2i\omega_1 \tag{A.2.53}$$

† In this case, it is simpler to use the facts that $Z_T(t)$ must vanish outside $(0-,T+)$ and be real in $(0-,T+)$ to determine $\gamma_2{}^{(\pm)}$.

and the desired solution is found to be specifically

$$Z_T(t) = \frac{2}{W_0\omega_0^4}\{[\omega_0^4 G - 2(\omega_F^2 - \omega_0^2)\ddot{G} + G^{(4)}]_{t-\epsilon} + A_0\delta(t-T) + A_1\delta'(t-T) + B_0\delta(t-0) + B_1\delta'(t-0)\} \qquad 0- < t < T+ \tag{A.2.54}$$

with

$$\begin{aligned} A_0 &= [-G^{(3)} + \omega_0^2\dot{G} + 2\omega_F\omega_0^2 G]_{T-\epsilon} \\ B_0 &= [G^{(3)} - \omega_0^2\dot{G} + 2\omega_F\omega_0^2 G]_{-\epsilon} \\ A_1 &= -[\ddot{G} + 2\omega_F\dot{G} + (4\omega_F^2 - \omega_0^2)G]_{T-\epsilon} \\ B_1 &= [\ddot{G} - 2\omega_F\dot{G} + (4\omega_F^2 - \omega_0^2)G]_{-\epsilon} \end{aligned} \tag{A.2.55}$$

These results hold for all $Q\ (\equiv \omega_0/2\omega_F \geq 0)$.

Two cases of particular interest are (1) *critical damping* ($\omega_F = \omega_0$) and (2) *high-Q*, or *narrowband, noise* ($\omega_0 \gg \omega_F$). For (1), we find that Eqs. (A.2.54), (A.2.55) reduce to

Critical damping

$$Z_T(t) = \frac{2}{W_0\omega_0^4}[(\omega_0^4 G + G^{(4)})_{T-\epsilon} + A_0'\delta(t-r) + A_1'\delta'(t-T) + B_0'\delta(t-0) + B_1'\delta'(t-0)] \qquad 0- < t < T+ \tag{A.2.56}$$

with

$$\begin{aligned} A_0' &= (-G^{(3)} + \omega_0^2\dot{G} + 2\omega_0^3 G)_{T-\epsilon} \qquad & B_0' &= (G^{(3)} - \omega_0^2\dot{G} + 2\omega_0^3 G)_{-\epsilon} \\ A_1' &= -(\ddot{G} + 2\omega_0\dot{G} + 3\omega_0^2 G)_{T-\epsilon} & B_1' &= (\ddot{G} - 2\omega_0\dot{G} + 3\omega_0^2 G)_{-\epsilon} \end{aligned} \tag{A.2.57}$$

This is an example of a limiting form of the exponential kernel mentioned earlier [cf. footnote referring to Eq. (A.2.4)]. Here Eq. (A.2.49) reduces, as $b_1 \to b_2$, to

$$K(\tau) = \psi(1 + \omega_0|\tau|)e^{-\omega_0|\tau|} \qquad \omega_F = \omega_0 \tag{A.2.58}$$

In the same way, we find for the narrowband cases that the solution (A.2.54) simplifies to the (approximate) result

Narrowband cases

$$Z_T(t) \doteq \frac{2}{W_0\omega_0^4}[(\omega_0^2 G + 2\omega_0^2\ddot{G} + G^{(4)})_{t-\epsilon} + (-G^{(3)} + \omega_0^2\dot{G})_{T-\epsilon}\delta(t-T) + (-\ddot{G} + \omega_0^2 G)_{T-\epsilon}\delta'(t-T) - (G^{(3)} - \omega_0^2\dot{G})_{-\epsilon}\delta(t-0) + (\ddot{G} - \omega_0^2 G)_{-\epsilon}\delta'(t-T)]] \qquad 0- < t < T+ \tag{A.2.59}$$

The Wiener-Hopf solution ($T \to \infty$) follows at once from Eqs. (A.2.54), (A.2.56), (A.2.59) if we drop the singular terms at $t = T$.

Example 3. Mixed Kernels; *RC* Kernels and White Noise. It frequently happens that the kernel $K(t,u)$ is the covariance function of the sum of two noise processes $N(t)$, $S(t)$, one of which, $N(t)$, is an (independent) white noise, so that our inhomogeneous integral equation takes the form of Eq. (A.2.8*c*). The general approach of Sec. A.2.3 is readily modified to handle this case, for the spectrum of this additive process is simply the sum of the two spectra. Accordingly, Eq. (A.2.3) can be written

$$\mathcal{W}_{S+N} = \mathcal{W}_S + W_{0N} = \prod_{n=0}^{N}(c_n^2 + \omega^2) \Big/ \prod_{n=0}^{N}(d_n^2 + \omega^2) \qquad M = N \tag{A.2.60}$$

where the order of numerator and denominator is now the same. Equations (A.2.30), (A.2.31) still apply, where now the steps for $X^{\pm}(p)$ [Eqs. (A.2.32) to (A.2.41)] are appropriately altered to account for the modified form of the kernel (e.g., the delta function that appears in addition to the sum of simple exponentials).

An alternative procedure is to observe that since for the *symmetrical* delta function

$$\delta(\tau - 0) = \lim_{\alpha_N \to \infty} \frac{\alpha_N}{2} e^{-\alpha_N|\tau|}$$

the white noise component of the mixed kernel may be treated as the limit of an *RC* kernel here, e.g.,

$$K_N(\tau) = W_{0N} \lim_{\alpha_N \to \infty} \frac{\alpha_N}{4} e^{-\alpha_N|\tau|} = \frac{W_{0N}}{2} \delta(\tau - 0) \tag{A.2.61}$$

Accordingly, we may consider first the solution for the case where the kernel K is entirely the sum of exponentials, using the approach of Sec. A.2.3 without modification, and then (formally) take the limit of this solution as $\alpha_N \to \infty$.

Let us illustrate these remarks in the case where $S(t)$ is an *RC* noise process, so that our integral equation (A.2.21) is specifically

$$\int_{0-}^{T+} \left[\psi_S e^{-\alpha|t-u|} + \frac{W_{0N}}{2} \delta(t - u) \right] Z_T(u) = G(t - \epsilon) \qquad 0- < t < T+ \tag{A.2.62}$$

Here we shall adopt the alternative procedure mentioned above and consider first the equation

$$\int_{0-}^{T+} \left(\psi_S e^{-\alpha|t-u|} + \frac{W_{0N}\alpha_N}{4} e^{-\alpha_N|t-u|} \right) Z_T(u) = G(t - \epsilon) \qquad 0- < t < T+ \tag{A.2.62a}$$

The kernel $K(t,u)$ contains two simple exponential terms, so that

$$\mathcal{W}\left(\frac{p}{2\pi i}\right) = \frac{4\psi_S\alpha}{\alpha^2 - p^2} + \frac{W_{0N}\alpha_N^2}{\alpha_N^2 - p^2} = C_0 \frac{c_1^2 - p^2}{(b_1^2 - p^2)(b_2^2 - p^2)} \tag{A.2.63}$$

with

$$M = 1, N = 2 \qquad a_1 = \psi_S \qquad a_2 = \frac{W_{0N}\alpha_N}{4} \qquad b_1 = \alpha \qquad b_2 = \alpha_N$$

$$c_1 = \alpha \left(\frac{1 + \gamma_0^2}{1 + \alpha^2\gamma_0^2/\alpha_N^2} \right)^{1/2} \tag{A.2.63a}$$

and with $\gamma_0^2 = 4\psi_S/\alpha W_{0N}$ and $C_0 = 4\psi_S + W_{0N}\alpha_N^2$. Putting this into Eqs. (A.2.42), (A.2.43) gives

$$\begin{aligned} \frac{2S_E(p)}{\mathcal{W}(p/2\pi i)} = 2C_0^{-1} \int_{0-}^{T+} G(\tau - \epsilon)\, d\tau \bigg[-p^2 + (b_1^2 + b_2^2 - c_1^2) \\ + \frac{b_1^2 b_2^2 - c_1^2(b_1^2 + b_2^2 - c_1^2)}{c_1^2 - p^2} \bigg] e^{-p\tau} \\ + 2C_0^{-1} \frac{[G^{(+)}e^{-pT}(b_1 - p) + G^{(-)}(b_1 + p)](b_2^2 - p^2)}{c_1^2 - p^2} \end{aligned} \tag{A.2.64}$$

$$H^{(\pm)}(p) \equiv 2\frac{X^{(\pm)}(p)}{\mathcal{W}(p/2\pi i)} = 2C_0^{-1}\gamma_2^{(\pm)} (e^{-pT} \text{ or } 1) \frac{(b_1 - b_2)(b_1 \mp p)(b_2 \mp p)}{(c_1 - p)(c_1 + p)} \tag{A.2.65}$$

Inverting Eqs. (A.2.64), (A.2.65) according to Eq. (A.2.31) then yields

$$\begin{aligned} \mathfrak{F}^{-1}\left\{\frac{2S_E(p)}{\mathcal{W}}\right\} = 2C_0^{-1} \bigg\{ \int_{0-}^{T+} G(\tau - \epsilon) \bigg[\frac{b_1^2 b_2^2 - c_1^2(b_1^2 + b_2^2 - c_1^2)}{-2c_1} e^{\pm c_1(t-\tau)} \bigg|_{t-\tau \lessgtr 0} \\ + (b_1^2 + b_2^2 - c_1^2)\delta(\tau - t) - \delta^{(2)}(\tau - t) \bigg] d\tau \\ + G^{(+)} \bigg[-\delta'(t - T) + b_1\delta(t - T) + \frac{(b_1 \mp c_1)(b_2^2 - c_1^2)}{2c_1} e^{\pm c_1(t-T)} \bigg|_{t \lessgtr T} \bigg] \\ + G^{(-)} \bigg[\delta'(t - 0) + b_1\delta(t - 0) + \frac{(b_1 \pm c_1)(b_2^2 - c_1^2)}{2c_1} e^{\pm c_1 t} \bigg|_{t \lessgtr 0} \bigg] \bigg\} \end{aligned} \tag{A.2.66}$$

Similarly, we get

$$\mathfrak{F}^{-1}\{H^{(+)}\} = 2C_0^{-1}\gamma_2^{(+)}(b_1 - b_2)\left[-\delta(t - T) + \frac{b_1b_2 + c_1^2 \mp c_1(b_1 + b_2)}{2c_1} e^{\pm c_1(t-T)}\Big|_{t \lessgtr T}\right] \tag{A.2.67}$$

$$\mathfrak{F}^{-1}\{H^{(-)}\} = 2C_0^{-1}\gamma_2^{(-)}(b_1 - b_2)\left[-\delta(t - 0) + \frac{b_1b_2 + c_1^2 \pm c_1(b_1 + b_2)}{2c_1} e^{\pm c_1 t}\Big|_{t \lessgtr 0}\right] \tag{A.2.68}$$

For $(0- < t < T+)$, the desired solution is accordingly

$$\begin{aligned} Z_T(t) = 2C_0^{-1}\Bigg(&\frac{b_1^2b_2^2 - c_1^2(b_1^2 + b_2^2 - c_1^2)}{-2c_1}\int_{0-}^{T+} G(\tau - \epsilon)e^{-c_1|t-\tau|}\,d\tau \\ &+ (b_1^2 + b_2^2 - c_1^2)G(t - \epsilon) - \ddot{G}(t - \epsilon) \\ &+ \frac{e^{c_1(t-T)}}{2c_1}\{G^{(+)}(b_1 - c_1)(b_2^2 - c_1^2) + \gamma_2^{(+)}(b_1 - b_2)[b_1b_2 + c_1^2 - c_1(b_1 + b_2)]\} \\ &+ \frac{e^{-c_1t}}{2c_1}\{G^{(-)}(b_1 - c_1)(b_2^2 - c_1^2) + \gamma_2^{(-)}(b_1 - b_2)[b_1b_2 + c_1^2 - c_1(b_1 + b_2)]\} \\ &+ \delta(t - T)\,[b_1G^{(+)} + \dot{G}(T - \epsilon) - (b_1 - b_2)\gamma_2^{(+)}] \\ &+ \delta(t - 0)[b_1G^{(-)} - \dot{G}(-\epsilon) - (b_1 - b_2)\gamma_2^{(-)}]\Bigg) \end{aligned} \tag{A.2.69}$$

where $Z_T = 0$ outside $(0-,T+)$, as usual.

The undetermined constants $\gamma_2^{(\pm)}$ are here most easily found from the requirement that Z_T vanish outside the interval $(0-,T+)$. After a little manipulation, we obtain

$$\gamma_2^{(+)} = \frac{D_3E_1 - E_3D_1}{\Delta} \qquad \gamma_2^{(-)} = \frac{D_1E_2 - D_2E_1}{\Delta} \tag{A.2.70}$$

with $\Delta = D_2E_3 - E_2D_3$, where specifically

$$\begin{aligned} D_1 &= (b_2^2 - c_1^2)[(b_1 + c_1)G^{(+)}e^{c_1T} + (b_1 - c_1)G^{(-)}] \\ &\qquad - [b_1^2b_2^2 - c_1^2(b_1^2 + b_2^2 - c_1^2)]\int_{0-}^{T+} e^{c_1\tau}G(\tau - \epsilon)\,d\tau \\ D_2 &= (b_1 - b_2)[b_1b_2 + c_1^2 + c_1(b_1 + b_2)]e^{c_1T} = E_3e^{c_1T} \\ D_3 &= (b_1 - b_2)[b_1b_2 + c_1^2 - c_1(b_1 + b_2)] = E_2e^{c_1T} \\ E_1 &= (b_2^2 - c_1^2)[(b_1 - c_1)G^{(+)}e^{-c_1T} + (b_1 + c_1)G^{(-)}] \\ &\qquad - [b_1^2b_2^2 - c_1^2(b_1^2 + b_2^2 - c_1^2)]\int_{0-}^{T+} e^{-c_1\tau}G(\tau - \epsilon)\,d\tau \end{aligned} \tag{A.2.71}$$

$$\Delta = (b_1 - b_2)^2[(b_1b_2 + c_1^2 + c_1b_1 + c_1b_2)^2e^{c_1T} - (b_1b_2 + c_1^2 - c_1b_1 - c_1b_2)^2e^{-c_1T}] \tag{A.2.72}$$

Note that, when $c_1 \to b_2$ (that is, $\gamma_0 \to \infty$, or, equivalently, $W_{0N} \to 0$), Eq. (A.2.69) reduces as expected to the earlier result [Eq. (A.2.48)] for the single RC kernel.

In the special case where the second term in the kernel represents white noise [cf. Eq. (A.2.62)], we let $\alpha_N \to \infty$ in the solutions (A.2.69) to (A.2.72), to get

$$\Delta \sim \alpha_N^4[(b_1 + c_1)^2e^{c_1T} - (b_1 - c_1)^2e^{-c_1T}] \equiv \Delta_0\alpha_N^4 \tag{A.2.73}$$

with

$$\gamma_2^{(+)} \to G^{(+)} + \frac{(b_1 + c_1)^2(b_1 - c_1)}{\Delta_0}\int_{0-}^{T+} e^{c_1\tau}G(\tau - \epsilon)\,d\tau - \frac{(b_1 - c_1)^2(b_1 + c_1)}{\Delta_0}\int_{0-}^{T+} e^{-c_1\tau}G(\tau - \epsilon)\,d\tau \tag{A.2.74a}$$

$$\gamma_2^{(-)} \to G^{(-)} + \frac{(b_1 + c_1)^2(b_1 - c_1)}{\Delta_0} e^{c_1 T} \int_{0-}^{T+} e^{-c_1\tau} G(\tau - \epsilon)\, d\tau - \frac{(b_1 - c_1)^2(b_1 + c_1)}{\Delta_0} e^{-c_1 T} \int_{0-}^{T+} e^{c_1\tau} G(\tau - \epsilon)\, d\tau \quad \text{(A.2.74}b\text{)}$$

(A.2.5), we

The solution (A.2.69) now reduces to

$$Z_T(t) = \frac{2}{W_{0N}} \left[\frac{b_1^2 - c_1^2}{-2c_1} \int_0^T G(\tau - \epsilon) e^{-c_1|t-\tau|}\, d\tau + G(t - \epsilon) + (G^{(+)} - \gamma_2^{(+)}) \frac{b_1 - c_1}{2c_1} e^{c_1(t-T)} + (G^{(-)} - \gamma_2^{(-)}) \frac{b_1 - c_1}{2c_1} e^{-c_1 t} \right] \quad 0 \le t \le T \quad \text{(A.2.75)}$$

where the $\gamma_2^{(\pm)}$ are given by Eqs. (A.2.74*a*), (A.2.74*b*) and $c_1 \to \alpha\sqrt{1 + \gamma_0^2}$, $b_1 = \alpha$, with $Z_T = 0$ outside the interval $(0-,T+)$ as before. Observe that, when $b_1 \to c_1$ (that is, $\gamma_0 \to 0$, or, equivalently, $\psi_S = 0$), we get here the expected result $Z_T(t) = (2/W_{0N})G(t - \epsilon)$ $(0 \le t \le T)$ for the simplest case of Eq. (A.2.48), when the kernel is the delta function $(W_{0N}/2)\delta(t - u)$ for white noise alone.

Example 4. A Mixed Kernel, with White Noise.† One integral equation of considerable interest to us in applications where the signal itself is a stationary, entirely random process is given by Eqs. (A.2.8), where $G(t - \epsilon)$ takes the specific form of Eq. (A.2.9) with $Q = DK_S$. Accordingly, we have

$$\int_{0-}^{T+} [K_S(t - u) + B\delta(t - u)] Z_T(u)\, du = D \int_{0-}^{T+} K_S(t - u) g_T(u)\, du \qquad 0- \le t \le T+ \quad \text{(A.2.76)}$$

The kernel K_S obeys the conditions 1 to 4 of Sec. A.2.1, and $g_T(t) = g(t)$ $(0- < t < T+)$, $g_T(t) = 0$ elsewhere, like $Z_T(t)$ [cf. Eqs. (A.2.20)].

To solve Eq. (A.2.76), we again use the continuity properties of the kernel and write as a modified form of Eqs. (A.2.24)

$$\int_{-\infty}^{\infty} K_S(t - u)[Z_T(u) - Dg(u)]\, du + BZ_T(t) = \begin{cases} \sum_{n=1}^{N} A_n^{(+)} e^{-b_n(t-T)} & t \ge T \\ 0 & 0 \le t \le T \\ \sum_{n=1}^{N} A_n^{(-)} e^{b_n t} & t \le 0 \end{cases} \quad \text{(A.2.77)}$$

where the $2N$ undetermined constants $A_n^{(\pm)}$ are to be found, as before, from the requirement that Z_T vanish outside $(0,T)$. Setting $p = i\omega$ and writing

$$S_T(p)_Z = \mathfrak{F}\{Z_T\} = \int_{0-}^{T+} e^{-p\tau} Z_T(\tau)\, d\tau \qquad S_T(p)_g = \mathfrak{F}\{g\} = \int_{0-}^{T+} e^{-p\tau} g(\tau)\, d\tau \quad \text{(A.2.78)}$$

[with their corresponding inverses $Z_T = \mathfrak{F}^{-1}\{S_T(p)_Z\}$, etc.], and using Eqs. (A.2.29), (A.2.31), we transform both members of Eqs. (A.2.77) in the usual way [cf. Eqs. (A.2.25) et seq.], to get

$$S_T(p)_Z = \left[2B + \mathcal{W}_S\left(\frac{p}{2\pi i}\right)\right]^{-1} \left[2 \sum_{n=1}^{N} \left(\frac{A_n^{(+)} e^{-pT}}{b_n + p} + \frac{A_n^{(-)}}{b_n - p}\right) + DS_T(p)_g \mathcal{W}_S\left(\frac{p}{2\pi i}\right)\right] \quad \text{(A.2.79)}$$

where $\mathcal{W}_S$ is the intensity spectrum $2\mathfrak{F}\{K_S\}$.

† See Middleton,[6] app. IV.

Our problem now is to determine the Fourier transform of $S_T(p)_Z$, to get Z_T, and to obtain the as yet unspecified constants $A_n^{(\pm)}$. We begin by observing that here there are $2N$ roots of

$$2B + \mathcal{W}_S(p_{\pm k}/2\pi i) = 0 \tag{A.2.80}$$

($k = 1, 2, \ldots, N$), and we assume that none of these roots is multiple, so that letting

$$H(p) \equiv [2B + \mathcal{W}_S(p/2\pi i)]^{-1} \tag{A.2.81}$$

we have

$$\int_{-\infty i}^{\infty i} e^{pt}\mathcal{W}_S\left(\frac{p}{2\pi i}\right) H(p)\,\frac{dp}{2\pi i} \equiv \sum_{n=1}^{N} h_n e^{-p_n|t|} = h(|t|) \tag{A.2.82}$$

Here the p_n ($n = 1, \ldots, N$) are the N roots of $H(p)^{-1}$ with *positive* real parts. That h is a function of $|t|$ follows directly from the fact that the real parts of the poles of $\mathcal{W}_S$ and $H(p)$ are symmetrically located on either side of the imaginary axis.

Since, by hypothesis above, none of the roots of Eq. (A.2.80) is multiple, it is convenient now to write

$$\mathcal{W}_S(p/2\pi i) \equiv M(p)/N(p) \tag{A.2.83a}$$

or

$$H(p) = N(p)/[2BN(p) + M(p)] \equiv N(p)/L(p) \tag{A.2.83b}$$

with

$$L(p) \equiv 2BN(p) + M(p) = L(-p) \tag{A.2.83c}$$

where $M(p)$ and $N(p)$ are both $O(N)$ in p^2, as is $L(p)$ also. In terms of Eq. (A.2.5), we have

$$\mathcal{W}_S(p/2\pi i) = 4\sum_{n=1}^{N} a_n b_n \prod_{j=1}^{N}{}' (b_j^2 - p^2) \Big/ \prod_{n=1}^{N} (b_n^2 - p^2) \tag{A.2.84}$$

(where the prime on the product Π indicates that the factor for $j = n$ is omitted), so that

$$M(p) = 4\sum_{n=1}^{N} a_n b_n \prod_{j=1}^{N}{}' (b_j^2 - p^2) \qquad N(p) = \prod_{n=1}^{N} (b_n^2 - p^2) \tag{A.2.85}$$

Writing c_n for p_n [cf. Eqs. (A.2.82) et seq.], let us define a new product by

$$L(p) \equiv A' \prod_{n=1}^{N} (c_n^2 - p^2) = 2B\prod_{n=1}^{N} (b_n^2 - p^2) + 4\sum_{n=1}^{N} a_n b_n \prod_{j=1}^{N}{}' (b_j^2 - p^2)$$

$$c_n \neq b_n,\ \mathrm{Re}\, c_n > 0 \tag{A.2.86}$$

The constant A' is found by comparing the second and third members of Eq. (A.2.86); It is in all cases $A' = 2B$. The various c_n's are similarly found, from the fact that Eq. (A.2.86) is an identity.

From Eqs. (A.2.81) to (A.2.83), straightforward integration gives

$$\int_{-\infty i}^{\infty i} e^{pt} M(p) L(p)^{-1}\,\frac{dp}{2\pi i} = \frac{2}{A'}\sum_{n=1}^{N} \frac{a_n b_n}{c_n} \prod_{j=1}^{N}{}' \frac{b_j^2 - c_n^2}{c_j^2 - c_n^2}\, e^{-c_n|t|} \tag{A.2.87}$$

from which it follows that

$$h_n = \frac{2}{A'}\frac{a_n b_n}{c_n} \prod_{j=1}^{N}{}' \frac{b_j^2 - c_n^2}{c_j^2 - c_n^2} = \frac{1}{B}\frac{a_n b_n}{c_n} \prod_{j=1}^{N}{}' \frac{b_j^2 - c_n^2}{c_j^2 - c_n^2} \tag{A.2.88}$$

At this point let us define still another function of p,

$$G(p) \equiv \mathcal{W}_S(p/2\pi i)H(p) = \int_{-\infty}^{\infty} h(|t|)e^{-pt}\,dt \qquad \mathrm{Re}\ p = 0 \tag{A.2.89}$$

[and continue $G(p)$ analytically, in the usual way]. The (inverse) Fourier transform of the last member of Eq. (A.2.79) is thus

$$\int_{-\infty i}^{\infty i} G(p)S_T(p)_g e^{pt}\,\frac{dp}{2\pi i} = \int_{-\infty}^{\infty} g_T(t-\tau)h(|\tau|)\,d\tau = \int_{0-}^{T+} h(|t-\tau|)g(\tau)\,d\tau \tag{A.2.90}$$

Since the zeros of $H(p)^{-1}$ are the same as the zeros of $L(p)$ [cf. Eqs. (A.2.83)], and since the poles of the expressions in the parentheses of Eq. (A.2.79) are the same as the corresponding zeros of $N(p)$, when we take the (inverse) Fourier transform of both members of Eq. (A.2.79) we note that these latter cancel and only the poles of the integrand occur at the zeros of $L(p)$. Using Eqs. (A.2.82) to (A.2.86), with the aid of Eq. (A.2.90), we get [remembering that $A' = 2B$; cf. Eq. (A.2.86)]

$$Z_T(t) = B^{-1}\int_{-\infty i}^{\infty i} \frac{dp}{2\pi i}\,e^{pt} \prod_{n=1}^{N} (b_n^2 - p^2)(c_n^2 - p^2)^{-1}\left[\sum_{n=1}^{N}\left(\frac{A_n^{(+)}e^{-pT}}{b_n + p} + \frac{A_n^{(-)}}{b_n - p}\right)\right]$$
$$+ D\sum_{n=1}^{N} h_n \int_{0-}^{T+} g(\tau)e^{-c_n|t-\tau|}\,d\tau \qquad 0- \le t \le T+ \tag{A.2.91}$$

with the requirement that $Z_T = 0$ outside $(0,T)$, determining the $2N$ constants $A_n^{(\pm)}$.

Let us now evaluate Eq. (A.2.91) and find the $A_n^{(\pm)}$. We observe that for $t > T$ the poles of the integrand are simple and occur at $p = -c_n$ and for $0 < t < T$ (all $t < T$) the poles are at $p = c_n$; for $t < 0$, they also occur at $p = c_n$. [Eqs. (A.2.92) to (A.2.107), (A.2.114) to (A.2.116) are incorrect. See Ref. 7 for the correct relations (D. M.).]

$t > T$:

$$0 = D\sum_{n=1}^{N}\left[h_n\int_{0-}^{T+} g(x)e^{c_n x}\,dx\right]e^{-c_n t}$$
$$+ B^{-1}\sum_{n=1}^{N}(A_n^{(+)}d_n^{(N)}e^{c_n T})e^{-c_n t} + B^{-1}\sum_{n=1}^{N}(A_n^{(-)}e_n^{(N)})e^{-c_n t} \tag{A.2.92a}$$

$0- < t < T+$:

$$Z_T(t) = D\sum_{n=1}^{N} h_n\int_{0-}^{T+} g(x)e^{-c_n|t-x|}\,dx + B^{-1}\sum_{n=1}^{N} A_n^{(+)}e_n^{(N)}e^{c_n(t-T)}$$
$$+ B^{-1}\sum_{n=1}^{N} A_n^{(-)}e_n^{(N)}e^{-c_n t} \tag{A.2.92b}$$

$t < 0$:

$$0 = D\sum_{n=1}^{N} e^{c_n t}\left[h_n\int_{0-}^{T+} g(x)e^{-c_n x}\,dx\right] + B^{-1}\sum_{n=1}^{N}(A_n^{(+)}e_n^{(N)}e^{-c_n T})e^{c_n t}$$
$$+ B^{-1}\sum_{n=1}^{N}(A_n^{(-)}d_n^{(N)})e^{c_n t} \tag{A.2.92c}$$

where

$$d_n^{(N)} = \prod_{j=1}^{N} (b_j + c_j) \prod_{\substack{j=1 \\ j \neq n}}^{N} (b_j - c_j) \Big/ \prod_{j=1}^{N} (c_j + c_n) \prod_{j=1}^{N}{}' (c_j - c_n) \quad \text{(A.2.93}a\text{)}$$

$$e_n^{(N)} = \prod_{\substack{j=1 \\ j \neq n}}^{N} (b_j + c_j) \prod_{j=1}^{N} (b_j - c_j) \Big/ \prod_{j=1}^{N} (c_j + c_n) \prod_{j=1}^{N}{}' (c_j - c_n) \quad \text{(A.2.93}b\text{)}$$

Equation (A.2.92b) is the desired solution, where the constants $A_n^{(\pm)}$ are found by the following argument:

Since Eqs. (A.2.92a), (A.2.92c) are each an identity, the coefficients of $e^{\mp c_n t}$ must vanish for each $n = 1, \ldots, N$, giving us $2N$ equations from which to determine the $2N$ unknown constants $A_n^{(\pm)}$. Letting

$$\begin{aligned} \alpha_n &\equiv Dh_n \int_{0-}^{T+} g(x)e^{c_n x}\,dx & \alpha_n' &\equiv Dh_n \int_{0-}^{T+} g(x)e^{-c_n x}\,dx \\ \beta_n &\equiv d_n^{(N)} e^{c_n T} B^{-1} & \beta_n' &\equiv e_n^{(N)} e^{-c_n T} B^{-1} \\ \gamma_n &\equiv e_n^{(N)} B^{-1} & \gamma_n' &\equiv d_n^{(N)} B^{-1} \end{aligned} \quad \text{(A.2.94)}$$

with $x_n \equiv A_n^{(+)}$, $y_n \equiv A_n^{(-)}$, we obtain for each n from Eqs. (A.2.92a), (A.2.92c) the N pairs of simultaneous linear equations

$$\alpha_n + \beta_n x_n + \gamma_n y_n = 0 \qquad \alpha_n' + \beta_n' x_n + \gamma_n' y_n = 0 \qquad n = 1, \ldots, N \quad \text{(A.2.95)}$$

for which the solutions are

$$x_n \equiv A_n^{(+)} = \frac{\alpha_n \gamma_n' - \alpha_n' \gamma_n}{\Delta_n} \qquad y_n = A_n^{(-)} = \frac{\alpha_n' \beta_n - \alpha_n \beta_n'}{\Delta_n} \quad \text{(A.2.96)}$$

with $\Delta_n = \beta_n' \gamma_n - \beta_n \gamma_n'$ ($\neq 0$). Specifically,

$$\Delta_n = B^{-2} \left[\prod_{j=1}^{N}{}' \frac{(b_j^2 - c_j^2)^2}{(c_j - c_n)^2} \prod_{j=1}^{N} (c_j + c_n)^{-2} \right] [(b_n - c_n)^2 e^{-c_n T} - (b_n + c_n)^2 e^{c_n T}] \quad \text{(A.2.97)}$$

with $b_j \neq c_j$, b_j, c_j distinct for all n. From Eqs. (A.2.93), (A.2.94), (A.2.96), (A.2.97), we obtain directly for the $2N$ constants $A_n^{(\pm)}$

$$A_n^{(+)} = F_n \left[\frac{(b_n + c_n) \int_{0-}^{T+} g(x)e^{c_n x}\,dx - (b_n - c_n) \int_{0-}^{T+} g(x)e^{-c_n x}\,dx}{(b_n - c_n)^2 e^{-c_n T} - (b_n + c_n)^2 e^{c_n T}} \right] \quad \text{(A.2.98)}$$

$$A_n^{(-)} = F_n \left[\frac{(b_n + c_n)e^{c_n T} \int_{0-}^{T+} g(x)e^{-c_n x}\,dx - (b_n - c_n)e^{-c_n T} \int_{0-}^{T+} g(x)e^{c_n x}\,dx}{(b_n - c_n)^2 e^{-c_n T} - (b_n + c_n)^2 e^{c_n T}} \right] \quad \text{(A.2.99)}$$

where

$$F_n = BDc_n h_n \prod_{j=1}^{N}{}' \frac{c_j^2 - c_n^2}{b_j^2 - c_j^2} = Da_n b_n \prod_{j=1}^{N}{}' \frac{b_j^2 - c_n^2}{b_j^2 - c_j^2} \quad \text{(A.2.100)}$$

These relations, in conjunction with Eq. (A.2.92b), complete our general solution of the integral equation (A.2.76), where K_S is a rational kernel that is the sum of simple exponentials. We remark once more that, when the corresponding spectrum $\mathcal{W}_S$ has multiple poles, an analogous procedure may be employed, or the desired results may be obtained from Eqs. (A.2.98) to (A.2.100) by a suitable passage to the limit [cf. Eq. (A.2.58)].

In the Wiener-Hopf case ($T \to \infty$), we find that

$$\lim_{T\to\infty} A_n^{(+)} = -F_n(b_n + c_n)^{-1} \lim_{T\to\infty} \left[e^{-c_nT} \int_{0-}^{T+} g(x)e^{c_nx}\, dx \right] < \infty \qquad \text{(A.2.101}a\text{)}$$

where $g(x)$ is assumed to be suitably bounded for all $t \geq 0$. We have also

$$\lim_{T\to\infty} A_n^{(-)} = -F_n(b_n + c_n)^{-1} \int_{0-}^{\infty} g(x)e^{-c_nx}\, dx < \infty \qquad \text{(A.2.101}b\text{)}$$

so that the solution (A.2.92b) in this instance becomes

$$Z_\infty(t) = \begin{cases} D \displaystyle\sum_{n=1}^{N} h_n \left[\int_{0-}^{\infty} g(x)e^{-c_n|t-x|}\, dx - \frac{b_n - c_n}{b_n + c_n} e^{-c_nt} \int_{0-}^{\infty} g(x)e^{-c_nx}\, dx \right] & 0- \leq t \\ 0 & t < 0- \end{cases} \qquad \text{(A.2.102)}$$

The integral equation

$$\int_0^T [K_S(t - u) + B\delta(t - u)]\rho_T(u,\tau)\, du = DK_S(t - \tau) \qquad 0 \leq t, \tau \leq T \qquad \text{(A.2.103)}$$

may be solved for $\rho_T(t,\tau)$ by these same methods, but as indicated earlier [cf. Eqs. (A.2.10) to (A.2.12)], since ρ_T and Z_T are linearly related by Eq. (A.2.12), we can obtain the desired results for ρ_T by a direct comparison of Eqs. (A.2.92b) and (A.2.12). The result is

$$\rho_T(t,\tau) = D \sum_{n=1}^{N} h_n e^{-c_n|t-\tau|} + B^{-1} \sum_{n=1}^{N} E_n^{(+)}(\tau) e_n^{(N)} e^{+c_n(t-T)} + B^{-1} \sum_{n=1}^{N} E_n^{(-)}(\tau) e_n^{(N)} e^{-c_nt}$$
$$0 \leq t, \tau \leq T \qquad \text{(A.2.104)}$$

where

$$E_n^{(+)}(\tau) = F_n \frac{(b_n + c_n)e^{c_n\tau} - (b_n - c_n)e^{-c_n\tau}}{(b_n - c_n)^2 e^{-c_nT} - (b_n + c_n)^2 e^{c_nT}} \qquad \text{(A.2.105)}$$

$$E_n^{(-)}(\tau) = F_n \frac{(b_n + c_n)e^{c_n(T-\tau)} - (b_n - c_n)e^{-c_n(T-\tau)}}{(b_n - c_n)^2 e^{-c_nT} - (b_n + c_n)^2 e^{c_nT}} \qquad \text{(A.2.106)}$$

and $\rho_T(t,\tau) = \rho_T(\tau,t)$, with $\rho_T = 0$ outside the square $(0,T)$. In the Wiener-Hopf case, Eq. (A.2.104) reduces to

$$\rho_\infty(t,\tau) = \begin{cases} D \displaystyle\sum_{n=1}^{N} h_n \left(e^{-c_n|t-\tau|} - \frac{b_n - c_n}{b_n + c_n} e^{-2c_nt} e^{c_n(t-\tau)} \right) \\ \hat{\rho}_\infty(t - \tau,\, t) \qquad t, \tau > 0 \end{cases} \qquad \text{(A.2.107)}$$

We illustrate these general results with some simple examples:

1. *RC noise signals.* Here, we let $B = D^{-1} = W_{0N}/2$ and find that

$$N = 1 \qquad a_1 = \psi_S \qquad b_1 = \alpha \qquad F_1 = c_1h_1 = 2\psi_S\alpha/W_{0N} \qquad h_1 = a_1b_1/c_1B \qquad \text{(A.2.108)}$$

The roots of Eq. (A.2.80) or Eq. (A.2.86) are obtained at once from

$$2B + 2\psi_S \left(\frac{1}{\alpha - p} + \frac{1}{\alpha + p} \right) = 0$$

Therefore,

$$c_1 = \alpha\sqrt{1+\gamma_0^2} \qquad \gamma_0^2 = \frac{W_{0S}}{W_{0N}} \qquad W_{0S} = \frac{4\psi_S}{\alpha}$$

$$h_1 = \frac{\alpha}{2}\gamma_0^2(1+\gamma_0^2)^{-1/2} \qquad F_1 = \frac{\alpha^2\gamma_0^2}{2} \tag{A.2.109}$$

$$d_1^{(1)} = \frac{b_1+c_1}{2c_1} = \frac{1+\sqrt{1+\gamma_0^2}}{2\sqrt{1+\gamma_0^2}} \qquad e_1^{(1)} = \frac{b_1-c_1}{2c_1} = \frac{1-\sqrt{1+\gamma_0^2}}{2\sqrt{1+\gamma_0^2}}$$

The desired solution is thus [from Eq. (A.2.92*b*)]

$$Z_T(t) = \frac{\gamma_0^2\alpha}{W_{0N}}(1+\gamma_0^2)^{-1/2}\int_{0-}^{T+} g(x)e^{-c_1|t-x|}\,dx + \frac{2}{W_{0N}}\,\frac{1-\sqrt{1+\gamma_0^2}}{\sqrt{1+\gamma_0^2}}\,(A_1^{(+)}e^{c_1(t-T)} + A_1^{(-)}e^{-c_1t}) \qquad 0- \le t \le T+ \tag{A.2.110}$$

with

$$A_1^{(+)} = \alpha\gamma_0^2\,\frac{(1+\sqrt{1+\gamma_0^2})\int_{0-}^{T+} g(x)e^{c_1x}\,dx - (1-\sqrt{1+\gamma_0^2})\int_{0-}^{T+} g(x)e^{-c_1x}\,dx}{(1-\sqrt{1+\gamma_0^2})^2e^{-c_1T} - (1+\sqrt{1+\gamma_0^2})^2e^{c_1T}} \tag{A.2.111a}$$

$$A_1^{(-)} = \alpha\gamma_0^2\,\frac{(1+\sqrt{1+\gamma_0^2})e^{c_1T}\int_{0-}^{T+} g(x)e^{-c_1x}\,dx - (1-\sqrt{1+\gamma_0^2})e^{-c_1T}\int_{0-}^{T+} g(x)e^{c_1x}\,dx}{(1-\sqrt{1+\gamma_0^2})^2e^{-c_1T} - (1+\sqrt{1+\gamma_0^2})^2e^{c_1T}} \tag{A.2.111b}$$

The Wiener-Hopf solutions (A.2.102) now become

$$Z_\infty(t) = \frac{\alpha\gamma_0^2}{W_{0N}}(1+\gamma_0^2)^{-1/2}\left[\int_{0-}^{\infty} g(x)e^{-c_1|t-x|}\,dx - 2\,\frac{1-\sqrt{1+\gamma_0^2}}{1+\sqrt{1+\gamma_0^2}}\,e^{-c_1t}\int_{0-}^{\infty} g(x)e^{-c_1x}\,dx\right] \qquad 0- < t \tag{A.2.112}$$

2. *LRC noise signals.* Here again we assume $B = D^{-1} = W_{0N}/2$. We also use the relations of Sec. A.2.3. The roots of $H(p)^{-1} = 0$ [cf. Eq. (A.2.81)] are found from

$$p^4 - 2(\omega_F^2 - \omega_1^2)p^2 + \omega_0^4(1+\gamma_0^2) = 0 \tag{A.2.113a}$$

to be

$$p_{1,2} = c_{1,2} = (a^2+b^2)^{1/4}e^{\pm(i/2)[\tan^{-1}(b/a)-\pi/2]} \qquad \text{Re } c_1, c_2 > 0 \tag{A.2.113b}$$

where $a \equiv \omega_1^2 - \omega_F^2$ (>0 here), $b \equiv \omega_0^2\sqrt{1+\gamma_0^2-(1-1/2Q^2)}$, $Q \equiv \omega_0/2\omega_F$, with $(a^2+b^2)^{1/4} = \omega_0(1+\gamma_0^2)^{1/4}$. Applying the above to Eqs. (A.2.88), (A.2.100), we get

$$F_2 = F_1^* \qquad F_1 = \frac{\omega_0^2\psi_S}{iW_{0N}\omega_1}\,\frac{b_2^2-c_1^2}{b_2^2-c_2^2} \qquad h_1 = h_2^* = \frac{\omega_0^2\psi_S}{iW_{0N}\omega_1c_1}\,\frac{b_2^2-c_1^2}{c_2^2-c_1^2} \tag{A.2.114}$$

and since $c_2^* = c_1$, we get finally [for $(0- < t < T+)$]

$$Z_T(t) = \text{Re}\left(\frac{b_2^2-c_1^2}{c_2^2-c_1^2}\,\frac{4\omega_0^2\psi_S}{iW_{0N}^2\omega_1c_1}\left\{\int_{0-}^{T+} g(x)e^{-c_1|t-x|}\,dx + (b_1-c_1)[(\quad)_{n=1}^{(+)}e^{-c_1(T-t)} + (\quad)_{n=1}^{(-)}e^{-c_1t}]\right\}\right) \tag{A.2.115}$$

where $(\quad)_{n=1}^{(\pm)}$ are, respectively, the quantities that are coefficients of F_1 in Eqs. (A.2.98),

(A.2.99) when $n = 1$. The Wiener-Hopf solution is

$$Z_\infty(t) = \mathrm{Re}\left\{\frac{b_2^2 - c_1^2}{c_2^2 - c_1^2}\frac{4\omega_0^2\psi_S}{W_{0N}^2\omega_1 c_1 i}\left[\int_{0-}^{\infty} g(x)e^{-c_1|t-x|}\,dx - \frac{b_1 - c_1}{b_1 + c_1}e^{-c_1 t}\int_{0-}^{\infty} g(x)e^{-c_1 x}\,dx\right]\right\} \tag{A.2.116}$$

Corresponding solutions of Eq. (A.2.103) for these *RC* and *LRC* cases follow by inspection from Eqs. (A.2.104) et seq.

REFERENCES

1. Lovitt, W. V.: "Linear Integral Equations," Dover, New York, 1950.
2. Morse, P. M., and H. Feshbach: "Methods of Theoretical Physics," vol. 1, McGraw-Hill, New York, 1953.
3. Davenport, W. B., Jr., and W. L. Root: "An Introduction to the Theory of Random Signals and Noise," McGraw-Hill, New York, 1958.
4. Laning, J. H., Jr., and R. H. Battin: "Random Processes in Automatic Control," McGraw-Hill, New York, 1956.
5. Reich, E., and P. Swerling, Jr.: Detection of a Sinewave in Gaussian Noise, *J. Appl. Phys.*, **24:** 289 (1953).
6. Middleton, D.: On the Detection of Stochastic Signals in Additive Normal Noise. Part I, *IRE Trans. on Inform. Theory*, **PGIT-3:** 86 (June, 1957).
7. ———, Remarks on and Revisions of Some Earlier Results in Two Recent Papers, *IRE Trans. on Info. Theory*, **PGIT-8** (October, 1962).

SUPPLEMENTARY REFERENCES AND BIBLIOGRAPHY

The references below supplement those appearing in the text. This list is by no means complete: it is limited to those books and papers which appear pertinent to the topics specifically treated in the preceding chapters. It is also unavoidably limited by considerations of space, accessibility to the reader, the actual availability and appositeness of the material, and the author's own judgment in these matters. For comprehensive lists, which in addition include applications to speech, linguistics, automata, biological mechanisms, group communications, and the like, the reader should consult the bibliographies of Stumpers,† Chessin,‡ and Green§ (the two Green bibliographies for what appears available¶ concerning Soviet efforts in the field). Other references may be found in the papers cited there and in the present study. Here, books are listed separately, followed by papers under eight convenient categories. The numbers after each category refer to the chapters in the text above, to which this reference material is mainly appropriate. The references themselves are ordered alphabetically by author; a complete list of authors cited here and in the text is available in the Name Index, starting on page 1121.

BOOKS

Bendat, J. S.: "Principles and Applications of Random Noise Theory," Wiley, New York, 1958.

Bunimovitch, V. I.: "Fluctuation Processes in Radio Receivers," Sovietskoe Radio, Moscow, 1951.

Cherry, E. C.: "Introduction to Communication Theory," Wiley, New York, 1954.

——— (ed.): "Information Theory," 3d London Symposium, September, 1955, Academic Press, New York, 1956.

Freeman, J. J.: "Principles of Noise," Wiley, New York, 1958.

Goldman, S.: "Frequency Analysis, Modulation, and Noise," McGraw-Hill, New York, 1948.

Jackson, W. (ed.): "Symposium on Information Theory—Report of Proceedings," 1st London Symposium, September, 1950, Ministry of Supply, London.

† F. L. H. M. Stumpers, A Bibliography of Information Theory—Communication Theory—Cybernetics, *IRE Trans. on Inform. Theory*, vol. PGIT-2, November, 1953, 1st suppl., **IT-1**(2): 31 (September, 1955), 2d suppl., **IT-3**(2): 150 (June, 1957).

‡ P. L. Chessin, A Bibliography on Noise, *IRE Trans. on Inform. Theory*, **IT-1**(2): 15 (September, 1955).

§ P. E. Green, Jr., A Bibliography of Soviet Literature on Noise, Correlation, and Information Theory, *IRE Trans. on Inform. Theory*, **IT-2**(2): 91 (June, 1956); Information Theory in the U.S.S.R., *IRE Wescon Conv. Record*, August, 1957, pt. 2, p. 67.

¶ Unfortunately, most of the Russian work listed in this supplementary bibliography was not available to the author until after the completion of the book.

Jackson, W.: "Communication Theory," 2d London Symposium, September, 1952, Academic Press, New York, 1953.

Khinchin, A. I.: "Mathematical Foundations of Information Theory," Dover, New York, 1957.

Kotelnikov, V. A.: "Theory of Optimum Noise Immunity," Government Power Engineering Press, Moscow-Leningrad, 1956; translated by R. A. Silverman, *MIT Lincoln Lab. Group Rept.* 34-67, March, 1958; McGraw-Hill, New York, 1960.

Levin, B. R.: "Theory of Random Processes and Its Application in Radio Engineering," Sovietskoe Radio, Moscow, 1957.

Proc. Symposium on Inform. Networks, April, 1954 (vol. 3, Polytechnic Institute of Brooklyn), Edwards, Ann Arbor, Mich., 1955.

Proc. Symposium on Modern Network Synthesis, April, 1955 (vol. 5, Polytechnic Institute of Brooklyn), Edwards, Ann Arbor, Mich., 1956.

Pugachev, V. S.: "Random Functions and Their Applications in Automatic Control Theory," Goztekhizdat, Moscow, 1957.

Rytov, S. M.: "Theory of Electric Fluctuations and Thermal Radiations," Akademiya Nauk Press, Moscow, 1953.

Solodovnikov, V. V.: "Introduction to Statistical Dynamics of Automatic Control Systems," Goztekhiteorizdat, Moscow-Leningrad, 1952.

PAPERS

I *Noise Statistics and Representations* (7, 8, 9, 17)

Arens, R.: Complex Processes for Envelopes of Normal Noise, *IRE Trans. on Inform. Theory*, **IT-3**: 204 (1957).

Bennett, W. R.: Distribution of the Sum of Randomly Phased Components, *Quart. Appl. Math.*, **5**: 385 (1948).

Bunimovitch, V. I., and M. A. Leontovich: On the Distribution of the Number of Large Deviations in Electrical Fluctuations, *Comptes rend. acad. sci. U.R.S.S.*, **53**(1): (1946).

Domb, C.: The Resultant of a Large Number of Events of Random Phase, *Proc. Cambridge Phil. Soc.*, **42**: 245 (1946).

Gold, B., and G. O. Young: Response of Linear Systems to Non Gaussian Noise, *IRE Trans. on Inform. Theory*, **PGIT-3**: 63 (March, 1954).

Kuznetsov, P. I.: Propagation of Random Functions through Nonlinear Systems, *Automatics and Telemechanics*, **15**: 200 (1954).

———, R. L. Stratonovich, and V. L. Tikhonov: Passage of Certain Random Functions through Linear Systems, *Automatics and Telemechanics*, **14**: 144 (1953).

———, ———, and ———: On the Duration of Excursions of a Random Function, *J. Tech. Phys.*, **24**: 103 (1954).

Marshall, J. S., and W. Hitschfeld: Interpretation of the Fluctuating Echo from Randomly Distributed Scatterers, Part I, *Can. J. Phys.*, **31**: 962 (1953).

Siegert, A. J. F.: On the First-passage-time Problem for a One-dimensional Markoffian Gaussian Random Function, *Phys. Rev.*, **70**: 449 (1946).

———: A Systematic Approach to a Class of Problems in the Theory of Noise and Other Random Phenomena. Part III, Examples, *IRE Trans. on Inform. Theory*, **IT-4**: 4 (1958).

Sponsler, G. C.: A Note on the Impulse-function Determination of Functional Probability-density Functions, *Inform. and Control*, **1**: 177 (May, 1958).

Stein, S., and J. E. Storer: Generating a Gaussian Sample, *IRE Trans. on Inform. Theory*, **IT-2**: 87 (1956).

Stumpers, F. L. H. M.: On a First-passage-time Problem, *Philips Research Repts.*, **5**: 270 (1950).

Tikhonov, V. I.: Distribution of Excursions of Normal Fluctuations According to Duration, *Radiotekh. i Elektron.*, **1**: 23 (1956).

Villars, F., and V. F. Weisskopf: The Scattering of Electromagnetic Waves by Turbulent Atmospheric Fluctuations, *Phys. Rev.*, **94**: 232 (1954).

Wallace, P. R.: Interpretations of the Fluctuating Echo from Randomly Distributed Scatterers, Part II, *Can. J. Phys.*, **31**: 995 (1953).

II *Spectra, Covariance Functions of Random Waves* (3, 4)

Cherry, E. C.: Quelques remarques sur le temps considéré comme variable complexe, *Onde élec.*, **34**: 7 (1954).

Fortet, R.: Average Spectrum of a Periodic Series of Identical Pulses, Randomly Displaced and Distorted, *Elec. Communs.*, **31**: 283 (1954); *Onde élec.*, **34**: 683 (1954).

Lampard, D. G.: A New Method of Determining Correlation Functions of Stationary Time Series, *J. IEE (London)*, vol. 104, August, 1954.

Macfarlane, G. G.: On the Energy-spectrum of an Almost Periodic Succession of Pulses, *Proc. IRE*, **37**: 139 (1949).

Meyer-Eppler, W.: Correlation and Autocorrelation in Communication Engineering, *Arch. elek. Übertragung*, **7**: 501, 531 (1953).

Siegert, A. J. F.: On the Evaluation of Noise Samples, *J. Appl. Phys.*, **23**: 737 (1952).

Turner, C. H. M.: On the Concept of an Instantaneous Power Spectrum and Its Relationship to the Autocorrelation Function, *J. Appl. Phys.*, **25**: 1347 (1954).

III *Noise Models and Mechanisms* (10, 11)

Burgess, R. E.: Fluctuation Noise in a Receiving Aerial, *Proc. Phys. Soc. (London)*, **58**: 313 (1946).

Callen, H. C., and T. A. Welton: On the Theory of Thermal Fluctuations or Generalized Noise, *Phys. Rev.*, **82**: 296 (1951).

Fürth, R.: On the Theory of Electrical Fluctuations, *Proc. Roy. Soc. (London)*, **192**: 593 (1948).

George, T. S., and H. Urkowitz: Fluctuation Noise in Microwave Superregenerative Amplifiers, *Proc. IRE*, **41**: 516 (1953).

Haeff, A. V.: On the Origin of Solar Radio Noise, *Phys. Rev.*, **75**: 1546 (1949).

Landon, V. D.: A Study of the Characteristics of Noise, *Proc. IRE*, **24**: 1514 (1936), **29**: 50 (1941), **30**: 425, 526 (1942).

MacDonald, D. K. C.: Spontaneous Fluctuations, *Progr. in Phys.*, **12**: 56 (1948, 1949).

Middleton, D., W. M. Gottschalk, and J. B. Wiesner: Noise in CW Magnetrons, *J. Appl. Phys.*, **24**: 1065 (1953).

Richardson, J. M.: The Linear Theory of Fluctuations Arising from Diffusional Mechanisms—An Attempt at a Theory of Contact Noise, *Bell System Tech. J.*, **29**: 117 (1950).

Rytov, S. M.: Electric Fluctuations and Thermal Radiations, *Uspekhi Fiz. Nauk*, **55**: 299 (1955).

———: Theory of Thermal Noise, Parts I, II, *Radiotekhnika*, **10**(2): 3, **10**(3): 3 (1955).

Schremp, E. J.: A Generalization of Nyquist's Thermal Noise Theorem, *Phys. Rev.*, **69**: 255 (1946).

Van der Ziel, A.: Thermal Noise at High Frequencies, *J. Appl. Phys.*, **21**: 399 (1950).

Wiener, N.: Harmonic Analysis of Irregular Motion, *J. Math. and Phys.*, **5**: 99 (1926).

IV *Noise and Signals in Nonlinear Systems; AM and FM Systems* (5, 13, 14)

Armstrong, E. H.: A Method of Reducing Disturbances in Radio—Signalling by a Method of Frequency Modulation, *Proc. IRE*, **24**: 689 (1936).

Baum, R. F.: The Correlation Function of Smoothly Limited Gaussian Noise, *IRE Trans. on Inform. Theory*, **IT-3**: 193 (1957).

Blachman, N. M.: The Output Signal-to-Noise Ratio of a Power-law Device, *J. Appl. Phys.*, **24**: 783 (1953).

Bunimovitch, V. I.: Conversion of Fluctuations with Non Linear Systems, *Tekh. Fiz.*, vol. 16, no. 6, 1946.

Campbell, L. L.: Rectification of Two Signals in Random Noise, *IRE Trans. on Inform. Theory*, **IT-2**: 119 (1956).

Deutsch, R.: Detection of Modulated Noise-like Signals, *IRE Trans. on Inform. Theory*, **PGIT-3**: 106 (March, 1954).

Francis, V. J., and E. G. James: Rectification of Signal and Noise, *Wireless Eng.*, **25**: 16 (1946).

Fränz, K., and T. Vellat: The Influence of Carrier Frequency on the Noise Following Amplitude Limiters and Linear Rectifiers, *Elek. Nachr.-Tech.*, **20**: 183 (1943).

Harrington, J. V., and T. F. Rogers: Signal-to-Noise Improvement through Integration in a Storage Tube, *Proc. IRE*, **38**: 1197 (1950).

MacDonald, D. K. C.: A Note on Two Definitions of Noise Figure in Radio Receivers, *Phil. Mag.*, **35**: 386 (1954).

MacFadden, J. A.: The Correlation Function of a Sine Wave plus Noise after Extreme Clipping, *IRE Trans. on Inform. Theory*, **IT-2**: 82 (1956).

Magness, T. A.: Spectral Response of a Quadratic Device to Non Gaussian Noise, *J. Appl. Phys.*, **25**: 1357 (1954).

Manasse, R., R. Price, and R. M. Lerner: Loss of Signal Detectability in Band-pass Limiters, *IRE Trans. on Inform. Theory*, **IT-4**: 34 (1958).

Miller, K. S., and R. I. Bernstein: An Analysis of Coherent Integration and Its Application to Signal Detection, *IRE Trans. on Inform. Theory*, **IT-3**: 237 (1957).

Montgomery, G. F.: A Comparison of Amplitude and Angle Modulation for Narrow-band Communication of Binary-coded Messages in Fluctuation Noise, *Proc. IRE*, **42**: 447, 1560 (1954).

Rudnick, P.: The Detection of Weak Signals by Correlation Methods, *J. Appl. Phys.*, **24**: 128 (1953).

Smith, R. A.: The Relative Advantages of Coherent and Incoherent Detectors, A Study of Their Output Noise Spectra under Various Conditions, *Proc. IEE (London)*, **98**(pt. IV): 43 (1951).

Stumpers, F. L. H. M.: On the Calculation of Impulse-noise Transients in Frequency-modulation Receivers, *Philips Research Repts.*, **2**: 468 (1947).

———: On a Nonlinear Noise Problem, *Philips Research Repts.*, **2**: 241 (1947).

Thomson, W. E.: The Response of a Non Linear System to Random Noise, *Proc. IEE (London)*, **102**(pt. C): 46 (1955).

Tucker, D. G., and J. W. R. Griffiths: Detection of Pulse Signals in Noise, *Wireless Eng.*, **30**: 264 (1953).

V *Coding and Information Measures* (6)

Bedrosian, E.: Weighted PCM, *IRE Trans. on Inform. Theory*, **IT-4**: 45 (1958).

Blachman, N. M.: A Comparison of the Informational Capacities of Amplitude and Phase-modulation Systems *Proc. IRE*, **41**: 748 (1953).

Davis, H.: Radar Problems and Information Theory, *IRE Conv. Record*, March, 1953, pt. 8, p. 39.

Dubarle, D.: Mécanique quantique et information, *Rev. questions sci.* (*Belg.-France*), **5:** 347 (1952).

Elias, P.: List Coding for Noisy Channels, *IRE Wescon Conv. Record*, 1957, pt. 2, pp. 19–94.

Feldman, C. B., and W. R. Bennett: Bandwidth and Transmission Performance, *Bell System Tech. J.*, **28:** 490 (1949).

Gelfand, I. M., and A. M. Yaglom: On Calculation of the Amount of Information of a Random Function Contained in Another Such Function, *Uspekhi Mat. Nauk*, vol. 12, no. 1, 1957.

Good, I. J., and K. Caj Doog: A Paradox Concerning Rate of Information, *Inform. and Control*, **1:** 113 (1958).

Hartley, R. V. L.: The Transmission of Information, *Bell System Tech. J.*, **7:** 535 (1928).

Jaynes, E. T.: Information Theory and Statistical Mechanics, *Phys. Rev.*, **106:** 620 (1957), **108:** 171 (1957).

Khinchine, A. I.: The Concept of Entropy in the Theory of Probability, *Uspekhi Mat. Nauk*, **8**(3): 3 (1953).

———: On the Principal Theorems of Information Theory, *Uspekhi Mat. Nauk*, **11**(1): 17 (1956).

Kullback, S.: An Application of Information Theory to Multivariate Analysis, *Ann. Math. Stat.*, **27:** 112 (1956).

McGill, W. J.: Multivariate Information Transmission, *IRE Trans. on Inform. Theory*, **PGIT-4:** 93 (September, 1954).

Muroga, S.: On the Capacity of a Noisy Continuous Channel, *IRE Trans. on Inform. Theory*, **IT-3:** 44 (1957). Corrections, *ibid.*, Vol. **IT-4:** 52 (1958).

Nyquist, H.: Certain Factors Affecting Telegraph Speed, *Bell System Tech. J.*, **3:** 324 (1924).

Rosenblatt-Roth, M.: Concept of Entropy in Probability Theory and Its Application in the Theory of Transmission over Communication Channels, 3d All-Union Mathematics Congress, Moscow, June, July, 1956.

Schmidt, K. O.: Frequenzbandbreite, Uebermittlungszeit und Amplituders Stufenzahl (Gerauschabstand) bei den verschiedenen Nachrichtenarten im Rechnen der Shannon Theorie, *Fern. Tech. Z.*, **6:** 555 (1953), **7:** 33 (1954).

Selfridge, O.: "Pattern Recognition and Learning," 3d London Symposium on Information Theory, 1955, Academic Press, New York, 1956.

Siforov, V. I.: On Noise Stability of a System with Error-correcting Codes, *IRE Trans. on Inform. Theory*, **IT-2:** 109 (1956).

Tuller, W. G.: Theoretical Limitations on the Rate of Transmission of Information, *Proc. IRE*, **37:** 468 (1949).

Watanabe, S.: A Study of Ergodicity and Redundance Based on Intersymbol Correlation of Finite Range, *IRE Trans. on Inform. Theory*, **PGIT-4:** 85 (1954).

———: A Theory of Multilevel Information Channel with Gaussian Noise, *IRE Trans. on Inform. Theory*, **IT-3:** 214 (1958).

White, W. D.: Information Losses in Regenerative Pulse-code Systems, *IRE Conv. Record*, 1954, pt. 4, p. 18.

Woodward, P. M.: Information Theory and the Design of Radar Receivers, *Proc. IRE*, **39:** 1521 (1951).

——— and I. L. Davies: A Theory of Radar Information, *Phil. Mag.*, **41:** 1001 (1950).

Young, G. O., and B. Gold: Effect of Limiting on the Information Content of Noisy Signals, *IRE Conv. Record*, 1954, pt. 4, p. 76.

Zadeh, L. A.: Some Basic Problems in Communication of Information, *Trans. N.Y. Acad. Sci.*, **14**: 2 (1952).

VI *Filtering, Prediction, and Signal-to-Noise Optimization* (16)

Bendat, J. S.: Exact Integral Equation Solutions and Synthesis for a Large Class of Optimum Time Variable Linear Filters, *IRE Trans. on Inform. Theory*, **IT-3**: 71 (1957).

Berkowitz, R. S.: Optimum Linear Shaping and Filtering Networks, *Proc. IRE*, **41**: 532 (1953).

Blum, M.: An Extension of the Minimum Mean Square Error Theory for Sampled Data, *IRE Trans. on Inform. Theory*, **IT-2**: 176 (1956).

———: On the Mean Square Noise Power of an Optimum Linear Discrete Filter Operating on Polynomial plus White Noise Input, *IRE Trans. on Inform. Theory*, **IT-3**: 225 (1957).

Bunimovitch, V. I.: Sensitivity of a Radiometer, *Zhur. Tekh. Fiz.*, **20**: 944 (1950).

Drenick, R. F., and P. Nesbeda: On a Class of Optimum Linear Predictors, *Ann. Math. Stat.*, **24**: 497 (1953).

Gerardin, L.: Le Signal minimum utilisable en réception radar et son améliorisation par certains procédés de corrélation, *Onde élec.*, **34**: 67 (1954).

Karlov, N. Y., and A. E. Salomonovich: The Automatic Null Radiometer of Centimeter Range for Studying Weak Noise Signals, *Radiotekh. i Elektron.*, **1**: 121 (1956).

Koschmann, A. H., and J. G. Truxal: Optimum Linear Filtering of Nonstationary Time Series, *Proc. Natl. Electronics Conf.*, **10**: 119 (1954).

Maximon, L. C., and J. P. Ruina: Some Statistical Properties of a Signal plus Narrowband Noise Integrated over a Finite Time Interval, *J. Appl. Phys.*, **27**: 1442 (1956).

Pike, E. W.: A New Approach to Optimum Filtering, *Proc. Natl. Electronics Conf.*, **8**: 407 (1952).

Pugachev, V. S.: General Theory of the Correlation of Stochastic Functions, *Izvest. Akad. Nauk S.S.S.R., Ser. Mat.*, **17**: 401 (1953).

Rochefort, J. S.: Matched Filters for Detecting Pulsed Signals in Noise, *IRE Conv. Record*, 1954, pt. 4, p. 30.

Troitskii, V. S.: The Sensitivity of Radiometers, *Zhur. Tekh. Fiz.*, **22**: 455 (1952).

Urkowitz, H.: Filters for Detection of Small Radar Signals in Clutter, *J. Appl. Phys.*, **24**: 1024 (1953).

Wax, N.: Signal-to-Noise Improvement and the Statistics of Track Populations, *J. Appl. Phys.*, **26**: 586 (1955).

Wold, H. A.: On Predictions in Stationary Time Series, *Ann. Math. Stat.*, **19**: 558 (1948).

Yaglom, A. M.: Introduction to Theory of Stationary Random Processes, *Uspekhi Mat. Nauk*, **7**: 3 (1952).

———: Correlation Theory of Processes with Random Stationary *n*th Increments, *Mat. Sbornik*, **37**(1): 141 (1955).

———: Extrapolation, Interpolation, and Filtration of Stationary Random Processes with Rational Spectral Density, *Trudy, Moscow Mat. Soc.*, **4**: 333 (1955).

Youla, D. C.: The Solution of a Homogeneous Wiener-Hopf Integral Equation Occurring in the Expansion of Second-order Stationary Random Functions, *IRE Trans. on Inform. Theory*, **IT-3**: 187 (1957).

Zadeh, L. A., and K. S. Miller: Generalized Ideal Filters, *J. Appl. Phys.*, **23**: 223 (1952).

VII *Detection Theory* (*Radar and Communications*) (19, 20, 23)

Drenick, R. F., S. Gartenhouse, and P. Nesbeda: Detection of Coherent and Non Coherent Signals, *IRE Conv. Record*, 1955, pt. 4, p. 114.

Dugundji, J., and E. Ackerlind: Automatic Bias Control for a Threshold Detector, *IRE Trans. on Inform. Theory*, **IT-3**: 65 (1957).

Green, B. A., Jr.: Radar Detection Probability with Logarithmic Detectors, *IRE Trans. on Inform. Theory*, **IT-4**: 50 (1958).

Heaps, H. S.: The Effect of a Random Noise Background upon the Detection of a Random Signal, *Can. J. Phys.*, **33**: 1 (1955).

Kaplan, E. L.: Signal Detection Studies with Applications, *Bell System Tech. J.*, **34**: 403 (1955).

Pachares, J.: A Table of Bias Levels Useful in Radar Detection Problems, *IRE Trans. on Inform. Theory*, **IT-4**: 38 (1958).

Sponsler, G. C., and F. L. Shader: PPI Light Spot Brightness and Probability Distributions, *J. Appl. Phys.*, **25**: 1271 (1954).

Stone, W. M.: On the Effect of Integration in a Pulsed Radar, Randomly Modulated Carrier, *J. Appl. Phys.*, **25**: 1543 (1954).

Swerling, P.: Detection of Fluctuating Pulsed Signals in the Presence of Noise, *IRE Trans. on Inform. Theory*, **IT-3**: 175 (1957).

VIII *Extraction Theory* (21)

Balakrishnan, A. V., and R. F. Drenick: On Optimum Nonlinear Extraction and Coding Filters, *IRE Trans. on Inform. Theory*, **IT-2**: 1661 (September, 1956).

Booton, R. C., Jr.: An Optimization Theory for Time-varying Linear System with Nonstationary Statistical Inputs, *Proc. IRE*, **40**: 977 (1952).

———: Nonlinear Control Systems with Random Inputs, *IRE Trans. on Circuit Theory*, **CT-1**: 9 (March, 1954).

Bose, A. G.: A Theory for the Experimental Determination of Optimum Nonlinear Systems, *IRE Conv. Record*, 1956, pt. 4, p. 21.

Duncan, D. B.: Response of Linear Time-dependent Systems to Random Noise, *J. Appl. Phys.*, **24**: 609, 1252 (1953).

Kay, I., and R. A. Silverman: On the Uncertainty Relation for Real Signals, *Inform. and Control*, **1**: 64 (1957).

SELECTED SUPPLEMENTARY REFERENCES (1996)

PART I AN INTRODUCTION TO STATISTICAL COMMUNICATION THEORY

Chapter 1 Statistical Preliminaries

[1]. A. T. Bharucha-Reid, *Elements of the Theory of Markoff Process and Their Applications*, McGraw-Hill, New York, 1960.
[2]. L. Brillouin, *Science and Information Theory*, 2nd ed., Academic Press, New York, 1962.
[3]. E. Parzen, *Stochastic Processes*, Holden Day, San Francisco, 1962.
[4]. W. Feller, *An Introduction to Probability Theory and Its Applications*, John Wiley, New York, 1962.
[5]. H. Cramér and M. R. Leadbetter, *Stationary and Related Stochastic Processes*, John Wiley, New York, 1967.
[6]. E. Wong, *Stochastic Processes in Information and Dynamical Systems*, McGraw-Hill, New York, 1971.
[7]. D. L. Snyder, *Random Point Processes*, John Wiley, New York, 1975.
[8]. J. W. Goodman, *Statistical Optics*, John Wiley, New York, 1985.
[9]. S. M. Rytov, Yu. A. Kravtsov, V. I. Tatarskii, *Principles of Statistical Radio Physics, Elements of Random Process Theory*, Springer-Verlag, New York, 1987.
[10]. C. W. Gardiner, *Handbook of Stochastic Methods*, 2nd ed.; Springer-Verlag, New York, 1990.
[11]. D. L. Snyder and M. I. Miller, *Random Point Processes in Time and Space*, Springer-Verlag, New York, 1991.
[12]. N. G. Van Kempen, *Stochastic Processes in Physics and Chemistry*, North Holland, New York, 1992.

Chapter 2 Operations in Ensembles

[1]. R. Deutsch, *Nonlinear Transformations of Random Processes*, Prentice Hall, Englewood Cliffs, NJ, 1962.

Chapter 3 Spectra, Covariances, and Correlation Functions

[1]. M. Rosenblatt, *Time Series Analysis*, John Wiley, New York, 1963.
[2]. A. M. Yaglom, *Correlation Theory of Stationary and Related Random Functions:* I. *Basic Results;* II. *Supplementary Notes and References*, Springer-Verlag, New York, 1987.
[3]. S. M. Rytov, Yu A. Kravtsov, V. I. Tatarskii, *Principles of Statistical Radio Physics: 2. Correlation Theory of Random Processes*, Springer-Verlag, New York, 1988.

Chapter 4 Sampling, Interpolation, and Random Pulse Trains

[1]. D. P. Petersen and D. Middleton, ''Sampling and Reconstruction of Wave-Number-Limited Functions in N-Dimensional Euclidean Spaces,'' *Information and Control,* Vol. 5, no. 4, pp. 279–320, December 1962.
[2]. ———, ''On Representative Observations,'' *Tellus*, Vol. 15, no. 4, pp. 387–405, December 1963.
[3]. ———, ''Reconstruction of Multidimensional Stochastic Fields from Discrete Measurements of Amplitude and Gradient,'' *Information and Control*, Vol. 7, no. 4, pp. 445–476, December 1964.

[4]. ________, "Linear Interpolation, Extrapolation, and Prediction of Random Space-Time Fields with a Limited Domain of Measurement," *IEEE Trans. Information Theory*, Vol. IT-11, no. 1, pp. 18-31, January 1965.

[5]. V. I. Belyayev, *Processing and Theoretical Analysis of Oceanographic Observations*, Nauka Dumka Pub. House, Kiev, 1973. See "Chapter" 19, "Chapter" 20, trans. Joint Publication Research Service (JPRS), U.S. Department of Commerce, JPRS #60803, December 17, 1973; JPRS #60808, December 19, 1973.

[6]. A. V. Oppenheim and R. W. Shafer, *Digital Signal Processing*, Prentice Hall, Englewood Cliffs, NJ, 1975.

[7]. R. J. Marks II, *Introduction to Shannon Sampling and Interpolation Theory*, Springer-Verlag, New York, 1991.

Chapter 5 Signals and Noise in Nonlinear Systems

[1]. R. B. Blackman and J. W. Tukey, *The Measurement of Power Spectra, Bell Systems Technical Journal*, Vol. 37, January and March 1958; also Dover, New York, 1959.

[2]. E. N. Skomal, *Man-Made Radio Noise*, D. Van Nostrand, New York, 1978.

Chapter 6 An Introduction to Information Theory

[1]. F. M. Reza, *An Introduction to Information Theory*, D. Van Nostrand, New York, 1961.

[2]. R. B. Ash, *Information Theory*, John Wiley, New York, 1965.

[3]. G. David Forney, Jr., *Concatenated Codes*, MIT Press, Cambridge, MA, 1966.

[4]. S. Kullback, *Information Theory and Statistics*, John Wiley, New York, 1959; Dover, New York, 1968.

[5]. R. G. Gallagher, *Information Theory and Reliable Communications*, John Wiley, New York, 1968.

[6]. T. Berger, *Rate Distortion Theory*, Prentice Hall, Englewood Cliffs, NJ, 1971.

[7]. I. F. Blake and C. Mullin, *An Introduction to Algebraic and Combinatorial Coding Theory*, Academic Press, New York, 1976.

[8]. R. J. McEliece, *The Theory of Information and Coding*, Addison-Wesley, Reading, MA, 1977.

[9]. A. J. Viterbi and J. K. Omura, *Principles of Digital Communications and Coding*, McGraw-Hill, New York, 1979.

[10]. B. Sklar, *Digital Communications*, Prentice Hall, Englewood Cliffs, NJ, 1988.

[11]. T. Cover and J. Thomas, *Elements of Information Theory*, John Wiley, New York, 1991.

PART II RANDOM NOISE PROCESSES

Chapter 7 The Normal Random Process: Gaussian Variates

Chapter 8 The Normal Random Process: Gaussian Functionals

Chapter 9 Processes Derived From the Normal

[1]. R. Deutsch, *Nonlinear Transformations of Random Processes*, Prentice Hall, Englewood Cliffs, NJ, 1962.

[2]. E. J. Wegman, S. C. Schwartz, J. B. Thomas (eds.), *Topics in Non-Gaussian Signal Processing*, Springer-Verlag, New York, 1989.

[3]. C. W. Gardiner, *Handbook of Stochastic Methods*, 2nd ed., Springer-Verlag, New York, 1990.

[4]. N. G. Van Kampen, *Stochastic Processes in Physics and Chemistry*, North Holland, New York, 1992.

[5]. M. Grigoriu, *Applied Non-Gaussian Processes*, Prentice Hall, Englewood Cliffs, NJ, 1995.

Chapter 10 The Equations of Langevin, Fokker-Planck, and Boltzmann

[1]. W. Bernard and Herbert B. Callen, "Irreversible Thermodynamics of Nonlinear Processes and Noise in Diverse Systems," *Reviews of Modern Physics* Vol. 31, no. 4, pp. 1017–1044, October 1959.

[2]. Melvin Lax, (I) "Fluctuations from the Nonequlibrum Steady State," *Review of Modern Physics,* Vol. 32, no. 1, pp. 25–64, January 1960; (II) "Influence of Trapping, Diffusion and Recombination on Carrier Concentration Fluctuations," *J. Phys. Chem. Solids* (Pergamon Press, Great Britain), 1960; (III) "Classical Noise III: Nonlinear Markoff Processes," *Reviews of Modern Physics*, Vol. 38, no. 2, pp. 358–379, April 1966; (IV) "Classical Noise IV: Langevin Methods," *Reviews of Modern Physics*, Vol. 38, no. 3, pp. 541–566, July 1966.

[3]. W. C. Lindsey, *Synchronization Systems*, Prentice Hall, Englewood Cliffs, NJ, 1972.

[4]. W. C. Lindsey and M. K. Simon, eds., *Phase-Locked Loops and Their Applications* (selected papers), IEEE Press, New York, 1978.

[5]. E. Jakeman and R. J. A. Tough, "Non-Gaussian Models for the Statistics of Scattered Waves," *Advances in Physics* (Great Britain), Vol. 37, no. 5, pp. 471–529, 1988.

[6]. S. M. Rytov, Yu. A. Kravtsov, and V. I. Tatarskii, *Principles of Statistical Radio Physics 3. Elements of Random Fields*, Springer-Verlag, New York, 1989.

[7]. H. Risken, *The Fokker-Planck Equation* 2nd ed., Springer-Verlag, New York, 1989.

[8]. C. W. Gardiner, *Handbook of Stochastic Methods*, 2nd ed., Springer-Verlag, New York, 1990.

Chapter 11 Thermal, Shot and Impulse Noise

[1]. W. Bernard and Herbert B. Callen, "Irreversible Thermodynamics of Nonlinear Processes and Noise in Diverse Systems," *Reviews of Modern Physics*, Vol. 31, no. 4, pp. 1017–1044, October 1959.

[2]. R. L. Stratonovich, *Topics in the Theory of Random Noise*, R. A. Silverman, trans., Gordon and Breach, New York, 1963.

[3]. Melvin Lax, (I) "Fluctuations from the Nonequilibrium Steady State," *Reviews of Modern Physics*, Vol. 32, no. 1, pp. 25–64, January 1960; (II) "Influence of Trapping, Diffusion and Recombination on Carrier Concentration Fluctuations," *J. Phys. Chem. Solids* (Pergamon Press, Great Britain), 1960; (III) "Classical Noise III: Nonlinear Markoff Processes," *Reviews of Modern Physics*, Vol. 38, no. 2, pp. 358–379, April 1966; (IV) "Classical Noise IV: Langevin Methods," *Reviews of Modern Physics*, Vol. 38, no., 3, pp. 541–566, July 1966.

[4]. D. L. Snyder, *Random Point Processes*, John Wiley, New York, 1975; also D. L. Snyder and M. I. Miller, *Random Point Processes in Time and Space*, Springer-Verlag, New York, 1991.

[5]. D. Middleton, "Statistical-Physical Models of Electromagnetic Interference," *IEEE Trans. Electromagnetic Compat.*, Vol. EMC-19, no. 3, pp. 106–127, August 1977.

[6]. E. N. Skomal, *Man-Made Radio Noise*, Chapter 5, Van Nostrand, New York, 1978.

[7]. D. Middleton, "Canonical and Quasi-Canonical Probability Models of Class A Interference," *IEEE Trans. Electromagn-Compat.*, Vol. EMC-25, no. 2, pp. 76–106, May 1983.

[8]. ———, "S. O. Rice and the Theory of Random Noise: Some Personal Recollections," *IEEE Trans. on Information Theory*, Vol. 34, no. 6, pp. 1367–1373, November 1988.

[9]. D. Middleton and A. D. Spaulding, ''Elements of Weak Signal Detection in Non-Gaussian Noise Environments,'' Chapter 5 in *Advances in Statistical Signal Processing*, Vol. 2, H. V. Poor and J. B. Thomas, eds., JAI Press, Greenwich, CT, 1993.

[10]. M. Grigoriu, *Applied Non-Gaussian Processes*, Prentice Hall, Englewood Cliffs, NJ, 1995.

[11]. D. Middleton, *Threshold Signal Processing*, Chapters 10, and Part III, American Institute of Physics, Press, New York, in press, 1997.

PART III APPLICATIONS TO SPECIAL SYSTEMS

Chapter 12 Amplitude Modulation and Conversion

[1]. M. Schwartz, W. R. Bennett, and S. Stein, *Communication Systems and Techniques*, McGraw-Hill, New York, 1966; IEEE Press, New York, 1996.

Chapter 13 Rectification of Amplitude Modulated Waves: Second Moment Theory

[1]. J. H. Van Vleck and D. Middleton, *The Spectrum of Clipped Noise, Proc. IEEE*, Vol. 54, no. 4, pp. 2–19, January 1966.

[2]. A. H. Haddad, ed., *Nonlinear Systems*, Benchmark Papers in Electrical Engineering and Computer Science, Halsted Press, John Wiley, New York, 1975.

Chapter 14 Phase and Frequency Modulations

[1]. J. J. Downing, *Modulation Systems and Noise*, Prentice Hall, Englewood Cliffs, NJ, 1964.

[2]. H. E. Rowe, *Signals and Noise in Communications Systems*, D. Van Nostrand, New York, 1965.

[3]. N. M. Blachman, *Noise and Its Effect on Communications*, Krieger, Malabar FL, 1st ed., 1966; 2nd ed., 1982.

Chapter 15 Detection of Frequency Modulated Waves: Second Moment Theory

[1]. S. O. Rice, ''Noise in FM Receivers,'' Chapter 25 of *Time Series Analysis*, M. Rosenblatt, ed., John Wiley, New York, 1963. See also D. Middleton, ''S. O. Rice and the Theory of Random Noise,'' *IEEE Trans on Information Theory*, Vol. 34, no. 6, pp. 1367–1373, November 1988.

[2]. N. M. Blachman, *Noise and Its Effect on Communications*, Krieger, Malabar FL, 1st ed., 1966; 2nd ed., 1982.

Chapter 16 Linear Measurements, Prediction, and Optimum Filtering

[1]. J. H. Van Vleck and D. Middleton, ''Theory of the Visual vs. Aural or Meter Reception of Radar Signals in the Process of Noise'', Radio Research Lab. Report No. 411–486, p. 721; ref. 6 of this book), May 1944 (Harvard University); see also page 149 of Vol. 24, of MIT Radiation Laboratory Series, McGraw-Hill, New York, 1950.

[2]. C. Eckart, ''The Theory of Noise Suppression by Linear Filters,'' Report 41–434 (p. 722, Ref. 23), Scripps Institute of Oceanography, La Jolla, CA, 1952.

[3]. D. O. North, ''An Analysis of the Factors Which Determine Signal/Noise Discrimination in Pulsed Carrier Systems,'' *Proc, IEEE*, Vol. 51, pp. 1016–1027, 1964.

[4]. J. S. Bendat and A. G. Piersol, *Random Data Analysis and Measurement Procedures*, John Wiley, New York, 1971.

[5]. A. V. Oppenheim and R. W. Schafer, *Digital Signal Processing*, Prentice Hall, Englewood Cliffs, NJ, 1975.
[6]. B. D. Steinberg, *Principles of Aperture and Array System Design*, John Wiley, New York, 1976.
[7]. N. M. Blachman, *Noise and Its Effect on Communications*, Krieger, Malabar FL, 1st ed., 1966; 2nd ed., 1982.
[8]. S. Haykin, ed., *Array Signal Processing*, Prentice Hall, Englewood Cliffs, NJ, 1985.
[9]. B. Widrow and S. D. Stearns, *Adaptive Signal Processing*, Prentice Hall, Englewood Cliffs, NJ, 1985.
[10]. J. V. Candy, *Signal Processing—The Model-Based Approach*, McGraw-Hill, New York, 1986.
[11]. S. Marple, Jr., *Digital Spectral Analysis with Applications*, Prentice Hall, Englewood Cliffs, NJ, 1987.
[12]. A. M. Yaglom, *Correlation Theory of Stationary Random Functions:* I. *Basic Results;* II. *Supplementary Notes and References*, Springer-Verlag, New York, 1987.
[13]. S. M. Rytov, Yu. A. Kravtsov, V. I. Tatarskii, *Principles of Statistical Radio Physics:* 2. *Correlation Theory of Random Processes*, Springer-Verlag, New York, 1988.
[14]. S. Haykin ed., *Advances in Spectrum Analysis and Array Processing I, II*, Prentice Hall, Englewood Cliffs, NJ, 1991.
[15.] C. L. Nikias and A. P. Petropulu, *Higher-Order Spectral Analysis,* Prentice Hall, Englewood Cliffs, NJ, 1993.

Chapter 17 Some Distribution Problems

[1]. P. I. Kuznetsov, R. L. Stratonovich, and V. I. Tikhanov, *Nonlinear Transformations of Stochastic Processes*, Trans. J. Wise and D. C. Cooper, Pergamon Press, New York, 1965.
[2]. R. Deutsch, *Nonlinear Transformations of Random Processes*, Prentice Hall, Englewood Cliffs, NJ, 1962.
[3]. C. W. Gardiner, *Handbook of Stochastic Methods*, 2nd ed., Springer-Verlag, New York, 1990.
[4]. N. G. Van Kampen, *Stochastic Processes in Physics and Chemistry*, North Holland, New York, 1992.

Chapter 18 Reception as a Decision Problem

[1]. R. D. Luce and H. Raiffa, *Games and Decisions,* John Wiley, New York, 1957.
[2]. S. S. Wilks, *Mathematical Statistics*, John Wiley, New York, 1962.
[3]. D. Middleton, *Topics in Communication Theory*, McGraw-Hill, New York, 1965; Reprint Edition, Peninsula, Los Altos, CA, Chapter 1, 1987.
[4]. E. T. Jaynes, *Papers on Probability, Statistics, and Statistical Physics*, R. D. Rosenkranz, ed., D. Reidel, Dordrecht, Holland, 1983.
[5]. J. O. Berger, *Statistical Decision Theory and Bayesian Analysis*, 2nd ed., Springer-Verlag, New York, 1985.

Chapter 19 Binary Detection Systems Minimizing Average Risk. General Theory

[1]. C. W. Helstorm, *Statistical Theory of Signal Detection*, 2nd ed., Pergamon Press, New York, 1968.
[2]. H. V. Poor, *An Introduction to Signal Detection and Estimation*, Springer-Verlag, New York, 1988.

Chapter 20 Binary Detection Systems Minimizing Average Risk. Examples

A. Radar

[1]. E. J. Kelly, I. S. Reed, and W. L. Root, "The Detection of Radar Echoes in Noise I, II," *Journal of SIAM*, Vol. 8, pp. 309–341, 481–507, 1960.

[2]. L. A. Wainstein and V. D. Zubakov, *Extraction of Signals from Noise*, R. A. Silverman, trans., Prentice Hall, Englewood Cliffs, NJ, 1962.

[3]. M. I. Skolnik, *Introduction to Radar Systems*, McGraw-Hill, New York, 1962.

[4]. C. E. Cook and Marvin Bernfeld, *Radar Signals*, Academic Press, New York, 1967.

[5]. H. Van Trees, *Detection, Estimation, and Modulation Theory*, I. *Detection, Estimation, and Linear Modulation Theory*, (1968); II. *Nonlinear Modulation Theory*, (1971); III. *Radar and Sonar Signal Processing and Gaussian Systems in Noise* (1971), John Wiley, New York.

[6]. A. W. Rihaczak, *Principles of High-Resolution Radar*, McGraw-Hill, New York, 1969.

[7]. R. O. Harger, *Synthetic Aperture Radar Systems*, Academic Press, New York, 1970.

[8]. S. Haykin and A. Steinhart, *Adaptive Radar Detection and Estimation*, John Wiley, New York, 1992.

[9]. C. W. Helstrom, *Elements of Signal Detection and Estimation*, Prentice Hall, Englewood Cliffs, NJ, 1995.

[10]. D. Middleton, *Threshold Signal Processing*, American Institute of Physics Press, New York, 1996.

B. Sonar

[11]. V. V. Ol'shevskii, *Statistical Methods in Sonar*, Consultants Bureau, Plenum, New York, 1978.

[12]. L. J. Ziomek, *Underwater Acoustics—A Linear Systems Approach*, Academic Press, New York, 1985.

[13]. D. Middleton, "Channel Modeling and Threshold Signal Processing in Underwater Acoustics: An Analytic Overview," *IEEE J. of Ocean Engineering*, Vol. OE-12, no. 1, pp. 4–28, January 1987.

(14]. ________, *Threshold Signal Processing*, American Institute of Physics Press, New York, late 1996.

C. Telecommunications Refs. [1–10, 14]. aforementioned and the following:

[15]. V. I. Kotel'nikov, *The Theory of Optimum Noise Immunity*, R. A. Silverman, trans., McGraw-Hill, New York, 1959.

[16]. J. M. Wozencroft and I. M. Jacobs, *Principles of Communication Engineering*, John Wiley, New York, 1965.

[17]. M. Schwartz, W. R. Bennett, and S. Stein, *Communication Systems and Techniques*, McGraw-Hill, New York, 1966; IEEE Press, New York, 1996.

[18]. A. J. Viterbi, *Principles of Coherent Communications*, McGraw-Hill, New York, 1966.

[19]. R. S. Kennedy, *Fading Dispersive Communication Channels*, John Wiley, New York, 1968.

[20]. W. C. Lindsey and Marvin K. Simon, *Telecommunications Systems Engineering*, Prentice Hall, Englewood Cliffs, NJ, 1973.

[21]. K. Brayer, *Data Communications via Fading Channels*, IEEE Press, New York, 1975.

[22]. A. J. Viterbi and J. K. Omura, *Principles of Digital Communications and Coding*, McGraw-Hill, New York, 1977.

[23]. R. C. Dixon, *Spread Spectrum Systems*, John Wiley, New York, 1975; see also R. C. Dixon ed., *Spread Spectrum Techniques*, IEEE Press, New York, 1977.

[24]. P. Stavroulakis, ed., *Interference Analysis of Communication Systems*, IEEE Press, New York, 1980.
[25]. J. E. Hershey and R. K. R. Yarlagadda, *Data Transportation and Protection*, Plenum, New York, 1986.
[26]. B. Sklar, *Digital Communications*, Prentice Hall, Englewood Cliffs, NJ, 1988.
[27]. H. V. Poor and J. B. Thomas, eds., *Advances in Statistical Signal Processing*. 2, *Detection*, JAI Press, Greenwich, CT, 1993.

Chapter 21 Extraction Systems Minimizing Average Risk—Signal Analysis

[1]. R. Deutsch, *Estimation Theory*, Prentice Hall, Englewood Cliffs, NJ, 1965.
[2]. D. Middleton and R. Esposito, "Simultaneous Optimum Detection and Estimation of Signals in Noise," *IEEE Trans. on Information Theory*, Vol. IT-14, no. 3, pp. 434–444; May 1968.
[3]. ———, "New Results in the Theory of Simultaneous Optimum Detection and Estimation of Signals in Noise," *Problemy Peredachi Informatsii*, Vol. 6, no. 2, pp. 6–20, April–June, 1970, Moscow, USSR; translation by Consultants Bureau, Plenum, New York, 1973.
[4]. S. Haykin, ed., *Detection and Estimation*, Benchmark Papers in Electrical Engineering and Computer Science, Vol. 13, Dowden, Hutchinson and Ross, Straudsbury, PA, 1976.
[5]. T. Kaileth, ed., *Linear Least-Squares Estimation*, Benchmark Papers in Electrical Engineering and Computer Science, Vol. 17, Dowden, Hutchinson and Ross, Straudsbury, PA, 1977.
[6]. H. V. Poor and J. B. Thomas, eds., *Advances in Statistical Signal Processing* 1. *Estimation*, JAI Press, Greenwich, CT, 1987.
[7]. H. V. Poor, *An Introduction to Signal Detection and Estimation*, Springer-Verlag, New York, 1988.
[8]. S. Haykin and A. Steinhardt, eds., *Adaptive Radar Detection and Estimation*, John Wiley, New York, 1992.

Chapter 22 Information Measures in Reception

[1]. L. Brillouin, *Science and Information Theory*, 2nd ed., Academic Press, New York, 1962.
[2]. S. Kullback, *Information Theory and Statistics*, John Wiley, New York, 1959; Dover, New York, 1968.
[3]. E. T. Jaynes, *Papers on Probability, Statistics, and Statistical Physics*, R. D. Rosenkranz, ed., Reidel, Dordrecht, Holland, 1983.

Chapter 23 Generalizations and Extensions See Refs. [8–10, 15–17], Chapter 20, above and:

[1]. J. D. Gibson and J. I. Melsa, *Introduction to Nonparametric Detection with Applications*, Academic Press, New York, 1975; IEEE Press, New York, 1996.
[2]. C. W. Helstrom, *Quantum Detection and Estimation*, Academic Press, New York, 1976.
[3]. R. M. Gagliardi and S. Karp, *Optical Communications*, John Wiley, New York, 1976.
[4]. R. O. Harger, *Optical Communication Theory*, Benchmark Papers in Electrical Engineering and Computer Science, Vol. 18, Dowden, Hutchinson and Ross, Straudsbury, PA, 1977.
[5]. P. J. Huber, *Robust Statistics*, John Wiley, New York, 1981.
[6]. J. W. Goodman, *Statistical Optics*, John Wiley, New York, 1985.
[7]. S. A. Kassam, *Signal Detection in Non-Gaussian Noise*, Springer-Verlag, New York, 1987.

[8]. E. J. Wegman, S. C. Schwartz, J. B. Thomas, *Topics in Non-Gaussian Signal Processing*, Springer-Verlag, New York, 1989.
[9]. F. C. Moon, *Chaotic and Fractal Dynamics*, John Wiley, New York, 1992.
[10]. A. V. Gapunov-Grekhov and M. I. Rabinovich, *Nonlinearities in Action*, Springer-Verlag, New York, 1992.
[11]. M. Basseville and I. V. Nikiforov, *Detection of Abrupt Changes*, Prentice Hall, Englewood Cliffs, NJ, 1993.
[12]. R. K Young, *Wavelet Theory and Its Applications*, Kluwer Academic Publishers, Norwell, MA, 1993.
[13]. S. Haykin, *Neural Networks*, IEEE Press, New York; Macmillan, New York, 1994.
[14]. C. L. Nikias and M. Shao, *Signal Processing with Alpha-Stable Distributions and Applications,* John Wiley, New York, 1995.
[15]. D. Middleton, *Threshold Signal Processing*, American Institute of Physics Press, New York, 1997.

Bayes Theory and Critiques

[16]. E. T. Jaynes, *Papers on Probability, Statistics, and Statistical Physics*, R. D. Rosenkranz, ed., D. Riedel, Dordrecht, Holland, 1983.
[17]. J. O. Berger, *Statistical Decision Theory and Bayesian Analysis; 2nd Edition*, Springer-Verlag, New York, 1985.
[18]. J. Skilling, ed., *Maximum Entropy and Bayesian Methods*, Kluwer Academic Publishers, Norwell, MA 02061, 1989.
[19]. W. H. Jefferys and J. O. Berger, ''Ockham's Razor and Bayesian Analysis,'' *American Scientist*, Vol. 80, pp. 64–72, 1992.
[20]. R. D. Cousins, ''Why Isn't Every Physicist a Bayesian?'' *American Journal of Physics*, Vol. 63, no. 5, pp. 398–409, May 1995. See References [9, 12] therein, among others.

NAME INDEX TO SELECTED SUPPLEMENTARY REFERENCES*

*Numbers in parentheses ( ) refer to Chapters; (P_2) refers to the 2nd Reprint Edition (IEEE Press).

GLOSSARY OF PRINCIPAL SYMBOLS

Following is a list of the principal symbols that appear in the preceding text, with their meaning or meanings. Part, chapter, section, or equation numbers where the symbol is defined are included, along with a brief description. In some instances, a symbol will possess several interpretations, an ambiguity that is removed in the text where these symbols are actually used. Many of the symbols appear throughout the book, while others (for the most part not listed here) are special to a single chapter or to one or two sections only.

A few conventions have been followed almost without exception:

1. Boldface roman is used for vectors and matrices.
2. Capital boldface script letters denote operations of various types.
3. The symbol $| \quad |$ usually means "magnitude of."
4. The abbreviation det means "determinant of."
5. The tilde $\sim$ signifies "transpose of."
6. The superscript asterisk (*) usually denotes the complex conjugate of a complex quantity, except in Part 4, where for the most part it signifies an optimum or optimized value.
7. $\mathcal{W}(f)$ is always a spectral intensity density.
8. The vinculum or, alternatively, the pointed brackets with subscripts $\langle \quad \rangle_{a,b\ldots}$ denote the statistical average; the pointed brackets $\langle \quad \rangle$ without subscripts are used in Parts 1 to 3 to indicate a *time* average and in Part 4, for the most part, as a statistical average, the significance being clear from the context.
9. Probability densities are represented by w, W; probabilities are denoted by lower case and capital p, P.
10. Σ, Π, $\int$ are the usual summation, product, and integral signs.

Some common abbreviations used in the text are:

CMLE = conditional maximum likelihood estimator (or estimate)
MLE = maximum likelihood estimator (or estimate)
UMLE = unconditional maximum likelihood estimator (or estimate)
c.f. = characteristic function
d.d. = distribution density
l.i.m. = limit in the mean (square)

A, A_0 = peak signal amplitudes (Sec. 1.3-7)
$A(t)$ = an amplitude modulation (Sec. 9.2)
a_n = a Fourier coefficient (Sec. 3.1-5)
a_0^2 = input signal-to-noise power ratio [Sec. 5.2-2, Eq. (13.21)]
$\mathbf{a}$ = a vector component of a signal process [Eqs. (20.24*a*)]
α = type I conditional error probability (Chap. 19); parameter of *RC* filter [Eqs. (3.55)]
$\alpha(t), \alpha_0(t)$ = components of a complex waveform [Eq. (2.66)]
α^* = type I error probability of optimum binary detection systems (Chap. 19)
arg = argument of [Eq. (2.75)]

B = bandwidth of white noise [Eq. (3.54*b*)]
$B_l(t)$ = envelope of *l*th spectral zone [Eqs. (2.103)]
B_n = a bias term (Chaps. 20, 21)
$B_n^{(2)}, B_n^{(4)}, B_T^{(2)}, B_T^{(4)}$ = bias terms in binary detection theory (Chap. 20)
B_0 = a band frequency (Sec. 4.2-4)
b_n = a Fourier coefficient (Sec. 3.1-5); spectral moment [Eqs. (3.66*b*)]
b_0 = a bias voltage (Chap. 13)
$\mathbf{b}$ = a vector component of a signal process [Eqs. (20.24*a*)]
β = type II (conditional) error probability (Chap. 19); a filter bandwidth (cf. Prob. 5.8); a covariance function [Eq. (9.32)]; a force constant [Eq. (10.3)]; a dynamic transconductance [Eq. (13.1)]
$\beta(t), \beta_0(t)$ = components of a complex waveform [Eqs. (2.66), (2.67)]
β_q = an eigenvalue in the modal equation for a linear network (Sec. 2.2-2)
β^* = type II error probability of optimum binary detection systems (Chap. 19)

C = channel capacity [Sec. 6.5-1, Eq. (6.85)]
$C(\mathbf{S},\boldsymbol{\gamma}), C(\boldsymbol{\theta},\boldsymbol{\gamma})$ = constant cost functions [Eq. (18.4), Part 4]
$C_\alpha, C_{1-\alpha}, C_\beta, C_{1-\beta}$ = constant costs in binary detection (Chap. 19)
C_E, C_C = constant costs (Sec. 21.2-1)
${}_nC_k$ = the binomial coefficient $n!/(n-k)!k!$
C_0 = a constant cost [Eqs. (6.98), (21.52*a*)]
$\mathbf{C}, \mathbf{C}_1, \mathbf{C}_2$ = contours of integration
c = velocity of light
c_n, C_n = Fourier coefficients (Sec. 3.1-5)

D = diffusion constant; spectral density (Sec. 10.4)
$D(x)$ = probability distribution of x (Sec. 1.2-1)
D_ϕ, D_F = phase- and frequency-modulation scale factor (Chap. 14)
$D_T(t,u;\boldsymbol{\theta})$ = a weighting function [Eq. (20.46*b*)]
$D_T(t,u;\boldsymbol{\theta})_0$ = a weighting function [Eq. (20.62*b*)]
$\mathfrak{D}_T(\gamma)$ = a Fredholm determinant (Chaps. 17, 20)
$\mathfrak{D}_y(f)$ = spectral distribution [Eq. (3.48)]
Δ = decision space (Part 4)
Δf = a bandwidth (Secs. 5.3-1, 15.3-2)
$\boldsymbol{\nabla}$ = the gradient operator in rectangular coordinates [Eq. (7.71)]
$\delta(x)$, $\delta(x - x_0)$ = Dirac delta functions (Sec. 1.2-4)
$\delta(\boldsymbol{\gamma}|\mathbf{V})$ = a decision rule (Chaps. 18, 19)
δ_m^n = Kronecker delta; $\delta_m^n = 1$ $(m = n)$; $\delta_m^n = 0$ $(m \neq n)$
$\partial/\partial t$, $\partial/\partial \xi$ = partial derivatives

E, E_N, $E_0(t)$, $E(t)$ = envelopes of (narrowband) waves
$Ei(-x)$ = exponential integral (Prob. 11.1, Sec. 15.2-2)
$E_{y^{(j)}}$ = energy in $y^{(j)}$ [Eq. (3.27)]
$\boldsymbol{E}\{\ \}$ = expectation operator [Eq. (1.21)]
$e(t;\boldsymbol{\theta})$ = error ensemble [Eq. (16.49)]
$\in$ = "belongs in" or "falls in" [Eqs. (19.5)]
ϵ, ϵ_0, ϵ_c = an epoch (Secs. 1.3-5, 16.3-1, Chaps. 19, 20)
ϵ_n = Neumann factor; $\epsilon_0 = 1$, $\epsilon_n = 2$ $(n \neq 0)$

F = (available) noise figure [Eq. (5.71)]
$F_n(\mathbf{V}|\mathbf{S})$ = conditional probability density of V, given S (Part 4)
$F_S(t;\boldsymbol{\theta})$ = a signal envelope [Eq. (20.22)]
$F_x(i\xi)$; $F_2(i\xi_1,i\xi_2;t)$; $F_{\mathbf{y}}(i\boldsymbol{\xi})$ = characteristic functions [Eqs. (1.25), (1.56)]
${}_1F_1(a;b;\pm x)$ = a confluent hypergeometric function (Sec. A.1.2)
${}_2F_1(a,b;c;x)$ = a hypergeometric function (Sec. A.1.3)
$\mathfrak{F}(\mathbf{S},\boldsymbol{\gamma})$ = a loss function (Chap. 18)
$\boldsymbol{\mathfrak{F}}\{\ \}$ = Fourier transforms [Eq. (1.28*a*)]
$\boldsymbol{\mathfrak{F}}^{-1}\{\ \}$ = inverse Fourier transform [Eq. (1.28*b*)]
f = frequency
$f(i\xi)$ = complex Fourier transform of $g(y)$ [Eqs. (2.89) (2.90)]

$G(f)$ = gain function (Sec. 5.3-1)
$g[y(t)]$ = output of a zero-memory nonlinear device, with $y(t)$ as input [Eqs. (1.131)]

$\mathbf{G}_0$, $\mathbf{G}$ = a structure matrix in detection and estimation theory (Chaps. 20, 21)

Γ = data space (Part 4)

$\Gamma(\nu + 1)$ = gamma function of $\nu + 1$ [Eq. (2.101*a*)]

γ = pulse density in time [Eq. (11.74)]

γ_σ, γ_θ, $\hat{\gamma}_\sigma$, γ_σ^* = estimators (Chap. 21)

γ_0, γ_1 = decisions (Chap. 19)

$\gamma_0(t)$ = phase factor of a narrowband linear filter, in the time domain [Eqs. (2.63)]

Υ = a set of decisions (Part 4)

$H(\sigma,\delta)$ = average information loss [Eqs. (18.9)]

$H(X)$ = average uncertainty; communication entropy (Sec. 6.3-1)

H_R = entropy rate [Eqs. (6.49)]

$H_0(x)$ = absolute entropy of x (Sec. 6.4-1)

$h(\mathbf{S},\delta)$ = conditional information loss [Eq. (18.8), Chap. 22]

$h(t)$ = weighting function of a linear invariant filter [Eq. (2.16)]

h_M = weighting function of a matched filter (Sec. 20.1-1)

$h_0(t)$ = slowly varying portions of the weighting function of a narrowband linear filter [Eqs. (2.63)]

$\mathcal{H}_{m,n,q}$, $H_{m,n}(t)$, $h_{m,n}$ = amplitude factors [Eqs. (5.14), (5.15), Sec. 15.4]

$\mathcal{H}\{X\}$ = Hilbert transform of X [Eq. (2.77)]; uncertainty in X [Eqs. (6.13)]

I = current

$I(x|y)$ = average information gain (Sec. 6.3-2)

$I_m(x)$ = modified mth-order Bessel function of the first kind (Sec. 5.1-3)

$\mathcal{I}(x|x)$ = gain of information [Eqs. (6.3)]

i = $\sqrt{-1}$, or index of summation: $i = 1, 2, \ldots,$ etc.

$\mathbf{i}_1$, $\mathbf{i}_2$, $\mathbf{i}_3$ = unit orthogonal vectors (Sec. 7.7)

Im { } = imaginary part of

J = jacobian (Sec. 1.2-14)

$J_m(x)$ = Bessel function, first kind, mth order

$J_\nu(x)$ = Bessel function, first kind, νth order (Sec. 13.4-4)

K_S, K_N, K_{SN} = covariance functions of signal and noise processes (Sec. 16.2-2)

K_{xy} = covariance of x and y

$K_y(t),\ K_y(t_2 - t_1)$ = covariance functions
$K_0(t)_N,\ k_0(t)_N,\ k_0$ = envelopes of covariance functions of a narrowband process (Sec. 3.3)
$K_0(z)$ = a zero-order modified Bessel function of the second kind [Eq. (9.30)]
$\mathbf{K},\ \mathbf{K}_{yz}$ = covariance matrices [Eqs. (1.53), (1.60), (3.143)]
$\mathbf{K}^{-1}$ = inverse covariance matrix
$\mathcal{K}$ = a threshold [Eqs. (19.15)]
$\boldsymbol{\mathcal{K}}$ = cumulant function [Eqs. (1.58a)]
$k,\ k(t),\ k_N(t)$ = normalized autovariance functions of the process $N(t)$
k_B = Boltzmann's constant
κ = an FM conversion factor [Eq. (15.12)]

L = inductance
$L_n(\mathbf{V},\boldsymbol{\theta}),\ l_n(\mathbf{V}|\boldsymbol{\theta})$ = likelihood functions (Sec. 21.1-2)
$\mathcal{L}(\mathbf{S},\delta)$ = loss rating [Eq. (18.2)]
$\boldsymbol{\mathcal{L}}\{\ \},\ \boldsymbol{\mathcal{L}}^{-1}\{\ \}$ = Laplace and inverse Laplace transforms [Eqs. (2.33b), (2.34)]
$\Lambda_n(\mathbf{V})$ = a generalized likelihood ratio (Chap. 19)
Λ_T = a likelihood ratio functional (Chaps. 19, 20)
$\boldsymbol{\Lambda}$ = a diagonal matrix [Eq. (7.20a)]
λ = a modulation index (amplitude modulation) [Eq. (1.110a), Chap. 13]
$\lambda^{(L)}_{l_1,\ldots,l_m}$ = Lth-order multivariate semi-invariant [Eq. (1.58)]
λ_m = mth semi-invariant [Eq. (1.35)]; the mth eigenvalue of a homogeneous integral equation (Sec. 8.2-2, Chap. 17)
$\lambda_0(t)$ = a component of the weighting function of a narrowband linear filter [Eq. (3.93b)]

$M_y(t)$ = second-moment function of the process y [Eq. (1.121b)]
$\mu,\ \mu_{kk}$ = second moments, variances [Eq. (7.82)]; modulation index (Chap. 14); the ratio p/q (Part 4)
$\mu_\phi,\ \mu_F$ = phase- and frequency-modulation indices (Chap. 14)

$N,\ N(t)$ = a noise process
$N_C,\ N_S$ = cosine and sine components of a narrowband noise process (Sec. 5.1-4)
$\mathbf{N}$ = noise vector (Part 4)
$\mathbf{n}$ = normalized noise vector (Part 4)
ν = power law of a zero-memory nonlinear device [Eq. (2.85c)]

ν_1 = average value (Sec. 7.1)

Ω = signal space (Part 4); angular frequency [Eq. (20.154a)]
$\Omega(t)$ = a covariance function (Chap. 14)
Ω_θ = (signal) parameter space (Chap. 21)
ω = $2\pi f$ = angular frequency
ω_c = angular carrier frequencies [Eq. (2.64)]
ω_D = an angular deviation frequency [Eqs. (2.65b), (15.3)]
ω_F = angular bandwidth or filter width
ω_N = a spectral bandwidth (Chap. 14)
ω_0 = $2\pi f_0$ = angular carrier frequencies (Sec. 1.3-7)

$P_n(x)$ = d.d. of $x = \log \Lambda_n$ with respect to H_1 (Chap. 19)
$P_y^{(j)}$ = average power [Eq. (3.30)]
$\mathcal{P}$ = principal part [Eq. (2.76)]
p = a probability
= iz = a complex variable (Sec. 2.2-2)
$\Phi_n(v,s)$, $\Phi_n(s,s)$, $\Phi_T(v,s)$, $\Phi_T(s,s)$ = quadratic forms and functionals (Chaps. 19, 20); also given as Φ_v, Φ_s
ϕ, $\phi(t)$, $\phi_{A,F}$, $\Phi(t)$, $\Psi(t)$ = phases and phase factors [Eq. (5.7)]; phase modulation [Eq. (14.1)]
$\phi_0(t)_N$ = phase of covariance function of a narrowband process (Sec. 3.3)
$\Psi_n(v)$, $\Psi_{0n}(v)$, $\Psi_T(v)$ = quadratic forms and functionals in signal detection and extraction theory (Chaps. 20, 21)
ψ = mean intensity of a stationary noise process
$\psi(z)$ = logarithmic derivative of the gamma function of z [Eqs. (13.117)]
$\psi_m(t)$, $\phi_m(t)$ = orthonormal functions (Sec. 8.2-1)

Q = the Q factor of a filter (Sec. 3.3-6); charge on a capacitor [Eq. (10.4a)]
$Q_n(x)$ = d.d. of $x = \log \Lambda_n$ with respect to H_0 (Chap. 19)
Q_T = a quadratic functional (Sec. 19.4-2)
$\mathbf{Q}$ = a discrete linear filter operation [Eq. (20.12)]; a linear transformation (Sec. 7.5-2)
$\mathbf{Q}_C$ = a discrete linear filter (Sec. 20.1-1)
$\mathbf{Q}_E$, $\mathbf{Q}_F$ = linear discrete filters (Chaps. 20, 21)
q = an a priori probability ($= 1 - p$) (Chap. 19)

R = information rate [Eq. (6.88)]; resistance (Chaps. 3, 10)

$R(\sigma,\delta)$ = average risk (Part 4)
$R^*(\sigma,\delta^*)$ = Bayes risk (minimum average risk) (Part 4)
R_M^* = Minimax average risk (Chap. 19)
$R_y^{(j)}(t)$ = autocorrelation function of $y^{(j)}(t)$ [Eq. (1.113)]
$R_{yz}^{(j)}(t)$ = cross-correlation function of $y^{(j)}$ and $z^{(j)}$ [Eq. (1.114)]
R_0 = a limiting level (Chaps. 13, 15)
$\mathbf{R}_{yz}^{(j)}(t)$ = coherence matrix (Sec. 3.4-1)
Re { } = real part of
$r(\mathbf{S},\delta)$ = conditional risk [Eq. (18.6)]
$r_y^{(j)}(t)$ = normalized autocorrelation function of $y^{(j)}$ (Sec. 3.1-3)
$\mathbf{r}_k$ = a displacement vector (Sec. 7.7-1)
$\rho(t)$ = autocorrelation function of a linear invariant filter [Eqs. (3.89)]
$\rho_G(t,u)$ = a structure factor (Chap. 20)
$\rho_x(t)$ = a normalized covariance function (Sec. 7.2)
$\rho_0(t)$ = a component of the weighting function of a narrowband filter [Eq. (3.93*a*)]

S, $S(t)$ = a signal process
S/N, $(S/N)_{out}$ = signal-to-noise ratios
$S_g(f)$, $S_g(i\omega)$, $S_g(i\omega)_T$ = amplitude spectral densities [Eqs. (2.43), (2.44*b*)]
$\mathbf{S}$ = signal vector (Part 4)
$s(t)$ = a normalized signal process (Part 4)
s_F = rms signal amplitude (Sec. 13.3)
$\mathbf{s}$ = normalized signal vector (Chap. 17, Part 4)
$\sigma(\mathbf{S})$, $\sigma(\boldsymbol{\theta})$ = a priori signal d.d.'s (Part 4)
σ_n^2, σ_z^2, σ^2 = variances

T = duration of an observation interval
T_0 = a period [Eq. (4.20)]; an absolute temperature (Sec. 5.3-1)
$\boldsymbol{T}\{\ \}$ = transformation (of an ensemble)
$\boldsymbol{T}_g\{\ \}$ = a nonlinear transformation (Sec. 2.3-1)
$\boldsymbol{T}_L\{\ \}$ = a linear transformation [Sec. 2.1-3(2)]
$\boldsymbol{T}_R^{(N)}$, $\boldsymbol{T}_T^{(N)}$, $\boldsymbol{T}_M^{(N)}$ = operations of reception, transmission, and the medium of propagation [Eq. (2.20)]
$\boldsymbol{T}_S\{\ \}$ = a discrete sampling operation (Sec. 4.1)
t, τ = time variables
t_λ = a smoothing or prediction time

$$\mathrm{tr}\,\mathbf{G} = \sum_{i=1}^{n} G_{ii}$$

Θ = error integral [Eq. (1.9)]
$\hat{\theta}^*$, $\hat{\theta}$ = estimators (Chap. 21)
$\theta(t)$ = a random phase (Sec. 9.1)
$\boldsymbol{\theta}$ = a set of signal parameters (Sec. 1.3-5)

U_E, U_H = electric and magnetic energy densities [Eqs. (11.8)]
$\{u\}$ = transmitted message set [Eqs. (2.20), (6.1)]
$u(t)$ = a sampling function (Sec. 4.2-1)
$\mathcal{U}$ = initial uncertainty [Eq. (6.2)]

$V(t)$; $\mathbf{V}$ = voltage process; sampled process
V_c, V_s = cosine and sine components of a narrowband $V(t)$ (Sec. 9.2)
$V_{in}(t)$ = an input voltage
V_{out} = an output voltage
v = a velocity
$\{v\}$ = received message set or set of final decisions [Eqs. (2.20), (6.1)]
$v(t)$; $\mathbf{v}$ = normalized voltage process; sampled process
$\mathbf{dv}$, $\mathbf{dV}$ = volume elements $dv_1 \cdots dv_n$, $dV_1 \, dV_2 \cdots dV_n$
$\mathbf{v}_F$ = a filtered voltage process [Eq. (20.12)]

$W_n(\mathbf{V})_N$ = nth-order d.d. of a noise process $\mathbf{N} = \mathbf{V}$ (Part 4)
W_0 = spectral intensity density of white noise [Eqs. (3.54)]
$\mathcal{W}_y^{(j)}$, $\mathcal{W}_\infty$, $\mathcal{W}_y$ = spectral intensity densities (Sec. 3.2-1)
$\mathcal{W}_{yz}$ = cross-spectral density [Eqs. (3.147), (3.148)]
$\mathcal{W}_z(f)_l$ = spectral density for lth zone [Eq. (5.28)]
$w(x)$, $w_n(x)$; $W(x)$, $W(\mathbf{x})$, etc. = probability densities
$w(x|y)$ = conditional probability density of x given y
$w_u(f)$ = spectral density of a sampling pulse [Eqs. (4.49)]

X, x = random variables (Part 4)
X_T = a weighting function; solution of an inhomogeneous integral equation (Chaps. 19, 20)
$\mathbf{x}$ = a random vector

$Y(i\omega)$ = a system function of a linear invariant network [Eq. (2.49)]; admittance function (Sec. 11.1-3)
$Y(i\omega)_T$ = system function of an optimum linear filter (Chap. 20)
$Y(p)$ = a transfer function of a linear network [Eq. (2.39)]

$Y_0(i\omega)$ = equivalent l-f representation of the system function of a narrowband filter [Eq. (2.62)]

$\dot{y}, \ddot{y}$ = first, second time derivatives of y

$y^{(k)}(t)$ = kth representation or member of the ensemble $y(t)$ (Sec. 1.3)

Z = a measurement (Chap. 16)

$Z(i\omega)$ = impedance function (Sec. 11.1-3)

NAME INDEX

SUBJECT INDEX

AUTHOR'S BIOGRAPHY

David Middleton, physicist, applied mathematician, educator, and author of An Introduction to Statistical Communication Theory, is an internationally recognized pioneer in statistical communication theory, a field in which he has been active for more than fifty years. Born in New York City on April 19, 1920, Dr. Middleton received his AB (summa cum laude) in 1942 from Harvard College and the AM degree in 1945 from Harvard University. During World War II (1942–1945) he was research assistant to Professor J. H. Van Vleck (Nobel Laureate in Physics) under whose aegis he took his Ph.D. in 1947. During the war, as Special Research Associate at Harvard's Radio Research Laboratory, directed by Professor Frederick Terman of Stanford, Dr. Middleton was engaged in various studies and applications of electronic countermeasures. In 1947 he became a Research Fellow at Harvard. A year of postdoctoral research there with Professor Leon Brillouin followed in 1948. In 1949, he became Assistant Professor of Applied Physics in the division of Engineering Sciences and Applied Physics, also at Harvard, where his investigations included analyses of AM and FM systems, random noise theory, and applications to system design and optimization.

Since 1954, in addition to his principal consulting activities, Dr. Middleton has served as Adjunct Professor in physics, mathematical sciences, and electrical engineering at various academic institutions, among them Columbia, Rensselaer, Denver, Johns Hopkins, Texas, Rice, and currently, the University of Rhode Island. He is the author of more than 160 papers and two books—(*Introduction* (1960–1972, 1987–1995, 1996–) and *Topics in Communication Theory* (1965–1977, 1987–)—the latter currently reprinted by IEEE Press and a third volume, *Threshold Signal Processing* (from the American Institute of Physics Press, estimated 1997). His main areas of research and applications have been and are currently (1) *Statistical Communication Theory*—random processes, detection, and estimation; (2) *Statistical Physics*—namely, the canonical channel, involving propagation and scattering in random media with ambient noise sources, with particular attention to underwater acoustics, sonar, and remote sensing, along with their physical counterparts in electromagnetic applications (e.g., telecommunications and radar); and (3) *Electromagnetic Compatibility*, with specific emphasis on nongaussian noise and interference models and, as in (1), including threshold signal transmission and reception for both man-made and natural electromagnetic and acoustic environments.

Dr. Middleton's research has emphasized the close and necessary relationship between the underlying physics and the intended engineering applications, whatever may be the particular physical environment. His approaches accordingly have focused first on the canonical channel, which is invariant of the particular physics, and have then proceeded to specific applications. Among the author's many contributions, from 1942 to the present, may be cited his early work with Professor Van Vleck establishing the "matched filter" concept (1944–), the theory of signals and noise through nonlinear devices (1943–), the formulation and development (1954–) of statistical comunication theory in Bayesian terms (with Van Matin, 1954, 1955), the development of statistical-physical models of nongaussian noise, and interference (1970–), the Bayesian theory (with Esposito) of joint detection and estimation of signals under a priori uncertainty (1968–), new approaches to scattering in random media (1970–), and the canonical theory of threshold detection and estimation in generalized noise (1965–).

Dr. Middleton is a consulting physicist and contractor with universities, industry, and the U.S. Departments of Energy, Defense, and Navy. He also has been and is currently a member of various advisory boards and panels. In addition, he is a Fellow of the Institute of Electrical and Electronic Engineers (IEEE), the American Physical Society (APS), the Acoustical Society of America (ASA), and the American Association for the Advancement of Science (AAAS) and is also the recipient of various prizes and awards. (See *Who's Who in America, Who's Who in Science and Engineering*, and other similar volumes for details.)